ESSENTIALS
OF ECOLOGY

ESSENTIALS
OF ECOLOGY
Third Edition

Colin R. Townsend
Department of Zoology, University of Otago, Dunedin, New Zealand

Michael Begon
Population Biology Research Group, School of Biological Sciences
The University of Liverpool, Liverpool, UK

John L. Harper
Professor Emeritus in the University of Wales
Visiting Professor in the University of Exeter, Exeter, UK

BLACKWELL PUBLISHING
350 Main Street, Malden, MA 02148-5020, USA
9600 Garsington Road, Oxford OX4 2DQ, UK
550 Swanston Street, Carlton, Victoria 3053, Australia

First edition published 2000 by Blackwell Publishing
Second edition published 2003
Third edition published 2008

2   2008

Library of Congress Cataloging-in-Publication Data

Townsend, Colin R.
    Essentials of ecology / Colin R. Townsend, Michael Begon, John L. Harper.—3rd ed.
        p.   cm.
    Includes bibliographical references and index.
    ISBN 978-1-4051-5658-5 (pbk. : alk. paper)
    1. Ecology.   I. Begon, Michael.   II. Harper, John L.   III. Title.

    QH541.T66 2008
    577—dc22   2007034694

A catalogue record for this title is available from the British Library.

Set in 9.5/12pt ClassGarmond
by Graphicraft Limited, Hong Kong
Printed and bound in Singapore by C.O.S. Printers Pte Ltd

The publisher's policy is to use permanent paper from mills that operate a sustainable forestry
policy, and which has been manufactured from pulp processed using acid-free and elementary
chlorine-free practices. Furthermore, the publisher ensures that the text paper and cover
board used have met acceptable environmental accreditation standards.

For further information on
Blackwell Publishing, visit our website:
www.blackwellpublishing.com

Short contents

SHORT CONTENTS

CONTENTS

PREFACE

By writing this book we hope to share with you some of our wonder at the complexity of nature, but we must all also be aware that there is a darker side: the fear that we are destroying our natural environments and the services they provide. All of us need to be ecologically literate so that we can take part in political debate and contribute to solving the ecological problems that we carry with us into the new millennium. We hope our book will contribute to this objective.

The genesis of this book can be found in the more comprehensive treatment of ecology in our big book *Ecology: from Individuals to Ecosystems* (Begon, Townsend & Harper, 4th edn, 2006). This is used as an advanced university text around the world, but many of our colleagues have called for a more succinct treatment of the essence of the subject. Thus, we were spurred into action to produce a distinctively different book, written with clear objectives for a different audience – those taking a semester-long beginners course in the essentials of ecology. We hope that at least some readers will be excited enough to go on to sample the big book and the rich literature of ecology that it can lead into.

In this third edition of *Essentials of Ecology* we have made the text, including mathematical topics, even more accessible. Ecology is a vibrant subject and this is reflected by our inclusion of literally hundreds of new studies. Some readers will be engaged most by the fundamental principles of how ecological systems work. Others will be impatient to focus on the ecological problems caused by human activities. We place heavy emphasis on both fundamental and applied aspects of ecology: there is no clear boundary between the two. However, we have chosen to deal first in a systematic way with the fundamental side of the subject, and we have done this for a particular reason. An understanding of the scope of the problems facing us (the unsustainable use of ecological resources, pollution, extinctions and the erosion of natural biodiversity) and the means to counter and solve these problems depend absolutely on a proper grasp of ecological fundamentals.

The book is divided into four sections. In the introduction we deal with two foundations for the subject that are often neglected in texts. Chapter 1 aims to show not only what ecology is but also how ecologists do it – how ecological understanding is achieved, what we understand (and, just as important, what we do not yet understand) and how our understanding helps us predict and manage. We then introduce 'Ecology's evolutionary backdrop' and show that ecologists need a full understanding of the evolutionary biologist's discipline in order to make sense of patterns and processes in nature (Chapter 2).

What makes an environment habitable for particular species is that they can tolerate the physicochemical conditions there and find in it their essential resources.

In the second section we deal with conditions and resources, both as they influence individual species (Chapter 3) and in terms of their consequences for the composition and distribution of multispecies communities, for example in deserts, rain forests, rivers, lakes and oceans (Chapter 4).

The third section (Chapters 5–11) deals systematically with the ecology of individual organisms, populations of a single species, communities consisting of many populations, and ecosystems (where we focus on the fluxes of energy and matter between and within communities). To understand patterns and processes at each of these levels we need to know the behavior of the level below. This section also includes a new Chapter 8 on 'Evolutionary ecology', responding to the feelings of some readers that, although evolutionary ideas pervade the book, there was still not sufficient evolution for a book at this level.

Finally, armed with knowledge and understanding of the fundamentals, the book turns to the applied questions of how to deal with pests and manage resources sustainably (whether wild populations of fish or agricultural monocultures) (Chapter 12), then to a diversity of pollution problems ranging from local enrichment of a lake by sewage to global climate change associated with the use of fossil fuels (Chapter 13) and lastly we develop an armory of approaches that may help us to save endangered species from extinction and conserve some of the biodiversity of nature for our descendants (Chapter 14).

A number of pedagogical features have been included to help you.

- Each chapter begins with a set of key concepts that you should understand before proceeding to the next chapter.
- Marginal headings provide signposts of where you are on your journey through each chapter – these will also be useful revision aids.
- Each chapter concludes with a summary and a set of review questions, some of which are designated challenge questions.
- You will also find three categories of boxed text:
  - 'Historical landmarks' boxes emphasize some landmarks in the development of ecology.
  - 'Quantitative aspects' boxes set aside mathematical and quantitative aspects of ecology so they do not unduly interfere with the flow of the text and so you can consider them at leisure.
  - 'Topical ECOncerns' boxes highlight some of the applied problems in ecology, particularly those where there is a social or political dimension (as there often is). In these, you will be challenged to consider some ethical questions related to the knowledge you are gaining.

An important further feature of the book is the companion internet web site, e.cology, accessed through www.blackwellpublishing.com and linked to the companion site of our big book, *Ecology*. This provides an easy-to-use range of resources to aid study and enhance the content of the book. Features include self-assessment multiple choice questions for each chapter in the book, an interactive tutorial to help students to understand the use of mathematical modeling in ecology, and high-quality images of the figures in the book that teachers can use in preparing their lectures or lessons.

Acknowledgments

ACKNOWLEDGMENTS

It is a pleasure to record our gratitude to the people who helped with the planning and writing of this book. Going back to the first edition, we thank Bob Campbell and Simon Rallison for getting the original enterprise off the ground and Nancy Whilton and Irene Herlihy for ably managing the project; and for the second edition, Nathan Brown (Blackwell, US) and Rosie Hayden (Blackwell, UK) for making it so easy for us to take this book from manuscript into print. For this third edition, we especially thank Nancy Whilton and Elizabeth Frank in Boston for persuading us to pick up our pens again (not literally) and Rosie Hayden, again, and Jane Andrew and Ward Cooper for seeing us through production. We are also grateful to the following colleagues who provided insightful reviews of early drafts of one or more chapters. For the first edition, Tim Mousseau (University of South Carolina), Vickie Backus (Middlebury College), Kevin Dixon (Arizona State University, West), James Maki (Marquette University), George Middendorf (Howard University), William Ambrose (Bates College), Don Hall (Michigan State University), Clayton Penniman (Central Connecticut State University), David Tonkyn (Clemson University), Sara Lindsay (Scripps Institute of Oceanography), Saran Twombly (University of Rhode Island), Katie O'Reilly (University of Portland), Catherine Toft (UC Davis), Bruce Grant (Widener University), Mark Davis (Macalester College), Paul Mitchell (Staffordshire University, UK) and William Kirk (Keele University, UK); and for the second, James Cahill (University of Alberta), Liane Cochrane-Stafira (Saint Xavier University), Hans deKroon (University of Nijmegen), Jake Weltzin (University of Tennessee at Knoxville) and Alan Wilmot (University of Derby, UK).

For this edition, our long-time mentor and collaborator John Harper has stepped from the treadmill to more fully enjoy his retirement. We owe him a special debt of gratitude that extends far beyond the past co-authorship of this book into all aspects of our lives as ecologists.

Last, and perhaps most, we are glad to thank our wives and families for continuing to support us, listen to us, and ignore us, precisely as required – thanks to Laurel, Dominic, Jenny, Brennan and Amelie, and to Linda, Jessi and Rob.

The publisher would like to thank Denis Saunders, from CSIRO, for use of the image in part 4 of the book.

PART ONE
Introduction

Chapter 1

Ecology and how to do it

Key concepts

In this chapter you will:

- learn how to define ecology and appreciate its development as both an applied and a pure science
- recognize that ecologists seek to describe and understand, and on the basis of their understanding, to predict, manage and control
- appreciate that ecological phenomena occur on a variety of spatial and temporal scales, and that patterns may be evident only at particular scales
- recognize that ecological evidence and understanding can be obtained by means of observations, field and laboratory experiments, and mathematical models
- understand that ecology relies on truly scientific evidence (and the application of statistics)

Nowadays, ecology is a subject about which almost everyone has heard and most people consider to be important – even when they are unsure about the exact meaning of the term. There can be no doubt that it is important; but this makes it all the more critical that we understand what it is and how to do it.

1.1 Introduction

the earliest ecologists

The question 'What is ecology?' could be translated into 'How do we define ecology?' and answered by examining various definitions of ecology that have been proposed and choosing one of them as the best (Box 1.1). But while definitions have conciseness and precision, and they are good at preparing you for an examination, they

1.1 *Historical landmarks*

Definitions of ecology

Ecology (originally in German, *Öekologie*) was first defined in 1866 by Ernst Haeckel, an enthusiastic and influential disciple of Charles Darwin. To him, ecology was 'the comprehensive science of the relationship of the organism to the environment'. The spirit of this definition is very clear in an early discussion of biological subdisciplines by Burdon-Sanderson (1893), in which ecology is 'the science which concerns itself with the external relations of plants and animals to each other and to the past and present conditions of their existence', to be contrasted with physiology (internal relations) and morphology (structure). For many, such definitions have stood the test of time. Thus, Ricklefs (1973) in his textbook defined ecology as 'the study of the natural environment, particularly the interrelationships between organisms and their surroundings'.

In the years after Haeckel, plant ecology and animal ecology drifted apart. Influential works defined ecology as 'those relations of *plants*, with their surroundings and with one another, which depend directly upon differences of habitat among plants' (Tansley, 1904), or as the science 'chiefly concerned with what may be called the sociology and economics of *animals*,

rather than with the structural and other adaptations possessed by them' (Elton, 1927). The botanists and zoologists, though, have long since agreed that they belong together and that their differences must be reconciled.

There is, nonetheless, something disturbingly vague about the many definitions of ecology that seem to suggest that it consists of all those aspects of biology that are neither physiology nor morphology. In search of more focus, therefore, Andrewartha (1961) defined ecology as 'the scientific study of the distribution and abundance of organisms', and Krebs (1972), regretting that the central role of 'relationships' had been lost, modified it to 'the scientific study of the *interactions* that determine the distribution and abundance of organisms', explaining that ecology was concerned with '*where* organisms are found, *how many* occur there, and *why*'. This being so, it might be better still to define ecology as:

the scientific study of the distribution and abundance of organisms and the interactions that determine distribution and abundance.

are not so good at capturing the flavor, the interest or the excitement of ecology. There is a lot to be gained by replacing that single question about definition with a series of more provoking ones: 'What do ecologists *do*?', 'What are ecologists *interested* in?' and 'Where did ecology emerge from in the first place?'

Ecology can lay claim to be the oldest science. If, as our preferred definition has it, 'Ecology is the scientific study of the distribution and abundance of organisms and the interactions that determine distribution and abundance' (Box 1.1), then the most primitive humans must have been ecologists of sorts – driven by the need to understand where and when their food and their (non-human) enemies were to be found – and the earliest agriculturalists needed to be even more sophisticated: having to know how to manage their living but domesticated sources of food. These early ecologists, then, were *applied* ecologists, seeking to understand the distribution and abundance of organisms in order to apply that knowledge for their own collective benefit. They were interested in many of the sorts of things that applied ecologists are still interested in: how to maximize the rate at which food is collected from natural environments, and how this can be done repeatedly over time; how domest-icated plants and animals can best be planted or stocked so as to maximize rates of return; how food organisms can be protected from their own natural enemies; and how to control the populations of pathogens and parasites that live on us.

In the last century or so, however, since ecologists have been self-conscious enough to give themselves a name, ecology has consistently covered not only applied but also fundamental, 'pure' science. A.G. Tansley was one of the founding fathers of ecology. He was concerned especially to understand, for understanding's sake, the processes responsible for determining the structure and composition of different plant communities. When, in 1904, he wrote from Britain about 'The problems of ecology' he was particularly worried by a tendency for too much ecology to remain at the descriptive and unsystematic stage (i.e. accumulating descriptions of communities without knowing whether they were typical, temporary or whatever), too rarely moving on to experimental or systematically planned, or what we might call a 'scientific', analysis.

a pure and applied science

His worries were echoed in the United States by another of ecology's founders, F.E. Clements, who in 1905 in his *Research Methods in Ecology* complained:

> The bane of the recent development popularly known as ecology has been a widespread feeling that anyone can do ecological work, regardless of preparation. There is nothing . . . more erroneous than this feeling.

On the other hand, the need of *applied* ecology to be based on its *pure* counter-part was clear in the introduction to Charles Elton's (1927) *Animal Ecology* (Figure 1.1):

> Ecology is destined for a great future . . . The tropical entomologist or mycologist or weed-controller will only be fulfilling his functions properly if he is first and foremost an ecologist.

In the intervening years, the coexistence of these pure and applied threads has been maintained and built upon. Many applied areas have contributed to the development of ecology and have seen their own development enhanced by ecological ideas and approaches. All aspects of food and fiber gathering, produc-tion and protection have been involved: plant ecophysiology, soil maintenance, forestry, grassland composition and management, food storage, fisheries, and control of pests and pathogens. Each of these classic areas is still at the forefront of

Figure 1.1

One of the great founders of ecology: Charles Elton (1900–1991).
Animal Ecology (1927) was his first book but *The Ecology of Invasions by Animals and Plants* (1958) was equally influential.

lots of good ecology and they have been joined by others. The biological control of pests (the use of pests' natural enemies to control them) has a history going back at least to the Ancient Chinese but has seen a resurgence of ecological interest since the shortcomings of chemical pesticides began to be widely apparent in the 1950s. The ecology of pollution has been a growing concern from around the same time and expanded further in the 1980s and 1990s from local to global issues. The closing decades of the last millennium also saw expansions both in public interest and ecological input into the conservation of endangered species and the biodiversity of whole areas, in the control of disease in humans as well as many other species, and in the potential consequences of profound human-caused changes to the global environment.

unanswered questions

And yet, at the same time, many fundamental problems of ecology remain unanswered. To what extent does competition for food determine which species can coexist in a habitat? What role does disease play in the dynamics of populations? Why are there more species in the tropics than at the poles? What is the relationship between soil productivity and plant community structure? Why are some species more vulnerable to extinction than others? And so on. Of course, unanswered questions – if they are *focused* questions – are a symptom of the health not the weakness of any science. But ecology is not an easy science, and it has particular subtlety and complexity, in part because ecology is peculiarly confronted by 'uniqueness': millions of different species, countless billions of genetically distinct individuals, all living and interacting in a varied and ever-changing world. The beauty of ecology is that it challenges us to develop an understanding of very basic and apparent problems – in a way that recognizes the uniqueness and complexity of all aspects of nature – but seeks patterns and predictions within this complexity rather than being swamped by it.

Summarizing this brief historical overview, it is clear that ecologists try to do a number of different things. First and foremost ecology is a science, and ecologists therefore try to *explain* and *understand*. There are two different classes of explanation in biology: 'proximate' and 'ultimate'. For example, the present distribution and abundance of a particular species of bird may be 'explained' in terms of the physical environment that the bird tolerates, the food that it eats and the parasites and predators that attack it. This is a *proximate* explanation – an explanation in terms of what is going on 'here and now'. However, we can also ask how this bird has come to have these properties that now govern its life. This question has to be answered by an explanation in evolutionary terms; the *ultimate* explanation of the present distribution and abundance of this bird lies in the ecological experiences of its ancestors (see Chapter 2).

understanding, description, prediction and control

In order to understand something, of course, we must first have a description of whatever it is we wish to understand. Ecologists must therefore *describe* before they explain. On the other hand, the most valuable descriptions are those carried out with a particular problem or 'need for understanding' in mind. Undirected description, carried out merely for its own sake, is often found afterwards to have selected the wrong things and has little place in ecology – or any other science.

Ecologists also often try to *predict* what will happen to a population of organisms under a particular set of circumstances, and on the basis of these predictions to control, exploit or conserve the population. We try to minimize the effects of locust plagues by predicting when they are likely to occur and taking appropriate action. We try to exploit crops most effectively by predicting when conditions will be favorable to the crop and unfavorable to its enemies. We try to preserve rare species by predicting the conservation policy that will enable us to do so. Some prediction and control can be carried out without deep explanation or understanding: it is not difficult to predict that the destruction of a woodland will eliminate woodland birds. But insightful predictions, precise predictions and predictions of what will happen in unusual circumstances can be made only when we can also explain and understand what is going on.

This book is therefore about:

1 How ecological understanding is achieved.
2 What we do understand (but also what we do not understand).
3 How that understanding can help us predict, manage and control.

1.2 Scales, diversity and rigor

The rest of this chapter is about the two 'hows' above: how understanding is achieved, and how that understanding can help us predict, manage and control. Later in the chapter we illustrate three fundamental points about doing ecology by examining a limited number of examples in some detail (Section 1.3). But first we elaborate on the three points, namely:

● ecological phenomena occur at a variety of scales;
● ecological evidence comes from a variety of different sources;
● ecology relies on truly scientific evidence and the application of statistics.

1.2.1 Questions of scale

Ecology operates at a range of scales: time scales, spatial scales and 'biological' scales. It is important to appreciate the breadth of these and how they relate to one another.

the 'biological' scale

The living world is often said to comprise a biological hierarchy beginning with subcellular particles and continuing through cells, tissues and organs. Ecology then deals with the next three levels:

- *individual organisms*;
- *populations* (consisting of individuals of the same species);
- *communities* (consisting of a greater or lesser number of populations).

At the level of the organism, ecology deals with how individuals are affected by (and how they affect) their environment. At the level of the population, ecology deals with the presence or absence of particular species, with their abundance or rarity, and with the trends and fluctuations in their numbers. Community ecology then deals with the composition or structure of ecological communities.

We can also focus on the pathways followed by energy and matter as these move among living and non-living elements of a fourth category of organization:

- *ecosystems* (comprising the community together with its physical environment).

With this level of organization in mind, Likens (1992) would extend our preferred definition of ecology (Box 1.1) to include 'the interactions between organisms and the transformation and flux of energy and matter'. However, we take energy/matter transformations as being subsumed in the 'interactions' of our definition.

a range of spatial scales

Within the living world, there is no arena too small nor one so large that it does not have an ecology. Even the popular press talk increasingly about the 'global ecosystem' and there is no question that several ecological problems can only be examined at this very large scale. These include the relationships between ocean currents and fisheries, or between climate patterns and the distribution of deserts and tropical rain forests, or between elevated carbon dioxide in the atmosphere (from burning fossil fuels) and global climate change.

At the opposite extreme, an individual cell may be the stage on which two populations of pathogens compete with one another for the resources that the cell provides. At a slightly larger spatial scale, a termite's gut is the habitat for bacteria, protozoans and other species (Figure 1.2) – a community whose diversity is comparable to that of a tropical rain forest in terms of the richness of organisms living there, the variety of interactions in which they take part, and indeed the extent to which we remain ignorant about the species identity of many of the participants. Between these extremes, different ecologists, or the same ecologist at different times, may study the inhabitants of pools that form in small tree-holes, the temporary watering holes of the savannas, or the great lakes and oceans; others may examine the diversity of fleas on different species of birds, the diversity of birds in different sized patches of woodland, or the diversity of woodlands at different altitudes.

a range of time scales

To some extent related to this range of spatial scales, and to the levels in the biological hierarchy, ecologists also work on a variety of time scales. 'Ecological succession' – the successive and continuous colonization of a site by certain species populations, accompanied by the extinction of others – may be studied over a period from the deposition of a lump of sheep dung to its decomposition (a

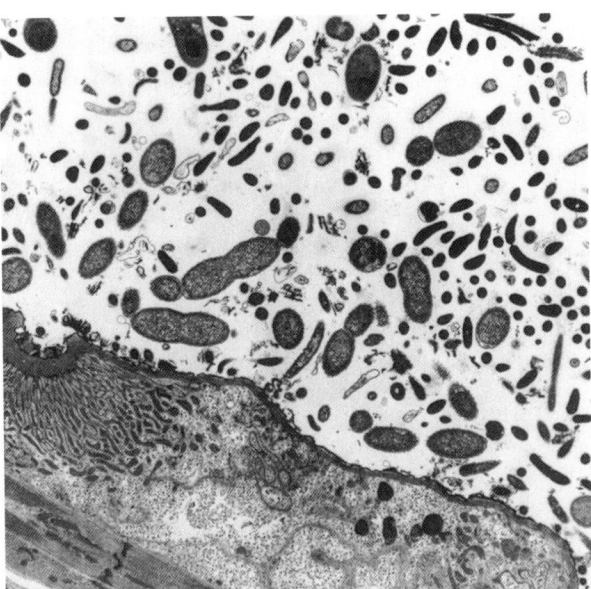

Figure 1.2

The diverse community of a termite's gut. Termites can break down lignin and cellulose from wood because of their mutualistic relationships (see Section 8.4.4) with a diversity of microbes that live in their guts.

matter of weeks), or from the change in climate at the end of the last ice age to the present day and beyond (around 14,000 years and still counting). Migration may be studied in butterflies over the course of days, or in the forest trees that are still (slowly) migrating into deglaciated areas following that last ice age.

Although it is undoubtedly the case that 'appropriate' time scales vary, it is also true that many ecological studies are not as long as they might be. Longer studies cost more and require greater dedication and stamina. An impatient scientific community, and the requirement for concrete evidence of activity for career progression, both put pressure on ecologists, and all scientists, to publish their work sooner rather than later. Why are long-term studies potentially of such value? The reduction over a few years in the numbers of a particular species of wild flower, or bird, or butterfly might be a cause for conservation concern – but one or more decades of study may be needed to be sure that the decline is more than just an expression of the random ups and downs of 'normal' population dynamics. Similarly, a 2-year rise in the abundance of a wild rodent followed by a 2-year fall might be part of a regular 'cycle' in abundance, crying out for an explanation. But ecologists could not be sure until perhaps 20 years of study has allowed them to record four or five repeats of such a cycle.

the need for long-term studies

This does not mean that all ecological studies need to last for 20 years – nor that every time an ecological study is extended the answer changes. But it does emphasize the great value to ecology of the small number of long-term investigations that have been carried out or are ongoing.

1.2.2 The diversity of ecological evidence

Ecological evidence comes from a variety of different sources. Ultimately, ecologists are interested in organisms in their natural environments (though for many organisms, the environment which is 'natural' for them now is itself manmade). Progress would be impossible, however, if ecological studies were limited to such natural environments. And, even in natural habitats, unnatural acts (experimental manipulations) are often necessary in the search for sound evidence.

observations and field experiments

Many ecological studies involve careful *observation* and monitoring, in the natural environment, of the changing abundance of one or more species over time, or over space, or both. In this way, ecologists may establish patterns; for example, that red grouse (birds shot for 'sport') exhibit regular cycles in abundance peaking every 4 or 5 years, or that vegetation can be mapped into a series of zones as we move across a landscape of sand dunes. But scientists do not stop at this point – the patterns require explanation. Careful analysis of the descriptive data may suggest some plausible explanations. But establishing what causes the patterns may well require *manipulative field experiments*: ridding the red grouse of intestinal worms, hypothesized to underlie the cycles, and checking if the cycles persist (they do not: Hudson et al., 1998), or treating experimental areas on sand dunes with fertilizer to see whether the changing pattern of vegetation itself reflects a changing pattern of soil productivity.

laboratory experiments

Perhaps less obviously, ecologists also often need to turn to laboratory systems and even mathematical models. These have played a crucial role in the development of ecology, and they are certain to continue to do so. Field experiments are almost inevitably costly and difficult to carry out. Moreover, even if time and expense were not issues, natural field systems may simply be too complex to allow us to tease apart the consequences of the many different processes that may be going on. Are the intestinal worms actually capable of having an effect on reproduction or mortality of individual grouse? Which of the many species of sand dune plants are, in themselves, sensitive to changing levels of soil productivity and which are relatively insensitive? *Controlled, laboratory experiments* are often the best way to provide answers to specific questions that are key parts of an overall explanation of the complex situation in the field.

simple laboratory systems . . .

Of course, the complexity of natural ecological communities may simply make it inappropriate for an ecologist to dive straight into them in search of understanding. We may wish to explain the structure and dynamics of a particular community of 20 animal and plant species comprising various competitors, predators, parasites and so on (relatively speaking, a community of remarkable simplicity). But we have little hope of doing so unless we already have some basic understanding of even simpler communities of just one predator and one prey species, or two competitors, or (especially ambitious) two competitors that also share a common predator. For this, it is usually most appropriate to construct, for our own convenience, *simple laboratory systems* that can act as benchmarks or jumping-off points in our search for understanding.

. . . and mathematical models

What is more, you have only to ask anyone who has tried to rear caterpillar eggs, or take a cohort of shrub cuttings through to maturity, to discover that even the simplest ecological communities may not be easy to maintain or keep free of unwanted pathogens, predators or competitors. Nor is it necessarily possible to construct precisely the particular, simple, artificial community that interests you; nor to subject it to precisely the conditions or the perturbation of interest. In many cases, therefore, there is much to be gained from the analysis of *mathematical models* of ecological communities: constructed and manipulated according to the ecologist's design.

On the other hand, although a major aim of science is to simplify, and thereby make it easier to understand the complexity of the real world, ultimately it is the real world that we are interested in. The worth of models and simple laboratory experiments must always be judged in terms of the light they throw on the working of more natural systems. They are a means to an end – never an end in themselves. Like all scientists, ecologists need to 'seek simplicity, but distrust it' (Whitehead, 1953).

1.2.3 Statistics and scientific rigor

For a scientist to take offence at some popular phrase or saying is to invite accusations of a lack of a sense of humor. But it is difficult to remain calm when phrases like 'There are lies, damn lies and statistics' or 'You can prove anything with statistics' are used, by those who should know better, to justify continuing to believe what they wish to believe, whatever the evidence to the contrary. There is no doubt that statistics are sometimes *mis*-used to derive dubious conclusions from sets of data that actually suggest either something quite different or perhaps nothing at all. But these are not grounds for mistrusting statistics in general – rather for ensuring that people are educated in at least the principles of scientific evidence and its statistical analysis, so as to protect them from those who may seek to manipulate their opinions.

In fact, not only is it not true that you can prove anything with statistics, the contrary is the case: you cannot *prove* anything with statistics – that is not what statistics are for. Statistical analysis is essential, however, for attaching a level of confidence to conclusions that can be drawn; and ecology, like all science, is a search not for statements that have been 'proved to be true' but for conclusions in which we can be confident.

ecology: a search for conclusions in which we can be confident

Indeed, what distinguishes science from other activities – what makes science 'rigorous' – is that it is based not on statements that are simply assertions, but that it is based (i) on conclusions that are the results of investigations (as we have seen, of a wide variety of types) carried out with the express purpose of deriving those conclusions; and (b) even more important, on conclusions to which a level of confidence can be attached, measured on an agreed scale. These points are elaborated in Boxes 1.2 and 1.3.

Statistical analyses are carried out after data have been collected, and they help us to interpret those data. There is no really good science, however, without forethought. Ecologists, like all scientists, must know what they are doing, and why they are doing it, *while* they are doing it. This is entirely obvious at a general level: nobody expects ecologists to be going about their work in some kind of daze. But it is perhaps not so obvious that ecologists should know how they are going to analyze their data, statistically, not only after they have collected it, not only while they are collecting it, but even before they begin to collect it. Ecologists must plan, so as to be confident that they have collected the right kind of data, and a sufficient amount of data, to address the questions they hope to answer.

ecologists must think ahead

Ecologists typically seek to draw conclusions about groups of organisms overall: what is the birth rate of the bears in Yellowstone Park? What is the density of weeds in a wheat field? What is the rate of nitrogen uptake of tree saplings in a nursery? In doing so, we can only very rarely examine every individual in a group, or in the entire sampling area, and we must therefore rely on what we hope will be a *representative* sample from the group or habitat. Indeed, even if we examined a whole group (we might examine every fish in a small pond, say),

ecology relies on representative samples

1.2 *Quantitative aspects*

Interpreting probabilities

P-values

The term that is most often used, at the end of a statistical test, to measure the strength of conclusions being drawn is a *P*-value, or probability level. It is important to understand what *P*-values are. Imagine we are interested in establishing whether high abundances of a pest insect in summer are associated with high temperatures the previous spring, and imagine that the data we have to address this question consist of summer insect abundances and mean spring temperatures for each of a number of years. We may reasonably hope that statistical analysis of our data will allow us either to conclude, with a stated degree of confidence, that there is an association, or to conclude that there are no grounds for believing there to be an association (Figure 1.3).

Null hypotheses

To carry out a statistical test we first need a *null hypothesis*, which simply means, in this case, that there is *no* association: that is, no association between insect abundance and temperature. The statistical test (stated simply) then generates a probability (a *P*-value) of getting a data set like ours if the null hypothesis is correct.

Suppose the data were like those in Figure 1.3a. The probability generated by a test of association on these data is $P = 0.5$ (equivalently 50%). This means that, if the null hypothesis really was correct (no association), then 50% of studies like ours should generate just such a data set, or one even further from the null hypothesis. So, if there was no association, there would be nothing very remarkable in this data set, and we could have no confidence in any claim that there *was* an association.

Suppose, however, that the data were like those in Figure 1.3b, where the *P*-value generated is $P = 0.001$ (0.1%). This would mean that such a data set (or one even further from the null hypothesis) could be expected in only 0.1% of similar studies if there was really no association. In other words, either something

very improbable has occurred, or there *was* an association between insect abundance and spring temperature. Thus, since by definition we do not expect highly improbable events to occur, we can have a high degree of confidence in the claim that there *was* an association between abundance and temperature.

Significance testing

Both 50% and 0.01%, though, make things easy for us. Where, between the two, do we draw the line? There is no objective answer to this, and so scientists and statisticians have established a convention in *significance testing*, which says that if *P* is less than 0.05 (5%), written $P < 0.05$ (e.g. Figure 1.3d), then results are described as statistically significant and confidence can be placed in the effect being examined (in our case, the association between abundance and temperature), whereas if $P > 0.05$, then there is no statistical foundation for claiming the effect exists (e.g. Figure 1.3c). A further elaboration of the convention often describes results with $P < 0.01$ as 'highly significant'.

'Insignificant' results?

Naturally, some effects are strong (for example, there is a powerful association between people's weight and their height) and others are weak (the association between people's weight and their risk of heart disease is real but weak, since weight is only one of many important factors). More data are needed to establish support for a weak effect than for a strong one. A rather obvious but very important conclusion follows from this: a *P*-value in an ecological study of greater than 0.05 (lack of statistical significance) may mean one of two things:

1. There really is no effect of ecological importance.
2. The data are simply not good enough, or there are not enough of them, to support the effect even though it exists, possibly because the effect itself is real but weak, and extensive data are therefore needed but have not been collected.

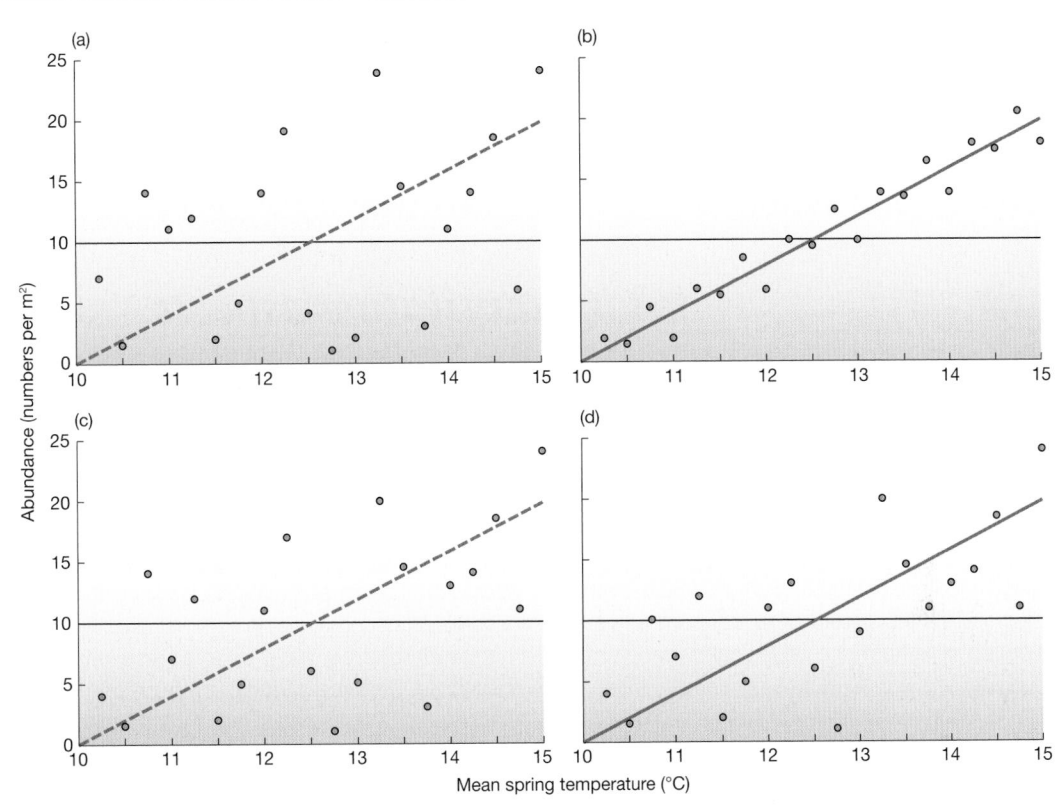

Figure 1.3
The results from four hypothetical studies of the relationship between insect pest abundance in summer and mean temperature the previous spring. In each case, the points are the data actually collected. Horizontal lines represent the *null hypothesis* – that there is no association between abundance and temperature, and thus the best estimate of expected insect abundance, irrespective of spring temperature, is the mean insect abundance overall. The second line is the *line of best fit* to the data, which in each case offers some suggestion that abundance rises as temperature rises. However, whether we can be confident in concluding that abundance does rise with temperature depends, as explained in the text, on statistical tests applied to the data sets. (a) The suggestion of a relationship is weak ($P = 0.5$). There are no good grounds for concluding that the true relationship differs from that supposed by the null hypothesis and no grounds for concluding that abundance is related to temperature. (b) The relationship is strong ($P = 0.001$) and we can be confident in concluding that abundance increases with temperature. (c) The results are suggestive ($P = 0.1$) but it would not be safe to conclude from them that abundance rises with temperature. (d) The results are not vastly different from those in (c) but are powerful enough ($P = 0.04$, i.e. $P < 0.05$) for the conclusion that abundance rises with temperature to be considered safe.

Quoting *P*-values
Furthermore, applying the convention strictly and dogmatically means that when $P = 0.06$ the conclusion should be 'no effect has been established', whereas when $P = 0.04$ the conclusion is 'there is a significant effect'. Yet very little difference in the data is required to move a *P*-value from 0.04 to 0.06. It is therefore far better to quote exact *P*-values, especially when they exceed 0.05, and think of conclusions in terms of shades of gray rather than the black and white of 'proven effect' and 'no effect'. In particular, *P*-values close to, but not less than, 0.05 suggest that something seems to be going on; they indicate, more than anything else, that more data need to be collected so that our confidence in conclusions can be more clearly established.

Throughout this book, then, studies of a wide range of types are described, and their results often have P-values attached to them. Of course, as this is a textbook, the studies have been selected because their results *are* significant. Nonetheless, it is important to bear in mind that the repeated statements $P < 0.05$ and $P < 0.01$ mean that these are studies where: (i) sufficient data have been collected to establish a conclusion in which we can be confident; (ii) that confidence has been established by agreed means (statistical testing); and (iii) confidence is being measured on an agreed and interpretable scale.

1.3 *Quantitative aspects*

1.3 QUANTITATIVE ASPECTS

Attaching confidence to results

Standard errors and confidence intervals

Following Box 1.2, another way in which the significance of results, and confidence in them, is assessed is through reference to standard errors. Again, simply stated, statistical tests often allow standard errors to be attached either to mean values calculated from a set of observations or to slopes of lines like those in Figure 1.3. Such mean values or slopes can, at best, only ever be estimates of the 'true' mean value or true slope, because they are calculated from data that are only a sample of all the imaginable items of data that could be collected. The standard error, then, sets a band around the estimated mean (or slope, etc.) within which the true mean can be expected to lie with a given, stated probability. In particular, there is a 95% probability that the true mean lies within roughly two standard errors (2 SE) of the estimated mean; we call this the *95% confidence interval*.

Hence, when we have, say, two sets of observations, each with its own mean value (for instance, the number of seeds produced by plants from two sites, Figure 1.4) the standard errors allow us to assess whether the means are significantly different from one another, statistically. Roughly speaking, if each mean is more than two standard errors from the other mean, then the difference between them is statistically significant with $P < 0.05$. Thus, for the study illustrated in Figure 1.4a, it would not be safe to conclude that plants from the two sites differed in their seed production. However, for the similar study illustrated in Figure 1.4b, the means are roughly the same as they were in the first study and are roughly as far apart, but the standard errors

Figure 1.4

The results of two hypothetical studies in which the seed production of plants from two different sites was compared. In all cases, the heights of the bars represent the mean seed production of the sample of plants examined, and the lines crossing those means extend 1 SE above and below them. (a) Although the means differ, the standard errors are relatively large and it would not be safe to conclude that seed production differed between the sites ($P = 0.4$). (b) The differences between the means are very similar to those in (a), but the standard errors are much smaller, and it can be concluded with confidence that plants from the two sites differed in their seed production ($P < 0.05$).

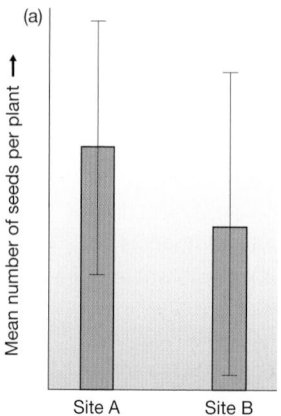
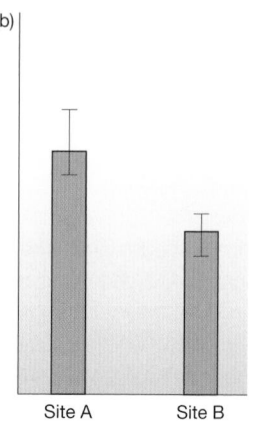

are smaller. Hence, the difference between the means is significant ($P < 0.05$), and we can conclude with confidence that plants from the two sites differed.

When are standard errors small?
Note that the large standard errors in the first study, and hence the lack of statistical significance, could have been due to data that were, for whatever reason, more variable; but they may also have been due to sampling fewer plants in the first study than the second. Standard errors are smaller, and statistical significance is easier to achieve, *both* when data are more consistent (less variable) *and* when there are more data.

we are likely to want to draw general conclusions from it: we might hope that the fish in 'our' pond can tell us something about fish of that species in ponds of that type, generally. In short, ecology relies on obtaining *estimates* from representative samples. This is elaborated in Box 1.4.

1.4 *Quantitative aspects*

Estimation: sampling, accuracy and precision

The discussion in Boxes 1.2 and 1.3 about when standard errors will be small or large, or when our confidence in conclusions will be strong or weak, not only has implications for the interpretation of data after they have been collected, but also carries a general message about planning the collection of data. In undertaking a sampling program to collect data, the aim is to satisfy a number of criteria:

1   That the estimate should be accurate or unbiased: that is, neither systematically too high nor too low as a result of some flaw in the program.
2   That the estimate should have as narrow confidence limits (be as precise) as possible.
3   That the time, money and human effort invested in the program should be used as effectively as possible (because these are always limited).

Random and stratified random sampling
To understand these criteria, consider another hypothetical example. Suppose that we are interested in the density of a particular weed (say wild oat) in a wheat field. To prevent bias, it is necessary to ensure that each part of the field has an equal chance of being selected for sampling. Sampling units should therefore be selected at random. We might, for example, divide the field into a measured grid, pick points on the grid at random, and count the wild oat plants within a 50 cm radius of the selected grid point. This unbiased method can be contrasted with a plan to sample only weeds from between the rows of wheat plants, giving too high an estimate, or within the rows, giving too low an estimate (Figure 1.5a).

Remember, however, that random samples are not taken as an end in themselves, but because random sampling is a means to truly representative sampling. Thus, randomly chosen sampling units may end up being concentrated, by chance, in a particular part of the field that, unknown to us, is not representative of the field as a whole. It is often preferable, therefore, to undertake *stratified random sampling* in which, in this case, the field is divided up into a number of equal-sized parts (*strata*) and a random sample taken from each. This way, the coverage of the whole field is more even, without our having introduced bias by selecting particular spots for sampling.

Separating subgroups and directing effort
Suppose now, though, that half the field is on a slope facing southeast and the other half on a slope facing

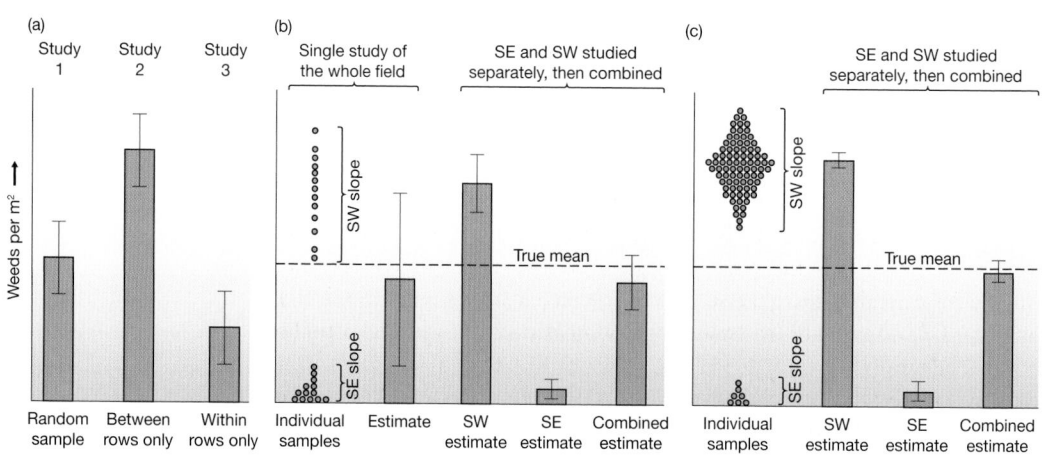

Figure 1.5

The results of hypothetical programs to estimate weed density in a wheat field. (a) The three studies have equal precision (95% confidence intervals) but only the first (from a random sample) is accurate. (b) In the first study, individual samples from different parts of the field (southeast and southwest) fall into two groups (left); thus, the estimate, although accurate, is not precise (right). In the second study, separate estimates for southeast and southwest are both accurate and precise – as is the estimate for the whole field obtained by combining them. (c) Following on from (b), most sampling effort is directed to the southwest, reducing the confidence interval there, but with little effect on the confidence interval for the southeast. The overall interval is therefore reduced: precision has been improved.

southwest, and that we know that aspect (which way the slope is facing) can affect weed density. Random sampling (or stratified random sampling) ought still to provide an unbiased estimate of density for the field as a whole, but for a given investment in effort, the confidence interval for the estimate will be unnecessarily high. To see why, consider Figure 1.5b. The individual values from samples fall into two groups a substantial distance apart on the density scale: high from the southwest slope; low (mostly zero) from the southeast slope. The estimated mean density is close to the true mean (it is accurate), but the variation among samples leads to a very large confidence interval (it is not very precise).

If, however, we acknowledge the difference between the two slopes and treat them separately from the outset, then we obtain means for each that have much smaller confidence intervals. What is more, if we average those means and combine their confidence intervals to obtain an estimate for the field as a whole, then that interval too is much smaller than previously (Figure 1.5b).

But has our effort been directed sensibly, with equal numbers of samples from the southwest slope, where there are lots of weeds, and the southeast slope, where there are virtually none? The answer is no. Remember that narrow confidence intervals arise from a combination of a large number of data points *and* little intrinsic variability (see Box 1.3). Thus, if our efforts had been directed mostly at sampling the southwest slope, the increased amount of data would have noticeably decreased the confidence interval (Figure 1.5c), whereas less sampling of the southeast slope would have made very little difference to that confidence interval because of the low intrinsic variability there. Careful direction of a sampling program can clearly increase overall precision for a given investment in effort. And generally, sampling programs should, where possible, identify biologically distinct subgroups (males and females, old and young, etc.) and treat them separately, but sample at random within subgroups.

1.3 Ecology in practice

In previous sections we have established in a general way how ecological understanding can be achieved, and how that understanding can be used to help us predict, manage and control ecological systems. However, the practice of ecology is easier said than done. To discover the real problems faced by ecologists and how they try to solve them, it is best to consider some real research programs in a little detail. While reading the following examples you should focus on how they illuminate our three main points: (i) ecological phenomena occur at a variety of scales; (ii) ecological evidence comes from a variety of different sources; and (iii) ecology relies on truly scientific evidence and the application of statistics. Every other chapter in this book will contain descriptions of similar studies, but in the context of a systematic survey of the driving forces in ecology (Chapters 2–11) or of the application of this knowledge to solve applied problems (Chapters 12–14). For now, we content ourselves with seeking an appreciation of how four research teams have gone about their business.

1.3.1 Brown trout in New Zealand: effects on individuals, populations, communities and ecosystems

It is rare for a study to encompass more than one or two of the four levels in the biological hierarchy (individuals, populations, communities, ecosystems). For most of the 20th century, physiological and behavioral ecologists (studying individuals), population dynamicists, and community and ecosystem ecologists tended to follow separate paths, asking different questions in different ways. However, there can be little doubt that, ultimately, our understanding will be enhanced considerably when the links between all these levels are made clear – a point that can be illustrated by examining the impact of the introduction of an exotic fish to streams in New Zealand.

Prized for the challenge they provide to anglers, brown trout (*Salmo trutta*) have been transported from their native Europe all around the world; they were introduced to New Zealand beginning in 1867, and self-sustaining populations are now found in many streams, rivers and lakes there. Until quite recently, few people cared about native New Zealand fish or invertebrates, so little information is available on changes in the ecology of native species after the introduction of trout. However, trout have colonized some streams but not others. We can therefore learn a lot by comparing the current ecology of streams containing trout with those occupied by non-migratory native fish in the genus *Galaxias* (Figure 1.6).

Mayfly nymphs of various species commonly graze microscopic algae growing on the beds of New Zealand streams, but there are some striking differences in their activity rhythms depending on whether they are in *Galaxias* or trout streams. In one experiment, nymphs collected from a trout stream and placed in small artificial laboratory channels were less active during the day than the night, whereas those collected from a *Galaxias* stream were active both day and night (Figure 1.7a). In another experiment, with another mayfly species, records were made of individuals visible in daylight on the surface of cobbles in artificial channels

the individual level – consequences for invertebrate feeding behaviour

(a)

(b)

Figure 1.6

(a) A brown trout and (b) a *Galaxias* fish in a New Zealand stream – is the native *Galaxias* hiding from the introduced predator?

placed in a real stream. Three treatments were each replicated three times – no fish in the channels, trout present and *Galaxias* present. Daytime activity was significantly reduced in the presence of either fish species, but to a greater extent when trout were present (Figure 1.7b).

These differences in activity pattern reflect the fact that trout rely principally on vision to capture prey, whereas *Galaxias* rely on mechanical cues. Thus,

Figure 1.7

(a) Mean number (± SE) of *Nesameletus ornatus* mayfly nymphs collected either from a trout stream or a *Galaxias* stream that were recorded by means of video as visible on the substrate surface in laboratory stream channels during the day and night (in the absence of fish). Mayflies from the trout stream are more nocturnal than their counterparts from the *Galaxias* stream. (b) Mean number (± SE) of *Deleatidium* mayfly nymphs observed on the upper surfaces of cobbles during late afternoon in channels (placed in a real stream) containing no fish, trout or *Galaxias*. The presence of a fish discourages mayflies from emerging during the day, but trout have a much stronger effect than *Galaxias*. In all cases, the standard errors were sufficiently small for differences to be statistically significant ($P < 0.05$).

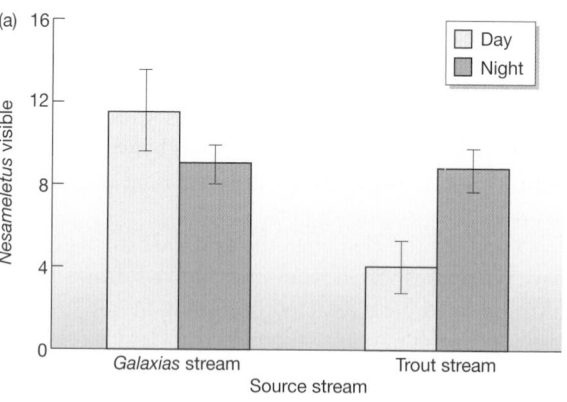

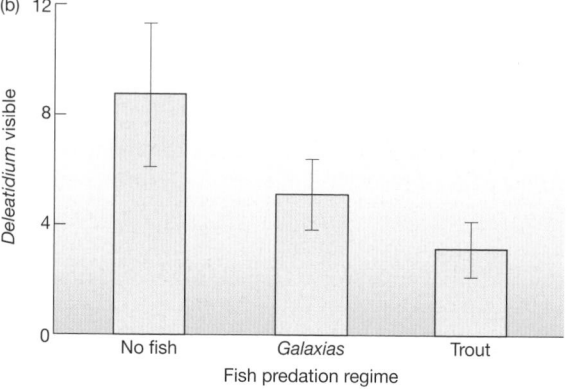

(a) AFTER MCINTOSH & TOWNSEND, 1994; (b) AFTER MCINTOSH & TOWNSEND, 1996

invertebrates in a trout stream are considerably more at risk of predation during daylight hours. And these conclusions are all the more robust because they derive both from the readily controlled conditions of a laboratory experiment and from the more realistic, but more variable, circumstances of a field experiment.

In the Taieri River in New Zealand, 198 sites were selected in a stratified manner by choosing streams of similar dimensions at random in each of three tributaries from each of eight subcatchments of the river. Care was taken not to succumb to the temptation of choosing sites with easy access (near roads or bridges) in case this biased the results. The sites were classified as containing: (i) no fish; (ii) *Galaxias* only; (iii) trout only; or (iv) both *Galaxias* and trout. At every site a variety of physical variables were measured (stream depth, flow velocity, phosphorus concentration in the stream water, percentage of the streambed composed of gravel, etc.). A statistical procedure called multiple discriminant functions analysis was then used to determine which physical variables, if any, distinguished one type of site from another. Means and standard errors of these key environmental variables are presented in Table 1.1.

Trout occurred almost invariably below waterfalls that were large enough to prevent their upstream migration; they tended to occur at low elevations because sites without waterfalls downstream tended to be at lower elevation. Sites containing *Galaxias* (or with no fish) were always upstream of one or several large waterfalls. The few sites that contained both trout and *Galaxias* were below waterfalls, at intermediate elevations, and in sites with cobble beds; the unstable nature of these beds may have promoted coexistence (at low densities) of the two species. This descriptive study at the population level therefore takes advantage of a 'natural' experiment (streams that happen to contain trout or *Galaxias*) to determine the effect of the introduction of trout. The most probable reason for the restriction of populations of *Galaxias* to sites upstream of waterfalls, which cannot be climbed by trout, is direct predation by trout on the native fish below the waterfalls (a single small trout in a laboratory aquarium has been recorded consuming 135 *Galaxias* fry in a day).

the population level – brown trout and the distribution of native fish

Table 1.1

Means and, in brackets, standard errors for important discriminating variables for fish assemblage classes in 198 sites in the Taieri River. In particular, compare the '*Galaxias* only' and 'brown trout only' classes. *Galaxias* are found on their own if there are large waterfalls downstream of the site (and at relatively high elevations where the stream bed has an intermediate representation of cobbles). Brown trout, on the other hand, generally occur where there are no downstream waterfalls (at slightly lower elevations and with a bed composition similar to the *Galaxias* class).

| | | VARIABLES | | |
SITE TYPE	NUMBER OF SITES	NUMBER OF WATERFALLS DOWNSTREAM	ELEVATION (m ABOVE SEA LEVEL)	% OF THE BED COMPOSED OF COBBLES
Brown trout only	71	0.42 (0.05)	324 (28)	18.9 (2.1)
Galaxias only	64	12.3 (2.05)	567 (29)	22.1 (2.8)
No fish	54	4.37 (0.64)	339 (31)	15.8 (2.3)
Trout + *Galaxias*	9	0.0 (0)	481 (53)	46.7 (8.5)

AFTER FLECKER & TOWNSEND, 1994

Figure 1.8

(a) Total invertebrate biomass and (b) algal biomass (chlorophyll *a*) (± SE) for an experiment performed in summer in a small New Zealand stream. In experimental replicates where trout are present, grazing invertebrates are rarer and graze less; thus, algal biomass is highest. G, *Galaxias* present; N, no fish; T, trout present.

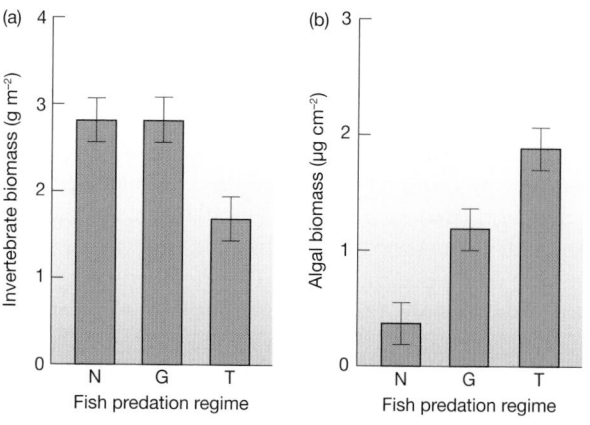

the community – brown trout cause a cascade of effects

That an exotic predator such as trout has direct effects on *Galaxias* distribution or mayfly behavior is not surprising. However, we can ask whether these changes have community consequences that cascade through to other species. In the relatively species-poor stream communities in the south of New Zealand, the plants are mainly algae that grow on the streambed. These are grazed by various insect larvae, which in turn are prey to predatory invertebrates and fish. As we have seen, trout have replaced *Galaxias* in many of these streams. An experiment involving artificial flow-through channels (several meters long, with mesh ends to prevent escape of fish but to allow invertebrates to colonize naturally) placed into a real stream was used to determine whether trout affect the stream food web differently from the displaced *Galaxias*. Three treatments were established (no fish, *Galaxias* present, and trout present, at naturally occurring densities) in each of several randomized blocks located in a stretch of a stream with each block separated by more than 50 m. Algae and invertebrates were allowed to colonize for 12 days before introducing the fish. After a further 12 days, invertebrates and algae were sampled (Figure 1.8).

A significant effect of trout reducing invertebrate biomass was evident ($P = 0.026$), but the presence of *Galaxias* did not depress invertebrate biomass from the no-fish control. Algal biomass, perhaps not surprisingly then, achieved its highest values in the trout treatment ($P = 0.02$). It is clear that trout do have a more pronounced effect than *Galaxias* on invertebrate grazers and, thus, on algal biomass. The indirect effect of trout on algae occurs partly through a reduction in invertebrate density, but also because trout restrict the grazing behavior of the invertebrates that are present (see Figure 1.7b).

the ecosystem – trout and energy flow

The sequence of studies above provided the impetus for a detailed energetics investigation of two neighboring tributaries of the Taieri River (with very similar physicochemical conditions), one being occupied by just trout and the other (because of a waterfall downstream) containing only *Galaxias*. No other fish were present in either stream. The hypothesis under examination was that the rate at which radiation energy was captured through photosynthesis by the algae would be greater in the trout stream because there would be fewer invertebrates and thus a lower rate of consumption of algae. Indeed, annual net 'primary' production (the rate of production of plant, in this case algal, biomass) was six times greater in the trout stream than in the *Galaxias* stream (Figure 1.9).

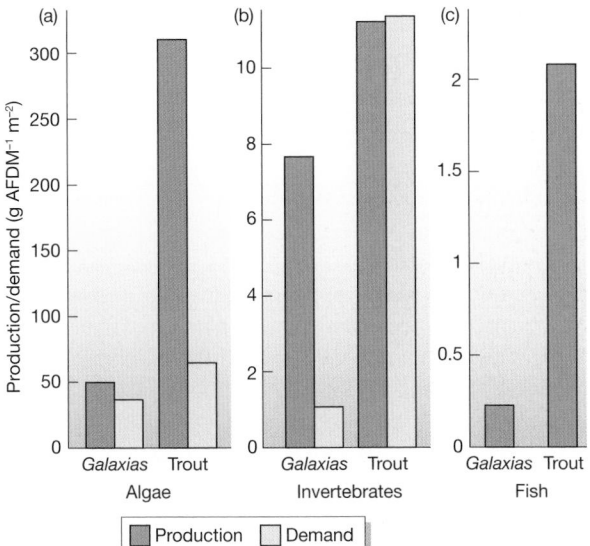

Figure 1.9

Annual estimates for 'production' of biomass at one trophic level, and the 'demand' for that biomass (the amount consumed) at the next trophic level, for (a) primary producers (algae), (b) invertebrates (which consume algae), and (c) fish (which consume invertebrates). Estimates are for a trout stream and a *Galaxias* stream. In the former, production at all trophic levels is higher, but because the trout consume essentially all of the annual invertebrate production (b), the invertebrates consume only 21% of primary production (a). In the *Galaxias* stream, these fish consume only 18% of invertebrate production, 'allowing' the invertebrates to consume the majority (75%) of annual primary production.

Moreover, the primary consumers (invertebrates that eat algae) produced new biomass in the trout stream at about 1.5 times the rate in the *Galaxias* stream, while trout themselves produced new biomass at roughly nine times the rate that *Galaxias* do (Figure 1.9).

Thus, the algae, invertebrates and fish are all 'more productive' in the trout stream than in the *Galaxias* stream; but *Galaxias* consume only about 18% of available prey production each year (compared to virtually 100% consumption by trout); while the grazing invertebrates consume about 75% of primary production in the *Galaxias* stream (compared to only about 21% in the trout stream) (Figure 1.9). Thus, the initial hypothesis appears to be confirmed: it is strong control by trout of the invertebrates that releases algae to produce and accumulate biomass at a fast rate.

A further ecosystem consequence ensues: in the trout stream, the higher primary production is associated with a faster rate of uptake by algae of plant nutrients (nitrate, ammonium, phosphate) from the flowing stream water (Simon et al., 2004).

This series of studies, therefore, illustrates some of the variety of ways in which ecological investigations may be pursued, and both the range of levels in the biological hierarchy that ecology spans and the way in which studies at different levels may complement one another. While it is necessary to be cautious when interpreting the results of an unreplicated study (only one trout and one *Galaxias* stream in the 'ecosystem study'), the conclusion that a trophic cascade is responsible for the patterns observed at the ecosystem level can be made with some confidence because of the variety of other corroborative studies conducted at the individual, population and community levels. Although brown trout are exotic invaders in New Zealand, and they have far-reaching effects on the ecology of native ecosystems, they are now considered a valuable part of the fauna, particularly by anglers, and generate millions of dollars for the nation. Many other invaders have dramatic negative economic impacts (Box 1.5).

AFTER HURYN, 1998

1.5 Topical ECOncerns

Invasions and homogenization of the biota: does it matter?

A recent analysis concluded that tens of thousands of invading exotic species in the United States cause economic losses totaling $137 billion each year (Pimentel et al., 2000). Table 1.2 breaks down the total into a variety of taxonomic groups.

Let us consider a few invaders with particularly dramatic consequences. The yellow star thistle (*Centaurea solstitalis*) now dominates more than 4 million hectares in California, resulting in the total loss of once productive grassland. Rats are estimated to destroy $19 billion of stored grains nationwide per year, as well as causing fires (by gnawing electric wires), polluting foodstuffs, spreading diseases and preying on native species. Introduced carp reduce water quality by increasing turbidity, while 44 native fish are threatened or endangered by fish invaders. The red fire ant (*Solenopsis invicta*) kills poultry, lizards, snakes and ground-nesting birds; in Texas alone, its estimated damage to livestock, wildlife and public health is put at about $300 million per year, and a further $200 million is spent on control. The zebra mussel (*Dreissena polymorpha*), which arrived in Michigan's Lake St. Clair in ballast water released from ships from Europe, has reached most aquatic

habitats in the eastern United States, and is expected to spread nationwide in the next 20 years. The large populations that develop threaten native mussels and other fauna, not only by reducing food and oxygen availability but by physically smothering them. The mussels also invade and clog water intake pipes, so that millions of dollars need to be spent clearing them from water filtration and hydroelectric generating plants. Overall, pests of crop plants, including weeds, insects and pathogens, engender the biggest economic costs. However, imported human disease organisms, particularly HIV and influenza viruses, cost $6.5 billion to treat and result in 40,000 deaths per year. (See Pimentel et al., 2000, for further details and references.)

Globalization has been the prevalent economic ideology in recent times. Globalization of the biota, in which successful invaders are moved around the world, often driving local species extinct, can be expected to lead to a general homogenization of the world's biota. [Lövei (1997) has colorfully referred to this as 'McDonaldization' of the biosphere.] Does biotic homogenization matter? Why?

Table 1.2

Estimated annual costs (billions of dollars) associated with invaders in the United States.

TYPE OF ORGANISM	NUMBER OF INVADERS	MAJOR CULPRITS	LOSS AND DAMAGE	CONTROL COSTS	TOTAL COSTS
Plants	5,000	Crop weeds	24.4	9.7	34.1
Mammals	20	Rats and cats	37.2	NA	37.2
Birds	97	Pigeons	1.9	NA	1.9
Reptiles and amphibians	53	Brown tree snake	0.001	0.005	10.006
Fishes	138	Grass carp, etc.	1.0	NA	1.0
Arthropods	4,500	Crop pests	17.6	2.4	20.0
Mollusks	88	Asian clams	1.2	0.1	1.3
Microbes (pathogens)	>20,000	Crop pathogens	32.1	9.1	41.2

NA, not available.
AFTER PIMENTEL ET AL., 2000

Yellow star thistle, *Centaurea solstitialis*.

Red fire ants, *Solenopsis*.

Zebra mussels, *Dreissena polymorpha*.

1.3.2 Successions on old fields in Minnesota: a study in time and space

'Ecological succession' is a concept that must be familiar to many who have simply taken a walk in open country – the idea that a newly created habitat, or one in which a disturbance has created an opening, will be inhabited, in turn, by a variety of species appearing and disappearing in some recognizably repeatable sequence. Widespread familiarity with the idea, however, does not mean that we understand fully the processes that drive or fine-tune successions; yet developing such understanding is important not just because succession is one of the fundamental forces structuring ecological communities, but also because human disturbance of natural communities has become ever more frequent and profound. We need to know how communities may respond to, and hopefully recover from, such disturbance, and how we may aid that recovery.

One particular focus for the study of succession has been the old agricultural fields of the eastern USA, abandoned as farmers moved west in search of 'fresh fields and pastures new'. One such site is now the Cedar Creek Natural History Area, roughly 50 km north of Minneapolis, Minnesota. The area was first settled by Europeans in

1856 and was initially subject to logging. Clearing for cultivation then began about 1885, and land was first cultivated between 1900 and 1910. Now there are agricultural fields that are still under cultivation and others that have been abandoned at various times since the mid-1920s. Cultivation led to depletion of nitrogen from soils that already were naturally poor in this important plant nutrient.

the use of natural experiments . . .

In the first place, studies at Cedar Creek illustrate the value of 'natural experiments'. To understand the successional sequence of plants that occur in fields in the years following abandonment we *could* plan an artificial manipulation, under our control, in which a number of fields currently under cultivation were 'forcibly' abandoned and the communities in them sampled repeatedly into the future. (We would need a number of fields because any single field might be atypical, whereas several would allow us to calculate mean values for, say, 'number of new species per year', and place confidence intervals around those means.) But the results of this experiment would take decades to accumulate. The natural experiment alternative, therefore, was to use the fact that records already exist of when many of the old fields were abandoned. This is what Tilman and his team did. Thus, Figure 1.10 illustrates data from a group of 22 old fields surveyed in 1983, having been abandoned at various times between 1927 and 1982 (i.e. between 1 and 56 years previously). Interpreted cautiously, these can be treated as 22 'snapshots' of the continuous process of succession in old fields at Cedar Creek in general, even though each field was itself only surveyed once.

A number of the shifting balances during succession are clear from the figure as statistically significant trends. Over the 56 years, the cover of 'invader' species (mostly agricultural weeds) decreased (Figure 1.10a) while the cover of species from nearby prairies increased (Figure 1.10b): the natives reclaimed their land. Of more general applicability, the cover of annual species decreased over time, while the cover of perennial species increased (Figure 1.10c, d). Annual species (those that complete a whole generation from seed to adult through to seeds again within a year) tend to be good at increasing in abundance rapidly in relatively empty habitats (the early stages of succession); whereas perennials (those that live for several or many years and may not reproduce in their early years) are slower to establish but more persistent once they do.

. . . in generating correlations

On the other hand, natural experiments like this, while frequently suggestive and stimulating (and too good an opportunity to miss), usually only generate *correlations*. They may therefore fail to establish what actually causes the observed patterns. In the present case, we can see the problem by noting, first, that field age is itself strongly correlated with nitrogen concentration in the soil – perhaps the single most important plant nutrient (Figure 1.10e). The question therefore arises: are the correlations in Figure 1.10a–d the result of an effect of field age itself? Or is the causal agent nitrogen, with which age is correlated?

artificial experiments: the search for causation

Manipulative field experiments can be used to help support – or refute – what so far is no more than a plausible explanation based on correlation. It seems to follow from the proposed explanation (time matters) that nitrogen itself has little role to play in driving these successions, and that manipulating nitrogen should do little to alter the species sequences that these fields have followed. To test this, Tilman's team selected a pair of fields (one abandoned for 46 years and the other for 14 years) and, over a 10-year period starting in 1982, subjected six replicate 4 m × 4 m plots in each field to one of two treatments: nitrogen added at rates of either 1 or 17 g m^{-2} yr^{-1} (Inouye & Tilman, 1995). Two questions in particular were being asked.

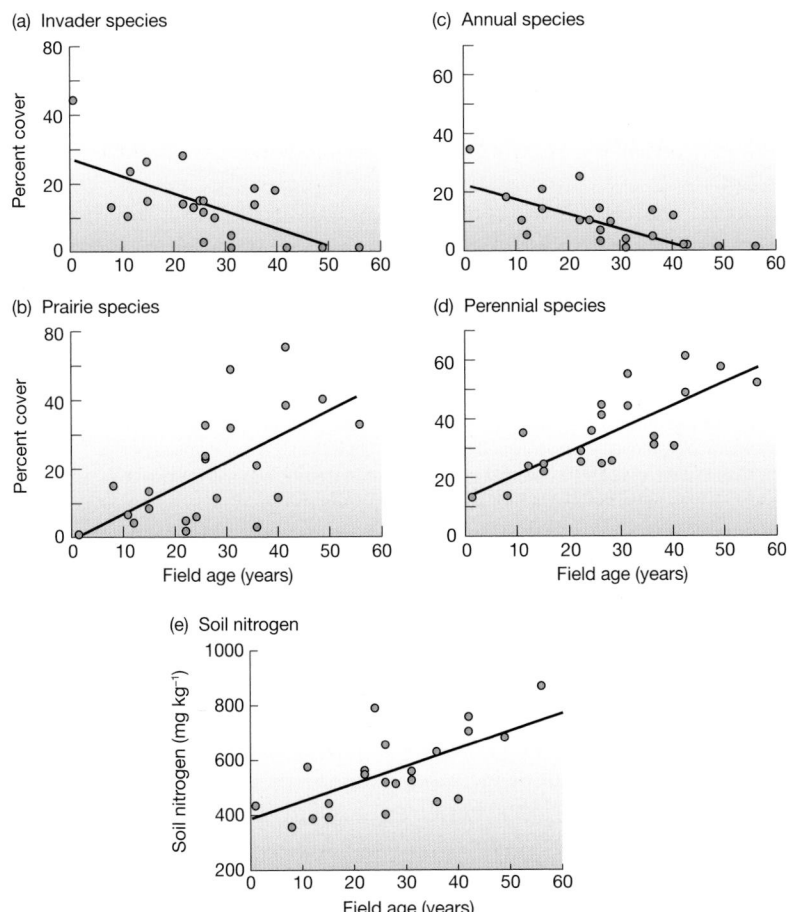

Figure 1.10

Twenty-two fields at different stages in an old-field succession were surveyed to generate the following trends with successional stage (field age): (a) invader species decreased, (b) native prairie species increased, (c) annual species decreased, (d) perennial species increased, and (e) soil nitrogen content increased. The best fit lines (see Box 1.2) are highly significant in every case ($P < 0.01$).

AFTER INOUYE ET AL., 1987

1 Do patches receiving different supply rates of nitrogen become less similar in species composition over time?

2 Do patches receiving similar supply rates of nitrogen become more similar in species composition over time?

The answer to the first question was clear: plots within a field were initially similar to one another but, 10 years later, plots receiving different amounts of nitrogen had diverged in species composition – and the greater the difference in nitrogen input, the greater the divergence (Inouye & Tilman, 1995).

The answer to the second question is illustrated in Figure 1.11. At the start of the experiment, the field abandoned for 46 was very different in species composition to the one only abandoned for 14 years. But 10 years later, plots within the two fields that had been subjected to similar rates of nitrogen input had become remarkably similar (Figure 1.11).

Thus, this experiment tends to refute the simplicity of our proposed explanation. Time itself is not the only cause of successional changes in species composition of these old fields. Differences in available nitrogen cause successions to diverge; similarities cause them to converge much more quickly than they would otherwise do. Time (= opportunity to colonize) and nitrogen are clearly intimately

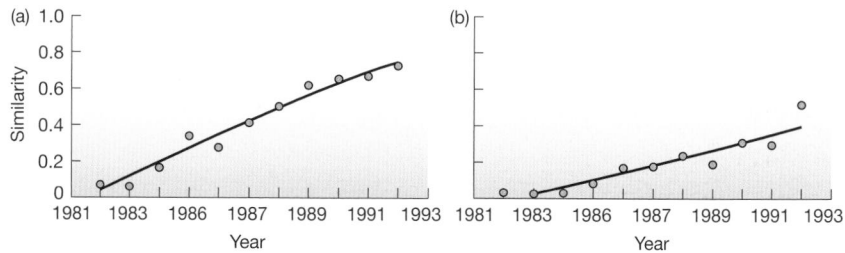

AFTER INOUYE & TILMAN, 1995

Figure 1.11

Results from an experiment in which plots within two old fields from Figure 1.10 were given artificial nitrogen addition treatments starting in 1982: one of the fields had been abandoned for 46 years and the other for 14 years. (a) Between 1982 and 1992, plots receiving 17 g of nitrogen m^{-2} yr^{-1} in the two fields became increasingly similar in composition. The similarity index measures the extent to which the species composition in the pair of fields is similar – identical compositions produce a similarity index of 1, entirely different compositions produce a similarity of 0. (b) Like (a) but with only 1 g of nitrogen m^{-2} yr^{-1}. Note in this case that there was still convergence in species composition between the two fields but to a lesser extent. In both cases the best fit lines are highly significant.

intertwined and further experiments will be required to disentangle their web of cause and effect – just one of many unanswered ecological questions.

Finally, experimental manipulations over extended periods like these may also provide important insights into the possible effects of more chronic human disturbances to natural communities. The lower rate of nitrogen addition in the experiment (1 g of nitrogen m^{-2} yr^{-1}) was similar to that experienced in many parts of the world as a result of increased atmospheric deposition of inorganic nitrogen (mainly derived from the burning of fossil fuels). Even these low levels apparently led to convergence of previously dissimilar communities over a 10-year period (Figure 1.11b). Experiments like this are crucial in helping us to predict the effects of pollutants, a point that is taken further in the next example.

insight into the effects of nitrogen pollution

1.3.3 Hubbard Brook: a long-term commitment of large-scale significance

The Cedar Creek study took advantage of a temporal pattern (a succession that takes decades to run its course) being reflected more or less accurately by a pattern in space (fields abandoned for different periods). The spatial pattern has the advantage that it could be studied within the time-bite of most research projects (3–5 years). It would have been better still to follow the ecological pattern through time but rather few researchers or institutions have risen to the challenge of designing research programs that last for decades.

A notable exception has been the work of Likens and associates at the Hubbard Brook Experimental Forest, an area of temperate deciduous forest drained by small streams in the White Mountains of New Hampshire in the USA. The researchers were pioneers with no precedents to follow. They decided to think big, and their work has shown the value of large-scale studies and long-term data records. The study commenced in 1963 and continues to the present. In the second edition of their classic book *Biogeochemistry of a Forested Ecosystem*, Likens and Bormann (1995) make poignant reference to three of their original collaborators who had died since the study began. Long term indeed.

Figure 1.12

The Hubbard Brook experimental forest. Note the experimental stream catchment from which all trees were removed – extending from the top left toward the center of the photograph.

The research team developed an approach called 'the small watershed technique' to measure the input and output of chemicals from individual catchment areas in the landscape. Because many chemical losses from terrestrial communities are channeled through streams, a comparison of the chemistry of stream water with that of incoming precipitation can reveal a lot about the differential uptake and cycling of chemical elements by the terrestrial biota. The same study can reveal much about the sources and concentrations of chemicals in the stream water, which in turn may influence the productivity of stream algae and the distribution and abundance of stream animals.

The catchment area (or watershed) – the extent of terrestrial environment drained by a particular stream – was taken as the unit of study because of the role that streams play in chemical export from the land. Six small catchments were defined and their outflows were monitored (Figure 1.12). A network of precipitation gauges recorded the incoming amounts of rain, sleet and snow. Chemical analyses of precipitation and stream water made it possible to calculate the amounts of various chemical elements entering and leaving the system. In most cases, the output of chemicals in streamflow was greater than their input from rain, sleet and snow (Table 1.3). The source of the excess chemicals was weathering of parent rock and soil, estimated at about 70 g m^{-2} yr^{-1}. The exception was nitrogen; less was exported in stream water than was added to the catchment in precipitation and by fixation of atmospheric nitrogen by microorganisms in the soil.

Likens had the brilliant idea of performing a large-scale experiment in which all the trees were felled in one of Hubbard Brook's six catchments. In terms of experimental design, statistical purists might argue the study was flawed because

the catchment area as a unit of study

insights from a large-scale field experiment

Table 1.3

Annual chemical budgets for forested catchment areas at Hubbard Brook (kg ha^{-1} yr^{-1}). Inputs are for dissolved materials in precipitation or in dryfall (gases or associated with particles falling from the atmosphere). Outputs are losses in stream water as dissolved material plus particulate organic material in the streamflow. The source of the excess chemicals (where outputs exceeded inputs) was weathering of parent rock and soil. The exception was nitrogen (as ammonium or nitrate ions) – less was exported than arrived in precipitation because of nitrogen uptake in the forest.

	NH_4^+	NO_3^-	SO_4^{2-}	K^+	Ca^{2+}	Mg^{2+}	Na^+
Input	2.7	16.3	38.3	1.1	2.6	0.7	1.5
Output	0.4	8.7	48.6	1.7	11.8	2.9	6.9
Net change*	+2.3	+7.6	−10.3	−0.6	−9.2	−2.2	−5.4

*Net change is positive when the catchment gains matter and negative when it loses it.

AFTER LIKENS ET AL., 1971

it was unreplicated. However, the scale of the undertaking rather precluded replication. In any case, it was the asking of a dramatically new question that made this study a classic rather than elegant statistical design.

Within a few months of felling all the trees in the drainage basin, the consequences were evident in the stream water. The overall export of dissolved inorganic substances from the disturbed catchment rose to 13 times the normal rate (Figure 1.13). Two phenomena were responsible. First, the enormous reduction in transpiring surfaces (leaves) led to 40% more precipitation passing through the ground water to be discharged to the streams, and this increased outflow caused greater rates of leaching of chemicals and weathering of rock and soil. Second, and more significantly, deforestation effectively broke the link between decomposition and nutrient uptake. In the spring, when the deciduous trees would normally have started production and taken up inorganic nutrients released by decomposer activity, these were instead available to be leached in the drainage water.

Likens knew from the beginning that the rain and snow at Hubbard Brook were quite acid but it was some years before the widespread nature of acid rain in North America became clear. In fact, Hubbard Brook is more than 100 km from the nearest urban industrial area, yet precipitation and stream water were both markedly acid as a result of atmospheric pollution from fossil fuels. The long-term records kept so meticulously since 1963 at Hubbard Brook have proved invaluable in monitoring progress in the war against acid rain and its long-term consequences. The value of such records of stream water concentrations can be seen for hydrogen, sulfate and nitrate, three ions associated with acid rain (which in simple terms is a mixture of dilute nitric and sulphuric acids; sulphuric acid is the dominant acid in the eastern USA). There have been statistically significant declines in average annual concentrations of H^+ and SO_4^{2-} since 1964/65, and also of NO_3^-, thought the latter is subject to much greater year to year variation (Figure 1.14). Of note, however, is the fact that the results for shorter periods suggest quite different trends. Consider the hydrogen ion graph where three periods of 4 years are highlighted in different colors. The first suggests an increasing trend, the second no change and the third a decreasing trend. In fact, no statistically significant, long-term trend was established until nearly two decades of data had been amassed (Likens, 1989).

for statistically significant trends to become evident, many years of data may be required

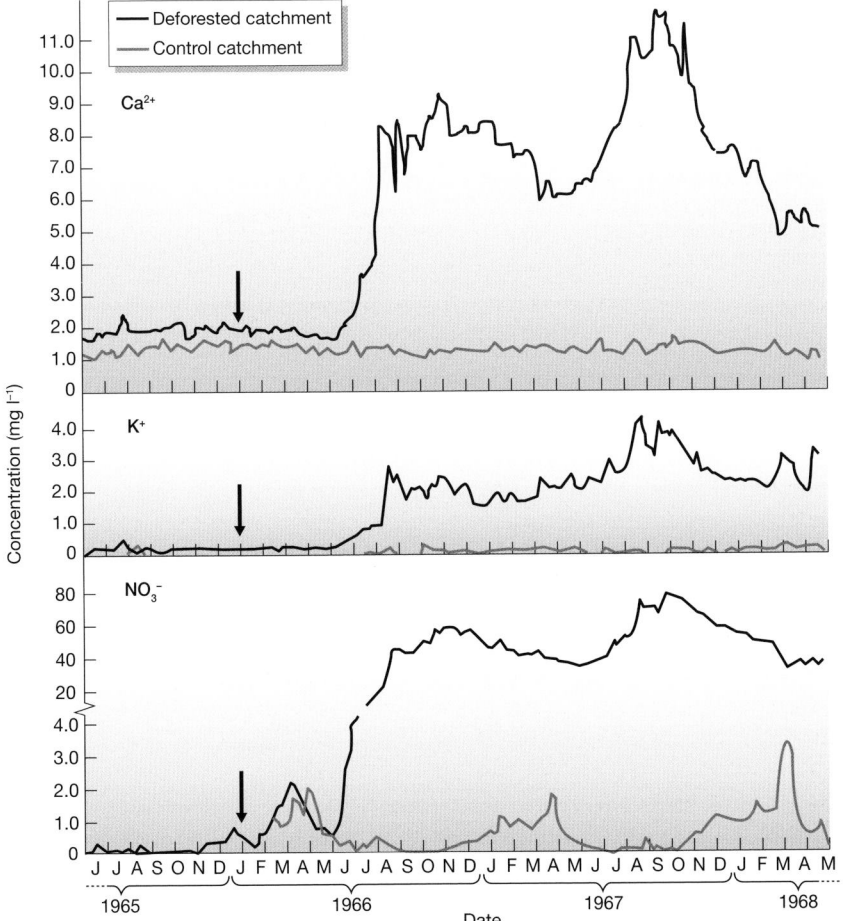

Figure 1.13

Concentrations of ions in stream water from the experimentally deforested watershed 2 and the control (unmanipulated) watershed 6 at Hubbard Brook. The timing of deforestation is indicated by arrows. In each case, there was a dramatic increase in export of the ions after deforestation. Note that the 'nitrate' axis has a break in it.

long data runs reveal the history of acid rain

It is thought that acid rain began in the USA in the early 1950s (before monitoring began at Hubbard Brook). After the passage of the Clean Air Act in 1970, emissions of SO_2 and particulates were reduced and this has been clearly reflected in stream water chemistry (Figure 1.14). Additional reductions in emissions have occurred as a result of the 1990 amendments to the Clean Air Act. However, critical questions remain – will forest and aquatic ecosystems recover from the effects of acid rain, and if so how long will it take (Likens et al., 1996)?

Using long-term data from Hubbard Brook and predictions of reductions to SO_2 emissions as a result of government legislation, Likens and Bormann (1995) estimated that by the turn of the millennium the sulfur loading in the atmosphere would still be three times higher than values recommended for protection of sensitive forests and aquatic communities (many plants, fish and aquatic invertebrates are intolerant of acid conditions). Moreover, declining inputs to Hubbard Brook of basic cations, such as calcium, may be causing the forests and streams to become even more sensitive to acidic inputs. Likens and Bormann (1995) hypothesized that a dramatic decline in forest growth rates during recent years may be related to a decline in calcium in the soil, a critical nutrient for tree growth. Acid rain may be responsible for the calcium deficiency.

Figure 1.14

Long-term changes in concentrations [microequivalents (μeq) l^{-1}] of H$^+$, NO$_3^-$, SO$_4^{2-}$ and Ca^{2+} in stream water from Hubbard Brook watershed 6 from 1963/64 to 1992/93. The declines are related to reductions in 'acid rain' affecting the Hubbard Brook area. The regression lines for all these ions have a probability of being significantly different from zero (no change) of $P < 0.05$; in other words there is a statistically significant pattern of decline in each. However, many years of data were needed before these patterns could be convincingly demonstrated. This is particularly marked for the hydrogen ion graph, where three periods of 4 years are highlighted in different colors. The first (in red) suggests an increasing trend, the second (in orange) no change and the third (in green) a decreasing trend.

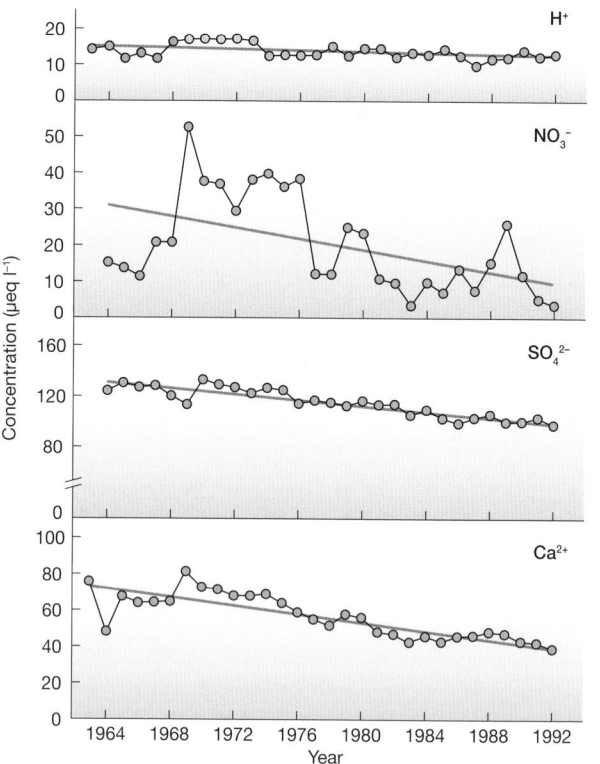

AFTER LIKENS & BORMANN, 1995

An associated reduction in bird populations in the forest may even be linked to this scenario. These unanswered questions are the subject of new phases of research at Hubbard Brook.

1.3.4 A modeling study: to discover why Asian vultures were heading for extinction

In 1997, vultures in India and Pakistan began dropping from their perches. Local people were quick to notice dramatic declines in numbers of the oriental white-backed vulture *Gyps bengalensis* (Figure 1.15) and the long-billed vulture *G. indicus*, but ecologists were puzzled. Repeated population surveys from 2000 to 2003 confirmed alarming rates of decline, defined technically as values of the 'population growth rate', λ (where the population size N in year t equals λ times the population size the previous year, $t - 1$; in other words $\lambda = N_t/N_{t-1}$). For the oriental white-backed vulture in India λ was 0.52 and in Pakistan it was 0.50, equating to a 48% and 50% decline per year, respectively. The state of affairs was a little less disastrous for the long-billed vulture in India where λ was 0.78, equating to a 22% decline per year.

These population crashes were of very great concern because of the crucial role vultures play in everyday life, disposing of the dead bodies of large animals, both wild and domestic. The loss of vultures enhanced carrion availability to wild dogs and rats, allowing their populations to increase and raising the probability of diseases such as rabies and plague being transmitted to humans. Moreover,

vulture populations in India and Pakistan were declining by 22–50% per year

contamination of nearby wells and the spread of disease by flies became more likely now that dead animals were not quickly picked clean by vultures. One group of people, the Parsees, were even more intimately affected because their religion calls for the dead to be taken in daylight to a special tower (dakhma) where the body is stripped clean by vultures within a few hours. It was crucial for ecologists to quickly determine the cause of vulture declines so that action could be taken.

It took a few years to find a common element in the deaths of otherwise healthy birds – each had suffered from visceral gout (accumulation of uric acid in the body cavity) followed by kidney failure. Soon a crucial piece in the jigsaw became clear: vultures dying of visceral gout contained residues of the drug diclofenac (Oaks et al., 2004). Then it was confirmed that carcasses of domestic animals treated with diclofenac were lethal to captive vultures. Diclofenac, a non-steroidal anti-inflammatory drug developed for human use in the 1970s, had only recently come into common use as a veterinary medicine in Pakistan and India. Thus, a drug that benefited domestic mammals proved lethal to the vultures that fed on their bodies.

The circumstantial evidence was strong, but given the relatively small numbers of diclofenac-contaminated dead bodies available to wild vultures, was the associated vulture mortality sufficient explanation for the population crashes? Or might other factors also be at play? This was the question addressed by Green and his team (2004) by means of a simulation population model. On the basis of their surveys of population declines and knowledge of birth, death and feeding rates, the researchers built a model to predict the behavior of the vulture populations. We show their model as a flow diagram (Figure 1.15); Green and his team developed mathematical formulae to predict changes in population size, but the details need not concern us here. The researchers posed the specific question: what proportion of carcasses (C) would have to contain lethal doses of diclofenac to cause the observed population declines? Their simulation model included the following assumptions:

... caused by drug-contaminated carcasses?

1 *Gyps* vultures do not breed (i.e. become adult) until they are 5 years old and then are capable of rearing only one juvenile per year, but only if both parents survive the breeding season of 160 days.

2 The fate of the population depends not only on rates of birth but also death. The pre-diclofenac 'baseline' survival rate of adult vultures (S) fell in the range 0.90–0.97, typical for large-bodied, long-lived birds. In other words, in the absence of diclofenac deaths, only 3–10% of adult vultures die each year.

3 Diclofenac poisoning reduces survival rate further. This depends on the probability an adult will eat from a diclofenac-affected carcass. In turn, this depends partly on the proportion of carcasses in the environment that contain diclofenac (C) and partly on how often vultures feed (F, the interval in days between feeding). Note that a single meal can sustain a vulture for 3 days and they do not feed every day; F ranges from 2 to 4 days. Vultures that feed more often (more times per year) are more likely to feed from a diclofenac-affected carcass and die.

4 The researchers had real estimates for population sizes in different years (N) and hence of λ (see above). In their modeling exercise they systematically varied the values for baseline survival S and feeding rate F. This is because

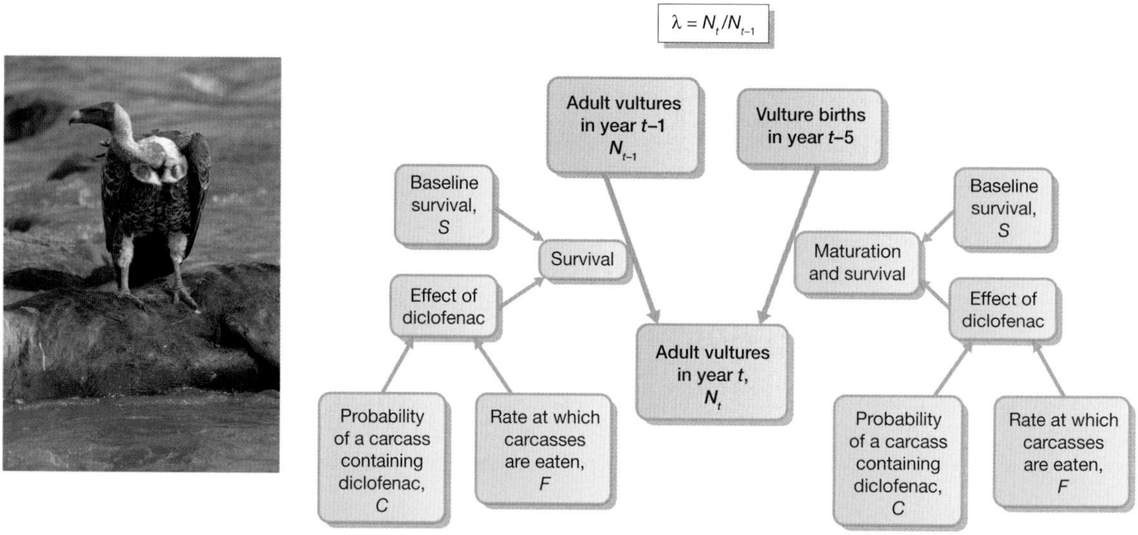

$$\lambda = N_t/N_{t-1}$$

Figure 1.15

Flow diagram showing the elements of a model of how the number of adult vultures in the population changes from one year (N_{t-1}) to the next (N_t). The oriental white-backed vulture, whose populations have shown disastrous declines in India and Pakistan, is shown in the inset. The number of adult vultures in year t depends on the number present the previous year ($t-1$), some of which die from natural causes (baseline survival) and others because of diclofenac poisoning. The number of adults in year t also depends on the number of vultures born 5 years previously ($t-5$), because vultures do not mature until they are 5 years old. Again, some newborn vultures die before maturity from natural causes and others because of diclofenac poisoning. The reduction in survival due to diclofenac depends on two things: the probability that a carcass contains diclofenac (C) and the rate at which carcasses are eaten (F).

they did not know precisely what the baseline survival or feeding rates were in these particular populations, although they did know the range in which the values fell. Thus, they ran the model for values of baseline survival of 0.90, 0.95 and 0.97, and with intervals between feeding of 2, 3 and 4 days.

5 Once all these parameters were entered into their model, the researchers could calculate the 'missing' parameter C – the proportion of carcasses that needs to be contaminated with diclofenac to account for the observed rate of population decline, λ (Table 1.4).

simulation models show that diclofenac-contaminated cattle are sufficient to explain vulture losses

Table 1.4 shows that at a maximum (for the Pakistani oriental white-backed vultures when adult survival is set at 0.97 and feeding interval is 4 days) only 0.743% or, in other words, 1 in 135 carcasses have to be dosed with diclofenac to cause the observed population decline. At a minimum (for Indian long-billed vultures when adult survival is set at 0.90 and feeding interval is 2 days) only 0.132% or 1 in 757 contaminated carcasses are required. The proportions of vultures found dead or dying in the wild with signs of diclofenac poisoning were closely similar to the proportions of deaths expected from the model if the observed population decline was due *entirely* to diclofenac poisoning. The researchers concluded, therefore, that diclofenac poisoning was a sufficient cause for the dramatic decline of wild vultures.

Clearly, urgent action is needed to prevent the exposure of vultures to live-stock carcasses contaminated with diclofenac and the Punjab government, for

Table 1.4

Modeled percentages of animal carcasses with lethal levels of diclofenac required to cause population declines at rates, λ, observed for long-billed vultures (LBW) or oriental white-backed vultures (OWBW) in India and Pakistan between 2000 and 2003. A value of 0.132%, for example, means that only 1 in 757 carcasses needs to be contaminated to cause the vulture decline. For each population, results are given for three feasible baseline adult survival rates, S (i.e. in the absence of diclofenac) and three values of the interval between vulture feeding bouts in days, F.

| | F | PERCENTAGE OF CARCASSES WITH LETHAL LEVEL | | |
		$S = 0.90$	$S = 0.95$	$S = 0.97$
LBV India	2	0.132	0.135	0.137
	3	0.198	0.202	0.205
	4	0.263	0.271	0.273
OWBV India	2	0.339	0.347	0.349
	3	0.508	0.521	0.526
	4	0.677	0.693	0.699
OWBV Pakistan	2	0.360	0.368	0.372
	3	0.538	0.551	0.558
	4	0.730	0.734	0.743

FROM GREEN ET AL., 2004

example, has now banned its use. Green and his colleagues also highlighted the need for research to identify alternative drugs that are effective in livestock and safe for vultures. Swan et al. (2006) have since tested a drug called meloxicam with promising results. Finally, given the depths to which the vulture populations have sunk, Green's team emphasize the importance of breeding vultures in captivity until diclofenac is under control. This is a sensible precaution to ensure long-term survival and to provide for future reintroduction programs.

This example, then, has illustrated a number of important general points about mathematical models in ecology:

1 Models can be valuable for exploring scenarios and situations for which we do not have, and perhaps cannot expect to obtain, real data (e.g. what would be the consequences of different baseline survival or feeding rates?).

2 They can be valuable, too, for summarizing our current state of knowledge and generating predictions in which the connection between current knowledge, assumptions and predictions is explicit and clear (given various values for S and F, and knowing λ, what values of C do these imply?).

3 In order to be valuable in these ways, a model does not have to be (indeed, cannot possibly be) a full and perfect description of the real world it seeks to mimic – all models incorporate approximations (the vulture model was, of course, a very 'stripped down' version of its true life history).

4 Caution is therefore always necessary – all conclusions and predictions are provisional and can be no better than the knowledge and assumptions on which they are based – but applied cautiously they can be useful (the vulture model prompted changes in management practices and research into new drugs).

5 Nonetheless, a model is inevitably applied with much more confidence once it has received support from real sets of data.

Summary
SUMMARY

Ecology as a pure and applied science

We define ecology as the scientific study of the distribution and abundance of organisms and the interactions that determine distribution and abundance. From its origins in prehistory as an 'applied science' of food gathering and enemy avoidance, the twin threads of pure and applied ecology have developed side by side, each depending on the other. This book is about how ecological understanding is achieved, what we do and do not understand, and how that understanding can help us predict, manage and control.

Questions of scale

Ecology deals with four levels of ecological organization: individual organisms, populations (individuals of the same species), communities (a greater or lesser number of populations) and ecosystems (the community together with its physical environment). Ecology can be done at a variety of spatial scales, from the 'community' within an individual cell to that of the whole biosphere. Ecologists also work on a variety of time scales. Ecological succession, for example, may be studied during the decomposition of animal dung (weeks), or during the period of climate change since the last ice age (millennia). The normal period of a research program (3–5 years) may often miss important patterns that occur over long time scales.

Diversity of ecological evidence

Many ecological studies involve careful observation and monitoring, in the natural environment, of the changing abundance of one or more species over time, or through space, or both. Establishing the cause(s) of patterns observed often requires manipulative field experiments. For complex ecological systems (and most of them are) it will often be appropriate to construct simple laboratory systems that can act as jumping-off points in our search for understanding. Mathematical models of ecological communities also have an important role to play in unraveling ecological complexity. However, the worth of models and simple laboratory experiments must always be judged in terms of the light they throw on the working of natural systems.

Statistics and scientific rigor

What makes the science of ecology rigorous is that it is based not on statements that are simply assertions, but on conclusions that are the results of carefully planned investigations with well thought-out sampling regimes, and on conclusions, moreover, to which a level of statistical confidence can be attached. The term that is most often used, at the end of a statistical test, to measure the strength of conclusions being drawn is a 'P-value' or probability level. The statements '$P < 0.05$' (significant) or '$P < 0.01$' (highly significant) mean that these are studies where sufficient data have been collected to establish a conclusion in which we can be confident.

Ecology in practice

Studies of the impacts of brown trout, introduced to New Zealand in the 20th century, have spanned all four ecological levels (individuals, populations, communities, ecosystems). Trout have replaced populations of native galaxiid fish below waterfalls. Laboratory and field experiments have established that grazing invertebrates in trout streams show an individual response, spending more time hiding and less time grazing. Trout cause a cascading community effect because the grazers impact less on the algae. Finally, a descriptive study revealed an ecosystem consequence: primary productivity by algae is higher in a trout stream than a galaxiid stream.

In the Cedar Creek Natural History Area are agricultural fields that are still under cultivation and others that have been abandoned at various times since the mid-1920s. This natural experiment was exploited to provide a description of the species sequence associated with succession on such abandoned fields. However, the fields differed not only in age but also in soil nitrogen. A set of field experiments, where soil nitrogen was augmented in a systematic way in fields of different age, showed that time and nitrogen interacted to cause the observed successional sequences.

The Hubbard Brook Experimental Forest study has been running since 1963. A large-scale experiment, involving the felling of all the trees in a single catchment area, resulted in a dramatic increase in chemical concentrations (particularly nitrate) in stream water. The loss of nitrate from the land and its increase in water can be expected to have consequences for the communities on both sides of the land–water interface. Monitoring of chemical concentrations for more than four decades in undisturbed catchments has revealed how acid rain has been diminishing as a result of the Clean Air Act. However, neither the forest nor the streams are immune from continuing effects of the pollution that caused acid rain.

Disturbing declines in vulture populations have profound implications for public health in India and Pakistan. A common element in the deaths was visceral gout, traced to an adverse effect of diclofenac used by veterinarians to treat domestic cattle, one source of food for vultures. Given the relatively small numbers of diclofenac-contaminated dead bodies available to wild vultures, a mathematical model was run to determine whether deaths due to diclofenac were a sufficient explanation for the population crashes, or whether other factors might also be at play. In fact, the proportion of vultures dying from diclofenac poisoning was very similar to that expected from the model if the decline was due *entirely* to diclofenac poisoning. Steps have now been taken to remedy the situation.

Review questions

Asterisks indicate challenge questions

1* Discuss the different ways that ecological evidence can be gained. How would you go about trying to answer one of ecology's unanswered questions, namely 'Why are there more species in the tropics than at the poles'?

2* The variety of microorganisms that live on your teeth have an ecology like any other community. What do you think might be the similarities in the forces determining species richness (the number of species present) in your oral community as opposed to a community of seaweeds living on boulders along the shoreline?

3 Why do some temporal patterns in ecology need long runs of data to detect them, while other patterns need only short runs of data?

4 Discuss the pros and cons of descriptive studies as opposed to laboratory studies of the same ecological phenomenon.

5 What is a 'natural field experiment'? Why are ecologists keen to take advantage of them?

6 Search the library for a variety of definitions of ecology: which do you think is most appropriate and why?

7* In a study of stream ecology, you need to choose 20 sites to test the hypothesis that brown trout have higher densities where the streambed consists of cobbles. How might your results be biased if you chose all your sites to be easy to access because they are near roads or bridges?

8 How might the results of the Cedar Creek study of old-field succession have been different if a single field had been monitored for 50 years, rather than simultaneously comparing fields abandoned at different times in the past?

9* When all the trees were felled in a Hubbard Brook catchment, there were dramatic differences in the chemistry of the stream water draining the catchment. How do you think stream chemistry would change in subsequent years as plants begin to grow again in the catchment area?

10 What are the main factors affecting the confidence we can have in predictions of a mathematical model?

Chapter 2

Ecology's evolutionary backdrop

Key concepts

KEY CONCEPTS

In this chapter you will:

- appreciate that Darwin and Wallace, who were responsible for the theory of evolution by natural selection, were both, essentially, ecologists
- understand that the populations of a species vary in their characteristics from place to place on both geographic and more local scales, and that some of the variation is heritable
- realize that natural selection can act very quickly on heritable variation – we can study it in action and control it in experiments
- understand that reciprocal transplanting of individuals of a species into each other's habitats can show a finely specialized fit between organisms and their environments
- appreciate that the origin of species requires the reproductive isolation of populations as well as natural selection forcing them to diverge

- realize that natural selection fits organisms to their past – it does not anticipate the future
- realize that the evolutionary history of species constrains what future selection can achieve
- understand that natural selection may produce similar forms from widely different ancestral lines (convergent evolution) or the same range of forms in populations that have become separated (parallel evolution)

As the great Russian-American biologist Dobzhansky said, 'Nothing in biology makes sense, except in the light of evolution.' But equally, very little in evolution makes sense except in the light of ecology: ecology provides the stage directions through which the 'evolutionary play' is performed. Ecologists and evolutionary biologists need a thorough understanding of each other's disciplines to make sense of key patterns and processes.

2.1 Introduction

The Earth is inhabited by a multiplicity of types of organism. They are distributed neither randomly nor as a homogeneous mixture over the surface of the globe. Any sampled area, even on the scale of a whole continent, contains only a tiny subset of the variety of species present on Earth. Why are there so many types of organism? Why are their distributions so restricted? Answering these ecological questions requires an understanding of the processes of evolution that have led to present-day diversity and distribution.

Until relatively recently, the emphasis with diversity was on using it (for example for medicine), exhibiting it in zoological and botanic gardens, and cataloging it in museums (Box 2.1). Without an understanding of how this diversity developed, such catalogs are more like stamp collecting than science. The enduring contribution of Charles Darwin and Alfred Russel Wallace was to provide ecologists with the scientific foundations to comprehend patterns in diversity and distribution over the face of the Earth.

all species are so specialized that they are almost always absent from almost everywhere

2.2 Evolution by natural selection

Darwin and Wallace (Figure 2.1) were both ecologists (although their seminal work was performed before the term was coined) who were exposed to the diversity of nature in the raw. Darwin sailed around the world as naturalist on the 5-year expedition of HMS *Beagle* (1831–36) recording and collecting in the enormous variety of environments that he explored on the way. He gradually developed the view that the natural diversity of nature was the result of a process of evolution

Darwin and Wallace were both ecologists

2.1 Historical landmarks

A brief history of the study of diversity

An awareness of the diversity of living organisms, and of what lives where, is part of the knowledge that the human species accumulates and hands down through the generations. Hunter–gatherer peoples needed (and still need) detailed knowledge of the natural history of their environments to obtain food successfully and to escape the hazards of being poisoned or eaten. The Arawaks of the South American equatorial forest know where to find and how to catch all the species of large animal around them and also the names of their trees and how they can be used.

Before 2000 BC, the Chinese emperor Shen Nung compiled what was perhaps the first written 'herbal' of useful plants and, by the first century AD, Dioscorides had described 500 species of medicinal plants and illustrated many of them.

Collections of living specimens in zoos and gardens also have a long history – certainly back to Greece in the seventh century BC. The urge to collect from the diversity of nature developed in the West in the 17th century when some individuals made their living by finding interesting specimens for other people's collections. John Tradescant the father (died 1638) and John Tradescant the son (1608–1662) spent most of their lives collecting plants and importing live specimens for the gardens of the British aristocracy. The father was the first English botanist to visit Russia (1618), bringing back many living plants; his son made three visits (1637, 1642 and 1654) to the New World to collect specimens in the American colonies.

Wealthy individuals built up vast collections into personal museums and traveled or sent travelers in search of novelties from new lands as they were discovered and colonized. Naturalists and artists (often the same people) were sent to accompany the major voyages of exploration to report and take home, dead or alive, collections of the diversity of organisms and artefacts that they found. The study of taxonomy and systematics developed and flourished – taxonomy gave names to the various types of organisms; systematics organized and classified them.

When big national museums were established (the British Museum in 1759 and the Smithsonian in Washington in 1846), they were largely compiled from the gifts of personal collections. Like zoos and gardens, the museums' main role was to make a public display of the diversity of nature, especially the new and curious and rare.

There was no need to explain the diversity – the biblical theory of the 7-day creation of the world sufficed. However, the idea that the diversity of nature had 'evolved' over time by progressive divergence from pre-existing stocks was beginning to be discussed around the turn of the 18th and 19th centuries. In 1844 an anonymous publication, *The Vestiges of Creation*, put the cat among the pigeons with a popular account of the idea that animal species had descended from other species.

in which *natural selection* favored some variants within species through a 'struggle for existence'. He developed this theme over the next 20 years through detailed study and an enormous correspondence with his friends as he prepared a major work for publication with all the evidence carefully marshalled. But he was in no hurry to publish.

In 1858, Wallace wrote to Darwin spelling out, in all its essentials, the same theory of evolution. Wallace was a passionate amateur naturalist. He had read Darwin's journal of the voyage of the *Beagle* and from 1847 to 1852, with his friend

(b)

(a)

Figure 2.1

Photographs of (a) Charles Darwin (lithograph by T. H. Maguire, 1849) and (b) Alfred Russel Wallace (1862).

H.W. Bates, he explored and collected in the river basins of the Amazon and Rio Negro, and from 1854 to 1862 made an extensive expedition in the Malay archipelago. He recalled lying on his bed in 1858 'in the hot fit of intermittent fever, when the idea [of natural selection] suddenly came to me. I thought it all out before the fit was over, and . . . I believe I finished the first draft the next day.'

Today, competition for fame and financial support would no doubt lead to fierce conflict about priority – who had the idea first. Instead, in an outstanding example of selflessness in science, sketches of Darwin's and Wallace's ideas were presented together at a meeting of the Linnean Society in London. Darwin's *On the Origin of Species* was then hastily prepared and published in 1859. *On the Origin of Species* may be considered the first major textbook of ecology, and aspiring ecologists would do well to read at least the third chapter.

Both Darwin and Wallace had read *An Essay on the Principle of Population*, published by Malthus in 1798. Malthus's essay was concerned with the human population, which, if its intrinsic rate of increase remained unchecked, would, he calculated, be capable of doubling every 25 years and overrunning the planet. Malthus realized that limited resources, as well as disease, wars and other disasters, slowed the growth of populations and placed absolute limits on their size. As experienced field naturalists, Darwin and Wallace realized that the Malthusian argument applied with equal force to the whole of the plant and animal kingdoms.

Darwin noted the great fecundity of some species – a single individual of the sea slug *Doris* may produce 600,000 eggs; the parasitic roundworm *Ascaris* may produce 64 million. But he realized that every species 'must suffer destruction

influence of Malthus's essay on Darwin and Wallace

during some period of its life, and during some season or occasional year, otherwise, on the principle of geometrical increase, its numbers would quickly become so inordinately great that no country could support the product.' In one of the earliest examples of population ecology, Darwin counted all the seedlings that emerged from a plot of cultivated ground 3 feet long and 2 feet wide: "Out of 357 no less than 295 were destroyed, chiefly by slugs and insects". Both authors, then, emphasized that most individuals die before they can reproduce and contribute nothing to future generations. Both, though, tended to ignore the important fact that those individuals that do survive in a population may leave different numbers of descendants.

The theory of evolution by natural selection, then, rests on a series of established truths:

fundamental truths of evolutionary theory

1 Individuals that form a population of a species are not identical.

2 Some of the variation between individuals is heritable – that is, it has a genetic basis and is therefore capable of being passed down to descendants.

3 All populations could grow at a rate that would overwhelm the environment; but in fact, most individuals die before reproduction and most (usually all) reproduce at less than their maximal rate. Hence, each generation, the individuals in a population are only a subset of those that 'might' have arrived there from the previous generation.

4 Different ancestors leave different numbers of descendants (descendants, *not* just offspring): they do not all contribute equally to subsequent generations. Hence, those that contribute most have the greatest influence on the heritable characteristics of subsequent generations.

Evolution is the change, over time, in the heritable characteristics of a population or species. Given the above four truths, the heritable features that define a population will inevitably change. Evolution is inevitable.

'the survival of the fittest'?

But which individuals make the disproportionately large contributions to subsequent generations and hence determine the direction that evolution takes? The answer is: those that were best able to survive the risks and hazards of the environments in which they were born and grew; and those who, having survived, were most capable of successful reproduction. Thus, interactions between organisms and their environments – the stuff of ecology – lie at the heart of the process of evolution by natural selection.

The philosopher Herbert Spencer described the process as 'the survival of the fittest', and the phrase has entered everyday language – which is regrettable. First, we now know that survival is only part of the story: differential reproduction is often equally important. But more worryingly, even if we limit ourselves to survival the phrase gets us nowhere. Who are the fittest? – those that survive. Who survives? – those that are fittest. Nonetheless, the term *fitness* is commonly used to describe the success of individuals in the process of natural selection. An individual will survive better, reproduce more and leave more descendants – it will be fitter – in some environments than in others. In a given environment, some individuals will survive better, reproduce more, and leave more descendants – they will be fitter – than other individuals.

Darwin had been greatly influenced by the achievements of plant and animal breeders: for example, the extraordinary variety of pigeons, dogs and farm animals that had been deliberately bred by selecting individual parents with exaggerated traits. He and Wallace saw nature doing the same thing; 'selecting' those individuals that survived from their excessively multiplying populations: hence the phrase 'natural selection'. But even this phrase can give the wrong impression. There is a great difference between human and natural selection. Human selection has an aim for the future – to breed a cereal with a higher yield, a more attractive pet dog or a cow that will yield more milk. But nature has no aim. Evolution happens because some individuals have survived the death and destruction of the past and reproduced more successfully in the past, not because they were somehow chosen or selected as improvements for the future.

natural selection has no aim for the future

Hence, past environments may be said to have selected particular characteristics of individuals that we see in present-day populations. Those characteristics are 'suited' to present-day environments only because environments tend to remain the same, or at least change only very slowly. We shall see later in this chapter that when environments do change more rapidly, often under human influence, organisms can find themselves, for a time, left 'high and dry' by the experiences of their ancestors.

2.3 Evolution within species

The natural world is not composed of a continuum of types of organism each grading into the next: we recognize boundaries between one sort of organism and another. In one of the great achievements of biological science, Linnaeus in 1735 devised an orderly system for naming the different sorts. Part of his genius was to recognize that there were features of both plants and animals that were not easily modified by the organisms' immediate environment, and that these 'conservative' characteristics were especially useful for classifying organisms. In flowering plants, the form of the flowers is particularly stable. Nevertheless, within what we recognize as species, there is often considerable variation, and some of this is heritable. It is on such intraspecific variation, after all, that plant and animal breeders work. In nature, some of this intraspecific variation is clearly correlated with variations in the environment and represents local specialization.

to understand the evolution of species we need to understand evolution within species

Darwin called his book *On the Origin of Species by Means of Natural Selection*, but evolution by natural selection does far more than create new species. Natural selection and evolution occur *within* species, and we now know that we can study them in action and within our own lifetime. Moreover, we need to study the way that evolution occurs within species if we are to understand the origin of new species.

2.3.1 Geographic variation within species

Since the environments experienced by a species in different parts of its range are themselves different (to at least some extent), we might expect natural selection to have favored different variants of the species at different sites. But evolution forces the characteristics of populations to diverge from each other (i) only if there is sufficient heritable variation on which selection can act; and (ii) provided that the forces of selection favoring divergence are strong enough to counteract the mixing and hybridization of individuals from different sites.

the characteristics of a species may vary over its geographic range

Two populations will not diverge completely if their members (or, in the case of plants, their pollen) are continually migrating between them, mating and mixing their genes.

The sapphire rockcress, *Arabis fecunda*, is a rare perennial herb restricted to calcareous soil outcrops in western Montana – so rare, in fact, that there are just 19 existing populations separated into two groups ('high elevation' and 'low elevation') by a distance of around 100 km. Whether there is local adaptation here is of practical importance: four of the low-elevation populations are under threat from spreading urban areas and may require reintroduction from elsewhere if they are to be sustained. Reintroduction may fail if local adaptation is too marked. Observing plants in their own habitats and checking for differences between them would not tell us if there was local adaptation in the evolutionary sense. Differences may simply be the result of immediate responses to contrasting environments made by plants that are essentially the same. Hence, high- and low-elevation plants were grown together in a 'common garden' (Figure 2.2a),

Figure 2.2

'Common garden' experiments (a) and reciprocal transplant experiments (b) compare the performance of organisms from different populations of the same species. In the former, organisms are taken from a variety of sources and reared under the same conditions. In the latter, organisms from two (or more) habitats are taken from their own habitat and reared alongside resident organisms in *their* own habitat, in a 'balanced' design such that all organisms are reared in their 'home' habitats and all 'away' habitats.

(a) Common garden experiments

(b) Reciprocal transplant experiments

FROM MCKAY ET AL. 2001

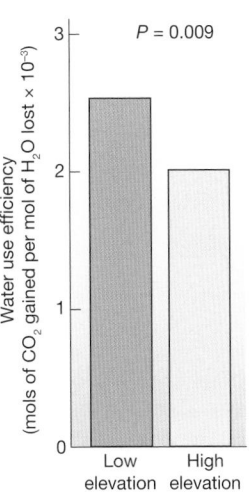
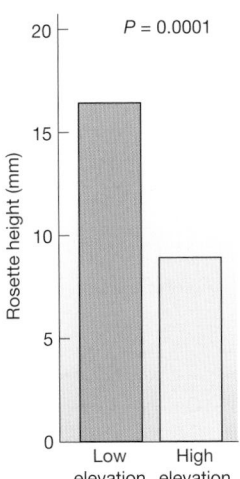
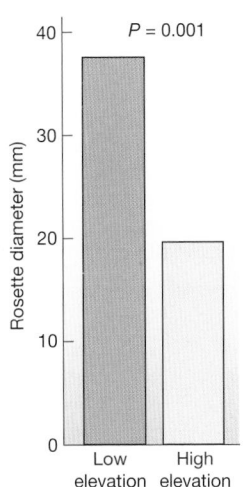

Figure 2.3

When plants of the rare sapphire rockcress from low elevation (drought-prone) and high elevation sites were grown together in a common garden, local adaptation was apparent: those from the low elevation site had significantly better water use efficiency as well as having both taller and broader rosettes.

eliminating any influence of contrasting immediate environments. The low-elevation sites were more prone to drought: both the air and the soil were warmer and drier; and the low-elevation plants in the common garden were indeed significantly more drought-tolerant: for example, they had significantly better 'water use efficiency' (their rate of water loss through the leaves was low compared to the rate at which carbon dioxide was taken in) as well as being much taller and 'broader' (Figure 2.3).

Differentiation over a much smaller spatial scale was demonstrated at a site called Abraham's Bosom on the coast of North Wales, UK. Here there was an intimate mosaic of very different habitats at the margin between maritime cliffs and grazed pasture, and a common species, creeping bent grass (*Agrostis stolonifera*) was present in many of the habitats. Figure 2.4 shows a map of the site and one of the transects from which plants were sampled; it also shows the results when plants from the sampling points along this transect were grown in a common garden. Each of four plants taken from each sampling point was represented by five rooted clonal replicates of itself. The plants spread by sending out shoots along the ground surface (stolons), and the growth of plants was compared by measuring the lengths of these. In the field, cliff plants formed only short stolons, whereas those of the pasture plants were long. In the experimental garden, these differences were maintained, even though the sampling points were typically only around 30 m apart – certainly within the range of pollen dispersal between plants. Indeed, the gradually changing environment along the transect was matched by a gradually changing stolon length, presumably with a genetic basis, since it was apparent in the common garden. Even over this small scale, the forces of selection seem to outweigh the mixing forces of hybridization.

On the other hand, it would be quite wrong to imagine that local selection always overrides hybridization – that all species exhibit geographically distinct variants with a genetic basis. For example, in a study of *Chamaecrista fasciculate*, an annual legume from disturbed habitats in eastern North America, plants were grown in a common garden that were derived from the 'home' site or were transplanted from distances of 0.1, 1, 10, 100, 1000 and 2000 km. Five characteristics were measured: germination, survival, vegetative biomass, fruit production

variation over very short distances

Figure 2.4

(a) Map of Abraham's Bosom, the site chosen for a study of evolution over very short distances. The green area is grazed pasture; the pale brown area represents cliffs falling to the sea. The numbers indicate sites from which the grass *Agrostis stolonifera* was sampled. Note that the whole area is only 200 m long. (b) A vertical transect across the study area showing gradual change in the numbered sites from pasture to cliff conditions. (c) The mean length of stolons produced in the experimental garden from samples taken from the transect.

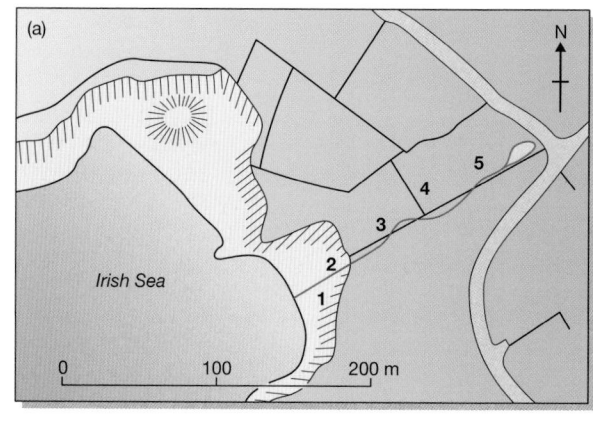

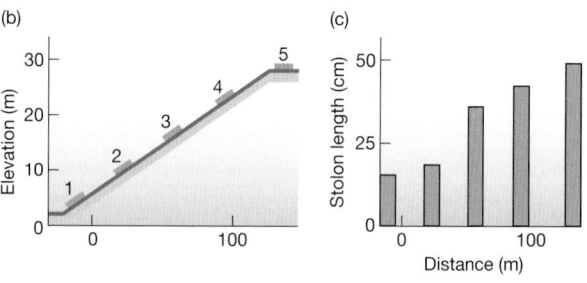

FROM ASTON & BRADSHAW, 1966

and the number of fruit produced per seed planted; but for all characters in all replicates there was little or no evidence for local adaptation except at the very farthest spatial scales (e.g. Figure 2.5). There is 'local adaptation' – but it's clearly not *that* local.

We can also test whether organisms have evolved to become specialized to life in their local environment in *reciprocal transplant* experiments (see Figure 2.2b): comparing their performance when they are grown 'at home' (i.e. in their original habitat) with their performance 'away' (i.e. in the habitat of others).

It can be difficult to detect the local specialization of animals by transplanting them into each other's habitat: if they do not like it, most species will run away.

reciprocal transplants test the match between organisms and their environment – e.g. sea anemones transplanted into each other's habitats

Figure 2.5

Percentage germination of local (transplant distance zero) and transplanted *Chamaecrista fasciculata* populations to test for local adaptation along a transect in Kansas. Data for 1995 and 1996 have been combined because they do not differ significantly. Populations that differ from the home population at $P < 0.05$ are indicated by an asterisk. Local adaptation occurs at only the largest spatial scales.

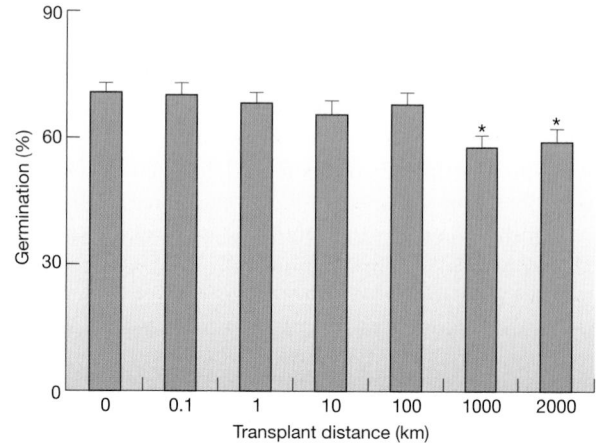

FROM GALLOWAY & FENSTER, 2000

Table 2.1

A reciprocal transplant experiment of the sea anemone *Actinia tenebrosa*. a, b and c are the three replicates in each colony. In each case the proportion of adults that were found brooding young is shown. Transplants back to the home sites are shown in bold print.

		TRANSPLANTED TO SITES AT:		
SITE OF ORIGIN		GREEN ISLAND	SALMON POINT	STRICKLAND BAY
Green island	a	**0.42**	0.68	0.78
	b	**0.80**	0.63	0.75
	c	**0.67**	0.62	0.61
Salmon Point	a	0.11	**0.42**	0.13
	b	0.18	**0.43**	0.28
	c	0.00	**0.50**	0.40
Strickland Bay	a	0.11	0.06	**0.33**
	b	0.00	0.06	**0.27**
	c	0.04	0.20	**0.27**

FROM AYRE, 1985

But invertebrates like corals and sea anemones are sedentary, and some can be lifted from one place and established in another. The sea anemone *Actinia tenebrosa* is found in pools on headlands around the coast of New South Wales, Australia. Ayre (1985) chose three colonies on headlands within 4 km of each other on which the anemone was abundant. Within each colony, he selected three transplant sites (each 3–5 m long) and at each he set aside three 1 m wide strips – two to receive anemones from the away sites and one to receive 'transplanted' individuals from the home site itself. Ayre cleared the experimental sites of all the anemones present and transplanted anemones into them. The number of juveniles brooded per adult was used as a measure of the performance of the anemones home and away.

The proportion of adults that were found brooding 11 months later is shown in Table 2.1. Anemones originally sampled from Green Island were rather successful in brooding young after being transplanted both home and away and did not show any specialization to their home environment. However, in all the other transplant experiments a greater proportion of anemones brooded young at home than at away sites: strong evidence of evolved local specialization. In later experiments, Ayre (1995) lifted anemones from a variety of sites as before, but he then kept them for a period to acclimate at a common site before transplanting them in a reciprocal experiment. This more severe test convincingly confirmed the results in Table 2.1.

Another reciprocal transplant experiment was carried out with white clover (*Trifolium repens*), which forms clones in grazed pastures. To determine whether the characteristics of individual clones matched local features of their environment, Turkington and Harper (1979) removed plants from marked positions in the field and multiplied them into clones in the common environment of a greenhouse. They then transplanted plants from each clone into the place in the vegetation from which it had originally been taken, and also to the places from

a reciprocal transplant experiment involving a plant

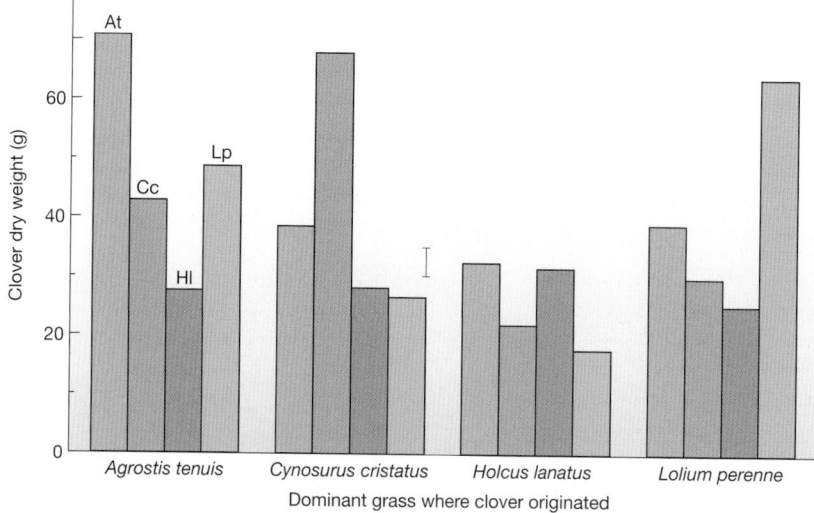

Figure 2.6

Plants of white clover (*Trifolium repens*) were sampled from a field of permanent grassland from local patches dominated by four different species of grass: *Agrostis tenuis* (At), *Cynosurus cristatus* (Cc), *Holcus lanatus* (Hl), and *Lolium perenne* (Lp). The clover plants were multiplied into clones and transplanted (in all possible combinations) into plots that had been sown individually with seeds of each of the four grass species. The histograms show the average weights of the transplanted white clover after 12 months' growth. The vertical bar indicates the difference between the height of any pair of columns that is statistically significant at $P < 0.05$. Note, in the panel of four histograms on the left, how clover that came originally from a patch of *Agrostis tenuis* grew significantly better in the presence of this grass (At) than any of the other species (Cc, Hl, Lp). Equivalent patterns are evident for clover that originated from patches of *Cynosurus cristatus* and *Lolium perenne* (strongest clover growth with Cc and Lp, respectively). Clover from *Holcus lanatus* patches did not follow the general trend, growing as well with At as with Hl.

where all the others had been taken. The plants were allowed to grow for a year before they were removed, dried and weighed. The mean weight of clover plants transplanted back into their home sites was 0.89 g but at away sites it was only 0.52 g, a statistically highly significant difference.

The clover plants had been chosen from patches dominated by four different species of grass. Hence, in a second experiment, samples from the different clones were planted into dense experimental plots of the four grasses (Figure 2.6). The mean yield of clovers grown with their original neighbor grass was 59.4 g; the mean yield with 'alien' grasses was 31.9 g, again a highly significant difference. Thus, clover clones in the pasture had evolved to become specialized such that they tended to perform best (make most growth) in their local environment and with their local neighbors.

In most of the examples so far, geographic variants of species have been identified, but the selective forces favoring them have not. This is not true of the next example. The guppy (*Poecilia reticulata*), a small freshwater fish from northeastern South America, has been the material for a classic series of evolutionary experiments. In Trinidad, many rivers flow down the northern range of mountains and are subdivided by waterfalls that isolate fish populations above the falls from those below. Guppies are present in almost all these water bodies, and

natural selection by predation: a controlled field experiment in fish evolution

FROM TURKINGTON & HARPER, 1979

Figure 2.7

Male and female guppies (*Poecilia reticulata*) showing two flamboyant males courting a typical, dull-colored female.

in the lower waters they meet various species of predatory fish that are absent higher up the rivers. The populations of guppies in Trinidad differ from each other in almost every feature that biologists have examined. Forty-seven of these traits tend to vary in step with each other (they *covary*) and with the intensity of the risk from predators. This correlation suggests that the guppy populations have been subject to natural selection from the predators. But the fact that two phenomena are correlated does not prove that one causes the other. Only controlled experiments can establish cause and effect.

Where guppies have been free or relatively free from predators, the males are brightly decorated with different numbers and sizes of colored spots (Figure 2.7). Females are dull and dowdy and (at least, to us) inconspicuous. Whenever we study natural selection in action, it becomes clear that compromises are involved. For every selective force that favors change, there is a counteracting force that resists the change. Color in male guppies is a good example. Female guppies prefer to mate with the most gaudily decorated males – but these are more readily captured by predators because they are easier to see.

This sets the stage for some revealing experiments on the ecology of evolution. Guppy populations were established in ponds in a greenhouse and exposed to different intensities of predation. The number of colored spots per guppy fell sharply and rapidly when the population suffered heavy predation (Figure 2.8a). Then, in a field experiment, 200 guppies were moved from a site far down the Aripo River where predators were common and introduced to a site high up the river where there were neither guppies *nor* predators. The transplanted guppies thrived in their new site, and within just 2 years the males had more and bigger spots of more varied color (Figure 2.8b). The females' choice of the more flamboyant males had dramatic effects on the gaudiness of their descendants, but this was only because predators were not present to reverse the direction of selection.

The speed of evolutionary change in this experiment in nature was as fast as that in artificial selection experiments in the laboratory. Many more fish were produced than would eventually survive (as many as 14 generations of fish occurred in the 23 months during which the experiment took place) and there was considerable genetic variation in the populations upon which natural selection could act.

Figure 2.8

(a) An experiment showing changes in populations of guppy *Poecilia reticulata* exposed to predators in experimental ponds. The graph shows changes in the number of colored spots per fish in ponds with different populations of predatory fish. The initial population was deliberately collected from a variety of sites so as to display high variability and was introduced to the ponds at time 0. At time S, weak predators (*Rivulus hartii*) were introduced to ponds R, a high intensity of predation by the dangerous predator *Crenicichila alta* was introduced into ponds C, while ponds K continued to contain no predators (the vertical lines show ± 2 SE). The number of spots per fish declined in treatments with the dangerous predator, but increased in the absence of fish or the presence of weak predators. (b) Results of a field experiment. A population of guppies originating in a locality with dangerous predators (*c*) was transferred to a stream having only the weak predator (*Rivulus hartii*) and, until the introduction, no guppies (*x*). Another stream nearby with guppies and *R. hartii* served as a control (*r*). The results shown are from guppies collected at the three sites 2 years after the introductions. Note how *x* and *r*, the sites with only weak predation, have converged and thus how *x* has changed dramatically from the source population with dangerous predators, *c*. In the absence of strong predators, the size, number and diversity of colored spots increased significantly within 2 years.

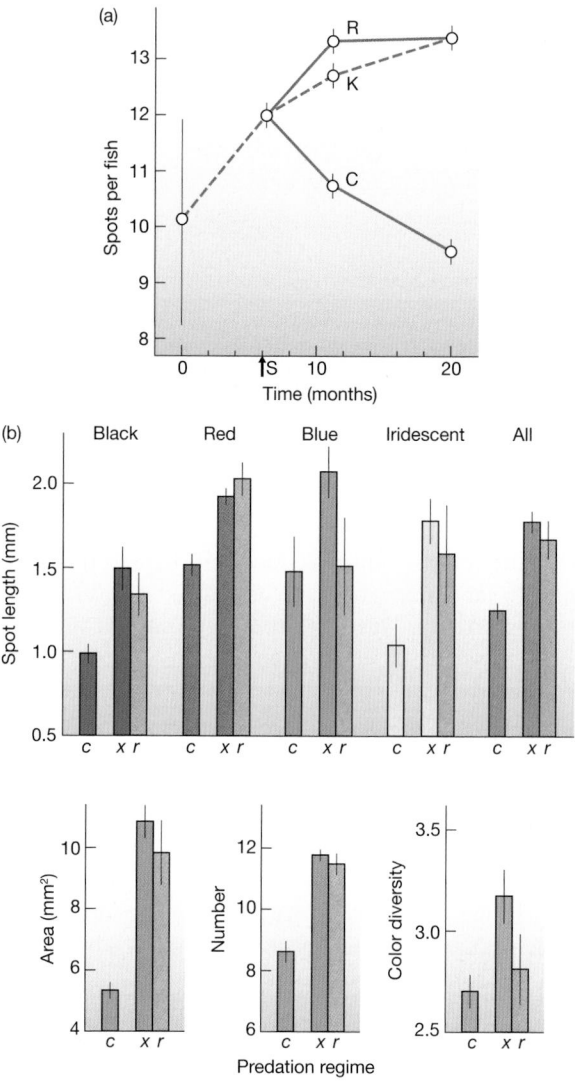

AFTER ENDLER, 1980

2.3.2 Variation within a species with manmade selection pressures

natural selection by pollution – the evolution of a melanic moth

It is not surprising that some of the most dramatic examples of natural selection in action have been driven by the ecological forces of environmental pollution – these can provide rapid change under the influence of powerful selection pressures. Pollution of the atmosphere in and after the Industrial Revolution has left evolutionary fingerprints in the most unlikely places. *Industrial melanism* is the phenomenon in which black or blackish forms of species of moths and other organisms have come to dominate populations in industrial areas. In the dark individuals, a dominant gene is responsible for producing an excess of the black pigment melanin. Industrial melanism is known in most industrialized countries, including some parts of the United States (e.g. Pittsburgh), and more than 100 species of moth have evolved forms of industrial melanism.

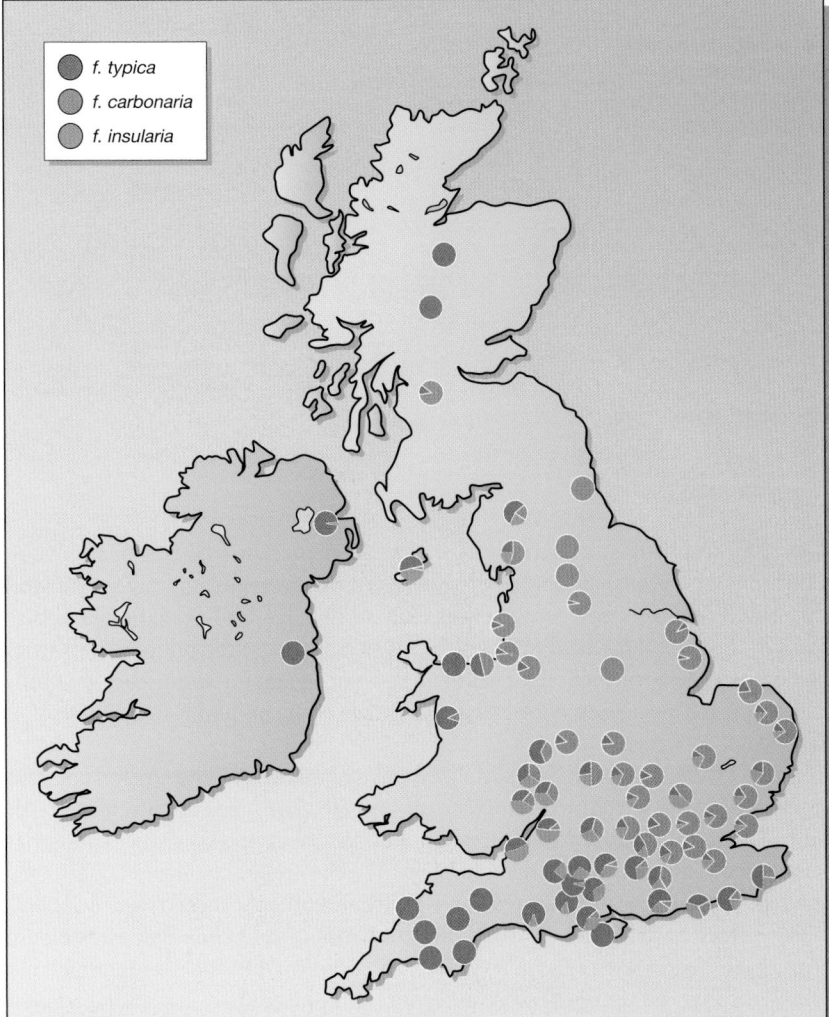

FROM FORD, 1975

Figure 2.9

Sites in Britain and Ireland where the frequencies of the pale (*forma typica*) and melanic forms of *Biston betularia* were recorded by Kettlewell and his colleagues. In all more than 20,000 specimens were examined. The principal melanic form (*forma carbonaria*) was abundant near industrial areas and where the prevailing westerly winds carry atmospheric pollution to the east. A further melanic form (*forma insularia*, which looks like an intermediate form but is due to several different genes controlling darkening) was also present but could not be detected where the genes for *forma carbonaria* were present.

The earliest recorded species to evolve in this way was the peppered moth (*Biston betularia*); the first black specimen was caught in Manchester, UK in 1848. By 1895, about 98% of the Manchester peppered moth population was melanic. Following many more years of pollution, a large-scale survey of pale and melanic forms of the peppered moth in Britain recorded more than 20,000 specimens between 1952 and 1970 (Figure 2.9). The winds in Britain are predominantly westerlies, spreading industrial pollutants (especially smoke and sulfur dioxide) toward the east. Melanic forms were concentrated toward the east and were completely absent from unpolluted western parts of England and Wales, northern Scotland, and Ireland.

The moths are preyed upon by insectivorous birds that hunt by sight. In a field experiment, large numbers of melanic and pale ('typical') moths were reared and released in equal numbers in a rural and largely unpolluted area of southern England. Of the 190 moths that were captured by birds, 164 were melanic and

Figure 2.10

Change in the frequency of the *carbonaria* form of the peppered moth *Biston betularia* in the Manchester area since 1950, covering the period where smoke pollution has been controlled and the frequency has declined dramatically. Vertical lines show standard errors.

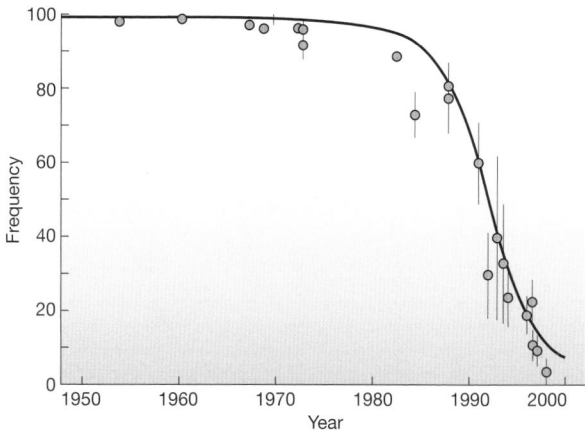

AFTER COOK ET AL., 1999

26 were typicals. An equivalent study was made in an industrial area near the city of Birmingham. Twice as many melanics as typicals were recaught. This showed that a significant selection pressure was exerted through bird predation, and that moths of the typical form were clearly at a disadvantage in the polluted industrial environment (where their light color stood out against a sooty background), whereas the melanic forms were at a disadvantage in the pollution-free countryside (Kettlewell, 1955).

In the 1960s, however, industrialized environments in Western Europe and the United States started to change as oil and electricity began to replace coal, and legislation was passed to impose smoke-free zones and to reduce industrial emissions of sulfur dioxide (see Chapter 13). The frequency of melanic forms then fell back to near preindustrial levels with remarkable speed (Figure 2.10).

The forces of selection at work, first in favor of and then against melanic forms, have clearly been related to industrial pollution, but the idea that melanic forms were favored simply because they were camouflaged against smoke-stained backgrounds may be only part of the story. The moths rest on tree trunks during the day, and non-melanic moths are well hidden against a background of mosses and lichens. Industrial pollution had not just blackened the moths' background; atmospheric pollution, especially SO_2, had also destroyed most of the moss and lichen on the tree trunks. Indeed the distribution of melanic forms in Figure 2.9 closely fits the areas in which tree trunks were likely to have lost lichen cover as a result of SO_2 and so ceased to provide such effective camouflage for the non-melanic moths. Thus SO_2 pollution may have been as important as smoke in selecting melanic moths.

natural selection by pollution – evolution of heavy-metal tolerance in plants

Some plants are tolerant of another form of pollution: the presence of toxic heavy metals such as lead, zinc, and copper, which contaminate the soil after mining. Populations of plants on contaminated areas may be tolerant, while at the edge of these areas a transition from tolerant to intolerant forms can occur over very short distances (Figure 2.11). In some cases it has been possible to measure the speed of evolution. Zinc-tolerant forms of two species of the grass *Agrostis capillaris* were found to have evolved under zinc-galvanized electricity pylons within 20–30 years of their erection (Al-Hiyaly et al., 1988).

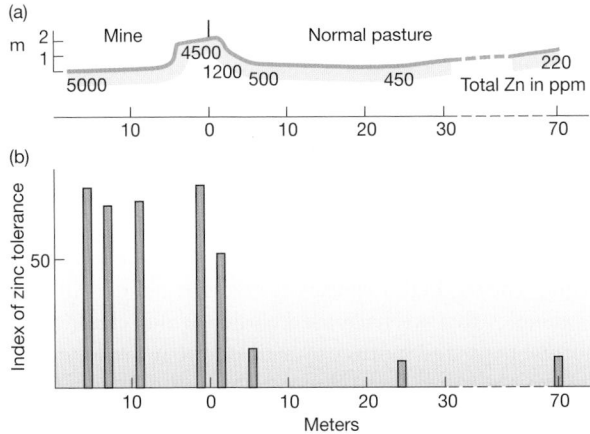

AFTER PUTWAIN, IN JAIN & BRADSHAW, 1966

Figure 2.11

The grass *Anthoxanthum odoratum* colonizes land heavily contaminated with zinc (Zn) on old mines. This is possible because the grass has evolved zinc-tolerant forms. (a) Samples of the grass were taken along a transect from a mine (at Trelogan in North Wales) into surrounding grassland (zinc concentrations in the soil are shown as parts per million, ppm) and were tested for zinc tolerance by measuring the length of roots that they produced when grown in a culture solution containing excess zinc. (b) The index of zinc tolerance falls off steeply over a distance of 2–5 m at the mine boundary.

2.3.3 *Evolution and coevolution*

It is easy to see that a population of plants faced with repeated drought is likely to evolve a tolerance of water shortage, and an animal repeatedly faced with cold winters is likely to evolve habits of hibernation or a thick protective coat. But droughts do not become any less severe as a result, nor winters milder. Physical conditions are not heritable: they leave no descendants, and they are not subject to natural selection. But the situation is quite different when two species interact: predator on prey, parasite on host, competitive neighbor on neighbor. Natural selection may select from a population of parasites those that are more efficient at infecting their host. But this immediately sets in play forces of natural selection that favor more resistant hosts. As they evolve, they put further pressure on the ability of the parasite to infect. Host and parasite are then caught in never-ending reciprocating selection: they *coevolve*. In many other ecological interactions, the two parties are not antagonists but positively beneficial to one another: *mutualists*. Pollinators and their plants, and leguminous plants and their nitrogen-fixing bacteria, are well-known examples. We consider coevolution in some detail when we return to more evolutionary aspects of ecology in Chapter 8.

2.4 The ecology of speciation

We have seen that natural selection can force populations of plants and animals to change their character – to evolve. But none of the examples we have considered has involved the evolution of a new species. Indeed Darwin's *On the Origin of Species* is about natural selection and evolution but is not really about the origin of species! 'Black' and 'typical' peppered moths are forms within a species, not different species. Likewise, the different growth forms of the grasses on the cliffs and pastures of Abraham's Bosom and the dull and flamboyant races of guppies are just local genetic classes. None qualifies for the status of distinct species. But when we ask just what criteria justify naming two populations as different species we meet real problems.

2.4.1 What do we mean by a 'species'?

Cynics have said, with some truth, that a species is what a competent taxonomist regards as a species. Darwin himself regarded species (like genera) as 'merely artificial combinations made for convenience'. On the other hand, in the 1930s, two American biologists, Mayr and Dobzhansky, proposed an empirical test that could be used to decide whether two populations were part of the same species or of two different species. They recognized organisms as being members of a single species if they could, at least potentially, breed together in nature to produce fertile offspring. They called a species tested and defined in this way a *biospecies*. In the examples that we have used earlier in this chapter we know that melanic and normal peppered moths can mate and that the offspring are fully fertile; this is also true of colored and dull guppies and of plants from the different types of *Agrostis*. They are all variations within species – not separate species.

biospecies do not exchange genes

In practice, however, biologists do not apply the Mayr–Dobzhansky test before they recognize every species: there is simply not enough time and resources. What is more important is that the test recognizes a crucial element in the evolutionary process. Two parts of a population can evolve into distinct species only if some sort of barrier prevents gene flow between them. If the members of two populations are able to hybridize and their genes are combined and reassorted in their progeny, then natural selection can never make them truly distinct.

orthodox speciation

The most orthodox scenario for speciation comprises a number of stages (Figure 2.12). First, two subpopulations become geographically isolated and natural selection drives genetic adaptation to their local environments. Next, as a *byproduct* of this genetic differentiation, a degree of reproductive isolation builds

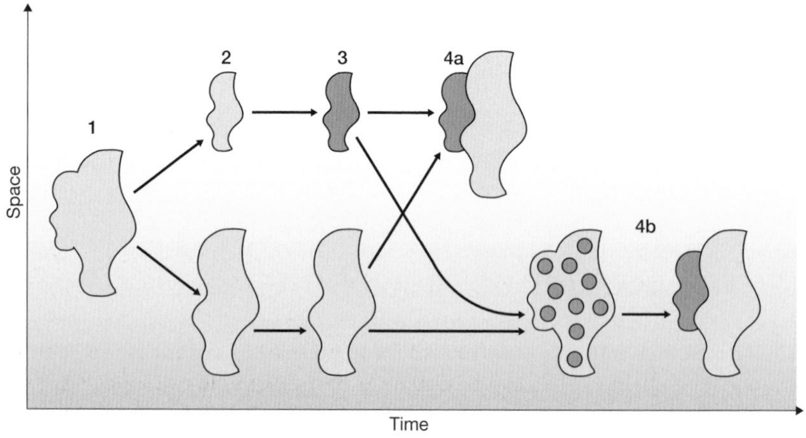

Figure 2.12

The orthodox picture of ecological speciation. A uniform species with a large range (1) differentiates into subpopulations (2; for example, separated by geographic barriers or dispersed onto different islands), which become genetically isolated from each other (3). After evolution in isolation they may meet again, when they are either already unable to hybridize (4a) and have become true biospecies, or they produce hybrids of lower fitness (4b), in which case evolution may favor features that prevent interbreeding between the 'emerging species' until they are true biospecies.

up between the two. This may be, for example, a difference in courtship ritual, tending to prevent mating in the first place. This is referred to as 'prezygotic' isolation. Alternatively, the offspring themselves may simply display a reduced viability. Then, in a phase of *secondary contact*, the two subpopulations re-meet. The hybrids between individuals from the different subpopulations are now of low fitness, because they are literally neither one thing nor the other. Natural selection will then favor any feature in either subpopulation that *reinforces* reproductive isolation, especially prezygotic characteristics, preventing the production of low-fitness hybrid offspring. These breeding barriers then cement the distinction between what have now become separate species.

It would be wrong, however, to imagine that all examples of speciation conform fully to this orthodox picture (Schluter, 2001). First, there may never be secondary contact. This would be pure 'allopatric' speciation (that is, with all divergence occurring in subpopulations in *different* places). This is especially likely for island species, which are examined further below.

allopatric and sympatric
speciation

Second, there has been increasing support for the view that a phase of physical isolation is not necessary: that is, '*sym*patric' speciation is possible (divergence occurring in subpopulations in the *same* place). One circumstance in which this seems likely to occur is where insects feed on more than one species of host plant, and where each requires specialization by the insects to overcome the plant's defenses. (Consumer-resource defense and specialization are examined more fully in Chapters 3 and 7.) Particularly persuasive in this is the existence of a continuum from populations of insects feeding on more than one host plant, through populations differentiated into 'host races' (coexisting subpopulations that specialize on different host plants but exchange genes at a rate of more than around 1% per generation), to distinct but closely related coexisting species, specializing on their particular hosts (Drès and Mallet, 2001). This continuum reminds us that the origin of a species, whether allopatric or sympatric, is a process, not an event. For the formation of a new species, like the boiling of an egg, there is some freedom to argue about when it is completed.

These same points are further illustrated by the extraordinary case of two species of sea gull. The lesser black-backed gull (*Larus fuscus*) originated in Siberia and colonized progressively to the west, forming a chain or *cline* of different forms, spreading from Siberia to Britain and Iceland (Figure 2.13). The neighboring forms along the cline are distinctive, but they hybridize readily in nature. Neighboring populations are therefore regarded as part of the same species and taxonomists give them only 'subspecific' status (e.g., *Larus fuscus graelsii*, *Larus fuscus fuscus*, the three words referring to genus, species and subspecies). Populations of the gull have, however, also spread east from Siberia, again forming a cline of freely hybridizing forms. Together, the populations spreading east and west encircle the northern hemisphere. They meet and overlap in northern Europe. There, the eastward and westward clines have diverged so far that it is easy to tell them apart, and they are recognized as two different *species*, the lesser black-backed gull (*Larus fuscus*) and the herring gull (*Larus argentatus*). Moreover, the two species do not hybridize: they have become true biospecies. We can see how two distinct species have evolved from one primal stock, and that the stages of their divergence remain frozen in the cline that connects them.

evolution in sea gulls

Figure 2.13

Two species of gull, the herring gull and the lesser black-backed gull, have diverged from a common ancestry as they have colonized and encircled the northern hemisphere. Where they occur together in northern Europe they fail to interbreed and are clearly recognized as two distinct species. However, they are linked along their ranges by a series of freely interbreeding races or subspecies.

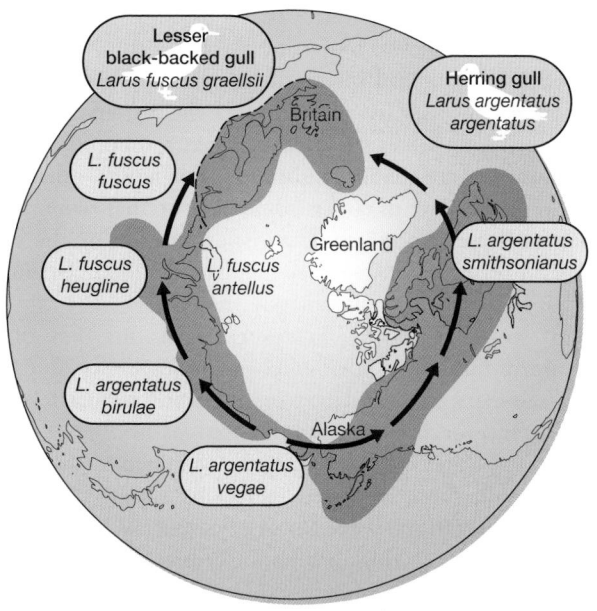

2.4.2 Islands and speciation

Darwin's finches

It is, though, when a population becomes split into completely isolated populations, dispersed onto different islands especially, that they most readily diverge into distinct species. The most celebrated example of evolution and speciation on islands is the case of Darwin's finches in the Galapagos archipelago. The Galapagos are volcanic islands isolated in the Pacific Ocean about 1000 km west of Equador and 750 km from the island of Cocos, which is itself 500 km from Central America. At more than 500 m above sea level the vegetation is open grassland. Below this is a humid zone of forest that grades into a coastal strip of desert vegetation with some endemic species of prickly pear cactus (*Opuntia*). Fourteen species of finch are found on the islands, and there is every reason to suppose that these evolved from a single ancestral species that invaded the islands from the mainland of Central America.

In their remote island isolation, the Galapagos finches have radiated into a variety of species in groups with contrasting ecologies (Figure 2.14). Members of one group, including *Geospiza fuliginosa* and *G. fortis*, have strong bills and hop and scratch for seeds on the ground. *Geospiza scandens* has a narrower and slightly longer bill and feeds on the flowers and pulp of the prickly pears as well as on seeds. Finches of a third group have parrot-like bills and feed on leaves, buds, flowers and fruits, and a fourth group with a parrot-like bill (*Camarhynchus psittacula*) has become insectivorous, feeding on beetles and other insects in the canopy of trees. A so-called woodpecker finch, *Camarhynchus* (*Cactospiza*) *pallida*, extracts insects from crevices by holding a spine or a twig in its bill. Yet a further group includes a species (*Certhidea olivacea*) that, rather like a warbler, flits around actively and collects small insects in the forest canopy and in the air. Populations of ancestor species became reproductively isolated, most likely after chance colonization of different islands within the archipelago, and evolved

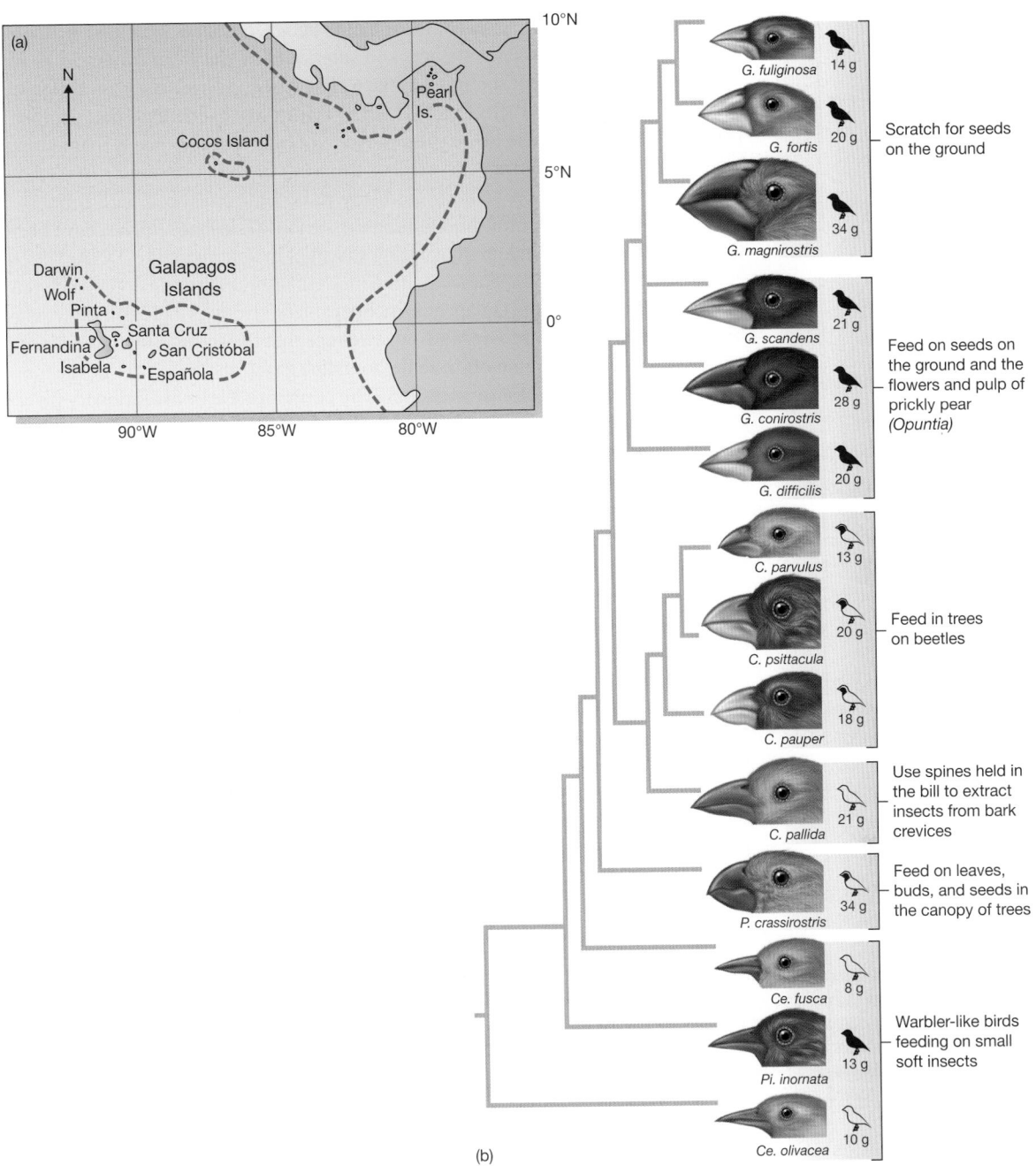

AFTER PETREN ET AL., 1999

Figure 2.14

(a) Map of the Galapagos Islands showing their position relative to Central and South America; on the equator 5° equals approximately 560 km.
(b) A reconstruction of the evolutionary history of the Galapagos finches based on variation in the length of microsatellite DNA. The *genetic distance* (a measure of the genetic difference) between species is shown by the length of the horizontal lines. Notice the great and early separation of the warbler finch (*Certhidea olivacea*) from the others, suggesting that it may closely resemble the founders that colonized the islands. The feeding habits of the various species are also shown. Drawings of the birds are proportional to actual body size. The maximum amount of black coloring in male plumage and the average body mass are shown for each species. *C., Camarhynchus; Ce., Certhidea; G., Geospiza; P. Platyspiza; Pi., Pinaroloxias*.

separately for a time. Subsequent movements between islands may have brought non-hybridizing biospecies together, and subsequently these have evolved to fill different niches. We will see in Chapter 6 that when individuals from different species compete, natural selection may act to favor those individuals that compete least with members of the other species. An expected consequence is that among a group of closely related species, such as Darwin's finches, differences in feeding and other aspects of their ecology are likely to become enhanced with time.

The evolutionary relationships among the various Galapagos finches have been traced by molecular techniques (analyzing variation in 'microsatellite' DNA; Petren et al., 1998) (Figure 2.14). These accurate modern tests confirm the long-held view that the family tree of the Galapagos finches radiated from a single trunk (i.e. was *monophyletic*) and also provides strong evidence that the warbler finch (*Certhidea olivacea*) was the first to split off from the founding group and is likely to be the most similar to the original colonist ancestors. The entire process of evolutionary divergence of these species appears to have happened in less than 3 million years.

island endemics

The flora and fauna of many other archipelagos show similar examples of great richness of species with many local *endemics* (i.e. species known only from one island or area). Lizards of the genus *Anolis* have evolved a kaleidoscopic diversity of species on the islands of the Caribbean; and isolated groups of islands, such as the Canaries off the coast of North Africa, are treasure troves of endemic plants. The endemics evolve, of course, because they are isolated from individuals of the original species, or other species, with which they might hybridize. An illustration of the importance of isolation in the evolution of endemics is provided by the animals and plants of Norfolk Island. This small island (about 70 km^2) is approximately 700 km from New Caledonia and New Zealand, but about 1200 km from Australia. Hence, the ratio of Australian species to New Zealand and New Caledonian species within a group can be used as a measure of that group's dispersal ability, and the poorer the dispersal ability the greater the isolation. As Figure 2.15 shows, the proportion of endemics on Norfolk Island is highest in groups with poor dispersal ability (more isolated) and lowest in groups with good dispersal ability.

Unusual and often rich communities of endemics may also pose particular problems for the applied ecologist (Box 2.2).

Figure 2.15

The evolution of endemic species on islands as a result of their isolation from individuals of an original species with which they might interbreed. Poorly dispersing (and therefore more 'isolated') groups on Norfolk Island have a higher proportion of endemic species and are more likely to contain species from either New Caledonia or New Zealand than from Australia, which is further away.

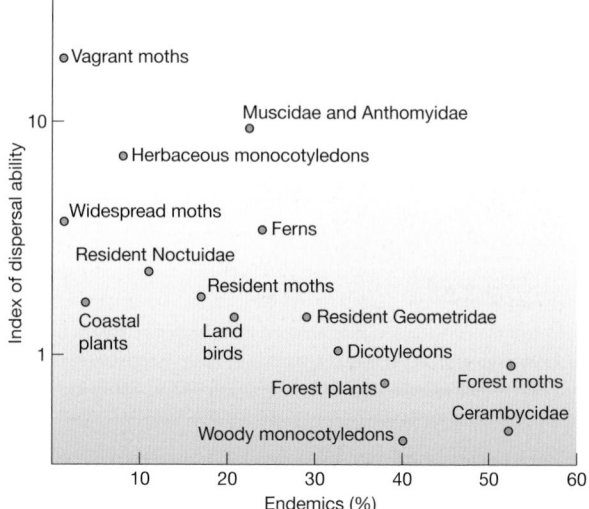

AFTER HOLLOWAY, 1977

2.2 *Topical ECOncerns*

2.2 TOPICAL ECONCERNS

Deep sea vent communities at risk

Deep sea vents are islands of warmth in oceans that are otherwise cold and inhospitable. As a consequence, they support unique communities, rich in endemic species. One of the latest controversies to pit environmentalists against industrialists concerns these deep sea vents, which are also now known to be sites rich in minerals. This newspaper article by William J. Broad appeared in the *San Jose Mercury News*, January 20, 1998.

> With miners staking claim to valuable metals lying in undersea lodes in the South Pacific, questions surface about how to prevent disasters in these fragile, little understood ecosystems.

> The volcanic hot springs of the deep sea are dark oases that teem with blind shrimp, giant tube worms and other bizarre creatures, sometimes in profusions great enough to rival the chaos of rain forests. And they are old.

> Scientists who study them say these odd environments, first discovered two decades ago, may have been the birthplace of all life on Earth, making them central to a new wave of research on evolution.

> Now, in a moment that diverse ranks of experts have feared and desired for years, miners are invading the hot springs, possibly setting the stage for the last great battle between industrial development and environmental preservation.

> The undersea vents are rich not just in life but in valuable minerals such as copper, silver and gold. Indeed, their smoky chimneys and rocky foundations are virtual foundries for precious metals. . . . The fields of undersea gold have long fired the imaginations of many scientists and economists, but no mining took place, in part because the rocky deposits were hard to lift from depths of a mile or more.

> Now, however, miners have staked the first claim to such metal deposits after finding the richest ores ever. The estimated value of copper, silver and gold at a South Pacific site is up to billions of dollars. Environmentalists, though, want to protect the exotic ecosystem by banning or severely limiting mining.

(Article written for the *New York Times*. Copyright Globe Newspaper Company; reprinted by permission.)

A deep sea vent community.
© WHOI, J. EDMOND, VISUALS UNLIMITED

Consider the following options and debate their relative merits:

1 *Allow the mining industry free access to all deep sea vents, since the wealth created will benefit many people.*

2 *Ban mining and other disruption of all deep sea vent communities, recognizing their unique biological and evolutionary characteristics.*

3 *Carry out biodiversity assessments of known vent communities and prioritize according to their conservation importance, permitting mining in cases that will minimize overall destruction of this category of community.*

2.5 Effects of climatic change on the evolution and distribution of species

Changes in climate, particularly during the ice ages of the Pleistocene (the past 2–3 million years), bear a lot of the responsibility for the present patterns of distribution of plants and animals. As climates have changed, species populations have advanced and retreated, have been fragmented into isolated patches, and may have then rejoined. Much of what we see in the present distribution of species represents phases in a recovery from past climatic change. Modern techniques for analyzing and dating biological remains (particularly buried pollen) are beginning to allow us to detect just how much of the present distribution of organisms is a precise, local, evolved match to present environments, and how much is a fingerprint left by the hand of history.

cycles of glaciation have occurred repeatedly

For most of the past 2–3 million years the Earth has been very cold. Evidence from the distribution of oxygen isotopes in cores taken from the deep ocean floor shows that there may have been as many as 16 glacial cycles in the Pleistocene, each lasting for up to 125,000 years (Figure 2.16a). Each cold (glacial) phase may have lasted for as long as 50,000–100,000 years, with brief intervals of only 10,000–20,000 years when the temperatures rose to, or above, those of today. In this case, present floras and faunas are unusual, having developed at the warm end of one of a series of unusual catastrophic warm periods.

Figure 2.16

(a) An estimate of the global temperature variations with time during glacial cycles over the past 400,000 years. The estimates were obtained by comparing oxygen isotope ratios in fossils taken from ocean cores in the Caribbean. The dashed line corresponds to the ratio 10,000 years ago, at the start of the present warming period. Periods as warm as the present have been rare events, and the climate during most of the past 400,000 years has been glacial. (b) Ranges in eastern North America, as indicated by pollen percentages in sediments, of spruce species (above) and oak species (below) from 21,500 years ago to the present. Note how the ice sheet contracted during this period.

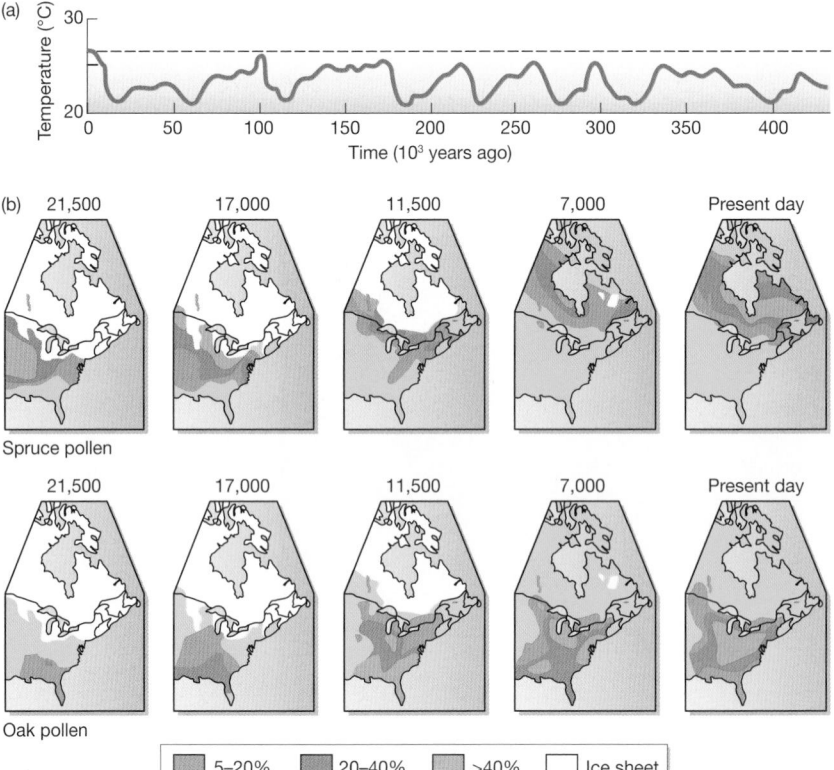

(a) AFTER EMILIANI, 1966; DAVIS, 1976; (b) AFTER DAVIS & SHAW, 2001

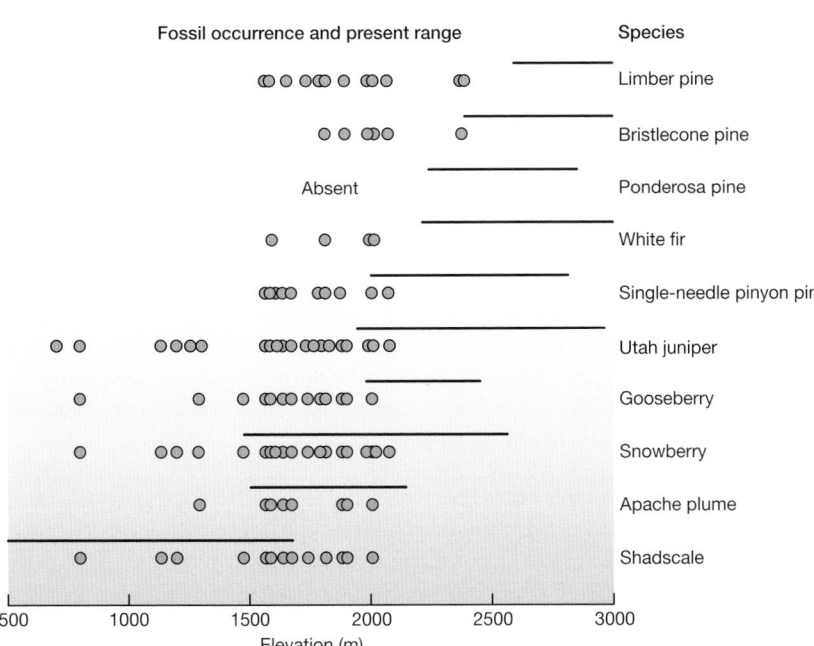

Fossil occurrence and present range | Species

Limber pine
Bristlecone pine
Ponderosa pine — Absent
White fir
Single-needle pinyon pine
Utah juniper
Gooseberry
Snowberry
Apache plume
Shadscale

Elevation (m)
500 1000 1500 2000 2500 3000

AFTER DAVIS & SHAW, 2001

Figure 2.17

The elevation ranges of 10 species of woody plant from the mountains of the Sheep Range, Nevada during the last glaciation (dots) and at present (solid line).

During the 20,000 years since the peak of the last glaciation, global temperatures have risen by about 8°C. The analysis of buried pollen – particularly of woody species, which produce most of the pollen – can show how vegetation has changed during this period (Figure 2.16b). As the ice retreated, different forest species advanced in different ways and at different speeds. For some, like the spruce of eastern North America, there was displacement to new latitudes; for others, like the oaks, the picture was more one of expansion.

We do not have such good records for the postglacial spread of the animals associated with the changing forests, but it is at least certain that many species could not have spread faster than the trees on which they feed. Some of the animals may still be catching up with their plants, and tree species are still returning to areas they occupied before the last ice age! It is quite wrong to imagine that our present vegetation is in some sort of equilibrium with (adapted to) the present climate.

Even in regions that were never glaciated, pollen deposits record complex changes in distribution: in the mountains of the Sheep Range, Nevada, for example, woody species show different patterns of change in elevational range (Figure 2.17). The species composition of vegetation has continually been changing and is almost certainly still doing so.

The records of climatic change in the tropics are far less complete than those for temperate regions. Many believe, though, that during cooler, drier glacial periods, the tropical forests retreated to smaller patches, surrounded by a sea of savanna. Support for this comes from the present-day distribution of species in the tropical forests of South America (Figure 2.18). There, particular 'hotspots' of species diversity are apparent, and these are thought to be likely sites of forest refuges during the glacial periods, and sites too, therefore, of increased rates of speciation (Ridley, 1993). On this interpretation, the present distributions of

the distribution of trees has changed gradually since the last glaciation

AFTER RIDLEY, 1993

Figure 2.18

(a) The present-day distribution of tropical forest in South America.
(b) The possible distribution of tropical forest refuges at the time when the last glaciation was at its peak, as judged by present-day hot spots of species diversity within the forest.

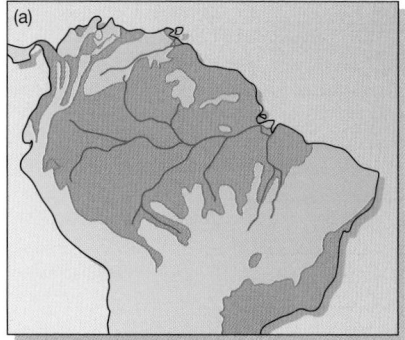
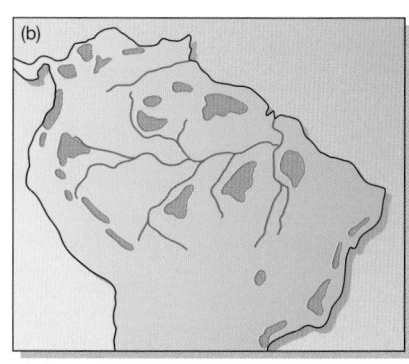

species may again be seen as largely accidents of history (where the refuges were) rather than precise matches between species and their differing environments.

Evidence of changes in vegetation that followed the last retreat of the ice hint at the likely consequences of the global warming (maybe 3°C in the next 100 years) that is predicted to result from continuing increases in 'greenhouse' gases in the atmosphere (see Chapter 13). But the scales are quite different. Postglacial warming of about 8°C occurred over around 20,000 years, and changes in the vegetation failed to keep pace even with this. But current projections for the 21st century require range shifts for trees at rates of 300–500 km per century compared to typical rates in the past of 20–40 km per century (and exceptional rates of 100–150 km). It is striking that the only precisely dated extinction of a tree species in the Quaternary period, that of *Picea critchfeldii*, occurred around 15,000 years ago at a time of especially rapid postglacial warming (Jackson & Weng, 1999). Clearly, even more rapid change in the future could result in extinctions of many additional species (Davis & Shaw, 2001).

predicted global warming by the 'greenhouse effect' is nearly 100 times faster than postglacial warming

2.6 Effects of continental drift on the ecology of evolution

land masses have moved . . .

The patterns of species formation that occur on islands appear on an even larger scale in the evolution of genera and families across continents. Many curious distributions of organisms between continents seem inexplicable as the result of dispersal over vast distances. Biologists, especially Wegener (1915), met outraged scorn from geologists and geographers when they argued that it must have been the continents that had moved rather than the organisms that had dispersed. Eventually, however, measurements of the directions of the Earth's magnetic fields required the same, apparently wildly improbable, explanation and the critics capitulated. The discovery that the tectonic plates of the Earth's crust move and carry the migrating continents with them reconciles geologist and biologist (Figure 2.19). While major evolutionary developments were occurring in the plant and animal kingdoms, their populations were being split and separated, and land areas were moving across climatic zones. This was happening while changes in temperature were occurring on a vastly greater scale than the glacial cycles of the Pleistocene episode.

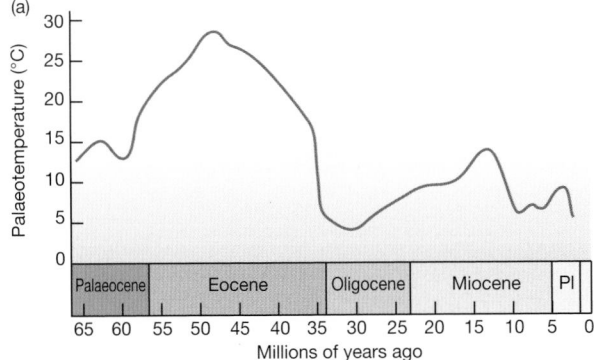

Figure 2.19

(a) Changes in temperature in the North Sea over the past 65 million years. During this period there were large changes in sea level that allowed dispersal of both plants and animals between land masses. (b–e) Continental drift. (b) The ancient supercontinent of Gondwanaland began to break up about 150 million years (Myr) ago. (c) About 50 Myr ago (early Middle Eocene) recognizable bands of distinctive vegetation had developed, and (d) by 32 Myr ago (early Oligocene) these had become more sharply defined. (e) By 10 Myr ago (early Miocene) much of the present geography of the continents had become established but with dramatically different climates and vegetation from today: the position of the Antarctic ice cap is highly schematic.

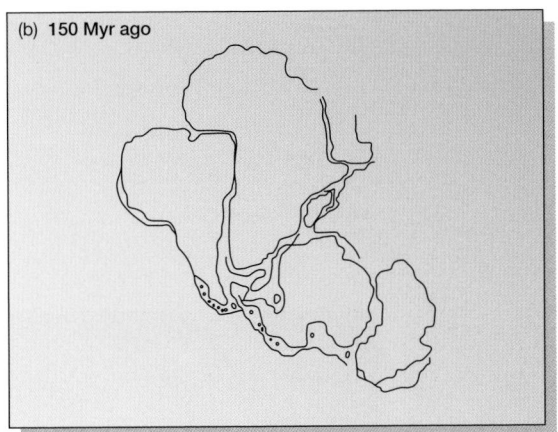

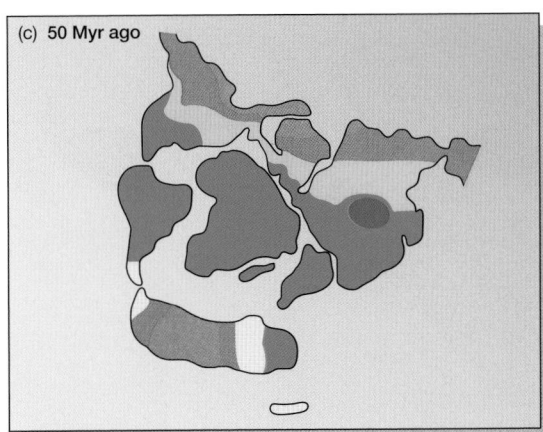

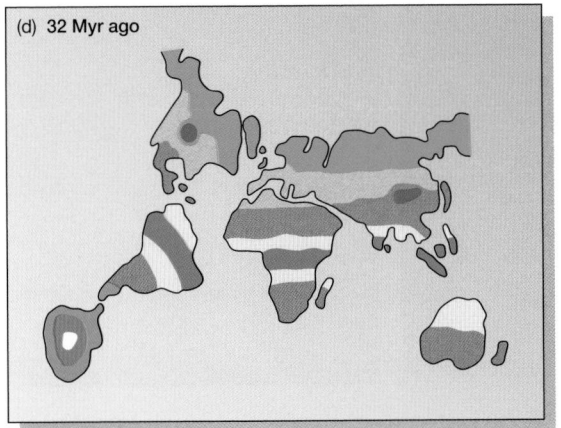

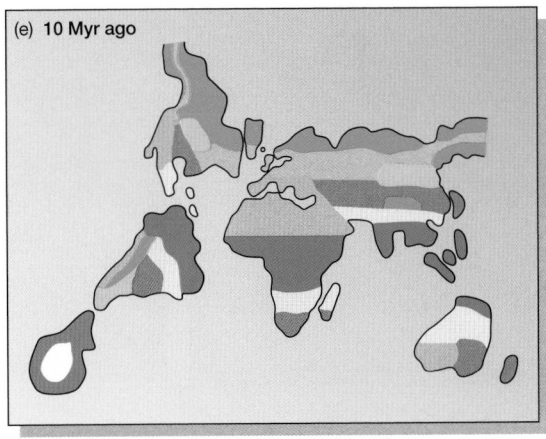

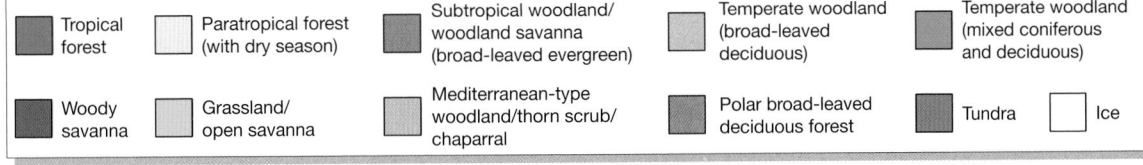

ADAPTED FROM NORTON & SCLATER, 1979; JANIS, 1993; AND OTHER SOURCES

. . . and divided populations that
have then evolved independently

The established drift of the continents answers many questions in the ecology of evolution. The curious world distribution of large flightless birds is one example (Figure 2.20a). The presence of the ostrich in Africa, the emu in Australia, and the very similar rhea in South America could scarcely be explained by dispersal

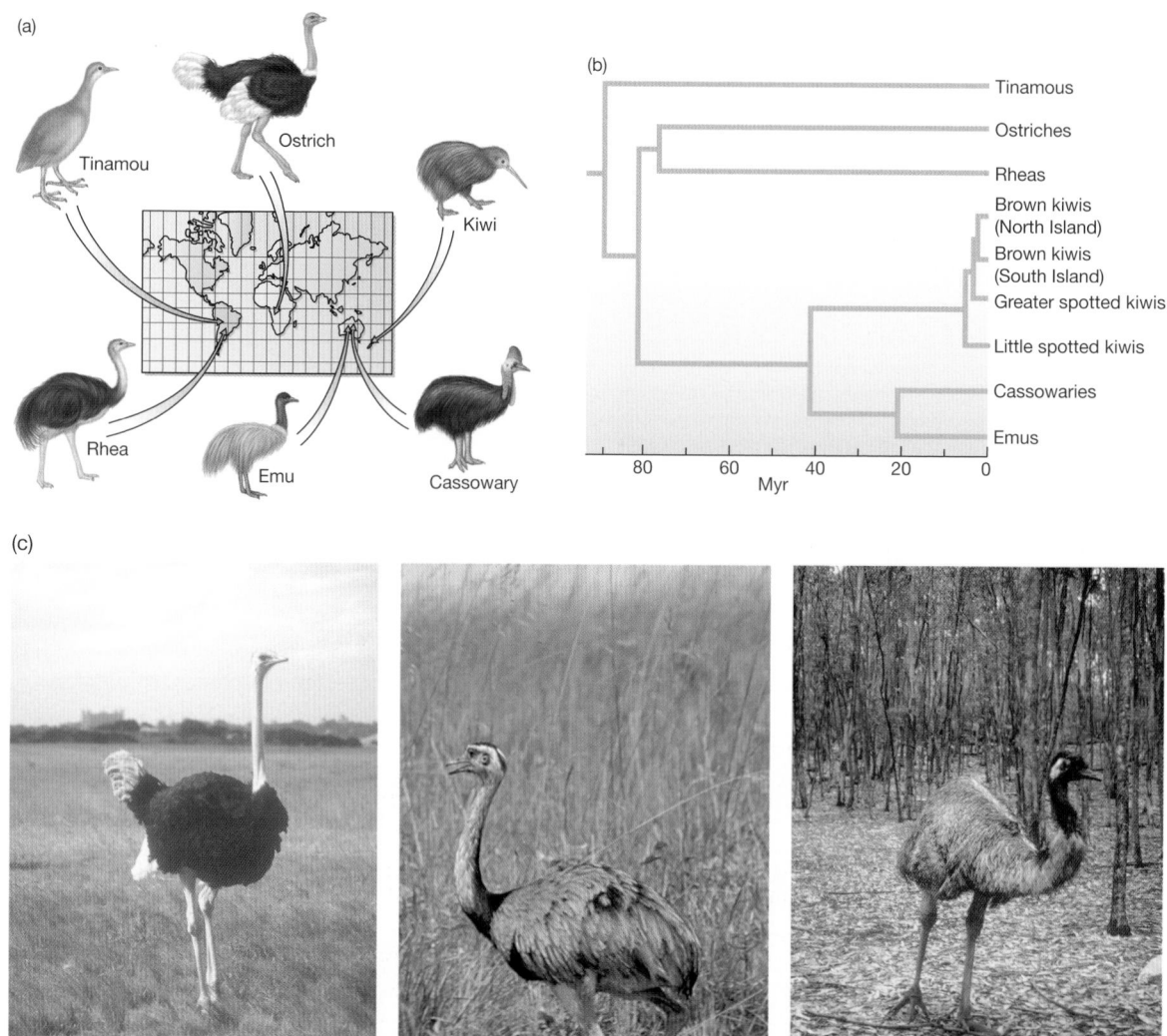

Figure 2.20

(a) The distribution of terrestrial flightless birds. (b) The phylogenetic tree of the flightless birds and the estimated times (million years, Myr) of their divergence. (c) Photos of large flightless birds found in three major continents: (left) the ostrich (*Struthio camelus*) is African and commonly occurs together with herds of zebra and antelope in savanna or steppe grasslands; (middle) the rhea (*Rhea americana*) is found in similar grasslands in South America (e.g. Brazil, Argentina), commonly together with herds of deer and guanacos; and (right) the emu (*Dromaius novaehollandiae*) inhabits equivalent habitats in Australia. Many other species of these very large, mainly herbivorous birds have been sought after by humans for food and have become extinct. The presence of these evolutionarily related and ecologically similar species in three widely separated continents is explained by the drifting apart of the continents from the time (150 Myr ago) when they were portions of the primitive continent of Gondwanaland (Figure 2.19).

of some common flightless ancestor. Now, techniques of molecular biology make it possible to analyze the time at which the various flightless birds started their evolutionary divergence (Figure 2.20b). The tinamous seem to have been the first to diverge and became evolutionarily separate from the rest, the *ratites*. Australasia next became separated from the other southern continents, and, from the latter, the ancestral stocks of ostriches and rheas were subsequently separated when the Atlantic opened up between Africa and South America. Back in Australasia, the Tasman Sea opened up about 80 million years ago and ancestors of the kiwi are thought to have made their way, by island hopping, about 40 million years ago across to New Zealand, where divergence into the present species happened relatively recently. The unraveling of this particular example implies the early evolution of the property of flightlessness and only subsequently the isolation of the different types between the emerging continents.

2.7 Interpreting the results of evolution: convergents and parallels

Flightlessness did not evolve independently on the different continents. However, there are many examples of organisms that have evolved in isolation from each other and then converged on remarkably similar forms or behavior. Such similarity is particularly striking when similar roles are played by structures that have quite different evolutionary origins – that is, when the structures are *analogous* (similar in superficial form or function) but not *homologous* (derived from an equivalent structure in a common ancestry). When this occurs, it is termed *convergent evolution*. Bird and bat wings are a classic example (Figure 2.21).

convergent evolution

Further examples show *parallels* in the evolutionary pathways of ancestrally related groups occurring after they were isolated from each other. The classic example is provided by placental and marsupial mammals. Marsupials arrived on what would become the Australian continent in the Cretaceous period (around 90 million years ago; see Figure 2.19), when the only other mammals present were the curious egg-laying monotremes (now represented only by the spiny anteaters and the duck-billed platypus). An evolutionary process of radiation then occurred among the Australian marsupials that in many ways accurately paralleled what was occurring among the placental mammals on other continents (Figure 2.22). It is hard to escape the view that the environments of placentals and marsupials contained ecological pigeonholes (niches) into which the evolutionary process has neatly 'fitted' ecological equivalents. In contrast to convergent evolution, however, the marsupials and placentals started to diversify from a common ancestral line, and both inherited a common set of potentials and constraints.

parallel evolution

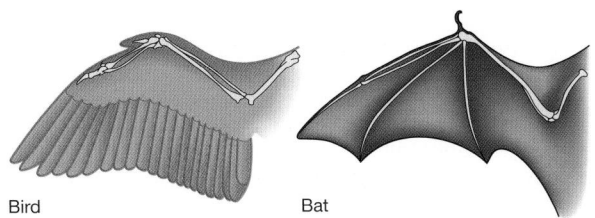

AFTER RIDLEY, 1993

Bird   Bat

Figure 2.21

Convergent evolution: the wings of bats and birds are analogous (not homologous). They are structurally different: the bird wing is supported by digit number 2 and covered with feathers; the bat wing is supported by digits 2–5 and covered with skin.

Figure 2.22

Parallel evolution of marsupial and placental mammals. The pairs of species are similar in both appearance and habit and usually (but not always) in lifestyle.

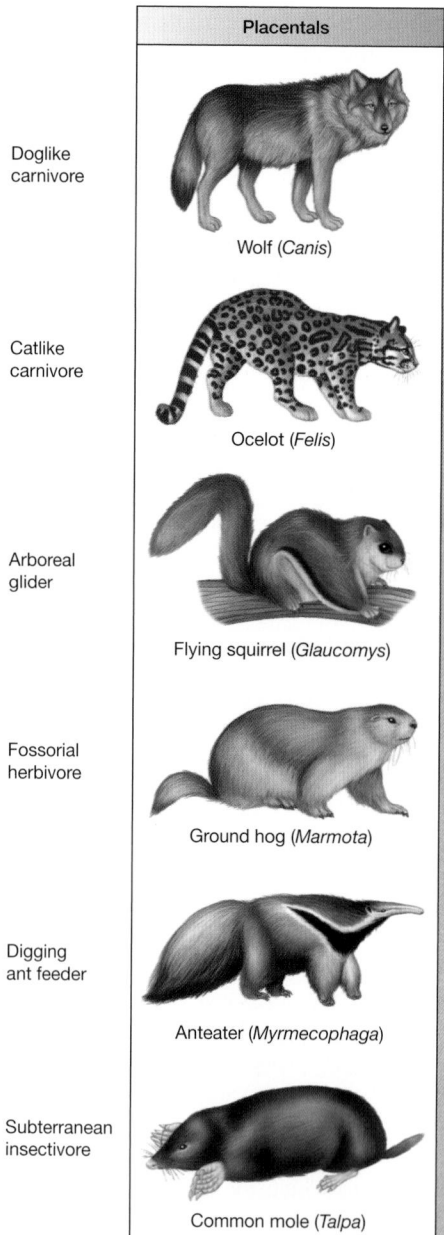
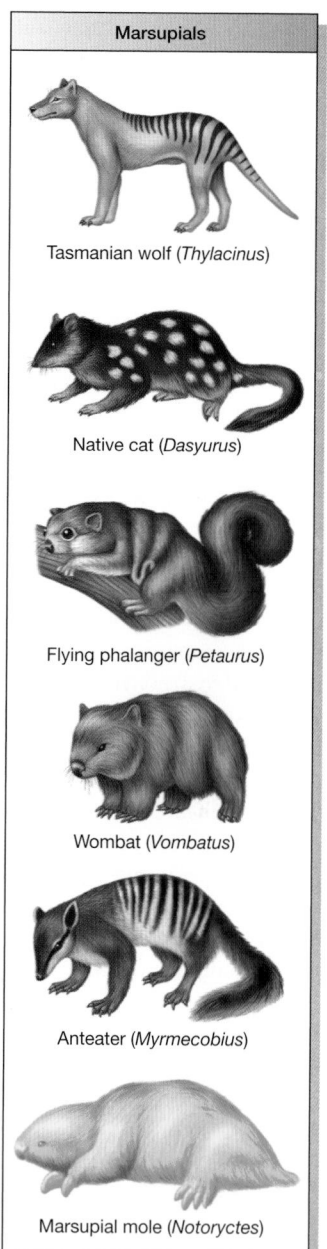

Doglike carnivore — Wolf (*Canis*) | Tasmanian wolf (*Thylacinus*)

Catlike carnivore — Ocelot (*Felis*) | Native cat (*Dasyurus*)

Arboreal glider — Flying squirrel (*Glaucomys*) | Flying phalanger (*Petaurus*)

Fossorial herbivore — Ground hog (*Marmota*) | Wombat (*Vombatus*)

Digging ant feeder — Anteater (*Myrmecophaga*) | Anteater (*Myrmecobius*)

Subterranean insectivore — Common mole (*Talpa*) | Marsupial mole (*Notoryctes*)

Placentals | Marsupials

interpreting the match between organisms and their environment

When we marvel at the diversity of complex specializations by which organisms match their varied environments there is a temptation to regard each case as an example of evolved perfection. But there is nothing in the process of evolution by natural selection that implies perfection. The evolutionary process works on the genetic variation that is available. It favors only those forms that are fittest from among the range of variety available, and this may be a very restricted choice. The very essence of natural selection is that organisms come to match their environments by being 'the fittest available' or 'the fittest yet': they are not 'the best imaginable'.

It is particularly important to realize that past events on Earth can have profound repercussions on the present. Our world has not been constructed by taking each organism in turn, testing it against each environment, and moulding it so that every organism finds its perfect place. It is a world in which organisms live where they do for reasons that are often, at least in part, accidents of history. Moreover the ancestors of the organisms that we see around us lived in environments that were profoundly different from those of the present. Evolving organisms are not free agents – some of the features acquired by their ancestors hang like millstones around their necks, limiting and constraining where they can now live and what they might become. It is very easy to wonder and marvel at how beautifully the properties of fish fit them to live in water – but just as important to emphasize that these same properties prevent them from living on land.

Having sketched out the evolutionary background for the whole of ecology in this chapter, we will return to some particular topics in evolutionary ecology in Chapter 8, especially aspects of coevolution, where interacting pairs of species play central roles in one another's evolution. However, since evolution does provide a backdrop to all ecological acts, its influence can of course be seen throughout the remainder of this book.

Summary
SUMMARY

The force of natural selection

Life is represented on Earth by a diversity of specialist species, each of which is absent from almost everywhere. Early interest in this diversity mainly existed among explorers and collectors, and the idea that the diversity had arisen by evolution from earlier ancestors over geological time was not seriously discussed until the first half of the 19th century. Charles Darwin and Alfred Russel Wallace (strongly influenced by having read Malthus's essay *An Essay on the Principle of Population*) independently proposed that natural selection constituted a force that would drive a process of evolution. The theory of natural selection is an ecological theory. The reproductive potential of living organisms leads them inescapably to compete for limited resources. Success in this competition is measured by leaving more descendants than others to subsequent generations. When these ancestors differ in properties that are heritable the character of populations will necessarily change over time and evolution will happen.

Darwin had seen the power of human selection to change the character of domestic animals and plants and he recognized the parallel in natural selection. But there is one big difference: humans select for what they want in the future, but natural selection is a result of events in the past – it has no intentions and no aim.

Natural selection in action

We can see natural selection in action within species in the variation within species over their geographic range and even over very short distances where we can detect powerful selective forces in action and recognize ecologically specialized races within species. Transplanting plants and animals between habitats reveals tightly specialized matches between organisms and their environments. The evolutionary responses of animals and plants to pollution demonstrate the speed of evolutionary change, as do experiments on the effects of predators on the evolution of their prey.

The origin of species

Natural selection does not normally lead to the origin of species unless it is coupled with the reproductive isolation of populations from each other – as occurs for example on islands and is illustrated by the finches of the Galapagos Islands. *Biospecies* are recognized when they have diverged enough to prevent them from forming fertile hybrids if and when they meet.

Climatic change and continental drift

Much of what we see in the present distribution of organisms is not so much a precise, locally evolved match to present environments as a fingerprint left by the hand of history. Changes in climate, particularly during the ice ages of the Pleistocene, bear a lot of the responsibility for the present patterns of distribution of plants and animals. On a longer time scale, many distributions make sense only once we realize that while major evolutionary developments were occurring, populations were being split and separated, and land areas were moving across climatic zones.

Parallel and convergent evolution

Evidence of the power of ecological forces to shape the direction of evolution comes from parallel evolution (in which populations long isolated from common ancestors have followed similar patterns of diversification) and from convergent evolution (in which populations evolving from very different ancestors have converged on very similar forms and behavior).

Review questions

Asterisks indicate challenge question

1* What do you consider to be the essential distinction between natural selection and evolution?

2 What was the contribution of Malthus to Darwin's and Wallace's ideas about evolution?

3 Why is 'the survival of the fittest' an unsatisfactory description of natural selection?

4 What is the essential difference between natural selection and the selection practiced by plant and animal breeders?

5 What are reciprocal transplants? Why are they so useful in ecological studies?

6 Is sexual selection, as practiced by guppies, different from or just part of natural selection?

7* Review the utility and applicability of the biospecies concept to a range of groups, including a common species of plant, a rare animal species of conservation interest and bacteria living in the soil.

8 What is it about the Galapagos finches that has made them such ideal material for the study of evolution?

9 What is the difference between convergent and parallel evolution?

10* The process of evolution can be interpreted as optimizing the fit between organisms and their environment or as narrowing and constraining what they can do. Discuss whether there is a conflict between these interpretations.

PART TWO
Conditions and Resources

Chapter 3

Physical conditions and the availability of resources

Key concepts

In this chapter you will:

- understand the nature of, and contrasts between, conditions and resources
- understand how organisms respond to the whole range of conditions like temperature, but also to 'extreme' conditions and to the timing of both variations and extremes
- appreciate how a plant's responses to, and its consumption of, the resources of solar radiation, water, minerals and carbon dioxide are intertwined
- appreciate the importance of contrasting body compositions in the consumption of plants by animals, and of overcoming defenses in the consumption of animals by other animals
- understand the effects of intraspecific competition for resources
- appreciate how responses to conditions and resources interact to determine ecological niches

For ecologists, organisms are really only worth studying where they are able to live. The most fundamental prerequisites for life in any environment are that the organisms can tolerate the local conditions and that their essential resources are being provided. We cannot expect to go very far in understanding the ecology of any species without understanding its interactions with conditions and resources.

3.1 Introduction

Conditions and resources are two quite distinct properties of environments that determine where organisms can live. Conditions are physicochemical features of the environment such as its temperature, humidity or, in aquatic environments, pH. An organism always alters the conditions in its immediate environment – sometimes on a very large scale (a tree, for example, maintains a zone of higher humidity on the ground beneath its canopy) and sometimes only on a microscopic scale (an algal cell in a pond alters the pH in the shell of water that surrounds it). But conditions are not consumed nor used up by the activities of organisms.

resources, unlike conditions, are consumed

Environmental resources, by contrast, *are* consumed by organisms in the course of their growth and reproduction. Green plants photosynthesize and obtain both energy and biomass from inorganic materials. Their resources are solar radiation, carbon dioxide, water and mineral nutrients. 'Chemosynthetic' organisms like many of the primitive Archaebacteria obtain energy by oxidizing methane, ammonium ions, hydrogen sulfide or ferrous iron; they live in environments like hot springs and deep sea vents using resources that were abundant during early phases of life on Earth. All other organisms use the bodies of other organisms as their food. In each case, what has been consumed is no longer available to another consumer. The rabbit eaten by an eagle is not available to another eagle. The quantum of solar radiation absorbed and photosynthesized by a leaf is not available to another leaf. This has an important consequence: organisms may *compete* with each other to capture a share of a limited resource.

In this chapter we consider, first, examples of the ways in which environmental conditions limit the behavior and distribution of organisms. We draw most of our examples from the effects of temperature, which serve to illustrate many general effects of environmental conditions. We consider next the resources used by photosynthetic green plants, and then we go on to examine the ways in which organisms that are themselves resources have to be captured, grazed or even inhabited before they are consumed. Finally we consider the ways in which organisms of the same species may compete with each other for limited resources.

Penguins do not find the Antarctic in the least bit 'extreme'.

3.2 Environmental conditions

3.2.1 What do we mean by 'harsh', 'benign' and 'extreme'?

It seems quite natural to describe environmental conditions as 'extreme', 'harsh', 'benign' or 'stressful'. But these describe how we, human beings, feel about them. It may seem obvious when conditions are extreme: the midday heat of a desert, the cold of an Antarctic winter, the salt concentration of the Great Salt Lake. What this means, however, is only that these conditions are extreme *for us*, given our particular physiological characteristics and tolerances. But to a cactus there is nothing extreme about the desert conditions in which cacti have evolved; nor are the icy fastnesses of Antarctica an extreme environment for penguins. But a tropical rain forest *would* be a harsh environment for a penguin, though it is benign for a macaw; and a lake is a harsh environment for a cactus, though it is benign for a water hyacinth. There is, then, a relativity in the ways organisms respond to conditions; it is too easy and dangerous for the ecologist to assume that all other organisms sense the environment in the way we do. Emotive words like harsh and benign, even relativities such as hot and cold, should be used by ecologists only with care.

3.2.2 Effects of conditions

Temperature, relative humidity and other physicochemical conditions induce a range of physiological responses in organisms, which determine whether the physical environment is habitable or not. There are three basic types of *response curve* (Figure 3.1). In the first (Figure 3.1a), extreme conditions are lethal, but between the two extremes there is a continuum of more favorable conditions. Organisms are typically able to survive over the whole continuum, but can grow actively only over a more restricted range and can reproduce only within an even narrower band. This is a typical response curve for the effects of temperature or pH. In the second (Figure 3.1b), the condition is lethal only at high intensities. This is the case for poisons. At low or even zero concentration

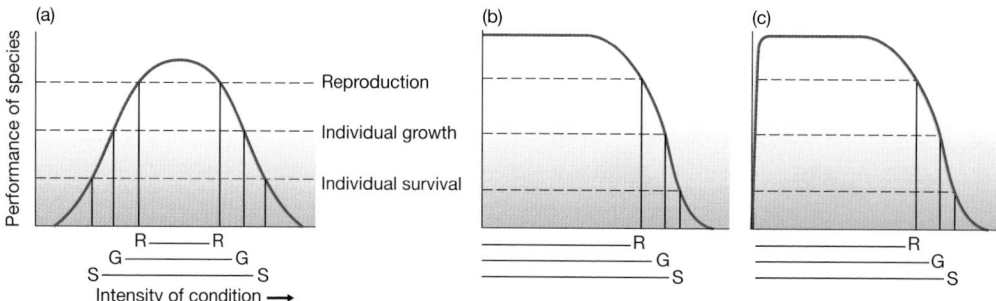

Figure 3.1

Response curves illustrating the effects of a range of environmental conditions on individual survival (S), growth (G), and reproduction (R). (a) Extreme conditions are lethal, less extreme conditions prevent growth, and only optimal conditions allow reproduction. (b) The condition is lethal only at high intensities; the reproduction–growth–survival sequence still applies. (c) Similar to (b), but the condition is required by organisms, as a resource, at low concentrations.

the organism is typically unaffected, but there is a threshold above which performance decreases rapidly: first reproduction, then growth, and finally survival. The third (Figure 3.1c), then, applies to conditions that are required by organisms at low concentrations but become toxic at high concentrations. This is the case for some minerals, such as copper and sodium chloride, that are essential resources for growth when they are present in trace amounts but become toxic conditions at higher concentrations.

effectively linear effects of temperature on rates of growth and development

Of these three responses, the first is the most fundamental. It is accounted for, in part, by changes in metabolic effectiveness. For each 10°C rise in temperature, for example, the rate of biological processes often roughly doubles, and thus appears as an exponential curve on a plot of rate against temperature (Figure 3.2a). The increase is brought about because high temperature increases the speed of molecular movement and speeds up chemical reactions. For an ecologist, however, effects on individual chemical reactions are likely to be less important than effects on rates of growth or development or on final body size, since these tend to drive the core ecological activities of survival, reproduction and movement (see Chapter 5). And when we plot rates of growth and development of whole organisms against temperature, there is quite commonly an extended range over which there are, at most, only slight deviations from linearity (Figure 3.2b, c). Either way, at lower temperatures (though 'lower' varies from species to species, as explained earlier) performance is likely to be impaired simply as a result of metabolic inactivity.

temperature and final size

Together, rates of growth and development determine the final size of an organism. For instance, for a given rate of growth, a faster rate of development will lead to smaller final size. Hence, if the responses of growth and development to variations in temperature are not the same, temperature will also affect final size. In fact, development usually increases more rapidly with temperature than does growth, such that, for a very wide range of organisms, final size tends to decrease with rearing temperature (Figure 3.3).

These effects of temperature on growth, development and size may be of practical rather than simply scientific importance. Increasingly, ecologists are called upon to predict. We may wish to know what the consequences would be, say, of a 2°C rise in temperature resulting from global warming. We cannot afford to assume

(a) AFTER MARZUSCH, 1952; (b) AFTER MONTAGNES ET AL., 2003; (c) AFTER HART ET AL., 2002

(a)

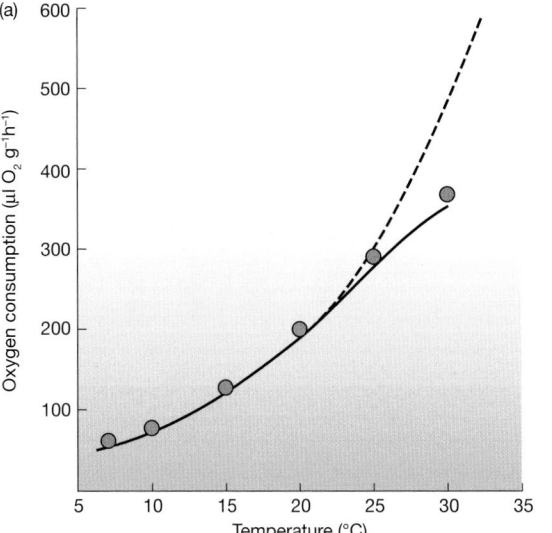

Figure 3.2

(a) The rate of oxygen consumption of the Colorado beetle (*Leptinotarsa decemlineata*), which increases non-linearly with temperature. It doubles for every 10°C rise in temperature up to 20°C but increases less fast at higher temperatures.
(b, c) Effectively linear relationships between rates of growth and development and temperature. The linear regression equations are shown. Both are highly significant. (b) Growth of the protist *Strombidinopsis multiauris*. (c) Egg-to-adult development in the mite *Amblyseius californicus*, where the vertical scale represents the proportion of total development achieved in 1 day at the temperature concerned.

(b)

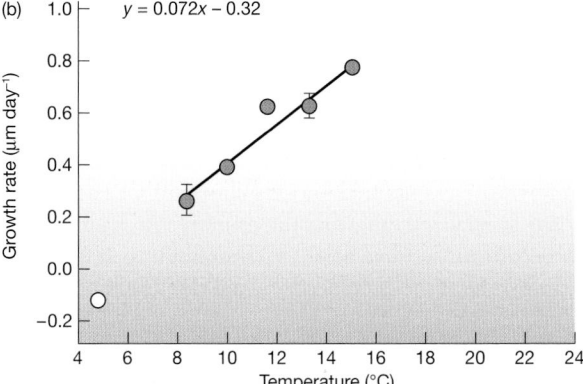

(c)

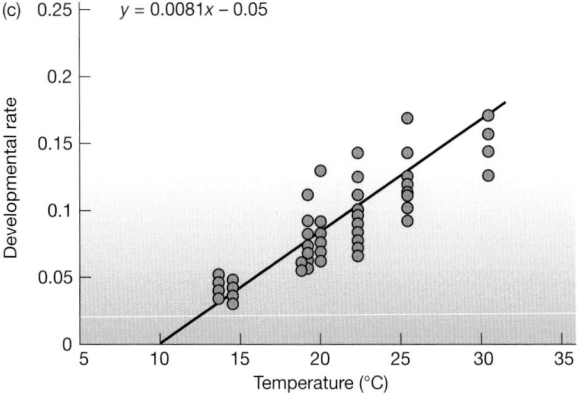

Figure 3.3

Final organism size decreases with increasing temperature, as illustrated in protists, single-celled organisms. Because the 72 data sets combined here were derived from studies carried out at a range of temperatures, both scales are 'standardized'. The horizontal scale measures temperature as a deviation from 15°C. The vertical scale measures size (cell volume, V) relative to the size at 15°C. The slope of the regression line is −0.025 (SE, 0.004; $P < 0.01$): cell volume decreased by 2.5% for every 1°C rise in rearing temperature.

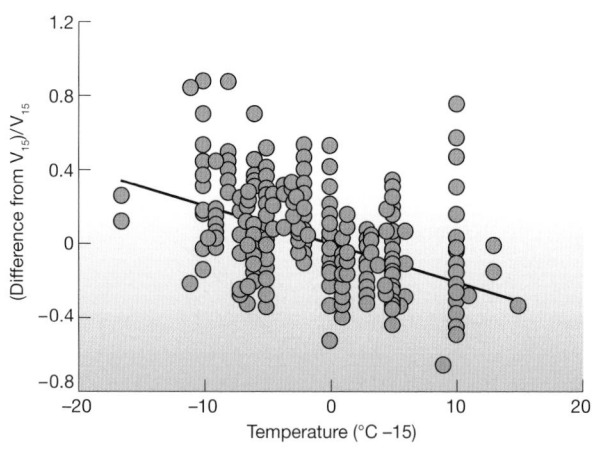

AFTER ATKINSON ET AL., 2003

exponential relationships with temperature if they are really linear, or to ignore the effects of changes in organism size on their role in ecological communities.

high and low temperatures

At extremely high temperatures, enzymes and other proteins become unstable and break down, and the organism dies. But difficulties may set in before these extremes are reached. At high temperatures, terrestrial organisms are cooled by the evaporation of water (from open stomata on the surfaces of leaves, or through sweating), but this may lead to serious, perhaps lethal, problems of dehydration; or, as water reserves run low, body temperature may rise rapidly. Even where loss of water is not a problem, for example among aquatic organisms, death is usually inevitable if temperatures are maintained for long above 60°C. The exceptions, *thermophiles*, are mostly specialized fungi and the primitive Archaebacteria. One of these, *Pyrodictium occultum*, can live at 105°C – something that is only possible because, under the pressure of the deep ocean, water does not boil at that temperature.

At temperatures a few degrees above zero, organisms may be forced into extended periods of inactivity and the cell membranes of sensitive species may begin to break down. This is known as *chilling injury*, which affects many tropical fruits. On the other hand, many species of both plants and animals can tolerate temperatures well below zero provided that ice does not form. If it is not disturbed, water can supercool to temperatures as low as −40°C without forming ice; but a sudden shock allows ice to form quite suddenly within plant cells, and this, rather than the low temperature itself, is then lethal, since ice formed within a cell is likely simply to disrupt and destroy it. If, however, temperatures fall slowly, ice can form between cells and draw water from within them. With dehydrated cells, the effects on plants are then very much like those of high-temperature drought.

the timing of extremes

The absolute temperature that an organism experiences is important. But the timing and duration of temperature extremes may be equally important. For example, unusually hot days in early spring may interfere with fish spawning or kill the fry but otherwise leave the adults unaffected. Similarly, a late spring frost might kill seedlings but leave saplings and larger trees unaffected. The duration and frequency of extreme conditions are also often critical. In many cases, a periodic drought or tropical storm may have a greater effect on a species' distribution than the average level of a condition. To take just one example: the saguaro cactus is

Saguaro cactus can only survive short periods at freezing temperatures.

liable to be killed when temperatures remain below freezing for 36 hours, but if there is a daily thaw it is under no threat. In Arizona, the northern and eastern edges of the cactus's distribution correspond to a line joining places where on occasional days it fails to thaw. Thus the saguaro is absent where there are occasionally lethal conditions – an individual need only be killed once.

3.2.3 Conditions as stimuli

Environmental conditions act primarily to modulate the rates of physiological processes. In addition, though, many conditions are important stimuli for growth and development and prepare an organism for conditions that are to come.

The idea that animals and plants in nature can anticipate, and be used by us to predict, future conditions ('a big crop of berries means a harsh winter to come') is the stuff of folklore. But there are important advantages to an organism that can predict and prepare for repeated events such as the seasons. For this, the organism needs an internal clock that can be used to check against an external signal. The most widely used external signal is the length of day – the photoperiod. On the approach of winter – as the photoperiod shortens – bears, cats and many other mammals develop a thickened fur coat, birds such as ptarmigan put on winter plumage, and very many insects enter a dormant phase (*diapause*) within the normal activity of their life cycle. Insects may even speed up their development as daylength decreases in the fall (as harsh winter conditions approach), but then speed up development again in the spring as daylength increases, once the pressure is on to have reached the adult stage by the start of the breeding season (Figure 3.4). Other photo-periodically timed events are the seasonal onset of reproductive activity in animals, the onset of flowering and seasonal migration in birds.

photoperiod is commonly used to time dormancy, flowering or migration

An experience of chilling is needed by many seeds before they will break dormancy. This prevents them from germinating during the moist warm weather

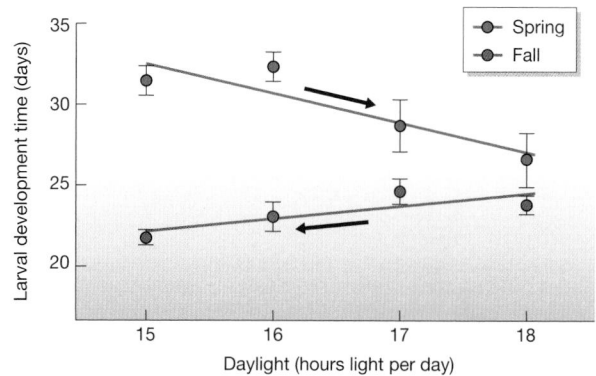

Figure 3.4

The effect of daylength on larval development time in the butterfly *Lasiommata maera* in the fall (third larval stage, before diapause) and spring. The arrows indicate the normal passage of time: daylength decreases through the fall (and development speeds up) but increases in the spring (development again speeds up). The bars are standard errors.

AFTER GOTTHARD ET AL., 1999

immediately after ripening and then being killed by the winter cold. As an example, temperature and photoperiod interact to control the seed germination of birch (*Betula pubescens*). Seeds that have not been chilled need an increasing photoperiod (indicative of spring) before they will germinate; but if the seed has been chilled, it starts growth without the light stimulus. Either way, growth should be stimulated only once winter has passed. The seeds of lodgepole pine, on the other hand, remain protected in their cones until they are heated by forest fire. This stimulus is an indicator that the ground has been cleared and that new seedlings have a chance of becoming established.

acclimatization

Conditions may themselves trigger an altered response to the same or even more extreme conditions: for instance, exposure to relatively low temperatures may lead to an increased rate of metabolism at such temperatures and/or to an increased tolerance of even lower temperatures. This is the process of *acclimatization* (called *acclimation* when induced in the laboratory). Antarctic springtails (tiny arthropods), for instance, when taken from 'summer' temperatures in the field (around 5°C in the Antarctic) and subjected to a range of acclimation temperatures, responded to temperatures in the range +2°C to −2°C (indicative of winter) by showing a marked drop in the temperature at which they froze (Figure 3.5); but at lower acclimation temperatures still (−5°C, −7°C), they

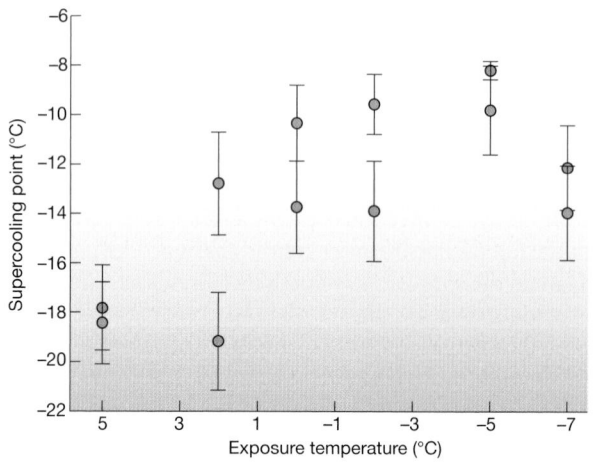

Figure 3.5

Acclimation to low temperatures. Samples of the Antarctic springtail *Cryptopygus antarcticus* were taken from field sites in the summer (ca. 5°C) on a number of days and their supercooling point (at which they froze) determined either immediately (controls, blue circles) or after a period of acclimation (brown circles) at the temperatures shown. The supercooling points of the controls themselves varied because of temperature variations from day to day, but acclimation at temperatures in the range +2°C to −2°C (indicative of winter) led to a drop in the supercooling point, whereas no such drop was observed at higher temperatures (indicative of summer) or lower temperatures (too low for a physiological acclimation response). Bars are standard errors.

AFTER WORLAND & CONVEY, 2001

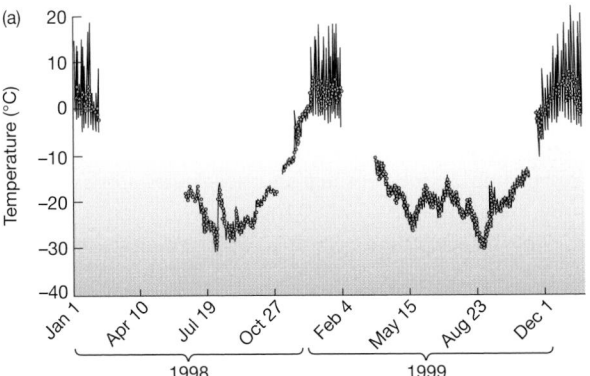

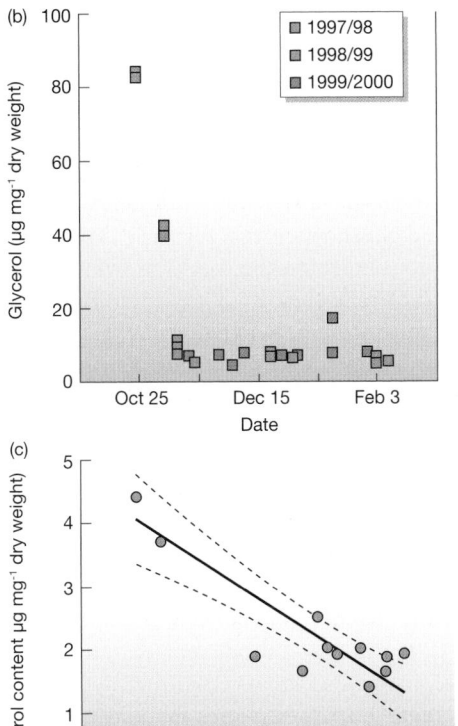

Figure 3.6

(a) Daily mean (points), maximum and minimum (tops and bottoms of lines, respectively) temperatures at Cape Bird, Ross Island, Antarctica. (b) Changes in the glycerol content of the springtail, *Gomphiocephalus hodgsoni*, from Cape Bird, which protect it from freezing (see (c)). This was extremely high over winter (as represented by the October value, the end of winter), but dropped to low values in the southern summer, when there was little need for any protection against freezing. (c) Confirmation that the supercooling point (at which ice forms) drops in the springtail as glycerol concentration increases.

AFTER SINCLAIR AND SJURSEN, 2001

showed no such drop because the temperatures were themselves too low for the physiological processes required to make the acclimation response. One way in which such increased tolerance is achieved is by forming chemicals that act as antifreeze compounds: they prevent ice from forming within the cells and protect their membranes if ice does form (Figure 3.6). Acclimatization in some deciduous trees (frost hardening) can increase their tolerance of low temperatures by as much as 100°C.

3.2.4 The effects of conditions on interactions between organisms

conditions may affect the availability of a resource, . . .

Although organisms respond to each condition in their environment, the effects of conditions may be determined largely by the responses of other community members. Temperature, for example, does not act on one species alone: it also acts on its competitors, prey, parasites and so on. Most especially, an organism will suffer if its food is another species that cannot tolerate an environmental condition. This is illustrated by the distribution of the rush moth (*Coleophora alticolella*) in England. The moth lays its eggs on the flowers of the rush (*Juncus squarrosus*) and the caterpillars feed on the developing seeds. Above 600 m, the moths and caterpillars are little affected by the low temperatures, but the rush, although it grows, fails to ripen its seeds. This, in turn, limits the distribution of the moth, because caterpillars that hatch in the colder elevations will starve as a result of insufficient food (Randall, 1982).

. . . the development of disease . . .

The effects of conditions on disease may also be important. Conditions may favor the spread of infection (e.g. winds carrying fungal spores), or favor the growth of the parasite, or weaken or strengthen the defenses of the host. For example, fungal pathogens of grasshopper, *Camnula pellucida*, in the United States develop faster at warmer temperatures, but they fail to develop at all at temperatures around 38°C and higher (Figure 3.7a), and grasshoppers that regularly experience such temperatures effectively escape serious infection (Figure 3.7b), which they do by 'basking', allowing solar radiation to raise their body temperatures by as much as 10–15°C above the air temperature around them (Figure 3.7c).

. . . or competition

Competition between species can also be profoundly influenced by environmental conditions, especially temperature. Two stream salmonid fishes, *Salvelinus malma* and *S. leucomaenis*, coexist at intermediate altitudes (and therefore intermediate temperatures) on Hokkaido Island, Japan, but only the former lives at higher altitudes (lower temperatures) and only the latter at lower altitudes. A reversal of the outcome of competition between the species, brought about by a change in temperature, appears to play a key role in this. For example, in experimental streams supporting the two species maintained at 6°C over a 191-day period (a typical high-altitude temperature), the survival of *S. malma* was far superior to that of *S. leucomaenis*; whereas at 12°C (typical low-altitude temperature), both species survived less well, but the outcome was so far reversed that by around 90 days all of the *S. malma* had died (Figure 3.8). Both species are quite capable, alone, of living at either temperature.

3.2.5 Responses by sedentary organisms

Motile animals have some choice over where they live: they can show preferences. They may move into shade to escape from heat or into the sun to warm up. Such choice of environmental conditions is denied to fixed or sedentary organisms. Plants are obvious examples, but so are many aquatic invertebrates such as sponges, corals, barnacles, mussels and oysters.

form and behavior may change with the seasons

In all except equatorial environments, physical conditions follow a seasonal cycle. Indeed, there has long been a fascination with organisms' responses to these (Box 3.1). Morphological and physiological characteristics can never be ideal for all phases in the cycle, and the jack-of-all-trades is master of none. One solution

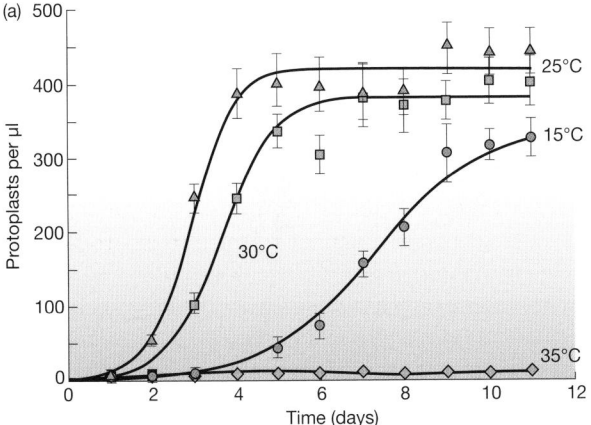

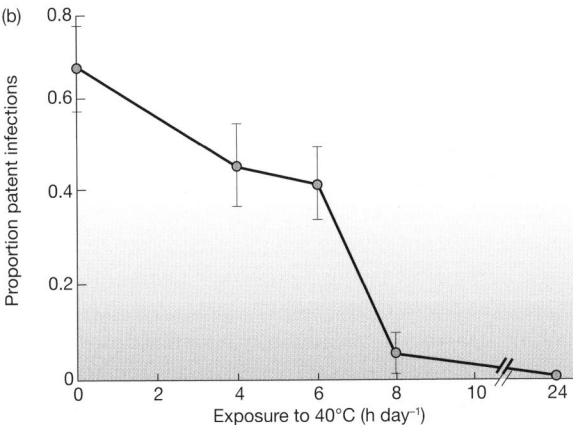

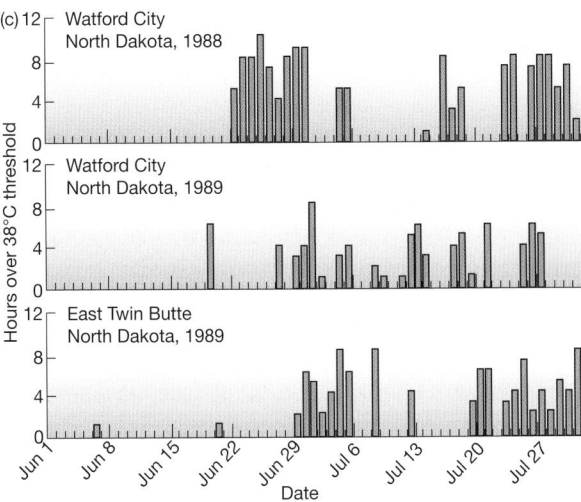

Figure 3.7

The effect of temperature on the interaction between the fungal pathogen, *Entomophaga grylli*, and the grasshopper, *Camnula pellucida*. (a) Growth curves over time of the pathogen (expressed as protoplasts per μl) at a range of temperatures: growth ceases at temperatures of around 38°C and higher. (b) The proportion of grasshoppers with patent (i.e. observable) infection with the pathogen drops sharply as grasshoppers spend more of their time at such higher temperatures. (c) Grasshoppers at two sites over 2 years did frequently raise their body temperatures to such high levels by basking.

AFTER CARRUTHERS ET AL., 1992

Figure 3.8

Changing temperature reverses the outcome of competition. At low temperature (6°C) on the left, the salmonid fish *Salvelinus malma* out-survives cohabiting *S. leucomaenis*, whereas at 12°C, right, *S. leucomaenis* drives *S. malma* to extinction. Both species are quite capable, alone, of living at either temperature.

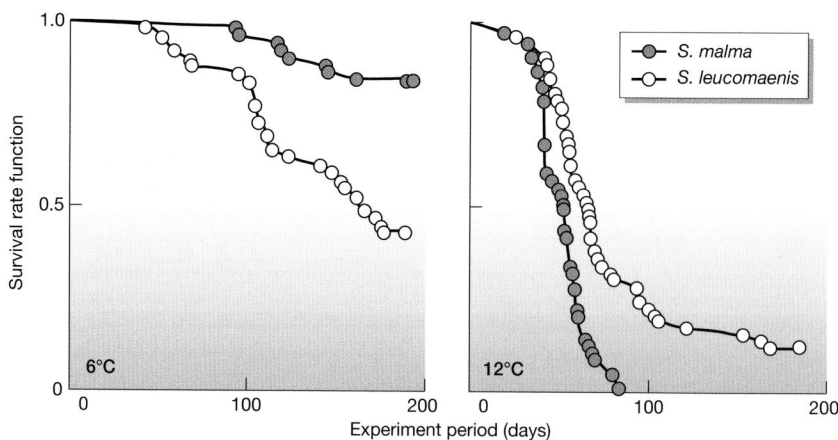

AFTER TANIGUCHI & NAKANO, 2000

3.1 Historical landmarks

3.1 HISTORICAL LANDMARKS

Recording seasonal changes

Recording the changing behavior of organisms through the season (*phenology*) was essential before agricultural activities could be intelligently timed. The earliest phenological records were apparently the Wu Hou observations made in the Chou and Ch'in (1027–206 BC) dynasties. The date of the first flowering of cherry trees has been recorded at Kyoto, Japan, since AD 812.

A particularly long and detailed record was started in 1736 by Robert Marsham at his estate near the city of Norwich, England. He called these records 'Indications of the spring'. Recording was continued by his descendants until 1947. Marsham recorded 27 phenological events every year: the first flowering of snowdrop, wood anemone, hawthorn and turnip; the first leaf emergence of 13 species of tree; and various animal events such as the first appearance of migrants (swallow, cuckoo, nightingale), the first nest building by rooks, croaking of frogs and toads, and the appearance of the brimstone butterfly.

Long series of measurements of environmental temperature are not available for comparison with the whole period of Marsham's records, but they are available from 1771 for Greenwich, about 160 km away. There is surprisingly close agreement between many of the flowering and leaf emergence events at Marsham and the mean January–May temperature at Greenwich (Figure 3.9). However, not surprisingly, events such as the time of arrival of migrant birds bears little relationship to temperature.

Analysis of the Marsham data for the emergence of leaves on six species of tree indicates that the mean date of leafing is advanced by 4 days for every 1°F increase in the mean temperature from February to May (Figure 3.10). Similarly, for the eastern United States, Hopkins' *bioclimatic law* states that the indicators of spring such as leafing and flowering occur 4 days later for every 1° latitude northward, 5° longitude westward or 400 feet (c. 120 m) of altitude.

Collecting phenological records has now been transformed from the pursuit of gifted amateurs to sophisticated programs of data collection and analysis. At least 1500 phenological observation posts are now maintained in Japan alone. The vast accumulations of data have suddenly become exciting and relevant as we try to estimate the changes in floras and faunas that will be caused by global warming.

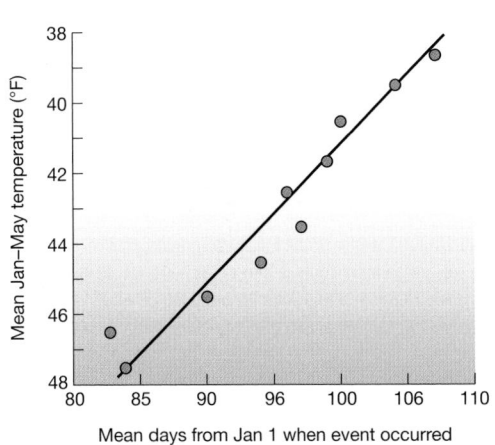

Figure 3.9

The relationships between mean January–May temperatures and the annual mean dates of 10 flowering and leafing events from the classic Marsham records started in 1736.

FROM REDRAWN FIGURES OF MARGARY, IN FORD, 1982

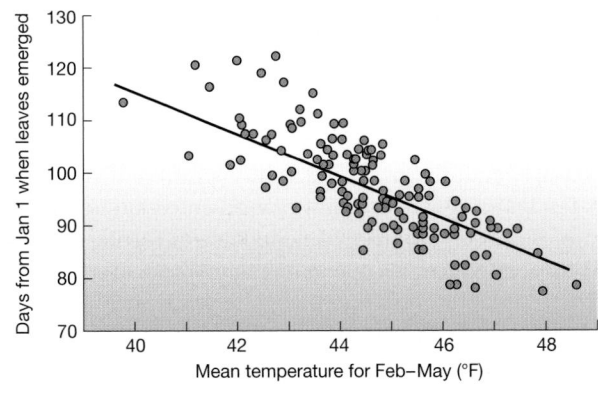

Figure 3.10

The relationship between the mean temperature in the 4-month period, February–May, and the average date of six leafing events. The correlation coefficient is −0.81.

FROM REDRAWN FIGURES OF KINGTON, IN FORD, 1982

is for the morphological and physiological characteristics to keep changing with the seasons (or even anticipating them, as in acclimatization). But change may be costly: a deciduous tree may have leaves ideal for life in spring and summer but faces the cost of making new ones every year. An alternative is to economize by having long-lasting leaves like those of pines, heathers and the perennial shrubs of deserts. Here, though, there is a cost to be paid in the form of more sluggish physiological processes. Different species have evolved different compromise solutions.

3.2.6 Animal responses to environmental temperature

Most species of animals are, like plants, *ectotherms*: they rely on external sources of heat to determine the pace of their metabolism. This includes the invertebrates and also fish, amphibians and lizards. Others, mainly birds and mammals, are *endotherms*: they can regulate their body temperature by producing heat within their body.

ectotherms and endotherms

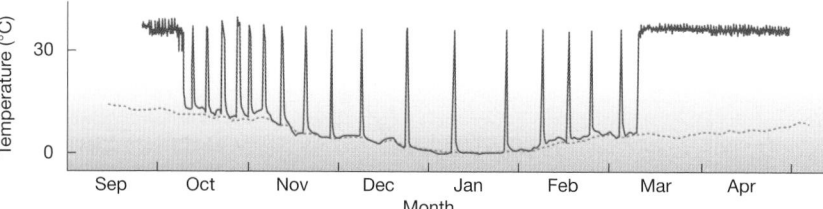

AFTER HUT ET AL., 2002

Figure 3.11

Changes in the body temperature over the 1996/97 winter of the European ground squirrel, *Spermophilus citellus* (solid line) compared to ambient soil temperature (dotted line) at the same depth at which it was hibernating. Note that during hibernation (early October to mid-March), body temperature was mostly indistinguishable from ambient temperature, apart from repeated brief periods of activity accompanied by 'normal' body temperatures.

The distinction between ectotherms and endotherms is not absolute. Some typical ectotherms, some insects for example, can control body temperature through muscle activities (e.g. shivering flight muscles). Some fish and reptiles can generate heat for limited periods of time, and even some plants can use metabolic activity to raise the temperature of their flowers. Some typical endotherms, on the other hand, such as dormice, hedgehogs and bats, allow their body temperature to fall and become scarcely different from that of their surroundings when they are hibernating (Figure 3.11).

Despite these overlaps, endothermy is inherently a different strategy from ectothermy. Over a certain narrow temperature range, an endotherm consumes energy at a basal rate. But at environmental temperatures further and further above or below that zone, endotherms expend more and more energy maintaining their constant body temperature. This makes them relatively independent of environmental conditions and allows them to stay longer at or close to peak performance. It makes them more efficient in both searching for food and escaping from predators. However, there is a cost – a high requirement for food to fuel this strategy.

The idea that organisms are harmed (and limited in their distributions) by environmental conditions not 'directly', but because of the energetic costs required to tolerate those conditions, is illustrated by a study examining the effect of a different condition: salinity. The freshwater shrimps *Palaemonetes pugio* and *P. vulgaris*, for example, co-occur in estuaries on the eastern coast of the USA at a wide range of salinities, but the former seems to be more tolerant of lower salinities than the latter, occupying some habitats from which the latter is absent. Figure 3.12 shows the mechanism likely to be underlying this. Over the low salinity range (though not at the effectively lethal lowest salinity) metabolic expenditure was significantly lower in *P. pugio*. *P. vulgaris* requires far more energy simply to maintain itself, putting it at a severe disadvantage in competition with *P. pugio*.

Endotherms have morphological modifications that reduce their energetic costs. In cold climates most have low surface area to volume ratios (short ears and limbs), and this reduces heat loss through surfaces. Typically, endotherms that live in polar environments are insulated from the cold with extremely dense fur (polar bears, mink, foxes) or feathers and extra layers of fat. In contrast, desert endotherms often have thin fur, and long ears and limbs, which help dissipate heat.

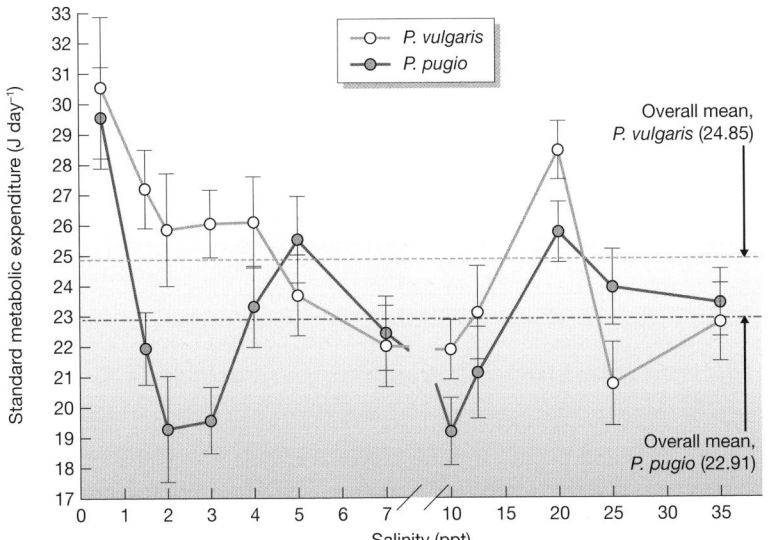

AFTER ROWE, 2002

Figure 3.12

Standard metabolic expenditure (estimated through minimum oxygen consumption) in two species of shrimp, *Palaemonetes pugio* and *P. vulgaris*, at a range of salinities. There was significant mortality of both species over the experimental period at 0.5 ppt (parts per thousand), especially *P. vulgaris* (75% compared to 25%).

Variability of conditions can set biological challenges as great as extremes. Seasonal cycles, for example, can expose an animal to summer heat close to its thermal maximum, and winter chill close to its thermal minimum. Responses to these changing conditions include the laying down of different coats in the fall (thick and underlain by a thick fat layer) and in the spring (a thinner coat and loss of the dense fat layer) (Figure 3.13). Some animals also take advantage of each other's body heat as a means to cope with cold weather by huddling together. Hibernation – relaxing temperature control – allows some vertebrates to survive periods of winter cold and food shortage (see Figure 3.11) by *avoiding* the difficulties of finding sufficient fuel over these periods. Migration is another avoidance strategy: the Arctic tern, to take an extreme example, travels from the Arctic to the Antarctic and back each year, experiencing only the polar summers.

temperatures that vary seasonally pose special problems

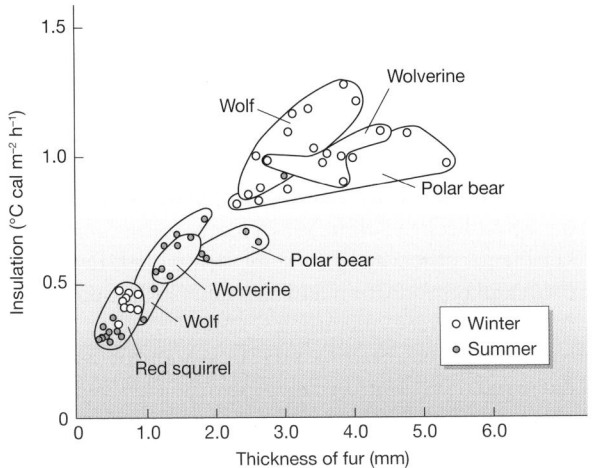

Figure 3.13

Seasonal changes in the thickness of the insulating fur coats of some Arctic and northern temperate mammals.

The thick, white winter coat and the thinner, browner summer coat of the Arctic fox.

3.2.7 Microorganisms in extreme environments

Microorganisms survive and grow in all the environments that are lived in or tolerated by animals and plants, and they show the same range of strategies – avoid, tolerate or specialize. Many microorganisms produce resting spores that survive drought, high temperature or cold. There are also some that are capable of growth and multiplication in conditions far outside the range of tolerance of higher organisms: they inhabit some of the most extreme environments on Earth. Temperatures maintained higher than 45°C are lethal to almost all plants and animals, but thermophilic ('temperature loving') microbes grow at much higher temperatures. Although similar in many ways to heat-intolerant microbes, the enzymes of these thermophiles are stabilized by especially strong ionic bonds.

Microbial communities that not only tolerate but grow at low temperatures are also known; these include photosynthetic algae, diatoms and bacteria that have been found on Antarctic sea ice. Microbial specialists have also been identified from other rare or peculiar environments: for example *acidophiles*, which thrive in environments that are highly acidic. One of them, *Thiobacillus ferroxidans*, is found in the waste from industrial metal-leaching processes and tolerates pH 1.0. At the other end of the pH spectrum, the cyanobacterium, *Plectonema nostocorum*, from soda lakes can grow at pH 13. As noted previously, these oddities may be relicts from environments that prevailed much earlier in Earth's history. Certainly, they warn us against being too narrow-minded when we consider the kind of organism we might look for on other planets.

3.3 Plant resources

Resources may be either biotic or abiotic components of the environment: they are whatever an organism uses or consumes in its growth and maintenance, leaving less available for other organisms. When a photosynthesizing leaf intercepts radiation, it deprives some of the leaves or plants beneath it. When a caterpillar eats a leaf, there is less leaf material available for other caterpillars. By their nature, resources are critical for survival, growth and reproduction and also inherently a potential source of conflict and competition between organisms.

resource requirements of non-motile organisms

If an organism can move about, it has the potential to search for its food. Organisms that are fixed and 'rooted' in position cannot search. They must rely on growing toward their resources (like a shoot or root) or catching resources that move to them. The most obvious examples are green plants, which depend on: (i) energy that radiates to them; (ii) atmospheric carbon dioxide that diffuses to them; (iii) mineral cations that they obtain from soil colloids in exchange for hydrogen ions; and (iv) water and dissolved anions that the roots absorb from the soil. In the following sections, we concentrate on green plants. But it is important to remember that many of the non-mobile animals, like corals, sponges and bivalve mollusks, depend on resources that are suspended in the watery environment and are captured by filtering the water or even just waiting for them open-mouthed.

3.3.1 Solar radiation

Solar radiation is a critical resource for green plants. We often refer to it loosely as 'light', but green plants actually use only about 44% of that narrow part of the spectrum of solar radiation that is visible to us between infrared and ultraviolet. The rate of photosynthesis increases with the intensity of the radiation that a leaf receives, but with diminishing returns; and this relationship itself varies greatly between species (Figure 3.14), especially between those that usually live in shaded habitats (which reach saturation at low radiation intensities) and those that normally experience full sunlight and can take advantage of it. Moreover, at high intensities, *photoinhibition* of photosynthesis may occur, such that the rate of fixation of carbon *decreases* with increasing radiation intensity. High intensities of radiation may also lead to dangerous overheating of plants. Radiation is an essential resource for plants, but they can have too much as well as too little.

The solar radiation that reaches a plant is forever changing. Its angle and intensity change in a regular and systematic way annually, diurnally and with depth within the canopy or in a water body (Figure 3.15). There are also irregular, unsystematic variations due to changes in cloud cover or shadowing by the leaves of neighboring plants. As light flecks pass over leaves lower in the canopy, they receive seconds or minutes of direct bright light and then plunge back into shade. The daily photosynthesis of a leaf integrates these various experiences; the whole plant integrates the diverse exposure of its various leaves.

There is enormous variation in the shapes and sizes of leaves. Most of the heritable variation in shape has probably evolved under selection not primarily for high photosynthesis, but rather for optimal efficiency of water use (photosynthesis achieved per unit of water transpired) and minimization of the damage done by foraging herbivores. Not all the variations in leaf shape are heritable, though:

sun and shade species

sun and shade leaves

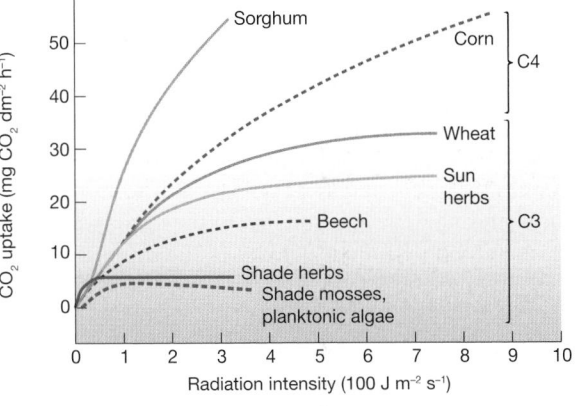

AFTER LARCHER, 1980, AND OTHER SOURCES

Figure 3.14

The response of photosynthesis by the leaves of various types of green plant (measured as carbon dioxide uptake) to the intensity of solar radiation at optimal temperatures and with a natural supply of carbon dioxide. (The different physiologies of C3 and C4 plants are explained later in Section 3.3.2.)

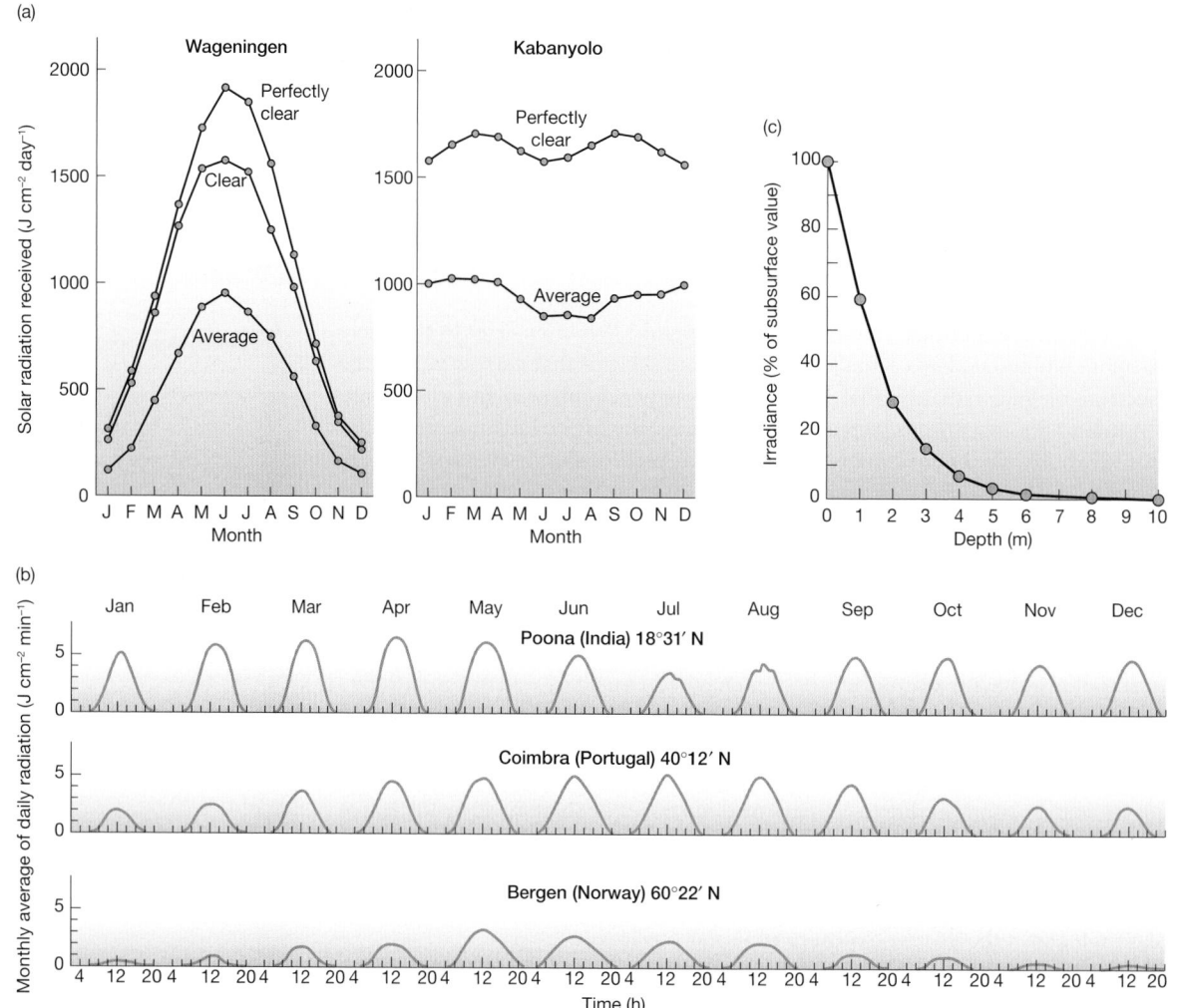

(a)

(A, B) AFTER DE WIT, 1965, AND OTHER SOURCES; (C) AFTER KIRK, 1994

Figure 3.15

(a) The daily totals of solar radiation received throughout the year at Wageningen (the Netherlands) and Kabanyolo (Uganda). (b) The monthly average of daily radiation recorded at Poona (India), Coimbra (Portugal) and Bergen (Norway). (c) Exponential diminution of radiation intensity with water depth in a freshwater habitat (Burrinjuck Dam, Australia).

many are responses by the individual to its immediate environment. Many trees, especially, produce different types of leaf in positions exposed to full sunlight ('sun leaves') and in places lower in the canopy where they are shaded ('shade leaves'). Sun leaves are thicker, with more densely packed chloroplasts (which process the incoming radiation) within cells and more cell layers. The more flimsy shade leaves intercept diffused and filtered radiation low in the canopy but may nonetheless supplement the main photosynthetic activity of the sun leaves high in the canopy.

sun and shade plants

Among herbaceous plants and shrubs, specialist 'sun' or 'shade' *species* are much more common. Leaves of sun plants are commonly exposed at acute angles to the midday sun and are typically superimposed into a multilayered canopy, where even the lower leaves may have a positive rate of net photosynthesis. The leaves of shade plants are typically arranged in a single-layered canopy and angled horizontally, maximizing their ability fully to capture the available radiation.

Other species develop as sun or shade plants, depending on where they grow. One such is the evergreen shrub, *Heteromeles arbutifolia*, which grows both in chaparral habitats in California, where shoots in the upper crown are regularly exposed to full sunlight and high temperatures, and also in shaded woodland habitats, where it receives around one-seventh as much radiation. A detailed study of this plant captures many of the points made above (Figure 3.16). As expected,

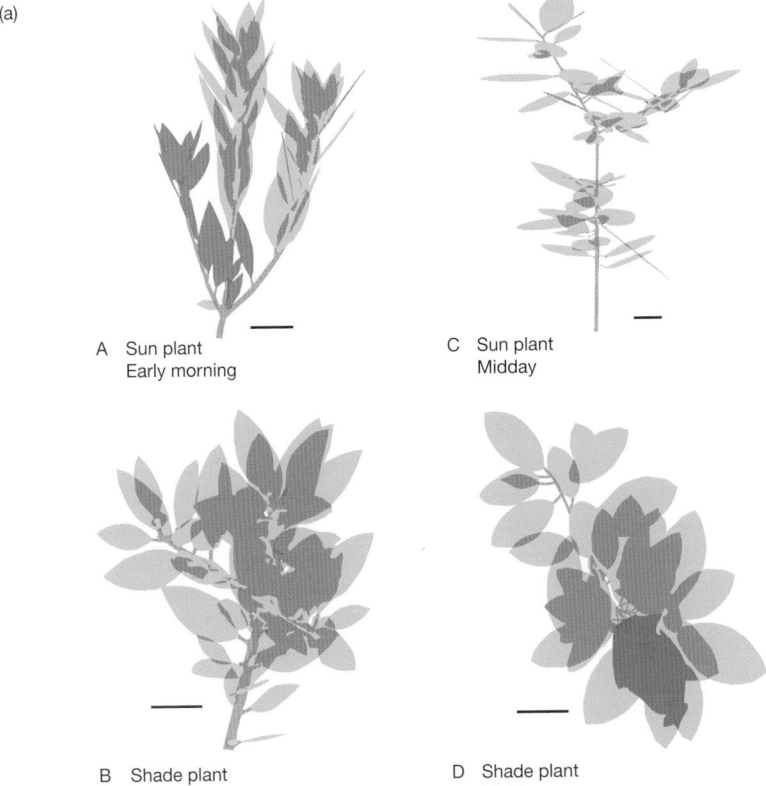

(a)

A Sun plant
Early morning

C Sun plant
Midday

B Shade plant
Early morning

D Shade plant
Midday

Figure 3.16

(a) Computer reconstructions of stems of typical sun (A, C) and shade (B, D) plants of the evergreen shrub *Heteromeles arbutifolia*, viewed along the path of the sun's rays in the early morning (A, B) and at midday (C, D). Darker tones represent parts of leaves shaded by other leaves of the same plant. Bars = 4 cm. (b) Observed differences in the leaves of sun and shade plants. Standard deviations are given in parentheses; the significance of differences are given following analysis of variance. (c) Consequent whole-plant properties of sun and shade plants. Letter codes indicate groups that differed significantly in analyses of variance ($P < 0.05$).

(b)

	Sun		Shade		P
Leaf angle (degrees)	71.3	(16.3)	5.3	(4.3)	<0.01
Leaf blade thickness (μm)	462.5	(10.9)	292.4	(9.5)	<0.01
Photosynthetic capacity, area basis (μmol CO_2 m^{-2} s^{-1})	14.1	(2.0)	9.0	(1.7)	<0.01
Chlorophyll content, area basis (mg m^{-2})	280.5	(15.3)	226.7	(14.0)	<0.01
Leaf nitrogen content, area basis (g m^{-2})	1.97	(0.25)	1.71	(0.21)	<0.05

(c)

	Sun plants		Shade plants	
	Summer	Winter	Summer	Winter
Fraction self-shaded	0.22[a]	0.42[b]	0.47[b]	0.11[a]
Display efficiency	0.33[a]	0.38[a,b]	0.41[b]	0.43[b]
Absorption efficiency	0.28[a]	0.44[b]	0.55[c]	0.53[c]

AFTER VALLADARES & PEARCY, 1998

the leaves of sun plants are thicker and have a greater photosynthetic capacity (more chlorophyll and nitrogen) per unit leaf area than those of shade plants (Figure 3.16b). As expected, too, sun-plant leaves are inclined at a much steeper angle to the horizontal, and they therefore absorb the direct rays of the overhead summer sun over a wider leaf area than the more horizontal shade-plant leaves. The more angled leaves of sun plants, though, are also less likely than shade-plant leaves to shade other leaves of the same plant from the overhead rays of the summer sun (Figure 3.16c). But in winter, when the sun is much lower in the sky, it is the shade plants that are much less subject to this 'self-shading'. The overall consequence of these differences is that 'display efficiency' – the proportion of incident radiation intercepted per unit area of leaf – is higher in shade than in sun plants, in summer because of the more horizontal leaves, but in the winter because of the relative absence of self-shading.

The properties of whole plants of *H. arbutifolia*, then, reflect both plant architecture and the morphologies and physiologies of individual leaves. The efficiency of light absorption per unit of biomass is massively greater for shade than for sun plants (Figure 3.16c), reflecting leaf angles, self-shading and leaf thickness. Overall, despite receiving only one-seventh of the radiation of sun plants, shade plants reduce the differential in their daily rate of carbon gain from photosynthesis to only a half. They successfully counterbalance their reduced photosynthetic capacity at the leaf level with enhanced light-harvesting ability at the whole-plant level. The sun plants, on the other hand, can be seen as striking a compromise between maximizing whole-plant photosynthesis while avoiding photoinhibition and overheating of individual leaves.

3.3.2 Water

water is lost from plants that photosynthesize

Most plant parts are largely composed of water. In some soft leaves and fruits, as much as 98% of the volume may be water. Yet this is a minute fraction of the water that passes from the soil through a plant to the atmosphere during plant growth. Photosynthesis depends on the plant absorbing carbon dioxide. This can only happen across surfaces that are wet – most notably the walls of the photosynthesizing cells in leaves. If a leaf allows carbon dioxide to enter, it is almost impossible to prevent water vapor from leaving. Likewise, any mechanism or process that slows down the rate of water loss, such as closing the stomata (pores) on the leaf surface, is almost bound to reduce the rate of carbon dioxide absorption and hence reduce the rate of photosynthesis.

wilting

Green plants serve as wicks that conduct water from the soil and release it to the atmosphere. If the rate of uptake falls below the rate of release, the body of the plant starts to dry out. The cells lose their turgidity and the plant wilts. This may just be temporary (though it may happen every day in summer), and they may recover and rehydrate at night. But if the deficit accumulates, the plant may die.

plant life in water deficit: avoiders and tolerators

Species of green plants differ in the ways in which they survive in dry environments. One strategy is to avoid the problems. *Avoiders* such as desert annuals, annual weeds and most crop plants have a short lifespan: their photosynthetic activity is concentrated during periods when they can maintain a positive water balance. For the remainder of the year, they remain dormant as seeds, a stage that requires neither photosynthesis nor transpiration. Some perennial plants shed their photosynthetic tissues during periods of drought. Some species then replace them with new leaf forms that are less extravagant of water or spend the driest season with no leaves at all – just green stems.

Other plants, *tolerators*, have evolved a different compromise, producing long-lived leaves that transpire slowly (for example, by having few and sunken stomata). They tolerate drought, but of course their photosynthesis is slower. These plants have sacrificed their ability to achieve rapid photosynthesis when water is abundant but gained the insurance of being able to photosynthesize throughout the seasons. This is not only a property of plants from arid areas but also of the pines and spruces that survive where water may be abundant but is usually frozen and therefore inaccessible.

The viability of alternative strategies to solve the problem of photosynthesizing in a dry environment is nicely illustrated by the trees of seasonally dry tropical forests and woodlands. These communities are found naturally in Africa, the Americas, Australia and India; but whereas, for example, the savannas of Africa and India are dominated by deciduous species (losing all leaves for at least 1 and usually 2–4 months each year), and the Llanos of South America are dominated by evergreens (a full canopy all year), in the savannas of Australia there are roughly equal numbers of deciduous and evergreen species (Figure 3.17). The deciduous species avoid drought in the dry season (April–November in Australia) as a result of their vastly reduced rates of transpiration, having shed their leaves (Figure 3.17a, b). The evergreens tolerate the threat of drought in the dry season (Figure 3.17b), but maintain a positive carbon balance throughout the year (Figure 3.17c), whereas the deciduous species make no net photosynthate at all for around 3 months.

coexisting alternative strategies in Australian savannas

The evaporation of water lowers the temperature of the body with which it is in contact. For this reason, if plants are prevented from transpiring they may overheat. This, rather than water loss itself, may be lethal. The desert honey-sweet (*Tidestromia oblongifolia*) grows vigorously in Death Valley, California despite the fact that its leaves are killed if they reach 50°C, a temperature that

water and overheating

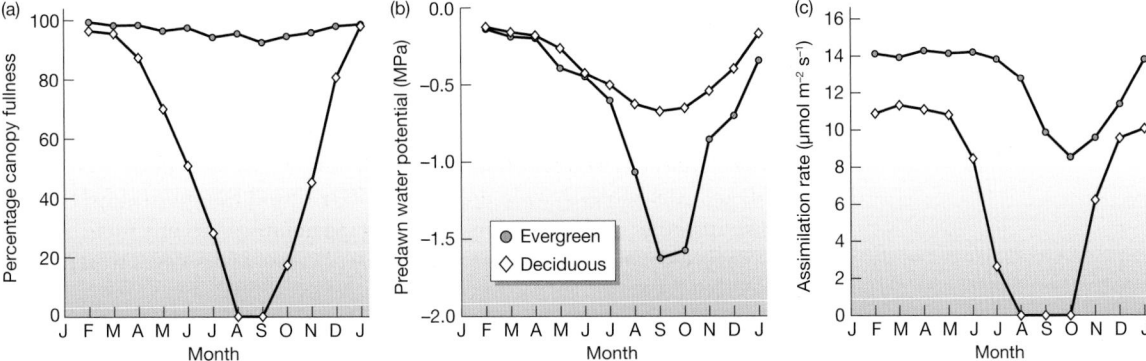

AFTER EAMUS, 1999

Figure 3.17

(a) Percentage canopy fullness for deciduous and evergreen trees in Australian savannas throughout the year. (Note that the southern hemisphere dry season runs from around April to November.) (b) Susceptibility to drought as measured by increasingly negative values of 'predawn water potential' for deciduous and evergreen trees. (c) Net photosynthesis as measured by the carbon assimilation rate for deciduous and evergreen trees.

is commonly reached in the surrounding air. Transpiration cools the surface of the leaf to a tolerable 40–45°C. Most desert plants bear hairs, spines and waxes on the leaf surface. These reflect a high proportion of incident radiation and help to prevent overheating. Other more general modifications in desert plants include the characteristic 'chunky' shape of succulents with few branches, giving a low surface area to volume ratio over which radiant heat is absorbed.

increasing the efficiency of water use: C4 and CAM

Specialized biochemical processes may increase the amount of photosynthesis that can be achieved per unit of water lost. The majority of plants on Earth photosynthesize using what is termed the *C3 pathway*. Although these plants are highly productive photosynthesizers, they are relatively wasteful of water, reach their maximum rates of photosynthesis at relatively low intensities of radiation, and are less successful in arid areas. Alternative pathways of photosynthesis – termed the *C4 pathway* and *CAM* (crassulacean acid metabolism) – are more economical in their water use. C4 plants have a particularly high affinity for carbon dioxide, and so absorb more per unit of water lost. CAM plants open their stomata at night and absorb carbon dioxide and fix it as malic acid. They close their stomata during the day and release the carbon dioxide internally for photosynthesis. C4 and CAM plants are most common in arid and, in particular, hot arid areas. They are restricted in range because the associated costs of their systems apparently make them less competitive under less arid conditions. For example, the photosynthesis of C4 plants is inefficient at low radiation intensities (see Figure 3.14) and so they are poor shade plants; while CAM plants must store their accumulated malic acid every night: most of them are succulents with extensive swollen water-storage tissues that cope with this problem.

obtaining water from the soil

Almost all falling water (rain, snow, etc.) bypasses the plants and passes to the soil. Some drains through the soil, but much is held against gravity by capillary forces and as colloids. Plants obtain virtually all their water from this stored reserve. Sandy soils have wide pores: these do not hold much water but what is there is held with weak forces and plants can withdraw it easily. Clay soils have very fine pores. They retain more water against the force of gravity, but surface tension in the fine pores makes it more difficult for the plants to withdraw it. The primary water-absorbing zone on roots is covered with root hairs that make intimate contact with soil particles (Figure 3.18). As water is withdrawn from the soil, the first to be released is from the wider pores, where capillary forces retain it only weakly. Subsequent water is withdrawn from narrower paths in which the water is more tightly held. Consequently, the more the soil around the roots is depleted of water, the more resistance there is to water flow. As a result of water withdrawal, roots create water depletion zones (or, more generally, resource depletion zones or RDZs) around themselves. The faster the roots draw water from the soil, the more sharply defined the RDZs, and the more slowly water will move into that zone. In a soil that contains abundant water, rapidly transpiring plants may still wilt because water does not flow fast enough to replenish the RDZs around their root systems (or because the roots cannot explore new soil volumes fast enough).

The shapes of root systems are much less tightly programmed than those of shoots. The root architecture that a plant establishes early in its life can determine its responsiveness to later events. Plants that develop under waterlogged conditions usually set down only a superficial root system. If drought develops later in the season, these same plants can suffer because their root system did not

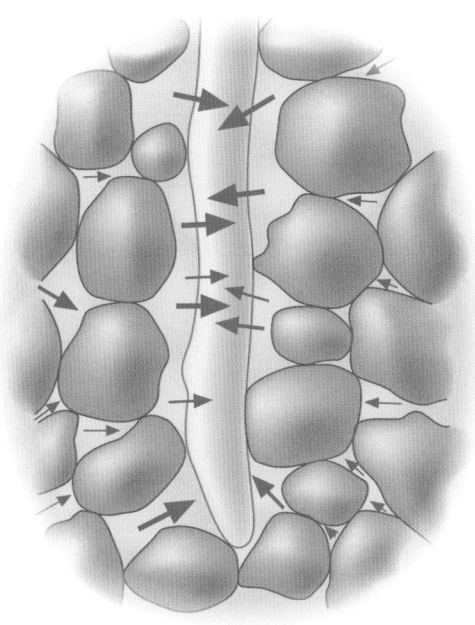

Figure 3.18
Highly diagrammatic picture of a root hair withdrawing water from pores in a very wet soil. Even the widest pores shown are full of water. As water is withdrawn, the wider pores become emptied and water flows only along the twisted pathways through narrower pores.

tap the deeper soil layers. A deep tap root, however, will be of little use to a plant in which most water is received from occasional showers on a dry substrate. Figure 3.19 illustrates some characteristic differences between the root systems of plants from damp temperate and dry desert habitats.

3.3.3 Mineral nutrients

Roots extract water from the soil, but they also extract key minerals. Plants require mineral resources of nitrogen (N), phosphorus (P), sulfur (S), potassium (K), calcium (Ca), magnesium (Mg) and iron (Fe), together with traces of manganese (Mn), zinc (Zn), copper (Cu) and boron (B). All of these must be obtained from the soil (or directly from the water in the case of free-living aquatic plants). Soils are patchy and heterogeneous, and as roots grow through them, they may meet regions that vary in nutrient and water content. They tend to branch profusely in the richer patches (Figure 3.20).

Root architecture is particularly important in this process, because different nutrients behave differently and are held in the soil by different forces. Nitrate ions diffuse rapidly in soil water, and rapidly transpiring plants may bring nitrates to the root surface faster than they are accumulated in the body of the plant. However, other key nutrients such as phosphate are tightly bound in the soil (have low diffusion coefficients). The phosphate RDZs of two roots 0.2 mm apart scarcely overlap, and the parts of a finely branched root system scarcely compete with each other. Consequently, if phosphate is in short supply, a highly branched surface root will greatly improve phosphate absorption. A more widely spaced extensive root system, in contrast, will tend to maximize access to nitrate.

the architecture of roots determines their foraging efficiency

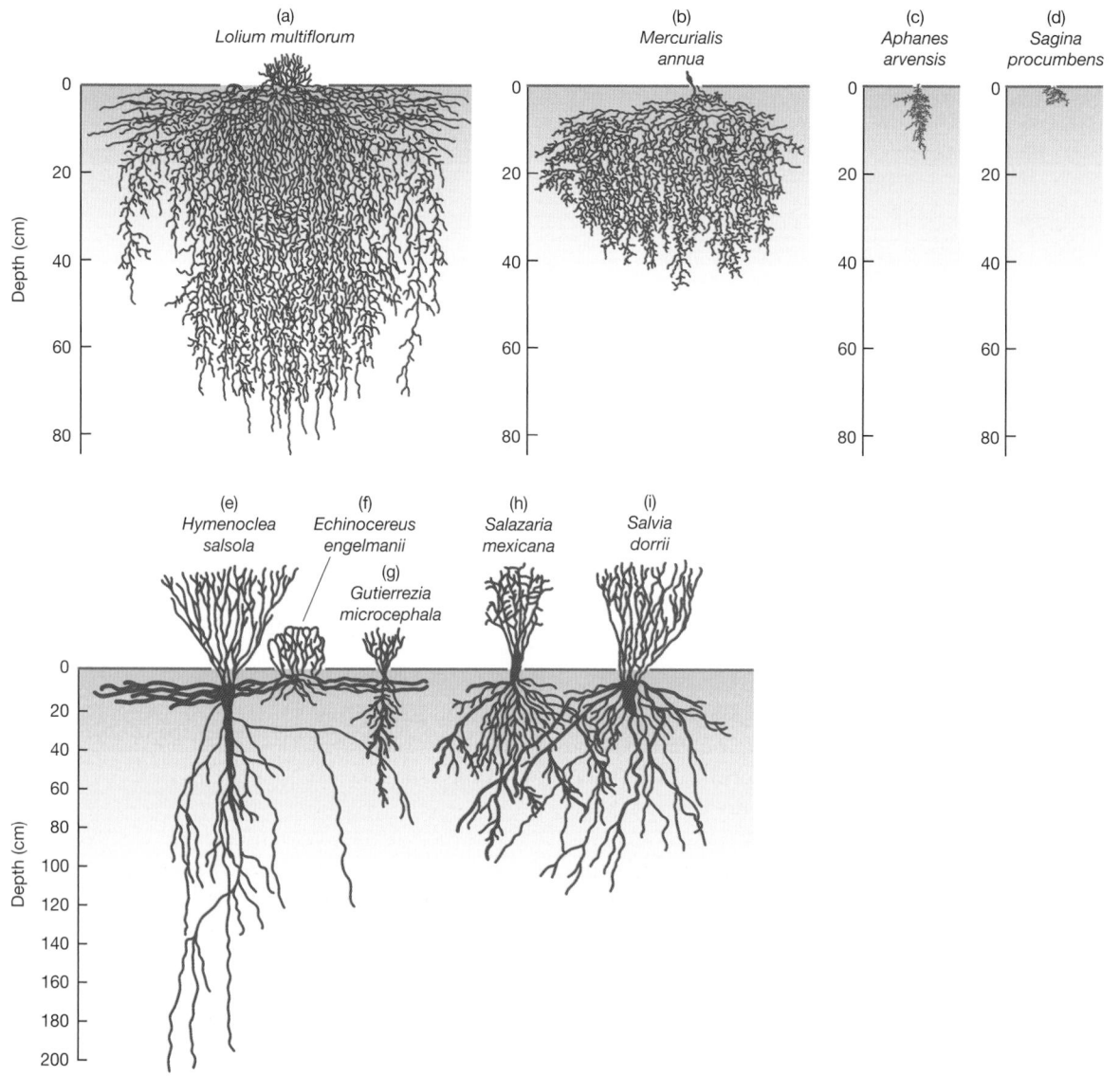

Figure 3.19

Profiles of root systems of plants from contrasting environments. (a–d) Northern temperate species of open ground: (a) *Lolium multiflorum*, an annual grass; (b) *Mercurialis annua*, an annual weed; and (c) *Aphanes arvensis* and (d) *Sagina procumbens*, both ephemeral weeds. (e–i) Desert shrub and semishrub species, Mid Hills, eastern Mojave Desert, California.

(a–d) FROM FITTER, 1991; (e–i) REDRAWN FROM A VARIETY OF SOURCES

3.3.4 Carbon dioxide

Plants take in carbon dioxide through the stomatal pores on leaf surfaces and, as we have seen, using the energy of sunlight, capture the carbon atoms during photosynthesis and release oxygen. Carbon dioxide varies in its concentration at a variety of scales. In 1750, atmospheric carbon dioxide concentrations were approximately 280 µl l^{-1}. Currently, the figure is over 370 µl l^{-1} and is increasing by 0.4–0.5% per year, mainly as a result of the burning of fossil fuels (Box 3.2).

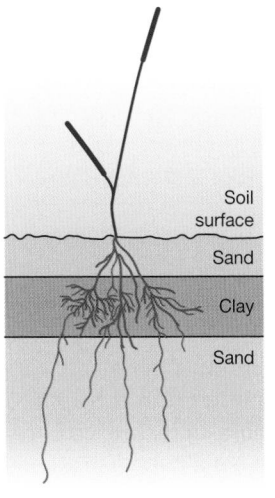

Figure 3.20

The root system developed by a young plant of wheat growing through a sandy soil with a layer of clay. Clays offer more nutrient resources and hold more water than sand and the roots respond by branching more intensively in the clay.

Soil surface

Sand

Clay

Sand

COURTESY OF J. V. LAKE

3.2 Topical ECOncerns

3.2 TOPICAL ECONCERNS

Global warming? Can we risk it?

Carbon dioxide is one of several global 'greenhouse gases' (see Section 13.3.1), whose increasing concentrations are believed by most scientists to be leading to rises in global mean temperatures, to a growth in the number of 'extreme' and 'record' weather events, and to the threat of the major biomes of Earth substantially changing their distribution (see Box 4.1).

The Intergovernmental Panel on Climate Change (IPCC) was established in 1988 by the World Meteorological Organization (WMO) and the United Nations Environment Programme (UNEP). Each report produced by the IPCC is written by some 200 independent scientists and other experts in approximately 120 countries around the world and is reviewed by another 400 independent experts.

A recent report (IPCC, 2001) describes the current state of understanding of the climate system, providing estimates of future change and highlighting areas of uncertainty. It concludes that an increasing body of observations point to a warming world – temperatures have risen during the past four decades in the lowest 8 km of the atmosphere, global average sea level has risen, snow cover and ice extent have decreased. These changes have occurred at the same time as atmospheric greenhouse gases continue to increase due to human activities and the panel points to new and stronger evidence that most of the observed warming over the past half century is attributable to human activities. Scientists now have greater confidence in the ability of models to project future climate, and all reasonable scenarios point to a substantial temperature increase. Globally averaged surface temperature is expected to increase by between 1.4°C and 5.8°C in the period 1990 to 2100, with complex consequences for weather patterns and sea level.

Policy-makers and law-makers are faced with different groups of scientific 'experts' offering different projections into the future, and with many interest groups, including a number of industries resisting attempts to force them to change their behavior in order to reduce emissions of greenhouse gases. Even though the majority of scientists believe the problem to be a very real one, the truth is that predictions of the future can never be made with absolute certainty.

Put yourself in the position of a politician. Would it be reasonable of you to demand major changes of significant sectors of the national economy, in order to avert a disaster that may never happen in any case? Or, since the consequences of the 'worst case' and even some of the 'middle of the road' scenarios are so profound, is the only responsible course of action to minimize risk – to behave as if disaster is certain if we do not change our collective behavior, even though it is not? One alternative might be to wait for better data. But suppose that by the time better data are available it is too late . . .

Plants have responded to even larger fluctuations of carbon dioxide over geological history. During the Triassic, Jurassic and Cretaceous periods, atmospheric concentrations of carbon dioxide were four to eight times greater than at present.

variations beneath a canopy

Concentrations can also vary in space and over short time scales. In a terrestrial community (Figure 3.21a), carbon dioxide concentration is highest (up to around $1800 \ \mu l \ l^{-1}$) very close to the ground in summer, where it is released rapidly from decomposing organic matter in the soil. Diffusion alone guarantees that the concentration quickly declines with increasing height, but in the daytime photosynthesizing plants also actively remove carbon dioxide from the air, whereas at night

Figure 3.21

(a) Average carbon dioxide concentrations for each hour of the day in a mixed deciduous forest (Harvard Forest, Massachusetts, USA) on November 21 and July 4 at three heights above the ground: ■, 0.05 m; ◇, 1 m; ◆, 12 m. (b) Variation in carbon dioxide concentration with depth in Lake Grane Langsø, Denmark in early July, and also in late August after the lake has become stratified with little mixing between the warm water at the surface and the colder water beneath.

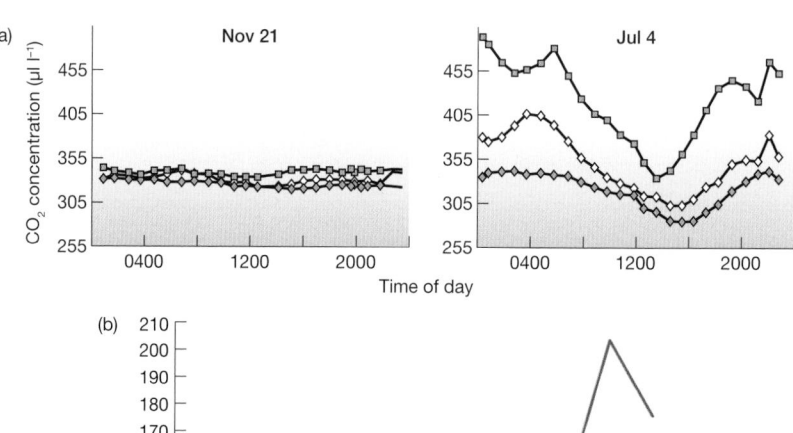

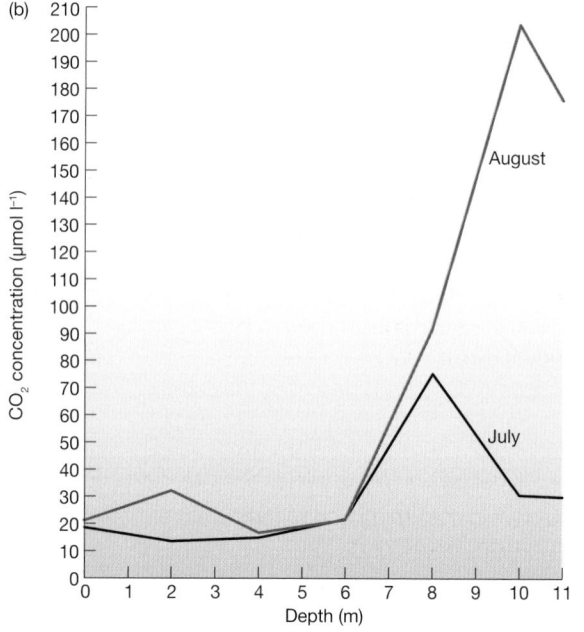

(a) AFTER BAZZAZ & WILLIAMS, 1991; (b) AFTER RIIS & SAND-JENSEN, 1997

concentrations increase as plants respire and there is no photosynthesis. During the winter, when low temperatures mean that rates of photosynthesis, respiration and decomposition are all slow, concentrations remain virtually constant through day and night at all heights. Thus, plants growing in different parts of a forest will experience quite different carbon dioxide environments: the lower leaves on a forest shrub will usually experience higher carbon dioxide concentrations than its upper leaves, and seedlings will live in environments richer in carbon dioxide than mature trees. In aquatic environments, variations in carbon dioxide concentration can be just as striking, especially when water mixing is limited, for example during the summer 'stratification' of lakes, with layers of warm water towards the surface and colder, carbon dioxide-rich layers beneath (Figure 3.21b).

Are higher concentrations of carbon dioxide better for plant growth? When other resources are present at adequate levels, additional carbon dioxide scarcely influences the rate of photosynthesis of C4 plants but increases the rate of C3 plants. Indeed, artificially increasing the carbon dioxide concentration in greenhouses is a commercial technique to increase crop (C3) yields. We might reasonably predict dramatic increases in the productivity of individual plants and of whole crops and natural communities as atmospheric concentrations of carbon dioxide continue to increase. However, there is also much evidence that the responses may be complicated. For example, when six species of temperate forest tree were grown for 3 years in a carbon dioxide-enriched atmosphere in a glasshouse, they were generally larger than controls, but the enhanced growth effect declined even within the relatively short time scale of the experiment (Bazzaz et al., 1993). Moreover, there is a general tendency for carbon dioxide enrichment to reduce the nitrogen concentration in above-ground plant tissues (Cotrufo et al., 1998), which may induce insect herbivores to eat 20–80% more foliage to maintain their nitrogen intake, effectively negating any growth enhancement.

what will be the consequences of current rises?

3.4 Animals and their resources

Green plants are *autotrophs*: their resources are quanta of radiation, ions and simple molecules. Plants assemble them into complex molecules (carbohydrates, fats, proteins) and then package them into cells, tissues, organs and whole organisms. It is these packages that form the food resources for virtually all other organisms, the *heterotrophs* (decomposers, predators, grazers, parasites). These consumers unpack the packages, metabolize and excrete some of the contents, and reassemble the remainder into their own bodies. They in turn may be consumed, unpacked and reconstituted in a chain of events in which each consumer becomes, in turn, a resource for some other consumer.

autotrophs and heterotrophs

Heterotrophs can generally be grouped as follows:

1 *Decomposers*, which feed on already dead plants and animals.
2 *Parasites*, which feed on one or a very few host plants or animals while they are alive but do not (usually) kill their hosts, at least not immediately.
3 *Predators*, which, during their life, eat many prey organisms, typically (and in many cases always) killing them.
4 *Grazers*, which, during their life, consume parts of many prey organisms, but do not (usually) kill their prey, at least not immediately.

The usual mental image of a predator–prey relationship is something akin to a lion eating a gazelle, but the relationship encompasses a much wider array of consumer–resource interactions. For example, a squirrel is a predator when it eats an acorn (it kills the acorn embryo); a whale is a predator as it feeds on krill; and even a fungus can be regarded as a predator when it feeds on and kills a growing seedling. In each case, the predator kills its food resource as it consumes all or part of it. Here, we concentrate on animal consumers (and take the subject further still in Chapter 7).

An important distinction between animal consumers is whether they are specialized or generalized in their diet. Generalists (*polyphagous* species) take a wide variety of prey species though they very often have clear preferences and a rank order of what they will choose when there are alternatives available. Specialists, on the other hand, may specialize on particular parts of their prey but range over a number of species. This is most common among herbivores because, as we shall see, different parts of plants are quite different in their composition. Thus, many birds specialize on eating seeds though they are seldom restricted to a particular species. Finally, a consumer may specialize on a single species or a narrow range of closely related species (when it is said to be *monophagous*). Examples are caterpillars of the cinnabar moth (which eat the leaves, flower buds and very young stems of a species of ragwort, *Senecio*) and many species of host-specific parasites.

monophagy and polyphagy

3.4.1 Nutritional needs and provisions

plants as (a variety of) foods

The various parts of a plant have very different compositions (Figure 3.22a) and so offer quite different resources. Bark, for example, is largely composed of dead

Figure 3.22

The composition of various plants (a) and animals (b) that may serve as food resources for herbivores or carnivores. Note that the different parts of a plant have very different compositions, whereas different species of animal (and their parts) are remarkably similar.

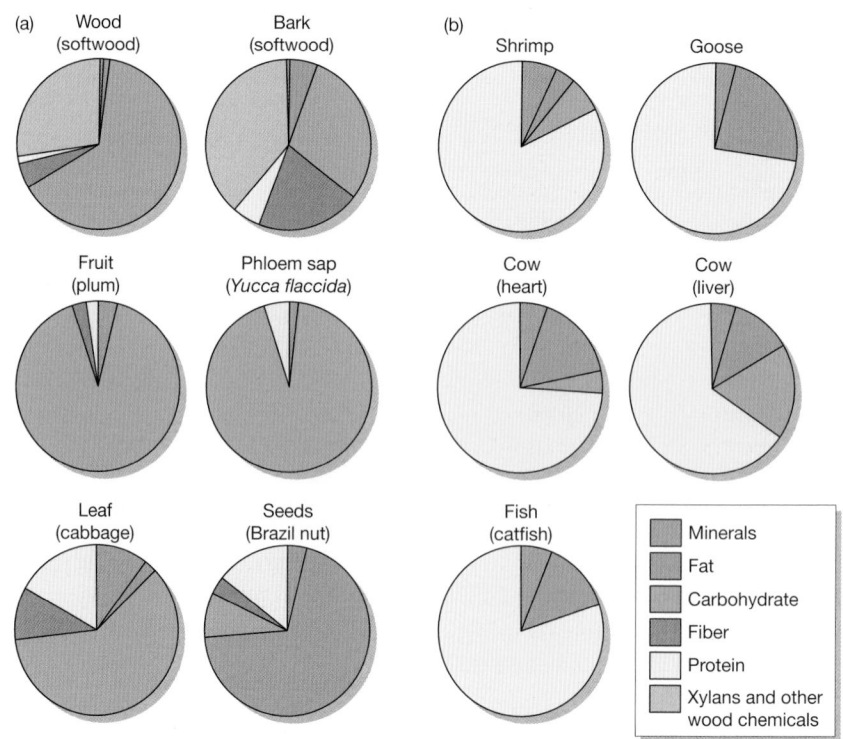

Figure 3.23
Examples of the variety of specialized mouthparts in herbivorous insects. (a) Honeybee with a long 'tongue' (glossia) for sucking. (b) Hawkmoth with an even longer sucking proboscis. (c) Leichhardt's grasshopper with large, plate-like chewing mandibles. (d) Acorn weevil with chewing mouthparts at the very end of its long rostrum. (e) Rose aphid with a piercing stylet.

cells with corky and lignified walls, packed with defensive phenolics, and is quite useless as a food for most herbivores (even species of 'bark beetle' specialize on the nutritious cambium layer just beneath the bark, rather than on the bark itself). The richest concentrations of plant proteins (and hence of nitrogen) are in the meristem in the buds at shoot tips and in leaf axils. Not surprisingly, these are usually heavily protected with bud scales and defended from herbivores by prickles and spines. Seeds are usually dried, packaged reserves rich in starch or oils as well as specialized storage proteins. And the very sugary and fleshy fruits are resources provided by the plant as 'payment' to animals that disperse the seeds. Very little of the plants' nitrogen is spent on these rewards.

The diversity of different food resources offered by plants is matched by the diversity of specialized mouthparts and digestive tracts that have evolved to consume them. This diversity is especially developed in the beaks of birds and the mouthparts of insects (Figure 3.23).

For a consumer, the body of a plant is a quite different package of resources from the body of an animal. First, plant cells are bounded by walls of cellulose, lignin and other structural carbohydrates that give plants their high fiber content and contribute to their high ratio of carbon to other elements. These large amounts of fixed carbon mean that plants are potentially rich sources of energy. Yet the overwhelming majority of animal species lack cellulolytic and other enzymes that

from plants into animals

can digest these compounds: they are quite useless as a direct energy resource for most herbivores. Moreover, the cell wall material of plants hinders the access of digestive enzymes to the plant cell contents. The acts of chewing by the grazing mammal, cooking by humans, and grinding in the gizzard of birds are necessary precursors to digestion of plant food because they allow digestive enzymes access to the cell contents. The carnivore, by contrast, can more safely gulp its food.

Many herbivores have made up for their own lack of cellulolytic enzymes by entering into a *mutualistic* (beneficial to both parties) association with cellulolytic bacteria and protozoa in their guts that do have the appropriate enzymes. The rumen (or sometimes the cecum) of many herbivorous mammals is a temperature-regulated culture chamber for these microbes into which already partially fragmented plant tissues flow continually (Figure 3.24). The microbes receive a home and a supply of food. The herbivorous 'host' benefits by absorbing many of the major byproducts of this microbial fermentation, especially fatty acids.

Unlike plants, animal tissues contain no structural carbohydrate or fiber component but are rich in fat and protein. The C : N ratio of plant tissues commonly exceeds 40 : 1, in contrast to ratios that rarely exceed 10 : 1 in bacteria, fungi and animals. Thus herbivores, which undertake the first stage of making animal

Figure 3.24

The digestive tracts of herbivores are commonly modified to provide fermentation chambers inhabited by a rich fauna and flora of microbes. The figure shows the digestive tracts of four different herbivorous mammals with the fermentation chambers highlighted in a darker shade. (a) Rabbit, with a fermentation chamber in the expanded cecum. (b) Zebra, with fermentation chambers in both cecum and colon. (c) Sheep, with foregut fermentation in an enlarged portion of the stomach, rumen and reticulum. (d) Kangaroo, with an elongate fermentation chamber in the proximal portion of the stomach.

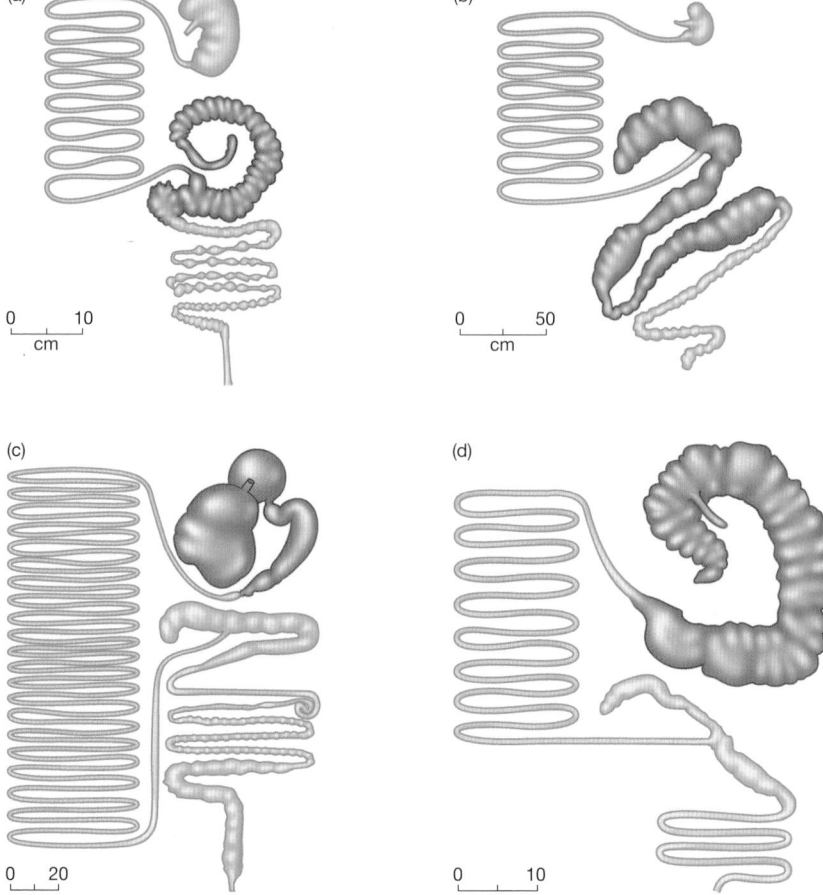

bodies out of plants, are involved in a massive burning off of carbon as the C : N ratio is lowered. The main waste products of herbivores are therefore carbon-rich compounds (carbon dioxide and fiber). Carnivores, on the other hand, get most of their energy from the protein and fats of their prey, and their main excretory products are consequently nitrogenous.

Even if the cell wall is excluded from consideration, the C : N ratio is high in plants compared with other organisms. Aphids, which gain direct access to cell contents by driving their stylets into the phloem transport system, gain a resource that is rich in soluble sugars (see Figure 3.22a). In their search for valuable nitrogen, they use only a fraction of this energy resource and excrete the rest in sugary rich honeydew that may drip as a rain from an aphid-infested tree. For most herbivores and decomposers, the body of a plant is a superabundant source of energy and carbon; it is other components of the diet, especially nitrogen, that are more usually limiting.

The bodies of different species of animal have remarkably similar composition (see Figure 3.22b). In terms of protein, carbohydrate, fat, water and minerals per gram, there is very little to choose between a diet of caterpillars or cod, or of earthworms, shrimps or venison. The packages may be differently assembled (and the taste may be different), but the contents are essentially the same. Moreover, the different parts of an animal have very similar nutritional content. Unlike herbivores, carnivores are not faced with difficult problems of digestion (and they vary very little in their digestive apparatus) but rather with difficulties in finding, catching and overcoming the defenses of their prey.

animals as food

3.4.2 Defense

The value of a resource to a consumer is determined not only by what it contains but by how well its contents are defended. Not surprisingly, organisms have evolved physical, chemical, morphological and behavioral defenses against being attacked. These serve to reduce the chance of an encounter with a consumer and/or increase the chance of survival in such encounters. The spiny leaves of holly are not eaten by larvae of the oak eggar moth, but if the spines are removed the leaves are eaten quite readily. No doubt similar results would be achieved in equivalent experiments with foxes as predators and de-spined hedgehogs or porcupines as prey. On a smaller scale, many plant surfaces are clothed in epidermal hairs (trichomes) that may keep the smaller predators (such as thrips and mites) away from the leaf surface (Figure 3.25; see also Figure 3.27a).

Any feature of an organism that increases the energy spent by a consumer in discovering or handling it is a defense if, as a consequence, the consumer eats less of it. The thick shell of a nut increases the time that an animal spends extracting a unit of effective food, and this may reduce the number of nuts that are eaten. We have already seen that most green plants are relatively overprovided with energy resources in the form of cellulose and lignin. It may therefore be cheap to build husks and shells around seeds (and woody spines on stems) if these defense tissues contain rather little protein, and if what is protected is far more valuable.

some resources are protected . . .

Both plants and animals have a battery of chemical defenses. The plant kingdom, in particular, is very rich in 'secondary' chemicals that apparently play no role in normal plant biochemical pathways. A defensive function is generally ascribed to these chemicals and a defensive role has been demonstrated

. . . or defended

Figure 3.25

A mite trapped in the protective trichomes (hairs) on the surface of a *Primula* leaf. The trichomes themselves support capsules of irritant volatile oils at their tip. Each white bar towards the foot of the image represents 10 μm.

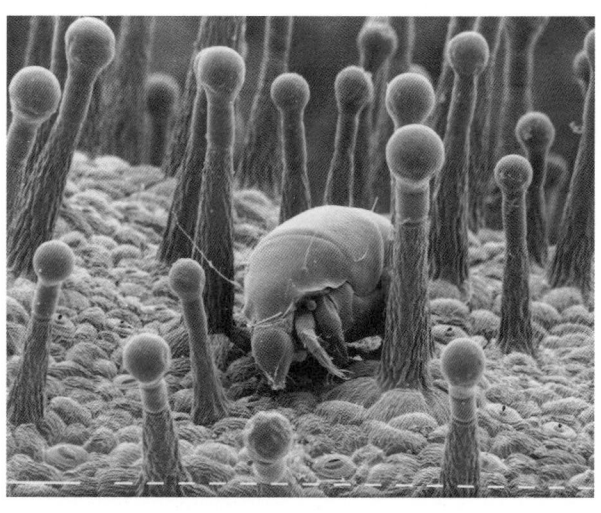

unequivocally in some cases. Populations of white clover, for example, commonly contain some individuals that release hydrogen cyanide when their tissues are attacked (*cyanogenic* forms) and others that do not; those that cannot are eaten by slugs and snails. The cyanogenic forms, however, are nibbled but then rejected (Table 3.1).

optimal defense theory: constitutive and inducible defenses

Noxious plant chemicals have been classified into two broad types. The first are *quantitative* chemicals (so-called because they are most effective at relatively high concentrations), which make the tissues that contain them, such as mature oak leaves, relatively indigestible. They are also often called *constitutive* chemicals, since they tend to be produced even in the absence of herbivore attack. The second type are toxic or *qualitative* chemicals, which are poisonous even in small quantities but can be produced relatively rapidly and are therefore commonly *inducible*: only produced in response to damage itself, and hence with lower fixed costs to the plants.

Plants differ in their chemical defenses from species to species and also from tissue to tissue within an individual plant. Broadly, relatively short-lived,

Table 3.1

Slugs (*Agriolimax reticulatus*) graze on the leaves of clover (*Trifolium repens*). There are forms of clover that release hydrogen cyanide when the cells are damaged. Slugs nibble clover leaves and reject cyanogenic forms but continue to consume the leaves of non-cyanogenic forms. Two plants, one of each form, were grown together in plastic containers and slugs were allowed to graze for seven successive nights. The table shows the numbers of leaves in different conditions after slug grazing. +/– indicate deviation from random expectation; the difference from random expectation is significant at $P < 0.001$.

	CONDITIONS OF LEAVES AFTER GRAZING			
	NOT DAMAGED	NIBBLED	UP TO 50% OF LEAF REMOVED	MORE THAN 50% OF LEAF REMOVED
Cyanogenic plants	160 (+)	22 (+)	38 (–)	9 (–)
Non-cyanogenic plants	87 (–)	7 (–)	30 (+)	65 (+)

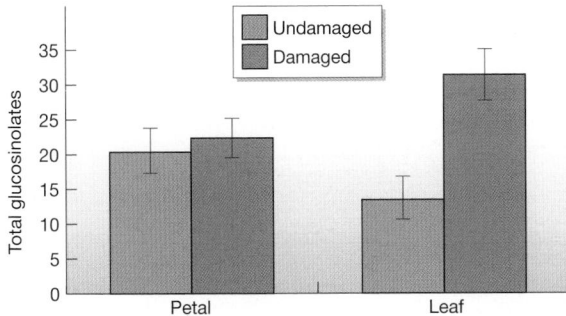

Figure 3.26

Concentrations of glucosinolates (μg mg^{-1} dry mass) in the petals and leaves of wild radish, *Raphinus sativus*, either undamaged or damaged by caterpillars of *Pieris rapae*. Bars are standard errors.

AFTER STRAUSS ET AL. 2004

ephemeral plants gain a measure of protection from consumers because of the unpredictability of their appearance in space and time. They therefore need to invest less in defense than predictable, long-lived species like forest trees. The latter, precisely because they are apparent for long periods to a large number of herbivores, tend to invest in constitutive chemicals that, while costly, afford them broad protection; whereas ephemeral plants tend to produce inducible toxins as required. Moreover, it may be predicted that, within an individual plant, the more important plant parts should be protected by costly, constitutive chemicals, whereas less important parts should rely on inducible toxins (McKey, 1979; Strauss et al., 2004). This is confirmed, for example, by a study of wild radish, in which plants were either attacked by caterpillars of the butterfly, *Pieris rapae*, or left as unmanipulated controls (Figure 3.26). Flower petals are known to be highly important to fitness in this insect-pollinated plant, and concentrations of toxic glucosinolates were twice as high in petals as in undamaged leaves: levels that were maintained constitutively, irrespective of whether the petals were damaged by the caterpillars. Leaves, on the other hand, have a much less direct influence on fitness: high levels of leaf damage can be sustained without any measurable effect on reproductive output. Constitutive levels of glucosinolates, as already noted, were low; but if leaves were damaged, the (induced) concentrations were even higher than in the petals.

Animals have more options than plants when it comes to defending themselves, but some still make use of chemicals. For example, defensive secretions of sulfuric acid of pH 1 or 2 occur in some marine gastropod groups, including the cowries. Other animals, which can tolerate the chemical defenses of their plant food, actually store the plant toxins and use them in their own defense. A classic example is the monarch butterfly, whose caterpillars feed on milkweeds, which contain cardiac glycosides, which are poisonous to mammals and birds. These caterpillars can store the poison, and it is still present in the adult. Thus, a bluejay will vomit violently after eating one, and, once it recovers, will reject all others on sight. In contrast, monarchs reared on cabbage are edible.

chemical defense in animals

Chemical defenses are not equally effective against all consumers. Indeed, what is unacceptable to some animals may be the chosen, even unique, diet of others. Many herbivores, particularly insects, specialize on one or a few plant species whose particular defense they have overcome. For example, females of the cabbage root fly, with eggs to lay, home in on a brassica crop from distances as far as 15 m downwind of the plants. It is probably hydrolyzed glucosinolates (toxic to many other species) that provide the attractive odor.

Figure 3.27

Lepidopterous caterpillars illustrate a range of defense strategies. (a) The irritating hairs of the gypsy moth. (b) Aposematism (advertizing distastefulness) in the black swallowtail. (c) A cryptic (camouflaged) noctuid, looking like bark. (d) Another swallowtail rearing and hence possibly startling a potential predator.

crypsis, aposematism, and mimicry

An animal may be less obvious to a predator if it matches its background, or possesses a pattern that disrupts its outline, or resembles an inedible feature of its environment. A straightforward example of such *crypsis* is the green coloration of many grasshoppers and caterpillars (Figure 3.27). Cryptic animals may be highly palatable, but their morphological traits and color (and their choice of the appropriate background) reduce the likelihood that they will be used as a resource. In contrast, noxious or dangerous animals often seem to advertize the fact by bright, conspicuous colors and patterns (*aposematism*; Figure 3.27b). The monarch butterfly (see earlier), for example, is aposematically colored. One attempt by a bird to eat an adult monarch is so memorable that others are subsequently avoided for some time. The adoption of memorable body patterns by distasteful prey, moreover, immediately opens the door for deceit by other species – there will be a clear advantage to a palatable prey if it *mimics* an unpalatable species. Thus, the palatable viceroy butterfly mimics the distasteful monarch, and a bluejay that has learned to avoid monarchs will also avoid viceroys.

behavior

By living in holes, animals (millipedes, moles) may avoid stimulating the sensory receptors of predators, and by 'playing dead' (opossum, African ground squirrel) animals may fail to stimulate a killing response. Animals that withdraw to a prepared retreat (rabbits and prairie dogs to their burrows, snails to their shells) or roll up and protect their vulnerable parts by a tough exterior (armadillos, hedgehogs) reduce their chance of capture. Other animals seem to try to bluff themselves out of trouble by threatening or startling displays (Figure 3.27d). Moths and butterflies that suddenly expose eye spots on their wings are one example. No doubt the most common behavioral response of an animal in danger of becoming a used resource is to run away.

3.5 Effects of intraspecific competition for resources

Resources are consumed. The consequence is that there may not be enough of a resource to satisfy the needs of a whole population of individuals. Individuals may then compete with each other for the limited resource. *Intraspecific competition* is competition between individuals of the same species.

In many cases, competing individuals do not interact with one another directly. Rather, they deplete the resources that are available to each other. Grasshoppers may compete for food, but a grasshopper is not directly affected by other grasshoppers so much as by the level to which they have reduced the food supply. Two grass plants may compete, and each may be adversely affected by the presence of close neighbors, but this is most likely to be because their resource depletion zones overlap – each may shade its neighbors from the incoming flow of radiation, and water or nutrients may be less accessible than they would otherwise be around the plants' roots. The data in Figure 3.28, for example, show the dynamics of the interaction between a single-celled aquatic plant, a diatom, and one of the resources it requires, silicate. As diatom density increases over time, silicate concentration decreases: there is then less available for the many than there had been previously for the few. This type of competition – in which competitors interact only indirectly, through their shared resources – is termed *exploitation*.

exploitation: competitors depleting each other's resources

On the other hand, competing individual vultures may fight one another over access to a newly found carcass. Individuals of other species may fight for ownership of a 'territory' and access to the resources that a territory brings with it. A barnacle that settles on a rock denies the space to another barnacle. This is called *interference* competition.

direct interference

Whether competition occurs through exploitation, interference or a combination of the two, its ultimate effect is on the *vital rates* of the competitors – their survival, growth and reproduction – compared with what they would have been if resources had been more abundant. Competition typically leads to decreased

competition and vital rates

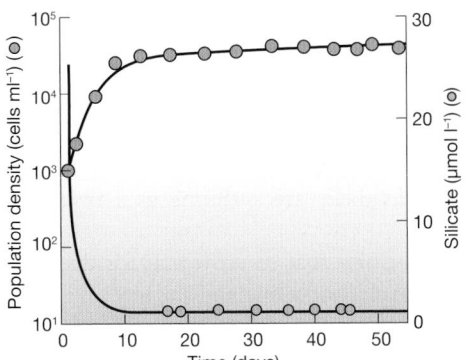

Figure 3.28

A population of the freshwater diatom *Asterionella formosa* was grown in flasks of culture medium. The diatom consumes silicate during growth and the population of diatoms stabilizes when the silicate has been reduced to a very low concentration.

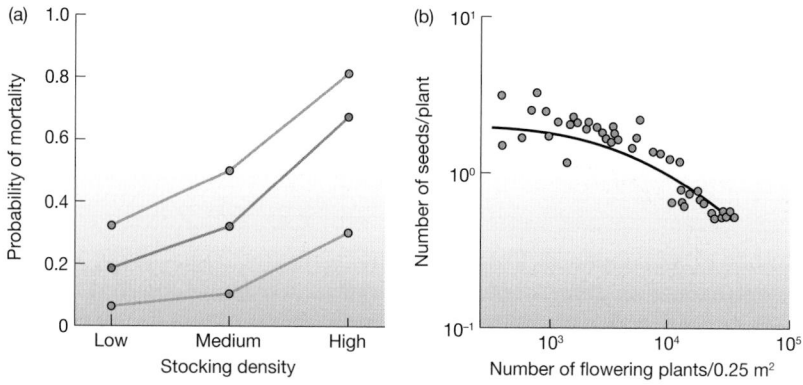

Figure 3.29

(a) The rate of mortality among steelhead trout (*Oncorynchus mykiss*) reared at a range of densities (32, 63 and 127 per m²) and at a range of food levels (1.4, 2.9 and 5.8 g of food pellets per day: yellow, maroon and blue lines, respectively). (b) The average number of seeds produced per plant of the dune grass *Vulpia fasciculata* growing at a range of densities.

(a) AFTER KEELEY, 2001; (b) AFTER WATKINSON & HARPER, 1978

rates of resource intake per individual, and thus to decreased rates of individual growth or development, perhaps to decreases in amounts of stored reserves or to increased risks of predation. Figure 3.29a shows how the mortality rate of steelhead trout increases, as the number of competing fish rises, at a range of food levels; Figure 3.29b shows how the birth rate of the sand dune grass *Vulpia* declines as individuals become increasingly crowded.

In practice, intraspecific competition is often a very one-sided affair: a strong early seedling will shade and suppress a stunted, late one; a large vulture is likely to fight off a smaller one. Some of the competitive strength of individuals is related to timing (the early seedling) or to random events (one seed may germinate in a depression where it obtains more water than its neighbors). Sometimes the winner and loser may be genetically different and then competition will be playing a role in natural selection.

density dependence

The effects of intraspecific competition on any individual are typically greater the more crowded the individual is by its neighbors – the more the resource depletion zones of other individuals overlap its own. This often translates into saying that the greater the density of a population of competitors the greater is the effect of competition. Hence, the effects of intraspecific competition are often said to be *density-dependent*. But it is doubtful that any organism has a way of detecting the density of its population. Rather, it responds to the effects of being crowded.

On the other hand, at low densities in the case of *Vulpia* (Figure 3.29b), the per capita birth rate or fecundity was *independent* of density (where per capita means literally 'per head' or 'per individual'). That is, the fecundity was effectively the same at a density of 1000 plants/0.25 m² as it was at a density of 500/0.25 m². Thus, there is no evidence at these densities that individuals are affected by the presence of other individuals and hence no evidence of intraspecific competition. But as density increases further, the per capita birth rate progressively decreases. These effects are now density-dependent, and this may be taken as an indication that at these densities, individuals are suffering as a result of intraspecific competition.

competition and the total number of survivors

The patterns in Figure 3.29 make the point that as crowding (or density) increases, the fecundity per individual is likely to decline and the mortality per individual likely to increase (which would mean that the survival rate per individual would *decrease*). But what can we expect to happen to the *total* number

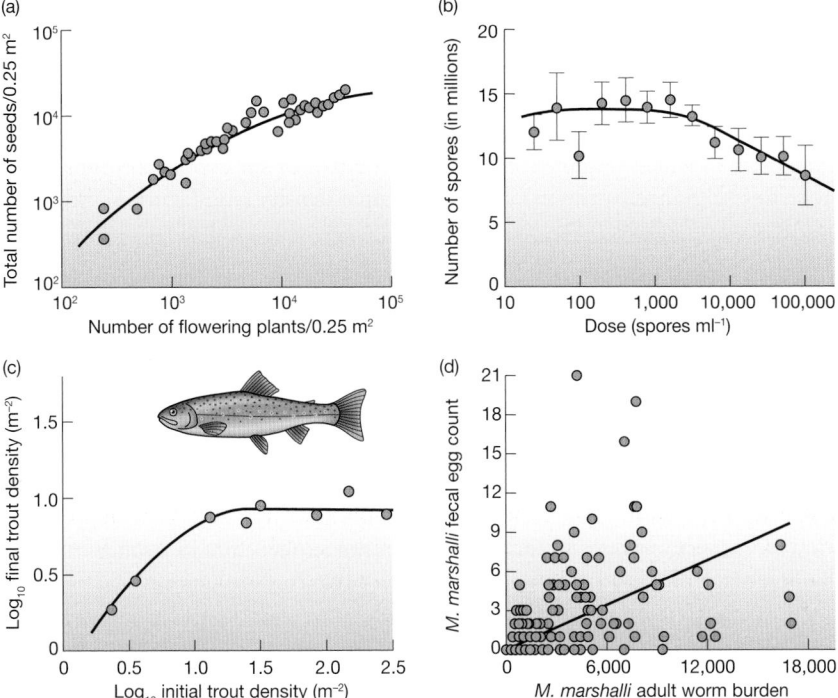

(a) AFTER WATKINSON & HARPER, 1978; (b) AFTER EBERT ET AL., 2000; (c) AFTER LE CREN, 1973; (d) AFTER IRVINE ET AL., 2001

Figure 3.30

Undercompensating, overcompensating and exactly compensating effects of intraspecific competition. (a) An undercompensating effect on fecundity: the total number of seeds produced by *Vulpia fasciculata* continues to rise as density increases. (b)When the planktonic crustacean *Daphnia magna* was infected with varying numbers of spores of the bacterium *Pasteuria ramosa*, the total number of spores produced per host in the next generation was independent of density (exactly compensating) at the lower densities, but declined with increasing density (overcompensating) at the higher densities. Standard errors are shown. (c) An exactly compensating effect on mortality: the number of surviving trout fry is independent of initial density at higher densities. (d) The total number of eggs of the parasitic nematode *Marshallagia marshalli* produced by infected reindeer (eggs per gram of feces) increased in direct proportion to the number of adult nematodes in the reindeer: there was no evidence of competition between the nematodes.

of seeds or eggs produced by populations at different densities – or to the total number of survivors? In some cases, although the rate per individual declines with increasing density, the total fecundity or total number of survivors in the population continues to increase. This can be seen (Figure 3.30a) to have been the case for the plant populations in Figure 3.29b – at least over the range of densities examined. In other cases, the rate per individual declines so rapidly with increasing density that the total fecundity or total number of survivors in the population actually gets smaller the greater the number of contributing individuals. This can be seen in Figure 3.30b for the highest densities of a bacterial parasite of the planktonic crustacean *Daphnia magna*.

In yet further cases, the mortality risk or fecundity per individual declines with increasing density such that the total number of survivors or total fecundity is the same irrespective of the number of contributing individuals. This is referred to as *exactly compensating density dependence*, and the competition leading to it is sometimes referred to as 'contest-like', since this is the pattern you would expect to see if there were a fixed number of winners and all the other competitors were doomed to lose. Examples are shown for fecundity in Figure 3.30b (at the lower densities) and for survivors in Figure 3.30c. Finally, of course, birth or mortality rates may be density-independent (no competition) throughout the range examined, in which case the total number of births or survivors will simply continue to rise in direct proportion to the original density (e.g. Figure 3.30d).

3.6 Conditions, resources and the ecological niche

Finally, many of the ideas in this chapter can be brought together in the concept of the ecological niche. The term *niche*, though, is frequently misunderstood and misused. It is often used loosely to describe the sort of place in which an organism lives, as in the sentence 'Woodlands are the niche of woodpeckers'. However, strictly, where an organism lives is not its niche but its *habitat*. A niche is not a place an idea: a summary of the organism's tolerances and requirements. The habitat of a gut microorganism would be an animal's alimentary canal; the habitat of an aphid might be a garden; and the habitat of a fish could be a whole lake. Each habitat, however, provides many different niches: many other organisms also live in the gut, the garden, or the lake – and with quite different lifestyles. The niche of an organism describes how, rather than just where, an organism lives.

the niche of an organism is defined by its needs and tolerances

The modern concept of the niche was proposed by Hutchinson in 1957 to address the ways in which tolerances and requirements interact to define the conditions and resources needed by an individual or a species in order to practice its way of life. Temperature, for instance, is a condition that limits the growth and reproduction of all organisms, but different organisms tolerate different ranges of temperature. This range is one *dimension* of an organism's ecological niche: Figure 3.31a shows how different species of plants vary in the temperature dimension of their niche. But there are many such dimensions for the niche of a species: its tolerance of various other conditions (relative humidity, pH, wind speed, waterflow, and so on), and its need for various resources (nutrients, water, food, and so on). Clearly the real niche of a species must be multidimensional.

It is easy to visualize the early stages of building such a multidimensional niche. Figure 3.31b illustrates the way in which two niche dimensions (temperature and salinity) together define a two-dimensional area that is part of the niche of a sand shrimp. Three dimensions, such as temperature, pH and the availability of a particular food, may define a three-dimensional niche volume (Figure 3.31c). It is hard to imagine (and impossible to draw) a diagram of a more realistic, multidimensional niche. (Technically, we now consider a niche to be an *n-dimensional hypervolume*, where *n* is the number of dimensions that make up the niche.) But the simplified three-dimensional version nonetheless captures the idea of the ecological niche of a species. It is defined by the boundaries that limit where it can live, grow and reproduce, and it is very clearly a concept rather than a place. The concept has become a cornerstone of ecological thought, as we shall see in later chapters.

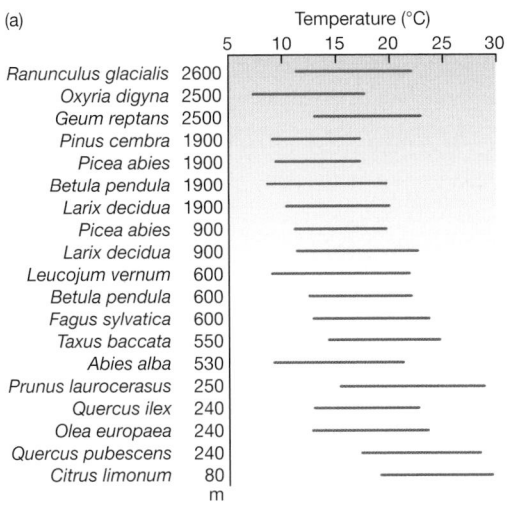

(a)

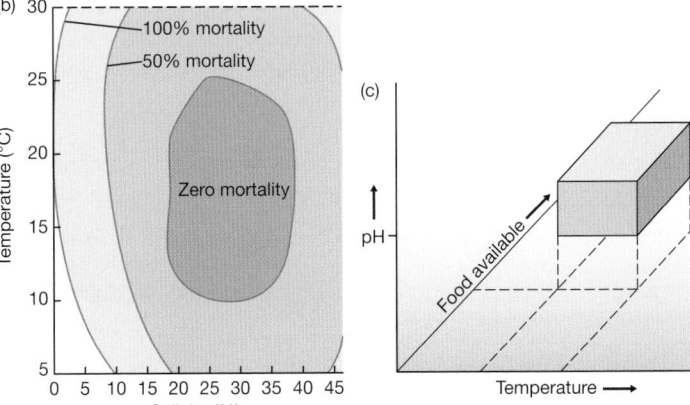

(b)

Figure 3.31

(a) A niche in one dimension. The range of temperatures at which a variety of plant species from the European Alps can achieve net photosynthesis at low intensities of radiation (70 W m^{-2}). (b) A niche in two dimensions for the sand shrimp (*Crangon septemspinosa*) showing the fate of egg-bearing females in aerated water at a range of temperatures and salinities. (c) A diagrammatic niche in three dimensions for an aquatic organism showing a volume defined by the temperature, pH and availability of food.

(a) AFTER PISEK ET AL., 1973; (b) AFTER HAEFNER, 1970

Summary

SUMMARY

Conditions and resources

Conditions are physicochemical features of the environment such as its temperature and humidity. They may be altered but are not consumed. Environmental resources are consumed by living organisms in the course of their growth and reproduction.

Environmental conditions

There are three basic types of response curve to conditions. Extreme conditions may be lethal with, between the two extremes, a continuum of more favorable conditions; or a condition may be lethal only at high intensities; or a condition may be required by organisms at low concentrations but become toxic at high concentrations.

These responses are accounted for, in part, by changes in metabolic effectiveness. But at extremely high temperatures, for example, enzymes and other proteins become unstable and break down, and the organism dies; and at high environmental temperatures, terrestrial organisms may encounter serious, perhaps lethal, problems of dehydration. At temperatures a few degrees above zero, organisms may be forced into extended periods of inactivity, or ice may form between cells and draw water from within them. The timing and duration of temperature extremes, however, may be as important as absolute temperatures.

In practice, the effects of conditions may be determined largely by the responses of other community members, through food consumption, disease or competition.

Many conditions are important stimuli for growth and development and prepare an organism for conditions that are to come.

Plant resources

Solar radiation, water, minerals and carbon dioxide are all critical resources for green plants. The shape of the curve that relates the rate of photosynthesis to the intensity of radiation varies greatly among species. The radiation that reaches a plant is forever changing; the plant integrates the diverse exposures of its various leaves.

Most variation in leaf shape has probably evolved under selection to optimize the photosynthesis achieved per unit of water transpired. Any mechanism or process that slows the rate of water loss, such as closing of the stomata, reduces the photosynthetic rate. If the rate of water uptake falls below the rate of release, the body of the plant starts to wilt. If the deficit accumulates, the whole plant may die. Plants may avoid or tolerate water shortage. Specialized biochemical processes may increase the amount of photosynthesis that can be achieved per unit of water lost in C4 and CAM (as opposed to C3) plants.

The primary water-absorbing zone on roots is covered with root hairs that make intimate contact with soil particles. Roots create water depletion zones around themselves. Root architectures are much less tightly programmed than those of shoots, and those established early in a plant's life can determine its responsiveness to later events. Roots also extract key minerals from the soil. Root architecture is particularly important here because different nutrients are held in the soil by different forces.

Animals and their resources

Green plants are autotrophs. Decomposers, predators, grazers and parasites are heterotrophs. The various parts of a plant have very different compositions and so offer quite different resources. This diversity is matched by the diversity of mouthparts and digestive tracts that have evolved to consume them. The body of a plant is a quite different package of resources from the body of an animal. To make better use of plant material, many herbivores enter into a mutualistic association with cellulolytic bacteria and protozoa in their alimentary canal.

The C : N ratio of plant tissues greatly exceeds those of bacteria, fungi and animals. Thus, herbivores typically have a superabundant source of energy and carbon, but nitrogen is often limiting; their main waste products are carbon dioxide and fiber. The bodies of different species of animal have remarkably similar compositions. Carnivores are not faced with problems of digestion, but rather with difficulties in finding, catching and overcoming the defenses of their prey. Carnivores' main excretory products are nitrogenous.

Effects of intraspecific competition for resources

Individuals may compete indirectly, via a shared resource, through exploitation, or directly, through interference. The ultimate effect of competition is on survival, growth and reproduction of individuals. Typically, the greater the density of a population of competitors, the greater is the effect of competition (density dependence). As a result, though, the total number of survivors, or of offspring, may increase, decrease or stay the same as initial densities increase.

Conditions, resources and the ecological niche

Where an organism lives is its habitat. A niche is a summary of an organism's tolerances and requirements. The modern concept, proposed by Hutchinson in 1957, is an *n*-dimensional hypervolume.

Review questions

1* Explain, referring to a variety of specific organisms, how the amount of water in different organisms' habitats may define either the conditions for those organisms, or their resource level, or both.

2 Discuss whether you think the following statement is correct: 'A layperson might describe Antarctica as an extreme environment, but an ecologist should never do so'.

3 In what ways do ectotherms and endotherms differ, and in what ways are they similar?

4* Drawing examples from a variety of both animals and plants, contrast the responses of tolerators and avoiders to seasonal variations in environmental conditions and resources.

5 Describe how plants' requirements to increase the rate of photosynthesis and to decrease the rate of water loss interact. Describe, too, the strategies used by different types of plants to balance these requirements.

6* Describe and account for the differences in both root and shoot architecture exhibited by different plants.

7 Account for the fact that the tissues of plants and animals have such contrasting C : N ratios. What are the consequences of these differences?

8 Describe the various ways in which animals use color to defend themselves against attacks by predators.

9 Explain, with examples, what exploitation and interference intraspecific competition have in common and how they differ.

10 What is meant when an ecological niche is described as an *n*-dimensional hypervolume?

Chapter **4**

Conditions, resources and the world's communities

Chapter contents

Key concepts

In this chapter you will:

- understand that conditions and resources interact to help determine the composition of whole communities
- appreciate that climatic patterns over the surface of the Earth are responsible for the large-scale pattern of distribution of terrestrial biomes (such as tropical rain forest, desert and tundra)
- recognize that biomes are not homogeneous because local topography, geology and soil influence the communities of plants and animals that occur
- appreciate that conditions and resources at a location may change over time scales ranging from hours to millennia, leading to parallel temporal patterns in the composition of communities
- understand that in most aquatic environments it is difficult to recognize anything comparable to terrestrial biomes: communities tend to reflect local conditions and resources rather than global patterns in climate

The interplay of conditions and resources profoundly influences the composition of the world's communities. At the global scale, patterns of climate circulation are largely responsible for distinctive terrestrial biomes, such as deserts and rain forests, with their characteristic assemblages of plants and animals. Distinct types of marine and freshwater communities can sometimes also be identified at a broad geographic scale. Within each biome or aquatic category, however, there are enormous variations in conditions and resources that are reflected in community patterns viewed at a smaller scale.

4.1 Introduction

Having examined in Chapter 3 the way individual organisms are affected by conditions and resources, we now turn to the larger question of how the interplay of conditions and resources influences whole communities (the assemblages of species that occur together). The answer to this question depends fundamentally on the scale at which we choose to study communities; this will be a pervasive theme throughout the chapter.

Not surprisingly, because of its influence on both conditions and resources, climate plays a major role in determining the large-scale distribution of different types of community across the face of the Earth. However, local factors, such as soil type in terrestrial environments and water chemistry in aquatic environments, are responsible for patchiness in community composition on much smaller scales. We discuss some of the causes of spatial patterns in community distribution in Section 4.2. Then, in Section 4.3, we turn to temporal patterns in conditions and resources that can change community composition over time scales from days to millennia. Section 4.4 describes the characteristics of the Earth's major terrestrial biomes and Section 4.5 deals with the diversity of aquatic communities.

scale and patchiness – central themes of this chapter

4.2 Geographic patterns at large and small scales

4.2.1 Large-scale climatic patterns

At the largest scale, the geography of life on Earth is mainly a consequence of the planet's movement through space. The tilt of the Earth as it orbits the sun causes solar radiation to strike the Earth's surface with different intensities at different latitudes (Figure 4.1). Because the equator is tilted toward the sun, equatorial and tropical areas receive more direct sunlight and are warmer than other latitudes. Warm air holds more moisture than cold air, increasing the water-holding capacity of air around the tropics. Solar radiation draws water from the vegetation by evaporation, but because the air is so moist, much of the water condenses and

solar radiation, . . .

Figure 4.1

The tilt of the Earth on its axis and its rotation around the sun define the amount of radiation striking the atmosphere around the Earth's surface. This, in combination with the daily spin of the Earth on its axis, is responsible for the large-scale patterns of rainfall and solar radiation that define the pattern of global climate. This diagram shows winter in the northern hemisphere with radiation falling almost vertically south of the equator, but the same amount of radiation is spread over greater areas north of the equator; less is therefore received, and there is less heating per unit area.

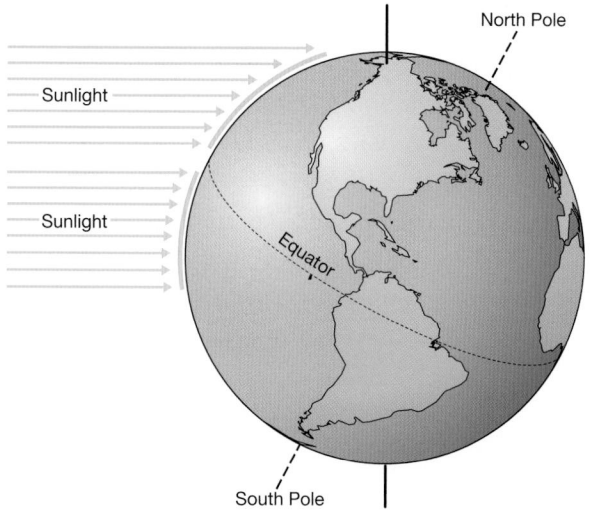

falls back as rain. Thus, the air that cycles to the atmosphere from the tropics is relatively dry, having lost most of its moisture as local rainfall before it ascends to the lower atmosphere.

The rotation of the Earth causes air masses from the tropics to curve to the north and south. Air that was warmed at the tropics (and which lost moisture as local rain) cools in the atmosphere and descends again at latitudes of approximately 30° (north and south). The air mass warms as it descends, increasing its capacity to hold water and causing the descending air mass to 'soak up' available water from the land. As a result, this is where most of the major deserts, including the Sahara, Kalahari, Mojave and Sonoran, are found. Another smaller evaporation/precipitation system occurs between 30° and 60° latitude, as warm air, now moist, rises and is blown further north or south, respectively. As it cools, the air descends again and rains, producing wetter environments.

. . . ocean currents . . .

Ocean currents have further powerful effects on climatic patterns. Southern waters circulate counterclockwise; they carry cold Antarctic waters up along the western coasts of continents and distribute warmer waters from the tropics along their eastern coasts (Figure 4.2). In the northern hemisphere, currents circulate clockwise, carrying cold Arctic waters along the eastern coasts of continents and returning warm tropical currents along western coasts. The cool, dry climate of eastern South America is an effect of the Antarctic Humboldt current; the relatively dry climate of California is a result of Arctic currents. Conversely, on the eastern side of North America the strong tropical Gulf Stream carries with it warm and moist air far into the Atlantic Ocean, affecting even the climate of Western Europe.

. . . and mountain ranges . . .

The topography of the land has consequences at an intermediate scale for the pattern of terrestrial climates. As winds meet mountain ranges they are forced up and become cooler as they rise. The cooler air holds less moisture so that water is released (as rain and snow) on the windward slopes of the mountains (the Rockies and Himalayas provide striking examples of this effect). As the air passes over to the leeward sides of the mountains it descends, becomes warmer and now absorbs water. This produces a desiccating effect and causes a *rain shadow* along the leeward slopes (Figure 4.3).

AFTER AUDESIRK & AUDESIRK, 1996

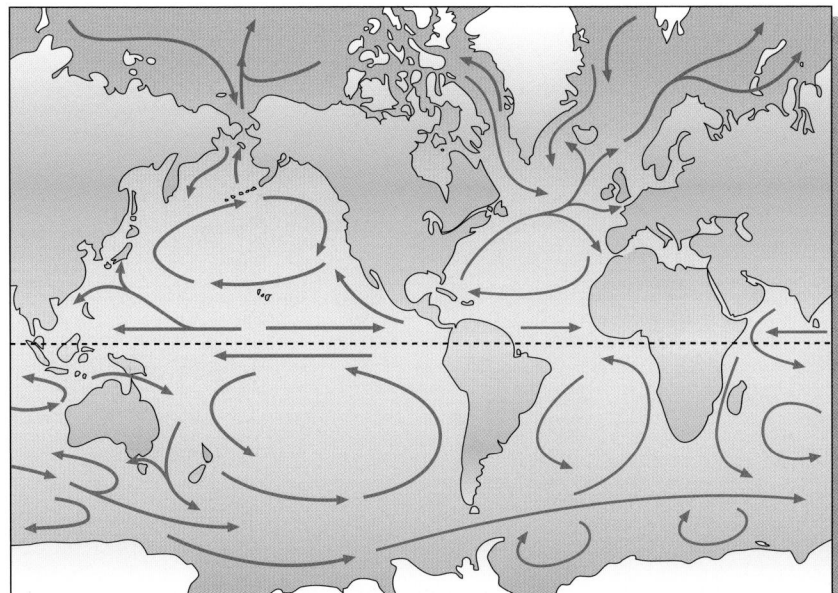

Figure 4.2

The movements of the major ocean currents. The general circulation in the northern hemisphere is clockwise, in the southern hemisphere counterclockwise, with consequences for continental climate patterns.

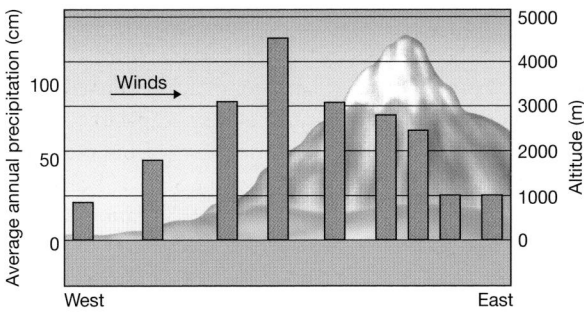

Figure 4.3

The typical influence of topography on rainfall (histogram bars) in the northern hemisphere. Moisture-laden westerlies are forced higher by a mountain range. As they rise they become cooler and release the moisture as rain or snow. This leaves a drier rain shadow on the eastern slopes.

The variety of influences on climate produces a mosaic of dry, wet, cool and warm climates over the surface of the globe. In the patches of this mosaic, distinctive terrestrial associations of vegetation and animals have formed. A world traveler sees repeatedly what can be recognized as characteristic types of vegetation, which ecologists call *biomes* (such as desert, savanna and rain forest). Figure 4.4 recognizes a set of biomes and shows their distribution as a global map. Figure 4.5 shows the ranges of rainfall and mean monthly minimum temperature that are critical in determining where the biomes are found. The characteristics of the communities inhabiting major biomes are described in Section 4.4.

> . . . produce a mosaic of dry, wet, cool and warm climates over the face of the Earth . . .

> . . . that, in turn, are responsible for the large-scale distribution of terrestrial biomes

4.2.2 Small-scale patterns in conditions and resources

It is easy to be seduced by cartographers who draw sharp lines on maps to show geographic boundaries. But neat pigeonholes, sharp categories and tidy boundaries are a convenience, not a reality of nature. Moreover, biomes are not homogeneous within their hypothetical boundaries; every biome has gradients

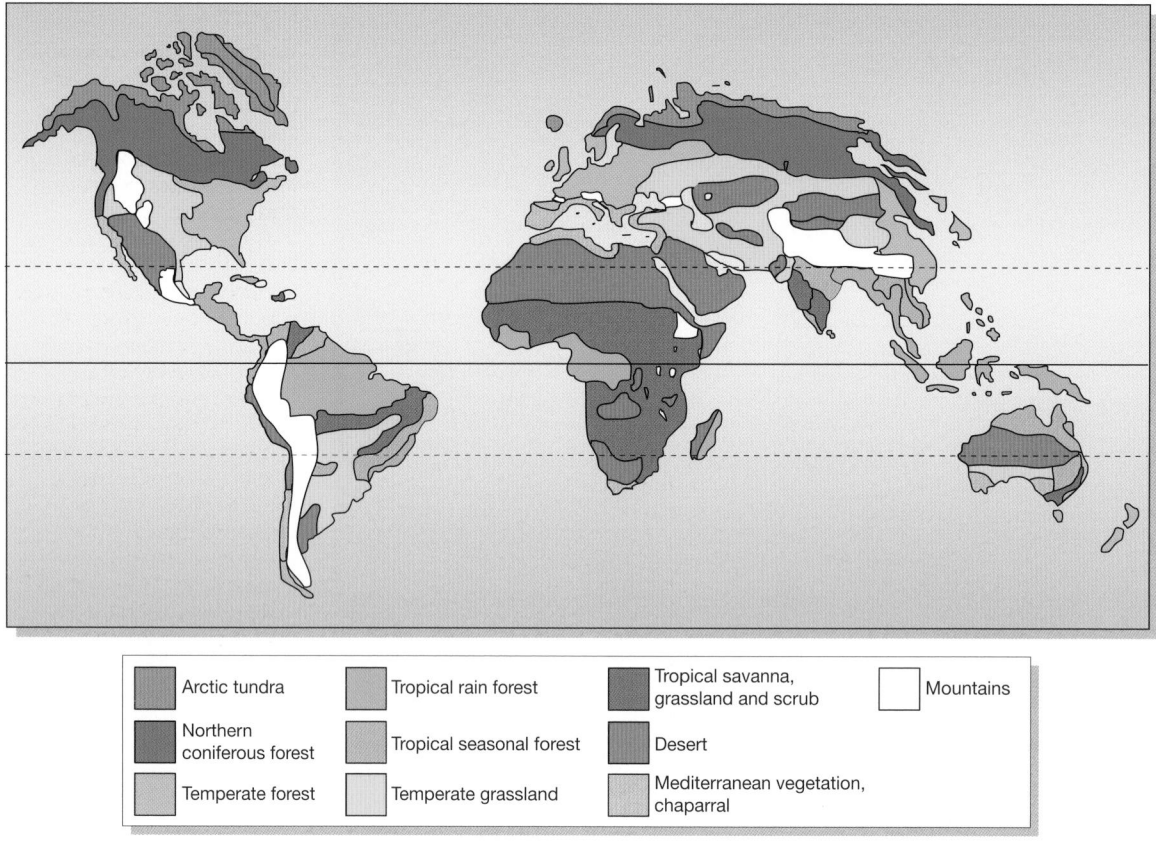

Figure 4.4

World distribution of the Earth's biomes. Their characteristic plant and animal communities are described in Section 4.4.

of physicochemical conditions related to local topography and geology. The communities of plants and animals that occur in different parts of this heterogeneous patchwork may be quite distinct.

local topography

Local variations in topography can override the broad climatic pattern described in Section 4.2.1. For example, temperature falls with increasing altitude and one effect is that vegetation high on a mountain in the tropics tends to resemble vegetation at low altitudes in northern latitudes. Traveling up a mountain in the tropics involves passing along a similar ecological gradient to that experienced when traveling northward from equator to pole (Figure 4.6).

local geology and soil

It is worth remembering that the Earth's surface would consist of a mosaic of different environments even if climate were identical everywhere. Geological history has provided a variety of rocks that differ in their mineral composition. When the surfaces of these rocks are decomposed by heat, frost and thaw, they give rise to a variety of types of soil that reflect their geological origin. Without soil, it is impossible for significant terrestrial vegetation to grow. Soils provide a source of stored water, a reserve of mineral nutrients, a medium in which atmospheric nitrogen can be fixed for plant use, and the support that allows plants to stand up and expose their leaves to the sunlight.

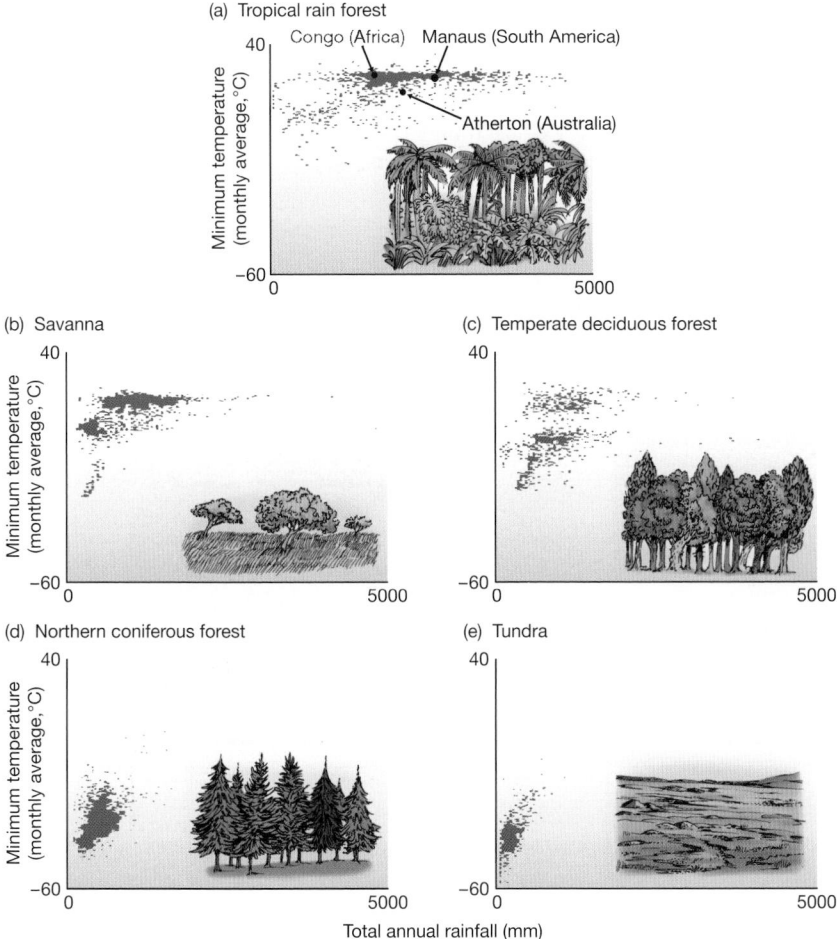

Figure 4.5

The variety of environmental conditions experienced in terrestrial environments can be described in terms of annual rainfall and mean monthly minimum temperatures. The diagrams show the range of conditions experienced in (a) tropical rain forest, (b) savanna, (c) temperate deciduous forest, (d) northern coniferous forest (taiga), and (e) tundra. Data points for a given biome come from different locations around the world. To illustrate this, data points for tropical rain forest on three different continents are shown in (a). Tropical rain forest has characteristically high mean monthly minimum temperatures and high rainfall. In contrast, tundra has both low temperatures and low precipitation. The other biomes occupy intermediate positions in this two-dimensional representation.

acidic and calcareous soils bear very different vegetation

Limestone rocks and chalk originated as marine deposits of calcium carbonate, often containing some magnesium and other carbonates. Where these deposits have been raised and exposed as land surfaces they become the basis for neutral or slightly alkaline *calcareous* soils, which bear a characteristic calcium-loving flora. On the other hand, plants normally found on more acid soils, such as *Rhododendron* and *Azalea*, are unsuccessful on calcareous soils. Strict calcium-lovers, in contrast, suffer on acidic soils, where they are intolerant of aluminum ions released at low pH. In the United States, for example, the calcium-loving yellow poplar (*Liriodendron tulipifera*) and northern white cedar (*Thuja occidentalis*) are found only on neutral or alkaline soils, whilst balsam fir (*Abies balsamea*) and eastern hemlock (*Tsuga canadensis*) are usually confined to highly acidic soils.

Variability in the organic matter component of soil also influences the biota that can occur. Organic matter accumulates at different rates in different soils and local variations in the balance between mineral and organic material in the soil contribute to the complexity of environmental mosaics. In extreme conditions, especially where the rocks are acidic, the temperatures are low and/or the soil is waterlogged, the decomposition of organic matter may be seriously impeded.

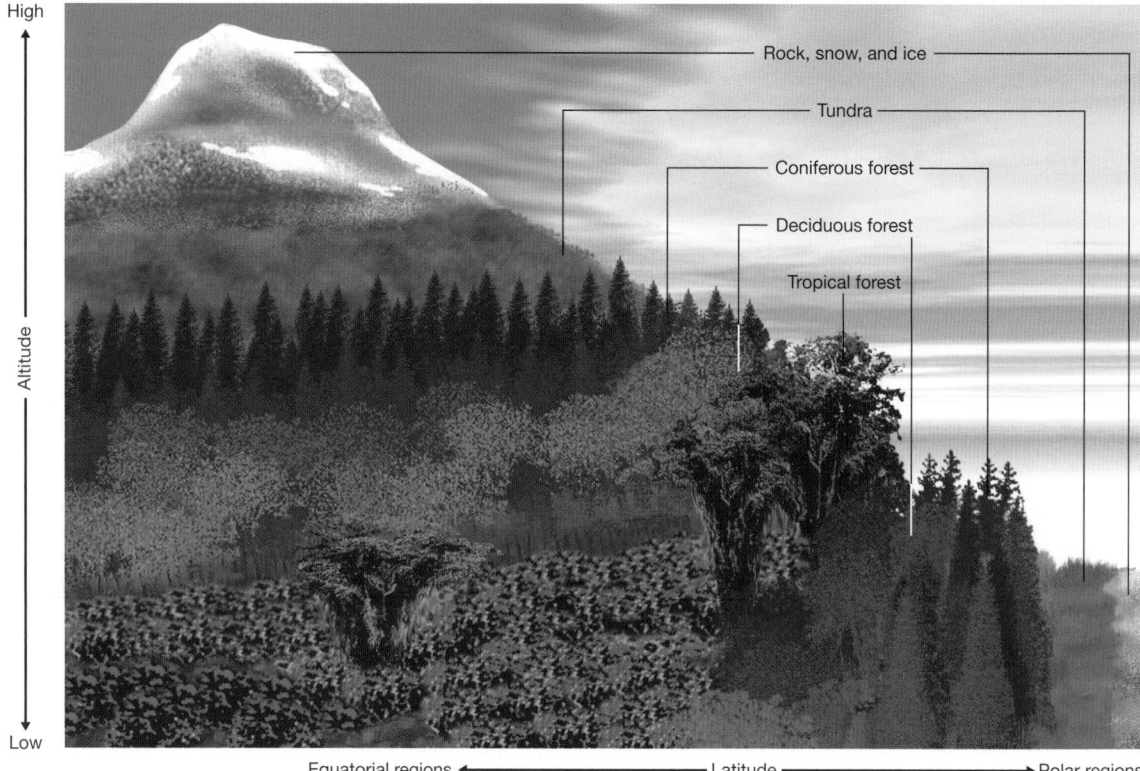

High

Altitude

Low

Rock, snow, and ice

Tundra

Coniferous forest

Deciduous forest

Tropical forest

Equatorial regions ◀——————— Latitude ———————▶ Polar regions

Figure 4.6

The effect of altitude and latitude on the distribution of biomes. Moving up in altitude is very similar to moving from equator to pole.

Then, peat bogs, with their very specialized plants and animals, form on the partially decomposed organic matter.

patchiness is in the eye of the observer . . .

To an ecologist, a *patch* in a community is an area in which a single variable distinguishes it from its surroundings. Thus, a fallen tree in a forest leaves a gap in the canopy and a patch on the forest floor where sufficient radiation may penetrate to allow seedlings to grow and eventually fill the gap. A tide pool is a patch on a rocky shore, but within that pool snails may graze and clear a patch free of algae. It is often useful to think of patches as the scale at which particular organisms experience the environment around them. For an aphid in a forest, an individual leaf of a particular species of tree is a patch – it provides both the conditions and the resources necessary for the insect. For a warbler feeding on caterpillars, the canopies of individual trees are patches that it encounters in its daily life. But owls or hawks hunt over a large part of the forest, and for them a patch may be the territory that each bird defends or perhaps even the whole forest over which it ranges.

. . . and all communities are patchy

4.2.3 Patterns in conditions and resources in aquatic environments

In most aquatic environments it is difficult to recognize anything comparable to terrestrial biomes. The exceptions occur at the ocean's edge; tropical mangrove swamps, coral reefs and temperate kelp forests have biotas that are as distinctive as any of the various terrestrial biomes, but this is largely due to their close relationship with major terrestrial climates. In contrast, the open oceans form a continuum in which there is flow of water and dissolved chemicals across the globe. We have seen how variation in the intensity of solar radiation from place to place and between the seasons has dramatic effects on the temperature and water relations of terrestrial environments. But this is not the case in the oceans. The high thermal capacity of water makes the oceans slow to heat and slow to cool. One effect is that the temperature of the water at one point on the globe is a better reflection of where the water has come from (along ocean currents) than of the local climate.

The world's large lakes can be distinguished and classified according to their physical conditions. For example, large lakes in lowland equatorial regions generally experience permanent stratification (distinct layers of water at particular temperatures), whereas seasonal patterns of stratification (in summer) and mixing (in fall) are the rule in temperate regions. Within the polar circles, permanent ice cover with no mixing is characteristic of large lakes. However, local geological conditions and basin size and shape have strong influences on conditions and resources in lakes, particularly in terms of water chemistry, a key determinant of lake flora and fauna. Consequently, a broad geographic classification of lake communities has only limited merit. In the case also of streams, rivers, estuaries and the open ocean, we will see that local conditions and resources are paramount in determining community patterns (see Section 4.5).

4.3 Temporal patterns in conditions and resources

The composition of communities can change over time scales ranging from hours to millennia, as conditions and resources themselves change. For example, the microbial community that colonizes and decomposes a dead mouse or fragment of a leaf may change from hour to hour. At the other extreme, we can trace patterns in community composition over tens of thousands of years. Thus, changes in climate during the Pleistocene ice ages bear much of the responsibility for present patterns of distribution of plants and animals. In the 20,000 years since the peak of the last glaciation, global temperatures have risen by about 8°C. Many tree species continue, even today, to migrate northward, following the retreat of the glaciers (Figure 4.7).

At intermediate temporal scales, predictable sequences of plant species may occur over periods ranging from years to centuries. For example, the successional sequence that occurs on cooled volcanic lava takes several centuries to run its course. This has been documented by comparing the plants living on lava flows from eruptions that occurred at different times on Miyake-jima Island, Japan (Figure 4.8). In the earliest stage of succession, conditions are harsh and soil is sparse and lacking in nitrogen-containing ions, an essential plant resource. Alders are first to colonize because they can fix atmospheric nitrogen into usable form. As nitrogen availability in the soil increases, many species of fern, herb, liana and tree enter the succession. After a century or two, late-successional trees (*Machilus* then *Castanopsis*) shade out many of the earlier arrivals. Succession – the predictable sequence of colonization and extinction after a disturbance – depends partly on changing conditions and resources, and partly on the differential competitive abilities of the plants themselves, a topic we return to in Chapter 9.

plant succession – the species sequence on volcanic lava flows

Figure 4.7

A map showing the spread of two species of forest tree in eastern North America after the retreat of the last ice age glaciation. Note that the two species of (a) eastern white pine (*Pinus strobes*) and (b) beech (*Fagus grandifolia*) have not followed the same invasion path. The lines on the maps (isochrones) define the time of arrival of each species at 1000-year intervals. The numbers on the map refer to thousands of years before present. The shaded brown areas show their present distributions.

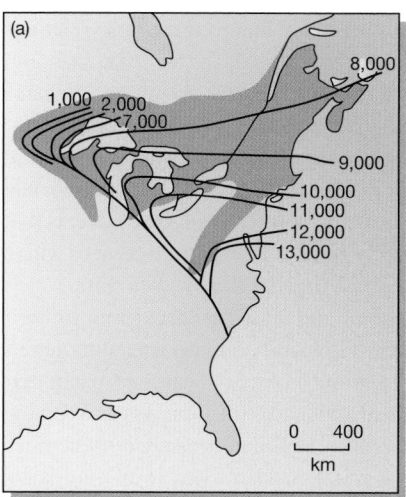
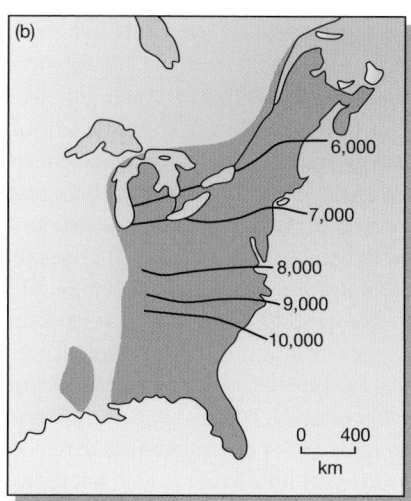

FROM DAVIS, 1976

Figure 4.8

(a) Locations of sampling sites (red dots) on 37 and 125-year-old lava flows on Miyake-jima Island, Japan. Sampling on 16-year-old lava was non-quantitative (no sampling sites shown). Sites outside these flows are at least 800 years old. Altitudinal contours are shown in meters.
(b) In the earliest stage of succession the only vegetation consists of a few small alder trees (*Alnus sieboldiana*). In the older plots (37–800 years old), 113 species were recorded, including ferns, herbs, lianas and trees. This succession consisted of:
(i) colonization of the bare lava by the nitrogen-fixing alder;
(ii) facilitation (through improved nitrogen availability) of mid-successional *Prunus speciosa* and the late-successional evergreen tree *Machilus thunbergii*;
(iii) establishment of a mixed forest in which *Alnus* and *Prunus* were shaded out; and (iv) competitive replacement of *Machilus* by the longer lived *Castanopsis sieboldii*.

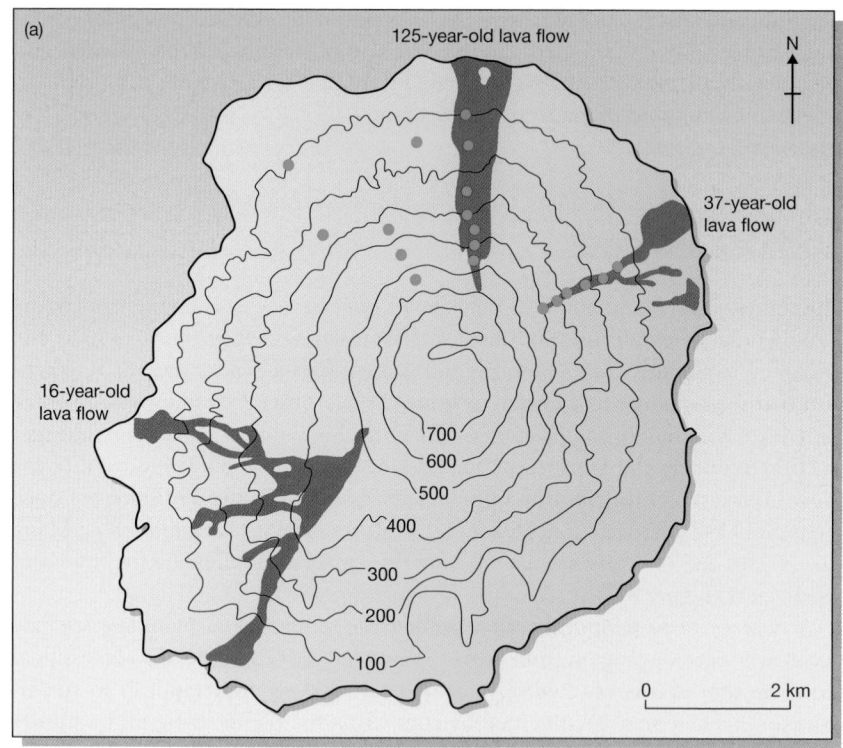
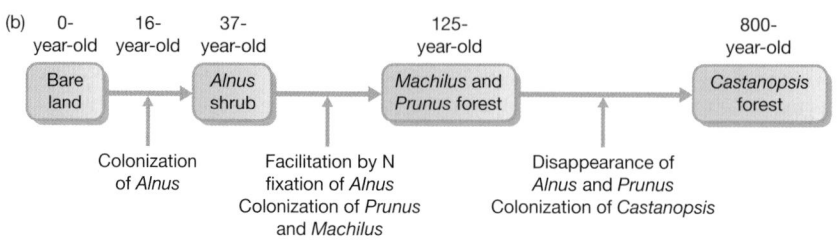

AFTER KAMIJO ET AL., 2002

4.4 Terrestrial biomes

Different biogeographers recognize different numbers of biomes; some make do with just five biomes and others find they need many more. The perspective of the scientist is as important as the system being studied; 'splitters' tend to distrust broad generalizations and emphasize the diversity of the natural world, whereas 'lumpers' force diversity into a minimum of easily mapped categories. The following are adequate for our purposes – tropical rain forest, savanna, temperate grassland, chaparral, desert, temperate deciduous forest, northern or boreal coniferous forest (taiga), and tundra.

the patterns that we recognize in nature depend on how we focus our attention

4.4.1 Describing and classifying biomes

We pointed out in Chapter 2 the crucial importance of geographic isolation in allowing populations to diverge under selection. The geographic distributions of species, genera, families and even higher taxonomic categories of plants and animals often reflect this geographic divergence. All species of lemurs, for example, are found on the island of Madagascar and nowhere else. Similarly, 230 species in the genus *Eucalyptus* (gum tree) occur naturally in Australia (and two or three in Indonesia and Malaysia). The lemurs and the gum trees occur where they do because they evolved there – not because these are the only places where they could survive and prosper. Indeed, many *Eucalyptus* species grow with great success and spread rapidly where they have been introduced to California or to Kenya. A map of the natural world distribution of lemurs tells us quite a lot about the evolutionary history of this group. But as far as its relationship with a biome is concerned, the most we can say is that lemurs happen to be one of the constituents of the tropical rain forest biome in Madagascar.

Another theme of Chapter 2 concerned the way species with quite different evolutionary origins have been selected to converge in their form and behavior. There were also examples of taxonomic groups that have radiated into a range of species with strikingly similar form and behavior (parallel evolution, as in the marsupial and placental mammals). Examples like these reveal much about the ways in which organisms have evolved to match the conditions and resources in their environments. But the different species need not characterize different biomes. Thus, particular biomes in Australia include certain marsupial mammals, while the same biomes in other parts of the world are home to their placental counterparts.

A map of biomes, then, is not usually a map of the distribution of species. Instead, it shows where we find areas of land dominated by plants with characteristic shapes, forms and physiological processes. These are the types of vegetation that can be recognized from an aircraft passing over them or from the windows of a fast car or train. It does not require a botanist to identify them. The scrubby chaparral vegetation characteristic of California provides a striking example. The spectrum of plant forms that gives this vegetation its distinctive nature also occurs in similar environments around the Mediterranean Sea and in Australia – but the species and genera of plants are quite different. We recognize different biomes from the types, not species identities, of organisms that live in them.

describing and classifying vegetation

When reading the brief descriptions of biomes that follow, it is important to bear in mind that the vegetation described is typical of the mature community that develops in different climatic regions (Figure 4.9). However, patchiness is always

(a)

(b)

(c)

(d)

Figure 4.9

Each biome is illustrated with two photographs, one focusing on the detail of the vegetation and the other providing a distant view and emphasizing the great structural variation to be found among the world's terrestrial communities. The animals found in each of these biomes also cannot be ignored; they are obvious in the savanna photo, but invertebrate and vertebrate animals are busy behind the scenes in all the biomes. (a) Above: Carrizo Badlands desert, Anza-Bonnego Desert State Park, California (© Doug Sokell); below: Red Rock Canyon, Las Vegas, Nevada (© Mark E. Gibson). (b) Above: Ozark Forest and Current River, Ozark National Scenic Riverways, Missouri (© Richard Thom); below: mature eastern deciduous

(e)

(f)

(g)

forest (© Bill Beatty). (c) Above: fir tree forest, Jasper National Park, Alberta, Canada (© Mark E. Gibson); below: foggy coniferous forest, Sierras (© Joe McDonald). (d) Above: Masai Mara Game Preserve at dawn (© Joe McDonald); below: African savanna with zebra and buffalo (© John Cunningham). (e) Above: rain forest, western slope of Andes, Ecuador (© C.P. Hickman); below: lake in mixed dipterocarp forest, Mulu National Park, Sarawak, Borneo (© Brian Rogers). (f) Above: a lone pronghorn antelope looks tiny in this vast mixed-grass prairie, Stanley County, central South Dakota (© Ron Spomer); below: view of prairie in flower with blazing star and black-eyed Susan (© Ann B. Swengel). (g) Above: green tundra with glacial moraine and Alaska mountain range, Denali National Park, Alaska (© Patrick J. Endes); below: wet summer tundra (© Doug Sokel).

present (based often on local topography and geology, Section 4.2.2) and small- and large-scale disturbances (caused by the death of individual trees, or by fires, storms or people) create a mosaic in which community successions are occurring (see Section 4.3).

4.4.2 Tropical rain forest

We have chosen to discuss tropical rain forest in greater depth than the other biomes because it represents the global peak of evolved biological diversity: all the other biomes suffer from a relative poverty of resources or more strongly constraining conditions.

Tropical rain forest is the most productive of the Earth's biomes with a photosynthetic productivity that can exceed 1000 g of carbon fixed per square meter per year (see Section 11.2.1). Such exceptional productivity results from the coincidence of high solar radiation received throughout the year and regular and reliable rainfall. The production is achieved, overwhelmingly, high in the dense forest canopy of evergreen foliage. It is dark at ground level except where fallen trees create gaps. A characteristic of this biome is that often many tree seedlings and saplings remain in a suppressed state from year to year and only leap into action if a gap forms in the canopy above them.

Almost all the action in a rain forest (not just photosynthesis but also flowering, fruiting, predation and herbivory) happens high in the canopy. Apart from the trees, the vegetation is largely composed of plant forms that reach up into the canopy vicariously, by climbing the trees (vines and lianas, including many species of fig) or growing as epiphytes, rooted on the damp upper branches. The epiphytes depend on the sparse resources of mineral nutrients that they extract from crevices and pockets of humus on the tree branches. The rich floras and faunas of the canopy are not easy to study; even to gain access to the flowers in order to identify the species of tree is difficult without the erection of tree walks. It is a measure of the problems of doing research in rain forest that botanists have trained monkeys to collect and throw down flowers and a research team has used hot air balloons to move over the canopy and work in it.

Most species of animals and plants in tropical rain forest are active throughout the year, though the plants may flower and ripen fruit in sequence. In Trinidad, for example, the forest contains at least 18 trees in the genus *Miconia*, whose combined fruiting seasons extend throughout the year; this contrasts with the situation in temperate latitudes (Figure 4.10).

Dramatically high species richness is the norm for tropical rain forest (see Section 10.5.2), and communities rarely if ever become dominated by one or a few species – a very different situation from the low biodiversity of northern coniferous forests. This raises some fundamental questions that have proved very difficult to resolve. First, what is it about the evolutionary history of tropical rain forest that has allowed such diversity to evolve? Part of the answer relates to the comparative stability of patches of rain forest during the ice ages. It is thought that during these periods, drought forced tropical rain forests to contract into 'islands' (in a 'sea' of savanna), and these expanded and coalesced again as wetter periods returned. This would have promoted genetic isolation of populations, a phenomenon that is so important for speciation to occur (see Section 2.4). We may also ask why it is that among the diversity of species in tropical rain forests,

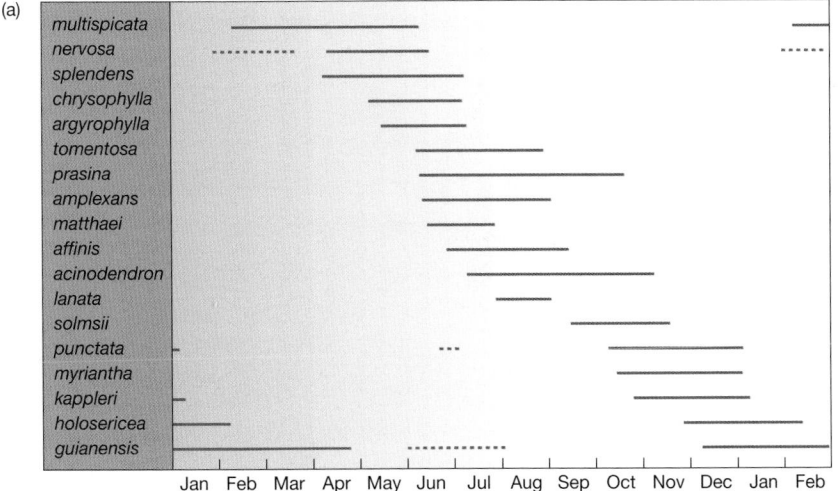

(a)

multispicata
nervosa
splendens
chrysophylla
argyrophylla
tomentosa
prasina
amplexans
matthaei
affinis
acinodendron
lanata
solmsii
punctata
myriantha
kappleri
holosericea
guianensis

Jan Feb Mar Apr May Jun Jul Aug Sep Oct Nov Dec Jan Feb

Figure 4.10

Contrasting patterns of fruit or seed production in tropical and temperate forests. (a) The fruiting seasons of 18 species of tree in the genus *Miconia* in the rain forest of Trinidad are spread throughout the year. (b) The seasonal production of fruit and seeds by herbs in a deciduous forest in Poland is concentrated in a relatively brief period of the year.

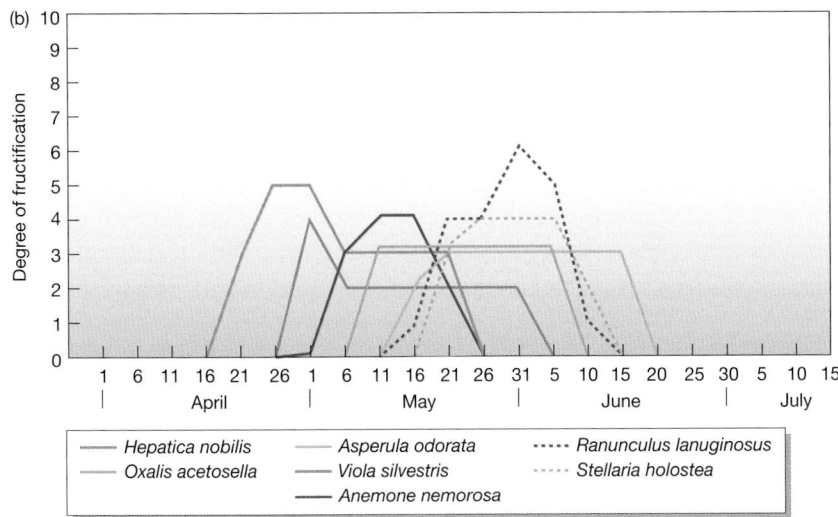

(b)

Degree of fructification

1 6 11 16 21 26 1 6 11 16 21 26 31 5 10 15 20 25 30 5 10 15
      April   May   June   July

—— *Hepatica nobilis*   —— *Asperula odorata*   ····· *Ranunculus lanuginosus*
—— *Oxalis acetosella*   —— *Viola silvestris*   ····· *Stellaria holostea*
                            —— *Anemone nemorosa*

AFTER HARPER, 1977

a few have not dominated and suppressed the rest in a struggle for existence. We will see later (Section 10.5.2) that at least part of the answer is that populations of specialized pathogens and herbivores develop near mature trees and attack new recruits of the same tree species nearby. Thus, the chance that a new seedling will survive can be expected to increase with its distance from a mature tree of the same species, reducing the likelihood of dominance by one or a few species in the forest.

The diversity of rain forest trees provides for a corresponding diversity of resources for herbivores (Figure 4.11). A variety of fresh young leaves are available throughout the year, and a constant procession of seed and fruit production provides reliable food for specialists such as fruit-eating bats. Moreover, a diversity of flowers, such as epiphytic orchids with their specialized pollinating mechanisms, require a parallel specialized diversity of pollinating insects. Rain forests are the center of diversity for ants – 43 species have been recorded

tropical rain forest is also associated with high animal diversity . . .

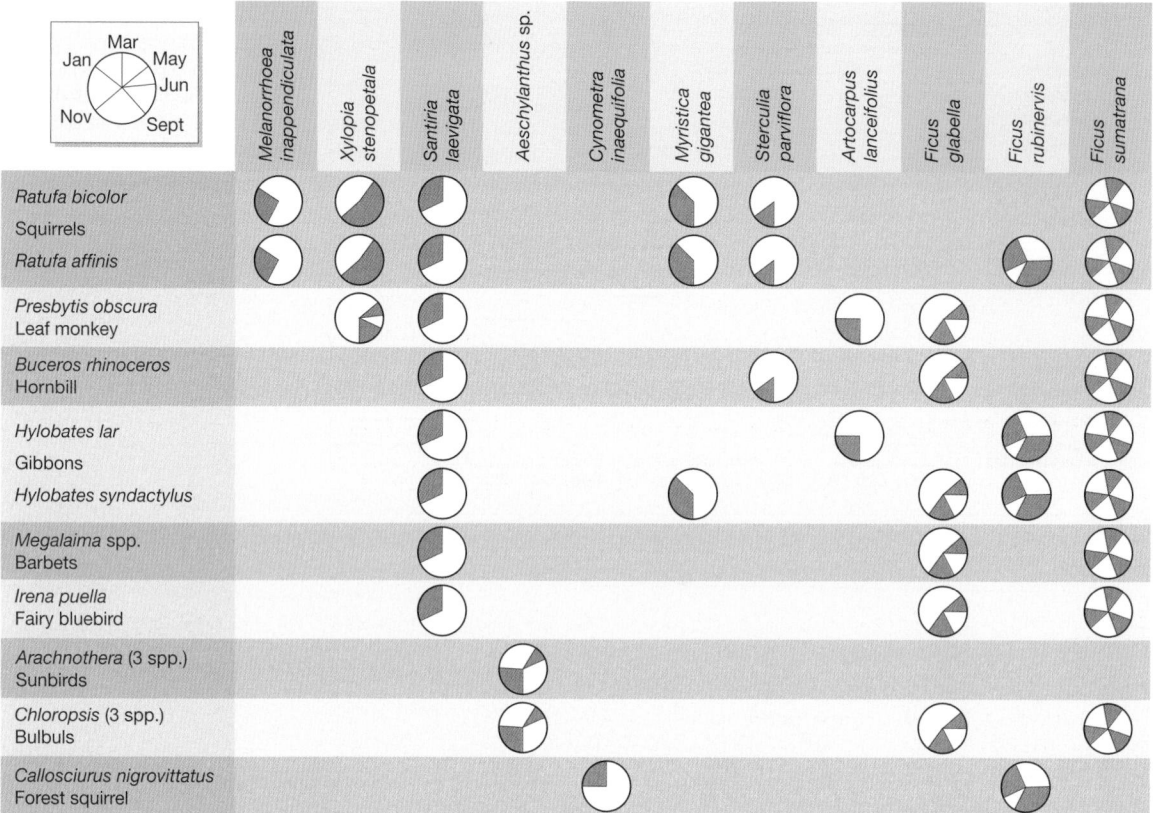

Figure 4.11

Animals (listed vertically) that feed on the fruit of trees (listed horizontally) at various times of the year at Selangor, Malaysia. Each circle is a calendar in which the feeding season is shown in dark brown. Each plant produces fruit only at certain times in the year, but there is fruit available for specialist fruit-eaters throughout the year.

on a single tree in a Peruvian rain forest. And there is even more diversity among the beetles; Erwin (1982) estimated that there are 18,000 species of beetle in 1 ha of Panamanian rain forest (compared with only 24,000 in the whole of the United States and Canada!).

... and intense soil activity

There is intense biological activity in the soil of tropical rain forests. Leaf litter decomposes faster than in any other biome and as a result the soil surface is often almost bare. The mineral nutrients in fallen leaves are rapidly released, and, as rainfall seeps down the soil profile, nutrients may be carried well below the levels at which roots can recover them. Almost all the mineral nutrients in a rain forest are held in the plants themselves, where they are safe from leaching. When such forests are cleared for agriculture, or the timber is felled or destroyed by fire, the nutrients are released and leached or washed away: on slopes the whole soil may go too. The full regeneration of soil and of a nutrient budget in a new forest may take centuries. Evidence of cultivated patches within rain forest can still be seen clearly from the air 40 years or more after they have been deserted.

All the other terrestrial biomes can be seen as the poor relations of tropical rain forest. They are all colder or drier and all are more seasonal. They have

AFTER HARPER, 1977

had prehistories that prevented the evolution of a diversity of animals and plants that approaches the remarkable species richness of tropical rain forest. Moreover, they are generally less suited to the lives of extreme specialists, both plant and animal.

4.4.3 Savanna

The vegetation of savanna characteristically consists of grassland with scattered small trees, but extensive areas have no trees. In the absence of other controlling factors, these tropical areas would be expected to be covered by forest. But forest development is kept in check by one of three factors, or a combination of these.

In some savannas, herds of grazing herbivores (e.g. zebra *Equus burchelli* and wildebeest *Connochaetes taurinus* in Africa) have a profound influence on the vegetation, favoring grasses (which protect their embryonic, actively dividing tissues in buds at or just below ground level) and hindering the regeneration of trees (because these same tissues are exposed to browsing animals and to fire).

In other cases, fire is the critical thing. Fire, whether natural or human-induced, can be a common hazard in the dry season and, like grazing animals, tips the balance in the vegetation against trees and favors perennial grasses, with their underground rhizomes and protected regenerating surfaces. In the savannas of Southeast Asia, palms are a feature because scorching of the outermost layer of the trunk does not kill these plants.

Finally, the advantage of grassland over forest in savannas, with their different regional names, may relate to unfavorable conditions, such as water-logging (Venezuelan *llanos*), severe drought (Central American *pine savannas*) or sparse soil nutrients (Brazilian *cerrado*).

Seasonal rainfall places the most severe restrictions on the diversity of plants and animals in savanna. Plant growth is limited for part of the year by drought, and there is a seasonal glut of food, alternating with shortage; as a consequence, the larger grazing animals suffer extreme famine (and mortality) in drier years. The strong seasonality of savanna ecology is well illustrated by its bird populations. An abundance of seeds and insects supports large populations of migrating birds, but only a few species can find sufficiently reliable resources to be resident year round.

seasonal glut and food shortage are characteristic of savanna

4.4.4 Temperate grasslands

Temperate grassland is the natural vegetation over large areas in every continent. These include the tall grass *prairie* of North America and *pampas* of South America, where rainfall is moderate and soils are rich, and the short grass *steppes* of Russia, typical of more semiarid conditions. These grasslands experience seasonal drought, but grazing animals also have a powerful impact. Populations of invertebrates, such as grasshoppers, are often very large and their biomass may exceed that of grazing vertebrates. The latter include bison (*Bison bison*), pronghorn antelope (*Antilocapra americana*) and gophers (*Thomomys bottae*) in North America, and saiga antelope (*Saiga tatarica*) and marmots (*Marmota bobac*) in Russia.

of all the biomes, temperate
grassland has been most
transformed by humans

Many of these natural grasslands have been cultivated and replaced by arable annual 'grasslands' of wheat, oats, barley, rye and corn. Such annual grasses of temperate regions, together with rice in the tropics, provide the staple food of human populations worldwide. In fact, the vast increase in the size of the human population in historical times (see Section 12.2) has depended on the domestication of grasses for human food or feed for domestic animals. At the drier margins of the biome, where cultivation is not economical, many of the grasslands are 'managed' for meat or milk production, sometimes requiring a nomadic human lifestyle. The natural populations of grazing animals, especially bison and pronghorns in North America and ungulates in Africa, have been driven back in favor of cattle, sheep and goats. Of all the biomes, this is the one most coveted, used and transformed by humans.

4.4.5 Desert

In their most extreme form, the hot deserts are too arid to bear any vegetation; they are as bare as the cold deserts of Antarctica. Where there is sufficient rainfall to allow plants to grow in arid deserts, its timing is always unpredictable.

contrasting patterns of behavior
of desert plants

Desert vegetation falls into two sharply contrasted patterns of behavior. Many species have an opportunistic lifestyle, stimulated into germination by the unpredictable rains (physiological 'internal' clocks are useless in this environment). They grow fast and complete their life history by starting to set new seed after a few weeks. These are the species that can occasionally make a desert bloom; the ecophysiologist Fritz Went called them 'belly plants' because only someone lying on the ground can appreciate their individual charm.

A different pattern of behavior of arid desert plants is to be long-lived with sluggish physiological processes. Cacti and other succulents, and small shrubby species with small, thick and often hairy leaves, can close their stomata (pores through which gas exchange takes place) and tolerate long periods of physiological inactivity. In arid deserts, freezing temperatures are common at night and tolerance of frost is almost as important as tolerance of drought.

animal diversity is low in deserts

The relative poverty of animal life in arid deserts reflects the low productivity of the vegetation and the indigestibility of much of it. Desert perennials including species of wormwood (*Artemisia*) and creosote plant (*Larrea mexicana*) in the southwestern United States, and mallee species of *Eucalyptus* in Australia, carry high concentrations of chemicals that are repellent to herbivores. Ants and small rodents rely on seeds as a relatively reliable perennial resource, whereas bird species are largely nomadic, driven by the need to find water. Only desert carnivores can survive on the water they obtain from their food. In the deserts of Asia and Africa, camels, donkeys and sheep are managed for transport and food by migrant groups of humans.

4.4.6 Temperate forest

Like all biomes, temperate forest includes, under one name, a variety of types of vegetation. At its low-latitude limits in Florida and New Zealand, winters are mild, frosts and droughts are rare, and the vegetation largely consists of broad-leaved evergreen trees. At its northern limits in the forests of Maine and the upper Midwest of the United States, the seasons are strongly marked, winter days are short and there may be 6 months of freezing temperatures. Deciduous trees, which dominate in most temperate forests, lose their leaves in the fall and become dormant after transferring much of their mineral content to the woody body of the tree. On the forest floor, diverse floras of perennial herbs often occur, particularly those that grow quickly in the spring, before the new tree foliage has developed.

All forests are patchy because old trees die, providing open environments for new colonists. This patchiness is on an especially large scale after hurricanes fell the older and taller trees or after fire kills the more sensitive species. In temperate forests the canopies are often composed of a mixture of long-lived species, such as red oaks (*Quercus rubra*) in the Midwest of the United States, and colonizers of gaps, such as sugar maple (*Acer saccharum*).

Temperate forests provide food resources for animals that are usually very seasonal in their occurrence (compare Figure 4.10b with 4.10a), and only species with short life cycles, such as leaf-eating insects, can be dietary specialists. Many of the birds of temperate forests are migrants that return in spring but spend the remainder of the year in warmer biomes.

Soils are usually rich in organic matter that is continually added to, decomposed and churned by earthworms and a rich community of other *detritivores* (organisms that feed on dead organic matter). Only waterlogging and low pH, in some locations, inhibit the decomposition of organic matter and force it to accumulate as peat.

temperate forest soils are rich in organic matter

Large swathes of deciduous forest in Europe and the United States have been cut down to provide for agriculture, but these have sometimes been allowed to regenerate as farmers abandoned the land (a conspicuous feature in New England).

4.4.7 Northern coniferous forest (taiga) grading into tundra

Northern (or boreal) coniferous forest (also known as taiga) and the treeless tundra occur in regions where the short growing season and the cold of winter limit the vegetation and its associated fauna.

Coniferous forest consists of a very limited tree flora. Where winters are less severe, the forests may be dominated by pines (*Pinus* species, which are all evergreens) and deciduous trees such as larch (*Larix*), birch (Betula) or aspens (*Populus*), often as mixtures of species. Farther north, these species give way to monotonous single-species forests of spruce (*Picea*) over immense areas of North America, Europe and Asia. This provides an extreme contrast to the biodiversity of tropical rain forests.

The areas of vegetation now occupied by tundra and northern coniferous forests (and much of northern deciduous forest) were occupied by the ice sheet during the last ice age, which only started to withdraw 20,000 years ago. Temperatures are now as high as they have ever been since that time, but the vegetation has not yet caught up with the changing climate and the forests are still spreading north. The very low diversity of northern floras and faunas is in part a reflection of a slow recovery from the catastrophes of the ice ages.

the low diversity of northern coniferous forest provides ideal conditions for pest outbreaks

Low-diversity communities provide ideal conditions for the development of disease and epidemics of pests. For example, the spruce budworm (*Choristoneura fumiferana*) lives at low densities in immature northern forests of spruce. As the forests mature, the budworm populations explode in devastating epidemics. These wreck the old forest, which then regenerates with young trees. This cycle takes about 40 years to run its course.

The overriding environmental constraint in northern spruce forests is the presence of permafrost: the water in the soil remains frozen throughout the year, creating permanent drought except when the sun warms the very surface. The root system of spruce can develop in the superficial soil layer, from which the trees derive all their water during the short growing season.

To the north of the spruce forest, the vegetation changes to tundra, with its low shrubs, grasses, sedges and small flowering plants, as well as mosses and lichens. In fact, forest and tundra often form a mosaic in the Low Arctic. In the colder areas, plants such as grasses and sedges disappear, leaving nothing rooted in the permafrost. High winds exaggerate the aridity of the environment, and ultimately vegetation that consists only of lichens and mosses gives way, in its turn, to the polar desert. The number of species of higher plants (i.e. excluding mosses and lichens) decreases from 600 species in the Low Arctic of North America to 100 species in the High Arctic (north of 83°) of Greenland and Ellesmere Island. In contrast, the flora of Antarctica contains only two native species of vascular plant and some lichens and mosses that support a few small invertebrates. The biological productivity and diversity of Antarctica are concentrated at the coast and depend almost entirely on resources derived from the sea.

dramatic animal population cycles are characteristic of northern biomes

The faunas of northern coniferous forests and tundra have intrigued ecologists because populations of lemmings, mice, voles and hares (herbivores), and the fur-bearing carnivores (e.g. lynx and ermine) that feed on them, pass through remarkable cycles of expansion and collapse (see Section 7.5.2). Lemmings (*Lemmus*) are famous for their population cycles and the role they play in the tundra. When the snow melts during a period when the lemming cycle is at a high point, the animals are exposed and they support large migratory populations of predatory birds (owls, skuas, gulls) and mammals such as weasels. Reindeer and caribou (they are the same species, *Rangifer tarandus*) occur in migrant herds capable of foraging on lichens of the tundra, which they can reach through the snow cover.

4.4.8 The future distribution of biomes

It is clear that the distribution of biomes has changed in the past in response to the ebb and flow of the ice ages. Nowadays, we are also acutely aware that their boundaries are probably on the move again. Predicted changes in global climate over the next few decades can be expected to result in dramatic changes to the distribution of biomes over the face of the Earth (Box 4.1). But the exact nature of these changes remains uncertain.

4.1 Topical ECOncerns

Predicted changes in the distribution of biomes as a result of global climate change

As a result of human activities, the atmosphere contains increasing concentrations of certain gases, particularly carbon dioxide, but also nitrous oxide, methane, ozone and chlorofluorocarbons (CFCs). These changes are predicted to lead to increased temperatures and altered patterns of climate over the face of the Earth (see Section 13.3.1). Given the controlling influence of climate on the distribution of biomes, ecologists expect the biome map of the world to change significantly as carbon dioxide concentrations double over the next 60–70 years.

It is no easy matter to predict the precise details of future climate or its consequences for biome distribution. Scientists have come up with a number of feasible scenarios, which differ according to the basic assumptions included in their models. The details of these need not concern us here. It is enough to note that the simulations shown in Figures 4.12 and 4.13 are based on a climate change model that assumes an effective doubling of carbon dioxide concentrations and takes into account the coupling of atmosphere and ocean in determining changes in patterns of temperature and rainfall. The model is known as MAPSS. This is translated into patterns in the distribution of biomes by simulating the potential mature vegetation that could live under the 'average' seasonal climate prevailing (see Neilson et al., 1998, for further details).

The distribution of biomes shown in Figure 4.12 is as simulated by the model for current climate conditions (Neilson et al., 1998). In other words, it is the model's picture of the way biomes are distributed now (and reflects reality well; note that the biome

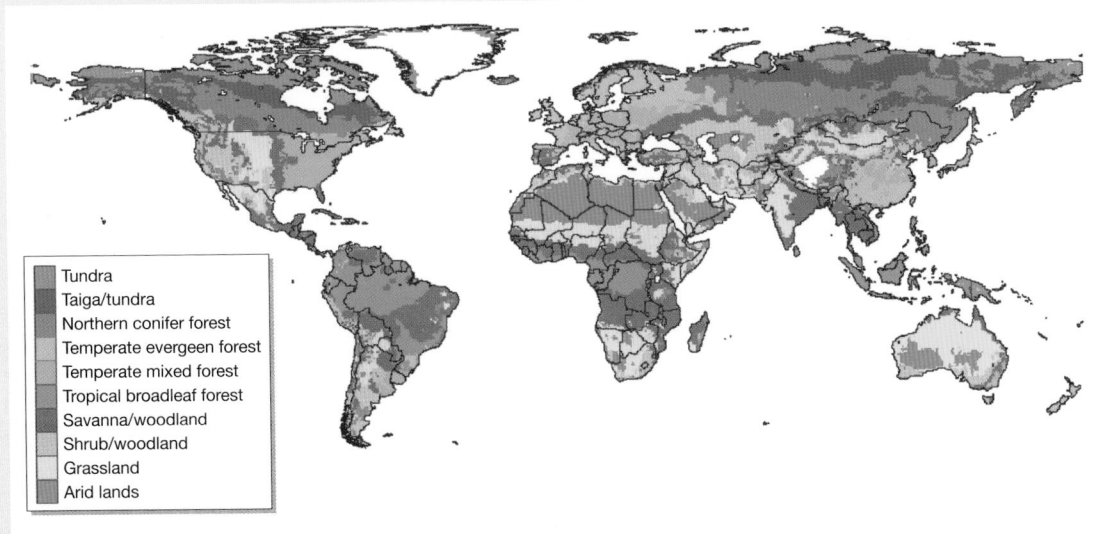

Tundra
Taiga/tundra
Northern conifer forest
Temperate evergeen forest
Temperate mixed forest
Tropical broadleaf forest
Savanna/woodland
Shrub/woodland
Grassland
Arid lands

Figure 4.12

The distribution of major biome types under the current climate, as simulated by the MAPSS biogeography model.

AFTER NEILSON ET AL., 1998

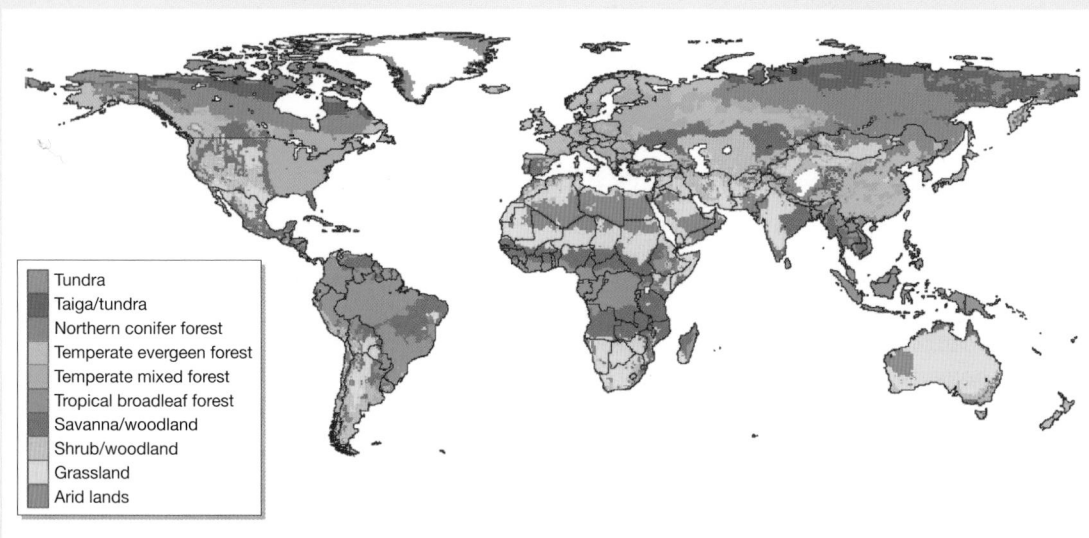

Figure 4.13

The potential distribution of major biomes resulting from climate changes associated with an effective doubling of carbon dioxide concentration, as simulated by the MAPSS biogeography model.

AFTER NEILSON ET AL., 1998

categories are not exactly the same as those we discuss elsewhere in the chapter). The map in Figure 4.13, by contrast, is the predicted distribution of biomes in 60–70 years' time (Neilson et al., 1998). This model predicts a reduction in area of the northern biomes of tundra and taiga/tundra (the open woodland that occurs between the treeless taiga and the dense northern coniferous forest). It also predicts a decrease in arid lands and an increase in temperate forest. These conclusions are in broad agreement with a variety of models that incorporate different starting assumptions.

4.5 Aquatic environments

The dominating characteristics of aquatic environments result from the physical properties of water. A water molecule is composed of an oxygen atom, which is slightly negatively charged, bonded with two hydrogen atoms, which are slightly positively charged. This dipolar structure enables water molecules to attract and dissolve more substances than any other liquid on Earth. Consequently, water can hold mineral ions in solution, providing the nutrient resources required for the growth of algae and higher plants.

the special properties of water as a medium in which to live

On the other hand, the solubility of oxygen, an essential resource for both plants and animals, decreases rapidly with increasing temperature, and oxygen diffuses only slowly in water. This problem can place major limits on life in water. Oxygen is rapidly used up when dead organic matter decomposes. In places where tree leaves accumulate or untreated sewage is discharged into a river or lake, decomposition can create anaerobic conditions, which are lethal for fish and other animals that have a high biological oxygen demand. Many aquatic animals maintain access to oxygen by forcing a continual flow of water over their respiratory surfaces (e.g. the gills of fish) or have very large surface areas relative to their body volume.

Water is viscous, and moving water transports whole living organisms, such as small plants and animals. It offers resistance to the movement of motile animals such as fish, otters and aquatic birds; not surprisingly, many motile aquatic animals are streamlined. Many plants that live in moving water depend on rooting in the substratum to hold them against water currents, and many smaller animals are attached to the plants or hide in crevices or under rocks where they are protected from the drag of moving water.

Water is unusual in remaining liquid over a wide range of temperatures. It requires a lot of energy to heat it (i.e. it has high thermal capacity), but retains heat efficiently. One consequence is that the temperature of large bodies of water (oceans and large lakes) varies little over the seasons. A further peculiar physical property of water is that it is less dense when frozen than when liquid. Like most liquids, water becomes denser and sinks as it cools. However, at temperatures below 4°C, water becomes less dense and when ice forms (at 0°C), it floats. Ice on a water surface insulates the water beneath; lakes and streams can remain liquid, free-flowing and inhabitable under a layer of ice.

4.5.1 Stream ecology

Streams and rivers contain a minute portion of the world's water (0.006%), but an enormous proportion of the fresh water that can be used by people. Consequently, they have been tapped, dammed, straightened, rerouted, dredged and polluted since the beginning of civilization. Understanding the impacts and sustainability of some of these practices begins with understanding the basics of stream ecology.

Streams and rivers are characterized by their linear form, unidirectional flow, fluctuating discharge and unstable beds. The narrow nature of river channels means that they are very intimately connected to the surrounding terrestrial environment. Thus, a proper understanding of river ecology requires us to consider the river and its drainage basin as a unit (see Section 1.3.3).

Oxygen concentration is often high in turbulent, upstream locations and low farther downstream, where higher temperatures cause reduced solubility. This is reflected in river fish communities, with active upstream species such as brown trout (*Salmo trutta*) having a high oxygen demand, whereas more sluggish species such as pike (*Esox lucius*) can tolerate the lower concentrations in their habitats downstream.

the importance of oxygen concentration, . . .

A variety of other chemical and physical conditions vary from stream to stream, or down the length of a given river. Figure 4.14 illustrates how the species composition of stream invertebrate communities varies with conditions. There were 30–40 species at each site (mainly the larvae of stoneflies, caddisflies and chironomid midges) with much overlap in the list of species present. The data were subjected to an analysis called *community classification*, which is conceptually similar to taxonomic classification. In taxonomy, similar individuals are grouped together in species, similar species in genera, and so on. In community classification, communities with similar species compositions are grouped together in

. . . pH and temperature . . .

Figure 4.14

The species composition of stream invertebrate communities varies with conditions such as pH, summer temperature and waterflow. (a) Classification of 34 stream communities. At each division, the communities are divided into classes with similar species compositions, and these divisions can be linked to particular differences in conditions, as shown. The classes are identified by the letters A–E. (b) The actual geographic distribution of community classes A–E in southern England. The classes associated with acid water conditions (D, E) occur typically in the headwaters of the streams.

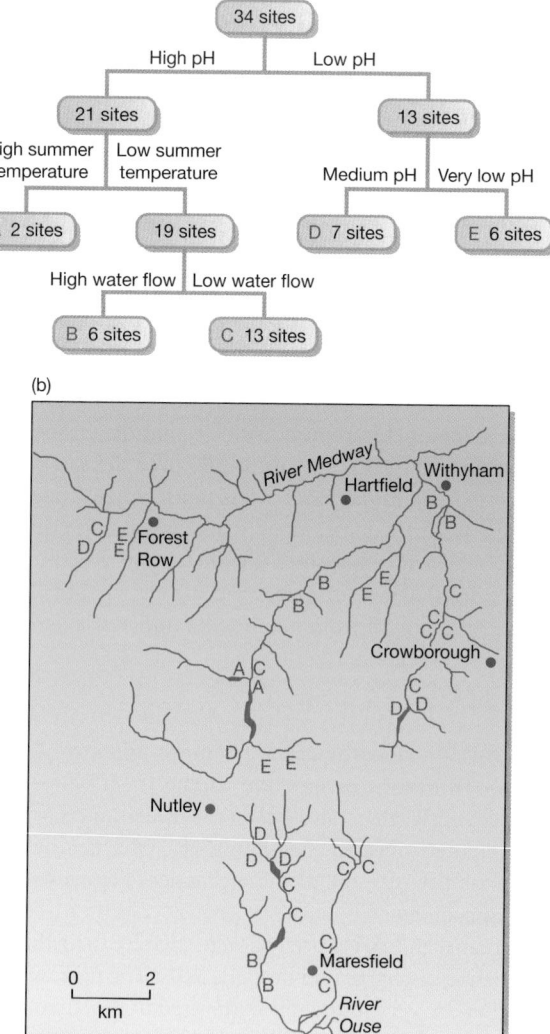

AFTER TOWNSEND ET AL., 1983

sets. These sets are then, in turn, grouped into more inclusive sets, and so on. In this case, the conditions that were most influential in determining the pattern of grouping – and thus were most influential in determining community composition – were pH, stream temperature and the volume of water flowing per unit time (discharge).

Because stream discharge responds to events such as thunderstorms and snowmelt, streams are highly disturbed systems. Stream ecologists have recently been looking at ways in which different regimes of disturbance of the streambed are reflected in the composition of the community. For example, the disturbance regimes of 54 stream sites in New Zealand were assessed by painting particles (pebbles, cobbles, boulders) representative of the streambeds and determining the percentage that moved during several periods; this varied from 10% to 85%. The insect inhabitants of the streams were categorized according to properties that might help them deal with highly disturbed conditions, including small size

. . . and disturbance of the streambed

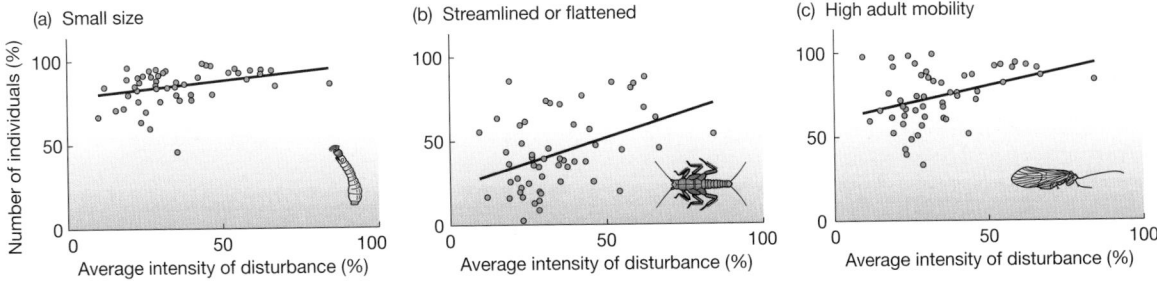

Figure 4.15

Disturbances play an important role in stream ecology, particularly of stream insects. Disturbed streams contained proportionately more larval insects that (a) were small, (b) had streamlined bodies, and (c) became adults that were strong fliers: characteristics that would enable these insects to withstand a disturbance and recolonize afterward. The best-fit lines (see Box 1.2) are very highly significant in every case ($P < 0.001$).

(small species generally have short life cycles and their populations can rapidly rebuild), a streamlined or flattened body (less prone to being dislodged) and good powers of flight of the adult insects that emerge from the stream to mate (more likely to recolonize after a disturbance). The representation of these traits was higher in the more disturbed streams, testifying to the ecological importance of disturbance regime (Figure 4.15).

The terrestrial vegetation surrounding a stream (the *riparian* vegetation) has two influences on the resources available to its inhabitants. First, by shading the streambed it may reduce primary production of attached algae and other plants. Second, by shedding leaves it can contribute directly to the food supply of animals and microorganisms. Rivers that begin their course in forested regions are often dominated by the external supply of organic matter, and many of the invertebrates have mouthparts that can handle large particles (*shredders*) (Vannote et al., 1980). Farther downstream, where the stream is wider and where shading is less intense, invertebrates that graze or scrape algae from stones (grazer–scrapers) may be more abundant. As a result of the shredding of large particles into small organic particles (and also physical processes that break up leaves), food for *collector–gatherers* and *collector–filterers* may also increase downstream (Figure 4.16).

interactions between the stream and surrounding land

When riparian vegetation is changed, for example when forest is converted to agriculture, there can be far-reaching effects. Less particulate organic matter enters the stream, but there is less shading and more nutrient runoff from farmland. Results are an increase in productivity of stream plants and a corresponding change in the stream food web. There may also be effects on discharge (increased when trees are removed), water temperature (higher if shading removed) and streambed characteristics (increased input of fine mineral particles). The more specific consequences of one particular interaction between human activity and stream ecology are described in Box 4.2.

may be disrupted by human activities

The intimate relationship between land and water is also obvious on the flood-plains of rivers such as the Amazon, where seasonal floods inundate huge areas of surrounding forest and provide massive inputs of nutrients and organic matter to the river. Many of the world's floodplains have been deliberately drained or cut off from their associated river channels, with profound consequences for river ecology.

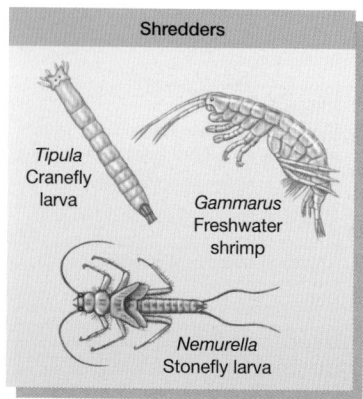
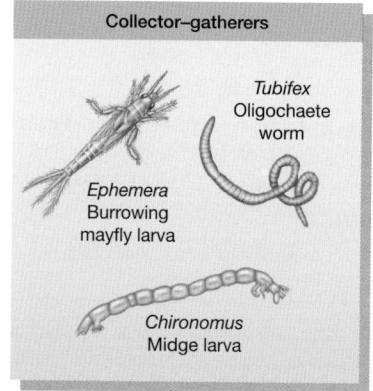
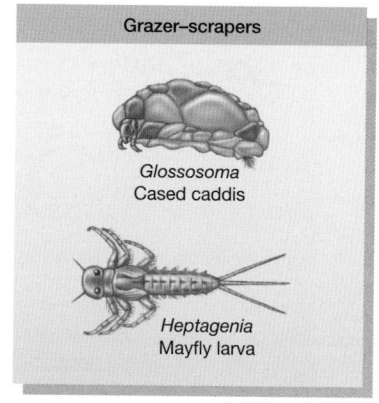
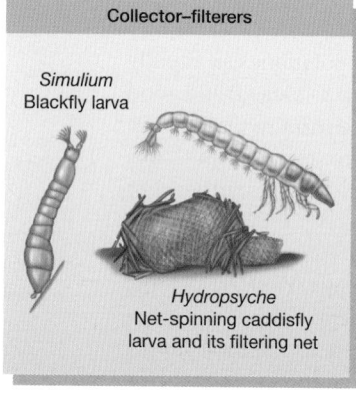
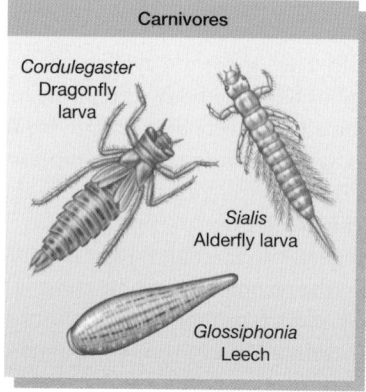

Figure 4.16

Examples of the various categories of invertebrate consumers in stream environments.

4.2 *Topical ECOncerns*

A tiny stream fish with big consequences for property development

Because streams are so intimately connected with their terrestrial catchment areas, human activities in their vicinity can have negative consequences for stream ecology. For example, landscaping or the construction of roads and buildings in the vicinity of waterways increases soil erosion and leads to silt runoff into streams. The Cherokee darter (*Etheostoma scotti*) lives in clear streams with beds made up of pebbles and cobbles. Streambeds covered in silt deny this species the ability to forage and spawn; it is now restricted to just a few streams.

COURTESY OF GEORGIA MUSEUM OF NATURAL HISTORY, E. SCOTTI_RICHLAND CREEK BJF0211

The following article by Clint Williams appeared in the *Atlanta Journal* on July 2, 2001.

Cherokee darter: tiny fish forces changes to project

While barely 2 inches long, the Cherokee darter has the power to move roads and redesign a golf course.

The tiny fish, protected under the federal Endangered Species Act, swims in the small, gravel-bottomed streams that wind through a planned 730-acre gated community straddling the Cobb-Paulding county line. And it's forcing the developer to reshape his plans in order to protect the fish.

'We have fine-tuned our layout in order to minimize our impact on the Cherokee darter', said Joe Horton, developer of the Governor's Club, a high-dollar golf course development. 'We're now on our sixth-generation site plan', Horton said.

The Cherokee darter, pale straw yellow with dark olive markings, was listed as threatened by the US Fish and Wildlife Service in 1994, not long after it was identified as a species distinct from the Coosa darter. The Cherokee darter is found only in roughly 20 small tributary systems of the Etowah River, according to a Fish and Wildlife report. But just a few streams have healthy populations.

'It's in a number of streams but that number is declining rapidly', said Seth Winger, a conservation ecologist at the Institute of Ecology at the University of Georgia.

The creeks running through the Governor's Club property are tributaries of Pumpkinvine Creek, which flows into the Etowah River below Allatoona Dam. There are 8000 feet of creeks on the tract, Horton said. A biological survey conducted before purchasing the property found four Cherokee darters . . . 'We're proud we have them', Horton said.

Having them will be a bit costly, however.

(Reproduced by permission of the PARS International Corp.)

1 *Is it reasonable that a small population of a species that occurs in about 20 other streams should disrupt economic development?*

2 *More specifically, how widespread would the species have to be (in how many streams, in how many states or countries) before developers should be allowed to ignore it?*

3 *Do you think is should be the responsibility of ecologists, such as the one quoted, simply to inform the public of the facts (as in this article)? Or is it reasonable for them to become involved in advocacy for a conservation cause?*

4.5.2 Lake ecology

Just as river ecology is defined by the unidirectional flow of water, lake ecology is defined by the relatively stationary nature of water within its basin. A critical component of lake ecology is the way in which water can stratify vertically in response to temperature (as mentioned in Section 4.2.3). As water sits in a lake basin, the upper layer is exposed to the sun and heats up. Because warm water is less dense than cold (and therefore tends to rise) the top layer stratifies – that is to say, it forms a layer that is quite separate from the colder water beneath. This layer, the *epilimnion*, is warm and well illuminated and has high oxygen content because surface waters exchange oxygen with the atmosphere. It is usually extremely productive, with high densities of plant and animal life.

In deeper lakes two further layers may form. Below the epilimnion is a transitional layer, the *thermocline*, in which temperature, oxygen concentration and light all decrease. The deepest layer, the *hypolimnion*, is cold and often poor in oxygen.

lakes may become thermally stratified, with major consequences for their ecology

It is here that sunken dead organic matter is decomposed and its mineral nutrients released. In temperate regions of the Earth, stratification of lake water breaks down in the fall when the upper layer cools. Currents then mix the water layers and the minerals released in the hypolimnion become available at the lake surface.

Lake ecologists are increasingly turning their attention to the larger spatial scale of whole lake districts. Lakes high in a landscape (such as those in northern Wisconsin) receive a greater proportion of their water from direct precipitation, whereas lakes at lower altitudes receive more water as an input from ground water (Figure 4.17). This is reflected in the higher concentrations of important ions in lakes low in the landscape. The contrasting ion concentrations can be expected, for example, to affect the ecology and distribution of freshwater sponges, whose skeletons require silica, and crayfish and snails, which have a particular need for calcium.

Nutrient-rich lakes may support a rich flora of microscopic, floating phytoplankton (microscopic plants), together with a diversity of invertebrates and fish species, but a rooted flora of flowering plants is confined to shallow waters near the shore, the *littoral zone*. This zone is usually rich in oxygen, light, food resources and hiding places. However, some fish and invertebrates specialize in the deeper colder waters of lakes. Lake trout (*Salvelinus namaycush*) and walleye (*Sander vitreus*) are two popular sport fish whose habitat is restricted to the colder regions of lakes.

saline lakes are common in some parts of the world

Many lakes in arid regions, lacking a stream outflow, lose water only by evaporation and become rich in sodium and other ions. These saline lakes should not be considered as oddities; globally, they are just as abundant as freshwater lakes. They are usually very fertile and have dense populations of blue-green algae, and some, such as Lake Nkuru in Kenya, support huge aggregations of plankton-filtering flamingoes (*Phoenicopterus roseus*).

4.5.3 The oceans

The oceans cover the major part of the Earth's surface and receive most of the Earth's income of solar radiation. However, much of this radiation is reflected at the water surface or absorbed by water itself and by particles in suspension. Even in clear water the intensity of radiation falls off exponentially with depth and photosynthesis is mainly restricted to the upper 100 m – the *euphotic zone*. In most waters the euphotic zone is much shallower, especially where water is more turbid close to coasts and estuaries.

The green plants that photosynthesize in the open oceans are planktonic, mainly single-celled algae that are capable of using solar radiation very efficiently. But, in the real world, many areas of ocean that receive the greatest intensity of solar radiation have the lowest biological activity – because plant productivity is limited by shortage of mineral nutrients. The great tropical parts of the Atlantic and Pacific Oceans have a biological productivity of less than 35 g carbon (g C) m^{-2} yr^{-1}. This compares with more than 800 g C m^{-2} yr^{-1} in terrestrial communities at the same latitudes.

The areas of greatest marine productivity (exceeding 90 g C m^{-2} yr^{-1}) occur where there is a reliable supply of minerals (especially nitrogen and phosphorus, and perhaps iron). This occurs via leaching from the land through rivers and estuaries or where deep currents in the oceans well up to the surface and bring dissolved

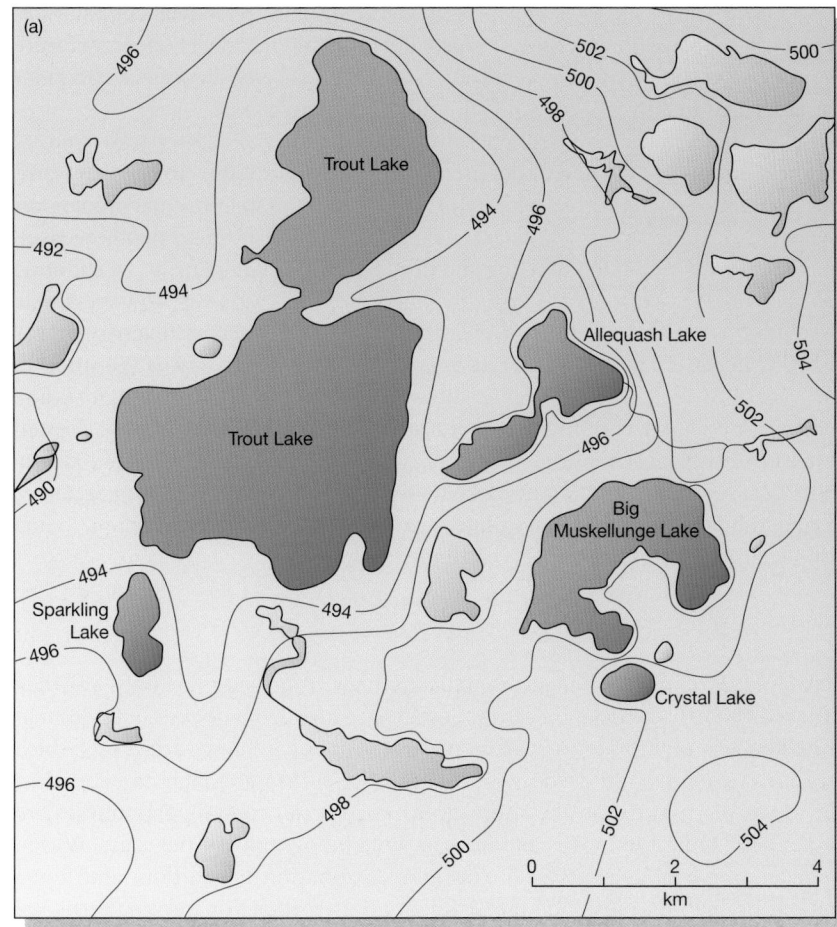

(a)

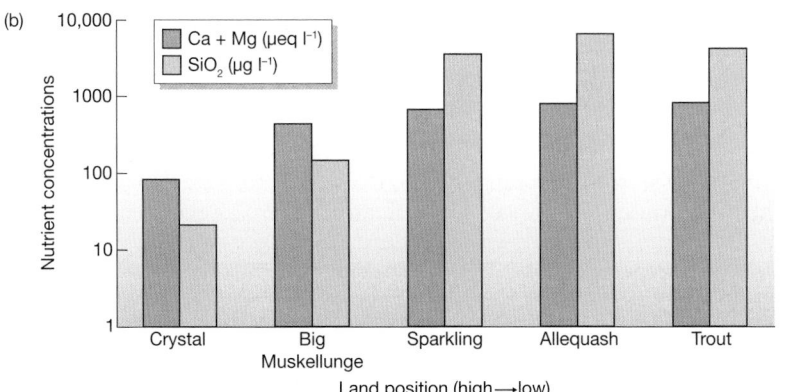

(b)

Land position (high→low)

Figure 4.17

Lakes at different positions in the landscape differ in the source of their water and the concentrations of chemicals important to their inhabitants. (a) Map of Wisconsin Lake District: study lakes are darkly shaded and contours are shown (meters above sea level). (b) Relationships between landscape position and concentrations of calcium and magnesium (Ca + Mg) and silica (SiO_2) in the five lakes. Lakes higher in the catchment area (Crystal and Big Muskellunge) have lower nutrient concentrations.

nutrients into the euphotic zone (see Section 11.2.2). In areas where upwellings occur, the ocean 'desert' becomes transformed to a productive environment, as, for example, off the coast of Peru. Dense populations of planktonic algae support small crustacea, which in turn are eaten by schools of anchovies (*Engraulis ringens*). The fish support sea lions, and flocks of cormorants, pelicans and gannets.

We saw earlier in this chapter that the distribution of terrestrial communities depends largely on the intensity of solar radiation and its effects on temperature and water availability. In complete contrast, variations between oceanic communities are ruled mainly by the availability of mineral nutrients.

unique communities occur in the abyssal depths of the oceans

Below the euphotic zone is increasing darkness, and the ocean floor is in total darkness, intensely cold and under great pressure. This abyssal environment supports the very slow biological activity of a community of extraordinary biological diversity (including worms, crustaceans, mollusks and fish found nowhere else), which depends on the rain of dying and dead organisms falling from the euphotic zone above. Many of the invertebrate animals are tiny, have very low metabolic rates and possess a lifespan that may last for decades. Yet further diversity occurs in hydrothermal vents that occur at a number of isolated places 2000–4000 m deep (see Box 2.2). In these remarkable environments, there are high sulfide concentrations and very high temperatures, up to 350°C, where superheated fluid emerges from 'chimneys', and there is a sharp gradient down to 2°C, the temperature normally encountered in abyssal depths close by. The vent areas are inhabited by productive thermophilic (heat-loving) bacteria and a unique fauna of polychaete worms, crabs and very large mollusks.

4.5.4 Coasts

Marine environments change dramatically near to coasts. Not only are they enriched by nutrients from the land; they are also affected by waves and tides that bring new physical forces to bear. In particular, there are now surfaces to which organisms can attach; indeed, if they do not do so they are liable to be washed out to sea or stranded on the shore. At a broad scale, coastal communities are strongly influenced by waves and tides and the topography of the coast. Within a single stretch of coast, we can recognize a zonation in the flora and fauna marked by high and low tide levels (Figure 4.18). Such zonation patterns are more obvious in sheltered situations where wave action is light, but become more fuzzy in very exposed situations.

waves and tides are key influences in coastal ecology

The extent of the littoral zone depends on the height of tides and the slope of the shore. Away from the shore, the tidal rise and fall are rarely greater than 1 m, but closer to shore, the shape of the land mass can funnel the ebb and flow of the water to produce extraordinary spring tidal ranges of, for example, nearly 20 m in the Bay of Fundy (between Nova Scotia and New Brunswick, Canada). In contrast, the shores of the Mediterranean Sea experience scarcely any tidal range.

On steep shores and rocky cliffs the littoral zone is very narrow and zonation is compressed. Both plants and animals are profoundly affected by the physical force of wave action. Anemones, barnacles and mussels attach themselves securely and permanently to the substrate and filter planktonic plants and animals from the water when the tides cover them. Other animals, such as limpets, move to graze, and crabs move with the tides and use rock crevices as refuges. The flora in a rocky infralittoral zone (Figure 4.18) is usually dominated by the large brown seaweeds (kelps), which fix themselves to the rock with specialized 'holdfasts'.

Environments are quite different on shallow sloping shores on which the tides deposit and stir up sand and mud. Here the dominant animals are mollusks

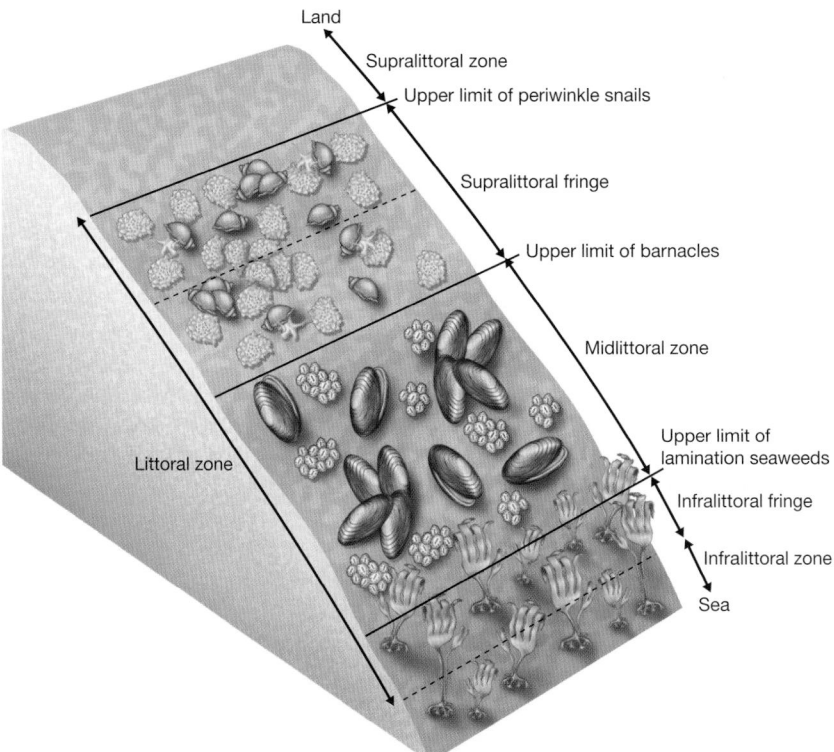

Land
Supralittoral zone
Upper limit of periwinkle snails
Supralittoral fringe
Upper limit of barnacles
Midlittoral zone
Littoral zone
Upper limit of lamination seaweeds
Infralittoral fringe
Infralittoral zone
Sea

AFTER RAFFAELLI & HAWKINS, 1996

Figure 4.18

A general zonation scheme for the seashore determined by relative lengths of exposure to the air and to the action of waves. At the top of the shore is the supralittoral zone (the splash zone above the high tide level). The littoral zone, between high and low tide levels, can be divided into a midlittoral zone together with a supralittoral fringe above and an infralittoral fringe below. The infralittoral zone proper lies below the low tidal limit. Characteristic communities of animals and plants occur in these different zonation bands.

and polychaete worms, living buried in the substrate and feeding by filtering the water when they are covered by the tides. This environment is completely free of large seaweeds, whose holdfasts can find no anchorage. Flowering plants are almost, but not completely, absent from intertidal environments. The exceptions occur where it is possible for them to be anchored by their roots and this requirement limits them to the more stable and muddy areas colonized by 'sea grasses' such as *Zostera* and *Posidonia* or tussocks of *Spartina*. In the tropics, mangroves occupy this kind of habitat, adding a shrubby, woody dimension to the marine littoral zone.

4.5.5 Estuaries

Estuaries occur at the confluence of a river (fresh water) and a tidal bay (salt water). They provide an intriguing mix of the conditions normally experienced in rivers, shallow lakes and tidal communities. Salt water, more dense than fresh water, tends to enter along the bottom of an estuary as a salt wedge. As it mixes with the outflowing fresh water, a brackish middle layer is created, then it returns downstream on the outgoing tide. The shape of the saltwater wedge is largely determined by the size of the discharge of the river flowing into the estuary; high discharge tends to create a smaller wedge of salt water and less mixing. The strong gradients in salinity, in both space and time, are reflected in a specialized estuarine fauna. Some animals cope through particular physiological mechanisms. Others avoid the variable salt concentrations by burrowing, closing protective shells or moving away when conditions do not favor them.

Summary
SUMMARY

Geographic patterns at large and small scales

The variety of influences on climatic conditions over the surface of the globe causes a mosaic of climates. This, in turn, is responsible for the large-scale pattern of distribution of terrestrial biomes. However, biomes are not homogeneous within their hypothetical boundaries; every biome has gradients of physicochemical conditions related to local topographic, geological and soil features. The communities of plants and animals that occur in these different locations may be quite distinct.

In most aquatic environments it is difficult to recognize anything comparable to terrestrial biomes; the communities of streams, rivers, lakes, estuaries and open oceans reflect local conditions and resources rather than global patterns in climate. The composition of local communities can change over time scales ranging from hours, through decades, to millennia.

Terrestrial biomes

A map of biomes is not usually a map of the distribution of species. Instead, it shows where we find areas of land dominated by plants with characteristic life forms.

Tropical rain forest represents the global peak of evolved biological diversity. Its exceptional productivity results from the coincidence of high solar radiation received throughout the year and regular and reliable rainfall.

Savanna consists of grassland with scattered small trees. Seasonal rainfall places the most severe restrictions on the diversity of plants and animals in savanna; grazing herbivores and fire also influence the vegetation, favoring grasses and hindering the regeneration of trees.

Temperate grassland occurs in the steppes, prairies and pampas. Typically, it experiences seasonal drought, but the role of climate in determining vegetation is usually overridden by the effects of grazing animals. Humans have transformed temperate grassland more than any other biome.

Many desert plants have an opportunistic lifestyle, stimulated into germination by the unpredictable rains; others, such as cacti, are long-lived and have sluggish physiological processes. Animal diversity is low in deserts, reflecting the low productivity of the vegetation and the indigestibility of much of it.

Temperate forests at lower latitudes experience mild winters, and the vegetation consists of broad-leaved, evergreen trees. Nearer the poles, the seasons are strongly marked, and vegetation is dominated by deciduous trees. Soils are usually rich in organic matter.

Northern coniferous forests have few tree species and contrast strongly with the biodiversity of tropical rain forests, reflecting a slow recovery from the catastrophes of the ice ages, and the overriding local constraint of frozen soil. Nearer the poles, the vegetation changes to tundra, and the two often form a mosaic in the Low Arctic. The mammal populations of the northern biomes often pass through remarkable cycles of expansion and collapse.

Aquatic environments

Streams and rivers are characterized by their linear form, unidirectional flow, fluctuating discharge and unstable beds. The terrestrial vegetation surrounding a stream has strong influences on the resources available to its inhabitants; the conversion of forest to agriculture can have far-reaching effects.

Lake ecology is defined by the relatively stationary nature of its water. Some lakes stratify vertically in response to temperature, with consequences for the availability of oxygen and plant nutrients. Lakes higher in a landscape may receive more of their water from rainfall; those at lower altitude receive more from ground water. Saline lakes in arid regions lack a stream outflow and lose water only by evaporation.

The oceans cover the major part of the Earth's surface and receive most of the solar radiation. However, many areas have very low biological activity because of a shortage of mineral nutrients. Below the surface zone is increasing darkness, but at the ocean floor there may be an abyssal environment that supports a diverse community with very slow biological activity.

Coastal communities are enriched by nutrients from the land, but they are also affected by waves and tides. Within a single stretch of coast, there is a zonation in the flora and fauna that differs between areas with heavy or light wave action.

Estuaries occur at the confluence of a river (fresh water) and a tidal bay (salt water). Strong gradients in salinity, in both space and time, are reflected in a specialized estuarine fauna.

Review questions

REVIEW QUESTIONS

Asterisks indicate challenge questions

1   Describe the various changes in climate that occur with changing latitude, including an explanation of why deserts are more likely to be found at around 30° latitude than at other latitudes.

2   How would you expect the climate to change as you crossed from west to east over the Rocky Mountains?

3* Biomes are differentiated by gross differences in the nature of their communities, not by the species that happen to be present. Explain why this is so.

4   The tropical rain forest is a diverse community supported by a nutrient-poor soil. Account for this.

5* Which of the Earth's biomes do you think have been most strongly influenced by people? How and why have some biomes been more strongly affected by human activity than others?

6   What is meant by the 'stratification' of water in lakes? How does it occur? And what are the reasons for variations in stratification from time to time and from lake to lake?

7   Describe how the logging of a forest may influence the community of organisms inhabiting a stream running through the affected area.

8   Why is much of the open ocean, in effect, a 'marine desert'?

9   Discuss some reasons why community composition changes as one moves (i) up a mountain, and (ii) down the continental shelf into the abyssal depths of the ocean.

10* Why are broad geographic classifications of aquatic communities less feasible than broad geographic classifications of terrestrial communities? What characteristics of aquatic ecosystems buffer the effects of climate?

PART THREE

Individuals, Populations, Communities and Ecosystems

Chapter 5

Birth, death and movement

Key concepts
Key concepts

In this chapter you will:

- appreciate the difficulties of counting individuals, but the necessity of doing so for understanding the distribution and abundance of organisms and populations
- appreciate the range of life cycles and patterns of birth and death exhibited by different organisms
- understand the nature and the importance of life tables and fecundity schedules
- understand the role and the importance of dispersal and migration in the dynamics of populations
- understand the impact of intraspecific competition on birth, death and movement, and hence on populations
- appreciate that life history patterns linking types of organism to types of habitat can be constructed but also recognize the limitations of those patterns

All questions in ecology – however scientifically fundamental, however crucial to immediate human needs and aspirations – can be reduced to attempts to understand the distributions and abundances of organisms, and the processes – birth, death and movement – that determine distribution and abundance. In this chapter, these processes, the methods of monitoring them and their consequences are introduced.

5.1 Introduction

what is a population?

As ecologists, we try to describe and understand the distribution and abundance of organisms. We may do so because we wish to control a pest or conserve an endangered species, or simply because we are fascinated by the world around us and the forces that govern it. A major part of our task, therefore, involves studying changes in the size of populations. We use the term *population* to describe a group of individuals of one species. What actually constitutes a population, though, varies from species to species and from study to study. In some cases, the boundaries of a population are obvious: the sticklebacks occupying a small lake are 'the stickleback population of the lake'. In other cases, boundaries are determined more by an investigator's purpose or convenience. Thus, we may study the population of lime aphids inhabiting one leaf, one tree, one stand of trees or a whole woodland. What is common to all uses of *population* is that it is defined by the number of individuals that compose it: populations grow or decline by changes in those numbers.

birth, death and movement change the size of populations

The processes that change the size of populations are birth, death and movement into and out of that population. Trying to understand the causes of changes in population size is important because the science of ecology is not just about understanding nature but often also about predicting or controlling it. We might, for example, wish to reduce the size of a population of rabbits that can do serious harm to crops. We might do this by increasing the death rate by introducing the myxomatosis virus to the population, or by decreasing the birth rate by offering them food that contains a contraceptive. We might encourage their emigration by bringing in dogs, or prevent their immigration by fencing.

Similarly, a nature conservationist may wish to increase the population of a rare endangered species. In the 1970s, the numbers of bald eagles, ospreys and other birds of prey in the United States began a rapid decline. This might have been because their birth rate had fallen, or their death rate had risen, or because the populations were normally maintained by immigration and this had fallen, or because individuals had emigrated and settled elsewhere. Eventually the decline was traced to reduced birth rates. The insecticide DDT (dichlorodiphenyltrichloroethane) was widely used at the time (it is now banned in the United States) and had been absorbed by many species on which the birds preyed. As a result, it accumulated in the bodies of the birds themselves and affected their physiological processes so that the shells of their eggs became so thin that the chicks often died in the egg. Conservationists charged with restoring the bald eagle population had to find a way to increase the birds' birth rate. The banning of DDT achieved this end.

5.1.1 What is an individual?

A population is characterized by the number of individuals it contains, but for some kinds of organism it is not always clear what we mean by an individual. Often there is no problem, especially for *unitary* organisms. Birds, insects, reptiles and mammals are all unitary organisms. The whole form of such organisms, and their program of development from the moment when a sperm fuses with an egg, is predictable and *determinate*. An individual spider has eight legs. A spider that lived a long life would not grow more legs.

But none of this is so simple for *modular* organisms such as trees, shrubs and herbs, corals, sponges and very many other marine invertebrates. These grow by the repeated production of modules (leaves, coral polyps, etc.) and almost always form a branching structure. Such organisms have an architecture: most are rooted or fixed, not motile (Figure 5.1). Both their structure and their precise program of development are not predictable but *indeterminate*. We could count the individual trees in a forest, but would this signify the 'size' of the tree population? Not unless we also noted whether the trees were young saplings (few leaves and branches each), or old individuals, each with many more such modules. Indeed, it may make more sense not to count the individual trees themselves but the total number of modules instead.

In modular organisms, then, we need to distinguish between the genet – the genetic individual – and the module. The *genet* is the individual that starts life as a single-celled zygote and is considered dead only when all its component modules have died. A *module* starts life as a multicellular outgrowth from another module and proceeds through a life cycle to maturity and death even though the form and development of the whole genet are indeterminate. We usually think of unitary organisms when we write or talk about populations, perhaps because we ourselves are unitary, and there are certainly many more species of unitary than of modular organisms. But modular organisms are not rare exceptions and oddities. Most of the living matter (biomass) on Earth and a large part of that in the sea is of modular organisms: the forests, grasslands, coral reefs and peat-forming mosses.

unitary and modular organisms

modular organisms are themselves populations of modules

5.1.2 Counting individuals, births and deaths

Even with unitary organisms, we face enormous technical problems when we try to count what is happening to populations in nature. A great many ecological questions remain unanswered because of these problems. For example, resources can only be focused on controlling a pest effectively if it is known when its birth rate is highest. But this can only be known by monitoring accurately either births themselves or rising total numbers – neither of which is ever easy.

If we want to know how many fish there are in a pond we might obtain an accurate count by putting in poison and counting the dead bodies. But apart from the questionable morality of doing this, we usually want to continue studying a population after we have counted it. Occasionally it may be possible to trap alive

the difficulties of counting

(a)

(b)

Figure 5.1

Modular plants (on the left) and animals (on the right), showing the underlying parallels in the various ways they may be constructed. (a) Modular organisms that fall to pieces as they grow: duckweed (*Lemna* sp.) (© John D. Cunningham) and *Hydra* sp. (© Larry Stepanowicz). (b) Freely branching organisms in which the modules are displayed as individuals on 'stalks': a vegetative shoot of a higher plant (*Lonicera japonica*) with leaves (feeding modules) and a flowering shoot (© Visuals Unlimited), and a hydroid colony (*Obelia*) bearing both feeding and reproductive modules (© Larry Stepanowicz).

all the individuals in a population, count them and then release them. With birds, for example, it may be possible to mark nestlings with leg rings and ultimately recognize every individual (except immigrants) in the population of a small woodland. It is not too difficult to count the numbers of large mammals such as deer on an isolated island. But it is very much more difficult to count the numbers of lemmings in a patch of tundra because they spend a large part of the year (and

(c)

(d)

(e)

Figure 5.1 (cont.)

(c) Stoloniferous organisms in which colonies spread laterally and remain joined by 'stolons' or rhizomes: a single plant of strawberry (*Fragaria*) spreading by means of stolons (© Science VU) and a colony of the hydroid *Tubularia crocea* (© John D. Cunningham). (d) Tightly packed colonies of modules: a tussock of the spotted saxifrage (*Saxifraga bronchialis*) (© Gerald and Buff Corsi) and a segment of the hard coral *Turbinaria reniformis* (© Dave B. Fleetham). (e) Modules accumulated on a long, persistent, largely dead support: an oak tree (*Quercus robur*) in which the support is mainly the dead woody tissues derived from previous modules (© Silwood Park) and a gorgonian coral in which the support is mainly heavily calcified tissues from earlier modules (© Daniel W. Gotshall).

may reproduce) under thick snow cover. And most other species are so small, or cryptic, or hidden, or fast moving that they are even more difficult to count.

Ecologists, therefore, are almost always forced to estimate rather than count. They may estimate the numbers of aphids on a crop, for example, by counting the number on a representative sample of leaves, then estimating the number of leaves per square meter of ground, and from this estimating the number of aphids per square meter. Sometimes more complex methods are used (Box 5.1), and at other times we may rely on indirect 'indices' of abundance. These can provide

5.1 *Quantitative aspects*

Mark–recapture methods for estimating population size

An estimate of the size of a population can sometimes be made by capturing a sample of individuals, marking them in some way (paint spots, leg rings) and then releasing them. Later, another sample is captured, and the proportion that is marked gives some estimate of the size of the whole population (Figure 5.2). For example, we might capture and mark 100 individuals from a population of sparrows and release them back into the population. If we later sample a further 100 individuals from the population and find half are marked, we could argue in the following way: half the sample are marked; the sample is representative of the whole population; therefore half the population are marked; 100 individuals were given a mark; therefore the whole population is composed of about 200 individuals. But this technique of mark and recapture is far less straightforward than it appears at first sight. There are many pitfalls in the sampling process and in interpretation of the data. Suppose, for example, that many of the individuals we marked died between our first and second visits. Modifications of the method would be needed to take account of this. For many organisms, however, it is the only technique that we have to estimate the size of a population.

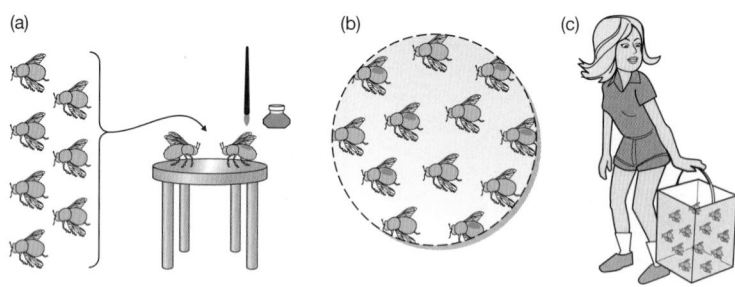

Figure 5.2

The mark and recapture technique for estimating the size of a population of mobile organisms (in simplified form). (a) On a first visit to a population of unknown total size N, a representative sample is caught (r individuals) and given a harmless mark. (b) These are released back into the population, where they remix with the unknown number of unmarked individuals. (c) On a second visit, a further representative sample is caught. Because it is representative, the proportion of marks in the sample (m out of a total sample of n) should, on average, be the same as that in the whole population (r out of a total of N). Hence N can be estimated.

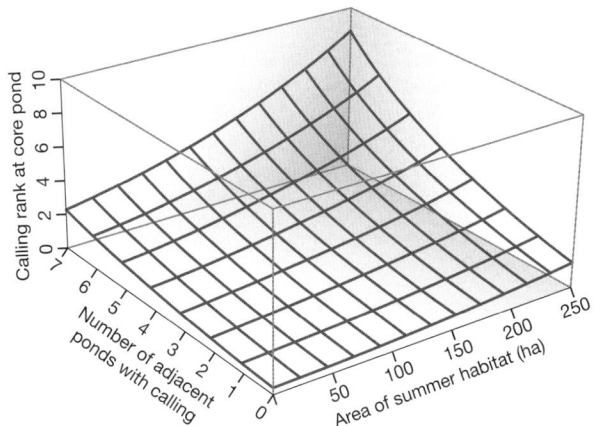

AFTER POPE ET AL. 2000

Figure 5.3

The abundance (calling rank) of leopard frogs (*Rana pipiens*) in ponds increases significantly with both the number of adjacent ponds that are occupied and the area of summer habitat within 1 km of the pond. Calling rank is the sum of an index measured on four occasions, namely: 0, no individuals calling; 1, individuals can be counted, calls not overlapping; 2, calls of <15 individuals can be distinguished with some overlapping; 3, calls of ≥15 individuals.

information on the relative size of a population, but usually give little indication of absolute size. As an example, Figure 5.3 shows how the abundance of Canadian leopard frogs was affected by the number of occupied ponds and the amount of summer (terrestrial) habitat in their vicinity. Here, frog abundance was estimated from the 'calling rank': whether there were no frogs, 'few', 'many' or 'very many' frogs calling on each of four occasions. Despite their shortcomings, even indices of abundance can provide valuable information.

Moreover, as we have already noted, for modular organisms it is often not even clear what it is we should be counting.

5.2 Life cycles

5.2.1 Life cycles and reproduction

If we wish to understand the forces determining the abundance of a population of organisms, we need to know the important phases of those organisms' lives: that is, the phases when these forces act most significantly. For this, we need to understand the sequences of events that occur in those organisms' life cycles.

There is a point in the life of any individual when, if it survives that long, it will start to reproduce and leave progeny. A highly simplified, generalized life history (Figure 5.4) comprises birth, followed by a pre-reproductive period, a period of reproduction, a post-reproductive period and then death as a result of senescence (though of course other forms of mortality may intervene at any time). The life histories of all unitary organisms can be seen as variations around this simple pattern, though a post-reproductive period (as seen in humans) is probably rather unusual.

Some organisms fit several or many generations within a single year, some have just one generation each year (annuals) and others (perennials) have a life cycle extended over several or many years. For all organisms, though, a period of growth occurs before there is any reproduction, and growth usually slows down (and in some cases stops altogether) when reproduction starts. Growth and reproduction both require resources and there is clearly some conflict between them. Thus, as the perennial plant *Sparaxis grandiflora* enters its reproductive

the conflict between growth and reproduction

Figure 5.4

An outline life history for a unitary organism. Time passes along the horizontal axis, which is divided into different phases. Reproductive output is plotted on the vertical axis.

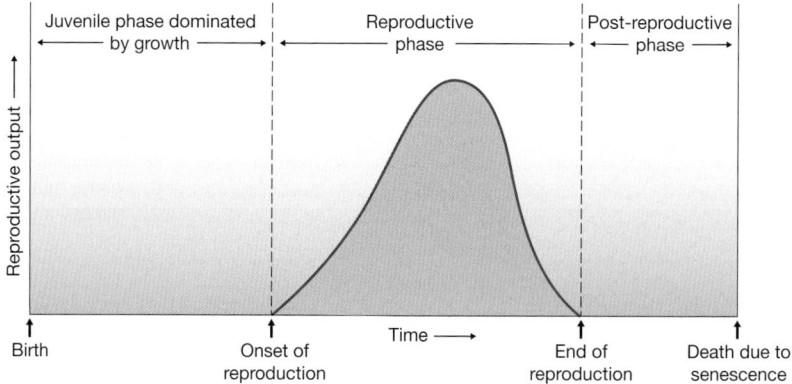

stage in the Southwestern Cape, South Africa, flowers, flower stalks and fruit (aspects of reproduction) can be seen to have been produced *at the expense* of roots and leaves (Figure 5.5). There are also many plants (e.g. foxgloves) that spend their first year in vegetative growth, and then flower and die in the second or a later year (called 'biennial' plants). But if the flowers of these species are removed before their seeds begin to set, the plants usually survive to the following year, when they flower again and set seed even more vigorously. It seems to be the cost of provisioning the offspring (seeds) rather than the flowering itself that is lethal. Similarly, pregnant women are advised to increase their caloric intake by as much as half their normal consumption: when nutrition is inadequate, pregnancy can harm the health of the mother.

iteroparous and semelparous species

Among both annuals and perennials, there are some – *iteroparous species* – that breed repeatedly, devoting some of their resources during a breeding episode not to breeding itself, but to survival to further breeding episodes (if they manage to live that long). We ourselves are examples. There are others, *semelparous species*, like the biennial plants already described, in which there is a single reproductive episode, with no resources set aside for future survival, so that reproduction is inevitably followed quickly by death.

Figure 5.5

Percentage allocation of the crucial resource nitrogen to different structures throughout the annual cycle of the perennial plant *Sparaxis grandiflora* in South Africa, where it sets fruit in the southern hemisphere spring (September–December). The plant grows each year from a corm, which it replaces over the growing season, but note the development of reproductive parts at the expense of roots and leaves toward the end of the growing season. The plant parts themselves are illustrated to the right for a plant in early spring.

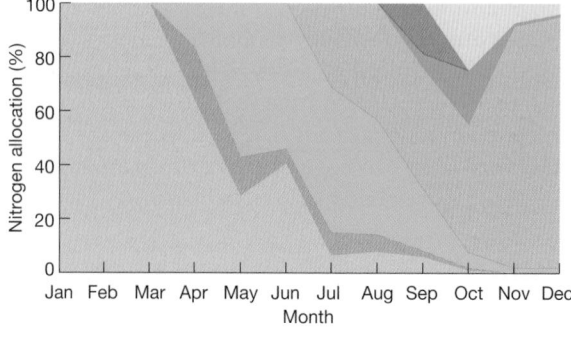
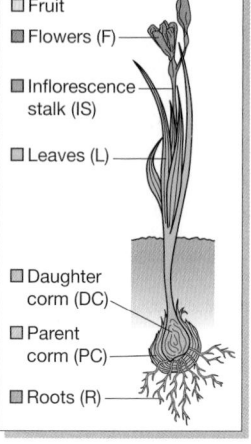

AFTER RUITERS & MCKENZIE, 1994

5.2.2 Annual life cycles

In strongly seasonal, temperate latitudes, most annuals germinate or hatch as temperatures start to rise in the spring, grow rapidly, reproduce and then die before the end of summer. The European common field grasshopper *Chorthippus brunneus* is an example of an annual species that is iteroparous. It emerges from its egg in late spring and passes through four juvenile stages of nymph before becoming adult in midsummer and dying by mid-November. During their adult life, the females reproduce repeatedly, each time laying egg pods containing about 11 eggs, and recovering and actively maintaining their bodies between the bursts of reproduction.

Many annual plants, by contrast, are semelparous: they have a sudden burst of flowering and seed set, and then they die. This is commonly the case among the weeds of arable crops. Others, such as groundsel, are iteroparous: they continue to grow and produce new flowers and seeds through the season until they are killed by the first lethal frost of winter. They die with their buds on.

Most annuals spend part of the year dormant as seeds, spores, cysts or eggs. In many cases these dormant stages may remain viable for many years; there are reliable records of seeds of the annual weeds *Chenopodium album* and *Spergula arvensis* remaining viable in soil for 1600 years. Similarly, the dried eggs of brine shrimps remain viable for many years in storage. This means that if we measure the length of life from the time of formation of the zygote, many so-called 'annual' animals and plants live very much longer than a single year. Large populations of dormant seeds form a *seed bank* buried in the soil: as many as 86,000 viable seeds per square meter have been found in cultivated soils. The species composition of the seed bank may be very different from that of the mature vegetation above it (Figure 5.6). Species of annuals that seem to have become locally extinct may suddenly reappear after the soil is disturbed and these seeds germinate.

seed banks

Dormant seeds, spores or cysts are also necessary to the many ephemeral plants and animals of sand dunes and deserts that complete most of their life cycle in less than 8 weeks. They then depend on the dormant stage to persist through the remainder of the year and survive the hazards of low temperatures in winter and the droughts of summer. In desert environments, in fact, the rare rains are not

ephemeral 'annuals' of deserts

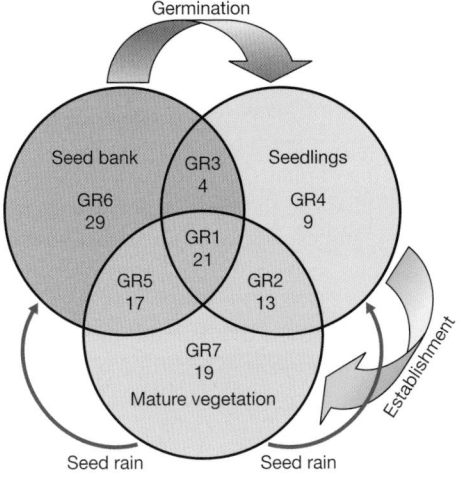

AFTER JUTILA, 2003

Figure 5.6

Species recovered from the seed bank, from seedlings and from the mature vegetation in a coastal grassland site on the western coast of Finland. Species may germinate from the buried seed bank into seedlings, and seedlings may establish themselves in the mature vegetation. Mature plants may contribute seeds (in the 'seed rain') that germinate into seedlings immediately or enter the buried seed bank. Seven species groups (GR1–GR7) are defined on the basis of whether they were found in only one, two or all three life stages. The marked difference in composition, especially between the seed bank and the mature vegetation, is readily apparent. Thirty-two species in the mature vegetation (19 + 13) were not represented in the seed bank; 33 species in the seed bank were not found in the mature vegetation, and 29 of these were not found as seedlings either.

A desert in bloom. In desert areas where rainfall is rare and seasonally unpredictable, a dense and spectacular flora of very short-lived annuals commonly develops after rain storms. They often complete their life cycle from germination to seed set in little more than a month.

© DOUG SOKELL, VISUALS UNLIMITED

necessarily seasonal, and it is only in occasional years that sufficient rain falls and stimulates the germination of characteristic and colorful floras of very small ephemeral plants.

5.2.3 Longer life cycles

repeated, seasonal breeders

There is a marked seasonal rhythm in the lives of many long-lived plants and animals, especially in their reproductive activity: a period of reproduction once per year (Figure 5.7a). Mating (or the flowering of plants) is commonly triggered by the length of the *photoperiod* – the light phase in the daily light–dark cycle, which varies continuously through the year – and usually makes sure that young are born, eggs hatch or seeds are ripened when seasonal resources are likely to be abundant.

In populations of perennial species, the generations overlap and individuals of a range of ages breed side by side. The population is maintained in part by survival of adults and in part by new births. A study of the great tit *Parus major*, for example, showed that of 50 eggs that were laid by a breeding population of 10 birds in one season, only 30 hatchlings survived to become fully fledged, and

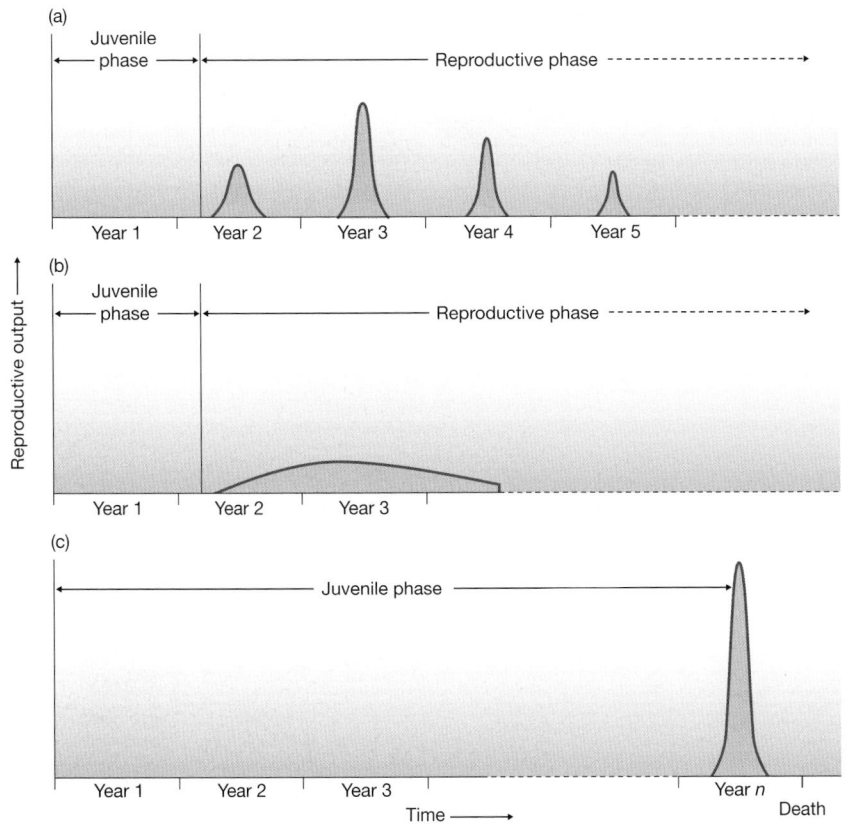

Figure 5.7

Simplified life histories for organisms living more than 1 year. (a) An iteroparous species breeding seasonally once per year. Death tends not to occur predictably after any given time, though a decline toward senescence is often observed. (b) An iteroparous species breeding continuously throughout the year. The pattern of death and decline is similar to that in (a). (c) A semelparous species passing several or many years in a pre-reproductive juvenile phase, followed by a burst of reproduction, followed in turn by inevitable death.

only three of these survived to adulthood the following year. These three 1-year-old birds were joined in that second year, though, by a further five birds aged between 2 and 5 years – the survivors of the previous year's 10 (Figure 5.8).

In wet equatorial regions, on the other hand, where there is very little seasonal variation in temperature and rainfall and scarcely any variation in photoperiod, we find species of plant that are in flower and fruit throughout the year – and continuously breeding species of animal that subsist on this resource (Figure 5.7b). There are several species of fig (*Ficus*), for instance, that bear fruit continuously and form a reliable year-round food supply for birds and primates. In more seasonal climates, humans are unusual in also breeding continuously throughout the year, though numbers of other species, cockroaches for example, do so in the stable environments that humans have created.

continuous breeders

Other plants and animals (Figure 5.7c) may spend almost all their lives in a long non-reproductive (juvenile) phase and then have one lethal burst of reproductive activity. We saw such semelparity earlier in biennial plants, but it is also characteristic of some species that live much longer than 2 years. The Pacific salmon is a familiar example. Salmon are spawned in rivers. They spend the first phase of their juvenile life in fresh water and then migrate to the sea, often traveling thousands of miles. At maturity they return to the stream in which they were hatched. Some mature and return to reproduce after only 2 years at sea; others mature more slowly and return after 3, 4 or 5 years. At the time of

semelparous species like salmon and bamboo

Figure 5.8

A diagrammatic life history for a population of great tits near Oxford, UK. Individuals typically live for several years; hence, the population in any one year is a combination of survivors from previous years and newborn individuals. Population sizes (in rectangles) are per hectare; the proportions surviving from one stage to the next are in triangles; the rate of egg production per female is shown in the diamond.

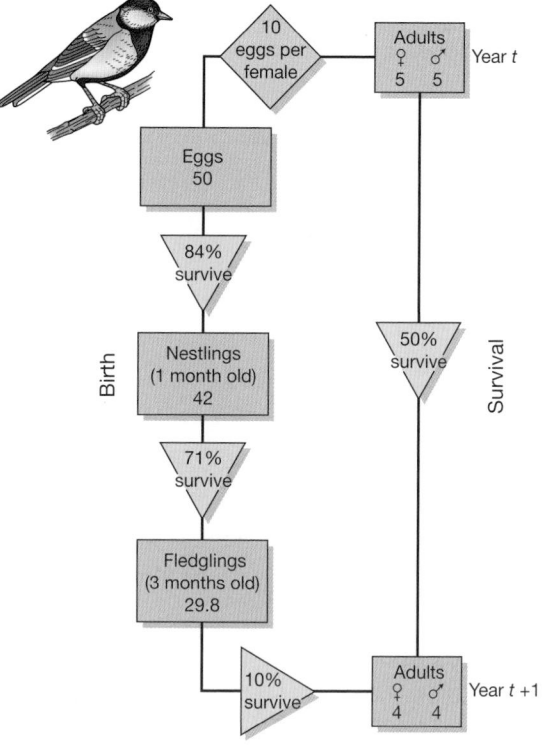

reproduction, the population of salmon is composed of overlapping generations of individuals. But all are semelparous: they lay their eggs and then die; their bout of reproduction is terminal.

There are even more dramatic examples of species that have a long life but reproduce just once. Many species of bamboo form dense clones of shoots that

Figure 5.9

The effect of plant age (years) and plant size (as measured by leaf area) on the probability of *Rhododendron lapponicum* shoots entering their reproductive phase. The relationships have been 'smoothed' by a statistical technique called 'logistic regression'. Note that the probability of reproduction increases with plant size at all ages. Also, older shoots are overall more likely to enter their reproductive phase because they tend to be bigger. However, at any given size, the probability of reproduction tends to *decrease* with age, making age itself a much poorer predictor of shoot fate than size.

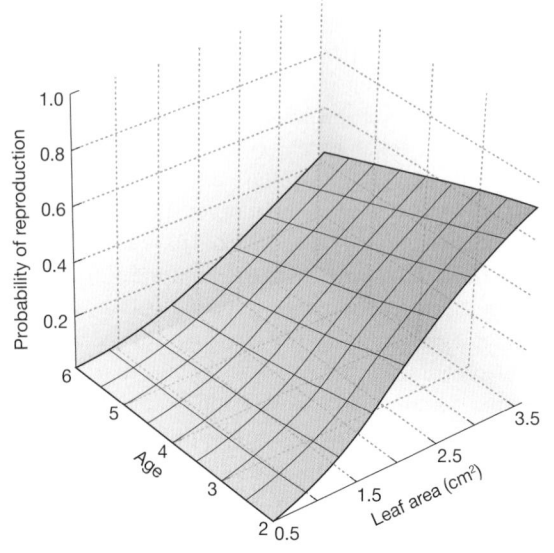

remain vegetative for many years: in some species, 100 years. The whole population of shoots then flowers simultaneously in a mass suicidal orgy. Even when shoots have become physically separated from each other, the parts still flower synchronously.

Organisms of long-lived species that are the same age, however, are not necessarily the same size – especially in modular organisms. Some individuals may be very old but have been suppressed in their growth and development by predators or by competition. Age, then, is often a particularly poor predictor of fecundity. An analysis that classifies the members of a population according to their size rather than their age (Figure 5.9) is often more useful in suggesting whether they will survive or reproduce.

size matters

5.3 Monitoring birth and death: life tables and fecundity schedules

The previous sections have outlined the different patterns of births and deaths in different species. But patterns are just a start. What are the *consequences* of these patterns in specific cases in terms of their effects on how a population might grow to pest proportions, say, or shrink to the brink of extinction? To determine these consequences, we need to monitor the patterns in a quantitative way.

There are different ways of doing so. To monitor and quantify survival, we may follow the fate of individuals from the same *cohort* within a population: that is, all individuals born within a particular period. A *cohort life table* then records the survivorship of the members of the cohort over time (Box 5.2). A different approach is necessary when we cannot follow cohorts but we know the ages of all the individuals in a population. We can then, at one time, describe the numbers of survivors of different ages in what is called a *static life table* (Box 5.2).

5.2 *Quantitative aspects*

The basis for cohort and static life tables

In Figure 5.10, a population is portrayed as a series of diagonal lines, each line representing the life 'track' of an individual. As time passes, each individual ages (moves from bottom left to top right along its track) and eventually dies (the dot at the end of the track). Here, individuals are classified by their age. In other cases it may be more appropriate to split the life of each individual into different developmental stages.

Time is divided into successive periods: t_0, t_1, etc. In the present case, three individuals were born (started their life track) prior to the time period t_0, four during t_0, and three during t_1. To construct a *cohort life table*, we direct our attention to a particular cohort (in this case, those born during t_0) and monitor what happens subsequently to the cohort. The life table is constructed by noting the number surviving to the start of each time period. Here, two of the four individuals survived to the beginning of t_1; only one of these was alive at the beginning of t_2; and none survived to the start of t_3. The first data column of the cohort life table

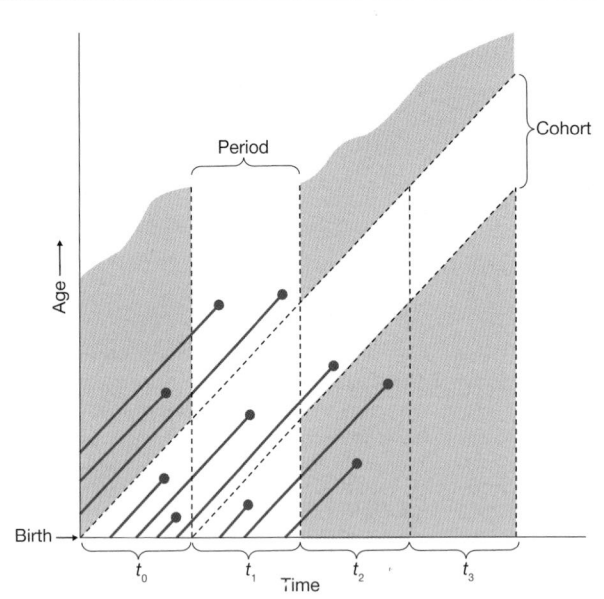

Figure 5.10

See text for details.

thus comprises the series of declining numbers in the cohort: 4, 2, 1, 0.

A different approach is necessary when we cannot follow cohorts but we know the ages of all the individuals in a population (perhaps from some clue such as the condition of the teeth in a species of deer). We can then, as the figure shows, direct our attention to the whole population during a single period (in this case, t_1) and note the numbers of survivors of different ages in the population. These may be thought of as entries in a life table *if* we assume that rates of birth and death are, and have previously been, constant – a very big assumption. What results is called a *static life table*. Here, of the seven individuals alive during t_1, three were actually born during t_1 and are hence in the youngest age group, two were born in the previous time interval, two in the interval before that, and none in the interval before that. The first data column of the static life table thus comprises the series 3, 2, 2, 0. This amounts to saying that over these time intervals, a typical cohort will have started with three and declined over successive time intervals to two, then two again, then zero.

The fecundity of individuals also changes with their age, and to understand properly what is going on in a population we need to know how much individuals of different ages contribute to births in the population as a whole: these can be described in *age-specific fecundity schedules*.

5.3.1 Cohort life tables

an annual life table for a plant

The most straightforward life table to construct is a cohort life table for annuals, because with non-overlapping generations it is indeed often possible to follow a single cohort from the first birth to the death of the last survivor. One such life table, for the annual plant *Phlox drummondii*, is shown in Table 5.1. An initial cohort of 996 seeds was followed from seed germination to the death of the last adult, with the life cycle broken down into successive periods of 14–63 days.

Table 5.1

A simplified cohort life table for the annual plant *Phlox drummondii*. The columns are explained in the text.

AGE INTERVAL (DAYS) $x–x'$	NUMBER SURVIVING TO DAY X a_x	PROPORTION OF ORIGINAL COHORT SURVIVING TO DAY X l_x	SEEDS PRODUCED IN EACH STAGE F_x	SEEDS PRODUCED PER SURVIVING INDIVIDUAL IN EACH STAGE m_x	SEEDS PRODUCED PER ORIGINAL INDIVIDUAL IN EACH STAGE $l_x m_x$
0–63	996	1.000	0.0	0.00	0.00
63–124	668	0.671	0.0	0.00	0.00
124–184	295	0.296	0.0	0.00	0.00
184–215	190	0.191	0.0	0.00	0.00
215–264	176	0.177	0.0	0.00	0.00
264–278	172	0.173	0.0	0.00	0.00
278–292	167	0.168	0.0	0.00	0.00
292–306	159	0.160	53.0	0.33	0.05
306–320	154	0.155	485.0	3.13	0.49
320–334	147	0.148	802.7	5.42	0.80
334–348	105	0.105	972.7	9.26	0.97
348–362	22	0.022	94.8	4.31	0.10
362–	0	0.000	0.0	0.00	0.00
Total			2408.2		2.41

$$R_0 = \Sigma\, l_x m_x = \frac{\Sigma F_x}{a_0} = 2.41.$$

Even when generations overlap, if individuals can be marked early in their life so that they can be recognized subsequently, it can be possible to follow the fate of each year's cohort separately. It is then possible to merge the cohorts from the different years so as to derive a cohort life table that combines information from the whole study period. An example is shown in Table 5.2: females from a population of the yellow-bellied marmot, *Marmota flaviventris*, which was live-trapped and marked individually from 1962 through to 1993 in the East River Valley of Colorado.

a cohort life table for marmots . . .

The first column in each life table is a list of the age classes (or in some cases, stages) of the organism's life: 14–63-day periods for *Phlox*, years for the marmots. The second column is then the raw data from each study, collected in the field. It reports the number of individuals surviving to the beginning of each age class (see Box 5.2).

Ecologists are typically interested not just in examining populations in isolation but in comparing the dynamics of two or more perhaps rather different populations (in the presence and absence of a pollutant, for instance). Hence, it is necessary to standardize the raw data so that comparisons can be made. This is done in the third column of the table, which is said to contain l_x values, where l_x is defined as the proportion of the original cohort surviving to the start of age class. The first value in the third column, l_0 (spoken: L zero), is therefore the proportion surviving to the beginning of this original age class. Obviously, in Tables 5.1 and 5.2, and in every life table, l_0 is 1.00 (the whole cohort is there at the start).

In the marmots, for example, there were 773 females observed in this youngest age class. The l_x values for subsequent age classes are then expressed as proportions

Table 5.2

A simplified cohort life table for female yellow-bellied marmots, *Marmota flaviventris*, in Colorado. The columns are explained in the text.

AGE CLASS (YEARS) x	NUMBER ALIVE AT THE START OF EACH AGE CLASS a_x	PROPORTION OF ORIGINAL COHORT SURVIVING TO THE START OF EACH AGE CLASS l_x	NUMBER OF FEMALE YOUNG PRODUCED BY EACH AGE CLASS F_x	NUMBER OF FEMALE YOUNG PRODUCED PER SURVIVING INDIVIDUAL IN EACH AGE CLASS m_x	NUMBER OF FEMALE YOUNG PRODUCED PER ORIGINAL INDIVIDUAL IN EACH AGE CLASS $l_x m_x$
0	773	1.000	0	0.000	0.000
1	420	0.543	0	0.000	0.000
2	208	0.269	95	0.457	0.123
3	139	0.180	102	0.734	0.132
4	106	0.137	106	1.000	0.137
5	67	0.087	75	1.122	0.098
6	44	0.057	45	1.020	0.058
7	31	0.040	34	1.093	0.044
8	22	0.029	37	1.680	0.049
9	12	0.016	16	1.336	0.021
10	7	0.009	9	1.286	0.012
11	3	0.004	0	0.000	0.000
12	2	0.003	0	0.000	0.000
13	2	0.003	0	0.000	0.000
14	2	0.003	0	0.000	0.000
15	1	0.001	0	0.000	0.000
Total			519		0.670

$$R_0 = \sum l_x m_x = \frac{\sum F_x}{a_0} = 0.67.$$

AFTER SCHWARTZ ET AL., 1998

of this number. Only 420 individuals survived to reach their second year (age class 1: between 1 and 2 years of age). Thus, in Table 5.2, the second value in the third column, l_1, is the proportion $420/773 = 0.543$ (that is, only 0.543 or 54.3% of the original cohort survived this first step). In the next row, $l_2 = 208/773 = 0.269$, and so on. For *Phlox* (Table 5.1), $l_1 = 668/996 = 0.671 = 67.1\%$ survived the first step.

In a full life table, subsequent columns would then use these same data to calculate the proportion of the original cohort that died at each stage and also the mortality rate for each stage, but for brevity these columns have been omitted here.

. . . and fecundity schedule . . .

Tables 5.1 and 5.2 also include fecundity schedules for *Phlox* and for the marmots (columns 4 and 5). Column 4 in each case shows F_x, the total number of the youngest age class produced by each subsequent age class: this youngest class being seeds for *Phlox* and, for the marmots, independent juveniles fending for themselves outside of their burrows. Thus, *Phlox* plants produced seed between around day 300 and day 350 in the year; while marmots produced young when they were between 2 and 10 years old.

The fifth column is then said to contain m_x values, fecundity: the mean number of the youngest age class produced per surviving individual of each subsequent class. For *Phlox*, it is apparent that fecundity, m_x, the mean number of seeds produced per surviving adult plant, reached a peak around day 340. For the marmots, fecundity was highest for 8-year-old females.

In the final column of a life table, the l_x and m_x columns are brought together to express the overall extent to which a population increases or decreases over time – reflecting the dependence of this on both the survival of individuals (the l_x column) and the reproduction of those survivors (the m_x column). That is, an age class contributes most to the next generation when a large proportion of individuals have survived and they are highly fecund, and it contributes least when few survive and/or they produce few (or no) offspring. The sum of all the $l_x m_x$ values, $\sum l_x m_x$, where the symbol $\sum$ means 'the sum of', is therefore a measure of the overall extent by which this population has increased or decreased in a generation. We call this the *basic reproductive rate* and denote it by R.

. . . combined to give the basic reproductive rate

For *Phlox* (Table 5.1), $R = 2.41$: this population set approximately 2.5 times more seed at the end of the generation (the end of the season) than was present at the beginning. For the marmots, $R = 0.67$: the population was declining to around two-thirds its former size each generation. However, whereas for *Phlox* the length of a generation is obvious, since, being an annual, there is one generation each year, for the marmots the generation length must itself be calculated. The details of that calculation are beyond our scope here, but its value, 4.5 years, matches what we can observe ourselves in the life table: that a 'typical' period from an individual's birth to giving birth itself (i.e. a generation) is around 4.5 years. Thus, Table 5.2 indicates that each generation, every 4.5 years, this particular marmot population was declining to around two-thirds its former size.

It is also possible to study the detailed pattern of decline in either the *Phlox* cohort or a cohort of marmots. Figure 5.11a, for example, shows the numbers surviving relative to the original population – the l_x values – plotted against the age of the cohort. However, this can be misleading. If the original population is 1000 individuals, and it decreases by half to 500 in one time interval, then this decrease looks more dramatic on a graph like Figure 5.11a than a decrease from 50 to 25 individuals later in the season. Yet the risk of death to individuals is the same on both occasions. If, however, l_x values are replaced by $\log(l_x)$ values, that is, the logarithms of the values, as in Figure 5.11b (or, effectively the same thing, if l_x values are plotted on a log scale), then it is a characteristic of logs that the reduction of a population to half its original size will always look the same. *Survivorship curves* are, therefore, conventionally plots of $\log(l_x)$ values against cohort age.

logarithmic survivorship curves

Figure 5.11b shows that there was a relatively rapid and constant decline in the size of the *Phlox* cohort over the first 6 months, but that the death rate thereafter remained steady and rather low until the very end of the season, when the survivors all died. For the marmots, Figure 5.11b shows an even more clearly constant decline until around the 10th year of life (when breeding ceased), followed by a brief period with effectively no mortality, after which the few remaining survivors died.

It is possible to see, therefore, even from these two examples, how life tables can be useful in characterizing the 'health' of a population – the extent to which it is growing or declining – and in identifying which stage in the life cycle (whether it is survival or birth) is apparently most instrumental in determining that rate of increase or decline. Either or both of these may be vital in determining how best to conserve an endangered species or control a pest.

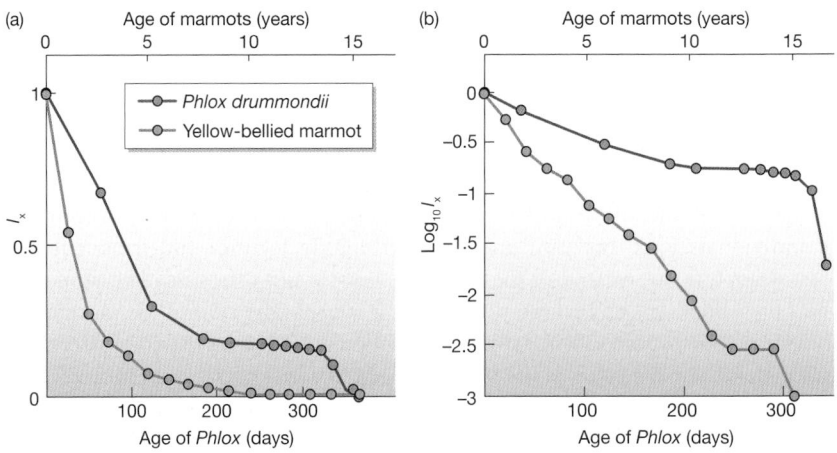

Figure 5.11

Following the survival of a cohort of *Phlox drummondii* (maroon, Table 5.1) and of the yellow-bellied marmot (yellow, Table 5.2). (a) When l_x is plotted against cohort age, it is clear that most individuals are lost relatively early in the lives of the cohorts, but there is no clear impression of the risk of mortality at different ages. (b) By contrast, a survivorship curve plotting $\log(l_x)$ against age shows, for *Phlox*, that an initial 6 months of moderate survivorship was followed by an extended period of higher survivorship (less risk of mortality) and then by very low survivorship in the final weeks of the annual cycle. For the marmots, there was virtually constant mortality risk until around age 10, followed by a brief period of low risk after which the remaining survivors died.

5.3.2 Life tables for populations with overlapping generations

Many of the species for which we have important questions, and for which life tables may provide an answer, have repeated breeding seasons like the marmots, or continuous breeding as in the case of humans, but constructing life tables here is complicated, largely because these populations have individuals of many different ages living together. Building a cohort life table is sometimes possible, as we have seen, but this is relatively uncommon. Apart from the mixing of cohorts in the population, it can be difficult simply because of the longevity of many species.

a static life table – useful if used
with caution

Another approach is to construct a 'population snapshot' in a static life table (see Box 5.2). Superficially, the data look like a cohort life table: a series of different numbers of individuals in different age classes. But great care is required: they can only be treated and interpreted in the same way if patterns of birth and survival in the population have remained much the same since the birth of the oldest individuals – and this will happen only rarely. Nonetheless, useful insights can sometimes be gained by combining the data from a static life table (an *age structure*: the numbers in different age classes) with corresponding background information. This is illustrated by a study of two populations of the long-lived tree *Acacia burkittii* in South Australia (Figure 5.12). Although differences in age structure between the populations are obvious, the reasons are not. Fortunately, background information provides important clues.

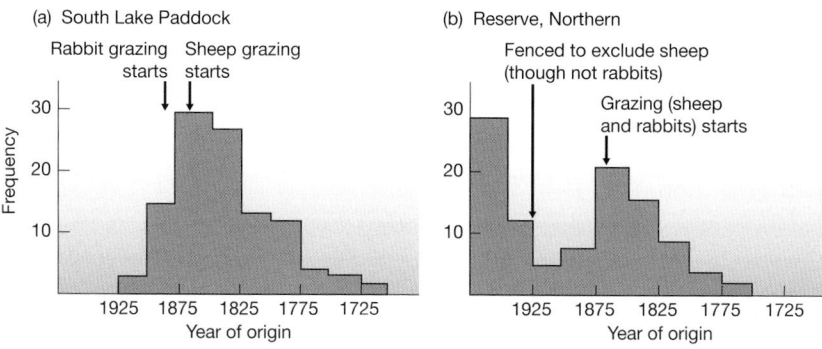

Figure 5.12

Age structures (and hence static life tables) of *Acacia burkittii* populations at two sites in South Australia. South Lake Paddock populations had been grazed by sheep from 1865 to 1970 and by rabbits from 1885 to 1970, whereas the Reserve population had been fenced in 1925 to exclude sheep (but did not exclude rabbits). With this information in hand, the effect of grazing from 1865 onward is evident in the decreased numbers of new recruits to both populations. However, the effects of fencing after 1925 are equally obvious in the Reserve population, where the proportion of new recruits increased dramatically. The effects of rabbit grazing on recruitment after fencing in the Reserve population can, however, still be detected, since, for example, the 1925–1940 age class was much smaller than the (pre-grazing) 1845–1860 class, even though the latter had survived an additional 75 years.

5.3.3 A classification of survivorship curves

Life tables provide a great deal of data on specific organisms. But ecologists search for generalities: patterns of life and death that we can see repeated in the lives of many species. Ecologists conventionally divide survivorship curves into three types, in a scheme that goes back to 1928, generalizing what we know about the way in which the risks of death are distributed through the lives of different organisms (Figure 5.13).

- In a type I survivorship curve, mortality is concentrated toward the end of the maximum lifespan. It is perhaps most typical of humans in developed countries and their carefully tended zoo animals and pets.

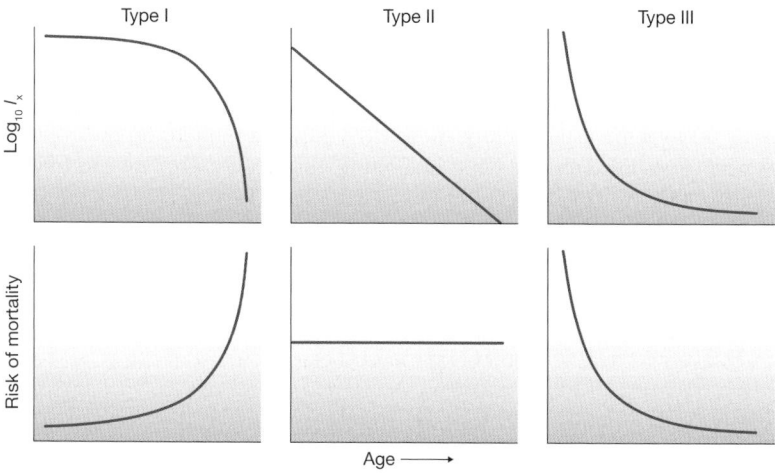

Figure 5.13

A classification of survivorship curves plotting log(l_x) against age, above, with corresponding plots of the changing risk of mortality with age, below. The three types are discussed in the text.

Figure 5. 14

Survivorship curves for the sand dune annual plant *Erophila verna* monitored at three densities: high (initially 55 or more seedlings per 0.01 m² plot), medium (15–30 seedlings per plot) and low (1–2 seedlings per plot). The horizontal scale (plant age) is standardized to take account of the fact that each curve is the average of several cohorts, which lasted different lengths of time (around 70 days on average).

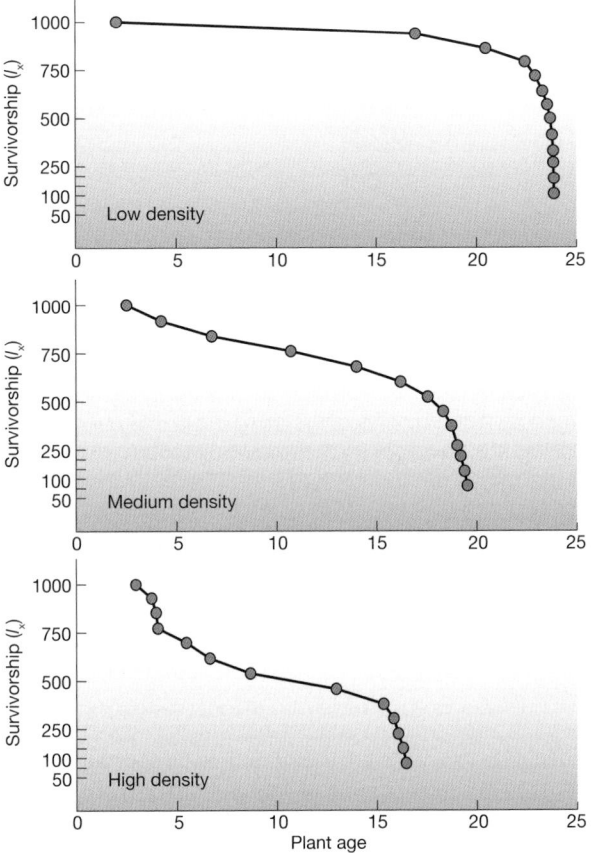

AFTER SYMONIDES, 1983

- A type II survivorship curve is a straight line signifying a constant mortality rate from birth to maximum age. It describes, for instance, the survival of buried seeds in a seed bank.
- In a type III survivorship curve there is extensive early mortality, but a high rate of subsequent survival. This is typical of species that produce many offspring. Few survive initially, but once individuals reach a critical size, their risk of death remains low and more or less constant. This appears to be the most common survivorship curve among animals and plants in nature.

These types of survivorship curve are useful generalizations, but in practice, patterns of survival are usually more complex. Thus, in a population of *Erophila verna*, a very short-lived annual plant inhabiting sand dunes, survival can follow a type I curve when the plants grow at low densities; a type II curve, at least until the last stages of life, at medium densities; and a type III curve in the early stages of life at the highest densities (Figure 5.14).

5.4 Dispersal and migration

patterns of distribution

Birth is only the beginning. If we were to stop there in our studies, many crucial ecological questions would remain unanswered. From their place of birth, all

 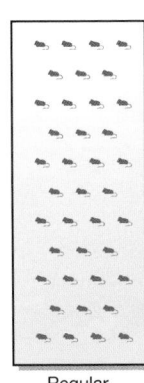 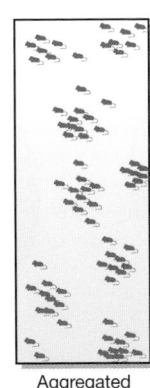

Random | Regular | Aggregated

Figure 5.15

Three generalized spatial patterns that may be exhibited by organisms across their habitat.

organisms move to locations where we eventually find them. Plants grow where their seeds fall, but seeds may be moved by the wind, water, animals or shifting soil. Animals move in search of food and safe havens, whether it is only to move 1 cm along a leaf from where their egg was deposited, or to move half-way around the globe. The effects of those movements are varied. In some cases they aggregate members of a population into clumps; in others they continually redistribute and shuffle them; and in still others they spread the individuals out. Three generalized spatial patterns that result from this movement – aggregated (clumped), random and regular (evenly) spaced – are illustrated in Figure 5.15. Clearly, movement and spatial distribution (the latter sometimes, confusingly, called 'dispersion') are intimately related.

Technically, the term *dispersal* describes the way individuals spread away from each other, such as when seeds are carried away from a parent plant or young lions leave the pride in search of their own territory. *Migration* refers to the mass directional movement of large numbers of a species from one location to another. Migration therefore describes the movement of locust swarms but also includes the smaller scale movements of intertidal organisms, back and forth twice a day, as they follow their preferred level of immersion or exposure.

Our view of dispersal and migration, and of the resulting distributions, is determined by the scale on which we are working. For example, consider the distribution of an aphid living on a particular species of tree in a woodland. On a large scale, the aphids appear to be aggregated in the woodlands and non-existent in the open fields. If the samples we took were smaller, and taken only in woodlands, the aphids would still appear to be aggregated, but now aggregated on their host trees rather than on trees in general. However, if samples were collected at an even smaller scale – the size of a leaf within a canopy – the aphids might appear to be randomly distributed over the tree as a whole. And on the scale experienced by the aphid itself (1 cm^2), the distribution might appear regular as individuals on a leaf spread out to avoid one another (Figure 5.16).

the perception of pattern depends on the spatial scale

This example also illustrates the difference between the 'average density' and the crowding experienced by individuals in a population. The *average density* is simply the total number of individuals divided by the total size of the habitat – but it depends very much on how we define the habitat. For the aphids, if it includes everything, woodland and non-woodland, then average density will be low. It will higher, but still quite low, if we include only woodland but every species of tree. It will be much higher, however, if we include only the aphids' host trees.

density and crowding

Figure 5.16

Are aphids distributed evenly, randomly or in an aggregated fashion? It all depends on the spatial scale at which they are viewed.

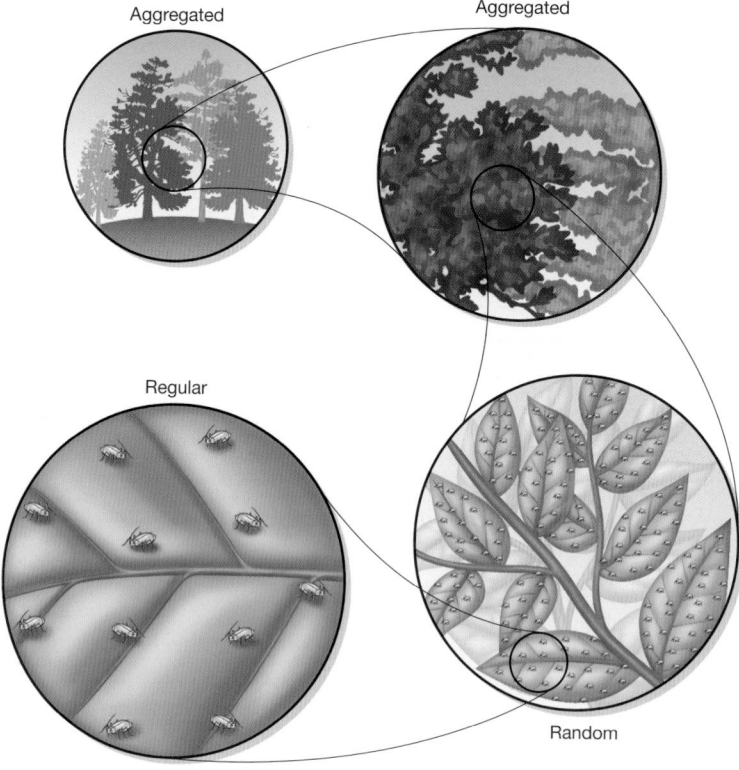

The average density of individuals in the United States is about 75 persons km^{-2}. Yet there are vast areas of the United States – rural and wilderness areas – within which the density is low, but also crowded cities and towns within which the density is much higher. And because the majority of people live in urban and suburban settings, the density actually experienced by people, on average, has been calculated at 3630 persons km^{-2}. There may be little impetus for dispersal, or migration, at the relatively low population pressure of 75 persons km^{-2}. At 3630 persons km^{-2}, however, individuals are much more likely to find ways to escape from their neighbors. Real measures of crowding as experienced by individuals are likely to be more important forces driving dispersal and migration than some average value of population density.

5.4.1 Dispersal determining abundance

dispersal: important but frequently neglected

Compared to birth and death, relatively few studies have examined the role of dispersal in determining the abundance of populations. However, studies that *have* looked carefully at dispersal have tended to bear out its importance. In a long-term and intensive investigation of a population of great tits, *Parus major*, near Oxford, UK, it was observed that 57% of breeding birds were immigrants rather than born in the population (Greenwood et al., 1978). And in a population of the Colorado potato beetle, *Leptinotarsa decemlineata*, in Canada, the average emigration rate of newly emerged adults was 97% (Harcourt, 1971). This makes the rapid spread of the beetle in Europe in the middle of the last century

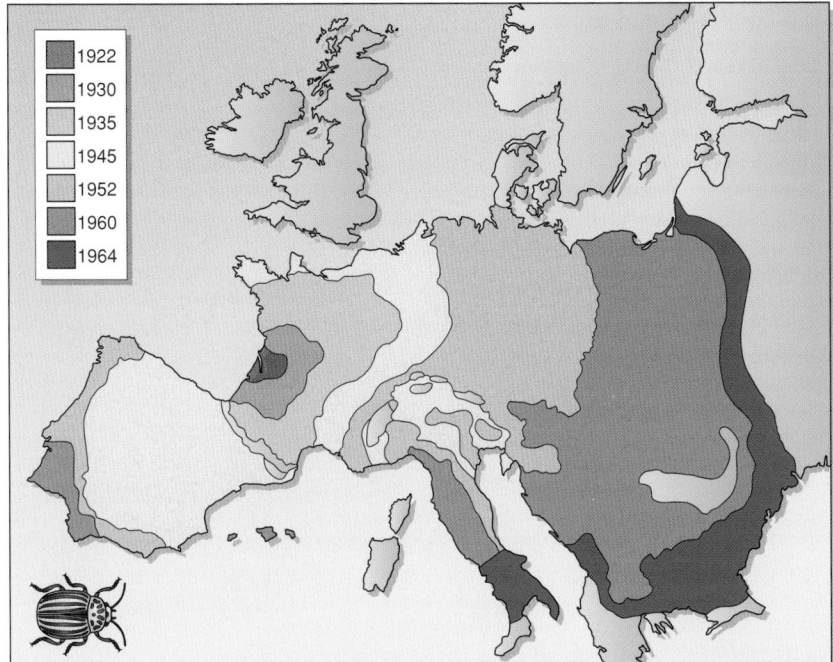

Figure 5.17

Spread of the Colorado beetle (*Leptinotarsa decemlineata*) in Europe, 1922–1964.

easy to understand (Figure 5.17). Indeed, most populations are more affected by immigration and emigration than is commonly imagined. Within the United States, for example, over 40% of US residents, over 100 million people, can trace their roots to the 12 million immigrants who entered the United States through the Ellis Island port from 1870 to 1920.

In fact, often the most important role played by dispersal in a population is to get the organisms there in the first place. For instance, the invasion of 116 patches of lowland heath vegetation in southern England by scrub and tree species was studied for the period from 1978 to 1987 (Figure 5.18). The most important factors accounting for such invasions were those describing the abundance of scrub and tree species in the vegetation bordering the heath patches. Invasions, and thus the subsequent dynamics of patches, were being driven by initiating acts of dispersal.

dispersal as invasion

One key force provoking dispersal is the more intense competition suffered by crowded individuals (see Section 3.5) and the direct interference between such individuals even in the absence of a shortage of resources. We frequently observe, therefore, that the highest rates of dispersal are away from the most crowded patches (Figure 5.19): emigration dispersal is commonly density-dependent.

density-dependent dispersal – and its converse

On the other hand, such density-dependent dispersal is by no means a general rule, and in some cases the converse pattern is observed – most dispersal at the lowest densities or *inverse* density dependence – a pattern often attributed to the avoidance of inbreeding between closely related individuals (and the lowered offspring fitness that would result), since on average, at low densities, a high proportion of those you grow up with are likely to be your close relatives. Furthermore, immigrants and emigrants not only influence the numbers in a population, they can also affect its composition. Dispersers are often the young, and males

age- and sex-biased dispersal

Figure 5.18

The invasion (i.e. increase in abundance) of most of the 116 patches of lowland heath in Dorset, UK by scrub and tree species between 1978 and 1987. The coastline is to the south and the county boundary to the east.

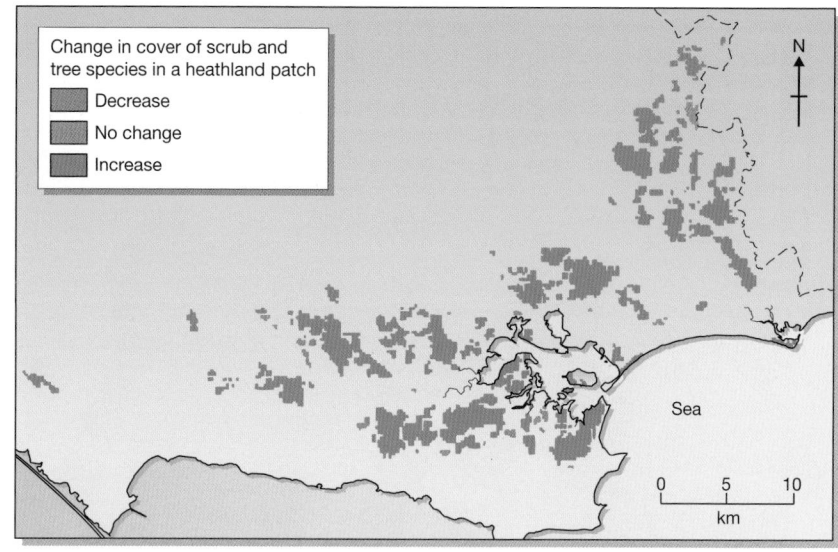

AFTER BULLOCK ET AL., 2002

Figure 5.19

Density-dependent dispersal. (a) The dispersal rates of newly hatched blackfly, *Simulium vittatum*, larvae increased with increasing density. (Data from Fonseca & Hart, 1996.) (b) The percentage of juvenile male barnacle geese, *Branta leucopsis*, dispersing from breeding colonies on islands in the Baltic Sea to non-natal breeding locations increased as density increased. (Data from van der Jeugd, 1999.)

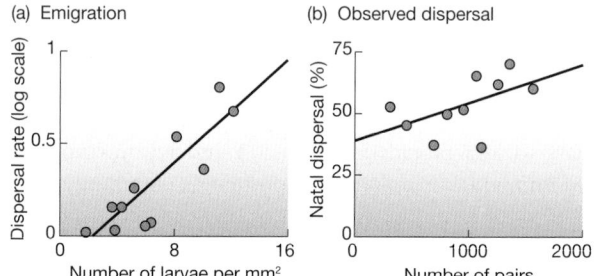

AFTER SUTHERLAND ET AL., 2002

frequently do more moving about than females. In mammal dispersal, for instance, age and sex biases, and the forces of inbreeding avoidance and competition avoidance, may all be tied intimately together. Thus, in an experiment with gray-tailed voles, *Microtus canicaudus*, 87% of juvenile males and 34% of juvenile females dispersed within 4 weeks of initial capture at low densities, but only 16% and 12%, respectively, dispersed at low densities (Wolff et al., 1997). There was massive juvenile dispersal; this was particularly pronounced in males; and the especially high rates at low densities argue in favor of inbreeding avoidance as a major force shaping the pattern.

5.4.2 The role of migration

The mass movements of populations that we call migration are (rather like density-dependent dispersal) almost always from regions where the food resource is declining to regions where it is abundant (or where it will be abundant for the progeny). By day, planktonic plants live in the upper layers of the water in lakes where the light needed for photosynthesis is brightest. At night they migrate to lower, nutrient-rich depths. Crabs migrate along the shore with the tides, following the movement of their food supply as it is washed up in the waves. At longer time scales, some shepherds still follow the ages-old practice of 'transhumance', moving their flocks of sheep and goats up to mountain pastures in summer and down again in the fall to track the seasonal changes in climate and food supply.

The long-distance migrations of terrestrial birds in many cases involve movement between areas that supply abundant food, but only for a limited time. They are areas in which seasons of comparative glut and famine alternate, and cannot support large all-year-round resident populations. For example, swallows (*Hirundo rustica*) migrate seasonally from northern Europe in the fall, when flying insects start to become rare, to South Africa when they are becoming common. In both areas the food supply that is reliable throughout the year can support only a small population of resident species. The seasonal glut supports the populations of invading migrants, which make a large contribution to the diversity of the local fauna.

5.5 The impact of intraspecific competition on populations

The concept of intraspecific competition was introduced in Section 3.5 because its intensity is typically dependent on resource availability. It re-emerges here because its effects are expressed through the focal topics of this chapter – rates of birth, death and movement. Competing individuals that fail to find the resources they need may grow more slowly or even die; survivors may reproduce later and less; or, as we have seen, if they are mobile, they may move farther apart or migrate elsewhere. Examples in which the dynamics of a species can be understood without a firm grasp of the effects of competition are rare.

The intensity of competition for limiting resources is often related to the density of a population, though, as we have seen, the straightforward density need not be a good measure of the extent to which its individuals are crowded. Modular sessile organisms are particularly sensitive to competition from their immediate neighbors: they cannot withdraw from each other and space themselves more evenly or escape by dispersal or migration. Thus, when silver birch trees (*Betula pendula*) were grown in small groups, there were more suppressed and dying branches on the sides of individual trees where their branches shaded each other than on the sides away from neighbors, where there was more vigorous growth (Figure 5.20).

crowding not density – especially in modular organisms

We saw in Section 3.5 that, over a sufficiently large density range, as density increases, competition between individuals generally reduces the per capita birth rate and increases the death rate, and that this effect is described as *density-dependent*. Thus, when birth and death rate curves are plotted against density on the same graph, and either or both are density-dependent, the curves must cross (Figure 5.21a–c). They do so at the density at which birth and death rates are equal, and because they are equal, there is no overall tendency at this density for the population either to increase or to decrease (ignoring, for convenience, both emigration and immigration). The density at the crossover point is called the *carrying capacity* and is denoted by the symbol K. At densities below K, births exceed deaths and the population increases. At densities above K, deaths exceed births and the population decreases. There is therefore an overall tendency for the density of a population under the influence of intraspecific competition to settle at K.

density-dependent birth and death and the carrying capacity

Figure 5.20

Mean relative bud production (new buds per existing bud) for silver birch trees (*Betula pendula*), expressed (a) as gross bud production and (b) as net bud production (birth minus death), in different interference zones (i.e. where they interfered to differing extents with their neighbors). (c) Plan of three trees, explaining these zones. ●, high interference; ●, medium; ●, low. Bars are standard errors.

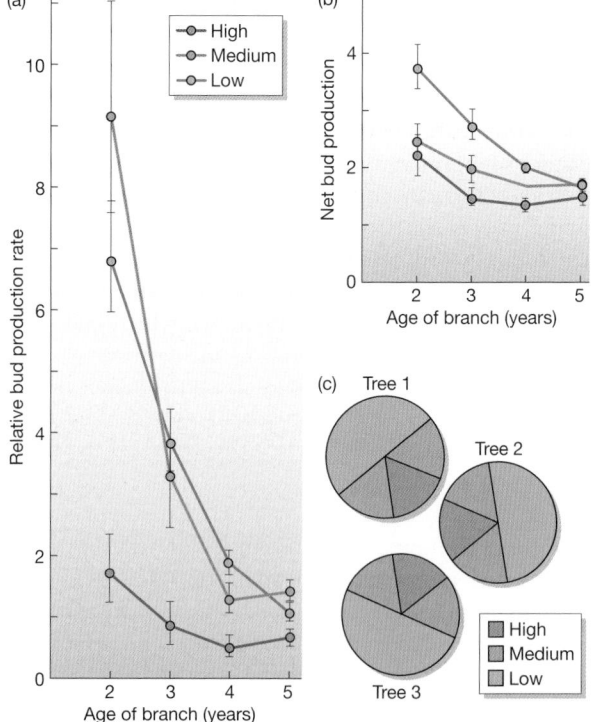

AFTER JONES & HARPER, 1987

Figure 5.21

Density-dependent birth and mortality rates lead to the regulation of population size. When both are density-dependent (a), or when either of them is (b, c), their two curves cross. The density at which they do so is called the *carrying capacity* (*K*). However, the real situation is closer to that shown by the thick lines in (d), where mortality rate broadly increases, and birth rate broadly decreases, with density. It is possible, therefore, for the two rates to balance not at just one density, but over a broad range of densities, and it is toward this broad range ('*K*') that other densities tend to move.

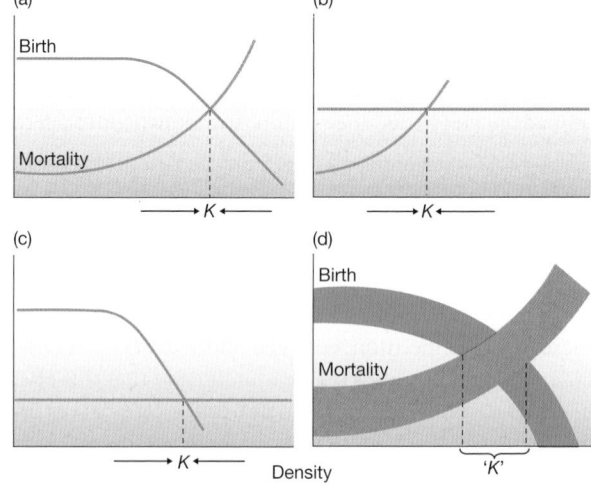

population regulation by competition – but not to a single carrying capacity

In fact, because of the natural variability within populations, the birth rate and death rate curves are best represented by broad lines, and *K* is best thought of not as a single density, but as a range of densities (Figure 5.21d). Thus, intraspecific competition does not hold natural populations to a single, predictable and unchanging level (*K*), but it may act upon a very wide range of starting densities and bring them to a much narrower range of final densities. It therefore tends to keep density within certain limits, and may thus be said to play a part in *regulating* the size of populations.

Of course, graphs like those in Figure 5.21 are generalizations on a grand scale. Many organisms, for example, have seasonal life cycles. For part of the year births vastly outnumber deaths, but later, after the period of peak births, there is likely to be a period of high juvenile mortality. Most plants, for example, die as seedlings soon after germination. Thus, although births may balance deaths over the year, a population that is 'stable' from year to year will often change dramatically over the seasons.

5.5.1 Patterns of population growth

When populations are sparse and uncrowded they may grow rapidly (and this can cause real problems – even with species that were previously endangered: Box 5.3). It is only as crowding increases that density-dependent changes in birth and death rates start to take effect. In essence, populations at these low

5.3 *Topical ECOncerns*

Sea otter populations on the rise

It is estimated that as many as 300,000 sea otters once populated the North Pacific, from Russia to Mexico. But hunting caused the population to plummet to a few thousand by 1911. Since then, numbers have shot back up to more than 100,000, although the animals have not returned everywhere. The following newspaper article by Craig Welch concerns the situation along the Washington coastline in the northwest United States. It appeared in the *Philadelphia Inquirer* on March 4, 2001.

© ALAMY IMAGES ACRN42

Sea otters colliding with fishing industry
Sea otters are rebounding in dramatic fashion along Washington's coast, and that is forcing marine biologists and wildlife managers to prepare for a potentially uncomfortable collision between the charismatic critters and some coastal fisheries.

'It's a classic recipe for political polarization', said Glenn VanBlaricom, an associate professor of marine ecology at the University of Washington. 'People love sea otters, but they could run right into shellfish harvesters whose livelihoods depend on their food sources.' Wiped out of Washington waters in the 19th century by pelt-hungry hunters, otters have staged a comeback since being reintroduced to the western shores of the Peninsula in the late 1960s.

The population has grown 30-fold in as many years, and their range is expanding so far and fast that some scientists suspect groups of otters may someday – for the first time – make Puget Sound home.

. . . While sea otters remain protected under Washington state law as an endangered species, their numbers are increasing by 10 percent a year. The population now hovers at 600 animals, roughly a quarter of what marine experts think the environment can sustain.

But such a healthy return comes with complications. Because they lack blubber, otters eat a quarter of their weight each day to fuel their supercharged metabolisms. Their munchies of choice include the seafood humans crave – sea urchins, Dungeness crabs, clams, abalone. And their recent travels toward rich harvest areas such as the Dungeness Spit put them on a direct route toward multimillion-dollar commercial, recreational and tribal shellfisheries.

Steven Jeffries, who heads marine-mammal investigations for the State Department of Fish and Wildlife, said it was tough to determine whether it would be a few years or a few decades before conflicts begin.

Consider the following options and debate their relative merits:

1   *Shellfisheries are of considerable importance to commercial, recreational and tribal fishers. How would you weigh up the competing demands of conservation and fishing? Should the sea otters remain absolutely protected or is there a case for culling or some other form of control of their spread?*

2   *The story in Washington is very different from that in parts of Alaska, where otter numbers are declining, or Los Angeles, where recent efforts have been made to reintroduce the species. Suggest some plausible reasons for the different population trajectories in different areas.*

densities grow by simple multiplication over successive intervals of time. This is *exponential* growth (Figure 5.22) and the rate of increase is the population's *intrinsic rate of natural increase* (denoted by r; Box 5.4). Of course, any population that behaved in this way would soon run out of resources, but as we have seen, the rate of increase tends to become reduced by competition as the population grows, and it falls to zero when the population reaches its carrying capacity (since birth rate then equals death rate). A steady reduction in the rate of increase as densities move toward the carrying capacity gives rise to population growth that is not exponential but S-shaped (Figure 5.22). The pattern is also often called *logistic* growth after the so-called logistic equation (Box 5.4).

Figure 5.22

Exponential (maroon line) and S-shaped or *sigmoidal* (blue line) increases in the size of a population (N) over time. These patterns describe the growth to be expected in general in populations in the absence (exponential) and under the influence (sigmoidal) of intraspecific competition, but are also generated, specifically, by the exponential and logistic equations shown (see also Box 5.4).

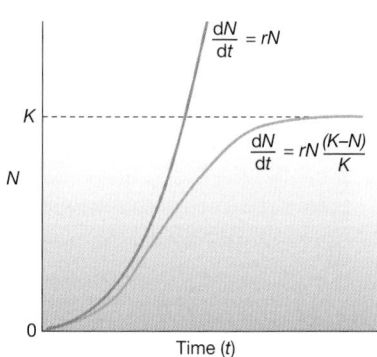

$$\frac{dN}{dt} = rN$$

$$\frac{dN}{dt} = rN\frac{(K-N)}{K}$$

5.4 Quantitative aspects

The exponential and logistic equations of population growth

In this box, simple mathematical models are derived for populations first in the absence of, and then under the influence of, intraspecific competition. These and other mathematical models play an important part in ecology (see Chapter 1). They help us to follow through the consequences of assumptions we may wish to make, and to explore the behavior of ecological systems that we may find it hard to observe in nature or construct in the laboratory. The particular models in this box themselves form the basis for more complex models of *interspecific* competition and predation: they are important building blocks. It is essential to appreciate, however, that a pattern generated by such a model – for example, the S-shaped pattern of population growth under the influence of intraspecific competition – is not of interest, or important, because it is generated by the model. There are many other models that could generate very similar (indistinguishable) patterns. Rather, the point about the pattern is that it reflects important, underlying ecological processes – and the model is useful in that it appears to capture the essence of those processes.

We start with a model of a population in which there is no intraspecific competition and then incorporate that competition later. Our models are in the form of *differential equations*, describing the net rate of increase of a population, which will be denoted by dN/dt (spoken: DN by DT). This represents the speed at which a population increases in size, N, as time, t, progresses.

The increase in size of the whole population is the sum of the contributions of the various individuals within it. Thus, the average rate of increase per individual, or the per capita rate of increase (per capita means 'per head') is given by $dN/dt \bullet (1/N)$. In the absence of intraspecific competition (or any other force that increases the death rate or reduces the birth rate) this rate of increase is a constant and as high as it can be for the species concerned. It is called the *intrinsic rate of natural increase* and is denoted by r. Thus:

$$dN/dt(1/N) = r$$

and the net rate of increase for the whole population is therefore given by:

$$dN/dt = rN$$

This equation describes a population growing *exponentially* (Figure 5.22).

Intraspecific competition can now be added. This we do by deriving the *logistic equation*, using the method set out in Figure 5.23. The net rate of increase per individual is unaffected by competition when N is very close to zero, because there is no crowding, nor a shortage of resources. It is still therefore given by r (point A). When N rises to K (the carrying capacity) the net rate of increase per individual is, by definition, zero (point B). For simplicity, we assume a straight line between A and B; that is, we assume a linear reduction in the per capita rate of increase, as a result of intensifying intraspecific competition, between $N = 0$ and $N = K$.

Thus, on the basis that the equation for any straight line takes the form $y = \text{intercept} + \text{slope } x$, where x and

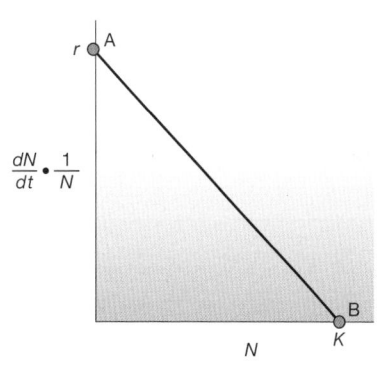

Figure 5.23

An ideal linear decline in the net rate of increase per individual with increasing population (N).

y are the variates on the horizontal and vertical axes, here we have:

$$dN/dt(1/N) = r - (r/K)N$$

or, rearranging,

$$dN/dt = rN[1 - (N/K)]$$

This is the logistic equation, and a population increasing in size under its influence is shown in Figure 5.22. It describes a sigmoidal or S-shaped growth curve approaching a stable carrying capacity, but it is only one of many reasonable equations that do this. Its major advantage is its simplicity. Nevertheless, it has played a central role in the development of ecology.

The S-shaped curve can best be seen in action in laboratory studies of micro-organisms or animals with very short life cycles (Figure 5.24a). In these kinds of experiment it is easy to have experimental control of environmental conditions and resources. In the real world, outside the laboratory and the mind of the mathematician, the world is less simple. The complex life cycles of organisms, changing conditions and resources through the seasons, and the patchiness of habitats introduce many complications. In nature, populations often follow a very bumpy ride along the path of perfect logistic growth (Figure 5.24b), though not always (Figure 5.24c).

Another way to summarize the ways in which intraspecific competition affects populations is to look at *net recruitment* – the number of births minus the number of deaths in a population over a period of time. When densities are low, net recruitment will be low because there are few individuals available either to give birth or to die. Net recruitment will also be low at much higher densities as the carrying capacity is approached. Net recruitment will be at its peak, then, at some intermediate density. The result is a 'humped' or dome-shaped curve (Figure 5.25). Again, of course, as with the ideal logistic curve, real data from

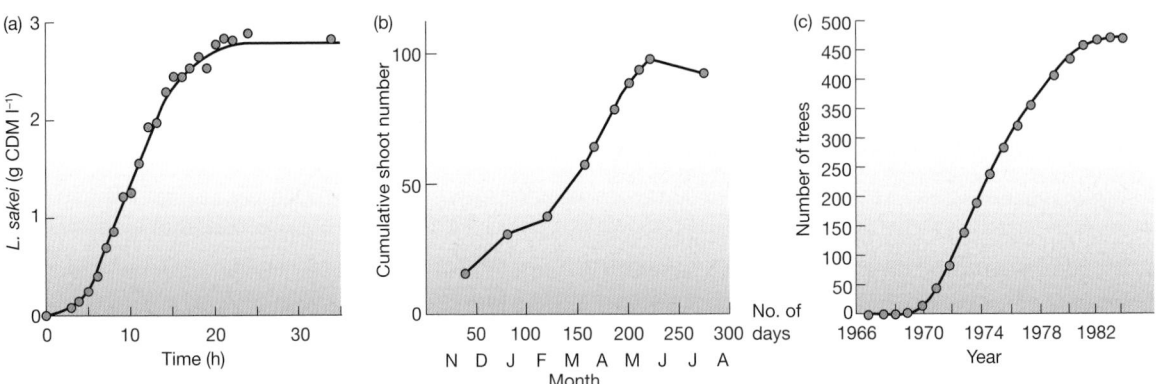

Figure 5.24

Real examples of S-shaped population increase. (a) The bacterium *Lactobacillus sakei* [measured as grams of 'cell dry mass' (CDM) per liter] grown in nutrient broth. (b) The population of shoots (i.e. modules – see Section 5.1.1) of the annual plant *Juncus gerardi* in a salt marsh habitat on the west coast of France. (c) The population of the willow tree (*Salix cinerea*) in an area of land after myxomatosis had effectively prevented rabbit grazing.

(a) AFTER LEROY & DE VUYST, 2001; (b) AFTER BOUZILLE ET AL., 1997; (c) AFTER ALLIENDE & HARPER, 1989

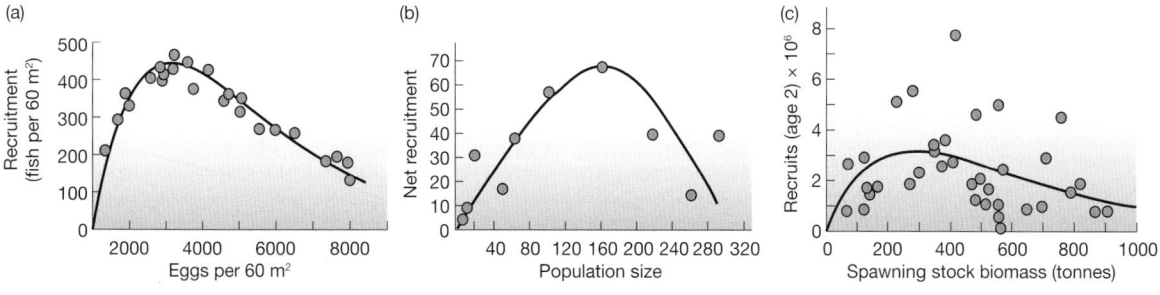

Figure 5.25

Some dome-shaped net recruitment curves. (a) Six-month-old brown trout, *Salmo trutta*, in Black Brows Beck, England between 1967 and 1989. (b) An experimental population of the fruitfly *Drosophila melanogaster*. (c) 'Blackwater' herring, *Clupea harengus*, from the Thames estuary, England between 1962 and 1997.

(a) AFTER MYERS, 2001; FOLLOWING ELLIOTT, 1994; (b) AFTER PEARL, 1927; (c) AFTER FOX, 2001

nature never fall on a single line. But the dome-shaped curve reflects the essence of net recruitment patterns when density-dependent birth and death are the result of intraspecific competition.

5.6 Life history patterns

One of the ways in which we can try to make sense of the world around us is to search for repeated patterns. In doing so, we are not pretending that the world is simple or that all categories are watertight, but we can hope to move beyond a description that is no more than a series of unique special cases. This final section of this chapter describes some simple, useful, though by no means perfect, patterns linking different types of life history and different types of habitat.

First, though, we return to a point made earlier: that in any life history there is a limited total amount of energy (or some other resource) available to an organism for growth and reproduction. Some trade-off may therefore be necessary: either grow more and reproduce less, or reproduce more and grow less. Specifically, there may be an observable cost of reproduction in that when reproduction starts, or increases, growth may slow or stop completely, as resources are diverted. We can, of course, look at this trade-off the other way around: an organism that makes vigorous growth, and so thrives in competition with its neighbors, may have to pay the price by reducing reproductive activity. In many forest trees, for example, growth rings in the trunk are conspicuously narrower in 'mast' years, when very heavy crops of seeds are produced (Figure 5.26a). Furthermore, as shown in Figure 5.26b, the diversion of resources to present reproduction may jeopardize subsequent survival (as also seen in the salmon and foxgloves described earlier), or simply reduce the capacity for future reproduction.

Yet it would be quite wrong to think that such negative, trade-off correlations abound in nature, only waiting to be observed. In particular, if there is variation between individuals in the amount of resource they have at their disposal, then there is likely to be a positive, not a negative, correlation between two apparently alternative processes – some individuals will be good at everything, others

the 'cost' of reproduction – a life history trade-off

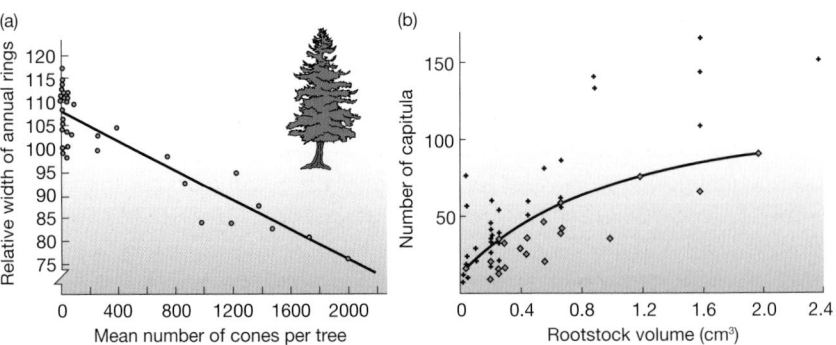

Figure 5.26

(a) The negative correlation between cone crop size and annual growth increment for a population of Douglas fir *Pseudotsuga menziesii*. There is a cost of reproduction: the more the trees reproduce, the less they grow. (b) The cost of reproduction in ragwort plants (*Senecio jacobaea*). The line divides plants that survive (◆) from plants that have died by the end of the season (+). There are no surviving plants above and to the left of the line. For a given size (measured as 'rootstock volume') only those that have made the smallest reproductive allocation (measured as 'number of capitula') survive, although larger plants are able to make a larger allocation and still survive.

(A) AFTER EIS ET AL., 1965; (B) AFTER GILMAN & CRAWLEY, 1990

consistently awful. For instance, in Figure 5.27, the snakes in the best condition produced larger litters but also recovered from breeding more rapidly, ready to breed again.

But early reproduction can yield some striking rewards, particularly because the progeny themselves start reproduction earlier. Populations of individuals that reproduce early in their life can grow extremely fast – even if this means producing many fewer total offspring over their life than they would otherwise. The effect is shown by considering the life cycle of fruitflies (*Drosophila*). The number of eggs produced by a female in her lifetime is about 780. Doubling that number would clearly boost the intrinsic rate of increase, but such a massive increase in reproductive output is asking a great deal of an individual. So, what other changes in the life history of *Drosophila* would have a similar effect? In fact, the same rise in the rate of increase would be attained simply by shortening the juvenile period from around 10 to around 8.5 days (reproducing sooner,

Figure 5.27

Female aspic vipers (*Vipera aspis*) that produced larger litters ('relative' litter mass because total female mass was taken into account) also recovered more rapidly from reproduction (not 'relative' because mass recovery was not affected by size) ($r = 0.43$; $P = 0.01$).

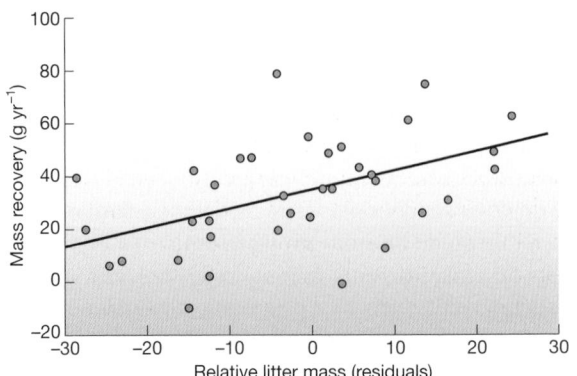

rather than growing longer). Conversely, the rate of growth of populations can be slowed by delaying the onset of reproduction. One very effective way in which the growth rate of human populations can be slowed down, for example (see Chapter 12), is by discouraging early marriage and childbearing.

We can now turn to the life history patterns themselves. The potential of a species to multiply rapidly is advantageous in environments that are short-lived, allowing the organisms to colonize new habitats quickly and exploit new resources. This rapid multiplication is a characteristic of the life cycles of terrestrial organisms that invade disturbed land (for example, many annual weeds), or colonize newly opened habitats such as forest clearings, and of the aquatic inhabitants of temporary puddles and ponds. These are species whose populations are usually found expanding after the last disaster or exploiting the new opportunity. They have the life cycle properties that are favored by natural selection in such conditions: the production of large numbers of progeny, early in the life cycle, rather than investing heavily in either growth or survival. They have been called *r* species, because they spend most of their life in the near-exponential, *r*-dominated phase of population growth (see Box 5.4), and the habitats in which they are likely to be favored have been called *r*-selecting.

r and *K* species

Organisms with quite different life histories survive in habitats where there is often intense competition for limited resources. The individuals that are successful in leaving descendants are those that have captured, and often held on to, the larger share of resources. Their populations are usually crowded and those that win in a struggle for existence do so because they have grown faster and/or larger (rather than reproducing) or have spent more of their resources in aggression or some other activity that has favored their survival under crowded conditions. They are called *K* species because their populations spend most of their lives in the *K*-dominated phase of population growth (see Box 5.4) – 'bumping up' against the limits of environmental resources – and the habitats in which they are likely to be favored have been called *K*-selecting.

A further common distinction between *r* and *K* species is whether they produce many small progeny (characteristic of *r* species) or few large progeny (characteristic of *K* species). This is another example of a life history trade-off: an organism has limited resources available for reproduction, and natural selection will influence how these are packaged. In environments where rapid population growth is possible, those individuals that produce large numbers of small progeny will be favored. The size of progeny can be sacrificed because they will usually not be in competition with others. However, in environments in which the individuals are crowded and there is competition for resources, those progeny that are well provided with resources by the parent will be favored. Producing progeny that are well endowed requires the trade-off of producing fewer of them (see, for example, Figure 5.28).

r, K and progeny size and number

The *r/K* concept can certainly be useful in describing some of the general differences among different organisms. For instance, among plants it is possible to describe a number of very broad and general relationships (Figure 5.29). Trees in a forest are splendid examples of *K* species. They compete for light in the canopy, and survivors are those that put their resources into early growth and overtopping their neighbors. They usually delay reproduction until their branches have an assured place in the canopy of leaves. Once established they hold on to their position and usually have a very long life, with a relatively low allocation

evidence for the *r/K* scheme?

Figure 5.28

Evidence for a trade-off between the number of offspring produced in a clutch by a parent and the individual fitness of those offspring: a negative correlation between the size of offspring (as measured by their snout–vent length, SVL) and the number of them in a litter in the Australian highland copperhead snake, *Austrelaps ramsayi* ($r^2 = 0.63$, $P = 0.006$). 'Residual' offspring and litter sizes have been used: these are the values arrived at after variations in maternal size have been allowed for, since both increase with maternal size.

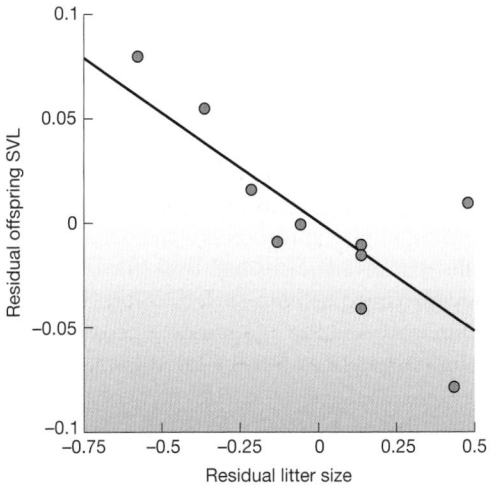

AFTER ROHR, 2001

Figure 5.29

Broadly speaking, plants show some conformity with the *r/K* scheme. For example, trees in relatively *K*-selecting woodland habitats: (a) have a relatively high probability of being iteroparous and a relatively small reproductive allocation; (b) have relatively large seeds; and (c) are relatively long-lived with relatively delayed reproduction.

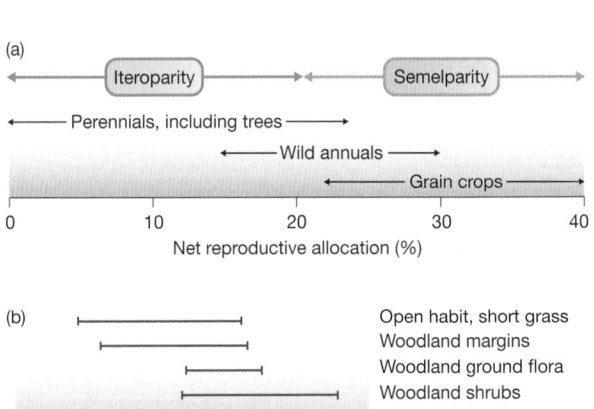

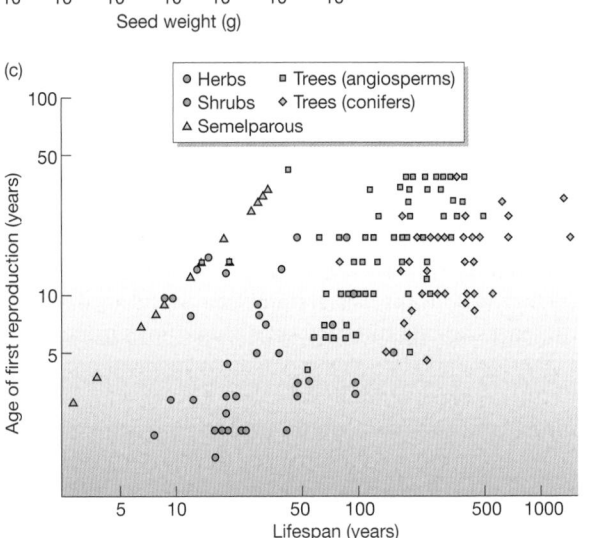

AFTER HARPER, 1977; FROM SALISBURY, 1942; OGDEN, 1968; HARPER & WHITE, 1974

to reproduction overall but large individual seeds. By contrast, in more disturbed, open, *r*-selecting habitats, the plants tend to conform to the general syndrome of *r* characteristics: a greater reproductive allocation, but smaller seeds, smaller size, earlier reproduction and a shorter life (Figure 5.29).

On the other hand, there seem to be about as many examples that fail to fit the *r/K* scheme as examples that correspond. One might regard this as a damning criticism of the *r/K* concept, since it undoubtedly demonstrates that the explanatory powers of the scheme are limited. But it is equally possible to regard it as very satisfactory that a relatively simple concept can help make sense of a large proportion of the multiplicity of life histories. Nobody, though, can regard the *r/K* scheme as the whole story. Like all attempts to classify species and their characteristics into pigeonholes, the distinction between *r* and *K* species has to be recognized as a convenient (and useful) human creation rather than an all-encompassing statement about the living world.

Summary

Counting individuals, births and deaths
Ecologists try to describe and understand the distribution and abundance of organisms. The processes that change the size of populations are birth, death and movement. A population is a number of individuals, but for some kinds of organism, especially modular organisms, it is not always clear what we mean by an individual.

Ecologists face enormous problems when they try to count what is happening to populations in nature. They almost always estimate rather than count. There are particular problems in counting modular organisms and the numbers of births and deaths.

Life cycles and reproduction
The life histories of all unitary organisms can be seen as variations around a simple, sequential pattern. Some organisms fit several or many generations within a single year, some breed predictably just once each year (annuals), and others (perennials) have a life cycle extended over several or many years. Some, iteroparous species, breed repeatedly; others, semelparous species, have a single reproductive episode followed quickly by death.

Most annuals germinate or hatch in spring, grow rapidly, reproduce and then die before the end of summer. Most spend part of the year dormant. There is a marked seasonal rhythm in the lives of many long-lived species. Where there is very little seasonal variation, some reproduce throughout the year; others have a long non-reproductive phase and then one lethal burst of reproductive activity.

Monitoring birth and death: life tables and fecundity schedules
Life tables can be useful in identifying what in a life cycle is apparently most instrumental in determining rates of increase or decline. A cohort life table records the survivorship of members of a single cohort. When we cannot follow cohorts, it may be possible to construct a static life table, but great care is required. The fecundity of individuals also changes with age, described in age-specific fecundity schedules.

Ecologists search for patterns of life and death that we can see repeated in the lives of many species. A useful set of survivorship curves (types I–III) has been developed, but in practice patterns of survival are usually more complex.

Dispersal and migration

Dispersal is the way individuals spread away from each other. Migration is the mass directional movement of large numbers of a species from one location to another. Movement and spatial distribution are intimately related. Dispersal and migration can have a profound effect on the dynamics of a population and on its composition.

The impact of intraspecific competition on populations

Over a sufficiently large density range, competition between individuals generally reduces the birth rate as density increases and increases the death rate (i.e. is density-dependent). Intraspecific competition therefore tends to keep density within certain limits and may thus be said to play a part in regulating the size of populations.

When populations are sparse and uncrowded they tend to exhibit exponential growth, but the rate of increase tends to become reduced by competition as the population grows, giving rise to population growth that is not exponential but S-shaped or logistic.

Intraspecific competition also affects net recruitment, typically resulting in a humped curve.

Life history patterns

There is typically a limited total amount of energy or some other resource available to an organism for growth and reproduction. There may be an observable cost of reproduction. But populations of individuals that reproduce early in their life can grow extremely fast.

The potential of a species to multiply rapidly is favored by natural selection in environments that are short-lived, allowing the organisms to colonize new habitats quickly and exploit new resources. Such species have been called *r* species. Where there is often intense competition for limited resources, the individuals that are successful in leaving descendants are those that have captured the larger share of resources, often because they were born larger and/or have grown faster (rather than reproducing): so-called *K* species. The *r/K* concept can be useful in interpreting many of the differences in form and behavior of organisms, but of course it is not the whole story.

Review questions

REVIEW QUESTIONS

Asterisks indicate challenge questions

1 Contrast the meaning of the word 'individual' for unitary and modular organisms.

2 In a mark–recapture exercise during which a population of butterflies remained constant in size, an initial sample provided 70 individuals, each of which was marked and then released back into the population. Two days later, a second sample was taken, totaling 123 individuals of which 47 bore a mark from the first sample. Estimate the size of the population. State any assumptions that you have had to make in arriving at your estimate.

3* Define annual, perennial, semelparous and iteroparous. Try to give an example of both an animal and a plant for each of the four possible combinations of these terms. In which cases is it difficult (or impossible) to come up with an example and why?

4 Contrast the derivation of cohort and static life tables and discuss the problems of constructing and/or interpreting each.

5 The following is an outline life table and fecundity schedule for a cohort of a population of sparrows. Fill in the missing values (wherever there is a question mark).

STAGE (X)	NUMBERS AT START OF STAGE (A_x)	PROPORTION OF ORIGINAL COHORT ALIVE AT START OF STAGE (l_x)	MEAN NO. OF EGGS PRODUCED PER INDIVIDUAL IN STAGE (m_x)
Eggs	173	?	0
Nestlings	107	?	0
Fledglings	64	?	0
1-year-olds	31	?	2.5
2-year-olds	23	?	3.7
3-year-olds	8	?	3.1
4-year-olds	2	?	3.5

$R = ?$

6 Describe what are meant by aggregated, random and regular distributions of organisms in space, and outline, with actual examples where possible, some of the behavioral processes that might lead to each type of distribution.

7* Why is the average density of people in the United States lower than the density experienced by people, on average, in the United States? Is a similar contrast likely to apply to most species? Why? Under what conditions might it not apply?

8* Compare unitary and modular organisms in terms of the effects of intraspecific competition both on individuals and on populations.

9 What is meant by the carrying capacity of a population? Describe where it appears, and why, in: (i) S-shaped population growth; (ii) the logistic equation; and (iii) dome-shaped net recruitment curves.

10 Explain why an understanding of life history trade-offs is central to an understanding of life history evolution. Explain the contrasting trade-offs expected to be exhibited by r-selected and K-selected species.

Chapter **6**

Interspecific competition

Chapter contents

Key concepts

In this chapter you will:

- appreciate the difficulty of distinguishing between the power and importance of interspecific competition in principle and in practice
- distinguish between fundamental and realized niches
- define the Competitive Exclusion Principle and understand its limitations
- appreciate the potential role of the evolutionary effects of competition in species coexistence and the difficulty of proving that role
- understand the nature and importance of niche complementarity
- appreciate the difficulties of determining the prevalence of current competition in nature, and of distinguishing between the effects of competition and mere chance

Interspecific competition is one of the most fundamental phenomena in ecology, affecting not only the current distribution and success of species but also their evolution. Yet the existence and effects of interspecific competition are often remarkably difficult to establish and demand an armory of observational, experimental and modeling techniques.

6.1 Introduction

Having been introduced to *intra*specific competition in previous chapters, it is not difficult to deduce what *inter*specific competition is. Its essence is that individuals of one species suffer a reduction in fecundity, survivorship or growth as a result of exploitation of resources or interference by individuals from another species. These competitive effects on individuals are likely to affect the population dynamics of the competing species. These, in turn, can influence the species' distributions and also their evolution. The distributions and abundances of species, of course, determine the compositions of the communities of which they are part. And evolution, in *its* turn, can influence the species' distributions and dynamics.

This chapter, then, is about both the ecological and the evolutionary effects of interspecific competition on individuals, on populations and on communities. But it also addresses a more general issue in ecology and indeed in science – that there is a difference between what a process can do and what it *does* do: a difference between what, in this case, interspecific competition is capable of doing and what it actually does in practice. These are two separate questions, and we must be careful to keep them separate.

two separate questions – the possible and actual consequences of competition

The way these different questions can be asked and answered will be different, too. To find out what interspecific competition is capable of doing is relatively easy. Species can be forced to compete in experiments, or they can be examined in nature in pairs or groups chosen precisely because they seem most likely to compete. But it is much more difficult to discover how important interspecific competition actually is. It will be necessary to ask how realistic our experiments were, how typical they were of the way species interact in nature, and how typical of pairs and groups of species generally were those singled out for special attention.

We begin, though, with some examples of what interspecific competition can do.

6.2 Ecological effects of interspecific competition

6.2.1 Competition between diatoms for silicate

Competition was investigated in the laboratory between two species of freshwater diatoms (single-celled plants), *Asterionella formosa* and *Synedra ulna*, both of which require silicate in the construction of their cell walls (see Section 3.5).

Figure 6.1

Competition between diatoms.
(a) *Asterionella formosa*, when grown alone in a culture flask, establishes a stable population and maintains a resource, silicate, at a constant low level. (b) When *Synedra ulna* is grown alone it does the same, but maintains silicate at an even lower level. (c) When grown together, in two replicates, *Synedra* drives *Asterionella* to extinction.

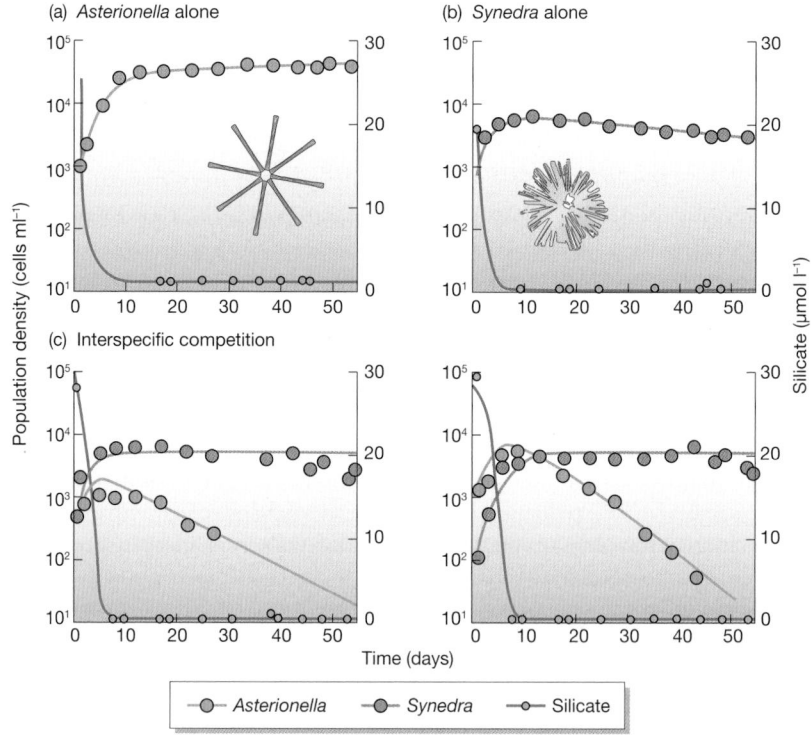

AFTER TILMAN ET AL., 1981

The population densities of the diatoms were monitored, but at the same time their impact on their limiting resource (silicate) was also being recorded. When either species was grown alone in a liquid medium to which resources were continuously being added, it established a steady population density while reducing the silicate to a constant low concentration (Figure 6.1a, b). However, in exploiting this resource, *Synedra* reduced the silicate concentration to a lower level than did *Asterionella*. Hence, when the two species were grown together, *Synedra* maintained the concentration at a level that was too low for the survival and reproduction of *Asterionella* and only *Synedra* survived (Figure 6.1c).

Thus, although both species were capable of living alone in the laboratory habitat, when they competed, *Synedra* excluded *Asterionella* because it was the more effective exploiter of their shared, limiting resource. A similar result has been obtained for the nocturnal, insectivorous gecko *Hemidactylus frenatus*, an invader of urban habitats across the Pacific basin, where it is responsible for population declines of the native gecko *Lepidodactylus lugubris* (Petren and Case, 1996). The diets of the two geckos overlap substantially and insects are a limiting resource for both. The invader is capable of depleting insect resources in experimental enclosures to lower levels than the native gecko, and the latter suffers reductions in body condition, fecundity and survivorship as a result.

more efficient exploiters exclude less efficient ones

6.2.2 Coexistence and exclusion of competing salmonid fishes

Salvelinus malma (Dolly Varden charr) and *S. leucomaenis* (white-spotted charr) are morphologically similar and closely related species of salmonid fish (see

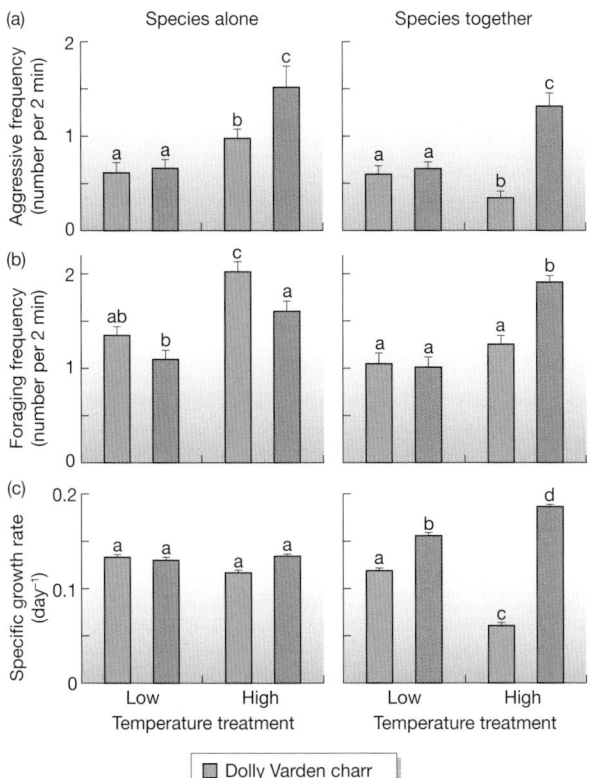

AFTER TANIGUCHI & NAKANO, 2000

Figure 6.2

(a) Frequency of aggressive encounters initiated by individuals of each fish species during a 72-day experiment in artificial stream channels with two replicates each of 50 Dolly Varden (blue histograms) or 50 white-spotted charr (maroon histograms) alone (allopatry) or 25 of each species together (sympatry); (b) foraging frequency; and (c) specific growth rate in length. Different letters indicate means are significantly different from each other.

Section 3.2.4). They are found together in many streams on Hokkaido Island in Japan, but Dolly Varden are distributed further upstream than white-spotted charr, with a zone of overlap at intermediate altitudes. In streams where one species is absent, the other expands its range. Water temperature, which has profound consequences for fish ecology, increases downstream.

In laboratory streams, higher temperatures (12°C as compared to 6°C) led to increased aggression in both species when they were tested alone. But this effect was reversed for Dolly Varden when white-spotted charr were also present (Figure 6.2a). Reflecting this, Dolly Varden charr were suppressed from obtaining favorable foraging positions and so foraged far less effectively when white-spotted charr were present at the higher temperature (Figure 6.2b). Also, when alone, neither species' growth rates were influenced by temperature, but when both species were present, growth of Dolly Varden charr decreased with increasing temperature, whereas that of white-spotted charr increased (Figure 6.2c), such that the growth rate of Dolly Varden was much lower than that of white-spotted charr at the higher temperature.

competitive advantage determined by temperature-dependent aggressive behaviour

These results are consistent with the hypothesis that the lower altitudinal boundary of Dolly Varden charr in the Japanese streams was due to temperature-mediated competition favoring white-spotted charr: they were more aggressive, foraged more effectively and grew far faster. But the results do not support the contention that the upper boundary of white-spotted charr is also due to temperature-mediated competitive difference; that is, Dolly Varden did not outcompete white-spotted charr in any of the experiments, even at the lower temperatures. Further work will be needed to determine why Dolly Varden exclude white-spotted charr upstream.

6.2.3 Some general observations

These two examples illustrate several points of general importance.

1 Competing species often coexist at one spatial scale but are found to have distinct distributions at a finer scale of resolution. Here, the fishes coexisted in the same stream, but each was more or less confined to its own altitudinal zone.

2 Species are often excluded by interspecific competition from locations at which they could exist perfectly well in the absence of interspecific competition. Here, Dolly Varden charr can live in the white-spotted charr zone – but only when there are no white-spotted charr there. Similarly, *Asterionella* can live in laboratory cultures – but only when there were no *Synedra* there.

fundamental and realized niches

3 We can describe this by saying that the conditions and resources provided by the white-spotted charr zone are part of the *fundamental niche* of Dolly Varden charr (see Section 3.6 for an explanation of ecological niches) in that the basic requirements for the existence of Dolly Varden charr are provided there. But the white-spotted charr zone does not provide a *realized niche* for Dolly Varden when white-spotted charr are present. Likewise, the laboratory cultures provided the requirements of the fundamental niches of both *Synedra* and *Asterionella*, but those of the realized niche for only *Synedra*.

4 Thus, a species' fundamental niche is the combination of conditions and resources that allow that species to exist, grow and reproduce when considered in isolation from any other species that might be harmful to its existence; whereas its realized niche is the combination of conditions and resources that allow it to exist, grow and reproduce in the presence of specified other species that might be harmful to its existence – especially interspecific competitors.

5 Competing species can therefore coexist when both are provided with a realized niche by their habitat (in the present case, the stream as a whole provided a realized niche for both fishes); but even in locations that provide a species with the requirements of its fundamental niche, that species may be excluded by another, superior competitor that denies it a realized niche there.

6 Finally, the fish study illustrates the importance of experimental manipulation if we wish to discover what is really going on in a natural population – 'nature' may need to be prodded to reveal its secrets.

6.2.4 Coexistence of competing diatoms

Another experimental study of competing diatoms looked at species coexisting on not one but two shared, limiting resources. The two species were *Asterionella formosa* (again) and *Cyclotella meneghiniana*, and the resources, which were both

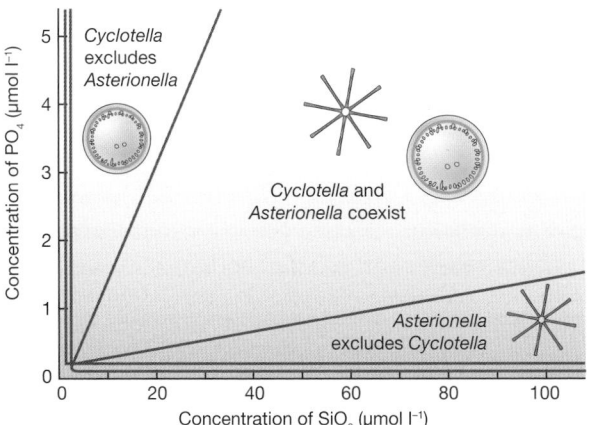

Figure 6.3

Asterionella formosa and *Cyclotella meneghiniana* coexist when there are roughly balanced supplies of silicate (SiO_2) and phosphate (PO_4), but *Asterionella* excludes *Cyclotella* when there are especially low supplies of phosphate, whereas *Cyclotella* excludes *Asterionella* when there are especially low supplies of silicate.

capable of limiting the growth of both diatoms, were silicate and phosphate. However, whereas *Cyclotella* was the more effective exploiter of silicate (reducing its concentration to a lower level), *Asterionella* was the more effective exploiter of phosphate. Thus, in cultures where there were especially low supplies of silicate, *Cyclotella* excluded *Asterionella* (Figure 6.3): such cultures failed to provide a realized niche for *Asterionella*, the inferior competitor there. Conversely, in cultures where there were especially low supplies of phosphate, *Asterionella* excluded *Cyclotella*. However, in cultures with relatively balanced supplies of silicate and phosphate, the two diatoms coexisted (Figure 6.3): with two species, both provided with sufficient supplies of a resource on which they were superior, there was a realized niche for both.

6.2.5 Coexistence of competing birds

It is not always so easy to identify the 'niche differentiation' or 'differential resource utilization' that allows competitors to coexist. Ornithologists, for example, are well aware that closely related species of birds often coexist in the same habitat. For example, five *Parus* species occur together in English broad-leaved woodlands: the blue tit (*Parus caeruleus*), the great tit (*P. major*), the marsh tit (*P. palustris*), the willow tit (*P. montanus*) and the coal tit (*P. ater*). All have short beaks and hunt for food chiefly on leaves and twigs, but at times on the ground; all eat insects throughout the year, and also seeds in winter; and all nest in holes, normally in trees. Yet, the closer we look at the details of the ecology of such coexisting species, the more likely we are to find ecological differences – for example, in precisely where within the trees they feed, in the size of their insect prey and the hardness of the seeds they take. We may be tempted to conclude that such species compete but coexist by eating slightly different resources in slightly different ways: 'differential resource utilization'. But in complex natural environments, such conclusions, while plausible, are difficult to prove.

Indeed, it is often not easy to prove even that the species compete. To do so, it is usually necessary to remove one or more of the species and monitor the responses of those that remain. This was done, for example, in a study of two very similar bird species: the orange-crowned warbler (*Vermivora celata*) and the virginia's warbler (*V. virginiae*), whose breeding territories overlap in central

coexistence through niche differentiation – and even competition – may be difficult to prove

Figure 6.4

Percentage difference in feeding rates (mean ± SE) at orange-crowned warbler and virginia's warbler nests on plots where the other species had been experimentally removed. Feeding rates (visits per hour to the nest with food) were measured during incubation (inc; rates of male feeding of incubating females on the nest) and during the nestling period (nstl; nestling feeding rates by both parents combined). *P*-values are from *t*-tests of the hypothesis that each species fed at higher rates on plots from which the other had been removed. This hypothesis was supported for virginia's warblers but not orange-crowned warblers.

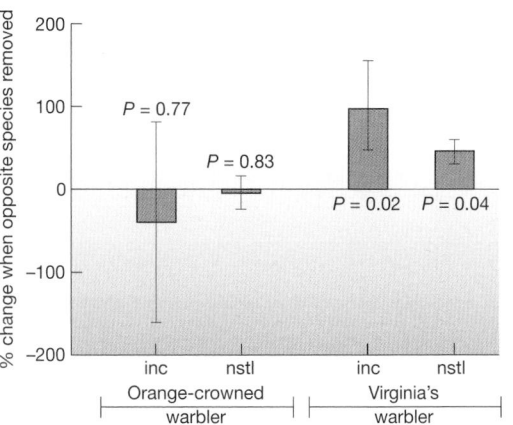

AFTER MARTIN & MARTIN, 2001

Arizona. On plots where one of the two species had been removed, the remaining species fledged between 78% and 129% more young per nest. The enhanced performance was due to improved access to preferred nest sites and consequent decreases in the loss of nestlings to predators. In the case of virginia's warblers, but not orange-crowned warblers, feeding rate also increased in plots from which the other species was removed (Figure 6.4).

6.2.6 Coexistence of competing rodents and ants

The examples described so far have all involved pairs of closely related species – diatoms, salmonid fish or birds. This is potentially misleading in at least two important respects. First, competition may occur amongst larger groups of species than just a pair – where it is sometimes, therefore, called 'diffuse' competition. And second, competition may occur between completely unrelated species.

competition between groups of unrelated species

Both points are illustrated by a study of interspecific competition involving seed-eating ants and seed-eating rodents in deserts of the southwestern United States. At the study sites, only two *guilds* (groups of species that feed on similar foods in a similar fashion; Root, 1967) fed on seeds: the rodents and the ants. By studying the size of the seeds harvested by each guild, it was apparent that the two exhibited significant overlap in the size of the seeds they ate (Figure 6.5). Ants did eat a larger proportion of the smallest seeds, but overall the potential for resource competition between them was very high.

As already noted, however, the only true test for whether competition occurs between them would be to manipulate the abundance of each competitor and observe the response of its counterpart. Consequently, eight plots were established in similar habitats. In two, rodents were trapped and excluded by fencing, to ensure that only ants now had access to the seeds. In another two, ants were eliminated by repeated applications of pesticides. In two further plots both ants and rodents were excluded, and finally two plots were maintained as unmanipulated controls.

When either rodents or ants were removed, there was a statistically significant increase in the numbers of the other guild: the depressive effect of interspecific competition from each guild on the abundance of the other was apparent. Also, when rodents were removed, the ants ate as many seeds as the rodents and ants

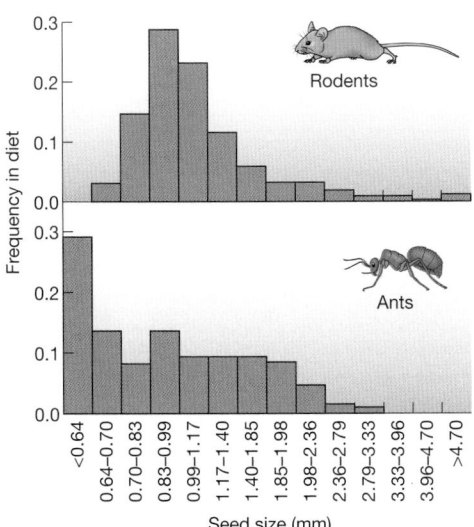

Figure 6.5

The diets of ants and rodents overlap: sizes of seeds harvested by coexisting ants and rodents near Portal, Arizona.

had previously eaten between them – as did the rodents when the ants were removed; only when both were removed did the amount of resource increase. In other words, under normal circumstances both guilds eat less and achieve lower levels of abundance than they would do if the other guild were absent. This clearly indicates that rodents and ants, although they coexist in the same habitat, compete interspecifically with one another.

6.2.7 The Competitive Exclusion Principle

The patterns that are apparent in these examples have also been uncovered in many others, and have been elevated to the status of a principle: the *Competitive Exclusion Principle* or Gause's Principle (named after an eminent Russian ecologist). It can be stated as follows:

- If two competing species coexist in a stable environment, then they do so as a result of niche differentiation, i.e. differentiation of their realized niches.

- If, however, there is no such differentiation, or if it is precluded by the habitat, then one competing species will eliminate or exclude the other.

Although the principle has emerged here from a contemplation of patterns evident in real sets of data, its establishment was – and many modern discussions of interspecific competition still are – bound up with a simple mathematical model of interspecific competition, usually known by the names of its two (independent) originators: Lotka and Volterra (Box 6.1).

There is no question that there is some truth in the principle that competitor species can coexist as a result of niche differentiation, and that one competitor species may exclude another by denying it a realized niche. But it is crucial also to be aware of what the Competitive Exclusion Principle does *not* say.

It does *not* say that whenever we see coexisting species with different niches it is reasonable to jump to the conclusion that this is the principle in operation. Each species, on close inspection, has its own unique niche. Niche differentiation

The Competitive Exclusion Principle – what it does and does not say

6.1 *Quantitative aspects*

The Lotka–Volterra model of interspecific competition

The most widely used model of interspecific competition is the Lotka–Volterra model (Volterra, 1926; Lotka, 1932). It is an extension of the logistic equation described in Box 5.4. Its virtues are (like the logistic) its simplicity, and its capacity to shed light on the factors that determine the outcome of a competitive interaction.

Within the logistic equation,

$$\frac{dN}{dt} = rN\frac{(K-N)}{K}$$

the particular term that models intraspecific competition is $(K - N)/K$. Within this term, the greater the value of N (the bigger the population), the greater is the strength of intraspecific competition. The basis of the Lotka–Volterra model is the replacement of this term by one that models both intra- and interspecific competition. In the model, we call the population size of the first species N_1, and that of a second species N_2. Their carrying capacities and intrinsic rates of increase are K_1, K_2, r_1 and r_2.

By analogy with the logistic, we expect the *total* competitive effect on, say, species 1 (intra- *and* interspecific) to be greater the larger the values of N_1 and N_2; but we cannot just add them together, since the competitive effects of the two species on species 1 are unlikely to be the same. Suppose, though, that 10 individuals of species 2 have, between them, only the same competitive effect on species 1 as does a single individual of species 1. The total competitive effect on species 1 will then be equivalent to the effect of $(N_1 + N_2 * 1/10)$ species 1 individuals. The constant (1/10 in the present case) is called a *competition coefficient* and is denoted by α_{12} (alpha one two). Multiplying N_2 by α_{12} converts it to a number of N_1 equivalents, and adding N_1 and $\alpha_{12}N_2$ together gives us the total competitive effect on species 1. (Note that $\alpha_{12} < 1$ means that individuals of species 2 have less inhibitory effect on individuals of species 1 than individuals of species 1 have on others of their own species, and so on.)

The equation for species 1 can now be written:

$$\frac{dN_1}{dt} = r_1N_1\frac{(K_1 - [N_1 + \alpha_{12}N_2])}{K_1}$$

and for species 2 (with its own competition coefficient, converting species 1 individuals into species 2 equivalents):

$$\frac{dN_2}{dt} = r_2N_2\frac{(K_2 - [N_2 + \alpha_{21}N_1])}{K_2}$$

These two equations constitute the Lotka–Volterra model.

The best way to appreciate its properties is to ask the question, 'Under what circumstances does each species increase or decrease in abundance?' In order to answer, it is necessary to construct diagrams in which all possible combinations of N_1 and N_2 can be displayed. This has been done in Figure 6.6. Certain combinations (certain regions in Figure 6.6) give rise to increases in species 1 and/or species 2, whereas other combinations give rise to decreases. It follows inevitably that there must also therefore be a so-called

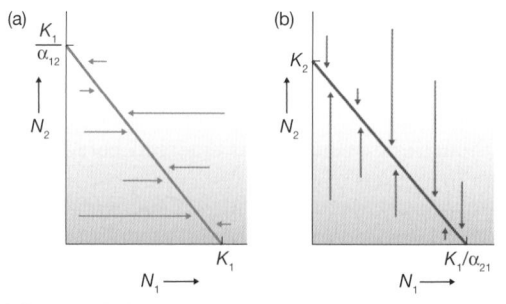

Figure 6.6

The zero isoclines generated by the Lotka–Volterra competition equations. (a) The N_1 zero isocline: species 1 increases below and to the left of it, and decreases above and to the right of it. (b) The equivalent N_2 isocline.

zero isocline for each species: that is, a line with combinations leading to increase on one side of it and combinations leading to decrease on the other, but along which there is neither increase nor decrease.

We can map out the regions of increase and decrease in Figure 6.6 for species 1 if we can draw its zero isocline, and we can do this by using the fact that *on* the zero isocline, $dN_1/dt = 0$ (the rate of change of species 1 abundance is zero, by definition). Rearranging the equation, this gives us, as the zero isocline for species 1:

$$N_1 = K_1 - \alpha_{21}N_2$$

Below and to the left of this, species 1 increases in abundance (arrows in the figure, representing this increase, point from left to right, since N_1 is on the horizontal axis). It increases because numbers of both species are relatively low, and species 1 is thus subjected to only weak competition. Above and to the right of the line, however, numbers are high, competition is strong and species 1 decreases in abundance (arrows from right to left). Based on an equivalent derivation, Figure 6.6b also shows the species 2 zero isocline, with arrows, like the N_2 axis, running vertically.

In order to determine the outcome of competition in this model, it is necessary to determine, at each point on a figure, the behavior of the joint species 1–species 2 population, as indicated by the pair of arrows. There are, in fact, four different ways in which the two zero isoclines can be arranged relative to one another, and these can be distinguished by the intercepts of the zero isoclines (Figure 6.7). The outcome of competition will be different in each case.

Looking at the intercepts in Figure 6.7a, for instance,

$$\frac{K_1}{\alpha_{12}} > K_2 \quad \text{and} \quad K_1 > \frac{K_2}{\alpha_{21}}$$

Rearranging these slightly gives us:

$$K_1 > K_2\alpha_{12} \quad \text{and} \quad K_1\alpha_{21} > K_2$$

The first inequality ($K_1 > K_2\alpha_{12}$) indicates that the inhibitory intraspecific effects that species 1 can exert on itself (denoted by K_1) are greater than the interspecific effects that species 2 can exert on species 1 ($K_2\alpha_{12}$). This means that species 2 is a weak interspecific competitor. The second inequality, however, indicates that species 1 can exert more of an effect

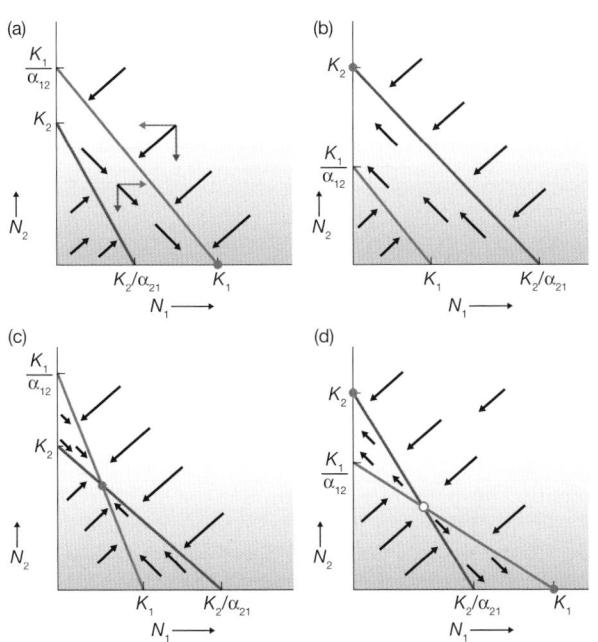

Figure 6.7

The outcomes of competition generated by the Lotka–Volterra competition equations for the four possible arrangements of the N_1 and N_2 zero isoclines. Black arrows refer to joint populations, and are derived as indicated in (a). The solid circles show stable equilibrium points. The open circle in (d) is an unstable equilibrium point. For further discussion, see box text.

on species 2 than species 2 can on itself. Species 1 is thus a *strong* interspecific competitor; and as the arrows in Figure 6.7a show, species 1 drives the weak species 2 to extinction and attains its own carrying capacity. The situation is reversed in Figure 6.7b. Hence Figure 6.7a and b describe cases in which the environment is such that one species invariably outcompetes the other, because the first is a strong interspecific competitor and the other weak.

In Figure 6.7c, by contrast:

$$K_1 > K_2\alpha_{12} \quad \text{and} \quad K_2 > K_1\alpha_{21}$$

In this case, both species have less competitive effect on the other species than those other species have on themselves; in this sense, both are weak competitors. This would happen, for example, if there were niche differentiation between the species – each competed mostly 'within' its own niche. The outcome, as Figure 6.7c shows, is that all arrows point towards a stable, equilibrium combination of the two species, which all joint populations therefore tend to approach: that is, the outcome of this type of competition is the stable coexistence of the competitors. Indeed, it is only this type of competition (both species having more effect on themselves than on the other

species) that does lead to the stable coexistence of competitors.

Finally, in Figure 6.7d:

$$K_2\alpha_{12} > K_1 \quad \text{and} \quad K_1\alpha_{21} > K_2$$

Thus individuals of both species have a greater competitive effect on individuals of the other species than those other species do on themselves. This will occur, for instance, when each species is more aggressive toward individuals of the other species than toward individuals of its own species. The directions of the arrows are rather more complicated in this case, but eventually they always lead to one or other of two alternative stable points. At the first, species 1 reaches its carrying capacity with species 2 extinct; at the second, species 2 reaches its carrying capacity with species 1 extinct. In other words, both species are capable of driving the other species to extinction, but which actually does so cannot be predicted with certainty. It depends on which species has the upper hand in terms of densities, either because they start with a higher density or because density fluctuations in some other way give them that advantage. Whichever species has this upper hand, capitalizes on that and drives the other species to extinction.

does not prove that there are coexisting competitors. The species may not be competing at all and may never have done so in their evolutionary history. We require proof of interspecific competition. In the examples above, this was provided by experimental manipulation – remove one species (or one group of species) and the other species increases its abundance or its survival. But most of even the more plausible cases for competitors coexisting as a result of niche differentiation have not been subjected to experimental proof. So just how important is the Competitive Exclusion Principle in practice? We return to this question in Section 6.5.

Part of the problem is that although species may not be competing now, their ancestors may have competed in the past, so that the mark of interspecific competition is left imprinted on the niches, the behavior or the morphology of their present-day descendants. This particular question is taken up in Section 6.3.

Finally, the Competitive Exclusion Principle, as stated above, includes the word 'stable'. That is, in the habitats envisaged in the principle, conditions and the supply of resources remain more or less constant – if species compete, then that competition runs its course, either until one of the species is eliminated or until the species settle into a pattern of coexistence within their realized niches. Sometimes this is a realistic view of a habitat, especially in laboratory or other controlled environments where the experimenter holds conditions and the supply of resources constant. However, most environments are not stable for long periods of time. How does the outcome of competition change when environmental heterogeneity in space and time are taken into consideration? This is the subject of the next section.

6.2.8 Environmental heterogeneity

As explained in previous chapters, spatial and temporal variations in environments are the norm rather than the exception. Environments are usually a patchwork of favorable and unfavorable habitats; patches are often only available temporarily; and patches often appear at unpredictable times and in unpredictable places. Under such variable conditions, competition may only rarely 'run its course', and the outcome cannot be predicted simply by application of the Competitive Exclusion Principle. A species that is a 'weak' competitor in a constant environment might, for example, be good at colonizing open gaps created in a habitat by fire, or a storm, or the hoofprint of a cow in the mud – or may be good at growing rapidly in such gaps immediately after they are colonized. It may then coexist with a strong competitor, as long as new gaps occur frequently enough. Thus, a realistic view of interspecific competition must acknowledge that it often proceeds not in isolation, but under the influence of, and within the constraints of, a patchy, impermanent or unpredictable world.

competition may only rarely 'run its course'

The following examples illustrate just two of the many ways in which environmental heterogeneity ensures that the Competitive Exclusion Principle is very far from being the whole story when it comes to determining the outcome of an interaction between competing species.

The first concerns the coexistence of a superior competitor and a superior colonizer: the sea palm *Postelsia palmaeformis* (a brown alga) and the mussel *Mytilus californianus* on the coast of Washington, USA (Paine, 1979) (Figure 6.8).

mussels, sea palms and the frequency of gap formation

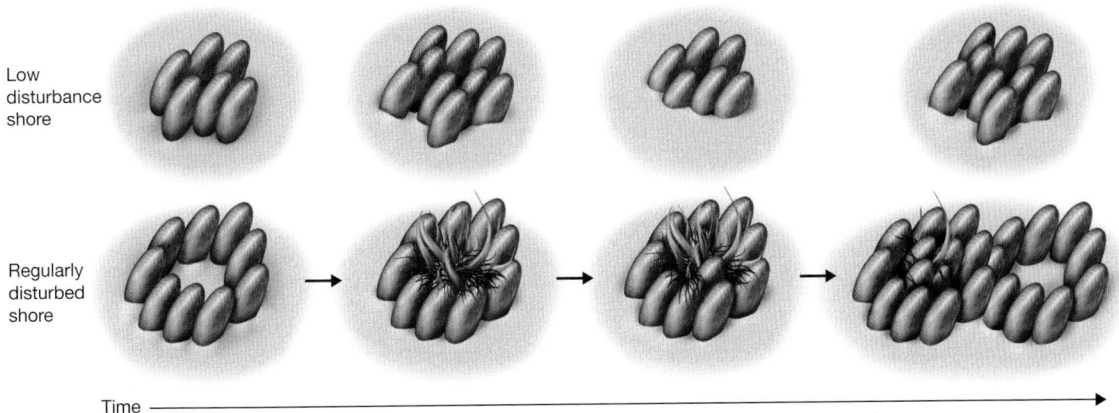

Low disturbance shore

Regularly disturbed shore

Time

Figure 6.8

On shores in which gaps are not created, mussels are able to exclude the brown alga *Postelsia*; but where gaps are created regularly enough the two species coexist, even though *Postelsia* is eventually excluded by the mussels from each gap.

Seashore with *Postelsia* and *Mytilus californianus*.

© GERALD AND BUFF CORSI, VISUALS UNLIMITED

Postelsia is an annual plant that must re-establish itself each year in order to persist at a site. It does so by attaching to the bare rock, usually in gaps in the mussel bed created by wave action. However, the mussels themselves slowly encroach on these gaps, gradually filling them and precluding colonization by *Postelsia*. In other words, in a stable environment, the mussels would outcompete and exclude *Postelsia*. But their environment is not stable – gaps are frequently being created. It turns out that these species coexist only at sites in which there is a relatively high average rate of gap formation (at least 7% of surface area per year), and in which this rate is approximately the same each year. Where the average rate is lower, or where it varies considerably from year to year, there is (either regularly or occasionally) a lack of bare rock for colonization by *Postelsia*. At the sites of coexistence, on the other hand, although *Postelsia* is eventually excluded from each gap, these are created with sufficient frequency and regularity for there to be coexistence in the site as a whole. In short, there is coexistence of competitors – but not as a result of niche differentiation.

coexistence as a result of aggregated distributions

A perhaps more widespread path to the coexistence of a superior and an inferior competitor is based on the idea that the two species may have independent, aggregated (i.e. clumped) distributions over the available habitat. This would mean that the powers of the superior competitor were mostly directed against members of its own species (in the high-density clumps), but that this aggregated superior competitor would be absent from many areas – within which the inferior competitor could escape competition. An inferior competitor may then be able to coexist with a superior competitor that would rapidly exclude it from a continuous, homogeneous environment.

That such aggregated distributions are indeed a reality is illustrated by a field study of two species of sand-dune plant, *Aira praecox* and *Erodium cicutarium*, in northwest England. Both species were aggregated, and the smaller plant, *Aira*, tended to be aggregated even at the smallest spatial scales (Figure 6.9a). The two species, though, were negatively associated with one another at these smallest scales (Figure 6.9b). Thus, *Aira* tended to occur in small single-species clumps and was therefore much less liable to competition from *Erodium* than would have been the case if they had been distributed at random.

The consequences of such aggregated distributions are illustrated by a study of experimental communities of four annual terrestrial plants – *Capsella*

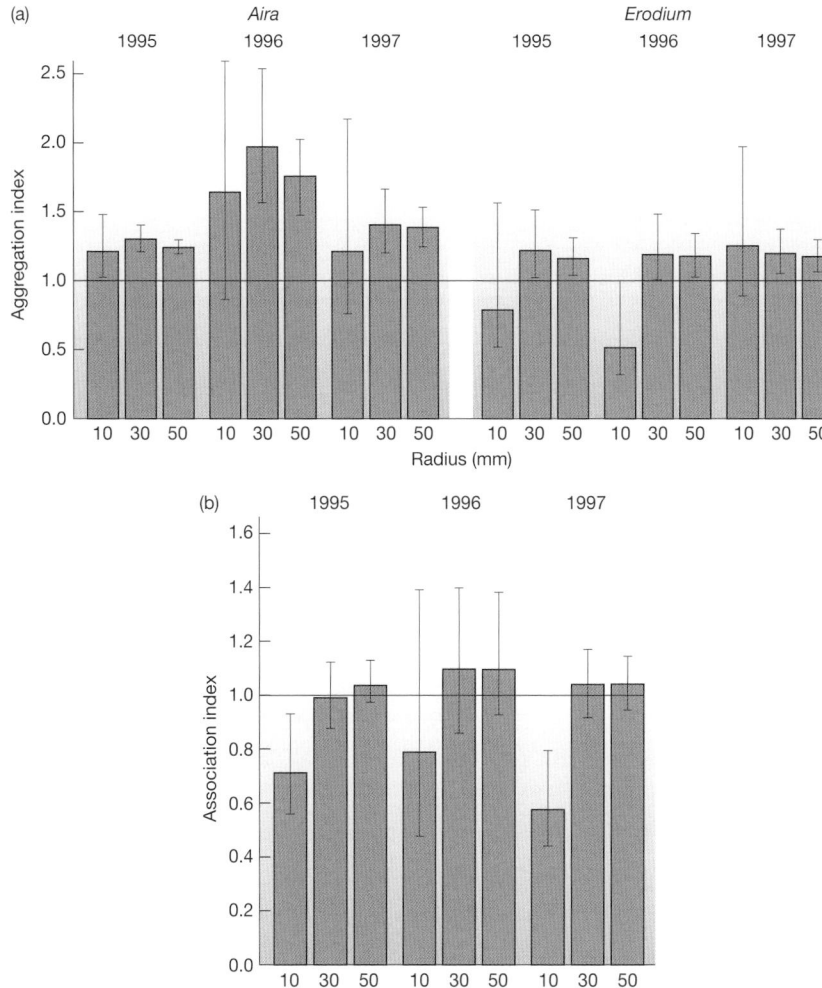

AFTER COOMES ET AL. 2002

Figure 6.9

(a) Spatial distribution of two sand-dune species, *Aira praecox* and *Erodium cicutarium* at a site in northwest England. An aggregation index of 1 indicates a random distribution. Indices greater than 1 indicate aggregation (clumping) within patches with the radius as specified; values less than 1 indicate a regular distribution. Bars represent 95% confidence intervals. (b) The association between *Aria* and *Erodium* in each of the 3 years. An association index greater than 1 would indicate that the two species tended to be found together more than would be expected by chance alone in patches with the radius as specified; values less than 1 indicate a tendency to find one species or the other. Bars represent 95% confidence intervals.

bursa-pastoris, *Cardamine hirsuta*, *Poa annua* and *Stellaria media* (Figure 6.10). *Stellaria* is known to be the superior competitor among these species. Replicate three- and four-species mixtures were sown at high density, and the seeds were either placed completely at random, or seeds of each species were aggregated in subplots within the experimental areas. Intraspecific aggregation harmed the performance of the superior *Stellaria* in the mixtures, whereas in all but one case aggregation improved the performance of the three inferior competitors. Again, coexistence of competitors was favored not by niche differentiation but simply by a type of heterogeneity that is typical of the natural world: aggregation ensured that most individuals competed with members of their own and not of another species.

These studies, and others like them, go a long way toward explaining the co-occurrence of species that in constant, homogeneous environments would probably exclude one another. The environment is rarely unvarying enough for competitive exclusion to run its course or for the outcome to be the same across the landscape.

Figure 6.10

The effect of intraspecific aggregation on above-ground biomass (mean ± SE) of four plant species grown for 6 weeks in three- and four-species mixtures (four replicates of each). The normally competitively superior *Stellaria media* (*Sm*) did consistently less well when seeds were aggregated than when they were placed at random (d). In contrast, the three competitively inferior species – *Capsella bursa-pastoris* (*Cbp*), *Cardamine hirsuta* (*Ch*) and *Poa annua* (*Pa*) – almost always performed better when seeds had been aggregated (a–c). Note the different scales on the vertical axes, and that the compositions of the mixtures are given only along the horizontal axis of (d).

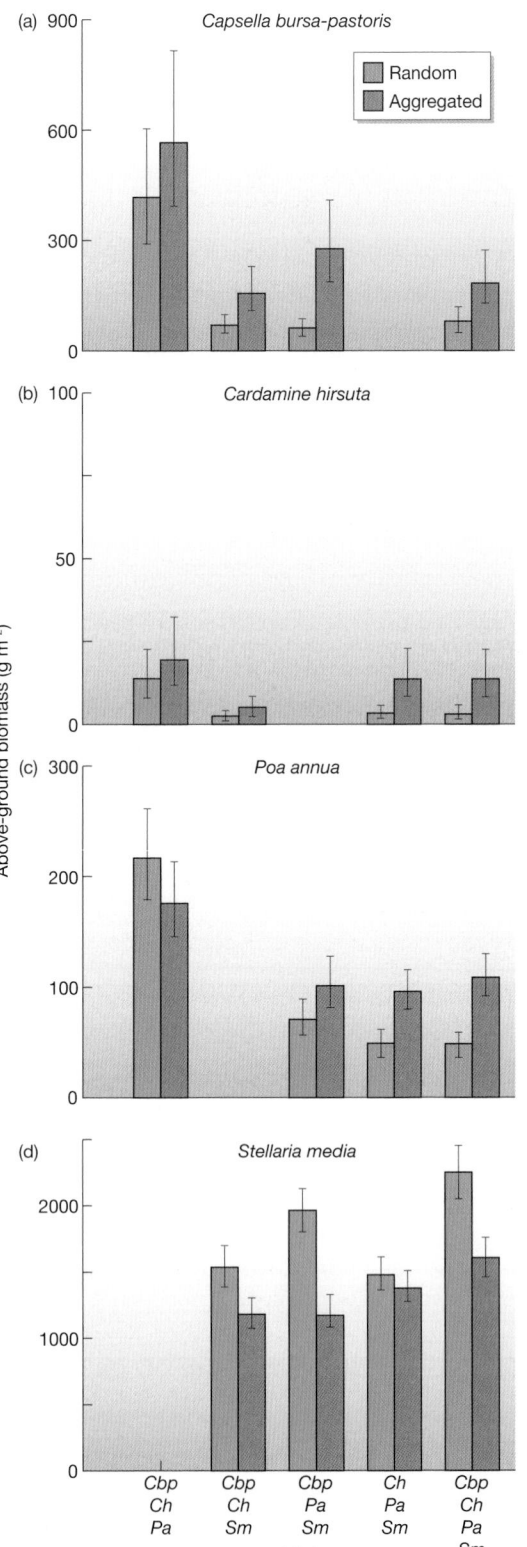

6.3 Evolutionary effects of interspecific competition

Putting to one side the fact that environmental heterogeneity ensures that the forces of interspecific competition are often much less profound than they would otherwise be, it is nonetheless the case that the potential of interspecific competition to adversely affect individuals is considerable. We have seen in Chapter 2 that natural selection in the past will have favored those individuals that, by their behavior, physiology or morphology, have avoided adverse effects that act on other individuals in the same population. The adverse effects of extreme cold, for example, may have favored individuals with an enzyme capable of functioning effectively at low temperatures. Similarly, in the present context, the adverse effects of interspecific competition may have favored those individuals that managed to avoid those competitive effects. We can, therefore, expect species to have evolved characteristics that ensure that they compete less, or not at all, with members of other species.

evolutionary avoidance of competition

How will this look to us at the present time? Coexisting species, with an apparent potential to compete, will exhibit differences in behavior, physiology or morphology that ensure that they compete little or not at all. Connell has called this line of reasoning 'invoking the ghost of competition past'. Yet the pattern it predicts is precisely the same as that supposed by the Competitive Exclusion Principle to be a prerequisite for the coexistence of species that still compete. Coexisting present-day competitors, and coexisting species that have evolved an avoidance of competition, can look the same.

invoking the ghost of competition past

The question of how important either past or present competition are as forces structuring natural communities will be addressed in the last section of this chapter (Section 6.5). For now, we examine some examples of what interspecific competition *can* do as an evolutionary force. Note, however, that by invoking something that cannot be observed directly (evolution), it may be impossible to prove an evolutionary effect of interspecific competition, in the strict sense of 'proof' that can be applied to mathematical theorems or carefully controlled experiments in the laboratory. Nonetheless, we consider some examples where an evolutionary (rather than an ecological) effect of interspecific competition is the most reasonable explanation for what is observed.

the difficulty of distinguishing ecological and evolutionary effects

6.3.1 Character displacement and ecological release in the Indian mongoose

In western parts of its range, the small Indian mongoose (*Herpestes javanicus*) coexists with one or two slightly larger species in the same genus (*H. edwardsii* and *H. smithii*), but these species are absent in the eastern part of its range (Figure 6.11a). The upper canine teeth are the mongoose's principal prey-killing organ, and these vary in size within and between species and between the sexes (female mongooses are smaller than males). In the east, where *H. javanicus* occurs alone (area VII in Figure 6.11a), both males and females have larger canines than in the western areas (III, V, VI) where it coexists with the larger species (Figure 6.11b). This is consistent with the view that where similar but larger mongoose species are present, the prey-catching apparatus of *H. javanicus* has

Figure 6.11

(a) Native geographic ranges of *Herpestes javanicus* (*j*), *H. edwardsii* (*e*) and *H. smithii* (*s*). (b) Maximum diameter (mm) of the upper canine (C^{sup}L) for *Herpestes javanicus* in its native range [data only for areas III, V, VI and VII from (a)] and islands on which it has been introduced. Symbols in blue represent mean female size and in maroon mean male size. Compared to area VII (*H. javanicus* alone), animals in areas III, V and VI, where they compete with the two larger species, are smaller. On the islands, they have increased in size since their introduction, but are still not as large as in area VII.

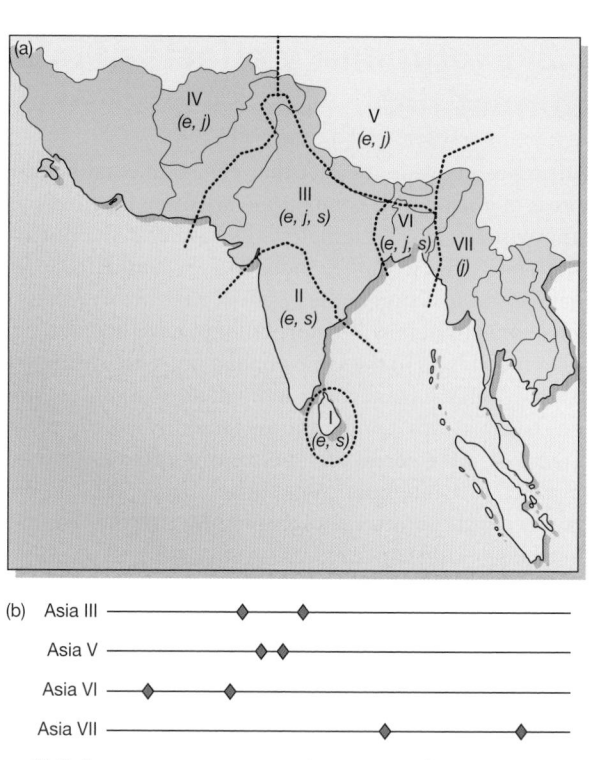

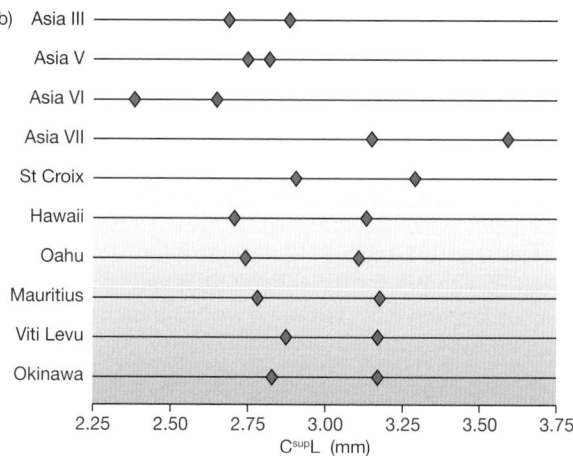

FROM SIMBERLOFF ET AL., 2000

been selected for reduced size (referred to as 'character displacement'), reducing the strength of competition with other species in the genus because smaller predators tend to take smaller prey. Where *H. javanicus* occurs in isolation, since no character displacement has occurred, its canine teeth are much larger. (Another strong candidate for the evolutionary effects of interspecific competition, especially because of its association with character displacement, is provided by Darwin's finches of the genus *Geospiza* living on the Galapagos islands, discussed in Section 2.4.2.)

smaller teeth in the small Indian mongoose when larger competitors are present

In fact, *H. javanicus* was introduced about a century ago to many islands outside its native range (often as part of a naive attempt to control introduced rodents). In these places, the larger competitor mongoose species were absent. Within 100–200 generations *H. javanicus* had increased in size (Figure 6.11b), so that the sizes of island individuals are now intermediate between those in the region of origin (where they coexisted with other species and were small) and those in the east where they occur alone. Their size on the islands is consistent with 'ecological release' from competition with larger species.

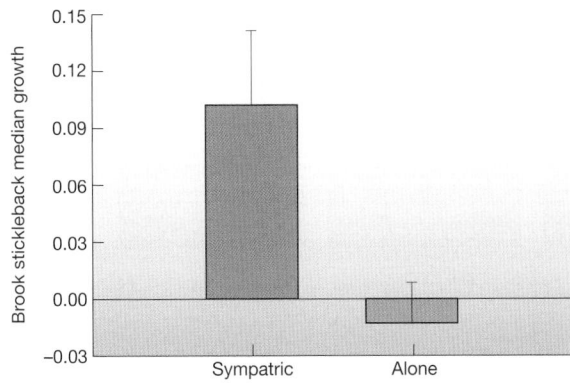

AFTER GRAY & ROBINSON, 2001

Figure 6.12

Means (with standard errors) of group median growth (natural log of the final mass of fish in each enclosure divided by the initial mass of the group) for sympatric brook sticklebacks, representing post-displacement phenotypes (maroon bar), and brook sticklebacks living alone, representing pre-displacement phenotypes (blue bar), both reared in the presence of ninespine sticklebacks. In competition with ninespine sticklebacks, growth was significantly greater for post-displacement vs pre-displacement phenotypes ($P = 0.012$).

6.3.2 Character displacement in Canadian sticklebacks

If character displacement has ultimately been caused by competition, then the effects of competition should decline with the degree of displacement. Brook sticklebacks, *Culaea inconstans*, coexist in some Canadian lakes with ninespine sticklebacks, *Pungitius pungitius* (the species are 'sympatric'), whereas in other lakes brook sticklebacks live alone. In sympatry, the brook sticklebacks possess significantly shorter gill rakers (more suited for foraging in open water), longer jaws and deeper bodies. We can consider the brook sticklebacks living alone as having pre-displacement morphology and the sympatric brook sticklebacks as post-displacement phenotypes. When each phenotype was placed separately in enclosures in the presence of ninespine sticklebacks, the pre-displacement brook sticklebacks grew significantly less well than their sympatric post-displacement counterparts (Figure 6.12). This is clearly consistent with the hypothesis that the post-displacement phenotype has evolved to avoid competition, and hence enhance fitness, in the presence of ninespine sticklebacks.

6.3.3 Evolution in action: niche-differentiated bacteria

The most direct way of demonstrating the evolutionary effects of competition within a pair of competing species is for the experimenter to induce these effects – impose the selection pressure (competition) and observe the outcome. Surprisingly perhaps, there have been very few successful experiments of this type. To find an example of niche differentiation giving rise to coexistence of competitors in a selection experiment, we must turn away from interspecific competition in the strictest sense to competition between three types of the same bacterial species, *Pseudomonas fluorescens*, which behave as separate species because they reproduce asexually. The three types are named 'smooth' (SM), 'wrinkly spreader' (WS) and 'fuzzy spreader' (FS), on the basis of the morphology of their colonies plated out on solid medium. In liquid medium they also occupy quite different parts of the culture vessel (Figure 6.13a), that is, they have separate niches. In vessels that were continually shaken, so that no separate niches could be established, an initially pure culture of SM individuals retained its purity (Figure 6.13b). But in the absence of shaking, WS and FS mutants arose in the SM population, increased in frequency and established themselves (Figure 6.13c): evolution had favored niche differentiation and the consequent avoidance of competition.

Figure 6.13

(a) Pure cultures of three types of the bacterium *Pseudomonas fluorescens* (smooth, SM; wrinkly spreader, WS; fuzzy spreader, FS) concentrate their growth in different parts of a liquid culture vessel. (b) In shaken culture vessels, pure SM cultures are maintained. Bars are standard errors. (c) But in unshaken, initially pure SM (●) cultures, WS (▲) and FS (■) mutants arise, invade and establish. Bars are standard errors.

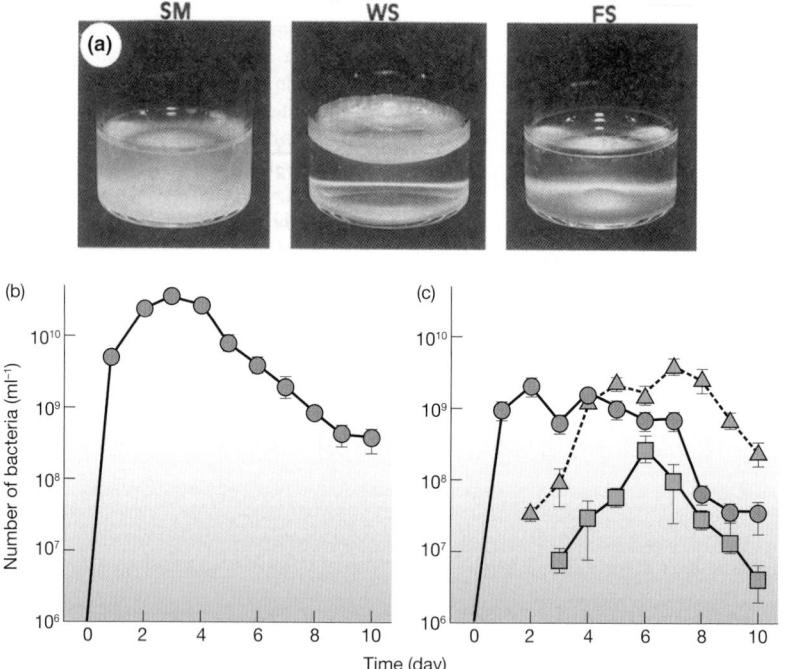

AFTER RAINEY & TREVISANO, 1998, BY PERMISSION OF NATURE

6.4 Interspecific competition and community structure

Interspecific competition, then, has the potential to either keep apart (Section 6.2) or drive apart (Section 6.3) the niches of coexisting competitors. How can these forces express themselves when it comes to the role of interspecific competition in molding the shape of whole ecological communities – who lives where and with whom?

6.4.1 Limiting resources and the regulation of diversity in phytoplankton communities

We begin by returning to the question of coexistence of competing phytoplankton species. In Section 6.2.4, we saw how two diatom species could coexist in the laboratory on two shared limiting resources – silicate and phosphate. In fact, theory predicts that the diversity of coexisting species should be proportional to the number of resources in a system that are at physiological limiting levels (Tilman, 1982): the more limiting resources, the more coexisting competitors. A direct test of this hypothesis examined three lakes in the Yellowstone region of Wyoming, USA using an index (Simpson's index) of the species diversity of phytoplankton there (diatoms and other species). If one species exists on its own, the index equals 1; in a group of species where biomass is strongly dominated by a single species, the index will be close to 1; when two species exist at equal biomass, the index is 2; and so on. According to the theory, therefore, this index

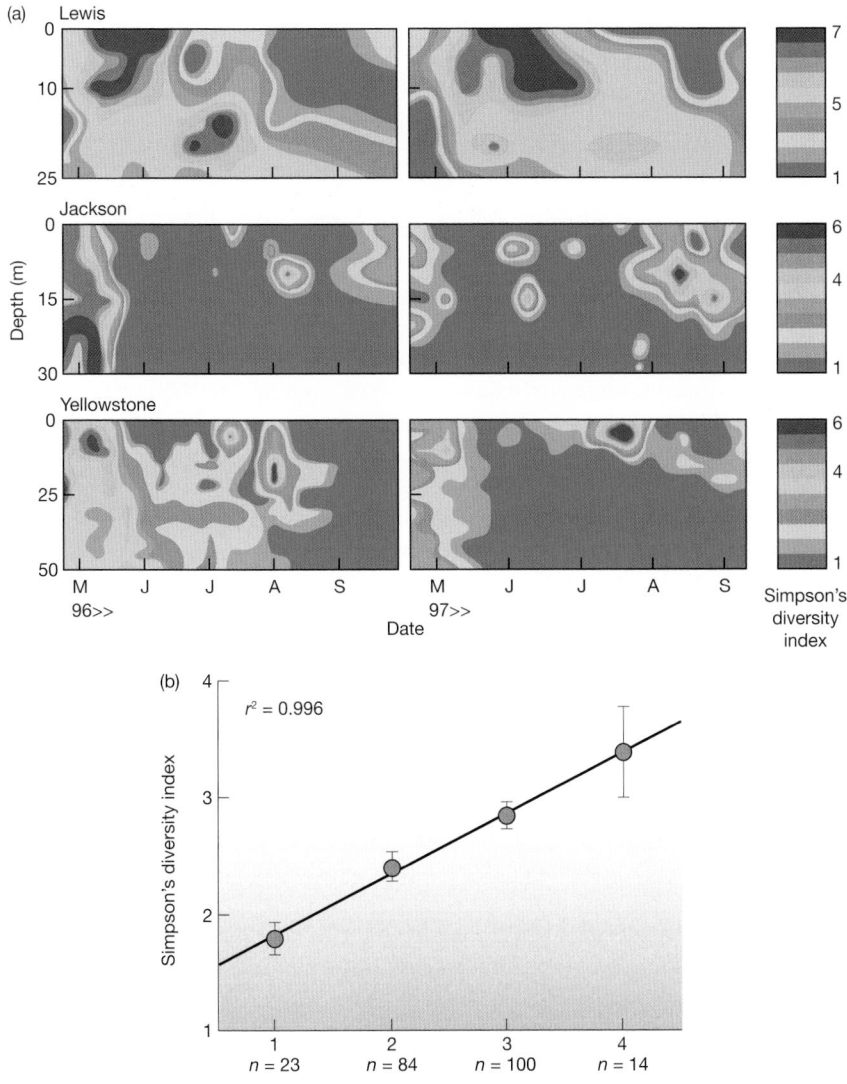

Figure 6.14

(a) Variation in phytoplankton species diversity (Simpson's index) with depth in 2 years in three large lakes in the Yellowstone region. Color indicates depth–time variation in a total of 712 discrete samples; maroon denotes high species diversity, blue denotes low species diversity. (b) Phytoplankton diversity (Simpson's index; mean ± SE) associated with samples with different numbers of measured limiting resources. It was possible to perform this analysis on 221 samples from those displayed in (a); the number of samples (n) in each limiting resource class is shown. Diversity clearly increases with the number of limiting resources.

should increase in direct proportion to the number of resources limiting growth. The spatial and temporal patterns in phytoplankton diversity in the three lakes for 1996 and 1997 are shown in Figure 6.14a.

The principal limiting resources for phytoplankton growth are nitrogen, phosphorus, silicon and light. These parameters were measured at the same depths and times that the phytoplankton were sampled, and it was noted where and when any of the potential limiting factors actually occurred at levels below threshold limits for growth. Consistent with the theory, species diversity increased as the number of resources at physiologically limiting levels increased (Figure 6.14b). This suggests that even in the highly dynamic environments of lakes, where equilibrium conditions are rare, resource competition plays a role in continuously structuring the phytoplankton community. It is heartening that the results

as predicted, highest phytoplankton diversity occurred where many resources were limiting of experiments performed in the artificial world of the laboratory (Section 6.2.4) are echoed here in the much more complex natural environment.

6.4.2 Niche complementarity amongst anemone fish in Papua New Guinea

In another study of niche differentiation and coexistence, a number of species of anemone fish were examined near Madang in Papua New Guinea. This region has the highest reported species richness of both anemone fish (nine) and their host anemones (10). Each individual anemone tends to be occupied by individuals of just one species of anemone fish, because the residents aggressively exclude intruders. However, aggressive interactions were less frequently observed between anemone fish of very different sizes. Anemones seem to be a limiting resource for the fish in that almost all anemones were occupied, and when some were transplanted to new sites they were quickly colonized. Surveys in four zones (nearshore, mid-lagoon, outer barrier reef and offshore: Figure 6.15a) showed that each anemone fish was primarily associated with a particular species of anemone; each also showed a characteristic preference for a particular zone (Figure 6.15b). Crucially, moreover, anemone fish that lived with the same anemone were typically associated with different zones. For example, *Amphiprion percula* occupied the anemone *Heteractis magnifica* in nearshore zones, while *A. perideraion* occupied *H. magnifica* in offshore zones. Finally, associated with the lowered level of aggression, small anemone fish species (*A. sandaracinos* and *A. leucokranos*) were able to cohabit the same anemone with larger species.

species similar in one dimension tend to differ in another dimension

Two important points are illustrated here. First, the anemone fish demonstrate *niche complementarity*; that is, niche differentiation involves several niche dimensions: species of anemone, zone on the shore and, almost certainly, some other dimension, perhaps food particle size, reflected in the size of the fish. Fish species that occupy a similar position along one dimension tend to differ along another dimension. Second, the fish can be considered to be a guild, in that they are a group of species that exploit the same class of environmental resource in a similar way, and insofar as interspecific competition plays a role in structuring communities, it tends to do so, as here, not by affecting some random sample of the members of that community, nor by affecting every member, but by acting within guilds.

6.4.3 Species separated in space or in time

In spite of the many examples where there is no direct connection between interspecific competition and niche differentiation, there is no doubt that niche differentiation is often the basis for the coexistence of species within natural communities. There are a number of ways in which niches can be differentiated. One, as we have seen, is resource partitioning or differential resource utilization. This can be observed when species living in precisely the same habitat nevertheless utilize different resources. In many cases, however, the resources used by ecologically similar species are separated spatially. Differential resource utilization will then express itself as either a microhabitat differentiation between the species (different species of fish, say, feeding at different depths) or even a difference in

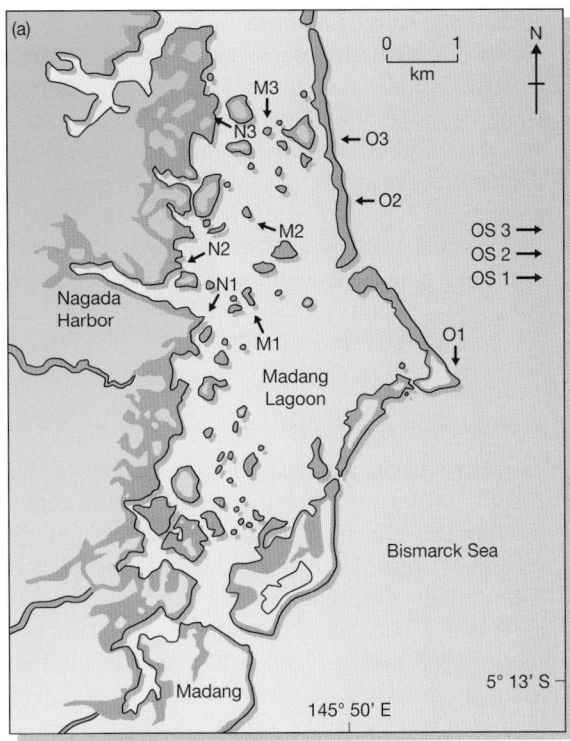

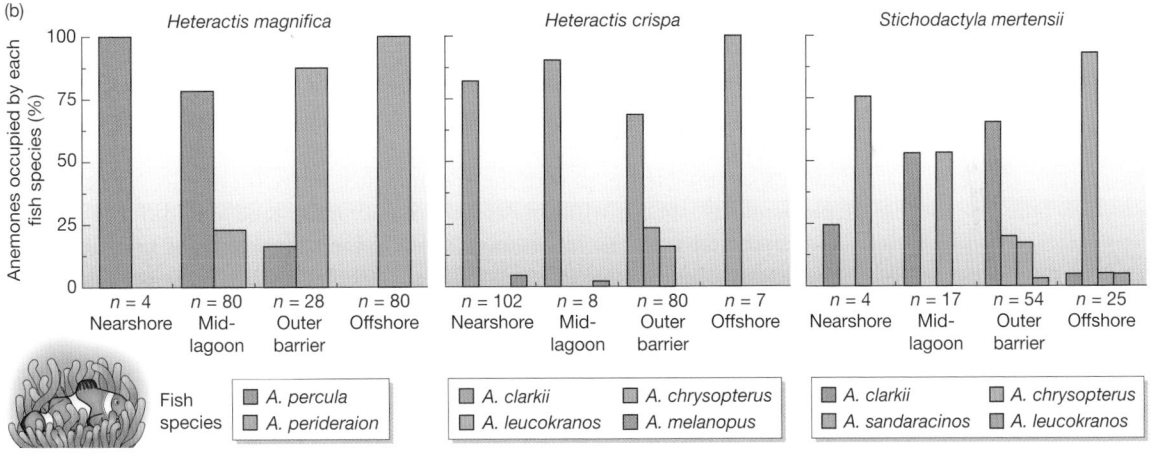

Figure 6.15

(a) Map showing the location of three replicate study sites in each of four zones within and outside Madang Lagoon (N, nearshore; M, mid-lagoon; O, outer barrier reef; OS, offshore reef). The blue areas indicate water, brown shading represents coral reef, and green represents land.
(b) The percentage of three common species of anemone (*Heteractis magnifica*, *H. crispa* and *Stichodactyla mertensii*) occupied by different anemone fish species (*Amphiprion* spp., in key below) in each of the four zones. The number of anemones censused in each zone is shown by *n*.

AFTER ELLIOTT & MARISCAL. 2001

geographic distribution. Alternatively, the availability of the different resources may be separated in time; that is, different resources may become available at different times of the day or in different seasons. Differential resource utilization may then express itself as a temporal separation between the species.

The other major way in which niches can be differentiated is on the basis of conditions. Two species may use precisely the same resources, but if their ability to do so is influenced by environmental conditions (as it is bound to be), and if they respond differently to those conditions, then each may be competitively superior in different environments. This too can express itself as either a micro-habitat differentiation, or a difference in geographic distribution, or a temporal separation, depending on whether the appropriate conditions vary on a small spatial scale, a large spatial scale or over time. Of course, it is not always easy to distinguish between conditions and resources, especially with plants (see Chapter 3). Niches may then be differentiated on the basis of a factor (such as water), which is both a resource and a condition.

6.4.4 Spatial separation in trees and tree-root fungi

trees in Borneo: height, depth, gaps and soil

Trees vary in their capacity to use resources such as light, water and nutrients. A study in Borneo of 11 tree species in the genus *Macaranga* showed marked differentiation in light requirements, from extremely light-demanding species such as *M. gigantea* to shade-tolerant species such as *M. kingii* (Figure 6.16a). Average light levels intercepted by the crowns of trees tended to increase as they grew larger, but the ranking of the species did not change. The shade-tolerant species were smaller (Figure 6.16b) and persisted in the understorey, rarely establishing in disturbed microsites (e.g. *M. kingii*), in contrast to some of the larger, high-light species that are pioneers of large forest gaps (e.g. *M. gigantea*). Others were associated with intermediate light levels and can be considered small-gap specialists (e.g. *M. trachyphylla*). The *Macaranga* species were also differentiated along a second niche gradient, with some species being more common on clay-rich soils and others on sand-rich soils (Figure 6.16b). This differentiation may be based on nutrient availability (generally higher in clay soils) and/or soil moisture availability (possibly lower in the clay soils because of thinner root mats and humus layers). Hence, as with the anemone fish, there is evidence of niche complementarity: species with similar light requirements tended to differ in terms of preferred soil textures. In addition, though, the apparent niche partitioning by *Macaranga* species was partly related to space horizontally (variation in soil types and in light levels from place to place) and partly to space vertically (height in the canopy, depth of the root mat).

separation with depth in ectomycorrhizal fungi

Differential resource utilization in the vertical plane has also been demonstrated for fungi intimately associated with plant roots (ectomycorrhizal fungi; see Section 8.4.5) in the floor of a forest of pine, *Pinus resinosa* (Figure 6.17). Until recently, it was not possible to study the distribution of ectomycorrhizal species in their natural environment. Now DNA analyses make this possible and allow their distributions to be compared. The forest soil has a well-developed litter layer above a fermentation layer (F layer) and a thin humified layer (H layer), with mineral soil beneath (B horizon). Of the 26 species separated by the DNA analysis, some were very largely restricted to the litter layer (group A in Figure 6.17), others to the F layer (group D), the H layer (group E) or the B horizon (group F). The remaining species were more general in their distributions (groups B and C). This is therefore an example of where a spatial (microhabitat) separation cannot simply be ascribed to one resource or condition: there are no doubt several that vary with the soil layers.

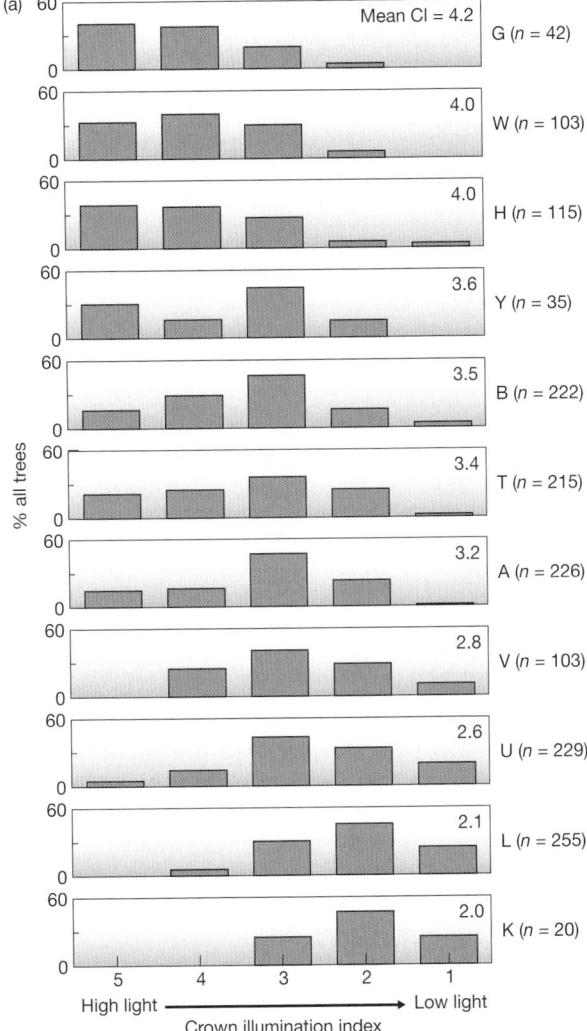

(a) 60

Mean CI = 4.2

G (n = 42)

60

4.0

W (n = 103)

60

4.0

H (n = 115)

60

3.6

Y (n = 35)

60

3.5

B (n = 222)

60

3.4

T (n = 215)

60

3.2

A (n = 226)

60

2.8

V (n = 103)

60

2.6

U (n = 229)

60

2.1

L (n = 255)

60

2.0

K (n = 20)

% all trees

5   4   3   2   1

High light ⟶ Low light

Crown illumination index

Figure 6.16

(a) Percentage of individuals in each of five crown illumination (CI) classes for 11 *Macaranga* species (sample sizes in parentheses). (b) Three-dimensional distribution of the 11 species with respect to maximum height, the proportion of stems in high light levels [class 5 in (a)] and proportion of stems in sand-rich soils. Each species of *Macaranga* is denoted by a single letter. G, *gigantean*; W, *winkleri*; H, *hosei*; Y, *hypoleuca*; B, *beccariana*; T, *triloba*; A, *trachyphylla*; V, *havilandii*; U, *hullettii*; L, *lamellate*; K, *kingii*.

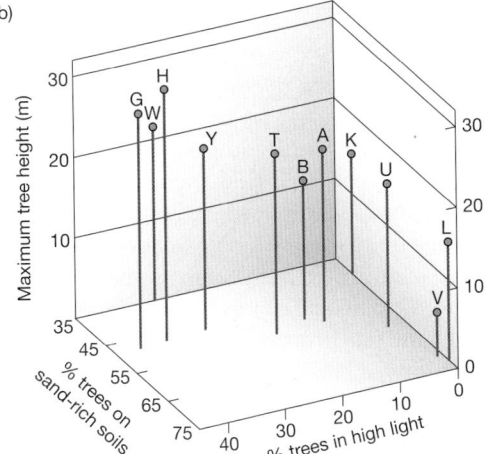

(b)

Maximum tree height (m)

% trees on sand-rich soils

% trees in high light

Figure 6.17

The vertical distribution of 26 ectomycorrhizal fungal species in the floor of a pine forest determined by DNA analysis. Most have not formally been named but are shown as a code. Vertical distribution histograms show the percentage of occurrences of each species in the litter (maroon), F layer (yellow), H layer (green) and B horizon (blue).

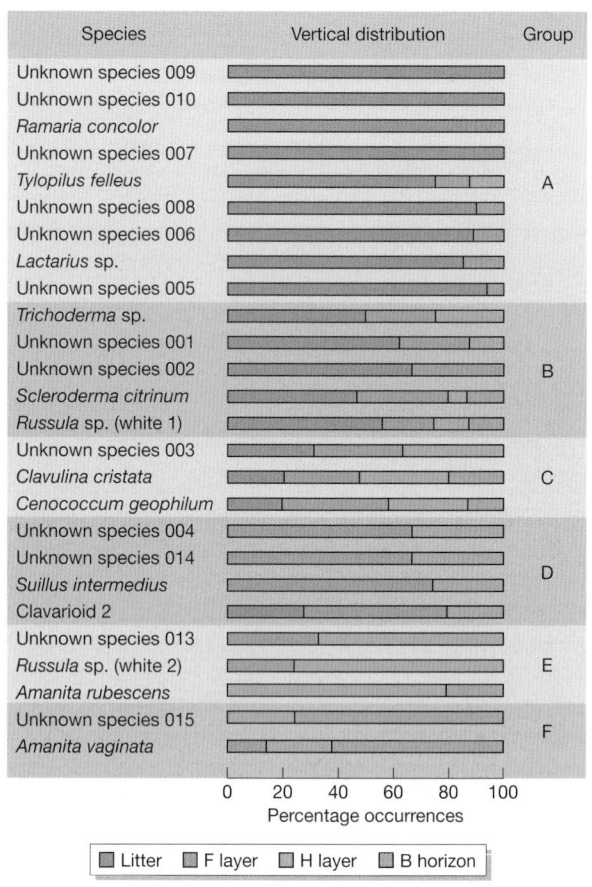

AFTER DICKIE ET AL., 2002

6.4.5 Temporal separation in mantids and tundra plants

staggered life cycles in mantids

One common way in which resources may be partitioned over time is through a staggering of life cycles through the year. It is notable that two species of mantids, which feature as predators in many parts of the world, commonly coexist both in Asia and North America. *Tenodera sinensis* and *Mantis religiosa* have life cycles that are 2–3 weeks out of phase. To test the hypothesis that this asynchrony serves to reduce interspecific competition, the timing of their egg hatch was experimentally synchronized in replicated field enclosures (Hurd & Eisenberg, 1990). *T. sinensis*, which normally hatches earlier, was unaffected by *M. religiosa*. In contrast, the survival and body size of *M. religiosa* declined in the presence of *T. sinensis*. Because these mantids are both competitors for shared resources and predators of each other, the outcome of this experiment probably reflects a complex interaction between the two processes.

nitrogen, depth and time in Alaskan plants

In plants too, resources may be partitioned in time. Thus, tundra plants growing in nitrogen-limited conditions in Alaska are differentiated in their timing of nitrogen uptake, as well as the soil depth from which it is extracted and the

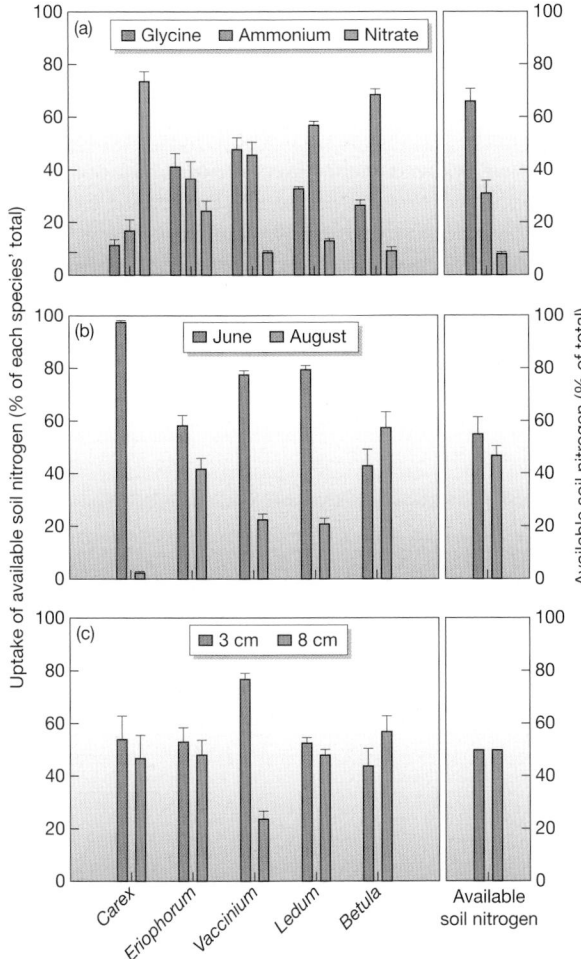

Figure 6.18

Mean uptake of available soil nitrogen (± SE) in terms of (a) chemical form, (b) timing of uptake and (c) depth of uptake by the five most common species in tussock tundra in Alaska. Data are expressed as the percentage of each species' total uptake (left panels) or as the percentage of the total pool of nitrogen available in the soil (right panels).

chemical form of nitrogen used. To trace how tundra species differed in uptake of different nitrogen sources, McKane et al. (2002) injected three chemical forms labeled with the rare isotope ^{15}N (ammonium, nitrate and glycine) at two soil depths (3 and 8 cm) on two occasions (June 24 and August 7). Concentration of the ^{15}N tracer was measured in each of five common tundra plants 7 days after application. The five plants proved to be well differentiated in their use of nitrogen sources (Figure 6.18). Cottongrass (*Eriophorum vaginatum*) and the cranberry bush (*Vaccinium vitis-idaea*) both relied on a combination of glycine and ammonium, but cranberry obtained more of these forms early in the growing season and at a shallower depth than cottongrass. The evergreen shrub *Ledum palustre* and the dwarf birch (*Betula nana*) used mainly ammonium, but *L. palustre* obtained more of this form early in the season while the birch exploited it later. Finally, the grass *Carex bigelowii* was the only species to use mainly nitrate. Here, niche complementarity can be seen along three niche dimensions: nitrogen source, depth and time.

6.5 How significant is interspecific competition in practice?

Competitors may exclude one another, or they may coexist if there is ecologically significant differentiation of their realized niches (Section 6.2). On the other hand, interspecific competition may exert neither of these effects if environmental heterogeneity prevents the process from running its course (Section 6.2.8). Evolution may drive the niches of competitors apart until they coexist but no longer compete (Section 6.3). All these forces may express themselves at the level of the ecological community (Section 6.4). Interspecific competition sometimes makes a high profile appearance by having a direct impact on human activity (Box 6.2). In this sense, competition can certainly be of practical significance.

6.2 *Topical ECOncerns*

Competition in action

When exotic plant species are introduced to a new environment, by accident or on purpose, they sometimes prove to be exceedingly good competitors and many native species suffer harmful consequences as a result. Some of them have even more far-reaching consequences for native ecosystems. This newspaper article by Beth Daley, published in the *Contra Costa Times* on June 27, 2001, concerns grasses that have invaded the Mojave Desert in the southern United States. Not only are the invaders outcompeting native wild flowers, they have also dramatically changed the fire regime.

Invader grasses endanger desert by spreading fire

The newcomers crowd out native plants and provide fuel for once-rare flames to damage the delicate ecosystem.

Charred creosote bushes dot a mesa in the Mojave Desert, the ruins of what was likely the first fire in the area in more than 1000 years.

Though deserts are hot and dry, they aren't normally much of a fire hazard because the vegetation is so sparse there isn't much to burn or any way for blazes to spread.

But, underneath these blackened creosote branches, the cause of the fire seven years ago has already grown back: flammable grasses fill the empty spaces between the native bushes, creating a fuse for the fire to spread again.

Tens of thousands of acres in the Mojave and other southwestern deserts have burned in the last decade, fueled by the red brome, cheat grass and Sahara mustard, tiny grasses and plants that grow back faster than any native species and shouldn't be there in the first place.

. . . The grasses brought to America from Eurasia more than a century ago have no natural enemies, and little can stop their spread across empty desert pavement. And, once an area is cleared of native vegetation by one or repeated fires, the grasses grow in even thicker, sometimes outcompeting native wildflowers and shrubs.

. . . 'These grasses could change the entire makeup of the Mojave Desert in short order', said William Schlesinger of Duke University, who has studied the Mojave Desert for more than 25 years. When he began his research in the 1970s, the grasses were in the Mojave, but there still were vast areas left untouched. Now, he said, the grasses are virtually everywhere and soon will be in concentrations large enough to fuel massive fires. 'This is not an easy problem to solve', he said.

. . . Despite the harsh conditions, a rainbow of wildflowers blooms regularly in the desert, sometimes carpeting the ground with blossoms after a rainstorm. Zebra-tailed lizards, rattlesnakes, desert tortoises and kangaroo rats are able to get by for long periods without water and bear up under the sun. But the innocuous-looking grasses threaten all these species by choking out wildflowers and killing off shelter and food that they rely on.

. . . Esque [of the US Geological Survey] has roped off 12 experimental sites, six of which he burned in 1999 to see how quickly invasive species re-establish themselves. But the result only showed the unpredictability of the desert:

the first year, the invasive red brome took hold, but this year, native wildflowers came back in force.

. . . Esque said 'It's not black and white with what is going on. We don't know if we are looking at coexistence or competition.'

(All content © 2001 *Contra Costa Times* (Walnut Creek, CA) and may not be republished without permission.)

1 *Some people have suggested bringing sheep into the desert to graze the invading grasses. Do you think this is a sensible idea? What further information would help you make a decision?*

2 *The US Geological Survey scientist found that red brome grass appeared to be outcompeting native flowers one year but not the next. Suggest some factors that may have changed the competitive outcome.*

In a broader sense, however, the significance of interspecific competition rests not on a limited number of high profile effects, but on an answer to the question 'How widespread are the ecological and evolutionary consequences of interspecific competition in practice?' We address this question in two ways. In the first, dealt with in Section 6.5.1, we ask 'How prevalent is current competition in natural communities?'. To demonstrate current competition requires experimental field manipulations, in which one species is removed from or added to the community and the responses of the other species are monitored. It is important to answer this question, because where current competition is demonstrable, neither the ghost of competition past nor spatial and temporal variation are likely to have a crucial role. And, if current competition is prevalent, then interspecific competition is likely to be an important structuring force in nature. However, even if current competition is *not* prevalent, past competition, and therefore competition generally, may still have played a significant role in structuring communities.

The second problem, dealt with in Section 6.5.2, is to distinguish between interspecific competition (past or present) and 'mere chance': species differ not as a reflection of interspecific competition but because they *are* different species. The many studies in which experimental field manipulations have not been possible can be examined to determine whether observed patterns provide strong evidence for a role for competition, or are open to alternative interpretations.

6.5.1 The prevalence of current competition

There have been two classic surveys of field experiments on interspecific competition. Schoener (1983) examined the results of all the experiments he could find – 164 studies in all. He found that approximately equal numbers of studies had dealt with terrestrial plants, terrestrial animals and marine organisms, but that studies of freshwater organisms amounted to only about half the number in the other groups. Amongst the terrestrial studies, however, he found that most were concerned with temperate regions and mainland populations, and that there were relatively few dealing with phytophagous (plant-eating) insects. Any conclusions were therefore bound to be subject to the limitations imposed by what ecologists had chosen to look at. Nevertheless, Schoener found that approximately 90% of the studies had demonstrated the existence of interspecific competition, and that the figures were 89%, 91% and 94% for terrestrial, fresh-water and marine organisms, respectively. Moreover, even if he looked at single species or small groups of species (of which there were 390) rather than at whole studies, which may have dealt with several groups of species, he found that 76% showed effects of competition at least sometimes, and 57% showed effects in all the conditions under which they were examined. Once again, terrestrial, freshwater and marine organisms gave very similar figures.

surveys of published studies of competition indicate that current competition is widespread . . .

Connell's (1983) review was less extensive than Schoener's: 72 studies, dealing with a total of 215 species and 527 different experiments. Interspecific competition was demonstrated in most of the studies, more than half of the species and approximately 40% of the experiments. In contrast to Schoener, Connell found that interspecific competition was more prevalent in marine than in terrestrial organisms, and also that it was more prevalent in large than in small organisms.

Taken together, Schoener's and Connell's reviews certainly seem to indicate that active, current interspecific competition is widespread. Its percentage occurrence amongst species is admittedly lower than its percentage occurrence amongst whole studies, but this is to be expected, since, for example, if four species were arranged along a single niche dimension and all adjacent species competed with each other, this would still be only three out of six (or 50%) of all possible pairwise interactions.

. . . but these surveys exaggerate to an unknown extent the true frequency of competition

Connell also found, however, that in studies of just one pair of species, inter-specific competition was almost always apparent, whereas with more species the prevalence dropped markedly (from more than 90% to less than 50%). This can be explained to some extent by the argument outlined above, but it may also indicate biases in the particular pairs of species studied, and in the studies that are actually reported (or accepted by journal editors). It is highly likely that many pairs of species are chosen for study because they are 'interesting' (because competition between them is suspected) and if none is found this is simply not reported. Judging the prevalence of competition from such studies is rather like judging the prevalence of debauched clergymen from the 'gutter press'. This is a real problem, only partially alleviated in studies on larger groups of species when a number of 'negatives' can be conscientiously reported alongside one or a few 'positives'. Thus the results of surveys, such as those by Schoener and Connell, exaggerate, to an unknown extent, the frequency of competition.

As previously noted, phytophagous insects were poorly represented in Schoener's data, but reviews of this group alone have tended to suggest either that competition is relatively rare in this group overall (Strong et al., 1984) or rare in at least certain types of phytophagous insects, for example 'leaf-biters' (Denno et al., 1995). On a more general level, it has been suggested that herbivores as a whole are seldom food-limited, and are therefore not likely to compete for common resources

(Hairston et al., 1960; Slobodkin et al., 1967). The bases for this suggestion are the observations that green plants are normally abundant and largely intact, they are rarely devastated, and most herbivores are scarce most of the time. Schoener found the proportion of herbivores exhibiting interspecific competition to be significantly lower than the proportions of plants, carnivores or detritivores.

Taken overall, therefore, current interspecific competition has been reported in studies on a wide range of organisms and in some groups its incidence may be particularly obvious, for example amongst sessile organisms in crowded situations. However, in other groups of organisms, interspecific competition may have little or no influence. It appears to be relatively rare among herbivores generally, and particularly rare amongst some types of phytophagous insect.

6.5.2 Competition or mere chance?

There is a tendency to interpret differences between the niches of coexisting species as confirming the importance of interspecific competition. But the theory of interspecific competition does more than predict 'differences'. It predicts not simply that the niches of competing species differ, but that they differ more than would be expected from chance alone. A more rigorous investigation of the role of interspecific competition, therefore, should address itself to the question: 'Does the observed pattern, even if it appears to implicate competition, differ significantly from the sort of pattern that could arise in the community even in the absence of any interactions between species?' This question has been the driving force behind analyses that seek to compare real communities with so-called *neutral models*. These are hypothetical models of actual communities that retain certain of the characteristics of their real counterparts, but reassemble or reconstruct some of the community components in a way that specifically excludes the consequences of interspecific competition. In fact, the neutral model analyses are attempts to follow a much more general approach to scientific investigation, namely the construction and testing of *null hypotheses*. The idea is that the data are rearranged into a form (the neutral model or null hypothesis) representing what the data would look like in the absence of interspecific competition. Then, if the actual data show a significant statistical difference from the null hypothesis, the null hypothesis is rejected and the action of interspecific competition is strongly inferred.

In fact, the approach has been applied to three different predictions of what a community structured by interspecific competition should look like: (i) potential competitors that coexist in a community should exhibit niche differentiation; (ii) this niche differentiation will often manifest itself as morphological differentiation; and (iii) within a community, potential competitors with little or no niche differentiation should not coexist, so each should tend to occur only where the other is absent ('negatively associated distributions'). The application of null hypotheses to community structure – that is, the reconstruction of natural communities with interspecific competition removed – has not been achieved to the satisfaction of all ecologists. But a brief examination of a study of niche differentiation in lizard communities shows the potential and rationale of the neutral model approach (Box 6.3). For these lizard communities, niches are more spaced out than would be expected by chance alone and interspecific competition therefore appears to play an important role in community structure.

neutral models

niche differentiation, morphological differentiation and negatively associated distributions

6.3 *Quantitative aspects*

6.3 QUANTITATIVE ASPECTS

Neutral models of lizard communities

Lawlor (1980) investigated differential resource utilization in 10 North American lizard communities, consisting of four to nine species. For each community, there were estimates of the amounts of each of 20 food categories consumed by each species. This pattern of resource use allowed the calculation, for each pair of species in a community, of an index of *resource use overlap*, which varied between 0 (no overlap) and 1 (complete overlap). Each community was then characterized by a single value: the mean resource overlap for all pairs of species present.

A number of 'neutral models' of these communities were then created. They were of four types. The first type, for example, retained the minimum amount of original community structure. Only the original number of species and the original number of resource categories were retained. Beyond that, species were allocated food preferences completely at random, such that there were far fewer species completely ignoring food in particular categories than in the real community. The niche breadth of each species was therefore increased. The fourth type, on the other hand, retained most of the original community structure: if a species ignored food in a particular category, then that was left unaffected, but among those categories where food was eaten, preferences were reassigned at random. These neutral models were then compared with their real counterparts in terms of their patterns of resource use overlap. If competition is a significant force in determining community structure, then the niches should be spaced out, and resource use overlap in the real communities should be less – and statistically significantly less – than that in the neutral models.

The results (Figure 6.19) were that in all communities, and for all four neutral models, the model mean overlap was higher than that observed for the real community, and that in almost all cases this was statistically significant. For these lizard communities, therefore, the observed low overlaps in resource use suggest that niches are more segregated than would be expected by chance alone, and that interspecific competition plays an important role in community structure.

A desert lizard of the southwestern United States.

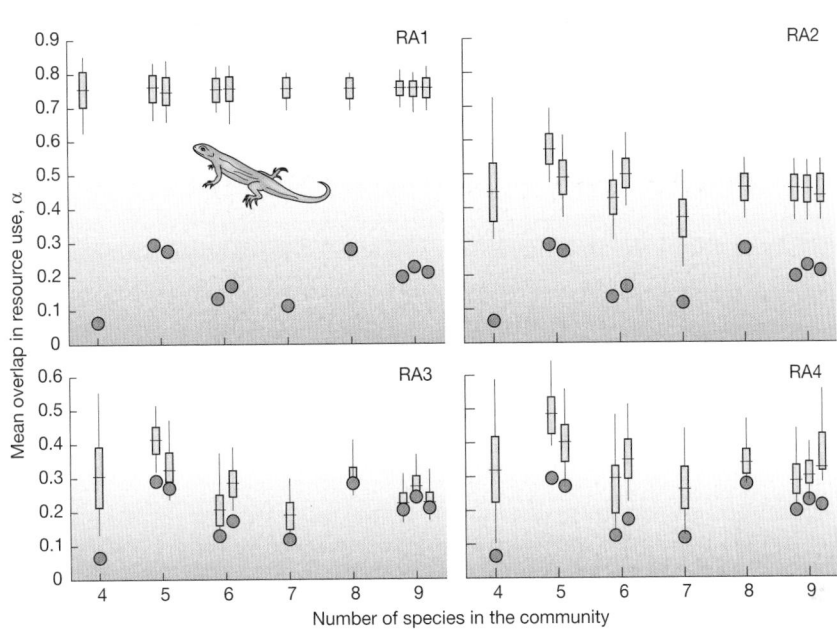

Figure 6.19

The mean indices of resource use overlap for each of 10 North American lizard communities are shown as solid circles. These can be compared, in each case, with the mean (horizontal line), standard deviation (vertical rectangle), and range (vertical line) of mean overlap values for the corresponding set of 100 randomly constructed communities. The analysis was performed using four different types of reorganization algorithms (RA1 to RA4).

AFTER LAWLOR, 1980

Where niche differentiation is manifested as morphological differentiation, the spacing out of niches can be expected to have its counterpart in regularity in the degree of morphological difference between species belonging to a guild. One example is shown in Figure 6.20 for four species of fossil strophomenide brachiopod (so-called 'lamp shells' that resemble bivalve mollusks) that appear from the fossil record to have coexisted. If successively sized species are compared, they have a consistent ratio for body outline length of around 1.5. Moreover, when Hermoyian et al. (2002) built 100,000 null models that each drew four species at random from the complete strophomenide brachiopod fossil fauna (74 taxa) and calculated size ratios between adjacent species, they rejected the null hypothesis that the observed ratios could have arisen from randomly selected taxa ($P < 0.03$), supporting the hypothesis that competition had played a key role in structuring this community.

morphological patterns

The null model approach to the analysis of distributional differences involves comparing the pattern of species co-occurrences at a suite of locations with what would be expected by chance. An excess of negative associations would then be consistent with a role for competition in determining community structure. Gotelli

negatively associated
distributions

Figure 6.20

Distributions of strophomenide body outline length (SOL) of samples of four coexisting species of brachiopods collected from a late Ordovician (ca. 448–438 million years before present) marine sediment in Indiana, USA. The species shown, from left to right, are *Eochonetes clarksvillensis*, *Leptaena richmondensis*, *Strophomena planumbona* and *Rafinesquina alternata*.

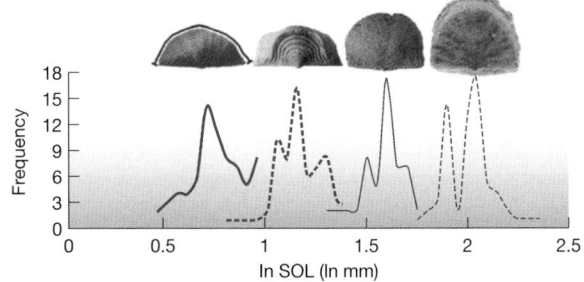

AFTER HERMOYIAN ET AL., 2002

and McCabe (2002) carried out a 'meta-analysis': an analysis of all the analyses of others that they could find (96 data sets in all) that had examined the distribution of species assemblages across sets of replicated sites. For every real data set, a 'checkerboard score', C, was computed. This is highest when every species-pair in a community forms a perfect checkerboard: sites are either 'black' or 'white' – the species never co-occur. It takes its lowest value when all species-pairs always co-occur. Next, 1000 randomized versions of each data set were simulated and C calculated each time. The observed C-value for each data set was then expressed as the number of standard deviations it was, C_s, from the mean of the simulations. The null hypothesis is that C_s should be zero (real communities not different from simulated communities), but in particular that a C_s-value greater than 2 indicates a significant negative association between species in the data set. The results, classified by taxonomic group, are shown in Figure 6.21. There was a significant excess of negative associations for plants and homeothermic vertebrates and for ants, but the excess was not significant for invertebrates (other than ants), fish, amphibians and reptiles.

This kind of pattern – sometimes a role for competition is confirmed, sometimes not – has been the general conclusion from the neutral model approach. What then should be our verdict on it? Perhaps most fundamentally, its aim is undoubtedly worthy. It concentrates the minds of investigators, stopping them from jumping to conclusions too readily; it is important to guard against the temptation to see competition in a community simply because we are looking for it. On the other hand, the approach can never take the place of a detailed understanding of the field ecology of the species in question, or of manipulative experiments designed to reveal competition by increasing or reducing species abundances. It, like so many other approaches, can only be part of the community ecologist's armory.

Figure 6.21

An analysis of data sets of species distributions across sites, classified by taxonomic group (mean ± SE) seeking evidence of an excess of negative associations, as measured by the standardized 'checkerboard score' (see text). The dashed line indicates an effects size of 2.0, which is the approximate 5% significance level.

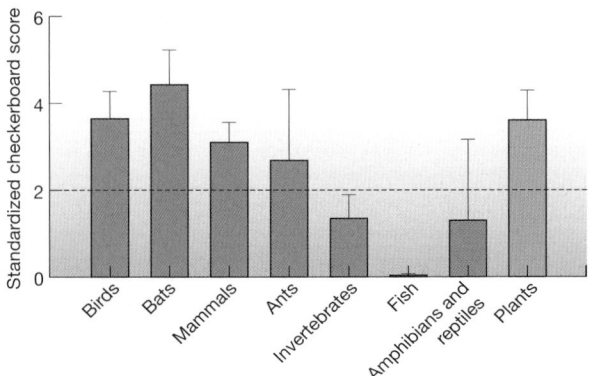

AFTER GOTELLI & MCCABE, 2002

Summary

Ecological effects of interspecific competition

The essence of interspecific competition is that individuals of one species suffer a reduction in fecundity, survivorship or growth as a result of exploitation of resources or interference by individuals of another species.

Species are often excluded by interspecific competition from locations at which they could exist perfectly well in the absence of interspecific competition.

With exploitation competition, the more successful competitor is the one that more effectively exploits shared resources. Two species exploiting two resources can compete but still coexist when each species holds one of the resources at a level that is too low for effective exploitation by the other species.

A fundamental niche is the combination of conditions and resources that allow a species to exist when considered in isolation from any other species. Whereas its realized niche is the combination of conditions and resources that allow it to exist in the presence of other species that might be harmful to its existence – especially interspecific competitors.

The Competitive Exclusion Principle states that if two competing species coexist in a stable environment, then they do so as a result of differentiation of their realized niches. If, however, there is no such differentiation, or if it is precluded by the habitat, then one competing species will eliminate or exclude the other. However, whenever we see coexisting species with different niches it is not reasonable to jump to the conclusion that this is the principle in operation.

The only true test for whether competition occurs between species is to manipulate the abundance of each competitor and observe the response of its counterparts.

Environments are usually a patchwork of favorable and unfavorable habitats; patches are often only available temporarily; and patches often appear at unpredictable times and in unpredictable places. Under such variable conditions, competition may only rarely 'run its course'.

Evolutionary effects of interspecific competition

Although species may not be competing now, their ancestors may have competed in the past. We can expect species to have evolved characteristics that ensure that they compete less, or not at all, with members of other species. Coexisting present-day competitors, and coexisting species that have evolved an avoidance of competition, can look, at least superficially, the same.

By invoking something that cannot be observed directly – 'the ghost of competition past' – it is impossible to prove an evolutionary effect of interspecific competition. However, careful observational studies have sometimes revealed patterns that are difficult to explain in any other way.

Interspecific competition and community structure

Interspecific competition tends to structure communities by acting within guilds – groups of species exploiting the same class of resource in a similar fashion.

Niche complementarity can be discerned in some communities, where coexisting species that occupy a similar position along one niche dimension tend to differ along another dimension.

Niches can be differentiated through differential resource utilization. In many cases, however, differential resource utilization expresses itself as either a microhabitat differentiation between the species or a difference in geographic distribution. Alternatively, differential resource utilization may express itself as a temporal separation between species. Niches can also be differentiated on the basis of conditions. This too can express itself as either a microhabitat differentiation, or a difference in geographic distribution, or a temporal separation.

▶

How significant is interspecific competition in practice?

Surveys of published studies of competition indicate that current competition is widespread but these exaggerate to an unknown extent the true frequency of competition.

The theory of interspecific competition predicts that the niches of competing species should be arranged regularly rather than randomly in niche space, that as a reflection of this they should be more distinct morphologically than expected by chance, and that competitors should be negatively associated in their distributions. Neutral models have been developed to determine what the community pattern would look like in the absence of interspecific competition. Real communities are sometimes structured in a way that makes an influence of competition difficult to deny.

Review questions

REVIEW QUESTIONS

Asterisks indicate challenge questions

1 Some experiments concerning interspecific competition have monitored both the population densities of the species involved and their impact on resources. Why is it helpful to do both?

2 Interspecific competition may be a result of exploitation of resources or of direct interference. Give an example of each and compare their consequences for the species involved.

3 Define fundamental niche and realized niche. How do these concepts help us to understand the effects of competitors?

4 With the help of one plant and one animal example, explain how two species may coexist by holding different resources at levels that are too low for effective exploitation by the other species.

5* Define the Competitive Exclusion Principle. When we see coexisting species with different niches is it reasonable to conclude that this is the principle in action?

6 Explain how environmental heterogeneity may permit an apparently 'weak' competitor to coexist with a species that might be expected to exclude it.

7* What is the 'ghost of competition past'? Why is it impossible to prove an evolutionary effect of interspecific competition?

8 Provide one example each of niche differentiation involving physiological, morphological and behavioral properties of coexisting species. How may these differences have arisen?

9 Define 'niche complementarity' and, with the help of an example, explain how it may help to account for the coexistence of many species in a community.

10* Discuss the pros and cons of the neutral model approach to evaluating the effects of competition on community composition.

Chapter **7**

Predation, grazing and disease

Key concepts

KEY CONCEPTS

In this chapter you will:

- distinguish the similarities and differences among 'true predators', grazers and parasites
- understand the subtleties of predation, including the capacity of prey to compensate
- appreciate the value of the optimal foraging approach for analyzing predator choices
- recognize the underlying tendency of populations of predators and prey to cycle and the 'damping' effect of crowding and patchy distributions
- understand the consequences of predation for community composition

Every living organism is either a consumer of other living organisms, or is consumed by other living organisms, or – in the case of most animals – is both. We cannot hope to understand the structure and dynamics of ecological populations and communities until we understand the links between consumers and their prey.

7.1 Introduction

predator: a term extending beyond the obvious examples

Ask most people to name a predator and they are almost certain to say something like lion, tiger or grizzly bear – something big, ferocious, instantly lethal. However, from an ecological point of view, a predator may be defined as any organism that consumes all or part of another living organism (its prey or host) thereby benefiting itself, but reducing the growth, fecundity or survival of the prey. Thus, this definition extends beyond the likes of lions and tigers by including organisms that consume all *or part* of their prey and also those that merely *reduce* their prey's growth, fecundity or survival. Predators are not all large, aggressive or instantly lethal – they need not even be animals. Here we examine these consumers together and try to understand the part they play in determining the structure and dynamics of ecological systems.

Within the broad definition, three main types of predator can be distinguished.

'true' predators, grazers and parasites

1 *'True' predators*:
  - invariably kill their prey and do so more or less immediately after attacking them;
  - consume several or many prey items in the course of their life.

  True predators therefore include lions, tigers and grizzly bears, but also spiders, baleen whales that filter plankton from the sea, zooplanktonic animals that consume phytoplankton, birds that eat seeds (each one an individual organism) and carnivorous plants.

2 *Grazers*:
  - attack several or many prey items in the course of their life;
  - consume only part of each prey item;
  - do not usually kill their prey, especially in the short term.

  Grazers therefore include cattle, sheep and locusts, but also, for example, blood-sucking leeches that take a small, relatively insignificant blood meal from several vertebrate prey over the course of their life.

3 *Parasites*:
  - consume only part of each prey item (usually called their host);
  - do not usually kill their prey, especially in the short term;
  - attack one or very few prey items in the course of their life, with which they therefore often form a relatively intimate association.

Parasites therefore include some obvious examples: animal parasites and pathogens such as tapeworms and the tuberculosis bacterium, plant pathogens like tobacco mosaic virus, parasitic plants like mistletoes, and the tiny wasps that form 'galls' on oak leaves. But aphids that extract sap from one or a very few plants with which they enter into an intimate association, and even caterpillars that spend their whole life on one host plant, are also, in effect, parasites.

On the other hand, these distinctions between 'true' predators, grazers and parasites, as with most categorizations of the living world, have been drawn in large part for convenience – certainly not because every organism fits neatly into one and only one category. We could, for example, have included a fourth class, the *parasitoids*, which are little known to non-biologists but are extensively studied by ecologists (and immensely important in the biological control of insect pests; see Chapter 12). Parasitoids are flies and wasps whose larvae consume their insect larva host from within, having been laid there as an egg by their mother. Parasitoids therefore straddle the 'parasite' and 'true predator' categories (only one host individual, which it always kills), fitting neatly into neither and confirming the impossibility of constructing clear boundaries.

parasitoids – and the artificiality of boundaries

There is, moreover, no satisfactory term to describe all the 'animal consumers of living organisms' to be discussed in this chapter. Detritivores and plants are also 'consumers' (of dead organisms, or of water, radiation, and so on); whilst the term 'predator' inevitably tends to suggest a 'true' predator even after we have defined it to encompass grazers and parasites too. But neither is it very satisfactory to be continually using the qualifier 'true' when discussing conventional predators. Thus, throughout this chapter, 'predator' will often be used as a shorthand term to encompass true predators, grazers and parasites, when general points are being made; but it will also be used to refer to predators in the more conventional sense, when it is obvious that this is what is being done.

A parasitoid wasp, which uses its long ovipositor to insert its eggs into the larvae of other insects, where they develop by consuming their host.

Figure 7.1

Relative growth rates (changes in height, with standard errors) of a number of different clones of the sand-dune willow, *Salix cordata*, in 1990 (a) and in 1991 (b), subjected either to no herbivory, low herbivory (four flea beetles per plant) or high herbivory (eight beetles per plant).

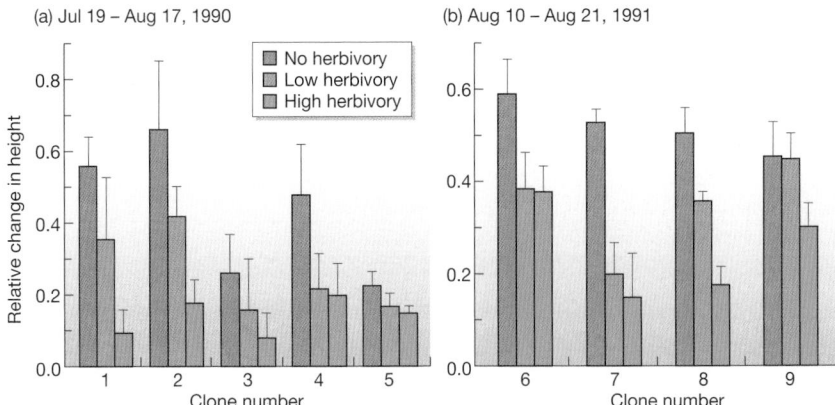

(a) Jul 19 – Aug 17, 1990   (b) Aug 10 – Aug 21, 1991

AFTER BACH, 1994

7.2 Prey fitness and abundance

predators reduce the fecundity and/or survival of individual prey

The fundamental similarity between predators, grazers and parasites is that each, in obtaining the resources it needs, reduces either the fecundity or the chances of survival of individual prey and may therefore reduce prey abundance. The effects of true predators on the survival of individual prey hardly need illustrating – the prey die. But the effects of grazers and parasites can be equally profound, if more subtle, as illustrated by the following two examples.

When the sand-dune willow, *Salix cordata*, was grazed by a flea beetle in two separate years – 1990 and 1991 – the reduction in the growth rate of the willow was marked in both years (Figure 7.1), but the consequences were rather different. Only in 1991 were the plants also subject to a severe shortage of water. Thus it was only in 1991 that the reduced growth rate was translated into plant mortality: 80% of the plants died in the high grazing treatment, 40% died in the low grazing treatment, but none of the ungrazed control plants died.

The pied flycatcher is a bird that migrates early each summer from tropical West Africa to Finland (and elsewhere in northern Europe) to breed. Males that arrive relatively early are particularly successful at finding mates. Late arrival therefore has a serious detrimental effect on the expected 'fecundity' of a male: the number of offspring that it can expect to father. Significantly, the later arrivals are disproportionately infected with the blood parasite *Trypanosoma* (Figure 7.2).

Figure 7.2

The proportion of males of male pied flycatchers (*Ficedula hypoleuca*) infected with *Trypanosoma* amongst groups of migrants arriving in Finland at different times.

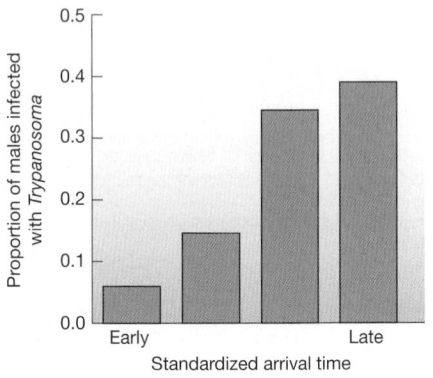

AFTER RATTI ET AL., 1993

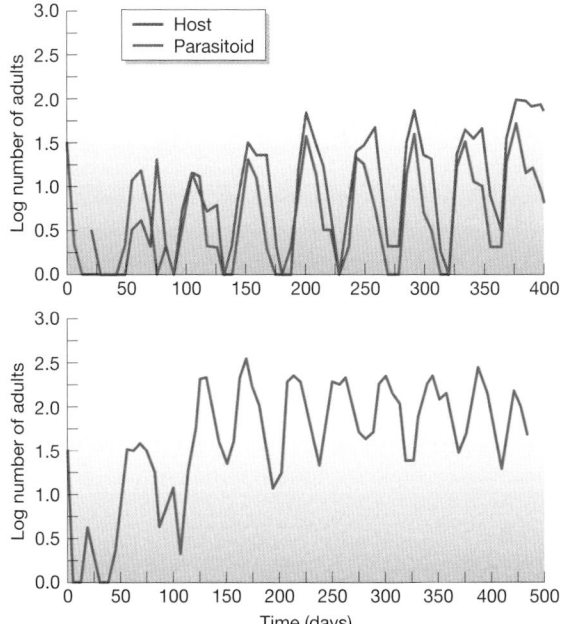

AFTER BEGON ET AL., 1995

Figure 7.3

Long-term population dynamics in laboratory population cages of a host (*Plodia interpunctella*), with and without its parasitoid (*Venturia canescens*). (a) Host and parasitoid, and (b) the host alone.

Infection with the parasite therefore has a profound effect on the reproductive output of individual birds.

It is not so straightforward, though, to demonstrate that reductions in the survival or fecundity of individual prey translate into reductions in prey abundance – we need to be able to compare prey populations in the presence and the absence of predators. As so often in ecology, we cannot rely simply on observation: we need experiments – either ones we set up ourselves, or natural experiments set up for us by nature.

For example, Figure 7.3 contrasts the dynamics of laboratory populations of an important pest, the Indian meal moth, with and without a parasitoid wasp, *Venturia canescens*. Ignoring the rather obvious regular fluctuations (cycles) in both moth and wasp, it is apparent that the wasp reduced moth abundance to less than one-tenth of what it would otherwise be (notice the logarithmic scale in the figure).

A particularly graphic example of the impact grazers can have is provided by the story of the invasion of Lake Moon Darra in North Queensland (Australia) by *Salvinia molesta*, a water fern that originated in Brazil. In 1978, the lake carried an infestation of 50,000 metric tons of the fern. In *Salvinia*'s native habitat in Brazil, the black long-snouted weevil (*Cyrtobagous* spp.) was known to graze only on *Salvinia*. Hence in June 1980, 1500 adults were released at an inlet to the lake and a further release was made in January 1981. By April 1981, *Salvinia* was dying throughout the lake, supporting an estimated population of one billion beetles. By August 1981, less than 1 metric ton of *Salvinia* remained. This was a 'controlled' experiment in that other lakes in the region continued to bear large populations of *Salvinia*.

All sorts of predators can cause reductions in the abundance of their prey. We shall see as this chapter develops, however, that they do not *necessarily* do so.

predators *can* reduce prey abundance – but do not *necessarily* do so

7.3 The subtleties of predation

There is much to be gained by stressing the similarities between different types of predators. On the other hand, it would be wrong to make this an excuse for over-simplification (there *are* important differences between true predators, grazers and parasites), or to give the impression that all acts of predation are simply a question of 'prey dies, predator takes one step closer to the production of its next offspring'.

7.3.1 Interactions with other factors

grazers and parasites may make prey more vulnerable to other forms of mortality

Grazers and parasites, in particular, often exert their harm not by killing their prey immediately like true predators, but by making the prey more vulnerable to some other form of mortality. For example, grazers and parasites may have a more drastic effect than is initially apparent because of an interaction with competition between the prey. This can be seen in a southern Californian salt marsh, where the parasitic plant, dodder (*Cuscuta salina*) attacks a number of plants including *Salicornia* (Figure 7.4). *Salicornia* tends to be the strongest competitor in the marsh, but it is also the preferred host of dodder. The distribution of plants in the marsh can therefore only be understood as a result of the interaction between competition and parasitism (Figure 7.4).

Infection or grazing may also make hosts or prey more susceptible to predation. For example, postmortem examination of red grouse (*Lagopus lagopus scoticus*) showed that birds killed by predators in the spring and summer carried significantly

Figure 7.4

The effect of dodder, *Cuscuta salina*, on competition between *Salicornia* and other species in a southern Californian salt marsh. (a) A schematic representation of the main plants in the community in the upper and middle zones of the marsh and the interactions between them (solid arrows: direct effects; dashed arrows: indirect effects). *Salicornia* (the relatively low growing plant in the figure) is most attacked by, and most affected by, dodder (which is not itself shown in the figure); but when uninfected, *Salicornia* competes strongly and symmetrically with *Arthrocnemum* at the *Arthrocnemum–Salicornia* border, and is a dominant competitor over *Limonium* and *Frankenia* in the middle (high *Salicornia*) zone. However, dodder significantly shifts the competitive balances. (b) Over time, *Salicornia* decreased and *Arthrocnemum* increased in plots infected with dodder. (b) Large patches of dodder suppress *Salicornia* and favor *Limonium* and *Frankenia*.

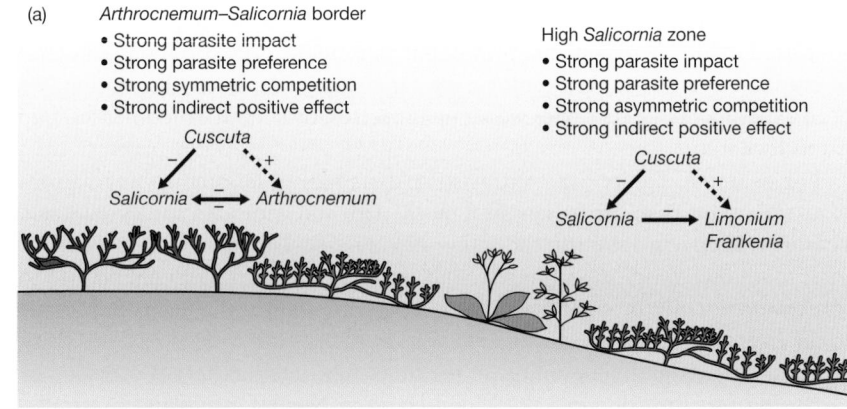

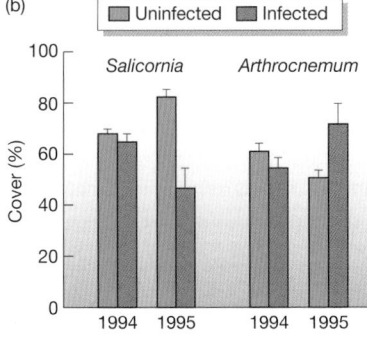
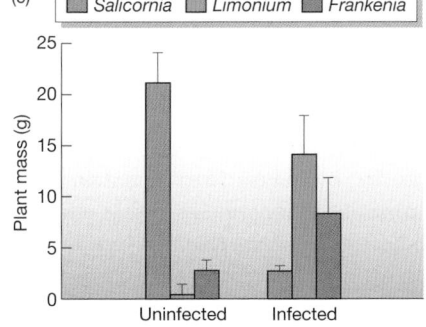

AFTER PENNINGS & CALLAWAY, 2002

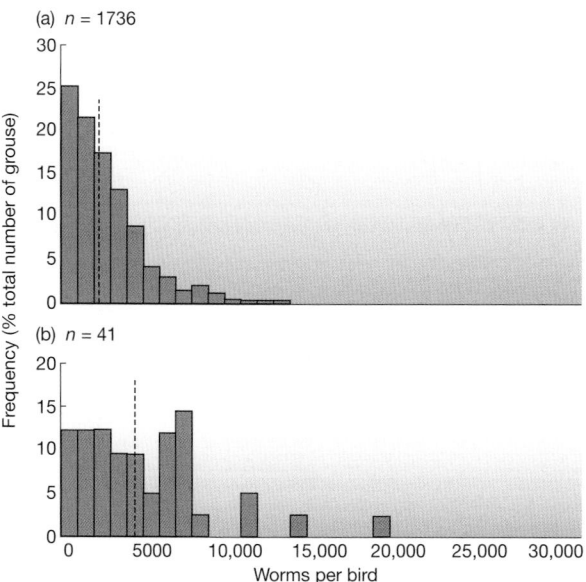

(a) *n* = 1736

(b) *n* = 41

Worms per bird

Frequency (% total number of grouse)

Figure 7.5

Infection with a nematode worm parasite makes red grouse more susceptible to predation. (a) Worm burdens of birds that are shot for 'sport', which may be taken as a representative sample of the whole population. (b) Worm burdens of those found killed by predators. The vertical line is the mean in each case, and the worm burdens of those caught by predators are clearly higher, typically, than those in the population as a whole.

greater burdens of the gut nematode parasite *Trichostrongylus tenuis* than the birds that remained in the fall (Figure 7.5).

7.3.2 Compensation and defense by individual prey

The effects of parasites and grazers, however, are not always more profound than they first seem. They are often *less* profound because, for example, individual plants can compensate in a variety of ways for the effects of herbivory (Strauss & Agrawal, 1999). The removal of leaves from a plant may decrease the shading of other leaves and thereby increase their rate of photosynthesis. Or, following herbivore attack, many plants compensate by utilizing stored reserves. Herbivory frequently alters the distribution of newly synthesized material within the plant, usually maintaining a balanced root : shoot ratio. When shoots are defoliated, an increased fraction of net production is channeled to the shoots themselves; when roots are destroyed, the switch is towards the roots. Often, there is compensatory regrowth of defoliated plants when buds that would otherwise remain dormant are stimulated to develop. There is also commonly a reduced subsequent death rate of surviving plant parts.

For example, when herbivory on the biennial plant field gentian (*Gentianella campestris*) was simulated by clipping to remove half its biomass (Figure 7.6a), subsequent production of fruits was increased (Figure 7.6b), but the outcome depended on the timing of clipping. Fruit production was much increased over controls if clipping occurred between July 12 and 20, but if clipping occurred later than this, fruit production was less in clipped plants than in unclipped controls. The period when the plants show compensation coincides with the time when damage by herbivores normally occurs.

Plants may also respond by initiating or increasing their production of defensive structures or chemicals. For example, a few weeks of grazing on the brown seaweed *Ascophyllum nodosum* by snails (*Littorina obtusata*) induced substantially increased

compensatory plant responses

defensive plant responses

Figure 7.6

(a) Clipping of field gentians to simulate herbivory causes changes in the architecture and numbers of flowers produced. (b) Production of mature (maroon histograms) and immature fruits (blue histograms) of unclipped control plants and plants clipped on different occasions from July 12 to 28, 1992. Means and standard errors are shown and all means are significantly different from each other ($P < 0.05$). Plants clipped on July 12 and 20 developed significantly more fruits than unclipped controls. Plants clipped on July 28 developed significantly fewer fruits than controls.

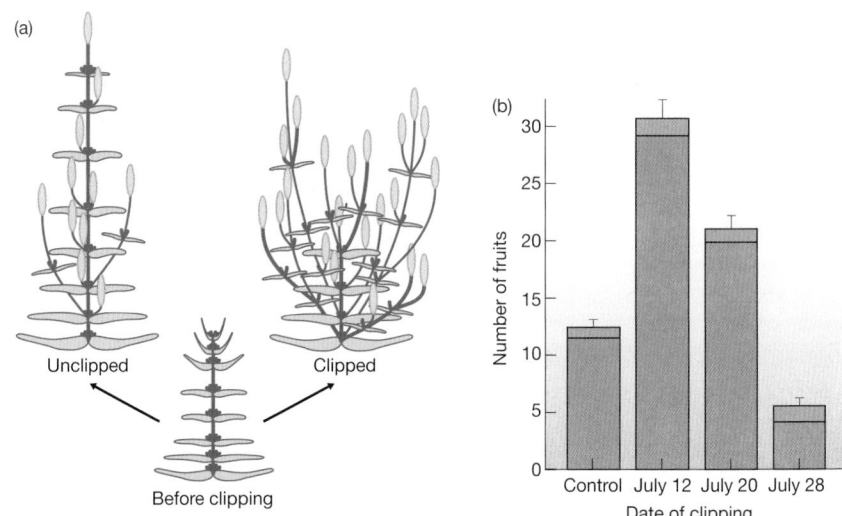

AFTER LENNARTSSON ET AL., 1998

concentrations of phlorotannins (Figure 7.7a), which reduce further snail grazing (Figure 7.7b). Interestingly, simple clipping of the plants did not have the same effect. The snails can stay and feed on the same plant for long time periods. Induced responses that take time to develop can still be effective in reducing damage.

The snails in Figure 7.7 suffer as a consequence of the seaweed's response (they eat less), and the plants benefit in that less of them is eaten. But that benefit comes to the plants at a cost (that of producing the chemicals), and it is therefore never straightforward to establish whether plants experience a net benefit in the longer term. One attempt to address this question looked at lifetime fitness of wild radish plants (*Raphanus sativus*) assigned to one of three treatments: (i) grazed by caterpillars, *Pieris rapae*; (ii) leaf-damage controls (an equivalent amount of biomass removed using scissors); and (iii) overall controls (undamaged). Earwigs (*Forficula* spp.) and other chewing herbivores caused 100% more leaf damage on control and damage control plants than on grazed plants, and there were 30% more phloem-sucking aphids on them (Figure 7.8a, b): the response induced

Figure 7.7

(a) Phlorotannin content of *Ascophyllum nodosum* plants after exposure to simulated herbivory (removing tissue with a hole punch) or grazing by the snail *Littorina obtusata*. Only the snail had the effect of inducing increased concentrations of the defensive chemical in the seaweed. Means and standard errors are shown. Different letters indicate that means are significantly different ($P < 0.05$). (b) In a subsequent experiment, the snails were presented with algal shoots from the control and the snail-grazed treatments in (a) – the snails ate significantly less of plants with high phlorotannin content.

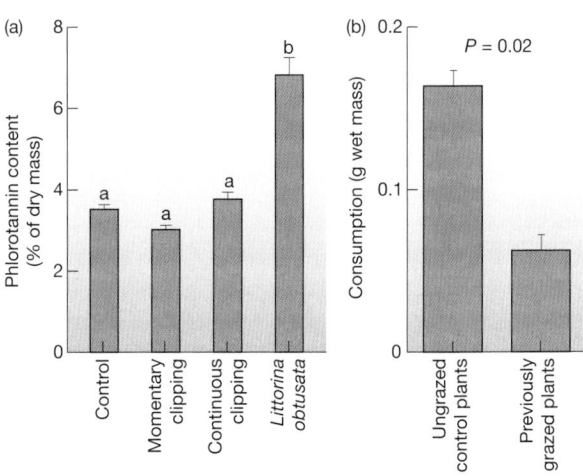

AFTER PAVIA & TOTH 2000

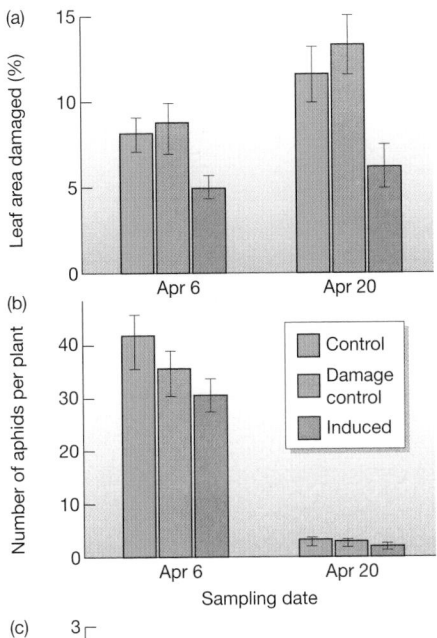

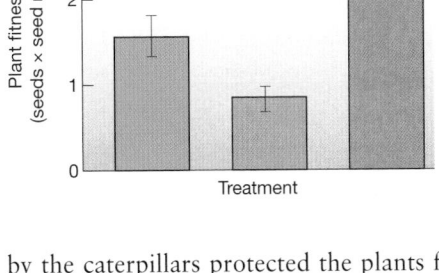

Figure 7.8

(a) Percentage leaf area consumed by chewing herbivores and (b) number of aphids per plant, measured on two dates (April 6 and 20) in three field treatments: overall control, damage control (tissue removed by scissors) and induced (caused by grazing of caterpillars of *Pieris rapae*). (c) Fitness of plants in the three treatments calculated by multiplying the number of seeds produced by the mean seed mass (in milligrams).

AFTER AGRAWAL, 1998

by the caterpillars protected the plants from additional herbivory. Moreover, despite any costs, this increased significantly (by more than 60%) the lifetime fitness of induced plants compared to the control plants. Plants cut with scissors, on the other hand, had 38% *lower* fitness than the overall controls, emphasizing the negative effect of tissue loss without the benefits of induction (Figure 7.8c). This fitness benefit occurred, however, only in environments containing herbivores. In their absence, the costs of producing the chemicals outweighed the benefits and plants suffered a reduction in fitness (Karban et al., 1999). Thus the benefits in the presence of herbivores were *net* benefits: benefits outweighed costs.

7.3.3 From individual prey to prey populations

In spite of these various qualifications, the general rule is that predators are harmful to individual prey. But the effects of predation on a population of prey are not always so predictable. The impact of predation is most commonly limited by compensatory reactions amongst the survivors as a result of reduced intraspecific competition. Outcomes of predation may, therefore, vary with food availability. When there is plenty of good food, and no competition, the effects of predation should be detectable. But when food is short and competition intense, predation

compensatory reactions amongst surviving prey . . .

Figure 7.9

Trajectories of numbers of grasshoppers surviving (mean ± SE) for fertilizer and predation treatment combinations in a field experiment involving caged plots in the Arapaho Prairie, Nebraska, USA.

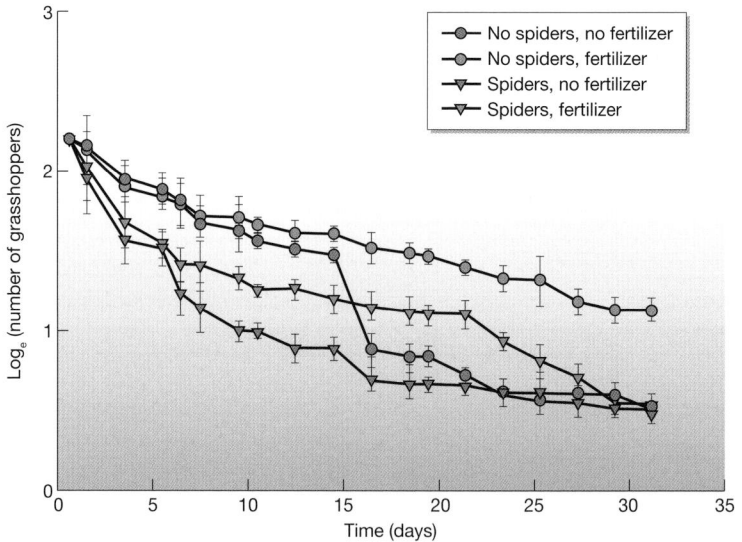

AFTER OEDEKOVEN & JOERN, 2000

may relieve competitive pressures and allow individuals to survive who would not otherwise do so. The results of an experiment that tested this are shown in Figure 7.9. The survival of grasshoppers (*Ageneotettix deorum*) was monitored in caged prairie plots with food (grass) that was either plentiful (fertilized) or limited (not fertilized), and in the presence or absence of predatory spiders. As predicted, with plentiful food, spider predation reduced the numbers surviving: a non-compensatory response. But with limited food, spider predation and food limitation were compensatory: the same numbers of grasshoppers were recovered at the end of the 31-day experiment.

Predation may also have a negligible impact on prey abundance if an increased loss of prey to predators at one stage of the prey's life simply leads to a decreased loss to predators at some other stage. If, for example, recruitment to a population of adult plants is not limited by the number of seeds produced, then insects that reduce seed production are unlikely to have an important effect on plant population dynamics. The point is illustrated by a study of the shrub, *Haplopappus venetus*, in California (Louda, 1982, 1983). The level of insect damage to the developing flowers and seeds was high. Experimental exclusion of flower and seed predators, therefore, caused a 104% increase in the number of developing seeds escaping damage. This led to an increase in the number of seedlings established. But subsequently this was followed by a much greater loss of seedlings, probably to vertebrate herbivores. As a consequence, original abundances were re-established in spite of the short-term importance of the seed predators.

... but compensation is often imperfect

Compensation, however, is by no means always perfect. Figure 7.10, for example, shows the results of an experiment in which Douglas fir seeds were sown both in open plots and in plots screened from rodents and birds. The immediate effect of this was an enormous reduction in the loss of seeds (though the screens were not totally effective). There were, however, compensatory *increases* in mortality from other causes. Nonetheless, in spite of this, the overall effect of screening was to more than double the number of seedlings still surviving 1 year after germination.

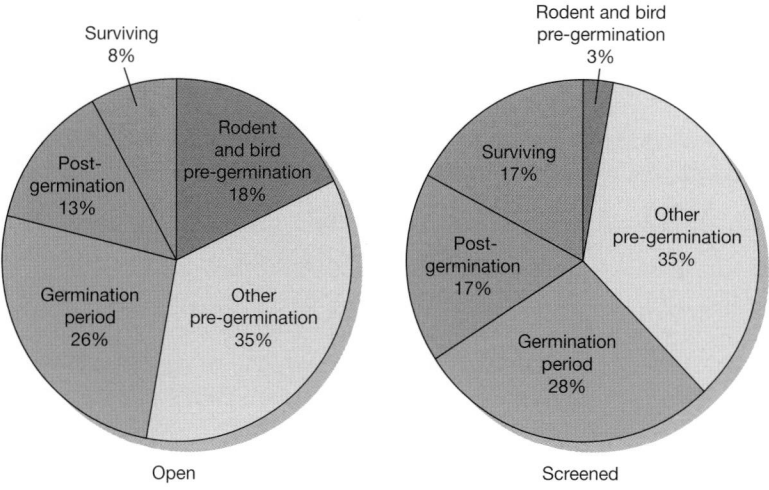

AFTER LAWRENCE & REDISKE, 1962

Open

Screened

Figure 7.10

When Douglas fir seeds are protected from vertebrate predation by screens, the lowered mortality is compensated for (but not *fully* compensated for) by increased mortality from other sources.

Predators may also have little impact on prey populations as a whole because of the particular individuals they attack. Many large carnivores, for example, concentrate their attacks on the old (and infirm), on the young (and naive) or on the sick. Thus, a study in the Serengeti found that cheetahs and wild dogs killed a disproportionate number from the younger age classes of Thomson's gazelles (Figure 7.11a) because: (i) these young animals were easier to catch (Figure 7.11b); (ii) they had lower stamina and running speeds; (iii) they were less good at out-maneuvering the predators (Figure 7.11c); and (iv) they may even have failed to recognize the predators. The effects of predation on the prey population will therefore have been less than would otherwise have been the case, because these young gazelles will have been making no present reproductive contribution to the population, and many would have died anyway, from other causes, before they were able to do so.

predators often attack the weakest and most vulnerable

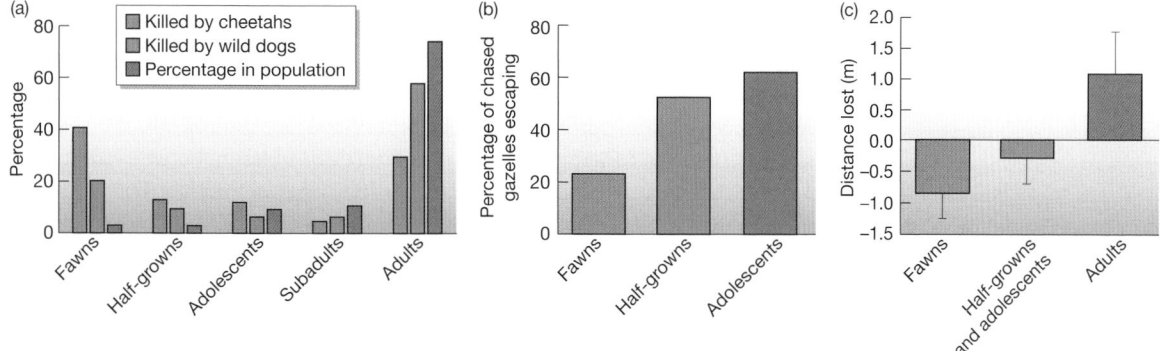

AFTER FITZGIBBON & FANSHAWE, 1989; FITZGIBBON, 1990

Figure 7.11

(a) The proportions of different age classes (determined by tooth wear) of Thomson's gazelles in cheetah and wild dog kills is quite different from their proportions in the population as a whole. (b) Age influences the probability for Thomson's gazelles of escaping when chased by cheetahs. (c) When prey (Thomson's gazelles) zigzag to escape chasing cheetahs, prey age influences the mean distance lost by the cheetahs.

Thomson's gazelle.

It is apparent, then, that the effects of a predator on an individual prey are crucially dependent on the response of the prey; and the effects on prey populations are equally dependent on which prey are attacked and on the responses of other prey individuals and other natural enemies of the prey. The effect of a predator may be more drastic than it appears, or less drastic. It is only rarely what it seems.

7.4 Predator behavior: foraging and transmission

So far, we have been looking, in effect, at what happens *after* a predator finds its prey. Now, we take a step back and examine how contact is established in the first place. This is crucially important, because this pattern of contact is critical in determining the predator's consumption rate, which goes a long way to determining its own level of benefit and the harm it does to the prey, which determines, in turn, the impact on the dynamics of predator and prey populations, and so on.

sit-and-wait predators

True predators and grazers typically 'forage'. Many move around within their habitat in search of their prey, and their pattern of contact is therefore itself determined by the predators' behavior – and sometimes by the evasive behavior of the prey (Figure 7.12a). This foraging behavior is discussed below. Other predators, web-spinning spiders for instance, 'sit and wait' for their prey, though almost always in a location they have selected (Figure 7.12b).

parasite transmission

With parasites and pathogens, on the other hand, we usually talk about transmission rather than foraging. This may be direct transmission between infectious and uninfected hosts when they come into contact with one another (Figure 7.12c), or free-living stages of the parasite may be released from infected hosts, so that it is the pattern of contact between these and uninfected hosts that is important (Figure 7.12d). The simplest assumption we can make for directly transmitted parasites – and one that often is made when attempting to understand their dynamics (discussed in Section 7.5) – is that transmission depends on infectious and uninfected hosts 'bumping into one another'. In other words, the overall rate of parasite transmission depends both on the density of uninfected, susceptible hosts (since these represent the size of the 'target') and on the density of infectious hosts (since this represents the risk of the target being 'hit') (Figure 7.12c).

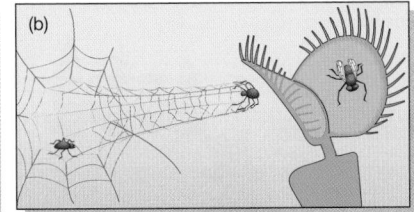

Figure 7.12

The different types of foraging and transmission. (a) Active predators seeking (possibly active) prey. (b) Sit-and-wait predators waiting for active prey to come to them. (c) Direct parasite transmission – infectious and uninfected hosts 'bumping into each other'. (d) Transmission between free-living stages of a parasite shed by a host and new, uninfected hosts.

7.4.1 Foraging behavior

There are many questions we might ask about the behavior of a foraging predator. Where, within the habitat available to it, does it concentrate its foraging? How long does it tend to remain in one location before moving on to another? And so on. Ecologists address all such questions from two points of view. The first is from the viewpoint of the *consequences* of the behavior for the dynamics of predator and prey populations. We turn to this in Section 7.5.

The second is the viewpoint of 'behavioral ecology' or 'optimal foraging'. The aim is to seek to understand why particular patterns of foraging behavior have been favored by natural selection. Most readers will be familiar with the general approach as applied, for example, to the anatomy of the bird's wing – we may seek to understand why a particular surface area, or a particular arrangement of feathers, has been favored by natural selection for the effectiveness they bring to the bird's powers of flight. Of course, this does not imply even a basic understanding of aerodynamics theory on the bird's part – only that those birds with the most effective wings have been favored in the past by natural selection and have passed their effectiveness on to their offspring. Likewise, in applying this approach to foraging behavior, there is no question of suggesting 'conscious decision-making' on the predator's part.

the evolutionary, optimal foraging approach

What, though, is the appropriate measure of 'effectiveness' in foraging behavior – the equivalent of flying ability as a criterion for a successful bird's wing? Usually, the *net* rate of energy intake has been used – that is, the amount of energy obtained per unit time, *after* account has been taken of the energy expended by the predator in carrying out its foraging. For many consumers, however, the efficient gathering of energy may be less critical than some other dietary constituent (e.g. nitrogen), or it may be of prime importance for the forager to consume a mixed and balanced diet. The predictions of optimal foraging theory do not apply to all the foraging decisions of every predator.

Figure 7.13

The types of foraging 'decisions' considered by optimal foraging theory. (a) Choosing between habitats. (b) The conflict between increasing input and avoiding predation. (c) Patch stay-time decisions. (d) The 'ideal free' decision – the conflict between patch quality and competitor density. (e) Optimal diets – to include or not to include an item in the diet (when something better might be 'round the corner').

applying the optimal foraging approach to a range of foraging behaviors

A range of the aspects of foraging behavior to which the optimal foraging approach has been applied is illustrated in Figure 7.13. These are elaborated on briefly here, before the whole approach is demonstrated by examining just one of them in detail.

- Where, within the habitat available to it, does a predator concentrate its foraging (Figure 7.13a)? Does it concentrate where the long-term expectation of net energy intake is highest *or* where the risk of extended periods of low intake is lowest?

- Does the location chosen by a predator reflect just the expected energy intake? Or does there appear to be some balancing of this against the risk of being preyed upon by its own predators (Figure 7.13b)?

- How long does a predator tend to remain in one location – one patch, say, of a patchy environment – before moving on to another (Figure 7.11c)? Does it remain for extended periods and hence avoid unproductive trips from one patch to another? Or does it leave patches early, before the resources there are depleted?

- What are the effects of other, competing predators foraging in the same habitat (Figure 7.11d)? The expected net energy intake from a location is now presumably a reflection of both its intrinsic productivity and the number of competing foragers. What is the expected distribution of the predators as a whole over the various habitat patches?

- The remaining 'question', in Figure 7.13e, and the one to which we now turn in Box 7.1 for a fuller illustration of the optimal foraging approach, is that of *diet width*. No predator can possibly be capable of consuming all types of prey. Simple design constraints prevent shrews from eating owls (even though shrews are carnivores) and prevent humming-birds from eating seeds. Even within their constraints, however, most animals consume a narrower range of food types than they are morphologically capable of consuming.

7.1 *Quantitative aspects*

Optimal diet width

Diet width is the range of food types consumed by a predator. In order to derive widely applicable predictions about when diets are likely to be broad or narrow, we need to strip down the act of foraging to its bare essentials. So, we can say that to obtain food, any predator must expend time and energy, first in searching for its prey, and then in handling it (i.e. pursuing, subduing and consuming it). While searching, a predator is likely to encounter a wide variety of food items. Diet width, therefore, depends on the responses of predators once they have encountered prey. Generalists, those with a broad diet, pursue a large proportion of the prey they encounter. Specialists, those with a narrow diet, continue searching except when they encounter prey of their specifically preferred type.

Generalists have the advantage of spending relatively little time searching – most of the items they find they pursue and, if successful, consume. But they suffer the disadvantage of including relative low-profitability items in their diet. That is, generalists enjoy a net intake of energy much of the time – but their *rate* of intake is often relatively low. Specialists, on the other hand, have the advantage of only including high-profitability items in their diet. But they suffer the disadvantage of spending a relatively large amount of their time searching for them. Thus, specialists spend

relatively long periods with a net expenditure of energy – but when they do take in energy it is at a relatively high rate. Determining the predicted optimal foraging strategy for a particular predator amounts to determining how these pros and cons should be balanced so as to maximize the *overall* net rate of energy intake, while searching for *and* handling prey (MacArthur & Pianka, 1966; Charnov, 1976).

We can start by taking it for granted that any predator will include the single most profitable type of prey in its diet: that is, the one for which the net rate of energy intake is highest. But should it include the next most profitable type of item too? Or, when it comes across such an item, should it ignore it and carry on searching for the *most* profitable type? And if it does include the second most profitable type, what about the third, and the fourth? And so on.

Consider first this 'second most profitable food type'. When will it pay a predator to include an item of this type in its diet (in energetic terms)? The answer is when, having found the item, its expected rate of energy intake over the time spent handling it exceeds its expected rate of intake if, instead, it continued to search for, and then handled, an item of the *most* profitable type. (The *expected* times are simply the average times for items of a particular type.) Expressing this in symbols, we call the expected searching and handling times for the most profitable type s_1 and h_1, and its energy content E_1, and the expected handling time for the second most profitable type h_2, and its energy content E_2. Then it pays the predator to increase the width of its diet if E_2/h_2 (i.e. the rate of intake, energy per unit time, if it handles the second-best type) is greater than $E_1/(s_1 + h_1)$ (the rate of intake if instead it searches for the most profitable type).

Suppose now that it did pay the predator to expand its diet. What about the third most profitable type? We argue in the same way as before: it will pay a predator to include this in its diet if, when it has found it, its expected rate of intake over the time spent handling it, h_3, exceeds the expected rate if it searches for and handles *either* of the two most profitable types, both already included in its diet. Thus, if we call $\bar{s}$, $\bar{h}$, and $\bar{E}$ the searching and handling times and energy content for items already in the diet, it will pay the predator to expand its diet if E_3/h_3 exceeds $\bar{E}/(\bar{s} + \bar{h})$, or, more generally, if E_n/h_n exceeds $\bar{E}/(\bar{s} + \bar{h})$, where n refers generally to the 'next' most profitable prey type (not already in the diet). The ecological implications of this rule are considered in the main text.

predictions of the optimal diet model

In summary, Box 7.1 suggests that a predator should continue to add increasingly less profitable items to its diet as long as this increases its overall rate of energy intake. This will serve to maximize its overall rate of energy intake. This 'optimal diet model', then, leads to a number of predictions.

1 Predators with handling times that are typically short compared to their search times should be generalists (i.e. have broad diets), because in the short time it takes them to handle a prey item that has already been found, they can barely begin to search for another prey item. This prediction seems to be supported by the broad diets of many insectivorous birds feeding in trees and shrubs. Searching is always moderately time-consuming, but handling the minute, stationary insects takes negligible time and is almost always successful. A bird, therefore, has something to gain and virtually nothing to lose by consuming an item once found, and overall profitability is maximized by a broad diet.

2 By contrast, predators with handling times that are long relative to their search times should be specialists: maximizing the rate of energy intake is achieved by including only the most profitable items in the diet. For instance, lions live more or less constantly in sight of their prey so that search time is negligible; handling time, on the other hand, and particularly pursuit time, can be long (and very energy-consuming). Lions consequently specialize on those prey that can be pursued most profitably: the immature, the lame and the old.

3 Other things being equal, a predator should have a broader diet in an unproductive environment (where prey items are relatively rare and search times relatively large) than in a productive environment (where search times are generally smaller). This prediction is supported by a study of brown and black bears (*Ursos arctos* and *U. americanus*) feeding on salmon in Bristol Bay in Alaska (Figure 7.14). When salmon availability was high, bears consumed less biomass per captured fish, targeting energy-rich fish (those that had not spawned) or energy-rich body parts (eggs in females, brain in males). That is, their diet became more specialized when prey were abundant.

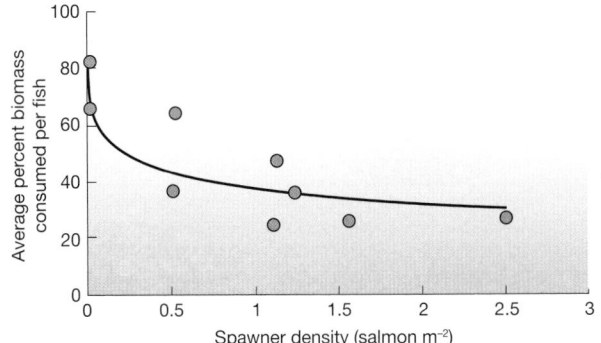

AFTER GENDE ET AL., 2001

Figure 7.14
As the spawning density (i.e. the abundance) of salmon increases, the average percentage of each salmon consumed by bears decreases: as prey abundance increases, the predators become more specialized.

Overall, then, we can see how an evolutionary, optimal foraging approach can help us make sense of predators' foraging behavior – how it makes predictions of what that behavior might be expected to be, and that these predictions may be supported by real examples.

7.5 Population dynamics of predation

What roles do predators play in driving the dynamics of their prey, or prey play in driving the dynamics of their predators? Are there common patterns of dynamics that emerge? The preceding sections should have made it plain that there are no simple answers to these questions. It depends on the detail of the behavior of individual predators and prey, on possible compensatory responses at individual and population levels, and so on. Rather than despair at the complexity of it all, however, we can build an understanding of these dynamics by starting simply and then adding additional features one by one to construct a more realistic picture.

building a picture from simple beginnings

7.5.1 Underlying dynamics of predator–prey interactions: a tendency to cycle

We begin by consciously oversimplifying – ignoring everything but the predator and the prey, and asking what underlying tendency there might be in the dynamics of their interaction. It turns out that the underlying tendency is to exhibit coupled oscillations – cycles – in abundance. With this established, we can turn to the many other important factors that might modify or override this underlying tendency. Rather than explore each and every one of them, however, Sections 7.5.4 and 7.5.5 examine just two of the more important ones: crowding and spatial patchiness. These two factors cannot, of course, tell the whole story; but they illustrate how the differences in predator–prey dynamics, from example to example, might be explained by the varying influences of the different factors with a potential impact on those dynamics.

Starting simply then, suppose there is a large population of prey. Predators presented with this population should do well: they should consume many prey and hence increase in abundance themselves. The large population of prey thus gives rise to a large population of predators (Figure 7.15). But this increasing

Figure 7.15

The underlying tendency for
predators and prey to display
coupled oscillations in abundance
as a result of the time delays in
their responses to each other's
abundance.

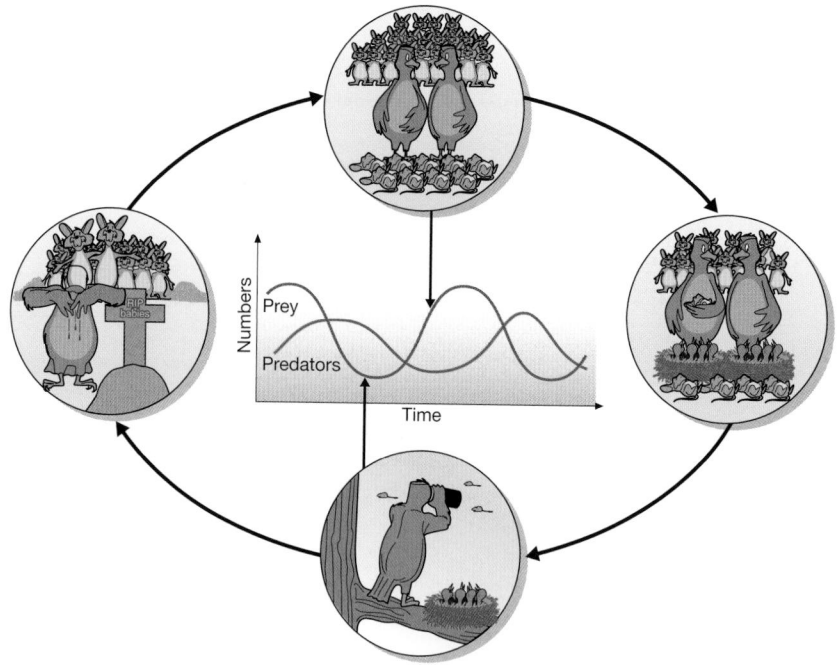

population of predators increasingly takes its toll of the prey. The large population of predators therefore gives rise to a small population of prey. Now the
predators are in trouble: large numbers of them and very little food. Their abundance declines. But this takes the pressure off the prey: the small population of
predators gives rise to a large population of prey – and the populations are back
to where they started. There is, in short, an underlying tendency for predators and
their prey to undergo coupled oscillations in abundance – population cycles
(Figure 7.15) – essentially because of the time delays in the response of predator
abundance to that of the prey, and vice versa. (A 'time delay' in response means,
for example, that a high predator abundance reflects a high prey abundance *in
the past*, but it *coincides* with declining prey abundance, and so on.) A simple
mathematical model – the Lotka–Volterra model – conveying essentially the same
message is described in Box 7.2.

7.5.2 Predator–prey cycles in practice

the 'expectation' of cycles is only
rarely fulfilled

This underlying tendency for predator–prey interactions to generate coupled
oscillations in abundance could produce an 'expectation' of such cycles in real
populations, but there were many aspects of predator and prey ecology that
had to be ignored in order to demonstrate this underlying tendency, and these
can greatly modify expectations. It is no surprise, then, that there are rather
few good examples of clear predator–prey cycles – albeit ones that have received
a great deal of attention from ecologists. Nonetheless, in trying to make sense of
predator–prey population dynamics, cycles – the underlying tendency – are a
good place to start.

7.2 Quantitative aspects

The Lotka–Volterra predator–prey model

Here, as in Boxes 5.4 and 6.1, one of the foundation-stone mathematical models of ecology is described and explained. The model is known (like the model of interspecific competition in Box 6.1) by the name of its originators: Lotka and Volterra (Volterra, 1926; Lotka, 1932). It has two components: P, the numbers present in a predator (or consumer) population, and N, the numbers or biomass present in a prey or plant population.

It is assumed that in the absence of consumers the prey population increases exponentially (see Box 5.4):

$$dN/dt = rN$$

But now we also need a term signifying that prey individuals are removed from the population by predators. They will do this at a rate that depends on the frequency of predator–prey encounters, which will increase with increasing numbers of predators (P) and prey (N). The exact number encountered and consumed, however, will also increase with the searching and attacking efficiency of the predator, denoted by a. The consumption rate of prey will thus be aPN, and overall:

$$dN/dt = rN - aPN \qquad (1)$$

Turning to predator numbers, in the absence of food these are assumed to decline exponentially through starvation:

$$dP/dt = -qP$$

where q is their mortality rate. But this is counteracted by predator birth, the rate of which is assumed to depend on: (i) the rate at which food is consumed, aPN; and (ii) the predator's efficiency, f, at turning this food into predator offspring. Overall:

$$dP/dt = faPN - qP \qquad (2)$$

Equations 1 and 2 constitute the Lotka–Volterra model.

The properties of this model can be investigated by finding zero isoclines (see Box 6.1). There are separate predator and prey zero isoclines, both of which are drawn on a graph of prey density (x-axis) against predator density (y-axis) (Figure 7.16). The prey zero isocline joins combinations of predator and prey densities that lead to an unchanging prey population, $dN/dt = 0$, while the predator zero isocline joins

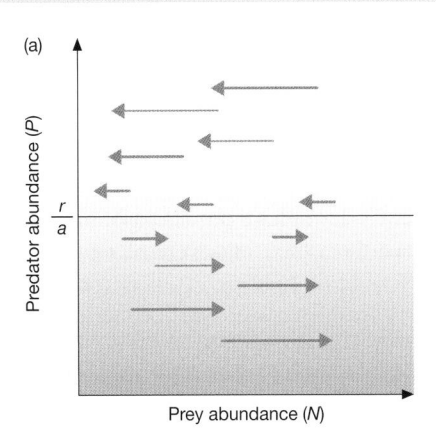

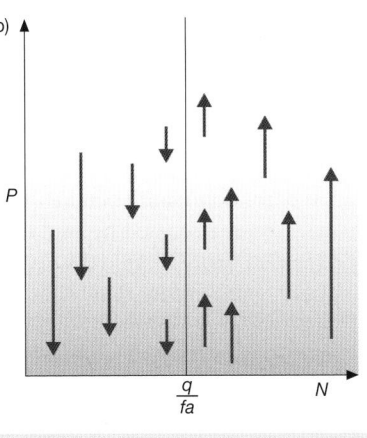

Figure 7.16

See box text for details.

combinations of predator and prey densities that lead to an unchanging predator population, $dP/dt = 0$.

In the case of the prey, we 'solve' for $dN/dt = 0$ in equation 1, giving the equation of the isocline as:

$$P = r/a$$

Thus, since r and a are constants, the prey zero isocline is a line for which P itself is a constant (Figure 7.16a): prey increase when predator abundance is low ($P < r/a$) but decrease when it is high ($P > r/a$).

Similarly, for the predators, we solve for $dP/dt = 0$ in equation 2, giving the equation of the isocline as:

$$N = q/fa$$

The predator zero isocline is therefore a line along which N is constant (Figure 7.16b): predators decrease when prey abundance is low ($N < q/fa$) but increase when it is high ($N > q/fa$).

Putting the two isoclines (and two sets of arrows) together in Figure 7.17 shows the behavior of joint populations. The various combinations of increases and decreases, listed above, mean that the populations undergo 'coupled oscillations' or 'coupled cycles' in abundance; 'coupled' in the sense that the rises and falls of the predators and prey are linked, with predator abundance tracking that of the prey (discussed biologically in the main text).

It is important to realize, however, that the model does not 'predict' the exact patterns of abundance that it generates. The world is much more complex than imagined by the model. But it does capture the essential tendency for coupled cycles in predator–prey interactions.

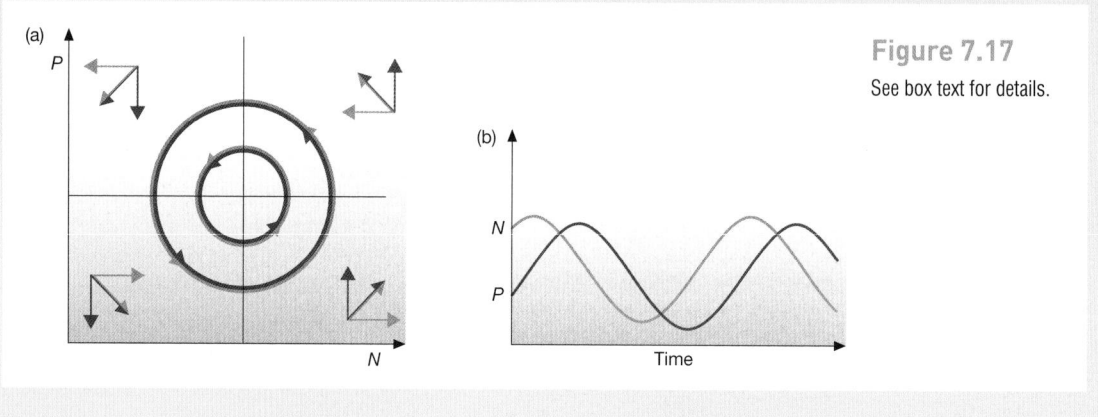

Figure 7.17

See box text for details.

plants, hares and lynx in
North America . . .

They do occur sometimes. It has been possible in several cases, for example, to generate coupled predator–prey oscillations, several generations in length, in the laboratory (Figure 7.18a; see also Figure 7.22c). Amongst field populations, there are a number of examples in which regular cycles of prey and predator abundance can be discerned. Cycles in hare populations, in particular, have been discussed by ecologists since the 1920s, and were recognized by fur trappers more than 100 years earlier. Most famous of all is the snowshoe hare, *Lepus americanus*, which in the boreal forests of North America follows a '10-year cycle' (although in reality this varies in length between 8 and 11 years; see Figure 7.18b). The snowshoe hare is the dominant herbivore of the region, feeding on the terminal twigs of numerous shrubs and small trees. A number of predators, including the Canada lynx (*Lynx canadensis*), have associated cycles

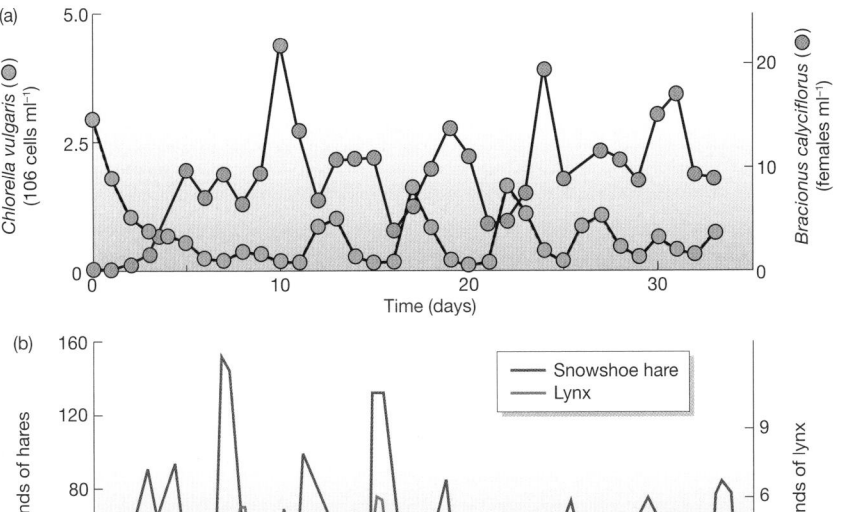

Figure 7.18

Coupled oscillations in the abundance of predators and prey. (a) Parthenogenetic female rotifers, *Bracionus calyciflorus* (predators, maroon circles), and unicellular green algae, *Chlorella vulgaris* (prey, blue circles), in laboratory cultures. (b) The snowshoe hare (*Lepus americanus*) and the Canada lynx (*Lynx canadensis*) as determined by the number of pelts lodged with the Hudson Bay Company.

The Canada lynx and the snowshoe hare – a predator and prey that may show coupled oscillations.

of similar length. The hare cycles often involve 10–30-fold changes in abundance, and 100-fold changes can occur in some habitats. They are made all the more spectacular by being virtually synchronous over a vast area from Alaska to Newfoundland.

But are the hare and lynx participants in a predator–prey cycle? This immediately seems less likely once one appreciates the number of other species with which both interact. Their food web (see Section 9.5) is shown in Figure 7.19. In fact, both experimental studies (Krebs et al., 2001) and statistical analyses of the population dynamics data (Stenseth et al., 1997) suggest that whereas the dynamics of the hares are driven by their interactions with both their food and their predators (especially lynx), the dynamics of the lynx are driven largely by their interaction with their hare prey, much as the food web might suggest. Both the hare–plant and the predator–hare interactions have some propensity

. . . but how are the cycles generated?

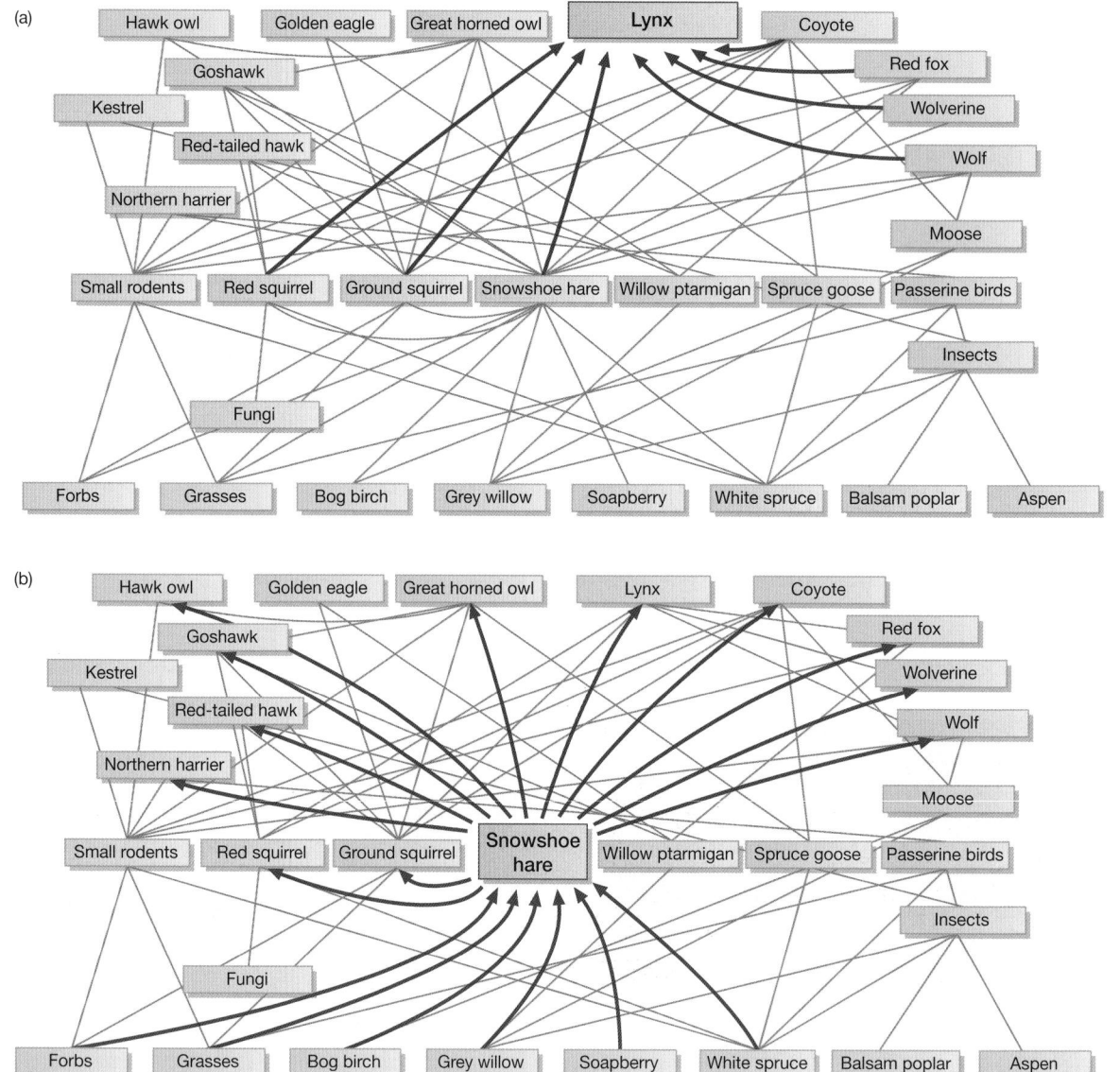

Figure 7.19

(a) The main species and groups of species in the boreal forest community of North America, with trophic interactions (who eats who) indicated by lines joining the species, and those affecting the Canada lynx shown as maroon arrows, pointing toward the consumer. (b) The same community, but with the interactions of the snowshoe hare shown as arrows.

AFTER STENSETH ET AL., 1997

to cycle on their own – but in practice the cycle seems normally to be generated by the interaction between the two. This warns us that even when we have a predator–prey pair both exhibiting cycles, we may still not be observing simple predator–prey oscillations.

Apparent instances of predator–prey cycles sometimes make the news – see Box 7.3 for an example.

7.3 *Topical ECOncerns*

7.3 TOPICAL ECONCERNS

A cyclical outbreak of a forest insect in the news

Large outbreaks of forest tent caterpillars occur about every 10 years, and each lasts for 2–4 years. During these outbreaks, massive damage is done to the foliage of forest trees over large tracks of land. This article appeared in the *Telegraph Herald* (Dubuque, Iowa) on June 11, 2001.

Caterpillars making a meal out of northern forests

Forest tent caterpillars have munched their way through much of northern Wisconsin, eating aspen, sugar maple, birch and oak from Tomahawk to southern Canada.

The insects move across roads in waves that make the pavement seem to crawl and hang from trees in large clumps. . . . 'One lady from Eagle River said they were on her house and on her driveway and on her sidewalk, and she was ready to move back to Oak Creek', said Jim Bishop, public affairs manager for the Department of Natural Resource's northern region.

Shane Weber, a DNR forest entomologist from Spooner, said the caterpillars on sidewalks, driveways and highways are a good sign. 'Whenever they start these mass overland moves, suddenly moving in waves across the ground, it means that they're starving, looking for another source of food', he said.

In Superior, customers have inundated Dan's Feed Bin [general store], looking for ways to rid their yards and homes of the insects. Employee Amy Connor said some customers held their telephones up to the window so Connor could hear the worms falling like hail. 'It's terribly gross', she said.

The caterpillars have eaten most of the leaves in the Upper Peninsula, said Jeff Forslund, of Hartland, who drove to Ramsey, Michigan. 'My grandfather has about 500 acres of aspen, and there isn't a leaf left', Forslund said.

Most of the trees will survive and the caterpillars should start spinning cocoons by mid-June, the DNR said. Forest entomologist Dave Hall said he expects the outbreak to peak this year. 'I can't imagine it getting much worse', he said. The last infestation of the native forest tent caterpillars in Wisconsin was in the late 1980s and early 1990s. . . . During the last tent caterpillar outbreak, several serious traffic collisions in Canada were blamed on slick roads from squashed tent caterpillars.

About 4 million of the fuzzy crawlers can be found per acre at the peak of the cyclical infestation, the DNR said.

1   *From what you have learnt about population cycles in this chapter, suggest an ecological scenario to account for the periodic outbreaks of these caterpillars.*

2   *Do you believe the comment attributed to a Department of Natural Resources (DNR) employee that the mass movement of the caterpillars is a good sign? How would you determine whether this behavior heralds an end to the peak phase of the cycle?*

7.5.3 Disease dynamics and cycles

basic reproductive rate and the transmission threshold

Cycles are also apparent in the dynamics of many parasites, especially microparasites (bacteria, viruses, etc.). To understand the dynamics of any parasite, the best starting point is its basic reproductive rate, conventionally called 'R nought', R_0. For microparasites, R_0 is the average number of new infected hosts that would arise from a single infectious host in a population of susceptible hosts. An infection will eventually die out for $R_0 < 1$ (each present infection leads to less than one infection in the future), but an infection will spread for $R_0 > 1$. There is therefore a 'transmission threshold' when $R_0 = 1$, which must be crossed if a disease is to spread. A derivation of R_0 for microparasites with direct transmission (see Figure 7.12c) is given in Box 7.4.

threshold population sizes and microparasite cycles

Box 7.4 provides us with a crucial insight into disease dynamics – for each directly transmitted microparasite there is a critical threshold population size that needs to be exceeded for a parasite population to be able to sustain itself. For example, measles has been calculated to have a threshold population size of around 300,000 individuals and is unlikely to have been of great importance until quite recently in human biology. However, it has generated major epidemics in the growing cities of the industrialized world in the 18th and 19th centuries, and in the growing concentrations of population in the developing world in the

7.4 *Quantitative aspects*

Transmission threshold for microparasites

Putting it simply, for microparasites with direct transmission, the basic reproductive rate, R_0, measures the average number of new infections arising from a single infected individual in a population of susceptible hosts. It increases with the average period of time over which an infected host remains infectious, L, since a long infectious period means plenty of opportunity to transmit to new hosts; it increases with the number of susceptible individuals in the host population, S, because more susceptible hosts offer more opportunities ('targets') for transmission of the parasite; and it increases with the transmission rate of the infection, β, because this itself increases first with the infectiousness of the parasite – the probability that contact leads to transmission – but also with the likelihood of infectious and susceptible hosts coming into contact as a reflection of the pattern of host behavior (Anderson, 1982). Thus, overall:

$$R_0 = S \cdot \beta L$$

We know that $R_0 = 1$ is a transmission threshold, in that below this the infection will die out but above it the infection will spread. But this in turn allows us to define a critical *threshold population size* S_T: the number of susceptibles that give rise to $R_0 < 1$. At that threshold, making $R_0 = 1$ in the equation means:

$$S_T = 1/\beta L$$

In populations with fewer susceptibles than this, the infection will die out ($R_0 < 1$), but with more than this, the infection will spread ($R_0 > 1$). The threshold population size is larger (more individuals are required to sustain an infection) when infectiousness (β) is low and/or infections themselves are short-lived (small L).

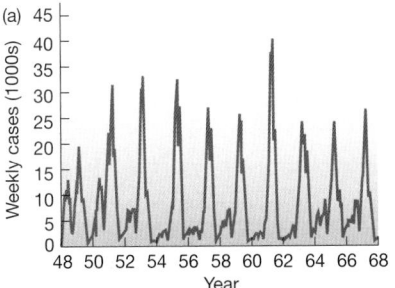

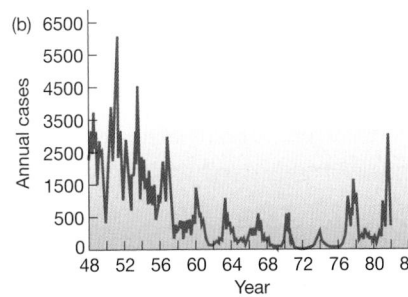

Figure 7.20

(a) Reported cases of measles in England and Wales from 1948 to 1968, prior to the introduction of mass vaccination. (b) Reported cases of pertussis (whooping cough) in England and Wales from 1948 to 1982. Mass vaccination was introduced in 1956.

20th century. Current estimates suggest that around 900,000 deaths occur each year from measles infection in the developing world (Walsh, 1983).

Moreover, the immunity induced by many bacterial and viral infections, combined with death from the infection, reduces the number of susceptibles in a population, reduces R_0, and therefore tends to lead to a decline in the incidence of the disease itself. In due course, though, there will be an influx of new susceptibles into the population (as a result of new births or perhaps immigration), an increase in R_0, an increase in incidence, and so on. There is thus a marked tendency with such diseases to generate a sequence from 'high incidence', to 'few susceptibles', to 'low incidence', to 'many susceptibles', to 'high incidence', etc. – just like any other predator–prey cycle. This undoubtedly underlies the observed cyclic incidence of many human diseases (especially prior to modern immunization programs), with the differing lengths of cycle reflecting the differing characteristics of the diseases: measles with peaks every 1 or 2 years, pertussis (whooping cough) every 3–4 years, diphtheria every 4–6 years, and so on (Figure 7.20).

7.5.4 Crowding

One fundamental feature that we have ignored so far is the fact that no predator lives in isolation: all are affected by other predators. The most obvious effects are competitive; many predators compete, and this results in a reduction in the consumption rate per individual as predator density increases (see Chapter 3). However, even when food is not limited, the consumption rate per individual can be reduced by increases in predator density by a number of processes known collectively as 'mutual interference'. For example, many predators interact behaviorally with other members of their population, leaving less time for feeding. Humming-birds actively and aggressively defend rich sources of nectar; parasitoid wasps will threaten and, if need be, fiercely drive away an intruder from their own area of tree trunk. Alternatively, an increase in consumer density may lead to an increased rate of emigration, or of consumers stealing food from one another (as do many gulls), or the prey themselves may respond to the presence of consumers and become less available for capture.

In all such cases, the underlying pattern is the same: the consumption rate per individual predator declines with increasing predator density. This reduction is likely to have an adverse effect on the fecundity, growth and mortality of individual predators, which intensifies as predator density increases. The predator population is thus subject to density-dependent regulation (see Chapters 3 and 5).

mutual interference amongst predators reduces the predation rate

AFTER ANDERSON & MAY, 1991

Figure 7.21

Host immune responses are necessary for density dependence in infections of the rat with the nematode *Strongyloides ratti*. Survivorship is independent of initial dose in mutant rats without an immune response (●; slope not significantly different from 0), but with an immune response (■) it declines (slope = −0.62, significantly less than 0; $P < 0.001$).

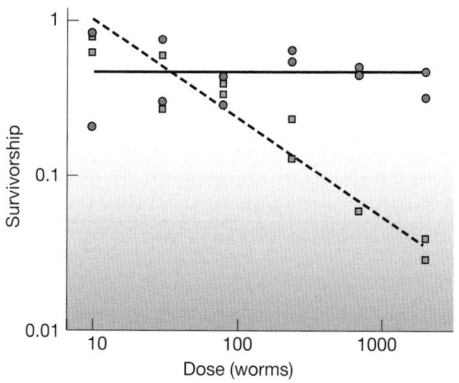

AFTER PATERSON & VINEY, 2002

competition or the immune response in parasites?

With parasites, too, it is to be expected that individuals will often interfere with each other's activities, and that there will be intraspecific competition between parasites and density dependence in their growth, birth and/or death rates. However, for vertebrate hosts at least, we need to remember that the intensity of the immune reaction elicited from a host also typically depends on the abundance of parasites. A rare attempt to disentangle these two effects utilized the availability of mutant rats lacking an effective immune response (Figure 7.21). These and normal, control rats were subjected to experimental infection with a nematode, *Strongyloides ratti*, at a range of doses. Any reduction in parasite fitness with dose in the normal rats could be due to intraspecific competition and/or an immune response that itself increases with dose; but clearly, in the mutant rats only the first of these is possible. In fact, there was no observable response in the mutant rats (Figure 7.21), indicating that at these doses, which were themselves similar to those observed naturally, there was no evidence of intraspecific competition, and that the pattern observed in the normal rats is entirely the result of a density-dependent immune response. Of course, this does not mean that there is never intraspecific competition amongst parasites within hosts, but it does emphasize the particular subtleties that arise when an organism's habitat is its reactive host.

Moreover, it is, of course, not only the predators that may be subject to the effects of crowding. Prey, too, are likely to suffer reductions in growth, birth and survival rates as their abundance increases and their individual intake of resources declines.

crowding tends to dampen or eliminate predator–prey cycles

The effect of either predator or prey crowding on their dynamics is, in a general sense, fairly easy to predict. Prey crowding prevents their abundance from reaching as high a level as it would otherwise do, which means in turn that predator abundance is also unlikely to reach the same peaks. Predator crowding, similarly, prevents predator abundance from rising so high, but also tends to prevent them from reducing prey abundance as much as they would otherwise do. Overall, therefore, crowding is likely to have a damping effect on any predator–prey cycles, reducing their amplitude or removing them altogether; not just because crowding chops off the peaks and troughs, but also because each peak in a cycle tends itself to generate the next trough (e.g. high prey abundance → high predator abundance → *low* prey abundance), so that the lowering of peaks in itself tends to raise troughs.

There are certainly examples that appear to confirm the stabilizing effects of crowding in predator–prey interactions. For instance, there are two groups of primarily herbivorous rodents that are widespread in the Arctic: the microtine rodents (lemmings and voles) and the ground squirrels. The microtines are renowned for their dramatic, cyclic fluctuations in abundance, but the ground squirrels have populations that remain remarkably constant from year to year, especially in open meadow and tundra habitats. There, significantly, they appear to be strongly self-limited by food availability, suitable burrowing habitat and their own spacing behavior (Karels & Boonstra, 2000).

7.5.5 Predators and prey in patches

The second feature that was ignored initially but will be examined here is the fact that many populations of predators and prey exist not as a single, homogeneous mass, but as a *metapopulation* – an overall population divided, by the patchiness of the environment, into a series of subpopulations, each with its own internal dynamics but linked to other subpopulations by movement (dispersal) between patches (a topic developed further in Section 9.3).

It is possible to get a good idea of the general effect of this spatial structure on predator–prey dynamics by considering the simplest imaginable metapopulation: one consisting of just two subpopulations. If the patches are displaying the same dynamics, and dispersal is the same in both directions, then the dynamics are unaffected: every 'lost' individual is counteracted by an equivalent gain. To put it simply, patchiness and dispersal have no effect in their own right. Differences between the patches, however, either in the dynamics within subpopulations or in the dispersal between them, tend, in themselves, to stabilize the interaction: to dampen any cycles that might exist. The reason is that any difference leads to asynchrony in the fluctuations in the patches. Inevitably, therefore, a population at the peak of its cycle tends to lose more by dispersal than it gains, a population at a trough tends to gain more than it loses, and so on. In addition, even with just two patches, if one subpopulation goes extinct, the other (asynchronous) subpopulation is unlikely to do so at the same time. Dispersers from it may therefore 'rescue' the first, allowing the population as a whole, the metapopulation, to persist. Dispersal and asynchrony together, therefore – and some degree of asynchrony is likely to be the general rule – tend to dampen fluctuations in predator–prey dynamics and make population persistence more likely.

Is it possible, though, to see the stabilizing influence of this type of meta-population structure in practice? One famous example is experimental work on a laboratory system in which a predatory mite *Typhlodromus occidentalis* fed on a herbivorous mite *Eotetranychus sexmaculatus*, which fed on oranges interspersed amongst rubber balls in a tray. In the absence of its predator, *Eotetranychus* maintained a fluctuating but persistent population (Figure 7.22a). However, if *Typhlodromus* was added during the early stages of prey population growth, it rapidly increased its own population size, consumed all of its prey and then became extinct itself (Figure 7.22b): the underlying predator–prey dynamics were unstable.

The interaction was altered, however, when the habitat was made more 'patchy'. The oranges were spread further apart and partially isolated from each other by placing a complex arrangement of petroleum jelly barriers in the tray,

dispersal and asynchrony dampen cycles

stabilizing metapopulation effects in Huffaker's mites . . .

Figure 7.22

Predator–prey interactions between the mite *Eotetranychus sexmaculatus* and its predator, the mite *Typhlodromus occidentalis*. (a) Population fluctuations of *Eotetranychus* without its predator. (b) A single oscillation of the predator and prey in a simple system. (c) Sustained oscillations in a more complex system.

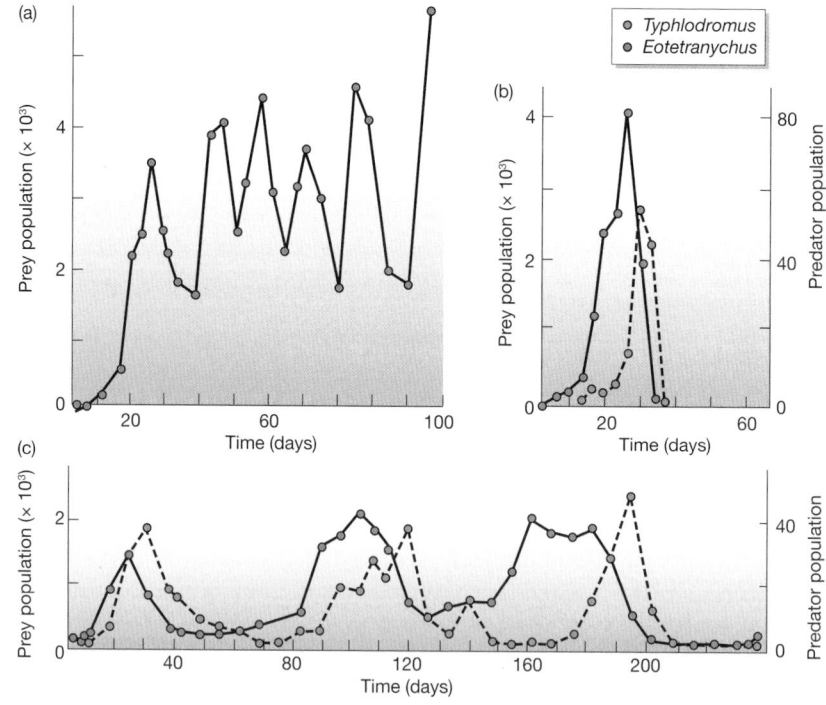

AFTER HUFFAKER, 1958

which the mites could not cross. The dispersal of *Eotetranychus* was facilitated, however, by inserting a number of upright sticks from which they could launch themselves on silken strands carried by air currents. Dispersal between patches was therefore much easier for prey than it was for predators. In a patch occupied by both, the predators consumed all the prey and then either became extinct themselves or dispersed (with a low rate of success) to a new patch. In patches occupied by prey alone, there was rapid, unhampered growth accompanied by successful dispersal to new patches. And in a patch occupied by predators alone, there was usually death of the predators before their food arrived. Predators and prey were therefore ultimately doomed to extinction in each patch – that is, the patch dynamics were unstable. But overall, at any one time, there was a mosaic of unoccupied patches, prey–predator patches heading for extinction, and thriving prey patches; and this mosaic was capable of maintaining persistent populations of both predators and prey (Figure 7.22c).

... and in starfish and mussels

A similar example, from a natural population, is provided by work off the coast of southern California on the predation by starfish of clumps of mussels (Murdoch & Stewart-Oaten, 1975). Clumps that are heavily preyed upon are liable to be dislodged by heavy seas so that the mussels die; the starfish are continually driving patches of their mussel prey to extinction. The mussels, however, have planktonic larvae that are continually colonizing new locations and initiating new clumps, whereas the starfish disperse much less readily. They aggregate at the larger clumps, but there is a time lag before they leave an area when the food is gone. Thus, patches of mussels are continually becoming extinct, but other clumps are growing prior to the arrival of the starfish.

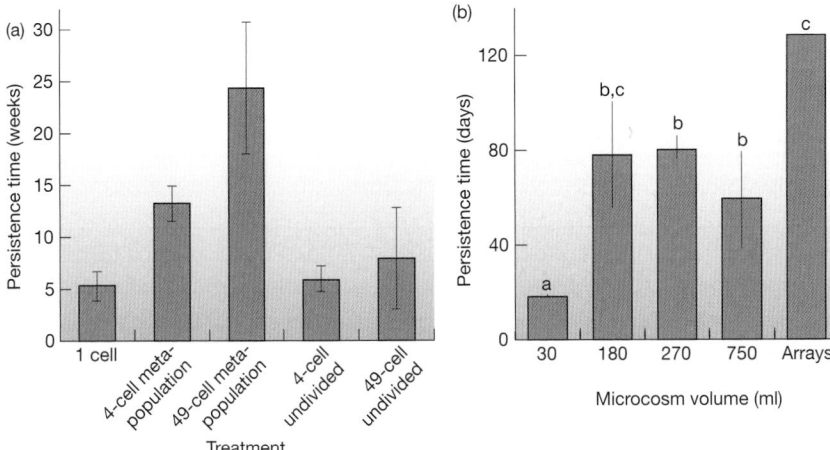

Figure 7.23

A metapopulation structure can increase the persistence of predator–prey interactions. (a) The parasitoid, *Anisopteromalus calandrae*, attacking its bruchid beetle host, *Callosobruchus chinensis*, lived on beans either in small single 'cells' (short persistence time, left), or in combinations of cells (4 or 49), which either had free access between them so that they effectively constituted a single population (persistence time not significantly increased, right), or had limited (infrequent) movement between cells so that they constituted a metapopulation of separate subpopulations (increased persistence time, center). Bars are standard errors. (b) The predatory ciliate, *Didinium nasutum*, feeding on the bacterivorous ciliate, *Colpidium striatum*, in bottles of various volumes (30–750 ml), where persistence time varied little, except in the smallest populations (30 ml) where times were shorter, and also in 'arrays' of 9 or 25 linked 30 ml bottles (metapopulations), where persistence was greatly prolonged: all populations persisted until the end of the experiment (130 days). Bars are standard errors; different letters above bars indicate treatments that were significantly different from one another ($P < 0.05$).

As with the mites, the combination of patchiness, the aggregation of predators in particular patches, and a lack of synchrony between the behavior of different patches appears capable of stabilizing the dynamics of a predator–prey interaction.

Others, too, have demonstrated the power of a metapopulation structure in promoting the persistence of coupled predator and prey populations when their dynamics in individual subpopulations are unstable. Figure 7.23a, for example, shows this for a parasitoid attacking its beetle host. Figure 7.23b shows similar results for prey and predatory ciliates (protists), where, in support of the role of a metapopulation structure, it was also possible to demonstrate the asynchrony in the dynamics of individual subpopulations and frequent local prey extinctions and recolonizations (Holyoak & Lawler, 1996).

A metapopulation structure, then, like crowding, can have an important influence on predator–prey dynamics. More generally, however, the message from this section is that predator–prey dynamics can take a wide variety of forms, but there are good grounds for believing that we can make sense of this variety through seeing it as a reflection of the way in which the different aspects of predator–prey interactions combine to play out variations on an underlying theme.

metapopulation effects in mites and ciliates

an explanation for the variety of predator–prey dynamics begins to emerge

(a) AFTER BONSALL ET AL., 2002; (b) AFTER HOLYOAK & LAWLER, 1996

7.6 Predation and community structure

What roles can predation play when we broaden our perspective from populations to whole ecological communities? Central to this is the notion that predation, in many of its effects, is just one of the forces acting on communities that can be described as a 'disturbance'. For example, the result of a predator opening up a gap in a community for colonization by other organisms is often indistinguishable from that of battering by waves on a rocky shore or a hurricane in a forest.

predation as an interruptor of competitive exclusion: predator-mediated coexistence

In fact, many of the effects of predation (and other disturbances) on community structure are the result of its interaction with competitive exclusion (taking up a theme introduced in Section 6.2.8). In an undisturbed world, the most competitive species might be expected to drive less competitive species to extinction. However, this assumes first that the organisms are actually competing. Yet there are many situations where predation may hold down the densities of competitors, so that resources are not limiting and individuals do not compete for them. When predation promotes the coexistence of species that might otherwise exclude one another, this is known as *predator-mediated coexistence*.

owls and tits on Scandinavian islands

For example, in a study of nine Scandinavian islands, pigmy owls (*Glaucidium passerinum*) occurred on only four of the islands, and the pattern of occurrence of three species of tit had a striking relationship with this distribution. The five islands without the predatory owl were home to only one species, the coal tit (*Parus ater*). However, in the presence of the owl, the coal tit was always joined by two larger tit species, the willow tit (*P. montanus*) and the crested tit (*P. cristatus*). Kullberg and Ekman (2000) argue that the coal tit is the superior competitor for food; but the two larger species are less affected than the coal tit by predation from the owl. It seems that the owl may be responsible for predator-mediated coexistence, by reducing the competitive dominance enjoyed by coat tits in its absence.

grazing by cattle can promote the coexistence of plants

In another example, grazing by local zebu cattle in natural pasture in the Ethiopian highlands was manipulated to provide a no-grazing control and four grazing intensity treatments in two sites. Figure 7.24 shows how the mean

Figure 7.24

Mean species richness of pasture vegetation in plots subjected to different levels of cattle grazing in two sites in the Ethiopian highlands in October. 0, no grazing; 1, light grazing; 2, moderate grazing; 3, heavy grazing; 4, very heavy grazing (estimated according to cattle stocking rates).

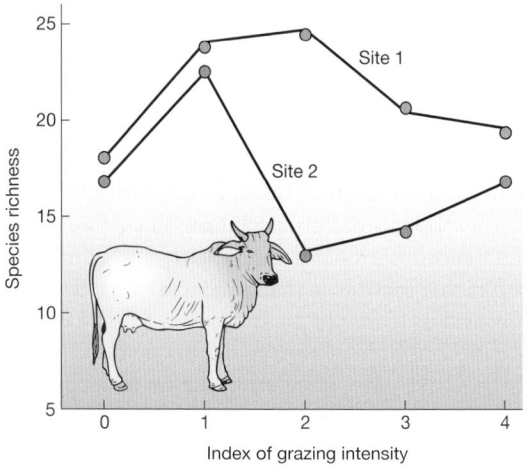

AFTER MWENDERA ET AL., 1997

number of plant species varied in the sites in October, the period when plant productivity was at its highest. Significantly more species occurred at intermediate levels of grazing than where there was no grazing or heavier grazing ($P < 0.05$).

In the ungrazed plots, several highly competitive plant species, including the grass *Bothriochloa insculpta*, accounted for 75–90% of ground cover. At intermediate levels of grazing, however, the cattle kept the dominant grasses in check and allowed a greater number of plant species to persist. But at very high intensities of grazing, cattle were forced to turn to less preferred species, driving some to extinction and allowing grazing-tolerant species such as *Cynodon dactylon* to become dominant, so that plant species numbers were again reduced (Figure 7.24). Overall, the number of species was greatest at intermediate levels of predation.

This suggests that, as a generalization, *selective* predation should favor an increase in species numbers in a community as long as the preferred prey are competitively dominant, although species numbers may also be low at very high predation pressures. To take another example, along the rocky shores of New England, the most abundant and important herbivore in mid and low intertidal zones is the periwinkle snail *Littorina littorea*. The snail will feed on a wide range of algal species but is relatively selective: it shows a strong preference for small, tender species and in particular for the green alga *Enteromorpha intestinalis*. The least-preferred foods are much tougher (e.g. the perennial red alga *Chondrus crispus* and brown algae).

Is *Enteromorpha*, the periwinkles' preferred food, a competitive dominant in their absence? Naturally, in a *Chondrus* pool, periwinkles feed on the young stages of many ephemeral algae that settle on *Chondrus*, including *Enteromorpha*. However, if periwinkles are artificially removed from a *Chondrus* pool, *Enteromorpha* and several other algae settle, grow and become abundant. *Enteromorpha* achieves competitive dominance, while *Chondrus* becomes bleached and then disappears. Conversely, adding periwinkles to *Enteromorpha* pools leads, in a year, to a decline in the percentage cover of *Enteromorpha* from almost 100% to less than 5%, as *Chondrus* colonizes and eventually comes to dominate. Clearly, periwinkles are responsible for the dominance of *Chondrus* in *Chondrus* pools.

The natural composition of tide pools in the rocky intertidal varies from almost pure stands of *Enteromorpha* to almost pure stands of *Chondrus*. Is grazing by the periwinkle responsible? A survey suggests that it is (Figure 7.25a). When periwinkles were absent or rare, *Enteromorpha* appeared to competitively exclude other species and the number of algal species was low. At very high densities of periwinkles, however, all palatable algal species were consumed to extinction, leaving almost pure stands of *Chondrus*. As with the cattle, therefore, it was at intermediate predation intensities that the abundance of *Enteromorpha* and other ephemeral algal species was reduced, competitive exclusion was prevented, and many species, both palatable and unpalatable, coexisted.

Why then do some pools contain periwinkles while others do not? Predation is again the answer. The periwinkle colonizes pools in its immature, planktonic stage. Planktonic periwinkles are just as likely to settle in *Enteromorpha* pools as *Chondrus* pools, but the crab *Carcinus maenas*, which can shelter in the

do most species occur at intermediate levels of predation?

selective predation on a rocky shore

Figure 7.25

The effect of *Littorina littorea* (periwinkle) density on species richness (a) in tide pools and (b) on emergent substrata. (c) The web of interactions giving rise to the relationship in tide pools shown in (a).

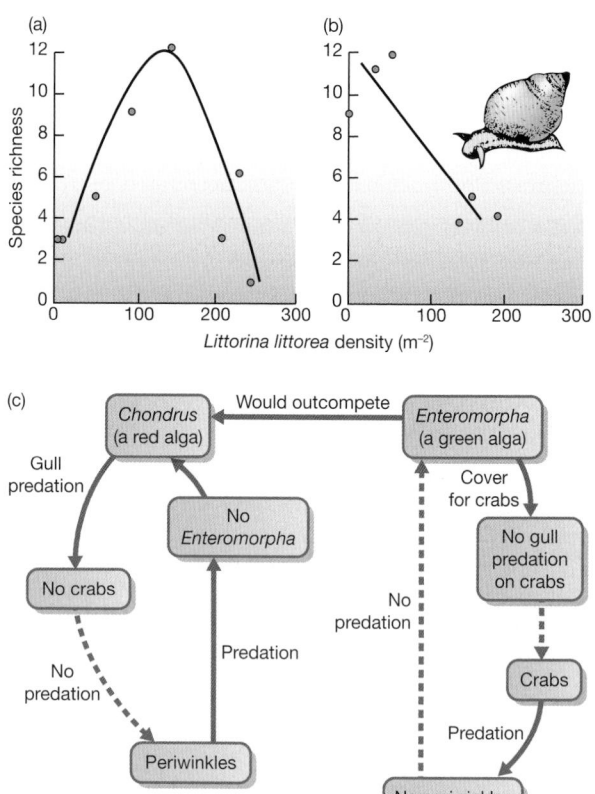

(a, b) AFTER LUBCHENCO, 1978

Enteromorpha canopy, feeds on the young periwinkles and prevents them from establishing. The final thread in this tangled web of predator–prey interactions is the effect of gulls, which prey on crabs where the dense green algal canopy is absent. Thus there is no bar to continuing periwinkle recruitment in *Chondrus* pools. These relationships, and the key roles of predation, are summarized in Figure 7.25c.

The picture is quite different, though, when the preferred prey species is not competitively dominant. Here, increased predation pressure should simply reduce the number of prey species in the community. This can also be illustrated on the rocky shores of New England, where the competitive dominance of the plants is more evenly balanced when they interact on emergent substrata rather than in tide pools. Any increase in the predation pressure, therefore, simply decreases the algal diversity, as the preferred, ephemeral species like *Enteromorpha* are consumed totally and prevented from re-establishing themselves (Figure 7.25b).

Overall, then, predation can have an important role in developing our understanding of the structure of ecological communities, not least in reminding us that the patterns we saw in Chapter 6 when we were focusing on interspecific competition may never get a chance to express themselves because communities in the real world rarely proceed smoothly to an equilibrium state.

Summary

Predation, true predators, grazers and parasites

A predator may be defined as any organism that consumes all or part of another living organism (its 'prey' or 'host') thereby benefiting itself, but, under at least some circumstances, reducing the growth, fecundity or survival of the prey.

'True' predators invariably kill their prey and do so more or less immediately after attacking them, and consume several or many prey items in the course of their life. Grazers also attack several or many prey items in the course of their life, but consume only part of each prey item and do not usually kill their prey. Parasites also consume only part of each host, and also do not usually kill their host, especially in the short term, but attack one or very few hosts in the course of their life, with which they therefore often form a relatively intimate association.

The subtleties of predation

Grazers and parasites, in particular, often exert their harm not by killing their prey immediately like true predators, but by making the prey more vulnerable to some other form of mortality.

The effects of grazers and parasites on the organisms they attack are often *less* profound than they first seem because individual plants can compensate for the effects of herbivory and hosts may have defensive responses to attack by parasites.

The effects of predation on a population of prey are complex to predict because the surviving prey may experience reduced competition for a limiting resource, or produce more offspring, or other predators may take fewer of the prey.

Predator behavior

True predators and grazers typically 'forage', moving around within their habitat in search of their prey. Other predators 'sit and wait' for their prey, though almost always in a selected location. With parasites and pathogens there may be direct transmission between infectious and uninfected hosts, or contact between free-living stages of the parasite and uninfected hosts may be important.

Optimal foraging theory aims to understand why particular patterns of foraging behavior have been favored by natural selection (because they give rise to the highest net rate of energy intake).

Generalist predators spend relatively little time searching but include relatively low-profitability items in their diet. Specialists only include high-profitability items in their diet but spend a relatively large amount of their time searching for them.

Population dynamics of predation

There is an underlying tendency for predators and prey to exhibit cycles in abundance, and cycles are observed in some predator–prey and host–parasite interactions. However, there are many important factors that can modify or override the tendency to cycle.

Crowding of either predator or prey is likely to have a damping effect on any predator–prey cycles.

Many populations of predators and prey exist as a 'metapopulation'. In theory, and in practice, asynchrony in population dynamics in different patches and the process of dispersal tend to dampen any underlying population cycles.

Predation and community structure

There are many situations where predation may hold down the densities of populations, so that resources are not limiting and individuals do not compete for them. When predation promotes the coexistence of species amongst which there would otherwise be competitive exclusion (because the densities of some or all of the species are reduced to levels at which competition is relatively unimportant) this is known as 'predator-mediated coexistence'.

The effects of predation generally on a group of competing species depend on which species suffers most. If it is a subordinate species, then this may be driven to extinction and the total number of species in the community will decline. If it is the competitive dominants that suffer most, however, the results of heavy predation will usually be to free space and resources for other species, and species numbers may then increase.

It is not unusual for the number of species in a community to be greatest at intermediate levels of predation.

Review questions

REVIEW QUESTIONS

Asterisks indicate challenge questions

1 With the aid of examples, explain the feeding characteristics of true predators, grazers, parasites and parasitoids.

2* True predators, grazers and parasites can alter the outcome of competitive interactions that involve their 'prey' populations: discuss this assertion using one example from each category.

3 Discuss the various ways that plants may 'compensate' for the effects of herbivory.

4 Predation is 'bad' for the prey that get eaten. Explain why it may be good for those that do not get eaten.

5* Discuss the pros and cons, in energetic terms, of (i) being a generalist as opposed to a specialist predator, and (ii) being a sit-and-wait predator as opposed to an active forager.

6 In simple terms, explain why there is an underlying tendency for populations of predators and prey to cycle.

7* You have data that shows cycles in nature among interacting populations of a true predator, a grazer and a plant. Describe an experimental protocol to determine whether this is a grazer–plant cycle or a predator–grazer cycle.

8 Define mutual interference and give examples for true predators and parasites. Explain how mutual interference may dampen inherent population cycles.

9 Discuss the evidence presented in this chapter that suggests environmental patchiness has an important influence on predator–prey population dynamics.

10 With the help of an example, explain why most prey species may be found in communities subject to an intermediate intensity of predation.

Chapter 8

Evolutionary ecology

Key concepts

KEY CONCEPTS

In this chapter you will:

- appreciate the range of molecular (DNA) markers that have been used in ecology
- understand how these markers can be put to work in determining the extent of subdivision within, and the degree of separation between, species
- recognize the importance of coevolutionary arms races in the dynamics of the component populations, especially of plants and their insect herbivores, and of parasites and their hosts
- understand the nature of mutualistic interactions in general and their crucial importance both for the species concerned and for almost all communities on the planet
- appreciate the particular contributions of mutualisms in diverse areas from farming, through the functioning of guts and roots, to the fixation of nitrogen by plants

We have noted previously that nothing in ecology makes sense, except in the light of evolution. But some areas of ecology are even more evolutionary than others. We may need to look within individuals to examine the details of the genes they carry, or to acknowledge explicitly the crucial and reciprocal role that species play in one another's evolution.

8.1 Introduction

In Chapter 2, we set the scene for the remainder of this book by illustrating how, to modify slightly Dobzhansky's famous phrase, 'nothing in ecology makes sense, except in the light of evolution'. But evolution does more than underpin ecology (and the whole of the rest of biology). There are many areas in ecology where evolutionary adaptation by natural selection takes center stage to the extent that the term 'evolutionary ecology' is often used to describe them. In several previous chapters, therefore, topics in evolutionary ecology have been dealt with, quite naturally, as integral parts of broader ecological questions. In Chapter 3, we examined the nature and importance of defenses that have evolved to protect plants and prey from their predators. In Chapter 5, we saw how patterns in life histories – schedules of growth, reproduction and so on – can only be understood in relation to corresponding patterns in the habitats in which they have evolved. In Chapter 6, we looked at interspecific competition as an evolutionary driving force, generating patterns in the coexistence and exclusion of competing species. And in Chapter 7, we discussed 'optimal foraging': the evolution of behavioral strategies that maximize predator fitness and thus mold their dynamic interactions with their prey.

This, of course, is not an exhaustive survey of topics in evolutionary ecology. In the present chapter, therefore, we deal with a number of others (though the final list will remain less than exhaustive). We focus especially on *coevolution*: pairs of species acting as reciprocal driving forces in one another's evolution. The question of coevolutionary 'arms races' between predators and their prey is taken up in Section 8.3, with a particular emphasis on host–pathogen interactions: each adaptation in the prey that fends off or avoids the attacks of a predator provoking a corresponding adaptation in the predator that improves its ability to overcome those defenses. However, not all coevolutionary interactions are antagonistic. Many species-pairs are 'mutualists': both parties benefiting, on balance at least, from the interactions in which they take part. Some of the most important of these mutualisms – pollination, corals and nitrogen fixation, for example – are discussed in Section 8.4. We begin, though, not with species interactions but with aspects of evolutionary differentiation within and between species, especially those detectable by modern techniques developed in molecular genetics and thus often described as aspects of 'molecular ecology'.

8.2 Molecular ecology: differentiation within and between species

For much of the time, it is entirely appropriate for ecologists to talk about 'populations' or 'species' as if they were singular, homogeneous entities: for example, we may talk of 'the distribution of Asian elephants', saying nothing about whether the species might be differentiated into distinct races or subgroups, as indeed it is (Figure 8.1). But for some purposes, knowing how much differentiation there is within species, or between one species and another, is critical for an understanding of their dynamics, and ultimately for managing those dynamics. Is a particular population derived largely from offspring born locally, or from immigrants from another, distinguishable population? Where exactly does the distribution of a particular species end and that of another, closely related species begin? In cases like these, being able to determine, at a variety of scales, who is most closely related to whom (and who is quite distinct from whom) may be essential.

the need to know who is most closely related to whom

Our ability to do this itself depends on the resolution with which we can differentiate individuals from one another and even determine where they came from or who their parents were. In the past, this was difficult and frequently impossible: reliance on simple, visual markers meant that all individuals within a species often looked the same, and even members of closely related species could often only be distinguished by experienced taxonomists looking down a microscope at, say, details of a male's genitalia. Now, though, molecular, genetic markers (albeit still requiring experts and expensive equipment) have massively

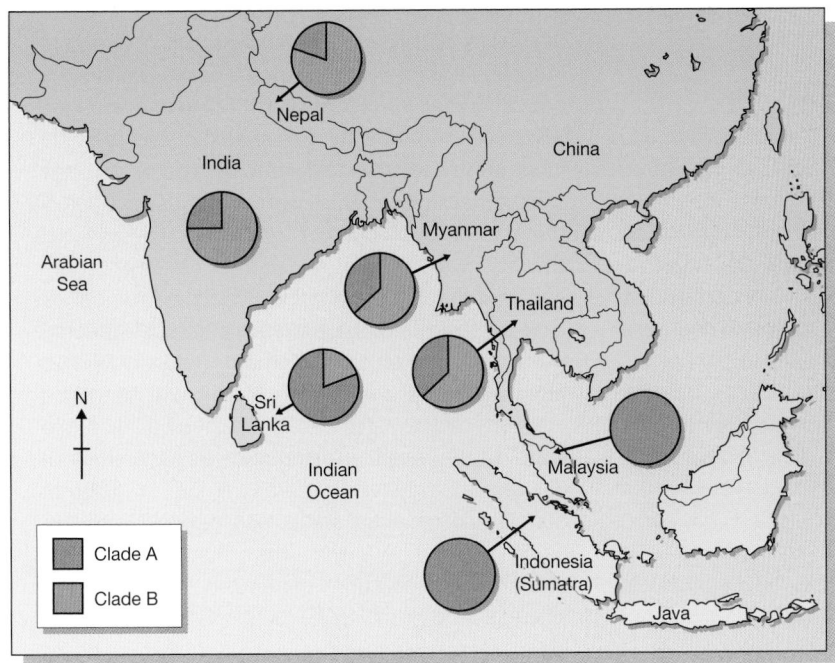

Figure 8.1

Distribution of two distinct 'clades' of the Asian elephant, *Elephas maximus* (groups with distinct evolutionary histories following their common origin), revealed only by an analysis of molecular markers. These clades coexist in many areas, though their distinctiveness itself suggests a degree of independence in their dynamics even when they do coexist.

increased the resolution at which we can differentiate between populations and even between individuals, and hence have vastly improved our ability to address these types of questions. We begin, therefore, in Box 8.1, with a brief survey of some of the most important of these molecular markers and their uses.

8.1 *Quantitative aspects*

Molecular markers

This is not the place for crash courses in either molecular biology or the laboratory methods used to extract, amplify, separate and analyze molecular markers, but it will nonetheless be useful to have some appreciation of their nature and key properties – and to be introduced to some of the technical terms and abbreviations that abound in this area. Most recent studies in ecology have used DNA of one type or other for molecular identification. We need, at the very least, to be aware that a length of DNA is characterized by the sequence of bases of which it is composed, adenine (A), cytosine (C), guanine (G) and thymine (T), and that in double-stranded DNA, these link across to one another in complementary base-pairs: A-T and G-C.

Choosing a molecular marker

The basis for all uses of molecular markers in ecology is that individuals can be differentiated from one another to greater or lesser extents as a result of molecular variation amongst them. The ultimate source of this variation is mutation in the sequence of bases, which, of course, occurs independently of its consequences for the organism concerned. What happens to the mutation, and the mutated organism, then depends essentially on the balance between selection and 'genetic drift' (random, undirected changes in gene frequency from generation to generation). If the mutation occurs in a region of DNA that is important because, say, it codes for a crucial part of an essential enzyme, then selection is likely to determine the outcome. An unfavorable mutation (the vast majority in important regions of DNA) will quickly be lost because the mutated organism is less fit than its

counterparts. Individuals will therefore differ relatively little in such regions, and if they do, differentiation is most likely to reflect 'adaptive' variation: different variants being favored in different individuals, perhaps because of where they live.

But there are also regions of DNA that appear not to code for important parts of enzymes or to perform any other function where the precise sequence is crucial. Variation in these regions is therefore said to be 'neutral', and mutations can accumulate there over time. Imagine two offspring of a single mating. They will be genetically very similar. But imagine now that each, literally, goes its own way. As each generation passes and mutations accumulate, the lineages derived from them will become increasingly divergent in those regions of their genome where variation is neutral, and lineages derived from those lineages will diverge in their turn. A snapshot taken in the future should allow us to determine, broadly, who has diverged most recently, and which groups have barely diverged at all, though our ability to do this will itself depend on the rate of mutation in the DNA region concerned: too slow and individuals will tend all to look the same; too fast and each individual sampled will tend to be so unique that its relationships to others will be hard to discern. Molecular markers are therefore chosen, ideally, such that the mutation rate matches the question being addressed. A study of differentiation between gerbils living in different burrow systems in the same, local population should use a region of DNA where the mutation rate is high (much divergence from generation to generation); whereas a study tracing the routes of colonization that have placed different populations of brown bears over the whole of Europe in the 10,000–12,000 years since the last glaciation should use a region where the mutation rate is relatively low.

Polymerase chain reaction (PCR)

As a practical point, most studies in molecular ecology, having extracted the DNA from the organism concerned, use the polymerase chain reaction (PCR) to amplify the amount of target material such there is sufficient available for analysis. By therefore being able to make use of small samples, this has revolutionized our ability to sample individuals 'non-invasively', using blood, hair, feces or wing clips. Very simply, PCR requires 'primers' that flank the particular sequence of DNA that is to be amplified. In the PCR reaction, nowadays fully automated, the originally double-stranded DNA is denatured to single strands, the primers anneal to the separated strands, and an enzyme, DNA polymerase, copies the sequence between the primers. This series of reactions is then repeated 30–40 times, and, since the process of repeated amplification is exponential, an originally small amount of target DNA in the midst of other, unwanted sequences becomes a large enough amount of target to be subjected to analysis. Note, though, that hidden within this brief description is the need to have identified not only informative target regions of DNA, but also the primers that characterize them.

Nuclear and mitochondrial DNA

In the past especially, many studies have used not nuclear DNA (inherited equally from both parents and holding the code for the vast majority of an organism's functions) but the relatively small lengths of mitochondrial DNA (mtDNA), found in the mitochondria in the cytoplasm of each of an organism's cells. The main advantages of mtDNA are that, almost always, it is inherited only from the mother (who contributes the cytoplasm to the fused egg) and does not undergo recombination. Thus, lineages can be more clearly traced from generation to generation. Also, the mutation rate is higher than for coding regions of nuclear DNA, allowing finer resolution differentiation. On the other hand, mtDNA offers only a small number of targets, and its maternal inheritance means that when disparate types meet in a population it is impossible to know whether any individuals are the result of

matings between them. Increasingly, therefore, studies are focusing on regions of nuclear DNA, though often in parallel with analyses of mtDNA genes, combining the advantages of both.

Microsatellites

Within the nuclear genome, sequences coding for proteins (i.e. genes) are by no means the only regions that have been utilized by molecular biologists. Microsatellites, for example, are regions of DNA in which the same two, three of four bases are repeated many times, preceded and followed in the sequence by flanking regions that uniquely identify each microsatellite (Figure 8.2a). The variability comes from the fact that the number of 'repeats' can vary, the resulting lengths of microsatellite DNA being measured by the speed at which they move through a semisolid medium (a 'gel') under the influence of an electric current (electrophoresis). Microsatellites may be highly polymorphic within a population. Thus, an appropriately chosen 'panel' of microsatellites for a species may effectively allow each individual in a population to be uniquely identified (a DNA 'fingerprint'), making microsatellites especially appropriate at the finer scales of differentiation.

Sequencing

As far as nuclear or mitochondrial genes are concerned, having chosen, extracted and amplified the target region from a sample of individuals, it is necessary to have some basis for differentiating individuals from one another, determining who is most similar to whom, and so on. Increasingly, as automation improves, and costs come down, the whole sequences of genes are being determined. As previously noted, regions of the same gene differ in terms of their functional importance (Figure 8.2b). Some regions are 'conserved' from individual to individual, from population to population, and often from species to species. These are (or are presumed to be) the regions of greatest functional importance, and they play effectively no part in differentiation. But there are other regions where far more variation is observed (and that can be presumed, therefore, to be neutral or at least subject to weaker selective constraints), and it is on the basis of this that individuals and populations can be differentiated.

Figure 8.2

(a) A 'locus', here, refers to the location of a region in the overall DNA sequence. An 'allele' is the particular variant of sequence that exists at that locus in a particular case. Remember that that sequence is of *two* strands of DNA, between which the bases are paired: G with C and A with T. This figure shows two contrasting alleles at a microsatellite locus, with its sequence of repeated bases (of differing length) in the two DNA strands (red) and exactly similar flanking regions at either end (black). (b) This figure, by contrast, shows the base sequence in just one DNA strand of a hypothetical gene (i.e. a sequence of DNA coding for a protein) from five individuals. Note the contrast between the conserved (unvarying) regions at either end, in black, and a variable region in red towards the center. Differentiation between individuals clearly depends on this variable region.

(a)

Allele 1 which has 10 repeats
```
...GCATTGCGATAACGTGTGTGTGTGTGTGTGTGCCATGCCGGATGA...
...CGTAACGCTATTGCACACACACACACACACACACGGTACGGCCTACT...
```
    Flanking region   Microsatellite   Flanking region

Allele 2 which has 8 repeats
```
...GCATTGCGATAACGTGTGTGTGTGTGTGTGCCATGCCGGATGA...
...CGTAACGCTATTGCACACACACACACACACGGTACGGCCTACT...
```

(b)

```
Individual 1..CGTAACGCTATTGCGCATTGTGATAACACCATGCCGGATGA..
Individual 2..CGTAACGCTATTGCGCCATCCGATCATATCATGCCGGATGA..
Individual 3..CGTAACGCTATTGCGCCTAGTCCTAGTGCCATGCCGGATGA..
Individual 4..CGTAACGCTATTGCGCCTAGCGAGAAAGTCATGCCGGATGA..
Individual 5..CGTAACGCTATTGCGCCTTACGATAACGTCATGCCGGATGA..
```

Restriction fragment length polymorphism (RFLP)

However, in the past especially, use was often made of 'restriction endonuclease' enzymes that cut DNA at specific recognition sites situated along its length and so split an original strand of DNA into fragments. Individuals differ, as a result of largely neutral mutations, in the location of these sites, and so they differ, too, in the lengths of the fragments generated, these lengths being monitored by electrophoresis. This variation, within a population, is known as restriction fragment length polymorphism, RFLP, and there are therefore separate polymorphisms for different restriction enzymes (because their recognition sites differ). Samples can thus be subjected in turn to a series of enzymes, and the most differentiated individuals will then differ in the greatest number of RFLPs. Its disadvantage, of course, is that it utilizes only a small part of the underlying sequence variation.

8.2.1 Differentiation within species

albatrosses

Albatrosses, wide ranging sea birds with the largest wingspans of any birds alive today, have achieved iconic status by virtue of their appearance in poems and stories, but of the 21 species normally recognized, 19 are regarded as 'threatened' with extinction and the other two as 'near threatened'. The black-browed albatross has recently been split by taxonomists into two species: *Thalassarche impavida*, found only on Campbell Island, between New Zealand and Antarctica, and *T. melanophris*, with breeding populations elsewhere in the sub-Antarctic, including the Falkland Islands, South Georgia and Chile (Figure 8.3a). The gray-headed albatross, *T. chrysostoma*, similar in size, also breeds on a number of sub-Antarctic islands, including South Georgia. The black-browed species usually remain associated with coastal shelf systems, whereas gray-headed albatrosses are far more 'oceanic' in their feeding grounds, but both, like all albatross species, are thought to return very close to their place of birth to breed (natal philopatry).

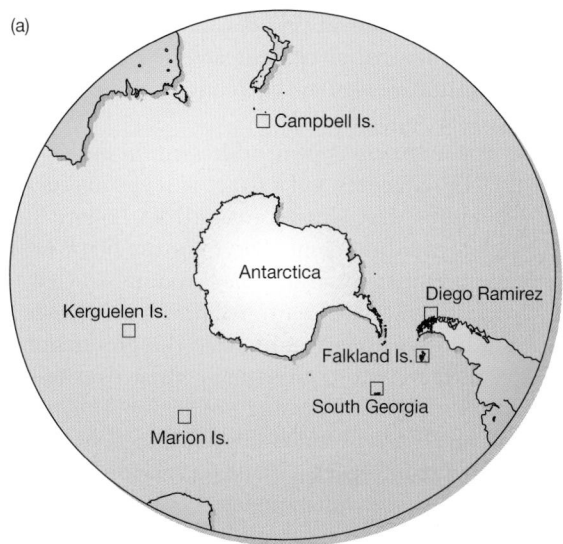

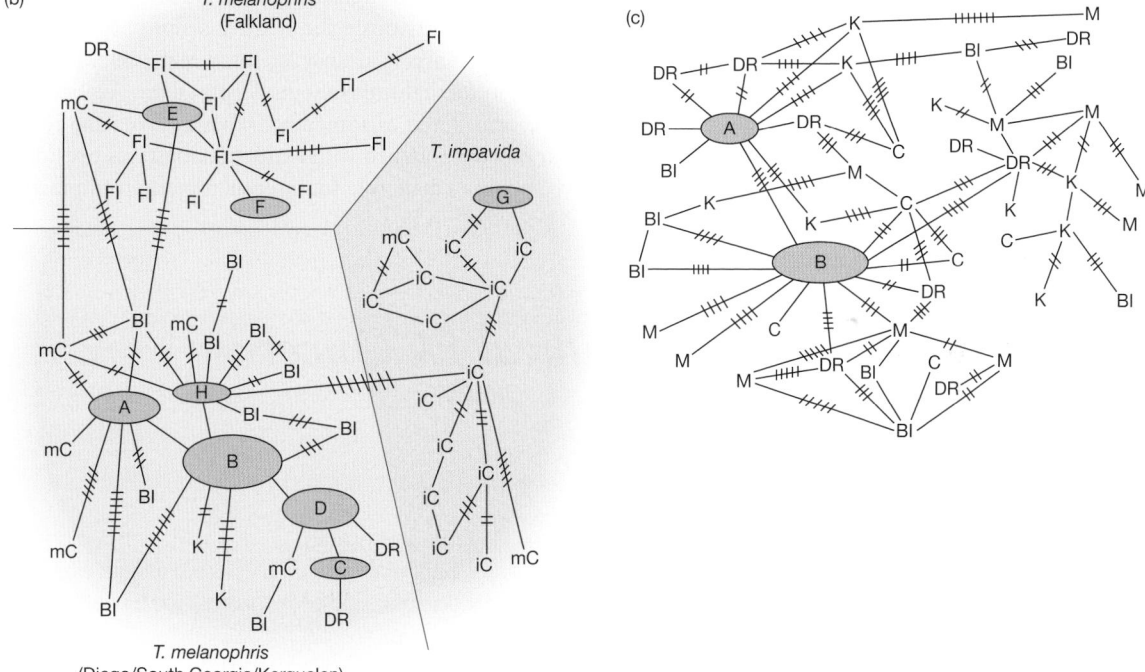

Figure 8.3

Population differentiation in albatrosses: black-browed albatrosses, *Thalassarche melanophris* and *T. impavida*, and the gray-headed albatross, *T. chrysostoma*. (a) Distribution of sites in the sub-Antarctic from which samples were taken. (b) The relationships amongst 73 black-browed albatrosses in the base sequence at a focal, variable site in their mtDNA. Where individuals from the same site shared exactly the same sequence, those individuals have been assigned a letter code (A, B, etc.) and placed in an oval proportional in size to the number of individuals. Individuals that do not fall into these groups, having a sequence unique within the data set, are identified as follows: BI, South Georgia, DR, Diego Ramirez (Chile), FI, Falkland Islands, K, Kerguelen Island (all *T. melanophris*); mC, *T. melanophris* from Campbell Island; and iC, *T. impavida* from Campbell Island. The cross-hatches represent the number of base differences between the individuals (or groups) they join. The samples fall into three 'clusters': *T. impavida*, *T. melanophris* from the Falkland Islands and *T. melanophris* from all other sites. Note though that the clustering is not perfect – as is normal, like the separation between the populations – and that some of the *T. melanophris* found on Campbell Island were identifiable as *T. melanophris–T. impavida* hybrids. (c) The relationships amongst 50 gray-headed albatrosses in the base sequence at a focal, variable site in their mtDNA. Coding is the same as in (b) except that M is Marion Island and C is Campbell Island. No separate clusters are discernable in this case.

AFTER BURG & CROXALL, 2001

With numbers in all sites declining year on year, therefore, the questions arise: 'How connected or separate are these populations? Should conservation efforts be directed at what are currently perceived to be whole species, or at particular breeding populations?'

These questions were addressed, in both species, by a study that used both mtDNA sequences and a panel of seven microsatellites (Burg & Croxall, 2001). The results were clearest for mtDNA (Figure 8.3b, c), but those for the microsatellites told the same story. For the black-browed species (Figure 8.3b), the molecular data confirmed the taxonomists' view that *T. impavida* was a separate species, but also demonstrated breeding between this species and *T. melanophris* on Campbell Island and indeed the production of hybrids between these two species there. More surprisingly, these data also demonstrated that the Falkland Islands support a breeding population of *T. melanophris* that is quite separate from an effectively indivisible population shared by Diego Ramirez (Chile), South Georgia and Kerguelen Island (in spite of the natal philopatry to these three sites). By contrast, the wider ranging gray-headed albatrosses, from all five of their sites, seemed to represent a single breeding population (Figure 8.3c) – again in spite of their natal philopatry.

molecular markers in conservation

From a conservation point of view, though, the most important conclusion relates to *T. melanophris*. Whereas previously the relative stability of the large Falkland Islands population was taken as insurance against a real vulnerability of the species to extinction, now, in the light of these molecular data, the Falkland Islands population should be considered as somewhat separate from the rest of the species, which itself is far more threatened with extinction than was previously appreciated. (A more active and immediate role for molecular markers in practical matters of conservation is described in Box 8.2.)

8.2 Topical ECOncerns

The forensic analysis of the origins of our food

As we shall discuss more fully in Chapter 12, there is an increasingly frequent conflict between exploiting natural populations as a necessary source of food and conserving those same populations, both as an end in itself and so that future generations have something to eat. In Canada, for example, Pacific salmonid fish are harvested from a large number of commercial (industrial) and sport fisheries, each managed in its own way in an attempt to ensure its continued viability. So, for instance, a fishery may be closed altogether at times when fish from other sources are readily available, in order to allow the stock to breed and recover. Nonetheless, threats to sustainability are very real: 2002 saw the first designation of a Canadian salmon stock, the Interior Fraser River coho salmon, as 'endangered', and many others require careful protection.

In an ideal world, policing, and hence management, of the different fisheries would be perfectly effective. But in reality, illegal fishing is bound to take place and cannot necessarily be countered simply by catching offenders 'in the act'. An alternative, then, or at least another weapon in the managers' armoury, is to be able to identify fish as having been illegally obtained at some other point in the chain from being caught to being eaten. Molecular markers make this possible.

Table 8.1

Species identification of salmonid samples obtained by fisheries officers in Canada because the material was believed to have been obtained illegally.

CASE (YEAR)	TISSUES	RESULT	LEGAL OUTCOME	FINE ($)
1 (1995)	Blood/scales/slime from containers	Coho	Conviction	1500
2 (1998)	Muscle	Chum	Conviction	1800
		Chinook		
		Coho		
3 (1998)	Muscle	Coho	Conviction	?
4 (1999)	Muscle	Atlantic	No charges	–
		Chinook		
		Coho		
5 (2000)	Muscle	Coho	Guilty plea	7500
6 (2000)	Muscle	Sockeye	Conviction	1000

AFTER WITHER ET AL., 2004

Table 8.2

Stock identification of salmonid samples obtained by fisheries officers in Canada because the material was believed to have been obtained illegally. IF&T refers to the Interior Fraser and Thompson tributaries.

CASE (YEAR)	SPECIES	RESULTS	OUTCOME	FINE ($)
1 (1998)	Sockeye	96.5% Fraser; 96.5% IF&T	Guilty plea	2,000
2 (1999)	Sockeye	100% Fraser; 100% IF&T	Conviction	15,000
3 (1999)	Chinook	91.4% Fraser	No conviction, under appeal	
4 (2000)	Sockeye	100% Fraser; 100% IF&T	Guilty plea	8,000
5 (2001)	Sockeye	97.8% Fraser; 97.8% IF&T	Guilty plea	3,000

AFTER WITHER ET AL., 2004

For example, the 10 species of Pacific salmon, *Oncorhynchus* spp., can be effectively distinguished from one another by RFLP profiling of targeted nuclear genes (Withler et al., 2004). Some results of applying such analyses to cases of suspected illegal possession of salmon are shown in Table 8.1. Case 2, for instance, involved a disaffected chef reporting a restaurant owner to the authorities. A fish was identified as a coho salmon, *O. kisutch*, which, because it showed no signs of having been frozen, could not have come from the previous years' legal harvest. The owner was duly fined.

Moreover, analyses based largely on microsatellites, with their finer scale of resolution, are able, even within a species, to tie a sample to a particular river – if not with certainty then at least with a very high probability. Some results of these analyses are shown in Table 8.2. In case 2 here, for instance, illegally sourced Fraser River sockeye salmon, *O. nerka*, were identified in an analysis of 50 cans of salmon and the defendant, fined $15,000, was found to be in possession of 100,000 cans with a 'street value' of $300,000–400,000.

What do you think of the level of the fines imposed? How does the seriousness of crimes like this compare to those of other crimes: street robbery or the possession of illegal drugs for personal use? Should those convicted be punished in proportion to the economic harm they may be doing to these particular fisheries, or should their fines be seen as a signal sent out to all those who ignore the need to restrain activity in exploited but vulnerable populations and to conserve them for future generations?

8.2.3 Differentiation between species: the red wolf

species or hybrid?

Issues in conservation surface again when we shift our focus from differentiation within to differentiation between species. The red wolf, *Canis rufus*, once had a widespread distribution in the southeastern United States (Figure 8.4a), but when, by the mid-1970s, that distribution had shrunk to a single population in eastern Texas, the US Fish and Wildlife Service instituted an emergency program to save it from extinction. Fourteen individuals were rescued from its final refuge and bred in captivity with a view to subsequent reintroduction in the wild. In the United States as a whole, the red wolf coexists with two other, closely related species, the gray wolf, *C. lupus*, and the coyote, *C. latrans*. Traditional analyses, based on morphological features, placed the red wolf as a genuine, separate species, intermediate in many ways between the gray wolf and the coyote (Nowak, 1979). However, as we shall see below, molecular markers suggest strongly that the red wolf is a hybrid arising from interbreeding between gray wolves and coyotes. A number of questions therefore suggest themselves (Wayne, 1996), including: 'Should the conservation status of the red wolf, and the amount of money spent on its conservation, be downgraded if it is acknowledged that it is 'only' a hybrid and not a full species?' And will attempts to save the red wolf by reintroduction be doomed, in any case, because of 'introgression' – the movement of genes from gray wolves or coyotes into the red wolf gene pool as a result of interbreeding?

mtDNA

The first molecular markers used to assess the degree of genetic isolation of red wolves from gray wolves and coyotes, albeit for a relatively small sample, were from mtDNA – both restriction fragment genotypes (RFLPs – see Box 8.1) and sequence variation within the cytochrome *b* gene. From the restriction site analysis carried out on contemporary captures (Figure 8.4b), it is clear, first, that the gray wolf and coyote samples were quite separate from one another; but also that samples from captive red wolves all fitted squarely within the cluster of coyote genotypes. And when sequence analysis was applied to museum pelts of red wolves from a variety of locations, and to a number of contemporary gray wolves and coyotes (Figure 8.4c), these too showed separate clusters for gray wolves and coyotes, and this time that red wolves had either gray wolf or coyote genotypes. Thus, the status of the red wolf as a separate species was called seriously into question, and its origin as a gray wolf–coyote hybrid was further supported by the observation of common, contemporary introgression of coyote genes into gray wolf populations throughout a region on the USA–Canadian border, where recent contact (the last 100 years) has been made as coyotes have moved north (Lehmann et al., 1991).

nuclear microsatellites

Investigations of microsatellites in the nuclear DNA have further clarified the red wolf story (Roy et al., 1994). First, studies on the USA–Canadian border confirmed the high frequency of contemporary coyote introgression into gray wolf gene pools (Figure 8.4d). Second, an analysis of 40 captive red wolves revealed that every one of the 53 microsatellite alleles they carried was also found in coyotes. Museum specimens of red wolves, too, failed to turn up unique red wolf alleles, and indeed, the historical and contemporary red wolf samples were themselves very similar. Finally, overall, red wolf samples, like contemporary gray wolf samples in the zone of hybridization, appear intermediate between coyotes and non-hybridizing gray wolves (Figure 8.4d). All of this argues in favor of the red wolf having its origins in gray wolf–coyote hybridization, with

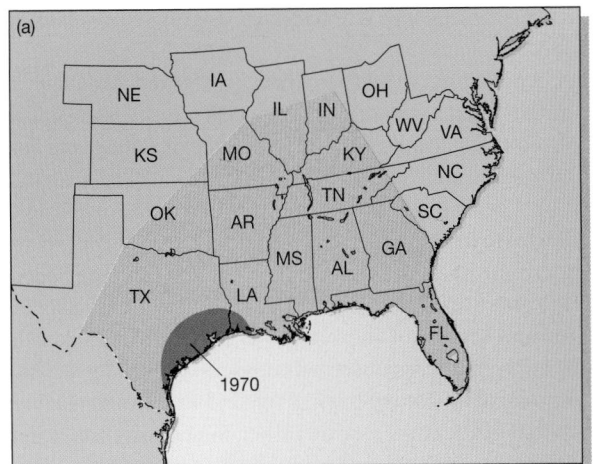

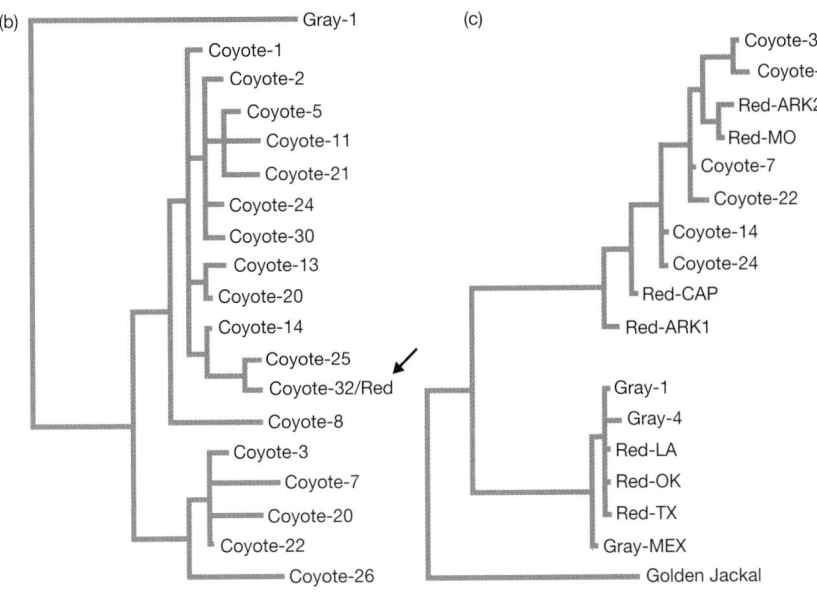

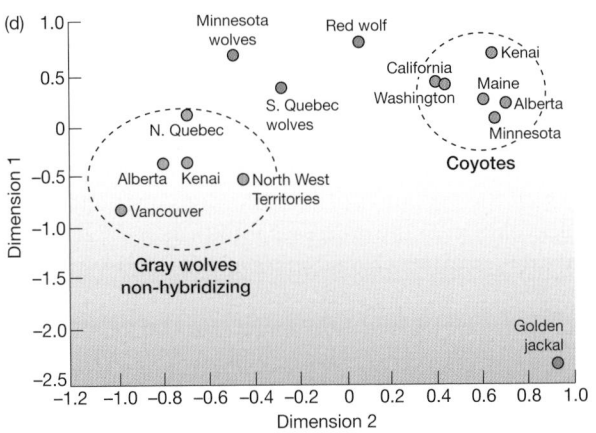

Figure 8.4

(a) The geographic range (light maroon) of the red wolf, *Canis rufus*, in the United States around 1700, and within that the smaller bounded area showing its range in southeastern Texas around 1970. (b) A 'phylogenetic tree' of coyote and red wolf mtDNA restriction-site genotypes (RFLPs). In a phylogenetic tree, the most similar (closely related) types are placed closest together, then linked to the type that is most similar to them, and so on, the lengths of the horizontal lines representing the degree of difference. The tree is 'rooted' (to give it context) by inclusion of a gray wolf (Gray-1). The numbers refer to different individuals. The arrow points to the single genotype shared by the eight captive red wolves that were sampled, which is clearly simply part of the coyote 'cluster'. (c) A phylogenetic tree constructed on similar principles but based on sequences of the cytochrome *b* gene in the mtDNA. Museum red wolf samples are from Arkansas (ARK), Missouri (MO), Louisiana (LA), Oklahoma (OK) and Texas (TX); CAP refers to a captive red wolf; and MEX refers to a gray wolf from Mexico. The tree is rooted by the inclusion of sequence data from the golden jackal, *C. aureus*. The red wolf genotypes are clearly parts of either the coyote or the gray wolf clusters. (d) The relationships between various coyote, gray wolf and red wolf populations at 10 nuclear DNA microsatellite loci, as demonstrated by an analysis that condenses the data from these 10 loci into two dimensions. The details of this analysis are unimportant here, as long as it is appreciated that the most similar populations are closest together in the figure. There are two clusters: coyotes and gray wolves from populations in which there is no hybridization with coyotes. Red wolves, and the populations of gray wolves from Minnesota and south Quebec where there is hybridization with coyotes, are located between these two clusters. Context, again, is provided by the location of the golden jackal.

(a–c) AFTER WAYNE & JENKS, 1991; (d) AFTER ROY ET AL., 1994

subsequent further hybridization with coyotes, as gray wolves became rare in the southeastern USA.

In answer to our original questions, then, (i) the red wolf seems, ultimately, to be a hybrid rather than a separate species with a more ancient origin, and (ii) any program of reintroduction clearly is in danger of failing as a result of introgression from coyotes, requiring sufficient densities of red wolves to minimize this possibility, and perhaps even barriers to the 'species' meeting (Fredrickson & Hedrick, 2006). However, whether biological status and practical difficulties combine to undermine even the desirability of reintroducing red wolves is not simply a scientific question. Public perception and opinion (in this case regarding the conservation importance of the red wolf) must also be taken into account. Similar remarks apply to most conservation issues, especially when public funds are involved. A molecular ecology perspective has been immensely informative – but information may sometimes muddy rather than clarify the waters.

8.3 Coevolutionary arms races

We turn now from evolution at the molecular level to evolution at the level of species interactions, starting with those in which species are 'in opposition' to one another. Following some general background, we turn first to interactions between insects and the plants they eat (Section 8.3.2) and then to those between parasites and their hosts (Section 8.3.3).

8.3.1 Coevolution

The dynamics of consumer resource pairs (see Chapter 7) are linked to the dynamics of whole webs of interacting species (see Chapter 9) by how specialized or generalized particular consumers are. Generalists draw the species of a community together into large interactive networks. Specialists divide communities into detached or semidetached compartments. Coevolution plays a vital part in determining how specialized or generalized particular consumers are.

It is not surprising, as we saw in Chapter 3, that many organisms have evolved defenses that reduce the chance of an encounter with a consumer and/or increase the chance of surviving such an encounter. But the interaction does not necessarily stop there. A better defended food resource (the 'prey') itself exerts a selection pressure on consumers to overcome that defense. A consumer that does so is likely to have invested in counteracting that defense as opposed to others, and will steal a march on its competitors, and so is likely to become relatively specialized on that prey type – which is then under particular pressure to defend itself against that particular consumer, and so on. A continuing interaction can therefore be envisaged in which the evolution of both the consumer and the prey depend crucially on the evolution of the other: what Ehrlich and Raven (1964) called a coevolutionary 'arms race', which, in its most extreme form, has a coadapted pair of species locked together in perpetual struggle.

one man's poison is another man's meat

Indeed, what is unacceptable to most animals may be the chosen, even unique, diet of others. It is, after all, an inevitable consequence of having evolved resistance to a prey's defenses that a consumer will have gained access to a resource unavailable to most (or all) other species. For example, the tropical legume *Dioclea metacarpa* is toxic to almost all insect species because it contains a non-protein amino acid, L-canavanine, which those insects incorporate (lethally) into their proteins in place of arginine. But a species of bruchid beetle, *Caryedes brasiliensis*, has evolved a modified enzyme that distinguishes between L-canavanine and arginine, and the larvae of these beetles feed solely on *D. metacarpa* (Rosenthal et al., 1976).

8.3.2 Insect–plant arms races

We discussed in Section 3.4.2 how attacks by herbivores select for plant-defensive chemicals. We also saw that these can be divided into 'qualitative' chemicals that are poisonous, can kill in small doses and tend to be induced by herbivore attacks, and 'quantitative' chemicals that are digestion-reducing, rely on an accumulation of ill effects and tend to be produced constitutively (i.e. all the time). These chemicals will select for adaptations in herbivores that can overcome them. It seems probable, however, that toxic chemicals, by virtue of their specificity, are likely to be the foundation of an arms race, requiring an equally specific response from a herbivore; whereas chemicals that make plants generally indigestible are much more difficult to overcome through any 'targeted' adaptation (Cornell & Hawkins, 2003). Put simply: plants relying on toxins are more prone to becoming involved in arms races with their herbivores (like the beetle and legume described above) than those relying on more 'quantitative' chemicals.

We can seek evidence for the toxin arms race hypothesis by asking whether specialist herbivores generally, locked in their coevolutionary arms races, perform better when faced with their plants' toxic chemicals than generalists; whereas generalists, having invested in overcoming a wide range of chemicals, perform better than specialists when faced with chemicals that have not provoked coevolutionary responses. Such evidence is provided by an analysis of a wide range of data sets for insect herbivores fed on artificial diets with added chemicals (892 insect–chemical combinations; Figure 8.5).

specialists are more prone to arms races

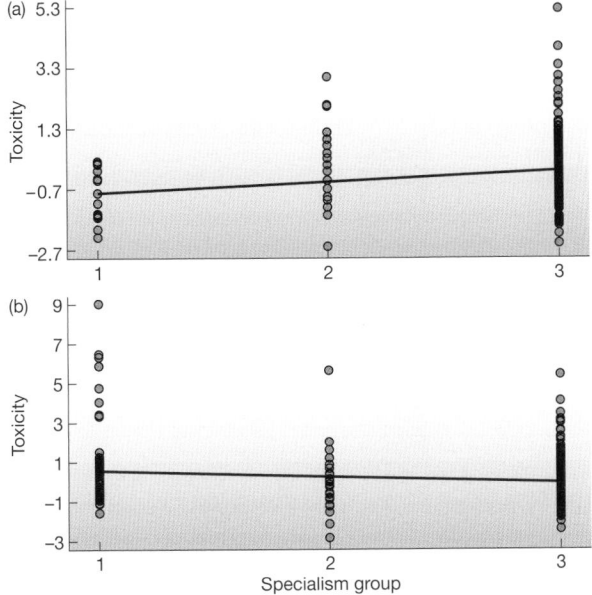

Figure 8.5

Combining data from a wide range of published studies, insect herbivores were split into three groups: 1, specialists (feeding from one or two plant families); 2, 'oligophages' (3–9 families); and 3, generalists (more than nine families). Chemicals were split into two groups: (a) those that are found in the normal hosts of the specialists and oligophages, and (b) those that are not. 'Toxicity' is measured from the mortality rates of insects on a standardized scale, since many studies have been combined. (a) It is apparent that more specialized insects suffered lower mortality on chemicals that have provoked a coevolutionary response from specialist herbivores. (b) It is apparent that more generalist insects suffered lower mortality on chemicals that have *not* provoked a coevolutionary response from specialist herbivores. $P < 0.005$ in both cases.

8.3.3 Coevolution of parasites and their hosts

The intimate association between parasites and their hosts makes them especially prone to coevolutionary arms races. Indeed, the specialization may go further than that between species. Within species, it is common to find a high degree of genetic variation in the virulence of parasites and/or in the resistance or immunity of hosts. Every few years, for example, as we are perhaps more aware than ever, a new strain of the influenza virus evolves of sufficient virulence and novelty to generate a widespread epidemic and mortality in human populations that had been relatively resistant to previously circulating strains. No strain has been more devastating – at the time of writing – than the worldwide epidemic (*pandemic*) of Spanish flu that followed World War I in 1918/19 and killed 20 million people – many more than died in the war itself. Human diseases can also provide examples of variation in host resistance. When the Native Americans of the Canadian Plains were forcibly settled onto reservations in the 1880s, their death rate due to tuberculosis (TB) initially exploded but then gradually declined (Figure 8.6). Environmental factors (inadequate diet, overcrowding, spiritual demoralization) undoubtedly played some part in this, but variation in resistance is also likely to have been significant. The mortality rate among the Native Americans was often 20 times that of the surrounding European colonist population, living in similar conditions but having been exposed previously to TB. Some native families had a particularly low mortality rate in the 1880s epidemic, and many of the survivors in the 1930s were descendants of those families (Ferguson, 1933; Dobson & Carper, 1996).

myxomatosis

It may seem straightforward that parasites in a population select for the evolution of more resistant hosts, which in turn select for more infective parasites: a classic arms race. In fact, the process is not necessarily so straightforward, though there are certainly examples where host and parasite drive one another's evolution. A most dramatic example involves the rabbit and the myxoma virus, which causes myxomatosis. The virus originated in the South American jungle rabbit *Sylvilagus brasiliensis*, where it causes a mild disease that only rarely kills

Figure 8.6

The mortality rate due to tuberculosis in three generations of Canadian Plains Native Americans after their forced settlement onto reservations.

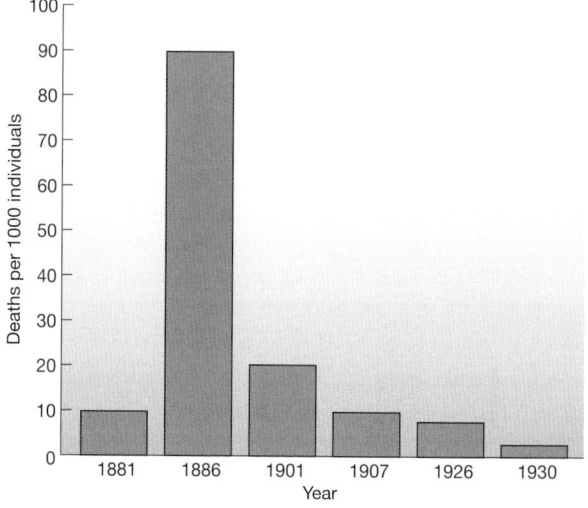

AFTER FERGUSON, 1933; DOBSON & CARPER, 1996

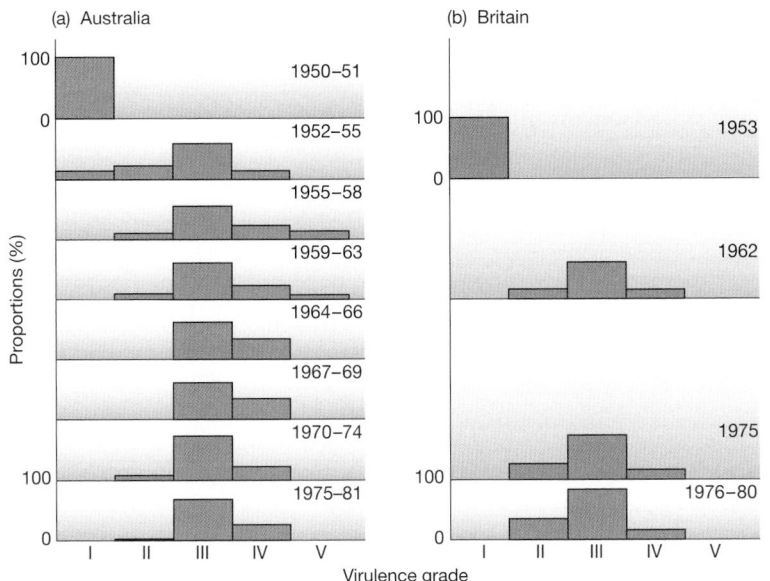

(a) Australia

(b) Britain

Proportions (%)

Virulence grade

FROM FENNER, 1983; AFTER MAY & ANDERSON, 1983

Figure 8.7

(a) The percentages in which various grades of myxoma virus have been found in wild populations of rabbits in Australia at different times from 1951 to 1981. Grade I is the most virulent. (After Fenner, 1983.) (b) Similar data for wild populations of rabbits in Great Britain from 1953 to 1980.

the host. The South American virus, however, is usually fatal when it infects the European rabbit *Oryctolagus cuniculus*. In one of the greatest examples of biological pest control, the myxoma virus was introduced into Australia in the 1950s to control the European rabbit, which had become a pest of grazing lands. The disease spread rapidly in 1950/51, and rabbit populations were greatly reduced – by more than 90% in some places. At the same time, the virus was introduced to England and France, and there too it resulted in huge reductions in the rabbit populations. The evolutionary changes that then occurred in Australia were followed in detail by Fenner and his associates, who had the brilliant foresight to establish baseline genetic strains of both rabbits and virus (Fenner, 1983). They used these to measure subsequent changes in the virulence of the virus and the resistance of the host as they evolved in the field.

When the disease was first introduced to Australia it killed more than 99% of infected rabbits. This 'case mortality' fell to 90% within 1 year and then declined further. The virulence of virus isolates was graded according to host survival time and the case mortality of control rabbits. The original, highly virulent virus was grade I, which killed > 99% of infected laboratory rabbits. Already by 1952, most of the virus isolates from the field were the less virulent grades III and IV. At the same time, the rabbit population in the field was increasing in resistance. When injected with a standard grade III strain, field samples of rabbits in 1950/51 had a case mortality of nearly 90%, which had declined to less than 30% only 8 years later (Figure 8.7).

This evolution of resistance is easy to understand: resistant rabbits are obviously favored by natural selection in the presence of the myxoma virus. The case of the virus, however, is subtler. The contrast between the virulence of the virus in the European rabbit and its lack of virulence in the American host with which it had coevolved, combined with the attenuation of its virulence in Australia and Europe after its introduction, fit a commonly held view that parasites evolve toward becoming benign to their hosts in order to prevent the parasite eliminating its host and thus eliminating its habitat. This view, however, is quite wrong. The parasites favored by natural selection are those with the greatest fitness (broadly, the greatest reproductive rate). Sometimes this is achieved through a decline in virulence, but sometimes it is not. In the myxoma virus, an initial decline in virulence was indeed favored – but further declines were not.

The myxoma virus is blood-borne and is transmitted from host to host by blood-feeding insect vectors. In the first 20 years after its introduction to Australia, the main vectors were mosquitoes, which feed only on live hosts. The problem for grade I and II viruses is that they kill the host so quickly that there is only a very short time in which the mosquito can transmit them. Effective transmission may be possible at very high host densities, but as soon as densities decline, it is not. Hence, there was selection against grades I and II and in favor of less virulent grades, giving rise to longer periods of host infectiousness. At the other end of the virulence scale, however, the mosquitoes are unlikely to transmit grade V of the virus because it produces very few infective particles. The situation was complicated in the late 1960s when an alternative vector of the disease, the rabbit flea *Spilopsyllus cuniculi* (the main vector in England), was introduced to Australia, apparently favoring more virulent strains than the mosquitoes had done. Overall, however, there has been selection in the rabbit–myxomatosis system not for decreased virulence as such, but for *increased transmissibility* (and hence increased fitness) – which happens in this system to be maximized at intermediate grades of virulence.

bacteria and bacteriophage

In other cases, host–parasite coevolution is more definitely antagonistic: increased resistance in the host and increased infectivity in the parasite. A classic example is the interaction between agricultural plants and their pathogens (Burdon, 1987), though in this case the resistant hosts are often introduced by human intervention. There may even be gene-for-gene matching, with a particular virulence allele in the pathogen selecting for a resistant allele in the host, which in turn selects for alleles other than the original allele in the pathogen, and so on. In fact, these detailed processes have proved difficult to observe, but this has been done with a system comprising the bacterium *Pseudomonas fluorescens* and its viral parasite, the bacteriophage (or phage) SBW25φ2, where such evolution is relatively easy to observe because generation times are so short. Changes in both host and parasite were monitored as 12 replicate coexisting populations of bacterium and phage were transferred from culture bottle to culture bottle. It is apparent that the bacteria became generally more resistant to the phage at the same time as the phage became generally more infective to the bacteria: each was being driven by the directional selection of an arms race (Figure 8.8).

This was only apparent, however, because each bacterial strain (from one of the 12 replicate pairs) was tested against all 12 phage strains, and each phage strain tested against all bacterial strains, and mean resistances and infectivities calculated. When, at the end of the experiment (Table 8.3), the resistance of each bacterial strain was tested against each phage strain in turn, it was clear that bacteria were almost always *most* resistant (and often wholly resistant) to the phage strain with which they coevolved. Clearly, the specific problems posed by particular phage strains had provoked equally specific evolutionary responses on the part of the bacterial strains.

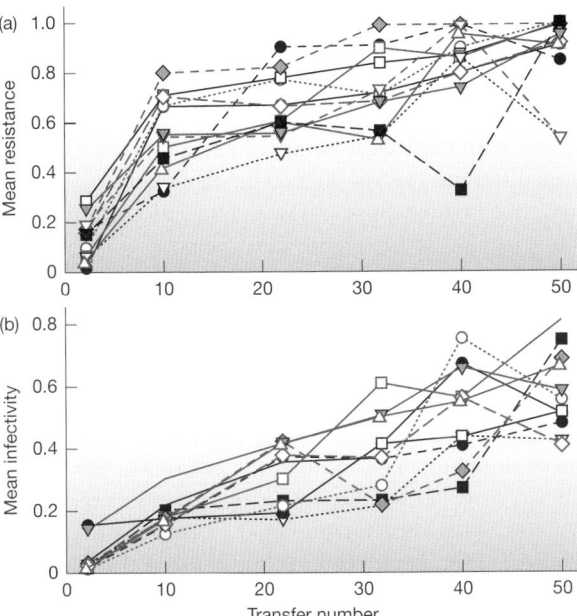

Figure 8.8

(a) Over evolutionary time (1 'transfer' ≈ 8 bacterial generations) bacterial resistance to phage increased in each of 12 bacterial replicates (designated by different symbols). 'Mean' resistance was the mean calculated over the 12 phage isolates from the respective time points. (b) Similarly, phage infectivity increased, where 'mean' infectivity was calculated over the twelve bacterial replicates.

Table 8.3

For each of 12 bacterial replicates (B1–B12) and their 12 respective phage replicates ($\phi1$–$\phi12$), entries in the table are the proportion of bacteria resistant to the phage at the end of a period of coevolution (50 transfers ≈ 400 bacterial generations). Coevolving pairs are shown along the diagonal in bold. Note that bacterial strains are usually most resistant to the phage strain with which they coevolved.

	BACTERIAL REPLICATES											
PHAGE REPLICATES	B1	B2	B3	B4	B5	B6	B7	B8	B9	B10	B11	B12
$\phi1$	**0.8**	0.9	1	1	1	1	1	1	0.85	0.85	0.75	0.65
$\phi2$	0.1	**1**	0.3	1	0.85	0.25	1	1	0.85	0.9	0.8	0.65
$\phi3$	0.75	0.75	**1**	1	1	0.9	1	1	0.85	0.9	0.9	0.65
$\phi4$	0.15	0.9	0.8	**1**	0.85	0.6	0.6	1	0.85	1	0.85	0.35
$\phi5$	0.25	0.9	1	1	**1**	0.9	1	0.8	0.85	1	0.8	0.65
$\phi6$	0.2	1	0.85	0.8	0.75	**0.8**	0.85	0.9	0.85	0.75	0.45	0.25
$\phi7$	0.2	0.75	0.6	1	0.4	0.45	**1**	0.9	0.85	1	0.75	0.35
$\phi8$	0	0.95	0.55	0.95	0.35	0.25	0.8	**1**	0.85	1	0.7	0.25
$\phi9$	0	0.7	0.55	0.45	0.7	0.35	1	1	**0.85**	1	0.5	0.1
$\phi10$	0	0.7	0.9	0.7	0.55	0.9	1	1	0.7	**1**	0.5	0.4
$\phi11$	0	0.5	0.9	0.75	0.7	1	1	0.95	0.75	1	**1**	0.35
$\phi12$	0	0.15	0	0.1	0.65	0.35	1	1	0.7	0.8	0.85	**0.4**

8.4 Mutualistic interactions

No species lives in isolation, but often the association with other species is especially close: for many organisms, the habitat they occupy is an individual of another species. Parasites live within the body cavities or even the cells of their hosts, nitrogen-fixing bacteria live in nodules on the roots of leguminous plants,

symbiosis and mutualism

and so on. *Symbiosis* ('living together') is the term that has been coined for such close physical associations between species, in which a 'symbiont' occupies a habitat provided by a 'host'. In fact, though, parasites are usually excluded from the category of symbionts, which is reserved instead for interactions where there is at least the suggestion of *mutualism*. A mutualistic relationship is simply one in which organisms of different species interact to their mutual benefit. Mutualism, therefore, need not involve close physical association: mutualists need not be symbionts. For example, many plants gain dispersal of their seeds by offering a reward to birds or mammals in the form of edible fleshy fruits, and many plants assure effective pollination by offering a resource of nectar in their flowers to visiting insects. These are mutualistic interactions but they are not symbioses.

It would be wrong, however, to see mutualistic interactions simply as conflict-free relationships from which nothing but good things flow for both partners. Rather, current evolutionary thinking views mutualisms as cases of reciprocal exploitation where nonetheless each partner is a *net* beneficiary (Herre & West, 1997).

Mutualisms themselves have often been neglected in the past compared to other types of interaction, yet mutualists compose most of the world's biomass. Almost all the plants that dominate grasslands, heaths and forests have roots that have an intimate mutualistic association with fungi. Most corals depend on the unicellular algae within their cells, many flowering plants need their insect pollinators, and many animals carry communities of microorganisms within their guts that they require for effective digestion.

The rest of this section is organized as a progression. We start with mutualisms in which no intimate symbiosis is involved; rather, the association is largely behavioral: that is, each partner behaves in a manner that confers a net benefit on the other. By Section 8.4.4, when we discuss mutualisms between animals and the microbiota living in their guts, we will have moved on to closer associations (one partner living within the other), and in Sections 8.4.5 and 8.4.6 we examine still more intimate symbioses in which one partner enters between or within another's cells.

mutualism: reciprocal exploitation not a cosy partnership

8.4.1 Mutualistic protectors

cleaner and client fish

'Cleaner' fish, of which at least 45 species have been recognized, feed on ecto-parasites, bacteria and necrotic tissue from the body surface of 'client' fish. Indeed, the cleaners often hold territories with 'cleaning stations' that their clients visit – and visit more often when they carry many parasites. The cleaners gain a food source and the clients are protected from infection. In fact, it has not always proved easy to establish that clients benefit, but experiments off Lizard Island on Australia's Great Barrier Reef were able to do this for the cleaner fish, *Labroides dimidiatus*, which eats parasitic gnathiid isopods from its client fish, *Hemigymnus melapterus*. Clients had significantly (3.8 times) more parasites 12 days after cleaners were excluded from caged enclosures (Figure 8.9a); but even in the short term (up to 1 day), although removing cleaners, which only feed during daylight, had no effect when a check was made at dawn (Figure 8.9b), this led to there being significantly (4.5 times) more parasites following a further day's feeding (Figure 8.9c).

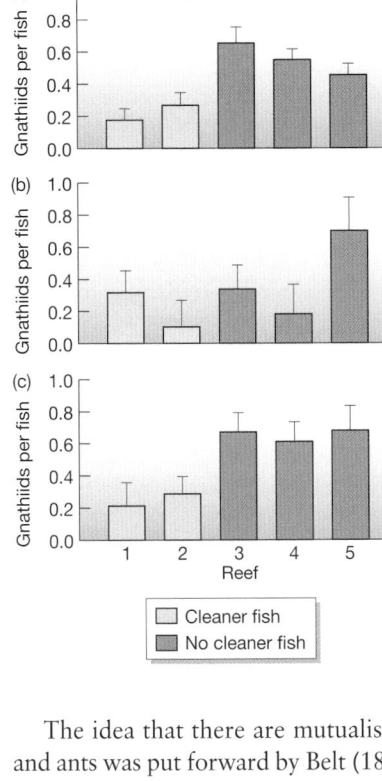

Figure 8.9

Cleaner fish really do clean their clients. The mean number of gnathiid parasites per client, *Hemigymnus melapterus*, at five reefs, from three of which cleaners, *Labroides dimidiatus*, were experimentally removed. (a) In a long-term experiment, clients without cleaners had more parasites after 12 days ($F = 17.6$, $P = 0.02$). (b) In a short-term experiment, clients without cleaners did not have significantly more parasites at dawn after 12 hours ($F = 1.8$, $P = 0.21$), presumably because cleaners do not feed at night. (c) However, the difference was significant after a further 12 hours of daylight ($F = 11.6$, $P = 0.04$). Bars are standard errors.

AFTER GRUTTER, 1999

The idea that there are mutualistic, 'protective' relationships between plants and ants was put forward by Belt (1874) after observing the behavior of aggressive ants on species of *Acacia* with swollen thorns in Central America. For example, the Bull's horn acacia (*Acacia cornigera*) bears hollow thorns that are used by its associated ant, *Pseudomyrmex ferruginea*, as nesting sites (Figure 8.10b); its

ant–plant mutualisms . . .

Figure 8.10

Structures of the Bull's horn acacia (*Acacia cornigera*) that attract its ant mutualist. (a) Protein-rich Beltian bodies at the tips of the leaflets. (b) Hollow thorns used by the ants as nesting sites.

(a) © MICHAEL FOGDEN, OXFORD SCIENTIFIC FILMS IHY360FOM00201; (b) © C. P. HICKMAN, VISUALS UNLIMITED

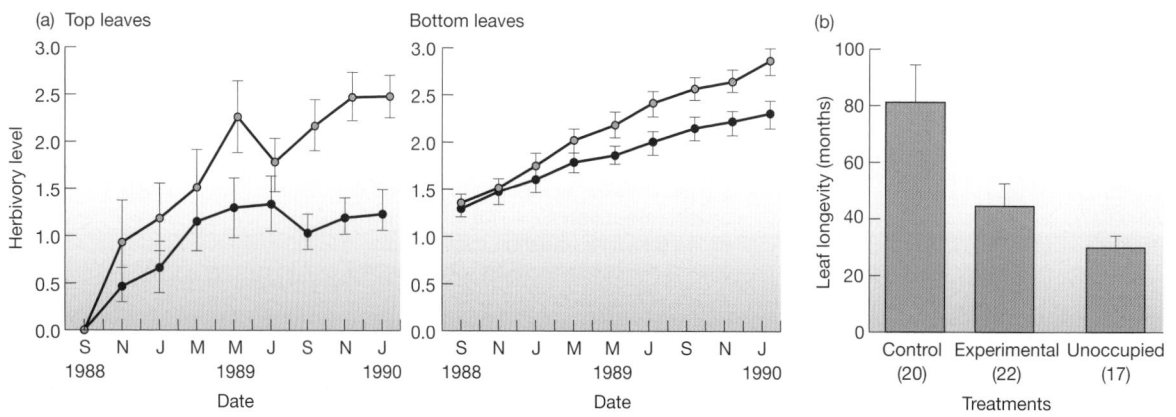

Figure 8.11

(a) The intensity of leaf herbivory (based on the cumulative proportion of leaf area removed) on plants of *Tachigali myrmecophila* naturally occupied by the ant *Pseudomyrmex concolor* (●, n = 22) and on plants from which the ants had been experimentally removed (○, n = 23). Bottom leaves were those present at the start of the experiment and top leaves were those emerging subsequently. (b) The longevity of leaves on plants of *T. myrmecophila* occupied by *P. concolor* (control) and from which ants were experimentally removed or from which ants were naturally absent. Error bars ± SE.

leaves have protein-rich 'Beltian bodies' at their tips (Figure 8.10a) which the ants collect and use for food; and it has sugar-secreting nectaries on its vegetative parts that also attract the ants. The ants, for their part, protect these small trees from competitors by actively snipping off shoots of other species and also protect the plant from herbivores – even large (vertebrate) herbivores may be deterred.

... but do the plants benefit?

In fact, ant–plant mutualisms appear to have evolved many times (even repeatedly in the same family of plants); and nectaries are present on the vegetative parts of plants of at least 39 families and in many communities throughout the world. Their precise role is not easy to establish. They clearly attract ants, sometimes in vast numbers, but carefully designed and controlled experiments are necessary to show that the plants themselves benefit, such as a study of the Amazonian canopy tree *Tachigali myrmecophila*, which harbors the stinging ant *Pseudomyrmex concolor* in specialized hollowed-out structures (Figure 8.11). The ants were removed from selected plants. These then bore 4.3 times as many phytophagous insects as control plants and suffered much greater herbivory, such that leaves on plants that carried a population of ants lived more than twice as long as those on unoccupied plants and nearly 1.8 times as long as those on plants from which ants had been deliberately removed.

8.4.2 The culture of crops or livestock

human agriculture

At least in terms of geographic extent, some of the most dramatic mutualisms are those of human agriculture. The numbers of individual plants of wheat, barley, oats, corn and rice, and the areas these crops occupy, vastly exceed what would have been present if they had not been brought into cultivation. The increase in the human population since the time of hunter–gatherers is some measure of the reciprocal advantage to *Homo sapiens*. Even without doing the experiment, we can easily imagine the effect the extinction of humans would have on the world population of rice plants or the effect of the extinction of rice plants on the

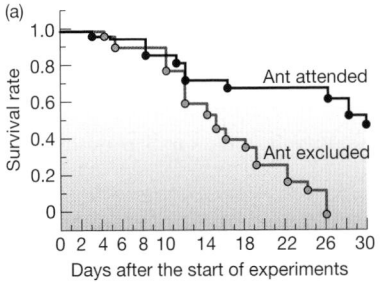
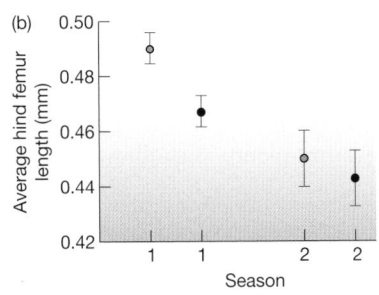
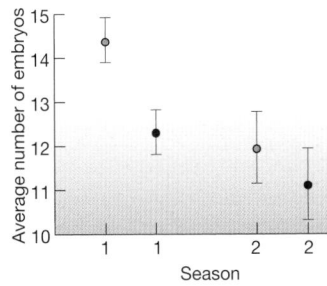

AFTER YAO ET AL., 2000

Figure 8.12

(a) Ant-excluded colonies of the aphid *Tuberculatus quercicola* were more likely to become extinct than those attended by ants ($\chi^2 = 15.9$, $P < 0.0001$). (b) But in the absence of predators (experimentally removed), ant-excluded colonies performed better than those attended by ants. Shown are averages for aphid body size (hind femur length; $F = 6.75$, $P = 0.013$) and numbers of embryos ($F = 7.25$, $P = 0.010$), $\pm$ SE, for two seasons (1: July 23 to August 11, 1998; 2: August 12 to August 31, 1998). Maroon circles, predator-free and ant-excluded treatment; black circles, predator-free and ant-attended treatment.

population of humans. Similar comments apply to the domestication of cattle, sheep and other mammals.

Similar 'farming' mutualisms have developed in termite and especially ant societies, where the farmers may protect individuals they exploit from competitors and predators and may even move or tend them. Ants, for example, farm many species of aphids (homopterans) in return for sugar-rich secretions of honeydew. The 'flocks' of aphids benefit through suffering lower mortality rates caused by predators, showing increased feeding and excretion rates, and forming larger colonies; but it would be wrong to imagine that this is a cosy relationship with nothing but benefits on both sides: the aphids are being manipulated – is there a cost to be entered on the other side of the balance sheet? This question has been addressed for colonies of the aphid *Tuberculatus quercicola* attended by the red wood ant *Formica yessensis* on the island of Hokkaido, northern Japan (Figure 8.12). As expected, in the presence of predators, aphid colonies survived significantly longer when attended by ants than when ants were excluded by smearing ant repellent at the base of the oak trees on which the aphids lived (Figure 8.12a). However, there *were* also costs for the aphids: in an environment from which predators were excluded, and the effects of ant attendance on aphids could thus be viewed in isolation, ant-attended aphids grew less well and were less fecund than those where ants as well as predators were excluded (Figure 8.12b).

aphids farmed by ants: do they pay a price?

8.4.3 The dispersal of seeds and pollen

Very many plant species use animals to disperse their seeds and pollen. About 10% of all flowering plants possess seeds or fruits that bear hooks, barbs or glues that become attached to the hairs, bristles or feathers of any animal that comes into contact with them. They are frequently an irritation to the animal, which often cleans itself and removes them if it can, but usually after carrying them some distance. In these cases the benefit is to the plant (which has invested resources in attachment mechanisms) and there is no reward to the animal.

seed dispersal

Quite different are the true mutualisms between higher plants and the birds and other animals that feed on fleshy fruits and disperse the seeds. Of course, for the relationship to be mutualistic, it is essential that the animal digests only

fruits

the fleshy fruit and not the seeds, which must remain viable when regurgitated or defecated. Thick strong defenses that protect plant embryos are usually part of the price paid by the plant for dispersal by fruit-eaters.

pollination

Many different kinds of animals have entered into pollination liaisons with flowering plants, including humming-birds, bats and even small rodents and marsupials (Figure 8.13). Most animal-pollinated flowers offer nectar, pollen or

Figure 8.13

Pollinators. (a) Honeybee (*Apis mellifera*) on raspberry flowers. (b) Cape sugarbird (*Promerops cafer*) feeding on *Protea eximia*.

(a)

(b)

both as a reward to their visitors. Floral nectar seems to have no value to the plant other than as an attractant to animals and it has a cost to the plant, because the nectar carbohydrates might have been used in growth or some other activity. Presumably, the evolution of specialized flowers and the involvement of animal pollinators have been favored because an animal may be able to recognize and discriminate between different flowers and so move pollen between different flowers of the same species but not to flowers of other species. Passive transfer of pollen, for example by wind or water, does not discriminate in this way and is therefore much more wasteful. On the other hand, where the vectors and flowers are highly specialized, as is the case in many orchids, virtually no pollen is wasted even on the flowers of other species.

The pollinators par excellence are, without doubt, the insects. Pollen is a nutritionally-rich food resource and in the simplest insect-pollinated flowers, pollen is offered in abundance and freely exposed to all and sundry. The plants rely for pollination on the insects being less than wholly efficient in their pollen consumption, carrying their spilt food with them from plant to plant. In more complex flowers, nectar (a solution of sugars) is produced as an additional or alternative reward. In the simplest of these, the nectaries are unprotected, but, with increasing specialization, nectaries are enclosed in structures that restrict access to the nectar to just a few visitor species. This range can be seen within the family Ranunculaceae. In the simple flower of *Ranunculus ficaria* the nectaries are exposed to all visitors, but in the more specialized flower of *R. bulbosus* there is a flap over the nectary, and in *Aquilegia* the nectaries have developed into long tubes and only visitors with long probosces (tongues) can reach the nectar. Unprotected nectaries have the advantage of a ready supply of pollinators, but because these pollinators are unspecialized they transfer much of the pollen to the flowers of other species. Protected nectaries have the advantage of efficient transfer of pollen by specialists to other flowers of the same species, but are reliant on there being sufficient numbers of these specialists.

> insect pollinators: from generalists to ultraspecialists

Charles Darwin (1859) recognized that a long nectary, as in *Aquilegia*, forced a pollinating insect into close contact with the pollen at the nectary's mouth. Natural selection may then favor even longer nectaries, and as an evolutionary reaction, the tongues of the pollinator would be selected for increasing length: reciprocal coevolution. Nilsson (1988) deliberately shortened the nectary tubes of the long-tubed orchid *Platanthera* and showed that the flowers then produced many fewer seeds – presumably because the pollinator was not forced into a position that maximized the efficiency of pollination.

8.4.4 Mutualistic gut inhabitants

Most of the mutualisms discussed so far have depended on patterns of behavior, where neither species lives entirely 'within' its partner. In many other mutualisms, one of the partners is a unicellular eukaryote or bacterium that is integrated more or less permanently into the body cavity or even the cells of its multicellular partner. The microbiota occupying parts of various animals' alimentary canals are the best known extracellular symbionts.

The crucial role of microbes in the digestion of cellulose by vertebrate herbivores has long been appreciated, but it now appears that the gastrointestinal tracts of all vertebrates are populated by a mutualistic microbiota. Protozoa and fungi are

> the vertebrate gut

usually present but the major contributors to these 'fermentation' processes are bacteria. Their diversity is greatest in regions of the gut where the pH is relatively neutral and food retention times relatively long. In small mammals (e.g. rodents, rabbits, hares), the cecum is the main fermentation chamber, whereas in larger non-ruminant mammals such as horses the colon is the main site. In ruminants, like cattle and sheep, and in kangaroos and other marsupials, fermentation occurs in specialized stomachs (see Figure 3.24).

The basis of the mutualism is straightforward. The microbes receive a steady flow of substrates for growth in the form of food that has been eaten, chewed and partly homogenized. They live within a chamber in which the pH and, in endotherms, temperature are regulated and anaerobic conditions are maintained. The vertebrate hosts, especially the herbivores, receive nutrition from food that they would otherwise find, literally, indigestible. The bacteria produce short-chain fatty acids (SCFAs) by fermentation of the host's dietary cellulose and starches and of the endogenous carbohydrates contained in host mucus and sloughed epithelial cells. SCFAs are often a major source of energy for the host: for example, they provide more than 60% of the maintenance energy requirements for cattle and 29–79% of those for sheep (Stevens & Hume, 1998). The microbes also convert nitrogenous compounds (amino acids that escape absorption in the midgut, urea that would otherwise be excreted by the host, mucus and sloughed cells) into ammonia and microbial protein, conserving nitrogen and water; and they synthesize B vitamins. The microbial protein is useful to the host if it can be digested – in the intestine by foregut fermenters and following coprophagy (eating their own feces) in hindgut fermenters – but ammonia is usually not useful and may even be toxic to the host.

8.4.5 Mycorrhizas

Most higher plants do not have roots, they have *mycorrhizas* – intimate mutualisms between fungi and root tissue. Plants of only a few families, such as the Cruciferae, are exceptions. Broadly, the fungal networks in mycorrhizas capture nutrients from the soil, which they transport to the plants in exchange for carbon. Many plant species can live without their mycorrhizal fungi in soils where neither nutrients nor water are ever limiting, but in the harsh world of natural plant communities, the symbioses, if not strictly obligate, are nonetheless 'ecologically obligate': that is, necessary if the individuals are to survive in nature (Buscot et al., 2000).

Generally, three major types of mycorrhiza are recognized. Arbuscular mycorrhizas are found in about two-thirds of all plant species, including most non-woody species and tropical trees. Ectomycorrhizal fungi form symbioses with many trees and shrubs, dominating boreal and temperate forests and also some tropical rain forests. Finally, ericoid mycorrhizas are found in the dominant plant species of heathland.

ectomycorrhizas

In ectomycorrhizas (ECMs), infected roots are usually concentrated in the litter layer of the soil. Fungi form a sheath of varying thickness around the roots. From there, hyphae radiate into the litter layer, extracting nutrients and water and also producing large fruiting bodies that release enormous numbers of wind-borne spores. The fungal mycelium also extends inward from the sheath, penetrating between the cells of the root cortex to give intimate cell-to-cell contact with the host and establishing an interface with a large surface area for the exchange of the products of photosynthesis, soil water and nutrients between the host plant and its fungal partner.

The ECM fungi are effective in extracting the sparse and patchy supplies of phosphorus and especially nitrogen from the forest litter layer. Carbon flows from the plant to the fungus, very largely in the form of simple hexose sugars: glucose and fructose. Fungal consumption of these may represent up to 30% of the plants' net rate of photosynthate production. The plants, though, are often nitrogen-limited, since in forest litter there are low rates of nitrogen mineralization (conversion from organic to inorganic forms), and inorganic nitrogen is itself mostly available as ammonia. It is therefore crucial for forest trees that ECM fungi can access organic nitrogen directly through enzymic degradation, and utilize ammonium as a preferred source of inorganic nitrogen. Nonetheless, the idea that this relationship between the fungi and their host plants is mutually exploitative rather than 'cosy' is emphasized by its responsiveness to changing circumstances. ECM growth is directly related to rate of flow of hexose sugars from the plant. But when the direct availability of nitrate to the plants is high, either naturally or through artificial supplementation, plant metabolism is directed away from hexose production (and export) and towards amino acid synthesis. As a result the ECM degrades: the plants seem to support just as much ECM as they appear to need.

Arbuscular mycorrhizas (AMs) do not form a sheath but penetrate *within* the roots of the host. Roots become infected from mycelium present in the soil or from germ tubes that develop from asexual spores, which are very large and produced in small numbers: a striking contrast with the ECM fungi. Initially, the fungus grows between host cells but then enters them and forms a finely branched intracellular 'arbuscule'.

arbuscular mycorrhizas

There has been a tendency to emphasize facilitation of the uptake of phosphorus as the main benefit to plants from AM symbioses (phosphorus is a highly immobile element in the soil, and is therefore frequently limiting to plant growth), but the truth appears to be more complex than this, with benefits demonstrated, too, in nitrogen uptake, pathogen and herbivore protection, and resistance to toxic metals (Newsham et al., 1995). Certainly, there are cases where the inflow of phosphorus is strongly related to the degree of colonization of roots by AM fungi. This has been shown for the bluebell, *Hyacinthoides non-scripta*, as colonization progresses during its phase of subterranean growth from August to February through to its above-ground photosynthetic phase thereafter (Figure 8.14a). Indeed, bluebells cultured without AM fungi are unable to take up phosphorus through their poorly branched system of roots (Merryweather & Fitter, 1995).

a range of benefits?

On the other hand, a set of experiments examined the growth of the annual grass *Vulpia ciliata* ssp. *ambigua* (Figure 8.14b) in which seedlings of *Vulpia* were grown with an AM fungus (*Glomus* sp.), with the pathogenic fungus *Fusarium oxysporum*, with both, and with neither. Growth was not enhanced by *Glomus* alone, but growth was harmed by *Fusarium* in the absence of *Glomus*. When both were present, growth returned to normal levels. Clearly, the mycorrhiza did not benefit the phosphorus economy of the *Vulpia*, but it did protect it from the harmful effects of the pathogen.

The key difference appears to be that *Vulpia*, unlike the bluebell, has a highly branched system of roots (Newsham et al., 1995). Plants with finely branched roots have little need for supplementary phosphorus capture, but development of that

it depends on the species

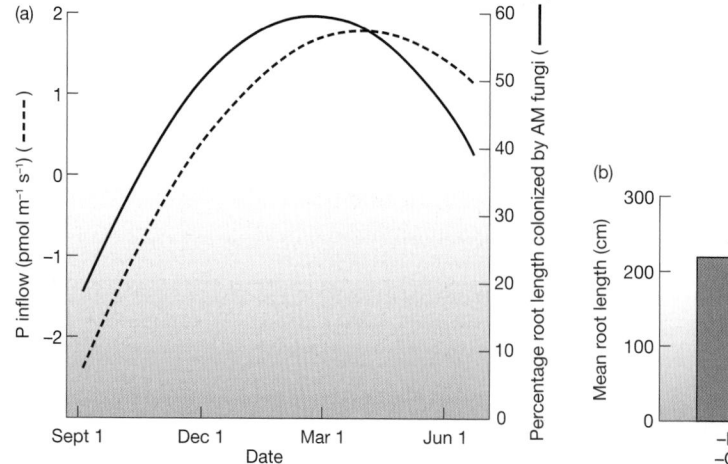
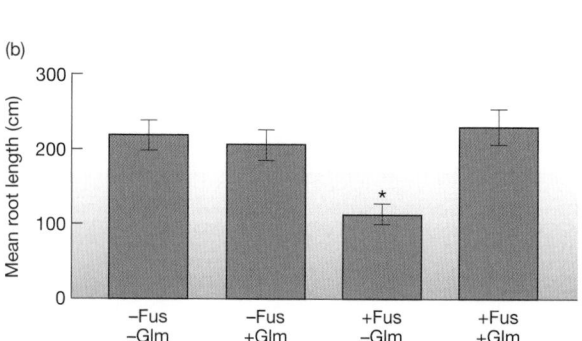

Figure 8.14

(a) Curves fitted to rates of phosphorus inflow (dashed line, left axis) and root colonization by arbuscular mycorrhiza (AM) fungi (solid line, right axis) in the bluebell, *Hyacinthoides non-scripta*, over a single growing season. Phosphorus uptake appears to be strongly linked to root colonization by the fungi. (b) The effects of a factorial combination of *Fusarium oxysporum* (Fus, a pathogenic fungus) and an AM fungus, *Glomus* sp. (Glm) on growth (root length) of *Vulpia* plants. Values are means of 16 replicates per treatment; bars are standard errors; the asterisk signifies a significant difference at $P < 0.05$ in a Fisher's pairwise comparison. In this case, the benefit provided by AM fungi seems not to be an improvement in nutrient uptake but protection against the pathogen.

(A) AFTER MERRYWEATHER & FITTER, 1995; NEWSHAM ET AL., 1995; (B) AFTER NEWSHAM ET AL., 1994, 1995

same root architecture provides multiple points of entry for plant pathogens. In such cases AM symbioses are therefore likely to have evolved with an emphasis on plant protection. By contrast, root systems with few lateral and actively growing meristems are relatively invulnerable to pathogen attack, but these root systems are poor foragers for phosphorus. Here, AM symbioses are likely to have evolved with an emphasis on phosphorus capture.

8.4.6 Fixation of atmospheric nitrogen in mutualistic plants

The inability of most plants and animals to fix atmospheric nitrogen is one of the great puzzles in the process of evolution, since nitrogen is in limiting supply in many habitats. However, the ability to fix nitrogen is widely though irregularly distributed amongst both the eubacteria ('true' bacteria) and the archaea (Archaebacteria), and many of these have been caught up in tight mutualisms with distinct groups of eukaryotes. The best known, because of the huge agricultural importance of legume crops, are the rhizobia, which fix nitrogen in the root nodules of most leguminous plants and just one non-legume, *Parasponia* (a member of the family Ulmaceae, the elms).

mutualisms of rhizobia and leguminous plants: several steps to a liaison

The establishment of the liaison between rhizobia and legume plants proceeds by a series of reciprocating steps. The bacteria occur in a free-living state in the soil and are stimulated to multiply by root exudates and cells that have been sloughed from roots as they develop. In a typical case, a bacterial colony develops on the root hair, which then begins to curl and is penetrated by the bacteria. The host responds by laying down a wall that encloses the bacteria and forms an 'infection thread', which grows within the host root cortex, and within which the rhizobia proliferate. Rhizobia in the infection thread cannot fix nitrogen, but some are released into host cells in a developing 'nodule', where, surrounded by a host-derived peribacteroid membrane, they differentiate into 'bacteroids' that can fix nitrogen. Meanwhile, a special vascular system develops in the host, supplying the products of photosynthesis to the nodule tissue and carrying away fixed-nitrogen compounds to other parts of the plant.

The costs and benefits of this mutualism need to be considered carefully. From the plant's point of view, we need to compare the energetic costs of alternative processes by which supplies of fixed nitrogen might be obtained. The route for most plants is direct from the soil as nitrate or ammonium ions. The metabolically cheapest route is the use of ammonium ions, but in most soils ammonium ions are rapidly converted to nitrates by microbial activity (nitrification). The energetic cost of reducing nitrate from the soil to ammonia is about 12 mol of adenosine triphosphate (ATP, the cell's energy currency) per mole of ammonia formed. The mutualistic process (including the maintenance costs of the bacteroids) is energetically slightly more expensive to the plant: about 13.5 mol of ATP. However, we must also add the costs of forming and maintaining the nodules, which may be about 12% of the plant's total photosynthetic output. It is this that makes nitrogen fixation energetically inefficient. Energy, though, may be much more readily available for green plants than nitrogen. A rare and valuable commodity (fixed nitrogen) bought with a cheap currency (energy) may be no bad bargain. On the other hand, when a nodulated legume is provided with nitrates (i.e. when nitrate is *not* a rare commodity) nitrogen fixation declines rapidly.

costs and benefits of rhizobial mutualisms

On the other hand, the mutualisms of rhizobia and legumes (and other nitrogen-fixing mutualisms) must not be seen as isolated interactions between bacteria and their own host plants. In nature, legumes normally form mixed stands in association with non-legumes. These are potential competitors with the legumes for fixed nitrogen (nitrates or ammonium ions in the soil). The nodulated legume sidesteps this competition by its access to its unique source of nitrogen. It is in this ecological context that nitrogen-fixing mutualisms gain their main advantage. Where nitrogen is plentiful, however, the energetic costs of nitrogen fixation often put the plants at a competitive *dis*advantage.

interspecific competition: a classic 'replacement series'

Figure 8.15, for example, shows the results of a classic experiment in which soybeans (*Glycine soja*, a legume) were grown in mixtures with *Paspalum*, a grass. The mixtures either received mineral nitrogen, or were inoculated with *Rhizobium*, or received both. The experiment was designed as a 'replacement series', which allows us to compare the growth of pure populations of the grass and legume with their performances in the presence of each other. In the pure stands of soybean, yield was increased very substantially *either* by inoculation with *Rhizobium*, *or* by application of fertilizer nitrogen, or by receiving both. The legumes can use either source of nitrogen as a substitute for the other. The grass, however, responded only to the fertilizer. Hence, when the species competed in the presence of *Rhizobium* alone, the legume contributed far more to the overall yield than did the grass: over a succession of generations, the legume would have outcompeted the grass. When they competed in soils supplemented

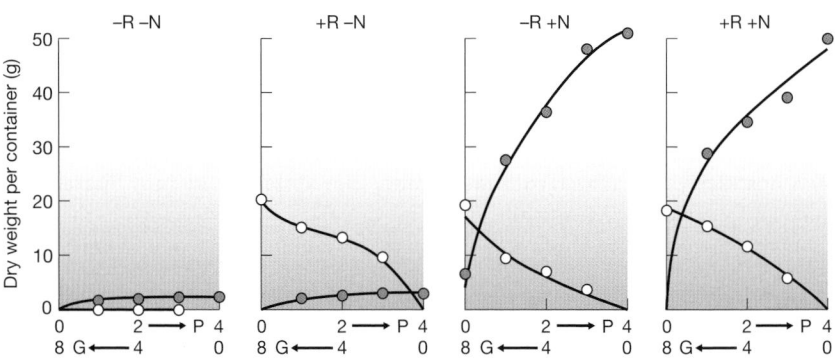

AFTER DE WIT ET AL., 1966

Figure 8.15

The growth of soybeans (*Glycine soja*, G, ○) and a grass (*Paspalum*, P, ●) grown alone and in mixtures with and without nitrogen fertilizer (N) and with and without inoculation with nitrogen-fixing *Rhizobium* (R). The plants were grown in pots containing 0–4 plants of the grass together with 0–8 plants of *Glycine*. Thus, moving left to right on the horizontal axis, the treatments are zero *Paspalum* (0P) and 8 *Glycine* (8G), 1P with 6G, 2P with 4G, 3P with 2G and, finally, 4P with 0G. The vertical scale on each figure shows the mass of plants of the two species in each container. –R –N, no *Rhizobium* and no fertilizer; +R –N, inoculated with *Rhizobium* but no fertilizer; –R +N, no *Rhizobium* but nitrate fertilizer was applied; +R +N, inoculated with *Rhizobium* and nitrate fertilizer was supplied. When the two species competed in the presence of nitrogen-fixing *Rhizobium* and without fertilizer, the soybeans (with their mutualistic relationship to *Rhizobium*) performed best, but in the presence of nitrogen fertilizer (with or without the *Rhizobium*) the grass outperformed the soybeans.

with fertilizer nitrogen, however, whether or not *Rhizobium* was also present, it was the grass that made the major contribution: long term, it would have out-competed the legume.

Quite clearly, then, it is in environments deficient in nitrogen that nodulated legumes have a great advantage over other species. But their activity raises the level of fixed nitrogen in the environment. After death, legumes augment the level of soil nitrogen on a very local scale with a 6–12-month delay as they decompose. Thus, their advantage is lost – they have improved the environment of their competitors, and the growth of associated grasses will be favored in these local patches. Hence, organisms that can fix atmospheric nitrogen can be thought of as locally suicidal. This is one reason why it is very difficult to grow repeated crops of pure legumes in agricultural practice without aggressive grass weeds invading the nitrogen-enriched environment. It may also explain why leguminous herbs or trees usually fail to form dominant stands in nature.

the shifting balance between nitrogen-fixers and non-fixers

Grazing animals, on the other hand, continually remove grass foliage, and the nitrogen status of a grass patch may again decline to a level at which the legume is once more at a competitive advantage. In a stoloniferous legume, such as white clover, the plant is continually 'wandering' through the sward, leaving behind it local grass-dominated patches, whilst invading and enriching with nitrogen new patches where the nitrogen status has become low. The symbiotic legume in such a community not only drives its nitrogen economy but also some of the cycles that occur within its patchwork (Cain et al., 1995).

We end this section, then, on a theme that has recurred repeatedly. To under-stand the ecology of mutualistic pairs, we must look beyond those species to the wider community of which they are part.

Summary

Molecular ecology: differentiation within and between species

For much of the time, it is entirely appropriate for ecologists to talk about 'populations' or 'species' as if they were singular, homogeneous entities, but for some purposes, knowing how much differentiation there is within species, or between one species and another, is critical for an understanding of their dynamics, and ultimately for managing those dynamics. Molecular genetic markers, of a variety of types, have massively increased the resolution at which we can differentiate between populations and even individuals.

Studies on albatrosses illustrate how even within a species of conservation importance, separate populations even more threatened with extinction may be hidden; while studies on salmon illustrate how molecular markers can be used to detect, and to prosecute, illegal fishermen. Molecular markers have also shown, for example, that a threatened 'species', the red wolf, may in fact be a hybrid between two other, relatively common species, with implications for both the desirability and the practicality of its conservation.

Coevolutionary arms races

A better defended food resource exerts a selection pressure on consumers to overcome that defense. A consumer that does so will steal a march on its competitors, and so is likely to become relatively specialized on that prey type – which is then under particular pressure to defend itself against that particular consumer, and so on: a coevolutionary 'arms race'. Plants relying on toxins are more prone to becoming involved in arms races with their herbivores than those relying on more 'quantitative' (digestion-reducing) chemicals.

The intimate association between parasites and their hosts makes them especially prone to coevolutionary arms races. However, the process is not necessarily so straightforward, as illustrated by the case of the myxoma virus and the European rabbit. The evolution of resistance in the rabbit is easy to understand, but the parasites favored by natural selection are those with the greatest reproductive rate.

In the myxoma virus, this occurs at intermediate levels of virulence because of increased transmissibility.

In other cases, host–parasite coevolution is more definitely antagonistic: increasing resistance in the host and increasing infectivity in the parasite. With bacteria and their viruses, this process can be observed in action, because generation times are so short.

Mutualistic interactions

No species lives in isolation, but often the association with other species is especially close: for many organisms, the habitat they occupy is an individual of another species – a symbiosis. A mutualistic relationship is one in which organisms of different species interact to their mutual benefit. Current evolutionary thinking views mutualisms as cases of reciprocal exploitation where nonetheless each partner is a *net* beneficiary. Mutualisms themselves have often been neglected in the past compared to other types of interaction, yet mutualists compose most of the world's biomass.

Pairs of species from many taxa take part in mutualistic associations in which one species protects the other from predators or competitors but gains privileged access to a food resource on the protected species.

Some of the most dramatic mutualisms are those of human agriculture, but similar 'farming' mutualisms have developed in termite and especially ant societies. Ants farm many species of aphids in return for sugar-rich secretions of honeydew. The aphids benefit through suffering lower mortality rates; but there are also costs: where aphid predators are excluded experimentally, aphids grow less well in the presence of ants.

Very many plant species use animals to disperse their seeds and pollen, and many different kinds of animals have entered into pollination liaisons with flowering plants. The pollinators par excellence, though, are the insects.

The gastrointestinal tracts of all vertebrates are populated by a mutualistic microbiota. The microbes receive a steady flow of substrates for growth in the form of food that has been eaten, and they live within a chamber in which pH and, in endotherms, temperature are regulated and anaerobic conditions are maintained. The vertebrate hosts receive nutrition from food that they would otherwise find, literally, indigestible.

Most higher plants do not have roots, they have mycorrhizas – intimate mutualisms between fungi and root tissue. In ectomycorrhizas (ECMs), fungi form a sheath of varying thickness around the roots. These fungi are effective in extracting the sparse and patchy supplies of phosphorus and especially nitrogen from the forest litter layer. Carbon flows from the plant to the fungus (mostly hexose sugars). However, ECM growth is directly related to the rate of flow of the sugars from the plant. When the direct availability of nitrate to the plants is high, plant metabolism is directed away from hexose production. As a result the ECM degrades: the plants seem to support just as much ECM as they appear to need. Arbuscular mycorrhizas (AMs) penetrate *within* the roots of the host. There has been

a tendency to emphasize facilitation of the uptake of phosphorus as the main benefit to plants from AM symbioses, but benefits have been demonstrated, too, in nitrogen uptake, pathogen and herbivore protection, and resistance to toxic metals.

The ability to fix nitrogen is widely distributed amongst both the eubacteria and the Archaebacteria, and many of these have been caught up in tight mutualisms with distinct groups of eukaryotes. The best known are the rhizobia, which fix nitrogen in the root nodules of most leguminous plants. Nitrogen fixation is often energetically inefficient, but energy may be much more readily available for green plants than nitrogen. On the other hand, when a nodulated legume is provided with nitrates, nitrogen fixation declines rapidly. The mutualisms of rhizobia and legumes (like other nitrogen-fixing mutualisms) must be seen in the context of competition between legumes and non-legumes.

Review questions

REVIEW QUESTIONS

Asterisks indicate challenge questions

1 Explain why molecular (DNA) markers have improved the ability of ecologists to study degrees of differentiation within and between species.

2* Review the range of molecular markers that have been used in molecular ecology, stressing their advantages and disadvantages at different scales of resolution.

3 Should the red wolf be conserved, or would that be a misguided waste of public money?

4 Why are some plants more likely than others to be involved in arms races with their insect herbivores?

5 Account for the decline in virulence of the myxomatosis virus in European rabbits after its initial introductions in Australia and Europe.

6 Compare and contrast the mutualistic associations of ants with plants they protect and aphids they farm.

7* Discuss the following propositions: 'Most herbivores are not really herbivores but consumers of the byproducts of the mutualists living in their gut' and 'Most gut parasites are not really parasites but competitors with their hosts for food that the host has captured'.

8 Compare the roles of fruits and nectar in the interactions between plants and the animals that visit them.

9 What are mycorrhizas and what is their significance?

10* Leguminous plants are a perfect example of a mutualistic association that can only be understood in the context of the ecological community within which it normally exists. Discuss.

Chapter 9

From populations to communities

Chapter contents

Key concepts

In this chapter you will:

- appreciate the variety of interacting abiotic and biotic factors that account for the dynamics of populations
- distinguish between the determination and regulation of population abundance
- understand how patchiness and dispersal between patches influence the dynamics of both populations and communities
- recognize the influence of disturbance on community patterns and understand the nature of community succession
- appreciate the importance of direct and indirect effects and distinguish between bottom-up and top-down control of food webs
- understand the relationship between the structure and stability of food webs

In previous chapters, we generally dealt with individual species or pairs of species in isolation, as ecologists often do. Ultimately, however, we must recognize that every population exists within a web of interactions with myriad other populations, across several trophic levels. Each population must be viewed in the context of the whole community, and we need to understand that populations occur in patchy and inconstant environments in which disturbance and local extinction may be common.

9.1 Introduction

Single-species populations have been the focus for many of the questions posed in previous chapters. In attempting to answer the most fundamental ecological question of all – what determines a species' abundance and distribution – we have chosen to ask separately about the role of conditions and resources, of migration, of competition (both intra- and interspecific), of mutualism, and of predation and parasitism. In reality, the dynamics of any population reflect a combination of these effects, though their relative importance varies from case to case. Now, therefore, we need to view the population in the context of the whole community, since each exists within a whole web of interactions (Figure 9.1), and each responds differently to the prevailing abiotic conditions.

In Section 9.2 we consider how abiotic and biotic factors combine to determine the dynamics of species populations. Then, in Section 9.3, we revisit one of the major themes of this book – the importance of patchiness and dispersal between patches in ecological dynamics – and discuss especially the importance of the concept of the metapopulation. Disturbances, such as forest fires and the storm battering of seashores, also play an important role in the dynamics of many populations and the composition of most communities. After each disturbance, there is a pattern of re-establishment of species that is played out against a background of changing conditions, resources and population interactions. We deal with temporal patterns in community composition, including community succession, in Section 9.4. Finally, in Section 9.5 we broaden our view further to examine food webs, like the one illustrated in Figure 9.1, with usually at least

Figure 9.1

Community matrix illustrating how each species may interact with several others in competitive interactions (among plant species 1, 2 and 3; or between grazers 4 and 5; or between predators 6 and 7) and predator–prey interactions (such as between 6 and 4, or 5 and 2).

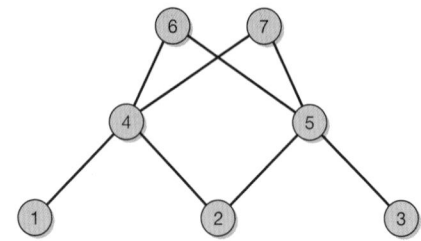

three trophic levels (plant–herbivore–predator), emphasizing the importance not only of direct but also of indirect effects that a species may have on others on the same trophic level or on levels below or above it.

9.2 Multiple determinants of the dynamics of populations

Why are some species rare and others common? Why does a species occur at low population densities in some places and at high densities in others? What factors cause fluctuations in a species' abundance? These are crucial questions when we wish to conserve rare species, or control pests, or manage natural, living resources, or when we wish simply to understand the patterns and dynamics of the natural world. To provide complete answers for even a single species in a single location, we need to know the physicochemical conditions, the level of resources available, the organism's life cycle and the influence of competitors, predators, parasites and so on – and how all these factors influence abundance through effects on birth, death, dispersal and migration. We now bring these factors together and consider how we might discover which actually matter in particular examples.

fluctuations in abundance are caused by a wide variety of biotic and abiotic factors

The raw material for the study of abundance is usually some estimate of the numbers of individuals in a population. However, a record of numbers alone can hide vital information. Picture three human populations, shown to contain identical numbers of individuals. One is an old people's residential area, the second is a population of young children and the third is a population of mixed age and sex. In the absence of information beyond mere numbers, it would not be clear that the first population was doomed to extinction (unless maintained by immigration), the second would grow fast but only after a delay and the third would continue to grow steadily. The most satisfactory studies, therefore, estimate not only the numbers of individuals (and their parts, in the case of modular organisms) but also those of different age, sex and size.

what total numbers can and cannot tell us

The data that accumulate from estimates of abundance may be used to establish correlations with external factors like food or weather. Correlations may be used to predict the future. For example, high intensities of the disease 'late blight' in potato crops usually occur 15–22 days after a period in which the minimum temperature is above 10°C and relative humidity is more than 75% for two consecutive days. Such a correlation may alert the potato grower to the need for protective spraying. Correlations may also suggest – but not prove – causal relationships. For example, a correlation may be demonstrated between the size of a population and its growth rate. But ultimately 'cause' requires a mechanism. When the population is large, many individuals may starve to death, or may fail to reproduce, or may become aggressive and drive out the weaker members. A correlation cannot tell us which. Nonetheless, correlations can be informative. Figure 9.2, for example, shows four examples in which population growth rate increases with the availability of food. It also suggests that in general, such relationships are likely to level off at the highest food levels where some other factor or factors place an upper limit on abundance.

what correlations can and cannot tell us

Figure 9.2

Increases in annual population growth rate with the availability of food, measured as pasture biomass (kg ha^{-1}) in (a) and (c), as vole abundance in (b) and as availability of food per capita in (d). (a) Red kangaroo (Bayliss, 1987). (b) Barn owl (modified from Taylor, 1994). (c) Wildebeest (Krebs et al., 1999). (d) Feral pig (Choquenot, 1998). Positive growth rates indicate increasing abundance; negative growth rates decreasing abundance.

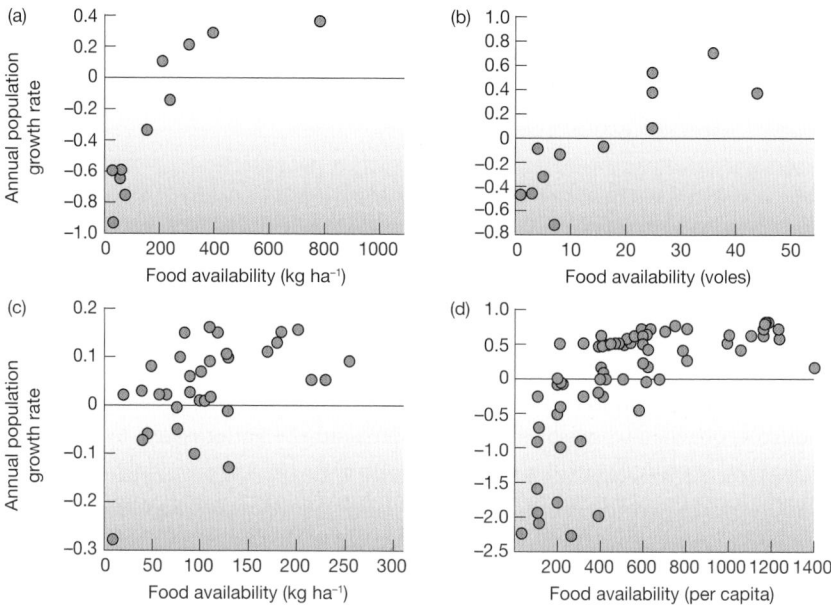

AFTER SIBLY & HONE, 2002

9.2.1 Fluctuation or stability?

many populations are very stable . . .

Some populations appear to change very little in size. One study that covered an extended timespan – though it was not necessarily the most scientific – examined swifts (*Micropus apus*) in the village of Selborne in southern England over more than 200 years. In one of the earliest published works on ecology, Gilbert White, who lived in the village, wrote in 1778 (see White, 1789):

> I am now confirmed in the opinion that we have every year the same number of pairs invariably. . . . The number that I constantly find are eight pairs, about half of which reside in the church, and the rest in some of the lowest and meanest thatched cottages.

More than 200 years later, Lawton and May (1984) visited the village and, not surprisingly, found major changes. Swifts are unlikely to have nested in the church for 50 years, and the thatched cottages have either disappeared or had their roofs covered with wire. Yet the number of breeding pairs of swifts regularly to be found in the village is now 12. In view of the many changes that have taken place in the intervening centuries, this number is remarkably close to the eight pairs so consistently found by White.

. . . but stability need not mean 'nothing changes'

But the stability of a population may conceal complex underlying dynamics. Another example of a population showing relatively little change in adult numbers from year to year is seen in an 8-year study in Poland of the small, annual sand-dune plant *Androsace septentrionalis* (Figure 9.3a). Each year, however, there was great flux within the population. Between 150 and 1000 new seedlings per square meter appeared, but subsequent mortality reduced the population by between 30% and 70%. Thus, the population appears to be kept within bounds. At least 50 plants always survived to fruit and produce seeds for the next season. By contrast, the mice in Figure 9.3b have extended periods of relatively low abundance interrupted by sporadic and dramatic irruptions.

(a)

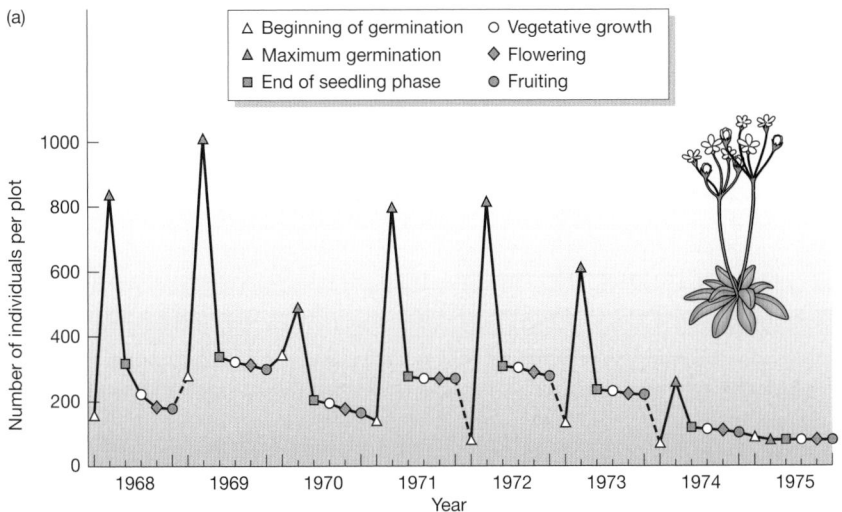

(b)

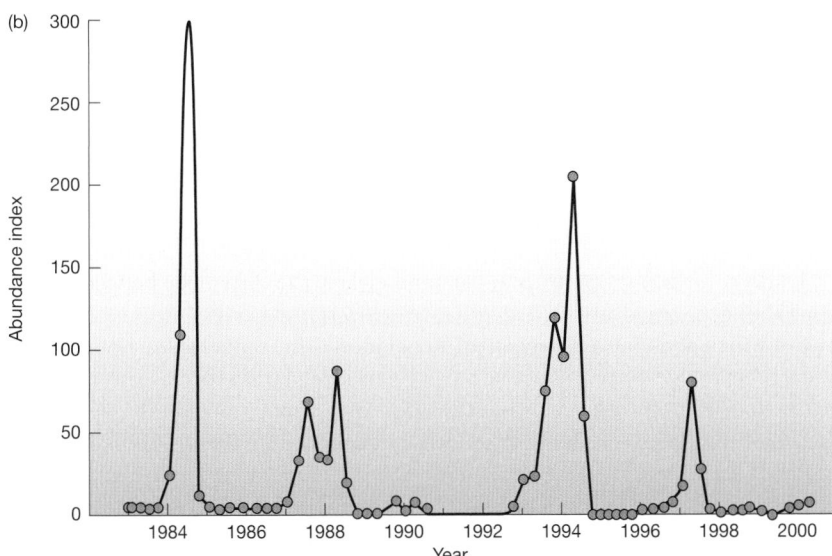

(a) AFTER SYMONIDES, 1979; (b) AFTER SINGLETON ET AL., 2001

Figure 9.3

(a) The population dynamics of *Androsace septentrionalis* during an 8-year study. (b) Irregular irruptions in the abundance of house mice (*Mus domesticus*) in an agricultural habitat in Victoria, Australia, where the mice, when they irrupt, are serious pests. The 'abundance index' is the number caught per 100 trap-nights. In the fall of 1984 the index exceeded 300.

9.2.2 Determination and regulation of abundance

Is the move from eight to 12 pairs of swifts over 200 years an indication of consistency or of change? Is the similarity between eight and 12 of most interest – or the difference between them? Some investigators have emphasized the apparent constancy of populations; others have emphasized the fluctuations.

Those who have emphasized constancy argue that we need to look for stabilizing forces within populations to explain why the populations do not exhibit unfettered increase or a decline to extinction (generally, density-dependent forces: for instance, competition between crowded individuals for limited resources). Those who have emphasized fluctuations often look to external factors, weather or disturbance, to explain the changes. Can the two sides be brought together to form a consensus?

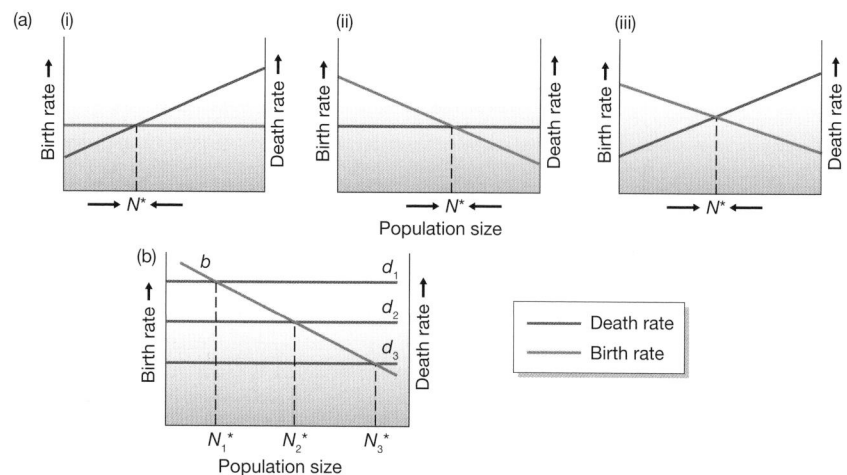

Figure 9.4

(a) Population regulation with: (i) density-independent birth and density-dependent death; (ii) density-dependent birth and density-independent death; and (iii) density-dependent birth and death. Population size increases when birth rate exceeds death rate and decreases when death rate exceeds birth rate. N^* is therefore a stable equilibrium population size. The actual value of the equilibrium population size is seen to depend on both the magnitude of the density-independent rate and the magnitude and slope of any density-dependent processes. (b) Population regulation with density-dependent birth, b, and density-independent death, d. Death rates are determined by physical conditions which differ in three sites (death rates d_1, d_2 and d_3). Equilibrium population size varies as a result (N_1^*, N_2^*, N_3^*).

the distinction between determination and regulation

To do so, it is important to understand clearly the difference between questions about the ways in which abundance is *determined* and questions about the way in which abundance is *regulated*. Regulation is the tendency of a population to decrease in size when it is above a particular level, but to increase in size when below that level. In other words, regulation of a population can, by definition, occur only as a result of one or more density-dependent processes (see Chapters 3 and 5) that act on rates of birth and/or death and/or movement (Figure 9.4a). Various potentially density-dependent processes have been discussed in earlier chapters on competition, predation and parasitism. We must look at regulation, therefore, to understand how it is that a population tends to remain within defined upper and lower limits.

On the other hand, the precise abundance of individuals will be determined by the combined effects of all the factors and all the processes that affect a population, whether they are dependent or independent of density (Figure 9.4b). We must look at the determination of abundance, therefore, to understand how it is that a particular population exhibits a particular abundance at a particular time, and not some other abundance.

In the past, certainly, some have believed that density-dependent, biotic interactions play the main role not only in regulating but also in determining population size, holding populations in a state of balance in their environments. Others have felt that most natural populations could be viewed as passing through a repeated sequence of setbacks and recovery. This view tends to reject any subdivision of the environment into density-dependent and density independent 'factors', preferring instead to see populations as sitting at the center of an ecological web, where various factors and processes interact in their effects on the population.

There is really no conflict between the two views. The first is preoccupied with what regulates population size and the second with what determines population size – and both are perfectly valid interests. No population can be absolutely free of regulation – long-term unrestrained population growth is unknown, and unrestrained declines to extinction are rare. Furthermore, any suggestion that density-dependent processes are rare or generally of only minor importance would be wrong. A very large number of studies have been made of various kinds of animals, especially of insects. Density dependence has by no means always been detected but it is commonly seen when studies are continued for many generations. For instance, density dependence was detected in 80% or more of studies of insects that lasted for more than 10 years (Hassell et al., 1989; Woiwod & Hanski, 1992).

both are perfectly valid interests

On the other hand, for many populations weather is typically the major determinant of abundance and other factors are of relatively minor importance. For instance, in one famous, classic study of a pest, apple thrips, weather accounted for 78% of the variation in the number of thrips (Davidson & Andrewartha, 1948); for predicting thrips' abundance, information on the weather is of paramount importance. So, what regulates the size of a population need not determine its size for most of the time. It would be wrong to give regulation or density dependence some kind of pre-eminence. It may be occurring only infrequently or intermittently, and it is likely that no natural population is ever truly at equilibrium: even when regulation is occurring, it may be drawing abundance toward a level that is itself changing in response to changing levels of resources. Thus, there are a range of possibilities: some populations in nature are almost always recovering from the last disaster (Figure 9.5a), others are usually limited by an abundant resource (Figure 9.5b) or by a scarce resource (Figure 9.5c), and others are usually in decline after sudden episodes of colonization (Figure 9.5d).

9.2.3 Key factor analysis

We can distinguish clearly between what regulates and what determines the abundance of a population, and see how regulation and determination relate to one another, by examining an approach known as *key factor analysis*. It has been applied to many insects and some other animals and plants and is based on calculating what are known as k-values for each phase of the life cycle. In fact, key factor analysis is poorly named, since it identifies key *phases* (rather than key factors) in the life of a study organism (those most important in determining abundance). Details are described in Box 9.1, but the approach can be understood simply by appreciating that the k-values measure the amount of mortality: the higher the k-value, the greater the mortality (k stands for 'killing power').

For a key factor analysis to be carried out, data are compiled in the form of a life table (see Chapter 5), such as that done for a Canadian population of the Colorado potato beetle (*Leptinotarsa decemlineata*) in Box 9.1. The sampling program in that case provided estimates of the population at seven stages: eggs, early larvae, late larvae, pupae, summer adults, hibernating adults and spring adults. One further category was included, females × 2, to take account of any unequal sex ratios among the summer adults.

Colorado potato beetles

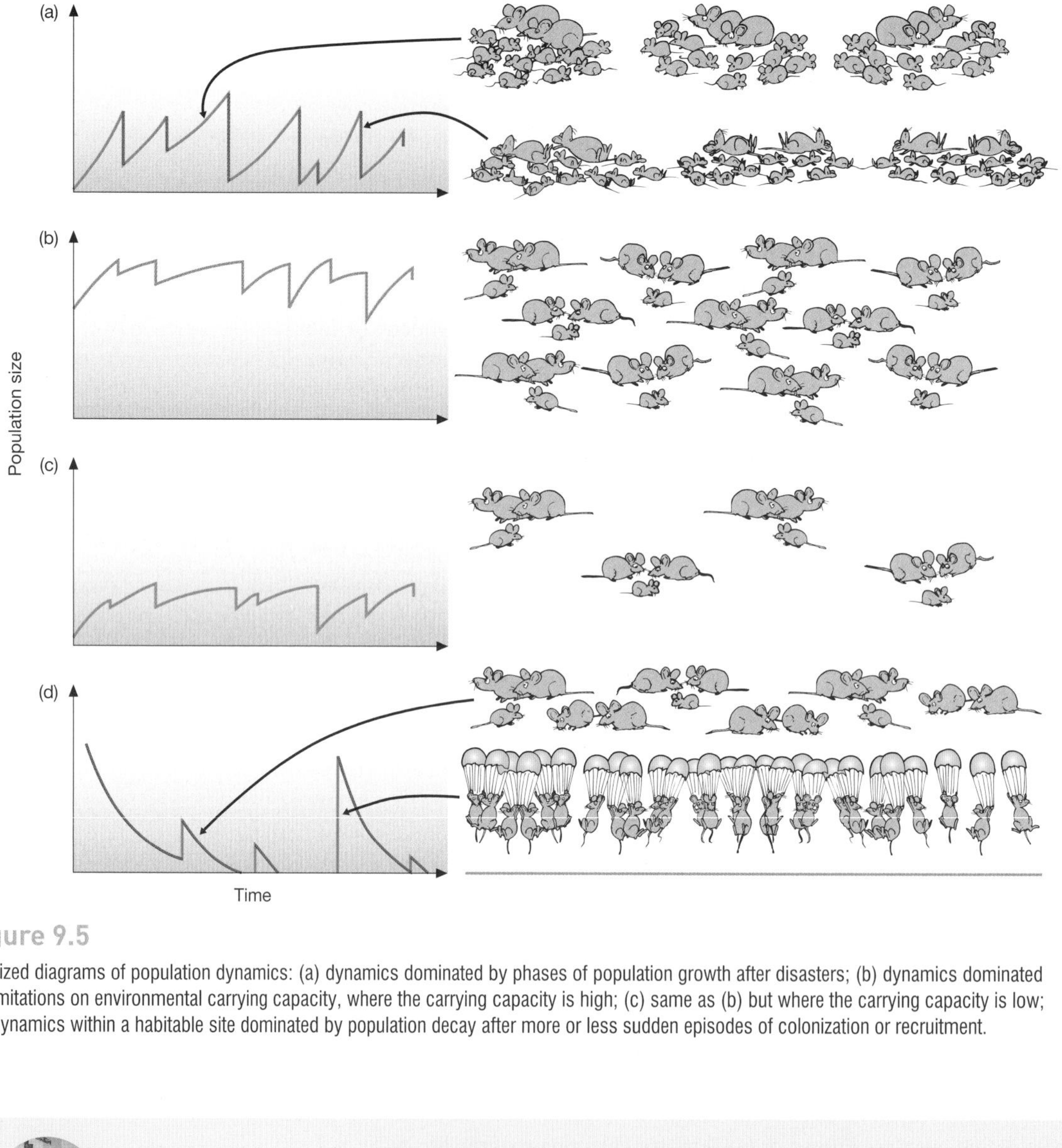

Figure 9.5

Idealized diagrams of population dynamics: (a) dynamics dominated by phases of population growth after disasters; (b) dynamics dominated by limitations on environmental carrying capacity, where the carrying capacity is high; (c) same as (b) but where the carrying capacity is low; (d) dynamics within a habitable site dominated by population decay after more or less sudden episodes of colonization or recruitment.

9.1 *Quantitative aspects*
9.1 QUANTITATIVE ASPECTS

Determining *k*-values for key factor analysis

Table 9.1 sets out a typical set of life table data, collected by Harcourt (1971) for the Colorado potato beetle, *Leptinotarsa decemlineata*, in Canada. The first column lists the various phases of the life cycle.

Spring adults emerge from hibernation around the middle of June, when potato plants are breaking through the ground. Within 3 or 4 days egg laying begins, and it continues for about 1 month. The eggs

Table 9.1

Life table data for the Canadian Colorado potato beetle.

AGE INTERVAL	NUMBERS PER 96 POTATO HILLS	NUMBERS DYING	MORTALITY FACTOR	FACTOR $LOG_{10}N$	k-VALUE	
Eggs	11,799	2,531	Not deposited	4.072	0.105	(k_{1a})
	9,268	445	Infertile	3.967	0.021	(k_{1b})
	8,823	408	Rainfall	3.946	0.021	(k_{1c})
	8,415	1,147	Cannibalism	3.925	0.064	(k_{1d})
	7,268	376	Predators	3.861	0.023	(k_{1e})
Early larvae	6,892	0	Rainfall	3.838	0	(k_2)
Late larvae	6,892	3,722	Starvation	3.838	0.337	(k_3)
Pupal cells	3,170	16	Parasitism	3.501	0.002	(k_4)
Summer adults	3,154	−126	Sex (52% ♀)	3.499	−0.017	(k_5)
Females × 2	3,280	3,264	Emigration	3.516	2.312	(k_6)
Hibernating adults	16	2	Frost	1.204	0.058	(k_7)
Spring adults	14			1.146		
					2.926	(k_{total})

are laid in clusters (approximately 34 eggs) on the lower leaf surface, and the larvae crawl to the top of the plant, where they feed throughout their development, passing through four stages. When mature,

An adult Colorado potato beetle (*Leptinotarsa decemlineata*) taking off from its host plant. Emigration by summer adults represents the key phase in the population dynamics of potato beetles.

they drop to the ground and form pupal cells in the soil. *Summer adults* emerge in early August, feed, and then re-enter the soil at the beginning of September to hibernate and become the next season's spring adults.

The next column lists the estimated numbers (per 96 potato hills) at the start of each phase, and the third column then lists the numbers dying in each phase, before the start of the next. This is followed, in the fourth column, by what were believed to be the main causes of deaths in each stage of the life cycle. The fifth and sixth columns then show how k-values are calculated. In the fifth column, the logarithms of the numbers at the start of each phase are listed. The k-values in the sixth column are then simply the differences between successive values in column 5. Thus, each value refers to deaths in one of the phases, and, similarly to column 3, the total of the column refers to the total death throughout the life cycle. Moreover, each k-value measures the rate or intensity of mortality in its own phase, whereas this is not true for the values in column 3 – there, values tend to be higher earlier in the life cycle simply because there are more individuals 'available' to die. These useful characteristics of k-values are put to use in *key factor analysis*.

Table 9.2

Summary of the life table analysis for Canadian Colorado beetle populations (see Box 9.1).

		MEAN	COEFFICIENT OF REGRESSION ON k_{TOTAL}
Eggs not deposited	k_{1a}	0.095	−0.020
Eggs infertile	k_{1b}	0.026	−0.005
Rainfall on eggs	k_{1c}	0.006	0.000
Eggs cannibalized	k_{1d}	0.090	−0.002
Egg predation	k_{1e}	0.036	−0.011
Larvae 1 (rainfall)	k_2	0.091	0.010
Larvae 2 (starvation)	k_3	0.185	0.136
Pupae (parasitism)	k_4	0.033	−0.029
Unequal sex ratio	k_5	−0.012	0.004
Emigration	k_6	1.543	0.906
Frost	k_7	0.170	0.010
	k_{total}	2.263	

AFTER HARCOURT, 1971

when does most mortality occur?

The first question we can ask is: 'How much of the total "mortality" tends to occur in each of the phases?' (Mortality is in inverted commas because it refers to all losses from the population.) The question can be answered by calculating the mean k-values for each phase, in this case determined over 10 seasons (that is, from 10 tables like the one in Box 9.1). These are presented in the third column of Table 9.2. Thus, here, most loss occurred amongst summer adults – in fact, mostly through emigration rather than mortality as such. There was also substantial loss of older larvae (starvation), of hibernating adults (frost-induced mortality), of young larvae (rainfall) and of eggs (cannibalization and 'not being laid').

the phases that determine abundance . . .

It is usually more valuable, however, to ask a second question: 'What is the relative importance of these phases as determinants of year-to-year *fluctuations* in mortality, and hence of year-to-year fluctuations in abundance?' This is rather different. For instance, a phase might repeatedly witness a significant toll being taken from a population (a high mean k-value), but if that toll is always roughly the same, it will play little part in determining the particular rate of mortality (and thus the particular population size) in any particular year. In other words, this second question is much more concerned with discovering what *determines* particular abundances at particular times, and it can be addressed in the following way.

Mortality during a phase that is important in determining population change – referred to as a *key phase* – will vary in line with total mortality in terms of both size and direction. It is a key phase in the sense that when mortality during it is high, total mortality tends to be high and the population declines – whereas when phase mortality is low, total mortality tends to be low and the population tends to remain large, and so on. By contrast, a phase with a k-value that varies quite randomly with respect to total k will, by definition, have little influence on changes in mortality and hence little influence on population size. We need therefore to measure the relationship between phase mortality and total mortality, and this is achieved by the *regression coefficient* of the former on the latter. The largest regression coefficient will be associated with the key phase causing population change, whereas phase mortality that varies at random with total mortality will generate a regression coefficient close to zero.

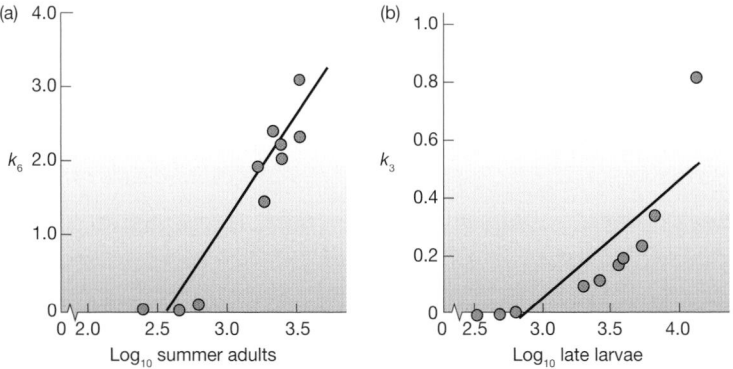

AFTER HARCOURT, 1971

Figure 9.6

(a) Density-dependent emigration by Colorado beetle summer adults (slope = 2.65).
(b) Density-dependent starvation of larvae (slope = 0.37).

In the present example (Table 9.2), the summer adults, with a regression coefficient of 0.906, are the key phase. Other phases (with the possible exception of older larvae) have a negligible effect on the changes in generation mortality.

What, though, about the possible role of these phases in the *regulation* of the Colorado beetle population? In other words, which, if any, act in a density-dependent way? This can be answered most easily by plotting k-values for each phase against the numbers present at the start of the phase. For density dependence, the k-value should be highest (that is, mortality greatest) when density is highest. For the beetle population, two phases are notable in this respect: for both summer adults (the key phase) and older larvae there is evidence that losses are density-dependent (Figure 9.6) and thus a possible role of those losses in regulating the size of the beetle population. In this case, therefore, the phases with the largest role in determining abundance are also those that seem likely to play the largest part in regulating abundance. But as we see next, this is by no means a general rule.

Key factor analysis has been applied to a great many insect populations, but to far fewer vertebrate or plant populations. Examples of these, though, are shown in Table 9.3 and Figure 9.7.

We start with populations of the wood frog (*Rana sylvatica*) in three regions of the United States (Table 9.3). The larval period was the key phase determining abundance in all regions, largely as a result of year-to-year variations in rainfall. In low-rainfall years, the ponds often dry out, reducing larval survival to catastrophic levels. Such mortality, however, was inconsistently related to the size of the larval population (only one of two ponds in Maryland, and only approaching significance in Virginia) and hence it played an inconsistent part in regulating the sizes of the populations. Rather, in two regions it was during the adult phase that mortality was clearly density-dependent (apparently as a result of competition for food) and, indeed, in two regions mortality was also most intense in the adult phase (first data column).

The key phase determining abundance in a Polish population of the sand-dune annual plant *Androsace septentrionalis* (Figure 9.7) were the seeds in the soil. Once again, however, mortality there did not operate in a density-dependent manner, whereas mortality of seedlings (not the key phase) was density-dependent.

Overall, therefore, key factor analysis (its rather misleading name apart) is useful in identifying important phases in the life cycles of study organisms, and useful too in distinguishing the variety of ways in which phases may be important:

... and the factors that regulate abundance

two further examples of key factor analysis

Table 9.3

Key factor (or key phase) analysis for wood frog populations in the United States: Maryland (two ponds, 1977–1982), Virginia (seven ponds, 1976–1982) and Michigan (one pond, 1980–1993). In each area, the phase with the highest mean k-value, the key phase and any phase showing density dependence are highlighted in bold.

AGE INTERVAL	MEAN k-VALUE	COEFFICIENT OF REGRESSION ON k_{TOTAL}	COEFFICIENT OF REGRESSION ON LOG (POPULATION SIZE)
Maryland			
Larval period	1.94	**0.85**	**Pond 1 : 1.03 ($P = 0.04$)**
			Pond 2 : 0.39 ($P = 0.50$)
Juvenile: up to 1 year	0.49	0.05	0.12 ($P = 0.50$)
Adult: 1–3 years	**2.35**	0.10	0.11 ($P = 0.46$)
Total	4.78		
Virginia			
Larval period	**2.35**	**0.73**	0.58 ($P = 0.09$)
Juvenile: up to 1 year	1.10	0.05	−0.20 ($P = 0.46$)
Adult: 1–3 years	1.14	0.22	**0.26 ($P = 0.05$)**
Total	4.59		
Michigan			
Larval period	1.12	**1.40**	1.18 ($P = 0.33$)
Juvenile: up to 1 year	0.64	1.02	0.01 ($P = 0.96$)
Adult: 1–3 years	**3.45**	−1.42	**0.18 ($P = 0.005$)**
Total	5.21		

AFTER BERVEN, 1995

Figure 9.7

Key factor analysis of the sand-dune annual plant *Androsace septentrionalis*. A graph of total generation mortality (k_{total}) and of various k-factors is presented. The values of the regression coefficients of each individual k-value on k_{total} are given in brackets. The largest regression coefficient signifies the key phase and is shown as a maroon line. Alongside is shown the one k-value that varies in a density-dependent manner.

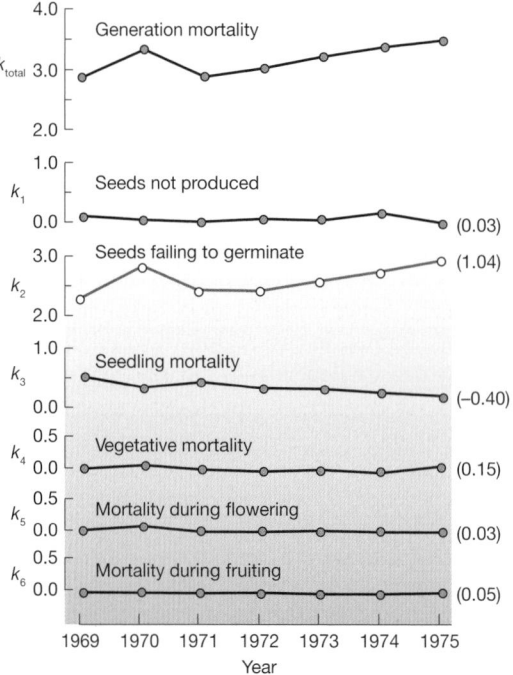
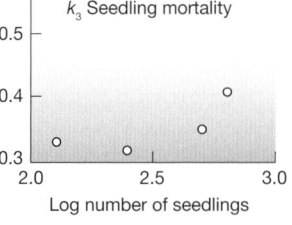

AFTER SYMONIDES, 1979; ANALYSIS IN SILVERTOWN, 1982

in contributing significantly to the overall sum of mortality; in contributing significantly to variations in mortality, and hence in *determining* abundance; and in contributing significantly to the *regulation* of abundance by virtue of the density dependence of the mortality. Box 9.2 presents an account of a topical problem, an understanding of which could benefit from key factor analysis.

9.2 *Topical ECOncerns*

9.2 TOPICAL ECONCERNS

Acorns, mice, ticks, deer and human disease: complex population interactions

Ecologists have been trying to uncover the complex interactions among acorn production, populations of mice and deer, parasitic ticks and, ultimately, a bacterial pathogen carried by the ticks that can affect people. It is clear that a thorough understanding of the abiotic factors that determine the size of the acorn crop and of the various population interactions can enable scientists to predict years when the risk of human disease is high. This is the topic of the following newspaper article in the *Contra Costa Times* on Friday, February 13, 1998, by Paul Recer.

More acorns may mean a rise in Lyme disease
A big acorn crop last fall could mean a major outbreak of Lyme disease next year, according

Female deer tick (*Ixodes dammini*), which carries Lyme disease (× 7).

© ROBERT CALANTINE, VISUALS UNLIMITED

to a study that linked acorns, mice and deer to the number of ticks that carry the Lyme disease parasite.

Based on the study, researchers at the Institute of Ecosystem Studies in Millbrook, New York, say that 1999 may see a dramatic upswing in the number of Lyme disease cases among people who visit the oak forests of the Northeast.

'We had a bumper crop of acorns this year, so in 1999, two years after the event, we should also have a bumper year for Lyme disease', said Clive G. Jones, a researcher at the Institute of Ecosystem Studies; '1999 should be a year of high risk for Lyme disease'.

Lyme disease is caused by a bacterium carried by ticks. The ticks normally live on mice and deer, but they can bite humans. Lyme disease first causes a mild rash, but left untreated can damage the heart and nervous system and cause a type of arthritis.

Jones, along with researchers at the University of Connecticut, Storrs, and Oregon State University, Corvallis, found that the number of mice, the number of ticks, the deer population and even the number of gypsy moths are linked directly to the production of acorns in the oak forest.

Jones said that in years following a big acorn crop, the number of tick larvae is eight times greater than in years following a poor acorn crop.

Additionally, he said, there are about 40 percent more ticks on each mouse.

The researchers tested the effect of acorns by manipulating the population of mice and the availability of acorns in forest plots along the Hudson River. Jones said the work, extended over several seasons, proved the theory that mice and tick populations rise and fall based on the availability of acorns.

How could a key factor analysis be used to pinpoint the phases of importance in determining risk of human disease?

9.3 Dispersal, patches and metapopulation dynamics

dispersal is ignored at the ecologist's peril

In many studies of abundance, the assumption has been made that the major events all occur within the study area, and that immigrants and emigrants can safely be ignored. But migration can be a vital factor in determining and/or regulating abundance. We have already seen, for example, that emigration was the predominant reason for the loss of summer adults of the Colorado potato beetle, which was both the key phase in determining population fluctuations and one in which loss was strongly density-dependent.

habitable sites and dispersal distance

Dispersal has a particularly important role to play when populations are fragmented and patchy – as many are. The abundance of patchily distributed organisms can be thought of as being determined by the properties of two features: the 'habitable site' and the 'dispersal distance' (Gadgil, 1971). Thus, a population may be small if its habitable sites are themselves small or short-lived or only few in number; but it may also be small if the dispersal distance between habitable sites is great relative to the dispersibility of the species, such that habitable sites that go extinct locally are unlikely to be recolonized.

To discover the limitations that the accessibility of habitable sites places on abundance, though, it is necessary to identify habitable sites that are not inhabited. This is possible, for example, for a number of butterfly species, because their larvae feed only on one or a few species of patchily distributed plants. Thus, by identifying habitable sites with these plants, whether or not they were inhabited, Thomas et al. (1992) found that the silver-studded blue butterfly *Plebejus argus* was able to colonize virtually all habitable sites less than 1 km from existing populations, but those further away (beyond the dispersal powers of the butterfly) remained uninhabited. The overall size of the population was determined as much by the accessibility of this patchy resource as by the total amount of the resource. Indeed, the habitability of some of these isolated sites was established when the butterfly was successfully introduced there (Thomas & Harrison, 1992). This, after all, is the crucial test of whether an uninhabited 'habitable' site is really habitable or not.

metapopulations

A radical change in the way ecologists think about populations has involved combining patchiness and dispersal in the concept of a *metapopulation*, the origins of which are described in Box 9.3. A population can be described as a

9.3 Historical landmarks

The genesis of metapopulation theory

A classic book, *The Theory of Island Biogeography*, written by MacArthur and Wilson and published in 1967, was an important catalyst in radically changing ecological theory. They showed how the distribution of species on islands could be interpreted as a balance between the opposing forces of extinctions and colonizations (see Chapter 10) and focused attention especially on situations in which those species were all available for repeated colonization of individual islands from a common source – the mainland. They developed their ideas in the context of the floras and faunas of real (i.e. oceanic) islands, but their thinking has been rapidly assimilated into much wider contexts with the realization that patches everywhere have many of the properties of true islands – ponds as islands of water in a sea of land, trees as islands in a sea of grass, and so on.

At about the same time as MacArthur and Wilson's book was published, a simple model of 'metapopulation dynamics' was proposed by Levins (1969). The concept of a *metapopulation* was introduced to refer to a subdivided and patchy population in which the population dynamics operate at two levels:

1 The dynamics of individuals within patches (determined by the usual demographic forces of birth, death and movement).
2 The dynamics of the occupied patches (or 'subpopulations') themselves within the overall metapopulation (determined by the rates of colonization of empty patches and of extinction within occupied patches).

Although both this and MacArthur and Wilson's theory embraced the idea of patchiness, and both focused on colonization and extinction rather than the details of local dynamics, MacArthur and Wilson's theory was based on a vision of mainlands as rich sources of colonists for whole archipelagos of islands, whereas in a metapopulation there is a collection of patches but no such dominating mainland.

Levins introduced the variable $p(t)$, the fraction of habitat patches occupied at time t. Note that the use of this single variable carries the profound notion that not all habitable patches are always inhabited. The rate of change in $p(t)$ depends on the rate of local extinction of patches and the rate of colonization of empty patches. It is not necessary to go into the details of Levin's model; suffice to say that as long as the intrinsic rate of colonization exceeds the intrinsic rate of extinction within patches, the total metapopulation will reach a stable, equilibrium fraction of occupied patches, even if none of the local populations is stable in its own right.

Perhaps because of the powerful influence on ecology of MacArthur and Wilson's theory, the whole idea of metapopulations was largely neglected during the 20 years after Levins's initial work. The 1990s, however, saw a great flowering of interest, both in underlying theory and in populations in nature that might conform to the metapopulation concept (Hanski, 1999).

metapopulation if it can be seen to comprise a collection of subpopulations, each one of which has a realistic chance both of going extinct and of appearing again through recolonization. The essence is a change of focus: less emphasis is given to the birth, death and movement processes going on within a single subpopulation; but much more emphasis is given to the colonization (= birth) and extinction (= death) of subpopulations within the metapopulation as a whole. From this

Figure 9.8

Comparison of the subpopulation sizes in June 1991 (adults) and August 1993 (larvae) of the Glanville fritillary butterfly (*Melitaea cinxia*) on Åland Island in Finland. Multiple data points are indicated by numbers. Many 1991 populations, including many of the largest, had become extinct by 1993.

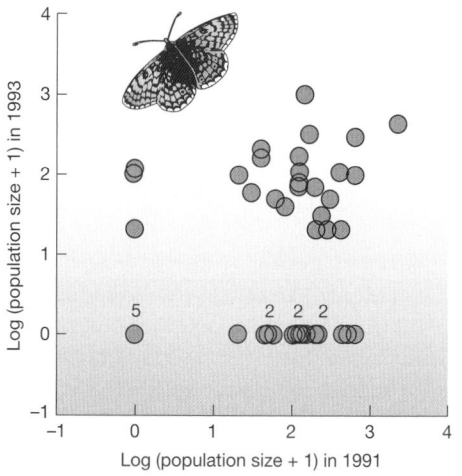

AFTER HANSKI ET AL., 1995

perspective, it becomes apparent that a metapopulation may persist, stably, as a result of the balance between extinctions and recolonizations, even though none of the local subpopulations is stable in its own right. An example of this is shown in Figure 9.8, where within a persistent, highly fragmented metapopulation of the Glanville fritillary butterfly (*Melitaea cinxia*) in Finland, even the largest subpopulations had a high probability of declining to extinction within 2 years.

metapopulation dynamics: the American pika

Aspects of the dynamics of metapopulations can be illustrated in a study of a small mammal, the American pika, *Ochotona princeps*, in California (Figure 9.9). The overall metapopulation could itself be divided into northern, middle and southern networks of patches, and the patch occupancy in each was determined on four occasions between 1972 and 1991. These data (Figure 9.9a) show that the northern network maintained a high occupancy throughout the study period, the middle network maintained a more variable and much lower occupancy, while the southern network suffered a steady and substantial decline.

The dynamics of individual subpopulations were not monitored, but these were simulated using models based on the principles of metapopulation dynamics and on general information on pika biology. When the three networks were simulated in isolation (Figure 9.9b), the northern network remained at a stable high occupancy (as observed in the data), but the middle network rapidly and predictably crashed, and the southern network eventually suffered the same fate. However, when the entire metapopulation was simulated as a single entity (Figure 9.9c), the northern network again achieved stable high occupancy, but this time the middle network was also stable, albeit at a much lower occupancy (again as observed), while the southern network suffered periodic collapses (also consistent with the real data).

This all suggests that within the metapopulation as a whole, the northern network acts as a net source of colonizers that prevent the middle network from suffering overall extinction. These in turn delay extinction in, and allow recolonization of, the southern network. The study therefore illustrates how whole metapopulations can be stable when their individual subpopulations are not. Moreover, the comparison of the northern and middle networks, both stable but at very different occupancies, shows how occupancy may depend on the size

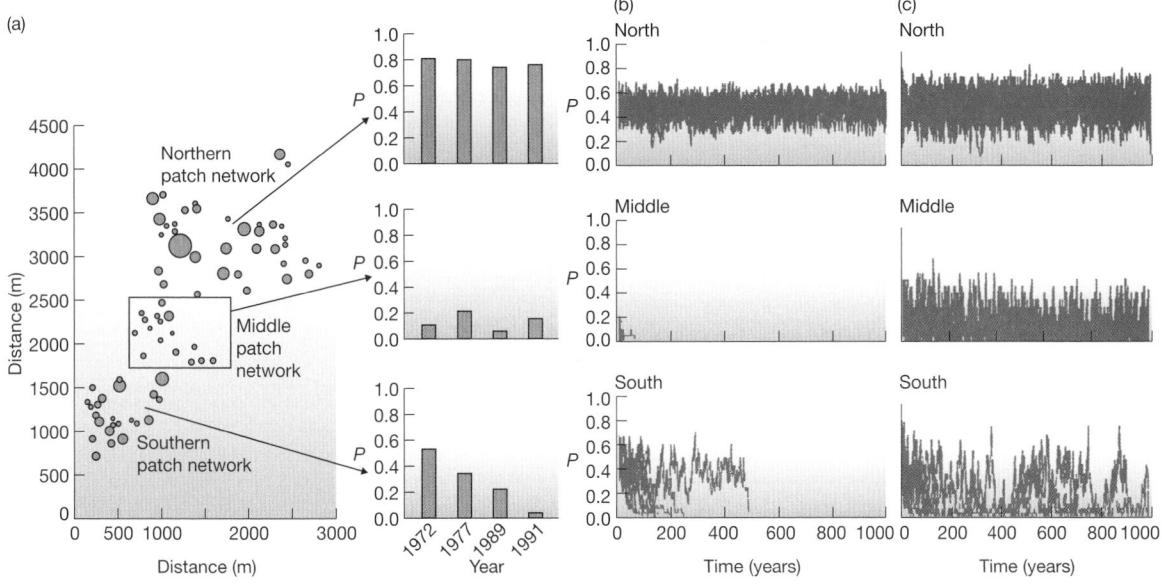

AFTER MOILANEN ET AL., 1998

Figure 9.9

The metapopulation dynamics of the American pika, *Ochotona princeps*, in Bodie, California. (a) The relative positions (distance from a point southwest of the study area) and approximate sizes (as indicated by the size of the dots) of the habitable patches, and the occupancies (as proportions, *P*) in the northern, middle and southern networks of patches in 1972, 1977, 1989 and 1991. (b) The simulated temporal dynamics of the three networks, with each of the networks simulated in isolation. Ten replicate simulations are shown, overlaid on one another, each starting with the actual data in 1972. (c) Equivalent simulations to (b) but with the entire metapopulation treated as a single entity.

of the pool of dispersers, which itself may depend on the size and number of the subpopulations.

Finally, the southern network in particular emphasizes that the observable dynamics of a metapopulation may have more to do with 'transient' behavior, far from any equilibrium. To take another example, the silver-spotted skipper butterfly (*Hesperia comma*) declined steadily in Great Britain from a widespread distribution over most calcareous hills in 1900, to 46 or fewer refuge localities (local populations) in 10 regions by the early 1960s (Thomas & Jones, 1993). The probable reasons were changes in land use – increased plowing of grasslands, reduced stocking with grazing animals – and the virtual elimination of rabbits by myxomatosis with its consequent profound vegetational changes. Throughout this non-equilibrium period, rates of local extinction generally exceeded those of recolonization. In the 1970s and 1980s, however, reintroduction of livestock and recovery of the rabbits led to increased grazing, and suitable habitats increased again. This time, recolonization exceeded local extinction, but the spread of the skipper remained slow, especially into localities isolated from the 1960s refuges. Even in southeast England, where the density of refuges was greatest, it is predicted that the abundance of the butterfly will increase only slowly – and remain far from equilibrium – for at least 100 years. Thus, it seems that around a century of 'transient' decline in the dynamics of the metapopulation is to be followed by another century of transient increase – except that the environment will no doubt alter again before the transient phase ends and the metapopulation reaches equilibrium.

transient dynamics may be as important as equilibria

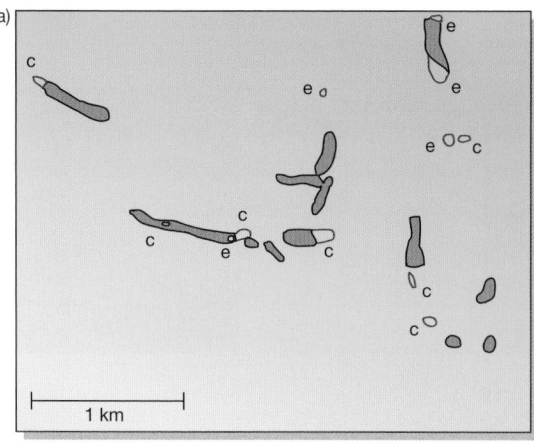 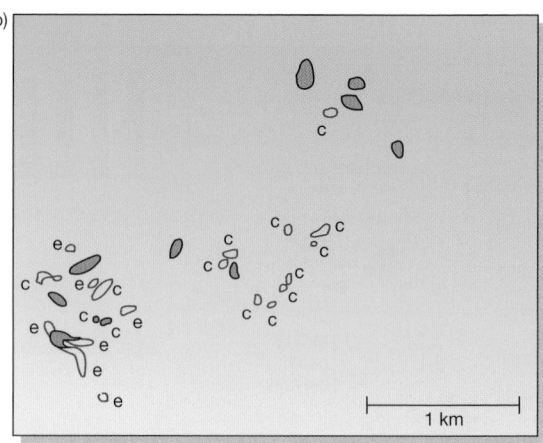

AFTER THOMAS & HARRISON, 1992

Figure 9.10

Two metapopulations of the silver-studded blue butterfly (*Plejebus argus*) in North Wales: filled outlines, present in both 1983 and 1990 ('persistent'); open outlines, not present at both times; e, present only in 1983 (presumed extinction); c, present only in 1990 (presumed colonization). (a) In a limestone habitat, where there was a large number of persistent (often larger) local populations among smaller, much more ephemeral local populations (extinctions and colonizations). (b) In a heathland habitat, where the proportion of smaller and ephemeral populations was much greater.

a continuum of metapopulation types

In reality, moreover, there is likely to be a continuum of types of metapopulation: from collections of nearly identical local populations, all equally prone to extinction, to metapopulations in which there is great inequality between local populations, some of which are effectively stable in their own right. This contrast is illustrated in Figure 9.10 for the silver-studded blue butterfly (*Plejebus argus*) in North Wales, UK.

metapopulations of plants? remember the seed bank

Finally, we must be wary of assuming that all patchy populations are truly metapopulations – comprising subpopulations, each one of which has a measurable probability of going extinct or being recolonized. The problem of identifying metapopulations is especially apparent for plants. There is no doubt that many plants inhabit patchy environments, and apparent extinctions of local populations may be common. This is shown in Figure 9.11 for the annual aquatic plant *Eichhornia paniculata*, living in temporary ponds and ditches in arid regions

Figure 9.11

Of 123 populations of the annual aquatic plant *Eichhornia paniculata* in northeast Brazil observed over a 1-year time interval, 39% went extinct, but the mean initial size of those that went extinct (dark bars) was not significantly different from those that did not (open bars). (Mann-Whitney U = 1925, $P > 0.3$).

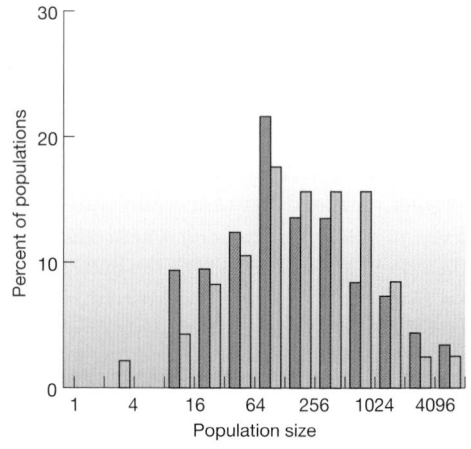

AFTER HUSBAND & BARRETT, 1996

in northeast Brazil. However, the applicability of the idea of recolonization following a genuine extinction is questionable in any plant species that has a buried seed bank (see Section 5.2.2). In *E. paniculata*, for instance, the heavy seeds almost always drop in the immediate vicinity of the parent rather than being dispersed to other patches. 'Extinctions', then, are typically the result of the catastrophic loss of habitat (note in Figure 9.11 that the chance of extinction has effectively nothing to do with the previous population size); and 'recolonizations' are almost always simply the result of the germination of seeds following habitat restoration. Recolonization by dispersal, a prerequisite for a true metapopulation, is extremely rare.

9.4 Temporal patterns in community composition

9.4.1 Founder-controlled and dominance-controlled communities

From the perspective of environmental patchiness, the metapopulation concept is important for our understanding of population dynamics, but when community organization is the focus of attention we usually refer to the *patch dynamics* concept. The concepts are closely related. Both accept that a combination of patchiness and dispersal between patches can give rise to dynamics quite different from those that would be observed if there was just one, homogeneous patch.

disturbances and the patch dynamics concept of community organization

Disturbances that open up gaps are common in all kinds of community. Gaps are simply patches within which many species suffer local extinction simultaneously. In forests, high winds, elephants or simply the death of a tree through old age may all create gaps. In grassland, agents include frost, burrowing animals and cattle dung. On rocky shores, gaps may be formed as a result of severe wave action during hurricanes, battering by moored boats or the action of predators.

Two fundamentally different kinds of community organization can be recognized (Yodzis, 1986). When all species are good colonists and essentially equal competitors, communities are described as *founder controlled*; when some species are strongly superior competitively, communities can be described as *dominance controlled*. The dynamics of the two are quite different, and we deal with them in turn.

In founder-controlled communities, species are approximately equivalent in their ability to invade gaps and can hold the gaps against all comers during their lifetime. Hence, the probability of competitive exclusion in the community as a whole may be much reduced where gaps are appearing continually and randomly. This can be referred to as a 'competitive lottery'. On each occasion that an organism dies (or is killed) a gap is opened for invasion. All conceivable replacements are possible, and species richness is maintained at a high level in the system as a whole. For example, three species of fish co-occur on the upper slope of Heron Reef, part of the Great Barrier Reef off eastern Australia: *Eupomacentrus apicalis*, *Plectroglyphidodon lacrymatus* and *Pomacentrus wardi*. Within rubble patches, the available space is occupied by a series of non-overlapping territories, which individuals hold throughout their juvenile and adult life, defending them against individuals of their own and other species.

founder-controlled communities: competitive lotteries

The Great Barrier Reef, Australia.

But there seems to be no particular tendency for space initially held by one species to be taken up, following mortality, by the same species. Nor is any sequence of ownership evident (Table 9.4). *Pomacentrus wardi* both recruited and lost individuals at a higher rate than the other two species, but all three species appear to have recruited at a sufficient level to balance their rates of loss and maintain a resident population of breeding individuals.

Indeed, communities of tropical reef fish in general may often conform to the founder-controlled model (Sale & Douglas, 1984). They are extremely rich in species. The number of fish species on the Great Barrier Reef ranges from 900 in the south to 1500 in the north, and more than 50 resident species may be recorded on a single patch of reef 3 m in diameter. Only a proportion of this species richness is likely to be attributable to resource partitioning of food and space – indeed the diets of many of the coexisting species are very similar. It is vacant living space that seems to be a crucial limiting factor, generated unpredictably in space and time when a resident dies or is killed. The lifestyles of the species match this state of affairs. They breed often, sometimes year-round, and produce

Table 9.4

For three species of reef fish, the numbers of each species observed occupying sites, or parts of sites, that had been vacated during the immediately prior period between censuses through the loss of residents of each species. The sites vacated through loss of 120 residents were reoccupied by 131 fish; the species of the new occupant is not dependent on the species of the previous resident ($\chi^2 = 5.88$; $P > 0.1$).

	REOCCUPIED BY:		
RESIDENT LOST	*E. APICALIS*	*P. LACRYMATUS*	*P. WARDI*
Eupomacentrus apicalis	9	3	19
Plectroglyphidodon lacrymatus	12	5	9
Pomacentrus wardi	27	18	29

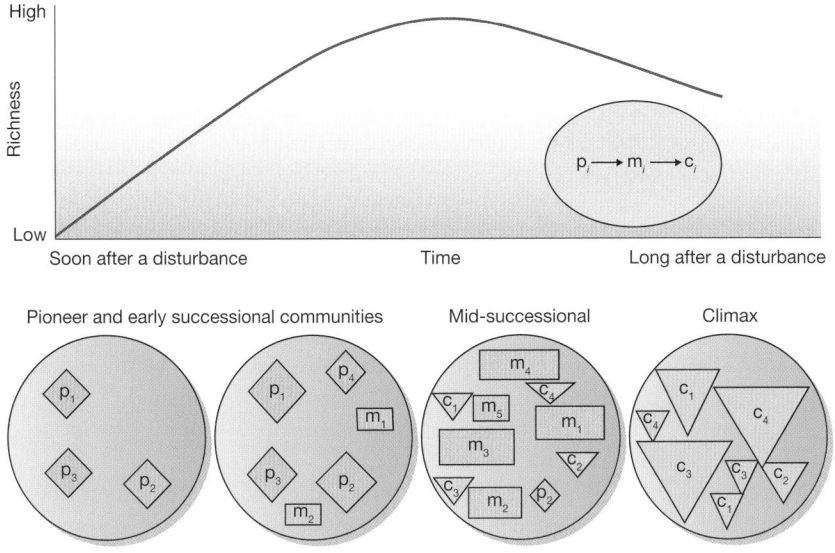

Figure 9.12

Hypothetical succession in a gap – an example of dominance control. The occupancy of gaps is reasonably predictable. Richness begins at a low level as a few pioneer (p_i) species arrive; reaches a maximum in midsuccession when a mixture of pioneer, mid-successional (m_i) and climax (c_i) species occur together; and drops again as competitive exclusion by climax species takes place.

numerous clutches of dispersive eggs or larvae. The species compete in a lottery for living space in which larvae are the tickets, and the first arrival at the vacant space wins the site, matures quickly and holds the space for its lifetime.

In dominance-controlled communities, by contrast, some species are competitively superior to others, and an initial colonizer of a patch cannot necessarily maintain its presence there. In these cases, disturbances that open up gaps lead to reasonably predictable sequences of species, because different species have different strategies for exploiting resources – early species are good colonizers and fast growers, whereas later species can tolerate lower resource levels and grow to maturity in the presence of early species, eventually outcompeting them. Such sequences are examples of community *successions*. An idealized view of a succession is shown in Figure 9.12. Open space is colonized by one or more of a group of opportunistic, early-succession species (p_1, p_2, etc., in Figure 9.12). As time passes, more species invade, often those with poorer powers of dispersal. These eventually reach maturity, dominating mid-succession (m_1, m_2, etc.) and many or all of the pioneer species are driven to extinction. Later still, the community reaches a *climax* stage when the most efficient competitors (c_1, c_2, etc.) oust their neighbors. In this sequence, if it runs its full course, the number of species first increases (because of colonization) then decreases (because of competition).

Some disturbances are synchronized over extensive areas. A forest fire may destroy a huge tract of a climax community. The whole area then proceeds through a more or less synchronous succession. Other disturbances are much smaller and produce a patchwork of habitats. If these disturbances are out of phase with one another, the resulting community comprises a mosaic of patches at different stages of succession.

dominance-controlled communities and community succession

9.4.2 Community succession

If an opened-up gap has not previously been influenced by a community, the sequence of species is referred to as a *primary succession*. Lava flows caused by volcanic eruptions, substrate exposed by the retreat of a glacier and freshly

primary and secondary successions

formed sand dunes are all examples. But where the species of an area has been partially or completely removed but seeds and spores remain, the subsequent sequence is termed a *secondary succession*. The loss of trees locally as a result of high winds may lead to secondary successions, as can cultivation followed by the abandonment of farmland (so-called old-field successions).

Primary successions often take several hundreds of years to run their course. However, on recently denuded rocks in the marine subtidal zone a primary succession may take only a decade or so. The research life of an ecologist is sufficient to encompass a subtidal succession but not that following glacial retreat. Fortunately, however, information can sometimes be gained over the longer time scale. Successional stages in time may be represented by community gradients in space. The use of historical maps, carbon dating or other techniques may enable the age of a community since initial exposure to be estimated. A series of communities currently in existence – a 'chronosequence' – can then be inferred to reflect succession.

a primary succession in duneland

An extensive chronosequence of dune-capped beach ridges occurs on the coast of Lake Michigan in the USA. Thirteen ridges of known age (30–440 years old) show a clear pattern of primary succession to forest. The dune grass *Ammophila breviligulata* dominates the youngest, still mobile, dune ridge. Within 100 years, these are replaced by evergreen shrubs such as *Juniperus communis* and by prairie bunch grass *Schizachyrium scoparium*. Conifers begin colonizing the dune ridges after 150 years, and a mixed forest of pine species develops between 225 and 400 years. Deciduous trees such as the oak and maple do not become important components of the forest until 440 years.

Experimental seed addition and seedling transplants have shown that later species are nonetheless capable of germinating in young dunes (Figure 9.13a). The more developed soil of older dunes may improve the performance of late-successional species, but their successful colonization of young dunes is mainly prevented by limited seed dispersal, together with seed predation by rodents (Figure 9.13b). Eventually, however, the early species are competitively excluded as trees establish and grow.

secondary succession on fields abandoned by farmers

Successions on old fields have been studied primarily in the eastern United States, where many farms were abandoned by farmers who moved west after the frontier was opened up in the 19th century. Most of the pre-colonial mixed conifer–hardwood forest had been destroyed, but regeneration was swift after the 'disturbance' caused by farmers came to an end. The early pioneers of the American West left behind exposed land that was colonized by pioneers of a very different kind. The typical sequence of dominant vegetation is: annual weeds → herbaceous perennials → shrubs → early successional trees → late successional trees. A particularly detailed study of old-field succession has been performed at the Cedar Creek Natural History Area in Minnesota on well-drained and nutrient-poor soil. This study is discussed in detail in Section 1.3.2.

Old-field succession has also been studied in the productive Loess Plateau in China, which for millennia has been affected by human activities so that few areas of natural vegetation remain. One study examined the vegetation at four plots abandoned by farmers for known periods of time: 3, 26, 46 and 149 years. Of a total of 40 plant species identified, different species were dominant (in terms of relative abundance and relative ground cover) in the different aged plots (Figure 9.14). The early-successional species were annuals and biennials with

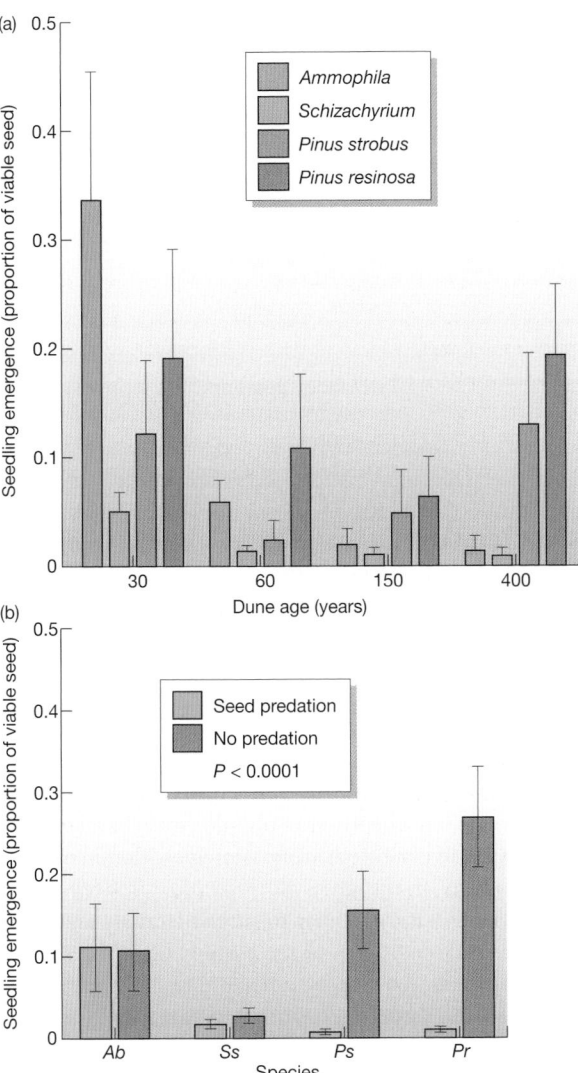

Figure 9.13

(a) Seedling emergence (means + SE) from added seeds of species typical of different successional stages on dunes of four ages.
(b) Seedling emergence of the four species (*Ab*, *Ammophila*; *Ss*, *Schizachyrium*; *Ps*, *Pinus strobes*; *Pr*, *Pinus resinosa*) in the presence and absence of rodent predators of seeds.

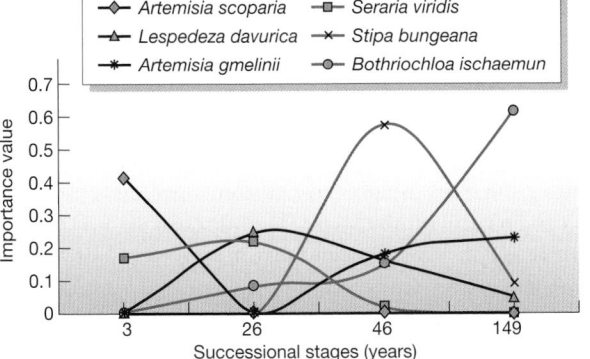

Figure 9.14

Variation in the relative importance of six species during an old-field succession on the Loess Plateau in China.

Table 9.5

Some representative photosynthetic rates (mg CO_2 dm^{-2} h^{-1}) of plants in a successional sequence. Late-successional trees are arranged according to their relative successional position.

PLANT	RATE	PLANT	RATE
Summer annuals		**Early-successional trees**	
Abutilon theophrasti	24	*Diospyros virginiana*	17
Amaranthus retroflexus	26	*Juniperus virginiana*	10
Ambrosia artemisiifolia	35	*Populus deltoides*	26
Ambrosia trifida	28	*Sassafras albidum*	11
Chenopodium album	18	*Ulmus alata*	15
Polygonum pensylvanicum	18		
Setaria faberii	38	**Late-successional trees**	
		Liriodendron tulipifera	18
Winter annuals		*Quercus velutina*	12
Capsella bursa-pastoris	22	*Fraxinus americana*	9
Erigeron annuus	22	*Quercus alba*	4
Erigeron canadensis	20	*Quercus rubra*	7
Lactuca scariola	20	*Aesculus glabra*	8
		Fagus grandifolia	7
Herbaceous perennials		*Acer saccharum*	6
Aster pilosus	20		

AFTER BAZZAZ, 1979

high seed production. By 26 years, the perennial herb *Lespedeza davurica*, with its ability to spread laterally by vegetative means and a well-developed root system, had replaced *Artemisia scoparia*. The 46-year-old plot was characterized by the highest species richness and diverse life history strategies, dominated by perennial lifestyles. The dominance of the grass *Bothriochloa ischaemun* at 149 years was related to its perennial nature, ability to spread clonally and high competitive ability. Unlike the abandoned fields of the eastern USA, the climax vegetation of the Loess Plateau appears to be steppe grassland rather than forest. But as in the idealized succession of Figure 9.12, an initial increase in species number as a result of colonization and a subsequent decrease as a result of competition are both apparent.

early and late successional species have different properties

Early-succession plants have a fugitive lifestyle. Their continued survival depends on dispersal to other disturbed sites. They cannot persist in competition with later species, and thus they must grow and consume the available resources rapidly. High growth and photosynthetic rates are crucial properties of the fugitive. Those of later successional plants are much lower (Table 9.5).

In contrast to the pioneer annuals, seeds of later successional plants can germinate in the shade, for example, beneath a forest canopy. They can continue to grow at these low light intensites, too – quite slowly but faster than the species they replace (Figure 9.15).

The early colonists among the trees usually have efficient seed dispersal; this in itself makes them likely to be early on the scene. They are usually precocious reproducers and are soon ready to leave descendants in new sites elsewhere. The late colonists are those with larger seeds, poorer dispersal and long juvenile phases. The contrast is between the lifestyles of the 'quickly come, quickly gone' and 'what I have, I hold'.

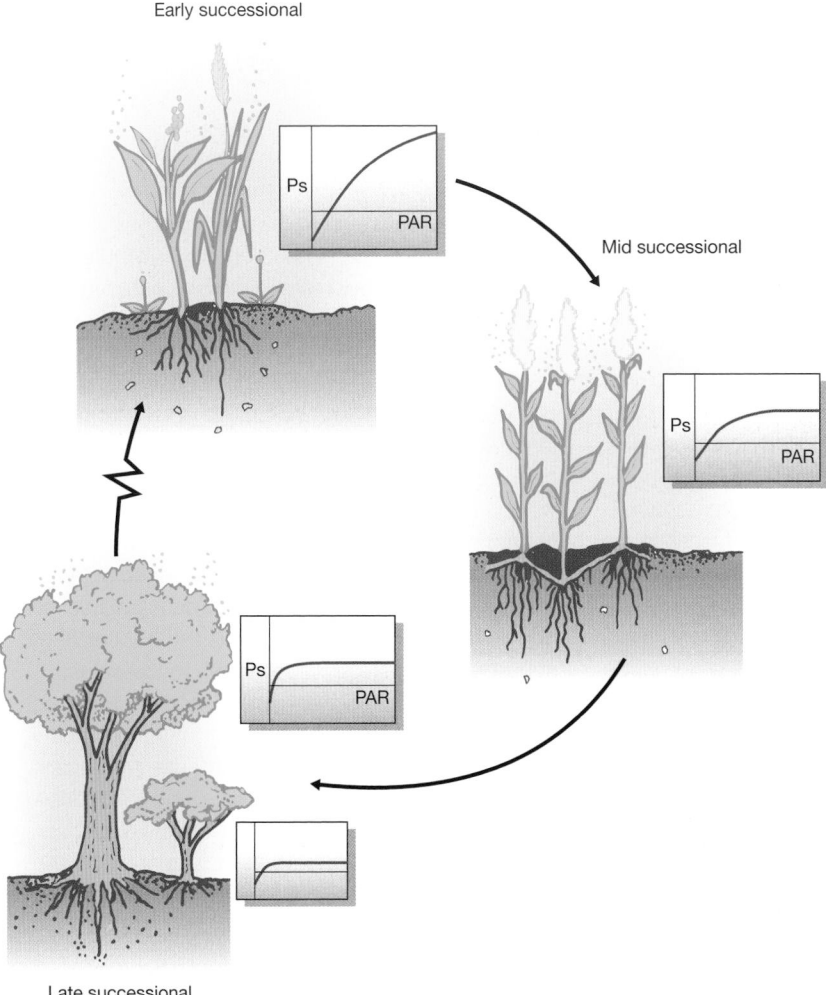

Early successional

Mid successional

Late successional

AFTER BAZZAZ, 1996

Figure 9.15

Idealized light saturation curves (photosynthetic rate, Ps, plotted against the quantity of photosynthetically active radiation, PAR) for early-, mid-, and late-successional plants.

The fact that plants dominate most of the structure and succession of communities does not mean that the animals always follow the communities that plants dictate. This will often be the case, of course, because the plants provide the starting point for all food webs and determine much of the character of the physical environment in which animals live. But it is also sometimes the animals that determine the nature of the plant community, for example, through heavy grazing or trampling (Box 9.4). More often, though, animals are passive followers of successions among the plants.

animals are often affected by, but may also affect, plant successions

Figure 9.12 was described as an idealized succession, and one respect in which it was idealized was in arriving at a climax community at the end. Do real successions reach a climax? Some may. The succession of seaweeds on an overturned boulder may reach a climax in only a few years. Old-field successions, on the other hand, might take 100–300 years to reach a climax, but in that time the probabilities of fire or severe hurricanes, which occur every 70 or so years in New England, are so high that a process of succession may never go to completion. Bearing in

the concept of a climax community

9.4 *Topical ECOncerns*

Conservation sometimes requires manipulation of a succession

Some endangered animal species are associated with particular stages of a succession. Their conservation then depends on a full understanding of the successional sequence, and intervention may be required to maintain their habitat at an appropriate successional stage.

An intriguing example is provided by a giant New Zealand insect, the weta *Deinacrida mahoenuiensis* (Orthoptera, Anostostomatidae). This species, which is believed to have been formerly widespread in forest habitat, was discovered in the 1970s in an isolated patch of gorse (*Ulex europaeus*). Ironically, in New Zealand gorse is an introduced weed that farmers spend much time and effort attempting to control. However, its dense, prickly sward provides a refuge for the giant weta against other introduced pests,

particularly rats but also hedgehogs, stoats and possums, which could readily capture wetas in their original forest home. Mammalian predation is believed to be responsible for weta extinction elsewhere.

New Zealand's Department of Conservation purchased this important patch of gorse from the landowner, who insisted that his cattle should still be permitted to overwinter in the reserve. Conservationists were unhappy about this, but the cattle subsequently proved to be part of the weta's salvation. By opening up paths through the gorse, cattle provided entry for feral goats that browse the gorse, producing a dense hedge-like sward and preventing the habitat from succeeding to a stage inappropriate to the wetas.

This story involves a single, endangered, endemic insect together with a whole suite of introduced pests (gorse, rats, goats, etc.) and introduced domestic animals (cattle). Before the arrival of people in New Zealand, the island's only land mammals were bats, and New Zealand's endemic fauna has proved to be extraordinarily vulnerable to the mammals that arrived with people. However, by maintaining gorse succession at an early stage, the grazing goats provide a habitat in which the wetas can escape the attentions of rats and other predators.

A giant weta on a gorse branch.
COURTESY OF GREG SHERLEY, DEPARTMENT OF CONSERVATION, WELLINGTON, NEW ZEALAND

Because of its economic cost to farmers, ecologists have been trying to find an appropriate biological control agent for gorse, ideally one that would eradicate it. How would you weigh up the needs of a rare insect against the economic losses associated with gorse on farms?

mind that forest communities in northern temperate regions, and probably also in the tropics, are still recovering from the last glaciation, it is questionable whether the idealized climax vegetation is often reached in nature.

In fact, the perception of whether a climax has been reached, like so much else in ecology, is likely to be a matter of scale. As mentioned previously, many successions take place in a mosaic of patches, with each patch, having been

successions in a patchwork – the size and shape of gaps

disturbed independently, at a different successional stage. Boulders on a rocky shore are a good example. Climax communities in such cases can then only occur, at best, on a very local scale. Moreover, when successions occur in a patchwork, the nature of the succession, both locally and overall, is likely to depend on the size and shape of the patches (gaps). The centers of very large gaps are most likely to be colonized by species producing propagules that travel relatively great distances. Such mobility is less important in small gaps, since most recolonization will be by propagules from, or simply lateral movement by, established individuals around the periphery.

Intertidal beds of mussels provide excellent opportunities to study the processes of formation and filling-in of gaps. In the absence of disturbance, mussel beds may persist as extensive monocultures. More often, they are an ever-changing mosaic of many species that inhabit gaps formed by the action of waves. The size of these gaps at the time of formation ranges from a single mussel space to hundreds of square meters. Gaps begin to fill as soon as they are formed. An experimental study of mussel beds of *Brachidontes solisianus* and *B. darwinianus* in Brazil aimed to determine the effects of patch size and location within a patch on the dynamics of succession (Figure 9.16).

High densities of the limpet *Collisella subrugosa* occurred in the smallest gaps in the first 6 months after gap formation, but not in medium or large gaps (Figure 9.16a). It was also much quicker to colonize the periphery than the center of the large gaps (Figure 9.16b). This association of the limpets with patch edges (and hence small patches) probably occurs because they are less vulnerable there to visually hunting predators. Small gaps were also most quickly colonized by lateral migration of the two mussel species (Figure 9.16a), but from around 6 months, *B. darwinianus* increasingly predominated and also built up its numbers in the medium and large gaps. In the absence of further disturbance, *B. darwinianus* would seem likely to outcompete *B. solisianus*. After around 6 months, too, the *Brachiodontes* mussels, which cannot be identified to species when they are small, recruited significantly from settled larvae in the central areas of the large gaps (Figure 9.16b). Finally, the barnacle *Chthamalus bisinuatus* also recruited from settled larvae, largely as a pulse after around 6 months, especially in the largest gaps (Figure 9.16a) and more in the center than at the periphery of the large gaps (Figure 9.16b).

Thus, the smaller the gap, the more the succession within it was dominated by lateral movement than by true migration, and even within a large gap, succession proceeded differently at the center and at the periphery. On the shore as a whole, as in any patchy and disturbed habitat, there was a mosaic of patches in different successional states – those states being determined by patch size, the time since the last disturbance and even on location within a patch.

9.5 Food webs

No predator–prey, parasite–host or grazer–plant pair exists in isolation. Each is part of a complex web of interactions with other predators, parasites, food sources and competitors within its community. Ultimately, it is these food webs that ecologists wish to understand. However, it has been useful to isolate groups of competitors as we did in Chapter 6, of predator–prey and parasite–host pairs as in Chapter 7,

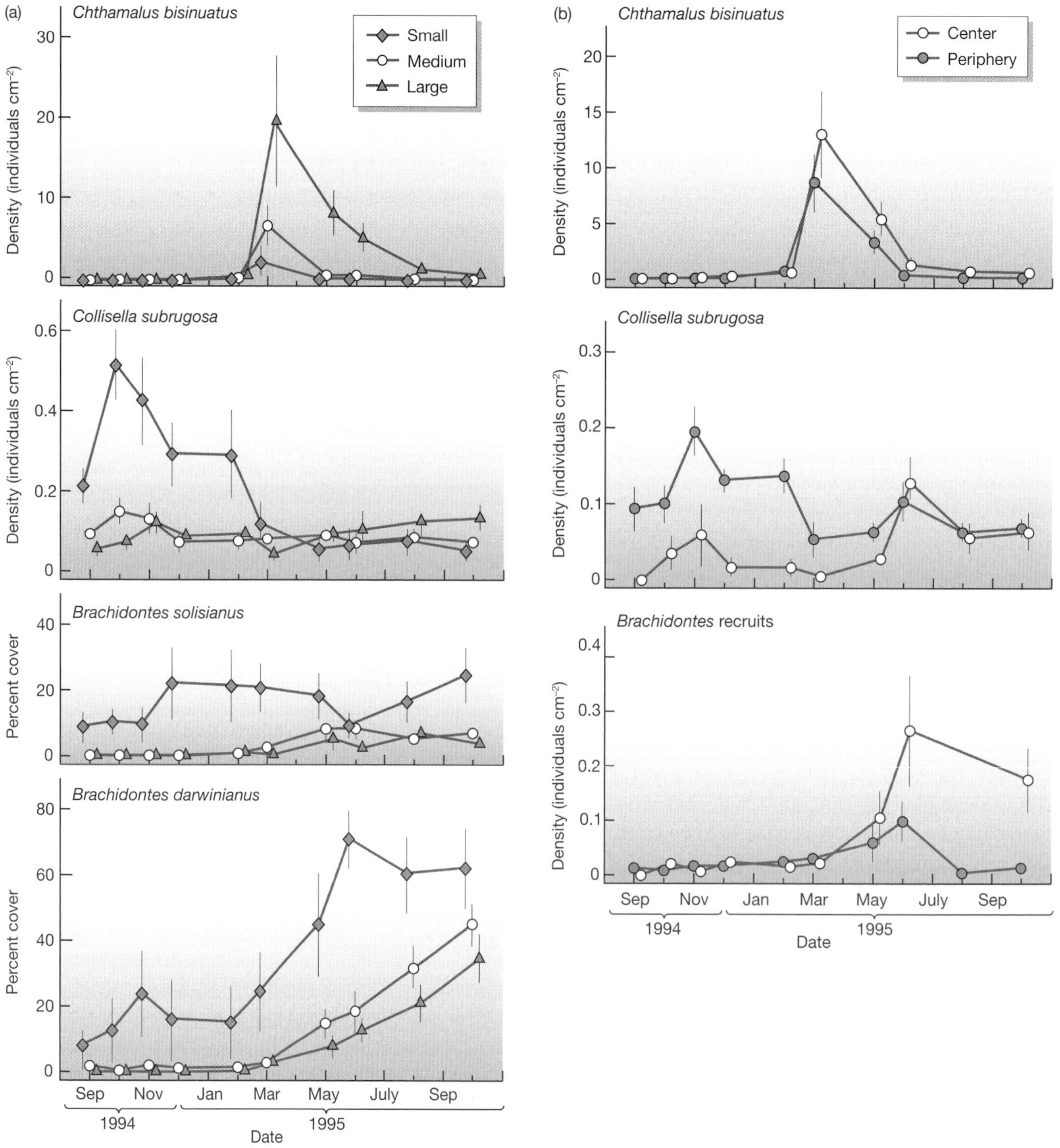

Figure 9.16

(a) Mean abundances (± SE) of four colonizing species in experimentally cleared small, medium and large gaps in intertidal mussel beds.
(b) Recruitment of three species at the periphery (within 5 cm of the gap edge) and in the centre of 400 cm² square gaps.

and of mutualists as in Chapter 8, simply because we have little or no hope of understanding the whole unless we have some understanding of the component parts. Toward the end of Chapter 7 (Section 7.6), our field of view was expanded to include the effects of predators on groups of competitors and to show, for example, the importance of predator-mediated coexistence.

AFTER TANAKA & MAGALHÃES, 2002

We now take this approach a stage further to focus on systems with at least three trophic levels (plant–herbivore–predator), and consider not only direct but also indirect effects that a species may have on others on the same or other trophic levels. The effects of a predator on individuals or even populations of its herbivorous prey, for example, are direct and relatively straightforward. But these effects may also be felt by any plant population on which the herbivore feeds, or by other predators of the herbivore, or other consumers of the plant, or competitors of the herbivore, or by the myriad species linked even more remotely in the food web.

food webs – shifting the focus to systems with at least three trophic levels

9.5.1 Indirect and direct effects

The deliberate removal of a species from a community can be a powerful tool in unraveling the workings of a food web. We might expect such removal to lead to an increase in the abundance of a competitor, or, if the species removed is a predator, to an increase in the abundance of its prey. Sometimes, however, when a species is removed, a competitor may actually decrease in abundance, and the removal of a predator can lead to a decrease in a prey population. Such unexpected effects arise when direct effects are less important than effects that occur through indirect pathways. For example, removal of a species might increase the density of one competitor, which in turn causes another competitor to decline.

These indirect effects are brought especially into focus when the initial removal is carried out for some managerial reason, since the deliberate aim is to solve a problem, not create further, unexpected problems. For example, there are many islands on which feral cats have been allowed to escape domestication and now threaten native prey, especially birds, with extinction. The 'obvious' response is to eliminate the cats (and conserve their island prey), but as a simple model shows (Figure 9.17), the programs may not have the desired effect, especially where, as is often the case, rats have also been allowed to colonize the island. The rats typically both compete with and prey upon the birds. The cats normally prey upon the rats as well as the birds. Hence, removal of the cats will relieve the pressure on the rats and is thus likely to increase not decrease the threat to the birds. For example, introduced cats on Stewart Island, New Zealand preyed upon an endangered flightless parrot, the kakapo, *Strigops habroptilus* (Karl & Best, 1982). But controlling cats alone would have been risky, since their preferred prey are three species of introduced rats, which, unchecked, could pose far more of a threat to the kakapo. In fact, Stewart Island's kakapo population was translocated to smaller offshore islands where exotic predators (like rats) were absent.

cats, rats and birds

The indirect effect within a food web that has probably received most attention is the so-called *trophic cascade*. It occurs when a predator reduces the abundance of its prey, and this cascades down to the trophic level below, such that the prey's own resources (typically plants) increase in abundance. Of course, it need not stop there. In a food chain with four links, a top predator may reduce the abundance of an intermediate predator, which may allow the abundance of a herbivore to increase, leading to a decrease in plant abundance.

trophic cascades – effects of shorebirds on limpet populations

One example of a trophic cascade, but also of the complexity of indirect effects, is provided by a 2-year experiment in which predation by birds was experimentally manipulated in an intertidal community on the northwest coast of the United States to determine the consequences for three limpet species and their algal food. Glaucous-winged gulls (*Larus glaucescens*) and oystercatchers

Figure 9.17

(a) Schematic representation of a model of an interaction in which a superpredator (such as a cat) preys both on mesopredators (such as rats, for which it shows a preference) and on prey (such as birds), while the mesopredator also attacks the prey. Each species also recruits to its own population, 'reproduction'. (b) The output of the model with realistic values for rates of predation and reproduction: with all three species present, the superpredator keeps the mesopredator in check and all three species coexist (left); but in the absence of the superpredator, the mesopredator drives the prey to extinction (right).

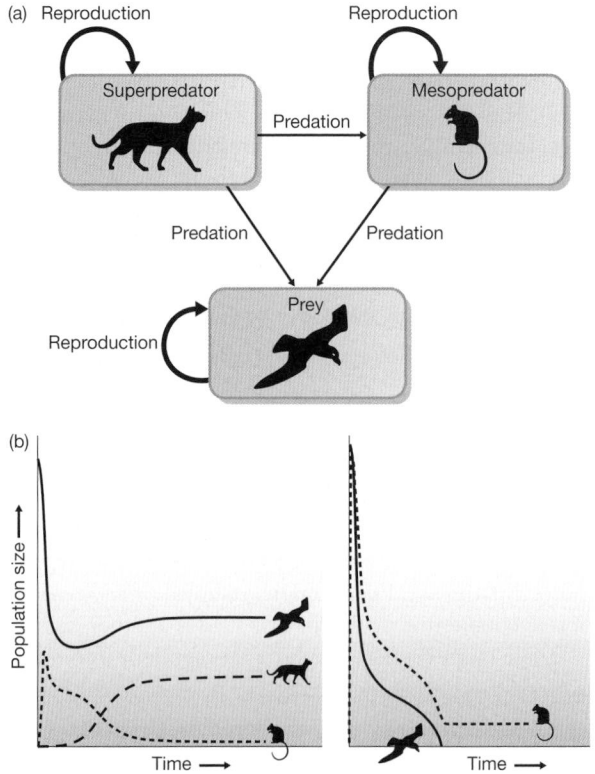

AFTER COURCHAMP ET AL., 1999

(*Haematopus bachmani*) were excluded by means of wire cages from large areas (each 10 m²) in which limpets were common. It became evident that excluding the birds increased the overall abundance of one of the limpet species, *Lottia digitalis*, as might have been expected, but a second limpet species (*L. strigatella*) became rarer, and the third, *L. pelta*, which was the one most frequently consumed by the birds, did not vary in abundance. The reasons are complex and go well beyond the direct effects of birds eating limpets (Figure 9.18).

L. digitalis, a light-colored limpet, tends to occur on light-colored goose barnacles (*Pollicipes polymerus*) where it is camouflaged, whereas the dark *L. pelta* occurs primarily on dark Californian mussels (*Mytilus californianus*). Predation by birds normally reduces the area covered by goose barnacles, and so excluding the birds increased goose barnacle abundance and also increased the abundance of *L. digitalis* (Figure 9.18). Increasing barnacle abundance also led to a decrease in the area covered by mussels, because they were now subject to more intense competition from the barnacles. This, one imagines, might have led to a decrease in the abundance of *L. pelta*, living predominantly on those mussels. However, the third limpet species, *L. strigatella*, is competitively inferior to the others, and the increase in abundance of *L. digitalis* therefore led to a decrease in the abundance of *L. strigatella*, which in turn released pressure on *L. pelta* such that overall its abundance remained effectively unchanged.

But the effects of bird predation also cascade down to the plant trophic level, because by consuming limpets, the birds normally reduce the grazing pressure of the limpets on fleshy algae, and by consuming goose barnacles, the birds

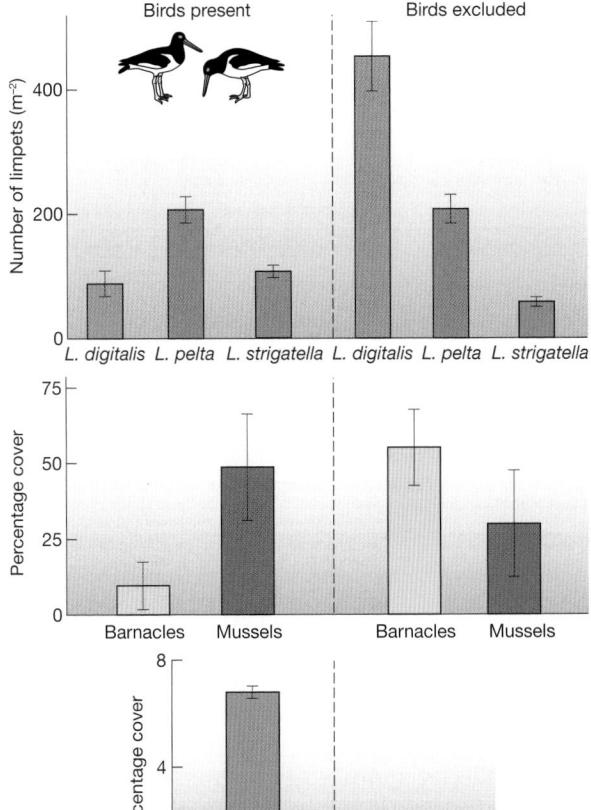

Figure 9.18
When birds are excluded from the intertidal community, barnacles increase in abundance at the expense of mussels, and three limpet species show marked changes in density, reflecting changes in the availability of cryptic habitat and competitive interactions as well as the easing of direct predation. Algal cover is much reduced in the absence of effects of birds on intertidal animals (means and standard errors are shown).

normally free up space for algal colonization. Hence, when the birds were excluded, algal cover decreased (Figure 9.18).

In a four-trophic-level system, if it is subject to trophic cascade, we might expect that as the abundance of a top carnivore increases, the abundances of primary carnivores in the trophic level below decrease, those of the herbivores therefore increase, and plant abundance decreases. This is what was found in a study, in the tropical lowland forests of Costa Rica, of *Tarsobaenus* beetles preying on *Pheidole* ants that prey on a variety of herbivores that attack ant-plants, *Piper cenocladum* (Figure 9.19a). These showed precisely the alternation of abundances expected in a four-trophic-level cascade: relatively high abundances of plants and ants associated with low levels of herbivory and beetle abundance at three sites, but low abundances of plants and ants associated with high levels of herbivory and beetle abundance at a fourth (Figure 9.19b). Moreover, when beetle abundance was manipulated experimentally at one of the sites, ant and plant abundance were significantly higher, and levels of herbivory lower, in the absence of beetles than in their presence (Figure 9.19c).

However, in another four-trophic-level community, in the Bahamas, consisting of sea grape shrubs, which were fed upon by herbivorous arthropods, and then web spiders (primary carnivores) and lizards (top carnivores), the results of

four trophic levels . . .

. . . that can act like three

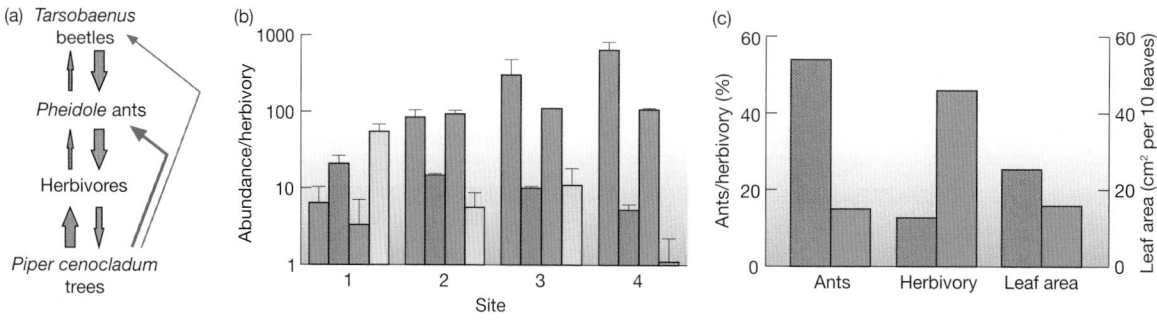

Figure 9.19

(a) Schematic representation of a four-level food chain in Costa Rica. Green arrows denote mortality and maroon arrows a contribution to the consumer's biomass; arrow breadth denotes their relative importance. Both (b) and (c) show evidence of a trophic cascade flowing down from the beetles: positive correlations between beetles and herbivores and between ants and trees. (b) At four sites, the relative abundance of ant-plants (blue bars), the abundance of beetles (maroon bars) and of ants (green bars) and the strength of herbivory (yellow bars) are shown. Means and standard errors are shown; the units of measurement are various and are given in the original references. (c) The results of an experiment at site 4 when replicate enclosures were established without beetles (maroon bars) and with beetles (green bars).

AFTER LETOURNEAU & DYER, 1998A, 1998B; PACE ET AL., 1999

experimental manipulations indicated a strong direct effect of the lizards on the herbivores but a weaker effect of the lizards on the spiders. Consequently, the net effect of top predators on plants was positive and there was less leaf damage in the presence of lizards. In essence, this four-trophic-level community functions as if it has only three levels.

We have seen that trophic cascades are normally viewed 'from the top', starting at the highest trophic level. So, in a three-trophic-level community, we think of the predators controlling the abundance of the herbivores and exerting top-down control. Reciprocally, the predators are subject to bottom-up control: abundance determined by their resources. The plants are also subject to bottom-up control, having been released from top-down control by the effects of the predators on the herbivores. Thus, in a trophic cascade, top-down and bottom-up controls alternate as we move from one trophic level to the next.

top-down or bottom-up control of food webs?

But suppose instead that we start at the other end of the food chain, and assume that the plants are controlled bottom-up by competition for their resources. It is still possible for the herbivores to be limited by competition for plants – *their* resources – and for the predators to be limited by competition for herbivores. In this scenario, all trophic levels are subject to bottom-up control, because the resource controls the abundance of the consumer but the consumer does not control the abundance of the resource. The question therefore arises: 'Are food webs – or are particular *types* of food web – dominated by either top-down or bottom-up control?'

why is the world green? . . .

The widespread importance of top-down control, foreshadowing the idea of the trophic cascade, was first advocated in a famous paper by Hairston et al. (1960), which asked 'Why is the world green?' They answered, in effect, that the world is green because top-down control predominates: green plant biomass accumulates because predators keep herbivores in check.

. . . or is it prickly and bad tasting?

Murdoch (1966), in particular, challenged these ideas. His view, described by Pimm (1991) as 'the world is prickly and tastes bad', emphasized that even if the world is green (assuming it is), it does not necessarily follow that the herbivores are failing to capitalize on this because they are limited, top down, by their predators.

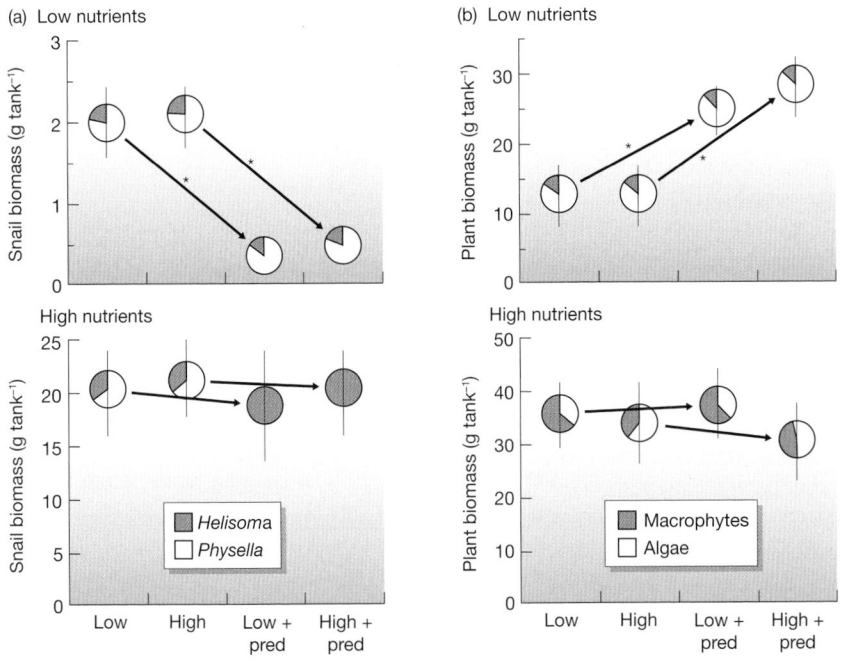

Figure 9.20

Top-down control, but only with low productivity. (a) Snail biomass and (b) plant biomass in experimental ponds with low or high nutrient treatments (vertical bars are standard errors). With low nutrients, the snails were dominated by the insect predator *Physella* (vulnerable to predation) and the addition of predators led to a significant decline (indicated by *) in snail biomass and a consequent increase in plant biomass (dominated by algae). But with high nutrients, *Helisoma* snails (less vulnerable to predation) increased their relative abundance, and the addition of predators led neither to a decline in snail biomass nor an increase in plant biomass (often dominated by macrophytes).

AFTER CHASE, 2003

Many plants have evolved physical and chemical defenses that make life difficult for herbivores. The herbivores may therefore be competing fiercely for a limited amount of palatable and unprotected plant material; and their predators may, in turn, compete for scarce herbivores. A world controlled from the bottom up may still be green.

That very little is required to switch control from one type to the other is emphasized by a study that examined the effect of nutrient concentrations on a freshwater web comprising an insect predator (*Physella gyrina*) feeding on two species of herbivorous snails feeding on water plants and algae (Figure 9.20). At the lowest nutrient concentrations, the snails were dominated by the smaller *P. gyrina* (they were vulnerable to predation), and the predator gave rise to a trophic cascade extending to the plants and algae. But at the highest nutrient concentrations, the snails were dominated by the larger *Helisoma trivolvis* (they were relatively invulnerable to predation), and no trophic cascade was apparent. This study, therefore, also lends support to Murdoch's proposition that 'the world tastes bad', in that invulnerable herbivores gave rise to a web with a relative dominance of bottom-up control. Overall, though, the elucidation of clear patterns in the predominance of top-down or bottom-up control remains a challenge for the future.

9.5.2 Population and community stability and food web structure

Of all the imaginable food webs in nature, are there particular types that we tend to observe repeatedly? Are some food web structures more stable than others? (We discuss what *stable* means in Box 9.5.) Do we observe particular types of

9.5 *Quantitative aspects*

9.5 QUANTITATIVE ASPECTS

What do we mean by 'stability'?

Among several, there are two important qualifications that can be made when we come to decide what we mean by stability. The first is the distinction between the resilience of a community and its resistance. A *resilient community* is one that returns rapidly to something like its former structure after that structure has been altered. A *resistant community* is one that undergoes relatively little change in its structure in the face of a disturbance.

The second distinction is between *fragile* and *robust stability*. A community has only fragile stability if it remains essentially unchanged in the face of a small disturbance but alters utterly when subjected to a larger disturbance, whereas one that stays roughly the same in the face of much larger disturbances is said to have stability that is dynamically robust.

To illustrate these distinctions by analogy, consider the following:

- a pool or billiard ball balanced carefully on the end of a cue

- the same ball resting on the table
- the ball sitting snugly in its pocket

The ball on the cue is stable in the narrow sense that it will stay there forever as long as it is not disturbed – but its stability is fragile, and both its resistance and its resilience are low: the slightest touch will send the ball to the ground, far from its former state (low resistance), and it has not the slightest tendency to return to its former position (low resilience).

The same ball resting on the table has a similar resilience: it has no tendency to return to exactly its former state (assuming the table is level), but its resistance is far higher: pushing or hitting it moves it relatively little. And its stability is also relatively robust: it remains 'a ball on the table' in the face of all sorts and all strengths of assault with the cue.

The ball in the pocket, finally, is not only resistant but resilient too – it moves little and then returns – and its stability is highly robust: it will remain where it is in the face of almost everything other than a hand that carefully plucks it away.

food web *because* they are stable (and hence persist)? Are populations themselves more stable when embedded in some types of food web than in others? These are important practical questions. We require answers if we are to determine whether some communities are more fragile (and more in need of conservation) than others; or whether there are certain 'natural' structures that we should aim for when we construct communities ourselves; or whether communities that have been restored are likely to stay 'restored'.

keystones in food web architecture

'Stability', of course, means stability in the face of a disturbance or perturbation, and most disturbances are, in practice, the loss of one or more populations from a community. What are the knock-on effects of such a loss? How profound are the consequences of the loss of that population for the rest of the community? Some species are more intimately and tightly woven into the fabric of a food web than others. A species whose removal would produce a significant effect (extinction or a large change in density) in at least one other species may be thought of as a strong interactor. The removal of some strong interactors leads to significant changes spreading throughout the food web – we refer to these as *keystone species*.

In building construction, a keystone is the wedge-shaped block at the highest point of an arch that locks the other pieces together. The removal of the keystone species, just like removal of the keystone in an arch, leads to collapse of the structure: it leads to extinction or large changes in abundance of several species, producing a community with a very different species composition. A more precise definition of a keystone species is one whose impact is 'disproportionately large relative to its abundance' (Power et al., 1996). This has the advantage of excluding from keystone status what would otherwise be rather trivial examples, especially species at lower trophic levels that may provide the resource on which a whole myriad of other species depend – for example, a coral, or the oak trees in an oak woodland.

Although the term was originally applied only to predators, it is now widely accepted that keystone species can occur at any trophic level. For example, lesser snow geese (*Chen caerulescens caerulescens*) are herbivores that breed in large colonies in coastal marshes along the west coast of Hudson Bay in Canada. At their nesting sites in spring, before growth of above-ground foliage begins, adult geese grub for the roots and rhizomes of plants in dry areas and eat the swollen bases of shoots of sedges in wet areas. Their activity creates bare areas (1–5 m^2) of peat and sediment. Few pioneer plant species are able to recolonize these patches, and recovery is very slow. Furthermore, in areas of intense summer grazing, 'lawns' of *Carex* and *Puccinellia* spp. have become established. Here, therefore, high densities of grazing geese are essential to maintain the species composition of the vegetation and its above-ground production (Kerbes et al., 1990). The lesser snow goose is a keystone species – the whole structure and composition of these communities are drastically altered by its presence.

For a long time, the conventional wisdom, arrived at largely through 'logical' argument, was that increased complexity within a community leads to increased stability (MacArthur, 1955; Elton, 1958); that is, more complex communities are more stable in the face of a disturbance such as the loss of one or more species. For example, it was argued that in more complex communities, with more species and more interactions, there were more possible pathways by which energy passed through the community. Hence, if there was a perturbation to the community (a change in the density of one of the species), this would affect only a small proportion of the energy pathways and would have relatively little effect on the densities of other species: the complex community would be resistant to change (Box 9.5).

a long-standing belief that complexity leads to stability . . .

However, as analyses of mathematical models of food webs have become more sophisticated, the conventional wisdom has by no means always received support (May, 1981; Tilman, 1999), and conclusions differ depending on whether we focus on individual populations within a community or on aggregate properties of the community such as their biomass or productivity. Briefly, the model food webs have been characterized by one or more of the following: (i) the number of species they contain; (ii) the *connectance* of the web (the fraction of all possible pairs of species that interact directly with one another); and (iii) the average interaction strength between pairs of species. At the level of the individual population, models have not always come to the same conclusion, but overall they suggest that increases in the number of species, increases in connectance and increases in average interaction strength – each representing an increase in complexity – all tend to *decrease* the tendency of individual populations within the community to return to their former state following a disturbance (their resilience, e.g. Figure 9.21). Thus these models suggest, if anything, that community complexity leads to population *instability*.

. . . that is not supported by mathematical models for individual populations

Figure 9.21

The effect of species richness (number of species) on the temporal variability (coefficient of variation, CV) of population size and aggregate community abundance in model communities in which all species are equally abundant and have the same CV. Thus, high values for CV equate to low levels of stability.

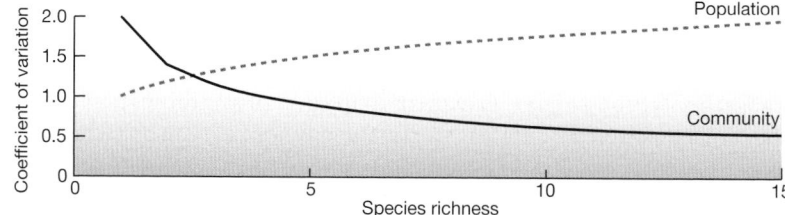

AFTER COTTINGHAM ET AL., 2001

but aggregate properties are more stable in richer model communities

However, the effects of complexity, especially species richness, on the stability of aggregate properties of model communities have been more consistent. Broadly, in richer communities, the dynamics of these aggregate properties are *more* stable (Figure 9.22). In large part, this is because, as long as the fluctuations in different populations are not perfectly correlated, there is an inevitable 'statistical averaging' effect when populations are added together – when one goes up, another is going down – and this tends to increase in effectiveness as richness (the number of populations) increases. Certainly, models indicate that there is no necessary, unavoidable connection linking stability to complexity.

complexity and stability in practice: populations

What is the evidence from real communities? Various studies have sought to build on the mathematical models by examining the relationships among number of species, connectance and interaction strength. The argument runs as follows. The only communities we can observe are those that are stable enough to exist. Hence, those with more species can only be sufficiently stable if there are compensatory decreases in connectance and/or interaction strength. But data on interaction strengths for whole communities are unavailable, so we assume, for simplicity, that average interaction strength is constant. Thus, communities with more species will only retain stability if there is an associated reduction in average connectance.

Figure 9.22

The relationships between connectance and species richness. (a) A compilation from the literature of 40 food webs from terrestrial, freshwater and marine environments. (b) A compilation of 95 insect-dominated webs from various habitats. (c) Seasonal versions of a food web for a large pond in northern England, varying in species richness from 12 to 32. (d) Food webs from swamps and streams in Costa Rica and Venezuela.

(a) AFTER BRIAND, 1983; (b) FROM SCHOENLY ET AL., 1991; (c) FROM WARREN, 1989; (d) FROM WINEMILLER, 1990; AFTER HALL & RAFFAELLI, 1993

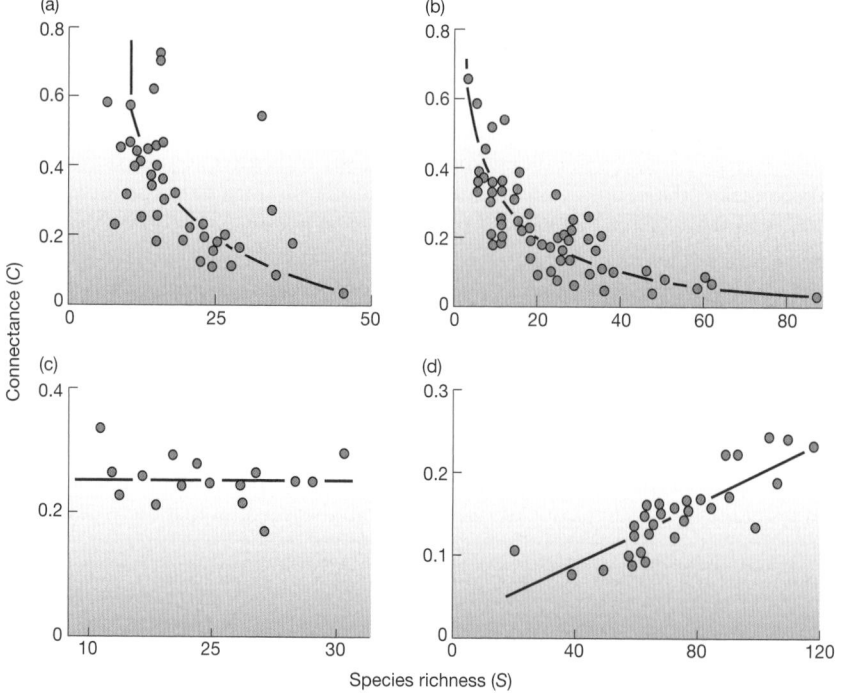

Early analyses of published food web data found, as predicted, that connectance decreased with species number (Figure 9.22a). These data, however, were not collected for the purpose of quantitative study of food web properties. In particular, the accuracy of identification varied substantially from web to web, and even in the same web components were sometimes grouped at the level of kingdom (e.g. 'plants'), sometimes as a family (e.g. Diptera) and sometimes as a species (polar bear) (see review by Hall & Raffaelli, 1993). More recent studies, in which food webs were more rigorously documented, indicate that connectance may decrease with species number (as predicted) (Figure 9.22b), or may be independent of species number (Figure 9.22c), or may even increase with species number (Figure 9.22d). Thus, the stability argument does not receive consistent support from food web analyses either.

The prediction that populations in richer communities are less stable when disturbed was also investigated by Tilman (1996), who pooled data for 39 common plant species from 207 grassland plots in Cedar Creek Natural History Area, Minnesota, collected over an 11-year period. He found that variation in the biomass of individual species increased significantly, but only very weakly, with the richness of the plots (Figure 9.23a). Thus, like the theoretical studies, empirical studies hint at decreased population stability (increased variability) in more complex communities, but the effect seems to be weak and inconsistent.

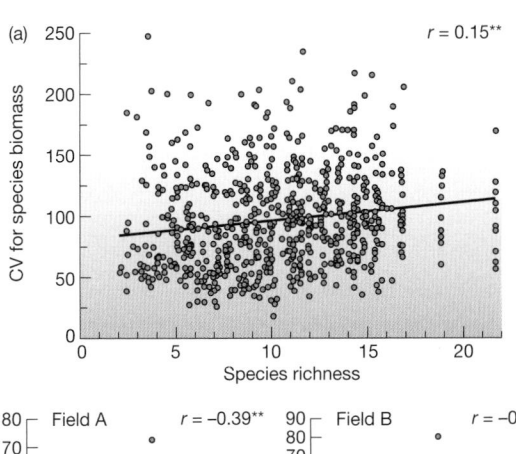

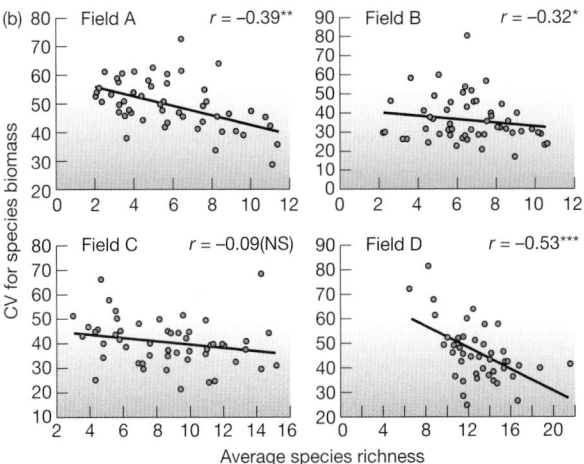

Figure 9.23

(a) The coefficient of variation (CV) of population biomass for 39 plant species from plots in four fields in Minnesota over 11 years (1984–1994) plotted against species richness in the plots. Variation increased with richness but the slope was very shallow. (b) The CV for community biomass in each plot plotted against species richness for each of the four fields (A–D). Variation consistently decreased with richness. In both cases, regression lines and correlation coefficients are shown (*, $P < 0.05$; **, $P < 0.01$; ***, $P < 0.001$).

Figure 9.24

Variation (i.e. 'instability') in productivity (standard deviation of carbon dioxide flux) declined with species richness in microbial communities observed over a 6-week period.

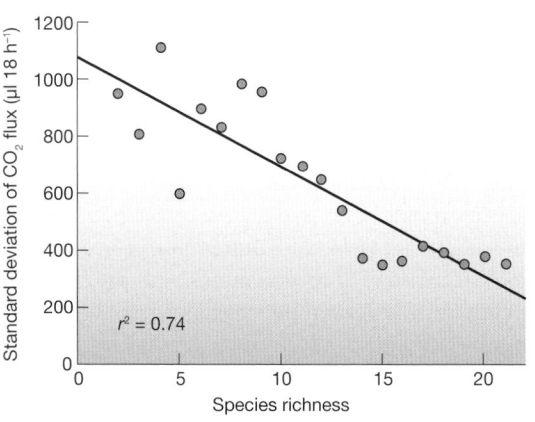

AFTER MCGRADY-STEED ET AL., 1997

complexity and stability in practice: whole communities

Turning to the aggregate, whole-community level, evidence is largely consistent in supporting the prediction that increased richness in a community increases stability (decreases variability). For example, in Tilman's (1996) Minnesota grasslands study, in contrast to the weak negative effect found at the population level, there was a strong positive effect of richness on the stability of community biomass (Figure 9.23b). Also, McGrady-Steed et al. (1997) manipulated richness in aquatic microbial communities (producers, herbivores, bacterivores, predators) and found that variation in another community measure, carbon dioxide flux (a measure of community respiration), also declined with richness (Figure 9.24).

importance of the nature of the community: keystones again

On the other hand, in an experimental study of small grassland communities perturbed by an induced drought, Wardle et al. (2000) found detailed community composition to be a far better predictor of stability than overall richness. Indeed, it is clear that the whole concept of a keystone species (see above) is itself a recognition of the fact that the effects of a disturbance on structure or function are likely to depend very much on the precise nature of the disturbance – that is, on *which* species are lost. Reinforcement of this idea is provided by a simulation study carried out by Dunne et al. (2002), in which they took 16 published food webs and subjected them to the sequential removal of species. Secondary extinctions followed most rapidly when the most connected species were removed, and least rapidly when the least connected were removed, with random removals lying between the two (Figure 9.25). This reminds us that the idiosyncrasies of individual webs are likely always to undermine the generality of any 'rules' even if such rules can be agreed on.

environmental predictability linked to community fragility?

In fact, even if complexity and instability are connected in models, this does not necessarily mean that we should expect to see an association between complexity and instability in real communities. Unstable communities will fail to persist when they experience environmental conditions that reveal their instability. But the range and predictability of environmental conditions will vary from place to place. In a stable and predictable environment, a community that is dynamically fragile may nevertheless persist. But in a variable and unpredictable environment, only a community that is dynamically robust will be able to persist. Hence, we might expect to see: (i) complex and fragile communities in stable and predictable environments, with simple and robust communities in variable and unpredictable environments; but (ii) approximately the same *observed* stability (in terms of population fluctuations,

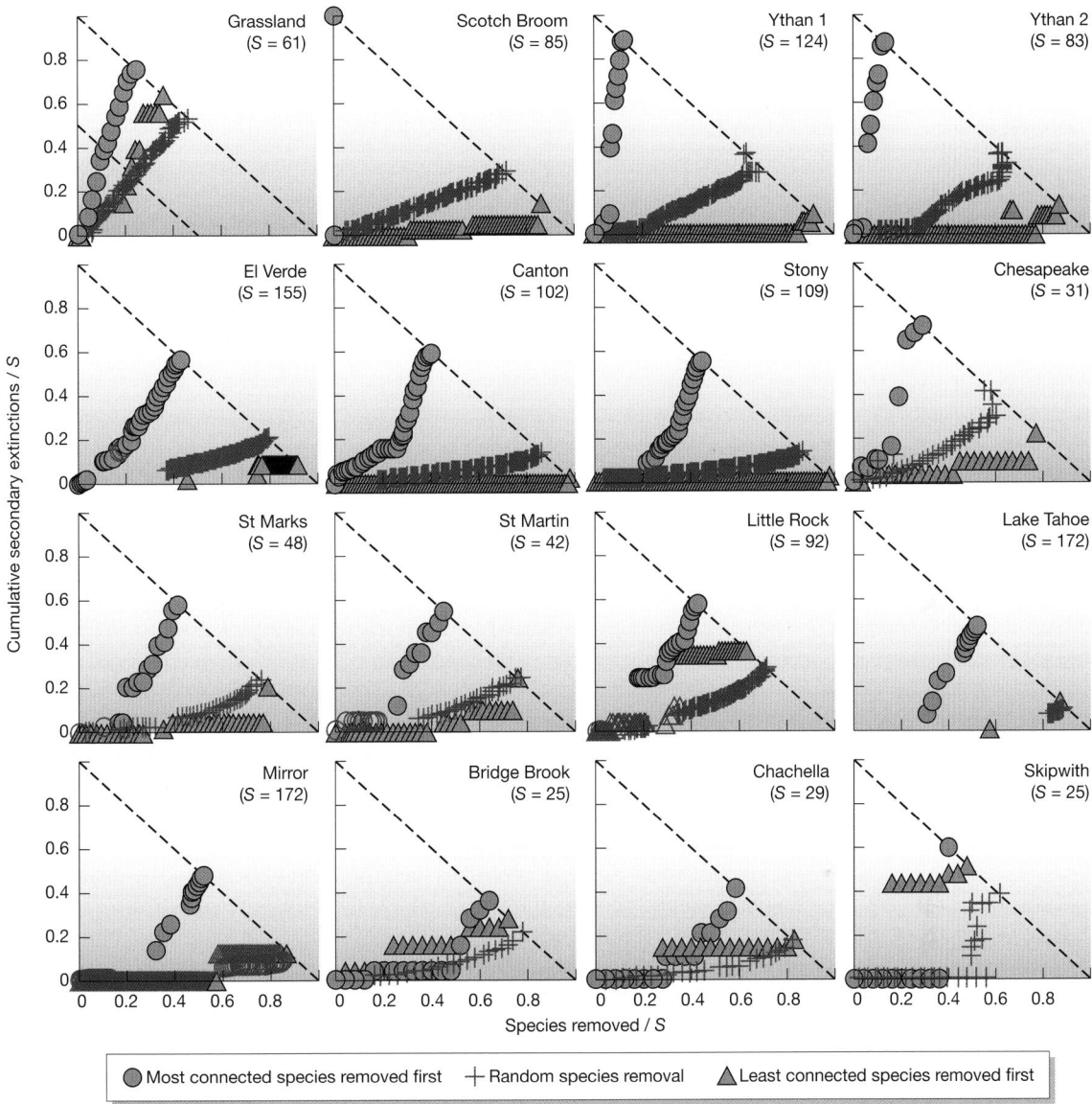

Figure 9.25

The results of a simulation study. The effect of sequential species removal on the number of consequential (secondary) species extinctions, as a proportion of the total number of species originally in the web, S, for each of 16 previously described food webs. The three different rules for species removal are described in the lower panel. Robustness of the webs (the tendency *not* to suffer secondary extinctions) was usually lowest when the most connected species were removed first and highest when the least connected were removed first.

and so forth) in all communities, since this will depend on the inherent stability of the community combined with the variability of the environment. One study tending to support this investigated 10 small streams in New Zealand that differ in the intensity and frequency of flow-related disturbances to their beds (Figure 9.26). Food webs in the more disturbed, 'unstable' streams were characterized by less complex communities: fewer species and fewer links between species.

Figure 9.26

In New Zealand streams, less disturbed sites support more 'complex' communities, with (a) more species (greater web size) and (b) greater connectance between species. The average number of feeding links per animal species (number of prey species in the diet) increases with the intensity of flow-related disturbances to the streambed.

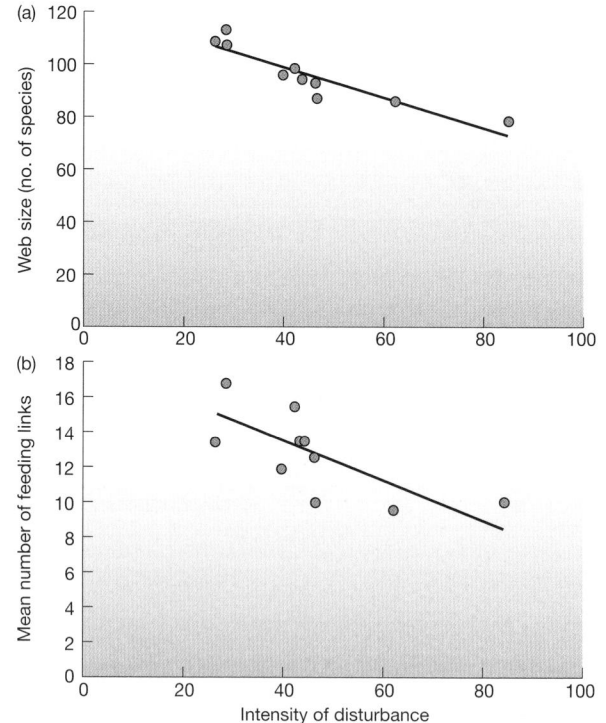

AFTER TOWNSEND ET AL., 1998

This line of argument, moreover, carries a further, very important implication for the likely effects of unnatural perturbations caused by humans on communities. We might expect these to have their most profound effects on the dynamically fragile, complex communities of stable environments, which are relatively unaccustomed to perturbations, and least effect on the simple, robust communities of variable environments, which have previously been subjected to repeated (albeit natural) perturbations.

Summary
SUMMARY

Multiple determinants of the dynamics of populations

To understand the factors responsible for the population dynamics of even a single species in a single location, it is necessary to have a knowledge of physicochemical conditions, available resources, the organism's life cycle, and the influence of competitors, predators and parasites on rates of birth, death, immigration and emigration.

There are contrasting theories to explain the abundance of populations. At one extreme, researchers emphasize the apparent stability of populations and point to the importance of forces that stabilize (density-dependent factors). At the other extreme, those who place more emphasis on density fluctuations may look at external (often density-independent) factors to explain the changes. Key factor analysis is a technique that can be applied to life table studies to throw light both on determination and on regulation of abundance.

Dispersal, patches and metapopulation dynamics

Movement can be a vital factor in determining and/or regulating abundance. A radical change in the way ecologists think about populations has involved focusing attention less on processes occurring within populations and more on patchiness, the colonization and extinction of subpopulations within an overall meta-population, and dispersal between subpopulations.

Temporal patterns in community composition

Disturbances that open up gaps (patches) are common in all kinds of community. Founder-controlled communities are those in which all species are approximately equivalent in their ability to invade gaps and are equal competitors that can hold the gaps against all comers during their lifetime. Dominance-controlled communities are those in which some species are competitively superior to others so that an initial colonizer of a patch cannot necessarily maintain its presence there.

The phenomenon of dominance control is responsible for many examples of community succession. Primary successions occur in habitats where no seeds or spores remain from previous occupants of the site: all colonization must be from outside the patch. Secondary successions occur when existing communities are disturbed but some at least of their seed, etc. remain. It can be very difficult to identify when a succession reaches a stable climax community, since this may take centuries to achieve and in the meantime further disturbances are likely to occur. The exact nature of the colonization process in an empty patch depends on the size and location of that patch. Many communities are mosaics of patches at different stages in a succession.

Food webs

No predator–prey, parasite–host or grazer–plant pair exists in isolation. Each is part of a complex food web involving other predators, parasites, food sources and competitors within the various trophic levels of a community.

The effect of one species on another (its herbivorous prey) may be direct and straightforward. But indirect effects may also be felt by any of the myriad species linked more remotely in the food web. One of the most common is a 'trophic cascade', in which, say, a predator reduces the abundance of a herbivore, thus increasing the abundance of plants.

Top-down control of a food web occurs in situations in which the structure (abundance, species number) of lower trophic levels depends on the effects of consumers from higher trophic levels. Bottom-up control, on the other hand, occurs in a community structure dependent on factors, such as nutrient concentration and prey availability, that influence a trophic level from below. The relative importance of these forces varies according to the trophic level under investigation and the number of trophic levels present.

Some species are more tightly woven into the food web than others. A species whose removal would produce a significant effect (extinction or a large change in density) in at least one other species may be thought of as a strong interactor. Removal of some strong interactors leads to significant changes that spread throughout the food web; we refer to these as keystone species.

The relationship between food web complexity and stability is uncertain (and care is needed in deciding what is meant by stability). Mathematical and empirical studies agree in suggesting that, if anything, population stability decreases with complexity, whereas the stability of aggregate properties of whole communities increases with complexity, especially species richness.

Review questions

REVIEW QUESTIONS

Asterisks indicate challenge questions

1* Construct a flow diagram (boxes and arrows) with a named population at its center to illustrate the wide range of abiotic and biotic factors that influence its pattern of abundance.

2 Population census data can be used to establish correlations between abundance and external factors such as weather. Why can such correlations not be used to prove a causal relationship that explains the dynamics of the population?

3 Distinguish between the determination and regulation of population abundance.

4* Imagine a number of species with patchy distributions: a plant, an insect and a mammal – or consider examples of such species with which you are familiar. How would you identify 'habitable patches' of these species that are not currently occupied by them?

5 What is meant by a 'metapopulation' and how does it differ from a simple 'population'?

6 Define founder control and dominance control as they apply to community organization. In a mosaic of habitat patches, how would you expect communities to differ if they were dominated by founder or dominance control?

7 What factors are responsible for changes in species composition during an old-field succession?

8* Draw up a food web of, say, six or seven species with which you are familiar and which spans at least three trophic levels. Take each species in turn and suggest the kind of community organization that would be necessary for this to be a keystone species.

9 What are meant by bottom-up and top-down control? How is the importance of each likely to vary with the number of trophic levels in a community?

10 Discuss what is understood about the relationship between the complexity and stability of food webs.

Chapter **10**

Patterns in species richness

Key concepts

In this chapter you will:

- understand the meanings of species richness, diversity indices and rank–abundance diagrams
- appreciate that species richness is limited by available resources, the average portion of the resources used by each species (niche breadth) and the degree of overlap in resource use
- recognize that species richness may be highest at intermediate levels of productivity, predation intensity or disturbance but tends to increase with spatial heterogeneity
- appreciate the importance of habitat area and remoteness in determining richness, especially with reference to the equilibrium theory of island biogeography
- understand richness gradients with latitude, altitude and depth, and during community succession, and the difficulties of explaining them
- appreciate how theories of species richness can also be applied to the fossil record

An accurate appreciation of the world's biological diversity is becoming increasingly important. For our conservation efforts to be effective we must understand why species richness varies widely across the face of the Earth. Why do some communities contain more species than others? Are there patterns or gradients in this biodiversity? If so, what are the reasons for these patterns?

10.1 Introduction

Why the number of species varies from place to place, and from time to time, are questions that present themselves not only to ecologists but to anybody who observes and ponders the natural world. They are interesting questions in their own right – but they are also questions of practical importance. It is clear that if we wish to conserve or restore the planet's biological diversity, then we must understand how species numbers are determined and how it comes about that they vary. We will see that there are plausible answers to the questions we ask, but these answers are not always clearcut. Yet this is not so much a disappointment as a challenge to ecologists of the future. Much of the fascination of ecology lies in the fact that many of the problems are blatant, whereas the solutions can be difficult to find. We will see that a full understanding of patterns in species richness must draw on our knowledge of all the areas of ecology discussed so far in this book.

determining species richness

The number of species in a community is referred to as its *species richness*. Counting or listing the species present in a community may sound a straightforward procedure, but in practice it is often surprisingly difficult, partly because of taxonomic inadequacies, but also because only a proportion of the organisms in an area can usually be counted. The number of species recorded then depends on the number of samples that have been taken, or on the volume of the habitat that has been explored. The most common species are likely to be represented in the first few samples, and as more samples are taken, rarer species will be added to the list. At what point does one cease to take further samples? Ideally, the investigator should continue to sample until the number of species reaches a plateau. At the very least, the species richness of different communities should be compared only if they are based on the same sample sizes (in terms of area of habitat explored, time devoted to sampling or, best of all, number of individuals included in the samples).

diversity indices and rank–abundance diagrams

An important aspect of the structure of a community is completely ignored, though, when its composition is described simply in terms of the number of species present – namely, that some species are rare and others common. Intuitively, a community of 10 species with equal numbers in each seems more diverse than another, again consisting of 10 species, with 91% of the individuals belonging to the most common species and just 1% in each of the other nine species. Yet, each community has the same species richness. *Diversity indices* are designed to combine both species richness and the evenness or equitability of the distribution of individuals among those species (Box 10.1). Moreover, attempts to describe a complex community structure by one single attribute, such as richness, or even diversity, can still be criticized because so much valuable information is lost. A more complete picture of the distribution of species abundance in a community is therefore sometimes provided in a *rank–abundance diagram* (Box 10.1).

10.1 *Quantitative aspects*

Diversity indices and rank–abundance diagrams

The measure of the character of a community that is most commonly used to take into account both species richness and the relative abundances of those species is known as the *Shannon* or the *Shannon–Weaver diversity index* (denoted by H). This is calculated by determining, for each species, the proportion of individuals or biomass (P_i for the ith species) that that species contributes to the total in the sample. Then, if S is the total number of species in the community (i.e. the richness), diversity (H) is:

$$H = -\sum P_i \ln P_i$$

where the summation sign $\sum$ indicates that the product ($P_i \ln P_i$) is calculated for each of the S species in turn and these products are then summed. As required, the value of the index depends on both the species richness and the evenness (equitability) with which individuals are distributed among the species. Thus, for a given richness, H increases with equitability, and for a given equitability, H increases with richness.

An example of an analysis using diversity indices is provided by the unusually long-term study that commenced in 1856 in an area of pasture at Rothamsted in England. Experimental plots received a fertilizer treatment once every year, and control plots did not. Figure 10.1 shows how species diversity (H) of grass species changed between 1856 and 1949. While the unfertilized area remained essentially unchanged, the fertilized area showed a progressive decline in diversity. This 'paradox of enrichment' is discussed in Section 10.3.1.

Rank–abundance diagrams, on the other hand, make use of the full array of P_i values by plotting P_i

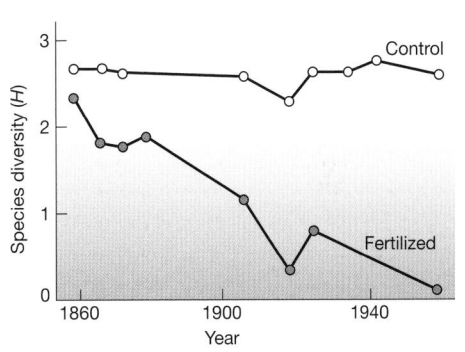

Figure 10.1

Species diversity (H) declined progressively in a plot of pasture that regularly received fertilizer in an experiment commencing in 1856 at Rothamsted in England. In contrast, species diversity remained constant in a control plot that received no fertilizer.

AFTER TILMAN, 1982

against 'rank'; i.e. the most abundant species takes rank 1, the second most abundant rank 2, and so on, until the array is completed by the rarest species of all. The steeper the slope of a rank–abundance diagram, the greater the dominance of common species over rare species in the community (a steep slope means a sharp drop in relative abundance, P_i, for a given drop in rank). Thus, in the case of the Rothamsted experiment, Figure 10.2 shows how the dominance of commoner species steadily increased (steeper slope) while species richness decreased over time.

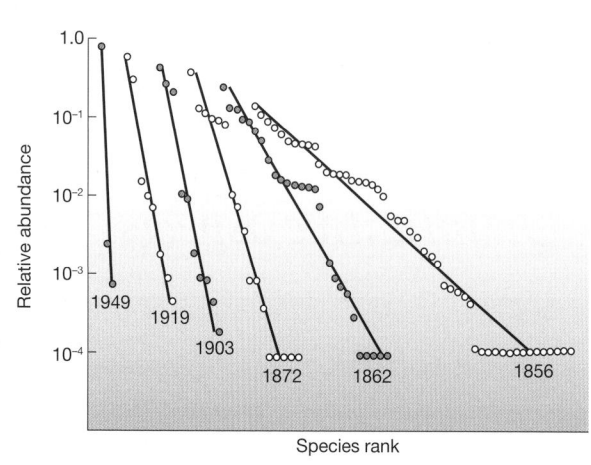

Figure 10.2

Change in the rank–abundance pattern of plant species in the Rothamstead fertilized plot from 1856 to 1949. Note how the slope of the regression line becomes progressively steeper with time since commencement of fertilizer addition. A steeper plot indicates that the commoner species comprise a greater proportion of the total community – in other words, this pasture community gradually became dominated by just a few species.

AFTER TOKESHI, 1993

Nonetheless, for many purposes, the simplest measure, species richness, suffices. In the following sections, therefore, we examine the relationships between species richness and a variety of factors that may, in theory, influence richness in ecological communities. It will become clear that it is not always easy to come up with unambiguous predictions and clean tests of hypotheses when dealing with something as complex as a community.

10.2 A simple model of species richness

To try to understand the determinants of species richness, it will be useful to begin with a simple model (Figure 10.3). Assume, for simplicity, that the resources available to a community can be depicted as a one-dimensional continuum, R units long. Each species uses only a portion of this resource continuum, and these portions define the *niche breadths* (n) of the various species: the average niche breadth within the community is $\bar{n}$. Some of these niches overlap, and the overlap between adjacent species can be measured by value o. The average niche overlap within the community is then $\bar{o}$. With this simple background, it is possible to consider why some communities should contain more species than others.

First, for given values of $\bar{n}$ and $\bar{o}$, a community will contain more species the larger the value of R, i.e. the greater the range of resources (Figure 10.3a). Second, for a given range of resources, more species will be accommodated if $\bar{n}$ is smaller, i.e. if the species are more specialized in their use of resources (Figure 10.3b). Alternatively, if species overlap to a greater extent in their use of resources (greater $\bar{o}$), then more may coexist along the same resource continuum (Figure 10.3c). Finally, a community will contain more species the more fully saturated it is; conversely, it will contain fewer species when more of the resource continuum is unexploited (Figure 10.3d).

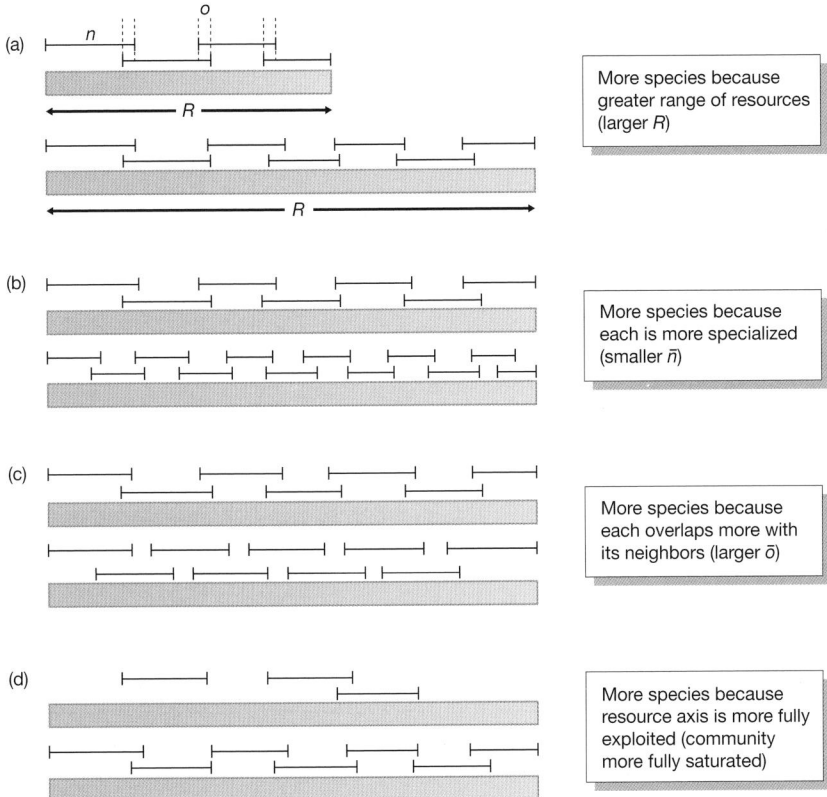

More species because greater range of resources (larger *R*)

More species because each is more specialized (smaller *n̄*)

More species because each overlaps more with its neighbors (larger *ō*)

More species because resource axis is more fully exploited (community more fully saturated)

AFTER MACARTHUR, 1972

Figure 10.3

A simple model of species richness. Each species utilizes a portion *n* of the available resources (*R*), overlapping with adjacent species by an amount *o*. More species may occur in one community than in another because: (a) a greater range of resources is present (larger *R*), (b) each species is more specialized (smaller average *n*), (c) each species overlaps more with its neighbors (larger average *o*), or (d) the resource dimension is more fully exploited.

We can now consider the relationship between this model and two important kinds of species interactions described in previous chapters: interspecific competition and predation. If a community is dominated by interspecific competition (see Chapter 6), the resources are likely to be fully exploited. Species richness will then depend on the range of available resources, the extent to which species are specialists and the permitted extent of niche overlap (Figure 10.3a–c). We will examine a range of influences on each of these three.

Predation, on the other hand, is capable of exerting contrasting effects (see Chapter 7). First, we know that predators can exclude certain prey species; in the absence of these species the community may then be less than fully saturated, in the sense that some available resources may be unexploited (Figure 10.3d). In this way, predation may reduce species richness. Second though, predation may tend to keep species below their carrying capacities for much of the time, reducing the intensity and importance of direct interspecific competition for resources. This may then permit much more niche overlap and a *greater* richness of species than in a community dominated by competition (Figure 10.3c).

competition and predation may influence species richness

The next two sections examine a variety of factors that influence species richness. To organize these, Section 10.3 focuses on factors that often vary spatially (from place to place): productivity, predation intensity, spatial heterogeneity and environmental 'harshness'. Section 10.4 then focuses on factors reflecting temporal variation: climatic variation, disturbance and evolutionary age.

10.3 Spatially varying factors that influence species richness

10.3.1 Productivity and resource richness

For plants, the productivity of the environment can depend on whichever nutrient or condition is most limiting to growth (dealt with in detail in Chapter 11). Broadly speaking, the productivity of the environment for animals follows the same trends as for plants, mainly as a result of the changes in resource levels at the base of the food chain.

increased productivity might be expected to lead to increased richness . . .

If higher productivity is correlated with a wider *range* of available resources, then this is likely to lead to an increase in species richness (Figure 10.3a). However, a more productive environment may have a higher rate of supply of resources but not a greater variety of resources. This might lead to more individuals per species rather than more species. Alternatively again, it is possible, even if the overall variety of resources is unaffected, that rare resources in an unproductive environment may become abundant enough in a productive environment for extra species to be added, because more specialized species can be accommodated (Figure 10.3b).

. . . and often does

In general, though, we might expect species richness to increase with productivity – a contention that is supported by an analysis of the species richness of trees in North America in relation to a crude measure of available environmental energy, *potential evapotranspiration* (PET). This is the amount of water that under prevailing conditions would evaporate or be transpired from a saturated surface (Figure 10.4a). However, while energy (heat and light) is

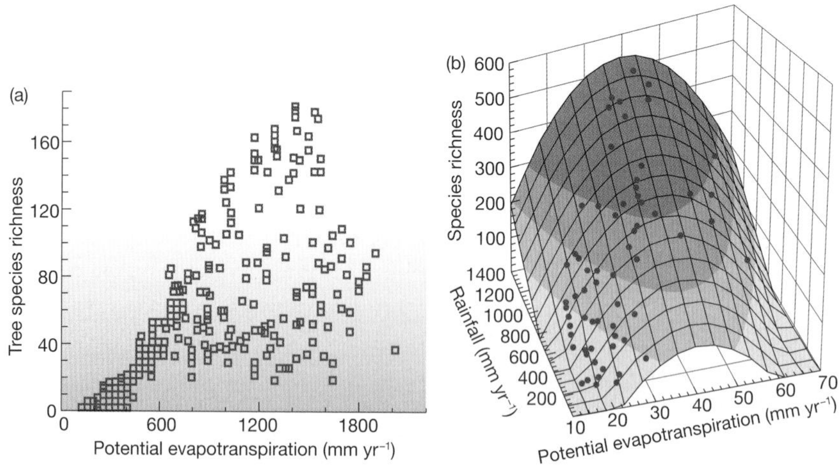

Figure 10.4

(a) Species richness of trees in North America (north of the Mexican border) in relation to potential evapotranspiration. For this analysis the continent was divided into 336 quadrats following lines of latitude and longitude. (b) Species richness of southern African trees (each dot represents a 25,000 km^2 map quadrat) in relation to both rainfall and potential evapotranspiration. The three-dimensional surface describes the regression relationship of species richness with rainfall and potential evapotranspiration. The surface is divided into zones of increasing depth of color representing increasing species richness.

(a) AFTER CURRIE & PAQUIN, 1987; CURRIE, 1991; (b) DATA FROM O'BRIEN, 1993; AFTER WHITTAKER ET AL., 2003

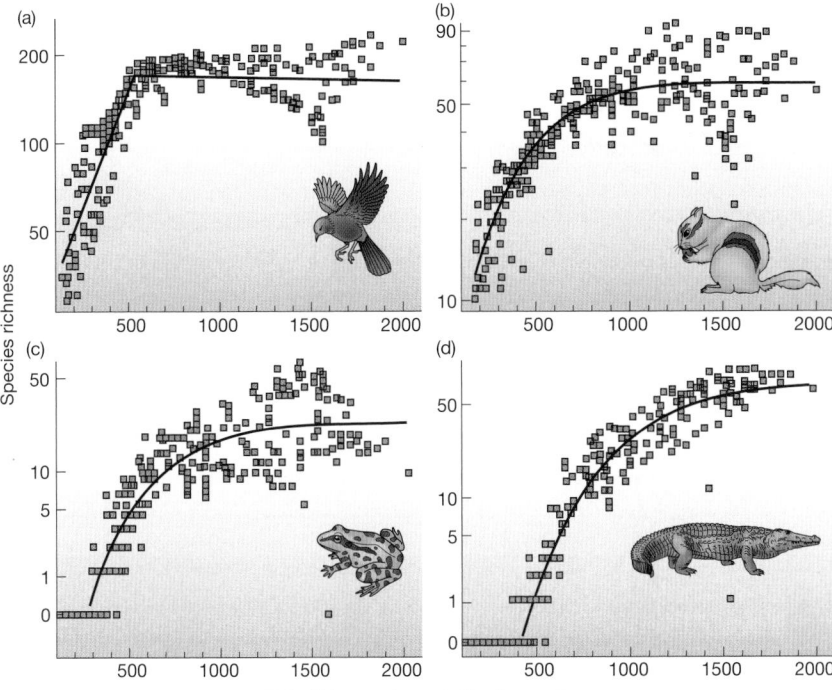

Figure 10.5

Species richness of: (a) birds, (b) mammals, (c) amphibians and (d) reptiles in North America in relation to potential evapotranspiration.

necessary for tree functioning, plants also depend critically on actual water availability. Indeed, energy and water availability inevitably interact, since higher energy inputs lead to more evapotranspiration and a greater requirement for water (Whittaker et al., 2003). Thus, in a study of southern African trees, species richness increased with water availability (annual rainfall), but first increased and then decreased with available energy (PET; Figure 10.4b). Such hump-shaped richness patterns will be a recurring feature in this chapter.

When the North American work (Figure 10.4a) was extended to four vertebrate groups, species richness correlated to some extent with tree species richness itself. However, the best correlations were consistently with PET (Figure 10.5). Why should animal species richness be positively correlated with crude atmospheric energy? The answer is not known with any certainty, but it may be because for an ectotherm, such as a reptile, extra atmospheric warmth would enhance the intake and utilization of food resources; while for an endotherm, such as a bird, the extra warmth would mean less expenditure of resources in maintaining body temperature and more available for growth and reproduction. In both cases, then, this could lead to faster individual and population growth and thus to larger populations. Warmer environments might therefore allow species with narrower niches to persist and such environments may therefore support more species in total (Turner et al., 1996) (see Figure 10.3b).

Sometimes there seems to be a direct relationship between animal species richness and plant productivity. Thus, there are strong positive correlations between species richness and precipitation for both seed-eating ants and seed-eating rodents in the southwestern deserts of the United States (Figure 10.6a). In such arid regions, it is well established that mean annual precipitation is closely related

Figure 10.6

Relationships between species richness and productivity. Where best-fit lines are shown (see Box 1.2), each is statistically significant. (a) The species richness of seed-eating rodents (triangles) and ants (circles) inhabiting sandy soils increased along a geographic gradient of increasing precipitation and, therefore, of increasing productivity. (b) Species richness of fish increased with primary productivity of phytoplankton in a series of North American lakes, while (c) species richness of the phytoplankton themselves showed a hump-shaped relationship, increasing with productivity when productivity was low, but declining at higher levels. (d) Species richness of desert rodents also showed a hump-shaped relationship when plotted against annual rainfall. (e) Percentage of published studies on plants and animals showing various patterns between species richness and productivity. All conceivable patterns have been detected, but hump-shaped and positive patterns, such as those shown in (a) to (d), are well represented. However, it is not uncommon for no pattern to be documented.

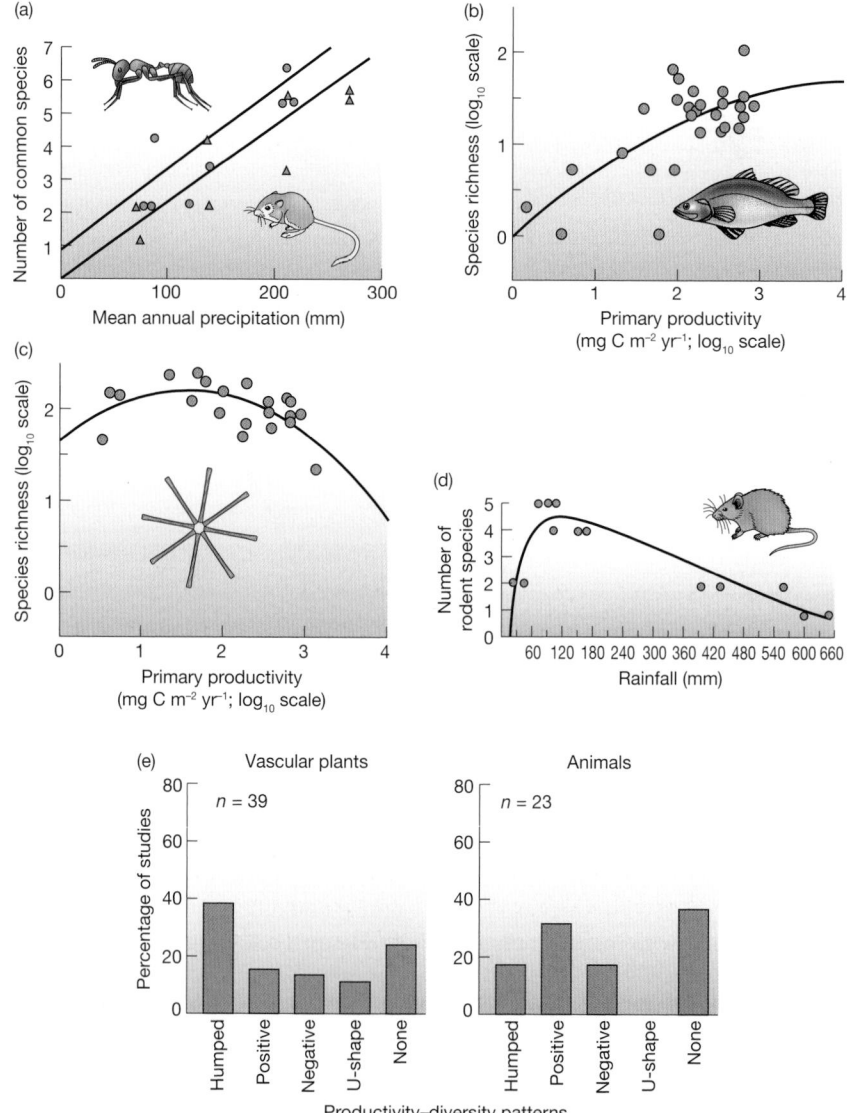

(a) AFTER BROWN & DAVIDSON, 1977; (b, c) AFTER DODSON ET AL., 2000; (d) AFTER ABRAMSKY & ROSENZWEIG, 1983; (e) AFTER MITTELBACH ET AL., 2001

to plant productivity and thus to the amount of seed resource available. It is particularly noteworthy that in species-rich sites, the communities contain more species of very large ants (which consume large seeds) and more species of very small ants (which take small seeds) (Davidson, 1977). It seems that either the range of seed sizes is greater in the more productive environments (Figure 10.3a) or the abundance of seeds becomes sufficient to support extra consumer species with narrower niches (Figure 10.3b). The species richness of fish in North American lakes also increases with an increase in productivity of the lake's phytoplankton (Figure 10.6b).

On the other hand, an increase in diversity with productivity is by no means universal, as shown for example, by the unique experiment that started in 1856 at Rothamsted in England (see Box 10.1). An 8 acre pasture was divided into

other evidence shows richness declining with productivity . . .

20 plots, two serving as controls and the others receiving a fertilizer treatment once a year. While the unfertilized areas remained essentially unchanged, the fertilized areas showed a progressive decline in species richness (and diversity).

Such declines have long been recognized. Rosenzweig (1971) referred to them as illustrating "the paradox of enrichment". One possible resolution of the paradox is that high productivity leads to high rates of population growth, bringing about the extinction of some of the species present because of a speedy conclusion to any potential competitive exclusion (see Section 6.2.7). At lower productivity, the environment is more likely to have changed before competitive exclusion is achieved. An association between high productivity and low species richness has been found in several other studies of plant communities (reviewed by Tilman, 1986). It can be seen, for example, where human activities lead to an increased input of plant resources like nitrates and phosphates into lakes, rivers, estuaries and coastal marine regions; when such 'cultural eutrophication' is severe, we consistently see a decrease in species richness of phytoplankton (despite an increase in their productivity).

It is perhaps not surprising, then, that several studies have demonstrated both an increase and a decrease in richness with increasing productivity – that is, that species richness may be highest at intermediate levels of productivity. Species richness declines at the lowest productivities because of a shortage of resources, but also declines at the highest productivities where competitive exclusions speed rapidly to their conclusion. For instance, there are humped curves when the number of lake phytoplankton species is plotted against overall phytoplankton productivity (Figure 10.6c; the decline at higher productivity is analogous to the cultural eutrophication mentioned above) and when the species richness of desert rodents is plotted against precipitation (and thus productivity) along a geographic gradient in Israel (Figure 10.6d). Indeed, an analysis of a wide range of such studies found that when communities differing in productivity but of the same general type (e.g. tallgrass prairie) were compared (Figure 10.6e), a positive relationship was the most common finding in animal studies (with fair numbers of humped and negative relationships), whereas with plants, humped relationships were most common, with smaller numbers of positives and negatives (and even some U-shaped curves – cause unknown!). Clearly, increased productivity can and does lead to increased or decreased species richness – or both.

. . . and further evidence suggests a 'humped' relationship

10.3.2 Predation intensity

The possible effects of predation on the species richness of a community were examined in Chapter 7: predation may increase richness by allowing otherwise competitively inferior species to coexist with their superiors (*predator-mediated coexistence*); but intense predation may reduce richness by driving prey species (whether or not they are strong competitors) to extinction. Overall, therefore, there may also be a humped relationship between predation intensity and species richness in a community, with greatest richness at intermediate intensities, such as that shown by the effects of cattle grazing (illustrated in Figure 7.24).

A classic example of predator-mediated coexistence is provided by a study that established the concept in the first place: the work of Paine (1966) on the influence of a top carnivore on community structure on a rocky shore (Figure 10.7).

predator-mediated coexistence by starfish on a rocky shore

Figure 10.7

Paine's rocky shore community. The profound influence of the predatory starfish could only be detected by removing them. In the absence of *Pisaster*, other species became dominant (first barnacles and then mussels) leading to an overall reduction in species richness. This is a classic case of predator-mediated coexistence.

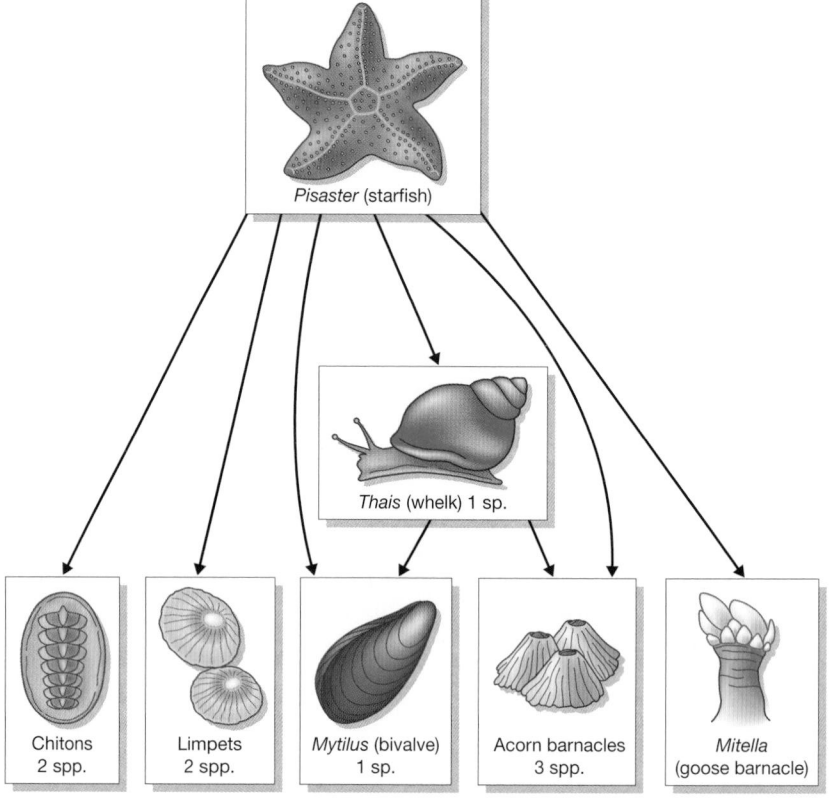

The starfish *Pisaster ochraceus* preys on sessile filter-feeding barnacles and mussels, and also on browsing limpets and chitons and a small carnivorous whelk. These species, together with a sponge and four macroscopic algae (seaweeds), form a typical community on rocky shores of the Pacific coast of North America. Paine removed all starfish from a stretch of shoreline about 8 m long and 2 m deep and continued to exclude them for several years. The structure of the community in nearby control areas remained unchanged during the study, but the removal of *Pisaster* had dramatic consequences. Within a few months, the barnacle *Balanus glandula* settled successfully. Later mussels (*Mytilus californicus*) crowded it out, and eventually the site became dominated by these. All but one of the species of alga disappeared, apparently through lack of space, and the browsers tended to move away, partly because space was limited and partly because there was a lack of suitable food. The main influence of the starfish *Pisaster* appears to be to make space available for competitively subordinate species. It cuts a swathe free of barnacles and, most importantly, free of the dominant mussels that would otherwise outcompete other invertebrates and algae for space. Overall, there is usually predator (starfish)-mediated coexistence, but the removal of starfish led to a reduction in number of species from 15 to eight. The concept of predator-mediated coexistence is not only intrinsically interesting; it also finds a surprising application in the field of restoration ecology (Box 10.2).

10.2 Topical ECOncerns

10.2 TOPICAL ECONCERNS

Using exploiter-mediated coexistence to assist grassland restoration

Species-rich meadows are now uncommon in agricultural landscapes in Europe because decades of intensive fertilizer application have allowed a few species to competitively exclude others, a pattern that echoes the results of the remarkable century-long Rothamsted experiment (see Figure 10.1). It is not uncommon nowadays for attempts to be made to restore the lost species richness of these pasture settings. One approach is to use what we know about predator-mediated coexistence or, more generally, exploiter-mediated coexistence. This occurs when one species 'exploits' as food a number of species in the community, reducing the dominance of the most competitively superior species and allowing less competitive species to maintain a foothold.

One example of exploiter-mediated coexistence occurs when parasites exert a leveling effect. *Rhinanthus minor*, an annual plant, is capable of its own limited photosynthesis but is known as a 'hemiparasite' because it typically taps into the photosynthetic products of other plants by building connections with their roots. Researchers reasoned that the presence of the hemiparasite might facilitate recovery to species-rich grassland via exploiter-mediated coexistence (Pywell et al., 2004). To test this hypothesis in an agriculturally impoverished grassland, they established experimental plots with various densities of *Rhinanthus minor*. After the hemiparasite populations had become established, the researchers sowed a mixture of seeds of 10 native wildflower species that had been lost from the grassland as a result of intensive agriculture. After 2 years the hemiparasite was found to have suppressed the growth of the parasitized plants and this led, the following year, to the desired increase in grassland species richness because competitive exclusion had been circumvented (Figure 10.8).

A species-rich flower meadow
© ALAMY IMAGES A4T6HC

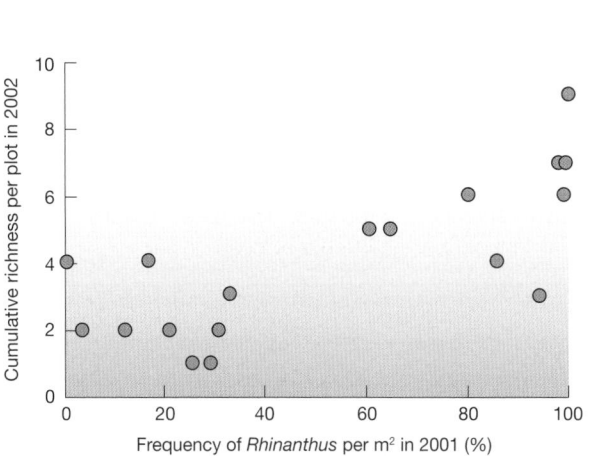

Figure 10.8

Relationship between frequency of occurrence of the hemiparasite *Rhinanthus minor* and species richness of plants per experimental plot of grassland. The presence of the hemiparasite leads to lower plant height, because of reduced success of the parasitized plants, and the following year to increased species richness because of suppression of competitive exclusion by the dominant species.

(LEFT) © ALAMY IMAGES A02Y49; (RIGHT) AFTER PYWELL ET AL., 2004

An understanding of exploiter-mediated coexistence holds promise for future meadow restoration efforts. Can you think of any other aspects of the theory of species richness that could be applied to the benefit of impoverished grasslands? (Clue – check out the 'intermediate disturbance hypothesis', described in Section 10.4.2. These intensively farmed landscapes have also been subject to regular and intensive disturbances caused by heavy mowing or grazing. What might the intermediate disturbance hypothesis have to offer in restoring grassland species richness?)

10.3.3 Spatial heterogeneity

Environments that are more spatially heterogeneous can be expected to accommodate extra species because they provide a greater variety of microhabitats, a greater range of microclimates, more types of places to hide from predators, and so on. In effect, the extent of the resource spectrum is increased (see Figure 10.3a).

richness and the heterogeneity of the abiotic environment

In some cases, it has been possible to relate species richness to the spatial heterogeneity of the abiotic environment. For instance, a study of plant species growing in 51 plots alongside the Hood River, Canada, revealed a positive relationship

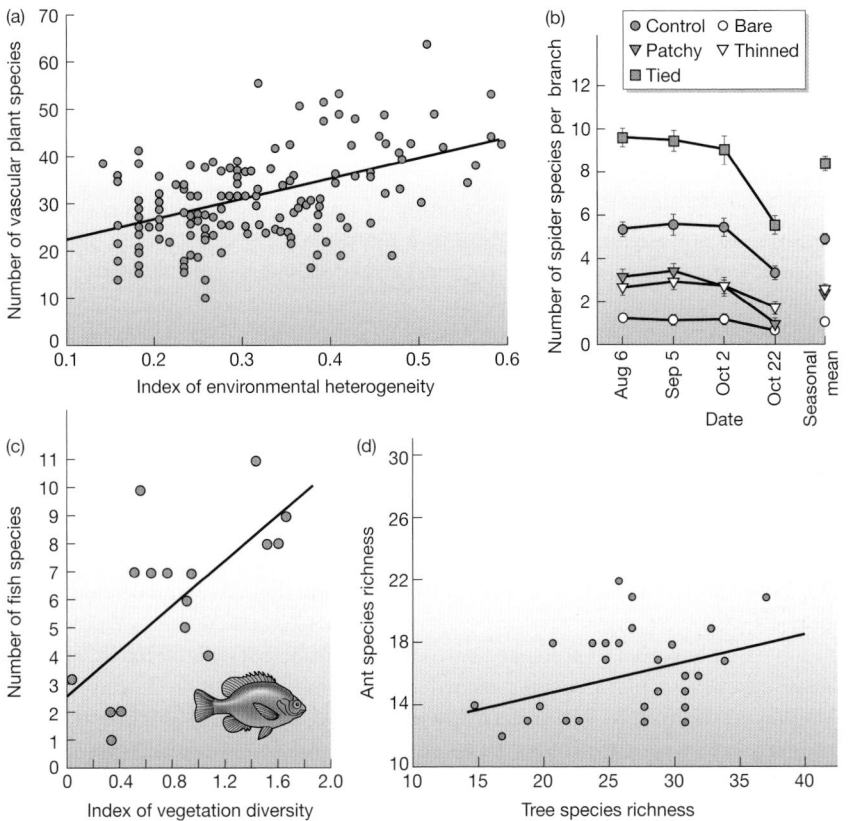

(a) AFTER GOULD & WALKER, 1997; (b) AFTER HALAJ ET AL., 2000; (c) AFTER TONIN & MAGNUSON, 1982; (d) AFTER RIBAS ET AL., 2003

Figure 10.9

(a) Relationship between the number of plants per 300 m^2 plot beside the Hood River, Northwest Territories, Canada, and an index (ranging from 0 to 1) of spatial heterogeneity in abiotic factors associated with topography and soil. (b) In an experimental study, the number of spider species living on Douglas fir branches increased with their structural diversity. Those 'bare', 'patchy' or 'thinned' were less diverse than normal ('control') by virtue of having needles removed; those 'tied' were more diverse because their twigs were entwined together. (c) Relationship between animal species richness and an index of structural diversity of vegetation for freshwater fish in 18 Wisconsin lakes. (d) Relationship between arboreal ant species richness in Brazilian savanna and the species richness of trees (a surrogate for spatial heterogeneity).

between species richness and an index of spatial heterogeneity (based, among other things, on the number of categories of substrate, slope, drainage regimes and soil pH present) (Figure 10.9a).

Most studies of spatial heterogeneity, though, have related the species richness of animals to the structural diversity of the plants in their environment, either as a result of experimental manipulation of the plants, as with the spiders in Figure 10.9b, but more commonly through comparisons of natural communities that differ in plant structural diversity (Figure 10.9c) or plant species richness (where higher species richness equates to greater spatial heterogeneity; Figure 10.9d).

animal richness and plant spatial heterogeneity

Whether spatial heterogeneity arises from the abiotic environment or is provided by biological components of the community, it is capable of promoting an increase in species richness.

10.3.4 Environmental harshness

Environments dominated by an extreme abiotic factor – often called *harsh* environments – are more difficult to recognize than might be immediately apparent. An anthropocentric view might describe as extreme both very cold and very hot habitats, unusually alkaline lakes and grossly polluted rivers. However, species have evolved and live in all such environments, and what is very cold and extreme for us must seem benign and unremarkable to a penguin.

We might try to get around the problem of defining environmental harshness by 'letting the organisms decide'. An environment may be classified as *extreme* if organisms, by their failure to live there, show it to be so. But if the claim is to be made – as it often is – that species richness is lower in extreme environments, then this definition is circular, and it is designed to prove the very claim we wish to test.

Perhaps the most reasonable definition of an extreme condition is one that requires, of any organism tolerating it, a morphological structure or biochemical mechanism that is not found in most related species and is costly, either in energetic terms or in terms of the compensatory changes in the biological processes of the organism that are needed to accommodate it. For example, plants living in highly acidic soils (low pH) may be affected directly through injury by hydrogen ions or indirectly via deficiencies in the availability and uptake of important resources such as phosphorus, magnesium and calcium. In addition, aluminum, manganese and heavy metals may have their solubility increased to toxic levels. Moreover, the activity of symbiotic fungi (mycorrhizas enhancing uptake of dissolved nutrients – see Section 8.4.5) or bacteria (fixation of atmospheric nitrogen – see Section 8.4.6) may be impaired. Plants can only tolerate low pH if they have specific structures or mechanisms allowing them to avoid or counteract these effects.

Environments that experience low pHs can thus be considered harsh, and the mean number of plant species recorded per sampling unit in a study in the Alaskan Arctic tundra was indeed lowest in soils of low pH (Figure 10.10a). Similarly, the species richness of benthic (bottom-dwelling) invertebrates in streams in southern England was markedly lower in the more acidic streams (Figure 10.10b). Further examples of extreme environments that are associated with low species richness include hot springs, caves and highly saline water bodies such as the Dead Sea. The problem with these examples, however, is that each is also characterized by other features associated with low species richness, such as low productivity and low spatial heterogeneity. In addition, many occupy small areas (caves, hot springs) or areas that are rare compared to other types of habitat (only a small proportion of the streams in southern England are acidic). Hence extreme environments can often be seen as small and isolated islands.

Figure 10.10

(a) The number of plant species in the Alaskan Arctic tundra increases with soil pH. (b) The number of taxa of invertebrates in streams in southern England increases with the pH of stream water.

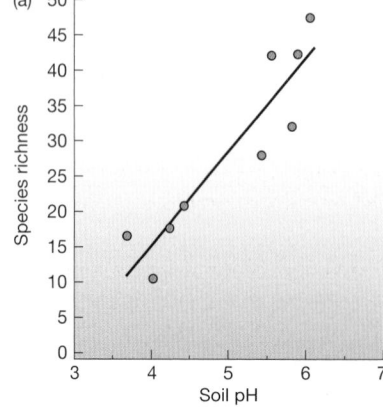

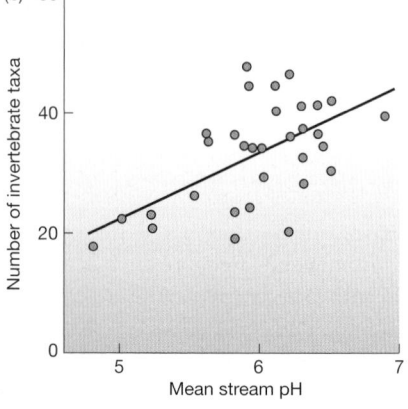

(a) AFTER GOUGH ET AL., 2000; (b) AFTER TOWNSEND ET AL., 1983

We will see in Section 10.5.1 that these features, too, are usually associated with low species richness. Although it appears reasonable that intrinsically extreme environments should as a consequence support few species, this has proved an extremely difficult proposition to establish.

10.4 Temporally varying factors that influence species richness

Temporal variation in conditions and resources may be predictable or unpredictable and operate on time scales from minutes through to centuries and millennia. All may influence species richness in profound ways.

10.4.1 Climatic variation

The effects of climatic variation on species richness depend on whether the variation is predictable or unpredictable (measured on time scales that matter to the organisms involved). In a predictable, seasonally changing environment, different species may be suited to conditions at different times of the year. More species might therefore be expected to coexist in a seasonal environment than in a completely constant one (see Figure 10.3a). Different annual plants in temperate regions, for instance, germinate, grow, flower and produce seeds at different times during a seasonal cycle; while phytoplankton and zooplankton pass through a seasonal succession in large, temperate lakes with a variety of species dominating in turn as changing conditions and resources become suitable for each.

temporal niche differentiation in seasonal environments

On the other hand, there are opportunities for specialization in a non-seasonal environment that do not exist in a seasonal environment. For example, it would be difficult for a specialist fruit-eater to persist in a seasonal environment when fruit is available for only a very limited portion of the year. But such specialization is found repeatedly in non-seasonal, tropical environments where fruit of one type or another is available continuously.

specialization in non-seasonal environments

Unpredictable climatic variation (climatic instability) could have a number of effects on species richness. On the one hand: (i) stable environments may be able to support specialized species that would be unlikely to persist where conditions or resources fluctuated dramatically (Figure 10.3b); (ii) stable environments are more likely to be saturated with species (Figure 10.3d); and (iii) theory suggests that a higher degree of niche overlap will be found in more stable environments (Figure 10.3c). All these processes could increase species richness. On the other hand, populations in a stable environment are more likely to reach their carrying capacities, the community is more likely to be dominated by competition, and species are therefore more likely to be excluded by competition ($\bar{o}$ is smaller, see Figure 10.3c).

Some studies seem to support the notion that species richness increases as climatic variation decreases. For example, there is a significant negative relationship between species richness and the range of monthly mean temperatures for birds, mammals and gastropods that inhabit the West coast of North America (from Panama in the south to Alaska in the north) (MacArthur, 1975). However, this correlation does not prove causation, since there are many other things that change between Panama and Alaska. There is no established relationship between climatic instability and species richness.

10.4.2 Disturbance

Previously, in Section 9.4, the influence of disturbance on community structure was examined, and it was demonstrated that when a disturbance opens up a gap, and the community is *dominance controlled* (strong competitors can replace residents), there tends in a community succession to be an initial increase in richness as a result of colonization, but a subsequent decline in richness as a result of competitive exclusion.

the intermediate disturbance hypothesis . . .

If the frequency of disturbance is now superimposed on this picture, it seems likely that very frequent disturbances will keep most patches in the early stages of succession (where there are few species) but also that very rare disturbances will allow most patches to become dominated by the best competitors (where there are also few species). This suggests an *intermediate disturbance hypothesis*, in which communities are expected to contain most species when the frequency of disturbance is neither too high nor too low (Connell, 1978). The intermediate disturbance hypothesis was originally proposed to account for patterns of richness in tropical rain forests and coral reefs. It has occupied a central place in the development of ecological theory because all communities are subject to disturbances that exhibit different frequencies and intensities.

. . . supported by studies of algae on boulders on a rocky shore . . .

Among a number of studies that have provided support for this hypothesis, we turn first to a study of green and red algae on different-sized boulders on the rocky shores of southern California (Sousa, 1979a, 1979b). Wave action disturbs small boulders more frequently than larger ones; thus, small boulders had a monthly probability of movement of 42%, intermediate-sized boulders a probability of 9%, and large boulders a probability of only 0.1%. After a disturbance clears space on a boulder, ephemeral green algae (*Ulva* spp.) are quick to colonize, but later in the year several species of perennial red alga feature in the succession, including *Gelidium coulteri*, *Gigartina leptorhinchos*, *Rhodoglossum affine* and *Gigartina canaliculata*. The last of these gradually takes over until within 2–3 years it dominates the community, tending to competitively exclude the early and mid-successional species. *G. canaliculata* then persists unless there is a further disturbance. Sousa found that algal species richness was lowest on the frequently (F) disturbed small boulders – these were dominated most often by *Ulva*. The highest levels of species richness were consistently recorded on the intermediate boulder class (I), most of which held mixtures of 3–5 abundant species from all successional stages. Finally, species richness on the rarely disturbed (R), largest boulders was lower than the intermediate class, with a monoculture of *G. canaliculata* on some of them (Figure 10.11a).

. . . and from studies of invertebrates in small streams and plankton in lakes

Disturbances in small streams often take the form of bed movements during periods of high discharge, and because of differences in flow regimes and in the substrata of stream beds, some stream communities are disturbed more frequently than others. This variation was assessed in 54 stream sites in the Taieri River in New Zealand. The pattern of richness of macroinvertebrate species conformed to the intermediate disturbance hypothesis (Figure 10.11b). Finally, in controlled field experiments, natural phytoplankton communities in Lake Plußsee (north

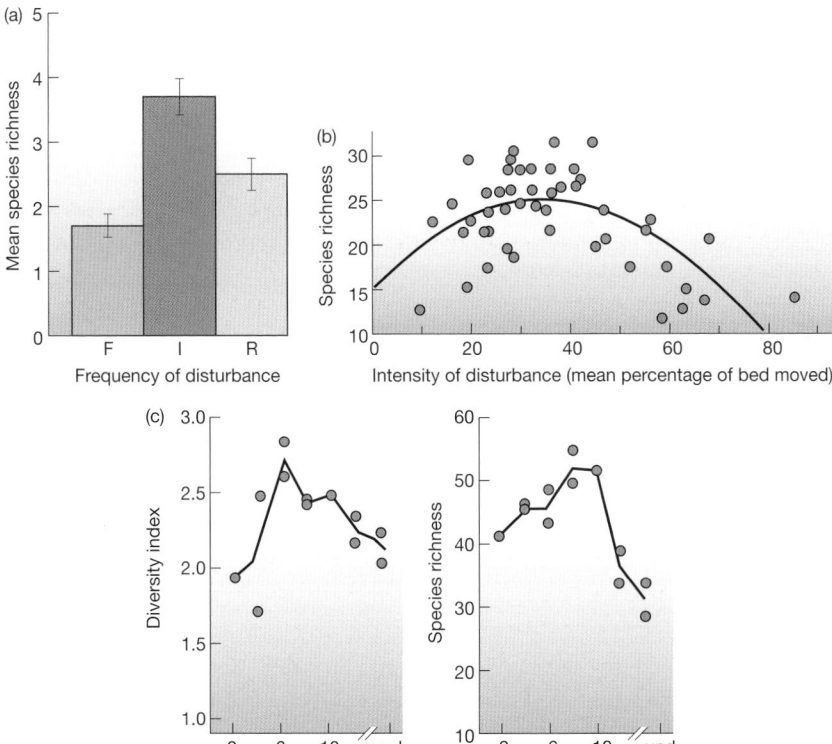

(a) AFTER SOUSA, 1979B; (b) AFTER TOWNSEND ET AL., 1997; (c) AFTER FLODDER & SOMMER, 1999

Figure 10.11

(a) Pattern in species richness (± SE) on rocky-shore boulders in each of three classes categorized according to the frequency with which they are disturbed: frequently disturbed (F), disturbed at an intermediate rate (I), and rarely disturbed (R). Species richness is highest at the intermediate level of disturbance. (b) Relationship between insect species richness and the intensity of disturbance, measured as the average percentage of the bed that moves during successive 2-month periods, in each of 54 stream sites in the Taieri River, New Zealand. Species richness is again highest at intermediate levels of disturbance. (c) Both species diversity (Shannon index) and species richness of phytoplankton communities are highest at intermediate frequencies of disturbance in controlled field experiments in Lake Plußsee in north Germany. 'und' represents species richness in the undisturbed state.

Germany) were disturbed at intervals of 2–12 days by disrupting the normal stratification in the water column with bubbles of compressed air. Again, both species richness and Shannon's diversity index were highest at intermediate frequencies of disturbance (Figure 10.11c).

10.4.3 Environmental age: evolutionary time

It has also often been suggested that communities that are 'disturbed' only on very extended time scales may nonetheless lack species because they have yet to reach an ecological or an evolutionary equilibrium. Thus communities may differ in species richness because some are closer to equilibrium and are therefore more saturated than others (see Figure 10.3d).

For example, many have argued that the tropics are richer in species than temperate regions at least in part because the tropics have existed over long and uninterrupted periods of evolutionary time, whereas the temperate regions are still recovering from the Pleistocene glaciations when temperate biotic zones shifted in the direction of the tropics. It now seems, however, that tropical areas were also disturbed during the ice ages, not directly by ice but by associated climatic changes that saw tropical forest contracting to a limited number of small refuges surrounded by grassland. Thus, although it seems likely that some communities, by virtue of disturbances in their distant past, are less saturated than others, we cannot pinpoint these communities with confidence.

An alternative explanation for lower species richness in temperate than tropical areas invokes the idea that species evolve faster in the tropics because of higher rates of mutation in these warmer climes. Wright et al. (2006) compared the rates of evolution of pairs of woody plant species, one each from tropical areas (e.g. *Eucalyptus deglupta*, *Clematis javana*, *Banksia dentate* and 42 others) and temperate areas (*Eucalyptus coccifera*, *Clematis paniculata*, *Banksia marginata*, etc., respectively). Evolution, as assessed by the rate of nucleotide substitution in a particular region of DNA, turns out to be more than twice as fast in the tropical species.

10.5 Gradients of species richness

Sections 10.3 and 10.4 have demonstrated how difficult explanations for variations in species richness are to formulate and test. It is easier to describe patterns, especially gradients, in species richness. These are discussed next. Explanations for these, too, however, are often very uncertain.

10.5.1 Habitat area and remoteness: island biogeography

species–area relationships on oceanic islands

It is well established that the number of species on islands decreases as island area decreases. Such a *species–area relationship* is shown in Figure 10.12a for plants on small islands east of Stockholm, Sweden.

habitat islands and areas of mainland

'Islands', however, need not be islands of land in a sea of water. Lakes are islands in a 'sea' of land, mountaintops are high-altitude islands in a low-altitude ocean, gaps in a forest canopy where a tree has fallen are islands in a sea of trees,

Figure 10.12

Species–area relationships: in each case the number of species increases with 'island' area. (a) For plants on small islands off the east coast of Sweden in 1999. (b) For birds inhabiting lakes ('islands' of water in a 'sea' of land) in Florida. (c) For bats inhabiting different-sized caves in Mexico. (d) For fish living in Australian desert springs connected to pools of different sizes. All regression lines are significant at $P < 0.05$; no line is shown in (b) because the regression is not significant.

(a) AFTER LOFGREN & JERLING, 2002; (b) AFTER HOYER & CANFIELD, 1994; (c) AFTER BRUNET & MEDELLÍN, 2001; (d) AFTER KODRIC-BROWN & BROWN, 1993

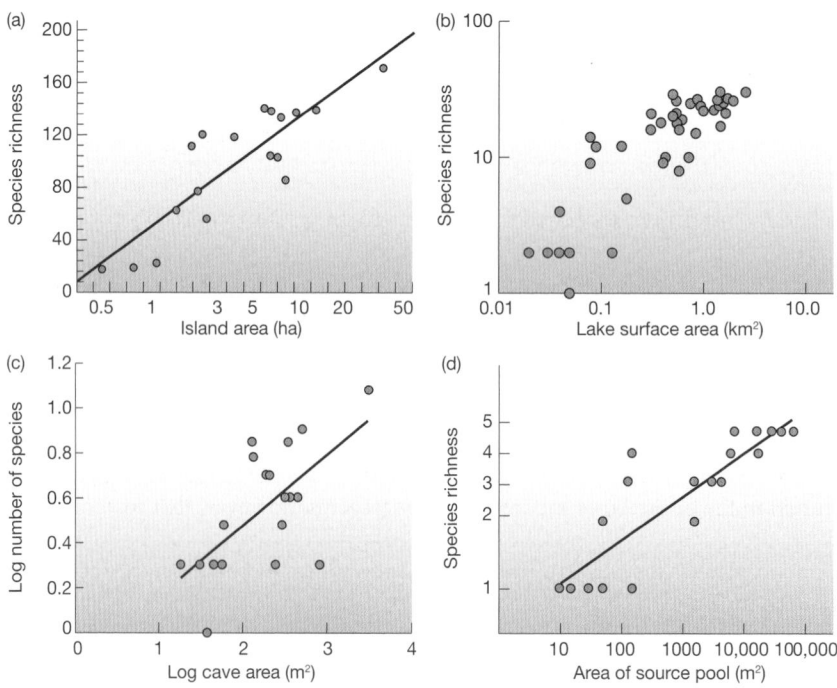

and there can be islands of particular geological types, soil types or vegetation types surrounded by dissimilar types of rock, soil or vegetation. Species–area relationships can be equally apparent for these types of islands (Figure 10.12b–d).

The relationship between species richness and habitat area is one of the most consistent of all ecological patterns. However, the pattern raises an important question: 'Is the impoverishment of species on islands more than would be expected in comparably small areas of mainland?' In other words, does the characteristic isolation of islands contribute to their impoverishment of species? These are important questions for an understanding of community structure since there are many oceanic islands, many lakes, many mountaintops, many woodlands surrounded by fields, many isolated trees and so on.

'island effects' and community structure

Probably the most obvious reason why larger areas should contain more species is that larger areas typically encompass more different types of habitat. However, MacArthur and Wilson (1967) believed this explanation to be too simple. In their *equilibrium theory of island biogeography* they argued that island size and isolation themselves played important roles: that the number of species on an island is determined by a balance between immigration and extinction; that this balance is dynamic, with species continually going extinct and being replaced (through immigration) by the same or by different species; and that immigration and extinction rates may vary with island size and isolation (Box 10.3).

10.3 Historical landmarks

MacArthur and Wilson's equilibrium theory of island biogeography

Taking immigration first, imagine an island that as yet contains no species at all. The rate of immigration of *species* will be high, because any colonizing individual represents a species new to that island. However, as the number of resident species rises, the rate of immigration of new, unrepresented species diminishes. The immigration rate reaches zero when all species from the source pool (i.e. from the mainland or from other nearby islands) are present on the island in question (Figure 10.13a).

The immigration graph is drawn as a curve, because immigration rate is likely to be particularly high when there are low numbers of residents and many of the species with the greatest powers of dispersal are yet to arrive. In fact, the curve should really be a blur rather than a single line, since the precise curve will depend on the exact sequence in which species arrive, and this will vary by chance. In this

sense, the immigration curve can be thought of as the *'most probable' curve*.

The exact immigration curve will depend on the degree of remoteness of the island from its pool of potential colonizers (Figure 10.13a). The curve will always reach zero at the same point (when all members of the pool are resident), but it will generally have higher values on islands close to the source of immigration than on more remote islands, since colonizers have a greater chance of reaching an island the closer it is to the source. It is also likely that immigration rates will generally be higher on a large island than on a small island, since the larger island represents a larger 'target' for the colonizers (Figure 10.13a).

The rate of species extinction on an island (Figure 10.13b) is bound to be zero when there are no species there, and it will generally be low when there are few species. However, as the number of resident

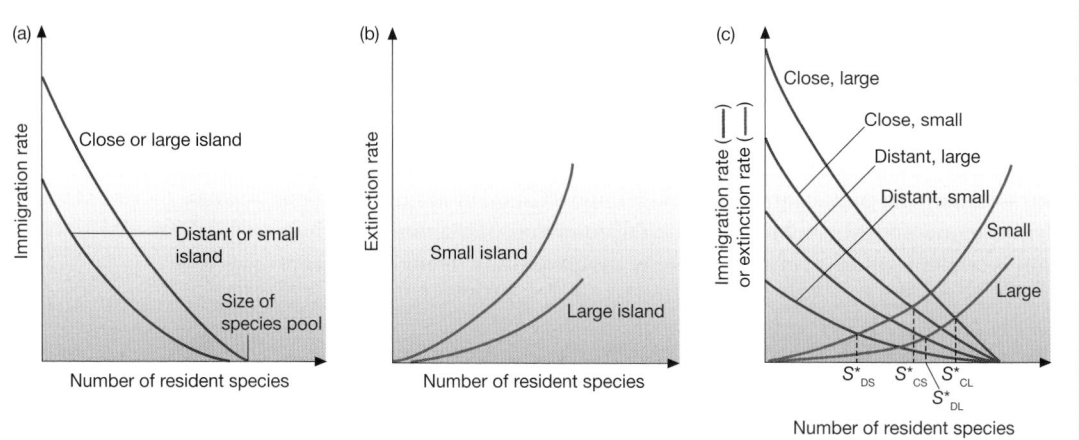

Figure 10.13

MacArthur and Wilson's (1967) equilibrium theory of island biogeography. (a) The rate of species immigration on to an island, plotted against the number of resident species on the island, for large and small islands and for close and distant islands. (b) The rate of species extinction on an island, plotted against the number of resident species on the island for large and small islands. (c) The balance between immigration and extinction on small and large islands and on close and distant islands. In each case, S^* is the equilibrium species richness; C, close; D, distant; L, large; S, small.

species rises, the extinction rate is assumed by the theory to increase, probably at a more than proportionate rate. This is thought to occur because with more species, competitive exclusion becomes more likely, and the population size of each species is on average smaller, making it more vulnerable to chance extinction. Similar reasoning suggests that extinction rates should be higher on small than on large islands – population sizes will typically be smaller on small islands (Figure 10.13b). As with immigration, the extinction curves are best seen as 'most probable' curves.

In order to see the net effect of immigration and extinction, their two curves can be superimposed (Figure 10.13c). The number of species where the curves cross (S^*) is a dynamic equilibrium and should be the characteristic species richness for the island in question. Below S^*, richness increases (immigration rate exceeds extinction rate); above S^*, richness decreases (extinction exceeds immigration). The theory, then, makes a number of predictions, described in the text.

MacArthur and Wilson's theory makes several predictions:

1 The number of species on an island should eventually become roughly constant through time.

2 This should be a result of a continual *turnover* of species, with some becoming extinct and others immigrating.

3 Large islands should support more species than small islands.

4 Species number should decline with the increasing remoteness of an island.

partitioning variation between habitat diversity and area itself

On the other hand, a higher richness on larger islands would be expected simply as a consequence of larger islands having more habitat types. Does richness increase with area at a rate *greater* than could be accounted for by increases in habitat

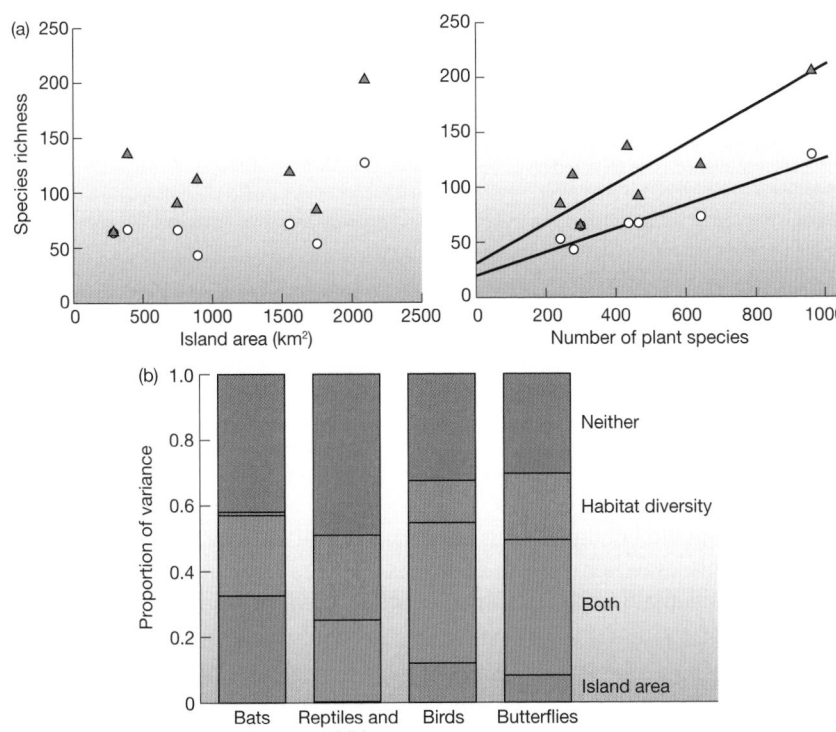

(A) AFTER BECKER, 1992; (B) AFTER RICKLEFS & LOVETTE, 1999

Figure 10.14

(a) The relationships between species richness of herbivorous (circles) and carnivorous (triangles) beetles of the Canary Islands and both island area (left) and plant species richness (right). (b) Proportion of variance in species richness, for four animal groups, among islands in the Lesser Antilles related uniquely to island area (blue), related uniquely to habitat diversity (orange), related to correlated variation between area and habitat diversity (green) and unexplained by either (maroon). Regression lines are significant at $P < 0.05$; no lines are shown in the left panel of (a) because the regression are not significant.

diversity alone? Some studies have attempted to partition species–area variation on islands into that which can be entirely accounted for in terms of habitat diversity, and that which remains and must be accounted for by island area in its own right. For beetles on the Canary Islands, the relationship between species richness and habitat diversity (as measured by plant species richness) is much stronger than that with island area, and this is particularly marked for the herbivorous beetles, presumably because of their particular food plant requirements (Figure 10.14a).

Contrasting with the Canary Island results, in a study of a variety of animal groups living on the Lesser Antilles islands in the West Indies, the variation in species richness from island to island was partitioned, statistically, into that attributable to island area alone, that attributable to habitat diversity alone, that attributable to correlated variation between area and habitat diversity (and hence not attributable to either alone) and that attributable to neither (Figure 10.14b). For reptiles and amphibians, like the beetles of the Canary Islands, habitat diversity was far more important than island area. But for bats, the reverse was the case; and for birds and butterflies, both area itself and habitat diversity had important roles to play. Overall, therefore, studies like this suggest a separate area effect (larger islands are larger targets for colonization; populations on larger islands have a lower risk of extinction) beyond a simple correlation between area and habitat diversity.

An example of species impoverishment on more remote islands can be seen in Figure 10.15 for non-marine, lowland birds on tropical islands in the southwest Pacific. With increasing distance from the large 'source' landmass of Papua New Guinea, there is a decline in the number of species, expressed as a percentage of the number present on an island of similar area but close to Papua New Guinea.

bird species richness on Pacific islands decreases with remoteness

Figure 10.15

The number of resident, non-marine, lowland bird species on islands more than 500 km from the large 'source' landmass of Papua New Guinea expressed as a percentage of the number of species on an island of equivalent area but close to Papua New Guinea – this can be thought of as the 'degree of saturation' of the bird community. It is plotted against island distance from Papua New Guinea.

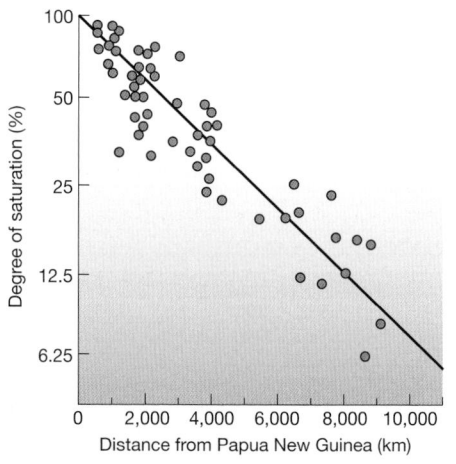

AFTER DIAMOND, 1972

species missing because of insufficient time for colonization

A more transient but nonetheless important reason for the species impoverishment of islands, especially remote islands, is the fact that many lack species that they could potentially support, simply because there has been insufficient time for the species to colonize. An example is the island of Surtsey, which emerged in 1963 as a result of a volcanic eruption. The new island, 40 km southwest of Iceland, was reached by bacteria and fungi, some sea birds, a fly and seeds of several beach plants within 6 months of the start of the eruption. Its first established vascular plant was recorded in 1965, the first moss colony in 1967 and the first bush (a dwarf willow, *Salix herbacea*) in 1998. An earthworm was found in 1993 and slugs in 1998, probably carried in by birds (Hermannsson, 2000). By 2004, more than 50 species of vascular plant, 53 mosses, 45 lichens and 300 species of invertebrate had been recorded, though not all persisted (Surtsey Research Society, website www.surtsey.is). Colonization by new species occurred both above and below the water line, with marine invertebrates, which disperse as larval stages in the ocean, accumulating faster than terrestrial plants (Figure 10.16).

evolution rates on islands may be faster than colonization rates

Finally, it is important to reiterate that no aspect of ecology can be fully understood without reference to evolutionary processes (see Chapter 2), and this is particularly true for an understanding of island communities. On isolated islands,

Figure 10.16.

Regular surveys of species richness of animals and plants have occurred since the emergence in 1963 of the volcanic Surtsey Island, near Iceland. Shown here are the results of standard surveys of coastal marine invertebrates up to 1992 (barnacles, isopods, decapods, mollusks, starfish, brittlestars, sea urchins and sea squirts; maroon circles) and of terrestrial vascular plants up to 2004 (open circles).

AFTER HERMANNSON, 2000; SURTSEY RESEARCH SOCIETY, WEBSITE WWW.SURTSEY.IS

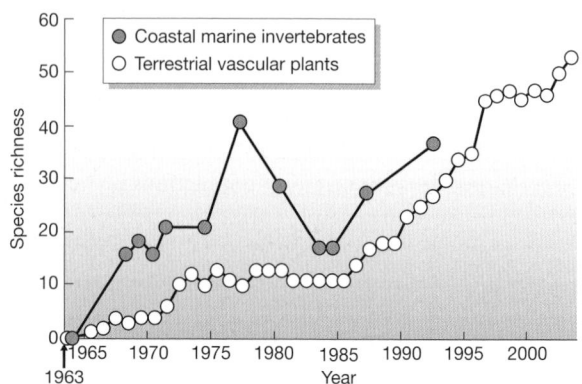

the rate at which new species evolve may be comparable to or even faster than the rate at which they arrive as new colonists. Clearly, the communities of these islands will be incompletely understood by reference only to ecological processes. Take the remarkable numbers of *Drosophila* species (fruitflies) found on the remote volcanic islands of Hawaii. There are probably about 1500 *Drosophila* species worldwide but at least 500 of these are found on the Hawaiian Islands; they have evolved, almost entirely, on the islands themselves. The communities of which they are a part are clearly much more strongly affected by local evolution and speciation than by the processes of invasion and extinction.

10.5.2 Latitudinal gradients

One of the most widely recognized patterns in species richness is the increase that occurs from the poles to the tropics. This can be seen in a wide variety of groups, including trees, marine invertebrates, butterflies and lizards (Figure 10.17). The pattern can be seen, moreover, in terrestrial, marine and freshwater habitats.

A number of explanations have been put forward for the general latitudinal trend in species richness, but not one of these is without problems. In the first place, the richness of tropical communities has been attributed to a greater intensity of predation and to more specialized predators. More intense predation could reduce the importance of competition, permitting greater niche overlap and promoting higher richness (see Figure 10.3c), but predation cannot readily be forwarded as

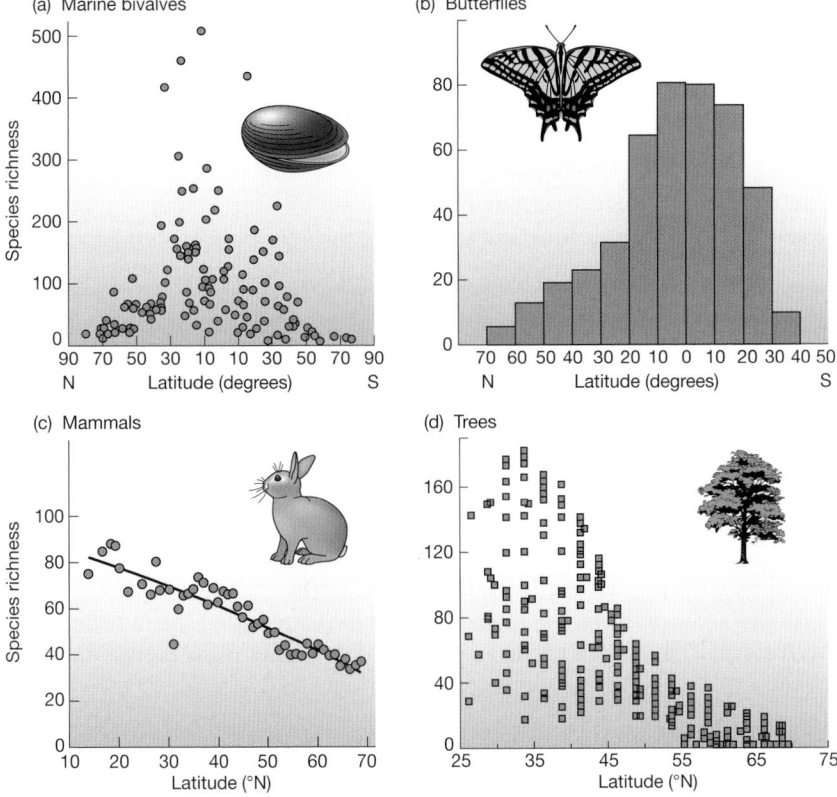

Figure 10.17

Latitudinal patterns in species richness for: (a) marine bivalves, (b) swallowtail butterflies, (c) mammals in North America, and (d) trees in North America. In each case there is a decline from low latitudes (the equator is at 0°) to high latitudes (the poles are at 90°).

(a) AFTER FLESSA & JABLONSKI, 1995; (b) AFTER SUTTON & COLLINS, 1991; (c) AFTER ROSENZWEIG & SANDLIN, 1997; (d) AFTER CURRIE & PAQUIN, 1987

the root cause of tropical richness, since this begs the question of what gives rise to the richness of the predators themselves.

productivity as an explanation?

Second, increasing species richness may be related to an increase in productivity as one moves from the poles to the equator. Certainly, on average, there is more heat and more light energy in increasingly tropical regions, and, as discussed in Section 10.3.1, both of these have tended to be associated with greater species richness, though increased productivity in at least some cases has been associated with reduced richness.

Moreover, light and heat are not the only determinants of plant productivity. Tropical soils tend, on average, to have lower concentrations of plant nutrients than temperate soils. The species-rich tropics might therefore be seen, in this sense, as reflecting their *low* productivity. In fact, tropical soils are poor in nutrients because most of the nutrients are locked up in the large tropical biomass. A productivity argument might therefore have to run as follows. The light, temperature and water regimes of the tropics lead to high biomass communities but not necessarily to diverse communities. This, though, leads to nutrient-poor soils and perhaps a wide range of light regimes from the forest floor to canopy far above. These in turn lead to high plant species richness and thus to high animal species richness. There is certainly no *simple* 'productivity explanation' for the latitudinal trend in richness.

climatic variation or evolutionary age as explanations?

Some ecologists have invoked the climate of low latitudes as a reason for their high species richness. Specifically, equatorial regions are generally less seasonal than temperate regions, and this may allow species to be more specialized (i.e. have narrower niches, see Figure 10.3b). The greater evolutionary 'age' of the tropics has also been proposed as a reason for their greater species richness, and another line of argument suggests that the repeated fragmentation and coalescence of tropical forest refugia promoted genetic differentiation and speciation, accounting for much of the high richness in tropical regions. And in a related context, we have already noted that the rate of evolution may be faster in the tropics (see Section 10.4.3). All these ideas are plausible too, but far from proven generalizations.

Overall, therefore, the latitudinal gradient lacks an unambiguous explanation. This is hardly surprising. The components of a possible explanation – trends with productivity, climatic stability and so on – are themselves understood only in an incomplete and rudimentary way, and the latitudinal gradient intertwines these components with one another, and with other, often opposing forces – isolation, harshness and so on.

10.5.3 Gradients with altitude and depth

A decrease in species richness with altitude, analogous to that observed with latitude, has frequently been reported in terrestrial environments (e.g. Figure 10.18a, b). On the other hand, some have reported an increase with altitude (e.g. Figure 10.18c), while about half the studies of altitudinal species richness have described hump-shaped patterns (e.g. Figure 10.18d) (Rahbek, 1995).

At least some of the factors instrumental in the latitudinal trend in richness are also likely to be important as explanations for altitudinal trends (though the problems in explaining the latitudinal trend apply equally to altitude). For example, declines in species richness have often been explained in terms of decreasing productivity associated with lower temperatures and shorter growing seasons at higher altitude, or physiological stress associated with climatic extremes near

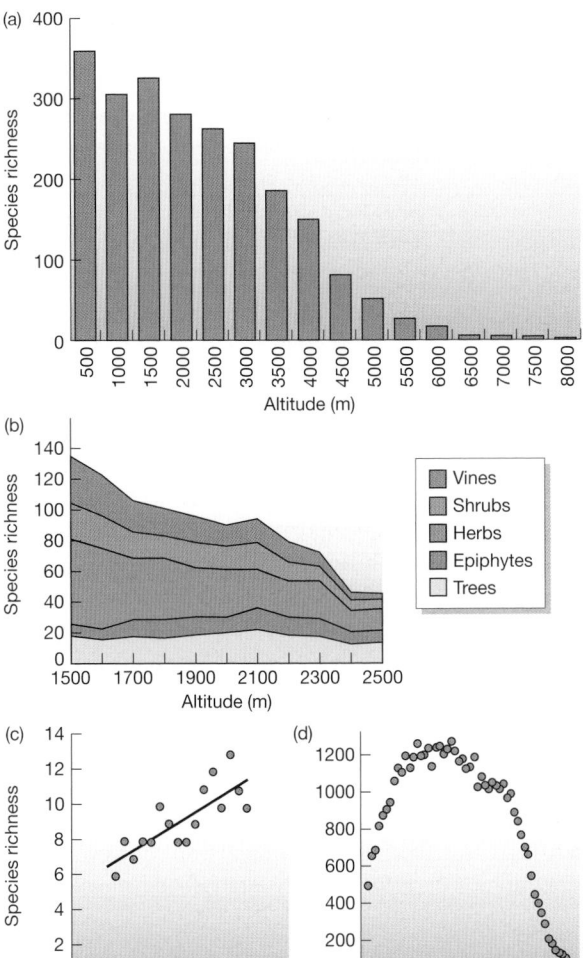

Figure 10.18

Relationships between species richness and altitude for:
(a) breeding birds in the Nepalese Himalayas, (b) plants in the
Sierra Manantlán, Mexico, (c) ants in Lee Canyon in the Spring
Mountains of Nevada, USA, and (d) flowering plants in the
Nepalese Himalayas. Species richness decreases with altitude
in (a) and (b), increases with altitude in (c) and shows a
hump-backed relationship in (d).

mountaintops. Indeed, the explanation for the converse, positive relationship
between ant diversity and altitude in Figure 10.18c, is that precipitation increased
with altitude in this case, resulting in higher productivity and less physiologically
extreme conditions at higher altitude. In addition, high-altitude communities
almost invariably occupy smaller areas than lowlands at equivalent latitudes, and
they will usually be more isolated from similar communities than lowland sites.
Therefore the effects of area and isolation are likely to contribute to observed
decreases in species richness with altitude.

In aquatic environments, the change in species richness with depth shows
some strong similarities to the terrestrial gradient with altitude. In larger lakes, the
cold, dark, oxygen-poor abyssal depths contain fewer species than the shallow
surface waters. Likewise, in marine habitats, plants are confined to the photic
zone (where light penetrates and they can photosynthesize), which rarely extends
below 30 m. In the open ocean, therefore, there is a rapid decrease in richness

Figure 10.19

Depth gradient in species richness
of bottom-dwelling vertebrates and
invertebrates (fish, decapods,
holothurians, asteroids) in the ocean
southwest of Ireland.

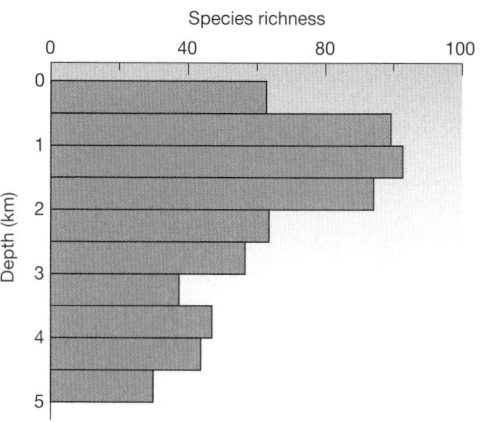

AFTER ANGEL 1994

with depth, reversed only by the variety of bizarre animals living on the ocean
floor. Interestingly, however, in coastal regions the effect of depth on the species
richness of benthic (bottom-dwelling) animals is to produce not a single gradient,
but a peak of richness at about 1000 m, possibly reflecting higher environmental
predictability there (Figure 10.19). At greater depths, beyond the continental
slope, species richness declines again, probably because of the extreme paucity of
food resources in abyssal regions.

10.5.4 Gradients during community succession

Section 9.4 described how, in community successions, if they run their full course,
the number of species first increases (because of colonization) but eventually
decreases (because of competition). This is most firmly established for plants, but
the few studies that have been carried out on animals in successions indicate, at
the least, a parallel increase in species richness in the early stages of succession.
Figure 10.20 illustrates this for birds following the cessation of shifting cultivation in tropical rain forest, and for insects associated with an old-field succession
in a temperate region.

To a certain extent, the successional gradient is a necessary consequence of
the gradual colonization of an area by species from surrounding communities
that are at later successional stages; that is, later stages are more fully saturated
with species (see Figure 10.3d). However, this is a small part of the story, since
succession involves a process of the replacement of species and not just the mere
addition of new ones.

a cascade effect?

Indeed, as with the other gradients in species richness, there is something of
a cascade effect with succession: one process that increases richness kick-starts a
second, which feeds into a third, and so on. The earliest species will be those that
are the best colonizers and the best competitors for open space. They immediately provide resources (and introduce heterogeneity) that were not previously
present. For example, the earliest plants generate resource depletion zones (see
Section 3.3.2) in the soil that inevitably increase the spatial heterogeneity of plant
nutrients. The plants themselves provide a new variety of microhabitats, and for
the animals that might feed on them they provide a much greater range of food
resources (see Figure 10.3a). The increase in herbivory and predation may then

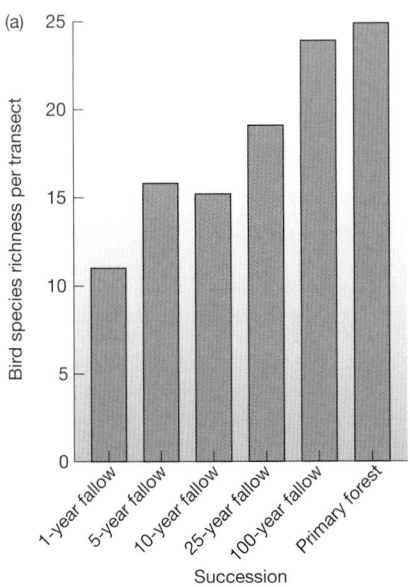

(a)

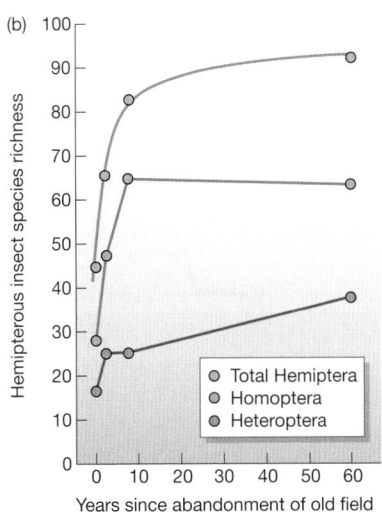

(b)

(a) AFTER SHANKAR RAMAN ET AL., 1998; (b) AFTER BROWN & SOUTHWOOD, 1983

Figure 10.20

Examples of increases in animal species richness during succession. (a) Bird species richness increased after shifting cultivation ceased in tropical rain forest in northeast India. Areas that were left fallow after being retired from cultivation for known periods were compared with the undisturbed primary forest. (b) The species richness of true bugs (insects in the suborders Homoptera and Heteroptera of the order Hemiptera) increased with time after an English farm field was taken out of cultivation.

feed back to promote further increases in species richness (predator-mediated coexistence, Figure 10.3c), which provides further resources and more heterogeneity, and so on. In addition, temperature, humidity and wind speed show much less temporal variation within a forest than in an exposed early successional stage, and the enhanced constancy of the environment may provide a stability of conditions and resources that permits specialist species to build up populations and persist (Figure 10.3b). As with the other gradients, the interaction of many factors makes it difficult to disentangle cause from effect. But with the successional gradient of richness, the tangled web of cause and effect appears to be of the essence.

10.6 Patterns in taxon richness in the fossil record

Finally, it is of interest to take the processes that are believed to be instrumental in generating present-day gradients in richness and apply them to trends occurring over much longer timespans. The imperfection of the fossil record has always been the greatest impediment to the paleontological study of evolution. Nevertheless, some general patterns have emerged, and our knowledge of six important groups of organisms is summarized in Figure 10.21.

Until about 600 million years ago, the world was populated virtually only by bacteria and algae, but then almost all the phyla of marine invertebrates entered the fossil record within the space of only a few million years (Figure 10.21a). We have seen that the introduction of a higher trophic level can increase richness at a lower level by 'exploiter-mediated coexistence'; thus, it can be argued that the first single-celled herbivorous protist was probably instrumental in the Cambrian explosion in species richness. The opening up of space by grazing on the algal monoculture, coupled with the availability of recently evolved eukaryotic cells, may have caused the biggest burst of evolutionary diversification in the planet's history.

the Cambrian explosion: exploiter-mediated coexistence?

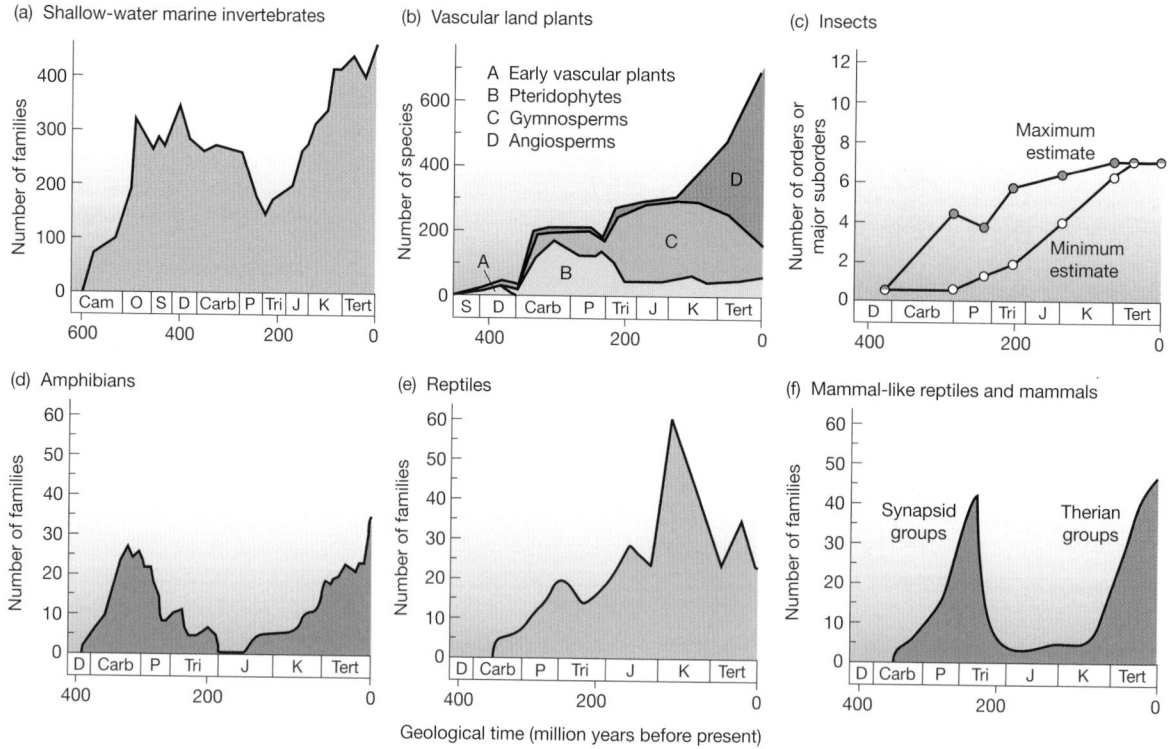

Figure 10.21

Patterns in taxon richness through the fossil record. (a) Families of shallow-water marine invertebrates. (b) Species of vascular land plants in four groups – early vascular plants, pteridophytes, gymnosperms and angiosperms. (c) Orders and major suborders of insects (minimum values are derived from definite fossil records; the maximum values include 'possible' records). (d) Families of amphibians, (e) families of reptiles and (f) families of 'mammal-like reptiles' (Synapsids) and Therian mammals (includes both marsupial and placental groups). Key to geological periods: Cam, Cambrian; O, Ordovician; S, Silurian; D, Devonian; Carb, Carboniferous; P, Permian; Tri, Triassic; J, Jurassic; K, Cretaceous; Tert, Tertiary.

the Permian decline: a species–area relationship?

In contrast, the equally dramatic decline in the number of families of shallow-water invertebrates at the end of the Permian (Figure 10.21a) could have been a result of the coalescence of the Earth's continents to produce the single super-continent of Pangaea; the joining of continents produced a marked reduction in the area occupied by shallow seas (which occur around the periphery of continents) and thus a marked decline in the area of habitat available to shallow-water invertebrates. Moreover, at this time the world was subject to a prolonged period of global cooling in which huge quantities of water were locked up in enlarged polar caps and glaciers, causing a widespread reduction of warm, shallow sea environments. Thus, a species–area relationship may be invoked to account for a reduction in taxon richness at this time.

competitive displacement among the major plant groups?

The analysis of fossil remains of vascular land plants (Figure 10.21b) reveals four distinct evolutionary phases: (i) a Silurian–mid-Devonian proliferation of early vascular plants; (ii) a subsequent late-Devonian–Carboniferous radiation of fern-like lineages (pteridophytes); (iii) the appearance of seed plants in the late Devonian and the adaptive radiation to a gymnosperm-dominated flora; and (iv) the appearance and rise of flowering plants (angiosperms) in the Cretaceous and Tertiary. It seems that after initial invasion of the land, made possible by

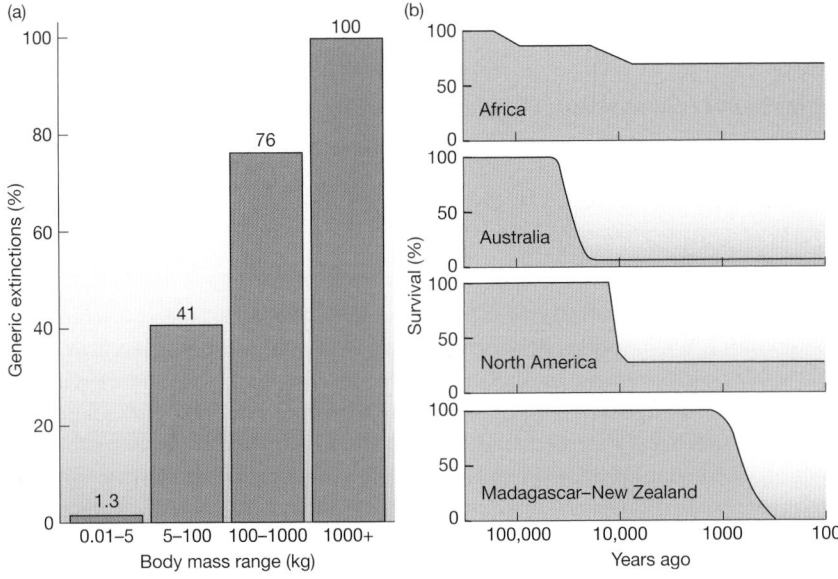

(a) AFTER OWEN-SMITH, 1987; (b) AFTER MARTIN, 1984

Figure 10.22

(a) The percentage of genera of large mammalian herbivores that have gone extinct in the last 130,000 years is strongly size-dependent (data from North and South America, Europe and Australia combined). (b) Percentage survival of large animals on three continents and two large islands (New Zealand and Madagascar). The dramatic declines in taxon richness in Australia, North America and the islands of New Zealand and Madagascar occurred at different times in history.

the appearance of roots, the diversification of each plant group coincided with a decline in species numbers of the previously dominant group. In two of the transitions (early plants to gymnosperms, and gymnosperms to angiosperms), this pattern may reflect the competitive displacement of older, less specialized taxa by newer and presumably more specialized taxa.

The first undoubtedly herbivorous insects are known from the Carboniferous. Thereafter, modern orders appeared steadily (Figure 10.21c) with the Lepidoptera (butterflies and moths) arriving last on the scene, at the same time as the rise of the angiosperms. Coevolution between plants and herbivorous insects (see Section 8.4.3) has almost certainly been, and still is, an important mechanism driving the increase in richness observed in both land plants and insects through their evolution.

Toward the end of the last ice age, the continents were much richer in large animals than they are today. For example, Australia was home to many genera of giant marsupials; North America had its mammoths, giant ground sloths and more than 70 other genera of large mammals; and New Zealand and Madagascar were home to giant flightless birds, the moas (Dinornithidae) and elephant birds (Aepyornithidae), respectively. During the past 30,000 years or so, a major loss of this biotic diversity has occurred over much of the globe. The extinctions particularly affected large terrestrial animals (Figure 10.22a), they were more pronounced in some parts of the world than others, and they occurred at different times in different places (Figure 10.22b). The extinctions mirror patterns of human migration. Thus, the arrival in Australia of ancestral aborigines occurred between 40,000 and 30,000 years ago; stone spear points became abundant throughout the United States about 11,500 years ago; and humans have been in both Madagascar and New Zealand for 1000 years. It can be convincingly argued, therefore, that the arrival of efficient human hunters led to the rapid overexploitation of vulnerable and profitable large prey. Africa, where humans originated, shows much less evidence of loss, perhaps because coevolution of

extinctions of large animals in the Pleistocene: prehistoric overkill?

large animals alongside early humans provided ample time for them to develop effective defenses (Owen-Smith, 1987).

The Pleistocene extinctions herald the modern age, in which the influence upon natural communities of human activities has been increasing dramatically.

10.7 Appraisal of patterns in species richness

richness patterns – generalizations and exceptions

There are many generalizations that can be made about the species richness of communities. We have seen how richness may peak at intermediate levels of available environmental energy or of disturbance frequency, and how richness declines with a reduction in island area or an increase in island remoteness. We find also that species richness decreases with increasing latitude, and declines or shows a hump-backed relationship with altitude or depth in the ocean. It increases with an increase in spatial heterogeneity but may decrease with an increase in temporal heterogeneity (increased climatic variation). It increases, at least initially, during the course of succession and with the passage of evolutionary time. However, for many of these generalizations important exceptions can be found, and for most of them the current explanations are not entirely adequate.

It also needs to be recognized that global patterns of species richness have been disrupted in dramatic ways by human activities, such as land-use development, pollution and the introduction of exotic species (Box 10.4).

10.4 Topical ECOncerns

The flood of exotic species

Throughout the history of the world, species have invaded new geographic areas, as a result of chance colonizations (e.g. dispersed to remote areas by wind or to remote islands on floating debris; see Section 10.5.1) or during the slow northward spread of forest trees in the centuries since the last ice age (see Section 2.5). However, human activities have increased this historical trickle to a flood, disrupting global patterns of species richness.

Some human-caused introductions are an accidental consequence of human transport. Other species have been introduced intentionally, perhaps to bring a pest under control (see Section 12.5), to produce a new agricultural product or to provide new recreational opportunities. Many invaders become part of natural communities without obvious consequences. But some have been responsible for driving native species extinct or changing natural communities in significant ways (see Section 14.2.3).

The alien plants of the British Isles illustrate a number of general points about invaders. Species inhabiting areas where people live and work are more likely to be transported to new regions, where they will tend to be deposited in habitats like those where they originated. As a result, more alien species are found in disturbed habitats close to human transport centers (docks, railways, cities) and fewer in remote mountain areas (Figure 10.23a). Moreover, more invaders to the British Isles are likely to arrive from nearby geographic locations (e.g. Europe) or

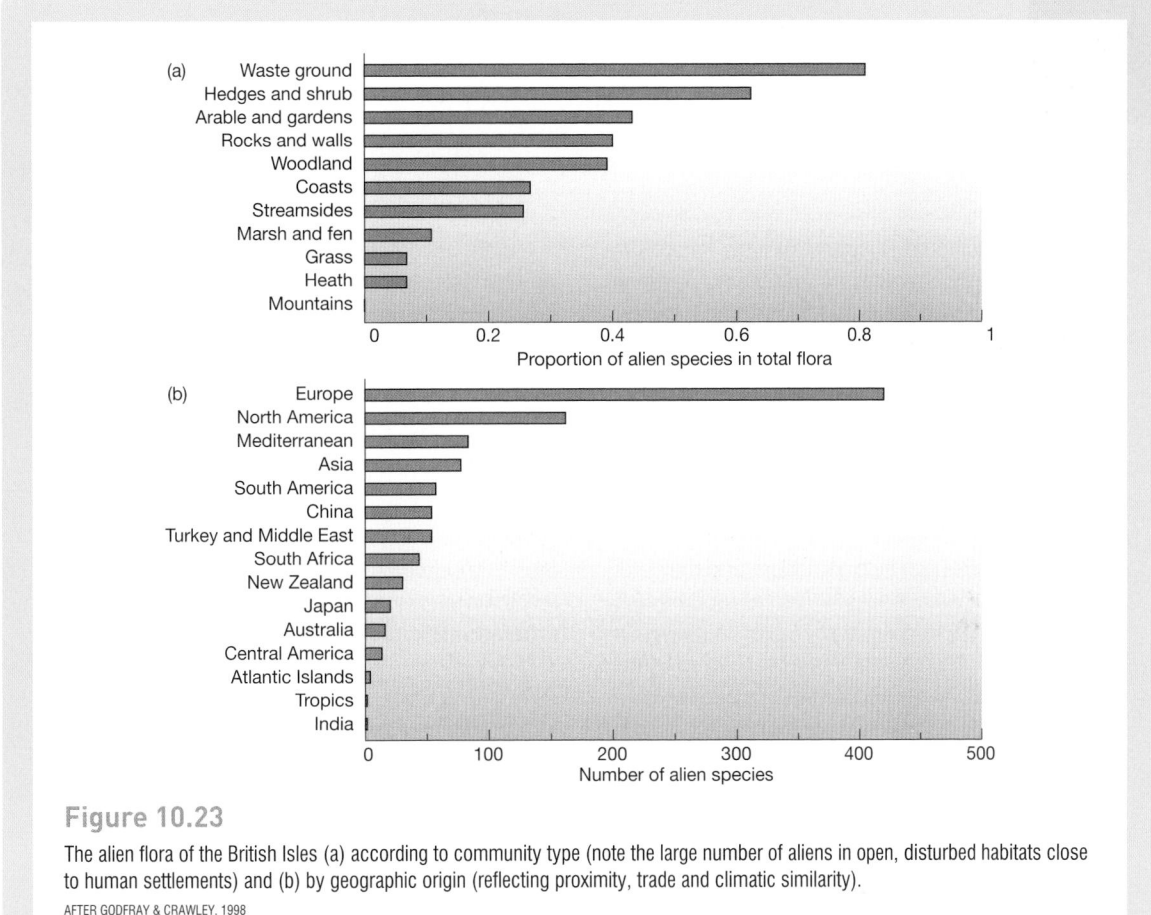

Figure 10.23

The alien flora of the British Isles (a) according to community type (note the large number of aliens in open, disturbed habitats close to human settlements) and (b) by geographic origin (reflecting proximity, trade and climatic similarity).

AFTER GODFRAY & CRAWLEY, 1998

from remote locations whose climate matches that of Britain (e.g. New Zealand) (Figure 10.23b). Note the small number of alien plants from tropical environments; these species usually lack the frost-hardiness required to survive the British winter.

Review the options available to governments to prevent (or reduce the likelihood) of invasions of undesirable alien species.

Unraveling richness patterns is one of the most difficult and challenging areas of modern ecology. Clear, unambiguous predictions and tests of ideas are often very difficult to devise and will require great ingenuity of future generations of ecologists. Because of the increasing importance of recognizing and conserving the world's biological diversity, though, it is crucial that we come to understand thoroughly these patterns in species richness. We will assess the adverse effects of human activities, and how they may be remedied, in Chapters 12–14.

Summary

Richness and diversity

The number of species in a community is referred to as its species richness. Richness, though, ignores the fact that some species are rare and others common. Diversity indices are designed to combine species richness and the evenness of the distribution of individuals among those species. Attempts to describe a complex community structure by one single attribute, such as richness or diversity, can still be criticized because so much valuable information is lost. A more complete picture is therefore sometimes provided in a rank–abundance diagram.

A simple model can help us understand the determinants of species richness. Within it, a community will contain more species the greater the range of resources, if the species are more specialized in their use of resources, if species overlap to a greater extent in their use of resources, or if the community is more fully saturated.

Productivity and resource richness

If higher productivity is correlated with a wider range of available resources, then this is likely to lead to an increase in species richness, but more of the same might lead to more individuals per species rather than more species. In general, though, species richness often increases with the richness of available resources and productivity, although in some cases the reverse has been observed – the paradox of enrichment – and others have found species richness to be highest at intermediate levels of productivity.

Predation intensity

Predation can exclude certain prey species and reduce richness or permit more niche overlap and thus greater richness (predator-mediated coexistence). Overall, therefore, there may be a humped relationship between predation intensity and species richness in a community, with greatest richness at intermediate intensities.

Spatial heterogeneity

Environments that are more spatially heterogeneous often accommodate extra species because they provide a greater variety of microhabitats, a greater range of microclimates, more types of places to hide from predators and so on – the resource spectrum is increased.

Environmental harshness

Environments dominated by an extreme abiotic factor – often called harsh environments – are more difficult to recognize than might be immediately apparent. Some apparently harsh environments do support few species, but any overall association has proved extremely difficult to establish.

Climatic variation

In a predictable, seasonally changing environment, different species may be suited to conditions at different times of the year. More species might therefore be expected to coexist than in a completely constant environment. On the other hand, opportunities for specialization (e.g. obligate fruit-eating) exist in a non-seasonal environment that are not available in a seasonal environment. Unpredictable climatic variation (climatic instability) could decrease richness by denying species the chance to specialize, or increase richness by preventing competitive exclusion. There is no established relationship between climatic instability and species richness.

Disturbance

The intermediate disturbance hypothesis suggests that very frequent disturbances keep most patches at an early stage of succession (where there are few species), but very rare disturbances allow most patches to become dominated by the best competitors (where there are also few species). Originally proposed to account for patterns of richness in tropical rain forests and coral reefs, the hypothesis has occupied a central place in the development of ecological theory.

Environmental age: evolutionary time

It has often been suggested that communities may differ in richness because some are closer to equilibrium and therefore more saturated than others, and that the tropics are rich in species in part because the tropics have existed over long and uninterrupted periods of evolutionary time. A simplistic contrast between the unchanging tropics and the disturbed and recovering temperate regions, however, is untenable.

Habitat area and remoteness: island biogeography

Islands need not be islands of land in a sea of water. Lakes are islands in a sea of land; mountaintops are high-altitude islands in a low-altitude ocean. The number of species on islands decreases as island area decreases, in part because larger areas typically encompass more different types of habitat. However, MacArthur and Wilson's equilibrium theory of island biogeography argues for a separate island effect based on a balance between immigration and extinction, and the theory has received much support. In addition, on isolated islands especially, the rate at which new species evolve may be comparable to or even faster than the rate at which they arrive as new colonists.

Gradients in species richness

Richness increases from the poles to the tropics. Predation, productivity, climatic variation and the greater evolutionary age of the tropics have been put forward as partial explanations.

In terrestrial environments, richness often (but not always) decreases with altitude. Factors instrumental in the latitudinal trend are also likely to be important in this, but so are area and isolation. In aquatic environments, richness usually decreases with depth for similar reasons.

In successions, if they run their full course, richness first increases (because of colonization) but eventually decreases (because of competition). There may also be a cascade effect: one process that increases richness kick-starts a second, which feeds into a third, and so on.

Patterns in taxon richness in the fossil record

The Cambrian explosion of taxa may have been an example of exploiter-mediated coexistence. The Permian decline may reflect a species–area relationship when the Earth's continents coalesced into Pangaea. The changing pattern of plant taxa may reflect the competitive displacement of older, less specialized taxa by newer, more specialized ones. The extinctions of many large animals in the Pleistocene may reflect the hand of human predation and hold lessons for the present day.

Review questions

REVIEW QUESTIONS

Asterisks indicate challenge questions

1 Explain species richness, diversity index and rank–abundance diagrams and compare what each measures.

2 What is the paradox of enrichment, and how can the paradox be resolved?

3 Explain, with examples, the contrasting effects that predation can have on species richness.

4* Researchers have reported a variety of hump-shaped patterns in species richness, with peaks of richness occurring at intermediate levels of productivity, predation pressure, disturbance and depth in the ocean. Review the evidence and consider whether these patterns have any underlying mechanisms in common.

5 Why is it so difficult to identify 'harsh' environments?

6   Explain the intermediate disturbance
    hypothesis.

7   Islands need not be islands of land in an ocean
    of water. Compile a list of other types of habitat
    islands over as wide a range of spatial scales
    as possible.

8* An experiment was carried out to try to
    separate the effects of habitat diversity and
    area on arthropod species richness on some
    small mangrove islands in the Bay of Florida.
    These consisted of pure stands of the
    mangrove species *Rhizophora mangle*, which
    support communities of insects, spiders,
    scorpions and isopods. After a preliminary
    faunal survey, some islands were reduced
    in size by means of a power saw and brute
    force! Habitat diversity was not affected, but
    arthropod species richness on three islands
    nonetheless diminished over a period of
    2 years (Figure 10.24). A control island, the size
    of which was unchanged, showed a slight
    increase in richness over the same period.
    Which of the predictions of island biogeography
    theory are supported by the results in the
    figure? What further data would you require to
    test the other predictions? How would you
    account for the slight increase in species
    richness on the control island?

9* A cascade effect is sometimes proposed
    to explain the increase in species richness

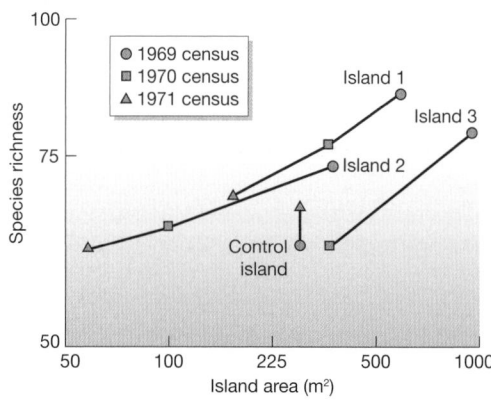

Figure 10.24

The effect on the number of arthropod species of artificially
reducing the size of three mangrove islands. Islands 1 and 2
were reduced in size after both the 1969 and 1970 censuses.
Island 3 was reduced only after the 1969 census. The control
island was not reduced, and the change in its species
richness was attributable to random fluctuations.

AFTER SIMBERLOFF, 1976

during a community succession. How might
a similar cascade concept apply to the
commonly observed gradient of species
richness with latitude?

10 Describe how theories of species richness that
    have been derived on ecological time scales
    can also be applied to patterns observed in
    the fossil record.

Chapter 11

The flux of energy and matter through ecosystems

Key concepts

KEY CONCEPTS

In this chapter, you will:

- recognize that communities are intimately linked with the abiotic environment by fluxes of energy and matter
- understand that net primary productivity is not evenly spread across the Earth
- appreciate that transfer of energy between trophic levels is always inefficient – secondary productivity by herbivores is approximately an order of magnitude less than the primary productivity on which it is based
- recognize that much more of a community's energy and matter passes through the decomposer system than the live consumer system
- appreciate that decomposition results in complex, energy-rich molecules being broken down by their consumers (decomposers and detritivores) into carbon dioxide, water and inorganic nutrients
- understand that in global geochemical cycles, nutrients are moved over vast distances by winds in the atmosphere and in the moving waters of streams and ocean currents

Like all biological entities, ecological communities require matter for their construction and energy for their activities. We need to understand the routes by which matter and energy enter and leave ecosystems, how they are transformed into plant biomass and how this fuels the rest of the community – bacteria and fungi, herbivores, detritivores and their consumers.

11.1 Introduction

All biological entities require matter for their construction and energy for their activities. This is true not only for individual organisms, but also for the populations and communities that they form in nature. The intrinsic importance of fluxes of energy and of matter means that community processes are particularly strongly linked with the abiotic environment. The term *ecosystem* is used to denote the biological community *together with* the abiotic environment in which it is set. Thus, ecosystems normally include primary producers, decomposers and detritivores, a pool of dead organic matter, herbivores, carnivores and parasites *plus* the physicochemical environment that provides living conditions and acts both as a source and a sink for energy and matter. It was Lindeman (1942) who laid the foundations for ecological energetics, a science with profound implications both for understanding ecosystem processes and for human food production (Box 11.1).

the standing crop and primary and secondary productivity

In order to examine ecosystem processes, it is important to understand some key terms.

- *Standing crop*. The bodies of the living organisms within a unit area constitute a standing crop of biomass.
- *Biomass*. By biomass we mean the mass of organisms per unit area of ground (or water) and this is usually expressed in units of energy (e.g. joules per square meter) or dry organic matter (e.g. tonnes per hectare). In practice we include in biomass all those parts, living or dead, that are attached to the living organism. Thus, it is conventional to regard the whole body of a tree as biomass, despite the fact that most of the wood is dead. Organisms (or their parts) cease to be regarded as biomass when they die (or are shed) and become components of dead organic matter.
- *Primary productivity*. The primary productivity of a community is the rate at which biomass is produced *per unit area* by plants, the primary producers. It can be expressed either in units of energy (e.g. joules per square meter per day) or of dry organic matter (e.g. kilograms per hectare per year).
- *Gross primary productivity*. The total fixation of energy by photosynthesis is referred to as gross primary productivity (GPP). A proportion of this, however, is respired away by the plant itself and is lost from the community as respiratory heat (R).

11.1 Historical landmarks

Ecological energetics and the biological basis of productivity and human welfare

A classic paper by Lindeman (1942) laid the foundations of a science of ecological energetics. He attempted to quantify the concept of food chains and food webs by considering the efficiency of transfer between trophic levels – from incident radiation received by a community through its capture by green plants in photosynthesis to its subsequent use by bacteria, fungi and animals.

Lindeman's paper was a major catalyst that stimulated the International Biological Programme (IBP for short). The subject of the IBP was 'the biological basis of productivity and human welfare'. Given the problem of a rapidly increasing human population, it was recognized that scientific knowledge would be required for rational resource management. Cooperative international research programs focused on the ecological energetics of areas of land, fresh waters and the seas. The IBP provided the first occasion on which biologists throughout the world were challenged to work together towards a common end.

More recently, another pressing issue has galvanized the ecological community into action. Deforestation, the burning of fossil fuels and other human influences are causing dramatic changes to global climate and atmospheric composition, and can be expected in turn to influence patterns of productivity and the composition of vegetation on a global scale. Among the prime objectives of the International Geosphere-Biosphere Programme (IGBP), established in the early 1990s, was to predict the effects of changes in climate and atmospheric composition on agriculture and food production. The Food and Agriculture Organization (FAO) of the United Nations reported recently that some of the predicted changes seemed to be advancing at a higher rate than anticipated, including:

1 A likely decline in precipitation in some food-insecure areas such as southern Africa and the northern region of Latin America.
2 Changes in seasonal distribution of rainfall, with less falling in the main crop-growing season.
3 Higher night-time temperatures, which may adversely affect grain production.
4 Disruption of food supply through more frequent and severe extreme weather events.

We will see in this chapter why changes to water availability and temperature, among other factors, can have such profound effects on productivity.

- *Net primary productivity*. The difference between GPP and R is known as net primary productivity (NPP) and represents the actual rate of production of new biomass that is available for consumption by heterotrophic organisms (bacteria, fungi and animals).
- *Secondary productivity*. The rate of production of biomass by heterotrophs is called secondary productivity.

A proportion of primary production is consumed by herbivores, which, in turn, are consumed by carnivores. These constitute the *live consumer system*. The fraction of NPP that is not eaten by herbivores passes through the *decomposer system*. We distinguish two groups of organisms responsible for the decomposition of dead organic matter (detritus): bacteria and fungi are called *decomposers* while animals that consume dead matter are known as *detritivores*.

live consumer systems and decomposer systems

11.2 Primary productivity

11.2.1 Geographic patterns in primary productivity

The functioning of the biota of the Earth, and of the communities across the surface of the planet, depend crucially on the levels of productivity that plants are able to achieve. The total NPP of the planet is estimated to be about 105 petagrams of carbon per year (1 Pg = 10^{15} g). Of this, 56.4 Pg C yr^{-1} is produced in terrestrial ecosystems and 48.3 Pg C yr^{-1} in aquatic ecosystems (Table 11.1). Thus, although oceans cover about two-thirds of the world's surface, they account for less than half of its production and most of the ocean is, in effect, a marine desert. On the land, tropical rain forests and savannas account between them for about 60% of terrestrial NPP, reflecting the large areas covered by these biomes and their high levels of productivity.

In the forest biomes of the world, there is a general latitudinal trend of increasing productivity from boreal (1019–1034 g C m^{-2} yr^{-1}), through temperate (1327–1499 g C m^{-2} yr^{-1}) to tropical (> 3000 g C m^{-2} yr^{-1}) forest (Falge et al., 2002). A similar latitudinal trend has been reported for tundra and grassland communities, various cultivated crops and lakes. Despite considerable variation, these general trends with latitude suggest that radiation (a resource) and temperature (a condition) may be the factors usually limiting the productivity of communities. Other factors can, however, constrain productivity within even narrower limits. In the sea, where no latitudinal trend has been reported, productivity is more often limited by a shortage of nutrients.

11.2.2 Factors limiting primary productivity

What, then, limits primary productivity? In terrestrial communities, solar radiation, carbon dioxide, water and soil nutrients are the resources required for primary production while temperature, a condition, has a strong influence on the

Table 11.1

Net primary production (NPP) per year summed for each of the major biomes and for the planet in total (in units of petragrams of carbon).

MARINE	NPP	TERRESTRIAL	NPP
Tropical and subtropical oceans	13.0	Tropical rain forests	17.8
Temperate oceans	16.3	Broadleaf deciduous forests	1.5
Polar oceans	6.4	Mixed broad/needleleaf forests	3.1
Coastal	10.7	Needleleaf evergreen forests	3.1
Salt marsh/estuaries/seaweed	1.2	Needleleaf deciduous forests	1.4
Coral reefs	0.7	Savannas	16.8
		Perennial grasslands	2.4
		Broadleaf shrubs with bare soil	1.0
		Tundra	0.8
		Desert	0.5
		Cultivation	8.0
Total	48.3	Total	56.4

FROM GEIDER ET AL., 2001

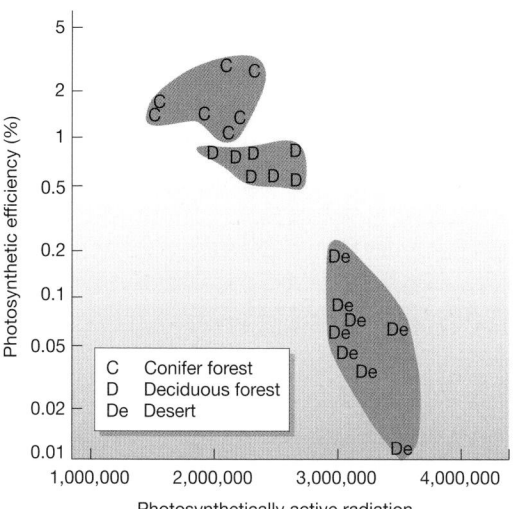

Figure 11.1

Photosynthetic efficiency (percentage of incoming photosynthetically active radiation converted to above-ground net primary production) for three sets of terrestrial communities in the United States. Desert ecosystems receive the greatest levels of radiation, but are much less efficient than forests in converting it to biomass.

rate of photosynthesis. Carbon dioxide is normally present at a level of around 0.03% of atmospheric gases and seems to play no significant role in determining differences between the productivities of different communities (although global increases in carbon dioxide concentration may bring big changes; Kicklighter et al., 1999). On the other hand, the intensity of radiation, the availability of water and nutrients, and temperature all vary dramatically from place to place. They are all candidates for the role of limiting factor. Which of them actually sets the limit to primary productivity?

Depending on location, something between 0 and 5 J of solar energy strike each square meter of the Earth's surface every minute. If all this were converted by photosynthesis to plant biomass (that is, if photosynthetic efficiency was 100%) there would be a prodigious generation of plant material, ten to a hundred times greater than recorded values. However, only about 44% of incident shortwave radiation occurs at wavelengths suitable for photosynthesis. Yet, even when this is taken into account, productivity still falls well below the maximum possible. For example, the conifer communities shown in Figure 11.1 had the highest net photosythetic efficiencies, but these were only between 1% and 3%. For a similar level of incoming radiation, deciduous forests achieved 0.5–1%, and, despite their greater energy income, deserts managed only 0.01–0.2%. These can be compared with short-term peak efficiencies achieved by crop plants under ideal conditions, when values from 3% to 10% can be achieved.

terrestrial communities use radiation inefficiently

There is no doubt that available radiation would be used more efficiently if other resources were in abundant supply. The much higher values of community productivity from agricultural systems bear witness to this. Shortage of water – an essential resource both as a constituent of cells and for photosynthesis – is often the critical factor. It is not surprising, therefore, that the rainfall of a region is quite closely correlated with its productivity (Figure 11.2a). There is also a clear relationship between NPP and mean annual temperature, but note that high temperature is associated with rapid transpiration, and thus higher temperatures increase the rate at which water shortage becomes important. Water shortage has direct effects on the rate of plant growth, but it also leads to less dense vegetation.

water and temperature as critical factors

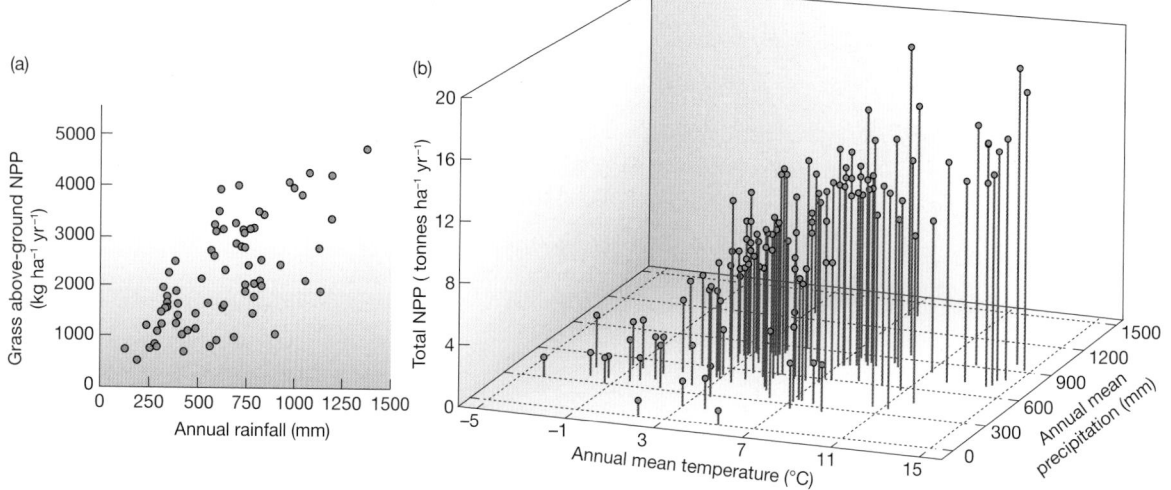

Figure 11.2

(a) Above-ground net primary productivity (NPP) of grass in savanna regions of the world in relation to annual rainfall. (b) Total NPP in relation to both annual precipitation and temperature on the Tibetan Plateau for ecosystems including forests, woodlands, shrublands, grasslands and desert.

(a) AFTER HIGGINS ET AL., 2000; (b) AFTER LUO ET AL., 2002

Vegetation that is sparse intercepts less radiation (much of which falls on bare ground), accounting for much of the difference in productivity between desert vegetation and forest in Figure 11.1. Figure 11.2b plots the NPP for a variety of ecosystem types against both temperature and annual rainfall – highest productivity occurs where temperature and rainfall are both high.

NPP increases with the length of the growing season

The productivity of a community can be sustained only for that part of the year when the plants bear photosynthetically active foliage. Deciduous trees have a self-imposed limit on the period of the year during which they bear foliage, while evergreen trees hold a canopy throughout the year. However, for much of the year conifer forest may barely photosynthesize at all, a pattern that is particularly marked in the colder boreal zones (Figure 11.3).

NPP may be low because appropriate mineral resources are deficient

No matter how brightly the sun shines, how often the rain falls and how equable the temperature, productivity must be low if there is no soil in a terrestrial community, or if the soil is deficient in essential mineral nutrients. Of all the mineral nutrients, the one with the strongest influence on community productivity is fixed nitrogen (in contrast to atmospheric nitrogen, which is not directly available for use in photosynthesis; fixed nitrogen occurs in inorganic ions such as nitrate). There is probably no agricultural or forestry system that does not respond to the application of nitrogen by increasing primary productivity, and this may well be true of natural vegetation as well. The deficiency of other elements, particularly phosphorus, can also hold the productivity of a community far below what is theoretically possible.

a succession of factors may limit primary productivity through the year

In fact, in the course of a year, the productivity of a terrestrial community may be limited by a succession of factors. The primary productivity of grasslands may be far below the theoretical maximum because the winters are too cold and the intensity of radiation is low, the summers are too dry, the rate of nitrogen supply is too slow, or because heavy grazing reduces the standing crop of photosynthetic leaves and much of the incident radiation falls on bare ground.

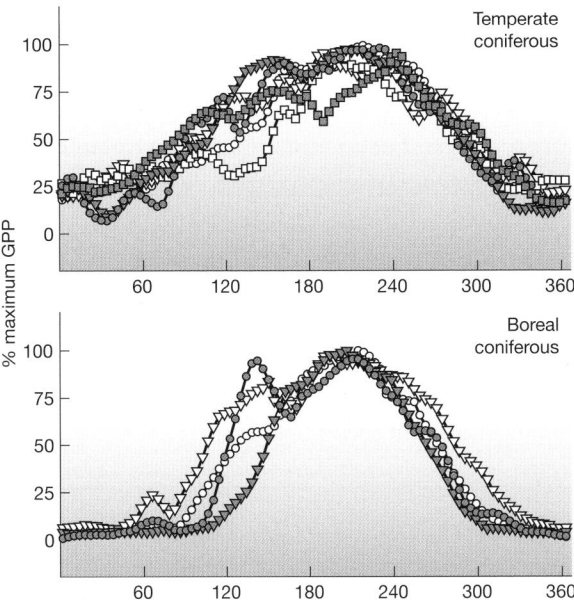

Figure 11.3

Seasonal development of maximum daily gross primary productivity (GPP) for conifer forests in temperate (Europe and North America) and boreal locations (Canada, Scandinavia and Iceland). The different symbols in each panel relate to different forests. Daily GPP is expressed as the percentage of the maximum achieved in each forest during the 365 days of the year. Note the extended periods with no photosynthesis in the colder boreal locations.

In aquatic communities, the factors that most frequently limit primary productivity are the availability of nutrients (particularly nitrate and phosphate) and the intensity of solar radiation that penetrates the column of water. Productive aquatic communities occur where, for one reason or another, nutrient concentrations are high (as for the lakes in Figure 11.4a). Lakes receive nutrients by the

productive aquatic communities occur where nutrient concentrations are high

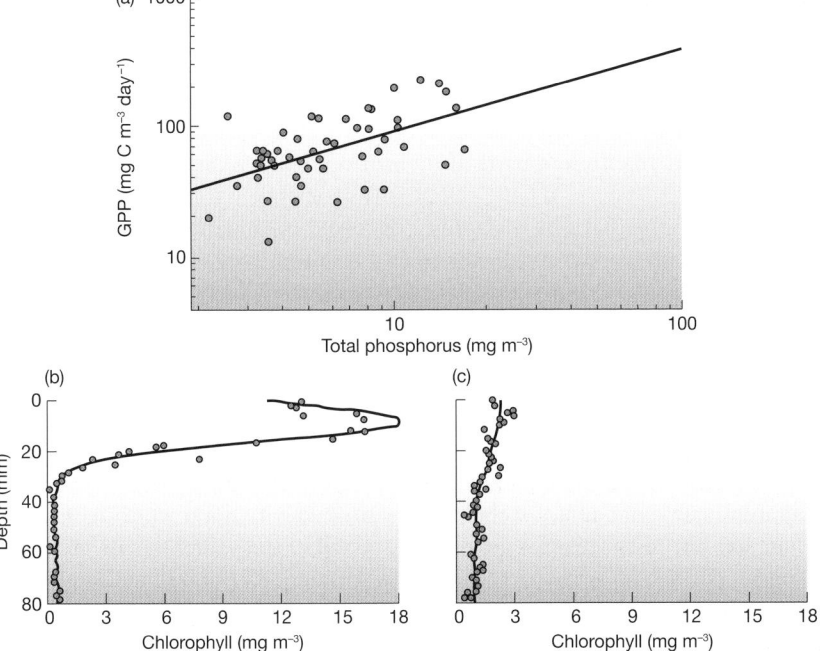

Figure 11.4

(a) The relationship between gross primary productivity (GPP) of phytoplankton (microscopic plants) and phosphorus concentration in some Canadian lakes. (b, c) Examples of vertical chlorophyll profiles recorded in the ocean off the coast of Namibia. The biomass of chlorophyll is an index of NPP of ocean phytoplankton. (b) A location associated with ocean upwelling: the nutrient-rich water fuels very high NPP by phytoplankton near the surface, but the dense phytoplankton cells reduce light penetration so that NPP is not detectable in deeper water. (c) A location where nutrient concentrations are much lower: NPP is thus low, but because light can penetrate more deeply, NPP can be detected to a greater depth. All regression lines are statistically significant.

weathering of rocks and soils in their catchment areas, in rainfall and as a result of human activity (fertilizers and sewage input; see Chapter 13); lakes vary considerably in nutrient availability.

In the oceans, locally high levels of primary productivity are associated with high nutrient inputs from two sources. First, nutrients may flow continuously into coastal shelf regions from estuaries. Productivity in the inner shelf region is particularly high because nutrient concentrations are high and the relatively clear water provides a reasonable depth within which net photosynthesis is positive (the *euphotic zone*). Closer to land, the water is richer in nutrients but highly turbid and its productivity is less. The least productive zones are in the open ocean where, although the water is clear and the euphotic zone is deep, there are generally extremely low concentrations of nutrients. Local regions of high productivity occur in the open ocean only where there are upwellings from deep, nutrient-rich water (compare Figure 11.4b and c).

11.3 The fate of primary productivity

Fungi, animals and most bacteria are heterotrophs: they derive their matter and energy either directly by consuming plant material or indirectly from plants by eating other heterotrophs. Plants, the primary producers, comprise the first trophic level in a community; primary consumers occur at the second trophic level; secondary consumers (carnivores) at the third, and so on.

11.3.1 The relationship between primary and secondary productivity

there is a general positive relationship between primary and secondary productivity

Since secondary productivity depends on primary productivity, we should expect a positive relationship between the two variables in communities. Figure 11.5 illustrates this general relationship in aquatic and terrestrial examples. Secondary productivity by zooplankton (small animals in the open water), whose main food is phytoplankton cells, is positively related to phytoplankton productivity in a range of lakes in different parts of the world (Figure 11.5a). The productivity of heterotrophic bacteria in lakes and oceans also parallels that of phytoplankton (Figure 11.5b); the bacteria metabolize dissolved organic matter released from intact phytoplankton cells or produced as a result of 'messy feeding' by grazing animals. Figure 11.5c shows how the abundance achieved by caterpillars (larvae of moths and butterflies) is tightly linked to annual rainfall (and thus primary productivity) on an island in the Galapagos Archipelago. One of Darwin's famous finches, the seed-eating *Geospiza fortis* (see Figure 2.14), also responds to increased plant production in wet years by raising significantly more broods of young (Grant et al., 2000).

most of the primary productivity does not pass through the grazer system

In both aquatic and terrestrial communities, secondary productivity by herbivores is approximately one-tenth of the primary productivity upon which it is based. Where has the missing energy gone? First, not all of the plant biomass produced is consumed alive by herbivores. Much dies without being grazed and supports a community of decomposers (bacteria, fungi and detritivorous animals). Second, not all the plant biomass eaten by herbivores (nor herbivore biomass eaten by

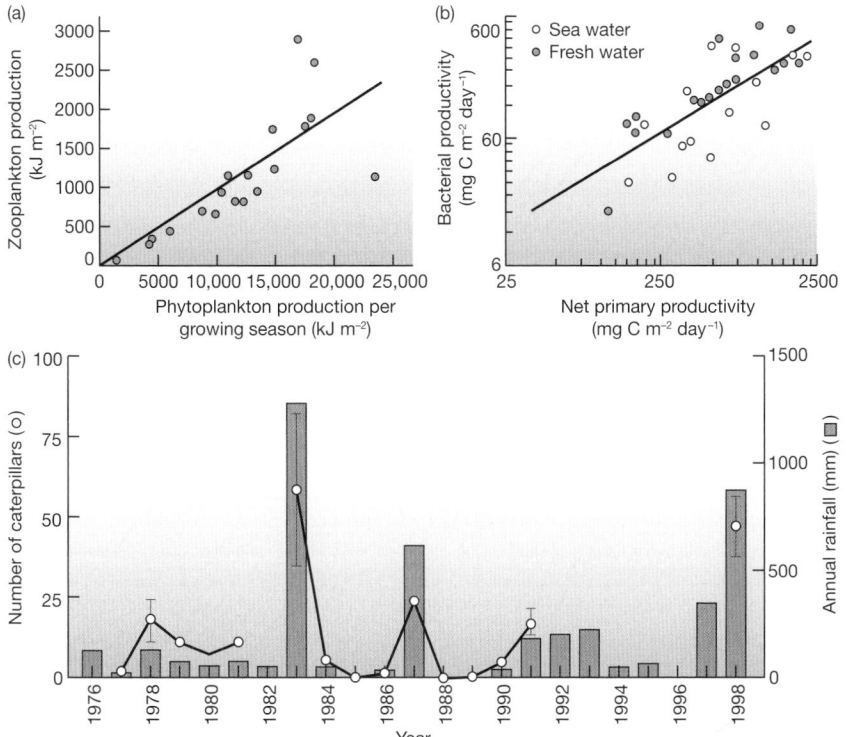

(a) AFTER BRYLINSKY & MANN, 1973; (b) AFTER COLE ET AL., 1988; (c) AFTER GRANT ET AL., 2000

Figure 11.5

The relationship between primary and secondary productivity for: (a) zooplankton in lakes, (b) bacteria in fresh and sea water, and (c) caterpillars (numbers and standard errors from a standard census) in relation to a histogram of annual rainfall on the Galapagos island of Daphne Major. Caterpillar numbers are an index of their annual secondary productivity; the primary productivity of plants, upon which the caterpillars feed, is closely correlated with annual rainfall. Regression lines are significant and caterpillar abundance is significantly correlated with annual rainfall at $P < 0.05$.

carnivores) is assimilated and available for incorporation into consumer biomass. Some is lost in feces, and this also passes to the decomposers. Third, not all the energy that has been assimilated is actually converted to biomass. A proportion is lost as respiratory heat. This occurs both because no energy conversion process is 100% efficient (some is lost as unusable random heat, consistent with the second law of thermodynamics) and also because the organisms do work that requires energy, again released as heat. These three energy pathways occur at all trophic levels and are illustrated in Figure 11.6.

11.3.2 The fundamental importance of energy transfer efficiencies

A unit of energy (a joule) may be consumed and assimilated by an invertebrate herbivore that uses part of it to do work and loses it as respiratory heat. Or it might be consumed by a vertebrate herbivore and later be assimilated by a carnivore that dies and enters the dead organic matter compartment. Here, what remains of the joule may be assimilated by a fungus and consumed by a soil mite, which uses it to do work, dissipating a further part of the joule as heat. At each consumption step, what remains of the joule may fail to be assimilated and pass in the feces to dead organic matter, or it may be assimilated and respired, or assimilated and incorporated into growth of body tissue (or the production of offspring). The body may die and what remains of the joule enters the dead organic matter compartment, or it may be captured alive by a consumer in the next trophic level where it meets a further set of possible branching pathways. Ultimately, each

possible pathways of a joule of energy through a community

Figure 11.6

The pattern of energy flow through a trophic compartment (represented as the maroon box).

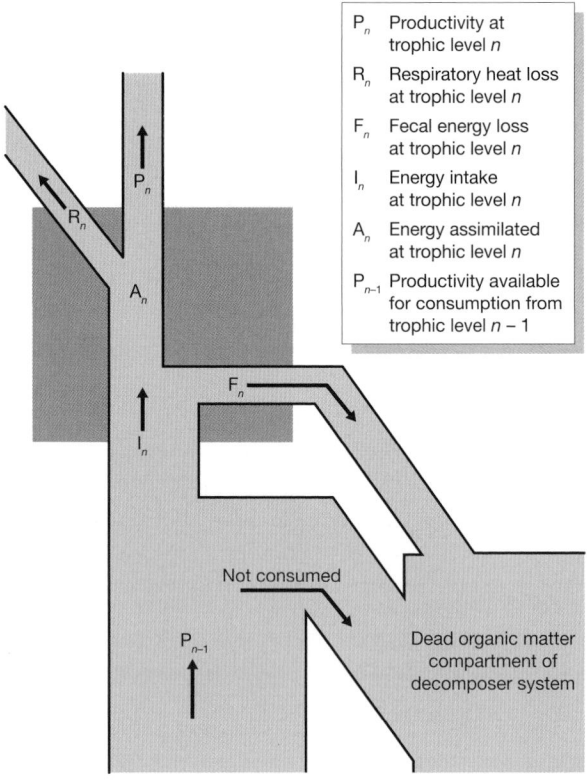

P_n	Productivity at trophic level n
R_n	Respiratory heat loss at trophic level n
F_n	Fecal energy loss at trophic level n
I_n	Energy intake at trophic level n
A_n	Energy assimilated at trophic level n
P_{n-1}	Productivity available for consumption from trophic level $n-1$

joule will have found its way out of the community, dissipated as respiratory heat at one or more of the transitions in its path along the food chain. Whereas a molecule or ion may cycle endlessly through the food chains of a community, energy passes through just once.

The possible pathways in the herbivore/carnivore (live consumer) and decomposer systems are the same, with one critical exception – feces and dead bodies are lost to the former (and enter the decomposer system), but feces and dead bodies from the decomposer system are simply sent back to the dead organic matter compartment at its base. Thus, the energy available as dead organic matter may finally be completely metabolized – and all the energy lost as respiratory heat – even if this requires several circuits through the decomposer system. The exceptions to this are situations: (i) where matter is exported out of the local environment to be metabolized elsewhere, for example detritus being washed out of a stream; and (ii) where local abiotic conditions have inhibited decomposition and left pockets of incompletely metabolized high-energy matter, otherwise known as oil, coal and peat.

consumption, assimilation and production efficiencies determine the relative importance of energy pathways

The proportions of net primary production flowing along each of the possible energy pathways depend on *transfer efficiencies* from one step to the next. We need to know about just three categories of transfer efficiency to be able to predict the pattern of energy flow. These are consumption efficiency (CE), assimilation efficiency (AE) and production efficiency (PE).

Consumption efficiency is the percentage of total productivity available at one trophic level that is consumed ('ingested') by the trophic level above. For primary consumers, CE is the percentage of joules produced per unit time as NPP that finds its way into the guts of herbivores. In the case of secondary consumers, it is the percentage of herbivore productivity eaten by carnivores. The remainder dies without being eaten and enters the decomposer system. Reasonable average figures for CE by herbivores are approximately 5% in forests, 25% in grasslands and 50% in phytoplankton-dominated communities. As far as carnivores are concerned, vertebrate predators may consume 50–100% of production from vertebrate prey but perhaps only 5% from invertebrate prey, while invertebrate predators consume perhaps 25% of available invertebrate prey production.

Assimilation efficiency is the percentage of food energy taken into the guts of consumers in a trophic level that is assimilated across the gut wall and becomes available for incorporation into growth or to do work. The remainder is lost as feces and enters the decomposer system. An 'assimilation efficiency' is much less easily ascribed to microorganisms, where food does not pass through a 'gut' and feces are not produced. Bacteria and fungi digest dead organic matter externally and, between them, typically absorb almost all the product: they are often said to have AEs of 100%. AEs are typically low for herbivores, detritivores and microbivores (20–50%) and high for carnivores (around 80%). The way that plants allocate production to roots, wood, leaves, seeds and fruits also influences their usefulness to herbivores. Seeds and fruits may be assimilated with efficiencies as high as 60–70%, and leaves with about 50% efficiency, while the AE for wood may be as low as 15%.

Production efficiency is the percentage of assimilated energy that is incorporated into new biomass – the remainder is entirely lost to the community as respiratory heat. PE varies according to the taxonomic class of the organisms concerned. Invertebrates in general have high efficiencies (30–40%), losing relatively little energy in respiratory heat. Amongst the vertebrates, ectotherms (whose body temperature varies according to environmental temperature; see Section 3.2.6) have intermediate values for PE (around 10%), whilst endotherms, which expend considerable energy to maintain a constant temperature, convert only 1–2% of assimilated energy into production. Microorganisms, including protozoa, tend to have very high PEs.

The overall *trophic transfer efficiency* from one trophic level to the next is simply $CE \times AE \times PE$. In the period after Lindeman's (1942) pioneering work (see Box 11.1), it was generally assumed that trophic transfer efficiencies were around 10%; indeed some ecologists referred to a 10% 'law'. However, there is certainly no law of nature that results in precisely one-tenth of the energy that enters a trophic level transferring to the next. For example, a compilation of trophic studies from a wide range of freshwater and marine environments revealed that trophic-level transfer efficiencies varied between about 2% and 24% – although the mean *was* 10.13% (standard error 0.49) (Pauly & Christensen, 1995).

11.3.3 The relative roles of the live consumer and decomposer systems

Given knowledge of NPP at a site, and CE, AE and PE for all the trophic groupings present (herbivores, carnivores, decomposers, detritivores), it is possible to map out the relative importance of different pathways. Figure 11.7 does this,

Figure 11.7

General patterns of energy flow for: (a) forest, (b) grassland, (c) a plankton community in the sea, and (d) the community of a stream or small pond. Relative sizes of boxes and arrows are proportional to the relative magnitude of compartments and flows. DOM, dead organic matter; LCS, live consumer system; NPP, net primary production.

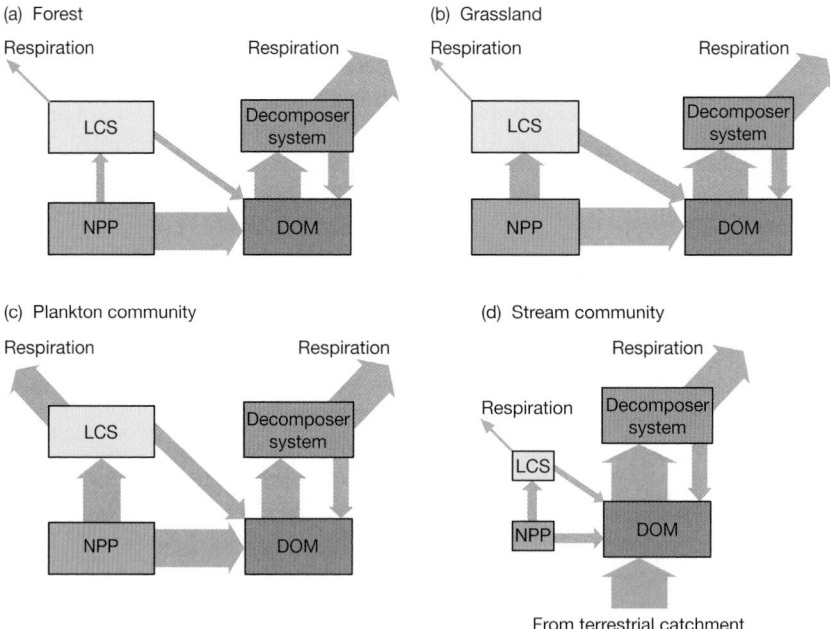

in a general way, for a forest, a grassland, a plankton community (of the ocean or a large lake) and the community of a small stream or pond. The decomposer system is probably responsible for the majority of secondary production, and therefore respiratory heat loss, in every community in the world (Figure 11.8). The 'live consumers' have their greatest role in open-water aquatic communities based on phytoplankton or in the beds of microalgae that occur in shallow water. In each case, a large proportion of NPP is consumed alive and assimilated at quite

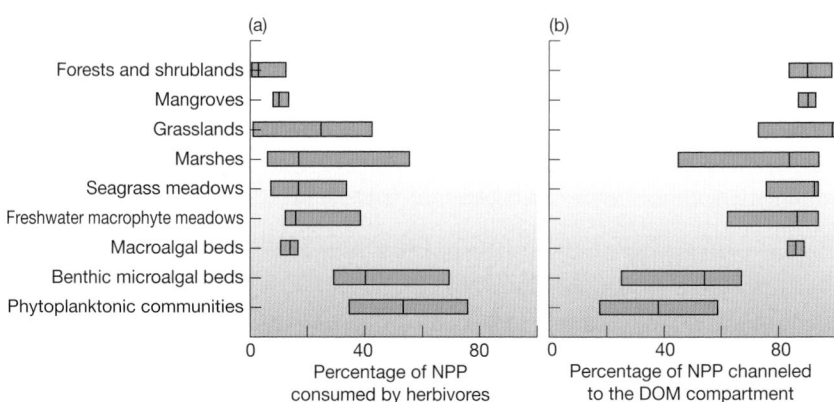

AFTER CEBRIAN, 1999

Figure 11.8

Box plots for a range of ecosystem types showing: (a) percentage of net primary production (NPP) consumed by herbivores and (b) percentage of NPP entering the dead organic matter (DOM) compartment. Boxes encompass 25% and 75% percentiles of published values and the central lines represent the median values. Phytoplankton and aquatic microalgal communities channel the largest proportions of NPP through herbivores and the smallest proportions through the DOM compartment.

a high efficiency (Figure 11.8a). In contrast, the decomposer system plays its greatest role where vegetation is woody – forests, shrublands and mangroves (Figure 11.8b). Grasslands and aquatic systems based on large plants [seagrasses, freshwater weeds and macroalgae (seaweeds)] occupy intermediate positions.

The live consumer system holds little sway in terrestrial communities because of low herbivore consumption efficiencies and assimilation efficiencies, and it is almost non-existent in many small streams and ponds simply because primary productivity is so low (Figure 11.7d). The latter often depend for their energy base on dead organic matter that falls or is washed or blown into the water from the surrounding terrestrial environment. The deep-ocean benthic community has a trophic structure very similar to that of streams and ponds. In this case, the community lives in water too deep for photosynthesis and energy is derived from dead phytoplankton, bacteria, animals and feces that sink from the autotrophic community in the euphotic zone above. From a different perspective, the ocean bed is equivalent to a forest floor beneath an impenetrable forest canopy.

11.4 The process of decomposition

Given the profound importance of the decomposer system, and thus of decomposers (bacteria and fungi) and detritivores, it is important to appreciate the range of organisms and processes involved in decomposition.

Immobilization is what occurs when an inorganic nutrient element is incorporated into organic form, primarily during the growth of green plants: for example, when carbon dioxide becomes incorporated into a plant's carbohydrates. Energy (coming, in the case of plants, from the sun) is required for this. Conversely, decomposition involves the release of energy and the *mineralization* of chemical nutrients – the conversion of elements from organic back to an inorganic form. Decomposition is defined as the gradual disintegration of dead organic matter (i.e. dead bodies, shed parts of bodies, feces) and is brought about by both physical and biological agencies. It culminates with complex, energy-rich molecules being broken down by their consumers (decomposers and detritivores) into carbon dioxide, water and inorganic nutrients. Ultimately, the incorporation of solar energy in photosynthesis, and the immobilization of inorganic nutrients into biomass, is balanced by the loss of heat energy and organic nutrients when the organic matter is mineralized.

decomposition defined

11.4.1 Decomposers: bacteria and fungi

If a scavenging animal, a vulture or a burying beetle perhaps, does not take a dead resource immediately, the process of decomposition usually starts with colonization by bacteria and fungi. Bacteria and fungal spores are always present in the air and the water, and are usually present on (and often in) dead material before it is dead. The early colonists tend to use soluble materials, mainly amino acids and sugars that are freely diffusible. The residual resources, though, are not diffusible and are more resistant to attack. Subsequent decomposition therefore proceeds more slowly, and involves microbial specialists that can break down structural carbohydrates (e.g. celluloses, lignins) and complex proteins such as suberin (cork) and insect cuticle.

bacteria and fungi are early colonists of newly dead material

11.4.2 Detritivores and specialist microbivores

specialist microbivores feed on bacteria and fungi, but most detritivores consume detritus too

The *microbivores* are a group of animals that operate alongside the detritivores, and which can be difficult to distinguish from them. The name microbivore is reserved for the minute animals that specialize at feeding on bacteria or fungi but are able to exclude detritus from their guts. In fact, though, the majority of detritivorous animals are generalist consumers, of both the detritus itself and the associated bacterial and fungal populations. The invertebrates that take part in the decomposition of dead plant and animal materials are a taxonomically diverse group. In terrestrial environments they are usually classified according to their size (Figure 11.9). This is not an arbitrary basis for classification, because size is

Figure 11.9

Size classification by body width of organisms in terrestrial decomposer food webs. Bacteria and fungi are decomposers. Animals that feed on dead organic matter (plus any associated bacteria and fungi) are detritivores. Carnivores that feed on detritivores include Opiliones (harvest spiders), Chilopoda (centipedes) and Araneida (spiders).

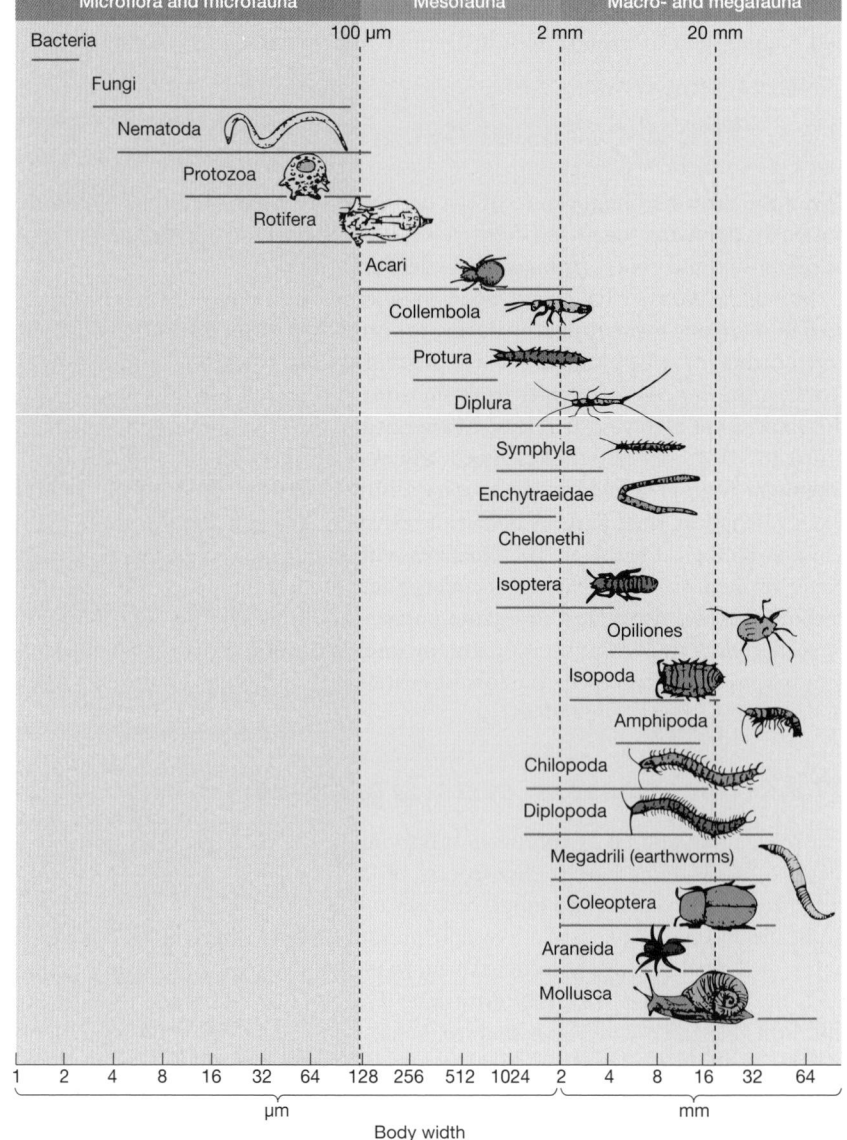

AFTER SWIFT ET AL. 1979

an important feature for organisms that reach their resources by burrowing or crawling among cracks and crevices of litter or soil.

In freshwater ecology, on the other hand, the study of detritivores has been concerned less with the size of the organisms than with the ways in which they obtain their food (refer back to Figure 4.16). For example, *shredders* are detritivores that feed on coarse particulate organic matter, such as tree leaves fallen into a river – these animals fragment the material into finer particles. On the other hand, *collector–filterers*, such as larvae of blackflies in rivers, consume the fine particulate organic matter that otherwise would be carried downstream. Because of very high densities (sometimes as many as 600,000 blackfly larvae per square meter of riverbed) a very large quantity of fine particulate matter is converted by the larvae into fecal pellets that settle on the bed and provide food for other detritivores (estimated at an amazing 429 tonnes dry mass of fecal pellets per day in a Swedish river; Malmqvist et al., 2001).

> aquatic detritivores are usually classified according to their feeding mode

11.4.3 Consumption of plant detritus

Two of the major organic components of dead leaves and wood are cellulose and lignin. These pose considerable digestive problems for animal consumers. Digesting cellulose requires *cellulase* enzymes but, surprisingly, cellulases of animal origin have been definitely identified in only one or two species. The majority of detritivores, lacking their own cellulases, rely on the production of cellulases by associated bacteria or fungi or, in some cases, protozoa. The interactions are of a range of types: (i) *obligate mutualisms* between a detritivore and a specific and permanent gut microflora (e.g. bacteria) or microfauna (e.g. termites); (ii) *facultative mutualisms*, where the animals make use of cellulases produced by a microflora that is ingested with detritus as it passes through an unspecialized gut (e.g. woodlice); or (iii) 'external rumens', where animals simply assimilate the products of the cellulase-producing microflora associated with decomposing plant remains or feces [e.g. springtails (Collembola)].

A variety of detritivores may be involved in fragmenting a single leaf. In experiments involving larvae of shredding stoneflies in streams, three different species were very similar in the efficiency with which they decomposed leaves of the alder tree, *Alnus incana*. However, average leaf loss was significantly greater when pairs of species were involved and was faster still when all three species were feeding on the leaf (Figure 11.10). The same number of stonefly larvae were

> the presence of more species of detritivore increases decomposition rate

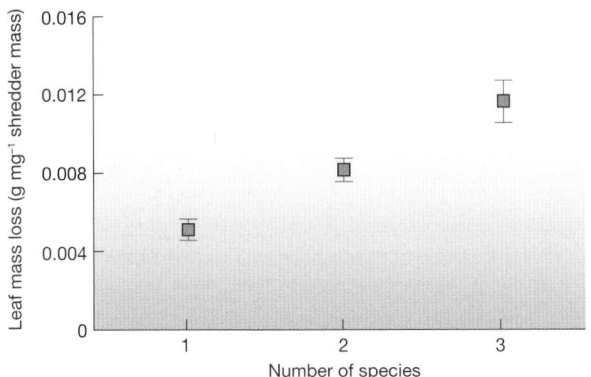

Figure 11.10

Variation in rate of loss of alder leaf mass in replicated stream experiments (per gram of leaf per milligram of shredder ± SE) caused by three species of shredder: larvae of the stoneflies *Protonemura meyeri*, *Nemoura avicularis* and *Taeniopteryx nebulosa*. The results are averaged for species acting on their own, for pairs of species in all possible combinations, and for all three species together (means ± SE). The decomposition rate was significantly faster when species operated in pairs, and was fastest of all when all three species were together.

AFTER JÓNSSON & MALMQVIST, 2000

included in every experiment (12 of a single species, six each in the species pairs, and four each when all three species were present) and the results were expressed in a standard way (leaf mass loss per gram of leaf per milligram of shredder in a 46-day experiment) so the result directly reflects the species richness present. These results are indicative of *complementarity* (each species feeds in a slightly different way so their combined effect is enhanced). Studies such as these have significant implications for the role that biological diversity plays in ecosystem functioning. Given current concerns about the extinction of species worldwide (see Chapter 14), we need to know whether diversity loss will have major consequences for the way ecosystems work. This is an important and controversial area (Box 11.2).

11.2 *Topical ECOncerns*

The importance of biological diversity in ecosystem functioning

Ecologists agree that some experimental evidence points to a significant role for biological diversity (biodiversity) in ecosystem functioning. Figure 11.10, for example, showed how decomposition rate is slower when fewer species are involved in the process. But some disagree about how much this matters – in other words, whether these kinds of result prove that biodiversity is critical to ecosystem health. This is a significant question at a time when global biodiversity is declining.

The following quotation comes from a commentary by Jocelyn Kaiser that appeared in 2000 in one of the major academic scientific journals, *Science* (289, 1282–1283).

Rift over biodiversity divides ecologists
A long-simmering debate among ecologists over the importance of biodiversity to the health of ecosystems has erupted into a full-blown war. Opposing camps are dueling over the quality of key experiments, and some are flinging barbs at meetings and in journals.

What lay behind such bellicose language? The disagreement began as part of the normal debate that should occur about any piece of research. To what extent are the conclusions justified from the results and how far can they be generalized from the special circumstances of the experiment to other situations in nature? Various studies around the world seemed to show that the loss of plant or animal species might adversely affect ecosystem function; for example, the productivity of grassland communities appears to be higher when more species are present. This could mean that biodiversity *per se* matters to productivity. But might variables other than species diversity have given rise to increased productivity? For example, perhaps such a result was a statistical artefact – higher productivity with higher species diversity might be explained simply by the addition of a more productive species to the list (and a more productive species is more likely to be present when more species are included in the experiment).

This kind of debate is healthy, but it took on a new dimension when one of the world's leading learned societies, the Ecological Society of America (ESA), published a pamphlet and sent copies to members of Congress. One of a series called 'Issues in Ecology', the pamphlet concerned the importance of biodiversity for ecosystem functioning. It summarized the results of several studies but with little discussion of doubts raised by skeptics in the ESA.

The commentator noted:

Other ecologists safely outside the fray say there is more at stake in this dispute than personalities and egos. Beyond the legitimate scientific question about how much can be learned from experiments is the nagging question – by no means limited to biodiversity – of when scientific data are strong enough to form the basis of policy decisions.

This debate was not really about the quality of the science (since every study has its limitations), but *rather the document that the ESA sent to Congress, which some said tended to present opinion as fact. Do you think scientists should remain entirely outside the political arena? If not, how would you ensure that balanced and generally accepted positions would be presented? Read the article by Hooper et al. (2005) 'Effects of biodiversity on ecosystem functioning: a consensus of current knowledge' in Ecological Monographs 75, 3–35. Decide whether the opposing factions have found an effective way forward – the list of authors includes people who were on different sides of the original debate.*

The decomposition of dead material is not simply due to the sum of the activities of decomposers and detritivores; it is largely the result of interaction between the two (Lussenhop, 1992). This can be illustrated by taking an imaginary journey with a leaf fragment through the process of decomposition, focusing attention on a part of the wall of a single cell. Initially, when the leaf falls to the ground, the piece of cell wall is protected from microbial attack because it lies within the plant tissue. The leaf is now chewed and the fragment enters the gut of a woodlouse. Here it meets a new microbial flora in the gut and is acted on by the digestive enzymes of the woodlouse. The fragment emerges, changed by passage through the gut. It is now part of the woodlouse's feces and is much more easily attacked by microorganisms, because it has been fragmented and partially digested. While microorganisms are colonizing the fecal pellet, it may again be eaten, perhaps by a springtail, and pass through the new environment of the springtail's gut. Incompletely digested fragments may again appear, this time in springtail feces, yet more easily accessible to microorganisms. The fragment may pass through several other guts in its progress from being a piece of dead tissue to its inevitable fate of becoming carbon dioxide and minerals.

11.4.4 Consumption of feces and carrion

The dung of carnivorous vertebrates is relatively poor-quality stuff. Carnivores assimilate their food with high efficiency (usually 80% or more is digested) and their feces retain only the least digestible components; their decomposition is probably caused almost entirely by bacteria and fungi. In contrast, herbivore dung still contains an abundance of organic matter and is sufficiently thickly spread in the environment to support its own characteristic fauna, consisting of many occasional visitors but with several specific dung-feeders. A good example is provided by elephant dung; within a few minutes of dung deposition the area is alive with beetles. The adult dung beetles feed on the dung but they also bury large quantities along with their eggs to provide food for developing larvae.

All animals defecate and die, yet feces and dead bodies are not generally very obvious in the environment. This is because of the efficiency of the specialist

Figure 11.11

The influence of woodlice on the rate of breakdown of feces of herbivorous caterpillars (*Operophthera fagata* – which feed on leaves of beech trees, *Fagus sylvatica*). After 6 weeks, twice as much of the fecal material had decomposed when woodlice were present.

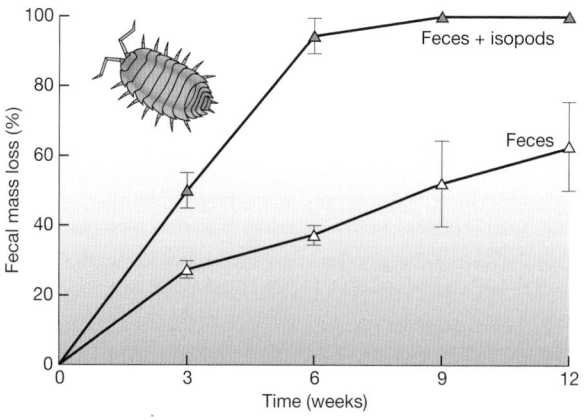

AFTER ZIMMER & TOPP, 2002

consumers of these dead organic products. On the other hand, where consumers of feces are absent, a build-up of fecal material may occur. Figure 11.11 shows how feeding by woodlice (*Porcellio scaber* and *Oniscus asellus*) speeds the breakdown of invertebrate feces. A more dramatic example is provided by the accumulation of cattle dung where these domestic animals have been introduced to locations lacking appropriate dung beetles. In Australia, for example, during the past 200 years, the cow population increased from just seven individuals (brought over by the first English colonists in 1788) to 30 million or so, producing 300 million cowpats per day. The lack of native dung beetles led to losses of up to 2.5 million hectares per year under dung. The decision was made in 1963 to establish in Australia beetles of African origin, able to dispose of bovine dung under the conditions where cattle are raised; more than 20 species have been introduced (Doube et al., 1991).

When considering the decomposition of dead bodies, it is helpful to distinguish three categories of organisms that attack carcasses. As before, decomposers (bacteria and fungi) and invertebrate detritivores have roles to play, but, in addition, scavenging vertebrates are often of considerable importance. Many carcasses of a size to make a single meal for one of a few of these scavenging detritivores will be removed completely within a very short time of death, leaving nothing for bacteria, fungi or invertebrates. This role is played, for example, by Arctic foxes and skuas in polar regions; by crows, gluttons and badgers in temperate areas; and by a wide variety of birds and mammals, including kites, jackals and hyenas, in the tropics.

11.5 The flux of matter through ecosystems

Chemical elements and compounds are vital for the processes of life. When living organisms expend energy (as they all do, continually), they do so, essentially, in order to extract chemicals from their environment, and hold on to them and use them for a period before they lose them again. Thus, the activities of organisms profoundly influence the patterns of flux of chemical matter.

The great bulk of living matter in any community is water. The rest is made up mainly of carbon compounds and this is the form in which energy is accumulated and stored. Carbon enters the food web of a community when a simple molecule, carbon dioxide, is taken up in photosynthesis. Once incorporated in NPP, it is available for consumption as part of a sugar, a fat, a protein or, very often, a cellulose molecule. It follows exactly the same route as energy, being successively consumed and either defecated, assimilated or used in metabolism, during which the energy of its molecule is dissipated as heat while the carbon is released again to the atmosphere as carbon dioxide. Here, though, the tight link between energy and carbon ends.

Once energy is transformed into heat, it can no longer be used by living organisms to do work or to fuel the synthesis of biomass. The heat is eventually lost to the atmosphere and can never be recycled: life on Earth is only possible because a fresh supply of solar energy is made available every day. In contrast, the carbon in carbon dioxide can be used again in photosynthesis. Carbon, and all other nutrient elements (nitrogen, phosphorus, etc.), are available to plants as simple organic molecules or ions in the atmosphere (carbon dioxide), or as dissolved ions in water (nitrate, phosphate, potassium, etc.). Each can be incorporated into complex carbon compounds in biomass. Ultimately, however, when the carbon compounds are metabolized to carbon dioxide, the mineral nutrients are released again in simple inorganic form. Another plant may then absorb them, and so an individual atom of a nutrient element may pass repeatedly through one food chain after another.

energy cannot be cycled and reused – matter can

Unlike the energy of solar radiation, moreover, nutrients are not in unalterable supply. The process of locking some up into living biomass reduces the supply remaining to the rest of the community. If plants, and their consumers, were not eventually decomposed, the supply of nutrients would become exhausted and life on Earth would cease.

We can conceive of pools of chemical elements existing in compartments. Some compartments occur in the *atmosphere* (carbon in carbon dioxide, nitrogen as gaseous nitrogen, etc.), some in the rocks of the *lithosphere* (calcium as a constituent of calcium carbonate, potassium in the rock called feldspar) and others in the waters of soil, streams, lakes or oceans – the *hydrosphere* (nitrogen in dissolved nitrate, phosphorus in phosphate, carbon in carbonic acid, etc.). In all these cases the elements exist in inorganic form. In contrast, living organisms (the biota) and dead and decaying bodies can be viewed as compartments containing elements in organic form [carbon in cellulose or fat, nitrogen in protein, phosphorus in adenosine triphosphate (ATP), etc.]. Studies of the chemical processes occurring within these compartments and, more particularly, of the fluxes of elements between them, comprise the science of biogeochemistry.

Nutrients are gained and lost by communities in a variety of ways (Figure 11.12). A nutrient budget can be constructed if we can identify and measure all the processes on the credit and debit sides of the equation.

biogeochemistry and biogeochemical cycles

11.5.1 Nutrient budgets in terrestrial ecosystems

Weathering of parent bedrock and soil, by both physical and chemical processes, is the main source of nutrients such as calcium, iron, magnesium, phosphorus and potassium, which may then be taken up via the roots of plants.

nutrient inputs

Atmospheric carbon dioxide is the source of the carbon content of terrestrial communities. Similarly, gaseous nitrogen from the atmosphere provides most

Figure 11.12

Components of the nutrient budgets of a terrestrial and an aquatic system. Inputs are shown in blue and outputs in black. Note how the two communities are linked by streamflow, which is a major output from the terrestrial system but a major input to the aquatic one.

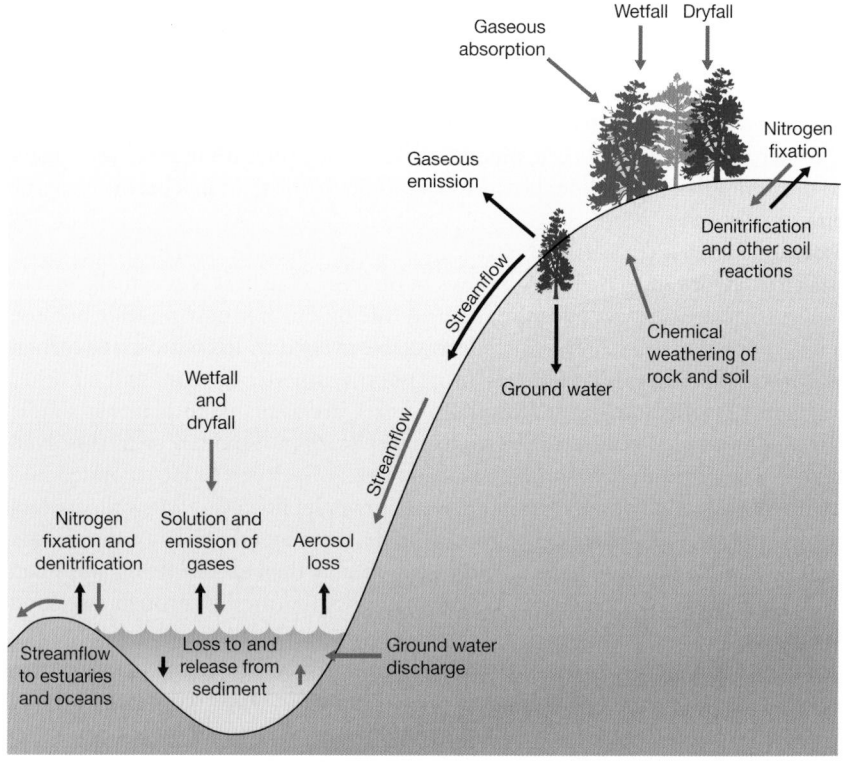

nutrient outputs

of the nitrogen content of communities. Several types of bacteria and blue-green algae possess the enzyme nitrogenase, which converts gaseous nitrogen to ammonium ions (NH_4^+) that can then be taken up through the roots and used by plants. All terrestrial ecosystems receive some available nitrogen through the activity of free-living, nitrogen-fixing bacteria, but communities containing plants such as legumes and alder trees (*Alnus* spp.), with their root nodules containing symbiotic nitrogen-fixing bacteria (see Section 8.4.6), may receive a very substantial proportion of their nitrogen in this way.

Other nutrients from the atmosphere become available to communities in *dryfall* (settling of particles during periods without rain) or *wetfall* (in rain, snow and fog). Rain is not pure water but contains chemicals derived from a number of sources: (i) trace gases, such as oxides of sulfur and nitrogen; (ii) aerosols, produced when tiny water droplets from the oceans evaporate in the atmosphere and leave behind particles rich in sodium, magnesium, chloride and sulfate; and (iii) dust particles from fires, volcanoes and windstorms, often rich in calcium, potassium and sulfate. Nutrients dissolved in precipitation mostly become available to plants when the water reaches the soil and can be taken up by plant roots.

Nutrients may circulate within the community for many years. Alternatively, the atom may pass through the system in a matter of minutes, perhaps without interacting with the biota at all. Whatever the case, the atom will eventually be lost through one of the variety of processes that remove nutrients from the system (Figure 11.12). These processes constitute the debit side of the nutrient budget equation.

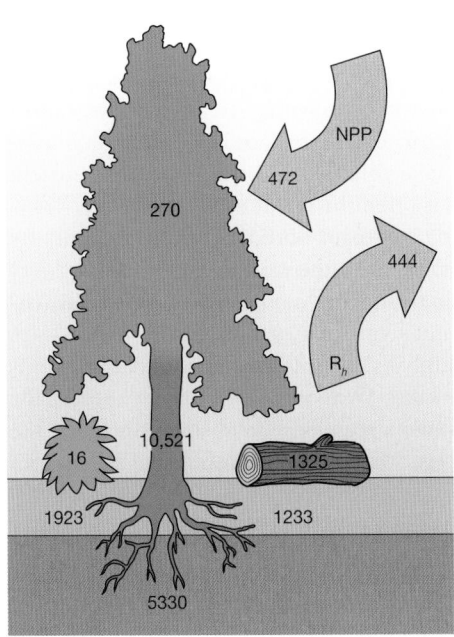

Figure 11.13

Annual carbon budget for a ponderosa pine (*Pinus pondersosa*) forest in Oregon, USA, where the trees are up to 250 years old. The numbers above ground represent the amount of carbon contained in tree foliage, in the remainder of forest biomass, in understorey plants and in dead wood on the forest floor. The numbers just below the ground surface represent tree roots (left) and litter (right). The lowest numeral is for soil carbon. The amounts of carbon stored in each of these elements of biomass are in g C m^{-2}. Values for net primary production (NPP) and for respiratory heat loss from heterotrophs (R_h) (i.e. microorganisms and animals) are in g C m^{-2} yr^{-1} (arrows). There is an approximate balance in the rate at which carbon is taken up in NPP and the rate at which it is lost as respiratory heat loss.

AFTER LAW ET AL., 2001

Release to the atmosphere is one pathway of nutrient loss. In many communities there is an approximate annual balance in the carbon budget; the carbon fixed by photosynthesizing plants is balanced by the carbon released to the atmosphere as carbon dioxide from the respiration of plants, microorganisms and animals (Figure 11.13). Plants themselves may be direct sources of gaseous and particulate release. For example, forest canopies produce volatile hydrocarbons (e.g. terpenes) and tropical forest trees appear to emit aerosols containing phosphorus, potassium and sulfur. Finally, ammonia gas is released during the decomposition of vertebrate excreta. Other pathways of nutrient loss are important in particular instances. For example, fire (either natural, or when, for instance, agricultural practice includes the burning of stubble) can turn a very large proportion of a community's carbon into carbon dioxide in a very short time, and the loss of nitrogen, as volatile gas, can be equally dramatic.

For many elements, the most substantial pathway of loss is in streamflow. The water that drains from the soil of a terrestrial community into a stream carries a load of nutrients that is partly dissolved and partly particulate. With the exception of iron and phosphorus, which are not mobile in soils, the loss of plant nutrients is predominantly in solution. Particulate matter in streamflow occurs both as dead organic matter (mainly tree leaves) and as inorganic particles.

It is the movement of water under the force of gravity that links the nutrient budgets of terrestrial and aquatic communities (see Figure 11.12). Terrestrial systems lose dissolved and particulate nutrients into streams and ground waters; aquatic systems (including the stream communities themselves, and ultimately the oceans) gain nutrients from streamflow and groundwater discharge. Refer to Section 1.3.3 for discussion of a study (at Hubbard Brook) that explored the chemical linkages at the land–water interface.

11.5.2 Nutrient budgets in aquatic communities

Aquatic systems receive the bulk of their supply of nutrients from stream inflow. In stream and river communities, and also in lakes with a stream outflow, export in outgoing stream water is a major factor. By contrast, in lakes without an outflow (or where this is small relative to lake volume), and also in oceans, nutrient accumulation in permanent sediments is often the major export pathway.

Many lakes in arid regions, lacking a stream outflow, lose water only by evaporation. The waters of these *endorheic* lakes (the word means 'internal flow') are thus more concentrated than their freshwater counterparts, being particularly rich in sodium but also in other nutrients such as phosphorus. Saline lakes should not be considered as oddities; globally, they are just as abundant in terms of numbers and volume as freshwater lakes (Williams, 1988). They are usually very fertile with dense populations of blue-green algae, and some, such as Lake Nakuru in Kenya, support huge aggregations of plankton-filtering flamingoes (*Phoeniconaias minor*).

The largest of all endorheic 'lakes' is the world ocean – a huge basin of water supplied by the world's rivers and losing water only by evaporation. Its great size, in comparison to the input from rain and rivers, leads to a remarkably constant chemical composition. The main transformers of dissolved inorganic carbon (essentially carbon dioxide dissolved from the atmosphere) are small phytoplankton cells, whose carbon is mainly recycled near the ocean surface via consumption by microzooplankton, release of dissolved organic substances and their mineralization by bacteria (Figure 11.14). In contrast, pathways involving larger phytoplankton and macrozooplankton are responsible for the majority

Figure 11.14

Pathways of carbon atoms in the ocean. Small phytoplankton, microzooplankton and bacteria recycle carbon in the mixed surface layer. Most of the carbon that moves to the deep ocean follows pathways involving larger phytoplankton and macrozooplankton, to be recycled again. A small proportion of remineralized inorganic carbon and particulate organic carbon is lost to the ocean sediment.

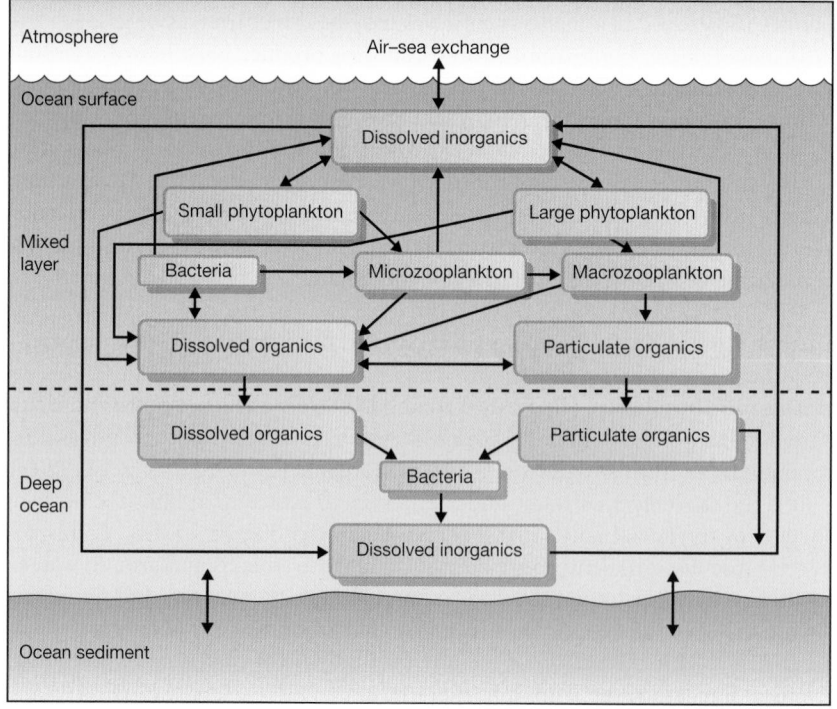

AFTER FASHAM ET AL., 2001

of carbon flux to the deep ocean floor. Some of this organic material is consumed by deep-sea animals, some is mineralized to inorganic form by bacteria and recirculated, and a small proportion becomes buried in the sediment. Figure 11.14 is essentially the ocean equivalent of the forest system in Figure 11.13. In contrast to the atmospheric source of carbon, nutrients such as phosphorus come from two sources – river inputs and water welling up from the deep. Phosphorus atoms in the surface water follow a similar set of pathways to carbon atoms, with about 1% of detrital phosphorus being lost to the deep sediment during each oceanic mixing cycle.

All water bodies receive nutrients, in inorganic and organic form, in the water draining from the land. It is no surprise, therefore, that human activities are responsible for dramatic changes in nutrient fluxes both locally (Box 11.3) and globally. We turn to global biogeochemical cycles in the next section.

11.3 *Topical ECOncerns*

Nutrient enrichment of aquatic ecosystems: a major problem for lakes and oceans

The excess input of nutrients from sources such as agricultural runoff and sewage has caused many 'healthy' *oligotrophic* lakes (low nutrients, low plant productivity with abundant water weeds, and clear water) to switch to a *eutrophic* condition where high nutrient inputs lead to high phytoplankton productivity (sometimes dominated by toxic bloom-forming species), making the water turbid, shading out large plants and, in the worst situations, leading to anoxia and fish kills. This process of *cultural eutrophication* of lakes has been understood for some time. But it was only recently that people noticed huge 'dead zones' in the oceans near river outlets, particularly those draining large catchment areas such as the Mississippi in North America and the Yangtze in China. The following extracts are from a news item posted by Associated Press on March 29, 2004.

Ocean dead zones on the increase
So-called 'dead zones', oxygen-starved areas of the world's oceans that are devoid of fish, top the list of emerging environmental challenges, the United Nations Environment Program [UNEP] warned Monday in its global overview.

The new findings tally nearly 150 dead zones around the globe . . . The main cause is excess nitrogen run-off from farm fertilizers, sewage and industrial pollutants. The nitrogen triggers blooms of microscopic algae known as phytoplankton. As the algae die and rot, they consume oxygen, thereby suffocating everything from clams and lobsters to oysters and fish.

'Human kind is engaged in a gigantic, global, experiment as a result of inefficient and often overuse of fertilizers, the discharge of untreated sewage and the ever rising emissions from vehicles and factories', UNEP Executive Director Klaus Toepfer said in a statement. 'Unless urgent action is taken to tackle the sources of the problem, it is likely to escalate rapidly.'

Suggest some 'urgent actions' that could be taken to alleviate the problem.

11.6 Global biogeochemical cycles

Nutrients are moved over vast distances by winds in the atmosphere and by the moving waters of streams and ocean currents. There are no boundaries, either natural or political. It is appropriate, therefore, to conclude this chapter by moving to an even larger spatial scale to examine global biogeochemical cycles.

11.6.1 The hydrological cycle

The principal source of water is the oceans; radiant energy makes water evaporate into the atmosphere, winds distribute it over the surface of the globe and precipitation brings it down to the Earth's surface (with a net movement of atmospheric water from oceans to continents), where it may be stored temporarily in soils, lakes and icefields (Figure 11.15). Loss occurs from the land through evaporation and transpiration or as liquid flow through stream channels and groundwater aquifers, eventually to return to the sea. The major pools of water occur in the oceans (97.3% of the total for the biosphere), the ice of polar icecaps and glaciers (2.06%), deep in the ground water (0.67%) and in rivers and lakes (0.01%) (Berner & Berner, 1987). The proportion that is in transit at any time is very small – water draining through the soil, flowing along rivers and present as clouds and vapor in the atmosphere – together this constitutes only about 0.08% of the total. However, this small percentage plays a crucial role, both by supplying the requirements for the survival of living organisms and for community productivity and because so many chemical nutrients are transported with the water as it moves.

The hydrological cycle would proceed whether or not a biota was present. However, terrestrial vegetation can modify the fluxes that occur. Vegetation can intercept water at two points on this journey, stopping some from reaching the ground water and moving it back into the atmosphere, by: (i) catching some in foliage from which it evaporates; and (ii) preventing some from draining from the soil water by taking it up via the roots into the plant's transpiration stream. We have seen earlier how cutting down the forest in a catchment in Hubbard Brook (see Section 1.3.3) increased the throughput of water to streams, along with its load of dissolved and particulate matter. It is small wonder that large-scale

Figure 11.15

The hydrological cycle, showing volumes of water in the 'reservoirs' of oceans, ice (polar and glacier), rivers and lakes, ground water and atmosphere (units of 10^6 km^3), and on the move as precipitation, runoff, evaporation and vapor transport (arrows: units of 10^6 km^3 yr^{-1}).

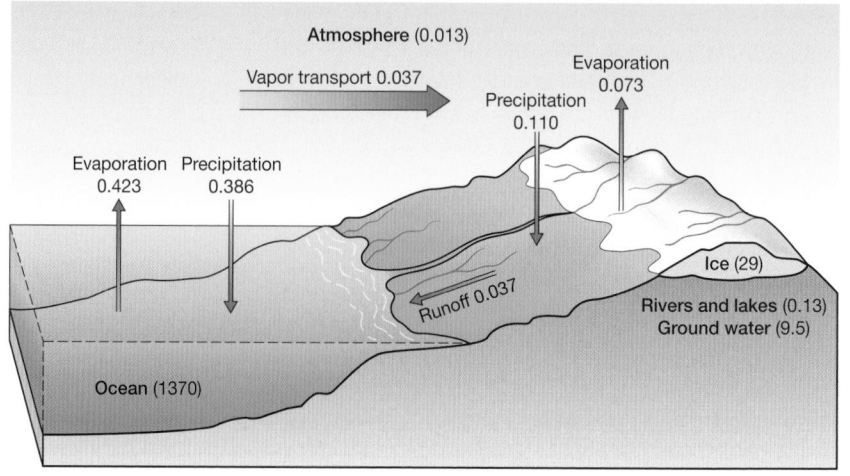

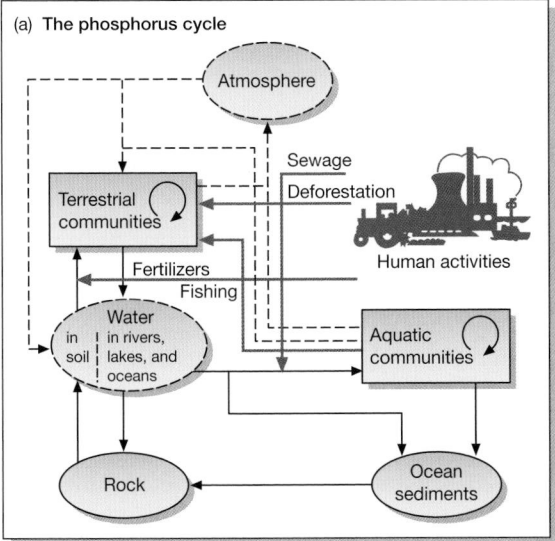

(a) The phosphorus cycle

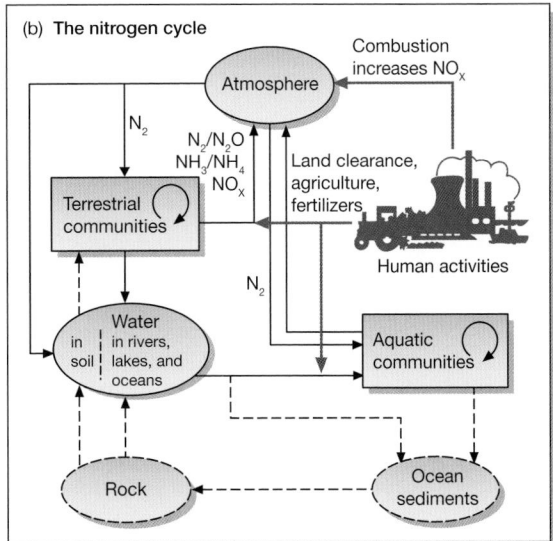

(b) The nitrogen cycle

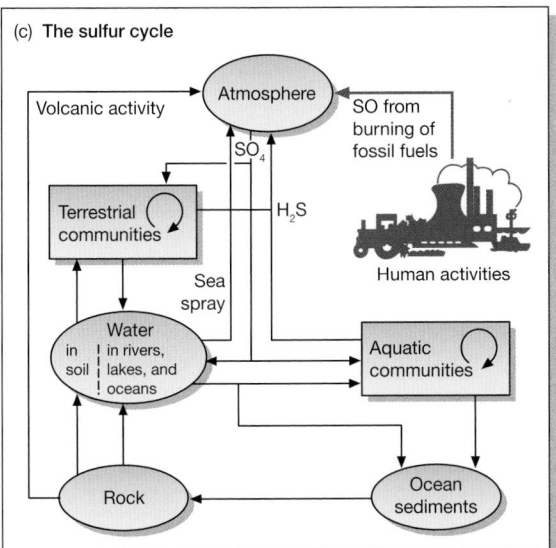

(c) The sulfur cycle

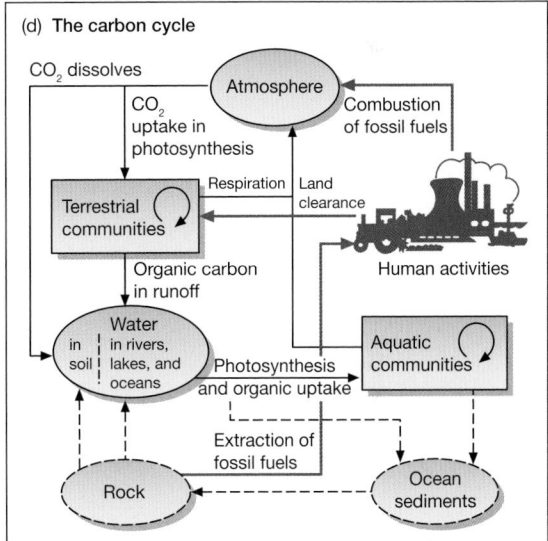

(d) The carbon cycle

Figure 11.16

The major global pathways of nutrients between the abiotic 'reservoirs' of atmosphere, water (hydrosphere) and rock and sediments (lithosphere), and the biotic 'reservoirs' constituted by terrestrial and aquatic communities. Human activities (maroon arrows) change nutrient fluxes in terrestrial and aquatic communities by releasing extra nutrients into both atmosphere and water. Cycles are presented for four important nutrient elements: (a) phosphorus, (b) nitrogen, (c) sulfur and (d) carbon. Insignificant compartments and fluxes are represented by dashed lines.

deforestation around the globe, usually to create new agricultural land, can lead to loss of topsoil, nutrient impoverishment and increased severity of flooding. Water is a very valuable commodity, and this is reflected in the difficult political exercise of dealing with competing demands – to divert river water for hydro-electric power generation or agricultural irrigation as opposed to maintaining the intrinsic values of an unmanipulated river.

The world's major abiotic reservoirs for nutrients are illustrated in Figure 11.16. We now consider these cycles in turn.

11.6.2 The phosphorus cycle

The principal stocks of phosphorus occur in the waters of soil, rivers, lakes and oceans and in rocks and ocean sediments. The phosphorus cycle may be described as a sedimentary cycle because of the general tendency for mineral phosphorus to be carried from the land inexorably to the oceans where ultimately it becomes incorporated in the sediments (Figure 11.16a).

the life history of a phosphorus atom

A 'typical' phosphorus atom, released from the rock by chemical weathering, may enter and cycle within the terrestrial community for years, decades or centuries before it is carried via the ground water into a stream. Within a short time of entering the stream (weeks, months or years), the atom is carried to the ocean. It then makes, on average, about 100 round trips between surface and deep waters, each lasting perhaps 1000 years. During each trip, it is taken up by surface-dwelling organisms, before eventually settling into the deep again. On average, on its 100th descent (after 10 million years in the ocean) it fails to be released as soluble phosphorus, but instead enters the bottom sediment in particulate form. Perhaps 100 million years later, the ocean floor is lifted up by geological activity to become dry land. Thus, our phosphorus atom will eventually find its way back via a river to the sea, and to its existence of cycle (biotic uptake and decomposition) within cycle (ocean mixing) within cycle (continental uplift and erosion).

11.6.3 The nitrogen cycle

the nitrogen cycle has an atmospheric phase of overwhelming importance

The atmospheric phase predominates in the global nitrogen cycle, in which nitrogen fixation and denitrification by microbial organisms are of particular importance (Figure 11.16b). However, nitrogen from certain geological sources may also be locally significant in fueling productivity in terrestrial and freshwater communities (Holloway et al., 1998, Thompson et al., 2001). The magnitude of the flux in streamflow from terrestrial to aquatic communities is relatively small, but it is by no means insignificant for the aquatic systems involved. This is because nitrogen is one of the two elements (along with phosphorus) that most often limit plant growth. Finally, there is a small annual loss of nitrogen to ocean sediments.

11.6.4 The sulfur cycle

Three natural biogeochemical processes release sulfur to the atmosphere: the formation of seaspray aerosols, anaerobic respiration by sulfate-reducing bacteria and volcanic activity (relatively minor) (Figure 11.16c). Sulfur bacteria release reduced sulfur compounds, particularly H_2S, from waterlogged bog and marsh communities and from tidal mudflats. A reverse flow from the atmosphere involves oxidation of sulfur compounds to sulfate, which returns to earth as both wetfall and dryfall.

the sulfur cycle has an atmospheric phase and a lithospheric phase of similar magnitude

The weathering of rocks provides about half the sulfur draining off the land into rivers and lakes, the remainder coming from atmospheric sources. On its way to the ocean, a proportion of the available sulfur (mainly dissolved sulfate) is taken up by plants, passed along food chains and, via decomposition processes, becomes available again to plants. However, in comparison to phosphorus and nitrogen, a much smaller fraction of sulfur takes part in internal recycling in terrestrial and aquatic communities. Finally, there is a continuous loss of sulfur to ocean sediments.

11.6.5 The carbon cycle

Photosynthesis and respiration are the two opposing processes that drive the global carbon cycle. It is predominantly a gaseous cycle, with carbon dioxide as the main vehicle of flux between atmosphere, hydrosphere and biota. Historically, the lithosphere played only a minor role; fossil fuels lay as dormant reservoirs of carbon until human intervention in recent centuries (Figure 11.16d).

Terrestrial plants use atmospheric carbon dioxide as their carbon source for photosynthesis, whereas aquatic plants use dissolved carbonates (i.e. carbon from the hydrosphere). The two subcycles are linked by exchanges of carbon dioxide between atmosphere and oceans. In addition, carbon finds its way into inland waters and oceans as bicarbonate resulting from weathering (carbonation) of calcium-rich rocks such as limestone and chalk. Respiration by plants, animals and microorganisms releases the carbon locked in photosynthetic products back to the atmospheric and hydrospheric carbon compartments.

> the opposing forces of photosynthesis and respiration drive the global carbon cycle

11.6.6 Human impacts on biogeochemical cycles

It goes almost without saying that human activities contribute significant inputs of nutrients to ecosystems and disrupt local and global biogeochemical cycles. For example, the amounts of carbon dioxide and oxides of nitrogen and sulfur in the atmosphere have been increased by the burning of fossil fuels and by car exhausts, and the concentrations of nitrate and phosphate in stream water have been raised by agricultural practices and sewage disposal. These changes have far-reaching consequences, which will be discussed in Chapter 13.

Summary

SUMMARY

Patterns in primary productivity
Primary production on land is limited by a variety of factors – the quality and quantity of solar radiation, the availability of water, nitrogen and other key nutrients, and physical conditions, particularly temperature. Productive aquatic communities occur where, for one reason or another, nutrient concentrations are unusually high and the intensity of radiation is not limiting.

The fate of primary productivity
Secondary productivity by herbivores is approximately an order of magnitude less that the primary productivity on which it is based. Energy is lost at each feeding step because consumption efficiencies, assimilation efficiencies and production efficiencies are all less than 100%. The decomposer system processes much more of a community's energy and matter than the live consumer system. The energy pathways in the live consumer and decomposer systems are the same, with one critical exception – feces and dead bodies are lost to the grazer system (and enter the decomposer system), but feces and dead bodies from the decomposer system are simply sent back to the dead organic matter compartment at its base.

The process of decomposition
Decomposition results in complex, energy-rich molecules being broken down by their consumers

(decomposers and detritivores) into carbon dioxide, water and inorganic nutrients. Ultimately, the incorporation of solar energy in photosynthesis, and the immobilization of inorganic nutrients into biomass, is balanced by the loss of heat energy and organic nutrients when the organic matter is decomposed. This is brought about partly by physical processes, but mainly by decomposers (bacteria and fungi) and detritivores (animals that feed on dead organic matter).

The flux of matter through ecosystems

Nutrients are gained and lost by communities in a variety of ways. Weathering of parent bedrock and soil, by both physical and chemical processes, is the dominant source of nutrients such as calcium, iron, magnesium, phosphorus and potassium, which may then be taken up via the roots of plants. Atmospheric carbon dioxide and gaseous nitrogen are the principal sources of the carbon and nitrogen content of terrestrial communities while other nutrients from the atmosphere become available as dryfall or in rain, snow and fog. Nutrients are lost again through release to the atmosphere or in the water that feeds into streams and rivers. Aquatic systems (including the stream communities themselves, and ultimately the oceans) gain nutrients from streamflow and groundwater discharge and from the atmosphere by diffusion across their surfaces.

Global biogeochemical cycles

The principal source of water in the hydrological cycle is the oceans; radiant energy makes water evaporate into the atmosphere, winds distribute it over the surface of the globe and precipitation brings it down to the Earth's surface. Phosphorus derives mainly from the weathering of rocks (lithosphere); its cycle may be described as sedimentary because of the general tendency for mineral phosphorus to be carried from the land inexorably to the oceans where ultimately it becomes incorporated in the sediments. The sulfur cycle has an atmospheric phase and a lithospheric phase of similar magnitude. The atmospheric phase is predominant in both the global carbon and nitrogen cycles. Photosythesis and respiration are the two opposing processes that drive the global carbon cycle, while nitrogen fixation and denitrification by microbial organisms are of particular importance in the nitrogen cycle. Human activities contribute significant inputs of nutrients to ecosystems and disrupt local and global biogeochemical cycles.

Review questions

Asterisks indicate challenge questions

1 A large proportion of the open ocean is, in effect, a marine desert. Why?

2* Describe the general latitudinal trends in net primary productivity. Suggest reasons why such a latitudinal trend does not occur in the oceans.

3* Table 11.2 presents the results of a study that contrasted the productivity of a deciduous beech forest (Fagus sylvatica) with that of a nearby evergreen spruce forest (Picea abies). The beech leaves photosynthesized at a greater rate (per gram dry weight) than those of spruce, and beech 'invested' a considerably greater amount of biomass in its leaves each year. But net primary productivity of the beech forest was lower than spruce forest. Why? If these species were grown together, which would you expect to come to dominate the forest? What factors other than productivity might influence the relative competitive status of the two species?

Table 11.2

Characteristics of representative trees of two contrasting species growing within 1 km of each other on the Solling Plateau, Germany.

	BEECH	NORWAY SPRUCE
Age (years)	100	89
Height (m)	27	25.6
Leaf shape	Broad	Needle
Annual production of leaves	Higher	Lower
Photosynthetic capacity per unit dry weight of leaf	Higher	Lower
Length of growing season (days)	176	260
Net primary productivity (metric tons of carbon per hectare per year)	8.6	14.9

AFTER SCHULZE, 1970; SCHULZE ET AL., 1977A, 1977B

4 What evidence suggests that the productivity of many terrestrial and aquatic communities is limited by nutrients?

5* In both aquatic and terrestrial communities, secondary productivity by herbivores is approximately one-tenth of the primary productivity upon which it is based. This has led some to suggest the operation of a 10% law. Do you subscribe to this view?

6 Account for the observation that in most communities much more energy is processed through the decomposer system than through the live consumer system.

7 Outline the role played by bacteria and fungi (decomposers) in the flux of energy and matter through a named ecosystem. Imagine what would happen if bacteria and fungi were magically removed – describe the resulting scenario.

8 Energy cannot be cycled and reused but matter can. Discuss this assertion and its significance for ecosystem functioning.

9 Is the ocean simply a large lake in terms of patterns of flux of energy and matter?

10 The hydrological cycle would proceed whether or not a biota was present. Discuss how the presence of vegetation modifies the flow of water through an ecosystem.

PART **FOUR**

Applied Issues in Ecology

Chapter **12**

Sustainability

Key concepts

In this chapter you will:

- appreciate the underlying dynamics of human population growth and its relationship to the sustainable (or unsustainable) use of resources
- understand the biological basis of sustainable harvesting of wild populations – particularly in fisheries
- recognize the benefits and costs of farming monocultures
- understand that much agricultural practice has not been sustainable because of loss and degradation of soil
- appreciate that water may be the least sustainable of global resources
- recognize the benefits and costs of different methods of pest control and the importance of devising integrated management practices

The sustainability of human activities, and of the size and distribution of the human population, have increasingly become preoccupations of the general public and of the politicians who represent them. But attaining or even approaching sustainability requires more than a will to do so – it requires ecological understanding, carefully acquired and even more carefully applied.

12.1 Introduction

what is 'sustainability'?

To call an activity 'sustainable' means that it can be continued or repeated for the foreseeable future. Concern has arisen, therefore, precisely because so much human activity is clearly unsustainable. We cannot go on increasing the size of the global human population; we cannot (if we wish to have fish to eat in future) continue to remove fish from the sea faster than the remaining fish can replace their lost companions; we cannot continue to harvest agricultural crops or forests if the quality and quantity of the soil deteriorates or water resources become inadequate; we cannot continue to use the same pesticides if increasing numbers of pests become resistant to them; we cannot maintain the diversity of nature if we continue to drive species to extinction.

Sustainability has thus become one of the core concepts – perhaps *the* core concept – in an ever-broadening concern for the fate of the Earth and the ecological communities that occupy it. In defining sustainability we used the words '*foreseeable* future'. We did so because, when an activity is described as sustainable, it is on the basis of what is known at the time. But many factors remain unknown or unpredictable. Things may take a turn for the worse (as when adverse oceanographic conditions damage a fishery already threatened by overexploitation), or some unforeseen additional problem may be discovered (resistance may appear to some previously irresistible pesticide). On the other hand, technological advances may allow an activity to be sustained that previously seemed unsustainable (new types of pesticide may be discovered that are more finely targeted on the pest itself rather than species that are innocent bystanders). However, there is a real danger that we observe the many technological and scientific advances that have been made in the past and act on the faith that there will always be a technological 'fix' coming along to solve our present problems, too. Unsustainable practices cannot be accepted simply from faith that future advances will make them sustainable after all.

sustainability 'comes of age'

The recognition of sustainability's importance as a unifying idea in applied ecology has grown gradually, but there is something to be said for the claim that sustainability really came of age in 1991. This was when the Ecological Society of America published 'The sustainable biosphere initiative: an ecological research agenda', 'a call-to-arms for all ecologists' with a list of 16 co-authors (Lubchenco et al., 1991). And in the same year, the World Conservation Union, the United Nations Environment Program and the World Wide Fund for Nature jointly published *Caring for the Earth. A Strategy for Sustainable Living*

(IUCN/UNEP/WWF, 1991). The detailed contents of these documents are less important than their existence. They indicate a growing preoccupation with sustainability, shared by scientists and pressure groups, and recognition that much of what we do is not sustainable.

The emphasis shifted more recently from a purely ecological perspective to one that incorporates economic and social conditions that influence sustainability (Milner-Gulland & Mace, 1998), a theme that has gathered pace in the new millennium. Thus, the Millennium Ecosystem Assessment, based on contributions from a large number of natural and social scientists, has as its aim providing both the general public and decision-makers with 'a scientific evaluation of the consequences of current and projected changes in ecosystems for human well-being' (Balmford & Bond, 2005; Millennium Ecosystem Assessment, 2005).

In this chapter, we first consider the size and rate of growth of the human population, a primary driver of the environmental problems that confront us (Section 12.2). Then we deal with two areas of applied ecology where sustainability is a particularly pressing issue – the harvesting of living resources from the wild (Section 12.3) and the production, in unnatural agroecosystems, of the food and fiber needs of humankind (Sections 12.4–12.7).

12.2 The human population 'problem'

12.2.1 Introduction

The root of most, if not all of the environmental problems facing us is the 'population problem', the effects of a large and growing population of humans. More people means an increased requirement for energy, a greater drain on non-renewable resources like oil and minerals, more pressure on renewable resources like fish and forests (Section 12.3), more need for food production through agriculture (Section 12.4) and so on. The issue is undoubtedly one of sustainability: things cannot go on the way they are. Yet it is not clear exactly what 'the problem' is (Box 12.1). Here, therefore, we examine first the size and growth rate of the global human population and how we reached our current state, then how successful we can expect to be in projecting forward into the future, before finally addressing 'the problem' more directly by asking the question, 'How many people can the Earth support?'

what is the human population problem?

12.2.2 Population growth up to the present

When the finger is pointed at human population *growth* as the key issue, it is often said that what is wrong is that the global population has been growing 'exponentially'. But in an exponentially growing population (see Chapter 5) the rate of increase per individual is constant. The population as a whole grows at an accelerating rate (a plot of numbers against time sweeps upwards), because the population growth rate is a product of the individual rate (constant) and the accelerating number of individuals. In Chapter 5, such exponential growth was contrasted with a population limited by intraspecific competition (such as one described by the 'logistic' equation), where the rate of increase per individual *decreases* as population size increases. In the case of the global human population

past population growth: 'more than exponential'

12.1 *Topical ECOncerns*

12.1 TOPICAL ECONCERNS

The human population problem

What is 'the human population problem'? This is not an easy question to pin down, but what follows are some possible versions of the answer (Cohen, 1995, 2003, 2005). The real problem, of course, may be a combination of these – or of these and others. There is little doubt, though, that there *is* a problem, and that the problem is 'ours', collectively.

- *The present size of the global human population is unsustainably high*. Around AD 200, when there were about a quarter of a billion people on Earth, Quintus Septimus Florens Tertullianus wrote that 'we are burdensome to the world, the resources are scarcely adequate to us'. By 2005 the total had risen to an estimated 6.5 billion (United Nations 2005).

- *It is not the size but the distribution over the Earth of the human population that is unsustainable*. The fraction of the population living, highly concentrated, in an urban environment has risen from around 3% in 1800 to 29% in 1950 and 47% in 2000. Each agricultural worker today has to feed her- or himself plus one city dweller; by 2050 that will have risen to two urbanites (Cohen 2005).

- *The present rate of growth in size of the global human population is unsustainably high*. Prior to the widespread agricultural revolution of the 18th century, the human population, very roughly, had taken 1000 years to double in size. The most recent doubling took just 39 years (Cohen 2001).

- *It is not the size but the age distribution of the global human population that is unsustainable*. In the 'developed' regions of the world, the percentage of the population that was elderly (over 65) rose from 7.6% in 1950 to 12.1% in 1990. This proportion will jump dramatically after 2010 when the large cohorts born after World War II pass 65.

- *It is not the size but the uneven distribution of resources within the global population that is unsustainable. In 1992, the 830 million people of the world's richest countries enjoyed an average income equivalent to US$22,000 per annum. The 2.6 billion people in the middle income countries received $1600. But the 2 billion in the poorest countries got just $400. These averages themselves hide enormous inequalities.*

1 *What role or responsibility does the individual, as opposed to government, have in responding to the human population problem?*

2 *Which of the variants of the problem, above, pose particular questions of the relationship between the developed and the developing parts of the world or between the 'haves' and the 'have nots'?*

Rich world

Poor world

(Box 12.2), however, the rate of increase per individual (and also therefore the annual percentage increase in size: the rate of increase per 100 individuals) has certainly not been decreasing – but neither has it remained constant (Cohen, 1995). Rather, the individual rate has itself been accelerating. Even exponential growth would be unsustainable; but the more-than-exponential growth that we have witnessed would, if continued, become unsustainable even sooner.

12.2 *Quantitative aspects*

The growth of human populations

Figure 12.1 shows estimates of the size of the global human population from 2000 years ago to the present. Apart from the occasional hesitation and even rarer downturn (such as that caused by the ravages of the Black Death towards the end of the 14th century) the overall picture has clearly been one of ever more rapid population growth: the slope of the curve gets steeper and steeper.

But is this exponential growth? The answer is a conclusive 'No'. Figure 12.1b shows this same graph

(black line), but also shows: (i) what an exponentially growing population would have looked like that started at the same point 2000 years ago and finished at the present population size; and (ii) for the sake of contrast, a population anchored at the same start and finish points but growing according to the logistic equation.

Disregarding the logistic as utterly unrealistic, it is also clear that exponential growth is much more 'gradual' than what has actually been observed. The crux of the difference between these three graphs is

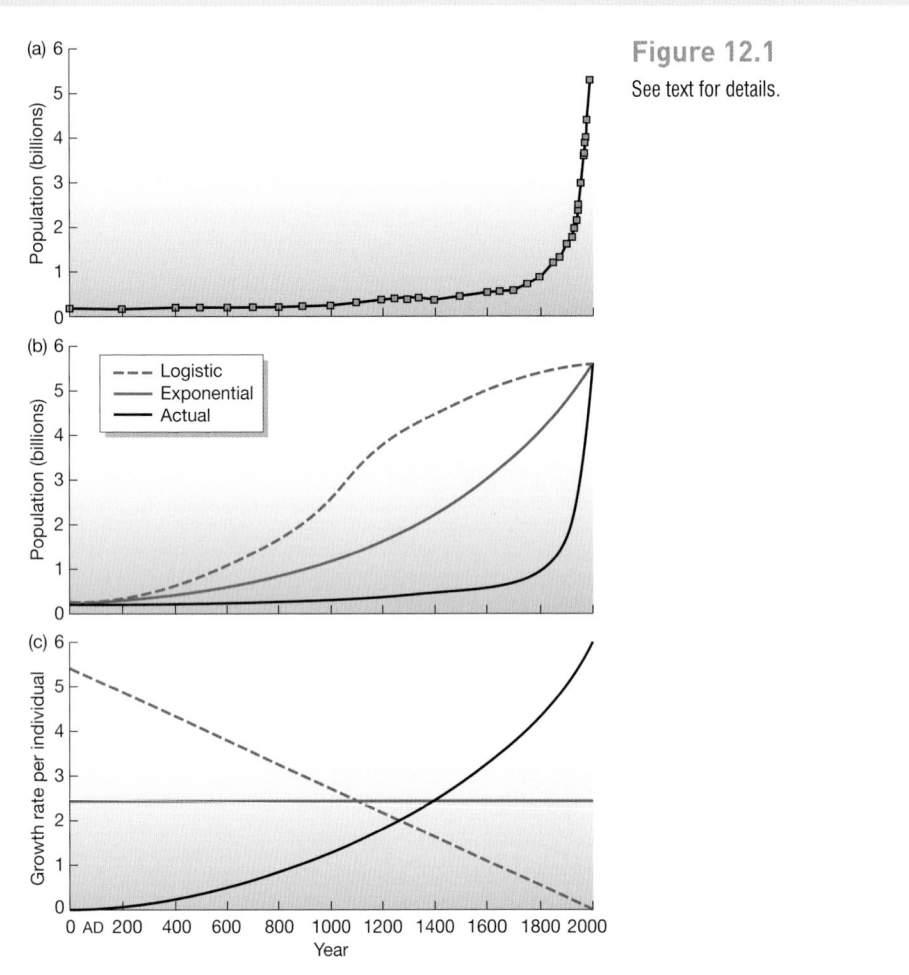

Figure 12.1
See text for details.

shown in Figure 12.1c, which uses the same information, but this time plots the changing *growth rate per individual* against time: the *per capita* rate. This parameter was introduced in Box 5.4, where it was described, formally, as $dN/dt \cdot (1/N)$ or, in words, as the rate of population growth, dN/dt, divided by the number of individuals. For the logistic, under the influence of increasingly intense intraspecific competition, the growth rate per individual declines in a straight line down to zero – as it always does for the logistic. For exponential growth, the rate is constant – again 'by definition'. But the actual growth curve gives rise to an individual rate which not only increases with time as the global population has increased – it increases more than linearly – it accelerates. The historical pattern of growth has been more than exponential.

12.2.3 *Predicting the future*

prediction is more than projection

It is interesting to see what has happened to the total human population in the past – and to do so alerts us to the scale of the problem we face – but the major practical importance of such a survey lies in the opportunity it might provide to

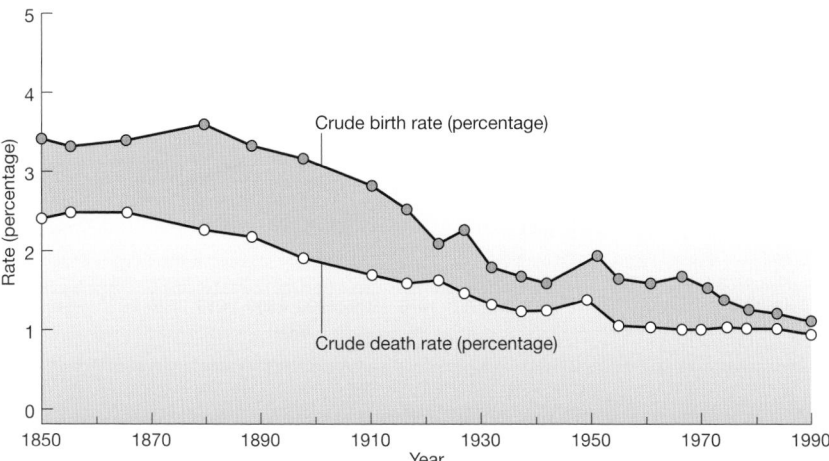

AFTER COHEN, 1995

Figure 12.2

The decline in the annual rate of population growth in Europe since 1850 has been associated with a decline in the death rate, followed by a decline in the birth rate, and an overall narrowing of the gap between the two.

predict future population sizes and rates of growth. There is an enormous difference, however, between projection and prediction. Simply to project forwards would be to make the almost certainly false assumption that things will go on in the future just as they have in the past.

Prediction, by contrast, requires an *understanding* of what has happened in the past, as well as how the present differs from the past, and finally how these differences might translate into future patterns of population growth. In particular, it is essential to recognize that the global population of humans is a collection of smaller populations, each with its own often very different characteristics. Like all ecological populations, the human population is heterogeneous.

the global population is heterogeneous

One common way in which subpopulations have been distinguished has been in terms of the 'demographic transition'. Three groups of nations can be recognized: those that passed through the demographic transition 'early' (pre-1945), 'late' (since 1945) or 'not yet' (pre-transition countries). The pattern, illustrated for the combined 'early transition' populations of Europe in Figure 12.2, is as follows. Initially, both the birth rate and the death rate are high, but the former is only slightly greater than the latter, so the overall rate of population increase is only moderate or small. (This is presumed to have been the case in all human populations at some time in the past.) Next, the death rate declines while the birth rate remains high, so the population growth rate increases. Subsequently, however, the birth rate also declines until it is similar to or perhaps even lower than the death rate. The population growth rate therefore declines again eventually (sometimes to become negative with the death rate higher than the birth rate), though at a population size far greater than before the transition began.

early, late and future demographic transitions

The hypothesis commonly proposed to explain this transition, put simply, is that it is an inevitable consequence of industrialization, education and general modernization leading, first, through medical advances, to the drop in death rates, and then, through the choices people make (delaying having children and so on) to the drop in birth rates.

Certainly, when all the regional populations of the world are considered together, there has been a dramatic decline from the peak population growth rate of about 2.1% per year in 1965–1970 to around 1.1–1.2% per year today (Figure 12.3). And, as Cohen (2005) points out, while population growth rate has

global population growth rate peaked before 1970 and has declined since then

Figure 12.3

Population growth rate averaged for the world as a whole from 1950 to 2050.

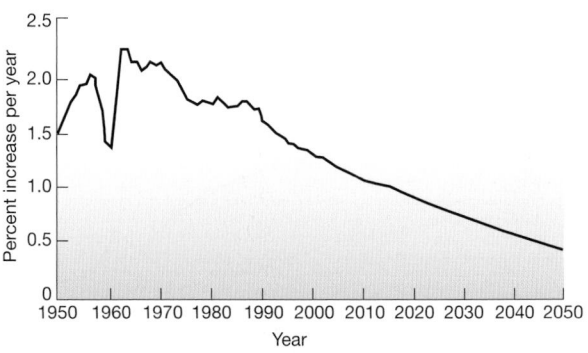

fallen at times in the past (during the plague and great wars), never before the 20th century has a fall in the global population growth rate been 'voluntary'.

The decade we are now passing through (2000–2010) has a very special place in human history because it will encompass three unique transitions:

the current decade is unique in the history of human population dynamics

1 Until now, young people (e.g. the 0–4 years class) have always outnumbered old people (e.g. the 60+ years class), but from 2000 the old will outnumber the young.

2 Until now, rural people have always outnumbered urban people, but from approximately 2007 urban people will predominate.

3 From 2003 onward, women, on average, worldwide have had, and will continue to have, too few or just enough children during their lifetime to replace themselves and the children's father in the next generation (Cohen 2005).

The first two transitions must be considered problematic from the point of view of sustainability – will the small population of workers be able to sustain a large body of senior citizens? And will the small population of agricultural workers be able to provide food for the rest of us? The third transition gives cause for some optimism – but the dramatic drop in population growth rate by no means provides an immediate fix to the population problem, as we will see in the next section.

12.2.4 Two future inevitabilities

unsustainable age structures?

Even if it were possible to effect some kind of demographic transition in all countries of the world, so that the birth rates were no more than the death rates (zero growth), would the 'population problem' be solved? The answer, sadly, is 'No', for at least two important reasons. First, there is a big difference in age structure between a population with equal birth and death rates in which both those rates are high and one in which both are low. When life tables were described in Chapter 5, we made the point that the net reproductive rate of a population was a reflection of the age-related patterns of survival and birth. A given net reproductive rate, though, can be arrived at through a literally infinite number of different birth and death patterns, and these different combinations them- selves give rise to different age structures within the population. If birth rates are

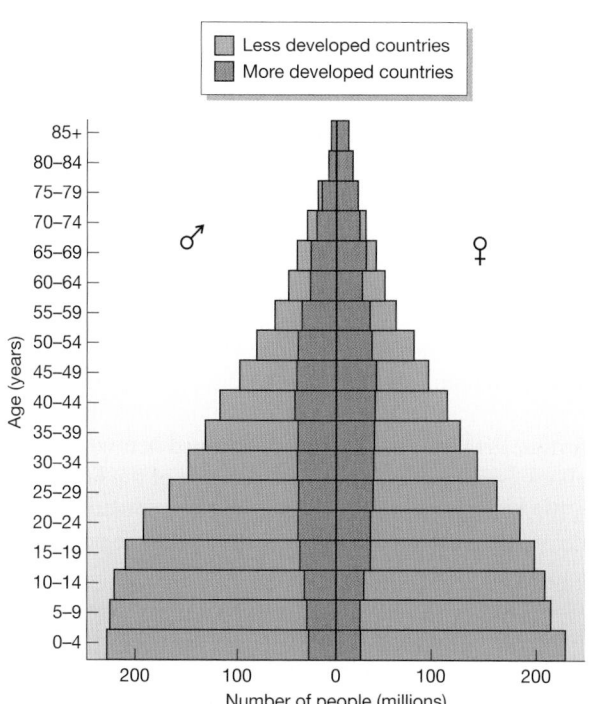

AFTER COHEN 2005

Figure 12.4

Predicted population size and age structure in 2050 for the less developed and more developed countries of the world. The horizontal scale is in millions of people (males to the left and females to the right), and the vertical scale shows age groups in 5-year increments. In the two centuries prior to 1950, Europe and the New World experienced the most rapid population growth, while the populations of most of Asia and Africa grew very slowly. But since 1950, rapid population growth has shifted from Western countries to Africa, the Middle East and Asia. Note the way the population of more developed countries becomes strongly biased towards older people, while that of less developed countries demonstrates a very much stronger representation of young people. China and the USA are excluded from the graph because they are exceptions in their categories: China's long-standing one-child policy will produce an age structure more like developed countries and the USA will retain a 'younger' age profile because of substantial immigration.

high but survival rates low ('pre-transition') then there will be many young, and relatively few old individuals in the population. But if birth rates are low and survival rates high – the 'ideal' to which we might aspire 'post-transition' – then relatively few young, productive individuals will be called upon to support the many who are old, unproductive and dependent (see Box 12.1). The size and growth rates of the human population are not the only problems: the age structure of a population adds yet another (Figure 12.4).

Moreover, suppose that our understanding was so sophisticated, and our power so complete, that we could establish equal birth and death rates tomorrow. Would the human population stop growing? The answer, once again, is 'No'. Population growth has its own momentum, which would still have to be contended with. Even with a birth rate matched to the death rate, there would be many years before a stable age structure was established, and in the mean time there would be considerable further population growth before numbers leveled off. According to a population projection prepared by the United Nations (the 'medium fertility variant'), the world's population is expected to grow from 6.3 billion today to peak at 8.9 billion in 2050 (Cohen, 2003). The reason, simply, is that there are, for example, many more babies in the world now than there were 25 years ago, and so even if birth rate per capita drops considerably now, there will still be many more births in 25 years' time, when these babies grow up, than at present; and these children, in turn, will continue the momentum effect before an approximately stable age structure is eventually established. As can be gauged from Figure 12.4 it is the populations in the developing regions of the world, dominated by young individuals, that will provide most of the momentum for further population growth.

the momentum of population growth

12.2.5 A global carrying capacity?

The current rate of increase in the size of the global population is unsustainable even though it is lower now than it has been: in a finite space and with finite resources, no population can continue to grow forever. What is an appropriate response to this? To suggest an answer, it is necessary to have some sense of a target, and thus it is interesting, and may be important, to know how large a population of humans could be sustained on the Earth. What is the global carrying capacity?

There is astonishing variation in the estimates that have been proposed over the last 300 or so years and even the estimates since 1970 span three orders of magnitude – from 1 to 1000 billion. To illustrate the difficulty in arriving at an estimate of global carrying capacity, a few examples are described here (see Cohen, 1995, 2005 for further details of the authors mentioned below).

some estimates of 'the global carrying capacity'

In 1679, van Leeuwenhoek estimated that the inhabited area of the Earth was 13,385 times larger than his home nation of Holland, whose population then was about 1 million people. He then assumed that all this area could be populated as densely as Holland, yielding an upper limit of roughly 13.4 billion.

In 1967, De Wit asked the question 'How many people can live on Earth if photosynthesis is the limiting process?' The answer he arrived at was roughly 1000 billion. He built into his calculation the fact that the length of the potential growing season varies with latitude, but assumed, amongst other things, that neither water nor minerals were limiting. He acknowledged that if people wanted to eat meat, or wanted what most of us consider a reasonable amount of living space, and so on, then the estimate would be much less.

By contrast, Hulett in 1970 assumed that levels of affluence and consumption in the United States were 'optimal' for the whole world, and that not only food but requirements for renewable resources like wood and non-renewable resources like steel and aluminum needed to be brought into the calculations. The figure he came up with was no more than 1 billion. Kates and others, in a series of reports from 1988, made similar assumptions but worked from global rather than United States averages and estimated a global carrying capacity of 5.9 billion people on a basic diet (principally vegetarian), or 3.9 billion on an 'improved' diet (about 15% of calories from animal products) or 2.9 billion on a diet with 25% of calories from animal products.

More recently, Wackernagel and his colleagues in 2002 sought to quantify the amount of land humans use to supply resources and to absorb wastes (embodied in their 'ecological footprint' concept). Their preliminary assessment was that people were using 70% of the biosphere's capacity in 1961 and 120% by 1999. They reasoned, in other words, that global carrying capacity was exceeded before the turn of the millennium – when our population was about 6 billion.

defining global carrying capacity is far from straightforward

As Cohen (2005) has pointed out, many estimates have been based on (or rely heavily on) a single dimension – biologically productive land area, water, energy, food and so on – and a difficulty with them all is the reality that the impact of one factor depends on the value of others. Thus, for example, if water is scarce and energy is abundant, water can be desalinated and transported to where it is in short supply, a solution that is not available if energy is expensive. Moreover, it is clear from the examples above that there is a difference between the number the Earth can support and the number that can be supported with an acceptable standard of living. The higher estimates come closer to the concept of a carrying capacity we normally apply to other organisms (see Chapter 5) – a number 'imposed' by the limiting resources of the environment. But it is unlikely that many of us would choose to live crushed up against an environmental ceiling or wish it on our descendants.

In any case, it is a big step to assume that the human population is limited 'from below' by its resources rather than 'from above' by its natural enemies. Infectious disease in particular, which not long ago was considered to be an enemy largely vanquished, is now once again perceived, for example by the World Health Organization, as a major threat to human welfare. Just consider the growing epidemics in tuberculosis and HIV and AIDS and the deaths caused by malaria. We saw in Chapter 7 that many infectious diseases thrive best in the densest populations.

Any suggestions we make about a global carrying capacity clearly depend on choices we make both for ourselves and for others. Most of us would choose to live at least as well as we do at present, but can the global population afford to choose for the whole world to live at least as well as those in developed countries do now? The answer to any question depends on what is meant by the question – defining what we mean by 'the global carrying capacity' is far from straightforward.

12.3 Harvesting living resources from the wild

A major limit to the number of people the Earth can support is the food that can be obtained. Populations of many species living freely in the wild are exploited for food by humans, who 'cull' or 'harvest' a proportion of the population, leaving some individuals behind to grow and reproduce for future harvests. Primitive human societies obtained all their resources like this, by hunting and gathering from nature, and humans continue to garner some resources in this way. The resources may be fish from the sea, deer from a moorland or timber from a forest. There is an important difference between resources obtained in this way and those that are farmed (Sections 12.4 and 12.6). Farmed resources are obtained by taking chosen species of plant or animal, domesticating them (often changing them genetically) and growing or rearing them in more or less controlled monocultures. These resources tend to be owned and managed by a farmer or organization. In contrast, most of the oceans and forests that are fished and hunted have at one time been common property, open to free-for-all unsustainable looting by all-comers. Recently, though, fishing and hunting have also come under increasing national and international regulation and national claims to 'ownership'. Many of our examples in this section are of fish or fisheries, but the principles apply to the harvesting of any natural resource.

12.3.1 Fisheries: maximum sustainable yields

Whenever a natural population is exploited there is a risk of overexploitation: too many individuals are removed and the population is driven into biological jeopardy or economic insignificance – perhaps even to extinction. Global catches of marine fish rose five-fold between 1950 and 1989 and many of the world's fish

aiming for the narrow path between over- and underexploitation

Figure 12.5

Changes in the contribution to global marine fish production made by fisheries in different phases of their exploitation. In the 1950s most of the catches were from undeveloped fisheries, but by 2000 most fisheries were fully exploited (near their maximum sustainable yield), or overexploited or had already collapsed.

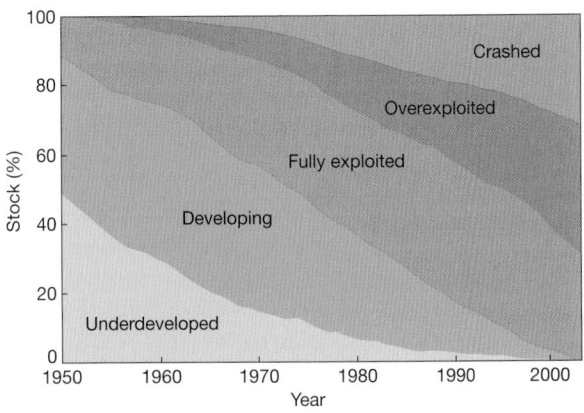

AFTER KHAN ET AL., 2006

stocks are now beyond the point of overexploitation (Figure 12.5). But harvesters also want to avoid underexploitation: if fewer individuals are removed than the population can bear, the harvested crop is smaller than necessary, potential consumers are deprived and those who do the harvesting are underemployed. It is not easy to tread the narrow path between under- and overexploitation. It is asking a great deal of a management policy to combine the well-being of the exploited species, the profitability of the harvesting enterprise, continuing employment for the workforce and the maintenance of traditional lifestyles, social customs and natural biodiversity.

population dynamics in the absence of exploitation – humped net recruitment curves

The most fundamental aspects of ecology that we need to understand here were introduced in Chapter 5 when the effects of intraspecific competition on populations were discussed. To determine the best way to exploit a population, it is necessary to know what the consequences will be of different exploitation 'strategies'. But in order to know these consequences, we first need some understanding of the dynamics of the population in the absence of, or prior to, exploitation. It is usual to assume that, before it is exploited, a harvestable population is crowded and intraspecific competition is intense. Summarizing from Chapter 5, and remembering that these are broad generalizations:

- populations in the absence of exploitation can be expected to settle around their carrying capacity, but exploitation will reduce numbers to less than this;
- exploitation, by reducing the intensity of competition, moves the population 'leftwards' along the humped net recruitment curve, increasing the net number of recruits to the population per unit time (Figure 12.6).

Figure 12.6

The humped relationship between the net recruitment into a population (births minus deaths) and the size of that population, resulting from the effects of intraspecific competition (see Chapter 5). Population size increases from left to right, but increasing rates of exploitation take the population from right to left.

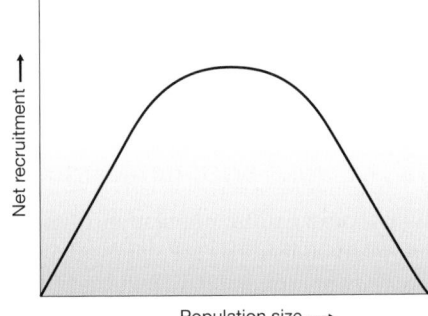

In fact, we can go further with Figure 12.6, since it is clear from the shape of the curve that there must be an 'intermediate' population size at which the rate of net recruitment is highest. Consider a time scale of years. The peak of the curve might be '10 million new fish recruited each year'. This is then also the highest number of new fish that could be removed from the population each year that the population itself could replenish. It is known as the *maximum sustainable yield* (MSY): the largest harvest that can be removed from the population regularly and indefinitely. It looks as though a fishery could tread the narrow path between under- and overexploitation if the fishers could find a way to achieve this MSY.

MSY – the narrow path?

The MSY concept has been the guiding principle in resource management for many years in fisheries, forestry and wildlife exploitation, but it is very far from being the perfect answer for a variety of reasons.

the MSY concept has shortcomings

1 By treating the population as a number of similar individuals, it ignores all aspects of population structure such as size or age classes and their differential rates of growth, survival and reproduction.

2 By being based on a single recruitment curve, it treats the environment as unvarying.

3 In practice, it may be impossible to obtain a reliable estimate of the MSY.

4 Achieving an MSY is by no means the only, nor necessarily the best, criterion by which success in the management of a harvesting operation should be judged. (It may, for example, be more important to provide stable, long-term employment for the workforce.)

12.3.2 Obtaining MSYs through fixed quotas

There are two simple ways of obtaining an MSY on a regular basis: through a 'fixed quota' and through a 'fixed effort'. With fixed quota MSY harvesting (Figure 12.7), the same amount, the MSY, is removed from the population every year. If (and it is a big if) the population stayed exactly at the peak of its net recruitment curve, then this could work: each year the members of the population, through their own growth and reproduction, would add exactly what the harvesting removed. But if by chance numbers fell even slightly below those at which the curve peaked, then the numbers harvested would exceed those recruited. Population size would then decline to below the peak of the curve,

the fragility of fixed quota harvesting . . .

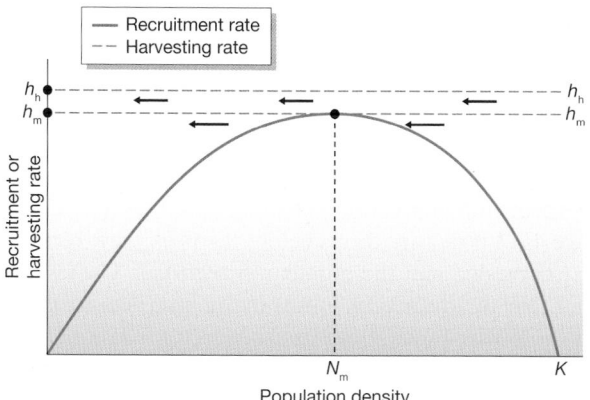

Figure 12.7

Fixed quota harvesting. The figure shows a single recruitment curve (solid line; recruitment in relation to density, N) and two fixed quota harvesting curves (dashed lines): high quota (h_h), and MSY quota (h_m). The arrows in the figure refer to changes to be expected in abundance under the influence of the harvesting rate to which the arrows are closest. The black dots indicate equilibria. At h_h the only 'equilibrium' is when the population is driven to extinction. The MSY is obtained at h_m because it just touches the peak of the recruitment curve (at a density N_m): populations greater than N_m are reduced to N_m, but populations smaller than N_m are driven to extinction. K is the carrying capacity, the density where the population is expected to settle in the absence of exploitation.

Figure 12.8

Landings of the Peruvian anchovy since 1950. Note the dramatic crash that resulted mainly as a result of overfishing. The stock has taken 20 years to rebuild.

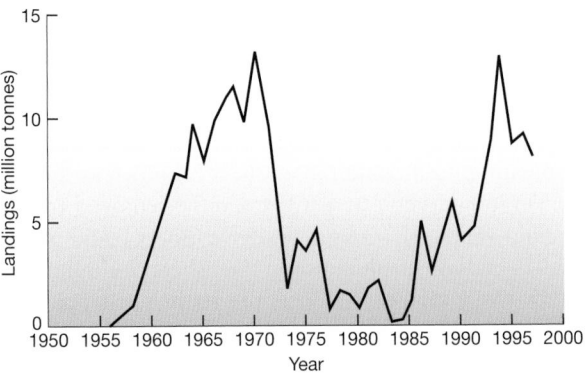

AFTER JENNINGS ET AL., 2001

and if a fixed quota at the MSY level were maintained the population would carry on declining until it was extinct (Figure 12.7). Furthermore, if the MSY was even slightly overestimated (and reliable estimates are hard to come by) then harvesting rate would always exceed the recruitment rate and extinction would again follow. In short, a fixed quota at the MSY level might be desirable and reasonable in a wholly predictable world about which we had perfect knowledge. But in the real world of fluctuating environments and imperfect data sets, these fixed quotas are open invitations to disaster.

... borne out in practice

Nevertheless, a fixed quota strategy has frequently been used – a management agency formulates an estimate of the MSY, which is then adopted as the annual quota. On a specified day in the year, the fishery is opened and the accumulated catch is logged. A fairly typical example is provided by the Peruvian anchovy (*Engraulis ringens*) fishery (Figure 12.8). From 1960 to 1972 this was the world's largest single fishery, and it constituted a major sector of the Peruvian economy. Fisheries experts advised that the MSY was around 10 million tonnes annually, and catches were limited accordingly. But the fishing capacity of the fleet expanded, and in 1972 the catch crashed. Overfishing seems, at the least, to have been a major cause of the collapse, although its effects were compounded with the influences of profound environmental fluctuations, discussed below. A moratorium on fishing would have been an ecologically sensible step, but this was not politically feasible: 20,000 people were dependent on the anchovy industry for employment. The Peruvian government therefore allowed fishing to continue. The stock took more than 20 years to recover.

12.3.3 Obtaining MSYs through fixed effort

relative robustness of fixed effort harvesting

An alternative to trying to maintain a constant harvest is to maintain a constant 'harvesting effort' (e.g. the number of 'trawler-days' in a fishery or the number of 'gun-days' with a hunted population). With such a regime the amount harvested should increase with the size of the population being harvested (Figure 12.9). Now, in contrast to Figure 12.7 if density drops below the peak, new recruitment exceeds the amount harvested and the population recovers. The risk of extinction is much reduced. The disadvantages, however, are first, because there is a fixed effort, the yield varies with population size (there are good, but, more to the point, bad years), and second, steps need to be taken to ensure that nobody makes a greater effort than they are supposed to. Nonetheless, there are many

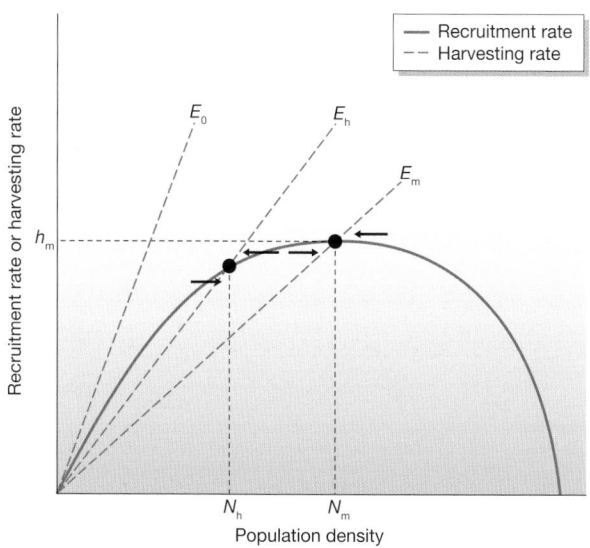

Recruitment rate
Recruitment rate
- - Harvesting rate

Figure 12.9

Fixed effort harvesting. Curves, arrows and dots as in Figure 12.7. The MSY is obtained with an effort of E_m, leading to a stable equilibrium at a density of N_m with a yield of h_m. At a somewhat higher effort (E_h), the equilibrium density and the yield are both lower than with E_m, but the equilibrium is still stable. Only at a much higher effort (E_0) is the population driven to extinction.

examples of harvests being managed by legislative regulation of effort. Harvesting of the important Pacific halibut (*Hippoglossus stenolepis*), for example, is limited by seasonal closures and sanctuary zones, though heavy investment in fisheries protection vessels is needed to control law breakers.

12.3.4 Beyond MSYs

There is no doubt that fishing pressure often exerts a great strain on populations. But the collapse of fish stocks in one year rather than any other is often the result of an occurrence of unusually unfavorable environmental conditions, rather than simply overfishing.

Harvests of the Peruvian anchovy (see Figure 12.8) collapsed from 1972 to 1973, but a previous steady rise in catches had already dipped in the mid-1960s as a result of an El Niño event: this happens when warm tropical water from the north reduces the upwelling, and hence the productivity, of the nutrient-rich cold Peruvian current coming from the south. By 1973, however, commercial fishing had so greatly increased that the subsequent El Niño event had even more severe consequences. There were some signs of recovery from 1973 to 1982, but a further collapse occurred in 1983 associated with yet another El Niño event. It is unlikely that the El Niño events would have had such severe effects if the anchovy had only been lightly fished. It is equally clear, though, that the history of the Peruvian anchovy fishery cannot be explained simply in terms of overfishing.

environmental fluctuations – the anchovy and El Niño

So far, this account has ignored population structure of the exploited species. This is a bad fault for two reasons. First, most harvesting practices are primarily interested in only a portion of the harvested population (mature trees, fish that are large enough to be saleable, etc.). Second, 'recruitment' is, in practice, a complex process incorporating adult survival, adult fecundity, juvenile survival, juvenile growth and so on, each of which may respond in its own way to changes in density and harvesting strategy. An example of a model that takes some of these variables into account was that developed for the Arcto-Norwegian cod fishery, the most northerly fish stock in the Atlantic Ocean. The numbers of fish

population structure and the Arctic cod (Gadus morhua)

Figure 12.10

Predictions for the stock of Arctic cod under three intensities of fishing and three different sizes of mesh in the nets. Larger meshes allow more and larger fish to escape capture. The largest effort (45%, bottom panel) is clearly unsustainable, regardless of the mesh size used. The largest sustainable catches are achieved with a low fishing effort (26%, upper panel) and a large mesh size.

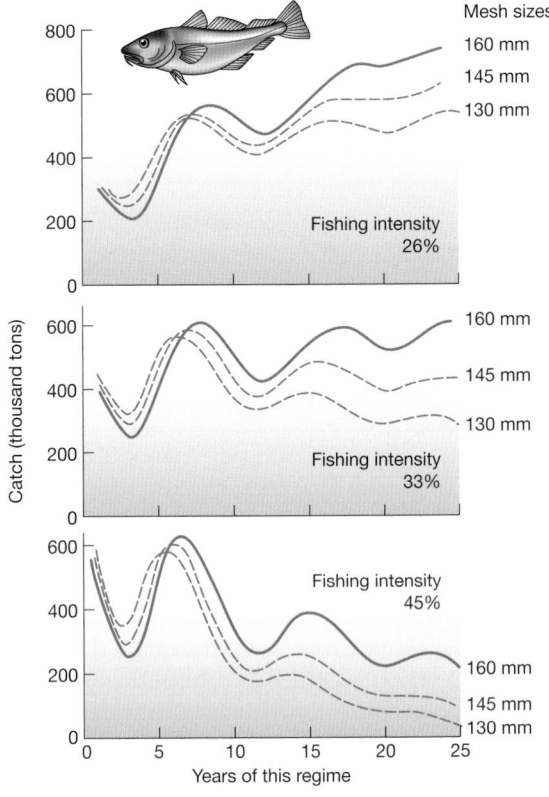

in different age classes were known for the late 1960s and this information was used to predict the tonnage of fish likely to be caught with different intensities of harvesting and with different net mesh sizes. The model predicted that the long-term prospects for the fishery were best ensured with a low intensity of fishing (less than 30%) and a large mesh size. These gave the fish more opportunity to grow and reproduce before they were caught (Figure 12.10). The recommendations from the model were ignored and, as predicted, the stocks of cod fell disastrously.

a strategy of taking only intermediate-sized fishes

Indigenous harvesters have long had their own 'regulations' to reduce the chance of overexploitation. In their harvesting of moi (*Polydactylus sexfilis*), Hawaiian fishermen, using traditional methods along the shore, take only intermediate-sized fishes, leaving both juveniles and large females. Thus they go a stage further than simply increasing net mesh size, which, while reducing the numbers of smaller individuals taken, nevertheless captures the largest individuals in the population. The good sense of the Hawaiian strategy has been reinforced by the discovery that large females of some fish not only produce exponentially more offspring but also that each of their offspring grows faster (Figure 12.11) and is more likely to reach adulthood. Protecting the largest individuals may give a great boost to sustainability.

precautionary management, closed areas and 'data-less' management

Managing most marine fisheries to achieve perfect, optimum yields is an unattainable dream. There are generally too few researchers to do the work and, in many parts of the world, no researchers at all. In these situations, a precautionary approach to fisheries management might involve locking away a proportion of a coastal or coral community in marine-protected areas (Hall, 1998). The term

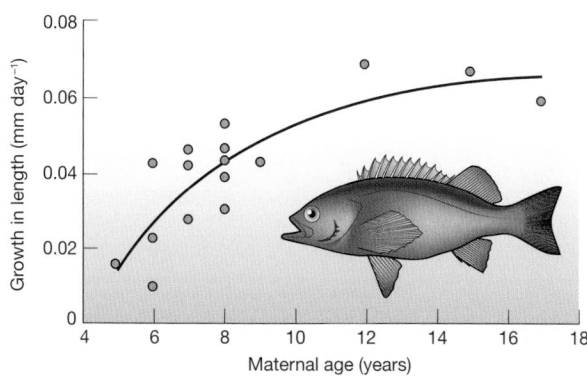

AFTER BOBKO & BERKELEY, 2004

Figure 12.11

The black rockfish (*Sebastes melanops*), off the coast of Oregon, USA, is a long-lived fish that produces live young. Not only do bigger fish produce more eggs to be fertilized, but the proportion of these that are in fact fertilized is itself greater in larger females. Futhermore, as shown in the graph, larvae produced by older (larger) females grow more than three times as fast as do larvae produced by their younger (smaller) counterparts.

data-less management has been applied to situations where local villagers follow simple prescriptions to make sustainability more likely – for example locals on the Pacific island of Vanuatu were provided with some simple principles of management for their trochus (*Tectus niloticus*) shellfishery: stocks should be harvested every 3 years and left unfished in between. The outcome has apparently been successful: continued economic viability (Johannes, 1998).

12.4 The farming of monocultures

Globally there is abundant food. Between 1961 and 1994 the per capita food supply in developing countries increased by 32% and the proportion of the world's population that was undernourished fell from 35% to 21%, though this is very unevenly distributed. Yet 800 million people remain hungry worldwide, and the rate of increase in per capita food production is falling.

Fishing and hunting (Section 12.3) have been human activities since our early history as hunter-gatherers. But the harvest that can be taken from nature was totally inadequate to support the main phases of growth of human populations. Increasingly, both animals and plants were domesticated and managed in ways that allowed much greater rates of production. The great bulk of the human food resource is now farmed – usually produced as dense populations of single species (*monocultures*). This allows them to be managed in specialized ways that can maximize their productivity, whether as immense monocultures of rice, corn or wheat (Figure 12.12), or as livestock factories producing beef, pork or poultry. Fish, indeed, are increasingly managed in the same way (aquaculture) – reared in enclosures, fed with controlled diets and harvested in mass production. Nearly a quarter of the fish supply in Asia is already produced in this way.

Only monoculture can maximize the rate of food production. This is because it allows the farmer to control and optimize with high precision the density of the populations (livestock or crop plants), the quantity and quality of their resources (food supplied to livestock; fertilizer and water to the crops) and often even the physical conditions of temperature and humidity. With many animals, the monocultures extend to segregating livestock or poultry into narrow age bands or age classes. If the only important criteria are economic ones, then there need be none of the uneconomic mixing of calves with cows or chickens with hens; fish eggs and fry can be segregated from potentially cannibalistic adults; the grossly

monoculture – and beyond

Figure 12.12
Agricultural monoculture:
wheat as far as the eye
can see.

uneconomic equality of the sex ratio that is common in nature can be distorted by culling to give efficient all-female dairy herds of cattle, or all-hen populations in batteries for egg production. This is a far cry from the ecology of the primitive human hunter-gatherers, who subsisted on their gleanings from the tangled web of wild nature!

but disease spreads in monocultures

To what extent, though, are modern farming methods sustainable? There is abundant evidence that a high price has to be paid to sustain the high rates of food production achieved by farmed monocultures. For example, they offer ideal conditions for the epidemic spread of diseases such as mastitis, brucellosis and swine fever among livestock and coccidiosis among poultry. Farmed animals are normally kept at densities far higher than their species would meet in nature with the result that disease transmission rates are magnified (see Chapter 7). In addition, high rates of transmission between herds occur as animals are sold from one farming enterprise to another, and it is easy for the farmers themselves, with mud on their boots and their vehicles, to act as vectors of pests and disease. The dramatic spread of foot and mouth disease in 2001 among British livestock provides a graphic example.

Crop plants, too, provide illustrations of the fragility of human dependence on monocultures. The potato, for example, was not introduced across the Atlantic to Europe until the second half of the 16th century, but three centuries later other foods had given way to it, and it had become the almost exclusive food crop of the poorer half of the population of Ireland. Dense monoculture, though, provided ideal conditions for the devastating spread of late blight (the fungal pathogen *Phytophthora infestans*) when it also crossed the Atlantic in the 1840s. The disease spread rapidly, dramatically reducing potato yields and also decomposing the tubers in storage. Out of the Irish population of about 8 million, 1.1 million died in the resulting famine and another 1.5 million emigrated to the UK and the USA.

In more modern history, an outbreak of southern corn leaf blight (caused again by a fungus, *Helminthosporium maydis*) developed in southeastern USA in the late 1960s and spread rapidly after 1970. Most of the corn grown in the area had been derived from the same stock and was genetically almost uniform. This extreme monoculture allowed one specialized race of the pathogen to have devastating consequences. The damage was estimated as at least $1 billion in the USA and had repercussions on grain prices worldwide. One of our favorite fruits is also at great risk of economic disaster (Box 12.3).

12.3 Topical ECOncerns

12.3 TOPICAL ECONCERNS

Can this fruit be saved? The banana as we know it is on a crash course toward extinction

In June 2005, Dan Koeppel filed the story below.

For nearly everyone in the US, Canada and Europe, a banana is a banana: yellow and sweet, uniformly sized, firmly textured, always seedless.

The Cavendish banana – as the slogan of Chiquita, the globe's largest banana producer, declares – is 'quite possibly the world's perfect food'. . . . It also turns out that the 100 billion Cavendish bananas consumed annually worldwide are perfect from a genetic standpoint, every single one a duplicate of every other. It doesn't matter if it comes from Honduras or Thailand, Jamaica or the Canary Islands – each Cavendish is an identical twin to one first found in Southeast Asia, brought to a Caribbean botanic garden in the early part of the 20th century, and put into commercial production about 50 years ago.

That sameness is the banana's paradox. After 15,000 years of human cultivation, the banana is too perfect, lacking the genetic diversity that is key to species health. What can ail one banana can ail all. A fungus or bacterial disease that infects one plantation could march around the globe and destroy millions of bunches, leaving supermarket shelves empty.

A wild scenario? Not when you consider that there's already been one banana apocalypse. Until the early 1960s, American cereal bowls and ice cream dishes were filled with the Gros Michel, a banana that was larger and, by all accounts, tastier than the fruit we now eat. Like the Cavendish, the Gros Michel, or 'Big Mike', accounted for nearly all the sales of sweet bananas in the Americas and Europe. But starting in the early part of the last century, a fungus called Panama disease began infecting the Big Mike harvest.

1 *Use a web search to discover the options that might be used to safeguard the banana industry.*

2 *How far fetched do you consider the risk of global economic terrorism by deliberate spread of a banana disease?*

12.4.1 Degradation and erosion of soil

A United Nations report (1998) stated:

Agricultural intensification in recent decades has taken a heavy toll on the environment. Poor cultivation and irrigation techniques and excessive use of pesticides and herbicides have led to widespread soil degradation and water contamination.

Around 300 million hectares are now severely degraded around the world and a further 1.2 billion hectares – 10% of the Earth's vegetated surface – can be described as moderately degraded. Clearly much of agricultural practice has not been sustainable.

Land without soil can support only very small primitive plants such as lichens and mosses that can cling onto a rock surface. The rest of the world's terrestrial

agriculture and forestry
requires soil

vegetation has to be rooted in soil, which gives it physical support. The soil also serves as a store of essential mineral nutrients and water that are extracted by the roots during plant growth. Soil develops by the accumulation of finely divided mineral products of rock weathering and decomposing organic residues from previous vegetation. The characteristics of the soil under natural vegetation in any particular climatic region and on any particular rock type depends on the balance between these processes of accumulation and forces that degrade and remove the soil.

soil forms . . . and is lost

The formation and persistence of soil in a region depend on local natural checks and balances. Soil may be lost by being washed or blown away, perhaps to be redeposited as an accumulation of fine-textured 'loess' somewhere else. Soil is best protected when it contains organic matter, is always wholly covered with vegetation, is finely interwoven with roots and rootlets and is on horizontal ground. Natural soil systems are probably *always* too fragile to be fully sustained when land is brought into cultivation. Dramatic evidence of unsustainable land use came from the 'dust bowl' disaster in the Great Plains of the United States and a similar disaster happening currently in China (Box 12.4).

12.4 *Historical landmarks*

Soil erosion, America's historical 'dust bowl' and China's current problem

Large areas of southeastern Colorado, southwestern Kansas and parts of Texas, Oklahoma and northeastern New Mexico were once used to support rangeland management of livestock. The vegetation consisted largely of native perennial grasses and had been neither ploughed nor sown with seed.

At the time of the First World War, much of the land was ploughed and annual crops of wheat were grown. There were poor crops in the early 1930s due to severe drought and the topsoil was exposed and carried away by the wind. Black blizzards of windblown soil blocked out the sun and piled the dirt in drifts. Occasionally the dust storms swept completely across the country to the East Coast. Thousands of families were forced to leave the region at the height of the Great Depression in the early and mid-1930s. The wind erosion was gradually halted with federal aid: windbreaks were planted and much of the grassland was restored. By the early 1940s the area had largely recovered.

The story is being repeated today in northwest China, where the need to feed 1.3 billion people has led to the raising of too many cattle and sheep, and the use of too many plows. This is more than the land can stand and 2300 km^2 are turning to desert each year. A huge dust storm blanketed areas from Canada to Arizona in April 2001 – the dust originated in China.

Dust bowl field and abandoned farm.

In an ideal sustainable world, new soil would be formed as fast as the old was lost. In Britain about 0.2 tonnes of new soil is produced naturally per hectare per year and it has been suggested that a tolerable (although not necessarily sustainable) rate of soil erosion might be about 2.0 t ha^{-1} yr^{-1}. However, rates of erosion have been recorded of up to 48 t ha^{-1} yr^{-1}!

Almost all (perhaps all) agricultural land will support higher yields if artificial fertilizers are applied to supplement the nitrogen, phosphorus and potassium supplied naturally by the soil. Fertilizers are cheap, easy to handle, of a guaranteed composition, allow even and accurate application, and higher and more predictable yields. When there is an overwhelming reliance on them, however, maintaining the organic matter capital of the soil tends to be neglected and has declined everywhere.

The degradation of soil by agriculture can be prevented, or at least slowed down, by: (i) incorporating farmyard manure, crop residues and animal wastes; (ii) alternating years under cultivation with years of fallow; or (iii) returning the land to grazed pasture or rangeland. Such practices conserve soil quality in technologically sophisticated agricultures in temperate regions.

But soil degradation is most serious and least easily prevented in less developed countries. The problems are greatest in high rainfall areas and on steeply sloping ground in the tropics where organic matter in the soil also decomposes more rapidly. The United Nations soil conservation strategy of 'Agenda 21' (formulated in Rio de Janeiro, 1992) recommended measures to prevent soil erosion and promote erosion control.

The most cost-effective technology used in reducing soil erosion is considered to be contour-based cultivation (Figure 12.13). In India, contour ditches have helped to quadruple the survival chances of tree seedlings and quintuple their early growth in height. Deeply rooted, hedge-forming 'vetiver' grass, planted in contour strips across hill slopes, slows water runoff dramatically, reduces erosion and increases the moisture available for crop growth. Currently 90% of soil conservation efforts in India are based on such biological systems. Simple technologies involving rock embankments constructed along contour lines for soil and water conservation have also been successful. Embanked fields in Burkina Faso (west Africa) yielded an average of 10% more crop production than traditional fields in a normal year and, in drier years, almost 50% more (United Nations, 1998).

soil maintenance

contour plowing and terracing – Agenda 21

Figure 12.13

Terracing of hill and mountainous land.

Such terracing provides a very high level of soil conservation but is possible only where labor is cheap. On lesser slopes, by ploughing and cultivating in strips along the contours, runoff of soil can be significantly reduced.

desertization and salinization

Agricultural land is also highly susceptible to degradation in arid and semiarid regions. Both overgrazing and excessive cultivation expose the soil directly to erosion by the wind and to rare but fierce rainstorms. In the process of '*desertization*', land that is arid or semiarid but has supported subsistence or nomadic agriculture gives way to desert. The process has often been slowed down for a time by irrigating the land. This gives a temporary remission but lowers the water table and salts accumulate in the topsoil (*salinization*). Once salts have started to accumulate, the process of salinization tends to spread and leads to an expansion of sterile, white salt deserts. This has been a particular hazard in irrigated areas of Pakistan.

forests protect . . . but not if harvested by clear felling

Forests protect soil from erosion because the canopy absorbs the direct impact of the rain on the soil surface, the perennial root systems bind the soil and leaf fall continually adds organic matter. But when forests are clear felled and then replanted, there is an open 'window of opportunity' for soil erosion until the forest canopy closes again. Cultivation and replanting along contours gives some control over soil erosion during this danger period, but the best precaution is to avoid clear felling and extract only a proportion of a forest stand at each harvest. This can often be technically difficult and more expensive.

12.4.2 The sustainability of water as a resource

water is a finite global resource

In the 1960s and 1970s, the main worry about the sustainability of global resources concerned energy supplies that were recognized to be finite and exhaustible. While energy resources remain finite, concern has shifted because exploration has revealed much larger reserves of oil, gas and even coal than had been entered into earlier environmental balance sheets. Water has now come into sharper focus. Fresh water, which is used in crop irrigation and for domestic consumption, is of crucial importance. On a global scale, agriculture is the largest consumer of fresh water, taking more than 70% of available supplies and more than 90% in parts of South America, central Asia and Africa.

water – the resource that future wars will be fought over?

There is a fixed stock of water on the globe and it is continually recycled as it evaporates from vegetation, land and sea and is then condensed and redistributed as precipitation. The human species now uses, directly or indirectly, more than half of the world's accessible water supply. The fresh water available per capita worldwide fell from 17,000 m^3 in 1950 to 7300 m^3 in 1995 and there is very considerable variation in availability from region to region (Figure 12.14). Many assessments of the problems of water supply suggest that countries with less than 1000 m^3 per person per year experience chronic scarcity. Water is widely thought to be the resource that future wars will be fought over. Even at a national level, the allocation of water resources can cause political problems, as occur for example in conflicts in California between urban and agricultural demands for water from the Colorado River. At the international level, conflict arises between countries that are upstream of their neighbors and are in a position to dam and divert water supplies. There are bitter cross-border disputes in South America, Africa and the Middle East between nations that share river basins.

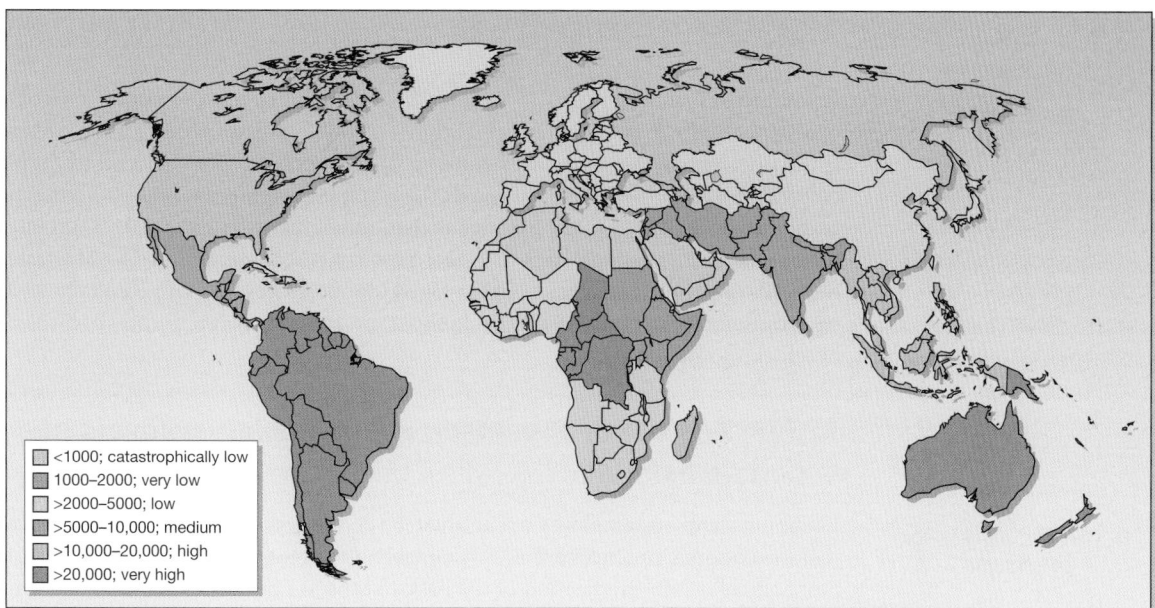

Legend:
- <1000; catastrophically low
- 1000–2000; very low
- >2000–5000; low
- >5000–10,000; medium
- >10,000–20,000; high
- >20,000; very high

Figure 12.14

Water availability per person from region to region of the globe in 2000. The units are in cubic meters per capita per year.

One response to chronic scarcity of water is to pump it from underground aquifers – but this often happens faster than the aquifers can be recharged. Such activity is clearly unsustainable. It is also the main cause of the loss of land from agriculture due to salinization. The demand for accessible supplies of water for both agriculture and domestic use has led to the plumbing of the Earth's river systems on a vast scale. The number of river dams more than 15 m high increased from about 5000 in 1950 to 38,000 in the 1990s.

In Chapter 13 we discuss the pollution of water by excreta, and by the pesticides and fertilizers applied in agriculture. Water that is uncontaminated by disease agents, nitrates or pesticides is an especially valuable commodity, but contamination occurs all too readily and removing contaminants (e.g. nitrates) is very expensive. Major dams built to control and conserve water in north and west Africa create large bodies of open water in which contamination spreads easily; one consequence has been the rapid spread along rivers of schistosomosiasis (a flatworm disease of humans), with infection rates rising from less than 10% to more than 98%.

contamination and conservation

Maintaining water supplies for human use also creates problems for the conservation of wildlife (see Chapter 14). The waterflow in many of the world's larger rivers is now very heavily controlled – in some cases little water now reaches the sea and wetlands have been lost or are at risk. Moreover, silt accumulates up-river instead of spreading into deltas and flood plains. The results may be catastrophic for wildlife areas and for human communities as well. For example, there is reason to believe that failure of silt deposition in the Nile delta (together with rising sea levels) may cause Egypt to lose up to 19% of its habitable land and displace 16% of its population within 60 years.

12.5 Pest control

what is a pest?

Pest control is another area in which the sustainability of agricultural practice may be threatened. A pest species is simply one that humans consider undesirable. Estimates suggest that there are around 67,000 species of pests that attack agricultural crops worldwide: 8000 weeds that compete with crops, and 9000 insects and mites and 50,000 plant pathogens that attack them (Pimentel, 1993). Here we consider the sustainability of insect pest control in agriculture to illustrate the types of problems that arise in managed monocultures. We could equally well have chosen the control of weeds or mollusks, or of the pests and diseases of farmed livestock, poultry or fish.

12.5.1 Aims of pest control: economic injury levels and action thresholds

EILs for pests, non-pests and potential pests

Economics and sustainability are intimately tied together. Market forces ensure that uneconomic practices are not sustainable. One might imagine that the aim of pest control is total eradication of the pest, but this is not the general rule. Rather, the aim is to reduce the pest population to a level at which it does not pay to achieve yet more control (the *economic injury level* or EIL). The EIL for a hypothetical pest is illustrated in Figure 12.15a. It is greater than zero (eradication is not profitable) but it is also below the typical, average abundance of the species. If the species was naturally self-limited to a density below the EIL, then it would never make economic sense to apply 'control' measures, and the species could

Figure 12.15

(a) The population fluctuations of a hypothetical pest. Abundance fluctuates around an 'equilibrium abundance' set by the pest's interactions with its food, predators, etc. It makes economic sense to control the pest when its abundance exceeds the economic injury level (EIL). Being a pest, its abundance exceeds the EIL most of the time (assuming it is not being controlled). (b) By contrast, a species that cannot be a pest always fluctuates below its EIL. (c) 'Potential' pests fluctuate normally below their EIL but rise above it in the absence of one or more of their natural enemies.

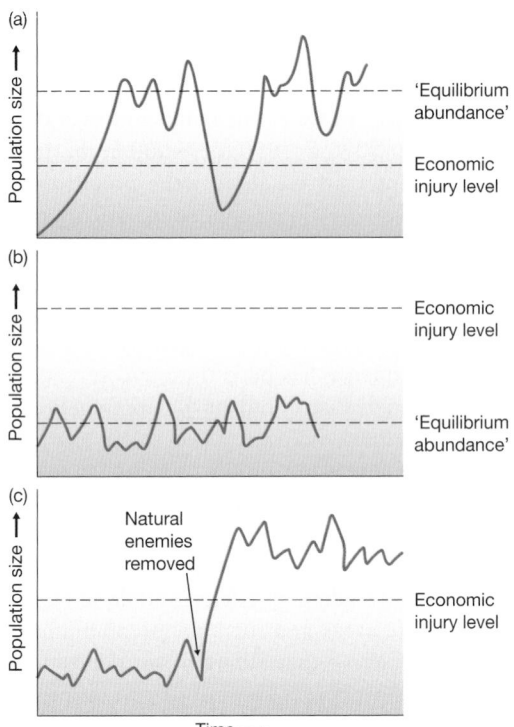

not, by definition, be considered a 'pest' (Figure 12.15b). There are other species, though, which have a carrying capacity (see Chapter 5) in excess of their EIL, but have a typical abundance that is kept below the EIL by natural enemies (Figure 12.15c). These are potential pests. They can become actual pests if their enemies are removed.

When a pest population has reached a density at which it is causing economic injury, however, it is generally too late to start controlling it. More important, then, is the *economic threshold* (ET): the density of the pest at which action should be taken to prevent it reaching the EIL. ETs are predictions based on detailed studies of past outbreaks or sometimes on correlations with climatic records. They may take into account the numbers not only of the pest itself but also of its natural enemies. As an example, in order to control the spotted alfalfa aphid (*Therioaphis trifolii*) on hay alfalfa in California, control measures have to be taken at specific times under certain circumstances:

1 In spring when the aphid population reaches 40 aphids per stem.
2 In summer and fall when the population reaches 20 aphids per stem, but the first three cuttings of hay are not treated if the ratio of ladybirds (beetle predators of the aphids) to aphids is one adult per 5–10 aphids, or three larvae per 40 aphids on standing hay, or one larva per 50 aphids on stubble.
3 During winter when there are 50–70 aphids per stem (Flint & van den Bosch, 1981).

12.5.2 Problems with chemical pesticides, and their virtues

A pesticide gets a bad name if, as is usually the case, it kills more species than just the one at which it is aimed. It may then become a pollutant (see Chapter 13). However, in the context of the sustainability of agriculture, the bad name is especially justified if it kills the pest's natural enemies and so contributes to undoing what it was employed to do. Thus, the numbers of a pest sometimes increase rapidly some time after the application of a pesticide. This is known as *target pest resurgence* and occurs when treatment kills both large numbers of the pest *and* large numbers of their natural enemies. Pest individuals that survive the pesticide or that migrate into the area later find themselves with a plentiful food resource but few, if any, natural enemies. The abundance of the pest population may then explode.

The after effects of applying a pesticide may involve even more subtle reactions. When a pesticide is applied, it may not be only the target pest that resurges. Alongside the target are likely to be a number of potential pest species that had been kept in check by their natural enemies (Figure 12.15c). If the pesticide destroys these, the potential pests become real ones – and are called *secondary pests*. A dramatic example concerns the insect pests of cotton in Central America. In 1950, when mass dissemination of organic insecticides began, there were two primary pests: the Alabama leafworm and the boll weevil (Smith, 1998). Organochlorine and organophosphate insecticides were applied fewer than five times a year and initially had apparently miraculous results – crop yields soared. By 1955, however, three secondary pests had emerged: the cotton bollworm, the cotton aphid and the false pink bollworm. The pesticide applications rose

target pest resurgence and secondary pest outbreaks

to 8–10 per year. This reduced the problem of the aphid and the false pink bollworm, but led to the emergence of five further secondary pests. By the 1960s, the original two pest species had become eight and there were, on average, 28 applications of insecticide per year. Clearly, such a rate of pesticide application is not sustainable.

Chemical pesticides lose their role in sustainable agriculture if the pests evolve resistance. The evolution of pesticide resistance is simply natural selection in action (see Chapter 2). It is almost certain to occur when vast numbers of a genetically variable population are killed. One or a few individuals may be unusually resistant (perhaps because they possess an enzyme that can detoxify the pesticide). If the pesticide is applied repeatedly, each successive generation of the pest will contain a larger proportion of resistant individuals. Pests typically have a high intrinsic rate of reproduction, and so a few individuals in one generation may give rise to hundreds or thousands in the next, and resistance spreads very rapidly in a population.

This problem was often ignored in the past, even though the first case of DDT (dichlorodiphenyltrichloroethane) resistance was reported as early as 1946 (houseflies in Sweden). The current scale of the problem is illustrated in Figure 12.16, which shows the exponential increase in the numbers of invertebrates that have evolved resistance and in the number of pesticides against which resistance has evolved. Resistance has been recorded in every family of arthropod pest (including dipterans such as mosquitoes and houseflies, as well as beetles, moths, wasps, fleas, lice, moths and mites) as well as in weeds and plant pathogens. Take the Alabama leafworm (see above), a moth pest of cotton, as an example. It has developed resistance in one or more regions of the world to aldrin, DDT, dieldrin, endrin, lindane and toxaphene.

If chemical pesticides brought nothing but problems, however – if their use was intrinsically and acutely unsustainable – then they would already have fallen out of widespread use. This has not happened. Instead, their rate of production

Figure 12.16

Global increases in the number of arthropod pest species reported to have evolved pesticide resistance and in the number of pesticide compounds against which resistance has developed. Each pest, on average, has evolved resistance to more than one pesticide, so there are now more than 2500 cases of evolution of resistance (pests × compounds).

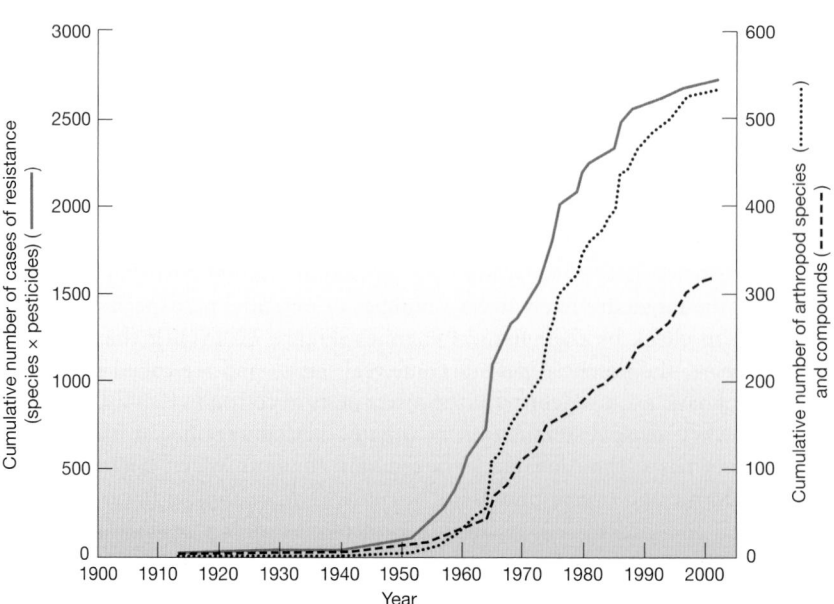

has increased rapidly. The ratio of cost to benefit for the individual producer has remained in favor of pesticide use: they do what is asked of them. In the USA, insecticides have been estimated to benefit the agricultural producer to the tune of around $5 for every $1 spent (Pimentel et al., 1978).

Moreover, in many poorer countries, the prospect of imminent mass starvation, or of an epidemic disease, are so frightening that the social and health costs of using pesticides have to be ignored. In general the use of pesticides is justified by objective measures such as 'lives saved', 'economic efficiency of food production' and 'total food produced'. In these very fundamental senses, their use may be described as sustainable. In practice, sustainability depends on continually developing new pesticides that keep at least one step ahead of the pests – pesticides that are less persistent, biodegradable and more accurately targeted at the pests.

12.5.3 Biological control

Outbreaks of pests occur repeatedly and so does the need to apply pesticides. But biologists can sometimes replace chemicals by other tools that do the same job and cost a great deal less. Biological control involves the manipulation of the natural enemies of the pests. There are three main types of biological control: importation, conservation and inoculation biological control.

three types of biological control

The first is the *importation* of a natural enemy from another geographic area – often the area where the pest originated. The objective is for the control agent to persist and thus maintain the pest below its economic threshold for the foreseeable future. This is a case of a desirable invasion of an exotic species and is often called *classical biological control*.

importation biological control

The most classic example of 'classical' biological control concerns the cottony cushion scale insect (*Icerya purchasi*), discovered as a pest of Californian citrus orchards in 1868. By 1886 it had brought the citrus industry to its knees. Species that colonize a new area may become pests because they have escaped the control of their natural enemies. Importation of some of these natural enemies is then, in essence, restoration of the status quo. A search for natural enemies led to the importation to California of two candidate species. The first was a parasitoid, a two-winged fly (*Cryptochaetum* sp.) that laid its eggs on the scale insect, giving rise to a larva that consumed the pest. The other was a predatory ladybird beetle (*Rodolia cardinalis*). Initially, the parasitoids seemed to have disappeared, but the predatory beetles underwent such a population explosion that, amazingly, all scale insect infestations in California were controlled by the end of 1890. Although the beetles have usually taken the credit, the long-term outcome has been that the beetles keep the scale insects in check inland, but *Cryptochaetum* is the main control near the coast (Flint & van den Bosch, 1981). The economic return on investment in biological control was very high in California and the ladybird beetles have subsequently been transferred to 50 other countries.

Another invasive scale insect was driving to extinction the national tree of the small South Atlantic island of St. Helena (the last home of another famous invader – Napoleon Bonaparte). Only 2500 St. Helena gumwoods (*Commidendrum robustum*) remained in 1991 as a result of attack by the South American scale insect *Orthezia insignis*. Fowler (2004) estimated that all remaining individuals of

Figure 12.17

Mean numbers (± SE, log scale), on continuously monitored 20 cm branchlets of 30 randomly selected gumwood trees, of the pest scale insect *Orthezia insignis* and its biological control agent, the ladybird *Hyperaspis pantherina*. Mean scale insect numbers dropped from more than 400 adults and nymphs (in September 1993) to fewer than 15 (in February 1995) when sampling ceased. Mean ladybird numbers increased from January to August 1994, coinciding with an obvious decline in scale insects, before ladybird numbers declined again. The highest recorded numbers of ladybirds were 1.3 adults and 3.4 larvae per 20 cm branchlet.

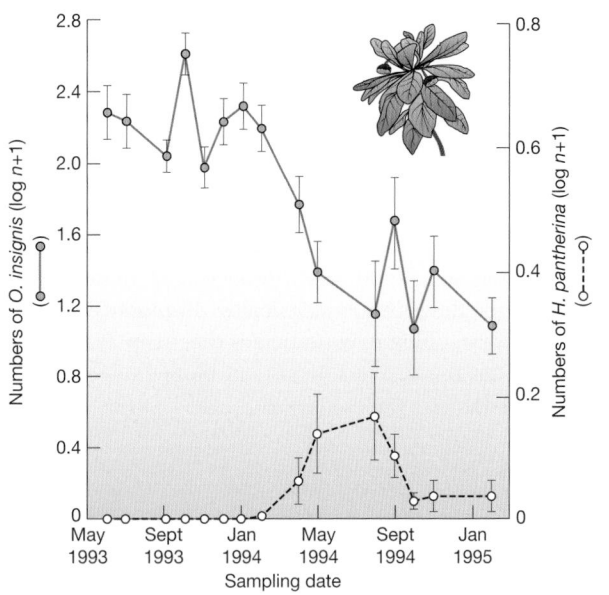

AFTER FOWLER 2004

this rare tree would have been dead by 1995. Another ladybird beetle saved the day. *Hyperaspis pantherina* was cultured and released on St. Helena in 1993 and as its numbers increased there was a corresponding 30-fold decrease in scale insect numbers (Figure 12.17). No scale outbreaks have been reported since 1995 and release of the ladybirds has been discontinued because the ladybird population is maintaining itself at low density in the wild, as good importation biocontrol agents should.

conservation biological control

In contrast to importation biological control, *conservation biological control* involves manipulations to increase the equilibrium density of natural enemies that are already native to the region where the pest occurs. In the case of the aphid pests of wheat (e.g. *Sitobion avenae*), predators that specialize on aphids include ladybirds and other beetles, heteropteran bugs, lacewings (Chrysopidae), fly larvae (Syrphidae) and spiders. Many of these natural enemies spend the winter in grassy boundaries at the edge of wheat fields, from where they disperse to reduce aphid populations around field edges. Farmers can protect grass habitat around their fields and even plant grassy strips in the interior to enhance these natural populations and the scale of their impact on the pests.

control by inoculation

A third class of biological control, *inoculation biological control* is widely practiced in the biological control of pests in glasshouses, where crops are removed, along with the pests and their natural enemies, at the end of the growing season. Two particularly important natural enemies used for inoculation are *Phytoseiulus persimilis*, a mite that preys on the spider mite *Tetranychus urticae*, a pest of roses, cucumbers and other vegetables, and *Encarsia formosa*, a chalcid parasitoid wasp of the whitefly *Trialeurodes vaporariorum*, a pest in particular of tomatoes and cucumbers.

biological control: excellent when it works . . .

Insects have been the main agents of biological control against both insect pests and weeds. Table 12.1 summarizes the extent to which they have been used and the proportion of cases where the establishment of an agent has greatly reduced or eliminated the need for other control measures.

Table 12.1

The record of insects as biological control agents against insect pests and weeds.

	INSECT PESTS	WEEDS
Control agent species	563	126
Pest species	292	70
Countries	168	55
Cases where agent has become established	1063	367
Substantial successes	421	113
Successes as a percentage of establishments	40	31

AFTER WAAGE & GREATHEAD, 1988

Biological control may appear to be a particularly environmentally friendly approach to pest control, but examples have come to light where even carefully chosen, and apparently successful, introductions of biological control agents have impacted on non-target species (Pearson & Callaway, 2003). *Cactoblastis* moths, which were introduced to Australia and were dramatically successful at controlling exotic cactuses, were accidentally introduced to Florida where they have been attacking several native cacti (Cory & Myers, 2000). Similarly, a seed-feeding weevil (*Rhinocyllus conicus*), introduced to North America to control exotic *Carduus* thistles, attacks several native thistles and has adverse impacts on populations of a native picture-winged fly (*Paracantha culta*) that feeds on the thistle seeds (Louda et al., 1997). Such ecological effects need to be better evaluated in future assessments of potential biocontrol agents.

. . . but sometimes non-target organisms are affected

12.6 Integrated farming systems

The desire for sustainable agriculture has increasingly led to more ecological approaches to food production, which are often given the label 'integrated farming systems'. Part of this, and something that preceded it historically, is a similar approach to pest control: *integrated pest management* (IPM).

IPM is a practical philosophy of pest management. It combines physical control (for example, simply keeping pests away from crops), cultural control (for example, rotating the crops planted in a field so pests cannot build up their numbers over several years), biological and chemical control and the use of resistant crop varieties. It has come of age as part of the reaction against the unthinking use of chemical pesticides in the 1940s and 1950s.

integrated pest management

IPM is ecologically based but uses all methods of control – including chemicals, where appropriate. It relies heavily on natural mortality factors, such as natural enemies and weather, and seeks to disrupt them as little as possible. It aims to control pests below an economically damaging level (the EIL), and it depends on monitoring the abundance of pests and their natural enemies and using various control methods as complementary parts of an overall program. IPM therefore calls for specialist pest managers or advisers. Broad-spectrum pesticides in particular, although not excluded, are used only very sparingly, and if chemicals are used at all it is in ways that minimize the costs and quantities used. The essence of the IPM approach is to make the control measures fit the pest problem, and no two pest problems are the same – even in adjacent fields.

Figure 12.18

Decision flow chart for the integrated pest management of potato tuber moths (PTM) in New Zealand. Boxed phrases are questions (e.g. 'Growth stage of crop?'), words in arrows are the farmer's answers to the questions (e.g. 'Pre tuber') and the recommended action is shown in the vertical boxes (e.g. 'Do not spray'). Note that February is late summer in New Zealand.

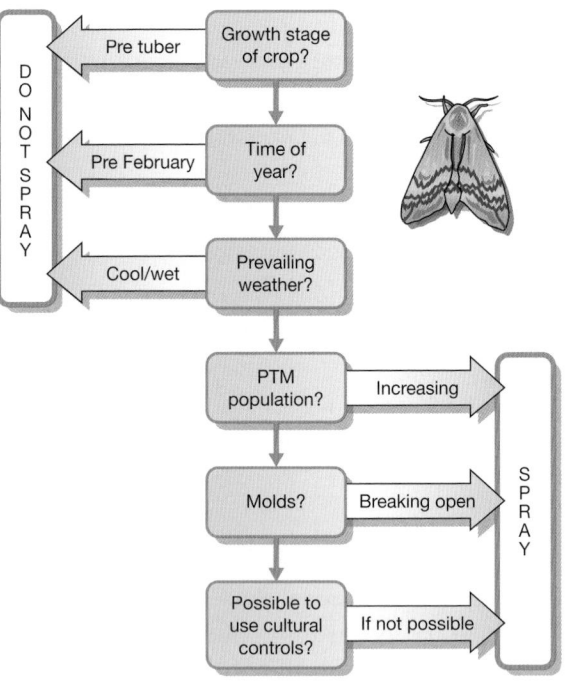

AFTER HERMAN, 2000

IPM for the potato tuber moth

The caterpillar of the potato tuber moth (*Phthorimaea operculella*) commonly damages crops in New Zealand. An invader from a warm temperate subtropical country, it is most devastating when conditions are warm and dry (i.e. when the environment coincides closely with its optimal niche requirements). There can be as many as 6–8 generations per year and different generations mine leaves, stems and tubers. The caterpillars are protected both from natural enemies (parasitoids) and insecticides when in the tuber, so control must be applied to leaf-mining generations. The IPM strategy for the potato tuber moth (Herman, 2000) involves monitoring (female pheromone traps, set weekly from midsummer, are used to attract males, which are counted), cultural methods (the soil is cultivated to prevent soil cracking, soil ridges are molded up more than once and soil moisture is maintained), and the use of insecticides, but only when absolutely necessary (most commonly the organophosphate, methamidophos). Farmers follow the decision tree shown in Figure 12.18.

integrated farming systems: LISA, IFS and LIFE

It has increasingly become apparent, in an agricultural context at least, that implicit in the philosophy of IPM is the idea that pest control cannot be isolated from other aspects of food production and is especially bound up with the means by which soil fertility is maintained and improved. Thus, a number of programs have been initiated to develop and put into practice sustainable food production methods that incorporate IPM, including not only IFS (integrated farming systems) but also LISA (low input sustainable agriculture) in the USA and LIFE (lower input farming and environment) in Europe (International Organisation for Biological Control, 1989; National Research Council, 1990). All share a commitment to the development of sustainable agricultural systems.

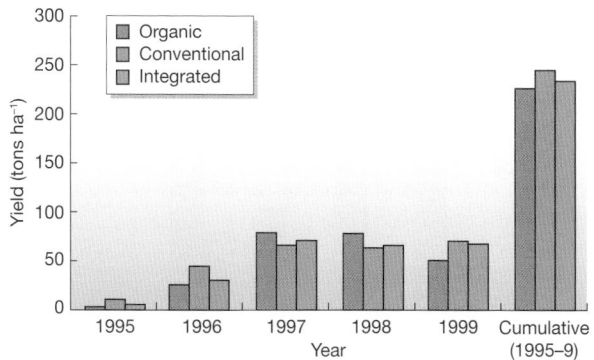

FROM REGANOLD ET AL., 2001

Figure 12.19

Fruit yields (metric tons per hectare) of three apple production systems.

These approaches have advantages in terms of reduced environmental hazard. Even so, it is unreasonable to suppose that they will be adopted widely unless they are also sound in economic terms. As we have already noted, in an industry such as agriculture, practices that are economically unsustainable are, ultimately, unsustainable overall. In this context, Figure 12.19 shows the yields of apples from organic, conventional and integrated production systems in Washington State from 1994 to 1999 (Reganold et al., 2001). Organic management excludes such conventional inputs as synthetic pesticides and fertilizers whilst integrated farming uses reduced amounts of chemicals by integrating organic and conventional approaches. All three systems gave similar apple yields but the organic and integrated systems had higher soil quality and potentially lower environmental impacts. When compared with conventional and integrated systems, the organic system produced sweeter apples, higher profitability and greater energy efficiency.

environmental and economic sustainability

12.7 Forecasting agriculturally driven global environmental change

Much attention has been focused on the predicted far-reaching consequences of global climate change caused by human activities such as the burning of fossil fuels. We deal with this in Chapter 13. However, very significant threats are also posed to ecosystems around the world by increasing agricultural development. In this chapter we have considered the problems of the more-than-exponential increase in the human population, and of the associated impacts of increased erosion, unsustainability of water supply, salinization and desertification, excess plant nutrients finding their way into waterways, and unwanted consequences of chemical pesticides. Model projections suggest that all these will increase over the next 50 years as more land is converted to grow crops and pasture (Tilman et al., 2001) (Figure 12.20). This can be expected to place biodiversity at high risk, particularly because population increases are predicted to be greatest in species-rich tropical areas. To control the environmental impacts of agricultural expansion, we will need scientific and technological advances as well as the implementation of effective government policies.

Figure 12.20

Projected increases in nitrogen (N) and phosphorus (P) fertilizers, irrigated land, pesticide use and total areas under crops and pasture by the years 2020 (maroon bars) and 2050 (green bars).

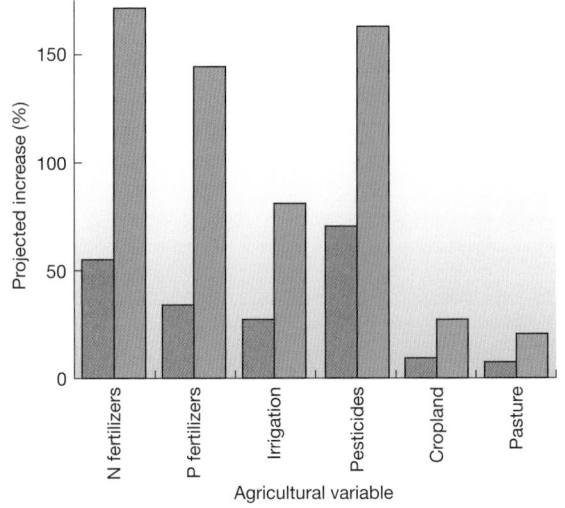

FROM LAURANCE 2001. BASED ON DATA IN TILMAN ET AL., 2001

Summary
SUMMARY

The human population 'problem'

Resource use by humans is defined as sustainable if it can be continued for the forseeable future. The root of most environmental problems is the 'population problem': a large human population that has been growing at a more-than-exponential rate.

Three groups of nations can be recognized: those that passed through the demographic transition 'early', 'late' or 'not yet'. Even if it were possible instantaneously to bring about the transition in all remaining countries of the world, the population problem would not be solved, partly because population growth has its own momentum.

The global carrying capacity for humans is variously estimated at between 1 and 1000 billion, depending mainly on what is deemed to constitute an acceptable standard of living.

Harvesting living resources from the wild

Whenever a natural population is exploited by harvesting there is a risk of overexploitation. But harvesters also want to avoid underexploitation – when potential consumers are deprived of resources and those who do the harvesting are underemployed.

The concept of the maximum sustainable yield (MSY) has been a guiding principle in the exploitation of natural populations. There are two simple ways of obtaining an MSY on a regular basis: through a 'fixed quota' and through a 'fixed effort'. Two limitations of the MSY approach are: (i) that it treats all individuals in the population as identical; and (ii) that it treats the environment as unvarying. Improved harvesting strategies correct both these oversights. Lack of knowledge of most fisheries around the world means that management is often based on the precautionary principle, often in the absence of data.

The farming of monocultures

Increasingly, animals and plants have been domesticated and managed in ways that allowed much larger harvests – usually as monocultures. But a high price may be paid to maintain these high rates of food production. Monocultures offer ideal conditions for the epidemic spread of diseases and lead to widespread degradation of land.

The sustainability of soil and of water supplies
In an ideal sustainable world, new soil would be formed as fast as the old was lost, but in most agricultural systems this is not achieved. When there is an overwhelming reliance on artificial fertilizers, maintaining the organic matter capital of the soil tends to be neglected and this has declined worldwide.

Soil degradation can be slowed down by incorporating manures and residues, alternating cultivation and fallow periods, or returning the land to grazed pasture. In tropical regions, terracing is widely practiced over hilly and mountainous terrain. In arid regions, overgrazing and excessive cultivation can lead to desertization and salinization.

Water is widely thought to be the resource that future wars will be fought over. On a global scale, agriculture is the largest consumer of fresh water. Pumping water from underground aquifers is the main cause of loss of agricultural land through salinization.

Pest control
The aim of pest control is to reduce the pest population to its economic injury level (EIL), but a so-called economic threshold may be of more immediate importance.

Pesticides may kill species other than their target and may give rise to target pest resurgence or secondary pest outbreaks. Pests may also evolve pesticide resistance.

Biologists may also manipulate the natural enemies of pests (biological control) via three forms of biological control – importation, conservation or inoculation – but even biocontrol agents can have unwanted effects on non-target species.

Integrated farming systems
Integrated pest management (IPM) is a practical philosophy of pest management that is ecologically based but uses all methods of control where appropriate. It relies heavily on natural mortality factors and calls for specialist pest managers or advisers.

Implicit in the philosophy of IPM is the idea that pest control cannot be isolated from other aspects of food production. A number of programs have been initiated to develop and put into practice sustainable food production methods that incorporate IPM. Evidence has been accumulating that this sustainable farming approach can yield improved economic returns too.

Agriculturally induced global change
It is clear that very significant threats are posed to ecosystems around the world by the increasing human population and concomitant increases to agricultural development. These are expected to have a particularly damaging effect on biodiversity because most agricultural growth is predicted to occur in the species-rich tropics.

Review questions

Asterisks indicate challenge questions

1* What is sustainability? Is it possible to have sustainable population growth? Sustainable use of fossil fuels? Sustainable use of forest trees? Justify your answers.

2 Describe what is meant by 'the demographic transition' in a human population. Explain why it might be important, for future management of human population growth, to discover whether the demographic transition is an academic ideal or a process through which all human populations necessarily pass.

3* The number of people that the Earth can support depends on their standard of living. Argue the case either for or against developing nations having the right to expect standards of living those in the developed world take for granted.

4 Contrast the ways in which 'fixed quota' and 'fixed effort' harvesting strategies seek to extract maximum sustainable yields from natural populations.

5 Discuss the pros and cons of agricultural monocultures.

6 One of the main bodies regulating the production of organic food (food produced without synthetic fertilizers or pesticides) in the United Kingdom is the Soil Association. Explain why you think it has adopted this name.

7 Explain the meaning and importance of the terms economic injury level and economic threshold.

8 Weigh up the advantages and disadvantages of the chemical and biological control of pests.

9 Explain why methods of pest control and methods of soil fertility maintenance need to be considered together in integrated farming systems.

10* Hilborn and Walters (1992) have suggested that there are three attitudes that ecologists can take when they enter the public arena. The first is to claim that ecological interactions are too complex, and our understanding and our data too poor, for definite pronouncements to be made (for fear of being wrong). The second possibility is for ecologists to concentrate exclusively on ecology and arrive at a recommendation designed to satisfy purely ecological criteria. The third is for ecologists to make ecological recommendations that are as accurate and realistic as possible, but to accept that these will be incorporated with a broader range of factors when management decisions are made – and may be rejected. Which of these do you favor, and why?

Chapter 13

Habitat degradation

Chapter contents

Key concepts

In this chapter you will:

- realize that *Homo sapiens* is just one species among many whose activities can reduce the quality of their environment – but to a dramatically greater extent
- understand that we have both physical effects (such as desertization and changes to riverflow) and chemical impacts (pollution by nitrates, carbon dioxide, chlorofluorocarbon, etc.)
- recognize that most pollutants produced on land ultimately affect the atmosphere or rivers, lakes and oceans
- understand that power generation is responsible for the most far-reaching environmental impacts when the carbon dioxide released contributes to global climate change
- appreciate the value to human welfare of ecosystem services lost when we degrade habitats

As the human population has grown and new technologies have been developed, we have had an ever-increasing impact on natural ecosystems. Physical degradation and chemical pollution associated with cultivation, power generation, urban life and industry have adversely affected human health and many 'ecosystem services' that were free and contributed greatly to human welfare. Our environmental problems have ecological, economic and sociopolitical dimensions, so a multidisciplinary approach will be needed to find solutions.

13.1 Introduction

13.1.1 Physical and chemical impacts of human activities

People destroy or degrade natural ecosystems to make way for agricultural, urban and industrial development. We physically damage the natural world when mining for non-renewable resources such as gold and oil, and even exploitation of a renewable resource can disrupt habitat when, for example, bottom trawling for fish damages deep-sea coral communities. The worldwide scale of damage is even greater as a result of chemical pollution produced by human activities such as defecation, cultivation, power generation and industry.

Homo sapiens – just another species?

Humans are not unique in degrading their environment. Feces, urine and dead bodies of animals are sometimes sources of pollution in their immediate environments – cattle avoid grass near their waste for several weeks, many birds carry away the fecal sacs of their nestlings and the 'undertaker' caste of honeybees removes dead bodies from the hive. Like us, many species also make profound physical changes to their habitats. Among the 'ecological engineers' of the natural world are beavers that construct dams, prairie dogs that build underground towns and freshwater crayfish that clear sediment from the riverbed. In each case, other species in the community are affected. And there are even species that, like farmers, increase plant nutrient concentrations in their habitats (leguminous plants – see Section 8.4.6), and others that produce 'pesticides' (certain plants produce allelochemicals, the function of which appears to be the inhibition of growth of neighbors).

the scale of human degradation reflects our population density and technology

When population density was low, and prior to our harnessing of non-food energy, humans probably had no greater impact than many other species. But now the scale of human effects is proportional to our huge numbers and the advanced technologies we employ.

physical degradation of habitats

Physical degradation of habitats includes soil loss and desertization caused by intensive agriculture (discussed in Section 12.4.1) and changes to river discharge as a result of water impoundment for hydropower generation or abstraction for irrigation of crops (Section 13.2.5).

chemical degradation – pollution

Chemical degradation has many causes. Pesticides are applied to land but find their way to places they were not intended to be – passing up food chains

(Box 13.1) and moving via ocean currents to the ends of the Earth. A plethora of other exotic chemicals enter the natural environment from a variety of industrial sources. But the most far-reaching kinds of chemical degradation result not from our production of exotic chemicals but rather the augmentation of simple compounds that already occur naturally. The heavy use on land of nitrogen fertilizer spills into rivers, lakes and oceans, where raised levels of nitrate severely disrupt ecosystem processes – with blooms of microscopic algae shading out waterweeds and, when the algae die and decompose, reducing oxygen and killing animals. Another pollutant route is via the atmosphere. Thus, hundreds of kilometers downwind of large population centers, acid rain (caused by emission of oxides of nitrogen and sulfur from power generation) kills trees and drives lake fish

13.1 Topical ECOncerns

13.1 TOPICAL ECONCERNS

Pollution and the thickness of birds' eggshells

The peregrine falcon (*Falco peregrinus*) is a particularly distinctive and beautiful bird of prey with an almost worldwide range. Until the 1940s, about 500 pairs bred regularly in the eastern states of the United States and about 1000 pairs in the west and in Mexico. In the late 1940s their numbers started a rapid decline, and by the mid-1970s the bird had disappeared from almost all the eastern states and its numbers had fallen by 80–90% in the west. Similar dramatic declines were occurring in Europe. Peregrine falcons were listed as an *endangered species* (at risk of extinction). The decline also occurred in many other birds of prey and was traced to failure to hatch normal broods. There was very high breakage of eggs in the nest.

The cause was eventually identified as the accumulation of DDT (dichlorodiphenyltrichloroethane) in the parent birds. The pesticide had apparently contaminated seeds and insects that had then been eaten by small birds and had accumulated in their tissues. In turn these had been caught and eaten by birds of prey and the pesticide interfered with their reproduction – in particular causing the eggs to have thin shells and be more likely to break.

The use of DDT was banned in the United States in 1972. Programs were developed to breed peregrines in captivity and at least 4000 peregrines were bred and released to the wild. Peregrines are now breeding successfully over much of the United States and are no longer considered an endangered species. In Britain, recovery has been so successful that the peregrine has become regarded as a pest by pigeon fanciers and lovers of the smaller songbirds.

It was possible to identify DDT pollution as a cause of eggshell thinning because eggshells had been collected as dated specimens in museums and private collections. A measure of eggshell thickness in collections of eggs of the sparrow hawk (*Accipiter nisus*) showed a sudden stepwise fall of 17% in 1947, when DDT began to be used widely in agriculture, and a steady increase in thickness after DDT was banned (Figure 13.1).

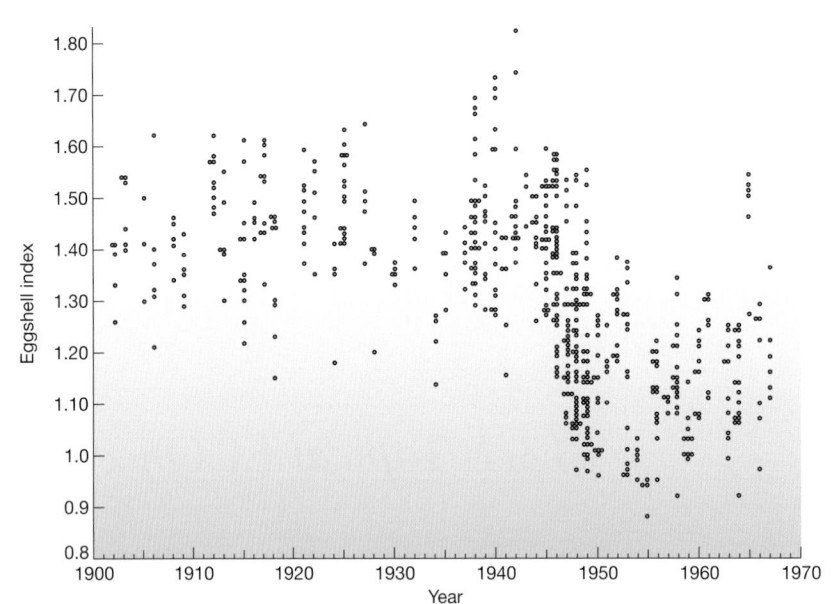

Figure 13.1

Graph showing the changes in sparrowhawk eggshell thickness (museum specimens) in Britain.
FROM RATCLIFFE, 1970

It was a surprise to ornithologists in Britain to find evidence of a decline in eggshell thickness of 2–10% in four species of thrush (*Turdus*) since the mid-19th century (Green, 1998). This seemed to have started long before the development of organic pesticides and there was no sudden change when DDT was introduced. Snails are an important part of the diet of thrushes; thrushes derive much of the calcium for their eggshells from snails. There is convincing evidence that acid rain, caused by release to the atmosphere of sulfur and nitrogen oxides from power generation and industry, has acidified leaf litter and reduced its calcium content, leading to a reduction in snail populations and in the calcium content of their shells. The shells of wild birds' eggs have therefore recorded two of the major, but quite different, forces of environmental pollution: pesticides (Section 13.2.5) and acid rain (Section 13.3.1).

extinct. And the biggest pollution problem of all involves the augmentation, via the burning of fossil fuels, of carbon dioxide in the atmosphere. The consequent global climate change has implications for every ecosystem in the world.

Our discussion of human degradation of habitats will first consider the consequences of cultivation (Section 13.2), before proceeding to an assessment of damage associated with the generation of power (Section 13.3), and then the ecological consequences of life in urban and industrial landscapes (Section 13.4). But first (Section 13.1.2) we will note how the cost of our activities can be tallied in relation to the free 'ecosystem services' that are lost when habitats are degraded. Discussion returns to this theme in the final section (13.5), which strikes a more optimistic note by discussing actions that can be taken to maintain or restore ecosystem services.

© ROB AND ANN SIMPSON, VISUALS UNLIMITED

Acid rain damage to spruce forest.

13.1.2 Economic costs of human impacts: lost ecosystem services

Biodiversity has intrinsic value. But there is also a utilitarian view of nature that focuses on the services that ecosystems provide for people to use and enjoy. *Provisioning services* include wild foods such as fish from the ocean and berries from the forest, medicinal herbs, fiber, fuel and drinking water, as well as the products of cultivation in agroecosystems. Nature also contributes the *cultural services* of esthetic fulfilment and educational and recreational opportunities. *Regulating services* include the ecosystem's ability to break down or filter out pollutants, the moderation by forests and wetlands of disturbances such as floods, and the ecosystem's ability to regulate climate (via the capture or 'sequestration' by plants of the greenhouse gas carbon dioxide). Finally, and underlying all the others, there are *supporting services* such as primary production, the nutrient cycling upon which productivity is based, and soil formation.

provisioning, cultural, regulating and supporting services

In the case of three important provisioning services – production of crops, livestock and aquaculture – human activities have had a positive effect. And because of increased tree planting in some parts of the world there has even been a global improvement in the sequestration of carbon by trees (a climate regulating service).

a few positive human effects on ecosystem services . . .

But we have degraded most of the other services (Millennium Ecosystem Assessment 2005). As discussed in Chapter 12, many fisheries are now overexploited (a negative effect on this provisioning service), while intensive agriculture has worked against the ecosystem's ability to replace soil lost to erosion (a regulating service). The continuing loss of forest in tropical regions has negative effects on the ability of the terrestrial ecosystem to regulate riverflow – deforestation increases flow during flooding and decreases it during dry periods. And, as we saw in Section 1.3.3, deforestation (or even just the loss of riverside vegetation) can diminish the terrestrial ecosystem's capacity to hold and recycle nutrients (another regulating service), releasing large quantities of nitrate and other plant nutrients

. . . but many negative effects

into waterways. Note that the modification of an ecosystem to enhance one service (e.g. intensification of agriculture to produce more crop per hectare – a provisioning service) generally comes at a cost to other services previously provided (loss of regulating services such as nutrient uptake and of cultural services such as sacred sites, streamside walks and valued biodiversity) (Townsend, 2007).

a valuation of ecosystem services . . .

The concept of ecosystem services is important because it focuses on how ecosystems contribute to human well-being, providing a counterpoint to the economic reasons that justify our degradation of nature in the first place (to produce food, fiber, fuel, housing and luxury products for a burgeoning human population).

Economists can put a value on nature in a variety of ways. A provisioning service for which there is a market is straightforward – values are easily ascribed to clean water for drinking or irrigation, to fish from the ocean and medicinal products from the forest. A more imaginative approach is required in other situations. Thus, the *travel cost* that tourists are willing to pay to visit a natural area provides a minimum value of the cultural service provided. To determine *contingent valuation*, surveys of the public assess their willingness to pay for each of a set of alternative land use scenarios; the answer is thus 'contingent' on a specific hypothetical scenario and description of the environmental service concerned. *Replacement cost* estimates how much would need to be spent to replace an ecosystem service with a man-made alternative, for example by substituting the natural waste disposal capacity of a wetland by building a treatment works. And when an ecosystem service has already been lost, the real costs become apparent. Take, for example, the largely deliberate burning of 50,000 km^2 of Indonesian vegetation in 1997 – the economic cost comprised US\$4.5 billion in lost forest products and agriculture, increased greenhouse gas emissions, reductions to tourism and healthcare expenditure on 12 million people affected by the smoke (Balmford & Bond 2005).

. . . adding up to a global total of \$38 trillion

Costanza et al. (1997) added up all ecosystem services worldwide, arriving at an estimate of US\$38 trillion (10^{12}) – more than the gross domestic product of all nations combined. This 'new economics' provides persuasive reasons for taking greater care of ecosystems and the biodiversity they contain.

13.2 Degradation via cultivation

When intensive livestock production forces animals to live the equivalent of urban life, their waste is produced faster than natural decomposers and detritivores can handle it (see Chapter 11). All the problems of human urban overpopulation then apply to domestic livestock. Intensive agriculture is also associated with an increase in the level of the nitrate and phosphate that runs into rivers and lakes (and into drinking water) and problems associated with the use of insecticides and herbicides. As we have already seen in Section 12.7, the environmental threats posed by agricultural intensification are expected to increase in the coming decades.

13.2.1 Intensive livestock management

excreta from cattle and pigs is bulky (and smelly) but poultry waste is more acceptable

Pigs, cattle and poultry are the three major contributors to pollution in industrialized agriculture feedlots. The waste from factory-farmed poultry is easily dried and forms a transportable, inoffensive and valuable crop and garden fertilizer. In contrast, the excreta from cattle and pigs are 90% water and have an unpleasant smell. A commercial unit for fattening 10,000 pigs produces as much pollution as a town of 18,000 inhabitants.

The law in many parts of the world increasingly restricts the discharge of agricultural slurry into watercourses. The simplest and often the most economically sound practice returns the material to the land as semisolid manure or as sprayed slurry. This dilutes its concentration in the environment to what might have occurred in a more primitive and sustainable type of agriculture and converts pollutant into fertilizer. Soil microorganisms decompose the organic components of sewage and slurry and most of the mineral nutrients become immobilized in the soil, available to be absorbed again by the vegetation.

Nitrogen is a special case: nitrate ions are not adsorbed in the soil and rainfall leaches them into drainage (and therefore potential drinking) water. The nitrate becomes a new pollutant and one of the biggest culprits is farm specialization where forage crops are grown in one area, but stock is fattened on the other side of the country. This means that fertilizer must be used to make up the shortfall when plants are reaped and transported to the stock, whose excreta can hardly be shipped all the way back to the farm of origin. In the USA, for example, only 34% of the nitrogen excreted in animal waste is returned to fields where the crops are grown (Mosier et al., 2002). Much of the rest eventually finds its way into streams and rivers. A change in practice to one where animal feed crops and stock fattening occur in the same area would certainly reduce nutrient loss to waterways.

13.2.2 Intensive cropping

Some of the nitrogen used in agricultural fertilizer is obtained by mining potassium nitrate in Chile and Peru, and some, as we have seen, comes from animal excreta, but the majority comes from the energy-expensive industrial process of nitrogen fixation, in which nitrogen is catalytically combined with hydrogen under high pressure to form ammonia and, in turn, nitrate. However, it is wrong to regard artificial fertilization as the only practice that leads to nitrate pollution; nitrogen fixed by crops of legumes such as alfalfa, clover, peas and beans also finds its way into nitrates that leach into drainage water.

most agricultural crops depend on fertilizer nitrogen – or nitrogen fixation by legumes

Excess nitrates in drinking water can be a health hazard – the Environmental Protection Agency in the United States recommends a maximum concentration of 10 mg l^{-1} in drinking water. Nitrates may contribute to the formation of carcinogenic nitrosamines and, in young children, may reduce the oxygen-carrying capacity of the blood. Public water systems are required to be monitored regularly and violations reported to the federal government. In 1998, for example, nearly 0.2% of children in the USA (117,000 children in all) lived in areas in which the nitrate standard was exceeded.

nitrates in drinking water are a hazard to health

There are a number of tools to minimize fertilizer loss from the land (thus saving money) to the water (where a useful resource becomes an irritating pollutant). Farmers might aim to maintain a ground cover of vegetation year-round, practice mixed cropping rather than monoculture and take care to return organic matter to the soil. The overriding objective should be to match nutrient supply to crop demand. Modern 'controlled release' fertilizers hold much promise in this regard (Mosier et al., 2002).

tools to minimize fertilizer loss from land

The excess input of nutrients, both nitrogen- and phosphorus-based, from agricultural runoff (and human sewage) has caused many 'healthy' *oligotrophic* lakes (low nutrient concentrations, low plant productivity with abundant water weeds, and clear water) to switch to a *eutrophic* condition where high nutrient inputs lead to high phytoplankton productivity (sometimes dominated by bloom-forming toxic species). This makes the water turbid, eliminates large plants and, in the worst situations, leads to anoxia and fish kills: so-called *cultural eutrophication*. Thus, important ecosystem services are lost, including the provisioning service of wild-caught fish and the cultural services associated with recreation.

The process of *cultural eutrophication* of lakes has been understood for some time. But only recently did scientists notice huge 'dead zones' in the oceans near river outlets, particularly those draining large catchment areas such as the Mississippi in North America and the Yangtze in China. The nutrient-enriched water flows through streams, rivers and lakes, and eventually to the estuary and ocean where the ecological impact may be huge, killing virtually all invertebrates and fish in areas up to 70,000 km^2 in extent. More than 150 sea areas worldwide are now regularly starved of oxygen as a result of decomposition of algal blooms, fueled particularly by nitrogen from agricultural runoff of fertilizers and sewage from large cities (UNEP, 2003). Oceanic dead zones are typically associated with industrialized nations and usually lie off countries that subsidize their agriculture, encouraging farmers to increase productivity and use more fertilizer.

13.2.3 Managing eutrophication

Lake eutrophication, where phosphorus is often the principal culprit, can be reversed by either chemical or biological means. Reduction of phosphorus inputs, by better managing fertilizer use, may be combined with an intervention such as chemical treatment to immobilize phosphorus in the sediment; recovery to a more oligotrophic state can occur within 10–15 years (Jeppesen et al., 2005). In essence, this is *bottom-up control* (see Section 9.5.1) of nutrient availability, reducing phytoplankton productivity and increasing water quality.

The aim of biological control – known as *biomanipulation* – is also to reduce phytoplankton density and increase water clarity, but via an increase in grazing by zooplankton resulting from the active reduction of the biomass of zooplanktivorous fish (by fishing them out or by increasing piscivorous fish biomass). The outcome is the same, but the process is now *top-down control* of a cascade in the food web.

Lathrop et al. (2002) biomanipulated Lake Mendota in Wisconsin, USA, by increasing the density of two piscivorous fish: walleye (*Stizostedion vitreum*) and northern pike (*Esox lucius*). More than 2 million fingerlings of the two species were stocked beginning in 1987 (Figure 13.2a) and total piscivore biomass stabilized at 4–6 kg ha^{-1}. The biomass of zooplanktivorous fish declined, as a result of increased predation by the piscivores, from 300–600 kg ha^{-1} prior to biomanipulation to 20–40 kg ha^{-1} in subsequent years. The consequent reduction in predation pressure on zooplankton (Figure 13.2b) led, in turn, to a switch from small zooplankton grazers (*Daphnia galeata mendotae*) to the larger and more efficient *Daphnia pulicaria*. The increased grazing pressure had the desired effect of reducing phytoplankton density and increasing water clarity (Figure 13.2c).

The only way to alleviate problems in the world's oceans is by careful management of terrestrial catchment areas to reduce agricultural runoff of nutrients and by treating sewage to remove nutrients before discharge (known as tertiary treatment – Section 13.4.1). The vegetation zones between land and water, such

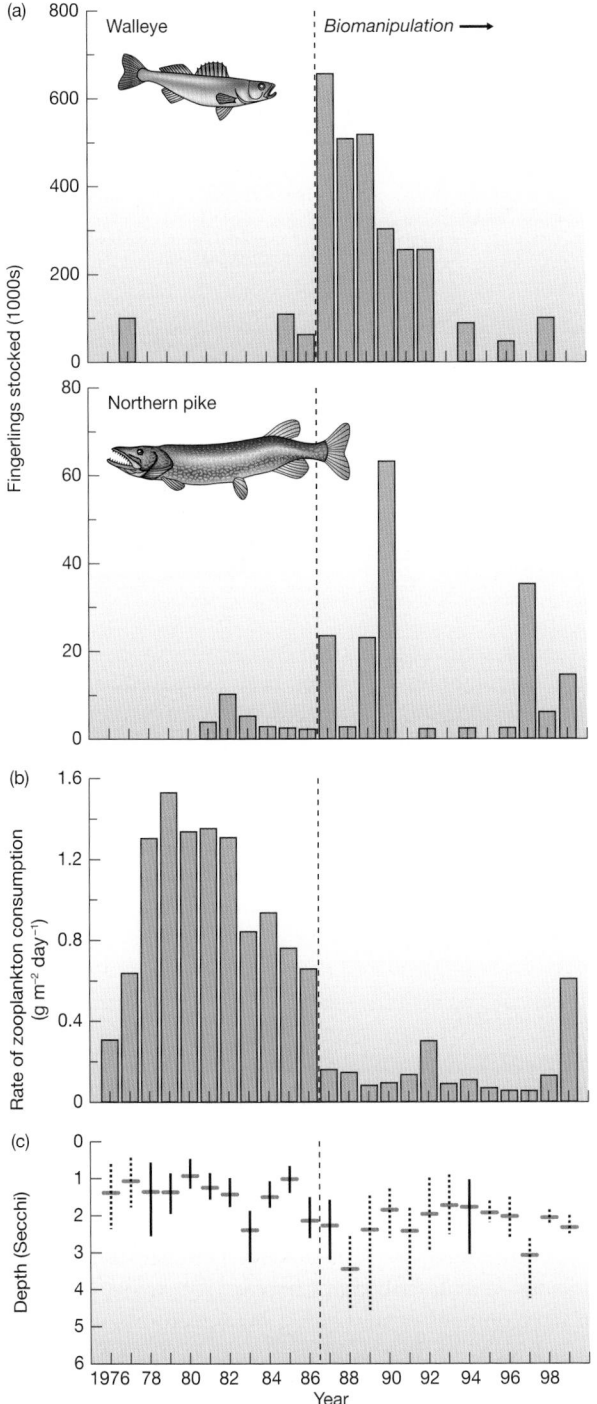

Figure 13.2

(a) Fingerlings of two piscivorous fish stocked in Lake Mendota; the major biomanipulation effort started in 1987 (vertical dashed line). (b) Estimates of zooplankton biomass consumed by zooplanktivorous fish per unit area per day. The principal zooplanktivore fish were *Coregonus artedi*, *Perca flavescens* and *Morone chrysops*. Note that the consumption of zooplankton was reduced because the piscivorous fish reduced densities of the zooplanktivorous fish. (c) Mean and range of the maximum depth at which a Secchi disk is visible (a measure of water clarity) during the summer from 1976 to 1999. The dotted vertical lines are for periods when the large and efficient grazer *Daphnia pulicaria* was dominant. This grazing zooplankton species was much more prominent after biomanipulation had allowed zooplankton to increase in density; *D. pulicaria* plays a large role in reducing the density of phytoplankton so that water clarity increases (Secchi disk visible at greater depth).

FROM BEGON ET AL., 2006; AFTER LATHROP ET AL., 2002

as wetland areas (consisting of swamps, ditches and ponds) and riparian forest along the banks of streams, can be particularly beneficial because the plants and microorganisms remove some of the dissolved nutrients as they filter through the soil. In this way, the riparian zone provides a regulating ecosystem service.

FROM VERHOEVEN ET AL., 2006, BASED ON ARHEIMER & WITTGREN, 2002

Figure 13.3

The locations of 148 wetlands under construction along tributaries of the Rönneå River in southern Sweden. If these are built to occupy 5% of the total land area, a 40% reduction can be expected in agricultural nitrogen input to the Baltic Sea.

But riparian and wetland communities have often been destroyed to provide a greater area for agricultural production. These ecosystems can sometimes be restored to a seminatural state. An alternative is 'treatment wetlands', which are constructed, planted and have water flow controlled to maximize the removal of pollutants from the water draining through them. Estimates for catchment areas in southern Sweden, which are a major source of nitrate enrichment of the Baltic Sea, indicate that to remove 40% of the nitrogen currently finding its way into the sea, a system of wetlands covering about 5% of the total land area would need to be recreated (Figure 13.3).

13.2.4 Pesticide pollution

Many of the manufactured chemicals that are used to kill pests have become important environmental pollutants. The most widely polluting pesticides are those used to control pests and weeds that damage crops in agriculture, horticulture and forestry or to kill pests that transmit diseases of livestock and humans. All are sprayed or dusted onto the areas in which the pests live, but only a very small proportion hits the target – most lands on the crop or on bare ground. Such pesticides are therefore used in much larger quantities than strictly necessary. The characteristics of the most widely used pesticides were described in Chapter 12.

In the early industrial development of pesticides, manufacturers were not much concerned with the specificity of their product. The potential for disaster is illustrated by the occasion when massive doses of the insecticide dieldrin were applied to large areas of Illinois farmland from 1954 to 1958 to 'eradicate' a grassland pest, the Japanese beetle. Cattle and sheep on the farms were poisoned, 90% of the cats on the farms and a number of dogs were killed, and among wildlife 12 species of mammals and 19 species of birds suffered losses (Luckman & Decker, 1960).

Chemical insecticides are generally intended to control particular target pests at particular places and times. Problems arise when they are toxic to many more species than just the target and particularly when they drift beyond the target areas and persist in the environment beyond the target time. The organochlorine insecticides have caused particularly severe problems because they are *biomagnified*. Biomagnification happens when a pesticide is present in an organism that becomes the prey of another and the predator fails to excrete the pesticide. It then accumulates in the body of the predator. The predator may itself be eaten by a further predator, and the insecticide becomes more and more concentrated as it passes up the food chain. Top predators in aquatic and terrestrial food chains, which were never intended as targets, can then accumulate extraordinarily high doses (Figure 13.4; see also Box 13.1).

pesticides are most polluting when they are unselective, persistent and if they 'biomagnify' in food chains

13.2.5 Physical degradation associated with cultivation

It hardly needs stating that one of the biggest impacts of cultivation is the physical loss of natural habitats, together with the species they contain. Sometimes, however, the impact is more subtle. A large proportion of the world's crops depend on insect pollinators and bees play a leading role. Farmers often rely on domesticated honeybees (*Apis mellifera*), importing hives when their crops are in flower. However, many wild bee species also pollinate crops (providing a free provisioning ecosystem service) and these species are much less abundant in landscapes that retain little natural vegetation.

loss of natural habitat to cultivation

Kremen et al. (2004) studied the role played by native bees in watermelon (*Citrullus lanatus*) fields on Californian farms that varied in the proportion of native and other habitats found nearby. Satellite imagery was used to quantify native upland habitat (woodland and chaparral), riparian woodland and highly modified land classes (agriculture, grassland dominated by non-native species, urban land) in the vicinity of each field. Kremen's team found that the proportion of upland native habitat within 1–2.4 km of the fields was strongly correlated with deposition of watermelon pollen by native bees, reflecting maximum flight distances of about 2.2 km for species that nest in these natural habitats. Next they calculated the proportion of surrounding land that must consist of upland native habitat to yield the 500–1000 pollen grains required per watermelon plant to produce marketable fruit. It turns out that 40% of habitat within 2.4 km of a field needs to be upland native habitat to provide sufficiently for melon pollination needs, providing a strong economic argument to conserve these natural habitats. For farms that are far from natural habitat, active restoration with native plants in hedgerows and ditches and around fields, barns and roads, might allow a target of about 10% native habitat to be achieved (equating to 20–40% of watermelon pollination needs).

Increasing agricultural intensity is usually associated with the removal of surface and ground water for irrigation. Coupled with impoundment of river water behind dams, this abstraction for irrigation can have dramatic physical consequences for patterns in riverflow. Thus, for example, the Nile in Africa, the Yellow River in China and the Colorado River in North America dry up for parts of the year before they reach the ocean. In many less dramatic cases, water abstraction for agricultural, industrial and domestic use changes the hydrographs (discharge patterns) of rivers both by reducing discharge (volume per unit time) and by altering daily and seasonal patterns of flow.

changes to river discharge via impoundment and irrigation

Figure 13.4

Organochlorines, applied as pesticides on land, are transported to the Arctic through river runoff and oceanic and atmospheric circulation. A study in the Barents Sea showed how two classes of pesticide are biomagnified during passage through the marine food chain. Concentrations in sea water are very low. Herbivorous copepods (that feed on phytoplankton) have higher concentrations (measured in nanograms per gram of lipid in the organisms), and predatory amphipods higher concentrations still. The polar cod (*Boreogadus saida*), which feeds on the invertebrates, and cod (*Gadus morhua*) which also includes polar cod in its diet, show further evidence of biomagnification. However, it is the higher steps in the food chain where biomagnification is most marked, because the sea birds that feed on the fish (black guillemots, *Cepphus grylle*) or on fish and other sea birds (glaucous gull, *Larus hyperboreus*) have much less ability to eliminate the chemicals than fish or invertebrates. Note how chlordanes are biomagnified to a lesser extent than polychlorinated biphenyls (PCBs). This results from the birds' greater ability to metabolize and excrete the former class of pesticide.

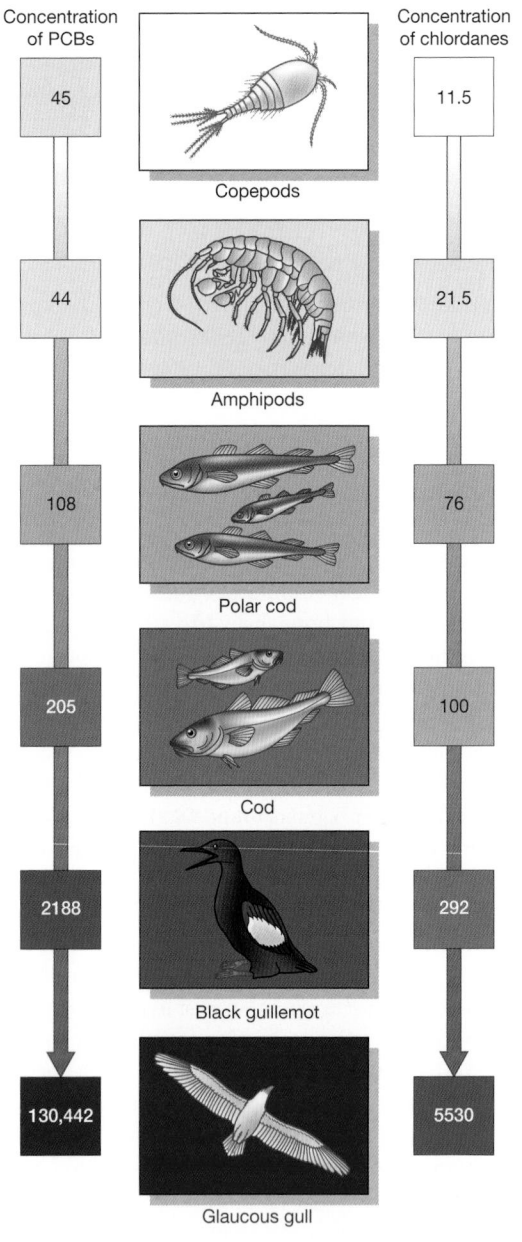

The rare Colorado pikeminnow (*Ptychocheilus lucius*), which eats other fish, is now restricted to the upper reaches of the Colorado River. Its present distribution is positively correlated with prey fish biomass, which in turn depends on the biomass of invertebrates upon which the prey fish depend, and this, in its turn, is positively correlated with algal biomass, the energy base of the food web (Figure 13.5a–c). Osmundson et al. (2002) argue that the rarity of pikeminnows can be traced to the accumulation of fine sediment on the riverbed, where it reduces algal productivity in downstream regions of the river. Historically, spring snowmelt often produced flushing discharges with the power to remove much of

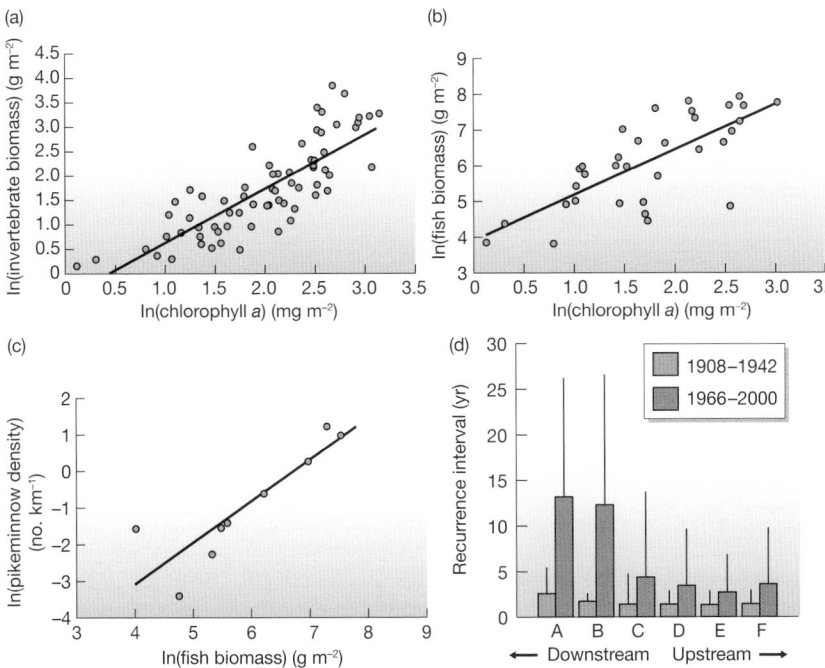

Figure 13.5

Interrelationships among biological parameters measured in a number of reaches of the Colorado River to determine the ultimate causes of the declining distribution of Colorado pikeminnows. (a) Invertebrate biomass versus algal biomass (chlorophyll *a*). (b) Prey fish biomass versus algal biomass. (c) Pikeminnow density versus prey fish biomass (from catch rate per minute of electrofishing). (d) Mean recurrence intervals in six reaches of the Colorado River (for which historical data were available) of discharges necessary to remove silt and sand that would otherwise accumulate, during recent (1966–2000) and pre-regulation periods (1908–1942). Lines above the histograms show maximum recurrence intervals.

FROM BEGON ET AL., 2006; AFTER OSMUNDSON ET AL., 2002

the silt and sand that would otherwise accumulate. As a result of river regulation, however, the mean recurrence interval of such discharges has increased from once every 1.3–2.7 years to only once every 2.7–13.5 years (Figure 13.5d), extending the period of silt accumulation. Managers must aim to incorporate ecologically influential aspects of the natural hydrograph of a river into restoration efforts if endangered (or valuable harvestable) species are to be sustained.

13.3 Power generation and its diverse effects

Since the industrial revolution of the 18th century, our use of fossil fuels has provided the power to transform much of the face of the planet through urbanization, industrial development, mining and highly intensive agriculture, forestry and fishing. In Section 13.3.1 we consider the far-reaching effects of chemical pollutants from fossil fuel use. Because fossil fuels are exhaustible, increasingly costly to extract, pollute the atmosphere and contribute to global warming, much recent emphasis has been placed on developing alternative energy sources that do not release carbon dioxide. The cleanest and safest technologies are expected to derive from hydropower schemes (already at a technologically advanced state in many parts of the globe), together with wind farms (rapidly developing) and solar and wave power. Nuclear power, whose popularity had declined because of concerns over security and radioactive waste disposal, is receiving renewed consideration because it does not release greenhouse gases. We discuss nuclear power in Section 13.3.2 and wind power in Section 13.3.3.

Figure 13.6

The concentration of atmospheric carbon dioxide measured at the Mauna Loa Observatory, Hawaii showing the seasonal cycle (dipping each northern summer when photosynthetic rates are maximal in the northern hemisphere) and, more significantly, the long-term increase that is due largely to the burning of fossil fuels.

COURTESY OF THE CLIMATE MONITORING AND DIAGNOSTICS LABORATORY (CMDL) OF THE NATIONAL OCEANIC AND ATMOSPHERIC ADMINISTRATION (NOAA)

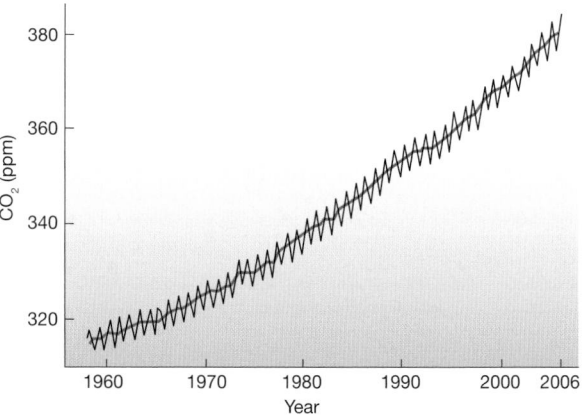

13.3.1 Fossil fuels and atmospheric pollution

The most profound and far-reaching consequences of the burning of fossil fuels, principally coal and oil, are those of atmospheric pollution. Thus, the concentration of carbon dioxide in the atmosphere has increased from about 280 parts per million (ppm) in 1750 to about 370 ppm today, and is projected to continue to rise to 700 ppm by the year 2100 unless there are rather drastic changes in human behavior. A remarkable census of atmospheric carbon dioxide was started in 1958 at Mauna Loa Observatory in Hawaii and this detected the extraordinary pattern shown in Figure 13.6. The principal cause of this increase has been the burning of fossil fuels, which in 1980, for example, released about 5.2×10^9 metric tons of carbon into the atmosphere (Table 13.1).

Table 13.1

Balancing the global carbon budget (in 10^9 metric tons per year) in 1980 to account for increases in atmospheric carbon caused by human activities. In the row labeled 'Missing' the minus sign indicates the need to identify an unknown uptake of carbon of the size shown. This has now been identified as fertilization of terrestrial vegetation by atmospheric carbon dioxide so that an increase of the order of what was estimated as 'missing' can be accounted for by an increase in the amount of carbon locked in extra vegetation biomass (Kicklighter et al., 1999).

	EXTREME LOW ESTIMATE	MEDIAN ESTIMATE	EXTREME HIGH ESTIMATE
Release to atmosphere			
Fossil fuel combustion	4.7	5.2	5.7
Cement production	0.1	0.1	0.1
Tropical forest clearance	0.4	1.0	1.6
Non-tropical forest clearance	−0.1	0.0	0.1
Total release	5.1	6.3	7.5
Accounted for			
Atmospheric increase	−2.9	−2.9	−2.9
Ocean uptake	−2.5	−2.2	−1.8
Missing?	−0.3	+1.2	+2.8

AFTER DETWILER & HALL, 1988

The clearing and burning of tropical forest to make way for agriculture or timber production and the decay of the residues make a further contribution to the increase in atmospheric carbon dioxide (Table 13.1). A considerable amount of this is recaptured in photosynthesis by the replacement vegetation (Kicklighter et al., 1999), but this is least when forest is converted to grassland, which has a much lower biomass. In total about 1.0×10^9 metric tons per year has been released through changes in tropical land use (Detwiler and Hall, 1988). This calculation was made for 1980, and the figure for tropical forest clearance must now be significantly greater as a result of the uncontrollable spread of forest fires in Indonesia and in South America following the droughts associated with the El Niño phenomenon of 1997/98.

atmospheric pollution due to the burning of fossil fuels and deforestation

The Earth's atmosphere behaves like a greenhouse. Solar radiation warms up the Earth's surface, which reradiates energy outward, principally as infrared radiation. Carbon dioxide – together with other gases whose concentrations have increased as a result of human activity (nitrous oxide, methane, ozone, chlorofluorocarbons) – absorbs infrared radiation. Like the glass of a greenhouse, these gases (and water vapor) prevent some of the radiation from escaping and keep the temperature high. The air temperature at the land surface is now $0.6 \pm 0.2°C$ warmer than in pre-industrial times. Given further predicted rises in greenhouse gases, temperatures will continue to rise by a global average of between $2.0°C$ and $5.5°C$ by 2100 (IPCC, 2001; Millennium Ecosystem Assessment, 2005), but to different extents in different places. Such changes will lead to a melting of glaciers and icecaps, a consequent rise in sea level, and large changes to global patterns of precipitation, winds, ocean currents and the timing and scale of storm events.

the greenhouse effect

In response to these changes, we can expect latitudinal and altitudinal shifts in species distributions and widespread extinctions as floras and faunas fail to track and keep up with the rate of change in global temperatures (Hughes, 2000). In addition, the global threats imposed by harmful invasive species will change. Take, for example, the Argentine ant (*Linepithema humile*), a native of South America. This is now established on every continent except Antarctica. It can achieve extremely high densities and has adverse consequences for biodiversity (eliminating native invertebrates) and for domestic life, swarming over human foodstuffs and even sleeping babies. A distributional model was developed for the ant, based on occurrences in its native and invaded ranges, and related both to climatic data (e.g. maximum, minimum and mean temperatures, precipitation, number of frost days, number of wet days) and topographic data (e.g. elevation, slope, aspect). The model provided a good fit with current distribution based on current climate. Next, predicted climate change was used to model the ant's future distribution. Figure 13.7 indicates in red those areas predicted to improve for the ant by 2050 (increased likelihood of ant occurrence) and in blue those areas expected to worsen. The species will retract its range in tropical areas but expand into higher latitudes. Ironically, the Argentine ant looks set to do less well in its native South America than in North America and Europe.

Efforts to eradicate Argentine ants have mostly been unsuccessful. The management response is therefore to increase biosecurity precautions in regions expected to become progressively more invadable in future.

Of the pollutants that humans release into the atmosphere, most are returned to Earth, about half as gases or particles and half dissolved or suspended in rain, snow and fog. They may be carried in the wind for hundreds of kilometers across

acid rain

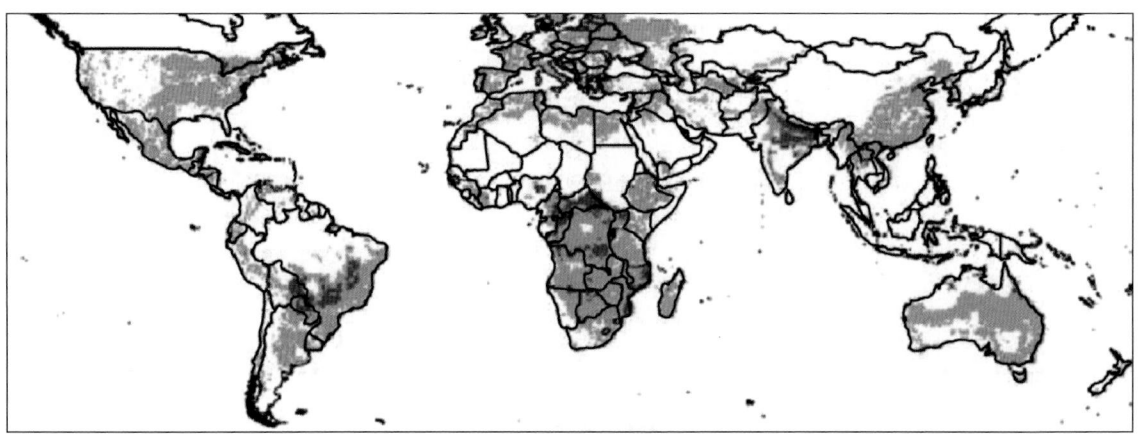

Figure 13.7

Predicted changes to the distribution of the Argentine ant between now and 2050. Red areas are those predicted to improve for Argentine ants, whereas blue areas are predicted to worsen for the species.

state and national borders, and when they cause harm they can be the source of bitter international dispute. Atmospheric pollutants sulfur dioxide (SO_2) and oxides of nitrogen (NO_x), contributed particularly by the burning of fossil fuels, interact with water and oxygen in the atmosphere to form dilute sulfuric and nitric acids, which fall as *acid rain*.

Rain water has a pH of about 5.6, but pollutants lower it to below 5.0 and values as low as 2.4 have been recorded in Britain, 2.8 in Scandinavia and even 2.1 in the United States. Many of the most visibly dramatic effects of acid rain have been observed in the forests of central Europe where industry depended on low-quality coal with a high sulfur content and forest dieback occurred on a massive scale. Even in the United States, high-elevation spruce forests have been affected, including the Shenandoah and Great Smoky Mountain national parks.

Further effects have occurred in lakes and streams, especially when the composition of the underlying soil and rock does not help to neutralize the acidity. A high concentration of hydrogen ions may itself be toxic, but changes in the availability of nutrients and other toxins are usually more important. At a pH below 4.0–4.5, the concentrations of aluminum (Al^{3+}), iron (Fe^{3+}) and manganese (Mn^{2+}) become toxic to most plants and to aquatic animals that expose delicate tissues directly to the water (such as the gills of fish). Acid rain is most damaging in water that is already naturally acidic: it may then lower the pH so far that it sterilizes the environment for many of the native species (e.g. Figure 13.8).

13.3.2 Nuclear power

When first developed, nuclear power was viewed as an almost ideal, long-term source of industrial and domestic power. However, the view that the release of radiation could be readily controlled faded rapidly. Some leakage occurs from nuclear power reactors, and it is doubtful whether the reprocessing of waste

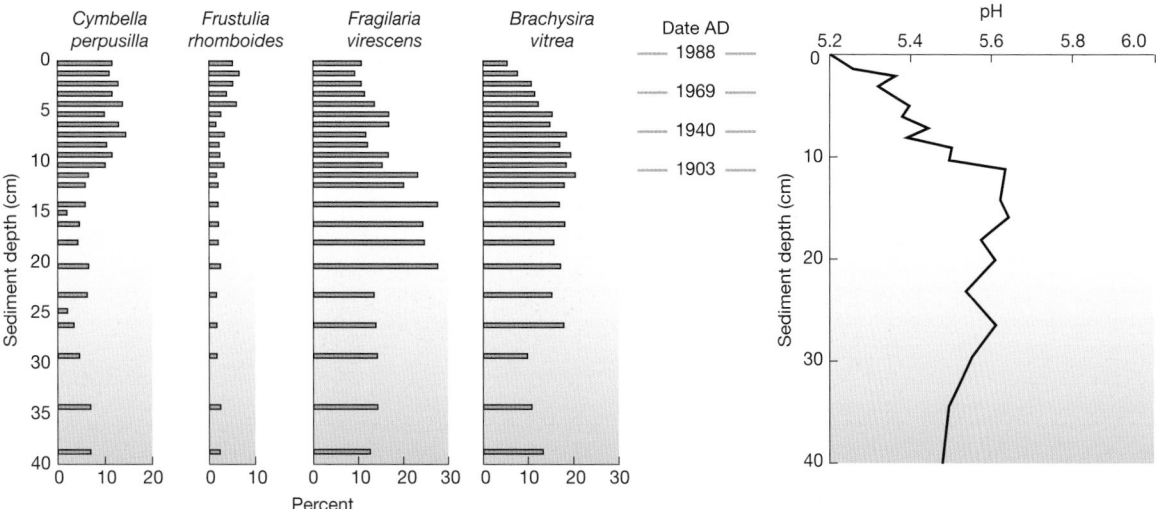

Figure 13.8

The history of the diatom flora of an Irish lake (Lough Maam, County Donegal) can be traced by taking cores from the sediment at the bottom of the lake. The percentage of various diatom species at different depths reflects the flora present at various times in the past (four species are illustrated). The age of layers of sediment can be determined by the radioactive decay of lead-210 (and other elements). We know the pH tolerance of the diatom species from their present distribution and this can be used to reconstruct what the pH of the lake has been in the past. Note how the waters have acidified since about 1900. The diatoms *Fragilaria virescens* and *Brachysira vitrea* have declined markedly during this period while the acid-tolerant *Cymbella perpusilla* and *Frustulia rhomboides* increased after 1900.

nuclear fuel can ever be made completely clean. Moreover, the polluting power of radioactive waste has a time scale that may be orders of magnitude greater than that of other human pollutants. For example, plutonium-239 has a half-life of about 25,000 years. Plutonium is separated and recovered from the spent fuel in nuclear reactors and stocks are expected to have risen to more than 100 metric tons by 2010. Ways have to be found to protect against risks of leakage over this sort of time scale, perhaps by burial in deep mines after incorporation in glass.

The radiation received by an organism arises from human activities (nuclear warfare, leakage from and accidents at nuclear plants, and medical use) together with a very similar sized contribution from 'background radiation' from cosmic rays and produced during the radioactive decay of materials such as radium and thorium in the Earth's crust. It is a sobering thought that the total radiation given to a cancer patient can be many thousand times greater than the total normal exposure from the combined natural and artificial background radiation.

> natural background radiation and that produced by human activities are of similar magnitude

A major accident in 1986 at a nuclear power station at Chernobyl in the Ukraine released 50–185 million curies of radionucleides into the atmosphere. Close to the explosion, 32 deaths occurred within a very short time. Farther away, individuals contracted radiation sickness and some died. Effects in the locality have continued to appear – livestock have been born deformed, and thousands of radiation-induced illnesses and deaths from cancer are expected in the longer term. Farther afield, wind-dispersed atmospheric pollution from Chernobyl was detected in Sweden 3 days after the accident. Fallout also reached the British

> Chernobyl – the worst nuclear pollution disaster so far

AFTER FLOWER ET AL. 1994

Figure 13.9

An example of long-distance environmental pollution: the distribution in 1988 in Great Britain of fallout of cesium-137 from the Chernobyl nuclear accident in the Soviet Union in 1986 (measured in Becquerels per kilogram). The contours show the persistence of the cesium on acidic upland soils where it is recycled through soil, plants and animals. On typical lowland soils, cesium does not persist in food chains.

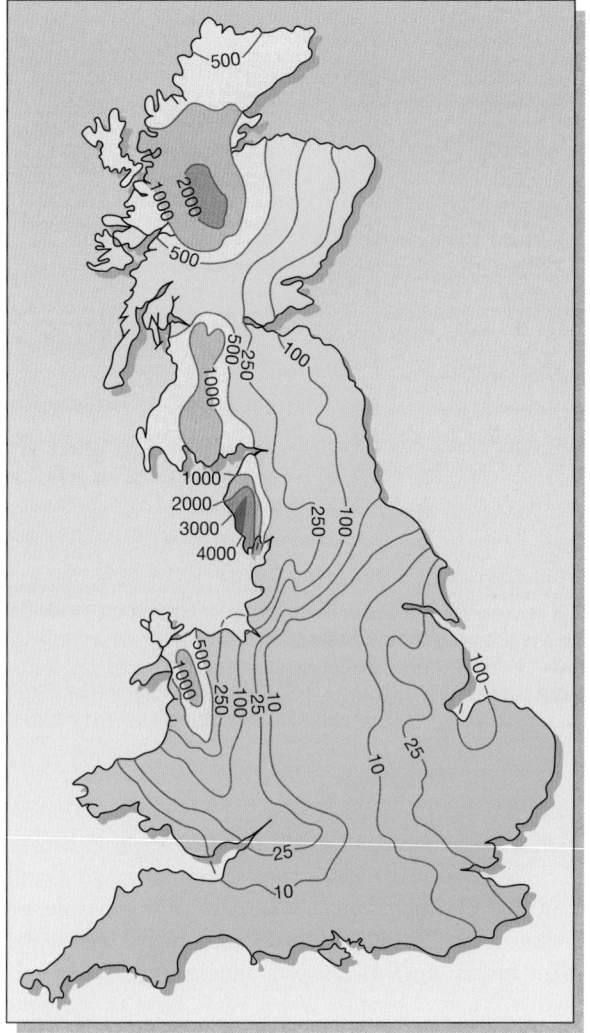

AFTER NERC, 1990

Isles. Figure 13.9 shows the persistence of cesium-137 in the acid soils of the northwest of Britain, where it was absorbed by plants and eaten by sheep. The sale of sheep for food was still banned more than 10 years after the accident because of persistence of the isotope at dangerous levels.

13.3.3 Wind power

In this time of global climate change, harnessing the power of the wind has much to commend it. But while this form of power generation does not release carbon dioxide, local communities often object to the massive structures that will appear in their localities. (This conundrum parallels the situation for hydropower stations, which produce clean power but at the cost of altered river discharge patterns and lost recreational opportunities downstream.) Wind farms also pose an ecological risk in terms of threats to migrating birds. On land, soaring birds such as falcons and vultures are at particular risk of colliding with the turbines (up to 100 m above the ground), particularly because the engineers often select their locations for the same wind-related reasons that birds select their routes (Barrios & Rodriguez, 2004). Many wind farms are also planned for marine settings – in Europe, for example, more than 100 applications have been submitted. Each may consist of as many as 1000 turbines, up to 150 m tall, as far offshore as 100 km and in water as deep as 40 m. The turbines may pose risks to migrating birds (from the smallest of song birds to cranes and birds of prey) as well as sea birds dispersing locally to find food.

Thousands of square kilometres of the marine environment off the German coast are planned for wind farming by 2030. To predict the possible consequences for bird populations, Garthe and Huppop (2004) developed a species sensitivity index (SSI) for 26 seabird species, combining their scores for a range of properties, including flight maneuverability (less agile species score highly because they are more likely to collide with turbines), flight altitude (species flying at 50–200 m score highly because they are more vulnerable to turbines than lower flyers), percentage time spent flying (those in the air for more of the time score highly) and conservation status ('vulnerable' or 'declining' score highly). The most sensitive species (highest SSI) include the non-maneuverable and 'vulnerable' black-throated diver (*Gavia arctica*) and the maneuverable but 'declining' sandwich tern (*Sterna sandvicensis*) that flies almost constantly and at perilous altitudes. The SSI for each species was then coupled with density distribution data (low-density species score highly because their populations are more at risk) to produce vulnerability maps (all bird species combined) for the German area of the North Sea. Three classes of vulnerability were assigned – 'major concern' (combined wind-farm sensitive data [WSI] > 43), 'less concern' (WSI < 24) and 'concern' (between these extremes) (Figure 13.10). Such ecological information should be taken into account when selecting wind farm locations.

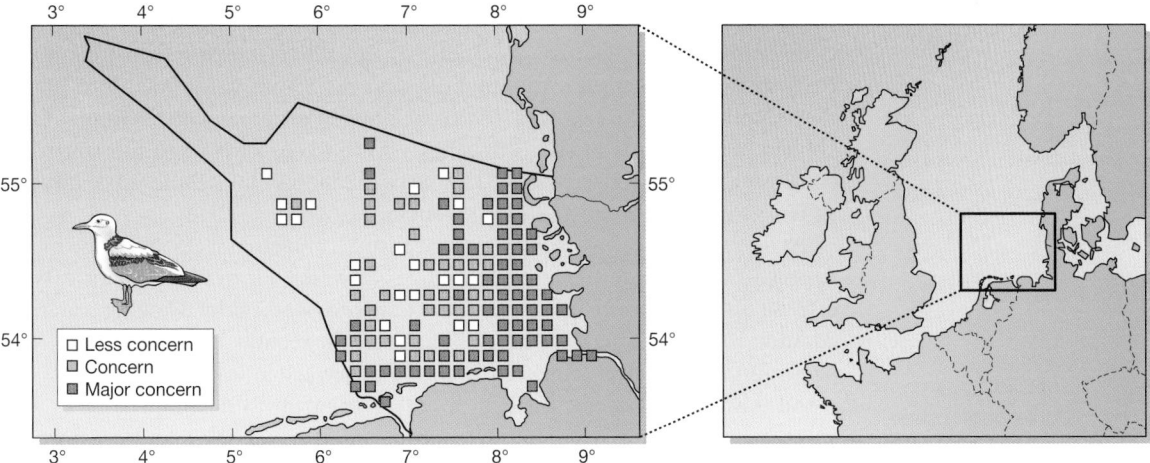

AFTER GARTHE & HUPPOP, 2004

Figure 13.10

Areas in the German sector of the North Sea (inset, right) where wind farm development is considered to be of 'less concern', 'concern' or 'major concern' on the basis of bird density patterns and species-specific sensitivity indexes (SSIs).

13.4 Degradation in urban and industrial landscapes

A wide range of habitat degradation occurs as a result of human activity in urban and industrial settings. Marked changes to riverflow result from the loss of permeable surfaces – roofs, pavements and roads are impermeable, in contrast to field and forest. Our feces, urine and dead bodies create large disposal problems in towns and cities because density is so high. Exotic industrial chemicals find their way into waterways and the atmosphere. And our mining activities, whether for fossil fuels or valuable jewels or ores, cause physical and chemical degradation to surrounding ecosystems. In this section, our examples encompass sewage disposal (Section 13.4.1), the industrial production of fluorocarbons and consequences for the ozone layer (Section 13.4.2) and the ecological problems associated with mining (Section 13.4.3).

13.4.1 Disposal of bodily waste

All human body products, but most notably feces and urine, can be regarded as pollutants. The Greeks were probably the first to control the accumulation of pollution within towns, and a law of 320 BC forbade dumping of waste in the streets. The Romans were also very pollution conscious, dumping city waste in pits outside the city walls. When Roman and Greek civilizations perished, their quite sophisticated control of urban pollution collapsed. Medieval castles, for example, were often designed with latrines projecting from castle walls that simply dumped waste at the base of the walls (the accumulated wastes give archeologists a direct record of historical diets and infestation with intestinal worms!). Until the 14th and 15th centuries the open streets again became the main, and often the only, destination for human and animal feces and urine. A special trade developed, that of the scavenger, who was paid to carry waste to dumps outside the cities; in 1714 every city in England had an official scavenger (the forerunner of the Environmental Protection Agency!). Even when water closets (invented by Thomas Crapper) began to be installed in some countries early in the 19th century, the underground reservoirs (cesspools) into which they emptied often overflowed and contaminated drinking water. Outbreaks of cholera in the middle of the 19th century were traced directly to this source of contamination, a discovery that led to the connection of household waste directly to sewers in both Britain and the United States.

At first glance, the easiest way to cope with accumulated feces and urine might appear to be to dilute them in large bodies of water. However, it is not easy to dispose of human waste and at the same time provide healthy drinking water. In addition to health issues, we have already seen in Section 13.2.2 how there can be profound ecological effects of disposing of sewage in water bodies.

when natural ecosystems cannot cope with human waste . . .

All natural ecosystems have an inherent capacity to decompose feces and up to a point natural decomposition processes in rivers, lakes and oceans may cope with increased organic matter from human sewage without obvious changes to the nature of the biological communities they contain. However, problems arise when the rate of sewage input exceeds this capacity. First, excessively high rates of decomposition of dead organic matter in rivers and lakes can lead to anaerobic conditions (causing the death of fish and invertebrates). This happens because oxygen is consumed by the decomposer microorganisms faster than it is replenished from photosynthesis by aquatic plants and diffusion from the air. Second, the supply of nutrients such as phosphate and nitrate that normally limit plant growth in water bodies may be increased to a level where algal growth is so great that it shades and kills other aquatic plants – the cultural eutrophication discussed in Section 13.2.2.

Modern sewage systems were developed as ecological devices for pollution management. They aim to capture pollutants from waste water and to clean it, usually in a drainage system separate from the one that carries heavy flows of storm water. Ideally, a sewage system cleans polluted water to a state suitable for drinking before discharging it back into rivers, lakes and the sea. The full treatment of sewage has three stages (Figure 13.11), though in many places only the first or first and second stages are actually used before discharge into the environment.

After paper, rags and plastic have been removed by passing the sewage through screens, *primary treatment* is a physical process in which much of the solid

. . . sewage treatment systems are needed

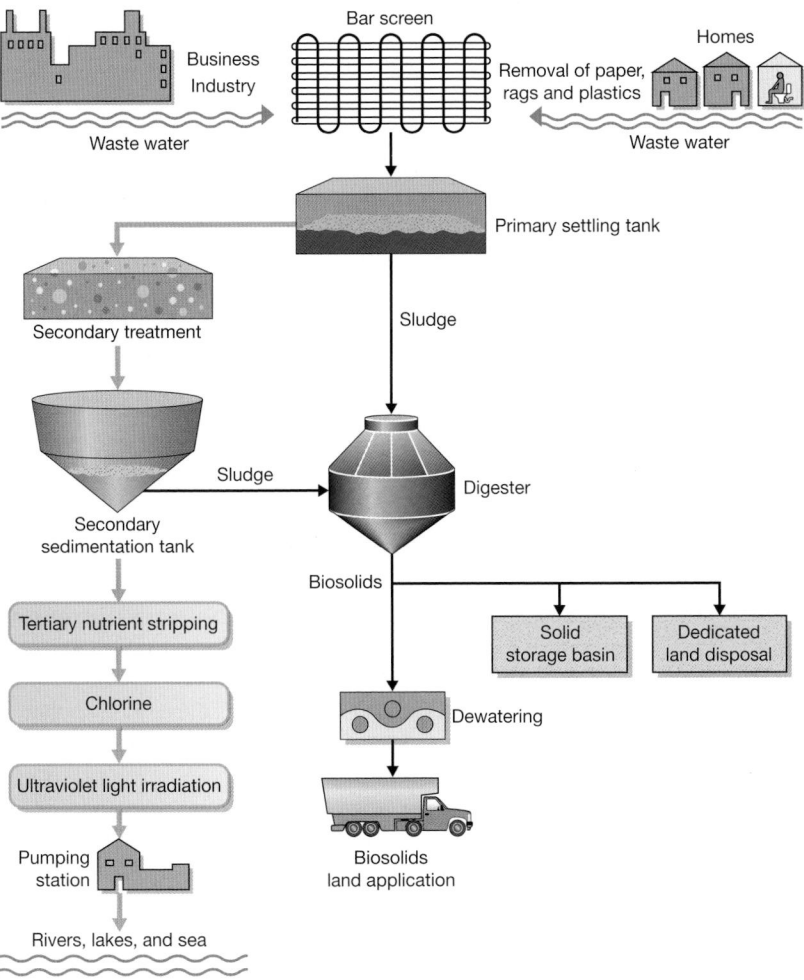

Figure 13.11

The sequence of treatments commonly applied to the sewage waste from a modern urban community.

organic sewage waste is allowed to settle to the bottom of settlement tanks, from which it is removed as sludge.

Secondary treatment is an engineered biological process designed to mimic (and indeed enhance) natural decomposition. In its simplest version, the partly cleaned water is sprayed onto a layer of crushed rock within which microorganisms have been encouraged to grow; as the water trickles down through these *percolating* or *trickling filters*, natural decomposition mineralizes much of the remaining organic matter, releasing carbon dioxide to the atmosphere. A more sophisticated and efficient method of secondary treatment is the *activated sludge method*, in which the sewage is passed into aerated tanks containing sludge that is activated, or seeded, with microorganisms. After secondary treatment the remaining solids are settled to yield more sludge. The waste water now appears clean, but it still contains two types of impurity, namely disease organisms and high concentrations of mineral nutrients, the latter having both health consequences (Section 13.2.2) and causing eutrophication if released into rivers and lakes.

A final 'polishing' stage usually includes chlorination, and sometimes ultra-violet (UV) light irradiation to kill bacteria. Full *tertiary treatment* involves the stripping of nutrients, largely by artificial and expensive chemical processes.

products of sewage treatment are themselves pollutants

Untreated sewage is obviously a pollutant, with adverse health and ecological consequences for water bodies into which it is discharged. However, discharge of sewage that has only been subject to primary treatment is still likely to cause eutrophication because it remains rich in organic matter and nutrients. Moreover, even secondary treatment removes only the organic matter, leaving waste water rich in plant nutrients. The sludge that accumulates in settling tanks is itself a pollutant that has to be disposed of, usually by dumping at sea or burying in landfill sites. Buried sludge decomposes anaerobically, sometimes taking more than 20 years to mineralize completely, and it produces methane, which is a greenhouse gas that contributes to global climate change (Section 13.3.1). Sludge can be more appropriately used as a fertilizer, either dried or as a liquid sprayed onto the land; in this way the nutrient cycle can be reconstituted by returning nutrients, assimilated from crops by people, to agricultural land to be taken up by future crops.

13.4.2 Chlorofluorocarbon compounds and thinning of the ozone layer

ozone can have adverse consequences locally . . .

Ozone is produced by the influence of sunlight on oxygen and during the oxidation of carbon monoxide and hydrocarbons such as methane. It has three very different roles in environmental pollution. The first two are negative, in the sense that undesirable polluting consequences occur as the concentration of ozone increases. First, in atmospheres polluted with methane, industrial hydrocarbons, oxides of nitrogen and carbon monoxide, ozone can reach concentrations that are toxic to plants and that contribute to smog. Second, ozone is also a greenhouse gas, though it is not particularly significant in this respect.

. . . but in the upper atmosphere it shields the Earth from damaging UV radiation

However, ozone also accumulates as a layer in the upper atmosphere. This 'ozone layer' is beneficial because it absorbs most of the UV radiation (wavelength 200–300 nm) incident on the Earth's upper atmosphere and so makes the Earth habitable for plants and animals. The increasing frequency of skin cancer among humans has focused attention on the damage caused by exposure to the sun and on the importance of stability of the ozone layer.

Evidence that nitric oxide produced by supersonic aircraft might contribute massively to reduce the level of atmospheric ozone led to the halting of their large-scale development. However, that has by no means been the end of the story. Chlorofluorocarbon compounds (CFCs) had been developed as aerosols and refrigerants and used on a very large, international scale. It became clear that these posed the threat that their chlorine content could interact with and destroy atmospheric ozone.

Ozone chemical processes are very complicated, and methane, nitrous oxide and carbon monoxide may all play a part in its decomposition. The upper atmosphere is not the easiest place in which to study the chemical characteristics of gases! But pollution of the upper atmosphere poses questions of the greatest significance for environmental scientists, especially since the discovery that the concentration of ozone in the atmosphere over Antarctica had started to decline by 1978 and was doing so very rapidly after 1982. The phenomenon happens at the start of the southern hemisphere spring (August to October). The size of the ozone hole over Antarctica on September 24, 2006 was one of the largest ever recorded, equivalent to more than the surface area of North America (Figure 13.12). It is

> chlorine compounds and other pollutants decompose ozone in the atmosphere and need to be phased out

(a)

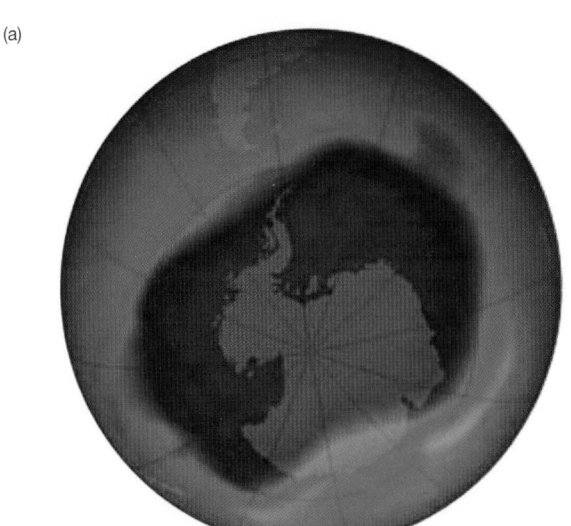

Figure 13.12

(a) An image of the ozone hole over Antarctica for September 24, 2006; the blue and purple colors are where there is least ozone (<220 Dobson units). (b) Average size of the ozone hole from September 7 to October 13 each year from 1980 to 2006. The vertical lines show the minimum and maximum areas during this period each year.

COURTESY OF THE US NATIONAL AERONAUTICS AND SPACE ADMINISTRATION, WWW.NASA.GOV clearly in the interests of humans and probably most other organisms that ozone concentrations should remain low close to the Earth's surface (e.g. minimizing smog) but high in the upper atmosphere, and that we should find out how to ensure this. International agreements to phase out CFCs are expected to lead to recovery of the ozone hole by about 2050.

13.4.3 Mining

physical disruption from the mining of fossil fuels

Our dependence on fossil fuels has effects that go beyond atmospheric pollution. The extraction and transport of coal and especially oil can also cause physical disruption of habitats. Thus, more than 1 million tonnes of oil enters the world's waterways every year from wells drilled into the seabed or from oil tankers. Oil in and on the sea affects wildlife in many ways. It reduces the level of aeration of the water, and it prevents light from penetrating the surface. Damage to invertebrates can be widespread, affecting chitons, mussels, crustaceans and bryozoans, as well as seaweeds and kelps. Feathers become choked with oil so that sea birds cannot fly and fish gills become coated and cease to function. The largest accident in the United States occurred on March 24, 1989, when the oil tanker *Exxon Valdez* ran aground in Prince William Sound, Alaska. It spilled nearly 50,000 tons of crude oil, which spread along the coast for nearly 1000 km, contaminating the shores of a national forest, five state parks, four state critical habitat areas and a state game sanctuary. The episode is believed to have killed 300 harbor seals, 2800 sea otters, 250,000 birds and possibly 13 killer whales. Many commercial fisheries were closed for a year or more because of the concern that fish caught in the area might find their way into the human food chain. By 1996, 28 species and resources were still listed as having failed to recover.

the mining and purification of copper

Metals were first used by humans in the late Stone Age, about 6500 years ago. Gold, silver and copper were the first metals used; they are easy to extract because they exist in nature as the metals themselves rather than as chemical compounds. Nuggets of pure metallic gold were found in riverbeds and were beaten and molded for decoration. Once such metals were valued it was an obvious step to dig and mine for them, and from that point almost every phase in the extraction and industrial use of metals involves a sequence of phases of environmental pollution.

Each type of metal has its own peculiarities. Here we use the mining and purification of copper to illustrate pollution through the extraction of metal. Copper is present in deposits either as the metal or as copper sulfide or oxide. Like most metal deposits, it usually exists in a mixture with other metals, some of which may be worth saving (e.g. gold), whereas others are discarded in more or less hazardous waste.

The mining industry may pollute at every stage of extraction, purification and disposal:

1 Mining and quarrying. Mining or quarrying exposes the metal and its ores. Many of the world's copper reserves are close to the surface and are easily extracted by open cast mining: the copper mines of Bougainville (Solomon Islands, Papua New Guinea) and of Utah are among the largest human scars on the Earth's surface (Figure 13.13).

Figure 13.13

Binyon Canyon Mine, Utah. A toxic and sterile environment created by the world's largest excavation.

2 Processing. The ores are crushed and finely ground. This processing immediately exposes ores to the elements, and even after the best has been extracted the residues are copper-rich and the metal leaches as toxic waste into rivers and lakes. Waters close to copper mines are commonly brilliantly blue-green with copper salts and quite sterile.

3 Concentration. The finely ground ore is agitated in water, and the metal becomes concentrated in the froth and dried to a cake. The remainder, which is still rich in copper, may be further concentrated to recover more of the metal. Ultimately water and solid 'tailings' have insufficient copper to warrant further extraction but contain sufficient copper to form a hazardous and polluting waste.

4 Purification through heat. The concentrate is then roasted to 1230–1300°C, polluting the atmosphere by the burning of the necessary fuel. The roasting drives off a host of pollutants such as arsenic, mercury and sulfur into the atmosphere.

5 Purification through electrolysis. The copper can now be purified by electrolysis, which leaves most of the other metals in a sludge that may be further purified (to remove gold, for example) but ultimately contributes yet more toxic waste.

The major role of some metals as environmental pollutants occurs after they have been purified and used industrially and are then released into the environment as industrial waste. Lead and mercury are particularly striking examples. Lead became an environmental pollutant from the moment that the Romans started to use it to make water pipes and so started to pollute their drinking water. It is ranked by the US Environmental Protection Agency as number 1 in their list of 275 hazardous substances, posing a particular risk for the development of the nervous system in young children and in the fetus. It is being phased out of many commercial uses.

lead and mercury can be especially dangerous pollutants

It is not clear whether lead pollution has significant consequences for wildlife on the land or in aquatic environments, but it does not appear to become concentrated along food chains. This is a major contrast with mercury.

Mercury is used in a variety of specialized applications in industry and medicine – in electric switches, batteries, fluorescent and mercury vapor lights, thermometers, barometers and dental amalgams. The main culprits in releasing mercury to the atmosphere are, in order of importance, coal-fired power plants, medical waste incinerators, municipal waste incinerators and industrial boilers. In the natural environment mercury can be converted by microbial activity to methylmercury, a form that is readily absorbed and accumulated up food chains, especially in lakes and estuaries. Fish, the top predators, may accumulate concentrations of mercury 10,000 to 100,000 times that in the surrounding water (Bowles et al., 2001). Native peoples who hunt and eat wildlife can accumulate even higher concentrations. Mercury is a serious poison that can cause permanent damage to the human brain and kidneys, and particularly to the developing fetus. It may also damage the immune system.

prospecting for plant species to restore contaminated sites

Land that has been damaged by mining is usually unstable, liable to erosion and devoid of vegetation. The simplest solution to land reclamation is the re-establishment of vegetation cover, because this will stabilize the surface, be visually attractive and self-sustaining (Bradshaw, 2002). Candidate plants for reclamation are those that are tolerant of the toxic heavy metals present. Of particular value are ecotypes – different genotypes, within a species, that fill different niches (see Section 2.3.1) – that have evolved resistance in mined areas. Thus, certain metal-tolerant grass genotypes (or cultivars) have been selected for commercial production in the UK for use on neutral to alkaline soils contaminated by acidic copper wastes (*Agrostis capillaris* cultivar 'Parys') or lead or zinc (*Festuca rubra* cultivar 'Merlin') (Baker, 2002).

In addition, many species characteristic of naturally metal-rich soils have evolved biochemical systems for nutrient acquisition, detoxification and the control of local geochemical conditions. *Phytoremediation* of metal-contaminated sites can take a variety of forms (Susarla et al., 2002). *Phytoaccumulation* occurs when the contaminant is taken up by the plants but is not degraded rapidly or completely; these plants, such as the zinc-accumulating herb *Thlaspi caerulescens*, are harvested to remove the contaminant and then replaced. *Phytostabilization*, on the other hand, takes advantage of the ability of root exudates to precipitate heavy metals and render them biologically harmless. Finally, *phytotransformation* involves elimination of a contaminant by the action of plant enzymes; for example, hybrid poplar trees *Populus deltoides* × *nigra* have the remarkable ability to degrade TNT (trinitrotoluene) and show promise for the restoration of munition dump areas.

13.5 Maintenance and restoration of ecosystem services

a triple bottom-line approach to natural resource management

We have now considered a range of examples of the many impacts of human activities on ecosystems, noting that these can often be measured in terms of lost 'ecosystem services' (Section 13.1.2). The concept of ecosystem services brings into focus three very different ways of looking at our effects on the natural world. First, there are the *environmental* outcomes – the realm of the ecologist. But there are also *economic* and *sociopolitical* perspectives. In this section we explore this triple bottom-line approach to sustainable natural resource use by considering two examples – one at a regional scale (Section 13.5.1) and the other global (Section 13.5.2).

13.5.1 Managing an agricultural landscape

When farm production becomes too intensive and widespread, biodiversity is lost because of the loss of species-rich habitat remnants and the impact of high levels of pesticides. At the same time there is an adverse effect on ecosystem services, such as the provision of water of high quality for drinking and contact recreation. Normally provided 'free' from a healthy landscape, these can be lost because of the input of large quantities of nitrogen and phosphorus, fine sediment from eroding land, and an increase in water-borne pathogens from farm animals that affect humans (such as the *Giardia* parasite).

The impact of agriculture depends on the proportion of the landscape that is used for production. One small farm – even if there is excessive use of plow, fertilizer and pesticide – will have little effect on biodiversity and water quality in the landscape as a whole. It is the cumulative effect of larger and larger areas of intensive agriculture that depletes the region's biodiversity and reduces the quality of water needed for other human activities. In other words, management of agricultural landscapes needs to be done at a regional scale.

Santelmann et al. (2004) integrated the knowledge of experts in environmental, economic and sociological disciplines into alternative visions of a particular landscape – the catchment area of Walnut Creek, in an intensively farmed part of Iowa, USA. They mapped the present pattern of land use and then created three future management strategies, assessing how farm income, water quality and biodiversity would be expected to change under each scenario. A *production* scenario imagines what the catchment will look like in 25 years if continued priority is given to corn and soybean production ('row' crops), following a policy that encourages extension of cultivation to all the highly productive soils available in the catchment. A *water quality* scenario envisions a new (hypothetical) federal policy that enforces chemical standards for river and ground water, and supports agricultural practices that reduce soil erosion. And a *biodiversity* scenario assumes a new (hypothetical) federal policy to increase the abundance and richness of native plants and animals – in this case a network of biodiversity reserves is established with connecting habitat corridors (including the riparian zones of rivers).

Figure 13.14 compares for the three scenarios the distribution of agricultural and 'natural' habitats in 25 years' time. Compared with the current situation, the 'production' scenario produces the most homogeneous landscape, with an increase in row crops and a decrease in the less profitable pasture and forage crops. The 'water quality' scenario leads to more extensive riparian strips of natural vegetation cover and more perennial crop cover (pasture and forage crops), which are conducive to both higher water quality and biodiversity. Finally, the 'biodiversity' scenario has even wider riparian strips together with prairie, forest and wetland reserves, and an increase in strip intercropping, a

comparing three scenarios for managing a catchment area

Figure 13.14

Present landscape (top left) and alternative future scenarios for the Walnut Creek catchment area in Iowa, USA. In comparison to the current situation, note how row crops increase at the expense of perennial cover in the 'production' scenario. In the 'water quality' scenario, note the increase in perennial cover (pasture and forage crops) and wider riparian buffers. In the 'biodiversity' scenario, note the increase in strip intercropping, the wide riparian buffers and the extensive prairie, forest and wetland restoration reserves.

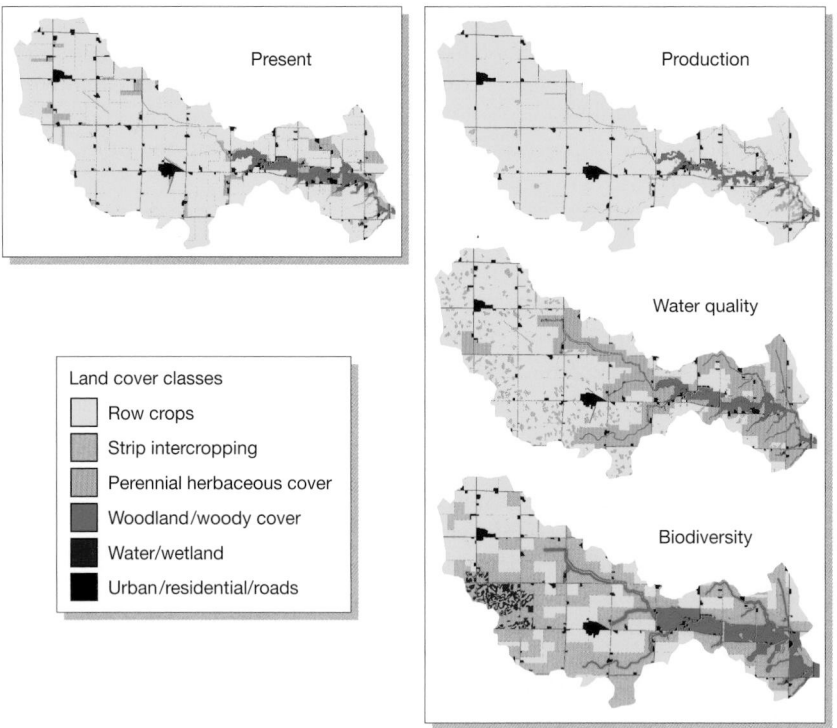

Land cover classes
- Row crops
- Strip intercropping
- Perennial herbaceous cover
- Woodland/woody cover
- Water/wetland
- Urban/residential/roads

FROM SANTELMANN ET AL., 2004

farming practice that benefits biodiversity because it increases connectivity between reserves.

The percentage change after 25 years in economic, water quality and biodiversity terms is shown for each scenario in Figure 13.15. Not surprisingly, the 'biodiversity' scenario ranks highest for improvements in plant and animal biodiversity. More unexpected is the finding that the land use and management practices required by the 'biodiversity' scenario are nearly as profitable to farmers as current practices. The 'biodiversity' scenario also ranks highest in terms of acceptability to farmers (based on farmer ratings of images of land cover under each scenario), and provides water quality improvements similar in magnitude to those in the 'water quality' scenario. Despite the slightly higher profitability of the 'production' scenario, it seems that the farmers would not be unhappy with a 'biodiversity' strategy that provides the greatest benefits to the community at large in terms of biodiversity and ecosystem services.

farmers accept a 'biodiversity' scenario despite lower productivity

13.5.2 Global environmental outcomes of different sociopolitical scenarios

Dealing with the diversity of views among neighbors in a farming region is difficult enough, but our biggest environmental problems require a multinational, global change to the way we deal with nature.

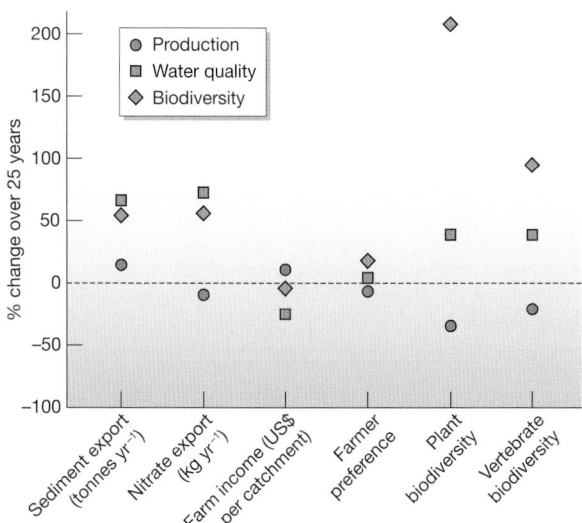

Figure 13.15

Percent change in the Walnut Creek catchment area for each scenario ('production', 'water quality' and 'biodiversity', compared to the current situation) in water quality measures (sediment, nitrate concentration), an economic measure (farm income in the catchment as a whole), a measure of farmer preference for each scenario (based on farmer ratings of images of what the land cover would look like under each scenario) and two biodiversity measures (plant and vertebrate). The 'biodiversity' and 'water quality' scenarios rank above the 'production' scenario in all but economic profitability.

AFTER SANTELMANN ET AL., 2004

An analysis of four contrasting sociopolitical scenarios in Table 13.2 explores likely trends in climate change, pollution problems and the state of ecosystem services. If there is little change in our sociopolitical outlook – that is, if our world remains regionalized and fragmented and mainly concerned with security and protection – the *order from strength* scenario is expected to apply, with poor economic growth, degradation of all ecosystem services and a large increase in global temperature. A more globally connected society (*global orchestration*) could produce higher economic growth and the biggest improvement for the poorest people, but at the cost of many ecosystem services and with the largest predicted temperature increase (particularly because of continued heavy fossil fuel use). The scenario *adapting mosaic*, of a world driven by local communities focusing on sound environmental management (such as our regional example in Section 13.5.1) would lead to the smallest economic growth, improvements to all ecosystem services and an intermediate rise in global temperature. Finally, the *technogarden* scenario, with its environmentally sound but highly managed ecosystems, and crucially with a climate change policy (stabilizing carbon dioxide at 550 ppm), leads to the smallest rise in temperature, reduces nutrient pollution of waterways and improves ecosystem services – except cultural ones, because many ecosystems are managed and unnatural.

Which of these, or other, scenarios comes to pass depends on a wide range of sociopolitical factors. Watch this space!

Table 13.2

Four scenarios that explore plausible futures for ecosystems and human well-being based on different assumptions about sociopolitical forces of change and their interactions. Greenhouse gas emissions [carbon dioxide (CO_2), methane (CH_4), nitrous oxide (N_2O) and 'Other'] are expressed as gigatons of carbon-equivalents (GtC-eq).

	GREENHOUSE GAS EMISSIONS TO 2050	PREDICTED TEMPERATURE RISE TO 2050 AND 2100	LAND USE CHANGES TO 2050	NITROGEN TRANSPORT IN RIVERS TO 2025	ECOSYSTEM SERVICES TO 2025
Global orchestration A globally connected society focused on global trade and economic liberalization. Assumes a reactive approach to ecosystem problems. Takes strong steps to reduce poverty and inequality and to invest in public goods such as infrastructure and education. Economic growth is the highest of the four scenarios, while population in 2050 is lowest (8.1 billion)	CO_2: 20.1 GtC-eq CH_4: 3.7 GtC-eq N_2O: 1.1 GtC-eq Other: 0.7 GtC-eq	2050: +2.0°C 2100: +3.5°C	Slow forest decline to 2025, 10% more arable land	Increased nitrogen in rivers	Provisioning services improved, regulating and cultural services degraded
Order from strength A regionalized and fragmented world, concerned with security and protection, emphasizing primarily regional markets, paying little attention to public goods and taking a reactive approach to ecosystem problems. Economic growth rate is the lowest (particularly in developing countries) while population growth is the highest of the scenarios (9.6 billion in 2050)	CO_2: 15.4 GtC-eq CH_4: 3.3 GtC-eq N_2O: 1.1 GtC-eq Other: 0.5 GtC-eq	2050: +1.7°C 2100: +3.3°C	Rapid forest decline to 2025, 20% more arable land	Increased nitrogen in rivers	All ecosystem services heavily degraded
Adapting mosaic River catchment-scale ecosystems are the focus of political and economic activity. Local institutions are strengthened and local ecosystem management strategies are common, with a strongly proactive (and learning) approach. Economic growth is low initially but increases with time. Population in 2050 is high (9.5 billion)	CO_2: 13.3 GtC-eq CH_4: 3.2 GtC-eq N_2O: 0.9 GtC-eq Other: 0.6 GtC-eq	2050: +1.9°C 2100: +2.8°C	Slow forest decline to 2025, 10% more arable land	Increased nitrogen in rivers	All ecosystem services improved
Technogarden A globally connected world relying on environmentally sound technology, using highly managed, often engineered, ecosystems to deliver ecosystem services, and taking a proactive approach to ecosystem management. Economic growth is relatively high and accelerating, while the 2050 population is midrange (8.8 billion). This is the only scenario to assume a climate policy (stabilizing CO_2 at 550 ppm)	CO_2: 4.7 GtC-eq CH_4: 1.6 GtC-eq N_2O: 0.6 GtC-eq Other: 0.2 GtC-eq	2050: +1.5°C 2100: +1.9°C	Forest increase to 2025, 9% more arable land	Decreased nitrogen in rivers	Provisioning and regulating services improved, cultural services degraded

FROM TOWNSEND, 2007; BASED ON MILLENNIUM ECOSYSTEM ASSESSMENT, 2005

Summary

SUMMARY

Physical and chemical impacts of human activities

People physically degrade or chemically pollute natural ecosystems when generating power or developing land for agricultural, urban and industrial purposes. Humans are not unique among species in degrading their environment, and when our population density was low, and prior to our harnessing of non-food energy, humans probably had no greater impact than many other species. But now the scale of human effects is proportional to our huge numbers and advanced technologies.

Habitat degradation has costs in terms of human health and lost ecosystem services, including provisioning services (such as wild foods and drinking water), cultural services (including educational and recreational opportunities), regulating services (such as the ecosystem's ability to break down pollutants or regulate climate) and supporting services (including primary production and soil formation).

Degradation via cultivation

The intensive production of livestock in factory farming is seriously polluting, and agricultural slurry may need to be thinly dispersed over extensive farmland to dilute it to a level that natural decomposers can deal with it. Intensive agriculture is associated with an increase in the nitrate and phosphate that runs into rivers, lakes and oceans. The consequent eutrophication may be counteracted by matching fertilizer supply to crop demand, restoring natural wetlands (or constructing artificial ones) to take up some of the excess nutrients before they enter rivers and, in lakes, by biomanipulating the level of grazing on phytoplankton to increase water clarity.

Many manufactured pesticides have become important environmental pollutants. Problems arise when pesticides are toxic to many more species than just the target and particularly when they drift beyond the target areas and persist in the environment. The organochlorine insecticides have been particularly problematic because they are progressively biomagnified in animals further up the food chain. Top predators in aquatic and terrestrial food chains, which were never intended as targets, can then accumulate very high doses.

Cultivation can also physically degrade a landscape through the loss of habitat diversity, while heavy irrigation depletes water in rivers and changes their patterns of flow, with adverse consequences for river inhabitants.

Power generation and its diverse effects

Our use of fossil fuels has provided the power to transform much of the face of the planet through intensive agriculture, urbanization and industrial development. The polluting effects of burning coal and oil include acid rain, which can affect lakes and forests in neighboring countries, and a dramatic increase in atmospheric carbon dioxide, which is responsible for climate change at the global level.

Recent emphasis has been placed on developing alternative energy sources that do not release carbon dioxide. The cleanest and safest technologies are expected to derive from hydropower schemes (already at a technologically advanced state in many parts of the globe), together with wind farms (rapidly developing, but with potential adverse consequences for migrating birds) and solar and wave power. Nuclear power, whose popularity had declined because of concerns over security and radioactive waste disposal, is receiving renewed consideration because it does not release greenhouse gases.

Degradation in urban and industrial landscapes

Our feces and urine create large disposal problems in towns and cities because density is so high. At its simplest, primary sewage treatment simply removes most of the solid organic matter. Secondary treatment mimics natural decomposition processes, eliminating organic matter but leaving high concentrations of nitrate and phosphate in the waste water. Tertiary treatment chemically removes these nutrients.

Exotic industrial chemicals also find their way into waterways and the atmosphere where they cause diverse problems. For example, chlorofluorocarbon compounds (CFCs), developed as aerosols and

▶

refrigerants and used on a very large international scale, were found to pose the threat that their chlorine content could interact with and destroy atmospheric ozone, which normally protects the worlds' biota from harmful UV radiation. International agreement to phase out CFCs is expected to solve the problem by 2050 (including recovery of the substantial ozone hole that forms annually over Antarctica).

Mining activities, whether for fossil fuels or metals, also cause physical and chemical degradation to surrounding ecosystems. For example, more than 1 million tonnes of oil enters the world's waterways every year from wells drilled into the seabed or from oil tankers, with adverse consequences for marine life. Mining for metals such as copper may also pollute at every stage of extraction, purification and disposal.

Land that has been damaged by mining is usually unstable, liable to erosion and devoid of vegetation. The simplest solution to land reclamation is the re-establishment of vegetation cover, because this will stabilize the surface, be visually attractive and self-sustaining. Candidate plants for reclamation are those that are tolerant of the toxic heavy metals present.

Maintenance and restoration of ecosystem services

The concept of ecosystem services brings into focus three very different ways of looking at our effects on the natural world – the triple bottom-line of environmental, economic and sociopolitical perspectives. Planning for sustainable use of natural resources usually needs to be carried out at regional or global scales.

The impact of agriculture depends on the proportion of the landscape that is used for production, and planning needs to be done at the regional scale and involve the knowledge of experts in environmental, economic and sociological disciplines. Dealing with the diversity of views among neighbors is difficult enough, but our biggest environmental problem – climate change due in large measure to the burning of fossil fuels – requires a multinational, global level of planning.

Review questions

REVIEW QUESTIONS

Asterisks indicate challenge questions

1 What are the features that distinguish human pollution of the environment from that by other social organisms?

2 Explain why it may be impossible to achieve increasing agricultural production without creating unacceptable levels of nitrate in drinking water.

3* Consider the toilet that you most frequently use. Find out where your sewage goes and how it is treated. What pollution problems are you contributing to as a result of your sewage disposal?

4 Describe the causes of acid rain and the way in which it damages terrestrial and aquatic communities.

5* Hydroelectric schemes provide one of the least polluting ways of generating power.

Nevertheless, they have a number of negative effects on natural systems. What are they?

6 Define the characteristics that make some pesticides particularly dangerous pollutants.

7 Describe the ways in which the use of metals by humans has created problems of environmental pollution.

8 Define the greenhouse effect and list the pollutants that contribute to it.

9* Review the case of the Asian vultures heading to extinction (see Section 1.3.4) and describe the ecosystem services that would be lost with the vultures. In outline, describe how economic value could be estimated for these services.

10* It is often argued that environmental pollution can be prevented only by 'making the polluter pay'. Discuss the ways in which this is, or might be, done.

Chapter **14**

Conservation

Key concepts

KEY CONCEPTS

In this chapter you will:

- recognize that in seeking to conserve the Earth's species and communities, we are often woefully ignorant of what there is to conserve
- appreciate that endangered species are usually rare, but not all rare species are endangered
- understand that some species are at risk for a single reason, such as overexploitation, habitat disruption or introduced species, but often a combination of factors is at work
- recognize that populations that become very small may experience genetic problems
- understand that conservation involves the development of species management plans but also often requires a broader, community perspective
- appreciate that global climate change further complicates conservation planning

Natural ecosystems have been placed at threat by a plethora of human influences, particularly in the face of a burgeoning human population. Conservation is the science concerned with increasing the probability that the Earth's species and communities (or, more generally, its biodiversity) will persist into the future. We need to appreciate the scale of the problem, understand the threats posed by human activities and consider how our knowledge of ecology can be brought to bear to provide remedies.

14.1 Introduction

what is biodiversity?

The term *biodiversity* makes frequent appearances in both the popular media and the scientific literature – but it often does so without an unambiguous definition. At its simplest, it is species richness, the number of species present in a defined geographic unit (see Chapter 10). Biodiversity, though, can also be viewed at scales smaller and larger than the species. For example, we may include genetic diversity within species, perhaps seeking to conserve genetically distinct subpopulations and subspecies (see Chapter 8). Above the species level, we may wish to ensure that species without close relatives are afforded special protection, so that the overall evolutionary variety of the world's biota is maintained as large as possible. At a larger scale still, we may include in biodiversity the variety of community types present in a region – swamps, deserts, early and late stages in a woodland succession and so on. Thus, 'biodiversity' may itself, quite reasonably, have a diversity of meanings. Yet it is necessary to be specific if the term is to be of any practical use. Ecologists must define precisely what it is they mean to conserve in their particular circumstances, and how to measure whether this has been achieved.

estimates of the number of species on Earth range from 3 to 30 million or more

Most often the focus of concern of conservation biologists is the rate of extinction of species in the face of human influence. To judge the scale of this problem, we need to know the total number of species that occur in the world, the rate at which these are going extinct and how this rate compares with that of pre-human times. Unfortunately, there are considerable uncertainties in our estimates of all these things. About 1.8 million species have so far been named (Figure 14.1), but the real number must be much larger. Estimates have been derived in a variety of ways. One approach, for example, uses information on the rate of discovery of new species to project forward, group by taxonomic group, to a total estimate of up to 6–7 million species in the world. However, the uncertainties in estimating global species richness are profound and our best guesses range from 3 to 30 million or more (Gaston, 1998).

modern extinction rates compared to historical extinction rates

An important lesson from the fossil record is that the vast majority of (probably all) species eventually become extinct – more than 99% of species that ever existed are now extinct. However, given that individual species are believed, on average, to have lasted about 1–10 million years, and if we estimate conservatively that the total number of species on Earth is 10 million, we would predict that only an average of between 100 and 1000 species (0.001–0.01%) would go extinct each

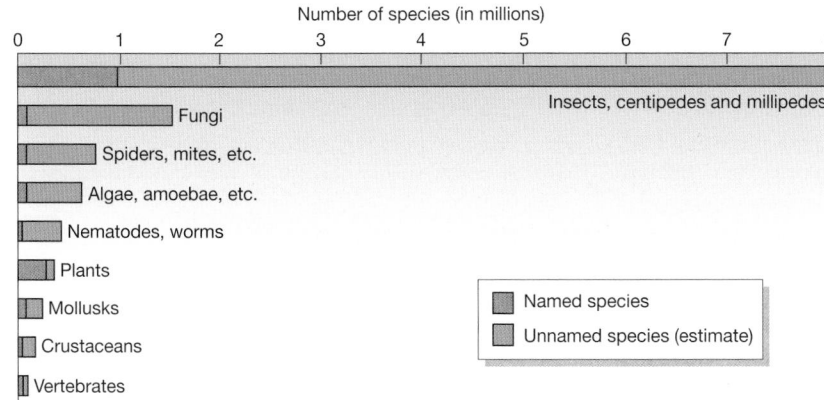

Figure 14.1

Numbers of species identified and named (maroon histograms) and estimates of unnamed species that exist (green histograms).

century. The current observed rate of extinction of birds and mammals of about 1% per century is 100–1000 times this 'natural' background rate. Furthermore, the scale of the most powerful human influence, habitat destruction, continues to increase.

The evidence, then, while inconclusive to a degree because of the unavoidable difficulty of making accurate estimates, suggests that our children and grand-children may live through a period of species extinction comparable to the 'natural' mass extinctions evident in the geological record (see Section 10.6). But should we care? To most, the answer is a resounding and unhesitating 'Yes'. Whether the answer seems obvious or debatable, however, it is important to consider *why* we should care – *why* biodiversity is valuable (Box 14.1).

14.1 *Topical ECOncerns*

What is the value of biodiversity?

To most people, biological diversity is undeniably of value but standard economics has generally failed to assign value to ecological resources. Thus, the costs of environmental damage or depletion of living resources have frequently been disregarded. A major challenge is the development of a new *ecological economics* (Costanza et al., 1997) in which the worth of species, communities and ecosystems can be assigned financial value to be set against the gains to be made in industrial and other human projects that may damage them. As we saw in Section 13.1.2, the value of biodiversity can be measured in terms of the 'free' *ecosystem services* it provides.

Many species have direct value and many more are likely to have a potential value that as yet remains untapped. For example, wild meat, fish and plants remain vital resources in many parts of the world, while most of the world's food is derived from plants that were originally domesticated from wild plants in tropical and semiarid regions. In future, wild strains of these species may be exploited for their genetic diversity, and quite different species of plants and animals may be found that are appropriate for domestication. Secondly, as we saw in Chapter 12, the potential benefits that might come from natural enemies if they could be used as biological control agents for pest species are enormous; most natural enemies of most pests remain unstudied and often unrecognized. Finally, about 40% of the prescription and non-prescription drugs used throughout the world have active ingredients extracted from plants and animals. Aspirin, probably the world's most widely used drug, was derived originally from the leaves of the tropical willow, *Salix alba*. The nine-banded armadillo (*Dasypus novemcinctus*) has been used to study leprosy and prepare a vaccine for the disease; the Florida manatee (*Trichechus manatus*), an endangered mammal, is being used to help understand hemophilia; while the rose periwinkle (*Catharanthus roseus*), a plant from Madagascar, has yielded two potent drugs effective in treating blood cancer. In all these cases, the species can be thought of as representing *provisioning ecosystem services* (see Section 13.1.2).

Other species have indirect economic value. For example, many wild insects are responsible for pollinating crop plants. This is another provisioning service. In a different context, the monetary value of ecotourism, which depends on biodiversity, is becoming ever more considerable. Each year, nearly 200 million adults and children in the USA take part in nature recreation and spend about $4 billion on fees, travel, lodging, food and equipment. Moreover, ecotourists, who visit a country wholly or partly to experience its biological diversity, spend approximately $12 billion a year worldwide on their enjoyment of the natural world (Primack, 1993). On a smaller scale, a multitude of natural history films, books and educational programs are 'consumed' annually without harming the wildlife upon which they are based. In these

contexts, biodiversity provides *cultural ecosystem services*. More ingenuity is required to find ways to measure the indirect economic benefits that accrue as a result of natural biodiversity; for example, biological communities can be of vital importance by maintaining the chemical quality of natural waters, in buffering ecosystems against floods and droughts, in protecting and maintaining soils, in regulating local and even global climate, and in breaking down or immobilizing organic and inorganic wastes. All of these are *regulating ecosystem services*.

It should be noted that many people point to ethical grounds for conservation, with every species being of value in its own right – a value that would still exist even if people were not here to appreciate or exploit them. From this perspective even species with no conceivable economic value require protection.

It would be wrong, though, to see things only from the point of view of conservation – not that there are really arguments *against* conservation as such, but there are arguments in favor of the human activities that make conservation a necessity: agriculture, the felling of trees, the harvesting of wild animal populations, the exploitation of minerals, the burning of fossil fuels, irrigation, the discharge of wastes and so on. To be effective, it is likely that the arguments of conservationists must ultimately be framed in cost–benefit terms because governments will always determine their policies against a background of the money they have to spend and the priorities accepted by their electorates.

A government conservation authority is considering a proposal to designate a marine reserve at a rocky promontory of great scenic beauty. The site is very diverse in species, including a few that are rare. Commercial and recreational fishers wish to continue fishing at this unusually productive site, local people have mixed feelings about an expected influx of tourists, while conservationists (who mostly live a long way from the site) believe that the conservation value is such that no fishing should be permitted and visitor numbers should be strictly controlled. Imagine that you are an arbitrator chairing a meeting of all interested parties. What arguments do you think they will put forward? What decision would you reach and why?

Conservation biology relies on an understanding of the threats facing biodiversity (Section 14.2). After presenting this background, we consider in Section 14.3 the options open to conservation biologists to maintain or restore biodiversity. Then, in Section 14.4 we consider some of the issues confronting conservation biologists in the face of global climate change. Section 14.5 provides the final word.

14.2 Threats to biodiversity

the classification of threat

A basic aim of conservation is to prevent species from becoming extinct either regionally or globally. But how do we define the risk of extinction that a species faces? A species can be described as:

- *critically endangered* if there is considered to be more than a 50% probability of extinction in 10 years or three generations, whichever is longer (Figure 14.2);
- *endangered* if there is more than a 20% chance of extinction in 20 years or five generations;
- *vulnerable* if there is a greater than 10% chance of extinction in 100 years;
- *near threatened* if a species is close to qualifying for a threat category or judged likely to qualify in the near future;
- of *least concern* if a species does not meet any of these threat categories (Rodrigues et al., 2006).

Based on the above criteria, for example, 12% of bird species, 20% of mammals and 32% of amphibians are threatened with extinction (being critically endangered, endangered or vulnerable; Rodrigues et al., 2006).

there are several ways of being rare

Species that are at high risk of extinction are almost always rare, but not all rare species are at risk. We need to ask what precisely we mean by rare. A species may be rare in the sense that its geographic range is small, or in the sense that its habitat range is narrow, or because local populations, even where they do occur, are small. Species that are rare on all three counts, such as the giant panda

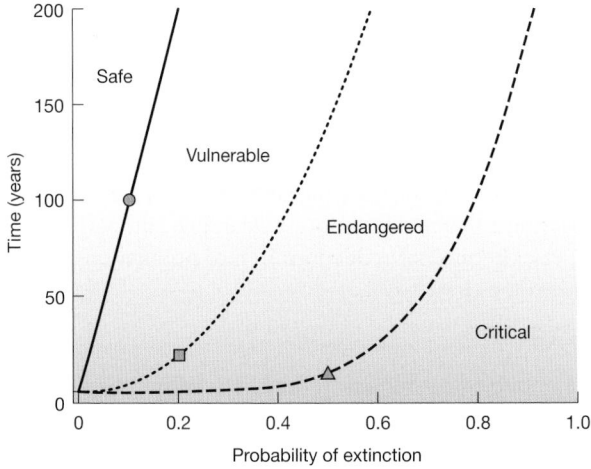

Figure 14.2

Levels of threat as a function of time and probability of extinction. The circle represents a 10% probability (i.e. 0.1) of extinction in 100 years (minimum criterion for a population to be designated 'vulnerable'). The square represents a 20% probability of extinction in 20 years (minimum criterion for the designation 'endangered'). The triangle represents a 50% probability of extinction in 10 years (minimum criterion for the designation 'critically endangered').

(*Ailuropoda melanoleuca*), are intrinsically vulnerable to extinction. However, species need only be rare in one sense in order to become endangered. For example, the peregrine falcon (*Falco peregrinus*) is broadly distributed across habitats and geographic regions, yet, because it exists always at low densities, local populations in the USA have become extinct and have had to be re-established with individuals bred in captivity (see Box 13.1).

some species have rarity thrust upon them

Nevertheless, rare species, just by virtue of their rarity, are not necessarily at risk of extinction. In fact it seems that many, probably most, species are naturally rare. We have already said (see Chapter 2) that almost everything is almost always absent from almost everywhere. Succinctly: many species are born rare – but others have rarity thrust upon them. Other things being equal, it will be easier to make a rare species extinct, simply because a localized effect may be sufficient to push it to the brink. Next, therefore, we deal with the various categories of human influence that increase the chances of species extinction.

14.2.1 Overexploitation

large animals are prone to overexploitation

The essence of overexploitation is that populations are harvested at a rate that is unsustainable, given their natural rates of mortality and capacities for reproduction (see Section 12.3). We have already discussed the idea that in prehistoric times humans were responsible for the extinction of many large animals, the so-called megaherbivores, by overexploiting them (see Section 10.6). In more recent times, the history of the great whales has followed a similar pattern, while today we are still taking our toll of other vulnerable giants. Sharks provide an interesting example. Among the most feared of species (although attacks are much rarer than held in the popular imagination), large numbers are taken for sport, many others to make shark fin soup, while a large proportion of the estimated annual 200 million shark kills are accidental by-catches of commercial fishing. Evidence is mounting that many species of shark have been declining in abundance, a trend that should come as no surprise given their late ages of maturity, slow reproductive cycles and low fecundities (Cortes, 2002). Sharks are among the most important predators in the marine environment, and their enforced rarity may have widespread repercussions in ocean communities.

the threat posed by collectors

A feature of animals that are collected for ornamentation, whether for their body parts or as exotic pets, is that their value to collectors goes up as they become more rare. Thus, instead of the normal safeguard of a density-dependent reduction in consumption rate at low density (see Section 7.5), the very opposite occurs. The phenomenon is not restricted to animals. New Zealand's endemic mistletoe (*Trilepidia adamsii*), for example, parasitic on a few forest understorey shrubs and small trees, was undoubtedly overcollected to provide herbarium specimens. Always a rare species, its extinction (recorded from 1867 to 1954 but not seen since) was due to overcollecting combined with forest clearance and perhaps an adverse effect on fruit dispersal because of reductions in bird populations.

14.2.2 Habitat disruption

Habitats may be adversely affected by human influence in three main ways. First, a proportion of the habitat available to a particular species may simply be destroyed, for urban and industrial development or for the production of food and other natural resources such as timber. Second, habitat may be degraded by pollution (see Chapter 13) to the extent that conditions become untenable for certain species. Third, habitat may be disturbed by human activities to the detriment of some of its occupants.

Forest clearance has been, and is still, the most pervasive of the forces of habitat destruction. Much of the native temperate forest in the developed world was destroyed long ago, while current rates of deforestation in the tropics are 1% or more per annum. As a consequence, more than half of the wildlife habitat has been destroyed in most of the world's tropical countries. The process of habitat destruction often results in the habitat available to a particular species being more fragmented than was historically the case. This can have several repercussions for the populations concerned, a point we take up again in Section 14.2.4.

Degradation by pollution can take many forms, from the application of pesticides that harm non-target organisms, to acid rain with its adverse effects on organisms as diverse as forest trees, amphibians in ponds and fish in lakes, to global climate change that may turn out to have the most pervasive influence of all. Aquatic environments are particularly vulnerable to pollution. Water, inorganic chemicals and organic matter enter from drainage basins, with which streams, rivers, lakes and continental shelves are intimately connected. Land use changes, waste disposal and water impoundment and abstraction can profoundly affect their patterns of waterflow and the quality of their water (Allan and Flecker, 1993).

Habitat disturbance is not such a pervasive influence as destruction or degradation but certain species are particularly sensitive. For example, diving and snorkeling on coral reefs, even in marine protected areas, can cause damage through direct physical contact with hands, body, equipment and fins. Often the disturbance is minor, but this can amount to cumulative damage and reduction in the populations of vulnerable branching corals. In one analysis of 214 divers in a marine park on Australia's Great Barrier Reef, 15% of divers damaged or broke corals, mostly by fin flicks (Rouphael & Inglis, 2001). Impacts were much more likely to be caused by male than female divers, whilst specialist underwater photographers caused more damage on average (1.6 breaks per 10 minutes) than divers without cameras (0.3 breaks per 10 minutes). Nature recreation, ecotourism and even ecological research are not without risk of disturbance and the decline of the populations concerned.

14.2.3 Introduced species

Invasions of exotic species into new geographic areas sometimes occur naturally and without human agency. However, human actions have increased this trickle to a flood. Human-caused introductions may occur either accidentally as a consequence of human transport, or intentionally but illegally to serve some private purpose, or legitimately to procure some hoped-for public benefit by bringing a pest under control, producing new agricultural products or providing novel recreational opportunities. Many introduced species are assimilated into communities without much obvious effect. However, some have been responsible for dramatic changes to native species and natural communities.

For example, the accidental introduction of the brown tree snake *Boiga irregularis* onto Guam, an island in the Pacific, has through nest predation reduced

habitat may be destroyed . . .

. . . or degraded . . .

. . . or disturbed

introduced predators

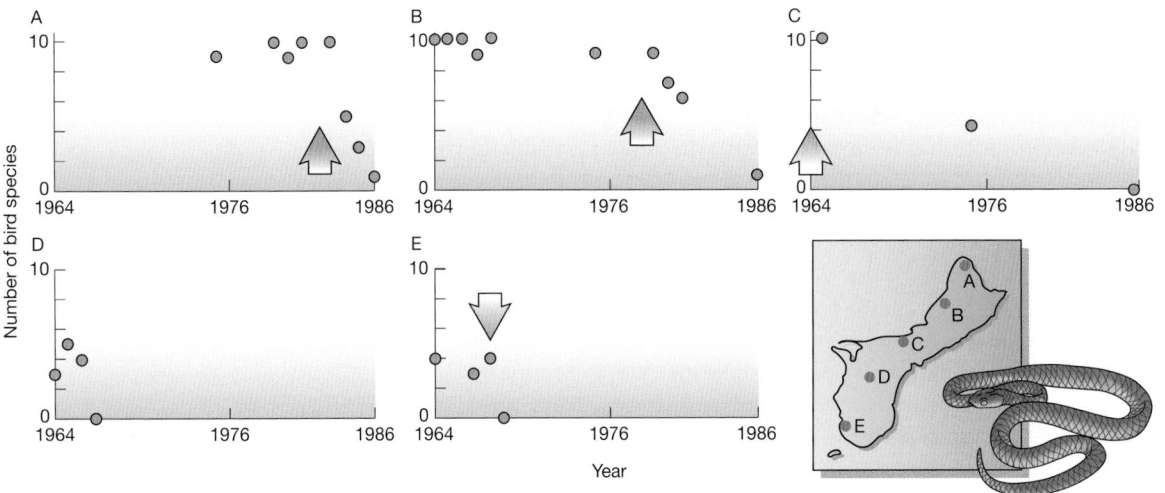

AFTER SAVIDGE, 1987

Figure 14.3

Decline in the number of forest bird species at five locations on the island of Guam. Large arrows indicate the first sightings of the brown tree snake at each location (in location D, the snake was first sighted in the early 1950s).

10 endemic forest bird species to the point of extinction. The gradual spread of the snake from its bridgehead population in the center of the island has been paralleled by the timing of the loss of bird species to the north and south (Figure 14.3). Similarly, the introduction as a source of human food of the predaceous Nile perch (*Lates nilotica*) to the enormously species-rich Lake Victoria in East Africa has driven most of its 350 endemic species of fish to extinction or near extinction (Kaufman, 1992).

introductions leading to homogenization

Conservation biologists are particularly concerned about the effects of introduced species wherever there are communities of native organisms that are largely endemic (that is, live nowhere else in the world). Indeed, one of the major reasons for the world's great biodiversity is the occurrence of centers of endemism so that similar habitats in different parts of the world are occupied by different groups of species that happen to have evolved there. If every species naturally had access to everywhere on the globe, we might expect a relatively small number of successful species to become dominant in each biome. The extent to which this homogenization can happen naturally is restricted by the limited powers of dispersal of most species in the face of the physical barriers that exist to dispersal. By virtue of the transport opportunities offered by humans, these barriers have been breached by an ever-increasing number of exotic species. The effects of introductions have been to convert a hugely diverse range of local community compositions into something much more homogeneous.

It would be wrong, however, to conclude that introducing species to a region will inevitably cause a decline in species richness there (Sax & Gaines, 2003). For example, there are numerous species of plants, invertebrates and vertebrates found in continental Europe but absent from the British Isles (many because they have so far failed to recolonize after the last glaciation). Their introduction would be likely to augment British biodiversity. The significant detrimental effect noted above arises where aggressive species provide a novel challenge to endemic biotas ill equipped to deal with them.

14.2.4 Demographic risks associated with small populations

Much of conservation biology is a crisis discipline. Thus, for example, the remaining population of giant pandas in China (or yellow-eyed penguins in New Zealand or spotted owls in North America) has become so small that if nothing is done the species may become extinct within a few years or decades. There is a pressing need to understand the dynamics of small populations.

These are governed by a high level of uncertainty – whereas large populations can be described as being governed by the law of averages (Caughley, 1994). Three kinds of uncertainty or variation are or particular importance to the fate of small populations.

1 *Demographic uncertainty*. Random variations in the number of individuals that are born male or female, or in the number that happen to die or reproduce in a given year, or in the genetic 'quality' of the individuals in terms of survival/reproductive capacities can matter very much to the fate of small populations. Suppose a breeding pair produces a clutch consisting entirely of females – such an event would go unnoticed in a large population but would be the last straw for a species down to its last pair.

2 *Environmental uncertainty*. Unpredictable changes in environmental factors, whether 'disasters' (such as floods, storms or droughts of a magnitude that occurs very rarely) or more minor (year to year variation in average temperature or rainfall), can also seal the fate of a small population. A small population is more likely than a large one to be reduced by adverse conditions to zero or to numbers so low that recovery is impossible.

3 *Spatial uncertainty*. Many species consist of an assemblage of subpopulations that occur in more or less discrete patches of habitat (habitat fragments). Since the subpopulations are likely to differ in terms of demographic uncertainty, and the patches they occupy in terms of environmental uncertainty, the dynamics of extinction and local recolonization can be expected to have a large influence on the chance of extinction of the overall metapopulation (see Section 9.3).

To illustrate some of these ideas, take the demise in North America of the heath hen (*Tympanuchus cupido cupido*). This bird was once extremely common from Maine to Virginia. Being highly edible and easy to shoot (and also susceptible to introduced cats and affected by conversion of its grassland habitat to farmland), it had by 1830 disappeared from the mainland and was only found on the island of Martha's Vineyard. In 1908 a reserve was established for the remaining 50 birds and by 1915 the population had increased to several thousand. However, 1916 was a bad year. Fire (a disaster) eliminated much of the breeding ground, there was a particular hard winter coupled with an influx of goshawks (environmental uncertainty) and finally poultry disease arrived on the scene (another disaster). At this point the remnant population was likely to have become subject to demographic uncertainty; for example, of the 13 birds remaining in 1928 only two were females. A single bird was left in 1930 and the species went extinct in 1932.

the case of the heath hen

Of the high risk factors associated with local extinctions of plant and animal species, having a small habitat area is probably the most pervasive. Figure 14.4

the importance of habitat area

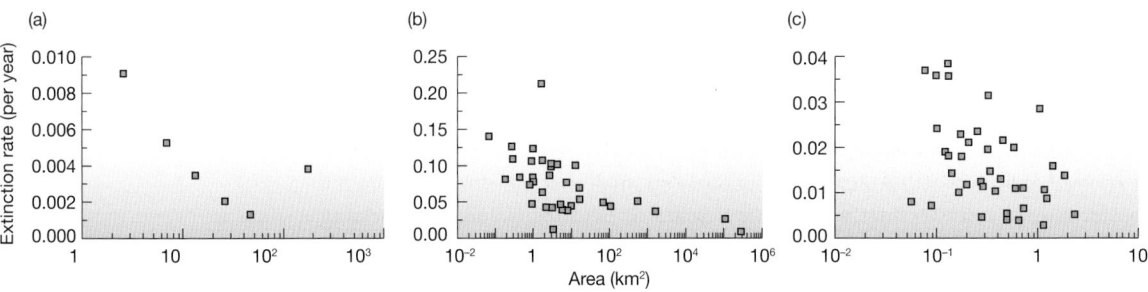

DATA ASSEMBLED BY PIMM, 1991

Figure 14.4

Percentage extinction rates as a function of habitat area for (a) zooplankton in lakes in northeastern USA, (b) birds on northern European islands, and (c) vascular plants in southern Sweden.

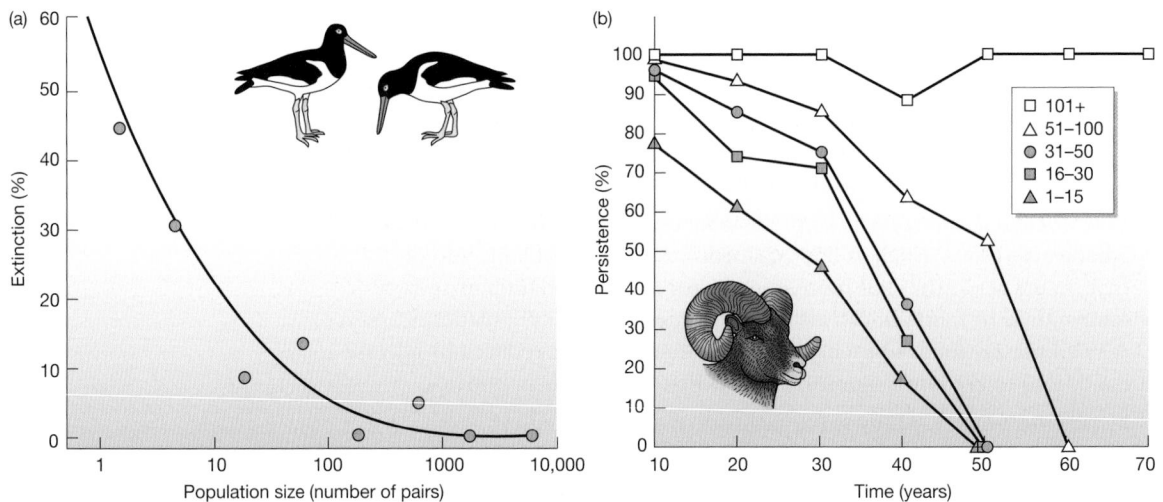

AFTER BERGER, 1990

Figure 14.5

(a) The extinction rate of island birds is higher for small populations. (b) The percentage of populations of bighorn sheep in North America that persists over a 70-year period is lowest where the initial population size was small (green triangles: 1–15 individuals) and is highest where initial populations size was large (open squares: >101 individuals). The regression line in (a) is statistically significant.

shows the negative relationships for a variety of taxa between annual extinction rate and area. No doubt the main reason for the vulnerability of populations in small areas is the fact that the populations themselves are small. This is illustrated in Figure 14.5 for bird species on islands and for bighorn sheep in various desert areas in southwest USA.

    In fact, loss of habitat frequently results not only in a reduction in the absolute size of a population but also the division of the original population into a meta-population of semi-isolated subpopulations. Further fragmentation can result in a decrease in the average size of fragments, an increase in the distance between them and an increase in the proportion of edge habitat (Burgman et al., 1993). A question of fundamental importance, then, is whether a species is more at risk simply because its population is subdivided. In other words, would a single population of a given size be less or more at risk than one divided into a number of subpopulations in habitat fragments?

habitat fragmentation

The answer lies in the balance between the connectedness of different sub-populations on the one hand, and the correlation between the dynamics of different subpopulations on the other. Thus, where the probability of dispersal between fragments (that is, connectedness) is high, metapopulations will tend to persist for longer than unfragmented populations. The reason is because when individual subpopulations go extinct, there is a good chance that they will be restarted by a colonist from another subpopulation. However, where extinction events in different subpopulations are strongly correlated (because environmental variation acts identically in all fragments), metapopulations will be more at risk than unfragmented populations. This is because the individual subpopulations, being small, are vulnerable to extinction, and when one goes extinct, they all tend to.

So far, attention has been focused on individual species, treating them as though they were largely independent entities and applying what we know about population dynamics. However, it hardly needs to be pointed out that conservation of biodiversity also requires a broader perspective in which we apply our knowledge of whole communities. If we ignore community interactions, a chain of extinctions may follow inexorably from the extinction of a particular native species, which therefore deserves special attention. Flying foxes in the genus *Pteropus*, which occur on South Pacific islands, are the major, and sometimes the only, pollinators and seed dispersers for hundreds of native plants (many of which are of considerable economic importance, providing medicines, fiber, dyes, prized timber and foods). Flying foxes are highly vulnerable to human hunters and there is widespread concern about declining numbers. On the island of Guam, for example, the two indigenous flying fox species are either extinct, or virtually so, and there are already indications of reductions in fruiting and dispersal (Cox et al., 1991).

chains of extinctions – taking a community perspective

14.2.5 Possible genetic problems in small populations

Theory tells conservation biologists to beware genetic problems in small populations that may arise through loss of genetic variation (Box 14.2). The preservation of genetic diversity is important in the first place because of the long-term evolutionary potential it provides. Rare forms of a gene (alleles), or combinations of alleles, may confer no immediate advantage but could turn out to be well suited to changed environmental conditions in the future. Small populations tend to have less variation and hence lower evolutionary potential.

loss of evolutionary potential

A more immediate potential problem is inbreeding depression. When populations are small there is a tendency for related individuals to breed with one another. All populations carry recessive alleles that can be lethal to individuals when homozygous (i.e. when the alleles provided by the mother and father are identical). Individuals that breed with close relatives are more likely to produce offspring where the harmful alleles are derived from both parents – so the deleterious effect is expressed. There are many examples of inbreeding depression – breeders of domesticated animals and plants, for example, have long been aware of reductions in fertility, survivorship, growth rates and resistance to disease.

the risk of inbreeding depression

In their study of 23 local populations of the rare plant *Gentianella germanica* in grasslands in the Jura Mountains (Swiss–German border), Fischer and Matthies (1998) found a negative correlation between reproductive performance and

14.2 *Quantitative aspects*

What determines genetic variation?

Genetic variation is determined primarily by the joint action of natural selection and genetic drift (where the frequency of genes in a population is determined by chance rather than evolutionary advantage). The relative importance of genetic drift is higher in small isolated populations that, as a consequence, are expected to lose genetic variation. The rate at which this happens depends on the effective population size (N_e). This is the size of the 'genetically idealized' population to which the actual population (N) is equivalent in genetic terms.

N_e is usually less, often much less, than N, for a number of reasons [detailed formulae can be found in Lande and Barrowclough (1987)]:

1 If the sex ratio is not 1:1; for instance, with 100 breeding males and 400 breeding females, $N = 500$ but $N_e = 320$.

2 If the distribution of progeny from individual to individual is not random; for instance, if 500 individuals each produce one individual for the next generation on average ($N = 500$), but the variance in progeny production is five (with random variation this would be one), then $N_e = 100$.

3 If population size varies from generation to generation, then N_e is disproportionately influenced by the smaller sizes; for instance, for the sequence 500, 100, 200, 900, 800, mean $N = 500$ but $N_e = 258$.

How many individuals are needed to maintain genetic variability? Franklin and Frankham (1998) suggest that an effective population size of 500–1000 might be needed to maintain longer term evolutionary potential.

Greater prairie chickens (*Tympanuchus cupido pinnatus*), closely related to the heath hens in Section 14.2.4, provide a good example of how genetic diversity may be related to population size. These birds were once widespread throughout the prairies of North America, but with the loss and fragmentation of this habitat many populations have become small and isolated. Johnson et al. (2003) used molecular biology techniques (see Section 8.2) to measure genetic diversity in both large (from 1000 to more than 100,000 individuals) and small prairie chicken populations (fewer than 1000 individuals). The mean number of alleles (per gene) ranged from 7.7 to 10.3 in the large populations, but was only 5.1–7.0 in the small populations. Prairie chicken populations were once linked by the 'gene flow' provided by migrants, keeping genetic diversity high. But current populations are isolated in their habitat fragments.

population size (Figure 14.6a–c). Furthermore, population size decreased between 1993 and 1995 in most of the studied populations, but population size decreased more rapidly in the smaller populations (Figure 14.6d). Seeds taken from small populations produced fewer flowers than seeds from large populations grown under identical conditions. We can conclude that genetic effects are of importance for population persistence in this rare species.

14.2.6 *A review of risks*

'drivers' of extinction

We have seen that extinction may be caused by one of a number of 'drivers', including overexploitation, habitat disruption and introduced species. The relative importance of different drivers for global bird biodiversity is illustrated in

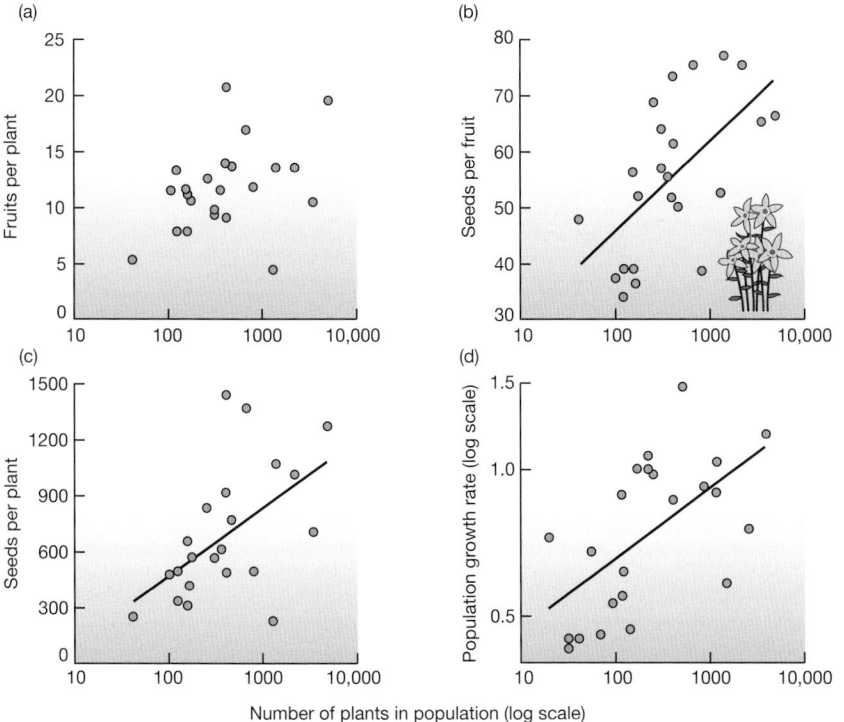

(a)

(b)

(c)

(d)

Number of plants in population (log scale)

Figure 14.6

Relationships for 23 populations of *Gentianella germanica* between population size and (a) mean number of fruits per plant, (b) mean number of seeds per fruit and (c) mean number of seeds per plant. (d) The relationship between population growth rate from 1993 to 1995 (ratio of population sizes) and population size (in 1994). All regression lines are significant at $P < 0.05$; no line is shown in (a) because the regression is not significant.

Figure 14.7. Bird extinctions during the last five centuries can be attributed, in roughly equal measure, to the effects of invasive species, overexploitation by hunters and habitat loss. Currently, habitat loss is the biggest problem facing threatened species (whether critically endangered, endangered or vulnerable).

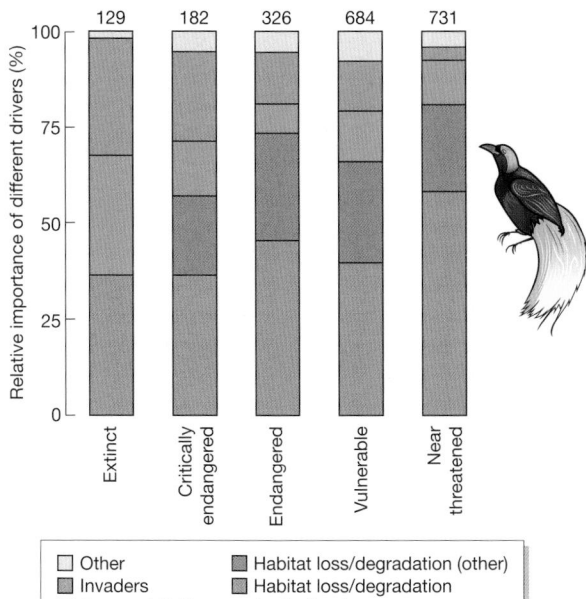

Figure 14.7

Relative importance of different 'drivers' responsible for the loss or endangerment of bird biodiversity. Patterns are shown for five categories of extinction threat (see Section 14.2). The values above each histogram are the numbers of species in each threat category globally. Habitat loss/degradation poses a much bigger risk now than in the past (compare histograms for endangered and vulnerable categories with extinct birds) and this is set to increase in the future, in particular via agricultural expansion (histogram for near threatened species).

Figure 14.8

Extinction vortices may progressively lower population sizes leading inexorably to extinction.

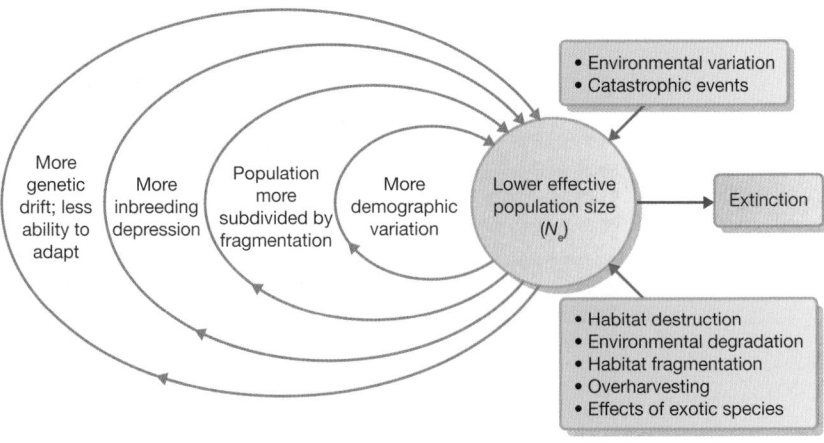

And in the case of 'near threatened' bird species, the ones that managers will have to attend to in future, habitat loss to agriculture is expected to be by far the most important driver.

extinction vortices

Some species are at risk for a single reason, but often, as in the case of the New Zealand mistletoe discussed earlier, a combination of factors is at work. It is interesting that no example of extinction due to genetic problems has so far come to light. Perhaps inbreeding depression has occurred, though undetected, as part of the 'death rattle' of some dying populations (Caughley, 1994). Thus, a population may have been reduced to a very small size by one or more of the processes described above, and this may have led to an increased frequency of matings among relatives and the expression of deleterious recessive alleles in offspring, leading to reduced survivorship and fecundity and causing the population to become smaller still – the so-called *extinction vortex* (Figure 14.8). The small populations of *Gentianella germanica* (Figure 14.6) may have entered an extinction vortex.

14.3 Conservation in practice

Given the environmental circumstances and species characteristics of a particular rare species, what is the chance it will go extinct in a specified period? Alternatively, how big must its population be to reduce the chance of extinction to an acceptable level? These are frequently the crunch questions in conservation biology and a tool, known as *population viability analysis*, is frequently called upon (Section 14.3.1). As a result of population modeling, managers determine the course of action most likely to prevent extinction. Sometimes, however, populations have become so small that their only chance of persistence involves translocation of individuals from viable populations elsewhere or from captive rearing programs. In these cases, managers can call on genetic theory to determine the best individuals to found or augment a population (Section 14.3.2). Conservation action often involves setting aside protected areas, sometimes designed for particular species (to provide a large enough area to accommodate the minimum viable population size) but often to protect biodiversity more generally. We discuss some of the principles of reserve design in Section 14.3.3.

14.3.1 Population viability analysis

Data sets such as those for bighorn sheep in desert areas in Figure 14.5b are unusual because they depend on a long-term commitment (in this case, by hunting organizations) to monitor a number of populations. If we set an arbitrary definition of the necessary *minimum viable population* (MVP) as one that will give at least a 95% probability of persistence for 100 years, we can explore data like these to provide an approximate estimate of MVP. Bighorn populations of fewer than 50 individuals all went extinct within 50 years, while only 50% of populations of 51–100 sheep lasted for 50 years. Evidently, for an MVP here we require more than 100 individuals; indeed, for these sheep such populations demonstrated close to 100% persistence over the maximum period studied of 70 years. The value for conservation of studies like this, however, is limited because they deal with species that are generally not at risk.

Simulation models known as *population viability analyses* (PVAs) provide an alternative, more specific way of gauging viability. Usually, these encapsulate survivorships and reproductive rates in age-structured populations (see Chapter 5). Random variations in these elements or in carrying capacity (K) can be employed to represent the impact of environmental variation, including that of disasters of specified frequency and intensity. Density dependence can be introduced where required. In the more sophisticated models, every individual is treated separately in terms of the probability, with its imposed uncertainty, that it will survive or produce a certain number of offspring in the current time period. The program is run many times, each giving a different population trajectory because of the random elements involved. The outputs, for each set of model parameters used, include estimates of population size each year and the probability of extinction during the modeled period (the proportion of simulated populations that go extinct).

Koalas (*Phascolarctos cinereus*) are regarded as 'near threatened' in Australia, with populations in different parts of the country varying from secure to vulnerable or extinct. Penn et al. (2000) used a widely available PVA tool (known as VORTEX; Lacy, 1993) to model two populations in Queensland, one thought to be declining (Oakey), the other secure (Springsure). Koala breeding commences at 2 years in females and 3 years in males. The other demographic values were derived from extensive knowledge of the two populations and are shown in Table 14.1. Note how the Oakey population had somewhat higher female mortality and fewer females producing young each year. The Oakey population was modeled from 1971 and the Springsure population from 1976 (when first estimates of density were available) and the model trajectories were indeed declining and stable, respectively (Figure 14.9). Over the modeled period, the probability of extinction of the Oakey population was 0.380 (i.e. 380 out of 1000 iterations went extinct) while that for Springsure was 0.063. Managers concerned with critically endangered species do not usually have the luxury of monitoring populations to check the accuracy of their predictions. In contrast, Penn et al. (2000) were able to compare the predictions of their PVAs with real population trajectories, because the koala populations have been continuously monitored since the 1970s (Figure 14.9). The predicted trajectories were close to the actual population trends, particularly for the Oakey population, and this gives added confidence to the modeling approach. The predictive accuracy of VORTEX and other simulation modeling tools has also been shown to be good for 21 other long-term animal data sets (Brook et al., 2000).

trying to determine the minimum viable population

simulation modeling – population viability analysis

koalas – identifying populations at particular risk

Table 14.1

Values used as inputs for simulations of koala populations at Oakey (declining) and Springsure (secure).
Values in brackets are standard deviations due to environmental variation; the model procedure involves
the selection of values at random from the range. Catastrophes are assumed to occur with a certain
probability; in years when the model 'selects' a catastrophe, reproduction and survival are reduced
by the multipliers shown (e.g. in a year with a catastrophe, reproduction is reduced to 55% of what it
would otherwise have been).

VARIABLE	OAKEY	SPRINGSURE
Maximum age	12	12
Sex ratio (proportion male)	0.575	0.533
Litter size of 0 (%)	57.00 (±17.85)	31.00 (±15.61)
Litter size of 1 (%)	43.00 (±17.85)	69.00 (±15.61)
Female mortality at age 0	32.50 (±3.25)	30.00 (±3.00)
Female mortality at age 1	17.27 (±1.73)	15.94 (±1.59)
Adult female mortality	9.17 (±0.92)	8.47 (±0.85)
Male mortality at age 0	20.00 (±2.00)	20.00 (±2.00)
Male mortality at age 1	22.96 (±2.30)	22.96 (±2.30)
Male mortality at age 2	22.96 (±2.30)	22.96 (±2.30)
Adult male mortality	26.36 (±2.64)	26.36 (±2.64)
Probability of catastrophe	0.05	0.05
Multiplier, for reproduction	0.55	0.55
Multiplier for survival	0.63	0.63
% males in breeding pool	50	50
Initial population size	46	20
Carrying capacity, K	70 (±7)	60 (±6)

AFTER PENN ET AL., 2000

Figure 14.9

Observed koala population trends (diamonds) compared with
predicted population performance (triangles, ± 1 SD) based on
1000 repeats of the VORTEX modeling procedure at (a) Oakey
and (b) Springsure. Real population censuses were not
performed every year.

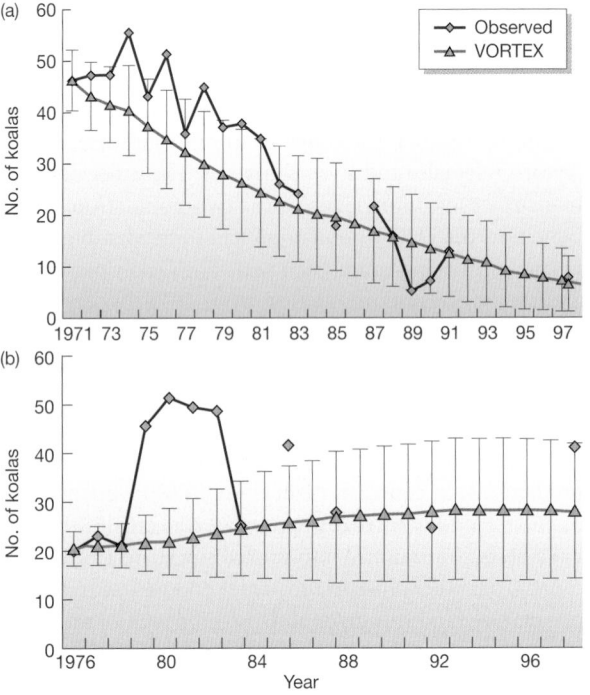

AFTER PENN ET AL., 2000

How can such modeling be put to management use? Local governments in New South Wales are obliged both to prepare comprehensive koala management plans and to ensure that developers survey for potential koala habitat when a building application affects an area greater than 1 ha. Penn et al. (2000) argue that PVA modeling can be used to determine whether any effort made to protect habitat is likely to be rewarded by a viable population.

The life histories of plants present particular challenges for simulation modeling, including seed dormancy, highly periodic recruitment of seedlings and clonal growth (Menges, 2000). However, as with endangered animals, different management scenarios can be simulated in PVAs. The royal catchfly, *Silene regia*, is a long-lived prairie perennial whose range has shrunk dramatically. Menges and Dolan (1998) collected demographic data for up to 7 years from 16 populations in the US Midwest. The populations, whose total adult numbers ranged from 45 to 1302, had been subject to different management regimes. This species, whose seeds do not show dormancy, has high survivorship and frequent flowering, but successful germination is very episodic – most populations in most years fail to produce seedlings.

the royal catchfly – management of an endangered plant

Simulation modeling made use of *population projection matrices*, which are particularly useful for analyzing species with overlapping generations. A population projection matrix acknowledges that most life cycles comprise a sequence of distinct classes with different rates of fecundity and survival. A matrix for one of the populations of *S. regia* is illustrated in Table 14.2. Such matrices were produced for each population in each year. Multiple simulations, each lasting 1000 years, were then run for every matrix to determine both the probability of extinction and the population's finite rate of increase, λ. This term has not yet been introduced, but note that it is related to the population's intrinsic rate of increase, r, discussed in Box 5.4. In fact, $r = \ln \lambda$. For now, all you need to appreciate is that a population will grow in size when $\lambda > 1$ and will decline when $\lambda < 1$; a value of $\lambda = 2$, for example, means that on average every individual in the population will give rise to two individuals in the next generation (either by producing one surviving offspring and staying alive itself, or by dying but producing two surviving offspring).

Figure 14.10 shows the median population growth rate λ for the 16 populations, grouped into cases where particular management regimes were in place.

Table 14.2

An example of a projection matrix (using the simulation modeling tool called RAMAS-STAGE) for a particular *Silene regia* population from 1990 to 1991, assuming successful germination of seedlings. The numbers represent the proportion changing from the stage in the column to the stage in the row (bold values represent plants remaining in the same stage). 'Alive undefined' represents individuals with no size or flowering data, usually as a result of mowing or herbivory. The numbers in the top row are seedlings produced by flowering plants. The finite rate of increase, λ, for this population is 1.67. (Note that a population will increase when $\lambda > 1$, and decrease when $\lambda < 1$.) The site is managed by regular burning.

	SEEDLING	VEGETATIVE	SMALL FLOWERING	MEDIUM FLOWERING	LARGE FLOWERING	ALIVE UNDEFINED
Seedling	–	–	5.32	12.74	30.88	–
Vegetative	0.308	**0.111**	0	0	0	0
Small flowering	0	0.566	**0.506**	0.137	0.167	0.367
Medium flowering	0	0.111	0.210	**0.608**	0.167	0.300
Large flowering	0	0	0.012	0.039	**0.667**	0.167
Alive undefined	0	0.222	0.198	0.196	0	**0.133**

Figure 14.10

Median rates of population increase (λ) of *Silene regia* populations in relation to management regime, for years with seedling recruitment (shaded circles) and without (open triangles). Unburned management regimes include just mowing, herbicide use or no management. All sites above the dashed line have values for λ of >1.0, indicating their capacity to grow in size. Those below the line are on paths to extinction.

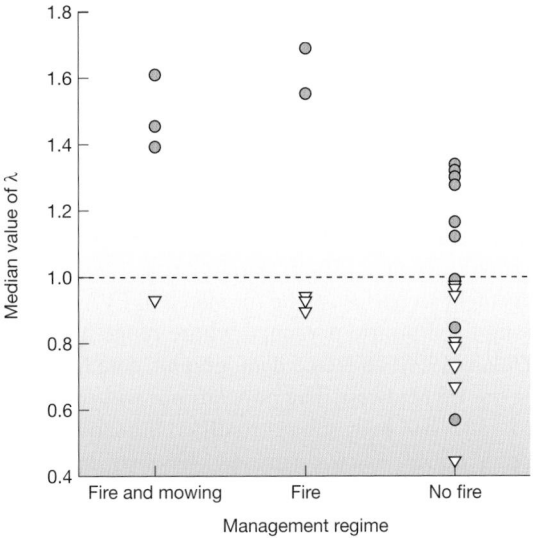

AFTER MENGES & DOLAN, 1998

This was done both for years when recruitment of seedlings occurred and for years when seedling recruitment did not occur. All sites where λ was greater than 1.35 when recruitment took place are managed by burning and some by mowing as well; none of these were predicted to go extinct during the modeled period. On the other hand, populations with no management regime, or whose management does not include fire, had lower values for λ and all except two had predicted extinction probabilities (over 1000 years) of from 0.10 to 1.00.

The obvious management recommendation is to use prescribed burning to provide opportunities for seedling recruitment. Low establishment rates of seedlings may be due to rodents or ants eating fruits or to competition for light with other plants – burnt areas probably reduce one or both of these negative effects.

14.3.2 Dealing with genetic issues

recovery of the pink pigeon

The pink pigeon (*Columba mayeri*), once widespread on the island of Mauritius, was down to only nine or 10 birds in 1990. As a result of the release of captive-bred individuals the population had swelled to 355 free-living individuals (plus more in captivity) by 2003. In captivity, the aim was to manage matings to retain high levels of genetic diversity and to minimize inbreeding. The captive population was originally descended from just 11 founder individuals, augmented in 1989–1994 by adding 12 more founder individuals (offspring of the remaining wild individuals).

Once captive-reared birds are released into the wild the incidence of inbreeding depression is not easy to control – the tactic of releasing a large number of birds provides the greatest chance of success. Between 1987 and 1997, 256 birds were reintroduced on Mauritius – wherever possible selecting birds with minimal inbreeding (based on records in breeding 'stud books') and releasing them in groups with good representation of the different founder ancestries. All birds were banded for unique identification.

The genetics and ecological success of both captive and wild populations have been carefully monitored. Thus, we can evaluate the impact of inbreeding

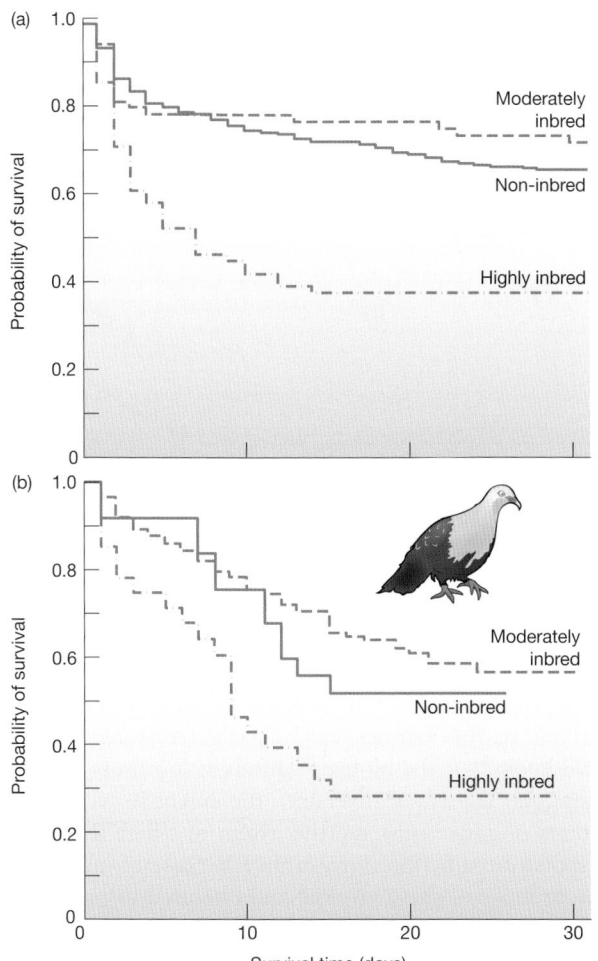

(a)

(b)

Probability of survival

Survival time (days)

Moderately inbred

Non-inbred

Highly inbred

Moderately inbred

Non-inbred

Highly inbred

AFTER SWINNERTON ET AL., 2004

Figure 14.11

Effect of inbreeding on probability of survival to 30 days of age of pink pigeon nestlings (a) in captivity and (b) in the wild population. Inbreeding is expressed as an index derived from known ancestry in relation to 23 founder individuals. The fewer founders in a bird's ancestry, the higher the index of inbreeding. Birds are grouped into three classes – non-inbred, moderately inbred and highly inbred. Only highly inbred birds show a powerful effect of inbreeding.

on survival and reproduction under the controlled situation of captive rearing and also in the more risky circumstances of the wild. Inbreeding reduced egg fertility and survival of nestlings (Figure 14.11), but effects were only strongly marked in the most inbred birds. The pink pigeon reintroduction success story has the added benefit of providing a rare quantification of the value of avoiding inbreeding when managing endangered populations.

14.3.3 Selecting conservation areas

Producing survival plans for individual species may be the best way to deal with species recognized to be in deep trouble and identified to be of special importance (e.g. keystone species described in Section 9.5.2, evolutionarily unique species or charismatic large animals that are easy to 'sell' to the public). However, there is no possibility that all endangered species could be dealt with one at a time. Conservation dollars are simply too limited for this. We can, though, expect to conserve the greatest biodiversity if we protect whole communities by setting aside protected areas. In fact, protected areas of various kinds (national parks,

Figure 14.12

Distribution of biodiversity hotspots, showing numbers of species of globally threatened birds plus amphibians mapped on an equal area basis (each grid cell is 3113 km²).

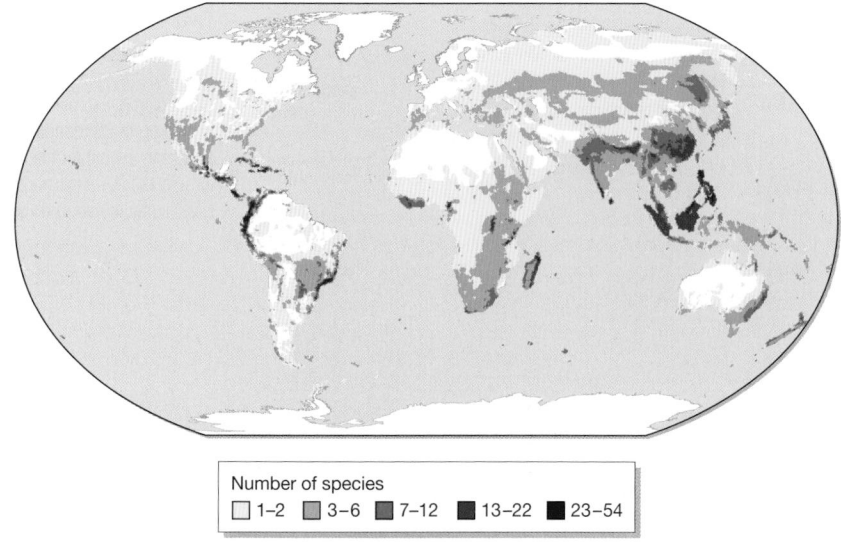

Number of species
☐ 1–2 ▨ 3–6 ▤ 7–12 ■ 13–22 ■ 23–54

AFTER RODRIGUES ET AL., 2006

nature reserves, sites of special scientific interest, etc.) grew both in number and area during the 20th century. Currently, about 7.9% of the world's land area is protected (and 0.5% of the sea area; Balmford et al., 2002).

biodiversity hotspots

It is important to devise priorities so that the restricted number of new protected areas, in terrestrial and marine settings, can be evaluated systematically and chosen with care. We know that the biotas of different locations vary in species richness (with particular centers of diversity), the extent to which the biota is unique (with centers of endemism) and the extent to which the biota is endangered (with hotspots of extinction, for example because of imminent habitat destruction). One or more of these criteria could be used to prioritize potential areas for protection (Figure 14.12).

the design of nature reserves

A perhaps rather surprising application of island biogeography theory (see Section 10.5.1) is in nature conservation. This is because many conserved areas and nature reserves are surrounded by an 'ocean' of habitat made unsuitable, and therefore hostile, by people. Can the study of islands in general provide us with design principles that can be used in the planning of nature reserves? The answer is a cautious 'Yes'; some general points can be made.

1 One problem that conservation managers sometimes face is whether to construct one large reserve or several small ones adding up to the same total area. If the region is homogeneous in terms of conditions and resources, it is quite likely that smaller areas will contain a subset of the species present in a larger area. In such as case it would be preferable to construct the larger reserve in the expectation of conserving more species in total (this recommendation derives from the species–area relationships discussed in Section 10.5.1).

2 On the other hand, if the region as a whole is heterogeneous, then each of the small reserves may support a different group of species and the total conserved might exceed that in one large reserve of the same size. In fact, collections of small islands tend to contain more species than a comparable area composed of one or a few large islands. The pattern

is similar for habitat islands and, most significantly, for national parks. Thus, several small parks contained more species than larger ones of the same area in studies of mammals and birds in East African parks, of mammals and lizards in Australian reserves, and of large mammals in national parks in the USA. It seems likely that habitat heterogeneity is a general feature of considerable importance in determining species richness.

3 A point of particular significance is that local extinctions are common events, and so recolonization of habitat fragments is critical for the survival of fragmented populations. Thus, we need to pay particular attention to the spatial relationships amongst fragments, including the provision of dispersal corridors. There are potential disadvantages – for example, corridors could increase the correlation among fragments of catastrophic effects, such as the spread of fire or disease – but the arguments in favor are persuasive. Indeed, high recolonization rates (even if this means conservation managers themselves moving organisms around) may be indispensable to the success of conservation of endangered metapopulations. Note especially that human fragmentation of the landscape, producing subpopulations that are more and more isolated, is likely to have had the strongest effect on populations with naturally low rates of dispersal. Thus, the widespread declines of the world's amphibians may be due, at least in part, to their poor potential for dispersal (Blaustein et al., 1994).

The basic approach in *complementarity* selection is to proceed in a stepwise fashion, selecting at each step the site that is most *complementary* to those already selected in terms of the biodiversity it contains. In the case of the coastal marine fishes around Western Australia, the results of a complementarity analysis showed that more than 95% of the total of 1855 species could be represented in just six, appropriately located, sections (each 100 km long) (see stars in Figure 14.13).

principles for selecting new reserves: 'complementarity' . . .

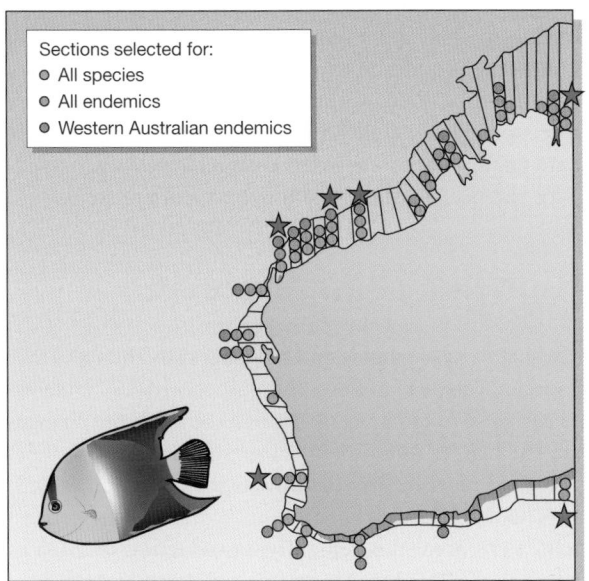

Figure 14.13

Coastline of Western Australia divided into 100 km lengths and showing the results of complementarity analysis to identify the minimum number of sites needed to include all the fish biodiversity for the region. Analyses were performed using all fish species, and separately for species endemic to Australia (found nowhere else) or those endemic to Western Australia. In the case of total fish biodiversity, 26 areas were needed if all 1855 fish species were to be incorporated (green circles) but only 6 areas (stars) would be needed to incorporate more than 95% of the total.

Sections selected for:
- All species
- All endemics
- Western Australian endemics

AFTER FOX & BECKLEY, 2005

Figure 14.14

Map of South Africa's Cape Floristic Region (CFR) showing site irreplaceability values for achieving a range of conservation targets in the 20-year conservation plan for the region. Irreplaceability is a measure, varying from 0 to 1, which indicates the relative importance of an area for the achievement of regional conservation targets. Existing reserves are shown in blue.

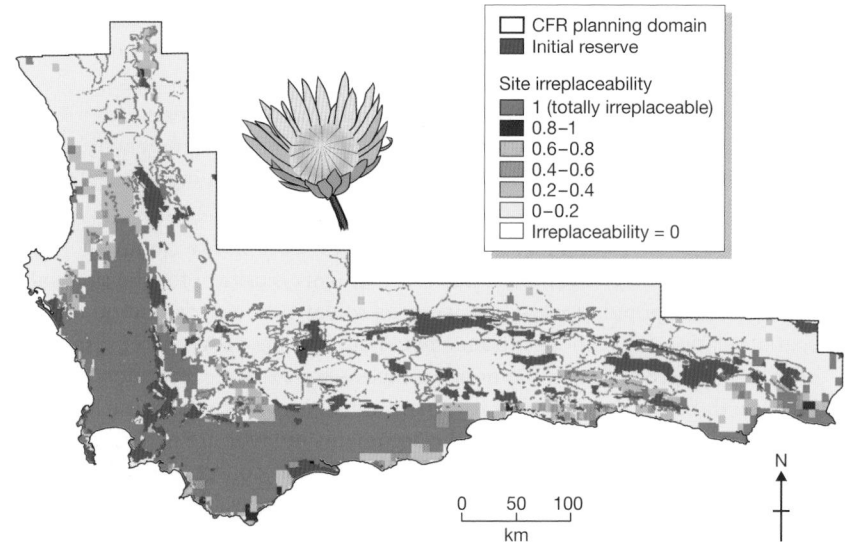

AFTER COWLING ET AL., 2003

. . . or 'irreplaceability'

An approach that contrasts subtly with complementarity analysis concerns the *irreplaceability* of each potential area. Irreplaceability is defined as the likelihood of an area being required to achieve conservation targets or, conversely, the likelihood of one or more targets not being achieved if the area is not included. Cowling et al. (2003) used irreplaceability analysis as part of their conservation plan for South Africa's Cape Floristic Province – a global hotspot with more than 9000 plant species. The research team identified a variety of conservation targets including, among others, the minimum acceptable number of species of *Protea* plants to be safeguarded (for which the region is famous), the minimum permissible number of ecosystem types and even the minimum permissible number of individuals (or populations) of large mammal species. They used an irreplaceability approach to guide the choice of areas to add to existing reserves that would best achieve the conservation targets (Figure 14.14). The ambitious aim is to achieve their overall goal by 2020 and they conclude that, in addition to areas that already have statutory protection, 42% of the Cape Floristic Province, comprising some 40,000 km², will need some level of protection. This includes all cases of high irreplaceability (>0.8) and some areas that are unimportant in terms of *Protea* and ecosystem types but are needed to provide for the needs of large mammals in lowland areas.

14.4 Conservation in a changing world

As we have seen, a basic idea that derives from island biogeography theory is that smaller areas contain fewer species. One way to assess the extinction risk of endemic species under global climate change is to estimate, on the basis of predicted changes to temperature and rainfall, the loss in area of key habitats. Thus, for example, the characteristic biota of the Cape Floristic Province, discussed in Section 14.3.3, is expected to lose 65% of its habitable area by 2050. On the basis of the general pattern relating species richness to area, this represents a reduction of 24% in number of species (Thomas et al., 2004). Moreover, this conclusion is based on the optimistic assumption that all *Protea* species are capable of dispersing to all currently uninhabited areas that become inhabitable (global climate change will also make some uninhabitable areas more hospitable). If no dispersal is assumed, and future ranges are simply those reduced parts of current ranges that remain inhabitable, 30–40% of species seem at risk of extinction. Similar fates could await diverse animal and plant taxa around the world (Box 14.3). In many cases, though, a suitable choice of protected areas can minimize the predicted losses.

14.3 *Topical ECOncerns*

The following article appeared in the *Boston Globe* on January 2, 2007.

The silence of the polar bears

Arctic polar bears are becoming canaries in the mine, warning of the consequences of global warming.

Even the Bush administration has been forced, grudgingly, to acknowledge this. Last week, it proposed to put the bears on the threatened species list because rising temperatures in the Arctic are depriving them of the ice platforms from which they hunt seals. But Interior Secretary Dirk Kempthorne acted only under pressure of a suit from environmental organizations, and has refused to admit that greenhouse gas emissions from vehicles and smokestacks are causing the ice loss and would have to be cut back to save the bears' habitat.

The administration still has a long way to go before it comes out of the denial that has left it on the sidelines as other nations take action to reduce greenhouse gases. If the United States does not quickly take a leadership role on this issue, polar bears will be only one of many species to suffer. So will human beings.

It is no surprise that one of the first species to be so affected by climate change is in the Arctic. In northern latitudes, temperatures are rising at twice the global rate and could rise by an additional 13 degrees Fahrenheit by the end of the century. Researchers say summer sea ice will decline by 50 to 100 percent, with a worst-case scenario from the National Center

© FLP 10140-00336-140

for Atmospheric Research predicting the ice could be all gone by 2040.

In areas where scientists have studied many of the world's 20,000 to 25,000 polar bears, they report thinner animals with lower female reproductive rates and lower survival rates for cubs. Bears have been seen cannibalizing one another and have drowned during ever-longer swims from ice floe to ice floe.

(Copyright 2007 Globe Newspaper Company; reprinted by permission.)

The proposal to declare polar bears 'threatened' was just the start of a year-long process in which the Department of Interior was due to call for comments before taking final action. The Department is also supposed to be working out a plan for the recovery of polar bears by limiting factors that harm them. What components would you expect to be included in a management plan? Do you think any plan can be effective unless it calls for measures to reduce emissions of carbon dioxide?

Temperature and rainfall also strongly influence the life cycle of butterflies. Beaumont and Hughes (2002) used predicted climate changes to model the future distributions of 24 Australian butterfly species. Under even a moderate set of future conditions (temperature increase of 0.8–1.4°C by 2050) the distributions of 13 of the species decreased by more than 20%. Most at risk are butterflies, such as *Hypochrysops halyetus*, that not only have specialized food plant requirements but also depend on the presence of ants for a mutualistic relationship (see Section 8.4) – this species is predicted to lose 58–99% of its current climatic range. Moreover, only one-fourth of its predicted future distribution occurs in locations that it currently occupies. This result highlights a general point for managers: regional conservation efforts and current nature reserves may turn out to be in the wrong place in a changing world.

Téllez-Valdés and Dávila-Aranda (2003) explored this issue for cacti, the dominant plant form in Mexico's Tehuacán-Cuicatlán Biosphere Reserve. From knowledge of the biophysical basis of current species distributions and assuming one of three future climate scenarios, they predicted future species distributions in relation to the location of the reserve. Table 14.3 shows how the potential ranges of species contracted or expanded in the various scenarios. Focusing on

Table 14.3

The core distributions (km^2) of cacti in Mexico under current conditions and as predicted for three climate change scenarios. Species in the first category of cacti are currently completely restricted to the 10,000 km^2 Tehuacán-Cuicatlán Biosphere Reserve. Those in the second category have a current range more or less equally distributed inside and outside the reserve. The current ranges of species in the final category extend widely beyond the reserve boundaries.

SPECIES CATEGORY	CURRENT	+1.0°C −10% RAIN	+2.0°C −10% RAIN	+2.0°C −15% RAIN
Restricted to the reserve				
Cephalocereus columna-trajani	138	27	0	0
Ferocactus flavovirens	317	532	100	55
Mammillaria huitzilopochtli	68	21	0	0
Mammillaria pectinifera	5,130	1,124	486	69
Pachycereus hollianus	175	87	0	0
Polaskia chende	157	83	76	41
Polaskia chichipe	387	106	10	0
Intermediate distribution				
Coryphantha pycnantha	1,367	2,881	1,088	807
Echinocactus platyacanthus f. grandis	1,285	1,046	230	1,148
Ferocactus haematacanthus	340	1,979	1,220	170
Pachycereus weberi	2,709	3,492	1,468	1,012
Widespread distribution				
Coryphantha pallida	10,237	5,887	3,459	2,920
Ferocactus recurvus	3,220	3,638	1,651	151
Mammillaria dixanthocentron	9,934	7,126	5,177	3,162
Mammillaria polyedra	10,118	5,512	3,473	2,611
Mammillaria sphacelata	3,956	5,440	2,803	2,580
Neobuxbaumia macrocephala	2,846	4,943	3,378	1,964
Neobuxbaumia tetetzo	2,964	1,357	519	395
Pachycereus chrysacanthus	1,395	1,929	872	382
Pachycereus fulviceps	3,306	5,405	2,818	1,071

the most extreme scenario (an average temperature increase of 2.0°C and a 15% reduction in rainfall), it is evident that more than half of the species that are currently restricted to the reserve are predicted to go extinct. A second category of cacti, whose current ranges are almost equally within and outside the reserve, are expected to contract their ranges, but in such a way that their distributions become almost completely confined to the reserve. A final category, whose current distributions are much more widespread, also suffer range contractions but in future they are still expected to be distributed within and outside the reserve. In the case of these cacti, then, the location of the reserve seems to cater adequately for potential range changes. But how many other nature reserves may turn out to be in the wrong place?

14.5 Finale

This final chapter has brought together a diversity of environmental problems (overexploitation, habitat disruption, introduced species, global climate change), which themselves require us to understand population, community and ecosystem dynamics. We have seen that the dynamics of endangered species are governed by a high level of uncertainty; despite this, our knowledge is sometimes sufficient to safeguard biodiversity.

Nevertheless, there is no room for complacency. We have insufficient knowledge and, just as important, insufficient financial resources to protect everything everywhere. In desperate times, painful decisions have to be made about priorities. Thus, wounded soldiers arriving at field hospitals in the First World War were subjected to a *triage* evaluation: priority 1, those who were likely to survive but only with rapid intervention; priority 2, those who were likely to survive without rapid intervention; priority 3, those who were likely to die with or without intervention. Conservation managers are often faced with the same kind of choices and need to demonstrate some courage in giving up on hopeless cases, and prioritizing those species and habitats where something can be done.

the 'triage' approach to setting priorities

The spectrum of opinions on conservation is complete. It ranges from the environmental terrorist, who is prepared to destroy property and put human life at risk for what is seen as unacceptable exploitation of animals, to the other extreme of the exploitational terrorist, who is prepared to destroy a rare habitat just as it is about to achieve protected status. There are zealots on both sides of the spectrum too. On the one hand, there are the industrialists, fishers, farmers and foresters who accept none of the conservationist case and are not prepared to look objectively at the scientific evidence, while, on the other, are the environmental zealots – preservationists who seem unwilling to accept any exploitation of the natural world, some even pronouncing that fishing or hunting or logging are intrinsically wrong. The middle ground is occupied by both exploiters and conservationists whose basic philosophy holds that natural resources can be used, but this should be in a sustainable and balanced manner. A thorough understanding of the principles and applications of ecological science should enable all to pay healthy regard to the scientific aspects of what, in its broader context, is very much an ethical, economic and sociopolitical problem. The task for the next generation of ecologists is to bring their understanding to bear in this challenging environment.

the challenge – taking a balanced view

Summary

SUMMARY

The scale of the problem

Conservation is the science concerned with increasing the probability that the Earth's species and communities (or, more generally, its biodiversity) will persist into the future. Biodiversity is, at its most basic, the number of species present, but it can also be viewed at smaller scales (e.g. genetic variation within populations) and larger scales (e.g. the variety of community types present in a region). About 1.8 million species have so far been named, but the real number is probably between 3 and 30 million. The current observed rate of extinction may be as much as 100–1000 times the background rate indicated by the fossil record.

Endangered species and rarity

A species may be rare in the sense that its geographic and/or habitat ranges are small or in the sense that local populations, even where they do occur, are small. Many species are naturally rare but just by virtue of their rarity species are not necessarily at risk of extinction. However, other things being equal, it will be easier to make a rare species extinct. Some species are born rare, others have rarity thrust upon them as a result of the actions of humans.

Threats to biodiversity

The principal causes of decline are overexploitation, habitat degradation and the introduction of exotic species. Overexploitation occurs when people harvest a population (for food or trophies) at a rate that is unsustainable. Humans adversely affect habitat in three ways – a proportion of available habitat may simply be destroyed, or it may be degraded by pollution, or it may be disturbed by human activities to the detriment of some of its occupants. Human-caused introductions of exotic species, which may occur accidentally or intentionally, have sometimes been responsible for dramatic changes to native species and natural communities.

Genetic problems

Rare alleles of a gene may confer no immediate advantage but could turn out to be well suited to changed environmental conditions in the future – small populations that have lost rare alleles through genetic drift have less potential to adapt. A more immediate potential problem is inbreeding depression – when populations are small there is a tendency for individuals breeding with one another to be related, and this may lead to reductions in fertility, survivorship, growth rates and resistance to disease.

The extinction vortex

A given population may have been reduced to a very small size by one or more of the processes described above, and this may have led to an increased frequency of matings among relatives and the expression of deleterious recessive alleles in offspring, leading to reduced survivorship and fecundity and causing the population to become smaller still – the so-called extinction vortex.

Conservation in practice

Much of conservation biology is a crisis discipline concerned with small populations in immediate danger of extinction. A high level of uncertainty governs the dynamics of small populations, whereas large populations can be described as being governed by 'the law of averages'. Three kinds of uncertainty or variation can be identified that are of particular importance to the fate of small populations: demographic uncertainty, environmental uncertainty and spatial uncertainty. Moreover, loss of habitat frequently results not only in a reduction in the absolute size of a population but also the division of the original population into a number of fragments.

Predicting minimum viable population size

Population viability analysis, a simulation modeling tool, can be used to estimate the minimum population size of a particular species that should ensure its persistence with an acceptable probability (e.g. greater than 90%) for a reasonable period (e.g. 100 years). Armed with such information, managers can work out the best approach to guard against extinction (supplementary feeding, predator control, one or more reserves of appropriate size, etc.)

Selecting protected areas

Given limited funds to purchase protected areas, it is important to devise priorities so that they can be evaluated systematically and chosen with care. We know that the biotas of different locations vary in species richness, the extent to which the biota is unique and the extent to which the biota is endangered; one or more of these criteria could be used to prioritize potential areas for protection. The principles of island biogeography theory provide some clues about the most appropriate shape and disposition of protected areas. The selection of a network of reserves to optimize the protection of biodiversity can be performed on the basis of 'complementarity' (selecting at each step the site that is most complementary to those already selected in terms of the biodiversity it contains) or 'irreplaceability' (defined in terms of the likelihood of an area being required to achieve specified conservation targets).

Global climate change and conservation

Predicted changes to patterns of temperature and rainfall around the world have important implications for conservation biology. Changes to environmental conditions will affect the size and location of habitable areas of species, whether or not they are currently at risk of extinction. Moreover, nature reserves may turn out to be in the wrong places. Models of global climate change can be used by ecologists to safeguard species and communities when planning for the conservation of individual species or designing reserve networks.

Review questions

REVIEW QUESTIONS

Asterisks indicate challenge questions

1* Of the estimated 3–30 million species on Earth, only about 1.8 million have so far been named. How important is it for the conservation of biodiversity that we can name the species involved?

2 Species may be 'rare' on three counts: what are these? From your own experience, provide examples of three 'rare' species and explain the nature of their rarity.

3* Researchers collected data on the relative abundance of 16 Peruvian mammal species in forest areas that contrasted in whether they were subject to light or heavy hunting by local people. As an index of vulnerability to hunting they used the reduction in relative abundance in the heavily versus lightly hunted areas. This was plotted against intrinsic rate of population increase (r_{max}), age of first reproduction and longevity (Figure 14.15). Provide explanations for the relationships shown in the figure. Would you expect the variables r_{max}, age of first reproduction and longevity to be intercorrelated? If so, how? Many species of large animals have gone extinct in the last 50,000 years. What light do the results of

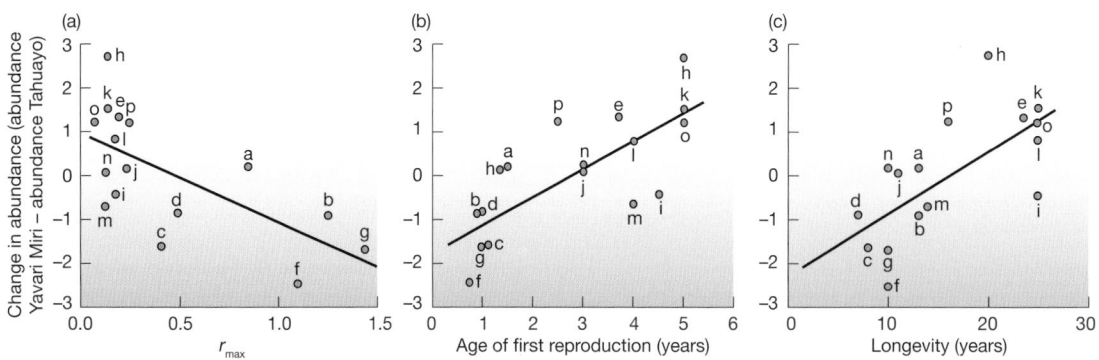

Figure 14.15

Relationships between (a) r_{max}, (b) age of first reproduction, and (c) longevity and the vulnerability of mammals to population declines measured as the change in abundance between lightly and heavily hunted areas of forest. The mammals are represented by the following letters: a, white-lipped peccary; b, collared peccary; c, red brocket deer; d, gray brocket deer; e, lowland tapir; f, black agouti; g, green acouchy; h, woolly monkey; I, howler monkey; j, red wakari monkey; k, brown capuchin; l, white-fronted capuchin; m, monk saki monkey; n, titti monkey; o, spider monkey; p, squirrel monkey.

AFTER BODMER ET AL., 1997

this study shed on the possible role of overexploitation by humans in historical extinctions? On the basis of these results, what advice would you give wildlife managers about conserving mammals in Peruvian forests?

4 Are there any circumstances where the intentional introduction of an exotic species can be considered a good thing because it enhances biodiversity?

5 Unpredictable temporal variability is a feature of most ecosystems. How can conservation biologists allow for such uncertainty when they devise species management plans?

6 Explain, with examples, how the loss or introduction of a single species can have conservation consequences throughout a whole ecological community.

7 In desperate times, painful decisions have to be made about priorities. Discuss the 'triage' approach to conservation assessment. List some highly endangered species of which you are aware and propose priorities for conservation action. Are any so hopeless that they should be allowed to go extinct?

8 Discuss the value of zoos and botanical gardens in nature conservation.

9 Discuss the advantages and limitations of using population viability analysis tools to devise species management plans.

10* The famous ecologist of the early 20th century, A.G. Tansley, when asked what he meant by nature conservation, said it was maintaining the world in the state he knew as a child. From your perspective, as we enter the new millenium, how would you define the aims of conservation biology?

References
REFERENCES

Abramsky, Z. & Rosenzweig, M.L. (1983) Tilman's predicted productivity–diversity relationship shown by desert rodents. *Nature*, 309, 150–151.

Agrawal, A.A. (1998) Induced responses to herbivory and increased plant performance. *Science*, 279, 1201–1202.

Akçakaya, H.R. (1992) Population viability analysis and risk assessment. In: *Proceedings of Wildlife 2001: Populations* (D.R. McCullough, ed.), pp. 148–157. Elsevier, Amsterdam.

Al-Hiyaly, S.A., McNeilly, T. & Bradshaw, A.D. (1988) The effects of zinc contamination from electricity pylons – evolution in a replicated situation. *New Phytologist*, 110, 571–580.

Allan, J.D. & Flecker, A.S. (1993) Biodiversity conservation in running waters. *Bioscience*, 43, 32–43.

Alliende, M.C. & Harper, J.L. (1989) Demographic studies of a dioecious tree. I. Colonization, sex and age-structure of a population of *Salix cinerea*. *Journal of Ecology*, 77, 1029–1047.

Anderson, R.M. (1982) Epidemiology. In: *Modern Parasitology* (F.E.G. Cox, ed.), pp. 205–251. Blackwell Scientific Publications, Oxford.

Anderson, R.M. & May, R.M. (1991) *Infectious Diseases of Humans: Dynamics and Control*. Oxford University Press, Oxford.

Andrewartha, H.G. (1961) *Introduction to the Study of Animal Populations*. Methuen, London.

Angel, M.V. (1994) Spatial distribution of marine organisms: patterns and processes. In: *Large Scale Ecology and Conservation Biology* (P.J. Edwards, R.M. May & N.R. Webb, eds), pp. 59–109. Blackwell Science, Oxford.

Arheimer, B. & Wittgren, H.B. (2002) Modelling nitrogen retention in potential wetlands at the catchment scale. *Ecological Engineering*, 19, 63–80.

Aston, J.L. & Bradshaw, A.D. (1966) Evolution in closely adjacent plant populations. II. *Agrostis stolonifera* in maritime habitats. *Heredity*, 21, 649–664.

Atkinson, D., Ciotti, B.J. & Montagnes, D.J.S. (2003) Protists decrease in size linearly with temperature: *ca.* 2.5%°C^{-1}. *Proceedings of the Royal Society of London, Series B*, 270, 2605–2611.

Audesirk, T. & Audesirk, G. (1996) *Biology: Life on Earth*. Prentice Hall, Upper Saddle River, NJ.

Ayre, D.J. (1985) Localized adaptation of clones of the sea anemone *Actinia tenebrosa*. *Evolution*, 39, 1250–1260.

Ayre, D.J. (1995) Localized adaptation of sea anemone clones: evidence from transplantation over two spatial scales. *Journal of Animal Ecology*, 64, 186–196.

Bach, C.E. (1994) Effects of herbivory and genotype on growth and survivorship of sand-dune willow (*Salix cordata*). *Ecological Entomology*, 19, 303–309.

Baker, A.J.M. (2002) The use of tolerant plants and hyperaccumulators. In: *The Restoration and Management of Derelict Land: Modern Approaches* (M.H. Wong & A.D. Bradshaw, eds), pp. 138–148. World Scientific Publishing, Singapore.

Balmford, A. & Bond, W. (2005) Trends in the state of nature and their implications for human well-being. *Ecology Letters*, 8, 1218–1234.

Balmford, A., Bruner, A., Cooper, P. et al. (2002) Economic reasons for conserving wild nature. *Science*, 297, 950–953.

Barrios, L. & Rodriguez, A. (2004) Behavioural and environmental correlates of soaring-bird mortality at on-shore wind turbines. *Journal of Applied Ecology*, 41, 72–81.

Bayliss, P. (1987) Kangaroo dynamics. In: *Kangaroos, their Ecology and Management in the Sheep Rangelands of Australia* (G. Caughley, N. Shepherd & J. Short, eds), pp. 119–134. Cambridge University Press, Cambridge.

Bazzaz, F.A. (1979) The physiological ecology of plant succession. *Annual Review of Ecology and Systematics*, 10, 351–371.

Bazzaz, F.A. (1996) *Plants in Changing Environments*. Cambridge University Press, Cambridge.

Bazzaz, F.A., Miao, S.L. & Wayne, P.M. (1993) CO_2-induced growth enhancements of co-occurring tree species decline at different rates. *Oecologia*, 96, 478–482.

Bazzaz, F.A. & Williams, W.E. (1991) Atmospheric CO_2 concentrations within a mixed forest: implications for seedling growth. *Ecology*, 72, 12–16.

Beaumont, L.J. & Hughes, L. (2002) Potential changes in the distributions of latitudinally restricted Australian butterfly species in response to climate change. *Global Change Biology*, 8, 954–971.

Becker, P. (1992) Colonization of islands by carnivorous and herbivorous Heteroptera and Coleoptera: effects of island area, plant species richness, and 'extinction' rates. *Journal of Biogeography*, 19, 163–171.

Begon, M., Sait, S.M. & Thompson, D.J. (1995) Persistence of a predator–prey system: refuges and generation cycles? *Proceedings of the Royal Society of London, Series B*, 260, 131–137.

Begon, M., Townsend, C.R. & Harper, J.L. (2006) *Ecology: from Individuals to Ecosystems*, 4th edn. Blackwell Publishing, Oxford.

Belt, T. (1874) *The Naturalist in Nicaragua*. J.M. Dent, London.

Berger, J. (1990) Persistence of different-sized populations: an empirical assessment of rapid extinctions in bighorn sheep. *Conservation Biology*, 4, 91–98.

Berner, E.K. & Berner, R.A. (1987) *The Global Water Cycle: Geochemistry and Environment*. Prentice Hall, Englewood Cliffs, NJ.

Berven, K.A. (1995) Population regulation in the wood frog, *Rana sylvatica*, from three diverse geographic localities. *Australian Journal of Ecology*, 20, 385–392.

Blaustein, A.R., Wake, D.B. & Sousa, W.P. (1994) Amphibian declines: judging stability, persistence, and susceptibility of populations to local and global extinctions. *Conservation Biology*, 8, 60–71.

Bobko, S.J. & Berkeley, S.A. (2004) Maturity, ovarian cycle, fecundity, and age-specific parturition of black rockfish (*Sebastes melanops*). *Fisheries Bulletin*, 102, 418–429.

Bodmer, R.E., Eisenberg, J.F. & Redford, K.H. (1997) Hunting and the likelihood of extinction of Amazonian mammals. *Conservation Biology*, 11, 460–466.

Bonnet, X., Lourdais, O., Shine, R. & Naulleau, G. (2002) Reproduction in a typical capital breeder: costs, currencies and complications in the aspic viper. *Ecology*, 83, 2124–2135.

Bonsall, M.B., French, D.R. & Hassell, M.P. (2002) Metapopulation structure affects persistence of predator–prey interactions. *Journal of Animal Ecology*, 71, 1075–1084.

Borga, K., Gabrielsen, G.W. & Skaare, J.U. (2001) Biomagnification of organochlorines along a Barents Sea food chain. *Environmental Pollution*, 113, 187–198.

Bouzille, J.B., Bonis, A., Clement, B. & Godeau, M. (1997) Growth patterns of *Juncus gerardi* clonal populations in a coastal habitat. *Plant Ecology*, 132, 39–48.

Bowles, K.C., Apte, S.C., Maher, W.A., Kawei, M. & Smith, R. (2001) Bioaccumulation and biomagnification of mercury in Lake Murray, Papua New Guinea. *Canadian Journal of Fisheries and Aquatic Science*, 58, 888–897.

Bradshaw, A.D. (2002) Introduction – an ecological perspective. In: *The Restoration and Management of Derelict Land: Modern Approaches* (M.H. Wong & A.D. Bradshaw, eds), pp. 1–6. World Scientific Publishing, Singapore.

Breznak, J.A. (1975) Symbiotic relationships between termites and their intestinal biota. In: *Symbiosis* (D.H. Jennings & D.L. Lee, eds), pp. 559–580. Symposium 29, Society for Experimental Biology. Cambridge University Press, Cambridge.

Briand, F. (1983) Environmental control of food web structure. *Ecology*, 64, 253–263.

Brook, B.W., O'Grady, J.J., Chapman, A.P., Burgman, M.A., Akçakaya, H.R. & Frankham, R. (2000) Predictive accuracy of population viability analysis in conservation biology. *Nature*, 404, 385–387.

Brookes, M. (1998) The species enigma. *New Scientist*, June 13, 1998.

Brown, J.H. & Davidson, D.W. (1977) Competition between seed-eating rodents and ants in desert ecosystems. *Science*, 196, 880–882.

Brown, V.K. & Southwood, T.R.E. (1983) Trophic diversity, niche breadth, and generation times of exopterygote insects in a secondary succession. *Oecologia*, 56, 220–225.

Brunet, A.K. & Medellín, R.A. (2001) The species–area relationship in bat assemblages of tropical caves. *Journal of Mammalogy*, 82, 1114–1122.

Brylinski, M. & Mann, K.H. (1973) An analysis of factors governing productivity in lakes and reservoirs. *Limnology and Oceanography*, 18, 1–14.

Buckling, A. & Rainey, P.B. (2002) Antagonistic coevolution between a bacterium and a bacteriophage. *Proceedings of the Royal Society of London, Series B*, 269, 931–936.

Bullock, J.M., Moy, I.L., Pywell, R.F., Coulson, S.J., Nolan, A.M. & Caswell, H. (2002) Plant dispersal and colonization processes at local and landscape scales. In: *Dispersal Ecology* (J.M. Bullock, R.E. Kenward & R.S. Hails, eds), pp. 279–302. Blackwell Publishing, Oxford.

Burdon, J.J. (1987) *Diseases and Plant Population Biology*. Cambridge University Press, Cambridge.

Burdon-Sanderson, J.S. (1893) *Inaugural address*. Nature, 48, 464–472.

Burg, T.M. & Croxall, J.P. (2001) Global relationships amongst black-browed and grey-headed albatrosses: analysis of population structure using mitochondrial DNA and microsatellites. *Molecular Ecology*, 10, 2647–2660.

Burgman, M.A., Ferson, S. & Akçakaya, H.R. (1993) *Risk Assessment in Conservation Biology*. Chapman & Hall, London.

Buscot, F., Munch, J.C., Charcosset, J.Y., Gardes, M., Nehls, U. & Hampp, R. (2000) Recent advances in exploring physiology and biodiversity of ectomycorrhizas highlight the functioning of these symbioses in ecosystems. *FEMS Microbiology Reviews*, 24, 601–614.

Cain, M.L., Pacala, S.W., Silander, J.A. & Fortin, M.-J. (1995) Neighbourhood models of clonal growth in the white clover *Trifolium repens*. *American Naturalist*, 145, 888–917.

Carignan, R., Planas, D. & Vis, C. (2000) Planktonic production and respiration in oligotrophic shield lakes. *Limnology and Oceanography*, 45, 189–199.

Carruthers, R.I., Larkin, T.S., Firstencel, H. & Feng, Z. (1992) Influence of thermal ecology on the mycosis of a rangeland grasshopper. *Ecology*, 73, 190–204.

Caughley, G. (1994) Directions in conservation biology. *Journal of Animal Ecology*, 63, 215–244.

Cebrian, J. (1999) Patterns in the fate of production in plant communities. *American Naturalist*, 154, 449–468.

Charnov, E.L. (1976) Optimal foraging: attack strategy of a mantid. *American Naturalist*, 110, 141–151.

Chase, J.M. (2003) Experimental evidence for alternative stable equilibria in a benthic pond food web. *Ecology Letters*, 6, 733–741.

Choquenot, D. (1998) Testing the relative influence of intrinsic and extrinsic variation in food availability on feral pig populations in Australia's rangelands. *Journal of Animal Ecology*, 67, 887–907.

Clements, F.L. (1905) *Research Methods in Ecology*. University of Nevada Press, Lincoln, NV.

Cohen, J.E. (1995) *How Many People Can the Earth Support?* W.W. Norton & Co., New York.

Cohen, J.E. (2001) World population in 2050: assessing the projections. In: *Seismic Shifts: the Economic Impact of Demographic Change* (J.S. Little & R.K. Triest, eds), pp. 83–113.

Conference Series No. 48. Federal Reserve Bank of Boston, Boston.

Cohen, J.E. (2003) Human population: the next half century. *Science*, 302, 1172–1175.

Cohen, J.E. (2005) Human population grows up. *Scientific American*, 293(3), 48–55.

Cole, J.J., Findlay, S. & Pace, M.L. (1988) Bacterial production in fresh and salt water ecosystems: a cross-system overview. *Marine Ecology Progress Series*, 4, 1–10.

Connell, J.H. (1978) Diversity in tropical rainforests and coral reefs. *Science*, 199, 1302–1310.

Connell, J.H. (1983) On the prevalence and relative importance of interspecific competition: evidence from field experiments. *American Naturalist*, 122, 661–696.

Cook, L.M., Dennis, R.L.H. & Mani, G.S. (1999) Melanic morph frequency in the peppered moth in the Manchester area. *Proceedings of the Royal Society of London, Series B*, 266, 293–297.

Coomes, D.A., Rees, M., Turnbull, L. & Ratcliffe, S. (2002) On the mechanisms of coexistence among annual-plant species, using neighbourhood techniques and simulation models. *Plant Ecology*, 163, 23–38.

Cornell, H.V. & Hawkins, B.A. (2003) Herbivore responses to plant secondary compounds: a test of phytochemical coevolution theory. *American Naturalist*, 161, 507–522.

Cortes, E. (2002) Incorporating uncertainty into demographic modeling: application to shark populations and their conservation. *Conservation Biology*, 16, 1048–1062.

Cory, J.S. & Myers, J.H. (2000) Direct and indirect ecological effects of biological control. *Trends in Ecology and Evolution*, 15, 137–139.

Costanza, R., D'Arge, R., de Groot, R. et al. (1997) The value of the world's ecosystem services and natural capital. *Nature* 387, 253–260.

Cotrufo, M.F., Ineson, P., Scott, A. et al. (1998) Elevated CO_2 reduces the nitrogen concentration of plant tissues. *Global Change Biology*, 4, 43–54.

Cottingham, K.L., Brown, B.L. & Lennon, J.T. (2001) Biodiversity may regulate the temporal variability of ecological systems. *Ecology Letters*, 4, 72–85.

Courchamp, F., Clutton-Brock, T. & Grenfell, B. (1999) Inverse density dependence and the Allee effect. *Trends in Ecology and Evolution*, 14, 405–410.

Cowling, R.M., Pressey, R.L., Rouget, M. & Lombard, A.T. (2003) A conservation plan for a global biodiversity hotspot – the Cape Floristic Region, South Africa. *Biological Conservation*, 112, 191–216.

Cox, P.A., Elmquist, T., Pierson, E.D. & Rainey, W.E. (1991) Flying foxes as strong interactors in South Pacific island ecosystems: a conservation hypothesis. *Conservation Biology*, 5, 448–454.

Crisp, M.D. & Lange, R.T. (1976) Age structure distribution and survival under grazing of the arid zone shrub *Acacia burkitii*. *Oikos*, 27, 86–92.

Currie, D.J. (1991) Energy and large-scale patterns of animal and plant species richness. *American Naturalist*, 137, 27–49.

Currie, D.J. & Paquin, V. (1987) Large-scale biogeographical patterns of species richness in trees. *Nature*, 39, 326–327.

Darwin, C. (1859) *On the Origin of Species by Means of Natural Selection*, 1st edn. John Murray, London.

Davidson, D.W. (1977) Species diversity and community organization in desert seed-eating ants. *Ecology*, 58, 711–724.

Davidson, J. & Andrewartha, H.G. (1948) The influence of rainfall, evaporation and atmospheric temperature on fluctuations in the size of a natural population of *Thrips imaginis* (Thysanoptera). *Journal of Animal Ecology*, 17, 200–222.

Davies, S.J., Palmiotto, P.A., Ashton, P.S., Lee, H.S. & Lafrankie, J.V. (1998) Comparative ecology of 11 sympatric species of *Macaranga* in Borneo: tree distribution in relation to horizontal and vertical resource heterogeneity. *Journal of Ecology*, 86, 662–673.

Davis, M.B. (1976) Pleistocene biogeography of temperate deciduous forest. *Geoscience and Management*, 13, 13–26.

Davis, M.B. & Shaw, R.G. (2001) Range shifts and adaptive responses to quarternary climate change. *Science*, 292, 673–679.

de Wit, C.T. (1965) Photosynthesis of leaf canopies. *Verslagen van Landbouwkundige Onderzoekingen*, 663, 1–57.

de Wit, C.T., Tow, P.G. & Ennik, G.C. (1966) Competition between legumes and grasses. *Verslagen van Landbouwkundige Onderzoekingen*, 112, 1017–1045.

Deevey, E.S. (1947) Life tables for natural populations of animals. *Quarterly Review of Biology*, 22, 283–314.

Denno, R.F., McClure, M.S. & Ott, J.R. (1995) Interspecific interactions in phytophagous insects: competition reexamined and resurrected. *Annual Review of Entomology*, 40, 297–331.

Detwiler, R.P. & Hall, C.A.S. (1988) Tropical forests and the global carbon cycle. *Science*, 239, 42–47.

Diamond, J.M. (1972) Biogeographic kinetics: estimation of relaxation times for avifaunas of south-west Pacific islands. *Proceedings of the National Academy of Science of the USA*, 69, 3199–3203.

Dickie, I.A., Xu, B. & Koide, R.T. (2002) Vertical niche differentiation of ectomycorrhizal hyphae in soil as shown by T-RFLP analysis. *New Phytologist*, 156, 527–535.

Dobson, A.P. & Carper, E.R. (1996) Infectious diseases and human population history. *Bioscience*, 46, 115–126.

Dodson, S.I., Arnott, S.E. & Cottingham, K.L. (2000) The relationship in lake communities between primary productivity and species richness. *Ecology*, 81, 2662–2679.

Doube, B.M., Macqueen, A., Ridsdill-Smith, T.J. & Weir, T.A. (1991) Native and introduced dung beetles in Australia. In: *Dung Beetle Ecology* (I. Hanski & Y. Cambefort, eds), pp. 255–278. Princeton University Press, Princeton, NJ.

Drès, M. & Mallet, J. (2001) Host races in plant-feeding insects and their importance in sympatric speciation. *Philosophical Transactions of the Royal Society of London, Series B*, 357, 471–492.

Dunne, J.A., Williams, R.J. & Martinez, N.J. (2002) Network structure and biodiversity loss in food webs: robustness increases with connectance. *Ecology Letters*, 5, 558–567.

Eamus, D. (1999) Ecophysiological traits of deciduous and evergreen woody species in the seasonally dry tropics. *Trends in Ecology and Evolution*, 14, 11–16.

Ebert, D., Zschokke-Rohringer, C.D. & Carius, H.J. (2000) Dose effects and density-dependent regulation in two microparasites of *Daphnia magna*. *Oecologia*, 122, 200–209.

Ehrlich, P. & Raven, P.H. (1964) Butterflies and plants: a study in coevolution. *Evolution*, 18, 586–608.

Eis, S., Garman, E.H. & Ebel, L.F. (1965) Relation between cone production and diameter increment of douglas fir (*Pseudotsuga menziesii* (Mirb). Franco), grand fir (*Abies grandis* Dougl.) and western white pine (*Pinus monticola* Dougl.). *Canadian Journal of Botany*, 43, 1553–1559.

Elliott, J.K. & Mariscal, R.N. (2001) Coexistence of nine anemonefish species: differential host and habitat utilization, size and recruitment. *Marine Biology*, 138, 23–36.

Elliott, J.M. (1994) *Quantitative Ecology and the Brown Trout*. Oxford University Press, Oxford.

Elton, C. (1927) *Animal Ecology*. Sidgwick & Jackson, London.

Elton, C.S. (1958) *The Ecology of Invasions by Animals and Plants*. Methuen, London.

Endler, J.A. (1980) Natural selection on color patterns in *Poecilia reticulata*. *Evolution*, 34, 76–91.

Erwin, T.L. (1982) Tropical forests: their richness in Coleoptera and other arthropod species. *Coleopterists Bulletin*, 36, 74–75.

Falge, E., Baldocchi, D., Tenhunen, J. et al. (2002) Seasonality of ecosystem respiration and gross primary production as derived from FLUXNET measurements. *Agricultural and Forest Meteorology*, 113, 53–74.

Fasham, M.J.R., Balino, B.M. & Bowles, M.C. (2001) A new vision of ocean biogeochemistry after a decade of the Joint Global Ocean Flux Study (JGOFS). *Ambio Special Report*, 10, 4–31.

Fenner, F. (1983) Biological control, as exemplified by smallpox eradication and myxomatosis. *Proceedings of the Royal Society, Series B*, 218, 259–285.

Ferguson, R.G. (1933) The Indian tuberculosis problem and some preventative measures. *National Tuberculosis Association Transactions*, 29, 93–106.

Fischer, M. & Matthies, D. (1998) Effects of population size on performance in the rare plant *Gentianella germanica*. *Journal of Ecology*, 86, 195–204.

FitzGibbon, C.D. (1990) Anti-predator strategies of immature Thomson's gazelles: hiding and the prone response. *Animal Behaviour*, 40, 846–855.

FitzGibbon, C.D. & Fanshawe, J. (1989) The condition and age of Thomson's gazelles killed by cheetahs and wild dogs. *Journal of Zoology*, 218, 99–107.

Flecker, A.S. & Townsend, C.R. (1994) Community-wide consequences of trout introduction in New Zealand streams. *Ecological Applications*, 4, 798–807.

Fleischer, R.C., Perry, E.A., Muralidharan, K., Stevens, E.E. & Wemmer, C.M. (2001) Phylogeography of the Asian elephant (*Elephus maximus*) based on mitochondrial DNA. *Evolution*, 55, 1882–1892.

Flessa, K.W. & Jablonski, D. (1995) Biogeography of recent marine bivalve mollusks and its implications of paleobiogeography and the geography of extinction: a progress report. *Historical Biology*, 10, 25–47.

Flint, M.L. & van den Bosch, R. (1981) *Introduction to Integrated Pest Management*. Plenum Press, New York.

Flower, R.J., Rippey, B., Rose, N.L., Appleby, P.G. & Battarbee, R.W. (1994) Palaeolimnological evidence for the acidification and contamination of lakes by atmospheric pollution in western Ireland. *Journal of Ecology*, 82, 581–596.

Fonseca, C.R. (1994) Herbivory and the long-lived leaves of an Amazonian ant-tree. *Journal of Ecology*, 82, 833–842.

Fonseca, D.M. & Hart, D.D. (1996) Density-dependent dispersal of black fly neonates is mediated by flow. *Oikos*, 75, 49–58.

Ford, E.B. (1975) *Ecological Genetics*, 4th edn. Chapman & Hall, London.

Ford, M.J. (1982) *The Changing Climate: Responses of the Natural Fauna and Flora*. George Allen & Unwin, London.

Fowler, S.V. (2004) Biological control of an exotic scale, *Orthezia insignis* Browne (Homoptera: Orthexiidae), saves the endemic gumwood tree, *Commidendrum robustum* (Roxb.) DC (Asteraceae) on the island of St Helena. *Biological Control*, 29, 367–374.

Fox, C.J. (2001) Recent trends in stock-recruitment of blackwater herring (*Clupea harengus* L.) in relation to larval production. *ICES Journal of Marine Science*, 58, 750–762.

Fox, N.J. & Beckley, L.E. (2005) Priority areas for conservation of Western Australian coastal fishes: a comparison of hotspot, biogeographical and complementarity approaches. *Biological Conservation*, 125, 399–410.

Franklin, I.R. & Frankham, R. (1998) How large must populations be to retain evolutionary potential? *Animal Conservation*, 1, 69–73.

Fredrickson, R.J. & Hedrick, P.W. (2006) Dynamics of hybridization and introgression in red wolves and coyotes. *Conservation Biology*, 20, 1272–1283.

Gadgil, M. (1971) Dispersal: population consequences and evolution. *Ecology*, 52, 253–261.

Galloway, L.F. & Fenster, C.B. (2000) Population differentiation in an annual legume: local adaptation. *Evolution*, 54, 1173–1181.

Garthe, S. & Huppop, O. (2004) Scaling possible adverse effects of marine wind farms on seabirds: developing and applying a vulnerability index. *Journal of Applied Ecology*, 41, 724–734.

Gaston, K.J. (1998) *Biodiversity*. Blackwell Science, Oxford.

Geider, R.J., Delucia, E.H., Falkowski, P.G. et al. (2001) Primary productivity of planet earth: biological determinants and physical constraints in terrestrial and aquatic habitats. *Global Change Biology*, 7, 849–882.

Gende, S.M., Quinn, T.P. & Willson, M.F. (2001) Consumption choice by bears feeding on salmon. *Oecologia*, 127, 372–382.

Gilman, M.P. & Crawley, M.J. (1990) The cost of sexual reproduction in ragwort (*Senecio jacobaea* L.). *Functional Ecology*, 4, 585–589.

Godfray, H.C.J. & Crawley, M.J. (1998) Introductions. In: *Conservation Science and Action* (W.J. Sutherland, ed.), pp. 39–65. Blackwell Science, Oxford.

Gotelli, N.J. & McCabe, D.J. (2002) Species co-occurrence: a meta-analysis of J.M. Diamond's assembly rules model. *Ecology*, 83, 2091–2096.

Gotthard, K., Nylin, S. & Wiklund, C. (1999) Seasonal plasticity in two satyrine butterflies: state-dependent decision making in relation to daylength. *Oikos*, 84, 453–462.

Gould, W.A. & Walker, M.D. (1997) Landscape-scale patterns in plant species richness along an arctic river. *Canadian Journal of Botany*, 75, 1748–1765.

Grant, P.R., Grant, B.R., Keller, L.F. & Petren, K. (2000) Effects of El Nino events on Darwin's finch productivity. *Ecology*, 81, 2442–2457.

Gray, S.M. & Robinson, B.W. (2001) Experimental evidence that competition between stickleback species favours adaptive character divergence. *Ecology Letters*, 5, 264–272.

Green, R.E. (1998) Long-term decline in the thickness of eggshells of thrushes, *Turdus* spp., in Britain. *Proceedings of the Royal Society of London, Series B*, 265, 679–684.

Green, R.E., Newton, I., Shultz, S., Cunningham, A.A., Gilbert, M., Pain, D.J. & Prakash, V. (2004) Diclofenac poisoning as a cause of vulture population declines across the Indian subcontinent. *Journal of Applied Ecology*, 41, 793–800.

Greenwood, P.J., Harvey, P.H. & Perrins, C.M. (1978) Inbreeding and dispersal in the great tit. *Nature*, 271, 52–54.

Grutter, A.S. (1999) Cleaner fish really do clean. *Nature*, 398, 672–673.

Grytnes, J.A. & Vetaas, O.R. (2002) Species richness and altitude: a comparison between null models and interpolated plant species richness along the Himalayan altitudinal gradient, Nepal. *American Naturalist*, 159, 294–304.

Hairston, N.G., Smith, F.E. & Slobodkin, L.B. (1960) Community structure, population control, and competition. *American Naturalist*, 44, 421–425.

Halaj, J., Ross, D.W. & Moldenke, A.R. (2000) Importance of habitat structure to the arthropod food-web in Douglas-fir canopies. *Oikos*, 90, 139–152.

Hall, S.J. (1998) Closed areas for fisheries management – the case consolidates. *Trends in Ecology and Evolution*, 13, 297–298.

Hall, S.J. & Raffaelli, D.G. (1993) Food webs: theory and reality. *Advances in Ecological Research*, 24, 187–239.

Hanski, I. (1999) *Metapopulation Ecology*. Oxford University Press, Oxford.

Hanski, I., Pakkala, T., Kuussaari, M. & Lei, G. (1995) Metapopulation persistence of an endangered butterfly in a fragmented landscape. *Oikos*, 72, 21–28.

Harcourt, D.G. (1971) Population dynamics of *Leptinotarsa decemlineata* (Say) in eastern Ontario. III. Major population processes. *Canadian Entomologist*, 103, 1049–1061.

Harper, J.L. (1977) *The Population Biology of Plants*. Academic Press, London.

Harper, J.L. & White, J. (1974) The demography of plants. *Annual Review of Ecology and Systematics*, 5, 419–463.

Hart, A.J., Bale, J.S., Tullett, A.G., Worland, M.R. & Walters, K.F.A. (2002) Effects of temperature on the establishment potential of the predatory mite *Amblyseius californicus* McGregor (Acari: Phytoseiidae) in the UK. *Journal of Insect Physiology*, 48, 593–599.

Hassell, M.P., Latto, J. & May, R.M. (1989) Seeing the wood for the trees: detecting density dependence from existing life-table studies. *Journal of Animal Ecology*, 58, 883–892.

Herman, T.J.B. (2000) Developing IPM for potato tuber moth. *Commercial Grower*, 55, 26–28.

Hermannsson, S. (2000) *Surtsey Research Report No. XI*. Museum of Natural History, Reykjavik, Iceland.

Hermoyian, C.S., Leighton, L.R. & Kaplan, P. (2002) Testing the role of competition in fossil communities using limiting similarity. *Geology*, 30, 15–18.

Herre, E.A. & West, S.A. (1997) Conflict of interest in a mutualism: documenting the elusive fig wasp–seed trade-off. *Proceedings of the Royal Society of London, Series B*, 264, 1501–1507.

Hilborn, R. & Walters, C.J. (1992) *Quantitative Fisheries Stock Assessment*. Chapman & Hall, New York.

Holloway, J.D. (1977) *The Lepidoptera of Norfolk Island, their Biogeography and Ecology*. Junk, The Hague.

Holloway, J.M., Dahlgren, R.A., Hansen, B. & Casey, W.H. (1998) Contribution of bedrock nitrogen to high nitrate concentrations in stream water. *Nature*, 395, 785–788.

Holyoak, M. & Lawler, S.P. (1996) Persistence of an extinction-prone predator–prey interaction through metapopulation dynamics. *Ecology*, 77, 1867–1879.

Hooper, D.U., Chapin, F.S., Ewel, J.J. et al. (2005) Effects of biodiversity on ecosystem functioning: a consensus of current knowledge. *Ecological Monographs*, 75, 3–35.

Hoyer, M.V. & Canfield, D.E. (1994) Bird abundance and species richness on Florida lakes: influence of trophic status, lake morphology and aquatic macrophytes. *Hydrobiologia*, 297, 107–119.

Hudson, P.J., Dobson, A.P. & Newborn, D. (1992) Do parasites make prey vulnerable to predation? Red grouse and parasites. *Journal of Animal Ecology*, 61, 681–692.

Hudson, P.J., Dobson, A.P. & Newborn, D. (1998) Prevention of population cycles by parasite removal. *Science*, 282, 2256–2258.

Huffaker, C.B. (1958) Experimental studies on predation: dispersion factors and predator–prey oscillations. *Hilgardia*, 27, 343–383.

Hughes, L. (2000) Biological consequences of global warming: is the signal already apparent. *Trends in Ecology and Evolution*, 15, 56–61.

Hunter, M.L. & Yonzon, P. (1992) Altitudinal distributions of birds, mammals, people, forests, and parks in Nepal. *Conservation Biology*, 7, 420–423.

Hurd, L.E. & Eisenberg, R.M. (1990) Experimentally synchronized phenology and interspecific competition in mantids. *American Midland Naturalist*, 124, 390–394.

Huryn, A.D. (1998) Ecosystem-level evidence for top-down and bottom-up control of production in a grassland stream system. *Oecologia*, 115, 173–183.

Husband, B.C. & Barrett, S.C.H. (1996) A metapopulation perspective in plant population biology. *Journal of Ecology*, 84, 461–469.

Hut, R.A., Barnes, B.M. & Daan, S. (2002) Body temperature patterns before, during and after semi-natural hibernation in the European ground squirrel. *Journal of Comparative Physiology B*, 172, 47–58.

Hutchinson, G.E. (1957) Concluding remarks. *Cold Spring Harbour Symposium on Quantitative Biology*, 22, 415–427.

Inouye, R.S., Huntly, N.J., Tilman, D., Tester, J.R., Stillwell, M. & Zinnel, K.C. (1987) Old-field succession on a Minnesota sand plain. *Ecology*, 68, 12–26.

Inouye, R.S. & Tilman, D. (1995) Convergence and divergence of old-field vegetation after 11 yr of nitrogen addition. *Ecology*, 76, 1872–1877.

Interlandi, S.J. & Kilham, S.S. (2001) Limiting resources and the regulation of diversity in phytoplankton communities. *Ecology*, 82, 1270–1282.

International Organisation for Biological Control (1989) *Current Status of Integrated Farming Systems Research in Western Europe* (P. Vereijken & D.J. Royle, eds). IOBC West Palaearctic Regional Service Bulletin No. 12(5). IOBC, Zurich.

IPCC (2001) *Third Assessment Report*. Working Group 1, Intergovernmental Panel on Climate Change. IPCC, Geneva.

Irvine, R.J., Stien, A., Dallas, J.F., Halvorsen, O., Langvatn, R. & Albon, S.D. (2001) Contrasting regulation of fecundity in two abomasal nematodes of Svarlbard reindeer (*Rangifer tarandus platyrynchus*). *Parasitology*, 122, 673–681.

IUCN/UNEP/WWF (1991) *Caring for the Earth. A Strategy for Sustainable Living*. World Conservation Union/United Nations Environmental Program/World Wide Fund, Gland, Switzerland.

Jackson, S.T. & Weng, C. (1999) Late quaternary extinction of a tree species in eastern North America. *Proceedings of the National Academy of Sciences of the USA*, 96, 13847–13852.

Jain, S.K. & Bradshaw, A.D. (1966) Evolutionary divergence among adjacent plant populations. I. The evidence and its theoretical analysis. *Heredity*, 21, 407–411.

Janis, C.M. (1993) Tertiary mammal evolution in the context of changing climates, vegetation and tectonic events. *Annual Review of Ecology and Systematics*, 24, 467–500.

Jennings, S., Kaiser, M.J. & Reynolds, J.D. (2001) *Marine Fisheries Ecology*. Blackwell Publishing, Oxford.

Jeppesen, E., Sondergaard, M., Jensen, J.P. et al. (2005) Lake responses to reduced nutrient loading – an analysis of contemporary long-term data from 35 case studies. *Freshwater Biology*, 50, 1747–1771.

Johannes, R.E. (1998) Government-supported village-based management of marine resources in Vanuatu. *Ocean Coastal Management*, 40, 165–186.

Johnson, C.G. (1967) International dispersal of insects and insect-borne viruses. *Netherlands Journal of Plant Pathology*, 73 (Suppl. 1), 21–43.

Johnson, J.A., Toepfer, J.E. & Dunn, P.O. (2003) Contrasting patterns of mitochondrial and microsatellite population structure in fragmented populations of greater prairie-chickens. *Molecular Ecology*, 12, 3335–3347.

Jones, M. & Harper, J.L. (1987) The influence of neighbours on the growth of trees. I. The demography of buds in *Betula pendula*. *Proceedings of the Royal Society of London, Series B*, 232, 1–18.

Jonsson, M. & Malmqvist, B. (2000) Ecosystem process rate increases with animal species richness: evidence from leaf-eating, aquatic insects. *Oikos*, 89, 519–523.

Jutila, H.M. (2003) Germination in Baltic coastal wetland meadows: similarities and differences between vegetation and seed bank. *Plant Ecology*, 166, 275–293.

Kaiser, J. (2000) Rift over biodiversity divides ecologists. *Science*, 89, 1282–1283.

Kamijo, T., Kitayama, K., Sugawara, A., Urushimichi, S. & Sasai, K. (2002) Primary succession of the warm-temperate broad-leaved forest on a volcanic island, Miyake-jima, Japan. *Folia Geobotanica*, 37, 71–91.

Karban, R., Agrawal, A.A., Thaler, J.S. & Adler, L.S. (1999) Induced plant responses and information content about risk of herbivory. *Trends in Ecology and Evolution*, 14, 443–447.

Karels, T.J. & Boonstra, R. (2000) Concurrent density dependence and independence in populations of arctic ground squirrels. *Nature*, 408, 460–463.

Karl, B.J. & Best, H.A. (1982) Feral cats on Stewart Island: their foods, and their effects on kakapo. *New Zealand Journal of Zoology*, 9, 287–294.

Karlsson, P.S. & Jacobson, A. (2001) Onset of reproduction in *Rhododendron lapponicum* shoots: the effect of shoot size, age, and nutrient status at two subarctic sites. *Oikos*, 94, 279–286.

Kerbes, R.H., Kotanen, P.M. & Jefferies, R.L. (1990) Destruction of wetland habitats by lesser snow geese: a keystone species on the west coast of Hudson Bay. *Journal of Applied Ecology*, 27, 242–258.

Kettlewell, H.B.D. (1955) Selection experiments on industrial melanism in the Lepidoptera. *Heredity*, 9, 323–342.

Khan, A.S., Sumaila, U.R., Watson, R., Munro, G. & Pauly, D. (2006) The nature and magnitude of global non-fuel fisheries subsidies. In: *Catching More Bait: a Bottom-up Re-estimation of Global Fisheries Subsidies* (U.R. Sumaila & D. Pauly, eds), pp. 5–37. Fisheries Centre Research Reports Vol. 14, No. 6. Fisheries Centre, University of British Columbia, Vancouver.

Kicklighter, D.W., Bruno, M., Donges, S. et al. (1999) A first-order analysis of the potential role of CO_2 fertilization to affect the global carbon budget: a comparison of four terrestrial biosphere models. *Tellus*, 51B, 343–366.

Kirk, J.T.O. (1994) *Light and Photosynthesis in Aquatic Ecosystems*. Cambridge University Press, Cambridge, UK.

Kodric-Brown, A. & Brown, J.M. (1993) Highly structured fish communities in Australian desert springs. *Ecology*, 74, 1847–1855.

Krebs, C.J. (1972) *Ecology*. Harper & Row, New York.

Krebs, C.J., Boonstra, R., Boutin, S. & Sinclair, A.R.E. (2001) What drives the 10-year cycle of snowshoe hares? *Bioscience*, 51, 25–35.

Krebs, C.J., Sinclair, A.R.E., Boonstra, R., Boutin, S., Martin, K. & Smith, J.N.M. (1999) Community dynamics of vertebrate herbivores: how can we untangle the web? In: *Herbivores: between Plants and Predators* (H. Olff, V.K. Brown & R.H. Drent, eds), pp. 447–473. Blackwell Science, Oxford.

Kremen, C., Williams, N.M., Bugg, R.L., Fay, J.P. & Thorp, R.W. (2004) The area requirements of an ecosystem service: crop pollination by native bee communities in California. *Ecology Letters*, 7, 1109–1119.

Kullberg, C. & Ekman, J. (2000) Does predation maintain tit community diversity? *Oikos*, 89, 41–45.

Lacy, R.C. (1993) VORTEX: a computer simulation for use in population viability analysis. *Wildlife Research*, 20, 45–65.

Lande, R. & Barrowclough, G.F. (1987) Effective population size, genetic variation, and their use in population management. In: *Viable Populations for Conservation* (M.E. Soulé, ed.), pp. 87–123. Cambridge University Press, Cambridge.

Larcher, W. (1980) *Physiological Plant Ecology*, 2nd edn. Springer-Verlag, Berlin.

Lathrop, R.C., Johnson, B.M., Johnson, T.B. et al. (2002) Stocking piscivores to improve fishing and water clarity: a synthesis of the Lake Mendota biomanipulation project. *Freshwater Biology*, 47, 2410–2424.

Laurance, W.F. (2001) Future shock: forecasting a grim fate for the Earth. *Trends in Ecology and Evolution*, 16, 531–533.

Law, B.E., Thornton, P.E., Irvine, J., Anthoni, P.M. & van Tuyl, S. (2001) Carbon storage and fluxes in ponderosa pine forests at different developmental stages. *Global Climate Change*, 7, 755–777.

Lawlor, L.R. (1980) Structure and stability in natural and randomly constructed competitive communities. *American Naturalist*, 116, 394–408.

Lawrence, W.H. & Rediske, J.H. (1962) Fate of sown douglas-fir seed. *Forest Science*, 8, 211–218.

Lawton, J.H. & May, R.M. (1984) The birds of Selborne. *Nature*, 306, 732–733.

Le Cren, E.D. (1973) Some examples of the mechanisms that control the population dynamics of salmonid fish. In: *The Mathematical Theory of the Dynamics of Biological Populations* (M.S. Bartlett & R.W. Hiorns, eds), pp. 125–135. Academic Press, London.

Lehmann, N., Eisenhawer, A., Hansen, K., Mech, L.D., Peterson, R.O., Gogan, P.J.P. & Wayne, R.K. (1991) Introgression of mitochondrial DNA into sympatric North American grey wolf population. *Evolution*, 45, 104–119.

Lennartsson, T., Nilsson, P. & Tuomi, J. (1998) Induction of overcompensation in the field gentian, *Gentianella campestris*. *Ecology*, 79, 1061–1072.

Leroy, F. & de Vuyst, L. (2001) Growth of the bacteriocin-producing *Lactobacillus sakei* strain CTC 494 in MRS broth is strongly reduced due to nutrient exhaustion: a nutrient depletion model for the growth of lactic acid bacteria. *Applied and Environmental Microbiology*, 67, 4407–4413.

Letourneau, D.K. & Dyer, L.A. (1998a) Density patterns of *Piper* ant-plants and associated arthropods: top-predator trophic cascades in a terrestrial system? *Biotropica*, 30, 162–169.

Letourneau, D.K. & Dyer, L.A. (1998b) Experimental test in a lowland tropical forest shows top-down effects through four trophic levels. *Ecology*, 79, 1678–1687.

Levins, R. (1969) Some demographic and genetic consequences of environmental heterogeneity for biological control. *Bulletin of the Entomological Society of America*, 15, 237–240.

Lichter, J. (2000) Colonization constraints during primary succession on coastal Lake Michigan sand dunes. *Journal of Ecology*, 88, 825–839.

Likens, G.E. (1989) Some aspects of air pollutant effects on terrestrial ecosystems and prospects for the future. *Ambio*, 18, 172–178.

Likens, G.E. (1992) *The Ecosystem Approach: its Use and Abuse*. Excellence in Ecology, Book 3. Ecology Institute, Oldendorf-Luhe, Germany.

Likens, G.E. & Bormann, F.G. (1975) An experimental approach to New England landscapes. In: *Coupling of Land and Water Systems* (A.D. Hasler, ed.), pp. 7–30. Springer-Verlag, New York.

Likens, G.E. & Bormann, F.H. (1994) *Biogeochemistry of a Forested Ecosystem*, 2nd edn. Springer-Verlag, New York.

Likens, G.E., Bormann, F.H., Pierce, R.S. & Fisher, D.W. (1971) Nutrient–hydrologic cycle interaction in small forested watershed ecosystems. In: *Productivity of Forest Ecosystems* (P. Duvogneaud, ed.), pp. 553–563. UNESCO, Paris.

Likens, G.E., Driscoll, C.T. & Buso, D.C. (1996) Long-term effects of acid rain: response and recovery of a forest ecosystem. *Science*, 272, 244–245.

Lindeman, R.L. (1942) The trophic–dynamic aspect of ecology. *Ecology*, 23, 399–418.

Lofgren, A. & Jerling, L. (2002) Species richness, extinction and immigration rates of vascular plants on islands in the Stockholm Archipelago, Sweden, during a century of ceasing management. *Folia Geobotanica*, 37, 297–308.

Lotka, A.J. (1932) The growth of mixed population: two species competing for a common food supply. *Journal of the Washington Academy of Sciences*, 22, 461–469.

Louda, S.M. (1982) Distributional ecology: variation in plant recruitment over a gradient in relation to insect seed predation. *Ecological Monographs*, 52, 25–41.

Louda, S.M. (1983) Seed predation and seedling mortality in the recruitment of a shrub, *Haplopappus venetus* (Asteraceae), along a climatic gradient. *Ecology*, 64, 511–521.

Louda, S.M., Kendall, D., Connor, J. & Simberloff, D. (1997) Ecological effects of an insect introduced for the biological control of weeds. *Science*, 277, 1088–1090.

Lövei, G.L. (1997) Global change through invasion. *Nature*, 388, 627–628.

Lubchenco, J. (1978) Plant species diversity in a marine intertidal community: importance of herbivore food preference and algal competitive abilities. *American Naturalist*, 112, 23–39.

Lubchenco, J., Olson, A.M., Brubaker, L.B. et al. (1991) The sustainable biosphere initiative: an ecological research agenda. *Ecology*, 72, 371–412.

Luckman, W.H. & Decker, G.C. (1960) A 5-year report on observations in the Japanese beetle control area of Sheldon, Illinois. *Journal of Economic Entomology*, 53, 821–827.

Lussenhop, J. (1992) Mechanisms of microarthropod–microbial interactions in soil. *Advances in Ecological Research*, 23, 1–33.

MacArthur, J.W. (1975) Environmental fluctuations and species diversity. In: *Ecology and Evolution of Communities* (M.L. Cody & J.M. Diamond, eds), pp. 74–80. Belknap, Cambridge, MA.

MacArthur, R.H. (1955) Fluctuations of animal populations and a measure of community stability. *Ecology*, 36, 533–536.

MacArthur, R.H. (1972) *Geographical Ecology*. Harper & Row, New York.

MacArthur, R.H. & Pianka, E.R. (1966) On optimal use of a patchy environment. *American Naturalist*, 100, 603–609.

MacArthur, R.H. & Wilson, E.O. (1967) *The Theory of Island Biogeography*. Princeton University Press, Princeton, NJ.

MacLulick, D.A. (1937) Fluctuations in numbers of the varying hare (*Lepus americanus*). *University of Toronto Studies, Biology Series*, 43, 1–136.

Malmqvist, B., Wotton, R.S. & Zhang, Y. (2001) Suspension feeders transform massive amounts of seston in large northern rivers. *Oikos*, 92, 35–43.

Malthus, T. (1798) *An Essay on the Principle of Population*. J. Johnson, London.

Martin, P.R. & Martin, T.E. (2001) Ecological and fitness consequences of species coexistence: a removal experiment with wood warblers. *Ecology*, 82, 189–206.

Martin, P.S. (1984) Prehistoric overkill: the global model. In: *Quaternary Extinctions: a Prehistoric Revolution* (P.S. Martin & R.G. Klein, eds), pp. 354–403. University of Arizona Press, Tuscon, AZ.

Marzusch, K. (1952) Untersuchungen über di Temperaturabhängigkeit von Lebensprozessen bei Insekten unter besonderer Berücksichtigung winter-schlafender Kartoffelkäfer. *Zeitschrift für vergleicherde Physiologie*, 34, 75–92.

May, R.M. (1981) Patterns in multi-species communities. In: *Theoretical Ecology: Principles and Applications*, 2nd edn (R.M. May, ed.), pp. 197–227. Blackwell Scientific Publications, Oxford.

McGrady-Steed, J., Harris, P.M. & Morin, P.J. (1997) Biodiversity regulates ecosystem predictability. *Nature*, 390, 162–165.

McIntosh, A.R. & Townsend, C.R. (1994) Interpopulation variation in mayfly antipredator tactics: differential effects of contrasting predatory fish. *Ecology*, 75, 2078–2090.

McIntosh, A.R. & Townsend, C.R. (1996) Interactions between fish, grazing invertebrates and algae in a New Zealand stream: a trophic cascade mediated by fish-induced changes to grazer behavior. *Oecologia*, 108, 174–181.

McKane, R.B., Johnson, L.C., Shaver, G.R. et al. (2002) Resource-based niches provide a basis for plant species diversity and dominance in arctic tundra. *Nature*, 415, 68–71.

McKay, J.K., Bishop, J.G., Lin, J.-Z., Richards, J.H., Sala, A. & Mitchell-Olds, T. (2001) Local adaptation across a climatic gradient despite small effective population size in the rare sapphire rockcress. *Proceedings of the Royal Society of London, Series B*, 268, 1715–1721.

McKey, D. (1979) The distribution of secondary compounds within plants. In: *Herbivores: their Interaction with Secondary Plant Metabolites* (G.A. Rosenthal & D.H. Janzen, eds), pp. 56–134. Academic Press, New York.

Menges, E.S. (2000) Population viability analyses in plants: challenges and opportunities. *Trends in Ecology and Evolution*, 15, 51–56.

Menges, E.S. & Dolan, R.W. (1998) Demographic viability of populations of *Silene regia* in midwestern prairies: relationships with fire management, genetic variation, geographic location, population size and isolation. *Journal of Ecology*, 86, 63–78.

Merryweather, J.W. & Fitter, A.H. (1995) Phosphorus and carbon budgets: mycorrhizal contribution in *Hyacinthoides non-scripta* (L.) Chouard ex Rothm. under natural conditions. *New Phytologist*, 129, 619–627.

Millennium Ecosystem Assessment (2005) *Ecosystems and Human Well-being: Biodiversity Synthesis*. World Resources Institute, Washington, DC.

Milner-Gulland, E.J. & Mace, R. (1998) *Conservation of Biological Resources*. Blackwell Science, Oxford.

Mittelbach, G.G., Steiner, C.F., Scheiner, S.M. et al. (2001) What is the observed relationship between species richness and productivity? *Ecology*, 82, 2381–2396.

Moilanen, A., Smith, A.T. & Hanski, I. (1998) Long-term dynamics in a metapopulation of the American pika. *American Naturalist*, 152, 530–542.

Montagnes, D.J.S., Kimmance, S.A. & Atkinson, D. (2003) Using Q_{10}: can growth rates increase linearly with temperature? *Aquatic Microbial Ecology*, 32, 307–313.

Mosier, A.R., Bleken, M.A., Chaiwanakupt, P. et al. (2002) Policy implications of human-accelerated nitrogen cycling. *Biogeochemistry*, 57/58, 477–516.

Murdoch, W.W. (1966) Community structure, population control and competition – a critique. *American Naturalist*, 100, 219–226.

Murdoch, W.W. & Stewart-Oaten, A. (1975) Predation and population stability. *Advances in Ecological Research*, 9, 1–131.

Mwendera, E.J., Saleem, M.A.M. & Woldu, Z. (1997) Vegetation response to cattle grazing in the Ethiopian Highlands. *Agriculture, Ecosystems and Environment*, 64, 43–51.

Myers, R.A. (2001) Stock and recruitment: generalizations about maximum reproductive rate, density dependence, and variability using meta-analytic approaches. *ICES Journal of Marine Science*, 58, 937–951.

National Research Council (1990) *Alternative Agriculture*. National Academy of Sciences, Academy Press, Washington, DC.

Neilson, R.P., Prentice, I.C., Smith, B., Kittel, T. & Viner, D. (1998) Simulated changes in vegetation distribution under global warming. Available as Annex C at www.epa.gov/globalwarming/reports/pubs/ipcc/annex/index.html.

NERC (1990) *Our Changing Environment*. Natural Environment Research Council, London. (NERC acknowledges the significant contribution of Fred Pearce to the document.)

Newsham, K.K., Fitter, A.H. & Watkinson, A.R. (1994) Root pathogenic and arbuscular mycorrhizal mycorrhizal fungi determine fecundity of asymptomatic plants in the field. *Journal of Ecology*, 82, 805–814.

Newsham, K.K., Fitter, A.H. & Watkinson, A.R. (1995) Multi-functionality and biodiversity in arbuscular mycorrhizas. *Trends in Ecology and Evolution*, 10, 407–411.

Niklas, K.J., Tiffney, B.H. & Knoll, A.H. (1983) Patterns in vascular land plant diversification. *Nature*, 303, 614–616.

Nilsson, L.A. (1988) The evolution of flowers with deep corolla tubes. *Nature*, 334, 147–149.

Norton, I.O. & Sclater, J.G. (1979) A model for the evolution of the Indian Ocean and the breakup of Gondwanaland. *Journal of Geophysical Research*, 84, 6803–6830.

Nowak, R.M. (1979) *North American Quaternary Canis*. Monograph No. 6, Museum of Natural History. University of Kansas, Lawrence, KA.

O'Brien, E.M. (1993) Climatic gradients in woody plant species richness: towards an explanation based on an analysis of southern Africa's woody flora. *Journal of Biogeography*, 20, 181–198.

Oaks, J.L., Gilbert, M., Virani, M.Z. et al. (2004) Diclofenac residues as the cause of vulture population decline in Pakistan. *Nature*, 427, 629–633.

Oedekoven, M.A. & Joern, A. (2000) Plant quality and spider predation affects grasshoppers (Acrididae): food-quality-dependent compensatory mortality. *Ecology*, 81, 66–77.

Ogden, J. (1968) *Studies on reproductive strategy with particular reference to selected composites*. PhD thesis, University of Wales, Bangor.

Osmundson, D.B., Ryel, R.J., Lamarra, V.L. & Pitlick, J. (2002) Flow–sediment–biota relations: implications for river regulation effects on native fish abundance. *Ecological Applications*, 12, 1719–1739.

Owen-Smith, N. (1987) Pleistocene extinctions: the pivotal role of megaherbivores. *Paleobiology*, 13, 351–362.

Pace, M.L., Cole, J.J., Carpenter, S.R. & Kitchell, J.F. (1999) Trophic cascades revealed in diverse ecosystems. *Trends in Ecology and Evolution*, 14, 483–488.

Paine, R.T. (1966) Food web complexity and species diversity. *American Naturalist*, 100, 65–75.

Paine, R.T. (1979) Disaster, catastrophe and local persistence of the sea palm *Postelsia palmaeformis*. *Science*, 205, 685–687.

Paterson, S. & Viney, M.E. (2002) Host immune responses are necessary for density dependence in nematode infections. *Parasitology*, 125, 283–292.

Pauly, D. & Christensen, V. (1995) Primary production required to sustain global fisheries. *Nature*, 374, 255–257.

Pavia, H. & Toth, G.B. (2000) Inducible chemical resistance to herbivory in the brown seaweed *Ascophyllum nodosum*. *Ecology*, 81, 3212–3225.

Pearl, R. (1927) The growth of populations. *Quarterly Review of Biology*, 2, 532–548.

Pearl, R. (1928) *The Rate of Living*. Knopf, New York.

Pearson, D.E. & Callaway, R.M. (2003) Indirect effects of host-specific biological control agents. *Trends in Ecology and Evolution*, 18, 456–461.

Penn, A.M., Sherwin, W.B., Gordon, G., Lunney, D., Melzer, A. & Lacy, R.C. (2000) Demographic forecasting in koala conservation. *Conservation Biology*, 14, 629–638.

Pennings, S.C. & Callaway, R.M. (2002) Parasitic plants: parallels and contrasts with herbivores. *Oecologia*, 131, 479–489.

Perrins, C.M. (1965) Population fluctuations and clutch size in the great tit, *Parus major* L. *Journal of Animal Ecology*, 34, 601–647.

Petren, K. & Case, T.J. (1996) An experimental demonstration of exploitation competition in an ongoing invasion. *Ecology*, 77, 118–132.

Petren, K., Grant, B.R. & Grant, P.R. (1999) A phylogeny of Darwin's finches based on microsatellite DNA variation. *Proceedings of the Royal Society of London, Series B*, 266, 321–329.

Pimentel, D. (1993) Cultural controls for insect pest management. In: *Pest Control and Sustainable Agriculture* (S. Corey, D. Dall & W. Milne, eds), pp. 35–38. Commonwealth Scientific and Research Organisation, East Melbourne, New South Wales.

Pimentel, D., Krummel, J., Gallahan, D. et al. (1978) Benefits and costs of pesticide use in U.S. food production. *Bioscience*, 28, 777–784.

Pimentel, D., Lach, L., Zuniga, R. & Morrison, D. (2000) Environmental and economic costs of nonindigenous species in the United States. *BioScience*, 50, 53–65.

Pimm, S.L. (1991) *The Balance of Nature: Ecological Issues in the Conservation of Species and Communities*. University of Chicago Press, Chicago.

Pitcher, T.J. & Hart, P.J.B. (1982) *Fisheries Ecology*. Croom Helm, London.

Pope, S.E., Fahrig, L. & Merriam, H.G. (2000) Landscape complementation and metapopulation effects on leopard frog populations. *Ecology*, 81, 2498–2508.

Power, M.E., Tilman, D., Estes, J.A. et al. (1996) Challenges in the quest for keystones. *Bioscience*, 46, 609–620.

Primack, R.B. (1993) *Essentials of Conservation Biology*. Sinauer Associates, Sunderland, MA.

Pywell, R.F., Bullock, J.M., Walker, K.J., Coulson, S.J., Gregory, S.J. & Stevenson, M.J. (2004) Facilitating grassland diversification using the hemiparasitic plant *Rhinanthus minor*. *Journal of Applied Ecology*, 41, 880–887.

Raffaelli, D. & Hawkins, S. (1996) *Intertidal Ecology*. Kluwer, Dordrecht.

Rahbek, C. (1995) The elevational gradient of species richness: a uniform pattern? *Ecography*, 18, 200–205.

Rainey, P.B. & Trevisano, M. (1998) Adaptive radiation in a heterogeneous environment. *Nature*, 394, 69–72.

Randall, M.G.M. (1982) The dynamics of an insect population throughout its altitudinal distribution: *Coleophora alticolella* (Lepidoptera) in northern England. *Journal of Animal Ecology*, 51, 993–1016.

Ratcliffe, D.A. (1970) Changes attributable to pesticides in egg breakage frequency and eggshell thickness in some British birds. *Journal of Applied Ecology*, 7, 67–107.

Rätti, O., Dufva, R. & Alatalo, R.V. (1993) Blood parasites and male fitness in the pied flycatcher. *Oecologia*, 96, 410–414.

Reganold, J.P., Glover, J.D., Andrews, P.K. & Hinman, H.R. (2001) Sustainability of three apple production systems. *Nature*, 410, 926–929.

Ribas, C.R., Schoereder, J.H., Pic, M. & Soares, S.M. (2003) Tree heterogeneity, resource availability, and larger scale processes regulating arboreal ant species richness. *Austral Ecology*, 28, 305–314.

Ricklefs, R.E. (1973) *Ecology*. Nelson, London.

Ricklefs, R.E. & Lovette, I.J. (1999) The role of island area *per se* and habitat diversity in the species–area relationships of four Lesser Antillean faunal groups. *Journal of Animal Ecology*, 68, 1142–1160.

Ridley, M. (1993) *Evolution*. Blackwell Science, Boston.

Riis, T. & Sand-Jensen, K. (1997) Growth reconstruction and photosynthesis of aquatic mosses: influence of light, temperature and carbon dioxide at depth. *Journal of Ecology*, 85, 359–372.

Risebrough, R. (2004) Fatal medicine for vultures. *Nature*, 427, 596–598.

Rodrigues, A.S.L., Pilgrim, J.D., Lamoreux, J.F., Hoffmann, M. & Brooks, T.M. (2006) The value of the IUCN Red List for conservation. *Trends in Ecology and Evolution*, 21, 71–76.

Rohr, D.H. (2001) Reproductive trade-offs in the elapid snakes *Austrelap superbus* and *Austrelap ramsayi*. *Canadian Journal of Zoology*, 79, 1030–1037.

Root, R. (1967) The niche exploitation pattern of the blue-grey gnatcatcher. *Ecological Monographs*, 37, 317–350.

Rosenthal, G.A., Dahlman, D.L. & Janzen, D.H. (1976) A novel means for dealing with L-canavanine, a toxic metabolite. *Science*, 192, 256–258.

Rosenzweig, M.L. (1971) Paradox of enrichment: destabilization of exploitation ecosystems in ecological time. *Science*, 171, 385–387.

Rosenzweig, M.L. & Sandlin, E.A. (1997) Species diversity and latitudes: listening to area's signal. *Oikos*, 80, 172–176.

Rouphael, A.B. & Inglis, G.J. (2001) 'Take only photographs and leave only footprints'? An experimental study of the impacts of underwater photographers on coral reef dive sites. *Biological Conservation*, 100, 281–287.

Roura-Pascual, N., Suarez, A.V., Gomez, C., Pons, P., Touyama, Y., Wild, A.L. & Townsend Peterson, A. (2004) Geographical potential of Argentine ants (*Linepithema humile* Mayr) in the face of global climate change. *Proceedings of the Royal Society of London, Series B*, 271, 2527–2534.

Rowe, C.L. (2002) Differences in maintenance energy expenditure by two estuarine shrimp (*Palaemonetes pugio* and *P. vulgaris*) that may permit partitioning of habitats by salinity. *Comparative Biochemistry and Physiology A*, 132, 341–351.

Roy, M.S., Geffen, E., Smith, D., Ostrander, E.A. & Wayne, R.K. (1994) Patterns of differentiation and hybridization in North American wolflike canids, revealed by analysis of microsatellite loci. *Molecular Biology and Evolution*, 11, 553–570.

Ruiters, C. & McKenzie, B. (1994) Seasonal allocation and efficiency patterns of biomass and resources in the perennial geophyte *Sparaxis grandiflora* subspecies *fimbriata* (Iridaceae) in lowland coastal Fynbos, South Africa. *Annals of Botany*, 74, 633–646.

Sale, P.F. (1979) Recruitment, loss and coexistence in a guild of territorial coral reef fishes. *Oecologia*, 42, 159–177.

Sale, P.F. & Douglas, W.A. (1984) Temporal variability in the community structure of fish on coral patch reefs and the relation of community structure to reef structure. *Ecology*, 65, 409–422.

Salisbury, E.J. (1942) *The Reproductive Capacity of Plants*. Bell, London.

Sanders, N.J., Moss, J. & Wagner, D. (2003) Patterns of ant species richness along elevational gradients in an arid ecosystem. *Global Ecology and Biogeography*, 12, 93–102.

Santelmann, M.V., White, D., Freemark, K. et al. (2004) Assessing alternative futures for agriculture in Iowa, USA. *Landscape Ecology*, 19, 357–374.

Savidge, J.A. (1987) Extinction of an island forest avifauna by an introduced snake. *Ecology*, 68, 660–668.

Sax, D.F. & Gaines, S.D. (2003) Species diversity: from global decreases to local increases. *Trends in Ecology and Evolution*, 18, 561–566.

Schluter, D. (2001) Ecology and the origin of species. *Trends in Ecology and Evolution*, 16, 372–380.

Schoener, T.W. (1983) Field experiments on interspecific competition. *American Naturalist*, 122, 240–285.

Schoenly, K., Beaver, R.A. & Heumier, T.A. (1991) On the trophic relations of insects: a food-web approach. *American Naturalist*, 137, 597–638.

Schulze, E.D. (1970) Dre CO_2-Gaswechsel de Buche (*Fagus sylvatica* L.) in Abhäbgigkeit von den Klimafaktoren in Freiland. *Flora, Jena*, 159, 177–232.

Schulze, E.D., Fuchs, M.I. & Fuchs, M. (1977a) Spatial distribution of photosynthetic capacity and performance in a mountain spruce forest in northern Germany. I. Biomass distribution and daily CO_2 uptake in different crown layers. *Oecologia*, 29, 43–61.

Schulze, E.D., Fuchs, M.I. & Fuchs, M. (1977b) Spatial distribution of photosynthetic capacity and performance in a mountain spruce forest in northern Germany. III. The significance of the evergreen habit. *Oecologia*, 30, 239–249.

Schwartz, O.A., Armitage, K.B. & Van Vuren, D. (1998) A 32-year demography of yellow-bellied marmots (*Marmota flaviventris*). *Journal of Zoology*, 246, 337–346.

Shankar Raman, T., Rawat, G.S. & Johnsingh, A.J.T. (1998) Recovery of tropical rainforest avifauna in relation to vegetation succession following shifting cultivation in Mizoram, north-east India. *Journal of Applied Ecology*, 35, 214–231.

Sibly, R.M. & Hone, J. (2002) Population growth rate and its determinants: an overview. *Philosophical Transactions of the Royal Society of London, Series B*, 357, 1153–1170.

Simberloff, D.S. (1976) Experimental zoogeography of islands: effects of island size. *Ecology*, 57, 629–648.

Simberloff, D.S., Dayan T., Jones, C. & Ogura, G. (2000) Character displacement and release in the small Indian mongoose, *Herpestes javanicus*. *Ecology*, 91, 2086–2099.

Simon, K.S., Townsend, C.R., Biggs, B.J.F., Bowden, W.B. & Frew, R.D. (2004) Habitat-specific nitrogen dynamics in New Zealand streams containing native or invasive fish. *Ecosystems*, 7, 777–792.

Sinclair, B.J. & Sjursen, H. (2001) Cold tolerance of the Antarctic springtail *Gomphiocephalus hodgsoni* (Collembola, Hypogastruridae). *Antarctic Science*, 13: 277–279.

Singleton, G., Krebs, C.J., Davis, S., Chambers, L. & Brown, P. (2001) Reproductive changes in fluctuating house mouse populations in southeastern Australia. *Proceedings of the Royal Society of London, Series B*, 268, 1741–1748.

Slobodkin, L.B., Smith, F.E. & Hairston, N.G. (1967) Regulation in terrestrial ecosystems, and the implied balance of nature. *American Naturalist*, 101, 109–124.

Smith, J.W. (1998) Boll weevil eradication: area-wide pest management. *Annals of the Entomological Society of America*, 91, 239–247.

Sousa, M.E. (1979a) Experimental investigation of disturbance and ecological succession in a rocky intertidal algal community. *Ecological Monographs*, 49, 227–254.

Sousa, M.E. (1979b) Disturbance in marine intertidal boulder fields: the nonequilibrium maintenance of species diversity. *Ecology*, 60, 1225–1239.

Stenseth, N.C., Falck, W., Bjornstad, O.N. & Krebs, C.J. (1997) Population regulation in snowshoe hare and lynx populations: asymmetric food web configurations between the snowshoe hare and the lynx. *Proceedings of the National Academy of Science of the USA*, 94, 5147–5152.

Stevens, C.E. & Hume, I.D. (1998) Contributions of microbes in vertebrate gastrointestinal tract to production and conservation of nutrients. *Physiological Reviews*, 78, 393–426.

Stoll, P. & Prati, D. (2001) Intraspecific aggregation alters competitive interactions in experimental plant communities. *Ecology*, 82, 319–327.

Strauss, S.Y. & Agrawal, A.A. (1999) The ecology and evolution of plant tolerance to herbivory. *Trends in Ecology and Evolution*, 14, 179–185.

Strauss, S.Y., Irwin, R.E. & Lambrix, V.M. (2004) Optimal defence theory and flower petal colour predict variation in the secondary chemistry of wild radish. *Journal of Ecology*, 92, 132–141.

Strong, D.R. Jr., Lawton, J.H. & Southwood, T.R.E. (1984) *Insects on Plants: Community Patterns and Mechanisms*. Blackwell Scientific Publications, Oxford.

Susarla, S., Medina, V.F. & McCutcheon, S.C. (2002) Phytoremediation: an ecological solution to organic chemical contamination. *Ecological Engineering*, 18, 647–658.

Sutherland, W.J., Gill, J.A. & Norris, K. (2002) Density-dependent dispersal in animals: concepts, evidence, mechanisms and consequences. In: *Dispersal Ecology* (J.M. Bullock, R.E. Kenward & R.S. Hails, eds), pp. 134–151. Blackwell Publishing, Oxford.

Sutton, S.L. & Collins, N.M. (1991) Insects and tropical forest conservation. In: *The Conservation of Insects and their Habitats* (N.M. Collins & J.A. Thomas, eds), pp. 405–424. Academic Press, London.

Swan, G., Naidoo, V., Cuthbert, R. et al. (2006) Removing the threat of diclofenac to critically endangered Asian vultures. *Public Library of Science Biology*, 4(3), e66. doi: 10.1371/journal.pbio.0040066.

Swift, M.J., Heal, O.W. & Anderson, J.M. (1979) *Decomposition in Terrestrial Ecosystems*. Blackwell Scientific Publications, Oxford.

Swinnerton, K.J., Groombridge, J.J., Jones, C.G., Burn, R.W. & Mungroo, Y. (2004) Inbreeding depression and founder diversity among captive and free-living populations of the endangered pink pigeon *Columba mayeri*. *Animal Conservation*, 7, 353–364.

Symonides, E. (1979) The structure and population dynamics of psammophytes on inland dunes. II. Loose-sod populations. *Ekologia Polska*, 27, 191–234.

Symonides, E. (1983) Population size regulation as a result of intra-population interactions. I. The effect of density on the survival and development of individuals of *Erophila verna* (L.). *Ekologia Polska*, 31, 839–881.

Tanaka, M.O. & Magalhaes, C.A. (2002) Edge effects and succession dynamics in *Brachidontes* mussel beds. *Marine Ecology Progress Series*, 237, 151–158.

Taniguchi, Y. & Nakano, S. (2000) Condition-specific competition: implications for the altitudinal distribution of stream fishes. *Ecology*, 81, 2027–2039.

Tansley, A.G. (1904) The problems of ecology. *New Phytologist*, 3, 191–200.

Taylor, I. (1994) *Barn Owls. Predator–Prey Relationships and Conservation*. Cambridge University Press, Cambridge.

Téllez-Valdés, O. & Dávila-Aranda, P. (2003) Protected areas and climate change: a case study of the cacti in the Tehuacán-Cuicatlán Biosphere Reserve, Mexico. *Conservation Biology*, 17, 846–853.

Thomas, C.D., Cameron, A., Green, R.E. et al. (2004) Extinction risk from climate change. *Nature*, 427, 145–148.

Thomas, C.D. & Harrison, S. (1992) Spatial dynamics of a patchily distributed butterfly species. *Journal of Applied Ecology*, 61, 437–446.

Thomas, C.D. & Jones, T.M. (1993) Partial recovery of a skipper butterfly (*Hesperia comma*) from population refuges: lessons for conservation in a fragmented landscape. *Journal of Animal Ecology*, 62, 472–481.

Thomas, C.D., Thomas, J.A. & Warren, M.S. (1992) Distributions of occupied and vacant butterfly habitats in fragmented landscapes. *Oecologia*, 92, 563–567.

Thompson, R.M., Townsend, C.R., Craw, D., Frew, R. & Riley, R. (2001) (Further) links from rocks to plants. *Trends in Ecology and Evolution*, 16, 543.

Tilman, D. (1982) *Resource Competition and Community Structure*. Princeton University Press, Princeton, NJ.

Tilman, D. (1986) Resources, competition and the dynamics of plant communities. In: *Plant Ecology* (M.J. Crawley, ed.), pp. 51–74. Blackwell Scientific Publications, Oxford.

Tilman, D. (1996) Biodiversity: population versus ecosystem stability. *Ecology*, 77, 350–363.

Tilman, D. (1999) The ecological consequences of changes in biodiversity: a search for general principles. *Ecology*, 80, 1455–1474.

Tilman, D., Fargione, J., Wolff, B. et al. (2001) Forecasting agriculturally driven global environmental change. *Science*, 292, 281–284.

Tilman, D., Mattson, M. & Langer, S. (1981) Competition and nutrient kinetics along a temperature gradient: an experimental test of a mechanistic approach to niche theory. *Limnology and Oceanography*, 26, 1020–1033.

Tokeshi, M. (1993) Species abundance patterns and community structure. *Advances in Ecological Research*, 24, 112–186.

Tonn, W.M. & Magnuson, J.J. (1982) Patterns in the species composition and richness of fish assemblages in northern Wisconsin lakes. *Ecology*, 63, 137–154.

Townsend, C.R. (2007) *Ecological Applications: Toward a Sustainable World*. Blackwell Publishing, Oxford.

Townsend, C.R. & Crowl, T.A. (1991) Fragmented population structure in a native New Zealand fish: an effect of introduced brown trout? *Oikos*, 61, 348–354.

Townsend, C.R., Hildrew, A.G. & Francis, J.E. (1983) Community structure in some southern English streams: the influence of physiochemical factors. *Freshwater Biology*, 13, 521–544.

Townsend, C.R., Scarsbrook, M.R. & Dolédec, S. (1997) The intermediate disturbance hypothesis, refugia and bio-diversity in streams. *Limnology and Oceanography*, 42, 938–949.

Townsend, C.R., Thompson, R.M., McIntosh, A.R. et al. (1998) Disturbance, resource supply, and food-web architecture in streams. *Ecology Letters*, 1, 200–209.

Turkington, R. & Harper, J.L. (1979) The growth, distribution and neighbour relationships of *Trifolium repens* in a permanent pasture. IV. Fine scale biotic differentiation. *Journal of Ecology*, 67, 245–254.

Turner, J.R.G., Lennon, J.J. & Greenwood, J.J.D. (1996) Does climate cause the global biodiversity gradient? In: *Aspects of the Genesis and Maintenance of Biological Diversity* (M. Hochberg, J. Claubert & R. Barbault, eds), pp. 199–220. Oxford University Press, London.

UNEP (2003) *Global Environmental Outlook Year Book 2003*. United Nations Environmental Program (UNEP), GEO Section, Nairobi, Kenya.

United Nations (1998) *Global Change and Sustainable Development: Critical Trends*. Report of the Secretary General, United Nations, New York. (Also available on the world wide web at www.un.org/esa/sustdev/trends.html.)

United Nations (2002) *Global Environmental Outlook 3.* Report of the United Nations Environmental Program (UNEP). UNEP, www.unep.org/GEO/geo3.

United Nations (2005) *The World Population Prospects: the 2004 Revision.* Analytical Report Vol. III. Department of Economic and Social Affairs, Population Division, United Nations. United Nations, New York.

Valentine, J.W. (1970) How many marine invertebrate fossil species? A new approximation. *Journal of Paleontology*, 44, 410–415.

Valladares, V.F. & Pearcy, R.W. (1998) The functional ecology of shoot architecture in sun and shade plants of *Heteromeles arbutifolia* M. Roem., a Californian chaparral shrub. *Oecologia*, 114, 1–10.

van der Juegd, H.P. (1999) *Life history decisions in a changing environment: a long-term study of a temperate barnacle goose population.* PhD thesis, Uppsala University, Uppsala.

Vannotte, R.L., Minshall, G.W., Cummins, K.W., Sedell, J.R. & Cushing, C.E. (1980) The river continuum concept. *Canadian Journal of Fisheries and Aquatic Sciences*, 37, 130–137.

Vázquez, G.J.A. & Givnish, T.J. (1998) Altitudinal gradients in tropical forest composition, structure, and diversity in the Sierra de Manantlán. *Journal of Ecology*, 86, 999–1020.

Verhoeven, J.T.A., Arheimer, B., Yin, C. & Hefting, M.M. (2006) Regional and global concerns over wetlands and water quality. *Trends in Ecology and Evolution*, 21, 96–103.

Volterra, V. (1926) Variations and fluctuations of the numbers of individuals in animal species living together. (Reprinted in 1931. In: *Animal Ecology* (R.N. Chapman, ed.), pp. 409–448. McGraw Hill, New York.)

Waage, J.K. & Greathead, D.J. (1988) Biological control: challeges and opportunities. *Philosophical Transactions of the Royal Society of London, Series B*, 318, 111–128.

Walsh, J.A. (1983) Selective primary health care: strategies for control of disease in the developing world. IV. Measles. *Reviews of Infectious Diseases*, 5, 330–340.

Wang, G.-H. (2002) Plant traits and soil chemical variables during a secondary vegetation succession in abandoned fields on the Loess Plateau. *Acta Botanica Sinica*, 44, 990–998.

Wardle, D.A., Bonner, K.I. & Barker, G.M. (2000) Stability of ecosystem properties in response to above-ground functional group richness and composition. *Oikos*, 89, 11–23.

Warren, P.H. (1989) Spatial and temporal variation in the structure of a freshwater food web. *Oikos*, 55, 299–311.

Watkinson, A.R. & Harper, J.L. (1978) The demography of a sand dune annual: *Vulpia fasciculata.* I. The natural regulation of populations. *Journal of Ecology*, 66, 15–33.

Wayne, R.K. (1996) Conservation genetics in the Canidae. In: *Conservation Genetics* (J.C. Avise & J.L. Hamrick, eds), pp. 75–118. Chapman & Hall, New York.

Wayne, R.K. & Jenks, S.M. (1991) Mitochondrial DNA analysis implying extensive hybridization of the endangered red wolf *Canis rufus. Nature*, 351, 565–568.

Webb, S.D. (1987) Community patterns in extinct terrestrial invertebrates. In: *Organization of Communities: Past and Present* (J.H.R. Gee & P.S. Giller, eds), pp. 439–468. Blackwell Scientific Publications, Oxford.

Webb, W.L., Lauenroth, W.K., Szarek, S.R. & Kinerson, R.S. (1983) Primary production and abiotic controls in forests, grasslands and desert ecosystems in the United States. *Ecology*, 64, 134–151.

Wegener, A. (1915) *Entstehung der Kontinenter und Ozeaner.* Samml. Viewg, Braunschweig. (English translation 1924. *The Origins of Continents and Oceans*, translated by J.G.A. Skerl. Methuen, London.)

White, G. (1789) *The Natural History and Antiquities of Selborne.* (Reprinted in 1977 as *The Natural History of Selborne* (G. White and R. Mabey). Penguin, London.)

Whitehead, A.N. (1953) *Science and the Modern World.* Cambridge University Press, Cambridge.

Whittaker, R.J., Willis, K.J. & Field, R. (2003) Climatic–energetic explanations of diversity: a macroscopic perspective. In: *Macroecology: Concepts and Consequences* (T.M. Blackburn & K.J. Gaston, eds), pp. 107–129. Blackwell Publishing, Oxford.

Williams, W.D. (1988) Limnological imbalances: an antipodean viewpoint. *Freshwater Biology*, 20, 407–420.

Winemiller, K.O. (1990) Spatial and temporal variation in tropical fish trophic networks. *Ecological Monographs*, 60, 331–367.

Withler, R.E., Candy, J.R., Beacham, T.D. & Miller, K.M. (2004) Forensic DNA analysis of Pacific salmonid samples for species and stock identification. *Environmental Biology of Fishes*, 69, 275–285.

Woiwod, I.P. & Hanski, I. (1992) Patterns of density dependence in moths and aphids. *Journal of Animal Ecology*, 61, 619–629.

Wolff, J.O., Schauber, E.M. & Edge, W.D. (1997) Effects of habitat loss and fragmentation on the behavior and demography of gray-tailed voles. *Conservation Biology*, 11, 945–956.

Wootton, J.T. (1992) Indirect effects, prey susceptibility, and habitat selection: impacts of birds on limpets and algae. *Ecology*, 73, 981–991.

Worland, M.R. & Convey, P. (2001) Rapid cold hardening in Antarctic microarthropods. *Functional Ecology*, 15, 515–524.

Wright, S., Keeling, J. & Gillman, L. (2006) The road from Santa Rosalia: a faster tempo of evolution in tropical climates. *Proceedings of the National Academy of Sciences of the USA*, 103, 7718–7722.

Yao, I., Shibao, H. & Akimoto, S. (2000) Costs and benefits of ant attendance to the drepanosiphid aphid *Tuberculatus quercicola. Oikos*, 89, 3–10.

Yodzis, P. (1986) Competiton, mortality and community structure. In: *Community Ecology* (J. Diamond & T.J. Case, eds), pp. 480–491. Harper & Row, New York.

Yoshida, T., Jones, L.E., Ellner, S.P., Fussmann, G.F. & Hairston, N.G., Jr. (2003) Rapid evolution drives ecological dynamics in a predator–prey system. *Nature*, 424, 303–306.

Zimmer, M. & Topp, W. (2002) The role of coprophagy in nutrient release from feces of phytophagous insects. *Soil Biology and Biochemistry*, 34, 1093–1099.

Continued on inside back cover

Understanding Medical-Surgical Nursing

UNDERSTANDING MEDICAL-SURGICAL NURSING

SECOND EDITION

LINDA S. WILLIAMS, MSC, RNC
Professor of Nursing
Jackson Community College
Registered Nurse Care Coordinator
W.A. Foote Memorial Hospital
Jackson, Michigan

PAULA D. HOPPER, MSN, RN
Associate Professor of Nursing
Jackson Community College
On-call Case Manager
W.A. Foote Memorial Hospital
Jackson, Michigan

F.A. DAVIS COMPANY / Philadelphia

F. A. Davis Company
1915 Arch Street
Philadelphia, PA 19103

Printed in the United States of America

Last digit indicates print number: 10 9 8 7 6 5 4 3 2 1

Acquisitions Editor: Lisa B. Deitch
Developmental Editors: Melanie J. Freely, Catherine Harold
Designer: Paul Fry
Cover Designer: Louis Forgione

As new scientific information becomes available through basic and clinical research, recommended treatments and drug therapies undergo changes. The author(s) and publisher have done everything possible to make this book accurate, up to date, and in accord with accepted standards at the time of publication. The authors, editors, and publisher are not responsible for errors or omissions or for consequences from application of the book, and make no warranty, expressed or implied, in regard to the contents of the book. Any practice described in this book should be applied by the reader in accordance with professional standards of care used in regard to the unique circumstances that may apply in each situation. The reader is advised always to check product information (package inserts) for changes and new information regarding dose and contraindications before administering any drug. Caution is especially urged when using new or infrequently ordered drugs.

ISBN 0-8036-1037-8

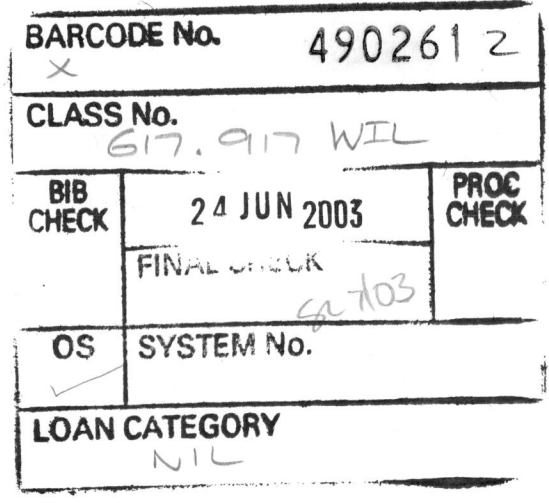

Continued on inside back cover

UNDERSTANDING MEDICAL-SURGICAL NURSING

UNDERSTANDING MEDICAL-SURGICAL NURSING

SECOND EDITION

LINDA S. WILLIAMS, MSC, RNC

Professor of Nursing
Jackson Community College
Registered Nurse Care Coordinator
W.A. Foote Memorial Hospital
Jackson, Michigan

PAULA D. HOPPER, MSN, RN

Associate Professor of Nursing
Jackson Community College
On-call Case Manager
W.A. Foote Memorial Hospital
Jackson, Michigan

F.A. DAVIS COMPANY / Philadelphia

F. A. Davis Company
1915 Arch Street
Philadelphia, PA 19103

Printed in the United States of America

Last digit indicates print number: 10 9 8 7 6 5 4 3 2 1

Acquisitions Editor: Lisa B. Deitch
Developmental Editors: Melanie J. Freely, Catherine Harold
Designer: Paul Fry
Cover Designer: Louis Forgione

As new scientific information becomes available through basic and clinical research, recommended treatments and drug therapies undergo changes. The author(s) and publisher have done everything possible to make this book accurate, up to date, and in accord with accepted standards at the time of publication. The authors, editors, and publisher are not responsible for errors or omissions or for consequences from application of the book, and make no warranty, expressed or implied, in regard to the contents of the book. Any practice described in this book should be applied by the reader in accordance with professional standards of care used in regard to the unique circumstances that may apply in each situation. The reader is advised always to check product information (package inserts) for changes and new information regarding dose and contraindications before administering any drug. Caution is especially urged when using new or infrequently ordered drugs.

ISBN 0-8036-1037-8

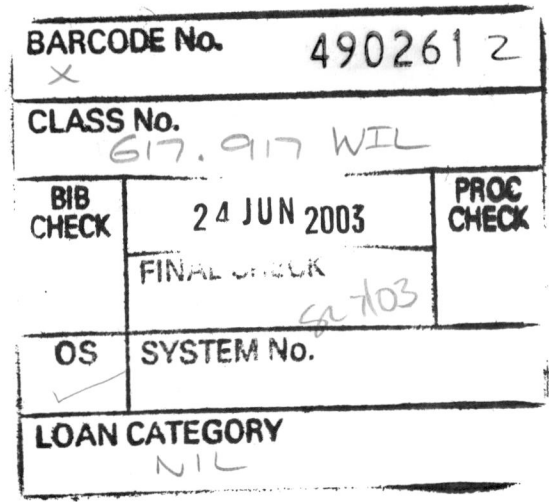

To **our students,** who provide us with inspiration, motivation,
and the joy of being a part of their learning experience.

To **Garland,**
with love, for his continued support and encouragement
and many trips to the mailbox with manuscripts.
To my **daughter, son-in-law,** and **grandchildren** who bring me joy.
To my parents, who have always been supportive
and encouraging through both editions of this text.

LINDA WILLIAMS

To **Dave,** with love,
who still loves me after 25 years and two editions of this text.
To my children, **Dan** and **Libby,** the two greatest young adults in the universe.
To **my mom,** who taught me strength and integrity.
I regret she did not live to see this second edition.
And to **my dad,** whose pride in this project keeps
me motivated when I get tired of writing.

PAULA HOPPER

Preface

Welcome to the second edition of *Understanding Medical-Surgical Nursing!* We are thrilled with the enthusiastic feedback we received on the first edition, and we have worked to make this second edition even more student-friendly. To those of you who gave us valuable suggestions on the first edition, thank you. We have incorporated much of what you told us into this edition while maintaining our original vision for the book.

We initially embarked on the first edition because we were unsatisfied with the materials available for practical/vocational nursing students. There were texts with lots of information, but very little attempt was made to help students *understand why* they were doing what they were doing. We have once again worked hard to provide a text written at an understandable level, with features that help students understand, apply, and practice the challenging content required to function as a practical/vocational nurse.

As in the first edition, this book uses the nursing process to provide a unifying framework. Within this framework, emphasis is placed on understanding, critical thinking, and application. We believe that a student who learns to think critically will be better able to apply information to new situations. We hope both students and instructors find this new book more valuable than ever for learning and understanding medical-surgical nursing.

▶ FEATURES OF THE BOOK

One of the most exciting features of this edition is a new design and full-color contents. Naturally, all of the text has been updated to reflect current health care and nursing practice. And, based in part on user feedback, we have also made a number of helpful changes and added exciting new features:

- A new chapter on alternative and complementary therapies has been added.
- A full chapter on caring for patients with AIDS has been added.
- *Questions to Guide Your Reading* have replaced Chapter Objectives. In our experience, the standard objectives found in many textbooks have little meaning and provide little assistance to students who have much reading to do in minimal time. The literature suggests that comprehension increases when students read guiding questions before reading the text. So we have provided a series of questions that students should keep in mind as they read. These questions can be translated easily back into objectives by instructors who prefer this format.

- Special features written by actual patients, called *Patient Perspectives,* have been added. These essays, focused on each patient's experience with illness, help to make patients' thoughts and feelings more meaningful and personal for students. If you have stories you feel would make valuable additions to the book, please share them with us as we prepare the third edition.
- Web links have been added to the text to help students do further research on topics of interest. Every effort was made to use only major established web sites that are unlikely to change in the near future.
- One of our most popular features—Critical Thinking Exercises—have been expanded to help students practice and think about what they are learning. We also added math calculations and documentation practice to Critical Thinking Exercises where applicable. These exercises are provided throughout the book to foster critical thinking and are followed by questions that require more than simple recall of material.
- Answers for the Critical Thinking Exercises and Review Questions have been included. Research supports the importance of immediate feedback to reinforce correct learning, so we feel strongly that students should have access to correct answers while they are studying, without having to wait for their next instructor contact. Obviously there can be many answers to some of the critical thinking questions. We have provided sample answers to help stimulate students' thinking.

We have also kept our most popular features from the first edition:

- Pronunciation key for new words at the beginning of each chapter
- Review of anatomy and physiology at the beginning of each unit
- Information on the effects of aging on body systems
- Learning tips throughout the text
- Word-building footnotes to break down complex words
- Comprehensive glossary of new words
- Nursing care plans with geriatric considerations
- Common laboratory and diagnostic tests
- Brief pathophysiology for each disorder
- Boxed presentations of Cultural Considerations, Gerontological Issues, Home Health Hints, and Nutrition Notes. (Ethical dilemmas have been moved to the Instructor Guide because we believe this material is best learned during instructor-guided discussion.)
- Critical thinking exercises throughout the text
- Review questions at the end of each chapter

As in the previous edition, we have included many figures because we believe they enhance understanding and readability. Most of the artwork and photographs are original and specifically designed to support the text. Additional figures, tables and boxed materials are used to clarify complex material.

Using the Features of This Book

As you begin each chapter, carefully read the section labeled *Questions to Guide your Reading*. Then, when you are finished reading each chapter, go back and make sure you can answer each question.

You will also find a list of new words and their pronunciations at the beginning of each chapter. These words appear in bold at their first use in the text, indicating that they also appear in the glossary at the end of the book. By learning the meanings of these words as you encounter them, you will increase your understanding of the material.

You also will encounter other devices to increase your understanding and retention of the material, such as mnemonics, acrostics, and learning tips. You may want to develop your own memory techniques in addition to those provided. (If you think of a good one, send it to us and you may find it in the next edition!) Many of the learning tips have been developed and used in our own classrooms. We find them helpful in fostering understanding of complex concepts or as memory aids. However, we want to stress that memorization is not the primary focus of the text but rather a foundation for understanding and thinking about more complex information. Understanding and application will serve you far better than memorization when dealing with new situations.

Each chapter includes one or more case studies designed to help you apply material that has been presented. A series of questions related to the case study will help you integrate the material with what you already know. These questions emphasize critical thinking, which is based on a foundation of recall and understanding of material. To enhance your learning, try to answer the questions before looking up the answers at the end of the chapter.

Review questions appear at the end of each chapter. These are written in a multiple-choice format to help you prepare for the NCLEX-PN. Again, to assess your learning, try to answer the questions before looking up the answers at the back of the book.

A bibliography at the end of each unit provides sources for additional reading material. Web sites have been included in many chapters. We believe it is important for you to interact with current technology to expand your informational resources.

Appendices are included for easy reference. They cover the following:

- Nursing diagnoses
- Lab values
- Abbreviations
- Common prefixes and suffixes to assist in learning word building techniques

Supplemental Materials

A *Student Workbook* is available to provide the student additional contact and practice with the material. Each chapter includes vocabulary practice, objective exercises, a case study or other critical thinking practice and review questions written in NCLEX-PN format. Answers are provided so students can solidify their understanding and learning with immediate feedback. Rationales are provided for review question answers.

An *Instructor's Guide* provides materials for use in the classroom. Each chapter has a chapter outline with suggested classroom activities. Also included are student activities suitable for duplicating and using for individual practice or for collaborative learning activities. These activities help the student to interact with the material, understand it, and apply it. Many of the activities are based on real patient cases and have been used with our own practical nursing students. Feedback from students has helped to refine the exercises. We believe the use of collaborative learning has greatly enhanced our students' success in achieving their educational and licensure goals. Another benefit is the sense of community the students develop as a result of working in groups. A brief introduction and guidelines for using collaborative learning techniques is included.

An expanded computerized test bank, available to instructors who adopt the textbook, provides test questions that assist students to prepare for State Board Examinations. These questions have been prepared according to test item writing protocols. The questions are in multiple choice format, and test recall and application of material. Many of the test questions have been developed, used and refined by the authors in their own medical-surgical courses for practical nursing students. The new test bank program for the second edition is easier to use and allows instructors to choose and modify the questions that best suit their classroom needs.

Finally, new for this edition is a comprehensive Power-Point program for classroom presentations. We have provided the basics; each unit presentation can be modified, reduced, or expanded by individual instructors to suit their needs.

► ACKNOWLEDGEMENTS

Many people helped us make this book a reality. First and foremost are our students, who provided us with the inspiration to undertake this project. We hope that they continue to find this text worth reading.

The F.A. Davis Company has been an exceptional publishing partner. We feel fortunate to have had their continued enthusiasm and confidence in our book. The staff at F.A. Davis has guided us through this project from its inception to publication to help us create a student-friendly book that truly promotes understanding of medical-surgical nursing.

F.A. Davis developmental editor Melanie Freely has been our guide, cheerleader and supporter. We appreciate all she has done to support our book. Lisa B. Deitch, acquisitions editor, also helped shepherd the process, as did Bette Haitsch and Bob Butler in Production. We thank them all for their assistance. Developmental editor Catherine Harold and her assistant Connie Warren helped us format the manuscript, meet deadlines and gave us helpful advice and guidance. They were a great mix of kind understanding and professionalism.

We thank the staff of Thomas Jefferson University Medical Center, in Philadelphia, and especially Ann Reynolds, for assisting us in the photo shoot originally done for the first edition. Thanks also to Nanine Hartzenbusch, our photographer, who provided a wonderful human touch to the photography.

Graphic World Publishing Services production editor Noelle Barrick did an outstanding job of coordinating the development of our manuscript into a book. Graphic World Illustration Studio's John Denk created the wonderful first edition artwork and updated it for the second edition.

Contributors from across the United States and Canada, including many well-known experts in their fields, brought expertise and diversity to the content. Their hard work is much appreciated. Reviewers from throughout the United States provided valuable insights that enhanced the quality of the text.

Many of our co-workers have contributed to this book and given us ongoing encouragement and validation of the worthiness of this project. Elizabeth Ackley, Marina-Martinez Kratz, Sharon Nowak, Kathy Walsh, Gail Ladwig, and Carroll Lutz were especially helpful in providing material, advice, and encouragement.

We wish to thank everyone who played a role, however large or small, in helping us to provide a tool to help students realize their dreams of becoming LPN/LVNs. We hope this book will help train nurses who can provide safe and expert care because they are able to think critically.

Contributors

Elizabeth J. Ackley, BSN, MSN, EdS
Professor of Nursing
Jackson Community College
Jackson, Michigan

Debra Aucoin-Ratcliff, BSN, MA
Programs Coordinator/Educator
Northern California Training Institute
Roseville, California

Cynthia Francis Bechtel, MS, RN, EMT-I
Associate Professor Practical Nursing
Mass Bay Community College
Framingham, Massachusetts

Virginia Birmie, RN, BScN, MSN
Nursing Instructor
Camosun College
Victoria, British Columbia
Canada

Linda Hopper Cook, BSN, MN
Instructor
University of Alberta
Collaborative Baccalaureate Nursing
    Program
Edmonton, Alberta
Canada

Rowena Elliott, MS, RN, CNN, C, CLNC
Assistant Professor
University of Mississippi
Jackson, Mississippi

Mary Friel Fanning, RN, MSN, CCRN
Director, Adult Cardiac Nursing Units
West Virginia University Hospitals
Morgantown, West Virginia

Sharon Gordon Dawson, RN, MSN, CNOR
Educator, Surgical Services
Swedish Medical Center
Englewood, Colorado

Paula D. Hopper, MSN, RN
Associate Professor of Nursing
Jackson Community College
Registered Nurse Care Coordinator
W.A. Foote Memorial Hospital
Jackson, Michigan

Cheryl L. Ivey, RN, MSN
Department Director
Emory University Hospital
Atlanta, Georgia

Rodney B. Kebicz, RN, BN, MN
Instructor
Assiniboine Community College
Winnipeg, Manitoba
Canada

Lynn Keegan, PhD, RN, HNC, FAAN
Director
Holistic Nursing Consultants
Port Angeles, Washington

Elaine Kennedy, EdD, RN
Professor of Nursing
Wor-Wic Community College
Salisbury, Maryland

Karen Kettleman-Hall, MS, RN
Director of Patient Care Services, Pain Services
Doctors Medical Center
Modesto, California

Carroll A. Lutz, MA, BSN
Associate Professor Emerita
Adjunct Faculty
Jackson Community college
Jackson, Michigan

Sharon D. Martin, MSN, BSN, APRN, BC
Associate Professor of Nursing
Saint Joseph's College
Standish, Maine

Maureen McDonald, RN, MS, CS
Associate Professor, Nursing Educations
Massasoit Community College
Brockton, Massachusetts

Cindy Meredith, MSN, RN
Director and Instructor of Nursing
Spring Arbor University
Spring Arbor, Michigan

Debbie Millar, MEd, BScN, RN, MBA candidate
Clinical Educator
Humber River Regional Hospital
Toronto, Ontario, Canada

Kathy Neeb, ADN, BA
RN Consultant
North Memorial Occupational Health Clinic
Robbinsdale, Minnesota

Sharon M. Nowak, MSN, RN, CCRN
Associate Professor of Nursing
Jackson Community College
Jackson, Michigan

Lazette V. Nowicki, BSN, MSN
Nursing Assistant Professor
Sacramento City College, Allied Health Department
Sacramento, California

Lynn Dianne Phillips, RN, MSN, CRNI
Director, Nursing Department
Butte Community College
Oroville, California

Winifred J. Ellenchild Pinch, EdD, MEd, MS, RN, BS
Professor
Creighton University
Omaha, Nebraska

Larry Purnell, PhD, MSN, BSN, FAAN
Professor
University of Delaware
Sudlersville, Maryland

Ruth Remington, PRD
Assistant Professor
University of Massachusetts, Lowell
Lowell, Massachusetts

Valerie C. Scanlon, PhD
Professor
College of Mount St. Vincent
Bronx, New York

MaryAnne Pietraniec Shannon, BSN, MSN
Doctoral Candidate at Michigan State University
Associate Professor of Nursing
Lake Superior State University
Sault Ste. Marie, Michigan

Patrick M. Shannon, JD, MPH, EdD
Attorney
Sault Ste. Marie, Michigan
Adjunct Faculty
Central Michigan University
Mt. Pleasant, Michigan

George B. Smith, MSN, BSN, ADN
Nursing Faculty
Hillsborough Community College
Tampa Florida

Rita Bolek Trofino, MNEd, RN, BSN
Continuing Education Program Manager
Cambria County Area Community College
Johnstown, Pennsylvania

Rose Utley, PhD, RN
Associate Professor
Southwest Missouri State University
Rogersville, Missouri

Linda S. Williams, MSN, RNC
Professor of Nursing
Jackson Community College
Jackson, Michigan

Bruce K. Wilson, PhD, MSN, BSN, LVN
Associate Professor
University of Texas—Pan America
Edinburg, Texas

Reviewers

Bonnie B. Anton, RN, MN
Cardiology Clinical Nurse
University of Pittsburgh Medical Center
Pittsburgh, Pennsylvania

Mary T. Bouchaud, MSN, CNS, RN, CRRN
Community Clinical Coordinator and Full Time Faculty
Thomas Jefferson University College of Health Professions
Philadelphia, Pennsylvania

Andrea' G. Bowden-Evans, MSN, RN, CRNP
Practical Nursing Instructor
Shelton State Community College
Tuscaloosa, Alabama

Teresa L. Bryan, RN, MSN, CFNP
Assistant Professor
Alcorn State University
Natchez, Mississippi

Joan R. Carnosso, RN, BSN
Instructor
Boise State University
Boise, Idaho

Barbara Chamberlain, MSN, RN, CNS, CCRN
Assistant Professor of Nursing
Drexel/MCP Hahnemann University
Philadelphia, Pennsylvania

Patricia G. Chichon, RN, APN-C, MSN
Private Practice—Pediatric/adult NP
The Chrysalis Center, Inc
Lambertville, New Jersey

Stephen R. Crumb, MSN
Nurse Practitioner, Cardiology
Fletcher Allen Health Care
Burlington, Vermont

Robin S. Culbertson, RN, BSN, MSN
Nursing Instructor
Okaloosa Applied Technology Center
Fort Walton Beach, Florida

Linda Dale, RN, BA, CDE
Diabetes Clinical Coordinator
Diabetes Center of Foote Hospital
Jackson, Michigan

R. Eric Doerfler, NP, MSN
President
Nightingale Health Centers, Inc.
Harrisburg, Pennsylvania

Mary Douglas, RN, BAN
Faculty
Hennepin Technical College
Eden Prairie, Minnesota

Kathleen Anne Fiato, RNC
Clinical Instructor
Questar III
Troy, New York

Andorra L. Foley, BSN, CEN
Staff RN, Emergency Dept.
University of Utah
Salt Lake City, Utah

Mary Jo Goolsby, EdD, MSN, APRN-C
Director of Research & Education
American Academy of Nurse Practitioners
Augusta, Georgia

Roger Green, ARNP, FNP, CS, MSN
Instructor
Seattle University School of Nursing
Seattle, Washington

Mary Elizabeth Haq, PhD, RN, FNP
Adjunct Associate Professor
Felician College
Lodi, New Jersey

B. Nicole Harder, BN, MPA
Faculty of Nursing
University of Manitoba
Winnipeg, Manitoba
Canada

Nancy B. Henry, RN, BSN, MSN
Faculty
Delaware Technical & Community College
Georgetown, Delaware

Anita L. Huse, BS, MSN, EdDC
Member Nursing Education and Professional Development
New England Medical Center
Boston, Massachusetts

Jo Kline, RN,C
Coordinator
Knox County Career Center
Mount Vernon, Ohio

Vanessa C. Kramasz, RN, MSN, APNP
Nursing Instructor
Gateway Technical College
Kenosha, Wisconsin

Pat S. Kupina, BSN, MSN
Professor
Joliet Junior College
Joliet, Illinois

Geneva Marie Lamm, LPN, ASN, BSN, MSN
Health Technologies Department Chair
Program Chair, Practical Nursing
Ivy Tech State College
Columbus, Indianapolis

Christina Lim, BAAN
Clinical Educator-Surgical Program
Humber River Regional Hospital
Downsview, Ontario
Canada

Cynthia A. Logan, BS, MS
Assistant Professor
Southeastern Louisiana University
Baton Rouge, Louisiana

Lynn Koch Lyons, BSN
Faculty
Greenville Technical College
Greenville, South Carolina

Cynthia W. McCoy, RN, PhD
Assistant Professor, School of Nursing
Troy State University
Troy, Alabama

Susan A. Moore, RN, PhD
Associate Professor of Nursing
HN Community Technical College
Manchester, New Hampshire

Carla Mueller, RN, PhD
Associate Professor, Director of Distance Education
University of Saint Francis
Fort Wayne, Indiana

Kim Penland, MSN, RN, APRN, BC
Assistant Professor of Nursing
University of Saint Francis
Fort Wayne, Indiana

Sandra Perkins, LPN
Charge Nurse in Long Term Care
Western Nebraska Community College
Scottsbluff, Nebraska

LuAnn J. Reicks, RNC, BS, BSN
Professor, PN Coordinator
Iowa Central Community College
Lake City, Iowa

Pat Reinhart, RN
Nursing Faculty and Health-care Representative for Customized
    Training and Continuing Education
Minneapolis Community and Technical College
Minneapolis, Minnesota

Anne W. Ryan, MSN, BSN, MPH
Associate Professor of Nursing
Chesapeake College
Wye Mills, Maryland

Alita K. Sellers, MSN, PhD
Chairperson, Health Sciences Division
West Virginia University
Parkersburg, West Virginia

Sheila M. Silsby, MS, RNCS, CCRN, CNN
Nurse Practitioner
Berkshire Medical Center
Pittsfield, Massachusetts

Kimberly J. Simmons, BSN, MSN, CS
Instructor
West Virginia University at Parkersburg
Parkersburg, West Virginia

Julie A. Slack, RN, MS
Faculty
Mohave Community College
Colorado City, Arizona

Joan Tilghman, MSN
Assistant Professor
Howard University
Washington, DC

Deborah L. Weaver, MSN, PhD
Associate Professor, College of Nursing
Valdosta State University
Valdosta, Georgia

Iris Winkelhake, RN, BSN, MEd
Program Chair, Practical Nurse Program
Southeast Community College
Lincoln, Nebraska

Patricia R. Yeargin, BSN, MN, MPH
Clinical Instructor
Southeast AIDS Training and Education Center, Emory
    University
Atlanta, Georgia

Contents

UNIT THIRTEEN
UNDERSTANDING THE NEUROLOGICAL
SYSTEM   801

45   Neurological Function, Assessment, and
Therapeutic Measures   803
George Byron Smith, Valerie C. Scanlon, and Sally Schnell

46   Nursing Care of Patients with Central Nervous
System Disorders   823
George Byron Smith and Sally Schnell

47   Nursing Care of Patients with Peripheral
Nervous System Disorders   873
George Byron Smith, Marsha A. Miles, and Deborah L. Roush

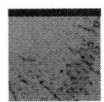
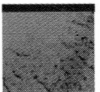

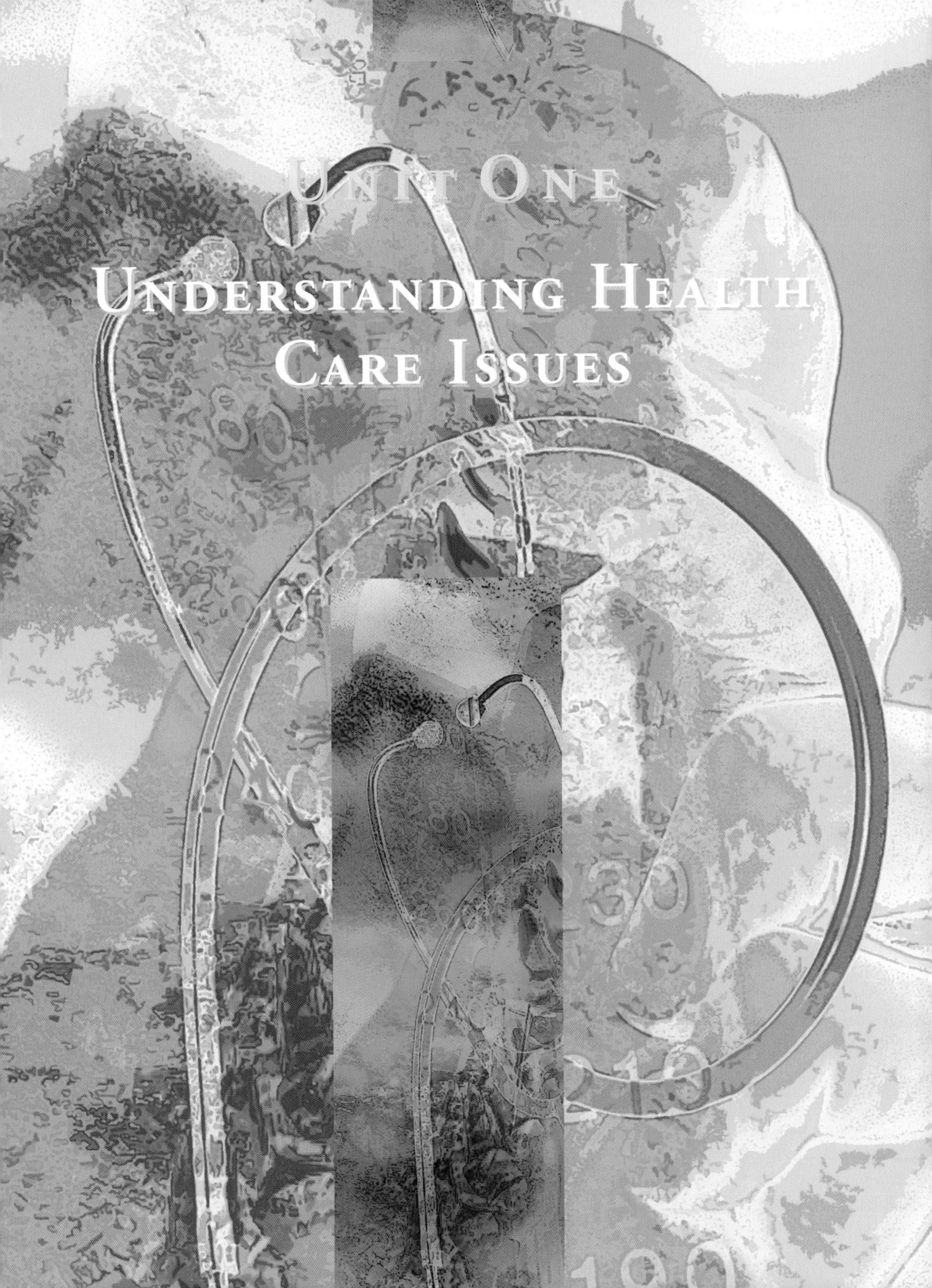

UNIT ONE

UNDERSTANDING HEALTH CARE ISSUES

1

CRITICAL THINKING AND THE NURSING PROCESS

Paula D. Hopper and Linda S. Williams

QUESTIONS TO GUIDE YOUR READING

1. How is the nursing process defined?
2. What are the characteristics of each step of the nursing process?
3. What are three characteristics of critical thinking?
4. How is critical thinking of value in the nursing process?
5. What is the role of the licensed practical nurse/licensed vocational nurse in using the nursing process?
6. How would you prioritize patient care based on Maslow's hierarchy of human needs?
7. How does critical thinking enhance the implementation of the nursing process?

Excellence in the delivery of nursing care requires good thinking. Each day nurses make many decisions that affect the care of their patients. For those decisions to be effective, the thought processes behind them must be sound.

CRITICAL THINKING

Nurses must learn to think critically. This means they must use their knowledge and skills to make the best decisions possible in patient care situations. Halpern[1] says that "**critical thinking** is the use of those cognitive (knowledge) skills or strategies that increase the probability of a desirable outcome." Critical thinking is sometimes called directed thinking because it focuses on a goal. Other terms used when talking about critical thinking include *reasoning, common sense, analysis,* and *inquiry.*

PROBLEM SOLVING

Problem solving is one type of critical thinking. Nurses solve problems on a daily basis. However, a problem can be handled in a way that may or may not help the patient.

For instance, consider Mr. Frank, who is in pain and requests pain medication. You check the medication record and find that his analgesic medication is not due for another 40 minutes. You can choose to manage this problem in a variety of ways. One obvious approach is to return to Mr. Frank and tell him that it is not time for the pain medica-

3

tion and that he will have to wait. This may solve your problem (you can move on to the next patient), but it does not solve the problem in an acceptable way for Mr. Frank.

An alternative approach is to use a problem-solving method:

1. Gather **data.** When Mr. Frank requests pain medication, the first thing you do is obtain more data. A common way to assess pain is to use a pain rating scale in which the patient rates pain on a scale of 0 (no pain) to 10 (the greatest pain possible). Mr. Frank states that his pain is in his back and rates it as an 8 on the 10-point scale. His history includes compression fractures of his spine. You return to the medication record and find that he has no alternative pain medications ordered.

2. Identify the problem. Mr. Frank is in acute pain, and the current medication orders may not be sufficient to provide pain relief.

3. Decide what outcome (sometimes called a goal) is desirable. The outcome should be determined by you (the nurse) and the patient, working together. The patient is intimately involved in this situation and deserves to be consulted. In this case you talk to Mr. Frank and determine that he needs pain relief now; he cannot wait until the next scheduled dose of medication. He states that he is able to tolerate a pain rating of 3 or less on a 10-point scale.

4. Plan what to do. Formulate and evaluate some alternative solutions. For example, you can decide to tell Mr. Frank that he has to wait 40 minutes; however, this will not help him reach his desired outcome of pain control. Giving the medication early might relieve his pain, but this would not be following the physician's orders and may have harmful effects for Mr. Frank. You could decide to try some alternative pain control methods, such as relaxation, distraction, or imagery. These might be helpful, but with a pain rating of 8, he may also need medication to reach his goal. Another alternative is to report to the physician that Mr. Frank's pain is not controlled with the current pain control regimen. Once you have several alternatives, decide which ones you think will best help the patient. Then you can discuss those options with the registered nurse (RN) and together decide the best thing to do; in this case you might decide to have the RN contact the physician while you work with the patient on relaxation exercises. You might decide to ask Mr. Frank if he would like to listen to some of the music his wife brought in for him. You can also tell Mr. Frank that the physician is being contacted. This would assure Mr. Frank that his pain relief needs are being pursued.

5. Implement the plan of care. Suppose the RN enters the room and informs you and the patient that the physician has changed the analgesic orders. You obtain and administer the first dose of the new analgesic. The RN also informs Mr. Frank that the physician has ordered a consultation with the institution's pain clinic.

6. Evaluate the plan of care. Did the plan work? As you reassess Mr. Frank 30 minutes later, he rates his pain level at 2. You think back to the desired outcome, compare it with the current **assessment,** and determine that your **interventions** were successful.

Can you see how using the problem-solving process led to a better outcome than simply choosing the first obvious option? You were able to affect a positive change: assisting a patient in achieving pain relief.

▶ NURSING PROCESS

You have just used the **nursing process** to solve a real problem. The nursing process is an organizing framework that links the process of thinking with actions in nursing practice. The nursing process can be used to assess patient needs; formulate nursing diagnoses; and plan, implement, and evaluate care. As a nursing student, you will consciously use the nursing process with each patient problem. With experience, you will internalize the nursing process and use it without as much conscious effort.

Role of the Licensed Practical Nurse and Licensed Vocational Nurse

The licensed practical nurse (LPN) or licensed vocational nurse (LVN) carries out a specific role in the nursing process, as described in Table 1–1. The LPN/LVN collects data, assists in formulating nursing diagnoses, assists in determining outcomes and planning care to meet patient needs, implements patient care interventions, and assists in evaluating the effectiveness of nursing interventions in achieving the patient's outcomes. It is the role of the LPN/LVN to provide direct patient care. This gives the LPN/LVN an opportunity to develop a relationship with the patient that aids in collecting valuable data. The LPN/LVN and the RN work as a team to analyze data and develop, implement, and evaluate the plan of care (Fig. 1–1).

Data Collection

The first step in the nursing process is data collection. The LPN/LVN assists the RN in collecting data from a variety of sources. Data is divided into two types: **subjective data** and **objective data.**

TABLE 1-1	ROLE OF THE LPN/LVN IN THE NURSING PROCESS
Steps of the Process	**Role of the LPN/LVN**
Assessment	Assists in collecting data
Nursing Diagnosis	Assists in choosing appropriate nursing diagnoses
Plan of Care	Assists in developing outcomes and planning care to meet outcomes
Implementation	Carries out those portions of the plan of care that are within the LPN/LVN's scope of practice
Evaluation	Assists in evaluation and revision of the plan of care

Figure 1-1   An LPN and an RN collaborating on a nursing care plan.

Subjective Data

Information that is provided verbally by the patient is called subjective data. Symptoms are subjective data. Subjective data are often placed in quotes, such as "I have a headache" or "I feel out of breath." You must listen carefully to the patient and understand that only the patient truly knows how he or she feels.

When collecting subjective data, first assess the chief complaint. Focus on the reason the patient is seeking health care. The question "What happened that made you decide to come to the hospital (clinic, office)?" can be helpful.

Once the patient has identified the main concern, further questioning can elicit more pertinent information. Use the letters of the "WHAT'S UP?" questioning format to remember questions to ask the patient (Table 1–2).

LEARNING TIPS

Practice assessing a symptom on a classmate. Ask the "WHAT'S UP?" questions.

Next, obtain a patient history. This is done by asking the patient and family questions about the patient's past and present health problems, including specific questions about each body system, family health problems, and risk factors for health problems. The patient's medical record may also be consulted for background history information.

In addition to assessment related to physiological functioning, ask the patient about personal habits that relate to health, such as exercise, diet, and the presence of stressors, per institutional assessment guidelines. Finally, assess the patient's family role, support systems, and cultural and spiritual beliefs.

Objective Data

Objective data are pieces of factual information obtained through physical assessment and diagnostic tests and are observable or knowable through the five senses. Objective

TABLE 1-2   WHAT'S UP? GUIDE TO SYMPTOM ASSESSMENT

W—Where is it?
H—How does it feel? Describe the quality.
A—Aggravating and alleviating factors. What makes it worse? What makes it better?
T—Timing. When did it start? How long does it last?
S—Severity. How bad is it? This can often be rated on a scale of 1 to 10.
U—Useful other data. What other symptoms are present that might be related?
P—Patient's perception of the problem. The patient often has an idea about what the problem is, or the cause, but may not believe that his or her thoughts are worth sharing unless specifically asked.

data are sometimes called *signs*. Examples of objective data include the following: respiratory rate 36, 3-cm red lesion, blood glucose 326 mg/dL, and patient moaning. Note that these are all observable or measurable by the nurse and do not require explanation by the patient.

Objective data are gathered through physical assessment. During the physical assessment, the nurse inspects, palpates, percusses, and auscultates (IPPA) to collect objective data. This is called the IPPA format. Special attention is given to areas that the patient has identified as potential problems.

INSPECTION. During the **inspection** phase of physical assessment, the nurse uses observation skills to systematically gather data that can be seen. This may include noting the patient's respiratory effort, observing skin color, or measuring a wound. This phase continues throughout the assessment.

PALPATION. **Palpation** involves use of the fingers or hands to feel something. The nurse might palpate the abdomen for firmness or use the back of the hand to palpate for heat, such as palpating the forehead for fever. Physicians and advanced practice nurses also use deep palpation to assess abdominal organs.

PERCUSSION. **Percussion** is a technique used by physicians and advanced practice nurses to determine the consistency of underlying tissues. The examiner taps on the patient to elicit a sound. Generally, the middle finger of the nondominant hand is placed on the area to be percussed, and the middle finger of the dominant hand taps on the nondominant one (Fig. 1–2). This prevents patient discomfort. A dull sound is heard over a fluid-filled area, such as the liver or urinary bladder. A flat sound is heard over a solid area, such as muscle. Tympany is a drumlike sound heard over air, such as gas in the stomach. Resonance is a hollow sound heard over air-filled lung tissue. Although this technique generally is not used by LPN/LVNs, it is helpful to understand what it is when assisting another care provider performing percussion during a physical examination.

AUSCULTATION. **Auscultation** is usually done with a stethoscope. The chest is auscultated for heart sounds. Lung sounds are listened to anteriorly (chest) and posteriorly

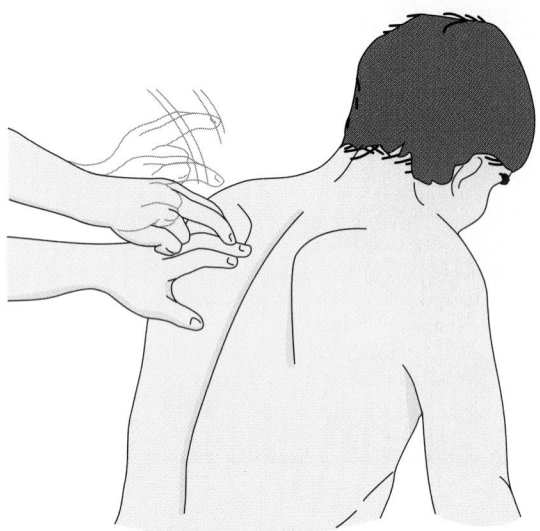

Figure 1-2 Percussion.

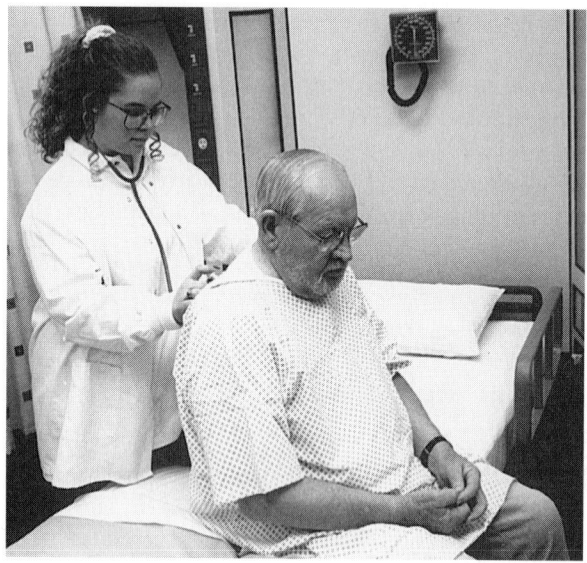

Figure 1-3 Auscultation of posterior lung sounds. (From Anderson, MA: Nursing leadership, management, and professional practice for the LPN/LVN. FA Davis, Philadelphia, 1997, p 29, with permission.)

(back) for normal and adventitious lung sounds or (Fig. 1–3).The abdomen is auscultated for bowel sounds. Major vessels may be auscultated for turbulent blood flow by a physician or RN.

During physical assessment, the IPPA format is followed in the given order, except when assessing the abdomen. The order for the abdomen assessment is IAPP, because percussion and palpation of the abdomen can alter the auscultation findings.

Data collection related to each body system is covered in individual chapters throughout the book.

Documentation of Data

Once the data have been collected, they are documented in the patient's medical record. Identification of a significant problem is reported immediately to an RN or physician and then documented. The recorded data should be accurate and concise. You should document exactly what was observed or stated by the patient, significant other, or health team members. Avoid interpreting the data and using words that have vague meanings in your documentation. For example, "nailbed color is pink" gives clearer information than "nailbed color is normal." "Capillary refill is 2 seconds" provides more precise data than "capillary refill is good." The statement "the wound looks better" is not meaningful unless the wound has been previously observed by the reader. Stating that "the wound is 1 by 2 inches, red, with no drainage or odor" provides data with which to compare the future status of the wound and determine whether it is responding to treatment. When documenting subjective data, direct quotations from the patient are desirable. Quotes accurately represent the patient's view and are least open to interpretation. Meaningful documentation promotes continuity of patient care.

LEARNING TIPS

Documenting exactly what is observed is more appropriate, easier, and less time consuming than seeking other words or ways to state observations. Nursing students and novices often search for elaborate phrases or words when simple, direct words are best. Stating exactly what is seen or heard usually provides the most clear and accurate information.

Nursing Diagnosis

Once data have been collected, the LPN/LVN assists the RN to compare the findings with what is considered "normal." Data are then grouped, or clustered, into sets of related information that identify problems.

According to the North American Nursing Diagnosis Association (NANDA), a **nursing diagnosis** is a clinical judgment about individual, family, or community response to actual or potential health problems or life processes.[2] Nursing diagnoses are standardized labels that make an identified problem understandable to all nurses. Nursing diagnoses are the foundation used to select interventions to achieve a desired outcome.

Nursing actions are either independent or collaborative. Independent nursing actions can be initiated by the nurse. Examples of independent nursing actions include teaching the patient deep breathing exercises, turning a patient every 2 hours, teaching about medications, and giving a back rub for comfort. Collaborative actions require a physician's order to perform them in response to both nursing and medical diagnoses. Examples of this are giving prescribed medications, applying antiembolism stockings, requesting a referral to physical therapy, and inserting a urinary catheter.

A diagnosis is considered "nursing" instead of "medical" if the interventions necessary to treat the problem are primarily independent nursing functions. Even with independent nursing functions, nurses often consult with physicians about plans of care. If, however, the physician directs most of the care related to a particular health problem, it is a medical diagnosis, rather than a nursing diagnosis. For example, a patient with pneumonia (a medical diagnosis) has many needs that depend on physician orders, such as respiratory treatments and antibiotics. The nurse, however, can provide important assessment findings related to the health problem and provide nursing measures such as encouraging fluid intake, coughing, and deep breathing. When the physician and nurse work closely together on a patient problem, it is referred to as a collaborative problem.

One of the NANDA nursing diagnoses is acute pain. In Mr. Frank's example the pain was assessed as a health problem, and a plan of care was developed to manage the pain. The physician was contacted for analgesic orders, and independent nursing actions were used, including relaxation and distraction. These independent nursing actions did not require a physician's order. See the inside cover of this book for a complete list of nursing diagnoses recommended by NANDA.

♥ CRITICAL THINKING

Nursing Diagnosis
Which of the following are NANDA nursing diagnoses? Which are medical diagnoses?

1. Impaired mobility
2. Ineffective coping
3. Herniated disk
4. Fractured femur
5. Diabetes
6. Impaired gas exchange
7. Appendicitis
8. Health-seeking behaviors

Answers at end of chapter.

A well-written nursing diagnosis helps guide development of a plan of care. The three parts to a diagnosis follow:

■ Problem—the nursing diagnosis label from the NANDA list
■ Etiology—the cause or related factor (often preceded by the words "related to")
■ Signs and symptoms—the subjective or objective data that are evidence that this is valid diagnosis (often preceded by the words "as evidenced by")

Assessment of problem, etiology, and signs and symptoms is called the PES format. Look again at the case study of Mr. Frank. A diagnosis using this format might read as follows: "Acute pain related to muscle spasms and nerve compression as evidenced by patient pain rating of 8." Note how the complete diagnosis gives you more helpful information than simply the label "pain." This additional information helps determine an appropriate outcome to guide intervention selection.

Plan of Care

Once nursing diagnoses are identified, an individualized plan of care to help the patient meet his or her care needs is de-

signed. It is important to include the patient in the development of the plan of care. The patient must be in agreement with the plan for it to be successful in meeting the desired outcomes. The first step in planning care after diagnoses are selected is to prioritize the diagnoses and develop outcomes, or goals, for each. Actions that will help the patient meet the desired outcomes can then be determined.

LEARNING TIPS

If you are developing a plan of care for a complex patient and are not sure where to start, go back to the assessment phase. Often additional information can help you better understand the patient's needs and develop a plan of care that is individualized to the patient's specific problem areas.

Prioritizing Care

Once you know what problems need to be addressed, you must decide which problem or intervention must be taken care of first. You and the patient decide together which problems take priority. Maslow's hierarchy of human needs is one commonly used psychological theory that can be used as a basis for determining priorities (Fig. 1–4). According to Maslow, humans must meet their most basic needs (those at the bottom of the triangle) first. They can then move up the hierarchy to meet higher-level needs.

Physiological needs are the most basic. For example, a person who is short of breath can't attend to higher-level needs because the physiological need for oxygen is not being met. Once physiological needs are met, the patient can concentrate on meeting safety and security needs. Love, belonging, and self-esteem needs are next; self-actualization needs are the last to be met.

Throughout life, individuals move up and down Maslow's hierarchy in response to life events. If a need occurs on a level below the patient's current level, the patient will move down to the level of that need. Once the need is fulfilled, the person can move upward on the hierarchy again.

In a nursing plan of care the patient's most urgent problem is listed first. According to Maslow's hierarchy of human needs, this usually involves a physiological need such as oxygen or water, because these are life-sustaining needs. If several physiological needs are present, life-threatening

---

♥ CRITICAL THINKING

Prioritizing Care
Based on Maslow's hierarchy of needs, list the following nursing diagnoses in order from highest (1) to lowest (5) priority. Give rationales for your decisions.

____ Deficient knowledge
____ Constipation
____ Disabled family coping
____ Anxiety
____ Ineffective airway clearance

Answers at end of chapter.

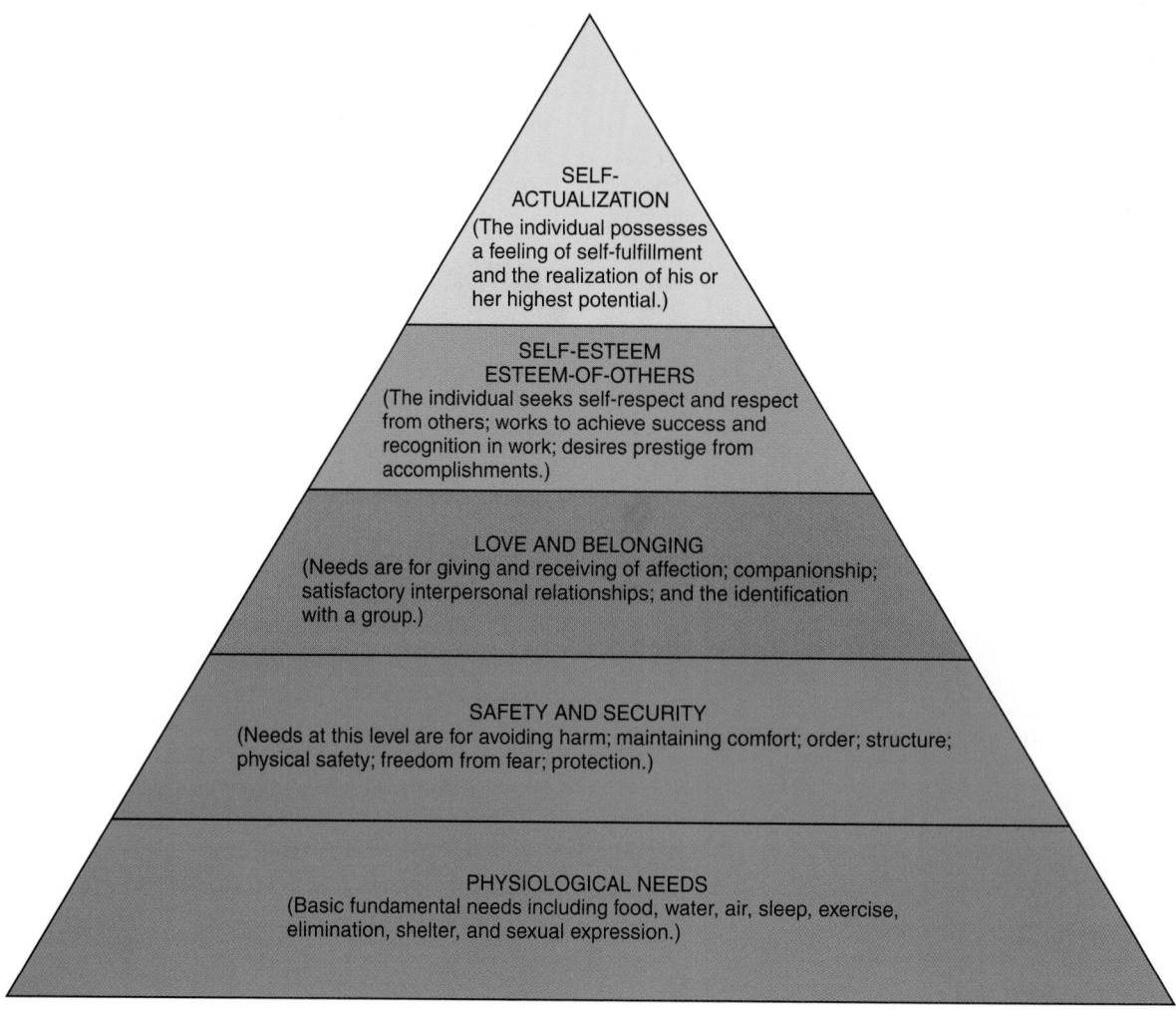

Figure 1-4 Maslow's hierarchy of human needs.

needs are ranked first, health-threatening needs are second, and health-promoting needs, although important, are last.

Once physiological needs are met, needs related to the next level of the hierarchy, safety and security, can be addressed. Remaining diagnoses are listed in order of urgency as they relate to the hierarchy. Needs can occur simultaneously on different levels and must be addressed in a holistic manner, with prioritization guiding the care provided.

Establishing Outcomes

An outcome is a statement that describes the patient's desired goal for a problem area. It should be measurable, be realistic for the patient, and have an appropriate time frame for achievement. *Measurable* means that the outcome can be observed, or is objective. It should not be vague or open to interpretation, with the use of subjective words such as *normal*, *large*, *small*, or *moderate*. Consider, for example, two outcomes:

■ The patient's shortness of breath will improve.
■ The patient will be less short of breath within 15 minutes as evidenced by patient rating the shortness of breath at less than 3 on a scale of 1 to 10, respiratory rate between 16 and 20, and relaxed appearance.

Although the first outcome seems appropriate, in reality it will be difficult to know when it has been met. There is nothing to objectively indicate when the problem has been resolved. The second outcome is objective. You can see that when the patient rates his or her shortness of breath less than 3, is breathing at a rate of 16 to 20, and appears relaxed, the desired outcome will have been met. The outcome is realistic, and the 15-minute time frame ensures that the patient's distress will be minimized. If the plan of care does not achieve the desired outcome in the given time frame, it should be evaluated and revised as needed.

When determining criteria for a measurable outcome, look at the signs and symptoms portion of the nursing diagnosis. The resolution or continuation of the signs and symptoms identified in the NANDA nursing diagnoses are evidence that nursing interventions were effective. If the desired outcome is not achieved, re-evaluation of the problem and interventions is needed. Look at another outcome example to see how criteria are used for measurement:

Nursing diagnosis—Ineffective airway clearance related to excess secretions as evidenced by coarse crackles and non-productive cough.

Outcome—Demonstrates clear lung sounds and productive cough within 8 hours.

Identifying Interventions

Interventions are the actions you take to help the patient meet the desired outcome. Therefore, interventions are considered goal directed. Any intervention that does not contribute to meeting the outcome should not be part of the plan of care.

One way to create a plan is to include interventions that can be categorized as "take, treat, and teach." In the first intervention category, "take," or identify, data that should be routinely collected related to the problem. Next, "treat" the problem by identifying deliberate actions to help reach the outcome. Last, identify what to "teach" the patient and family for the patient to learn to care for himself or herself.

Look again at the nursing diagnosis of impaired airway clearance. A plan of care for this problem using the take, treat, and teach method might look like the following:

Take: Auscultate lung sounds every 4 hours and prn.
       Assess respiratory rate every 4 hours and prn.
Treat: Provide 2 L of fluids every 24 hours.
       Offer expectorant as ordered.
       Provide cool mist vaporizer in room.
Teach: Teach the patient the importance of fluid intake.
       Teach the patient to cough and deep breathe every 1 to 2 hours.

In addition to identifying interventions, it is important to understand how and why they will work. This is called identifying rationales. For example, assess lung sounds and respiratory rate every 4 hours because increased crackles and respiratory rate indicate retained secretions. Fluids are provided to help liquefy secretions and ease their removal. Sound rationale that is evidence (research) based should guide the selection of each nursing intervention.

Implementation

Once the plan of care has been identified, it must be communicated to the patient, family, and health team members and then implemented. One way a plan of care is communicated is by writing it as a nursing care plan. The nursing care plan is documented on the patient's medical record and lets other nurses know the patient's priority problems, the desired outcomes, and the plan for meeting the outcomes. In this way all nurses can be provide consistent care for the patient.

When implementing the plan of care, the actions listed as interventions are performed. The patient's response to each intervention is noted and documented. This documentation provides the basis for **evaluation** and revision of the plan of care.

Evaluation

The last step of the process is evaluation. Evaluation examines both outcomes and interventions. The nurse continuously evaluates the patient's progress toward the desired outcomes and the effectiveness of each intervention. If the outcomes are not reached within the given time frame, or if the interventions are ineffective, the plan of care is revised. Any part of the plan of care can be revised, from the diagnosis or desired outcome to the interventions. Acute care institutions require review and updating of the plan of care every 24 hours.

▶ SUMMARY

Although the nursing process is presented in discrete steps in this chapter, in reality it is a continuous process. The nurse is constantly using critical thinking skills to assess, evaluate, and revise the plan of care to meet the patient's changing needs.

In most chapters throughout this book, sample nursing care plans are provided. These are general guidelines for patient care. Any plan of care must be individualized based on a patient's unique characteristics and needs.

Answers to Critical Thinking

Nursing Diagnosis

1. Impaired mobility = nursing
2. Ineffective coping = nursing
3. Herniated disk = medical
4. Fractured femur = medical
5. Diabetes = medical
6. Impaired gas exchange = nursing
7. Appendicitis = medical
8. Health-seeking behaviors = nursing

Prioritizing Care

4	Deficient knowledge, safety and security need
2	Constipation, physiological need
5	Disabled family coping, love and belonging need
3	Anxiety, safety and security need
1	Ineffective airway clearance, physiological need

1. Which of the following statements best defines the nursing process?
   a. The process that nurses use to write nursing care plans
   b. The process used by nurses to evaluate nursing care
   c. A framework that links the process of thinking with nursing actions
   d. A framework that promotes collaboration with other members of the health team

2. Which one of the following parts of the nursing process can be carried out by the LPN?
   a. Implementation of interventions
   b. Nursing diagnosis
   c. Analysis of data
   d. Evaluation of outcomes

3. Which of the following pieces of information is considered objective data?
   a. Patient is short of breath.
   b. Patient states, "I feel short of breath."
   c. Patient's respiratory rate is 28.
   d. Patient is feeling panicky.

4. An LPN is collecting data on a newly admitted patient. He has an ulcerated area on his left hip that is 2 inches in diameter and 1 inch deep, with yellow exudate. Which of the following statements best communicates findings in the patient's database?
   a. Wound on left hip, 2 inches diameter, 1 inch deep, infected
   b. Left hip wound is large, deep, and has yellow drainage
   c. Pressure ulcer on left hip, yellow drainage
   d. Wound on left hip 2 inches in diameter, 1 inch deep, yellow exudate

5. A 34-year-old mother of three children is admitted to a respiratory unit with pneumonia. Based on Maslow's hierarchy of needs, which of the following patient problems should the nurse address first?
   a. Frontal headache from stress of hospital admission
   b. Shortness of breath from newly diagnosed pneumonia
   c. Anxiety related to concern about leaving children
   d. Deficient knowledge about treatment plan

6. Which of the following is a nursing diagnosis?
   a. Stroke
   b. Renal failure
   c. Fracture
   d. Acute pain

7. You have just completed teaching a patient the importance of stopping smoking. Which of the following patient responses best evaluates patient teaching effectiveness?
   a. "I have a brother who died of lung cancer. I know smoking is bad."
   b. "I tried to quit 5 years ago, and I really would like to, but it is very hard."
   c. "Thank you for the information. I will call the Smoke Stoppers organization today."
   d. "I know you are right; I should stop smoking."

REFERENCES

1. Halpern, D: Thought and knowledge: an introduction to critical thinking, ed 3. Lawrence Erlbaum Associates, NJ, 1996, p 5.

2. North American Nursing Diagnosis Association (NANDA): Nursing diagnosis: definitions and classification, 1999–2000. Philadelphia, 1999.

2

ETHICAL AND LEGAL ISSUES FOR NURSING

Winifred Ellenchild Pinch and Patrick Shannon

KEY TERMS

administrative laws
(ad-MIN-i-**STRAY**-tive LAWZ)

beneficence (buh-**NEF**-i-sens)

civil law (**SIV**-il **LAW**)

code of ethics (**KOHD** OF **ETH**-icks)

confidentiality (**KON**-fi-den-she-**AL**-i-tee)

criminal law (**KRIM**-i-nuhl **LAW**)

deontology (DA-on-**TOL**-o-gee)

distributive justice
(dis-**TRIB**-yoo-tiv **JUS**-tiss)

empathy (**EM**-puh-thee)

ethical (**ETH**-i-kuhl)

feminist (**FEM**-uh-nist)

fidelity (fi-**DEL**-i-tee)

law (**LAW**)

liability (LYE-uh-**BIL**-i-tee)

limitation of liability
(LIM-i-**TAY**-shun OF LYE-uh-**BIL**-i-tee)

malpractice (mal-**PRAK**-tiss)

morality (muh-**RAL**-i-tee)

negligence (**NEG**-li-jense)

nonmaleficence (NON-muh-**LEF**-i-sens)

paternalism (puh-**TER**-nuhl-izm)

respondeat superior
(res-**POND**-ee-et sue-**PEER**-ee-or)

standard of best interest
(**STAND**-erd OF **BEST IN**-ter-est)

summons (**SUM**-muns)

torts (**TORTS**)

utilitarian (yoo-TILL-i-**TAR**-i-en)

values (**VAL**-use)

veracity (vuh-**RAS**-i-tee)

welfare rights (**WELL**-fare **RIGHTS**)

QUESTIONS TO GUIDE YOUR READING

1. Why are ethics important in health care?

2. What roles do values and rights have in making ethical decisions?

3. What is an example of a character trait and how does it relate to nursing?

4. What are the definitions of the four major principles in ethics and how would you apply one to an ethical dilemma?

5. What is one ethical theory and how would you apply it to an ethical dilemma?

6. What are the steps of the ethical decision-making model?

7. What are three legal concepts that relate to nursing practice?

8. How can you provide quality care and limit your liability?

9. How would you define malpractice insurance?

▶ ETHICS AND VALUES

One of the most exciting and challenging areas of nursing practice today is bioethics. Bioethics means the **ethical** principles and **values** of a particular culture or group that are applied to life. In the current practice of nursing, bioethics has come to be most closely associated with ethics in health care: life as affected by health, sickness, disease, or trauma. Ethics is the study of traditions, values, and beliefs as they relate to persons and their relationships with one another. Defining what is "good" and "bad," what is "right" and "wrong," and what values are used to make those judgments are all part of ethics. What *should* we do? and What *ought* we to do? are also ethical questions. We are responsible for knowing *why* we might take the intended action. Morals or **morality** are also related to ethics. Some individuals use the terms *ethics* and *morals* interchangeably; however, *morals* more specifically refers to our personal values, the standards set by our own conscience, and our personal choices of good and bad, right and wrong.

Values are ideals or concepts that give meaning to an individual's life. Values are most commonly derived from societal norms, religion, and family traditions. They serve as a guide for making decisions and taking certain actions in everyday life. Values are not usually written down, but it can be helpful to make a list of your values and attempt to rank them by priority. Value conflicts often occur in everyday life and can force an individual to select a higher-priority value over a lower-priority value. For example, a nurse who values both her career and her family may be forced to decide between going to work or staying home with a sick child.

Values exist on many different levels. Individuals have personal values that govern their lives and actions. Many groups and organizations have values that may or may not be identical to personal values. When a person becomes a member of a group or organization, he or she agrees to accept the values of the group. Examples of groups include clubs, churches and religious organizations, and professions. The values of a profession are usually outlined in a **code of ethics.** Failure to adhere to a code of ethics of a profession may be an indication that the individual really does not want to belong to that group. Society as a whole has values. As a member of a society or country, an individual accepts the values of that culture.

Ethical issues surround us throughout our entire lifetime. Bioethical issues are particularly prevalent in our professional lives for several reasons. Health care has the potential to affect us at every milestone of human development. Life and death issues can be especially difficult for health care professionals, who may be faced with these issues every day. Today's sophisticated technology and complex treatments, which continue to become more advanced, add another challenge. When sophisticated interventions are available, the ethical question is should we use them? Ethical issues are raised when the organization of health care delivery systems becomes more complex as mergers and alliances come and go. Despite the large percentage of our country's wealth dedicated to health care, financial resources are not unlimited. How those resources are divided can be troublesome, with the needs and desires of many different groups and organizations competing for a slice of the pie.

Although bioethical issues abound, not all issues are problems and not all ethical problems make headline news. Nurses are involved in some ethical decisions every day that may not become problems. A promise might be extended to return to the patient and assess whether or not the pain medication was effective. Patients might share information that they want to remain confidential. A family agrees to a do not resuscitate (DNR) order for their loved one when they believe there is no hope for recovery. A young man in the prime of life is given cardiopulmonary resuscitation (CPR) when he suddenly goes into respiratory failure on the day of his proposed discharge. These bioethical examples usually are not problems but, depending on the situation, can develop into ethical dilemmas.

An ethical dilemma is created when the required moral action is not clear or individuals in the situation do not agree on the proposed solution. Potential solutions may appear to be equally good or, worse, equally risky: A promise cannot be kept; information cannot remain confidential; DNR orders may not be acceptable for some individuals. Not every patient should be the recipient of CPR endeavors, even those who are very young. When patients are conscious, their choices are usually respected, but on occasion even that premise can be difficult to apply. Often groups of individuals must resolve a conflict that can occur if there is disagreement between physicians and families, nurses and physicians, or among family members.

A basic mastery of several elements will enhance your ability to perform competently when bioethical issues and decision making are the focus. Understanding the ethical component of your nursing role is a first step. Acquiring knowledge about relevant ethical material is also essential. An ethical decision-making process is a useful tool for examining ethical dilemmas. Together these elements provide a foundation from which you can begin to explore the meaning of bioethics in nursing practice today. For more information about bioethics, visit the America Journal of Bioethics Online at http://www.ajobonline.com and the National Institutes of Health at http://www.nih.gov/sig/bioethics.

Ethical Obligations and Nursing

As a nurse you are an invaluable member of the health care team. Members of the team contribute to patient care according to their educational preparation and assigned responsibilities. A common goal among all health team members is to provide holistic care to the patient, because the individual's mind, body, and spirit are all affected by disease, illness, and trauma. Part of that holistic care encompasses your responsibility to practice ethically.

In addition to practicing within the **law,** nurses have ethical obligations related to the law. First, if the law is unethical or has serious limitations, a basic moral obligation of the nurse is to make an effort to change that law. Laws can unfairly affect health care in relation to accessing health care, subsidizing health care, or allocating health care dollars. Laws can be unethical because they are enacted based on power, authority, and political influences rather than high moral standards. Laws are made to reflect common ethical beliefs (do not kill, do not steal). However, in a democratic society with a diverse population such as in the United States, there are serious questions about the degree to which society should enforce moral behavior through the law.

Nursing Code of Ethics

Some of the major ethical obligations of nursing practice are addressed within the nursing code of ethics. As a professional guide for ethical practice, the National Association for Practical Nurse Education and Service (NAPNES) developed a code of ethics for the licensed practical nurse/licensed vocational nurse (LPN/LVN). Like all codes of ethics, this code is an important document because it is a public statement of the basic ethical principles and standards for LPN/LVNs. A code should contain those basic ethical premises with which most members agree and for which the public can hold the professional morally accountable. A code does not dictate a particular action, nor is it a legal document, although the code should not be in conflict with the law. The code is not enforced by any organization, and no punishment exists if a nurse fails to adhere. A code must be interpreted because it usually contains broad statements, but it does serve as a general guideline for the most important professional ethical issues. From time to time codes are revised and updated. Although an early medical code, such as the Hippocratic oath from the fourth century B.C., contains good moral advice, the passage of time and the occurrence of notable events make it incomplete for medical practice today. Changes in society prompt us to examine the profession's ethical standards and update them appropriately.

Virtues

Particular ethical obligations can be summarized in the form of virtues. Each individual is endowed with specific character traits. As we grow and develop we acquire certain other character traits. Virtues are one kind of character trait most often associated with one's values and morality or conscience. Virtues are different from the skills that we acquire as nurses. Skills are used to implement various actions, something we do, whereas virtues define who we are. Examples of virtues include **fidelity, veracity,** integrity, compassion, discernment, trustworthiness, and respectfulness. These are all ideal traits, and although we may strive to act as virtuous persons, we may not always succeed. These are also the sort of traits that the profession encourages nurses to develop. For example, faithfulness and respect for patient beliefs are specifically included in the code. Virtues are related to roles and responsibilities and are therefore time dependent. For example, in the early decades of organized nursing in the United States, nurses occupied a handmaiden role relative to the physician. Obedience to the physician was a virtue, but such obedience is no longer held as a virtue by nurses.

FIDELITY. Fidelity is the obligation to be faithful to commitments made to self and others. In health care, fidelity includes faithfulness or loyalty to agreements and responsibilities accepted as part of the practice of nursing. Fidelity is the main support for the concept of accountability, although conflicts in fidelity might arise because of obligations owed to different individuals or groups. For example, nurses have an obligation of fidelity to the patients they care for to provide the highest quality care possible, as well as an obligation of fidelity to their employing institution to follow its rules and policies. Nurses can have an ethical dilemma when a hospital's policy on staffing creates a situation that does not allow the nurses to provide the quality of care they feel is necessary.

Maintaining a patient's privacy and **confidentiality** is related to fidelity (Fig. 2–1). Privacy and confidentiality may or may not be explicit promises. For example, nurses are obligated to only discuss the patient under circumstances in which it is necessary to deliver high-quality health care. Care should be taken to hold change-of-shift reports and case conferences in settings where the discussion will remain private.

Confidentiality also includes the communication of information through the posting of unit census and various schedules for tests, procedures, or special examinations (operating room, physical therapy, radiology), storage and access of patient information through computers, and the transmission of patient information via fax machines, for

Figure 2-1 Maintaining privacy is a patient right and conveys caring to the patient.

example. Many individuals have legitimate access to the patient's chart in addition to direct caregivers: faculty members in the course of making student assignments, accrediting agencies, risk managers, quality assurance personnel, insurance companies, and researchers. Each is obligated to maintain patient confidentiality to the extent that concealing information does not compromise mandated reports (communicable diseases or gunshot wounds), releases already granted by the patient (such as when insurance coverage was obtained), or gathering data in the aggregate without identification of specific patients (research or institutional statistics).

VERACITY. Veracity is the virtue of truthfulness. It requires health care providers to tell the truth and not intentionally deceive or mislead patients. As with other rights and obligations, there are limitations to this virtue. The primary limitation occurs when telling patients the truth would seriously harm (principle of **nonmaleficence**) their ability to recover, or when the truth would produce greater illness. Another difficult situation is created in relation to diagnostic information. Although giving diagnostic information is the responsibility of the physician or registered nurse (RN), LPN/LVNs sometimes find themselves caught in situations in which they must deal with patients' questions. If LPN/LVNs feel uncomfortable about reinforcing physician or RN explanations about unpleasant information, they may avoid answering patients' questions directly by using half-truths or claiming ignorance. However, patients do have a right to know this information.

INTEGRITY. Integrity is the holistic, moral sense of self sustained over time. Each individual has a cluster of values, beliefs, and traditions that form the basis for moral decision making—in a sense, the conscience. This sense of self can be compromised when the nurse is requested to act in a manner that requires setting aside or acting against values, belief, or traditions. The nurse who believes in sanctity of life and that human life begins at conception will not be able to maintain integrity if required to participate in abortion procedures. At the same time, this nurse also has the responsibility not to accept a professional position where abortion will be an issue—the nurse cannot abandon a patient in need of nursing services. The nurse with integrity is faithful to professional responsibilities and obligations.

COMPASSION. Compassion or caring is a central virtue in nursing. Some label this **empathy** and connect it to the ability of a nurse to identify with a patient's suffering, pain, or disability. Patients are comforted by the compassionate nurse, and this emotion can assist in the healing process. A nurse without any emotion robs patients of the full potential for healing that is only possible when all parties are actively engaged in all aspects of the relationship. Compassion should not be so dominant that it clouds judgment and prevents effective and efficient provision of nursing services. Compassion needs to be tempered with rationality.

DISCERNMENT. Discernment has been described as practical wisdom or common sense. This is the ability to understand, to have insight into the hidden, as well as the obvious, elements of a situation. A nurse who has discernment is one who is sensitive to the patient's actions and responses and does not necessarily accept what is seen at face value. Verbal and nonverbal communication, as well as concrete signs and symptoms, all contribute to the overall evaluation of the patient by the discerning nurse. This is sometimes translated as practical wisdom because this nurse has a depth of understanding of patients that leads to the selection of the appropriate action, which in turn is implemented in a caring manner.

TRUSTWORTHINESS. Trustworthiness is essential for patients to be able to rely on those professionals who care for them. Certain nurses may be preferred by patients because they convey a sense of confidence with the care they provide. In many situations, patients' lives are truly in the nurse's hands. With a trustworthy nurse, stress and anxiety over the illness are reduced and coping is improved. In fact it is the erosion of trust that is cited as a major factor in the escalating lawsuits that have been initiated against health professionals.

RESPECTFULNESS. Respectfulness is an attitude of a nurse toward the patient that indicates valuing that patient as a unique individual. All patients should be treated with dignity and their autonomy acknowledged. Nurses are obligated to make an effort to identify their own biases and prejudices and work to avoid stereotyping and "isms" in patient relationships—classism, sexism, racism, ageism, and ethnocentrism—as well as eliminate discrimination based on religion, sexual orientation, or disability. Attentiveness to bias and discrimination is not limited to private interactions, but nurses are also responsible for drawing attention to unacceptable statements or negative actions that reflect disrespect for race, religion, gender, class, age, culture, sexual orientation, or disability in any health care situation.

Rights

Rights represent at least two ways to think about what we are owed or what we deserve. First, some rights make claims about goods and services (right to clean water, right to food). These are positive rights, and because something needs to be provided, there is a responsibility for someone to furnish these items. Second, other rights can be determined to protect us (right to privacy, right to self-determination). These are negative rights preventing some action that would intrude on our lives or prevent us from acting as we choose.

Laws guarantee some rights. Others are moral rights based on values and ethical principles but are not enforceable by law. "Basic human rights" is a common phrase that we hear when discussing the condition of various people around the world, especially when those rights are compromised. The United Nations has a document, "The Universal Declaration of Human Rights," that serves to represent what all people should be provided with or protected from.

In health care, discussions of rights often include the concept of patients' rights. For many years, patients were seen simply as passive recipients of whatever treatments or actions professionals determined necessary for their conditions. Now, professionals recognize patient autonomy and patients' active participation in health care. The American Hospital Association (AHA) first devised a patients' bill of rights in 1973, which formally began a recognition of what patients are entitled to but may not always receive. The revised bill includes statements on confidentiality, informed consent, and the right to refuse treatment. Many health care organizations followed with other more specific bills of rights, such as those developed by nursing homes and veterans' hospitals.

In bioethics many issues can be framed within a rights context. An important rights issue is whether people have a "right" to health care. Such a right is discussed at every level of society, from local governments that determine services they will provide in city clinics and public schools to the federal government, which periodically grapples with the debate on national health insurance. Another prominent dispute is the "right to die" with dignity, which arouses more interest as a larger segment of society reaches the later decades of life. The threat of the loss of autonomy and the possibility of being subjected to endless, painful technological interventions while dying helps fuel this concern. "Right to life" is another central concept in our society as groups organize politically to prevent abortions and overturn the *Roe v. Wade* decision of the Supreme Court. This right also extends to discussions of reproductive rights and the health care of pregnant women. These are but a few examples of rights issues and conflicts. Others can be identified as various areas in medical-surgical nursing are explored.

Building Blocks of Ethics

The discipline of ethics itself provides us with knowledge that can be applied to situations in health care. The professional's understanding of these basic concepts, presented here in the form of ethical principles and ethical theories, helps specifically target the ethical aspect of the problem. Principles and theories provide some insight into the ethical dimension. Knowledge about ethics cannot in itself solve the problem or dilemma. However, knowledge of ethics helps focus on the ethical aspect and aid in grounding ethical decision making in ethical rationale. Presenting a particular position relative to a situation or describing values, beliefs, and traditions using these common ethical terms helps clarify discussions and possibly prevents escalating arguments. Problems with communication, management of the unit, and legislation or the law can more easily be separated from the ethical dimension and resolved in their own problem-solving session.

Ethical Principles

The ethical principles that are widely used when examining bioethical dilemmas include autonomy, **beneficence,** nonmaleficence, and justice. Given these ethical principles' prominence in the bioethical literature, a basic understanding of them is necessary.

AUTONOMY. Autonomy is the right of self-determination, independence, and freedom. *Autonomy* refers to patients' rights to make health care decisions for themselves, even if health care providers do not agree with those decisions. Preventing patients from making autonomous decisions or deciding for patients without regard for their preferences is **paternalism.** Autonomy also encompasses the professional's self-determination and freedom.

Autonomy, as with most rights, is not an absolute right, and under certain conditions, limitations can be imposed on autonomy. Generally these limitations occur when an individual's autonomy interferes with the rights, health, or well-being of others. For example, patients generally have an autonomous right to refuse any or all treatments. This autonomous right is guaranteed by federal legislation known as the Patient's Self-Determination Act. However, in the case of contagious diseases that affect society, such as tuberculosis (TB), an individual can be forced by the health care and legal systems to take medications to cure the disease. Individuals can also be quarantined to prevent the spread of disease.

BENEFICENCE. Beneficence views the primary goal of health care as doing good for patients. In general, the term good includes more than just technically competent care for patients. Good care requires a holistic approach to patients, including respect for their beliefs, feelings, and wishes, as well as those of their family and significant others. A problem sometimes encountered in implementing the principle of beneficence is in determining what exactly is good for another and who can best make the decision about this.

NONMALEFICENCE. Nonmaleficence is the requirement that health care providers do no harm to their patients, either intentionally or unintentionally. This is one of the oldest obligations in health care, dating back to the Hippocratic oath (fourth century B.C.). In a sense it is the opposite side of the coin of beneficence, and it is difficult to speak of one concept without mentioning the other. In current health care practice the principle of nonmaleficence is often violated in the short run to produce a greater good in the long-term treatment of the patient. For example, a patient may undergo a painful and debilitating or mutilating surgery to remove a cancerous growth, thereby avoiding death and prolonging life.

By extension, the principle of nonmaleficence also requires a nurse to protect from harm those who cannot protect themselves. This protection from harm includes groups such as children, the mentally incompetent, the unconscious, and those who are too weak or debilitated to protect themselves.

JUSTICE. Justice includes both fairness and equality. Concerns for justice encompass both material resources (**distributive justice**) and the fair treatment of individuals, groups, and communities (psychologically, socially, legally, politically). When the patient makes an appointment for 9 A.M. at the outpatient clinic and his or her neighbor makes a 10 A.M. appointment, each expects to be seen by the primary care provider at the designated time unless some emergency prevails. Unequal treatment would result if a walk-in who has no pressing problem is seen by the provider in place of the 9 A.M. appointment, forcing each subsequent appointment to be delayed beyond the pre-arranged time. Distribution of material resources can be complex because it involves not only benefits (what we receive) but also burdens (what we may be taxed for but then do not receive). Burdens are not just monetary, but also include such factors as the unequal participation of individuals in medical research and the sacrifices family members make when caring for disabled individuals in the home.

One of the most serious limitations of these principles is the lack of any built-in priority when applying them to an ethical dilemma. Autonomy is not automatically prioritized over justice, or beneficence over nonmaleficence. However, these principles are helpful in categorizing various preferences and positions when examining a dilemma, and their categorizing can lead to a clarification of various disagreements among participants in the ethical discussion. Working with principles moves the discussion to a focus on ethics rather than a particular personal viewpoint or feeling. Such a strategy can also avoid a power struggle between participants who simply want to win the argument. An example of an ethical dilemma is the following scenario: A nurse attempts to support a patient's refusal of surgery, while the physician claims that the patient must have surgery or she will lose her leg. Realizing that the nurse is arguing the case from the perspective of autonomy, whereas the physician's desire to perform surgery is attributable to beneficence, moves the discussion to one based on conflicting principles rather than conflict between individuals. From this point of clarification, the discussion can focus on autonomy and beneficence and the rationale for the possible prioritizing of one principle over the other. This strategy does not resolve the dilemma, but it makes it less personal and forces participants to develop sound, ethical rationales for their solutions.

Ethical Theories

Another helpful device for examining ethical dilemmas is the use of various ethical theories. Ethical theories are concepts that are more complete than principles for analyzing ethical dilemmas. However, only a brief description is included here; a more in-depth understanding can be obtained from other resources, such as the Web links.

Two major theories used in bioethics are defined first, and then a subcategory of theories collectively known as feminist theories are introduced; finally, recognition of the relationship of theology or religion to bioethics completes this section.

UTILITARIANISM. In **utilitarian** theory the consequences or the outcomes of the dilemma are the most important elements to consider in decision making. Actions are morally preferred if they produce more happiness or greater benefits than unhappiness or burdens. The cost-benefit analysis is a very common approach used by institutions and organizations. A hospital that has the responsibility to care for hun-

dreds of patients is not as concerned with the individual patient who is caught in the bureaucracy of its functioning. This is not to say that all institutions operate on this theory at all times, but in general, rules, policies, and procedures are developed with the majority in mind. The sacrifice of the individual for the good of the majority is a significant criticism of this theory.

DEONTOLOGY. **Deontology** is a philosophical theory in which rules and guides for behavior and decisions regarding moral action are formed based on a search for the definition of duty. Ethical decisions are determined irrespective of the outcomes or consequences. For example, health professionals might operate by a rule that indicates that a moral person never lies. No matter how much the truth might hurt, the truth is told. Another rule might be to never use people as a means to an end. Translated this means that regardless of the benefits, individuals cannot be forced to participate in medical research studies to benefit others. An individual's right to voluntarily participate in research must be respected. Research does not have to benefit the individual participant as long as this is understood by the individual. However, the individual cannot be used simply to meet the investigator's needs. Acting morally only because one has a duty to do so, without any consideration of the outcome, is a serious limitation of this theory.

FEMINIST THEORIES. **Feminist** approaches focus on gender bias, discrimination, and prejudice in addition to the more serious actions of oppression and violence against women. These are important because of power differentials among the providers and recipients of health care, as well as the previous treatment of normal development, injury, illness, disease, and dying in women. Feminist theories are really a group of various approaches to ethical problems and decision making rather than one uniform approach. Feminists do use utilitarian and deontological theories, as well as various other approaches, to develop a systematic organization of ethical rules and principles. Some feminist theories focus on feminine attributes (emotional attachment, caring, prominence of relationships, and connection with people), emphasizing their importance in ethical decision making. In general, those who support this approach observe the devaluing of these attributes in society. Caring, as an example of one attribute, is important personally, but supporters recognize that caring is not systematically rewarded in our society.

Other theories are more politically oriented and concentrate their energies on developing approaches to ethical decision making that minimize the subordination and marginalization of women, whatever other principles they support. For example, in a liberal approach to ethical decision making, human beings have a political responsibility to organize and maintain society based on freedom and the various rights that we have. Rules protect our privacy. From a feminist perspective, the liberal approach focuses on making sure that women are included in the political processes when rules are made, helping to develop legislation, which serves to shape the communities in which we live. In health care a liberal feminist approach focuses on providing women with educational and promotional opportunities for leadership positions in health care systems. In leadership positions, women have the power and authority to make the health care system more accommodating for women.

THEOLOGICAL PERSPECTIVES. Theological perspectives include the many religious traditions represented in our culture. Religious teachings are key concepts for ethical decision making for some individuals. In fact, for many individuals these teachings are the primary source of values and morals. Jehovah's Witness believers' rejection of blood transfusions is a common example of how religious beliefs affect health care decision making. This religious group has collected a large amount of information about blood substitutes and alternative therapies. Leaders of Jehovah's Witnesses are prepared to provide education for health professionals about their beliefs and acceptable interventions. In another area, a number of religious traditions oppose abortion, which affects both health professionals and patients. Euthanasia is an issue addressed by numerous organized religious groups, which in turn affects how patients make decisions regarding end-of-life issues. One of the difficulties with religious traditions is that it is not simply the official church teaching that is involved, but the individual member's interpretation of that teaching. Assessment of the importance of this dimension of the patient's life is important in the ethical analysis.

Ethical Decision Making

A variety of models and frameworks for ethical decision making are available. The steps listed in this section are a combination of several ideas that have been suggested. In its simplest form, ethical decision making is a problem-solving process. Similar to the nursing process, as discussed in the first chapter, the steps described in this section take the user through a set of strategies that assist in approaching a problem in an organized and systematic manner. Nurses applying these steps use critical thinking skills in order to be as logical and objective as possible. This is not to say that feelings and emotions are to be put aside. Bonding, love, commitment, and other sentiments define our human condition. However, a balanced perspective is the goal, which includes respecting these feelings without allowing them to overshadow the process and the outcome. Although making the final decision may not be easy, the user knows that once these steps have been completed, the situation has been thoroughly examined and all reasonable aspects explored before the decision was made.

Step 1: Gather and Verify the Information

Many kinds of information can contribute to an in-depth understanding of the situation, which provokes an ethical dilemma. Basically one needs to know who is involved, what is involved, and the context of the problem. Foremost is information about the patient, what the patient says when conscious and competent. Determining competency can sometimes be a challenging process. The chart is an obvious resource for clinical data, but so are the health professionals who are assigned to the patient, because not every

individual records all that is known about the patient. When appropriate, health care records from previous hospitalizations may be helpful, as well as charts from other institutions and agencies. External factors also need to be included, such as legal aspects, governing policies and procedures of the institution, and available resources. The amount of information gathered depends on the time frame and urgency of the decision.

Step 2: Clarify the Values of All the Participants

The values, beliefs, and traditions of all participants are important, especially the patient's, but more so when the patient is unconscious or incompetent and a group is attempting to reach a consensus based on some understanding of the patient. Advance directives or a living will prepared by the patient are often helpful as guides when the patient can no longer express personally held ethical beliefs. Determining the ethical orientation of all participants clarifies their perspectives and moves decision making to a more objective level. Decision makers must be especially aware of assumptions, stereotyping, biases, and prejudices related to socioeconomic status, ethnic identification, gender, religious preference, and other common characteristics of the participants.

Step 3: Identify the Ethical Dilemma and the Conflicts in Values

Separating the ethical aspects from the nonethical aspects facilitates the decision-making process. For example, a conflict can arise as a result of gaps in communication rather than a conflict of ethical principles. When the communication problem is solved, the dilemma is resolved, and as it turns out, there is no ethical dilemma. Sometimes both nonethical and ethical aspects need to be addressed in parallel fashion to reach a resolution of the situation. The importance of various ethical values for each participant should be explored.

Step 4: Examine Possible Actions and the Consequences of Each

There is no magic number of "possible actions," but it is easy to pose solutions at the extremes of possible actions. Take informing the patient as an example: The two obvious solutions are to tell and not to tell. But clinical practice is not that simple, and it is more useful to also pose solutions that are between the extremes. Suppose the issue is a critically injured patient who wants to know the condition of her spouse and child, who were also in the automobile accident. Her spouse was killed, and her child is in critical condition. At the moment her own status is unstable, and telling her might have a deleterious effect on her own health. Telling her is certainly an option, just one with certain limitations at the moment. Not saying anything in response to her question is a possibility but seems to show a lack of respect for her dignity as a person. What might be some possible actions in between those extremes? Is there some response that tells part of the information while avoiding the worst part, her spouse's death? Even if the partial truths seem unacceptable at first glance, they ought to be added to the list of possible actions. Examination of a range of possible actions aids in identifying where the line is drawn for determining morally acceptable choices versus those not morally acceptable or for creating a list of each choice's risks and benefits.

Step 5: Determine the Ethical Foundation for Each Action

Each action should be based on selected ethical values, principles, or theories. Perhaps the code of ethics lends some support as ethical rationale. One strategy that is proposed on occasion, especially for an incompetent or unconscious patient, is an application of the standard of best interest. The best interest standard involves the determination of what action is in the best interest of the patient, given the information that is known about the patient and the situation. Family members together with health care providers usually make the best interest determination. Ideally this decision is made in an objective manner, setting aside any special interests of the family or health professionals. For example, it may be in the best interest of the patient to be discharged and go home so that recovery can be optimized in that setting. Family members themselves may have other goals that make placement in a long-term care facility better for them. Health professionals may be pressured by administrative personnel to simply arrange a discharge, regardless of where the patient might be discharged to or what might be best for the patient, because of utilization review and financial reimbursement considerations. Under the best interest standard, the optimal placement is in the home because it is best for the patient.

Step 6: Determine the Best Action with the Strongest Ethical Support

Each action is judged based on its risks, benefits, and supporting rationale. The actions are then ranked in order of their priority. Strong ethical support for the first priority is required, as well as a reasonable potential for the action to be implemented. For example, if placement in a nursing home after discharge from the acute care facility is ethically the best solution for a patient, but there is no room available in the potential facilities, that action must be set aside. If the preferred action fails, it is possible to move to choice number 2. The designation of these priorities does not necessarily mean that their supporting ethical rationale always results in the same ranking, nor does it mean that principles or theories are always in opposition to one another.

Utilitarians and deontologists may disagree about the important principles in decision making. However, it should not be assumed that different theorists cannot reach a mutually agreeable decision. Although their rationale for the decision may differ, the final solution each proposes can be identical. A case in point involves the relief of pain and suffering. The deontologist might argue that there is a duty among health professionals to provide pain medications and make patients comfortable. This action shows respect for the dignity of the individual. A utilitarian might argue that pain and suffering should be relieved because doing so creates a positive outcome: the patient is more comfortable without pain. In an even broader context, the more pain and suffering that is relieved, the greater the number of patients who benefit. So both theorists support the same action, but for different reasons.

Even theorists, such as utilitarians, can disagree among themselves. A classic case involving confidentiality was solved based on utilitarian concepts, yet both the ruling vote and the dissenting vote were based on utilitarian principles.[1] The majority opinion of the court indicated that more good would be created by disregarding confidentiality obligations because overriding the patient's confidentiality would have saved the life of a third person who the patient threatened to murder. The minority opinion of the court stated that confidentiality should not be breached because it gave society the message that psychiatrists would not keep patient confessions secret and therefore would inhibit most individuals who have psychiatric problems from seeking professional help.

Step 7: Implement the Action

Needless to say, the selected action needs to be carried out. Responsibilities for the professionals and others in the situation can be assigned. Assigning a coordinator for the implementation of the action, especially if several individuals have responsibilities for portions of the action, enhances the potential for success, ensuring that someone will check on the process of implementation and that the action is evaluated.

Step 8: Evaluate the Outcome

We can learn from success and failure. Successful resolution of an ethical dilemma provides us with knowledge for the next ethical dilemma. Although no two dilemmas are exactly alike, selected features of a particular dilemma may parallel a subsequent situation and assist in solving the new dilemma. Failures also teach us a lesson. Retracing all the steps of the ethical decision-making process may provide insight and greater understanding of what steps of the process could have been better implemented. Evaluation is also important because the next time we confront a similar dilemma we may not have the luxury of examining the new dilemma in as much detail. Prior experience enables us to face new dilemmas more confidently and to determine the better action more quickly. One caution, however: It is very important to evaluate the outcome when decisions are made and the action taken within a limited time frame.

CRITICAL THINKING

Ethical Decisions

1. Identify a health care–related ethical dilemma you have encountered as a student. How did you solve the dilemma? What expert resources did you use?
2. Apply the ethical decision-making model to the ethical dilemma. How are your decision-making process and proposed actions different when using the model?

Summary

What lies ahead? We cannot predict exactly, but as a nurse you are sure to encounter ethical issues on a regular basis. Now that a selected knowledge base has been acquired, the next step is to practice decision making skills.

▶ LEGAL CONCEPTS

We are all members of a society. To promote harmony, safety, and productivity, members of society create rules. The rules of society can be informal or formal. An informal social rule, for example, is opening a door for someone. A criminal statute (law) is an example of a formal rule of society. Social rules or codes promote our social well-being. It would be unsafe to live in a community that existed without rules. All societies must require minimum standards of conduct for their members. Laws are the governmental mandates of a society that define individuals' duties to themselves, their neighbors, and the government. The failure to adhere to laws can result in punishment that may include imprisonment or monetary fines.

Regulation of Nursing Practice

Nursing is a licensed health care profession. Nurses must be licensed by their state to practice nursing. The rationale for state licensure is to improve the quality of health care services and to protect the public. As such, state governments have created licensing boards to establish the entry-level requirements for nurses. These licensing boards also establish regulations that define the scope of appropriate nursing practice for licensed nurses. These licensing regulations are found in state nursing practice laws and within regulations that are made by the licensing agency.

The state nursing practice laws and the attendant nursing regulations establish the parameters within which nurses must practice to obtain and maintain state license. These regulations are referred to as **administrative laws.** These considerations and mandates can be the basis for disciplinary actions by the licensing body. The failure to adhere to the regulatory mandates of the nursing licensing body can result in the loss of the privilege to practice nursing. Unprofessional conduct and conviction of a crime are examples of possible violations of nursing regulations.

Nursing Liability and the Law

Liability refers to the level of responsibility that society places on individuals for their actions. In recent years this responsibility has been interpreted to mean the financial responsibility owed to those who are injured by wrongful actions. Laws establish liability or responsibility for these wrongful actions. Following the law is a major part of the practice of nursing.

Administrative laws establish the licensing authority of the state to create, license, and regulate the practice of nursing. **Criminal law** regulates behaviors for citizens within this country. **Civil law** provides the rules by which individuals seek to protect their personal and property rights.

Criminal and Civil Law

All individuals, regardless of their occupations, are required to obey the criminal laws of the government. Criminal laws establish the rules of social behavior and define the punishment for the breaking of those rules, which can result in imprisonment or monetary fines.

Criminal law is different from civil law in the nature of remedies that are used for punishment. A crime is viewed as an action taken by an individual against society that the government will prosecute and punish. The breaking of a criminal law may result in criminal punishment and civil liability. For instance, an intoxicated driver may go to jail for a crime and also be held civilly liable for any personal injury that resulted. Examples of criminal acts are assault, battery, rape, murder, and larceny.

Civil laws dictate how disputes are settled among individuals and how liability is assigned for wrongful actions. For health care workers, civil liability is a constant concern. The potential for civil liability is demonstrated by an increased use of health care procedures such as diagnostic testing, which results in higher health care costs. Civil liability is a method by which a patient can seek financial recovery for injuries and losses caused by the wrongful action or lack of action by a health care worker.

A civil liability suit begins with the filing of a complaint with a court. A copy of the complaint must be given or served by the plaintiff to the defendant. A **summons,** which is a notice to defendants that they are being sued, is attached to the complaint. The complaint describes the claim being made by the plaintiff, and the summons instructs the defendant that the complaint must be answered within a specified period, usually 20 to 30 days. Nurses served with a summons relating to work should notify their employers. These important documents must be taken seriously. The nurse must make sure that the summons is answered. If the employer does not answer the summons, the nurse must seek legal counsel to answer the summons within the specified time. If the nurse fails to answer the summons and complaint, it may result in a default judgment, which is acknowledgment of liability.

Civil wrongs caused by the act or omission of a health care worker can be physical, emotional, and financial in nature. The person claiming a civil cause of action and injury is the plaintiff, and the person alleged to have caused the injury is the defendant. Lawsuits involving civil wrongs are called **torts.** The institution that employs the worker may also become liable for the acts or omissions of its employees. This theory of law is called **respondeat superior.** It is important for employees to understand that their work may result in civil liability for their employers.

Civil or tort liability for health care workers can be based on intentional actions, unintentional actions, and even the omission of action. **Malpractice** may be defined as a breach of the duty that arises out of the relationship that exists between the patient and the health care worker. This term includes liability that may arise from intentional torts and unintentional torts. Intentional torts are lawsuits wherein the defendant is accused of intentionally causing injury to the plaintiff. Examples of intentional torts are assault, battery, defamation, false imprisonment, outrage, invasion of privacy, and wrongful disclosure of confidential information (Table 2–1).

Negligence

An unintentional tort is known as **negligence.** Negligence occurs when injury results from the failure of the wrongdoer to exercise care. This failure to follow due care in the protection of the person injured is referred to as a breach of duty. Professionals owe a higher duty of care to their patients. The failure of a health care professional to follow a prescribed duty of care is called malpractice. Professional negligence, therefore, is referred to as malpractice (Box 2–1). All professionals, including LPN/LVNs, are responsible for their own actions, whether they are intentional or negligent in nature. Although the employing agency is also responsible for the actions of its employees, employees always remain responsible for their own actions as well.

Limitation of Liability

All professions are concerned with the **limitation of liability.** This means that health care professionals can do some things to limit their individual and institutional liability. Some examples of liability limitations are ensuring patient rights, accurately documenting procedures, following institutional policies, acquiring individual malpractice or liability insurance, pursuing continuing education, and practicing in accordance with the current standards of the nursing profession. Several states have enacted tort reform legislation. Much of this legislation is directed at limiting liability for health care professionals and institutions. Examples of this reform legislation are limitations on the dollar amount allowable for a patient's damages, shortening the time in which a patient can file a lawsuit, and requiring stringent expert medical evaluation of a claim before a lawsuit can be filed.

BOX 2-1

Components Necessary for a Finding of Negligence
1. A duty of care owed to patients
2. A breach of duty to exercise care
3. Injury and damages occurring from this breach of duty

All patients are entitled to quality care and treatment with dignity and respect. To provide quality care and limit liability, understand and provide the rights your patient is entitled to and question directions that are controversial, given verbally, concern situations of high liability, or involve a discrepancy between the direction and standard policy. Rights are defined as something due an individual according to just claims, legal guarantees, or moral and ethical principles. **Welfare rights,** also called legal rights, are rights that are based on a legal entitlement to some good or benefit. These rights are guaranteed by laws such as the Bill of Rights and if violated can come under the powers of the legal system. For example, citizens of the United States have a right to equal access to employment regardless of race, sex, or religion. The American Hospital Association drafted a patients' bill of rights in 1973. The type of treatment and care a patient has the right to expect is outlined. Congress is currently involved with federal legislation related to a patients' bill of rights. Stay informed on the status of this legislation, which you will be expected to follow. Knocking before entering a patient's room and introducing oneself to the patient are examples of a patient's rights.

Documentation is a legal record of your actions. Document nursing actions based on orders given, as well as the name and title of the person who gave the verbal direction. Documentation must be clear, honest, and accurate. Always practice at a level that is generally accepted by the nursing profession. Failure to adhere to acceptable practice standards is cause for concern and can create potential liability for both you and the health care institution.

It is important to understand that some employers do not provide malpractice insurance for their nursing employees. Always ask an employer exactly who is covered under the employer's liability insurance. If nursing employees are covered, the employer's insurance will provide coverage from liability only as long as the employee follows the employer's work policies. For this reason, employer-provided liability insurance is not personal liability insurance. As a result, LPN/LVNs often carry personal liability insurance.

TABLE 2-1	**INTENTIONAL TORTS**
Assault	Unlawful conduct that places another in immediate fear of an unlawful touching or battery; real threat of bodily harm
Battery	Unlawful touching of another
Defamation	Wrongful injury to another's reputation or standing in a community; may be written (libel) or spoken (slander)
False Imprisonment	Unlawful restriction of a person's freedom
Outrage	Extreme and outrageous conduct by a defendant in the care of the client or the body of a deceased individual
Invasion of Privacy and Wrongful Disclosure of Confidential Information	Liability when a client's privacy is invaded physically or when records are released without authority

REVIEW QUESTIONS

1. As the nurse provides care to patients, it is important to have an understanding of ethics for which of the following reasons?

   a. Health care resources are not limited.

   b. Technological interventions are always helpful.

   c. Health care systems are becoming more simplified.

   d. Human development has the potential for health crisis at every stage.

2. The nurse is planning quality care for a patient without regard to race, ethnicity, and gender. Which of the following ethical obligations does this exhibit?

   a. Fidelity

   b. Integrity

   c. Respectfulness

   d. Compassion

3. The nurse is caring for a patient whose family requests that a feeding tube be inserted. The patient has an advance directive indicating that a feeding tube is not to be inserted. As the nurse and physician consider what is best for the patient in this ethical dilemma, which of the following principles is represented?

   a. Autonomy

   b. Beneficence

   c. Nonmaleficence

   d. Justice

4. The nurse has a question about how his nursing license is regulated. In which of the following documents will the nurse find this information?

   a. An institutional policy

   b. Nursing ethics code

   c. State nursing practice laws

   d. National nursing standards

5. In providing professional nursing care, the nurse understands that the law of negligence requires which of the following to create liability?

   a. A crime

   b. Assault

   c. Ethical violations

   d. A breach of duty

REFERENCES

1. *Tarasoff v. Regents of the University of California*, 17 Cal.3d 425, 1976.

3

CULTURAL INFLUENCES ON NURSING CARE

Larry Purnell

KEY TERMS

beliefs (bee-**LEEFS**)

cultural (**KUL**-chur-uhl)

cultural awareness (**KUL**-chur-uhl
a-**WARE**-ness)

cultural competence (**KUL**-chur-uhl
KOM-pe-tens)

cultural diversity (**KUL**-chur-uhl
di-**VER**-si-tee)

cultural sensitivity (**KUL**-chur-uhl
SEN-si-**TIV**-i-tee)

cultures (**KUL**-churs)

customs (**KUS**-tums)

ethnic (**ETH**-nick)

ethnocentrism (**ETH**-noh-SEN-trizm)

generalizations (**JEN**-er-al-i-ZAY-shuns)

stereotype (**STER**-ee-oh-**TIGHP**)

traditions (tra-**DISH**-uns)

values (**VAL**-use)

worldview (**WERLD**-vyoo)

QUESTIONS TO GUIDE YOUR READING

1. What are the meanings of the basic terms and concepts common to culture and ethnicity?

2. What are your own cultural characteristics, values, beliefs, and practices?

3. What are attributes of culturally diverse patients and their families and how do they influence nursing care?

4. What data should you collect from culturally diverse patients and their families?

5. How would you provide a holistic approach to patient care according to cultural characteristics and attributes?

Cultural diversity in the United States is increasing. As a result, cultural and ethnic differences between nurses and their patients are becoming more evident and must be recognized. Thus there is a need for nurses to become knowledgeable about cultures other than their own. This chapter provides you with the basics of culture and its impact on health promotion and wellness.

▶ CULTURE DEFINED

Culture refers to the socially transmitted behavior patterns, beliefs, values, customs, arts, and all other characteristics of people that guide their worldview. Cultural beliefs, val-

ues, customs, and traditions are primarily learned within the family on an unconscious level. They are also learned from the community in which one lives, in religious organizations, and in schools. All individuals and groups have the right to maintain their cultural practices as they deem appropriate. Because culture has powerful influences on a patient's interpretation of health and responses to nursing care, valuing diversity in nursing practice enhances the delivery and effectiveness of care (Fig. 3–1). As you accumulate knowledge about specific ethnic and cultural groups, you will be challenged to look at the differences and similarities across cultures.

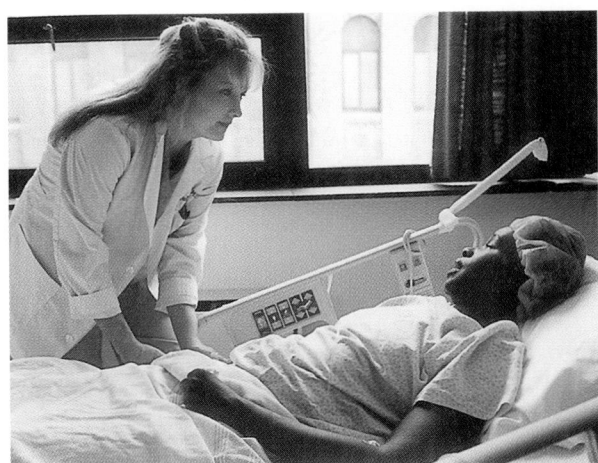

Figure 3-1 The nurse must assess patients' special needs related to their cultural backgrounds.

The terms **cultural sensitivity, cultural awareness,** and **cultural competence** have different meanings. Cultural sensitivity is knowing politically correct language and not making statements that may be offensive to another person's cultural beliefs. Cultural awareness focuses on history and ancestry and emphasizes an appreciation for and attention to arts, music, crafts, celebrations, foods, and traditional clothing. Cultural competence includes the skills and knowledge required to provide nursing care and has at least four components. The first component is having an awareness of your own culture and not letting it have an undue influence on the patient. The second is having specific knowledge about the patient's culture. The third is accepting and respecting cultural differences. The fourth is adapting nursing care to the patient's culture.

Even though you may have knowledge about another culture, **ethnocentrism** can still pervade your attitudes and behavior. Ethnocentrism is the universal tendency for human beings to think that their ways of thinking, acting, and believing are the only right, proper, and natural ways. Ethnocentrism perpetrates an attitude that beliefs that differ greatly are strange or bizarre and therefore wrong. Ethnocentrism can be a major barrier to providing culturally competent care. Additionally, you must be careful not to **stereotype** a patient. A stereotype is an opinion or belief ascribed to an individual. For example, the statement "all Chinese people prefer traditional Chinese medicine" is a stereotype. This stereotype is not true. Although many Chinese people may prefer traditional Chinese medicine for some health conditions, not all Chinese people prefer traditional Chinese medicine. Some Chinese people prefer the Western medicine that is practiced in the United States.

However, one can still make **generalizations** about an ethnic individual without stereotyping. Whereas a general-

ization, or assumption, may be true for the group, it does not necessarily fit the individual. Therefore you must seek additional information to determine whether the generalization fits the individual. The challenge is to understand the patient's cultural perspective. If you have specific cultural knowledge, you can improve therapeutic interventions by becoming a co-participant with patients and their families. To do this, you must develop a personal, open style of communication and be receptive to learning from patients from cultures other than your own.

▶ CHARACTERISTICS OF DIVERSITY

Primary and secondary characteristics of diversity affect how people view their culture. The primary characteristics of diversity include nationality, race, skin color, gender, age, and religious affiliation. Secondary characteristics include socioeconomic status, education, occupation, military experience, political beliefs, length of time away from the country of origin, urban versus rural residence, marital status, parental status, physical characteristics, sexual orientation, and gender issues.

Cultural Groups in the United States

This chapter describes selected attributes of some of the cultural groups in the United States. These groups are European-Americans, Native-Americans, African Americans, Spanish/Hispanics/Latinos, Asian/Pacific Islander–Americans, Arab-Americans, and Appalachians.

> **Cultural Self-Assessment Exercise**
> - Identify your primary and secondary cultural characteristics. How do they affect your worldview?
> - Share these views with others in your class.

Attributes presented for each group include communication styles, family organization, nutrition practices, death and dying issues, health care beliefs, and traditional health care practitioners. Traditional health care practitioners are those practitioners from the patient's native culture, such as shamans, herbalists, and other traditional healers. Racial and ethnic biologic variations, susceptibility to disease, and genetic diseases are covered elsewhere in this textbook. (See Ethical Consideration Box 3–1 and Gerontological Issues Box 3–2.)

▶ COMMUNICATION STYLES

Communication styles include verbal and nonverbal variations. Verbal communication includes spoken language, dialects, and voice volume. Dialects are variations in grammar, word meanings, and pronunciation of spoken language. Nonverbal communication includes the use and degree of eye contact, the perception of time, and physical closeness when talking with peers and perceived superiors.

In some societies, people are expected to maintain eye contact without staring, which denotes that they are listening and

---

ethnocentrism: ethnos–race + centr–center, self + ism–condition

When dealing with patients from cultural backgrounds that are different from their own, nurses are sometimes faced with the possible risk of imposing their own values and standards in situations that they do not fully understand. The following is an example of that possible hazard.

A woman of Iranian cultural background delivered a still-born baby. After the delivery, she was transferred to the medical-surgical unit for postpartum care, a practice commonly followed when stillbirths occur. Sharon, an experienced licensed practical nurse (LPN), was assigned to care for the patient. While reading through her chart, Sharon noticed that the woman's husband refused to let his wife see the baby. The chart also contained a photograph of the dead baby, an ultrasound picture, and the footprints and hand-prints taken after the birth.

As expected, the woman appeared very depressed, and Sharon felt a great deal of sympathy for her. While caring for this woman, Sharon mentioned that there was a photograph and other mementos of the child and asked if she would like to see them. The woman's husband, who was visiting at the time, called Sharon out of the room and indicated his very strong displeasure that Sharon let his wife know about these pictures without first asking him. He expressed his desire to see the pictures but refused to let his wife see them. He also admonished Sharon to stop talking with his wife and emphasized the fact that Sharon had no understanding at all of their culture.

Sharon completed the shift without additional communication with the woman, who was discharged before Sharon returned the next day for her shift. Sharon was plagued by a feeling that she had made a major error in caring for this woman and had failed to meet the patient's emotional needs.

It is evident from this story that the American conception of the ideal marital relationship is not the same for other cultures, including this patient and her husband from Iran. It would be easy to judge the couple based on American values of individual independence and autonomy. Instead of judging, nurses need to assess the family's values, as well as their role expectations in that particular situation. Nurses should respect each patient's cultural origins and be sensitive to their cultural preferences. These values and preferences cannot be assumed, even when nurses can correctly identify the culture of the family. Cultural stereotyping is a possible mistake. The role of the woman in this Iranian family is different from that in the American culture; roles can also reflect the socioeconomic status and other characteristics of the patient or family. Connecting with the family's culture and strengths increases the nurse's ability to communicate with and help that family.

Aging, Ethnicity, Health, and Illness

In the United States, ethnic minority elders have serious health and well-being issues. Compared with white or European-American elders, ethnic minority elders are more likely to:

- Live in poverty.
- Have a shorter life expectancy.
- Experience debilitating disease processes or functional disability at a higher rate and at an earlier age.
- Have difficulty accessing health care services.

Remember that older adults need to be assessed within their personal cultural context. Avoid generalizing cultural practices to individuals or families without first assessing whether this practice or belief is true for them. For example, it would be wrong to assume that an older Mexican-American woman who lives with her extended family will get the family's support for assistance with bathing and other activities of daily living. If an older Chinese woman uses herbs and folk treatments for common complaints, it does not mean that she will not use the services, treatments, or medications of Western medicine.

events may be flexible, whereas medical appointments and business engagements start on time.

The second dimension of time relates to whether the culture is predominantly concerned with the past, present, or future. Past-oriented individuals maintain traditions that were meaningful in the past and may worship ancestors. Present-oriented people accept the day as it comes, with little regard for the past; the future is unpredictable. Future-oriented people anticipate a bigger and better future and place a high value on change. However, some individuals balance all three views—they respect the past, enjoy living in the present, and plan for the future.

Nursing Assessment and Strategies

Ask the following questions:

- By what name do you prefer to be called?
- What language do you speak at home?

Cultural Self-Assessment Exercise

Whether we realize it or not, everyone has a cultural or ethnic background. This exercise is to help identify your personal beliefs and values, which are largely passed on to you through your family.

- How do you identify yourself in terms of racial, cultural, or ethnic background? From what country did your ancestors originate? Were your parents from the same or similar ethnic backgrounds?
- What stories do you remember that your parents, grandparents, or other relatives told about relocating in the United States? Do you know why they originally came to America?
- How do these stories compare with those of others from similar backgrounds?
- How do these stories compare with those of others from different backgrounds?

Remember, one's values and beliefs are not better than another's—they are just different.

can be trusted. However, in other societies, as a sign of respect, people should not maintain eye contact with superiors such as teachers and those in positions of higher status.

The perception of time has two dimensions. The first dimension is related to clock versus social time. For example, some cultures have a flexible orientation to time and events, and appointments take place when the person arrives. An event scheduled for 2 P.M. may not begin until 2:30 or when a majority of the people arrive. For others, time is less flexible, and appointments and social events are expected to start at the agreed-on time. For many, social

■ Are you normally on time for appointments?

Be sure to do the following:

■ Take cues from the patient for voice volume.
■ Be an active listener, and become comfortable with silence.
■ Avoid appearing rushed.
■ Be formal with greetings until told to do otherwise.
■ Take greeting cues from the patient.
■ Speak slowly and clearly. Do not speak loudly or with exaggerated mouthing.
■ Explain why you are asking specific questions.
■ Give reasons for treatments.
■ Repeat questions if necessary.
■ Provide written instructions in the patient's preferred language.
■ Obtain an interpreter if necessary.

Cultural Self-Enrichment Exercise

• How many languages do you speak? Do you speak a dialect of your dominant language? Does it interfere with communication with your patients?
• Do you speak in a soft, medium, or loud tone of voice? Does this tone change in different situations? How close do you stand when you speak with close friends? Does this distance change when you converse with your teacher, your religious leader, or a politician?
• Identify characteristics from your worldview in terms of being present, past, and future oriented.
• By what name do you prefer to be called? Why? Does this change in different situations?

▶ FAMILY ORGANIZATION

Family organization includes the perceived head of the household, gender roles, and roles of the elderly and extended family members. The head of the household may be patriarchal (male dominated), matriarchal (female dominated), or egalitarian (shared equally between men and women). An awareness of the family dominance pattern is important for determining with which family member to speak when health care decisions have to be made.

In some cultures, specific roles are outlined for men and women. Men are expected to protect and provide for the family, manage finances, and deal with the outside world. Women are expected to maintain the home environment, including child care and household tasks. You must accept that not all societies share or even desire an egalitarian family structure.[1]

Roles for the elderly and extended family vary among culturally diverse groups. In some cultures the elderly are seen as being wise, are deferred to for decision making, and are held in high esteem. Their children are expected to provide for them when they are no longer able to care for themselves. In other cultures, although the elderly may be loved by family members, they may not be given such high regard and may be cared for outside the home when self-care becomes a concern.

The extended family is very important in some groups, and a single household may include several generations living together out of desire rather than out of necessity. The extended family may include both blood-related and non-blood-related individuals who are provided with family status. For others, each generation lives separately and has its own living space.

Cultural Self-Enrichment Exercise

• Who is considered the head of the household in your family?
• Are there specified gender roles for family members?
• What are the roles of the elderly in your family?
• Do you identify with an extended family? Are they all blood relatives? What roles do they play?
• What kind of decisions do men make and what kind of decisions do women make?

Nursing Assessment and Strategies

Ask the following questions:

■ Who makes the decisions in your household?
■ Who takes care of money matters, does the cooking, or is responsible for child care?
■ Who decides when it is time to see a health care provider?
■ Who lives in your household? Are they all blood related?

Be sure to do the following:

■ Observe the use of touch between family members.
■ Allow family members to decide where they want to stand or sit for comfort.

▶ NUTRITION PRACTICES

Nutrition practices include the meaning of food to individuals, food choices and rituals, food taboos, and how food and food substances are used for health promotion and wellness. Cultural beliefs influence what people eat or avoid. Food offers security and acceptance; is necessary as a means of survival and relief from hunger; plays a significant role in socialization; has symbolic meaning for peaceful coexistence; is used to promote healing; denotes caring, closeness, and kinship; and may be used as an expression of love or anger.[2]

Culturally congruent dietary counseling, such as changing amounts and preparation practices and including ethnic food choices, can reduce health risks. Whenever possible, you should determine a patient's dietary practices during the intake interview. Culturally diverse patients may refuse to eat on a schedule of American mealtimes or eat American foods. Counseling about food group requirements, intake restrictions, and exercise that respects cultural behaviors is essential.

Most cultures have their own nutritional practices for health promotion and disease prevention. For many, a balance of different types of foods is important for maintaining health and preventing illness. Common folk practices recommend specific foods during illness and for prevention of illness or disease. Therefore a thorough history and assessment of dietary practices can be an important diagnostic tool to guide health promotion.

Cultural Self-Enrichment Exercise

- What is the meaning of food in your culture?
- Are there any dietary deficiencies or limitations in food for you?
- What cultural or ethnic foods do you prepare at home?
- When you eat out for lunch or dinner, what are your favorite ethnic foods? Which ethnic foods do you not like? Why?
- What are some of the dietary practices you engage in when you are ill?
- What kinds of foods do you eat to stay healthy?
- Are there any taboo or restricted foods in your culture or personal belief system?

Nursing Assessment and Strategies

Ask the following questions:

- What do you eat to stay healthy?
- What do you eat when you are ill?
- Are there certain foods that you do not eat? Why?
- Do certain foods cause you to become ill? What are they?
- Who prepares the food in your household?
- Who purchases the food in your household?

▶ DEATH AND DYING ISSUES

Death rituals of cultural groups are the least likely to change over time. To avoid cultural taboos, you must become knowledgeable about rituals surrounding death and bereavement. For some, the body should be buried whole. Therefore an amputated limb may be buried in a future grave site, and organ donation would probably not be acceptable. Cremation may be preferred for some, whereas for others it is taboo and burial is the preferred practice. Views on autopsy vary accordingly. Some cultural groups have elaborate ceremonies that last for days in commemoration of the dead. To some individuals these rituals appear to be a celebration, and in a sense they are—a celebration of the person's life rather than a mourning of the person's death.

The expression of grief in response to death varies within and among cultural and ethnic groups. For example, in some cultures, loved ones are expected to suffer the grief of death in silence, with little display of emotion. In other cultures, loved ones are expected to demonstrate an elaborate display of emotions to show that they cared for the individual. These variations in the grieving process may cause confusion if you perceive some individuals as overreacting and others as not caring. You must accept that there are culturally diverse behaviors associated with the grieving process. Bereavement support strategies include being physically present, encouraging reality orientation, openly acknowledging the family's right to grieve as they need to, assisting the family to express their feelings, encouraging interpersonal relationships, promoting interest in a new life, and making referrals to other staff and spiritual leaders as appropriate.

Nursing Assessment and Strategies

Ask the following questions:

- What are the usual burial practices in your family?
- Do you believe in autopsy?

Be sure to do the following:

- Observe expressions of grief. Support the family in their expression of grief.
- Observe for differences in the expression of grief among family members.
- Offer to obtain a religious counselor/spiritual leader if the family wishes.

Cultural Self-Enrichment Exercise

- What are the usual burial practices in your family?
- What is expected of family members and friends after a loved one dies?
- How is grief expressed in your family? Are there different expectations for men and women?
- Are any specific rituals associated with death?

▶ HEALTH CARE BELIEFS

The focus of health care includes prevention versus acute care practices and traditional, religious, and biomedical beliefs. Additionally, individual responsibility for health, self-medicating practices, views toward mental illness, and the patient's response to pain and the sick role are shaped by one's culture.

Most societies combine biomedical health care with traditional, folk, and religious practices, such as praying for good health and wearing amulets to ward off diseases and illnesses. The health care system abounds with individual and family folklore practices for curing or treating specific illnesses. Many times folk therapies are handed down from family members and may have their roots in religious beliefs. Examples of folk medicines include covering a boil with axle grease, wearing copper bracelets for arthritic pain, mixing wild turnip root and honey for a sore throat, and drinking herbal teas. As an adjunct to biomedical treatments, many people use complementary therapies, such as acupressure, acumassage, reflexology, and other traditional therapies specific to the cultural group.

Often folk practices are not harmful and should be incorporated into the patient's plan of care. However, some may conflict with prescription medications, intensify the treatment regimen, or cause an overdose. It is essential to inquire about the full range of therapies being used, such as food items, teas, herbal remedies, nonfood substances, over-the-counter medications, medications prescribed by others, and medications borrowed from others. If patients perceive that you do not accept their beliefs and practices, they may be less open to sharing information and less compliant with prescribed treatment.

Mental illness may be seen by many as being unimportant compared with physical illness. Mental illness is culture bound, and what may be perceived as a mental illness in one society may not be considered a mental illness in another society. Among some cultures, having a mental illness or an emotional difficulty is considered a disgrace and is taboo. As a result, the family is likely to keep the mentally ill or handicapped person at home as long as possible.[3] (See the Home Health Hints.)

HOME HEALTH HINTS

The effect of the patient's cultural beliefs and practices related to health care are more evident when care is provided in the home. The nurse must adapt care to the patient's environment, rather than the patient adapting to the nurse's hospital environment. The nurse is a guest in the patient's house.

Cultural responses to pain and the sick role vary among cultures. For example, some individuals are expected to openly express their pain. Others are expected to suffer their pain in silence. For some, the sick role is readily accepted, and any excuse is accepted for not fulfilling daily obligations. Others minimize their illness and make extended efforts to fulfill their obligations.

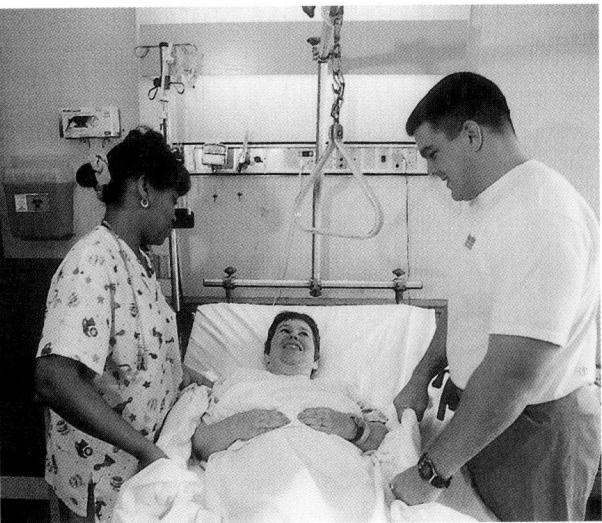

Figure 3-2 Patients and health care workers may come from a variety of cultural backgrounds.

Cultural Self-Enrichment Exercise

- How do you define health for yourself?
- How do you define illness for yourself?
- Identify preventive health care practices that you use.
- When you see a health care provider for a minor illness, what do you expect the health care provider to do for you?
- Identify your self-medicating behaviors.
- What home remedies do you use when you are ill?
- What meaning does pain have to you? What measures do you use when you are in pain?
- What are your personal views toward autopsy, organ donation, organ transplantation, and receiving blood or blood products?

Nursing Assessment and Strategies

Ask the following questions:

- ■ What do you usually do to maintain your health?
- ■ What do you usually do when you are sick?
- ■ What kind of home treatments do you use when you are sick?
- ■ Who is the first person you see when you are sick?
- ■ What do you do when you have pain?
- ■ Do you wear charms or bracelets to ward off illness?
- ■ Do you take herbs or drink special teas when you are sick? What are they?
- ■ Do you practice special rituals or prayers to maintain your health?

▶ HEALTH CARE PRACTITIONERS

Health care practitioner choices are made based on the patient's perceived status and previous use of traditional, religious, and biomedical health care providers. In Western societies, educated health care providers are treated with great respect (Fig. 3–2). However, some people prefer traditional

healers because they are known to the individual, family, and community.[1]

Because some patients may be especially modest, you need to respect differences in gender relationships when providing care. Some people may be especially modest because of their religion, seeking out same-gender nurses and physicians for intimate care. Respect these patients' modesty by providing privacy and assigning a same-gender care provider when possible.

Cultural Self-Enrichment Exercise

- What complementary health care practitioners have you used? Were they successful?
- Identify complementary health care practitioners used by your friends. Were they successful?
- When you are ill and need to see a health care provider, do you prefer a same-gender provider? Why or why not?

Nursing Assessment and Strategies

Ask the following questions:

- ■ What health care providers besides physicians and nurses do you see when you are ill?
- ■ Do you object to male or female health care providers giving physical care to you?

Be sure to do the following:

- ■ Observe for alternative care providers who may visit the patient in the health care facility.

▶ ETHNIC AND CULTURAL GROUPS

As of the 2000 census, the federal government has initiated new terminology for classifying people of diverse racial and ethnicity. The terminology used in this chapter reflects this new classification.

American Indians/Alaskan Natives

There are more than 400 American Indian/Alaskan Native tribes in the United States, totaling 0.9% of the population.[4] Although there are similarities among Native-Americans, each tribe has its own unique perspective on health and illness. Many traditional American Indians/Alaskan Natives live on reservations; others live in urban areas and practice few of their traditions. Many American Indians/Alaskan Natives have a strong belief that illness is caused by an imbalance with nature and the universe.

Tribal identity is maintained through powwows, ceremonial events, and arts and crafts that are taught to children at a young age. Communicating with nature is important for maintaining life forces.[1] American Indians/Alaskan Natives are the original inhabitants of North America.

American Indians/Alaskan Natives are underrepresented in all the health professions. They are consistently identified as the most underrepresented minority group in institutions of higher learning.[5] See Cultural Consideration Box 3–3 for cultural attributes of American Indians/Alaskan Natives.

BOX 3-3 CULTURAL CONSIDERATION

Cultural Group	Communication Styles	Family Organization	Nutrition
American Indians/ Alaskan Natives	Each tribe has its own language. Many speak English, Spanish, or both. Talking loudly may be considered rude. Touch is not acceptable from strangers. Pointing with a finger may be considered rude. Direct eye contact may be considered rude, even with friends. Appointments may not be kept. Most are present oriented.	Most tribes are matriarchal, but some are egalitarian. Gender roles are usually flexible. Elderly persons are highly respected. Family bonds are strong. Extended family members are very important.	Herbs are used to cleanse the body of evil spirits and poison. Nontraditional diets tend to be high in fat. Diet commonly lacks fruits and vegetables on reservations.
European-Americans	Primary language is English. Dialects vary by region of the country but are mutually understandable. May also speak language of native country. Loud voice tone is the norm. Eye contact should be maintained without staring. Value change, are future oriented, and emphasize accomplishments. Punctual in social and business situations. Many concerned with status. Touch infrequently.	Goal is egalitarian family relationships, but some are nominally patriarchal or matriarchal. Goal is gender role flexibility and equality. Have impersonal relationships with outsiders. Elderly persons are loved but may not be respected for wisdom or sought for advice. Extended family not usually important on a daily basis. Stress individualism over the group.	Diets tend to be high in fat and sodium. The only group that uses the food pyramid. Believe a balanced diet promotes a healthy body. Eating and drinking may be social rituals. Culture stresses thinness to be attractive.
African-Americans/ blacks	Primary language is English. Many speak "black English" occasionally, depending on the situation. Usually loud voice volume. Punctuality is flexible. Many are present oriented, believe there is little control over their destiny. Place emphasis on nonverbal behaviors. Direct eye contact may be interpreted as aggression. Are comfortable with close personal space. Touch frequently with friends, less so with strangers.	Usually matriarchal or egalitarian household. Flexible gender roles. Augmented families with non–blood relatives sharing space. Group goals more important than individual goals. Strong family ties. Extended family assists in crisis. Elderly persons respected, especially maternal grandmothers. Church commonly plays a central role, and church members may be seen as family.	Diet commonly high in fat and sodium. Many have lactose intolerance. Food selections may vary according to socioeconomic class and rural versus urban residence. Being overweight is seen as positive. Food is a symbol of health and wealth.

BOX 3-3 CULTURAL CONSIDERATION

Cultural Group	Communication Styles	Family Organization	Nutrition
Spanish/Hispanics/Latinos	Primary language either English or Spanish. Numerous Spanish dialects, which are regionally specific and may not be mutually understandable. Present oriented, believe there is no control over destiny. Flexible with time and appointments. Some believe that direct eye contact can cause illness ("evil eye"). Touch with close same-gender friends but not with opposite-gender friends or strangers.	Patriarchal, matriarchal, or egalitarian households with women taking a more active role in health care decisions. Usually strict gender roles. Children are highly valued and taken everywhere, rarely left with babysitters. Elderly persons are highly respected. Extended family members, including godparents, are important.	Important for food to be served warm—not too hot or too cold. Many subscribe to the hot-and-cold theory, the concept that illness is caused when the body is exposed to an imbalance of hot and cold substances. Foods are classified as hot and cold, and their consumption must be balanced or illness will occur. Food choices vary by country of origin. Many adults have lactose intolerance.
Asians	Language specific to each country. Many dialects, with 55 in China alone. Loud talking is considered rude. Much communication is nonverbal. May be reluctant to maintain eye contact with elders or superiors. Touch acceptable only between members of the same gender. Present oriented. Flexible with social engagements, punctual in business appointments. "Yes" may not mean "I agree" or "I understand." It may mean "I hear you." Reluctant to disclose personal information.	Patriarchal households. Gender roles are well defined. Children are extremely important. Elderly persons are seen as wise and are respected. A child's bad behavior is reflected on the entire family. Multigenerational and extended family important and may share the same living space. Responsibility of the family is to care for elderly, handicapped, and mentally ill at home. Group takes priority over the individual.	Foods are balanced between the yin and yang. Foods are used for health maintenance and treatment of illness. Diet is high in salt. Many have lactose intolerance. Food is a fundamental form of socialization. Food choices and preparation vary by specific country.
Arab-Americans	Primary language is Arabic. Most speak some English. Exaggerated, spirited, and loud voice is used. May be reluctant to disclose personal information. Maintain intense eye contact. Stand very close when talking. Touch only between the same gender. Flexible time schedules socially and for appointments. Present to future oriented according to socioeconomic status.	Patriarchal household. Well-defined gender roles. Very protective of children. Physical punishment is common. Loyalty to family over self. Elders are respected. Family obligated to care for elders. Extended family members important, freely offer advice. Extended family live in close proximity.	Food is a symbol of love, friendship, and generosity. Food preparation takes a long time; only fresh meats and vegetables are used when possible. Lunch is the main meal. Some do not eat pork or drink alcohol (Islam). Believe fasting cures disease. Believe in the hot-and-cold theory of foods. Many fast from sunrise to sunset during the holy month of Ramadan.
Appalachians	English language spoken with dialectal variations—"ing" endings become "in," some plurals formed by adding "es" instead of "s" (e.g., "breastes"). Individualism and self-reliance idealized. May deny anger and tend to not complain. Direct eye contact from strangers may be perceived as aggression or hostility. Comfortable with silence. Relaxed, enjoying body rhythms instead of strong adherence to clock time.	Traditional home was patriarchal but is currently changing to egalitarian. Women are providers of emotional strength. Women and men marry earlier than in many other cultures. Have larger families than the rest of the white American groups. Most believe in physical punishment as a form of discipline. Elders are respected. Usually care for elderly persons in the home.	Wealth means having plenty of food. High-cholesterol organ meats are commonly used. Use lard for frying and baking. Diet frequently deficient in iron, calcium, and vitamin A.

continued

BOX 3-3 CULTURAL CONSIDERATION—CONT'D

Cultural Group	Death and Dying Issues	Health Care Beliefs	Health Care Practitioners
American Indians/ Alaskan Natives	Believe body should go into the afterlife whole. Some engage in a cleansing ceremony after touching a dead body. Home has to be cleaned if person dies at home. Taboo to talk directly about death or a grave diagnosis. Openly express grief. Tribal laws may dictate cremation versus burial.	Many practice acute care only. Use herbs, corn, dances, and prayers to cure and balance life forces with nature. Combine Western medicine with traditional practices. Have a fear of witchcraft. Inanimate objects are believed to ward off evil. Theology and medicine are interwoven. Promote harmony with nature. Pain is something to be endured. Sick role is not usually supported.	Traditional healers include shamans, singers, diviners, crystal gazers, hand tremblers. Elderly may request same-gender direct care provider.
European-Americans	Autopsy and burial or cremation usually connected with religious practices or are individual decisions. Have varied expression of grief. Men are expected to be in more control of grief than women.	Believe humans can control nature. Have strong belief and value in technology. Practice primary acute care practices, but recent trend is moving toward prevention. Value individual responsibility for health. May use folk remedies or over-the-counter medicines before seeing a health care practitioner. Use prayers and religious symbols for good health. Have controlled expression of pain. Need little encouragement for pain relief. Sick role not well accepted except with a major illness.	Primarily use Western-educated health care providers. Recent trend is to use complementary therapists, such as chiropractors, reflexologists, and acu-massage. Most accept physical care from either gender.
African-Americans/ blacks	Death does not end connection between people. Body is kept intact after death; prefer no autopsy. Relatives may communicate with the dead person. Offer eulogy at burial with religious songs. Usually prefer burial, rarely cremation. Express grief openly with emotional catharsis.	Believe diseases may be natural, caused by cold or bad air, food, or water. Believe diseases may be unnatural, caused by voodoo, witchcraft, or a hex. Believe serious illness may be sent from God. May resist preventive care because illness is God's will. Folk and herbal medicine prevalent. Use prayers for prevention and health recovery. Pain is seen as a sign of illness. Sick role not seen as a burden.	A respected elderly female community member commonly sought for initial health care. Use spiritualists who receive their gift from God. Some seek care from voodoo priest or priestess. May seek care from root doctor who uses herbs, oils, candles, and ointments.
Spanish/Hispanics/ Latinos	Burial is the usual practice, rarely cremation. May resist autopsy; body should be buried whole. May have elaborate ceremonial burial. Women very expressive with grief; men are expected to maintain control of their grief.	Health beliefs strongly affected by religion. Believe health and illness largely God's will. May have shrines or statues in the home to pray for good health. Frequently use over-the-counter medicines and medicines brought from home country. Theory of hot and cold foods used for health maintenance and treatment of disease.	Santero in Cuba practices Santeria, a religion with animal sacrifice and worship of many gods. Curandero uses herbs. Espiritulsta uses prayers and amulets. Sobador manipulates muscles and performs massage. High degree of modesty, so may prefer same-gender provider for intimate care.

BOX 3-3 CULTURAL CONSIDERATION

Cultural Group	Death and Dying Issues	Health Care Beliefs	Health Care Practitioners
Spanish/Hispanics/ Latinos—cont'd		Many are fearful and suspicious of hospitals. Expressive with pain. Easily enter the sick role. Primarily receive acute care with little prevention. Believe children and women are more susceptible to the evil eye.	
Asians	Autopsy not understood by many. Cremation acceptable, but burial is also common depending on the specific population. Extended grieving time (7 to 30 days) for the more traditional. Expression of grief is highly varied between men and women and among specific countries.	Good health is a gift from ancestors. Imbalances in the yin and yang cause illness. Believe blood is the source of life and is not replenished. Amulets are worn to ward off disease. Cleanliness is highly valued. Prayers are used for healing. Over-the-counter medicine use is common. Use of herbal teas and medicine is common. Commonly stoical with pain. Many delay seeking help until very ill.	Acupuncturist uses needle insertions at specific points to control pain. Practice acumassage: deep tissue and muscle massage. Moxibustion therapy uses heat instead of needles or massage. May practice coining, in which a heated coin is rubbed over body surfaces, causing dermabrasion. Cupping: involves heated cups applied over the body, which causes suction, leaving large ecchymotic areas when the cup is removed.
Arab-Americans	Believe death is God's will. At the time of death the bed should face the holy city of Mecca. Perform ritual washing of the body after death. No cremation or autopsy allowed. May weep with grief, but limited.	Primarily receive acute care with little prevention. Little responsibility for self-care. Pray five times each day for health. Combine prayers with meditation. Freely use over-the-counter medication. Illness is a punishment for sins. Considered rude and cruel to communicate a grave diagnosis. Organs may be purchased for transplantation. Will display overt expression of pain. Sick dependency role readily accepted. Injections preferred over pills.	Accept and prefer Western practitioners. May use cupping and phlebotomy. Primarily same-gender care only, especially among elderly persons.
Appalachians	Funerals are plain and an important social function. Family and friends remain at bedside when a death is expected. Deceased usually buried in his or her best clothes. Elaborate meals are served after the funeral. Funeral services are accompanied by singing. Clergy are good at helping family through the grieving process.	Prayer is a primary source of strength. Believe good health is largely a result of God's will. Biomedical care may be used as a last resort. Use many folk remedies, primarily passed on by "granny" practitioners who use poultices, herbs, and teas. Self-medicate first before seeking a biomedical practitioner. A major health concern is the state of the blood.	Physicians are commonly seen as outsiders. Family folk practitioners used as a primary source of health care. Nurses are highly respected in the culture. Practitioner must ask the patient what he or she thinks is the problem to be effective.

European-Americans

European-American is the term used to describe people living in the United States whose heritage is from the countries of England, Scotland, Wales, Ireland, Norway, Switzerland, Sweden, the Netherlands, and other northern European countries. European-American groups include the white ethnic groups. Many of the descendants of these original European immigrants practice the unique attributes of the subcultures from which they originate. There is much diversity in the primary and secondary characteristics of diversity within this cultural group.

Many European-Americans maintain the value of individualism over group norms and are activity oriented. Most European-Americans practice Western medicine that uses high technology and emphasizes scientific discovery.[1] See Cultural Consideration Box 3–3 for common cultural attributes of European-Americans.

African-Americans/Blacks

African-Americans/blacks are the largest ethnic group in the United States and represent more than 100 racial strains.[1] They make up 12.3% of the population.[4] Although African-Americans/blacks live in all 50 states, more than half live in the South. It is important to understand that not all people with black skin identify themselves as African-American. Many black-skinned people from the Caribbean prefer terms more specific to their identity, such as *Haitian, Jamaican,* or *West Indian.*

African-Americans/blacks have been called by many names. Their ancient African name is Nehesu or Nubian. During slavery days in America, they were called *Negro,* a Spanish-Portuguese word meaning "black." After emancipation in 1863, they were called *colored,* a term adopted by the First Colored Men's Convention in the United States in 1831. The United States Bureau of the Census adopted the word *Negro* in 1880. During the civil rights movements in the 1960s, the term *black* was used to signify a philosophy of life instead of color. In the 1970s these ethnic peoples referred to themselves as African-Americans because they were proud of both their African and American heritages. In 1988 the term *African-American* was widely adopted in the United States by individuals whose ancestry originated from Africa. These terms continue to cause confusion when people attempt to use the "politically correct" term for this group in the United States. Some individuals prefer terms more appropriate to their individual heritage instead of *African-American/black.* Additionally, titles such as the National Black Nurses Association and the National Association for the Advancement of Colored People still exist.[1]

African-Americans/blacks are underrepresented in colleges and universities, managerial and administrative positions, and the health care professions. They are overrepresented in high-risk, hazardous occupations such as the steel and tire industries, construction industries, and high-pollution factories. See Cultural Consideration Box 3–3 for cultural attributes of African-Americans/blacks.

Spanish/Hispanics/Latinos

The term *Spanish/Hispanics/Latinos* is used to describe people whose cultural heritage has a strong Spanish influence. However, many people in this group prefer to identify themselves as Chicano or with terms that provide a country of origin, such as *Mexican, Peruvian, Puerto Rican,* and *Cuban.*[1] The largest Spanish/Hispanics/Latinos populations in the United States are Mexican-Americans (64%), Puerto Ricans (11%), Central Americans (7%), South Americans (7%), Cubans (4%), and Caribbean (3%). Hispanics also come from Venezuela, Colombia, Ecuador, Peru, Bolivia, Paraguay, Chile, Argentina, Uruguay, Dominican Republic, Guatemala, El Salvador, Costa Rica, Honduras, Nicaragua, and Panama. Thus there is much diversity in the Spanish/Hispanics/Latinos population in the United States.[1]

Some Spanish/Hispanics/Latinos speak only Spanish, some speak only English, some speak both Spanish and English, and some speak neither Spanish nor English but rather an Indian dialect, depending on individual circumstances and the length of time spent in the United States. Spanish is the second most common spoken language in the United States.

Spanish/Hispanics/Latinos compose 12.5% of the U.S. population and live in all 50 states. More than 90% live in cities. Four out of every five Spanish/Hispanics/Latinos are born and raised in the United States.

The majority of Spanish/Hispanics/Latinos practice adaptations from the Roman Catholic religion. Their close relationship with God makes it acceptable for people to experience visions and dreams in which God or the saints speak directly to them. Thus health care providers must be careful not to attribute these culture-bound visions to hallucinations that indicate a need for psychiatric services.[1] See Cultural Consideration Box 3–3 for cultural attributes of Spanish/Hispanics/Latinos.

Asian-Americans

This large group, with more than 5,575,000 people, is far from homogeneous. The term *Asian,* as used in most references, includes 32 different groups. These groups include Asians, Pacific Islanders, Indochinese, and other Asian groups. Asians include people from Korea, Japan, and 54 ethnic groups from China. Pacific Islanders include Hawaiians, Polynesians, Filipinos, Malaysians, and Guamanians. Indochinese populations include Cambodian, Vietnamese, Hmong, and Laotian. Other Asian groups include Asian Indian, Pakistani, and Thai.

Although it is difficult to determine exact numbers of Asians from specific countries because of the method of keeping population statistics, they are a significant and fast-growing population in the United States. It is important for many Asian patients to "save face." Individual shame is shared with the family and community. Most Asians see the nurse as an authority figure. Although known by different names, most Asian cultures practice the *yin* and *yang* balance of forces for illness prevention and maintaining health. Yin is considered female and represents cold and weakness. Yang is considered male and represents strength and warmth. Foods and all forces are classified as yin or yang and must be balanced or illness occurs. Yin and yang forces are major components of traditional Chinese medicine, which includes acupressure, acumassage, cupping, and moxibustion.[1] See Cultural Consideration Box 3–3 for cultural attributes of Asian-Americans.

Arab-Americans

Arab-Americans are a large and diverse population, with more than 3,000,000 in the United States. Some common bonds include the Arabic language and the Islamic religion. Arab-Americans include people from Morocco, Algeria, Tunisia, Libya, Sudan, and Egypt and the western Asian countries of Lebanon, occupied Palestine, Syria, Jordan, Iraq, Iran, Kuwait, Bahrain, Qatar, United Arab Emirates, Saudi Arabia, Oman, and Yemen. Many early Arab immigrants were Christians from Lebanon and Syria.

Although many Arab-Americans favor professional occupations, many are underemployed, have their own businesses, and work in a variety of other occupations. Arab-Americans, whether born in the United States or in Arab countries, are more educated than the average American.[1,6] They are more likely to be in managerial and professional specialty occupations than any other ethnic group in America.[6] However, a significant number of Arab-Americans, primarily foreign born, are unemployed and live in poverty. See Cultural Consideration 3–3 for cultural attributes of Arab-Americans.

Appalachians

The term *Appalachian* is derived from the Apalache Indians, who inhabited what is today known as the Appalachian Mountain region. This chapter uses the term *Appalachian* to describe people born in this region and their descendants. Most Appalachians trace their heritage to Scotland, Wales, Ireland, England, or Germany. This large, mountainous region consists of part or all of the states of Georgia, Alabama, Mississippi, Virginia, West Virginia, North Carolina, South Carolina, Kentucky, Tennessee, Ohio, Maryland, Pennsylvania, and New York. Parts of Appalachia have insufficient roads, public transportation systems, and airports, creating a disparity in educational and health care facilities.

Appalachians are loyal, caring, family oriented, religious, hardy, independent, honest, patriotic, and resourceful. They value home, which is a connectedness to the land rather than a physical structure. Additional characteristics ascribed to Appalachians are avoidance of aggression and assertiveness, refusal to interfere with others' lives, avoidance of dominance over others, and avoidance of arguments and desire for agreement. They are private people who want to offend no one and may not easily trust outsiders.[1] See Cultural Consideration 3–3 for cultural attributes of Appalachians.

► SUMMARY

Rarely do practicing nurses have the luxury to assess each patient comprehensively on a first encounter. The essentials for culturally competent interventions are obtained as needed. The patient database can be added to as time permits. Astute observations, an openness to diversity, and a willingness to learn from patients are requirements for effective cross-cultural competence in clinical practice. Through these avenues, you can provide culturally competent nursing care. Cultural competence is not a luxury; it is a necessity.

You can obtain additional information from the Transcultural Nursing Society at http://www.tcns.org, or go to http://nursing.about.com and click on "Transcultural."

⁓Ⅳ⁓ CRITICAL THINKING

Mr. Bautista

Mustafa Bautista, age 62, and his wife, Mina, age 60, immigrated to the United States 3 years ago from Saudi Arabia. During a celebration of Ramadan, Mustafa collapsed and was taken to the nearest emergency room. He was admitted to the observation unit for diagnostic testing. He does not speak or understand English well, and when the nurse asks him a question, he stares at her, makes elaborate hand gesture, and speaks in a very loud voice. Mustafa is a devoted adherent of Islam, and his wife brings him the Koran, from which he reads several times each day.

Even though Mustafa is prescribed bedrest, he gets out of bed several times each day, kneels on a small carpet, and prays. He refuses to let the female nurses bathe him. Yesterday he refused his breakfast and lunch. Today he refused his dinner and insisted on keeping the tray at his bedside. At 8:00 P.M. he took the pork chops from his tray and placed them in the hall. Additionally, he refuses his oral medicines and insists on injections instead.

1. What nonverbal communication characteristics does this patient display that are common among people of Arabic descent?
2. Does his loud voice mean that he is angry?
3. What might you do when Mr. Bautista gets out of bed to pray?
4. Why do you think Mr. Bautista refused his breakfast and lunch?
5. Why did Mr. Bautista remove the pork chops from his tray and leave them in the hall?
6. What might you do to improve Mr. Bautista's nutrition while he is in the hospital?
7. What is Ramadan?
8. Why does Mr. Bautista insist on injections rather than oral medication?
9. What might you do to get Mr. Bautista to take oral medications rather than injections?

Answers at end of chapter.

1. The nurse is explaining the medication schedule to Mrs. Bing Bing, a 45-year-old Chinese woman with minimal English language skills. Which of the following will best help ensure compliance?
    a. Provide written instructions in English.
    b. Speak slowly and clearly, using the same phrase each time.
    c. Speak loudly to make sure she understands you.
    d. Speak slowly and exaggerate the words.

2. A 12-year-old Mexican child needs an appendectomy. Who should sign the informed consent for the operation?
    a. The mother
    b. The father
    c. Both parents
    d. The parents decide

3. Your 42-year-old male Islamic patient in a long-term care facility needs to have a urinary catheter inserted to obtain a specimen. Who is the best person to perform the procedure?
    a. A female registered nurse
    b. A male nurse
    c. An orderly
    d. The patient's wife

4. A 22-year-old Osage American Indian woman is slow at giving responses and does not maintain eye contact with you when you are doing her intake interview. Which of the following interpretations of her behavior is most likely accurate?
    a. Direct eye contact may be interpreted as rude in her culture.
    b. She does not want to answer personal questions.
    c. She does not understand you.
    d. She does not want to talk with a nurse and prefers a physician.

5. An 82-year-old African-American woman is in the coronary care unit to rule out a myocardial infarction. Several church members come to visit her on a daily basis. Which of the following actions by the nurse is correct?
    a. Carefully explain to them that only family members are allowed to visit in the coronary care unit.
    b. Allow two church members to visit and have them represent the entire group.
    c. Set up a specific time during which several of them can visit for brief periods.
    d. Report the situation to her physician.

6. A 42-year-old Puerto Rican man has been admitted for reconstructive orthopedic surgery on his knee. His wife brings jars of special blends of spices that he wants to put on his food because the hospital food is too bland. He is on a general diet. What action should you take?
    a. Allow him to use them.
    b. Carefully explain that family cannot bring food items to the hospital.
    c. Have the dietitian speak with the family.
    d. Report the situation to the physician.

R E F E R E N C E S

1. Purnell, L, and Paulanka, B (eds.): Transcultural health care: a culturally competent approach. FA Davis, Philadelphia, 1998.
2. Leininger, ME: Transcultural eating patterns and nutrition: transcultural nursing and anthropological perspectives. Holistic Nursing Practice, 3(1), 16, 1988.
3. Louie, KB: Providing health care to Chinese patients. Topics in Clinical Nursing, 7(3), 18, 1985.
4. United States Bureau of the Census: United States Census 2000. Author, 2001. Available at http://www.census.gov.
5. Preito, D: American Indians in medicine: The need for Indian healers. Academic Medicine 64, 388, 1989.
6. Zogby, J: Arab American today: A demographic profile of Arab-Americans. Washington, DC: Arab American Institute, 1990.

Answers to CRITICAL THINKING

Mr. Bautista

1. Culture-bound communication patterns that Mr. Bautista displays include speaking in a loud voice, liberal use of hand gesturing, and maintaining intense eye contact.
2. A loud voice does not necessarily mean that Mr. Bautista is angry.
3. The nurse should instruct Mr. Bautista to call someone to help him get out of bed when he is ready to pray and when he is ready to return to bed. The danger of injuring himself while praying is minimal once he is out of bed and kneeling on his prayer rug.
4. During Ramadan, Islamic people fast from sunup to sundown. Breakfast and lunch are served during daylight hours. Therefore Mr. Bautista refused his meals.
5. Mr. Bautista removed the pork chops from his room because Islamic people do not eat pork.
6. Because Islamic people fast from sunup until sundown during Ramadan, serve Mr. Bautista breakfast before sunup and hold his dinner until sundown. Provide snacks between dinner and breakfast as necessary.
7. Ramadan is a holy month for adherents of Islam. It involves special religious observances.
8. Many Arab-Americans prefer injections over oral medications because it is believed that injections are stronger than oral medicine. Additionally, fasting may include not taking medications.
9. Administering medications after sundown may improve compliance. Explaining to Mr. Bautista that fasting is not required during times of illness may also improve compliance.

4

ALTERNATIVE AND COMPLEMENTARY THERAPIES

Lynn Keegan

KEY TERMS

acupuncture (ak-yoo-**PUNGK**-chur)
allopathic (**AL**-oh-**PATH**-ik)
Ayurvedic (**AY-YUR-VAY**-dik)
chiropractic (ky-roh-**PRAK**-tik)
homeopathy (**HO**-mee-**AH**-pa-thee)
naturopathy (**NAY**-chur-**AH**-pa-thee)
osteopathic (**AHS**-tee-ah-**PATH**-ik)

QUESTIONS TO GUIDE YOUR READING

1. What is the difference between an alternative and complementary therapy?
2. What are some systems of health care that have contributed to the development of new therapies?
3. How can the different types of therapies be classified?
4. What are some safety issues in alternative and complementary therapies?
5. What is the role of the licensed practical nurse/licensed vocational nurse/registered nurse (LPN/LVN/RN) in assisting a patient with alternative and complementary therapies?

Health care in the twenty-first century requires that nurses recognize the shift of thinking toward the incorporation of alternative and complementary approaches to care. Nurses at all levels and in every area of practice are answering the call to use new methods to care for the ill and to enhance the health of those who are well.

Holistic nursing was a precursor to many of the now popular alternative and complementary therapies. It was introduced in the 1970s and has been growing ever since. Holistic nursing is simply defined as caring for the whole person—body, mind, and spirit—in an ever-changing environment.

► ALTERNATIVE OR COMPLEMENTARY: WHAT'S THE DIFFERENCE?

The words *alternative* and *complementary* are sometimes used interchangeably, but they are not the same. *Alternative* therapy, sometimes called "unconventional" therapy, refers to a therapy used instead of conventional or mainstream therapy. An example is using **acupuncture** instead of analgesics. *Complementary therapy* refers to a therapy used in addition to a conventional therapy. For example, a nurse might use guided imagery, music, and relaxation techniques for pain control in addition to prescribed drug therapy.

A good all-around site for alternative and complementary medicine is http://www.healthy.net. To learn about holistic nursing and complementary therapies, visit http://www.ahna.org.

► INTRODUCTION OF NEW SYSTEMS INTO TRADITIONAL AMERICAN HEALTH CARE

There are many new and different philosophies within the scope of expanded medical and nursing practice. These systems reflect cultures and attitudes in healing that range from East to West and from ancient to modern.

In the United States the primary system of medicine is just called *medicine,* although some people refer to it as **allopathic** medicine. Upon closer look, you will find that a number of other schools of thought and philosophies are also being increasingly used. The most frequently seen new systems include **Ayurvedic,** Chinese, **chiropractic,**

---

acupuncture: acus—needle + punctura—puncture

allopathic: allos—other + pathic—disease or suffering
ayurvedic: ayu—life + veda—knowledge or science
chiropractic: cheir—hand + pracktos—to do

35

naturopathic, American Indian, and **osteopathic** medicine. Each philosophical system can stand alone or, as in most instances in the United States, be used in combination with other systems. Most of these recently introduced systems use alternative and complementary therapies.

Allopathic/Western Medicine

The most common name for allopathic medicine is *Western medicine*. Other commonly used terms for Western medicine are *conventional medicine* and *mainstream medicine*. Many people have not heard the term *allopathic* because most doctors and nurses do not refer to themselves or their practice with this name. Practitioners of other systems of medicine more often use the term when referring to what most of us consider mainstream medicine. Allopathy is a method of treating disease with remedies that produce effects different from those caused by the disease itself. For example, when a patient has a bacterial infection, a Western medical practitioner prescribes an antibiotic to eliminate the invading pathogen.

Practitioners of Western medicine are medical doctors, nurses, and allied health personnel. This system of medicine uses scientific data to determine the validity of a diagnosis and the effectiveness of treatment. In other words, it is evidence-based medicine. This means that peer-reviewed medical literature is very important. In scientific investigations, results can be verified and reproduced through various types of studies and statistical analyses. Practitioners use a variety of therapies, including drugs, surgery, and radiation therapy. Western medicine practitioners have made most of the significant advances and developments in modern medicine.

For further information see the American Medical Association site at http://www.ama-assn.org.

Ayurvedic Medicine

Ayurveda is the ancient Hindu system of medicine, which originated in India. Its main goals are to maintain the health of healthy people and cure the illnesses of sick people. Ayurveda maintains that illness is the result of falling out of balance with nature. Diagnosis is based on three metabolic body types call *doshas*. An Ayurvedic doctor determines the dosha type as vata, pitta, or kapha. Treatment usually involves prescribing a diet, herbal remedies, breath work, physical exercise, yoga, meditation, massage, and a rejuvenation or detoxification program.

Ayurveda is rapidly becoming more popular in America. The books and videos of Deepak Chopra are an example of this increasingly popular system of therapy. An introduction to Ayurveda can be found at http://www.ayurveda.org.

Traditional Chinese Medicine

Traditional Chinese medicine is thousands of years old and involves such techniques and practices as acupuncture, acu-

pressure, herbs, massage, and qi gong. The diagnosis and treatment of disturbances of qi (pronounced "chee")—or vital energy—is a distinctive characteristic of Chinese medicine.

Acupuncturists, one segment of traditional Chinese medicine practitioners, claim to be able to tell much about a patient's state of health by checking pulses, looking at the color of the tongue, checking facial color, assessing voice and smell, and asking a variety of questions. To treat patients, acupuncturists insert one or more needles along the meridians (pathways) where qi flows (Fig. 4–1). Many acupuncturists also prescribe herbs.

Chiropractic Medicine

Daniel David Palmer founded chiropractic in 1895. Chiropractic holds that illness is a result of nerve dysfunction. The main treatment modality of chiropractors is manual adjustment and manipulation of the vertebral column and the extremities. They use direct hand contact and mechanical and electrical treatment methods to manipulate joints. The goal is to remove the interference with nerve function so the body can heal itself. Chiropractors do not perform surgery, nor do they prescribe drugs. See the American Chiropractic Association site at http://www.amerchiro.org.

Homeopathic Medicine

Homeopathy was developed by Samuel Hahnemann in Germany in the early nineteenth century. Homeopathy is based on Hahnemann's principle that "like cures like"; in other words, that tiny doses of a substance that create the symptoms of disease in a healthy person will relieve those symptoms in a sick person.

Although there are schools and courses to train homeopaths, no diploma or certificate from any school or program is a license to practice homeopathy in the United States. Medical doctors and doctors of osteopathy are granted certificates of competency to practice homeopathy. Other health care practitioners may be allowed to use homeopathy within the scope of their state licenses.

See The National Center for Homeopathy site at http://www.homeopathic.org. This site has a research section with annotated listings from the medical literature discussing the value of homeopathy. In addition, there is a good review of licensing laws regarding homeopathy.

Naturopathic Medicine

Naturopathy primarily uses natural therapies such as nutrition, botanical medicine (herbs), hydrotherapy (water-based therapy), counseling, physical medicine, and homeopathy to treat disease, promote healing, and prevent illness. Naturopathic physicians have a doctor of naturopathy (ND) degree and can be licensed in 11 states. There are three schools of naturopathic medicine in the United

osteopathic: osteo—bone + pathy—disease

homeopathy: homeo—like + pathos—disease
naturopathy: naturo—nature + pathy—disease

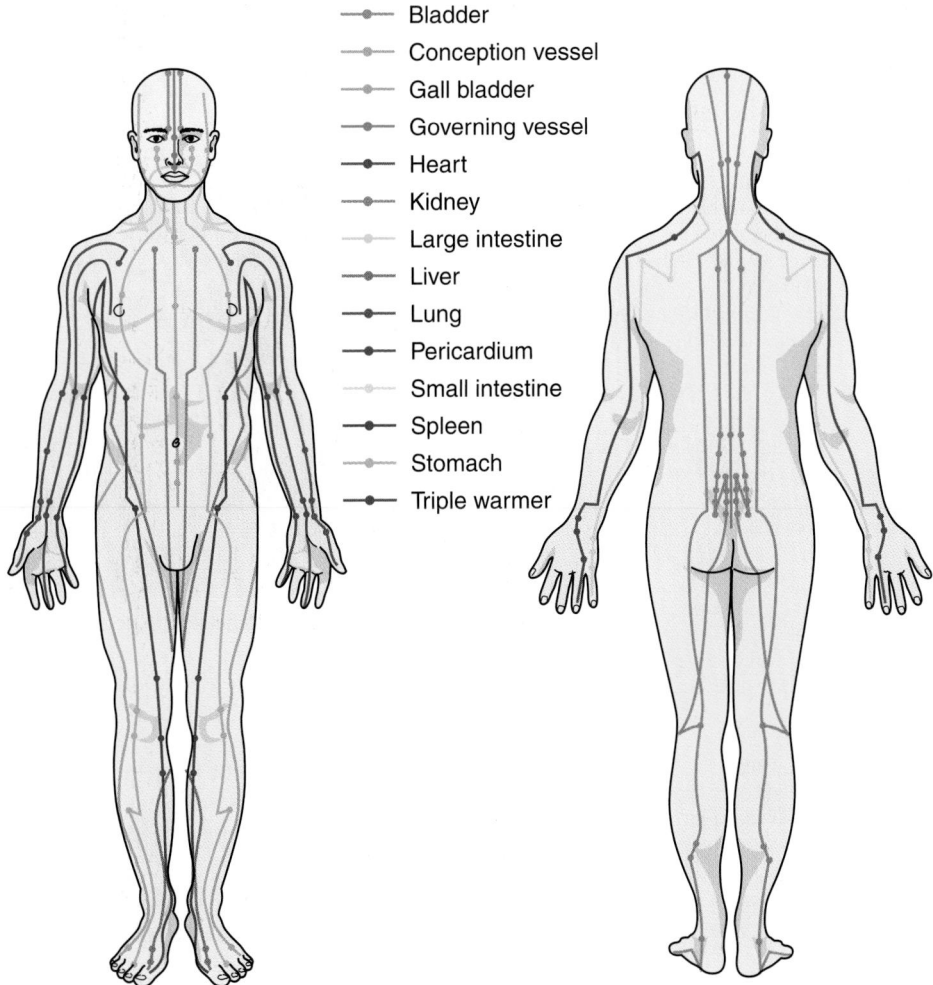

- Bladder
- Conception vessel
- Gall bladder
- Governing vessel
- Heart
- Kidney
- Large intestine
- Liver
- Lung
- Pericardium
- Small intestine
- Spleen
- Stomach
- Triple warmer

Figure 4-1 Qi meridians are used in the Chinese medicine techniques of acupressure and acupuncture.

States. For more information about naturopathy, visit http://www.naturopathic.org.

American Indian Medicine

American Indian medical practices vary from tribe to tribe. In general, American Indian medicine is a community-based system with rituals and practices such as the sweat lodge, herbal remedies, the medicine wheel, the sacred hoop, the "sing," and shamanistic healing. For example, an ill person may be placed in a small, enclosed sweat lodge while singing or chanting is done outside the lodge. It is believed that toxic substances are drawn out in the sweat of the person inside the lodge. After the ceremony, the ill person may be placed on a cot outside and be prayed over. You can learn more about American Indian Medicine at the Association of American Indian Physicians' site at http://www.aaip.com.

Osteopathic Medicine

Osteopathic medicine was founded in the United States in 1874 by Andrew Taylor Still, a frontier physician who was dissatisfied with the state of medicine at that time. This practice of medicine emphasizes the interrelationship of the body's nerves, muscles, bones, and organs. The osteopathic philosophy involves treating the whole person; recognizes the body's ability to heal itself; and stresses the importance of diet, exercise, and fitness with a focus on prevention. For more information about osteopathy, visit the American Osteopathic Association site at http://www.aoa-net.org.

► ALTERNATIVE AND COMPLEMENTARY THERAPIES

See Table 4-1 for a summary of the most common alternative and complementary therapies. Discussion of all therapies is beyond the scope of this book. Therapies are listed to make you aware of others you might want to research.

Herb Use

Many people use herbs for healing. However, when doing so, the patient should only take herbs under the supervision of a health care provider. Herbs can aid in healing, but they also can harm. Some of the more common herbs are described in Table 4–2. Figure 4–2 shows echinacea, an herb commonly prepared for use as an immune system booster.

It is important to note that herbal remedies are not foods. They have potent medicinal effects and can interact with prescribed medications and even surgery. This can be

TABLE 4-1	CATEGORIES AND TYPES OF ALTERNATIVE AND COMPLEMENTARY THERAPIES
Category of therapy	**Types of individual therapies**
Herbal medicine and nutritional supplements	Herbal medicine
	Nutrition and special diet therapies
	Nutritional supplements
Mind-body therapies	Art therapy
	Music/sound therapies
	Guided imagery
	Hypnosis and hypnotherapy
	Meditation and relaxation
Posture and mobility therapies	Movement therapies
	Tai chi and qi gong
	Yoga
Touch therapies and bodywork	Massage and related massage therapies
	Acupressure
Energetic therapies	Biofeedback
	Healing touch
	Magnet therapy
	Polarity therapy
	Reiki
	Spiritual healing
	Therapeutic touch
Miscellaneous therapies	Aquatherapy/hydrotherapy
	Aromatherapy
	Chanting
	Chelation therapy
	Colon therapy
	Kinesiology
	Light therapy
	Pet therapy

TABLE 4-2	COMMON HERBS AND THEIR INTENDED PROPERTIES AND USES
Herb	**Purported properties/uses**
Aloe vera	Soothing agent, used for skin lesions; toxin absorber
Bee pollen	Increases energy, stamina, and strength
Capsaicin	For tenderness and pain of osteoarthritis, fibromyalgia, diabetic neuropathy, and shingles
Chamomile	For anxiety, stomach distress, and infant colic
Echinacea	Antiviral; used for colds, flu, and other infections
Feverfew	Antiinflammatory; used for migraine headaches and as an appetite stimulant; promotes menstruation, eliminates worms, suppresses fever
Ginger	For nausea and vomiting, hypertension, and high cholesterol
Ginkgo	May improve memory and help cognitive function in Alzheimer's disease
Kava	For anxiety, insomnia, low energy, muscle tension
St. John's wort	For mild to moderate depression; viral infections, including human immunodeficiency virus (HIV); and herpes

Figure 4-2   Echinacea is a commonly used herb for colds and flu.

problematic, both because herbs are readily available in health food stores and drugstores and because many patients do not tell their doctors about their herb use. Be sure to assess patients' use of herbs and supplements, and educate them about the need to inform primary care providers when using herbs.

Safety and Effectiveness of Alternative Therapies

Safety generally means that the benefits outweigh the risks of a treatment or therapy. If your patient is interested in using an alternative or complementary therapy, first counsel the patient to talk with the primary care provider. The patient also should ask the practitioner of the therapy about its safety and effectiveness. Patients should tell their primary care providers and alternative practitioners about all therapies they are receiving, because this information may used to consider the safety of their entire treatment plan.

The patient should be as informed as possible and continue gathering information even after a practitioner has been selected. You can find information about specific therapy or therapies in journal articles and books. See Box 4–1

BOX 4-1	Questions Patients Should Ask Before Starting an Alternative or Complementary Therapy

1. What will this therapy do for me?
2. What are the advantages of this therapy?
3. What are the disadvantages?
4. What are the side effects?
5. What risks are associated with this therapy?
6. How much will it cost?
7. How long will it take? How many treatments will be necessary?
8. How will it interact with my other therapies?
9. What research has been done on this therapy?

for questions patients should ask when considering alternative therapies.

ROLE OF THE LPN/LVN

Patients may ask you about the use of an alternative or a complementary therapy. Because the safety and effectiveness of many therapies is still unknown, advising patients presents a challenge. The following steps may help you to help your patient. You can advise patients to do the following:

1. Take a close look at the background, qualifications, and competence of the proposed practitioner. Check credentials with a state or local regulatory agency with authority over the area of practice they seek. Are they licensed or certified? By whom?
2. Visit the practitioner's office, clinic, or hospital. Evaluate the conditions of the setting.
3. Talk with others who have used this practitioner.
4. Consider the costs. Are the treatments covered by insurance, or is it direct pay?
5. Discuss use of an alternative or complementary therapy with their primary care provider.

You can help direct your patient's questions and concerns to the primary care provider. See Home Health Hints for tips.

HOME HEALTH HINTS

- When taking a health history, ask the patient or caregiver about the use of complementary or alternative therapies, because these may influence the effect or side effects of some prescription medications.
- Be mindful of the importance of complementary and alternative therapies to the patient's health care belief system.
- Consider the alternative practitioner as part of the patient's health care team.

NURSING APPLICATIONS
Familiarization Strategies

There are ways to gain confidence with alternative and complementary therapies:

1. Begin by trying one or two of these therapies yourself. Start by choosing a basic therapy such as massage, music, or guided imagery. Follow the guidelines listed in Box 4–1 to make sure it is a safe strategy. Not only will you encounter the benefits firsthand, but you will also come away with a better understanding of what patients experience.
2. Ask your patients if they use any alternative or complementary therapies and what their responses to them are. Try to eliminate any preconceived notions you might have. Your patients will feel more comfortable mentioning them to you if they feel you understand the treatment and why they decided to use it.
3. If you decide to become involved, get adequate instruction in the therapies before you administer them. Many

universities and agencies offer continuing education courses on these therapies, and some nursing schools incorporate alternative and complementary therapies in their skills courses.

If you wish to incorporate alternative and complementary therapies into your practice, be sure to check your state's nurse practice act for any regulations. Discuss them with the patient and his or her primary care provider before using them. If you work for a hospital or other health care institution, also check institutional policy.

As the public understands more about alternative and complementary therapies, you can expect to see even greater demand for these treatments. Nurses have been in the forefront of developing the holistic philosophy that has now become an accepted standard of care.

Relaxation Therapies
Progressive Muscle Relaxation

Progressive muscle relaxation is a simple technique that anyone can learn. It is the process of alternately tensing and relaxing muscle groups. The purpose of the technique is to help the participant identify subtle levels of mental and physical tension that accompany mental and emotional stress. When our conscious awareness of the tensions increases, we can learn to relax and thus reduce the effects of stress and tension. The idea is to become aware of the ability to fine-tune the relaxation response.

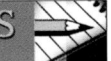

LEARNING TIPS

Try using progressive muscle relaxation the next time you are anxious during a nursing exam.

Guided Imagery

Guided imagery is another example of a complementary therapy that many nurses use. Guided imagery involves using mental images to promote physical healing or changes in attitudes or behavior. Practitioners may lead patients through visualization exercises or offer instruction in using imagery as a self-help tool. Guided imagery is often used to alleviate stress and to treat stress-related conditions such as insomnia and high blood pressure. People with cancer, acquired immunodeficiency syndrome (AIDS), chronic fatigue syndrome, and other disorders can use specific images to boost the immune system. Guided imagery is often an

CRITICAL THINKING

Mr. Jones
Scenario: Mr. Jones asks you whether he should stop his chemotherapy and try magnet therapy for his prostate cancer. How do you respond?

Answers at end of chapter.

added layer of therapy for someone who has learned the progressive relaxation skill discussed previously.

A common guided imagery technique begins with a general relaxation process. For those new to the process, it is good to begin with the progressive muscle relaxation as cited above. Box 4–2 gives an example of guided imagery. Guided imagery works best when all the senses are used. This exercise is very basic but gives an idea of how the technique works. When used for healing, many more steps are involved. For additional information read journal articles, books, and Internet sites. It is important to note that with this and any of the alternative and complementary therapies, one must have some training and skill before using the technique with patients. One of a number of good sites on guided imagery is http://www.healthy.net/agi.

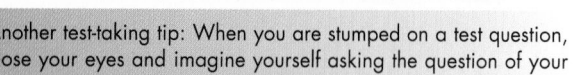

Another test-taking tip: When you are stumped on a test question, close your eyes and imagine yourself asking the question of your favorite instructor. Imagine what his or her answer would be.

Biofeedback

Biofeedback might be considered the third tier of learning of progressive relaxation. It is a technique used especially for stress-related conditions such as asthma, migraines, insomnia, and high blood pressure. Biofeedback is a way of monitoring and controlling tiny metabolic changes in one's body with the aid of sensitive machines.

► SUMMARY

The use of alternative and complementary therapies is growing in the United States. Nurses must be aware of their use and ask about them in their patient assessments. Nurses who have additional training in specific therapies can use many of them when caring for patients. When patients use complementary therapies, nurses should view it as an attempt to increase potential benefit and not as a sign of distress or dissatisfaction.

BOX 4–2   Guided Imagery

Assist your patient to progress through the following steps:
- Assume a comfortable position in a quiet environment.
- Close your eyes and keep them closed until the exercise is completed.
- Breathe in and out deeply to the count of four, repeating this step four times.
- When relaxed, think of a favorite peaceful place and prepare to take an imaginary journey there.
- Picture what this place looks like and how comfortable you feel being there.
- Listen to all the sounds; feel the gentle, clean air; and smell the pleasant aromas.
- Continue to breathe deeply, and appreciate the feeling of being in this special place.
- Feel the sense of deep relaxation and peace of this place.
- As you continue to breathe deeply, slowly and gently bring your consciousness back to the setting in this room.
- Slowly and gently open your eyes, stretch, and think about how relaxed you feel.

Answers to CRITICAL THINKING

Mr. Jones

As with all medical treatments, it is important to support the established therapy the physician has prescribed. Therefore a good response might be the following:

"Mr. Jones, chemotherapy is an established medical treatment for your condition. There is a lot of evidence for its effectiveness in the medical literature. If you want to supplement your therapy, there may be some other treatments you can add. I suggest that you discuss your feelings about seeking some additional treatments with your doctor."

REVIEW QUESTIONS

1. Which of the following statements best defines a complementary therapy?
   a. An alternative treatment that is used in place of a conventional treatment
   b. A treatment that is often dangerous and should be avoided
   c. A treatment that can be used in addition to a conventional treatment
   d. A treatment that is used after conventional treatments have failed

2. Which of the following therapies is most likely to use research-based interventions?

   a. Naturopathy
   b. Osteopathy
   c. Allopathy
   d. Homeopathy

3. Mr. Flynn is admitted for heart failure. He has been taking a ginger supplement in addition to his prescribed medications at home. What is the best response by the nurse when he shares this information during the admission interview?
   a. "Nonprescription supplements can interact with prescription medications. You should not take it any longer."

b. "Ginger can be effective for hypertension. Be sure to monitor your blood pressure while you are taking it."

c. "Ginger is a safe supplement because it is a food. It should not interact with your medications."

d. "You should check with your physician to make sure the ginger doesn't interact with your other medications before you continue to take it."

4. According to recent research, biofeedback-assisted relaxation training may be used to promote healing of foot ulcers by:

a. increasing peripheral perfusion.

b. eliminating bacteria in the wounds.

c. enhancing growth of new tissue.

d. reducing wound drainage.

5. Which of the following statements best describes the most important role of the LPN in alternative and complementary therapies?

a. The LPN should become familiar with and practice at least one alternative or complementary therapy.

b. The LPN should become adept at collecting and reporting data related to patients' use of alternative and complementary therapies.

c. The LPN should discourage use of alternative and complementary therapies because they can interact negatively with conventional therapies.

d. The LPN does not need to become involved in alternative and complementary therapies.

U N I T   1   B I B L I O G R A P H Y

Baker, C: Reflective learning: A teaching strategy for critical thinking. J Nurs Educ 35(1):19, 1996.

Colbath, JD, and Prawlucki, PM (eds): Holistic Nursing Care (collection of 12 articles). The Nursing Clinics of North America. 36(1). WB Saunders, Philadelphia, 2001.

Green, C: Critical Thinking in Nursing: Case Studies Across the Curriculum. Prentice-Hall, Upper Saddle River, NJ, 2000.

Dossey, B, Keegan, L, and Guzzetta, C: Holistic Nursing: A Handbook for Practice, ed 2. Aspen, Gaitherburg, MD, 2000.

Keegan, L: Healing with Alternative and Complementary Therapies. Albany, NY: Delmar, 2001.

Maynard, C: Relationship of critical thinking ability to professional nursing competence. J Nurs Educ 35(1):13, 1996.

Purnell, L, and Paulanka, B (eds): Transcultural Health Care: A Culturally Competent Approach. FA Davis, Philadelphia, 1998.

Rane-Szostak, D, and Fisher Robertson, J: Issues in measuring critical thinking: Meeting the challenge. J Nurs Educ 35(1):5, 1996.

Reich, WT: Encyclopedia of Bioethics. Macmillan, New York, 1995.

Rubenfeld, MG, and Scheffer, BK: Critical Thinking in Nursing: An Interactive Approach. JB Lippincott, Philadelphia, 1995.

Sherwin, S: No Longer Patient: Feminist Ethics and Health Care. Temple University, Philadelphia, 1992.

Tedesco, P, and Cicchetti, J: Like cures like: Homeopathy. Am J Nurs, 101:9, 2001.

Tong, R: Feminine and Feminist Ethics. Wadsworth, Belmont, CA, 1993.

Waddell, DL, Hummel, ME, and Sumners, AD: Three herbs you should get to know. Am J Nurs, 101(4), 2001.

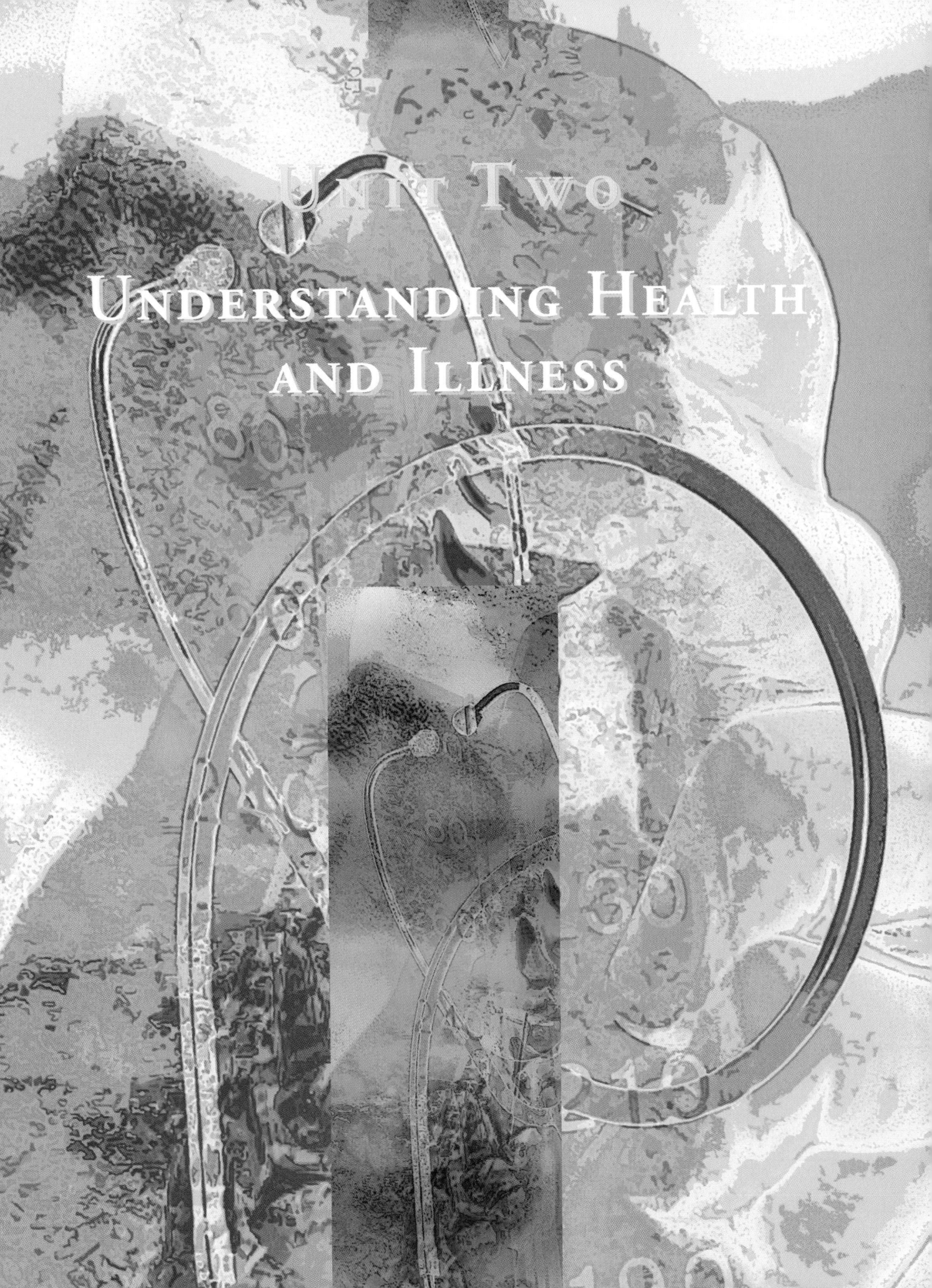

UNIT TWO

UNDERSTANDING HEALTH AND ILLNESS

5

Fluid, Electrolyte, and Acid-Base Balance and Imbalance

Donna D. Ignatavicius and Bruce K. Wilson

KEY WORDS

acidosis (ass-i-**DOH**-sis)

alkalosis (al-ka-**LOH**-sis)

anion (**AN**-eye-on)

antidiuretic (**AN**-ti-**DYE**-yoo-**RET**-ik)

cation (**KAT**-eye-on)

dehydration (**DEE**-high-**DRAY**-shun)

diffusion (di-**FEW**-zhun)

dysrhythmia (dis-**RITH**-mee-yah)

edema (e-**DEE**-ma)

electrolytes (ee-**LEK**-troh-lites)

extracellular (**EX**-trah-**SELL**-yoo-ler)

filtration (fill-**TRAY**-shun)

hydrostatic (**HIGH**-droh-**STAT**-ik)

hypercalcemia
(**HIGH**-per-kal-**SEE**-mee-ah)

hyperkalemia
(**HIGH**-per-kuh-**LEE**-mee-ah)

hypermagnesemia
(**HIGH**-per-**MAG**-nuh-**ZEE**-mee-ah)

hypernatremia
(**HIGH**-per-nuh-**TREE**-mee-ah)

hypertonic (**HIGH**-per-**TAHN**-ik)

hyperventilation
(**HIGH**-per-**VEN**-ti-**LAY**-shun)

hypervolemia
(**HIGH**-poh-voh-**LEE**-mee-ah)

hypocalcemia
(**HIGH**-poh-kal-**SEE**-mee-ah)

hypokalemia
(**HIGH**-poh-kuh-**LEE**-mee-ah)

hypomagnesemia
(**HIGH**-poh-**MAG**-nuh-**ZEE**-mee-ah)

hyponatremia
(**HIGH**-poh-nuh-**TREE**-mee-ah)

hypotonic (**HIGH**-poh-**TAHN**-ik)

hypovolemia
(**HIGH**-poh-voh-**LEE**-mee-ah)

interstitial (**IN**-ter-**STISH**-uhl)

intracellular (**IN**-trah-**SELL**-yoo-ler)

intracranial (**IN**-trah-**KRAY**-nee-uhl)

intravascular (**IN**-trah-**VAS**-kyoo-lar)

isotonic (**EYE**-so-**TAHN**-ik)

osmosis (ahs-**MOH**-sis)

osteoporosis (**AHS**-tee-oh-por-**OH**-sis)

semipermeable
(**SEM**-ee-**PER**-mee-uh-bull)

transcellular (trans-**SELL**-yoo-lar)

QUESTIONS TO GUIDE YOUR READING

1. What are the purposes of fluids and electrolytes in the body?
2. What are the signs and symptoms of common fluid imbalances?
3. Which patients are at the highest risk for dehydration and fluid overload?
4. What data should you collect in patients with fluid and electrolyte imbalances?
5. What is the current medical treatment for patients with fluid and electrolyte disturbances?
6. What are the education needs of patients with fluid imbalances?
7. What are the common causes, signs and symptoms, and treatments for sodium, potassium, calcium, and magnesium imbalances?
8. Which foods have high sodium, potassium, and calcium content?
9. What are common causes of acidosis and alkalosis?
10. How do ABGs change for each type of acid-base imbalance?

The body undergoes continuous dynamic change. The proper amount of fluid is needed to support these changes and to transport building and waste materials. Approximately 60 percent of a young adult's body weight is water. Elderly people are less than 50 percent water, and infants are between 70 percent and 80 percent water. Women have less body water because they have more fat than men. Fat cells do not contain water.

In addition to water, body fluids also contain solid substances that dissolve, called solutes. Some solutes are **electrolytes** and some are nonelectrolytes. Electrolytes are chemicals that can conduct electricity when dissolved in water. Examples of electrolytes are sodium, potassium, calcium, magnesium, acids, and bases; these are discussed later in this chapter. Nonelectrolytes do not conduct electricity. Examples are glucose and urea.

▶ **FLUID BALANCE**

Fluids are located both inside the cells (**intracellular** fluid [ICF]) and outside the cells (**extracellular** fluid [ECF]). As seen in Figure 5–1, ECF can be further divided into three types: **interstitial** fluid, **intravascular** fluid, and **transcellular** fluid.

Interstitial fluid is the water that surrounds the body's cells and includes lymph. Fluids and electrolytes move between the interstitial fluid and the intravascular fluid, which is the plasma of the blood. Transcellular fluids are those in specific compartments of the body, such as cerebrospinal fluid, digestive juices, and synovial fluid in joints.

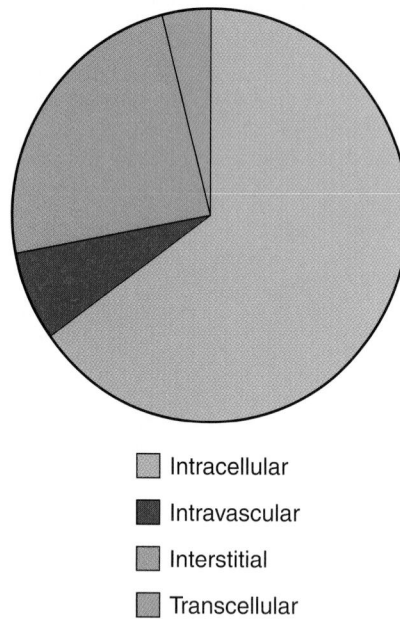

□ Intracellular
■ Intravascular
□ Interstitial
□ Transcellular

Figure 5-1 Normal distribution of total body water.

Control of Fluid Balance

The primary control of water in the body is through pressure sensors in the vascular system, which stimulate or inhibit the release of **antidiuretic** hormone (ADH) from the pituitary gland. A diuretic is a substance that causes the kidneys to excrete more fluid. ADH works in just the opposite way. ADH causes the kidneys to retain fluid. If fluid pressures within the vascular system decrease because there is less fluid, more ADH is released. If fluid pressures increase, less ADH is released and the kidneys eliminate more fluid.

Movement of Fluids and Electrolytes in the Body

Fluids and electrolytes move in the body by active and passive transport systems. Active transport depends on the presence of adequate cellular adenosine triphosphate for energy. The most common example of active transport is the sodium-potassium pump. This pump, located in the cell membrane, causes sodium to move out of the cell and potassium to move into the cell when needed.

In passive transport, no energy is expended specifically to move the substances. General body movements aid passive transport. The three passive transport systems are **diffusion, filtration,** and **osmosis.**

Diffusion is a process in which the substance moves from an area of higher concentration to an area of lower concentration. If you pour cream into a cup of coffee, the movement of the molecules causes the cream to eventually be dispersed throughout the beverage. If you stir the coffee, this process occurs at a faster rate. Body movement assists passive transport, like stirring the coffee. It causes the diffusion to occur at a faster rate.

Filtration is the movement of both water and smaller molecules through a **semipermeable** membrane. The semipermeable membrane works like a screen that keeps the larger substances on one side and only permits the smaller molecules to filter to the other side of the membrane. Filtration is promoted by **hydrostatic** pressure differences between areas.

Hydrostatic pressure is the force that water exerts, sometimes called water pushing pressure. In the body, filtration is important for the movement of water, nutrients, and waste products in the capillaries. The capillaries serve as semipermeable membranes allowing water and smaller substances to move from the vascular system to the interstitial fluid, but larger molecules and red blood cells are kept inside the capillary walls.

---

electrolyte: electro—electricity + lyte—dissolve
intracellular: intra—within + cellular—cell
extracellular: extra—outside of + cellular—cell
interstitial: inter—between + stitial—tissue
intravascular: intra—within + vascular—blood vessel
transcellular: trans—across + cellular—cell

---

antidiuretic: anti—against + diuretic—urination
diffusion: diffuse—spread, scattered
filtration: filter—strain through
osmosis: osmo—impulse + osis—condition
semipermeable: semi—half or part of + permeable—passing through
hydrostatic: hydro—water + static—standing

Osmosis is the movement of water from an area of lower substance concentration to an area of higher concentration. The substances exert an osmotic pressure, sometimes called water pulling pressure. The term *osmolarity* refers to the concentration of the substances in body fluids. The normal osmolarity of the blood is between 270 and 300 milliosmoles per liter (mOsm/L).

Another term for osmolarity is *tonicity*. Fluids or solutions can be classified as **isotonic, hypotonic,** or **hypertonic.** A fluid that has the same osmolarity as the blood is called isotonic. For example, a 0.9 percent saline solution (normal saline) is isotonic to the blood and is often used as a solution for intravenous (IV) therapy. A solution that has a lower osmolarity than blood is called hypotonic. When a hypotonic solution is given to a patient, the water leaves the blood and other ECF areas and enters the cells. Hypertonic solutions exert greater osmotic pressure than blood. When a hypertonic solution is given to a patient, water leaves the cells and enters the bloodstream and other ECF spaces.

Fluid Gains and Fluid Losses

Water is very important to the body for cellular metabolism, blood volume, body temperature regulation, and solute transport. Although people can survive without food for several weeks, they can survive only a few days without water.

Water is gained and lost from the body every day. In addition to liquid intake, some fluid is obtained from solid foods. When too much fluid is lost, the brain's thirst mechanism tells the individual that more fluid intake is needed. Older adults are more prone to fluid deficits because they have a diminished thirst reflex and their kidneys do not function as effectively. An adult loses as much as 2500 mL of sensible and insensible fluid each day. Sensible losses are those of which the person is aware, such as urination. Insensible losses may occur without the person recognizing the loss. Perspiration and water lost through respiration and feces are examples of insensible losses.

▶ FLUID IMBALANCES

Fluid imbalances are common in any clinical setting. Elderly people are at the highest risk for life-threatening complications that can result from either fluid deficit, more commonly called **dehydration,** or fluid overload. Infants are at risk for fluid deficit because they take in and excrete a large proportion of their total body water each day.

Dehydration

Although there are several types of dehydration, only the most common type is discussed in this chapter. Dehydration occurs when there is not enough fluid in the body, especially in the blood (intravascular area).

Pathophysiology and Etiology

The most common form of dehydration results from loss of fluid from the body, resulting in decreased blood volume. This decrease is referred to as **hypovolemia.** Hypovolemia occurs when the patient is hemorrhaging or when fluids from other parts of the body are lost. For example, severe vomiting and diarrhea, severely draining wounds, and profuse diaphoresis (sweating) can cause dehydration.

Hypovolemia may also occur when fluid from the intravascular space moves into the interstitial fluid space. This process is called third spacing. Examples of conditions in which third spacing is common include burns, liver cirrhosis, and extensive trauma. Box 5–1 lists the common causes of dehydration.

As described earlier in this chapter, the body initially attempts to compensate for fluid loss by a number of mechanisms. If the cause of dehydration is not resolved or the patient is not able to replace the fluid, a state of dehydration occurs.

Prevention

You can help prevent dehydration by identifying patients who have the highest risk for developing this condition. High-risk patients include the elderly, infants, children, and any patient who has one of the conditions listed in Box 5–1.

Adequate hydration is another important intervention to help prevent dehydration. You should encourage patients to drink adequate fluids. If the patient is unable to take enough fluid by mouth, check with the physician for an order for IV therapy to replace fluid in the body.

Signs and Symptoms

Thirst is the initial symptom experienced by otherwise healthy adults in response to hypovolemia. As the percentage of water in the blood goes down, the percentage of other

BOX 5–1 Common Causes of Dehydration

Long-term nothing by mouth (NPO) status
Hemorrhage
Profuse diaphoresis (sweating)
Diuretic therapy
Diarrhea
Vomiting
Gastrointestinal suction
Draining fistulas
Draining abscesses
Severely draining wounds
Systemic infection
Fever
Frequent enemas
Ileostomy
Cecostomy
Diabetes insipidus

isotonic: iso—equal + tonic—strength
hypotonic: hypo—less than + tonic—strength
hypertonic: hyper—more than + tonic—strength
dehydration: de—down + hydration—water

hypovolemia: hypo—less than + vol—volume + emia—blood

substances goes up, resulting in the thirst response. As the blood volume decreases, the heart pumps the remaining blood faster but not as powerfully, resulting in a rapid, weak pulse and a low blood pressure. The body pulls water into the vascular system from other areas, resulting in decreased tear formation, dry skin, and dry mucous membranes.

The individual with dehydration has poor skin turgor. Turgor is poor if the skin is pinched and a small "tent" remains (called *tenting*). Temperature increases because the body is less able to cool itself through perspiration. Temperature may not appear elevated in an elderly person because an elder's normal body temperature is often several degrees lower than a younger person's. Urine output decreases and the urine becomes more concentrated as water is conserved. Dehydration should be considered in any adult with a urine output of less than 30 mL per hour. The urine may appear darker because it is less diluted. The patient becomes constipated as the intestines absorb more water from the feces. A major method of evaluating dehydration is weight loss. A pint of water weighs approximately 1 pound. Symptoms of dehydration in the elderly may be atypical (Gerontological Issues 5–2).

BOX 5-2 GERONTOLOGICAL ISSUES

As a person ages, total body water decreases from 60 percent to 50 percent of total body weight. The age-related decrease in total body water is secondary to an increase in body fat and decreased thirst sensation. These factors increase the risk of developing dehydration.

Manifestations of dehydration in an older adult are different from typical manifestations in a younger person. Atypical manifestations include altered mental status, light-headedness, and syncope. These occur because a patient with hypovolemia has an inadequate blood supply, and therefore oxygen supply, to the brain.

Complications

If dehydration is not treated, lack of sufficient blood volume causes organ function to decrease and eventually fail. The brain, kidneys, and heart must be adequately supplied with blood (perfused) to function properly. The body protects these organs by decreasing blood flow to other areas. When these organs no longer receive their minimum requirements, death results.

Diagnostic Tests

The patient with dehydration usually has an elevated blood urea nitrogen (BUN) level and elevated hematocrit. Both values are increased because there is less water in proportion to the solid substances being measured. The specific gravity of the urine also increases as the kidneys attempt to conserve water, resulting in a more concentrated urine.

Medical Treatment

The goals of medical treatment are to replace fluids and resolve the cause of dehydration. In a patient with moderate or severe dehydration, IV therapy is used. Isotonic fluids that have the same osmolarity as blood are typically administered.

Nursing Process

You can play a major role in identifying patients who are dehydrated.

ASSESSMENT. Assess the patient for signs and symptoms of dehydration. All the classic signs and symptoms may not be present.

When assessing an elderly patient for skin turgor (tenting), assess the skin over the forehead or sternum. The skin over these areas usually retains elasticity and is therefore a more reliable indicator of skin turgor.

Weight is the most reliable indicator of fluid loss or gain. A loss of 1 to 2 pounds or more per day suggests water loss rather than fat loss. The patient in the hospital setting should be weighed every day. The patient in the nursing home or home setting should be weighed three times a week if the patient is at risk for fluid imbalance. The nursing staff should weigh the patient before breakfast using the same scale each time.

NURSING DIAGNOSES. The primary nursing diagnoses for the patient with dehydration include the following:

- Deficient fluid volume related to fluid loss or inadequate fluid intake
- Decreased cardiac output related to insufficient blood volume
- Impaired oral mucous membrane related to inadequate oral secretions
- Ineffective (cerebral) tissue perfusion related to insufficient blood volume
- Constipation related to decreased body fluids

PLANNING. The health care team works together to restore fluids in the patient. The expected outcome is that the patient will be adequately hydrated and not experience further episodes of dehydration.

IMPLEMENTATION. Patients can become dehydrated in any setting. For the patient who is mildly dehydrated, interventions can be implemented in the home or nursing home setting. For moderate or severe dehydration, the patient may require hospitalization to treat the underlying cause and replace fluid losses. Make sure the patient does not receive too much fluid replacement, which could cause fluid overload, discussed in the next section of this chapter. Fluid intake and output (I&O) should be carefully measured and monitored (Cultural Consideration 5–3).

BOX 5-3 CULTURAL CONSIDERATION

Muslims who celebrate Ramadan fast for 1 month from sunup to sundown. Although the ill are not required to fast, the pious may still insist on doing so. Fasting may include not taking fluids and medications during daylight hours. Therefore the nurse may need to alter times for medication administration, including intramuscular medication. Special precautions may need to be taken to prevent dehydration in Muslim patients.

EVALUATION. As part of evaluation, determine whether the patient has been adequately rehydrated, using the assessment techniques described earlier.

Patient Education

The patient, family, and significant others need to be taught the importance of reporting early signs and symptoms of dehydration to a physician or other health care provider. At home or in the nursing home, infections often cause fever and sepsis, a serious condition in which the infection invades the bloodstream. The body attempts to decrease the temperature through perspiration. The patient becomes dehydrated as a result and can become increasingly ill.

CRITICAL THINKING

Mrs. Levitt

Mrs. Levitt is a 92-year-old widow who has been in a nursing home for 4 years. Today she complains that her urine smells bad and that her heart feels like it is beating faster than usual. You suspect that she is becoming dehydrated. You check her urine and find that it is a dark amber color and that it smells strong. Her heart rate is 98, blood pressure 126/74, respiratory rate 20, and temperature 99.2.

1. What other data should you collect, and what results do you expect?
2. Which interventions should you provide at this time?
3. How should you document your findings?

Answers at end of chapter.

Fluid Overload

Fluid overload, sometimes called overhydration, is a condition in which a patient has too much fluid in the body. Most of the problems related to fluid overload result from too much fluid in the bloodstream or from dilution of electrolytes and red blood cells.

Pathophysiology and Etiology

The most common result of fluid overload is **hypervolemia,** in which there is excess fluid in the intravascular space. Healthy adult kidneys can compensate for mild to moderate hypervolemia. The kidneys increase urinary output to rid the body of the extra fluid.

The causes of fluid overload are related to excessive intake of fluids or inadequate excretion of fluids. Conditions that can cause excessive fluid intake are poorly controlled IV therapy, excessive irrigation of wounds or body cavities, and excessive ingestion of water. Conditions that can result in inadequate excretion of fluid include renal failure, heart failure, and syndrome of inappropriate antidiuretic hormone. These conditions are discussed elsewhere in this book.

Prevention

One of the best ways to prevent fluid overload is to avoid excessive fluid intake. For example, you should monitor the patient receiving IV therapy for signs and symptoms of fluid overload. In at-risk patients an electronic infusion pump or a quantity-limiting device, such as a burette, should be used to control the rate of infusion.

Also monitor the amount of fluid used for irrigations. For example, when a patient's stomach is being irrigated (gastric lavage), be sure an excess amount of fluid is not absorbed.

Signs and Symptoms

The vital sign changes seen in the patient with fluid overload are the opposite of those found in patients with dehydration. The blood pressure is elevated, pulse is bounding, and respirations are increased and shallow. The neck veins may become distended, and pitting **edema** in the feet and legs may be present. The skin is pale and cool. The kidneys increase urine output, and the urine appears diluted, almost like water. The patient rapidly gains weight. In severe fluid overload the patient develops moist crackles in the lungs, dyspnea, and ascites (excess peritoneal fluid).

Complications

Acute fluid overload typically results in congestive heart failure. As the fluid builds up in the heart, the heart is not able to properly function as a pump. The fluid then backs up into the lungs, causing a condition known as pulmonary edema. Other major organs of the body cannot receive adequate oxygen, and organ failure can lead to death.

Diagnostic Tests

In the patient experiencing fluid overload, the BUN and hematocrit levels tend to decrease from hemodilution. The plasma content of the blood is proportionately increased when compared with the solid substances. The specific gravity of the urine also diminishes as the urinary output increases.

Medical Treatment

Once the patient's breathing has been supported, the goal of treatment is to rid the body of excessive fluid and resolve the underlying cause of the overload. Drug therapy and diet therapy are commonly used to decrease fluid retention.

POSITIONING. To facilitate ease in breathing, the head of the patient's bed should be in semi-Fowler's or high Fowler's position. These positions allow greater lung expansion and thus aid respiratory effort. Once the patient has been properly positioned, oxygen therapy may be necessary (Fig. 5–2).

OXYGEN THERAPY. Oxygen therapy is typically used to ensure adequate perfusion of major organs and to minimize dyspnea. If the patient has a history of chronic obstructive

---

hypervolemia: hyper—more than + vol—volume + emia—blood

edema: swelling

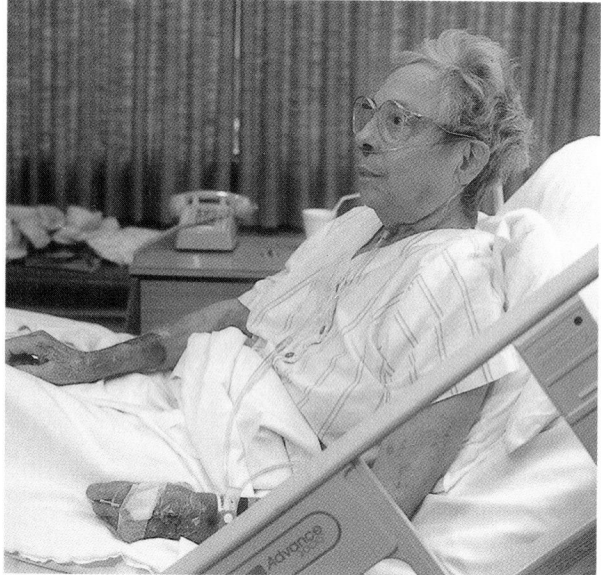

Figure 5-2   Patient in high Fowler's position with oxygen.

pulmonary disease, such as emphysema or chronic bronchitis, do not administer more than 2 L per minute of oxygen. At higher oxygen doses the patient may lose the stimulus to breathe and may suffer respiratory arrest.

DRUG THERAPY. Diuretics are frequently administered to rapidly rid the body of excess water. A diuretic is a drug that increases elimination by the kidneys. The drug of choice for fluid overload when the patient has adequately functioning kidneys is usually furosemide (Lasix). Furosemide is a loop (high-ceiling) diuretic that causes the kidneys to excrete sodium and water. Sodium (Na^+) and water tend to move together in the body. Potassium (K^+), another electrolyte, is also lost, which can lead to a potassium deficit, discussed later in this chapter.

Furosemide may be given by the oral, intramuscular, or IV route. The oral route is used most commonly for mild fluid overload. IV furosemide is administered by a registered nurse (RN) or physician for severe overload. The patient should begin diuresis within 30 minutes after receiving IV furosemide. If not, another dose is given.

DIET THERAPY. Mild to moderate fluid restriction may be necessary, as well as a sodium-restricted diet. In collaboration with the dietitian, a physician prescribes the specific restriction necessary, usually a 1 to 2 g Na^+ restriction for severe overload. Different diuretics result in differing electrolyte elimination. Specific diet therapy depends on the medications the patient is receiving and the patient's underlying medical problems.

Nursing Process

The nurse plays a pivotal role in the care of a patient with fluid overload. Prompt action is needed to prevent life-threatening complications.

ASSESSMENT. Observe a patient who is at high risk for fluid overload and monitor fluid I&O carefully. If the pa-

tient is drinking adequate amounts of fluid (1500 mL per day or more) but is voiding in small amounts, the fluid is being retained by the body.

Assess for edema, which is fluid that accumulates in the interstitial tissues. If the edema is pitting, a finger pressed against the skin over a bony area such as the tibia leaves a temporary indentation. For patients in bed, check the sacrum for edema. For patients in the sitting position, check the feet and legs.

As mentioned earlier, weight is the most reliable indicator of fluid gain. A gain of 1 to 2 pounds or more per day indicates fluid retention, even though other signs and symptoms may not be present.

NURSING DIAGNOSES. Common nursing diagnoses for a patient with fluid overload are as follows:

- Excess fluid volume related to excessive fluid intake or inadequate excretion of body fluid
- Decreased cardiac output related to excess work on the heart from fluid retention
- Ineffective tissue perfusion related to dependent edema
- Risk for impaired gas exchange related to fluid in the lungs

PLANNING. The expected outcome for the patient is that fluid balance is achieved and that the patient does not experience life-threatening cardiac or pulmonary failure.

IMPLEMENTATION. The patient at risk for fluid overload may be in the hospital, nursing home, or private home. Severe overload requires hospitalization, possibly to a critical care unit, and placement on a ventilator if the patient has severe pulmonary edema. Mild or moderate overload may be treated outside the hospital, but the patient must follow the treatment plan carefully to avoid worsening the condition.

If the patient is placed on a fluid restriction, you can work with the RN to determine how it should be implemented. For example, if a patient is on a 1000 mL per day fluid restriction, you might plan for 150 mL with each meal, 450 mL to be given to the patient to use as he or she likes during the day and 100 mL to be used during the night. Be sure to include the patient in your planning, and remember to reserve enough fluid for swallowing pills. Post a sign in the patient's room so other caregivers know how much fluid the patient can have.

EVALUATION. During treatment, evaluate whether the expected outcomes are achieved. Many patients must remain on drug and diet therapy after hospital discharge to prevent the problem from recurring.

Patient Education

In collaboration with the dietitian, the patient, family, or other caregiver should be instructed about the fluid and sodium restrictions to prevent further problems (Nutrition Notes Box 5-4). High-sodium foods to avoid are listed in Box 5-5.

Teaching caregivers about diuretic therapy is essential to prevent electrolyte imbalances. If a potassium-losing diuretic is prescribed, teach which foods are high in potas-

BOX 5-4 Nutrition Notes

Reducing Sodium Intake

Many foods, such as dairy products, grain products, and some vegetables, are naturally high in sodium, but the major sources of sodium in the diet are salted and processed foods, including baked goods and condiments. For example, compare the sodium content of cheddar and American cheese and of fresh and cured pork in Box 5–2. Drinking water may contain significant amounts of sodium, particularly if it is a softened or mineral water. Because of the numerous "hidden" sources of sodium, a dietitian's services are often necessary.

Specific definitions for reduced sodium food products have been adopted. Note that serving size is an important variable. The amount of daily intake of sodium recommended for adults is 2400 mg.

- Salt or sodium free: <5 mg sodium per serving
- Very low sodium: <35 mg sodium per serving (per 100 g if main dish)
- Low sodium: <140 mg sodium per serving (per 100 g if main dish)

BOX 5-5 Common Food Sources of Sodium*

Food Source	Amount (mg)
Table salt (1 tsp)	2000
Cheddar cheese (1 oz)	176
Cottage cheese (4 oz)	457
American cheese (1 oz)	439
Whole milk (8 oz)	120
Skim milk (8 oz)	126
Butter (1 tsp)	123
White bread (1 slice)	123
Whole-wheat bread (1 slice)	159
Soy sauce (1 tbsp)	1029
Ketchup (1 tbsp)	156
Mustard (1 tbsp)	188
Beef, lean (4 oz)	60
Pork, lean, fresh (4 oz)	60
Pork, cured (4 oz)	850
Chicken, light meat (4 oz)	70
Chicken, dark meat (4 oz)	70

Source: Ignativicius, DD, Workman, ML, and Mishler, MS: Medical-Surgical Nursing: A Nursing Process Approach, ed 2. WB Saunders, Philadelphia, 1995, with permission. (Data from Pennington, J: Bowe's and Church's Food Values of Portions Commonly Used, ed 16. JB Lippincott, Philadelphia, 1992.)
*U.S. Department of Agriculture recommended daily allowance for adults: 1100–3300 mg.

sium, such as oranges and other citrus fruits, melons, bananas, and potatoes. The patient's serum potassium level needs to be periodically monitored by a physician or home care nurse. If it becomes too low, an oral potassium supplement is needed.

The family or other caregiver also needs to be taught common signs and symptoms of fluid overload that should be reported to a physician or other health care provider. Of special importance is weight gain. The patient should be weighed at least three times a week in the home or nursing home if he or she is at high risk for fluid overload.

CRITICAL THINKING

Mr. Peters

Mr. Peters is a 32-year-old man with a congenital heart problem. He has been recovering from congestive heart failure and fluid overload. Today his blood pressure is higher than usual and his pulse is bounding. He is having trouble breathing and presses the call light for your assistance.

1. What should you do first when you assess Mr. Peters's condition?
2. What questions should you ask him?
3. What other assessments should you perform?

Answers at end of chapter.

▶ ELECTROLYTE BALANCE

Natural minerals in food become electrolytes or ions in the body through digestion and metabolism. Electrolytes are usually measured in milliequivalents per liter (mEq/L) or in milligrams per deciliter (mg/dL).

Electrolytes are one of two types: **cations,** which carry a positive electrical charge, and **anions,** which carry a negative electrical charge. Although there are many electrolytes in the body, this chapter discusses the most important ones, including sodium (Na^+), potassium (K^+), calcium (Ca^{2+}), and magnesium (Mg^{2+}). These electrolytes are maintained in different concentrations inside the cell and outside the cell because of pumps in the cell wall (Fig. 5–3).

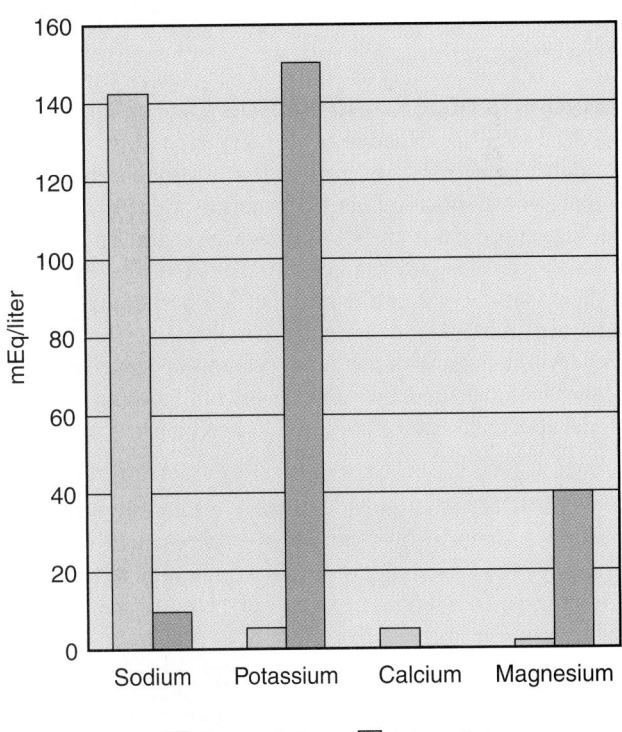

Figure 5-3 Extracellular and intracellular electrolytes.

cation: cat—descending + ion—carrying
anion: an—without + ion—carrying

ELECTROLYTE IMBALANCES

The two types of electrolyte imbalances are deficit and excess. In general, if a patient experiences a deficit of an electrolyte, the electrolyte is replaced either orally or intravenously. If the patient experiences an excess of the electrolyte, treatment focuses on getting rid of the excess, often by the kidneys. The underlying cause of the imbalance must also be treated.

The most important aspect of nursing care is preventing and assessing electrolyte imbalances. High-risk patients should be identified and monitored carefully. Serum electrolytes are measured on a regular basis. As a general rule, patients should be checked for electrolyte imbalance when there is a change in their mental state (either increased irritability or decreased responsiveness) or when their muscles start cramping. Patient education is another important nursing role for patients with electrolyte imbalances.

Sodium Imbalances

The normal level of serum sodium is 135 to 145 mEq/L. Because sodium is the major cation in the blood, it helps maintain serum osmolarity. Therefore sodium imbalances are often associated with fluid imbalances, described earlier in this chapter. Sodium is also important for cell function, especially in the central nervous system. The two sodium imbalances are **hyponatremia** (sodium deficit) and **hypernatremia** (sodium excess).

Hyponatremia

Hyponatremia occurs when the serum sodium level is less than 135 mEq/L.

PATHOPHYSIOLOGY AND ETIOLOGY. Many conditions can lead to either an actual or a relative decrease in sodium. In an actual decrease the patient has inadequate intake of sodium or excessive sodium loss from the body. As the percentage of sodium in the ECFs decreases, water is pulled by osmotic pressure into the cells. In a relative decrease the sodium is not lost from the body but leaves the intravascular space and moves into the interstitial tissues (third spacing). Another cause of a relative decrease occurs when the plasma volume increases (fluid overload), causing a dilutional effect. The percentage of sodium compared with the fluid is diminished.

PREVENTION. Additional sodium is commonly administered to patients at high risk for hyponatremia (Box 5–6), usually by the IV route. Individuals who have high fevers or who engage in strenuous exercise or physical labor, especially in the heat, need to replace both sodium and water. Hyponatremia is especially dangerous for the elderly patient.

SIGNS AND SYMPTOMS. Unfortunately, the signs and symptoms of hyponatremia are vague and depend somewhat

---

hyponatremia: hypo—less than + natr—sodium + emia—blood

hypernatremia: hyper—more than + natr—sodium + emia—blood

BOX 5-6	Patients at High Risk for Hyponatremia

Nothing by mouth (NPO)
Excessive diaphoresis (sweating)
Diuretics
Gastrointestinal suction
Syndrome of inappropriate antidiuretic hormone
Excessive ingestion of hypotonic fluids
Freshwater near-drowning
Decreased aldosterone

on whether a fluid imbalance accompanies the hyponatremia. The patient with sodium and fluid deficits has signs and symptoms of dehydration (discussed earlier). The patient with a sodium deficit and fluid overload has signs and symptoms associated with fluid overload.

In addition, the patient experiences mental status changes, including disorientation, confusion, and personality changes, caused by cerebral edema (fluid around the brain). Weakness, nausea, vomiting, and diarrhea may also occur.

COMPLICATIONS. In severe hyponatremia, respiratory arrest or coma can lead to death. The patient who also has fluid overload may develop pulmonary edema, another life-threatening complication.

DIAGNOSTIC TESTS. The primary diagnostic test is to obtain the serum sodium level, which is lower than the normal value when hyponatremia is present. The serum osmolarity also decreases in patients with hyponatremia. Other laboratory tests may be affected if the patient experiences an accompanying fluid imbalance. Serum chloride (Cl^-), an anion, is often depleted when sodium decreases because these two electrolytes commonly combine as NaCl (salt in solution, or saline).

MEDICAL TREATMENT. Medical treatment focuses on resolving the underlying cause of hyponatremia and replacing the lost sodium. The physician orders IV saline for patients who have hyponatremia without fluid overload. The saline solution given may be isotonic (0.9 percent) or hypertonic (3 percent), depending on the severity of the problem.

For patients who have a fluid overload, a fluid restriction is often ordered. Diuretics that rid the body of fluid but do not cause sodium loss may also be needed. For patients with cerebral edema, steroids may be prescribed to reduce **intracranial** swelling. I&O are strictly monitored, and the patient is weighed daily.

Hypernatremia

Hypernatremia occurs when the serum sodium level is above 145 mEq/L.

PATHOPHYSIOLOGY AND ETIOLOGY. Serum sodium increase may be an actual increase or a relative increase. In an

---

intracranial: intra—within + cranial—cranium (skull)

actual increase the patient receives too much sodium or is unable to excrete sodium, as seen in renal failure. In a relative increase the amount of sodium does not change but the amount of fluid in the intravascular space decreases. The percentage of sodium (solid) is increased in relationship to the amount of plasma (water).

In mild hypernatremia, most excitable tissues, such as muscle and neurons of the brain, become more stimulated. The patient becomes irritable and has tremors. In severe cases, these tissues fail to respond.

PREVENTION. Prevention of hypernatremia is not as simple as prevention of hyponatremia. Most patients have a sodium excess as a result of an acute or chronic illness. Patients with a potential for electrolyte imbalance must always have their IV fluids carefully regulated.

SIGNS AND SYMPTOMS. Thirst is usually one of the first symptoms to appear. If you eat salty foods, such as potato chips, the amount of sodium in the body increases and you become thirsty. Other signs and symptoms of hypernatremia are vague and nonspecific until severe excess is present. Like the patient with a sodium deficit, the patient experiencing sodium excess has mental status changes, such as agitation, confusion, and personality changes. Seizures may also occur.

At first, muscle twitches and unusual contractions may be present. Later, skeletal muscle weakness occurs that can lead to respiratory failure. If fluid deficit or fluid overload accompanies the hypernatremic state, the patient also has signs and symptoms associated with these imbalances.

COMPLICATIONS. The patient experiencing severe hypernatremia may become comatose or have respiratory arrest as skeletal muscles weaken.

DIAGNOSTIC TEST. The most reliable diagnostic test is the serum sodium level, which indicates an increase above the normal level. Serum osmolarity may also increase. If the patient has a fluid imbalance, other laboratory values, such as BUN, hematocrit, and urine specific gravity, are affected (see earlier discussion).

MEDICAL TREATMENT. If a fluid imbalance accompanies hypernatremia, it is treated first. For example, fluid replacement without sodium in a patient with dehydration should correct a relative sodium excess. If the kidneys are not excreting adequate amounts of sodium, diuretics may help if the kidneys are functional. If the kidneys are not functioning properly, dialysis may be ordered. I&O and daily weights are strictly monitored.

The cause of hypernatremia is also treated in an attempt to prevent further episodes of this imbalance. For some patients, a sodium-restricted diet is prescribed.

Potassium Imbalances

Potassium is the most common electrolyte in the ICF compartment. Therefore only a small amount, 3.5 to 5 mEq/L, is found in the bloodstream. Minimal changes in this laboratory value cause major changes in the body.

Potassium is especially important for cardiac muscle, skeletal muscle, and smooth muscle function. As the serum potassium level falls, the body attempts to compensate by moving potassium from the cells into the bloodstream.

The two potassium imbalances are **hypokalemia** (potassium deficit) and **hyperkalemia** (potassium excess). Hypokalemia is the most commonly occurring imbalance.

Hypokalemia

Hypokalemia occurs when the serum potassium level falls below 3.5 mEq/L.

PATHOPHYSIOLOGY AND ETIOLOGY. Most cases of hypokalemia result from inadequate intake of potassium or excessive loss of potassium through the kidneys. Hypokalemia most often occurs as a result of medications. Potassium-losing diuretics (e.g., furosemide [Lasix]), digitalis preparations (e.g., digoxin [Lanoxin]), and corticosteroids (e.g., prednisone [Deltasone]) are examples of drugs that cause increased excretion of potassium from the body. Potassium may also be lost through the gastrointestinal (GI) tract, which is rich in potassium and other electrolytes. Severe vomiting, diarrhea, and prolonged GI suction cause hypokalemia. Major surgery and hemorrhage can also lead to potassium deficit.

PREVENTION. Most patients having major surgery receive potassium supplements in their IV fluids to prevent hypokalemia. For patients receiving drugs known to cause hypokalemia, foods high in potassium may prevent a deficit (Box 5–7). Patients receiving digitalis must be closely monitored because hypokalemia can enhance the action of digitalis and cause digitalis toxicity.

SIGNS AND SYMPTOMS. Many body systems are affected by a potassium imbalance. Muscle cramping occurs with either a deficit or an excess of potassium. Vital signs change because the respiratory and cardiovascular systems need potassium to function properly. Skeletal muscle activity diminishes, resulting in shallow, ineffective respirations. The pulse is typically weak, irregular, and thready because the heart muscle is depleted of potassium. A major danger is an irregular heartbeat (**dysrhythmia**), which can lead to a cardiac arrest. Orthostatic (postural) hypotension may also be present.

The nervous system is usually affected as well. The patient experiences changes in mental status, followed by lethargy. The motility of the GI system is slowed, causing nausea, vomiting, abdominal distention, and constipation. Vomiting may further increase potassium loss.

COMPLICATIONS. If not corrected, hypokalemia can result in death from dysrhythmia, respiratory failure and arrest, or

hypokalemia: hypo—less than + kal—potassium + emia—blood
hyperkalemia: hyper—more than + kalpotassium + emia—blood
dysrhythmia: dys—bad or disordered + rhythmia—measured motion

BOX 5-7 **Common Food Sources of Potassium***

Food Source	Amount (mg)
Corn flakes (1¼ c)	26
Cooked oatmeal (¾ c)	99
Egg (1 large)	66
Codfish, raw (4 oz)	400
Salmon, pink, raw (3½ oz)	306
Tuna fish (4 oz)	375
Apple, raw with skin (1 medium)	159
Banana (1 medium)	451
Cantaloupe (1 c pieces)	494
Grapefruit (½ medium)	175
Orange (1 medium)	250
Raisins (½ c)	700
Strawberries, raw (1 c)	247
Watermelon (1 c pieces)	186
White bread (1 slice)	27
Whole-wheat bread (1 slice)	44
Beef (4 oz)	480
Beef liver (3½ oz)	281
Pork, fresh (4 oz)	525
Pork, cured (4 oz)	325
Chicken (4 oz)	225
Veal cutlet (3½ oz)	448
Whole milk (8 oz)	370
Skim milk (8 oz)	406
Avocado (1 medium)	1097
Carrot (1 large)	341
Corn (4-inch ear)	196
Cauliflower (1 c pieces)	295
Celery (1 stalk)	170
Green beans (1 c)	189
Mushrooms (10 small)	410
Onion (1 medium)	157
Peas (¾ c)	316
Potato, white (1 medium)	407
Spinach, raw (3½ oz)	470
Tomato (1 medium)	366

Sources: Pennington, J: Bowe's and Church's Food Values of Portions Commonly Used, ed 16. JB Lippincott, Philadelphia, 1992. Ignatavicius, DD, et al: Medical-Surgical Nursing: A Nursing Process Approach, ed 2. WB Saunders, Philadelphia, 1995, with permission.

*U.S. Department of Agriculture recommended daily allowance for adults: 1875–5625 mg.

coma. The patient must be treated promptly before these complications occur.

DIAGNOSTIC TESTS. The primary laboratory test is to obtain a serum potassium level. The patient's electrocardiogram (ECG) may show cardiac dysrhythmias associated with potassium deficit. In addition to a decrease in serum potassium level, the patient may have an acid-base imbalance known as metabolic **alkalosis,** which commonly accompanies hypokalemia. In metabolic alkalosis, the serum potential of hydrogen (pH) of the blood increases (above 7.45) so that the blood is more alkaline than usual. Acid-base imbalances are discussed later in this chapter.

alkalosis: alka—alkaline + losis—condition

MEDICAL TREATMENT. The goal of treatment is to replace potassium in the body and resolve the underlying cause of the imbalance. For mild to moderate hypokalemia, oral potassium supplements are given.

For severe hypokalemia, IV potassium supplements are given. Because the kidneys eliminate excess potassium, physicians frequently add potassium to the IV fluids only after the patient has voided. Potassium is a potentially dangerous drug, especially when administered intravenously. In too high a concentration, it causes cardiac arrest. For this reason, most health care facilities do not permit concentrated potassium in patient care areas. The IV container should be changed to one premixed with the proper potassium concentration instead of adding concentrated potassium to the existing container. Potassium is *never* given by IV push. The patient's laboratory values must be monitored carefully to prevent giving too much potassium.

Teach the patient about the side effects of oral potassium and precautions associated with potassium administration. Box 5–8 summarizes the precautions you need to be aware of when giving oral potassium supplements.

Hyperkalemia

Hyperkalemia is a condition in which the serum potassium level exceeds 5 mEq/L. It is rare in a person with healthy kidneys.

PATHOPHYSIOLOGY AND ETIOLOGY. Hyperkalemia may result from an actual increase in the amount of total body potassium or from the movement of intracellular potassium

BOX 5-8 **Education Tips for Patients Receiving Oral Potassium Supplements**

- Do not substitute one potassium supplement for another.
- Dilute powders and liquids in juice or other desired liquid to improve taste and to prevent gastrointestinal irritation. Follow manufacturer's recommendations for the amount of fluid to use for dilution, most commonly 4 oz per 20 mEq of potassium.
- Do not drink diluted solutions until mixed thoroughly.
- Do not crush potassium tablets, such as Slow-K or K-tab tablets. Read manufacturer's directions regarding which tablets can be crushed.
- Administer slow-release tablets with 8 oz of water to help them dissolve.
- Do not take potassium supplements if taking potassium-sparing diuretics such as spironolactone or triamterene.
- Do not use salt substitutes containing potassium unless prescribed by the physician.
- Take potassium supplements with meals.
- Report adverse effects, such as nausea, vomiting, diarrhea, and abdominal cramping, to the physician.
- Have frequent laboratory testing for potassium levels as recommended by the physician.

Source: Adapted from Lee, CA, Barrett, CA, and Ignatavicius, DD: Fluids and Electrolytes: A Practical Approach, ed 4. FA Davis, Philadelphia, 1996.

into the blood. Overuse of potassium-based salt substitutes or excessive intake of oral or IV potassium supplements can cause hyperkalemia. Use of potassium-sparing diuretics (e.g., spironolactone [Aldactone]) may also contribute to hyperkalemia. Patients with renal failure are at risk for hyperkalemia because the kidneys cannot excrete potassium.

Movement of potassium from the cells into the blood and other ECF is common in massive tissue trauma and metabolic **acidosis.** Metabolic acidosis is an acid-base imbalance commonly seen in patients with uncontrolled diabetes mellitus. Acid-base imbalances are discussed later in this chapter.

PREVENTION. For patients receiving potassium supplements, hyperkalemia can be prevented by monitoring serum electrolyte values and the patient's signs and symptoms.

SIGNS AND SYMPTOMS. Most cases of hyperkalemia occur in patients who are hospitalized or those undergoing medical treatment for a chronic condition. The classic manifestations are muscle twitches and cramps, later followed by profound muscular weakness; increased GI motility (diarrhea); slow, irregular heart rate; and decreased blood pressure.

COMPLICATIONS. Cardiac dysrhythmias and respiratory failure can occur in severe hyperkalemia, causing death.

DIAGNOSTIC TESTS. In addition to an elevated serum potassium level, an irregular ECG is associated with hyperkalemia. If the patient also has metabolic acidosis, the serum pH falls below 7.35.

MEDICAL TREATMENT. For mild, chronic hyperkalemia, dietary limitation of potassium-rich foods may be helpful. Potassium supplements are discontinued, and potassium-losing diuretics are given to patients with healthy kidneys. For patients with renal problems, a cation exchange resin, such as sodium polystyrene sulfonate (Kayexalate), is administered either orally or rectally. This drug releases sodium and absorbs potassium for excretion through the feces and out of the body.

In cases in which cellular potassium has moved into the bloodstream, administration of glucose and insulin can facilitate the movement of potassium back into the cells. During treatment of moderate to severe hyperkalemia, the patient should be in the hospital on a cardiac monitor.

Calcium Imbalances

Calcium is a mineral that is primarily stored in bones and teeth. A small amount is found in ECF. The normal value for serum calcium is 9 to 11 mg/dL, or 4.5 to 5.5 mEq/L. Minimal changes in serum calcium levels can have major negative effects in the body.

Calcium is needed for the proper function of excitable tissues, especially cardiac muscle. It is also needed for adequate blood clotting. The two calcium imbalances are **hypocalcemia** and **hypercalcemia.**

Hypocalcemia

Hypocalcemia occurs when the serum calcium falls below 9 mg/dL or 4.5 mEq/L.

PATHOPHYSIOLOGY AND ETIOLOGY. Although calcium deficit can be acute or chronic, most patients develop hypocalcemia slowly as a result of chronic disease or poor intake. The woman who is postmenopausal is most at risk for hypocalcemia. As a woman ages, calcium intake typically declines. The parathyroid glands recognize this decrease and stimulate bone to release some of its stored calcium into the blood for replacement. The result is a condition known as **osteoporosis,** in which bones become porous and brittle and fracture easily. The woman who is postmenopausal has a decreased level of estrogens, hormones that help prevent bone loss in the younger woman. Immobility or decreased mobility also contributes to bone loss in many patients. The patients at highest risk for osteoporosis are thin, petite, Caucasian women.

Hypocalcemia can also result from inadequate absorption of calcium from the intestines, as seen in patients with Crohn's disease, a chronic inflammatory bowel disease. An insufficient intake of vitamin D prevents calcium absorption as well. Conditions that interfere with the production of parathyroid hormone, such as partial or complete surgical removal of the thyroid or parathyroids, can also cause hypocalcemia.

Finally, patients with hyperphosphatemia (usually those with renal failure) often experience hypocalcemia. Calcium and phosphate have an inverse relationship. When one of these electrolytes increases, the other decreases, and vice versa.

PREVENTION. In the United States the typical daily calcium intake is less than 550 mg. The adequate intake (AI) of calcium for adults ages 19 to 50 is 1000 mg; the AI for adults over age 50 is 1200 mg.

Hypocalcemia can be prevented in premenopausal and postmenopausal women by consuming calcium-rich foods and by taking calcium supplements. These supplements can be purchased over the counter in any pharmacy or large food store. An inexpensive source of calcium for patients who do not require vitamin D supplementation is calcium carbonate (Tums), which provides 240 mg of elemental calcium in each tablet.

---

acidosis: acid—acidic + osis—condition

hypocalcemia: hypo—less than + calc—calcium + emia—blood
hypercalcemia: hyper—more than + calc + calcium + emia—blood
osteoporosis: osteo—bone + porosis—porous

Vitamin D supplementation may be required in addition to calcium for homebound or institutionalized patients who have no exposure to the sun. The ultraviolet light causes the skin to manufacture vitamin D.

SIGNS AND SYMPTOMS. Chronic hypocalcemia is usually not diagnosed until the patient breaks a bone, usually a hip. Acute hypocalcemia, which can occur after surgery or in patients with acute pancreatitis, has several signs and symptoms. These include increased and irregular heart rate, mental status changes, hyperactive deep tendon reflexes, and increased GI motility, including diarrhea and abdominal cramping. Two classic signs that can be used to assess for hypocalcemia are Trousseau's sign and Chvostek's sign.

To test for Trousseau's sign, inflate a blood pressure cuff around the patient's upper arm for 1 to 4 minutes. In a patient with hypocalcemia the hand and fingers become spastic and go into palmar flexion (Fig. 5–4). A positive Chvostek's sign test also indicates calcium deficit. To test for this sign, tap the face just below and in front of the ear. Facial twitching on that side of the face indicates a positive test (Fig. 5–5).

COMPLICATIONS. In severe hypocalcemia, seizures, respiratory failure, or cardiac failure can occur and lead to death if not aggressively treated. The patient may have a sudden laryngospasm that will stop air from entering the patient's lungs.

DIAGNOSTIC TESTS. The patient with hypocalcemia has a lowered serum calcium and an abnormal ECG. The parathyroid hormone level may be increased because it stimulates bone to release more calcium into the blood.

MEDICAL TREATMENT. In addition to treating the cause of hypocalcemia, calcium is replaced. For mild or chronic hypocalcemia, oral calcium supplements with or without vitamin D are given. Calcium supplements should be administered 1 to 2 hours after meals to increase intestinal absorption.

For patients with acute or severe hypocalcemia, IV calcium gluconate or calcium chloride is given. When a pa-

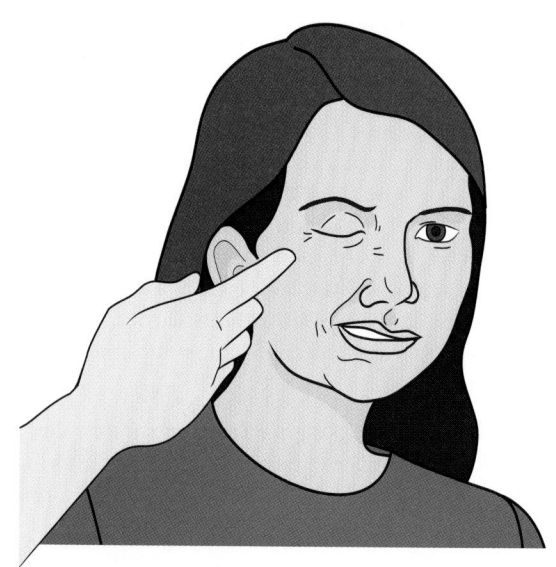

Figure 5-5 Chvostek's sign.

tient has had thyroid or parathyroid surgery, this medication must be readily available for emergency use.

For patients with hyperphosphatemia, usually those with renal failure, aluminum hydroxide is used to bind the excess phosphate for elimination via the GI tract. As the phosphate decreases, the serum calcium begins to increase closer to normal levels.

Diet therapy is an important part of treatment. Teach the patient, family, or other caregiver which foods are high in calcium (Table 5–1). Many foods today are fortified with calcium. Vitamin D foods are also encouraged, especially milk and other dairy products.

Hypercalcemia

Hypercalcemia occurs when the serum calcium is above 11 mg/dL or 5.5 mEq/L.

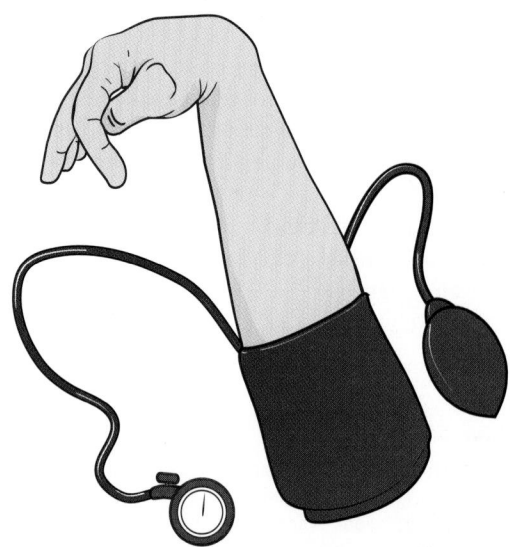

Figure 5-4 Trousseau's sign.

TABLE 5-1	QUANTITIES OF FOOD CONTAINING CALCIUM EQUAL TO 1 CUP OF MILK, IN ORDER OF ENERGY CONTENT	
Food	**Amount**	**Kilocalories**
Skim milk	1 c	86
Grated Parmesan cheese	4.3 tbs	99
Plain low-fat yogurt	0.7 c	101
Swiss cheese	1.1 oz	118
2% milk	1 c	121
Whole milk	1 c	150
Cheddar cheese	1.5 oz	171
Processed American cheese	1.7 oz	180
Low-fat yogurt with fruit	0.9 c	199
Blue cheese	2 oz	200
Vanilla milkshake	0.9 c	273
2% low-fat cottage cheese	2 c	410
Hard ice cream, vanilla	1.7 c	459
Soft ice cream	1.3 c	479
Cottage cheese, creamed, large curd	2.25 c	529
Sherbet	2.9 c	786

Source: Adapted from Lutz, CA, and Przytulski, KR: Nutrition and Diet Therapy, ed 2. FA Davis, Philadelphia, 1997, p 172.

CRITICAL THINKING

Mrs. Wright

Mrs. Wright is a 77-year-old petite Caucasian woman who lives alone at home. She is on a fixed income and rarely eats calcium-rich foods. She recently fell and broke her hip. After surgery she returned home under the care of a home health agency.

1. What made the patient at high risk for a fracture?
2. What would you expect her serum calcium level to have been before the fall?
3. What patient teaching related to diet and calcium supplements should the home health care nurse include during his or her home visits?

Answers at end of chapter.

PATHOPHYSIOLOGY AND ETIOLOGY. Chronic hypercalcemia can result from excessive intake of calcium or vitamin D, renal failure, hyperparathyroidism, cancers, and overuse or prolonged use of thiazide diuretics, such as hydrochlorothiazide (HydroDiuril). Acute hypercalcemia can occur as an emergency in patients with invasive or metastatic cancers.

PREVENTION. Although many causes of increased calcium cannot be prevented, a person receiving calcium supplements should be monitored carefully. Some women believe that if 2 or 3 tablets a day is helpful, consuming twice that much will help even more. The result can be serum calcium excess. Educating the public about the proper amount of calcium needed each day and the danger of too much calcium is very important.

SIGNS AND SYMPTOMS. Patients who have mild hypercalcemia or a slowly progressing calcium increase may have no obvious signs and symptoms. However, acute hypercalcemia is associated with increased heart rate and blood pressure, skeletal muscle weakness, and decreased GI motility. The patient also has a decreased blood clotting capability.

COMPLICATIONS. In some cases the patient may experience renal or urinary calculi (stones) resulting from the buildup of calcium. In more severe cases of acute hypercalcemia, the patient may experience respiratory failure caused by profound muscle weakness or heart failure caused by dysrhythmias.

MEDICAL TREATMENT. Patients with severe hypercalcemia should be hospitalized and placed on a cardiac monitor. Unless contraindicated by other conditions, the primary treatment is to give large amounts of fluids and promote diuresis. Saline infusions are the most useful solutions to promote renal excretion of calcium.

The physician also discontinues thiazide diuretics if the patient was receiving them and prescribes diuretics that promote calcium excretion, such as furosemide (Lasix). Other drugs that bind with calcium to lower calcium levels may also be used, such as plicamycin (Mithramycin, Mithracin) and D-penicillamine (Cuprimine).

If hypercalcemia is so severe that cardiac problems are present, hemodialysis, peritoneal dialysis, or ultrafiltration may be necessary to cleanse the blood of excess calcium. (See Chapter 35 for discussion of these procedures.)

Magnesium Imbalances

Magnesium and calcium work together for the proper functioning of excitable cells, such as cardiac muscle and nerve cells. Therefore an imbalance of magnesium is usually accompanied by an imbalance of calcium.

The normal value for serum magnesium is 1.5 to 2.5 mEq/L. The magnesium imbalances are called **hypomagnesemia** and **hypermagnesemia.**

Hypomagnesemia

Hypomagnesemia occurs when the serum magnesium level falls below 1.5 mEq/L. It results from either a decreased intake or an excessive loss of magnesium. Causes of inadequate intake include malnutrition and starvation diets. Patients with severe diarrhea and Crohn's disease are unable to absorb magnesium in the intestines.

One of the major causes of hypomagnesemia is alcoholism, which causes both a decreased intake and an increased renal excretion of magnesium. Certain drugs, such as loop (high-ceiling) and osmotic diuretics, aminoglycosides (e.g., gentamicin), and some anticancer agents (e.g., cisplatin), can increase renal excretion of magnesium.

The signs and symptoms of hypomagnesemia are similar to those for hypocalcemia, including positive Trousseau's and Chvostek's signs, described earlier in this chapter.

The goal of management is to treat the underlying cause and replace magnesium in the body. Magnesium sulfate is administered intravenously. If the serum calcium is also low, calcium replacement is prescribed. The patient is placed on a cardiac monitor because of magnesium's effect on the heart. Life-threatening dysrhythmias can lead to cardiac failure and arrest.

Hypermagnesemia

Hypermagnesemia results when the serum magnesium level increases above 2.5 mEq/L. The most common cause of hypermagnesemia is increased intake coupled with decreased renal excretion caused by renal failure.

Signs and symptoms are usually not apparent until the serum level is greater than 4 mEq/L. Then the signs and symptoms include bradycardia and other dysrhythmias, hypotension, lethargy or drowsiness, and skeletal muscle weakness. If not treated, the patient experiences coma, respiratory failure, or cardiac failure.

When kidneys are functioning properly, loop diuretics such as furosemide (Lasix) and IV fluids can help increase magnesium excretion. For patients with renal failure, dialysis may be the only option.

---

hypomagnesemia: hypo—less than + magnes—magnesium + emia—blood
hypermagnesemia: hyper—more than + magnes—magnesium + emia—blood

ACID-BASE BALANCE

The cells of the body function best when the body fluids and electrolytes are within a very narrow range. Hydrogen (H^+) is another ion that must stay within its normal limits. The amount of hydrogen determines whether a fluid is an acid or base.

An acid is a substance that releases a hydrogen ion. The stronger the acid, the more hydrogen ions are released. A common acid in the body is hydrochloric acid (HCl), which is found in the stomach. A base is a substance that binds hydrogen. A common base in the body is bicarbonate (HCO_3). *Alkali* is another word for base.

Sources of Acids and Bases

Acids and bases are formed in the body as part of normal metabolic processes. Acids are formed as end products of glucose, fat, and protein metabolism. These are called fixed acids because they do not change once they are formed. A weak acid, carbonic acid, can be formed when the carbon dioxide resulting from cellular metabolism combines with water. This acid can again change to bicarbonate (a base) and hydrogen and therefore is not a fixed acid.

The ECF maintains a delicate balance between acids and bases. The strength of the acids and bases can be measured by pH. The pH of a solution can vary from 0 to 14, with 7 being neutral, 0 to 6.99 being acid, and 7.01 to 14 being base, also called alkaline. The normal serum pH is 7.35 to 7.45, or slightly alkaline. It must remain in an extremely narrow range to sustain life. A pH lower than 6.9 or higher than 7.8 is usually fatal.

Control of Acid-Base Balance

As discussed in the sections on fluid and electrolyte balance, the body has several ways in which it tries to compensate for changes in the serum pH. Three major mechanisms are used: cellular buffers, the lungs, and the kidneys.

Cellular buffers are the first to attempt a return of the pH to its normal range. Examples of cellular buffers are proteins, hemoglobin, bicarbonate, and phosphates. These buffers act as a type of sponge to "soak up" extra hydrogen ions if there are too many (too acidic) or release hydrogen ions if there are not enough (too alkaline).

The lungs are the second line of defense to restore normal pH. When the blood is too acidic (pH is decreased), the lungs "blow off" additional carbon dioxide through rapid, deep breathing. This reduces the amount of carbon dioxide available to make carbonic acid in the body. If the blood is too alkaline (pH is increased), the lungs try to conserve carbon dioxide through shallow respirations.

The kidneys are the slowest to respond to changes in serum pH, taking as long as 24 to 48 hours to assist with compensation. The kidneys help in a number of ways, including regulating the amount of bicarbonate (base) that is kept in the body. If the serum pH lowers and becomes too acidic, the kidneys reabsorb additional bicarbonate rather than excreting it so that it can help neutralize the acid. If

the serum pH increases and becomes too alkaline, the kidneys excrete additional bicarbonate to get rid of the extra base. The kidneys also buffer pH by forming acids and ammonium (a base).

Acidosis or alkalosis that is corrected for by the body is referred to as compensated. The pH is returned to normal or near normal, but the PCO_2 and HCO_3 are abnormal.

ACID-BASE IMBALANCES

Most acid-base imbalances are caused by a number of acute and chronic illnesses or conditions. The primary treatment for each of the imbalances is to manage the underlying cause, which corrects the imbalance. The role of the nurse is to identify patients at risk and monitor laboratory test values for significant changes.

The laboratory tests that are used are called arterial blood gases (ABGs). As the name implies, the blood sample that is analyzed must be from an artery, rather than a vein. The femoral, brachial, and radial arteries are most often used to obtain the sample. Table 5–2 lists these major tests and normal values.

The two broad types of acid-base imbalance are acidosis and alkalosis. Each of these types can occur suddenly, which is called an acute imbalance, or develop over a long period, referred to as a chronic imbalance.

When the serum pH falls below 7.35, the patient has acidosis because the blood becomes more acidic than normal. Too much acid in the body or too little base causes acidosis. Acidosis can be divided into two types: respiratory and metabolic. Respiratory acidosis is caused by problems occurring in the respiratory system. Metabolic acidosis is the result of problems in the rest of the body.

When the serum pH increases above 7.45, the patient has alkalosis because the blood becomes more alkaline or basic. Alkalosis is caused by too little acid in the body or too much base. It can also be divided into two types: respiratory alkalosis and metabolic alkalosis.

Respiratory Acidosis

As the name indicates, the primary cause of this type of acidosis is respiratory problems. Carbon dioxide is not adequately "blown off" during expiration, causing a buildup of carbon dioxide in the blood. As mentioned earlier, carbon dioxide mixes with water to create a weak acid in the body, thus increasing the acidity of the blood.

Acute respiratory acidosis is caused by hypoventilation, usually as a result of an acute flare up of chronic respiratory disease, drugs, or neurological problems that depress breathing. Patients with chronic respiratory disease may have chronic respiratory acidosis.

The signs and symptoms of respiratory acidosis involve the central nervous system and the musculoskeletal system. As carbon dioxide increases, mental status is altered, progressing from confusion and lethargy to stupor and coma if not treated. The lungs are not able to get rid of excess carbon dioxide. Instead respirations become more depressed and shallow as muscle weakness worsens.

TABLE 5-2	Arterial Blood Gas Values and Changes in Acid-Base Imbalances		
	pH	Pco₂	Hco₃
Normal Values	7.35–7.45	32–45 mm Hg	2026 mEq/L
Respiratory acidosis	Decreased	Increased	Normal
Respiratory acidosis with compensation	Nearly normal	Increased	Increased
Respiratory alkalosis	Increased	Decreased	Normal
Respiratory alkalosis with compensation	Nearly normal	Decreased	Decreased
Metabolic acidosis	Decreased	Normal	Decreased
Metabolic acidosis with compensation	Nearly normal	Decreased	Decreased
Metabolic alkalosis	Increased	Normal	Increased
Metabolic alkalosis with compensation	Nearly normal	Increased	Increased

The treatment of respiratory acidosis is aggressive management of the underlying respiratory problem, discussed in the respiratory unit of this text.

Metabolic Acidosis

Metabolic acidosis can result from too much acid in the body (usually fixed acids) or too little bicarbonate in the body. Uncontrolled diabetes mellitus and end-stage renal failure are the two most common causes of metabolic acidosis resulting from increased fixed acids.

The GI tract is rich in bicarbonate. Patients experiencing severe diarrhea or prolonged nasointestinal suction are at high risk for metabolic acidosis as a result of bicarbonate (base) loss. As seen in Table 5–2, the serum pH decreases and the bicarbonate level decreases. As mentioned earlier in the discussion on hyperkalemia, serum potassium tends to increase in the presence of metabolic acidosis. Excess hydrogen in the ECF moves into the cells in exchange for potassium, which leaves the cells and enters the blood. In a sense, this is a way of compensating for the acidotic state.

The signs and symptoms are similar to those associated with respiratory acidosis, with the exception of the respiratory pattern. To help compensate for the acidotic state, the lungs get rid of extra carbon dioxide through Kussmaul's respirations. Kussmaul's respirations are deep and rapid and can occur only in patients with healthy lungs.

The treatment for the patient with metabolic acidosis is management of the underlying disease or condition. Information about disease management, such as diabetes, is found elsewhere in this book.

Respiratory Alkalosis

Respiratory alkalosis is probably the least common acid-base imbalance. It occurs when there is excessive loss of carbon dioxide through **hyperventilation.** Patients may hyperventilate when they are severely anxious or fearful. Patients who hyperventilate have rapid shallow respirations, are light-headed, and may become confused. The heart rate increases and the pulse becomes weak and thready. The serum pH is increased and the PaCO₂ is very low. Mechanical ventilation can also cause respiratory alkalosis, and it can occur as a result of being at high altitudes.

Respiratory alkalosis is treated by having patients rebreathe their own carbon dioxide with the use of either a rebreathing mask or a plain paper bag. The underlying cause must also be treated.

Metabolic Alkalosis

Metabolic alkalosis results from excessive ingestion of bicarbonate or other bases into the body or loss of acids from the body. Overuse or abuse of antacids or baking soda (sodium bicarbonate) can lead to metabolic alkalosis. Because the stomach contains hydrochloric acid, prolonged vomiting or nasogastric suction can cause loss of acid and also lead to metabolic alkalosis.

The serum pH is increased, as is bicarbonate. As discussed under potassium imbalances, the serum potassium decreases. Hydrogen from the ICF moves into the blood in exchange for potassium, which moves from the blood into the cells. This is one way that the body works to keep an acid-base balance. Hypocalcemia may also accompany hypokalemia.

The signs and symptoms of metabolic alkalosis are related to hypokalemia and hypocalcemia rather than the alkalotic state, itself. Treatment involves identifying the underlying cause and managing it as quickly as possible.

hyperventilation: hyper—more than + ventilation—air

REVIEW QUESTIONS

1. Mrs. Rodriguez is a 93-year-old patient with diarrhea and dehydration. She has been admitted to the hospital from an extended care facility. Which of the following symptoms of dehydration do you expect to see?
   a. Pale-colored urine, bradycardia
   b. Disorientation, poor skin turgor
   c. Decreased hematocrit, hypothermia
   d. Lung congestion, abdominal discomfort

2. Which of the following is the *most* reliable way to monitor Mrs. Rodriguez's fluid status?
   a. I&O
   b. Skin turgor
   c. Daily weights
   d. Lung sounds

3. When caring for a patient with fluid overload, which of the following interventions will help relieve respiratory distress?
   a. Elevate the head of the bed.
   b. Encourage the patient to cough and deep breathe.
   c. Increase fluids to promote urine output.
   d. Perform percussion and postural drainage.

4. Mr. Janes is being treated for hypokalemia. When evaluating his response to potassium replacement therapy, which of the following changes in his assessment should you observe for?
   a. Improving visual acuity
   b. Worsening constipation
   c. Decreasing serum glucose
   d. Increasing muscle strength

5. Which of the following organ systems is most at risk for life-threatening complications when a patient has hyperkalemia?
   a. Cardiovascular
   b. Renal
   c. Nervous
   d. Musculoskeletal

6. Which of the following adaptive responses occurs when a patient is in metabolic acidosis?
   a. Respiratory depression
   b. Kussmaul's respirations
   c. Increased urine output
   d. Highly concentrated urine

Answers to CRITICAL THINKING

Mrs. Levitt

1. Check her weight and compare it with her previous weights. Dehydration is associated with weight loss. Monitor mental status for disorientation. Check skin turgor for tenting. Continue to monitor vital signs.
2. Encourage increased fluid intake; notify the RN or physician if Mrs. Levitt is unable to take in additional fluids or if the fluids do not normalize assessment findings.
3. S: "My urine smells bad, and my heart is beating fast."
   O: Pt's urine is dark amber and strong smelling. VS P 98, BP 126/74, RR 20, T 99.2. Fluids encouraged. RN notified.

Mr. Peters

1. Raise the head of the bed to assist breathing.
2. Questions to ask might include the following: When did your symptoms begin? Have you had these symptoms before? (If the patient is too dyspneic to answer, do not ask many questions.)

3. Check breath sounds for crackles, observe for dependent edema and ascites, observe for distended neck veins, assess skin for color and temperature, check weight and compare with previous weight, and monitor I&O.

Mrs. Wright

1. The patient is at high risk for osteoporosis, and thus fracture, because she is an elderly, petite, Caucasian woman.
2. Her serum calcium levels should have been lower than normal.
3. Teach her about consuming foods high in calcium, teach her about the need to be compliant with taking her calcium supplements, and teach her to take them 1 to 2 hours after meals for best absorption by the body.

6

Nursing Care of Patients Receiving Intravenous Therapy

Lynn Phillips and Jill Secord

KEY TERMS

bolus (**BOH**-lus)

cannula (**KAN**-yoo-lah)

extravasation (eks-**TRA**-vah-**ZAY**-shun)

hypertonic (**HIGH**-per-**TAW**-nick)

hypotonic (**HIGH**-poh-**TAW**-nick)

intravenous (**IN**-trah-**VEE**-nus)

isotonic (**EYE**-so-**TAW**-nick)

phlebitis (fla-**BYE**-tis)

QUESTIONS TO GUIDE YOUR READING

1. What is the definition of intravenous therapy?

2. How is the practice of intravenous therapy regulated?

3. What are the indications for intravenous therapy?

4. What are the preferred sites for intravenous therapy?

5. What factors influence the condition, size, and long-term use of veins?

6. What steps are used for insertion of an intravenous catheter?

7. How will you know if your nursing interventions to prevent complications of intravenous therapy are effective?

8. How will you calculate a drip rate for a patient receiving a parenteral solution?

9. What is the difference between isotonic, hypertonic, and hypotonic solutions?

10. How would you explain the basic differences between central venous access devices: percutaneous catheters, peripherally inserted central catheters, tunneled catheters, and implanted ports?

Intravenous (IV) therapy is the administration of fluids or medication via a needle or catheter (**cannula**) directly into the bloodstream. The practice of IV therapy is governed by state nurse practice acts as statutory laws. Some states now include IV therapy within the licensed practical nurse (LPN) and licensed vocational nurse (LVN) roles. The practice acts define the parameters within which individuals are qualified and licensed to practice nursing in

a particular state and serve to codify the nursing obligation to act in the best interest of society.

Various specialty organizations, such as the Infusion Nurses Society (INS), the Centers for Disease Control and Prevention (CDC), the Occupational Safety and Health Administration, and the American Society for Parenteral and Enteral Nutrition (ASPEN), set forth guidelines for standards of practice for infusion therapy.

See http://www.ins1.org/ for more information about the INS. See http://www.cdc.gov/niosh/2000-108.html for information on preventing needlestick injuries. See http://www.clinnutr.org for information on ASPEN. See http://www.oshasafety.com for information about workplace safety.

---

intravenous: intra—within + venous—vein
cannula: tube or sheath

▶ INDICATIONS FOR INTRAVENOUS THERAPY

Patients receive a variety of substances via IV therapy, including fluids, electrolytes, nutrients, blood products, and medications. Patients can receive life-sustaining fluids, electrolytes, and nutrition when they are unable to eat or drink adequate amounts. The IV route also allows rapid delivery of medication in an emergency. Many medications are faster acting and more effective when given via the IV route. Other medications can be administered continuously to maintain a therapeutic blood level. Patients with anemia or blood loss receive lifesaving IV transfusions. Patients who are unable to eat for an extended period can have their nutritional needs met with total parenteral nutrition (TPN).

When a patient needs intermittent rather than continuous IV therapy, access to the venous system can be provided by a locking device whereby an IV catheter is inserted and capped with a port or cap that is resealable. (See the intermittent infusion section later in this chapter.) This is to provide access to the bloodstream for intermittent or emergency medications.

▶ INTRAVENOUS ACCESS

Intravenous therapy can be administered into the systemic circulation via the peripheral or central veins. Peripheral veins lie beneath the epidermis, dermis, and subcutaneous tissue of the skin. They usually provide easy access to the venous system. Central veins are located close to the heart. Special catheters that end in a large vessel near the heart are called central lines. This chapter primarily discusses peripheral catheters. The definitions of the various central venous access devices are discussed briefly at the end of the chapter.

▶ ADMINISTERING PERIPHERAL INTRAVENOUS THERAPY

Starting a Peripheral Line

The Phillips 15-step approach to starting a peripheral line offers an organized and thorough method, as described in Table 6–1. Remember to always check your institution's policy before performing any procedure.

Check Physician's Order

A physician's order is necessary to initiate IV therapy. According to the INS, a prescriber's verbal order written by a nurse in the medical record in a hospital setting should be signed by the prescriber within an appropriate time. The order should include solution, volume, rate, and route. If medication is ordered, the order should also include the medication, dosage, and frequency.

Wash Hands

Before beginning the procedure, wash your hands for 15 to 20 seconds. Wear gloves when inserting the needle and any time you have a risk of exposure to body fluid.

TABLE 6–1	PHILLIPS 15-STEP METHOD FOR STARTING A PERIPHERAL LINE
Phase	**Step**
Precannulation (preparation)	1. Check physician's order. 2. Wash your hands for 15 to 20 seconds. 3. Prepare the equipment. 4. Assess the patient. 5. Select the site and dilate the vein.
Cannulation (venipuncture)	6. Select the needle (catheter). 7. Put on gloves. 8. Prepare the site. 9. Enter the vein using the direct or indirect method. 10. Stabilize the catheter with tape, and apply a dressing.
Postcannulation (cleanup)	11. Label the site, tubing, and bag. 12. Properly dispose of used equipment. 13. Educate the patient. 14. Calculate the drip rate, if applicable. 15. Document the procedure.

Source: Phillips, L: Manual of IV therapeutics (3rd ed.). FA Davis, Philadelphia, 2001.

Gather Equipment

Obtain the following equipment and inspect it for integrity:

- Clean gloves
- Prepping solution (70 percent isopropyl alcohol, povidone-iodine, tincture of iodine 2 percent, or chlorhexidine)
- Sterile 2-inch by 2-inch gauze pads
- ½-inch or 1-inch tape
- Disposable latex tourniquet
- Cannulas (over-the-needle sizes 18, 20, 22, and 24 are the most common)
- Appropriate administration set
- IV solution (inspected for puncture holes, visible contamination, and expiration date)
- PRN device (locking device) if the catheter is maintained as a saline lock
- IV pole if needed

Some institutions have IV start kits that contain a tourniquet, gloves, alcohol, bandages, and prepping solution.

Once the solution is verified and inspected for integrity, the administration set is spiked to puncture the solution bag or bottle, taking care to keep the spike and the bag opening sterile. The administration set is then primed with the IV solution ordered by the physician.

Assess and Prepare Patient

Several factors should be considered before venipuncture. The type of solution, condition of vein, duration of therapy, cannula size needed, patient age, patient activity, presence of disease or previous surgery, presence of a dialysis shunt or graft, medications being taken by the patient (such as anticoagulants), and allergies must be assessed before a venipuncture. Provide privacy for the procedure, explain the procedure to the patient, and evaluate the patient's knowledge of the procedure by talking with the patient before assessing the upper arms for suitable venipuncture sites.

Select Site and Dilate Vein

Proper vein selection is important to accommodate the prescribed therapy and to minimize potential complications (Box 6–1). Avoid use of an arm on the side where the patient has had a mastectomy, has a dialysis access site, or is scheduled for a surgical procedure. The patient's condition and diagnosis, age, vein condition, size, location, and type and duration of therapy should be considered before initiation of intravenous therapy (Box 6–2). The vein should be able to accommodate the gauge and length of catheter used.

Hand veins are used first if long-term intravenous therapy is expected (Cultural Consideration 6–3). This allows each successive venipuncture to be made proximal to the site of the previous one, which eliminates the passage of irritating fluids through a previously injured vein and discourages leakage through old puncture sites. Hand veins can be used successfully for most hydrating solutions, but they are best avoided when irritating solutions of potassium or antibiotics are anticipated.

Vein size must also be considered. Small veins do not tolerate large volumes of fluid, high infusion rates, or irritating solutions. Large veins should be used for these purposes. Figure 6–1 shows peripheral veins that may be used for IV therapy.

If veins are constricted, venipuncture is more difficult. Fever, anxiety, and cold temperatures can cause veins to constrict. Smoking before the insertion of an IV line also causes veins to constrict.

BOX 6-1   **Considerations for Vein Selection**

- Age of patient
- Availability of sites
- Size of catheter to be used
- Purpose of infusion therapy
- Osmolarity of solution to be infused
- Volume, rate, and length of infusion
- Degree of mobility desired

BOX 6-2   **General Considerations When Initiating Intravenous Therapy**

1. Use veins in the upper part of the body.
2. When multiple sticks are anticipated, make the first venipuncture distally and work proximal with subsequent punctures.
3. If therapy will be prescribed for longer than 3 weeks, a long-term access device should be considered.
4. Avoid using venipunctures in affected arms of patients with radical mastectomies or a dialysis access site.
5. If possible, avoid taking a blood pressure on the arm receiving an infusion because the cuff interferes with blood flow and forces blood back into the needle. This may cause a clot or cause the vein or catheter to rupture.
6. No more than two attempts should be made at venipuncture before getting help.
7. Immobilizers should not be placed on or above an infusion site.

BOX 6-3   **CULTURAL CONSIDERATION**

Among the Vietnamese, the head is considered sacred. Thus the practice of starting IV lines in the scalp may cause a Vietnamese patient significant anxiety. If the patient must have an IV line in the scalp, carefully explain the necessity for it.

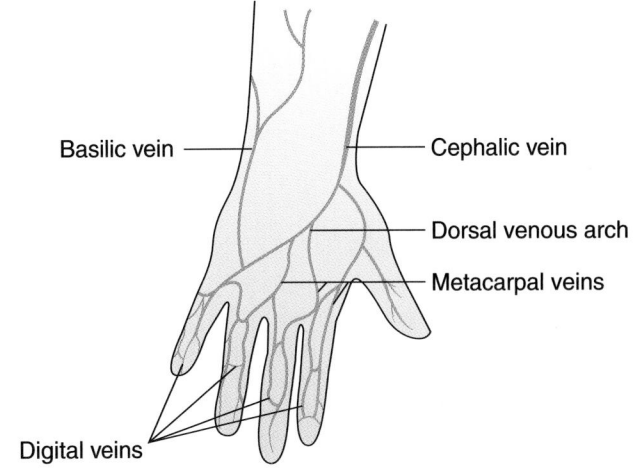

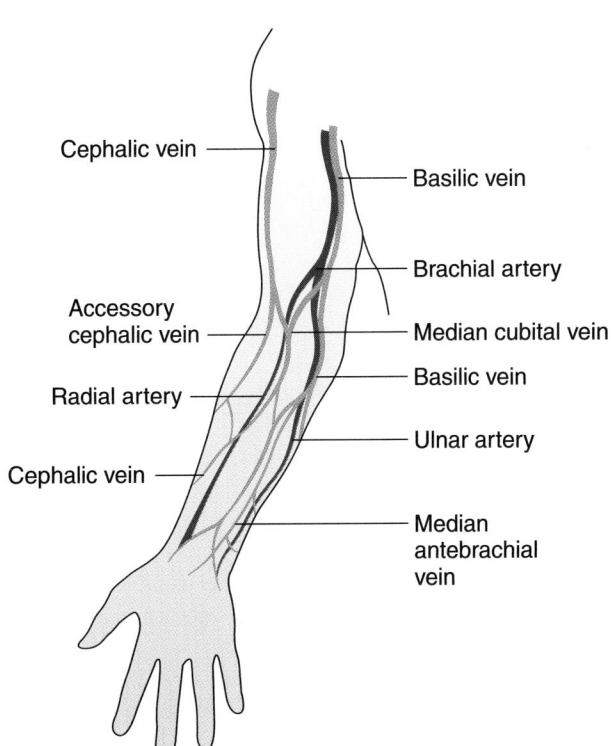

Figure 6-1   Peripheral veins used for IV therapy. (Modified from Phillips, L: Manual of IV therapeutics [3rd ed.]. FA Davis, Philadelphia, 2001.)

DILATE VEIN. A tourniquet helps to dilate and stabilize the vein, easing venipuncture and threading of the catheter. Place the tourniquet 6 to 8 inches above the insertion site. If the tourniquet is too close to the insertion site, it will create too much pressure and cause the vein to burst. The tourniquet should be tight enough to impede venous flow while maintaining arterial flow. A tourniquet should be at least 1 inch wide and should not be left on for more than 3 minutes to prevent impaired blood flow to the extremity.

Occasionally, additional techniques are necessary to distend the vein. Placing the arm in a dependent position or placing a warm towel over the site for several minutes before applying the tourniquet help to dilate a vein. The whole extremity must be warmed to improve blood flow to the area. Opening and closing the fist pumps blood to the extremity and increases blood flow to help dilate the vein. A blood pressure cuff inflated to 30 mm Hg is an appropriate method for vein dilation, especially with fragile veins in the elderly.

Choose the Cannula

Needles have been largely replaced with flexible plastic catheters, or cannulas, that are inserted over a needle (Fig. 6–2). The needle (or stylet) is removed after the catheter is in place. These are available in a variety of sizes (gauges) and lengths. For patient comfort, choose the smallest-gauge catheter that will work for the intended purpose. Use smaller-gauge needles (20 to 24 gauge) for fluids and slow infusion rates. Use larger needles (18 gauge) for rapid fluid administration and viscous solutions such as blood. Also consider vein size when choosing a catheter gauge. Refer to institution policy and equipment stock for specific recommendations. Keep in mind that the INS recommends that short peripheral catheters be removed every 72 hours and immediately upon suspected contamination.

Gloves

The CDC recommends following standard precautions whenever exposure to blood or body fluids is likely. Wearing latex or vinyl gloves provides basic protection from blood and body fluids.

Prepare the Site

Clean the peripheral insertion site with an antimicrobial solution before cannula placement. If the skin is dirty, it should be washed with soap and water before application of the antimicrobial solution. If the patient has excess hair, it can be clipped with scissors. Be sure to follow institution policy when choosing a solution. Most institutions use an alcohol- or iodine-based product. Avoid using alcohol after an iodine preparation because alcohol negates the effect of iodine.

Apply the solution in a circular motion, starting at the intended site and working outward to clean an area 2 to 3 inches in diameter. If alcohol is used, it should be applied with friction for at least 30 seconds or until the final applicator is visually clean. Blotting of excess solution at the insertion site is not recommended. Allow the solution to air dry completely.

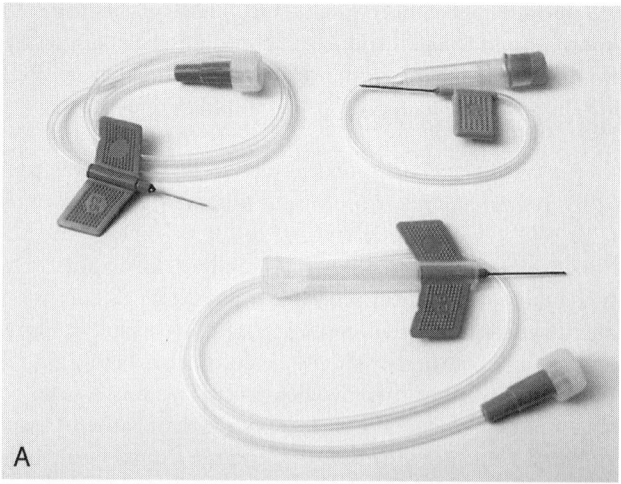

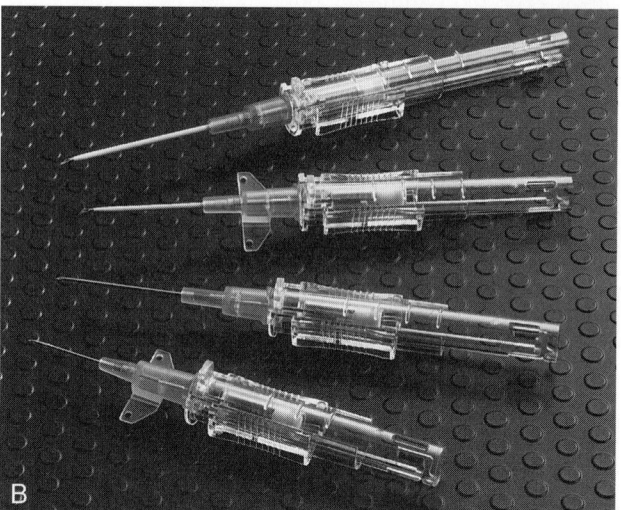

Figure 6-2 *(A)* Types of scalp vein needles. (Courtesy Becton Dickinson, Franklin Lakes, N.J.) *(B)* Shielded needle system PROTECTIV catheters. (Courtesy Johnson & Johnson, Arlington, Tex.)

Insert the Cannula

Venipuncture can be performed using a direct (one-step) or indirect (two-step) method. The direct method is appropriate for small-gauge needles, fragile hand veins, or rolling veins. The indirect method can be used for all venipunctures.

Hold the catheter with the bevel (slanted opening) of the needle facing up. With the tourniquet in place, enter the vein using either the direct or indirect approach. When using the direct entry approach, hold the needle at a 30- to 45-degree angle directly above the vein and then penetrate the skin and vein in one motion (Fig. 6–3). When using some newer catheters, the angle of insertion is minimal.

The indirect approach may help decrease vein collapse. To use it, hold the needle at a 30- to 45-degree angle over the skin next to (not over) the vein. Once the skin is punctured, lower the needle angle and locate and puncture the vein. Depending on the type of device used, a small flash of blood may be seen in the tubing or at the hub of the catheter when the needle is in the vein. The angle of the

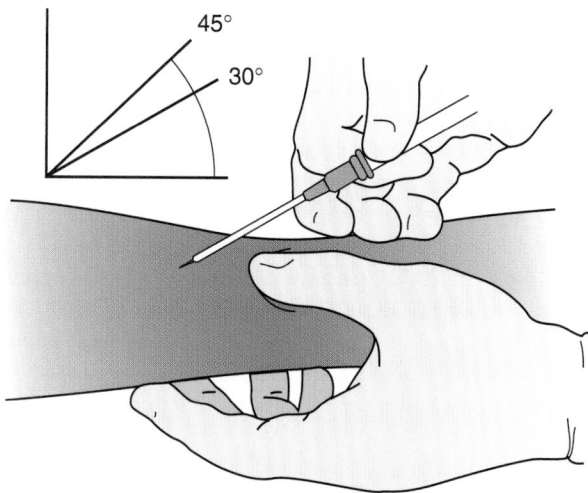

Figure 6-3 Insert the needle of choice bevel up at a 30-to 45-degree angle, depending on the vein location and catheter.

needle is then lowered so that it is parallel with the skin as it is threaded into the lumen of the vein. If a catheter-over-needle device is used, the needle is advanced one-fourth inch and then the catheter is advanced for its remaining length as the metal needle (stylet) is withdrawn.

The tourniquet is then released, and the IV solution or injection cap is connected to the hub of the catheter. Blood may ooze from the hub at this time. If an injection cap is being used, the catheter is flushed with 0.9 percent sodium chloride solution to check for patency. A smooth, easy flush and no signs of infiltration indicate that the catheter is patent and that the prescribed solution can be administered.

LEARNING TIPS

Use traction (a downward pulling motion to make the skin taut below the puncture site) before venipuncture to stabilize the skin and prevent the vein from rolling during venipuncture (Fig. 6–4).

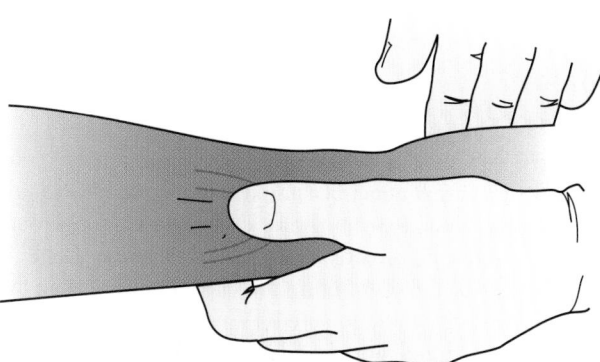

Figure 6-4 Pull skin below the intended puncture site using a downward motion to stabilize the skin and prevent the vein from rolling.

Stabilize the Cannula and Dress the Site

A common problem in IV therapy is dislodgement of the cannula. Secure taping keeps the catheter in place and stable, thus preventing complications caused by damage to the intima of the vein. There are several different techniques for taping a cannula securely, including the U, H, and chevron methods (Fig. 6–5). Take care to apply tape in a manner that does not constrict blood flow to the extremity.

A transparent, semipermeable membrane dressing allows the nurse to stabilize the cannula and monitor the venipuncture site for redness or swelling and provides an occlusive dressing for the site. Another acceptable method of dressing management is the use of sterile 2-inch by 2-inch gauze over the venipuncture site and a piece of 1-inch tape over the gauze. Band-Aids are not acceptable dressings over catheters.

Arm boards are not used routinely. However, if a confused patient places the IV site in danger, the extremity can be immobilized as a last resort; this requires a physician's order.

Label the Site

The IV setup should be labeled in three areas: the insertion site, the tubing, and the solution container. Once the venipuncture procedure is completed, label the setup with the date, time, catheter type and size, and your initials.

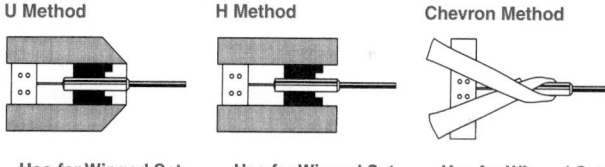

U Method	H Method	Chevron Method
Use for Winged Set	**Use for Winged Set**	**Use for Winged Set**
1. Cut three strips of 1/2-in tape. With sticky side up, place one strip under tubing.	1. Cut three strips of 1-in tape.	1. Cover thr venipuncture with transparent dressing or 2 x 2 gauze dressing.
2. Bring each side of the tape up, folding it over the wings of the needle. Press it down, parallel with the tubing.	2. Place one strip of tape over each wing, keeping the tape parallel with the needle.	2. Cut a long 5- to 6-in strip of 1/2-in tape. Place one strip of tape, sticky side under hub, parallel with the dressing.
3. Loop the tubing and secure it with a piece of 1-in tape.	3. Place another strip of tape perpendicular to the first two. Place over the wings to stabilize wings and hub.	3. Cross the end of the tape over the opposite side of the needle so that the tape sticks to the patient's skin.
		4. Apply a piece of 1-in tape across the wings of the chevron. Loop the tubing and secure it with another piece of 1-in tape.

*For all methods, include on the last piece of tape the date, time of insertion, size of gauge, length of needle or catheter, and your initials.

Figure 6-5 Taping techniques for IV access devices. (From Phillips, L: Manual of IV therapeutics [3rd ed.]. FA Davis, Philadelphia, 2001.)

Dispose of Equipment

All needles and blood-contaminated equipment should be disposed of according to institution policy in a tamper-proof, nonpermeable container.

Educate Patient

Patients have the right to receive information on all aspects of their care in a manner they can understand. They also have the right to accept or refuse treatment.

Calculate Drip Rate

All IV infusions should be monitored frequently for accurate flow rates and complications associated with infusion therapy. (See the section on calculating drip rates later in this chapter.)

Document

Documentation is noted in the medical record according to institution policy. All IV solutions are also documented on the medication administration record in the medical record. Document the following:

- Date and time of insertion
- Manufacturer's brand name and style of device
- Gauge and length of the device
- Location of the accessed vein
- Solution infusing and rate of flow
- Method of infusion (gravity or pump)
- Number of attempts needed for a successful IV start
- Patient's specific comments related to the procedure
- Signature

CRITICAL THINKING

Insertion of a Peripheral Line
A patient is admitted with a diagnosis of symptomatic anemia and has an atrioventricular (AV) shunt in the left arm. An IV line is ordered for administration of two units of packed red blood cells. What must be taken into consideration when assessing the patient for an appropriate venipuncture site?

Answers at end of chapter.

▶ TYPES OF INFUSIONS

Continuous

In a continuous infusion, the physician orders the infusion in milliliters to be delivered over a specific amount of time. The infusion is kept running constantly until discontinued by the physician. An IV controller or roller clamp allows the solution to infuse at a constant rate.

Intermittent Infusion

Intermittent IV lines are "capped off" with an injection port and used only periodically. Thus intermittent IV therapy is administered at prescribed intervals. Patency must be ensured before an intermittent site is used to inject a drug

or solution. Check for backflow of blood in the syringe before injection.

Sites that are capped with an injection cap are called saline or heparin locks. A diluted solution of heparin, an anticoagulant, may used to flush the needle after each use and every 8 hours or according to institution policy. However, many institutions are changing this practice and specifying saline solution for flushes, which is believed to be safer and less costly. In *2000 Standards of Practice,*[2] the INS recommends the use of saline (0.9 percent sodium chloride) solution for maintaining peripheral locking devices, whereas heparin is recommended for central venous access devices. Check your institution's policy and guidelines.

LEARNING TIPS

Always check for catheter patency before injecting any substance into the circulatory system.

Regular flushing ensures cannula patency. Flushing also prevents the mixing of incompatible medications and solutions. Positive pressure must be maintained in the lumen of the catheter during the administration of the flush solution to prevent a backflow of blood into the cannula lumen. This is accomplished by continuing to slowly inject the saline or heparin solution even as the needle is withdrawn from the cap. Intermittent cannulas should be flushed after administration of IV medication, blood sampling, and conversion from continuous to intermittent IV therapy.

If resistance is met while a cannula is being flushed, a clot may be occluding the cannula. Do not exert pressure in an attempt to restore patency because doing so may dislodge the clot into the vascular system or rupture the catheter.

Some medications that are given intermittently are not compatible with heparin and therefore are given by the **s**odium chloride **a**dministration (of drug) **s**odium chloride **h**eparin (SASH) method:

1. Flush with **S**aline solution.
2. **A**dminister the medication.
3. Flush with **S**aline solution.
4. Flush with **H**eparin solution.

Bolus

A **bolus** drug (sometimes called an IV push [IVP] drug) is injected slowly via a syringe into the IV site or tubing port. It provides a rapid effect because it is delivered directly into the patient's bloodstream. Bolus drugs can be dangerous if they are given incorrectly, and a drug reference should always be checked to determine the safe amount of time over which the drug can be injected. IV push drugs are administered by registered nurses (RNs) and are not within the

bolus: mass or lump

scope of practice of the LPN/LVN. However, you should be aware of the drugs being given and should observe the patient for desired or adverse effects.

► METHODS OF INFUSION

Gravity Drip

Gravity is often used to drip a solution into a vein (Fig. 6–6). The solution is positioned about 3 feet above the infusion site. If it is positioned too high above the patient, the infusion may run too fast. Positioned too low, it may run too slowly. Flow is controlled with a roller, screw, or slide clamp. A mechanical flow device can be added to achieve accurate delivery of fluid with minimal deviation.

Calculating Drip Rates

When using a gravity set, the nurse must calculate the drops required per minute to deliver fluid at the ordered rate. Commercial parenteral administration sets vary in the number of drops delivering 1 mL. Sets typically deliver 10, 15, 20, or 60 drops per mL of fluid. For example, to deliver 100 mL per hour using a set with 10 drop factor tubing, a flow rate of 17 drops per minute is necessary. To administer the same amount using a set with 15 drop factor tubing, a flow rate of 25 drops per minute is necessary. Check the manufacturer's instructions on the administration set to determine how many drops per milliliter (drop factor) are delivered by the set. Sets delivering 60 drops per milliliter are called minidrip or microdrip sets and are used for solutions that need to be infused slowly.

To determine drops per minute of an IV solution, the nurse needs to know the amount of fluid to be given in a specified time interval and the drop factor of the administration set to be used. The formula for determining drops per minute is as follows:

$$\frac{\text{mL per hour} \times \text{drops per mL (drop factor)}}{60 \text{ (minutes per hour)}} = \text{drops per minute}$$

Sample Problem: A physician orders 1000 mL of 5 percent dextrose in water to be infused at 125 mL per hour. You have on hand an administration set that delivers 10 drops per 1 mL.

Formula:

$$\frac{125 \text{ mL} \times \text{(drop factor)}}{60} = \text{drops per minute}$$

$$\text{Step 1:} \quad \frac{125 \times 10}{60} = \text{drops per minute}$$

$$\text{Step 2:} \quad \frac{125}{6} = 21 \text{ drops per minute}$$

Factors Affecting Flow Rates

CHANGE IN CATHETER POSITION. A change in the catheter's position may push the bevel either against the wall of the vein, which will decreased the flow rate, or away from the wall of the vein, which may increase the flow rate. Careful taping and avoidance of joint flexion above the site minimizes this problem.

HEIGHT OF THE SOLUTION. Because infusions flow by gravity, a change in the height of the infusion bottle or a change in the level of the bed can increase or decrease the flow rate. The flow rate increases as the distance between the solution and the patient increases. A patient may alter the flow rate greatly simply by standing up. The ideal height for a solution is 3 feet above the level of the heart.

PATENCY OF THE CATHETER. A small clot or fibrin sheath may occlude the needle lumen and decrease or stop the flow rate. Clot formation can result from irritation, increased venous pressure, or backup of blood into the line. Avoid use of a blood pressure cuff on the affected extremity because of the resulting transient increase in venous pressure. A regular flush schedule helps maintain patency.

Electronic Control Devices

Electronic pumps and controllers regulate the rate of infusion (Fig. 6–7). Controllers measure the amount of solution delivered and depend on gravity to deliver the infusion. Pumps use positive pressure to deliver the solution. Pumps are often used for central lines to help overcome the high pressure of the central circulation.

Pumps and controllers are used for the infusion of precise volumes of solution. Institution policy often dictates use of controllers for infusion of potent medications, such as heparin, concentrated morphine, and chemotherapy solutions,

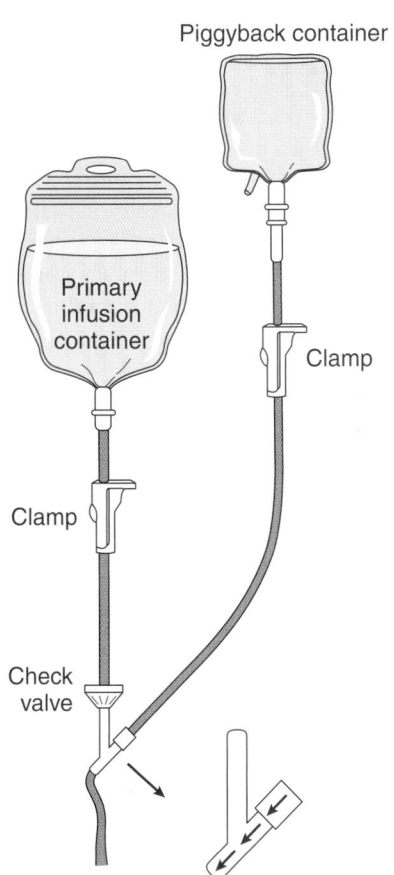

Figure 6-6 Gravity drip setup with piggyback infusion.

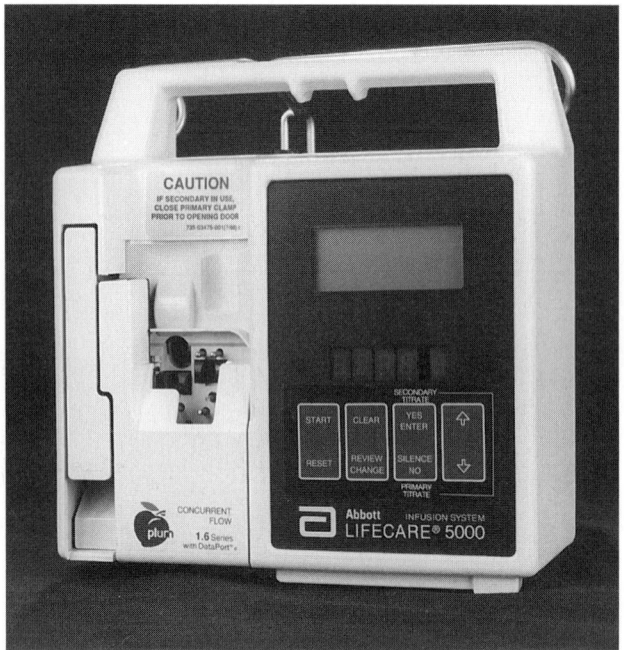

Figure 6-7 Infusion pump: LIFECARE 5000. (Courtesy Abbott Laboratories Hospital Products Division, North Chicago, Ill.)

and for very fast or slow rates. Some electronic infusion devices are portable and are designed to be worn on the body. These are called ambulatory infusion devices. It is important to know the type of pump being used and its manufacturer's guidelines.

Filters

It is recommended by the Infusion Nursing Standards of Practice that filters be used routinely for the delivery of IV therapy. Filters fit onto the IV tubing between the solution and the insertion site and remove contaminants from the IV fluid. A 0.22-micron filter removes bacteria and fungi from IV fluids. Check institution policy and manufacturers' guidelines for use of filters.

▶ TYPES OF FLUIDS

Fluids and electrolytes administered intravenously pass directly into the plasma space of the extracellular fluid compartment. They are then absorbed based on the characteristics of the fluid and the hydration status of the patient. The most commonly infused fluids are dextrose and sodium solutions. These are called crystalloid solutions.

Dextrose Solutions

Dextrose in water is available in many concentrations and provides carbohydrates in a readily usable form. Solutions of 2.5 percent, 5 percent, and 10 percent dextrose in water are used for peripheral infusions. Concentrations of 20 percent and above must be given into a large vein and are infused via a central line. These high concentrations can be used for treating hypoglycemia or in combination with TPN because they supply a large number of calories.

Sodium Chloride Solutions

Sodium chloride solutions are available in concentrations of 0.25 percent, 0.33 percent, 0.45 percent, 0.9 percent (normal saline solution), 3 percent, and 5 percent. Combination dextrose and sodium chloride solutions, such as 5 percent dextrose with 0.45 percent sodium chloride (often referred to as "D5 and a half"), are commonly used.

Electrolyte Solutions

Electrolyte solutions are used to replace lost fluids and electrolytes. Lactated Ringer's solution is an example of a premixed electrolyte solution. Potassium is an electrolyte that is commonly added to a solution to replace deficits. Potassium is limited to 10 to 20 mEq per hour and is never administered as an IV bolus because of the risk of cardiac complications.

Tonicity of IV Solutions

Intravenous fluids may be classified as **isotonic, hypotonic,** or **hypertonic.** (See Chapter 5 to review these concepts.) Isotonic fluids have the same concentration of solutes to water as body fluids. Hypertonic solutions have more solutes (i.e., are more concentrated) than body fluids. Hypotonic solutions have fewer solutes (i.e., are less concentrated) than body fluids. Water moves from areas of lesser concentration to areas of greater concentration. Therefore hypotonic solutions send water into areas of greater concentration (cells), and hypertonic solutions pull water from the more highly concentrated cells.

Isotonic Solutions

Normal saline (0.9 percent sodium chloride) solution is an isotonic solution that has the same tonicity as body fluid. When administered to a patient requiring water, it neither enters cells nor pulls water from cells; it therefore expands the extracellular fluid volume. A solution of 5 percent dextrose in water (D5W) is also isotonic when infused, but the dextrose is quickly metabolized, making the solution hypotonic.

Hypotonic Solutions

Hypotonic fluids are used when fluid is needed to enter the cells, as in the patient with cellular dehydration. They are also used as fluid maintenance therapy. An example of a hypotonic solution is 0.45 percent sodium chloride solution.

Hypertonic Solutions

Examples of hypertonic solutions include 5 percent dextrose in 0.45 percent sodium chloride, 5 percent dextrose in 0.9 percent sodium chloride, and 5 percent dextrose in lactated Ringer's solution. Hypertonic solutions are used to ex-

---

isotonic: iso—equal + tonic—tension or tone
hypotonic: hypo—deficient + tonic—tension or tone
hypertonic: hyper—excessive + tonic—tension or tone

pand the plasma volume, as in the hypovolemic patient. They are also used to replace electrolytes.

▶ NURSING PROCESS FOR THE PATIENT RECEIVING IV THERAPY

Assessment

A patient receiving IV therapy is assessed routinely. IV therapy is a medical intervention, and the nurse is responsible primarily for appropriate monitoring, documenting, and reporting related to the therapeutic goals. Some institution policies require assessment as often as every hour. Assessment should be systematic and thorough. It begins with observation and evaluation of the patient for signs of fluid imbalance. This is especially important when caring for an older patient (Gerontological Issues Box 6–4). Daily weighing and measuring of intake and output help determine whether the patient is retaining too much fluid. Skin turgor, mucous membrane moisture, vital signs, and level of consciousness also indicate hydration status. New onset of fine crackles in the lungs can indicate fluid retention. Table 6–2

BOX 6-4	GERONTOLOGICAL ISSUES

Care of the Patient with Intravenous Therapy

When an older patient is receiving fluids intravenously, the nurse must regularly assess the patient for potential fluid volume excess. Symptoms of fluid volume excess include:
• elevated blood pressure.
• full bounding pulse.
• shallow but rapid respirations.
• jugular-venous distention.
• increased urine output.
• lung sounds—moist crackles related to pulmonary edema.
If these signs are present:
• immediately turn down IV to a minimum drip rate (1 mL per minute); do not discontinue or shut IV off because the physician may want to administer IV diuretics.
• Position the patient to maximize lung expansion.
• Check peripheral oxygen perfusion with an oximeter.
• Start emergency oxygen per mask or nasal cannula if indicated.
• Closely monitor patient's vital signs, level of consciousness, and oxygen perfusion along with fluid output.
• Assist the physician or RN with IV push administration of diuretic medication such as furosemide if ordered.

extravasation: extra—outside + vas—vessel + tion—condition

TABLE 6-2	COMPLICATIONS OF INTRAVENOUS THERAPY

Local Complications	Symptoms	Prevention	Treatment
Phlebitis (inflammation of vein)	Pain and erythema at insertion site	Anchor cannula well. Avoid insertion near joint. Dilute medication. Use large veins. Use an in-line filter.	Remove cannula. Restart in new site.
Infiltration (solution leaks out of the vein into tissues)	Insertion site puffy and cool	Monitor patency of intravenous line. Check for blood return if unsure.	Stop the infusion. Apply warm compresses. Elevate extremity. Restart in new site.
Extravasation—infiltration of a vesicant (drug that causes tissue necrosis)	Redness, edema, exudate at site	Strict aseptic technique during insertion and site care. Inspect fluid before hanging.	Discontinue infusion. Replace in new site.
Infection	Serious local tissue necrosis or death of the surrounding tissue; pain	Check that IV line is patent with blood return before and during vesicant drug infusion.	Stop infusion immediately. Follow institution policy. Many have extravasation kits. Pharmacist can provide information related to specific drug.
Pain (can occur with some medications even if IV is patent)	Complaint of discomfort at or above infusion site	Check manufacturer recommendations for specific drugs.	Obtain physician order to slow infusion, dilute medication, or add lidocaine.

Systemic Complications	Symptoms	Prevention	Treatment
Circulatory overload (fluid administered too fast or too much)	Dyspnea, new-onset crackles, bounding pulse, intake greater than output	Monitor flow rate, lung sounds, and intake and output. Use controller in elderly patients.	Raise head of bed. Reduce flow rate and contact physician.
Infection (septicemia)	Fever, chills, thready pulse, tachycardia	Follow strict aseptic technique at all times. Inspect all fluids and equipment before infusion.	Discontinue infusion. Be prepared to send catheter and tubing to laboratory for culture. Culture IV site if ordered by physician.
Pulmonary embolism (clot or particle in pulmonary artery)	Shortness of breath, feeling of panic, chest pain, bloody sputum	Never irrigate a plugged cannula. Inspect all fluids for particles.	Notify physician for emergency treatment. Administer IV anticoagulants per order.

TABLE 6-2 COMPLICATIONS OF INTRAVENOUS THERAPY—CONT'D

Systemic Complications	Symptoms	Prevention	Treatment
Air embolism (air in bloodstream)	Cyanosis; hypotension; weak, rapid pulse; loss of consciousness	Inspect tubing for cracks, poor connections, other places where air could enter. Highest risk in central lines.	Clamp tubing. Administer oxygen. Place patient on left side with head down to allow air to rise into the right atrium and be excreted via pulmonary circulation.
Speed shock (shock symptoms from infusing drug too rapidly)	Syncope, flushing, headache, chest pain, hypotension, respiratory distress, cardiac arrest	Check drug manufacturer recommendations for proper dilution and rate of administration. Use IV controllers.	Discontinue infusion. Contact physician or pharmacist for treatment for specific drug.
Incompatibility (two drugs do not dissolve together)	Symptoms of embolism	Check drug reference or pharmacist before mixing medications or infusing via same IV tubing. If a precipitate is noted in tubing, do not infuse.	Notify physician and pharmacist for drug-specific remedy.

lists other symptoms of complications, along with prevention and treatment strategies.

Inspection of the site for redness or swelling and evaluation of the integrity of the dressing should be documented. Inspect the tubing to ensure tight connections and the absence of kinks or defects. Inspect the solution container and compare it with the physician's order for type, amount, and rate. Report complications to the RN or physician.

Nursing Diagnosis, Planning, and Implementation

All patients receiving IV therapy have a risk for complications. Monitor each patient carefully for the onset of complications, and report them promptly to the RN. Patients also may experience some anxiety related to IV therapy. Patient education may help alleviate anxiety. The RN is responsible for explaining the actions of and rationale for therapy.

Evaluation

The RN is responsible for evaluation and thus monitors the patient for evidence that the goals of therapy are being met and that complications are avoided. The LPN/LVN collects data that contribute to the evaluation. For example, if antibiotic therapy is administered, monitor the patient's temperature and other signs that the infection is resolving. If IV therapy is ordered to correct dehydration, monitor skin turgor, vital signs, and other appropriate signs of improved fluid balance. Document all findings and report them to the RN.

CRITICAL THINKING

Blood in IV Tubing
You have walked into the room of a patient whose IV has blood backed up in the tubing. When you open the clamp to increase the flow, nothing happens. What should you do?

Answer at end of chapter.

► COMPLICATIONS OF IV THERAPY

See Table 6–2 for complications, their prevention, and treatment. Any complication or unusual incident should be reported to the physician, and an incident report should be prepared according to institution policy. This applies to the hospital and the home situation.

CRITICAL THINKING

Complications of IV Therapy
A patient is receiving 5 percent dextrose in water at 83 mL per hour. One hour after the infusion starts, the patient complains of pain at the site. The site is cool to the touch and swollen, and the infusion rate is sluggish.

1. What might be happening?
2. How do you further assess the patient?
3. What action do you take?
4. How should you document your findings?

Answers at end of chapter.

► ALTERNATIVE ACCESS ROUTES

The role of the LPN/LVN in central venous access in most states is limited to assisting the RN with assessments. It is important for you to be able to recognize the different central venous access devices.

Central Venous Catheters

Central venous catheters terminate in the superior vena cava near the heart (Fig. 6–8). They are used when peripheral sites are inadequate or when large amounts of fluid or irritating medication must be given. Central catheter devices include a percutaneous catheter, peripherally inserted central catheter (PICC), tunneled catheter, and implanted port. These devices can have one, two, or three lumens in the catheter or one or more port chambers. Each lumen exits the site in a separate line, called a tail. Multilumen

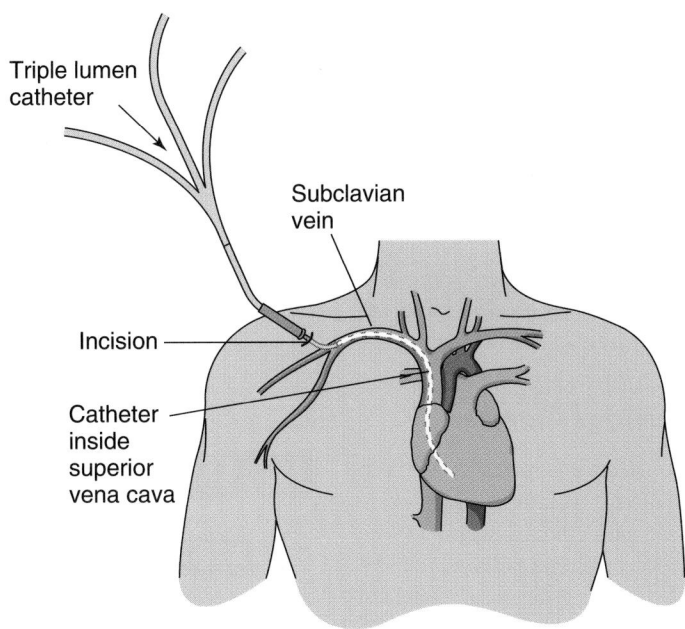

Triple lumen
catheter

Subclavian
vein

Incision

Catheter
inside
superior
vena cava

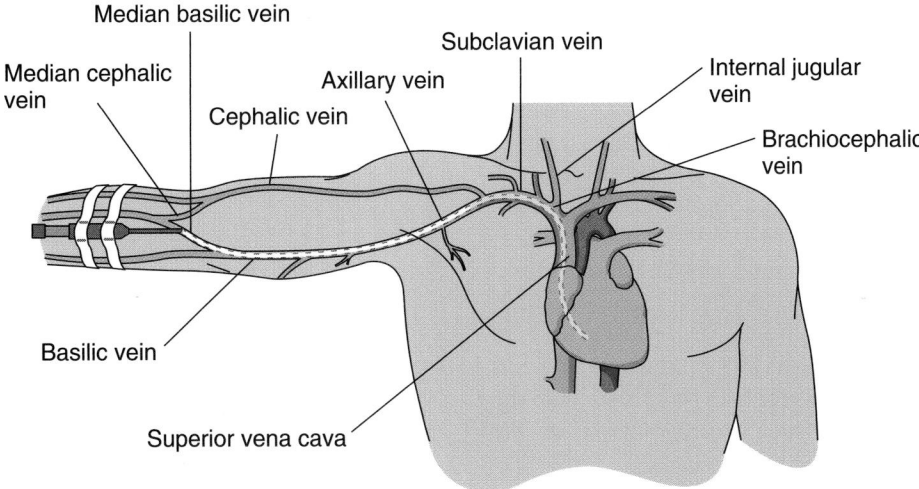

Median basilic vein

Median cephalic
vein

Cephalic vein

Axillary vein

Subclavian vein

Internal jugular
vein

Brachiocephalic
vein

Basilic vein

Superior vena cava

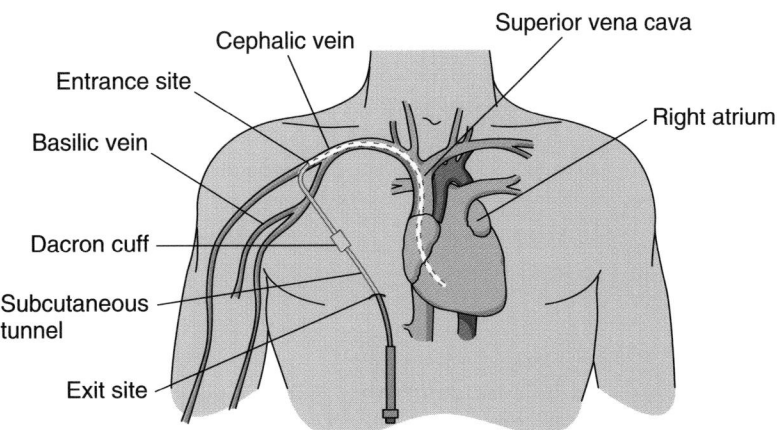

Cephalic vein

Superior vena cava

Entrance site

Right atrium

Basilic vein

Dacron cuff

Subcutaneous
tunnel

Exit site

Figure 6-8   Central lines. *(A)* Triple-lumen subclavian catheter. *(B)* PICC line. *(C)* Tunneled catheter. *(B and C* modified from Phillips, L: Manual of IV therapeutics [3rd ed.]. FA Davis, Philadelphia, 2001, pp 529, 549.)

catheters allow for the administration of incompatible solutions at the same time.

Percutaneous Catheter

A percutaneous catheter is inserted by a physician into the jugular or subclavian vein. After insertion, correct placement is determined by x-ray before the catheter is used. These short-term central venous catheters may remain in place up to several weeks, but usual placement time is 7 days. These catheters are inserted at the bedside and are cost effective for short-term central venous access in the acute care setting.

Peripherally Inserted Central Catheter

A PICC line is a long catheter that is inserted in the arm and that terminates in the central circulation. This device is used when therapy will last more than 2 weeks or the medication is too caustic for peripheral administration. Specially trained RNs insert PICC lines. They can be left in place for long periods, minimizing the trauma of frequent IV insertions. Consult with a physician if long-term therapy is anticipated.

It is important to follow the manufacturer's recommended guidelines for flushing the catheter and to be aware of your institution's policy. An RN removes the PICC or midline catheter when therapy is terminated. An LPN/LVN may assist the RN with this procedure if the state nursing practice act permits.

Ports

A port is a reservoir that is surgically implanted into a pocket created under the skin, usually in the upper chest. An attached catheter is tunneled under the skin into a central vein. An advantage of a port is that, when not in use, it can be flushed and left unused for long periods. Because the port is under the skin, the patient can swim and shower without risk of contaminating the site.

Ports come in a variety of sizes and styles and are now being used in many areas of the body. Ports can be used to administer chemotherapeutic agents and antibiotics that are toxic to tissues and are suitable for long-term therapy. Ports should be accessed only by specially trained RNs. Most ports require the use of special noncoring needles that are specifically designed for this purpose. Regular needles can cut a "core" in the septum, which can travel through the catheter and cause an embolism.

Nursing Management of Central Access Devices

Central lines have sterile, occlusive dressings to protect the site. Sterile gloves and mask should be worn during the dressing change. Whenever tubing is disconnected for changing, the patient is instructed to perform Valsalva's maneuver if able. This maneuver increases thoracic pressure and prevents air from entering the central line. Institution policy directs specific dressing and tubing change procedures.

▶ NUTRITION SUPPORT

Total parenteral nutrition is complete IV nutrition that is administered to patients who are unable to take adequate nutrients via the enteral route (mouth or tube feeding). TPN may be used to promote wound healing or to help the patient achieve optimal weight before surgery, or it may be used to avoid malnutrition from chronic disease or after surgery. Patients with ulcerative colitis, trauma, or cancer cachexia are candidates for TPN. Every effort should be made to return a patient on TPN to oral or tube feedings as soon as possible (Nutrition Notes Box 6–5).

BOX 6–5   *Nutrition Notes*

Attending to Patients Receiving Total Parenteral Nutrition

Serious metabolic complications of total parenteral nutrition can occur quickly. Rapid shifts in potassium, magnesium, phosphorus, and glucose among the fluid compartments of the body can become life-threatening. Glucose in excess of the body's need can produce carbon dioxide retention, hyperlipoproteinemia, and fatty deposits in the liver. Therefore administration of TPN requires careful monitoring. The goal is to provide sufficient nutrients but not excessive amounts that create physiological stress.

To avoid complications, TPN infusions are started slowly at low concentrations. Adjustments are made cautiously, changing one component (amount, concentration, or rate) at a time and observing the patient's reaction.

In a patient whose only nutrition is obtained from TPN, the gastrointestinal cells will atrophy. Therefore the patient must be weaned from parenteral feeding while oral or tube feedings are slowly introduced. As gastrointestinal intake increases, TPN is decreased. Sometimes, when circumstances permit, gastrointestinal function can be maintained throughout TPN administration by giving the patient a liquid or light diet along with TPN.

TPN provides and maintains the essential nutrients required by the body. Solutions contain carbohydrates, amino acids, lipid emulsions, electrolytes, trace elements, and vitamins in varied amounts according to the patient's needs. Parenteral nutrition requires filtration and an electronic infusion device for administration. In the home setting an ambulatory infusion device is used to allow the patient more mobility.

Initial assessment includes the patient's height, daily weight, nutritional status, and current laboratory values. Because of the high glucose concentration of TPN, the patient is at risk for infection and blood glucose disturbances. Insulin therapy may be necessary during TPN administration. Ongoing assessments include blood glucose levels according to institution policy and monitoring for signs and symptoms of infection, hyperglycemia, and hypoglycemia. When TPN therapy is begun, the rate is increased gradually to the prescribed rate to help prevent hyperglycemia. When it ends, the rate is gradually decreased to prevent hypoglycemia.

When nutritional solutions contain final concentrations exceeding 10 percent dextrose or 5 percent protein, they must be administered via a central catheter. When final concentrations are 10 percent dextrose or lower or 5 percent protein, they may be administered through a peripheral vein. Peripheral therapy is a short-term intervention because it does not provide adequate nutrition over an extended period.

It is essential that the entire health care team be involved in TPN therapy because TPN is total replacement for food. The pharmacist, dietitian, physician, and nurse communicate in a team conference to discuss the assessment, plan, and outcome criteria. Many institutions have nutrition teams that assess the appropriateness of TPN for individual patients.

▶ HOME INTRAVENOUS THERAPY

As health care costs continue to escalate, patients are using more alternatives to hospitalization. Subacute care, skilled nursing care in nursing homes, and home health care are growing. Home IV therapy allows many patients the benefit of early discharge and the ability to accomplish health care in the privacy and comfort of their own homes. Some home health agencies employ nurses to instruct patients and their families in the administration of home IV therapy (see Home Health Hints).

HOME HEALTH HINTS

If IV therapy is to be continued after hospital discharge, assist the RN in teaching the patient and family or caregiver the skills to oversee the IV therapy. A home care nurse will continue teaching after discharge.

In addition to basic IV care, teach the caregiver the effects and side effects of the medications and signs and symptoms that should be reported to the nurse or doctor.

Instruct the home care patient to refrain from smoking for at least 30 minutes before IV insertion to prevent vasoconstriction and ensure successful venipuncture.

Home IV antibiotic therapy is becoming the method of choice in the long-term treatment of a number of infections, including bacterial endocarditis, osteomyelitis, and septic arthritis. Other patients with chronic diseases choose to receive TPN at home. The health team can assess patients and their families for their ability to manage home IV therapy.

REVIEW QUESTIONS

1. Which of the following is the best resource for the nurse who has a question about implementation of IV therapy at a specific institution?
   a. An experienced nurse
   b. Institution policy
   c. The physician
   d. INS standards

2. Which of the following sites is the best choice for initiating a peripheral IV catheter?
   a. In the antecubital fossa
   b. In a lower extremity
   c. In the hand or forearm
   d. Distal to previous insertion sites

3. Which of the following techniques helps dilate a vein for IV catheter insertion?
   a. Apply a cool compress.
   b. Elevate the extremity.
   c. Apply a tourniquet 6 to 8 inches above the insertion site.
   d. Teach the patient a relaxation technique.

4. Which of the following is the drop factor for minidrip tubing?
   a. 10
   b. 15
   c. 30
   d. 60

5. The patient is to receive 1000 mL of 5 percent dextrose in water at 75 mL per hour. The tubing has a drop factor of 10. The nurse adjusts the drip rate to how many drops per minute?
   a. 13
   b. 19
   c. 25
   d. 75

6. A client receiving IV therapy via a central line develops hypotension, cyanosis, and tachycardia. You note a crack in the tubing. Which of the following actions should you take first?
   a. Have the RN call the physician.
   b. Clamp the tubing and administer oxygen.
   c. Raise the head of the bed.
   d. Slow the infusion and lay the patient flat.

BIBLIOGRAPHY

1. Phillips, L.D: Manual of IV therapeutics (3rd ed.). FA Davis, Philadelphia, 2001.
2. Infusion Nurses Society: Intravenous nursing standards of practice. Author, Norwood, Mass, 2000.
3. Centers for Disease Control and Prevention, Division of HIV/AIDS Prevention, National Center for HIV, STD and TB Prevention: Surveillance Report, 9 (1), 1997.

Answers to CRITICAL THINKING

Insertion of a Peripheral Line

Consider the following when assessing the patient for an appropriate venipuncture site:
1. An 18-gauge catheter should be used for blood administration whenever possible.
2. The catheter should not be inserted in the arm that has the shunt.
3. The catheter should be placed in the forearm. Hand veins are too small to accommodate the delivery of blood.

Blood in IV Tubing

Your patient's IV line is clotted. If it has been so for a long time, it will not be salvageable. Do not flush it because doing so can dislodge the clot into the circulation. Discontinue the IV and insert a new catheter.

Complications of IV Therapy

1. The IV fluid may be leaking at the insertion site and flowing into the subcutaneous tissue, a problem known as extravasation.
2. Consider whether the pain could be caused by the buildup of fluid under the skin.
3. Compare the insertion site with the opposite limb. If the IV solution has infiltrated, stop the infusion, discontinue the cannula, and restart the needle in a new site.
4. "Patient complains of pain at IV site in right arm; area is cool to touch and edematous in 45 cm area around site. Flow rate sluggish. Infusion discontinued; IV restarted in left arm with 22-gauge catheter. Infusing well with no signs of infiltration."

7

NURSING CARE OF PATIENTS WITH INFECTIONS

Sharon Martin and Elizabeth Chapman

KEY TERMS

aerobic (air-**O**-bick)

anaerobic (ann-air-**OH**-bick)

antibodies (**AN**-ti-baw-dees)

antigen (**AN**-tih-jen)

asepsis (ah-**SEP**-sis)

bacteria (back-**TEER**-e-ah)

colonization (COLLIN-i-**ZAY**-shun)

dormant (**DOOR**-mant)

flora (**FLOOR**-a)

fungi (**FUNG**-guy)

host (**HOE**-st)

morbidity (more-**BID**-it-ee)

mortality (more-**TAL**-it-ee)

nosocomial infection
(no-zoh-**KOH**-mee-uhl in-**FECK**-shun)

pathogen (**PATH**-o-jen)

phagocytosis (fay-go-sigh-**TOH**-sis)

protozoa (pro-tow-**ZOH**-ah)

reservoir (**REZ**-er-VWAR)

rickettsia (ra-**KET**-see-ah)

sepsis (**SEP**-sis)

standard precautions
(**STAN**-derd pre-**KAW**-shuns)

Staphylococcus (STAFF-il-oh-**KOCK**-uss)

vector (**VECK**-tur)

virulence (**VEER**-you-lence)

virus (**VIGH**-rus)

QUESTIONS TO GUIDE YOUR READING

1. What types of organisms cause infectious disease?

2. How can you interrupt the routes of transmission of infectious disease?

3. How can you assist the body's defense mechanisms to fight infectious disease?

4. What are the signs and symptoms of a localized versus a generalized infection?

5. How would you apply the principles of medical and surgical asepsis?

6. How will you use standard precautions to protect yourself from infection?

7. What nursing care will you provide for a patient with an infectious disease?

8. How will you know if your nursing care has been effective?

The authors acknowledge the contribution to this chapter by Betty Ackley.

► THE INFECTIOUS PROCESS

Events in the chain of infection are shown in Figure 7–1. To prevent an infection, the chain must be broken. If infection occurs, treatment focuses on breaking the chain of infection to prevent the spread of infection to others (Cultural Consideration Box 7–1).

A **pathogen** is able to cause disease. When pathogenic microbes are present in the body without causing symptomatic infection, it is referred to as **colonization.** When an infection occurs without producing symptoms, it is known as a subclinical infection. Identification of a subclinical infection is made from an increased antibody level for the mi-

---

BOX 7-1   CULTURAL CONSIDERATION

One concern when caring for refugees, migrant workers, and immigrants is treating infectious conditions that jeopardize both the patient and the resident population. Some immigrants may suffer from malaria; dengue fever; gastrointestinal parasites; tuberculosis; hepatitis A, B, or C; and other infectious diseases. The hepatitis B virus is hyperendemic in Vietnam, Cambodia, and Laos. Most people are infected during childhood and then bring it with them when they emigrate to the United States. Standard precautions are a must when working with any patient.

Among the Native Americans, infectious health problems include the plague, tick fever, and the Muerto Canyon *Hantavirus*. Many of these illnesses are due to the reservation's rodent population, which includes prairie dogs and deer mice.

Some Appalachians live in rural areas that lack electricity, plumbing, and running water, putting them at increased risk of infection. Teach all patients to thoroughly wash their hands after toileting and coming in contact with raw food or contaminated water.

The drug isoniazid (INH), used in the treatment of tuberculosis, may be metabolized differently by some ethnic and racial groups. Half of European-Americans and a smaller percentage of Asian persons are slow eliminators of INH. These individuals may have high blood concentrations of this medicine, leading to harmful reactions. Carefully monitor these patients for reactions to INH.

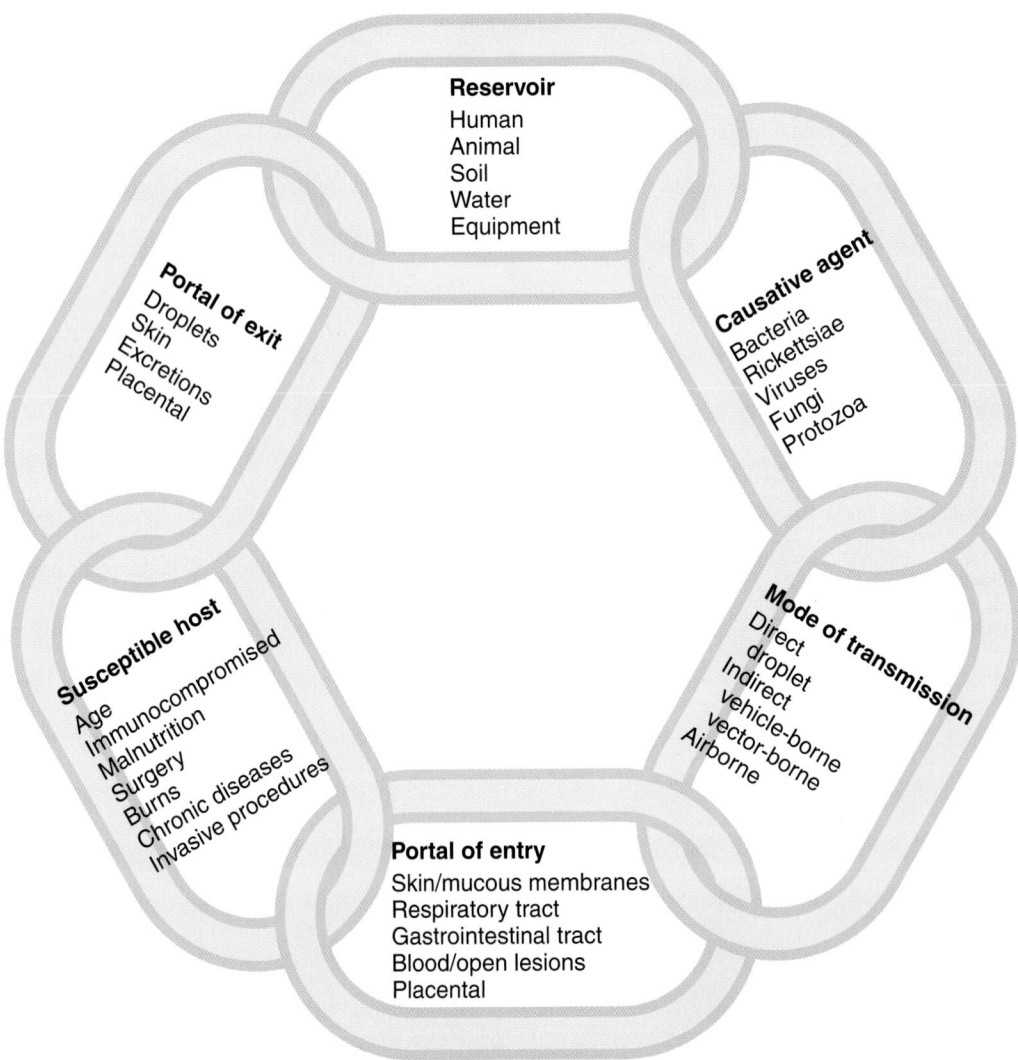

Figure 7-1   Chain of events in the infectious process.

crobe. An infection results in apparent signs and symptoms and injury to the **host.**

Reservoir

A **reservoir** is the place in the environment where infectious agents live, multiply, and reproduce so they can be transmitted to a susceptible host. A reservoir can be animate, such as people, insects, animals, and plants, or inanimate, such as water, soil, or medical devices.

Causative Agents

Microorganisms that cause infection include **bacteria, virus**es, **fungi, protozoa,** helminths, and a newly discovered agent called a prion (Table 7–1). Microbes that occur naturally in or on a particular body part are known as normal **flora.** They are usually harmless, or nonpathogenic, because they do not normally produce disease in a healthy person. Normal flora are helpful to the human host. For example, intestinal flora (bacterial) assist in vitamin K production, a necessary nutrient for normal blood clotting. However, if

---

protozoa: proto–first + zoon–animal

these bacteria get into another area of the body, such as the blood, they may produce disease and are then referred to as pathogens.

Bacteria

Bacteria are single-celled organisms that usually reproduce by simple cellular division. Bacteria may depend on a host or may live and reproduce outside a host. Most bacteria produce cell walls that are susceptible to antibiotics. However, bacteria can mutate to survive.

Bacteria are named according to their shape (spherical [coccus], rod [bacillus], and spiral [spirillum]) and classified according to their staining properties (Gram's method, acid-fast staining). Bacteria respond to stains in one of three ways: gram-positive bacteria stain purple; gram-negative bacteria lose purple stain when exposed to alcohol but stain red with a second dye; and acid-fast bacteria keep purple stain when an acid is applied.

Bacterial growth depends on oxygen, nutrition, light, temperature, and humidity. **Aerobic** bacteria, such as those found on the skin, need oxygen to live. **Anaerobic** bacteria, such as bacteria in the gastrointestinal tract, live without oxygen. Most bacteria that inhabit humans grow best at body temperature, 98.6° F (37° C).

TABLE 7–1	COMMON INFECTIONS
Microorganism	**Type or Site of Infection**
Gram-Positive Bacteria	
Staphylococcus aureus	Pneumonia, cellulitis, peritonitis, and toxic shock
Staphylococcus epidermidis	Postoperative bone/joints, IV line–related phlebitis
Staphylococcus pneumoniae	Pneumonia, meningitis, otitis media, sinusitis, septicemia
Gram-Negative Bacteria	
Escherichia coli	Urinary tract, pyelonephritis, septicemia, and gastroenteritis
Klebsiella pneumoniae	Pneumonia and wounds
Legionella pneumophila	Pneumonia
Neisseria gonorrhoeae	Gonorrhea
Pseudomonas aeruginosa	Wounds, urinary tract, pneumonia, and IV lines
Salmonella enteritidis	Gastroenteritis, food poisoning
Viruses	
Herpes virus group	Cold sores/fever blisters, genital herpes
Epstein-Barr	Infectious mononucleosis
Varicella zoster	Skin (chickenpox and shingles)
Hepatitis (A, B, C, D, E)	Liver
Human immunodeficiency virus	Acquired immunodeficiency syndrome
Influenza (A, B, C)	Bronchiolitis, pneumonia
Rubella	German measles
Rubeola	Measles
Protozoa	
Giardia lamblia	Gastroenteritis
Trichomonas vaginalis	Trichomoniasis
Dientamoeba fragilis	Diarrhea, fever
Entamoeba histolytica	Amebic dysentery
Toxoplasma gondii	Toxoplasmosis
Plasmodium falciparum	Malaria
Fungi	
Candida albicans	Nailbed, thrush, vaginitis
Histoplasma capsulatum	Pneumonia

Rod-shaped bacteria form spores that are thick walled and hard to kill. Spores remain in a resting state until favorable conditions exist that allow the organism to resume normal function. Prolonged exposure to high temperature destroys spores on surgical equipment. Bleach is used in patient rooms to kill spores from *Clostridium difficile* (C. diff).

RICKETTSIAE. Rickettsiae are a type of bacteria that must be inside living cells to reproduce. Rickettsiae **vectors** (living organisms that transmit disease) are infected fleas, ticks, mites, and lice that bite humans. Several diseases are caused by **rickettsia**. Rocky Mountain spotted fever, caused by *Rickettsia rickettsii*, whose reservoirs (the places in nature where the organism usually lives and multiplies without causing disease) are rodents and dogs, is transmitted to humans by tick bite.

Viruses

Viruses are organisms smaller than bacteria that depend on host cells to live and reproduce (see Table 7–1). Invaded host cells make more virus material. The new virus particles are then released either by destroying the host cell or by forming small buds that break away to infect other cells.

When a virus enters a cell, it may immediately trigger disease or remain **dormant** (inactive) for years without causing illness. An example of this is *Human herpesvirus 3* (varicella zoster virus), which can cause disease quickly (chickenpox) or remain dormant for years, eventually erupting in the disease called shingles. Antibiotics are not effective against viruses. Newer antiviral drugs are used to decrease symptoms caused by viruses and to decrease the viral load (the number of virus cells in the patient's blood).

Fungi

Fungi are a group of organisms that includes yeasts, molds, and mushrooms and can produce highly resistant spores (see Table 7–1). Because fungi do not contain chlorophyll, they must obtain food from living organisms or dead organic matter. Normal flora of the mouth, skin, vagina, and intestinal tract include many fungi. Most fungi are not pathogenic, and serious fungal infections are rare. Antifungal medications treat fungal infections.

Protozoa

Protozoa are single-celled parasitic organisms with flexible membranes that live in the soil and obtain nourishment from dead or decaying organic material (see Table 7–1). Protozoa infect humans through fecal-oral contamination or through ingestion of food or water contaminated with cysts or spores, through host-to-host contact, or by the bite of a mosquito or other insect that has previously bitten an infected person.

Helminths

Helminths are wormlike parasitic animals: roundworms, flatworms, tapeworms, pinworms, hookworms, and flukes. Disease transmission occurs through skin penetration of larva or ingestion of helminth eggs. Trichinellosis (caused by the roundworm *Trichinella spiralis*) is a disease caused by eating raw or undercooked meat of pigs or wild animals that contain *Trichinella* larvae.

Prions

Prions are recently identified organisms or agents thought to be unique proteins with long incubation periods. How they reproduce is unknown. Research will shed light on these unusual organisms or protein particles, which are thought to cause mad cow disease and the human dementia Jakob-Creutzfeldt syndrome.

Mode of Transmission

Once the causative agent exits the reservoir, a means of transfer to a susceptible host is needed. Transmission methods for microorganisms include direct contact, indirect contact, and through the air.

Direct Contact

Direct transmission occurs through touching, biting, kissing, sexual contact, or droplet spray into the eyes or mucous membranes while sneezing, coughing, spitting, singing, or talking. Droplet spread is usually limited to 3 feet or less. Illnesses spread by direct transmission may include influenza, impetigo, scabies, conjunctivitis, pediculosis, herpes, C. *difficile*, and all sexually transmitted diseases, including human immunodeficiency virus (HIV). Protect yourself and your patients from direct transmission with hand washing, aseptic technique, gloves, surgical masks, goggles, gowns, and booties.

Indirect Contact

Indirect transmission is either vehicle-borne or vector-borne. Vehicle-borne transmission is the spread of an infectious organism by contact with a contaminated object, such as a toy, soiled bedding, dressings from a wound, surgical instruments, water, food, and biologic products such as blood, serum, plasma, tissues, and organs. Vehicle-borne illnesses include conjunctivitis, trichinellosis, HIV, and hepatitis A, B, C, D, and E. Vehicle transmission can be avoided through proper hand washing, excellent cleaning of the patient environment, and provision of clean water and food supplies. Vector-borne transmission is the spread of infectious organisms through a living source other than humans, such as an insect, flea, mouse, or rat. Diseases spread through vectors include malaria, plague, and Lyme disease. Vector transmission can be reduced with insect repellants, avoidance of infested areas, and rodent control.

Airborne

Airborne transmission is different from droplet transmission (see Direct Contact earlier) because the particles floating in the air are much smaller, remain suspended in the air for a long time, and may travel large distances. Airborne organisms can be inhaled or deposited on the mucus membrane of a susceptible host. Measles, chickenpox, and tuberculosis are transmitted by airborne transmission. Airborne transmission is prevented with the use of high-efficiency particulate air (HEPA) respirators (also known as a tuberculosis [TB] mask). HEPA respirators filter the tiniest particles from the air, unlike surgical masks, which can allow such particles to pass into the respiratory system of a host. Institutions provide individual fit testing and training for HEPA respirator use for each health care worker.

L E A R N I N G   T I P S

If you provide care for patients with suspected or confirmed diseases that are spread through airborne transmission, such as TB, be sure you have your own fit-tested HEPA mask to wear. Do not use other masks because they do not provide adequate protection. If you cannot obtain your own fit-tested mask, you should not enter the patient's room.

Multiple Modes of Transmission

Many diseases have multiple modes of transmission requiring a variety of protective techniques. For example, chickenpox is transmitted by direct contact, indirect contact, and airborne transmission. It is no wonder 80 percent to 90 percent of susceptible persons exposed develop the disease. Understanding the modes of transmission of a particular disease allows you to use the appropriate means of protection without using unnecessary supplies that increase costs.

Portal of Entry

To produce disease, organisms must gain entry into a susceptible host. Routes of entry into a susceptible host include the respiratory tract, skin (usually nonintact), mucous membranes, gastrointestinal tract, genitourinary tract, and placenta. Once the organism enters the host, it *may* lead to disease, depending on the condition of the host and many other factors, such as the **virulence** of the organism.

Susceptible Host

The body has many defense mechanisms to prevent infection, such as intact skin and mucous membranes, and a functioning immune system. A breakdown in these defenses increases the possibility of infection. Factors that increase susceptibility to infection are very young age, old age, malnourishment, immunocompromise, chronic disease, stress, and invasive procedures (Gerontological Issues Box 7–2).

BOX 7–2   GERONTOLOGICAL ISSUES

Infection and Older Patients
Often an older patient may not have typical symptoms of an infection. For example, a serious bacterial infection may cause no elevation in temperature. In fact, a fever is not a common complaint or sign of infection for an older adult.

This difference among elderly patients may cause significant delay in providing appropriate treatment and care. Be alert for the following in the elderly patient, which may indicate an infection:

- Behavioral change, such as pacing or irritability.
- Masking of the symptoms of infection by a chronic disease. For example, the inflammation and pain of degenerative joint disease may make it difficult for a patient to recognize an infection in an affected joint.

Portal of Exit

The portal of exit is the route by which the infectious agent leaves the host, who has become a reservoir for infection: respiratory tract, skin, mucous membranes, gastrointestinal tract, genitourinary tract, blood, open lesions, or placenta.

▶ THE HUMAN BODY'S DEFENSE MECHANISMS

Skin and Mucous Membranes

Intact skin and mucous membranes are the body's first line of defense against infection. Oral mucous membranes have many layers, making it difficult for organisms to enter the body. The skin has acidic (pH <7.0) properties that render some organisms unable to produce disease. For example, many bacteria prefer an alkaline (pH >7.0) environment for reproduction. There is also an abundance of normal flora that impairs the growth of pathogens both on the skin and in the gastrointestinal (GI) tract.

Mucociliary Membranes

Cilia are hairlike structures lining upper respiratory tract mucous membranes that protect the lungs. Cilia trap mucus, pus, dust, and foreign particles to prevent them from entering the lungs. Then the cilia push the trapped particles up to the pharynx with wavelike movements for expectoration.

Gastric Juices

Gastric juices inside the stomach are very acidic (pH 1.0 to 5.0). This acidic environment destroys most organisms that enter the stomach.

Immunoglobulins

Immunoglobulins are proteins found in serum and body fluids that may act as **antibodies** to destroy invading organisms and prevent the development of infectious disease. Antibodies are proteins that are produced by B-lymphocytes when foreign **antigens** of invading cells are detected. Antigens are markers on the surface of cells that identify cells as being the body's own cells (autoantigens) or as being foreign cells (foreign antigens). Antibodies combine with specific foreign antigens on the surface of the invading organisms, such as bacteria or viruses, to control or destroy them. Antigens are neutralized or destroyed by antibodies in several ways. Antibodies can initiate destruction of the antigen, neutralize toxins released by bacteria, promote antigen clumping with the antibody, or prevent the antigen from adhering to host cells.

Leukocytes and Macrophages

Leukocytes (white blood cells) are the primary cells that protect against infection and tissue damage. There are five types of leukocytes: neutrophils, which are phagocytic cells

---

antigen: anti—against + gennan—to produce

focusing on bacteria and small particles; monocytes, which become macrophages and are mainly phagocytic on tissue debris and large particles; lymphocytes, whose functions include antigen recognition and antibody production; basophils, which respond to inflammation from injury; and eosinophils, which destroy parasites and respond in allergic reactions.

After recognizing a foreign antigen, neutrophils and macrophages engulf and digest it, a process known as **phagocytosis.** The macrophages move the antigen fragments to their surface to be recognized by T lymphocytes to further stimulate action of the immune system. Phagocytes ingest and destroy bacteria, damaged or dead cells, cellular debris, and foreign substances.

Lysozymes

Lysozymes are bactericidal enzymes present in white blood cells and most body fluids, such as tears, saliva, and sweat. These enzymes dissolve the walls of bacteria, destroying them.

Interferon

If an invading organism is a virus, white blood cells and fibroblasts release interferon (a group of antiviral proteins). Interferon aids in the destruction of infected cells and inhibits production of the virus within infected cells. Tumor cell growth may also be inhibited by interferon.

Inflammatory Response

The inflammatory response occurs as a result of any bodily injury. This response can be caused by pathogens, trauma, or other events causing injury to tissues. Infection may or may not be present.

Vascular Response

The first step of the inflammatory process is local vasodilation, which increases blood flow to the injured area. Increased blood flow creates redness and heat at the injury. Pathogenic organisms can trigger the first step of the inflammatory process.

Inflammatory Exudate

Next, increased permeability of the blood vessels allows plasma to move out of the capillaries and into the tissues. Swelling occurs resulting in pain from pressure on nearby nerve endings.

Phagocytosis and Purulent Exudate

The final step of the inflammatory process is the destruction of pathogenic organisms and their toxins by leukocytes. During this process, a purulent exudate (pus) may form that contains protein, cellular debris, and dead leukocytes.

Immune System

The immune system is the body's final line of defense against infection. Immune cells and lymphoid tissue work with the body's other defense mechanisms. Immune cells include lymphocytes (T cells, B cells, and natural killer cells), which have protective functions related to specific antigens. Macrophages assist T and B lymphocytes.

The lymphoid organs are the thymus, which is vital to the development of the immune system, and the bone marrow, which produces leukocytes. Other lymphoid structures include the spleen, tonsils, intestinal lymphoid tissue, and lymph nodes, where immune cells grow and whose filtering of foreign materials prevents them from entering the bloodstream. The spleen destroys old or damaged red blood cells and contains large amounts of lymphocytes.

The immune system is a finely tuned network that functions together to protect the body from invasion by pathogenic organisms. When this network breaks down, infectious disease can result.

▶ INFECTIOUS DISEASE
General Clinical Manifestations of Infections
Localized Infection

Localized infection is caused by an increase of microbes in one area that triggers the inflammatory response. Manifestations of a local infection include pain, redness, swelling, and warmth at the site. Pain is most severe when the infection occurs in closed cavities. Redness and swelling are seen when surface structures are involved. Warmth may be felt at the site. Temperature may rise.

Generalized Infection

Generalized infections occur when there is systemic or whole body involvement. Symptoms of generalized infection may include headache, malaise, muscle aches, fever, and anorexia. As the infection progresses, there can be an increase in fever, elevated white blood cell count, decreased blood pressure, mental confusion, tachycardia, and shock. **Sepsis** is the term used for an infection that has spread to the bloodstream.

Laboratory Assessment

Several methods are used to identify pathogens. One method is performing a microscopic examination, such as Gram's method of staining. Another method is culture examination, in which the organism is grown on a laboratory plate and identified within 24 to 48 hours. A sensitivity examination is then done, which exposes the organism to many antibiotics to determine which the pathogen is sensitive to and which it is resistant to for treatment.

A serum antibody test measures reaction to a certain antigen. A positive result on this test does not always mean an active infection is present. It can simply mean there has been an exposure to the antigen, so it is not as accurate as a culture.

phagocytosis: phagein—to eat + dytos—cell + osis—condition

A complete blood cell count with differential (CBC with diff) is usually obtained when an infectious disease is suspected. The five different types of leukocytes and their levels are identified. Elevations in specific leukocytes occur based on the type and severity of the pathogen.

Erythrocyte sedimentation rate (ESR, sed rate) is an early screening test for inflammation but not a definitive test for infection. During the inflammatory process, red blood cells become heavier and during the test settle to the bottom of a tube. The ESR measures in millimeters per hour the speed at which the red blood cells settle in the tube. The faster the settling, the greater the inflammation.

Other tests such as x-rays, computed tomography (CT), and magnetic resonance imagery (MRI) are helpful in identifying abscesses (walled-off infections). Skin tests diagnose infections, such as the Mantoux test (commonly called a PPD), which screens for tuberculosis. (See Chapter 28.)

Immunity

Immunity is the ability of the body to protect itself from disease. (See Chapter 52.) There are several types of immunity:

- Natural immunity occurs in species and prevents one species from contracting illnesses found in another species.
- Innate immunity is genetic, hereditary immunity a person is born with.
- Acquired immunity is obtained either actively or passively through exposure to an organism, from a vaccine, or from an injection of immunoglobulins (antibodies) or is passed from mother to baby.

Mononucleosis

Infectious mononucleosis (referred to as mono or the kissing disease) is an infection that is usually caused by the Epstein-Barr virus (a herpes virus). Mononucleosis is a contagious disease that anyone may develop. However, it is mainly diagnosed in young adults. Most adults have been exposed to the virus and have antibodies to it but never develop the disease. The virus remains in the body for a lifetime but rarely causes another infection.

Mononucleosis is primarily spread through person-to-person contact, mainly through saliva. Sharing utensils, food, or beverages can transmit the disease. Coughing or sneezing of small droplets of infected saliva or mucus into the air allows the virus to be inhaled by others.

The incubation period for mononucleosis is 4 to 8 weeks after exposure. Symptoms during the first 3 days include extreme fatigue, loss of appetite, and chills. Then a severe sore throat, headache, high fever, reddened throat and tonsils with a white coating, generalized lymphadenopathy (enlarged lymph nodes in two different sites other than inguinal nodes), or diarrhea occur. The spleen enlarges 50% of the time. Occasionally a rash develops similar to the measles.

Signs and symptoms, as well as diagnostic tests, confirm mononucleosis. Lymphocyte levels are elevated. The monospot and heterophile antibody tests confirm mononucleosis.

Usually no specific treatment is needed. The illness runs its course as other viral illnesses do. Antiviral drugs are not effective. Symptoms are treated as needed with supportive care. Fatigue may last for months. Rest is important. If the spleen is enlarged, contact sports, lifting, and straining are avoided to prevent trauma and rupture. Mild inflammation of the liver or hepatitis may occur, but no treatment is usually required. Other complications are rare.

▶ INFECTION CONTROL IN THE COMMUNITY

Many levels of organizations work closely together to control communicable diseases. The World Health Organization (WHO) and the Centers for Disease Control and Prevention (CDC) teach standards to prevent and control diseases and monitor disease outbreaks. Local health departments teach how to prevent and control the spread of disease. The community health nurse collects health data, sets up immunization programs within the community, and educates the public about health and illness.

Community immunization programs have helped reduce infectious diseases. Although requirements vary from state to state, most elementary schools require some proof of childhood immunization. Many colleges also require proof of immunization to help control the outbreak of diseases such as measles. Vaccination has played a major role in preventing many infectious diseases and controlling the spread of existing disease outbreaks. In addition, educating the public about the importance of hand washing, immunization, clean water, safe food handling techniques, and safer sex precautions in preventing the spread of disease cannot be overemphasized.

Both the hospital and home health care nurse have a responsibility to provide infection control teaching for the patient and family. Such techniques may include use of disposable dishes and utensils and disposable gloves or proper disposal of contaminated items. Techniques used should be specific to the interruption of the transmission of the particular disease.

▶ BIOTERRORISM

Preparing for a bioterrorism attack is difficult. The Global Outbreak Alert and Response Network monitors for global outbreaks of infectious diseases. Staying informed and understanding how to respond if an emergency or disaster occurs is the best way to be prepared. Participating in your agency's disaster drills can increase your confidence if a disaster occurs. Infectious agents that might be used in a biologic attack include anthrax, botulism, plague, and smallpox.

For more information on public health emergency preparedness and response and facts about bioterrorism agents, visit www.bt.cdc.gov. For more information on infectious diseases, visit www.cdc.gov/ncidod/eid. For the Federal Emergency Management Agency, visit www.fema.gov.

Anthrax

Anthrax, a bacterial disease caused by *Bacillus anthracis*, is found around the world. Anthrax occurs in animals (such as goats, sheep, cattle, deer, horses). Anthrax spores are found in the environment and can survive for many years in water and soil. Exposure to infected animals and animal products (such as infected wool, hides, body fluids, feces, or bone meal) can result in infection. Anthrax is not transmitted person to person, so isolation is not needed.

There are three kinds of anthrax: skin anthrax (most common), occurring when a spore enters the skin through a cut or an abrasion; inhalation anthrax, resulting from inhalation of airborne anthrax spores; and gastrointestinal tract anthrax, acquired from eating infected meat. All forms of anthrax can be fatal if not treated promptly.

Symptoms of skin anthrax occur 1 to 7 days after contact. The affected skin itches with papular and vesicular lesions, edema, ulceration, and sloughing of the skin. Inhalation anthrax exposure results in flulike symptoms after 48 hours, followed 2 to 4 days later by severe respiratory distress and shock; there is a high **mortality** rate within 24 hours of onset of respiratory distress and shock.

Standard precautions are used during patient care for all forms of anthrax. Vaccines for anthrax are available and given to people primarily in high-risk situations. To treat anthrax, antibiotic therapy is beneficial if given before or just after onset of illness. Penicillin, ciprofloxacin, and doxycycline are effective antibiotics. Antibiotic therapy is also given as prophylaxis for exposure to anthrax spores. Chlorine destroys spores. Boiling contaminated articles in water for 30 minutes destroys anthrax.

▶ INFECTION CONTROL IN HEALTH CARE AGENCIES

If on admission to a hospital or health care agency a patient already has an infection, it is referred to as a community-acquired infection. An infection that develops as a result of the stay in the hospital or health care agency is called a **nosocomial infection.** The host's condition plays a major role in whether or not an infection is acquired. Patients in the hospital are commonly debilitated, malnourished, or immunocompromised. Multiple antibiotic therapy also increases susceptibility to other types of infection and promotes the resistance of pathogens to antibiotics. Therefore the risk of developing a nosocomial infection is very high. Some areas within an institution tend to have an increased number of nosocomial infections, such as intensive care, neonatal, dialysis, oncology, and burn units. Patients in these areas tend to undergo more invasive procedures and are debilitated, increasing susceptibility to infection.

Several pathogens are commonly responsible for causing nosocomial infections:

- *Escherichia coli* is the most common pathogen causing nosocomial urinary tract infections. *E. coli* normally lives in the healthy intestinal tract of humans. *E. coli* can be spread by the patient, by the unwashed hands of a health care worker, or through contaminated food and water.
- *Staphylococcus aureus* (commonly known as staph) is the most common pathogen causing nosocomial surgical wound infections and nosocomial septicemia. Staph usu-

ally lives in the nose and on the skin of healthy people. (Some highly pathogenic strains of staph are also emerging in the world.)
- *Pseudomonas aeruginosa* is the most common pathogen in nosocomial pneumonia. It is found in soil, around water, and in the health care setting around sinks, water, irrigating solutions, and nebulizers on respiratory equipment.

The single most effective way to prevent and control the spread of infection is by effective hand washing. Most organisms in the institutional setting are transmitted via the hands of health care workers. Hands *must* be washed before and after every patient contact to prevent the direct transmission of organisms (Fig. 7–2). Patients may also transmit organisms with inadequate hand washing. Teach patients the importance of hand washing after handling their own secretions. The use of gloves decreases the transmission of organisms, but CDC guidelines also require hand washing before and after glove use.

Asepsis

The concept of **asepsis** (freedom from organisms) is important for all health care workers who have direct or indirect patient contact. For hospitalized patients, the most common sites for infectious diseases are the genitourinary tract, respiratory tract, bloodstream, and surgical wounds. Be aware of patients at risk of developing these infections and protect them with aseptic techniques.

Medical Asepsis

Medical asepsis is commonly referred to as clean technique. The goal is to reduce the number of pathogens or prevent

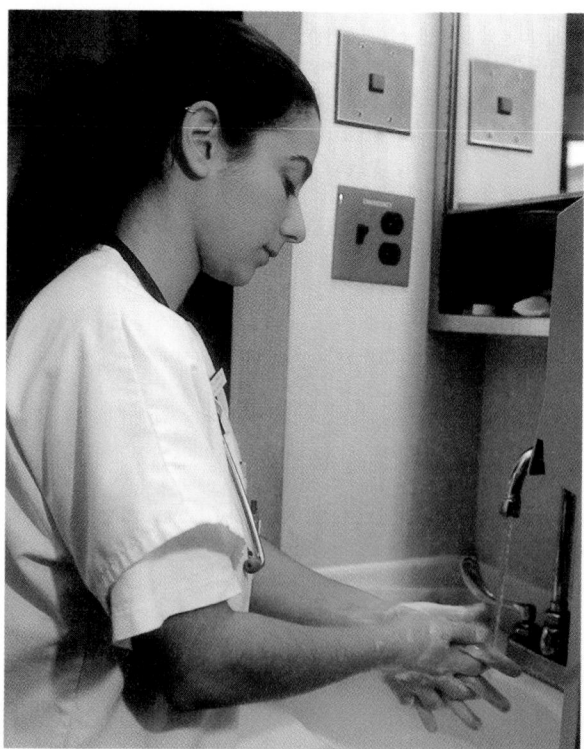

Figure 7-2 Frequent hand washing by health care workers helps reduce the spread of microorganisms.

CRITICAL THINKING

Mrs. Sampson

Mrs. Sampson has neutropenia from chemotherapy treatments.

1. Why is hand washing the most important intervention you can do to help prevent infection for Mrs. Sampson?
2. What would be a priority nursing diagnosis for Mrs. Sampson?
3. What type of isolation could be beneficial to Mrs. Sampson?

Answers at end of chapter.

LEARNING TIPS

Proper hand washing requires wetting the hands, soaping, and lathering, with at least 10 seconds of rubbing your hands together under running water with fingertips pointed downward. Interlace your fingers to cleanse between them, rub your nails against your palms to clean under the nails, and then rinse your hands with fingertips pointed downward. Dry your hands with clean paper towels. Use the paper towel to turn off the faucet. Apply lotion to your hands to prevent drying and possible cracking or open areas in which infection could develop.

the transmission of pathogens from one person to another. Frequent, proper hand washing is one of the best ways to achieve this goal. Gowns, gloves, masks, and protective eyewear or rooms with special ventilation may also be helpful (Fig. 7–3). Disinfectants and precautions as defined by the CDC are also crucial tools. Techniques used should be appropriate to interrupt the spread of the known pathogen. As part of medical asepsis you should keep your own body and clothing clean to prevent spread of infection to patients, yourself, and your family (Box 7–3).

Surgical Asepsis

Surgical asepsis (sterile technique) refers to an item or area that is free of all microorganisms and spores. Surgical asepsis is used in surgery and to sterilize equipment. Articles can be subjected to intense heat or chemical disinfectants to destroy all organisms. The use of pressurized steam sterilizers, called autoclaves, kills even the most powerful organisms.

 CRITICAL THINKING

Which of the following patients is at greatest risk for infection and why?

1. Mr. Ashland, age 55, is hospitalized for a hernia repair. He is overweight and has adult-onset diabetes.
2. Mrs. Burrows, age 72, is hospitalized for a broken hip. She is thin, frail, has dementia, and has undergone placement of a urinary catheter.
3. Jackson Dent, age 2, is hospitalized for minor surgery. Jackson is thin and small for his age.

Answers at end of chapter.

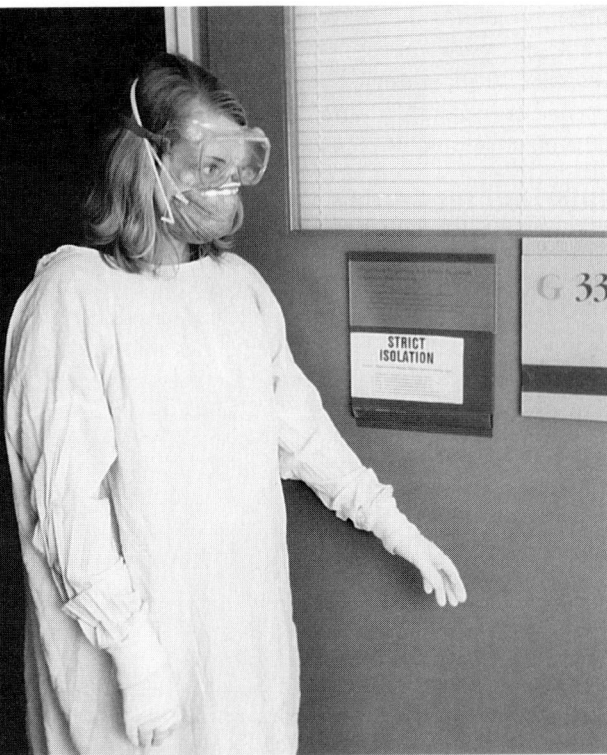

Figure 7-3 Gloves, gowns, masks, goggles, and face shields help prevent the spread of infection to health care workers and patients.

BOX 7-3 **Guidelines to Prevent Spread of Infection to the Patient, Self, and Family**

- Bathe daily and wear a clean uniform/clothing every day.
- Keep your fingernails short, piano playing length, and do not wear acrylic nails. Both long nails and acrylic nails have been associated with spread of infection to patients and can be colonized with harmful bacteria. Multiple studies have demonstrated that long fingernails and artificial nails harbor bacteria and have caused infections for patients that sometimes have resulted in death.
- Avoid wearing rings and bracelets that harbor organisms.
- Cleanse your stethoscope at least daily and in between patient use with alcohol. VRE bacteria have been cultured from stethoscopes in a hospital setting.
- Wash your hands between each patient contact. Hand washing is recognized as the single most important action to take to prevent spread of infection. If you are unable to use soap and water, use a nondrying alcohol hand gel, which has shown to be effective in cleansing hands and preventing drying of the skin.
- Follow prescribed isolation precautions for your protection, as well as that of the patient.
- Remove your uniform and bathe/shower when you come home from work. This will decrease the spread of antibiotic-resistant bacteria to your home and your family. Keep your nursing shoes clean and put away from the rest of the family.

Some equipment cannot be exposed to moist heat, so gas sterilizers are used instead. Once these articles are sterilized, they are dated, packaged, and sealed. Once a package is opened or outdated, it is no longer considered sterile. Sterile technique is rarely required in the home care setting.

Infection Prevention Guidelines

The CDC guidelines for infection control and isolation precautions are used in hospitals and health care agencies. Health care agencies use these guidelines to establish their own policies. CDC and agency guidelines are continuously updated and should be followed for your protection, as well as your patients'. Current CDC guidelines for isolation precautions in hospitals include two tiers of precautions:

standard precautions and transmission-based precautions (Table 7–2).

Standard Precautions

Standard precautions require you to assume that all patients are infectious regardless of their diagnosis. Standard precautions combine reducing the spread of blood-borne pathogens found in blood and body fluids with body substance isolation to decrease transmission of moist body substance pathogens. Standard precautions apply to blood, secretions, excretions, open skin, mucous membranes, and all body fluids, excluding sweat. All patients with draining wounds or secretions of body fluids are considered infectious until an infection is confirmed or ruled out. Using gloves,

---

TABLE 7–2   STANDARD PRECAUTIONS AND TRANSMISSION-BASED PRECAUTIONS

Standard Precautions

Use standard precautions for all patient care. Combine standard precautions with transmission-based precautions as needed based on the patient's illness.

Hand washing	Wash hands with nonmicrobial soap unless specially contraindicated before and after using gloves, between patients, and between procedures on same patient.
Gloves	Wear gloves before contact with any body fluids or substances.
Mask, eye protection, face shield	Use protective equipment for patient care if splashes or sprays of blood or body fluids are likely.
Gown	Wear gown to protect skin/prevent soiling of clothing for patient care if splashes or sprays of blood or body fluids are likely.
Occupational health and blood-borne pathogens	Dispose of sharps properly; do not recap needles.
Patient care equipment	Clean reusable equipment before reuse. Discard single-use items properly.
Linen	Handle linen to avoid clothing contamination.
Patient placement	Use private room for infectious patients.

Transmission-based Precautions

Airborne Precautions

Examples: measles, tuberculosis, varicella (chickenpox, shingles)

Patient placement	Provide private room with regulated air flow. Keep door closed.
Respiratory protection	Do not enter room if susceptible to measles or chickenpox unless no immune caregivers available. If susceptible, wear fit-tested N95 or HEPA respirator.
Patient transport	Limit patient transport to essential purposes. Place surgical mask on patient.

Droplet Precautions

Examples: diphtheria (pharyngeal), *Haemophilus influenzae* (epiglottitis, meningitis, pneumonia, sepsis), influenza, mumps, mycoplasma pneumonia, *Neisseria meningitidis* (meningitis, pneumonia, sepsis), pertussis, pneumonic plague, rubella, streptococcal pharyngitis, pneumonia

Patient placement	Provide private room or separation of at least 3 feet between the infected patient and other patients.
Mask	Wear mask for patient care if within 3 feet of patient.
Patient transport	Limit patient transport to essential purposes. Place surgical mask on patient.

Contact Precautions

Examples: cellulitis, *Clostridium difficile*, skin infections (diphtheria, herpes simplex virus, impetigo, pediculosis, scabies), viral/hemorrhagic conjunctivitis, viral hemorrhagic infections (Ebola, Lassa, or Marburg), herpes zoster

Patient placement	Provide private room or place with patient with same infection and no other infection.
Hand washing, gloves, gown	Protect self and others from contaminated items.
Patient transport	Limit patient transport.
Patient care equipment	Dedicate the use of noncritical patient care equipment to a single patient.

Data from Public Health Service, US Department of Health and Human Services, Centers for Disease Control and Prevention, Atlanta, Georgia; and Garner JS, Hospital Infection Control Practices Advisory Committee: Guideline for isolation precautions in hospitals. Infect Control Hosp Epidemiol 17:53, 1996; and Am J Infect Control 24:24, 1996.

gowns, masks, goggles, face shields, and, most important, hand washing helps prevent the spread of infection to health care workers and other patients.

Transmission-based Precautions

Transmission-based precautions are used for patients with specific communicable diseases that can be transmitted to others. Transmission-based precautions are used with standard precautions to control the spread of infection.

Prevention of Respiratory Tract Infections

Nosocomial pneumonia has been linked with the highest infection mortality rate in hospitalized patients. Patients who are at highest risk for pneumonia are those with endotracheal, nasotracheal, or tracheostomy tubes because these invasive tubes bypass the normal defenses of the upper respiratory tract. To prevent infection in these patients in the agency setting, care of respiratory tubes usually requires sterile technique. If patients are on ventilators or receiving respiratory treatments, the tubing and equipment are usually changed every 48 hours. Protecting debilitated patients or those with an ineffective cough from aspirating also helps prevent pneumonia.

Prevention of Genitourinary Tract Infections

The most common hospital-acquired infection is a urinary tract infection. Patients with urinary catheters are at the greatest risk. The urinary tract is normally a sterile tract, but insertion of a catheter into the bladder may allow organisms to enter. Institutional policy on appropriate use of urinary catheters differs, and you should follow the policy of your particular institution. Appropriate uses of catheters include use in patients with urinary obstructions and neurogenic bladder conditions and in those in shock.

LEARNING TIPS

Catheters should be used only when necessary because of the **morbidity** and mortality associated with infections that develop from them.

Indwelling urinary catheters should be removed as soon as possible. For patients requiring long-term use of urinary catheters, intermittent catheterization is preferred because it has been shown to significantly reduce the risk of infection. Using strict aseptic technique while inserting and caring for the catheter in the health care agency is imperative. The catheter tubing must be securely anchored to the patient's leg, according to agency protocol, so it does not move in and out of the urethra. Movement can encourage organisms to enter the urinary tract.

The closed urinary drainage system seal should never be opened. If intermittent irrigation is ordered, sterile technique must be used to protect both ends from contamination. The drainage bag should be positioned so that it is never higher than the level of the bladder to prevent backflow of urine into the bladder, which could contaminate the sterile urinary tract. If an indwelling urinary catheter and drainage system is used long term, the catheter and the entire system should be changed regularly using sterile technique. Standards in home care differ from institutional care because patients are generally at lower risk of infection within their own environment.

All long-term indwelling urinary catheters are considered colonized, but only a few will cause infection in the patient. Remember that the most crucial point at which bacteria may enter the patient is during the insertion of the catheter, so excellent technique is required. Another point to remember is that the urinary tract is highly vascular (many blood vessels close to the surface), so that an infection in this tract can easily result in bacteremia (bacteria in the blood), which can then progress to septicemia (infection in the blood), a potentially life-threatening condition.

CRITICAL THINKING

Mr. Carson
While working in an extended care facility you see Mrs. Brandt's nursing assistant wheeling Mr. Carson to activities. He has a long-term urinary catheter. The urine bag is hung on the arm of the wheelchair with a cloth partially covering it. What is your responsibility in this situation?

Answers at end of chapter.

Prevention of Surgical Wound Infections

The initial dressing for surgical wounds is applied in the operating room using sterile aseptic technique. Postoperative orders indicate when to change the dressing. Sterile technique should be used. The wound is monitored with every dressing change for signs of infection.

Protection from Septicemia (Sepsis)

Septicemia (commonly called blood poisoning) is a blood infection with a variety of causes, including infection in another body site, invasive catheters and solutions (central lines, arterial lines, pulmonary artery catheters, urinary catheters). Insertion and care of these catheters require sterile technique and careful observation for infection signs. All solutions should be examined for expiration date, signs of contamination, cloudiness, particles, or discoloration before use. Indications of sepsis such as fever, tachypnea, tachycardia, hypotension, and elevated white blood cell count should be reported promptly to the physician for immediate treatment. Blood cultures may be ordered. Antibiotics are used to treat sepsis. A new intravenous (IV) drug, drotrecogin alfa (activated) (Xigris), is available to treat severe sepsis when death is likely. Drotrecogin alfa (activated) is given for 96 hours. It is associated with a risk of bleeding. Drotrecogin alfa (activated) has reduced mortality in patients that may be due to anti-inflammatory effects.

▶ ANTIBIOTIC-RESISTANT INFECTIONS

Antibiotic-resistant infections are on the rise and are creating significant treatment problems. These infections result in increased health care costs, morbidity (death), and mortality (illness). Preventive measures to reduce the occurrence of antibiotic-resistant infections, include avoiding the use of antibiotics in animal feed, teaching patients to take all prescribed medications exactly as ordered, and avoiding the use of antibiotics for viral infections (common cold or flu). Newer American College of Physicians/American Society of Internal Medicine guidelines suggest that most upper respiratory infections do not require antibiotic treatment. For healthy adults with the symptoms of bronchitis, sinusitis, pharyngitis, and other upper respiratory infections, it is suggested to use over-the-counter cold symptom remedies and saltwater gargles. Two infections that are on the rise include methicillin-resistant *Staphylococcus aureus* (MRSA) and vancomycin-resistant enterococci (VRE). For more information on antibiotic resistance, visit the World Health Organization at www.who.int/emc/amr.html.

LEARNING TIPS

Antibiotics are not effective against viral infections such as colds or the flu. People need to be educated that antibiotics do not work on viral infections. The misuse of antibiotics is creating antibiotic-resistant "superbugs." People should not ask health care providers for antibiotics to treat viral infections. Use of antibacterial products such as antibacterial soaps may also contribute to the superbug problem. You play a vital role in educating people of this growing problem.

MRSA

A serious antibiotic-resistant infection is caused by methicillin-resistant *S. aureus* (commonly called MRSA). MRSA is difficult to treat and has a high mortality rate. MRSA affects mainly the elderly and the chronically ill. Vancomycin hydrochloride, a potent and expensive antibiotic, can be used intravenously to treat MRSA. Fear exists that the bacteria will further mutate to become resistant to all currently available antibiotics. A few isolated cases of this have been documented worldwide. The risk that these resistant organisms will spread is real, which will return us to the preantibiotic days when *S. aureus* was a killer.

VRE

Vancomycin-resistant enterococci infections are common. Although enterococci are normal flora in the GI and female genital tracts, VRE are a new pathogenic strain. VRE are transmitted via direct or indirect contact. Patients at risk for VRE infections include those with indwelling urinary or central venous catheters, the immunocompromised or critically ill, those receiving multiple antibiotics or vancomycin therapy, surgical patients, and those with extended hospital stays. Preventive VRE measures focus on proper hand washing, education of health care workers, aggressive infection-control methods and restricting use of vancomycin. Patients with VRE should be isolated, and current CDC and institutional isolation policies should be strictly followed (Fig. 7–4). Treatment is difficult, involving combination antibiotic therapy, but VRE may also be resistant to other antibiotics.

New drugs have been developed for treating antibiotic-resistant infections. Quinupristin/dalfopristin (Synercid) is an intravenous drug for VRE treatment. When using quinupristin/dalfopristin, the IV line can only be flushed before and after with 5 percent dextrose in water (D5W), not 0.9 percent normal saline or heparin. Linezolid (Zyvox) can treat both VRE and MRSA and should ideally only be used for these serious infections, so that resistance does not develop to it. Research is ongoing to develop more antibiotics for resistant organisms, but antibiotic-resistant bacteria remain a very serious threat to the health of the world population.

▶ TREATMENT OF INFECTIOUS DISEASES

Treatment is begun once an infectious organism and the affected body system are identified. Many factors play a part in determining the drug to be used. The type of organism is very important in selecting an appropriate medication.

- Antibiotics treat bacterial infections, not viruses, fungi, helminthes, or prions.
- Antiviral medications treat viral infections, but their use is aimed at symptom control rather than cure.
- Antifungal drugs are available for fungal infections, but cure may require their extended use.

The drug of choice must be able to destroy (or control) the pathogen while preserving the patient's healthy cells (Relevent Research Box 7–4). Cost-effectiveness is another concern when selecting a medication. Newer antibiotics can be very expensive and therefore are not available to everyone.

Numerous antibiotics are available on the market today to treat bacterial infections. When giving an antibiotic, several things must be considered. First, consider any allergies the patient may have to a particular group or class of antibiotics. Patients may have allergies to one antibiotic group that prevents the use of chemically similar drugs. Therefore all allergies should be reported to the prescribing health care provider.

The patient's kidney and liver function should be known. Many antibiotics are metabolized by the liver and excreted by the kidneys. Diseases of these organs may require a lower dose. Antibiotic levels fluctuate greatly de-

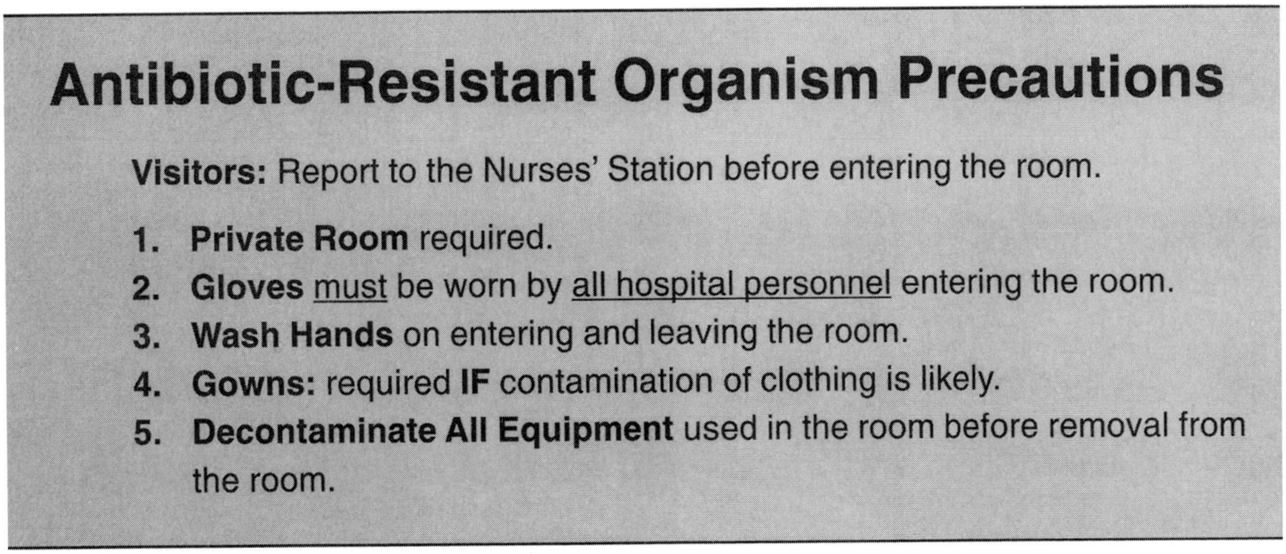

Antibiotic-Resistant Organism Precautions

Visitors: Report to the Nurses' Station before entering the room.

1. **Private Room** required.
2. **Gloves** <u>must</u> be worn by <u>all hospital personnel</u> entering the room.
3. **Wash Hands** on entering and leaving the room.
4. **Gowns:** required **IF** contamination of clothing is likely.
5. **Decontaminate All Equipment** used in the room before removal from the room.

Figure 7-4 Antibiotic-resistant organism precautions for Vancomycin resistant enterococci (VRE).

RELEVANT RESEARCH BOX 7–4

Use of Probiotics to Prevent Infection

Probiotics are live microorganisms that interact with the human flora, improving the intestinal microbial balance with normal bacterial flora. An example of a probiotic is yogurt that contains live cultures. Probiotics can also be taken as a capsule, but there is only preliminary research to demonstrate what bacteria is helpful for different conditions.

Some probiotics have shown the ability to prevent the diarrhea associated with antibiotic use, including decreasing the incidence of *Clostridium difficile*.[1] Clinically significant benefit has been shown in the use of probiotics to treat acute infectious diarrhea in children.[2] Some studies have shown the benefit of probiotics to prevent and treat vaginal infections, and also prevent urinary tract infections.[2,3] Research is ongoing to prove that probiotics may be useful to control inflammatory diseases and some allergies and to stimulate the immune system, which may reduce the incidence of respiratory infections.[4] It should be noted that the severely immunosuppressed patient should avoid use of probiotics because they contain live bacteria. Probiotics is a new and exciting area of research in infection control.

1. Elmer, GW: Probiotics: "Living drugs." *Am J Health Sys Pharm* 58:12, 2001.
2. Szajewska, H, and Mrukowicz, JZ: Probiotics in the treatment and prevention of acute infectious diarrhea in infants and children: a systematic review of published randomized, double-blind, placebo-controlled trials. *J Pediatr Gastroenterol Nutr* 33:4, 2001.
3. Reid, G, and Bruce A: Selection of Lactobacillus strains for urogenital probiotic applications. *J Infect Dis* (suppl)183:1, 2001.
4. Vanerhoof JA: Probiotics: future directions. *Am J Clin Nutr* 73:6, 2001.

pending on organ function, age, sex, health, and many other factors. Antibiotic peak (highest blood level) and trough levels (lowest blood level) need to be monitored to prevent toxicity and damage to major organs. Site of infection is another factor that determines the choice of antibi-

otic. Possible routes of administration include oral, parenteral, intravenous, or instillation into a body cavity.

Antibiotics can be classified as either bactericidal or bacteriostatic (Table 7–3). Bactericidal agents kill bacteria, whereas bacteriostatic agents inhibit or retard bacterial growth, leaving the final destruction of the bacteria up to the infected host's immune system. Bacteriostatic agents therefore may be less helpful for the patient who is immunocompromised.

General Principles

Consult drug references for further information on anti-infective medications. You are responsible for administering medications correctly and for teaching the patient the importance of taking these medications properly. Certain nursing and patient teaching responsibilities relate to *all* anti-infective medications. These general principles include the following.

Nursing Responsibilities

■ All patient allergies are noted, especially any documented allergy to the specific drug ordered or cross-allergies to the drug ordered, and the primary provider is notified before treatment is started.

■ Monitor and report any side effects or signs of allergic response, especially anaphylactic reactions.

■ Observe and report any signs of superinfection (one that occurs as a result of antibiotic use). For example, thrush may develop as antibiotics disrupt the normal flora of the GI tract.

Patient Education

■ Stress to the patient and family the need to take *all* the medication exactly as prescribed. Explain that stopping treatment when the patient feels better results in potential relapse with more resistant organisms.

■ Explain what signs and symptoms to watch for (allergic and nonallergic side effects) and what to do about them, especially when to call the primary provider.

TABLE 7–3	Drugs for Treating Infectious Disease	
Type and Examples	**Nursing Considerations**	**Patient Teaching**
Bactericidal Antibiotics		
Penicillins amoxicillin (Amoxil), ticarcillin (Ticar), penicillin G, ampicillin (Omnipen)	• Most widely used antibiotic. • Most effective against gram-positive organisms. • Monitor patient for allergic reaction (rash, hives, itching) or anaphylactic shock (fever, chills, trouble breathing, lower blood pressure, tight throat). Keep epinephrine available. • If signs of allergic reaction occur, stop parenteral drug and notify primary care provider immediately. • Watch for superinfection.	• Review signs of allergic reactions. • Tell patient and family to stop drug and call primary provider if they occur. • Tell patient to notify primary care provider if white patches appear in mouth or vagina becomes irritated.
Cephalosporins cephalothin (Keflin), cefazolin (Ancef), cefaclor (Ceclor), ceftriaxone (Rocephin)	• First-generation drugs are most effective against gram-positive organisms. • Second- and third-generation are more effective against gram-negative organisms. • Patient with penicillin allergy may have cross-allergy to cephalosporins. • Common side effects include GI disturbance, phlebitis, pain at injection site, rash, and hives. • Less common but serious side effects include kidney and liver damage and superinfection. • Monitor blood urea nitrogen, creatinine, lactic dehydrogenase, aspartate aminotransferase, and alanine aminotransferase to detect kidney or liver damage.	• Tell patient to take drug on an empty stomach, 1 hour before or 2 hours after meals, to increase absorption. • Tell patient to take drug at specified intervals over 24-hour period.
Aminoglycosides amikacin (Amikin), gentamicin (Garamycin), tobramycin (Nebcin)	• Used to treat gram-negative organisms. • Usually given parenterally because they are not absorbed well in GI tract. • Monitor peak and trough levels to keep drug in therapeutic range. • Assess patient's kidney function. • Check for ototoxicity (ringing in ears, deafness). • Don't mix or infuse with any other drug.	• Tell patient to report signs of allergy, tinnitus, vertigo, or hearing loss. • Tell patient to obtain daily weights because weight gain may indicate kidney problems.
Fluoroquinolones ciprofloxacin (Cipro), levofloxacin (Levaquin), norfloxacin (Noroxin), ofloxacin (Floxin), trovafloxacin (Trovan)	• Used to treat a variety of infections, such as bronchitis, bone and joint infection, pneumonia, tuberculosis, sexually transmitted disease, and urinary tract infection. • Give drug on an empty stomach. • Do not give with antacids that contain aluminum, calcium, or magnesium. • Monitor liver function and report signs of dysfunction (fatal hepatitis may occur).	• Tell patient to take with a full glass of water; encourage fluids. • Tell patient to report side effects: swelling of face/throat; trouble swallowing; shortness of breath; itching; hives; pain in shoulder, hands, or heel tendons. • Explain that drug may cause drowsiness and increase sensitivity to light. Advise sun protection.
Bacteriostatic Antibiotics		
Tetracyclines tetracycline HCl, doxycycline (Vibramycin), minocycline HCl (Minocin)	• Treat most gram-positive and gram-negative organisms. • Tetracycline HCl comes only in oral form. The others are usually given IV because intramuscular route leads to poor absorption. • Give 1 hour before or 2 hours after meals. • Do not give with milk, milk products, or antacids because they impair absorption. • GI disturbances are the most common side effect. • Tetracyclines increase anticoagulant activity. Monitor prothrombin time, international normalized ratio, partial prothrombin time as ordered.	• Suggest eating crackers and juice (but not a full meal) to reduce GI upset. • Tell patient to avoid prolonged sun exposure during therapy.
Erythromycin E-Mycin, E.E.S., Erythrocin	• A broad-spectrum antibiotic effective against many gram-negative and gram-positive organisms. • Give on an empty stomach 1 hour before or 2 hours after meals. • Give with a full glass of water (not with acidic fruit juices, such as orange juice or grapefruit juice). • When giving drug IV, administer slowly to decrease vein irritation.	• Urge patient to take drug around the clock and complete entire course of treatment. • Explain that gastric distress is common but not a reason to stop the drug. • Suggest that patient contact primary care provider if side effects are intolerable.

TABLE 7-3	DRUGS FOR TREATING INFECTIOUS DISEASE — CONT'D	
Type and Examples	**Nursing Considerations**	**Patient Teaching**
Bacteriostatic Antibiotics		
Sulfonamides sulfamethoxazole (Gantanol), sulfasalazine (Azulfidine), sulfisoxazole (Gantrisin), trimethoprim-sulfamethoxazole (Bactrim, Septra)	• Effective against most gram-positive and many gram-negative organisms. • Commonly used with urinary tract infections, *Pneumocystis carinii* pneumonia, and otitis media. • Common side effects include rash, pruritus, nausea, vomiting, phlebitis, signs of bleeding. • IV drug should be given over 1 hour. • Monitor intake and output. • Fluid intake should be at least 1500 mL daily. • Bleeding time may increase, so use caution with anticoagulants. • Phenytoin (Dilantin) toxicity may be increased. • The risk of hypoglycemia may be increased.	• Instruct patient to take drug on an empty stomach 1 hour before or 2 hours after meals. • Tell patient to take with a full glass of water; encourage fluids. • Advise against prolonged exposure to the sun. • Tell patient to stop drug and call primary care provider if signs of allergic reaction or bleeding occur.
Antifungal Agents		
Amphotericin B	• Interferes with the cell wall structure of the fungus, causing it to die. • Given parenterally and reserved for life-threatening fungal infections. Monitor patient during first hour of infusion for febrile reaction. • Monitor injection site often because drug is very irritating to tissues. • Report side effects: nausea, vomiting, diarrhea. • Monitor intake and output, blood urea nitrogen (BUN), and creatinine levels for signs of kidney damage. • Obtain daily weight because fluid retention follows kidney damage. • Encourage 2000 to 3000 mL of fluid daily to help flush drug through kidneys.	• Explain the purpose of treatment and need for long-term IV therapy. • Instruct patient and family on side effects and possible discomfort at IV site.
Fluconazole (Diflucan)	• Used to treat candidal and urinary tract infections, as well as cryptococcal meningitis. • Can be given either parenterally or orally. • Cultures are obtained before giving drug. • Monitor BUN and creatinine. Use cautiously if patient has renal impairment. • Monitor liver function. • Monitor side effects: hepatotoxicity, nausea, vomiting, diarrhea, abdominal discomfort.	• Tell patient to take drug at same time each day and not to double a dose if one is missed. • Teach patient and family to notify the primary provider at the first sign of yellow skin, dark urine, or pale stools (signs of liver damage).

CRITICAL THINKING

Mr. Cheevers

Mr. Cheevers is admitted to the hospital for IV antibiotic therapy. He states that he has no allergies. One hour after the infusion begins you happen to meet the nursing assistant coming down the hall with a blanket. He casually says, "Mr. Cheevers is very cold. I'm bringing him a blanket." It is a warm mid-June day. What is your responsibility in this situation?

Answers at end of chapter.

▶ NURSING PROCESS

Assessment

It is important to recognize the earliest signs and symptoms of infection. Early detection can help provide early treatment to prevent major complications and reduce costs.

Signs of Respiratory Tract Infections

Patients with respiratory tract infections may have a cough, a congested or runny nose, a sore throat, chest congestion, or chest pain. The throat may be reddened, or there may be white patches in the back of the throat. Lung sounds can include crackles, rhonchi, or wheezing. Ask patients if they have a productive cough and the amount, frequency, and color of the sputum. A sputum culture is obtained to identify the presence of pathogenic organisms for appropriate treatment.

Signs of Gastrointestinal Tract Infections

The symptoms of GI tract infections may include nausea, vomiting, diarrhea, cramping, and anorexia. Patients may have frequent episodes of emesis and diarrhea and need to be monitored for signs of dehydration resulting from the loss of fluid. Stool cultures may be ordered.

Signs of Genitourinary Tract Infection

Symptoms of a urinary tract infection can include voiding urgency, frequency, burning, flank pain, change in color of urine, foul odor, or discharge. Monitor frequency, amount, color, and odor of the urine. Urinalysis and urine cultures are ordered to detect organisms. Collect urine cultures using sterile (via catheter) or clean (via clean catch) technique. Urinary tract infections need to be treated because those left untreated can lead to serious kidney problems.

Nursing Diagnosis

Nursing diagnoses for a patient with an infection may include the following:

- Risk for infection related to external factors
- Pain related to the infectious process
- Imbalanced nutrition: less than body requirements related to problems eating or digesting food
- Ineffective protection related to suppressed immune function
- Deficient knowledge related to disease process and treatment

Planning

The patient's goals are to remain free of infection symptoms, have pain relieved, consume adequate nourishment, remain free of infection, and verbalize knowledge of disorder and therapy.

Implementation

It is important to follow agency policy and procedures for infection control. Understanding how infections are contracted and spread help control infectious disease. Ensuring that the appropriate medication is given on time and understanding how to care for patients with infections can help the patient recover quickly. All members of the health care team must be aware of the patient's infection and their role in helping the patient recover. Providing emotional support to the patient is also important (see Patient Perspective Box 7–5).

Patients, families, and all members of the health care team must be aware of the events that cause infections and understand treatment plans for infections. Patients who are prone to infection because of immunosuppression should take special precautions to prevent infection (Box 7–6). Education should be directed toward prevention of infectious diseases. Teaching patients how to participate in their care and having them assist in the development of their plan of care promote compliance with treatment. Patients' understanding of how infections occur helps them in controlling their risk for developing an infection.

BOX 7-5 *Patient Perspective*

Emotions of Chronic Infection

Edie

It was back! I wasn't sure I could deal with it one more time. I've been hospitalized four times with this same infection (cellulitis) in my leg. It feels like I've lost my life. I can't count on being able to do anything and go anywhere because the infection just keeps coming back.

My left leg is now all swollen, reddened, discolored, and very painful. I can tell when I'm infected by more than the pain in the leg. I feel weak, kind of spacey, and once I passed out. I'm so sick of going into the emergency room—waiting forever to get admitted, all the IV starts and blood draws. With these infections, I've had a PICC [peripherally inserted central catheter] line twice. I've been sent home on IV antibiotics, sometimes for weeks at a time. I've learned how to hang my own IV antibiotics; in fact, I've learned much more than I ever wanted to.

My cellulitis is associated with chronic lymphedema, causing swelling in my legs. I'm working hard to keep the swelling down, so that the infection doesn't reoccur. Wish me luck and keep giving me psychosocial support during your nursing care. I'm not sure how I will be able to deal with this much longer.

BOX 7-6 **Teaching Points to Prevent Infection in the Elderly, Debilitated, or Immunocompromised Patient**

- Wash your hands frequently, using proper technique.
- Avoid crowds or anyone with an infection.
- Stay well nourished because food helps keep the immune system healthy.
- Have a flu shot yearly and a pneumonia shot as recommended by your primary care provider.
- Wash raw fruits and vegetables thoroughly, cook food thoroughly, and store food safely to prevent food poisoning. (NOTE: with severe immunocompromise, raw foods, soft cheeses, and yogurt may be contraindicated.)
- If your immune system is depressed, notify your primary care provider about any elevated temperature, even in the absence of other symptoms. People with depressed immune function cannot mount the usual immune response to infection, and a low-grade fever may be the only sign of infection.

Evaluation

Patient outcomes are met if the patient:

- has laboratory data within normal limits, indicating infection is controlled and no symptoms of infection are present.
- reports pain is relieved.
- maintains ideal body weight for height and weight and eats a balanced diet.
- remains free of infection.
- describes therapy and carries out treatment.

REVIEW QUESTIONS

1. Which of the following nursing actions should the nurse include in the plan of care to help maintain the body's first line of defense against infection?
   a. Help the patient cough and deep breathe.
   b. Give an antibiotic as ordered.
   c. Apply lotion to clean skin.
   d. Help the patient void.

2. Which of the following is the most important nursing action for the nurse to use to prevent a hospital-acquired urinary tract infection in a patient with an indwelling urinary catheter?
   a. Ensure an adequate intake of IV and oral fluids.
   b. Use clean technique for catheter insertion.
   c. Position drainage bag higher than bladder level.
   d. Maintain a closed urinary drainage system.

3. The nurse is to give a newly ordered antibiotic to a patient with a wound infection. Which of the following should the nurse do first?
   a. Check all patient allergies.
   b. Check the patient's temperature.
   c. Change dressing and note wound appearance.
   d. Give antibiotic at meal time.

4. Which of the following is the most important technique for the nurse to use during patient care to prevent infection transmission?
   a. Wear gloves
   b. Wear gown
   c. Wash hands
   d. Wear mask

5. Which of the following statements suggests that the patient understands the general principles of appropriate antibiotic use?
   a. "I'll take this until I feel better."
   b. "I have some pills from the last time that I had this infection that I will take."
   c. "I'll take all of this as it says on the label."
   d. "I take only half a pill to reduce the cost."

Answers to CRITICAL THINKING

Mrs. Sampson

1. Hand washing reduces the microorganisms on the nurse's hands to help reduce their transmission from patient to patient. This helps prevent exposure to pathogens and infection.
2. Risk for infection.
3. Reverse isolation, the goal of which is to protect the patient from exposure to organisms rather than to protect others from exposure to the patient.

Who Is at Greatest Risk of Infection?

1. Mr. Ashland has two risk factors: chronic disease and probably stress.
2. Mrs. Burrows has many risk factors, including old age, debilitated condition, probable malnourishment, probable stress, and an invasive procedure. This patient is at greatest risk.
3. Jackson has four risk factors: his young age, probable stress, possible malnourishment, and an invasive procedure.

Mr. Carson

1. Ask the nursing assistant about the Foley bag placed above the patient's bladder. Explain that the bag should always stay below the level of the bladder both for proper drainage and for infection control.
2. Assist the nursing assistant in repositioning the bag properly.
3. Let Mr. Carson's primary nurse know about the potential backflow of urine so that he can be assessed in the next few days for signs of a bladder infection.

Mr. Cheevers

Mr. Cheevers may be experiencing a sign of allergic reaction to the medication. The fact that he has no history of allergy is no guarantee that he is not experiencing one now. He needs to be evaluated immediately, and if allergy is suspected, the IV should be stopped and the primary provider notified. Epinephrine should be on hand if the patient develops an anaphylactic response. The primary nurse needs to be alerted to the situation in the event that it worsens. Later the nursing assistant can be instructed about the signs of allergic response.

8

NURSING CARE OF PATIENTS IN SHOCK

Cindy Bechtel and Ruth Remington

KEY TERMS

acidosis (ass-i-**DOH**-sis)

acute pulmonary hypertension (ah-**KEWT PULL**-muh-**NAIR**-ee **HIGH**-per-**TEN**-shun)

anaerobic (AN-air-**ROH**-bik)

anaphylaxis (AN-uh-fi-**LAK**-sis)

bronchospasm (**BRONG**-koh-spazm)

cardiac output (**KAR**-dee-ack **OWT**-put)

cardiogenic (KAR-dee-oh-**JEN**-ick)

cyanosis (SIGH-uh-**NOH**-sis)

distributive (dis-**TRIB**-yoo-tiv)

dysrhythmia (dis-**RITH**-mee-yah)

epinephrine (EP-i-**NEFF**-rin)

extracardiac (EX-trah-**KAR**-dee-ack)

hypotension (HIGH-poh-**TEN**-shun)

hypovolemic (HIGH-poh-voh-**LEEM**-ick)

ischemia (iss-**KEY**-me-ah)

lactic acid (LAK-tik **ASS**-id)

laryngeal edema (lah-**RIN**-jee-uhl uh-**DEE**-muh)

myocardium (MY-oh-**KAR**-dee-um)

myocarditis (MY-oh-kar-**DYE**-tis)

neurogenic (NEW-roh-**JEN**-ik)

norepinephrine (NOR-ep-i-**NEFF**-rin)

oliguria (AWH-li-**GYOO**-ree-ah)

perfusion (per-**FEW**-zhun)

pericardial tamponade (PER-ih-**KAR**-dee-uhl TAM-pon-**AID**)

sepsis (**SEP**-sis)

tachycardia (TAK-ih-**KAR**-dee-yah)

tachypnea (TAK-ip-**NEE**-ah)

tension pneumothorax (**TEN**-shun NEW-moh-**THOR**-raks)

thrombi (**THROM**-bye)

toxemia (tock-**SEE**-me-ah)

trauma (**TRAW**-mah)

Trendelenburg (tren-**DELL**-en-berg)

urticaria (**UR**-ti-**CARE**-ee-ah)

QUESTIONS TO GUIDE YOUR READING

1. What is the definition of shock?
2. How would you explain the pathophysiology of shock?
3. What are the etiologies, signs, and symptoms of the four categories of shock?
4. What is the current medical treatment for shock?
5. Which data should you collect when caring for patients with shock?
6. What nursing care should you provide for patients with shock?
7. How will you know if your nursing interventions have been effective?

A patient in shock is in a state of circulatory collapse that results in organ damage and death without immediate treatment. Massive bleeding, overwhelming infection, severe allergic reactions, and cardiac failure are examples of conditions that may lead to shock. No matter what its source, shock is a medical emergency that requires rapid, comprehensive nursing assessment and intervention in collaboration with the health care team.

The commonly used definition of shock is "inadequate tissue **perfusion**," in which there is insufficient delivery of oxygen and nutrients to the body's tissues and inadequate removal of waste products from the tissues. All body systems are affected by reduced oxygen supplies. The resulting injury to the body can be treated in the early stages of shock, but if shock is prolonged, it leads to irreversible cell damage and death. By the time the blood pressure drops, cellular and tissue damage have already occurred. Therefore it is important to identify patients at risk for shock and carefully assess them for early symptoms.

▶ PATHOPHYSIOLOGY OF SHOCK

Tissue perfusion and blood pressure are maintained in the body by three mechanisms: adequate blood volume, an effective cardiac pump, and effective blood vessels. The body is able to compensate for failure of one of these mechanisms by making a change in one or both of the other two. Shock occurs when compensatory mechanisms fail and inadequate tissue perfusion occurs. Common causes include inadequate **cardiac output** caused by heart failure, a sudden loss of blood volume resulting from hemorrhage, or a sudden decrease in peripheral vascular resistance caused by **anaphylaxis, sepsis,** and neurological alterations.

Metabolic and Hemodynamic Changes in Shock

When blood pressure falls, the body responds by activating the sympathetic nervous system (SNS). The SNS secretes **epinephrine** and **norepinephrine** to increase cardiac output by causing the heart to beat faster and stronger. Blood is shunted away from the skin, kidneys, and intestines to preserve blood flow to the brain, liver, and heart. Epinephrine, cortisol, and glucagon raise blood glucose levels to supply cells with fuel. Stimulation of the renin-angiotensin-aldosterone system (RAAS) from decreased cardiac output causes vasoconstriction and retention of sodium and water to decrease further fluid loss. Respiratory rate increases to deliver more oxygen to the tissues. Together these compensatory responses produce the classic signs and symptoms of the initial phase of shock: **tachycardia, tachypnea, oliguria,** restlessness, anxiety, and cool, clammy skin with pallor. If oxygen delivery remains inadequate, signs and symptoms of moderate and irreversible shock phases are seen (Table 8–1).

LEARNING TIPS

Tachycardia is a compensatory mechanism that is usually the first sign of shock. When a patient develops sustained tachycardia, it is a signal that the patient's condition is changing. Be aware that elderly patients cannot tolerate tachycardia very long because their ability to adapt to stress is reduced.

Consider the cause of the tachycardia. For example, a surgical patient who develops tachycardia may be hemorrhaging and should be assessed for bleeding. Be aware that with internal hemorrhaging there may not be any visible signs of bleeding. Changes in vital signs may be the only evidence.

Provide prompt intervention, such as applying direct pressure to an area of hemorrhage, and implement the physician's orders immediately.

Inadequate tissue blood flow causes an important change in cellular metabolism. When cells are deprived of oxygen, they shift to **anaerobic** metabolism to continue to receive nutrition and energy. As you may recall, anaerobic metabolism is an inefficient form of metabolism that can supply the energy needs of the cell for a few minutes only. After that, the body's metabolic rate and temperature begin to fall as a result of reduced energy production. Anaerobic metabolism results in the production of **lactic acid** as an unwanted byproduct. Unless the lactic acid can be circulated to the liver and thus removed from the bloodstream, the blood will become increasingly acidic. **Acidosis,** which is a fall in blood pH below 7.35, is one of the classic signs of shock.

CRITICAL THINKING

Anaerobic Metabolism
Why is anaerobic metabolism necessary and helpful if it produces the complication of metabolic acidosis?

Answer at end of chapter.

Effect on Organs and Organ Systems

Prolonged shock causes extensive damage to the organs and organ systems (Table 8–2). Inadequate blood flow results in tissue **ischemia** and injury. Because blood is shunted away from the kidneys early in shock to save fluid and provide oxygen to vital organs, the kidneys commonly are injured first. The kidneys can tolerate reduced blood flow for about 1 hour before sustaining permanent damage. Cells in the kidneys die when there is a lack of oxygen and nutrients. If there is widespread damage to the kidneys, complete renal failure is likely. Renal failure resulting from inadequate blood flow to the kidneys can be prevented and treated by replacing lost fluids.

Several organs of the gastrointestinal system may be injured early in shock. Inadequate circulation to the intestines may result in injury of the mucosa and may even cause paralytic ileus. **Toxemia** may result when the body absorbs

---

anaphylaxis: an—without + phylaxis—protection
tachycardia: tachy—fast + cardia—heart condition
tachypnea: tachy—fast + pnea—breathing
oliguria: olig—few + uria—urine condition

---

anaerobic: an—without + aerobic—presence of oxygen
acidosis: acid—sour + osis—condition

TABLE 8-1   SIGNS AND SYMPTOMS OF SHOCK PHASES

	Phases		
Signs and Symptoms	Mild/Compensating	Moderate/Progressive	Severe/Irreversible
Heart Rate	Tachycardia	Tachycardia	Slowing
Pulses	Bounding	Weaker, thready	Absent
Blood Pressure			
Systolic	Normal	Below 90 mm Hg In hypertensive 25% below baseline	Below 60 mm Hg
Diastolic	Normal	Decreased	Decreasing to 0
Respirations	Elevated	Tachypnea	Slowing
Depth	Deep	Shallow	Irregular, shallow
Temperature	Varies	Decreased May elevate in septic shock	Decreasing
Level of consciousness	Anxious, restless, irritable, alert, oriented	Confused, lethargy	Unconscious, comatose
Skin and Mucous Membranes	Cool, clammy, pale	Cold, moist, clammy, pale	Cyanosis, mottled cold, clammy
Urine Output	Normal	Decreasing to less than 20 mL/h	15 mL/h decreasing to anuria
Bowel Sounds	Normal	Decreasing	Absent

TABLE 8-2   EFFECT OF SHOCK ON ORGANS AND ORGAN SYSTEMS

Lungs	Acute respiratory failure Acute respiratory distress syndrome
Renal	Renal failure
Heart	Dysrhythmias, myocardial ischemia, and myocardial depression
Liver	Abnormal clotting; decreased production of plasma proteins; elevated serum levels of ammonia, bilirubin, and liver enzymes
Immune System	Depletion of defense components
Gastrointestinal System	Mucosal injury, paralytic ileus, pancreatitis, absorption of endotoxins and bacteria
Central Nervous System	Ischemic damage, necrosis, brain death

into the circulation normally occurring bacteria and endotoxins from inside the bowel. The liver may be injured both by ischemia and by toxins created by the shock state as blood is circulated through it for cleansing. Signs and symptoms of liver injury include decreased production of plasma proteins; abnormal clotting, because clotting factor production by the liver is impaired; and elevated serum levels of ammonia, bilirubin, and liver enzymes.

The immune system is also affected by shock. Many of the body's defenses become depleted from shock, leaving the body vulnerable to infection. Also, if the liver has been damaged, it is unable to assist the immune system in providing defense.

The body attempts to preserve blood supply to the heart and brain because these are vital organs that require a continuous supply of oxygen. Shock places extra demands on the heart itself, creating a situation in which the heart is in extra need of oxygen at a time when oxygen supplies are already low. When the **myocardium** receives inadequate oxygenation, cardiac output decreases and shock worsens. The pumping ability of the heart can be further depressed by acidosis, toxins released into the blood from ischemic tissues,

or ischemia-induced **dysrhythmias.** If the brain is deprived of circulation for more than 4 minutes, brain cells die from a lack of oxygen and glucose. Lengthy shock may result in brain death.

▶ COMPLICATIONS FROM SHOCK

Acute respiratory distress syndrome (ARDS), disseminated intravascular coagulation (DIC), and multisystem organ failure are three especially grave conditions that may follow a prolonged episode of shock. Patients with ARDS usually develop respiratory failure despite high levels of supplemental oxygen and mechanical ventilation. DIC results from ischemic damage to the endothelial lining of blood vessels. The formation of multiple tiny **thrombi,** microscopic debris, and depletion of tissue clotting factors cause abnormal bleeding and additional tissue damage. DIC itself may cause shock and death. Multisystem organ failure is a major cause of death following shock. It usually begins with respiratory failure, followed by failure of the kidneys, heart, liver, and finally cerebral and gastrointestinal function.

LEARNING TIPS

To understand what *disseminated intravascular coagulation* means, define each of the words and then put the definitions together.

   Disseminated = scattered or widespread
   Intravascular: intra = inside + vascular = vessels
   Coagulation = clotting

   These definitions put together tell you that DIC is scattered, widespread clotting inside the vessels.

   At first, hemorrhage does not seem reasonable in light of a clotting problem, but if you think about what is occurring, it does make sense. When many clots form throughout the body in response to stressors, few clotting factors remain available to form the clots needed to prevent hemorrhage. As a result, hemorrhage is a risk in DIC.

dysrhythmia: dys—difficult or abnormal + rhythmia—rhythm

► CLASSIFICATION OF SHOCK

The different forms of shock are classified by their cardiovascular characteristics. The four shock categories are: (1) **hypovolemic** shock caused by a decrease in the circulating blood volume, (2) **cardiogenic** shock caused by cardiac failure, (3) **extracardiac** obstructive shock caused by a blockage of blood flow in the cardiovascular circuit outside the heart, and (4) **distributive** shock caused by excessive dilation of the venules and arterioles (Table 8–3). Most cases of clinical shock show only some components of each of these categories. However, this classification system is helpful in understanding shock. The hallmark characteristic, which all forms of shock exhibit, is a decrease in blood pressure. The blood pressure usually falls below the level required to provide an adequate supply of blood to the tissues.

Hypovolemic Shock

Any severe loss of body fluid may lead to hypovolemic shock. Hypovolemic shock can be caused by dehydration; internal or external hemorrhage; fluid loss from burns, vomiting, or diarrhea; or loss of intravascular fluid into the interstitium as a result of sepsis or **trauma.** Clinical signs and symptoms include pale, cool, clammy skin; tachycardia; tachypnea; flat, nondistended peripheral veins; decreased jugular vein circumference; decreased urine output; and altered mental status. With a 10 percent loss of blood volume,

tachycardia is the only obvious sign. At 20 to 25 percent blood loss, tachycardia and mild to moderate **hypotension** are present. With a loss of 40 percent or greater, all clinical signs and symptoms of shock are present.

Cardiogenic Shock

Cardiogenic shock results when the heart fails as a pump. It occurs in 5 to 10 percent of patients with acute myocardial infarction (AMI). In most cases, approximately 40 percent of the myocardium must be lost to produce cardiogenic shock. Patients with cardiogenic shock have signs and symptoms similar to hypovolemic shock, except that they may display distended jugular and peripheral veins, as well as other symptoms of heart failure, such as pulmonary edema. Other causes of cardiogenic shock include rupture of heart valves, acute **myocarditis,** end-stage heart disease, severe dysrhythmias, or traumatic injury to the heart.

Obstructive Shock

Extracardiac obstructive shock occurs when there is a blockage of blood flow in the cardiovascular circuit outside the heart. Several conditions may cause obstructive shock. **Pericardial tamponade,** which is the filling of the pericardial sac with blood, compresses the heart and limits its filling capacity. **Tension pneumothorax** compresses the heart from an abnormal collection of air in the pleural space and

TABLE 8-3	CATEGORIES OF SHOCK	
Category	**Causes**	**Signs and Symptoms**
Hypovolemic Shock	Any severe loss of body fluid; dehydration, internal or external hemorrhage, fluid loss from burns or from vomiting or diarrhea, or loss of intravascular fluid into the interstitium.	Tachycardia, tachypnea, hypotension, cyanosis, oliguria, flat nondistended peripheral veins, decreased jugular veins, and altered mental status.
Cardiogenic Shock	Myocardial infarction, myocarditis, end-stage cardiomyopathy, severe dysrhythmias, valvular disease, severe electrolyte imbalance, and drug overdoses.	Dysrhythmias, labored respirations, hypotension, cyanosis, oliguria, altered mental status, possibly distended jugular and peripheral veins, and symptoms of congestive heart failure.
Obstructive Shock	Any block to the cardiovascular flow. Pericardial tamponade, tension pneumothorax, intrathoracic tumors, massive pulmonary emboli, and large systemic emboli.	Tachycardia, tachypnea, hypotension, cyanosis, oliguria, and altered mental status; jugular veins may be distended.
Distributive Shock	Any condition causing massive vasodilation of the peripheral circulation. Subcategories include anaphylactic, septic, and neurogenic shock.	See subcategories below.
Anaphylactic shock	Insect stings, antibiotics, anesthetics, contrast dye and blood products are typical allergens.	Tachycardia, tachypnea, hypotension, cyanosis, oliguria, and altered mental status. May also have urticaria, laryngeal edema, and severe bronchospasm. If conscious, may be extremely apprehensive and complain of a metallic taste.
Septic shock	Massive release of chemical mediators and endotoxins causes loss of vascular autoregulatory control and loss of fluid into the interstitium. Bacteria, especially gram-negative strains, protozoans, and viruses.	Early or warm phase: blood pressure, urine output, and neck veins may be normal. Skin warm and flushed with full veins. Fever usually present, although temperature may be subnormal. Late phase: tachycardia, tachypnea, hypotension, oliguria, flat jugular and peripheral veins, and cool clammy skin. Normal or subnormal temperature.
Neurogenic shock	Dysfunction or injury to the nervous system. Spinal cord injury, general anesthesia, fever, metabolic disturbances, and brain injuries.	Early phase: hypotension and altered mental status, bradycardia, and skin that is warm and dry. Late phase: tachycardia, tachypnea, and cool clammy skin.

interferes with normal cardiac functioning. **Acute pulmonary hypertension,** a sudden abnormally elevated pressure in the pulmonary artery, increases resistance for blood flowing out the right side of the heart. All these conditions decrease cardiac output, which can lead to shock. Tumors or large emboli may also cause shock. Signs and symptoms of obstructive shock are similar to those of hypovolemic shock, except that jugular veins are usually distended.

Distributive Shock

Distributive shock occurs when peripheral vascular resistance is lost because of massive vasodilation of the peripheral circulation. Distributive shock includes anaphylactic, septic, and **neurogenic** shock.

Anaphylactic Shock

Anaphylactic shock occurs when the body has an extreme hypersensitivity reaction. It occurs most commonly from insect stings, antibiotics (especially penicillins), anesthetics, contrast dye, and blood products. The signs and symptoms are similar to those seen in hypovolemic shock. Additionally, patients may have symptoms specific to allergic reactions, including **urticaria,** wheezing, **laryngeal edema,** and severe **bronchospasm.** If conscious, patients may be extremely apprehensive and short of breath, and they may complain of a metallic taste.

Septic Shock

Septic shock is caused by systemic infection and inflammation. Extensive release of chemical mediators and endotoxins causes dilation of blood vessels and loss of fluid into the interstitial space. Most cases of sepsis are caused by gram-negative bacteria, although other bacteria and viruses may be the cause. Septic shock is the leading cause of death among critical care patients. Predisposing conditions include trauma, diabetes mellitus, corticosteroid therapy, immunocompromise (e.g., as seen in patients with human immunodeficiency virus [HIV] and in those undergoing chemotherapy treatment for cancer), and an indwelling Foley catheter.

During the early, or warm, phase of septic shock (which may be referred to as "pink" shock), blood pressure, urine output, and neck vein size may be normal, but the skin is warm and flushed. Fever is present in the majority of patients, although some may have a subnormal temperature. Untreated, septic shock progresses to a second phase with signs and symptoms similar to hypovolemic shock: hypotension, oliguria, tachycardia, tachypnea, flat jugular and peripheral veins, and cold, clammy skin. Body temperature may be normal or subnormal.

Neurogenic Shock

Neurogenic shock occurs when dysfunction or injury to the nervous system causes extensive dilation of peripheral blood vessels. It is a rarer form of shock, occurring most commonly as a result of injury to the spinal cord (referred to as spinal shock). Other causes include general anesthesia, fever, metabolic disturbances, and brain contusions and concussions. Signs and symptoms include hypotension and altered mental status and, during the early phases, bradycardia and warm, dry skin. As shock progresses, however, tachycardia and cool, clammy skin develop.

► MEDICAL-SURGICAL MANAGEMENT OF SHOCK

Because of the emergency nature of shock, life-threatening symptoms must be treated immediately (Table 8–4). The exact nature of the shock must be assessed while interven-

TABLE 8-4	MEDICAL-SURGICAL MANAGEMENT OF SHOCK
Respiratory support	Oxygen (nasal cannula, face mask, non-rebreather mask, assisted ventilations with bag-valve-mask, ventilator)
	SPo₂ > 90%
	Venous lactic acid < 2.2 mmol/L
Cardiovascular support	Vasopressor medication (dopamine) if fluid resuscitation not effective
	Revascularization of heart in cardiogenic shock via angioplasty with or without stent or fibrinolytic therapy (alteplase)
	Antidysrhythmics
	Positive inotropes
Adequate circulatory volume	1–3 L IV fluids
	Crystalloid fluids: normal saline (0.9% sodium chloride) or lactated Ringer's solution
	3 mL crystalloid solution for every 1 mL fluid lost
	Blood or blood products
	Urine output > 30 mL/h
	Hemoglobin > 10 g/dL
Control of bleeding	Pressure dressings
	Surgical intervention
Treatment of life-threatening injuries	Surgical interventions
	Medications
Determination and treatment of causes of shock	Septic shock: antibiotics
	Cardiogenic shock: morphine, diuretics, nitrates
	Anaphylactic shock: epinephrine, diphenhydramine (Benadryl), methylprednisolone (Solu-Medrol), aminophylline

TABLE 8-5	ASSESSMENT OF THE PATIENT IN SHOCK
Signs and Symptoms	Tachycardia, tachypnea, hypotension, oliguria, cyanosis, and altered mental status
Laboratory Tests	Complete blood count, serum osmolarity, blood chemistries, prothrombin time, partial thromboplastin time, blood typing and crossmatch, serum lactate, arterial blood gases, cardiac isoenzymes, urinalysis
Imaging	Chest x-ray, spinal films, computed tomography, echocardiogram
Monitoring	Electrocardiogram, arterial pressure monitor, central venous pressure monitor, pulmonary artery catheter, gastric pH

tions such as ventilatory and circulatory support are being implemented (Table 8–5). The order of interventions and testing is guided by the stability of the patient. Intervention priorities are as follows:

1. Airway and respiratory support
2. Cardiovascular support
3. Maintenance of circulatory volume
4. Control of bleeding if present
5. Assessment of neurological status
6. Treatment of life-threatening injuries
7. Determination and treatment of the cause of shock

▶ NURSING PROCESS

Assessment

Assessment of the patient in shock must be carried out quickly and should always start with the ABCDs: airway, breathing, circulation, and disability.

Airway is assessed for patency and opened as necessary. A compromised airway must be treated immediately with the head-tilt/chin-lift method, an oral or nasal airway, or endotracheal intubation.

Breathing is assessed for rate, depth, and symmetry of chest movement. The patient is observed for use of accessory muscles. Lung sounds are auscultated. Crackles may be found in the patient with cardiogenic shock or in the patient who has received too much intravenous (IV) fluid.

Circulation is assessed with blood pressure. A narrowing pulse pressure may be present before a drop in systolic pressure and indicates a decrease in cardiac stroke volume and peripheral vasoconstriction. Peripheral pulses are palpated. Tachycardia is the first sign of shock. The pulse is assessed for quality; commonly it is weak and thready in a patient with shock. As shock progresses, the peripheral pulses become be bradycardic or absent. The radial pulse is be present if the patient's systolic blood pressure is below 80 mm Hg. A capillary refill greater than 2 seconds indicates inadequate circulation. Capillary refill has been found to be an unreliable indicator of shock in adults, especially elderly people. Other observations regarding circulation include

distended neck veins, which may be collapsed or full; skin that may be cool, pale, and diaphoretic; presence of **cyanosis;** mucous membranes that may be pale and dry; and thirst. Rapidly scan the entire body for evidence of bleeding or other injuries. Palpate the abdomen for signs of internal bleeding, such as a tender, distended, boardlike abdomen.

Disability is assessed by determining the patient's level of consciousness (LOC).

LEARNING TIPS

To assess level of consciousness, determine if the patient is alert by asking his or her name, the date, and the location. If the patient can answer all three questions, he or she is "alert and oriented ✕ 3 (person, place, time)." You can also use these indicators:

A: alert
V: responds to verbal commands
P: responds to painful stimuli
U: unresponsive

All four limbs are assessed for circulation, sensation, and mobility. Bilateral responses are compared for equality. **Circulation** is assessed by palpating pulses for presence and quality. **Sensation** is determined by touching the patient's hands and feet and asking what the patient feels and if there is any numbness or tingling. **Mobility** (motor ability) is assessed by having the patient move all four limbs and wiggle the fingers and toes. Have the patient push with his or her feet against your hands and squeeze two of your fingers to assess strength.

A "head-to-toe" approach can follow the primary ABCD assessment. The presence, severity, and location of pain or nausea and vomiting are assessed. Bowel sounds are auscultated to determine whether they are normal, absent, hyperactive, or hypoactive. When an indwelling urinary catheter has been placed, the color of the urine and the rate of urine output are noted. Body temperature should be noted.

CRITICAL THINKING

Classic Signs of Shock

What is the cause and compensatory purpose of each of the classic signs of shock: tachycardia, tachypnea, oliguria, pallor, and cool, clammy skin?

Answer at end of chapter.

CRITICAL THINKING

Beta Blockers

You are assessing a patient admitted with sepsis. The patient has been taking a beta blocker. You are assessing the patient for septic shock. What sign of shock will not be present in a patient taking a beta blocker?

Answer at end of chapter.

Nursing Diagnosis

The major nursing diagnoses for shock include but are not limited to the following:

■ Ineffective tissue perfusion (renal, cerebral, cardiopulmonary, gastrointestinal, peripheral) related to hypovolemia or inadequate cardiac output or mechanical reduction of venous and/or arterial blood flow
■ Fear related to severity of condition and unknown outcome
■ Deficient knowledge related to unfamiliar condition

See Nursing Care Plan 8–1 for the Patient in Shock.

Planning

The patient's goals are to (1) demonstrate adequate tissue perfusion, (2) verbalize reduced fear, and (3) understand the condition and its treatment.

Nursing Interventions

The cardiovascular status of patients in shock must be carefully monitored based on their stability. IV fluids, medications, and oxygen are given as ordered. The **Trendelenburg** position is avoided, but a modified Trendelenburg position (raising legs only) may be used to increase venous return to the heart except in patients with cardiogenic shock. If there is no neck or spinal trauma, the patient is placed supine with the trunk horizontal and the legs elevated 20 to 45 degrees, knees straight. The modified Trendelenburg position prevents gravity from stimulating cardiac reflexes, compressing coronary arteries, and causing the abdominal organs to press against the diaphragm. In cardiogenic shock, any position that could might increase blood flow to the compromised heart could be fatal and therefore is avoided.

Reducing patient fear is important in controlling shock and preventing complications. The patient should be allowed to express fears and ask questions. Family members should be allowed to provide support to the patient.

The patient and family are included in education sessions. Simple explanations of what has occurred and what is happening can be followed by more complete explanations as the patient's condition improves. Interventions should include explanation of all diagnostic and treatment procedures, rationales, and outcome goals.

BOX 8-1   NURSING CARE PLAN FOR THE PATIENT IN SHOCK

 Ineffective tissue perfusion (renal, cerebral, cardiopulmonary, gastrointestinal, peripheral) related to hypovolemia or inadequate cardiac output or inadequate vascular tone

PATIENT OUTCOME
Patient demonstrates adequate tissue perfusion

EVALUATION OF OUTCOME
Is patient's skin warm/dry, peripheral pulses present/strong? Are vital signs within patient's normal range? Are lung sounds normal, intake/output balanced, edema absent, pain/discomfort absent? Is patient alert/oriented?

INTERVENTION	RATIONALE	EVALUATION
Monitor vital signs.	Changes in vital signs, which indicate change in condition, can be detected early and treated promptly.	Is heart rate between 60 and 100 beats per minute? Is heart rhythm regular? Are peripheral pulses strong? Is systolic blood pressure (BP) > 100 mm Hg? Is patient alert and oriented × 3?
Maintain airway and provide oxygenation.	Ensures adequate oxygenation and tissue perfusion.	Is SPo₂ > 90%? Is skin pink? Are respirations between 12 and 20?
Monitor intake and output. Provide adequate fluid intake.	Provides adequate cardiac output to perfuse tissues. Assesses renal function. Urine output is an indicator of renal function.	Is urinary output > 30 mL/h? Are mucous membranes moist? Is skin turgor < 3 seconds?
Position patient appropriately (head elevated for patients with shortness of breath, increased intracranial pressure).	Proper positioning promotes circulation and helps prevent skin breakdown.	Is edema noted? Is skin breakdown noted?
Provide quiet, restful environment. Maintain body temperature with warmed IV fluids, room temperature, blankets.	Conserves energy and lowers tissue oxygen demands. Recovery is aided by normal body temperature.	Is patient resting comfortably without anxiety? Is body temperature within normal limits?
Assess for pain and provide pain relief measures.	Pain increases tissue demands for blood and oxygen.	Is patient pain free?

GERIATRIC

Change positions slowly.	Age-related losses of cardiovascular reflexes can result in hypotension.	Is systolic BP > 100 mm Hg?

⑆ CRITICAL THINKING

Modified Trendelenburg Position
Why would the use of modified Trendelenburg positioning be life threatening in a patient in cardiogenic shock?

Answer at end of chapter.

Evaluation

The ultimate goal of nursing care for the patient in shock is to restore normal tissue perfusion quickly enough to ensure recovery without complications. Short-term goals may include the following:

1. Reduce blood and fluid loss
2. Maintain adequate cardiac output
3. Restore adequate circulating volume
4. Reduce fear
5. Maintain optimal body temperature
6. Maintain normal baseline LOC

   Once these goals are met, discharge planning for home care or rehabilitation can begin based on the patient's individualized needs.

⑆ CRITICAL THINKING

Mr. Hall
Mr. Hall, who is 55 years old, has suffered an acute myocardial infarction. He is complaining of chest pain (rated 10 out of 10 on the pain rating scale) and difficulty breathing. His SPO_2 is 89 percent. Crackles are heard on auscultation of breath sounds. The electrocardiogram shows an irregular and rapid heartbeat. He is restless and apprehensive.

1. Name three nursing priorities for Mr. Hall's care.
2. What type of IV fluid and rate is appropriate for Mr. Hall?
3. What signs and symptoms indicate Mr. Hall is in cardiogenic shock?

Answers at end of chapter.

⑆ CRITICAL THINKING

Mrs. Neal
Mrs. Neal, who is 45 years old, came to the emergency room in severe hypovolemic shock after sustaining several bleeding wounds in an automobile accident. Her shock is resolving after receiving several transfusions and surgical repair of her injuries. She has just been admitted to the surgical unit for postoperative care.

1. What postoperative nursing assessments should be performed first?
2. Mrs. Neal's family is very alarmed by her condition. What interventions can you provide to decrease their anxiety?
3. What postoperative complications may develop in Mrs. Neal?
4. What documentation is appropriate for Mrs. Neal?

Answers at end of chapter.

REVIEW QUESTIONS

1. Which of the following mechanisms does the body use to compensate for shock?
   a. Peripheral nervous system depression
   b. Central nervous system depression
   c. Sympathetic nervous system stimulation
   d. Parasympathetic nervous system stimulation

2. Which of the following findings is seen specifically with anaphylactic shock?
   a. Tachycardia
   b. Hypotension
   c. Laryngeal edema
   d. Oliguria

3. Which of the following conditions causes the decreased level of consciousness commonly found in patients experiencing shock?
   a. Severe pain
   b. Endotoxins
   c. Cerebral hypoxia
   d. Cerebral edema

4. Which of the following nursing diagnoses is most appropriate to include in the plan of care for a patient experiencing shock?
   a. Fatigue
   b. Ineffective tissue perfusion
   c. Ineffective health maintenance
   d. Hopelessness

5. You enter a patient's room and find the patient lying in a pool of blood from a leg incision that has opened. The patient is restless and confused. You call for help and take vital signs. Which of the following treatments for shock should be started first?
   a. IV fluids
   b. Oxygen
   c. Vasopressor medications
   d. Antibiotics

Answers to Critical Thinking

Anaerobic Metabolism

Anaerobic metabolism is the source of nutrition and energy for the cell that prevents cellular death when oxygen is not available. It is a short-term compensatory mechanism to save the cell until oxygen becomes available again.

Classic Signs of Shock

Tachycardia is caused by decreased cardiac output and reduced tissue oxygenation. Its purpose is to increase cardiac output and oxygen delivery by causing more heartbeats to pump out blood from the heart.

Tachypnea is caused by decreased tissue oxygenation. Its purpose is to increase respirations so more oxygen is available for delivery to tissues.

Oliguria is caused by a reduced blood flow to the kidneys. Its purpose as a compensatory mechanism is to conserve as much fluid as possible to help maintain a normal blood pressure.

Pallor is caused by reduced blood volume or flow. When pallor results from compensation, it is due to peripheral vasoconstriction that occurs to shunt blood volume to the vital organs.

Cool, clammy skin is the result of decreased blood flow to the skin and the release of moisture (sweat) from the skin. The sympathetic nervous system causes these compensatory mechanisms; peripheral vasoconstriction shunts blood to the vital organs, and sweating cools the body in anticipation of the fight or flight response, which generates body heat when it occurred.

Beta Blockers

Tachycardia will not be present. Beta blockers block the response of the sympathetic nervous system, which is activated in shock.

Modified Trendelenburg Position

The modified Trendelenburg position increases venous blood return to the heart, which is helpful in improving cardiac output in shock. In cardiogenic shock, any position that increases blood flow to the compromised heart increases the heart's workload. This could overwhelm or flood the heart so that it is unable to keep up with the blood volume returning to it. The result could be death.

Mr. Hall

1. Nursing priorities for Mr. Hall include relief of chest pain and anxiety, stabilization of cardiac rhythm and vital signs, and adequate tissue oxygenation.
2. Because Mr. Hall has crackles, which indicate fluid in the lungs, he should not be given IV fluids. He should have an IV access to give IV medications as needed. He already has too much fluid for his heart to handle, and giving him IV fluids could be life threatening.
3. Signs of cardiogenic shock include decreased blood pressure; increased heart and respiratory rates; cyanosis; decreased urine output; cool, pale skin; and decreased mental status.

Mrs. Neal

1. Assessment of respiratory status and cardiovascular status, inspection of surgical wounds for bleeding, assessment of mental status, and need for pain relief should be performed first.
2. Explain the cause of shock and all interventions, rationales, and desired outcomes. Keep the environment calm, provide for privacy, and answer all questions in a matter-of-fact and reassuring manner. Allow Mrs. Neal's family to visit.
3. Unrelieved pain, bleeding, infection, and respiratory complications are possible.
4. Airway: rate, depth, regularity of respirations; breath sounds, SPO_2. Vital signs: cardiac rhythm, quality of pulses, skin color, blood pressure, body temperature. Urine output: oral and IV intake, fluid balance. Pain: measures to relieve pain and evaluation of those measures. Dressings: any bleeding. Bowel sounds.

9

NURSING CARE OF PATIENTS IN PAIN

Karen P. Kettelman-Hall

KEY TERMS

addiction (uh-**DIK**-shun)

adjuvant (ad-**JOO**-vant)

agonist (**AG**-un-isst)

analgesic (AN-uhl-**JEE**-zik)

antagonist (an-**TAG**-on-ist)

breakthrough (**BRAYK**-THROO)

ceiling effect (**SEE**-ling e-**FEKT**)

endorphins (en-**DOR**-fins)

enkephalins (en-**KEF**-e-lins)

equianalgesic (EE-kwee-AN-uhl-**JEE**-zik)

neuropathic (NEW-roh-**PATH**-ik)

nociceptive (NOH-see-**SEP**-tiv)

opioid (**OHP**-ee-OYD)

pain (PAYN)

patient-controlled analgesia (**PAY**-shunt kon-**TROHLD** AN-uhl-**JEE**-zee-ah)

physical dependence (**FIZ**-ik-uhl dee-**PEN**-dens)

prostaglandins (**PRAHS**-tah-GLAND-ins)

pseudoaddiction (sue-doh-ah-**DICK**-shun)

psychological dependence (SY-ko-**LAW**-jick-al dee-**PEN**-dens)

serotonin (SER-ah-**TOH**-nin)

suffering (**SUFF**-er-ing)

titration (tigh-**TRAY**-shun)

tolerance (**TALL**-er-ens)

transdermal (trans-**DER**-mal)

QUESTIONS TO GUIDE YOUR READING

1. How is pain defined?
2. What are the common myths and barriers to the effective management of pain?
3. What are the differences between addiction, physical dependence, and tolerance?
4. What is current knowledge of the basic physiology of the pain response?
5. What are characteristics that help to define acute, chronic nonmalignant, and cancer pain?
6. What are the primary components of a pain assessment?
7. How is the World Health Organization analgesic ladder used for the treatment of pain?
8. What are the three categories of analgesics and their uses?
9. What are commonly used pain treatment modalities and when are they used?
10. How are nondrug pain management techniques utilized?
11. How does ethical decision making play a role in the care of the patient in pain?

▶ THE PAIN PUZZLE

Pain is the most common reason patients seek medical advice. However, despite the widespread nature of the problem, pain is often untreated or undertreated. The care of patients with pain is challenging and requires a systematic approach to assessment and treatment.

Decisions about pain management require careful assessment of the patient's condition and attention to the ethical principles that influence patient care. Providing information and giving the patient choices help maintain autonomy. When patients are not involved in the process of choosing their own pain management options, they cannot be autonomous. Just as risks, benefits, and alternatives to surgery and anesthesia are discussed with the patient, so should pain management options be discussed in the process of obtaining informed consent. Nurses often worry about overmedicating patients, thinking that they are "doing good" (beneficence) or "doing no harm" (nonmaleficence) by withholding medication from a patient they do not believe is in pain (Ethical Consideration Box 9–1). It is important to learn as much as you can about pain and pain management, so you can effectively advocate for your patients, assist with patient education, and provide appropriate resources.

The Joint Commission for Accreditation of Healthcare Organizations (JCAHO) published pain management standards in 2000 and began assessing the management of pain in hospitals based on these standards in January 2001. These standards have added to the importance of pain management. They address assessment and appropriate pharmacological management of pain, as well as patient and family teaching, postoperative pain, management of side effects, discharge planning, and continuous quality improvement. These guidelines can be found at the JCAHO Web site.

For more information on pain management, visit the following Web sites. For some sites, you may need to type "pain" in the search window.

http://www.ahrq.gov
http://www.ampainsoc.org
http://www.cancer.org
http://www.iasp-pain.org
http://www.jcaho.org/standards_frm.html
http://www.who.org

Cultural differences must be considered when planning care for the patient in pain. People from various cultures have different ways of expressing pain (Cultural Consideration Box 9–2). Some may be dramatic and emotional; others tend to be stoic and quiet. Knowledge of widely accepted information about different ethnic and cultural groups can be useful in understanding a patient's experience and what care might be considered acceptable. It is important, however, to assess a patient's pain care needs individually and not make assumptions based on culture or ethnicity alone.

Because of the importance of controlling health care costs, the entire health team must provide care in the most cost-effective manner possible while continuing to provide the best quality of care. Effective pain management can help reach those goals by enhancing comfort, minimizing

---

BOX 9-1 ETHICAL CONSIDERATION

Case Study-Controlling Pain

A client was admitted to the medical unit 2 days after his 83rd birthday with a diagnosis of metastatic cancer of the pancreas. He had several full-course treatments of chemotherapy and radiation during the previous year with only temporary remissions of the disease. His condition deteriorated rapidly during the previous month, and his family was no longer able to care for him at home. The patient's physician expected him to die within a few weeks, if not sooner, and admitted him to the hospital primarily for pain control. On admission, the physician ordered morphine sulfate (MS) 5 mg IV to be given every 2 hours around the clock.

Although this seemed like a relatively large dose of a strong narcotic medication to be given this frequently, the patient tolerated the treatment for the first 36 hours, and reported a significant reduction in his pain. On the third morning after his admission, he was difficult to arouse for his morning vital signs and breakfast. When his nurse, Kathy P., finally did manage to awaken the patient, he was confused, and his blood pressure was 82/50 with shallow respirations at 10/min. He still complained of severe generalized pain and asked for "more of that medication that helped so much the day before."

Kathy did not give the patient his scheduled 0800 dose of MS. She phoned the physician and reported her assessment of the patient, expressing her belief that continuing the medication at the previously prescribed dose and frequency would be fatal to this patient. The physician, who had been up most of the night with an emergency patient, stated sarcastically, ". . . and your patient has such a productive life to look forward to—give the #$@& medication like I ordered it! And don't call me any more about this—I'll be in after lunch!" and slammed down the phone.

What should Kathy do? What are her ethical obligations in this particular case? How should the nurse weigh the risks and benefits of providing adequate pain medication given the possible side effects of hastening the patient's death? If the nurse has a moral objection to carrying out a physician's order, what options are available to that nurse? Use the ethical decision-making model to resolve this ethical dilemma. What information can assist in solving the problem if the material about virtues is applied to this situation? Would it make a difference if the nurse thought the patient should receive the pain medication and the patient refused on the basis of stoicism or from a religious perspective in which suffering is seen as a way to grow spiritually?

---

side effects of **opioids** and complications related to inadequate pain control, and reducing the lengths of hospital stays.

It is not possible to record a patient's pain on a machine or measure it with a blood or urine sample. It is difficult to objectively gauge the effect of a nursing intervention on pain, whether the intervention is medication or nondrug therapy. Pain and its treatment can be likened to a difficult puzzle that requires many pieces to solve.

In this chapter the many challenges of pain assessment and treatment are discussed. Some of the tools needed to effectively deal with these challenges are presented. Common myths and barriers that continue to affect nursing practice are clarified.

BOX 9-2   CULTURAL CONSIDERATION

The pain experience may differ between and among ethnocultural individuals. Remember that individuals within groups vary, and not all fit the general descriptions and previous experience with pain. (See Chapter 3.)

Culture	Expression and Meaning of Pain	Patient Preferences	Assessment	Interventions
Native-American	Frequently do not request pain medicine and are undertreated. May not realize that they can ask for pain medicine. Many believe pain is something that must be endured. May describe pain in general terms such as "not feeling good." The word for pain varies according to the tribal language.	Many prefer traditional herbal medicines. May complain to family member or visitor, who relays message to caregiver.	Frequently ask patient and family members or visitors if patient has pain. Observe for nonverbal clues of pain.	Explain that the control of pain can promote healing. Offer pain medicine as needed. Allow adequate time for response; silence is valued. Maintain a calm, relaxing environment. Incorporate traditional practices for pain relief if not harmful.
European-American	Strong sense of stoicism, especially men. Fear of being dependent may decrease use of pain medicine. Many have fear of addiction. May continue to work and carry out daily activities and minimize pain.	Prefer relaxation and distractions as means of pain control.	Observe for nonverbal signs of pain. Use visual analog or numerical pain scales to assess severity of pain.	Encourage use of pain medicine as needed. Incorporate distraction and relaxation techniques.
African-American	Usually openly and publicly display pain, but this is highly variable. Many, especially the elderly, fear that medication may be addictive. Pain is seen as a sign of illness or disease. Many believe that suffering and pain are inevitable and should be endured.	May focus on spirituality and religious beliefs to endure pain. Prayers and the laying on of hands are thought to relieve pain if the client has enough faith.	Observe for verbal and nonverbal expressions of pain. Use of pain scales is helpful.	Offer pain medication as needed. Allow meditation and prayer along with pain medication. Support patient's spiritual practices.
Hispanic-American	Puerto Ricans tend to be expressive of pain and discomfort. Moaning, groaning, and crying are culturally accepted ways of dealing with and reducing pain. Mexicans may bear pain stoically because it is "God's will." Many feel that pain and suffering are a consequence of immoral behavior. For men, expressing pain shows weakness and a possible loss of a respect. The Spanish word for pain is *dolar*.	Prefer oral or intravenous medication for pain. Heat, herbal teas, and prayer are used to manage pain.	Visual analog and numerical scales may be helpful. Observe and compare verbal and nonverbal behaviors indicating pain.	Do not censor verbal expression of pain. Incorporate traditional practices as permitted. For individuals who are stoical with pain, encourage pain medicine frequently. Explain that pain control can hasten healing.
Asian-American	Chinese and Koreans tend to be stoical, although there is a wide variation, and describe pain in terms of diverse body symptoms instead of locally. Filipinos may view pain as a part of living an honorable life. Some view this as an opportunity to reach a fuller life and to atone for past transgressions. Frequently stoic and tolerate pain to a high degree. Some moan as an expression of pain.	Prefer oral or intravenous pain medications. Like warm compresses. For Koreans, intramuscular injections may be seen as an invasion of privacy. Vietnamese maintain self-control as a means of pain relief.	Observe for nonverbal signs of pain. Vietnamese may not understand numerical scale of rating pain. Observing facial expression may be a good indicator of pain.	Incorporate traditional healing methods as much as possible. Offer and encourage pain medicines to promote healing.

CULTURAL CONSIDERATION—CONT'D

Culture	Expression and Meaning of Pain	Patient Preferences	Assessment	Interventions
Asian-American—cont'd	For the Japanese, bearing pain is a virtue and a matter of family honor. Some, especially older individuals, may fear addiction. For Vietnamese, enduring pain is an indication of strong character. Family may be very attentive and request pain medication for the patient.	Vietnamese maintain self-control as a means of pain relief.		
Arab-American	See pain as something to be controlled. May express pain openly to family with elaborate verbal expressions, less so with caregivers. Expect prompt interventions for pain control. Tend to describe pain as diffuse rather than locally. May use terms such as fire, hot, and cold.	Intramuscular or intravenous usually preferred over oral medications.	Compare verbal and nonverbal characteristics of pain to determine degree of pain.	Engage family to help with distraction and relaxation techniques. Administer medication promptly.
Appalachian	Pain is something to be endured, and many respond stoically, especially men. May continue activities until pain forces cessation of work.	Some may place a knife or axe under the bed or mattress to "cut pain."	Observe for verbal and nonverbal signs of pain. Many do not like visual analog or numerical pain scales.	Offer pain medication frequently. Encourage relaxation techniques and distraction. Talking about the pain may have pain-relieving results.

▶ DEFINITIONS OF PAIN

According to Margo McCaffery, a well-known consultant in the care of patients with pain, "Pain is whatever the experiencing person says it is, existing whenever the experiencing person says it does."[1] This is a reminder to nurses to accept the patient's report of pain.

In 1979 pain was defined by the International Association for the Study of Pain (IASP) as an unpleasant sensory and emotional experience associated with actual or potential tissue damage or described in terms of such damage.[2] This definition indicates that pain is complex and has not only a physical component but an emotional one as well.

What is pain, and why does it exist? Pain is a protective mechanism or a warning. Pain in the presence of injury may help to prevent further injury. Consider the patient who experiences a fracture and holds it still to prevent further damage or a child who touches a hot stove and pulls his or her hand away before a serious burn occurs.

Why is untreated or undertreated pain a bad thing? Basic complications can occur when pain is experienced. The body produces a stress response in the presence of pain during which harmful substances are released from injured tissue. Reactions include breakdown of tissue, increased metabolic rate, impaired immune function, and negative emotions. In addition, pain prevents the patient from participating in self-care activities such as walking, deep breathing, and coughing. Consider the patient who has had chest surgery and then has to cough and deep breathe. It

hurts! So the patient tries not to cough, turn, or even move. Retained pulmonary secretions and pneumonia can develop. If the patient does not move around, return of bowel function is delayed and an ileus can result. When pain is well controlled, patients are able to do what they need to do to get well and go home from the hospital or continue with recovery activities.

Suffering often accompanies pain, but not all pain has the element of suffering. Suffering can exist without pain and pain without suffering. Suffering represents a threat to one's self-image or life. According to the latest cancer pain guidelines from the Agency for Health Care Policy and Research (now called the Agency for Health Care Research and Quality), suffering is defined as the state of severe distress associated with events that threaten the intactness of the person.[3] Pain can be a constant reminder for patients with cancer that they have a life-threatening illness. Suffering can often be relieved if patients believe that their pain can be relieved.

▶ MYTHS AND BARRIERS TO EFFECTIVE PAIN MANAGEMENT

Treatment of patients in pain is influenced by a number of factors, including the way the nurse was treated when in pain as a child. Why aren't patients believed when they report pain? Why do nurses and other health care team members often insist that patients behave a certain way before they are believed? Common myths about pain may impair the nurse's ability to be objective about pain and create barriers to effective treatment. Because there is no objective measure for pain, nurses may rely on what is comfortable rather than what has been proven to be effective.

Myth: A person who is laughing and talking is not in pain.

Fact: A person in pain is likely to use laughing and talking as a form of distraction. This can be very effective in the management of pain, especially when used in conjunction with appropriate drug therapies.

Myth: If morphine is given too early to the patient with cancer pain, it will not work when the patient really needs it, toward the end, when the pain is worse.

Fact: Morphine is an opioid **agonist.** Opioid doses can be escalated (titrated upward) indefinitely as needed as the patient's pain increases. There is no **ceiling effect,** or maximum effective dose. Side effects such as sedation or clinically significant respiratory depression may temporarily limit the dose or the rate at which the dose can be increased.

Myth: Respiratory depression is common in patients receiving opioid pain medications.

Fact: Respiratory depression is uncommon in patients receiving opioid pain medications. If patients are monitored carefully when they are at risk, such as with the first dose of an opioid or when a dose is increased, respiratory depression is preventable. A patient's respiratory status and level of sedation (LOS) should be routinely monitored using an LOS scale.

Myth: Pain medication is more effective when given by injection.

Fact: Intramuscular (IM) injections are not recommended because they are painful, have unreliable absorption from the muscle, and have a lag time to peak effect and rapid falloff compared with oral administration. Oral administration is the first choice if possible; the intravenous (IV) route has the most rapid onset of action and is the preferred route for postoperative administration.

Myth: Teenagers are more likely to become addicted than older patients.

Fact: **Addiction** to opioids is very uncommon in all age groups when taken for pain by patients without a prior drug abuse history.

CRITICAL THINKING

Smithers and Barnett

Mrs. Smithers had an abdominal hysterectomy and is sitting up in bed the morning after surgery, putting on her makeup. On morning rounds she is smiling but reports that her pain is at 6 on a scale of 0 to 10. Mr. Barnett has just been transferred from the surgical intensive care unit the day after surgery for multiple injuries. He is moaning and reports his pain at 6 on a scale of 0 to 10. Which of these patients is really having as much pain as they say they are? How can you make that judgment?

Answers at end of chapter.

▶ MORE DEFINITIONS

Nurses often express concern about patients who require large amounts of pain medication or know exactly when their next dose of pain medication is due. Nurses may say that such patients are addicted or that they are "clock watchers," but do we really know what that means? Patients are expected to be informed about their medications and involved in their care, but when they know when their medications are due, we may become suspicious. In truth, if a patient is watching the clock, the most likely reason is because he or she is in pain. The most common reason that patients ask for more pain medicine is because they have increased pain. It is important to understand the differences between addiction, **physical dependence,** and **tolerance.** When talking with patients and teaching them about their medications, it is important to help them understand these differences as well. It is something many patients worry about.

Tolerance simply means that it takes a larger dose to provide the same level of pain relief. Physical dependence is a physiological phenomenon that most people experience after a few weeks of continuous opioid use. If an opioid is abruptly discontinued after a few weeks of use, the patient may experience a withdrawal syndrome, exhibiting symptoms such as sweating, tearing, runny nose, restlessness, irritability, tremors, dilated pupils, sleeplessness, nausea, vomiting, and diarrhea. These symptoms can be prevented by slowly weaning a patient from an opioid, rather than stopping it suddenly.

Psychological dependence is another term for addiction and is defined as a pattern of compulsive drug use characterized by a continued craving for an opioid and the need to use the opioid for effects other than pain relief. **Pseudoaddiction** has been described in patients who are prescribed opioid doses that are too low or spaced too far apart to relieve their pain, and certain behavioral characteristics resembling psychological dependence, such as drug-seeking behaviors, have developed. Educating patients and families about these concerns helps alleviate their fears and increases their satisfaction with the pain relief measures that are employed.

▶ MECHANISMS OF PAIN TRANSMISSION

Many theories of how pain is transmitted are in the literature. The specificity theory, developed by Descartes in 1644, proposed that body trauma sends a message directly to the brain, causing a sort of "bell" to ring, prompting a response from the brain. In 1965 Melzack and Wall proposed the gate control theory, which describes the dorsal horn of the spinal cord as a gate, allowing impulses to go through when there is a pain stimulus and closing the gate when those impulses are inhibited. The gate control theory stimulated massive research on the physiology of pain.

---

pseudoaddiction: pseudo—false + addiction—psychological dependence

Much more is known about the transmission of pain today. **Endorphins** are endogenous (naturally occurring) chemicals that act like opioids to inhibit pain impulses in the spinal cord and brain. Unfortunately they degrade too quickly to be considered effective **analgesics.** Endorphins are the chemicals that stimulate the long-distance runner's "high." **Enkephalins** are one type of endorphin.

Pain is transmitted through the dorsal horn of the spinal cord and other points in the central nervous system to higher centers of the brain with the influence of chemicals known as neurotransmitters, which are released during pain or trauma. These chemicals include **serotonin, prostaglandins,** and others. Many treatments and analgesics are designed, based on known principles, to inhibit the release of these chemicals at different points along the pain pathway.

Mechanisms of pain transmission are **nociceptive** and **neuropathic.** *Nociception* refers to the body's reaction to noxious stimuli, such as tissue damage, with the release of pain-producing substances. Nociceptive pain in the visceral organs may be referred to other parts of the body. See Figure 9–1 for sites of referred pain. Neuropathic pain is pain associated with injury to either the peripheral or central nervous system. Unlike nociceptive pain, neuropathic pain is not usually localized, and it may spread to involve other areas along the nerve pathway.

▶ TYPES OF PAIN

Pain is often categorized according to whether it is acute, cancer related, or chronic nonmalignant. Acute pain is described as pain that follows injury to the body and subsides when healing takes place. It may be associated with objective physical signs such as increased heart rate and elevated blood pressure. Examples of acute pain include pain related to fractures, burns, or other trauma. Cancer pain may be acute, chronic, or intermittent and often has a definable cause such as tumor invasion or neuropathy caused by the cancer treatment. Chronic nonmalignant pain persists beyond the time when healing usually takes place. Examples include low back pain, arthritic pain, and phantom limb pain. Chronic nonmalignant pain may have nociceptive, as well as neuropathic, components and may require a variety of medications and nondrug treatments. Because of the body's ability to adapt, patients with chronic nonmalignant pain or chronic cancer pain may not appear to be in pain. The physiological responses that accompany acute pain, such as elevated heart rate and blood pressure, cannot be sustained without harm to the body, so the body adapts and the vital signs return to normal.

▶ OPTIONS FOR TREATMENT OF PAIN

Analgesics

Medications that relieve pain are called analgesics. Analgesics are the biggest pieces of the pain management puzzle. There are three main categories of analgesics: opioids, nonopioids, and **adjuvants.** Opioids are classified by their ability to bind to opioid receptors in the brain and spinal cord, as well as other areas of the body, inhibiting the perception of pain. Nonopioids include nonsteroidal anti-inflammatory drugs (NSAIDs) and acetaminophen (Tylenol). Adjuvants include classes of drugs that were originally developed for a different purpose but have been found to have pain-relieving properties in certain painful conditions.

---

nociceptive: noci—pain + ceptive—reception
neuropathic: neuro—nerves + pathy—disease, suffering

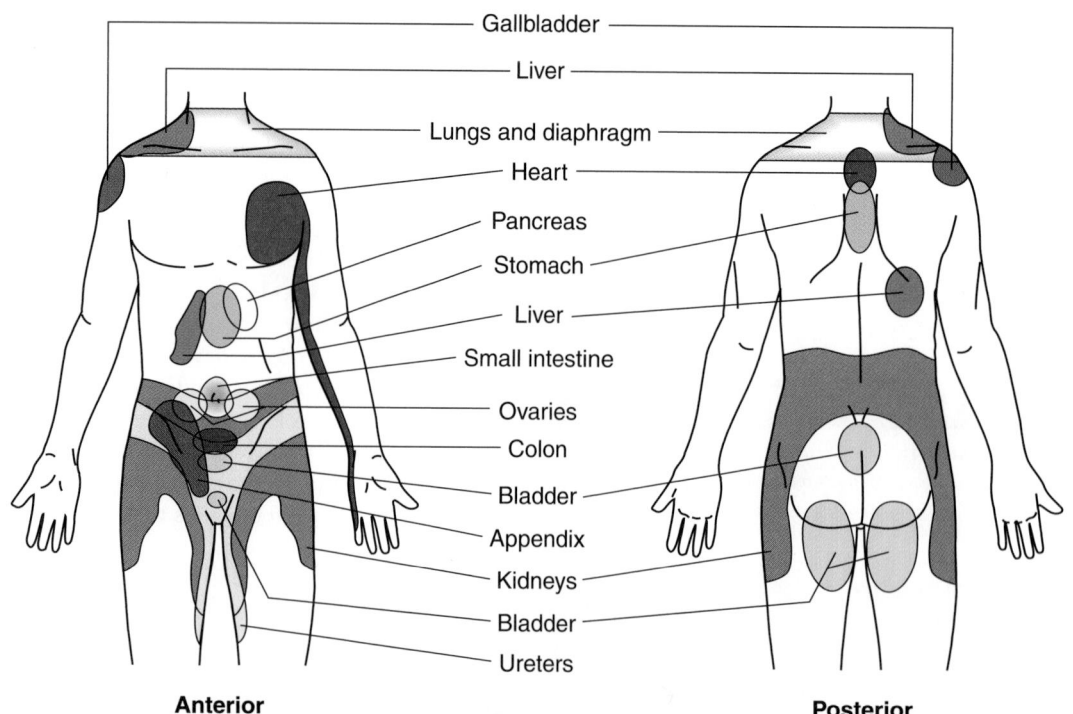

Gallbladder
Liver
Lungs and diaphragm
Heart
Pancreas
Stomach
Liver
Small intestine
Ovaries
Colon
Bladder
Appendix
Kidneys
Bladder
Ureters

Anterior   **Posterior**

Figure 9-1   Sites of referred pain.

Nonopioids

Nonopioids are generally the first class of drugs used for treatment of pain. They can be useful for acute and chronic pain from a variety of causes, such as surgery, trauma, arthritis, and cancer. These medications are limited in their use because they have a ceiling effect to analgesia. A ceiling effect indicates that there is a dose beyond which there is no improvement in the **analgesic** effect and there may be an increase in side effects. When used in combination with opioids, care must be taken to ensure that the dose of the nonopioid drug does not exceed the maximum safe dose for a 24-hour period. For example, if a patient receiving two aspirin with codeine tablets every 4 hours continues to experience pain, the dose cannot be increased because of the potential toxic effects of the aspirin (Table 9–1). Nonopioids do not produce tolerance or physical or psychological dependence. They do have antipyretic (fever-reducing) effects. This class of drugs works primarily at the site of injury, or peripherally, rather than in the central nervous system like opioids do. NSAIDs block the synthesis of prostaglandin, one of many chemicals necessary for pain transmission. In general it is helpful to include a nonopioid in any analgesic regimen, even if the pain is severe enough to require the addition of an opioid.

Opioids

Opioids are added to nonopioids for pain that cannot be managed effectively by nonopioids alone. Opioids are classified as full agonists, partial agonists, or mixed agonists and **antagonists.** Full agonists have a complete response at the opioid receptor site; the partial agonist has a lesser response. The mixed agonist and antagonist activates one type of opioid receptor while blocking another.

Opioids alone have no ceiling effect to analgesia. Doses can safely be increased to treat increasing pain if the patient's respiratory status and level of sedation are stable. See Box 9–3 for side effects of opioids. Controlled-release opioids such as oxycodone (Oxycontin) and morphine (MS Contin) are effective for prolonged periods.

Controlled- or time-release medication should never be crushed, but always taken whole. Whenever a controlled-release preparation is used, it is important to have an immediate-release medication available for **breakthrough** pain, such as oral morphine solution (OMS) or other immediate-release opioid.

Morphine is the drug of choice for the treatment of moderate to severe pain. It is the drug used as a standard against which all other analgesics are compared. (See Table 9–2 for **equianalgesic** doses of medications.) Morphine is long acting (4 to 5 hours) and available in many forms, making it convenient, as well as affordable, for the patient. It has a slower onset than the other opioid agonists.

Hydromorphone (Dilaudid) is commonly used for moderate to severe pain as well. It is shorter acting than morphine and has a somewhat faster onset. It is a good option for pain management in most patients.

Meperidine (Demerol), also an opioid agonist, should be reserved for healthy patients requiring opioids for a short period or for those who have unusual reactions or allergic responses to other opioids. When broken down in the body,

---

equianalgesic: equi—equal + analgesic—relieving pain

BOX 9-3	Common Side Effects of Opioids

Sedation
Respiratory depression*
Constipation
Nausea, vomiting
Itching
Constricted pupils*

*These effects are not common, but they alert the nurse to possible overdose.

TABLE 9–1	SIDE EFFECTS OF NONOPIOIDS
Drug	**Side Effects**
NSAIDs (including aspirin)	Gastrointestinal (GI) irritation and bleeding
	Inhibition of platelet aggregation, increasing risk of GI bleeding
	Renal insufficiency in some patients, especially the elderly
	Patients with asthma at risk for hypersensitive reactions
Acetaminophen	Necrosis of the liver with overdose

TABLE 9-2	EQUIANALGESIC CHART		
Approximate doses of medications in milligrams to equal same amount of pain relief.			
Drug	**Parenteral (IM/IV/SQ)**	**Oral Dose**	**Conversion Factor (Parenteral to PO)**
Morphine	10 mg	30 mg	3
Codeine	130 mg	200 mg NR	1.5
Hydromorphone (Dilaudid)	1.5 mg	7.5 mg	5
Methadone (Dolophine)	10 mg	20 mg	2
Meperidine (Demerol)	75 mg	300 mg NR	4

Source: Adapted from McCaffery, M: Pain: Assessment and Use of Analgesics (conference handouts). Fall 1996.
NR—not recommended at that dose.

it produces a toxic metabolite called normeperidine. Normeperidine is a cerebral irritant that can cause side effects ranging from dysphoria and irritable mood to seizures. It has a long half-life even in healthy patients, so patients with impaired renal function are at increased risk. Meperidine use should be avoided in patients over the age of 65, in those with impaired renal function, and in those receiving monoamine oxidase inhibitor (MAOI) antiepressants. Additionally, the effective dose of oral meperidine is three to four times the parenteral dose, a toxic dose that is not recommended.

Fentanyl (Sublimaze, Duragesic) can be administered parenterally, intraspinally, or by **transdermal** patch (Duragesic patch). Fentanyl is commonly used intravenously with anesthesia for surgery and also for relief of postoperative pain via the epidural route. Administration can also be effective via IV **patient-controlled analgesia** (discussed later in this chapter) for postoperative pain management, but other drugs may be preferred. IV fentanyl is very short acting and must be administered frequently. The fentanyl patch is useful for the patient with stable cancer pain and requires dosing only every 3 days.

Methadone (Dolophine) is a potent analgesic that has a longer duration of action than morphine. It has a very long half-life and accumulates in the body with continued dosing. Dosing intervals should be lengthened after pain relief has been achieved. Methadone is well absorbed from the gastrointestinal tract and is very effective when given orally at doses similar to the parenteral dose. Methadone is also used in drug treatment programs during detoxification from heroin and other opioids. Patients on methadone maintenance can present a unique challenge when admitted to the hospital. It is important to continue the maintenance dose even if other pain medications are given after surgery or trauma.

LEARNING TIPS

It is important to monitor a patient's level of sedation and respiratory status whenever administering opioids. Increased sedation, decreased respiratory effort, and constricted pupils can be signs of opioid overdose. Careful monitoring and **titration** of opioids can prevent opioid-induced respiratory depression.

Opioid Antagonists

Naloxone (Narcan) is a pure opioid antagonist that counteracts, or antagonizes, the effect of opioids. It is often used in the emergency department setting for treatment of opioid overdose. Caution must be used when naloxone is given to the patient receiving opioids for the treatment of pain. If too much naloxone is given too fast, it can reverse not only the respiratory depression and sedation but the analgesia as well.

transdermal: trans—across + dermal—the skin

CRITICAL THINKING

Mrs. Shepard

Mrs. Shepard is 92 years old and has undergone an open cholecystectomy. Her continuous epidural infusion of analgesic has been discontinued. The physician has ordered oral hydrocodone with acetaminophen (Vicodin) every 3 to 4 hours as required for pain. This is her second postoperative day, and she refuses to get out of bed because her pain is 7 on a scale of 0 to 10.

The medication record shows that Mrs. Shepard has not received any pain medication since the continuous epidural infusion was stopped 3 hours ago.

1. Why is Mrs. Shepard in so much pain?
2. What complications can occur as a result of her pain?
3. Each Vicodin tablet contains 500 mg of acetaminophen. The maximum daily dose of acetaminophen is 4 g. If she takes one Vicodin every 3 hours, is her dose safe?
4. What can be done to relieve her pain and better prevent it in the future?

Answers at end of chapter.

Antagonists are generally shorter acting than the opioid that is being used. If the antagonist is given because of respiratory depression, the dose may need to be repeated because its effect may wear off before that of the opioid.

Some analgesics are classified as combined agonist and antagonist. These drugs bind with some opioid receptors and block others. The most commonly used agonist-antagonist drugs are butorphanol (Stadol) and nalbuphine (Nubain).

How does this information translate into nursing practice? Consider, for example, a patient who is taking sustained-release morphine every 12 hours to control metastatic bone pain and is experiencing breakthrough pain between doses. You observe that butorphanol has been ordered for pain by another doctor and administer it. The butorphanol will antagonize, or counteract, the effects of the morphine, and the patient may develop acute pain. It is important to be informed about the actions of all drugs that are administered and to be aware of possible drug interactions that may interfere with patient care.

Agonist-antagonist drugs are also used sometimes to counteract opioid side effects. Nalbuphine (Nubain) can be used to treat itching and nausea that may accompany the administration of opioids. A smaller dose can be given so that the analgesia is not reversed completely along with the reversal of the side effect.

Analgesic Adjuvants

Adjuvants are classes of medications that may potentiate the effects of opioids or nonopioids, have analgesic activity themselves, or counteract unwanted effects of other analgesics. Adjuvants are especially important when treating pain that does not respond well to traditional analgesics alone.

Steroids can be used to treat a variety of pain conditions, including acute and chronic cancer-related pain. They may be used as part of actual cancer treatment because of their toxicity to some cancer cells, or they may reduce pain by decreasing inflammation and the resultant compression of healthy tissues. Their use is standard emergency practice in the treatment of suspected spinal cord compression. Patients with pain caused by malignant lesions pressing on nerves such as the brachial or lumbosacral plexus may receive large doses of steroids.

Benzodiazepines such as midazolam (Versed) or diazepam (Valium) are effective for the treatment of anxiety or muscle spasms associated with pain. These drugs do not provide pain relief except in the treatment of muscle spasms. Benzodiazepines may cause sedation, which limits the amount of opioid that may be given safely.

Tricyclic antidepressants such as amitriptyline, imipramine, desipramine, and doxepin have been shown to relieve pain related to neuropathy and other painful nerve-related conditions. These medications must be taken for a time before they are fully effective, and patients must be instructed about this so that they do not stop taking the medication after a few days. Additional benefits of this class of medications may include mood elevation and improved ability to sleep.

Anticonvulsants such as phenytoin (Dilantin) and carbamazepine (Tegretol) are often used to relieve the sharp or cutting pain caused by peripheral nerve syndromes. Again, these medications must be taken regularly before full benefit is realized.

Amphetamines such as methylphenidate hydrochloride (Ritalin) may be used to counteract the sedating effects of opioids in some patients.

OTHER INTERVENTIONS. Other pain treatments include the use of radiation therapy or antineoplastic chemotherapy to help shrink tumors that are causing pain for a patient with cancer. Chemotherapy is also used for treating pain associated with connective tissue disorders such as rheumatoid arthritis or systemic lupus erythematosus. Bowel regimens, laxatives, enemas, or antigas medication to decrease abdominal fullness may be considered pain management if it is the primary source of discomfort for the patient. In patients with osteoporosis, drugs that result in calcium uptake by the bones can aid in relief of pain. These may include hormonal agents and medications that decrease calcium resorption from bone, such as etidronate (Didronel).

The IV administration of strontium-89 chloride (Metastron) by a qualified physician helps ease the bone pain of some patients with metastatic bone cancer. It is most effective for patients who have hormone-related cancers, such as cancer of the breast or prostate.

Lidocaine/prilocaine (EMLA, a eutectic mixture of local anesthetics) cream is a topical local anesthetic that decreases pain associated with procedures such as venipunctures and lumbar punctures. It is most effective if left in place for 1 hour before the procedure and covered with a semipermeable membrane dressing such as Tegaderm or Opsite before the needle stick.

PLACEBOS. Use of placebos involves the administration of an inactive substitute such as normal saline in place of a real medication. In the past, placebos were sometimes given in an attempt to determine whether a patient's pain was "real." Placebos are also given in drug studies (clinical trials) to compare a new drug with an inactive substance. The use of placebos is not justified in the treatment of pain. Placebos should only be used if the patient has agreed to be a part of a research study and has consented to the possibility that he or she may receive a placebo. The use of placebos is a denial of the patient's report of pain. If a placebo is ordered for a patient, discuss your concerns with the physician and nurse supervisor.

Routes for Medication Administration

ORAL. Medication administration by the oral (PO) route is preferred whenever possible. It is convenient, flexible, and inexpensive and provides consistent blood levels when given around the clock (ATC).

RECTAL. When the patient is not able to take medications by the oral route, rectal administration is an alternative. Many medications may be given by the rectal route. Often oral preparations (tablets or capsules) can be administered rectally, as well as intrastromally, intravaginally, or buccally.

INHALATION. Butorphanol (Stadol) can be administered via the nasal route and is sometimes used for treatment of migraine headaches. It is an agonist-antagonist, and caution must be used if the patient is also taking opioids, as discussed earlier. The use of aerosolized morphine is now being used successfully in some patients with diseased lungs.

TRANSDERMAL. Fentanyl patches are used in the treatment of chronic pain and are dosed in micrograms (mcg) per hour. It may seem easy enough to apply a patch, but there are some special considerations when using this route.

- There is about a 12-hour delay until an effective level is reached, and the patient may require an immediate-release form of pain medication until that time.
- The patches may not be as effective in patients who smoke.
- Absorption may be reduced if the patch is placed in an area with little or no fatty tissue.
- Absorption may be increased in patients with body temperatures of 101°F or greater. Increased or erratic absorption may require patches to be replaced more frequently.
- When applying the patch, use caution to prevent touching the membrane covering the medication to prevent exposure. If contact occurs, rinse with plain water.
- Patches should not be used with other sustained-release analgesics.
- Multiple patches may be used.
- When the patch is discontinued, the residual medication in the skin continues to be absorbed for about 17 hours. During this time the patient must be monitored for pain control and overdose if another pain medication has been administered.
- Caution should be exercised when disposing of the patch because active drug remains even after 72 hours.

INTRAMUSCULAR. The IM route is appropriate only if the medication cannot be delivered by another route. IM injections are painful and inconvenient and absorption is unpredictable. IM injections are never recommended for treatment of chronic pain.

INTRAVENOUS. The IV route is the preferred route of opioid administration for postoperative pain and when the patient is unable to tolerate oral medications. IV infusions of opioids are preferred over intermittent doses of IM or subcutaneous medication for chronic cancer pain. Continuous infusion of an opioid provides a steady blood level, which tends to give the most effective analgesia with the fewest side effects. Routes other than IV require lag time for absorption of the drug into the circulation. Self-administration of an opioid by the patient is called patient-controlled analgesia (PCA). PCA is most commonly administered intravenously, although it can be administered orally or via epidural. This chapter focuses on IV PCA.

IV PCA is a safe method for postoperative pain management that many patients prefer over intermittent injections. A special pump is programmed to administer the prescribed dose of opioid by pushing a patient-controlled button. A lockout interval is also programmed into the pump, which regulates the frequency with which the patient can receive a dose of medication to protect against overdose. A typical dose of morphine is 1 mg, with a lockout interval of 5 to 10 minutes. All PCA pumps can also deliver a continuous, or basal, infusion in addition to the patient-controlled dose. One advantage of the basal infusion is that the patient can have intervals of uninterrupted sleep. Basal infusions should be used cautiously in patients who are newly receiving opioids. An hourly limit is also prescribed by the physician and programmed into the pump, which keeps track of the number of doses a patient uses (doses actually received by the patient) and the number of attempts made (times the button was pushed). This helps you know how much medication the patient is using and if the use of the pump is understood. If the patient has a record of many attempts and very few injections, the patient may not be waiting the full interval between doses. This occurs because the patient is in severe pain and cannot wait or because he or she does not understand the instructions. Patient education should include information about the medication that is being administered, safety features and the use of the pump, potential side effects, and what to report to the nurse. Refer to institutional policy for safe PCA administration guidelines. It is important that patients, families, and caregivers understand that no one should push the button except the patient. This modality is safe if it is patient controlled and appropriately monitored. Family members can help by reminding their loved one to use the PCA if they think the person is in pain.

SUBCUTANEOUS. If IV access is problematic, the subcutaneous route can be used as an alternative for opioid therapy in patients with chronic cancer pain. A needle is placed into the subcutaneous tissue and a small volume of a high concentration of opioid can be infused for effective pain management.

INTRASPINAL. The intraspinal (epidural or subarachnoid) route is an appropriate pain relief modality for some patients. Patients with traumatic injuries, including rib fractures or orthopedic injuries of the pelvis and lower extremities, and patients undergoing chest, abdominal, or lower extremity orthopedic surgical procedures may benefit from the epidural route of opioid administration. Advantages to the intraspinal route of administration include the ability to provide superior pain relief while administering overall lower doses of opioid. There is less systemic effect of the opioid because it is administered close to the site of the nerves serving the area of injury or surgical incision and not into the systemic circulation. Intraspinal opioids can be safely administered in a variety of hospital and home care settings if protocols for the care and monitoring of the patient are in place and enforced.

Nondrug Therapies

Nondrug treatments are usually classified as cognitive-behavioral interventions or physical agents. The goals of these two groups of treatments are different. Cognitive-behavioral interventions can help patients understand and cope with pain and take an active part in its assessment and control. The goals of physical agents may include providing comfort, correcting physical dysfunction, altering physiological response, and reducing fear that might be associated with immobility.

Cognitive-Behavioral Interventions

Included in this group are interventions such as educational information, relaxation exercises, guided imagery, distraction (e.g., music, television), and biofeedback. These treatments require extra time for detailed instruction and demonstration.

Providing patients with educational information about what to expect and how patients can participate in their own care has been shown to decrease patients' reports of postoperative pain and analgesic use. Relaxation can be accomplished through a variety of methods. The patient may prefer a relaxation exercise with a script that can be practiced and used the same way each time or simply the use of a favorite piece of music that will allow a state of muscle relaxation and freedom from anxiety. Guided imagery uses the patient's imagination to take the patient away from the pain to a favorite place, such as a beach in Tahiti. The success of guided imagery does not mean that the pain is in any way imaginary. The use of distraction is something that many of us do without thinking about it. We focus our attention on something other than the pain. Patients watch a favorite television program or laugh with visitors when they are in pain. When the program is over or the visitors leave, the patient may notice the pain again and ask for a dose of pain medication.

Biofeedback is commonly used in behavioral chronic pain programs to show patients how to teach their bodies to respond to different signals. Biofeedback has been very useful in patients with migraine headaches. When patients experience the aura that many migraine sufferers get before the headache, they begin the exercise that relaxes them and prevents the migraine.

Physical Agents

Physical agents may contribute directly to the patient's comfort. Applications of heat or cold, massage, exercise and immobilization, and transcutaneous electrical nerve stimulation (TENS) are commonly used physical agents.

The application of heat to sore muscles and joints is effective for pain relief. Heat works to increase circulation, induce muscle relaxation, and decrease inflammation when applied to a painful area. Heat can be applied using dry or moist packs or wraps or in a bath or whirlpool. Heat is contraindicated in conditions that would be worsened by its use, such as in an area of trauma, because of the possibility of increased swelling caused by vasodilation. To prevent burns, heat should not be applied over areas of decreased sensation.

Cold can reduce swelling, bleeding, and pain when used to treat to a new injury. Cold can be applied by a variety of methods, such as cold wraps and cold packs, as well as localized ice massage. Patients often choose heat over cold if they have the choice. Cold is better tolerated over a small area. Alternating heat and cold therapies is most effective if not contraindicated.

Massage and exercise are used to stretch and regain muscle and tendon length and probably work by relaxing the muscles. Massage pressure can be superficial or deep. It is important that massage is acceptable and not offensive to the patient. Immobilization is used following a variety of orthopedic procedures, as well as fractures and other injuries worsened by movement.

Physical agents are readily available, are inexpensive, and require little preparation or instruction. It is important to use nondrug treatments to enhance appropriate drug treatments, not as a substitute.

▶ NURSING PROCESS

Assessment

Accurate assessment of pain is essential to effective treatment. Without appropriate assessment, it is not possible to intervene in a way that meets the patient's needs. The American Pain Society recommends that nurses consider pain the "fifth vital sign."[4] That way it will be routinely assessed whenever other vital signs are assessed. The WHAT'S UP? format found in Chapter 1 can assist you in performing a complete and effective assessment (Table 9–3). Following are some additional key points for assessing pain and putting together more pieces of the pain puzzle.

Believe the Patient

Pain is what the patient says it is, not what the nurse or physician thinks it should be. When a member of the health

TABLE 9-3	WHAT'S UP? FORMAT FOR ASSESSMENT OF PAIN

W—Where is the pain? Be specific. Use drawing of body if necessary.
H—How does the pain feel? Is it shooting, burning, dull, sharp?
A—Aggravating and alleviating factors. What makes the pain better? Worse?
T—Timing. When did the pain start? Is it intermittent? Continuous?
S—Severity. How bad is the pain on a 0 to 10 (0 to 5; faces scale)
U—Useful other data. Are you experiencing any other symptoms associated with the pain or pain treatment? Itching, nausea, sedation, constipation?
P—Perception. What is the patient's perception of what caused the pain?

care team distrusts the patient's report of pain, the patient can usually sense that he or she is not believed. The patient may compensate by either underreporting pain or, less commonly, anxiously overreporting.

Take a Pain History

Information is obtained from the patient about the pain that he or she is experiencing. Allowing and encouraging the patient to describe the pain in his or her own words helps establish a trust relationship between you and the patient and also helps discover the effects the pain is having on the patient's lifestyle. Is the pain keeping the patient from eating, sleeping, or participating in work or family activities? The patient's emotional and spiritual distress and coping abilities should be assessed to individualize the interventions to the patient's needs. The patient can tell you how he or she has coped with pain previously and what treatment measures have been effective in the past.

A variety of tools are available to assist in accurate and complete pain assessment. You should become familiar with the tool used in your setting and use it consistently. It is of utmost importance that all health care personnel caring for the patient use the same pain rating scale, whether it is a numerical scale (e.g., 0 to 5 or 0 to 10), a visual analog scale (a 10-cm line on which the level is marked and then measured from the beginning of the line to indicate the amount of pain), or the Wong-Baker faces scale (Figs. 9–2 and 9–3). Whichever scale is used, it must be one that has been validated with research. Some scales are cute and interesting

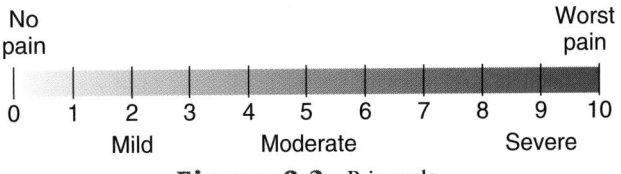

Figure 9-2 Pain scale.

1) Explain to the child that each face is for a person who feels happy because he or she has no pain (hurt, or whatever word the child uses) or feels sad because he or she has some or a lot of pain.
2) Point to the appropriate face and state, "This face is . . .":
   0—"very happy because he doesn't hurt at all."
   1—"hurts just a little bit."
   2—"hurts a little more."
   3—"hurts even more."
   4—"hurts a whole lot."
   5—"hurts as much as you can imagine, although you don't have to be crying to feel this bad."
3) Ask the child to choose the face that best describes how he or she feels. Be specific about which pain (e.g., "shot" or incision) and what time (e.g., now? earlier before lunch?).

Figure 9-3 Wong-Baker faces. (From Wong, DL: Whaley & Wong's Essentials of Pediatric Nursing, ed 5. Mosby, St Louis, 1997, with permission.)

but may not be accurate or even useful. The best tools are simple and easy to use. Longer questionnaires require more time and may cause distress for the patient in acute pain but may be helpful when doing a complete pain history (Fig. 9–4). A scale should also be used to monitor the patient's level of sedation following opioid administration (Fig. 9–5). Any unexpected decrease in the patient's level of sedation should be reported promptly to the registered nurse (RN) or physician.

It is important to use the patient's own descriptions and words when taking the pain history, such as *aching*, *knifelike*, or *throbbing*. This is also true when the patient is experiencing neuropathic pain. Neuropathic pain is often difficult to define. Terms commonly used to describe neuropathic pain include *burning*, *shocklike*, and *tingling*.

Do a Complete Physical Assessment

A good physical assessment is necessary to determine the effect of the pain and pain treatments on the body. It helps identify all the pain sites and helps you to prioritize the seemingly overwhelming task of helping the patient achieve acceptable pain management and good qualify of life. As discussed previously, the patient with acute pain may exhibit signs such as grimacing and moaning or elevated pulse and blood pressure, but these signs cannot be relied on to "prove" that the patient is in pain. The only reliable source of pain assessment is the patient's self-report (Gerontological Issues Box 9–4).

Planning and Implementation

Establish a pain control goal during the planning phase. The patient must be asked to determine an acceptable level of pain if complete freedom from pain is not possible. It is also important to identify the patient's activity goals. After

---

BOX 9-4 GERONTOLOGICAL ISSUES

The older patient may have different manifestations of pain than a younger patient. Older patients who are confused may be unable to tell the nurse that they are feeling pain. The nurse should consider incidents of restlessness and confusion as possible signs of pain. Pulling at dressings, tugging at IV sites, and trying to climb over the side rails to get out of bed can also be symptoms of discomfort.

The nurse can anticipate pain and provide relief measures to prevent severe pain. Pain medications and basic comfort care can be administered routinely if pain is likely. Nagging achiness in hands and feet is often noted as a reason for decreased activity, inability to sleep, and altered functional ability. A hand or foot massage using lotion and gentle massage strokes is often a very relaxing comfort measure.

CRITICAL THINKING

Mr. Sebastian

Mr. Sebastian is a 75-year-old gentleman who has been diagnosed with lung cancer and is anxious about leaving the hospital to return home following a thoracotomy. The nursing assessment reveals the need for home health care for dressing changes and teaching about the medications he will need at home. While in the hospital, Mr. Sebastian has required 5 mg morphine IV every 4 hours around the clock.

What discharge instructions must be given to Mr. Sebastian and his wife before sending him home? How might his pain be managed at home to prevent unnecessary readmissions to the hospital?

Answers at end of chapter.

---

surgery, goals may include the ability to ambulate and sleep without pain. For patients with chronic pain, the goals may be different. For example, if a patient with terminal cancer wants to be able to attend her granddaughter's wedding, you can assist the patient in reaching that goal. She can be

Pain Assessment Chart (For Admission and/or Follow-up)

1. Patient _____ 2. DX _____

Assessment on Admission

Date _____/_____/_____   Pain ☐   No Pain ☐   Date of Pain Onset _____/_____/_____

1. Location of Pain (indicate on drawing)

2. Description of Predominant Pain (in patient's words) _____

_____

_____

3. Intensity [Scale 0 (no pain) — 10 (most intense)] _____   Right   Left   Left   Right

_____

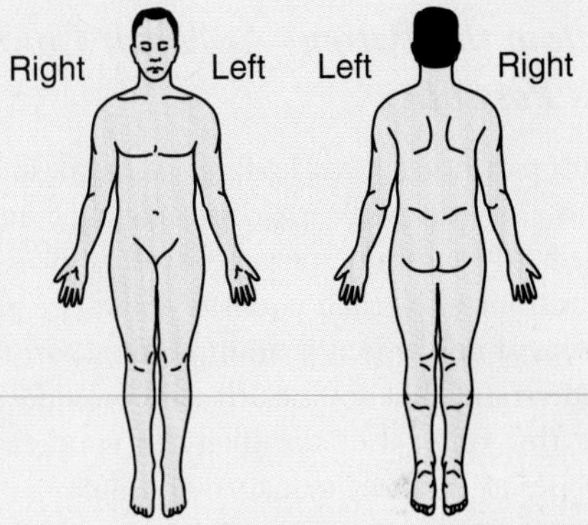

4. Duration and when occurs _____

_____

5. Precipitating Factors _____

_____

6. Alleviating Factors _____

_____

7. Accompanying Symptoms

   GI: Nausea ☐   Emesis ☐   Constipation ☐   Anorexia ☐

   CNS: Drowsiness ☐   Confusion ☐   Hallucinations ☐

   Psychosocial: Mood _____   Anger _____

   Anxiety _____   Depression _____

   Relationships _____

_____

8. Other Symptoms

   Sleep _____   Fatigue _____

   Activity _____   Other _____

9. Present Medications _____

_____

_____

   Doses and times medicated last 48 hours _____

_____

10. Breakthrough Pain _____

_____

Signature: _____

Figure 9-4 Pain assessment chart. (Modified from The Purdue Frederik Company, Norwalk, Conn, with permission.)

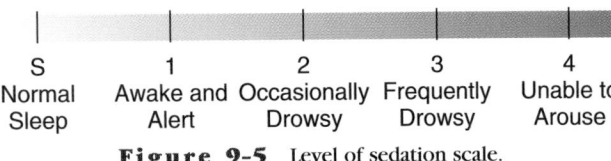

S	1	2	3	4
Normal Sleep	Awake and Alert	Occasionally Drowsy	Frequently Drowsy	Unable to Arouse

Figure 9-5 Level of sedation scale.

taught to reserve energy that day for the activity that is most important to her. Instructing her in optimal timing of her pain medication will also assist her in reaching a good comfort level for the activity.

Some additional principles to consider during the planning phase follow.

Allow the Patient As Much Control As Possible

Pain can bring forth feelings of helplessness and hopelessness. Giving patients pain management options allows them to maintain some control. It is also the nurse's responsibility to teach patients about the goals of pain management and why it is an important part of care. If patients understand that the health care provider's goals and theirs are the same, they are likely to cooperate with and contribute to the pain management plan.

Pain Affects the Whole Family

It is important to include the whole family in the plan. Understanding family dynamics helps the nurse in implementing an effective pain management plan. Cultural influences are important to consider when planning pain treatment (see Chapter 3). It is difficult for family members to see loved ones in pain. Including them in the planning helps them feel that they may be helping to make the patient more comfortable. (See Home Health Hints.)

HOME HEALTH HINTS

- Emotional or spiritual distress and fear related to dependence on family caregivers may alter the patient's perception or report of pain.

- Several alternative measures are easily taught to patients and caregivers in the home. For example, ice can be made in paper cups and used for a cold massage of a painful area.

- Repositioning can alleviate some aches, and soothing music can serve as a distraction from the pain.

Pain is Exhausting

Pain may keep the patient from sleeping. This cycle of sleeplessness and pain must be interrupted to help the patient. The need for adequate rest must not be ignored. This is more often an ongoing problem for the patient with chronic nonmalignant pain or chronic cancer pain, and it is perhaps more difficult to manage. The patient must get at least 4 to 6 hours of uninterrupted sleep to be relaxed enough to break the cycle. Controlled-release opioid pain medications may help maintain pain relief, allowing the patient to sleep. If controlled-release medications are not used, it may be necessary to wake a patient to administer pain medication so that the pain does not get out of control. The addition of a sedative may be necessary to allow the patient to sleep.

A Team Approach to Pain Management

A plan must be developed using an interdisciplinary approach, including the patient and family, the nurse, and the physician. Others, such as the occupational and physical therapist, chaplain, social worker, and pharmacist, should be included as appropriate. Communication is the important link allowing this team to be effective in creating a plan that works for the patient. Plans must be individualized to meet the special needs of each patient and vary greatly depending on the type of pain the patient is experiencing.

In 1990 the World Health Organization (WHO) developed the WHO analgesic ladder, which involves choosing among three levels of treatments based on intensity of pain (Fig. 9-6). The ladder, which helps direct the interventions required when using medications to treat pain, was developed for the treatment of cancer pain but can be used when treating other types of pain as well.

When experiencing mild pain (level I on the WHO ladder), the patient can usually sleep, perform activities of daily living, and even work. The first level of the ladder addresses the use of nonopioid analgesics. When pain is unre-

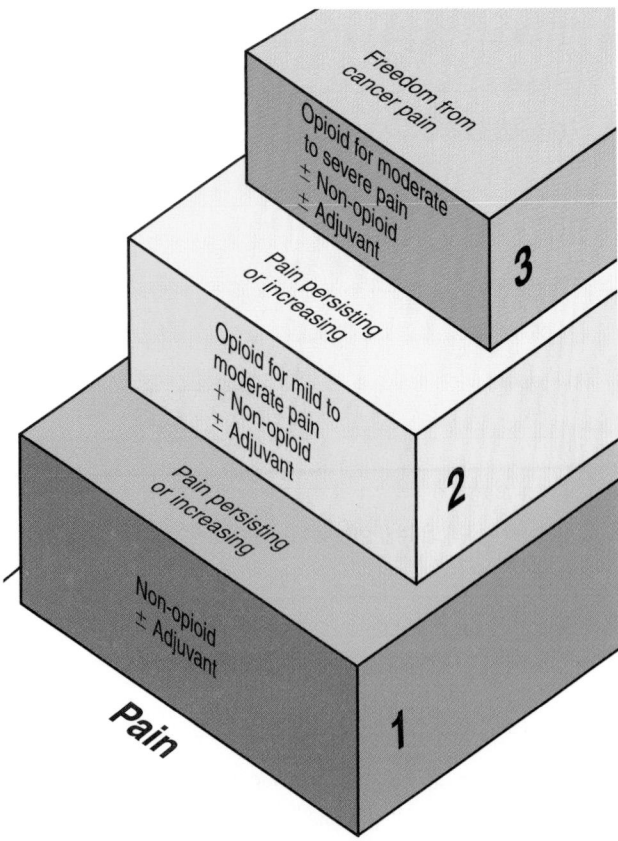

Figure 9-6 World Health Organization three-step analgesic ladder.

lieved by maximum ATC dosing, the treatment moves up the ladder to level II (mild to moderate pain) and adds an opioid analgesic. The patient with mild to moderate pain may not be able to sleep or may have trouble working and staying focused. If pain increases beyond that which is controlled by the level II analgesics, it is time to move on to level III. At this level the pain is moderate to severe and is affecting the quality of the patient's life, and he or she may not be able to perform activities of daily living. At level III an adjuvant analgesic may be added.

Analgesics should be given on an ATC basis to prevent breakthrough pain, especially for cancer and chronic benign pain. For patients experiencing surgical or traumatic pain, analgesics should be given ATC until the pain decreases to a level that allows medications to be given as required (prn). When using the WHO ladder it is important to keep in mind that it may not be necessary to start at level I if the patient is having severe pain. Analgesics from level III on the WHO ladder may be the starting point for some patients.

CRITICAL THINKING

Mrs. Zales

Mrs. Zales, a 32-year-old woman, was admitted for a hysterectomy after being treated for painful endometriosis for 12 months. After her surgery she had a PCA pump with hydromorphone, which was effective in relieving her pain. Forty-eight hours after surgery the surgeon discontinued the PCA pump and ordered oral hydrocodone with acetaminophen (Lortab). It was ineffective, so an order for hydromorphone, 2 to 4 mg orally every 3 to 4 hours as needed, was added. The nurses gave only one dose of the hydromorphone, then, thinking that her pain should be lessening, switched Mrs. Zales back to the hydrocodone with acetaminophen. By the next morning she was in severe pain, and the on-call physician ordered IM meperidine and promethazine (Phenergan). Mrs. Zales's discharge was delayed until her pain could be controlled.

What do you think happened? How could the delayed discharge have been avoided?

Answers at end of chapter.

Patient Education

Patients must be informed about the medications they are taking for pain management so they can take an active role in their care. Patients informed about the goals of pain management are more likely to report unrelieved pain so that they can receive prompt and effective treatment. Goals include a satisfactory comfort level with minimal side effects and complications of pain and its treatment, as well as a reduced period of recovery.

It is important to provide the patient with information about the drug's common side effects, the frequency of the dose and its duration of action, and potential drug-drug and food-drug interactions if indicated. Education must be presented at a level that the patient can understand. Informed patients use their medications more effectively and safely.

CRITICAL THINKING

Ms. Jackson

Ms. Jackson had abdominal surgery 2 days ago. She has been receiving morphine via IV PCA at an average of 2.5 mg per hour for the last 6 hours. She rates her pain at 3 on a scale of 0 to 10. She is to be discharged today. Her physician has ordered codeine 30 mg with acetaminophen (Tylenol with codeine No. 3), 1 or 2 tablets every 4 hours as needed for pain at home.

Will Ms. Jackson be comfortable at home? Why or why not?

Answers at end of chapter.

LEARNING TIPS

Many medication interventions are available for the treatment of pain. Whenever possible, administer analgesics by the mouth, by the WHO ladder, and by the clock.

Evaluation

The final phase of the nursing process is evaluation. Once the plan of care has been implemented, evaluate whether the patient's goals have been met. Evaluation is an ongoing process that continues as the plan of care is carried out to be sure that pain management is effective and that changes are made as needs are identified. Has the patient's identified goal for an acceptable level of pain been met? How well were the pain treatments tolerated? Were there any adverse effects to the medications that were given? Were the side effects managed appropriately? Poorly managed side effects can be as problematic as poorly managed pain. For example, when nausea has been identified as something that is always a problem for the patient after surgery, it is imperative to include in the plan how it will be addressed. Was monitoring appropriate at each point in the pain management plan for the prevention of uncontrolled pain and clinically significant respiratory depression? Was the patient able to participate in activities that he or she identified as important? The plan should be continuously updated based on the evaluation. (See Nursing Care Plan 9–5 for the Patient in Pain.)

BOX 9-5   NURSING CARE PLAN FOR THE PATIENT IN PAIN

PATIENT OUTCOME
Pain is at a level that is acceptable to the patient. Patient is able to participate in activities that are important to him or her.

EVALUATION OF OUTCOME
Is pain at a level that is acceptable to the patient? Is the patient able to participate in activities that he or she has identified as important?

INTERVENTION	RATIONALE	EVALUATION
Assess pain based on patient report. Use the WHAT'S UP format.	Patient's pain is defined as what the patient says it is, when the patient says it is occurring.	Does the patient verbalize his or her pain? Does the patient use verbal or nonverbal messages that imply trust in nurse's belief of pain report?
Teach the patient to use a pain rating scale. Use the same scale consistently.	A rating scale is the most reliable method for assessing pain severity.	Does the patient understand the use of the scale and use it to report pain?
Determine with patient what is an acceptable pain level.	Only the patient can decide what pain level is tolerable.	Is the patient's pain at an acceptable level?
Assess whether pain is acute, chronic, or both.	Acute and chronic pain may present differently and may require different interventions.	Has acute versus chronic pain been identified? Are treatments appropriate?
Assess need for and offer emotional and spiritual support for the experience of pain and suffering.	Pain, as well as disease processes, can be accompanied by feelings of powerlessness and distress.	Does the patient appear emotional, angry, or withdrawn? Does the patient have difficulty making decisions? Is the patient-nurse relationship therapeutic?
Give analgesics before pain becomes severe. For moderate to severe pain, give analgesic ATC.	Severe pain is more difficult to relieve.	Is analgesic schedule effective?
Assess for pain relief approximately 1 hour after administration of oral analgesics, or 30 minutes after IV analgesics.	If pain is not relieved, additional measures will be needed.	Does patient report acceptable level of relief?
Observe for anticipated side effects of pain medication.	Many pain medications cause nausea and constipation. The nausea usually subsides after several days, but the constipation does not.	Are side effects occurring? Can they be managed? Does medication regimen need to be adjusted?
If opioids are being used, assess for respiratory depression and reduced level of consciousness at regular intervals.	High doses or sudden increases in dose of an opioid can result in respiratory depression and reduced level of consciousness.	Is the patient's respiratory rate greater than 8 per minute or above the parameter ordered by the physician? Is the patient alert and oriented?
Institute measures to prevent constipation: 8 to 10 glasses of fluid daily (unless contraindicated), fiber in meals, fiber or bulk laxatives, exercise as tolerated.	Opioids cause constipation.	Are the patient's bowels moving according to his or her usual pattern?
Teach patient alternative pain relief interventions, such as relaxation and distraction, to be used with medication.	Alternative interventions can help the patient feel in control and may help reduce the perception of pain.	Does the patient use alternative interventions effectively?
Assess whether patient is taking pain medications appropriately, and if not, assess reasons. Instruct in how to manage pain interventions.	Pain medications must be taken appropriately to be effective.	Is the patient able to manage the pain control regimen? Are adjustments necessary?

1. Mr. White is walking up and down the hall and visiting with other patients. He is laughing and joking. He approaches the nurse's station, asks for his pain shot, and reports that his pain is 6 on a scale of 0 to 10. Based on McCaffery's definition of pain, which of the following assumptions by the nurse is most likely correct?
   a. Mr. White is not really in pain but just wants his medication.
   b. Mr. White is having pain at a level of 6 on a scale of 0 to 10.
   c. Mr. White is in minimal pain and should receive a pill instead of a shot.
   d. Mr. White is in pain but does not need his pain medication yet.

2. Ms. Williams has incisional pain following total hip replacement surgery. Which of the following types of pain is she experiencing?
   a. Nociceptive
   b. Neuropathic
   c. Chronic

3. Which of the following methods is the most reliable way to assess a patient's pain?
   a. Obtain a pain history from the patient.
   b. Observe the patient for signs of pain.
   c. Ask the patient to rate his or her pain using a valid assessment scale.
   d. Ask a family member to rate the patient's pain.

4. Jim is hospitalized following a motor vehicle accident. He has multiple orthopedic injuries and is in acute pain. He has an order for morphine 6 mg IV every 2 to 3 hours as needed. Assuming his respiratory rate and level of sedation are acceptable and his pain is controlled, which of the following analgesic schedules will be most effective for relieving his pain?
   a. Offer the analgesic every 2 to 3 hours.
   b. Tell him to put on his light when he feels pain, and give the drug immediately when he requests it.
   c. Give the IV analgesic every 2 to 3 hours ATC.
   d. Alternate the IV analgesic with an oral analgesic.

5. Mrs. Edwards has terminal cancer and has been requiring 5 mg of IV morphine every 1 to 2 hours to control her pain, yet she is laughing and enjoying her visitors. Which of the following explanations of her behavior is most likely correct?
   a. Denial of pain is common in patients with cancer.
   b. Mrs. Edwards's cancer is improving.
   c. Mrs. Edwards is hiding her pain when visitors are present.
   d. Distraction can be an effective treatment for pain when used with appropriate drug treatments.

6. Mr. Lawrence is an 88-year-old gentleman admitted with a broken hip after a fall. He has an order for meperidine 50 to 75 mg IM q4–6h prn pain. As his nurse, which of the following actions should you take?
   a. Give the meperidine every 4 hours ATC.
   b. Offer the meperidine every 6 hours because you know that his liver and kidney function may be diminished.
   c. Administer an NSAID with the meperidine for added pain relief.
   d. Talk to the RN or physician about getting an order for a different analgesic.

REFERENCES

1. McCaffery, M, and Pasero, C: Pain: Clinical Manual, ed 2. Mosby, St Louis, 1999.
2. International Association for the Study of Pain: http://www.iasp-pain.org, December 2001.
3. Agency for Health Care Policy and Research: AHCPR Clinical Practice Guideline: Management of Cancer Pain. US Department of Health and Human Services, Rockville, Md, 1994.
4. American Pain Society: Pain: The Fifth Vital Sign. http://www.ampainsoc.org, December 2001.

Answers to CRITICAL THINKING

Smithers and Barnett

It is important to accept both patients' pain reports. Assessment should be based on what the patient says rather than what is observed. Each patient copes with his or her pain in a unique way, and the nurse cannot judge whether one is in more pain than the other.

Mrs. Shepard

1. Pain medication is most effective when given on a routine schedule around the clock to avoid breakthrough pain. Mrs. Shepard's epidural infusion should continue to relieve her pain for a time, up to several hours after it is discontinued, depending on the medication used. The oral medication is most effective when given at the time the epidural is stopped so that it is taking effect as the epidural effects wear off.
2. Pain prevents patients from moving freely. Postoperative complications such as retained pulmonary secretions and ileus can occur when patients are immobile. Effective pain management can help prevent these complications.
3. If she takes a dose every 3 hours, then she will receive 8 doses in 24 hours; 500 mg × 8 = 4000 mg or 4 g, which is the maximum safe dose. Recall that elderly patients metabolize and excrete medications more slowly than younger patients. If she will need the hydrocodone/acetaminophen more than a few days, it would be wise to consult with the physician about giving the opioid and acetaminophen separately.
4. Mrs. Shepard should be instructed what her role will be when her pain management regimen is altered. Does she have to ask for the pain medication or will it just be brought to her? Patient and family education is vital to success in management of a patient's pain.

Mr. Sebastian

Home instruction regarding ATC administration of pain medication is indicated. MS Contin, a long-acting form of morphine, may be an option for Mr. Sebastian, along with an immediate-release preparation for breakthrough pain. Also, information about what to do and who to contact if pain becomes unmanageable is necessary to help prevent readmissions to the hospital.

Mrs. Zales

Mrs. Zales was probably tolerant to opioids because of her need for medication for chronic pain over the last year. For this reason, she needed more medication than a nontolerant patient who does not usually use opioids. Also, the belief that promethazine and other phenothiazines potentiate opioids is a myth. They do cause increased levels of sedation and may limit the amount of opioid that can be given safely. IM injections are not recommended because they are painful, absorption is not predictable, and there is a delay between injection and relief. Nurses often base the treatment of a patient's pain on what they usually do or what they think should be effective rather than on sound pain management practices and principles. A more rational approach to Mrs. Zales would have been regular pain assessment with ATC treatment until pain began to subside. If her pain level had been better controlled, she might have been discharged on oral analgesics without the delay.

Ms. Jackson

Using an equianalgesic conversion, we can determine whether Ms. Jackson is likely to have good pain relief based on her requirement with the PCA. Her current pain level of 3 shows that the morphine has been effective. Remember that the pump keeps a history of what the patient uses, which is the best indicator of what the patient needs. Ms. Jackson has used 15 mg of morphine during the past 6 hours. An equianalgesic dose of Tylenol with codeine No. 3 would be almost 200 mg of codeine, but only 30 to 60 mg has been ordered. In addition, if Ms. Jackson takes enough Tylenol with codeine No. 3 to get 200 mg of codeine, she will receive a dangerous dose of both the codeine and the acetaminophen. The physician needs to be contacted for different analgesic orders.

10

NURSING CARE OF PATIENTS WITH CANCER

Martha Spray, Vera Dutro, and Debbie Millar

KEY TERMS

alopecia (AL-oh-**PEE**-she-ah)

anemia (uh-**NEE**-mee-yah)

anorexia (AN-oh-**REK**-see-ah)

benign (bee-**NINE**)

biopsy (**BY**-ahp-see)

cancer (**KAN**-sir)

carcinogen (kar-**SIN**-oh-jen)

chemotherapy (KEE-moh-**THER**-uh-pee)

cytotoxic (SIGH-toh-**TOCK**-sick)

in situ (in-**SIT**-yoo)

leukopenia (LOO-koh-**PEE**-nee-yah)

malignant (muh-**LIG**-nunt)

metastasis (muh-**TASS**-tuh-sis)

mucositis (MYOO-koh-**SIGH**-tis)

neoplasm (**NEE**-oh-PLAZ-uhm)

oncology (on-**CAW**-luh-gee)

oncovirus (**ON**-koh-VIGH-russ)

palliation (pal-ee-**AY**-shun)

radiation therapy
(RAY-dee-**AY**-shun **THER**-uh-pee)

stomatitis (STOH-mah-**TIGH**-tis)

thrombocytopenia
(THROM-boh-SIGH-toh-**PEE**-nee-ah)

tumor (**TOO**-mur)

vesicant (**VESS**-i-kant)

xerostomia (ZEE-roh-**STOH**-mee-ah)

QUESTIONS TO GUIDE YOUR READING

1. What are the normal structure and functions of the cell?

2. What changes occur in the cell when it becomes malignant?

3. Which medications are commonly used as chemotherapy agents?

4. What are the special nursing needs of the patient receiving chemotherapy or radiation therapy?

5. Which data should you collect when caring for a patient with cancer?

6. What are nursing interventions for common oncological emergencies?

7. How will you know if your nursing interventions have been effective?

8. What is the role of hospice in providing care for patients with advanced cancer?

► REVIEW OF NORMAL ANATOMY AND PHYSIOLOGY OF CELLS

Cells are the smallest living structural and functional subunits of the body. Although human cells vary in size, shape, and certain metabolic activities, they have many characteristics in common.

Cell Structure

Human cells have a cell membrane, cytoplasm, cell organelles, and, with the exception of mature red blood cells, a nucleus. In the mature red blood cell the nucleus has been lost. Each cell structure has a specific and vital function. The cell membrane forms the outer boundary of the cell and is made up of protein, phospholipids, and cholesterol. Proteins serve three different purposes: (1) Some are pores or enzymes to permit transport of materials, (2) some are receptor sites for hormones to trigger a cell's activity, and (3) some are antigens to identify the cell as belonging in the body.

A cell membrane is selectively permeable, meaning that not all substances pass through. The lipids permit the diffusion of lipid-soluble materials into or out of the cell. Materials may enter or leave a cell in one of several ways. These transport mechanisms (with examples of their importance in the body) are summarized in Table 10–1.

Cytoplasm and Cell Organelles

Cytoplasm is a watery solution of minerals, gases, and organic molecules that is found between the cell membrane and the nucleus. Chemical reactions (such as the synthesis of adenosine triphosphate [ATP]) take place in the cytoplasm. Cell organelles are subcellular structures with specific functions, which are summarized in Table 10–2 and shown in Figure 10–1. Many cell organelles are found in the cytoplasm.

Nucleus

The nucleus of a cell is surrounded by a double-layered nuclear membrane with many pores. Inside the nucleus are one or more nucleoli and the chromosomes of the cell.

A nucleolus is a small sphere made of deoxyribonucleic acid (DNA), ribonucleic acid (RNA), and protein. The nucleoli form a type of RNA called ribosomal RNA, which is part of the cell organelle called the ribosome and is involved in protein synthesis.

The nucleus is the control center of the cell because it contains the chromosomes. The 46 chromosomes of a human cell are made of DNA and protein. DNA is the genetic code for the characteristics and activities of the cell. Specific regions of DNA are called genes; a gene is the code for one protein. Not all the genes in any cell are active, only those relative few needed for the proteins to carry out their specific functions. These proteins may be structural, such as the collagen of connective tissue, or functional, such as the hemoglobin of red blood cells. Important functional proteins are the enzymes that catalyze the specific reactions characteristic of each type of cell.

TABLE 10-1	CELLULAR TRANSPORT MECHANISMS	
Mechanism	**Definition**	**Example in the Body**
Diffusion	Movement of molecules from an area of greater concentration to an area of lesser concentration	Exchange of gases in the lungs or body tissues
Osmosis	The diffusion of water	Absorption of water by the intestines or kidneys
Facilitated Diffusion	Carrier enzymes move molecules across cell membranes	Intake of glucose by most cells
Active Transport	The use of ATP to move molecules from an area of lesser concentration to an area of greater concentration	Absorption of glucose and amino acids by the small intestine
Filtration	Movement of water and dissolved materials from an area of higher pressure to an area of lower pressure (blood pressure)	Formation of tissue fluid; the first step in the formation of urine
Phagocytosis	A moving cell engulfs something	White blood cells engulf bacteria
Pinocytosis	A stationary cell engulfs something	Cells of the kidney tubules reabsorb small proteins

TABLE 10-2	FUNCTIONS OF CELL ORGANELLES
Organelle	**Function(s)**
Endoplasmic Reticulum	Passageway for transport of materials within the cell
	Synthesis of lipids
Ribosomes	Site of protein synthesis
Golgi Apparatus	Synthesis of carbohydrates
	Packaging of materials for secretion from the cell
Mitochondria	Site of aerobic cell respiration, synthesis of ATP
Lysosomes	Contain enzymes to digest ingested material or damaged tissue
Centrioles	Organize the spindle fibers during cell division
Cilia	Sweep materials across the cell surface, as in the respiratory passages
Flagellum	Enables a sperm cell to move

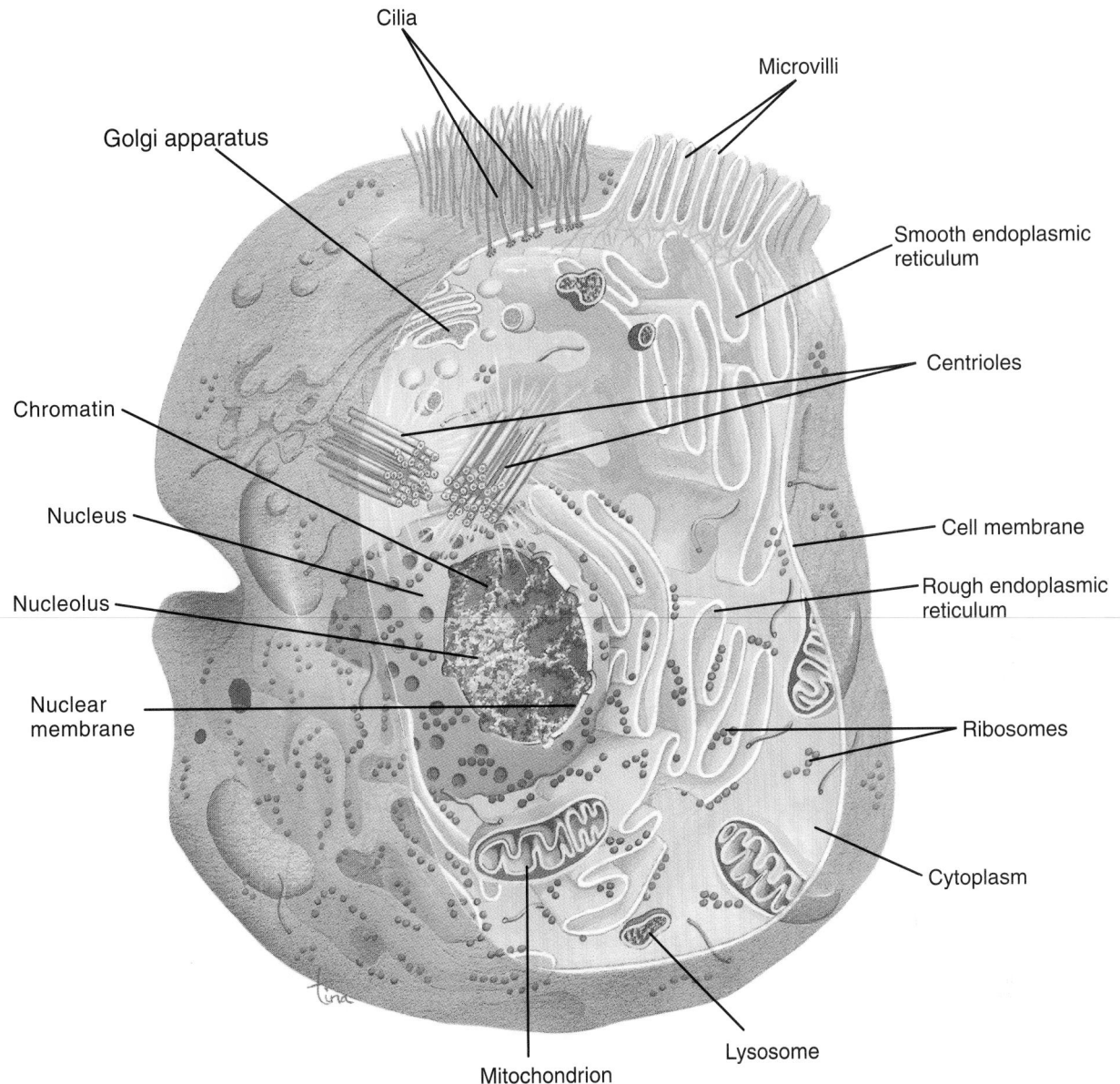

Cilia

Microvilli

Golgi apparatus

Smooth endoplasmic reticulum

Centrioles

Chromatin

Nucleus

Cell membrane

Nucleolus

Rough endoplasmic reticulum

Nuclear membrane

Ribosomes

Cytoplasm

Mitochondrion

Lysosome

Figure 10-1 Schematic diagram of a typical human cell. (From Scanlon, V, and Sanders, T: Essentials of Anatomy and Physiology, ed 3. FA Davis, Philadelphia, 1999.)

Genetic Code and Protein Synthesis

The genetic code of DNA is the code for the amino acid sequences needed to synthesize a cell's proteins. This process is shown in Figure 10–2 and may be described simply as follows. A complementary copy of the DNA's gene is made by a molecule called messenger RNA (mRNA). The mRNA then moves to the cytoplasm of the cell and attaches to the ribosomes. Transfer RNA (tRNA) molecules bring the necessary amino acids to the proper places on the mRNA molecule, and enzymes of the ribosomes catalyze the formation of peptide bonds to link the amino acids into a finished protein.

As with any complex process, mistakes are possible. Should there be a mistake in the DNA code, the process of protein synthesis may go on anyway, but the resulting protein will not function normally; this is the basis for genetic diseases. DNA mistakes acquired during life are called mu-

tations. A mutation is any change in the DNA code. Ultraviolet rays or exposure to certain chemicals may cause structural changes in the DNA code. These changes may kill the affected cells or may irreversibly alter their function. Such altered cells may become **malignant,** unable to function normally but very active; this is the basis of some forms of **cancer.**

Mitosis

Mitosis is the process by which a cell reproduces itself. One cell, after its 46 chromosomes have duplicated themselves, divides into two cells, each with membrane, cytoplasm, and organelles from the original cell and a complete set of chromosomes. Mitosis is necessary for the growth of the body and the replacement of dead or damaged cells. Some cells are capable of mitosis and others are

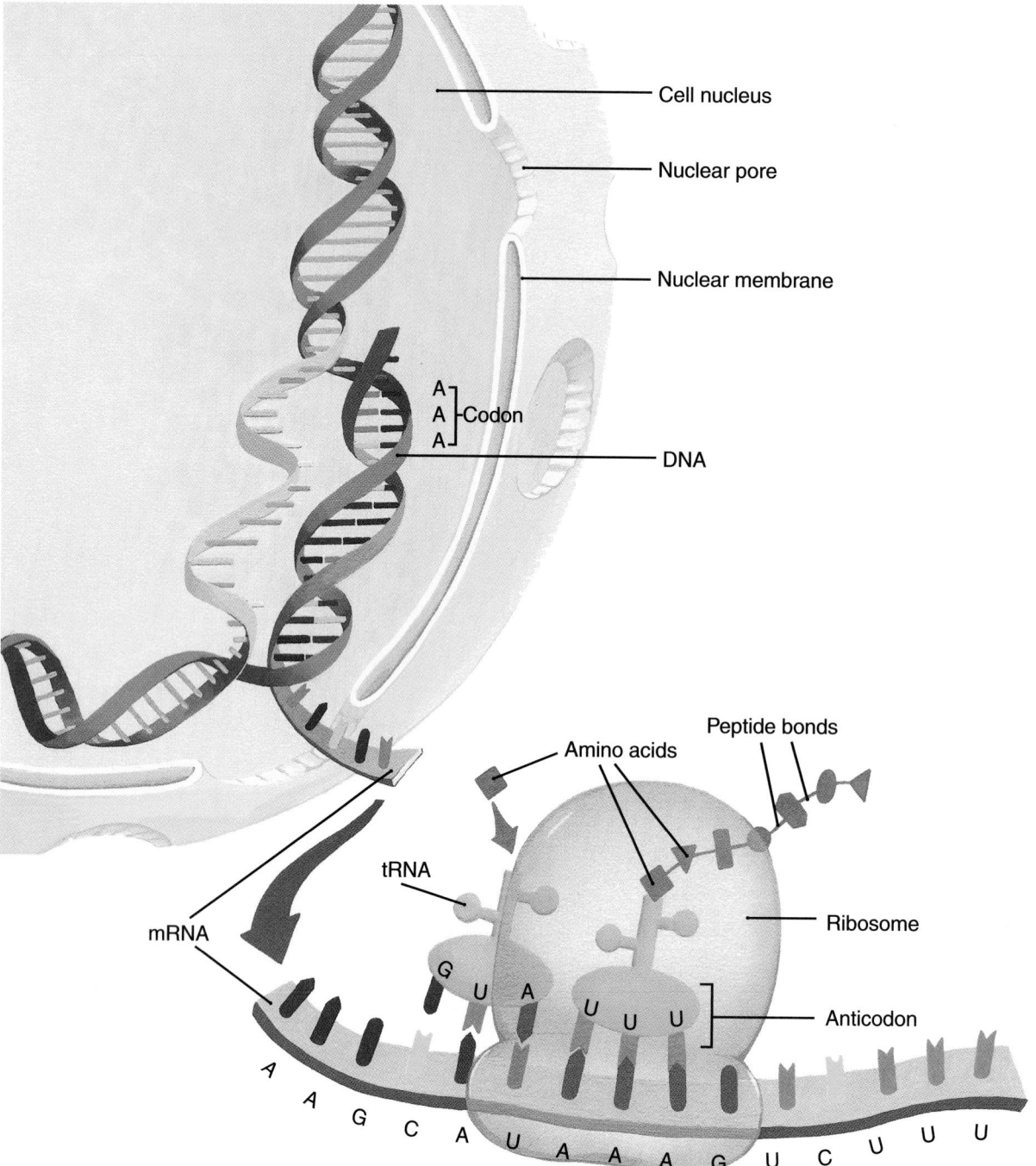

Figure 10-2   Schematic diagram of the process of protein synthesis. (From Scanlon, V, and Sanders, T: Essentials of Anatomy and Physiology, ed 3. FA Davis, Philadelphia, 1999.)

not. Cells of the epidermis of the skin undergo mitosis continuously to replace the superficial cells that are constantly worn off the skin surface. The same is true of cells that line the stomach and intestines. Cells in the red bone marrow also divide frequently; red blood cells have a fixed life span (about 120 days) and must be replaced. Some cells seem to be capable of only a limited number of divisions, and when that limit has been reached, the cells die and are not replaced.

Other cells do not undergo mitosis to any great extent after birth. Nerve cells (neurons) are unable to divide, and

muscle cells have very limited mitotic capability. When such cells are lost through injury or disease, the loss of their function in the individual is usually permanent.

Cell Cycle

The cell cycle involves a series of changes through which a cell progresses, starting from the time it develops until it reproduces itself. The duration of the cell's life, the time it takes for mitosis to occur, the growth ratio (percentage of cycling cells), the frequency of cell loss, and the doubling time (time for a **tumor** to double its size) are important concepts related to tumor growth and treatment strategies.

Cells can occupy three functions in the cell cycle: cells that are actively dividing, cells that leave the cycle after a certain point and die, and cells that temporarily leave the cycle and remain inactive until reentry into the cycle. Inactive cells continue to synthesize RNA and protein (Fig. 10–3).

Cells and Tissues

A tissue is a group of cells with similar structure and functions. The four groups of human tissues are epithelial, connective, muscle, and nerve.

Epithelial tissues form coverings and linings throughout the body. Often the cells are capable of mitosis, and damage to the tissue may be repaired. The healing of a cut to the skin is a typical example.

There are many types of connective tissues with varied functions. For example, blood is a connective tissue involved in the transportation of materials throughout the body. Fibrous connective tissue, made mostly of the protein collagen, forms strong membranes such as those around muscles, attaching structures such as ligaments and tendons and the dermis of the skin. Bone and cartilage are supporting connective tissues. Adipose tissue, another form of connective tissue, stores fat as potential energy. Many kinds of connective tissue cells are capable of mitosis.

There are three kinds of muscle tissue: skeletal muscle, smooth muscle, and cardiac muscle. Skeletal muscle tissue makes up the voluntary muscles attached to the skeleton. Smooth muscle is found in viscera such as the stomach and intestines; the walls of arteries and veins; the walls of the bronchial tubes; and, in women, the uterus. Cardiac muscle forms the walls of the chambers of the heart. As mentioned, the cells of muscle tissue have little ability to reproduce themselves.

Nerve tissue is made of neurons and supporting cells; in the central nervous system these supporting cells are called neuroglia. Although mature nerve cells are not capable of mitosis, many neuroglia are capable of mitosis. It is the neuroglia, not neurons, that usually form the tumors that develop in the central nervous system.

► INTRODUCTION TO CANCER CONCEPTS

Oncology is the branch of medicine dealing with tumors. Oncology nursing is also called cancer nursing; it is an important component of medical-surgical nursing care. A list of cancer resources is provided in Box 10–1. Cancer is second only to heart disease in mortality rates in the United States. The American Cancer Society reports that an estimated 10 million Americans alive today have a history of cancer.

Early accounts of cancer date back to the seventeenth century B.C. Documentation of the benefits of early cancer detection and its impact on treatment exist from the beginning of the nineteenth century. Today microscopic technology and genetic engineering provide physicians with a better understanding of tumor growth and cell activity and a means for early cancer detection and intervention.

► BENIGN TUMORS

Normal cells that reproduce abnormally result in **neoplasms,** or tumors. *Neoplasm* is a term that combines the Greek word *neo,* meaning "new," and *plasia,* meaning "growth," to suggest new tissue growth. A neoplasm is an enlargement of tissue and the formation of an abnormal mass. A neoplasm develops as cells multiply. Not all neoplasms contain cancer cells; however, a neoplastic cell is responsible for producing a tumor and shows a lively growing cell. New neoplastic growth is very difficult to detect until it contains about 500 cells and is approximately 1 cm.

A **benign** tumor is defined as a cluster of cells that is not normal to the body but is noncancerous. Benign tumors grow more slowly and have cells that are the same as the original tissue. An organ containing a benign tumor usually continues to function normally, whereas an organ affected with a cancerous tumor eventually ceases to function. *Malignant,* a term often used as a synonym for cancer, is defined as a growth that resists treatment. A comparison of benign and malignant tumors is found in Table 10–3.

---

oncology: onco—mass + logy—word, reason
neoplasm: neo—new + plasm—form

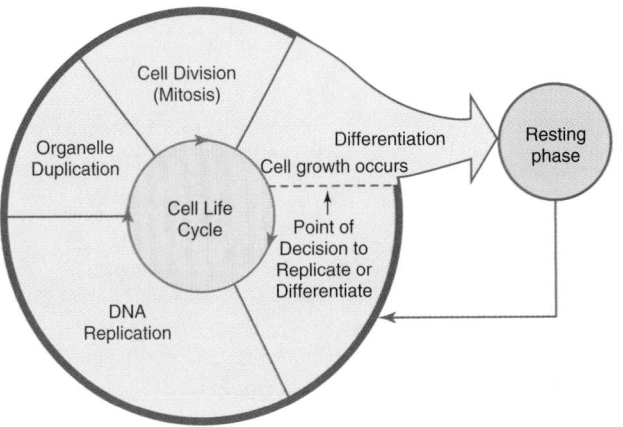

Figure 10–3 Cell cycle.

BOX 10-1 **Cancer Resources**

American Cancer Society
1-800-ACS-2345
http://www.cancer.org

National Cancer Institute
301-496-8531
http://cancernet.nci.nih.gov/
http://www.nci.nih.gov

Oncology Nursing Society
412-921-7373
http://www.ons.org/

TABLE 10-3 BENIGN AND MALIGNANT TUMORS		
	Benign	**Malignant**
Growth Rate	Typically slow expansion	Often rapid with cell numbers doubling normal cell growth; malignant cells infiltrate surrounding tissue
Cell Features	Typical of the tissue of origin	Atypical in varying degrees of the tissue or origin; altered cell membrane; contains tumor-specific antigens
Tissue Damage	Minor	Often causes necrosis and ulceration of tissue
Metastasis	Not seen; remains localized at origin site	Often spreads to form tumors in other parts of the body
Recurrence after Treatment	Seldom recurrence after surgical removal	Recurrence can be seen after surgical removal and following radiation and chemotherapy
Related Terminology	Hyperplasia, polyp, and benign neoplasia	Cancer, malignancy, and malignant neoplasia
Prognosis	Not injurious unless location causes pressure or obstruction to vital organs	Death if uncontrolled

► CANCER

Cancer is a group of cells that grows out of control, taking over the function of the affected organ. Cancer cells are described as poorly constructed, loosely formed, and without organization. A simplistic definition is "confused cell."

LEARNING TIPS

Cancer is not contagious.

Pathophysiology

Cancer is not one disease, but many diseases with different causes, manifestations, treatments, and prognoses. It is a disease of more than 100 different types, caused by mutation of cellular genes. Cancer takes on the characteristics of the cell it mutates and then takes on characteristics of the mutation. Growth-regulating signals in the cell's surrounding environment are ignored as the abnormal cell growth increases. Normal cells are limited to about 50 to 60 divisions before they die. Cancer cells do not have a division limit and are considered immortal.

The progression from a normal cell to a malignant cell follows a pattern of mutation, defective division and abnormal growth cycles, and defective cell communication. Cell mutation occurs when a sudden change affects the chromosomes, causing the new cell to differ from the parent. The malignant cell enzymes destroy the gluelike substance found between normal cells, which disrupts the transfer of information used for normal cell structure.

LEARNING TIPS

An individual's cancer risk is viewed as the balance between exposure and susceptibility to **carcinogens.**

Cancer Classification

Cancers are identified by the tissue affected, speed of cell growth, cell appearance, and location. Neoplasms occurring in the epithelial cells are called carcinomas. Carcinoma is the most common type of cancer and includes cells of the skin, gastrointestinal system, and lungs (Figs. 10-4 and 10-5). Cancer cells affecting connective tissue, including fat, the sheath that contains nerves, cartilage, muscle, and bone, are called sarcomas. *Leukemia* is the term used to describe the abnormal growth of white blood cells. Cancers involving cells of the lymphatic system, lymph nodes, and spleen are called lymphomas (Table 10-4).

---

carcinogen: karkinos—cancer, crab + genesis—birth

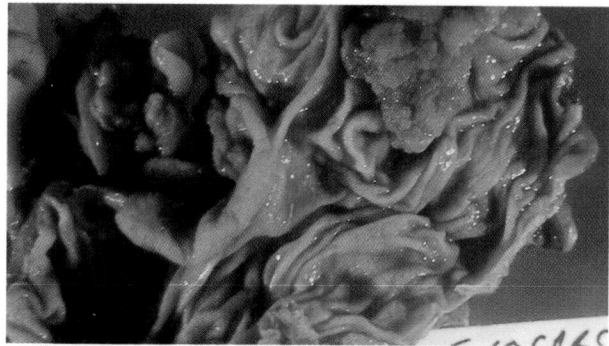

Figure 10-4 Adenocarcinoma of the caecum. (Photo courtesy Dinesh Patel, MD. Medical Oncology, Internal Medicine, Zanesville, Ohio.)

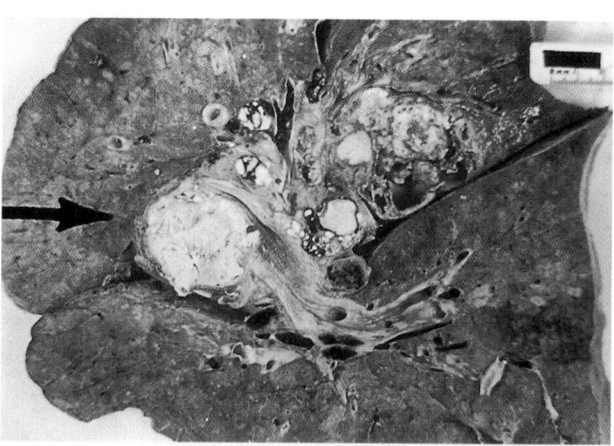

Figure 10-5 Lung cancer. (Photo courtesy Dinesh Patel, MD. Medical Oncology, Internal Medicine, Zanesville, Ohio.)

TABLE 10-4	TUMOR DESCRIPTION	
Tumor Type	**Character**	**Origin**
Fibroma	Benign	Connective tissue
Lipoma	Benign	Fat tissue
Carcinoma	Cancerous	Tissue of the skin, glands, and digestive, urinary, and respiratory tract linings
Sarcoma	Cancerous	Connective tissue, including bone and muscle
Leukemia	Cancerous	Blood, plasma cells, and bone marrow
Lymphoma	Cancerous	Lymph tissue
Melanoma	Cancerous	Skin cells

Spread of Cancer

Neoplastic cells that remain in one area are considered localized, or **in situ,** cancers. These tumors may be difficult to visualize on clinical examination and are detected through microscopic cell examination. In situ tumors are often removed surgically and require no further treatment. **Metastasis** is the term used to describe the spread of the tumor from the primary site into separate and distinct areas.

Metastasis is the stage at which cancer cells acquire invasive behavior characteristics and cause the surrounding tissue to change (Fig. 10–6). Metastasis occurs primarily because cancer cells break away more easily than normal cells and can survive for a time independently from other cells. There are three steps in the formation of a metastasis. Cancer cells are able to (1) invade blood or lymph vessels, (2) move by mechanical means, and (3) lodge and grow in a new location.

Metastatic tumors carry with them the cell characteristics of the original or primary tumor site. As a result, surgeons are able to determine the original tumor site based on metastatic cell characteristics. For example, lung tissue found in the brain suggests a primary lung tumor with metastasis to brain tissue. Common sites of metastasis are the lungs, liver, bones, and brain.

---

in situ: in—in + situ—position
metastasis: meta—beyond + stasis—stand

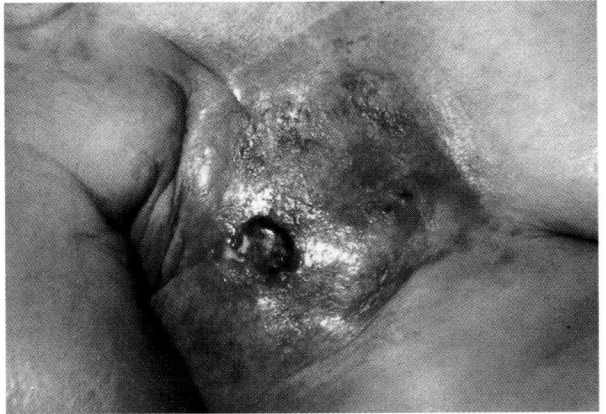

Figure 10-6 Invasive metastasis to skin area following mastectomy for breast cancer. (Photo courtesy Dinesh Patel, MD. Medical Oncology, Internal Medicine, Zanesville, Ohio.)

Incidence of Cancer

Cancer affects all age groups, although the incidence is higher in people ages 60 to 69. The second highest age group is ages 70 to 79. Men have a higher incidence of cancer than women. Cancer in people over age 60 is thought to occur from a combination of exposure to carcinogens and weakening of the body's immune system.

Some cancers, such as Wilms' tumor of the kidney and acute lymphocytic leukemia, occur more commonly in young people. The cause of tumors in young people is not well understood, but genetic predisposition tends to be a major factor.

The most common type of cancer in adults is skin cancer; it is also considered to be the most preventable. Exposure to ultraviolet radiation (sunlight) increases the risk of skin cancer. Wearing protective clothing and sunscreen can greatly reduce the risk of skin cancer.

Lung cancer is responsible for the highest mortality rate in both men and women and also is commonly preventable. Cigarette smoking is the main cause, along with air pollution and exposure to chemical agents.

Men have a high incidence of prostate cancer between ages 60 and 79. Cancer of the colon and rectum has been linked to the consumption of high-fat, low-fiber diets and ranks as the third highest cancer in men (Fig. 10–7).

The highest incidence of cancer in women is in the lungs, and the second highest is in the breast. Women with a family history of breast cancer have a greater risk than those with no family history. Commercial testing for the oncogene linked with breast cancer is available and marketed for high-risk women, especially those in the Ashkenazi Jewish population. Genetic testing is done through genetic counseling programs, and the cost ranges from $700 to $2400, depending on the geographic region.

Mortality Rates

Cancer survival rates have improved over the past 30 years. The American Cancer Society reports a gain in patient survival rates from 33 percent in the 1960s to 40 percent currently. A 5-year period is used to monitor cancer patients' progress following diagnosis and treatment. Survival statistics are based on persons living 5 years in remission.

For more information about cancer incidence and mortality data, visit the National Cancer Institute Web site at http://www.nci.nih.gov or http://www.seer.ims.nci.nih.gov,

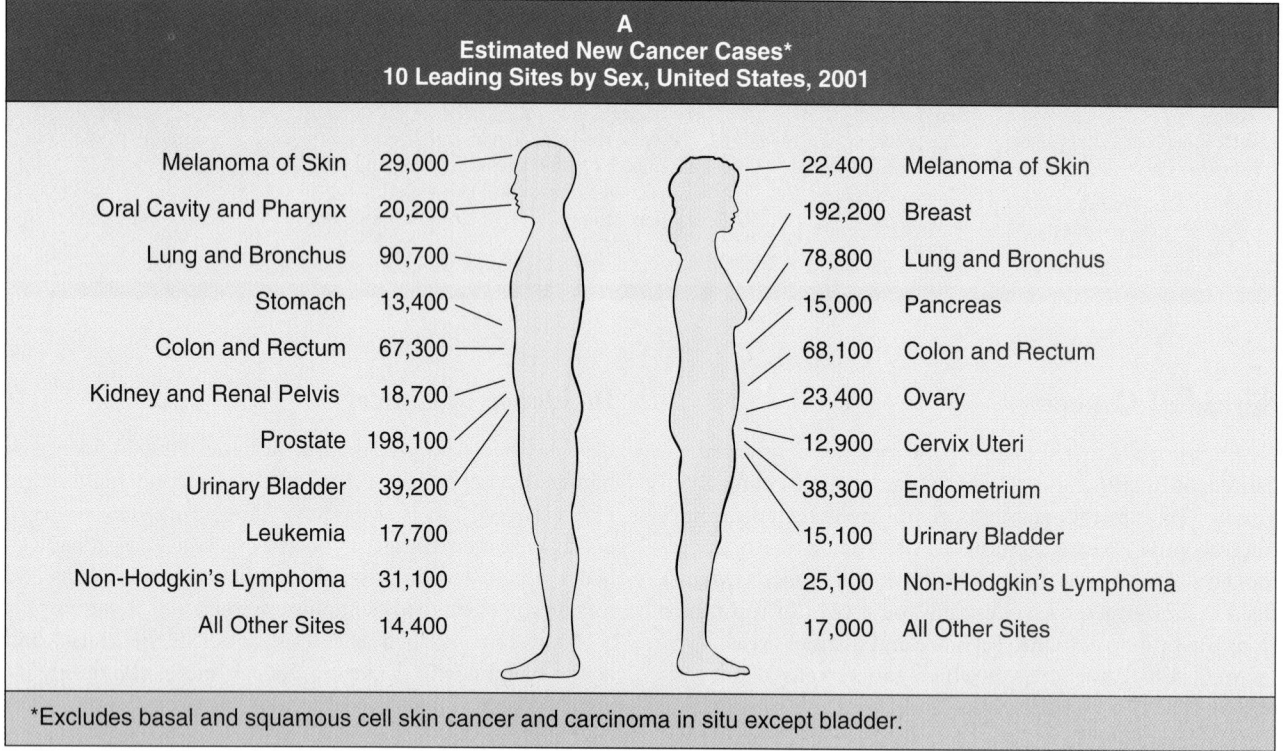

A
Estimated New Cancer Cases*
10 Leading Sites by Sex, United States, 2001

Melanoma of Skin	29,000	22,400	Melanoma of Skin
Oral Cavity and Pharynx	20,200	192,200	Breast
Lung and Bronchus	90,700	78,800	Lung and Bronchus
Stomach	13,400	15,000	Pancreas
Colon and Rectum	67,300	68,100	Colon and Rectum
Kidney and Renal Pelvis	18,700	23,400	Ovary
Prostate	198,100	12,900	Cervix Uteri
Urinary Bladder	39,200	38,300	Endometrium
Leukemia	17,700	15,100	Urinary Bladder
Non-Hodgkin's Lymphoma	31,100	25,100	Non-Hodgkin's Lymphoma
All Other Sites	14,400	17,000	All Other Sites

*Excludes basal and squamous cell skin cancer and carcinoma in situ except bladder.

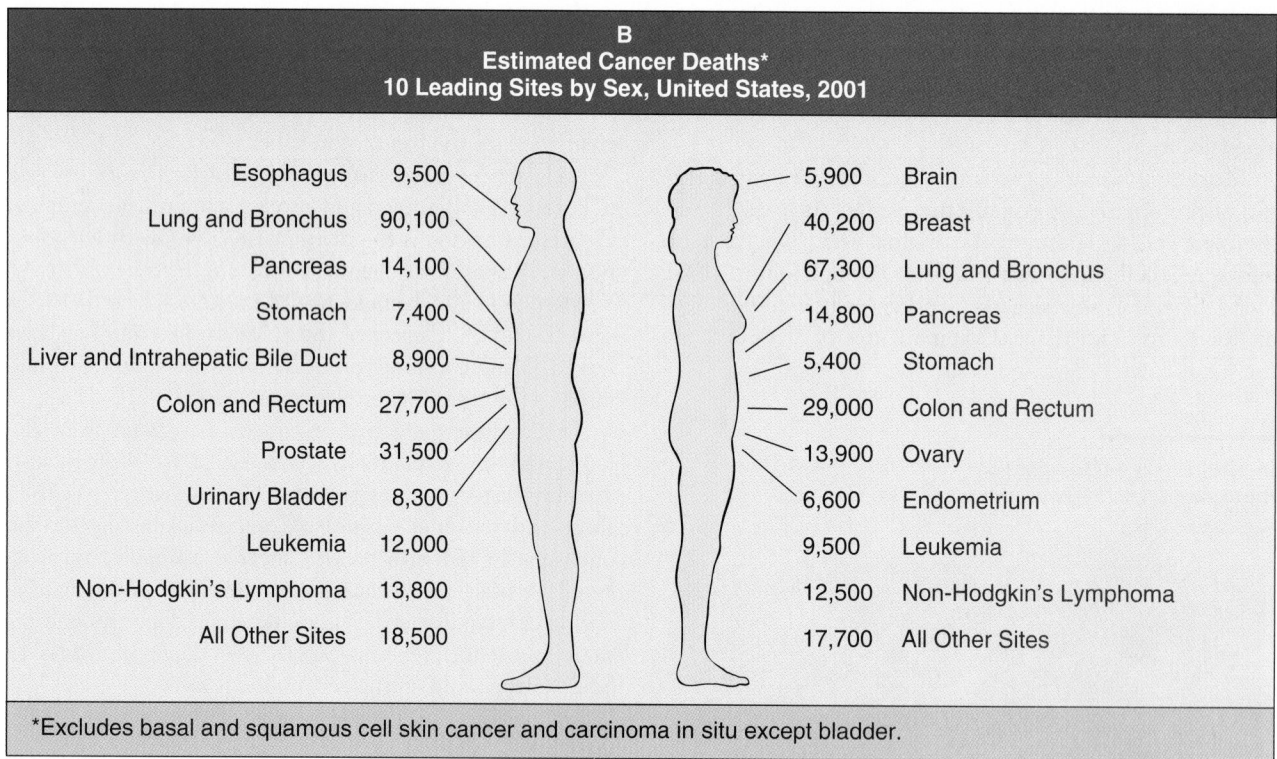

B
Estimated Cancer Deaths*
10 Leading Sites by Sex, United States, 2001

Esophagus	9,500	5,900	Brain
Lung and Bronchus	90,100	40,200	Breast
Pancreas	14,100	67,300	Lung and Bronchus
Stomach	7,400	14,800	Pancreas
Liver and Intrahepatic Bile Duct	8,900	5,400	Stomach
Colon and Rectum	27,700	29,000	Colon and Rectum
Prostate	31,500	13,900	Ovary
Urinary Bladder	8,300	6,600	Endometrium
Leukemia	12,000	9,500	Leukemia
Non-Hodgkin's Lymphoma	13,800	12,500	Non-Hodgkin's Lymphoma
All Other Sites	18,500	17,700	All Other Sites

*Excludes basal and squamous cell skin cancer and carcinoma in situ except bladder.

Figure 10-7   *(A)* Estimated new cancer cases, 10 leading sites by sex, United States, 2001. *(B)* Estimated cancer deaths, 10 leading sites by sex, United States, 2001. (Modified from Cancer Facts and Figures, American Cancer Society, 2001.)

or visit the National Coalition for Cancer Survivorship Web site at http://www.cansearch.org.

Etiology

Cancer cell growth and reproduction involves a two-step process. The first step in cancer growth is called initiation.

Initiation causes an alteration in the genetic structure of the cell (DNA). Cell alteration is associated with exposure to a carcinogen. The cellular change primes the cell to become cancerous.

Promotion is the second type of cancer cell growth. It occurs after repeated exposure to carcinogens causes initiated cells to mutate. During the promotion step, a tumor forms from mutated cell reproduction.

A healthy immune system can often destroy cancer cells before they replicate and become a tumor. It is important to remember that any substance that weakens or alters the immune system puts the individual at risk for cell mutation. Medical researchers support the theory that cancer is a symptom of a weakened immune system.

Risk Factors

Increased risk of cancer is linked to several environmental factors. An evaluation of cancer begins with assessment of well-known risk factors such as specific viruses; exposure to radiation, chemicals, and irritants; genetics; diet; and general immunity. Certain racial and ethnic groups also are at higher risk for some types of cancer (Cultural Consideration Box 10–2).

BOX 10-2 CULTURAL CONSIDERATION

Many racial and ethnic groups in the United States have high rates of cancer. Although risk factors for the development of specific cancers are similar, barriers to prevention and nursing strategies to reduce risk factors vary among ethnicities.

Foreign-born and first-generation white men from Norway, Sweden, and Germany have an increased risk of stomach cancer. This suggests an interrelation among ethnic, geographic, and dietary risk factors as the cause of this high incidence of stomach cancer. Assessing for this data among these populations may assist in the diagnostic process.

Recent Eastern European immigrants may be at risk for thyroid cancer and leukemia because of the current industrial pollution and radiation exposure from the Chernobyl nuclear disaster in 1986. Some contamination occurred in Estonia, Latvia, Lithuania, and Poland. This may constitute a health hazard and may affect both recent immigrants and visitors to these countries. It is essential that health care providers carefully screen individuals for these cancers.

The Appalachian region has the highest rate of cervical cancer in the United States. Thus the nurse needs to encourage Appalachian women to have gynecological examinations on a yearly basis, and be flexible with appointments.

Cancer sites among African-Americans include the prostate, breast, lung, colon, rectum, cervix, pancreas, and esophagus. Because African-Americans are overrepresented in the working class, they experience increased exposure to hazardous occupations. For example, African-American men are at a higher risk for developing cancer related to their high representation in the steel and tire industries and in factories manufacturing chemicals and pesticides. They have the highest overall cancer rate, the highest overall mortality rate, and their 5-year survival rate is 30% lower than that of European-Americans. In general, African-Americans report later for treatment than European-Americans. Colon tumors are deeper within African-Americans, making detection on digital examination more difficult. Poverty, a diet high in fat and low in fiber, and lower levels of thiamine, riboflavin, vitamins A and C, and iron may increase cancer risk among African-Americans. Additionally, cigarette smoking, inner city living with pollution, obesity, and alcohol consumption increase African-Americans' risk for developing cancer.

Lack of medical care access acts as a barrier to prevention among African-Americans. Survival, not prevention, is the priority for some. Additional barriers include a lack of cancer risk teaching and detection in some African-American communities, lack of health insurance, and little stigma attached to alcohol consumption, and smoking. Strong family ties encourage seeking health care from family members before professionals.

Primary strategies for cancer prevention and increasing survival among African-Americans include using African-American professionals as speakers in community activities, using church-based information dissemination, providing forums in African-American communities, and addressing smoking advertisements in African-American communities. Additional strategies include involving granny healers and ministers, changing food preparation practices and amounts rather than changing cultural food habits, involving extended family members in educational campaigns, and using African-American sports leaders in media campaigns.

Hispanic populations in the United States have an increased incidence for some types of cancer. Cervical cancer is increased among Central and South American women. Pancreatic, liver, and gallbladder cancer is increased among Mexican-Americans. Many Mexican-Americans are less aware of the early warning signs of cancer; many are more fearful of getting cancer than the general public; and many work in mining, factories using chemicals, and farming using pesticides.

Barriers to preventive health care among many Hispanics include high poverty rates, low educational rates, a preference for health care providers who understand Spanish, a preference for health care information presented in Spanish, a delay in seeking treatments for symptoms, and using lay healers as a first choice in health care. Additionally, many have a fear of surgical intervention with a body cavity left open to air and have decreased access to health care. For some, an undocumented immigration status creates a fear of reprisal.

Nursing approaches effective among Hispanics include educating lay healers regarding cancer prevention and early warning signs of cancer, using bilingual health care providers, using Hispanic health care providers whenever available, using respected Hispanic community leaders in educational programs, presenting videos in Spanish using Hispanic actors, educating the entire family because of close family networks, and connecting with Hispanic community churches, restaurants, and stores. Additionally, the nurse can use the 1-800-4-CANCER telephone number for Spanish translation and counseling, become involved with Hispanic community movements, and provide information in community and regional Hispanic newspapers and community publications.

Cervical, liver, lung, stomach, multiple myeloma, esophageal, pancreatic, and nasopharyngeal cancer are higher among Chinese-Americans. Chinese-American women have a 20% higher rate of pancreatic cancer. High rates of stomach and liver cancer in Korea predispose recent immigrants to these conditions. Thus the nurse needs to assess and teach newer immigrants regarding these types of cancer.

High rates of stomach, breast, colon, and rectal cancer common among Japanese people may be related to the high sodium content of the Japanese diet, a genetic predisposition, consumption of salted fish and contaminated grain, hepatitis B, smoking, vitamin A deficiency, low vitamin C intake, chronic esophagitis, and pulmonary sequelae of cigarette smoking. Barriers to prevention include the following: prevention models are not native to their culture; there may be a lack of trust in Western medicine; they have decreased access

BOX 10-2 CULTURAL CONSIDERATION—CONT'D

to health care; some are unable to speak the English language; and for some, an undocumented immigration status creates a fear of reprisal.

Nursing approaches to improve cancer risk prevention among Asians and Pacific Islanders include education about prevention versus acute care practice, educating native healers, involvement in community with respected native leaders, videos and literature in native language, and incorporating native healing practices such as traditional Chinese medicine.

Native-American populations have an increased risk for skin, pancreatic, gallbladder, liver, and prostate cancer. Risk factors for the development of cancer include obesity, a diet high in fat, high rates of alcohol consumption, and high rates of smoking. Barriers to preven-

tion include a lack of Native-American health care providers, health care providers' unfamiliarity with Native-American cultures, lack of financial resources, and a lack of integration of Native-American healing practices into prevention practices. Nursing approaches to decrease cancer risk prevention among Native-American populations include the following: incorporate prevention into Native-American healing practices; educate Native-American lay healers regarding cancer prevention practices; work with tribal community leaders; respect modesty, gender roles, and tribal customs; work with the Indian Health Service and Bureau of Indian Affairs; encourage traditional customs of physical fitness and exercise; and encourage dietary portion control and healthy food preparation practices instead of changing cultural food habits.

VIRUSES. Certain viruses, such as the **oncoviruses** (RNA-type viruses), are linked to cancer in humans. Retrovirus is an enzyme produced by RNA tumor viruses and is found in human leukemia cells.

The Epstein-Barr virus (EBV), which causes infectious mononucleosis, is also associated with Burkitt's lymphoma. Herpes simplex virus II has been associated with cervical and penile cancers. Papillomavirus associated with genital warts is considered one cause of cervical cancer in women. Chronic hepatitis B is linked with liver cancer.

RADIATION. There is an increased incidence of cancer in persons exposed to prolonged or large amounts of radiation. Ionizing radiation involving x-rays; alpha, beta, and gamma rays; and ultraviolet rays such as sunlight play a major role in promoting leukemia and skin cancers, primarily melanomas.

Persons exposed to radioactive materials in large doses, such as a radiation leak or an atomic bomb, are at risk for leukemia, breast, bone, lung, and thyroid cancer. Controlled **radiation therapy** is used to treat cancer patients by destroying rapidly dividing cancer cells. Radiation can also damage normal cells. The decision to use radiation is made after careful evaluation of the tumor's location and vulnerability to other treatments.

CHEMICALS. Chemicals are present in air, water, soil, food, drugs, and tobacco smoke. Chemical carcinogens are implicated as triggering mechanisms in malignant tumor development. Length of exposure time and degree of exposure intensity to chemical carcinogens are associated with risk for cancer development.

Smoking accounts for 87 percent of lung cancer worldwide. Chemical agents, as in tobacco, are more toxic when used with alcohol. Alcohol and tobacco are the most frequent causes of cancers of the mouth and throat. Chemicals used in manufacturing, such as vinyl chloride, are associated with liver cancer.

IRRITANTS. Chronic irritation or inflammation caused by irritants such as snuff or pipe smoke often cause cancer in

local areas. Nevi (moles) that are chronically irritated by clothing, especially clothing contaminated by chemical residue, may become malignant. Asbestos found in temperature and sound insulation has been proven to cause a particularly virulent type of lung cancer.

GENETICS. Genetics plays a large part in cancer formation. Certain breast cancers are linked to a specific gene mutation. Skin and colon cancers have a genetic tendency. People with Down syndrome (a chromosomal abnormality) have a higher risk of developing acute leukemia.

DIET. Diet is a large factor in both cause and prevention of malignancies. People who eat high-fat, low-fiber diets are more prone to develop colon cancers. Diets high in fiber reduce the risk of colon cancer. High-fat diets are linked to breast cancer in women and prostate cancer in men. Consumption of large amounts of pickled, smoked, and charbroiled foods has been linked with esophageal and stomach cancers. A diet low in vitamins A, C, and E is associated with cancers of the lungs, esophagus, mouth, larynx, cervix, and breast.

HORMONES. Hormonal agents that disturb the balance of the body may also promote cancer. Long-term use of the female hormone estrogen is associated with cancer of the breast, uterus, ovaries, cervix, and vagina. It has been found that children born of mothers who took diethylstilbestrol (DES) during pregnancy have an increased incidence of reproductive cancers. DES is a synthetic hormone with estrogen-like properties used in the past to prevent miscarriage.

Tumors of the breast and uterus are tested for estrogen or progesterone influence. If a breast tumor is malignant, the tumor is tested and treatment varies depending on whether it is positive for estrogen or progesterone dependence.

IMMUNE FACTORS. A healthy immune system destroys mutant cells quickly on formation. An individual with altered immunity is more susceptible to cancer formation when exposed to small amounts of carcinogens compared with someone with a healthy immune system. Immune system suppression allows malignant cells to develop in large numbers.

---

oncovirus: onco—bulk + virus

Altered immunity is noted in persons with chronic illness and stress. An increased risk of cancer follows a traumatic, stressful period in life, such as the loss of a mate or a job. Failure to decrease stress productively contributes to a higher incidence of chronic illnesses. Thus a cycle of stress, illness, and increased cancer risk develops. Individuals with acquired immunodeficiency syndrome (AIDS) have a compromised immune system and an increased risk for certain cancers. A decline in the immune system is also noted as the body ages. A weaker immune system contributes to chronic illnesses and cancers associated with the elderly population.

Detection and Prevention

Nurses play an important role in preventing and detecting cancer. They help educate patients about risk factors, self-examination, and cancer screening programs (Nutrition Notes Box 10–3). Early diagnosis and treatment provide time to stop the progression of cancer.

BOX 10-3  *Nutrition Notes*

Reducing Cancer Risk

Encourage patient to consume these foods:

- Fruits and vegetables, especially those rich in vitamin C or carotene
- Cruciferous vegetables (cabbage, broccoli, brussels sprouts)
- Whole grains

Encourage patient to limit these foods:

- Excessive meat, especially when smoked, salted, charbroiled, or cooked at high temperature
- Excessive fat (more than 30% of daily calories)
- Excessive calories
- Alcohol

An annual physical examination helps medical personnel detect the seven warning signals of cancer promoted by the American Cancer Society. The warning signals can be remembered with the mnemonic CAUTION:

- **C**hange in bowel or bladder habits
- **A** sore that fails to heal
- **U**nusual bleeding or discharge
- **T**hickening or lump in breast or other tissue
- **I**ndigestion or swallowing difficulties
- **O**bvious change in wart or mole
- **N**agging cough or hoarseness

Monthly breast self-examination is recommended after puberty for both men and women, and monthly self-testicular examination is recommended for men. (See Chapter 39 for information on how to do self-examinations.) Mammography (a specific x-ray of breast tissue used to detect a mass too small for palpation) is recommended once between ages 35 and 39 to provide a baseline and then annually after age 40. Routine pelvic examination and Papanicolaou testing (Pap smear) are recommended annually

after age 18 to screen for cervical cancer. Cytological examination of cervical cells increases the chance of diagnosing cervical cancer in situ.

Promotion of healthy lifestyles, including proper diet and exercise, helps strengthen the immune system and reduce cancer risks. The American Cancer Society promotes smoking cessation campaigns and supports the effort by stating that smoking is the most preventable cause of death from lung cancer. Secondhand smoke contributes to an increased risk of lung cancer in nonsmokers as well.

The American Cancer Society recommends one of the following five options to screen for colorectal cancer, beginning at age 50:

1. An annual stool test for blood
2. A flexible sigmoidoscopy every 5 years
3. A yearly stool test for blood and flexible sigmoidoscopy every 5 years
4. A double-contrast barium enema every 5 years
5. A colonoscopy every 10 years

Option 3 is preferred. Screening should begin earlier and take place more frequently in high-risk people.

Detection of prostate cancer is significant for men beginning around age 50. Digital rectal examination and prostate-specific antigen (PSA) blood testing is recommended annually for men older than 50 who have a life expectancy of at least 10 years and for younger men who are at higher risk. See Gerontological Issues Box 10–4 for specific screening recommendations for older adults.

BOX 10-4 GERONTOLOGICAL ISSUES

Screening Guidelines

Cancer is more common in older adults. Therefore screening for early detection is important. Use the following American Cancer Society guidelines for patients over age 65.

Colorectal cancer:

- Sigmoidoscopy every 3 to 5 years
- Fecal occult blood test yearly
- Digital rectal exam yearly

Breast cancer:

- Breast self-exam monthly
- Breast clinical exam yearly
- Mammogram yearly

Prostate cancer:

- Prostate exam (digital rectal exam) yearly
- Prostatic surface antigen yearly

Currently, much attention is directed toward genetic testing and identification of persons at risk for cancer. Genetic testing technology poses both legal and ethical questions concerning confidentiality and insurance cost issues. The cooperation of family members is important because genetic testing is done after a family member has been diagnosed with cancer. Family members may experience a variety of emotions surrounding the increased risk for themselves and their guilt over the role they may have played in increasing the risk for their children.

Preventive cancer vaccines are being developed for cancers associated with specific viruses. At present, most cancer vaccines are therapeutic rather than prophylactic and are used to stimulate the patient's immune system to destroy cancer cells. Vaccine therapy for malignant melanoma is being tested.

Diagnosis of Cancer

A diagnosis of cancer can be a very frightening experience (see Patient Perspective Box 10–5). Often people try to mask symptoms because they are so frightened of the disease. Exploring patient attitudes and perceptions about the disease helps you construct an effective teaching plan. A careful and thorough assessment of the patient's present and past medical and surgical history and pertinent family history are obtained. A complete physical examination provides both objective and subjective data. The most conclusive information about the health of tissue is acquired by examining cell activity through biopsy.

Biopsy

Accurate identification of a cancer can only be done by **biopsy** (the surgical removal of tissue cells). Microscopic examination of a piece of suspected tissue or aspirated body fluid can confirm the presence of mutant cells. A biopsy is commonly done in a physician's office or outpatient surgery department.

Incisional biopsy is an invasive procedure that involves the surgical removal of a small amount of tissue for inspection. Tissue can also be removed during endoscopic procedures (insertion of a tube to observe the inside of a hollow organ or cavity), such as a lung biopsy done during bronchoscopy. Excisional biopsy is used to remove an entire tissue mass. Needle aspiration biopsy involves insertion of a needle into tissue for fluid or tissue aspiration (Fig. 10–8).

BOX 10-5   *Patient Perspective*

Robyn

I am a 43-year-old woman with three children and in the prime of my life, at least I thought. That was before I was diagnosed with cancer in my left breast. I was breast-feeding at the time I felt the lump and, although I went for a biopsy, I felt sure the lump resulted from a blocked milk duct or fibroid cyst. But, unbelievably, the biopsy came back positive for cancer. My whole life flipped upside down. I was devastated.

I was scheduled for surgery within a week, and my emotions were in complete turmoil. I felt nauseated all the time, vomited almost every morning, and had diarrhea daily. My stomach felt like it had a pot of bees inside. I never cried so much in my life. Thinking about all the tests, the surgery, and the untold ways my life could be affected made me a nervous wreck. Finally, I got down on my knees and turned this whole crisis over to God. I couldn't handle it anymore, so I asked God to give me peace, and I placed all my trust and faith in Him. It worked, and I was finally able to get control and face this thing head on.

As I went for further testing, the nurses and technicians I met were all very helpful and informative. Some actually broke down with me because they had endured this same disease. We would hold each other and then exchange phone numbers just to "talk" if I needed it. It was very encouraging to know these women had made it through, and I could too.

The surgery went smoothly, and I was released the next day. There wasn't much discomfort, and I felt good physically. The nerve was removed and a scar runs from the center of my chest down under my armpit. I was able to return to work in 3 weeks. The doctor gave me a prescription for a prosthesis as soon as the drains were removed and I began healing. We went to a specialty place to be fitted, and, although the prosthesis was nothing like the real thing, I looked normal and it helped to build my confidence.

The impact on my family was one of complete bewilderment because I had no family history of this type of cancer. Everyone tried to help with positive sentiments like we caught it early, breast cancer has a high cure rate, periodic followup can keep you cancer free, and so on. My husband and children supported and comforted me. I tried to focus on them because I want to be there for them when they graduate, get married, etc. My mother and sister helped get me to all my appointments and filled my prescriptions.

Chemo was advised as a follow-up treatment, and I was scheduled for four rounds, one every 3 weeks. This was undoubtedly the worst thing I have ever endured. Not even giving birth can compare to the way chemo makes you feel. I had a very bad experience the first round and was extremely sick and unable to eat for 5 days. I wondered why I didn't just die from the cancer because I felt that this was killing me. Before the second round, I told the doctor how violently ill I'd been and she adjusted the dosages of some of the drugs. Still, I was very groggy; although I didn't vomit, I still wasn't feeling myself. For the third round, they changed a medication and I withstood the side effects a lot better—although I was still nauseated, light-headed, fatigued, and unable to focus, eat, or taste anything. At times it was hard just to put one foot in front of the other. They prepared me for the loss of my hair, but you really don't know how hard that is until it starts coming out in globs. Not just the loss, but then you have such a long time to wait for it to grow back. When the chemo is over, it's hard to look back and feel the way you did then, but when you look in the mirror and your hair is still gone, it's a hard reminder.

All through being diagnosed and dealing with breast cancer I have felt a tremendous outpouring of love and caring, not only from my immediate family but also from my church family. I was never so well taken care of. All the hugs, cards, calls, food, and flowers brought to the house encouraged me tremendously. It makes it a little easier to cope when you know you have so many people who care and are concerned enough to take time out of their daily lives to give you support.

I'm lucky because my sister is an RN and prepared me for many of the side effects and difficulties. She also was there to help ask questions and get information from other survivors that kept me in a positive frame of mind. I know that without her help and God's grace and peace, my recovery would not have been so easy. Looking back I can't really feel all those terrible emotions and symptoms, but I still am afraid of the unknown. It is not easy when it is you and not someone else this happens to.

Now that I'm through the worst part of this, I take positive steps every day to enjoy the little things in life. I feel that the more you keep involved in everyday activities and become educated in the disease and its treatments, the easier it is to deal with. I am taking a drug called Tamoxifen now and will be for 5 years. Two of the side effects are hot flashes and sweats. If this is all I have to deal with, however, praise God. My prognosis is very good, and I am expecting a complete cure because I am a survivor.

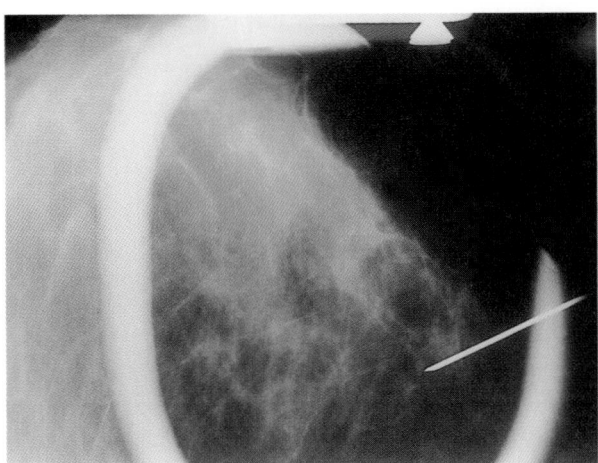

Figure 10-8 Fine needle breast biopsy. (Photo courtesy Dinesh Patel, MD. Medical Oncology, Internal Medicine, Zanesville, Ohio.)

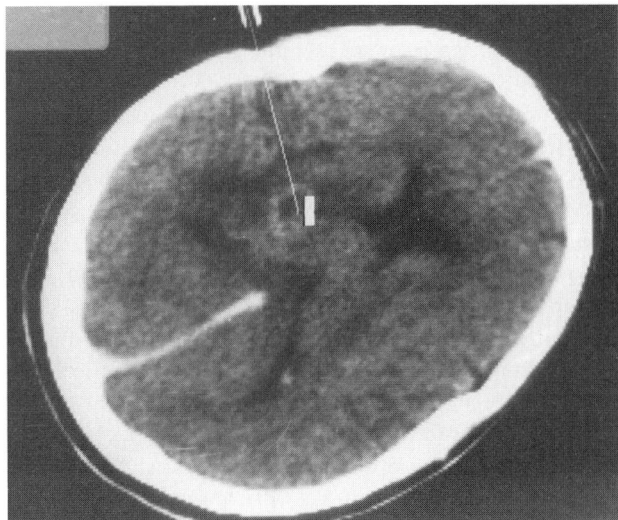

Figure 10-9 Stereotactic biopsy of a brain lesion. (Photo courtesy Dinesh Patel, MD. Medical Oncology, Internal Medicine, Zanesville, Ohio.)

This procedure is considered less invasive than incisional or excisional biopsy. Transcutaneous aspiration involves the insertion of a fine needle into tissue such as breast, prostate, or salivary gland and is used for diagnosing metastatic cancers. Frozen section biopsy provides immediate evaluation of the tissue sample during a surgical procedure. By freezing the tissue sample for microscopic examination, a quick analysis is possible, which helps direct additional surgical intervention. Frozen section biopsy is useful in the diagnosis and surgical intervention of breast cancer.

Stereotactic biopsy is a safe and efficient procedure for evaluating lesions in the brain and breast. The procedure is done by a specially trained radiologist. The biopsy site must be firmly immobilized. The lesion is scanned for location, and a small incision is made for easy insertion of a small fiberoptic instrument (Fig. 10–9). Stereotactic biopsy of the brain involves a local anesthetic because a small hole in the skull is made. Breast stereotactic biopsy uses pressure exerted by a mammogram machine to secure the breast; anesthesia may not be necessary.

Laboratory Tests

Blood, serum, and urine tests are important in establishing baseline values and general health status. Laboratory values are used with other assessment findings. An elevated white blood cell (WBC) count is expected if the patient has evidence of infection; however, an increase in WBCs without infection raises suspicion of leukemia. Fifty percent of patients with liver cancer have increased levels of bilirubin, alkaline phosphatase, and glutamic-oxaloacetic transaminase.

Bone marrow aspiration is done to learn the number, size, and shape of red and white blood cells and platelets. Bone marrow aspiration is a major tool for diagnosis of leukemia. (See Chapter 23 for a description of this test and related nursing care.)

Tumor markers, also called biochemical markers, are proteins, antigens, genes, hormones, and enzymes produced and secreted by tumor cells. Tumor markers help confirm a diagnosis of cancer, detect cancer origin, monitor the effect

of cancer therapy, and determine cancer remission. Tumor markers include the following:

- Prostatic acid phosphatase (PAP)—high levels noted in prostate cancer
- PSA—elevated levels associated with prostate cancer
- Cancer antigen (CA) 15-3—elevated levels noted in breast cancer; useful in monitoring patient response to therapy for metastatic breast cancer
- CA 125—increased levels in ovarian, cervical, liver, and pancreas cancers
- CA 19-9—used to diagnose and evaluate pancreatic and hepatobiliary cancer; levels elevated in cancer
- Carcinoembryonic antigen (CEA)—increased levels suggest tumor activity

Cytological Study

Cytology is the study of the formation, structure, and function of cells. Cytological diagnosis of cancer is obtained primarily through Pap smears of cells shed from a mucous membrane (e.g., cervical or oral smear). Test results are based on the degree of cell abnormality. Normal results reflect no cellular changes. Slight cellular changes are considered normal, with a possible link to abnormal cells seen in infection. Severe cellular changes reflect a higher probability of precancerous or cancerous cellular activity. Infection causes cellular changes and contributes to an increase in abnormal cells detected.

Radiological Procedures

X-ray examination is a valuable diagnostic tool in detecting cancer of the bones and hollow organs. Routine chest x-ray examination is one diagnostic test used in detecting lung cancer. Mammography is a reliable and noninvasive low-radiation x-ray procedure for detecting breast masses. Breast tissue is compressed to allow better visualization of the soft tissue. You must alert patients that soft tissue compression causes a degree of discomfort, but the compression is necessary to obtain an accurate picture. The discomfort is brief (Figs. 10–10 and 10–11).

Contrast media x-ray studies are used to detect abnormalities of bone and the gastrointestinal and urinary systems. Contrast media can be given by various methods. Barium is given orally for visualization of the esophagus and stomach or rectally for visualization of the colon (e.g., a barium enema). Intravenous injection of contrast media is used for lung and brain scans.

Computed tomography (CT) provides a three-dimensional, cross-sectional, computerized picture of the body. CT scans are important in the diagnosis and staging of malignancies and can detect minor variations in tissue thickness. The use of a contrast medium enhances the accuracy of an abdominal CT scan. CT scans are also used to improve the accuracy of inserting a fine needle for biopsy.

Nuclear Imaging Procedures

Nuclear medicine imaging involves camera imaging of organs or tissues containing radioactive media. Radioactive compounds are given intravenously or by ingestion. These studies are highly sensitive and can detect sites of abnormal cell growth months before changes are seen on an x-ray.

Positron emission tomography (PET) scanning provides information about cellular biochemical and metabolic activity. Patients are given biochemical compounds, and im-

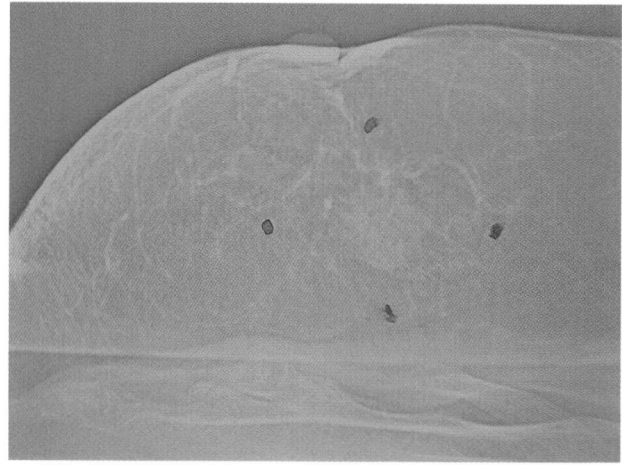

Figure 10-11 Mammogram. (Photo courtesy Dinesh Patel, MD. Medical Oncology, Internal Medicine, Zanesville, Ohio.)

ages are made of the tissue through gamma-camera tomography. PET scans have been useful in brain imaging.

Ultrasound Procedures

Ultrasonography uses high-frequency sound waves to provide images of deep soft-tissue structures in the body. The procedure is noninvasive and does not use x-rays. Echoes from high-frequency sound waves outline tissue density and masses. This technology helps detect tumors of the pelvis and breast. Ultrasound may also be used to distinguish between benign and malignant breast tumors.

Magnetic Resonance Imaging

Magnetic resonance imaging (MRI) creates sectional images of the body. MRI can be done with or without contrast dye and does not use radiation. The patient is placed in a cylinder-shaped magnetic field. The magnetic field aligns the nuclei of body cells in one direction. The magnetized cells are then excited by radiofrequency pulses. Images are made as cell nuclei change their alignment. MRI is valuable in the detection, localization, and staging of malignant tumors in the central nervous system, spine, head, and musculoskeletal system. MRI cannot be used in patients with pacemakers, implanted pumps, surgical clips, metal knees or hips, or tattooed eyeliner.

Endoscopic Procedures

An endoscopic examination allows the direct visualization of a body cavity or opening. The procedure involves the insertion of a flexible endoscope containing fiberoptic glass bundles that transmit light and can produce an image. Endoscopy enables the surgeon to biopsy abnormal tissue and is used to detect lesions of the throat, esophagus, stomach, colon, and lungs.

Oral endoscopic procedures require patient preparation to reduce the risk of aspirating stomach secretions. The patient is given nothing to eat or drink before and immediately after the examination. A local anesthetic is used during the examination to anesthetize the throat. Following the procedure, oral food and fluids are withheld until the

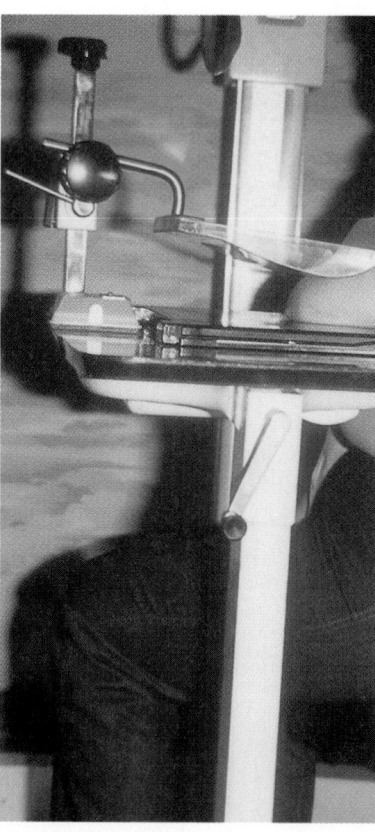

Figure 10-10 Mammogram procedure. (Photo courtesy Dinesh Patel, MD. Medical Oncology, Internal Medicine, Zanesville, Ohio.)

gag reflex returns, to prevent aspiration. The gag reflex is assessed by placing a tongue blade on the back of the tongue or by using a cotton-tipped swab at the back of the throat to stimulate the gag reflex after the procedure.

Staging and Grading

Tumor staging is used to determine the stage of solid-tumor masses, providing valuable information about the potential success of treatment plans. Tumor staging is important in the development of an international system that can compare statistics among cancer centers. The most common system used for staging tumors is called tumor, node, metastasis (TNM).

This staging system classifies solid tumors by size and tissue involvement. TNM stages are T0 (no tumor), Tis (tumor in situ), and T1 through T4 (progressive increase in tumor size or involvement). Extent of lymph node involvement ranges from N0, no nodes, to N4, a large amount of lymph node involvement. Metastasis is described as M0, no metastasis, to M1, metastasis to some area (Table 10–5).

There is also a rating or grading system to define the cell types of tumors. Tumors are classified according to the percentage of cells that are differentiated (mature). If the tissue of a neoplastic tumor closely resembles normal tissue, it is called well differentiated. A poorly differentiated tumor is a malignant neoplasm that contains some normal cells, but most of the cells are abnormal. The better defined or differentiated the tumor, the easier it is to treat.

Treatment for Cancer

There are three main types of treatment for cancer: surgery, radiation therapy, and **chemotherapy.** To find out more about cancer treatment options, go to the American Cancer Society Web site at http://www.cancer.org.

---

chemotherapy: chemo—chemistry + therapy—treatment

TABLE 10-5	TUMOR, NODE, METASTASIS CLASSIFICATION, STAGING, AND TISSUE INVOLVEMENT	
Classification	**Staging**	**Tissue Involvement**
Primary tumor (T)		
T_{is}	Stage I	Tumor in situ, indicates no invasion of other tissues
T_1, T_2, T_3, T_4	Stage II	Ranges indicate progressive increase in tumor size with local metastasis
Regional lymph node involvement (N)		
N_0	No nodes	
N_1, N_2, N_3	Stage III	Metastasis to regional lymph nodes
Metastasis (M)		
M_0		No metastasis
M_1	Stage IV	Distant metastasis

Surgery

Surgery can be used to cure cancer when it is possible to remove the entire tumor. Skin cancers and well-defined tumors without metastasis can be removed without any additional intervention.

Prophylactic surgery is used to remove moles or lesions that have the potential to become malignant. Colon polyps are often removed to prevent malignancies from developing, especially if the polyps are considered premalignant. An extreme example of prophylactic surgery is a woman who elects to have a mastectomy (surgical removal of the breast) because of a high incidence of breast cancer in her family.

Surgery may also be done for **palliation** (symptom control). Surgical removal of tissue to reduce the size of the tumor mass is helpful, especially if the tumor is compressing nerves or blocking the passage of body fluids. The goals of palliative surgery are less discomfort and an improved quality of life.

Reconstructive surgery can be done for cosmetic enhancement or for return of function of a body part. Facial reconstruction is important for a patient's self-image after removal of head or neck tumors. Women can elect to have breast implants after mastectomy.

Surgical intervention for cancer treatment is typically not an emergency intervention, which allows patients and medical personnel time for planning. Autologous blood donation (blood donated by the patient before surgery) has become popular and reduces the risk of exposure to bloodborne infections. The American Red Cross and many hospitals have programs specifically designed for autologous blood donation and accept donations from 30 days to 72 hours before surgery.

You can play a major role in reducing the patient's fears about postoperative pain. Patient-controlled analgesia (PCA) provides patients with some control over their pain. Therapies such as deep relaxation, imagery, and hypnosis can be used with traditional pain control measures.

It is important to encourage patients to express and discuss their fears. Patients with a limited understanding of cancer may fear that tissues will not heal postoperatively. Provide information about wound care, including dressing changes and drainage tubes, to increase the patient's knowledge base and sense of control. Visual aids concerning tumor site and surgical procedures are valuable teaching tools.

Patients who are undernourished are poor surgical candidates and require intervention such as enteral or parenteral nutrition before and after surgery. Patients with cancer are also at increased risk for postoperative deep vein thrombosis (DVT). Preoperative teaching includes the importance of leg movement, early ambulation, wearing antiembolism stockings, and recognizing symptoms of DVT, such as a cramping sensation in the calf muscle and pain when the foot is dorsiflexed.

Radiation

Radiation is used commonly in the treatment of cancer for control or palliation, and it can be curative if the disease is localized. The decision to use radiation is commonly based on cancer site and size. Radiation destroys cancer cells by affecting cell structure and the cell environment. It is used in fractionated (divided) doses to prevent destructive side effects; however, side effects can occur in the area being treated because of damage to normal cells.

The size of a large tumor can be decreased with radiation before surgery, making surgical intervention more effective and less dangerous. Palliative radiation is used to reduce the size of a large cancerous lesion and consequently reduce pressure and pain. Radioisotopes inserted into cancerous tissue during surgery help destroy the cancerous cells without removing the organ.

NURSING CARE OF THE PATIENT RECEIVING RADIATION TREATMENT. Symptoms of tissue reaction to radiation treatment can be expected about 10 to 14 days after the start of the treatment program and continue up to 2 weeks after treatment is completed. Typical reactions and appropriate nursing interventions include the following:

- Fatigue. Encourage the patient to nap frequently and prioritize activities. Reassure the patient that the feeling will go away when the treatments are completed.
- Nausea, vomiting, and **anorexia.** Encourage the patient to take prescribed medication for nausea and vomiting. Anorexia can be eased by giving small amounts of high-carbohydrate, high-protein foods and avoiding foods high in fiber.
- **Mucositis** (inflammation of the mucous membranes, especially of the mouth and throat). Encourage the patient to avoid irritants such as smoking, alcohol, acidic food or drinks, extremely hot or cold foods and drinks, and commercial mouthwash. Advise the patient to perform mouth care before meals and every 3 to 4 hours. A neutral mouthwash is appropriate and can be made by using 1 ounce of diphenhydramine hydrochloride (Benadryl) elixir diluted in 1 quart of water or normal saline solution. Agents that coat the mouth, such as Maalox, are sometimes used. Lidocaine hydrochloride 2 percent viscous has an anesthetic effect on the mouth and throat.
- **Xerostomia** (dry mouth). Encourage frequent mouth care. Saliva substitute is available over the counter and is helpful, especially at night when patients complain of a choking sensation from extreme dryness.
- Skin reactions. These vary from mild redness to raw moist lesions similar to a second-degree burn. Skin surfaces that are especially warm and moist, such as the groin, perineum, and axillae, have poor tolerance to radiation. Prophylactic skin care includes keeping skin dry; keeping it free from irritants, such as powder, lotions, deodorants, and restrictive clothing; and protecting it against exposure to direct sunlight.
- Bone marrow depression. This reaction occurs with both radiation and chemotherapy. Weekly blood cell counts are done to detect low levels of WBCs, red blood cells, and platelets. Transfusions of whole blood, platelets, or other blood components may be necessary.

SAFETY CONSIDERATIONS. Radiation may be administered externally or internally. External radiation is given by a trained medical specialist in a designated area in the hospital or clinic. Internal radiation is administered to patients admitted to a health care facility.

Safety guidelines must be followed when caring for a patient with radioactive materials implanted into tissue or body cavities or administered orally or intravenously. Nursing responsibilities include knowledge about the following:

1. Radiation source being used
2. Method of administration
3. Start of treatment
4. Length of treatment
5. Prescribed nursing precautions

Personnel involved with radiation therapy must recognize the three primary factors in radiation protection: time, distance, and shielding. These three factors depend on the type of radiation used. *Time* involves the time spent administering care, *distance* involves the amount of space between the radioisotope and the nurse, and *shielding* involves the use of a barrier such as a lead apron.

You must work efficiently when caring for patients who are receiving radioisotopes that are releasing gamma rays. Your exposure to radiation is proportionate to the time spent and the distance from the radiation source. For example, you will receive less exposure standing at the foot of the bed of a patient with radioisotopes inserted into the head than if you stand at the head of the bed (Fig. 10–12). Time and distance are used to protect the nurse, visitors, and other personnel.

It is important to teach the patient and family members the reason nursing care focuses on providing only essential care. Speedy nursing encounters and visitor restrictions are better accepted and less likely to promote feelings of isolation when patients understand the reasons behind them.

Drainage from the site of a radioactive colloid injection is considered radioactive, and the physician must be informed immediately. Dressings contaminated with radioactive seepage must be removed with long-handled forceps. Radioactive materials must never be touched with unprotected hands; shielding is required to prevent exposure to radiation. Contamination from radioisotope applicators or interstitial implants cannot occur when the capsule is intact; contamination occurs when the capsule is broken.

---

anorexia: an—not + orexis—appetite
mucositis: muco—mucous (membrane) + itis—inflammation
xerostomia: xero—dry + stoma—mouth

L E A R N I N G TIPS

Remember to use the principles of time, distance, and shielding to protect yourself from radiation exposure.

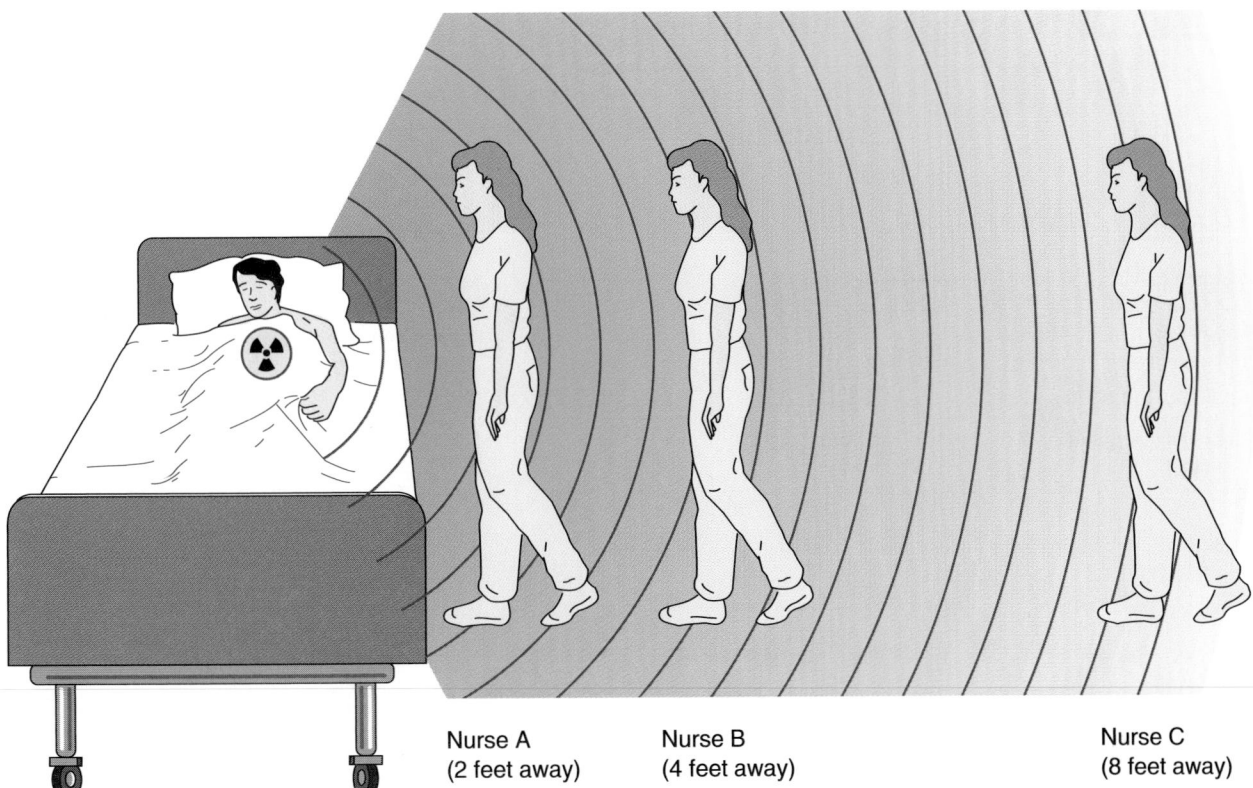

Figure 10-12 Radiation distancing. Nurse B receives less radiation than Nurse A, and Nurse C receives less radiation than Nurse B.

Chemotherapy

Chemotherapy is chemical therapy that uses **cytotoxic** drugs to treat cancer. Cytotoxic drugs can be used for cure, control, or palliation of cancerous tumors and are described according to how they affect cell activity. For example, alkylating agents bind with DNA to stop the production of RNA; antimetabolites substitute for nutrients or enzymes in the cell life cycle; mitotic inhibitors interfere with cell division; antibiotics inhibit DNA and RNA synthesis; and hormonal agents alter the hormonal structure of the body. Chemotherapy is usually more effective when multiple drugs are given in multiple doses. Examples of specific drugs and their adverse effects are provided in Table 10–6.

The effects of chemotherapy are systemic unless used topically for skin lesions. Chemotherapy is used preoperatively to shrink tumors and postoperatively to treat residual tumors. Factors influencing the effectiveness of chemotherapy are tumor type, available chemotherapeutic drugs, and genetics.

ROUTES OF ADMINISTRATION. Drugs may be given via oral, intramuscular, intravenous, or topical routes. The dosage of medication is regulated by the size of the individual and the toxicities of the drug. The administration of intravenous chemotherapeutic drugs requires specialized training and knowledge of antineoplastic drugs.

Vesicant drugs are given only by the intravenous route. These drugs cause blistering of tissue that eventually leads to necrosis if they infiltrate, or leak out of the blood vessel, into soft tissue (Fig. 10–13). Skin grafts may be necessary if tissue damage is extensive.

Central Lines. Central lines are intravenous catheters that terminate in the superior vena cava near the right atrium of the heart. This is a large vessel that allows for dilution of vesicant drugs and reduces the risk of infiltration. Central lines may be external, with the distal end of the catheter exiting the skin, or internal, with the distal catheter ending in an implanted port. See Chapter 6 for additional information on central lines.

SIDE EFFECTS OF CHEMOTHERAPY. Toxicities in patients receiving chemotherapy vary according to the medications given; however, some general side effects are commonly associated with chemotherapy drugs. Fast-growing epithelial cells, such as those of the hair, blood, skin, and gastrointestinal tract, are generally affected by both chemotherapy and radiation.

Bone Marrow. Chemotherapy is toxic to the bone marrow, where the blood cells are produced. Patients may develop low white blood cell counts **(leukopenia),** increasing their susceptibility to infection and sepsis. A reduction in

---

cytotoxic: cyt—cell + toxic—poison

vesicant: vesicate—to blister
leukopenia: leuko—white cells + penia—lack

TABLE 10-6   CHEMOTHERAPY MEDICATIONS

Medication	Classification	Method of Administration	Common Adverse Effects
Bleomycin	Antibiotic	IM or IV	Fever and chills, cough, shortness of breath; in severe cases, pulmonary fibrosis.
Busulfan	Alkylating	PO	Unusual bleeding or bruising, diarrhea, fatigue, nausea, vomiting
Carboplatin (Paraplatin)	Alkylating	IV	Abdominal pain, nausea and vomiting, electrolyte imbalances
Carmustine (BCNU)	Alkylating	IV	Fever and chills, nausea and vomiting
Cisplatin	Alkylating	IV	Difficulty hearing, fever and chills, ringing in ears, nausea and vomiting.
Cyclophosphamide (Cytoxan)	Alkylating	IV or PO	Nausea and vomiting, blood in the urine, loss of hair
Cytarabine (Cytosar)	Antimetabolic	IV	Fever and chills, unusual bleeding or bruising, sore throat, tiredness, nausea and vomiting
Dactinomycin (Cosmegen)	Antibiotic (vesicant)	IV	Loss of hair, nausea and vomiting, tiredness
Daunorubicin	Antibiotic (vesicant)	IV	Red urine, nausea and vomiting, loss of hair
Docetaxel (Taxotere)	Toxoid	IV	Fatigue, edema, nausea and vomiting, stomatitis, anemia, thrombocytopenia, myalgia, hair loss
Doxorubicin (Adriamycin)	Antibiotic (vesicant)	IV	Red urine, nausea and vomiting, loss of hair, cardiac damage
Estramustine (Emcyt)	Alkylating	PO	Nausea, vomiting, anorexia, estrogenic, edema, breast tenderness or enlargement
Etoposide (VP-16)	Alkylating (mild vesicant)	IV	Nausea and vomiting, loss of hair, numbness and tingling in fingers and toes
Floxuridine (FUDR)	Antimetabolic	IV	Diarrhea, nausea and vomiting, loss of appetite, sores in mouth
Fludarabine (Fludara)	Antimetabolic	IV	Bone marrow depression, nausea, vomiting, diarrhea, neurotoxicity
Fluorouracil (5-FU)	Antimetabolic	IV	Diarrhea, loss of appetite, loss of hair, nausea and vomiting
Gemcitabine (Gemzar)	Nucleoside analogue	IV	Dyspnea, edema, nausea, vomiting, diarrhea, stomatitis, hematuria, hair loss, bone marrow suppression
Hydroxyurea (Hydrea)	Antimetabolic	PO	Fever and chills, sore throat, drowsiness, diarrhea, nausea and vomiting
Ifosfamide (Ifex)	Alkylating	IV	CNS toxicity, nausea, vomiting, cystitis, hair loss
Irinotecan (Camptosar)	Enzyme inhibitor	IV	Dizziness, headache, insomnia, dyspnea, edema, nausea, vomiting, constipation, stomatitis, hair loss, bone marrow suppression, weight loss
Melphalan (Alkeran)	Alkylating	PO	Nausea and vomiting
Methotrexate	Antimetabolite	PO, IV, intrathecal	Blood in urine, sun sensitivity, diarrhea, sores in mouth, jaundice, nausea and vomiting
Mitomycin-C (Mutamycin)	Antibiotic (vesicant)	IV	Blood in urine, nausea and vomiting, loss of appetite
Mitotane (Lysodren)	Miscellaneous	PO	Dizziness, drowsiness, GI distress
Mitoxantrone (Novantrone)	Antibiotic	IV	Headache, dyspnea, diarrhea, nausea, vomiting, stomatitis, hair loss, fever, bone marrow suppression
Paclitaxel (Taxol)	Miscellaneous	IV	Nausea, vomiting, muscle aches, cardiac toxicities
Procarbazine	Alkylating	PO	MAO inhibitor, drowsiness, nausea and vomiting
Rituximab (Rituxan)	Miscellaneous	IV	Nausea, vomiting, bone marrow suppression, bronchospasm
Streptozocin (Zanosar)	Antibiotic	IV	Anxiety, chills, nausea and vomiting
Tamoxifen (Nolvadex)	Hormone	PO	Hot flashes, weight gain, nausea, bone pain
Topotecan (Hycamtin)	Enzyme inhibitor	IV	Headache, dyspnea, nausea, vomiting, diarrhea, hair loss, bone marrow suppression
Vinblastine (Velban)	Vinca alkaloid	IV	Muscle pain, nausea and vomiting, loss of hair
Vincristine (Oncovin)	Vinca alkaloid	IV	Constipation, difficulty walking, tingling in fingers and toes
Vinorelbine (Navelbine)	Vinca alkaloid (vesicant)	IV	Fatigue, constipation, nausea, hair loss, bone marrow suppression, neurotoxicity

CNS = Central nervous system; GI = gastrointestinal; IM = intramuscular; IV = intravenous; MAO = monoamine oxidase inhibitor.

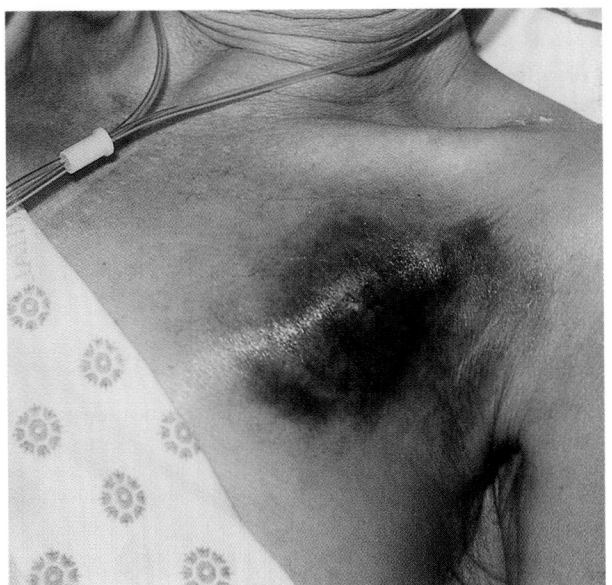

Figure 10-13 Necrosis of skin tissue resulting from administration of a vesicant chemotherapy drug. (Photo courtesy Dinesh Patel, MD. Medical Oncology, Internal Medicine, Zanesville, Ohio.)

platelets **(thrombocytopenia)** increases the risk of bruising and bleeding and can require platelet transfusions. Increased risk of **anemia** occurs with the reduction of red blood cells and may require blood transfusions.

Gastrointestinal Tract. The gastrointestinal tract is susceptible to the toxicity of chemotherapy drugs. Patients often become nauseated and vomit or experience diarrhea. **Stomatitis** is a common complaint and is discussed under side effects of radiation. These side effects can be controlled with medication.

Hair. Alopecia (hair loss) is common with many chemotherapy drugs. This is a temporary condition, and growth of the new hair usually starts when the chemotherapy medication is stopped. Alopecia involves the entire body and includes eyebrows, eyelashes, and axillary and pubic hair. Hair regrowth may be of different color or texture than the original hair. It is not uncommon for individuals who originally had straight hair to regrow curly hair.

Reproductive System. The effects of chemotherapy or radiation can cause temporary or permanent alterations of the reproductive system. Occasionally, patients are rendered sterile because of the treatment. Issues concerning fertility should be discussed with the patient before treatment. Measures such as freezing ova and the use of sperm banks provide options for the patient and his or her partner.

Neurological System. Drugs may affect the neurological system. An adverse reaction to vincristine is neurotoxicity,

---

thrombocytopenia: thrombo—clot + cyte—cell + penia—lack
anemia: an—not + emia—blood
stomatitis: stoma—mouth + itis—inflammation

which may result in tingling or numbness in the extremities and in severe cases may cause footdrop from muscle weakness.

Less common complications include renal toxicities, such as pain and burning on urination, and hematuria. Doxorubicin (Adriamycin) has been associated with permanent heart damage, and bleomycin can cause pulmonary fibrosis.

Severe toxic side effects can be controlled by carefully limiting the amount of medication given and constantly monitoring the patient for complications.

New Treatments Being Researched

Laetrile is a medication made from the pits of peaches or apricots. The main ingredient of this medication is cyanide. The drug is given with very strict dietary guidelines, and conditions are very difficult for the patient to manage.

Hyperthermia has been used with radiation and chemotherapy. It has been beneficial in some types of cancer but is seldom used except in investigational studies.

Biologic response modifiers (such as interferons) are drugs used to stimulate the immune system. These drugs are used commonly for specific types of cancer and have produced some beneficial results. They are also being used in many investigational studies.

▶ NURSING PROCESS FOR THE PATIENT WITH CANCER
Assessment

Patients are assessed for many different problems associated with cancer and its treatment (Nursing Care Plan Box 10–6). Assessment of patient knowledge concerning the disease, treatment, and expected outcomes helps you provide adequate information about patient rights and responsibilities. See Box 10–7 for the cancer survivor's bill of rights.

Assess home safety management and coping skills of the patient and family. Determine patient and caregiver strengths and weaknesses related to therapeutic care.

Assess for side effects of chemotherapy or radiation, as described earlier. Infection and septicemia can occur quickly, especially when the patient's white blood cell count is low. Monitor body temperature frequently to detect infection. Examine the patient daily for signs of inflammation or purulent drainage at potential infection sites, such as old aspirate sites, old and new venipuncture sites, mouth and rectal mucosa, perineal area, axilla, incisions, earlobes if ears are pierced, under breasts, and between toes. Monitor the patient for signs of respiratory infection, such as a sore throat, cough, shortness of breath, and purulent sputum. Signs of urinary infection include burning, pain, urgency, frequency, and presence of blood or pus in the urine.

Bleeding can occur when the platelet count is low. Report bleeding gums, easy bruising, tarry stools, and bloody urine to the physician immediately.

Monitor the patient's weight, and note complaints of nausea, vomiting, and diarrhea related to either the disease or treatment. Patients with cancer are at risk for dehydration and wasting syndrome. Pain may also keep the patient from eating.

BOX 10-6   NURSING CARE PLAN FOR THE PATIENT WITH CANCER

> **Risk for ineffective coping related to the diagnosis and treatment of cancer as evidenced by behaviors such as denial, isolation, anxiety, and depression**

OUTCOMES

Patient will cope effectively as evidenced by identifying stressors related to illness and treatment; communicating needs, concerns, and fears; and use of appropriate resources to support coping.

EVALUATION OF OUTCOMES

Is patient able to identify stressors and communicate concerns? Does patient have and appropriately use support systems?

INTERVENTIONS	RATIONALE	EVALUATION
Assess effective coping mechanisms used in the past and currently available to the patient.	Coping mechanisms that worked in the past may be helpful again, and the nurse can support appropriate choices.	Is the patient able to identify and draw on past coping mechanisms?
Use active listening skills to encourage the patient to express feelings and fears.	The patient must identify fears to be able to cope effectively with them.	Does the patient identify fears and concerns?
Assess the meaning of quality of life to the patient.	Once identified, the nurse can assist the patient to achieve quality-of-life goals.	Is the patient able to identify the meaning of quality of life? Are there ways the nurse can assist the patient to reach quality-of-life goals?
Assess for suicide risks.	A patient who feels hopeless may be at risk for suicide.	Is the patient at risk? Are suicide precautions necessary?
Explore outlets that promote feelings of personal achievement.	Personal achievement promotes self-esteem.	Does the patient have creative outlets that promote feelings of achievement? Can the nurse assist in implementing these activities?
Promote the use of humor.	Humor can be distracting and therapeutic.	Does the patient use humor? Does it provide temporary distraction from concerns?

> **Pain related to tissue injury from tumor invasion or surgical intervention**

OUTCOME

Patient will be pain free as evidenced by patient statement of comfort on pain scale.

EVALUATION OF OUTCOME

Does patient state pain is controlled?

INTERVENTIONS	RATIONALE	EVALUATION
Ask patient to rate pain on a scale from 0 to 10 (0 = absence of pain; 10 = worst pain).	A complete pain assessment should be done before and after any interventions for pain.	Does patient use pain assessment scale effectively? Is patient in pain?
Educate patient on use of patient-controlled analgesic (PCA).	PCA allows the patient to be in control of own pain relief.	Does PCA keep patient pain free and able to participate in desired activities?
Monitor pain relief every 2 to 4 hours.	Alternative medications may be necessary for breakthrough pain.	Is PCA or epidural analgesic effective?
Explain and encourage use of relaxation techniques and other complementary techniques.	Relaxation techniques can reduce pain intensity by reducing skeletal muscle tension.	Does patient use relaxation techniques effectively?

The patient's pain may be slight to severe, with both physical and emotional components. Pain assessment includes location, onset, intensity, pattern, duration, mental state, and known relief measures. This may be accomplished with the WHAT'S UP? format presented in Table 1–2. Assessment of intensity is best done by using a pain scale, such as a 0 to 10 scale with 0 indicating no pain and 10 indicating the worst pain imaginable. The patient's perceptions about pain management and the impact of pain on performance of activities of daily living are also assessed.

LEARNING TIPS

When assessing patients with possible side effects of chemotherapy and radiation, use the pneumonic BITES:

  B—Bleeding suggests low platelet count.

  I—Infection suggests low WBC count and a risk for septicemia.

  T—Tiredness suggests anemia.

  E—Emesis places the patient at risk for altered nutrition and fluid and electrolyte imbalance.

  S—Skin changes may be evidence of radiation reaction or skin breakdown.

BOX 10-7 *Cancer Survivor's Bill of Rights*

The American Cancer Society promotes the following Survivor's Bill of Rights to promote cancer care.

1. Survivors have the right to assurance of lifelong medical care, as needed. the physicians and other professionals involved in their care should continue their constant efforts to be:
   - sensitive to the cancer survivor's lifestyle choices and need for self-esteem and dignity;
   - careful, no matter how long their patients have survived, to take symptoms seriously, and not dismiss aches and pains, for fear of recurrence is a normal part of survivorship;
   - informative and open, providing survivors with as much or as little candid medical information as they wish, and encouraging their informed participation in their own care;
   - knowledgeable about counseling resources, and willing to refer survivors and their families as appropriate for emotional support and therapy, which will improve the quality of individual lives.
2. Survivors will have the right to the pursuit of happiness. This means they have the right:
   - to talk with their families and friends about their cancer experience if they wish, but to refuse to discuss if that is their choice and not to be expected to be more upbeat or less blue than anyone else;
   - to be free of the stigma of cancer as a "dread disease" in all social relations;
   - to be free of blame for having gotten the disease and of guilt for having survived it.
3. In the workplace, survivors have the right to equal job opportunities. They have the right:
   - to apply for jobs worthy of their skills, and for which they are trained and experienced;
   - to be hired, promoted, and accepted on return to work, according to their individual abilities and qualifications, and not according to "cancer" or "disability" stereotypes;
   - to privacy about their medical histories.
4. Every effort should be made to assure all survivors adequate health insurance, whether public or private. This includes:
   - survivors have the right to be included in group coverage at the place of employment;
   - physicians, counselors, and other professionals must keep themselves and survivors informed and up to date on available group or individual health policy options;
   - social policy makers, both in government and in the private sector, must seek to broaden insurance programs like Medicare to include diagnostic procedures and treatment to help prevent recurrence and lessen survivor anxiety.

Source: From Cancer Survivor's Bill of Rights, American Cancer Society.

Nursing Diagnosis

Individual diagnoses must be determined based on specific patient assessment. Possible nursing diagnoses include the following:

- Chronic pain related to disease process and cancer treatment

- Risk for infection related to diminished immunity and bone marrow suppression as a result of chemotherapy or radiation
- Risk for injury related to bleeding tendencies associated with chemotherapy and radiation
- Nutrition, imbalanced: less than body requirements related to anorexia, nausea, or vomiting associated with disease, pain, and treatment
- Deficient self-care related to weakness and fatigue
- Anticipatory grieving related to potential disease outcome
- Caregiver role strain related to patient care and anticipated outcome
- Social isolation related to changing relationships
- Ineffective sexuality pattern related to change in body functions
- Disturbed body image related to surgical procedures such as an ostomy or loss of hair associated with chemotherapy.

Planning and Implementation

Short- and long-term goals must be set and mutually agreed upon by the nurse and the patient. You must advocate for active patient participation in the manner that the patient requires. Short-term goals are centered around control of pain and side effects of therapy. Long-term goals are determined by the patient's lifestyle, quality of life, and outcome of therapy.

Pain

Pain intervention involves a multidimensional nursing approach. Cancer pain is associated with (1) surgical interventions, postoperative recovery, and painful medical or nursing procedures; (2) pressure from a large tumor mass on nerves and blood vessels; (3) psychological depression and anxiety about mortality; and (4) cultural views and perceptions concerning the cancer process.

Pain affects quality of life for the patient with cancer. Chronic pain places the patient at risk for depression, hopelessness, delirium, loss of control, exhaustion, and suicide. Patients need reassurance that measures will be taken to control their pain and that they will not become addicted to pain medication (Ethical Consideration Box 10–8).

Obstacles to effective pain management include a lack of understanding about pain, inadequate pain assessment, ineffective treatment with analgesics, and fears of addiction, sedation, and respiratory depression.

Cancer pain is treated first with analgesics. The goal is effective prevention of pain, rather than treatment of pain once it occurs. Use of medications for pain is discussed in Chapter 9.

Procedures such as administration of anesthetics or nerve blocks are used for intractable pain. Nursing responsibilities associated with these procedures include teaching the patient why and how the procedure will be done, potential complications, and expected benefits from the procedure.

Noninvasive therapies may be used in addition to pain medication to increase the patient's control over the pain.

Pain Medications for the Patient with Cancer

Ms. Parker, LPN, returned to work on the oncology unit after a 2-week vacation. During report, she could hear moans of pain emanating from the room of Mr. Stipe, a 32-year-old male suffering from metastatic bone cancer. She had liked Mr. Stipe when he was admitted for diagnosis and initial treatment several months ago, and now he was back to die. The metastatic growths in his spine were causing intolerable pain, while the metastatic growths in his liver were threatening death.

The goal the oncology team had established for Mr. Stipe was to keep him as pain free as possible, and correspondingly large doses of morphine sulfate had been ordered both by continuous drip and PCA pump boosters. In reviewing Mr. Stipe's chart for the last shift, Ms. Parker noted that he had received 800 mg by drip with 20 mg boosters every hour. She became very concerned that this much morphine would cause respiratory depression and arrest.

The head nurse verified the dose with Ms. Parker and also explained that Mr. Stipe had a very high tolerance to the medication due to his severe pain and long-term use of the medication for pain control. When Mr. Stipe screamed in pain again, the head nurse instructed Ms. Parker to give another 20 mg booster. "Our goal is to keep him comfortable," she said, "and the only way to do that is to give him his medication." Ms. Parker agreed that the pain must be relieved, but she felt very uncomfortable giving more medication on top of the medication he had already been given. What if he went into respiratory arrest after the next dose of medication? Would Ms. Parker be responsible for his death? What effect does the patient's dying status have on the ethical obligation to provide pain medication? What should a nurse do when the treatment ordered for the patient has the potential to compromise personal values and conscience?

These may include (1) cutaneous stimulation, which eliminates or decreases pain by using massage and vibration, heat and cold, and menthol application; (2) distraction, such as music therapy or coping statements; (3) relaxation through deep breathing exercises and humor therapy; (4) imagery; and (5) comfort measures, such as good body positioning, a comfortable mattress and pillows, and so on.

Risk for Infection

Leukopenia occurs when the WBC count is decreased to 2000 (normal range is 5000 to 10,000/mm³), increasing the patient's risk for infection. Patients with leukopenia must be protected against sources of infection, such as people with transmissible illnesses; bird, cat, or dog feces; stagnant water in flower vases, denture cups, irrigating containers, respiratory equipment, or soap dishes; and fruit peelings.

Report signs and symptoms of infection to a physician immediately. Antibiotics and analgesics are administered as ordered. The patient is advised to avoid intercourse while the WBC count is low to prevent secondary infection.

Risk for Injury Related to Bleeding

Thrombocytopenia increases the risk for bleeding or hemorrhage. The normal platelet level is 150,000 to 300,000/mm³. Potential for bleeding exists when the platelet count is 50,000; risk for spontaneous bleeding occurs when the count is less than 20,000. Bleeding precautions are necessary when the platelet count falls below 50,000.

Bleeding precautions include monitoring platelet counts daily and observing for signs of bleeding such as bruising, petechiae, bleeding gums, tarry stools, and black emesis. Intramuscular and subcutaneous injections are avoided to prevent bleeding into the injection site. Safety measures are instituted to prevent cuts, bruises, and falls. The patient is advised to use an electric razor, avoid blowing the nose, and avoid intercourse to prevent bleeding for the duration of the thrombocytopenia.

Imbalanced Nutrition

Patients are instructed about the importance of maintaining proper nutrition. Nutrition Notes Box 10–9 presents criteria for determining whether a patient needs nutritional support. Consult with a dietitian and physician for dietary supplements and medications used to control nausea, vomiting, and diarrhea. Keep the environment free of odors that might induce nausea, such as strong disinfectants, perfumes, deodorizers, and body wastes. Monitor intake and output every 8 hours, and weigh the patient daily.

BOX 10-9 *Nutrition Notes*

Assessing the Need for Nutritional Support

Any two of the following findings indicate a need for nutritional support:

- Weight loss of 10% or more of body weight
- Serum albumin less than 3.4 g/dL
- Serum transferrin less than 190 mg/dL

Source: *Daly, HM, and Shinkwin, M: Nutrition and the cancer patient. In Holleb, AI, Fink, DJ, and Murphy, GP (eds): American Cancer Society Textbook of Clinical Oncology. American Cancer Society, Atlanta, 1991.*

Room-temperature foods or cold foods and clear liquids may help reduce vomiting. Sour foods such as hard candy, lemon, and pickles might be helpful. Music or relaxation exercises may help distract the patient from the nausea. Adding nutmeg to foods may help slow down motility of the gastrointestinal tract. Provide mouth care before meals to make eating more pleasant. Small, high-calorie meals are better tolerated than large meals. Administer pain medication before meals to help reduce the impact of pain on the appetite. Instruct the patient to avoid fluids with meals to prevent premature feelings of fullness and to avoid exercise before meals to prevent fatigue. See Nutrition Notes Box 10–10 for additional nutrition interventions.

Self-Care Deficit

Psychosocial issues related to cancer are as varied as the persons afflicted with the disease. You can help the patient ex-

BOX 10-10  *Nutrition Notes*

Treating Problems Related to Nutrition

Try these ideas if a cancer patient has early satiety or anorexia, a bitter or metallic taste, local oral effects, nausea and vomiting, diarrhea, or an altered immune response.

Early Satiety and Anorexia

- Select nutrient-dense foods. For example, fortify puddings and milkshakes with dry skim milk powder.
- Encourage appropriate exercise.
- Present food attractively.
- Remove covers from food containers away from the bedside if strong odors annoy the patient.
- Offer small, frequent meals.
- Provide home-cooked food.
- If meals are rejected, offer 1 oz of a complete nutritional supplement every hour.

Bitter or Metallic Taste

- Cook in glass containers in a microwave oven.
- Serve food cold or at room temperature.
- See if the patient prefers eggs, fish, poultry, and dairy products to beef and pork.
- Experiment with sauces and seasonings. Sweet sauces and marinades may improve the palatability of meats.

Local Oral Effects

- Ulcerations—Offer soft, mild foods; cream sauces, gravies, and dressings for lubrication; cold foods for numbing; and soda straws for liquids. Topical vitamin E oil may be effective for chemotherapy-induced lesions. Avoid hot items, salty or spicy foods, and acidic juices. If an anesthetic mouthwash is prescribed, caution the patient to chew carefully to avoid biting the lips, tongue, or cheeks.
- Dry mouth—Offer frequent sips of water or an artificial saliva. Lubricate with gravies, butter, margarine, milk, cream, or bouillon. Sugarless hard candy, chewing gum, or popsicles may stimulate saliva production.

- Dysphagia—Teach the patient to make swallowing a conscious act (inhale, swallow, exhale) and to experiment with head position. Offer foods with a smooth, even consistency. Thick liquids are easier to swallow than thin. Encourage dunking breads in a beverage to soften.

Nausea and Vomiting

- Administer antiemetics on a regular prophylactic schedule.
- Suggest dry crackers.
- Offer liquids between meals to reduce stomach volume and low-fat meals to facilitate stomach emptying.
- Instruct the patient to chew thoroughly, eat slowly, and rest afterward.
- Arrange unconventional meal schedule to take advantage times when patient feels better.
- Avoid serving favorite foods when the patient is nauseous to avoid an association between those foods and vomiting.

Diarrhea

- Suggest a low-residue diet. (Citrotein and Enlive are clear liquid complete nutritional supplements.)
- Try a lactose-free diet for temporary lactose intolerance.
- Propose pectin-containing foods (apples, strawberries, citrus fruits) to absorb water in the bowel.
- Recommend active cultures of yogurt to repopulate intestine.

Altered Immune Response

- Restrict fresh fruits and vegetables that cannot be peeled or adequately disinfected.
- Consider avoiding yogurt to prevent translocation of the bacteria to the bloodstream.

Ganley, BJ: Effective mouth care for head and neck radiation therapy patients. Med-Surg Nurs 4:133, 1995.
Wadleigh, RD, et al: Vitamin E in the treatment of chemotherapy-induced mucositis. Am J Med 92:481, 1992.

---

plore perceptions about quality of life. Help the patient rank quality-of-life issues based on an understanding of the disease process, treatment, and specific limitations. Culture and age affect cancer perceptions (e.g., in a culture in which life expectancy is short, possible death from cancer in the later years is not a significant threat).

It is helpful for patients with cancer to anticipate a time when they will no longer be able to care for their own needs. You can assist patients to determine resources that can be called on at that time. If patients live with family members, the family can be instructed in how to assist in daily care. Home health nurses and hospice care can also be invaluable.

Anticipatory Grieving

To many people, cancer is synonymous with death. Although some cancers are very aggressive and death occurs quickly after diagnosis, many cancers are controllable and allow years of living after diagnosis and treatment. Make sure that the patient has received accurate information about the prognosis, but never give false hope. Use thera-

peutic communication techniques to help the patient talk about an anticipated death. Encourage family members to spend time with the patient. Contact the patient's minister or clergy if the patient agrees.

Risk for Caregiver Role Strain

The emotional issues of the disease can be devastating for patients, family members, and other support persons. Income is lost and lifestyles are changed when the disease affects wage-earning members of the family. When a patient's caregiver works outside the home, the stress of dual responsibility takes its toll. Family growth needs are often replaced with meeting basic survival needs such as providing food, clothing, and shelter. This can create many guilt feelings for both the patient and family. You can help by providing information about local community resources.

Role changes are sometimes needed and may be difficult for family members to make. If the dominant person is the patient, the less dominant partner may have to assume new responsibilities that contribute to stress. Your ability to actively listen to family members' concerns is important.

Social Isolation

Isolation can be either self-imposed or imposed by friends and family as terminal illness issues are confronted. It can be very frustrating to see a loved one decline with cancer; often people say they are "afraid of saying or doing the wrong thing" so they "just stay away." Patients may engage in self-blame, and their anger may lead to depression and further isolation. It is important to recognize signs of depression and suicidal tendencies.

Ineffective Sexuality Patterns

Maintaining healthy sexuality is important to all patients, including those with cancer. Individual sexual attitudes and practices vary, making it difficult to define what is "normal." Normal can be best described as what the patient and partner consider pleasurable.

A decline in sexual desire is not uncommon during cancer treatment. Anxiety about sexual intercourse includes fears concerning contracting cancer from the patient and fears that sexual intercourse will make the cancer worse.

It is important to stress that cancer is not contagious. It cannot be passed from person to person, even through contact as close as kissing, intercourse, or oral and genital sex. Sexual activity is usually safe during and after cancer treatment. Patients who are advised to abstain from sexual intercourse because of low blood counts are temporarily at risk for secondary infections and bleeding until blood counts return to normal. They do not put their partners at risk. Closeness and intimacy are often desired even if sexual intercourse is not.

Pain is the most common problem for both male and female patients during intercourse. Pain is associated with pelvic surgery, radiation therapy, or treatment that affects hormone levels, and pain can prevent the male patient from achieving an erection. Most obstacles can be overcome if the couple are open and honest about their feelings. Be prepared to address sexuality issues in a timely fashion.

Disturbed Body Image

The loss of any body part can be traumatic to a person's self-image. You can provide information about plastic surgery and prosthetics and help the patient through the grieving process related to the loss.

Cancer treatment is costly, and financial aid may be needed. Provide information about community assistance programs or services. Social service personnel are available in most health care facilities.

Evaluation

If interventions have been effective, the patient can describe known facts about the disease and the prescribed treatment plan. The patient verbalizes relief or a reduction in pain intensity. Pain management should provide the patient with periods of rest and an ability to appropriately focus on living. The patient or family demonstrates knowledge of risk factors associated with bleeding and infection. The patient remains infection free and maintains intact skin as evidenced by an absence of reddened or ulcerated areas. The patient has no signs of bleeding or anemia. The patient maintains optimal nutritional status, as reflected by caloric intake adequate to meet body requirements, balanced intake and output, absence of nausea and vomiting, and good skin turgor. The patient feels free to share feelings about his or her disease and coping abilities and achieves or maintains control of his or her body.

CRITICAL THINKING

Mrs. Jones

Mrs. Jones is admitted to your unit following a simple mastectomy for breast cancer. The tumor was staged as a T2, N0, M0. Estrogen and progesterone receptors were negative. A bone scan was negative for metastasis. She is scheduled for four chemotherapy treatments, 3 weeks apart. The medications prescribed are high doses of doxorubicin and cyclophosphamide (Cytoxan). A central line is inserted for chemotherapy.

1. What does the staging of Mrs. Jones's tumor mean?
2. What major side effects of her medications should you observe for?
3. Why was a central line inserted?
4. What nursing diagnoses are appropriate for Mrs. Jones?

Answers at end of chapter.

► HOSPICE CARE OF THE PATIENT WITH CANCER

Patients who are considered terminal and have a life expectancy of 6 months or less are eligible for hospice care, which provides humanistic care for dying people and their families. The dying person is provided care in a home or homelike setting that promotes comfort and quality of life until death. Hospice care is offered as an inpatient or outpatient service (see Home Health Hints).

HOME HEALTH HINTS

- The home health or hospice nurse helps manage cancer pain in the home. Intramuscular dosing of pain medication should be avoided because of pain associated with injections and the burden this places on the caregiver. Oral or intravenous analgesics are preferred. For moderate to severe pain, medication dosing should be around-the-clock with prn doses for breakthrough pain. The nurse should anticipate constipation from opioid administration and treat prophylactically.
- Home health nurses are in key positions for making timely referrals for hospice care. Eligible clients are those who have a life expectancy of 6 months or less, have a desire for supportive care rather than continued treatments, and have a friend or relative who is willing to coordinate the care.

Inpatient services are used for symptom control and respite care for the family. Family and pets are allowed to stay with the patient. Hospice care deals with the family in crisis and continues after the patient dies, with follow-up counseling, listening, nurturing, and referral.

Outpatient care is given in the home with family members providing the primary care. Support care is given by the hospice staff. Medications and supplies are furnished by the hospice service. At home, the patient can enjoy loved ones, pets, plants, music, and other personal possessions for as long as possible.

► ONCOLOGICAL EMERGENCIES
Superior Vena Cava Syndrome

Superior vena cava syndrome occurs in patients with lung cancer when the tumor or enlarged lymph nodes block the circulation in the vena cava. This results in edema of the head and neck and may lead to seizures. Radiation therapy can be used to shrink the tumor and allow for circulation to resume naturally.

Spinal Cord Compression

Spinal cord compression may develop in patients with bone metastasis when the bones collapse. This is a very painful problem and requires pain management while radiation is given to relieve the symptoms. Patients may develop some motor loss when this occurs. Often a myelogram or bone scan is used for diagnosis.

Hypercalcemia

Hypercalcemia occurs when the serum calcium level exceeds 11 mg/dL. In patients with cancer, hypercalcemia is associated with the release of calcium into the blood from bone deterioration. It is common in patients with bone metastasis, especially metastasis from breast cancer. It can be treated with intravenous medication to lower the calcium levels.

Pericardial Effusion/Cardiac Tamponade

Pericardial effusion, or cardiac tamponade, is a condition usually caused by direct invasion of the cancer, causing the pericardial sac to fill with fluid. Treatment involves draining the fluid from the heart sac by pericardiocentesis and using sclerosing agents to keep the pericardial sac from refilling with fluid.

Disseminated Intravascular Coagulation

Disseminated intravascular coagulation (DIC) involves an abnormal activation of the clot formation and fibrin mechanisms of the blood, resulting in the consumption of coagulation factors and platelets. Patients with DIC are at high risk for thrombus formation, infarctions, and bleeding. Treatment includes fresh frozen plasma and cryoprecipitates with heparin.

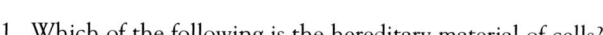

1. Which of the following is the hereditary material of cells?
   a. Protein in the ribosomes
   b. DNA in the chromosomes
   c. RNA in the nucleus
   d. Ribosomes in the cytoplasm

2. Mr. Michaels, age 56, has been recently diagnosed with prostate cancer. He asks you, "How do malignant tumors differ from benign tumors?" Which of the following is the most accurate statement?
   a. Malignant tumors invade surrounding cells.
   b. Malignant tumors are encapsulated.
   c. Malignant tumors remain localized.
   d. Malignant tumors always require surgical intervention.

3. A patient asks the meaning of alopecia. Which of the following is the most appropriate response?
   a. Itching of skin
   b. Frequent bowel movements
   c. Blood vessel constriction
   d. Hair loss

4. Why is the cell cycle important in chemotherapy?
   a. Dividing cells tend to be more vulnerable to chemotherapy.
   b. Chemotherapeutic agents act on only the last phase of the cell cycle.
   c. Chemotherapeutic agents are unable to destroy all cancer cells.
   d. Most normal cells are not affected by chemotherapeutic agents.

5. Mr. West is receiving radiation therapy and notes that the skin over the treated area is slightly reddened. What is the best care for this area?
   a. Applying moist normal saline compresses
   b. Lubricating the skin with water-soluble lubricant
   c. Keeping the skin clean and protected from the sun
   d. Applying a heat lamp to the area two times a day

6. Mr. Waist, age 70, admitted for diagnostic tests to confirm cancer of the colon, is restless and uncomfortable. He states that he will need a pain pill soon. Which of the following nursing actions is best before administering pain medication?
   a. Assess the patient's anxiety level.
   b. Assess the patient's understanding of the side effects of pain medication.
   c. Determine the patient's pain tolerance.
   d. Assess the success of past pain management measures.

REFERENCES

1. American Cancer Society: Cancer prevention and early detection worksheet. http://www.cancer.org, 2001.
2. American Cancer Statistics: www.cancer.org, 2002.
3. Genovese, L, and Wholihan, D: The "VANAC team": Establishing a cancer prevention team. Cancer Nurs 18(6):421, 1995.
4. Groenwald, S, et al: Cancer Nursing, ed 3. Jones & Bartlett, Boston, 1993.

Answers to CRITICAL THINKING

Mrs. Jones

1. Mrs. Jones's tumor is beginning to invade surrounding tissue. There is no lymph node involvement and no metastasis.
2. Doxorubicin is commonly associated with red urine. Cyclophosphamide can cause blood in the urine. Both medications can cause nausea, vomiting, and alopecia. Both are vesicants.
3. Because the drugs are vesicants, it is important to inject them into a large vein.
4. Many diagnoses are appropriate, including acute pain related to surgical incision, disturbed body image related to alopecia and loss of a breast, imbalanced nutrition related to nausea and vomiting, risk for injury related to medication side effects, and deficient knowledge about cancer treatment and management of side effects.

NURSING CARE OF PATIENTS HAVING SURGERY

Linda S. Williams

KEY TERMS

adjunct (**ADD**-junkt)

anesthesia (AN-es-**THEE**-zee-uh)

anesthesiologist
(an-es-**THEE**-zee-uhl-la-just)

aseptic (ah-**SEP**-tik)

atelectasis (AT-e-**LEK**-tah-sis)

debridement (day-breed-**MAHNT**)

dehiscence (dee-**HISS**-ents)

evisceration (E-**VIS**-sir-a-shun)

hematoma (HEE-muh-**TOH**-mah)

hypothermia (HIGH-poh-**THER**-mee-ah)

induction (in-**DUCK**-shun)

intraoperative (IN-trah-**AHP**-er-uh-tiv)

perioperative (PER-ee-**AHP**-er-uh-tiv)

postoperative (post-**AHP**-er-uh-tiv)

preoperative (pre-**AHP**-er-uh-tiv)

purulent (**PURE**-u-lent)

serosanguineous
(SEER-oh-**SANG**-gwin-ee-us)

surgeon (**SURGE**-on)

QUESTIONS TO GUIDE YOUR READING

1. How would you describe the three phases of perioperative nursing?

2. What is your role in each perioperative phase?

3. What are six factors influencing surgical outcomes?

4. What is the purpose of preoperative patient data collection?

5. How would you enhance learning for the elderly preoperative patient?

6. What are some nursing interventions for common postoperative patient needs?

7. How will you know if your nursing interventions have been effective?

8. What are signs and symptoms of common postoperative complications?

9. What are the criteria for ambulatory discharge?

10. How would you help reinforce patient discharge instructions?

11. What is the postoperative role of the home health nurse?

Surgery is the use of instruments during an operation to treat injuries, diseases, and deformities. Physicians perform surgical procedures, including **surgeons**, family practice physicians, or other physicians trained to do certain surgical procedures. Surgery is performed in clinics, physicians' offices, ambulatory surgical centers, and hospitals. Laser and scope technology continues to lead to new procedures that offer less risk, less invasion, faster recovery, and reduced hospitalization or ambulatory surgery. Today surgery is a safe, effective treatment option because of medications such as antibiotics and anesthetics that allow a quicker recovery.

▶ SURGERY CATEGORIES

Surgical procedures are named according to the involved body organ, part, or location and are given a suffix that describes what is done during the procedure (Table 11–1). The urgency levels to have a successful outcome for the patient determine the scheduling time frame for a surgery (Table 11–2). The various indications for surgery are listed in Table 11–2.

The author acknowledges the contributions to this chapter by Linda Nabozny and Deanna Casler.

TABLE 11-1 SURGICAL PROCEDURE SUFFIXES

Suffix	Meaning	Word-Building Examples
-ectomy	Removal by cutting	crani (skull) + ectomy = craniectomy
		appen (appendix) + ectomy = appendectomy
-orrhaphy	Suture of or repair	colo (colon) + orrhaphy = colorrhaphy
		herni (hernia) + orrhaphy = herniorrhaphy
-oscopy	Looking into	colon (intestine) + oscopy = colonoscopy
		gastr (stomach) + oscopy = gastroscopy
-ostomy	Formation of a permanent artificial opening	ureter + ostomy = ureterostomy
		colo (colon) + ostomy = colostomy
-otomy	Incision or cutting into	oust (bone) + otomy = osteotomy
		thoro (thorax) + otomy = thoracotomy
-plasty	Formation or repair	oto (ear) + plasty = otoplasty
		mamm (breast) + plasty = mammoplasty

TABLE 11-2 SURGERY CATEGORIES

Type	Definition	Examples
Urgency Level		
Emergent	Immediate surgery needed to save life or limb.	Ruptured aortic aneurysm or appendix, traumatic limb amputation, loss of extremity pulse from emboli
Urgent	Surgery needed within 24–30 hours.	Fracture repair, infected gallbladder
Elective	Planned/scheduled, with no time requirements	Joint replacement, hernia repair, skin lesion removal
Indications		
Aesthetic	Requested by patient for improvement	Blepharoplasty, breast augmentation
Diagnostic	To obtain tissue samples, make an incision, or use a scope to make a diagnosis	Biopsy
Exploratory	Confirmation or measurement of extent of condition	Exploratory laparotomy
Preventive	Removal of tissue before it causes a problem	Mole or polyp removal
Curative	Removal of diseased or abnormal tissue	Inflamed appendix, tumor, benign cyst
Reconstructive	Correction of defects of body parts	Scar repair, total knee replacement
Palliative	Alleviation of symptoms without curing disease	Rhizotomy (cuts nerve root to relieve pain), partial tumor removal to relieve pain or pressure, gastrostomy tube for swallowing problem

PERIOPERATIVE PHASES

There are three phases in the surgical process: **preoperative, intraoperative,** and **postoperative.** They are referred to collectively as **perioperative,** which is the time surrounding and during surgery. Each of the perioperative surgical phases has a defined time frame in which specific events related to surgery occur (Table 11–3). Throughout each phase, nurses interact closely with the patient, physician, and other members of the health care team.

TABLE 11-3 PERIOPERATIVE SURGICAL PHASES

Perioperative	All three phases surrounding and during surgery
Preoperative	Begins with decision for surgery and ends with transfer to the operating room
Intraoperative	Begins with transfer to operating room and ends with admission to postanesthesia care unit (PACU)
Postoperative	Begins with admission to PACU and continues until recovery is complete

PREOPERATIVE PHASE

Your primary roles in the preoperative phase are to:

- assist in data collection for developing the patient's plan of care.
- reinforce explanations and instructions given to the patient and family by the physician and registered nurse.
- provide emotional and psychological support for patients and their families.

Studies show that a patient's family experiences anxiety during the surgical procedure even if the patient's surgery is elective and of short duration. You can make a major contribution by reducing the family's anxiety. If families are less anxious, they are more helpful in assisting patients as they recover.

Other health team members assist in preparing the patient for surgery. The physician obtains a medical history, performs a physical examination, and orders diagnostic testing. Registered nurses perform a baseline preoperative assessment, provide explanations and instructions, offer patients and families emotional and psychological support to ease anxiety, develop a plan of care, and then verify the patient's name, surgical site, allergies, and related information when the patient arrives in the surgical waiting area.

Factors Influencing Surgical Outcomes

When preparing a patient for surgery and developing a nursing care plan, the goal is to identify and implement actions that reduce surgical risk factors. Preoperative care focuses on helping the patient achieve the best possible surgical outcome by being in the healthiest possible condition for surgery.

Emotional Responses

The word *surgery* causes a common emotional reaction in patients and their families. You need to be aware of these reactions to assist the patient in coping with them. If any of the patient's fears are extreme, such as a fear of dying or not waking after surgery, the physician should be informed.

Surgical patients may experience various fears related to **anesthesia:** possible brain damage; feeling sensation during surgery; feeling loss of control; a fear of not waking up. The patient should discuss these concerns with the **anesthesiologist.** Listening to music or using guided imagery before surgery may help reduce patients' anxiety and calm them.

A concern of being in pain is normal among surgical patients. Patients are informed about pain control methods. During the operation, the anesthesia provider uses medications to control pain. Analgesics are given for pain relief after surgery. Complementary techniques can also be used to help reduce pain, such as guided imagery or focused breathing.

Changes in body image may be a great fear for some patients. The thought of disfigurement, mutilation, bleeding, or having a scar causes great anxiety for some patients.

Age

Surgery can be a positive experience that promotes quality of life for many elderly patients. For healthy older patients, age alone does not mean that they are a greater surgical risk. Complications can occur, however, related to previous health status, immobilization occurring from surgery, normal aging changes reducing the effectiveness of deep breathing and coughing, and the effects of administered medications (Gerontological Issues Box 11–1). Older patients may require a longer time to recover from anesthetic agents because of aging changes in drug metabolism and elimination. To reduce complications the elderly patient's health problems need to be controlled before surgery.

BOX 11-1 **GERONTOLOGICAL ISSUES**

Surgical Considerations for the Older Adult

Older adults usually have limited physiological reserve, resulting in decreased ability to compensate for changes that occur during surgery. There is increased risk for hemorrhage, anemia, fluid/electrolyte imbalance, and infection. Increased risk for complications is secondary to age-related loss of blood vessel elasticity and decreased cardiac, respiratory, and renal reserves. Nursing interventions should be aimed at these age-related changes before, during, and after the surgical procedure to help reduce complications.

Preoperatively:

- Reassure the patient and family.
- Pad bony prominences to protect against pressure ulcers and muscle and bone discomfort.
- Teach what to expect before, during, and after surgery, diet changes, description and length of surgical procedure, activities in the recovery room, pain management, coughing and deep breathing exercises, procedures, and treatments (e.g., dressings, catheters).
- Ensure preoperative screening: blood work, radiographic studies, nutritional assessments, pulmonary function tests, electrocardiogram.

Intraoperatively:

- Assess patient for hypothermia (cool temperature in operating room, medications that slow metabolism).
- Assess patient for hypoxia (older adult may exhibit restlessness).
- Assess patient for hemorrhage.
- Assess patient's output (urine, drainage, bleeding, vomitus).

Postoperatively:

Pain Control—Provide adequate pain relief so required postoperative activities, such as deep breathing, coughing, position changes, and exercise, can be performed more effectively.

Respiratory Function—Reduce respiratory complications by encouraging deep breathing and coughing:

- Do deep breathing and coughing after pain medication has begun to take effect because this will increase the older patient's ability to take deeper breaths. Assess the patient carefully when giving narcotic analgesics because they may cause respiratory depression.
- Use a pillow and instruct the patient to hold it firmly over abdominal or chest incisions to support the incision. Taking a deep breath increases chest expansion, as well as abdominal pressure, which may pull or stretch an incision.
- Elderly patients perform deep breathing and coughing exercises better if the nurse performs the exercises with them. For example, say the following:

   "Let's take a deep breath in through the nose, hold it and count to three, then slowly blow it out completely through the mouth. When you blow the air out, shape your lips like they are going to whistle. Great, let's do it again."

Mobility—Encourage mobility through the following nursing actions and observations:

- Use pillows to support the patient's body alignment; assist the patient to ambulate as soon as possible after surgery; and regularly help the patient with passive or active range-of-motion exercises, along with flexion and extension exercises, for legs and feet.
- Deep vein thrombosis is a risk related to venous pooling in the lower extremities. This risk is increased with postoperative inactivity.
- If patients lay in one position too long, pressure ulcers can develop. When tissues are compressed between bones and the bed surface, blood supply is reduced to the tissue and cells begin to die. This results in painful open wounds.

BOX 11-1 GERONTOLOGICAL ISSUES—CONT'D

Bowel Function—Assess bowel sounds. It is common for patients to feel bloated after surgery. Increasing activity, such as walking—not just sitting in a chair—stimulates peristaltic action of the bowel. This helps expel flatus and reduce discomfort.

Urinary Function—Be aware of the following aspects of urinary function:

- It is common for individuals to have difficulty emptying their bladder after surgery. Patients who are sleeping but restless should be evaluated for bladder distention. It is often difficult to void on a bed pan or in a urinal in a supine position.
- Older men with an enlarged prostate may have even greater difficulty voiding if they have received medications that have urinary retention side effects.
- Assisting patients to sit or stand to use urinals, use a bedside commode, or ambulate to the bathroom promotes bladder emptying and helps avoid the use of urinary catheters.

- Measure urine output that is voided or from a catheter. Note the color and odor of the urine. The elderly are prone to dehydration, and this provides an indication of their hydration status for intervention.

Delirium—Perform the following nursing actions to minimize delirium:

- Monitor level of consciousness routinely. Provide a calm environment and orient the patient to their environment. Restraints should not be used because they can worsen delirium.
- Recognize that urinary catheter presence can contribute to delirium, so methods to avoid the need for a catheter should be tried.

Hydration and Nutrition

A normal fluid and electrolyte balance decreases complications. Patients should be well nourished to adequately heal and recover from surgery. Higher levels of protein (tissue repair and healing), vitamin C (collagen formation), and zinc (tissue growth, skin integrity, and cell-mediated immunity) are required. Obese or underweight patients may not heal as well and may have complications. Obese patients have more respiratory problems and wound healing difficulties, such as delayed healing and wound **dehiscence** (opening of the incision). Emaciated individuals may have more infections and delayed wound healing because they lack the nutrients needed for tissue healing. Identifying patients at risk and providing preoperative intervention reduce the risks of surgery for these patients (Nutrition Notes Box 11–2).

Smoking and Alcohol

The use of tobacco and alcohol increases the surgical patient's risks. Smoking thickens and increases the amount of lung secretions and reduces the action of cilia that remove the secretions. Patients should be encouraged to avoid smoking for 24 hours before surgery. A patient who smokes and has a chronic lung disorder should be encouraged to stop smoking 3 to 4 weeks before surgery. Cessation of smoking allows an increase in the action of the lungs' defense mechanisms and makes more hemoglobin available to carry oxygen during surgery.

Long-term alcohol use may cause nutritional deficiencies and liver damage, which can create bleeding problems, fluid volume imbalances, and drug metabolism alterations. In addition, alcohol interacts with medications and should be avoided before surgery.

Diseases

Chronic disorders may increase the patient's surgical risk unless they are well controlled. A preoperative assessment and clearance for surgery by the patient's physician may be needed. For diabetics, the stress of surgery can alter blood

BOX 11-2 Nutrition Notes

Screening and Nourishing the Preoperative Patient

In a recent study, preoperative protein energy malnutrition was identified in 38% of patients undergoing abdominal surgery for benign conditions. Of these malnourished patients, 75% developed postoperative complications, compared with 22% of well-nourished patients. Mild to severe malnutrition affected 39% of patients undergoing gastrointestinal or orthopedic surgical procedures, but only about two-thirds of them received nutritional support in another study. History of a gain or loss of 10 pounds within the past 6 months should trigger a nutritional assessment; however, just 59% of the patients in this study even had a weight recorded on their chart.

Before elective surgery, the patient may have time to correct some nutritional deficiencies. Many patients are instructed to lose weight to reduce the risk of surgery. If they are anemic, an iron preparation can be administered. At least 2 to 3 weeks are required for objective evidence of the effectiveness of nutritional therapy. In patients identified as high risk, 7 days of preoperative total parenteral nutrition (TPN) produced a sixfold reduction in major sepsis. Before surgery on the gastrointestinal tract, a low-residue diet may be given for 2 to 3 days to minimize the feces in the bowel.

Many botanical products sold in the United States over the counter as nutritional supplements profoundly affect body systems during surgery. So many people are taking botanical products that the American Society of Anesthesiologists issued a warning to consumers of herbal medicine to stop taking the products 2 to 3 weeks before scheduled surgery. Possible interactions cited were an unintended deepening of anesthesia and problems with bleeding and blood pressure. In addition to inquiring about weight loss, the preoperative assessment should include questions about herbal or traditional medicines. Many patients do not regard these products as medicines or are reluctant to tell their doctors that they are using them. For more information about the American Society of Anesthesiologists' warning, visit the public education section of www.asahq.org.

glucose levels. Patients with chronic lung disorders may have pulmonary complications from anesthesia. To prepare patients with lung disorders for surgery, teach them to deep breathe and cough and to use an incentive spirometer (Fig. 11–1).

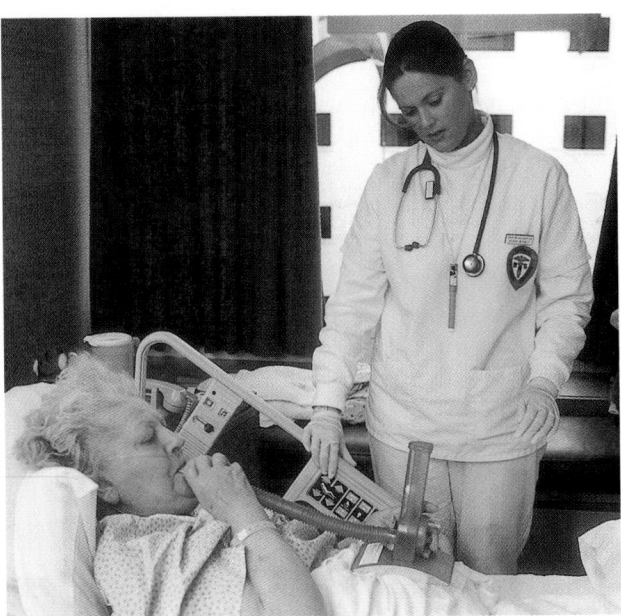

Figure 11-1 An incentive spirometer aids lung expansion.

Preoperative Patient Assessment

Nonemergent surgical patients are assessed before admission to a surgical facility or hospital in the preadmission testing (PAT) department. PAT nurses and anesthesia department members obtain a health history, identify risk factors, begin patient and family teaching, and make necessary referrals to social work, support groups, and educational programs, such as a total joint education program for patients undergoing a total joint replacement. Laboratory tests, electrocardiograms (ECGs), chest x-rays, and other diagnostic testing are done based on the patient's needs. A urine or serum pregnancy test as appropriate for female patients may be done to prevent fetal exposure to anesthetics. Health information and

diagnostic testing results are reviewed. Interventions are ordered for abnormalities. Then preoperative, intraoperative, and postoperative care is planned.

Federal law dictates that patients must be asked before surgery if they have a signed advance directive. An advance directive (e.g., health care durable power of attorney or living will) indicates a person's wishes for medical care if they become unable to speak for themselves. A copy of the advance directive is entered into the medical record. If there is no advance directive, written information on advance directives is provided. A health care power of attorney allows patients to place someone of their own choosing, such as a relative or friend, in control of their medical decisions if they are unable to make them. A living will instructs the physician when to provide, withhold, or withdraw treatment that prolongs life and specifies types of treatment the patient wishes, such as comfort care only.

Preoperative Patient Admission

Upon admission for surgery, subjective and objective patient data are collected (Table 11–4). It is important for the patient to have access to their contact lenses, glasses, or hearing aids whenever explanations or instructions are given to ensure effective communication. During the interview, the patient's emotional reaction to surgery is noted. If the patient is anxious, you should explore the cause of the anxiety and allow the patient to express any concerns. Past experiences of patients or their significant others can impact the surgical experience.

Subjective Data

During data collection, it is important to ask the patient if there have been any personal or family problems with anesthesia. A rare hereditary muscle disease known as malignant hyperthermia can predispose the patient to a serious

TABLE 11-4 NURSING ASSESSMENT OF THE PREOPERATIVE PATIENT
Subjective Data: Health History Questions
Demographic information: Name, age, marital status, occupation, roles?
History of condition for which surgery is scheduled: Why are you having surgery?
Medical history: Any allergies, acute or chronic conditions, current medications, pain, or prior hospitalizations?
Surgical history: Any reactions or problems with anesthesia? Previous surgeries?
Tobacco use: How much do you smoke? Pack-year history (number of packs per day 3 number of years)?
Alcohol use: How often do you drink alcohol? How much?
Coping techniques: How do you usually cope with stressful situations? Support systems?
Family history: Hereditary conditions-diabetes, cardiovascular, anesthesia problems?
Female patients: Date of last menses and obstetrical information?
Objective Data: Body System Review
Vital signs, oxygen saturation
Height and weight
Emotional status: calm, anxious, tearful, affect
Neurological: ability to follow instructions
Skin: color, warmth, bruises, lesions, turgor, dryness, mucous membranes
Respiratory: infection: cough; breath sounds; chronic obstructive pulmonary disease; respiratory rate, pattern, and effort; barrel chest
Cardiovascular: angina, myocardial infarction, heart failure, hypertension, valvular heart disease, mitral valve prolapse, heart rate and rhythm, peripheral pulses, edema, jugular vein distention
Gastrointestinal: bowel sounds, date of last bowel movement, abdominal distention
Musculoskeletal: deformities, weakness, decreased range of motion, crepitation, gait, artificial limbs, prostheses

life-threatening reaction to certain anesthetic agents (discussed later). Prior surgeries are recorded.

Medications

All medications (herbs and over-the-counter, prescription, and recreational drugs) that the patient is taking must be reviewed. Alterations in drug dosages and routes of administration may be required. You should ensure that the patient clearly understands all medication instructions. For example, patients taking an anticoagulant such as warfarin (Coumadin) may be told by their physician to decrease or stop it several days before surgery to avoid bleeding problems during surgery. Many herbs can interfere with medications used during surgery or increase bleeding times. Patients may be instructed to stop certain herbs several days or weeks before surgery.

Diabetic patients on insulin are usually given instructions by the physician to either hold their insulin or take half of their normal dose of insulin the day of surgery. On the day of surgery, blood glucose monitoring is done every 4 hours to ensure that blood glucose levels are maintained within a desired range.

Patients on chronic oral steroid therapy cannot abruptly stop their medication if they are told to take nothing by mouth (NPO) before or after surgery. Serious complications, including circulatory collapse, can develop if steroids are stopped abruptly. The physician should order a patient's steroid therapy to be given by a parenteral route if the patient is NPO, so that it is not interrupted. You should ensure that the therapy is ordered and continued for the patient by an alternate route.

Patients should be asked about the use of drugs such as cocaine, marijuana, or opioids because these drugs can interact with anesthesia or other medications. To obtain accurate information, patients should be told of this potential interaction. Information and questions should be stated in a nonjudgmental manner. For example, you should ask, "How much alcohol do you drink daily or weekly?" instead of "Do you drink alcohol?" The first statement assumes that people drink alcohol. This allows the patient who does not drink to indicate none and the patient who does to state an amount rather than having to say yes and then give an amount with further questioning. More accurate responses are given because this approach is viewed more positively by the patient who consumes alcohol. Another example would be to ask the patient, "What roles do drugs or alcohol play in your life?"

Objective Data

A physical assessment of body systems is performed. This information can highlight risk factors for surgery, determine the type of anesthesia to be used, and assist in planning interventions to reduce risk factors. A cough, cold, or fever is reported to the physician because surgery may be delayed until the patient recovers from an acute infection. Dentures, bridges, capped teeth, and loose teeth are documented because they can become dislodged during intuba-

tion (insertion of endotracheal breathing tube) for general anesthesia, causing complications.

SKIN ASSESSMENT. The skin is observed for color, bruises, and open areas. Palpation of the skin is done to check turgor, warmth, and dryness. Mucous membranes are inspected for color and moistness, especially for elderly patients, who are at greater risk for hydration problems.

SENSORY ASSESSMENT. The patient's ability to see and hear are assessed. Assistive devices (e.g., contact lenses, glasses, or hearing aids) are documented. Patients must have access to these devices whenever explanations or instructions are given for effective communication.

NERVOUS SYSTEM ASSESSMENT. Assessing a patient's ability to understand and follow directions is important because preoperative instructions such as being NPO after a specified time are given. Instructions must be followed to avoid delaying surgery. If cognitive impairments are found, the physician is informed. A family member may need to be included in teaching sessions to assist the patient with following instructions.

RESPIRATORY ASSESSMENT. Respiratory rate and pattern and breathing effort are noted. Oxygen saturation is obtained using an electronic oximeter attached to the finger, nose, or ear that measures the oxygen saturation of the blood flowing past the sensor. The thorax is inspected for use of accessory muscles. A barrel chest (increase in anteroposterior diameter of the chest that looks like the shape of a barrel) is noted. Patients who develop barrel chests usually have chronic obstructive lung disease and may be at greater risk for lung problems. Breath sounds are auscultated, and abnormalities are documented and reported to the physician. Pulmonary function tests or arterial blood gas measurements are done based on the patient's history of lung disease and assessment findings, such as shortness of breath, inability to tolerate activity without dyspnea (feeling out of breath and laboring to breathe), or abnormal breath sounds such as crackles or wheezes.

LEARNING TIPS

Crackles are produced by air flowing over secretions or the snapping open of collapsed airways. They sound like the crackling of Velcro being opened and can be fine (soft, short, high-pitched) or coarse (loud, long, low-pitched). Wheezes are a continuous musical sound caused by airway narrowing.

CARDIOVASCULAR ASSESSMENT. Apical heart rate and rhythm are auscultated. Pulses are palpated for presence and quality. Edema or jugular vein distention is documented.

MUSCULOSKELETAL ASSESSMENT. The patient's joint range of motion, muscular strength, gait, and mobility are observed. The presence of deficits, artificial limbs, prostheses, or pain is documented. Joints are inspected and palpated for warmth, redness, swelling, or crepitation (grating with movement). Positioning in surgery may be affected by any abnormal findings.

ABDOMINAL ASSESSMENT. The abdomen is inspected for distention and firmness. Bowel sounds are auscultated. Ostomies are noted.

DIAGNOSTIC TESTING. Patient preoperative diagnostic testing is based on the patient's age, medical history, and assessment findings. Institutional protocols for patients of specified age ranges or with certain conditions may guide testing (Table 11–5). Abnormal findings are reported to the physician.

Nursing Process for Preoperative Patients
Nursing Assessment

The nursing assessment findings are used to plan care for the surgical patient. (See Table 11–4.)

Nursing Diagnosis

Common preoperative nursing diagnoses can include the following:

- Anxiety related to potential change in body image, hospitalization, pain, loss of control, and uncertainties surrounding surgery
- Fear related to expectation of pain and surgical risk factors
- Deficient knowledge related to lack of prior experience with surgical routines and procedures.

Planning

The patient's goals are to have decreased anxiety, reduced fear, and increased knowledge of surgical routines and procedures.

Nursing Interventions

ANXIETY OR FEAR. Anxiety is a feeling of apprehension or uneasiness resulting from the uncertainties and risks associated with surgery. Fear, a feeling of dread from a source known to the patient, is an extreme reaction to surgery. Anxiety and fear can be reduced by informing patients about procedures and surgical routines. Patients should be allowed to express their concerns. If the patient expresses extreme anxiety or fear, inform the physician because complications or even death could result. When fear is excessive, the physician may reschedule the surgery until the patient is better able to cope.

DEFICIENT KNOWLEDGE. Deficient knowledge occurs from a lack of information and experience with surgery. Providing information to patients decreases their anxiety, promotes informed choice, and increases self-care abilities. Teaching is caring in action and empowers patients to be a participant in their care. Teaching can be done in preadmission testing, by telephone for outpatient surgery, and upon admission.

Teaching sessions include the patient, family members, or others assisting the patient through the surgical experience. Patient anxiety levels should be considered in planning teaching sessions because learning can be affected by high anxiety levels. Knowledge deficiencies should be identified by both you and the patient so that he or she is motivated to learn. Use a variety of teaching methods (discussion, written materials and instructions, models, and videos) to allow for different learning styles and to reinforce learning. Individualize explanations so the patient is given enough information without being overwhelmed. Gerontological Issues Box 11–3 describes methods to provide a positive learning experience for the elderly patient by adapting

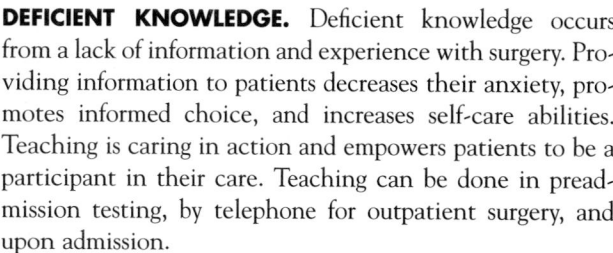

TABLE 11-5 PREOPERATIVE DIAGNOSTIC TESTS	
Diagnostic Test	**Purpose**
Chest x-ray	Detect pulmonary and cardiac abnormalities
Oxygen saturation	Obtain baseline level and detect abnormality
Serum Tests	
Arterial blood gases	Obtain baseline levels and detect pH and oxygenation abnormalities
Bleeding time	Detect prolonged bleeding problem
Blood urea nitrogen	Detect kidney problem
Creatinine	Detect kidney problem
Complete blood cell count	Detect anemia, infection, clotting problem
Electrolytes	Detect potassium, sodium, chloride imbalances
Fasting blood glucose	Detect abnormalities, monitor diabetes control
Pregnancy	Detect early, unknown pregnancy
Partial thromboplastin time	Detect clotting problem
Prothrombin time, INR	Detect clotting problem, monitor warfarin therapy
Type and cross-match	Identify blood type to match blood for possible transfusion
Urine Tests	
Pregnancy	Detect early, unknown pregnancy
Urinalysis	Detect infection, abnormalities

INR = International normalized ratio.

BOX 11-3 GERONTOLOGICAL ISSUES

Considerations for Elderly Patient Teaching Session

Environmental considerations:

- Comfortable: anxiety free, quiet, appropriate temperature
- Correctly lit: small, intense, nonglare, soft white light (not fluorescent)
- Private: no distractions, no background noise

Presentation considerations:

- Assess readiness to learn.
- Assess comfort and safety needs.
- Use past experience and relate to new learning.
- Base learning on assessment data and current knowledge base.
- Use preferred learning style, using understandable words.
- Use legible audiovisual materials: large print, black print on white nonglare paper.
- If using colors, remember that elders see red, orange, and yellow best; blue, violet, and green are difficult for elders to see.
- Perform ongoing assessment of energy level of patient.
- Answer questions as they occur.

Presenter considerations:

- Have positive attitude and belief in self-care promotion for elderly.
- Earn trust by being viewed as credible, positive role model.
- Maintain professional appearance.
- Use knowledge of aging changes in presentation:
  - Speak slowly in low tone.
  - Sit near patient for best visibility.
  - Ensure that prostheses are in place: glasses, hearing aids.
  - Allow patient increased response time and use memory aids.
  - Use touch appropriately to convey caring.
- Teach most important information first.
- Present one idea at a time.
- Provide instruction using multiple senses (vision and hearing).
- Emphasize concrete versus abstract information.
- Provide repetition.
- Ask for feedback to ensure comprehension.
- Provide feedback and positive reinforcement.

---

to aging changes that may affect learning. All teaching and levels of patient understanding is documented.

PREOPERATIVE ROUTINES. Preoperative teaching provides information about common surgical preparation procedures and routines:

- Date and time of admission and surgery
- Admission procedures: arrive about 2 hours before surgery to allow preparation time
- Length of stay, items to bring and wear
- Recovery
- Family information: where to wait during surgery and who communicates patient's status to them
- Discharge criteria: after outpatient surgery, a responsible adult must take the patient home

Preoperative Instructions. To reduce the risk of aspiration when anesthesia is started, as well as postoperative nausea and vomiting, the anesthesiologist orders fluid and food restrictions. The patient is told when to stop fluid and food intake (NPO), usually after midnight the night before surgery. If surgery is scheduled for the afternoon, clear liquids in the early morning may be allowed by the anesthesiologist. Patients may brush their teeth or rinse their mouth if no water is swallowed. Cancellation of surgery may result if the patient has not been NPO as ordered.

Any medications the patient is to take the morning of surgery, with an ounce of water, are explained. Special preparations, such as an enema, are also described. For abdominal or intestinal surgery, enemas are ordered to empty the bowel in order to reduce fecal contamination preoperatively and straining or distention postoperatively.

Instructions for postoperative care are given before surgery so the patient is alert when being taught and has time to learn. Patients should be told that active participation in postoperative care aids in their recovery. Teach patients how to report their pain level using a pain rating scale (such as a 0 [none] to 10 [severe] rating scale, a color-based rating scale, or a scale using pictures of faces showing varying degrees of frowning or smiling that indicate a certain pain level) so that prompt pain relief can be provided (see Chapter 9). Pain relief methods are described, such as analgesic injections, an epidural catheter, or patient-controlled analgesia (PCA). Anticipated dressings, tubes, casts, or special equipment, such as a continuous passive motion machine for total knee replacement, are also described. If needed, crutches are fitted to the patient, and their proper use is explained and demonstrated.

Postoperative exercises are taught to decrease complications. They include deep breathing and coughing, use of incentive spirometry, leg exercises, turning, and how to get out of bed. After an exercise is taught, the patient should perform a return demonstration so understanding and ability to perform the exercise correctly can be evaluated.

Deep breathing helps prevent the development of **atelectasis** (collapse of the lung caused by hypoventilation or mucus obstruction preventing some alveoli from opening and being fully ventilated) by expanding and ventilating the lungs. The patient is taught to sit up, exhale fully, take in a deep breath through the nose, hold the breath and count to three, and then exhale completely through the mouth. The patient is told to repeat this hourly while awake, in sets of five, for 24 to 48 hours postoperatively.

Incentive spirometry may also be ordered postoperatively to prevent atelectasis by increasing lung volume, alveoli expansion, and venous return. (See Fig. 11–1.) Any patient can benefit from incentive spirometry, especially the elderly and those at increased risk for developing lung complications. The spirometer stays at the patient's bedside for hourly use while awake. Offer the spirometer to the patient

---

atelectasis: ateles—imperfect + ektasis—expansion

each hour to ensure that it is used. Teach patients to do the following:

- Sit upright, at 45 degrees minimum, if possible.
- Take two normal breaths. Place mouthpiece of spirometer in mouth.
- Inhale until target, designated by spirometer light or rising ball, is reached, and hold breath for 3 to 5 seconds.
- Exhale completely.
- Perform 10 sets of breaths each hour.

Coughing moves secretions to prevent pneumonia. Teach patients how to cough effectively (Table 11–6). Give pain medication before coughing and offer reassurance that coughing should not harm the incision. Several sets of coughing are performed, if not contraindicated by the patient's condition (such as hernia repair or head injury), every 1 to 2 hours while the patient is awake.

Leg exercises, if not contraindicated, improve circulation and help prevent complications related to stasis of blood, such as emboli formation. Patients are told to lie down, raise the leg, bend it at the knee, flex the foot, extend the leg, and lower it to the bed. Each leg is exercised in sets of five. Foot circles are also done every hour while awake. Teach the patient to raise a leg slightly off the bed with toes pointed. Draw a circle in the air with the great toe, rotating to the right four times, then to the left. Repeat this five times and then do the same with the other foot.

Patients are taught that turning from side to side in bed is aided by bending the leg that is to be on top and placing a pillow between the legs to support the top leg. They are told, unless contraindicated, to use the bed rail to pull themselves over to the side. They are encouraged to deep breathe while turning instead of holding their breath to promote comfort.

To reduce the strain on the incision and to make it easier for patients to get out of bed, patients are instructed to turn to their side, without pillows between their knees. Then they should place their hands flat against the bed and push up while swinging their legs out of bed and into a sitting position. Patients should be told to sit for a few minutes after changing position to avoid dizziness and falling. They should also deep breathe while sitting up to promote lung expansion.

Evaluation

The goal of decreased anxiety is achieved if the patient states and demonstrates that anxiety is relieved. If the patient is able to learn during teaching sessions, anxiety is not a barrier to learning.

The goal for relieved fear is accomplished if the patient verbalizes fears or states that fear is decreased or gone.

The goal for correcting deficient knowledge is reached if the patient states understanding of the information presented and accurately performs return demonstrations.

Preoperative Consent

Before performing surgery, it is the physician's responsibility to obtain voluntary, written, informed consent from the patient. The consent gives legal permission for the surgery and has two purposes. It protects the patient from unauthorized procedures, and it protects the physician, anesthesiologist, hospital, and hospital employees from claims of performing unauthorized procedures. A consent is needed for all invasive procedures, anesthesia, blood administration, and radiation or cobalt therapy.

Informed consent involves three elements. (1) The physician must tell the patient in understandable terms about the diagnosis, the proposed treatment and who will perform it, the likely outcome, possible risks and complications of treatment, alternative treatments, and the prognosis without treatment. If the patient has questions before signing the consent, the physician must be contacted to provide further explanation to the patient. It is not within the nurse's scope of practice to provide this information. (2) The consent must be signed before analgesics or sedatives are given because patients must demonstrate that they are informed and understand the surgery. (3) Consent must be given voluntarily. No persuasion or threats can be used to influence the patient. The patient can withdraw consent at any time, even after the consent form has been signed.

It is often your role to obtain and witness the patient's or authorized person's signature on the consent form. You should ensure that the person signing the consent form understands its meaning and that it is being signed voluntarily. If the patient is unable to read, the entire consent must be read to the patient before it is signed. Patients are unable to give consent if they are unconscious, are mentally incompetent, are minors, or have received analgesics or drugs that alter central nervous system function within time frames specified by agency policy. Consent can be obtained in any of these cases from parents, next of kin, or legal guardians.

In a medical emergency, the patient may not be able to give consent. In this case the next of kin or legal guardian may give telephone consent, or a court order can be

TABLE 11-6	TEACHING PATIENTS COUGHING TECHNIQUES
Procedure	**Rationale**
Have patient sit up and forward.	Promotes lung expansion and ability to generate forceful cough
Show patient how to splint incision with hands, pillow, or blanket.	Reduces incision pressure so it does not feel as if incision is opening
Have patient inhale and exhale deeply three times through mouth.	Helps expand lungs
Take in deep breath and cough out the breath forcefully with three short coughs using diaphragmatic muscles. Take in quick deep breath through mouth, cough deeply, and deep breathe.	Generates forceful cough and expands lungs to help move secretions

obtained. If time does not permit this, the physician documents the need for treatment in the chart as necessary to save the patient's life or avoid serious harm, according to state law and institutional policy.

LEARNING TIPS

Witnessing a Consent

As the patient's advocate you should ensure, before the consent is signed, that the patient is informed about the surgery and has no further questions for the physician.

If the patient does have questions, the consent should not be signed and the physician should be contacted to answer the patient's questions.

Your signature as a witness on a consent form indicates that you observed the informed patient or patient's authorized representative voluntarily sign the consent form. It does not mean that you informed the patient about the surgical procedure; that is the responsibility of the physician.

Preparation for Surgery
Preoperative Checklist

A preoperative checklist is completed and signed by the nurse (per agency policy) before the patient is transported from the unit to surgery (Fig. 11–2). The checklist provides guidance for preoperative preparation of the patient:

■ An identification band is placed on the patient. A hospital gown is given to the patient to wear. Underwear is removed, depending on the type of surgery.
■ Vital signs are taken and recorded as baseline information and to assess patient status.

■ Makeup, nail polish, and one artificial nail (if applicable) are removed to allow assessment of natural color and pulse oximetry for oxygenation status during surgery.
■ Removal of hair pins, wigs, and jewelry prevents loss or injury. Rings are taped in place if the patient does not want to take them off, unless the ring is on the operative side (arm or chest surgery), in case edema occurs.
■ Dentures, contact lenses, and prostheses are removed to prevent injury. Some patients are concerned about body image and do not want family members to see them without dentures or makeup. Remove dentures after the family goes to the waiting room and insert them before the family sees the patient postoperatively.
■ Glasses and hearing aids go with patients to surgery if they are unable to communicate without them. Label them with the patient's name and document where they go.
■ All orders, diagnostic test results, consents, and history and physical (required on the chart) are reviewed for completion and documented on the checklist.
■ Patient valuables are recorded and given to a family member or locked up per institutional policy by the nurse.
■ Antiembolism devices are applied if ordered.
■ Patients are asked to void before sedating preoperative medications are given, unless a urinary catheter is present, to prevent injury to the bladder during surgery.

Preoperative Medications

The final preparation before surgery is giving preoperative medications (Table 11–7). The medications may be given alone or in combination to achieve desired effects. Preoperative medications may be ordered at a specific time, usually 1 hour before, or on call to surgery (surgery calls you to instruct you that it is time to give the drugs). All medications

TABLE 11-7 PREOPERATIVE MEDICATIONS

Category	Medication	Purpose
Narcotics	morphine sulfate	Analgesia
	meperidine (Demerol)	Enhancement of postoperative pain relief
	fentanyl (Sublimaze)	
Antianxiety and sedative hypnotics	diazepam (Valium)	Sedation
	hydroxyzine hydrochloride (Vistaril)	Anxiety reduction
	lorazepam (Ativan)	
	midazolam (Versed)	
	phenobarbital sodium (Luminal sodium)	
Anticholinergic	atropine sulfate	Secretion reduction
	glycopyrrolate (Robinul)	
	scopolamine hydrobromide	
Antiemetic	droperidol (Inapsine)	Control nausea and vomiting; may be effective into the postoperative period
	ondansetron (Zofran)	
	metoclopramide (Reglan)	
	promethazine hydrochloride (Phenergan)	
Histamine (H₂) antagonist	cimetidine (Tagamet)	Reduction of acidic gastric secretions in case aspiration occurs either silently or with vomiting
	famotidine (Pepcid)	
	ranitidine (Zantac)	
Alkalinizing agent	Sodium citrate and citric acid (Bicitra)	Prevention of an asthmalike attack, pneumonitis, pulmonary edema, or severe hypoxia
Antibiotic	cefazolin (Ancef)	Prevention of postoperative infection
	cefoxitin (Cefotan)	
	ampicillin (Omnipen)	

Pre-op Surgical Checklist   **Client Name**

_____ I.D. BAND ON   _____

_____ NPO AS ORDERED

_____ PRE-OP TEACHING COMPLETED

_____ INFORMED CONSENT SIGNED

_____ HISTORY AND PHYSICAL ON CHART

_____ ALLERGIES

_____ LAB RESULTS

_____   CBC: HGB _____ HCT _____ WBC _____ PLATELETS _____

_____   POTASSIUM _____

_____   URINALYSIS _____

_____   PREGNANCY TEST   SERUM _____ URINE _____

_____   PT _____ PTT _____ BLEEDING TIME _____

_____   TYPE AND SCREEN _____ CROSSMATCH _____-___ UNITS

_____ ECG ON CHART

_____ CHEST X-RAY REPORT ON CHART

_____ SHOWERED/BATHED

_____ HOSPITAL GOWN ON

_____ PREPS COMPLETED AS ORDERED

_____ ANTIEMBOLISM STOCKINGS

_____ JEWELRY TAPED/REMOVED: DISPOSITION _____

_____ VALUABLES: DISPOSITION _____

_____ DENTURES, PROSTHESIS REMOVED

_____ HAIR PINS, WIGS, MAKE UP, NAIL POLISH, ONE ACRYLIC NAIL REMOVED

_____ CONTACT LENSES REMOVED

_____ VOIDED

_____ VITAL SIGNS: T _____ P _____ R _____ BP _____

_____ PRE-OP MEDICATIONS GIVEN _____ SIDE RAILS UP _____

_____ IV STARTED _____

_____ EYE GLASSES AND HEARING AID(S) TO OR

_____ OLD CHART TO OR

_____ X-RAYS TO OR

_____ FAMILY LOCATION _____

_____ NEXT OF KIN _____

_____ CLIENT READY FOR SURGERY _____

         TIME _____   (NURSE SIGNATURE)

         COMMENTS:

Figure 11–2   Sample preoperative checklist form.

TABLE 11–8 POSTOPERATIVE PATIENT HOSPITAL ROOM PREPARATION

After patient transfer to surgery, the nurse should prepare the patient's room for the patient's postoperative care needs to be ready for the patient's return from the postanesthesia care unit.

Preparation	Rationale
Bed	
Bed linens should be clean and are changed if used before surgery by patient.	Reduces contamination of surgical wound
Place disposable, absorbent, waterproof pads on bottom sheet if drainage is expected.	Protects linen from wetness and soiling so a patient in pain does not have to be disturbed for linen change.
Apply lift sheet on bed of patient needing assistance with repositioning.	Makes lifting and turning easier for patient and nurse.
Have extra blankets available.	Patient may be cold or have a low temperature.
Fanfold top cover to end of bed or to side of bed away from patient transfer side.	Bed ready to receive patient on transfer from cart to bed. Allows covers to be easily pulled up over patient.
Obtain extra pillows as needed for positioning, elevating extremities, splinting during coughing.	Pillows help maintain position when patient is turned, elevate operative extremities for comfort and swelling reduction, or splint an incision during coughing.
Equipment	
Obtain IV pole.	Surgical patients have IV infusions postoperatively.
Have emesis basin at bedside.	Nausea or vomiting may occur, especially after movement during transfer.
Have tissues and washcloths in room.	Promotes comfort: washing face or a cool cloth on forehead.
Have urinal or bedpan available in room.	Patients may be unable to get out of bed for first voiding.
Prepare suction setup for tracheostomy, nasogastric tube, or drains as ordered.	Suction may be ordered related to surgical procedures: Sterile suction: tracheostomy; Nasogastric tube: thoracic, abdominal, gastrointestinal surgery; T-tube: cholecystectomy
Have oxygen set up as needed.	After tracheostomy, patients wear humidified oxygen mask.
Obtain special equipment as indicated by the surgical procedure.	Institutional policy and physician orders may require specialized equipment. Examples: Jaw surgery: suction, wire cutters, tracheostomy tray; Tracheostomy: suction, extra tracheostomy set, tracheostomy care supplies; Transurethral resection of prostate: irrigation supplies.
Have vital sign equipment available.	Promotes ability to promptly obtain vital signs.
Documentation Forms	
Obtain agency postoperative documentation forms and place in room	Promotes timely and accurate documentation of patient data.

IV = Intravenous.

administered are documented. The bed rails are raised for safety and the patient is instructed not to get up alone after medications are given.

Transfer to Surgery

When the surgery department is ready, the patient is taken to the surgical holding area on a stretcher. The patient's chart, inhaler medications for those with asthma, and glasses or hearing aids for those who need them to communicate also go to the surgical holding area. The patient can be accompanied by family members.

During surgery, the family waits in the surgical waiting area, which is a communication center where the family is kept informed regarding the patient's status. The physician calls the family there when surgery is over. Families may be given beepers so that they can walk outside or to other areas of the hospital and still be reached.

After Transfer

After the patient goes to surgery, prepare the patient's room and necessary equipment so it is ready for the patient's return (Table 11–8).

▶ INTRAOPERATIVE PHASE

When the patient is transferred to the operating table, the next phase of the perioperative period, the intraoperative phase, begins. Surgery may take place in a hospital operating room (OR) or freestanding ambulatory or outpatient surgical centers. Additionally, surgery is performed in physician's offices, cardiac catheterization laboratories, radiology centers, emergency rooms, and specialized units that perform endoscopy procedures.

Gloves are worn as part of the surgical sterile field by the OR team. The OR team members must perform a sterile surgical hand scrub to reduce the amount of microorganisms on their hands and arms. An antimicrobial soap scrub is used to reduce the chances of microorganism contamination. Jewelry (e.g., watches, rings, bracelets) is also removed. Fingernails are kept short and clean. Artificial nails and colored nail polish are not permitted. Artificial nails can harbor microorganisms and colored nail polish prevents seeing if nails are clean.

The OR is designed to enhance **aseptic** technique. Clean and contaminated areas are separated. Special ventilation systems control dust and prevent air from flowing into the OR from hallways. The temperature and humidity in the room are controlled to discourage bacterial growth. Everyone entering the OR wears surgical scrubs, shoe covers, caps, masks, and goggles to protect the patient from infection and themselves from blood-borne pathogens. Traffic in and out of the OR is limited. Strong disinfectants are used to clean the OR after each surgical case, and instruments are sterilized.

Before the patient arrives in surgery, a nursing plan of care is developed from preadmission assessment data. The OR is prepared based on the plan of care (Nursing Care Plan Box 11–4). Attention is given to safety needs of the patient. Special needs are addressed, such as a tall patient's need for longer table length or an elderly patient's positioning needs resulting from osteoporosis. A surgical case cart containing sterile instruments required for the patient's case is prepared ahead of time. Items such as needles and sponges are counted before surgery and again before closure of the incision to account for all of them and ensure that none have been left in the patient. To ensure safety, electrical equipment is checked for proper functioning.

Health Care Team Member Roles

The licensed practical nurse/licensed vocational nurse's (LPN/LVN's) major role in this phase is as a scrub nurse who assists the physician during the surgery. Scrub nurses scrub their hands and arms and dress in sterile garb. They maintain the sterile instrument field and hand the sterile surgical instruments to the physician. The scrub nurse creates a sterile field and removes the sterile instruments from the cart.

Other health care team members involved in this phase may include the following:

- Physician (medical doctor [MD], doctor of osteopathy [DO], oral surgeon, or podiatrist)
- Surgical (first) assistant: assists the physician and is either another physician, a specially trained registered nurse, or a physician's assistant

- Anesthesiologist: physician who specializes in anesthesia and supervises certified registered nurse anesthetists in the operating room
- Certified registered nurse anesthetist (CRNA): registered nurse trained and certified in the administration of anesthesia, usually at the master's degree level
- Registered nurse (RN): circulates in OR; roles include being patient's advocate, planning care, protecting patient safety, monitoring patient positioning, checking vital signs and making patient assessment, reducing patient's anxiety, monitoring sterility during surgery, preparing skin before incision, managing equipment such as by making sponge counts, documenting the procedure, and aiding health team communications
- Surgical (second assistant) technician: assists physician, such as scrub nurse (may be an RN, LPN/LVN, or surgical technologist)

Patient Arrival in Surgery

The nurse greets the patient; verifies the patient's name, age, allergies, surgeon performing the surgery, consent, surgical procedure (especially right or left when applicable), and medical history; answers questions; and alleviates anxiety. The patient is introduced to the anesthesiologist and CRNA, who also verify patient information and explain the type of anesthesia that is to be used. All surgical patients have intravenous (IV) fluids started, and the patient may receive prophylactic antibiotics.

As a patient enters the OR, he or she should be told what to expect:

- "The room may feel cool, but you can request extra blankets."
- "There is a lot of equipment in the room, including a table and large, bright overhead lights."
- "Several health care team members will introduce themselves to you."
- "Your physician will greet you."

The patient is assisted onto the operating table, and a safety strap is carefully applied. Monitoring equipment is applied and readings recorded. Then the anesthesia provider begins anesthesia. When the anesthesia provider gives permission, the patient is carefully positioned to prevent pressure points that could cause tissue or nerve damage. Any necessary tubes that are not already in place, such as a nasogastric tube or urinary catheter, are inserted by the RN.

BOX 11-4 **NURSING CARE PLAN FOR INTRAOPERATIVE NURSING DIAGNOSES AND OUTCOMES**

- Risk for perioperative-positioning injury related to positioning, chemicals, electrical equipment, and effect of being anesthetized
  Free from injury
- Risk for impaired skin integrity related to chemicals, positioning, and immobility
  Skin integrity is maintained
- Risk for deficient fluid volume related to NPO status and blood loss
  Maintains blood pressure, pulse, and urine output within normal limits
- Risk for infection related to incision and invasive procedures
  Is free of symptoms of infection
- Pain related to positioning, incision, and surgical procedure
  Reports pain is relieved to satisfactory level

Patient allergies are checked. Then a skin prepping solution that the patient is not allergic to, such as povidone iodine, is used to cleanse the skin. A large area surrounding the operative site is scrubbed to allow for extension of the incision. The scrub is completed in a circular motion from inside to outer edge. If an allergic reaction to the solution occurs, it can cause skin redness and blistering wherever the solution was used. After the skin is scrubbed, a sterile drape is applied with the incisional area left exposed.

Anesthesia

Anesthesia is used for surgery to prevent pain and allow the procedure to be done safely. The type of anesthesia and the anesthetic agents are ordered by the anesthesia provider with input from the patient and physician (Cultural Consideration Box 11–5).

BOX 11-5	CULTURAL CONSIDERATION

Care of the Patient Having Surgery

Chinese people show a greater increase in heart rate in response to atropine than Caucasian persons. Thus the nurse needs to carefully monitor the pulse after atropine is given as a preoperative medication. Additionally, Chinese people are more sensitive to the sedative effects of diazepam (Valium) and require lower doses. Thus the nurse needs to carefully monitor the Chinese patient for untoward effects of diazepam (Valium).

There are two types of anesthesia: general and local (regional). General anesthesia (GA) causes the patient to lose sensation, consciousness, and reflexes. GA acts directly on the central nervous system. Local anesthesia results in the loss of sensation to a region of the body without the loss of consciousness. Local anesthesia works by blocking the nerve impulses along the nerve where it is injected.

General Anesthesia

General anesthesia is commonly given by an IV or inhalation route. GA is chosen when patients are very anxious or do not want local anesthesia, when the surgical procedure will take a long time and there is a need for muscle relaxation, or when the patient is unable to cooperate, as in head injury, muscle disorders, or impaired cognitive function.

INTRAVENOUS AGENTS. To begin most general anesthesia, the patient is induced (i.e., anesthesia is caused) with a short-acting IV agent that provides a rapid, smooth **induction** (the period from when anesthetic is first given until full anesthesia is reached). Because these agents last only a few minutes, they are used along with inhalation agents, which maintain anesthesia during surgery. After induction, the patient is intubated with an endotracheal (ET) tube to provide mechanical ventilation and anesthesia (Fig. 11–3).

INHALATION AGENTS. Maintenance of anesthesia is accomplished by using inhalation agents. These agents are de-

induction: inductio—to lead in

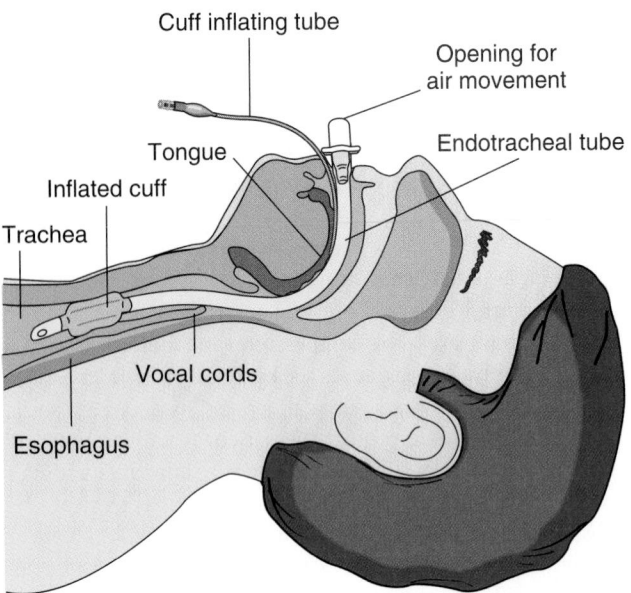

Figure 11-3 Endotracheal tube with cuff inflated.

livered, controlled, and excreted through mechanical ventilation. Inhalation agents and the ET tube can be irritating to the respiratory tract. Complications that can occur from their use include laryngospasm (sudden violent contraction of the vocal cords), laryngeal edema, irritated throat, or injury to the vocal cords. When the tube is removed, the nurse should closely monitor the patient and be prepared to provide respiratory support and assist with reintubation if complications arise.

ADJUNCT AGENTS. An **adjunct** agent is a medication used along with the primary anesthetic agents. These medications can include narcotics to control pain, muscle relaxers to avoid movement of muscles during surgery, antiemetics to control nausea or vomiting, and sedatives to supplement anesthesia.

Malignant Hyperthermia

Malignant hyperthermia (MH) is a rare hereditary muscle disease that can be triggered by some types of general anesthesia agents. Obtaining a history of anesthesia problems in the patient or family members can help detect the potential for development of this condition so that precautions can be taken. A muscle biopsy can also diagnose this problem. Patients with this condition can undergo surgery safely with careful planning and choice of anesthetic agents by the anesthesia provider.

With MH there is increased metabolism in muscles, which produces a very high fever and muscle rigidity, as well as tachycardia, tachypnea, hypertension, dysrhythmias, hyperkalemia, metabolic and respiratory acidosis, and cyanosis. MH is life threatening, so immediate treatment is required or death results. Surgery is stopped and anesthesia discontinued immediately. Oxygen at 100% is given. The patient must be cooled with ice and infusions of iced solutions. Dantrolene sodium (Dantrium), a muscle relaxant that relieves the muscle spasms, is the most effective medication for treating malignant hyperthermia. Dantrolene sodium is kept readily available in the OR and administered according to the treatment protocol of the Malignant Hyperthermia Association of the United States. For more information about malignant hyperthermia, visit www.mhaus.org.

Local (Regional) Anesthesia

Local anesthesia is selected when the patient is not anxious, can tolerate the local agent, and is not required by the surgical procedure to be unconscious or relaxed. It is a good choice for some outpatient procedures or when the patient has not been NPO. The anesthesia provider, or sometimes the physician, administers local anesthesia.

Local anesthetic agents may include bupivacaine hydrochloride (Marcaine), lidocaine (Xylocaine), and dibucaine (Nupercainal). Topical administration places the agent directly on the surgical area. Local infiltration is achieved by injecting the medication into the tissue where the incision is to be made. A regional block is done by injecting the local agent along a nerve that carries impulses in the region where anesthesia is desired. There are several types of regional blocks. A nerve block is the injection of a nerve at a specific point. A Bier block is done by placing a tourniquet on an extremity to remove the blood and then injecting the local agent into the extremity. A field block is a series of injections surrounding the surgical area. A spinal or epidural block is injection of a local agent into an area around the spinal nerves.

SPINAL AND EPIDURAL BLOCKS. Injection of a local agent into the subarachnoid space produces spinal block (Fig. 11–4). Epidural block occurs when the local agent is injected

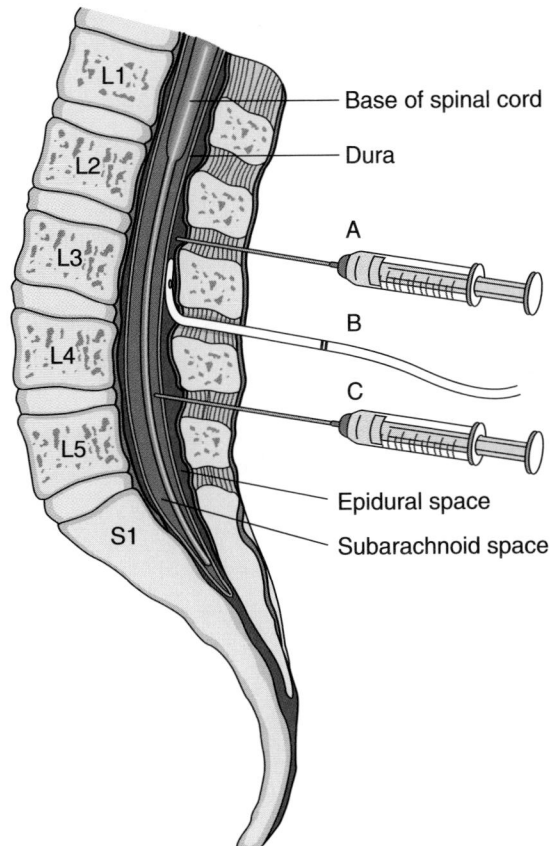

Figure 11-4 Injection of spinal anesthesia. *(A)* Epidural anesthesia. *(B)* Epidural catheter. *(C)* Spinal anesthesia.

Base of spinal cord
Dura
A
B
C
Epidural space
Subarachnoid space

into the epidural space. Spinal and epidural blocks are used mainly for lower extremity and lower abdominal surgery. Both motor and sensory function is blocked. The patient must be carefully monitored for complications. Hypotension results from sympathetic blockade causing vasodilation, which reduces venous return to the heart and therefore reduces cardiac output. Respiratory depression results if the block travels too far upward. As the block wears off, patients feel as if their legs are very heavy and numb. This is normal, and reassurance should be offered to the patient that this type of feeling does not last after the block wears off.

Complications. A postdural puncture headache may occur in 10 to 40 percent of patients caused by leakage of cerebrospinal fluid (CSF) from the needle hole in the dura that does not close when the needle is withdrawn. The use of a small-gauge spinal needle (less than 25 gauge) helps prevent headaches. Photophobia or double vision may also be present. Postanesthesia orders usually include methods to help reduce the pain of a headache, including positioning the patient flat and forcing fluids.

If a spinal headache develops, it may be severe and can last for weeks. Lying flat and prone (on abdomen) may help relieve the pain of the headache. Adequate fluid intake and analgesics may be ordered, and steroids may be helpful. If the headache continues, a blood patch treatment can be used to stop the CSF leakage. This treatment can be done by the anesthesiologist at the bedside or in the postanesthesia care unit (PACU). To create a blood patch, approximately 10 mL of the patient's own blood is injected into the epidural space at the previous puncture site. The injected blood forms a clot that "patches" the dura hole to prevent further CSF leakage. Pain relief should occur quickly if the patch is successful. Blood patch treatment can be repeated.

Adjunct Anesthesia Techniques

Several techniques may be used with anesthesia to improve the outcome of surgery. Cryoanesthesia uses cooling or freezing in a localized area to block pain impulses. Acupuncture produces loss of sensation. Controlled hypotension decreases blood loss by lowering the blood pressure.

Conscious Sedation

Conscious sedation is purposeful, minimal sedation that does not cause the complete loss of consciousness during selected medical, dental, or diagnostic procedures. IV medications such as sedatives, hypnotics, and opioids are given to produce conscious sedation. Selection of patients who are eligible for conscious sedation is based on the procedure, the patient's general health, patient preference, and physician preference. Examples of short procedures for which conscious sedation is used are dental procedures, endoscopy, cardiac catheterization, cardioversion, and closed fracture reduction. During conscious sedation, patients are comfortable, respond purposefully, and maintain their own patent airway. Medications are ordered by the physician and usually administration by a specially trained registered nurse, as defined by agency and state scope of nursing practice.

A signed surgical consent is obtained. Then an IV for medications and fluids is inserted before the procedure. Patient monitoring is done every 5 minutes to check vital signs, electrocardiogram, and oxygen saturation. Changes are reported to the physician. Oxygen may be given by nasal cannula or mask. Emergency equipment (e.g., airway suction, defibrillator, drugs) is on standby, according to advanced cardiac life support (ACLS) protocols.

After the procedure, the patient awakens easily and quickly. The patient is monitored every 15 to 30 minutes for response to the procedure and the drugs, until the patient is fully awake and stable. The patient is ready for discharge when vital signs return to baseline and are stable, oral fluids are retained, he or she has voided (if applicable), and written and oral discharge teaching is given to both the patient and the responsible adult to whom the patient is being discharged. The responsible adult and the patient must sign the instructions. Instructions include the following: an adult must drive the patient home and provide a safe environment; the patient must not and will not drive or operate heavy machinery or sign legal documents for 24 hours.

Transfer from Surgery

When surgery is completed and anesthesia stopped, the patient is stabilized for transfer. After local anesthesia, the patient may return directly to a nursing unit. After general and spinal anesthesia, the patient goes to the PACU or in some cases the intensive care unit (ICU).

Patient safety, which is always a priority, is an important concern at this time. The patient is never left alone. Ensuring a patent airway and preventing falls and injury from uncontrolled movements are priorities. The anesthesia provider and OR nurses transfer the patient to the PACU and monitor the patient until the PACU nurse is able to receive the report and assume care of the patient. This begins the final patient perioperative phase, the postoperative period.

▶ POSTOPERATIVE PHASE

The postoperative phase begins when the patient is admitted to the PACU or a nursing unit and ends with the patient's postoperative evaluation in the physician's office. The family is updated on the patient's status by the physician as the patient is admitted to the PACU.

Admission to the Postanesthesia Care Unit

The goal of the postanesthesia nurse is to promote recovery from anesthesia. The nurse's role in the PACU begins by receiving a patient report from the OR nurse and anesthesia provider. When the patient is admitted to the PACU, an admission assessment is done. The priority areas of patient assessment are the following:

- Respiratory status and patency of airway
- Vital signs
- Level of consciousness
- Surgical site incision/dressing
- Pain level

Oxygen by nasal cannula or mask is given if the patient has had general anesthesia, or as ordered. Some patients who are still intubated may require mechanical ventilation. Continuous monitoring is done on all patients for ECG, pulse oximetry, and blood pressure measurements. The surgical site incision or dressing is assessed. Drainage and **hematoma** formation are documented and reported. Urinary catheter, drains, and nasogastric tubes are checked for function and patency as applicable.

The patient's body temperature is measured on admission, and then blankets heated in blanket warmers are applied. If the patient's temperature is below normal, a warming blanket that can be set to a desired temperature is applied. Body temperature may be decreased as a result of a cool OR environment, anesthesia, cool IV solutions, and incisional openings, which allow heat loss. Patient recovery is aided by avoiding a decrease in body temperature during surgery. Using forced warm air before surgery to warm the patient, rather than warm blankets, has been shown to maintain body temperature best upon arrival in PACU. Be aware that the elderly are at increased risk of **hypothermia.** Temperature is measured again before PACU discharge because a normal body temperature is usually one of the discharge criteria.

Shivering may occur from anesthesia or as a result of being cold. It is important to control shivering because it increases oxygen consumption 400 to 500 percent. Meperidine (Demerol) is effective in relieving shivering when anesthesia is the cause. If the patient is cold, raising the body temperature is helpful to decrease shivering.

PACU nursing responsibilities are listed in Box 11–6. Vital signs and assessment are done at least every 15 min-

---

BOX 11–6   Postanesthesia Care Unit Nursing Responsibilities

Airway maintenance
Vital signs
Respiratory assessment
Neurological assessment
Surgical site status
General assessment
Patient safety
Monitoring anesthetic effects
Pain relief
Assessing PACU discharge readiness

---

hematoma: heimatos—blood + oma—tumor

utes, IV fluid infusion is maintained, and IV analgesics are given for pain as needed. Antiemetics are administered for nausea or vomiting. Deep breathing and coughing, if not contraindicated by the surgical procedure, are encouraged. Coughing increases pressure, which could cause harm to some surgical areas. Surgical procedures that prohibit coughing include hernia repair; eye, ear, intracranial, and plastic surgery; and jaw surgery. If the patient is no longer NPO, ice chips or sips of water may be offered when the patient is fully awake to promote comfort for a dry mouth. If ordered, postoperative therapies such as patient-controlled analgesia, which allows patients to administer their own pain medication (see Chapter 9), or continuous passive motion machines for joint replacement surgeries are begun in the PACU.

Nursing Process for Postoperative Patients in PACU

Many postoperative complications may occur in the PACU or later in the postoperative phase. The cause of these complications may be the surgical procedure, anesthesia, blood and fluid loss, immobility, unrelieved pain, or other diseases the patient may have. Nursing care focuses on preventing, detecting, and caring for these complications.

Respiratory Function

ASSESSMENT. Normal respiratory function can be altered in the immediate postoperative period by airway obstruction, hypoventilation, secretions, laryngospasm, or decreased swallowing and cough reflexes. Respiratory function assessment includes respiratory rate, depth, ease, and pattern. Breath sounds, chest symmetry, accessory muscle use, and sputum are also observed.

NURSING DIAGNOSIS. Respiratory-related nursing diagnoses for the PACU include the following:

■ Ineffective airway clearance related to obstruction, anesthesia medications, and secretions
■ Ineffective breathing pattern related to anesthesia medications and pain
■ Risk for aspiration related to depressed cough and gag reflexes and reduced level of consciousness

PLANNING. The patient's goals are to have a patent airway at all times, breathe comfortably and maintain normal arterial blood gases, and have clear lung sounds.

NURSING INTERVENTIONS. The priority nursing responsibility in the PACU is to ensure that patients maintain a patent airway. Airway obstruction is usually caused from relaxed muscles allowing the tongue to block the pharynx in the patient who has a decreased level of consciousness. It may be necessary to manually open a patient's airway to prevent obstruction if the patient has snoring respirations and has not completely emerged from anesthesia. The jaw-thrust method is used to open the airway.

Patients are positioned on their side, unless contraindicated, to protect the airway until they are awake. Then when they are awake, elevating the head in a supine position assists respiration.

Suction equipment is always readily available to clear secretions or emesis. Aspiration is a risk because of the effects of anesthesia, which depresses cough and gag reflexes and reduces level of consciousness. As a result, secretions may accumulate and not be effectively controlled.

Oxygen therapy is started on all general anesthesia patients. Hypoventilation can be an effect of anesthesia medications or analgesics, decreased level of consciousness, or an incision in the thorax causing painful respirations.

Deep breathing is encouraged to expand the lungs. Pain is controlled with carefully administered analgesics to promote deep breathing but avoid respiratory depression. Respiratory depression is reported to the anesthesiologist for prompt treatment.

EVALUATION. The goal for ineffective airway clearance and aspiration is achieved if the patient's airway remains patent and lung sounds remain clear.

The goal for ineffective breathing pattern is met if the patient's respiratory rate is within normal limits, no dyspnea is reported, and arterial blood gases are within normal limits.

Cardiovascular Function

ASSESSMENT. Alterations in cardiovascular function can include hypotension, dysrhythmias, and hypertension. Hypotension can be the result of blood and fluid volume loss or cardiac abnormalities. Shock can result from significant blood and fluid volume loss (see Chapter 8). Dysrhythmias may occur from hypoxia, altered potassium levels, hypothermia, pain, stress, or cardiac disease. New-onset hypertension can develop from pain, a full bladder, or respiratory distress. Cardiovascular function assessment includes heart rate, blood pressure, ECG, and skin temperature, color, and moistness. Vital signs are compared with baseline readings to determine if they are normal. Tachycardia, hypotension, pale skin color, cool, clammy skin, and decreased urine output indicate hypovolemic shock, which requires prompt treatment. Abnormal findings are promptly reported to the physician for intervention.

LEARNING TIPS

Tachycardia: An Early Warning Sign
Tachycardia is a compensatory mechanism designed to provide adequate delivery of oxygen in times of altered function. It is usually the earliest warning sign that an abnormality is occurring. It should be a red flag to assess the patient and ask yourself what this particular patient is likely to be experiencing that is compromising oxygenation, allowing you to begin prompt treatment.

Patient Condition	Possible Causes of Compromised Oxygenation
Postoperative patient	Hemorrhage, respiratory depression, pain
Myocardial infarction patient	Cardiogenic shock, pain
Respiratory patient	Respiratory distress
Trauma patient	Hemorrhage, severe pain

NURSING DIAGNOSIS. The PACU cardiovascular-related nursing diagnoses include the following:

- Deficient fluid volume related to blood and fluid loss or NPO status
- Decreased cardiac output related to volume loss, medication effects, or cardiac disease

PLANNING. The patient's goals are to maintain blood pressure, pulse, and urine output within normal limits; and to maintain blood pressure, pulse, and rhythm within normal limits.

NURSING INTERVENTIONS

- Dressings and incisions are checked for color and amount of drainage.
- IV fluids are maintained at the ordered rate to replace lost fluids but avoid fluid overload.
- Intake and output are monitored to detect imbalances.
- Providing pain relief, warming the patient, and preventing bladder distention help prevent hypertension.

EVALUATION. The goal for deficient fluid volume is met if vital signs and urine output are within normal limits.

The goal for decreased cardiac output is achieved if vital signs and rhythm are within normal limits.

Neurological Function

Until its effects wear off, anesthesia can alter neurological function. Patients may arrive in the PACU either awake, arousable, or sleeping. If patients are sleeping, they should become more alert during their stay in the PACU. As emergence from anesthesia occurs, patients may become wild or agitated for a short period. This is referred to as emergence delirium. Amnesia effects can be caused by anesthesia. Movement, sensations, and perceptions may also be altered by anesthesia. Movement is the first function to return after spinal anesthesia.

For geriatric patients, it is important to review their history to understand if they have any cognitive or neurological deficits. Confused patients may be agitated or frightened when they awaken. It is helpful to know how caregivers normally communicate with the patient. You should understand that the patient may not be able to report pain or follow commands. If possible, it may be helpful to have a familiar relative or caregiver with the patient in the PACU to calm them and help them communicate. You should watch for nonverbal pain cues such as moaning, grimacing, rubbing an area, and restlessness. If patients have limited movements or sensations before surgery, you should know this to obtain an accurate assessment of anesthesia effects.

ASSESSMENT. A neurological assessment includes level of consciousness; orientation to person, place, and time; pupil size and reaction to light; and motor and sensory function.

NURSING DIAGNOSIS. The PACU neurological-related nursing diagnoses include the following:

- Risk for injury related to anesthesia effects causing decreased level of consciousness, sensation, and movement.

- Disturbed sensory perception related to decreased level of consciousness, amnesiac effects of anesthesia, or spinal anesthesia.

PLANNING. The patient's goals are to maintain safety and be free from injury.

NURSING INTERVENTIONS

- All patient data are verified until patients are awake and can communicate.
- Patients' safety is maintained until they are fully awake or movement and sensation return following spinal anesthesia. Side rails are kept up. Extremities are positioned in proper alignment and protected from injury until sensation and movement return.
- Tubes, dressings, and IVs are secured and monitored.
- As patients wake up, orientation explanations are provided and repeated until amnesiac anesthesia effects are no longer present. Examples of explanations include the following: "Mr. Smith, surgery is over; you are in the recovery room." "Your family is waiting for you and knows you are in the recovery room." "The doctor spoke with your family about how you are doing."

EVALUATION. The goals for risk for injury and disturbed sensory perception are met if the patient remains free from injury.

Pain

ASSESSMENT. If patients are awake, they are asked to rate the presence of pain using a scale, such as 0 to 10, a pain scale using color, or pictures that rate pain. The location and character of the pain are documented. When patients are not fully awake, vital signs and nonverbal indications of pain should be monitored. Nonverbal indications of pain can include abnormal vital signs, restlessness, moaning, grimacing, rubbing, or pulling at specific areas or equipment.

PLANNING. The patient's goal for pain is to report that pain is relieved at a satisfactory level.

NURSING INTERVENTIONS

- Patient pain is monitored and promptly relieved. IV narcotic analgesics are used for their rapid onset. If PCA is ordered, it is started in PACU. Pain can result from the surgical procedure, movement, deep breathing, anxiety, or a full bladder. Positioning during surgery and the presence of devices such as nasogastric tubes, catheters, IVs, or ET tubes can also be a source of pain.
- Repositioning the patient, providing warmth, and emptying a full bladder may alleviate pain.
- Playing music in the PACU, such as nature sounds or Mozart, dimming the lights, and reducing noise can reduce pain.

EVALUATION. After pain medication is given, patients are asked to rate their pain level. The goal for pain is met if they report a decreased level of pain that is satisfactory to them. For example, the patient reports pain of 10 on a scale of 0 to 10. You medicate the patient, and 30 minutes later the patient rates pain as 2 on a scale of 0 to 10.

CHAPTER 11 NURSING CARE OF PATIENTS HAVING SURGERY **163**

Discharge from the Postanesthesia Care Unit

The length of stay in the PACU if the patient remains stable is normally 1 hour. A postanesthesia recovery scale is used to score the patient's readiness to be discharged. The scale rates categories such as respiration, oxygen saturation, level of consciousness, activity, and circulation. The anesthesiologist orders the patient to be transferred to a nursing unit or discharged home when discharge criteria are met (Table 11–9). The patient may be transferred to the ICU if frequent monitoring is needed.

Transfer to Nursing Unit

The PACU nurse gives a report of the patient's condition to the unit nurse when the patient is transferred to the nursing unit. The patient is moved into bed on the nursing unit with assistance to prevent dislodging IVs, tubes, drains, or dressings. After patients are placed in bed, the following safety interventions are performed:

- The bed is placed in its lowest position, with the side rails raised.
- The nurse's call button is placed within easy patient reach and answered promptly.
- Patients are instructed that they should be assisted with ambulation when they get up.

When patients get up postoperatively, especially for the first time, they may be weak or dizzy. One or two health care workers should assist the patient and allow the patient to dangle before standing (Fig. 11–5). These precautions can help prevent falls.

Nursing Process for Postoperative Patients

A complete patient assessment is performed after transfer to the nursing unit. Respiratory status, vital signs (including temperature), level of consciousness, surgical site, dressings, and pain assessment are especially noted. IV site and patency are assessed. The IV solution and infusion rate are verified and monitored. Nasogastric tubes are hooked to suction or clamped as ordered. Drains and catheters are positioned to promote proper functioning.

TABLE 11-9	DISCHARGE CRITERIA FOR POSTANESTHESIA CARE UNIT OR AMBULATORY SURGERY

Vital signs stable
Patient awake or at baseline level of consciousness
Drainage or bleeding not excessive
Respiratory function not depressed
Oxygen saturation above 90%

Additional Criteria for Ambulatory Surgery

No nausea or vomiting
No IV narcotics within last 30 minutes
Voided if required by surgical procedure or ordered
Is ambulatory or has baseline mobility
Understands discharge instructions
Provides means of contact for follow-up telephone assessment
Released to responsible adult

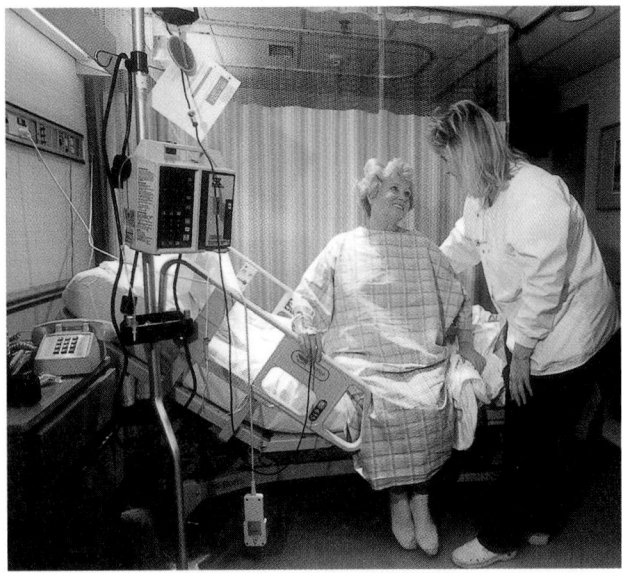

Figure 11-5 Patient dangling.

After patients are discharged from the PACU, your role is to provide interventions to promote recovery. These interventions may include monitoring for complications, providing postoperative care, educating patients and their significant others, making necessary referrals, and providing home health care (Nursing Care Plan Box 11–7 for the Postoperative Patient). The patient is watched for possible problems such as respiratory depression, hemorrhage, and shock. This is especially important during the first 24 hours postoperatively.

Respiratory Function

ASSESSMENT. The patient's respiratory status is assessed and monitored to ensure that a patent airway is maintained. Chest symmetry is noted. If the patient's airway is compromised, immediate action is taken to support the airway and the physician is notified. Breath sounds are auscultated for adequate air exchange and absence of abnormal sounds in all lobes. The patient's cough is assessed if it is not contraindicated by the type of surgery, such as hernia repair or eye, ear, intracranial, jaw, or plastic surgery.

NURSING DIAGNOSES. Postoperative respiratory-related nursing diagnoses include the following:

- Ineffective airway clearance related to ineffective cough and secretion retention
- Ineffective breathing pattern related to analgesic medications and pain

PLANNING. The patient's goals are to have a patent airway and clear breath sounds and normal arterial blood gases. Factors that may inhibit this goal should be identified and discussed with the patient.

NURSING INTERVENTIONS

- Regular monitoring of the patient's respiratory rate, depth, and effort; cough strength; and breath sounds should be done. Postoperative patients are at risk for developing atelectasis and pneumonia. They may have a weak cough as

BOX 11-7   NURSING CARE PLAN FOR THE POSTOPERATIVE PATIENT

 Ineffective airway clearance related to ineffective cough and secretion retention

PATIENT OUTCOMES

Patient maintains a patent airway at all times. Breath sounds remain clear.

EVALUATION OF OUTCOMES

Is patient able to clear own secretions? Are breath sounds clear?

INTERVENTION	RATIONALE	EVALUATION
Monitor breath sounds.	Abnormal breath sounds such as crackles or wheezes can indicate retained secretions.	Are breath sounds clear?
Encourage deep breathing and coughing and use of incentive spirometer hourly while awake.	Lung expansion helps prevent atelectasis and keeps lungs clear of secretions.	Does patient perform breathing and coughing and use incentive spirometer?
Ensure patient pain is relieved before activity.	Movement can cause or increase pain.	Does patient state pain is controlled before activity?
Encourage movement by turning every 2 hours and ambulating as able.	Movement promotes lung expansion and movement of secretions.	Is patient moving?

 Acute pain related to surgery, nausea, and vomiting

PATIENT OUTCOMES

Patient reports pain management relieves pain satisfactorily and describes pain management plan.

EVALUATION OF OUTCOMES

Does patient report satisfactory pain relief? Is patient able to describe pain management plan?

INTERVENTION	RATIONALE	EVALUATION
Assess pain using grating scale such as 0 to 10.	Self-report is the most reliable indicator of pain.	Does patient report pain using scale?
Provide analgesics prn.	Analgesics relieve pain.	Is patient's pain less after medication?
Provide antiemetics prn.	Antiemetics relieve nausea and vomiting.	Is patient's nausea and vomiting less after medication?
Position patient comfortably.	Incisions, drains, tubing, equipment, and bed rest can cause discomfort, which positioning can relieve.	Does patient report positioning is comfortable?

	GERIATRIC	
When assessing pain speak clearly and slowly so elderly patient can hear.	If elderly patient does not hear or misunderstands, pain may not be reported accurately to ensure appropriate intervention provided.	Does patient hear and report pain and relief accurately using pain scale?
Assess elderly patients' pain level regularly, observing nonverbal pain cues (restlessness, grimacing, moaning), especially for those cognitively impaired.	The pain of elderly patients is often underreported and undertreated, especially if cognitively impaired, and noting nonverbal cues can aid in pain treatment.	Are nonverbal cues present in elderly patients, especially those cognitively impaired?

 Risk for infection related to inadequate primary defenses from surgical wound

PATIENT OUTCOMES

Patient remains free from infection.

EVALUATION OF OUTCOMES

Does patient remain free from infection?

INTERVENTION	RATIONALE	EVALUATION
Observe incision for signs and symptoms of infection.	Redness, warmth, fever, and swelling indicate infection.	Are signs and symptoms of infection present?
Monitor drainage and maintain drains.	Drains remove fluid from the surgical site to prevent infection development.	Are drainage amount and color normal for procedure? Are drains functioning?
Maintain sterile technique for dressing changes.	Sterile technique reduces infection development.	Is incision free of signs and symptoms of infection?

a result of being drowsy from anesthesia or analgesics. If fine crackles are heard in the lung bases, the patient should be encouraged to deep breathe or cough. Afterward you should listen again to see if the crackles have cleared.

■ Preoperative teaching of deep breathing and coughing is reinforced and encouraged every hour while the patient is awake. Deep breathing and coughing, especially through the first postoperative day, help prevent atelectasis and keep lungs clear of secretions. If secretions are retained, mucus plugs can develop and block bronchioles, causing alveoli to collapse. Infection can develop from the stasis of mucus, resulting in pneumonia.

■ An incentive spirometer also helps prevent atelectasis by encouraging patients to inspire a deep breath, which opens their airways. After the patient is taught to use the incentive spirometer, it is placed within the patient's reach and encouragement to use it is given frequently.

■ Pain should be controlled so the patient does not guard against deep respirations or coughing. An incision near the diaphragm makes deep breathing or coughing painful, so patients avoid deep breathing or coughing.

■ Turning the patient at least every 2 hours helps expand the lungs and moves secretions. Bedrest results in decreased movement of secretions, so ambulating the patient as soon as possible also aids in keeping the airway clear.

EVALUATION. If the patient's breath sounds are clear and arterial blood gases remain normal, the goals are met. The patient's ability and willingness to deep breathe or cough and use an incentive spirometer every hour while awake are important in achieving the goal.

Circulatory Function

ASSESSMENT. The primary concern in monitoring the patient's circulatory status is the detection and prevention of hemorrhage, shock, and thrombophlebitis. Vital signs and skin temperature, color, and moistness are observed and compared with baseline data. The incision or dressing is checked for drainage or hematoma formation. Drainage may leak down the patient's side and pool underneath the patient. While wearing gloves, you should feel underneath the patient or turn the patient to check for bleeding. Any signs of hemorrhage or shock are promptly reported to the physician.

Institutional policy is followed for frequency of patient monitoring. A typical protocol for monitoring vital signs and oxygen saturation is every 15 minutes for 4 hours, then every 30 minutes for 2 hours, hourly for 4 hours, and then every 4 hours if stable.

The lower extremities of surgical patients are observed. Tenderness or pain in the calf may be the first indication of a deep vein thrombosis. Leg swelling, warmth, and redness, as well as fever, may also be present. Bilateral calf and thigh measurement is done daily if thrombophlebitis is suspected or diagnosed. Peripheral pulses and capillary refill are also checked.

NURSING DIAGNOSIS. Postoperative cardiovascular-related nursing diagnoses include the following:

■ Deficient fluid volume related to blood and fluid loss or NPO status
■ Ineffective tissue perfusion: peripheral or pulmonary related to interruption of blood flow

PLANNING. The patient's goals are to maintain blood pressure and pulse within normal limits and maintain normal tissue perfusion.

NURSING INTERVENTIONS. Vital signs are monitored and abnormal trends noted. Dressings, incisions, drains, and tubes are checked for color and amount of drainage. Bright red drainage or excessive drainage amounts are reported immediately to the physician. Intake and output are monitored to detect imbalances. IV fluids are maintained at the ordered rate.

Slower blood flow during surgery, dehydration, leg straps, and positioning may contribute to venous injury or thrombosis development. Preventive interventions to decrease the development of thrombophlebitis and pulmonary embolism should be used (see Chapter 17). Leg exercises that were taught preoperatively are encouraged hourly while the patient is awake. Early postoperative ambulation is a major preventive technique for thrombosis. Patients' pain should be controlled to facilitate their ability to participate in early ambulation. Knee-length or thigh-length antiembolism elastic stockings or intermittent pneumatic compression may be ordered. (See Fig. 15–15.) Low-dose heparin, low-molecular-weight heparin (enoxaparin [Lovenox]), warfarin, and plasma expanders such as dextran 40 and dextran 70 reduce clot formation. It is important to avoid pressure under the knee from pillows, rolled blankets, or prolonged bending of the knee. Leg elevation is helpful in preventing venous stasis.

EVALUATION. The goal for deficient fluid volume is met if vital signs and urine output are within normal limits.

The goal for ineffective tissue perfusion is met if tissue blood flow remains normal.

Postoperative Pain

It is common for patients to experience pain after surgery, although each patient's pain experience varies. In addition to incisional pain, painful muscle spasms can occur. Nausea and vomiting, ambulation, coughing, deep breathing, and anxiety can cause discomfort and increase postoperative pain. Unrelieved pain has negative physiological effects. It also impairs deep breathing and coughing and hinders early ambulation, which may increase complications, length of hospital stay, and health care costs. Nurses should understand that providing pain relief not only reduces suffering, it also has positive benefits for a quicker recovery. It is important for nurses to stay informed of advances in pain management and ensure that they make pain relief a priority in providing patient care. (See Chapter 9.)

ASSESSMENT. Nurses must be proactive and diligent in their provision of pain relief. Studies show that commonly used intramuscular (IM) pain medication orders may not provide adequate pain relief in many patients and can result in the patient's waiting for periods of up to 20 minutes for requested pain injections. Anticipating postoperative patients' pain by regularly monitoring their pain level instead of waiting until it is the time for the next dose of pain medication is essential to provide quality nursing care for pain relief.

Elderly patients, including the very oldest, experience postoperative pain. Pain is not a normal part of aging. Careful assessment of elderly patients' unique aging changes, chronic diseases, and pain relief needs is required to appropriately treat their pain. Cognitively impaired adults are at risk for undertreatment of their postoperative pain. Pain relief needs remain even when patient communication is impaired. You must ensure that you assess and treat pain for these patients.

If patients are not fully awake on transfer, vital signs and nonverbal indications of pain should be monitored. Nonverbal indicators of pain include abnormal vital signs (usually elevated blood pressure, although hypotension can occur in some patients), restlessness, moaning, grimacing, and rubbing or pulling at specific body areas or equipment. Patients who are awake are asked the location of the pain, to rate the presence of pain and describe the pain quality, such as sharp, aching, throbbing, or burning, which you should document.

NURSING DIAGNOSIS. Postoperative pain-related nursing diagnoses include pain related to tissue damage from surgery, muscle spasms, nausea, or vomiting.

PLANNING. The patient's goal for pain is to report pain relief at a satisfactory level. Patients should be involved in setting the goal for the level of pain relief that is satisfactory to them using the pain rating scale.

NURSING INTERVENTIONS

Patients are monitored for pain and provided prompt interventions for relief. If analgesics are used IV initially, such as in the PACU, and then ordered as IM injections on the unit, you should know the length of action of the IV analgesic. IV analgesics usually have a shorter duration than IMs. You should consult the physician or pharmacist for appropriate timing intervals for IV to IM doses of analgesics.

LEARNING TIPS

- For their first dose of an IM analgesic, patients in pain should not have to wait the ordered time interval of the IM dose after an intravenous analgesic dose (i.e., 3 hours if the IM order is morphine 10 mg IM q3h prn). Having to wait when the IV analgesic is no longer effective can cause needless pain.
- PCA may be started in the PACU. The patient's ability to use PCA, the patient's response to the medication, and the relief obtained from it are monitored. If PCA is not effective or if side effects occur, the physician should be notified.
- Comfortable positioning, warming the patient, and relieving a patient's full bladder can also alleviate pain. Attention to environmental factors such as bright overhead lighting, excessive noise or visitors, and extreme room temperatures also helps promote comfort.
- Antiemetics should be given as ordered to relieve the discomfort of nausea and vomiting. If vomiting occurs, patients should be turned onto one side to aid emesis removal and prevent aspiration. A nasogastric tube may be ordered to help control vomiting.

EVALUATION. Thirty minutes after pain medication is given, patients are asked to rate their pain level. If patients are sleeping at this time, you should allow them to sleep. When they awaken, you can ask them to rate their pain level. The goal for pain is met when patients report a decreased level of pain, as compared with their previous level of pain, that is satisfactory to them. If the patient does not report satisfactory pain relief, the physician should be promptly notified of the inadequate pain relief.

CRITICAL THINKING

Mrs. Wood

Mrs. Wood, following a hysterectomy, returns to the unit from the PACU. Mrs. Wood's postoperative vital signs and assessment findings are normal. You ask Mrs. Wood her pain level, noting that she moans occasionally, moves her legs, and pulls at her covers near her abdominal incision. She is drowsy but says it hurts. In the PACU, she received 10 mg of morphine IV 55 minutes ago. Morphine 5 to 10 mg IM is ordered every 3 hours as needed.

1. What nonverbal pain cues does Mrs. Wood display?
2. How should you document Mrs. Wood's pain?
3. What action should you take to relieve Mrs. Wood's pain?
4. When should you next monitor Mrs. Wood's pain level?
5. If Mrs. Wood indicates that her pain is unrelieved after 30 minutes, what action should you take?
6. You are to give Mrs. Wood morphine 8 mg IM now. You have available morphine 10 mg/mL. How many milliliters will you give?

Answers at end of chapter.

Urinary Function

ASSESSMENT. The patient's urinary status is monitored to ensure that normal function is maintained because urinary retention can occur after anesthesia. For outpatient surgery, patients may be required to void before being discharged. If the patient has a urinary catheter, the amount, color, and consistency of the urine are noted. Otherwise, monitor the patient's first postoperative voiding to prevent bladder distention. Patients should void within 8 hours of their last voiding. For patients having urinary or gynecologic procedures, the patient may need to void within 4 to 6 hours to prevent increased pressure on the surgical site. Catheterization may be necessary if the patient is unable to void.

If patients report the inability to void, the bladder is palpated for distention. You should be aware that restlessness can be caused by discomfort from a full bladder. A distended bladder requires intervention to empty it. Efforts are made to promote voiding before inserting a urinary catheter because of the risk of infection.

The amount, color, and consistency of the patient's urine is noted. The body's stress response to the surgical experience stimulates the sympathetic nervous system ("fight or flight" response), which saves fluid by reducing urine output. Therefore initially urine output may be reduced and concentrated. Then it should gradually increase, becoming less concentrated and lighter in color.

NURSING DIAGNOSIS. Postoperative urinary-related nursing diagnoses include the following:

■ Impaired urinary elimination related to surgery, pain, anesthesia, altered positioning
■ Urinary retention related to surgery, pain, anesthesia, altered positioning

PLANNING. The patient's goals are to completely and regularly empty the bladder and to remain free from pain during voiding.

NURSING INTERVENTIONS

■ Output is measured and recorded on most postoperative patients, especially those undergoing major procedures or urological surgery, elderly patients, and those with an IV or urinary catheter.
■ If the patient's urinary output from the catheter is less than 30 mL in 1 hour, it is reported to the physician.
■ Patients who void small, frequent amounts (30 to 50 mL every 20 to 30 minutes) or dribble may have retention overflow and may not be emptying their bladder. This is not normal and may require catheterization to empty the bladder and prevent complications.
■ To promote voiding, patients should be assisted to the bathroom or bedside commode, and men should be allowed to stand or sit to void if possible. If bedpans are used, they should be warmed to prevent reflexive sphincter tightening.
■ If the patient is unable to void, techniques to promote voiding should be used before catheterization. Running water, pouring warm water over a female patient's perineum, or drinking a hot beverage may stimulate voiding. The patient is given privacy to void after safety is ensured. Having patients place their feet solidly on the floor relaxes the pelvic muscles to aid voiding.
■ If the patient is uncomfortable, has a distended bladder, or has not voided within the specified time frame, the physician is notified. After other measures have been unsuccessful, an order to catheterize the patient, either with a straight or an indwelling catheter, is usually received.

EVALUATION. The goal for impaired urinary elimination is met if the patient is able to void completely and regularly.

The goal for urinary retention is met if the patient is able to void without pain or complications.

Surgical Wound Care

ASSESSMENT. A wound is a break in the skin. An incision is a wound made by a physician with a sharp instrument such as a scalpel. A puncture wound has a small opening and may be made with a scalpel to insert a tube or drain. Incisions are closed with sutures or staples, or surgical glue, which is painless and produces less scarring, is rapidly applied (Fig. 11–6). Wound healing occurs in phases (Table 11–10). As the wound heals, sutures or staples are removed in 7 to 10 days, and Steri-Strips may be applied to continue supporting the wound as it heals.

When a wound occurs, it disrupts the integrity of the skin, giving bacteria an entry point into the body. Wounds can be clean or dirty. Clean wounds are surgical wounds that are not infected. Contaminated wounds include accidental wounds or surgical incisions exposed to gastrointestinal (GI) contents or unsterile conditions. Infected wounds and dirty wounds contain microorganisms from trauma, ruptured organs, or infection. Necrotic and infected tissue is removed before infected wounds are closed. This is known as **debridement.**

NURSING DIAGNOSIS. Postoperative skin-related nursing diagnoses include the following:

■ Impaired skin integrity related to surgical incision
■ Risk for infection related to inadequate primary defenses from surgical wound

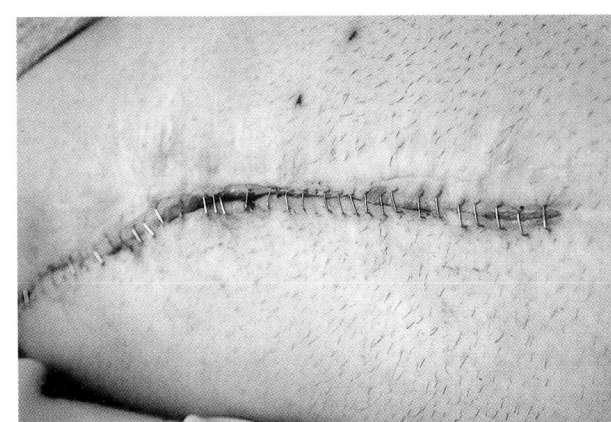

Figure 11-6 A stapled incision. *(A)* Note wound edges not approximated at arrows. *(B)* Arrows indicate puncture sites where drains were inserted.

TABLE 11-10	WOUND HEALING		
Wound Healing Processes			
First intention	Wound edges well approximated, little tissue loss		
Second intention	Large tissue loss, granulation tissue fills in area over time		
Third intention	Granulation tissue fills in area with scar formation		
Wound Healing Phases			
Phase	**Time Frame**	**Wound Healing**	**Patient Effect**
Phase I	Incision to second postoperative day	Inflammatory response	Fever, malaise
Phase II	Third to fourteenth postoperative day	Granulation tissue forms	Feeling better
Phase III	Third to sixth postoperative week	Collagen deposited	Raised scar formed
Phase IV	Months to 1 year	Collagen deposited	Flat, thin scar

PLANNING. The patient's goals are to regain skin integrity and to remain free from infection. The patient should know the signs and symptoms indicating infection that should be reported to the physician.

NURSING INTERVENTIONS

Drains. Drains are inserted into wounds during surgery to prevent accumulation of blood, lymph, or necrotic tissue in wounds that can lead to infection or delayed healing. Drains may work by gravity or suction. Penrose drains are open, soft, flat, rubberlike drains that carry drainage out of the wound. Moderate **serosanguineous** drainage is expected from a Penrose drain and may require frequent dressing changes. Examples of drains that use suction to gently enhance drainage include the Jackson-Pratt, Hemovac, and Mini-Snyder Hemovac (Fig. 11–7). These drains are closed systems that may require periodic emptying and reapplication of the suction by compressing the drain. Output is recorded when the drainage is emptied. The amount of drainage expected varies with the type of surgery. Specialized drainage systems allow the autotransfusion of bloody drainage back to the patient to maintain hemoglobin levels without the risks associated with blood transfusions, such as transfusion reactions or transmission of infections.

Surgical wound drainage initially is sanguineous (red) and changes to serosanguineous (pink) and then serous (pale yellow) after a few hours to days. Drainage that is bright red, remains sanguineous after a few hours, or is profuse should be promptly reported to the physician because the patient may be hemorrhaging.

Dressings. Dressings are applied to surgical wounds for several reasons. They protect the wound, absorb drainage, prevent contamination from body fluids, provide comfort, and apply pressure to reduce swelling or bleeding as in a pressure dressing. The initial dressing is applied in surgery and then is usually removed by the physician approximately 24 hours postoperatively. If drainage appears on the initial dressing,

you will usually reinforce it with another dressing, according to physician orders or institution policy, and document the color, amount, and consistency of the drainage.

After the initial dressing is removed, if the wound is dry and the edges intact (approximated), the physician may not order the dressing to be replaced. This allows easy observation of the wound and avoidance of applying tape to the skin. Draining wounds are dressed with several layers that are changed as needed. The Centers for Disease Control and Prevention's standard precautions are used when changing dressings. When the old dressing is removed, it should be done carefully to prevent dislodging of tubes or drains. The condition of the wound is documented with each dressing change. It is normal for the incision to be puffy and red from the inflammatory response. The surrounding skin should be the patient's normal color and temperature. Sterile technique is used to reapply the new dressing. Correct tape application over the dressing is done by gently laying the tape over the dressing and applying even pressure on each side of the wound. Pressure should not be applied on top of the wound by pulling on the tape from one side of the wound to the other side.

EVALUATION. The goal for impaired skin integrity is met if the patient's wound heals and skin integrity is regained without complications.

The goal for risk for infections is achieved if the patient remains free from signs and symptoms of infection.

WOUND COMPLICATIONS. Wound problems can include hematoma, infection, dehiscence, and **evisceration.** A hematoma occurs from bleeding in the wound and into the tissue around the wound. A clot forms from the bleeding. If the clot is large with swelling, the clot may need to be removed by the physician.

Infected wounds may be warm, reddened, and tender and have **purulent** (pus) drainage. The drainage may have a foul odor. A fever and elevated white blood cell (WBC) count

---

serosanguineous: sero—whey + sanguineous—bloody

evisceration: e—out + viscera—body organs

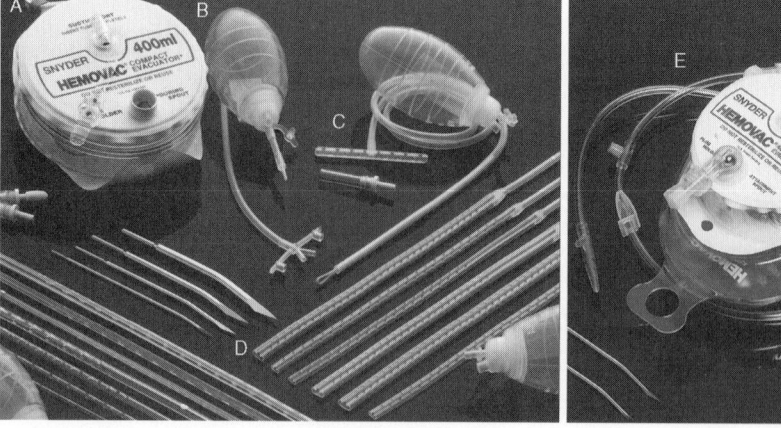

Figure 11-7   Surgical drains. *(A)* Hemovac. *(B)* Mini-Synder Hemovac. *(C)* T drain. *(D)* Flat drains. *(E)* Hemovac autotransfusion system. (From Zimmer Patient Care Products, Dover, Ohio, with permission.)

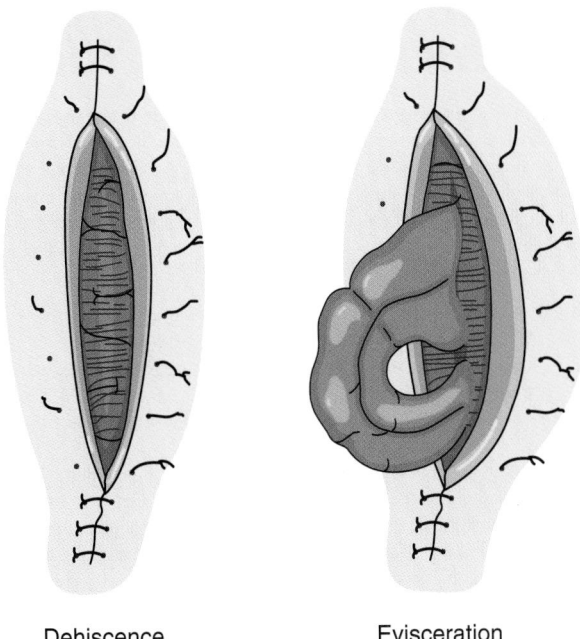

Dehiscence   Evisceration

Figure 11-8 *(A)* Wound dehiscence. *(B)* Wound evisceration.

may be present. Antibiotics are used to treat the infection.
Careful use of sterile technique for incisional care can help
prevent infection.

Dehiscence and evisceration are serious wound complications (Fig. 11–8). Wound dehiscence is the sudden burst-
ing open of a wound's edges, which may be preceded by an
increase in serosanguineous drainage. Evisceration is the
viscera spilling out of the abdomen. Dehiscence and evisceration often occur with abdominal incisions in patients
who are malnourished, obese, elderly, or who have poor
wound healing. Supporting the wound during coughing and
other activities that pull on the incision or applying an abdominal binder on patients who are at risk help prevent dehiscence and evisceration. When evisceration occurs, the
patient may have pain and vomiting and may report that
"something let loose" or "gave way."

If dehiscence or evisceration occurs, place the patient in
low Fowler's position with flexed knees. Cover the wound
with sterile dressings or towels moistened with warm sterile
normal saline. Notify the physician immediately of this surgical emergency. Apply gentle pressure over the wound and
keep the patient still and calm. Monitor vital signs for evidence of shock (e.g., tachycardia, tachypnea, dyspnea, hypotension). IV fluids are infused as ordered. Prepare the patient for immediate surgery to close the wound.

For dehisced surgical incisions that resist healing, a
newer device, vacuum-assisted closure (VAC), is available
to aid in healing the incision. The device can be used on
other types of wounds. Wound closure is aided by application of local negative pressure to wound edges by VAC.

Gastrointestinal Function

Nutritional intake and bowel functioning can be affected by
surgery and anesthesia. Application of NPO status and un-

dergoing bowel preparation often occur preoperatively. Postoperatively, intestinal handling during surgery, the need to
rest the GI tract following a procedure, immobility, lack of
peristalsis, and complications such as nausea and vomiting,
paralytic ileus (from peristalsis stopping), constipation, or
obstruction can interfere with normal GI function.

ASSESSMENT. Postoperatively, bowel sounds are auscultated every 4 hours. If no bowel sounds are heard initially,
you should listen for 5 minutes in each quadrant before documenting that they are absent. Normally, 5 to 30 bowel
sounds are heard per minute. Bowel sounds can be absent,
hypoactive, normal, or hyperactive. The abdomen is
checked to see if it is soft or firm and flat or distended. When
the patient begins passing flatus or stool, document it. Any
abnormal findings are reported to the physician.

After abdominal surgery, peristalsis and bowel sounds
usually stop for 24 to 72 hours. Flatus is usually absent for 24
to 72 hours postoperatively. Patients do not have bowel
movements until peristalsis returns. The patient is kept
NPO until flatus and bowel sounds return or as ordered by
the physician. A nasogastric or gastrointestinal tube may be
inserted and attached to suction to remove secretions and
flatus until peristalsis returns. The patient's abdominal girth
is measured if distention occurs. Drainage from the decompression tube is observed for amount, color, and consistency.
Intake and output are measured. Removal of gastric secretions can cause electrolyte imbalances. Signs and symptoms
of electrolyte imbalance can include new-onset confusion
or weakness, which should be reported.

NURSING DIAGNOSIS. Postoperative gastrointestinal-
related nursing diagnoses include the following:

- Imbalanced nutrition: less than body requirements related
  to NPO, pain, nausea
- Constipation related to decreased peristalsis, immobility,
  altered diet, narcotic side effect
- Impaired oral mucosal membranes related to NPO status

PLANNING. The patient's goals are to resume normal dietary intake and maintain weight within normal limits; to
return to normal bowel elimination patterns and report
freedom from gas pains and constipation; and to maintain
intact, moist mucous membranes.

NURSING INTERVENTIONS. Your primary focus is to help
restore normal GI functioning. When normal GI function is
achieved, dietary intake can be resumed. Until the patient
can resume a normal dietary intake, as determined by the surgical procedure, IV fluids, total parenteral nutrition, or enteral feedings may be required (Nutrition Notes Box 11–8).

EVALUATION. The goal for imbalanced nutrition: less than
body requirements is met if patients are able to maintain
their baseline weight and resume a normal dietary intake.

The goal for constipation is met if patients are free from
discomfort and establish a regular bowel elimination pattern.

The goal for impaired oral mucous membranes is met if
patients maintain intact, moist mucous membranes.

BOX 11-8  *Nutrition Notes*

Nourishing the Postoperative Patient

After surgery, intravenous 5% glucose in water is commonly prescribed. Two liters of this solution contain only 340 calories, insufficient to meet the patient's energy needs but enough to prevent ketosis from breakdown of adipose tissue. Previously well-nourished adults generally have nutrient reserves for 3 to 4 days of semistarvation. To prevent excessive muscle protein from being used for energy, adequate nourishment should be delivered to the patient within 3 days. Complete nutrition is available in clear liquid formulations such as Citrotein and Enlive.

To avoid abdominal distention, oral feedings are delayed until peristalsis returns, as evidenced by the patient's passing flatus or the health care worker's auscultating bowel sounds. Patients usually progress from clear liquids to a regular diet as soon as possible. If "diet as tolerated" is prescribed, the patient should be asked what sounds good. Sometimes a full dinner tray when the patient doesn't feel well "turns off" the appetite. After gastrointestinal surgery, oral food and fluids are deferred longer than with other surgeries to allow healing. When particular amounts are prescribed, those limits should be strictly implemented to preserve the suture lines. One other precaution is often taken after surgery on the mouth and throat: After tonsillectomy, for example, no red liquids are given so that bleeding can be seen and vomitus is not mistaken for blood.

Specific nutrients necessary for healing are as follows:

- Vitamin C for collagen formation
- Vitamin K for blood clotting
- Zinc for tissue growth, skin integrity, and cell-mediated immunity
- Protein for controlling fluid balance and edema and for manufacturing antibodies and white blood cells, as well as for building scar tissue

---

LEARNING TIPS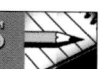

The primary purpose of most IV fluids is to provide hydration. Most IV solutions do not provide enough nutrients or calories to prevent malnutrition. A 1000-mL IV solution containing 5% dextrose provides only about 170 calories. This does not meet an adult's daily caloric needs, especially if healing is occurring. You should ensure that early consideration of other nutritional methods is made to meet the patient's dietary needs.

- If nausea and vomiting occur, antiemetics may be given.
- Early ambulation, exercise, and diet promote restoration of GI functioning. The patient should be encouraged to be as active as possible. If gas pains occur, techniques to relieve the pain, as tolerated, include ambulation, having the patient lie prone, and pulling the knees up to the chest.
- If the patient is NPO, oral care is provided frequently. Good oral hygiene helps promote appetite. A dry mouth is uncomfortable and can cause increased bacterial accumulation and mucous membrane breakdown. Products such as dry mouth toothpaste, moisturizing gel, or mouthwash can be used to promote good oral hygiene and decrease dryness. If the patient is allowed occasional ice chips, they can be made from normal saline to prevent depletion of electrolytes. Alcohol-containing mouthwashes increase dryness and should be avoided. Lemon glycerin swabs may also increase dryness.
- If a paralytic ileus develops, abdominal distention, absent bowel sounds, and pain may result. If severe peristalsis occurs, nausea and vomiting result. As ordered, a nasogastric tube can be used to decompress the GI tract until peristalsis returns.
- When the patient is able to resume oral intake, food is introduced gradually to tolerance. Water and clear liquids are usually started first. Then the diet is advanced to soft and finally solid foods.

Mobility

ASSESSMENT. It is important for the patient to move as much as possible to prevent complications and promote healing. Pain, incisions, tubes, drains, dressings, and other equipment may make movement difficult. You should determine the patient's ability to move in bed, to get out of bed, and to walk. Pain levels that may interfere with movement are assessed. The patient's tolerance to activity is observed. Understanding of how to perform exercises is noted.

NURSING DIAGNOSIS. Postoperative mobility-related nursing diagnoses include the following:

- Impaired physical mobility related to surgery, decreased strength, and movement restriction
- Activity intolerance related to immobility and weakness

PLANNING. The patient's goals are to resume normal physical activity and to demonstrate increased activity tolerance.

NURSING INTERVENTIONS. The patient is assisted with positioning in bed. Pillows can be used to support the body

in good alignment. The patient should be turned at least every 2 hours. Positioning should be alternated from supine to side to side. Patients should move themselves as much as possible for several reasons. Movement increases circulation and promotes lung expansion to prevent complications. Also, patients know where it hurts and are less likely to increase their pain when moving.

Encouraging exercises in bed is important, especially if ambulation is not possible. Exercises should be done hourly while awake to promote normal functioning and prevent complications.

Deep breathing, range of motion of all joints, and isometric exercises of the abdominal, gluteal, and leg muscles should be performed by the patient. Passive joint range of motion is done by the nurse if the patient is unable to do active range of motion.

As patients are assisted to sit up, the head of the bed is raised slowly to allow the circulatory system to adjust to the position change. If patients report dizziness or feeling faint, the head of the bed is lowered. Vital signs and skin color are noted. Patients should rest about 1 hour, as tolerated, before sitting up again. Once patients tolerate a sitting position, they can be dangled on the side of the bed in preparation for ambulation.

If dangling is tolerated, patients can be ambulated. Before getting up, patients should pedal their feet to "wake up" the muscles controlling the arteries. To rise, they should keep their eyes forward and move slowly until they feel adjusted to being up. Usually the patient ambulates a short distance the first time and increases the distance as tolerated. One or two health care workers should assist the patient and use a gait (walking) belt for safety. Walkers with wheels and seats may also be used for support and for resting if the patient becomes dizzy or tired. If patients feel faint or dizzy or their vital signs change, they should be assisted back to bed. A wheelchair may be needed to transport them safely back to their room.

EVALUATION. The goal for impaired physical mobility is met if patients are able to increase ambulation and resume normal activities.

The goal for activity intolerance is met if patients are able to increase activity to desired level and maintain vital signs within normal limits.

Postoperative Patient Discharge

Discharge planning begins in preadmission testing and continues after admission to ensure that the patient is ready for a timely discharge. When the patient meets discharge criteria, the physician discharges the patient from either the ambulatory setting or the hospital.

Ambulatory Surgery

DISCHARGE CRITERIA. Generally the patient can be considered a candidate for discharge 1 hour after surgery if discharge criteria are met (see Table 11–9). Discharge criteria include stable vital signs, no bleeding, no nausea or vomiting, and controlled pain that is not severe. Depending on the type of surgical procedure, such as urological, gynecologic, or hernia surgery, the patient may be required to void before discharge. The patient should be able to sit up without dizziness before discharge. When patients meet discharge criteria and are discharged by the physician, they are released to a responsible adult. They are not permitted to drive themselves home because of the effects of anesthesia and medications they have received.

DISCHARGE INSTRUCTIONS. Patients and their families are given written discharge instructions before discharge. Elderly patients should have a caregiver participate in the discharge instruction session to understand what observations to make and what to do if complications develop. The instruction form is signed by the patient or an authorized representative to indicate understanding. Prescriptions and a copy of the instructions, to provide a reference for later, are sent with the patient. The patient is encouraged to rest for 24 to 48 hours. The patient is to avoid operating machinery, driving, drinking alcoholic beverages, and making major decisions for 24 hours because the effects of undergoing surgery can alter energy levels and thinking ability. The physician orders any fluid, dietary, activity, or work restrictions.

Patients are taught wound care, medication information (including side effects), and signs and symptoms of complications to report to the physician. Phone numbers for the physician, surgical facility, and emergency care are provided. Patients are informed of the date for their follow-up visit to the physician and told to call and make an appointment. They are also told that a nurse will call the next day to check on their progress and answer any questions.

Inpatient Surgery

DISCHARGE CRITERIA. The physician determines the patient's readiness for discharge from the hospital. Postoperative lengths of stay vary based on the surgical procedure and the patient's individual needs. Before discharge, a complete assessment of the patient is performed and documented.

DISCHARGE INSTRUCTIONS. Patients and their families are given prescriptions and a copy of written instructions that are signed by the patient to indicate understanding before discharge. All necessary teaching is completed before discharge. If more teaching or reinforcement is needed, a referral to a home health nurse can be requested.

The physician orders any fluid, dietary, activity, or work restrictions. Patients are taught wound care, medication information (including side effects), and signs and symptoms of complications to report to the physician. Patients are informed of the date for their follow-up visit to the physician.

Home Health Care

The role of the home health nurse is to assist the patient in the recovery process (Home Health Hints).

A referral to the home health nurse is made when the patient requires the following:

- Continued assistance with skilled nursing interventions, such as wound care, IV medications, or ostomy care
- Additional teaching to be able to perform self-care, such as diabetic teaching for a patient with newly diagnosed diabetes or ostomy care
- Assessment of the recovery process
- Assistance needed because of weakness, lack of social support, or development of complications; care provided in the home is adapted to the patient's resources and environment to facilitate compliance

Home health care workers can include assistants who help with activities of daily living and household chores, LPNs/LVNs who provide basic nursing care, and registered nurses who perform patient assessments, teaching, and complex nursing care. The frequency of visits is determined by the patient's needs. The RN can contact the physician from the patient's phone for orders for abnormal findings.

HOME HEALTH HINTS

After a patient comes home from surgery, the home care nurse can help give direction to the family to prepare the room where the patient will be staying. Some preparations include the following:

- It is helpful if the room can be on the same floor with the bathroom, kitchen, and living space.
- If an extended recovery period or illness is expected, the den or living room might be considered as the primary living space to provide room for equipment and make companionship easier. The patient can see activity in the home and be included in family activity. Also, caregivers can be more attentive to the patient's needs and save countless footsteps.

Special equipment may be needed, including the following:

- For the patient on bedrest, a hospital bed with full side rails helps with a variety of position changes and better height for the caregiver.
- Draw sheets made of folded twin sheets are needed, as well as extra pillows for positioning.
- A bedside stand is needed for personal and toilet articles.
- A bedside commode can be placed near the bed if the patient cannot walk to the bathroom. A bedpan or urinal may be needed. A functional female urinal is easier to use than a bedpan.
- A flexible tube with a shower head that connects to the bathtub faucet is convenient and allows the patient more independence in bathing.
- Installation of grab bars and tub stools and skid-proofing of a shower or tub are important safety measures to help prevent falls.
- If the patient is eligible for insurance coverage for durable medical equipment, a physician's order must be obtained.

It is helpful for caregivers to keep a notebook in the hospital and continue it at home. Treatments, medicines, observations, procedures, doctor and nurse visits, instructions, and therapies with dates and times can be recorded. This helps prevents confusion, prepares the next caregiver, and affords better organization of time and resources for everyone. It can also be nice to see who visited the patient.

Families of patients recovering from surgery should provide items to keep the patient occupied and comfortable: talking books, inspirational reading material, pictures, and their favorite pajamas, robe and slippers, and coverlet.

Answers to CRITICAL THINKING

Mrs. Wood

1. Moaning occasionally, moving legs restlessly, and pulling covers near abdominal incision are nonverbal pain cues.
2. Document pain levels by actual observations: occasional moaning, restless leg movements, and pulling of covers near abdominal incision. By patient's statement: "It hurts." Because Mrs. Wood is too drowsy to use the pain scale, other data are used. When Mrs. Wood is more awake, explanation of the pain scale should be reinforced and used.
3. Review pain medication orders to determine if analgesics can be given. Noting that an IV analgesic was given 50 minutes ago and the IM analgesic is ordered every 3 hours, you should request that the physician or pharmacist be consulted to determine appropriate time intervals. If the consultation indicates it is time to give the analgesic, verify that vital signs are still stable and then give the analgesic. You should also consider other pain relief measures such as patient warmth, positioning, or environmental issues such as bright lighting, room temperature, and noise.
4. After administration of the analgesic, Mrs. Wood's pain level should be assessed in at least 30 minutes to determine pain relief. If Mrs. Wood is asleep, she should not be awakened unless you determine it is necessary. Nonverbal cues should be observed and respirations counted and documented. If no indication of pain is noted, Mrs. Wood's pain level should be monitored by you at least hourly or as needed.
5. Document pain level on scale of 0 to 10 and have the physician notified of inadequate pain relief. The patient should not have to wait the 3-hour interval in pain. Consider providing other pain relief measures while the physician is being notified.
6. Method 1: $\dfrac{\text{Desired}}{\text{Available}} \times \text{Volume} = \dfrac{8}{10} \times 1 \text{ mL} = \dfrac{8}{10} = 0.8 \text{ mL}$

   Method 2:   H : V :: D : X
   On hand Vehicle   Desired dose   Amount to give
   10 mg : 1 mL :: 8 mg : X mL
   $10 x = 8$
   $x = 0.8 \text{ mL}$

1. Which of the following is a LPN/LVN patient care role in the preoperative phase?
    a. Assisting in data collection
    b. Explaining the surgical procedure
    c. Obtaining preoperative orders
    d. Providing informed consent

2. When teaching the elderly preoperative patient, which of the following is a teaching strategy that improves learning?
    a. Sit near a window with bright sunlight
    b. Use large black-on-white printed materials
    c. Sit beside patient
    d. Use blue and green materials

3. Which of the following related to the patient providing consent for surgery is within the LPN/LVN's scope of practice?
    a. Obtaining informed consent
    b. Providing informed consent
    c. Answering surgical procedure questions
    d. Requesting patient questions be referred to physician

4. The nurse is assessing a postoperative patient. Which of the following findings would the nurse correctly evaluate as being the earliest indicator of hemorrhage or shock that should be reported to the physician?
    a. Tachycardia
    b. Polyuria
    c. Nausea
    d. Fever

5. Which one of the following would the nurse evaluate as indicating that the patient's diet can be resumed as ordered?
    a. Absence of flatus
    b. Bowel sounds every 8 seconds
    c. Excessive thirst
    d. Absent bowel sounds

6. Which of the following interventions should the nurse include in the plan of care to help prevent atelectasis in a postoperative patient?
    a. Coughing and deep breathing
    b. Holding breath while moving
    c. Restricting fluids
    d. Leg exercises

7. Which of the following is one criterion the nurse uses to evaluate patient readiness for discharge from ambulatory surgery?
    a. Ability to drive an automobile
    b. Ability to ambulate 50 feet
    c. Being pain free
    d. Absence of nausea or vomiting

12

NURSING CARE OF PATIENTS WITH EMERGENT CONDITIONS

Rose Utley

KEY TERMS

abrasion (a-**BRAY**-zhun)

avascular necrosis
(a-**VAS**-cue-lur nah-**CROW**-sis)

amputation (am-pew-**TAY**-shun)

anaphylactic shock
(an-uh-fah-**LAK**-tik **SHAHK**)

asphyxia (as-**FIX**-ee-a)

capillary permeability
(**KAP**-ih-lar-ee **PER**-me-a-**BILL**-i-tee)

capillary refill (**KAP**-ih-lar-ee **RE**-fill)

cardiac tamponade
(**KAR**-dee-ack tam-pon-**AID**)

cardiogenic (kar-dee-o-**JEN**-ick)

distributive (dis-**TRIB**-u-tive)

flail chest (**FLAY**-ul chest)

full-thickness burn
(**FUL THICK**-ness **BERN**)

gastric lavage (**GAS**-trick la-**VAHJ**)

heatstroke (**HEET**-strohk)

hypoproteinemia
(**HIGH**-poh-pro-teen-**EE**-mee-ah)

hypovolemic shock
(**HIGH**-poh-voh-**LEEM**-ik **SHAHK**)

laceration (lass-ur-**A**-shun)

obstructive shock
(ahb-**STRUK**-tive **SHAHK**)

partial-thickness burn
(**PAR**-shul **THICK**-ness **BERN**)

rule of nines (**ROOL** of nines)

shock (**SHAHK**)

tetanus (**TET**-nus)

triage (**TREE**-ahj)

QUESTIONS TO GUIDE YOUR READING

1. What are common signs and symptoms of shock?

2. How would you describe hypovolemic, distributive, and cardiogenic shock?

3. What are the components of the primary survey?

4. How would you prioritize interventions for a multisystem trauma victim?

5. What is the difference between a partial-thickness and full-thickness burn?

6. What are the symptoms of inhalation injury?

7. What are the stages of hypothermia and hyperthermia?

8. What are priorities of care for poison overdose?

9. What is your role in crisis situations and psychiatric emergencies?

The ability to recognize an emergent condition, prioritize, and provide quick assessments and interventions is essential in nursing. In emergent situations, initial assessment and intervention is guided by the ABCs of the primary survey. The primary survey of the patient's airway, breathing, and circulation allows recognition, prioritization, and treatment of life-threatening situations. The secondary survey, a rapid head-to-toe assessment, identifies additional serious injuries throughout the body. This chapter presents components of the primary and secondary surveys followed by specific emergent conditions with application of the nursing process.

▶ PRIMARY SURVEY

To recognize life-threatening conditions and determine priorities of care, an intitial assessment of the patient's airway, breathing, and circulation is conducted. This process is known as the primary survey. The components of the primary survey are listed in Table 12–1.

A = Airway

The airway is the most important component of the primary survey. The airway is opened using the chin lift or jaw thrust maneuver (Fig. 12–1). It is essential to maintain alignment of the cervical spine if there is any possibility of a cervical spine injury. The neck should not be hyperextended, flexed, or rotated until spinal injury is ruled out because any movement may worsen an existing cervical spine injury. The airway is inspected for obstruction, including loose teeth, foreign objects, bleeding, and vomitus. Next, any visible airway obstructions are removed using a finger sweep or suction if available.

Airway adjuncts, such as nasopharyngeal or oropharyngeal airways, may be used to keep the airway open. When additional airway support and mechanical ventilation are required, advanced airway adjuncts, such as endotracheal intubation or cricothyroidotomy, may be performed by specially trained emergency personnel or physicians.

B = Breathing

After the patency of the airway is ensured, the patient is assessed for spontaneous breathing and respiratory rate and depth. If the patient is not breathing, interventions are conducted before proceeding. The patient may be ventilated with a mouth-to-face mask or a bag-valve-face mask. Endotracheal intubation is the preferred method of maintaining an airway in an unconscious patient because it ensures air-

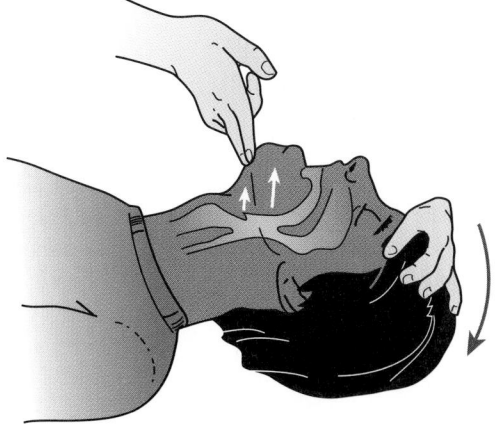

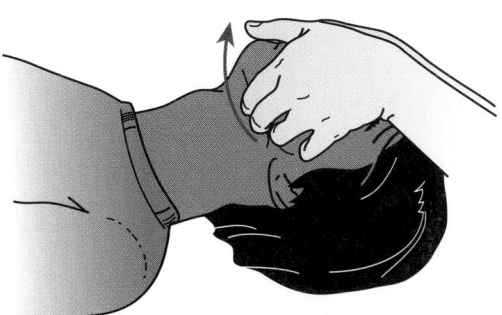

Figure 12-1 *(A)* Chin lift maneuver is used to open the airway. *(B)* Jaw thrust maneuver is used to open the airway if the patient may have a head or neck injury.

way patency and protects the lungs from aspiration. (See Chapter 26.) The nurse observes whether the patient's chest rises and falls spontaneously and auscultates for breath sounds bilaterally.

C = Circulation and Consciousness

The carotid pulse is palpated for quality and rate. The skin is inspected for color and temperature. External bleeding is controlled by external pressure and elevation when possible. Any life-threatening conditions that may compromise circulation are assessed and interventions provided before proceeding. Other conditions that may compromise circulation include internal bleeding, **shock** resulting from hemorrhage, or major burns. Large-gauge intravenous (IV) cannulas (16 gauge or 18 gauge) are initiated for fluid resuscitation. If the patient does not have a pulse, cardiopulmonary resuscitation must be initiated. If a pulse can be palpated, vital signs are taken and recorded.

After assessment of airway, breathing, and circulation, a brief neurological assessment is conducted to determine the level of consciousness, which may range from alert and talking to unresponsive. If the patient is unresponsive, a painful stimulus is applied, such as rubbing the sternum, pressing a pen against the base of the nail, or applying periorbital

TABLE 12-1	COMPONENTS OF THE PRIMARY SURVEY
A	Airway
B	Breathing
C	Circulation and consciousness

pressure. The patient is observed for any response to the pain, and the response is recorded.

▶ SECONDARY SURVEY

For victims of severe trauma, a secondary survey is conducted. Major body areas that may sustain serious injury, such as the head, spine, chest, abdomen, and musculoskeletal system, are quickly examined to detect additional injuries. To adequately perform a head-to-toe assessment, the patient's clothing is removed. Each major body area is inspected and palpated for deformity, bruising, opens wounds, bleeding, and pain. The chest is palpated for subcutaneous emphysema, a crackling feeling in the skin caused by air escaping from the lungs into the soft tissue. The chest is auscultated for equal breath sounds and heart sounds. The abdomen is inspected for open wounds, bruising, and distention, and then auscultated for bowel sounds. Tenderness and rigidity of the abdomen are noted and reported. Next the extremities are inspected and palpated for injury, deformity, peripheral pulses, temperature, and capillary refill. Movement and sensation in the fingers and toes are noted. If the patient is conscious and lucid, a history is taken. Details of the accident or illness and information regarding the patient's medical history are obtained if possible.

Subcutaneous emphysema feels similar to the sensation of squeezing a bubble in the bubble wrap used for packaging.

▶ SHOCK

Shock is a condition of acute peripheral circulatory failure, causing inadequate and progressively failing tissue perfusion that results in cell death if not treated. During the initial phases of shock, compensatory adjustments allow the body to adapt to the circulatory changes. Eventually, however, these compensatory mechanisms fail and cellular perfusion decreases, causing cell death. There are four types of shock: hypovolemic, **cardiogenic,** obstructive, and **distributive.** (See Chapter 8.)

Nursing Process for Shock

Assessment

When assessing a patient at risk for shock, be aware of the signs and symptoms common to all types of shock (Box 12–1). It is important to note the patient's initial level of consciousness and monitor the patient for any subsequent changes. A progressive decline in level of consciousness indicates an urgent need for intervention. Pulses indicate the strength of the heart's contractions. Because a pulse is an immediate indicator of the patient's condition, it should be taken frequently during any emergency condition. Changes in blood pressure may also indicate changes in blood vol-

---

BOX 12-1   **Common Signs and Symptoms of Shock**

- Restlessness and anxiety
- Weak, rapid, thready pulse
- Cold and clammy skin
- Pale skin color
- Shallow, rapid, labored breathing
- Gradually and steadily falling blood pressure
- Alteration in consciousness in severe shock state
- Thirst

ume. Blood pressure changes can occur rapidly but usually not as swiftly as pulse changes.

Skin temperature and color changes may be seen with shock. Severe blood loss activates the "fight-or-flight" response in the sympathetic nervous system, which causes the skin to become cool and clammy. This occurs when peripheral blood vessels constrict to shunt blood to vital organs. Skin color depends on the presence of circulating blood in the vessels of the skin. Pale, white, or ashen skin indicates insufficient circulation. In patients with deeply pigmented skin, color changes may be apparent in the nailbeds, conjunctiva of the eye, or mucous membranes of the mouth.

Capillary refill is checked on nailbeds to evaluate arterial circulation to an extremity. The nailbed is compressed to produce blanching (lighter color change), released, and the seconds counted until color returns to the blanched area. Normally, nail color should return within 3 seconds after the pressure is released. Patients in shock may have delayed or absent capillary refill.

Gently squeeze your own nailbed. Do you see the color change? Count the seconds until the color returns. That is your capillary refill time.

Nursing Diagnosis

Examples of applicable nursing diagnoses include the following:

■ Decreased cardiac output related to hypovolemia
■ Ineffective tissue perfusion: cerebral, related to hypovolemia

Planning

The patient's goals are to maintain vital signs and level of consciousness within normal limits.

Interventions

Patients who have signs or symptoms of shock must be treated immediately. Ongoing data collection helps to iden-

BOX 12-2 Guiding Principles for Treating Shock

- Maintain an open airway and give oxygen as ordered.
- Control external bleeding by direct pressure.
- Keep the patient supine if possible.
- Accurately record vital signs.
- Give IV fluids as ordered.
- Give the patient nothing to eat or drink until surgery is ruled out.

tify the probable cause of shock so that appropriate treatment can be given.

Guiding principles for treating patients in shock are listed in Box 12–2. External bleeding is controlled by applying direct pressure to stop the flow of blood and allow normal coagulation to occur. If bleeding continues after a dressing is in place, additional manual pressure may be indicated. If additional dressings are needed, they are applied over the initial dressing. Elevation of a bleeding extremity helps stop venous bleeding and should be combined with direct pressure. When direct pressure and elevation do not control hemorrhage, pressure-point control should be attempted (Fig. 12–2). The chosen artery for pressure-point control must be proximal to the injury site and must overlie a bony structure.

Vital signs are continually taken, recorded, and reported to the physician if changes occur. The legs may be elevated to heart level to promote venous return to the heart (contraindicated in cardiogenic shock). A blanket can be used to help keep the patient from getting cold; however, the patient also should not be allowed to overheat because this causes peripheral blood vessels to dilate, which draws blood away from vital organs. IV fluids are administered to increase circulating volume.

Tourniquets are used only as a last resort to control bleeding. If a tourniquet is required, a blood pressure cuff, inflated proximal to the bleeding point, is used. Material that is narrow or cuts into the skin should never be used. Tourniquets should not be placed below the knee or elbow or covered with a bandage. Neurovascular damage to the underlying tissues can result from an improperly applied tourniquet.

Evaluation

Criteria indicating a positive outcome for the treatment of shock include a rising blood pressure; a strong pulse; warm, dry skin; and a calmer patient. If signs of shock persist, the physician must be notified. Persistent shock can lead to irreversible shock and death.

Irreversible shock occurs when attempts fail to increase cardiac output and tissue perfusion to vital organs. The body's compensatory mechanisms are no longer helpful, and medical interventions to alleviate shock are unsuccessful. Severely decreased cardiac output and acidosis with multisystem organ failure leads to cardiac arrest and death.

▶ ANAPHYLAXIS

Anaphylaxis is a severe allergic hypersensitivity reaction. The reaction may occur suddenly after initial contact with an allergen or after any subsequent exposure. Signs and symptoms result from a massive release of chemical mediators from mast cells and basophils throughout the body. Chemical mediators lead to vasodilation and capillary leaking that results in hypotension and possible vascular collapse. Signs and symptoms of an allergic reaction are listed in Box 12–3.

Pathophysiology

Anaphylactic shock is a form of distributive shock. There is no loss of blood, but there is excessive vasodilation. Bronchi constrict, and air movement into the lungs becomes increasingly difficult. Increased fluid and mucus are secreted into the bronchial passages. Fluid in the air passages and constricted bronchi cause wheezing. The body is rapidly deprived of needed oxygen by this respiratory system reaction. Signs of severe anaphylaxis include hypotension, decreased level of consciousness, and respiratory distress

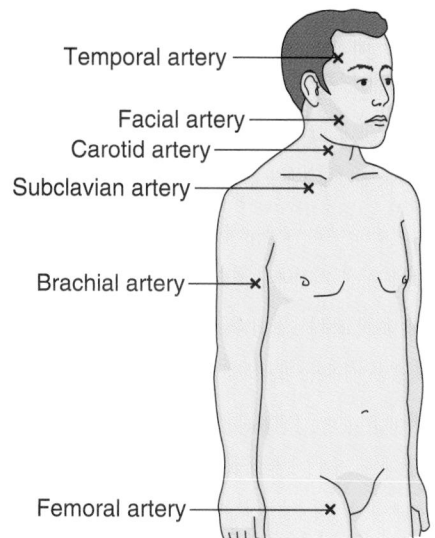

Figure 12-2 Arterial pressure points to control bleeding.

Temporal artery
Facial artery
Carotid artery
Subclavian artery
Brachial artery
Femoral artery

BOX 12-3 Signs and Symptoms of an Allergic Reaction

- Generalized itching and burning
- Urticaria (hives)
- Swelling about the lips and tongue
- Dyspnea
- Bronchospasm and wheezing
- Chest tightness and cough
- Anxiety
- Hypotension

with stridor and cyanosis. One of the increasing causes of anaphylactic shock is latex allergy reaction. This type of allergy is on the rise among health care workers as a result of exposure to health care products such as gloves that are made of latex. Using latex-free products limits exposure and reduces the risk of developing this type of allergy.

Nursing Process for Anaphylaxis

Nursing Diagnosis

Examples of applicable nursing diagnoses include the following:

■ Ineffective breathing patterns related to airway constriction
■ Decreased cardiac output related to vasodilation

Planning

The patient's goals are to maintain vital signs and level of consciousness within normal limits.

Interventions

When anaphylaxis occurs, airway compromise is an immediate threat to life and must be treated quickly. The patient is given oxygen for respiratory distress. Epinephrine, the drug of choice for treating anaphylaxis, is used to decrease edema, dilate bronchial smooth muscle, and increase blood pressure by vasoconstriction. The standard adult dose of epinephrine is 0.2 mg to 0.5 mg of a 1:1000 solution given subcutaneously. Injections can be repeated every 10 to 15 minutes until the desired effect is achieved or significant side effects occur. Antihistamines are used to control the allergic rash and pruritus. Steroids are given in gradually tapered doses to prevent return of symptoms.

Evaluation

When treatment is effective, the patient should show an immediate reversal of shock symptoms. Breathing is easy, and blood pressure and pulse return to the normal range. Breath sounds become clear, and hives and pruritus subside.

▶ MAJOR TRAUMA

Major trauma is the fifth leading cause of death in the United States. It mainly affects perons under age 36. Young men have the highest overall incidence of traumatic injury. Victims of major trauma may receive injury to an isolated vital organ or to multiple body systems. To recognize life-threatening conditions and determine priorities of care, a primary survey followed by a secondary survey is conducted to detect additional serious injuries.

Mechanism of Injury

When assessing a victim of major trauma, it is important to determine the mechanism of injury whereby energy is transferred from the environment to the person (Geronotological Issues Box 12–4). Injuries resulting from the transfer of mechanical energy are either penetrating or blunt. Penetrating, or open, injuries may be caused by any sharp object, such as broken glass or a knife, or by projectiles traveling at high speed, such as bullets or fragments from an explosion.

BOX 12-4   GERONTOLOGICAL ISSUES

Injuries and Older Adults

Older adults are at a high risk for falls that put them at risk for bruises, abrasions, cuts, and fractures. Nurses who initially assess older adults with injuries requiring treatment must ask questions and perform assessments that would identify if the patient is a victim of abuse or neglect.

Injuries Caused by Falls versus Battery or Assault

Any unexplained bruises, burns, abrasions, cuts, fractures, evidence of old injuries or bruises, burns, and cuts that are in different stages of healing suggest abuse. The pattern of an injury can also suggest abuse—for example, cigarette burns in areas covered with clothing; bruises or friction burns in a ring around the neck, ankles, or wrists; welts, burns, or bruises in the outline of a hand or belt buckle; multiple similar injuries in an area, such as whip marks across the buttocks or back of the legs; defensive injury pattern of bruising; and trauma to the hands and forearms.

Injuries related to falls have a predictable injury pattern related to the history and report of the fall. When an older adult falls there is bruising of the hands and knees caused when the person attempts to break the fall. Additional bruising or injuries to the front of the body, arms, and head could be caused by hitting furniture or other items during the fall. Skin tears on the arms are common with a fall. Often, a friend or family member sees the older adult starting to fall and tries to steady the person by grabbing the area, tearing the skin. Ask questions to be sure that the report of the fall incident is consistent with the presenting injuries.

Any form of abuse or suspicion of abuse must be reported to the state agency that investigates reports of suspected abuse. It is not the nurse's responsibility to prove that there has been abuse or neglect, only to report incidents or cases of possible abuse.

In blunt, or closed, injuries the skin surface is intact. The injury from blunt trauma usually extends beyond the point of impact to surrounding and underlying structures, such as a broken rib that results from a blow to the chest.

The mechanism of injury from firearms is related to the energy created and dissipated by the bullet into the surrounding tissues. Damage caused by a gunshot wound and the trajectory of the bullet depends on the projectile mass, the type of tissue struck, the striking velocity, and the range. Entrance wounds are round or oval and may be surrounded by an **abrasion** rim. Powder burns are visibile if the firearm was discharged at close range. Exit wounds are usually larger than entrance wounds and commonly produce a starburst or stellate wound. Documentation of these wounds should include a clear description of their appearance but should not include the words *entry* or *exit*. Patients with gunshot wounds near the level of the diaphragm should be evaluated for both abdominal and thoracic injuries.

Surface Trauma

Surface trauma is an injury that does not break the skin (closed wound) or an open wound in which the skin surface is broken. Types of closed wounds include contusions (bruising) and hematomas (collection of blood under the skin). Types of open wounds include abrasions, **lacerations,** avulsions, **amputations,** and punctures.

Abrasions are a scratching of the epidermal and dermal layers of the skin. They bleed very little but can be extremely painful because of inflamed nerve endings. Dirt may be ground into abrasions and can pose an infection threat when large areas of skin are involved.

Lacerations are open wounds resulting from snagging or tearing of tissue. Skin tissue may be partly or completely torn away. Lacerations can cause significant bleeding if blood vessels or arteries are involved.

Avulsions involve a full-thickness skin loss in which wound edges cannot be approximated. This type of injury is often seen in machine operators or in lawn mower and power tool accidents.

An amputation is a partial or complete severing of a body part. In cases of complete amputation the arteries usually spasm and retract into the tissue, resulting in less bleeding than does a partial amputation, in which the lacerated arteries continue to bleed.

Puncture wounds result from sharp, narrow objects such as knives, nails, or high-velocity bullets. They can often be deceptive because the entrance wound may be small with little or no bleeding. It is difficult to estimate the extent of damage to underlying organs as a result. Puncture wounds usually do not bleed profusely unless they are located in the chest or abdomen.

Nursing Process for Surface Trauma

NURSING DIAGNOSIS. Examples of applicable nursing diagnoses include the following:

- Acute pain related to tissue trauma
- Impaired skin integrity related to trauma
- Risk for infection related to tissue trauma

PLANNING. The patient's goals are to maintain vital signs within normal limits, remain free of infection, restore skin integrity, and be pain free.

INTERVENTIONS. The management of closed wounds includes application of ice, elevation, and immobilization of the affected part to decrease swelling and relieve pain. Deformity, pain, and inability of the patient to voluntarily move the affected part may indicate an underlying fracture. The management of open wounds includes application of direct pressure to control bleeding. Open wounds are irrigated with sterile saline solution to thoroughly remove dirt and debris and clean exposed tissue to prevent infection. Injuries that could potentially cause shock are identified and treated accordingly.

Open wounds are treated with a **tetanus** immunization if it has been more than 5 years since one was last given. Tetanus is a disease caused by the bacillus *Clostridium tetani,* which enters the body through an open wound. Tetanus causes seizures, muscle spasms, stiffness of the jaw, coma, and death. Tetanus vaccinations should begin at 2 months of age and be followed by a series of pediatric immunizations

until age 15. Thereafter, booster vaccinations are recommended every 10 years in the absence of an open wound.

If the patient has sustained an amputation, bleeding is controlled with direct pressure and elevation. A tourniquet is applied only as a last resort. If a tourniquet is necessary, it should be made of wide material such as a blood pressure cuff, which is less damaging to nerves and blood vessels. A dressing is applied to the amputated extremity, referred to as the stump. The stump is covered with sterile saline–moistened gauze followed by dry gauze, which is held in place with an elastic bandage for pressure. Amputated parts are taken to the hospital with the patient for possible reattachment. At the hospital, the amputated part is rinsed with saline solution, wrapped in sterile gauze, and placed in a sealed plastic bag, which is then placed in slushy ice water.

For a patient with an injury caused by an impaled object, it is imperative that the object not be removed unless it is obstructing the airway. Removing an impaled object may cause additional trauma. Impaled objects are never cut off, broken off, or shortened unless transportation to the emergency department is otherwise impossible. A bulky dressing is applied around the object to stabilize it and reduce motion.

EVALUATION. When treatment is effective for surface trauma, the patient reports a satisfactory pain level and the patient's wound heals without infection.

Head Trauma

Sharp blows to the head can cause shifting of intracranial contents and lead to brain tissue contusion. The pathophysiology of head trauma can be divided into two phases. The first phase is the initial injury that occurs at the time of the accident and cannot be reversed. The second phase involves intracerebral bleeding and edema from the initial injury, which causes increased intracranial pressure (ICP). Management of head trauma is directed at the second phase and involves decreasing ICP. Early and late signs and symptoms of increased ICP are listed in Box 12–5.

BOX 12-5 **Signs and Symptoms of Increased Intracranial Pressure**

Early Signs and Symptoms of Increased ICP

- Headache
- Nausea and vomiting
- Amnesia
- Altered level of consciousness
- Changes in speech
- Drowsiness

Late Signs and Symptoms of Increased ICP

- Dilated nonreactive pupils
- Unresponsiveness
- Abnormal posturing
- Widening pulse pressure
- Decreased pulse rate
- Changes in respiratory pattern

Nursing Process for Head Trauma

ASSESSMENT. The mechanism of injury is determined to identify the extent of injury. Loss of consciousness immediately after the injury indicates that a concussion has occurred. The Glasgow Coma Scale (GCS) is used to rate a patient's level of consciousness (Fig. 12–3). The highest score is 15, indicating that the patient is alert and only needs observation. Scores lower than 13 may indicate the need for immediate treatment. Morbidity and mortality are highest for patients with GCS scores of 8 or lower. Pupil size and reaction are monitored and recorded. Dilated or nonreactive pupils indicate increased ICP and a need for immediate intervention.

NURSING DIAGNOSIS. An example of an applicable nursing diagnosis includes the following:

■ Ineffective tissue perfusion: cerebral, related to cerebral edema

PLANNING. The patient's goal is to remain alert and oriented.

INTERVENTIONS. Oxygen is administered to the patient who has sustained head trauma. If the patient has an altered level of consciousness or deteriorating respiratory effort, endotracheal intubation is performed. Oxygen delivery and ventilation improve cerebral tissue oxygenation and perfusion. The head of the patient's bed is elevated 15 to 30 degrees, if possible, to reduce ICP. The patient's head position should remain at midline to ensure unobstructed venous

drainage. Intravenous access is established to maintain hemodynamic stability and access for medications. Mannitol IV, an osmotic diuretic, may be ordered to decrease cerebral edema. If the patient is agitated, calming the patient is important because agitation increases ICP.

EVALUATION. Criteria for evaluating a positive outcome in the treatment of head trauma include a GCS of 14 to 15. The goal is met if the patient is alert, oriented, and able to follow verbal commands; vital signs are within normal limits; and pupils are equal in size, shape, and reactivity to light.

Spinal Trauma

Spinal cord injury most often results from motor vehicle crashes, sports injuries, falls, and assaults, with most cases occurring in men ages 16 to 30. The cervical spine is especially vulnerable to traumatic injury. Patients who have sustained severe multiple injuries should be suspected of having a spinal cord injury, especially when they have signs of head trauma. All trauma patients should be treated as though they have a spinal cord injury until proven otherwise. Stabilization of the neck and back with a cervical collar and backboard is essential until spinal cord injury is ruled out (Fig. 12–4).

Nursing Process for Spine Trauma

ASSESSMENT. Spinal nerves are located in the spinal cord and transmit motor and sensory impulses to the body. The higher a traumatic lesion is on the spinal column, the more extensive will be the loss of muscle and sensory function. The patient's muscle functions correlate with the level of spinal injury (Table 12–2). A spinal cord injury at the level of C5 or above interferes with diaphragmatic function and affects respiratory effort, which must be carefully assessed. The patient's level of muscle control and ability to feel each extremity is noted and recorded.

NURSING DIAGNOSIS. Examples of applicable nursing diagnoses include the following:

■ Ineffective breathing pattern related to neck injury
■ Ineffective airway clearance related to neck injury
■ Impaired mobility related to neck injury

PLANNING. Goals for the patient are to maintain arterial blood gases within normal limits and maintain or increase mobility.

INTERVENTIONS. During initial treatment of a patient with head or neck trauma, it is imperative that the neck remain immobilized. A cervical collar and backboard must be in place. If the cervical spinal cord has been traumatized, the effectiveness of breathing may be altered. If signs of respiratory distress are present, the patency of the airway should be maintained using the jaw thrust or chin lift maneuver, along with suction and airway adjuncts as needed. Oxygen is administered to improve tissue oxygenation. Advanced adjunct airway equipment, including an endotracheal tube, must be readily available.

GLASGOW COMA SCALE	
Areas of Response	**Points**
Eye Opening	
Eyes open spontaneously	4
Eyes open in response to voice	3
Eyes open in response to pain	2
No eye opening response	1
Best Verbal Response	
Oriented (e.g., to person, place, time)	5
Confused, speaks but is disoriented	4
Inappropriate, but comprehensible words	3
Incomprehensible sounds but no words	
are spoken	2
None	1
Best Motor Response	
Obeys command to move	6
Localizes painful stimulus	5
Withdraws from painful stimulus	4
Flexion, abnormal decorticate posturing	3
Extension, abnormal decerebrate posturing	2
No movement or posturing	1
Total Possible Points	**3–15**
Major Head Injury	**≤8**
Moderate Head Injury	**9–12**
Minor Head Injury	**13–15**

Figure 12-3 The Glasgow Coma Scale is used to determine level of consciousness.

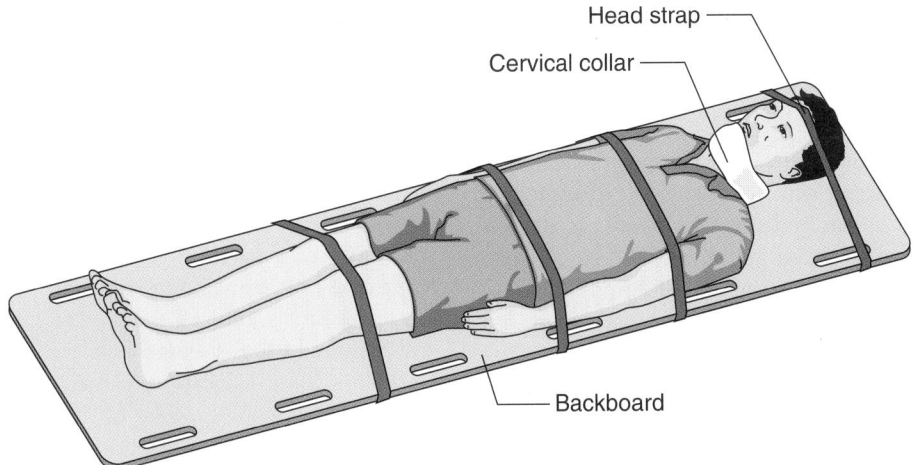

Figure 12-4   Immobilization of a patient suspected of having a spinal cord injury using a backboard and cervical collar.

TABLE 12-2	CORRELATING SPINAL INJURY WITH IMPAIRMENT OF MOTOR FUNCTION
Injury Level	**Impairment**
S3–S5 or above	Patient unable to tighten anus
L4–L5 or above	Patient unable to flex foot and extend toes
L2–L4 or above	Patient unable to extend and flex legs
C5–C7 or above	Patient unable to extend and flex arms

EVALUATION. Outcome criteria for effectively treating cervical spine trauma include maintaining a regular rate, rhythm, and pattern of breathing; maintaining clear lung sounds by protecting the airway from aspiration; and maintaining intactness of mobility by keeping the spine immobile until injury is ruled out.

Chest Trauma

Chest trauma can damage the heart and lungs and cause life-threatening injuries, including pericardial tamponade, hemothorax, tension pneumothorax, and **flail chest**. Potentially life-threatening injuries include pulmonary and myocardial contusion, aortic and tracheobronchial disruption, and diaphragmatic rupture.

Chest trauma can result in laceration of lung tissue and cause a change in the negative intrapleural pressure. Air or blood leaking into the intrapleural space collapses the lung, resulting in a hemothorax and ineffective ventilation. In a tension pneumothorax, air is trapped in the pleural space during exhalation, resulting in increased pressure on the unaffected lung. The heart, great vessels, and trachea shift toward the unaffected side of the chest. As a result, blood flow to and from the heart is greatly reduced, causing a decrease in cardiac output. An uncorrected tension pneumothorax is fatal.

Chest trauma can also injure the heart and great vessels and reduce the amount of circulating blood volume. The heart may be bruised (myocardial contusion) or may sustain direct trauma. **Cardiac tamponade** occurs when blood accumulates in the pericardial sac and increases pressure around the heart. The increased pericardial pressure prevents the heart chambers from filling and contracting effectively. A patient with cardiac tamponade exhibits hypotension, tachycardia, and neck vein distention and requires immediate intervention to reduce the pressure in the pericardial sac and restore normal filling and contraction of the heart chambers.

Nursing Process for Chest Trauma

ASSESSMENT. Patients with major chest injuries can have dramatic symptoms. They may exhibit classic signs of shock with cyanosis, dyspnea, and restlessness. The patient's breathing pattern and effectiveness of respirations are assessed. The rise and fall of the chest is observed, as well as symmetrical chest movement. Any bruising on the chest or upper abdomen is noted. Seat belts and restraint systems can cause significant bruising in high-impact crashes.

NURSING DIAGNOSIS. Examples of applicable nursing diagnoses include the following:

- Ineffective breathing pattern related to unstable chest wall segment or lung collapse
- Deficient fluid volume related to hemorrhage
- Decreased cardiac output related to compression of heart and great vessels

PLANNING. Goals for the patient are to maintain arterial blood gases and vital signs within normal limits.

INTERVENTIONS. Patients with major chest injuries require immediate treatment. Supplemental oxygen is administered to promote tissue oxygenation. A chest tube may be inserted by the physician to relieve a pneumothorax or a hemothorax. IV fluids are initiated with 18- or 16-gauge cannulas. The patient's vital signs and oxygen saturation are continuously monitored to detect signs of shock. Respiratory status is observed to detect ineffective ventilation.

Patients with unstable vital signs need immediate surgical intervention in the operating room. Patients with stable vital signs undergo radiographic studies to determine the extent of cardiac or pulmonary injury.

EVALUATION. Outcome criteria for evaluating the achievement of goals for a patient with chest trauma include maintenance of a patent airway and effective breathing pattern. Respirations are of normal rate and depth with equal chest expansion and no dyspnea or cyanosis. Vital signs and oxygen saturation are within normal limits. Jugular vein distention is absent, and the trachea is midline. Urinary output is 30 to 50 mL/hour. Capillary refill is less than 3 seconds, and skin is warm and dry.

Abdominal Trauma

The organs of the abdomen are vulnerable to injury because there is limited bony protection. Injury to organs such as the spleen and liver, which have a rich blood supply, can result in rapid loss of blood volume and **hypovolemic shock.** If the urinary bladder ruptures, urine leaks into the abdomen and blood may be detected at the urinary meatus or perineum. Penetrating trauma can cause lacerations to abdominal organs, resulting in rapid blood loss and hypovolemic shock.

Nursing Process for Abdominal Trauma

ASSESSMENT. Vital signs are taken to detect tachycardia and hypotension from shock. The shape of the abdomen is observed to detect distention from intra-abdominal hemorrhage. Skin color, bruising, open wounds, and penetrating trauma are noted. The abdomen is auscultated for bowel sounds. The perineum is inspected for blood from the urethra.

NURSING DIAGNOSIS. Examples of applicable nursing diagnoses include the following:

■ Deficient fluid volume related to abdominal organ injury
■ Impaired urinary elimination related to urethral or renal trauma

PLANNING. Goals for the patient are to maintain vital signs and urinary output within normal limits.

INTERVENTIONS. Abdominal organs may be injured as a result of severe blunt or penetrating trauma. If hypotension is present, intra-abdominal hemorrhage may exist. IV fluids are administered via 16- or 18-gauge cannulas to restore circulating volume. A peritoneal lavage may be performed to detect intra-abdominal hemorrhage. An indwelling urinary catheter may be ordered unless blood is present at the urethra. A nasogastric tube may be inserted to decompress the stomach. Open abdominal wounds are covered with a sterile dressing. If abdominal organs are exposed, they are covered with sterile saline–soaked dressings to prevent tissue necrosis.

EVALUATION. Expected outcome criteria for patients with abdominal trauma include effective circulating volume as evidenced by vital signs within normal limits. Skin or mucous membrane color is pink, and skin is warm and dry. Urine output is 30 to 50 mL/hour with no hematuria.

Orthopedic Trauma

Fractured bones can result in blood loss, compromised circulation, infection, and immobility. Unstable pelvic fractures can cause injury to the genitourinary system or disrupt the veins in the pelvis. Fractures of large bones such as the femur and tibia can cause significant blood loss. For example, a fractured femur can cause up to 1500 mL of blood loss and fractured tibia or humerus can cause up to 750 mL of blood loss. Joint dislocations can cause neurovascular compromise by applying pressure to the nerves and blood vessels. Delayed fracture reduction (realignment or setting) can cause **avascular necrosis,** which leads to death of the affected tissue and bone.

Nursing Process for Orthopedic Trauma

ASSESSMENT. Vital signs and pain level are assessed to detect abnormalities. A respiratory assessment is done to detect a pulmonary embolism as a result of a long bone fracture. The injured extremity is inspected, and skin color and capillary refill are noted. Skin integrity, protruding bone, or deformity are noted. Pulses distal to the injury are palpated to assess circulation to the area. Motor function and sensation are assessed to determine the extent of nerve injury.

NURSING DIAGNOSIS. Examples of applicable nursing diagnoses include the following:

■ Acute pain related to trauma
■ Deficient fluid volume related to hemorrhage
■ Impaired physical mobility related to bone injury

PLANNING. Goals for the patient are to maintain vital signs within normal limits, be pain free, and maintain motor and sensory function at baseline level.

INTERVENTIONS. The extremity is splinted and immobilized if there is severe pain or deformity. Splinting promotes comfort and prevents further damage to surrounding tissue by preventing movement of broken bone ends. The joints above and below the affected area are immobilized using a folded towel or a pillow until the patient is evaluated by a physician. Because the extremity may swell after injury, all jewelry is removed before applying a splint. Skin color, temperature, distal pulses, capillary refill, movement, and sensation of the extremity are re-evaluated after splint application. The extremity is elevated and ice applied to reduce edema and relieve pain. Deformed extremities from fractures or dislocations are splinted in the position they are found unless the distal circulation is severely compromised. A patient with a deformed extremity and an absent distal pulse requires immediate medical intervention by a physician to realign the bones. Bleeding from an injured extremity may be severe enough to cause hypovolemic shock. Patients should be monitored for signs of shock and treated accordingly.

LEARNING TIPS

If an extremity is fractured, splint it as it lies to prevent further damage. If the distal circulation is severely compromised, the patient needs immediate medical intervention.

EVALUATION. Outcome criteria for patients with orthopedic trauma include the following:

■ Effective circulating volume as evidenced by strong and palpable pulses, normal blood pressure, normal skin color, skin that is warm and dry, and capillary refill that is less than 3 seconds
■ Pain controlled to a satisfactory level
■ Normal motor function and sensation in the extremity

▶ BURNS

Burn injuries are acutely painful events that may be dramatic in appearance. Nursing care depends on the extent and depth of the burn injury and the presence of any associated factors such as smoke inhalation, blunt trauma, or fractures. The skin protects the body by preventing bacterial or viral invasion, enhancing temperature regulation, and conserving body fluids and electrolytes. These functions are impaired with a burn injury and can lead to multisystem alterations. The more extensive the burn injury, the greater the potential for complications and mortality. The patient's age may contribute to the risk of mortality as well. Infants under age 2 and elderly patients over age 60 have the highest mortality rates from major burns.

Types of Burns

Burns can be thermal, chemical, or electrical. The most common are thermal burns, caused by steam, scalds, fire injuries, contact with hot substances, or excessive exposure to the sun or radiation. Chemical burns are caused by acids or alkalis. Alkali burns are usually more serious than acid burns because alkalis penetrate deeper and burn longer. The concentration of the chemical agent and the duration of exposure determine the extent and depth of damage. Electrical burns can be caused by low-voltage (alternating) current or high-voltage (alternating or direct) current.

Burn Pathophysiology

Three factors influence the extent of the burn injury: the intensity of the energy source, the duration of exposure to the energy source, and the conductance of the tissue exposed. Increased intensity with increased exposure causes greater tissue damage. Initially there is decreased blood flow to the local burn area, followed by vasodilation and increased **capillary permeability,** which causes fluid and electrolytes to leak from cells into the interstitial space. Fluid can also be lost directly through the burn wound. These responses cause intravascular fluid loss and can place the patient at risk for hypovolemic shock.

Burn Classification

Burns are classified as either major or minor depending on the depth of the burn, the percentage of body surface area burned, the age of the patient, medical history, and the part of the body burned.

Burn depth is described as either partial thickness or full thickness, based on the appearance of the wound (Fig. 12–5). Partial-thickness burns can be either superficial or deep. Superficial **partial-thickness burns** primarily involve the epidermis of the skin. These wounds appear red and cause severe local pain. Common examples of this type of burn include sunburns and minor steam burns. This type of burn typically heals in 2 to 7 days.

Deep partial-thickness burns involve the entire epidermal layer and part of the dermis. These burns appear red or mottled and are more painful. The epidermis is blistered or broken (Fig. 12–6). Deep partial-thickness burns can be-

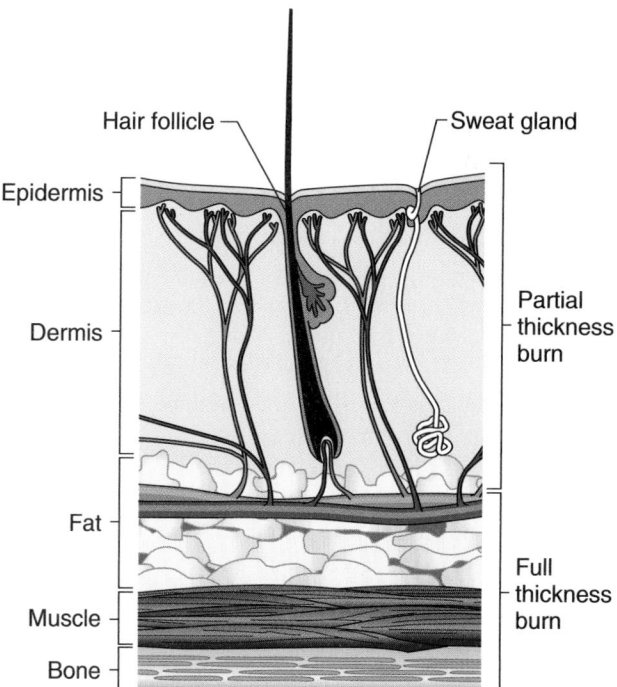

Figure 12-5 Partial- and full-thickness burns and structures affected.

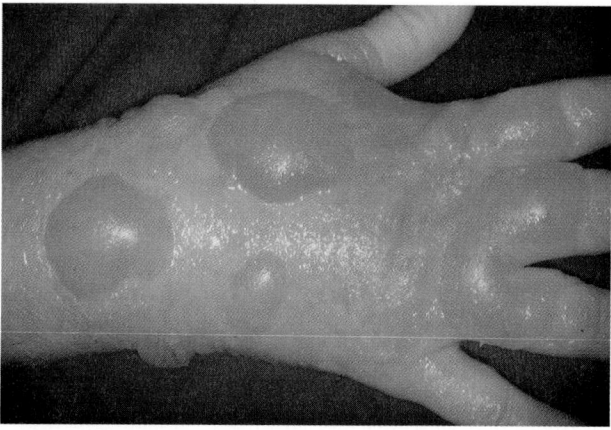

Figure 12-6 A blistered partial-thickness thermal burn.

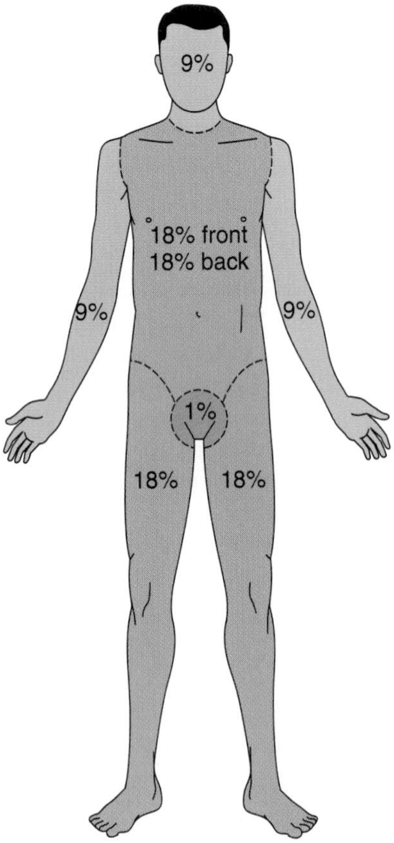

Figure 12-7 The rule of nines is used to estimate the amount of body surface burned using areas approximating 9 percent and multiples of 9 percent.

come full-thickness injuries if they become infected or if there is further trauma to the site.

Full-thickness burns involve all the layers of the skin and the subcutaneous tissue. The burn appears pale white or charred, red or brown, and leathery. Nerve endings have been destroyed, so the severely burned area may be painless; however, the surrounding areas may be very painful and sensitive to air.

The percent of total body surface area (TBSA) burned can be estimated using the **rule of nines,** which divides the body surface into areas equal to 9 percent or a multiple of 9 percent (Fig. 12–7). Deep partial-thickness burns of the face, neck, hands, feet, and genitalia are considered severe even though they do not involve a large percentage of the TBSA. Superficial partial-thickness burns are not included in the estimates of TBSA because the skin does not lose its functioning ability.

Chemical Burns

Chemical burns differ from thermal burns in that the burning process continues until the agent is inactivated, neutralized, or diluted with water. The degree of damage caused by a chemical depends on its concentration and quantity, its mechanism of action, and the duration of contact. Most chemical burns are caused by strong acids or alkalis that get on the skin or clothing. The fumes of strong chemicals can cause pulmonary burns.

To stop the burning process of most chemical burns, all saturated clothing must be removed and the burn copiously irrigated with water. The stream of water should be gentle to avoid further injury to the burned skin. Strong alkalis may cause more severe burns than strong acids because they penetrate more deeply into the tissues. Dry chemicals such as lime must be brushed off before flushing with water because they may be activated by contact with water and cause more damage to the skin.

Eyes involved in chemical burns can be seriously damaged. Eyes should be immediately flushed under running water for at least 5 minutes for an acid burn and 10 to 20 minutes for an alkali burn. Immediately after flushing the eye, the patient should be taken for evaluation and treatment of the injury.

Electrical Burns

Electrical burns frequently are much more serious than they appear. As the electrical current passes through the body, it damages the inner tissues, leaving little evidence of a burn on the skin surface. The type and voltage, resistance, path of transmission through the body, and duration of contact determine the amount of damage sustained. From the point of entry, an electrical current follows the path of least resistance, causing one or more tracks of damage. Energy from a high-voltage electrical current may disrupt the normal electrical rhythm of the heart, causing cardiac arrest. The electrical current also may cause violent muscle contractions, resulting in fractures or dislocations.

Lightning contains massive amounts of energy (up to 50 million volts) and can reach temperatures in excess of 50,000° F. Injuries that result from lightning include cardiac arrest, neurological injuries, blunt trauma, and superficial burns. Surface burns from lightning have a spidery, feathery, branching appearance.

Patients with minor electrical burns may have lost consciousness briefly and are often confused. The skin protects the body by preventing bacterial or viral invasion, enhancing temperature regulation, and conserving body fluids and electrolytes. These functions are impaired with a burn injury and can cause multisystem physiological alterations.

Moderately injured persons show more obvious altered mentation and may be combative or comatose. The more extensive the burn injury, the greater the potential for complications and mortality.

Nursing Process for Burns
Assessment

The assessment of the burn patient begins with the ABCs of the primary survey (Nursing Care Plan Box 12–6). The history should include the mechanism and time of the injury and a description of the surrounding environment, including the presence of noxious chemicals and inhalation of smoke in an enclosed space. The greatest threat to life in a patient with a major burn injury is smoke or heat inhalation, which causes edema in the respiratory passages. Continuous assessment of respiratory status is essential when

BOX 12-6 NURSING CARE PLAN FOR THE PATIENT WITH BURNS

 Ineffective airway clearance related to edema from burn injury

OUTCOME
Patient has patent airway.

EVALUATION OF OUTCOME
Is respiratory rate within normal limits? Is pulse oximetry reading above 90 percent? Can the patient talk without a hoarse voice?

INTERVENTION	RATIONALE	EVALUATION
Assess for burns to the face, singed nasal hair, or soot around the mouth and nose.	Burns to the face swell rapidly and can compromise the airway. Evidence of soot and singed nasal hair can be a sign of smoke inhalation and pulmonary injury.	Is respiratory rate between 10 and 20? Is pulse oximetry reading above 90 percent?
Elevate the head of bed 30 degrees.	Elevation can minimize edema by dependent gravity.	Can patient talk without a hoarse voice?

 Deficient fluid volume related to fluid loss secondary to increased capillary permeability

OUTCOME
Maintains normal vital signs and has urine output of more than 30 mL/hour.

EVALUATION OF OUTCOME
Are blood pressure, pulse, and respiratory rate within normal limits? Is urinary output 30 mL/hour?

INTERVENTION	RATIONALE	EVALUATION
Monitor vital signs and report abnormalities.	Changes in heart rate and blood pressure can be readily detected.	Is pulse rate less than 120? Is blood pressure within normal limits?
Give intravenous fluids as ordered.	Fluid volume is replaced by infusing intravenous fluids.	Is patient's urinary output 30 mL/hr?

Acute pain related to major burn

OUTCOME
Patient states pain is at satisfactory level.

EVALUATION OF OUTCOME
Does patient state pain is relieved?

INTERVENTION	RATIONALE	EVALUATION
Administer narcotics as ordered.	Burns cause damage or exposure of the nerve endings.	Does patient state pain is relieved?
Keep burned areas covered with sterile dressings.	Air movement over exposed nerve endings increases pain.	Does patient report comfort?

 Risk for infection related to altered skin integrity

OUTCOME
Patient remains free from signs of infection.

EVALUATION OF OUTCOME
Does the patient have a normal temperature? Is the white blood cell (WBC) count less than 10,000? Is the burn wound absent of purulent drainage?

INTERVENTION	RATIONALE	EVALUATION
Wash hands frequently. Wear sterile gloves, gown, and mask when providing care. Maintain sterile dressings.	Patient's first line of defense (the skin) can no longer provide protection from bacterial or viral invasion.	Are signs of infection present: fever, purulent wound drainage, elevated WBCs?
Monitor patient's temperature and WBC results.	Abnormal findings are detected and can be reported to physician.	Is WBC count less than 10,000? Is patient's core temperature normal?

you observe burns or soot on the face, singed nasal hairs, a hoarse voice, or restlessness.

Nursing Diagnosis

Examples of applicable nursing diagnoses include the following:

- Deficient fluid volume related to increased capillary permeability
- Pain related to major burn
- Risk for infection related to impaired skin integrity

Planning

The patient's goals are to maintain vital signs and oxygen saturation within normal limits, have pain relieved, remain free of infection, and restore skin integrity.

Interventions

The first responsibility when caring for a burn victim is to stop any further burning. The victim is removed from the burning area. If clothing is on fire, the victim is placed on the ground and rolled in a blanket. All jewelry is removed as soon as possible because swelling begins soon after injury. The burned area is covered with a dry sterile dressing or clean linen. If the TBSA affected appears to be less than 10%, cool saline-moistened sterile dressings may be applied within the first 10 minutes to help relieve pain and reduce the heat content of the tissue. Generally, wet dressings should not be used if the TBSA is greater than 10 percent because they provide a pathway for infection and cause hypothermia. Blisters provide a protective covering to underlying tissue and should not be ruptured. Cold packs and ice are never placed over the burn area because they can cause further thermal tissue damage.

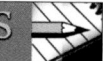

L E A R N I N G   TIPS

Over-the-counter ointments, lotions, butter, and antiseptics are never used on a major burn because they may promote infection, retain heat, and cause more pain.

Burns of the face may swell rapidly and can compromise the airway. Facial and respiratory burns should be suspected if nasal hair, eyebrows, eyelashes, or scalp hair are singed, if soot is on the face or neck, or if the patient has a hoarse voice or is coughing. Facial burns are treated by elevating the head of the bed to 30 degrees to minimize edema. Oxygen is administered to the patient with potential pulmonary injury. Equipment for endotracheal intubation should be readily available. Because large fluid losses occur in burn injuries, an IV infusion with large-bore cannulas should be started. The patient's weight and the extent of the burn determine fluid resuscitation needs.

The patient is kept warm because when skin is lost, a burn victim cannot maintain body heat. IV narcotics are administered for pain. Partial-thickness burns that involve a small area are cleaned with sterile saline solution, covered with a $\frac{1}{8}$-inch layer of an anti-infective cream such as silver sulfadiazine (Silvadene, Flamazine), and covered with bulky, fluffed dressings. Major full-thickness wounds are covered with dry, sterile dressings or linen. Patients with major burns are transferred to a specialized burn unit.

Evaluation

When treatment is effective for deficient fluid volume, vital signs remain normal. The goal for pain is met if the patient reports satisfactory relief. The goal for impaired skin integrity and risk for infection is met if the patient's wounds heal without infection.

CRITICAL THINKING

Mr. Smith

Mr. Smith is a 28-year-old man who was welding close to a natural gas line. The flame of the welder caused the gas line to explode, throwing Mr. Smith 50 feet. He landed on his back. He is brought to the emergency department by the rescue squad. Mr. Smith is awake, alert, and oriented. He has soot around his mouth and nose. He sustained deep partial-thickness burns to his neck, upper chest, and both forearms. He is complaining of pain from his burns and also thoracic back and hip pain. His pulse rate is 100. His blood pressure is 160/90. His respiratory rate is 20.

1. What is the first priority of care for Mr. Smith?
2. Is Mr. Smith at risk for respiratory burns? Why?
3. Are Mr. Smith's vital signs within normal limits?
4. Would wet or dry dressings be preferable for his large areas of deep partial-thickness burns?
5. Mr. Smith is wearing a neck chain and a wedding ring. Should they be removed immediately, or should you wait until Mr. Smith's wife arrives to take them? Why?
6. Mr. Smith continues to complain of hip and back pain. In reviewing his mechanism of injury, what other injuries could Mr. Smith have?

Answers at end of chapter.

▶ HYPOTHERMIA

Normally the body maintains its temperature in a narrow range on either side of 98.6° F (37° C) to allow chemical reactions to work most efficiently. Body heat escapes to the environment through conduction, convection, radiation, and evaporation. Heat loss is inversely proportional to body size and body fat. Fat insulates because it has less blood flow and consequently has less ability to vasodilate and lose heat.

Nursing Process for Hypothermia
Assessment of Stages of Hypothermia

Hypothermia occurs when the core body temperature falls below 95° F (35° C). As the core temperature falls below 95° F, the body is less able to regulate its temperature and generate body heat, causing progressive loss of body heat to occur.

In cases of mild hypothermia (core temperature between 90° F [32° C] and 95° F [35° C]) the patient is usually alert, shivering, and may appear clumsy, apathetic, or irritable (Table 12–3). Hypoglycemia can occur because glucose and glycogen stores are depleted by long-term shivering. Respiratory rate, heart rate, and cardiac output decrease.

More severe hypothermia occurs between 85° F and 90° F (29° C and 32° C). Shivering stops and muscle activity decreases. Initially, fine muscle coordination ceases. Then, as core body temperature continues to drop, all muscle activity stops and the muscles become rigid. The patient becomes lethargic and less interested in combating the cold environment. The patient's level of consciousness begins to markedly decrease at 89.6° F (32°C); the patient becomes lethargic and disoriented and begins to hallucinate. The pupils become dilated. As the core body temperature falls to 82° F (28° C), the patient becomes apneic, the pulse becomes slower and weaker, and cardiac dysrhythmias occur. The profoundly hypothermic patient has a core temperature of less than 80° F (27° C) and usually appears dead, with no obtainable vital signs. Determination of death should be made only after aggressive core rewarming to at least 90° F (32° C).

Nursing Diagnosis

An example of an applicable nursing diagnosis includes the following:

■ Hypothermia related to cold exposure

Planning

The patient's goal is to maintain a body temperature within normal limits (Nursing Care Plan Box 12–7).

Interventions

Initial treatment of the hypothermic patient consists of rewarming the patient, stabilizing vital functions, and preventing further heat loss. The patient is removed from the cold environment. All wet clothing is removed to prevent further heat loss. The patient's core body temperature guides treatment. If the body temperature is above 82.4° F (28° C), passive rewarming is preferred. The room temperature is set to 70° F to 75° F. The patient is wrapped in

TABLE 12–3	DEFINING CHARACTERISTICS AND OUTCOME CRITERIA FOR HYPOTHERMIA
Core Body Temperature	**Defining Characteristics**
Below 95° F (35° C)	• Skin cold to touch • Lack of coordination • Slurred speech
Below 91.4° F (33° C)	• Cardiac dysrhythmias • Cyanosis
Below 89.6° F (32° C)	• Shivering replaced by muscle rigidity • Hypotension • Dilated pupils
Below 82.4° F (28° C)	• Absent deep tendon reflexes • Hypoventilation (3–4 breaths per minute) • Ventricular fibrillation possible
Below 80.6° F (27° C)	• Coma • Flaccid muscles • Fixed, dilated pupils • Ventricular fibrillation to cardiac standstill • Apnea

Outcome Criteria
• Core body temperature is greater than 95° F (35° C).
• Patient is alert and oriented.
• Cardiac dysrhythmias are absent.
• Acid-base balance is normal.
• Pupils react normally.

warm, dry blankets. Heat loss from the head is reduced by covering the head with warm towels.

For a patient with a core body temperature below 82.4° F (28° C), active rewarming is needed. A heating blanket (carbon-fiber) and radiant heat lights are used. Warm, humidified oxygen is administered. Warm IV fluids are administered. Body temperature is constantly monitored using a rectal probe. Heated **gastric lavage,** heated peritoneal lavage, or cardiopulmonary bypass may be performed for profound hypothermia. Cardiac drugs are given sparingly because as the body warms, peripheral vasodilation occurs. Drugs that were trapped in the peripheral circulation are

BOX 12-7 **NURSING CARE PLAN FOR THE PATIENT WITH HYPOTHERMIA**

 Hypothermia related to exposure to cold environment

OUTCOME
Body temperature and vital signs are within normal limits.

EVALUATION OF OUTCOME
Is patient's body temperature greater than 95° F (35° C)? Is patient alert and oriented? Is cardiac rhythm normal?

INTERVENTION	RATIONALE	EVALUATION
Monitor patient's core body temperature.	Abnormal body temperature can be detected and treated.	Is body temperature greater than 95° F (35° C)?
Monitor pulse and electrocardiogram (ECG) rhythm.	Cardiac dysrhythmias may occur at temperatures below 91.4° F (33° C).	Is pulse rate and ECG rhythm normal?
Monitor patient's level of consciousness.	Level of consciousness becomes markedly decreased at temperatures of 32° C (89.6° F).	Is the patient alert?
Institute rewarming passively or actively as ordered.	Rewarming is necessary to return body temperature to desirable range.	Is body core temperature rising to normal range?

then suddenly released during rewarming, leading to a bolus effect that may cause fatal dysrhythmias.

Evaluation

Desired outcome criteria for the patient with hypothermia is a core body temperature higher than 95° F (35° C), no cardiac dysrhythmias, pulse and blood pressure within normal limits and an alert and oriented status.

► FROSTBITE

The extremities are vulnerable to cold injury. Frostnip occurs when exposed parts of the body become very cold but not frozen. This condition usually is not painful. The skin becomes pale and blanched. Contact with a warm object such as someone's hand may be all that is needed to rewarm the part. During rewarming, the affected part may tingle and become red.

Frostbite occurs when body parts become frozen. The extremities are at increased risk because blood shunts away from them to maintain core body temperature. The affected tissue feels hard and frozen. Most frostbitten parts are white, yellow-white, or blue-white. When rewarmed, the skin appears deep red, hot, and dry to touch. The severity of a cold injury is determined by the duration of the exposure, the temperature to which the body part was exposed, and the wind velocity during exposure.

Interventions for frostbite include protecting the affected area from further trauma. To prevent additional damage, the frostbitten part is handled gently and never rubbed. The injured part is loosely covered with a dry, sterile dressing. The patient is not allowed to stand or walk on a frostbitten foot. The affected extremity is elevated to heart level to minimize edema and promote blood flow.

► HYPERTHERMIA

Usually the body's heat-regulating mechanisms work very well, allowing people to tolerate significant temperature changes. The body's most efficient mechanisms to decrease body heat are sweating and dilation of blood vessels in the skin. When blood vessels dilate, blood comes to the skin surface to increase the rate of radiation of heat from the body. However, when these mechanisms become overwhelmed, the consequences can be disastrous and irreversible. Those at greatest risk for heat illnesses include children, elderly people, and patients with cardiac disease.

Hyperthermia results when thermoregulation breaks down because of excess heat generation, an inability to dissipate heat, overwhelming environmental heat, or a combination of these factors. Unlike a fever, in which the thermal

set point is elevated, in heat illness the thermal set point remains normal and hyperthermia occurs because of an inability to dissipate heat. Antipyretics are of no use in hyperthermia and in fact may contribute to complications.

Stages of Hyperthermia
Assessment

Illness from heat exposure can take three forms: heat cramps, heat exhaustion, and **heatstroke** (Box 12–8). As heat illness progresses, circulating blood volume decreases, causing dehydration. Fluid intake is crucial in the prevention of heat illness.

Heat Cramps

Heat cramps, the mildest form of heat illness, involve painful muscle spasms, usually in the legs or abdomen, that occur after strenuous exercise. Large amounts of salt and water can be lost as a result of excessive sweating, causing stressed muscles to spasm. When heat cramps occur, the patient should be removed from the hot environment. Tight clothing should be loosened. The patient should sit or lie down until the cramps subside. The patient should be given water or a diluted (half-strength) balanced electrolyte solution to drink. With adequate rest and fluid replacement, the body adjusts the distribution of electrolytes and the cramps disappear.

Heat Exhaustion

Heat exhaustion occurs when the body loses so much water and electrolytes through heavy sweating that hypovolemia occurs. Heat exhaustion is largely a manifestation of the strain placed on the cardiovascular system attempting to maintain normothermia. Cerebral function is unimpaired, although the patient may show minor irritability and poor

---

BOX 12-8 **Defining Characteristics and Outcome Criteria for Environmental Hyperthermia**

Defining Characteristics

Early signs:
- Core body temperature 100.4° F to 102.2° F (38° C to 39° C)
- Diaphoresis
- Cool, clammy skin
- Dizziness
- Pulse rate > 100

Late signs:
- Increasing body core temperature of 106° F (41° C) or more
- Hot, dry, flushed skin
- Altered mental status
- Coma or seizures possible
- Hypotension

Outcome Criteria
- Core body temperature less than 101° F (38.3° C)
- Patient alert and oriented
- Skin warm and dry to touch

LEARNING TIPS

Older adults are vulnerable to hyperthermia. In times of extreme summer temperatures, elderly people who live alone should be checked to make sure they are not experiencing hyperthermia. If they do not have fans or air conditioning available, they should be taken to a cooler environment.

judgment. The ability to sweat remains. The skin is usually cold and clammy and the face gray. Sodium and water loss cause the patient to become dehydrated. The body temperature is usually normal or slightly elevated, from 100.4° F to 102.2° F (38° C to 39° C). The patient may complain of feeling dizzy, weak, or faint, with nausea or a headache. Vomiting and diarrhea may also be present.

The patient should be removed from the hot environment and tight clothing loosened. If the patient is fully alert, oral fluids (water or diluted balanced salt solution) are administered. If symptoms do not resolve promptly, the patient will need IV fluids.

Heatstroke

If symptoms of heat exhaustion are not treated, heatstroke can develop. Altered mental status and an inability to sweat are key symptoms in heatstroke. Some patients show confusion, irrational behavior, or psychosis; others develop a coma or seizures. Because the sweating mechanism has been overwhelmed, many heatstroke victims have hot, dry, flushed skin. The body temperature rises rapidly to 106° F (41° C) or more, and the patient's level of consciousness decreases. If heatstroke is not treated, death results.

Nursing Process for Hyperthermia
Nursing Diagnosis

Examples of applicable nursing diagnoses include the following:

- Hyperthermia related to exposure to hot environment
- Deficient fluid volume related to hypervolemia

Planning

The patient's goal is to maintain the body temperature within normal limits.

Interventions

Emergency treatment of heatstroke consists of reducing the body temperature and rapidly cooling the victim. The patient should be undressed and allowed to cool. Evaporative cooling is the most efficient method of cooling. Tepid water can be used as a mist spray over the patient, with a strong continual breeze from electric fans to enhance evaporation. If the patient is hypotensive, IV fluids are administered.

Patients suffering from heatstroke are admitted to the intensive care unit because late complications can appear suddenly and require immediate management. Relatively common occurrences include seizures, cerebral ischemia, renal failure, late cardiac decompensation, and gastrointestinal bleeding. Long-term prognosis is variable, depending on the patient's previous state of health and length of time under heat stress.

Evaluation

Desired outcome criteria for the hyperthermic patient include a core body temperature that is below 101° F (38.3° C), warm and dry skin, a strong pulse, blood pressure within normal limits, and an alert and oriented status.

▶ POISONING AND DRUG OVERDOSE

Poisons are introduced into the body by ingestion, inhalation, injection, absorption, or venom bites. Poisons act by changing cellular metabolism, causing damage to structures, or disturbing function. Many toxins and poisons alter the patient's mental status, making it difficult to obtain an accurate history.

Nursing Process for Poisoning
Assessment

The primary nursing responsibility is to recognize that a poisoning has occurred and then attempt to determine the nature of the poison. The method of exposure is established so that removal or interruption of the toxin can begin. Most ingested poisons are drugs, but about one-third of poisonings are caused by cleaners, soaps, insecticides, acids, or alkalis. Many household plants are poisonous if they are accidentally ingested. Some plants cause local irritation of the skin, and others can affect the circulatory system, gastrointestinal tract, or central nervous system.

Objects at the scene, such as empty bottles, scattered pills, or chemicals, should be sent with the patient to the hospital to help establish the identity of the substance. The patient's physical appearance may also give a clue to the type of substance taken. Intravenous needle tracks, burns, erythema, and flushed skin may help identify the poison or toxic exposure. Poison control centers have access to information concerning virtually all poisonous substances, available antidotes, and appropriate emergency treatment.

Nursing Diagnosis

An example of an applicable nursing diagnosis includes the following:

- Risk for injury related to absorption of poisoning agent

Planning

The patient's goal is to maintain normal vital signs and be free of injury.

Interventions for Ingested Poisons

Emergency treatment for the patient who has ingested poison includes rapid removal of the poison from the gastrointestinal tract and dilution of the remainder. Syrup of ipecac can be used to induce vomiting if the patient is fully alert and if the ingested substance is nonerosive and not petroleum based. The usual dose of syrup of ipecac is 15 mL for children ages 1 to 5 and 30 mL for ages 6 and up. The syrup is followed by several glasses of water. Vomiting usually occurs within 30 minutes. The dose may be repeated once if vomiting does not occur after 30 minutes.

Gastric lavage may be indicated to flush ingested poisons from the stomach. A large gastric tube is inserted via the nose or mouth into the patient's stomach. Water is instilled via the tube in 60-mL amounts and withdrawn to evacuate any remaining poison. This is repeated until 2 L of water have been lavaged or until the gastric return is clear of any pill fragments or substance. Before removing the gastric lavage tube, activated charcoal with sorbitol can be administered to absorb toxins and facilitate rapid transit of the poison through the intestinal tract.

Evaluation

A desired outcome is that the patient remains free from injury and has stable vital signs.

Inhaled Poisons

Inhaled poisons include natural gas, pesticides, carbon monoxide, chlorine, and other gases. Carbon monoxide is odorless and can produce profound hypoxia by combining with hemoglobin molecules (and displacing oxygen) in red blood cells. The patient's carboxyhemoglobin levels are monitored to direct appropriate therapy. Inhalation of chlorine is very irritating to the respiratory system and can produce airway obstruction and pulmonary edema.

When an inhalation injury occurs, the patient must be moved into fresh air and away from the toxin. Supplemental oxygen is given as ordered. A patient exposed to prolonged inhalation of a poison may experience lung damage. Respiratory status must be closely monitored to detect complications.

Injected Poisons

Injected poisons pose compelling problems because they are difficult to remove or dilute. Usually they result from drug overdose, but they can also result from the bites and stings of insects or animals. Local swelling and tissue destruction may occur at the injection site. All jewelry is removed because swelling may occur. A cold pack is applied to decrease local pain and swelling around the injection site. The identity of the injected drug or toxin must be established so that adverse effects can be anticipated and managed.

Insects

Insect stings or bites cause anaphylaxis in a small percentage of people; however, symptoms are typically limited to localized pain, swelling, heat, and redness. Potentially dangerous stings or bites may come from bees, wasps, yellow jackets, hornets, certain ants, scorpions, and some spiders. Treatment involves applying ice to the site and elevating the affected part. Cellulitis can occur hours later and may require medical treatment.

When a patient has sustained a bee or wasp sting, examine the area for the stinger and remove it by gently scraping it off the skin. Tweezers or forceps are not used to remove the stinger because squeezing the stinger can inject more venom into the patient. Placing ice over the injury site may help slow the rate of toxin absorption.

Two types of spiders—the black widow and the brown recluse—can inflict serious and sometimes life-threatening bites. Both species are found throughout the United States, and antivenin is available for treatment. Black widow spiders are glossy black and have a distinctive, bright red-

orange marking in the shape of an hourglass on the abdomen. They are found in dry, dim places around buildings, in woodpiles, and among debris. Their venom is neurotoxic and causes systemic symptoms, including cramping of large muscle groups, dyspnea, weakness, sweating, nausea, vomiting, and rash. Death is uncommon, and symptoms generally subside in 48 hours.

The brown recluse spider is dull brown and has a dark violin-shaped mark on its back. It tends to live in dark areas, under rocks, in woodpiles, and in old abandoned buildings. The venom of the brown recluse causes severe local tissue damage. The area becomes red, swollen, and tender and develops a pale, mottled, cyanotic center. A large ulcer can develop within 48 hours if not treated promptly. Systemic symptoms include fever, chills, nausea, vomiting, arthralgia, and weakness.

Snake Bites

Only a small percentage of snake bites are caused by poisonous snakes. The most prevalent poisonous snakes are the coral snake and the pit vipers, which include rattlesnakes, copperheads, and cottonmouth moccasins. Envenomation occurs when the snake's hollow fangs puncture the skin and inject venom, which is stored in sacs located at the back of the snake's head. A poisonous snake bite has two small puncture wounds with surrounding discoloration, swelling, and pain. Envenomation by any of the pit viper snakes produces burning pain at the site of the injury. Swelling and discoloration occur within 5 to 10 minutes after the bite.

Interventions are focused on decreasing the circulation of venom throughout the patient's system by keeping the patient calm and immobilizing the affected part. Venous tourniquets placed above and below the fang marks help limit the spread of venom through the veins of the extremity. The tourniquets should not stop arterial flow. The patient's pulse should be palpable below the tourniquets after they are applied. The site of the bite is cleaned with soap and water. The patient is kept calm until antivenin can be given. Medical treatment of the patient with a poisonous snake bite should be directed by an experienced toxicologist.

▶ NEAR-DROWNING

Drowning is death from **asphyxia** after submersion in water. Near-drowning is submersion with at least temporary survival of the victim. Life-threatening complications of near-drowning are respiratory failure and ischemic neurological injury from hypoxia and acidosis. When submersion occurs, conscious victims hold their breath until reflex inspiratory efforts override breath holding. As water is aspirated, laryngospasm occurs, producing severe hypoxia. In wet drowning the laryngospasm is less prolonged and fluid enters the lungs after the vocal cords relax. In dry drowning the laryngospasm is prolonged and prevents fluid from entering the lungs. Most successfully resuscitated victims experience dry drowning.

If a person survives submersion, acute respiratory failure may follow. The incidence of serious pulmonary complications is high in this group. Symptoms of impaired gas exchange may be delayed as long as 24 hours after the incident. Contaminants in the water can irritate the pulmonary system and cause additional complications. Cardiovascular complications can occur secondary to hypoxia and metabolic acidosis, resulting in dysrhythmias.

Nursing Process for Near-Drowning
Assessment

Vital signs are assessed to detect abnormal readings. Respiratory rate and pattern are observed. Any dyspnea or signs of airway obstruction are noted. Skin color or cyanosis is noted. The patient's level of consciousness may be altered due to anoxia.

Nursing Diagnosis

An example of an applicable nursing diagnosis includes the following:

■ Ineffective tissue perfusion related to severe anoxia

Planning

The patient's goal is to maintain level of consciousness and vital signs within normal range, with clear breath sounds that are equal bilaterally.

Interventions

Resuscitative efforts should always start with the ABCs of the primary survey. Supplemental oxygen is administered to increase tissue oxygenation. Adjunct airway equipment should be available. Endotracheal intubation and insertion of a nasogastric tube to decompress the stomach may be needed.

Aggressive resuscitative efforts should be used on victims of cold water drowning when submersion time is 1 hour or less. Hypothermia can decrease the metabolic needs of the brain and contribute to neurological recovery after prolonged submersion.

Evaluation

Factors that influence the outcome of near-drowning include the temperature of the water, length of time submerged, cleanliness of the water, and age of the victim. The younger the patient, the better the chance for survival. Expected outcome criteria to evaluate the near-drowning patient include normal respiratory rate and pattern and normal vital signs, an alert and oriented status, and skin that is warm and dry to touch with capillary refill of less than 3 seconds.

▶ PSYCHIATRIC EMERGENCIES

A psychiatric emergency exists when people no longer possess the coping skills necessary to maintain their usual level of functioning. The patient's moods, thoughts, or actions may be so disordered that the patient has the potential to endanger or harm self or others if the situation is not quickly controlled. If acute psychiatric episodes are not managed, they can result in life-threatening, suicidal, violent, or psychologically damaging behavior. If an emotional trauma is not managed successfully, a condition known as posttraumatic stress disorder may result in which tension,

anxiety, guilt, and fear concerning the traumatic event produce cognitive, affective, and behavioral responses to memories of the event long after the event has passed.

Nursing Process for Psychiatric Emergencies
Assessment

Causes of psychiatric emergency symptoms are varied and require thorough assessment of the patient's history and mental status. Information from the patient's medical history may produce possible organic causes contributing to the patient's presenting symptoms. Endocrine dysfunction, electrolyte abnormalities, and head trauma are examples of medical conditions that may cause changes in mental status. A medication history is obtained to determine compliance with medication regimens and any recent changes in medications. Information regarding recent use of alcohol or illicit drugs should be obtained because these substances can heighten psychiatric emergencies. A brief mental status examination is conducted. The patient's appearance, behavior, cognitive function, thought content, and thought processes are noted. The nurse determines whether the patient is having problems concentrating, following instructions, or recalling his or her medical history.

A crisis occurs when people enter a sudden state of emotional turmoil and are unable to resolve the situation with their own resources. Common emotional or behavioral manifestations of psychiatric crises include responses to stressful events, anxiety, depression, psychosis, and mania. Anxiety may range in severity from mild to panic. Panic evolves into complete disorganization and loss of control. The patient in panic is terrified and needs external controls to avoid harm. Depression is an affective disorder most commonly characterized by physical ailments and somatizations. An assessment of the person's suicide risk is important. Antidepressants are used to restore the balance of brain neurochemicals and diminish the symptoms of depression. Psychotic patients experience impaired thought processes and thought content characterized by hallucinations, delusions, ideas of reference, thought broadcasting, and thought insertion. Psychotic thinking and abnormal speech patterns interfere with the patient's attempt to communicate rationally. Manic behavior is most commonly the result of manic-depressive (bipolar) disorder. Manic persons typically exhibit bizarre, extreme, and hyperactive behaviors. Manic persons are also at high risk for injuring themselves or others.

Nursing Diagnosis

Examples of applicable nursing diagnoses include the following:

■ Anxiety related to situational stress
■ Risk for injury related to impaired judgment
■ Fear related to alteration in thought content

Planning

The patient's goals are to reduce anxiety, remain free from injury, and reduce fear.

Interventions

During crisis intervention, strategies are designed to reduce the negative impact of a distressing event. It is important to establish an atmosphere of trust so the patient feels free to discuss problems. The patient's physical and emotional complaints are acknowledged by using active listening. The environment is made safe, and external sources of stimulation are reduced. When speaking to the patient, the nurse speaks directly and truthfully and never promises unachievable things. Trusted supportive members of the patient's family may be involved to calm the person and encourage cooperation. Bystanders or adversive family members who could create further complications are restricted. Do not threaten, challenge, or argue with a disturbed patient. Correct misconceptions, but not in an argumentative manner. Be firm but unthreatening and show respect for the patient by not laughing or joking.

Interventions involve reducing fear and potential harm to self or others. Sometimes physical restraint is necessary. Restraint equipment is not shown to the patient until sufficient help is available to use it. The patient's behavior is documented on the chart, along with written physician orders for the use of physical restraints. The restraints must not restrict circulation. Pulses and capillary refill are assessed frequently while the patient is restrained.

Antipsychotic medications are administered as ordered, and their response is monitored. Haloperidol (Haldol) is used if a patient needs rapid tranquilization.

Evaluation

The patient's goals are met if the patient reports reduced anxiety and fear and remains free from injury.

▶ DISASTER RESPONSE

Disaster is defined as any event that overwhelms existing personnel, facilities, equipment, and capabilities of a responding agency, institution, or community. Potential sources of disaster include internal events such as fires and explosions; external events such as floods, storms, fires, earthquakes, and tornados; and created events such as motor vehicle accidents, plane crashes, and acts of terrorism.

External disasters involve a communitywide response of several different agencies, including emergency medical system (EMS) providers, fire agencies, law enforcement, and hospitals. These agencies work together to coordinate search, rescue, transportation, communication, and treatment of multiple victims. Hospitals serve as the major treatment area for victims of a disaster, referred to as casualties. When a disaster occurs, the hospital activates its disaster plan, which outlines specific duties for each nursing unit

and the staff for each nonnursing department as well. Typically each nursing unit prepares for the influx of casualties by calling all available off-duty staff to report to work and by discharging noncritical patients. In a hospital disaster plan, each nursing unit is usually designated to receive specific types of casualties, such as major trauma, burns, medical, pediatric, or psychiatric. The emergency department serves as the **triage** and stabilization area for the casualties. To facilitate the triage (sorting for the purpose of assigning priorities), stabilization, and transportation of numerous causalities, additional emergency department staff may be called in to work. In addition, the hospital disaster plan may require assigning one or more staff from each nursing unit and each nonnursing department to a specific area or task within the emergency department, such as triage, first aid, critical care, burn treatment area, family room, or transportation.

During a disaster, decision making and prioritization of patient care are guided by the resources and personnel available. Patients who are seriously injured and have the greatest chance of full recovery are treated first. Each hospital, as well as each agency involved in responding to a disaster, follows a disaster response plan that outlines the roles and responsibilities of the agency staff and the procedures to follow when interacting with the media, families, other agencies, and causalities. Disaster drills are conducted on a regular basis to evaluate and rework plans. You should be familiar with your agency's disaster plans and policies and know your role and responsibilities in a disaster.

Answers to CRITICAL THINKING

Mr. Smith

1. The airway is the first priority because edema from inhalation burns can occlude the airway.
2. You know that Mr. Smith is at risk for respiratory burns because of the soot near his mouth and nose. He should be closely monitored. Assessment should include respiratory rate and pattern and the patient's ability to speak without a hoarse voice. Abnormal breathing sounds such as wheezing indicate partial upper airway occlusion.
3. The vital signs are within normal limits.
4. Deep partial-thickness burns should be covered with dry dressings. Because the skin can no longer protect the patient, wet dressings provide a medium for bacterial invasion. Wet dressings can also cause a decrease in body temperature because the skin can no longer maintain thermoregulation.
5. Jewelry should always be immediately removed before edema formation begins.
6. Mr. Smith was involved in an explosive incident and thrown 50 feet. He could have sustained fractures of the pelvis or back. He also may have internal organ injuries from blunt trauma.

REVIEW QUESTIONS

1. You are assessing a patient who is hypovolemic. Which of the following signs and symptoms indicate that the patient is experiencing profound shock?
   - a. Sacral edema
   - b. Jugular vein distention
   - c. Palpable, bounding pulse
   - d. Decreasing blood pressure

2. You are caring for a patient who is hemorrhaging from a puncture wound. Which one of the following interventions would you use first to control the arterial bleeding?
   - a. Application of a tourniquet
   - b. Application of pressure
   - c. Pressure point massage
   - d. Pressure dressing

3. Which one of the following is an immediate threat to life during acute anaphylaxis?
   - a. Airway obstruction
   - b. Generalized itching
   - c. Hypotension
   - d. Tachycardia

4. When interacting with a psychotic patient, which of the following interventions is helpful to gain the patient's trust?
   - a. Play along
   - b. Make promises
   - c. Avoid eye contact
   - d. Show respect

5. A patient whose home has burned is admitted to the emergency room. Which of the following symptoms would alert you to the potential for inhalation injury?
   - a. Peripheral edema
   - b. Singed nasal hairs
   - c. Jugular vein distention
   - d. Increased capillary refill time

REFERENCES

1. National Center for Health Statistics: Top 10 causes of death. http://www.CDC.gov/nchs.html, 2001.

2. Sheehy, SB, et al: Manual of Clinical Trauma Care: The First Hour. Mosby, St. Louis, 1999.

Acute Pain Management Guideline Panel. Acute Pain Management: Operative or Medical Procedures and Trauma. Clinical Practice Guideline. AHCPR Pub No 92-0032. Agency for Health Care Policy and Research, Public Health Service, US Department of Health and Human Services, Rockville, MD, 1992.

American Cancer Society: *Guide to pain control: powerful methods to overcome cancer pain.* American Cancer Society, Atlanta, 2001.

American Society of Anesthesiologists. Anesthesiologists warn: if you're taking herbal products, tell your doctor before surgery. Press release 5/26/99. Accessed 5/16/00 at http://www.asahq.org/PublicEducation/herbal.html.

Arroll, B, and Kenealy, T: Antibiotics for the common cold (Cochrane Review). In: *The Cochrane Library,* 1, Update Software, Oxford, 2002.

Asuncion, MM, and Koushik VS: Shock state in the elderly. Clinical Geriatrics, 8(8), 40, 2000.

Atassi, KA, and Harris, ML: Disseminated intravascular coagulation. Nursing 2001, 31(3), 64, 2001.

Barnes, S: Patient preparation: the physical assessment. Journal of Perianesthesia Nursing 17:1, 2002.

Barton, MD: Antibiotic use in animal feed and its impact on human health. Nutrition Research Reviews 2000, 13:2, 2000.

Bruun, LI, et al: Prevalence of malnutrition in surgical patients: evaluation of nutritional support and documentation. Clinical Nutrition, 18:141, 1999.

Burney KY: Tips for timely management of febrile neutropenia. Oncology Nursing Forum 27(4):617-618, 2000.

Buss, H, and Melderis, K: PACU pain management algorithm. Journal of Perianesthesia Nursing 17:1, 2002.

Campbell, M, and Pruitt, J: Radiation therapy: protecting your patient's skin. RN 59:46, 1996. Cancer Survivor's Bill of Rights. American Cancer Society, Atlanta, 1990.

Carroll, A: Handwashing for health-care workers in domestic care settings. British Journal of Community Nursing, 6:5, 2001.

Casella-Gordon, V: Acid-base disorders and arterial blood gases. In Schell, HM, and Puntillo KA (eds.): Critical care nursing secrets. Hanley and Belfus, Inc., Philadelphia, 2001, 124-130.

Centers for Disease Control and Prevention. Guideline for prevention of intravenous therapy-related infection. U.S. Department of Health and Human Services, Atlanta, GA, 1995.

Chin, J: Control of communicable diseases manual, 17th ed. American Public Health Association, Washington, DC, 2000.

Davinder, J: Perioperative cardiac management, Emergency Medicine Journal, 2:7, 2001.

Doenges, ME, and Moorhouse, MF: Nurses's pocket guide: diagnoses, interventions, and rationales, 7th ed. FA Davis, Philadelphia, 2000.

Donegan, W: Tumor-related prognostic factors for breast cancer. CA: A Cancer Journal for Clinicians, 47:1, 1997.

Edel, E, et al: Impact of a 5-minute scrub on the microbial flora found on artificial, polished, or natural fingernails of operating room personnel. Nursing Research, 47:1, 1998.

Fink, K: The research column. Is Trendelenburg a wise choice? Journal of Emergency Nursing, 25(1):60, 1999.

Fossum, S, et al: A comparison study on the effects of prewarming patients in the outpatient surgery setting. Journal of Perianesthesia Nursing, 16:3, 2001.

Gebbie, K, and Qureshi, K: Emergency and disaster preparedness: core competencies for nurses. American Journal of Nursing, 102:1, 2002.

Gonzales, R, et al: Infections due to vancomycin-resistant *enterococcus faecium* resistant to linezolid. Lancet, 357:9263, 2001.

Gonzales, R, et al: Principles of appropriate use for treatment of acute respiratory tract infections in adults: background, specific aims, and methods. Annals of Internal Medicine, 134:6, 2001.

Graydon, J, Bubela, N, Irvine, D, and Vincent, L: Fatigue-reducing strategies used by patients receiving treatment for cancer. Cancer Nursing, 18:23, 1995.

Greisinger, AJ, et al: Terminally ill cancer patients: their most important concerns. Cancer Practice, 5:147, 1997.

Hallatt, SA: Electrolyte disturbances. In Schell, HM, and Puntillo, KA (eds.): Critical care nursing secrets. Hanley and Belfus, Inc., Philadelphia, 2001, 318-333.

Hasdai D, et al: Cardiogenic shock complicating acute coronary syndromes. Lancet, 356(9231):749, 2000.

Hughes, MK: Sexuality and the cancer survivor: a silent co-existence. Cancer Nursing, 23:6, 2000.

Iggulden, H: Dehydration and electrolyte disturbance. Nursing Standard, 13(19):48-56, 1999.

Isolauri, E: Probiotics in human disease. American Journal of Clinical Nutrition, 73(6):1142S-1146S, 2001.

Jacobs, V: Informational needs of surgical patients following discharge. Applied Nursing Research, 13:1, 2000.

Janowski, M: Managing cancer pain. RN, 58:30, 1995.

Jones RD, et al: Moisturizing alcohol hand gels for surgical hand preparation. AORN Journal, 71(3), 584-590, 2000.

Kee, C, and Miller, V: Perioperative care of the older adult with auditory and visual changes. AORN Journal, 70:6, 1999.

Kershner, K: Comedy and cancer. Frontiers, Cleveland Clinic, Cleveland, 1996.

Kornfeld, H: Co-meditation: guiding patients through the relaxation process, RN 58:57, 1995.

Lipson, J, and Haifizi, H: Iranians. In Purnell, L, and Paulanka, B (eds.): Transcultural health care: a culturally competent approach. FA Davis, Philadelphia, 1997.

McCaffery, M: Stigmatizing patients as addicts. American Journal of Nursing, 101(5), 2001.

McCaffery, M: Using the 0-10 pain rating scale. American Journal of Nursing, 101(10), 2001.

McCaffery, M, and Pasero, C: Pain: Clinical manual, 2nd ed. Mosby, St Louis, 1999.

Metheny, N: Fluid and electrolyte balance: nursing considerations, 4th ed. Lippincott, Williams & Wilkins, Philadelphia, 2000.

Moolenaar, R, et al: A prolonged outbreak of *Pseudomonas aeruginosa* in a neonatal intensive care unit: did staff fingernails play a role in disease transmission? Infection Control and Hospital Epidemiology, 21:2, 2000.

Mower-Wade, DM, Bartley, MK, and Chiari-Allwein, JL: How to respond to shock. Dimensions of Critical Care Nursing, 20(2):22, 2001.

Mower-Wade, DM, Bartley, MK, and Chiari-Allwein, JL: Shock: do you know how to respond? Nursing 2000, 30(7):34, 2000.

Ostrow, CL: Use of the Trendelenberg position by critical care nurses: Trendelenberg survey. American Journal of Critical Care, 6(3):172, 1997.

Ostrow CL, Hupp E, and Topjian D: The effect of Trendelenburg and modified Trendelenburg positions on cardiac output, blood pressure and oxygenation: a preliminary study. American Journal of Critical Care, 3:382, 1994.

Pace, B: Preventing dehydration from diarrhea. Journal of the American Medical Association, 285(3):362, 2001.

Phillips, LD: Manual of IV Therapeutics, ed 3. FA Davis, Philadelphia, 2001.

Phillips, LD: Teaching intravenous therapy using innovative strategies. Journal of Intravenous Nursing, 17:40, 1994.

Purnell, L, and Paulanka, B (eds.): Transcultural health care: a culturally competent approach. FA Davis, Philadelphia, 2001.

Roberts, M: Basic pain management. South Mississippi Home Health, Hattiesburg, MS, 1997.

Sainio, C, et al: Patient participation in decision making. Cancer Nursing, 24:3, 2001.

Sandlin, D: Herbal supplements: healthy or harmful? Journal of Perianesthesia Nursing, 17:1, 2002.

Schmidt, C: The basics of therapeutic touch. RN, 58:50, 1996.

Schweid, L, Etheredge, C, and McCullough, M: Will you recognize these oncological crises? RN, 57:23, 1994.

Sheff, B: Taking aim at antibiotic-resistant bacteria. Nursing 2001, 31:11, 2001.

Shertzer, KE, et al: Music and the PACU environment. Journal of Perianesthesia Nursing 16:2, 2001.

Tappen, R, et al: Preoperative assessment and discharge planning for older adults undergoing ambulatory surgery. AORN Journal, 73:2, 2001.

Towle, C; Turkish Americans. In Purnell, L, and Paulanka, B (eds): Transcultural health care: a culturally competent approach. FA Davis, Philadelphia, 1997.

Tremblay, LN, Ritozoli, SB, and Brenneman, FD: Advances in fluid resuscitation of hemorrhagic shock. Canadian Journal of Surgery, 44(3):172, 2001.

U.S. Food and Drug Administration: FDA issues health advisory regarding the safety of Sporanox products and Lamisil tablets to treat fungal nail infections, May 17, 2001. Accessed at www.fda.gov/bbs.topics/answers/2001/ans01083.html.

Vermett, E: Malignant hyperthermia. American Journal of Nursing, 98:4, 1998.

Walton, J: Helping high risk surgical patients beat the odds. Nursing 2001, 31:3, 2001.

Wong, FWH: A new approach to ABG interpretation. American Journal of Nursing, 99(8):34-36, 1999.

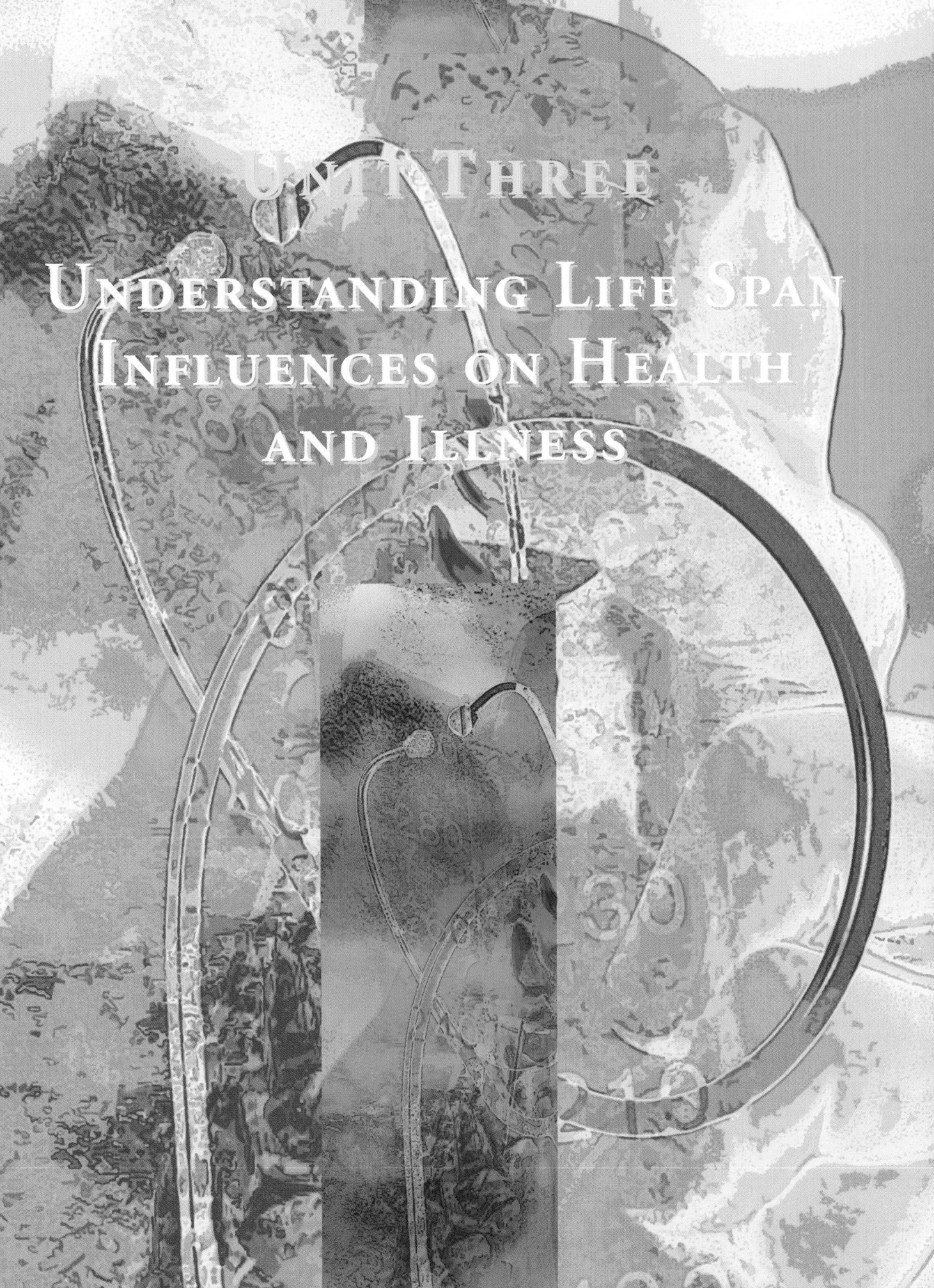

UNIT THREE

UNDERSTANDING LIFE SPAN INFLUENCES ON HEALTH AND ILLNESS

13

INFLUENCES ON HEALTH AND ILLNESSS

Linda S. Williams

KEY TERMS

chronic illness (KRAH-nick **ILL**-nes)

developmental stage (DEE-vell-up-**MEN**-tal STAYJ)

hopelessness (**HOHP**-less-nes)

health (HELLTH)

illness (**ILL**-ness)

powerlessness (**POW**-er-less-nes)

respite care (**RES**-pit CARE)

spirituality (SPIHR-it-u-**AL**-it-tee)

terminal illness (**TERM**-in-al **ILL**-nes)

QUESTIONS TO GUIDE YOUR READING

1. How would you define health and illness?
2. What are the eight developmental stages?
3. How would you define chronic illness?
4. What are the effects of chronic illness?
5. What special needs do caregivers have?
6. What are the benefits of respite care?
7. What are health promotion methods?
8. What nursing interventions would you use for a chronically ill patient?
9. What nursing care would you provide for terminally ill patients and their significant others?

▶ HEALTH, WELLNESS, AND ILLNESS

Nurses generally care for people who are ill and need assistance to meet their basic needs, so it may be difficult to visualize or imagine patients as healthy and productive people. Developing an understanding of health, wellness, and illness can positively affect nursing care. People are made up of physical, psychological, cultural, sociological, and spiritual aspects. All these aspects need to be considered when studying health, wellness, and illness.

LEARNING TIPS

To foster understanding of how an ill patient, especially an older patient, was once healthy and active, ask family members to bring in photos showing the patient at various ages or engaged in favorite activities. Displaying these photos in the patient's room or on a hallway bulletin board allows caregivers to see the patient in times not associated with illness.

The concept of illness is one of imbalance or disharmony resulting from a problem that causes one to be sick. Physical causes of illness are usually easily recognized, such as exercise that induces an asthma attack in an asthmatic person or a fall that breaks a bone. Illness can also result from a psychological, sociological, cultural, or spiritual imbalance. After the loss of a spouse, one may experience loneliness and depression and a loss of balance in the social and psychological aspects of life. A hospitalization may increase disharmony if cultural beliefs and practices are not understood or upheld by health care providers. A person faced with a terminal diagnosis may lose hope and direction in life, causing anxiety and despair. So rather than being exclusive concepts, **health** and **illness** are dynamic and ever-changing states of being. A health crisis such as a myocardial infarction (MI) overwhelms a patient's ability to maintain a normal level of wellness. Two months after the MI, however, this patient could be enjoying a higher level of wellness than before the MI if he has lost weight, is walking daily, and is eating a nutritious low-fat diet.

► THE NURSE'S ROLE IN SUPPORTING AND PROMOTING WELLNESS

The goal of nursing care can best be defined as helping patients achieve their highest possible level of wellness. To do this, consider the patient's strengths, assets, and resources, as well as weaknesses, liabilities, and disabilities. Encourage the patient to take a personal inventory and recognize what is required to attain wellness. Working together, the patient, family, and members of the health care team set wellness goals and develop a plan of action that will help meet those goals. The plan of care focuses on the following six main areas:

- Mobilizing resources
- Providing a safe and adaptable environment
- Assisting the patient to learn about his or her health problem and treatment
- Performing and teaching the patient to perform health care procedures
- Anticipating problems and recognizing potential crises
- Evaluating the plan and progress toward the goals with the patient and family

► DEVELOPMENTAL STAGES

The **developmental stages** of life focus on the balance a person must achieve for high-level wellness within that stage. Across the life span at various ages and stages of life there are recognized potential threats to health. Risk factors are related to growth and development, physical attributes, family, behavior, social interactions, environment, lifestyle choices, and ethnic background.

Developmental stages begin at birth and continue through death. Erik Erikson proposed eight stages of psychological development: 0 to 1 year—trust versus mistrust; 2 to 3 years—autonomy versus shame and doubt; 4 to 5 years—initiative versus guilt; 6 to 12 years—industry versus inferiority; 13 to 18 years—identity versus role confusion; young adult—intimacy versus isolation; middle-aged adult—generativity versus self-absorption; old adult—ego integrity versus despair.[1,2] Developmental stages for the young adult, middle-aged adult, and older adult are discussed here.

The Young Adult

Developing intimacy versus isolation is the sixth psychological development task. The young adult's task is to develop relationships with a spouse, family, or friends that are warm, affectionate, and developed through fondness, understanding, caring, or love. When this stage is not successfully resolved, the individual often experiences isolation from others. This stage encompasses ages 18 to 45. Physically, growth is usually completed by age 20. Socially, young adults begin to move away from their parents to develop their own families. The young adult begins to attempt to develop a place in society through school, work, and social activities. This is the stage in which intimacy or closeness develops with partners and friends. The decision to marry, have children, or have a pet is demonstration of the person's striving for intimacy. Challenges to intimacy are tasks that must be overcome in this stage. Within a marriage, communication, financial issues, and the needs of children must be successfully met to maintain the marriage. There are cultural issues facing the young adult. Melding one's traditions and customs with the traditions and customs of a spouse, family, or friends is a major responsibility, as is the passing on of culture to children. Values and beliefs, which arise from an individual's culture or conscience, serve as guidelines for behavior.

Common Health Concerns

The lifestyle choices of young adults may place their health at risk. Health promotion for this age group is mainly focused on preventing or limiting possible risk factors. Young adults need to understand the importance of diet and exercise in maintaining health. They are also in the position of teaching these lifelong habits to their children. Positive health practices in young adulthood help prevent long-term complications. Maintaining an aerobic exercise program and following a diet that is low in fat help keep weight down and avoid obesity, as well as promoting cardiovascular health. Blood cholesterol should be kept below 200 mg/dL. Avoiding sun exposure and using sunscreen are important to avoid sunburn and permanent sun damage to the skin. Years of sun exposure may cause skin cancer. Tobacco use started in the teen years is often carried on throughout young adulthood and is linked to chronic bronchitis, emphysema, and oral, throat, and lung cancer in later life.

In the early part of young adulthood, the individual is in the workforce or is preparing for the work world with a college or vocational education. Being a novice in the work world and accepting new independence, freedom, and responsibilities can introduce stressors into the young adult's life. Unfortunately, overeating, alcohol use, drug use, cigarette smoking, and family violence are all negative lifestyle choices and poor coping mechanisms for stress. Young adults need to be aware of their individual stress and be encouraged to develop positive coping mechanisms for stress. Exercise, support groups, music, and meditation are just a few positive ways to cope with stress.

Although marriage commonly occurs within this age group, this group also has the highest rate of divorce. Trying to make a marriage work is a hard task. The blending of two people into a couple requires a lot of creative communication and loving care. When stressors overwhelm the couple's coping mechanisms or coping strategy, the marriage relationship is in trouble. Sometimes because of the high rate of divorce many couples choose to live together without being married. However, the avoidance of making a commitment may set these relationships up for failure.

If young adults are sexually active with multiple partners, they are at risk for sexually transmitted diseases. Safer sex guidelines and information on birth control should be available for the sexually active young adult.

Pregnancy is a common health occurrence for women in this age group. Because research indicates that a mother's health practices directly affect the health of the developing fetus, nutrition, drug and alcohol use, physical health, and effective stress coping mechanisms are lifestyle issues that need to be discussed with every pregnant woman. Prenatal care should be encouraged and readily available to pregnant women.

CRITICAL THINKING

Rita

Rita, age 25, and her husband are trying to start a family. Rita visits her physician, who confirms that she is pregnant. Which information and health practices should Rita and her husband be instructed in during a prenatal health examination?

Answers at end of chapter.

The Middle-Aged Adult

People ages 45 to 65 are considered to be in the middle adult years. The psychological developmental task of this age group is developing generativity versus self-absorption. Generativity is demonstrated by a concern and support for others, along with a vision for future generations. The unresolved conflict would be preoccupation with personal needs or self-absorption. Physically, middle adults start to notice signs of decreased endurance and intolerance for physical exercise if they have not maintained healthy lifestyle choices. Socially, their children are adolescents or young adults who require extra attention and assistance with entering adulthood. Along with this, middle-aged adults may also be faced with the challenging demands of caring for aging parents. Middle adults also look over their lives and assess accomplishments versus unrealized goals. Midlife crisis may occur as this self-inspection leads to a desire to change work, social, or family situations to try to meet unrealized goals. Planning for retirement by developing meaningful pastimes and interests outside of work and preparing for financial security is another important task during this period.

Common Health Concerns

Adverse health choices, such as smoking, use of alcohol, drug use, sedentary lifestyle, diet high in saturated fat, and overeating, often have serious consequences for middle adulthood. Hypertension and heart disease are major health concerns, as are chronic bronchitis, emphysema, and lung cancer. Cardiovascular disease and cancer cause most of the deaths in this age group. However, middle adulthood is not too late to begin lifestyle changes that positively affect health. Replacing high-risk habits with lifestyle choices such as regular exercise, healthful eating, weight reduction, and positive stress-coping mechanisms are positive changes. Helping adults recognize the benefits of these lifestyle choices and empowering them to change is the major challenge for health care professionals for this age group.

CRITICAL THINKING

Mr. Paul

Fifty-four-year-old Mr. Paul calls the doctor's office for the fourth time this month complaining of severe indigestion and requesting a medication that will work to fix it. He has refused to have an x-ray examination or other diagnostic tests because he "can't fit them into" his "busy schedule." Mr. Paul has his own insurance business. His wife quit her job to monitor the activities of their 13-year-old son, who was not going to school every day. The couple's twin daughters are both in college out of state.

1. Why might Mr. Paul be experiencing health problems?
2. What is affecting the developmental tasks Mr. Paul needs to perform?

Answers at end of chapter.

The Older Adult

Development and *aging* seem to be terms that are directly opposite because aging is commonly viewed as deterioration or a "downhill slide" to inevitable death. This negative view of aging is supported by cultures that idolize and strive to maintain youthfulness. Being old is often equated with outliving one's usefulness. However, people are living productive, fulfilling lives into their 80s, 90s, and even 100s. Many of these older adults are likely to be found working in their gardens, exercising, or socializing (Fig. 13–1).

Figure 13-1 Socialization helps older adults maintain integrity.

The developmental goal for people age 65 or older is integrity versus despair. Health state, environment, relationships, and lifestyle choices influence the diversity found in this group. Physical health is often a concern for older adults. Chronic health problems that require medication and treatment often also require lifestyle changes or adaptations. Because psychological development for the older adult is characterized by striving for integrity, older adults often spend time reflecting on their lives. Integrity is measured by feeling satisfied with having lived a full and productive life. When this life reflection is characterized by misgivings, missed opportunities, dissatisfaction, and disappointment, the older adult feels hopeless despair.

Coping with loss is a major challenge for older adults. Life events such as retirement, illness, or death of a spouse and changes such as decreased physical ability are losses facing older adults. Coping with aging is also influenced by the individual's cultural beliefs. Cultural viewpoints on the social role and value of aged members affect the health of older adults. Sometimes the greatest loss for older adults is their lack of connection with the world and a lack of being part of a greater purpose.

Common Health Concerns

As a result of changes in older people's health, the focus of care is on assisting older adults to meet their physical, psychological, cultural, sociological, and spiritual needs. Encouraging the use of community services for seniors and promoting self-care are important. Most older adults continue to live in their own homes or apartments, but impairment in mobility and the ability to carry out instrumental activities of daily living, which include shopping for groceries, preparing meals, and cleaning and maintaining a home, threaten their independence. Having to ask or pay others to perform tasks that they formerly were able to do themselves is seen as a significant loss by many older adults. Adding the loss of a spouse, the death of friends, or the lack of social contacts may further isolate an older person, leading to depression and **hopelessness.** The accumulation of losses can overwhelm an older adult's resources and coping mechanisms and is related to a high rate of suicide, especially for older men. Suicide is the ultimate expression of hopelessness.

Older adults need to be encouraged to remain active and to continue to pursue interests. Most communities have transportation services such as buses and vans that operate to meet the needs of older adults. Senior centers offer diverse programs and services to older adults. Some senior groups are focused on community service; others mainly plan trips or sponsor activities such as dances or bowling leagues (Fig. 13–2). Seniors also have opportunities to continue to work in areas of interest as volunteers. Schools, hospitals, nursing facilities, parks, museums, zoos, community theaters, and youth groups all welcome older adult volunteers. Colleges and universities offer discounted tuition for senior citizens, and there are elder hostel programs across the country, which are programs especially designed for older adults. Elder hostels offer a variety of programs

Figure 13-2 Older adults can remain active by participating in activities such as dances. (From Anderson, M: Nursing Leadership, Management and Professional Practice for the LPN/LVN. FA Davis, Philadelphia, 1997, p 200, with permission.)

such as photography, Civil War history, nature survival, bird watching, and painting.

As people age, chronic illnesses and disability increase. Unfortunately, chronic diseases commonly limit an older person's ability to be independent in self-care and activities of daily living. Hypertension is common, as are heart disease and strokes in this age group. Managing blood pressure, losing weight, eating a low-fat diet, not smoking, enhancing effective stress-coping strategies, and regular exercise decrease the potential for cardiovascular disease.

Changes in mobility and chronic pain may limit an older person's activity and impede an active lifestyle. Falls are a serious concern for older adults. Falls and accidents can be prevented by in-home safety assessments and altering the home environment to ensure the safety of the older adult. Bathrooms should be equipped with grab bars and nonskid mats. Bath chairs or benches make getting into a bath or shower less risky. Removing clutter, throw rugs, small furniture, and electrical cords decreases the risk that an older adult will trip and fall.

Social stigmas related to memory changes such as forgetfulness, dementia, and senility are a serious worry for many older adults. The elderly often confuse their own depression with senility and attempt to hide their symptoms rather than seek treatment. Sensory impairments can further isolate the patient. Impaired vision can be caused by decreased peripheral vision, macular degeneration, cataracts, or glaucoma. One of the most dramatic losses for many older adults is not being able to safely drive a car. This is usually associated with a loss of independence and free will. Many older adults continue to drive during the day but not at night because of night vision problems. Decreased hearing is also common in older adults. Loss of high-pitch discrimination and reduced ability to filter background noises cause older adults to hear the background noise more clearly than a one-to-one conversation when in a crowded room.

CRITICAL THINKING

Mrs. Riccardi

Mrs. Riccardi, age 87, lives alone in her small apartment. Her daughter and son are both retired and live in the same community. Mrs. Riccardi had a dizzy spell, so her daughter brought her to the hospital emergency department (ED). Upon admission, Mrs. Riccardi's blood pressure was 208/128 and she had blurred vision in the left eye that resolved after 1 hour in the ED. She was diagnosed with hypertension, which had possibly contributed to a small stroke or transient ischemic attack (TIA). Mrs. Riccardi was started on furosemide (Lasix) 20 mg twice a day and digoxin (Lanoxin) 0.125 mg every day. She was discharged to her home. The plan of care addressed safety issues.

1. Why might Mrs. Riccardi be at increased risk of falling?
2. What nursing interventions would help promote Mrs. Riccardi's independence and safety?

Answers at end of chapter.

CHRONIC ILLNESS

Chronic illness is defined as an illness that is long-lasting or that recurs. Chronic illnesses usually interfere with the patient's ability to perform activities of daily living. Medical care and hospitalization, commonly at least once a year, are often required on an ongoing basis. A **terminal illness,** however, could or is expected to cause death within 6 months.

When caring for those with chronic illness, the goal of nursing care is to maintain and improve the patient's quality of life. A chronic illness also affects the patient's family's quality of life. Therefore, when planning patient care consider the family's needs for adapting to the chronic illness.

Fostering hope is an important intervention that should be a primary foundation of care planning for the chronically ill. A chronic illness may appear to be a hopeless situation if no cure is possible. If recovery from an illness is not possible, it might be thought that nothing can be done for the patient. However, whenever there is life, there is potential for growth in areas such as developmental tasks, health promotion, knowledge, or spirit. Individuals have developmental tasks to perform even as they cope with illness or prepare for a peaceful death.

Incidence of Chronic Illness

The incidence of chronic illness is rising for several reasons. First, people are living longer, in part because of better hygiene, nutrition, vaccinations, antibiotic development, and exercise. As a result, a larger elderly population is living long enough to develop many chronic illnesses. Second, medical advances have resulted in reduced mortality from some chronic illnesses, so that patients live longer with these illnesses. Third, today's technology and modern lifestyles affect the development of some chronic illnesses. Examples include a sedentary lifestyle; exposure to air and water pollution, chemicals, and carcinogens; substance abuse; and stress.

Some of the most common chronic conditions include chronic sinusitis, arthritis, hypertension, orthopedic dysfunction, decreased hearing, heart disease, bronchitis, asthma, and diabetes. Preventive measures can be taught to reduce the incidence of these conditions.

Types of Chronic Illnesses

There are a variety of chronic illnesses resulting from several different causes (Box 13–1). These illnesses can have varying degrees of severity and can affect length of life. A chronic illness can lead to the development of other illnesses, such as hypertension that then causes chronic renal failure. Chronic illnesses can have onsets at various ages, but the elderly are most commonly affected, often having several chronic illnesses at one time (Box 13–2).

BOX 13-1 **Examples of Chronic Illnesses by Causes**

Genetic
Cystic fibrosis
Huntington's disease
Muscular dystrophy
Sickle cell anemia

Congenital
Heart defects
Malabsorption syndromes
Spina bifida

Acquired
Acquired immunodeficiency syndrome (AIDS)
Arthritis
Cancer
Cataracts
Chronic obstructive pulmonary disease
Diabetes
Head or spinal cord injury
Multiple sclerosis
Peripheral vascular disease

BOX 13-2 **Examples of Chronic Illnesses in the Elderly**

Arthritis
Cataracts
Cerebrovascular accident
Chronic lung disease
Diabetes
Hearing impairments
Heart disease
Hypertension
Peripheral vascular disease
Visual impairments

Gerontological Influence

The elderly are one of the largest age groups living with chronic illness. As people live longer, elderly spouses or older family members are increasingly being called on to care for a chronically ill family member. Children of elders who themselves are reaching their 60s are being expected to care for their parents. These elderly caregivers may also be experiencing a chronic illness themselves. In this case it is usually the less ill spouse that provides care to the other spouse. The elderly family unit is at great risk for ineffective coping or further development of health problems. Assess all members of the elderly family to ensure that their health needs are being met.

Elderly adults are very concerned about becoming dependent on and a burden to others. They may become depressed and give up hope if they feel that they are a burden to others. Establishing short-term goals or self-care activities that allow them to participate or have small successes are important nursing actions that can increase their self-esteem (Cultural Consideration Box 13–3).

BOX 13-3 CULTURAL CONSIDERATION

Traditional Appalachians believe that disability is natural with aging and is inevitable. This belief discourages the use of rehabilitation as an option. Thus to promote rehabilitation efforts among Appalachians, the nurse may need to stress self-help and a return to physical function.

Barriers to caring for a chronically ill elderly patient include a lack of information about treatments, medications, or special diets and being unfamiliar with supportive services in the community such as meal programs or **respite care.** You should be aware of these and provide this information, as well as a resource number for questions.

Effects of Chronic Illness

For the patient to live as normally as possible with a chronic illness, many adjustments are usually necessary. Lifelong routines and habits may need to be changed. Daily living patterns are affected by routines that are established to cope with the illness. Treatment needs such as going to therapy sessions, performing peritoneal dialysis exchanges, or monitoring blood sugars can interrupt daily life.

Chronic Sorrow

Chronic sorrow is a normal response felt by those affected by a chronic illness. It is an intermittently occurring permanent sadness in response to the loss caused by a chronic condition. It can be felt by the patient or the patient's significant others. It is a common feeling among those with chronic illness. The nursing diagnosis chronic sorrow may apply to those experiencing chronic illness. When this sadness occurs, nursing care should focus on being comforting and supportive. Providing information and assisting with coping strategies such as fostering support systems are great interventions to help those with chronic sorrow.

Spiritual Distress

Patients with chronic illness can experience spiritual distress when faced with the limitations of their illness. Maintaining patients' quality of life includes assisting patients with their spiritual needs. Religious and spiritual needs are important to most people whose lives have been disrupted with new challenges from chronic illness. Patients must be helped to find meaning in the illness and realistic hope. Interventions that address **spirituality** may need to be performed first to allow success of subsequent nursing care.

Develop a comfortable approach in assessing and meeting patients' spiritual needs. Several factors may make one uncomfortable in caring for the patient's spiritual needs. These factors include a lack of training, a lack of understanding of one's own spiritual needs and beliefs, and not recognizing or believing that this is your role. Examine your own spiritual needs to define a personal spiritual view. By doing this, you will develop insight into others' spiritual needs and resources, as well as gain a greater understanding of issues surrounding your patients' spiritual needs.

Many people use spirituality to cope with chronic illness. It helps give them a sense of wholeness, hope, and peace during a time filled with uncertainty and anxiety. Spirituality plays an important role in empowering patients to handle their condition. It is a source of inner strength that allows the patient to experience a sense of unity. Hospital interventions may include use of a meditation room for quiet reflection or prayer, chaplain visits, or worship services. To help meet the patient's spiritual needs, assist the patient with transportation to the meditation room or worship services.

Accreditation agencies require the spiritual needs of patients to be addressed and documented by nurses. Nursing diagnoses related to spiritual needs include spiritual distress, risk for spiritual distress, and readiness for enhanced spiritual well-being.

LEARNING TIPS

Spiritual needs should not be thought of in only a religious focus. Spirituality is feeling connected with a higher power. Everyone has spiritual needs that involve hope, peace, and wholeness. Spiritual care goes beyond simply asking the patient's religion. It is assessing patients' perceptions of spirituality and then devising ways to assist them to meet their spiritual needs.

Powerlessness

Those with chronic illness often feel powerless because they are uncertain of their ability to control what may happen to them (Fig. 13–3). A chronic illness can take an unknown course in relation to its seriousness and controllability. This leaves the patient vulnerable to the many phases of a

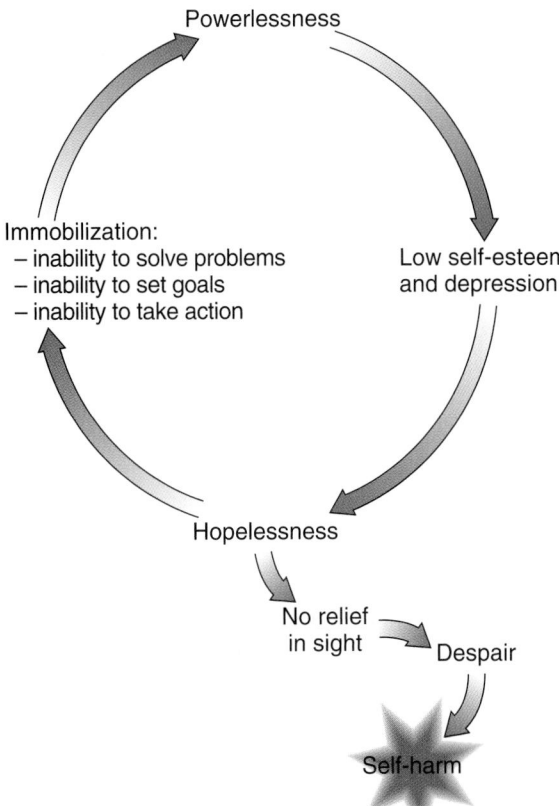

Powerlessness

Immobilization:
– inability to solve problems
– inability to set goals
– inability to take action

Low self-esteem
and depression

Hopelessness

No relief
in sight

Despair

Self-harm

Figure 13-3 Powerlessness-hopelessness cycle. (Modified from Miller, J: Coping with Chronic Illness. Overcoming Powerlessness. FA Davis, Philadelphia, 2000, p 526.)

chronic illness (the diagnosis, the instability phase, an acute illness or crisis, remissions, and a terminal phase).

Treatments that the patient undergoes can be a new experience that is painful, frightening, and invasive. If patients do not understand what is happening, they can feel overwhelmed and alone. This contributes to a feeling of **powerlessness** because the patient cannot control the outcome. Patients with chronic illness are faced with a lack of control throughout their illness. It influences the reactions of the patient to the illness. The nursing diagnoses of powerlessness or risk for powerlessness may apply to many chronically ill patients.

COPING. Patients can be helped to feel more in control of their illness if you remember to include them in their care; listen to their feelings, values, and goals; and explain all procedures first. Complex medical language should be avoided when talking with patients to increase their understanding and feeling of being included in their care instead of isolated. In addition, coping with a chronic illness can be aided if the patient develops a positive attitude toward the illness. This can be accomplished if the patient gains knowledge, uses a problem-solving approach to difficulties, and becomes motivated to continue adapting to the illness and not succumb to a defeatist attitude.

A variety of coping techniques can be useful. Ask the patient's perception of the illness and identify coping techniques that the patient has previously used successfully.

New coping resources may need to be added to effectively deal with the patient coping tasks that are associated with chronic illness. Support services in the community should be made available to the patient and family. To cope effectively, the patient should become comfortable with the newly defined person he or she is to become. The nursing diagnoses of ineffective coping, compromised family coping, disabled family coping, and readiness for enhanced family coping may apply to those dealing with chronic illness.

HOPE. Before coping resources can be used, hope must be established by the patient. False hope is not beneficial and should be replaced with realistic hope. Providing patients with accurate knowledge regarding their fears helps do this. Hope should not be directed toward a cure that may not be possible but rather at living a quality life with the functional capacity that the patient has. Over the course of the illness, hope needs to be maintained for both the patient and family. Periodically assess if the patient is maintaining hope. Many studies have shown that patients adapt better when hope is high.

Many nursing interventions may increase hope. The use of humor helps patients be lighthearted and hopeful. Patients should be encouraged to live each moment to the fullest and experience the joy of being alive. Awakening the senses to appreciate the environment can bring a feeling of

CRITICAL THINKING

Mr. Soloman

Mr. Soloman, age 88, still lives in his own home with his wife of 60 years. He is in good health except for poor vision, which developed slowly over the last 8 years. He cares for his yard, which is his pride and joy, and grows prize-winning tomatoes each year. He plays golf three times a week and walks every day to the neighborhood grocery store or bank. The employees know him and cheerfully assist him with his vision limitations.

Mr. Soloman's wife had always been the homemaker, whereas he was the family provider. Now his wife is in the early stages of Alzheimer's disease, exhibiting confusion, fatigue, and paranoia. She is unable to perform activities of daily living, so he has willingly assumed the homemaker and caregiver roles. They complement each other's limitations because she can still see and is helpful to Mr. Soloman when she is not confused.

Over time, Mr. Soloman's wife's health declines and she must be institutionalized. Mr. Soloman remains in his home alone. His family is concerned about him being alone and eventually convinces him to move into a studio apartment in senior housing. He is very reluctant to leave his home and does not actively participate in moving and selling his home. Mr. Soloman rarely leaves his apartment, sleeps 14 hours a day, and eats one daily meal. He tries to visit his wife by taking the bus but finds it difficult because of his limited vision, so he rarely goes to see her. Within a year, Mr. Soloman develops pneumonia and after a brief illness dies in his sleep.

1. Why did Mr. Soloman exhibit the described behaviors after he moved?
2. What interventions could have been used to empower Mr. Soloman?
3. Why might Mr. Soloman have developed pneumonia and died?

Answers at end of chapter.

hope and peace. Simple things, such as the smell of baking bread, the clean scent of the air after a rain, or the scent of pine trees, can make one appreciate the beauty of nature and inspire hope. Family members need to be encouraged to help foster hope for the patient. By doing this, family members may gain hope as well. During times of acute illness, the patient needs to maintain as much control as possible and be told that any loss of control related to treatments is usually temporary. This prevents a continual feeling of loss of power. The use of music or inspirational reading material can reduce stress and help the patient find meaning in life. This in turn fosters hope. Hopeful patients are empowered and no longer feel powerless. The nursing diagnosis of hopelessness may apply to the chronically ill.

Sexuality

Chronic illness can affect a patient's sexuality. Sexuality includes femininity or masculinity, as well as sexual activity. Body image changes affect the way patients view themselves and are viewed by others. If patients have a negative body image perception, they may withdraw and become depressed. When interacting with patients, be aware of your facial expressions, nonverbal cues such as appearing hurried or keeping a distance, use of or lack of touch, and amount of time spent with the patient. When patients believe they have lost their femininity or masculinity, their self-worth decreases. Interventions to enhance sexuality should be used, such as obtaining a wig from the wig bank for patients undergoing chemotherapy.

There are many forms of sexual expression. Sexual intimacy can include touching, hugging, or sharing time together. Provide patients with the opportunity to discuss sexuality concerns or questions. Assume a professional and confidential approach to this topic, which is usually considered a private matter by patients. Chronically ill patients can be referred to sex counselors for information on ways to cope with sexual issues in relation to their illness. Support groups can also be helpful.

Patients in extended care facilities should be given private time with their significant other, if appropriate. Elderly patients need to have their sexuality needs met just as younger patients do. Because sexuality is a part of a person's lifelong identity, ensure that elderly patients' sexuality is addressed in their plans of care. Grooming methods can increase a patient's self-esteem and sexual identity. Women may get their hair and nails done; men can be shaved or get a haircut.

Roles

Chronically ill patients are usually faced with altering their accustomed roles in life. Common roles that may be affected for the adult patient include that of being a spouse, grandparent, parent, provider, homemaker, or friend. Not only is the patient faced with dealing with these role alterations, the family must also adapt to these changes. Family members may have to take on new roles themselves to compensate for roles the patient is no longer able to perform. The nursing diagnosis of ineffective role performance should be included in the plan of care for the patient and family.

The patient is faced with giving up aspects of old roles at the same time that new roles related to being chronically ill need to be assumed. Grieving accompanies the loss of old roles. If a patient is no longer able to participate in social events such as being a golf team member or a committee member, grief work occurs to help the patient accept the loss and maintain dignity. With other roles, only certain aspects of the role may change. For example, in the parenting role, patients may still be there as support systems for the child, although they are no longer able to be the disciplinarian. Whatever the role loss, the patient needs to be allowed to grieve the loss. The nursing diagnosis of grieving may help in planning care for the patient.

The new roles the patient may have to assume related to chronic illness include dependency, ongoing health care consumer, and self-care agent, and being chronically ill. Patients need to learn how to cope with these new roles. They need to gather knowledge and be given understanding as they become familiar with these roles. For patients used to being independent before the illness, being dependent on others to meet activities of daily living can cause a loss in self-esteem. Navigating the complex health care and financial reimbursement systems can be overwhelming. Transportation needs and waiting times for medical appointments can be difficult for the patient who must deal with them on an ongoing basis. Becoming a self-care agent requires assuming responsibility for meeting one's own care needs. Handling chronic illness covers many areas, such as living with pain, having altered mobility, or complying with daily treatments. Deficient knowledge is a nursing diagnosis helpful for fostering learning for these new roles.

As patients live with chronic illness over time, they become experts on their own illness. However, today's health care system tends to assume control over patients and does not respect the patient's own knowledge. Patients who are not given this respect take charge of caring for themselves by seeking knowledge and trying complementary healing methods. Be sensitive to the patient's knowledge and respect it, because respect increases patient's self-esteem.

Family and Caregivers

Families are affected by the chronic illness of a family member in many ways. Chronic illness care is usually provided in the home so that families become involved in the management of the illness (Home Health Hints). Family members may have to take on new family roles or assume the role of caregiver. Decreased socialization, lost income, and increased medical expenses can increase family stress and tension. For more information, visit www.caregiver.org or contact Family Caregiver Alliance, 425 Bush St., Suite 500, San Francisco, CA, 94108, (415) 434-3508.

Families must learn to cope with the stress of illness and its often unpredictable course. Most families develop ways to cope with the patient's illness the majority of the time and may become closer as a family unit. Children adapt better to a parent's illness when they receive parental support. Families often deal with the illness on a day-by-day basis and take a passive approach for most problems to let them

HOME HEALTH HINTS

- Home care nurses can strengthen a patient's self-care capacity by (1) saying "Let me assist you" instead of "Let me do this for you"; (2) being a partner in caring instead of being a caregiver; (3) empowering the patient instead of doing it all for him or her.

- Using humor can be helpful during a visit, unless the patient is distraught, anxious, or angry. Comics or jokes from magazines can be read. Humor can relieve the patient's and caregiver's stress and humanize the care that must be performed. Keep in mind, however, that humor may be irritating if it is taken as downplaying the seriousness of a situation.

- When patients have the use of only one hand or arm, provide a sponge for personal grooming instead of washcloth; it is easier to use and hold.

- Encourage the family of bed-confined patients to purchase an inexpensive portable intercom (such as a nursery monitor) to give the caregiver freedom to move about the house and hear the patient if help is needed.

- The home health nurse has an opportunity to increase a patient's self-concept by emphasizing the patient's existing abilities, joys, and talents. This can be done by observing the home environment for clues: photographs, trophies, hobby paraphernalia, flowers, or homemade items. Each clue observed can be a springboard to discussing triumphs and losses. It can help the nurse understand how the patient copes to assist in planning care.

- Being attentive to any of the patient's efforts or accomplishments, such as number of steps taken, shaving without help, interest in a book or TV show, or decisions on fixing up the house or cleaning, provides opportunities to offer support and praise for accomplishments.

- The home health social worker should be informed of any patient concerns regarding cost of medicines or equipment. There are many programs to the assist the chronically ill to obtain resources.

- Always carry a pocket mask for cardiopulmonary resuscitation (CPR).

- Encourage discussion regarding advance directives with the patient and family/caregiver early in the care process. Decisions regarding resuscitation and the use of technology to prolong life are more difficult when the patient is in crisis.

- Refer families to community CPR classes as desired. Families can be empowered and the patient may feel more secure when families are taught CPR.

- Patients and families should be taught that if an ambulance is called they should turn on an outside light, open the door, and move furniture to enable the emergency medical technicians to get to the patient more easily if possible.

work themselves out. During times of exacerbation or crisis, however, the family may need coping assistance.

Patients are often concerned about being a burden to their families. It is important to determine both the family's and the patient's feelings about the care required by the patient. The family's ability to provide this care adequately must also be considered in care planning. If the family lacks the desire, skills, or resources to adequately care for the patient, alternative care options must be explored such as home health care, adult foster care, or extended care facilities.

Patient's caregivers often have certain ideas about the care that the patient should receive. This may come into conflict with the views of health care providers. Caregiver input into the patient's plan of care should be sought so that everyone has a clear understanding of goals and expectations for the patient's care.

Caregivers commonly experience depression, role strain, guilt, powerlessness, and grieving caused by caregiving. Be aware of this to detect indications that caregivers are in need of help in dealing with these feelings. Chronic care coaches are resources available to caregivers who can provide insight, encouragement, and support for caring for the chronically ill. Nursing diagnoses for caregivers include risk for caregiver role strain or caregiver role strain.

RESPITE CARE. When caregivers are required to provide 24-hour care for a patient, they can experience burnout, fatigue, and stress, which, if extreme, in some cases can lead to patient abuse. Caregivers may not be able to leave patients alone even briefly because of wandering behaviors, confusion, or safety issues. They may not ever be able to get a normal night's sleep and suffer from sleep deprivation because of the patient's wandering or around-the-clock treatment needs.

Caregivers must be given periodic relief from their responsibilities of caregiving to reduce the stress of always having to be responsible. Everyone requires private time for reflection or pursuing favorite hobbies. Caregivers may need to get away overnight or for a weekend simply to sleep soundly and be refreshed.

Respite care is designed to provide caregivers with a much-needed break from caregiving by providing someone else to assume the caregiver role. Know your community's respite care services to share with caregivers. Unfortunately, there are often not enough respite care services available to meet the needs of caregivers. Most respite care is provided by volunteers who receive training. As the number of chronically ill persons grows, more respite care programs must be developed to promote the health of the caregiver and in turn the patient.

CRITICAL THINKING

Mrs. Burden

Mrs. Burden, age 64, is caring for her husband, who has Alzheimer's disease. He exhibits wandering behaviors. He gets up at night and in freezing winter weather is found walking down the street in only his pajamas. He attempts to cook and burns the pans. He is unable to express his needs. He disrobes frequently and is incontinent. Mrs. Burden quit her job to care for him. She no longer goes to lunch weekly with her friends. Her children live out of town. She places a chair and tin cans in front of the home's doors as an alarm in case her husband tries to open the doors.

1. What is Mrs. Burden at risk of developing?
2. What nursing diagnoses should be included in a plan of care for Mrs. Burden?
3. What nursing interventions would be beneficial for Mrs. Burden?

Answers at end of chapter.

Finances

Managing a chronic illness can be expensive. Income can be lost if the patient is unable to work or caregivers are forced to stay home. Family savings can quickly be wiped out. If the patient is covered by insurance, it may not cover all the patient's expenses or it may have caps on lifetime coverage amounts. Expenses may involve medications, medical equipment or supplies, therapy, acute care, and home care. Inadequate funds can place a strain on families. This can lead to the nursing diagnosis of compromised family coping, disabled family coping, or readiness for enhanced family coping. Nurses may need to refer patients to a social worker or sources of financial aid to help them meet their financial needs.

Health Promotion

Health promotion is possible and necessary at all levels of age or disability. With the increase in the elderly population, it is essential to understand the role of health promotion for the elderly who have chronic illness. Patients with chronic illness make daily lifestyle choices that affect their health. For example, the patient with chronic lung disease who smokes can make a choice to smoke or to quit smoking. Patients with degenerative joint disease can choose whether or not to keep their weight within ideal weight ranges to reduce wear and tear on their joints. Arthritic persons can reduce their fatigue levels by pacing their activities and scheduling daily rest periods.

Those with chronic illness consider health promotion important, so encourage health promotion efforts. Patients need to be assisted to strive toward high-level wellness. This can be achieved by looking at the patient's strengths and weaknesses holistically to develop a plan of care. Determine the patient's risk factors to plan methods of promoting health. Providing patients with knowledge to make informed decisions empowers them to take control of their lives and reach for their greatest potential. The nursing diagnosis of health-seeking behaviors can be used to assist patients in promoting their health.

Nursing Care

Nursing care is primarily devoted to caring for patients with chronic illness. Develop an understanding that the wishes of the patient must be respected even if you do not agree with them. Patients have the right to establish their own goals in partnership with the health care team. Because of the nature of chronic illness, understand the unique needs of patients and families experiencing chronic illness. These needs differ from acute care as far as depth of knowledge needs and the compounding problems that are usually faced by the patient.

Most chronic illness care occurs in the home and community rather than the acute care setting. Therefore family members and caregivers, even more so than in acute care, must be assessed and included in the plan of care. As the numbers of those with chronic illness grow, community support for the chronically ill and their caregivers needs to grow. Training programs for caregivers must be available and offered affordably.

A major focus of nursing care for the chronically ill is teaching. These patients and their families have tremendous educational needs if they are to successfully learn to cope with a long-term illness. The following are primary tasks that the chronically ill need to perform:

- Be willing and able to carry out the medical regimen.
- Understand and control symptoms.
- Prevent and manage crises.
- Reorder time to meet demands caused by the illness, such as treatments, medication schedules, and pacing of activities.
- Adjust to changes in the disease over the course of time, whether positive or negative.
- Prevent social isolation from physical limitations or an altered body image.
- Compensate for symptoms and limitations in order to be treated as normally as possible by others.

Explain individualized interventions to deal with these tasks during teaching sessions. Unique approaches are needed to positively assist chronically ill patients and their families on their long-term journey.

▶ TERMINAL ILLNESS

Death is a natural occurrence after many illnesses, especially chronic illness. Reactions to death vary by culture and individuals' responses. Elizabeth Kübler-Ross described five stages of emotional reactions that people have to dying: denial and isolation, anger, bargaining, depression, and acceptance.[3] These emotional stages assist the patient and family to deal with the news of a terminal illness. Each person experiences these stages uniquely and may move back and forth among them.

Denial allows hope to continue after being informed that death may occur and prevents the individual from being psychologically overwhelmed. Anger is reflected in the question, "Why me?" In bargaining, a deal is made with God if the patient is allowed to live just a little longer, usually to attend a special event such as a graduation. Depression indicates that the patient is sad and feels that nothing more can be done. With acceptance, the patient is at peace and simply seeks a comfortable and dignified death.

The main focus of terminal nursing care is providing the patient and family with a pain-controlled and dignified death that allows participation by family members. This care is given in many locations: the patient's home, a hospice facility, an extended care facility, or a hospital. Determine the goals of the patient and family in dealing with the terminal illness. Nursing diagnoses that may apply at this time include anticipatory grieving or grieving for either the patient or family. Families should be taught what to expect and do as death approaches, so they do not panic when signs of impending death occur (Table 13–1). If the patient becomes comatose and no further medical treatment is desired, comfort care is usually provided until death occurs. At this time, the primary focus may become the patient's significant others, who need assistance with emotional support and grief work.

TABLE 13-1   SIGNS AND SYMPTOMS OF APPROACHING DEATH

Signs and Symptoms	Cause	Nursing Care
Coolness, mottling	Vasoconstriction	Use blankets.
Sleeping	Metabolism changes	Sit quietly; remember that hearing is last sense to leave.
Disorientation	Metabolism changes	Orient if helpful.
Incontinence	Muscles relaxing	Keep clean and comfortable.
Chest congestion	Weakened cough to remove secretions	Give oral care to keep mouth moist.
Restlessness	Decreased brain oxygen	Calm and distract patient with music, massage, reading.
Oliguria	Less fluid intake	Consider if catheter is needed (often it is not).
Decreased fluid and food intake	Body is saving energy, may cause natural analgesia	Do not force intake; give oral care, sips of ice if desired.
Breathing pattern changes	Less circulation to organs	Position in semi-Fowler's or on side.

REVIEW QUESTIONS

1. The nurse is assessing a 68-year-old patient's developmental stage and finds that the patient is retired and that the patient's spouse died 4 months ago. The nurse identifies the patient as being in which one of the following developmental stages?
   a. Generativity versus self-absorption
   b. Identity versus role confusion
   c. Intimacy versus isolation
   d. Integrity versus despair

2. The nurse is planning care for a patient with diabetes. Which one of the following effects does the nurse understand is most likely to occur with a chronic illness?
   a. Powerfulness
   b. Hopefulness
   c. Increased socialization
   d. Spiritual distress

3. The nurse is providing care for a chronically ill patient. Which one of the following is an appropriate nursing intervention for a chronically ill patient?
   a. Decreasing educational information
   b. Limiting visiting hours for family members
   c. Including family members in teaching sessions
   d. Setting goals for the patient

4. The nurse is assisting with goal setting for a patient with a diabetes. Which one of the following goals should the nurse focus on as the primary goal of chronic illness treatment?
   a. Ensuring a cure
   b. Providing comfort care
   c. Maintaining quality of life
   d. Undergoing new experimental treatments

5. The nurse is caring for a patient who is comatose and in respiratory failure. Which one of the following would the nurse correctly interpret as a sign of impending death?
   a. Warmth of extremities
   b. Increased body temperature
   c. Insomnia during the night
   d. Mottling of lower extremities

REFERENCES

1. Erikson, EH: Childhood and Society, ed 2. WW Norton, New York, 1963.
2. Erikson, EH: Identity and the Life Cycle. WW Norton, New York, 1980.
3. Kübler-Ross, E: On Death and Dying. Macmillan, New York, 1969.

Answers to CRITICAL THINKING

Mr. Paul

1. Mr. Paul's physical health is being affected by poor diet choices, excessive stomach acid secretion or other gastrointestinal problems, and stress.
2. It is easy to recognize the psychological stress related to parenting skills when a child is in trouble. Decreased family income with increased family expenses (two children in college) can cause financial strains and more economic pressure on Mr. Paul's business. With family problems or health problems, Mr. Paul may be questioning why things are happening to him and his family, causing him spiritual distress.

Rita

Prenatal education information should be offered to Rita and her husband. This education should include an overview of Rita's health needs, what to expect during pregnancy, ways Rita's husband can be supportive, and information on prenatal classes. Rita's physical examination should include a vaginal examination, blood pressure, and blood work. It is important to be aware of any sexually transmitted diseases that may be transferred to the fetus or during birth. A rubella titer (a test for immunity to rubella or measles) is important because of potential birth defects if the mother has rubella while pregnant. Elevated blood glucose may be a sign of diabetes, and low red blood cell counts and low hemoglobin are related to anemia. To prepare a woman for pregnancy, prenatal vitamins or vitamins with iron and folic acid (necessary for effective neural tube development in the first 3 months of pregnancy) are recommended. Because of the increased workload of the heart during pregnancy, blood pressure needs to be closely monitored. A balanced diet, maintaining an exercise program, and continuing to develop effective and positive ways to deal with stress are very important for pregnant women. In preparation for pregnancy, Rita also needs information on the negative effects that cigarette smoking, alcohol use, and drug use can have on the developing fetus.

Mrs. Riccardi

1. Falls could be caused by environmental problems, such as throw rugs that may move or cause tripping, clutter, electrical cords in walking paths, lack of hand grips in the bathroom, or lack of nonskid mats in the shower or tub. Poor vision and altered depth perception can result in missing a stair step or obstacles. Weakness or orthostatic hypotension can cause an unsteady gait or fall. Furosemide can cause urinary urgency and incontinence. If incontinence occurs, falls may result from a wet, slippery floor.

2. It is important for the nurse to instruct the patient and family about home safety. Mrs. Riccardi may even benefit by using a cane or a walker if she is unsteady. Because Mrs. Riccardi lives alone, an emergency alert system, such as a small transmitter that is worn around the neck or wrist with a button that can be activated in emergencies, would be beneficial. When activated, the transmitter alerts an answering service to contact designated individuals to check on the patient. Safety with medications is also an important consideration. Patients who take diuretics or other medications that lower blood pressure must be aware of the potential for orthostatic hypotension. Orthostatic hypotension is a drop in blood pressure that happens when a person moves from a lying to sitting or sitting to standing position. It is often accompanied by dizziness or light-headedness. Some people may even faint, causing a fall.

Mr. Soloman

1. Mr. Soloman had lost control of his world and felt powerless. His environment, both home and outdoors, was shrinking. He had to give up his daily routines and interactions with others. His purpose in life was gone when he was no longer caring for his wife. He was separated from his loved one. His visual limitations made his new environment unfamiliar and frightening.
2. Options to keep him safely in his home could have been explored with his input. After the move, he should have been thoroughly oriented to his environment. He should have been asked to explain what he wanted his life to be like as he adapted to this new developmental stage. Hobbies and interests should have been continued. Visual support services should have been contacted for ideas. Transportation should have been arranged to allow him to visit his wife. It should have been determined whether phone calls to his wife were possible.
3. He was depressed and slept from a lack of any interests. His lungs were at risk for pneumonia because of his long periods of immobility. He lost hope and gave up on living, which decreased his ability to fight the pneumonia.

Mrs. Burden

1. Mrs. Burden is at risk for sleep deprivation, fatigue, stress, and burnout.
2. Nursing diagnoses include disturbed sleep pattern, fatigue, social isolation, risk for caregiver role strain, and deficient knowledge.
3. Beneficial nursing interventions would include teaching about Alzheimer's, a chronic care coach, respite care referral, alarm devices for wandering, and stress management techniques.

14

NURSING CARE OF ELDERLY PATIENTS

MaryAnne Pietraniec-Shannon

KEY TERMS

activities of daily living
(ack-**TIV**-i-tees of **DAY**-lee **LIV**-ing)

arrhythmias (uh-**RITH**-mee-yahs)

aspiration (ASS-pi-**RAY**-shun)

cataract (**KAT**-uh-rackt)

constipation (KON-sti-**PAY**-shun)

contractures (kon-**TRACK**-churs)

dementia (dee-**MEN**-cha)

depression (dee-**PRESS**-shun)

edema (uh-**DEE**-muh)

expectorate (eck-**SPECK**-tuh-RAYT)

extrinsic factors (eks-**TRIN**-sik **FAK**-ters)

glaucoma (glaw-**KOH**-mah)

homeostasis (HOH-mee-oh-**STAY**-sis)

intrinsic factors (in-**TRIN**-sik **FAK**-ters)

macular degeneration
(**MACK**-you-lar dee-JEN-uh-**RAY**-shun)

nocturia (nock-**TYOO**-ree-ah)

optimum level of functioning (**OP**-teh-mum **LEV**-uhl of **FUNK**-shun-ing)

osteoporosis (AHS-tee-oh-por-**OH**-sis)

perception (per-**SEP**-shun)

pressure ulcer (press-sure **ULL**-sir)

range of motion (RANJE of **MOH**-shun)

reality orientation
(ree-**AL**-i-tee OR-ee-en-**TAY**-shun)

sensory deprivation
(**SEN**-suh-ree DEP-ri-**VAY**-shun)

sensory overload
(**SEN**-suh-ree **OH**-ver-lohd)

urinary incontinence
(**YOOR**-i-NAR-ee in-**KON**-ti-nents)

QUESTIONS TO GUIDE YOUR READING

1. How would you define aging?
2. What are basic physiological changes associated with advancing age?
3. How would you describe psychological and cognitive changes associated with advancing age?
4. What are nursing implications for the physiological and psychological changes associated with advancing age?
5. What nursing practices promote safety for the older patient?

▶ WHAT IS AGING?

Over time it is easy to visually notice changes that occur in the human body. Both physical structures and body functions undergo changes and declines with advancing age. Although there is not one commonly accepted definition or theory to explain these declines, there is an understanding that aging is a universal and normal process that starts at conception and continues until death.

Older adults are increasing in number. In 2000, 34.7 million adults were older than age 65.1 By 2025 this number is expected to be 61.9 million. Within this group, those older than 84 are the fastest-growing segment. There were 4.9 million adults older than 84 in 2000, and by 2025 it is projected there will be more than 7 million. For more population data, visit www.census.gov.

For this chapter, aging is defined as a maturational process that creates the need for individual adaptation because of physical and psychological declines that occur during a lifetime. Even though aging truly begins at conception, the focus in this chapter is on the maturational process that is experienced after age 65 (older adult). Those older than age 84 are usually the frailest. The changes discussed do not always occur in the sixth decade of life, so chronological age should not be the basis for determining health issues. For some people, aging effects go unnoticed in their daily functioning; for others these effects cause varying degrees of impairment. Functional age (health, independence, and functional abilities) should be used as the basis of individual care needs. It is important to understand that most elderly people function independently in the home and community. Many are still working. With supportive, educative care as needed, these people are able to maintain their independent abilities. This chapter discusses aging changes and resulting disabilities that may require more intensive nursing care than independent, healthy older adults need.

L E A R N I N G TIPS

Placing older adults into age categories overlooks their unique aging experience. The concept of functional age recognizes that aging is individual and promotes individualized nursing assessment and development of plans of care for the older adult.

Although aging is universal, it remains a unique experience for each individual. Factors that contribute to this process can be grouped into two categories. **Intrinsic factors** focus on genetic theories of aging, such as biological clock theory or programmed aging theory, and some aspects of physiological theories of aging, such as wear-and-tear theory or stress adaptation theory. **Extrinsic factors** focus on environmental influences, such as pollutants, free radical theory, and stress-adaptation theory. Regardless of which factors have the greatest influence on this process of aging, **perception** and attitude also play key roles in how the changes over time affect the individual. It is through the filter of perception and attitude that the individual identifies, defines, and adapts to the changes that occur in structure and function over time. These factors have implications not only for older patients but also for their families and the health care providers working with them.

▶ PHYSIOLOGICAL CHANGES

Over time cells change, and they do not function as efficiently as in earlier years (Table 14–1). The physical changes that are seen when looking at the older patient are

TABLE 14-1 PHYSIOLOGICAL AGING CHANGES

Body System	Aging Change	Effect of Change
Cardiovascular	Increased conduction time	Heart rate slows, unable to increase quickly
	Decreased cardiac output	Less oxygen delivered to tissues
	Decreased blood vessel elasticity	Increased blood pressure increases cardiac workload
	Irregular heartbeats	Poor heart oxygenation, decreased cardiac output, heart failure
	Leg veins dilate, valves less efficient	Varicose veins, fluid accumulation in tissues
Endocrine and Metabolism	Basal metabolic rate slows	Possible weight gain
	Altered adrenal hormone production	Decreased ability to respond to stress
	Decreased insulin release	Hyperglycemia
Gastrointestinal	Reduced taste and smell	Appetite may be reduced
	Decreased saliva	Dry mouth, altered taste
	Decreased gag reflex, relaxation of lower esophageal sphincter	Increased **aspiration** risk
	Delayed gastric emptying	Reduced appetite
	Reduced liver enzymes	Reduced drug metabolism and detoxification
	Decreased peristalsis	Reduced appetite, constipation
Genitourinary	Kidney size decreases	Able to live with 10% renal function
	Decreased bladder size, tone, changes from pear to funnel shaped	Frequency of urination increased
	Muscles weaken	Incontinence
	Decreased concentrating ability	Nocturia
	Less sodium saved	Risk for dehydration
	Reduced renal blood flow	Decreased renal clearance of all medications

TABLE 14-1	PHYSIOLOGICAL AGING CHANGES—CONT'D	
Body System	**Aging Change**	**Effect of Change**
Immunological	Decreased function	Infection and cancer risk greater
	Increased autoimmune response	Increased autoimmune diseases
Integumentary	Reduced cell replacement	Healing slower
	Water loss	Dryness of the skin
	Increased pigmentation	Aging spots
	Thinning of skin layers	Skin more fragile
	Decreased subcutaneous fat	Less insulation and protective cushioning
	Decreased sebaceous and sweat glands	Dryness and decreased temperature regulation
	Hard, dry nails	Brittle nails
	Thinning scalp hair	Baldness
	Decreased melanin	Gray hair
	Decreased skin elasticity	Wrinkle development
Musculoskeletal	Decreased muscle mass	Reduced strength
	Decreased muscle tone	Muscles look flabbier
	Decreased elasticity of tendons and ligaments	Movements are restricted
	Muscle responses slowed	Response time increased
	Bone thinning, softening	Decreasing bone density
	Joint stiffening	Decreased flexibility
	Vertebral disk water loss	Decreased height
Neurological		
Central nervous system	Loss of brain cells	Able to maintain function with remaining cells
	Decreased brain blood flow	Short-term memory loss
	Decreased regulation of body temperature	Hypothermia, hyperthermia risk
	Decreased endorphins	Increased depression
Peripheral nervous system	Decreased sensation	Risk for injury, burns
	Increased reaction times	Slow response, injury risk
	Decreased motor coordination	Unsteady, fall risk
Respiratory	Decreased lung capacity	Dyspnea with activity
	Decreased cough and gag reflexes	Aspiration, infection risk
	Reduced lung tissue tone	Shallow, faster respirations
	Reduced lung emptying on exhalation	CO_2 retention
	Decreased fluid and ciliary action	Mucous obstruction, infection risk
Sensory		
Eye	Lens less elastic	Decreased near and peripheral vision
	Lens opaque, yellows	Cataracts
	Cornea more translucent	Blurry vision
	Smaller pupil	Decreased dark adaptation
	Decreased violet, blue, green color vision	See red, orange, yellow colors better
	Arcus senilus—blue or milky lipid ring on iris edge	No effect on vision
Ear	Degeneration of auditory nerve	Lose high-frequency tones, deafness
	Excess bone impairs sound conduction	Deafness
Nose	Decreased smell	Decreased ability to smell substances such as smoke, gas causing safety risk; appetite reduced
Sexuality		Lack of sexual expression, suppression of desires
	Availability of partner or privacy decreases	Increased time needed for sexual stimulation
	Slower sexual arousal time	Psychologically causes concern
Men	Decreased erection, slower ejaculation	Painful intercourse
Women	Less vaginal lubrication	Increased vaginal infection risk
	Vaginal acidity reduced	

slight as compared with the cellular changes that occur. Cellular decline in structure and function increases in severity and extent over time. Although the body works hard to maintain **homeostasis,** it is often unable to fully adapt to many of the declines that result from aging. Cells that die cannot regenerate themselves. As a result, structures are altered, and the body tries to adapt to make the revised structure meet functional demands.

homeostasis: homios—similar + stasis—standing

Common Physical Changes in Older Patients and Their Implications for Nursing

Key Changes in the Muscular System with Aging

Key aging-related changes in the muscular system include the following:

■ Decrease in muscle mass, so muscles look smaller
■ Decrease in muscle tone, so muscles look flabbier
■ Slower muscle responses, so response time is increased
■ Decrease in elasticity of tendons and ligaments, restricting movements

NURSING IMPLICATIONS. Changes in the muscular system have implications for movement, strength, and endurance. Restricted movements are most commonly seen in the arms, legs, and neck of the older patient, who may demonstrate limited **range of motion** (ROM) in these areas. Because muscle response abilities are slowed, it will take longer for the older patient to move. This increased response time has implications on the older individual's confidence level regarding personal ability to perform routine tasks.

Key Changes in the Skeletal System with Aging

Key aging-related changes in the skeletal system include the following:

■ Eroding cartilage
■ Exaggerated bony prominence
■ Joint stiffening and decreased flexibility
■ **Osteoporosis,** a thinning and softening of the bone
■ Shortening in height caused by water loss in the intervertebral disks of the spinal column, flexion of the spine, and stooped posture

NURSING IMPLICATIONS. Because muscles and bones work together for movement, aging skeletal changes are most obvious when the older patient is moving. **Contractures** of the fingers and hands can limit the individual's ability to perform self-care tasks called **activities of daily living** (ADL). It is important to assist the patient with ROM exercises if help is needed to prevent the long-term disabilities that contractures bring (Fig. 14–1). Performing ROM exercises in warm water helps the patient who experiences discomfort with these exercises. If the individual has arthritis, anti-inflammatory medications should be given so that their action peaks when the exercises begin. Older patients on anti-inflammatory medicines should be monitored closely for gastrointestinal upset or bleeding and taught the symptoms of bleeding to report.

Decreased bone density is influenced by diet and weight-bearing exercise, so balanced diets rich in calcium and vita-

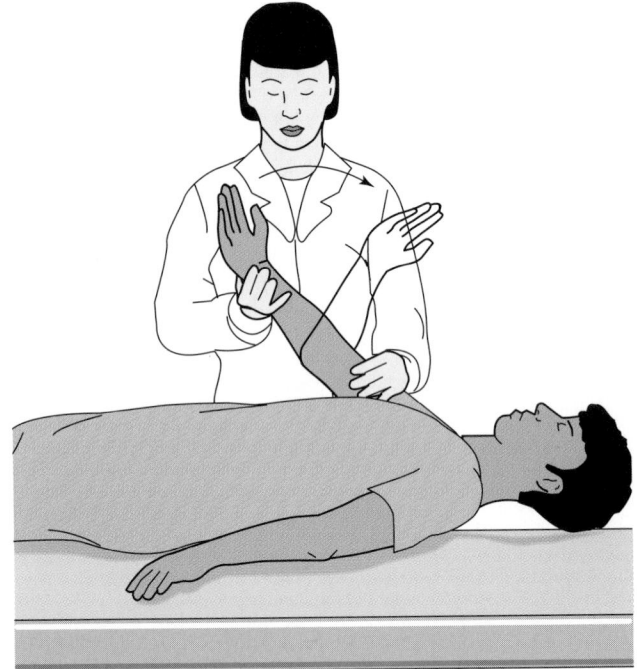

Figure 14-1   Nurse assists patient in range-of-motion exercises to prevent the development of contractures.

min D and safe and sensible exercise programs should be promoted. Patients should be encouraged to ambulate whenever possible, using sensible shoes that have nonskid soles and sturdy assistive devices if needed, such as handrails, canes, or walkers (Home Health Hints). When the decreasing density of older bones is considered, it is easy to understand that broken bones do not always result from a fall but may indeed be the reason for the fall in many older patients.

Key Changes in the Integumentary System with Aging

Key aging-related changes in the integumentary system include the following:

■ Increased dryness of the skin
■ Increased pigmentation, causing liver or aging spots
■ Thinning in the layers of the skin, which makes the skin more fragile
■ Decreased elasticity of the skin, causing wrinkles to develop
■ Decreased subcutaneous fat layer of skin, so older patients have less insulation and less protective cushioning
■ Hardness and dryness of nails, making them more brittle
■ Decrease in nail growth rate and strength
■ Thinning of scalp hair (primarily men)
■ Increased growth and coarseness of nose, ear, and facial hair
■ Decrease in melanin, which results in gray hair
■ Decreased sebaceous and sweat glands, which has implications for dryness and decreased temperature regulation

---

osteoporosis: osteon—bone + poros—a passage + osis—condition

HOME HEALTH HINTS

- Try to schedule therapy visits and nurse visits on the same day, to decrease fatiguing the older patient.

- Because many older patients keep their homes warm, the home health nurse should wear layers of clothing, removing a layer as needed, rather than adjusting the heat in the older patient's home.

- Pagers should be placed on a silent mode if possible to avoid startling or confusing the older patient.

- Nurses should not assume that the older patient remembers them when they answer the door. Nurses should state their name and why they are there. A large-letter name tag should be in clear view.

- To enhance the effectiveness of a visit, the older patient should be asked to visit in a quiet room of the home with the primary caregiver invited in at the appropriate time. This may help the older patient stay more focused, offer privacy, and assist hearing.

- Making a sign for the door of the home giving visitors instructions, such as ringing the doorbell several times, knocking loudly, or using the other door, may be helpful to the older patient. Do not provide information that would inform a stranger that an older patient lives in the home.

- Suggesting that limiting visitors or not allowing persons with colds to visit can be helpful in preventing illness for the older patient.

- Stressors in a patient's life, such as annoying visitors, chastisement by caregivers, and harassment by bill collectors, are often experienced firsthand by the home health nurse. Document and share those with the home care team, so a coordinated approach can be taken.

- When auscultating lungs, ask the older patient to take deep breaths in and out through the mouth. This may stimulate coughing, which is an opportune time to teach deep breathing and coughing exercises.

- If the older patient seems fatigued early in the day, ask about sleeping patterns and things that disturb it, such as barking dogs, traffic noise, and visitors. Recommend ear plugs or changing rooms to obtain a good night's rest. Check medications for insomnia listed as a side effect and suggest dosing of these medications early in the day if appropriate. A 15- to 30-minute nap in the early afternoon can be helpful.

- One of the first signs of infection in the older patient is confusion.

- One of the first signs of dehydration is tachycardia. Instruct the patient regarding adequate hydration.

- Assess the older patient's environment for safety hazards on each visit and promote safety.

- If the older patient has dentures, assess if they are worn for eating. If not, assess why they are not worn (e.g., sores, improper fit from weight loss) and discuss solutions.

- Checking the refrigerator for outdated food is helpful. The older patient is often on a limited budget and has been taught not to waste food. These factors along with a decreased sense of smell and taste can lead to food poisoning in the older patient.

- Encourage the uses of spices and herbs, such as parsley, oregano, lemon, garlic, and basil, instead of salt and sugar. Suggest keeping pared apples and slices of oranges in the refrigerator for snacks.

- If a Meals on Wheels program is available, ask if the older patient would like to be placed on the service.

- Use a warming tray when feeding an older patient who takes a longer time to eat.

- When swallowing is difficult, freezing liquids helps, so they can be eaten with a spoon or like a popsicle. Milkshakes, high-protein drinks, instant breakfast mix, or eggnog are thicker liquids that are easier to swallow.

- If an older woman wears perineal pads or adult briefs, ask how many are used in a 24-hour period to assess the degree of incontinence or amount of output. Have the patient keep a voiding diary to further assess the degree of incontinence.

- Suggest a bedside commode when a weakened older patient is on diuretics or has a history of falling or confusion. Placing it next to the bed at night helps reduce the risk of falls and eases the caregiver's burden.

- If the older patient reports constipation, review the diet and make suggestions regarding adequate fluid and fiber. A mixture of equal parts of applesauce, bran, and prune juice is often helpful to prevent or relieve constipation. Discourage the use of mineral oil because it can make vitamins less effective.

- When drawing blood from the hand of an older patient, use the smallest needle possible.

- Hold light pressure for at least 2 minutes after the needle is removed. Do not use a Band-Aid on the fragile skin of the older patient if the bleeding has stopped with pressure.

- When teaching, it is important to acknowledge the patient's knowledge and life experiences. When given a chance, patients tell how they have maintained their health over the years. Use open-ended scenarios for teaching, such as, "What would you do if you were alone and fell?" Teaching should occur with patients, not to them. The nurse is in the patient's home, which is a personal place. Patient dignity should always remain intact during home care visits and teaching sessions.

NURSING IMPLICATIONS. The skin, which is the first line of defense against infection and injury, does not work as effectively in older patients (Fig. 14–2). In the elderly, skin injuries take longer to heal, and those longer healing times are usually complicated by the fact that many older patients have multiple chronic diseases, such as diabetes and circulatory ailments.

The older patient with limited mobility is especially prone to developing pressure ulcers (Fig. 14–3). These ulcers usually develop over a bony prominence of the body (e.g., ears, shoulders, elbows, tip of the spine, pelvic bone ridges, knees, heels, or ankles). Pressure ulcers are caused by ischemia that results from continuous pressure on an area of the body. Ischemia from unrelieved pressure can develop in 20 to 40 minutes. Early signs of **pressure ulcer** formation are warmth, redness, tenderness, and a burning sensation at the potential ulcer site. These potential ulcer sites are aggravated by lack of activity and the weight of the body. For this reason it is especially important to take the time to assess skin integrity daily.

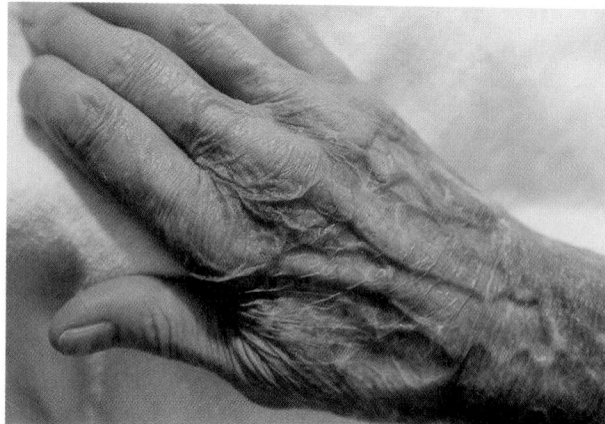

Figure 14-2   Thin, fragile skin of older patient.

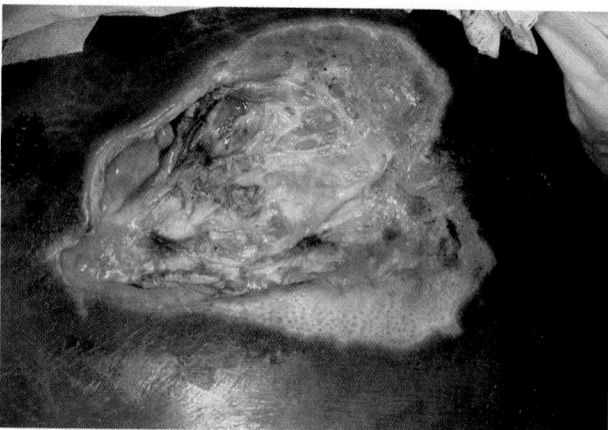

Figure 14-3   Pressure ulcer. (From Goldsmith, LA, Lazarus, GS, and Tharp, MD: Adult and Pediatric Dermatology: A Color Guide to Diagnosis and Treatment. FA Davis, Philadelphia, 1997, p 445, with permission.)

Skin care includes gently stimulating nonreddened intact skin sites with massage, moisturizing with creams regularly, avoiding the use of hot water and a complete daily bath, and limiting the use of soap. If able, patients should be taught to shift their weight every 15 minutes when sitting. For immobile patients, consistent repositioning is essential. Repositioning every 30 minutes is the most beneficial to help ensure ulcer prevention. Keeping bed linens clean, dry, and wrinkle free also aids in the prevention of pressure ulcer formation.

As with care of the skin, nail care is important for older individuals. Soaking in warm water helps soften nails to ease in their trimming while encouraging blood flow to the peripheral areas of the body. Filing the nails with an emery board is safer than cutting the nails. To prevent accidental injury to the feet, patients should be instructed not to walk barefooted. Potential pressure points of the feet should be identified and closely monitored, and the patient should be referred to a podiatrist for treatment should there be any concerns. Diabetics should assess their feet daily because they may have decreased sensation, causing lack of awareness of foot irritation or injury.

Key Changes in the Cardiovascular System with Aging

Key aging-related changes in the cardiovascular system include the following:

- Slowed heart rate
- Decreased cardiac output from less effective functioning of the heart and blood vessels, yielding less oxygen to body tissues
- Decreased elasticity of the blood vessels, so the circulatory system is less efficient
- Reduced ability of the heart to quickly increase its rate in response to an emergency because of thickening of the heart valves, left ventricle, and aorta (when rate finally does increase, the heart takes longer to return to resting rate)
- More irregular heartbeats, **arrhythmias,** which lead to poor oxygenation of the heart
- Commonly, a lack of classic symptoms of cardiac emergencies
- Increased peripheral vascular resistance in blood vessels, yielding increased blood pressure
- More visible superficial blood vessels of the legs
- Less efficiency of leg blood vessel valves, creating the risk for an accumulation of excess fluids in the leg tissues

NURSING IMPLICATIONS. Be a good observer when it comes to caring for older patients because many early symptoms related to circulatory problems are subtle. Cardiovascular disease, which is separate from the process of aging, accounts for half of all deaths in people older than 65 years. Older adults must be educated about prevention practices that promote healthy circulation and encouraged to take prescribed medications as ordered and maintain a balanced intake and output of fluids. Older patients receiving intravenous therapy should be monitored closely because fluid overload can occur quickly in older people.

Special care should be taken to maintain good skin integrity and provide appropriate stimulation. If **edema** is present in the legs, they should be elevated to assist the fluid return to the upper body and supportive, nonrestrictive stockings should be worn as ordered. Concerns regarding leg circulation should be identified and reported early to the physician before problems develop. It is important to identify whether the circulation problem is arterial or venous before treatment is ordered.

Because quick changes in body position can make the older patient feel weak and dizzy, it is important to stand next to older patients as they dangle their legs over the side of the bed before rising to stand. Changes in body positioning from lying to sitting to standing need to occur gradually to accommodate the less efficient circulatory systems of older patients. Older patients may find comfort in the security of an ambulatory belt or walker if they fear unsteadiness when in an upright position. Falls continue to rank as a

edema: oldema—swelling

leading cause of accidental death in older patients. Every effort should be made to decrease the risk of falling for the older patient. The fall history of patients should be assessed to identify whether they are at high risk of falling, so that preventive measures can be used.

Key Changes in the Respiratory System with Aging

Key aging-related changes in the respiratory system include the following:

- Decreased lung capacity
- Weaker cough or gag reflex, increasing risk for upper respiratory infections
- Reduced tone of lung tissue, so respirations increase to 16 to 25 per minute and are more shallow
- Reduced tone of the diaphragm muscle
- Less complete emptying of the lungs, with greater CO_2 retention
- Decreased blood flow to the lungs, contributing to cardiac arrhythmias

NURSING IMPLICATIONS. Because the respiratory system is less efficient with advancing age, the older patient has a decreased tolerance for activity in general. Nurses should therefore pace activities for older patients instead of letting them confine themselves to bed. It is important to prevent overexertion, so rest periods should be scheduled. However, the rest periods should not outnumber the activity sessions planned throughout the course of the patient's day.

Cough, marked fatigue, and confusion may be early signs of an inadequate oxygen uptake. Respiratory rates greater than 25 per minute may be an early indication of a lower respiratory tract infection. Because overall muscle strength is reduced, the older patient performs the O_2–CO_2 exchange in the less efficient upper lobes of the lung instead of the larger lower lobes specifically designed for this purpose. Because lung recoil strength is decreased, mucus may be more difficult for the older patient to **expectorate.** This situation is compounded by the fact that older individuals also have less effective cough and gag reflexes, which creates greater potential for lung problems. Because of the normal changes in the respiratory system that occur with aging, it is important to include coughing, deep breathing, and position changes in the exercise program designed to stimulate all lobes of the older patient's lungs. At the prevention level, encourage the older individual to receive a pneumonia vaccine and an annual flu shot. This is important because influenza and pneumonia combine to be the fourth leading cause of death in people over age 65. Be aware that lifelong habits, such as smoking and respiratory pollutant exposure in employment settings, secondhand smoke, or paints and glues used in hobbies, are cumulative over time and can contribute to respiratory sensitivity for the older patient.

Key Changes in the Gastrointestinal System with Aging

Key aging-related changes in the gastrointestinal system include the following:

- Changes in taste and smell, which affect the enjoyment of eating
- Decreased saliva production
- Decreased gag reflex and relaxation of lower esophageal sphincter, increasing the risk of aspiration
- Delayed gastric emptying
- No functional changes in the small intestine
- Decreased tone in external sphincter
- Marked decline in liver enzymes, which affects drug metabolism and detoxification
- Decreased peristalsis from generalized weakness of muscle activity
- Alteration in bowel habits

NURSING IMPLICATIONS. Many factors can alter appetite, ingestion, digestion, and absorption of nutrients in food, regardless of an individual's age. However, the structural and functional changes that occur with advancing age put the older patient at greater risk for not obtaining the nutrients needed to sustain a healthy body (Nutrition Notes Box 14–1).

There is little that can be done to change the physical alterations in the older body that make getting needed nutrients more difficult. Being knowledgeable and committed to providing the support necessary to meet nutritional goals is important. Patients should be assisted with toileting before helping them eat. An appropriate amount of time for the patient to accomplish the task of eating needs to be provided. If the patient needs assistance with eating, be sensitive to the patient's pace while allowing the patient to have as much control as possible.

Additional forms of support can focus on the need to recognize that for some cultures eating has a strong social component. Encourage the patient to eat out of bed and with others as much as possible, while respecting the patient's right of refusal to eat in a designated social setting. Some patients may not eat as well when seated next to agitated or confused residents in a common dining hall.

Offer continuing support by maintaining a calm and comfortable environment that aids digestion. Certain food combinations enhance nutrient absorption, such as vitamin C foods taken with plant foods high in iron, which increase the iron's absorption rate. Offer these selections together whenever possible. Family members can be asked to bring in familiar seasonings patients used at home. If acceptable, familiar seasonings in shakers that patients apply themselves can be used to stimulate a lagging appetite. Work closely with the dietitian or nutritionist, who may provide other ideas to promote healthful eating for older patients.

Although many older patients wear dentures or partial plates, it is important not to stereotypically assume that all older individuals wear dentures. Tooth loss is not a normal change of aging. With proper lifelong dental care, teeth

---

expectorate: ex—out + pectus—breast

BOX 14-1 — Nutrition Notes

Meeting the Nutritional Needs of the Older Adult

Energy needs decrease 5% for every decade after age 40, partly because a 70-year-old has 40% less skeletal muscle than in younger days. Adipose tissue requires fewer calories to sustain than skeletal muscle does, so special attention must be given to maximizing nutritional content in fewer calories if a healthy weight is to be maintained. A modified food guide pyramid for adults age 70 and older adds eight servings of water to the base of the pyramid, increases milk servings to three, and recommends calcium, vitamin D, and vitamin B_{12} supplements. New recommended daily allowances for vitamin B_{12} specify that fortified foods or supplements furnish most of this vitamin for people over age 50.

Decreased visual acuity and impaired dexterity may make shopping for food and preparing it difficult or even hazardous. Arthritis affects not only mobility, but also jaw movements, so chewing may be problematic. By age 65, about 40% of U.S. residents have lost all their teeth and have to develop skill in the use of dentures. It is recommended that an individual learn to drink with them first, then learn to manipulate soft foods, and lastly learn to bite and chew with the dentures. Older persons produce less saliva than younger people. The sense of taste declines in most but not all aging patients, who as a result may increase their intake of salt and sugar to the detriment of a prescribed diet plan.

Achlorhydria may occur as a result of aging or from chronic ingestion of antacids. In either case, protein digestion and absorption of iron and vitamins B_{12} and C can be impaired by the lack of gastric acid. Although anemia can be caused by poor nutrition, older patients should be evaluated for hidden blood loss just as younger ones are. High-protein supplements should be offered to older individuals with caution, because normal aging reduces kidney function significantly and excretion of nitrogenous products of excessive protein intake could stress the urinary system. In addition, a high-protein intake increases calcium excretion. The vitamin necessary to metabolize calcium, vitamin D, may be deficient in older persons who neither drink milk nor spend time outdoors.

Many mature patients are at increased risk for food, nutrient, and drug interactions. Situations associated with these interactions are the use of many drugs, including alcohol; the need for long-term drug therapy in chronic illness; and poor or marginal nutritional status. Identification of any of these factors should prompt a thorough nutritional assessment or referral to a registered dietitian.

should last a lifetime. If the older patient does have dentures, be aware that any significant change in body weight will affect their fit and comfort and the patient's nutritional intake. Because of this, it is important to conduct regular assessments of the mouth when assisting the older patient during oral care.

Medications may cause taste disturbances or problems with dry mouth that may also affect the older patient's ability to meet nutritional needs. Some medications create problems with bowel motility, resulting in **constipation.** Constipation can also result from a change in routine, stress, and anxiety. To gain a greater understanding of the situation, talk with the patient about previous assistive procedures and establish an expectation baseline for bowel

elimination. Educate the older patient about the important relationship between intake of fiber and water and exercise in the promotion of effective bowel evacuation. Enemas, suppositories, and medications should be considered for use only after dietary management is found to be ineffective.

Key Changes in the Endocrine-Metabolic System with Aging

Key aging-related changes in the endocrine-metabolic system include the following:

- Slowing in the basal metabolic rate, requiring a 5 percent reduction in calorie consumption to maintain weight
- Alteration in hormone production, including changes in estrogen, progesterone, and adrenal secretions
- Decreased pancreatic insulin release and peripheral sensitivity
- Decreased glucose tolerance with advancing age

NURSING IMPLICATIONS. Increased incidence of metabolic disease, such as diabetes, occurs with advancing years. Because of this, older patients should be encouraged to participate in screening programs for early detection of metabolic problems.

Because there is a notable decrease in the effectiveness and interaction of all hormones as one ages, it becomes especially difficult for the older body to respond appropriately to any stressful situations. Nurses should spend time addressing the psychological needs of patients by recognizing and addressing those actual and perceived stressors that affect their care. Preventive care to keep patients free from the stress of illness is also important.

Key Changes in the Genitourinary System with Aging

Key aging-related changes in the genitourinary system include the following:

- Decreased kidney size
- Decreased kidney function, urinary output, and adaptability
- Reduction of blood flow to the kidneys because of decreased cardiac output and increased peripheral resistance
- Diminished kidney filtration rate and tubular function, which cause a decrease in the renal clearance of all medications
- Decreased bladder size and tone
- Increased incidence of urinary tract infection with age, especially in women
- Longer correction times for fluid and electrolyte imbalances

NURSING IMPLICATIONS. Many older people have an urge to urinate at night, which is referred to as **nocturia.** Changes in the kidneys, lack of gravity influence when in the recumbent position, fluid retention, and medication use

---

nocturia: nocte—night + ouron—urine

contribute to nocturia. Approximately 30 minutes after lying down many older people need to urinate, because fluid that was held in the legs by gravity is circulated through the kidneys and urine is produced, causing the urge to void. Plan safe toileting access for older patients during the night. Emptying the bladder becomes more difficult to control from weakening of bladder and perineal muscles and also from a change in brain sensation regarding the need to void. As a result, the older patient may have difficulty controlling urination either knowingly or unknowingly, resulting in **urinary incontinence.**

L E A R N I N G TIPS

With age, the urge to void is not felt as early as it once was. This aging change combined with other changes in the urinary system, such as the bladder becoming funnel shaped, contribute to the older patient's voiding urgency. If the patient is not assisted to void promptly on request, incontinence can result.

Incontinence is socially embarrassing and can occur in both male and female patients. It is said to be one of the major reasons that older people are admitted to nursing homes. The incidence in men is most often associated with prostate enlargement; women are more likely candidates because of weakened perineal muscles. Older women are more likely to experience urinary incontinence because they have short urethras that are not supported by strong perineal muscles and more urinary tract infections than older men.

Management of urinary incontinence needs to be tailored to the particular need of the patient. Bladder training programs have been effective when patients are reminded on a regular basis that it is time to urinate. If incontinence results from problems that affect the toileting task itself, such as clothing removal or distance to the bathroom, steps need to be taken to eliminate those obstacles that stand in the way of continence. Clothing with Velcro fasteners can replace buttons or zippers that are difficult for the patient to manipulate. Urinary briefs can help instill confidence in older patients who previously restricted activities because of fear of urine leakage. With use of the supportive incontinence brief, assess for early signs of perineal skin breakdown and proper application of the proper brief to meet individual needs.

Older patients who are aware of their incontinence may try to inappropriately decrease the chance for leakage by severely limiting fluid intake. This approach often results in dehydration, which disturbs the acid-base and electrolyte balance in the body. Over time, dehydrated patients may have problems with vomiting, diarrhea, weakness, and confusion. Because fluid intake needs to be encouraged in older patients, focus educational efforts on topics such as liquid intake timing and beverage selection (e.g., teaching that caffeine and alcoholic beverages should be avoided because they normally increase urinary output).

In addition, review patient medications and personal medication practices, such as taking diuretics late in the

day, which then causes nighttime urination. Teaching perineal muscle support exercises and techniques is successful for some. All patients with unexplained urinary incontinence should be referred to a specialist for further evaluation because it is not a normal condition of aging.

CRITICAL THINKING

Mr. Jones

Mr. Jones, age 72, lives at home. His home has wood floors with throw rugs. The bathroom is located in the hall outside his bedroom. Mr. Jones has nocturia and is occasionally incontinent from urgency to void. He takes bumetanide (Bumex) daily.

1. What additional assessment data should be obtained about Mr. Jones regarding his urinary status and home environment?
2. Are safety concerns present in the home environment?
3. What nursing diagnoses should be included in Mr. Jones's nursing care plan?
4. What should be included in a teaching plan for Mr. Jones?

Answers at end of chapter.

Key Changes in the Immunological System with Aging

Key aging-related changes in the immune system include the following:

- Decreased immune response
- Decreased number and function of T cells, leading to impaired ability to produce antibodies to fight disease

NURSING IMPLICATIONS. Older patients tend to have more chronic diseases that may increasingly depress their immune responses over time. Although older patients have fewer colds, they are at higher risk for influenza and other complications once they get a cold. It takes longer to recover from infections, so patients and their family members need to be told that a prolonged recovery is to be expected. It is also important to screen visitors for illness before they visit a recuperating older patient.

Efforts at prevention should focus on teaching the older patient about the importance of obtaining current immunizations, reducing stress, eating right, exercising, and maintaining a healthful lifestyle. Individuals on medications for conditions that put them at risk for immunosuppression, such as steroids, should be aware of the need to take additional safety precautions around others who are ill. Nurses can help patients help themselves by encouraging techniques such as proper hand washing that support universal precautions.

Key Changes in the Neurological System with Aging

Key aging-related changes in the neurological system include the following:

- Progressive loss of brain cells
- Decreased blood flow and oxygen utilization to the brain

■ Decreased protein synthesis
■ Decrease in sensitivity and the sensation pathways
■ Increased reaction times
■ Decreased motor coordination
■ Decreased equilibrium
■ Decreased ability of the hypothalamus to regulate body temperature
■ Change in neurotransmitter secretion levels

NURSING IMPLICATIONS. Changes in the nervous system of the older patient can be seen in both the peripheral and central systems. These changes have significant meaning for the older individual, especially in the area of safety. With normal aging there is a slowed response to stimuli and a marked decrease in the speed of the psychomotor response to that stimuli. Because stronger stimuli are required to elicit any neurological response, the older patient is unable to perceive early signs of danger. To protect from accidental skin burns, the thermostat on water heaters should be lowered and electrical heating devices such as heating pads, electric blankets, and mattress pads should not be used.

Another safety issue of concern focuses on changes in balance. It is more difficult to maintain balance as one ages, especially when musculoskeletal changes are considered. Special caution should be taken to assist older patients with transferring and ambulation activities so optimum levels of safety can be maintained.

Fine tremors of the hand are a normal finding with advancing age and tend to increase when the older patient is cold, excited, hungry, or active. Assistive devices provided by an occupational therapist, such as handle grips or anchored equipment, can help eliminate the unsteadiness created when fine tremors make it more difficult to accomplish activities of daily living (Fig. 14–4). Accompanied by generalized muscle weakness, normal neurological changes that arise with advancing age usually occur on both sides of the body at the same time. One-sided weakness, sensory problems, and performance problems should always be referred for further evaluation. Coarse tremors of the finger, forearm, head, eyelids, or tongue that occur when the body part is at rest may be a sign of a neurological problem such as Parkinson's disease. These generally occur on one side of the body first and should always be referred for further evaluation.

Key Changes in the Sensory System with Aging

Key aging-related changes in the sensory system include the following:

■ Decreased visual perception
■ Decreased elasticity of the eardrum
■ Decreased sense of smell
■ Decreased taste perception
■ Decreased touch sensation

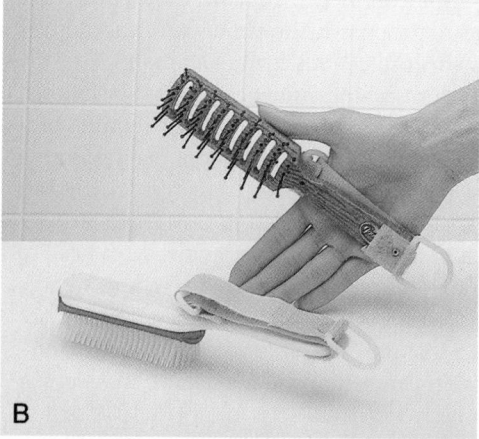

Figure 14-4 Assistive devices for activities of daily living. *(A)* Sock and stocking aid. *(B)* Easy-pull hair brush. *(C)* Food guard. (Courtesy Sammons Preston, Inc., Bollingbrook, Ill, with permission.)

NURSING IMPLICATIONS. Problems that occur with normal changes in the aging sensory system are not inevitable with advancing age. Early identification and prompt referral can minimize sensory loss, which has a strong impact on the older person's psychological health.

Normal aging changes in the eye affect the focus ability of the lens and the maneuverability of the eye muscles to meet the needs of near and far vision. It is especially difficult for older patients to read fine print, although it is usually manageable with reading glasses or a bifocal lens. When caring for patients who wear glasses, it is important to remind patients to consistently wear their glasses as needed and to help the patient keep them clean and in good repair. When glasses are removed by the patient, keep them accessible and in a protective, labeled case.

The older patient has a more difficult time adjusting to changes between light and dark settings. Seeing in the dark can be enhanced with the use of a red night-light because red lighting is more easily detected by the cones and rods in the older patient's eye. Make every effort to reduce glare from bright sunlight because sensitivity to glare is enhanced with the normal changes that occur in the aging eye (Fig. 14–5). The glare from car headlights may impair vision in older patients and reduce their ability to drive at night.

Several eye disorders can affect the aging eye. (See Chapter 49.) **Cataracts,** one of the most common pathological problems affecting the aging eye, cloud the lens and impair vision. **Glaucoma** is a chronic disease characterized by increased intraocular pressure that can damage the optic nerve. Another common visual condition that occurs later in life is **macular degeneration,** which results in a loss of central vision. Regardless of the means, any visual loss in older patients puts them at risk for developing psychological problems with disorientation, withdrawal, or self-imposed isolation caused by **sensory deprivation.**

Hearing loss is a common condition found in older individuals. Although the severity of the hearing loss caused by aging is variable, the stigma it carries is the same. For most older patients, the first difficult sounds to discriminate are the high-pitched tones. Therefore it is often more effective to whisper when communicating with the hearing-impaired individual, because whispering decreases the pitch of the sounds. Shouting is not helpful because both volume and pitch are increased. It is best to speak to a hearing-impaired person in a moderate volume and a lower tone. It is also helpful to stand in front of hearing-impaired patients so they can see the speaker's face during communication (Box 14–2).

Two types of hearing loss can occur separately or in combination in the hearing-impaired patient. Conduction loss is due to blockage or damage to the mechanisms that transmit sound to the middle ear, where it is carried by the acoustic

---

BOX 14-2 **Communicating with Hearing-Impaired Patients**

- Ensure that hearing aids are on with working batteries, if applicable.
- Face patient so the speaker's face is visible to patient.
- Speak toward patient's best side of hearing.
- Speak in a clear, moderate-volume, low-pitched tone.
- Do not shout because this distorts sounds.
- Recognize that high-frequency tones and consonant sounds are lost first—s, z, sh, ch, d, g.
- Eliminate background noise because it can distort sounds.

Figure 14-5 *(Left)* Normal view. *(Right)* The effects of glare as seen by an older patient. (From Matteson, MA, and McConnell, ES: Gerontological Nursing. WB Saunders, Philadelphia, 1988, p 314, with permission.)

nerve to the brain. Conduction hearing loss can be due to a variety of factors, some as simple as ear wax buildup, and in most cases can be successfully managed. The other type of hearing loss, sensorineural loss, is due to damage to the structures in the inner ear and is much more difficult to manage. This type of loss can result from illness, medication use, or long-term abusive noise pollution that damages the sensitive structures involved with sensation and interpretation of sound. It is important to use hearing protection throughout life because damage to the ear is usually not reversible.

New-technology hearing aids can help most people with either type of hearing loss. Patients should be referred to an audiologist for further evaluation if hearing loss is suspected. Hearing aids are not well accepted by all older persons or their family members because they are a visual sign of a loss. Because of this, maintain patient privacy if a referral for further evaluation is made. If the older patient already has a hearing aid, it should be kept accessible to the patient at all times. The hearing aid should be kept clean with soap and water. A cotton swab removes wax buildup in tiny areas of the hearing aid. When not in use, the hearing aid should be stored in a labeled protective container and the battery turned off to conserve it. Extra batteries should be readily available.

As discussed in the gastrointestinal section of this chapter, the senses of taste and smell work closely together. Like other sensory losses, a declining ability to taste is not a problem exclusive to the older individual. Both ill-fitting dentures and poor oral care can contribute to an alteration in the sense of taste, as can certain medications, tobacco substances, and oral disease.

Although they are often treated together as one sensory loss, the incidence of loss of the sense of smell is more common than the disorder that focuses on the loss of taste. In some cases the loss of smell may be due to sinus problems, nasal obstruction, or allergies. If olfactory receptors are the primary cause for losing the sense of smell, there is little that can be done to treat the loss. Assess patients' medication charts, as well as their oral cavity, if they report a recent loss of smell or taste sensation.

Nurses also need to be aware of the care environment they provide for the older patient. Overstimulation produced by **sensory overload,** commonly experienced in intensive care hospital settings, can create psychological and physical strains that can make it difficult for the older patient to cope. As a patient advocate, be sensitive to the patient's care environment and changes in the patient's health care status, identifying and minimizing overload situations early in the recuperation period.

Key Changes in the Sexuality System with Aging

Key age-related changes in sexuality include the following:

- Chronic illnesses, which may affect sexual functioning
- Functional sexual changes in older men, commonly alteration in ability to obtain or maintain an erection and to ejaculate

- Functional sexual changes in older women, such as decreased vaginal lubrication
- Sexual problems caused by psychological factors, which are more common than physical ones in elderly patients
- Longer sexual arousal time in the older patient
- Increase in sexual activities that focus on nonintercourse intimacy behaviors, including various forms of touch

NURSING IMPLICATIONS. The older patient's sexuality is an important aspect to consider when providing care. It is essential to accept that sexuality is one of the basic physiological needs identified in Maslow's hierarchy for all individuals regardless of age. Be aware of personal attitudes and values about sexuality, sex, and aging, being careful that personal stereotypes and beliefs do not interrupt the older person's attempt to maintain sexual identity. Because privacy remains a common problem for individuals in health care settings, provide the patient some scheduled private time for sexual expression. As new sexual enhancement treatments become more available to patients, there may be a greater openness to discuss sexual expression in conversations with their older patients. Still, there remain many older patients who may have difficulty overcoming barriers to sexual expression because of cultural or religious beliefs, lack of a suitable partner, fear of failure, fear of consequences, illness, side effects of medications, and certain chronic diseases that may alter libido or sexual functional abilities. Because sexual discussions with these patients may be more difficult, it is important to address the sexual history section as sensitively, professionally, and completely as other sections of the patient assessment form (Table 14–2).

TABLE 14–2	**NURSING ASSESSMENT OF SEXUALITY**

As with any nursing skill, practice will increase your comfort in collecting sexuality data for the older patient so these needs are not ignored. Assessment of sexuality concerns may lead to the discovery of other health issues and is a valuable part of data collection. As you collect data, be aware of problems (e.g., pain, medication side effects, mobility, dexterity, elimination, angina, surgery) identified in the history that may impact sexual functioning. Also, establish rapport with the patient by collecting other data first and turning to sexuality data at the end of the history. Older patients often respond to sexuality questions but will not initiate the topic.

Naturally, you should use a professional approach and manner, and tell the patient that you will now ask questions about sexuality. Start with questions that are likely to be more comfortable for the patient, as follows:

- Address roles first. For example: What concerns (do you have or has your illness created) in carrying out your roles with your (spouse or sexual partner)?
- Address relationship issues next. For example: What effect on your relationship with your (spouse or sexual partner) has this (illness, symptom, chronic illness) had?
- Address body image. For example: What effect has your (illness, surgery [mastectomy, prostate, ostomy], chronic illness, aging-related changes) had on your self-concept of being a (man or woman)?
- Address sexually transmitted diseases. For example: Do you have any discharge or open sores (vaginal, penile)?

COGNITIVE AND PSYCHOLOGICAL CHANGES IN THE OLDER PATIENT

Cognition

In addition to those physiological changes that occur with advancing age, older patients are also experiencing changes that have an effect on cognition. Cognition includes abilities related to intelligence, memory, orientation, judgment, calculation abilities, and learning. Cognition focuses on intake of information and storage, processing, and retrieval of stored information. For the most part, older patients store information without much conscious effort. If dealing with information becomes difficult, the older person may begin to worry. Unless this worry is addressed, the patient's concern may result in psychological problems and fears.

Many factors can affect cognition. Sensory changes and diseases associated with age can cause misinterpretation of information being collected. Pain from chronic diseases, such as arthritis, can limit cognition as pain takes over the body and mind. Sleep deprivation caused by worry or fear can make it more difficult to perform routine tasks. Medications that cause drowsiness as a side effect can also impair cognition.

Long-term memory retrieval is easier to accomplish in old age than short-term memory retrieval. Assist the patient having short-term memory problems by using written lists, visual cues, and other memory-enhancing systems to aid in strengthening short-term memory skills.

Intelligence does not decline as one ages if it is measured using an appropriate instrument that focuses on accuracy and not on speed of response. Although cognitive abilities do tend to slow down with advancing age, they are not lost. Most of the subtle declines that do occur with information processing and retrieval do not need to interfere with the older patient's abilities in performing activities of daily living.

Nurses need to assist in the identification of individual problem situations and work with older patients and their families in developing strategies to better address their needs. It is common knowledge that health, good nutrition, and adequate sleep are important factors for brain functioning. These factors should be the cornerstones in planning care for the older patient.

Coping Abilities

How the individual chooses to adapt to a change in functional ability over time has a significant impact on how that individual will work through the entire maturational process called aging. In addition to normal aging changes, many older patients are coping with compounding changes that occur because of chronic disease. Nurses must not lose sight of the fact that in addition to changes caused by some decline in physical and cognitive functioning, older persons simultaneously work to deal with many societal and culturally perceived losses associated with advancing age. Changes in employment status and societal and family roles and shifts from independence to dependence may have a strong psychological impact on both older patients and significant others. With such a combination of "losses," an alteration in the confidence level of the older patient may be seen, requiring encouragement of self-care behaviors.

The older person's personality, attitude, past life experiences, and desire to adapt to change are all intrinsic influencing factors that assist the older patient in coping with changes brought on by advancing age. Extrinsic factors include things such as financial status, family support, and support provided by those who directly care for the patient. If older patients have the energy, desire, determination, and support of those who care for them, they will be better able to optimize use of their cognitive functions toward health.

Depression

There are times when the psychological impact of change is too difficult to cope with, and loneliness, grief, or sadness does not easily allow the older person to cognitively focus on health. When this happens, **depression** can result, with the potential to disable the older individual's mind and body. Depression is the most common psychiatric problem among older adults. This psychological condition, which causes a disturbance in mood, increases the risk for suicide, physical health complaints, and sleep disturbances.

It is important to understand that the frequency and intensity of depression generally increase with advancing age. Depression can result from physical changes in the brain caused by medications or conditions that affect the neurotransmitters in the brain or from psychological changes at an emotional level, such as maladaptive coping from a perceived loss. Regardless of the cause, depression is a condition that has the potential to be reversed with prompt identification and treatment. It is important, therefore, to be sensitive to what the older individual is and is not saying during communications. Patients should be referred for treatment if depression is suspected before maladaptive behaviors occur.

Dementia

Unlike depression, **dementia** involves a more permanent progressive deterioration of mental functioning. Dementia is often characterized by confusion, forgetfulness, impaired judgment, and personality changes. There are two main types of dementia: multi-infarct dementia and Alzheimer's disease.

The multi-infarct classification results from repeated strokes that affect the brain tissue. The onset for this condition can be sudden or gradual, but its course is marked by a cyclical worsening and lessening of signs and symptoms that vary with the intensity and location of the brain damage.

The cause of Alzheimer's disease is unknown, but symptoms gradually begin with impaired memory that progresses to language and motor function losses. The course of this disease is marked by stages that can occur up to 14 years before death. It is only on autopsy of the brain that a definitive diagnosis of Alzheimer's disease can be made. Until then, Alzheimer's is suspected after other types of dementia have been ruled out with testing.

With any dementia, help the older patient maintain an **optimum level of functioning** in an atmosphere that provides for physical and emotional safety. In efforts to help ground the confused patient, incorporate **reality orientation** or validation into all nursing interventions. Sensory overload should be decreased for confused patients. Speak calmly and slowly and provide nonthreatening therapeutic touch if accepted by the patient.

In addition, it is important to address the education and support needs of the patient's family as they learn to deal with changes in behavior demonstrated by the older patient. Refer any patient with confusion for a mental status examination to help in determining whether the person suffers from depression or dementia. It is essential that a sensory status examination be conducted on the older patient before a mental status examination is conducted to help ensure the accuracy of the test results.

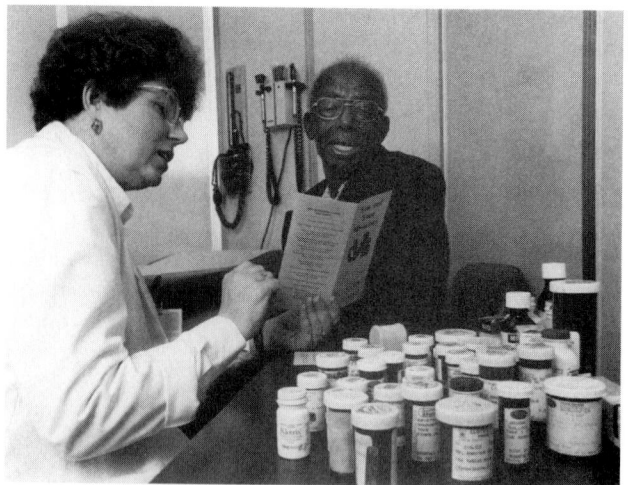

Figure 14-6 Older patients use many medications and must understand their correct use. (From Matteson, MA, and McConnell, ES: Gerontological Nursing. WB Saunders, Philadelphia, 1988, p 588, with permission.)

Sleep and Rest Patterns

The need for sleep in elderly patients does not decrease with age, but the pattern usually varies from earlier times in their lives. As in any age group, the lack of sleep leads to fatigue, irritability, increased sensitivity to pain, and increased likelihood of accidental behaviors. This is why it is important to obtain a baseline sleep and rest history when the older patient is admitted to the health care unit. Consider sleep patterns, bedtime rituals, rest and nap patterns, daily exercise patterns, stress level, dietary intake patterns, and lifestyle issues such as caffeine, alcohol, and nicotine intake when completing the sleep and rest assessment of an older patient.

Circulatory problems may disrupt normal sleep patterns for the older person and may be the only clue of an impending health problem. The anxious patient who is unable to sleep may be calmed with back rubs, foot rubs, a warm bath, warm milk, or a glass of wine, if not contraindicated, when patterned bedtime rituals prove to be unsuccessful. Sleep medications should be used only as a last resort because they can affect the quality or depth of the sleep the older person receives and may produce unwanted side effects.

Medication Management

One of the most difficult tasks for older patients, their family members, and the health care provider caring for them centers around the topic of medication management. Because older patients are more susceptible to drug-induced illness and adverse side effects for a variety of reasons already addressed in this chapter, nurses need to be especially aware of what the older patient is taking, how it is being taken, and what effect it is having.

Older patients use many medicines for various ailments (Fig. 14–6). Most have more than one chronic illness for which they take medications. Sometimes these different medications interact and produce side effects that can be dangerous. Not only do health care providers need to be concerned about prescribed medicines, they also need to look at the types of over-the-counter medicines older patients take, as well as the self-prescribed extracts, elixirs, herbal teas, cultural healing substances, and other home remedies commonly used by individuals of their age cohort.

Health care providers not only need to be concerned about overuse and combinations of medications, but also misuse of them. If an older patient crushes a large enteric-coated pill so that it can be taken in food and easily swallowed, it destroys the enteric protection and can inadvertently cause damage to the stomach and intestinal system. Because the U.S. health care system does not have a universal medication program for older individuals, some patients intentionally skip prescribed doses in efforts to save money. When prescribed doses are not being taken as expected, problems do not clear up as quickly and new problems may result.

Work closely with older patients and their families on medication education. Patients need to know what each prescribed pill is for, when it should be taken, how it should be taken, and side effects to report. As a patient advocate, work with the pharmacist to remedy administration concerns so that the prescribed dose can be taken as directed. Write down early medication side effects, educating patients or designated care providers to be proactive in their care.

Medication use, abuse, and misuse need to be addressed regularly with older persons. Concerns need to be closely monitored and addressed before they evolve into problem situations. Helping older patients consistently adhere to a prescribed medication routine with visual and verbal supports helps all those involved in the process of their medication management. This method also encourages self-care and supportive independence, as able, making for easier and safer medication use.

HEALTH PROMOTIONAL ROLE IN NURSING CARE OF THE OLDER PATIENT

Overall, the nurse who works with older patients must recognize the importance of developing strong skills in the areas of listening and observation. With competence in these skill areas, patterns of change can be readily identified and concerns can be addressed before they become problems for the elderly. Although this is important to consider for all patients who receive nursing care, it becomes especially important when working with older patients. This is because older persons often do not present with the typical symptom patterns associated with health problems seen in younger individuals, and early signs of health problems are often not evident in older patients.

By learning about normal changes expected with advancing age, older patients and their families can be helped to recognize commonly held societal myths associated with aging. In efforts to help older patients meet their optimum level of functioning, it is important to assess the older patient holistically, remembering that nursing assessment is an ongoing process. To assist in focusing on health promotion with older patients, look to health education activities, screening efforts for disease detection, immunization programs, and specific safety practices (Table 14–3).

TABLE 14-3 NURSING CARE FOCUS ON SAFETY ALPHABET FOR OLDER PATIENTS	
A is for ABILITIES	• Know your abilities. • Know patient's abilities. • Base nursing actions on your abilities. • Seek out assistance when needed.
B is for BODY MECHANICS AND ALIGNMENT	• Use proper body mechanics. • Use appropriate assistive devices. • Ensure patient is in proper body alignment.
C is for COMFORT	• Ensure physical and emotional comfort during care. • Use pain scale during each assessment.
D is for DELIBERATE MOVEMENTS	• Plan ahead and communicate plans to patient. • Demonstrate confidence during care. • Alert patient to planned movements by saying, "Moving on three. One, two, three." • Ensure patient assists with moves as able.
E is for ENVIRONMENT	• Always place call light within reach. • Keep environment uncluttered and safe. • Ask patient's permission before moving items. • Put items back as patient prefers.
F is for FALLS	• Remember falls are a primary concern for older patients. • Use interventions to prevent them: Assist patients with ambulation, use assistive devices, answer call lights promptly, provide accessible toileting facilities, use night-lights, avoid use of throw rugs.
G is for GIVING YOUR TIME	• Allow more time to perform actions. • Do not rush older patients. • Provide time for listening and observing, so concerns are addressed before becoming problems.
H is for HANDWASHING	• Use correct handwashing protocols to protect yourself and older patients. • Use standard precautions to protect yourself and older patients.

Answers to CRITICAL THINKING

Mr. Jones

1. Does he live alone? Does he have a fall history? If so, does he wear a device to signal for help? What type of night-light is used? How far is it to the bathroom? Does he take his bumetanide early in the day rather than at night? Does he void before going to bed? Does he anticipate needing to void 30 minutes after lying down?
2. Wood floors that are slippery when wet from incontinence are a safety hazard, as well as throw rugs that may slide or cause tripping. An appropriate night-light should be available.
3. Nursing diagnoses include functional incontinence related to distance to bathroom; deficient knowledge related to safety, medication administration, and nocturia; and risk for injury related to slippery floors from incontinence, use of throw rugs, lighting.

4. A teaching plan should include the following:
   • Safety: Place urinal at bedside to prevent incontinence on way to bathroom. Use red night-light to improve vision and prevent falls. Use easily cleaned floor covering that is secure and absorbent to avoid falls. Consider the need for wearing device that sends a signal for help.
   • Medication administration: Take diuretics early in the day to avoid having to get up frequently at night.
   • Nocturia: Void before lying down. Anticipate need to void after lying down by reclining in chair for 30 minutes before going to bed and then void on way to bed.

REVIEW QUESTIONS

1. The nurse is assessing a patient who says "I am shorter now." The nurse responds based on the understanding that which of the following commonly contributes to individuals becoming shorter with age?
   a. Contractures
   b. Bone degeneration in the legs
   c. Hyperextension of the cervical spine
   d. Water loss from spinal intervertebral disks

2. The nurse is caring for a patient on bedrest. The nurse is contributing to the plan of care based on the understanding that pressure ulcers occur at sites of ischemia, which can begin to develop in which of the following time frames?
   a. 5 to 10 minutes
   b. 20 to 40 minutes
   c. 60 minutes
   d. 120 minutes

3. The nurse is teaching a patient who has hypertension about the disease. Which of the following does the nurse understand contributes to increased blood pressure with age?
   a. Increased peripheral vascular resistance
   b. Decreased peripheral vascular resistance
   c. Increased cardiac output
   d. Decreased cardiac output

4. The nurse is planning care for a patient at risk of aspiration. Which one of the following does the nurse understand increases the older patient's risk for aspiration?
   a. Increased lung capacity
   b. Decreased lung capacity
   c. Decreased gag reflex
   d. Increased gag reflex

5. The nurse is assessing a patient's sexuality. The nurse understand that which one of the following is the most common factor related to sexual problems in older patients?
   a. Social
   b. Financial
   c. Physical
   d. Psychological

REFERENCES

1. U.S. Bureau of the Census: Demographic Data, 2000. Retrieved August 4, 2001, from http//:www.census.gov.

UNIT THREE BIBLIOGRAPHY

Abrams, W, and Berkow, R (eds.): The Merck Manual of Geriatrics. Merck Sharp and Dome Research Laboratories, Rahway, NJ, 2000.

Burke, M, and Walsh, M: Gerontologic Nursing: Wholistic Care of the Older Adult. Mosby, St. Louis, 1997.

do Rozario, L: Spirituality in the lives of people with disability and chronic illness: A creative paradigm of wholeness and reconstitution. Disabil Rehabil 19(10):427, 1997.

Ebersole, P, and Hess, P: Geriatric Nursing and Healthy Aging. Mosby, St. Louis, 2001.

Frankenfield, D, Cooney, RN, Smith, JS, and Rowe, WA: Age-related differences in the metabolic response to injury. J Trauma 48(1):49-56, 2000.

Hayes, M: A phenomenological study of chronic sorrow in people with type 1 diabetes. Pract Diabetes Int 18(2):65, 2001.

Kubusch, S, and Wichowski, H: Restoring power through nursing interventions. Nurs Diagn 8(1):7, 1997.

Lewis, KS: Emotional adjustment to a chronic illness. Lippincott's Primary Care Practice, 2(1):38, 1998.

Medaliem, JH: The patient and family adjustment to chronic disease in the home. Disabil Rehabil 19(4):163, 1997.

Miller, J: Coping with Chronic Illness. Overcoming Powerlessness. FA Davis, Philadelphia, 2000.

National Academy of Sciences: Dietary Reference Intakes. National Academy Press, Washington, DC, 1998.

Olshanshky, S: Chronic sorrow: A response to having a mentally defective child. Soc Casework 43:191-193, 1962.

Przytulski, K: Nutrition and Diet Therapy. FA Davis, Philadelphia, 2001.

Purnell, L, and Paulanka, B (eds.): Transcultural Health Care: A Culturally Competent Approach. FA Davis, Philadelphia, 2001.

Rankin, S, and Weekes, D: Life-span development: A review of theory and practice for families with chronically ill members. Scholar Inq Nurs Pract 14(4):355, 2000.

Russell RM, Rasmussen, H, and Lichtenstein, AH: Modified food guide pyramid for people over 70 years of age. J Nutr 129:751, 1999.

Thorson, J: Aging in a Changing Society, ed 2. Brunner/Mazel, Ann Arbor, MI, 2000.

Tremethick, MJ: Thriving, not just surviving. The importance of social support among the elderly. J Psychosoc Nurs Ment Health Serv 35(9):27, 1997.

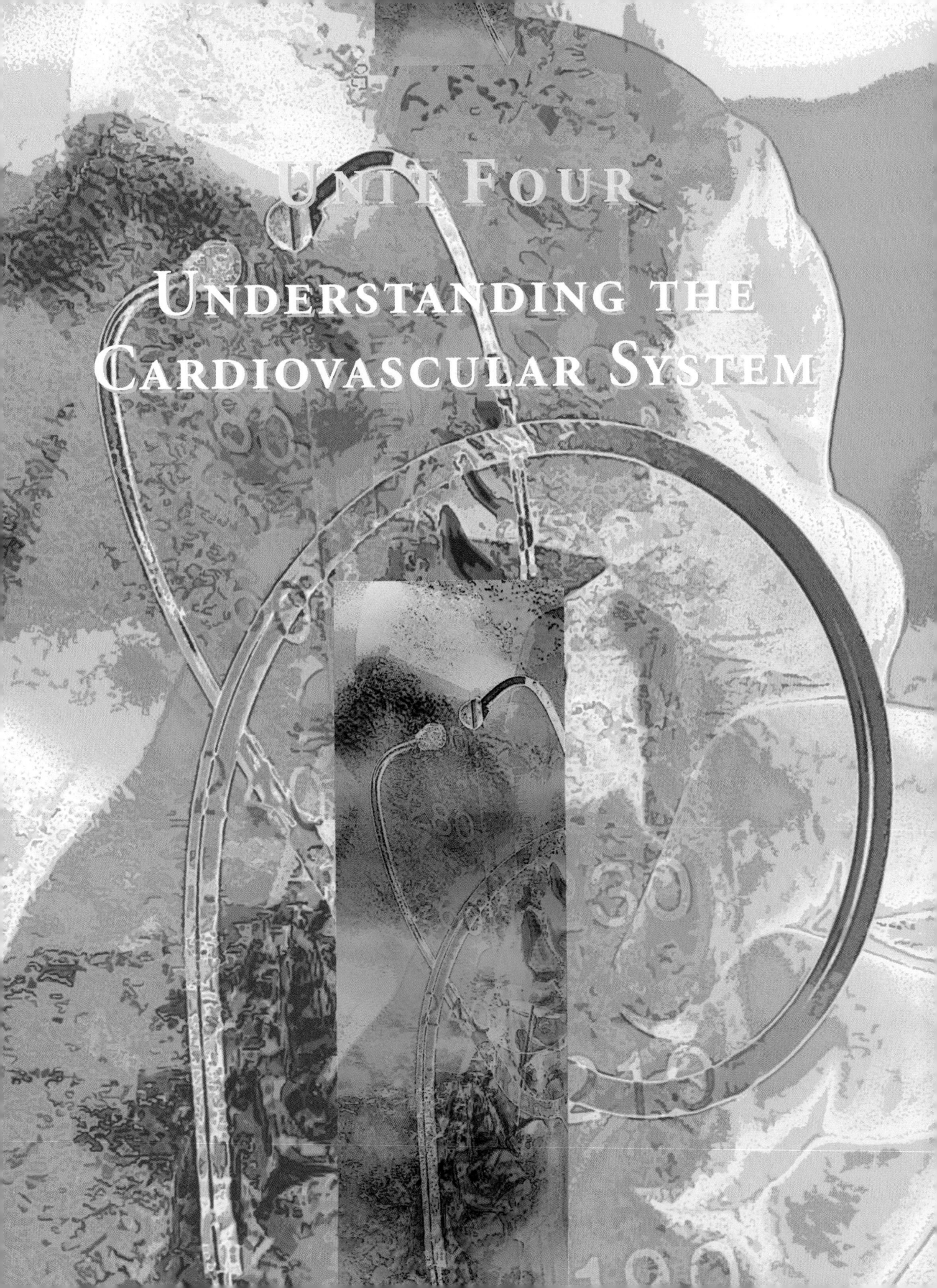

UNIT FOUR

UNDERSTANDING THE CARDIOVASCULAR SYSTEM

15

CARDIOVASCULAR SYSTEM FUNCTION, ASSESSMENT, AND THERAPEUTIC MEASURES

Linda S. Williams

KEY TERMS

arteriosclerosis
(ar-TIR-ee-oh-skle-**ROH**-sis)

atherosclerosis (ATH-er-oh-skle-**ROH**-sis)

bruit (brew-E)

claudication (KLAW-di-**KAY**-shun)

clubbing (**KLUB**-ing)

dysrhythmias (dis-**RITH**-mee-yahs)

Homans' sign (**HOH**-manz SIGHN)

ischemic (iss-**KEY**-mik)

murmur (**MUR**-mur)

pericardial friction rub
(PER-ee-**KAR**-dee-uhl **FRIK**-shun RUB)

poikilothermy (POY-ki-loh-**THER**-mee)

point of maximum impulse
(POYNT OF **MAKS**-i-muhm **IM**-puls)

preload (**PREE**-lohd)

pulse deficit (PULS **DEF**-i-sit)

thrill (THRILL)

QUESTIONS TO GUIDE YOUR READING

1. What is the normal anatomy of the cardiovascular system?

2. What is the normal function of the cardiovascular system?

3. What data should you collect when caring for a patient with a disorder of the cardiovascular system?

4. What are the diagnostic tests commonly performed to diagnose disorders of the cardiovascular system?

5. What nursing care should you provide for patients undergoing each of the diagnostic tests?

6. What are common therapeutic measures used for patients with disorders of the cardiovascular system?

▶ REVIEW OF NORMAL ANATOMY AND PHYSIOLOGY

The cardiovascular system consists of the heart and blood vessels, including arteries, capillaries, and veins. Its function is to pump and distribute blood throughout the body.

Heart
Cardiac Location and Pericardial Membranes

The heart is located in the mediastinum, the area between the lungs in the thoracic cavity. It is enclosed by three pericardial membranes that make up the pericardial sac. The outermost of these membranes is the fibrous pericardium, which forms a loose-fitting sac around the heart. The second, or middle, layer is the parietal pericardium, a serous membrane that lines the fibrous layer. The third and innermost layer, the visceral pericardium or epicardium, is a serous membrane on the surface of the heart muscle. Between the parietal and visceral layers is serous fluid, which prevents friction as the heart beats.

Cardiac Structure and Vessels

The walls of the four chambers of the heart are made of cardiac muscle (myocardium) and are lined with endocardium, which is smooth epithelial tissue that prevents abnormal

227

clotting. The endocardium also covers the valves of the heart and continues into blood vessels as the lining. The coronary vessels include the arteries and capillaries that circulate oxygenated blood throughout the myocardium and the veins that return unoxygenated blood to the heart. The two main coronary arteries are the first branches of the ascending aorta, just outside the left ventricle (Fig. 15–1).

The upper chambers of the heart are the thin-walled right atrium and left atrium, which are separated by the interatrial septum. The lower chambers are the thicker-walled right and left ventricles, which are separated by the interventricular septum. Each septum is made of myocardium that forms a common wall between the two chambers.

The right atrium receives deoxygenated blood from the upper body by way of the superior vena cava and from the lower body by way of the inferior vena cava. (See Fig. 15–1.) This blood flows from the right atrium through the tricuspid valve into the right ventricle. Backflow during ventricular systole (contraction and emptying) is prevented by the tricuspid or right atrioventricular (AV) valve (Fig. 15–2). The right ventricle pumps blood through the pulmonary semilunar valve to the lungs by way of the pulmonary artery. The pulmonary semilunar valve prevents backflow of blood into the right ventricle during ventricular diastole (relaxation and filling).

The left atrium receives oxygenated blood from the lungs by way of the four pulmonary veins. This blood flows through the mitral or left AV valve into the left ventricle.

The mitral valve prevents backflow of blood into the left atrium during ventricular systole. The left ventricle pumps blood through the aortic valve to the body by way of the aorta. The aortic valve prevents backflow of blood into the left ventricle during ventricular diastole.

The tricuspid and mitral valves consist of three and two cusps, respectively. These cusps, or flaps, are connective tissue covered by endocardium and are anchored to the floor of the ventricle by the chordae tendineae and papillary muscles. The papillary muscles are columns of myocardium that contract along with the rest of the ventricular myocardium. This contraction pulls on the chordae tendineae and prevents inversion of the AV valves during ventricular systole. (See Fig. 15–2.)

Although each ventricle pumps the same amount of blood, the much thicker walls of the left ventricle pump with approximately six times the force of the right ventricle to distribute the blood throughout the body. This difference in force is reflected in the great difference between systemic and pulmonary blood pressure.

Cardiac Conduction Pathway and Cardiac Cycle

The cardiac conduction pathway is the pathway of electrical impulses that generates a heartbeat. The sinoatrial (SA) node in the wall of the right atrium is a specialized mass of cardiac muscle that depolarizes rhythmically and most rap-

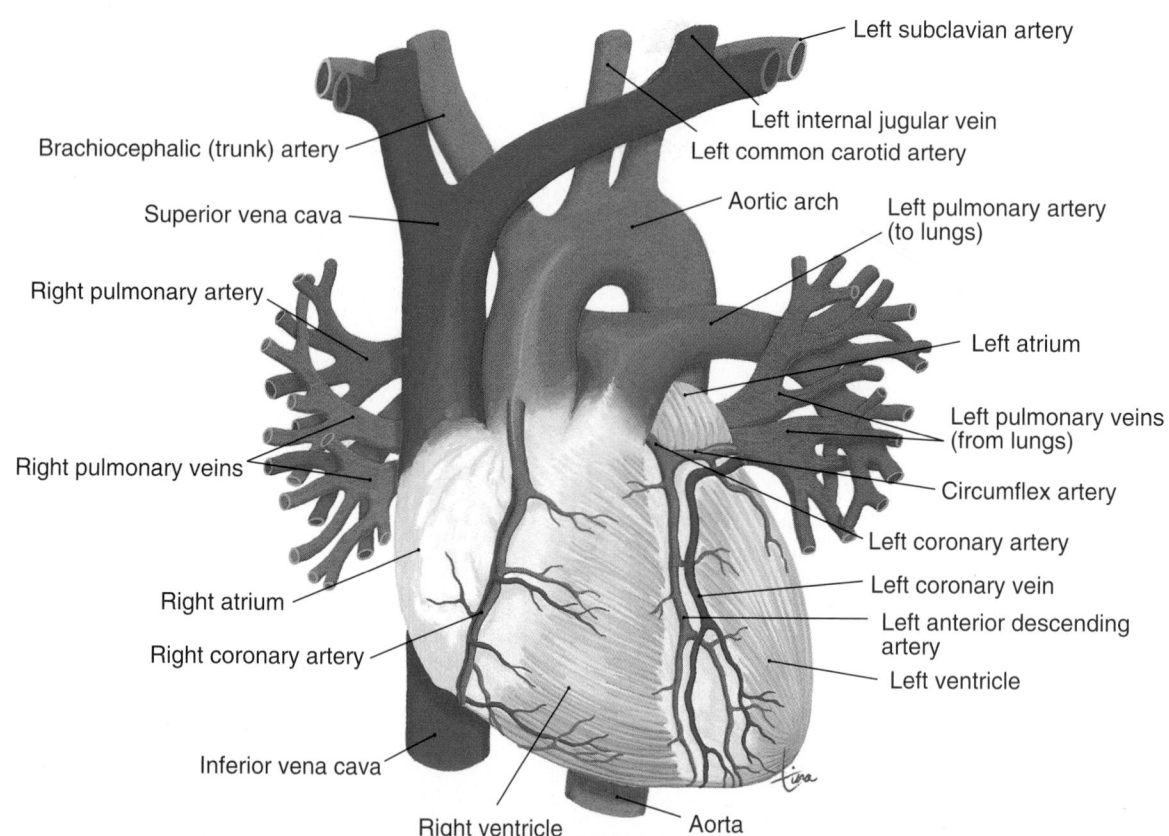

Figure 15-1 Anterior view of the heart and major blood vessels. (From Scanlon, V, Sanders, T: Essentials of Anatomy and Physiology, ed 3. FA Davis, Philadelphia, 1999, p 260.)

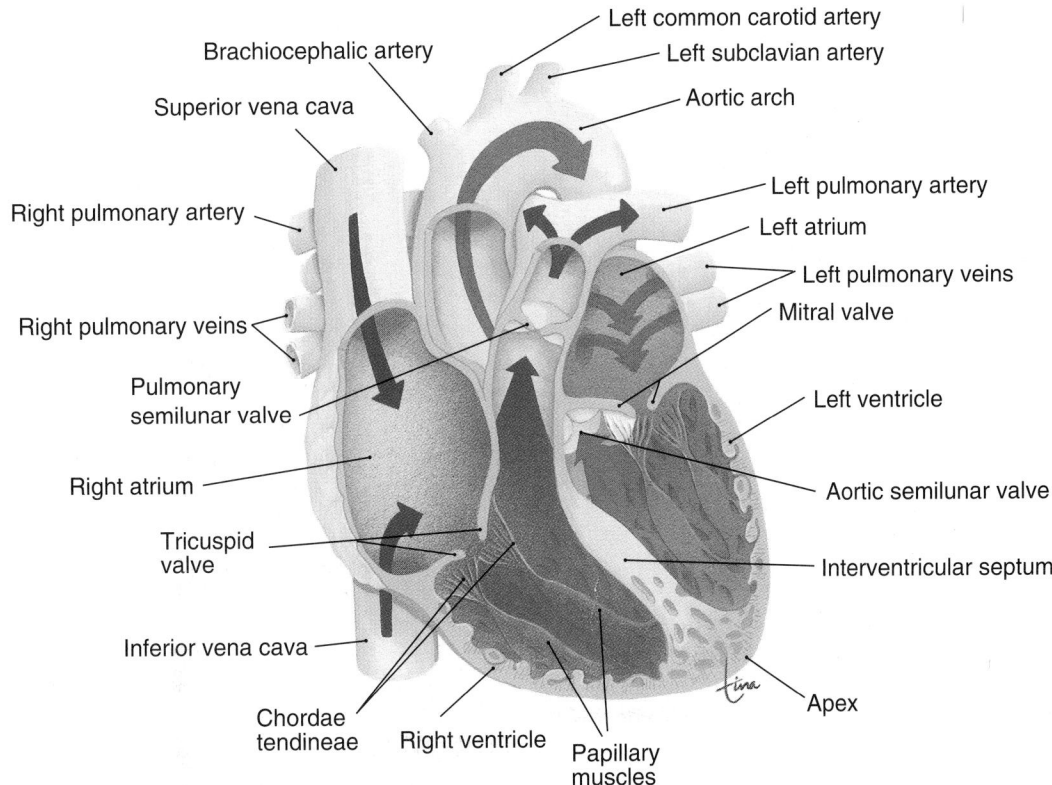

Figure 15-2 Frontal section of the heart showing internal structures and cardiac blood flow. (From Scanlon, V, Sanders, T: Essentials of Anatomy and Physiology, ed 3. FA Davis, Philadelphia, 1999, p 260.)

idly, 60 to 80 times per minute, and therefore initiates each heartbeat. For this reason it is sometimes called the pacemaker, and a normal heartbeat is called a normal sinus rhythm. From the SA node, impulses travel to the AV node located in the lower interatrial septum, to the bundle of His in the upper interventricular septum, to the right and left bundle branches in the septum, and to the Purkinje fibers in the rest of the ventricular myocardium. If the SA node becomes nonfunctional, the AV node can initiate each heartbeat, but at a slower rate of 40 to 60 beats per minute. The bundle of His is capable of generating the beat of the ventricles, but at the much slower rate of 15 to 40 beats per minute.

A cardiac cycle is the sequence of mechanical events during one heartbeat. Simply stated, the two atria contract simultaneously, followed a fraction of a second later by the simultaneous contraction of the two ventricles. The contraction, or systole, of each set of chambers is followed by relaxation, or diastole, of the same set of chambers.

The atria in diastole continually receive blood from the veins. As pressure in the atria increases, the AV valves are forced open, causing most of the blood to flow passively into the ventricles. Atrial systole, referred to as atrial kick, pumps the remaining blood into the ventricles, and then the atria relax. Ventricular systole follows. The pressure in the ventricles causes the AV valves to close and forces the semilunar valves to open. Blood is then pumped into the aorta and pulmonary artery. There is no passive blood flow. Any blood leaving the ventricles must be pumped. Toward the very end of ventricular systole, as the pressure drops, the

blood tends to flow backward. It is this backflow of blood that closes the semilunar valves. The ventricles and atria are then all in diastole; the atria continue to fill until pressure opens the AV valves again, and the cycle is repeated.

The events of the cardiac cycle create the normal heart sounds. The first of the two major sounds (the lubb of lubb-dupp) is caused by the closure of the AV valves during ventricular systole. The second sound is created by the closure of the aortic and pulmonary semilunar valves.

Cardiac Output

Cardiac output is the amount of blood pumped by the left ventricle in 1 minute (the right ventricle pumps a similar amount); it is determined by multiplying stroke volume by pulse. Stroke volume is the amount of blood pumped by a ventricle in one beat and averages 60 to 80 mL. With an average resting heart rate of 70 beats per minute, an average resting cardiac output is 5 to 6 L (approximately the total blood volume of an individual that is pumped within 1 minute). During exercise, venous return increases and stretches the ventricular myocardium, which in response contracts more forcefully. This is known as Starling's law of the heart, and the result is an increase in stroke volume. More blood is pumped with each beat, and at the same time, the heart rate increases, causing cardiac output to increase to as much as four times the resting level, and even more for athletes.

The ejection fraction is a measure of ventricular efficiency and is usually about 60 percent. It is the stroke volume divided by total blood in the ventricle (also known as the end-diastolic volume, which is approximately 120 to 130 mL). Lower values indicate that the ventricle is not pumping as forcefully and that more blood remains in the ventricle at the end of systole. The normal end-systolic volume is about 50 to 60 mL.

Regulation of Heart Rate

The heart generates its own electrical impulse, which begins at the SA node. The nervous system, however, can change the heart rate in response to environmental circumstances (Fig. 15–3). In the brain the medulla contains the cardiac centers: the accelerator center and the inhibitory center. Sympathetic nerve impulses—along sympathetic nerves from the thoracic spinal cord to the SA node, AV node, and ventricular myocardium—increase rate and force of contraction. Parasympathetic impulses—along the vagus nerve to the SA node and AV node—decrease heart rate.

The information for changes necessary in the heart rate comes to the medulla from pressoreceptors and chemoreceptors located in the internal carotid arteries and the aortic arch. The pressoreceptors, specialized cells in the carotid and aortic sinuses, detect changes in blood pressure. The chemoreceptors are located in the carotid and aortic bodies and are cells specialized to detect changes in the oxygen content of the blood. In response to either a drop in blood pressure or a decrease in blood oxygen level, the heart receives sympathetic impulses and beats faster in an attempt to provide sufficient oxygenation for tissues.

Hormones and the Heart

The hormone epinephrine, secreted by the adrenal medulla in stressful situations, is sympathomimetic in that it increases the heart rate and force of contraction and it dilates the coronary vessels. This in turn increases cardiac output and systolic blood pressure.

Aldosterone, a hormone produced by the adrenal cortex, is important for cardiac function because it helps regulate blood levels of sodium and potassium, both of which are needed for the electrical activity of the myocardium. The blood level of potassium is especially critical because even a small deficiency or excess impairs the rhythmic contractions of the heart.

The atria of the heart secrete a hormone of their own called atrial natriuretic peptide (ANP) or atrial natriuretic hormone (ANH). As its name suggests, ANP increases the excretion of sodium by the kidneys, perhaps by inhibiting secretion of aldosterone by the adrenal cortex or renin by the kidneys. Atrial natriuretic peptide is secreted when a higher blood pressure or greater blood volume stretches the walls of the atria. The loss of sodium is accompanied by the loss of more water in urine, which decreases blood volume and perhaps blood pressure as well.

Blood Vessels
Arteries and Veins

Arteries and arterioles carry blood from the heart to capillaries. Their walls are relatively thick and consist of three layers. Arteries carry blood under high pressure, and the outer layer

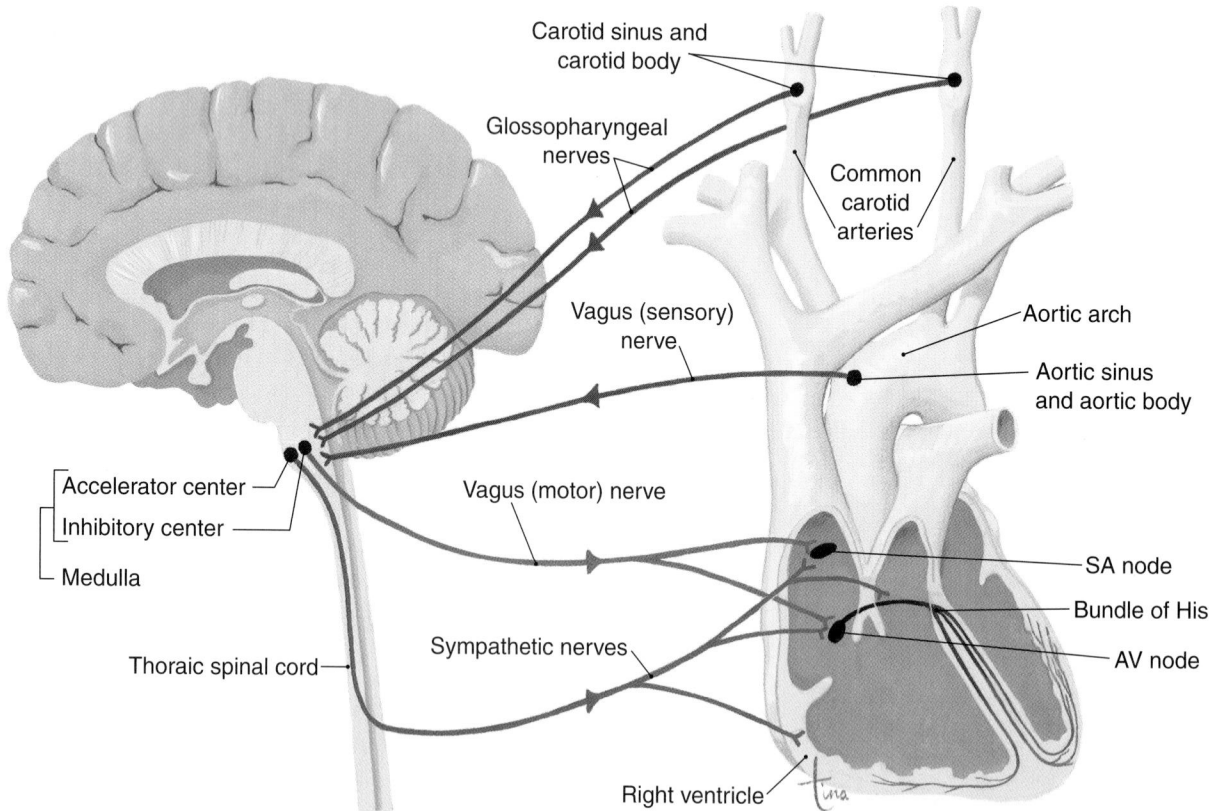

Figure 15-3 Nervous system regulation of the heart. (From Scanlon, V, Sanders, T: Essentials of Anatomy and Physiology, ed 3. FA Davis, Philadelphia, 1999, p 269.)

of fibrous connective tissue prevents rupture of the artery. The middle layer of smooth muscle and elastic connective tissue contributes to the maintenance of normal blood pressure (BP), especially diastolic BP, by changing the diameter of the artery. The diameter of arteries is regulated primarily by the sympathetic division of the autonomic nervous system. The lining is simple squamous epithelium, called endothelium, which is very smooth to prevent abnormal clotting.

Veins and venules carry blood from capillaries to the heart. Their walls are relatively thin because there is less smooth muscle (veins do not have as important a role in the maintenance of BP as arteries). Sympathetic impulses can bring about extensive constriction of veins, however, and this becomes important in situations such as severe hemorrhage. The lining of veins is endothelium that prevents abnormal clotting; at intervals it is folded into valves to prevent backflow of blood. Valves are most numerous in the veins of the extremities, especially the legs, where blood must return to the heart against the force of gravity.

Capillaries

Capillaries carry blood from arterioles to venules and form extensive networks in most tissues. The exceptions are cartilage, the epidermis, and the lens and cornea of the eye. Their walls, a continuation of the lining of arteries and veins, are one cell thick to permit the exchanges of gases, nutrients, and waste products between the blood and tissues (Fig. 15–4). Blood flow through a capillary network is regulated by a precapillary sphincter, a smooth muscle cell that contracts or relaxes in response to tissue needs. In an active tissue such as exercising skeletal muscle, for example, the rapid oxygen uptake and carbon dioxide production causes dilation of the precapillary sphincters to increase blood flow. At the same time, precapillary sphincters in less active tissues constrict to reduce blood flow. This is important because there is not enough blood in the body to fill all the capillaries at once; the fixed volume must constantly be shunted or redirected to where it is needed most.

The blood pressure in capillaries is 30 to 35 mm Hg at the arterial end of the network, and it drops to about 15 mm Hg at the venous end. This pressure is low enough to prevent rupture of the capillaries but high enough to permit filtration. Tissue fluid is formed from the plasma in capillaries by the process of filtration. Because capillary blood pressure is higher than the pressure of the surrounding tissue fluid, plasma and dissolved materials such as nutrients are forced through the capillary walls to become tissue fluid. Some of this tissue fluid returns to the capillaries, and some is collected in lymph capillaries. Now called lymph, it too is returned to the blood, by the system of lymph vessels. Should blood pressure within the capillaries increase, more tissue fluid than usual is formed, which is too much for the lymph vessels to collect. This is called edema.

Blood Pressure

Blood pressure is the force of the blood against the walls of the blood vessels and is measured in mm Hg, systolic over diastolic. The normal range of systemic arterial pressure is 90 to 135/60 to 85 mm Hg. Blood pressure decreases in the arterioles and capillaries, and the systolic and diastolic pressures merge into one pressure. As blood enters the veins, BP decreases further and approaches zero in the caval veins. As mentioned previously, the blood pressure in the capillaries is of great importance, and normal blood pressure is high enough to permit filtration for nourishment of tissues but low enough to prevent rupture.

The arteries and veins are usually in a state of slight constriction that helps to maintain normal blood pressure, especially diastolic pressure. This is called peripheral resistance; it is regulated by the vasomotor center in the medulla, which generates impulses along sympathetic vasoconstrictor nerves to all vessels with smooth muscle to maintain slight constriction. More impulses per second increase vasoconstriction and raise blood pressure; fewer impulses per second bring about vasodilation and lower blood pressure. The information for changes needed in the vessel diameter comes to the medulla from the pressoreceptors and chemoreceptors located in the internal carotid arteries and aortic arch.

Blood pressure is also affected by many other factors. If heart rate and force increase, blood pressure increases up to a point. If the heart is beating very fast, the ventricles are not filled before they contract, cardiac output decreases, and blood pressure drops. The strength of the heart's contractions depends on adequate venous return, which is the amount of blood that flows into the atria. Decreased venous return results in weaker contractions.

Venous return depends on several factors: constriction of the veins so that blood does not pool in them, the skeletal muscle pumping to squeeze the deep veins of the legs, and the muscles of respiration compressing and expanding the veins in the chest cavity. The valves in the veins prevent backflow of blood and thus contribute to the return of blood to the heart.

The elasticity of the large arteries also contributes to normal blood pressure. When the left ventricle contracts, the blood stretches the elastic walls of the large arteries, which absorb some of the force. When the left ventricle relaxes, the walls recoil or snap back and put pressure on the blood. Normal elasticity, therefore, lowers systolic pressure, raises diastolic pressure, and maintains normal pulse pressure. Pulse pressure is the difference between the systolic and diastolic pressures. The usual ratio of systolic to diastolic to pulse pressure is 3:2:1.

Renin-Angiotensin-Aldosterone Mechanism

The kidneys are of great importance in the regulation of blood pressure. If blood flow through the kidneys decreases, renal filtration decreases and urinary output decreases to preserve blood volume. Decreased blood pressure stimulates the kidneys to secrete renin, which initiates the renin-angiotensin mechanism (Fig. 15–5). Renin splits the plasma protein angiotensinogen (from the liver) to form angiotensin I, which is changed to angiotensin II by a converting enzyme found primarily in lung tissue. Angiotensin

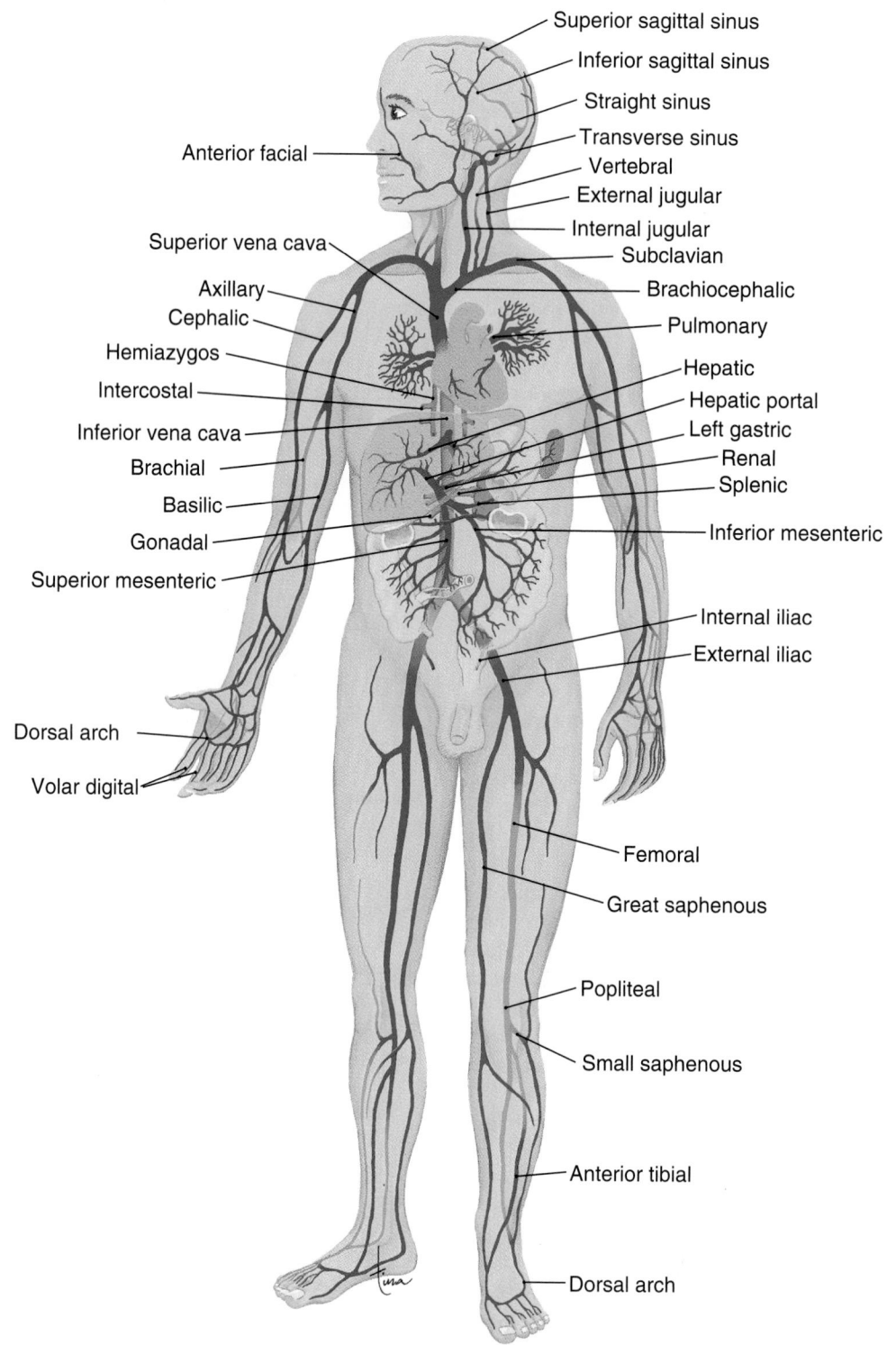

Figure 15-4 Structure of an artery, arteriole, capillary network, venule, and vein. (From Scanlon, V, Sanders, T: Essentials of Anatomy and Physiology, ed 3. FA Davis, Philadelphia, 1999, p 277.)

II causes vasoconstriction and stimulates secretion of aldosterone, both of which raise blood pressure.

Aldosterone, secreted by the adrenal cortex, increases the reabsorption of sodium ions by the kidneys. Water follows the sodium back to the blood; this increases blood volume and blood pressure. Other hormones that affect BP include those of the adrenal medulla: norepinephrine, which causes vasoconstriction throughout the body, and epinephrine, which increases cardiac output and causes vasoconstriction in skin and viscera. Antidiuretic hormone (ADH), from the posterior pituitary, directly increases water reabsorption by the kidneys, thus increasing blood volume and blood pressure. Atrial natriuretic peptide, secreted by the atria of the heart, is believed to inhibit aldosterone and

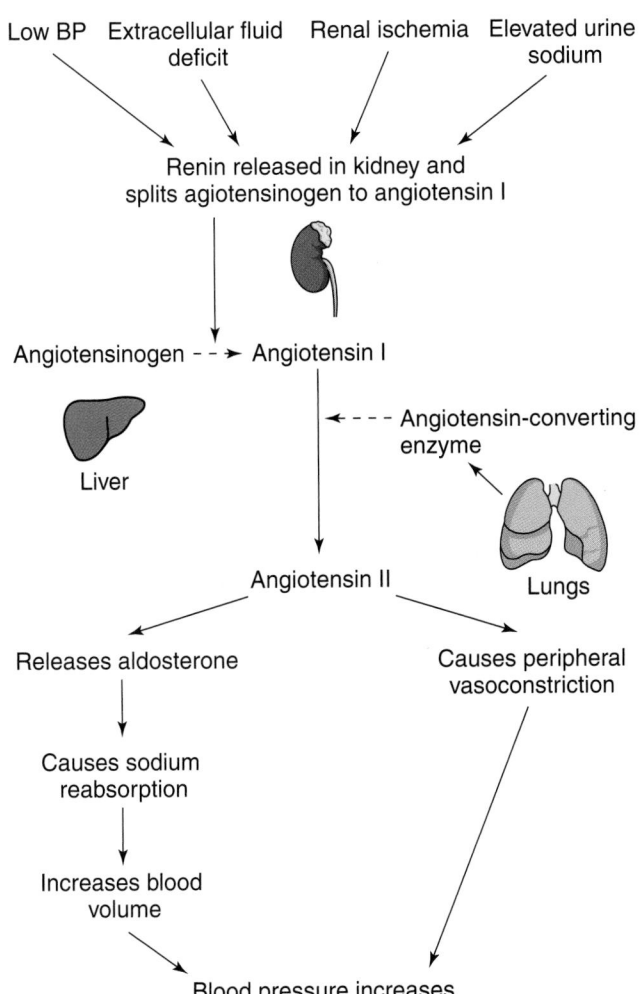

Low BP   Extracellular fluid   Renal ischemia   Elevated urine
         deficit   sodium

Renin released in kidney and
splits agiotensinogen to angiotensin I

Angiotensinogen - - ▸ Angiotensin I

Liver

◂ - - - Angiotensin-converting
         enzyme

Lungs

Angiotensin II

Releases aldosterone   Causes peripheral
                                        vasoconstriction

Causes sodium
reabsorption

Increases blood
volume

Blood pressure increases

Figure 15-5 The renin-angiotensin aldosterone mechanism.

renin secretion and thereby increase renal excretion of sodium ions and water, which decreases blood volume and subsequently blood pressure.

Pathways of Circulation

The two pathways of circulation are pulmonary and systemic. (See Fig. 15–2.) Pulmonary circulation begins at the right ventricle, which pumps deoxygenated blood into the pulmonary artery. The pulmonary artery branches into two arteries, one to each lung. The pulmonary capillaries around the alveoli of the lungs are the site of gas exchange. Oxygenated blood returns to the left atrium by way of the pulmonary veins. The blood pressure in the pulmonary circulation is always low because the right ventricle pumps with only about one-sixth the force of the left ventricle. The arterial pressure is approximately 20 to 25/8 to 10 mm Hg, and the pulmonary capillary pressure is lower still. This is important to prevent filtration in pulmonary capillaries, which keeps tissue fluid from accumulating in the alveoli of the lungs, causing pulmonary edema.

Systemic circulation begins in the left ventricle, which pumps oxygenated blood into the aorta, the many branches of which give rise to capillaries within the tissues. Deoxy-

genated blood returns to the right atrium by way of the superior and inferior vena cava. The hepatic portal circulation is a special part of the systemic circulation in which blood from the capillaries of the digestive organs and spleen flows through the portal vein and into the capillaries (sinusoids) in the liver before returning to the heart. This pathway permits the liver to regulate the blood levels of nutrients such as glucose, amino acids, and iron and to remove potential toxins such as alcohol or medications from circulation.

Aging and the Cardiovascular System

It is believed that the "aging" of blood vessels, especially arteries, begins in childhood, although the effects are not apparent until later in life. **Atherosclerosis** is the deposition of lipids on and in the walls of arteries over a period of years, which narrows their lumens, decreases blood flow, and forms rough surfaces that may stimulate intravascular clot formation. **Arteriosclerosis** is the gradual deterioration of the walls of arteries as a result of the pressure they have withstood over many decades. Average resting blood pressure tends to increase with age and may contribute to stroke or left-sided heart failure. The thinner-walled veins, especially those of the legs, may also weaken and stretch, making their valves incompetent.

With age, the heart muscle becomes less efficient, and there is a decrease in both maximum cardiac output and heart rate, although resting levels may be more than sufficient (Gerontological Issues Box 15–1). The health of the

atherosclerosis: athere—porridge + sklerosis—hardness
arteriosclerosis: arteriaartery + sklerosis—hardness

BOX 15-1   GERONTOLOGICAL ISSUES

Cardiovascular changes that occur in the older adult include the following:

- Decrease in heart size
- Thickening of left ventricular wall
- Increased collagen in the cardiac muscle
- Decreased elastin in the cardiac muscle
- Stiffer and thicker cardiac valves
- Fibrosis of the SA node
- Decreased number of pacemaker cells
- Calcification of blood vessels
- Loss of arterial distensibility
- More tortuous vessels
- Decreased response to baroreceptors

These age-related changes increase the risk of developing the following:

- Sinus bradycardia
- Atrial fibrillation
- Atherosclerotic heart disease
- Elevated systolic blood pressure
- Decreased cardiac output and cardiac reserve
- Hypertension
- Peripheral vascular disease
- Postural hypotension
- Vagal syncope

myocardium depends on its blood supply, and with age there is greater likelihood that atherosclerosis will narrow the coronary arteries. Hypertension causes the left ventricle to work harder, and it may hypertrophy and outgrow its blood supply. The heart valves may become thickened by fibrosis, leading to heart **murmur. Dysrhythmias** are more common in the elderly as the cells of the conduction pathway become less efficient.

▶ CARDIOVASCULAR DISEASE

Cardiovascular disease is a major health concern in the United States according to statistics of the American Heart Association. In 1998 the single leading cause of death in America was coronary heart disease (CHD), which was responsible for one of every five U.S. deaths. About 220,000 people a year die of sudden death without ever reaching the hospital, usually because of ventricular fibrillation. The greatest cause of death in women is cardiovascular disease, mainly from CHD.

With cardiovascular disease a significant health problem in the United States, efforts to reduce it that target those especially at risk are greatly needed. The prevalence of CHD is greatest for Mexican-American men and African-American women, followed by African-American men and Mexican-American women, and then non-Hispanic white men and women. The incidence of angina is greater in women than in men, with African-American women having the greatest incidence. High blood pressure, the silent killer, occurs in one in four adults, with approximately one-third unaware that they have it. Non-Hispanic blacks are more likely to suffer from high blood pressure than are non-Hispanic whites.

Lifestyle plays a major role in risk factors for cardiovascular disease. Smoking contributes to approximately one in five cardiovascular diseases deaths. Dietary fat intake accounts for approximately 34 percent of the American diet, which increases cholesterol. Americans continue to be sedentary. Exercise needs to be promoted for all, including children, to reduce cardiovascular disease. For more information on cardiovascular disease statistics visit www. americanheart.org.

▶ NURSING ASSESSMENT OF THE CARDIOVASCULAR SYSTEM

Nursing assessment of the cardiovascular system includes a patient health history and physical examination (Gerontological Issues Box 15–2). If the patient is experiencing an

---

dysrhythmia: dys—difficult or abnormal + rhythmia—rhythm

BOX 15-2	GERONTOLOGICAL ISSUES

Older patients commonly have signs and symptoms atypical of a disorder. For example, the only symptom of myocardial infarction in an older patient may be dyspnea. Chest pain, a typical symptom, may not be present.

---

acute problem, the nursing assessment should focus on the most serious signs and symptoms and physical assessment data until the patient is stabilized (Table 15–1). An in-depth nursing assessment can be completed when the patient is stable. For stable or chronic cardiac conditions, a complete nursing assessment is done on admission. The nurse provides privacy and makes the patient as comfortable as possible before beginning the nursing assessment.

Subjective Data

To understand the patient's cardiovascular problems, the nurse asks about past and current symptoms, use of prescribed and over-the-counter medications, use of recreational drugs, surgeries, treatments, and such risk factors as diet, activity, tobacco use, and recent stressors. Assessment of symptoms includes asking questions for WHAT'S UP?: where it is, how it feels, aggravating and alleviating factors, timing, severity, useful data for associated symptoms, and perception by the patient of the problem.

Health History

The entire body can be affected by cardiovascular problems. Some symptoms can have more than one cause. For example, shortness of breath can be the result of either heart failure or chronic obstructive pulmonary disease. The health history can help determine the cause of the symptom. For cardiovascular problems, the assessment focuses on the areas listed in Table 15–2.

TABLE 15-1	ACUTE CARDIOVASCULAR NURSING ASSESSMENT
History	**Significance**
Allergies	For medication administration, diagnostic dyes
Smoking history	Risk factor for cardiovascular disorders
Medications	Toxic levels; influencing symptoms
Pain	Location—chest, calf; radiation—arms, jaw, neck; description—pressure, indigestion, tightness, burning, angina, myocardial infarction, thrombus, embolism
Dyspnea	Left-sided heart failure; pulmonary edema or embolism
Fatigue	Decreased cardiac output
Palpitations	Dysrhythmias
Dizziness	Dysrhythmias
Weight gain	Right-sided heart failure
Physical Assessment	**Possible Abnormal Findings**
Vital signs	Bradycardia, tachycardia, hypotension, hypertension, tachypnea, apnea, shock
Heart rhythm	Dysrhythmias
Edema	Right-sided heart failure
Jugular vein distention	Right-sided heart failure
Breath sounds	Crackles, wheezes with left-sided heart failure
Cough, sputum	Acute heart failure—dry cough, pink frothy sputum

TABLE 15-2 CARDIOVASCULAR HISTORY ASSESSMENT

Question	Rationale
Pain: WHAT'S UP? Format	
• Where is pain? Does it radiate?	• Cardiac pain may radiate to shoulders, neck, jaw, arms, or back. Vascular disorders cause extremity pain.
• How does it feel? Discomfort, burning, aching, indigestion, squeezing, pressure, tightness, heaviness, numbness in chest area? Fullness, heaviness, sharpness, throbbing in legs?	• Pain can be associated with angina or MI. The quality of pain varies. Venous pain is a fullness or heaviness. Sharpness or throbbing is arterial pain.
• Aggravating/alleviating factors that increase/relieve the pain?	• Activity may cause or increase angina. Rest or medications may relieve angina. Leg activity pain, intermittent claudication, results from decreased perfusion that is aggravated by activity. Rest pain, from severe arterial occlusion, increases when lying. Dangling reduces the pain because blood flow is increased by gravity.
• Timing of pain: onset, duration, frequency?	• Pain may be continuous, intermittent, acute, or chronic. Arterial occlusion causes acute pain.
• Severity of pain?	• Rate pain on a scale of 0 to 10.
• Useful data for associated symptoms?	• Accompanying symptoms and their characteristics guide diagnosis and treatment.
• Perception of patient about problem?	• Patient's insight to problem is helpful in planning care.
Level of Consciousness (LOC)	
• What is your name? What is the month? Year? Where are you now?	• A lack of oxygen caused by cardiac disease can decrease LOC.
Dyspnea	
• Are you short of breath? What increases your shortness of breath? What relieves your shortness of breath?	• Dyspnea can be present with heart failure that reduces cardiac output, on exertion in angina pectoris or from a pulmonary embolus resulting from thrombophlebitis, heart failure, or dysrhythmias.
Palpitations	
• Are you having palpitations or irregular heartbeat? Does your heart ever race, skip beats, or pound?	• Palpitations can occur from dysrhythmias resulting from ischemia, electrolyte imbalance, or stress. Dizziness can be associated with dysrhythmias.
Fatigue	
• Have you noticed a change in your energy level?	• Fatigue occurs from reduced cardiac output resulting from heart failure.
• Are you able to perform activities that you would like to?	• Functional abilities can be limited from fatigue.
Edema	
• Have you had any swelling in your feet, legs, or hands?	• Right-sided heart failure can cause fluid accumulation in the tissues.
• Have you gained weight?	• Fluid retention causes weight gain.
Paresthesia/Paralysis	
• Any numbness, tingling, or other abnormal sensations in extremities?	• Numbers and tingling, pins and needles, and crawling sensations are paresthesia.
• Can you move your extremity?	• Paralysis is inability to move extremity. Reduced nerve conduction from decreased oxygen supply causes paresthesia and paralysis.

MEDICAL HISTORY. If previous medical records are available, they can provide objective patient data that can be supplemented with patient responses. If medical records are not available, the patient is asked about previous conditions that could affect the cardiovascular system. A history of childhood illnesses that can lead to heart disease, such as rheumatic fever or scarlet fever, is noted. Other conditions include pulmonary disease, hypertension, kidney disease, cerebral vascular accident or brain attack, transient **ischemic** attack, renal disease, anemia, streptococcal sore throat, congenital heart disease, thrombophlebitis, and alcoholism. Patient allergies, previous hospitalizations, and surgeries are documented. Baseline diagnostic tests are helpful for comparison with current tests. Functional limitations that are related to cardiovascular problems, such as performing activities of daily living (ADLs), walking, climbing stairs, or completing household tasks, are also assessed.

MEDICATIONS. Medication use is noted. This includes prescription drugs, over-the-counter medications such as aspirin that can prolong clotting time, and recreational drugs. The medication history includes the patient's understanding of

---

ischemic: ischein—hold back + haima—blood

the medication and the medication name, dosage, reason for taking, last dose, and length of use.

FAMILY HISTORY. A family history of cardiovascular conditions is assessed because many cardiac problems are hereditary. Health histories of close relatives, such as parents, siblings, and grandparents, are the most significant.

HEALTH PROMOTION. Risk factors such as diet, activity, tobacco use, and recent stressors for the patient are assessed in the health history. The patient's health promotion activities are noted, especially for risk factors that are modifiable through changes in lifestyle.

Objective Data
Physical Assessment

The patient's general appearance is observed. The patient's level of consciousness, which is an indicator of oxygenation of the brain, is assessed. Height and weight are recorded. Vital signs are measured.

BLOOD PRESSURE. The average adult reading is 120/80 with a normal range of 100/60 to 140/90. For accurate measurement, the correct size cuff for the patient is used. Readings in both arms are done for comparison. A difference in the readings is reported to the physician. The arm with the higher reading is used for ongoing measurements. The leg may be used if necessary, with a larger blood pressure cuff. The reading in the leg is normally 10 mm Hg higher than in the arm.

Blood pressure measurements are done with the patient lying, sitting, and standing to detect abnormal variations with postural changes. When the patient sits or stands, a drop in the systolic pressure of up to 15 mm Hg and either a drop or slight increase in the diastolic pressure of 3 to 10 mm Hg is normal. In response to the drop in blood pressure, the pulse increases 15 to 20 beats per minute to maintain cardiac output. Orthostatic hypotension, also referred to as postural hypotension, is a greater than normal change in these pressures and indicates a problem that should be investigated by the physician. The patient may experience dizziness when changing positions, so fall prevention methods should be used during blood pressure measurement.

PULSES. The apical pulse is auscultated for 1 minute to assess rate and rhythm. Normal heart rate is 60 to 100 beats per minute. In athletic people the heart rate is often slower, around 50 beats per minute, because the well-conditioned heart pumps more efficiently. Apical pulse rhythm is documented as regular or irregular. The apical rate can be compared with the radial rate to assess equality. If there are fewer radial beats than apical beats, a **pulse deficit** exists and should be reported to the physician.

Arterial pulses are palpated for volume and pressure quality. They are palpated bilaterally and compared for equality. A normal vessel feels soft and springy. A sclerotic vessel feels stiff. The quality of the pulses is described on a 4-point scale as follows: 0 is absent; 1+ is weak, thready; 2+ is normal; and 3+ is bounding. An absent pulse is not pal-

---

LEARNING TIPS

Anticipate potential drops in blood pressure with position changes. Orthostatic or postural hypotension is a drop in systolic blood pressure greater than 15 mm Hg, a drop or slight increase in the diastolic blood pressure greater than 10 mm Hg, and an increase in heart rate greater than 20 beats per minute in response to the drop in blood pressure. It can be found in patients of any age but is most commonly found in the older patient. The patient often reports light-headedness or syncope because the drop in pressure decreases the amount of oxygen-rich blood traveling to the brain. The change from a lying to a sitting or a sitting to a standing position may cause the drop in pressure. This blood pressure drop increases the risk of fainting and falling. Factors that can contribute to orthostatic hypotension include fluid volume deficit, diuretics, analgesics, and pain.

To assess orthostatic hypotension, do the following:

1. Take patient's lying blood pressure and heart rate.
2. Assist patient to sitting position. Ask if dizzy or light-headed with each position change. If yes, ensure safety from fainting or falling. A gait or walking belt can be used. With any position change, if patient experiences additional symptoms with the dizziness and decreased blood pressure and increased heart rate, assist the patient into bed, take blood pressure, and notify the physician. Consider the possible cause of the orthostatic hypotension—hemorrhaging, dehydration, diuretics—to plan patient care.
3. Wait at least 1 minute, and then take patient's sitting blood pressure and heart rate. If patient remains dizzy or light-headed, continue sitting position for 5 minutes if tolerated. Do not attempt to bring the patient to standing. Repeat sitting blood pressure. If blood pressure has increased and patient is no longer dizzy, assist patient to stand.
4. Assist patient to stand. Wait at least 1 minute, and then take patient's standing blood pressure and heart rate. If blood pressure drops and patient is dizzy or light-headed, continue standing position for 5 minutes, if tolerated. Do not attempt to ambulate patient. Repeat standing blood pressure and heart rate.
5. Document all heart rate and blood pressure measurements, including extremity used and patient position when reading was obtained (e.g., right arm lying 132/78, sitting 118/68, standing 110/60). Also document patient tolerance, symptoms, and nursing interventions if symptomatic.
6. Report abnormal findings to physician.

---

pable. A thready pulse is one that disappears when slight pressure is applied and returns when the pressure is removed. The normal pulse is easily palpable. The bounding pulse is strong and present even when slight pressure is applied. When the normal vessel is palpated, a tapping is felt. In the abnormal vessel that has a bulging or narrowed wall, a vibration is felt, which is called a **thrill.** When auscultating an abnormal vessel, a humming is heard that is caused by the turbulent blood flow through the vessel. This is referred to as a **bruit.**

RESPIRATIONS. The rate and ease of respirations are observed. Breath sounds are auscultated. Sputum characteristics such as amount, color, and consistency are noted. Pink, frothy sputum is an indicator of acute heart failure. A dry cough can occur from the irritation caused by the lung congestion resulting from heart failure.

Clubbing—early

160°

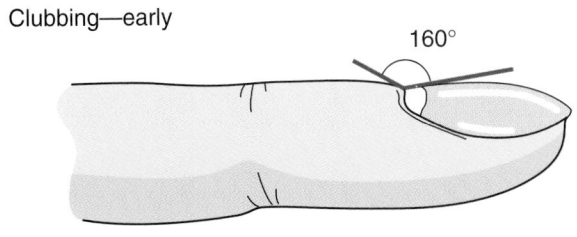

Clubbing—severe

Greater than 180°

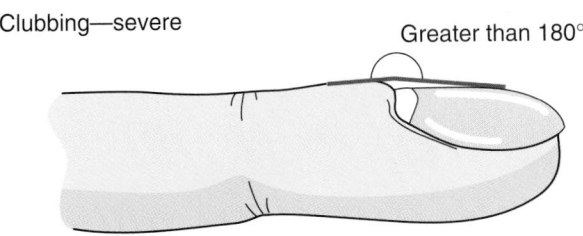

Figure 15-6 Clubbing of the fingers.

INSPECTION. During the health history, inspection begins by noting any shortness of breath when the patient speaks or moves. The patient's skin is noted for oxygenation status through the color of skin, mucous membranes, lips, earlobes, and nailbeds. Pallor may indicate anemia or lack of arterial blood flow. Cyanosis shows an oxygen distribution deficiency. A reddish-brown discoloration (rubor) found in the lower extremities occurs from decreased arterial blood flow. A brown discoloration and cyanosis when the extremity is dependent may be seen in the presence of venous blood flow problems. Hair distribution on the extremities is observed. Decreased hair distribution, thick, brittle nails, and shiny, taut, dry skin occur from reduced arterial blood flow. Venous blood return is assessed by inspecting extremities for varicose veins, stasis ulcers, or scars around the ankles and signs of thrombophlebitis such as swelling, redness, or a hard, tender vein.

The internal and external jugular neck veins are observed for distention. Normally, in the upright position, the veins are not visible. Distention of the veins in an upright position of 45 to 90 degrees indicates an increase in the venous volume. This is most commonly caused by right-sided heart failure. To observe for this, the patient is gradually elevated to a 45- to 90-degree position and any distention noted.

Capillary refill time assesses arterial blood flow to the extremities. The patient's nailbed is briefly squeezed, causing blanching, and then released. The amount of time that it takes for the color to return to the nailbed after release of the squeezing pressure is the capillary refill time. Normal capillary refill time is 3 seconds or less. Longer times indicate anemia or a decrease in blood flow to the extremity.

Clubbing of the nailbeds occurs from oxygen deficiency over time. It is often caused by congenital heart defects or the long-term use of tobacco. The distal ends of the fingers and toes swell and appear clublike. With clubbing, the normal 160-degree angle formed between the base of the nail and the skin is lost, causing the nail to be flat (Fig. 15–6).

Later, the nail base elevates, the angle exceeds 180 degrees, and the nail feels spongy when squeezed.

PALPATION. In addition to palpating the arteries, the thorax can be palpated at the **point of maximum impulse** (PMI). The PMI is palpated by placing the right hand over the apex of the heart. If palpable, a thrust is felt when the ventricle contracts. An enlarged heart may shift the PMI to the left of the midclavicular line.

The temperature of the extremities is palpated bilaterally for comparison. Palpation begins proximally and moves distally along the extremity. In areas of decreased arterial blood flow, the ischemic area feels cooler than the rest of the body because it is blood that warms the body. In the absence of sufficient arterial blood flow, the area becomes the temperature of the environment. This is called **poikilothermy.** A warm or hot extremity indicates a venous blood flow problem.

Six Ps characterize peripheral vascular disease:

- Pain
- Poikilothermia
- Pulselessness
- Pallor
- Paralysis
- Paresthesia (decreased sensation)

Edema is palpated in the extremities and dependent areas such as the sacrum for the supine patient (Fig. 15–7). Edema can occur from right-sided heart failure, gravity, or altered venous blood return. The nurse assesses the severity of the edema by pressing with a finger for 5 seconds over a

poikilothermy: poikilos—varied + therme—heat

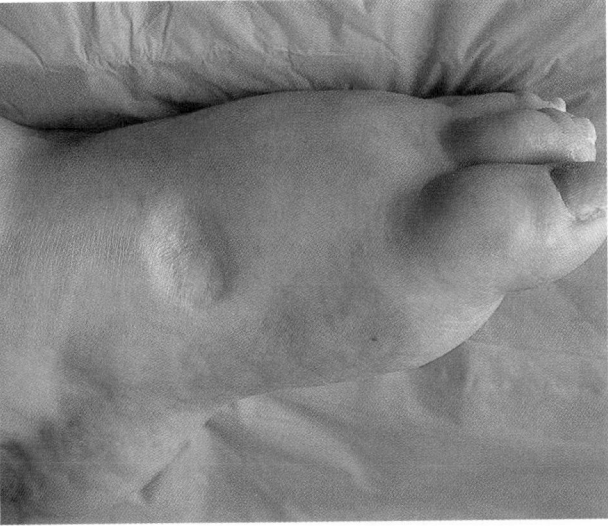

Figure 15-7 Pitting edema. Application of pressure over a bony area displaces the excess fluid, leaving an indentation or pit.

bone, the medial malleolus or tibia, in the area of edema. If the finger imprint or indentation remains, the edema is pitting. Measuring the leg circumference is an accurate method for monitoring the edema.

Homans' sign is an assessment for venous thrombosis; however, in less than 50 percent of patients with thrombosis the test is not positive. A positive Homans' sign is pain in the patient's calf or behind the knee when the foot is quickly dorsiflexed with the knee in a slightly flexed position (Fig. 15–8). Homans' sign should not be performed if a positive diagnosis of thrombosis has been made.

PERCUSSION. Percussion is performed by a physician to detect cardiac enlargement. Usually only the left border of the heart can be percussed. The heart is heard as dullness, in contrast to the resonance heard over the lungs.

AUSCULTATION. The normal heart sounds heard with a stethoscope placed on the wall of the chest are produced by the closing of the heart valves. These areas indicate where the sounds are best heard because sound in blood-flowing vessels is transmitted in the direction of the blood flow. The first heart sound (S1) is heard at the beginning of systole as lubb when the tricuspid and mitral (AV) valves close (Fig. 15–9). The second heart sound (S2) is heard at the start of diastole as dupp when the aortic and pulmonic semilunar valves close. The diaphragm of the stethoscope is used to hear the high-pitched sounds of S1 and S2. Extra heart sounds, usually indicating a pathologic condition, may be heard with practice. Normally no other sounds are heard between S1 and S2. With the bell of the stethoscope placed at the apex, a third heart sound (S3) or a fourth heart sound (S4) may be heard. Having patients lean forward or lie on their left side can make the heart sounds easier to hear by

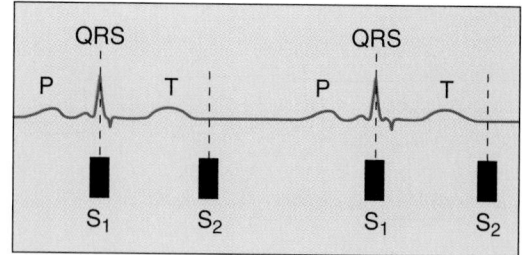

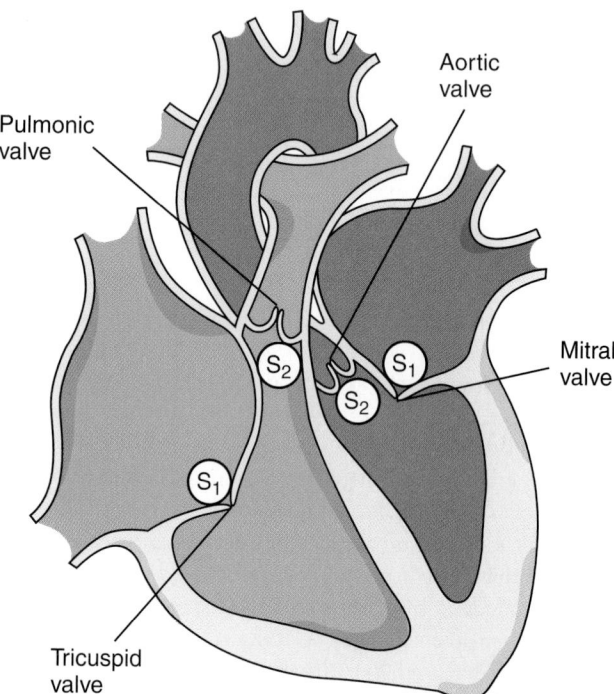

Figure 15-9   Heart sounds shown on electrocardiogram. S1 is heard at the beginning of systole, and S2 is heard at the beginning of diastole.

bringing the area of the heart where the sound may be heard closer to the chest wall. S3 is normal for children and younger adults. It sounds like a gallop and is a low-pitched sound heard early in diastole. S3 may be heard with left-sided heart failure, fluid volume overload, and mitral valve regurgitation. S4 is also a low-pitched sound, similar to a gallop but heard late in diastole. It occurs with hypertension, coronary artery disease, and pulmonary stenosis.

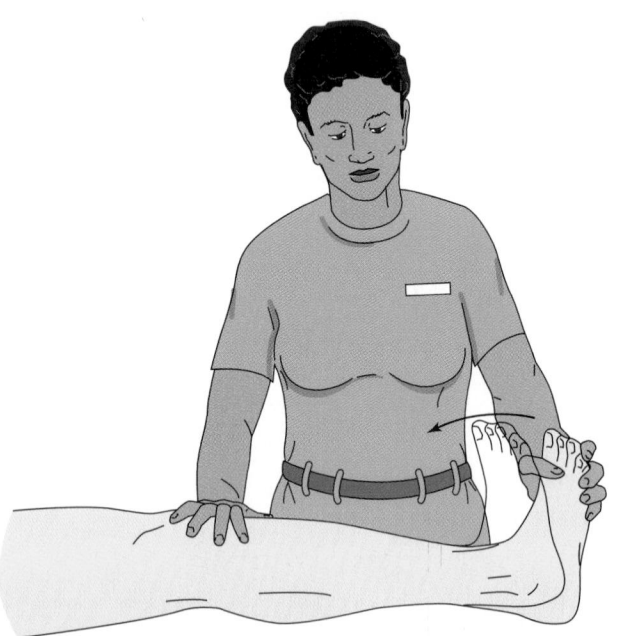

Figure 15-8   Assessment of Homan's sign for venous thrombosis. The foot is quickly dorsiflexed with the knee flexed. Calf or knee pain is noted. This assessment should not be performed if a positive diagnosis of thrombosis has been made.

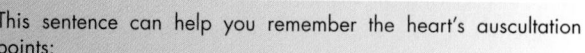

L E A R N I N G   TIPS

This sentence can help you remember the heart's auscultation points:

All	(aortic)
People	(pulmonic)
Eat	(Erb's point)
Three	(tricuspid)
Meals	(mitral)

Murmurs are caused by turbulent blood flow through the heart and major blood vessels. A murmur is a prolonged sound caused by a narrowed valve opening or a valve that does not close tightly. A swishing sound that ranges in intensity from faint to very loud is produced. The intensity of the murmur is graded by the physician on a scale of 1 to 6. The timing of the murmur in relationship to the cardiac cycle of systole or diastole is also documented by the physician.

A **pericardial friction rub** occurs from inflammation of the pericardium. The intensity of a rub can range from soft and faint to loud enough to be audible without a stethoscope. A rub has a grating sound like sandpaper being rubbed together that occurs when the pericardial surfaces rub together during the cardiac cycle. (See the Learning Tip on pericardial friction rub in Chapter 17.) Having the patient sit and lean forward allows a rub to be heard more clearly. The rub is best heard to the left of the sternum using the diaphragm of the stethoscope. A pericardial friction rub may occur after a myocardial infarction or chest trauma.

Diagnostic Studies

Diagnostic test results provide valuable assessment information for the nurse (Table 15–3). These data are combined

---

⫙ CRITICAL THINKING

Mrs. Smith

Mrs. Smith, 78, is admitted to the hospital with shortness of breath. Initial assessment findings are BP 152/88, pulse 104, respiration 26, temperature 99.4° F, short of breath at rest, shortness of breath increases with activity, ankles swollen, heart tones sound far away, nailbeds are very light pink, no pain, has not eaten well for 2 weeks, 6-pound weight gain in 1 week, sleeps on three pillows, veins in neck are visible on both sides. A diagnosis of acute myocardial infarction with heart failure is made by a physician.

1. Why is Mrs. Smith not reporting chest pain with a diagnosis of acute myocardial infarction?
2. How should swollen ankles be assessed to provide complete and measurable data?
3. What should be documented for the assessment of the swollen ankles?
4. How should the assessment findings for the swollen ankles be documented?
5. How should the assessment findings be documented for the additional symptoms Mrs. Smith has?

Answers at end of chapter.

TABLE 15-3	**DIAGNOSTIC TESTS AND PROCEDURES**	
Procedure	**Definition**	**Nursing Management**
	Noninvasive	
Chest x-ray film	Anterior-posterior and left lateral views of chest taken to show heart size and contour and lungs.	Assess x-ray history and whether pregnant. Remove metal items. Teaching—no discomfort.
Magnetic resonance imaging (MRI)	Three-dimensional image of heart.	Assess for metallic items and claustrophobia. Give antianxiety medication as ordered before MRI. Teaching—must lie still in long, small cylinder with loud, pounding sounds. Can talk to technician, listen to music.
Electrocardiogram (ECG)	Electrodes on skin carry electrical activity of heart from different views to show rhythm of heart, size of chambers, and heart damage.	Teaching—no discomfort. Explain procedure.
Holter monitor	Recording of ECG for up to 48 hours to match abnormalities with symptoms recorded in client's diary.	Apply electrodes and leads. Teaching—keep accurate diary; push event button for symptoms. No showers or baths. Return visit.
Transtelephonic event recorder	Records ECG events infrequently occurring for transmission over phone to interpreting center for analysis. Pacemaker checks done.	Teaching—explain use of recorder and how to transmit. Ensure good skin contact for best ECG tracing.
Pressure measurement	Blood pressures taken at several sites along extremity to show area of occlusion or decreased blood flow at rest and with exercise.	Teaching—no discomfort. Explain procedure.
Exercise treadmill test (ETT)	Evaluates effects of exercise on heart and vascular circulation; ECG and vital signs are continuously monitored. Test stopped if symptoms develop.	Monitor vital signs and ECG before, during, and after test until stable. Teaching—explain procedure, wear walking shoes and comfortable clothes.
Echocardiogram	Transducer transmits sound waves that bounce off heart to produce heart images and show blood flow. Provides audio and graphic data.	May be done at bedside. Patient lies on left side. Teaching—no discomfort, gel applied.
Transesophageal echocardiogram	Probe with transducer on end inserted into esophagus, depth and angle directed by physician. Shows clearer images of heart because no lung or rib tissue is crossed. Dye injected for blood flow study.	Monitor vital signs and oxygen saturation. Encourage client to relax. Suction continually during procedure. Teaching—NPO 6 hours before test. Sedation and local throat anesthetic given.

TABLE 15-3 DIAGNOSTIC TESTS AND PROCEDURES—CONT'D

Procedure	Definition	Nursing Management
Radioisotopes		
	IV injection of radioactive isotopes, which are taken up by heart and scanned with scintillation camera to show cardiac contractility, injury, and perfusion.	Assist patient to lie supine with arms over head for about 30 minutes. Teaching—explain procedure, inform that radioactivity is small and gone within a few hours.
Thallium Imaging	IV injection of thallium 201 to evaluate cardiac blood flow. Cold spots show infarcted areas. With exercise, thallium given 1 minute before end of test to circulate the thallium. Scan done within 10 minutes and repeated in 2–4 hours for comparison.	Teaching—explain procedure, inform that radioactivity is small and gone within a few hours. Light meal only between scans.
Dipyridamole thallium imaging	Dipyridamole (Persantine) IV is a vasodilator given to increase blood flow to coronary arteries; test is same as thallium imaging.	Teaching—explain procedure, instruct no caffeine or aminophylline 12 hours before. Same as thallium imaging.
Technetium pyrophosphate imaging	Technetium 99m pyrophosphate IV given. Shows hot spot in heart injury area. Scanned 2 hours later.	Teaching—explain procedure, inform that radioactivity is small and gone within a few hours.
Blood pool imaging	Technetium 99m pertechnetate IV. Studies effects of drugs, recent MI, and congestive heart failure. Serial studies are done over several hours. May be done at bedside.	Teaching—explain procedure.
Positron emission tomography (PET)	Nitrogen-13-ammonia IV given and scanned for cardiac perfusion. Then fluoro-18-deoxyglucose IV given and scanned for cardiac metabolic function. In normal heart, scans match; in injured heart, they differ. Exercise may also be used.	Patient's blood glucose must be 60 to 140 mg/dL for accuracy. Teaching—explain procedure. Must lie still during scan. If exercise used, NPO and no tobacco use.
Doppler ultrasound	Sound waves bounce off moving blood, producing recordings. Evaluates PVD.	Teaching—explain procedure.
Plethysmography	After the leg being tested is raised 30 degrees with the patient supine, a pressure cuff is inflated on the leg to distend the veins. Blood flow is measured with electrodes. Cuff is then rapidly deflated and venous volume changes recorded. Thrombi detected by less venous volume.	Teaching—explain procedure. No discomfort. Takes 30–45 minutes.
Technetium 99m Sestamibi	Technetium 99m sestamibi is given IV and scanned 1.5 to 2 hours later. Areas of myocardial cell damage take up the radioisotope; when scanned, these areas appear as hot spots.	Explain procedure.
Serum Enzymes		
Creatine kinase (CK)	Heart, brain, skeletal muscle contain CK enzymes. Damaged cells release CK. With MI, CK elevates in 6 hours and returns to baseline in 48–72 hours. Normal: male 5–55 U/mL; female 5–25 U/mL.	Avoid IM injections and take baseline CK before inserting IVs to avoid elevating CK from muscle cell damage. Serial sampling done.
CK-MB	Only heart muscle contains MB isoenzyme, which rises with MI in 6 hours and returns to baseline in 72 hours. Normal: 0–7 IU/L.	Same as CK.
Lactic dehydrogenase (LDH)	Intracellular enzyme found in many tissues, including heart. Elevates with MI after 8 hours and returns to baseline in 7 days. Normal: 80–120 U.	Serial sampling done.
LDH_1 and LDH_2	LDH contains five isoenzymes. Normally $LDH_2 > LDH_1$. In MI, ratio reverses: $LDH_1 > LDH_2$.	Same as LDH.
Serum aspartate aminotransferase (AST)	Nonspecific for cardiac damage.	No special care.
Myoglobin	Rises in 8 hours after MI and returns to baseline in 4–8 days; 99% indicative of MI. Rises in 1 hour after MI and peaks in 4–12 hours so must be drawn within 18 hours of chest pain onset. Normal: 250–450 ng/mL.	No special care.

TABLE 15-3 DIAGNOSTIC TESTS AND PROCEDURES—CONT'D

Procedure	Definition	Nursing Management
Serum Enzymes		
Cardiac troponin I	Cardiac cell protein. Elevated levels sensitive indicator of MI. Levels elevated up to 7 days.	No special care.
Serum Lipids		
Cholesterol	Measure CAD risk. Normal: 140–200 mg/dL.	Fasting not required.
Triglycerides	Elevated in cardiovascular disease. Normal: 40–190 mg/dL.	Eat normal diet for 2 weeks before test. Fasting required for 12 hours. Water allowed but no alcohol.
Phospholipids	May elevate in cardiovascular disease. Normal: 125–380 mg/dL.	Same as triglycerides.
Lipoproteins	Electrophoresis done to separate lipoproteins: VLDL, LDL, HDL. HDL protects against CAD; LDL increases CAD risk. Normal lipoproteins: 400–800 mg/dL. Desirable: LDL less than HDL (values vary with age).	Same as triglycerides.
Invasive		
Angiography	Dye injected into vessels to make them visible on x-rays to assess patency, injury, or aneurysms. Coronary—coronary arteries via cardiac catheter. Peripheral—peripheral arteries or veins.	Precare: Informed consent required. NPO 4–18 hours before test. Assess dye allergies. Teaching—sedative and local anesthesia may be used; burning sensation from dye; monitored continuously. Postcare: assess vital signs, circulation mobility sensation, catheter insertion site, or puncture site for hemorrhage, hematoma every 15 minutes for 1 hour, then every 30 minutes to 1 hour. Apply insertion site pressure (dressing, sandbag) as needed. Immobilize extremity for several hours as ordered.
Cardiac catheterization	Catheter inserted into heart for data on oxygen saturation and chamber pressures. Dye may be injected to visualize structures.	Same as angiography. Sensory teaching—table is hard, cool cleansing solution used, sting felt from local anesthetic, hear monitor beeping, feel pressure of catheter insertion, dye warm, burning feeling, headache, chest pain briefly, hear camera, feel table move.
Hemodynamic monitoring	Continuous readings of arterial BP, cardiac and pulmonary pressures, CO and SvO obtained with catheter attached to transducer and monitor to diagnose and guide treatment. Normal: BP 120/80; right atrial pressure 2–6 mm Hg; pulmonary artery systolic pressure/pulmonary artery diastolic pressure 20–30/0–10; pulmonary artery wedge pressure 4–12 mm Hg; CO 4–8 L/min; SvO$_2$ 60–80%.	Informed consent signed for insertion. Continuous assessment to ensure proper placement of catheter is maintained and a permanent catheter wedge does not go undetected. Recording of readings and monitoring of insertion site for signs of infection.
Central venous pressure (CVP)	Catheter inserted into a vein and threaded into vena cava. Monitors CVP, which reflects fluid volume status. Normal: 2–6 mm Hg.	Informed consent signed for insertion. Recording of readings and monitoring of insertion site for signs of infection.

with the health history and physical assessment to plan care for the patient.

NONINVASIVE

Pulse Oximetry. A transdermal clip or patch placed on a finger, a toe, an ear, or the nose is attached to a monitor and used to assess arterial oxygen saturation. A light in the clip or patch passes through the artery and provides data to measure the oxygen saturation. Normal levels are 95 percent or greater.

Chest X-ray Examination. A chest x-ray examination shows the size, position, contour, and structures of the heart

(Fig. 15–10). It shows heart enlargement, calcifications, and fluid around the heart. Heart failure can be confirmed with a chest x-ray examination. Correct placement of pacemaker leads and pulmonary artery catheters in the heart can be confirmed.

Fluoroscopy uses a luminescent x-ray screen to show cardiac structures and pulsations. It is used as a guide when placing cardiac catheters or pacemaker leads.

Magnetic Resonance Imaging. A three-dimensional image of the heart is produced by magnetic resonance imaging (MRI). The patient must lie still in a long, small-diameter

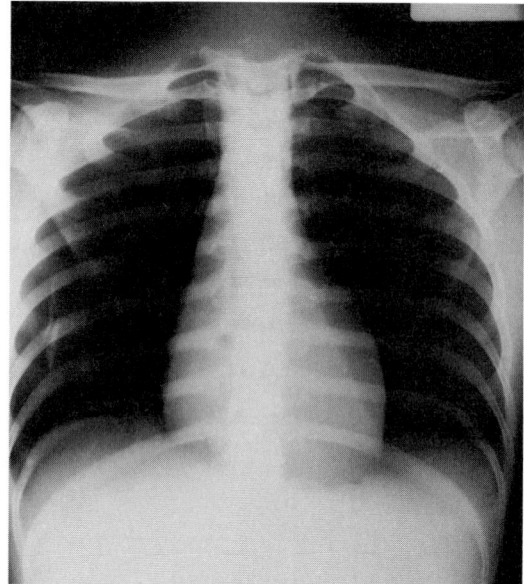

Figure 15–10 Normal chest x-ray film. Note white outline of heart borders in center. (From McKinnis, LN: Fundamentals of Orthopedic Radiology. FA Davis, Philadelphia, 1997, p 15.)

cylinder with a strong magnetic field. Because of the small cylinder's close proximity to the patient, many experience claustrophobia. Patients are asked if they are claustrophobic before the test, and antianxiety medication is given 1 hour before the MRI as ordered. Music may help to ease the patient's anxiety as well. No metallic items are permitted in the machine, so some patients, such as those with pacemakers or metal implants, shavings, or shrapnel, are not candidates for this test.

Electrocardiogram. The electrocardiogram (ECG) assesses the electrical activity of the heart from different views. The ECG shows abnormalities related to conduction, rate, rhythm, heart chamber enlargement, myocardial ischemia, myocardial infarction, and electrolyte imbalances. Abnormalities in cardiac function can be detected and the area of abnormality pinpointed with the aid of the different views on the ECG.

When an ECG is requested, information that aids in its interpretation is provided, including the patient's sex, age, height, weight, blood pressure, and cardiac medications. The patient requires no special preparation but is given an explanation of the procedure and told that the ECG is painless.

To obtain an ECG, electrodes are placed on the skin to transmit electrical impulses to the ECG machine for recording. The electrical impulses from the heart appear as waves on graph paper. (See Chapter 20.) One electrode is placed on each limb and several across the chest. The right leg is a ground electrode. One view of the heart using a combination of the electrodes to obtain the view is called a lead. The standard 12-lead ECG, using a combination of the electrodes, provides 12 views of the heart.

Signal-averaged ECG. The signal-averaged ECG is used to diagnose whether a patient is at risk of developing ventricu-

lar tachycardia and possible sudden death. A computer records low level signals not detected by a regular ECG. These electrical signals, referred to as late potentials, occur at the end of the QRS and into the ST segment. Late potentials place the patient at risk for ventricular dysrhythmias.

Ambulatory Electrocardiogram Monitoring. Continuous monitoring of the ambulatory patient is possible with the use of tape recorders.

Holter Monitoring. A Holter monitor, which weighs 2 pounds, continuously records one lead for up to 48 hours. The patient wears loose-fitting clothing and may only sponge bathe while wearing the monitor. The patient records a diary of activities and symptoms and pushes the event button if symptoms occur. Symptoms are documented for later correlation with the ECG recordings. Dysrhythmias or myocardial ischemia that occurs infrequently can be detected. The recordings are scanned by a computer and interpreted by a physician.

Transtelephonic Event Recorders. This recorder is used for dysrhythmic events that occur so infrequently that they would not likely be captured in 48 hours, or for follow-up evaluation of permanent pacemakers. With this recorder the patient has greater flexibility because the device is worn when needed or when symptoms occur. The disadvantage is that if the event is brief it may be missed before the recorder is on. When an event is recorded, the patient can transmit it at a convenient time over the telephone for printout and analysis. The telephone mouthpiece is placed over the signal transmitter box for transmission. After transmission the recording can be erased and the recorder reused.

Pressure Measurement. Pressure readings are done to assess areas of occlusion or narrowing in vessels. Blood pressure readings are taken at intervals along the extremity. Reduced readings are found in areas with blood flow problems.

Tilt Table Test. The tilt table test is used to help diagnose the cause of syncope ("fainting spells"). Heart rate and blood pressure are monitored during a change in position from lying down to standing up.

Exercise Tolerance Testing. The exercise tolerance test or stress test measures cardiac function or peripheral vascular disease during a defined exercise protocol (Fig. 15–11). Before the test, patients are given an explanation of the test and told not to smoke, eat, or drink for 2 to 4 hours before the test. They are also instructed to wear comfortable walking shoes, a loose top, and for women a supportive bra. After the test, patients should rest and wait to eat. They should also avoid eating or drinking stimulants such as caffeine and temperature extremes such as going out into cold weather for a few hours after the test.

Cardiac Stress Test. This test simulates sympathetic nervous system (fight or flight) stimulation. It shows the heart's response to increased oxygen needs. Before the test, baseline vital signs are obtained. Then, while the patient exercises on a treadmill, on a stationary bicycle, or by climbing

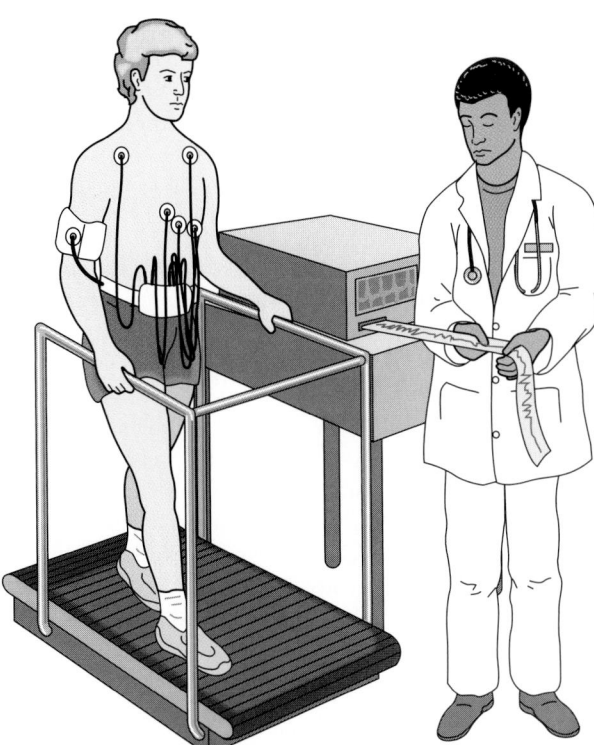

Figure 15-11 Performance of stress test.

stairs, vital signs, oxygen saturation, skin temperature, physical appearance, chest pain, and ECG are monitored to help ensure patient safety. The test is completed when the patient reaches peak heart rate (patient's age subtracted from 220), experiences chest pain, is unable to exercise further, or develops vital sign or ECG changes. Vital signs and ECG continue to be monitored after the test until they return to baseline.

The cardiac stress test is used to evaluate coronary artery disease. It aids in diagnosing ischemic heart disease, the cause of chest pain and dysrhythmias. The functional capacity of the heart can also be measured after a cardiac event or to plan a physical fitness or rehabilitation program.

Peripheral Vascular Stress Test. The patient walks for 5 minutes at 1.5 miles per hour on the treadmill. At certain intervals, pulse volume measurements are taken, including baseline resting, during the test, and final resting after the test. This test assesses response to activity. If **claudication** occurs, the test is stopped.

Echocardiogram. An echocardiogram is an ultrasound test that records the motion of the heart structures, including the valves, as well as the heart size, shape, and position. No preparation is required for a cardiac ultrasound. This test transmits ultrasonic sound waves into the heart so that the returned echoes can be recorded on videotape as audio and visual information. An ECG is recorded at the same time

for comparison purposes. Abnormalities that may be seen on the echocardiogram include heart enlargement, valvular abnormalities, thickened cardiac walls or septum, and pericardial effusion.

Exercise echocardiography diagnoses CAD during exercise-induced cardiac ischemia by detecting cardiac wall motion abnormalities.

Transesophageal Echocardiogram. Transesophageal echocardiogram provides a clearer picture than transthoracic echocardiography. It produces images by using a transducer on a probe that is placed in the esophagus. The images are clearer because lung and rib tissue does not have to be penetrated by the sound waves. The physician controls the position of the probe and takes pictures as it travels within the esophagus. Patients take nothing by mouth (NPO) for about 6 hours before the test, receive a sedative, and have their throat locally anesthetized. After the procedure, patients remain NPO until their gag reflex returns.

Radioisotope Imaging. For this type of imaging, small amounts of radioisotopes are given intravenously. The patient is then scanned with a gamma camera and a radionuclide image is produced. Radiation exposure is similar to that of other x-ray examinations. These tests can provide information about myocardial ischemia or infarction, cardiac blood flow, and ventricle size and motion.

Thallium Imaging. Thallium 201, a radioactive analog of potassium, is used to detect impaired myocardial perfusion. It is injected intravenously (IV), and muscle cells absorb it. After 10 to 15 minutes the heart is scanned to see where the thallium has concentrated. Four hours later the scan is repeated to look for changes. Healthy myocardial cells with good blood flow take up the thallium. Areas in which the thallium is not seen are referred to as cold spots and indicate ischemia or infarction. The patency of a coronary artery graft may also be assessed with this test. This test is used often because the short half-life of thallium results in lower radiation exposure.

Exercise testing may be combined with thallium injection to detect blood flow changes with activity and after rest. The patient exercises and about 2 minutes before stopping is given thallium. Scans are taken immediately and again in 2 to 4 hours. Cold spots on initial images indicate ischemia. If the cold spots are gone in later images, it indicates exercise-induced ischemia. If the cold spots are still present in later images, they show scarred areas.

If patients are unable to participate in exercise for the thallium stress test, dipyridamole (Persantine) or adenosine, coronary vasodilators, can be given. These drugs simulate the increased blood flow to healthy myocardial cells that occurs with exercise.

Technetium Pyrophosphate Scan. Technetium 99m pyrophosphate is injected for this test. Areas of ischemia or myocardial cell damage take up the radioisotope, and when scanned these areas appear as hot spots. Acute myocardial infarction (MI) size and location can be detected, but old MIs cannot be detected.

---

claudication: claudicare—to limp

Technetium 99m Sestamibi. Technetium 99m sestamibi is given IV and the patient is scanned 1.5 to 2 hours later. Areas of myocardial cell damage take up the radioisotope, and when scanned these areas appear as hot spots.

Blood Pool Imaging. Technetium 99m pertechnetate is injected IV and remains in the bloodstream; it is not taken up by myocardial cells. A camera follows the flow of the radioactivity, which shows ventricular function and wall motion and the ejection fraction of the heart.

Positron Emission Tomography. Positron emission tomography (PET) shows myocardial perfusion and viability with three-dimensional images. Nitrogen-13-ammonia is injected IV first and then scanned to show myocardial perfusion. Next, fluoro-18-deoxyglucose is given intravenously and then scanned to show myocardial metabolic function. If there is ischemia or heart damage, the two scans are different. For example, in ischemia of viable cells, blood flow is decreased but metabolism elevated. Treatment to increase blood flow improves cardiac function in this case. Before the test the patient's blood glucose should be normal, and caffeine and tobacco should be avoided for 4 hours before the test.

Doppler Ultrasound. In this test, sound waves are transmitted to an artery or vein to assess blood flow problems. The sound waves bounce off moving blood cells and return a sound frequency in relationship to the amount of blood flow. With decreased blood flow the sounds are reduced. This test requires no patient preparation, takes about 20 minutes to complete, and is painless.

Plethysmography. With this test, blood volume and changes in blood flow are measured to diagnose deep vein thrombosis and pulmonary emboli and to screen patients for peripheral vascular disease. The leg being tested is raised 30 degrees with the patient supine. A pressure cuff is then inflated on the leg to distend the veins. Blood flow is measured with electrodes, and the cuff is then rapidly deflated and venous volume changes are recorded. Thrombi are detected by reduction in venous volume.

BLOOD STUDIES

Cardiac Troponin. Cardiac muscle contains proteins called troponins, which control the muscle fibers that contract or squeeze the heart muscle. Troponin I and Troponin T are highly sensitive indicators of myocardial damage, which is helpful in diagnosing MI. They are proteins found only in cardiac cells. When injured or dead, cardiac cells release these proteins, which results in elevated levels within 4 to 6 hours of damage. These levels peak in 10 to 24 hours and remain elevated for up to 7 days after injury. Troponin T appears slightly earlier than Troponin I after cardiac damage. Troponin T is better at showing slight cardiac damage and predicting 30-day mortality for cardiac patients.

Cardiac Enzymes. When heart cells are damaged or die, they rupture and release their enzymes into the bloodstream. Levels of these enzymes rise in the serum as a result. Common cardiac enzyme tests are creatine kinase (CK), also referred to as creatine phosphokinase (CPK), and lactic dehydrogenase (LDH). However, these enzymes are also found in body cells other than the heart, so a more organ-specific test, the enzymes' isoenzymes (or different forms), is measured. Serum aspartate aminotransferase (AST) is a nonspecific cardiac test formerly known as serum glutamic-oxaloacetic transaminase (SGOT).

Creatine Kinase. Creatine kinase (CK) is found in three types of tissue: brain, skeletal muscle, and heart muscle. Isoenzymes of CK contained in these tissues are CK-BB (brain), CK-MM (skeletal muscle), and CK-MB (heart muscle). Levels of CK-MB rise within 4 to 6 hours after cardiac cells are damaged, peak in 12 to 24 hours, and return to normal in 48 to 72 hours. Serial CK levels are drawn at intervals to track trends. It is important to avoid invasive procedures such as IVs and intramuscular (IM) injections before drawing the first CK to prevent elevation in the CK levels from cell trauma caused by the procedure. Medications are often given IV rather than IM to prevent this elevation.

Lactic Dehydrogenase. Another intracellular enzyme that is found in a variety of body cells is LDH. There are five types of isoenzymes for LDH: LDH_1 and LDH_2 are found in the heart, kidneys, and red blood cells (RBCs); LDH_3 is specific to the lungs; and LDH_4 and LDH_5 are found in skeletal muscle and the liver. Normally LDH_2 is higher than LDH_1. After an MI, the pattern reverses and LDH_1 rises higher than LDH_2. Levels of LDH rise 8 to 12 hours after an MI, peak at 24 to 48 hours, and return to normal in 5 to 7 days.

Myoglobin. Myoglobin is a protein found in skeletal and cardiac muscle and is released into the bloodstream when cell damage occurs. Because myoglobin is not site specific, it can only provide an estimate of damage and is used with other more specific tests such as troponin to diagnosis an MI. Myoglobin levels elevate within 1 hour of an acute MI. Peak levels are reached 4 to 12 hours after an MI, and levels return to normal within 18 hours after the onset of chest pain, so it is a test that must be done early when MI is suspected.

Blood Lipids. Lipids include triglycerides, cholesterol, and phospholipids. Lipoproteins carry these lipids attached to proteins. Triglycerides are found in very low density lipoproteins (VLDL). Cholesterol is mainly found in low-density lipoproteins (LDL). High-density lipoproteins (HDL) are a mixture of one-half protein and one-half phospholipids and cholesterol.

A lipid profile can screen for increased risk for coronary artery disease. Patients must fast for 12 hours before the test, and, although water is not withheld, alcohol is restricted for 24 hours before the test. High levels of LDL are linked to an increase in CAD because they circulate cholesterol in the arteries. High-density lipoproteins play a protective role against CAD because they carry cholesterol to the liver to be metabolized. Controlling lipids is very important in reducing CAD (Cultural Consideration Box 15–3).

BOX 15-3 CULTURAL CONSIDERATION

Among French-Canadians, familial chylomicronemia (hyperlipoproteinemia type I), an autosomal recessive disorder, occurs with the highest frequency worldwide. Familial hypercholesterolemia can lead to coronary thrombosis. Thus the nurse can improve the health of French-Canadians by encouraging early diagnostic workups for familial chylomicronemia and encouraging healthful lifestyles.

INVASIVE STUDIES

Angiography. Arteriography and venography are the two types of angiography (Fig. 15–12). Arteriography examines arteries. Venography studies veins. Angiography uses dye injected into the vascular system to visualize the vessels on radiographs. This test is used to assess blood clot formation, to assess peripheral vascular disease (PVD), and to test vessels for potential grafting use.

The patient must be assessed for allergies, give informed consent, be NPO for about 4 hours before the test, and be informed that the dye produces a hot, burning feeling when injected. After the procedure the patient is assessed for several hours. Vital signs, allergic reaction signs, hemorrhage at the injection site, and pulses are monitored.

Cardiac Catheterization. Cardiac catheterization allows the study of the heart's anatomy and physiology. It is an invasive diagnostic procedure that measures pressures in the heart chambers, great blood vessels, and coronary arteries and provides information on cardiac output and oxygen saturation. Fluoroscopy is used, and dye can be injected once the catheter is in place to visualize the heart chambers and vessels. This procedure is often done before heart surgery.

Right-sided Catheterization. In right-sided catheterization, a catheter with or without a fiberoptic tip is inserted into the basilic or cephalic vein or the femoral vein and advanced into the vena cava. It is then moved through the right chambers of the heart and into the pulmonary artery. The catheter can be wedged momentarily in the artery by inflating the balloon at the tip of the catheter. This position provides the pulmonary artery wedge pressure (PAWP),

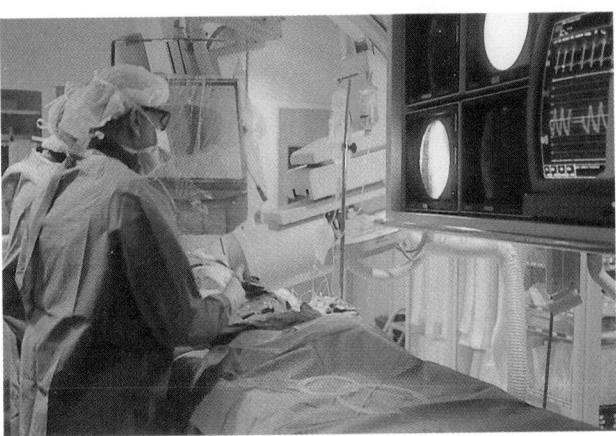

Figure 15-12 Coronary angiography and cardiac catheterization.

which reflects pressures in the left side of the heart. Other pressures obtained with right-sided cardiac catheterization are right atrial pressure, which reflects central venous pressure, pulmonary artery systolic and diastolic pressures, and cardiac output, and mixed venous oxygen saturation (SvO_2) if a fiberoptic catheter is used.

Left-sided Catheterization. The left side of the heart can be directly assessed by inserting a catheter into the brachial or the femoral artery. It is advanced against the flow of blood into the aorta, through the aortic valve, and into the left ventricle. Coronary angiography, which visualizes the coronary arteries with dye, can be done with this approach. The catheter is inserted into the opening of the coronary arteries, the dye is injected, and x-ray films are taken. Coronary artery disease can be assessed with coronary angiography.

An informed consent must be obtained. The patient is assessed for allergies to iodine, seafood, and shellfish and kept NPO before the procedure. Patients should be told that during the test they are awake and a warm, flushing sensation may be felt when the dye is injected; the room has a lot of equipment; a movable table is used; the patient's vital signs and ECG are monitored constantly; and the length of the procedure is 2 to 3 hours.

Complications of the procedure can be allergic reaction, breaking of the catheter, hemorrhage, thrombus formation, emboli of air or blood, dysrhythmias, MI, cerebrovascular accident (CVA), and puncture of the heart chambers or lungs.

After the catheter is removed, firm pressure must be applied to the insertion site for several minutes to prevent hemorrhage or hematoma formation. A pressure dressing or sandbag may be applied to the site when bleeding is stopped and is removed in several hours. Vital signs are assessed according to the physician's orders and the institution's policies. During vital sign checks, the puncture site is assessed and peripheral pulses are verified. The extremity used for insertion must not be moved or flexed for several hours after the procedure. Patients usually may eat and are instructed to drink fluids to help eliminate the dye from the body. If the patient is stable and no significant findings are found, the patient may be discharged.

Hemodynamic Monitoring. Bedside monitoring can be done to monitor the pressures in the blood vessels or heart. A catheter attached to a transducer and monitor, called an arterial line, can be inserted into the radial or femoral artery to measure continuous arterial blood pressure.

Ongoing monitoring of cardiac pressures, cardiac output, and central venous pressure (CVP) can be done with either a central catheter or a pulmonary artery catheter. Central venous pressure is measured directly with a central catheter inserted into the vena cava via the brachial, femoral, subclavian, or jugular vein (Fig. 15–13). It is measured indirectly with the pulmonary artery catheter (Fig. 15–14). The right atrial pressure measurement obtained from the pulmonary artery catheter reflects the pressure in the vena cava. Central venous pressure measures **preload** or fluid

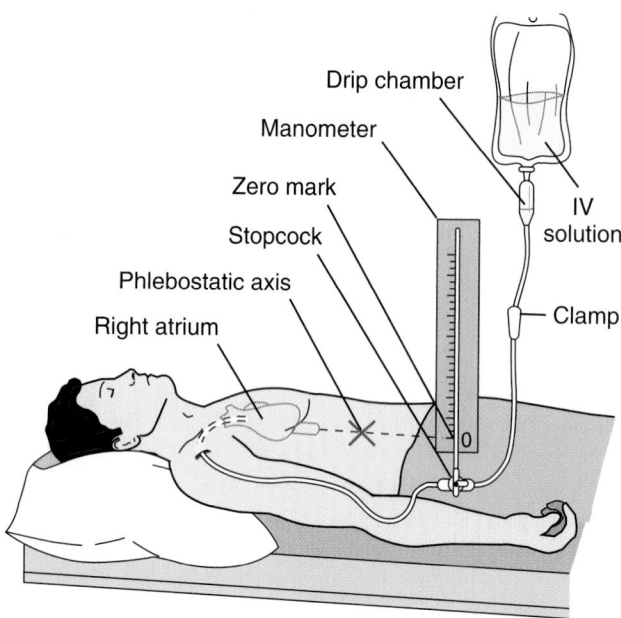

Figure 15-13   CVP measurement.

volume status; CVP readings used in fluid or diuretic therapy have been primarily replaced by pulmonary artery catheter measurements.

Electrophysiology Study. To study the heart's electrical system, one or more catheters with electrodes are inserted via the femoral vein into the right side of the heart. The electrical impulses are recorded. Dysrhythmias can be triggered to help the physician diagnose problems. A consent is obtained and the patient is NPO 6 to 8 hours before the test.

▶ THERAPEUTIC MEASURES FOR THE CARDIOVASCULAR SYSTEM

Exercise

A prescribed walking program helps promote blood flow by contracting the skeletal muscles and may reduce symptoms of peripheral vascular disease. For patients recovering from cardiac surgery or a myocardial infarction, activity is gradually increased. Exercise is very important for optimum cardiac functioning. A cardiac rehabilitation program is usually prescribed, and individualized exercise goals are determined. After discharge from the hospital, exercise three times a week for 20 to 30 minutes is encouraged. Mild stretching should be done before and after the exercise.

Smoking Cessation

Smoking causes vasoconstriction that can last up to 1 hour after the smoking of one cigarette. For patients with cardiac or vascular disease, blood flow is reduced, which can exacerbate symptoms. Patients should be encouraged to stop smoking and should be provided with support information such as cessation programs and support groups. For more information on smoking cessation, visit www.americanheart.org.

Diet

Teaching the patient to eat a healthy, balanced diet is important to help reduce the risk for coronary artery disease. Weight reduction, if needed, is encouraged, as well as increasing physical activity. In May 2001 the National Heart, Lung, and Blood Institute released therapeutic lifestyle changes (TLC) that are recommended to help control CHD risk factors. The TLC diet is low in cholesterol (<200 mg/day) and saturated fats (<7 percent of calories). Soluble fiber (10 to 25 g/day) and plant stanols/sterols (2 g/day) are also recommended to reduce LDL levels. (See Chapter 18.) For more information on the therapeutic lifestyle changes, visit http://hin.nhlbi.nih.gov/cvd_frameset.htm.

A diet low in sodium may be prescribed if excess fluid volume is a problem. (See Chapter 5). If diuretics are used that promote potassium loss, potassium levels must be monitored and adequate amounts of potassium included in the diet. To reduce fat, red meats, fried foods, whole milk, and cheese should be limited or avoided. Cholesterol can be reduced by avoiding egg yolks, organ meats, animal fats, and shellfish. Five to six servings of fruits and vegetables should be eaten daily. Increasing fish intake and eating poultry without skin are parts of a healthy diet.

Oxygen

Supplemental oxygen is administered to patients with chest pain to help ensure that the heart receives sufficient oxygen

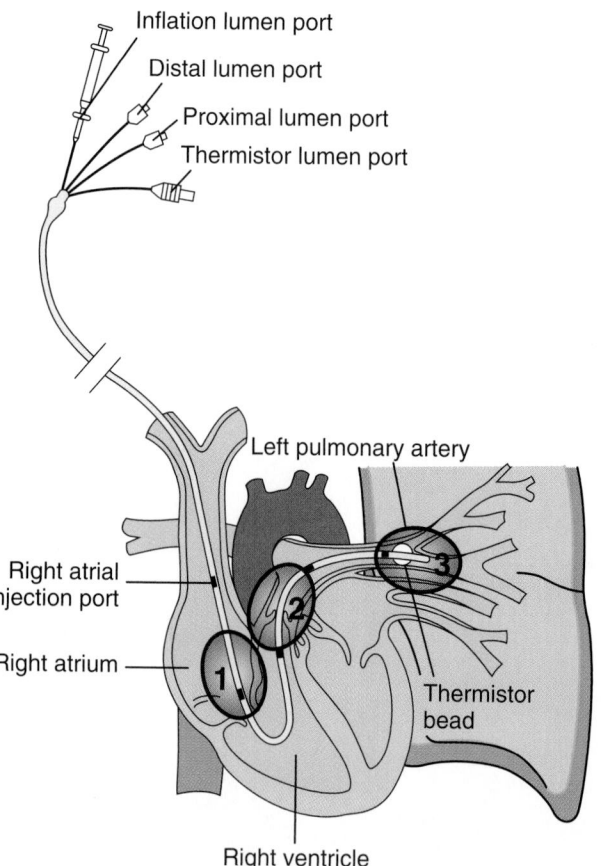

Figure 15-14   Pulmonary artery catheter (above) and placement of the inflated balloon in the pulmonary artery.

to function. Oxygen may be delivered via a nasal cannula or face mask. The patient must be taught safety precautions necessary for home use of oxygen if it is ordered, such as avoiding open flames and not smoking when the oxygen is in use.

Medications

The primary cardiovascular drugs are cardiac glycosides, vasodilators, antihypertensives, antidysrhythmics, antianginals, anticoagulants, and thrombolytics. They are discussed in further detail where the disorders they are used to treat are explained. (See Chapters 16 to 22.)

Antiembolism Devices

Antiembolism devices improve arterial blood flow and venous return to prevent the formation of blood clots. They are used for patients with peripheral vascular disease, on bedrest, or after surgery or trauma.

Elastic Stockings

Antiembolism stockings apply pressure over the leg to promote the movement of fluid and prevent stasis of fluid. These stockings may be knee or thigh length. They must be applied correctly so that a tourniquet effect is not produced by the stockings. For ease in application, the stocking is turned inside out to the heel, the foot portion is placed on the patient up to the heel, and then the remaining stocking is pulled up over the leg. The tops of the stockings should be 1 to 2 inches below the bottom of the kneecap. They should not roll down or they will cause stasis rather than prevent it. The stockings are removed for 20 minutes twice a day and the skin inspected for irritation. Elderly patients may require assistance in applying the stockings if they have impaired manual dexterity.

Intermittent Pneumatic Compression Devices

An intermittent pneumatic compression device consists of plastic inflatable stockings that are filled intermittently with air by an attached motor (Fig. 15–15). This device promotes fluid movement by simulating the contraction of the leg muscles to prevent thrombosis development. The compartments in the stockings inflate to 35 to 55 mm Hg of pressure, beginning in the ankle compartment and progressing next to the calf compartment and finally the thigh compartment. The nurse should monitor the device for proper pressure inflation.

Lifestyle and Cardiac Care

To reduce risk factors or promote recovery from cardiovascular disease, lifestyle changes are often needed. Long-standing habits are difficult to change. Support groups can offer encouragement that is helpful in promoting a healthy lifestyle. Patients should be referred to community support groups as needed.

Patients recovering from cardiac disorders are often anxious about resuming sexual activity but are embarrassed to discuss it. This is an area that is often overlooked when caring for patients. Sexual counseling should be offered to patients and their partners. Patients often have misconceptions that are unfounded but interfere with resuming their sexual activity. If patients have angina, nitroglycerin can be taken prophylactically before sexual activity. After a myocardial infarction, sexual activity can be resumed in 1 to 2 months or when the patient can climb two flights of stairs without symptoms, as ordered by the physician. Sexual activity is a form of physical activity. Patients should be given information to make an informed decision on when they are ready to resume this activity.

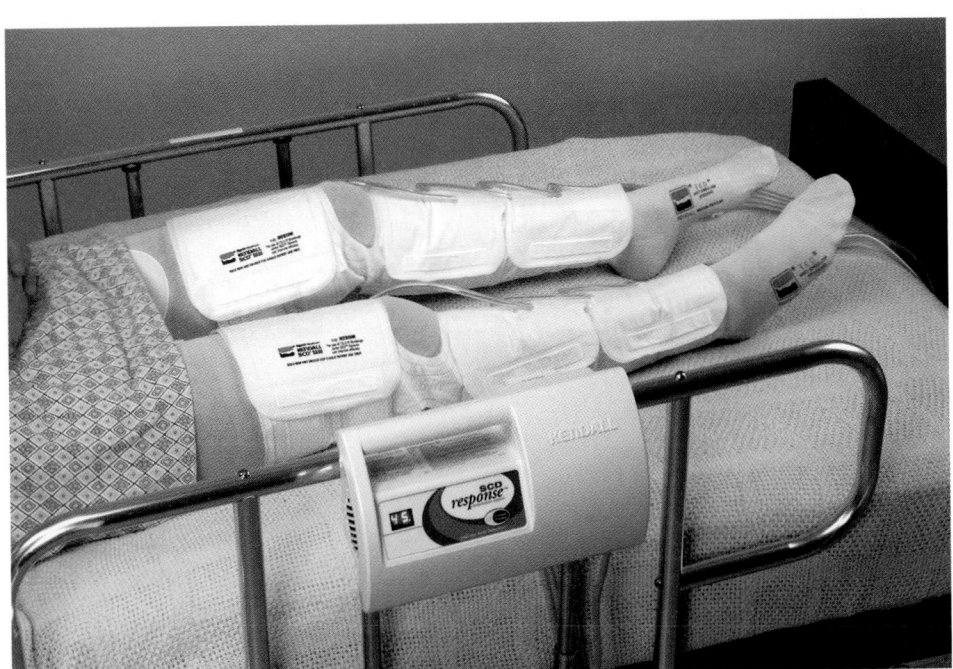

Figure 15-15 The Kendall SCD Response Compression system provides sequential personalized compression cycles that minimize stasis and maximize blood flow to prevent thrombosis and pulmonary embolus development. (Courtesy Kendall, Mansfield, Mass.)

REVIEW QUESTIONS

1. The mitral and tricuspid valves prevent backflow of blood from which of the following?
   a. Ventricles to atria when the ventricles contract
   b. Atria to ventricles when the ventricles relax
   c. Ventricles to atria when the atria contract
   d. Atria to ventricles when the atria contract

2. Which of the following describes the purpose of the endocardium of the heart?
   a. Cover the heart muscle and prevent friction.
   b. Support the coronary blood vessels.
   c. Line the chambers of the heart and prevent abnormal clotting.
   d. Prevent backflow of blood from atria to ventricles.

3. The function of the coronary blood vessels is to do which of the following?
   a. Prevent abnormal clotting within the heart.
   b. Bring oxygenated blood to the myocardium.
   c. Carry deoxygenated blood to the lungs.
   d. Carry oxygenated blood to the lungs.

4. Which of the following is the location of the cardiac centers in the nervous system?
   a. Cerebrum
   b. Hypothalamus
   c. Spinal cord
   d. Medulla

5. The functions of angiotensin II are to increase which of the following?
   a. Vasodilation and ADH secretion
   b. Vasoconstriction and aldosterone secretion
   c. Heart rate and vasodilation
   d. Heart rate and ADH secretion

6. The increase of resting blood pressure with age may contribute to which of the following?
   a. Dysrhythmias
   b. Thrombus formation
   c. Left-sided heart failure
   d. Peripheral edema

7. Which one of the following is a modifiable cardiovascular risk factor that should be noted during patient data collection?
   a. Age
   b. Gender
   c. Ethnic origin
   d. Tobacco use

8. If it takes longer than 3 seconds for the color to return when assessing capillary refill, it may indicate which of the following?
   a. Decreased arterial flow to the extremity
   b. Increased arterial flow to the extremity
   c. Decreased venous flow from the extremity
   d. Increased venous flow from the extremity

9. Which one of the following is an important safety intervention that should be used while assessing a patient for orthostatic hypotension?
   a. Reality orientation
   b. Gait or walking belt
   c. Liquids at bedside
   d. Standing patient quickly

10. You are caring for a patient who is on bedrest. In which area will you assess for the presence of edema?
    a. Arms
    b. Ankles
    c. Sternum
    d. Sacrum

11. Which of the following should be included in patient teaching for coronary angiography with femoral catheter insertion site?
    a. Dye injection causes hot, flushing sensation.
    b. General anesthesia is administered.
    c. Claustrophobia may be experienced.
    d. Ambulation is possible immediately after procedure.

12. A high-fiber diet for cardiac patients is recommended to do which of the following?
    a. Increase absorption of nutrients
    b. Reduce cardiac workload
    c. Reduce edema development
    d. Reduce appetite

REFERENCES

1. American Heart Association: 2001 Heart and Stroke Statistical Update. American Heart Association, Dallas, 2000.
2. Ma, Y, et al: A mutation in the human lipoprotein lipase gene as the most common cause of familial chylomicronemia in French Canadians. N Engl J Med 324(25):176, 1991.
3. Summary of the Third Report of the NCEP Expert Panel on Detection, Evaluation, and Treatment of High Blood Cholesterol in Adults (Adult Treatment Panel III), NIH Publication No. 01-3305, May 2001.

Answers to CRITICAL THINKING

Mrs. Smith

1. An older patient commonly does not experience typical disorder symptoms. Chest pain is often not present because of reduced nerve sensitivity with aging for an MI. Dyspnea is the classic symptom of MI in the older patient.
2. Inspect both legs to determine edematous areas. Determine location and severity of edema by pressing finger for 5 seconds over the medial malleolus and moving up the leg along the tibia until no edema is found. Assess bilaterally. Measure leg circumference.

3. Document location of edema and whether edema is nonpitting or pitting. Document findings for both legs.
4. Documentation should state "Bilateral pitting ankle edema" with leg circumference measurement number.
5. Additional symptoms should be documented as follows: Dyspnea at rest that increases with exertion, heart tones clear and distant, nailbeds pale, pain free, poor appetite for 2 weeks, 6-pound weight gain in 1 week, three-pillow orthopnea, bilateral jugular vein distention.

16

NURSING CARE OF PATIENTS WITH HYPERTENSION

Mary Friel Fanning and Diane Lewis

KEY TERMS

cardiac output (**KAR**-dee-yak **OWT**-put)

diastolic blood pressure
(dye-ah-**STAH**-lik BLUHD **PRE**-shure)

essential hypertension
(e-**SEN**-shul HIGH-per-**TEN**-shun)

hypertension (HIGH-per-**TEN**-shun)

hypertensive crisis
(HIGH-per-**TEN**-siv **CRY**-sis)

hypertrophy (high-**PER**-truh-fee)

isolated systolic hypertension (**EYE**-suh-lay-ted sis-**TAL**-lik HIGH-per-**TEN**-shun)

normotensive (nor-mo-**TEN**-siv)

peripheral vascular resistance (puh-**RIFF**-uh-ruhl **VAS**-kyoo-lar ree-**ZIS**-tense)

plaque (PLAK)

primary hypertension
(**PRY**-mare-ee HIGH-per-**TEN**-shun)

secondary hypertension
(**SEK**-un-DAR-ee HIGH-per-**TEN**-shun)

systolic blood pressure
(sis-**TAL**-ik BLUHD **PRE**-shure)

viscosity (vis-**KAH**-si-tee)

QUESTIONS TO GUIDE YOUR READING

1. How would you define hypertension and hypertensive crisis?

2. How would you explain the pathophysiology of hypertension?

3. What are the causes and risk factors, including ethnicity, for hypertension?

4. What are the signs and symptoms of hypertension?

5. What is the current medical treatment, including lifestyle modifications and medications, for hypertension?

6. What are four classifications of hypertension in adults and recommendations for treatment based on blood pressure?

7. What are common complications of hypertension?

8. What nursing care teaching and learning strategies will you provide for patients with hypertension in adjusting to a chronic illness?

9. How will you know if your nursing interventions have been effective?

Hypertension, also known as high blood pressure, is a condition in which the blood pressure, on at least two or more readings on different dates after an initial screening, is found to be higher than normal. If the **systolic blood pressure** is above 140 mm Hg or the **diastolic blood pressure** is above 90 mm Hg, an adult age 18 or older should be evaluated for hypertension by a health care provider. As we age, the incidence of hypertension increases. Therefore we can expect to see more cases of hypertension as our population ages and life spans increase. For more information on hypertension, visit www.americanheart.org.

► PATHOPHYSIOLOGY OF HYPERTENSION

Normally the heart pumps blood through the body to meet the cells' needs for oxygen and nutrients. As it pumps, it forces blood through the blood vessels to the vital organs and tissues. The pressure exerted by blood on the walls of the blood ves-

---

hypertension: hyper—excessive + tensio—tension
systolic: systole—concentration
diastolic: diastole—expansion

sels is measured as blood pressure. Blood pressure is determined by **cardiac output** (CO), **peripheral vascular resistance** (PVR; the ability of the vessels to stretch), the **viscosity** (thickness) of the blood, and the amount of circulating blood volume. Decreased stretching ability and increased viscosity and fluid volume increase blood pressure. (See Chapter 15.)

Several processes influence blood pressure by controlling CO and PVR. These processes include nervous system regulation, arterial baroreceptors and chemoreceptors, the renin-angiotensin-aldosterone mechanism, and balancing of body fluids. One way blood pressure is influenced is through adjustment of the CO, which is the amount of blood that the heart pumps out each minute. The heart rate rises to increase CO in response to either physical or emotional activities that require more oxygen for the organs and tissue. PVR also influences blood pressure; it is the opposition that blood encounters as it flows through vessels. Anything causing blood vessels to become more narrow increases PVR. Any time PVR is increased, more pressure is needed to push the blood through the vessel, so blood pressure is increased as a result. If PVR is decreased, less pressure is needed. Increased arteriolar PVR is the main mechanism that elevates blood pressure in hypertension.

Factors that impair normal regulation of blood pressure may lead to hypertension. Many of these factors are not well understood. Sympathetic nervous system overstimulation, which causes vasoconstriction, can contribute to hypertension. Alterations in baroreceptors and chemoreceptors may also influence the development of hypertension. For example, baroreceptors may become less sensitive from prolonged increases in vessel pressure and subsequently fail to stimulate vasodilation through vessel stretching. Additionally, increases in hormones that cause sodium retention, such as aldosterone, lead to increased fluid retention. Changes in kidney function that alter the excretion of fluid also result in an increase in overall body fluid that may contribute to hypertension.

Primary Hypertension

Primary or **essential hypertension** is the chronic elevation of blood pressure from an unknown cause. These unknown causes influence the factors that control blood pressure, resulting in a hypertensive state.

Secondary Hypertension

Secondary hypertension has a known cause. In other words, it is a sign of another problem, such as a kidney abnormality, a tumor of the adrenal gland, or a congenital defect of the aorta. When the cause of secondary hypertension is treated before permanent structural changes occur, blood pressure usually returns to normal. Treatment may include surgery or medication.

Isolated Systolic Hypertension

Isolated systolic hypertension (ISH) is a systolic pressure of 160 mm Hg or greater and a normal diastolic pressure of

90 mm Hg or less. This is an abnormal finding. This type of hypertension occurs mainly in elderly people, although it can occur at any age. For people with a systolic pressure higher than 160 mm Hg and a normal diastolic pressure found on two separate visits, a referral to a physician for further evaluation is recommended. Treatment of ISH is recommended to decrease cardiovascular disease, especially risk of stroke. Lifestyle modifications are usually tried first if the systolic elevation is not too severe. If lifestyle modifications fail to reduce the systolic pressure, antihypertensive medication is added.

▶ SIGNS AND SYMPTOMS OF HYPERTENSION

Often hypertension causes no signs or symptoms other than elevated blood pressure readings. As a result, hypertension is referred to as the "silent killer." Patients with hypertension are often first diagnosed when seeking health care for reasons unrelated to hypertension. In a small number of cases, a patient with hypertension may complain of a headache, bloody nose, or blurred vision, although it is usually impossible for a patient to correlate the absence or presence of symptoms with the degree of blood pressure elevation.

Most signs and symptoms of hypertension stem from long-term effects on the large and small blood vessels of the heart, kidneys, brain, and eyes. These effects are known as target organ disease.

▶ DIAGNOSIS OF HYPERTENSION

Diagnosis is based on a health history to assess a patient's risk factors for hypertension, any previous diagnosis of hypertension, presence of any signs and symptoms, history of kidney or heart disease, and current use of medications. Although there are no diagnostic studies specifically for hypertension, there are diagnostic tests that can be helpful in identifying related information, such as damage to organs or blood vessels. The types of diagnostic tests performed depend on the stage of the hypertension (Table 16–1) or other medical conditions that may be present at the time of evaluation.

The Joint National Committee (JNC) on Prevention, Detection, Evaluation, and Treatment of High Blood Pressure was created by the National Heart, Lung, and Blood Institute of the National Institutes of Health. The sixth report of the JNC (called the JNC VI) defines normal and abnormal blood pressures for adults age 18 years and older and reclassified blood pressure measurement and treatment guidelines for physicians, clinicians, nurses, and community programs to follow. (See Table 16–1.) For more information on the JNC guidelines, visit www.nhlbi.nih.gov.

Diagnostic Tests

Laboratory tests such as urinalysis, blood urea nitrogen, and creatinine may indicate kidney damage from high blood pressure. Serum levels of sodium, calcium, chloride, potassium, magnesium, and phosphate are essential to evaluate the patient's fluid, electrolyte, and acid-base balance. A

viscosity: viscous—sticky

TABLE 16-1   THE STAGES OF HYPERTENSION AND RECOMMENDATIONS FOR FOLLOW-UP

Category	Systolic Blood Pressure (mm Hg)	Diastolic Blood Pressure (mm Hg)	Recommended Follow-up
Normal	<130	<85	2 years
High normal	130–139	85–89	1 year
Hypertension			
Stage 1	140–159	90–99	2 months
Stage 2	160–179	100–109	1 month
Stage 3	≥180	≥110	1 week-immediately

Adapted from The National Heart, Lung, and Blood Institute, The Sixth Report of the Joint National Committee on Prevention, Detection, Evaluation, and Treatment of High Blood Pressure. Arch Intern Med 157(21):2413, 1997.

complete blood cell count detects blood disorders. Tests of blood glucose, uric acid, cholesterol, and triglyceride levels help determine possible causes of cardiovascular and hypertensive disease. An electrocardiogram or chest x-ray examination may assist in recognizing abnormal heart function. These diagnostic tests are also used to monitor the effects of prescribed treatment.

▶ RISK FACTORS FOR HYPERTENSION

A combination of genetic (nonmodifiable) and environmental (modifiable) risk factors are thought to be responsible for the development of hypertension, although the cause remains unknown. Nonmodifiable risk factors—those that cannot be changed—include a family history of hypertension, age, ethnicity, and diabetes mellitus. Modifiable risk factors—those that can be changed—include stress levels, blood glucose levels, activity levels, smoking, and salt and alcohol intake. Weight reduction, improved meal planning, reduced salt and alcohol intake, increased physical activity, not smoking, and managing stress can all help to decrease blood pressure.

Nonmodifiable Risk Factors
Family History of Hypertension

Hypertension is seen more commonly among people with a family history of hypertension. Indeed, people with a family history have almost twice the risk of developing hypertension as those with no family history. People with a family history of hypertension should be encouraged to have their blood pressure checked regularly.

Age

People age differently because of their genetic and environmental risk factors and lifestyle habits. Thus the results of the aging process may be reflected in wide variations of blood pressure among elderly people. As a person ages, **plaque** builds up in the arteries and the blood vessels become stiffer and less elastic, causing the heart to work harder to force blood through the vessels. These vessel changes increase cardiac output to maintain blood flow into the circulation and subsequently raise blood pressure in the elderly.

Race and Ethnicity

Cultural Consideration Box 16–1 discusses hypertension among various ethnic groups.

Diabetes Mellitus

Two-thirds of adults who have diabetes mellitus also have hypertension. The risk of developing hypertension with a family history of diabetes and obesity is two to six times greater than when there is no family history. Approximately 80 percent of people with type 2 diabetes mellitus (non–insulin dependent) are overweight. Lifestyle modifications

BOX 16-1   CULTURAL CONSIDERATION

Hypertension continues to be the most serious health problem for African-Americans in the United States. More than 5 million of the 26 million African-Americans living in the United States are hypertensive. They suffer higher mortality and morbidity rates related to hypertension and at an earlier age. African-Americans from lower socioeconomic backgrounds have higher blood pressure than African-Americans from higher socioeconomic backgrounds. Additionally, African-Americans are 3.2 times more likely to develop kidney failure related to hypertension than European-Americans.

Hypertension among African-Americans is usually caused by increased renin activity resulting in greater sodium and fluid retention. Thus African-Americans respond better to diuretics such as furosemide (Lasix) and hydrochlorothiazide (HydroDIURIL) than to beta blockers such as propranolol (Inderal). Hypertension among European-Americans is more often caused by chemical imbalances; thus they respond better to beta blockers.

Chinese people are more sensitive than Caucasians to the effects of propranolol on heart rate and blood pressure, requiring only half the blood level of European-Americans to achieve a therapeutic effect. Propranolol is eliminated from the bodies of many Chinese persons at double the rate of European-Americans. They are more likely to suffer fatigue as a side effect. Thus the nurse must carefully monitor the Chinese patient for therapeutic and side effects.

Hypertension among Japanese-Americans is primarily related to the high sodium content of the Japanese diet, stress, and a high rate of cigarette smoking.

High rates of hypertension among Koreans and Filipinos are due to the stress of immigration, preserving foods in salt, and using condiments high in sodium.

and adherence to therapy are crucial to prevent the heart attacks, strokes, blindness, and kidney failure associated with high blood glucose and blood pressure levels.

Modifying Risk Factors

The JNC VI suggests advising patients with hypertension to use lifestyle modifications. These modifications include weight reduction, stress management, moderation of dietary sodium and alcohol intake, increased physical activity, and smoking cessation (Box 16–2). Lifestyle modifications are most often used with antihypertensive drugs to control hypertension and enhance the drug effects (Nutrition Notes Box 16–3).

Weight Reduction

There is a strong relationship between excess body weight and increased blood pressure. Weight reduction is one of

---

BOX 16–2 **Lifestyle Modifications for Hypertension**

- Lose weight.
- Limit alcohol intake.
- Get regular aerobic exercise.
- Decrease amount of salt intake.
- Include daily allowances of potassium, calcium, and magnesium.
- Stop smoking.
- Reduce dietary saturated fat and cholesterol.
- Manage stress.

Adapted from The National Heart, Lung, and Blood Institute, The Sixth Report of the Joint National Committee on Prevention, Detection, Evaluation, and Treatment of High Blood Pressure. Arch Intern Med 157(21):2413, 1997.

BOX 16–3  *Nutrition Notes*

Reducing Blood Pressure with Diet
In studies the Dietary Approaches to Stop Hypertension (DASH) diet reduced blood pressure significantly in **normotensive** people and produced even greater reductions in hypertensive people. Rather than emphasizing restriction of foods, the DASH diet increases the intake of certain commonly available, not specialty, foods. So that weight loss would not confound the results, participants in the clinical trial were given additional calories if they began to lose weight. All the tested diets contained about 3 grams of sodium. Also, everyone was instructed to consume no more than three caffeinated beverages and no more than two standard alcoholic beverages per day to minimize the effects of those substances on the results. On a 2000-calorie diet, an individual following the DASH diet would consume the following:

7 to 8 servings of grains
4 to 5 servings of vegetables
4 to 5 servings of fruits
2 to 3 servings of low-fat or nonfat dairy foods
2 or fewer servings of meats, poultry, and fish
4 to 5 servings per week of nuts, seeds, and legumes
2.5 servings of fats and oils

---

the most important, if not the most important, lifestyle modification to lower blood pressure. The health care provider and dietitian should be consulted to help the patient develop a weight-reduction diet and other methods of weight loss.

Stress Management

Reducing stress can play a major role in the treatment of patients with hypertension. Stress stimulates the sympathetic nervous system (fight-or-flight response). This stimulation causes the vessels to constrict and activates the renin-angiotensin mechanism. Those who have high stress levels tend to develop hypertension more often than those who do not. Additionally, studies have shown that an increase in stress can also raise the body's production of cholesterol, which can lead to cardiovascular disease. It is important for patients to learn how to deal with stress. For many patients, stress management techniques, such as exercise, relaxation therapies, yoga, meditation, and biofeedback, may be useful in controlling their response to stress and lowering their blood pressure.

Groups who are economically deprived typically have an increased incidence of hypertension. Factors such as poor nutritional habits, low-status jobs, frustration, discontent, and suppression of hostility contribute to stress-related hypertension. Other factors that may affect health include reduced access to quality health care and poor living conditions.

Meal Planning

SALT INTAKE. Research has shown that some people may develop high blood pressure by eating a diet high in salt. Patients whose blood pressure can be lowered by restricting dietary sodium are called salt sensitive. This sensitivity is particularly common among African-Americans, elderly persons, and patients with diabetes and obesity. Patients with hypertension should be instructed not to add salt while cooking and not to add table salt to their food. Processed foods or foods in which salt can be easily tasted (e.g., canned soups, ham, bacon, salted nuts) should also be avoided.

INTAKE OF POTASSIUM, CALCIUM, AND MAGNESIUM. Recent studies are inconclusive as to the role that low dietary potassium, calcium, and magnesium intake play in the development of high blood pressure. A balanced diet that ensures adequate intake of these nutrients is important in maintaining general health. Foods rich in potassium include oranges, bananas, and broccoli. Milk, yogurt, and spinach are rich in calcium. Vegetables such as spinach, garbanzo beans, and lima beans are good sources of magnesium. Whenever possible, fresh or frozen foods should be selected rather than canned foods to increase intake of these nutrients.

Alcohol Consumption

The regular consumption of three or more drinks per day can increase the risk of hypertension and cause resistance to antihypertensive therapy. The nurse should counsel hypertensive patients who drink alcohol to avoid it or at least limit their daily intake to 1 oz of alcohol per day (i.e., 2 oz of 100-proof whiskey, 8 oz of wine, or 24 oz of beer). Blood pressure may decrease or return to normal when alcohol consumption is limited or eliminated.

Exercise

People with sedentary lifestyles have an increased risk of hypertension compared with people who exercise regularly. Exercise helps prevent and control hypertension by reducing weight, decreasing peripheral resistance, and decreasing body fat. Moderate activity, such as 30 to 45 minutes of brisk walking three to five times weekly, is recommended by JNC VI. Patients with hypertension should be evaluated by a health care provider before starting any exercise program.

Smoking

Smoking is a major risk factor for cardiovascular disease and is associated with a high incidence of stage 3 hypertension.

CRITICAL THINKING

Mrs. Miller

Mrs. Miller, age 54, visits a health clinic because she has a headache every morning. The nurse collects data on Mrs. Miller and finds that she is an office manager, smokes a pack of cigarettes a day, eats fast food for lunch at her desk, has two adult children, and is recently divorced. Mrs. Miller has been in good health and takes two aspirin for her headaches daily.

1. What are Mrs. Miller's risk factors for hypertension?
2. What is the most significant patient information identified? Why?
3. Why is hypertension referred to as the silent killer?
4. Why should Mrs. Miller be told of the need for lifelong therapy if she is diagnosed with hypertension?

Answers at end of chapter.

Patients who smoke may show an increase in blood pressure because nicotine constricts the blood vessels. The nurse should instruct patients with hypertension to quit or decrease smoking to reduce the risk of myocardial infarction and stroke. A referral by the nurse to a smoking cessation program can be helpful.

▶ HYPERTENSION TREATMENT

The JNC VI has guidelines for selecting therapy based on severity of blood pressure risk factors and the presence of target organ disease or cardiovascular disease. If the no- or low-risk hypertensive patient's blood pressure remains at or above 140/90 mm Hg during the 6- to 12-month period of lifestyle modifications, the JNC VI advises the practitioner to add antihypertensive medications to the patient's antihypertensive therapy. The medication is selected by the physician or health care provider in collaboration with the patient. For patients with severe hypertension, high-risk factors, or target organ disease, drug therapy is started immediately together with lifestyle modifications.

The goal of drug therapy is to decrease diastolic pressure to 90 mm Hg or less and systolic pressure to 160 mm Hg or less. For patients with stage 1 or 2 hypertension, diuretics and beta blockers are recommended by the JNC VI for initial drug therapy. If the response is inadequate to achieve the blood pressure goal, the dosage may be increased, another drug may be substituted, or a drug from a different class may be added. There are eight categories of medications to treat hypertension: diuretics, alpha blockers, beta blockers, calcium channel blockers, angiotensin-converting enzyme (ACE) inhibitors, central agents, peripheral agents, and vasodilators. Examples of these medications are given in Table 16–2.

The treatment plan of lifestyle modifications and medications is effective only when patients accept the diagnosis

---

TABLE 16-2 ANTIHYPERTENSIVE MEDICATIONS

Medication	Action	Side Effects	Nursing Considerations
Diuretics			
Thiazide and thiazide-like diuretics Chlorothiazide (Diuril) Chlorthalidone (Hygroton) Metolazone (Zaroxolyn) Indapamide (Lozol) Hydrochlorothiazide (HydroDIURIL)	Inhibits sodium reabsorption in the distal tubule to remove sodium, extra-cellular fluid, and potassium, reducing cardiac output	Hypokalemia, hyponatremia, dehydration, muscle weakness, dry mouth, hypotension	Monitor electrolytes. Teach need for increased dietary or supplemental potassium. **Geriatric:** Teach about postural hypotension, especially in hot weather and need to get up slowly. Teach to take in morning to avoid getting up in night.
Loop diuretics Bumetanide (Bumex) Furosemide (Lasix)	Rapid action, block sodium and water reabsorption; diuresis profound	Same as thiazides	Monitor electrolytes. Must replace electrolytes. **Geriatric:** Same as thiazides.
Potassium-sparing diuretics Amiloride (Midamor) Spironolactone (Aldactone) Triamterene (Dyrenium)	Remove sodium and extracellular fluid but keep potassium	Hyperkalemia, hyponatremia, nausea, vomiting, diarrhea	Monitor electrolytes. Give after meals to decrease nausea. **Geriatric:** Mental confusion and unsteady gait may occur, so ensure safety.

TABLE 16-2   ANTIHYPERTENSIVE MEDICATIONS—CONT'D

Medication	Action	Side Effects	Nursing Considerations
Adrenergic Inhibitors			
Beta blockers Atenolol (Tenormin) Propranolol (Inderal) Pindolol (Visken) Carvedilol (Coreg) Metoprolol (Lopressor, Toprol XL) Labetalol (Normodyne)	Block beta₁ heart receptors to reduce heart rate and blood pressure; some reduce peripheral vascular resistance	Heart failure Bradycardia, shortness of breath, fatigue, insomnia, numb hands, dizzy, weakness, hyperglycemia	Assess for signs of heart failure. Teach not to stop agent abruptly. **Geriatric:** Assess for toxicity. Teach about postural hypotension and need to get up slowly.
Alpha blockers Prazosin (Minipress) Terazosin (Hytrin)	Vasodilate to reduce peripheral vascular resistance	Dizziness, hypotension, headache, nausea, palpitations	Monitor for hypotension. Teach to make position changes slowly. **Geriatric:** May be weak and fatigued, so ensure safety.
Central-acting adrenergic inhibitors Clonidine (Catapres) Methyldopa (Aldomet) Guanfacine (Tenex) Guanabenz (Wytensin)	Suppress central nervous system	Drowsiness, sedation, dry mouth, constipation Weakness	Caution patient not to take alcohol. Suggest gum or hard candy. Encourage a high-fiber diet. **Geriatric:** Assess bowel routine and monitor for constipation.
Adrenergic Inhibitors			
Peripheral-acting adrenergic inhibitors Guanadrel (Hylorel) Doxazosin (Cardura) Terazosin (Hytrin)	Decrease production of norepinephrine to lower blood pressure	Dizziness, headache Orthostatic hypotension, lethargy, nasal stuffiness, edema	Monitor for first dose orthostatic hypotension. Inform patient to report signs of depression: mood swings, insomnia, anorexia. **Geriatric:** Postural hypotension common. Assess for depression.
Angiotensin-converting Enzyme Inhibitors			
Benazepril (Lotensin) Captopril (Capoten) Enalapril maleate (Vasotec) Lisinopril (Prinivol, Zestril) Fosinopril (Monopril) Quinapril (Accupril)	Block conversion of angiotensin I to angiotensin II	Cough, rash	**Geriatric:** Assess for toxicity and recognize need for reduced dose with renal impairment.
Calcium Channel Blockers			
Amlodipine (Norvasc) Nifedipine (Procardia) Diltiazem (Cardizem) Verapamil (Calan, Isoptin) Felodipine (Plendil) Isradipine (DynaCirc, DynaCirc CR) Nicardipine (Cardene, Cardene SR) Nisoldipine (Sular)	Block entry of calcium into smooth muscles to reduce afterload with vasodilation	Headache, dizziness, peripheral edema, heart failure, arrhythmias	Monitor for hypotension. Treat headache with acetaminophen. **Geriatric:** Decreased dosage may need to be used.
Vasodilators			
Hydralazine (Apresoline) Minoxidil (Loniten) Nitroprusside (Nipride)	Vasodilate arteries, arterioles to reduce afterload	Headache, palpitation	Treat headache with acetaminophen.

of hypertension and include lifelong treatment in their daily routine. Patients should be instructed that antihypertensive therapy usually must be continued for the rest of their lives. Patients should be reminded that although they may be feeling better with the modifications and medications, the hypertension is still present even if it is well controlled. The patient should be told not to discontinue medications unless a physician or advanced practice nurse instructs the patient to do so.

Antihypertensive medications can have unpleasant side effects. Patients should be told what these side effects are and to report them if they do occur, so that alterations in the medications can be made if possible. Impotence can be one of the side effects of these medications. Male patients may be reluctant to discuss this side effect and instead choose to discontinue the medication. The nurse should be proactive and inform male patients about this side effect, so they will understand that if impotence occurs and is reported, the physician can make adjustments in the medication regimen.

L E A R N I N G   TIPS

Hypertension Lifestyle Modifications
L—Limit salt and alcohol.
I—Include daily potassium, calcium, and magnesium.
F—Fight fat and cholesterol.
E—Exercise regularly.
S—Stress management.
T—Try to quit smoking.
Y—Your medications are to be taken daily.
L—Lose weight.
E—End-stage complications will be avoided!

CRITICAL THINKING

Mrs. Bell

Mrs. Bell, 80 years old and a widow for 15 years, visits her physician. She lives alone in her own home with a bathroom down the hall from the bedroom. She has wood floors with throw rugs in the hall and a tile floor in the bathroom. Her son lives in the same city. She has a 10-year history of hypertension for which she is taking bumetanide (Bumex) and propranolol (Inderal), when she remembers them. She wears glasses and has a cataract. She has an unsteady gait and nocturia. She is 40 pounds overweight and is sedentary.

1. What are Mrs. Bell's modifiable and nonmodifiable risk factors for hypertension?
2. Why is Mrs. Bell taking bumetanide and propranolol to treat her hypertension?
3. What teaching methods could be used to help ensure that Mrs. Bell will understand and follow her treatment plan?
4. Why should safety needs be addressed in the nursing care plan?
5. What safety interventions should the patient and family be taught?

Answers at end of chapter.

► COMPLICATIONS OF HYPERTENSION

Common complications of hypertension include coronary artery disease, atherosclerosis, myocardial infarction (MI), stroke, and kidney or eye damage. The severity and duration of the increase in blood pressure determine the extent of the vascular changes causing organ damage. High blood pressure levels may also result in an increase in the size of the left ventricle, referred to as **hypertrophy.** Elevated blood pressure damages the small vessels of the heart, brain, kidneys, and retina. The results are a progressive functional impairment of these organs, or target organ disease.

► SPECIAL CONSIDERATIONS

Blood pressure should be well controlled before any invasive procedure. Hypertensive patients are at greater risk for strokes, MI, kidney failure, and pulmonary edema. These patients should be instructed to continue their blood pressure medications until the time of the procedure, unless otherwise directed by their physician or health care provider. They should resume their antihypertensive medications as soon as possible after the procedure, unless they are given new instructions by the physician.

► NURSING PROCESS

Nursing Assessment

Assessment of a patient with hypertension includes the patient's history, medications, and physical assessment. Assessing what hypertensive patients and their families know about hypertension and associated risk factors is essential for planning patient and family education and subsequent lifestyle modification needs.

Nursing Diagnosis

Based on the data and defining characteristics, a nursing diagnosis should be chosen in collaboration with the health care team and the patient (Nursing Care Plan Box 16–4). Because the risk of hypertension depends on the number and severity of modifiable risk factors, several nursing diagnoses may be applicable. Common diagnoses may include but are not limited to the following:

■ Deficient knowledge related to lack of exposure to hypertension information.
■ Ineffective therapeutic regimen management related to complexity of therapy, cost of medications, lack of symptoms, side effects of medications, need to alter long-term lifestyle habits, normal blood pressure controlled by therapy.

Planning

After nursing diagnoses have been identified, specific goals for blood pressure control should be set by the physician, nurse, patient, and family. Any barriers to meeting these goals should be discussed with the patient and family so that the plan can be carried out more effectively. The patient's goals are to (1) explain hypertension and needed medications, and (2) verbalize the ability to follow prescribed therapy.

Nursing Interventions

To meet their goals, patients may need more information, guidance, and support from the health care team. Referrals to other resources such as the dietitian, social worker, pharmacist, and home health nurse should be included (Home Health Hints). The patient and family should be allowed to maintain a sense of control, make informed decisions regarding care, and develop the skills necessary to make lifestyle modifications. Behavioral changes are the most difficult for the patient to initiate and maintain. The nurse plays a major role in therapy and treatment compliance for hypertensive patients. Elderly patients may have unique

BOX 16-4   NURSING CARE PLAN FOR THE PATIENT WITH HYPERTENSION

▶ **Ineffective therapeutic regimen management related to complexity of therapy, cost of medications, lack of symptoms, side effects of medications, need to alter long-term lifestyle habits, normal blood pressure controlled by therapy**

OUTCOMES
Verbalizes ability and willingness to comply with treatment.

EVALUATION OF OUTCOMES
Is patient able to state how lifestyle will include therapy? Does patient identify and problem solve barriers for therapy?

INTERVENTION	RATIONALE	EVALUATION
Identify patient's modifiable risk factors and lifestyle modification needs.	Identifying risk factors is the first step in planning therapy. Patient must understand the relationship of these risk factors with hypertension and complication development.	Can patient state rationale for modifying risk factors to prevent complication development?
Identify factors that are barriers to patient complying with therapy.	Factors such as finances, transportation, aging changes, patient motivation, habits, and reading and educational level can be barriers for therapy.	Are barriers present for patient?
Develop plan to overcome barriers. Make referrals as needed.	Identified barriers can be overcome with planning and intervention, such as referral to support groups or for financial assistance or prescription delivery service and instructions provided at level of patient's learning ability.	Have barriers been eliminated? Is patient willing to use referrals?

GERIATRIC

INTERVENTION	RATIONALE	EVALUATION
Assess ability to take medications daily: financially, obtaining refills, understanding directions.	Elderly patients may be on a fixed income, lack transportation, or lack ability to take several medications several times a day. Simplifying this process, to one medication if possible, can increase compliance.	Is patient able to obtain medications? Can patient self-administer medications accurately on daily basis?
Teach patient to take medications as prescribed and not to skip dosages.	Elderly patients may skip dosages to save money, reduce side effects, or reduce need to void.	Does patient take dosages as prescribed? Does patient express concern over cost, side effects, or frequent voiding?
Teach patient to change positions slowly to prevent falls.	Antihypertensive medications can cause hypotension, resulting in dizziness and weakness and possibly leading to falls.	Does patient understand how to change positions slowly? Does patient experience dizziness or weakness?

HOME HEALTH HINTS

- For meal planning, most patients eat fast foods occasionally. Assist them in choosing foods that are low in fat, sugar, and salt (e.g., choose chicken salads with low-fat dressing or fajitas without sour cream and guacamole).

- Because medication and electrolyte interaction can occur with salt substitute, which often contains potassium, the physician should be consulted before the patient uses it.

- Teach patients how to read labels for fat and salt content. If patients are on a 2- to 3-g sodium diet, instruct them on eating breads or cereals that contain 200 mg or less of sodium per serving or canned vegetables of 150 mg of sodium per serving. Fresh vegetables are better, but cost and storage must be considered. Providing written suggestions for the caregiver who does the grocery shopping increases compliance with diet therapy.

- Home exercise using weights can be improvised using canned goods and bags of sugar as weights. The amount of weight being used is easily identified for documentation by the labeling on the food item.

- The following suggestions may help a patient decrease or stop smoking: Use cinnamon mouthwash on arising; put away all ashtrays but one and keep it in a place not normally used for smoking; find ways to keep hands busy at times when usually holding cigarette, such as when drinking coffee or alcohol.

- Encourage patients to put "No Smoking" signs on their door to avoid passive smoking.

- Medication compliance can be a challenge for the elderly hypertensive patient. Instruct patients to take medication as prescribed even if they are feeling well or if side effects, which they should report, are present. If the medicines are too expensive for the patient, check with the physician and pharmacist for less expensive alternatives.

- During home visits count the amount of pills in a bottle to assess compliance. Remind the patient to get refills and keep physician appointments by writing them on the calendar.

- Because many of the antihypertensive medicines can cause bradycardia, teach the patient or caregiver to take the patient's pulse and to call the nurse if it is below 60 or the parameters defined by the physician or agency.

- Monitor carefully for symptoms of congestive heart failure when the patient is on beta blockers. This is a side effect that needs to be caught early and reported to the physician.

- Encourage the patient to obtain a home blood pressure monitoring device. Instruct the patient caregiver on proper use and logging the date, time, and reading obtained. The home health nurse should review the log on each visit.

- Patients should be instructed to weigh themselves every morning after voiding, to wear the same amount of clothing each time, and to keep a log for the nurse to review.

needs related to hypertension therapy (Gerontological Issues Box 16–5).

Evaluation

One of the best outcome criteria for evaluation of the plan of care for reducing hypertension is reaching a lowered blood pressure with minimal side effects and no evidence of target organ damage. Other indicators of success in implementing the plan of care are weight loss, maintaining a low-sodium diet, decreasing alcohol intake, decreasing or not smoking, and managing stress on a daily basis. If the patient is satisfied and comfortable with the quality of life after the modifications, another goal has been met. Ongoing evaluation is necessary, because this is a lifelong condition that may require periodic teaching and reinforcement of knowledge as deficiencies are identified.

BOX 16-5 GERONTOLOGICAL ISSUES

Managing Antihypertensive Therapy
For safety, teach older adults who take antihypertensive drugs to rise slowly to prevent the effects of orthostatic hypotension. Dizziness may result, increasing the risk of falling. Deficiencies in fluid volume can be a common problem for older people as well, and diuretics can contribute to fluid volume deficiencies. Careful fluid balance assessment is important to prevent dehydration. Older adults may be more sensitive to medications, so monitor them carefully for adverse effects. Some older people may need lower dosages.

▶ PATIENT EDUCATION

The use of educational programs in hospitals, clinics, churches, and health fairs helps increase patient motivation to adhere to antihypertensive therapy. Instructions provided by the nurse are directed toward helping patients control their blood pressure through self-care measures, as well as the prescribed medical regimen. To effectively control hypertension, patients must be knowledgeable about their condition. They need to be taught about hypertension, its treatment, and the need for a lifelong commitment to controlling it. (See Nursing Care Plan Box 16–4.)

▶ HYPERTENSIVE EMERGENCY

Hypertensive crisis is a severe type of hypertension characterized by rapidly progressive elevations in blood pressure with diastolic values above 110 mm Hg. Patients who are untreated, fail to comply with antihypertensive therapy, or stop their medication abruptly are at risk for hypertensive crisis. A patient with hypertensive crisis may experience morning headaches, blurred vision, dizziness, nosebleeds, and dyspnea. A diminished level of consciousness, weakness, paralysis, palpitations, or complaints of chest pain may also indicate a hypertensive crisis and should be reported immediately. Patients with hypertensive crises are admitted to the critical care unit. In some cases the blood pressure may need to be reduced within 1 hour to prevent organ damage. An intravenous medication such as nitroprusside (Nipride) may be given to quickly reduce blood pressure during this crisis.

REVIEW QUESTIONS

1. Which one of the following is true of the cause of primary hypertension?
   a. It is caused by a tumor of the adrenal gland.
   b. It is caused by renal artery stenosis.
   c. It is caused by coarctation of the aorta.
   d. There is no known cause.

2. Which one of the following is often the only sign of hypertension?
   a. Sacral edema
   b. Elevated blood pressure
   c. Tachycardia
   d. Neck vein distention

3. The nurse should give which one of the following instructions to a patient receiving a thiazide diuretic?
   a. Eliminate salt in your diet.
   b. Change positions slowly.
   c. Take your medication before bed.
   d. Empty your bladder after taking the first dose.

4. For which one of the following blood pressure readings should a follow-up visit be recommended?
   a. 128/70
   b. 136/76
   c. 140/94
   d. 142/86

5. Which one of the following is the most important lifestyle modification for the hypertensive patient who is obese?
   a. Reduce weight.
   b. Restrict salt intake.
   c. Quit smoking.
   d. Decrease alcohol intake.

REFERENCES

1. The Sixth Report of the Joint National Committee on Prevention, Detection, Evaluation, and Treatment of High Blood Pressure. Arch Intern Med 157(21):2413, 1997.

2. Appel, LJ, et al: A clinical trial of the effects of dietary patterns on blood pressure. N Engl J Med 336:1117, 1997.

Answers to CRITICAL THINKING

Mrs. Miller

1. Risk factors include gender; age; smoking cigarettes; a diet high in fat, salt, and calories; recent divorce; a stressful occupation; and morning headaches.
2. Morning headaches. Mrs. Miller may be experiencing an episode of hypertensive crisis and should be evaluated immediately by a health care provider.
3. "Silent killer" refers to the fact that there are often no signs or symptoms associated with hypertension.
4. Lifelong therapy is required because there is no cure for hypertension and complications need to be prevented.

Mrs. Bell

1. Nonmodifiable risk factors include age, gender, and history of hypertension. Modifiable risk factors include weight and compliance with antihypertensive therapy.
2. Diuretics remove excess salt and water to decrease blood volume and lower blood pressure. Beta blockers stop the beta receptors from receiving the message from the brain for the heart to work harder. Therefore the heart rate and blood pressure decrease.
3. Assess patient's reading level and primary language. Provide patient with written instructions in large letters about medications. Include family members and encourage their support in reinforcing the importance of adhering to the treatment plan.
4. Patient is 80 years old, makes frequent trips to the bathroom related to diuretics, and a side effect of propranolol is weakness and fatigue.
5. Make arrangements for a bedside commode to reduce the distance and urgency to get to the bathroom. Encourage the family to place night-lights in the bedroom, hall, and bathroom. Explain that throw rugs increase the risk of falling and that wood or tile floors can be slippery when wet and hard if a fall occurs. Encourage removal of throw rugs, and suggest carpeting these areas if possible. Suggest the use of safety bars in hall and bathroom for support or other walking aids as needed. If incontinence is a concern, suggest wearing an adult brief to prevent a wet, slippery floor. Suggest discussing with the physician an exercise program to increase strength, such as lifting small, lightweight objects (e.g., canned foods), squeezing a rubber ball, or riding an exercise bike if able. These exercises can be done while sitting so they are not a fall-risk activity.

17

NURSING CARE OF PATIENTS WITH INFLAMMATORY AND INFECTIOUS CARDIOVASCULAR DISORDERS

Linda S. Williams

KEY TERMS

beta-hemolytic streptococci (**BAY**-tuh-HEE-moh-**LIT**-ick STREP-toh-**KOCK**-sigh)

cardiac tamponade (**KAR**-dee-yak TAM-pon-**AYD**)

cardiomegaly (KAR-dee-oh-**MEG**-ah-lee)

cardiomyopathy (KAR-dee-oh-my-**AH**-pah-thee)

chorea (kaw-**REE**-ah)

Dressler's syndrome (**DRESS**-lers **SIN**-drohm)

emboli (**EM**-boh-li)

infective endocarditis (in-**FECK**-tive EN-doh-kar-**DYE**-tis)

international normalized ratio (IN-ter-**NASH**-uh-nul **NOR**-muh-lized **RAY**-she-oh)

myectomy (my-**ECK**-tuh-mee)

myocarditis (MY-oh-kar-**DYE**-tis)

pericardial effusion (PER-ee-**KAR**-dee-uhl ee-**FYOO**-zhun)

pericardial friction rub (PER-ee-**KAR**-dee-uhl **FRICK**-shun RUB)

pericardiectomy (PER-ee-kar-dee-**ECK**-tuh-mee)

pericardiocentesis (PER-ee-KAR-dee-oh-sen-**TEE**-sis)

pericarditis (PER-ee-kar-**DYE**-tis)

petechiae (pe-**TEE**-kee-ee)

rheumatic carditis (roo-**MAT**-ick kar-**DYE**-tis)

rheumatic fever (roo-**MAT**-ick **FEE**-ver)

thrombophlebitis (THROM-boh-fle-**BYE**-tis)

QUESTIONS TO GUIDE YOUR READING

1. What are the pathophysiology, etiology, signs and symptoms, diagnostic tests, and medical treatment for rheumatic carditis?

2. What nursing care would you provide for a patient with rheumatic carditis?

3. What are the pathophysiology, etiology, signs and symptoms, diagnostic tests, and medical treatment for infective endocarditis, myocarditis, and pericarditis?

4. What nursing care would you provide for infective endocarditis, myocarditis, and pericarditis?

5. What are the pathophysiology, etiology, signs and symptoms, complications, diagnostic tests, and medical treatment for dilated, hypertrophic, and restrictive cardiomyopathy?

6. What nursing care would you provide for dilated, hypertrophic, and restrictive cardiomyopathy?

7. What are the pathophysiology, etiology, signs and symptoms, complications, diagnostic tests, and medical treatment for thrombophlebitis?

8. What are risk factors, prevention measures, and nursing care for thrombophlebitis?

▶ INFLAMMATORY AND INFECTIOUS CARDIAC DISORDERS

The entire heart (**rheumatic carditis**) or layers of the heart (endocarditis, **myocarditis,** and **pericarditis**) can become inflamed or infected (Fig. 17–1).

Rheumatic Carditis
Pathophysiology and Etiology

A serious complication of **rheumatic fever** is rheumatic carditis. Rheumatic fever occurs as an autoimmune reaction to an upper respiratory (throat) group A **beta-hemolytic streptococci** infection (Table 17–1). Two to three weeks after the streptococcal infection, rheumatic fever occurs. Although rheumatic fever can occur at any age, it typically occurs between ages 5 and 15, and can recur along with rheumatic carditis.

With carditis from rheumatic fever, all layers of the heart become inflamed. Pericardial layers are covered with an ex-

---

myocarditis: myo—muscle + kardia—heart + itis—inflammation
pericarditis: peri—around + kardia—heart + itis—inflammation

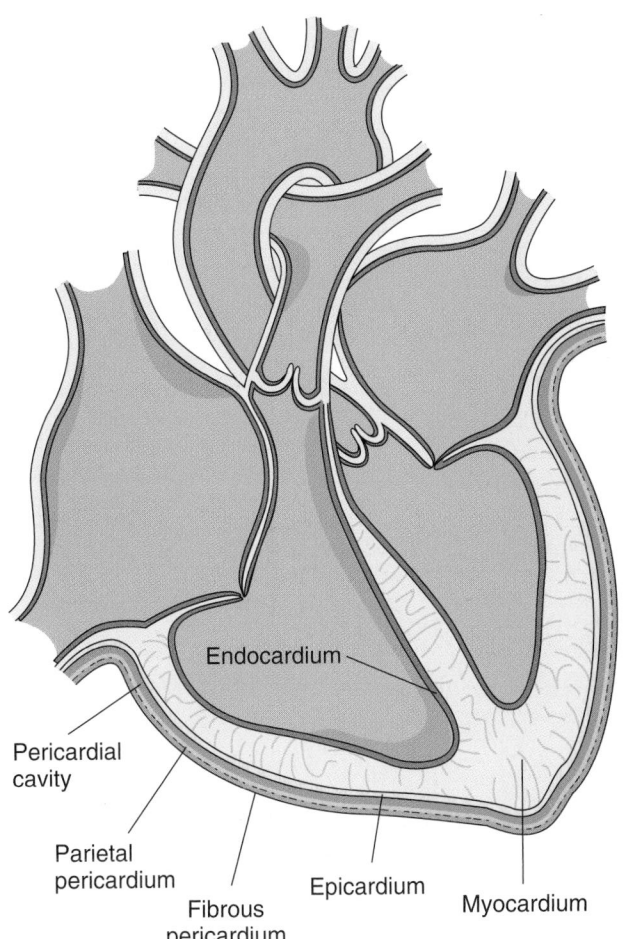

Figure 17-1 Layers of the heart.

TABLE 17-1	RHEUMATIC FEVER SUMMARY
Signs and symptoms	Fever
	Polyarthritis
	Subcutaneous nodules
	Chorea with rapid, uncontrolled movements
	Rheumatic carditis
	Arthralgia
	Pneumonitis
Diagnostic tests	Throat culture (identifies streptococcal infection)
	Antistreptolysin O titer level greater than 250 IU/mL
	Elevated ESR and WBC
Medical treatment	Antimicrobial medication
	Anti-inflammatory medication
	Control of symptoms
	Prophylactic antibiotics
Nursing diagnoses	Acute pain related to joint or cardiac inflammation
	Anxiety related to disease process
	Decreased cardiac output related to valvular damage and carditis

ESR = erythrocyte sedimentation rate.

---

udate and become thickened. As healing takes place, the pericardial sac can be damaged or destroyed by fibrosis. Nodules (Aschoff's bodies) form in myocardial tissue that become scar tissue over time. The endocardium, specifically the mitral valve, is the most seriously affected. Tiny, pinhead-size vegetations from blood and fibrin form on the valve leaflets. This can lead to thickening, fibrosis, and calcification of the valve leaflets and support structures. If the valve leaflets do not close completely, regurgitation of blood can occur. If the valve leaflets do not open fully (valvular stenosis), blood movement is impaired and severe heart failure may result.

Signs and Symptoms

Rheumatic carditis signs and symptoms include tachycardia, heart murmur, **pericardial friction rub,** chest pain, heart enlargement, electrocardiogram changes (e.g., PR interval lengthening), and evidence of heart failure.

Prevention

Preventing rheumatic fever by detecting and treating streptococcal infections promptly with penicillin (or erythromycin [Erythromid] in the case of a patient with a penicillin allergy) is important to prevent rheumatic carditis. Signs and symptoms of a throat streptococcal infection include sudden sore throat, fever of 101° F to 104° F, chills, throat redness with exudate, sinus or ear infection, and lymph node enlargement. A throat culture provides an accurate diagnosis. For those who have had rheumatic fever, lifelong prophylactic antibiotics before dental or invasive procedures are given because of the risk of developing **infective endocarditis** (discussed later).

Medical Management

Antibiotics are started promptly. Activity is limited based on the severity of cardiac involvement. Other treatment for cardiac involvement is based on symptoms.

Nursing Management

A history of recent illnesses (e.g., sore throat, streptococcal infection, or scarlet fever) is obtained from the patient, including past episodes of rheumatic fever, heart disease, joint pain, and current medications. A physical assessment detects murmurs, pericardial friction rub, and heart failure signs (e.g., jugular vein distension, edema, dyspnea, crackles, cough, and fatigue). Vital signs are documented, noting fever and tachycardia.

Nursing care focuses on relieving the patient's pain and anxiety, maintaining normal cardiac function, and educating the patient about the disease. Pain is relieved with analgesics, aspirin, or corticosteroids as ordered. Vital signs are monitored and symptoms of heart failure are reported to the physician. An explanation of the disease and its treatment is provided to promote understanding of acute and lifelong prophylactic treatment. A patient statement of willingness to maintain health supports goal achievement.

Infective Endocarditis

Infective endocarditis (IE) is an infection of the endocardium. Males develop IE more often than females. Even with antibiotic treatment, this infection can be fatal.

Pathophysiology

Cardiac defects result in turbulent blood flow that erodes the endocardium. Infective endocarditis begins when the invading organism (such as bacteria) attaches to eroded endocardium where platelets and fibrin deposits have formed a vegetative lesion. Then more platelets and fibrin cover the multiplying organism. This covering protects the microbes, reducing the ability to destroy them. Damage to valve leaflets occurs as the vegetations grow. As blood flows through the heart, these vegetations may break off and become **emboli.**

With bacteremia, bacteria attach mainly to the valves of the heart, although any heart endothelial surface can be infected. Damaged valves from conditions such as mitral valve prolapse with regurgitation, rheumatic heart disease, congenital defects, and valve replacements are especially prone to bacterial invasion. The mitral valve is the valve most commonly infected, with the aortic valve the second most commonly infected. Heart failure may result from valve damage, especially of the aortic valve.

Etiology

Portals of entry for organisms into the bloodstream result from intravenous (IV) drug use; surgery; dental or invasive procedures; and infections of the skin and gastrointestinal or genitourinary tracts. Risk factors include the following:

- IV drug use
- Compromised immune system
- Congenital or valvular heart disease (e.g., mitral valve prolapse with regurgitation, valvular replacement surgery, or rheumatic carditis)
- Gingival gum disease

Prevention

Because dental disease is a common cause of IE, oral care is an important preventative measure. Additionally, antibiotic therapy before invasive or dental procedures is recommended for patients with cardiac disease or prior endocarditis.[1] It is important that these patients are taught to inform health care team members (including their dentists) of their history before a procedure. Ensure that this history is reported to the physician, so that prophylactic therapy can be given if needed.

Signs and Symptoms

Fever (99° F to 103° F [37.2° C to 39.4° C]) is a common sign, although the elderly may be afebrile (Table 17–2). A

TABLE 17-2	INFECTIVE ENDOCARDITIS SUMMARY
Signs and symptoms	Fever Murmur Night sweats Fatigue Weight loss Weakness Pain in abdomen, joints, muscles, back Nailbed splinter hemorrhages Petechiae
Diagnostic tests and findings	Blood cultures (identify causative organism) Transesophageal echocardiography (identifies vegetations on heart valves) WBC count with differential (identifies elevation)
Medical treatment and purpose	Acute therapy: IV antimicrobial medications such as penicillin, vancomycin, amphotericin B (to cure infection) Antipyretics (to reduce fever) Rest (to decrease cardiac workload) Surgical valve replacement (to restore normal valve function) Prophylactic antibiotic therapy (to prevent infection)
Complications	Emboli Heart failure
Nursing diagnoses	Acute pain related to fever from cardiac infection Activity intolerance related to reduced oxygen delivery from decreased cardiac output Decreased cardiac output related to impaired valvular function or heart failure Deficient diversional activity related to restricted mobility from prolonged IV therapy Ineffective tissue perfusion related to emboli

new murmur is usually heard as valvular damage occurs. Splinter hemorrhages may be seen in the distal nailbed (black or red-brown longitudinal short lines). **Petechiae** (tiny red flat spots) resulting from microembolization of the vegetation may occur on mucous membranes, conjunctivae, or skin (Fig. 17–2). Janeway lesions (small, painless red-blue lesion on palms and soles) are an acute finding. Osler's nodes (small, painful nodes on fingers and toes) from cardiac emboli are a late finding (Fig. 17–3).

Complications

Vegetative emboli can be a major complication of IE. If organ embolization occurs, signs and symptoms that reflect

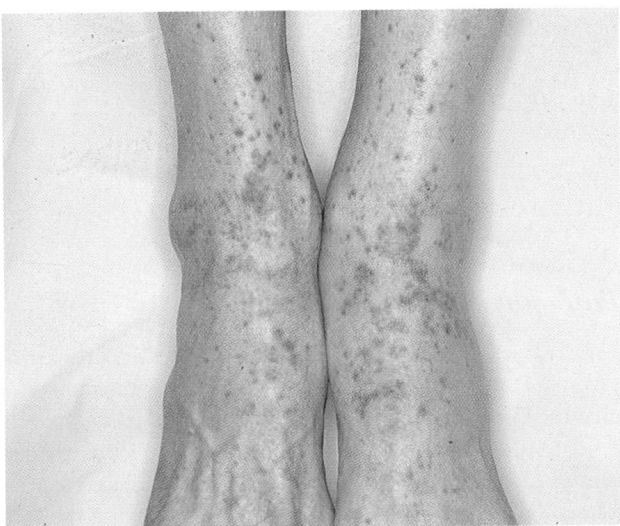

Figure 17-2 Petechiae. (From Goldsmith, L: Adult & Pediatric Dermatology. FA Davis, Philadelphia, 1997, p 61.)

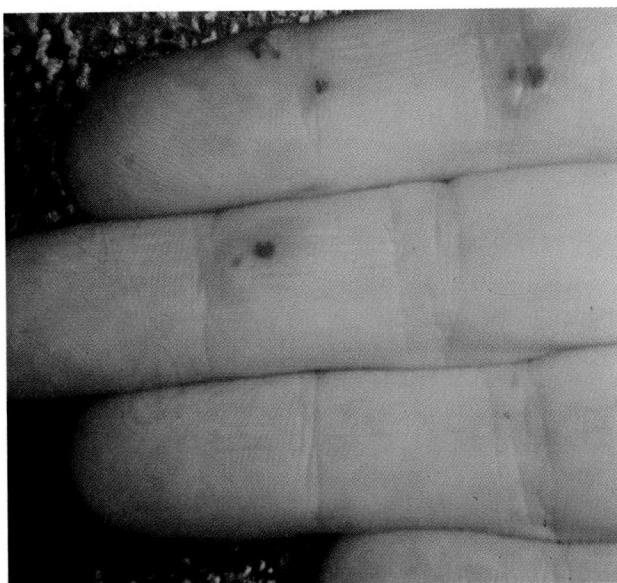

Figure 17-3 Osler's nodes. (From Goldsmith, L: Adult & Pediatric Dermatology. Philadelphia, FA Davis, 1997, p 188.)

---

petechiae: petecchia—skin spot

the organ that was affected by the emboli are seen. Brain emboli may produce changes in level of consciousness or stroke. Kidney emboli cause pain in the flank area, hematuria, or renal failure. Emboli in the spleen cause abdominal pain. Emboli in the small blood vessels can impair circulation in the extremities. Pulmonary emboli result in sudden dyspnea, cough, and chest pain. (See Chapter 28.)

Heart structures can be damaged or destroyed by IE. Stenosis or regurgitation of a heart valve may also result. As the infection progresses and causes more damage to heart structures, heart failure may occur. (See Chapter 21.)

Diagnostic Tests

Table 17–2 shows diagnostic tests for IE. Positive blood cultures identify the causative organism of IE, and echocardiography shows cardiac effects.

Medical Treatment

Initial treatment begins with hospitalization. An antimicrobial drug is selected that will destroy the organism identified by the blood culture. For bacterial infections, penicillin (or vancomycin for those allergic to penicillin) is commonly used. These medications are given intravenously over a period of 4 to 6 weeks, often once a day. A lengthy course of high-dose antibiotics is needed to penetrate the vegetations and reach the microbes. Rest and supportive symptom care are also used. If afebrile and without complications, the patient is discharged to continue IV antibiotic therapy at home. The patient's response to the drug is monitored via the home care nurse and through laboratory testing. Changes in antibiotics may be made in response to side effects, allergies, organism resistance to the drug, or relapses.

Surgical replacement or repair of valves is usually required for patients with severely damaged heart valves, prosthetic valve infection, recurrent infection, multiple emboli from damaged valves, or heart failure. (See Chapter 22.) Recovery from the disease can be greatly improved with surgery. Antimicrobial therapy continues after surgery.

Nursing Management

A patient history is obtained that includes risk factors for IE and recent infections or invasive procedures (Table 17–3). Vital signs are measured and recorded, and heart sounds are auscultated to detect murmurs. Signs of heart failure and emboli are noted. The physician is notified immediately if circulatory impairment, such as cold skin, decreased capillary refill, cyanosis, or absent peripheral pulses in an extremity, or symptoms of organ-related emboli are detected.

Nursing care aims to maintain the patient's normal cardiac function, monitor symptoms and complications, provide medications as ordered, and teach the patient about the disease, medications, treatments, and prevention techniques. See Nursing Care Plan Box 17–1 for the Patient with Infective Endocarditis for specific nursing interventions.

Teaching about the disease provides patients and families with the ability to provide IV antibiotics at home and promotes health maintenance to prevent IE. Good hygiene

TABLE 17-3 NURSING ASSESSMENT FOR PATIENTS WITH INFECTIVE ENDOCARDITIS

Subjective Data

Health History
Infections (rheumatic fever, previous endocarditis, streptococcal or staphylococcal, syphilis)?
Cardiac disease (valvular surgery, congenital)?
Childbirth?
Invasive procedures (surgery, dental, catheterization, IV therapy, cystoscopy, gynecologic)?
Malaise?
Anorexia?

Medications
Steroids, immunosuppressants, prolonged antibiotic therapy, IV drug use, alcohol abuse?

Respiratory
Dyspnea on exertion or when lying (orthopnea)?
Cough?

Cardiovascular
Palpitations, chest pain, fatigue, or activity intolerance?

Musculoskeletal
Weakness, arthralgia, myalgia?

Knowledge of Condition

Objective Data

Fever, Diaphoresis

Respiratory
Crackles, tachypnea

Cardiovascular
Murmurs, tachycardia, dysrhythmias, edema, headache

Integumentary
Nailbed splinter hemorrhages, petechiae on lips, mouth, conjunctivae, feet, or antecubital area

Renal
Hematuria

Diagnostic Test Findings
Anemia, elevated WBC count, elevated ESR, positive blood cultures, ECG showing conduction problems, echocardiogram showing valvular dysfunction and vegetations, chest x-ray exam show heart enlargement (cardiomegaly) and lung congestion

ECG = electrocardiogram; ESR = erythrocyte sedimentation rate; WBC = white blood cell.

CRITICAL THINKING

Mrs. Jones

Mrs. Jones, 28 years old, is admitted to the hospital with a fever of 100° F (37° C), chills, fatigue, anorexia, and pain in her joints. A physical assessment reveals splinter hemorrhages in left index finger nailbed and petechiae on her chest. She is diagnosed with a heart murmur and infective endocarditis.

1. Why is a heart murmur heard with endocarditis?
2. What do splinter hemorrhages look like?
3. What do petechiae indicate?
4. How would you document Mrs. Jones's assessment findings?
5. What type of medication does the nurse expect to be ordered to treat the infection?
6. Why does Mrs. Jones have chills if her temperature is elevated?
7. What signs and symptoms might occur if the complications of heart failure develop?
8. Why does Mrs. Jones need to be taught that she needs prophylactic antibiotics before dental or invasive procedures?

Answers at end of chapter.

Pericarditis
Pathophysiology and Etiology

Pericarditis is an acute or chronic inflammation of the pericardium (the sac surrounding the heart). The inflammation creates a problem for the heart as it tries to expand and fill. As a result, ventricular filling is reduced, which then decreases cardiac output and blood pressure. Acute pericarditis can be caused by a variety of factors, including the following:

■ Infections: viruses, bacteria, fungi, or Lyme disease
■ Drug reactions
■ Connective tissue disorders: systemic lupus erythematosus, rheumatic fever, or rheumatoid arthritis
■ Neoplastic disease
■ Postpericardiotomy (i.e., after cardiac surgery)
■ Postmyocardial infarction, 1 to 12 weeks (**Dressler's syndrome**)
■ Renal disease or uremia
■ Trauma from chest injury or invasive thoracic procedures

Acute pericarditis resolves usually in less than 6 weeks. Recurrence is possible.

Chronic constrictive pericarditis is the result of fibrous scarring of the pericardium. The heart becomes surrounded by a thickened, stiff sac that limits the stretching ability of the heart's chambers for filling. Heart failure may result. Chronic constrictive pericarditis results from neoplastic disease and metastasis, radiation, or tuberculosis.

Signs and Symptoms

Chest pain is the most common symptom of pericarditis (Table 17–4). The pain is located substernally and over the heart and may radiate to the clavicle, neck, left scapula, or epigastric area. The intense, sharp, grating pain increases with deep inspiration, coughing, moving the trunk, or lying

is essential, including dental care. Skin care includes bathing, using proper handwashing technique with soap, avoiding nailbiting, not popping pimples or lancing boils, and washing and applying antibiotic ointment to cuts. Brushing with a soft-bristle toothbrush (prevents gum trauma) twice a day reduces plaque formation, which traps bacteria. Twice yearly dental cleaning using prophylactic antibiotics is important. It is essential that patients understand the need to request and take prophylactic antibiotics as needed before invasive procedures. Patients are also taught symptom recognition (e.g., fever, chills, sweats) and seeking of prompt medical care. The patient is educated on the importance of having blood cultures drawn before antibiotics are started. The patient's statement of understanding and a willingness to follow lifestyle changes supports goal achievement.

BOX 17-1   NURSING CARE PLAN FOR THE PATIENT WITH INFECTIVE ENDOCARDITIS

 Decreased cardiac output related to impaired valvular function or heart failure

PATIENT OUTCOMES
Has adequate cardiac output as evidenced by vital signs within normal limits, no dyspnea or fatigue.

EVALUATION OF OUTCOMES
Are patient's vital signs within normal limits with no dyspnea or fatigue?

INTERVENTIONS	RATIONALE	EVALUATION
Assess vital signs, murmurs, dyspnea, and fatigue.	Vital signs, dyspnea, and fatigue are indicators of cardiac output decline.	Are vital signs within normal limits with no dyspnea or fatigue?
Give oxygen as ordered.	Supplemental oxygen provides more oxygen to the heart.	Are breathing pattern and oxygen saturation within normal limits?
Provide rest as ordered.	Cardiac workload and oxygen needs are reduced with rest.	Are vital signs within normal limits and no fatigue reported?
Elevate head of bed 45 degrees.	Venous return to heart is reduced and chest expansion improved.	Are vital signs within normal limits and respirations easy?

 Activity intolerance related to reduced oxygen delivery from decreased cardiac output

PATIENT OUTCOMES
Patient will state less fatigue in response to activity.

EVALUATION OF OUTCOMES
Does patient report less fatigue? Is patient able to participate in desired activities?

INTERVENTIONS	RATIONALE	EVALUATION
Assist with activities of daily living (ADLs) prn.	Assistance conserves energy.	Are ADLs completed?
Provide rest and space activities.	Cardiac workload and oxygen needs are reduced with rest.	Does patient report less fatigue?

 Deficient diversional activity related to restricted mobility from prolonged intravenous therapy

PATIENT OUTCOMES
States participation in satisfying diversional activities.

EVALUATION OF OUTCOMES
Does patient participate in diversional activities? Does patient state satisfaction with activities?

INTERVENTIONS	RATIONALE	EVALUATION
Assess patient's preferred activities and hobbies.	Activity preference should be known to plan satisfactory diversional activities.	Are patient's preferred activities known?
Plan patient's schedule around relaxing and fun activities.	Self-esteem is fostered with increased patient control.	Does patient offer input into scheduled care? Is input followed?
Use pet therapy.	Individuals who can interact with pets live longer and healthier.	Does patient state enjoyment of pet therapy?
Provide a mix of physical, mental, and social activities on a rotating schedule.	Rotating stimulating activities and visitors will keep patient interested and avoid fatigue.	Does patient state satisfaction in activities with no fatigue?

flat. The pain may be relieved by sitting up and leaning forward. Other symptoms depend on the cause of the pericarditis and may include dyspnea, low-grade fever, and cough. Dyspnea occurs as a result of decreased cardiac output and reduced oxygenation.

The classic sign of pericarditis is a pericardial friction rub, a grating, scratchy, high-pitched sound that is heard when a rub is present. The rub is a result of friction from the inflamed pericardial and epicardial layers rubbing together as the heart fills and contracts. Depending on the severity of the pericarditis, the rub may be faint when auscultated or loud enough to be audible without auscultation. The rub may be heard intermittently or continuously. It is usually heard over the lower left sternal border of the chest during each heartbeat. It is present in approximately 50 percent of those with pericarditis.

Chronic constrictive pericarditis produces the signs and symptoms of right-sided heart failure. Atrial fibrillation may also be seen in some patients with chronic constrictive pericarditis.

TABLE 17–4	PERICARDITIS SUMMARY
Signs and symptoms	Chest pain
	Dyspnea
	Low-grade fever
	Cough
	Pericardial friction rub
Diagnostic tests	Complete blood cell count
	Electrocardiogram
	Echocardiogram
	MRI
	CT
Medical treatment	Anti-inflammatory medication
	Corticosteroids
	Pericardiocentesis
	Pericardial window
Complications	Pericardial effusion
	Cardiac tamponade
Nursing diagnoses	Acute pain related to inflammation of pericardium
	Anxiety related to disease process
	Decreased cardiac output related to cardiac constriction

Diagnostic Tests

Table 17–4 lists diagnostic tests. The electrocardiogram reveals ST-T wave elevation in all leads. Echocardiogram results show **pericardial effusion**s. Serum laboratory tests focus on causes of the pericarditis, such as an elevated white blood cell (WBC) count, indicating a bacterial or viral infection, or elevated blood urea nitrogen or creatinine levels, indicating uremia. Fluid obtained during **pericardiocentesis** is examined to diagnose the cause. In chronic constrictive pericarditis, computed tomography (CT) or magnetic resonance imaging (MRI) may show a thickened pericardium.

Medical Treatment

If the patient is unstable, prompt intervention is required, such as an emergency pericardiocentesis. When the patient is stable, the cause is determined so that appropriate treatment can be administered, such as antibiotics for bacterial infections. Bedrest is used to reduce the heart's workload during acute symptoms. Nonsteroidal anti-inflammatory drugs (NSAIDs) such as indomethacin (Indocin) are given to resolve inflammation and reduce pain. Corticosteroids may be used when NSAIDs are not effective. Hemodialysis is used to treat uremic pericarditis.

L E A R N I N G   TIPS

To simulate the sound of a pericardial friction rub, hold the diaphragm of a stethoscope against the palm of one hand; listen through the stethoscope as you rub the index finger of the opposite hand over the knuckles of the hand holding the diaphragm. The sound you hear is similar to that of a pericardial friction rub.

Chronic effusive pericarditis can be treated with a pericardial window to allow continuous drainage of pericardial fluid into the pleural space. A pericardial window is created surgically by removing a portion of the outer pericardial layer.

Chronic constrictive pericarditis is treated with **pericardiectomy,** which is the surgical removal of the entire tough, calcified pericardium. Pericardiectomy relieves constriction of the heart and allows normal filling of the ventricles.

Complications

A pericardial effusion (buildup of fluid in pericardial space) is the most common complication of pericarditis. A rapidly developing effusion, such as one occurring from trauma, can produce symptoms at smaller amounts of fluid than slowly developing effusions, such as pericarditis from tuberculosis, with larger amounts of fluid. The increasing fluid presses on nearby tissue. Pressure on lung tissue can produce dyspnea, cough, and tachypnea. The heartbeat sounds distant. The body's compensatory mechanisms attempt to maintain blood pressure.

As the fluid accumulation grows, **cardiac tamponade,** another complication of pericarditis, can occur. Cardiac tamponade is a life-threatening compression of the heart by fluid accumulated in the pericardial sac. Cardiac output drops and then blood pressure falls as compensatory mechanisms fail. The patient shows symptoms of decreased cardiac output, such as restlessness, confusion, tachycardia, and tachypnea. Jugular vein distention is present from increased venous pressure, and heart sounds are distant. Cardiac tamponade requires immediate treatment with pericardiocentesis to puncture the pericardium with a 16-gauge needle and remove the excess fluid in the pericardial sac (Fig. 17–4). After the procedure, the patient is monitored for complications, such as dysrhythmias, laceration of a coronary artery, or laceration of the myocardium or pneumothorax.

Nursing Management

A patient history is obtained that includes any cardiac disease, recent infections, and current medications. Chest pain, pericardial friction rub, heart sounds, and signs of heart failure are noted. Vital signs are documented, noting fever and tachycardia.

Nursing care focuses on relieving the patient's pain and anxiety and maintaining normal cardiac function. Symptoms are monitored to detect complications. Pain, which may be severe, is relieved by giving NSAIDs or corticosteroids as ordered. Allowing the patient to assume a position of comfort by sitting up and leaning forward also re-

pericardiocentesis: peri—around + kardia—heart + centesis—puncture
cardiac tamponade: kardia—heart + tamponade—plug

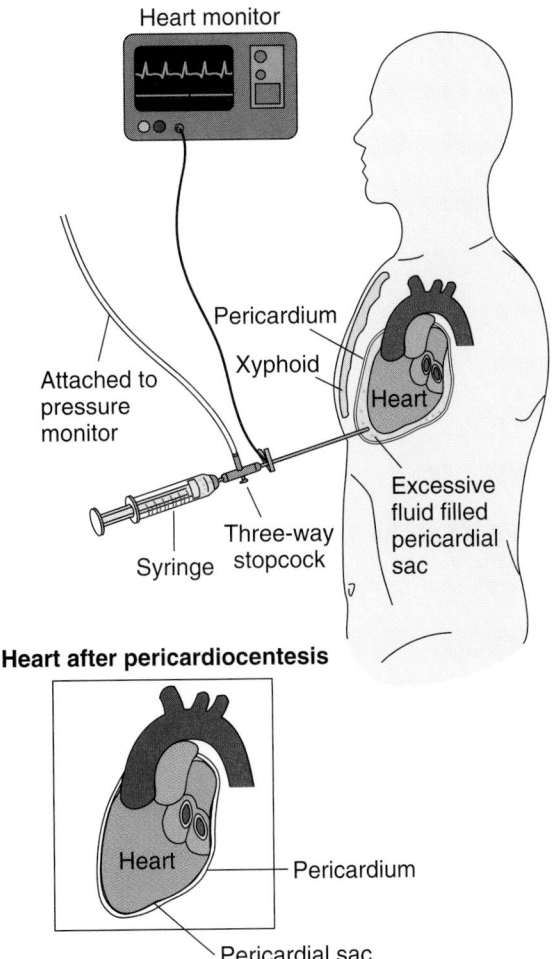

Figure 17-4 Pericardiocentesis.

lieves pain. Maintenance of normal cardiac function includes monitoring vital signs and observing for the presence of symptoms of cardiac tamponade or heart failure. The detection of these symptoms is immediately reported to the physician. Teaching the patient about pericarditis and its treatment relieves anxiety and allows a feeling of control by allowing the patient to make knowledgeable health care decisions.

Myocarditis
Pathophysiology and Etiology

In myocarditis, inflammation of the myocardium occurs. The amount of muscle destruction and necrosis that occurs as a result of myocarditis determines the extent of damage to the heart. The heart may enlarge in response to the damaged muscle fibers, although most cases of myocarditis are benign, with few signs or symptoms.

Myocarditis is a rare condition that most commonly develops following a virus. Other causes are bacteria, parasites, fungi, rickettsiae, spirochetes, medications, lead toxicity, autoimmune factors, human immunodeficiency virus (HIV), rheumatic fever, systemic lupus erythematosus, pericarditis or infective endocarditis, or cardiac transplant rejection.

Signs and Symptoms

Signs and symptoms of myocarditis vary from none to severe cardiac manifestations. Fatigue, fever, pharyngitis, malaise, dyspnea, palpitations, muscle aches, gastrointestinal (GI) discomfort, and enlarged lymph nodes may occur early from a viral infection. Cardiac manifestations such as chest discomfort, pain, or tachycardia may occur about 2 weeks after a viral infection. Occasionally, sudden death may occur.

Diagnostic Tests

A percutaneous endomyocardial biopsy during the first 6 weeks of inflammation is the preferred diagnostic test for myocarditis, although it is positive only about 30 percent of the time. MRI and gallium-67 scanning are helpful. An electrocardiogram (ECG) shows dysrhythmias, commonly sinus tachycardia.

Medical Treatment

Treatment is aimed at the cause, if known, such as antibiotics for bacterial infections. Interventions to reduce the heart's workload, such as bedrest, limited activity, and administration of oxygen, are used. Heart failure is treated with medication to strengthen the heart's contractility and slow the heart's rate, which reduces the heart's workload and oxygen needs. Digoxin is often used to treat heart failure. With myocarditis, the heart is sensitive to digoxin, which may be used to treat heart failure, and toxicity may occur even with small doses. The patient should be monitored closely for signs of digoxin toxicity, which may include anorexia, nausea, vomiting, bradycardia, dysrhythmias, or malaise.

Nursing Management

Recent illnesses, toxin exposure, cardiac diseases, activity tolerance, and current medications are documented. Vital signs and signs of heart failure, such as jugular vein distention, peripheral edema, crackles, and dyspnea are noted.

Nursing care is aimed at the patient's maintenance of normal cardiac function by monitoring vital signs and symptoms and administering medications as ordered. Reducing the patient's anxiety and increasing the patient's knowledge can be achieved through teaching about the disease. Determining diversional activities with the patient for times when activity is restricted further reduces anxiety.

Interventions to reduce fatigue include providing assistance as needed, having frequent rest periods, and teaching energy conservation methods.

Cardiomyopathy

Cardiomyopathy is an enlargement of the heart muscle. There are three types of cardiac structure and function

---

cardiomyopathy: kardia—heart + myo—muscle + pathy—disease

abnormalities in cardiomyopathy: (1) dilated, (2) hypertrophic, and (3) restrictive (Fig. 17–5). A consequence of all types of cardiomyopathy can be heart failure (Fig. 17–6). There is no cure for cardiomyopathy.

Dilated Cardiomyopathy

In dilated cardiomyopathy the size of the ventricular cavity enlarges. Contractile function decreases as the myocardial tissue is destroyed. Blood moves more slowly through the left ventricle, which often results in blood clot formation. Dilated cardiomyopathy is the most frequent type. Dilated cardiomyopathy may be hereditary, follow infectious myocarditis, or be caused by chronic alcohol or cocaine use, HIV, thiamine or zinc deficiencies, or infections.

Hypertrophic Cardiomyopathy

Hypertrophic cardiomyopathy is enlargement of the muscle walls of the left ventricle. The ventricular walls are rigid, which decreases ventricular filling. If the enlarged septum obstructs the outflow of blood through the aortic valve, it is known as obstructive hypertrophic cardiomyopathy. Hyper-

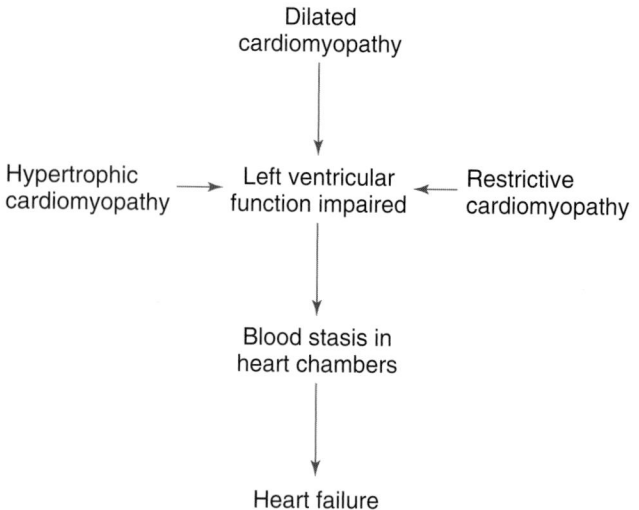

Figure 17-6 All types of cardiomyopathies can lead to heart failure.

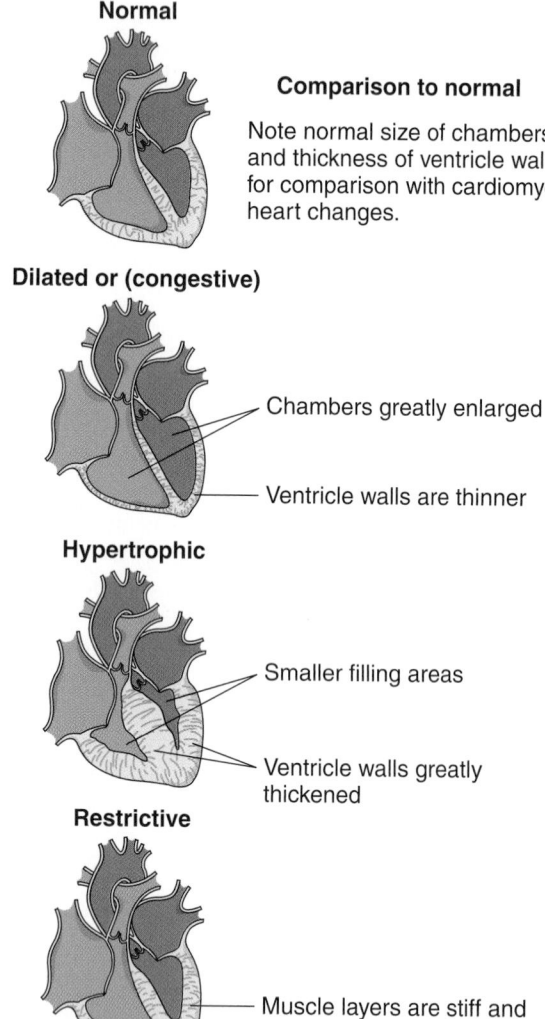

Figure 17-5 Comparison of the normal heart structure with each type of the cardiomyopic heart structure.

trophic cardiomyopathy can be a hereditary disorder that is transmitted as a dominant trait. Sudden death in young athletes may be due to this condition and is often the first sign of the disease.

Restrictive Cardiomyopathy

Restrictive cardiomyopathy impairs ventricular stretch and limits ventricular filling. Cardiac muscle stiffness is present, although systolic emptying of the ventricle remains normal. Restrictive cardiomyopathy is the rarest form of cardiomyopathy. It may be caused by infiltrative diseases such as amyloidosis that deposit the protein amyloid within the myocardial cells, which makes the muscle stiff. Visit the American Heart Association at www.americanheart.org for more cardiac information.

Signs and Symptoms

Manifestation of cardiomyopathy depend on the type of abnormality. Most patients show varying degrees of signs and symptoms of heart failure (Table 17–5). With dilated cardiomyopathy, left ventricular and then right-sided heart failure with a poor prognosis is seen. Dyspnea on exertion, orthopnea, fatigue, and sometimes atrial fibrillation occur. In hypertrophic cardiomyopathy, exertional dyspnea related to the obstruction of cardiac output is the most common symptom. Atypical chest pain occurs at rest and is not relieved with nitrates. With restrictive cardiomyopathy, exertional dyspnea occurs because of the heart's inability to increase filling when needed during exertion. Syncope, angina, and palpitations are common manifestations related to reduced cardiac output.

Diagnostic Tests

Cardiomegaly is seen on a chest x-ray examination. Echocardiography shows muscle thickness and chamber size

---

cardiomegaly: kardia—heart + mega—large

TABLE 17–5	CARDIOMYOPATHY SUMMARY
Signs and symptoms	Angina
	Exertional dyspnea
	Fatigue
	Syncope
Diagnostic tests	Cardiac catheterization
	Echocardiography
Medical treatment	Anticoagulants
	Antidysrhythmics
	For dilated cardiomyopathy: vasodilators, cardiac glycosides, cardiomyoplasty, heart transplant
	For hypertrophic cardiomyopathy: beta blockers, calcium channel blockers, myectomy
	For restrictive cardiomyopathy: vasodilators, heart transplant
Complications	Heart failure
Nursing diagnoses	Decreased cardiac output related to impaired myocardial function
	Activity intolerance related to cardiac insufficiency
	Anxiety related to disease process

to differentiate between the types of cardiomyopathy. Changes related to enlarged chamber size, tachycardia, and dysrhythmias can be seen on the ECG. Cardiac catheterization with angiocardiography may also be useful.

Medical Treatment

Treatment is palliative and aimed at managing heart failure and the underlying cause if known. (See Chapter 21.) In dilated or restrictive cardiomyopathy, vasodilators may be given to reduce the heart's workload and digoxin may be given to increase cardiac output. Anticoagulants are given to prevent emboli formation in patients with atrial fibrillation. Antidysrhythmics or cardioversion are used for dysrhythmias.

For hypertrophic cardiomyopathy, beta blockers or calcium channel blockers are used. In obstructive hypertrophic cardiomyopathy, digoxin and vasodilators are not given because they can increase the obstruction. Exercise is often restricted to prevent sudden death.

If medical management is not successful, surgery is considered. (See Chapter 22.) For hypertrophied muscle, surgery to remove part of the ventricular septum (**myectomy**) is done to allow greater outflow of blood. Septal ablation that uses alcohol to reduce heart wall size is being studied. Cardiomyoplasty is used for dilated cardiomyopathy when heart transplant is not an option. (See Fig. 21–6.) The heart's contraction strength is aided by the patient's latissimus dorsum muscle, which has been wrapped around the heart and contracts with each heartbeat. For severe heart failure, primarily in those with dilated cardiomyopathy and sometimes those with restrictive cardiomyopathy, a heart transplant may be the only hope for survival. A ventricular

myectomy: myo—muscle + ectomy—cutting out

assist device may be used until a donor is found. Many patients die while waiting for a donated heart because donated organs are limited.

Nursing Management

A patient history is obtained that includes signs and symptoms and assessment of family support systems because of the chronic nature of the disease. A physical assessment is done, noting vital signs and any signs or symptoms of heart failure.

Nursing care focuses on maintaining normal cardiac function, increasing activity tolerance, relieving anxiety, and educating the patient about the disease and its treatment. Patients with cardiomyopathy can be very ill. Careful monitoring is done to detect complications, such as heart failure, emboli, or dysrhythmias. The physician is immediately notified of problems. Home health care is often used for these patients to maintain their functional ability and reduce hospitalizations.

Maintenance of normal cardiac function includes monitoring vital signs and symptoms of heart failure. Increasing activity tolerance includes planning rest periods, scheduling activities in small amounts, avoiding tiring activities, and providing small meals that require less energy to digest than large meals. Patients are taught to avoid alcohol because it decreases cardiac function.

Reducing anxiety is important and can be accomplished by providing explanations for procedures, as well as educating the patient about the disease and its treatment. This may allow patients to feel in control of their lives by being able to make knowledgeable decisions about their health care. Methods to incorporate necessary lifestyle changes such as avoiding fatigue and scheduling rest periods can be helpful. Emotional support is greatly needed by these patients and their families because of the chronic nature of this disease.

Patient and Significant Other Education

Patients need to be taught the importance of medication compliance to prevent heart failure. The patient and family should have emergency telephone numbers readily available. Families should learn cardiopulmonary resuscitation (CPR). In terminal stages of the disease, the patient and family should be informed about the availability of hospice care and be emotionally supported during the grieving process.

▶ VENOUS DISORDERS
Thrombophlebitis

Thrombophlebitis is the formation of a clot and inflammation within a vein. The clot usually forms first and then inflammation occurs. Thrombophlebitis is the most common disorder of veins, with the legs being most often affected. Any superficial or deep vein can be involved. Deep vein thrombosis (DVT) is the most serious form of thrombophlebitis because pulmonary emboli can result if the thrombus detaches. (See Chapter 28.)

Pathophysiology

A venous thrombus is made up of platelets, red blood cells, white blood cells, and fibrin. Platelets attach to a vein wall and then a tail forms as more blood cells and fibrin collect. As the tail grows, it drifts in the blood flowing past it. The turbulence of the blood flow can cause parts of the drifting thrombus to break off and become emboli that travel to the lungs.

Etiology

Three factors are involved in the formation of a thrombus: stasis of blood flow, damage to the lining of the vein wall, and increased blood coagulation (Table 17–6). Venous stasis occurs when blood flow is reduced, veins are dilated, muscle contractions are decreased, or vein valves are faulty. When the wall of a vein is damaged, it provides a site for a thrombus to form. Intravenous therapy and venipuncture cause trauma to the vein, and IV catheters in place longer than 48 to 72 hours increase the risk of inflammation and thrombus. Increased coagulation of the blood promotes thrombus formation. Patients on oral anticoagulants that are abruptly stopped experience increased clotting of the blood. Smoking, oral contraceptive use, and estrogen therapy also increase blood coagulation. Hematological disorders can also lead to altered blood coagulation and increased risk of thrombus formation.

Prevention

Identification of risk factors and patient education is important to allow the use of preventive interventions for thrombosis. Because the elderly are at increased risk for thrombus formation, a family member should be instructed along with the elderly person in techniques that may be difficult for the elderly person to perform. Dehydration, which is common in the elderly population, should be avoided to reduce thrombus risk.

thrombophlebitis: thromb—lump (clot) + phleb—vein + itis—inflammation

IMMOBILITY. People with sedentary jobs that require long periods of sitting, standing, or traveling long distances should change positions, perform knee and ankle flexion exercises, or walk at regular intervals to prevent stasis of blood. Patients on bedrest should have legs elevated above the level of the heart if possible to prevent pooling of blood. Postoperatively or in times of bedrest, active or passive range-of-motion exercises should be done to increase blood flow. Postoperatively, early ambulation is a major preventative technique for thrombosis. Patients' pain should be controlled to facilitate their ability to participate in early ambulation. Deep breathing aids in improving blood flow in the large thoracic veins. Smoking should be avoided because nicotine causes vasoconstriction.

PROPHYLACTIC ANTIEMBOLISM DEVICES. Patients with peripheral vascular disease, those on bedrest, and those who have had surgery or trauma may use antiembolism devices to improve blood flow. Knee- or thigh-length elastic stockings apply pressure to the leg. They must be applied correctly to avoid a tourniquet effect. Older patients with decreased manual dexterity may need assistance. The stockings are removed for 20 minutes twice a day and the skin is inspected for irritation. Intermittent pneumatic compression (IPC) devices fill plastic inflatable stockings intermittently with air to move venous blood in the legs by simulating contraction of the leg muscles. (See Figure 15–15.) They may be used in combination with elastic stockings. Research that compares the various preventive measures for DVT and rates of DVT in surgical patients has shown that the lowest incidence of DVT occurs with elastic stockings and IPC devices used together.

PROPHYLACTIC MEDICATION. Low-molecular-weight heparin (LMWH) is given postoperatively to prevent thrombosis. Enoxaparin (Lovenox), given subcutaneously, is an example of a LMWH. Subcutaneous heparin may also be used postoperatively to prevent thrombosis. A common prophylactic dose of heparin is 5000 IU subcutaneously every 8 to 12 hours. Partial thromboplastin time (PTT) is not usually monitored with prophylactic doses of LMWH or heparin.

TABLE 17-6 PREDISPOSING CONDITIONS FOR THROMBOPHLEBITIS

Condition	Type	Examples
Venous stasis	Reduction of blood flow	Shock, heart failure, myocardial infarction, atrial fibrillation
	Dilated veins	Vasodilators
	Decreased muscle contractions	Immobility, sitting for long periods as in traveling, fractured hip, paralysis, anesthesia, surgery, obesity, advanced age
	Faulty valves	Varicose veins, venous insufficiency
Venous wall injury		Venipuncture, venous cannulation at same site for > 48 hours, venous catheterization, surgery, trauma, burns, fractures, dislocation, IV medications (potassium, chemotherapy drugs, antibiotics, IV hypertonic solutions), contrast agents, diabetes, cerebrovascular disease
Increased coagulation of blood		Anemia, malignancy, antithrombin III deficiency, oral contraceptives, estrogen therapy, smoking, discontinuance of anticoagulant therapy, dehydration, malnutrition, polycythemia, leukocytosis, thrombocytosis, sepsis, pregnancy

Oral anticoagulants such as warfarin (Coumadin) can be used in the high-risk patient to decrease thrombosis. Today the preferred indicator of warfarin effectiveness is the **international normalized ratio** (INR) value. The INR is preferred because it uses a standardized testing reagent. This means that it can be used around the world with no variation in results as occurs with prothrombin time (PT). Prothrombin time was previously the most commonly used test; however, PT control values can vary from laboratory to laboratory. Therefore a standardized test was developed, so that warfarin dosage variations would not occur as a result of differing laboratory control values. The INR value is reported along with the prothrombin time. These values correspond in meaning (i.e., normal, subtherapeutic, therapeutic, toxic). Some practitioners still order PT, so a discussion of how to interpret the PT is presented for understanding. For more information on these medications, visit www.fda.gov/cder/drug/default.htm.

LEARNING TIPS

Before administering anticoagulants, laboratory values must be assessed to ensure patient safety. INR normal and desired therapeutic values for the patient's disorder are provided on the laboratory report. These INR values do not require calculation of a therapeutic range because the values are given on the report. Compare the patient's INR value with the desired INR value to determine if it is safe to give the warfarin.

Understanding that INR is the preferred test for warfarin effectiveness, you may still want to know how to calculate a therapeutic range for prothrombin time. Prothrombin times are measured in seconds. The normal value range gives the seconds required for a fibrin clot to form during the test. If a patient is on warfarin, the purpose is to increase the time (seconds) it takes the blood to clot. It is therefore expected that the prothrombin time will be elevated by warfarin therapy.

Because a therapy, warfarin, is being given, a prothrombin time range that safely considers the expected effects of the warfarin is needed. This is called the therapeutic range (i.e., a high and a low value). Warfarin's therapeutic range is 1.5 to 2 times the normal prothrombin time range. To monitor the patient's therapeutic prothrombin time, compare the patient's result with the therapeutic range that you calculate. For example:

Patient's value on warfarin: 16 seconds
Normal prothrombin time range: 9–12 seconds
To calculate therapeutic range,
multiply	1.5	2
	× 9 seconds	×12 seconds
The therapeutic range is:	13.5 seconds to	24 seconds

Compare the patient's value of 16 seconds with the therapeutic range of 13.5 to 24 seconds to determine that the patient is safely within the therapeutic range.

INTRAVENOUS THERAPY. Monitoring of venous IV sites should be performed according to institutional policy time frames to detect signs of thrombophlebitis. Venous cannula sites should be changed regularly according to institutional

guidelines (e.g., every 48 to 72 hours) to prevent thrombus formation.

Signs and Symptoms

Up to 50 percent of patients have no symptoms with thrombophlebitis in the legs. For others the symptoms vary according to the size and location of the thrombus (Table 17–7). If adequate collateral circulation is present near the involved area, symptoms may be reduced. For some patients a pulmonary embolus is the only evidence of a DVT.

SUPERFICIAL VEINS. Thrombophlebitis in a superficial vein may produce redness, warmth, swelling, and tenderness in the area around the site of the thrombus. The vein feels like a firm cord, which is referred to as induration. The saphenous vein is the most commonly affected vein in the leg. Varicosity of the vein is usually the cause. In the arm, IV therapy is the most common cause.

DEEP VEINS. In a deep vein thrombus of the leg, swelling, edema, pain, warmth, venous distention, and tenderness with palpation of the calf may be present in the affected leg. Obstruction of blood flow from the leg causes edema and varies with the location of the thrombus. An elevated temperature may also be present. Pain in the calf with sharp dorsiflexion of the foot, a classic indication known as a positive Homans' sign, is present in less than 40 percent of those with thrombophlebitis and is not specific to DVT. Once a DVT is positively diagnosed, it is important to avoid performing Homans' sign because it may cause the clot to become dislodged. Cyanosis and edema may occur if the large veins such as the vena cava are involved.

TABLE 17-7	THROMBOPHLEBITIS SUMMARY
Signs and symptoms	Superficial veins: redness, warmth, swelling, and tenderness
	Deep veins: swelling, edema, pain, warmth, venous distention, and tenderness
Diagnostic tests	D-dimer and coagulation tests
	Duplex ultrasound
	Impedance plethysmography
	MRI
	Venography
Medical treatment	Superficial veins: warm, moist heat; analgesics; NSAIDs; compression stockings
	Deep veins: LMWH; heparin; warfarin; bedrest with extremity elevation above the level of the heart for 5–7 days; warm, moist heat; compression stocking therapy; thrombolytic therapy; thrombectomy; vena cava filter
Complications	Pulmonary embolism
	Chronic venous insufficiency
Nursing Diagnoses	Acute pain related to inflammation of vein
	Impaired skin integrity related to venous stasis
	Anxiety related to uncertain prognosis of disease

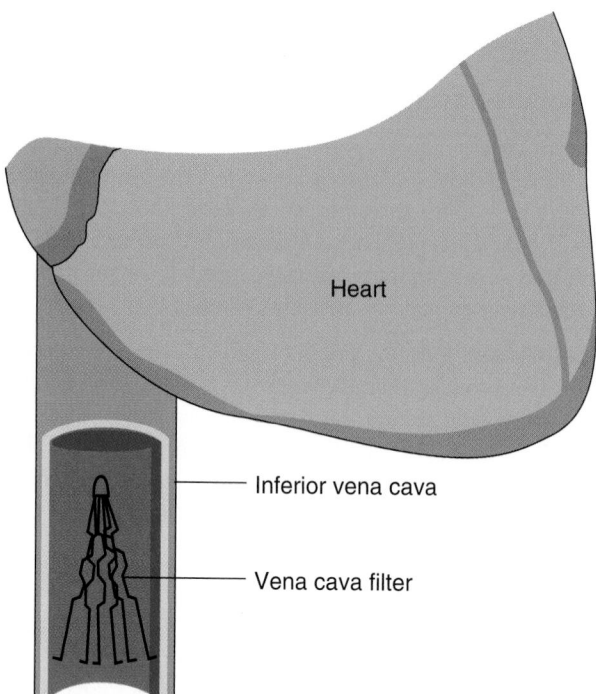

Figure 17-7 Vena cava filter placed in the inferior vena cava to prevent emboli from reaching the lungs.

Complications

The most serious complication of deep vein thrombophlebitis is pulmonary embolism, which is a life-threatening emergency. (See Chapter 28.) Another complication, chronic venous insufficiency, results from damage to the valves in the vein and causes venous stasis. Signs and symptoms from venous insufficiency that may appear years after a thrombus include edema, pain, brownish discoloration and ulceration of the medial ankle, venous distention, and dependent cyanosis of the leg. This condition can be difficult to treat.

Medical and Surgical Treatment

Diagnostic tests are done to guide treatment. (See Table 17-7.) The goals of treatment are to relieve pain and prevent pulmonary emboli, thrombus enlargement, and further thrombus development. Superficial thrombophlebitis is treated with warm, moist heat; analgesics; NSAIDs; and compression stockings.

Patients with a proximal DVT may be treated at home if they do not have pulmonary embolism, cardiovascular or pulmonary disease, obesity, or renal failure and are able to comply with follow-up care. A LMWH (e.g., enoxaparin, dalteparin, or ardeparin) is given subcutaneously daily or twice a day while oral warfarin is started. Both are taken until the INR is within therapeutic range (about 5 days); the LMWH is then stopped.

For some DVTs a hospital stay may be required. Treatment includes bedrest with leg elevation above heart level for 5 to 7 days; warm, moist heat; elastic stocking (initially on unaffected leg only until acute symptoms are gone on affected leg); and anticoagulants. A continuous heparin IV infusion is usually started for up to 10 days to prevent fur-

ther enlargement of the thrombus and development of new thrombi; it has little effect on the existing clot, which the body dissolves over time. An oral anticoagulant, warfarin, is begun 4 to 5 days before the heparin is stopped. When the therapeutic INR goal is reached, the heparin is stopped. Warfarin is continued for several months. In some cases of massive DVT, thrombolytic therapy (alteplase, tissue plasminogen activator [TPA], streptokinase, urokinase) may be used to dissolve the clot, but there is a high risk of bleeding with these drugs.

Surgical treatment is used to prevent pulmonary emboli or chronic venous insufficiency when anticoagulant therapy cannot be used or the risk of pulmonary emboli is great. Venous thrombectomy removes the clot through a venous incision. In some cases a vena cava filter is placed into the inferior vena cava through the femoral or right internal jugular vein (Fig. 17-7). Once in place, it is opened and attaches to the vein wall. The filter traps clots traveling toward the lungs without hindering blood flow.

Nursing Management

A patient history is obtained that includes recent IV therapy or use of contrast dyes, surgery, extremity trauma, childbirth, bedrest, recent long trip, cardiac disease, recent infections, and current medications. A physical assessment is done noting pain, fever, tenderness, positive Homans' sign, redness, warmth, swelling, edema, and a firm, cordlike vein in the affected extremity. Daily measurements are taken of bilateral thighs and calves and recorded to monitor swelling. Coagulation tests are monitored.

See Nursing Care Plan Box 17-2 for specific nursing interventions. Pain is relieved by giving analgesics and NSAIDs as ordered and applying warm, moist soaks. Legs are elevated to reduce edema. (See Home Health Hints.)

 HOME HEALTH HINTS

- Home health nurses can assist patients to develop energy-conserving techniques by being observant of their lifestyle. For instance, notice the room and chair that they spend most of the day in. Television trays and baskets can be used to hold items they may need or want, such as water pitcher and glass, TV remote, reading material, paper and pen, mail, medicines, telephone and phone book, snacks, washcloth, or tissues. Answering the door should be done by the caregiver. When the patient is alone, a note can be placed on the door with instructions; however, the instructions should not convey that the patient is alone.

- Other techniques to conserve energy are putting a carrying pouch on the front bar of a walker to carry items such as a portable phone or tissues and, if the house has stairs, putting a chair at the top and bottom of the stairs so the patient can rest.

- When a patient with venous circulation problems is sitting in a recliner with the leg rest up, the nurse should note whether pressure is being applied on the popliteal area or calf muscle. The angle of the recliner and the patient's height affect the position of the pressure. A small, flat pillow is placed underneath the knees and lower legs to open the angle and relieve the pressure.

BOX 17-2 NURSING CARE PLAN FOR THE PATIENT WITH THROMBOPHLEBITIS

 Acute pain related to inflammation of vein

PATIENT OUTCOME

Reports satisfactory pain relief.

EVALUATION OF OUTCOME

Does patient report satisfactory pain relief?

INTERVENTIONS	RATIONALE	EVALUATION
Assess pain using rating scale such as 0–10.	Self-report is the most reliable indicator of pain.	Does patient report pain using scale?
Provide analgesics and NSAIDS as ordered.	Pain is reduced when inflammation is decreased.	Is patient's rating of pain lower after medication?
Apply warm, moist soaks.	Heat relieves pain and vasodilates, which increases circulation to reduce swelling. Moist heat penetrates more deeply.	Does patient report increased comfort with warm, moist soaks? Is swelling reduced?
Maintain bedrest with leg elevation above heart level.	Elevation decreases swelling, which reduces pain.	Is swelling reduced?

 Impaired skin integrity related to venous stasis

PATIENT OUTCOME

Skin remains intact without edema.

EVALUATION OF OUTCOME

Does patient's skin remain intact? Is edema present?

INTERVENTIONS	RATIONALE	EVALUATION
Assess skin for edema, skin color changes, and ulcers. Measure extremities.	Assessment will detect signs of skin integrity impairment and extremity swelling.	Are skin changes seen? Do daily measurements show a change in swelling?
Elevate legs.	Elevation decreases swelling.	Is swelling reduced?
Fit and apply elastic stockings after edema reduced as ordered.	Elastic stockings are fitted after edema is reduced to avoid constriction. They increase venous blood flow to reduce swelling.	Is swelling reduced?
Teach patient to avoid crossing legs or wearing constricting clothes.	Crossing legs and constrictive clothes impair venous return.	Does patient state understanding of teaching?

Elastic stockings are fitted and applied when edema is reduced. The presence of dyspnea, tachycardia, tachypnea, blood-tinged sputum, chest pain, or changes in level of consciousness are immediately reported to the physician.

Teaching the patient about the disease and treatment is important to reduce anxiety about complications and to enhance compliance with treatment to prevent complications.

REVIEW QUESTIONS

1. The nurse evaluates the patient as understanding how to prevent rheumatic fever if the patient states that rheumatic fever can be prevented by treating streptococcal infections with which of the following?
    a. Penicillin
    b. Prednisone
    c. Cortisone
    d. Cyclosporine

2. For patients recovering from infective endocarditis, discharge teaching to prevent recurrence should include which of the following?
    a. Proper use of isolation techniques
    b. Keeping vaccinations up to date
    c. Need to obtain annual flu injection
    d. Need for antibiotics before invasive procedures

3. The nurse is planning care for a patient with cardiomyopathy. For which of the following complications of cardiomyopathy should the nurse assess?
    a. Thrombophlebitis
    b. Heart failure
    c. Rheumatic fever
    d. Pulmonary embolism

4. Which of the following signs and symptoms indicates to the nurse the presence of a deep vein thrombus in the patient's leg?
    a. Calf swelling
    b. Crackles
    c. Jugular vein distention
    d. Negative Homans' sign

5. The nurse is to give warfarin (Coumadin). Which of the following laboratory tests should the nurse review before giving the medication?
    a. International normalized ratio
    b. Partial thromboplastin time
    c. Plasma fibrinogen level
    d. Thrombin clotting time

REFERENCES

1. Dajani, A, et al: Prevention of bacterial endocarditis: Recommendations by the American Heart Association. JAMA 277(22):1794, 1997.

Answers to CRITICAL THINKING

Mrs. Jones

1. A heart murmur is heard from damaged heart valves.
2. Splinter hemorrhages appear as black or red lines in the nails.
3. Petechiae indicate that tiny pieces of a lesion on the endocardium or valves have broken off and become microemboli.
4. Subjective assessment findings might include patient statements such as, "I have pain in my joints and am chilled" or "I am fatigued and have no appetite."
    Objective findings are as follows: temperature 100° F (37° C), red splinter hemorrhages in left index finger nailbed, many petechiae on chest.
5. Expected medications include IV antibiotics.

6. Removing blankets to decrease fever results in chills and shivering, which further increases body temperature from the heat generated by muscular activity during shivering. Therefore Mrs. Jones should be kept covered to prevent chills.
7. For left-sided heart failure, crackles, wheezes, cough, or dyspnea might be seen. In right-sided heart failure, peripheral edema or jugular vein distention could be present.
8. Prophylactic antibiotics are needed to prevent another episode of infective endocarditis, which can result from bacteria entering the circulation during the invasive procedure, attaching to areas of the endocardium damaged from the current infection, and growing.

18

NURSING CARE OF PATIENTS WITH OCCLUSIVE CARDIOVASCULAR DISORDERS

Maureen McDonald and Elizabeth Chapman

KEY TERMS

acute coronary syndromes
(A-cute **KOR**-uh-na-ree Sin-dr-**OME**s)

aneurysm (**AN**-yur-izm)

angina pectoris
(an-**JIGH**-nah **PEK**-tuh-riss)

arteriosclerosis
(ar-**TIR**-ee-oh-skle-ROH-sis)

atherosclerosis (**ATH**-er-oh-skle-ROH-sis)

atheroma (ATH-er-**OMA**)

collateral circulation
(koh-**LA**-ter-al SIR-kew-**LAY**-shun)

coronary artery disease
(**KOR**-uh-na-ree **AR**-tuh-ree di-ZEEZ)

embolism (**EM**-buh-lizm)

high-density lipoprotein
(HIGH **DEN**-si-tee LIP-oh-**PROH**-teen)

hyperlipidemia
(HIGH-per-LIP-i-**DEE**-mee-ah)

intermittent claudication
(IN-ter-**MIT**-ent KLAW-di-**KAY**-shun)

lymphangitis (lim-FAN-je-**EYE**-tis)

myocardial infarction
(MY-oh-**KAR**-dee-yuhl in-**FARK**-shun)

peripheral arterial disease
(puh-**RIFF**-uh-ruhl ar-**TIR**-ee-uhl di-ZEEZ)

plaque (PLAK)

Raynaud's disease (ra-**NOHZ** di-ZEEZ)

thrombosis (throm-**BOH**-sis)

varicose veins (**VAR**-i-kohz VAINS)

venous stasis ulcers
(VEE-nus **STAY**-sis UL-sers)

QUESTIONS TO GUIDE YOUR READING

1. What are the etiologies, signs, and symptoms of coronary artery disease, angina pectoris, and myocardial infarction?

2. What are the current medical treatments for coronary artery disease, angina pectoris, and myocardial infarction?

3. What data should you collect when caring for patients with coronary artery disease, angina pectoris, and myocardial infarction?

4. What nursing care will you provide for patients with coronary artery disease or myocardial infarction?

5. What should you include in the education plan for the patient with coronary artery disease?

6. How will you know if your nursing interventions have been effective?

7. What are the etiologies, signs, and symptoms for each of the peripheral vascular disorders?

8. What is the current medical treatment for each peripheral vascular disorder?

9. What nursing care will you provide for patients with each peripheral vascular disorder?

Cardiovascular disorders are the leading cause of disability and death in the United States. Diseases of the heart and peripheral vessels can affect quality of life and alter the ability of the individual to perform tasks of everyday living. Long-term disability can place a burden on families and businesses and can create high insurance costs. Many factors leading to cardiovascular diseases can be controlled or modified. Education is important in preventing and treating occlusive cardiovascular diseases.

▶ ARTERIOSCLEROSIS AND ATHEROSCLEROSIS

Arteriosclerosis is a term used to describe conditions that affect arteries and may lead to occlusive cardiovascular disease. The lining of the artery and arteriole walls become thickened and hardened and lose elasticity. Arteriosclerosis is referred to as hardening of the arteries. Atherosclerosis, a type of **arteriosclerosis,** is the formation of **plaque** within the arterial wall. Arteriosclerosis and **atherosclerosis** are conditions that develop over a long period (beginning in early childhood). Arteriosclerosis and atherosclerosis usually occur together.

Pathophysiology

Atherosclerosis affects the inner lining of the artery. The first step in the development of atherosclerosis is injury to the endothelial cells that line the walls of the arteries. This injury causes inflammation and immune reactions. Lipids, platelets, and other clotting factors accumulate. Scar tissue replaces some of the arterial wall. An early indication of injury is a fatty streak on the lining of the artery. This buildup of fatty deposits is known as plaque. Plaque has irregular, jagged edges that allow blood cells and other material to adhere to the wall of the artery. Over time this buildup becomes calcified and hardened (arteriosclerotic), causing turbulence that damages cells and increases the buildup within the vessel. As the vessel becomes stenosed (narrowed), partial or total occlusion of the artery may occur, resulting in reduced blood flow. The area distal to the occlusion may become ischemic as a result.

Causes

There are many theories as to causes of arteriosclerosis and atherosclerosis:

- Genetics related to predisposition for **hyperlipidemia** (increase in blood lipids)
- Diabetes mellitus related to hyperglycemia
- Hypertension
- Smoking
- Obesity
- Sedentary lifestyle
- Increased serum homocysteine levels (an amino acid)
- Increased serum iron levels
- Infection

Signs and Symptoms

Arteriosclerosis or atherosclerosis are not usually symptomatic until later stages of development. If symptoms do occur, they may include chest pain or dizziness caused by decreased blood supply and oxygen to the heart. Reduced blood flow in the extremities is reflected by pallor in the nailbeds, a reddish-purple color in the lower extremities, thickened nails, dry skin, or loss of hair on the extremities. Peripheral pulses may be diminished or absent. Skin temperature in the extremities is cooler, and there is prolonged capillary refill (greater than 3 seconds).

Diagnostic Tests

Cholesterol and triglycerides are often elevated in patients with atherosclerosis. (See Table 15-3.) Total cholesterol levels above 200 mg/dL increase risk of myocardial infarction. Low-density lipoproteins (LDL) increase **coronary artery disease** (CAD) risks, but **high-density lipoproteins** (HDL) are protective against CAD. Blood glucose levels should also be checked because elevated levels may increase the risk for atherosclerosis. Radiological studies of the arteries can be performed to show narrowed or occluded vessels. (See Chapter 15.)

Treatment

A healthy lifestyle, cholesterol screening, frequent checkups, and medications are helpful in controlling arteriosclerosis and atherosclerosis.

Diet

Saturated fats are the primary cause of increased cholesterol levels, which promote arteriosclerosis and atherosclerosis. Because the formation of plaque within arteries is primarily caused by fatty deposits, an adherence to a low-fat diet is recommended (Nutrition Notes Box 18–1). The American Heart Association has complete guidelines and diets for decreasing fat and cholesterol intake. For more information, visit http://americanheart.org.

Smoking

Smoking contributes to a loss of high-density lipoprotein (HDL). These proteins are considered the best cholesterol to have in the body to decrease the risk of cardiovascular disorders. The rate of progressive damage to blood vessels is increased with smoking. Smoking also causes vasoconstriction, which leads to **angina pectoris** and cardiac dysrhythmias. Education on the risks of smoking and exposure to secondhand smoke should be presented to patients. The American Cancer Society has many programs to help patients quit smoking. For more smoking cessation information, visit www.cancer.org.

Exercise

Increased activity raises HDL levels (the desirable cholesterol). Over time, exercise also leads to the development of **collateral circulation,** which allows blood to flow around

---

hyperlipidemia: hyper—above + lipos—fat + emia—blood

angina pectoris: angina—to choke + pectora—chest

BOX 18-1  Nutrition Notes

Controlling Blood Cholesterol with Diet

Two-thirds of the body's cholesterol is produced by the liver and intestines. The ability to reduce the amount of cholesterol manufactured in response to an increased dietary intake of it is believed to be genetically determined. Although only foods of animal origin contain cholesterol, some vegetable products contain transfatty acids, potent risk factors for cardiovascular disease.

The National Cholesterol Education Program introduced a new diet called the therapeutic lifestyle changes (TLC) diet based on the following:

Monounsaturated fat	Up to 20% of total calories
Polyunsaturated fat	Up to 10% of total calories
Saturated fat	Less than 7% of total calories
Total fat	25 to 35% of total calories
Carbohydrate	50 to 80% of total calories
Fiber	20 to 30 grams per day
Plant stanols/sterols*	2 grams per day

*These compounds in plants, which structurally resemble cholesterol but are not absorbed in the human body to any extent, actually inhibit the absorption of cholesterol. Plant sterols have been useful, either alone or along with low-fat diets, in reducing low-density lipoprotein cholesterol. Table spreads and salad dressings containing plant sterols are commercially available. Plant sterols are contraindicated in a rare inherited metabolic disease (40 reported cases) called sitosterolemia, in which plant sterols are absorbed at a high rate and are not removed effectively by the liver, resulting in premature atherosclerosis.

occluded sites. Before beginning an exercise program, the patient should consult a physician.

Medications

Lowering lipid levels is the major treatment for atherosclerosis (Table 18–1). When dietary control is not effective, medication is also used. It may take 4 to 6 weeks before lipid levels respond to drug therapy. If one drug does not control the lipids, another drug can be added.

► CORONARY ARTERY DISEASE

Coronary artery disease is a term applied to obstructed blood flow through the coronary arteries to the heart muscle. The primary cause of coronary artery disease is atherosclerosis. If blood flow reduction resulting from CAD is severe and prolonged, a **myocardial infarction** (heart attack) can occur, causing irreversible damage.

Pathophysiology and Etiology

As discussed in the sections on atherosclerosis and arteriosclerosis, an accumulation of fatty deposits and minerals in the coronary arteries, called an **atheroma** or plaque, leads to stenosis and eventually occlusion of the artery. In CAD, blood flow to the myocardium is reduced. If myocardial oxygen demands are not met, ischemia results, which can lead to chest pain. The pain associated with CAD occurs from a lack of oxygen to the myocardium as a result of the CAD and is referred to as angina pectoris.

Risk Factors

Risk factors can be broken down into two components: those that cannot be changed and those that can be changed (Table 18–2). If coronary artery disease is not prevented or treated early, it can progress to more serious cardiac disorders. These include angina, myocardial infarction, heart failure, cardiac dysrhythmias, and even sudden death. Altering risk factors may help protect the patient from developing cardiovascular disease. To take a heart attack risk assessment, visit www.americanheartassociation.com/risk/quiz.html.

Medical Management

Prevention is the goal for patients with CAD. Educating patients on dietary changes, cessation of smoking, controlling hypertension, and diabetes can decrease their risk of CAD.

TABLE 18-1 LIPID-LOWERING DRUGS

Class	Drugs	Side Effects	Nursing Implications
Statins First-line drugs to reduce low-density lipoprotein by reducing cholesterol synthesis	Atorvastatin (Lipitor) Lovastatin (Mevacor) Pravastatin (Pravachol) Simvastatin (Zocor)	Impaired liver function Rhabdomyolysis (lethal breakdown of skeletal muscle)	• Tell patient to take in evening when cholesterol synthesis is highest. • Teach patient to report muscle pain. • Monitor liver function studies.
Fibrates Reduce triglycerides	Fenofibrate (Tricor) Gemfibrozil (Lopid)	Heartburn, gallstones	• Tell patient to take 30 minutes before morning and evening meals.
Bile acid sequestrants Lower cholesterol by binding bile acids, so stored cholesterol is used to make more bile acids	Colestipol (Colestid) Colesevelam HCl (WelChol, Sankyo) Cholestyramine (Questran)	Headache, heartburn, constipation, gas May interfere with absorption of fat-soluble vitamins A, D, E, K	• Fruits and vegetables high in fiber should be added to diet to reduce constipation and other GI effects noted with bile acid sequestrants.
Niacin Prevents conversion of fats into very-low-density lipoproteins. First antilipid agent used.	Niacin (Nicotinic acid)	Gastritis Gout Flushing	• Tell patient to take with liquids and meals and to take other drugs 1 hour before or 4 hours after. • Tell patient to take with meals or milk to avoid GI upset. Used with diet, exercise, and smoking cessation therapy to lower lipids.

GI = gastrointestinal.

TABLE 18-2	RISK FACTORS FOR CORONARY ARTERY DISEASE
Risk Factors That Cannot Be Changed	
Heredity	CAD risk factors can run in families.
Ethnicity	African-Americans have a higher incidence of atherosclerosis.
Gender	Men have more risk factors and higher incidence of CAD.
Age	Men have increased incidence after age 50. Women have increased incidence after menopause.
Risk Factors That Can Be Changed or Controlled	
Smoking	Causes vasoconstriction and increases myocardial oxygen demand. Decreases HDL.
Hypertension	Vasoconstriction increases myocardial oxygen demand.
Elevated serum cholesterol	Level above 240 mg/dL increases the risk of developing CAD.
Diabetes	Increases the risk of hypertension, obesity, and elevated blood lipids
Obesity	Increases heart workload and risk of hypertension, diabetes, glucose intolerance, hyperlipidemia.
Stress	Increases heart workload and risk for hypertension.
Elevated serum homocysteine	Increases CAD risk. Foods that contain folic acid (fruits, green leafy vegetables) reduce homocysteine levels.
Sedentary lifestyle	Increases obesity, hypertension, hyperlipidemia.

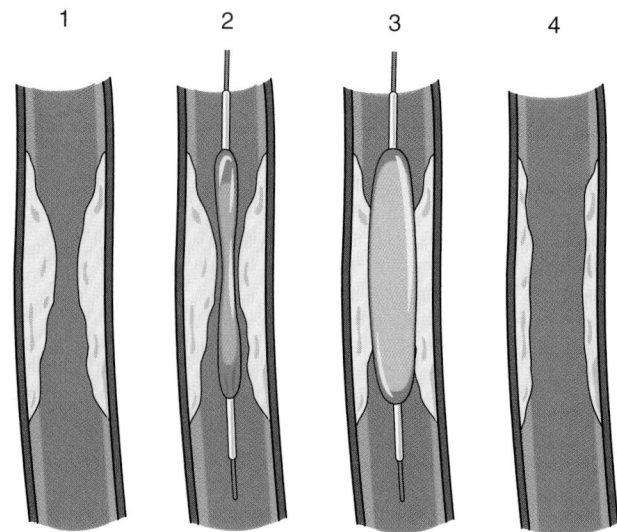

Figure 18-1 Percutaneous transluminal coronary angioplasty.

Dietary changes are made to reduce saturated fats to less than 10 percent of daily food intake. Cholesterol intake on the step 1 diet is less than 300 mg per day and less than 200 mg per day on the step 2 diet. Medication may be given to reduce cholesterol levels. Anticoagulants are used to prevent the formation of a thrombus.

Surgical Management
Percutaneous Transluminal Coronary Angioplasty

Percutaneous transluminal coronary angioplasty (PTCA) is a minimally invasive procedure that helps reduce symptoms of CAD (Fig. 18–1). A catheter is inserted via the femoral or brachial artery and is advanced into the heart. Once the blocked coronary artery is entered, the balloon on the catheter is inflated. This procedure compresses the plaque against the wall of the artery, thus restoring the opening of the artery. The symptoms of CAD are usually reduced, but the underlying progression of atherosclerosis continues. Reocclusion of the artery often occurs within a few months. Over time, PTCA may need to be repeated.

Coronary Atherectomy

Coronary atherectomy is used to cut and remove plaque from atherosclerotic coronary arteries. The catheter has a central rotating blade that shaves off the plaque and contains it for removal and pathological analysis. Calcium channel blockers are given before the procedure to prevent vasospasms from the vibrating cutter. To prevent clot formation, an antiplatelet agent is given after the procedure.

Coronary Artery Stents

Coronary artery stents are used to prevent closure of a coronary artery from an atherosclerotic lesion. Stents are placed in a procedure similar to PTCA. A stent provides support to a coronary artery wall at the area of stenosis to keep blood flowing through the artery. Different types of materials, such as bioabsorbable materials or stainless-steel mesh, and designs, such as self-expanding or balloon expandable, are used to make stents (Fig. 18–2). Complications associated

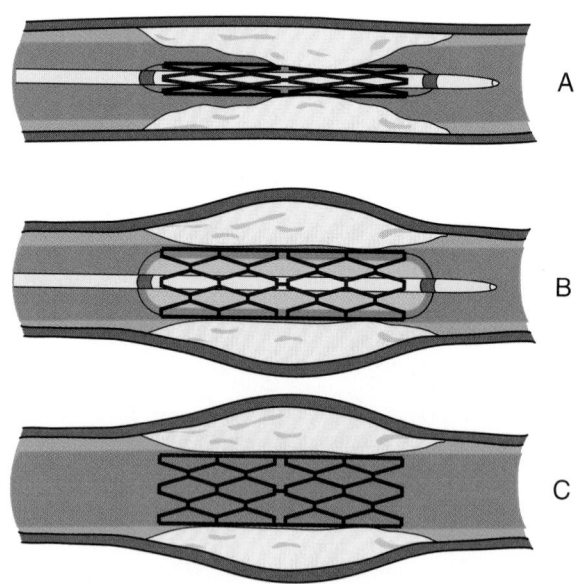

Figure 18-2 Insertion of a coronary artery stent: *(A)* A balloon catheter with a collapsed stent is advanced to the location of a coronary artery lesion. *(B)* The balloon is inflated, which expands the stent and compresses the lesion to increase the artery opening. *(C)* The balloon is then deflated and removed, leaving the expanded stent in place to prevent the artery from closing.

with stent placement include **thrombosis,** bleeding from anticoagulation, stent occlusion, or coronary artery dissection. An antiplatelet (ticlopidine [Ticlid]) and anticoagulant (warfarin [Coumadin]) are prescribed after the procedure. The antiplatelet agent is taken long term, but the anticoagulant is often only needed up to 6 weeks after the procedure.

Coronary Artery Bypass Graft

Coronary artery bypass graft (CABG) surgery involves bypassing one or more blocked coronary arteries. Patients who have severe CAD or have had myocardial infarctions may be candidates for a CABG. Patients usually have a cardiac catheterization to identify blocked coronary arteries and determine the need for surgery. (See Chapter 22.)

Transmyocardial Laser Revascularization

Transmyocardial laser revascularization is a newer treatment option for patients unable to have angioplasty or coronary artery bypass surgery. Channels are made through the myocardium into the ventricles in areas of ischemia with a fiberoptic catheter. However, it is not known why this procedure decreases symptoms of ischemia and improves quality of life.

Nursing Process

Caring for a patient with CAD includes monitoring blood pressure and heart rate and documenting abnormal values or reports of chest pain and any symptoms that may point to the presence of complications from CAD. The care of the patient is directed at teaching and working with the patient to change or modify risk factors. Teaching the patient how to control risk factors and manage symptoms is the major focus of nursing care for the patient with CAD.

► ACUTE CORONARY SYNDROMES

Acute coronary syndromes are a group of conditions that are caused by a lack of oxygen to the heart muscle. These conditions include unstable angina, non–Q wave myocardial infarction, and ST segment elevation myocardial infarction. Patients with acute coronary syndromes are at high risk for myocardial infarction and death.

► ANGINA PECTORIS
Pathophysiology

Angina pectoris (chest pain) is a symptom of ischemia. When an increased workload is placed on the heart, as in exercise or strenuous activity, there is an increased demand for oxygen. Normally, when the heart needs more oxygen, the coronary arteries dilate to carry more blood. However, with CAD, the narrowed vessels are unable to dilate and supply the heart with this extra blood and oxygen. This inability to supply more blood and oxygen causes myocardial ischemia. Chest pain results from the ischemia but usually lasts only for a few minutes, especially if activity is stopped. If adequate blood supply to the myocardium is restored with rest, no myocardial damage usually occurs.

Signs and Symptoms

Anginal pain manifests itself in several ways. Patients often describe the pain as a heaviness, tightness, or viselike or crushing pain in the center of the chest. The pain can radiate down one or both arms, with the left arm being more common, into the shoulder, neck, jaw, or back. Patients may also describe a heaviness in their arms or a feeling of impending doom. During the episode of pain, the patient may be pale, diaphoretic, or dyspneic. The pain is usually brought on by exertion and subsides with rest. It can be relieved with a vasodilator medication such as nitroglycerin (NTG). Episodes of chest pain may increase in frequency and severity over time. If patients do not heed this warning to stop their activity and rest, they may be at risk for a myocardial infarction or sudden death.

Types
Stable Angina

Stable angina occurs when the atherosclerotic arteries cannot dilate to increase blood flow to the myocardium. When increased physical activity and stress place an added demand on the heart, the patient develops midsternal chest pain. The pain of stable angina usually subsides when the activity is stopped or with the use of a vasodilating drug.

Variant Angina (Prinzmetal's Angina)

The pain of variant angina is the same as in stable angina, except it has a longer duration and can occur at rest. The pattern of occurrence is often cyclical, with pain presenting about the same time each day. This type of angina is often caused by coronary artery spasms and usually does not cause damage to the myocardium.

Unstable Angina

Unstable angina occurs in patients with worsening CAD and is noted by its changing pattern. Rest does not decrease the chest pain of unstable angina. This pain may even occur when the patient is at rest. The episodes of chest pain with unstable angina increase in frequency and severity, placing the patient at risk for myocardial damage or sudden death.

Diagnostic Tests

Diagnostic tests are discussed in Chapter 15.

Electrocardiogram

An electrocardiogram (ECG) may help with the diagnosis of angina pectoris, although it remains normal in 20 to 30 percent of those with angina. ECGs obtained during an episode of chest pain may identify which coronary artery is involved. The ECG changes that may be seen in patients with angina may include a depressed T wave, which indicates myocardial ischemia, or an elevated ST segment, indicating myocardial injury.

Exercise Electrocardiogram

In an exercise ECG (stress test), the patient gets on a treadmill or stationary bicycle. Exercise activity is increased until the patient has reached 85 percent of his or her maximum heart rate. If the patient experiences chest pain or if noticeable ECG or vital sign changes occur during the test, it may be an indication of ischemia.

Another type of stress test, the dipyridamole (Persantine) thallium stress test, uses an intravenous medication in place of exercise. This test is used for patients who cannot exercise. Dipyridamole dilates coronary arteries and produces the effect of exercise. The collateral vessels also dilate, taking blood away from the coronary arteries, which may cause ischemia if CAD is present. Radiological imaging is then done to determine the area of the heart that is ischemic.

Radioisotope Imaging

In radioisotope imaging, the injection of a radioisotope is performed during a stress test. When the maximum heart rate is reached or ECG changes or anginal symptoms, occur the test is stopped. The patient is taken to a room with special imaging equipment to identify myocardial areas that are poorly perfused (ischemic).

Coronary Angiography

Coronary angiography uses an intravenous dye to view the coronary arteries. Coronary arteries that are partially or totally blocked are identified.

Treatments

Treatment for angina is directed at relieving and preventing anginal episodes that could lead to a myocardial infarction. The risk factors identified for the patient determine the course of treatment. Weight reduction; a low-fat, low-cholesterol diet; and stress reduction may help decrease the number of attacks.

Most patients with angina are placed on medication that reduces oxygen demand and increases oxygen supply to the myocardium. The three major groups of medication used for angina are vasodilators, calcium channel blockers, and beta blockers (Table 18–3).

Vasodilators

Nitroglycerin (a nitrate) is the drug of choice for acute anginal attacks. NTG can be administered sublingually, orally, transdermally, intravenously, or as a lingual spray. When administered sublingually, NTG may relieve chest pain within 1 to 2 minutes (Box 18–2).

Long-acting nitrates can be given orally, by ointment, or by transdermal patches. Nitrates all act to maintain coronary artery vasodilation. Isosorbide dinitrate (Isordil) and isosorbide mononitrate (ISMO) are examples of long-acting nitrates. A problem with long-acting nitrates is the devel-

---

TABLE 18-3 PHARMACOLOGICAL TREATMENT FOR ANGINA PECTORIS

Class	Drug	Route	Side Effects
Vasodilators	Nitrates/nitroglycerin (Nitrostat)	Sublingual/spray	Headaches, light-headedness, postural hypotension, tachycardia, flushing
	Isosorbide dinitrate (Isordil)	Oral	
	Isosorbide monotrate (ISMO)	Oral	
	Nitroglycerin (Transderm Nitro, Nitro-Bid)	Topical	
Calcium channel blockers	Diltiazem (Cardizem)	Oral	Headache, peripheral edema, dysrhythmias, flushing, dizziness, atrioventricular blocks
	Nifedipine (Procardia)	Oral	
	Verapamil (Calan, Isoptin)	Oral	
Beta blockers	Propranolol (Inderal)	Oral	Dizziness, bradycardia, hypotension, nausea, confusion, fatigue, agranulocytosis, laryngospasm
	Metoprolol (Lopressor)	Oral	
	Atenolol (Tenormin)	Oral	

Nursing Implications

Vasodilators	Monitor blood pressure and heart rate before and after giving.
	Caution patient to rise slowly because of orthostatic hypotension, especially with sublingual nitroglycerin.
	Take up to three nitroglycerin tablets at 5-minute intervals. If pain is not relieved, call a doctor or emergency medical assistance immediately.
	Do not remove tablets from bottle. Vasodilators become inactive when exposed to heat, air, light, and moisture.
	Tell patient a burning or tingling sensation may be felt under the tongue with sublingual nitroglycerin.
Calcium channel blockers	Monitor blood pressure and heart rate before and after administering.
	Hold if blood pressure is less than 90 systolic or heart rate is less than 50 and call physician.
Beta blockers	Assess blood pressure and heart rate before and after giving.
	Administer with food.
	Propranolol should not be given to patients with asthma.
	Do not stop abruptly.

BOX 18-2 Key Points for Using Sublingual Nitroglycerin

- Carry NTG at all times.
- Keep NTG tightly sealed in the original container.
- Replace NTG every 6 months for maximum effect.
- Take NTG before an activity known to cause chest pain.
- Take one NTG tablet and repeat every 5 minutes up to three doses if pain is not relieved. If pain is unrelieved after three doses, call for emergency medical care or have someone else drive to the hospital.
- Tingling may be felt under the tongue when NTG is used.
- NTG may cause a headache.
- NTG may cause light-headedness; rise slowly to prevent falls.

L E A R N I N G TIPS

To help you identify beta blockers, remember that their generic names end with –ol.

opment of a tolerance to the drug. To prevent tolerance, the patch or ointment is usually removed at bedtime and reapplied in the morning, giving the patient an 8- to 12-hour nitrate-free period.

Calcium Channel Blockers

Calcium is required for electrical excitability of cardiac cells and contraction of the myocardium. Calcium channel blockers relax vascular smooth muscle, which leads to decreased peripheral vascular resistance (afterload) and decreased myocardial oxygen demand. These drugs dilate main coronary arteries, increasing the myocardial oxygen supply. Nifedipine (Procardia) and verapamil (Calan, Isoptin) are potent inhibitors of coronary artery spasms and are used to treat variant (Prinzmetal's) angina. Calcium channel blockers are also used to decrease systolic and diastolic blood pressures and to slow the heart rate. These drugs are commonly given in conjunction with other vasodilators and beta blockers. Because these drugs are slow acting, they are ineffective in relieving acute anginal attacks.

Beta Blockers

Beta blockers decrease the workload on the heart and block the effects of epinephrine and norepinephrine on the heart (increased heart rate and vasoconstriction), thus preventing anginal attacks. There is decreased myocardial oxygen demand secondary to decreased heart rate, decreased myocardial contractility, and decreased blood pressure. Because of these decreased effects, beta blockers should be avoided in patients with any degree of heart failure, because the heart failure would become worse. There are nonselective and selective types of beta-adrenergic blockers. Patients with asthma or chronic obstructive pulmonary disease (emphysema, bronchitis, and bronchiectasis) should avoid nonselective beta-adrenergic blockers because they cause bronchoconstriction. Examples of beta-blocking agents used to treat chronic angina include propranolol (Inderal), metoprolol (Lopressor), and atenolol (Tenormin). Metoprolol and atenolol are more cardioselective and can be used in patients with asthma and chronic obstructive pulmonary disease (COPD). Beta blockers are not effective for coronary artery spasms.

Nursing Process

Obtaining a thorough history on a patient admitted with a diagnosis of angina pectoris is important in developing a plan of care and should include the following:

- How long the patient has had angina
- Risk factors
- Triggering activities
- How pain is relieved

If the patient reports chest pain, areas of assessment and documentation should include the following:

- Type, location, and pain radiation to other areas of the body
- Vital signs and skin color and temperature
- Presence of dyspnea, labored respirations, diaphoresis, or nausea

Oxygen and sublingual NTG are given to the patient with chest pain as ordered. Blood pressure and heart rate are assessed before and after NTG is given. The nurse must promote rest and decrease anxiety for the patient with chest pain. A patient who is experiencing chest pain should never be left alone. Emotional support is important because patients and their families are often afraid that the patient may die.

▶ MYOCARDIAL INFARCTION

A myocardial infarction (MI), commonly known as a heart attack, results in the death of heart muscle. An MI occurs from a partial or complete blockage of a coronary artery, which decreases the blood supply to the cells of the heart supplied by the blocked coronary artery. The extent of the cardiac damage varies depending on the location and amount of blockage in the coronary artery. This is a potentially devastating condition. The ability of the heart to contract, relax, and propel blood throughout the body requires healthy cardiac muscle. When the patient has an MI, part of the heart muscle no longer functions as it should. Cardiac conduction, blood flow, and function can be dramatically altered by an MI.

Those with MIs are typically men over 40 with atherosclerosis development. Although MIs can occur at any age in men or women, women who smoke and use oral contraceptives are at greater risk for MI.

Pathophysiology

Myocardial infarction does not happen immediately. Ischemic injury evolves over several hours before complete necrosis and infarction take place. The ischemic process affects the subendocardial layer, which is most sensitive to hypoxia. This process leads to depressed myocardial contractility. The body's attempt to compensate for decreased cardiac function triggers the sympathetic nervous system to increase heart rate. The change in heart rate increases myocardial oxygen demand, further depressing the myocardium.

Prolonged ischemia can produce severe cellular damage and necrosis of cardiac muscle. Once necrosis takes place, the contractile function of the muscle is permanently lost. The heart has a zone of ischemia and injury around the necrotic area (Fig. 18–3). The zone of injury is next to the necrotic area and is susceptible to becoming necrosed. If treatment is initiated within the first hour of symptoms of the MI, the area of damage can be minimized. Around the injury zone is an area of ischemia and viable tissue. If the heart responds to treatment, this area can rebuild and maintain collateral circulation. If prolonged ischemia takes place, the size of the infarction can be quite large. The size of the infarction depends on how quickly the blood supply from the blocked artery can be restored.

The area that is affected by an MI depends on the coronary artery involved and the extent of occlusive coronary disease (Fig. 18–4). Being familiar with the anatomy of the heart and the area of the MI helps the nurse anticipate dysrhythmias, conduction disturbances, and heart failure, which are the major complications of MIs (Table 18–4).

The anterior interventricular branch of the left coronary artery is the area that feeds the anterior wall of the heart, which also includes most of the left ventricle. An occlusion in this area causes an anterior wall MI. When the left ventricle is affected, there can be severe loss of left ventricular function, leading to severe changes in the hemodynamic status of the patient.

The right coronary artery (RCA) feeds the inferior wall and parts of the atrioventricular node and the sinoatrial node. An occlusion of the RCA leads to an inferior MI and abnormalities in impulse formation and conduction. Serious dysrhythmias can occur early in an inferior MI that may be life threatening.

The left circumflex coronary artery feeds the lateral wall of the heart and part of the posterior wall of the heart. A lesion in the circumflex leads to a lateral wall infarction of the left ventricle.

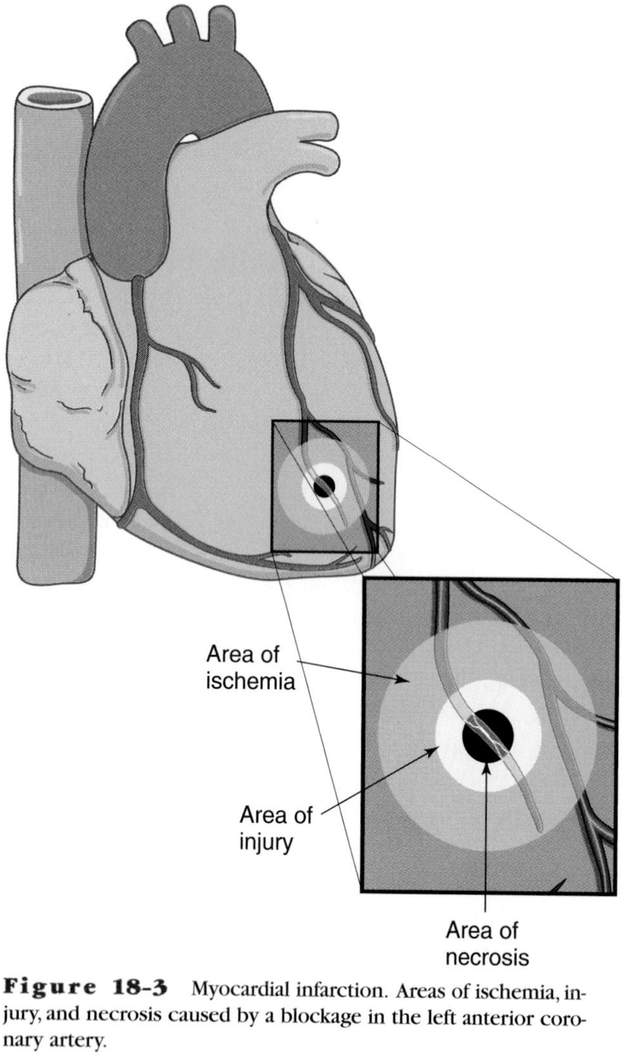

LEARNING TIPS

To remember what coronary artery occlusion results in a specific MI location, use coast-to-coast U.S. location initials such as those given below. You can personalize the locations with initials of landmarks familiar to you.

Location	Coronary Artery	Resulting MI Location
Los **A**ngeles	**L**eft anterior descending	**A**nterior
Cedar **P**oint	**C**ircumflex	**P**osterior
Rhode **I**sland	**R**ight	**I**nferior

Signs and Symptoms

Chest pain is a classic symptom of an MI. The pain begins suddenly and continues without relief with rest or NTG. The pain in the center of the chest is usually described as crushing, viselike, or as if an elephant is standing on the chest. The pain may radiate to the back, one or both arms and shoulders, neck, or jaw. The pain can imitate indigestion or a gallbladder attack with abdominal pain and vomiting. Other classic MI symptoms include shortness of breath, dizziness, nausea, and sweating (Box 18–3). When listening to lung sounds, crackles or wheezing may be heard. The pulse may be rapid or irregular, and an extra heart sound (referred to as S3 or S4) may be present. The presence of an extra heart sound can mean ventricular failure is imminent.

Women may not experience the classic symptoms of an MI. Research has only recently focused on understanding women and cardiac disease. Atypical symptoms reported by women may include extreme fatigue, epigastric pain, shortness of breath, or cramping in the chest. Women also often do not associate their symptoms with a heart attack because they believe it is a male disease.

Area of
ischemia

Area of
injury

Area of
necrosis

Figure 18-3 Myocardial infarction. Areas of ischemia, injury, and necrosis caused by a blockage in the left anterior coronary artery.

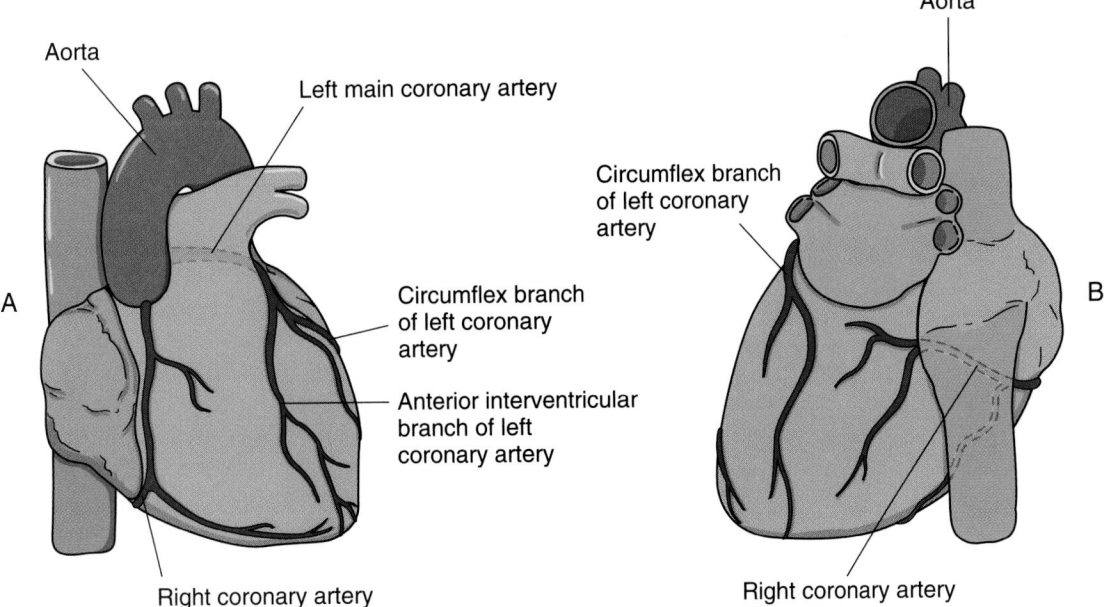

Figure 18-4 *(A)* Coronary arteries, frontal view. *(B)* Coronary arteries, posterior view.

People often deny or fail to recognize that they are having an MI because they experience atypical MI symptoms or their symptoms are similar to other mild conditions such as indigestion. Patients have reported that the symptoms of an MI that they experienced were not what they expected. If people expect to have the dramatic heart attack symptoms seen on television (which are usually not the same as those in real life) and they do not, they are likely to wait to seek treatment. People often wait 2 to 24 hours before seeking medical care, yet the first hour after symptom onset is crucial for seeking the newer reperfusion treatments that restore blood flow, minimize tissue damage, and save lives.

Because so few patients arrive at the emergency room quickly enough to benefit from newer treatments, the National Heart, Lung, and Blood Institute (NHLBI) and the American Heart Association launched a major new heart attack education campaign on 9-11-01 called "Act in Time

to Heart Attack Signs." The purpose of this campaign is to educate people on the importance of recognizing heart attack symptoms, working with a physician to create a heart attack survival plan, and calling 911 as soon as symptoms begin. For more information on the Act in Time to Heart Attack Signs, visit www.nhlbi.nih.gov, www.redcross.org, and www.americanheart.org.

The National Heart Attack Alert Program (NHAAP) is another public education program developed by the NHLBI. The message of this program is symptom recognition and "60 minutes to treatment" to improve survival and reduce tissue damage. For more information, visit www.nhlbi.nih.gov/nhlbi/othcomp/opec/nhaap/nhaapage.htm.

Gerontological Implications

Some myocardial infarctions occur without the presence of pain. This is referred to as a silent MI and occurs most often

TABLE 18-4	COMPLICATIONS OF MYOCARDIAL INFARCTION	
Complication	**Types or Symptoms**	**Interventions**
Dysrhythmias	Premature ventricular contractions, ventricular tachycardia, ventricular fibrillation, heart block	Continuous cardiac monitoring Protocols for treatment of dysrhythmias (see Chapter 20)
Cardiogenic shock	Decreased blood pressure; increased heart rate; diaphoresis; cold, clammy, gray skin	Immediate initiation of treatment to decrease infarct size, control pain and dysrhythmias Thrombolytic therapy Dopamine and dobutamine
Heart failure/pulmonary edema	Dizziness, orthopnea, weight gain, edema, enlarged liver, jugular venous distention, crackles	Correct underlying cause, relieve symptoms, increase cardiac contractility, administer furosemide (Lasix) and digoxin
Ventricular aneurysms, rupture of muscles or valves of the heart, septal rupture	Signs of cardiogenic shock, death	Mortality rate high; immediate treatment of MI to limit extent of damage
Pericarditis (inflammation of the heart muscle)	Chest pain, increased with movement, deep inspiration, or cough; pericardial friction rub (fine grating sound)	Relieved when sits up and leans forward Anti-inflammatory drugs (aspirin, indomethacin [Indocin])

BOX 18-3 Myocardial Infarction Summary

Signs and Symptoms

Classic
Crushing, viselike chest pain with radiation to arm, shoulder, neck, jaw, or back with a combination of the following:
Shortness of breath
Dizziness
Nausea
Sweating

Aytpical
Absence of classic pain with any one or a combination of the following:
Epigastric or abdominal pain
Cramping in the chest
Fatigue
Anxiety
Falling

Diagnostic Tests
ECG
Serum cardiac troponin I or T
Serum myoglobin
Serum CK-MB
Serum magnesium

Medical Treatment
Vital signs, oxygen saturations, intake and output monitored
Oxygen
Medications:
Aspirin
Morphine sulfate
Thrombolytics
Nitrates
Vasodilators
Beta blockers
Anticoagulants
Antidysrhythmics
Low-sodium clear liquids advanced to low-fat, low-cholesterol, and low-sodium diet; no caffeine
Fluid restriction
Bedrest with a bedside commode
Daily weight checks

Complications
Dysrhythmias
Heart failure

Nursing Diagnoses
Acute pain related to decreased coronary blood flow causing myocardial ischemia
Decreased cardiac output related to myocardial ischemia, heart rate changes, or dysrhythmias
Activity intolerance related to fatigue
Deficient knowledge related to lack of knowledge about myocardial infarction and its treatment

in the elderly patient. It may also occur in those with diabetes regardless of age. When pain is not present, the only symptom may be a sudden onset of shortness of breath or fainting, restlessness, or a fall experienced by the patient. Atypical presentation of MI symptoms is normal in the elderly patient, especially in those older than age 85. Because the elderly have had more time to develop collateral circulation than younger people, they often do not have as many complications with an MI.

Diagnostic Tests

Patients with a strong familial history of MI should be considered at risk until an MI is ruled out. The most useful indicators of an MI are patient history, ECG, and serum cardiac troponin I or T, myoglobin, and CK-MB levels. (See Chapter 15.) Magnesium levels are also checked, especially for those on diuretic therapy. Before thrombolytic or heparin therapy, prothrombin time (PT) and partial thromboplastin time (PTT) are determined. The ECG usually shows the area that has infarcted, as well as the ischemic areas of the heart. Myocardial damage is seen as ST segment elevation, the presence of a Q wave, or T-wave abnormalities (Fig. 18–5). Serial ECGs are done to monitor changes indicating damage or ischemia.

Treatment

Treatment should be sought within 15 minutes for any unrelieved chest pain. The American Heart Association recommends taking one uncoated adult aspirin at the onset of chest pain. Delays in seeking care can limit treatment options and result in more cardiac damage (Box 18–4). Patients need to be educated that "time is muscle." As time passes during an MI, more muscle is lost. Patients should not drive themselves to the hospital if they are having chest pain. Emergency medical care (911 or local emergency services number) should be called.

The presence of chest pain indicates a lack of oxygen to the myocardium. Patients reporting chest pain are treated as if they have an MI until it has been proven otherwise through testing. Treatment is given to increase oxygen supply to the heart muscle.

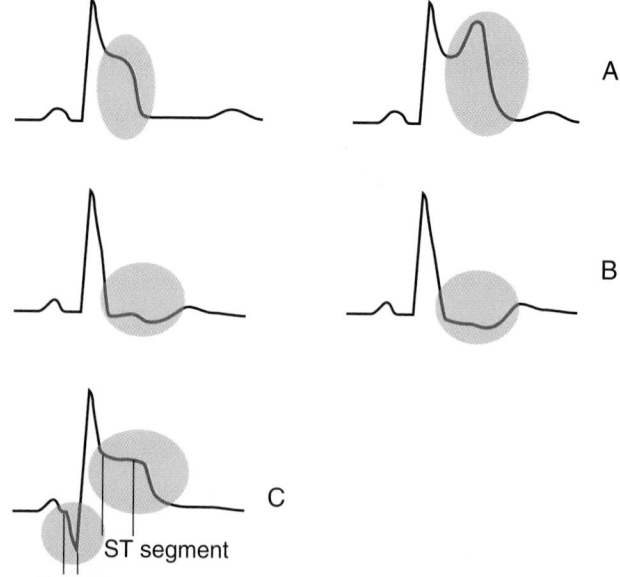

Figure 18-5 ECG changes during myocardial infarction. *(A)* Injury: ST segment elevation. *(B)* Ischemia: ST segment inversion. *(C)* Necrosis: large Q wave and ST segment elevation.

Preventing Delays in Myocardial Infarction Treatment

- Understand symptoms and "time is muscle" principle.
- Develop an action plan and rehearse it.
- Understand normal emotional responses of anxiety, denial, or embarrassment.
- Educate family to follow action plan.
- Establish protocols in workplaces for employees experiencing myocardial infarction.
- Establish emergency room policies that reduce delays, such as having equipment and medication readily available.

Oxygen

Oxygen is administered immediately, usually at 2 L per minute via nasal cannula. Arterial blood gases (ABGs) are drawn to determine the patient's oxygen needs. Oxygen can be administered via mask if higher concentrations are needed. Mechanical ventilation can be provided when indicated by ABGs.

Medication

Table 18–5 summarizes pharmacological treatment of myocardial infarction.

ANALGESICS. Analgesics are given for relief of chest pain. Morphine sulfate is the most commonly used narcotic for several reasons. In addition to pain relief, it helps decrease anxiety, opens bronchioles, and reduces preload and afterload, which can help increase blood supply and oxygen to the myocardium. It is given intravenously in small doses that are titrated to meet the patient's pain relief needs.

VASODILATORS. NTG sublingually, topically, or by intravenous (IV) drip can be administered for vasodilation to supply more blood to the myocardium to reduce pain and the workload of the heart. In the acute phase the IV route is usually used.

THROMBOLYTICS. Thrombolytic therapy is used to dissolve a blood clot that is occluding a coronary artery. Many communities allow initiation of thrombolytic therapy by paramedics in the field. Studies have revealed a decreased incidence of mortality and morbidity and less extensive tissue damage when thrombolytic treatment is used. Thrombolytic therapy must be started within a specified time range from the onset of symptoms, usually within 1 to 6 hours, before necrosis results.

Glycoprotein 11b/111a inhibitors (abciximab, tirofiban) may be used as an adjunct to thrombolysis or PTCA in patients with acute myocardial infarction. These drugs work by inhibiting platelet aggregation.

Activity

Initially patients are kept on bedrest with a bedside commode for bowel movements to decrease myocardial oxygen demand. Then activity is advanced gradually as tolerated.

Diet

During the acute phase of an MI, small, easily digested meals are served. Caffeine is usually restricted because it increases heart rate and causes vasoconstriction. Fluids may be restricted if the patient is in heart failure as well. Initially a low-sodium clear liquid diet may be ordered. Then a low-fat, low-cholesterol, and low-sodium diet may be ordered.

TABLE 18-5 PHARMACOLOGICAL TREATMENT FOR MYOCARDIAL INFARCTION

Drug	Purpose
Oxygen by nasal cannula, face mask, or face tent	Decreases myocardial oxygen demand and improves cellular oxygenation
Aspirin: one adult tablet PO. First dose not enteric coated. Daily dose enteric coated.	Decreases platelet aggregation to prevent clots
Analgesics: Morphine sulfate, IV push	Relieves pain, decreases anxiety, slows respirations, decreases preload and afterload
Nitrates: nitroglycerin by sublingual route, lingual spray, topical, IV infusion	Dilates coronary arteries to increase blood flow to the heart to relieve chest pain
Thrombolytics IV: alteplase (Ativase, tissue plasminogen activator [TPA]), anistreplase (Eminase), reteplase (Retavase), streptokinase (Streptase), urokinase (Abbokinase)	Lyses clots and prevents others from forming
Anticoagulants: heparin sodium by IV push or IV infusion, enoxaparin (Lovenox) subcutaneously	Prevents more clots from forming but has no effect on clots already present
Beta blockers IV or PO: propranolol (Inderal), metoprolol (Lopressor), atenolol (Tenormin)	Decreases heart rate and force of contraction and decreases oxygen requirements of the heart
Angiotensin-converting enzyme inhibitors PO: captopril (Capoten), lisinopril (Prinivil, Zestril), ramipril (Altace), trandolapril (Mavik)	Reduces death and heart failure after MI
Antidysrhythmics IV bolus or infusion: lidocaine (Xylocaine)	Used to control ventricular dysrhythmias (premature ventricular contractions, ventricular tachycardia)
Additional medications as needed: **Antiemetic**	Controls nausea and vomiting
Antianxiety	Reduces anxiety
Antacid	Relieves heartburn and indigestion
Stool softener/laxative	Reduces straining with bowel movement and prevents constipation

The number of grams of sodium is prescribed by the physician. If the patient is obese, a dietitian may work with the patient and family to devise a diet that is suitable and palatable for the patient.

Smoking

Smoking should be avoided, and patients are instructed on the hazards of continuing to smoke. Referral to a tobacco cessation program can be made. The nurse needs to work with patients to help them understand and accept lifestyle changes.

Nursing Process

Assessment

A thorough history is obtained to identify risk factors that may contribute to a myocardial infarction. All patients admitted with chest pain are treated as having a possible MI until it has been ruled out. Continuous cardiac monitoring, serial ECGs, and laboratory values help identify life-threatening dysrhythmias and determine the degree of cardiac damage. Controlling chest pain immediately helps diminish anxiety and the negative physiological effects pain has on the body.

Nursing Diagnosis

Nursing diagnoses may include but are not limited to the following:

■ Pain related to reduced coronary artery blood flow and increased myocardial oxygen needs

■ Decreased cardiac output related to ischemia or infarction, changes in heart rate and rhythm, and decreased contractility
■ Anxiety related to threat of death, changes in lifestyle, and chest pain
■ Activity intolerance related to imbalance between oxygen supply and demand, weakness, and fatigue
■ Deficient knowledge related to disease process, lifestyle changes, and medication

See Nursing Care Plan Box 18–5 for the Patient with Myocardial Infarction for more information.

Planning

The patient and family should be included in planning care. They need to understand that the purpose of treatment is to prevent further myocardial damage. The plan of care should focus on reducing factors that contribute to an increased workload of the heart. The patient's goals are to (1) report that the pain management regimen relieves pain; (2) maintain palpable pulses, warm and dry skin, and vital signs within normal range; (3) state that anxiety is relieved; (4) maintain desired activities; and (5) understand the disease and its treatment.

Nursing Interventions

Nursing care focuses on relieving pain, reducing the workload of the heart, and relieving patient and family anxiety. The patient should be positioned for comfort and ease in breathing. Semi-Fowler's position is usually preferred by patients having respiratory distress. When patients are sitting

---

BOX 18-5   NURSING CARE PLAN FOR THE PATIENT WITH MYOCARDIAL INFARCTION

 Acute pain related to decreased coronary blood flow causing myocardial ischemia

PATIENT OUTCOMES
Patient will exhibit signs of decreased pain. Patient will exhibit signs of relaxation.

EVALUATION OF OUTCOMES
Does patient state pain is reduced?

INTERVENTIONS	RATIONALE	EVALUATION
• Assess for location, duration, intensity, and radiation; use a scale of 0 to 10.	Identifies type and severity of pain.	What is pain level, location, duration, intensity, and radiation?
• Assess blood pressure, pulse, and respiration.	Vital signs may elevate episodes of pain.	Are vital signs within normal limits?
• Obtain ECG as ordered.	Identifies location of infarction or ischemia.	Is ECG normal?
• Administer oxygen as ordered.	Helps prevent hypoxia.	Are ABGs within normal limits? Is oxygen saturation greater than 90 percent?
• Instruct patient to report pain at first onset.	Helps control pain quickly to prevent further ischemia.	Does patient report pain?
• Instruct patient to rest during pain.	Activity increases oxygen demand and can increase chest pain.	Does patient remain quiet and relaxed?
• Remain with patient during chest pain until it is relieved.	Provides comfort and reassurance to decrease anxiety and fear.	Are anxiety and fear decreased?
• Explain and assist with alternative pain relief measures: • Positioning • Diversional activities • Relaxation techniques	These measures help decrease painful stimuli, allowing the patient to focus on other things.	Does patient express relief and decreased stress?
• Medicate as ordered.	Helps eliminate pain.	Is pain relieved?

BOX 18-5   Nursing Care Plan for the Patient with Myocardial Infarction—cont'd

Decreased cardiac output related to ischemia or infarction, changes in heart rate and rhythm, and decreased contractility

PATIENT OUTCOMES

Patient will maintain adequate cardiac output and tissue perfusion. Patient will exhibit signs of improved cardiac output and tissue perfusion.

EVALUATION OF OUTCOMES

Does patient have heart rate greater than 60 and less than 100, blood pressure greater than 90/60 and less than 140/90, and urine output greater than 30 mL/h?

INTERVENTIONS	RATIONALE	EVALUATION
• Monitor blood pressure greater than 90/60, heart rate greater than 60 and less than 100, urine output greater than 30 mL/h.	Indirect indicators of cardiac output.	Are indicators within normal limits?
• Listen to lung sounds.	Crackles indicate heart failure.	Are lungs clear?
• Monitor peripheral circulation, pulses, capillary refill, edema, color, and temperature.	Indicators of adequate tissue perfusion.	Does patient have strong peripheral pulses, capillary refill less than 3 seconds, no edema, warm skin, pink nailbeds?
• Monitor ECG.	Identifies dysrhythmias.	Is patient's ECG within normal limits?
• Administer medications as ordered by physician, such as vasodilators, beta blockers, calcium channel blockers, and cardiac glycosides.	Helps improve contractility, cardiac output, and tissue perfusion.	Does patient show signs of improved contractility, increased cardiac output, and tissue perfusion?
• Promote and provide for adequate rest, quiet environment, bedrest; place in semi-Fowler's position.	Decreases cardiac workload and stress and allows for improved breathing.	Is patient relaxed?

upright in bed, supporting their arms on pillows reduces the workload of the heart by eliminating the force of gravity on unsupported arms. The patient is taught to avoid Valsalva's maneuver, which occurs when the patient bears down against a closed glottis, mouth, and nose. Valsalva's maneuver increases intrathoracic pressure, slows the heart rate, and decreases blood return to the heart. Valsalva's maneuver can occur when the patient moves or turns in bed or strains with bowel movements. Patients must be taught not to hold their breath when moving in bed and also not to strain with bowel movements. Stool softeners are usually ordered by the physician to prevent straining with bowel movements. A record of the patient's bowel movements is kept, and intake and output are monitored.

Emotional support should be provided. The patient should be allowed to verbalize fears. Procedures should be explained. Family support should also be offered. Spouses also experience anxiety and need support, including ongoing information and explanations, being allowed to stay with the patient, and being involved in the patient's care. Spouses or significant others often ignore their own needs to focus on the patient. Nurses need to help spouses or significant others meet their own needs, so that they are better able to support the patient.

Teaching about the therapeutic regimen includes information about the disease, medications, diet, activity, and rehabilitation needs that may require lifestyle changes. Diet, stress reduction, a regular exercise program, cessation

of smoking if necessary, and following a medication schedule require extensive patient and family teaching. This disease can affect all aspects of a patient's lifestyle. Issues about family and job roles and sexual activities need to be addressed. Patients need time to understand information that has been presented and should be encouraged to express any questions, needs, or fears.

Evaluation

Patient goals are met if the following occur:

1. Satisfactory pain relief is reported using a rating scale of 0 to 10
2. Vital signs are within normal limits, no pain is reported, and no alteration in level of consciousness is observed
3. Anxiety is reduced
4. Fatigue is reduced, and the patient gains the ability to complete tasks and engage in desired activities
5. Understanding of the disease, medications, and reasons for lifestyle changes and a willingness to follow the regimen are stated

Cardiac Rehabilitation

Cardiac rehabilitation is begun when the patient's acute symptoms are relieved. The purpose of cardiac rehabilitation is to improve cardiac function and assist the patient to return to as normal a life as possible. Cardiac rehabilitation protocols are used in many institutions. The first two phases of rehabilitation occur in the hospital. Activities for each hospital day, such as types and amounts of self-care and activity, are specified in protocols. The third phase begins with hospital discharge and focuses on returning to prior levels of activity and function. Outpatient programs are often ordered for patients in this phase. In this phase, patients are encouraged to maintain optimum physical fitness and to continue healthy lifestyles.

CRITICAL THINKING

Mrs. Sims

Mrs. Sims, 43, is admitted to the intensive care unit with a diagnosis of atypical chest pain. She has a history of midsternal chest cramping. The pain is radiating to her left and right shoulders and down her left arm. Her pain increases with activity and decreases with rest. She smokes one and a half packs of cigarettes per day and is 50 pounds overweight. The cardiac monitor shows normal sinus rhythm without dysrhythmias. She has NTG sublingual ordered prn for chest pain.

One hour after admission, Mrs. Sims reports midsternal chest pain radiating to left neck and jaw. The cardiac monitor shows sinus tachycardia with occasional premature ventricular contractions (PVCs). Her blood pressure is 100/70, respirations are 20 and unlabored, and skin is warm and dry.

1. What actions should you take?
2. What is happening to Mrs. Sims?
3. How is angina differentiated from an MI?
4. What are four indicators of an MI?
5. What medical interventions can be used for an MI?
6. What education is indicated for Mrs. Sims?

Answers at end of chapter.

▶ PERIPHERAL VASCULAR SYSTEM

Peripheral vascular disease (PVD) may be either arterial or venous in origin. PVD is very common in people who are elderly or diabetic. You should understand if the origin of the problem is arterial or venous to prevent serious complications from occurring. Typically with arterial disease the patient reports leg pain with leg elevation or calf pain during activity that disappears with rest. Venous disease has a slower onset and is not associated with activity. Pain occurs when the legs are positioned dependently (down), causing pooling of blood at the ankles.

▶ ARTERIAL THROMBOSIS AND EMBOLISM
Pathophysiology

A thrombus (blood clot) adheres to the vessel wall. Acute arterial thrombi occur where there is injury to an arterial wall, sluggish flow, or plaque formation secondary to atherosclerotic changes. Other causes of arterial thrombosis are polycythemia, dehydration, and repeated arterial needle sticks. If a thrombus breaks off and travels, it becomes an **embolism** that occludes an arterial vessel that is too small to allow it to pass. Some of the causes of an arterial embolism are dysrhythmias, prosthetic heart valves, and rheumatic heart disease.

Signs and Symptoms

Usually there is an abrupt onset of symptoms with acute arterial occlusion. If a patient also has chronic arterial insufficiency, the symptoms may not occur as rapidly because collateral circulation has developed and can supply some blood to the occluded area. Symptoms depend on the artery occluded, the tissue supplied by that artery, and whether collateral circulation is present.

There are six clinical signs of acute arterial occlusion, known as the six *P*s: pain, pallor, pulselessness, paresthesia (numbness), paralysis, and poikilothermia (temperature). The patient experiences pain, numbness, and decreased movement in the extremity, which is pale and without pulses distal to the occlusion. The extremity feels cold because blood normally provides warmth. If treatment is not initiated immediately, ischemia occurs and can progress to tissue necrosis and gangrene development within hours.

Medical Treatment

Early treatment is necessary to protect and save the affected limb. Anticoagulant therapy is started immediately. Intravenous heparin is the treatment of choice to prevent further clotting. Heparin has no effect on existing clots. An initial IV bolus of heparin, usually 5000 IU, is given. An IV infusion is then started as ordered. The patient remains on heparin therapy for several days. Daily PTTs are monitored to maintain therapeutic heparin levels. After 3 to 7 days, warfarin (Coumadin) is added. Warfarin, an oral anticoagulant, takes 3 to 5 days to reach therapeutic levels. The heparin is continued until a therapeutic warfarin level is reached. To monitor warfarin's effects, international normalized ratios (INRs) and PTs are done daily, and adjustments in warfarin doses are made based on the results.

For patients with severe occlusions, especially if the risk of limb loss is imminent, surgery or thrombolytic agents are used to save the extremity. During an emergency embolectomy or thrombectomy, the artery is cut open, the emboli or thrombus is removed, and the vessel is sutured closed. Thrombolytic agents dissolve the thrombus or embolus.

CRITICAL THINKING

Mrs. May

Mrs. May is admitted with severe rheumatoid arthritis, which has left her immobile for 7 months. She is returning to her room following a whirlpool treatment when she suddenly reports severe pain in her left groin.

1. What is your first action?
2. After assessing Mrs. May, what action should be taken next?
3. What are the possible causes of these sudden symptoms?
4. How would you document Mrs. May's symptoms?
5. What would the immediate interventions be?
6. What medical interventions would you anticipate?
7. What surgical procedure may need to be done if the risk of losing the limb is imminent?

Answers at end of chapter.

► PERIPHERAL ARTERIAL DISEASE

Peripheral arterial disease (PAD) causes chronic, progressive narrowing of arterial vessels that leads to obstruction. PAD usually affects the lower extremities. Atherosclerosis is the leading cause of occlusive disease.

Pathophysiology

The purpose of the arterial system is to delivery oxygen-rich blood to the vascular beds. Anything that impedes this flow causes an imbalance in supply and demand for oxygen. Decreased nutrition, cellular waste accumulation, and the development of ischemia occur at the area distal to the obstruction. With the increased debris and sluggish flow, thrombosis and embolism become major problems.

The body has several mechanisms to attempt to compensate for reduced blood flow, including peripheral vasodilation, anaerobic metabolism, and development of collateral circulation. However, these mechanisms are not intended to meet the ongoing blood supply needs of the body. It takes time for collateral circulation to develop, blood vessels eventually reach their limit of dilation, and anaerobic metabolism is only a very short term compensatory mechanism. Eventually this lack of blood supply produces signs of ischemia, and if not corrected, ulceration, gangrene, and necrosis of the extremity occur; amputation of the limb may then become necessary.

Signs and Symptoms

Symptoms usually occur late in the course of PAD. It is only when diminished blood flow begins to produce changes in the extremities that symptoms occur. Pain in the calves of the lower extremities associated with activity or exercise, called **intermittent claudication,** is a common symptom of arterial occlusive disease. When blood supply to the muscles is decreased, the muscles are unable to receive adequate oxygen and ischemia develops. As ischemia increases, the muscle develops a cramping-type pain that usually subsides when the activity is stopped. As PAD progresses, the pain is present even at rest, thus indicating severe arterial occlusion.

Skin color changes are associated with decreased blood supply. The extremity is pale when the leg is elevated. If the leg is in a dependent position, it becomes reddish-purple or cyanotic. The extremity is cool to touch even in warm environments. As occlusion of the arteries progresses, there are diminished or absent arterial pulses. The loss of circulation leads to tissue death and gangrene. See Box 18–6 for stages of peripheral arterial disease.

Diagnostic Tests

Noninvasive studies can be used to diagnose occlusive disorders. The ankle-brachial (arm) blood pressure index is used to determine pressures in the upper and lower extremities. Normally, pressures in the legs and arms should be equal. When an occlusion occurs in the lower extremities, the pressures between the upper and lower extremities become unequal. A Doppler ultrasound measures the velocity

---

BOX 18-6 **Stages of Peripheral Arterial Disease**

Stage I: Asymptomatic
1. No claudication is present.
2. Bruit (swishing sound over artery or vein) or aneurysm may be present.
3. Physical examination may rarely reveal decreased pulses.

Stage II: Claudication
1. Muscle pain, cramping, or burning is exacerbated by exercise and relieved by rest.
2. Symptoms can be produced by exercise.

Stage III: Rest Pain
1. Pain while resting commonly wakes the patient at night.
2. Pain is described as numbness, burning, or toothache-type pain.
3. Pain usually occurs in the distal portion of the extremity and only rarely in the calf or ankle.
4. Pain is relieved by placing the extremity in a dependent position.

Stage IV: Necrosis/Gangrene
1. Ulcers and blackened tissue occur on the toes, forefoot, and heel.
2. A distinctive gangrenous odor is present.

Source: Ignatavicius, DD, and Hausman, KA: Pocket Companion for Medical-Surgical Nursing, ed 2. WB Saunders, Philadelphia, 1995, p 93, with permission.

---

of the blood flow. Magnetic resonance imaging (MRI) can give definitive images of blood vessels and degrees of arterial closure.

L E A R N I N G   T I P S

Arterial insufficiency risk assessment: ankle-brachial index

1. Take brachial blood pressure.
2. Take ankle blood pressure (use arm blood pressure cuff placed above the ankle).
3. Divide ankle systolic pressure by brachial systolic pressure (e.g., 98 ÷ 130 = 0.75).
4. Arterial disease = 0.8 or less.
5. Do not treat arterial disease like venous disease!

Arteriography is an invasive procedure in which an x-ray examination of an artery is taken after dye has been injected. This invasive procedure requires a patient's consent and identification of any patient allergies to iodine. Arteriography is usually done before surgical intervention is considered.

Medical Treatment

Conservative medical treatment is initiated with mild to moderate occlusive disease. This includes patients who experience pain on activity that ceases with rest. This type of patient usually receives medication for vasodilation and diet management if necessary. Surgical intervention is used for the patient who experiences pain at rest or who has leg ulcers that do not heal. Surgical treatment includes endarterectomy to remove the atherosclerotic lesion or grafting to bypass the occluded area. (See Chapter 22.)

Diet

To help control atherosclerosis development, the diet should be low fat, low cholesterol, and low calorie if the patient is overweight. Teaching the patient to avoid red meats, fried foods, whole milk, and cheese is important. Avoiding high-cholesterol foods, such as egg yolks, organ meats, animal fats, and shellfish, helps lower lipid levels.

Medications

Drug therapy is geared toward the symptoms and causes of the occlusive disease. The same drugs used to decrease cholesterol and lipid levels in atherosclerosis are used with occlusive disease. Vasodilators can be used, but their effectiveness is not the same for all patients. Pentoxifylline (Trental) is commonly the drug of choice for those patients with occlusive disorders who experience intermittent claudication. This drug makes red blood cells more flexible to improve perfusion. The major side effect is gastrointestinal upset, so it should be taken with meals. Thrombolytic therapy is used when an occlusion is caused by a thrombus or an embolus.

Invasive Therapies

Percutaneous transluminal angioplasty (PTA) can be used to dilate a narrowed peripheral vessel. It is similar to PTCA, which was discussed earlier. Complications that can occur with this procedure are a ruptured artery as a result of the stretching caused by the balloon and clot formation. Peripheral atherectomy is another invasive procedure used to remove plaque from atherosclerotic arteries. Intravascular stents can also be used to maintain patency of the artery. After stent placement, patients are given platelet aggregation inhibitors.

Nursing Process for Peripheral Arterial Diseases

Assessment

When dealing with patients having arterial occlusive disorders, monitoring peripheral circulation is important. Careful assessment of pulses, capillary refill, temperature, color, and presence of edema helps identify patients at risk for complications.

Nursing Diagnosis

The nursing diagnoses for PAD may include but are not limited to the following:

- Ineffective tissue perfusion related to decreased arterial perfusion
- Pain related to decreased blood supply to lower extremities
- Activity intolerance related to activity pain
- Deficient knowledge: peripheral arterial disease related to complications, medications, or postoperative care.

See Nursing Care Plan Box 18–7 for the Patient with Peripheral Arterial Occlusive Disorders.

Planning

Arterial occlusive disease is a progressive disorder that can be controlled or at least slowed. Including the patient and family in the development of the plan of care will help ensure their understanding of the reasons for the current therapy. The patient's goals include the following:

- Maintain palpable pulses and warm and dry skin
- Report that pain is relieved
- Maintain desired activities
- Understand the disease and its treatment

Nursing Interventions

The patient should be able to describe symptoms and complications associated with arterial occlusive disorders. Medication cards can be developed to help the patient understand the different medications being used and any special considerations required. The patient should be seen by a dietitian before discharge for instructions on a low-fat, low-cholesterol diet. Proper leg positioning should be taught to prevent complications and the potential loss of the extremity. Patients should not elevate their legs, especially if pain is increased. Legs should be kept in a dependent position to encourage blood flow with the assistance of gravity. Regular 30-minute walks have been shown to be helpful in maintaining circulation and reducing the incidence of blood clots for those with PAD.

Evaluation

The best indicators that outcomes for the patient with PAD have been achieved are that the patient is experiencing less pain with activity and that there are no complications with activity. If the patient states understanding of the disease and the need to comply with treatment, the goal for deficient knowledge is achieved. This disorder requires constant monitoring of its status and progression.

▶ ANEURYSMS

An **aneurysm** is a bulging or dilation at a weakened point of an artery. The artery diameter is often increased by 50 percent. The cause is unknown, but anything that weakens the artery wall or causes loss of elasticity in the artery can cause an aneurysm. Atherosclerosis, hypertension, smoking, trauma, and congenital abnormalities are risk factors for an aneurysm. Heredity may also play a role. Aneurysms can occur in any artery in the body but are common in the abdominal aorta, which is the focus of the rest of this discussion.

Abdominal aortic aneurysms (AAA) are often silent if they are less than 4 cm. Most people do not even know that they have an AAA. Men older than age 50 are at the highest risk of death from an AAA. The incidence of AAA increases with age.

| BOX 18-7 | **NURSING CARE PLAN FOR THE PATIENT WITH PERIPHERAL ARTERIAL OCCLUSIVE DISORDERS** |

 Acute pain related to impaired circulation to extremities causing intermittent or continuous pain

PATIENT OUTCOMES
Patient will report that pain is controlled.

EVALUATION OF OUTCOMES
Does patient report relief from pain by nonpharmacological or pharmacological methods?

INTERVENTIONS	RATIONALE	EVALUATION
• Monitor for intermittent claudication or pain at rest.	Helps determine degree of occlusive disease.	Does patient have pain during activity or at rest?
• Note peripheral circulation, pulses, color, temperature, presence of edema, and skin breakdown.	Determines the degree of tissue perfusion and complications.	Does patient have pulses, warm skin, capillary refill less than 3 seconds, no evidence of skin breakdown?
• Administer medication as ordered:		Does patient show signs of increased circulation and relief of pain following administration of medications?
• Analgesics	Relieves chronic or acute pain.	
• Vasodilators	Increases blood flow to extremities.	
• Calcium channel blockers	Decrease vasospastic episodes.	
• Promote bedrest if pain is present.	Rest decreases muscle contraction and prevents further ischemia in extremities.	Is patient able to rest?
• Position lower extremities below heart level.	Increases arterial flow to lower extremities.	Are pulses strong, capillary refill less than 3 seconds, extremities pink and warm?
• Protect extremities from cold or trauma.	Extremities with decreased circulation have decreased sensation, which increases risk of injury.	Are extremities injury free?

▶ **Ineffective tissue perfusion related to interruption of arterial flow in arms and legs**

PATIENT OUTCOMES
Patient will exhibit signs of increased arterial blood flow and tissue perfusion.

EVALUATION OF OUTCOMES
Does patient have strong peripheral pulses, capillary refill less than 3 seconds, warm skin, absence of edema?

INTERVENTIONS	RATIONALE	EVALUATION
• Check peripheral pulses, capillary refill, color, temperature, and presence of edema every 4 hours.	Indication of adequate tissue perfusion.	Are peripheral pulses strong, nailbeds pink, capillary refill less than 3 seconds with no edema noted?
• Check skin for intactness, healed areas, signs of ulceration or infection.	Chronic arterial occlusion leads to decreased blood flow, resulting in tissue damage and poor wound healing.	Is skin intact?
• Place extremities lower than heart, feet on floor in sitting position, head of bed elevated on blocks.	Increases blood flow to the legs and feet.	Does patient have adequate tissue perfusion signs?
• Avoid bending knees, pillows under knees, prolonged sitting or crossing legs.	These activities impede blood flow to extremities.	Does patient exhibit understanding of improving blood flow?
• Inspect lower extremities frequently. Clean feet with mild soap; dry carefully. Protect from injury.	Cleaning prevents trauma to feet, protecting feet from things that can lead to ulcerations.	Is patient free from trauma or breaks in skin of the lower extremities?

Figure 18-6 Types of aneurysms.

Types

There are different types of aneurysms (Fig. 18–6). A fusiform aneurysm is the dilation of the entire circumference of the artery. A saccular aneurysm is a bulging on only one side of the artery wall. A dissecting aneurysm occurs when a cavity is formed from a tear in the artery wall, usually the intimal (inner) layer. The layers of the artery are then separated as blood is pumped into the tear with each heartbeat, expanding the cavity, which is then prone to rupturing.

Signs and Symptoms

Usually there are few if any symptoms. As the AAA grows, symptoms may develop. Back or flank pain is the classic symptom; the pain is caused by the aneurysm pressing against nerves of the vertebrae. Depending on the location and size of the aneurysm, there may be complaints of abdominal pain, a feeling of fullness, or nausea caused by pressure on the intestines. Changing positions may temporarily relieve the symptoms. Because the symptoms are vague, they are often not associated with an AAA. There may be a pulsating mass in the abdomen caused by an

AAA that is only discovered during routine physical or x-ray examination.

Severe, sudden back, flank, or abdominal pain and a pulsating abdominal mass can indicate that the aneurysm may be about to rupture. With rupture, the patient's blood pressure may drop and signs of shock may be present. Immediate surgery is needed for a ruptured AAA. The mortality rate is high with a ruptured aneurysm.

Diagnostic Tests

Computed tomography (CT) and ultrasound are the most common diagnostic tools used to confirm the presence of an aneurysm. Small aneurysms may be watched over time to see if they enlarge. Aortography can be performed when surgical intervention is considered to identify the size and exact location of the aneurysm.

Medical Treatment

Medical treatment consists of medication to maintain lower blood pressures because patients with aneurysms often have hypertension. If the blood pressure is allowed to get too high, it can cause the arterial wall to rupture. Surgical treatment—a bypass graft—is performed when the patient is experiencing pain or showing signs of circulatory compromise. An aneurysm that is larger than 5 cm requires surgery. The risk of rupture is greatest when the aneurysm reaches 5 cm or greater.

A conventional open surgical repair or newer endovascular grafting may be done for an AAA. Endovascular grafting involves the transluminal placement (through the femoral artery) and attachment of a sutureless aortic endograft or stent-graft prosthesis at the site of the AAA. In the endograft procedure a balloon catheter positions and opens the graft. The graft remains attached to the inner wall of the aorta when the catheter is withdrawn. Blood flow continues through the aorta, bypassing the aneurysm. Another method uses a stent-graft that opens to fit the diameter of the aorta to reduce pressure on the aneurysm. Endovascular surgery requires less hospitalization time and a quicker recovery.

Nursing Care

Careful monitoring of those with an AAA is necessary. Patients must understand their medications and the importance of taking antihypertensives as prescribed. Stress may be a risk factor that should be addressed. Lifting heavy objects can increase pressure within the artery and may be restricted even in the individual being treated with more conservative measures. Postoperatively, the patient should avoid lifting heavy objects (Home Health Hints).

▶ RAYNAUD'S DISEASE

A vasoconstrictive response causing ischemia from exposure to cold and stress is known as Raynaud's disease. It occurs more often in women who live in cold climates. Raynaud's primarily affects the hands but can also occur in the

HOME HEALTH HINTS

Cardiac

- When a patient reports cardiac symptoms, the nurse should ask about possible precipitating factors, such as emotional stress, alcohol or drug abuse, or if a hot bath was taken.

- The elderly patient does not usually describe angina as pain but may say it feels like pressure, or burning, or a butterfly. Ask what the patient was doing when the angina began, such as yard work, eating, receiving emotional news, or engaging in sexual activity. Symptoms of accompanying nausea or sweating are common with an MI.

- Post-MI patients usually come home with a schedule of graduated activities. Review them with the patient and caregivers and post it in the home.

- Encourage the caregiver to walk with the patient outdoors when the weather is nice, avoiding extreme cold or heat. Caution them to avoid uneven sidewalks or areas with no sidewalks.

- When a patient is enrolled in a formal cardiac rehabilitation program, be sure to reinforce the exercises within the formal guidelines, incorporating them into the home health care plan. The nurse should inspect the layout of the house to identify those areas or the daily activities of the patient that cause more exertion than the patient realizes. Examples are the number of steps or stairs inside and outside and the amount of reaching, pulling, or pushing that the patient actually does.

- It may be appropriate to have a home health social worker visit a post-MI or postsurgical cardiac patient and provide input into the plan of care (anxiety and environmental stress reduction, relaxation techniques).

- When assessing the patient and the medication record, ask if the patient has regular bowel movements without straining. Reinforce that straining should be avoided. Ask the physician to prescribe a stool softener or laxative as needed.

Vascular

- When a patient wears elastic compression stockings and the home health nurse has to check the appearance of the lower extremities, the patient should lie down to help remove them. The stockings should be put back on while the patient is still lying down.

- Remind the patient to put the elastic stockings on before getting out of bed in the morning when edema is decreased and to leave the stockings on all day.

- Compression stockings should be discarded when they become too easy to put on. Because they are expensive, many patients keep them even after they are ineffective.

- Instruct the home health patient on these easy exercises to enhance venous return: going up and down on toes when standing and pedaling (like using car brake and gas pedals) when sitting.

- Teach patients with varicose veins not to jump, lift weights, wear high heels, or use hot tubs. Legs should be elevated on a regular basis.

- When assessing edema of a bedridden patient, check all dependent areas, especially the sacral area.

- Remember that if a patient goes to bed with lower leg edema, that fluid returns to the circulatory system and eventually to the kidneys, requiring extra trips to the bathroom during the night. Have patients with edema recline for 30 minutes before bed and then void before going to bed. They will then have less interrupted sleep during the night.

---

feet, ears, or nose. To be diagnosed with Raynaud's disease, the patient must experience intermittent attacks of ischemia for at least 2 years.

Pathophysiology

Raynaud's disease is characterized by spasms of small arteries in the digits. These spasms prevent arterial blood from perfusing the fingertips and sometimes the toes. The spasms can occur unilaterally and in one or two digits, but most often they occur bilaterally and in all digits.

Signs and Symptoms

The hands, when exposed to cold, exhibit vascular spasms and a marked decrease in blood flow to the tissues. The resulting effect in the tissues is ischemic pain. After several minutes of ischemia, hyperemia occurs. Hyperemia is intense reddening of the hands from dilation of all the vessels of the hands. Pain becomes more intense at this time. Patients with Raynaud's disease goes through phases of blanching of the skin, pain, and reddening of the skin. This disease can progress over time; the vessels remain constricted and the severe decrease in blood flow can lead to fingers becoming gangrenous and necrotic.

Medical Treatment

Conservative treatment is attempted first. The patient is instructed to keep the hands warm. Gloves should be worn when going outside, cleaning a refrigerator, or preparing cold foods. Patients are instructed in the importance of protecting the hands from injury and avoiding things that contribute to vasoconstriction, such as smoking, alcohol, and caffeine. Reducing stress levels can also help prevent vasoconstriction. Vasodilators are sometimes prescribed to help the patient avoid peripheral vasoconstriction. Low doses of nifedipine or long-acting nitrates can be used.

To treat Raynaud's disease surgically, the sympathetic reflex must be blocked. This is accomplished by interrupting the sympathetic nerve impulses from the spinal cord to the hand, which is known as sympathectomy.

Nursing Care

Education is the primary goal for patients with Raynaud's disease. Teaching the patient to protect the hands is very important. Stressing the use of gloves in cold climates, reducing vasoconstrictive activity, and decreasing stress levels helps reduce the number and severity of attacks.

► THROMBOANGIITIS OBLITERANS (BUERGER'S DISEASE)

Buerger's disease is a recurring inflammation of small and medium arteries and veins of the lower extremities. The disease is usually the result of occlusion of the vessels by thrombus formation. The cause is unknown, but heavy cigarette smoking is a major contributing factor. Some studies indicate an autoimmune response to tobacco products as a possible cause.

The inflammation and irritation of the vessels contribute to the development of vasospasms. These vasospasms lead to an obstruction in blood flow. The tissues become hypoxic, and the development of ischemic pain can occur. This ischemia, left untreated, can lead to ulceration and gangrene.

Intermittent claudication and other symptoms of occlusive disease are common in patients with Buerger's disease. Other symptoms include numbness or decreased sensation and cool extremities. Lower extremities can be red or cyanotic when in a dependent position, and pulses may be diminished. Depending on the degree of ischemia, ulceration or gangrene may be present.

Because the primary contributing factor is smoking, there is an urgency in helping the patient to cease smoking. The patient must be made aware of the effect smoking has on the body and that the disease will progress and further damage other vessels. Therapy and nursing care for Buerger's disease is the same as those used for other arterial occlusive diseases. The use of calcium channel blockers such as diltiazem (Cardizem) promotes vasodilation and may help with intermittent claudication. Careful inspection of the lower extremities for signs of breakdown is important, so early treatment can begin.

► VARICOSE VEINS

Varicose veins are elongated, tortuous, dilated veins. The exact cause is unknown. Varicose veins are divided into primary and secondary varicosities.

Pathophysiology

Primary varicosities are believed to be caused by a structural defect in the vessel wall. Along with the defect, the dilation of the vessel can lead to incompetent venous valves. The valves help prevent blood from refluxing. If reflux occurs, it can cause further dilation of the vessel. The superficial veins are the vessels most often involved in primary varicosities.

Secondary varicosities are caused by an acquired or congenital pathological condition of the deep venous system. This produces dilation of collateral and superficial veins. As a result, there is an interference of blood return to the heart, which leads to stasis, or pooling, of the blood in the deep venous system. This increases the pressure within the system, pushing blood into the collateral vessels and producing varicosities in the superficial veins.

Causes

A number of factors can lead to varicose veins. The wall defects have been identified as a familial tendency and may be inherited. Any factor that may contribute to increasing hydrostatic pressure within the leg, such as prolonged standing, pregnancy, and obesity, may promote venous dilation. Incompetent valves within the veins can cause blockage of blood flow and lead to dilated veins.

Signs and Symptoms

The most common manifestation is the disfigurement of the lower extremity with primary varicosities. There may be dull pain, especially after prolonged standing. This usually can be relieved by walking or elevating the extremity. With secondary varicosities, the pain and disfigurement may be more severe. There can be development of edema or ulceration if circulation is severely compromised.

Medical Treatment and Nursing Care

The primary goals are to improve circulation, relieve pain, and avoid complications. Treatment is usually not indicated if the problem is only cosmetic. Conservative treatment is geared to reduction of factors that contribute to varicose veins. Patients are taught the importance of assessing peripheral circulation and avoiding trauma to the lower extremities. Weight reduction, elevation of the extremities, walking, and exercise help increase muscle strength and contraction. Tight-fitting clothes at tops of legs or waist should not be worn. Wearing support hose assists in blood flow return to the heart. Elastic compression stockings should be used as ordered. (See Home Health Hints.) Injection sclerotherapy is used to treat superficial varicosities. Surgical intervention involves stripping the vein to remove incompetent valves. Surgery is performed when venous insufficiency cannot be controlled or prevented with conservative treatment.

► VENOUS INSUFFICIENCY

Venous insufficiency is a chronic condition. Damaged or aging valves within the veins interfere with blood return to the heart, causing pooling of blood in the lower extremities.

Venous Stasis Ulcers

Venous stasis ulcers are the end result of chronic venous insufficiency. Dysfunctional valves in the venous system prevent or reduce venous blood return. As venous pressure increases, venous stasis occurs. Over time the congestion and decreased venous circulation lead to changes in the lower extremities. There may be edema and a brownish discoloration of the leg and foot, with the surrounding skin hardened and leathery in appearance. The brown color occurs when veins rupture, releasing red blood cells into the tissues; the red blood cells then break down and stain the tissue brown.

Stasis ulcers develop from the increased pressure and rupture of small veins. Signs of skin breakdown are most commonly seen at the medial malleolus of the ankle. Stasis ulcers are a serious complication of venous insufficiency that are difficult to cure and can affect the patient's quality of life.

Medical Treatment

The focus of treatment is to decrease edema and heal the skin ulcerations. Compression wraps such as elastic stockings or bandage wraps are necessary to decrease edema. Elastic wraps should be started at the foot, with greater tension applied there, and wrapped up the leg. Rewrapping the elastic bandage twice a day is necessary. It is important to ensure that the wraps are not too tight at the top, which prevents return of blood to the heart.

Bedrest and elevation of legs and feet above the heart are important to assist with drainage of lower extremities. Patients are advised not to keep legs dependent and to avoid long periods of standing or sitting. The foot of the bed should be elevated 5 to 6 inches. Additionally, patients should be encouraged to exercise and walk often during nonacute episodes. Patients should be taught not to cross their legs or wear constrictive clothing that would decrease venous blood return to the heart.

Skin ulcers are usually cultured and treated with topical antibiotics if needed. Wound care can be chronic and challenging. (See Chapter 51.) An Unna boot, which is a gauze dressing coated with zinc oxide, calamine, and glycerin, may be used to promote healing in severe ulcers. Zinc promotes wound healing and can be soothing. The Unna boot is applied snugly and provides compression therapy as well. It is changed every 2 to 7 days. Skin grafting may be necessary if ulcerations are severe or do not heal.

Nursing Interventions

The patient's leg is protected from injury. Heating devices are avoided because of decreased sensitivity. The patient is taught treatment measures to reduce edema and pain. Emotional support is provided to patients with chronic ulcers that affect quality of life and require ongoing wound care.

▶ LYMPHATIC SYSTEM

The lymphatic system returns fluid from other tissues in the body to the bloodstream. It is a pumpless system with one-way valves that return the fluid to the heart. Any interruption in the flow of lymph results in edema.

Lymphangitis

Lymphangitis is an inflammation of the lymphatic channels. The infection can occur in the arms or legs and is commonly caused by *Staphylococcus* or *Streptococcus* bacteria.

Signs and Symptoms

Symptoms include pain at the site of the infected area, which may appear reddish. There may also be a red streak that follows the lymphatic channel. The patient may experience chills and fever. Lymph nodes in the area of infection can be enlarged and painful.

Medical Treatment

Medical therapy is initiated with a broad-spectrum antibiotic as the drug of choice. The use of heat on the extremity, as well as elevating it, can help improve circulation. Physicians may order the use of pneumatic pressure devices to help alleviate congestion.

Nursing Care

Frequent assessment and meticulous wound care, if needed, are performed. The nurse monitors the size of the extremity and notifies the physician of any increase in size or possible spread of infection.

REVIEW QUESTIONS

1. Which of the following is a risk factor that can be controlled to prevent the development of cardiovascular disease?
   a. Family history of cardiovascular disease
   b. Hypertension
   c. Ethnicity
   d. Family history of diabetes mellitus

2. The nurse is assessing a long-term care resident, 84 years old, who is not feeling well. Which of your assessment findings is an atypical symptom of a myocardial infarction when chest pain is not present?
   a. Shortness of breath
   b. Chest pain
   c. Sweating
   d. Nausea

3. You are scheduled to give a patient who had a myocardial infarction aspirin, 5 gr PO, now. How many milligrams of aspirin will you give?
   a. 5 mg
   b. 60 mg
   c. 150 mg
   d. 300 mg

4. Which one of the following is a classic symptom of peripheral arterial occlusive disease?
   a. Angina
   b. Edema
   c. Intermittent claudication
   d. Stasis ulcers

5. The nurse is caring for a patient, 64, who has had an MI and has peripheral arterial disease. Which of the following statements by the patient indicates understanding of how to manage the pain of peripheral arterial disease?
   a. "I will lie down frequently."
   b. "I will use a reclining chair."
   c. "I will sit with my legs down."
   d. "I will do knee flexion exercises."

REFERENCES

1. Miettinen, TA, et al: Reduction of serum cholesterol with sitostanol-ester margarine in a mildly hypercholesterolemic population. N Engl J Med 333:1308, 1995.
2. Hallikainen, MA, and Uusitupa, MIJ: Effects of 2 low-fat stanol ester-containing margarines on serum cholesterol concentrations as part of a low-fat diet in hypercholesterolemic subjects. Am J Clin Nutr 69: 403, 1999.
3. Plat, J, and Mensink, RP: Safety aspects of dietary plant sterols and stanols. Postgraduate Medicine: A Special Report. 32. November, 1998.

Answers to CRITICAL THINKING

Mrs. Sims

1. Place on bedrest, administer oxygen via nasal cannula at 2 L per minute, assess blood pressure and pulse, administer nitroglycerin sublingual as ordered, obtain ECG, and notify physician.
2. She may be having an anginal attack versus acute MI.
3. Nitroglycerin usually stops chest pain associated with angina. Rest may also alleviate chest pain. Neither nitroglycerin nor rest will relieve the pain of an acute MI.
4. Indicators of an MI include patient history, ECG changes with ST segment elevation, elevated troponin I, and CK-MB elevation.
5. Medical interventions include nitroglycerin drip, morphine, anticoagulant therapy (heparin), and thrombolytic agents to dissolve the clot. A cardiac catheterization can determine which coronary artery is blocked. Angioplasty or a coronary artery bypass graft may be done to reroute blood.
6. Educate Mrs. Sims about the risks of smoking and being overweight.

Mrs. May

1. Assess the patient's left leg for color, temperature, capillary refill, and pulses: femoral, popliteal, dorsalis pedis, and posterior tibial. Compare findings with findings in the right leg.
2. If unable to palpate pulses, use a Doppler ultrasound that enhances sound to locate pulses.
3. The patient's symptoms could be caused by an embolism above left femoral artery.
4. To document finding, you would obtain more assessment data. A sample of SOAP charting for your additional findings is given:
   S: "I have a severe pain in my left groin that just started. It is at 9."
   O: Grimacing, moaning, and holding left upper leg. Left leg cool, color pale, nailbeds pale, capillary refill 10 seconds, unable to palpate pulses. Faint femoral and popliteal pulse, no dorsalis pedis or posterior tibial pulse heard with Doppler. Right leg warm, pink, capillary refill 3 seconds, with all pulses palpable.
   A: Ineffective tissue perfusion
   P: Notify physician stat.
5. Immediate interventions include complete bedrest, protecting the leg, and notifying the physician.
6. Medical interventions could include medication for pain and use of an anticoagulant, such as heparin. If no pulses are present, a thrombolytic agent, such as streptokinase or tissue plasminogen activator, which dissolves already formed clots, may be ordered. Surgery is possible.
7. Thrombolectomy or embolectomy may be necessary to save the limb.

19

NURSING CARE OF PATIENTS WITH CARDIAC VALVULAR DISORDERS

Linda S. Williams

In the normal heart, blood flows in one direction because of the presence of heart valves. There are four valves in the heart: mitral, tricuspid, pulmonic, and aortic (see Fig. 15–2). The chordae tendineae and papillary muscles are attachment structures for both the mitral and tricuspid valves. They ensure that the valves close tightly. The pulmonic and aortic valves do not have these attachment structures.

Damage to the valves or their surrounding structures can result in abnormal valvular functioning (Fig. 19–1). The valves of the left side of the heart are most commonly affected and are discussed in this chapter. Forward blood flow can be hindered if the valve is narrowed, or stenosed, and does not open completely. If the valve does not close completely, blood backs up, which is referred to as **regurgitation** or insufficiency. The abnormal blood flow increases the workload of the heart and increases the pressures in the affected heart chamber.

Valvular damage may occur from congenital defects, rheumatic fever, or infections. Congenital defects occur mainly in children, and rheumatic heart disease occurs mainly in adults. Prophylactic antibiotic therapy helps prevent rheumatic fever and subsequent rheumatic heart disease and is recommended to prevent valvular disease.

---

regurgitation: re—again + gurgitare—to flood

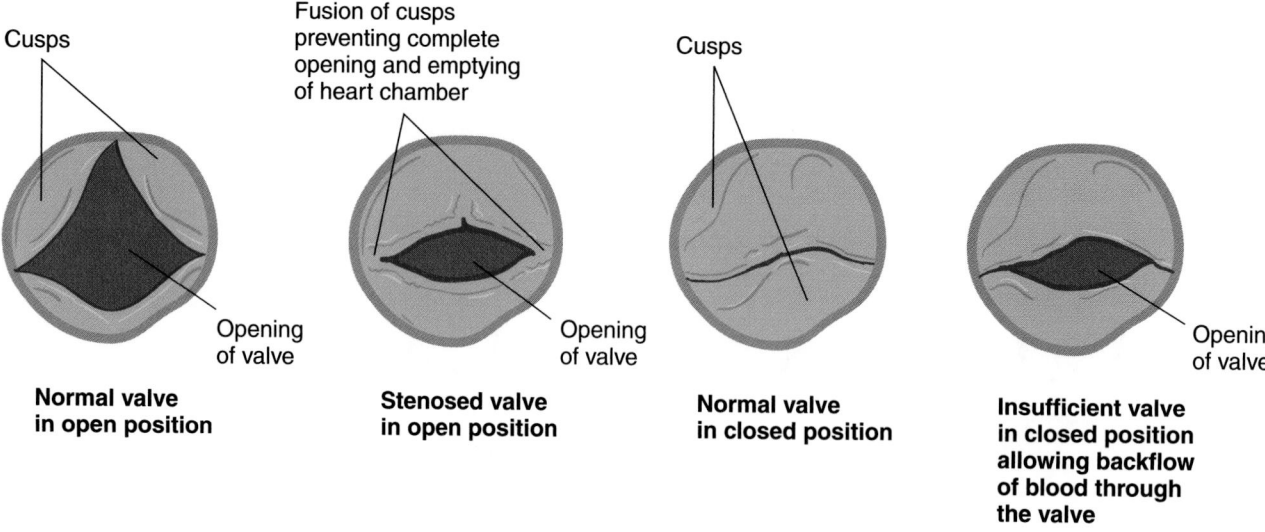

Figure 19-1 Openings of stenosed and insufficient valves compared with a normal valve.

LEARNING TIPS

The opening of a stenosed valve and an insufficient valve look very similar, and the results of extra blood building up in a chamber are the same (see Fig. 19–1). However, the problem is different. Remember what the defect is in each disorder to understand why the blood is building up in a chamber.

A valve that does not open fully (stenosed) does not allow a chamber to empty normally. Blood builds up in that chamber as a result. For example, mitral **stenosis** does not allow the left atrium to empty easily, so blood builds up in the left atrium.

A valve that does not close fully (insufficient) allows blood to flow back into the chamber that emptied. Blood builds up in that chamber as a result. For example, mitral **insufficiency** allows blood to backflow from the left ventricle into the left atrium after the left atrium has emptied, so blood builds up in the left atrium.

▶ VALVULAR DISORDERS
Mitral Valve Prolapse
Pathophysiology

During ventricular systole, when pressure in the left ventricle rises, the flaps of the mitral valve normally remain closed and stay within the atrioventricular junction. In mitral valve prolapse (MVP), however, one or both flaps bulge backward into the left atrium during systole. This is due to one flap being too large or a defect in the chordea tendineae that secures the valve to the heart wall. If the bulging flaps do not fit together, mitral regurgitation can occur with varying degrees of severity.

Etiology

MVP tends to be hereditary, although the cause is unknown. Infections that damage the mitral valve may be a contributing factor. It is the most common form of valvular

---

heart disease and typically occurs in women 20 to 55 years of age who are thin and have slight chest deformities.

Signs and Symptoms

Most patients with MVP do not have symptoms. If symptoms do occur, they may include chest pain, dysrhythmias, palpitations, fatigue, dyspnea, anxiety, dizziness, or syncope.

Diagnostic Tests

Auscultation for a click or **murmur** is the first diagnostic step for MVP. Several other diagnostic tests can be used when MVP is suspected (Box 19–1). A normal electrocardiogram (ECG) is usually seen with MVP. A two-dimensional or Doppler echocardiogram that shows valve abnormalities is the best noninvasive test to detect MVP. For more severe cases, a cardiac catheterization can show the bulging flaps of the mitral valve on coronary angiogram.

Medical Treatment

Unless patients have severe mitral regurgitation, MVP is a benign disorder. No treatment is needed unless symptoms are present (Box 19–2). The severity of MVP and symptoms produced determine the treatment used. A healthy lifestyle, including a good diet, exercise, and avoidance of stimulants and caffeine, is important to prevent symptoms. Stress management may reduce the occurrence of symptoms. Beta blockers may be helpful for those with rapid heart rates to reduce the heart rate and the stress placed on the mitral

---

BOX 19-1	Diagnostic Tests for Cardiac Valvular Disorders

History and physical examination
Electrocardiogram
Chest x-ray examination
Echocardiography
Cardiac catheterization

---

stenosis: stenos—narrow
insufficiency: in—not + sufficiens—sufficient

Medical Treatments for Cardiac Valvular Disorders

Rheumatic fever prophylaxis
Prophylactic antibiotic therapy considered
Anticoagulant therapy
Medication therapy
   Digitalis
   Diuretics
   Antidysrhythmics
Low-sodium diet
Percutaneous balloon valvuloplasty
Surgical
   Valvuloplasty
      Closed commissurotomy
      Open commissurotomy
      Annuloplasty
   Valve replacement

valve. Surgical repair or replacement of the valve can be done for severe cases of MVP.

Complications

Rare complications include dilation of the left side of the heart, heart failure, bacterial endocarditis, and emboli. Preventive antibiotic therapy before invasive procedures may be needed for some patients with thickened mitral valves to prevent endocarditis.[1] Aspirin or anticoagulants may be ordered to help prevent emboli.

CRITICAL THINKING

Sue Tepley

Sue Tepley, 32, has mitral valve prolapse and reports palpitations whenever she experiences stress. She drinks three cups of coffee daily. Today she is admitted for an outpatient cystoscopy.

1. What might you hear when auscultating Sue's heart sounds?
2. Why does Sue experience palpitations?
3. What medication might you expect to be ordered preoperatively for Sue?
4. Why does Sue need to be informed that she may need prophylactic antibiotics before invasive procedures?
5. What other information does Sue need to manage her MVP?

Answers at end of chapter.

Mitral Stenosis
Pathophysiology

Mitral stenosis results from thickening of the mitral valve flaps and shortening of the chordae tendineae, causing narrowing of the valve opening. Older patients with mitral stenosis usually have calcification and fibrosis of the mitral valve flaps. The narrowed opening obstructs blood flow from the left atrium into the left ventricle. The left atrium enlarges to hold the extra blood volume caused by the obstruction. As a result of this increased blood volume, pressure rises in the left atrium. Pressures then rise in the pulmonary circulation and the right ventricle as blood volume

backs up from the left atrium. The right ventricle dilates to handle the increased volume. Eventually the right ventricle fails from this excessive workload, reducing the blood volume delivered to the left ventricle and subsequently decreasing cardiac output.

Etiology

The major cause of mitral stenosis is rheumatic fever, with symptoms often taking two to four decades to appear after the illness. It is a continuous and progressive disease. Less common causes include congenital defects of the mitral valve, tumors, rheumatoid arthritis, systemic lupus erythematosus, calcium deposits, and rheumatic endocarditis.

Signs and Symptoms

At first, patients may be asymptomatic. A click or low-pitched murmur may be heard. Then mild symptoms progressing to more severe symptoms develop. Pulmonary symptoms are most commonly seen. Dyspnea, cough, and hemoptysis from pulmonary congestion are the major symptoms. Fatigue and intolerance to activity result from decreased cardiac output. Palpitations from atrial flutter or fibrillation caused by atrial enlargement and chest pain from decreased cardiac output may be experienced. Complications from emboli formed from the stasis of blood in the left atrium include stroke and seizures. If the right ventricle fails, symptoms related to heart failure are seen. (See Chapter 21.)

Diagnostic Tests

Mitral stenosis is diagnosed with data from the patient history and physical examination and findings from diagnostic tests. (See Box 19–1.) The ECG shows enlargement of the left atrium and right ventricle and changes in the P waveform. Atrial flutter or fibrillation may be seen. A chest x-ray examination confirms enlargement of the affected heart chambers. Two-dimensional and Doppler echocardiography are becoming the noninvasive gold standard for evaluation of valvular disease. They show the narrowed mitral valve opening and decreased motion of the valve. Cardiac catheterization, an invasive test, measures the pressures in the heart chambers and pulmonary vessels, as well as cardiac output. Elevated pressures and a reduced cardiac output are found. Angiography during the catheterization reveals the severity of the stenosis.

Medical Treatment

Individualized prophylactic antibiotic therapy may be given to prevent endocarditis.[1] Anticoagulants are given to patients with atrial fibrillation to prevent development of emboli from stasis of blood in the atrium. If heart failure develops, symptoms are treated with medications such as digitalis and diuretics and other therapies used for heart failure. (See Chapter 21.)

For less severe cases, percutaneous balloon **valvuloplasty,** which uses a balloon to dilate the stenosed heart valve, is done in a cardiac catheterization laboratory (Fig. 19–2). One approach for valvoplasty is to insert a balloon

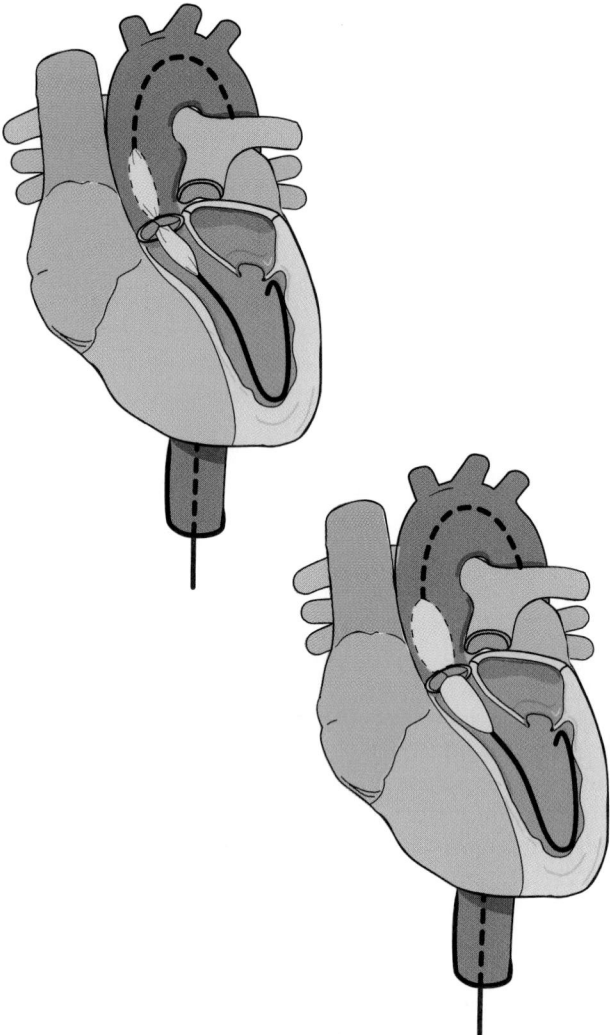

Figure 19-2   Percutaneous balloon valvuloplasty.

catheter via the venous circulation into the right atrium. Then the catheter is threaded through a small hole pierced in the atrial septum into the left atrium through the mitral valve. Inflation of the balloon opens the stenosed valve flaps. Complications may include dysrhythmias, emboli, hemorrhage, and cardiac tamponade. There are fewer complications with balloon valvuloplasty than with surgery.

Surgical treatment can include **commissurotomy** or **annuloplasty,** both forms of valvular repair (valvuloplasty) or valve replacement. Mitral valve replacement is used for more severe cases and when symptoms of ventricular failure develop. These procedures may require the use of cardiopulmonary bypass. In commissurotomy the valve flaps that have adhered to each other and closed the opening between them, known as the commissure, are separated to enlarge the valve opening. Annuloplasty is the repair or reconstruction of the valve flaps or annulus. It may involve the use of prosthetic rings. Valve replacement with prosthetic valves has

---

commissurotomy: commissura—joining together + tome—
incision
annuloplasty: annulus—ring + plasty—formed

been done since 1952. Valve replacement uses mechanical or biologic valves from animal or human tissue (Fig. 19-3). Valvular surgery is discussed in Chapter 22.

Mitral Regurgitation
Pathophysiology

Mitral regurgitation, or insufficiency, is the incomplete closure of the mitral valve. It allows some backflow of blood into the left atrium with each contraction of the left ventricle. Factors that prevent the mitral valve flaps from fitting together tightly may include tearing, shortening or rigidity of a flap, or shortening of the chordae tendineae.

When the mitral valve does not close completely, backflow of blood into the left atrium occurs. This blood is then extra volume that is added to the incoming blood from the lungs. With chronic mitral regurgitation, the increase in blood volume dilates and increases pressure in the left atrium. In response to the extra blood volume delivered by the left atrium, the left ventricle compensates by dilating. If the compensatory mechanism of dilation is inadequate, pressures rise in the pulmonary circulation and then in the right ventricle as blood volume backs up from the left atrium. The left ventricle and eventually the right ventricle may fail from this increased strain.

Etiology

The major cause of mitral regurgitation is rheumatic heart disease. Other causes include endocarditis, rupture or dysfunction of the chordae tendineae or papillary muscle, MVP, or congenital defects.

Figure 19-3   Valve prosthesis. An SJM Masters Series valve. (Courtesy St. Jude Medical, Inc., St. Paul, Minn.)

Signs and Symptoms

Initially patients may be asymptomatic. The symptoms of chronic mitral regurgitation are similar to those of mitral stenosis. Dyspnea and cough occur from increased pulmonary congestion. Palpitations and an irregular pulse from atrial fibrillation may result. Weakness and fatigue from decreased cardiac output occur if the left ventricle begins to fail.

If acute mitral regurgitation develops, such as in papillary muscle rupture following myocardial infarction, pulmonary edema and shock symptoms will be exhibited.

Diagnostic Tests

The ECG shows enlargement of the left atrium and left ventricle and changes in the P waveform. Atrial flutter or fibrillation may be seen. A chest x-ray examination confirms hypertrophy of the affected heart chambers. Two-dimensional or Doppler echocardiography shows left atrial enlargement and regurgitation of blood. Cardiac catheterization with dye also shows regurgitation.

Medical Treatment

Without symptoms, there is no general medical treatment. Prophylactic antibiotic therapy may be given to prevent infectious endocarditis before invasive procedures.[1] If atrial fibrillation with rapid heart rate is present, it can be controlled with digitalis, calcium channel blockers, or beta blockers. Emboli occur less frequently, but anticoagulation is still prescribed. Symptoms of heart failure are treated with therapies for heart failure. (See Chapter 21.) When symptoms develop, surgical treatment includes mitral valve repair or replacement (Fig. 19–4).

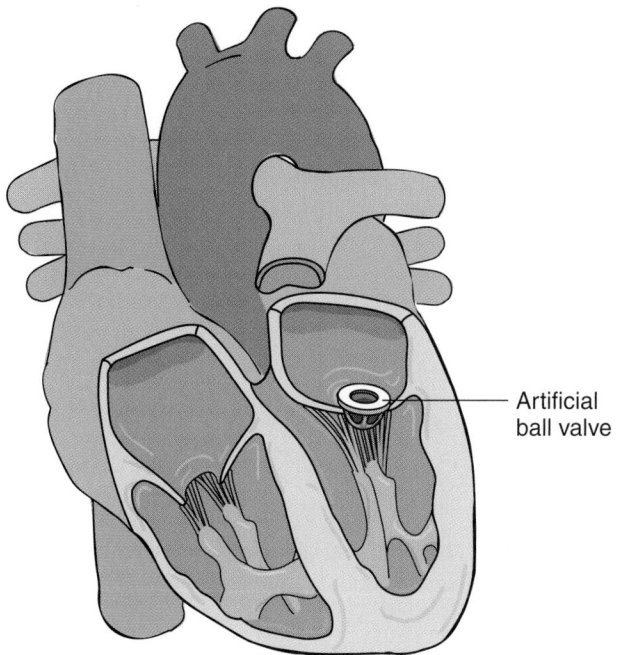

Artificial ball valve

Figure 19-4 Mitral valve replacement with ball valve prosthesis.

Aortic Stenosis
Pathophysiology

Blood flow from the left ventricle into the aorta is obstructed through the stenosed aortic valve. The opening of the aortic valve may be narrowed from thickening, scarring, calcification, or fusing of the valve's flaps. To compensate for the difficulty in ejecting blood into the aorta, the left ventricle contracts more forcefully. In chronic stenosis the left ventricle hypertrophies to maintain normal cardiac output. With increased narrowing of the valve opening, the compensatory mechanisms are unable to continue and the left ventricle fails to move blood forward. This results in decreased cardiac output and heart failure.

Etiology

The major causes of aortic stenosis are congenital defects or rheumatic heart disease. Mitral valve stenosis is often also present when the cause is rheumatic heart disease. Calcification of the aortic valve can be related to aging and is seen more with an increase in the elderly population.

Signs and Symptoms

It may take many years or decades before signs or symptoms of aortic stenosis are observed. When symptoms do occur, evaluation is essential because the disease can progress dramatically. If the mitral valve is diseased, signs and symptoms may appear earlier, allowing the detection of aortic stenosis.

Angina pectoris is a primary symptom that occurs as a result of the increased oxygen needs of the hypertrophied myocardium. The extra workload of the left ventricle and the hypertrophy of the cardiac muscle require more oxygen. Angina results if these oxygen needs are not met. In the young patient, angina indicates severe obstruction.

Other signs and symptoms include a murmur, syncope from dysrhythmias or decreased cardiac output, and heart failure signs and symptoms. Orthopnea, dyspnea on exertion, and fatigue are indicators of left ventricular failure. Progressive heart failure can result in pulmonary edema and right-sided heart failure.

Diagnostic Tests

ECG shows enlargement of the left ventricle and left atrium. A chest x-ray examination confirms hypertrophy of the left ventricle. Left atrial enlargement may be seen but occurs primarily when mitral stenosis is also present. Two-dimensional and Doppler echocardiography show thickening of the left ventricular wall, impaired movement of the aortic valve, and the severity of the disease. Serial echocardiography (e.g., annually or less often), may be recommended for those with moderate or severe disease. Cardiac catheterization will show elevated left ventricular pressure and decreased cardiac output.

Medical Treatment

Generally the treatment of choice is valve replacement because of the risk of sudden death when severe symptoms are present. Valvotomy may be considered for young adults. Prophylactic antibiotics are considered before invasive procedures.[1] Heart failure symptoms are treated carefully. Medications that reduce the contractility of the heart and subsequently cardiac output are avoided to prevent further failure.

CRITICAL THINKING

Mrs. Hesche

Mrs. Hesche, 72, has aortic stenosis and is admitted to the hospital with angina. She had an episode of syncope 2 days ago. She reports that she tires easily.

1. Mrs. Hesche asks what aortic stenosis is. What will you tell her and how will you document it?
2. Why might Mrs. Hesche be experiencing angina?
3. What nursing care related to safety needs are important to include in Mrs. Hesche's plan of care?
4. What nursing diagnosis and care are relevant for Mrs. Hesche's report of being tired?
5. Digoxin (Lanoxin) 0.25 mg is prescribed for Mrs. Hesche. It is time to give her digoxin and you have available digoxin 0.125 mg tablets. How many tablets will you give?

Answers at end of chapter.

Aortic Regurgitation
Pathophysiology

The aortic valve cusps may be scarred, thickened, or shortened in chronic aortic regurgitation. A backflow of blood from the aorta into the left ventricle occurs if the aortic valve cusps do not close completely. The left ventricle's blood volume increases with this backflow of blood that is in addition to the normal flow of blood from the left atrium. To handle the increased volume, the left ventricle compensates with dilation and hypertrophy to deliver a stronger contraction. This stronger contraction ejects more blood volume with each beat to maintain cardiac output. Over time the heart's contraction is not effective and the left ventricle fails, causing a cardiac output drop and pulmonary edema.

Etiology

Rheumatic heart disease is the usual cause of aortic regurgitation. Other causes include congenital defects, syphilis, endocarditis, severe hypertension, and rheumatoid arthritis. An acute cause of aortic regurgitation is aortic dissection.

Signs and Symptoms

Exertional dyspnea and fatigue are the first symptoms of chronic aortic regurgitation. They appear after years of progressive valvular dysfunction. The patient may report feeling a forceful heartbeat. The palpated pulse is forceful and then quickly collapses (Corrigan's pulse). The diastolic blood pressure decreases to widen the pulse pressure. This compensates for an increase in systolic blood pressure. Angina pectoris may occur late. The angina is atypical, often happening at night, when a lower pulse rate results in delivery of less oxygen to the myocardium. Eventually heart failure symptoms develop if the left ventricle fails.

In acute dysfunction, profound symptoms of pulmonary distress, chest pain, and shock symptoms occur. The prognosis is poor.

Diagnostic Tests

The ECG shows left ventricle hypertrophy. A chest x-ray examination confirms hypertrophy of the left ventricle and aorta. With severe regurgitation, left atrial enlargement may be seen. An echocardiogram shows an enlarged left ventricle. Cardiac catheterization reveals elevated left ventricular diastolic pressure and, with dye injection, shows the regurgitation of blood into the left ventricle.

Medical Treatment

Vasodilator therapy such as nifedipine may be useful for some patients to reduce systolic blood pressure and subsequently cardiac workload. Prophylactic antibiotic therapy may be given before invasive procedures.[1] Surgical valve replacement is the treatment of choice for most patients, especially if symptoms or left ventricular dysfunction are present. (See Box 19–2.)

Nursing Process for Valvular Disorders
Nursing Assessment

A history that includes information presented in Table 19–1 is obtained. Vital signs are measured and recorded. Heart sounds are auscultated to detect murmurs. Any signs and symptoms of heart failure are noted. (See Chapter 21.)

Nursing Diagnoses

The major nursing diagnoses for all valvular disorders are the same and include those for heart failure as well if heart failure symptoms are present. (See Nursing Care Plan Box 19–3 for the Patient with Cardiac Valvular Disorders.) The diagnoses may include but are not limited to the following:

- Pain related to reduced coronary artery blood flow and increased myocardial oxygen needs
- Decreased cardiac output related to valvular stenosis or insufficiency or heart failure
- Activity intolerance related to decreased oxygen delivery from decreased cardiac output
- Excess fluid volume related to heart failure and the secondary reduction in renal blood flow for filtration
- Ineffective therapeutic regimen management related to lack of knowledge about disorder

Planning

The patient and family should be included in planning care. The patient's goals include the following:

- Report that the pain management regimen relieves pain.
- Maintain vital signs and oxygen saturation within normal range.
- Maintain desired activities.
- Remain free of edema and maintain clear lung sounds.
- Understand the disease and its treatment.

TABLE 19-1	NURSING ASSESSMENT FOR PATIENTS WITH CARDIAC VALVULAR DISORDERS
Subjective Data	
Health history	Infections—rheumatic fever, endocarditis, streptococcal or staphylococcal, syphilis
	Congenital defects
	Cardiac disease—myocardial infarction, cardiomyopathy
Respiratory	Dyspnea at rest, on exertion, when lying, or that awakens patient?
	Cough or hemoptysis?
Cardiovascular	Any palpitations, chest pain, dizziness, fatigue, activity intolerance?
Medications	
Knowledge of condition	
Coping skills	
Objective Data	
Respiratory	Crackles, wheezes, tachypnea
Cardiovascular	Murmurs, extra heart sounds, dysrhythmias, edema, jugular vein distention, Corrigan's pulse, increased or decreased pulse pressure
Integumentary	Clubbing; cyanosis; diaphoresis; cold, clammy skin; pallor
Diagnostic test findings	

BOX 19-3	NURSING CARE PLAN FOR THE PATIENT WITH CARDIAC VALVULAR DISORDERS

 Decreased cardiac output related to valvular stenosis or insufficiency or heart failure

PATIENT OUTCOMES

Patient has adequate cardiac output as evidenced by vital signs within normal limits (WNL), no dyspnea or fatigue.

EVALUATION OF OUTCOMES

Does patient have vital signs WNL with no dyspnea or fatigue?

INTERVENTIONS	RATIONALE	EVALUATION
Assess vital signs, chest pain, and fatigue.	Vital signs, chest pain, and fatigue are indicators of cardiac output decline.	Are vital signs WNL with no chest pain or fatigue?
Give oxygen as ordered.	Supplemental oxygen provides more oxygen to the heart.	Is breathing pattern normal?
Provide bedrest or rest periods as ordered.	Cardiac workload and oxygen needs are reduced with rest.	Are vital signs WNL and no fatigue reported?
Elevate head of bed 45 degrees.	Venous return to heart is reduced and chest expansion improved.	Are vital signs WNL and respirations easy?
GERIATRIC		
Assess for cardiac medication side effects and teach patient side effects to report.	Toxic side effects are more common due to altered metabolism and excretion of medications in the elderly.	Are side effects present for medications patient is taking?
		Does patient understand side effects to report?

 Activity intolerance related to decreased oxygen delivery from decreased cardiac output

PATIENT OUTCOMES

Patient will show normal changes in vital signs with less fatigue in response to activity.

EVALUATION OF OUTCOMES

Does patient have normal changes in vital signs with activity? Does patient report decreased fatigue with activity?

INTERVENTIONS	RATIONALE	EVALUATION
Assist as needed with activities of daily living (ADLs).	Energy is conserved with ADL assistance.	Are all ADLs completed?
Provide rest and space activities.	Cardiac workload and oxygen needs are reduced with rest.	Are vital signs WNL with activity?
		Is patient able to perform activities when allowed extra time?
GERIATRIC		
Slow pace of care and allow patient extra time to perform activities.	Elderly patients can often perform activities if allowed time to slowly perform them and rest at intervals.	Does blood pressure remain WNL when changing position?
Ensure safety when mobilizing elderly patient.	Orthostatic hypertension is common in the elderly.	Does patient ambulate without injury?

Nursing Interventions

Nursing care is aimed at relieving patients' pain, maintaining patients' normal cardiac function, improving patients' ability to participate in activity, maintaining fluid balance, educating patients to understand the therapeutic management of their valvular disorder, promoting prevention techniques, monitoring symptoms, and providing preoperative and postoperative care as needed. See Nursing Care Plan Box 19–3 for the Patient with Cardiac Valvular Disorders for specific nursing interventions.

Pain management is achieved by assessing pain on an ongoing basis using a rating scale such as 0 to 10 with 10 being the most severe pain. Providing pain medication such as nitroglycerin as needed can relieve the pain. (Before nitroglycerin is given, blood pressure is checked, because nitroglycerin lowers blood pressure). Teaching the patient to pace activities with frequent rest periods can reduce the workload of the heart. This helps prevent ischemia and subsequently angina.

Maintenance of normal cardiac function includes monitoring of vital signs, intake and output, and daily weights if heart failure is present or diuretics are given. Sodium may be restricted to reduce fluid retention. Smoking cessation information is provided. Medications are given as prescribed.

Interventions that improve the quality of life include activities of daily living that reduce fatigue. Assistance as needed, frequent rest periods, and energy conservation techniques are planned. (See Chapter 21.) Exercise tolerance is assessed and daily exercise is planned according to this tolerance.

Ongoing monitoring for signs or symptoms of excess fluid volume are important to allow early detection and treatment. Daily weights provide the most accurate assessment of weight gain caused by excess fluid. The weights should be obtained at the same time of day, using the same scale and type of clothing, for accuracy. Assessment for edema, jugular vein distention, lung crackles or wheezes, dyspnea, and orthopnea should be done. Restricting sodium intake and administering diuretics as ordered assist in decreasing fluid volume excess. Monitoring intake and output is also important to detect fluid imbalances. Potassium is also monitored because some diuretics (e.g., furosemide [Lasix]) decrease potassium levels. It is important to understand that fluid volume excess develops more easily in the elderly patient and can be very serious. Therefore assessing and monitoring risk factors for fluid volume excess in the elderly patient is critical.

PATIENT EDUCATION. Education, an important nursing intervention, promotes understanding of the disorder, as well as health maintenance, prevention of complications, and early recognition of symptoms, so medical care can be sought. Teaching is provided for any medications the patient is taking. If the patient is on anticoagulants for atrial fibrillation or valve replacement, a Medic-Alert bracelet should be worn and monthly appointments to check prothrombin time and international normalized ratio (INR) values should be kept. For elderly patients it is important to include caregivers or family members in teaching sessions to assist with understanding of the information being taught.

Information on endocarditis prevention is essential for patients with most valvular problems. Damaged cardiac valves are prone to developing infection from organisms such as *Streptococcus viridans* or *Staphylococcus epidermidis*. During invasive procedures in which bleeding is possible, these organisms can enter the circulation, attach to damaged valves, and multiply. Patients should be taught the possible need for prophylactic antibiotics before invasive procedures, including dental work or surgery, to prevent endocarditis.[1] Patients should consult their physician about the need for prophylactic antibiotics.

Evaluation

The goal for the nursing diagnosis of pain is met if the patient reports satisfactory pain relief. Satisfactory pain relief is measured by comparing the patient's stated level of pain on a 0 to 10 rating scale with the patient's predetermined goal for acceptable level of pain.

The goal for the nursing diagnosis of decreased cardiac output is met if the patient's vital signs are within normal range and no symptoms of heart failure are present.

The goal for the nursing diagnosis of activity intolerance is met if the patient reports reduced fatigue and the ability to complete tasks and engage in desired activities.

The goal for the nursing diagnosis of excess fluid volume is met if the patient remains free of edema, maintains appropriate weight, maintains clear lung sounds without dyspnea, has normal vital signs, and has no evidence of neck vein distention.

The goal for the nursing diagnosis of ineffective therapeutic regimen management is met if the patient verbalizes understanding of completed teaching and does not have recurrence of symptoms.

REVIEW QUESTIONS

1. The nurse is evaluating patient teaching for mitral valve prolapse. The patient shows understanding of the prognosis of MVP if she states which of the following?
   a. "The prognosis is poor."
   b. "This is usually a benign condition."
   c. "Heart failure often occurs."
   d. "Symptoms begin mildly and then quickly progress."

2. The nurse is assessing a patient with aortic stenosis. Which of the following symptoms would the nurse expect a patient with aortic stenosis to report?
   a. Peripheral edema
   b. Angina
   c. Headache
   d. Weight loss

3. A patient just diagnosed with aortic regurgitation asks what this means. Which of the following is the nurse's best response for describing what occurs in aortic regurgitation?
   a. "Backflow of blood into the right ventricle"
   b. "Backflow of blood into the left ventricle"
   c. "Impaired emptying of the right ventricle"
   d. "Impaired emptying of the left ventricle"

4. A patient with mitral stenosis develops atrial fibrillation. Which one of the following medications does the nurse understand is given to the patient to prevent embolitic complications from atrial fibrillation?
   a. Bumetanide (Bumex)
   b. Furosemide (Lasix)
   c. Penicillin (Bicillin)
   d. Warfarin (Coumadin)

5. A patient is scheduled for cardiac valve testing. Which of the following tests would the nurse explain will show the cardiac valves and how they function?
   a. ECG
   b. Chest x-ray examination
   c. Echocardiogram
   d. Cardiac catheterization

REFERENCES

1. Dajani, A, et al: Prevention of bacterial endocarditis: recommendations by the American Heart Association. JAMA 277(22):1794, 1997.

Answers to CRITICAL THINKING

Sue Tepley

1. You might hear a murmur.
2. Stress and caffeine increase the occurrence of palpitations.
3. Prophylactic antibiotics might be ordered preoperatively.
4. She might need prophylactic antibiotics to prevent infective endocarditis, which can result from bacteria entering the circulation during invasive procedures, attaching to the damaged valve, and growing.
5. To help manage her condition, Sue needs a definition of MVP, stress management techniques, to know she should reduce caffeine intake (e.g., with decaffeinated coffee), and to understand symptoms of endocarditis to report to her physician.

Mrs. Hesche

1. In aortic stenosis the valve is narrowed, which makes it more difficult for blood to leave the left ventricle and go into the aorta. This means there can be less blood flow to the body.

Documentation: Asked "What is aortic stenosis?" Listened attentively during explanation that in aortic stenosis the valve is narrowed making it more difficult for blood to leave left ventricle to go to aorta. This means there can be less blood flow to the body. Stated interested in learning more about diagnosis.

2. Angina results if the heart's oxygen needs are not met because of reduced cardiac output.
3. Nursing care should include fall precautions due to syncope and fatigue.
4. Diagnoses and care include the following: Self-care deficits related to fatigue, so plan for meeting ADL needs. Activity intolerance related to fatigue, so plan rest periods between activity and monitor vital signs with activity.
5. You should give two tablets. An example of solving this problem is:

$$\frac{\text{Desired dose:}}{\text{Dose on hand:}} \quad \frac{0.25 \text{ mg}}{0.125 \text{ mg}} = 2 \text{ tablets}$$

20

NURSING CARE OF PATIENTS WITH CARDIAC DYSRHYTHMIAS

Linda S. Williams and Elizabeth Chapman

KEY TERMS

ablation (uh-**BLAY**-shun)

atrial depolarization (**AY**-tree-uhl DE-poh-lahr-i-**ZAY**-shun)

atrial systole (**AY**-tree-uhl **SIS**-tuh-lee)

atrioventricular node (**AY**-tree-oh-ven-**TRICK**-yoo-lar NOHD)

bigeminy (bye-**JEM**-i-nee)

bradycardia (BRAY-dee-**KAR**-dee-yah)

bundle of His (**BUN**-duhl of HISS)

cardioversion (KAR-de-oh-**VER**-zhun)

defibrillation (dee-**FIB**-ri-lay-shun)

dysrhythmia (dis-**RITH**-mee-yah)

electrocardiogram (ee-LECK-troh-**KAR**-dee-oh-GRAM)

fluoroscopy (fluh-**RAHS**-kuh-pee)

hyperkalemia (HIGH-per-kuh-**LEE**-mee-ah)

hypomagnesemia (**HIGH**-poh-MAG-nuh-**ZEE**-mee-ah)

isoelectric line (EYE-so-e-**LEK**-trick LINE)

multifocal (MUHL-tee-**FOH**-kuhl)

nodal or junctional rhythm (**NOHD**-uhl or **JUNGK**-shun-uhl **RITH**-uhm)

sinoatrial node (SIGH-noh-**AY**-tree-al NOHD)

trigeminy (try-**JEM**-i-nee)

unifocal (YOO-ni-**FOH**-kuhl)

ventricular diastole (ven-**TRICK**-yoo-lar dye-**AS**-tuh-lee)

ventricular escape rhythm (ven-**TRICK**-yoo-lar es-**KAYP RITH**-uhm)

ventricular repolarization (ven-**TRICK**-yoo-lar RE-pol-lahr-i-**ZAY**-shun)

ventricular systole (ven-**TRICK**-yoo-lar **SIS**-tuh-lee)

ventricular tachycardia (ven-**TRICK**-yoo-lar TACK-ee-**KAR**-dee-yah)

QUESTIONS TO GUIDE YOUR READING

1. What are the functions of the sinoatrial node, atrioventricular node, and Purkinje fibers?
2. How does electrical activity flow through the heart?
3. What are the six components of the cardiac cycle and what do they represent?
4. What are the five steps used for dysrhythmia interpretation?
5. What are current medical treatments for each of the cardiac dysrhythmias?
6. What are types and uses of cardiac pacemakers?
7. What nursing care would you provide for patients with dysrhythmias or a pacemaker?

▶ CARDIAC CONDUCTION SYSTEM

The heart's electrical conduction system (Fig. 20–1) initiates an impulse to stimulate the cardiac muscle to contract. The conduction system's electrical activity can be viewed on a cardiac monitor or recorded on an **electrocardiogram** (ECG) tracing. The activity seen on the ECG does not mean that the heart has contracted in response to the electrical impulse. To verify that contraction occurs, the patient's vital signs and pulses are monitored.

Located in the upper posterior wall of the right atrium is the **sinoatrial** (SA) **node.** The SA node is the primary pacemaker of the heart. It normally fires at a rate of 60 to 100 beats per minute (bpm). As a protective mechanism if the SA node does not function properly, other areas of the heart can initiate impulses to keep the heart beating. If the SA node fails, the **atrioventricular** (AV) **node** initiates an impulse at 40 to 60 bpm. The rhythm produced when the AV node initiates the impulse is called a **nodal or junctional rhythm.** The body can usually function adequately with this rhythm. If the AV node is unable to initiate an impulse, then the ventricles take over at 20 to 40 bpm. When the ventricles initiate the impulse, it is referred to as complete heart block or a **ventricular escape rhythm.** These rhythms are the heart's last attempt to compensate for loss of SA and AV node conduction. The ventricular rate of 20 to 40 bpm is not adequate to meet the body's oxygen needs, so the patient begins to show signs of inadequate cardiac output such as dyspnea, abnormal vital signs, and changes in level of consciousness. Treatment is usually necessary to reestablish a normal heart rate as soon as possible when the SA node is not functioning normally.

After the SA node fires, the impulse travels across both atria, stimulating them to contract. This is known as **atrial systole.** This atrial contraction propels blood out of the atria and into the relaxed ventricles during **ventricular diastole.** The impulse travels down the atria to the AV node where it is delayed briefly. It then travels down the **bundle of His,** which divides into right and left bundle branches. From there the impulse quickly travels through the Purkinje fibers, stimulating both ventricles to contract. This is known as **ventricular systole.**

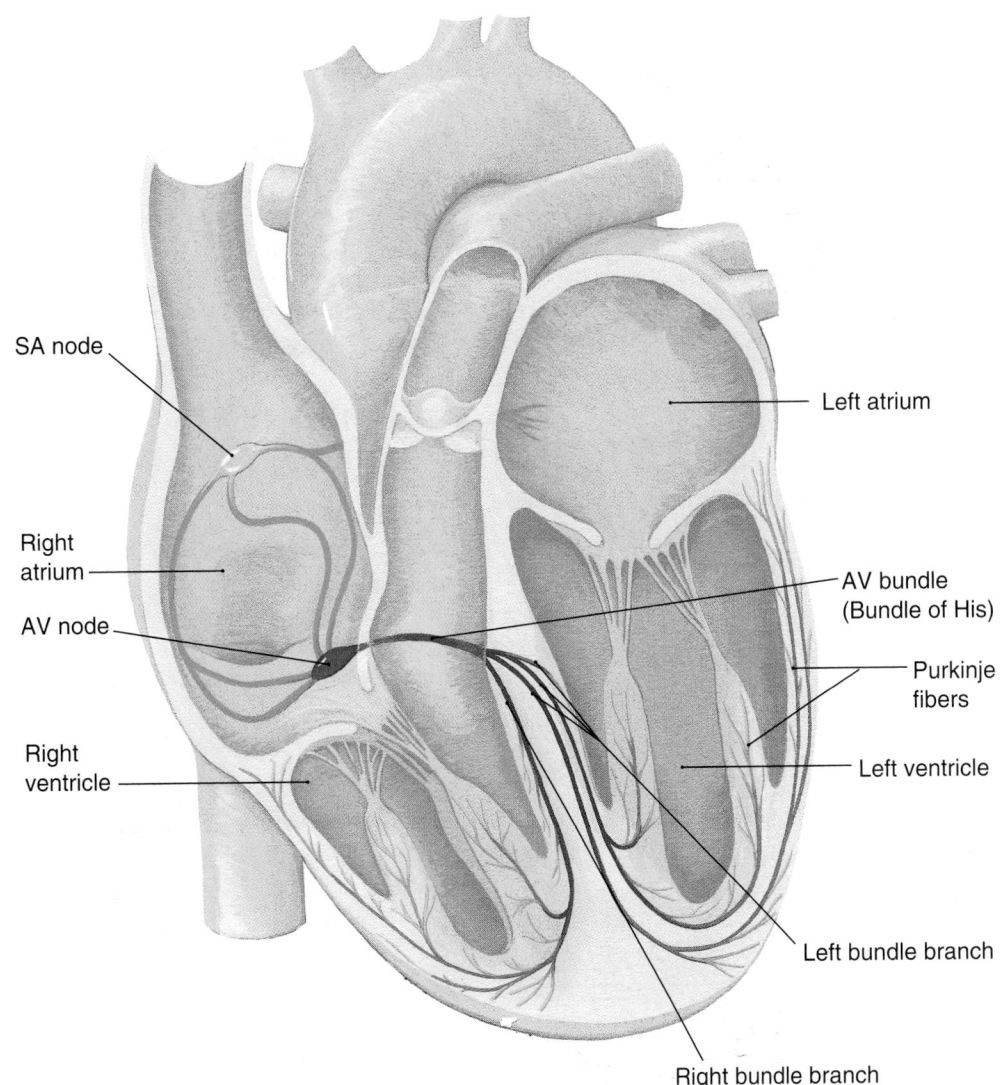

Figure 20-1 Conduction pathway of the heart. (From Scanlon, V, and Sanders, T: Essentials of Anatomy and Physiology, ed 3. FA Davis, Philadelphia, p 266, with permission.)

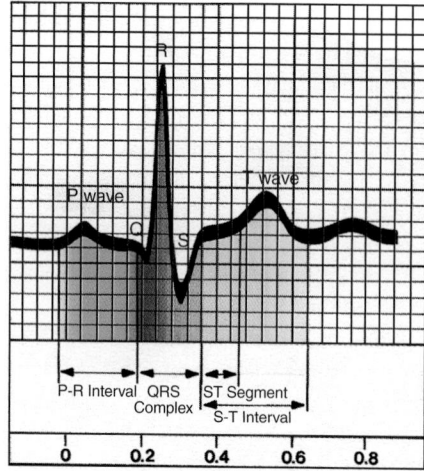

Figure 20-2   Components of the cardiac cycle. (From Scanlon, V, and Sanders, T: Essentials of Anatomy and Physiology, ed 3. FA Davis, Philadelphia, p 266, with permission.)

Cardiac Cycle

To interpret rhythms, a cardiac cycle must be identified. A cardiac cycle is the period from the beginning of one heartbeat to the beginning of the next. It is the electrical representation of the impulse that stimulates contraction and then relaxation of the atria and ventricles. Within the cardiac cycle, there is a P wave, a QRS complex, and a T wave. To be considered normal, a cycle must consist of each of these components. Figure 20–2 shows the components of a cardiac cycle.

► ELECTROCARDIOGRAM

A cardiac monitor or ECG tracing shows the electrical activity of the heart. It is used to verify normal heart function and detect abnormal heart function. Specialized training is required to interpret ECG abnormalities; this is usually done by physicians. You can learn characteristics of a normal heart rhythm and rules for common **dysrhythmia**s so that you will be able to report rhythm changes to your supervisor or physician.

ECG monitoring provides a view of the heart's electrical activity. Leads that are placed on the patient allow different views of the heart to be seen. A 12-lead ECG provides 12 different perspectives of the heart. Waveforms change in appearance in different leads. For continuous monitoring, lead II is the most commonly used. In lead II the waveforms can be expected to be upright.

Electrocardiogram Graph Paper

The intervals of each of the components of a cardiac cycle can be measured on the ECG graph paper on which the rhythm is recorded. The graph paper is calibrated in a grid with small squares divided into heavy lined blocks of 25 (five squares wide and five squares high; Fig. 20–3). Each small box is 0.04 seconds wide. There are five small squares, which equal 0.20 seconds of time, between two horizontally

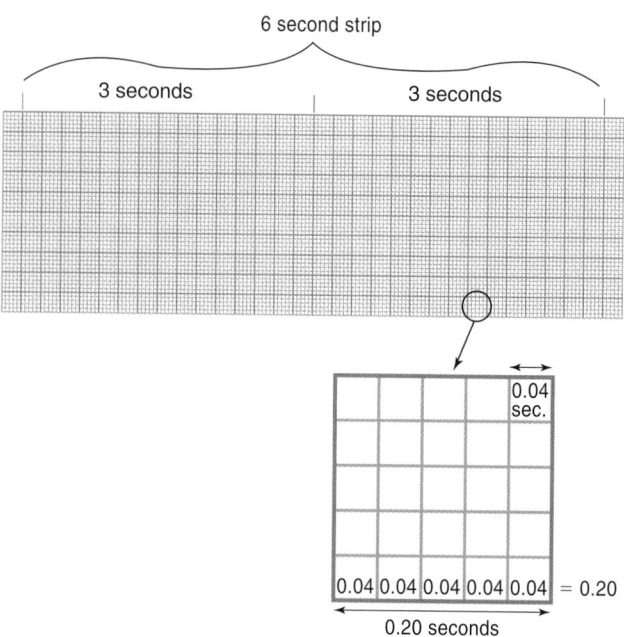

Figure 20-3   Electrocardiogram recording paper time intervals.

heavy vertical black lines. (See Fig. 20–3.) The height of waveforms (amplitude) is measured vertically. Each small square is 1 mm. Ten millimeters is the standard height that the ECG uses to measure patient waveforms.

When the ECG is on but there is no electrical activity detected, a straight line is produced. Known as the **isoelectric line,** this occurs when there are no positive or negative electrical wave deflections. Cardiac cycle impulses (seen as waves), depending on how they travel through the heart, are either upright (positive) or downward (negative) from the isoelectric line on the ECG graph paper.

► COMPONENTS OF A CARDIAC CYCLE

P Wave

The P wave is the first wave of the cardiac cycle and represents **atrial depolarization.** When the SA node fires, it normally appears rounded and symmetrical. There is one P wave in a normal cardiac cycle. Disorders that change atrial size cause alterations in P wave shape and size.

PR Interval

The PR interval represents the time it takes the electrical impulse to travel down the atrium to the AV node. It starts at the beginning of the P wave and ends at the beginning of the QRS complex. Counting the number of small boxes

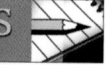

LEARNING   TIPS

To make measuring waves easier, try to find one that starts at the beginning of one small box. (See Fig. 20–4.) If the wave starts or ends in the middle of a box, count it as one-half of a box, which is 0.02 seconds.

---

dysrhythmia: dys—difficult or abnormal + rhythm—rhythm

horizontally that the interval covers determines the length of the PR interval (Fig. 20–4). The normal PR interval is 0.12 to 0.20 seconds.

QRS Complex

The QRS complex represents ventricular depolarization and is composed of three waves, the Q, R, and S. The Q wave is the first downward deflection after the P wave but before the R wave. The R wave is the first upward deflection after the P wave. The last part of the QRS complex is the S wave, which is the second negative deflection after the P wave when a Q wave is present or the first negative deflection after the R wave (see Fig. 20–2). The S wave ends when it returns to the isoelectric line. All three waves are not always present in every QRS complex. Even with absent waves, the QRS is still known as a QRS complex (Fig. 20–5).

Atrial repolarization occurs during the interval of the QRS but is not seen because of the more powerful ventricular activity. The presence of the P wave in the next cardiac cycle indicates that atrial repolarization has occurred.

QRS Interval

To measure the QRS interval, count the number of boxes from the wave that begins the QRS complex to the end of the wave that completes the QRS complex. For example, when a Q, R, and S are present, measure from the beginning of the Q wave to the end of the S wave (Fig. 20–6). The normal QRS interval is 0.06 to 0.10 seconds.

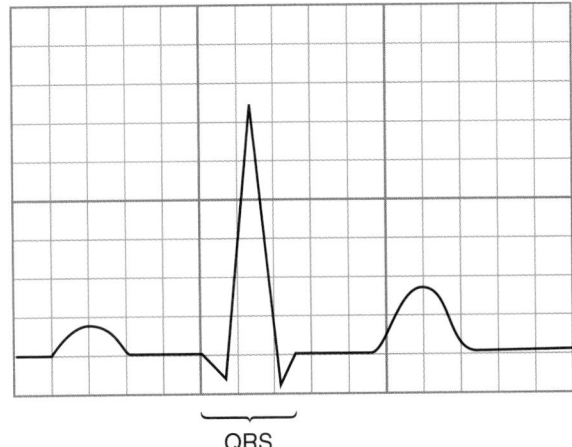

Figure 20-6 QRS interval. This QRS interval covers 2½ boxes. Each full box is 0.04 seconds. One-half box is 0.02 seconds; 2.5 × 0.04 = 0.10 seconds.

T Wave

The T wave represents **ventricular repolarization,** the resting state of the heart, when the ventricles are filling with blood and preparing to receive the next impulse. The T wave starts at the next upward (positive) deflection, after the QRS complex, and ends with a return to the isoelectric line. The T wave can be a downward deflection after the QRS complex in some ECG leads, or it can indicate ischemia of the heart (Fig. 20–7).

U Wave

The U waveform is usually not present. It is seen in patients with hypokalemia, which is a low serum potassium level. It occurs shortly after the T wave and can distort the configuration of the T wave (Fig. 20–8).

ST Segment

The ST segment reflects the time from completion of a contraction (depolarization) to recovery (repolarization) of myocardial muscle for the next impulse. The ST segment starts at the end of the QRS and ends at the beginning of the T wave (Fig. 20–9). The ST segment is examined if a

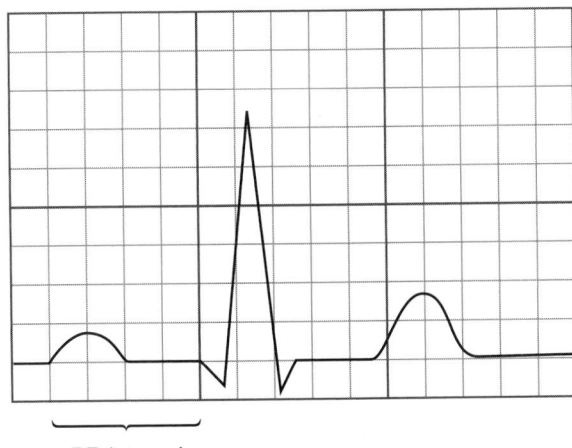

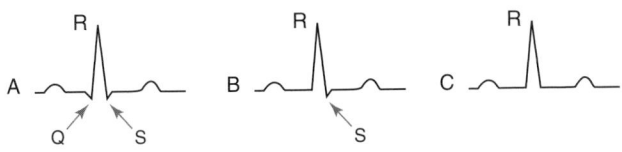

Figure 20-4 PR interval.

Figure 20-5 (A) QRS complex with a Q wave. (B) QRS complex without a Q wave. (C) QRS complex without a Q or an S wave.

Figure 20-7 (A) T wave with positive deflection. (B) T wave with inverted, negative inflection, indicating ischemia.

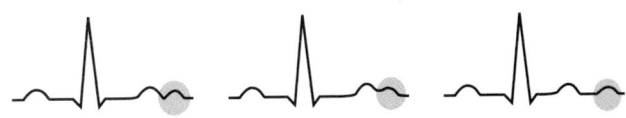

Figure 20-8 Different locations where U waves appear.

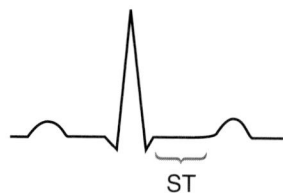

Figure 20-9   ST segment.

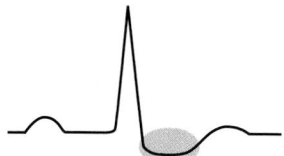

Figure 20-10   ST segment inverted or depressed.

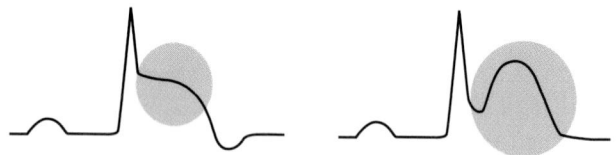

Figure 20-11   ST segment elevated.

TABLE 20-1	FIVE-STEP PROCESS FOR DYSRHYTHMIA INTERPRETATION	

After assessing the topics listed here, you should be able to name the patient's dysrhythmia.

Step	Topic	Assessment Questions
1	Regularity of rhythm	• Is the rhythm regular? Irregular? • Is there a pattern to the irregularity?
2	Heart rate	• What is the heart rate?
3	P waves	• Is there one P wave for every QRS complex? • Are the P waves regular and constant? • Do the P waves look alike? • Are the P waves upright and in front of every QRS complex?
4	PR interval	• Is the PR interval normal? • Is the PR interval constant or varying?
5	QRS interval	• Is the QRS interval normal? • Is the QRS interval constant? • Do the QRS complexes all look alike?

patient is experiencing chest pain. Changes in the ST segment can indicate the presence of ischemia or an injury pattern suggestive of myocardial damage. If a patient is experiencing ischemia, the ST segment can be inverted or depressed (Fig. 20–10). If the patient is experiencing an injury pattern, the ST segment elevates (Fig. 20–11).

▶ INTERPRETATION OF CARDIAC RHYTHMS

Five-Step Process for Dysrhythmia Interpretation

An orderly, systematic method for interpreting ECG rhythms should be used. Following this process, in order, increases understanding of items to examine and ensures that nothing is omitted. Five steps are examined in this process (Table 20–1). The findings of these five steps are then used to interpret the ECG rhythm according to the rules for each dysrhythmia. A 6-second ECG strip is used when interpreting rhythms. (See Fig. 20–3.)

1. Regularity of the Rhythm

The regularity or rhythm of the heartbeat can be determined by looking at the R-R interval on the ECG (Fig. 20–12). The same spacing between each R-R interval, with a variation of no greater than two small boxes, is seen in a normal rhythm. One way to determine the regularity of a rhythm is to count the number of small boxes between every R wave, which normally should remain the same. A more common way is to use a caliper to measure the spacing of the R-R interval.

A caliper is a small, two-sided, movable metal instrument with a sharp point at the end of each side. It is V-shaped when spread apart to measure various distances. To use a caliper for measuring, one point is placed at the top of an R wave and the other point is spread apart until it rests on the top of the next R wave. Then, without changing the distance between the points, the caliper is moved from one R wave to the next across the whole ECG tracing (also known as a strip) to see if the distance remains the same for each R-R interval. If the distance is the same, the rhythm is regular. If the distance changes, the rhythm is irregular. An irregular rhythm can be regularly irregular, which means it has a predictable pattern of irregularity, or irregularly irregular, without any pattern of irregularity.

LEARNING TIPS

If a caliper is not available, a piece of paper can be placed on the ECG strip. A mark can be made at the top of one R wave and another mark made at the top of the next R wave. The marks on the paper can then be moved across the R-R intervals on the strip (just as caliper points would be) to determine whether the rhythm is regular or irregular.

2. Heart Rate

After the components of a cardiac cycle are identified and the rhythm regularity determined, the heart rate is counted. There are two methods to calculate the heart rate.

1. Six second method: At the top of the ECG graph paper there are vertical marks at 3-second intervals. (See Fig. 20–3.) Count the number of R waves in a 6-second strip and multiply the total by 10 (the number of 6 seconds in a minute) to obtain the beats per minute (6 seconds × 10 =

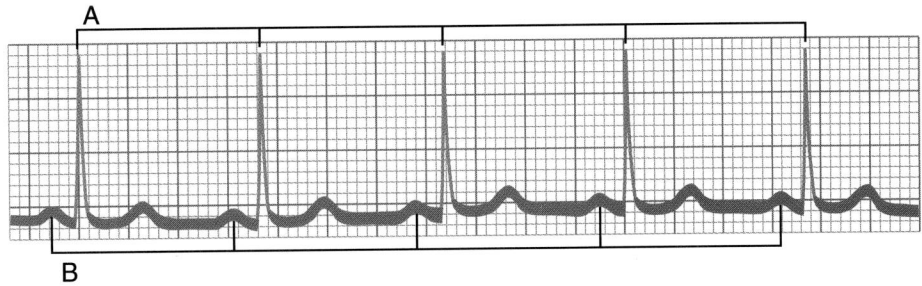

Figure 20-12 Normal cardiac waves are equal distances apart. *(A)* R-R waves. *(B)* P-P waves.

60 seconds or 1 minute) (Fig. 20–13). The 6-second method is used for irregular rhythms. It may also be used when a rapid estimate of a regular rhythm is needed, because it is not the most accurate method for regular rhythms.

2. Count the number of small (0.04-second) boxes between two R waves and divide that number into 1500. This gives the bpm, because 1500 small boxes equal 1 minute (Fig. 20–14). This method is used only for regular rhythms and is very accurate.

3. P Wave

The P waves on the ECG strip are examined to see if (1) there is one P wave in front of every QRS; (2) they are regular; and (3) the P waves all look alike. (See Fig. 20–12.) If the P waves all meet these criteria, they are considered normal. If they do not, further examination of the strip is necessary to determine the dysrhythmia.

4. PR Interval

All PR intervals are measured to determine whether they are normal and constant. If the PR is found to vary, it is important to note whether there is a pattern to the variation.

5. QRS Complex

The QRS intervals are measured to determine whether they are all normal and constant. Then the QRS complexes are examined to see if they all look alike.

▶ NORMAL SINUS RHYTHM
Description

Normal sinus rhythm is the normal cardiac rhythm. It originates in the SA node and represents a series of complete and regular cardiac cycles with a normal heart rate. The heart rate must be 60 to 100 bpm and have a regular rhythm. Each cycle must consist of a P wave, QRS complex, and a T wave. The measurements of the PR intervals and QRS intervals must fall within normal limits. If all these elements exist, the rhythm is interpreted as normal sinus rhythm (Fig. 20–15).

Normal Sinus Rhythm Rules

The rules for normal sinus rhythm are as follows:

1. Rhythm: regular
2. Heart rate: 60 to 100 bpm

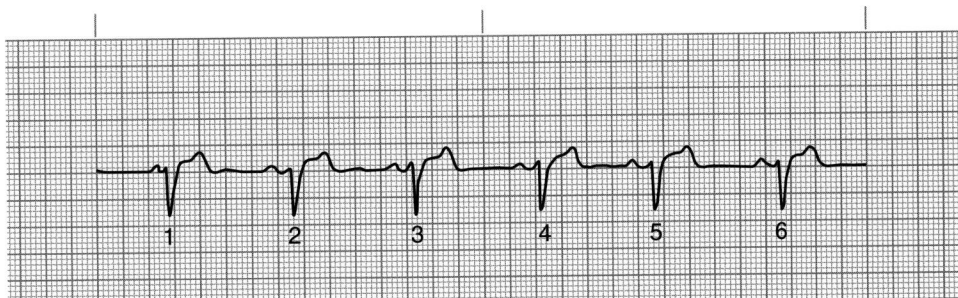

Figure 20-13 Counting R waves in a 6-second strip. There are six R waves in this 6-second strip, and 6 × 10 = 60 beats per minute.

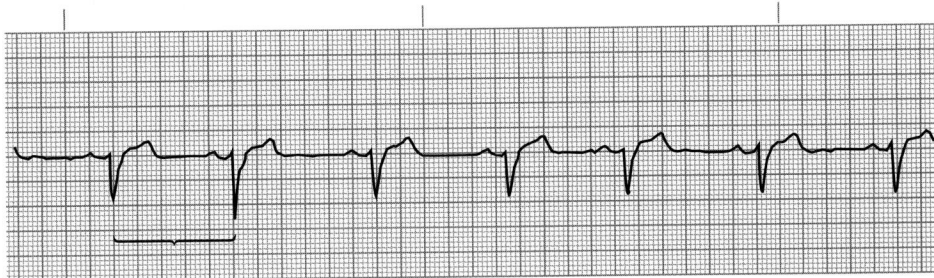

Figure 20-14 Heart rate. Counting large boxes and dividing into 300. Count number of large boxes between two R waves and divide into 300; five large boxes, 300/5 = 60 beats per minute.

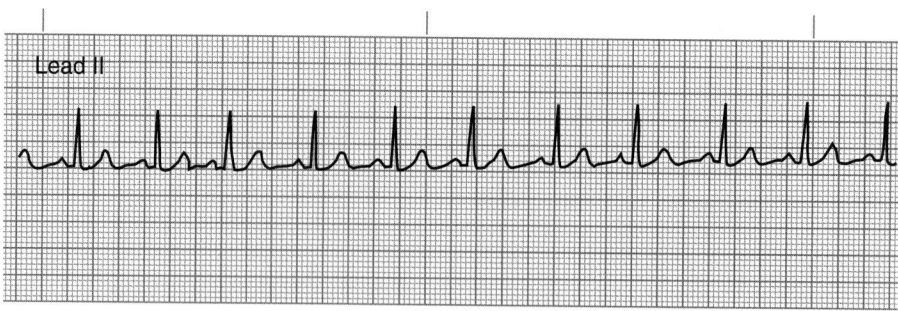

Figure 20-15   Normal sinus rhythm.

3. P waves: rounded, precede each QRS complex, alike
4. PR interval: 0.12 to 0.20 seconds
5. QRS interval: 0.06 to 0.10 seconds

▶ DYSRHYTHMIAS

Two terms are used for rhythm disturbances: arrhythmia and dysrhythmia. An arrhythmia is an irregularity or loss of rhythm of the heartbeat, and a dysrhythmia is an abnormal, disordered, or disturbed rhythm. These two terms are used interchangeably, but dysrhythmia is the most accurate term for the discussion of abnormal rhythms. The American Heart Association provides cardiac information and guidelines for cardiac care for dysrhythmias. For more information about cardiac care for dysrhythmias, visit the American Heart Association Web site at www.american-heart.org/arrhythmia.

Several mechanisms can cause irregularity or a dysrhythmia. Two examples of these mechanisms are a disturbance in the formation of an impulse and a disturbance in the conduction of the impulse. When impulse formation is disturbed, the impulse may arise from the atria, the AV node, or the ventricles. This disturbance can be seen as an increased or decreased heart rate, early or late beats, or atrial or ventricular fibrillation. With a disturbance in conduction there may be normal formation of the impulse, but it becomes blocked within the electrical conduction system, resulting in abnormal conduction (as in heart block or bundle branch blocks).

Dysrhythmias Originating in the Sinoatrial Node

Rhythms arising from the SA node are referred to as sinus rhythms. The SA node, being the pacemaker of the heart, fires normally at 60 to 100 bpm. Disturbances in conduction from the SA node can cause irregular rhythms or abnormal heart rates. Dysrhythmias arising from the SA node are rarely dangerous. Patients, especially those with heart, lung, or kidney disease, who cannot tolerate rapid or slow heart rates as shown by symptoms may require treatment.

LEARNING TIPS

The origin and type of problem, such as sinus **bradycardia**, are used to name a dysrhythmia. A dysrhythmia originating in the SA node is a sinus rhythm, but it is not normal. Therefore the term *normal* is not used; the origin *sinus* is used, and the type of problem, *bradycardia*, is identified.

Sinus Bradycardia

DESCRIPTION. Bradycardia is a slower-than-normal heart rate. Sinus bradycardia has the same cardiac cycle components as a normal sinus rhythm. The only difference is a slower heart rate caused by fewer impulses originating from the SA node (Fig. 20–16). Can you see that the name sinus bradycardia tells you this difference? The name says the impulse is coming from the sinus node (sinus) but at a slower rate than normal (bradycardia). It's easy to understand what is happening in the dysrhythmia when you look at what the name is telling you!

CAUSES. Medications, myocardial infarction (MI), and electrolyte imbalances can cause bradycardia. Well-conditioned athletes also can have slower heart rates because their hearts work more efficiently.

SINUS BRADYCARDIA RULES. The rules for sinus bradycardia are as follows:

1. Rhythm: regular
2. Heart rate: less than 60 bpm
3. P waves: smoothly rounded, precede each QRS complex, alike
4. PR interval: 0.12 to 0.20 seconds
5. QRS interval: 0.06 to 0.10 seconds

SIGNS AND SYMPTOMS. Sinus bradycardia rarely produces symptoms unless it is so slow that it reduces cardiac output. Symptoms consist of fatigue or fainting episodes. Patients are monitored for their responses to bradycardia.

TREATMENT. Treatment is usually not required if the patient is asymptomatic. The patient is observed, and the underlying cause is determined for correction. Oxygen and intravenous (IV) access may be started. If bradycardia is due to a heart block dysrhythmia, insertion of a cardiac pacemaker may be required. If the patient is symptomatic, treatment includes atropine sulfate, transcutaneous pacing, and dopamine, epinephrine, or isoproterenol. The medications are given intravenously by a registered nurse for immediate effect. These treatments are used to increase heart rate for a short time until the cause can be determined and treated.

bradycardia: bradys—slow + kardia—heart

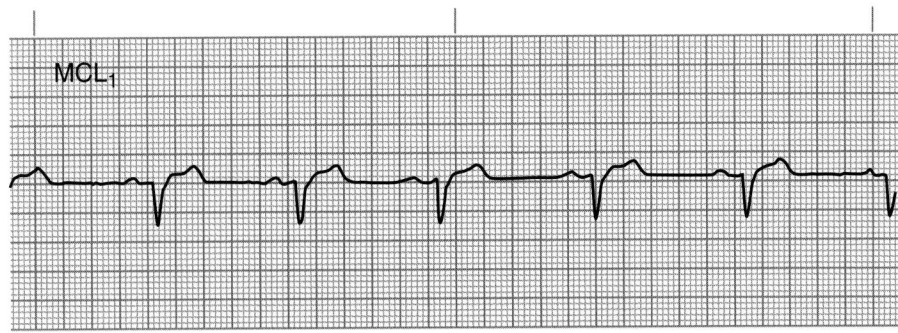

Figure 20-16 Sinus bradycardia.

Sinus Tachycardia

DESCRIPTION. Tachycardia is defined as a heart rate greater than 100 beats per minute. Sinus tachycardia has the same components as a normal sinus rhythm except the heart rate is faster (Fig. 20–17). This is due to more impulses originating from the SA node than normal.

CAUSES. Causes of sinus tachycardia include physical activity; hemorrhage; shock; medications such as epinephrine, atropine, or nitrates; dehydration; fever; MI; electrolyte imbalance; fear; and anxiety. Tachycardia occurs as a compensatory mechanism when hypoxia is present and more cardiac output is needed to deliver oxygen to organs and tissues.

SINUS TACHYCARDIA RULES. The rules for sinus tachycardia are as follows:

1. Rhythm: regular
2. Heart rate: 101 to 180 bpm
3. P waves: rounded, precede each QRS complex, alike
4. PR interval: 0.12 to 0.20 seconds
5. QRS interval: 0.06 to 0.10 seconds

SIGNS AND SYMPTOMS. Sinus tachycardia may not produce symptoms. If the heart rate is very rapid and sustained for long periods, the patient may experience angina or dyspnea. Elderly patients may become symptomatic more rapidly than younger patients (Geronotological Issues Box 20–1). Patients with MI may not tolerate a rapid heart rate because it increases cardiac workload; therefore more severe symptoms may occur.

TREATMENT. Treatment depends on the cause and the patient's symptoms. Treating the underlying cause usually cor-

BOX 20-1   GERONTOLOGICAL ISSUES

Factors that increase the risk of dysrhythmias in elderly people include the following:

- Digitalis toxicity (most common)
- Hypokalemia
- Acute infection
- Hemorrhage
- Angina
- Coronary insufficiency (exercise, stress)

Dysrhythmias that occur most often in elderly people include the following:

- Atrial fibrillation (atria beating 400 to 700 times per minute)
- Sick sinus syndrome (alternating episodes of bradycardia, normal sinus rhythm, tachycardia, and periods of long sinus pause)
- Heart block (delayed or blocked impulses to the atria or ventricles)

Age-related effects of dysrhythmias include the following:

- Weakness
- Fatigue
- Forgetfulness
- Palpitations
- Dizziness
- Hypotension
- Bradycardia
- Syncope

Elderly people are especially sensitive to changes in heart rate that increase the heart's workload. Whenever the heart works harder, as in tachycardia, the cardiac cells require more oxygen to function properly. Elderly patients have less ability to adapt to sudden changes or stressors, and they may not be able to tolerate tachycardia for very long. Any new-onset tachycardia in an elderly patient should be reported promptly.

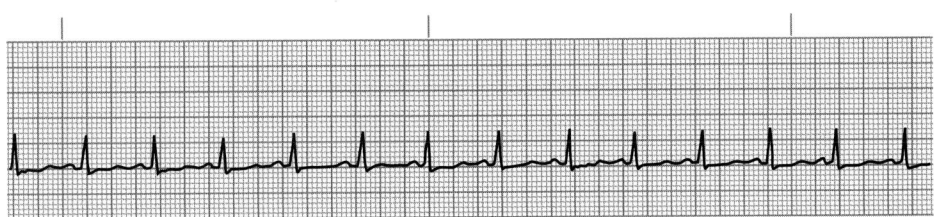

Figure 20-17 Sinus tachycardia.

rects the tachycardia. For example, if the patient is hemorrhaging, immediate intervention is needed to stop the bleeding and restore normal blood volume. The treatment goal is to decrease the heart's workload and resolve the cause. Medications such as digoxin, calcium channel blockers (verapamil), or beta blockers (propranolol) may be given to slow the heart rate. Oxygen may also be prescribed to help ensure an adequate supply for the heart.

LEARNING TIPS

Tachycardia is often the first sign of hemorrhage. It is a compensatory mechanism to maintain cardiac output. If a patient develops sudden tachycardia, consider whether hemorrhage could be the cause, such as in postoperative patients, patients with gastrointestinal bleeding or cancer, or trauma patients. The bleeding may be external, or it may be internal and therefore not visible. Apply pressure to the site if the bleeding is obvious. Monitor the patient and report the tachycardia and any obvious bleeding promptly.

Dysrhythmias Originating in the Atria

As previously discussed, all areas of the heart can initiate an impulse. The SA node is the primary pacemaker, but if the atria initiate impulses faster than the SA node, they become the primary pacemaker. Atrial rhythms are usually faster than 100 bpm and can exceed 200 bpm. When an impulse originates outside the SA node, the P waves look different from the rounded P waves produced by the SA node (flatter, notched, or peaked), which indicates that the SA node is not controlling the heart rate. The atrial impulse travels to the ventricles to initiate a normal QRS complex after each P wave.

Premature Atrial Contractions

DESCRIPTION. The term *premature* refers to an "early" beat. When the atria fire an impulse before the SA node fires, a premature beat results. If the underlying rhythm is sinus rhythm, the distance between R waves is the same except where the early beat occurs. When looking at the ECG strip, a shortened R-R interval is seen where the premature beat occurs. The R wave preceding the premature atrial contraction (PAC) and the PAC's R wave are close together, followed by a pause, with the next beat being regular (Fig. 20–18).

CAUSES. Causes of PACs include hypoxia, smoking, stress, myocardial ischemia, enlarged atria in valvular disorders, medications (such as digoxin), electrolyte imbalances, atrial fibrillation onset, and heart failure.

PREMATURE ATRIAL CONTRACTIONS RULES. The rules for premature atrial contractions are as follows:

1. Rhythm: premature beat interrupts underlying rhythm where it occurs
2. Heart rate: depends on the underlying rhythm; if normal sinus rhythm (NSR), 60 to 100 bpm
3. P waves: early beat is abnormally shaped
4. PR interval: usually appears normal, but premature beat could have shortened or prolonged PR interval
5. QRS interval: 0.06 to 0.10 seconds (indicates normal conduction to ventricles)

SIGNS AND SYMPTOMS. Premature atrial contractions can occur in healthy individuals, as well as in the patient with a diseased heart. No symptoms are usually present. If many PACs occur in succession, the patient may report the sensation of palpitations.

TREATMENT. PACs are usually not dangerous, and often no treatment is required other than correcting the cause. Frequent PACs indicate atrial irritability, which may worsen into other atrial dysrhythmias. Quinidine or procainamide can be given to a patient having frequent PACs to slow the heart rate.

Atrial Flutter

DESCRIPTION. In atrial flutter the atria contract, or flutter, at a rate of 250 to 350 bpm. The very rapid P waves appear as flutter, or F waves, on ECG and appear in a saw-tooth pattern. Some of the impulses get through the AV node and reach the ventricles, resulting in normal QRS complexes. There can be from two to four F waves between QRS complexes. If impulses pass through the AV node at a consistent rate, the rhythm is regular (Fig. 20–19). The classic characteristics of atrial flutter are more than one P wave before a QRS complex, a saw-tooth pattern of P waves, and an atrial rate of 250 to 350 bpm.

CAUSES. Causes of atrial flutter include rheumatic or ischemic heart diseases, congestive heart failure (CHF), hypertension, pericarditis, pulmonary embolism, and postoperative coronary artery bypass surgery. Many medications can also cause this dysrhythmia.

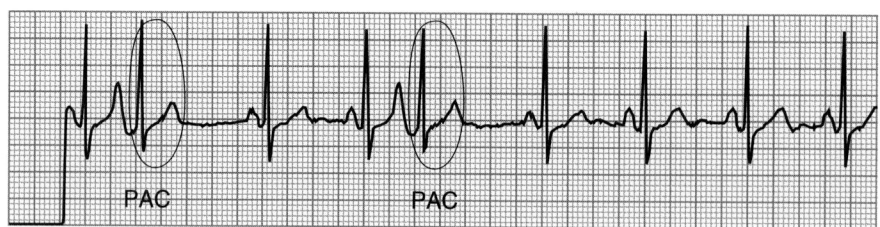

Figure 20-18 Premature atrial contractions (PAC).

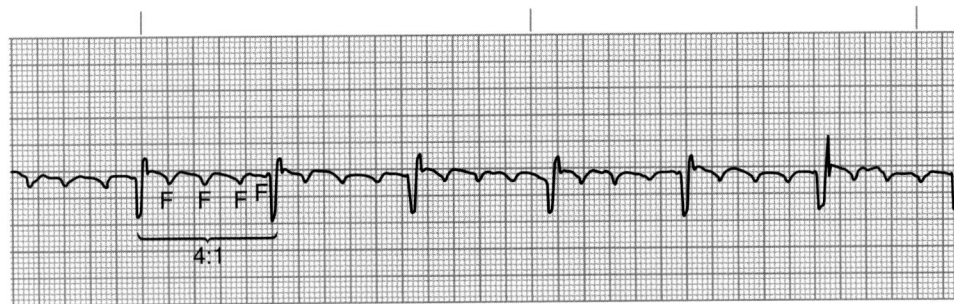

Figure 20-19   Atrial flutter.

ATRIAL FLUTTER RULES. The rules for atrial flutter are as follows:

1. Rhythm: atrial rhythm regular; ventricular rhythm regular or irregular depending on consistency of AV conduction of impulses
2. Heart rate: ventricular rate varies
3. P waves: flutter or F waves with saw-tooth pattern
4. PR interval: none measurable
5. QRS complex: 0.06 to 0.10 seconds

SIGNS AND SYMPTOMS. The presence of symptoms in atrial flutter depends on the ventricular rate. If the ventricular rate is normal, usually no symptoms are present. If the rate is rapid, the patient may experience palpitations, angina, or dyspnea.

TREATMENT. The ventricular rate and cardiac output guide treatment. The goal is to control the ventricular rate and convert the rhythm. A rapid ventricular rate or symptoms of decreased cardiac output require **cardioversion** (electrical shock). If the rate is greater than 150 bpm, immediate cardioversion is needed. Medications that may be used to control the rate include calcium channel blockers and beta blockers. For rhythm conversion, digoxin can be used to slow conduction through the AV node and increase cardiac contractility. Other medications, such as quinidine, procainamide, or propranolol, can also be used to slow the heart rate.

Atrial Fibrillation

DESCRIPTION. In atrial fibrillation the atrial rate is extremely rapid and chaotic. An atrial rate of 350 to 600 bpm can occur. However, the AV node blocks most of the impulses, so the ventricular rate is much lower than the atrial rate. There are no definable P waves because the atria are fibrillating, or quivering, rather than beating effectively. No P waves can be seen or measured. A wavy pattern is produced on the ECG. Because the atrial rate is so irregular and only a few of the atrial impulses are allowed to pass through the AV node, the R waves are irregular. The ventricular rate varies from normal to rapid.

Atrial fibrillation can be self-limiting, persistent, or permanent. A complication of this dysrhythmia is an increased risk of thrombus formation in the atria from blood stasis caused by poor emptying of blood from the quivering atria (Fig. 20–20). This can result in stroke or pulmonary emboli.

CAUSES. Causes of atrial fibrillation include aging, rheumatic or ischemic heart diseases, heart failure, hypertension, pericarditis, pulmonary embolism, and postoperative coronary artery bypass surgery. Medications can also cause this dysrhythmia.

ATRIAL FIBRILLATION RULES. The rules for atrial fibrillation are as follows:

1. Rhythm: grossly or irregularly irregular
2. Heart rate: atrial rate not measurable; ventricular rate under 100 is controlled response; greater than 100 is rapid ventricular response
3. P waves: no identifiable P waves
4. PR interval: none can be measured because no P waves are seen
5. QRS complex: 0.06 to 0.10 seconds

SYMPTOMS. With atrial fibrillation, most patients feel the irregular rhythm. Many describe it as palpitations or a skipping heartbeat. When checking a patient's radial pulse, it may be faint because of a decreased stroke volume (volume of blood ejected with each contraction). If the ventricular rhythm is rapid and sustained, the patient can go into left ventricular failure.

TREATMENT. Treatment is based on the patient's stability. If the patient is unstable, cardioversion is done immediately to try to return the heart to normal sinus rhythm. If the patient is stable, medications to restore and maintain a normal sinus rhythm and control the ventricular rate may be used. The ventricular rate may be controlled with such medications as digoxin, beta blockers, or calcium channel blockers. Medications approved by the Food and Drug Administration to convert atrial fibrillation and maintain a normal sinus rhythm include dofetilide, quinidine, flecainide, propafenone, and ibutilide IV. Dofetilide is a newer drug that has proven to be very effective. Anticoagulant therapy (aspirin for low-risk patients, warfarin for those at high risk), which can be long term or lifelong, is given to reduce thrombi. International normalized ratio (INR) and pro-

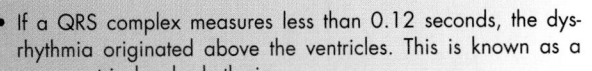

L E A R N I N G TIPS

- If a QRS complex measures less than 0.12 seconds, the dysrhythmia originated above the ventricles. This is known as a supraventricular dysrhythmia.
- Ventricular dysrhythmias produce wide QRS complexes that are greater than 0.11 seconds.

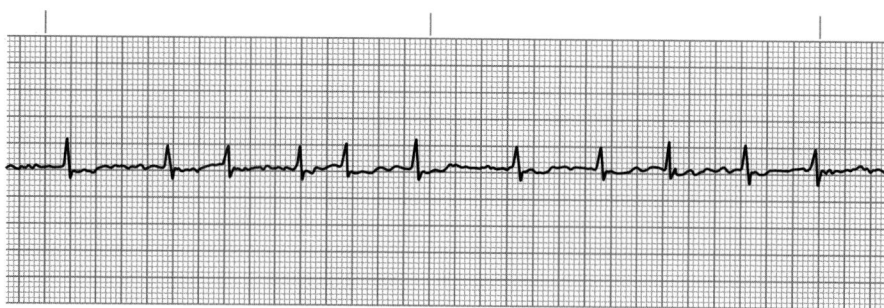

Figure 20-20   Atrial fibrillation.

thrombin time levels must be carefully monitored for patients on warfarin. Chemical or electrical cardioversion may be performed to convert the rhythm after sufficient anticoagulation (about 3 weeks). If known, the underlying cause of the atrial fibrillation should also be treated.

There are new advances in therapy for atrial fibrillation. Dual-chamber pacing for those with sinus node problems or biatrial pacing, as well as pacemaker recognition of atrial fibrillation, helps to prevent this dysrhythmia. Implantable cardioverter defibrillators (ICDs) can deliver a shock activated by the physician or patient to end the atrial fibrillation. Because this is a planned event, medications for comfort can be taken by the patient before the shock.

For patients with atrial fibrillation who do not respond to medications, **ablation** procedures may be performed. (These are discussed later.) A surgical procedure can be performed if other treatments fail. The maze procedure was first done as open heart surgery, but a percutaneous, nonsurgical catheter maze procedure is being investigated to eliminate the risks of open heart surgery (See Chapter 22.) In the open heart maze procedure incisions are made in the atria that create a "maze," or route, for electrical impulses to

---

ablation: ab—away from + lat—carry

---

travel through to the AV node. These impulses cannot go off-course because scar tissue surrounds the incision sites. In many cases this cures atrial fibrillation. However, some patients may need continued medications or a pacemaker following this procedure.

Ventricular Dysrhythmias
Premature Ventricular Contractions

DESCRIPTION. Premature ventricular contractions (PVCs) originate in the ventricles from an ectopic focus (a site other than the SA node). The ventricles are irritable and fire prematurely, before the SA node. When the ventricles fire first, the impulses are not conducted normally through the electrical pathway. This results in a wide (greater than 0.11 seconds), bizarre QRS complex on an ECG (Fig. 20–21).

PVCs can occur in different shapes. The shape of the PVC is referred to as **unifocal** (one focus) if all the PVCs look the same because they come from the same irritable ventricular area. **Multifocal** PVCs do not all look the same because they are originating from several irritable areas in the ventricle. There can be several repetitive cycles or patterns of PVCs:

■ **Bigeminy** is a PVC that occurs every other beat (a normal beat and then a PVC) (Fig. 20–22).

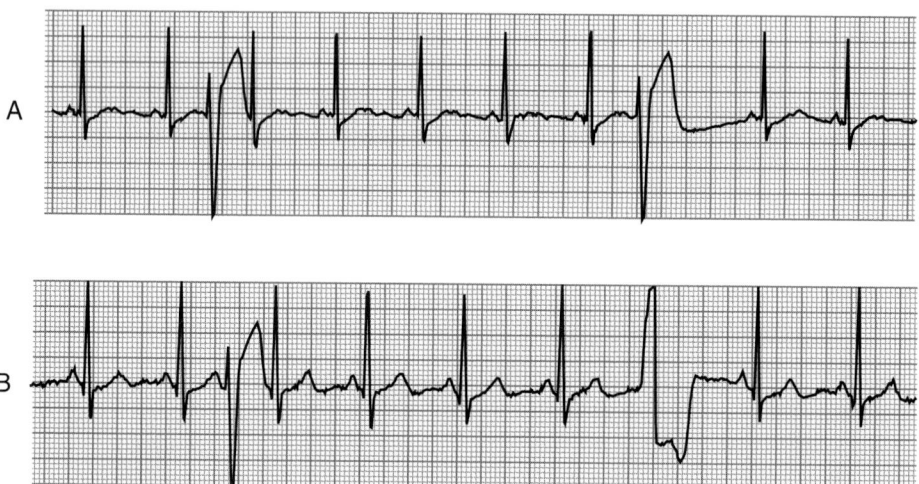

Figure 20-21   Premature ventricular contractions. *(A)* Unifocal PVCs arise from one area and look the same. *(B)* Multifocal PVCs arise from different foci and may look different.

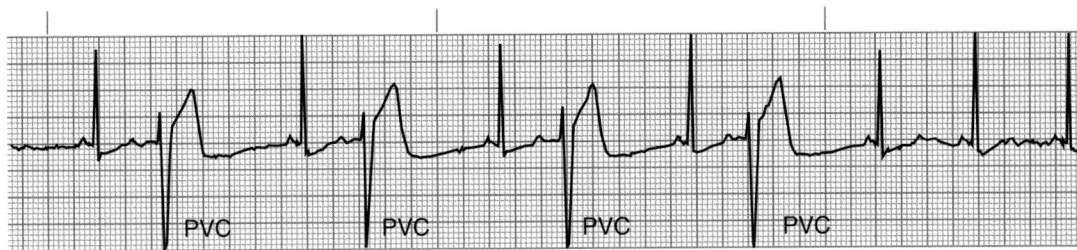

Figure 20-22 Bigeminal premature ventricular contractions.

- **Trigeminy** is a PVC that occurs every third beat (two normal beats and then a PVC).
- **Quadrigeminy** is a PVC that occurs every fourth beat (three normal beats and then a PVC).
- When two PVCs occur together, they are referred to as a couplet (pair).
- If three or more PVCs occur in a row, it is referred to as a run of PVCs or ventricular tachycardia.

CAUSES. Use of caffeine or alcohol, anxiety, hypokalemia, cardiomyopathy, ischemia, and MI are common causes of PVCs.

PREMATURE VENTRICULAR CONTRACTION RULES. The rules for premature ventricular contraction are as follows:

1. Rhythm: depends on the underlying rhythm; PVC usually interrupts rhythm
2. Heart rate: depends on underlying rhythm
3. P waves: absent before PVC QRS complex
4. PR interval: none for PVC
5. QRS complex: if PVC is greater than 0.11 seconds; T wave is in the opposite direction of QRS complex (i.e., QRS upright, T downward or QRS downward, T upright)

SIGNS AND SYMPTOMS. PVCs may be felt by the patient and are described as a skipped beat or palpitations. With frequent PVCs, cardiac output can be decreased, leading to fatigue, dizziness, or more severe dysrhythmias.

TREATMENT. Treatment depends on the type and number of PVCs and whether symptoms are produced. A few PVCs do not usually require treatment. However, if the PVCs are more than six per minute, regularly occurring, multifocal, falling on the T wave (known as "R-on-T phenomenon," which can trigger life-threatening dysrhythmias), or caused by an acute MI, they can be dangerous. Antidysrhythmic drugs that depress myocardial activity are used to treat PVCs. Examples of drugs that may be given intravenously by the registered nurse and then followed by a continuous IV drip are procainamide (Procan) and lidocaine (Xylocaine).

Ventricular Tachycardia

DESCRIPTION. The occurrence of three or more PVCs in a row is referred to as **ventricular tachycardia** (VT) (Fig. 20–23). VT results from the continuous firing of an ectopic ventricular focus. During VT, the ventricles rather than the SA node become the pacemaker of the heart. The pathway

---

CRITICAL THINKING

Mrs. Mae

Mrs. Mae, age 70, is 5 days post-MI without complications. You assist her back to bed at 1400 after she ambulates. Her oxygen is on at 2 L/min via nasal cannula. Her vital signs are 126/78, 82, 18. She has no pain and says she feels good after walking. The cardiac monitor shows normal sinus rhythm. Five minutes later you see that the monitor shows sinus rhythm with PVCs of less than six per minute. Her vital signs are now 132/84, apical 92 irregular, 22. She reports no pain but says, "I can feel my heart skipping, it takes my breath away." You call the RN while staying with the patient for reassurance.

1. What should you do first?
2. What should you do regarding the dysrhythmia?
3. What might be some of the causes for this dysrhythmia?
4. What symptoms, if any, would you expect to be present?
5. What would you do if symptoms were present?
6. What type of orders would you anticipate from the physician?
7. How would you document your findings?

Answers at end of chapter.

---

of the ventricular impulses is different from normal conduction, producing a wide (greater than 0.11 seconds), bizarre QRS complex.

CAUSES. Myocardial irritability, MI, and cardiomyopathy are common causes of VT. Respiratory acidosis, hypokalemia, digoxin toxicity, cardiac catheters, and pacing wires can also produce VT.

VENTRICULAR TACHYCARDIA RULES. The rules for ventricular tachycardia are as follows:

1. Rhythm: usually regular, may have some irregularity
2. Heart rate: 150 to 250 ventricular bpm; slow VT is below 150 bpm
3. P waves: absent
4. PR interval: none
5. QRS complex: greater than 0.11 seconds

SIGNS AND SYMPTOMS. The seriousness of ventricular tachycardia is determined by the duration of the dysrhythmia. Sustained VT compromises cardiac output. Patients are aware of a sudden onset of rapid heart rate and can experience dyspnea, palpitations, and light-headedness. Angina commonly occurs. The severity of symptoms can increase rapidly if the left ventricle fails and complete cardiac arrest results.

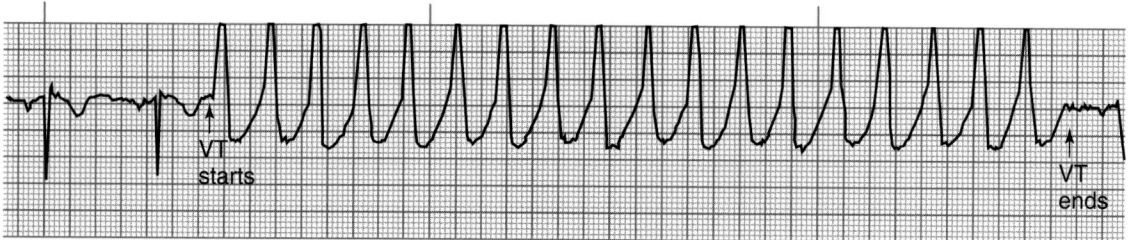

Figure 20-23 Ventricular tachycardia.

TREATMENT. If the patient is pulseless or not breathing, cardiopulmonary resuscitation (CPR) and immediate defibrillation are required, followed by antiarrhythmic drugs. Current advanced cardiac life support (ACLS) protocols for VT treatment should be used by individuals certified in ACLS.

If the patient is stable, medications can be tried first, such as amiodarone, procainamide, sotalol, lidocaine, phenytoin, or beta blockers following ACLS protocols. Magnesium can be used to help stabilize ventricular muscle excitability if the patient's magnesium level is low.

CRITICAL THINKING

Mrs. Parker

You are caring for Mrs. Parker, age 66, on the cardiac medical unit. She had an MI and several episodes of ventricular tachycardia while she was in the ICU before transferring to your unit. At 1600 you find her unresponsive, with no palpable pulses and shallow respirations and in VT on the ECG. Vital signs are BP 80/40, P 150, R 6.

1. Why are there no palpable pulses?
2. What is occurring to the heart when it is in VT?
3. What action should you take?
4. How will you document your findings?

Answers at end of chapter.

Ventricular Fibrillation

DESCRIPTION. Ventricular fibrillation occurs when many, many ectopic ventricular foci fire at the same time. Ventricular activity is chaotic with no discernible waves (Fig. 20-24). The ventricle quivers and is unable to initiate a contraction. There is a complete loss of cardiac output. If this rhythm is not terminated immediately, death ensues.

CAUSES. **Hyperkalemia, hypomagnesemia,** electrocution, coronary artery disease, and MI are all possible causes of ventricular fibrillation. Placement of intracardiac catheters and cardiac pacing wires can also lead to ventricular irritability and then ventricular fibrillation.

VENTRICULAR FIBRILLATION RULES. The rules for ventricular fibrillation are as follows:

1. Rhythm: chaotic and extremely irregular
2. Heart rate: not measurable
3. P waves: none
4. PR interval: none
5. QRS complex: none

SIGNS AND SYMPTOMS. Patients experiencing ventricular fibrillation lose consciousness immediately. There are no heart sounds, peripheral pulses, or blood pressure. These are all indicative of circulatory collapse. Additionally, respiratory arrest, cyanosis, and pupil dilation occur.

TREATMENT. Immediate defibrillation is the very best treatment for terminating ventricular fibrillation. Each minute that passes without defibrillation reduces survival. CPR is started until the defibrillator is available. Automatic external defibrillators (AEDs) provide quick access to easily used technology for defibrillation (see defibrillation section). Endotracheal intubation and ventilation support respiratory function. Medications are given according to ACLS protocols and may include epinephrine, vasopressin, amiodarone, lidocaine, magnesium, and procainamide.

---

hyperkalemia: hyper—above + kalium—potassium + emia—blood
hypomagnesemia: hypo—below + magnes—magnesium + emia—blood

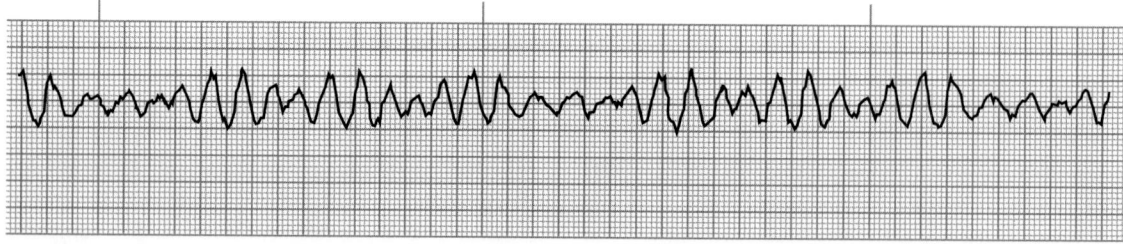

Figure 20-24 Ventricular fibrillation.

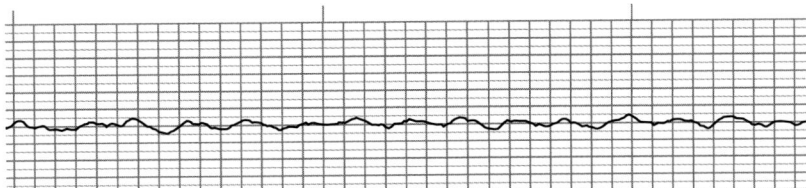

Figure 20-25 Asystole.

Asystole

DESCRIPTION. Asystole (the silent heart) is the absence of electrical activity in the cardiac muscle. It is referred to as cardiac arrest. A straight line appears on an ECG strip (Fig. 20–25). Ventricular fibrillation usually precedes this rhythm and must be reversed immediately to help prevent asystole.

CAUSES. Ventricular fibrillation and a loss of a majority of functional cardiac muscle due to an MI are common causes of asystole. Hyperkalemia is another cause of asystole.

ASYSTOLE RULES. The rules for asystole are as follows:

1. Rhythm: none
2. Heart rate: none
3. P waves: none
4. PR interval: none
5. QRS complex: none

TREATMENT. CPR is started immediately. ACLS protocols for asystole are used. Endotracheal intubation to support respirations is performed. Transcutaneous pacing is considered, then epinephrine and atropine are administered.

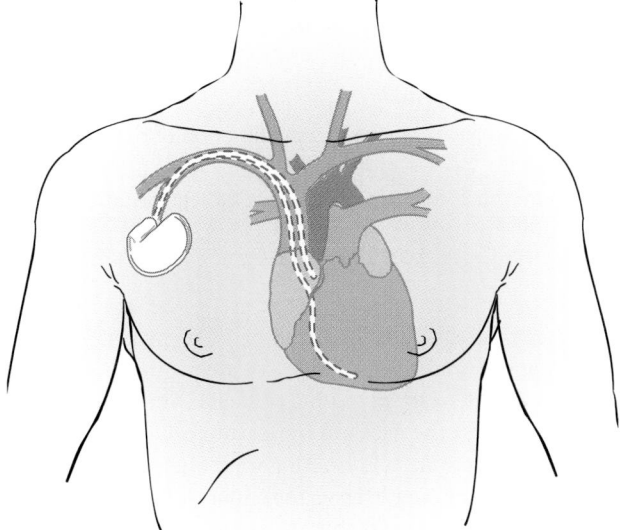

Figure 20-26 Insertion of dual-chamber permanent pacemaker.

CRITICAL THINKING

Mr. Peet

You are making rounds. When you enter Mr. Peet's room, you note that he is having difficulty breathing and is unresponsive.

1. What are your initial actions?
2. What should you do after assessing and finding no pulse or respirations?
3. What is your responsibility during a cardiac/respiratory arrest code?

Answers at end of chapter.

▶ CARDIAC PACEMAKERS

Pacemakers can be external and temporary or internal and permanent (Fig. 20–26). They are used to override dysrhythmias or to generate an impulse when the heart is beating too slowly. Transcutaneous pacemakers are used in emergency situations because they are quick and easy to apply. Impulses are delivered to the heart through the skin from the external generator via electrodes that are attached to the chest and back.

Temporary pacemakers are used for bradycardias or tachycardias that do not respond to medications or cardioversion. They may also be used after an MI to allow the heart time to heal when the diseased myocardium is unable to respond to or is not receiving electrical conduction because of damage within the system. The temporary pacemaker becomes the electrical conduction system and stimulates the atria and ventricles to contract to maintain cardiac output. Temporary pacemakers can be inserted during valve or open heart surgery or in the cardiac catheterization laboratory or critical care unit at the bedside as emergency treatment until surgery can be scheduled to insert a permanent pacemaker.

Permanent pacemaker insertion is a surgical procedure in which **fluoroscopy,** a screen that shows an image similar to a radiograph, is used. The pacemaker generator is implanted subcutaneously and attached to one or two leads (insulated conducting wires) that are inserted via a vein into the heart. The lead can then deliver the impulse directly to the heart wall. A single-lead pacemaker paces either the right atrium or right ventricle depending on its chamber placement. Dual-chamber pacemakers have two leads, with one in the right atrium and the other in the right ventricle. This allows pacing of both chambers. Usually pacemakers are set at a prescribed rate of 72 bpm. Activity-responsive pacemakers provide a rate range (e.g., 60 to 115 bpm) in response to a person's activity level. This provides the patient with greater flexibility for increasing cardiac output when needed, such as during exercise.

When a patient is in a paced rhythm, a small spike is seen on the ECG before the paced beat. This spike is the electrical stimulus. It can precede the P wave, QRS complex, or both depending on what is being paced (Fig. 20-27). Patients may have all paced beats (100-percent paced), a mixture of their own beats and paced beats, or all of their own beats. Pacemakers should not fire when patients have their own beats.

Problems that can occur with pacemakers include the following:

- Failure to sense the patient's own beat
- Failure to pace because of a malfunction of the pulse generator
- Failure to capture, which is a lack of depolarization

Nursing Care for Pacemakers

Patients are placed on a cardiac monitor and strict bedrest for 12 to 24 hours after insertion of a pacemaker. The patient's apical pulse is monitored frequently to detect changes in the heart rhythm. Irregular heart rhythms or a rate slower than the pacemaker's set rate can indicate pacemaker malfunction. The dressing at the pacemaker insertion site is monitored every 2 to 4 hours for signs of bleeding. Any change in heart rhythm, complaints of chest pain, or changes in vital signs must be reported immediately. Patients may have a sling on the operative side arm for 24 to 48 hours to help prevent dislodgement of the pacemaker lead from the cardiac wall.

Patient education for pacemaker care before discharge includes the following:

- Incision care. The patient should check the incision daily and report evidence of inflammation or infection (redness, swelling, warmth, tenderness, pain, fever, or discharge) to a physician.
- Methods for taking a radial pulse. The patient should call a physician if the pulse is slower than the pacemaker's set rate.
- The patient should report symptoms of dizziness, fainting, irregular heartbeats, or palpitations.
- The patient should understand the importance of wearing medical alert jewelry and carrying a pacemaker information card.
- The patient should avoid radiation, magnetic fields (e.g., magnetic resonance imaging [MRI], industrial magnets), high voltage (e.g., power plant, arc welding, high-tension wires), antitheft devices, and large running motors (e.g., distributor coil of running engine).
- The patient will need to tell airport security about the pacemaker, because it may trigger metal detectors (they do not harm the pacemaker).

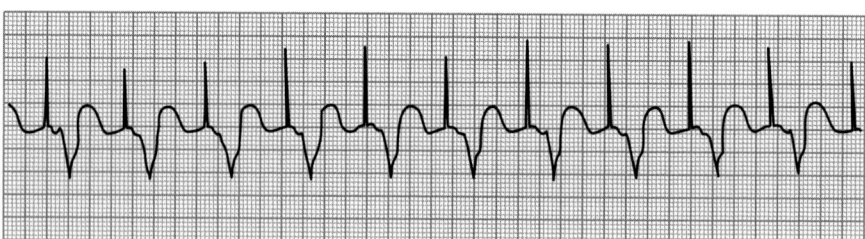

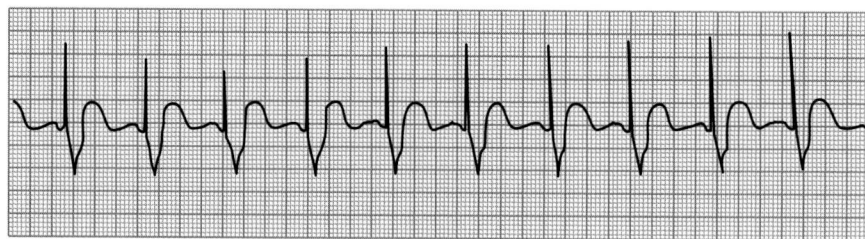

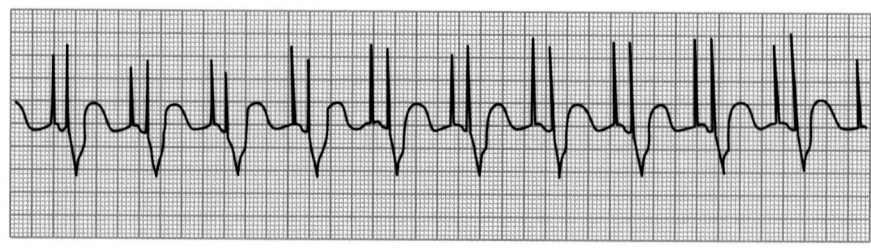

Figure 20-27 *(A)* Atrial-only pacemaker. *(B)* Ventricular-only pacemaker. *(C)* AV sequential pacemaker that paces both chambers.

- Grounded appliances (usually includes microwave ovens) and office equipment are safe to use.
- The patient must avoid lifting more than 10 pounds, making major arm movements, or participating in contact sports for 6 weeks after surgery. Normal activity is usually resumed after 6 weeks.
- The patient must keep scheduled appointments with the physician. Periodic pacemaker checks will be done by the physician or over the telephone. Reprogramming of the pacemaker can be done by the physician if needed.

⌇⋏⋀ CRITICAL THINKING

Mr. Treacher

Mr. Treacher, age 58, underwent pacemaker placement 6 days ago and is being transferred to the medical floor. After transfer, his vital signs are 138/72, 72 bpm, and 100-percent paced rhythm. Thirty minutes later, he says that he feels weak and tired. His vital signs are now 100/60, 60 bpm, and irregular.

1. What is your first action?
2. What actions should be taken next?
3. What might be happening to Mr. Treacher?
4. What interventions should you anticipate next?

Answers at end of chapter.

▶ DEFIBRILLATION

Defibrillation is a lifesaving procedure used for lethal dysrhythmias. It delivers an electrical shock to reset the heart's rhythm. It is used to terminate pulseless ventricular tachycardia or ventricular fibrillation. Self-adhesive pads, conductive jelly, or saline pads are placed on the patient's chest to prevent electrical burns from the defibrillator and promote conduction of the electrical charge. After the defibrillator is charged, the paddles are pressed firmly and evenly against the chest wall at the second intercostal space, right of the sternum and on the anterior axillary line at the fifth intercostal space (Fig. 20–28).

If the paddles are not pressed firmly against the chest wall during defibrillation, burns or electrical arcing can result when the shock is delivered. For safety, the person defibrillating must announce "clear." The phrase "One. I'm clear. Two. You're clear. Three. All clear" is suggested. No one, including the person defibrillating, should touch the bed or patient during this time to avoid also being shocked. ACLS protocols specify guidelines for resuscitation. If the first shock is unsuccessful, a total of three shocks can be given initially, at increasing energy levels (200, 300, 360 Joules).

LEARNING TIPS

Atrial fibrillation is easy to identify based on its two classic characteristics: a lack of identifiable P waves and an irregularly irregular rhythm (R waves).

---

defibrillation: de—from + fibrillation—quivering fibers

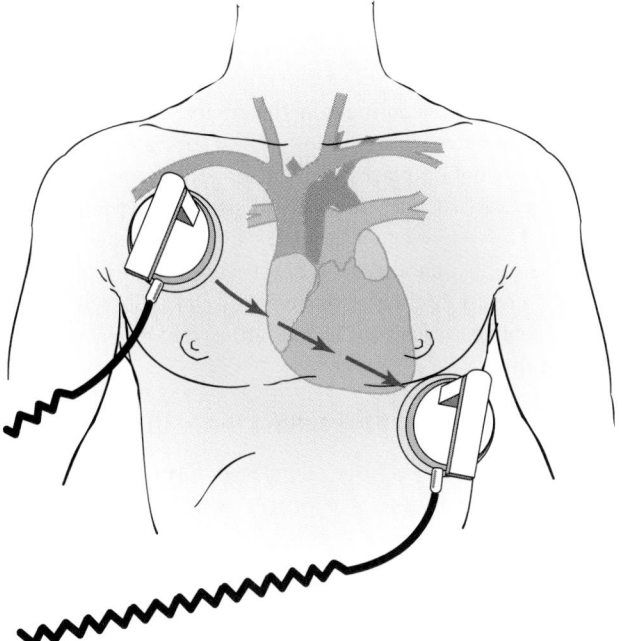

Figure 20-28 Placement of defibrillator paddles on chest.

A defibrillator may have a synchronization mode that allows R waves, if present, to be sensed so that an electrical shock can be delivered at an appropriate time in the cardiac cycle. This timing is important to prevent a more life-threatening dysrhythmia from developing. When defibrillation of a pulseless ventricular dysrhythmia is desired, the synchronize mode must be off (unsynchronized cardioversion). In this mode the charge is immediately released when the trigger is pressed. In the synchronized mode (synchronized cardioversion) there is a delay in the release of the charge while the R wave is sensed for appropriate timing.

After successful defibrillation, the patient is assessed for a pulse and adequate tissue perfusion. Vital signs, peripheral pulses, and level of consciousness should be noted. The patient is treated in the critical care unit after successful resuscitation.

Emotional support for the patient having experienced cardiac arrest and defibrillation is a very important aspect of nursing care. This can be an extremely frightening event for the patient. It is important to explain what happened to the patient and to listen and allow the patient to express any concerns. The patient is reassured that continuous monitoring is done in the critical care unit. Families also require emotional support during resuscitation of a loved one.

▶ OTHER METHODS TO CORRECT DYSRHYTHMIAS

Automatic External Defibrillators

An AED is an external device that automatically analyzes rhythms and either automatically delivers or prompts operators to deliver an electrical shock if a shockable rhythm (ventricular fibrillation or VT) is detected. Minimally trained laypersons or hospital and rescue personnel can use these devices with little risk of injury to the patient because the analyzes the rhythm instead of the operator. The patient is connected to the AED with adhesive sternal-apex pads attached to cables coming from the device. This connection allows hands-free defibrillation. AEDs increasingly are being kept in public places such as shopping malls, airports, stadiums, casinos, golf courses, and airplanes for immediate access. The American Heart Association encourages increasing the availability of AEDs because defibrillation attempts must occur within minutes of cardiac arrest to increase chance of survival. In the future, AEDs may even be kept in the home.

Implantable Cardioverter Defibrillator

An implanted ICD is surgically placed during a minor procedure into the chest of a patient who experiences life-threatening dysrhythmias or is at risk for sudden cardiac death (Fig. 20-29). ICDs have decreased the number of deaths from these dysrhythmias by analyzing and treating these heart rhythms. When an abnormal rhythm is detected that could cause death (ventricular fibrillation), it automatically delivers an electrical shock. If the dysrhythmia does not convert on the initial shock, more shocks are delivered sequentially.

If the device detects VT, it cardioverts the rhythm using lower energy. ICDs also have antitachycardia pacing ability if a tachycardic rhythm is detected. Battery life depends on usage, but a battery may last up to 7 years. A physician can tell when the battery is getting low and that the entire unit needs to be changed within a few months. New technology, such as smaller defibrillators that are easier to implant and have a longer battery life, is continually being developed for ICDs.

Patients with ICDs are extremely anxious about having another cardiac arrest and receiving shocks from the ICD. Defibrillator or cardioversion shocks may feel like a kick in

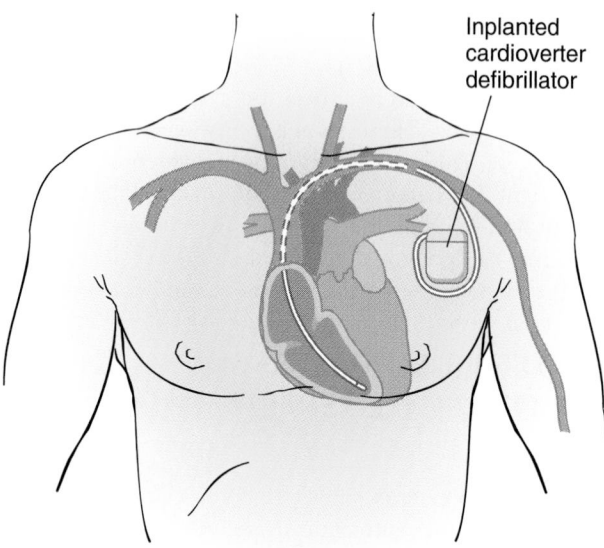

Figure 20-29 An implanted cardioverter defibrillator in place.

Inplanted cardioverter defibrillator

the chest. Reinforcement of patient and family education is very important in preparing the patient for discharge. Those with ICDs should take precautions to prevent problems with the ICD by taking the following precautions:

- Avoiding MRIs
- Avoiding metal detectors and standing near security gates or store entrances
- Avoiding equipment with strong electrical or magnetic fields (e.g., amusement rides, slot machines, remote-control toys, stereo speakers)
- Keeping cell phones 6 inches from the ICD

The nurse provides emotional support, answers all questions, and ensures that any misunderstood information is corrected before discharge.

Cardioversion

Elective cardioversion is used for dysrhythmias such as atrial fibrillation, atrial flutter, and supraventricular tachycardias that are not responsive to drug therapy. Conscious sedation is often used. The patient is given a sedative and monitored by anesthesia personnel during the procedure. Cardioversion is performed with a defibrillator set in the synchronize mode. The defibrillator is attached to the patient by electrode wires, which enables the ECG to be viewed on a screen. On the defibrillator there is a switch labeled "synchronize," which is on during cardioversion. When the defibrillator is in the synchronized mode, it marks a highlighted area on the patient's R waves, which must be recognized to deliver a shock.

The number of joules delivered with each shock is determined by a physician but usually ranges from 25 to 50 joules. Self-adhesive pads or conduction jelly is placed on the chest. The paddles must be firmly and evenly pressed against the pads or jelly on the chest until the shock is delivered to prevent skin burns and electrical arcing. When the discharge trigger is pressed, the shock is released when the machine senses it is safe to do so. The cardiac monitor screen is observed to see the discharge of the shock and to note the patient's ECG response. The patient's pulse and vital signs are checked.

If cardioversion is successful, there should be a return to normal sinus rhythm. If the rhythm does not immediately convert, more cardioversion attempts can be made as determined by the physician. After the procedure, the patient is monitored for skin burns, rhythm disturbances, respiratory problems, hypotension, and changes in the ST segment.

Ablation

When medications or other treatments are not successful in treating a dysrhythmia, ablation of cardiac conduction pathways may be used to stop the dysrhythmia. Intracardiac echocardiography and intracardiac mapping with fluoroscopy are done before ablation to determine the area of the heart requiring treatment. A newer mapping technique, tridimensional mapping, which may be more accurate and does not need fluoroscopy, is being investigated.

Forms of ablation include mechanical, chemical, and radiofrequency. Mechanical ablation destroys the involved tissue with cryosurgery or through surgical removal. Chemical ablation inserts alcohol or phenol through an angioplasty catheter into the area of the heart producing the unwanted beats. Radiofrequency ablation delivers high-frequency energy via a catheter to necrose selected conduction pathway areas. Following any of these ablation procedures, a temporary or permanent pacemaker may be needed if normal conduction tissue is damaged. Postprocedural care is similar to postangioplasty or postcardiac catheterization care. (See Chapter 15.)

▶ NURSING PROCESS FOR THE PATIENT WITH DYSRHYTHMIAS

Assessment

Assessment of the cardiac system, respiratory rate, breath sounds, and urinary output is important. Monitoring apical and radial pulses at frequent intervals helps detect dysrhythmias. Most dysrhythmias are not life threatening. Patients at risk for dysrhythmias require careful monitoring so that any dysrhythmias are detected and treated. A patient's complaints of dizziness, chest pain, or palpitations should always be reported to the physician.

Nursing Diagnoses

The major nursing diagnoses for dysrhythmias may include but are not limited to the following:

- Decreased cardiac output related to dysrhythmias or response to medications
- Ineffective tissue perfusion related to decreased cardiac output
- Activity intolerance related to decreased cardiac output
- Anxiety related to fear of dying, knowledge deficit, and diagnostic procedures
- Deficient knowledge related to dysrhythmias, diagnostic procedures, medications, and treatment

Planning

The goal of therapy is to identify patients at risk for dysrhythmias and promote adequate cardiac output. You need to identify factors that may contribute to increased cardiac workload. Careful assessment of all systems will focus on areas that might be affected by decreased cardiac output.

The patient's goals include the following:

- Maintain vital signs and oxygen saturation within normal range.
- Maintain warm, dry skin; palpable pulses; and normal vital signs.
- Maintain desired activities.
- Decrease anxiety.
- Understand the disease and its treatment.

Implementation

To implement the plan of care, the patient and family should be included. Assist them in understanding the plan and the reasons for the prescribed interventions. The patient and family need to be given time to understand the plan of care and to express their needs and fears. See Nursing Care Plan Box 20–2 for the Patient with Dysrhythmias for nursing interventions.

BOX 20-2   Nursing Care Plan for the Patient with Dysrhythmias

 Decreased cardiac output related to dysrhythmias

PATIENT OUTCOMES
(1) Patient cardiac status stabilizes. (2) Patient tolerates activities of daily living (ADL).

EVALUATION OF OUTCOMES
(1) There is an absence of dysrhythmias. (2) Patient is able to perform ADL without tachycardia, chest pain, or weakness.

INTERVENTIONS	RATIONALE	EVALUATION
Assess apical and radial pulses every 2 to 4 hours. Assess blood pressure and urinary output. Monitor mental status every 2 to 4 hours. Assess lung sounds every 2 to 4 hours. Administer O₂ as ordered.	Monitors for dysrhythmias, impending cardiac arrest, or shock. Blood pressure, pulse, and urinary output are indicators of cardiac output. Dizziness, confusion, and restlessness may indicate decreased cerebral blood flow. Dysrhythmias can cause heart failure. Increases oxygenation to the heart and brain.	Is the patient free of dysrhythmias with vital signs within normal limits? Does patient show signs of decreased cerebral perfusion, such as confusion? Are lungs clear with no report of dyspnea?
Ensure that patient gets adequate rest and does not exceed activity tolerance.	Reduces dyspnea and decreases O₂ demand on the myocardium.	Is patient free of chest pain, confusion, and light-headedness? Does patient rest and tolerate activity without dyspnea or chest pain?

GERIATRIC

Administer medications as ordered and observe for adverse reactions.	Older patients may have decreased renal and liver function that may lead to rapid development of toxicity.	Does patient have signs of toxicity?

BOX 20-2 **NURSING CARE PLAN FOR THE PATIENT WITH DYSRHYTHMIAS—CONT'D**

Anxiety related to situational crisis

PATIENT OUTCOMES

(1) Patient is able to effectively manage anxiety. (2) Patient will report decreased anxiety.

EVALUATION OF OUTCOMES

(1) Patient uses effective coping mechanisms to manage anxiety. (2) Patient expresses decreased anxiety.

INTERVENTIONS	RATIONALE	EVALUATION
Assess level of anxiety.	Establishes a baseline.	What is patient's level of anxiety?
Encourage patient and family to verbalize fears.	Helps correct and clarify their concerns.	What are patient's feelings or fears?
Explain procedures to patient and family.	Lack of knowledge increases anxiety. Also will help with compliance of therapy.	Does patient express understanding of therapy with decreased anxiety?
Identify and reduce as many environmental stressors as possible.	Anxiety often results from lack of trust in the environment.	Can patient describe two situations that increase tension?
Teach patient relaxation techniques to be performed every 4 to 6 hours, such as guided imagery, muscle relaxation, and meditation.	These measures can restore psychological and physical equilibrium and help decrease anxiety.	Is patient successful in demonstrating relaxation methods?
Medicate with antianxiety agents as ordered.	Aids the patient in decreasing anxiety.	Does patient show decreased anxiety?

Family members should be taught CPR or given information on local CPR classes. This training gives the patient and family a sense of control and hope. In the event the patient requires CPR, the family can take action instead of simply standing by and feeling helpless. The patient will feel more secure in knowing that immediate help from family members is available at home until medical help arrives (Home Health Hints).

Evaluation

Evaluation of the outcome stems from regulation of dysrhythmias. Compliance with medication and therapy is the best indicator of understanding. Involvement of the patient and family in asking questions and being involved in the plan of care leads to positive outcomes and control of dysrhythmias.

HOME HEALTH HINTS

- The nurse should have a pocket mask for CPR available at all times.

- A car phone or portable phone affords a home health nurse safety, convenience, and efficiency, especially if emergency help is needed, because some patients do not have phones.

- Patients prone to dysrhythmias should avoid straining with bowel movements. If the patient reports straining, request a laxative or stool softener order from the physician.

- Patients who come home with a pacemaker should be instructed to wear loose tops. Women should not wear tight bras.

- Symptoms of infection to watch for after a pacemaker is implanted are redness, swelling, warmth, and pain at the site.

- Instruct patients with a pacemaker to take their pulse once a day, in the morning, for a full minute. Assist them in setting up a log to record date, time, and pulse reading. Instruct them to call if the pulse varies outside parameters set by the physician.

- Patients on beta blockers need to know how to take their pulse, because bradycardia is a major side effect. For pulse below 50, call the nurse or physician.

- Advise patients who are leaving home for weekend or holidays to refill medicines ahead of time. Also, the physician may write a prescription for patients to keep in their wallet for emergencies.

REVIEW QUESTIONS

1. Which of the following is the correct sequence for normal electrical impulse movement through the cardiac conduction system?
   a. Vena cavae, right atrium, right ventricle, pulmonary artery
   b. SA node, AV node, bundle of His, Purkinje fibers
   c. Pulmonary veins, left atrium, left ventricle, aorta
   d. Purkinje fibers, AV node, bundle of His, SA node

2. If a patient is in pulseless ventricular tachycardia, which of the following is the first choice of treatment?
   a. Synchronized cardioversion
   b. Pacemaker
   c. Defibrillation
   d. Antiarrhythmic medication

3. The nurse is ambulating a patient who is recovering from an MI and the patient develops chest pain with an irregular pulse. What is the safest way to return the patient to his bed?
   a. Ambulate to room with one assistant
   b. With assistance by stretcher
   c. With assistance by a wheelchair
   d. After completion of ambulation

4. You are to give a patient amiodarone 800 mg/day PO in two divided doses. You have available 200-mg tablets. How many tablets will you give for each dose?
   a. 2
   b. 4
   c. 6
   d. 8

5. A patient has a radial pulse of 58 bpm. Which of the following is the appropriate term for documenting this rhythm?
   a. Normal
   b. Bradycardia
   c. Tachycardia
   d. Asystole

Answers to CRITICAL THINKING

Mrs. Mae

1. Assess the patient's vital signs and heart sounds; note symptoms; obtain an ECG per agency protocol.
2. Report the patient findings to the RN or physician.
3. Possible causes include hypokalemia or ischemia leading to irritability of the heart.
4. Symptoms might include light-headedness, feeling of heart skipping, chest pain, or fatigue.
5. To alleviate symptoms, elevate head of bed to comfort, monitor vital signs, and maintain oxygen at 2 L/min via nasal cannula per agency protocol. Remain with the patient to help alleviate anxiety. Notify the RN.
6. Orders might include ECG, oxygen, potassium, or electrolytes.
7. Documentation should include the following:
   1400: Ambulated 20 feet with one assist. Vital signs stable. Stated "feel good. Pain zero."
   Tolerated well. Assisted to bed. Oxygen at 2 L/min via nasal cannula.
   1405: See ECG strip with intermittent PVCs. Vital signs: 132/84, apical 92 irregular, 22.
   "Pain zero. I can feel my heart skipping, it takes my breath away."
   RN notified.

Mrs. Parker

1. A heart in VT has an ectopic focus firing. The heart is unable to maintain adequate cardiac output with such a rapid heart rate. The rapid and irregular heart rhythm does not allow the heart chambers time to adequately fill and empty, thereby reducing the blood volume with each beat. This in turn affects the peripheral circulation, causing the absence of palpable pulses.
2. In VT one or more sites in the ventricle may be initiating impulses. The rapid rate of VT overrides the normal pacemaker of the heart. The rhythm can be regular or irregular. The inability of the heart to conduct impulses along normal pathways prevents the chambers from emptying and filling properly. This leads to a decreased cardiac output and can lead to cardiac arrest if the rhythm is not converted.
3. Call a code and begin CPR. Report findings to code team upon their arrival.
4. Documentation should include the following:
   1600: Patient found in bed unresponsive to verbal and tactile stimuli. Respirations shallow. No palpable pulses. BP 80/40, P 150, R 6. Monitor shows VT (see strip). Code called from room. CPR started. Code team arrived at 1602. Report given to code team leader.

Mr. Peet

1. Initially you should assess responsiveness and the presence of a carotid pulse. Check for breathing.
2. Open the airway. Call for assistance or use the patient's phone to report a cardiac arrest. Initiate CPR until help arrives.
3. Once help or the code team arrives, the licensed practical nurse/licensed vocational nurse (LPN/LVN) reports the patient's status. The code team leader delegates responsibilities. Many facilities have protocols for each team member in a code. The LPN/LVN assists in the code as delegated by the RN in charge.

Mr. Treacher

1. Your first actions should be to obtain an ECG per agency protocol and to notify the RN and physician.
2. You should keep the head of the bed elevated and administer oxygen at 2 L/min via nasal cannula per protocol. Turn the patient onto his side because this may help float the pacemaker wire to the chamber wall for better contact. Monitor the patient's ECG, vital signs, and symptoms, and remain with patient to provide emotional support.
3. Mr. Treacher could be experiencing pacemaker malfunction.
4. Interventions could include transfer to a step-down unit or intensive care unit (ICU), reprogramming of the pacemaker, or a return to surgery for manipulation or replacement of the pace-

21

NURSING CARE OF PATIENTS WITH HEART FAILURE

Linda S. Williams

KEY TERMS

afterload (**AFF**-ter-lohd)

cor pulmonale (**KOR** PUL-mah-**NAH**-lee)

cyanosis (SIGH-an-**NOH**-sis)

hepatomegaly (HEP-uh-toh-**MEG**-ah-lee)

orthopnea (or-**THOP**-knee-a)

paroxysmal nocturnal dyspnea (PEAR-ox-**IS**-mall knock-TURN-al DISP-knee-a)

peripheral vascular resistance (puh-**RIFF**-uh-ruhl **VAS**-kyoo-lar ree-**ZIS**-tense)

preload (**PREE**-lohd)

pulmonary edema (**PULL**-muh-NAIR-ee uh-**DEE**-muh)

splenomegaly (SPLEE-noh-**MEG**-ah-lee)

QUESTIONS TO GUIDE YOUR READING

1. How would you describe the pathophysiology of left- and right-sided heart failure?

2. What is acute heart failure?

3. What are causes of acute and chronic heart failure?

4. What are signs and symptoms of acute and chronic heart failure?

5. What nursing care would you provide for diagnostic tests for heart failure?

6. What is the medical treatment for acute and chronic heart failure?

7. What nursing care would you provide for acute and chronic heart failure?

8. What would you include in your teaching plan for patients with heart failure and their families?

▶ HEART FAILURE

Heart failure is a syndrome that occurs as a result of the progressive inability of the heart to pump enough blood to meet the body's oxygen and nutrient needs. It can cause decreased tissue perfusion, fatigue, fluid volume overload in the intravascular and interstitial spaces, and reduced quality and length of life. Causes of heart failure include coronary artery disease, myocardial infarction, cardiomyopathy, heart valve problems, and hypertension. In the elderly the most common cause of heart failure is cardiac ischemia. It may develop rapidly (acute), as with cardiogenic shock and **pulmonary edema,** or over time (chronic) as a result of another disorder, such as hypertension or pulmonary disease.

The incidence of heart failure is increasing as the elderly population and patient survival rates increase. Heart failure is the most common reason for hospital admission in the elderly. Quality of life is often impaired. The patient may experience many functional limitations and symptoms, and there is a high mortality rate. Readmission rates to hospitals soon after discharge for heart failure treatment are high and

pose a challenge for health care providers. For more information, visit the American Heart Association at www.americanheart.org or the Heart Failure Society of America at www.hfsa.org.

Pathophysiology

The heart is divided into two separate pumping systems. The right side of the heart forms one pump. The left side of the heart forms the other pump. Normally these pumps work together to ensure that equal amounts of blood enter and leave the heart.

Blood flow through the heart begins in the right atrium. (See Chapter 15.) Unoxygenated blood from the body's venous system enters the right atrium from the inferior and superior venae cavae. Next the blood enters the right ventricle to be pumped into the pulmonary artery and into the lungs for oxygenation. After receiving oxygen in the lungs, the blood is returned to the left atrium via the four pulmonary veins. The oxygenated blood then enters the left ventricle and is pumped out into the aorta and the systemic circulation.

Proper cardiac functioning requires each ventricle to pump out equal amounts of blood over time. If the amount of blood returned to the heart becomes more than either ventricle can handle, the heart can no longer be an effective pump. Conditions that cause heart failure may affect one or both of the heart's pumping systems. Therefore heart failure can be classified as right-sided heart failure, left-sided heart failure, or biventricular heart failure. The ventricle is the area of the heart's pumping system that commonly fails. Of the two ventricles, the left ventricle is typically the one to weaken first because it has the greatest workload. The right and left sides of the heart's pumping system work together in a closed system to continuously move blood forward, so failure of one side eventually leads to failure of the other side.

LEARNING TIPS

To visualize and understand the effects of heart failure, trace the flow of blood backward from each ventricle. Along the backward path from the failing ventricle, congestion develops and produces the signs and symptoms seen in heart failure. If you understand the backward path of congestion, you can identify the signs and symptoms specifically associated with right- or left-sided heart failure.

LEARNING TIPS

To understand heart failure, compare it to a dam in a river:
In a river without a dam, the water flows freely; in the normal circulatory system, blood flows freely.

- In a river with a dam, the water is blocked by the dam and builds up behind it; in heart failure, the failing ventricle acts like the dam in the river, causing blood to back up behind it.
- When the dam on the river malfunctions, too much water builds up behind it and the riverbanks flood; in heart failure, if too much blood builds up behind the failing ventricle, the lungs or peripheral tissues are flooded (edema).
- Heart failure can be the result of systolic (contractile) dysfunction, diastolic (relaxation) dysfunction, or a mixed systolic and diastolic dysfunction. Systolic dysfunction is a contractile problem in which the ventricle is unable to generate enough force to pump blood from the ventricle. Diastolic dysfunction is a problem with the ventricle's ability to relax and fill. Mixed systolic and diastolic dysfunction is a combination of the two defects.

Left-Sided Heart Failure

A certain amount of force must be generated by the left ventricle during a contraction to eject blood into the aorta through the aortic valve. This force is referred to as **afterload.** The pressure within the aorta and arteries influences the force needed to open the aortic valve to pump blood into the aorta. This pressure is called **peripheral vascular resistance** (PVR).

Hypertension is one of the major causes of left-sided heart failure because it increases the pressure within arteries. Increased pressure in the aorta makes the left ventricle work harder to pump blood into the aorta. Over time the

strain caused by the increased workload causes the left ventricle to weaken and fail. Other conditions that can lead to left-sided heart failure are described in Table 21-1. Among these conditions are disorders that (1) restrict the outflow of blood from the left ventricle, as in aortic valve stenosis or coarctation of the aorta, which is a malformation causing narrowing; (2) impair contractility of the heart, as in myocardial infarction or cardiomyopathy; and (3) allow blood to flow backward into the left atrium, as in valvular disorders.

With left-sided heart failure, blood backs up from the left ventricle into the left atrium and then into the four pulmonary veins and lungs (Fig. 21-1). This increases pulmonary pressure, causing movement of fluid first into the interstitium and then the alveoli. Alveolar edema is more serious because it reduces gas exchange across the alveolar capillary membrane. Shortness of breath and **cyanosis** may

TABLE 21-1	CAUSES OF LEFT-SIDED HEART FAILURE	
Cause	**Primary Effect on Left Ventricular Workload**
Hypertension	Resistance increased from elevated pressure
Coarctation of the aorta	Resistance increased from elevated pressure
Myocardial infarction	Increased workload from poor contractility
Cardiomyopathy	Increased workload from poor contractility
Aortic stenosis	Increased volume to pump
Mitral regurgitation	Increased volume to pump

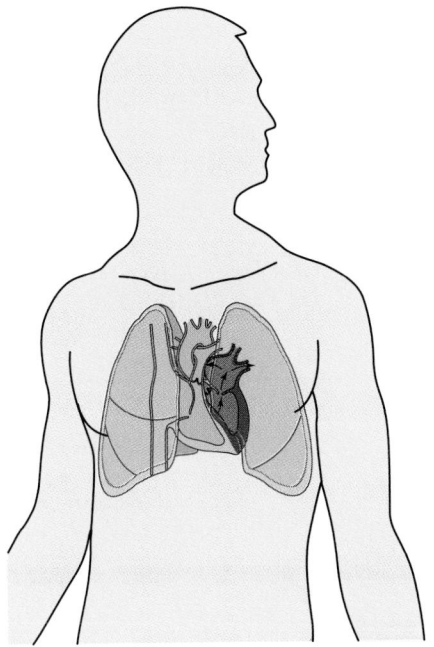

Figure 21-1 Left-sided heart failure. Shaded areas indicate areas of congestion from blood backup caused by the failing left side of the heart.

result from the decreased oxygenation of the blood leaving the lungs. If the fluid buildup is severe, pulmonary edema occurs, which requires immediate medical treatment.

Right-Sided Heart Failure

Causes of right-sided heart failure are described in Table 21-2. The major cause of right-sided heart failure is left-sided heart failure. When the left side fails, fluid backs up into the lungs and pulmonary pressure is increased. The right ventricle must continually pump blood against this increased fluid and pressure in the pulmonary artery and lungs. Over time this additional strain eventually causes it to fail.

Conditions causing right-sided heart failure increase the work of the right ventricle. They increase the amount of contractile force needed or they require pumping of excess blood volume **(preload).** Among these conditions are disorders that (1) increase pulmonary pressures, such as emphysema or congenital heart defects; (2) restrict the outflow of blood from the right ventricle, as in pulmonary valve stenosis; and (3) allow left atrial blood to flow into the right atrium, thereby increasing blood volume in the right ventricle, as in septal defects. When the right ventricle hypertrophies or fails because of increased pulmonary pressures, it is referred to as **cor pulmonale.**

When the right ventricle fails, it does not empty normally and there is a backward buildup of blood in the systemic blood vessels. As the blood backs up from the right ventricle, right atrial and systemic venous blood volume increases. The jugular neck veins, which are not normally visible, become distended and can be seen when the person is in a 45-degree upright position. Edema may occur in the peripheral tissues, and the abdominal organs can become engorged (Fig. 21-2). Congestion in the gastrointestinal tract causes anorexia, nausea, and abdominal pain. As the failure progresses, blood pools in the hepatic veins and the liver becomes congested, known as **hepatomegaly.** Pain in the right upper quadrant and impaired liver function are caused by this liver congestion. Systemic venous congestion also leads to engorgement of the spleen, known as **splenomegaly.**

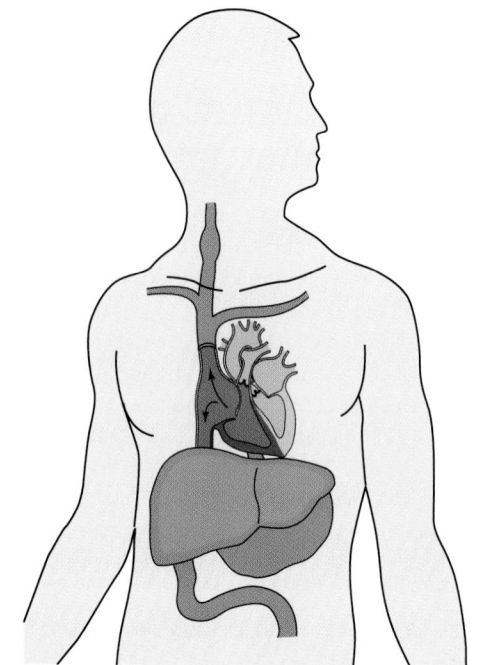

Figure 21-2 Right-sided heart failure. Shaded areas indicate areas of congestion from blood backup due to the failing right side of the heart.

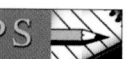

LEARNING TIPS

To understand the signs and symptoms of left-sided versus right-sided heart failure, remember that left-sided signs and symptoms are found in the lungs. *Left* begins with L, as does *Lung*:
Left = Lungs, L = L
Any signs and symptoms not related to the lungs (L) are caused by right-sided failure.

▶ COMPENSATORY MECHANISMS TO MAINTAIN CARDIAC OUTPUT

Compensatory mechanisms help ensure that an adequate amount of blood is being pumped out of the heart. Although these mechanisms are designed to maintain cardiac output, they can also contribute to heart failure and create a cycle that instead of being helpful leads to further heart failure.

When the sympathetic nervous system detects low cardiac output, it speeds up the heart rate by releasing epinephrine and norepinephrine. Although this raises cardiac output (cardiac output = heart rate × stroke volume), the increased heart rate also increases the oxygen needs of the heart. In response to low renal blood flow, the kidneys activate the renin-angiotensin-aldosterone system, and antidiuretic hormone is released from the pituitary gland to conserve water, causing decreased urine output. This adds to the fluid retention problem already found in heart failure.

Over time the heart responds to the increased workload by enlarging its chambers (dilation) and increasing its muscle mass (hypertrophy). In dilation the heart muscle fibers stretch to increase the force of myocardial contractions,

TABLE 21-2	CAUSES OF RIGHT-SIDED HEART FAILURE
Cause	**Primary Right Ventricular Workload Effect**
Pulmonary hypertension	Resistance increased from elevated pressure
Cor pulmonale	Resistance increased from elevated pressure
Pulmonary stenosis	Increased volume to pump
Atrial septal defect	Increased volume to pump

cor pulmonale: cor—heart + pulm—lung
hepatomegaly: hep—liver + mega—large
splenomegaly: splen—spleen + mega—large

which is known as the Frank-Starling phenomenon. In hypertrophy the muscle mass of the heart increases, creating more contractile force. Both these compensatory mechanisms also increase the heart's oxygen needs.

▶ PULMONARY EDEMA

Pulmonary edema, also known as acute heart failure, is severe fluid congestion in the alveoli of the lungs and is life threatening. Pulmonary edema occurs in an acute event such as a myocardial infarction (MI) or when the heart is severely stressed, causing the left ventricle to fail. Complications of pulmonary edema include dysrhythmias and cardiac arrest.

Pathophysiology

First, pressure rises in the lung's venous blood vessels and blood builds up. As pressures continue to rise, fluid moves into the interstitial spaces. Then, with continued pressure increases, fluid containing red blood cells moves into the alveoli. Finally, the alveoli and airways become filled with fluid, reducing gas exchange and oxygen levels.

Signs and Symptoms

Signs and symptoms of pulmonary edema are listed in Table 21-3. Pink, frothy sputum is a classic symptom of pulmonary edema caused by the increased lung congestion and pres-

TABLE 21-3	ACUTE HEART FAILURE SUMMARY
Signs and symptoms	Rapid respirations with accessory muscle use
	Severe dyspnea, orthopnea
	Crackles and wheezes
	Coughing
	Pink, frothy sputum
	Anxiety, restlessness
	Pale skin and mucous membranes
	Clammy, cold skin
Diagnostic tests	Chest x-ray examination
	Arterial blood gases
	Electrocardiogram
	Hemodynamic monitoring
Medical treatment	Oxygen via cannula, mask, or mechanical ventilation
	Positioning in high or semi-Fowler's position
	Bedrest
	Drug therapy
	Morphine IV
	Diuretics IV
	Inotropic agents IV
	Vasodilators IV
	Human B-type natriuretic peptide
	Frequent vital signs, urinary output
	Pulmonary pressures
	Daily weights
	Treatment of underlying cause
Nursing diagnoses	Impaired gas exchange
	Decreased cardiac output
	Excess fluid volume
	Acute pain
	Anxiety

IV = intravenous.

sures that allow leaking of fluid into the alveoli. Compensatory mechanisms increase the heart rate and blood pressure; however, as pulmonary edema worsens, the blood pressure may fall.

Diagnosis

Diagnostic studies are listed in Table 21-3. The congested pulmonary system can be seen on x-ray examination. Arterial blood gases (ABGs) will show a decrease in PaO_2 that continues as the edema worsens and an increase in $PaCO_2$, causing respiratory acidosis (pH < 7.35). The pulmonary artery catheter will show elevated pulmonary pressures and a decreased cardiac output.

Medical Management

Immediate treatment is necessary to prevent patients from drowning in their own secretions. (See Table 21-3.) The goal of therapy is to reduce the workload of the left ventricle in order to improve cardiac output and reduce the patient's anxiety. Care for the patient is usually provided in an intensive care unit. Treatment for the underlying cause occurs at the same time that the patient is being treated for the pulmonary edema.

Treatment includes positioning the patient upright to make breathing easier. In Fowler's position, the lungs can more easily expand. Ask the patient what position is preferred. Oxygen is given usually by mask to provide higher amounts. In severe cases of pulmonary edema, endotracheal intubation and mechanical ventilation may be necessary. Medications are given intravenously to reduce anxiety, relax airways, and increase peripheral blood pooling to decrease preload (morphine); reduce fluid congestion; reduce preload; strengthen heart contractions; reduce arterial pressure and sodium and water retention to relieve dyspnea (nesiritide [Natrecor]). Nesiritide is part of a newer drug class called human B-type natriuretic peptide (hBNP). hBNP is normally secreted by the ventricular myocardium in response to heart failure.

▶ CHRONIC HEART FAILURE

Signs and Symptoms

The signs and symptoms of chronic heart failure are influenced by the patient's age, the underlying cause and severity of the heart disease, and the ventricle that is failing. Chronic heart failure is a progressive disorder, so signs and symptoms may worsen over time. Signs and symptoms caused by a specific failing ventricle are listed in Table 21-4.

Fatigue and Weakness

Fatigue and weakness are the earliest symptoms of heart failure. They occur from the decreased amount of oxygen reaching the tissues. Throughout the day the fatigue worsens, especially with activity.

Dyspnea

A failing left ventricle produces prominent respiratory effects. Dyspnea is a common symptom of left-sided heart

TABLE 21–4 CHRONIC HEART FAILURE SUMMARY

Signs and Symptoms

Right-sided Heart Failure	Left-sided Heart Failure
Jugular vein distension	Dyspnea
Dependent peripheral edema	Dry cough
Ascites	Crackles, wheezing
Weight gain	Orthopnea
Splenomegaly	Paroxysmal nocturnal dyspnea
Hepatomegaly	Cheyne-Stokes respirations
GI discomfort	Cyanosis
Fatigue, weakness	Fatigue, weakness
Tachycardia	Tachycardia
Nocturia	Nocturia

Diagnostic Tests	Complications
History and physical examination	Hepatomegaly
Electrocardiogram	Splenomegaly
Chest x-ray examination	Pleural effusion
Exercise stress test	Left ventricular thrombus and emboli
Nuclear imaging studies	Cardiogenic shock
Echocardiography	
Coronary angiography	
Cardiac catheterization	
Serum laboratory tests: ABGs, electrolytes, liver enzymes, BUN, creatinine	
Hemodynamic monitoring	

Medical Treatment

Noninvasive	Invasive
Identification and treatment of underlying cause	Synchronizing pacemaker
Oxygen by cannula or mask	Mechanical assistive devices
Drug therapy:	Intra-aortic balloon pump
ACE Inhibitors	Left ventricular assist device
Diuretic	Total artificial heart
Beta blockers	Surgery
Inotropic agents	Cardiomyoplasty
Aldosterone antagonist	Cardiac transplant
Anticoagulants	
Antidysrhythmic agents	
Individualized activity plan	
Dietary sodium restriction	
Fluid restriction	
Daily weights	

Nursing Diagnoses

Impaired gas exchange
Decreased cardiac output
Excess fluid volume
Activity intolerance
Disturbed sleep pattern
Powerlessness

ACE, angiotensin-converting enzyme; BUN = blood urea nitrogen.

---

failure. It occurs from the pulmonary congestion that impairs gas exchange between the alveoli and capillaries. Dyspnea stimulates compensatory mechanisms that produce short, rapid respirations. Dyspnea is classified in several ways:

■ Exertional dyspnea is shortness of breath that increases with activity.
■ Orthopnea is dyspnea that increases when lying flat. In an upright position, gravity holds fluid in the lower extremities.

In a supine position, gravitational forces are removed, allowing fluid to move from the legs to the heart, which overwhelms the already congested pulmonary system. When orthopnea is present, two or more pillows are often used for sleeping, and the documentation should state the number of pillows used. For example, use of three pillows would be three-pillow orthopnea.

■ Paroxysmal nocturnal dyspnea (PND) is sudden shortness of breath that occurs after lying flat for a time. PND results from excess fluid accumulation in the lungs. The sleeping person awakens with feelings of suffocation and anxiety. Relief is obtained by sitting upright for a short

---

orthopnea: orth—straight + pnea—to breathe

time, which reduces the amount of fluid returning to the heart.

Cough

A chronic, dry cough is common in heart failure. The coughing increases when lying down from increased irritation of the lung mucosa. This irritation is due to the increase in pulmonary congestion that occurs when gravity no longer keeps fluid in the legs and more fluid returns to the heart and lungs.

Crackles and Wheezes

Pulmonary congestion causes abnormal breath sounds such as crackles and wheezes. These sounds indicate the presence of increased fluid in the lungs. Crackles are produced from fluid buildup in the alveoli resulting from increased pressure in the pulmonary capillaries. Wheezes occur from bronchiolar constriction caused by the increased fluid.

LEARNING TIPS

To simulate the sound of crackles, open a piece of Velcro or rub hair together next to your ear. These sounds are similar to the sound of crackles heard with a stethoscope.

Tachycardia

The sympathetic nervous system compensates for the decreased cardiac output in heart failure by releasing epinephrine and norepinephrine to increase the heart rate. Normally this is helpful because the increased heart rate increases the amount of blood ejected by the heart to maintain an adequate cardiac output. However, whenever the heart works faster it also requires more oxygen, which a failing heart finds it difficult to supply.

Chest Pain

Chest pain may occur from ischemia in the patient with heart failure. Decreased cardiac output results in decreased oxygen delivery to the heart itself, via the coronary arteries. Compensatory mechanisms designed to maintain cardiac output increase the workload and oxygen needs of the heart and are counterproductive in heart failure. Tachycardia increases the oxygen needs of the heart. The kidneys compensate by retaining sodium and fluid, which increases the fluid volume returning to the heart (preload) and therefore the heart's workload and oxygen needs. Pain also increases oxygen requirements, adding further to the cycle of heart failure.

Cheyne-Stokes Respiration

A breathing pattern of shallow respirations building to deep breaths followed by a period of apnea characterizes Cheyne-Stokes breathing. The apneic period occurs because the deep breathing causes carbon dioxide levels to drop to a level that does not stimulate the respiratory center. This ap-

nea may last up to 30 seconds and is then followed by the shallow-to-deeper respiratory pattern of Cheyne-Stokes as carbon dioxide levels rise again.

Edema

Edema occurs in heart failure as a result of (1) systemic blood vessel congestion and (2) sympathetic compensatory mechanisms that cause the kidneys to activate the renin-angiotensin-aldosterone system, in which antidiuretic hormone is released from the pituitary gland, causing sodium and water to be retained. Systemic or pulmonary edema can occur in heart failure. The effect of backward buildup of pressure in the systemic blood vessels is seen with distention of the jugular veins, swelling of the legs and feet, sacral edema in the individual on bedrest, and increased fluid within the abdominal cavity and organs. An acute buildup of fluid in the lungs produces pulmonary edema.

Nocturia

Nocturia is an increase in urine output at night. After lying down, fluid in the lower legs returns to the circulatory system. Renal blood flow and filtration is increased, resulting in greater urine production and the need to urinate frequently during the night. Nocturia may occur up to six times per night, contributing to the patient's fatigue from lack of sleep.

LEARNING TIPS

Patients often have to get up to void shortly after going to bed. This is due to fluid in the legs returning to the heart and then the kidneys for filtering after a person lays down. To help patients get as much undisturbed rest as possible, teach them to recline with legs at or above heart level for at least 30 minutes before going to bed. Then they can void before going to bed, instead of soon after going to bed.

Cyanosis

The skin, nailbeds, or mucous membranes may appear blue, or cyanotic, from decreased oxygenation of the blood. Cyanosis is a late sign of heart failure. It is associated primarily with left-sided heart failure.

Altered Mental Status

Less cardiac output decreases the amount of oxygen delivered to the brain. As a result, restlessness, insomnia, confusion, and impaired memory may occur. A decrease in level of consciousness may occur.

Malnutrition

Several factors contribute to malnutrition in the person with chronic heart failure. Altered mental status, dyspnea, and fatigue interfere with the ability to eat. Anorexia and gastrointestinal (GI) upset occur from pressure exerted by excess fluid surrounding the GI structures. Absorption of food may also be impaired by this pressure.

CRITICAL THINKING

Mr. Shepard (1)

Mr. Shepard, 66, has a family history of cardiac disease. He has been hypertensive for 10 years and takes captopril daily. His baseline vital signs are blood pressure 122/78, pulse 80, respiration 18, height 66 in, and weight 170 lb. During a visit to his physician, he states that he has been short of breath during his daily 2-mile walk and has been using two pillows at night for sleep. As he talks, the physician notes that he has an intermittent dry cough. His physical examination shows blood pressure 140/86, pulse 106, respiration 24, weight 178 lb, and bilateral crackles in the lung bases.

1. What signs and symptoms of heart failure does Mr. Shepard have?
2. Do the signs and symptoms reflect right- or left-sided heart failure?
3. Why are each of the signs and symptoms occurring?
4. Why is Mr. Shepard using two pillows for sleeping?

Answers at end of chapter.

Complications of Heart Failure

Complications of heart failure are listed in Table 21-4. The liver and spleen enlarge from the fluid congestion, which causes impaired function, cellular death, and scarring. Pleural effusion, a leakage of fluid from the capillaries of the lung into the pleural space, can occur. The elevated pressures in the capillaries of the lung cause this leakage. Thrombosis and emboli can occur as a result of poor emptying of the ventricles, which leads to stasis of blood. Aspirin or anticoagulants are often prescribed to prevent thrombus formation in patients with heart failure. Cardiogenic shock, often caused by a myocardial infarction that damages the left ventricle, occurs when the left ventricle is unable to supply the tissues with enough oxygen and nutrients to meet their needs. Cardiogenic shock is a life-threatening condition that requires immediate treatment. (See Chapter 8.)

Diagnostic Tests

Diagnostic tests are done to identify the cause of the failure and determine the degree of failure present (see Table 21-4):

- A chest x-ray examination shows the size, shape, and enlargement of the heart and congestion in the pulmonary vessels.
- Cardiac dysrhythmias that precipitate and contribute to heart failure can be diagnosed with electrocardiogram (ECG; see Chapter 20). Chamber enlargement in the atrium from heart failure is shown by P-wave changes and in the left ventricle by increased voltage and deeper S waves in some V leads.
- Exercise stress testing and nuclear imaging studies provide information on activity tolerance, which is usually limited in heart failure.
- Echocardiography measures the size of the heart chambers to detect enlargement and assess valvular function and motion of the ventricles.

- Cardiac catheterization and angiography are used to detect underlying heart disease that may be the cause of heart failure.
- Direct assessment of the heart's pressures is done with hemodynamic monitoring. A catheter is inserted into the heart and pulmonary artery to transmit pressures to a cardiac monitor. These cardiac and pulmonary pressures are then used to guide medical therapy.
- Serum laboratory tests may show elevated serum blood urea nitrogen (BUN) and elevated serum creatinine from renal failure and elevated liver enzymes from liver damage.

CRITICAL THINKING

Mr. Shepard (2)

Mr. Shepard's chest x-ray examination shows an enlarged heart (cardiomegaly).

1. Why is Mr. Shepard's heart enlarged?
2. What is the significance of an enlarged heart?

Answers at end of chapter.

Medical Management

The overall goal of medical treatment for chronic heart failure is to improve the heart's pumping ability and decrease the heart's oxygen demands. Treatment of heart failure focuses on (1) identifying and correcting the underlying cause, (2) increasing the strength of the heart's contraction, (3) maintaining optimum water and sodium balance, and (4) decreasing the heart's workload (Fig. 21-3). Heart failure management requires a team approach that may involve physicians, case managers, nurses, dietitians, physical therapists, occupational therapists, pharmacists, social workers, and clergy. Heart failure critical pathways (treatment guidelines), as well as heart failure clinics, are being used to ensure quality-based outcomes while reducing treatment costs.

The severity of heart failure determines the individualized therapy selected. Noninvasive approaches are usually tried first. (See Table 21-4.) If noninvasive treatment is not effective, invasive approaches may be used.

Oxygen Therapy

One of the major problems caused by heart failure is a reduction in oxygen delivered to the tissues. The heart failure signs and symptoms of this are fatigue, dyspnea, altered mental status, and cyanosis. Oxygen therapy assists in supplying the oxygen needs of the tissues. In mild heart failure, oxygen may be delivered via nasal cannula. For more severe cases, arterial blood gas values guide oxygen delivery, either via masks that provide high concentrations of oxygen or with mechanical ventilation.

Activity

Activity tolerance depends on the severity of heart failure signs and symptoms. Severe symptoms may require bedrest

Balancing act

Figure 21-3 Management of heart failure is a balancing act. (Modified from Stanley, M, and Beare, P: Gerontological Nursing. FA Davis, Philadelphia, 1995, p 198.)

with restricted activity until treatment reduces the symptoms. For stable heart failure, a regular exercise program can be helpful in improving cardiac function and reducing heart failure effects. Patients should be encouraged to stay as active as possible within the parameters the physician has pre-

scribed. An individualized walking program that increases activity over time is often prescribed. Patients should be taught how to exercise safely without causing symptoms and to understand that overexertion can produce fatigue the next day. Referral to a cardiac rehabilitation program can be helpful.

Nutrition

Dietary sodium is often restricted to decrease fluid retention. Restricting sodium is challenging. (See Nutrition Notes Box 5-4.) Compliance is often low because low-sodium foods are not very appealing. A referral to a dietitian is important. Diet counseling helps the patient and family understand the need for dietary compliance and the need to provide menus that are appealing and easy to use. Salt substitutes often use potassium in place of sodium, so the patient and physician should discuss their use. Spices, herbs, and lemon juice may be suggested to flavor unsalted foods. Eating should remain pleasurable for the patient to avoid malnutrition. Focus on showing patients the foods that they like and can still have rather than talking about only the foods they cannot have.

LEARNING TIPS

In severe heart failure with abdominal discomfort present, malnutrition is a concern. The patient can be anorexic, but the weight gain that occurs with fluid retention can mask the weight loss occurring from the anorexia. Monitor food intake to ensure weight gain from fluid retention does not allow malnutrition to be undetected.

Drug Therapy

The major classifications of oral drugs used to treat heart failure are listed in Table 21-5. Potassium supplements,

TABLE 21-5 ORAL MEDICATIONS FOR HEART FAILURE			
Class	**Drugs**	**Side Effects**	**Nursing Implications**
ACE inhibitor plus a diuretic is first-line therapy to decrease afterload. Decreases cardiac hypertrophy.	captopril (Capoten) enalapril (Vasotec) fosinopril (Monopril) lisinopril (Prinivil, Zestril) moexipril (Univasc) perindopril (Aceon) quinapril (Accupril) ramipril (Altace) trandolapril (Mavik)	Cough, hypotension, altered taste, rash, proteinuria, hyperkalemia, angioedema (dyspnea, facial swelling), neutropenia (with captopril)	Check blood pressure before giving. Monitor WBC with captopril. Teach patient to change positions slowly to reduce orthostatic hypotension. Take captopril and moexipril on an empty stomach (1 hour before or 2 hours after meals). Teach to report development of cough, so drug can be changed. Teach to check blood pressure weekly and report changes to physician. Teach to report rash, sore throat/mouth, fever, swelling of hands/feet/face/tongue, difficulty breathing/ swallowing, chest pain or irregular heartbeat.
Angiotensin II receptor inhibitor decreases cardiac hypertrophy and apoptosis.	losartan (Cozaar) valsartan (Diovan)	Headache, dizziness, angioedema (dyspnea, facial swelling)	Assess blood pressure and pulse before giving. Correct dehydration before initiating therapy. Teach to change positions slowly to reduce orthostatic hypotension.

ACE = angiotensin-converting enzyme; WBC = white blood cell.

TABLE 21-5 Oral Medications for Heart Failure—cont'd

Class	Drugs	Side Effects	Nursing Implications
May be used if ACE inhibitor not tolerated. Survival effects currently being studied.			Teach to report rash, sore throat/mouth, fever, swelling of hands/feet/face/tongue, difficulty breathing/ swallowing, chest pain or irregular heartbeat.
Diuretics			
Loop diuretics (potassium wasting) decrease fluid overload.	bumetanide (Bumex) furosemide (Lasix) torsemide (Demadex)	Hypokalemia, hypochloremia, hypomagnesemia, hyponatremia, dehydration, hypotension	Do not give to sulfa-allergic patients. Assess blood pressure and pulse before giving. Monitor electrolyte levels (especially potassium level and for those on digitalis), and fluid status (daily weight, intake, output, thirst, dry mouth, weakness, oliguria) throughout therapy. Administer in A.M. to avoid nocturia.
Thiazide diuretics (potassium wasting) decrease fluid overload.	chlorothiazide (Diuril) chlorthalidone (Hygroton, Thalitone) hydrochlorothiazide (HydroDiuril, HCTZ, Microzide)	Hypokalemia, dizziness, hypotension, photosensitivity	Assess blood pressure, intake, output, daily weight, and edema. Do not give to sulfa-allergic patients. Administer in A.M. to avoid nocturia. Monitor potassium level throughout therapy. Teach to report signs of hypokalemia: weakness, fatigue, arrhythmias. Teach about potassium supplements and, if on digitalis, about increased risk of toxicity with hypokalemia. Teach to monitor weight daily and report changes. Teach use of sunscreen to prevent photosensitivity reaction.
Beta blockers reduce sympathetic nervous system input, which decreases heart rate and improves cardiac output. Also reduce cardiac remodeling.	bisoprolol (Zebeta) carvedilol (Coreg), specific for heart failure, contains beta blocker plus antioxidant, effective in slowing progression of heart failure metoprolol tartrate (Toprol XL, Lopressor)	Dizziness, hypotension, hyperglycemia, fluid retention, diarrhea, bradycardia, heart blocks, impotence, bronchospasm (carvedilol)	Avoid carvedilol in patients with asthma (causes bronchospasm). Check apical heart rate and blood pressure. If heart rate is below 50 or blood pressure below 100 systolic, contact physician before giving. Teach to take pulse daily and blood pressure biweekly and to contact physician before taking drug if pulse rate is below 60. Teach to change positions slowly to reduce orthostatic hypotension. Diabetic patients should closely monitor blood sugar. Teach to monitor for worsening of heart failure symptoms.
Aldosterone antagonist blocks effects of aldosterone and has positive effect on sodium and potassium balance.	spironolactone (Aldactone)	Hyperkalemia, nausea, vomiting, anorexia, diarrhea, headache, clumsiness, gynecomastia	Do not give if hyperkalemic. Administer in A.M. to avoid nocturia. Monitor potassium level and fluid status throughout therapy. Teach signs of hyperkalemia: weakness, fatigue, confusion, dyspnea, arrhythmias, confusion. Elderly patients are at higher risk for electrolyte imbalances.
Inotropes—cardiac glycoside (positive inotrope and negative chronotrope) increase the force and contraction of the myocardium, which increases cardiac output. Slows heart rate to reduce workload of heart and controls atrial fibrillation if present.	digoxin (Lanoxicaps, Lanoxin)	Fatigue, nausea, vomiting, anorexia, headache, bradycardia, cardiac arrhythmias. Toxicity: abdominal pain, anorexia, nausea, vomiting, visual changes (blurred, yellow-green halos, photophobia, diplopia) bradycardia, dysrhythmias	Take apical pulse for 1 minute; if below 60, contact physician before giving. Instruct patient to take medication exactly as directed, at the same time each day. Teach patient to take pulse and to contact health care professional before taking medication if pulse rate is below 60 or above 100. Therapeutic digoxin levels: 0.5 to 2 mg/mL. Elderly are more susceptible to toxicity. Teach signs and symptoms of digitalis toxicity to patient and family. Periodically monitor drug level and electrolytes (hypokalemia, hypomagnesemia, hypercalcemia make patients more susceptible to toxicity). Antidote for toxicity is digoxin immune FAB (Digibind).

TABLE 21-5 ORAL MEDICATIONS FOR HEART FAILURE—CONT'D			
Class	**Drugs**	**Side Effects**	**Nursing Implications**
Vasodilators decrease afterload, which increases cardiac output and reduces cardiac workload. Used for patients who cannot take ACE inhibitors.	isosorbide dinitrate (ISDN, Isorbid, Isordil, Isonate) hydralazine (Apresoline)	Headache, dizziness, hypotension, tachycardia	Assess blood pressure and pulse before giving. Teach to change positions slowly to reduce orthostatic hypotension. Isosorbide dinitrate: Administer 1 hour before or 2 hours after meals for faster absorption. Chewable tablets should be chewed well before swallowing and held in the mouth for 2 minutes. Extended-release tablets and capsules should be swallowed whole. Do not chew, crush, or break. Sublingual tablets should be held under tongue until dissolved; no eating, drinking, or smoking until tablet is dissolved; replace sublingual tab if swallowed. Headache is common initially and is treated with aspirin. Hydralazine: Take with meals to enhance absorption. Teach to avoid sudden changes in position to avoid orthostatic hypotension. Teach to immediately report fatigue, fever, aching, rash, chest pain, sore throat, numbness, tingling, or weakness of hands/feet.

anticoagulants, and antidysrhythmics may be given on an individualized basis.

ANGIOTENSIN-CONVERTING ENZYME INHIBITORS. Angiotensin-converting enzyme (ACE) inhibitors are now considered a standard in therapy for heart failure. ACE inhibitors are used for vasodilation. It is now known that ACE inhibitors also offer additional benefit by preventing some structural cardiac changes that occur in heart failure. These remodeling changes lead to progressive cardiac deterioration, so inhibiting these changes is of great benefit in treating the condition.

LEARNING TIPS

To help you identify ACE inhibitors, remember that their generic names end with *-pril*.

DIURETICS. Diuretics are used to reduce fluid volume and decrease pulmonary venous pressure, so the heart does not have to work so hard. Diuretics act on various areas of the kidneys to promote the excretion of edema fluid. A combination of diuretics may be used to achieve the desired effect. Electrolytes (especially potassium levels, to prevent hypokalemia) and fluid balance (to prevent dehydration) should be carefully monitored during therapy. Potassium supplements are often given with potassium-wasting diuretics.

LEARNING TIPS

Always check potassium levels before giving a potassium-wasting diuretic such as furosemide (Lasix) or bumetanide (Bumex) or before giving a potassium supplement, which the patient may be taking due to diuretic therapy. Do not give diuretic if potassium levels are low or potassium supplement if potassium is high.

BETA BLOCKERS. Beta-blocker therapy is a newer recommendation for some heart failure patients along with standard therapy. Beta blockers are started at low doses and gradually increased. Improved cardiac output and activity tolerance are benefits with this therapy.

LEARNING TIPS

To help you identify beta blockers, remember that their generic names end with *-ol*.

INOTROPIC AGENTS. Inotropic drugs strengthen ventricular contraction to increase cardiac output. Inotropic agents include the cardiac glycosides (digitalis, digoxin), sympathomimetics (dopamine, dobutamine), and phosphodiesterase inhibitors (amrinone, milrinone). The sympathomimetics and phosphodiesterase inhibitors are usually used short term.

Digitalis. In addition to improving contraction strength, digitalis preparations decrease conduction time within the heart, which slows the heart rate to allow more complete emptying of the ventricles. Digitalis may increase myocardial oxygen needs, so it is being used more cautiously. Monitoring of serum drug levels is necessary to detect toxic levels of the drug. If toxic levels are present, the drug is stopped to allow digitalis levels to decrease over time. If life-threatening dysrhythmias occur, a digoxin antibody, digoxin immune FAB (Digibind), can be administered.

ALDOSTERONE ANTAGONIST. An aldosterone antagonist, spironolactone (Aldactone), blocks the effects of aldosterone and has a positive effect on sodium and potassium balance. The use of this drug in combination with other drugs (ACE inhibitor and loop diuretic) has been shown to reduce morbidity and mortality for heart failure. Potassium must be monitored carefully as spironolactone is a potassium-sparing agent.

Pacemaker for Congestive Heart Failure

The InSync biventricular cardiac pacing system is a new device similar to a pacemaker to reduce symptoms of heart failure. The device paces both ventricles at the same time so that they beat together. In heart failure, one of the problems is that the ventricles do not always beat together. This results in a slight delay between the beats of each ventricle that impairs cardiac function. Resetting the heartbeat reduces symptoms and improves quality of life.

Mechanical Assistive Devices

Mechanical assistive devices can provide temporary support to patients in cardiogenic shock or awaiting cardiac transplant. These devices increase the cardiac output of the patient. Research is ongoing into the development of mechanical pumps that may become heart replacements. These mechanical assistive devices are used primarily in critical care settings.

INTRA-AORTIC BALLOON PUMP. The intra-aortic balloon pump (IABP) increases circulation to the coronary arteries. More oxygen is then available to the myocardium, which reduces the workload of the heart. The IABP is inserted into the femoral artery and positioned in the descending aortic arch (Fig. 21-4). The IABP is attached to a machine that senses ventricular contraction. During diastole the balloon inflates, forcing increased blood into the coronary arteries. During systole the balloon deflates to allow blood to flow past it.

VENTRICULAR ASSIST DEVICES. A ventricular assist device (VAD) temporarily supports cardiac output and allows the failing ventricle to rest. There are several types of VADs (Fig. 21-5). VADs can pump blood directly from either the right atrium to the pulmonary artery (right ventricular failure) or the left atrium to the aorta (left ventricular failure). Two devices can be used for biventricular failure. A newer VAD (Novacor Left Ventricular Assist System) is an electromechanical pump, about the size of a human heart, which is implanted into the abdominal wall. This pump is used for patients awaiting transplant and is being studied for long-term use.

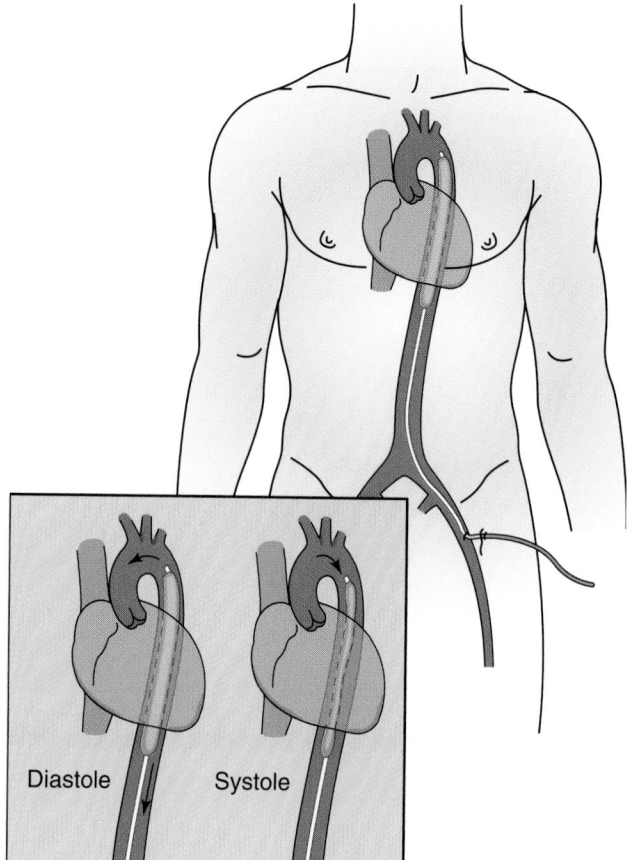

Diastole   Systole

Figure 21-4 Intra-aortic balloon pump.

Surgical Management

CARDIOMYOPLASTY. Cardiomyoplasty is a surgical procedure that uses the patient's own skeletal muscle (latissimus dorsi) to enhance the function of the heart and improve circulation (Fig. 21-6). The muscle is positioned around the aorta or heart, and a cardiomyostimulator and leads are implanted to stimulate muscle contraction. The muscle is rested for 2 weeks after surgery and then programmed to gradually start stimulation. After 4 months, the muscle can be stimulated either at a 1:1 or 1:2 heartbeat ratio.

For patients with end-stage heart failure, cardiac transplantation is an option. (See Chapter 22.) Mechanical assistive devices may be used to support cardiac function while the patient awaits a donor heart.

CRITICAL THINKING

Mr. Shepard (3)
During Mr. Shepard's visit, the physician tells him to continue the ACE inhibitor, the diuretic, and a 2-g sodium diet.

1. Why is the ACE inhibitor continued?
2. Will the ACE inhibitor affect preload or afterload?
3. Why is the diuretic ordered?
4. Why is a 2-g sodium diet ordered?
5. What is the overall goal of the ordered treatment?

Answers at end of chapter.

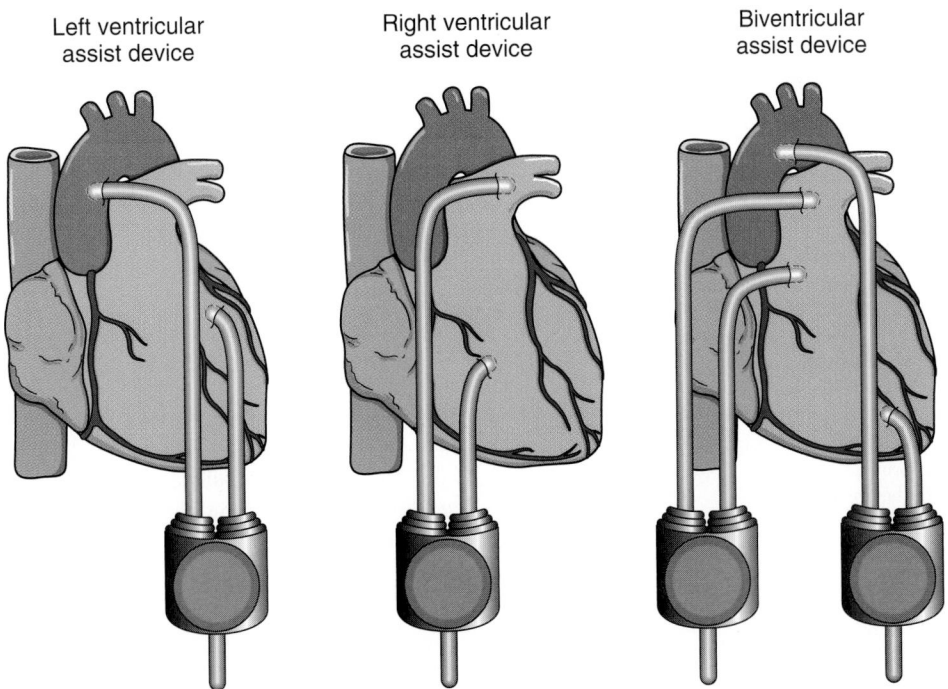

Left ventricular
assist device

Right ventricular
assist device

Biventricular
assist device

Figure 21-5   Different types of ventricular assist devices. (Modified from Ruppert, S, Kernicki, J, and Dolan, J: Dolan's Critical Care Nursing. FA Davis, Philadelphia, 1996, p 336.)

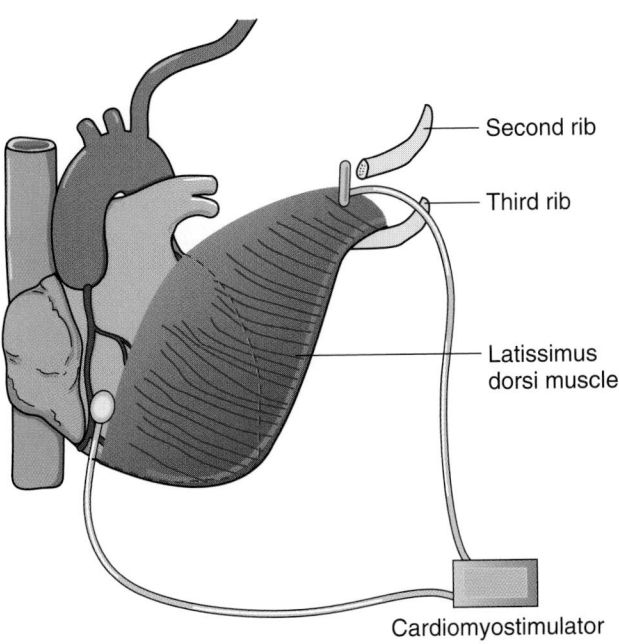

Second rib

Third rib

Latissimus
dorsi muscle

Cardiomyostimulator

Figure 21-6   Cardiomyoplasty. (Modified from Ruppert, S, Kernicki, J, and Dolan, J: Dolan's Critical Care Nursing. FA Davis, Philadelphia, 1996, p 329.)

Nursing Process for Chronic Heart Failure

Nursing Assessment

A nursing history and physical examination are done to gather subjective and objective data. While obtaining data, focus on areas that might indicate the presence of heart failure (Table 21-6).

Planning

The patient and family should be included in planning care. The patient's goals include the following:

■ Maintain desired activities
■ Remain free of edema and dyspnea
■ Awaken refreshed with less fatigue
■ Reduce anxiety
■ Follow a mutually agreed on health maintenance plan

Nursing Diagnosis

See Nursing Care Plan 21-1 for the Patient with Chronic Heart Failure for common nursing diagnoses.

Nursing Interventions

The major focus of nursing care for chronic heart failure patients includes the following:

■ Improving oxygenation
■ Educating patients about their condition and health maintenance needs
■ Promoting self-care with energy conservation techniques
■ Facilitating patient and family coping

Oxygenation can be improved with oxygen therapy and a decrease in the body's need for oxygen. Decreased metabolic oxygen needs can be achieved with rest, positioning, medications, fluid balance, and oxygen consumption control.

OXYGEN. Oxygen therapy is ordered by the physician and guided by blood gas analysis. Before starting oxygen therapy, explain the therapy to the patient. For chronic heart failure, oxygen is administered at 2 to 6 L/min via nasal cannula.

TABLE 21-6	NURSING ASSESSMENT FOR THE PATIENT WITH CHRONIC HEART FAILURE

Subjective Data

History

Respiratory
Lung disease?
How many flights of stairs can be climbed without dyspnea?
How many pillows used for sleeping?
Dyspnea at rest or that awakens from sleeping?

Cardiovascular
Any cardiac disease history: hypertension, myocardial infarction, valvular problem, anemia, dysrhythmias, palpitations
Chest pain—precipitating factors, severity, relieving factors
Can activities of daily living be performed?
Can activities performed 6 months, 4 months, 2 months, 2 weeks ago still be done?
Any dizziness (vertigo) or fainting (syncope)?

Fluid retention
Daily sodium intake?
Weight gain?
Are shoes tight? Do ankles swell?

Gastrointestinal
Is appetite good?
Any nausea, vomiting, or abdominal pain?

Urinary
Decrease in daytime urine output?
How often does patient go to the bathroom at night (nocturia)?

Neurological
Any change in behavior?

Medications
Knowledge of condition
Coping skills

Objective Data

Respiratory
Tachypnea, crackles, wheezing, respiratory effort, dyspnea with exertion

Cardiovascular
Tachycardia, dysrhythmias, jugular vein distention, peripheral edema—degree of pitting

Gastrointestinal
Abdominal distention, ascites, hepatomegaly, splenomegaly

Neurological
Confusion, decreased level of consciousness, restlessness, impaired memory

Integumentary
Cold, clammy skin; pallor; cyanosis

General
Weight

Diagnostic test findings

The effects of the oxygen should be monitored carefully. Oxygen should be used cautiously in all patients, so that their stimulus to breathe is not diminished (especially if the patient has chronic obstructive pulmonary disease [COPD]).

Home Oxygen Therapy. If a patient will be using home oxygen therapy, instructions must be given on the proper use of the oxygen and safety precautions for oxygen use.

The family must also understand oxygen safety precautions and be willing to comply with them. Smoking is prohibited when oxygen is in use.

REST AND ACTIVITY. Reduction of the body's oxygen demands decreases the workload of the heart. A balance of rest and activity that does not produce signs or symptoms of oxygen deprivation is essential. The activity level of the patient is determined by the severity of the heart failure. During times of exertion, monitor the patient's vital signs and respiratory effort for oxygen deprivation. If activity intolerance develops, the activity should be stopped.

POSITIONING. Semi-Fowler's or high Fowler's position makes breathing easier. In upright positions the lungs are able to expand more fully and gravity decreases the amount of fluid returned to the heart, thereby reducing the heart's workload.

FLUID RETENTION. Monitoring daily weights for weight gain is important in detecting fluid retention. Edema usually is not observed until there is 5 to 10 pounds of extra fluid present. A baseline weight should be obtained when heart failure is diagnosed. Daily weights should be measured on the same scale, at the same time of day, and with the same type of clothing worn to ensure accuracy. A good time to obtain a daily weight is in the morning after the bladder is emptied. Documentation of daily weights should include the date and time of the weight, the scale used, the clothing worn, and the weight measurement. A weight journal can be kept by the patient. Tell patients to report weight gains of 2 to 3 pounds over 1 to 2 days.

OXYGEN CONSUMPTION. Increased oxygen consumption by the heart should be avoided. Tachycardia increases the oxygen needs of the heart and should be reported promptly to the physician for treatment. Elderly patients are especially vulnerable to the effects of tachycardia because of their decreased reserves. Constipation should be prevented because straining during defecation, Valsalva's maneuver, increases the heart's workload by increasing venous return to the heart. Stool softeners should be administered, as ordered, to prevent straining.

Patients should be taught methods of saving energy while performing activities of daily living (ADL). Activities should be alternated with periods of rest. Fatigue should be avoided. A referral to occupational therapy and physical therapy can be helpful in developing techniques that allow the patient to conserve energy during self-care. Some suggestions for conserving energy include placing frequently used objects at waist level to avoid reaching overhead, planning bathing activities to include rest periods, and using Velcro fasteners to make dressing easier.

MEDICATIONS. Because heart failure is a progressive, chronic condition, patients may require lifetime medications. Combination drug therapy is often needed. Taking many pills each day can be challenging. Financial resources, compliance, and ongoing monitoring are issues that must be considered.

BOX 21-1 NURSING CARE PLAN FOR THE PATIENT WITH CHRONIC HEART FAILURE

 Activity intolerance related to fatigue caused by oxygen imbalance

PATIENT OUTCOMES
Patient will show increased activity tolerance with vital signs within normal limits (WNL) in response to activity.

EVALUATION OF OUTCOMES
Does the patient participate in activities and maintain vital signs WNL?

INTERVENTIONS	RATIONALE	EVALUATION
Provide rest, space activities, and conserve energy.	Myocardial oxygen need is decreased with rest and energy conservation.	Does patient participate in activity with no pulse rate or ECG changes?
Assist as needed with activities of daily living (ADL).	Conserve energy by assisting with ADL.	Are patient's ADLs met?
Teach use of assistive devices and lifestyle changes.	Assistive devices can overcome limitations to increase activity.	Does patient incorporate assistive devices into lifestyle changes?
GERIATRIC		
Increase time allowed to complete activities.	Independence and participation are increased if extra time is allowed for tasks.	Does patient report greater ability to complete activities with fewer symptoms?

 Excess fluid volume related to heart failure and the secondary reduction in renal blood flow for filtration

PATIENT OUTCOMES
Patient remains free from edema and dyspnea, has clear lung sounds, and maintains baseline weight.

EVALUATION OF OUTCOMES
Does patient have clear lung sounds with baseline weight maintained?

INTERVENTIONS	RATIONALE	EVALUATION
Monitor for edema, weight gain, jugular vein distention (JVD), lung crackles.	Excess fluid is indicated by edema, sudden weight gain, JVD, and crackles in the lungs.	Is edema, weight gain, JVD, or crackles present?
Decrease sodium intake as ordered.	Sodium retains fluid.	Does patient restrict sodium intake?
Administer diuretics or inotropics as ordered.	Diuretics promote fluid excretion. Inotropics increase cardiac contraction strength.	Is output increased and edema or dyspnea reduced?
Monitor intake and output.	Intake and output will show imbalances.	Are intake and output balanced for 24 hours?

 Disturbed sleep pattern related to nocturia and inability to lie down and sleep comfortably

PATIENT OUTCOMES
Patient awakens refreshed and is less fatigued during the day.

EVALUATION OF OUTCOMES
Does patient wake up less frequently during the night and feel more refreshed with less fatigue during the day?

INTERVENTIONS	RATIONALE	EVALUATION
Identify barriers to sleep.	Anxiety, nocturia, diuretics, orthopnea, or paroxysmal nocturnal dyspnea can make sleep difficult.	Does patient identify sleep barriers?
Assist patient in identifying positions of comfort for sleeping.	Use of pillows or a recliner can decrease orthopnea.	Can patient identify a position of comfort?
Teach patient cause of dyspnea at night.	Anxiety about falling asleep and waking up short of breath is reduced.	Can patient explain cause of dyspnea?
Encourage patient to recline for 30 to 60 minutes before bedtime.	Reclining before bedtime redistributes fluid and increases voiding that can occur before going to sleep instead of after going to sleep.	Is patient awakened to void after going to bed less often?
GERIATRIC		
Encourage patient to take diuretics early in the day.	Nocturia is reduced if diuretics are taken earlier in the day.	Does patient take diuretics early and report less nocturia?

Diuretics require monitoring of the patient's potassium levels and blood pressure. To prevent hypokalemia, potassium supplements may be prescribed during diuretic therapy, and a diet with high-potassium foods is encouraged. If too much fluid is removed, the patient may become hypotensive and orthostatic hypotension can develop. The patient may then be dizzy and at risk of falling. Caution the patient to change positions slowly to prevent falls during diuretic therapy.

Inotropic agents such as digitalis strengthen heart contractions and slow the heart rate. Before administration of a digitalis drug, the patient's apical pulse should be counted for 1 minute. If the pulse is below 60 beats per minute, notify the physician to determine if the drug should be given. Some patients are given digitalis even if their heart rates are between 50 and 60 beats per minute, as long as their heart's conduction system is normal or if it is due to other medications such as a beta blocker. When giving digitalis, be aware that hypokalemia increases the heart's sensitivity to digitalis. A patient can become toxic on a normal dose of digitalis when hypokalemia is present. This is important to note because many people on digitalis also take diuretics, which may lower potassium levels. Monitoring for signs and symptoms of digitalis toxicity should be done routinely during patient assessment. Early symptoms of digitalis toxicity are anorexia, nausea, and vomiting. Other signs and symptoms include bradycardia or other dysrhythmias, visual problems, and mental changes. The elderly are especially prone to toxic effects of this drug and may exhibit confusion when levels are toxic.

Medications with vasodilating effects reduce the heart's workload by decreasing vascular pressure. Blood pressure is monitored when administering vasodilators.

Medication Teaching. Patients and their families are taught the purpose, side effects, and precautions for prescribed medications. Teach them to report side effects to the physician. If dizziness occurs from drugs that reduce blood pressure, the drugs can be staggered so that they are not all taken at the same time. Patients taking a digitalis preparation should be taught to take their pulse. They should hold the medication and call their physician if their heart rate is less than 60 beats per minute or below the lower-limit heart rate set by their physician. Patients on diuretics should be taught the following:

1. Take them during the day before 4 P.M. to decrease being awakened at night to void (if desired).
2. Have a readily available and obstacle-free bathroom or commode to prevent incontinence and falls.
3. Eat high-potassium foods if taking a potassium-wasting diuretic.
4. Weigh themselves daily and report weight gains.

Patients should understand the importance of taking their medication as prescribed, even if they do not have symptoms. A schedule should be developed so patients remember to take their medications.

LOW-SODIUM DIET AND WEIGHT CONTROL. A diet assessment should be completed to help establish an appealing diet plan for the patient with heart failure. Factors such as the patient's food likes and dislikes, cultural influences, economic status, and food preparation resources (e.g., shopping ability, storage, refrigeration, and cooking equipment) influences compliance with any restricted diet and therefore should be assessed. It is helpful to include a dietitian in the planning process, as well as the food preparer or a support person, to increase diet compliance.

Excess weight influences heart failure by increasing the heart's workload. For overweight patients, weight reduction may help eliminate the underlying cause of heart failure. Diet counseling and support should be given to the obese patient to encourage weight loss. If anorexia occurs in the later stages of heart failure, the patient's intake should be evaluated. Several small meals rather than three large meals will decrease the heart's workload. If the patient's nutritional needs are not being met, the physician should be informed for a referral to a dietitian.

Patients should be taught which foods are high and low in sodium content. With this knowledge, patients can help design a daily meal plan using low-sodium foods that are appealing to them. Food preparers should be taught not to salt food during cooking, and table salt should be eliminated. Explaining seasoning alternatives such as spices, herbs, and lemon juice is helpful in making food taste better.

EDUCATION. Chronic management of heart failure requires patient and family understanding of the disease process, management of home oxygen therapy, diet and weight control, and medications (Home Health Hints). The patient and family must recognize the importance of each of these factors in order to foster a productive life for the patient with chronic heart failure. A discussion of the pathology of heart failure using simple terms should be included in the teaching plan for the patient. Signs and symptoms that the patient is to report to the physician should also be highlighted (Box 21-2).

COPING. Living with a chronic illness can be frustrating for both patients and their families. An assessment of coping skills used by patients and their families can be used to develop a plan for coping with this current illness. Available support systems are explained to patients. A referral to a social worker can be helpful in providing resources that may make living with heart failure easier. Understanding the chronic na-

---

BOX 21-2   Patient and Family Education

Heart Failure Signs and Symptoms to Report to the Physician

Shortness of breath
Fatigue
Dry cough
Shortness of breath when lying down (orthopnea)
Episodes of sudden awakening with shortness of breath (PND)
Weight gain of 2 to 3 pounds over 1 to 2 days
Ankle or foot edema
Nocturia
Anorexia

ture of heart failure is important for patients, families, and caregivers to positively deal with the emotions and feelings that can result. Referral to a nurse managed heart failure clinic has been shown to decrease hospitalization rates and increase effective management of the therapeutic regimen.

Evaluation

Goals are met if the patient:

■ reports reduced fatigue and the ability to complete tasks and engage in desired activities.
■ has less edema and no dyspnea and maintains a baseline weight.
■ is able to sleep and wake up refreshed without being fatigued during the day.
■ reports anxiety levels are reduced.
■ follows the health maintenance plan without complication development.

CRITICAL THINKING

Mr. Shepard (4)
The nurse meets with Mr. Shepard after the physician orders the ACE inhibitor to be continued, a diuretic, and 2-g sodium diet.

1. What information should the nurse teach Mr. Shepard based on the prescribed treatment?
2. What types of foods should be included in Mr. Shepard's diet?
3. Why does the nurse instruct Mr. Shepard to weigh himself daily?
4. Why does the nurse tell Mr. Shepard to weigh himself at the same time of day, on the same scale, and with the same type of clothing?

Answers at end of chapter.

HOME HEALTH HINTS

- Some patients may not have a scale in their home. It may be necessary to assist them in obtaining one or to leave an agency scale in the home for daily weights.

- The most objective way to document edema is to use a tape measure on the abdominal girth, thigh, calf, and ankle in centimeters. Measure at the same place each visit, such as measuring the girth of the calf at a specified distance above the medial malleolus.

- The sacrum, back, and sides of a bedridden patient should be checked to note edema. These are dependent areas in the bedridden patient, so fluid accumulates in these areas instead of the ankles.

- Blood drawn for potassium levels needs to be transported to the lab within 1 hour. Ice should not be put directly on the tube because this can cause destruction of the cells and a false elevation in the potassium level.

- Blood drawn for digoxin levels should be taken to the lab within 2 to 3 hours.

- Patients on sodium-restricted diets who already have canned vegetables in their home can still use them even though they are not the low-sodium type. They should be instructed to pour off the liquid and rinse the vegetables before heating them for serving. The use of herbs and spices helps make them more flavorful.

- For the patient on a low-sodium diet, an effective diet teaching technique is to have the patient name the foods highest in sodium. Asking the patient to rename the list on each visit helps knowledge retention and compliance.

- If patients have a poor appetite, ask their caregiver if they eat well when eating with others. Anorexia could be a sign of loneliness and depression if they eat well with others instead of an effect from the heart failure.

- Assist patients in taking medications at times that fit their lifestyle. A morning dose of a diuretic may limit what they can do for the next few hours. An afternoon dose might encourage compliance. Lack of compliance is a major factor in the rehospitalization of patients with heart failure.

- A dose of diuretic too late in the day may cause frequent awakenings during the night to void.

- The home health nurse should periodically check the contents of medicine bottles. If pills have been cut in half, ask about this. Often it is an attempt by the patient to "stretch" the medicine to decrease expenses. There are community or drug company programs that help purchase medicines for patients with financial need. Eligible Medicaid patients can apply for medication cards.

- Visual disturbances can occur from digitalis toxicity. If the patient sees halos around lights or red-green tinting on everything, report this to the physician.

- Troublesome side effects of an ACE inhibitor such as captopril (Capoten) are an intractable cough and hypotension. It should be noted how much coughing the patient is doing. The physician sets parameters for the blood pressure and reporting of abnormal findings.

- Oxygen concentrators are widely used in home care. Long tubing allows the patient ease in moving about the home. Patients need to be cautioned about keeping the tubing out of their way and not kinking it. if the patients also has chronic obstructive pulmonary disease (COPD), the oxygen flow rate is generally limited to 2 L/min, so the patient's drive to breathe is not decreased. A note stating this should be placed near the gauge as a reminder to the patients.

- As the home health nurse becomes acquainted with the patient, it is easier to pick up on signs of oxygen deprivation and hypoxia, such as confusion, combativeness, or unusual expressions of anger.

- For patients with orthopnea, a foam wedge can be obtained from a medical equipment company to use under their head when sleeping, instead of pillows.

- As patients with heart failure feel better, they may go back to the old habits that cause an increase in fluid. The home health nurse can help by providing information about the disease and help patients foster their own independence and ways of coping with the condition. Each home health visit is a teaching opportunity that empowers patients with a knowledge base to empower them to take control of their health.

REVIEW QUESTIONS

1. A patient asks the nurse what heart failure is. Which of the following is the nurse's best response?
   a. "In heart failure the heart pumps too much blood into the pulmonary veins."
   b. "In heart failure the heart is unable to pump enough blood for the body's oxygen needs."
   c. "Heart failure is a buildup of blood in the aorta from the heart's left ventricle."
   d. "With a failing heart, the heart stops beating, so blood is not pumped out."

2. Which of the following does the nurse understand is a major cause of left-sided heart failure?
   a. Hypertension
   b. Congenital heart defects
   c. Pulmonary valve stenosis
   d. Septal defects

3. A patient who has been treated for heart failure is being discharged from the hospital. He is taking 20 mg furosemide (Lasix) daily. Which of the following statements by the patient would indicate understanding of instructions for his medications?
   a. "I will take the Lasix in the morning."
   b. "I will take the Lasix in the evening."
   c. "I will drink lots of fluids with the Lasix."
   d. "I will take it with meals."

4. The nurse is caring for a patient receiving bumetanide (Bumex) to reduce preload for heart failure. While assessing the patient the nurse notes the patient has less ankle edema and jugular vein distention than earlier. The next dose of bumetanide is scheduled in 1 hour. Which of the following actions should the nurse take next?
   a. Notify the physician.
   b. Hold the bumetanide.
   c. Give the bumetanide as scheduled.
   d. Give the bumetanide early.

5. Which of the following assessments should the nurse teach the patient to perform to monitor fluid status at home?
   a. Weigh daily.
   b. Weigh weekly.
   c. Weigh biweekly.
   d. Weigh monthly.

Answers to CRITICAL THINKING

Mr. Shepard (1)

1. Signs and symptoms of heart failure include shortness of breath, two-pillow orthopnea, dry cough, tachycardia (pulse 106), tachypnea (respiration 24), and bilateral crackles.
2. Left-sided heart failure is indicated by the findings.
3. Shortness of breath: fluid in the lungs impairs gas exchange; orthopnea: lying flat increases fluid accumulation in the lungs, causing dyspnea; dry cough: fluid in the lungs irritates the mucosal lining of the lungs; tachycardia: sympathetic compensation to increase cardiac output; tachypnea: sympathetic compensation to increase blood oxygenation; bilateral crackles: fluid trapped in the lungs.
4. The two pillows help prevent orthopnea by using a more upright position, which allows gravity to decrease fluid accumulation in the lungs.

Mr. Shepard (2)

1. Mr. Shepard's heart is enlarged to compensate for the strain caused by increased peripheral vascular resistance from hypertension in order to maintain an adequate cardiac output.
2. An enlarged heart requires more oxygen, which often cannot be supplied in heart failure.

Mr. Shepard (3)

1. The ACE inhibitor is needed for vasodilation, which reduces peripheral vascular resistance and decreases the heart's workload, and to prevent cardiac remodeling.
2. The ACE inhibitor will affect afterload.
3. The diuretic is ordered to decrease fluid volume, which reduces preload and decreases the heart's workload.
4. The diet is ordered to reduce water retention, which decreases preload and decreases the heart's workload.
5. The goal is to decrease the heart's workload and increase its efficiency by reducing preload and peripheral vascular resistance.

Mr. Shepard (4)

1. After assessment of Mr. Shepard's knowledge base, medication teaching should be given on the ACE inhibitor and diuretic that includes their purpose, side effects, and precautions. A schedule for taking the medications can be planned. An explanation of the purpose of a low-sodium diet and menu planning based on Mr. Shepard's likes and dislikes should be done.
2. Low-sodium foods should be selected to prevent fluid retention, and high-potassium foods should be included to prevent hypokalemia from the diuretic if appropriate. Low-sodium foods include puffed rice, wheat cereals, fruits, chicken, beef, eggs, and potatoes. High-sodium foods include tomato juice, sauerkraut, softened water, buttermilk, cheese, smoked meats, canned tuna, canned soup, pickles, instant rice, and instant potatoes. High-potassium foods include bran products, avocado, bananas, prunes, oranges, baked potato, sweet potato, spinach (cooked), chocolate, nuts, and molasses.
3. Daily weighing is necessary to detect a rapid weight gain that indicates fluid retention (2 to 3 lb over 2 days) and to measure weight loss resulting from the diuretic.
4. These instructions ensure accuracy of the weight so that comparison to the baseline weight detects a weight gain or loss.

22

Nursing Care of Patients Undergoing Cardiovascular Surgery

Sharon M. Nowak

KEY TERMS

anastomose (uh-NAS-tuh-**MOS**)

annuloplasty (**AN**-yoo-loh-PLAS-tee)

atelectasis (AT-e-**LECK**-tah-sis)

cardioplegia (KAR-dee-oh-**PLEE**-jee-ah)

commissurotomy (KOM-i-shur-**AHT**-oh-mee)

encephalopathy (en-SEFF-uh-**LAHP**-ah-thee)

endarterectomy (end-AR-tur-**ECK**-tuh-mee)

gastroepiploic (GAS-troh-EP-i-**PLOH**-ick)

hemolysis (he-**MAHL**-e-sis)

hypothermia (HIGH-poh-**THER**-mee-ah)

hypoxemia (HIGH-pock-**SEE**-mee-ah)

hypoxia (high-**POCK**-see-ah)

leukocytosis (LOO-koh-sigh-**TOH**-sis)

mediastinum (ME-dee-ah-**STYE**-num)

nephrotoxic (NEFF-roh-**TOCK**-sick)

pancreatitis (PAN-kree-uh-**TIGH**-tis)

paresthesia (PAR-es-**THEE**-zee-ah)

pericardiocentesis (PER-ee-KAR-dee-oh-sen-**TEE**-sis)

pericardiotomy (PER-ee-KAR-dee-**AH**-tah-mee)

pericarditis (PER-ee-kar-**DYE**-tis)

sternotomy (stir-**NAH**-tuh-mee)

tachydysrhythmia (TACK-ee-dis-**RITH**-mee-yah)

transmyocardial (TRANS-my-o-**KAR**-dee-ul)

valvotomy (val-**VAH**-tuh-mee)

QUESTIONS TO GUIDE YOUR READING

1. What are the preoperative and postoperative routines and procedures for cardiac surgery?
2. What is the purpose and complications of cardiopulmonary bypass?
3. How would you describe the myocardial revascularization procedures?
4. What are the differences between commissurotomy, annuloplasty, and valve replacement?
5. What are the differences in postoperative care for the two types of cardiac valves?
6. What are the complications of cardiac and valve surgery?
7. What are an embolectomy and endarterectomy?
8. What nursing care will you provide for patients having cardiac or vascular surgery?
9. How will you know if your nursing interventions have been effective?

► CARDIAC SURGERY

As heart disease symptoms increase in severity and frequency or the disease process worsens, cardiac surgery may be used as treatment. Although cardiac surgery has become commonplace, it is still a major surgery with numerous complications, as well as physical, emotional, and social stressors.

Preparation for Surgery

For elective heart surgery, patients may be admitted to the hospital on the morning of surgery or 1 to 3 days before the surgery based on their medical history. A nursing assessment is important to provide baseline data that can be used for postoperative comparison and early discharge planning. In addition to routine admission testing, patients with chronic obstructive pulmonary disease (COPD) may have pulmonary function tests and baseline arterial blood gases (ABGs) done (Box 22–1). Patients with carotid bruits have carotid studies to determine the amount of occlusion in the carotid artery. If the occlusion is significant, a carotid **endarterectomy,** which removes the plaque on the lining of the blocked or diseased carotid artery, is performed, usually several weeks before having cardiac surgery.

Medications that may increase bleeding or reduce fluid volume may be ordered by the physician to be held before surgery. Drugs that increase bleeding include aspirin, often stopped 3 to 7 days preoperatively; warfarin (Coumadin), often stopped 4 to 5 days preoperatively; and heparin, stopped 4 hours preoperatively. During surgery fluid volume and blood pressure may be decreased by blood loss or medications. Therefore diuretics, which could further reduce fluid volume and blood pressure, are withheld up to 2 days before surgery. Because the patient takes nothing by mouth (NPO) 8 to 12 hours before surgery, insulin and oral hypoglycemic agents are reduced or withheld the morning of surgery.

Patients recover more quickly and have less postoperative stress when they have thorough preoperative teaching. Explanations of pain management, endotracheal tube (ETT), methods of communicating, ventilator, chest tubes, coughing and deep breathing exercises, intravenous (IV) lines, urinary catheter, incision care, and various equipment alarms are provided to the patient and family. It should be emphasized that patients are not able to talk while the ETT

---

endarterectomy: end—inside + arter—artery + ectomy—excision

| BOX 22–1 | **Routine Admission Testing** |

12-lead electrocardiogram (ECG)
Chest x-ray exam
Complete blood cell count (CBC)
Coagulation studies
Chemistry profile
Crossmatched for blood

is in place. Additionally, a preoperative family tour of the patient's initial postoperative unit and the waiting area helps prepare them for the surgical experience. A referral to pastoral care, if desired, can be comforting to the patient and family.

The anesthesiologist assesses the patient before surgery and orders preoperative medications. An antiseptic scrub shower is taken the night before and the morning of surgery. The patient is NPO after midnight the night before surgery.

Cardiopulmonary Bypass

The majority of cardiac surgeries use a cardiopulmonary bypass (CPB) pump in which blood is temporarily diverted away from the heart and lungs to the special pump. This diversion allows for a bloodless and motionless surgical field while the function of the heart and lungs is maintained by the pump. After the **sternotomy** is made, the vena cava and ascending aorta are cannulated (a small tube with multiple holes is placed into each of them). The aorta is cross-clamped between the heart and the cannula by clamping two large hemostats over the aorta in opposite directions. The cannulas are then attached to the pump tubing. Blood then flows from the body through the vena cava cannula to the CPB pump for oxygenation. After the pump oxygenates and removes carbon dioxide from the blood, the blood is returned to the body. The blood flows through the cannula into the ascending aorta, where it then circulates through the body (Fig. 22–1).

Using the CPB pump can have a unique set of complications. Before going on the pump, the patient is anticoagulated with heparin until the partial thromboplastin time (PTT) is five to six times greater than normal. Immediately before the patient comes off the pump, the effects of the heparin are reversed with protamine sulfate (antidote for heparin). Heparin is absorbed and stored in organs and tissue and can be sporadically released hours after surgery. As a result, the patient may have excessive bleeding. The risk of an air embolism is minimized by priming the pump with lactated Ringer's solution and maintaining careful observation. The priming solution increases circulating volume, which then results in a shifting of fluid into the interstitial tissue and edema formation. These fluid shifts can continue up to 6 hours after surgery and can cause hypotension.

Methods for providing closed-chest cardiopulmonary bypass and **cardioplegia** are being studied. With closed-chest methods, the chest is not opened to provide CPB or cardioplegia. Instead, small incisions are made in the chest and a video-assisted thoracoscope is inserted and used to perform the entire surgery. This technique of closed-chest surgery, with or without CPB, is termed minimally invasive direct coronary artery bypass (MIDCAB) surgery. When CPB is not used, mechanical devices are used to stabilize the portion of the heart being bypassed while the remainder of the heart continues to beat. Unfortunately, these techniques are not

---

sternotomy: stern—sternum + otomy—incision into
cardioplegia: cardio—heart + plegia—paralysis

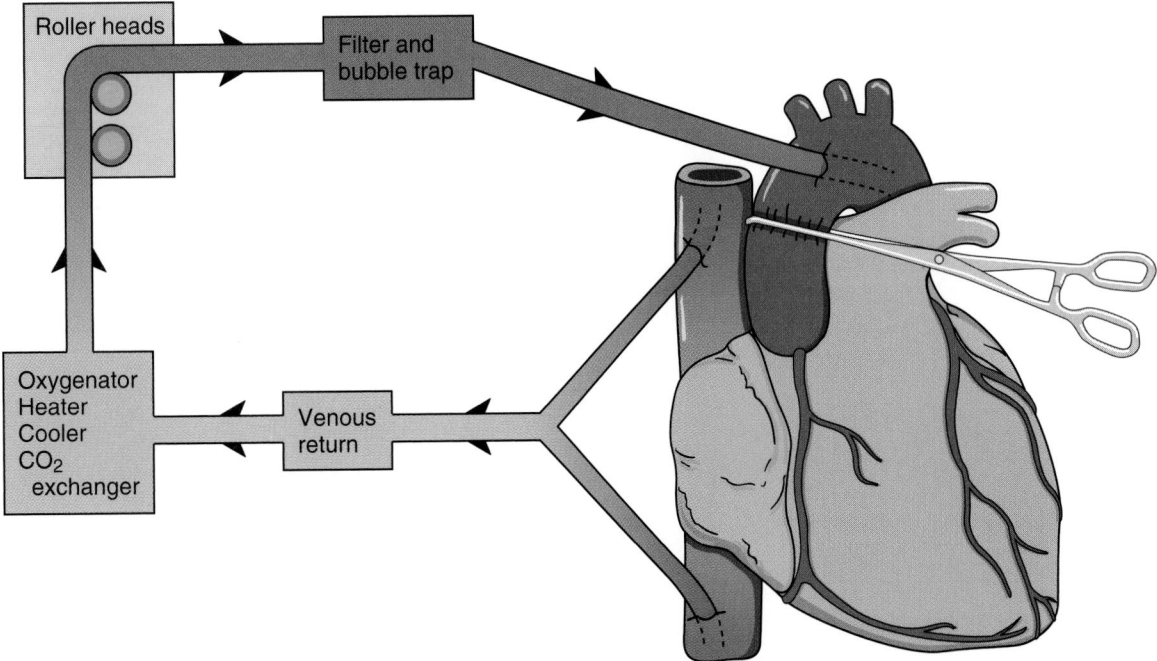

Figure 22-1 Cardiopulmonary bypass pump components.

options for all patients. Risk for complications associated with MIDCAB are much lower than with the traditional CABG procedure, and the recovery time is often weeks less. It is hoped that MIDCAB performed on a beating heart can become a common surgical option for qualified patients.

General Procedure for Cardiac Surgery

After the patient is placed on CPB, a cardioplegia solution is infused into the aortic root along with iced saline. This solution is placed around the heart to cause cardiac standstill. When the surgery is completed, the patient's blood is warmed in the CPB circuit and the patient is slowly weaned from CPB. The heart starts beating again after it is warmed and defibrillated. Temporary pacing wires are attached to the heart before the CPB pump is discontinued, so an external temporary pacemaker can be used if bradycardia develops. Once the heart is beating, CPB is stopped. Mediastinal chest tubes are placed to drain remaining blood and fluid from the chest. The sternotomy is closed with wires through the sternum and then sutures for the layers of tissue and skin. While still under anesthesia, the patient is transferred to an intensive care unit (ICU). Patients usually stay in ICU for 1 to 2 days, although those undergoing MIDCAB may not need an ICU stay.

Surgical Procedures
Myocardial Revascularization

CORONARY ARTERY BYPASS GRAFT. Coronary artery bypass graft (CABG) surgery is a procedure used to increase blood flow and oxygen to the myocardium and alleviate anginal symptoms. Significant occlusions in the coronary arteries are bypassed with vein or artery grafts (Fig. 22–2).

One or more vessel bypasses can be performed during the procedure. The saphenous vein from the leg or an internal mammary artery from the chest wall is generally used. The right **gastroepiploic** artery (RGEA), a branch of the gastroduodenal artery, and the inferior epigastric artery (IEA) have also been used for repeat CABG operations.

While the sternotomy is made, the vein graft is being removed from the body. The graft is flushed with a heparinized solution to check for leaks, then set aside for use during the surgery. The patient is then placed on CPB. After cardiac standstill occurs, one end of the graft is **anastomosed** (joined to the coronary artery distal to the occlusion) while the proximal end of the graft is anastomosed, usually to the ascending aorta. The surgery then continues as described in the general surgery procedure section.

Resecting the mammary arteries for grafting is more difficult and time consuming than resecting the saphenous vein, but their patency is longer. The proximal end of the artery is left attached to its origin, and the distal end is anastomosed to the coronary artery distal to the occlusion. See Patient Perspective Box 22–2.

TRANSMYOCARDIAL LASER REVASCULARIZATION. For patients who are not candidates for either MIDCAB or traditional CABG revascularization procedures, there is an alternative procedure called **transmyocardial** laser revascularization (TMR) or percutaneous transluminal myocardial laser revascularization (PTMR). A laser is used to make 20 to 30 small channels into the myocardium to reduce angina.

---

gastroepiploic: gastro—stomach + epiploic—pertaining to
transmyocardial: trans—across + myo—muscle + cardial—heart

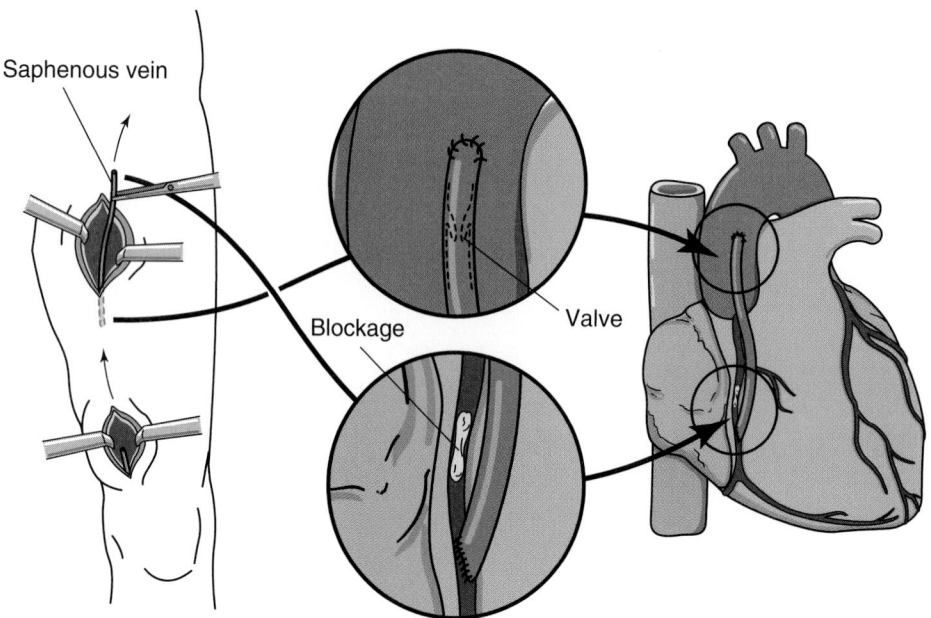

Figure 22-2 Myocardial revascularization: coronary artery bypass graft surgery.

Keith: Coronary Artery Bypass Graft

I was 72 when I had coronary artery bypass surgery. I had some very minor symptoms, and a heart catheterization showed significant coronary artery blockage. I ended up having five bypass grafts performed. It was a long surgery. I felt disoriented for several days afterward—mostly while I was in the intensive care unit.

I felt absolutely no pain of any type after surgery. However, I have had a strange sensation in my legs at the incision sites ever since surgery, but not pain! I also have experienced no depression and have assumed an attitude that I am in better condition than before surgery, so why should I worry?

I was provided with excellent care at home after discharge from the hospital. The nursing and rehab personnel were outstanding. I had a capable visiting nurse, physical therapist, or occupational therapist practically every day. Physical therapy was difficult at first but shortly became routine and easy.

My experience during cardiac rehab was outstanding. I had three sessions per week for 12 weeks. I have always exercised a lot, so I had no problems attaining the exercise levels suggested. It was a very positive experience, and I enjoyed the association with others in the same situation. We even staged a graduation when we finished. I wore a tuxedo jacket with my gym shorts! Since completing my rehab, I have religiously maintained a minimum 40-minute exercise schedule three times per week. I do this because I want to stay healthy for a long time.

The only negative aspect I experienced was a minor stroke during surgery. As a result, my balance is not as good as I would like. I need to be more careful, particularly when running or bike riding. Perhaps this is just older age, but it started just after surgery, so that is the reason I assume it is due to the minor stroke.

The impact on my life and family has been minimal. The primary change is being more conscious of my diet and an emphasis on exercise. Otherwise, I live pretty much the same as before.

An advantage of TMR over PTMR is that the heart is exposed, which allows channels to be made through the entire thickness of the heart, not just the inner surface as in PTMR. Future treatments may combine coronary bypass surgery with TMR for anginal relief when widespread blockages exist.

Heart Valve Repairs

A **commissurotomy** is a heart valve repair for a stenosed valve. For commissurotomy, the patient is placed on CPB and a left atriotomy (incision into the atrium) is made to expose the valve. The valve cusps are either incised with a knife or broken apart with a dilator. The atrium is sewn closed, CPB is discontinued, and surgery continues as described in the general surgery procedure section. Commissurotomy is most commonly performed on the mitral valve.

A balloon **valvotomy** is a procedure that repairs a stenosed heart valve. The balloon catheter is inserted through a diseased valve and then inflated to open the stenosed valve. This has been found to have effects similar to the commissurotomy, but the valvotomy must be performed blindly.

Another type of valve repair is **annuloplasty,** which repairs the annulus of a valve. The mitral valve is the most common valve repaired in this way. Sutures or a ring may be placed in the valve annulus to improve closure of the leaflets. Similar procedures are used on the tricuspid valve; however, the aortic valve is not readily repaired in this manner.

---

commissurotomy: commissur—connecting band of tissue + otomy—incision into
valvotomy: valv—valve + otomy—incision into
annuloplasty: annulo—annulus + plasty—repair

Heart Valve Replacement

Valves used for cardiac valve replacement may be either mechanical or biologic. There are three types of mechanical valves: caged ball, monoleaflet, and bileaflet (Fig. 22–3). Mechanical valves are durable but create turbulent blood flow. The turbulent flow can lead to clot formation, requiring lifelong anticoagulant therapy. Biologic (tissue) valves come from three sources: porcine (pig), bovine (cow), or allografts (human). Tissue valves have a very low incidence of thrombus formation and do not require lifelong anticoagulant therapy, but they may not last as long as mechanical valves (Cultural Consideration Box 22–3).

BOX 22-3 **CULTURAL CONSIDERATION**

Because the pig is considered a dirty animal to religious Jews and Muslims, only bovine, synthetic, or human valves should be used for these patients. Likewise, because the cow is sacred among Hindus, only porcine, synthetic, or human valves should be used for Hindu patients.

For mitral valve replacement (MVR), a left atriotomy is made after the patient is on CPB. For an aortic valve replacement (AVR), an incision is made above the right coronary artery in the aorta. Then in either valvular procedure, the diseased valve is excised and the new valve sutured in place. The incision is closed, and surgery then continues as described in the general surgery procedure section.

Just as coronary artery revascularization procedures are becoming less invasive, so are surgical interventions for cardiac valves. These minimally invasive techniques continue to use CPB but involve small incisions, which increase pa-

A. Caged ball valve

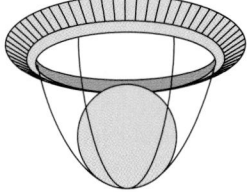

B. Monoleaflet

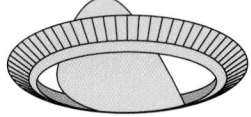

C. Bileaflet

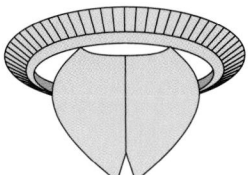

Figure 22-3 Types of mechanical heart valves.

tient comfort and reduce recovery time while decreasing costs.

Ventricular Aneurysm Repair

Indications that a ventricular aneurysm requires surgery include persistent angina, symptoms of heart failure, large aneurysms impeding the function of the heart, left ventricular failure, or **tachydysrhythmias.** Once the patient is on CPB, an incision is made exposing the heart and then the aneurysm is resected (cut out), leaving a fibrous border (Fig. 22–4). The ventricle is wiped clean and irrigated to ensure all thrombi are removed. The opening is then closed with sutures or patched with a graft. Air that entered the ventricle while it was open is aspirated, and the surgery continues as described in the general surgery procedure section.

Cardiomyoplasty

Cardiomyoplasty uses the patient's own muscle (latissimus dorsi) to improve the contractile function of the heart. (See Chapter 21.)The muscle is wrapped around the aorta or heart, and a cardiomyostimulator and leads are implanted to stimulate muscle contraction. The muscle is rested for 2 weeks after surgery and then programmed to gradually start stimulation. After 4 months, the muscle can be stimulated at either a 1:1 or 1:2 heartbeat ratio. Benefits from cardiomyoplasty include the improved contractility of the failing heart and reduced changes in heart failure remodeling from the wrapping or enclosing effects of the muscle. Cardiomyoplasty may reduce the need for heart transplantation by delaying or preventing end-stage heart failure.

Cardiac Trauma

Two types of cardiac trauma can occur: nonpenetrating and penetrating. Nonpenetrating injuries, or contusions, occur from blunt trauma such as motor vehicle accidents or contact sports in which direct compression or force is applied to the upper torso. Contusions may vary from small bruises to hemorrhage.

There may be few or no external injuries indicating traumatic cardiac injury. The patient may be asymptomatic or exhibit signs and symptoms identical to a myocardial infarction. In severe contusions, laboratory results may show elevated creatine kinase MB (CK-MB) or troponin-I levels.

If bleeding occurs into the pericardial sac, cardiac tamponade (compression of the heart from the blood collecting in the sac) may occur. When the heart is compressed, it cannot expand to fill with blood. As a result, cardiac output drops, which can be life threatening. If signs of shock occur, a **pericardiocentesis** must be performed. (See Chapter 17.) A needle is inserted into the pericardial sac to aspirate the collecting fluid. With its own pressure, the tamponade may seal the area of bleeding, so no cardiac decompensation oc-

tachydysrhythmias: tach—fast + dys—bad + rhythm—regularity + ia—condition
pericardiocentesis: peri—around + cardio—heart + centesis—surgical puncture

Ventricular
aneurysmectomy

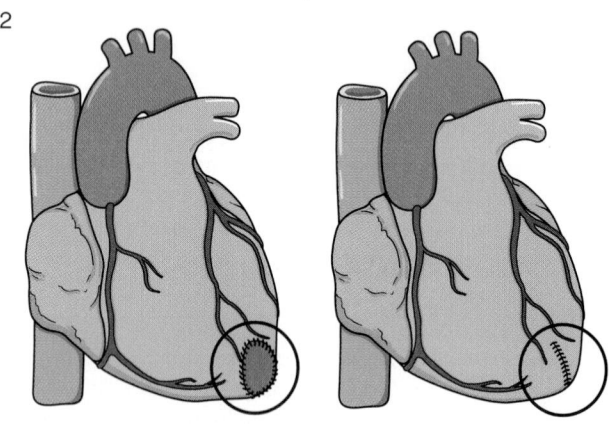

Ventricular
aneurysmorrhaphy
Dacron patch

1: Aneurysm is cut and closed.
2: Patch inserted inside incision, then closed.

Figure 22-4 Ventricular aneurysm repair.

curs. In this case only bedrest and observation are required. There are no long-term effects with most contusions. With severe contusions, however, scarring and necrosis of the myocardium may decrease cardiac output and increase the risk for cardiac rupture.

Penetrating traumas may be an external injury to the chest, such as a stab or gunshot wound, or an internal injury, such as invasive lines that penetrate the cardiac muscle. Complications vary depending on the size, location, and cause of injury. Tamponade occurs from bleeding into the pericardial sac if the pericardium is sealed off by clot formation. A hemothorax develops if blood drains into the pleural space in the chest. A pneumothorax occurs if air collects in the pleural space. Signs and symptoms of hemorrhage and myocardial ischemia may be noted. Cardiac trauma treatment usually requires surgery to repair the damage so that hemostasis can be regained.

Cardiac Tumors

Cardiac tumors are either primary or secondary in origin and can be benign or malignant. Primary cardiac tumors are less common than secondary (metastatic) cardiac tumors. Benign primary tumors are commonly myxomas.

A myxoma is a soft, gelatinous, intracavitary mass that is attached to the endocardium by a narrow stalk (peduncle). Myxomas are found in any of the four chambers of the heart but are more common in the left atrium. Symptoms vary depending on the location of the tumor and its effect on valve function. There is a high risk of emboli formation with this type of tumor because it breaks apart easily. Surgical excision of the tumor offers the patient a good prognosis.

The majority of malignant primary cardiac tumors are sarcomas. Secondary cardiac tumors are generally carcinomas that have metastasized. The symptoms depend on the location and size of the mass. Treatment consists of palliative measures such as surgery, radiation, and chemotherapy. For **pericarditis,** a pericardiocentesis may be performed. If pericarditis recurs, a window (opening) is cut into the pericardium for fluid drainage to prevent fluid buildup and cardiac tamponade. The prognosis is poor for malignant or metastatic cardiac tumors.

Complications of Cardiac Surgery
Cardiovascular Complications

Several cardiovascular complications can occur from cardiac surgery, including postoperative bleeding, cardiac tamponade, dysrhythmias, perioperative myocardial infarction, pericarditis, and myocardial depression.

Postoperative bleeding into the **mediastinum** is a common complication. Bleeding may come from oozing blood vessels within the chest, graft anastomoses, or myocardial incisions. These oozing areas may result from increased blood pressure, which may occur if patients are shivering until they are rewarmed after surgery, or coagulation problems occurring from platelet, fibrin, and other coagulation factors being destroyed in the CPB circuit. Inadequate reversal of the heparin used for CPB or a delayed release of heparin stored in organs and fatty tissue can add to postoperative bleeding. Conservative treatment is instituted before taking the patient back to surgery to find the cause of bleeding.

Cardiac tamponade can occur if clots block the chest tubes, preventing blood from draining out of the pericardial sac. This heart constriction can lead to cardiac arrest without prompt treatment.

Dysrhythmias are fairly common and have many causes: **hypoxia, hypoxemia,** acidosis, electrolyte imbalance (especially hypokalemia), pain, anxiety, inflammation from the **pericardiotomy,** presence of the CPB cannulas, or gastric distention. Varying degrees of heart block are associated

---

mediastinum: medi—middle + stinum—sternum
hypoxia: hypo—below normal + ox—oxygen + ia—condition

with valve replacement due to the close proximity of the atrioventricular node to all the valves.

A perioperative myocardial infarction can occur from graft thrombosis, embolus, inadequate myocardial preservation during aortic cross-clamping, or inadequate revascularization.

Pericarditis, an inflammation of the pericardial sac, is due to the handling of the pericardium. This inflammation may occur immediately or months later. Should pericarditis recur numerous times, it is termed postpericardiotomy syndrome.

Myocardial depression, which results in decreased cardiac output, is usually a temporary occurrence caused by hypoxemia, acidosis, interstitial fluid accumulation, or electrolyte imbalance. Hemodilution and diuretics can cause hypomagnesemia (low serum magnesium) and hypocalcemia (low serum calcium), which also depress myocardial function.

Hemodilution occurs when excess fluid is present in the vascular system. This excess fluid dilutes the particles normally found in the fluid, such as minerals, electrolytes, and blood cells. So even though there may have been a normal amount of a particle before the excess fluid developed, the excess fluid dilutes or makes this normal amount less concentrated. As a result the values of the particles are low.

Try this example: Stir 1 drop of food coloring into 1/4 cup of water. In a second container, stir 1 drop of food coloring into 1 cup of water. Compare the color of the two containers. The color of the first container simulates a normal value for the coloring particles. When more fluid is added (hemodilution), there is less color, indicating that the particles have been diluted and are now at a lower value than normal.

Pulmonary Complications

The most common postoperative problem is **atelectasis,** which is the collapse of airways caused by hypoventilation. Atelectasis can occur from pain, exhaustion, prolonged anesthesia, shallow breathing, poor cough effort, and decreased ambulation and movement. Atelectatic areas contribute to hypoxemia and become sites for infections such as pneumonia.

Deep breathing and coughing are important for most surgical patients to perform to prevent atelectasis.

Pleural effusions occur when fluid accumulates within the pleural lining. Heart failure, inflammation, and hypoalbuminemia (low serum albumin) cause changes in cell wall

permeability and pressures that lead to an imbalance in the production and reabsorption of pleural fluid. Treatment involves the insertion of a chest tube to drain the fluid and allow lung re-expansion.

The phrenic nerve conducts the motor impulses to the diaphragm. Phrenic nerve paralysis can occur from direct injury during the resection of the internal mammary artery or from the cold saline used for CPB. Symptoms include dyspnea and paradoxical (conflicting) abdominal breathing. On chest x-ray examination, the diaphragm is elevated on inspiration. The paralysis may be unilateral or bilateral. Patients with pre-existing pulmonary disease require continued ventilator support. Phrenic nerve paralysis resolves itself, but the time required varies.

Renal Complications

Low arterial pressure on CPB activates the renin-angiotensin-aldosterone system, thereby creating vasoconstriction, which decreases renal perfusion and function. Other factors that decrease renal function are **nephrotoxic** antibiotics, **hypothermia,** and **hemolysis** from CPB. If the kidneys do not recover, the patient may be started on hemodialysis until renal functioning improves.

Gastrointestinal Complications

Postoperative complications in the gastrointestinal (GI) tract can range from the minor annoyance of medication-induced diarrhea (quinidine or various antibiotics) to a life-threatening bowel infarction. Gastric distention can occur if the nasogastric tube is not properly functioning while the patient is on the ventilator. This distention can be uncomfortable and may impede the ability of the lungs to expand, causing decreased ventilation. As with any surgery, a paralytic ileus can develop, necessitating a longer gastric decompression time with a nasogastric tube and NPO status.

Bowel infarction secondary to ischemia of the mesenteric vasculature is rare yet life threatening. Bowel infarction can be caused from low blood flow while the patient is on CPB, myocardial depression, or emboli. Patients with bowel infarction are extremely ill and may have sepsis, respiratory distress, severe abdominal pain, metabolic acidosis, or diarrhea. Surgery to resect the necrotic bowel may be indicated. This low blood flow state can also cause a necrotic injury to the pancreas, leading to **pancreatitis.**

Immune and Humoral System Complications

As blood passes along the foreign surfaces of the CPB circuit, the complement system (an immune response involving plasma proteins and immunoglobulins) is activated. This immune response leads to fever, **leukocytosis,**

---

nephrotoxic: nephro—kidney + tox—poison + ic—pertaining to
hypothermia: hypo—below normal + therm—temperature + ia—pertain to
hemolysis: hemo—blood + lysis—destruction
pancreatitis: pancreat—pancreas + itis—inflammation
leukocytosis: leuko—white + cyto—cell + sis—condition

atelectasis: atel—imperfect + ectasis—stretching

vasodilation, and increased vascular permeability, resulting in interstitial fluid accumulation.

Neurological Complications

Several neurological complications can develop with cardiac surgery, including peripheral nerve injury, central nervous system deficits, and pain. Peripheral nerve injury can be associated with improper or prolonged positioning on the operating table. Examples of peripheral nerve injury include the following:

- Brachial plexus injury—weakness, numbness, and burning in the hand
- Ulnar nerve injury—numbness and tingling in the fourth and fifth fingers
- **Paresthesia** and burning in the leg from which the saphenous vein was taken for a coronary artery bypass graft

Central nervous system deficits include **encephalopathy** (brain dysfunction) and brain embolism. Encephalopathy involves the entire brain, unlike an embolism. Cause of encephalopathy include hypoxia, hypoperfusion, and volume overload resulting in brain tissue edema. An embolism results from dysrhythmias or aortic cross-clamping. Either of these conditions can lead to altered level of consciousness and confusion.

Pain from cardiac surgery is subjective and varies from patient to patient (Patient Perspective). Chest pain is usually due to the sternotomy, during which there is retracting of the ribs and stretching of the chest muscles and ligaments. With internal mammary artery grafts, anterior or posterior chest pain may be considerable resulting from manipulation and resection of the mammary artery.

Other Complications

Patients may experience varying degrees of memory loss, depression, immobilizing fear, frustration in role adjustments, hallucinations, cognitive dysfunction, or sexual dysfunction. Sleep disturbances such as insomnia, nightmares, and frequent waking are common and tend to prolong the symptoms.

Valve Replacement Complications

Tissue valves, over time, can undergo degenerative changes and calcification, leading to valve failure. The most common complication associated with mechanical valves is thrombus formation. An embolism can occur when thrombi form on the mechanical valve and then break off. Embolism rarely occurs with tissue valves and is prevented with anticoagulant therapy for mechanical valves. When the patient is on warfarin, ongoing monitoring of international normalized ratio (INR) and prothrombin time is important.

Other complications include anemia and endocarditis. Anemia is due to hemolysis of red blood cells (RBCs) as they come in contact with mechanical valve structures. Endocarditis is an infection of the inner lining of the heart. Microorganisms tend to grow on the valve leaflets or the sewing ring of mechanical valves. These growths can make valves incompetent or break off to become emboli.

CRITICAL THINKING

Mr. Jones

Mr. Jones is transferred to the nursing unit 12 hours after a quadruple CABG. Preoperative vital signs were blood pressure 164/88, apical pulse 62 regular, respiratory rate 18, temperature 98.4° F. Assessment findings are blood pressure 100/56, apical pulse 105 and irregular, respiratory rate 28 and shallow, temperature 99.8° F, lung sounds diminished with crackles in bilateral bases, pedal pulses weak bilaterally, chest and leg dressings dry and intact, and no urinary catheter. Mr. Jones is being monitored for first postoperative voiding.

1. Which findings may indicate pulmonary problems?
2. List four nursing interventions for the altered pulmonary status.
3. What are three reasons why the apical pulse could be elevated?
4. What are two reasons why the blood pressure could be low?

Answers at end of chapter.

► CARDIAC TRANSPLANTATION

Cardiac transplantation is reserved for patients with end-stage cardiac disease. Strict criteria for the selection of recipients and donors are applied to optimize survival (Box 22–4). Preoperative teaching is done once the recipient is accepted into the transplant program. For more information on heart transplantation, visit http://www.nlm.nih.gov/medlineplus/hearttransplantation.html.

Surgical Procedure

Once a donor heart is found, the recipient is notified, admitted to the hospital, and immediately prepared for surgery. The general procedures for this surgery are similar to those described in the cardiac surgery section. Two types of cardiac transplant procedures are performed: orthotopic and heterotopic. In the orthotopic procedure, once the patient is on CPB, the recipient's diseased heart is removed, leaving the posterior wall of the atria, superior vena cava and inferior vena cava, and pulmonary vein (Fig. 22–5). The aorta and pulmonary artery are cut. The donor's atria, aorta, and pulmonary artery are then anastomosed to the recipient's atria, aorta, and pulmonary artery. Surgery then continues as previously described. The heterotopic procedure joins the donor heart and vessels to the recipient's heart and vessels without removing the recipient's heart. The donor heart rests in the right side of the chest.

paresthesia: para—abnormal + esthes—sensation + ia—pertaining to
encephalopathy: en—inside + cephalo—head + pathy—disease

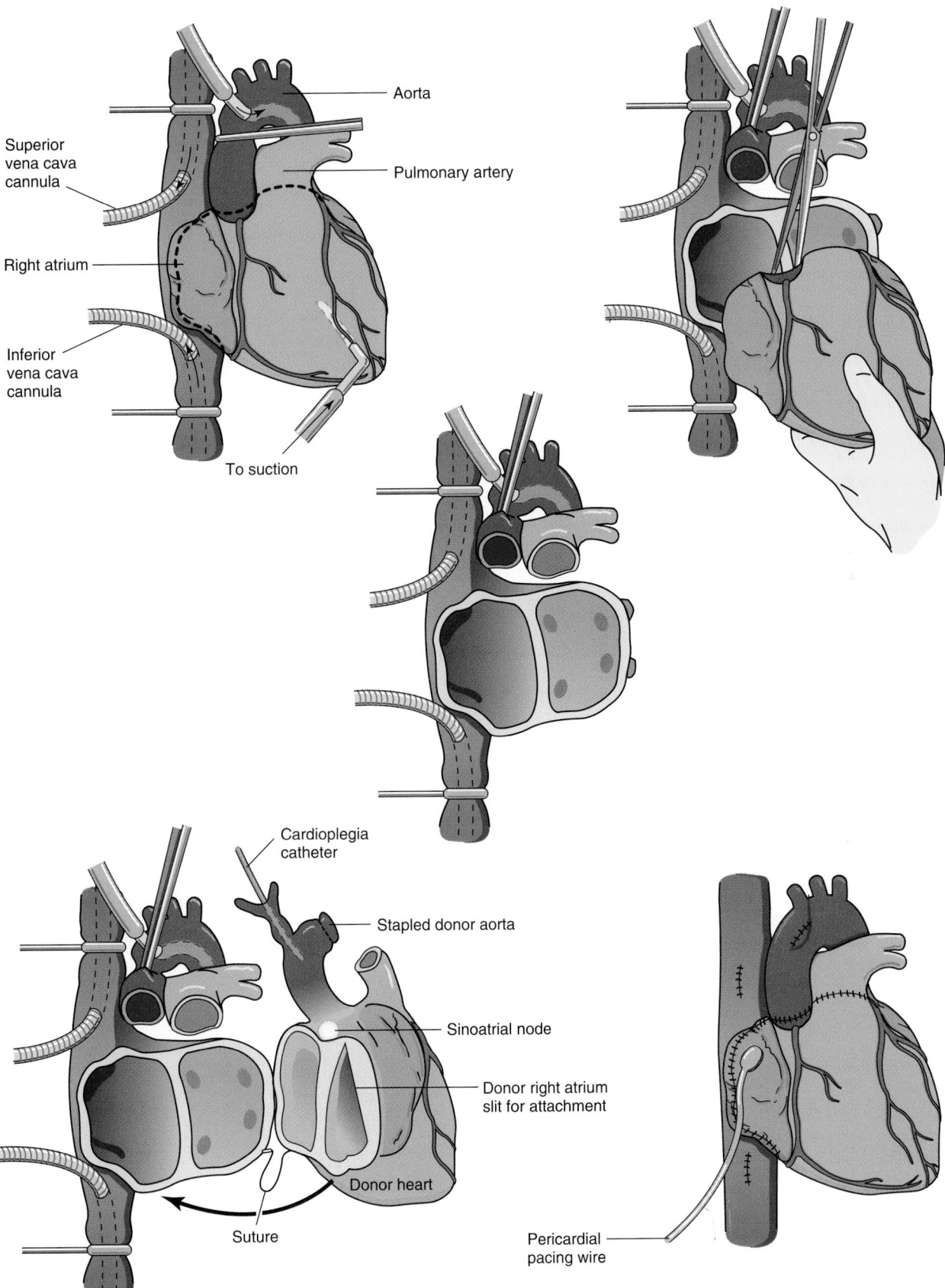

Figure 22-5 Heart transplantation.

BOX 22-4 **Cardiac Transplant Criteria**

Donor Criteria	Recipient Criteria
Younger than 40 years of age	Younger than 55 years of age
No significant cardiac or malignant disease	Class IV cardiac disease (not treatable with other medical or surgical treatment, less than 6–12 months survival)
No active infections	No irreversible pulmonary hypertension
No severe hypertension or diabetes mellitus	No unresolved pulmonary infarcts
Only ± 20 lb difference in weight between donor and recipient	No systemic disease limiting survival
	No drug addiction or peptic ulcer disease

CRITICAL THINKING

Mrs. Eden

Mrs. Eden, 45, a single mother of two, is transferred to a surgical unit 5 days after a cardiac transplant. She is withdrawn and has a poor appetite. Her vital signs are stable. However, on ambulating to the bathroom, she is very weak, requiring two nurses to help her. Her respiratory rate increases from 20 to 32 and is slightly labored, and her apical pulse increases from 88 to 103.

1. Is Mrs. Eden tolerating this activity? Why or why not?
2. List four reasons why Mrs. Eden has a poor appetite.
3. Give four nursing interventions for Mrs. Eden's poor appetite.
4. Give three reasons why Mrs. Eden is withdrawn.

Answers at end of chapter.

Immunosuppressive therapy begins preoperatively with high loading doses of cyclosporine, azathioprine, steroids, or other medications. The risk for rejection is highest immediately after surgery and decreases with time, so doses of immunosuppressive medication are also highest initially after surgery and decrease with time (Nutrition Notes Box 22–5). Lifelong antirejection therapy is required.

Complications

Heart transplantation complications may include all those stated for cardiac surgery, as well as heart rejection, which is the major cause of death within the first year. Immunosuppressive therapy is required to prevent rejection of the heart. To detect rejection, biopsies of cardiac muscle are commonly performed weekly during the first 3 to 6 weeks after surgery, then every 3 months for 1 year, and then yearly. If a biopsy shows damaged cells, indicating rejection, antirejection drug therapy may be changed.

In addition, infection and malignancies may occur as a result of the immunosuppressive therapy. The medications used for immunosuppressive therapy also may cause adverse reactions. Cyclosporine is nephrotoxic and hepatotoxic. Azathioprine is also hepatotoxic and may cause bone marrow depression or skin sensitivity to the sun. Steroids can cause osteoporosis, hyperglycemia, sodium and water retention, gastritis, hypertension, and cataract formation. muromonab-CD3 (Orthoclone OKT3), used to treat rejection reactions in spite of immunosuppressive therapy, can cause fever, malaise, and headaches.

Medical Management

The patient receives a diuretic when CPB is stopped to aid in excretion of excessive circulating fluid. Intake and output are monitored hourly, and the patient is observed for fluid overload. Lung sounds are monitored frequently for crackle, and weight and electrolyte levels are checked daily.

Postcardiotomy syndrome (PCS) may occur from day 2 to 5 after surgery and last a few weeks. Patients may arouse normally and be oriented but exhibit mild confusion or psychosis. Pupillary reaction and motor response are assessed. The safety of the patient is maintained with side rails up, bed in low position, and nursing call light within reach. The patient is given as much rest and as little sensory stimulation as possible. The family is kept informed and involved in the patient's recovery.

Sleeping is difficult because of postoperative pain and the continuous level of activity in the ICU. Sleep is promoted in 90-minute intervals by dimming lights and decreasing all sensory stimulation near the patient. Additionally, listening to a favorite soothing tape recording with earphones or the use of ordered narcotics for pain may also help sedate and relax the patient.

Temperature is monitored every 4 hours and complete blood cell count (CBC) and white blood cell (WBC) results are monitored for indications of infection. If oral thrush (white patches) develops, an antifungal agent is ordered. A urine culture to diagnose a urinary tract infection is ordered if cloudy urine or urinary tract burning occur.

▶ ARTIFICIAL HEART AND VENTRICULAR ASSIST DEVICES

Newer technology is being used to aid the failing heart. The purpose of these artificial hearts or ventricular assist devices is to be a bridge to transplantation or a long-term solution when other options are not available to treat the failing heart. Because there are not enough donor hearts, this technology may benefit many patients. Several trials of these devices are occurring now. For more information about artificial hearts and ventricular assist devices, visit the National Institutes of Health or a manufacturer's Web site at the following addresses:

BOX 22-5  *Nutrition Notes*

The potency of antirejection medications may be reduced if patients drink grapefruit juice with them. Grapefruit increases body metabolism.

Individuals who drink wine may not get the desired therapeutic effects of certain immunosuppressive medications.

▶ NURSING PROCESS: PREOPERATIVE CARDIAC SURGERY OR CARDIAC TRANSPLANT

Preoperative and postoperative needs for the patient undergoing cardiac surgery are discussed here and in Chapter 11.

Assessment

A baseline assessment is important for postoperative comparison and to begin discharge planning (Box 22–6). Pain control needs and circulatory status are essential items. Results of diagnostic laboratory tests, x-ray examinations, and other studies are reported if abnormal. Typing and cross-matching of ordered units of blood are placed on hold.

Nursing Diagnosis

The nursing diagnoses for preoperative cardiac surgery or transplant may include but are not limited to the following:

- Anxiety related to threat of death, pain, powerlessness, or lifestyle changes
- Deficient knowledge: knowledge of preoperative and postoperative procedures related to unfamiliar process.

▶ NURSING PROCESS: POSTOPERATIVE CARDIAC SURGERY OR CARDIAC TRANSPLANT

Assessment

The patient is accompanied to the ICU by the anesthesiologist, who gives the nurse a report of the procedure, complications, and hemodynamic and ventilatory management of the patient. The patient is connected to a cardiac monitor and mechanical ventilator. The mechanical ventilator is used for 4 to 24 hours. A temporary pacemaker is connected to the epicardial pacing wires if they were placed during surgery as a precaution to treat bradycardia. The patient is

---

BOX 22–6 **Preoperative Nursing Assessment**

Past and present medical conditions and surgeries
Present medications
Allergies: medications, foods, other
Description of anginal pain and symptoms
Vital signs with blood pressure taken in both arms
Patient's usual functioning level
Patient's and family members' current knowledge of procedure
Family roles and support
Financial, transportation, or home life concerns

---

placed under a warming device, such as a light or blanket. The chest tubes are monitored, the nasogastric tube is placed to suction, and the urinary catheter is placed for gravity drainage. A head-to-toe assessment of the patient, including dressings, tubes, and IV lines, is performed. Of importance are signs of awakening, shivering, pain, lung and heart sounds, and palpation of the entire chest and neck to detect crepitus (air in the subcutaneous tissue from opening the chest). Blood is drawn for a CBC, electrolytes, coagulation studies, and arterial blood gases. Cardiac transplant patients may be in isolation for their own protection, depending on the agency's policy.

After the initial transfer assessment, vital signs, oxygen saturation, and cardiac pressures are monitored and recorded every 15 to 30 minutes, with decreasing frequency as the patient stabilizes. Body temperature is monitored continuously while warming measures are used. Intake and output are measured and vital signs are checked. A 12-lead electrocardiogram (ECG) is done to detect perioperative myocardial infarction. A chest x-ray examination is done to check central line and endotracheal tube placement and to detect a pneumothorax or hemothorax, diaphragm elevation, or mediastinal widening from bleeding. At this point the family may see the patient, and patient care is explained.

Nursing Diagnosis

The nursing diagnoses for the patient who has had cardiac surgery may include but are not limited to the following:

- Acute pain related to surgical incision or reperfusion of tissue
- Ineffective airway clearance related to intubation, anesthesia, pain, or sedation
- Impaired gas exchange related to volume overload, atelectasis, phrenic nerve injury, or preexisting pulmonary dysfunction
- Decreased cardiac output related to myocardial depression, hypothermia, or dysrhythmias
- Risk for infection related to surgical incisions or immunosuppression
- Deficient knowledge: therapeutic regimen related to lack of previous exposure

Nursing diagnoses for postoperative cardiac surgery or transplant are discussed in Nursing Care Plan Box 22–7 for the Postoperative Patient Undergoing Cardiac or Transplant Surgery.

Planning

The patient's goals include the following:

- Maintain pain relief.
- Maintain a patent airway and clear lung sounds.
- Maintain oxygen saturation greater than 95 percent.
- Maintain vital signs within normal limits.
- Remain free from infection.
- Understand the surgery and the therapeutic regimen.

Nursing Interventions

Pain

After cardiac surgery, pain is monitored in relation to the patient's preoperative anginal or infarction-associated pain. Chest pain after surgery can be frightening for patients. Knowing that chest pain can occur from the surgical incision rather than the heart is important to the patient. Oth-erwise the patient may not associate surgical chest pain with the incision and instead think the pain is anginal or infarction pain.

Ineffective Airway Clearance

Coughing and deep breathing and using incentive spirome-try every hour while awake are important to prevent respi-ratory complications. Elevating the head of the bed, sitting

---

BOX 22-7   NURSING CARE PLAN FOR THE POSTOPERATIVE PATIENT UNDERGOING CARDIAC OR TRANSPLANT SURGERY

▶ **Acute pain related to sternotomy, leg incisions, internal mammary artery resection, or pericarditis**

PATIENT OUTCOMES

Patient will state pain is relieved or tolerable. Patient is able to rest and perform respiratory treatments.

EVALUATION OF OUTCOMES

Does patient state pain is within acceptable levels? Is patient able to rest and perform respiratory therapies?

INTERVENTIONS	RATIONALE	EVALUATION
• Assess characteristics of pain with each episode.	A thorough description is needed to deter-mine cause and plan actions.	Does patient describe pain on scale of 0 to 10?
• Splint chest incision with all movement and coughing and deep breathing.	Stabilizes sternum and incision to increase comfort.	Can patient splint chest incision independ-ently?
• Encourage patient to report pain even when pain is mild.	It is easier to keep pain under control when mild.	Does patient report pain when mild?
• Turn, reposition every 2 hours.	Changes muscle position, relieving stiffness.	Is patient comfortable without stiffness?
• Offer back rubs frequently.	Relaxes tense muscles retracted during opera-tion.	Is patient able to rest in comfort?
• Instruct patient to take a deep breath before movement and exhale slowly during move-ment.	Keeps muscles relaxed, minimizing tension with guarding and pain.	Can patient perform coughing and deep breathing techniques as instructed?

▶ **Decreased cardiac output related to myocardial depression, hypothermia, bleeding, unstable dysrhythmias, or hypoxemia**

PATIENT OUTCOMES

Patient will remain free of major side effects of pharmacological support. Patient will maintain vital signs within normal limits (WNL), palpable peripheral pulses, urine output greater than 30 mL/h, and normal sinus rhythm.

EVALUATION OF OUTCOMES

Is patient free of major side effects? Are vital signs WNL?

INTERVENTIONS	RATIONALE	EVALUATION
• Monitor vital signs.	Trends reflect problems.	Are vital signs WNL?
• Assess peripheral circulation.	Mottling or weak pulses may indicate poor cardiac output (CO).	Do peripheral pulses remain strong with nor-mal skin color, temperature, capillary refill?
• Monitor intake and output.	Fluid deficit or excess can alter CO.	Does total intake equal output?
• Assess lung sounds and charac-ter of sputum.	Wet lung sounds may indicate heart failure or pulmonary edema.	Are lungs clear?
• Monitor temperature closely while rewarming the patient.	Febrile state increases heart rate and myocar-dial oxygen consumption.	Does temperature remain less than or equal to 98.6° F (37° C)?
• Assess for shivering.	Shivering increases the blood pressure, de-creasing CO and increasing risk for bleeding.	Is patient's shivering controlled?
• Monitor chest tube drainage for increase or sudden decrease.	Drainage >200 mL/h may lead to hypovo-lemia and a decrease in CO.	Is patient free from tamponade and hypovo-lemia?
• Monitor ECG.	Premature ventricular contractions and atrial fibrillation decrease CO.	Does patient remain in normal sinus rhythm or controlled dysrhythmia?
• Monitor electrolytes.	Low calcium and magnesium and high potas-sium decrease contractility and CO.	Are electrolytes WNL?
• Monitor ABGs.	Acidosis decreases heart function, and a low CO may lead to further acidosis.	Are ABGs WNL?

| BOX 22-7 | **NURSING CARE PLAN FOR THE POSTOPERATIVE PATIENT UNDERGOING CARDIAC OR TRANSPLANT SURGERY—CONT'D** |

 Risk for infection related to inadequate primary defenses from surgical wound or immunosuppression (transplants)

PATIENT OUTCOMES

Patient will remain free from infection.

EVALUATION OF OUTCOMES

Does patient remain free from infection?

INTERVENTIONS	RATIONALE	EVALUATION
• Observe incision for signs and symptoms of infection.	Redness, warmth, fever, and swelling indicate infection.	Are signs and symptoms of infection present?
• Monitor drainage and maintain drains.	Drains remove fluid from the surgical site to prevent infection development.	Are drainage amount and color normal for procedure? Are drains functioning?
• Maintain sterile technique for dressing changes.	Sterile technique reduces infection development.	Is incision free of signs and symptoms of infection?
• Monitor and report abnormal findings for temperature, lung sounds, sputum, and urine consistency.	Low-grade (immunosuppressed) or high-grade fever, crackles, yellow-green sputum color, or cloudy urine can indicate infection.	Is the patient's temperature WNL and are lung sounds, sputum, and urine clear?
• Encourage coughing and deep breathing and incentive spirometer use.	Lung infections can be prevented with lung expansion and secretion removal.	Does patient perform coughing and deep breathing and use incentive spirometer?

upright, splinting the chest incision with a pillow or blanket roll, and relieving pain increase the patient's ability to cough and deep breathe. Elderly patients have a weakened cough force. These patients should be encouraged and assisted to cough as effectively as possible.

Impaired Gas Exchange

Daily chest x-ray examination results are monitored for diaphragm elevation, atelectasis, pulmonary effusions, and pneumothorax or hemothorax. The patient is given oxygen and suctioned as needed. Chest tubes are maintained. Respiratory treatments with bronchodilators may be given while the patient is on the ventilator and after extubation. Diuretics and steroids are given to reduce excess fluids and improve air exchange.

Decreased Cardiac Output

Trends in the patient's cardiac output are monitored. Body temperature is continuously monitored until warming measures are discontinued, which occurs when the core body temperature nears 37° C (98.6° F). While patients are being rewarmed, they are assessed for shivering, which may be felt as a fine vibration at the mandibular angle of the jaw. Shivering greatly increases cardiac oxygen needs. Paralyzing agents given with narcotics eliminate shivering.

Risk for Infection

Hand washing is extremely important to prevent infection. Dressings are kept dry and intact. Sterile technique is used to change dressings, and incisions are assessed with every dressing change. All IV solutions and tubing are changed according to the agency's protocol for immunosuppressed patients (often every 24 hours). IVs and invasive lines

should be discontinued as early as possible. The cardiac transplant patient may be in reverse (protective) isolation.

For the immunosuppressed patient, the risk of infection is great. Observation is the key to early detection of infection. A fever is usually low grade in an immunosuppressed patient because the suppressed immune system cannot produce a normal response to an infection. Sputum color is noted and a sputum culture is requested if changes occur. Urine appearance is noted, and a request for a urine culture is made if changes occur. The patient's oral mucosa is checked every shift for sores or thrush resulting from immunosuppression.

Deficient Knowledge

Awakening with many questions, strange auditory and tactile sensations, and the inability to speak are very frightening and frustrating to the patient. Keeping eye contact with the patient and using touch appropriately can be very soothing. If lip reading is unsuccessful, using simple closed-ended questions, nonverbal gestures, communication boards, and magic slates may be effective. Explanations regarding procedures are given in simple terms.

The family is taught to promote self-care by the patient. Cardiac transplant patients commonly have memory deficits, cognitive dysfunction, and short attention spans resulting from long-term decreased cerebral perfusion. Information must be given in small increments. Family involvement in teaching sessions is important to promote understanding and retention.

Discharge teaching should include treatment, complications, activity, medications, and how to enhance one's quality of life. Sexual functioning questions and concerns should also be discussed with the patients and their partners. In addition to written materials, audiotapes can be used to provide discharge information.

Additional Nursing Care for Cardiac Transplant Patients

Cardiac transplant patients may have feelings of sadness and grief for the donor and his or her family. These feelings may be offset by great elation, relief, and hope after a long wait for the transplant. Patients should be told that these feelings are normal. They should be allowed to express their feelings when they are ready. Emotional support may be needed.

Transplant rejection is a possible complication of this surgery. Patients need to understand the importance of following instructions regarding medications and testing that are related to preventing or detecting rejection.

Cardiac transplant patients are followed in an exercise rehabilitation program that closely monitors their activity progression in relation to myocardial oxygen consumption and signs of activity intolerance. Progression may be slow depending on the patient's condition before surgery. Most patients reach an activity level allowing them to participate in many recreational sports.

▶ VASCULAR SURGERY

Vascular impairments requiring surgery may be acute or chronic and involve arteries, veins, or lymphatic vessels.

Embolectomy and Thrombectomy

When an artery becomes completely occluded by an embolus or thrombus, it is considered a surgical emergency. Surgical removal to restore blood flow and oxygenation to the tissue distal to the occlusion is imperative to decrease ischemia and necrosis. Fogarty catheters (long narrow catheters with an inflatable balloon tip) are used to remove the thrombus or embolus. A small incision is made in the blood vessel near the occlusion, and the catheter tip is inserted past the occlusion. The balloon is inflated, and the catheter with the occlusive material is drawn back through the incision. The blood vessel and wound are sutured as circulation is assessed.

Vascular Bypasses and Grafts

Vascular bypass surgery involves the use of either autografts, such as the patient's own saphenous vein, or a synthetic graft material. The graft is anastomosed to the artery proximal to the occlusion and tunneled past the occlusion, where the distal end of the graft is anastomosed to the artery (Fig. 22–6). The graft is assessed for hemostasis and function, and then the wound is sutured closed.

Repair of a diseased area of a blood vessel, such as an aortic abdominal aneurysm, is performed with resection of the diseased area and replacement with a graft (aortic aneurysmectomy). This is usually an elective procedure. However, if an aneurysm is dissecting or ruptures, it is a surgical emergency.

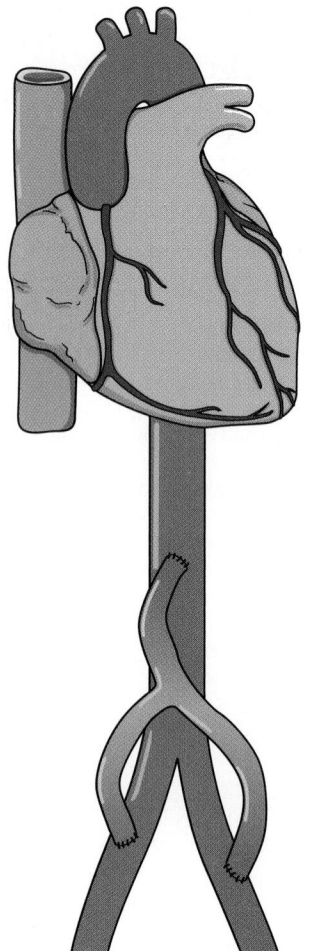

Figure 22-6 Aortic-femoral bypass.

Newer techniques are being used for bypass surgery. One technique uses video-assisted aortofemoral bypass without laparotomy. This method reduces length of hospital stay and patient recovery times.

Arteriovenous (AV) grafts or fistulas are created when permanent circulatory access is needed, such as for hemodialysis. Grafts are created by attaching one end of a synthetic graft to an artery and the other end to a nearby vein. An AV fistula is formed when a selected artery and vein are directly anastomosed to each other. The radial artery and cephalic vein in the forearm are most commonly used for these procedures. Local anesthesia is usually used.

Endarterectomy

Arteriosclerotic plaques are dissected from the lining of the arterial wall and removed in a procedure called an endarterectomy. This is most commonly performed on the carotid artery. To control blood flow, the artery is clamped on both sides of the occlusion, and an incision is made into the artery. The plaque within the artery is removed with forceps. The artery is irrigated to remove any further debris and then closed with sutures. The clamps are removed, and the skin incision is closed. A drain may be placed to help prevent hematoma formation.

Angioplasty

Minimally invasive techniques can also be used to open plaque-blocked arteries. These techniques include balloon or laser angioplasty. A flexible laser-tipped catheter is inserted into an artery and advanced to the site of the blockage. The laser sends out pulsating beams of light, which vaporize the plaque.

Stents

Stents are placed inside an artery to provide support to the artery walls and keep them open. (See Chapter 18.) Stents are placed in a procedure similar to percutaneous transluminal coronary angioplasty (PTCA). Stents may also be used in combination with other procedures such as angioplasty.

Complications of Vascular Surgeries

Bleeding and hemorrhage can occur with all vascular surgeries. Drainage can be expected with most surgeries. Drainage is usually small when peripheral vessels are involved. But with involvement of the great vessels, drainage is usually heavier and drains are often placed to prevent swelling and hematoma formation. If hemorrhage occurs, manual pressure is applied to the site of bleeding and the physician notified immediately.

Reocclusion is possible with any vascular surgery. If thrombi or emboli develop and block blood flow, it is a surgical emergency.

Hematoma formation is a complication of carotid surgery that can compress and compromise the airway. Monitoring the operative site visually for a hematoma is essential.

Neurological dysfunction may occur if plaque embolizes to the brain during surgery.

Extensive surgeries may also result in significant blood loss, leading to fluid volume deficit or shock.

Medical Management

Frequent assessments are ordered postoperatively. Neurological checks following carotid surgery are initially ordered hourly. Neurovascular checks of extremities every 1 to 4 hours are usually ordered. The incisional area is monitored for hematoma formation. Abdominal girth measurements (for aortic aneurysm repair) are ordered. An increase in abdominal girth may indicate hemorrhage. Any abnormal change is reported immediately to the physician. A loss of a pulse can indicate that circulation has been impaired in the vessel. The patient's need to return to surgery is anticipated if signs of impaired circulation are found.

CBC, prothrombin time (PT), PTT, and electrolytes may be ordered daily. Intake and output monitoring may be initially ordered hourly, then every 4 to 8 hours. Imbalances in fluid status should be reported. IV crystalloid solutions, volume expanders, or blood may be ordered for fluid deficits.

▶ NURSING PROCESS: PREOPERATIVE VASCULAR SURGERY

Assessment

A baseline assessment is important for postoperative comparison and discharge planning. Pain control needs and circulatory status are assessed. Diagnostic test (CBC, electrolytes, PT, PTT, and bleeding time) results are reviewed, and typing and crossmatching of blood to be placed on hold is performed.

Nursing Diagnosis

The nursing diagnoses for preoperative vascular surgery may include but are not limited to the following:

- Acute or chronic pain related to ischemia of tissue distal to occlusion or aneurysm
- Anxiety related to unknown outcome, pain, powerlessness, or threat of death
- Deficient knowledge: preoperative and postoperative procedures related to unfamiliar process.

▶ NURSING PROCESS: POSTOPERATIVE VASCULAR SURGERY

Assessment

On transfer postoperatively to either the ICU or surgical unit, the patient is positioned comfortably and a head-to-toe assessment is performed and documented, with abnormal findings reported to the physician. Once a patent airway is ensured, vital signs are monitored according to institutional policy, or more frequently if they are unstable. The patient's pain level is rated on a scale of 0 to 10. All IVs and drains are monitored. Patients with extensive surgery may have a nasogastric tube. Measurement of intake and output is usually done hourly. Laboratory tests are also monitored.

Arteriovenous grafts and fistulas are assessed for patency by palpating for a thrill (a tremor) and auscultating for a bruit (swishing sound) at the site of the graft or fistula. If a thrill or bruit is not present, the physician is notified immediately. Dressings or incisions are checked, and any drainage or hematoma development is noted and marked.

For carotid surgeries, airway and neurological checks are performed hourly or as ordered. Airway checks involve assessment of respirations and tracheal position. Any respiratory distress is reported for prompt intervention. If the trachea is not midline, the shift to either side is noted and reported immediately. Impaired respiratory function may occur rapidly, with tracheal shifting from bleeding or hematoma development in the neck. Neurological checks include level of consciousness; orientation; pupil size and reaction; cranial nerve function, such as eye and tongue movements; movement, strength, and sensation of extremities; and speech. Temporal pulses are also assessed (located at the hairline near the outer end of the eyebrow).

Initially, neurovascular checks are performed hourly for aortic or extremity vascular surgery. Neurovascular checks include extremity movement and sensation, presence of numbness or tingling, pulses, temperature, color, and capillary refill (less than 3 seconds normally). Peripheral pulses are palpated, or assessed with Doppler ultrasound if not palpable, marked, and compared with the unaffected extremity to detect deficits. If a pulse is absent or weak or the extremity is cool or dusky, the physician is notified immediately. A return to surgery for an embolectomy or other procedure is anticipated.

LEARNING TIPS

Neurovascular checks refer to the assessment of an extremity. (*Neurological checks* refer to assessment of the central nervous system). The following are areas to examine on an extremity when doing neurovascular checks. They are identified under the category for which they provide information:

Neurological	Vascular
Movement	Pulses
Sensation	Capillary refill
Numbness	Color (nailbed or skin)
Tingling	Temperature

Nursing Diagnosis

The nursing diagnoses for postoperative vascular surgery may include but are not limited to the following (Nursing Care Plan Box 22–8):

■ Acute pain related to surgical incision or reperfusion of tissue
■ Ineffective airway clearance related to intubation, anesthesia, pain, or sedation
■ Risk for infection related to surgical incisions
■ Ineffective tissue perfusion related to hypotension, hypothermia, emboli, or vascular reocclusion
■ Deficient knowledge: surgical recovery needs related to lack of prior exposure

Planning

The patient and family should be included in planning postoperative care goals. The patient's goals include the following:

■ Have pain relieved.
■ Maintain vital signs and oxygen saturation within normal range.
■ Remain free from infection.
■ Maintain peripheral pulses and tissue perfusion.
■ Remain free of edema and maintain clear lung sounds.
■ Understand the disease and its treatment.

Nursing Interventions
Acute Pain

Pain is usually mild and can be easily controlled after peripheral surgery. If pain is severe, it may indicate ischemia caused by spasm or reocclusion. With great vessel surgery, severe pain may be experienced, requiring more than one pain management method for relief. All characteristics of the pain are assessed to compare with previous pain episodes.

Ineffective Airway Clearance

After carotid endarterectomy, the head of the bed should be elevated 30 to 45 degrees. Elevation minimizes venous oozing and helps prevent hematoma formation in the neck, which can compromise respiratory function. Patients are encouraged to deep breathe to expand the alveoli. Depending on the procedure, coughing may need to be avoided to prevent increased pressure on the surgical site.

Risk for Infection

Hand washing is important to prevent infection. Temperature is monitored every 4 hours. Dressings are kept dry and intact and incisions are assessed with every dressing change. Sterile technique is used to change dressings.

Ineffective Tissue Perfusion

Circulatory parameters should be double-checked between nurses changing shifts, so that the oncoming nurse has a baseline for comparison. The presence and quality (strength) of a pulse and how it was assessed (palpated or via Doppler ultrasound) is documented.

Deficient Knowledge

The patient's and family's understanding regarding the postoperative recovery period is assessed. Physician's orders and routines regarding ambulation, exercise or restrictions, pain management, incision care, coughing and deep breathing, medications (heparin or warfarin), prevention, and early signs and symptoms of thrombus or embolus formation are explained. Written, as well as verbal, information should be given for future reference by the patient (Home Health Hints). Audiotapes are also a helpful method of providing discharge information. The patient can listen to the tapes in the hospital and have questions answered by the nurse; then the tapes can be taken home for review as needed or to share with other family members who may not have been able to participate in hospital teaching sessions.

Evaluation

For evaluation of nursing diagnoses for postoperative vascular surgery, see Nursing Care Plan Box 22–8.

BOX 22-8	NURSING CARE PLAN FOR THE POSTOPERATIVE PATIENT UNDERGOING VASCULAR SURGERY

▶ **Acute pain related to surgical incision and reperfusion of tissue**

PATIENT OUTCOMES

Patient will state that the pain is relieved or is tolerable. Patient will rest comfortably, perform respiratory treatments as necessary, and perform activities of daily living (ADLs).

EVALUATION OF OUTCOMES

Does patient state pain is relieved or acceptable? Is patient able to rest and participate in respiratory treatments and ADLs?

INTERVENTIONS	RATIONALE	EVALUATION
• Assess severity of pain, as well as all other qualities.	Peripheral vascular surgery pain is usually mild, and severe pain may indicate reocclusion. Major vascular surgery pain is severe. Pain relief is individualized.	Does patient state pain is at a tolerable level with a patent vessel?
• Ask patient to rate pain after analgesic is given.		Does patient state pain is controlled at a tolerable level?
• Notify physician if pain is unrelieved.	Different analgesic may be needed to give relief.	Are patient's pain relief needs met?

GERIATRIC		
• Ensure that elderly patient's pain is relieved.	Pain is not a normal part of aging, and elderly patients need and are entitled to adequate pain relief.	Does patient rate pain as none or at a tolerable level using a scale of 0 to 10?
• Use opioid pain medications cautiously. Consider reducing frail elderly patient's first opioid dose by 25%-50%, and increase as safe and needed and as ordered.	Elderly patients are more susceptible to peak effects and duration of analgesia of opioids.	Are patient's vital signs and sedation levels within normal limits (WNL)?

▶ **Ineffective tissue perfusion related to hypotension, hypothermia, emboli, vascular spasm, or reocclusion**

PATIENT OUTCOMES

Patient will have palpable peripheral pulses: adequate capillary refill; and normal color, temperature, motor, and sensory function of extremities. Patient will have reactive pupils and baseline cognitive function.

EVALUATION OF OUTCOMES

Is patient's circulatory status WNL? Does patient have reactive pupils and baseline cognitive function intact?

INTERVENTIONS	RATIONALE	EVALUATION
• Assess circulation, movement, and sensation to extremities every 1 to 4 hours.	Early detection of spasm or reocclusion minimizes risk of ischemia and necrosis.	Does graft or vessel remain patent?
• Mark location of pulses on affected extremity.	Allows for quick location of pulses.	Are pulses located easily?
• Perform neurological checks every 2 to 4 hours (carotid).	Allows early detection of complications.	Are major neurological or circulatory problems detected?
• Perform circulation or neurological check between nurses when changing caregiver.	Subtle changes can be detected and new caregiver has baseline for comparison.	Is baseline assessment done?
• Measure abdominal girth every shift (abdominal aortic surgery).	Increasing girths may indicate bleeding into abdomen.	Does abdominal girth remain unchanged?
• Take temperature every 4 hours.	May indicate infection or hypothermia with need for further warming.	Does patient remain normothermic?
• Monitor CBC as ordered.	RBC count, hemoglobin, and hematocrit decrease with insidious bleeding into abdomen or significant hematoma formations.	Is CBC WNL?
• Avoid constricting measures on affected extremity: knee gatch of bed, adhesive tape, tight dressings.	Prevent further decrease in blood flow to compromised extremity.	Is blood flow to affected extremity maintained?
• Auscultate AV shunts and grafts for bruits and palpate for thrills.	Any decrease or cessation of bruit or thrill indicates occlusion.	Does AV shunt or fistula remain patent?

HOME HEALTH HINTS

- A copy of the discharge instructions should be reviewed with the patient and family to ensure consistent understanding between the nurse and the family of the plan of care.

- A home health social worker visit to a postsurgical cardiac patient may be useful to help the patient and family plan for lifestyle changes to reduce anxiety and environmental stress.

- The ability of the patient caregiver to provide necessary care should be assessed. The support and resources available to the caregiver should be explored to prevent caregiver role strain.

- The need for caregiver respite care should be assessed, especially over time. If respite care is needed, assist the caregiver in identifying respite care resources in the family or community.

- Assist the caregiver to identify a plan to distribute the care workload among family members, if possible.

- Teach the caregiver stress management techniques to use.

- The importance of taking medications as prescribed needs to be reinforced, especially for immunosuppressive medications for cardiac transplant patients.

- Advise patients who are leaving home for the weekend or holidays to refill medicines ahead of time to ensure that they do not run out. The physician can write a prescription for the patient to have for emergency refills.

- Patients with cardiac transplants should not be exposed to people with illnesses such as colds or the flu. This should be explained to family members and visitors.

- Monitor the patient following cardiovascular surgery for complication development such as incisional infection, pneumonia, thrombophlebitis, or pulmonary emboli. Report any abnormal findings.

- Chest pain from esophageal reflux can mimic cardiovascular symptoms. The home health nurse should ask if the pain is related to consuming large meals, lying down, or bending over, or if it is relieved with antacids or food. Inform the physician of these findings.

CRITICAL THINKING

Mr. Smith

Mr. Smith has just returned to the surgical unit from the postanesthesia recovery room after undergoing a right carotid endarterectomy. His assessment findings include sleepy but arousable; oriented to person, place, and time; pupils equal and reactive to light; moving all extremities spontaneously and in response to commands equally and strongly; right neck dressing dry and intact; blood pressure 146/82, apical pulse 74, and respiratory rate 16.

1. How should Mr. Smith be positioned?

One hour later Mr. Smith is very sleepy but arouses to verbal and tactile stimulation. His pupils are equal, but the right pupil reacts more slowly than the left. He is only moving all extremities with repeated commands, with the left side moving less than the right. There is a 2-cm shadow of drainage on the right neck dressing. His vital signs are BP 176/88, P 64, R 26.

2. What are seven areas of concern found during this assessment of Mr. Smith?
3. What should your first action be?
4. List four priority items you should continue assessing as the physician is contacted.

Answers at end of chapter.

CRITICAL THINKING

Mr. Jangles

You are caring for Mr. Jangles, age 63, who has just returned from surgery after a embolectomy of the right lower leg. He has a history of insulin-dependent diabetes mellitus, hypertension, renal insufficiency, and a myocardial infarction 2 weeks ago. He is 6 feet tall and weighs 316 pounds.

1. What are four priority assessment areas for you to perform for Mr. Jangles?
2. What other information might you want to know regarding Mr. Jangles's medical history?
3. List three priority nursing diagnoses for Mr. Jangles.
4. State one outcome for each nursing diagnosis.
5. What are nursing interventions for each nursing diagnosis that you will perform?

Answers at end of chapter.

REVIEW QUESTIONS

1. A patient is scheduled for vascular surgery. The patient is taking digoxin, furosemide, potassium, warfarin, and famotidine. Which medication may be stopped several days before surgery?
   a. Digoxin (Lanoxin)
   b. Furosemide (Lasix)
   c. Warfarin (Coumadin)
   d. Famotidine (Pepcid)

2. Which one of the following is the purpose of CABG surgery?
   a. Cure coronary artery disease
   b. Increase blood flow to the myocardium
   c. Prevent spasms of the coronary arteries
   d. Decrease blood flow to the coronary arteries

3. The nurse is planning care for a patient having a cardiac valve replacement. With which type of valve will the patient be on anticoagulant therapy to prevent thrombus formation?

   a. Mechanical valves
   b. Porcine valves
   c. Allograft valves
   d. Bovine valves

4. Which one of the following can cause postoperative bleeding in a patient who has had an abdominal aortic aneurysm repair?
   a. Coronary artery spasm
   b. Hypotension
   c. Hypertension
   d. Heparin reversal

5. The nurse is assessing a patient who has had an AV fistula created. How is the patency of the fistula assessed?
   a. Check the pulse proximal to the fistula.
   b. Palpate the skin temperature proximal to the fistula.
   c. Check skin color and capillary refill proximal to the fistula.
   d. Palpate the thrill and auscultate the bruit over the fistula.

Answers to CRITICAL THINKING

Mr. Jones

1. Apical pulse 105, respirations shallow, lung sounds diminished with crackles, respiratory rate 28, temperature 99.8° F, and irregular apical pulse all could indicate pulmonary problems.
2. Nursing interventions include the following: Have patient turn, cough, and deep breathe; provide incentive spirometry; elevate head of bed; culture sputum; administer pain medication as ordered; splint the incision; administer diuretics if ordered; assess lungs every 2 to 4 hours.
3. Hypovolemia, hypoxemia, and a febrile state could cause apical pulse to be elevated.
4. Hypovolemia and irregular rhythm could cause low blood pressure.

Mrs. Eden

1. No, Mrs. Eden is not tolerating this activity, as evidenced by her increased respiratory rate and apical rate.
2. Steroids, immunosuppressive therapy, depression, and fatigue could be causing her poor appetite.
3. Nursing interventions related to Mrs. Eden's poor appetite could include the following: offering small, frequent meals; having family bring favorite foods from home; allowing the patient to rest before meals; providing oral hygiene before meals; administering antiemetics before meals; giving a high-calorie meal at peak appetite.
4. Mrs. Eden could be withdrawn because of changes in her lifestyle as a result of her transplant, extreme fatigue and concerns regarding how she will raise her children, grieving for the donor, and fear that she will reject her new heart.

Mr. Smith

1. The head of the bed should be elevated 30 to 45 degrees.
2. Areas of concern include increased blood pressure, increased respiratory rate, decreased apical pulse, drainage on dressing, more difficult to arouse, unequal movement, unequal pupillary reaction.

3. Have the registered nurse notify the physician and stay with the patient.
4. Priority assessments include respiratory status (rate, depth, stridor), bleeding (increased drainage, swelling around neck, tracheal shift to left), neurological status (decreased level of consciousness, movement of extremities), and vital signs (increased blood pressure, decreased apical pulse).

Mr. Jangles

1. Priority assessments include respiratory status, circulatory status of right leg and foot, vital signs, and pain level.
2. A medical history should include Mr. Jangles's usual blood sugar values, insulin dose, ambulation aids, gait, knowledge base regarding his various disease processes, and what led to this hospitalization.
3. (1) Priority nursing diagnoses include pain related to surgery of right lower leg; (2) ineffective tissue perfusion related to embolectomy of right lower leg, renal insufficiency; (3) risk for injury related to leg surgery, diabetes, obesity.
4. Outcomes include (1) verbalizes relief of pain; (2) maintains adequate tissue perfusion as evidenced by palpable peripheral (pedal) pulses, warm and dry skin; (3) remains free from injury.
5. Nursing interventions include the following: (1) Position (especially right leg) for comfort; keep the right leg slightly elevated; educate the patient regarding the need to ask for pain medication before pain is too severe; educate the patient regarding the need to take pain medication to minimize the negative physiological effects of pain; assess pain on a pain scale; evaluate the effectiveness of medication using the same pain scale, report ineffective pain measures. (2) Check pedal pulses, surgical dressing, pedal sensation and movement, and color, initially every hour; report changes; check capillary refill; assess for pain in extremities; assess for edema in extremities; keep leg elevated slightly. (3) Make sure the nursing call light is within reach; provide assistance with ambulation; use walking aids.

REFERENCES

1. Peters, W, et al: Closed-chest cardiopulmonary bypass and cardioplegia: Basis for less invasive cardiac surgery. Ann Thorac Surg 63(6):1748, 1997.

UNIT FOUR BIBLIOGRAPHY

Ackley, BJ, and Ladwig, GB: Nursing diagnosis handbook. Mosby, St. Louis, 2001.

Aikat, S, and Ghaffari, S: A review of pericardial diseases: Clinical, ECG and hemodynamic features and management. Cleve Clin J Med 67(12):903-14, 2000.

Ammon, S: Managing patients with heart failure. Am J Nurs 101:12, 2001.

Anderson, LA: Abdominal aortic aneurysm. J Cardiovasc Nurs 15(4):1, 2001.

Ansell, J, et al: Managing oral anticoagulant therapy. Chest 119:22S, 2001.

Artinian, N: Perceived benefits and barriers of eating heart healthy. MedSurg Nurs 10(3):129, 2001.

Bither, C, and Apple, S: Home management of the failing heart. Am J Nurs 101:12, 2001.

Bonow, RO, et al: ACC/AHA guidelines for the management of patients with valvular heart disease: A report of the American College of Cardiology/American Heart Association Task Force on Practice Guidelines (Committee on Management of Patients With Valvular Heart Disease). J Am Coll Cardiol 32:1486-588, 1998.

Bouknight, D: Current management of mitral valve prolapse. Am Fam Physician 61(11):3343, 2000.

Burkhoff, D, et al: Transmyocardial laser revascularization compared with continued medical therapy for treatment of refractory angina pectoris: A prospective randomized trial. Lancet, 354(9182):885, 1999.

Carelock, J, and Clark, A: Heart failure: Pathophysiologic mechanisms. Am J Nurs 101:12, 2001.

Chambers, J: The clinical and diagnostic features of mitral valve disease. Hosp Med 62(2):72, 2001.

Chavey, WE, et al: Guideline for the management of heart failure caused by systolic dysfunction, Part I. Guideline development, etiology and diagnosis. Am Fam Physician 64:5, 2001.

Chavey, WE, et al: Guideline for the management of heart failure caused by systolic dysfunction, Part II. Treatment. Am Fam Physician 64:6, 2001.

Corsetti AL, and Perry, D: A comprehensive approach to facilitating the recovery of the cardiac surgery patient. J Cardiovasc Nurs 12:82, 1998.

Crumlish, C, et al: When time is muscle. Am J Nurs 100(1):26, 2000.

Deglin, JH, and Vallerand, AH: Davis's Drug Guide for Nurses. FA Davis, Philadelphia, 2001.

Demirag, M, et al: Mechanical versus biological valve prosthesis in the mitral position: A 10-year follow up of St. Jude Medical and Biocor valves. J Heart Valve Dis 10(1):78, 2001.

Etoch, SW, Koenig, SC, and Laureano, MA: Results after partial left ventriculectomy versus heart transplantation for idiopathic cardiomyopathy. J Thorac Cardiovasc Surg 117(5):952, 1999.

Fan, R, and Viccellio, P: Restrictive cardiomyopathy. eMedicine Journal 2(7), 2001.

Feldman, AM, and McNamara, D: Myocarditis. N Engl J Med 343(19):1388, 2000.

Finkelmeier, B: Cardiothoracic Surgical Nursing. Lippincott Williams & Wilkins, 2000.

Fleury, J, and Moore, S: Family-centered care after acute myocardial infarction. J Cardiovasc Nurs 13(3):73, 1999.

Franco-Cereceda, A, et al: Partial left ventriculectomy for dilated cardiomyopathy: Is this an alternative to transplantation? J Thorac Cardiovasc Surg 121(5):879, 2001.

Gylys, K, and Gold, M: Acute coronary syndromes: New developments in pharmacological treatment strategies. Crit Care Nurse 20(2):202, 2000.

Hamm, CW: Cardiac biomarkers for rapid evaluation of chest pain. Circulation 104:1454, 2001.

Heart Failure Society of America (HFSA) Practice Guidelines. HFSA guidelines for management of patients with heart failure caused by left ventricular systolic dysfunction—Pharmacological approaches. J Card Fail 5:4, 1999. Available online: www.hfsa.org/pdf/lvsd heart failure.pdf.

Howes, D: Myocarditis. eMedicine Journal 2(5), 2001.

Indik, JH, and Alpert, JS. Post-myocardial infarction pericarditis. Curr Treat Options Cardiovasc Med 2(4):351, 2000.

Jones, M, Hoffman, L, and Makaroun, M: Endovascular grafting for repair of abdominal aortic aneurysm. Crit Care Nurse 20(4):204, 2000.

Knox, D, and Mischke, L: Implementing a congestive heart failure disease management program to decrease length of stay and cost. J Cardiovasc Nurs 14:1, 1999.

Kyngas, H, and Lahdenpera, T: Compliance of patients with hypertension and associated factors. J Adv Nurs 29(4):832, 1999.

Ledoux, D: Acquired valvular heart disease. In Woods, SL, Sivarajan Froelicher, ES, and Motzer, SA: Cardiac Nursing, 4th ed. Lippincott, Philadelphia, 2000.

Ledoux, D, and Luikart, H: Cardiac surgery. In Woods, SL, Sivarajan Froelicher, ES, and Motzer, SA: Cardiac Nursing, 4th ed. Lippincott, Philadelphia, 2000.

Lewis, R, et al: Ethanol-induced therapeutic myocardial infarction to treat hypertrophic obstructive cardiomyopathy. Crit Care Nurs 21(2):20, 2001.

Lutz, CA, and Przytulski, KR: Nutrition and Diet Therapy, 2nd ed. FA Davis, Philadelphia, 2001.

McAvoy, J: Cardiac pain: Discover the unexpected. Nursing 2000 30(3):34, 2000.

McCormick, J, and Deeg, M: Pharmacologic treatment of dyslipidemia. Am J Nurs 100(2):55, 2000.

Metra, M, Giubbini, R, and Nodari, S: Differential effects of beta-blockers in patients with heart failure: A prospective, randomized, double-blind comparison of the long-term effects of metoprolol versus carvedilol. Circulation 102(5):546, 2000.

Murphy-Lavoie, H, Preston, C: Dilated cardiomyopathy. eMedicine Journal 2(3), 2001.

Navuluri, R: Nursing implications of anticoagulant therapy. Am J Nurs 101:12, 2001.

Pitt, B, et al: The effect of spironolactone on morbidity and mortality in patients with severe heart failure. Randomized aldactone evaluation study investigators. N Engl J Med 341(10):709, 1999.

Popovic, AD, and Stewart, WJ: Echocardiographic evaluation of valvular stenosis: The gold standard for the next millennium? Echocardiography 18(1):59-63, 2001.

Rabbani, LE: Acute coronary syndromes: Beyond myocyte necrosis. N Engl J Med 345:1057, 2001.

Samama, MM, et al: A comparison of enoxaparin with placebo for the prevention of venous thromboembolism in acutely ill medical patients. Prophylaxis in medical patients with enoxaparin study group. N Engl J Med 341(11):793, 1999.

Strom, B, et al: Dental and cardiac risk factors for infective endocarditis: A population-based, case-control study. Ann Intern Med 129(10), 1998.

Schreiber, D: Deep venous thrombosis and thrombophlebitis. eMedicine Journal 2(6), 2001.

Schwetz, B: Pacemaker for congestive heart failure. JAMA 286:17, 2001.

Shipton, B, and Wahba, H: Valvular heart disease: Review and update. Am Fam Physician 63(11):2201, 2001.

Siomko, A: Demystifying cardiac markers. Am J Nurs 100(1):36, 2000.

Slaughter, MS, and Ward, HB: Surgical management of heart failure. Clin Geriatr Med 16:3, 2000.

Thornton, S: Differential diagnosis of infective endocarditis. J Am Acad Nurse Pract 12(5), 2000.

Todd, B: Physicians' recommendations to patients for use of antibiotic prophylaxis to prevent endocarditis. JAMA 284(1), 2000.

Urbano, F: Peripheral signs of endocarditis. Hosp Physician 36(5), 2000.

Valley, V: Pericarditis and cardiac tamponade. eMedicine Journal 2(6), 2001.

Whelton, PK, et al: Sodium reduction and weight loss in the treatment of hypertension in older persons: A randomized controlled trial of nonpharmacologic interventions in the elderly (TONE). TONE Collaborative Research Group. JAMA 279(11):839, 1998.

Yau, TM, et al: Mitral valve repair and replacement for rheumatic disease. J Thorac Cardiovasc Surg 119(1):53-60, 2000.

Yoshikawa, T: State of infectious diseases health care in older persons. Clin Geriatr 7(5), 1999.

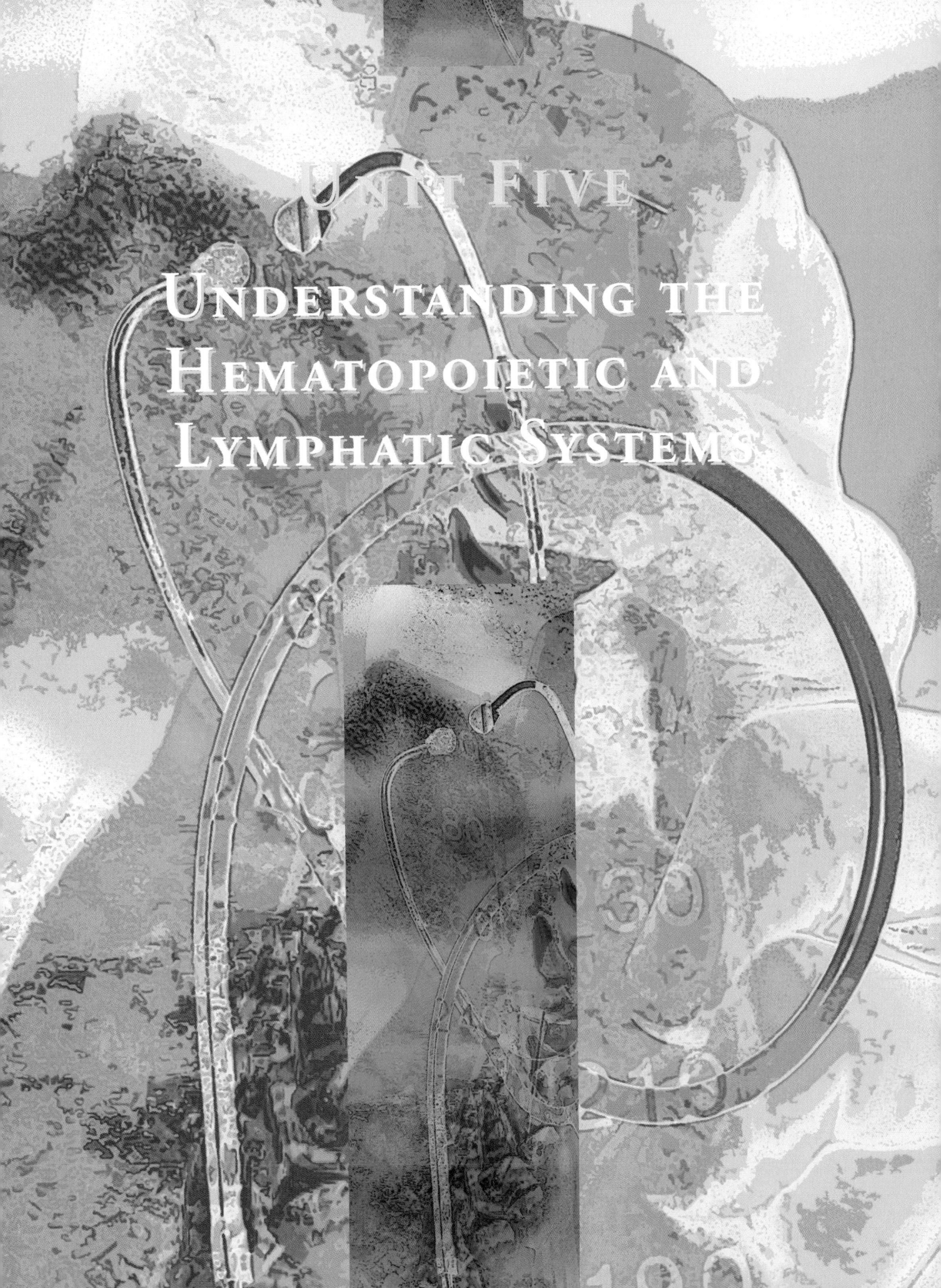

Unit Five

Understanding the Hematopoietic and Lymphatic Systems

23

HEMATOPOIETIC AND LYMPHATIC SYSTEM FUNCTION, ASSESSMENT, AND THERAPEUTIC MEASURES

Valerie C. Scanlon and Cheryl L. Ivey

KEY TERMS

ecchymoses (ECK-ih-**MOH**-sis)

lymphedema (LIMPF-uh-**DEE**-mah)

petechiae (puh-**TEE**-kee-eye)

purpura (**PUR**-pur-uh)

thrombocytopenia
(THROM-boh-SIGH-toh-**PEE**-nee-ah)

QUESTIONS TO GUIDE YOUR READING

1. What are the components of blood?
2. How are changes in the blood or blood-producing processes manifested as disease processes?
3. What is the sequence of events in the process of blood clotting?
4. What data should you collect when caring for a patient with a disorder of the hematological or lymphatic system?
5. What laboratory and diagnostic studies are used when evaluating the hematological and lymphatic systems?
6. What nursing care should you provide for patients undergoing each of the diagnostic tests?
7. What are common therapeutic measures for patients with hematological and lymphatic disorders?
8. What is the role of the licensed practical vocational nurse (LVN) in administering blood products?

► NORMAL ANATOMY AND PHYSIOLOGY

Blood

Hematology is the study of blood and its parts, functions, and abnormalities. The lymphatic system includes the lymph nodes and nodules that destroy pathogens. The lymphatic system produces many of the white blood cells that fight infection, and it returns lymph back to the blood via the lymph vessels.

The general functions of blood are transportation of oxygen, nutrients, and cellular waste products; regulation of body temperature, pH, and fluid balance; and production of cells that offer the body protection. Specific aspects of these functions are discussed with the particular part of the blood that is responsible for each.

A human body holds between 4 and 6 L of blood; 52 to 62 percent is plasma, and 38 to 48 percent is cells. The blood cells are the red blood cells (RBCs or erythrocytes), white blood cells (WBCs or leukocytes), and platelets (thrombocytes). All these blood cells are produced by the red bone marrow (RBM), a hematopoietic (blood-producing) tissue found in flat and irregular bones. The red bone marrow contains the undifferentiated stem cells that are the precursor cells for all blood cells. The other hematopoietic tissue is the lymphatic tissue of the lymph nodes, lymph nodules, spleen, and thymus, which produces only lymphocytes and monocytes. Table 23–1 shows normal cell counts.

Plasma

Plasma is the liquid portion of the blood and is about 91 percent water. It is the transporting medium for nutrients,

TABLE 23-1 REVIEW OF BLOOD CELL VALUES AND DISORDERS

Blood Cells Tested	Normal Value	Sample Disorders
Red Blood Cells		
Number of circulating RBCs	4.1–6.0 million/mm³	Increased in chronic hypoxia; decreased in anemia or blood loss
Hematocrit (cellular portion of blood)	38–48%	Increased in dehydration or chronic hypoxia; decreased in anemia or blood loss
Hemoglobin (reflects oxygen carrying capacity of blood)	12–18 g/100 mL	Increased in chronic hypoxia; decreased in blood loss or anemia
Reticulocytes (number of circulating immature RBCs)	0–1.5%	Increased in hypoxia or anemia; decreased in RBC maturation defect
White Blood Cells		
Number of circulating WBCs	5000–10,000/mm³	Increased in infection
Neutrophils	55–70%	Increased in infection
Eosinophils	1–3%	Increased in allergic response, some leukemias
Basophils	0.5–1%	Increased in hyperthyroidism, some bone marrow disorders, ulcerative colitis
Lymphocytes	20–35%	Increased in viral infections, chronic bacterial infection, some leukemias
Monocytes	3–8%	Increased in chronic inflammatory disorders, some leukemias
Platelets		
Number of circulating thrombocytes	150,000–300,000/mm³	Increased from trauma; decreased with blood disorders; low platelet count causes risk for bleeding

wastes, hormones, antibodies, and carbon dioxide. Plasma proteins include clotting factors, albumin, and globulins. Clotting factors such as prothrombin and fibrinogen are synthesized by the liver and circulate until activated in the clotting mechanism. Albumin is also synthesized by the liver, and it helps maintain blood volume and blood pressure by pulling tissue fluid into the venous ends of the capillary networks. Alpha and beta globulins are synthesized by the liver to be carrier molecules for substances such as fats, and gamma globulins are the antibodies produced by lymphocytes.

Plasma is also important in maintaining temperature because blood carries heat. The water of plasma is warmed by passage through active organs, such as the liver or skeletal muscles, and this heat is distributed as blood circulates throughout the body. This process may be visible in people with light skin. The flush of fever or vigorous exercise is caused by vasodilation in the dermis, allowing blood to circulate near the body surface and lose heat. A person in a cold environment may be pale, as vasoconstriction in the dermis keeps blood circulating in the core of the body to preserve heat.

The normal pH range of blood is 7.35 to 7.45, which is slightly alkaline. Chemical buffer systems in the blood prevent sudden fluctuations in pH and contribute to the body's acid-base balance.

Red Blood Cells

Mature RBCs are biconcave disks without nuclei; they carry oxygen bonded to the iron in hemoglobin (Hgb). Oxyhemoglobin is formed in the pulmonary capillaries where the hemoglobin combines with the oxygen in the lungs. Once hemoglobin gives up its oxygen to the cells of the body, it becomes reduced hemoglobin. The amount of hemoglobin in RBCs and the amount of iron in that hemoglobin are (in addition to the number of RBCs) the determining factors for the amount of oxygen the blood can carry. A lack of iron, hemoglobin, or RBCs is a form of anemia, which causes symptoms such as shortness of breath and weakness.

The rate of RBC production by the red bone marrow is most influenced by blood oxygen level. Hypoxia stimulates the kidneys to secrete erythropoietin, which increases the rate of RBC production and thus the oxygen-carrying capacity of the blood. At such times, immature RBC stages may be found in greater abundance in peripheral blood. A normoblast is the last RBC stage with a nucleus; a reticulocyte has visible fragments of its endoplasmic reticulum. These cells usually remain in the red bone marrow until mature; their presence in large numbers in peripheral blood indicates an insufficient amount of mature RBCs to meet the oxygen demands of the body.

Also required for normal production of RBCs is dietary intake of sufficient protein and iron to synthesize hemoglobin. The vitamins folic acid and vitamin B_{12} are necessary for DNA synthesis in the stem cells of the red bone marrow. The continuous mitosis of these cells depends on their ability to produce new sets of chromosomes. Vitamin B_{12} is also called extrinsic factor because its source is food. The parietal cells of the stomach lining produce intrinsic factor, which is a chemical that combines with vitamin B_{12} to prevent its digestion and promote absorption in the small intestine.

Red blood cells live for about 120 days and then become fragile and are phagocytized by fixed macrophages in the liver, spleen, and red bone marrow. The iron is returned to the red bone marrow for synthesis of new hemoglobin or is stored in the liver. The heme portion of the hemoglobin is converted to bilirubin, a bile pigment that the liver excretes into bile for elimination in the feces. Diseases such as malaria and sickle cell anemia cause an accelerated destruction of red blood cells. This hemoglobin release may cause the blood level of bilirubin to rise. When the bilirubin level is elevated, it stains the body fluids bright yellow to dark orange, depending on the bilirubin levels. This is known as jaundice.

Each person has a hereditary blood type, which refers to the antigens present on the RBCs. The two most important types are the ABO group and the Rh factor. The ABO type (A, B, O, or AB) indicates the antigens present (or not present, as in the case of type O) on the RBCs. In the plasma are antibodies for antigens that are not present in the blood; these antibodies can interact with antigens in transfused blood if the donor's blood does not match the recipient's blood (Table 23–2). To be Rh positive means that the D antigen is present on the RBCs; being Rh negative means that the antigen is not present. Rh-negative people do not have natural antibodies to the D antigen but will produce them if given Rh-positive blood.

White Blood Cells

White blood cells are larger than RBCs and have nuclei when mature. The granular WBCs (neutrophils, eosinophils, and basophils) are produced only in the red bone marrow. The agranular WBCs (lymphocytes and monocytes) are produced in the lymphatic tissue, as well as the red bone marrow. Table 23–1 shows normal values and percentages for each type in a differential count. WBCs carry out their functions in tissue fluid, as well as the blood, and all are involved in immunity or inflammation (response to injury).

Monocytes become macrophages, which phagocytize pathogens and dead tissue; neutrophils are more numerous but phagocytize only pathogens. Eosinophils detoxify foreign proteins during allergic reactions and parasitic infections. Basophils release histamine as part of inflammatory reactions. There are two groups of lymphocytes: T cells and B cells. T cells may be helper, suppressor, killer, or memory T cells. B cells become plasma cells, which produce antibodies to foreign antigens and also become memory cells.

Platelets

Platelets are formed in the red bone marrow; they are pieces of large cells called megakaryocytes. Platelets are involved in all mechanisms of hemostasis: vascular spasm, platelet plugs, and chemical clotting.

When a blood vessel is damaged, platelets release serotonin, which promotes contraction of smooth muscle and thereby vasoconstriction of an artery or a vein. Such constriction makes the break smaller, perhaps small enough to be covered by a clot. The clot is more likely to stay in place and stop any continued bleeding because it covers a smaller area. Capillaries have no smooth muscle and cannot constrict but are so small that breaks can be closed by platelet plugs. Platelets become sticky, adhering to the rough edges of the broken capillary and to one another, eventually forming a platelet plug that stops the bleeding.

Platelets also produce platelet factors, chemicals whose release is stimulated by contact of blood with a rough surface (either a break or damaged vessel lining), which are necessary for the first of the three stages of chemical clotting. In stage 1, platelet factors, clotting factors from the liver, tissue thromboplastin, and calcium ions react to form prothrombin activator (also called prothrombinase). In stage 2, prothrombin activator converts prothrombin (synthesized by the liver) into thrombin. In stage 3, thrombin converts soluble fibrinogen (also from the liver) to insoluble fibrin, strands of which form the clot. Calcium ions are also required for stages 2 and 3.

Excessive clotting in the vascular system is prevented in several ways. The very smooth endothelial lining of blood vessels repels platelets so that they do not stick to intact vessel walls. Heparin produced by mast cells inhibits the clotting mechanism. Antithrombin (synthesized by the liver) inactivates excess thrombin to prevent the clotting mechanism from becoming a vicious cycle.

The Lymphatic System

The lymphatic system consists of lymph, the system of lymph vessels, the lymph nodes and nodules, the spleen, and the thymus. Functions of the lymph system are to return tissue fluid to maintain blood volume and to protect the body against pathogens and other foreign material (the latter is called immunity and is covered in Chapter 52).

Lymphatic Vessels

Lymph is tissue fluid that has entered lymph capillaries (tissue fluid is formed from plasma by filtration in blood capillaries); it must be returned to the blood to maintain blood volume and blood pressure. Lymph capillaries are found in most tissue spaces; they anastomose, forming larger and larger lymph vessels, which have valves to prevent backflow of lymph. Lymph from the lower body and the upper left quadrant enters the thoracic duct (in front of the vertebral column) and is returned to the blood in the left subclavian vein (Fig. 23–1). Lymph from the upper right quadrant enters the right lymphatic duct and is returned to the blood in the right subclavian vein.

Lymph Nodes and Nodules

Lymph nodes are masses of lymphatic tissue (producing lymphocytes and monocytes) along the pathways of the lymph vessels. As lymph flows through the nodes, the WBCs produced enter the lymph, foreign materials are phagocytized by fixed macrophages, and fixed plasma cells produce antibodies to foreign antigens. The major paired groups of lymph nodes are the cervical, axillary, and inguinal nodes. These areas are located at the junction of the head (cervical) and

TABLE 23-2	ABO BLOOD TYPES	
Type	Antigens Present on RBCs	Antibodies Present in Plasma
A	A	Anti-B
B	B	Anti-A
AB	Both A and B	Neither anti-A nor anti-B
O	Neither A nor B	Both anti-A and anti-B

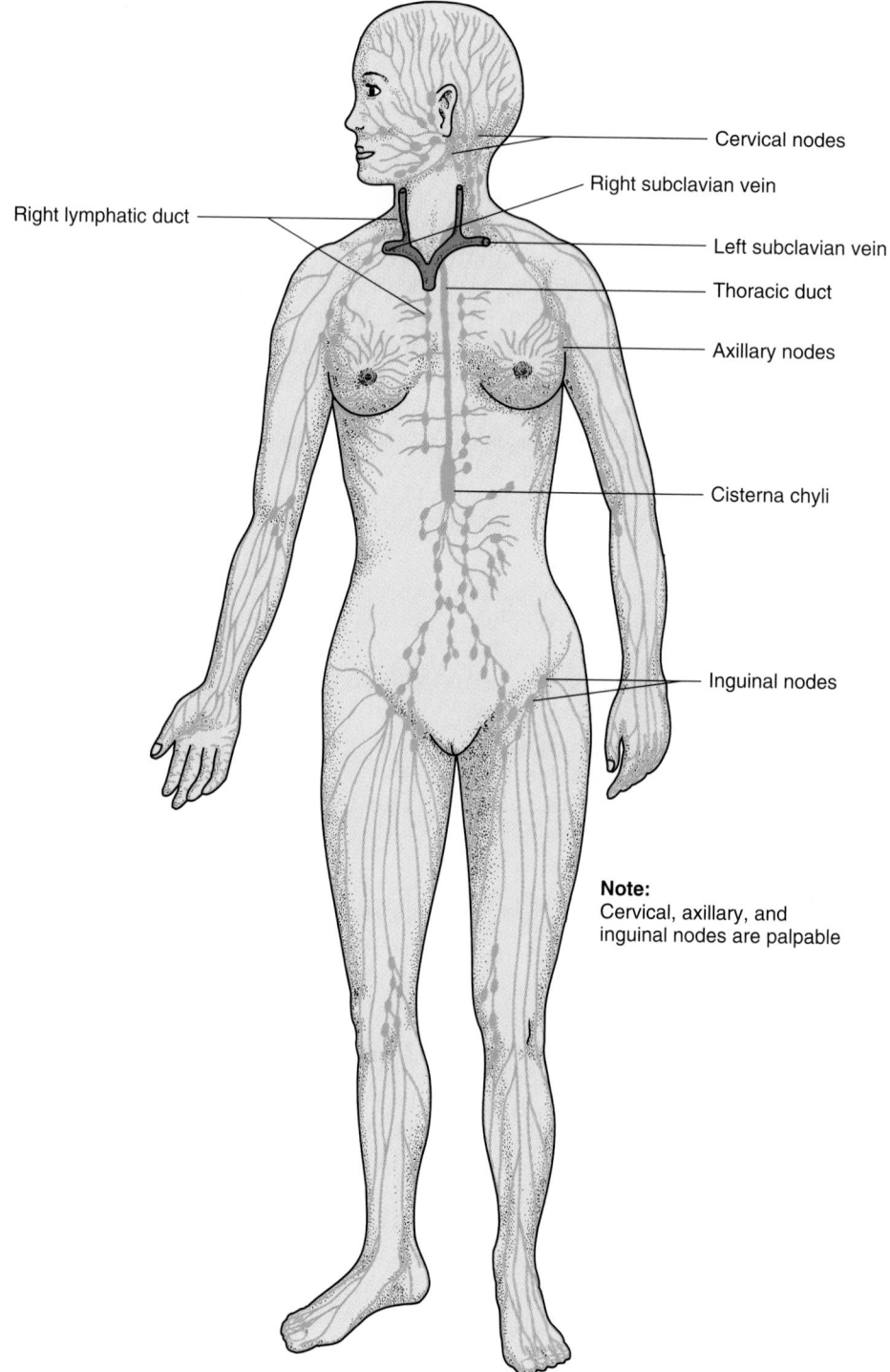

Cervical nodes

Right subclavian vein

Right lymphatic duct

Left subclavian vein

Thoracic duct

Axillary nodes

Cisterna chyli

Inguinal nodes

Note:
Cervical, axillary, and
inguinal nodes are palpable

Figure 23-1   System of lymph vessels and major groups of lymph nodes. (From Scanlon, V, Sanders, T: Essentials of Anatomy and Physiology, ed 3. FA Davis, Philadelphia, 1999, p 307, with permission.)

extremities (axillary and inguinal) with the trunk to remove pathogens in the lymph from the extremities before the lymph is returned to the blood.

Lymph nodules are small masses of lymphatic tissue found just beneath the epithelium of all mucous membranes. The body tracts lined with mucous membranes are those that have openings to the environment: the respiratory, digestive, urinary, and reproductive systems. Any natural body opening is a potential portal of entry for

pathogens; any pathogens that penetrate the epithelium usually are destroyed by the macrophages in the lymph nodules. The tonsils, which protect the oral and nasal portions of the pharynx, are familiar examples of lymph nodules, although most lymph nodules do not have names.

Spleen

The spleen is located in the upper left quadrant of the abdominal cavity, just below the diaphragm, behind the stomach. The lower ribcage protects the spleen from mechanical injury. In the fetus, the spleen produces red blood cells, a function assumed by the red bone marrow after birth. The spleen has several functions after birth.

The spleen produces lymphocytes and monocytes, which enter the blood. It also contains fixed plasma cells that produce antibodies to foreign antigens. It also contains fixed macrophages that phagocytize pathogens or other foreign material in the blood. These macrophages also phagocytize old RBCs and form bilirubin. By way of portal circulation, the bilirubin is sent to the liver for excretion in the bile.

The spleen is not considered a vital organ because other organs compensate for its functions if the spleen must be removed (although a person without a spleen is somewhat more susceptible to certain bacterial infections, such as pneumonia and meningitis). The liver and red bone marrow remove old RBCs from circulation, and the many lymph nodes and nodules produce lymphocytes and monocytes and phagocytize pathogens (as does the liver).

Thymus

The thymus is located below the thyroid gland on the front of the trachea; in the fetus or infant it extends under the sternum. With increasing age, the thymus shrinks, and relatively little thymus tissue is found in adults. The thymus produces T lymphocytes, or T cells, and hormones that contribute to the maturation of the immune system; that is, the ability to destroy foreign material. This ability is usually established by age 2. Immunity is covered in Chapter 52.

Aging and the Hematopoietic and Lymphatic Systems

Elderly people undergo a number of changes in these systems. Erythrocytes are produced more slowly. Iron deficiency is common, which leads to low hemoglobin levels. Plasma volume decreases, which increases the risk for dehydration. Immune responses become less efficient, making the elderly person more susceptible to infection, such as flu or pneumonia, and even cancer. Elders should take advantage of both flu and pneumonia vaccines. Autoimmune diseases such as rheumatoid arthritis are also more common in the elderly.

▶ NURSING ASSESSMENT
History

A thorough nursing assessment starts with an in-depth patient history (Table 23–3). Specific problems usually seen in patients with hematological disorders include abnormal bleeding, **petechiae** (small purplish hemorrhagic spots under the skin), **ecchymoses** (larger areas of discoloration from hemorrhage under the skin), and **purpura** (hemorrhage into the skin, mucous membranes, and organs), as well as fatigue, weakness, shortness of breath, and fever.

Begin by obtaining the patient's biographical data, marital status, occupation, religion, age, sex, and ethnic background. This information can give you valuable clues to risk factors. For example, even though hemophilia almost always occurs in males, females may carry the gene. Sickle cell anemia occurs mostly in African-Americans but also affects people of Mediterranean or Asian ancestry. Pernicious anemia occurs most often in people of northern European ancestry. By carefully collecting this information, you may be obtaining important clues that will help pinpoint the patient's problem. Finally, focus on the assessment of symptoms by using the WHAT'S UP? format presented in Chapter 1.

A complete review of past illnesses and family history is always indicated and can provide some valuable information. The social history is also useful. After developing good rapport with the patient, explore dietary and alcohol intake habits, any drug use or abuse, and sexual habits, all of which may cause changes in the hematological system.

An occupational review may reveal exposure to some hazardous substances that can cause bone marrow dysfunc-

TABLE 23-3	QUESTIONS ASKED DURING ASSESSMENT OF THE HEMATOLOGICAL AND LYMPHATIC SYSTEMS	
Question	**Rationale**
"Why are you seeking health care?"	Signs and symptoms of hematological/lymphatic disorders may be nonspecific. Any body system can be involved.
"How is the health of your blood relatives?"	Some blood and immune disorders are hereditary.
"Describe your usual diet."	Dietary deficiencies can lead to anemia or altered immune response.
"How much alcohol do you drink each day?"	Excess alcohol intake can lead to folic-acid deficiency anemia.
"What herbs or alternative therapies do you use?"	Herbs are drugs and can cause adverse reactions in the blood and immune system.
"What is your occupational and military history?"	Exposure to certain hazardous substances can lead to leukemias, other cancers, or anemias.
"Have you noticed any change in your energy level?"	Anemia and many cancers are associated with fatigue.
"Have you experienced nosebleeds or any other unusual bleeding? Have you had bloody or black bowel movements?"	Bleeding may indicate low platelet levels or a clotting factor deficiency.
"Do you experience shortness of breath or faintness?"	These are symptoms of anemia.
"Have you noticed any changes in your skin?"	Bleeding into the skin or mucous membranes can indicate a bleeding disorder.
"Have you noticed swelling in your neck, armpits, or groin?"	Swollen lymph nodes may indicate inflammation, infection, or some cancers.

tion. Certain occupations, such as working in a paint factory, tool and dye processing, and even dry cleaning are sometimes implicated in the formation of some hematological cancers. Military information may also reveal sources of exposure that can help during the diagnostic phase for hematological and lymphatic disorders.

Physical Examination

Hematological disorders can involve almost every body system, so each sytem must be carefully assessed. First, assess vital signs, which can reveal important clues. For example, frequent fevers indicate a poorly functioning immune system. Markedly subnormal temperatures may indicate an overwhelming gram-negative infection when combined with other abnormal assessment data. Heart and respiratory rate abnormalities may indicate decreased blood volume or decreased oxygen supply. Finally, check the level of consciousness; changes may occur with hypoxia, fever, and intracranial bleeding.

An inspection of physical structures most relevant to the blood and lymph systems includes the skin, mucous membranes, fingernails, eyes, lymph nodes, liver, and spleen. Observe the patient's skin color, noting pallor (indicating anemia), cyanosis (indicating poor oxygenation of RBCs), or jaundice (indicating liver disease or hemolysis). Observe the face, neck, and hands for areas of local inflammation, enlargements, or obvious sites of infection. Check the entire skin surface for purpura or other signs of bleeding, and examine the oral mucosa for color changes and petechiae or ecchymoses. These findings may indicate a bleeding disorder. Other important findings can be obtained after careful examination of the skin, looking for dryness and coarseness, which can indicate anemia. Inquire about itching, which may also indicate blood or lymph disorders.

Fingernails can give important clues about the patient's health. For instance, long striations (lines) on the nails or spoon-shaped nails may indicate anemia. Clubbed fingertips and raised nailbeds may indicate long-term hypoxia, which can be caused by anemia or heart disease.

Abdominal assessment should include the use of auscultation, or listening with the stethoscope. Listen intently and try to identify high-pitched, tinkling sounds, which might indicate intestinal obstruction, versus the regular, gurgling, and lower-pitched noises normally heard in the intestines. In some cases you will measure the abdominal girth and record the measurement in the nurse's notes. This baseline measurement might be useful if the patient begins to exhibit abdominal enlargement secondary to ascites or bleeding.

Next, progress to palpation, making sure that the patient is comfortable and that the patient's privacy is protected. Make sure to warm your hands. You can start with an examination of the lymph nodes, although this is more commonly done by the registered nurse or physician. The nodes of the neck can be gently palpated while the patient is sitting by simply running the hands over the neck and axillary surfaces. When examining the axillary nodes, the patient

may lie down. Any nodes that are palpable, with or without **lymphedema,** indicate some type of lymphatic disorder. Note the location, size, tenderness, texture, and fixation of the node groups. Enlarged or tender nodes indicate current or previous inflammation. Sternal tenderness may or may not be present. If present, this finding indicates that the bone marrow is being "packed" with an abnormal number and type of cells.

▶ DIAGNOSTIC TESTS

A number of diagnostic tests can help rule out or confirm a suspected diagnosis based on the analysis of the formed elements of the blood and the bone marrow. Specific studies include a complete blood cell count (CBC), coagulation studies, agglutination studies, bone marrow aspiration, and needle biopsy.

Blood Tests

Examples of laboratory studies routinely done for patients with hematological disorders include CBC, total hemoglobin concentration (Hgb), hematocrit levels (Hct), and platelet levels. (See normal values in Table 23–1.)

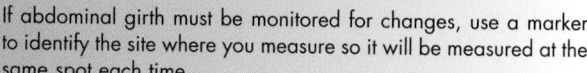

If abdominal girth must be monitored for changes, use a marker to identify the site where you measure so it will be measured at the same spot each time.

Coagulation Tests

Tests in this category include bleeding times, capillary fragility test, prothrombin time (PT), international normalized ratio (INR), partial thromboplastin time (PTT), and thrombin clotting time (TCT) (Table 23–4). Agglutination tests include ABO blood typing, Rh typing, crossmatching blood samples, and direct antiglobulin tests (also known as a Coombs' test).

⁀ CRITICAL THINKING

Mrs. Brown

Mrs. Brown is on warfarin (Coumadin) therapy because of a blood clot in her leg. She has a PT drawn at the lab, and the result is 12 seconds. Will the physician most likely increase her daily dose of warfarin, decrease it, or leave it the same? (Use Table 23–4 to figure out the answer.)

Answer at end of chapter.

Bone Marrow Biopsy

Biopsy information may be obtained through bone marrow aspiration, which in most states is obtained by a physician.

---

lymphedema: lymph—fluid found in lymphatic vessels + edema—swelling

TABLE 23-4 COAGULATION STUDIES

Test	Normal Value	Comments
Prothrombin time (affected by activity of clotting factors V, VII, X, prothrombin, and fibrinogen)	Men 9.6–11.8 seconds Women 9.5–11.3 seconds	Desired results 1.5 to 2 times longer for patient on warfarin (Coumadin) therapy
International normalized ratio	2.0–3.0 seconds for patient using anticoagulants (3.0–4.5 for recurrent problems)	Monitored with prothrombin time for warfarin therapy Standardized test adopted by World Health Organization
Partial thromboplastin time (affected by activity of clotting factors, prothrombin, and fibrinogen)	30–45 seconds	Desired results 1.5 to 2 times longer for patient on heparin therapy
Thrombin clotting time (measures time for fibrin clot to form after addition of thrombin)	10–15 seconds	Desired results 1.5 to 2 times longer for patient on heparin therapy
Bleeding time (measures time for small puncture wound to stop bleeding)	2.5–9.5 minutes	Indicates platelet function
Capillary fragility test	Fewer than 10 petechiae appearing in a 2-inch circle after application of a blood pressure cuff at 100 mm Hg for 5 minutes	Tests ability of capillaries to resist rupture under pressure Excessive fragility may be associated with thrombocytopenia

Aspiration of the bone marrow is done to obtain a specimen that can be viewed under the microscope. Purposes of this test include the diagnosis of hematological disorders; monitoring the course of treatment; discovery of other disorders, such as primary and metastatic tumors, infectious diseases, and certain granulomas; and isolation of bacteria and other pathogens by culture.

An accurate bone marrow specimen in an adult can be obtained from the sternum, the spinous processes of the vertebrae, or the anterior or posterior iliac crests. If a biopsy is also required, the latter are the preferred sites. Bone marrow biopsy, as well as aspiration, is considered a minor surgical procedure. Both are carried out under aseptic conditions. For iliac crest aspiration, the patient is placed comfortably on the side with the back slightly flexed. The posterior iliac crest is cleansed and covered with antiseptic solution. The skin, the subcutaneous tissue, and the periosteum are anesthetized using 1- or 2-percent lidocaine (Xylocaine). A 2- to 3-mm incision is made to facilitate penetration with a 14-gauge, 2- to 4-cm bone marrow needle. The incision is made to avoid introducing a skin plug into the marrow cavity, which can cause infection.

The nurse's role in bone marrow biopsy is multifaceted. First, it is often the nurse who helps coordinate between the laboratory and the physician, establishing a time to do the procedure and determining who obtains the supplies, such as the disposable bone marrow aspiration tray or other specialized needles, from the central supply area. The nurse may also assist in the following before the procedure takes place: administering analgesics according to the physician's order, helping to position the patient before and during the procedure, helping the patient maintain the necessary position, and observing the aspiration site for bleeding and infection. The nurse should also provide emotional support to the patient before, during, and after the procedure.

Lymphangiography

Disorders of the lymph system can be evaluated using lymphangiography. This procedure involves injection of a dye into the lymphatic vessels of the hand or foot. Various x-ray views are then taken to determine lymph flow or blockages. X-ray examinations are repeated in 24 hours to assess lymph node involvement.

Following the procedure, the physician may order a pressure dressing and immobilization of the injected limb to prevent leakage at the site. Continue to monitor the limb for swelling, circulatory status, and changes in sensation. Warn the patient that the skin, urine, or feces may be blue-tinged from the dye for about 2 days.

Lymph Node Biopsy

If a lymph node is enlarged, it may be biopsied to determine whether the cause is infection or malignancy. A biopsy may be done with a needle aspiration or surgical incision. A small dressing or Band-Aid is applied. Following the procedure, review signs of bleeding and infection with the patient and recommend that they be reported to the physician.

▶ THERAPEUTIC MEASURES

Therapeutic nursing measures are based on information obtained during the nursing assessment. A nursing diagnosis commonly used in patients with hematological disorders is fatigue related to decreased oxygen supply, secondary to decreased hemoglobin level. A patient with this nursing diagnosis should be taught to conserve energy through careful planning of rest and activity cycles, reducing stress, structuring the environment, enhancing dietary intake of iron and minerals, and avoiding highly emotional situations that can aggravate the fatigue level.

A second nursing diagnosis is ineffective protection: risk for bleeding, which may be associated with the medical diagnoses of hemophilia, **thrombocytopenia,** or disseminated intravascular coagulation. Monitor vital signs every 4 hours and assess sites of bleeding for change. Assess for changes in mental status, hypotension, and tachycardia, which may indicate bleeding and impending shock. See Chapter 24 for nursing interventions for patients at risk for bleeding.

Blood Administration

Blood is administered by a registered nurse. The licensed practical nurse/licensed vocational nurse (LPN/LVN) may be called on to assist with proper identification procedures and monitoring of vital signs during the transfusion. Table 23–5 lists blood components that may be ordered. The main goal is to give them safely and avoid mistakes. Make sure to use proper identifying information to ensure that the right patient is receiving the right blood products. A careful system most often used in health care institutions is outlined next.

Safety Steps

IDENTIFICATION. Safety is the first priority. This includes checking and double-checking the patient's identity. Great care is taken in the blood bank to match the donor's blood type with the recipient's blood type. After obtaining the unit of blood from the blood bank but before hanging the unit, two nurses (at least one of them a registered nurse) will check the following information at the patient's bedside (Fig. 23–2). Do not take this identification step for granted, because if the nurse makes an error and accidentally transfuses blood that does not match the patient's blood type, the results can be fatal.

■ Ask the patient to state his or her name aloud, if alert and able to speak.
■ Use the patient's identification band to confirm the identity and compare it with the information on the paperwork obtained from the blood bank.

TABLE 23-5	BLOOD PRODUCTS
Product	**Use**
Packed red blood cells	Severe anemia or blood loss
Frozen red blood cells	Autotransfusion (blood taken from patient and saved for future surgery), prevention of febrile reactions
Platelets	Bleeding caused by thrombocytopenia
Albumin	Hypovolemia caused by hypoalbuminemia
Fresh frozen plasma	Provides clotting factors for bleeding disorders, occasionally used for volume replacement
Cryoprecipitates	Bleeding caused by specific missing clotting factors

thrombocytopenia: thrombocyte—platelet + penia—lack

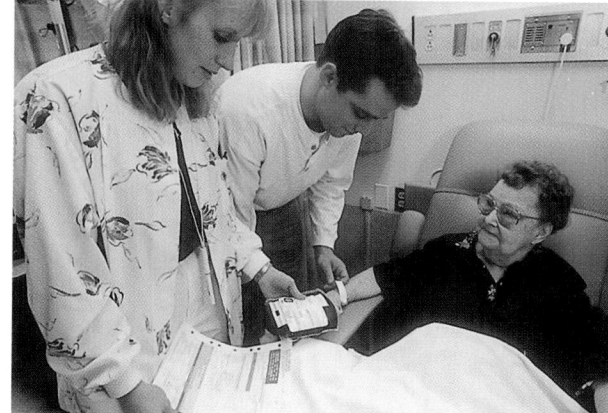

Figure 23-2 Two nurses check a patient's identification before administering a unit of blood.

■ Examine the blood bag and verify that the patient information and any other information, such as the ABO type, Rh type, and unit number, all match.
■ Finally, check the expiration date on the blood bag.

Do not give the unit of blood if any of the information does not match. Notify the blood bank immediately of any discrepancies, and delay the transfusion until the differences are cleared up. Remember, there is no room for error when transfusing blood products.

FILTERING. Filters are used with the blood administration tubing to prevent potentially harmful particles from entering the patient. Most often, the filter that comes with the transfusion tubing is sufficient for each unit of packed RBCs. In some situations, special filters may be needed to remove leukocytes or microaggregates. The blood bank can advise in these situations.

WASHED OR LEUKOCYTE DEPLETED. There are instances when packed RBCs (PRBCs) are ordered as "washed," often arriving from the blood bank in a round bag. The washing process removes almost all the plasma and can decrease the occurrence or severity of a febrile reaction. In addition, leukocyte filters may be used to completely remove all WBCs. This removal process is used in cases in which many transfusions are anticipated and decreases the chance of antigen sensitization, as well as transmission of certain viruses, such as cytomegalovirus.

WARMED. If the patient has had a severe bleeding episode and the nurses are helping to give replacement therapy through rapid, multiple transfusions, the physician may consider ordering a blood warmer. It works just as the name implies, warming the cold blood from the blood bank to the standard body temperature of 98.6° F. This warming helps prevent hypothermia, which can cause heart dysrhythmias, or shivering, which destroys blood cells and platelets.

Administration

GUIDELINES. It is important to use the correct intravenous needle size to infuse blood. The best sizes are 18- and 20-gauge needles. In a pinch a 22-gauge cannula can be used, but an infusion pump may be needed to help the thick blood move through the smaller lumen. Make sure to use only normal saline solution to help dilute the blood and to flush the intravenous lines before and after the transfusions. Any other type of fluid or medication may cause the blood product to clump, clot, or not infuse at all. Generally, 2 hours is a good time frame to transfuse each unit of packed cells. If it must transfuse more slowly because of the patient's condition, make sure that the unit does not hang longer than 4 hours. After that time, the blood is too warm and begins to deteriorate.

MONITORING. Careful monitoring of the patient's response to the transfusion is done to prevent complications or to detect and treat them quickly if they occur. Vital signs are taken and documented before starting the transfusion, after the blood has begun to infuse, and after the infusion. Some institutions require vital sign monitoring every 15 to 30 minutes during the earliest part of the transfusion, and then slightly less often for the duration of the infusion. Always follow institution guidelines. During the transfusion, the patient is assessed for signs and symptoms of transfusion reactions.

Complications

You can assist the registered nurse in the quick detection of complications. Many health care workers think of transfusing blood components as a routine procedure because it is a common activity. Do not be fooled. It is a serious procedure that can be life threatening if errors occur. Complications may include febrile reactions, hypersensitivities, hemolytic reactions, anaphylaxis, circulatory overload, and even death. Regular monitoring according to institution policy can help detect complications early, when treatment is most effective.

FEBRILE REACTION. By far the most common reaction is fever (febrile reaction), occurring up to 2 percent of the time. The risk of a febrile reaction goes up with each unit of blood product given to the patient. Many times, febrile reactions occur after the transfusion is completed, but they can occur at any time. This is the reason for obtaining a set of baseline vital signs, including the patient's temperature. Once a febrile reaction begins, the most common signs are an increasing fever and shaking chills, which can be severe. Other symptoms may include chest pain, headache, hypotension, and nausea or vomiting. If these symptoms occur, stop the transfusion and notify the physician. The physician usually treats the reaction symptomatically with acetaminophen. Often the physician orders the transfusion to continue once the patient is more comfortable. Make sure that the 4-hour hang rule is not violated.

URTICARIAL REACTION. Urticarial (hive) reactions are usually associated with antigens in the plasma accompanying the transfusion. There may be a fever, but the cardinal sign is the appearance of urticaria or a hivelike rash. On discovery of this reaction, notify the physician immediately. Expect that the patient will be given a dose of diphenhydramine (Benadryl) and then the transfusion will be restarted. Again, make sure the 4-hour hang rule is not violated.

HEMOLYTIC REACTION. The most deadly and, fortunately, the rarest of the possible reactions is an acute hemolytic reaction. The cause of this reaction is transfusion of incompatible blood. The result is hemolysis (destruction) of RBCs. Generally this type of serious reaction is noticed within minutes of starting the transfusion. The patient may report back pain, chest pain, chills, fever, shortness of breath, nausea, vomiting, or a feeling of impending doom. As the reaction progresses, the patient begins to show signs of shock, hypotension, oliguria, and decreased consciousness. Late signs and symptoms include those associated with disseminated intravascular coagulation: uncontrollable bleeding from many different sites at the same time, usually ending in death. At the first sign of this type of reaction, immediately stop the transfusion and stay with the patient. Institute emergency procedures to notify the charge nurse, the physician, and the blood bank. The vein is kept open with normal saline using a new administration set (ensuring that no more incompatible blood is administered) so that emergency drugs can be administered. High volumes of fluids are administered to decrease shock and hypotension, and high doses of diuretics are given to promote urine flow, because the kidney is the most likely organ system to be damaged. Dialysis may be instituted.

LEARNING TIPS

When a patient has a bacterial infection, the neutrophils, which are one type of WBC, rise in number to help fight it. There are two forms of neutrophils: segmented (mature) and bands (immature). Initially the number of segmented neutrophils goes up. Then, as the infection becomes more severe, the number of bands begins to rise.

An easy way to remember this is that the WBCs are part of the body's defenses, just like the military is part of a country's defenses. When needed, sargents who are fully trained or mature are called to battle first. If they are unable to fight off the invading enemy, new recruits being trained in boot camp are called in to help.

So segmented neutrophils (segs) are like the sergeants, fully mature and ready to fight. The bands are like boot camp recruits, immature and not fully trained. However, in an acute infection bands are necessary to prevent the body from being overwhelmed by the infection and losing the battle.

As you look at the differential WBC count, if the segs are elevated but the bands are normal, it is probably a new infection. If the bands are also elevated, the infection is worsening. The more elevated they are, the more severe the infection.

Lymphocytes fight viral infections and are elevated during a virus. A common pattern in the WBCs is produced for either a bacterial or viral infection. If the infection is acute bacterial:

$$\text{Segs} \uparrow \text{Bands} \uparrow \text{Lymphocytes} \downarrow$$

If the infection is viral:

$$\text{Segs} \downarrow \text{Bands} \downarrow \text{Lymphocytes} \uparrow$$

This can be remembered as the bone marrow producing the cells most needed during the time of viral infection and reducing production of those cells least needed. When the infection is resolved, all the cells should return to their normal production levels.

ANAPHYLACTIC REACTION. Anaphylactic reactions are not common but may be seen more often in patients who have received many transfusions or have had many pregnancies. Usually the source of the anaphylaxis is from sensitization to immune globulins passed from the donor's unit of blood product. In this type of reaction the very first milliliters of blood containing the allergens to pass into the patient's system may be enough to cause the patient to develop respiratory or cardiovascular collapse. Other more common symptoms include severe gastrointestinal cramping, instant vomiting, and uncontrollable diarrhea. If the patient exhibits these signs and symptoms, stop the transfusion at once and stay with the patient. Have someone else notify the registered nurse and the physician, using institutional emergency procedures. Emergency resuscitation measures must be instituted until the code team arrives, up to and including cardiopulmonary resuscitation. Expect the patient to be intubated and receive oxygen, steroids, and other drugs as needed for life support. After the emergency has passed, this patient will likely need to receive transfusions from frozen, deglycerolized blood cells.

CIRCULATORY OVERLOAD. Circulatory overload is caused by rapid transfusion in a short period, particularly in elderly and debilitated patients. Usual signs and symptoms include chest pain, cough, frothy sputum, distended neck veins, crackles and wheezes in the lung fields, and increased heart rate. If symptoms occur, stop the transfusion and notify the physician. Anticipate administration of diuretics, which help get rid of the excess fluid. Later the transfusion may be restarted, but at a much slower rate (Gerontological Issues Box 23–1).

BOX 23-1   GERONTOLOGICAL ISSUES

Elderly patients have less cardiac and renal ability to adapt to changes in blood volume, so they have a much higher risk of fluid overload when receiving blood transfusions. Carefully monitor lung sounds and vital signs both before and during a transfusion. New onset of dyspnea, crackles, hypertension, or bounding pulse should be reported to the registered nurse or physician immediately.

Answers to CRITICAL THINKING

Mrs. Brown

The physician will most likely increase Mrs. Brown's warfarin dose. Note in Table 23–4 that the PT for a patient on warfarin therapy should be 1.5 to 2 times longer than normal. That is the reason the warfarin is ordered—to prolong the time it takes for blood to clot. If a normal PT is 9.5 to 11.3 seconds, a therapeutic PT for Mrs. Brown would be 14.25 (or 9.5 × 1.5) to 22.6 (or 11.3 × 2). Her result of 12 seconds is not therapeutic.

REVIEW QUESTIONS

1. Clotting factors such as prothrombin are produced by which of the following?
   a. Red bone marrow
   b. Liver
   c. Spleen
   d. Lymph nodes

2. Which of the following best describes the function of erythropoietin?
   a. Increase production of platelets to promote clotting
   b. Decrease production of platelets to prevent abnormal clotting
   c. Increase RBC production to correct hypoxia
   d. Decrease RBC production to prevent hypoxia

3. The return of tissue fluid to the blood is important to maintain normal functioning in which of the following?
   a. Blood clotting
   b. Blood volume
   c. White blood cell formation
   d. Red blood cell formation

4. Which of the following refers to the portion of the blood in which cellular elements are suspended?
   a. Cytoplasm
   b. Platelets
   c. Plasma
   d. Hemoglobin

5. Mrs. Lee has a platelet count of 23,000/mm³. Which of the following actions should you take?
   a. Request an order for an anticoagulant.
   b. Protect Mrs. Lee from injury.
   c. Encourage Mrs. Lee to drink plenty of fluids.
   d. No action is necessary. This is a normal level.

6. You are assessing Ken and find small red-purple dots over most of his skin surfaces. He says that he has not noticed them before. Which action should you take first?
   a. Report your findings immediately to the registered nurse or physician.
   b. Document your findings objectively in the medical record.
   c. Assist Ken to apply a soothing lotion.
   d. Administer an antihistamine as needed.

24

NURSING CARE OF PATIENTS WITH HEMATOPOIETIC DISORDERS

Cheryl L. Ivey

KEY TERMS

anemia (uh-**NEE**-mee-yah)

disseminated intravascular coagulation (dis-**SEM**-i-NAY-ted IN-trah-**VAS**-kyoo-lar koh-AG-yoo-**LAY**-shun)

glossitis (glah-**SIGH**-tis)

hemarthrosis (HEEM-ar-**THROH**-sis)

hemolysis (he-**MAHL**-e-sis)

hemophilia (HEE-moh-**FILL**-ee-ah)

idiopathic thrombocytopenic purpura (ID-ee-oh-**PATH**-ik THROMB-boh-SIGH-toh-**PEE**-nik **PUR**-pew-rah)

leukemia (loo-**KEE**-mee-ah)

pancytopenia (PAN-sigh-toh-**PEE**-nee-ah)

panmyelosis (PAN-my-e-**LOH**-sis)

pathological fracture (PATH-uh-**LAH**-jik-uhl **FRAHK**-chur)

phlebotomy (fle-**BAH**-tuh-mee)

polycythemia (PAH-lee-sigh-**THEE**-me-ah)

thrombocytopenia (THROM-boh-SIGH-toh-**PEE**-nee-ah)

QUESTIONS TO GUIDE YOUR READING

1. How would you explain the pathophysiology of each of the hematological disorders discussed in this chapter?
2. What are the etiologies, signs, and symptoms of each of the disorders?
3. What is current medical treatment for each of the disorders?
4. What data should you collect when caring for patients with disorders of the hematological system?
5. What nursing care will you provide for patients with each of the covered disorders?
6. How will you know if your nursing interventions have been effective?
7. What care should you provide for the patient with sickle cell anemia? For patients with cancers of the hematological system?
8. What precautions should you institute to prevent bleeding in patients with clotting disorders?

Patients with hematopoietic disorders have problems related to their blood. Some problems are caused by too many cells, others by too few or defective cells. When red blood cells (RBCs) are affected, oxygen transport is also affected, causing symptoms of poor oxygenation. When white blood cells (WBCs) are affected, the patient is unable to effectively fight infections. If platelets or clotting factors are affected, bleeding disorders occur.

▶ DISORDERS OF RED BLOOD CELLS

Anemias

The term **anemia** describes a condition in which there is a deficiency of RBCs, hemoglobin, or both in the circulating blood. Because hemoglobin carries oxygen, this results in a reduced capacity to deliver oxygen to the tissues, producing symptoms such as weakness and shortness of breath, which lead the patient to seek medical help.

Pathophysiology

A decrease in the numbers of RBCs can be traced to three different conditions: (1) impaired production of RBCs, as in aplastic anemia and nutrition deficiencies; (2) increased destruction of RBCs, as in hemolytic or sickle cell anemia; or (3) massive or chronic blood loss. Some anemias are related to genetic problems in certain cultures (Cultural Consideration Box 24–1). It is important to remember that the general term *anemia* refers to a symptom or a condition secondary to another problem and is not a diagnosis in itself. Different types of anemia are discussed later in this chapter.

BOX 24–1	CULTURAL CONSIDERATION

In the past, Iranian cross-cousin marriages have resulted in an increased incidence of several forms of anemia and hemophilia. These marriages are now being addressed through genetic counseling and premarital screening for carriers. People are also tested for vitamin B_{12} or folic acid deficiencies linked to an enzyme deficiency.[1]

A sex-linked genetic disease common in the Chinese is glucose-6-phosphate dehydrogenase (G6PD) deficiency, an enzyme deficiency affecting the person's red blood cells and resulting in anemia. Mediterranean G6PD is common, causing a hemolytic crisis when fava beans are eaten, when aspirin or certain other drugs are taken, or in acidotic or hypoxemic states. Mediterranean-type G6PD deficiency is an inherited disorder most fully expressed in males, with a carrier state in females.

Among Asian Indians, sickle cell disease is highly prevalent; the gene is detected in 16.5 percent of selected populations. Sickle cell anemia is the most common genetic disorder among African-American populations. Sickle cell anemia also is found in individuals who live in areas where malaria is endemic, such as the Caribbean, the Middle East, the Mediterranean region, and Asia.

Etiologies

DIETARY DEFICIENCIES. Iron, folic acid, and vitamin B_{12} are all essential to production of healthy RBCs. A deficiency of any of these nutrients can cause anemia. Pernicious anemia is associated with a lack of intrinsic factor in stomach secretions, which is necessary for absorption of vitamin B_{12}. See Nutrition Notes Box 24–2 for more information.

HEMOLYSIS. Hemolysis is the destruction, or lysis, of RBCs. Destruction of RBCs leads to anemia and is termed hemolytic anemia. This may be a congenital disorder, or it may be caused by exposure to certain toxins.

OTHER CAUSES. Thalassemia anemia is a hereditary anemia found in persons from Southeast Asia, Africa, Italy, and the Mediterranean Islands. Individuals with thalassemia do not synthesize hemoglobin normally. Individuals with chronic disease also develop anemia (Gerontological Issues Box 24–3). Additional causes of anemia are discussed under the separate headings of aplastic and sickle cell anemias.

Signs and Symptoms

Symptoms of anemia include pallor, tachycardia, tachypnea, irritability, fatigue, and shortness of breath (Table 24–1). In addition to these symptoms, the patient with pernicious (vitamin B_{12}) anemia may experience numbness of the hands or feet and weakness, because vitamin B_{12} is necessary for normal neurological function. Pernicious anemia is also associated with a sore, beefy red tongue. Patients with iron deficiency may also have fissures at the corners of the mouth, an inflamed tongue (**glossitis**), and spoon-shaped fingernails.

Diagnostic Tests

A complete blood cell count (CBC) is done to determine the number of RBCs and WBCs per cubic millimeter. The size, color, and shape of the blood cells are determined by microscopic examination. Hemoglobin and hematocrit are below normal in anemia. Serum iron, ferritin, and total iron-binding capacity measurements are done to diagnose iron deficiency anemia. Serum folate is measured if folic acid deficiency is suspected. A bone marrow analysis may also be done.

Patients with pernicious anemia have low gastric acid levels, and many have antibodies to intrinsic factor. Both abnormalities are associated with poor absorption of vitamin B_{12}. If blood loss is suspected, additional tests are done to determine the source of bleeding.

Medical Treatment

Treatment begins with elimination of the contributing causes. Intake of the deficient nutrient can sometimes be increased in the diet or by administration of a supplement. Changing cooking habits, taking dietary supplements, decreasing

---

anemia: a—not + emia—blood

hemolysis: heme—blood + lysis—dissolution
glossitis: glos—tongue + itis—inflammation

BOX 24-2 *Nutrition Notes*

Understanding Common Nutritional Anemias

Although not its only cause, nutritional deficiencies can produce anemia. Nutrients vital to the construction of red blood cells include iron, folic acid, and vitamin B_{12}. Even if the cause of the anemia is dietary, other therapies may be employed in addition to nutritional interventions.

Iron deficiency anemia, the most common nutrient deficiency in the world, is characterized by smaller than normal RBCs. Insufficient intake of iron, excessive blood loss, or lack of stomach acid can lead to iron deficiency anemia. Individuals at greatest risk of iron deficiency are women of childbearing age and young children. Even before frank anemia is seen, cognitive abilities can be impaired. In early iron deficiency, serum transferrin, a blood protein that carries iron, rises in an attempt to increase iron-carrying capacity. Later the hemoglobin and hematocrit levels drop. A new measure of iron status is plasma transferrin receptor. It increases even in mild deficiency, making it a useful diagnostic aid for patients with inflammatory diseases.*

Relatively good sources of iron that are commonly included in Western diets are red meat; dark green leafy vegetables; dried fruits; and enriched, fortified, or whole grain products. Foods rich in vitamin C can be used to enhance absorption of iron from nonmeat sources. Iron supplements are often given to treat iron deficiency. This therapy should be continued for several months after hemoglobin and hematocrit levels return to normal to enable the body to rebuild iron stores.

Folic acid or vitamin B_{12} deficiencies produce anemias characterized by larger than normal RBCs. Both these vitamins are necessary for normal RBC production. Folic acid aids in the formation of DNA and heme, the iron-containing portion of hemoglobin. Conditions that increase the metabolic rate increase the need for folic acid. Many drugs, including alcohol,

anticonvulsants, and oral contraceptives, interfere with its absorption, metabolism, or excretion and can contribute to the development of anemia. Good food sources of folic acid include liver, green leafy vegetables, legumes, and enriched grain products. Because folic acid markedly decreases the occurrence of fetal neural tube defects such as spina bifida, women capable of becoming pregnant are advised to consume 400 µg of synthetic folic acid from fortified foods or supplements in addition to the folic acid furnished by a varied, balanced diet.

Vitamin B_{12} is essential for the manufacture of RBCs and for synthesis and maintenance of myelin, the fatty covering of nerves that facilitates rapid transmission of impulses. Vitamin B_{12} requires a highly specific protein-binding factor called intrinsic factor, secreted by glands in the stomach, to be absorbed. Intrinsic factor and vitamin B_{12} (also called extrinsic factor) combine in the proximal small intestine to form a complex to transport vitamin B_{12} to the ileum, where it is absorbed. Extrinsic factor (vitamin B_{12}) is found in foods from animal sources such as meat, fish, shellfish, poultry, and milk. Anyone eating these foods regularly is not at risk of vitamin B_{12} deficiency, but strict vegetarians are at risk. Because the deficiency here is dietary, a dietary supplement is the treatment.

In contrast, pernicious anemia is a disease caused by lack of intrinsic factor. Pernicious anemia occurs more frequently in older persons and is attributed to antibodies against gastric parietal cells and intrinsic factor. Extensive gastric resection can also result in insufficient intrinsic factor. Because the deficiency is not dietary, neither is the treatment. Pharmaceutical vitamin B_{12} is usually given by injection to circumvent the absorption problem. Symptoms of vitamin B_{12} deficiency are, in usual order of appearance, numbness and tingling in hands and feet, followed by RBC changes. Moodiness, confusion, depression, delusions, and overt psychosis appear next. Lastly, irreparable nerve damage occurs, and eventually, death. Because of the neurological damage, vitamin B_{12} deficiency should be considered in a person being evaluated for dementia.

*Connor, JR, and Beard, JL: Dietary iron supplements in the elderly: To use or not to use? Nutrition Today 32:102, 1997.

BOX 24-3 **GERONTOLOGICAL ISSUES**

Anemia of chronic disease is often diagnosed in an older patient who has an underlying medical condition that causes altered iron metabolism, deficiency of erythropoietin, or shortened life span of red blood cells. Unfortunately, anemia of chronic disease is often mistaken for iron deficiency anemia. Nutritional deficiencies and blood loss are common causes of iron deficiency anemia.

TABLE 24-1 **CLINICAL MANIFESTATIONS OF ANEMIA**

Body System	Mild (Hgb = 10–14 g/dL)	Moderate (Hgb = 6–10 g/dL)	Severe (Hgb < 6 g/dL)
Skin	None	None	Pallor, jaundice, pruritis
Eyes	None	None	Jaundiced conjunctiva and sclera, retinal hemorrhages, blurred vision
Mouth	None	None	Glossitis, smooth tongue
Cardiovascular	Palpitations	Increased palpitations	Tachycardia, increased pulse pressure, systolic murmurs, angina, congestive heart failure, myocardial infarctions
Lungs	Exertional dyspnea	Frank dyspnea	Tachypnea, orthopnea, dyspnea at rest
Neurological	None	None	Headache, vertigo, irritability, depression, impaired thought processes
Gastrointestinal	None	None	Anorexia, hepatomegaly, splenomegaly
Musculoskeletal	None	None	Bone pain
General	None	Fatigue	Sensitivity to cold, weight loss, lethargy

Hgb = hemoglobin.

alcohol intake, and controlling chronic diarrhea can help correct folic acid deficiency. If symptoms of anemia are acute, a blood transfusion may be necessary.

Nursing Process for the Patient with Anemia

ASSESSMENT. Monitor hemoglobin and hematocrit levels and other laboratory studies as ordered and report any downward trend. Monitor responses to therapy. Assess the patient's fatigue level and ability to ambulate safely and perform activities of daily living (ADLs). Monitor degree of dyspnea. Assess for pallor in the skin and conjunctivae.

NURSING DIAGNOSIS. Nursing diagnoses are based on assessment data and may include the following:

- Activity intolerance related to tissue hypoxia and dyspnea
- Imbalanced nutrition: less than body requirements related to disease, treatment, and lack of knowledge of adequate nutrition
- Risk for injury: falls related to weakness and dizziness
- Impaired oral mucous membranes

PLANNING AND INTERVENTION
Activity Intolerance. Plan care to conserve energy after periods of activity. Assist the patient with self-care activities as needed. Place articles within easy reach of the patient to reduce physiological demands on the body. Encourage the patient to limit visitors, telephone calls, and unnecessary interruptions. Allow rest periods between activities.

Monitor vital signs to evaluate tolerance to activity. If the pulse or respiratory rate increases more than 20 percent from baseline during activity, the activity is too strenuous. Administer oxygen as ordered to relieve dyspnea. Blood transfusions may be ordered if hemoglobin levels are very low or symptoms are severe.

Imbalanced Nutrition. If the anemia is caused by a dietary deficiency, consult the dietitian to provide diet modifications and instruction. Teach the patient with folic acid deficiency that daily requirements can be met by including foods from each food group at every meal. If the patient has a severe deficiency, dietary folic acid will not be enough; supplementation is the only way to correct the imbalance. Instruct the patient to continue taking the supplements until advised to stop by the physician.

Vitamin B_{12} is administered by intramuscular (IM) injection. The patient with pernicious anemia needs B_{12} injections for life.

Instruct the patient with iron deficiency in the use of iron supplements and side effects, which include nausea, diarrhea or constipation, and dark stools. If liquid supplements are used, give them with a straw to avoid staining the teeth. Iron is sometimes given as an intramuscular injection (Imferon). It should be given by the Z-track method to avoid staining at the site. Foods high in iron should be included in the diet; vitamin C enhances absorption of iron. (See Nutrition Notes Box 24–2.)

Risk for Injury. Assist the patient to change positions slowly to decrease dizziness and risk of falls. Assist with ambulation as needed. Protect the patient with pernicious anemia from injuries resulting from decreased sensation. Take special care with heating pads, turning and positioning, and other potential sources of injury because the patient may not feel pain.

Impaired Oral Mucous Membranes. If the patient has glossitis, provide good oral hygiene and soft, bland foods until healing occurs. Instruct the patient to use a very soft toothbrush and perform oral care after each meal and at bedtime.

Evaluation. When successfully treated, the patient will tolerate a normal level of activity without shortness of breath or excess fatigue. The patient should be able to explain the correct treatment plan and therapeutic measures for long-term prevention.

Aplastic Anemia

PATHOPHYSIOLOGY. Aplastic anemia differs from other types of anemia in that the bone marrow becomes fatty and incapable of production of the necessary numbers of RBCs. Also known as hypoplastic anemia, the cells that are produced are normal in size and shape, but there are not enough of them to sustain life. The resulting **pancytopenia** (reduced numbers of all the formed elements from the bone marrow—RBCs, platelets, and WBCs) is the indicator that something is wrong with the bone marrow. Left untreated, aplastic anemia is almost always fatal.

CAUSES. Aplastic anemia may be congenital—that is, the person is born with bone marrow incapable of producing the correct number of cells. In addition, it may be due to exposure to toxic substances such as industrial chemicals (e.g., benzenes and insecticides), chemotherapy medications, or use of cardiopulmonary bypass during surgery. Other causes include some bacterial and viral infections, such as tuberculosis and hepatitis.

SIGNS AND SYMPTOMS. The clinical features of aplastic anemia vary with the severity of the bone marrow failure. As with other anemias, early symptoms include progressive weakness, fatigue, pallor, shortness of breath, and headaches. As the disease progresses and the anemia and pancytopenia worsen, other symptoms, such as tachycardia and heart failure, may appear. Ecchymoses and petechiae appear on the skin surface because of the reduced platelet count (Fig. 24–1; also see Fig. 24–5). Blood may ooze from mucous membranes. Puncture sites may progress from oozing to frank bleeding. Often there is overt bleeding into vital organs. When aplastic anemia is left untreated, most patients die from infection or bleeding secondary to the lack of production of WBCs and platelets.

DIAGNOSTIC TESTS. The diagnosis of aplastic anemia begins with a CBC. Usually all values are reported as very low, with the occasional exception of the red blood cell count,

---

pancytopenia: pan—all + cyto—cell + penia—poverty

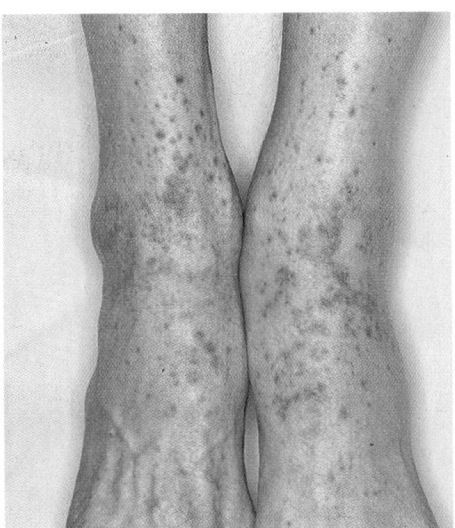

Figure 24–1 Petechiae on the skin from thrombocytopenia. (From Goldsmith, LA, et al: Adult and Pediatric Dermatology. FA Davis, Philadelphia, 1997, p 61, with permission.)

in part because of the longer life span of RBCs. Eventually the RBCs are also depleted. If the patient is having gross bleeding internally or externally, the RBC level drops rapidly and dramatically. The most definitive test is the bone marrow biopsy. Because the bone marrow is essentially dead, the result is often described as a "dry tap," in which pale, fatty, yellow, fibrous bone marrow is extracted instead of the red, gelatinous bone marrow normally seen. Not surprising, the more fatty and pale the marrow is at the bone marrow biopsy, the more dysfunctional the bone marrow is. Other diagnostic tests include total iron-binding capacity (TIBC) and serum iron level. It is common to find both these levels elevated because the RBCs are not being produced to use up the stores of iron in the production of hemoglobin.

TREATMENT. Early identification of the cause of the anemia and correction of the underlying problem are important to survival. Unfortunately, it is often difficult to determine the cause, and there is no way to reverse the damage already done. Aggressive supportive measures may be the only treatment. Most of these measures are aimed at prevention of infection and bleeding.

Today the most effective treatment for aplastic anemia is bone marrow transplantation (Patient Perspective Box 24–4). Another common therapy is the administration of steroids to stimulate production of cells in the weakened bone marrow. Occasionally the administration of hormones may work to increase the viability of the marrow. Steroid and other hormone treatments may be tried before attempting a bone marrow transplant.

A new line of therapy is also available. In many treatment institutions, limited success is being obtained with the use of colony-stimulating factors, natural elements now being produced synthetically. For example, erythropoietin (Epogen) stimulates the production of RBCs and filgrastim (granulocyte colony stimulator [Neupogen]) stimulates the

production of WBCs. The major drawback to this type of therapy is the high cost. Many of the pharmaceutical manufacturers have patient access programs that help with the costs of these medications.

NURSING INTERVENTIONS. Nursing care of patients with symptoms of reduced hemoglobin levels is presented under Nursing Process for the Patient with Anemia. If the patient's platelet count is low (usually less than 20,000), the patient is also placed on bleeding precautions (Box 24–5). If the white count is low, the patient must be protected from infection (Box 24–6).

Sickle Cell Anemia

PATHOPHYSIOLOGY. Sickle cell anemia is an inherited anemia in which the RBCs have a specific mutation that makes the hemoglobin in the red cells very sensitive to oxygen changes. Any time a decrease in the oxygen tension is sensed, the cells begin an observable physical change process from their usual spherical shape to a sickle or crescent shape (Fig. 24–2). Sickled cells are very rigid and easily cracked and broken. The abnormal shape also causes the

---

BOX 24–4 *Patient Perspective*

Bone Marrow Transplant

In June of 1997 I took my daughter to the doctor for her sports physical. Later that day, I received a call telling me to take her to the university hospital immediately because she had a serious life-threatening illness. I kept telling myself and my husband that our small-town hospital must have made some sort of error. As it turned out, they had not. My daughter was diagnosed with aplastic anemia and needed a bone marrow transplant. I became obsessed with the illness, poring over every tidbit of medical information I could find. Sometimes I found myself out in the car unable to remember where I was going; sometimes I had to pull over because my eyes were filled with tears and I could no longer see.

My daughter was 16 at the time of her illness, yet it is the parents who sign consent forms and make the choices in care. When the chemotherapy was started and running through the IV tubing, I felt like grabbing the tubing and pinching it off, yelling "I need more time to think about this decision," but time was running out. Without the transplant, she had about 8 months to live.

After transplantation, my daughter was in an isolation room for a month. I stayed with her every day, and at night I stayed at the inn that was attached to the hospital. If I was needed, I wanted to be no more than a minute away. I was one of the luckier parents because I had the financial means to manage this process. I thought about how horrible it would be if I had other children at home. Sometimes I would have such an urge to run away and escape from it all. I attended support groups that were held on the hospital unit. I got to know a lot of other parents with sick kids, and it became very upsetting to me at times. One day parents told me how well their child was doing; the next day I saw the child's room empty and thought he must have gone home, only to find out later that he had died during the night. I wondered if my daughter would be next.

I look at my daughter now, 4 years later, alive and perfectly healthy, and I tell myself that I made the right choices for her. But she tells me that, if it happens again, she will not go through chemotherapy. I wonder, is chemo worse than death?

BOX 24-5 **Interventions to Prevent Bleeding in the Patient with Thrombocytopenia**

1. Use an electric razor instead of a safety razor for shaving.
2. Use a soft toothbrush or gauze to clean the teeth.
3. Avoid invasive procedures as much as possible, including enemas, douches, suppositories, and rectal temperatures.
4. Avoid intramuscular injections.
5. To avoid injury when checking blood pressure, pump cuff up only until pulse is obliterated.
6. Avoid blood draws whenever possible. Use established access sites or group draws into once daily.
7. Maintain pressure on intravenous (IV), blood draw, and other puncture sites for 5 minutes.
8. Encourage use of shoes or slippers when out of bed.
9. Keep area clutter free to prevent bumps and bruises.
10. Avoid use of drugs that interfere with platelet function, such as aspirin products and nonsteroidal anti-inflammatory drugs.
11. Administer stool softeners as ordered to prevent straining to have a bowel movement.
12. Move and turn patient gently to avoid bruising.
13. Instruct patient to avoid blowing the nose.

BOX 24-6 **Interventions for the Patient at Risk for Infection**

1. The patient should be in a private room.
2. All personnel and visitors should wash hands before entering the room.
3. The patient should be taught to wash hands before and after using the toilet and before and after eating.
4. The patient and family should be instructed to wash hands before touching.
5. Staff or visitors with known infections should not enter the patient's room.
6. The patient should not handle flowers or plants brought into the room.
7. Raw fruits, vegetables, and milk products should be avoided.
8. Foley catheters and other invasive devices should be avoided.
9. If invasive procedures are necessary, strict aseptic technique should be used.
10. Use acetaminophen if an antipyretic is necessary; aspirin may induce bleeding.

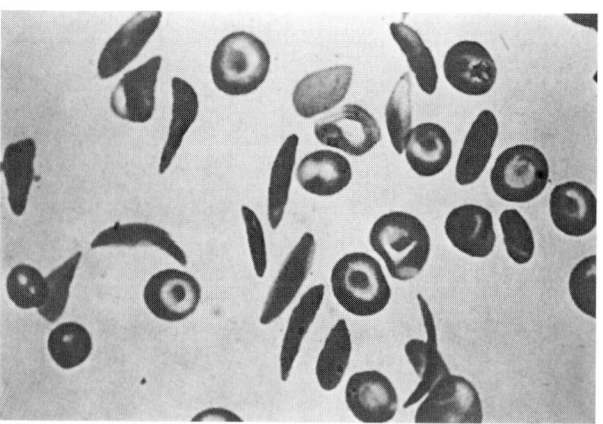

Figure 24-2 Sickled cells in sickle cell disease.

with sickle cell anemia. Normal red cells live about 120 days. Sickled cells survive only about 15 to 20 days, an 80- to 90-percent decrease in cell survival.

ETIOLOGY. Sickle cell disease is an autosomal recessive hereditary disorder. This means that if both parents pass on the abnormal hemoglobin, the child will have the disease. If only one parent passes on the abnormal hemoglobin, the child will have the sickle cell trait and will be able to pass the trait (or the disease if the other parent is also affected) on to his or her child.

In the United States, sickle cell anemia is most often found in those of African or Eastern Mediterranean origin. Worldwide, many persons residing in Asia, the Caribbean, the Middle East, and Central America are affected. Nearly 10 percent of African-Americans have the sickle cell trait; 1 out of every 400 African-American infants born has inherited the two sets of abnormal genes necessary to have the disease. Symptoms do not appear in infants until after the age of 6 months, because up to that age the infant is using hemoglobin manufactured during fetal life, which is not affected by the sickling process.

SIGNS AND SYMPTOMS. The sickling changes just described are a daily occurrence. The rapid return of the oxygen level to normal returns the cells to their normal shape for the most part.

Occasionally the sickling process cannot be reversed and the problem continues unabated. This sudden and severe sickling is called a sickle cell crisis. As more and more sickling occurs, the blood becomes sluggish and does not flow easily. It tends to collect in the capillaries and veins of the organs of the chest and abdomen, as well as joints and bones, and cause infarction (tissue necrosis resulting from lack of blood supply). Tissue necrosis results in pain, fever, and swelling.

Factors that contribute to the development of a sickle cell crisis include those related to decreased oxygenation. Some examples include pneumonia with hypoxia, exposure to cold, diabetic acidosis, and severe infection. Sickle cell anemia presents problems for the patient who needs surgery. Anesthesia and blood loss during surgery and postoperative dehydration can trigger a crisis.

cells to become tangled in the vessels, veins, and organs. The result is congestion, clumping, and clotting.

As red cells are broken, the cellular contents spill out into the general circulation. The resulting increase in the bilirubin level causes jaundice. Gallstones (cholelithiasis) may develop because of the increased amounts of bile pigments. The spleen and liver may enlarge because of the increase in retained cells and cellular materials.

Because of the differences in shape and texture, there is a significant decrease in life span of the RBCs in patients

Common symptoms produced during sickle cell crises include severe pain and swelling in the joints, especially of the elbows and knees, as the sickled cells impede circulation. Abdominal pain is common with swelling of the spleen and engorgement of the vital organs. Hypoxia occurs as fever and pain increase, causing the patient to breathe rapidly. The male patient may have a continuous, painful erection (priapism) from impaired blood flow through the penis. Symptoms of renal failure are common as circulation is slowed and the kidneys become clogged with cellular debris.

Repeated crises and infarctions lead to chronic manifestations such as hand-foot syndrome, an unequal growth of fingers and toes from infarction of the small bones in the hands and feet (Fig. 24–3). Additional manifestations of sickle cell disease are shown in Figure 24–4.

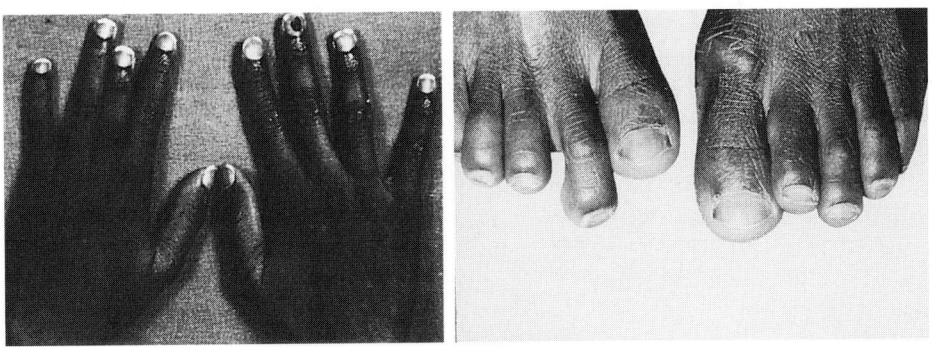

Figure 24-3 Hand-foot syndrome. Note different lengths of fingers and toes. (Courtesy Sandoz Pharmaceutical Corp., East Hanover, NJ.)

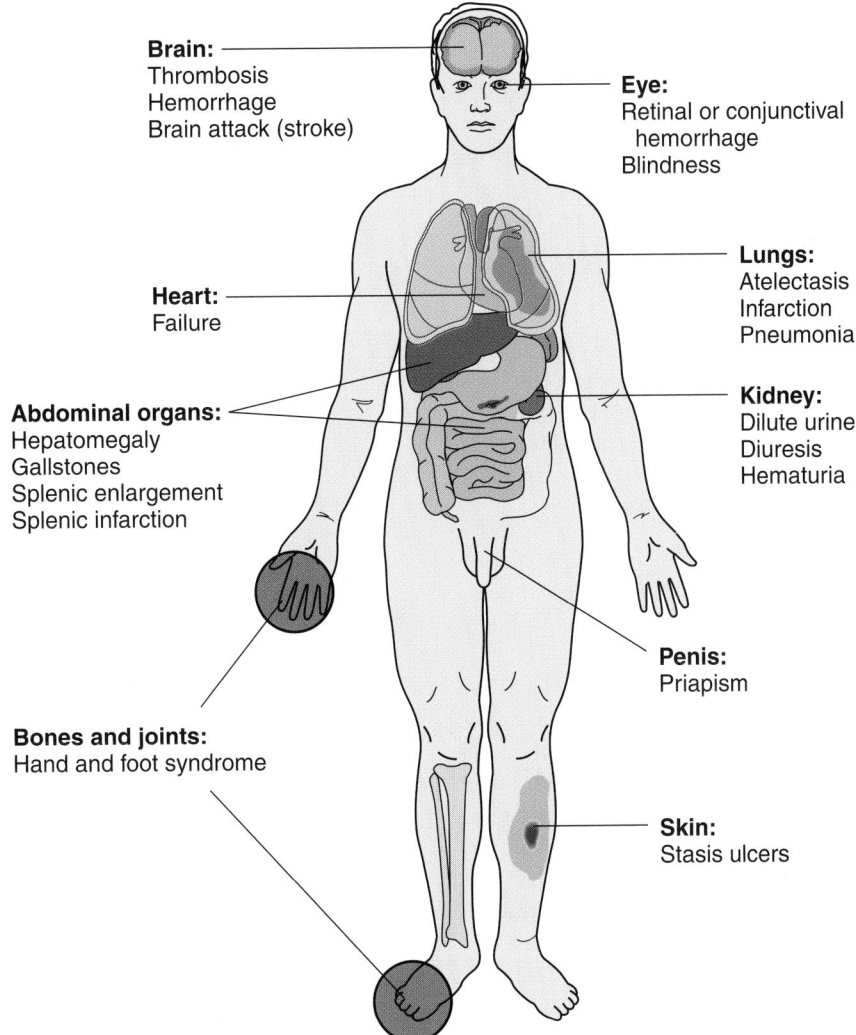

Brain:
Thrombosis
Hemorrhage
Brain attack (stroke)

Eye:
Retinal or conjunctival
    hemorrhage
Blindness

Lungs:
Atelectasis
Infarction
Pneumonia

Heart:
Failure

Abdominal organs:
Hepatomegaly
Gallstones
Splenic enlargement
Splenic infarction

Kidney:
Dilute urine
Diuresis
Hematuria

Penis:
Priapism

Bones and joints:
Hand and foot syndrome

Skin:
Stasis ulcers

Figure 24-4 Clinical manifestations of sickle cell anemia.

The patient with sickle cell anemia has impaired quality of life. Often, strenuous exercise or more exotic activities, such as scuba diving, are impossible because of the risk of crisis. Dehydration exacerbates the symptoms of sickle cell disease. Crises may occur without any apparent cause. In general, crises last from 4 to 6 days. They may occur in cycles close together for a time and then may become dormant for months to years. The cause of death in patients with sickle cell anemia is usually infection, stroke, or organ involvement.

DIAGNOSTIC TESTS. The most telling feature of sickle cell disease is a blood smear that shows sickle-shaped RBCs in circulation. The Sickledex test is a screening test that shows sickling of RBCs when oxygen tension is low. Hemoglobin electrophoresis is a test used to determine the presence of hemoglobin (Hgb) S, the abnormal form of hemoglobin. Also, there is a decreased amount of hemoglobin, a lowered RBC count, an elevated WBC count, and a decreased erythrocyte sedimentation rate.

TREATMENT. No cure is available for sickle cell anemia. Treatment is aimed at continual patient education to prevent crises and supportive care when crises occur. Some patients may be placed on low-dose oral penicillin to help prevent infections, decreasing the risk of crises.

During acute crises, the patient is admitted to the hospital for 5 to 7 days. The nurse can anticipate that the patient will require sedation and analgesia for severe pain and blood transfusions to replace the sickled red cells lost by being caught, crushed, and destroyed. Oxygen therapy decreases the dyspnea caused by the anemia, and large amounts of oral and intravenous fluids are given to flush the kidneys of the byproducts of the many broken cells' debris. Antibiotics are used to treat infection that may have triggered the crisis.

New treatments are being developed to treat sickle cell disease. Frequent blood transfusions, often monthly, are one of the newest treatment recommendations. A drug, hydroxyurea, has been shown to decrease crises, but it can cause life-threatening side effects. Bone marrow transplantation is also being investigated as a potential cure.

NURSING CARE. In the patient in crisis, assess circulation in the extremities every 2 hours, including pulse oximetry, capillary refill, peripheral pulses, and temperature. Frequent pain assessment is also necessary.

Encourage oral fluids, and assist the registered nurse (RN) to monitor intravenous (IV) fluids. Administer opioid analgesics such as morphine as ordered for acute pain. They are often given intravenously by the RN or through use of patient-controlled analgesia (PCA). Warm compresses to the painful areas, covering the patient with a blanket, and keeping the room temperature above 72° F reduce the vasoconstricting effects of cold. Cold compresses are contraindicated because they decrease circulation and increase the number of sickled cells caught in the painful area. Avoid restrictive clothing and raising the knee gatch of the bed, because these can also restrict circulation. Administer acetaminophen (Tylenol) to control fever; avoid aspirin because

it may increase acidosis, which can worsen the crisis. Encourage bedrest during the acute phase of the crisis.

PATIENT TEACHING. During remission, teach the patient how to prevent acute episodes. Advise the patient to avoid tight-fitting clothing that restricts circulation. Also encourage the patient to avoid strenuous exercise, which increases oxygen demand, and cold temperatures and smoking, which cause vasoconstriction. Alcoholic beverages can also trigger a crisis and should be avoided. Patients should never fly in unpressurized aircraft or undertake mountain climbing or other sports that can cause hypoxia. Encourage patients to get a pneumococcal vaccine and yearly flu vaccines. Encourage fluids to maintain hydration and reduce blood viscosity. Genetic counseling is important to prevent passing on the trait or disease to children. For more information, visit www.sicklecelldisease.org.

Polycythemia
Pathophysiology and Etiology

Polycythemia is really two separate disorders that are easily recognizable by similar characteristic changes in the RBC count. In both forms of polycythemia, the blood becomes so thick with an overabundance of RBCs that it closely resembles sludge. This thickness does not allow the blood to move easily. Laboratory tests show a hemoglobin greater than 18 mg/dL, the RBC mass greater than 6 million, and the hematocrit more than 55 percent.

Polycythemia vera (PV) is known as primary polycythemia. Its cause is unknown. Because the RBCs, platelets, and WBCs are overproduced, the bone marrow becomes packed with too many cells. As this overabundance of cells spills out into the general circulation, the organs become congested with cells and the tissues become packed with blood. The skin takes on a ruddy appearance from the buildup of red cells. PV is usually found in patients over age 50.

In contrast, secondary polycythemia is the result of long-term hypoxia. Common coexisting conditions that may predispose a patient to develop secondary polycythemia include pulmonary diseases such as chronic obstructive pulmonary disease (COPD), cardiovascular problems such as chronic heart failure, living in high altitudes, and smoking. The body makes more red cells in response to the low oxygenation associated with these conditions. Secondary polycythemia is a compensatory mechanism rather than an actual disorder.

Signs and Symptoms

The patient with PV commonly presents with hypertension, visual changes, headache, vertigo, tinnitus, and dizziness. Laboratory results show an increased level of all the bone marrow components (RBCs, WBCs, platelets), which is called **panmyelosis.** The patient may exhibit nosebleeds and bleeding gums, retinal hemorrhages, exertional dyspnea, and chest pains because of the increased pressure exerted by the excess cells. The patient with PV usually has a ruddy (or reddish) complexion and abdominal pain with an early feeling of fullness because of the enlarged liver and spleen. Nearly all the symptoms in PV are due to the major problems of hypervolemia, hyperviscosity, and engorgement of capillary beds. Without treatment, patients with PV die of thrombosis or hemorrhage.

Treatment

Treatment of PV takes place in two stages. The first stage is to decrease the hyperviscosity problem. The most common first-line treatment is therapeutic **phlebotomy.** Phlebotomy involves withdrawal of blood, which is then discarded. From 350 to 500 mL of blood are removed each time on an every-other-day basis, with the goal being a hematocrit of 40 to 45 percent. This reduces the RBC levels, and the patient usually feels more comfortable quickly. Repeated phlebotomies eventually cause iron deficiency anemia, which in turn stabilizes RBC production; phlebotomies can then be reduced to every 2 to 3 months.

The problem that remains is the increased white blood cell and platelet counts, because phlebotomy does very little to correct these overloads. Chemotherapeutic agents or radiation therapy may be used to suppress production of these cells in some patients. **Leukemia** is a side effect of this therapy, so it is used only if the benefits outweigh the risks.

Nursing Care

Explain the phlebotomy procedure and reassure the patient that the treatment will relieve the most distressing symptoms. The procedure is the same as that used for donating blood. The patient should be active and ambulatory to help prevent thrombus formation. When bedrest is necessary, passive and active range-of-motion exercises should be implemented. Monitor the patient for complications such as hypovolemia and bleeding. Advise the patient to report any signs or symptoms of bleeding immediately.

If the patient has more advanced manifestations, such as an enlarged liver or spleen, offer several small meals each day so that the patient will be more comfortable while still receiving adequate nutrition. In addition, encourage the patient to increase his or her daily intake of fluids to reduce the viscosity of the blood. A dietitian can be consulted to discuss ways to maintain good nutrition. If the patient is on drug therapy, monitor CBC and platelet counts.

Teaching should emphasize the need for continued follow-up with the physician once the patient begins to feel better. Instruct the patient on specific symptoms to report, such as chest pain, increased joint pain, decreased activity tolerance, and fever.

Patient Education

Instruct the patient to drink at least 3 L of water daily to reduce blood viscosity. Encourage avoidance of tight or restrictive clothing and elevation of feet when resting to prevent impairment of circulation. Use of support hose when active also promotes circulation. Teach the patient to watch for and report any signs of iron deficiency anemia, such as pallor, weight loss, and dyspnea. If anticoagulants or antiplatelets are ordered, instruct the patient about side effects

to watch for and the importance of routine laboratory tests. Routine bleeding precautions are implemented (Box 24–5). Warn the patient to stop activities at the first sign of chest pains.

► HEMORRHAGIC DISORDERS
Disseminated Intravascular Coagulation

Disseminated intravascular coagulation (DIC) involves a series of events that result in hemorrhage.

Pathophysiology

As its name implies, this syndrome is a catastrophic, overwhelming state of accelerated clotting throughout the peripheral blood vessels. In a short period, all the clotting factors and platelet supplies are exhausted and clots can no longer be formed. This results in bleeding from nearly every route possible. DIC is not a disease but is a syndrome that develops secondary to some other severe physical problem. Once this deadly syndrome develops, the progression of symptoms is rapid.

Massive clotting in blood vessels leads to organ and limb necrosis. Organs most often affected include the kidneys and the brain, but other blood-engorged organs, such as the lungs, the pituitary and adrenal glands, and the gastrointestinal mucosa, are commonly involved. DIC is usually acute in onset, although in some patients it becomes a chronic condition. The prognosis depends on early diagnosis and intervention and the severity of the hemorrhaging. DIC has a very high mortality rate.

Etiology

DIC can develop after any condition in which the body has sustained major trauma. The sources of trauma are varied and can include an overwhelming infection and sepsis or an overt viral infection. Obstetric complications such as abruptio placentae, amniotic fluid embolism, or a retained dead fetus can trigger the onset of DIC. Cancer-related causes of DIC include acute leukemia and widely metastatic cancers of the lung. Massive tissue necrosis found in severe crush or burn injuries may increase the risk for development of DIC. Tissue necrosis secondary to extensive abdominal surgery with leakage of the intestinal contents can be related to DIC onset. Rarer causes of this condition have included heatstroke, shock, and poisonous snakebites, as well as fat embolism secondary to broken long bones.

Signs and Symptoms

Abnormal bleeding without a history of a serious hemorrhagic disorder is a cardinal sign of DIC. Early signs of bleeding include petechiae, ecchymoses (Fig. 24–5), and bleeding from venipuncture sites. Bleeding may progress to IV sites, skin tears, surgical sites, incisions, and the gastrointestinal tract and oral mucosa. Pain and enlargement of joints develop if bleeding into the joints occurs. All these signs and symptoms may occur at the same time. Massive bleeding may also be accompanied by nausea, vomiting, dyspnea, oliguria, convulsions, coma, shock, major organ system failure, and severe muscle, back, and abdominal pain.

Diagnostic Tests

Initial laboratory findings in DIC include a prolonged prothrombin time (PT) and partial thromboplastin time (PTT), decreased platelet count, and increased evidence of fibrin degradation products. A decrease in hemoglobin is the result of spilled hemoglobin from the increased numbers of broken red cells. Blood urea nitrogen (BUN) and serum creatinine levels may also be increased. See Table 24–2 for additional findings.

Medical Treatment

Effective treatment of DIC depends on early recognition of the condition. Treatment is first aimed at correcting the underlying cause. Additional treatment consists of supportive interventions, including administration of blood, fresh frozen plasma, and platelets and the infusion of cryoprecipitates to support hemostasis. Some health care organizations include the use of intravenous heparin to help prevent the

initial microembolization, but this practice is controversial. Additional therapies are being investigated.

Nursing Care

Care of the patient with DIC is a nursing challenge. Early intervention requires early recognition and reporting of signs of bleeding. In addition to supportive care, focus on the prevention of further bleeding episodes. Care should be taken to avoid any trauma that might cause bleeding. Be careful not to dislodge clots from any site because another clot may not form and the patient will hemorrhage. See Box 24–5 for bleeding precautions.

Patient Teaching

Because a patient with DIC is often placed in the intensive care unit, there are many chances for patient and family teaching. Explain all diagnostic tests to the patient if he or she is alert. If not, keep the family informed. A large part of family education is preparing the family for what the patient may look like in terms of bleeding and bruising, as well as specific equipment that may be in place, such as IV lines, nasogastric (NG) tube, and Foley catheter. It may be helpful to enlist the aid of social workers, chaplains, and other members of the health care team to help support the family.

TABLE 24-2	LABORATORY ABNORMALITIES IN DISSEMINATED INTRAVASCULAR COAGULATION

Screening Test	Finding
Prothrombin time (PT)	Prolonged
Partial thromboplastin time (PTT)	Prolonged
Activated partial thromboplastin time (APTT)	Prolonged
Thrombin time (TT)	Prolonged
Fibrinogen	Reduced
Platelets	Reduced
Fibrin split products (FSP; also known as fibrin degradation products [FDP])	Elevated
Protamine sulfate	Strongly positive
Dimers (cross-linked fibrin fragments)	Elevated
Antithrombin III	Reduced
Factor assays (for factors V, VII, VIII, X, and XIII)	Reduced

CRITICAL THINKING

Mrs. Johns
Mrs. Johns is admitted to your unit with DIC following the difficult delivery of her new baby.

1. What will you assess as you care for Mrs. Johns?
2. What treatment do you anticipate?
3. What concerns is Mrs. Johns likely to have?

Answers at end of chapter.

Idiopathic Thrombocytopenic Purpura
Pathophysiology and Etiology

This disease results from increased platelet destruction by the immune system. Any time platelets are reduced, the risk for bleeding increases. Acute **idiopathic thrombocytopenic purpura** (ITP) usually affects children between the ages of 2 and 6, whereas chronic ITP mainly affects adults younger than age 50. The major population affected is women between ages 20 and 40.

Acute ITP usually occurs after an acute viral illness such as rubella or chickenpox. It may also be drug induced or associated with pregnancy. ITP is generally thought to be related to an immune system dysfunction. Antibodies responsible for platelet destruction have been found in nearly all diagnosed patients.

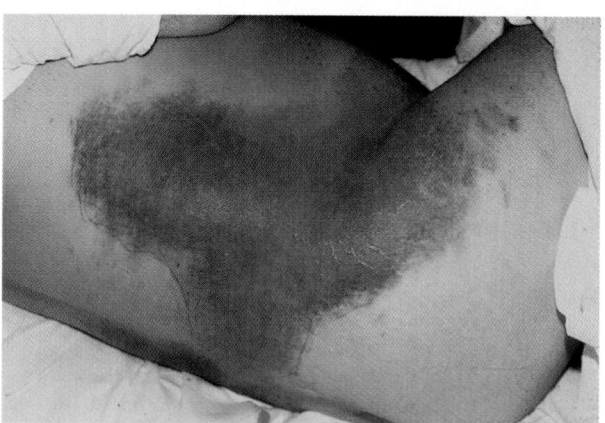

Figure 24-5 Extensive hemorrhage into the skin in DIC. Note how the area is outlined in pen so the nurse can assess if the area is spreading. (From Harmening, DM: Clinical Hematology and Fundamentals of Hemostasis, ed 3. FA Davis, Philadelphia, 1997, p 520, with permission.)

idiopathic thrombocytopenic purpura: idio—unknown + pathic—disease + thrombo—clot + cyto—cell + penic— lack + purpura—hemorrhage in the skin

Signs and Symptoms

ITP produces clinical changes that are common to all forms of **thrombocytopenia:** petechiae, ecchymoses, and bleeding from the mouth, nose, or gastrointestinal (GI) tract. Bleeding may occur in vital organs, such as the brain, which may prove fatal. In the acute type, onset may be sudden and without warning, causing easy bruising, nosebleeds, and bleeding gums. Onset of chronic ITP is usually insidious.

Diagnostic Tests

A platelet count of less than 20,000/mm^3 and a prolonged bleeding time suggest ITP. The greatly decreased platelet level places the patient at serious risk for hemorrhage. Examination of the platelets under the microscope shows them to be small and immature. Anemia may be present if there has been a bleeding episode. If a bone marrow aspiration is performed, the results show an adequate amount of the precursor cells for platelets, the megakaryocytes. However, instead of the 7- to 10-day life span that platelets usually have, these immature platelets have a life span of just a few hours.

Treatment

Most cases of acute ITP resolve spontaneously without treatment. Initial treatment, if necessary, is often the administration of steroids. The purpose of the steroids is to prolong the life of the platelets and strengthen the capillaries, making them less likely to break and cause bleeding. Some physicians order the use of chemotherapy drugs. The spleen may be removed, because it is the primary site involved in platelet destruction. Often the patient undergoing splenectomy has tried all other courses of treatment unsuccessfully and may be experiencing bleeding episodes. Acute bleeding episodes are treated with transfusions of blood, platelets, and vitamin K.

Nursing Interventions

Care for the patient with ITP is the same as any patient with a bleeding disorder. See Box 24–5 for bleeding precautions. Teach the patient to watch for and report signs and symptoms of bruising and bleeding (Box 24–7). The patient should avoid trauma and restrict activity during severe episodes.

---

thrombocytopenia: thrombo—clot + cyto—cell + penia—lack

BOX 24–7 **Patient Teaching: Signs and Symptoms of Bleeding**

Notify your health care provider if the following occur:
- Easy bruising of skin
- Petechiae (small red spots on skin)
- Blood in urine
- Black tarry stools
- Bleeding from nose or gums
- New onset of painful joints

Hemophilia

Hemophilia is a group of hereditary bleeding disorders that result from a severe lack of specific clotting factors. The two most common are hemophilia A (classic hemophilia) and hemophilia B (Christmas disease). Von Willebrand's disease is another related bleeding disorder, but it represents a minority of cases and is not discussed in this chapter.

Pathophysiology

Recall that many different clotting factors make up the clotting mechanism. Hemophilia A accounts for 80 percent of all types of hemophilia and results from a deficiency of factor VIII. Hemophilia B is a factor IX deficiency; approximately 15 percent of persons with hemophilia have this type. The severity and prognosis of hemophilia depend on the degree of deficiency of the specific clotting factors. Mild hemophilia has the best prognosis because it does not cause spontaneous bleeding and joint deformities as does severe hemophilia.

After an injury, the person with hemophilia forms a platelet plug (which differs from a clot) at the site of an injury as would normally be expected, but the clotting factor deficiency keeps the patient from forming a stable fibrin clot. Continued bleeding washes away the platelet plug that initially formed. Contrary to popular myths, people with hemophilia do not bleed "faster" and are not at risk from small scratches.

Etiology

Hemophilia A and B are inherited as X-linked recessive traits. This means that the female carrier (daughter of an affected father) has a 50 percent chance of transmitting the gene to each son or daughter. Daughters who receive the gene are carriers, and sons who receive the gene are born with hemophilia. It is technically possible for daughters to be affected with hemophilia, although it is very rare.

Signs and Symptoms

Bleeding occurs as a result of injury or, in severe cases, spontaneously (unprovoked by injury). Bleeding into the muscles and joints (**hemarthrosis**) is common. Severe and repeated episodes of joint hemorrhage cause joint deformities, especially in the elbows, knees, and ankles, which decreases the patient's range of motion and ability to walk.

In mild hemophilia, excessive bleeding is usually associated only with surgery or significant trauma. However, once a person with mild hemophilia begins to bleed, the bleeding can be just as serious as that of the patient with a more severe form.

The patient with moderate hemophilia has an occasional bout of spontaneous bleeding. In severe hemophilia, spontaneous bleeding occurs more frequently. It would be possible for the patient to develop hemarthrosis or bleeding

---

hemophilia: hemo—blood + philia—to love
hemarthrosis: hem—bleeding + arthr—joint + osis—condition

into the brain without any precipitating trauma. Severe episodes may produce large subcutaneous and deep intramuscular hematomas. Major trauma can cause bleeding so severe that it becomes life threatening.

Another unfortunate problem related to hemophilia treatment is the frequent need to replace clotting factors and other blood products. Before 1986, blood banks and other centers did not routinely test for human immunodeficiency virus (HIV) antibodies. Depending on the patient's age and frequency of treatment, many patients may have been exposed to HIV or hepatitis. Blood banks and pharmacies have checked their blood supplies for the presence of HIV since 1986. Today the plasma proteins are artificially created or thoroughly cleansed to prevent transmission of disease.

Diagnostic Tests

Laboratory data reveal a prolonged PTT. The various factor levels are then measured to determine which is missing. Once the missing factor is identified, the type of hemophilia is determined and necessary treatments can be implemented.

Medical Treatment

Hemophilia is not curable. However, treatment advances have improved outcomes and many patients can now live a normal life span. Treatment is aimed at the prevention of crippling deformities and at increasing life expectancy. Treatment involves stopping bleeding episodes by increasing the blood levels of the missing clotting factors. Hemophilia A is treated with factor VIII; hemophilia B is treated with factor IX. Each is available in a freeze-dried powder that is reconstituted with water and administered intravenously. The newest treatment is factors made using recombinant DNA technology, without the use of any human blood products. Blood transfusions are uncommon but may be necessary after severe trauma or surgery.

Complications related to therapy usually occur therapy is started too late. Minor trauma generally needs to be treated with at least 72 hours of added clotting factors; major traumas and surgeries may require up to 14 days of added factors to prevent sudden bleeding. Health care workers should pay careful attention to the patient who says that bleeding is starting, even when no outward signs are evident. The patient usually knows from experience if bleeding is starting. If treatment is delayed at this time, the results can be disastrous. Some patients with severe disease are treated prophylactically to prevent bleeding.

Nursing Process

ASSESSMENT. Because one goal is prevention of bleeding episodes, assess the patient and family for knowledge of the disease and its treatment and understanding of preventive measures. Most patients care for themselves at home, starting their own IVs and administering treatment independently. Hospitalization is necessary only for surgery or major trauma. During an acute episode of bleeding, the hemoglobin and hematocrit are carefully monitored. Factor VIII or IX levels are monitored to determine if factor replacement has reached adequate levels. Vital signs are monitored for falling blood pressure and rising pulse rate, which are signs of hypovolemic shock. All body systems are assessed for signs of bleeding (see Box 24–7). A pain assessment is done using the WHAT'S UP? format.

NURSING DIAGNOSIS. Priority nursing diagnoses include the following:

- Pain related to bleeding into tissues
- Risk for injury related to effects of bleeding
- Risk for ineffective management of therapeutic regimen related to knowledge deficit

PLANNING AND IMPLEMENTATION. Goals of treatment are to prevent bleeding and to access quick treatment and prevent complications if bleeding occurs.

Pain. Pain during acute bleeding episodes is managed with opioids as prescribed. Use of patient-controlled analgesia is helpful, especially because this uses the intravenous route. Intramuscular injections are contraindicated because of the risk of causing bleeding into the muscle.

Risk for Injury. Routine precautions to prevent bleeding are summarized in Box 24–5. In the event of an acute episode of bleeding, the patient is transfused with factor concentrates (or in rare circumstances, fresh frozen plasma, cryoprecipitate, blood, or a combination of these). See Chapter 23 for transfusion of blood products. Ice or pressure on bleeding sites may help slow bleeding. Special preventive care is exercised if the patient requires surgery or dental procedures; these can be life-threatening events for the patient with hemophilia.

Risk for Ineffective Management of Therapeutic Regimen. You can assist the RN to instruct the patient and family in care of the disorder. They must learn to prevent bleeding episodes, recognize signs and symptoms of bleeding, and obtain emergency care in the event that bleeding does occur. Many patients are taught to administer treatment at home. A nationwide system of comprehensive hemophilia treatment centers coordinates care for these patients. Periodic multidisciplinary clinics are held, at which social service, dental, rehabilitation, nursing, financial, and medical needs can be addressed.

EVALUATION. If nursing care is effective, the patient will be comfortable and bleeding will be prevented or complications minimized. The patient and family will be able to state appropriate measures to prevent and treat bleeding episodes.

▶ DISORDERS OF WHITE BLOOD CELLS
Leukemia

The term *leukemia* literally means "white blood." It was first identified in 1845 when the blood of victims was examined and found to have an excess of "colorless" cells.

Pathophysiology

Leukemia is a malignant disease of the WBCs that affects all age groups. The immature WBCs (blast cells) generate in an explosive fashion in the bone marrow, lymph tissue, and spleen. These cells are abnormal and unable to effectively fight infection. There are so many abnormal cells developed and dumped into the peripheral circulation that they tend to collect in the body organs and tissues, especially where circulation is sluggish. Areas especially prone to infiltration with these immature WBCs are the oral mucosa, anus, sinuses, and lungs. At the time of diagnosis, these areas are often inflamed, painful, and infected. It is common for some patients to be diagnosed only after suffering with an infection that does not clear up easily.

As the disease progresses, the bone marrow continues to produce large numbers of the useless cells, the peripheral circulation is filled with the abnormal cells, and the bone marrow is packed with blast cells; production of most other normal cells is impossible. The patient becomes anemic because of the lack of RBC production, and bleeding becomes a problem as fewer and fewer platelets are manufactured. But most importantly, even though the WBC count is very high, there are very few normal, mature, and active white cells with which to fight infection. Thus the patient often begins to have raging infections that do not respond to antibiotics. Without treatment, the leukemia leaves the patient unable to fight infection, unable to control bleeding, and in a downward spiral of fatigue and anorexia. Leukemia, untreated, is almost always fatal.

Classifications

Leukemias are classified as either acute or chronic and either lymphoid or myeloid. Symptoms of the acute leukemias begin very suddenly and the patient is very sick, whereas chronic leukemias develop very slowly and patients can be surprised by the diagnosis because they feel well. Lymphoid leukemias affect the lymphocytes. Myeloid leukemias originate in the stem cells of the bone marrow that develop into monocytes, granulocytes, erythrocytes, and platelets. The most common leukemias are discussed next.

ACUTE LEUKEMIAS. Acute lymphocytic leukemia (ALL) commonly affects children younger than age 15 and involves abnormal growth of the lymphocyte precursors (lymphoblasts). Acute myelogenous (myeloblastic) leukemia (AML) usually affects persons older than age 20 and has a poor prognosis. The patient with acute leukemia may present with sudden onset of high fever, abnormal bleeding from the mucous membranes, petechiae, ecchymoses, and easy bruising after minor trauma. Death usually results from infection.

CHRONIC LEUKEMIAS. Chronic lymphocytic leukemia (CLL) predominantly affects the B and T lymphocytes and usually occurs in adults older than age 40. Chronic myelogenous leukemia (CML) is characterized by the Philadelphia chromosome and occurs most often between the ages of 40 and 45.

Chronic leukemia usually develops in a three-stage process. First is the insidious phase, characterized by anemia and mild bleeding abnormalities. During this stage, the patient often feels well and is not even aware that he or she is sick. After a time, generally years, the disease progresses to

the accelerated and acute phases, in which the scenarios are similar to the events seen in acute leukemias. Chronic leukemia is almost always fatal; the average survival time is 3 to 4 years after onset of the chronic phase and 3 to 6 months after onset of the acute phase. With advances in treatments, it is not uncommon to encounter patients who have been living with chronic leukemia for 10 years or more.

Etiology

The cause of leukemia is unknown. Risk factors are thought to include certain viruses, because remnants of viruses have been found in leukemic cells. Often there are genetic and immunological factors involved. For example, persons with Down syndrome are more likely to develop leukemia. Other authorities point to exposure to radiation, in part because radiologists have been found to have a higher-than-average development of leukemia. Some patients have developed leukemia after being treated for another unrelated malignancy using radiation or chemotherapy. Researchers note the higher occurrence rate in persons who lived through the Hiroshima and Nagasaki bombings during World War II. Water polluted with benzenes and other chemicals may be a factor. There is no single clear-cut cause for the development of leukemia.

Signs and Symptoms

Symptoms are similar for all types of leukemia and include low-grade fever caused by infection, pallor, weakness, lassitude, shortness of breath, and malaise caused by anemia. These symptoms may be present weeks or months before the appearance of other symptoms. The patient may also have dyspnea, fatigue, tachycardia, palpitations, and abdominal pain. Sternal pain and rib tenderness may result from crowding of bone marrow. If the leukemia has invaded the central nervous system, the patient may experience confusion, headaches, and personality changes. During the acute phase the patient may exhibit high fevers from infection. Ecchymosis or petechiae may result from thrombocytopenia.

Diagnostic Tests

Although a simple CBC often points toward the diagnosis, only bone marrow aspiration can show the degree of proliferation of the malignant WBCs and confirm the diagnosis of leukemia. The complete blood cell count may also show a decrease in the numbers of platelets, RBCs, and mature WBCs. A lumbar puncture helps determine if the central nervous system is involved. Genetic analysis of the peripheral blood and bone marrow components may show the presence of the Philadelphia chromosome in patients with CML.

Medical Treatment

Systemic chemotherapy aims to eradicate the leukemic cells and induce a remission. Remission means that the bone marrow is free to produce normally occurring cells in normal proportions without production of the immature

WBCs. Chemotherapy types vary with the types of leukemia and the level of involvement. Radiation therapy is also sometimes used for initial treatment of leukemia.

The overall goal of the first treatments is to get the patient to a state of remission. Occasionally, partial remission is achieved, when everything looks good except for an occasional leukemic cell seen in the bone marrow. Remission is not the same as cure.

There are four phases to the treatment of leukemia: induction, intensification, consolidation, and maintenance. Induction is the period in which an attempt to get the patient into remission is made. This first phase is difficult because chemotherapy is given in extremely high doses and on an aggressive timetable. Often the patient becomes quite ill from complications of the treatment. The patient may become depressed because the treatment seems worse than the disease at this stage. The nurse must help the patient deal with the side effects of anemia, thrombocytopenia, and leukopenia (see also Leukemia Summary, Table 24–3). (See Boxes 24–5 and 24–6 and Chapter 10.)

If the first remission is accomplished, the other phases of treatment are begun. Intensification is similar to the initial induction phase, using the same drugs at even higher doses. The next phase, consolidation, is used to ensure that all leukemic cells have been eradicated from the body. Finally the patient graduates to maintenance therapy, in which the

patient is kept free of leukemic cells (and in remission) for a period of years (and hopefully a lifetime). This requires years of continued chemotherapy treatments, often on a monthly basis. Radiation therapy may be used throughout this course of treatment to decrease the size of the liver or spleen or to decrease the numbers of leukemic cells in the central nervous system.

Bone marrow transplantation (BMT) is sometimes used to treat leukemia. Often the precursors to BMT include total body irradiation and high-dose chemotherapy. The goal of these seemingly radical treatments is to destroy all the patient's malignant bone marrow and then, at the last possible moment, replace it with a donor's clean and healthy bone marrow (allogenic transplant). One other type of bone marrow transplant, known as an autologous transplant, uses the patient's own diseased bone marrow, which is harvested, chemically treated and cleaned, stored, and later reinfused. This stage of bone marrow replacement is sometimes called bone marrow rescue. Transplanted bone marrow is given to the patient like a blood transfusion—generally through a central line placed in the chest. Once infused into the bloodstream, the new marrow travels to the bones, where, under the best of circumstances, it begins to grow and function normally.

Bone marrow transplants are being performed at more and more centers across the United States. Common reasons for BMT include leukemia, lymphoma, breast cancer, aplastic anemia, and congenital immunodeficiency disorders.

A new and promising treatment for leukemia is peripheral blood stem cell transplantation. Hematopoietic stem cells can be collected from the patient during remission, then reinfused at a later time. Donor stem cells are also sometimes used if a good match can be found.

Nursing Care

The patient with leukemia is at risk for many problems, including fatigue, bleeding, infection, and other complications of the disease and its treatment. The patient must understand his or her disease process and treatment regimen in order to participate in self-care. See Nursing Care Plan Box 24–8 for the Patient with Leukemia for interventions to

TABLE 24-3	**Leukemia Summary**
Symptoms	Fever (related to infection)
	Pallor
	Weakness, malaise
	Tachycardia
	Dyspnea
	Bone pain
	Headaches, confusion
Diagnostic tests	CBC
	Bone marrow aspiration
	Lumbar puncture
Therapeutic management	Chemotherapy
	Radiation therapy
	Bone marrow transplant

BOX 24-8   Nursing Care Plan for the Patient with Leukemia

 **Risk for injury from infection or bleeding related to pancytopenia**

PATIENT OUTCOMES

The patient is free from infection and bleeding. Signs and symptoms of infection or bleeding are reported promptly.

EVALUATION OF OUTCOMES

Is the patient free from infection and bleeding, or are problems reported so that quick intervention can prevent further complications?

INTERVENTIONS	**RATIONALE**	**EVALUATION**
Monitor vital signs every 4 hours and as needed.	Elevated temperature is a sign of infection. Falling blood pressure and elevated pulse rate may indicate sepsis or blood loss.	Are vital signs stable?
Monitor for swelling, redness, purulent drainage.	These are signs of infection and should be reported promptly.	Are signs of infection present?

BOX 24-8   NURSING CARE PLAN FOR THE PATIENT WITH LEUKEMIA—CONT'D

Protect patient from sources of infection (see Box 24–6).	Patient is at risk for infection because of ineffective WBCs.	Are precautions being observed to prevent infection?
Observe for tarry stools, petechiae, ecchymosis (see Box 24–7).	These are signs of bleeding and should be reported promptly.	Are signs of bleeding present?
Protect patient from injury that could cause bleeding (see Box 24–5).	Patient is at risk for bleeding because of reduced platelet count.	Are precautions being observed to prevent injury and bleeding?

 Fatigue related to decreased red cell count and oxygenation and effects of treatments

PATIENT OUTCOMES
Patient is able to participate in activities that are important to him or her.

EVALUATION OF OUTCOMES
Is patient able to identify and participate in activities that are important to him or her?

INTERVENTIONS	RATIONALE	EVALUATION
Assess fatigue using the WHAT'S UP? format.	A good assessment establishes a baseline and aids in planning.	Is fatigue present? To what degree?
Assist patient to identify activities that are important to him or her (e.g., activities of daily living [ADLs] attending a child's wedding, taking a trip). Assist in setting goals to work toward the desired activity.	If the patient cannot do everything he or she wishes, it may help to focus on the most important things.	Can patient identify important activities? What are they? How can the nurse assist the patient to reach activity goals?
Encourage a balanced diet. Contact dietitian as needed.	Poor nutrition contributes to fatigue.	Is patient eating a balanced diet? Is weight stable?
Allow periods of rest between activities.	Any activity (ADL, x-rays, even talking) can increase fatigue.	Is patient able to rest?
Ensure adequate sleep. Obtain order for sleeping aid if indicated.	Lack of sleep worsens fatigue.	Does patient state feeling rested on awakening? Is medication needed?
Provide for ADLs when patient is unable to do so independently.	Extreme fatigue may prevent the patient from participating in self-care.	Does patient need total assistance?

 Impaired oral mucous membranes related to chemotherapy and pancytopenia

PATIENT OUTCOMES
The patient's oral mucous membranes will remain intact.

EVALUATION OF OUTCOMES
Are oral mucous membranes intact, without lesions?

INTERVENTIONS	RATIONALE	EVALUATION
Assess mouth daily for redness, edema, and lesions.	Routine assessment helps identify problems early so treatment can be implemented.	Are mucous membranes intact?
Encourage adequate nutrition and fluids.	Poor nutrition and dehydration increase the risk of oral lesions.	Is patient eating and drinking?
Encourage patient to brush teeth after meals with a soft toothbrush. If irritation is severe or if the patient is at risk for bleeding, use swabs or sponge Toothettes instead of a toothbrush.	Brushing the teeth controls tooth and gum disease; a toothbrush may be too harsh if the patient is at risk for bleeding.	Is mouth care being provided after meals? Is mouth care irritating? Are alternative methods needed?
Avoid use of lemon-glycerin swabs for mouth care.	Lemon-glycerin swabs are drying to oral mucosa.	Are products used appropriate?
Obtain an order for a mouthwash such as 1 oz of diphenhydramine (Benadryl) elixir diluted in 1 qt of water or normal saline. Obtain an order for a topical anesthetic if mouth is very inflamed and painful.	Diphenhydramine reduces inflammation; anesthetics reduce pain.	Does mouthwash soothe pain?

BOX 24-8 **NURSING CARE PLAN FOR THE PATIENT WITH LEUKEMIA—CONT'D**

Encourage the patient to avoid smoking, alcohol, acidic food or drinks, extremely hot or cold foods and drinks, and commercial mouthwash.	These things can be irritating to the mucosa.	Does patient state understanding of things to avoid?

GERIATRIC

Advise patient to remove dentures for cleaning and at bedtime.	Dentures left in for long periods can impair circulation and increase risk of lesions.	Are dentures removed for cleaning and at night?

deal with these problems. Additional diagnoses include knowledge deficit and anxiety. The following Web sites provide resources for patients and families with leukemia. See also Chapter 10 for general care of the patient with cancer.

For more information, visit the following organizations:
American Cancer Society at www.cancer.org
Leukemia and Lymphoma Society at www.leukemia.org
National Cancer Institute at www.nci.nih.gov

CRITICAL THINKING

Mr. Washington

Mr. Washington is on your unit undergoing initial treatment for leukemia. You enter his room and find it full of visitors.

1. What concerns do you have?
2. What do you do?

Answers at end of chapter.

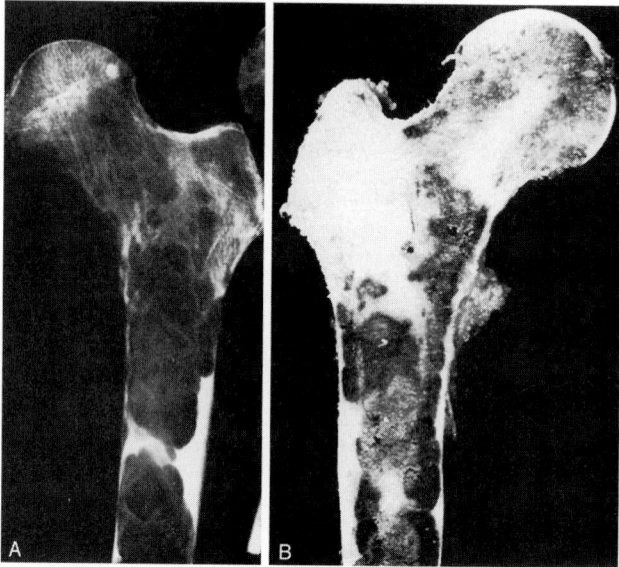

Figure 24-6 X-ray of bone destruction in multiple myeloma. (From Huether, SE, and McCance, KL: Understanding Pathophysiology. Mosby, St Louis, 1996, p 548, with permission.)

▶ MULTIPLE MYELOMA

Multiple myeloma is a deadly cancer of the plasma cells in the bone marrow. When the disease is caught in its early stages, treatment can prolong life by 3 to 5 years. More important, early detection can decrease the amount of pain and disability due to bony destruction and **pathological fractures.** Unfortunately, almost half the patients die within the first 3 months after diagnosis because of the silent and deadly nature of the disease. Another 40 percent of patients die within 2 years after diagnosis. Because early diagnosis is not often made, only 10 percent of patients can expect to live to the 5-year mark. Multiple myeloma most often affects men ages 50 to 70.

Pathophysiology

In this disorder, cancerous plasma cells in the bone marrow begin reproducing uncontrollably. These cells infiltrate bone tissue all over the body and produce hundreds of tumors that begin to devour the bone tissue. X-ray examination may show holes in the bones, forming a Swiss cheese pattern (Fig. 24–6). As more and more of these holes are formed, the bone integrity becomes compromised and weak. Multiple myeloma usually affects the bones of the skull, pelvis, ribs, and vertebrae.

As the disease continues, the plasma cells infiltrate the major organs, including the liver, spleen, lymph nodes,

lungs, adrenal glands, kidneys, skin, and GI tract. Because the diagnosis is usually made only after widespread invasion of the bones is well underway, the overall prognosis of patients with this disease is poor. Although the overall result of the disease is the devastating destruction of the bone and widespread osteoporosis, death is often from sepsis.

Etiology

The cause of multiple myeloma is unknown, although it is being researched. Some authorities believe this disease to be related to chronic allergies and hypersensitivity reactions. This line of thought stems from the fact that plasma cells are the first line of defense and are the producers of the immunoglobulins that help fight foreign bodies. For some reason these defenders get out of control and begin to attack the host, as well as foreign invaders. People who work in rubber, leather, farming, and petroleum industries are more likely to develop multiple myeloma. Radiation and chemical exposure may also be factors.

Signs and Symptoms

Skeletal pain is the most common complaint. The patient may describe the pain as constant severe back pain that increases with exercise or movement. The patient may complain about pain in the ribs. Other signs and symptoms include achiness of the long bones, joint swelling and tenderness, low-grade fever, and general malaise. Sometimes there is evidence of early peripheral neuropathy secondary to vertebral collapse and mild spinal cord compression. The patient may be unable to feel the true temperature of bath water and be burned or may be unable to feel wounds and infections on the feet. In more severe cases of cord compression, the patient may lose control of bladder and bowels. This is a true oncological emergency. Prompt emergency treatment is necessary to keep the patient from becoming paralyzed.

Occasionally the patient will have pathological fractures of the long bones. These are fractures that occur with no trauma, such as the person who breaks a leg just turning over in bed or breaks a rib while sneezing. In advanced disease there is anemia, weight loss, thoracic spinal deformities from multiple rib destruction, and a loss of height because of pathological fractures and compacting of the vertebrae.

Because calcium is mobilized from the bones and into the blood, the patient is at risk for hypercalcemia. Signs and symptoms include anorexia, nausea, vomiting, mental changes (especially confusion), seizures, and weakness and fatigue. Kidney stones may result as the excess calcium passes through the kidneys.

Patients are susceptible to infection because of compromised immune function. Pneumonia is a common finding in patients with multiple myeloma. They may develop anemia because of bone marrow dysfunction and reduced erythropoietin formation by diseased kidneys.

Patients often develop kidney failure as the filtering capacity of the kidney is blocked with calcium. Other factors include recurrent infections and deposits of myeloma cells in the kidneys.

Diagnostic Tests

A CBC shows moderate to severe anemia. Examination of the WBC count may show an increase in the number of white cells secondary to infection. X-ray examinations may show changes in the lungs and diffuse osteoporosis in the bones not already riddled with holes. Urine studies are positive for the M-type globulins (Bence-Jones proteins) in 40 percent of patients. Bone marrow biopsy is done to confirm the diagnosis and determine the disease's stage.

Blood chemistries often show an increased amount of calcium in the blood. Hypercalciuria results as the calcium released out of the bones is flushed out in the urine. An intravenous pyelogram may be done to see how much calcium is blocking the kidneys. A 24-hour urine collection is done to evaluate protein excretion.

Medical Treatment

Long-term treatment of multiple myeloma consists of a two-pronged approach: (1) managing the disease and (2) managing the symptoms. To manage the disease, high-dose steroids (prednisone) and oral or intravenous chemotherapy agents are given. The goal of drug therapy is to suppress the plasma cell proliferation, which then helps decrease the amount and speed of bone destruction.

The second approach is control of symptoms. The nurse monitors the patient for signs and symptoms of hypercalcemia, hyperuricemia, dehydration, respiratory infection, renal problems, and pain. External beam irradiation may be given to especially painful areas of bone involvement. Fortunately this treatment is quite effective, usually decreasing pain intensity in just a few days. The patient can expect to have a daily (or perhaps a twice-daily) therapy treatment over a course of 10 to 14 days, delivered directly to the painful bony areas. Vigorous attention to administering pain medications during the early course of treatment greatly reduces the patient's pain levels.

The patient may need a laminectomy if vertebral collapse occurs. Because of demineralization of the bone, with resulting large amounts of calcium in the blood and urine, surgery for kidney stones and eventual dialysis for acute or chronic kidney failure may be necessary.

A newer treatment involves high-dose chemotherapy combined with stem cell transplantation. The patient's own peripheral stem cells can be removed and reinfused. These stem cells can then differentiate into new, healthy cells. Methods of cleaning the cells to prevent contamination with malignant cells are being researched.

Nursing Care

Assess for fever or malaise that can signal the onset of infections. Other conditions to be alert for include anemia, hypercalcemia, fractures, and renal complications. Monitor intake and output, and strain urine for stones. Elevated BUN and creatinine levels will alert you to possible renal failure. Report back pain, leg weakness, sensory loss, or loss of bowel or bladder function, because these might indicate spinal cord compression.

Keeping the patient mobile is very important. The physical therapist and occupational therapist can help the patient continue to be active. Bones in use are strongest, so the patient should remain up and moving as much as possible to help stimulate calcium resorption and decrease demineralization. Assist the patient with walking because of the risk of pathological fractures of the long bones. If the patient is unsteady, use a walker or a support belt. Keeping the patient up and active also decreases the risk of respiratory complications. Urination also is enhanced in the patient who does not need to rely on the use of a bedpan or urinal.

If the patient is bedridden, reposition him or her every 2 hours to prevent complications related to immobility; use a lift sheet to move the patient gently and decrease the risk of pathological fractures. Provide passive range-of-motion exercises and encourage deep breathing.

Teach the patient the importance of good hydration at all times to minimize complications of hypercalcemia. Administer fluids so that daily output is never less than 1500 mL. Depending on time of year and the type and level of patient activities, the patient may need to have an intake of more than 4 L daily.

If hypercalcemia occurs, the physician will order an IV of normal saline to infuse at a high rate followed by regular administration of diuretics. The goal is to get the serum calcium level below 10 mg/dL. Oral compounds are also available to help keep the calcium level within normal limits. See Home Health Hints for additional suggestions for patients being cared for at home.

HOME HEALTH HINTS

- Patients who are at risk for infection can place a sign on the front door of their homes to limit visitors or ask persons with colds to come back when they are well. The patient may appreciate the home nurse's giving permission to be assertive in such circumstances.

- To prevent bruising, have the patient cut the feet off of white sport socks and wear them on the arms. They can be hidden under long-sleeve shirts and blouses and provides a cushion when doing housework.

- Patients with sickle cell anemia usually have lower blood pressures. It is important to report even mild hypertension for these patients.

REVIEW QUESTIONS

1. Which assessment finding would you expect to find in the patient who has anemia?
    a. Pain
    b. Dyspnea
    c. Vision changes
    d. Skin rash

2. Which of the following activities is contraindicated for the patient with sickle cell anemia?
    a. Riding in an elevator
    b. Taking a long car trip
    c. Running in a marathon
    d. Listening to a concert

3. Which explanation for bleeding should you give to the wife of a patient with DIC?
    a. "He is bleeding because he does not have enough RBCs."
    b. "He is bleeding because his white cells are depleted."
    c. "He is bleeding because his blood pressure is so high that it forces blood from mucous membranes."
    d. "He is bleeding because his body's clotting factors have all been used up."

4. Which instruction will help the mother of a child with hemophilia prevent bleeding episodes?
    a. "Your son should avoid contact sports."
    b. "Your son will have to avoid all potentially irritating foods."

    c. "Your son must never shave."
    d. "Your son should always live near a major hospital system."

5. Which family member should not be permitted to visit a patient with newly diagnosed leukemia?
    a. The one who has a new baby at home.
    b. The one who has a history of asthma.
    c. The one who has received recent radiation treatment for cancer.
    d. The one who has a runny nose.

6. Which of the following assessment findings is most frequently encountered in patients with multiple myeloma?
    a. Pathological fractures
    b. Mental status changes
    c. Raging infections
    d. Bleeding tendencies

7. Which of the following nursing interventions is most appropriate for a patient with thrombocytopenia?
    a. Avoid intramuscular injections.
    b. Keep visitors who are ill away from the patient.
    c. Encourage 4 L of fluid daily.
    d. Allow rest between activities.

REFERENCES

1. Purnell, L, and Paulanka, B (eds): Transcultural Health Care: A Culturally Competent Approach. FA Davis, Philadelphia, 1998.

Answers to CRITICAL THINKING

Mrs. Johns

1. Monitor Mrs. Johns's vital signs and report falling blood pressure and rising pulse immediately. Inspect her skin for petechiae and ecchymoses. Outline ecchymotic areas with an ink pen in order to see if the area is increasing in size. Monitor urine for signs of blood. Test stools for occult blood. Monitor vaginal discharge for increasing bleeding. Report any changes promptly.
2. Anticipate assisting the RN with administration of blood or blood products. Instruct Mrs. Johns in the importance of preventing injury that could cause further bleeding. Other care will be supportive.
3. Mrs. Johns will be concerned for her new baby, who is most likely on another unit or already discharged home. Allow Mrs. Johns to talk about her concerns. Arrange visits with her family and baby if permitted by her condition and her physician.

Mr. Washington

1. Because of his leukemia and his treatment, Mr. Washington is at risk for infection. If he develops an infection, he will have great difficulty getting over it. With so many visitors in the room, it is likely that one or more has a cold or virus. They may not be aware of the risk this poses to Mr. Washington. Mr. Washington is probably also fatigued because of his disease and treatment, and visiting requires energy.
2. You should kindly explain that he is very susceptible to catching colds or other illnesses and that it would be best to limit visitors to one or two at a time. Point out that persons with symptoms of colds or flu should not enter the room at all. Visits should also be brief to prevent overtiring the patient.

25

NURSING CARE OF PATIENTS WITH LYMPHATIC DISORDERS

Cheryl L. Ivey

KEY TERMS

lymphoma (lim-**FOH**-mah)
splenectomy (sple-**NEK**-tuh-mee)
splenomegaly (SPLEE-noh-**MEG**-ah-lee)

QUESTIONS TO GUIDE YOUR READING

1. What is the most distinguishing feature of Hodgkin's disease?
2. What are the differences between non–Hodgkin's lymphoma and Hodgkin's disease in terms of age of onset, symptoms, and curability?
3. What data should you collect when caring for patients with lymphatic disorders?
4. What nursing care will you provide for patients with lymphatic disorders?
5. What nursing care will you provide for patients undergoing a splenectomy?
6. What will you teach the patient undergoing therapeutic splenectomy?

Lymphatic disorders include Hodgkin's disease and the non–Hodgkin's **lymphomas.** Because the spleen is part of the lymph system, this chapter also discusses **splenectomy.**

▶ HODGKIN'S DISEASE

Despite its name, Hodgkin's disease (HD) is a lymphoma, which is a cancer of the lymph system. Its distinguishing feature is Reed-Sternberg cells, which make it different from all the other forms of lymphoma. HD is more prevalent in men than in women and occurs most often in young adults ages 15 to 40. After a decrease in incidence in persons ages 40 to 55, the incidence peaks again in adults older than age 55. Of all the lymphomas, HD is the most curable type, even when the disease is widely spread at the time of diagnosis.

Pathophysiology

Lymph nodes are made of tightly bound fibers and cells that serve as filtering devices for the body's immune system. Most often, HD begins as a single changed lymph node, usually in the cervical lymph nodes of the neck. As the disease progresses, the cancer invades the lymph node chains, node by node. The path of cancer infiltration is usually the same as the path for lymph fluid flow. Left untreated, other lymphoid tissues such as the spleen become infiltrated with HD. The major organs eventually become involved with Hodgkin's disease. Common complaints of patients with organ involvement may include shortness of breath, feelings of fullness, weakness, and malaise. These organ-related symptoms usually motivate the patient to seek medical help.

A tentative diagnosis of HD is based on one or more painlessly enlarged nodes in the cervical, axillary, or inguinal areas. A biopsy of several of the enlarged nodes is performed with the intent of searching for the presence of Reed-Sternberg cells, which confirms the diagnosis.

Etiology

The exact cause of HD is unknown. A possible viral origin has been proposed; it is more common in people who have had mononucleosis. Sometimes it occurs in families, suggesting a genetic link. Patients with impaired immune function, such as those with acquired immunodeficiency

---

lymphoma: lymph + oma—tumor

syndrome (AIDS) or those taking immunosuppressive drugs after organ transplant are also at higher risk.

Signs and Symptoms

Painless swelling in one or more of the common lymph node chains is a usual presentation (Fig. 25–1). This swelling can range from barely perceptible to a size similar to that of a softball, occasionally even larger. The patient may complain of generalized pruritus. One other curious event, alcohol-induced pain, is occasionally present. With just a few sips of any type of alcohol-containing beverage (beer, wine, or liquor), the patient may complain of intense pain at the site of disease. Because the lymph nodes in the upper chest and neck are often involved, the patient may have symptoms of obstruction, such as cough, dysphagia, and even stridor.

Other common symptoms may include persistent low-grade fever, night sweats, fatigue, weight loss, and malaise. When these additional symptoms are present, the prognosis is worse. In older adults there may be no enlarged lymph nodes visible and these secondary symptoms may be the only presenting symptoms. Other symptoms associated with late-stage disease include edema of the neck and face, possible jaundice, nerve pain, enlargement of the retroperitoneal nodes, and infiltration of the spleen; liver and bones may also be involved.

Diagnostic Tests and Staging

Diagnosis usually begins with a lymph node biopsy of the easiest lymph node to access. Lymph node biopsies are done to check for abnormal histiocyte proliferation, nodular fibrosis, and necrosis. Other tests include bone marrow biopsy and aspiration, liver and spleen biopsies, routine chest x-ray examination, abdominal computed tomography (CT) scan to check for presence of disease in the liver and spleen, lung scan, and bone scan. In some larger cancer centers, lymphangiography may be performed to view the flow of lymph in the lymph network. This test is accomplished by a skilled physician injecting dye into the lymph tracts located between the toes. The dye is observed as it migrates up the lymph chains, signaling blockages where present.

Hematological tests (e.g., complete blood cell count [CBC]) may show wide variability of red blood cells, indicating mild to severe anemia. The white blood cell (WBC) count is often abnormal and extreme (either very high or very low) because of bone marrow infiltration by disease. This places the patient at increased risk for infection.

These same tests are used for staging. HD is staged based on the Ann Arbor Clinical Staging Classification and is as follows:

- Stage I disease is limited to a single lymph node or site.
- In stage II disease, two or more nodes are involved on the same side of the diaphragm. Limited organ involvement may or may not be present.
- Stage III disease is characterized by nodes on both sides of the diaphragm, with or without organ involvement.
- Stage IV, the most serious form of the disease and the least curable, includes widely disseminated disease in several organs or tissues with or without associated lymph node involvement.

Medical Treatment

Appropriate therapy includes the use of radiation and chemotherapy, depending on the stage of the disease. Radiation therapy, administered on an outpatient basis over a 4- to 6-week period, can cure most patients with stage I or stage II disease. Combinations of chemotherapy and radiation therapy are used for patients with stage III and stage IV disease, with results based on the location and the stage of disease.

Nursing Care

Most nursing interventions are aimed at symptom management. If the patient is experiencing pruritis or night sweats, nursing interventions are aimed at alleviation of discomfort. These may include changing the gown and bed linens

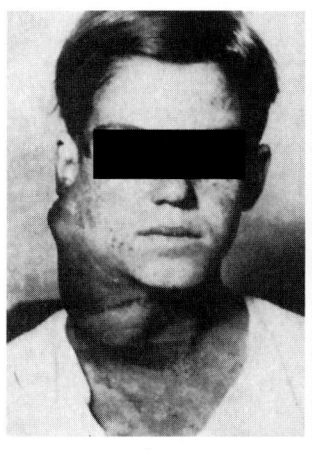 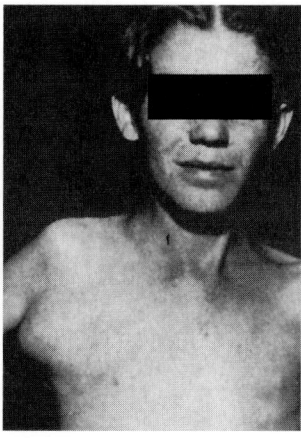 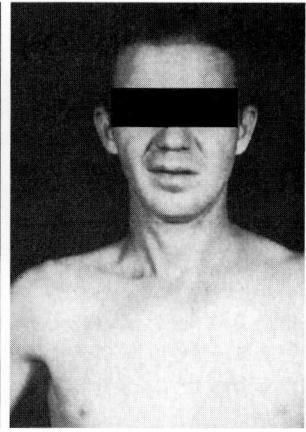

A   B   C

Figure 25-1 Cervical Hodgkin's disease. *(A)* Young boy with extensive cervical Hodgkin's disease. *(B)* Appearance several years later, when axillary manifestation developed. *(C)* Appearance 23 years after initial treatment with radiation. (From del Regato, JA, Spjut, HJ, and Cox, JD: Cancer: Diagnosis, Treatment, and Prognosis, ed 6. Mosby, St Louis, 1985, with permission.)

several times a night and helping the patient remain clean and dry. Keeping the patient and family involved in the plan of care may relieve anxiety.

After the diagnosis is established, the licensed practical nurse/licensed vocational nurse (LPN/LVN) collaborates with the registered nurse (RN) to teach the patient and the family about the treatments required. Specific information regarding chemotherapy and radiation therapy is necessary. Most oncology units have helpful pamphlets or other printed material. Type and length of therapy is based on the stage of the cancer at the time of diagnosis.

Later, nursing interventions are tailored to alleviate problems that arise secondary to chemotherapy and radiation therapy. See Chapter 10 for nursing interventions for these problems. Also see Nursing Care Plan 25–1 for the Patient with Lymphoma.

Patient Teaching

In addition to the teaching needs outlined above, make sure that the patient and the family know about local chapters of the American Cancer Society and the Leukemia and Lymphoma Society. Both of these organizations have information, financial assistance, and counseling referral sources, which most patients find valuable. Another source for chemotherapy and radiation therapy information is the National Cancer Institute (NCI). The NCI can send as much information as the patient requests regarding all aspects of treatment at no charge. For more information about Hodgkin's lymphoma, visit www.cancer.org. Or visit the Leukemia and Lymphoma Society at www.leukemia.org or the National Cancer Institute at www.nci.nih.gov.

BOX 25–1   NURSING CARE PLAN FOR THE PATIENT WITH LYMPHOMA

 Activity intolerance related to fatigue and anemia

PATIENT OUTCOMES
Patient will have activities of daily living (ADLs) needs met by self or caretaker.

EVALUATION OF OUTCOMES
Is patient able to carry out ADLs or are ADL needs met by a caretaker?

INTERVENTIONS	RATIONALE	EVALUATION
Assess amount of activity that causes fatigue.	Assessment helps guide plan of care.	How much can patient do before becoming fatigued?
Assist patient with activities as necessary.	The patient may need assistance with ADLs if fatigue is extreme.	Does the patient need assistance? Can family members assist?
Provide oxygen therapy as ordered.	Oxygen therapy can increase oxygen levels and activity tolerance.	Does patient tolerate activity better with oxygen therapy?
Instruct patient to space rest with activities.	Rest periods decrease oxygen need and allow patient to conserve energy for next activity.	Is patient able to tolerate activity better after a rest period?

 Risk for infection related to bone marrow involvement and side effects of treatment

PATIENT OUTCOMES
Patient will have no signs or symptoms of infection.

EVALUATION OF OUTCOMES
Are signs and symptoms of infection absent? Is temperature within normal limits?

INTERVENTIONS	RATIONALE	EVALUATION
Assess patient for risk factors for infection.	The WBC count may be very high or very low, placing the patient at risk for infection.	Is the patient at risk? Are additional interventions indicated?
Monitor patient for signs and symptoms of infection, such as cough, fever, malaise, erythema, pain, or drainage and report immediately.	Early detection and treatment of infection provides the best results.	Are signs and symptoms of infection present?
Teach patient and significant other signs and symptoms of infection to watch for and report.	The patient must be involved in monitoring for infection when at home.	Does patient verbalize understanding of signs and symptoms of infection and importance of reporting?
Teach the patient to avoid exposure to others with influenza or other infections.	Exposure increases risk for infection, especially with compromised immune function.	Does patient verbalize understanding of sources of infection to avoid?
Teach patient proper hand washing and good oral and personal hygiene.	These activities reduce risk of infection.	Does patient demonstrate proper hand washing and hygiene?

BOX 25-1   NURSING CARE PLAN FOR THE PATIENT WITH LYMPHOMA—CONT'D

▶ **Risk for ineffective coping related to new diagnosis and potential lifestyle changes to accommodate treatments**

PATIENT OUTCOMES
The patient states ability to manage lifestyle changes and medical management of condition.

EVALUATION OF OUTCOMES
Does patient carry out self-care necessary to manage treatment?

INTERVENTIONS	RATIONALE	EVALUATION
Assess patient's level of distress related to uncertainty of the future, bothersome symptoms, changes in self-concept, and past coping mechanisms.	Obtaining information regarding past experiences helps the nurse identify and correct misconceptions. The nurse can support effective coping mechanisms that worked in the past.	Is patient able to identify sources of anxiety? Are past coping mechanisms effective?
Assess for signs of maladaptive behaviors that interfere with responsible health practices, such as missed appointments or failure to attend to symptoms.	Long-term survival depends on keeping therapy schedule. The ability to manage and report symptoms early keeps the patient out of the hospital and in control of his or her own life.	Does the patient keep appointments? Does the patient participate in self-care activities and report symptoms promptly?
Assist the patient to identify support systems and resources. Refer to social worker or other community resources as needed.	Resources can assist with participation in treatment plan, home care, or financial assistance.	Are resources identified and helpful?
Refer patient and family to a cancer survivors' support group.	Others who have been through treatment themselves can be a good support for patients with cancer.	Does the patient state that the support group is helpful?

✎ CRITICAL THINKING

Jeanine

Jeanine is a 60-year-old nurse diagnosed with stage II Hodgkin's disease. She wishes to continue working at her job on a respiratory unit at the local hospital while she undergoes treatment. What concerns do you have about this?

Answers at end of chapter.

▶ NON–HODGKIN'S LYMPHOMAS

All of the other types of lymphomas are clumped into a diverse classification known as the non–Hodgkin's lymphomas (NHLs). It is possible to sort these other types of lymphomas into different categories based on the degree of malignancy. Non–Hodgkin's lymphomas arise in the lymphoid tissues of the body, just as Hodgkin's disease does, but they differ in several ways. See Table 25–1 for a comparison of Hodgkin's and non–Hodgkin's lymphomas.

Pathophysiology

The most distinguishing difference is the absence of the Reed-Sternberg cells in non–Hodgkin's lymphomas. Instead, many of these lymphomas arise from the B cells and T cells. The B cells are involved in recognizing and destroying specific antigens. Cells specifically involved include the memory B cells and the plasma cells. The T cells also are involved in registering antigens, but there are many more kinds of T cells. These include the amplifier T cells, helper T cells, suppressor T cells, memory T cells, cytotoxic T cells, and delayed hypersensitivity T cells. An abnormality in any of these cells can result in a type of NHL. Most cases of NHL are of B-cell origin.

Etiology

The cause of non–Hodgkin's lymphomas is unclear, but some viruses, such as Epstein-Barr virus and herpesvirus, are thought to play a role in their development. Immune prob-

---

TABLE 25-1   HODGKIN'S DISEASE VERSUS NON-HODGKIN'S LYMPHOMA

	Hodgkin's Disease	Non-Hodgkin's Lymphomas
Age	Younger	Older
Degree of debilitation (overall)	Less	More
Presence of fever, night sweats (indicating more advanced disease)	More likely	Less likely
Spread to other areas at time of diagnosis	Local to regional area of spread	Advanced cancer—spread to many different areas
Types	Just one	Many different types

lems such as AIDS increase risk. Exposure to nuclear waste and some toxic chemicals may also increase risk.

Signs and Symptoms

Clinical features of malignant lymphomas include enlarged, painless, rubbery nodes in the cervical and supraclavicular areas; enlarged tonsils and adenoids; and occasional symptoms of dyspnea and cough. As the disease progresses, the patient may report fatigue, malaise, weight loss, and night sweats, similar to Hodgkin's disease. NHL usually progresses more rapidly than HD.

Diagnostic Tests

Diagnosis is confirmed by histological evaluation of biopsied lymph nodes, tonsils, bone marrow, liver, bowel, skin, or any other affected tissues. Other relevant tests include bone scans, chest x-ray examination, lymphangiography, liver and spleen scan, CT of the abdomen, and intravenous pyelogram to determine the extent of the disease. Laboratory tests include a CBC, which often indicates anemia; serum uric acid level; and liver function studies. Serum calcium level may be elevated if bone lesions are present.

Medical Treatment

Treatment usually involves multimodal therapy, including the use of chemotherapy and radiation therapy given in combination. Chemotherapy treatments are given on a set schedule of approximately once every month for about 6 months. These treatments may be performed on an inpatient or an outpatient basis. Radiation therapy is given to affected areas in advanced stages of NHL. Stem cell transplant may be tried in patients with advanced disease.

Nursing Care

The nurse provides emotional support by keeping the patient and family informed during the testing phase. Symptoms such as night sweats can be managed with frequent linen and gown changes. If the patient has severe anemia and is symptomatic, arrange his or her daily schedule to conserve energy. Transfuse packed red blood cells as ordered.

Give antibiotics as ordered if the patient has an infection or if the WBC count is especially low. Observe standard precautions and teach the patient and family about the need for hand washing and avoiding large crowds. Teach the patient to avoid others with colds and flu to decrease the chance of infection.

Help the patient maintain nutrition with attractively prepared meals. Involve other professionals in the care of the patient. Include the chaplain if available and if the patient is exhibiting spiritual distress. Spend time listening to the patient's concerns. Refer the patient and family to the resources listed earlier for more information.

▶ SPLENIC DISORDERS

The spleen is involved in a number of disorders, including cancers of the blood, lymph, and bone marrow; hereditary conditions such as sickle cell disease; and acquired problems such as idiopathic thrombocytopenia. Under normal circumstances, the spleen is not paid much attention; it generally performs its functions without much fanfare.

If the spleen enlarges markedly, the condition is referred to as **splenomegaly.** Other times, the spleen may or may not be enlarged, but the function is out of control so that too many red blood cells and platelets are removed from the peripheral circulation. Sometimes the spleen is not able to perform its job because of bleeding into the pulp of the organ, which renders it useless. Bleeding into the spleen can occur from various illness or from trauma. Regardless of the nature of the malfunction, an occasional treatment option is splenectomy.

Splenectomy

Splenectomy is the surgical removal of the spleen. This is sometimes used to treat select hematological disorders and is also used, under different circumstances, to stage (or determine spread) in lymphomas. Splenectomy is performed fairly often in the United States, but, like any surgery, it is not without risk.

Patient Teaching

Explain to patients that this surgery removes the spleen, usually under general anesthesia. Inform patients that they can live a normal life after the surgery but that they may be more prone to infection and that they should receive the influenza vaccine each year, preferably in the early fall.

Preoperative Care

Before the surgery, ensure that the CBC and coagulation profile are completed and reported to the physician. If ordered, you may assist the RN in the transfusion of blood to correct the underlying anemia and to prepare for the loss of a great deal of blood stored in the spleen. Vitamin K is often ordered to correct clotting factor deficiencies.

Take the patient's vital signs and perform a baseline respiratory assessment. Note especially any signs of respiratory infections such as fever, chills, crackles, wheezes, or cough. If any of these are noted, make sure that the physician is aware of them because surgery may need to be delayed. Teach the patient routine coughing and deep breathing techniques to help prevent postoperative respiratory complications.

Postoperative Care

During the early postoperative period, watch carefully for bleeding, either external or internal. Be prepared to administer narcotics for pain, usually on an around-the-clock schedule so the patient is comfortable enough to deep breathe, cough, and ambulate. After narcotic administration be sure to observe for side effects, which may include incomplete pain relief or hypoventilation. Monitor for fever every 4 hours, and expect a mild, low-grade, transient fever postoperatively. A persistent fever may indicate abscess or hematoma formation.

If the surgery was performed to decrease the numbers of cells being removed from the peripheral circulation, monitor the platelet count. Often the count begins to rise in just a few days, but it may take up to 2 weeks for the platelets to normalize.

Complications

A splenectomy can cause complications such as bleeding, pneumonia, and atelectasis. Respiratory problems occur because of the spleen's position close to the diaphragm. This placement requires the need for a high surgical incision that is very painful. Often the patient tries to restrict lung expansion after surgery to keep from hurting, but this splinting behavior may leave the patient at risk for pneumonia and respiratory problems. In addition, splenectomy patients are usually more vulnerable to infection, especially influenza, because the spleen's role in the immune response is no longer filled.

Other possible complications from splenectomy include the development of pancreatitis and fistula formation. This is due to the fact that the tail of the pancreas is very close to the spleen, and irritation may have occurred.

Another serious complication is that of overwhelming postsplenectomy infection (OPSI). The causative agents in OPSI include streptococci, *Neisseria* spp., and influenza bacteria. Patients at greatest risk of OPSI may include patients who had a splenectomy secondary to a cancer condition or during childhood.

Early symptoms of OPSI include fever and malaise that seem unremarkable. However, the infection may progress within a few hours to sepsis and death. Unfortunately, OPSI has a mortality rate as high as 70 percent. Be sure to include the signs and symptoms of OPSI in presplenectomy patient education. Also stress the need to promptly obtain medical attention for the patient at the first signs and symptoms of OPSI. The patient should be directed to continue to receive lifetime vaccinations against these bacteria.

Answers to CRITICAL THINKING

Jeanine

Jeanine will probably be fatigued from her disease, and fatigue may increase further from side effects of treatment. Staff nursing jobs can be tiring even for healthy nurses. In addition, she will be around patients with respiratory diseases, many of which are contagious. Because of risk for infection secondary to the disease process and the treatment regimen, Jeanine might want to take a leave of absence during treatment or ask to be reassigned to a different area that is less demanding and away from direct patient care until her treatments have been completed.

REVIEW QUESTIONS

1. Which of the following signs is most commonly present at diagnosis with the lymphomas?
    a. High fever
    b. Painful lymph nodes
    c. Enlarged lymph nodes
    d. Abdominal pain

2. Stage III Hodgkin's disease is defined as which of the following?
    a. Lymphatic involvement on both sides of the diaphragm
    b. Localized involvement of more than two adjacent or nonadjacent regions on one side of the diaphragm
    c. Diffuse involvement of one or more extralymphatic organs or tissues such as the bone marrow or liver
    d. Localized involvement of a single lymph node site, usually located in the cervical or supraclavicular area

3. Which of the following interventions is appropriate for the patient at home who is at risk for infection?
    a. Advise the patient to stay indoors.
    b. Encourage the patient to avoid crowds.
    c. Maintain strict isolation.
    d. Instruct the patient to avoid blood draws.

4. Which of the following circumstances places the patient at most risk for respiratory complications following a splenectomy?
    a. Disturbance of clotting factors
    b. Nothing by mouth (NPO) status
    c. Need for frequent dressing changes
    d. Location of surgical incision

5. Which of the following postsplenectomy symptoms should the patient report immediately?
    a. Fever and malaise
    b. Pain at the incision site
    c. Discomfort with coughing
    d. Fatigue

UNIT FIVE BIBLIOGRAPHY

Atassi, KA, and Harris, ML: Actionstat: Disseminated intravascular coagulation. Nursing 2001 31(3):64, 2001.

Barrick, MC, and Mitchell, SA: Multiple myeloma: Recent advances for this common plasma cell disorder. American Journal of Nursing Apr, Suppl: 6–12, 2001.

Gutaj, DA: Oncology today: Lymphoma. RN 63(8):32–38, 2000.

Harmening, DM: Clinical hematology and fundamentals of hemostasis ed 3. FA Davis, Philadelphia, 1997.

Holcomb, SS: Anemia: Pointing the way to a deeper problem. Nursing 2001 31(7):36–43, 2001.

Lutz, CA, and Przytulski, KR: Nutrition and diet therapy ed 3, FA Davis, Philadelphia, 2001.

Maxson, JH: Management of disseminated intravascular coagulation. Critical Care Nursing Clinics of North America 12(3):341–352, 2000.

Medoff, E: Oncology today: New horizons. Leukemia. RN 63(9):42–50, 2000.

Sulton, LL: What's wrong with this patient? Idiopathic thrombocytopenic purpura. RN 61(3):35–39, 1998.

Test your knowledge: Assessment of febrile neutropenic patients. Clinical Journal of Oncology Nursing 4(4):182, 184, 2000.

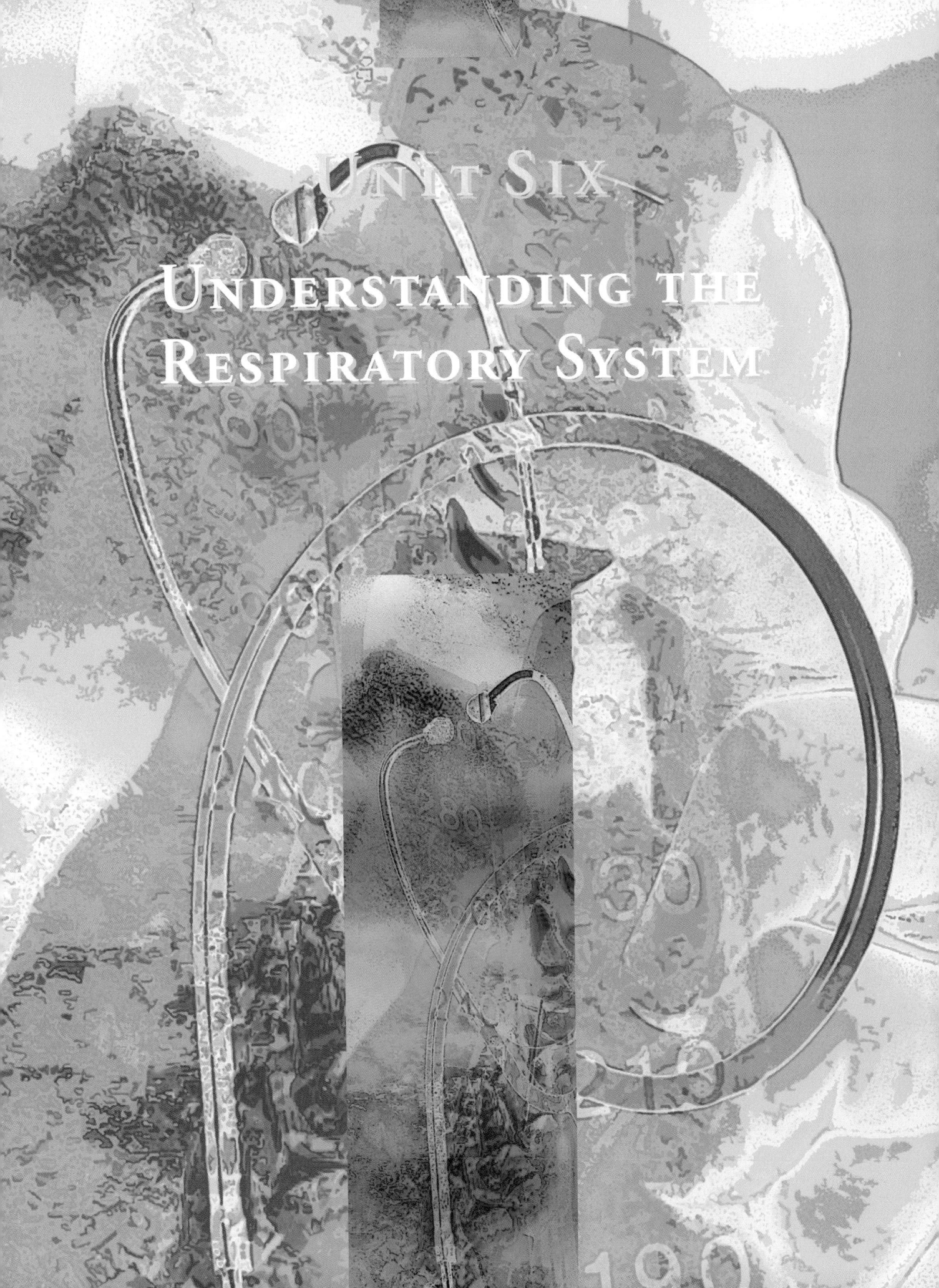

UNIT SIX

UNDERSTANDING THE RESPIRATORY SYSTEM

26

RESPIRATORY SYSTEM FUNCTION AND ASSESSMENT

Paula D. Hopper

KEY TERMS

adventitious (ad-ven-**TI**-shus)

apnea (**AP**-nee-ah)

crepitus (**KREP**-ih-tuss)

cyanosis (SIGH-uh-**NOH**-sis)

dyspnea (**DISP**-nee-ah)

respiratory excursion (**RES**-pi-rah-TOR-ee eks-**KUR**-zhun)

thoracentesis (THOR-uh-sen-**TEE**-sis)

tidaling (**TIGH**-dah-ling)

tracheostomy (TRAY-key-**AHS**-tuh-me)

tracheotomy (TRAY-key-**AH**-tuh-me)

QUESTIONS TO GUIDE YOUR READING

1. What are the structures of the respiratory system, and what is the function of each?
2. How does aging affect the respiratory system?
3. What questions should you ask when you take a history for a patient with a respiratory problem?
4. What findings do you expect when you inspect, palpate, percuss, and auscultate the chest?
5. What are the common diagnostic tests performed to diagnose disorders of the respiratory system?
6. What nursing care should you provide for patients undergoing each of these diagnostic tests?
7. What are common therapeutic measures used for patients with respiratory disorders?

► NORMAL ANATOMY AND PHYSIOLOGY

The respiratory system consists of the nose, nasal cavities, pharynx, larynx, trachea, bronchial tree, lungs, and respiratory muscles. The parts outside the chest cavity are collectively called the upper respiratory tract, and those within the chest cavity make up the lower respiratory tract (Fig. 26–1). The lungs are the site of gas exchange between the air and the blood; the rest of the system moves air into and out of the lungs.

Nose and Nasal Cavities

The nose is made of bone and cartilage covered with skin; hairs inside the nostrils block the entry of dust and other particles. The two nasal cavities are inside the skull and are separated by the nasal septum, which is made of the vomer and ethmoid bones. The nasal mucosa is ciliated epithelium that is highly vascular; it warms and moistens inhaled air. Dust and microorganisms are trapped on mucus produced by goblet cells and swept backward and down to the pharynx by the cilia.

The paranasal sinuses are air cavities in the maxillae and the frontal, sphenoid, and ethmoid bones that open into the nasal cavities. They are lined with ciliated epithelium, and the mucus produced usually drains into the nasal cavities. Their functions are to make the skull lighter in weight and to provide resonance for the voice.

Pharynx

The pharynx is posterior to the nasal and oral cavities and has three parts: nasopharynx, oropharynx, and laryngopharynx. The nasopharynx is an air passage above the level of the soft palate, which rises to block the nasopharynx during swallowing. The eustachian tubes from the middle ear cavities open into the nasopharynx; the adenoid is a lymph nodule on its posterior wall. The oropharynx is behind the oral cavity and is both an air and a food passage. The palatine tonsils are on the lateral walls, and together with the lingual tonsils on the base of the tongue and the adenoid, they form a ring of lymphatic tissue around the pharynx and destroy pathogens that penetrate the mucosa. The laryngopharynx is both an air and a food passage; it

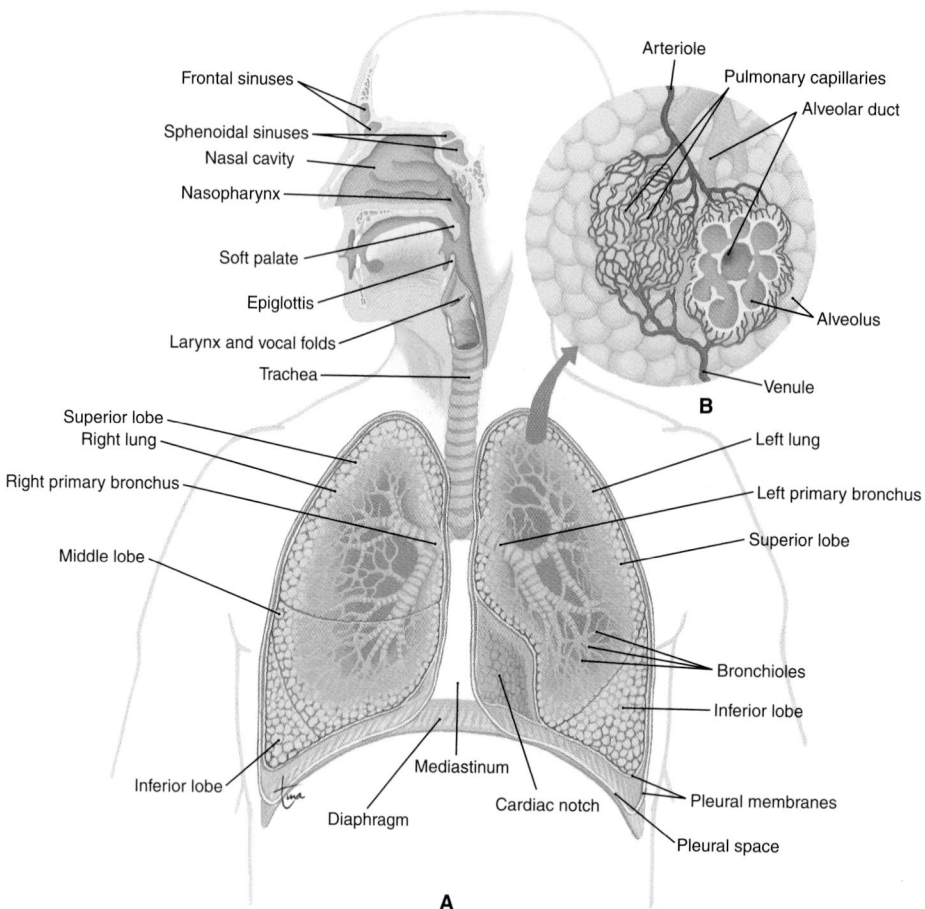

Figure 26-1 Respiratory system, anterior view, with microscopic view of alveoli and pulmonary capillaries. (Modified from Scanlon, VC, Sanders, T: Essentials of Anatomy and Physiology, ed 3. FA Davis, Philadelphia, 1999, p 330, with permission.)

opens anteriorly into the larynx and posteriorly into the esophagus.

Larynx

The larynx is the voice box and the airway between the pharynx and trachea. It is made of nine pieces of cartilage, a firm yet flexible tissue that keeps the airway open, and is lined with ciliated epithelium. The thyroid cartilage, commonly called the Adam's apple, is the largest of these cartilage pieces and is palpable on the front of the neck. The epiglottis is the uppermost cartilage and covers the larynx like a flap when the larynx is elevated during swallowing. On either side of the glottis (the airway) are the vocal cords. When pulled together across the glottis and vibrated by exhaled air, the vocal cords produce sounds that may be turned into speech. The vagus and accessory cranial nerves are the motor nerves to the larynx.

Trachea and Bronchial Tree

The trachea is a tube 4 to 5 inches long that extends from the larynx to the primary bronchi. C-shaped pieces of cartilage in the wall keep the trachea open. The mucosa is ciliated epithelium; mucus with trapped dust and microorganisms is swept upward toward the pharynx and is usually swallowed.

The bronchial tree is the series of air passages within the lungs, a succession of progressively smaller tubes that terminate in the alveoli. The right and left primary bronchi are branches of the trachea. Each gives rise to secondary bronchi; their structure is like that of the trachea. The bronchioles, however, have no cartilage in the walls to maintain patency, and they may be closed completely by contraction of their smooth muscle.

Lungs and Pleural Membranes

The lungs fill the chest cavity on either side of the heart, extending from the clavicles to the diaphragm. On the medial surface of each lung is an indentation called the hilus, where the primary bronchus and the pulmonary artery and veins enter the lung.

The pleural membranes are the serous membranes of the thoracic cavity. The visceral pleura is the membrane that covers the lungs; the parietal pleura lines the chest cavity. A small amount of serous fluid between these membranes prevents friction and keeps the membranes together during breathing.

The functional units of the lung are the millions of alveoli, the air sacs that are the site of gas exchange. Both the alveoli and the surrounding pulmonary capillaries (see Fig. 26–1) are made of simple squamous epithelium; that is, their walls are only one cell in thickness to permit diffusion of gases.

Each alveolus is lined with a thin layer of tissue fluid that is essential for the diffusion of gases, but the surface tension of the fluid tends to make the walls of an alveolus stick together internally. Certain alveolar cells secrete pulmonary surfactant, a lipoprotein that mixes with the tissue fluid and decreases surface tension to permit inflation. Also in the alveoli are the alveolar macrophages, which phagocytize pathogens or bits of particulate matter (such as air pollution) that have not been trapped and swept out by the cilia.

Between clusters of alveoli is elastic connective tissue; it is capable of recoil when stretched (during inhalation) and contributes significantly to normal exhalation. The recoil of this tissue ensures that normal exhalation is a passive process that does not require the expenditure of energy.

Mechanism of Breathing

Ventilation is the term for the movement of air into and out of the alveoli. Air moves from high-pressure to low-pressure areas (pressure gradients), some of which are created by the respiratory muscles, which in turn are controlled by the nervous system. The respiratory centers are in the medulla and pons. The main respiratory muscles are the diaphragm below the lungs and the external and internal intercostal muscles between the ribs. Accessory muscles of respiration are used during exercise and times of respiratory distress; these include the sternocleidomastoid and scalene muscles.

Pressures important to breathing include atmospheric pressure, intrapleural pressure, and intrapulmonic pressure. Atmospheric pressure is the pressure of the air around us, which at sea level is 760 mm Hg; atmospheric pressure decreases as altitude increases. Intrapleural pressure is in the potential pleural space between the pleural membranes. Serous fluid causes the two membranes to adhere to each other, and because the elastic lungs are always tending to collapse and pull the visceral pleura away from the parietal pleura, the pressure in this potential space is always below atmospheric pressure (about 756 mm Hg). This is called a negative pressure. Intrapulmonic pressure is the pressure inside the alveoli and bronchial tree. This pressure fluctuates below and above atmospheric pressure during each cycle of breathing.

Inhalation

Inhalation, also called inspiration, occurs when motor impulses from the medulla cause contraction of the respiratory muscles. Impulses along the phrenic nerves cause the diaphragm to contract and move downward (its dome shape flattens). Impulses along the intercostal nerves cause the external intercostal muscles to pull the ribcage upward and outward, expanding the chest cavity in all directions. This then expands the pleural membranes. Intrapleural pressure becomes even more negative, but the serous fluid keeps the membranes together and the lungs expand as well. As the lungs expand, intrapulmonic pressure falls below atmospheric pressure and air enters the nose and respiratory passages. Entry of air continues until intrapulmonic pressure equals atmospheric pressure; this is a normal inhalation. A deeper inhalation requires a more forceful contraction of the respiratory muscles to expand the chest cavity and lungs even further and permit the entry of more air.

Exhalation

Normal exhalation is a passive process that begins when motor impulses from the medulla decrease and the diaphragm and external intercostal muscles relax. The lungs are compressed as the chest cavity becomes smaller and the recoil of the elastic lung tissue compresses the alveoli. Intrapulmonic pressure rises above atmospheric pressure, and air is forced out of the lungs until the two pressures are again equal. Under normal circumstances, energy is not required for exhalation (as it is for inhalation) because the elasticity of the lungs causes recoil and forces air out. A forced exhalation beyond the normal amount is an active process that requires contraction of the internal intercostal muscles to pull the ribcage downward and inward and contraction of the abdominal muscles to force the dome of the diaphragm upward, increasing compression of the lungs.

Transport of Gases in the Blood

Oxygen is carried in the blood by iron in the hemoglobin (Hgb) of red blood cells (RBCs). The iron-oxygen bond is formed in the lungs, where the partial pressure of oxygen (PO_2) is high. In tissues where the PO_2 is low, hemoglobin releases much of its oxygen.

Most carbon dioxide is carried in the blood in the form of bicarbonate ions in the plasma. These ions are formed when carbon dioxide enters RBCs and is converted to carbonic acid (H_2CO_3), which ionizes to bicarbonate ions (HCO_3^-) and hydrogen ions (H^+). The bicarbonate ions leave the RBCs for the plasma, and the remaining hydrogen ions are buffered by the hemoglobin in the RBCs. When the blood reaches the lungs, an area of lower partial pressure of carbon dioxide (PCO_2), these reactions are reversed—carbon dioxide is reformed and diffuses into the alveoli to be exhaled.

Regulation of Respiration

Respiration is regulated by both nervous and chemical mechanisms. The medulla contains an inspiration center and an expiration center. The inspiration center generates impulses that bring about contraction of the respiratory muscles; the result is inhalation. Sensory impulses from baroreceptors in the inflating lungs to the medulla depress the inspiration center; this is called the Hering-Breuer inflation reflex and helps prevent overinflation of the lungs. In the pons the apneustic center prolongs inhalation and the pneumotaxic center helps bring about exhalation. These centers provide a normal breathing rhythm, 12 to 20 breaths per minute with exhalation slightly longer than inhalation. When there is a need for more forceful exhalations, the inspiration center activates the expiration center, which brings about contraction of the internal intercostal muscles.

Normal breathing is essentially a reflex, but because the respiratory muscles are skeletal (or voluntary) muscles, it is possible to make changes. The cerebral cortex may override the medulla to permit voluntary changes in breathing, such as faster or slower breathing, holding one's breath, or singing. Eventually, however, the medulla resumes control and breathing again becomes a reflex.

Chemical regulation of respiration involves the blood levels of oxygen and carbon dioxide. Decreased blood oxygen is detected by chemoreceptors in the carotid body and aortic body; the response by the medulla is to increase respiration to take more air into the lungs. Increased blood carbon dioxide is detected by central chemoreceptors in the medulla (actually responding to the lowered pH); the response is increased respiration to exhale more carbon dioxide, which raises the pH back toward normal.

Carbon dioxide is usually the major regulator of respiration because even small changes in its blood level change the pH. Fluctuations in oxygen level have no effect on pH, and an adequate oxygen level in the blood can be maintained even if breathing ceases for a few minutes. Residual air in the lungs is a contributing factor, as is the fact that air contains much more oxygen than we usually use (exhaled air is 16 percent oxygen), with the excess available if necessary. Oxygen becomes the major regulator only when its blood level is very low, as may occur with severe, chronic pulmonary disease.

Respiration and Acid-Base Balance

Because of its role in regulating the amount of carbon dioxide in body fluids, the respiratory system is important in the maintenance of acid-base balance, measured by blood pH. Any decrease in the rate or efficiency of respiration permits excess carbon dioxide to accumulate in the blood. Accumulation of excess hydrogen ions lowers pH. This is called respiratory acidosis and can occur as a consequence of pulmonary disease or any impairment of gas exchange in the lungs.

Respiratory alkalosis occurs when the rate of respiration increases and carbon dioxide is very rapidly exhaled. Less carbon dioxide in the blood means that fewer hydrogen ions are formed and the pH rises. Although not a common condition, respiratory alkalosis may occur during states of anxiety accompanied by hyperventilation or when accommodating to a high altitude, before RBC production increases to provide sufficient oxygenation of tissues.

The respiratory system may also help compensate for pH changes that are said to be metabolic; that is, of any cause other than respiratory. Metabolic acidosis occurs when the concentration of hydrogen ions in body fluids is above normal. Common causes include kidney disease, untreated diabetes mellitus, and severe diarrhea. Respiratory compensation involves an increase in the rate and depth of respiration to exhale more carbon dioxide, which decreases hydrogen ion formation and raises the pH toward normal. Metabolic alkalosis may be caused by overingestion of antacid medications or by vomiting of stomach contents only. Respiratory compensation involves a decrease in the breathing rate to retain carbon dioxide in the body, increasing the formation of hydrogen ions, which lowers the pH toward normal.

Respiratory compensation for an ongoing metabolic pH imbalance (such as kidney failure) cannot be complete because there are limits to the amounts of carbon dioxide that may be exhaled or retained. At most, respiratory compensation is only about 75 percent effective.

Acid-base balance is discussed further in Chapter 5.

Effects of Aging on the Respiratory System

The respiratory muscles, like all skeletal muscles, weaken with age. Lung tissue loses its elasticity, and even alveoli are lost as their walls deteriorate. These changes result in decreased ventilation and lung capacity and a corresponding decrease in blood oxygen level. Chronic alveolar hypoxia, from diseases such as emphysema or chronic bronchitis, may lead to pulmonary hypertension, which in turn overworks the right ventricle of the heart. The cilia of the respiratory mucosa deteriorate with age, and the alveolar macrophages are not as efficient, which makes elderly people more prone to respiratory infections. See Table 26–1 for more information.

▶ NURSING ASSESSMENT
Health History

Many factors in a patient's personal and family history affect respiratory function. Questions to ask while assessing the patient with a history of respiratory dysfunction are presented in Table 26–2. If at any time while you're taking the history the patient relates a specific symptom, redirect the line of questioning to further assess that symptom. One such line of questioning, as presented in Chapter 1, is the WHAT'S UP? format. For example, if the patient admits to shortness of breath (SOB), respond with the following questions (*Where is it?* doesn't apply to shortness of breath, so it may be skipped):

TABLE 26-1 EFFECTS OF AGING ON RESPIRATION	
Aging change	**Effect**
Weakened respiratory muscles	Decreased force of cough, increased work to breathe
Decreased elasticity of lung tissue	Decreased ventilation
Fewer alveoli	Decreased gas exchange, reduced activity tolerance
Deterioration of cilia	Reduced ability to raise secretions
Reduced efficiency of alveolar macrophages	Increased risk for infection

TABLE 26-2	QUESTIONS ASKED DURING NURSING ASSESSMENT OF THE RESPIRATORY SYSTEM

Question	Rationale
Do you often have headaches or sinus tenderness?	These may indicate sinusitis.
Do you often experience nosebleeds?	A history of nosebleeds may indicate an abnormality that can predispose to future nosebleeds.
Has your voice changed?	A voice change may indicate a variety of disorders of the nose or throat. Further investigation is necessary.
Do you ever feel short of breath, like you can't get enough air?	Many respiratory and cardiac problems result in shortness of breath.
Do you have a cough? Is it productive? What does the sputum look like?	A cough indicates respiratory irritation or excessive secretions. Yellow or green sputum may accompany an infection. Blood in the sputum may occur with tuberculosis, pulmonary embolism, or cancer.
Have you recently experienced night sweats, chills, fever?	These are symptoms of tuberculosis.
Do you ever feel confused, light-headed, or restless?	These symptoms might indicate a low PO_2, reducing oxygen to the brain.
Have you had any chest surgeries?	This may reveal problem areas the patient has not yet mentioned.
Do you have any allergies that cause respiratory symptoms? How do you treat them?	The patient may take over-the-counter medications for allergies that affect respiratory function or interact with prescribed medications.
Do you smoke? How many packs per day? For how many years? Are you exposed to secondhand smoke?	Many respiratory disorders are caused or aggravated by exposure to tobacco smoke.
Are you or have you been exposed to airborne pollutants at work?	Pollutants such as asbestos, coal dust, or chemicals can cause lung disease.
Do you take any medications or use inhalers (prescribed or over-the-counter) for your respiratory problems?	Information about medications gives further information about disorders, severity, and treatment. The nurse should also consider drug interactions and side effects.
Do you use home oxygen or other home respiratory treatments?	This helps the nurse determine the severity of disease and the treatment.
Do any of your blood relatives have emphysema, asthma, or tuberculosis?	Some respiratory disorders have a hereditary tendency. Tuberculosis is contagious.

- **H**ow does it feel? Does breathing feel tight, gasping, suffocating?
- **A**ggravating and alleviating factors? How much activity causes the SOB? Does anything else aggravate it? What do you do to lessen your SOB?
- **T**iming? When did you first experience SOB? Does it happen more at any particular time of day or year?
- **S**everity? Rate your SOB on a scale of 0 to 10, with 0 being easy breathing and 10 being the worst shortness of breath you can imagine.
- **U**seful other data? Do you have any other symptoms that occur along with the shortness of breath?
- **P**atient's perception? What do you think is causing your shortness of breath?

You must also be aware of cultural influences on the patient's health (Cultural Consideration Box 26–1).

Physical Assessment
Inspection

Inspection begins during the nursing history and continues during the physical assessment. Start with the nose, observing for symmetry, swelling, or other abnormalities. Note whether the patient is short of breath (dyspneic) while speaking or moving. If the patient is very dyspneic, he or she may speak in short sentences.

Observe the patient for use of accessory muscles of breathing. Use of the sternocleidomastoid muscles causes the shoulders to raise during labored inspiration. During forced expiration, the abdominal and intercostal muscles

BOX 26-1	CULTURAL CONSIDERATION

Pulmonary diseases associated with Japanese people include asthma related to dust mites in the straw mats that cover floors in Japanese homes, air pollution from living in urban areas, and cardiac and pulmonary sequelae of cigarette smoking, which is prevalent among Japanese.[1] The nurse should encourage patients who have straw mats and who wish to keep them to have them sterilized.

Patients from Poland, Ireland, or other countries where mining is a primary occupation may have an increased incidence of respiratory disease. Mining and heavy industry are health risk factors related to the development of pulmonary diseases. It is essential that health care providers carefully screen Polish and Irish immigrants for respiratory conditions.

Health care practitioners should be aware of the variations among ethnic peoples of color when assessing for cyanosis. Cyanosis and decreased blood hemoglobin levels in darker-skinned individuals gives the skin an ashen color instead of a bluish color. Thus the nurse must examine the sclera, conjunctiva, buccal mucosal, tongue, lips, nailbeds, and palms and soles of the feet to assess for lowered oxygen levels.

Smoking is deeply ingrained in the Arab-American culture. Offering cigarettes is a rite of Arab hospitality. Arabs may have difficulty stopping smoking because of these cultural rituals.[2]

Populations living in inner cities are at increased risk for respiratory diseases related to pollution.[3] Strategies to increase the effectiveness in getting African-Americans to stop smoking include working with community and church groups in African-American communities.

Appalachians, with a preponderance of jobs in pottery making, mining, furniture making, textiles, and fabricated materials, are at increased risk for respiratory diseases.[4] Nurses can improve the health status of their Appalachian patients by encouraging the use of face masks when working in these industries.

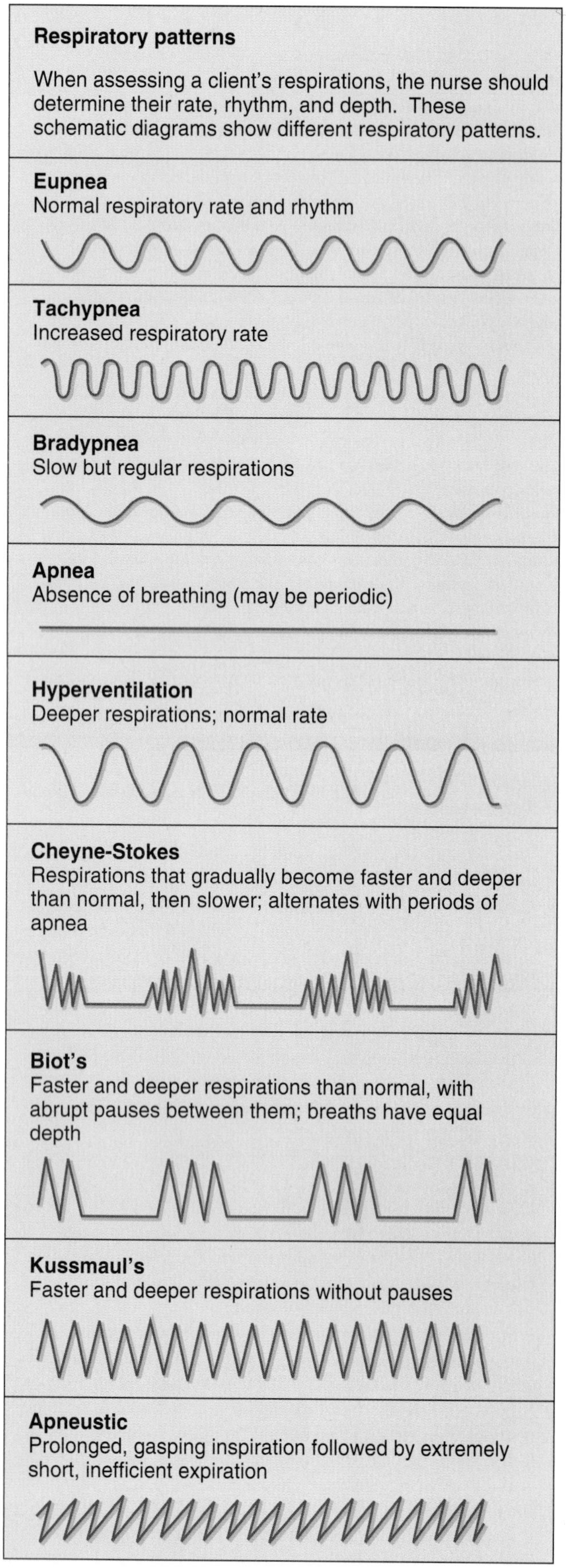

Respiratory patterns

When assessing a client's respirations, the nurse should determine their rate, rhythm, and depth. These schematic diagrams show different respiratory patterns.

Eupnea
Normal respiratory rate and rhythm

Tachypnea
Increased respiratory rate

Bradypnea
Slow but regular respirations

Apnea
Absence of breathing (may be periodic)

Hyperventilation
Deeper respirations; normal rate

Cheyne-Stokes
Respirations that gradually become faster and deeper than normal, then slower; alternates with periods of apnea

Biot's
Faster and deeper respirations than normal, with abrupt pauses between them; breaths have equal depth

Kussmaul's
Faster and deeper respirations without pauses

Apneustic
Prolonged, gasping inspiration followed by extremely short, inefficient expiration

Figure 26-2 Abnormal respiratory patterns. (Modified from Morton, PG: Health Assessment in Nursing, ed 2. FA Davis, Philadelphia, 1993, p 98, with permission.)

contract. The use of accessory muscles for breathing indicates respiratory distress. Retraction of the chest wall between the ribs can indicate serious distress.

Note color of skin, lips, mucous membranes, and nailbeds. A bluish color is called **cyanosis** and is a late sign of oxygen deprivation. Observe the trachea and chest for symmetry. Count respirations per minute, noting depth and rhythm. Irregular respirations, or periods of **apnea,** can indicate a pathological condition and are described in Figure 26–2. Observe the shape of the chest. Normally the chest is about twice as wide (side to side) as it is deep (front to back). If it is more rounded, it is called a barrel chest, which is associated with air trapping.

Palpation

Palpate the frontal and maxillary sinuses if sinus inflammation is suspected (Fig. 26–3). Use your thumbs to palpate gently below the eyebrows and below each cheekbone. Tenderness may indicate sinus inflammation or infection.

Respiratory excursion can be palpated. This is a rough measurement of chest expansion on inspiration. It is not necessary to palpate expansion on every patient, but it may be helpful if hypoventilation or asymmetry is suspected. See Figure 26–4 for how to palpate respiratory excursion. You can also palpate for **crepitus** (also called subcutaneous emphysema). Crepitus feels like Rice Krispies under the skin when felt with the fingers. It occurs when

Sinus locations

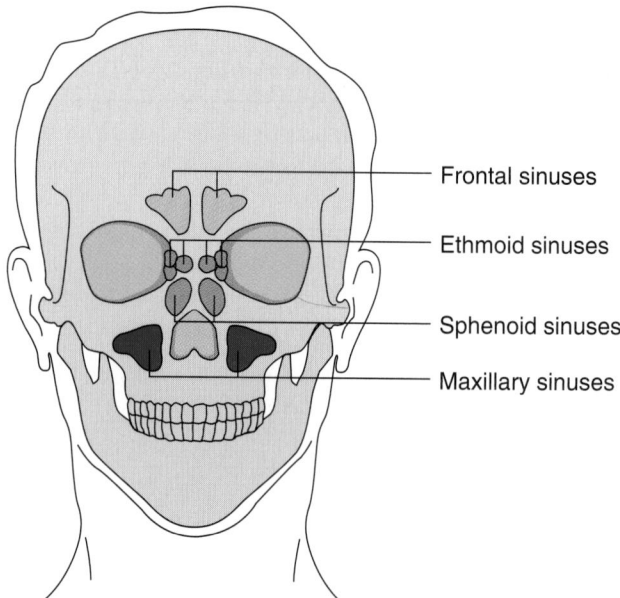

Frontal sinuses

Ethmoid sinuses

Sphenoid sinuses

Maxillary sinuses

Figure 26-3 Paranasal sinuses. (Modified from Morton, PG: Health Assessment in Nursing, ed 2. FA Davis, Philadelphia, 1993, p 194, with permission.)

cyanosis: cyan—dark blue + osis—condition
apnea: a—not + pnea—breath

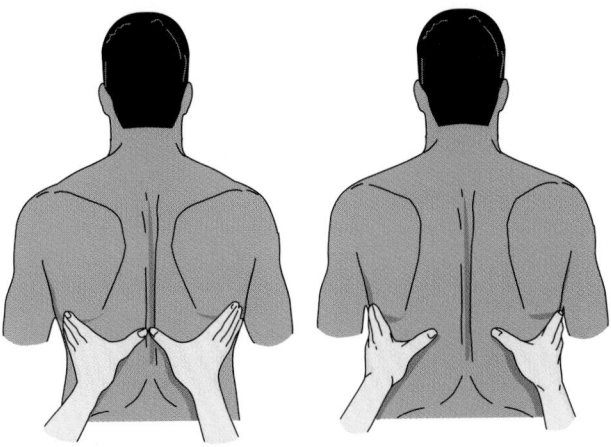

Posterior respiratory excursion
Stand behind the patient and place your thumbs beneath the scapulae area on either side of the spine at the level of the tenth rib. Grasp the lateral rib cage and rest your palms gently over the lateroposterior surface. Avoid applying excessive pressure to prevent restricting the patient's breathing.
As the patient inhales, the posterior chest should move upward and outward and your thumbs should move apart. When the patient exhales, your thumbs should return to midline and again touch.

Figure 26-4 Palpation of respiratory excursion. (Modified from Morton, PG: Health Assessment in Nursing, ed 2. FA Davis, Philadelphia, 1993, p 267, with permission.)

air leaks into subcutaneous tissues because of pneumothorax or a leaking chest tube site. Palpation for crepitus is not done routinely, but rather when the possibility of an air leak exists.

Percussion

Percussion is done by the experienced nurse. It involves tapping on the anterior and posterior chest, in each interspace, and comparing sounds from side to side. A normal chest sounds resonant and is the same on both the right and left sides, except over the heart. If other percussion notes are heard, they may indicate a pathological condition and should be reported. See Chapter 1 for additional information on percussion.

Auscultation

Auscultation provides valuable information about respiratory status. Use the diaphragm of the stethoscope to listen to the anterior and posterior chest during an entire inspiration and expiration at each interspace (Fig. 26–5). Auscultation of the posterior chest is easiest if the patient is sitting, but if necessary it may be done with the patient in a side-lying position. Asking the patient to breathe deeply through the mouth can help enhance the sounds. Allow the patient to rest at intervals to prevent hyperventilation. Regular and frequent practice helps you learn to distinguish normal from abnormal breath sounds. Abnormal extra sounds (another term is **adventitious**) indicate a pathological condition and are described in Table 26–3.

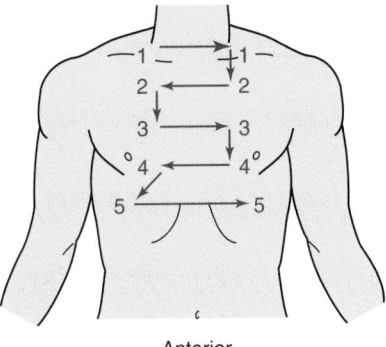

Anterior

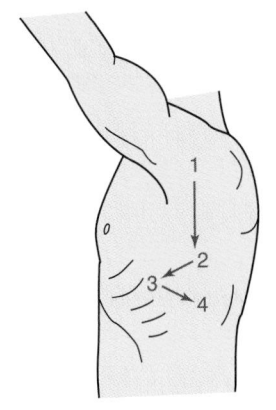

Lateral

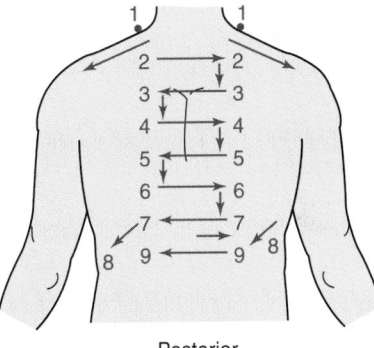

Posterior

Figure 26-5 Auscultation of the chest. Use a systematic approach to auscultate the chest, comparing sounds from side to side. (Modified from Morton, PG: Health Assessment in Nursing, ed 2. FA Davis, Philadelphia, 1993, p 268, with permission.)

🅦⎯ CRITICAL THINKING

Timothy
Timothy is a 16 year old brought into the emergency room by his mother because of an asthma attack. He says he feels short of breath, but when you listen to his lungs you hear no wheezing.

1. Does Timothy really need to be in the emergency room?
2. What should you do?
3. What do you think could be happening?

Answers at end of chapter.

TABLE 26-3   ABNORMAL LUNG SOUNDS

Abnormal (Adventitious) Sound	Cause of Sound	Description	Associated Disorders
Coarse crackles (sometimes called rales)	Fluid in airways	Moist bubbling sound, heard on inspiration or expiration	Pulmonary edema, bronchitis, pneumonia
Fine crackles (rales)	Alveoli popping open on inspiration	Velcro being torn apart, heard at end of inspiration	Heart failure, atelectasis
Wheezes	Narrowed airways	Fine high-pitched violins mostly on expiration	Asthma
Stridor	Airway obstruction	Loud crowing noise heard without stethoscope	Obstruction from tumor or foreign body
Pleural friction rub	Pleura rubbing together	Sound of leather rubbing together, grating	Pleurisy, lung cancer, pneumonia, pleural irritation
Diminished	Decreased air movement	Faint lung sounds	Emphysema, hypoventilation, obesity, muscular chest wall
Absent	No air movement	No sounds heard	Pneumothorax, pneumectomy

▶ DIAGNOSTIC TESTS

Laboratory Tests

Blood Tests

Measurement of red blood cells and hemoglobin can give information about the oxygen-carrying capacity of the blood. **Dyspnea** can be caused by a reduction in RBCs or Hgb. See Table 26-4 for normal values.

Sputum Culture and Sensitivity

A sputum culture identifies pathogens present in the sputum. The sensitivity test determines which antibiotics will be effective against those pathogens. To obtain a sputum specimen, first obtain a sterile container. Some institutions have special containers for sputum that help prevent transmission of infection to the health care worker (Fig. 26–6). Instruct the patient to take several deep breaths and cough sputum into the sputum container. It is important that the patient does not simply spit saliva or sinus drainage into the cup. The specimen must come from the lungs. It may be easiest to obtain a specimen first thing in the morning (after mouth care) because secretions build up during the night.

---

dyspnea: dys—bad + pnea—breathing

TABLE 26-4   COMMON LABORATORY TESTS

Test	Normal Values	Associated Conditions
Red blood cell count	Male: 4.5–6.2 million Female: 4.2–5.4 million cells/mm³ venous blood	↑ in chronic lung disease, dehydration ↓ in anemia, hemorrhage, overhydration with intravenous fluids
Hemoglobin	Male: 13.5–18 g/dL Female: 12–16 g/dL	Same as RBC count
White blood cell count	5000–10,000 cells/ mm³ venous blood	↑ in infection

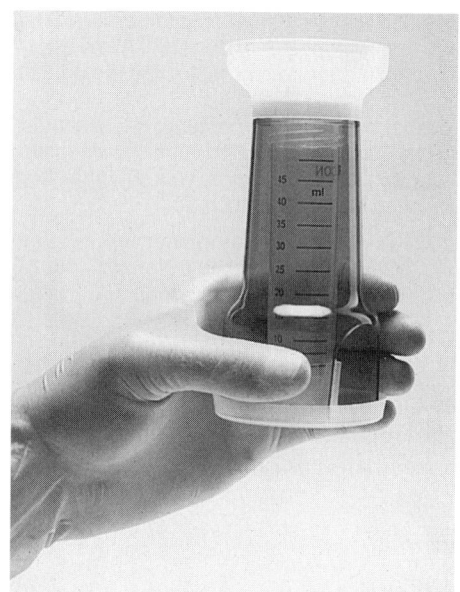

Figure 26-6 A special container that helps prevent transmission of infection is often used to collect sputum for culture. (Photo courtesy Becton Dickinson Microbiology Systems.)

The specimen is sent to the laboratory immediately. If the patient is unable to cough up sputum, extra fluids or a bedside humidifier may help. A respiratory therapist (RT) may be able to help obtain a specimen with a nebulized mist treatment or with a special suction catheter with a sputum trap. A physician order may be needed for these procedures.

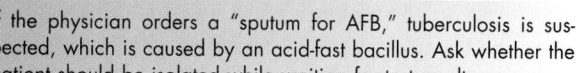

LEARNING TIPS

If the physician orders a "sputum for AFB," tuberculosis is suspected, which is caused by an acid-fast bacillus. Ask whether the patient should be isolated while waiting for test results.

Throat Culture

A throat culture is done to determine the presence of viral or bacterial pathogens in the pharynx. Use a culture tube with an attached swab (Culturette) to reach into the pharynx (without touching the patient's mouth) and rub the red area or lesions. Use a tongue blade to help hold the tongue down while obtaining the culture. Once the culture is obtained, squeeze the end of the culture tube to release the culture medium and return the swab to the tube. Send it immediately to the laboratory for analysis.

Arterial Blood Gas Analysis

Arterial blood gases (ABGs) are measured to determine the effectiveness of gas exchange. The "Pa" portion of the ABG results refer to the partial pressure of the gas in arterial blood. See Table 26–5 for a basic interpretation of ABGs. The blood sample is usually taken from the radial artery in the wrist by a physician or laboratory technician specially trained to do this. This can be painful for the patient. You may be asked to place pressure on the site for 5 minutes after the test to prevent bleeding.

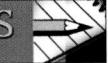

L E A R N I N G TIPS

Remember 50. If the PaO_2 falls below 50 and the $PaCO_2$ is above 50, the patient is in trouble and the physician should be notified. This is a crude analysis, but it is helpful when a quick assessment is needed.

Oxygen Saturation

The oxygen saturation test (also called pulse oximetry, O_2 sat, and SaO_2) is simple and noninvasive. A sensor is placed on the patient's finger or ear that measures the percentage of hemoglobin that is saturated with oxygen. Oxygen saturation can be measured at rest or while the patient is walking to determine the patient's exercise tolerance. It is also often done with and without supplemental oxygen to determine the patient's need for oxygen supplementation at home. See Table 26–5 for normal values. An SaO_2 value below 95 percent should be reported to the physician. If the SaO_2 is less than 75 percent, prepare for emergency intervention.

Other Tests
Chest X-ray Examination

A chest x-ray examination may be ordered to help diagnose a variety of pulmonary disorders. Usually, posterior-anterior (PA) and side views (lateral) are taken. If a hospitalized patient is too ill to go to the radiology department, a portable chest x-ray machine can be used to obtain a PA view.

Ventilation-Perfusion Scan

During a ventilation-perfusion scan (also called a lung scan or VQ scan), a radioactive substance is injected intravenously and a scan is done to view blood flow to the lungs (perfusion). Another radioactive substance is inhaled, and scanning shows how well oxygen is distributed in the lungs (ventilation). If an area of the lungs is well ventilated but has no blood supply, a pulmonary embolism is suspected. Chronic lung disease may cause poor ventilation and perfusion.

Pulmonary Function Studies

Pulmonary function studies are a series of tests done to determine lung volume, capacity, and flow rates. These are commonly used to help diagnose and monitor restrictive or obstructive lung disease. The patient is asked to use a special mouthpiece to blow into a cylinder that is connected to a computer. A computer printout is generated to show the results. See Table 26–6 for normal values. Some patients use handheld peak expiratory flow rate (PEFR) meters at home to monitor asthma symptoms. They might notice changes in PEFR before symptoms occur, allowing them to begin treatment before the problem becomes more serious.

Pulmonary Angiography

Pulmonary angiography involves an x-ray examination of the pulmonary vessels after intravenous (IV) administration of a radiopaque dye. A catheter is inserted into the femoral, brachial, or jugular vein and threaded through the heart to the pulmonary artery, where the dye is injected. Pulmonary angiography is used to help diagnose pulmonary embolism or other pulmonary vessel disorders. Patients receive nothing by mouth (NPO) for 4 to 8 hours before the procedure. Question the patient about allergies to x-ray dyes before scheduling the test. As with any procedure involving dye, the patient is informed that the dye may cause a warm feeling when injected. A signed consent form is required before any invasive procedure is performed. Medications may be administered before or during the test for patient comfort.

After angiography, place the patient flat in bed for 3 to 8 hours, as ordered by the physician. Monitor vital signs and observe the injection site for bleeding. A sandbag may be used to place pressure on the site. Encourage fluid intake to promote excretion of the dye.

TABLE 26–5	ARTERIAL BLOOD GAS ANALYSIS	
	Normal Values	**Interpretation**
PaO_2	75–100 mm Hg	↑ in hyperventilation ↓ in impaired respiratory function
$PaCO_2$	35–45 mm Hg	↑ in impaired gas exchange ↓ in hyperventilation
PH	7.35–7.45	↑ in respiratory alkalosis with low $PaCO_2$ ↓ in respiratory acidosis with high $PaCO_2$
HCO_3^-	22–26	↑ to buffer $PaCO_2$ in acidosis ↓ to buffer $PaCO_2$ in alkalosis
Oxygen saturation	95–100%	↑ in hyperventilation ↓ in impaired respiratory function

TABLE 26-6	PULMONARY FUNCTION VALUES	
Test	**Definition**	**Normal Values***
Tidal volume (TV)	Air inspired and expired in one breath	400–600 mL at rest
Residual volume (RV)	Air remaining in lungs after maximum exhalation	1000–1500 mL
Functional residual capacity (FRC)	Air remaining in lungs after normal expiration	2300 mL
Vital capacity (VC)	Maximum amount of air expired after maximum inspiration	4600 mL
Inspiratory reserve	Amount of air beyond tidal volume that can be taken in with the deepest possible inhalation	2000 to 3000 mL
Expiratory reserve	Amount of air beyond tidal volume in the most forceful exhalation	1000 to 1500 mL
Forced vital capacity (FVC)	Maximum amount of air expired forcefully after maximum inspiration	3000–5000 mL
Peak expiratory flow rate (PEFR)	Maximum flow of air expired during FVC (this is a rate rather than a volume)	450 L/min

*Normal values are approximate—computer determines normal values based on patient's height and weight.

Bronchoscopy

A bronchoscopy involves the use of a flexible endoscope to examine the larynx, trachea, and bronchial tree. A bronchoscopy can be used diagnostically for visualization or to obtain a biopsy specimen for examination. It can also be used therapeutically to remove an obstruction, foreign body, or thick secretions. The patient is told that he or she will be able to breathe through the nose and that oxygen can be administered through the tube if necessary. A signed consent is necessary for this procedure.

The patient is NPO for 6 to 8 hours before the procedure. Be prepared to administer a sedative, and commonly an injection of atropine to dry excess secretions, before the procedure. An anesthetic spray may be used to numb the throat. After the test, monitor vital signs and watch for signs of laryngeal edema. Sputum may be blood tinged. The patient is NPO until the gag reflex returns. Check for the gag reflex by touching the pharynx with a cotton swab. Once the gag reflex is positive, ask the patient to swallow a sip of water before offering foods or fluids. A sore throat may be relieved with lozenges.

▶ THERAPEUTIC MEASURES

Smoking Cessation

Probably the most important intervention for preventing and treating respiratory disease is smoking cessation. Many respiratory disorders are caused or aggravated by smoking, and stopping can prevent disease from occurring or slow its progression significantly. See Table 26-7 for interventions used to help patients stop smoking. Remind patients that if they have tried quitting before and failed, that does not mean that they will never be able to quit. Many patients try several times before quitting successfully. Formal smoking cessation programs and support groups can be helpful.

Many internet sites have information to help people stop smoking. Among these are the American Lung Association site at www.lungusa.org and the National Lung Health Education Program at www.nlhep.org. Or simply type *smoking cessation* into any search engine.

Deep Breathing and Coughing

Effective coughing can keep the airways clear of secretions. An ineffective cough is exhausting and fails to bring up secretions. Instruct the patient to take two or three deep breaths, using the diaphragm. This helps get the air behind the secretions. After the third deep inhalation, tell the patient to hold the breath and cough forcefully. This is repeated as necessary. Good hydration can facilitate this process.

Breathing Exercises

Breathing exercises are essential for patients with chronic lung disease. Diaphragmatic and pursed-lip breathing increase the effectiveness of breathing and help reduce panic when dyspnea occurs.

Diaphragmatic Breathing

The diaphragm is the major muscle of breathing, but patients often use less efficient accessory muscles when they are short of breath. Conscious use of the diaphragm during breathing can be relaxing and conserves energy. With practice, the patient should be able to use diaphragmatic breathing all the time without thinking about it. The patient is taught to do the following:

TABLE 26-7	INTERVENTIONS TO STOP SMOKING
Intervention	**Rationale**
Counseling	Counseling alone can result in a 3–5% quit rate.
Setting a quit date	The cold turkey method is more effective than slow tapering, although the patient may choose to taper before the quit date.
Nicotine replacement	Nicotine gum and patches are available without a prescription and can reduce withdrawal symptoms.
Drug therapy (bupropion [Zyban], buspirone [BuSpar])	Better success rates are achieved with a combination of nicotine and bupropion.

Source: Modified from Petty, TL: COPD: Interventions for smoking cessation and improved ventilatory function. Geriatrics 55(12):30–39, 2000.

1. Place one hand on the abdomen and the other on the chest.
2. Concentrate on pushing out the abdomen during inspiration and relaxing the abdomen on expiration. The chest should move very little.

Pursed-Lip Breathing

This technique can be used any time the patient feels short of breath. It helps keep airways open during exhalation, which promotes carbon dioxide excretion. It should be done with diaphragmatic breathing. Counting during breathing also distracts the patient, reducing panic. The patient is taught to do the following:

1. Inhale slowly through the nose to the count of two.
2. Exhale slowly through pursed lips to the count of four.

Positioning

The patient who is short of breath should be positioned to conserve energy while allowing for maximum lung expansion. The patient in bed can use a Fowler's or semi-Fowler's position. Most respiratory patients do not tolerate lying flat. Some patients prefer to sit in a chair while leaning forward and placing their elbows on their knees. Others may sit at a desk or use an over-bed table with a pillow on it to lean on.

Recent research shows that in patients with unilateral (one-sided) lung disease, oxygen saturation is increased in the "good lung down" lateral position. This is a side-lying position with the good lung in the dependent position. Researchers believe that gravity causes greater blood flow to the dependent, "good" lung, thereby increasing oxygenation.

Oxygen Therapy

Oxygen therapy is ordered by the physician when the patient is unable to maintain oxygenation. Many patients are placed on supplemental oxygen when their oxygen saturation is less than 90 percent on room air. The physician's order should include the method of administration and the flow rate. A variety of delivery methods are described in the following sections. The role of the nurse in oxygen therapy includes monitoring the flow rate, ensuring that the cannula and tubing or other device remain properly placed, and monitoring the patient's response to treatment. If the patient becomes short of breath while on oxygen therapy, an RT or a physician should be notified. Instruct the patient to avoid smoking, using electrical equipment, and performing other activities that can cause fire in the presence of oxygen. The RT is knowledgeable about oxygen therapy and is an excellent resource when questions arise.

Low-Flow Devices

NASAL CANNULA. The nasal cannula is the most common method of oxygen administration. Oxygen is delivered through a flexible catheter that has two short nasal prongs (Fig. 26–7). For the nasal cannula to be most effective, the patient must breathe through his or her nose. The cannula allows the patient to eat and talk, and it is generally more comfortable than other methods of administration. If the nasal mucous membranes become dry, an RT can place a water source on the system to humidify the oxygen. Oxygen can be delivered at 1 to 6 L/minute via a nasal cannula, according to the physician's order.

MASKS. Masks are used when a higher oxygen concentration is needed (Fig. 26–8). A disadvantage to masks is that they make some patients feel claustrophobic. Also, a mask must be replaced by a cannula for eating.

Simple Face Mask. A rate of 5 to 10 L/minute can deliver oxygen concentrations from 40 to 60 percent with a simple face mask.

Partial Rebreather Mask. A partial rebreather mask uses a reservoir to capture some exhaled gas for rebreathing. Vents on the sides of the mask allow room air to mix with oxygen. It can deliver oxygen concentrations of 50 percent or greater.

Nonrebreather Mask. A nonrebreather mask has one or both side vents closed to limit the mixing of room air with oxygen. The vents open to allow expiration but remain closed on inspiration. The reservoir bag has a valve to store oxygen for inspiration but does not allow entry of exhaled air. It is used to deliver oxygen concentrations of 70 to 100 percent.

When a patient is using a partial rebreather or nonrebreather mask, ensure that the reservoir is never allowed to collapse to less than half full.

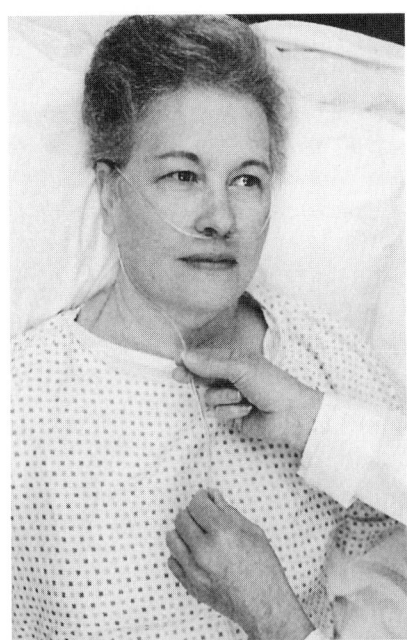

Figure 26-7 Nasal cannula for oxygen delivery.

LEARNING TIPS

If a patient suddenly becomes confused, check the oxygen delivery system. The patient may have taken off the cannula, or the tubing may be kinked or disconnected, resulting in hypoxia and confusion.

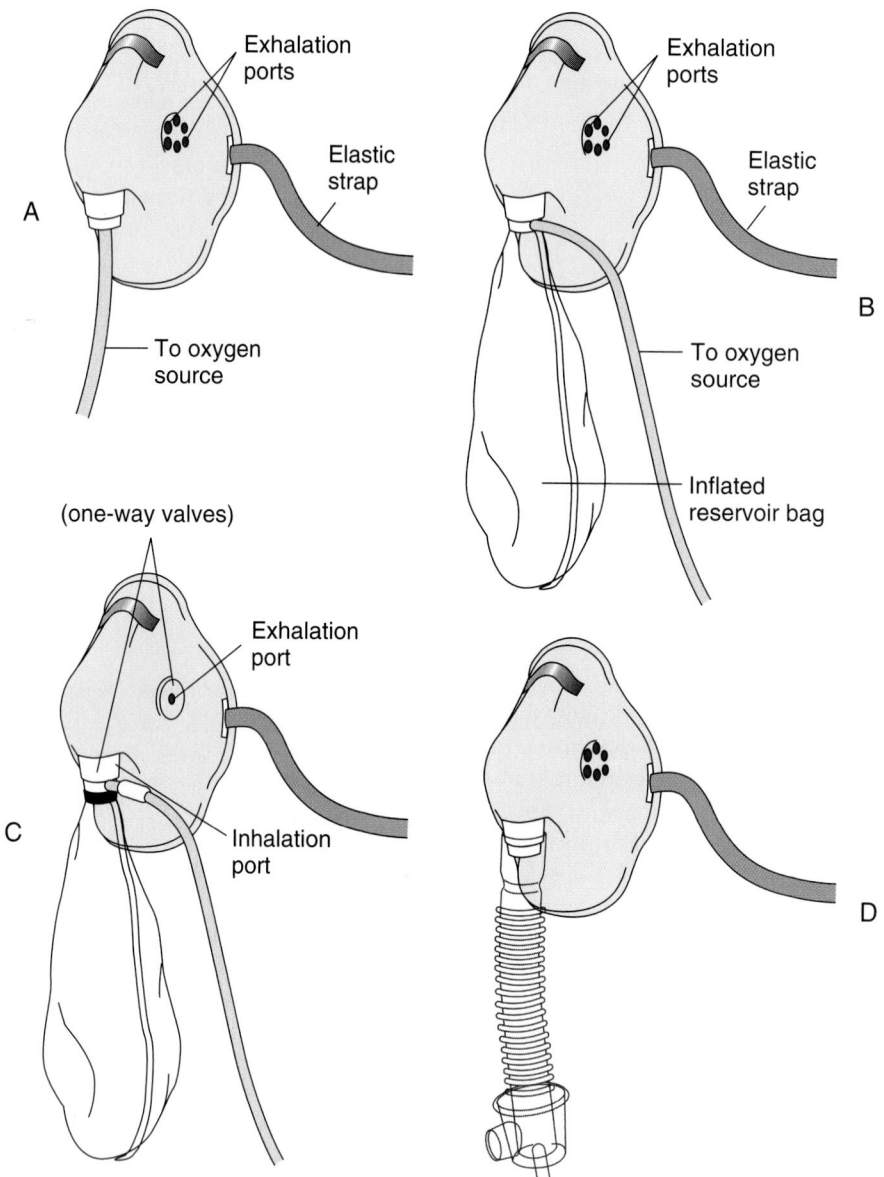

Figure 26-8   Oxygen masks. *(A)* Simple mask. *(B)* Partial rebreather mask. *(C)* Nonrebreathing mask. *(D)* Venturi mask.

High-Flow Devices

VENTURI MASK. A Venturi mask is used for the patient who requires precise percentages of oxygen, such as the patient with chronic lung disease with CO_2 retention. A combination of valves and specified flow rates determines oxygen concentration.

Transtracheal Catheter

A transtracheal catheter is a small tube that is surgically placed through the base of the neck directly into the trachea to deliver oxygen (Fig. 26–9). This is an attractive alternative for some patients who are on long-term oxygen therapy at home, because it does not obstruct the nose or mouth and can be easily covered with a loose scarf or collar. The patient is taught to remove and clean the catheter two or three times a day to prevent mucus obstruction. Check

institution policy and procedure and the respiratory care department for specific care instructions.

Risks of Oxygen Therapy

Patients with chronic airflow limitation (CAL) have chronically high $PaCO_2$ levels. Therefore they depend on low PaO_2 levels to stimulate breathing, and high supplemental oxygen flow rates can depress respirations. Patients with CAL should be maintained on no more than 1 to 2 L of oxygen per minute.

In addition, any patient can suffer lung damage from high oxygen concentrations delivered for more than 24 hours. If a patient exhibits symptoms of dry cough, chest pain, numbness in the extremities, lethargy, or nausea, the physician should be contacted. A PaO_2 greater than 100 mm Hg should also be reported.

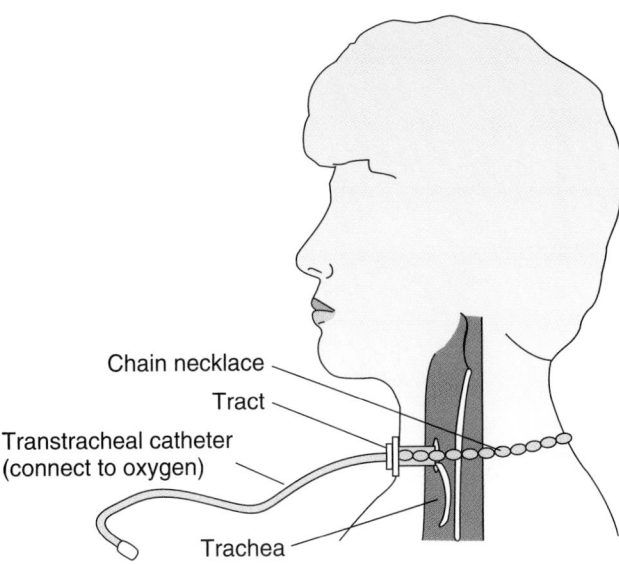

Figure 26-9 Transtracheal oxygen catheter.

Nebulized Mist Treatments

Nebulized mist treatments (NMTs) use a nebulizer to deliver medication directly into the lungs (Fig. 26–10). Such topical use of medication reduces systemic side effects. Bronchodilators such as albuterol or metaproterenol, mixed with normal saline solution and sometimes with supplemental oxygen, are most commonly administered. Other medications, including corticosteroids, mucolytics, and antibiotics, may also be given. An RT or a specially trained nurse administers the NMT. The patient uses a handheld reservoir with tubing and a mouthpiece to breathe in the medication. NMTs are commonly ordered every 4 to 6 hours and as needed. You may call for an NMT as needed (prn) when a patient with chronic pulmonary disease becomes acutely dyspneic. Some patients are taught to administer their own NMTs at home.

Metered Dose Inhalers

Metered dose inhalers (MDIs) are another way to administer topical medication directly into the lungs, minimizing systemic side effects. Medications delivered in this way in-

clude corticosteroids, bronchodilators, and mast cell inhibitors (cromolyn sodium). Figure 26–11 shows one way to use an MDI. An alternate method is the open-mouth technique, in which the patient keeps the mouth open in step 3 rather than closing the lips around the inhaler. It is important that the RT or nurse carefully instruct the patient, because improper use can reduce the effectiveness of the medication. It is also important to teach the patient to never overuse adrenergic bronchodilator inhalers. Patients with chronic disease tend to use extra puffs when they feel short of breath. Adrenergic bronchodilators, when used too often, can cause severe bronchoconstriction and even death.

Incentive Spirometry

Incentive spirometers are devices used to encourage deep breathing in patients at risk for collapse of lung tissue, a condition called atelectasis (Fig. 26–12). These devices are

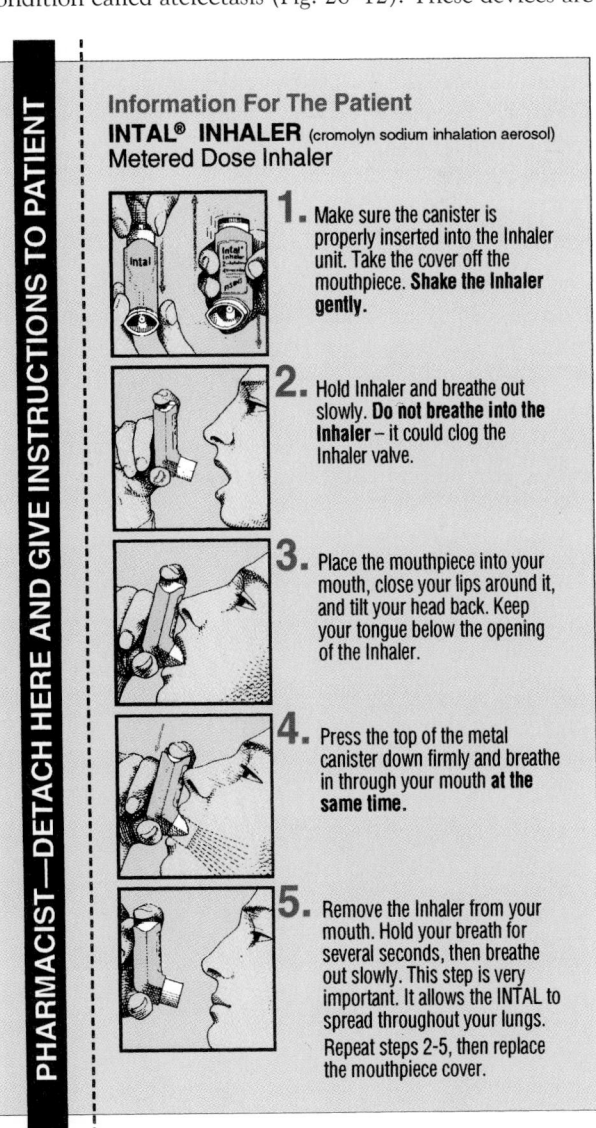

Figure 26-11 Sample instructions for use of a metered dose inhaler. Patient instructions for Intal inhaler. (Reprinted with permission from Rhône-Poulenc Rorer Pharmaceuticals, Inc. © 1996, Fisons Corporation, a Rhône-Poulenc Rorer company. All rights reserved. See package insert for specific product instructions.)

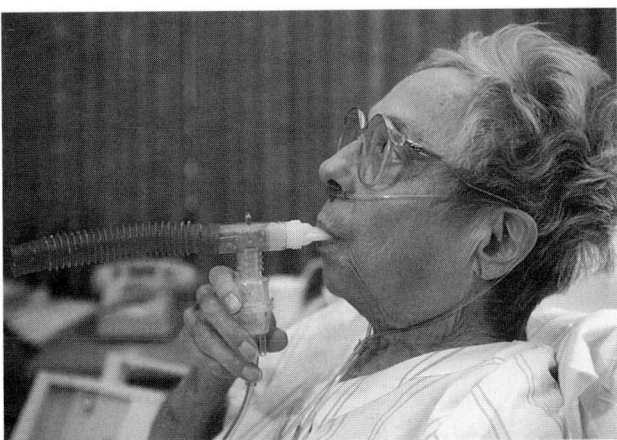

Figure 26-10 Patient receiving nebulized mist treatment.

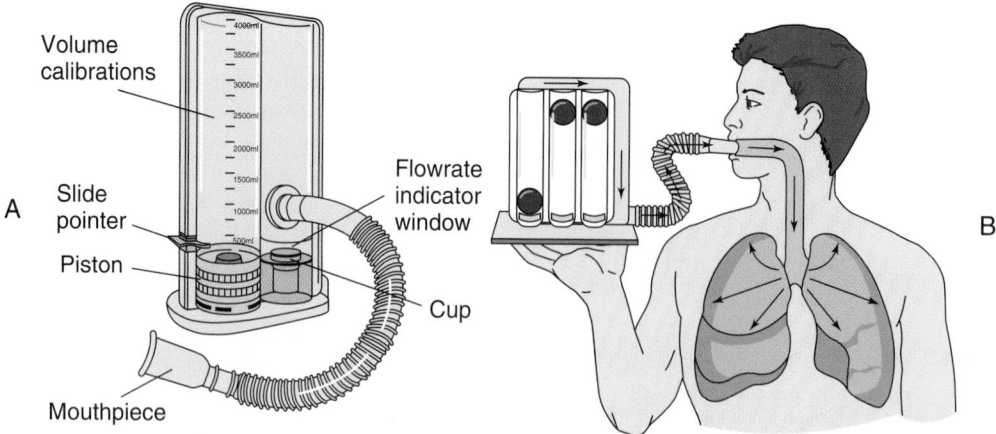

Figure 26-12 Incentive spirometers. *(A)* Voldyne volumetric deep breathing exerciser. *(B)* Triflow II incentive breathing exerciser. (Modified from Barnes, TA: Respiratory Care Principles. FA Davis, Philadelphia, 1991, p 434, with permission.)

commonly ordered for postoperative patients. Patients are instructed to use the spirometer 10 times each hour they are awake. Because a variety of spirometers are available, consult with an RT and read package inserts for specific directions for use.

Chest Physiotherapy

Chest physiotherapy (CPT), which includes postural drainage, percussion, and vibration, helps move secretions from deep inside the lungs (Fig. 26–13). It is indicated for the patient who has a weak or ineffective cough and is therefore at risk for retaining secretions. Patients with chronic obstructive pulmonary disease, cystic fibrosis, or bronchiectasis and patients on ventilators benefit from CPT.

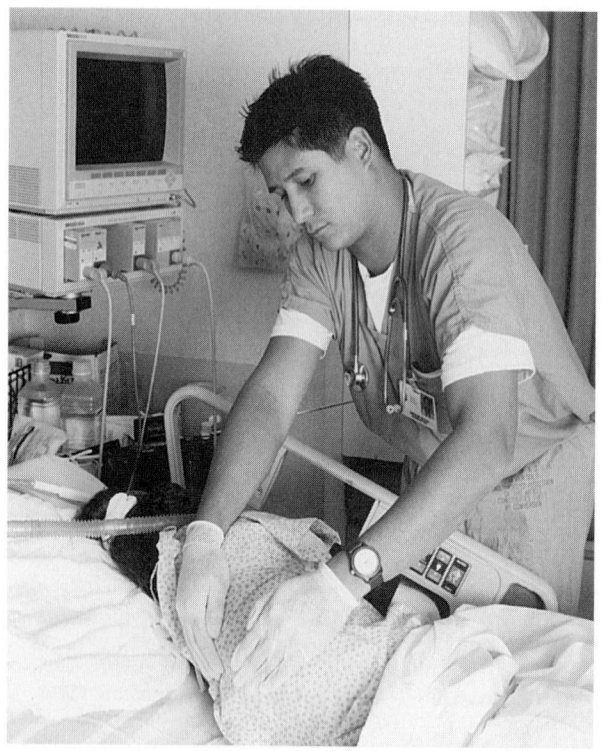

Figure 26-13 Patient receiving chest physiotherapy.

CPT is performed by an RT or specially trained nurse. For postural drainage, the patient is placed in various positions (head down, to help drain secretions) and turned periodically during the treatment so all lobes of the lungs are drained. The therapist uses cupped hands to strike the chest repeatedly (percussion), producing sound waves that are transmitted through the chest, loosening secretions. The therapist may also apply vibration to the patient's chest, using the hands or a vibrator, to loosen secretions. A nebulizer treatment should be given before CPT to humidify secretions. The patient is instructed to cough and deep breathe at intervals during and after the treatment.

Flutter Mucus Clearance Device

An alternative to chest physiotherapy is a small handheld device called the Flutter mucus clearance device. When the patient blows into the mouthpiece, it makes a heavy steel ball inside bounce around in its chamber, which then sends vibrations back into the airways. This helps to loosen mucus and open the airways.

Thoracentesis

Thoracentesis involves the insertion of a needle into the pleural space. It is commonly done to aspirate fluid in patients with pleural effusion (fluid trapped in the pleural space). The procedure may be diagnostic, to determine the source of fluid, or therapeutic, to remove fluid and reduce respiratory distress. It may also be performed to aspirate blood or air or to inject medication.

When assisting a physician with a thoracentesis, first verify that the patient understands the procedure and that written consent has been obtained if required by institution policy. Have the patient void before the procedure. The patient should be aware that a sensation of pressure may be felt, but that severe pain is rare. An analgesic may be administered if ordered before the procedure. A special procedure tray is obtained that has the equipment needed by the

thoracentesis: thoraco—chest + centesis—puncture

physician. The patient is placed in a sitting position, bending over the bedside table, or in a side-lying position if unable to sit. You can position yourself in front of the patient and encourage relaxation during the procedure. If you are asked to hand equipment to the physician, be sure to keep everything sterile. The physician uses a local anesthetic before inserting a needle into the patient's back through the desired interspace. Specimens are withdrawn through the needle, labeled, and sent to the laboratory. If the thoracentesis is being done for therapeutic reasons, a sterile container is used to collect the remaining fluid. As much as 2 L can be removed, sometimes more, and the patient will report immediate reduction of dyspnea.

After the procedure, the physician may apply a petroleum jelly dressing to prevent air leakage into the wound. Assess vital signs, breath sounds, and the puncture site according to the physician's orders (e.g., every 15 minutes times two, every 30 minutes times two, then every 4 hours for 24 hours). The patient is usually maintained on bedrest for 1 hour after the procedure. Label and send specimens to the laboratory as ordered. The physician may order a post-procedure x-ray examination to ensure that the lung was not punctured, causing a pneumothorax.

Chest Drainage

Continuous chest drainage involves insertion of one or two chest tubes by the physician into the pleural space to drain fluid or air. The tubes are connected to a chest drainage system that collects the fluid or allows escape of the air.

Indications

Chest tubes and a chest drainage system are used when fluid or air has collected in the pleural space. This can occur with a pneumothorax, pleural effusion, or penetrating chest injury or during chest surgery. These conditions are covered in Chapter 28.

Chest Tube Insertion

The physician inserts drainage tubes through the chest wall into the pleural space, either in surgery or at the bedside. If removal of air is the goal, the tube is inserted into the upper anterior chest, in the second to fourth intercostal space. If removal of fluid is the goal, the tube is inserted in the lower lateral chest, in the eighth or ninth intercostal space. If a patient has both air and fluid to drain, two tubes are inserted and may be joined with a Y connector before connecting to the tubing that leads to a drainage system. You can assist the physician by obtaining a chest tube insertion tray and chest drainage system and preparing it according to the manufacturer's directions. Ensure that the patient understands the procedure and that written consent has been obtained according to institutional policy. An analgesic is administered if ordered. Chest tube insertion is sometimes an emergency intervention, which necessitates that the nurse help prepare the patient quickly.

Once the tube has been inserted and the system is in place, ensure that each connection is securely taped with adhesive tape, to prevent a break in the system. Vaseline gauze and a sterile occlusive dressing are applied over the insertion site to prevent air leakage.

Obtain two padded clamps to keep at the bedside. These are used to clamp the chest tube if the chest drainage system becomes accidentally disconnected from the tubing, for changing the drainage system, or for a trial period before chest tube removal. The tubes are never clamped for more than a few seconds, however, because this prevents air escape and causes buildup of air in the pleural space. This creates a tension pneumothorax, which is a life-threatening emergency. (See Chapter 28.)

Chest Drainage System

The drainage system has evolved from a set of glass bottles to a one-piece molded plastic system with chambers that correspond to the bottles. Some newer one-piece systems use special valves to eliminate the need for water. Some physicians, however, continue to use the bottle system. Studying the bottle system helps to understand the one-piece system (Fig. 26–14). One, two, or three bottles can be used. Study the pictures as you read the following sections.

WATER SEAL BOTTLE OR CHAMBER. Each time the patient exhales, air trapped in the pleural space travels through the chest tube to the water seal bottle or chamber, under the water, then bubbles up and out of the bottle. The water acts as a seal, allowing air to escape from the pleural space but preventing air from getting back in during the negative pressure of inspiration. Water in the tube fluctuates up with each inspiration and down with each expiration, as much as 5 to 10 cm. This is called **tidaling.** When the lung is reinflated, tidaling stops. If tidaling stops before the lung is reinflated, the tubing should be checked for a

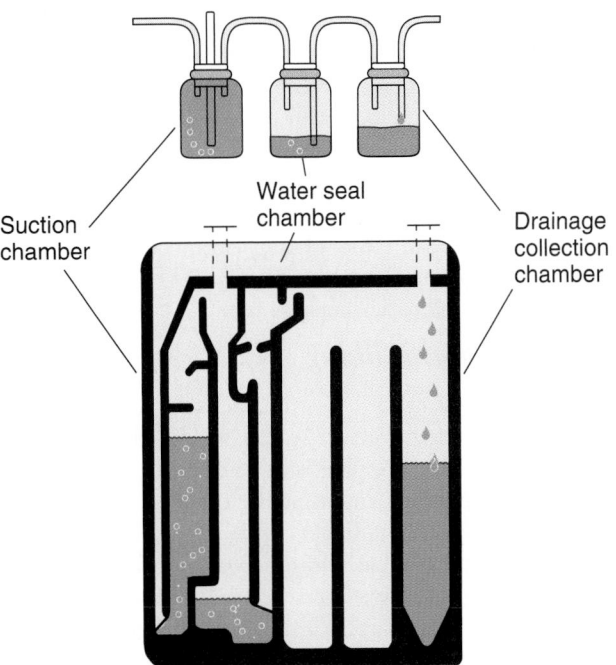

Figure 26-14 Pleur-Evac chest drainage system. (Courtesy Deknatel Snowden Pencer, Inc., Tucker, Ga.)

kink or occlusion. If constant bubbling occurs in the water seal chamber, the system should be checked immediately for leaks.

SUCTION BOTTLE OR CHAMBER. Sometimes a suction source is used to speed lung reinflation. A separate bottle with tubing attached to suction is used. The amount of suction depends on the level of water in the bottle, not the amount of suction set on the machine. The suction level is ordered by the physician and is almost always negative 20 cm of water. The suction should be turned on far enough to cause gentle bubbling in the suction bottle or chamber. Vigorous bubbling causes water evaporation and alter the amount of suction. If water evaporates, more must be added to maintain the correct amount of suction.

DRAINAGE BOTTLE OR CHAMBER. Sometimes a third bottle is needed to drain fluid from the pleural space. Drainage may be from pleural effusion, chest trauma, or surgery. Sometimes a small amount of drainage occurs because of the insertion of the chest tube. The drainage chamber is not emptied to measure drainage. Rather, the drainage level in the bottle or chamber is marked and timed each shift to monitor the amount. It is documented as output on the intake and output record. If drainage suddenly increases or becomes very bloody, notify the physician.

Nursing Care

Nursing care of a patient with a chest tube involves regular assessment of the patient and the drainage system. See Box 26–2 for specific assessment and care. If permitted by the physician, patients can be free to move around with the chest tube and drainage system. The drainage system must always be kept upright and below the level of the chest. If the patient must be transported, the drainage system is transported with the patient. Ask the physician if the patient can be safely transported without suction. The suction control chamber is then left open to allow air to escape. Tubing is not clamped for transport.

If a chest tube is accidentally pulled out before the pneumothorax is resolved, air can reenter the pleural space. Some physicians want an occlusive dressing placed over the site to prevent air from reentering. However, an occlusive dressing increases the risk of trapped air building up and placing pressure on the heart. You should be aware of the preferences of the patient's physician.

Stripping and Milking

In the past it was routine to strip and milk the tubing from the patient to the drainage system, to dislodge clots and maintain patency. Stripping is done by holding the proximal end of the tubing and using the other hand to squeeze the tubing between two fingers while sliding the fingers toward the drainage system. This is repeated on small sections of tubing until all have been stripped. It is now known, however, that this process can create negative pressure at the openings in the tubing that are within the pleural space, which can suck lung tissue in and cause damage. Stripping should be done only if it is ordered by the physician.

BOX 26-2   **Care of the Patient with a Chest Drainage System**

Assess the patient according to institution policy. Start with the patient and move toward the drainage system.
 1. Observe respiratory rate, effort, and symmetry.
 2. Assess shortness of breath, pain, or other discomforts.
 3. Auscultate lung sounds (lung sounds may initially be muffled or absent on the side of a collapsed lung but should gradually return to normal as the lung reinflates).
 4. Confirm that dressing is intact; observe for drainage. If necessary, reinforce the dressing and notify the physician. Do *not* change the dressing.
 5. Palpate around insertion sites for crepitus.
 6. Check all tubing for kinks, breaks, or broken connections. Verify that all connections are securely taped.
 7. Ensure that there are no dependent loops of tubing. Excess tubing should be coiled on the bed.
 8. Verify that drainage system is below level of patient's chest at all times.
 9. Check drainage system or bottles for cracks or leaks.
10. Check water seal chamber for correct water level and for tidaling (unless lung reinflated). Add water if evaporation has decreased level. If continuous bubbling is present, check entire system for leaks and notify physician.
11. Check suction control chamber for gentle bubbling (or open to air). Confirm correct amount of water if indicated. Add water if needed.
12. Check and mark amount of drainage in collection chamber every 8 hours and prn or as ordered. Report any marked increase in bloody drainage. Record drainage as output.
13. Document findings.
Notify physician if
• *The patient suddenly complains of increasing dyspnea.*
• *The drainage chamber is full and needs to be changed.*

Milking is done by gently squeezing portions of tubing from the patient to the system, without any sliding motion. This is somewhat safer for the patient but is still not done routinely. If tubing appears to be occluded, consult with the physician for specific orders.

Removal of Chest Tube

When the reason for the chest tube is resolved, the physician removes it and places a petroleum jelly gauze and a sterile occlusive dressing over the site. Continue to watch for development of crepitus and to monitor the dressing site and respiratory status.

CRITICAL THINKING

Miss Israel
Miss Israel has a chest tube in place for a spontaneous pneumothorax. You note that the water seal chamber is bubbling vigorously. What could cause bubbling in the water seal chamber? What should you do?

Answers at end of chapter.

Tracheostomy

A **tracheotomy** is a surgical opening through the base of the neck into the trachea. It is called a **tracheostomy** when it is more permanent and has a tube inserted into the opening to maintain patency (Fig. 26–15). The patient breathes through this opening, bypassing the upper airways. A tracheostomy is performed for a variety of reasons, such as in patients who have had a laryngectomy for cancer, patients with airway obstruction caused by trauma or tumor, patients who have difficulty clearing secretions from the airway, and patients who need prolonged mechanical ventilation.

The tracheostomy tube consists of three parts: an outer cannula, an inner cannula, and an obturator (Fig. 26–16). The obturator is a guide that is used only during insertion of the tube. After insertion, the obturator is immediately removed and kept at the bedside (commonly taped to the wall above the bed) for emergency use if the tracheostomy tube is accidentally removed. The outer cannula remains in place at all times and is secured by ties to prevent dislodging. The inner cannula is removed at intervals, usually every 8 hours and as needed for cleaning. Some newer tracheostomy tubes eliminate the need for an inner cannula. The tube may be metal or plastic. Plastic tubes generally have disposable inner cannulas, which can be replaced rather than cleaned. Plastic tubes may also have balloonlike cuffs that are inflated to prevent air escape during mechanical ventilation and to prevent aspiration of food or secre-

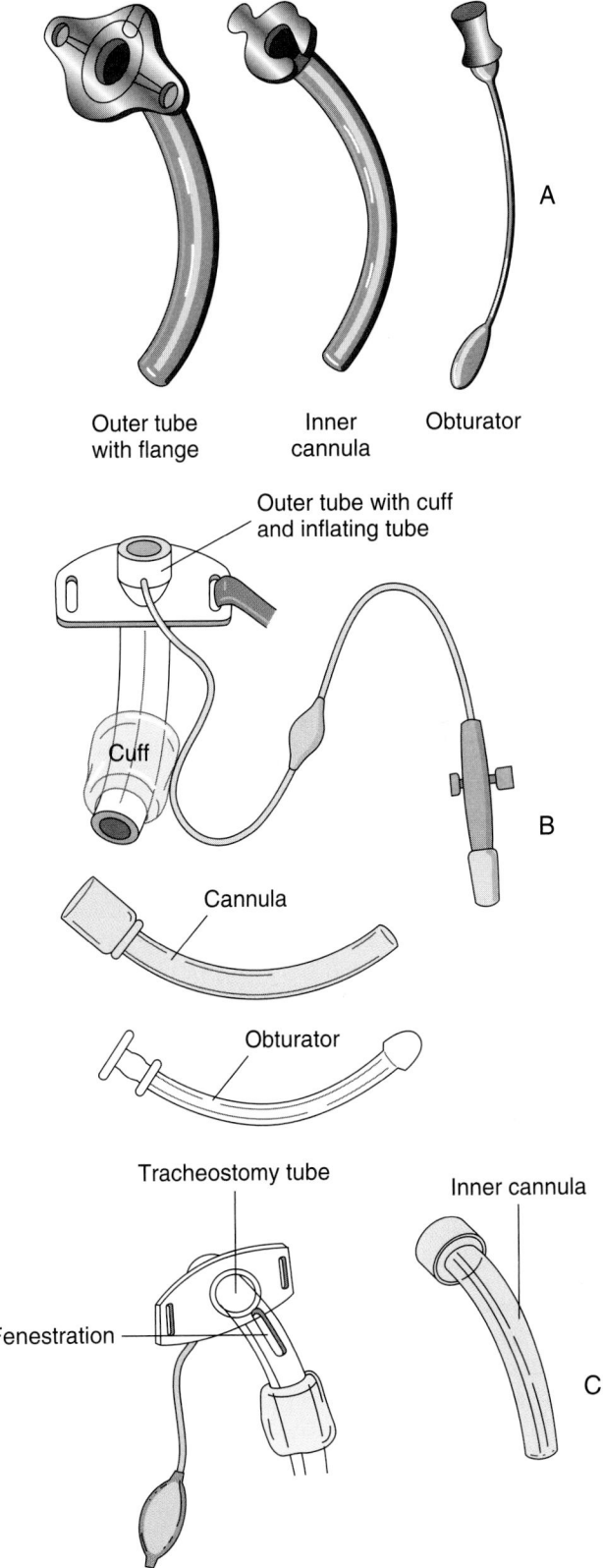

Figure 26-16 Tracheostomy tube. *(A)* Metal tube. *(B)* Cuffed plastic tube. *(C)* Fenestrated tube.

tions. Institution policy may dictate that cuffs be deflated routinely to prevent tissue damage. See Box 26–3 for routine tracheostomy cleaning.

Communication is problematic for the patient with a tracheostomy tube, because air is diverted out the tube rather

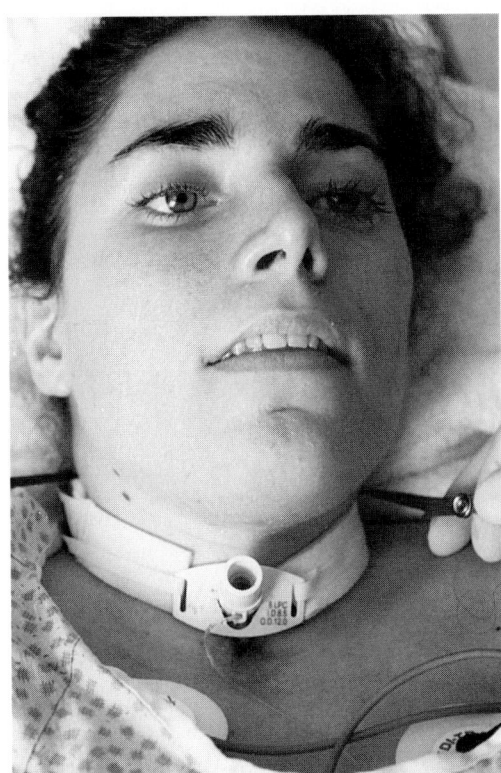

Figure 26-15 Patient with tracheostomy.

---

tracheotomy: trach—trachea + otomy—incision
tracheostomy: trache—trachea + ostomy—opening or mouth

BOX 26-3   Tracheostomy Cleaning Procedure

1. Assemble equipment: tracheostomy care kit, sterile water or saline, suction equipment, hydrogen peroxide.
2. Explain the procedure to the patient.
3. Suction inner cannula if necessary. See Box 26-2.
4. Open and prepare the kit, keeping all equipment sterile. Fill one side of basin with half peroxide and half saline and the other with saline.
5. Don clean gloves.
6. Remove old tracheostomy dressing.
7. Remove inner cannula from tracheostomy tube and place it in peroxide solution.
8. While inner cannula is removed, the patient may be suctioned if necessary.
9. Don sterile gloves.
10. Use brush and pipe cleaners to clean inner cannula. Place in water or saline to rinse. Dry inside of cannula with pipe cleaner. Reinsert into tracheostomy tube.
11. Use cotton swab and sterile gauze with sterile peroxide and saline to clean around tracheostomy site. Rinse with saline to prevent skin irritation.
12. Replace ties. Remove old ties after new ties are securely in place.
13. Apply sterile tracheostomy dressing (drain sponge or "trach pants"). Use precut or folded dressing. Cutting gauze creates fibers that can enter the tracheostomy.

NOTE: A procedure manual should be consulted for more detailed instruction.

than past the vocal cords and out the mouth. Fenestrated tubes are tubes with openings (fenestra) in the cannula to allow air to flow up into the larynx for speaking (see Fig. 26–16). The patient is taught to plug the opening of the tube while speaking to divert air through the fenestra. A newer option is the Passy-Muir tracheostomy speaking valve. This is a special valve that allows air into the tracheostomy during inspiration but closes and redirects air up and out the nose and mouth on expiration, allowing the patient to speak. For the valve to work, the tracheostomy tube must be small enough for air to flow around it or it must be fenestrated.

Some tracheostomies are permanent. However, some patients can be weaned from the tracheostomy tube when their condition has improved enough to allow breathing without it. The physician may replace the tube with a smaller tube to prepare the patient for its removal. This allows a plug to be inserted into the tracheostomy tube at intervals to force the patient to breathe around the tube through the nose and mouth. When the tracheostomy tube has been removed, the opening may be taped shut and covered with gauze until it is healed. The gauze often becomes saturated with secretions and is changed as needed.

Nursing Process for the Patient with a Tracheostomy

See Nursing Care Plan Box 26–4 for the Patient with a Tracheostomy.

CRITICAL THINKING

Mr. Smith

Mr. Smith had a plastic cuffed tracheostomy tube that was small enough to allow airflow around it for talking when the cuff was deflated. The cuff was inflated for lunch to prevent him from aspirating. A friend stopped by for a chat and assisted Mr. Smith to plug his tracheostomy so he could talk. Mr. Smith's face turned dark red, and his expression showed extreme anxiety.

1. What happened?
2. How could you help prevent this in the future?
3. How would you document this occurrence?

Answers at end of chapter.

BOX 26-4   NURSING CARE PLAN FOR THE PATIENT WITH A TRACHEOSTOMY

 Risk for ineffective airway clearance related to increase in secretions

PATIENT OUTCOME
Airway is free of secretions.

EVALUATION OF OUTCOME
Is airway free of secretions?

INTERVENTION	RATIONALE	EVALUATION
• Assess lung sounds every 4 hours and prn.	Coarse crackles or wheezes may indicate secretions in airways.	Are coarse crackles or wheezes present?
• Encourage patient to deep breathe and cough as able.	Patients may be able to clear own secretions without suctioning.	Is patient able to cough up secretions effectively?
• Encourage fluids if not contraindicated.	Fluids help hydrate secretions, making them easier to cough up.	Is patient taking adequate fluids? Are secretions thin?
• Encourage ambulation as able, or turn every 2 hours.	Movement helps mobilize secretions.	Is patient mobilized as much as possible?
• Suction patient using sterile technique prn. Suction only when necessary.	Suctioning clears secretions from airways. Unnecessary suctioning irritates airways.	Is suction necessary? Is airway free of secretions after suctioning?
• Monitor and document amount, color, and character of secretions. Report increase in secretions accompanied by fever.	Purulent sputum accompanied by fever may indicate pneumonia.	Is sputum clear or white, and scant in amount? Does purulent sputum need to be reported?

BOX 26-4   NURSING CARE PLAN FOR THE PATIENT WITH A TRACHEOSTOMY—CONT'D

 ### Risk for infection related to bypass of normal respiratory defense mechanisms

PATIENT OUTCOME

Patient is free from symptoms of infection.

EVALUATION OF OUTCOME

Is patient free from symptoms of infection?

INTERVENTION	RATIONALE	EVALUATION
• Use good hand-washing practice.	Hand-washing is important in preventing infection.	Do all caregivers use good hand-washing technique?
• Monitor and report signs and symptoms of infection: fever, increased respiratory rate, purulent sputum, elevated white blood cell (WBC) count.	Early recognition and treatment of infection enhances outcome.	Are signs of infection present?
• Protect tracheostomy opening from foreign material: food, sprays, powders.	Foreign materials in the tracheostomy may cause pneumonia.	Is tracheostomy adequately protected?
• Use meticulous sterile technique for all tracheostomy care and suctioning.	Use of nonsterile technique may introduce microorganisms into the respiratory tract.	Is sterile technique used by all caregivers?
• Encourage a well-balanced diet. Consult dietitian prn.	A well-balanced diet enhances immune function.	Is patient eating a balanced diet or receiving adequate supplementation?
• Take measures to prevent aspiration of food or secretions (see Chapter 46).	Aspiration can cause pneumonia.	Is there evidence that the patient is aspirating?

 ### Impaired verbal communication related to presence of tracheostomy tube

PATIENT OUTCOMES

Patient uses alternate methods of communication effectively. Patient expresses satisfaction with ability to communicate needs.

EVALUATION OF OUTCOME

Is patient able to use alternative methods to express needs? Does patient express satisfaction with ability to do so?

INTERVENTION	RATIONALE	EVALUATION
• Take time to allow patient to communicate needs.	Communication takes time; patient may become frustrated if hurried.	Does patient feel he or she is given adequate time for communication of needs?
• Watch for patient's nonverbal cues.	Gestures and facial expression can provide valuable cues.	Is the patient attempting to communicate with nonverbal cues?
• Offer pen and paper or magic slate (if patient is literate).	The patient may be able to write out his or her needs/concerns.	Is patient able to write out needs?
• Use picture board (see sample in Chapter 46).	The patient can point to a picture (water, toileting) that indicates his or her need.	Is patient able to point appropriately to needs?
• Teach patient with fenestrated or small tracheostomy tube how to cover opening with a plug or clean finger in order to talk.	Covering opening diverts air into larynx and allows speech.	Is patient able to communicate in this manner?
• Consult with speech therapist.	Speech therapist may have additional methods for communicating with patient.	Did speech therapist provide alternative communication techniques?

Disturbed body image related to presence of tracheostomy

PATIENT OUTCOMES

Patient verbalizes acceptance of tracheostomy. Patient is willing to participate in tracheostomy care.

EVALUATION OF OUTCOMES

Does patient verbalize acceptance of tracheostomy? Does patient participate in learning to care for tracheostomy?

INTERVENTION	RATIONALE	EVALUATION
• Assess patient's feelings about tracheostomy.	Assessment provides basis for care.	Are the patient's feelings within an expected range for such a change in body image?

BOX 26-4   NURSING CARE PLAN FOR THE PATIENT WITH A TRACHEOSTOMY—CONT'D

- Approach patient with an accepting attitude.
- Allow patient opportunity to verbalize concerns about tracheostomy.
- Refer patient to support group if available.
- Assist patient in finding attractive ways to conceal tracheostomy if desired.

The patient will be aware of the nurse's nonverbal body language.
Verbalizing concerns helps the patient to sort out feelings and problem solve.

The patient may benefit from talking with others with tracheostomies.
Loose scarves or collars can help conceal and protect the tracheostomy.

Does the patient indicate a feeling of acceptance from the nurse?
Does the patient verbalize feelings as needed? (Note: Some patients do not wish to share feelings and should not be forced to do so.)
Is patient receptive to a support group referral?

Is patient satisfied with appearance of tracheostomy?

 Deficient knowledge related to care of new tracheostomy

PATIENT OUTCOMES

Patient and significant other will verbalize understanding of self-care, demonstrate tracheostomy self-care procedures, and state resources for help after discharge.

EVALUATION OF OUTCOMES

Are patient and significant other able to verbalize self-care actions and rationale? Are patient and significant other able to correctly demonstrate care procedures? Is patient able to state how to obtain help after discharge?

INTERVENTION	RATIONALE	EVALUATION
• Assess patient's and significant other's baseline knowledge of self-care.	Teaching should only be initiated if a knowledge deficit exists.	Does patient exhibit knowledge of self-care?
• Instruct patient and significant other in the following (see text for specific instruction): • Tracheostomy cleaning • Deep breathing and coughing • Suctioning • Prevention of infection and symptoms to report to health care provider • Protection of tracheostomy from pollutants, water (no swimming, careful showering)	The patient will need to care for self after discharge.	Is patient able to verbalize understanding of self-care and demonstrate all procedures correctly?
• Provide follow-up with home health nurse after discharge.	A home health nurse can provide reinforcement of instruction at home.	Is patient receptive to having a home health nurse assist?

Suctioning

Suctioning involves the use of a flexible catheter to remove secretions from the respiratory tract of a patient who is unable to cough effectively. This may be the patient with a tracheostomy or endotracheal tube, or one with overwhelming secretions.

The procedures for suctioning are presented in Boxes 26-5 and 26-6. A procedure manual should be consulted for more detailed instruction. Remember that suctioning is both frightening and uncomfortable for a patient. Patients sometimes feel as though oxygen is being "vacuumed" from their lungs. Suctioning can cause hypoxia, vagal stimulation with resulting bradycardia, and even cardiac arrest. It is done only when necessary, rather than on a routine basis. Coughing is the most effective way to clear secretions and should be encouraged if the patient is capable. Each step should be explained to the patient, even if he or she is unresponsive.

LEARNING TIPS

Hold your own breath while suctioning your patient. It will keep you mindful of how the patient feels, because suctioning removes not only secretions but also oxygen from the respiratory tract.

Intubation

Some patients are unable to breathe effectively because of airway obstruction or respiratory failure. These patients are intubated with a special endotracheal (ET) tube through the nose or mouth and into the trachea (Fig. 26-17). Patients in cardiopulmonary arrest are intubated during advanced cardiac life support, and patients undergoing general anesthesia during surgery are intubated and mechanically ventilated. Most intubated patients are also mechanically ventilated. Some patients have advance di-

BOX 26-5 **Oropharyngeal or Nasopharyngeal Suction Procedure**

1. Gather equipment: sterile suction catheter, sterile gloves, sterile container (these items may be found in a single "cath and glove" kit); sterile water or saline, suction machine with tubing.
2. Explain procedure to patient.
3. Connect catheter to suction tubing, keeping catheter inside sterile sleeve. Turn on suction to level specified by institution policy (usually 80–120 mm Hg for wall suction).
4. Pour saline into sterile container.
5. Put on sterile gloves. Keep dominant hand sterile at all times.
6. Suction small amount of saline into catheter to rinse catheter and test suction.
7. Have patient take several deep breaths.
8. With thumb control uncovered to stop suction, insert suction catheter through mouth or nose into the trachea until resistance is met or patient coughs.
9. Slowly withdraw catheter, suctioning intermittently while rotating it. The entire procedure should take no more than 10 to 15 seconds.
10. After allowing patient to rest, repeat steps 6 through 9 two more times if needed.

NOTE: A procedure manual should be consulted for more detailed instruction.

BOX 26-6 **Suctioning Procedure for the Patient with a Tracheostomy**

1. Gather equipment: sterile suction catheter, sterile gloves, sterile container (these items may be found in a single "cath and glove" kit); sterile water or saline, suction machine with tubing, manual resuscitation bag.
2. Explain procedure to patient.
3. Connect catheter to suction tubing, keeping catheter inside sterile sleeve. Turn on suction to level specified by institution policy (usually 80–120 mm Hg for wall suction). Connect oxygen source to manual resuscitation bag.
4. Pour saline into sterile container.
5. Put on sterile gloves. Keep dominant hand sterile at all times.
6. Suction small amount of saline into catheter.
7. Oxygenate patient with three ventilations using a manual resuscitation bag connected to an oxygen source, using the nonsterile hand. If the patient is mechanically ventilated, use manual sigh.
8. With thumb control uncovered to stop suction, insert suction catheter through tracheostomy tube until patient coughs or resistance is met.
9. Slowly withdraw catheter, suctioning intermittently while rotating it. The entire procedure should take no more than 15 seconds.
10. Allow patient to rest.
11. Repeat steps 6 through 10 two more times if needed.
Some older sources recommend instilling sterile saline into the tracheostomy to loosen secretions. *This should be avoided.* It is now known that this procedure is not effective and may actually cause a drop in the patient's SaO_2.

NOTE: A procedure manual should be consulted for more detailed instruction.

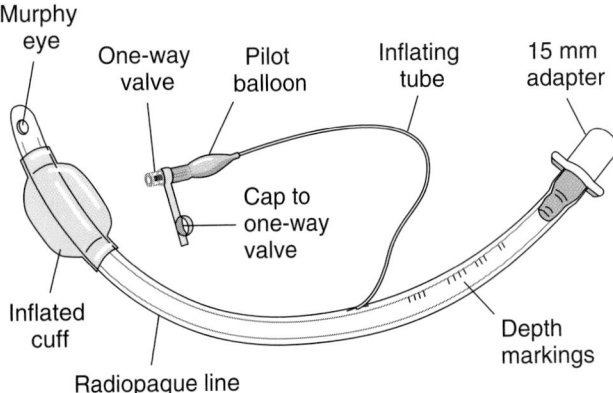

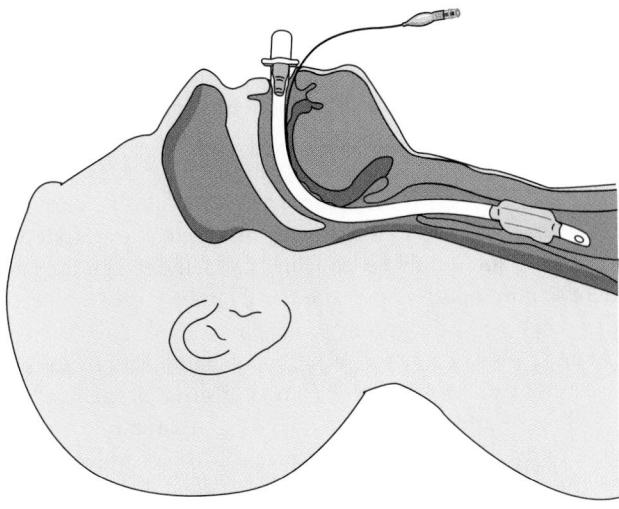

Placement of tube in airway

Figure 26-17 Endotracheal tube. (Modified from Barnes, TA: Respiratory Care Principles. FA Davis, Philadelphia, 1991, p 425, with permission.)

rectives that indicate that they do not wish to be intubated. You should be familiar with the patient's wishes and bring them to the attention of the physician if necessary.

Because intubation can damage the vocal cords and surrounding tissues, it is usually a short-term intervention. Patients who need long-term support have a tracheostomy tube placed.

Nursing care of the intubated patient includes regular assessment of the patient's respiratory status and tube placement. Lung sounds are auscultated bilaterally to ensure that the tube has not been displaced into one bronchus. The tube is carefully secured with tape or a Velcro holder to avoid dislodging. Oral tubes are repositioned and resecured to the opposite side of the mouth every 24 hours or according to institution policy to prevent tissue damage. An adhesive skin barrier should be applied under the tape to protect the skin. If the patient is alert, he or she is instructed to be careful not to pull on the tube. You may need to obtain an order for soft wrist restraints if absolutely necessary for the confused patient. Restraints can be avoided if a family member is available to sit with the patient. Some nursing interventions for the patient with a tracheostomy are also appropriate for the intubated patient. (See Nursing Care Plan 26–4 for the Patient with a Tracheostomy.)

Endotracheal tubes have a cuff (a balloonlike area around the tube) to help maintain proper placement and to prevent leakage of air around the tube. An RT usually inflates the cuff and maintains a specific cuff pressure and should be consulted for assistance with this activity.

Patients with ET tubes may need suctioning if they are unable to cough effectively. Visible secretions in the tube, crackles or wheezes heard with or without the stethoscope, or a drop in SaO_2 without another obvious cause are signs that suctioning is necessary. The ET tube suctioning procedure is sterile and is the same as suctioning a tracheostomy tube. Some institutions have in-line suctioning devices, which are connected to the ET tube within a sterile sleeve. This maintains sterility, protects the nurse, and simplifies the suctioning procedure. Oral suction keeps the mouth free of secretions. Meticulous mouth care is done every 4 hours and as needed. The intubated patient is often extremely anxious, especially if he or she is alert. Explain the purpose of all care activities. Suctioning is particularly anxiety-producing and should be explained carefully, even if the patient is unresponsive. Because the ET tube passes between the vocal cords, the patient is unable to speak. Provide paper and pencil or a picture board for communication. Yes/no questions can be answered by a nod or shake of the head.

Monitor arterial blood gas and oxygen saturation values and notify the physician of changes. If oxygen values drop or the patient becomes confused or agitated, the patient should be immediately inspected for a disconnected oxygen source or excessive secretions.

If the physician determines that the patient can breathe effectively without the tube, the tube will be removed. Before removal, the patient's mouth and tube are suctioned and the cuff is deflated. After removal, the patient is observed closely for laryngeal edema or respiratory distress. The patient is maintained in high Fowler's position to maximize chest expansion.

Mechanical Ventilation

Ventilators are devices that provide ventilation (respirations) for patients who are unable to breathe effectively on their own (Fig. 26–18). Ventilators use positive pressure to push oxygenated air via a cuffed ET or tracheostomy tube into the lungs at preset intervals. Patients may need mechanical ventilation after some surgeries, after cardiac or respiratory arrest, for declining arterial blood gases related to worsening respiratory disease, or for neuromuscular disease or injury that affects the muscles of respiration.

Ventilator Modes

Ventilators can control ventilation or assist the patient's own respirations. See Table 26–8 for terms that aid understanding of ventilator function. There are many types and models of ventilators. Consult with the respiratory care department for an explanation of a patient's ventilator and how to troubleshoot alarms that may sound.

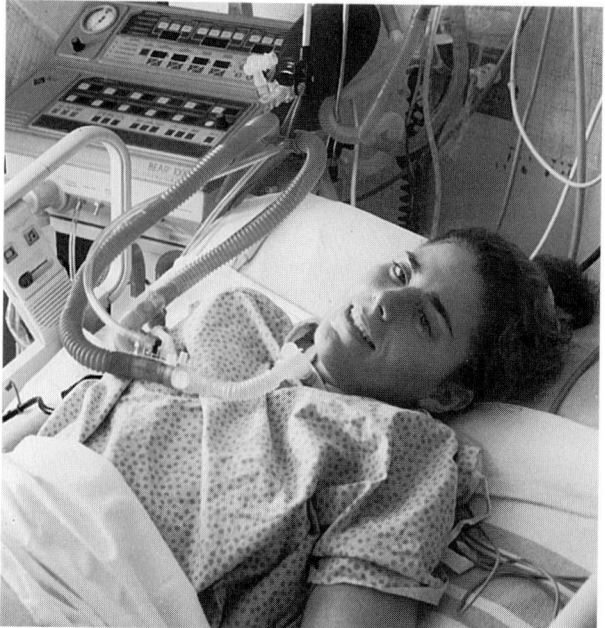

Figure 26-18   Patient on ventilator.

Ventilator Alarms

Several types of alarms are found on ventilators. Low-pressure alarms sound if the ventilator senses reduced pressure in the system. This can be caused by disconnected tubing, leaks in tubing or around the ET tube, or an underinflated cuff. A low-pressure alarm may also sound if the patient has attempted to remove the tube.

TABLE 26-8	**VENTILATOR TERMINOLOGY**
FIO_2	Fraction of inspired oxygen.
Tidal volume	Amount of air delivered with each breath.
Rate	Frequency of breaths delivered.
Assist control mode (AC; also called continuous mechanical ventilation, or CMV)	Ventilator delivers a breath each time patient begins to inspire. If patient does not breathe, the machine continues to deliver a preset number of breaths per minute.
Synchronized intermittent mandatory ventilation (SIMV)	Allows patient to breathe independently but delivers a minimum number of ventilations per minute as necessary. Synchronized to patient's own respiratory pattern.
Pressure support (PS)	Provides positive pressure on inspiration to decrease the work of breathing.
Continuous positive airway pressure (CPAP)	Provides positive pressure on inspiration and expiration to keep alveoli open in a spontaneously breathing patient.
Positive end-expiratory pressure (PEEP)	Provides positive pressure on expiration to help keep small airways open.

High-pressure alarms sound for higher-than-normal resistance to air flow. This might occur if the patient needs to be suctioned; if the patient is biting on the tube, coughing, or trying to talk; if tubing is kinked or otherwise obstructed; or if worsening respiratory disease causes decreased lung compliance. In addition, the high-pressure alarm may be triggered if the patient is anxious and is unable to time his or her breaths with those of the ventilator. Water in the tubing might also cause a high-pressure alarm. Consult with the respiratory care department for guidance in disconnecting and draining the tubing.

A loss of power alarm may signal a power failure or a disconnected plug. Be aware of emergency power sources and be prepared to manually ventilate if necessary.

Volume and frequency alarms sound when tidal volume or number of breaths per minute fall outside preset parameters.

When an alarm sounds, always check the patient first. If the patient is stable, the machine may then be checked. Determine why the alarm is sounding and correct the problem quickly. If no cause can be found, disconnect the patient from the ventilator and call for help. Use a manual resuscitation bag until an RT arrives.

NURSING RESPONSIBILITIES. Before initiating mechanical ventilation, it is important that the health care team be aware of advance directives and consult with the patient and family, because many patients do not wish to be intubated and mechanically ventilated. Some patients accept mechanical ventilation if it is a temporary measure but not if it might be a permanent intervention.

Until recently, ventilators were used only in intensive care units. Now ventilators are seen on medical-surgical units, in nursing homes, and even in patients' homes. It is important that a team approach be used when caring for a mechanically ventilated patient. The social worker; RT; physical, occupational, and speech therapists; dietitian; nurse; and physician all work together to provide the comprehensive care needed by this patient. Respiratory therapists usually take responsibility for routine monitoring and equipment maintenance. The nurse is responsible for monitoring the patient, ensuring that ventilator settings are maintained as prescribed, providing initial response to alarms, keeping tubing free from water accumulation, and keeping the patient's airway free from secretions. In addition, the nurse keeps a manual resuscitation bag at the bedside for emergencies.

Patients who are mechanically ventilated are unable to talk and can become very uncomfortable and anxious with no easy way to communicate. See Box 26–7 for one nurse's tips for making ventilated patients feel more secure. These tips were developed after the author interviewed 12 patients who had been intubated. They shared their fears, anxieties, and physical discomforts.

Noninvasive Positive-Pressure Ventilation

Noninvasive positive-pressure ventilation (NIPPV) is an alternative to intubation and mechanical ventilation for patients who are able to breathe on their own but are un-

---

| **BOX 26-7** | **Tips for Caring for Mechanically Ventilated Patients** |

- Introduce yourself to the patient each time you enter the room. Make sure he or she can see you.
- Explain everything you are about to do.
- Check ventilator settings regularly.
- Give sedatives or antianxiety drugs as ordered. Request an order if necessary. Find out cause of unexplained anxiety (patient may be hypoxemic).
- Reassure the patient that anxiety is normal and that relaxing will help the ventilator to work with him or her.
- Assess for comfort and reposition at regular intervals. Be careful not to pull on the ventilator tubing. (Pulling hurts.)
- Suction quickly and smoothly, without jabbing. Avoid the use of saline with suctioning.
- Provide good oral care, moistening the lips with a cool washcloth and water-based lubricant. (Patients get thirsty.)
- Use restraints only as a last resort.
- Take the time to communicate with the patient. Talk to him or her, and provide a magic slate or pen and paper for the patient to talk to you. Make sure the call light is within reach at all times.
- Answer patient's call light and ventilator alarms promptly.

Source: Modified from Jablonski, RAS: If ventilator patients could talk. RN, Feb 1995, p 32.

---

able to maintain normal blood gases. Patients with severe respiratory disease, sleep apnea, or neuromuscular diseases such as amyotrophic lateral sclerosis (ALS) that weaken respiratory muscles can benefit from this treatment. Instead of the invasive endotracheal or tracheostomy tube, NIPPV uses an external masklike device that fits over the nose or mouth and nose (Fig. 26–19). It can be successful in patients who are alert, able to cooperate, do not have excessive secretions, and are able to breathe on their own for periods of time. It can be used with or without supplemental oxygen. In an acutely ill patient, oxygen saturations are monitored.

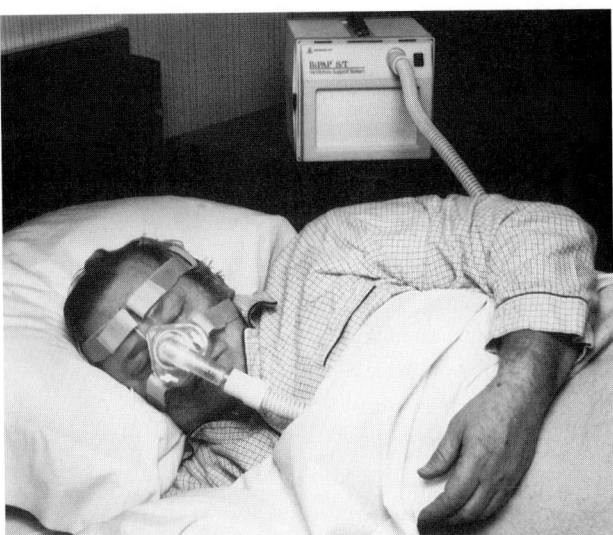

Figure 26-19 Noninvasive positive-pressure ventilation. (Courtesy Respironics, Inc.)

Two basic types of NIPPV are available: continuous positive airway pressure (CPAP) and bilevel positive airway pressure (BiPAP, Respironics, Inc.). In CPAP the same amount of positive pressure is maintained throughout inspiration and expiration to prevent airway collapse. In BiPAP a different level of positive pressure is used on inspiration and expiration.

Problems to be alert for in patients receiving NIPPV include skin irritation from the mask and gastric distention from swallowing air. Apply an adhesive skin barrier to the areas that come in contact with the mask to prevent irritation. To prevent gastric distention, place the patient in semi-Fowler's position and consult with the RT to adjust air delivery pressure if necessary. Topical saline or a special humidifier on the machine can reduce nose and mouth dryness.

An air leak around the mask can cause air to blow in the patient's eyes, which can be irritating. If this happens, remove the mask and reposition it. Another problem is patient acceptance of NIPPV. Many patients do not like the tight mask covering their nose or mouth. Be patient in explaining the reason for this treatment and check the patient frequently to help control anxiety. Be sure to assess the patient's goals for therapy. Some patients may choose not to use NIPPV, but they must be fully aware of possible consequences.

Patients can use NIPPV nearly continuously, removing it to eat or use the bathroom. Other patients use it only when they are sleeping and are able to breathe effectively on their own during the day. Some use it for a few days until an acute exacerbation of disease is resolved, and others continue its use indefinitely at home.

REVIEW QUESTIONS

1. During inhalation, which of the following muscle contractions takes place to enlarge the chest cavity from top to bottom?
   a. Diaphragm moves down.
   b. External intercostal muscles move down.
   c. Diaphragm moves up.
   d. Internal intercostal muscles move up.

2. Deteriorating cilia in the respiratory tract predispose the elderly to which of the following?
   a. Chronic hypoxia
   b. Pulmonary hypertension
   c. Respiratory infection
   d. Decreased ventilation

3. Which of the following terms is used to describe violinlike sounds heard on chest auscultation?
   a. Crackles
   b. Wheezes
   c. Friction rub
   d. Stridor

4. Which of the following is a normal value for oxygen saturation?
   a. Less than 60 percent
   b. 61 to 85 percent
   c. 86 to 95 percent
   d. More than 95 percent

5. The purpose of pursed-lip breathing is to promote which of the following?
   a. Carbon dioxide excretion
   b. Carbon dioxide retention
   c. Oxygen excretion
   d. Oxygen retention

6. Which of the following instructions is correct when teaching a patient how to use a metered dose inhaler?
   a. "Inhale deeply, place canister in mouth, depress top of canister, exhale."
   b. "Exhale, place canister in mouth, depress canister and inhale at the same time."
   c. "Cough, place canister in mouth, inhale deeply, cough again."
   d. "Exhale, depress canister, place in mouth, inhale deeply."

7. Which of the following actions by the nurse is appropriate when vigorous bubbling is noted in the suction control chamber of a chest drainage system?
   a. Check the tubing for leaks.
   b. Notify the physician.
   c. Reduce the level of wall suction.
   d. Clamp the chest tube.

8. You hear a high-pressure alarm sounding on a mechanically ventilated patient. Which of the following should you check first?
   a. Tubing
   b. Power to the ventilator
   c. Ventilator settings
   d. The patient

REFERENCES

1. Sharts-Hopko, N: Japanese-Americans. In Purnell, L, and Paulanka, B (eds): Transcultural Health Care: A Culturally Competent Approach. FA Davis, Philadelphia, 1998.
2. AbuGharbieh, P: Arab-Americans. In Purnell, L, and Paulanka, B (eds): Transcultural Health Care: A Culturally Competent Approach. FA Davis, Philadelphia, 1998.
3. Campinha-Bacote, J: African-Americans. In Purnell, L, and Paulanka, B (eds): Transcultural Health Care: A Culturally Competent Approach. FA Davis, Philadelphia, 1998.
4. Purnell, L, and Counts, M: In Purnell, L, and Paulanka, B (eds): Transcultural Health Care: A Culturally Competent Approach. FA Davis, Philadelphia, 1998.

Answers to CRITICAL THINKING

Timothy

1. There is no way to know whether Timothy needs to be in the emergency room without further assessment. Remember that shortness of breath is very subjective and must be evaluated before discharge.
2. Collect further data. Have Timothy rate his shortness of breath. Look at his color and use of accessory muscles. Check his vital signs, peak expiratory flow rate, and oxygen saturation.
3. If Timothy is having an asthma attack, one explanation for the absence of wheezing on auscultation is that he is not moving enough air to generate the wheezing sound. If his airways are extremely tight, breath sounds may be so diminished that wheezing is not heard. This is a bad sign rather than a good one. If you suspect that this is happening, call for help. The physician may want to begin treatment quickly before further evaluation is done.

Miss Israel

Bubbling in the water seal chamber indicates a leak in the system. Vigorous bubbling may indicate a large leak, and the physician should be contacted immediately. After you check the patient, check the entire system for cracks or leaks and correct any problems discovered.

Mr. Smith

1. Mr. Smith plugged his tracheostomy while the cuff was still inflated, so no air could get to his lungs. If the plug is not removed immediately he will be totally unable to breathe. Whenever the plug is in place, air must be able to travel around the tracheostomy tube or through the opening of a fenestrated tube for the patient to breathe.
2. To prevent this from happening in the future, Mr. Smith should be taught how his tracheostomy tube works and how to care for it.
3. "Trach tube cuff inflated for lunch as ordered. Answered call for help at 1230, found patient dark red in color, unable to breathe, trach plugged. Trach unplugged, respirations restored, vital signs stable. Patient stated he plugged trach so he could talk to his friend. Function of trach cuff explained to patient and friend. Both verbalize understanding to only plug trach when cuff is deflated or to call for nurse if unsure."

27

NURSING CARE OF PATIENTS WITH UPPER RESPIRATORY TRACT DISORDERS

Paula D. Hopper

KEY TERMS

dysphagia (dis-**FAYJ**-ee-ah)

epistaxis (EP-iss-**TAX**-iss)

exudate (**EKS**-yoo-dayt)

laryngectomee (lare-in-**JEK**-tah-mee)

laryngitis (lare-in-**JIGH**-tiss)

myalgia (my-**AL**-jee-ah)

nasoseptoplasty (NAY-zoh-**SEP**-toh-plass-tee)

pharyngitis (fair-in-**JIGH**-tiss)

rhinitis (rye-**NIGH**-tiss)

rhinoplasty (**RYE**-noh-plass-tee)

sinusitis (SINE-u-**SIGH**-tiss)

QUESTIONS TO GUIDE YOUR READING

1. How would you explain the pathophysiology of the disorders of the upper respiratory tract?
2. What are the etiologies, signs, and symptoms of disorders of the upper respiratory tract?
3. What is current medical treatment for disorders of the upper respiratory tract?
4. What nursing care should you be prepared to provide for the patient with an upper respiratory disorder?
5. How will you know if your care has been effective?
6. What are the special needs of the patient who has undergone a laryngectomy?

Disorders of the upper respiratory tract include problems occurring in the nose, sinuses, pharynx, larynx, and trachea. Many of these problems are minor illnesses that can be cared for at home. Others can become serious if they are not recognized and treated in a timely manner.

▶ DISORDERS OF THE NOSE AND SINUSES

Epistaxis

Pathophysiology

Epistaxis is more commonly known as a nosebleed. The nose can bleed either from the anterior or posterior region. Anterior bleeds are much more common and originate from a group of vessels called the Kiesselbach plexus. Anterior bleeds are easier to locate and treat than posterior bleeds, because the blood vessels of the posterior nose are larger and bleeding can be severe and difficult to control.

Etiology

The most common cause of epistaxis is dry, cracked mucous membranes. Trauma, forceful nose blowing, nose picking, and increased pressure on fragile capillaries from hypertension are also factors. Anything that reduces the blood's ability to clot, such as hemophilia or leukemia, regular aspirin use, anticoagulant therapy, or chemotherapy, can also predispose a patient to nosebleeds.

Treatment

Instruct the patient with a nosebleed to sit in a chair and lean forward slightly to avoid aspirating or swallowing blood. If the patient swallows blood, it will be difficult to assess the extent of bleeding. Be sure to wear gloves and follow standard precautions. Place pressure on the nares for 5 to 10 minutes to stop bleeding. However, avoid placing pressure on the nose if a fracture is suspected, to avoid further trauma. Ice packs to the nose and eye area may be used to constrict the vessels that are bleeding. Once the bleeding has stopped, instruct the patient to avoid blowing the nose for several hours to prevent further bleeding.

If first aid measures are ineffective in stopping bleeding, a physician may attempt more invasive treatment. Local application of a vasoconstrictive agent might be used to constrict the bleeding vessels. If the bleeding vessel can be located, the physician may cauterize it by use of an electrical cauterizing device, or by application of silver nitrate.

Gauze may be used to pack the anterior or posterior nasal cavity. The anterior cavity is packed firmly but gently, usually with half-inch petroleum gauze. To pack the posterior cavity, the physician must use a catheter and string via the nose to draw the packing through the mouth and into the posterior nasal cavity (Fig. 27–1). The strings are then brought out the mouth and taped to the patient's face so they can be used 2 to 4 days later to remove the packing. Placement and removal of packing can be very uncomfortable for the patient. If there is time, administration of an analgesic before the procedure is helpful. Petroleum jelly on the packing helps prevent gauze from adhering to the nasal mucosa. If the packing is to remain in place for a prolonged period, it is coated with an antibiotic ointment to reduce the risk of infection.

Another method to stop bleeding is the use of a nasal balloon catheter. This device employs a catheter with a balloon on the end that is inflated after placing it near the bleeding vessels in the nasal cavity. The inflated balloon places pressure on the bleeding vessels to stop the bleeding.

If the patient has lost a significant amount of blood, a transfusion may be necessary. Nosebleeds rarely cause death, because blood loss lowers blood pressure, which in turn slows the bleeding. Ultimately the cause of the epistaxis is determined and corrected if possible.

Nursing Care

Monitor bleeding, noting the amount and color of drainage. Monitor vital signs and hemoglobin level for signs of excessive blood loss. If the patient swallows repeatedly, inspect the back of the throat for bleeding. If bleeding does not stop or if it worsens, notify a registered nurse (RN) or physician immediately.

If posterior packing has been used, monitor the patient for airway obstruction from slipped packing. Know how to remove the packing in case of emergency. Institute comfort measures, and maintain the placement of the strings that will be used to remove the packing. Once bleeding is controlled, caution the patient not to blow the nose for several hours and to avoid nose picking. The packing will be removed by the physician. If the cause of the bleeding is known, teach the patient how to prevent recurrence.

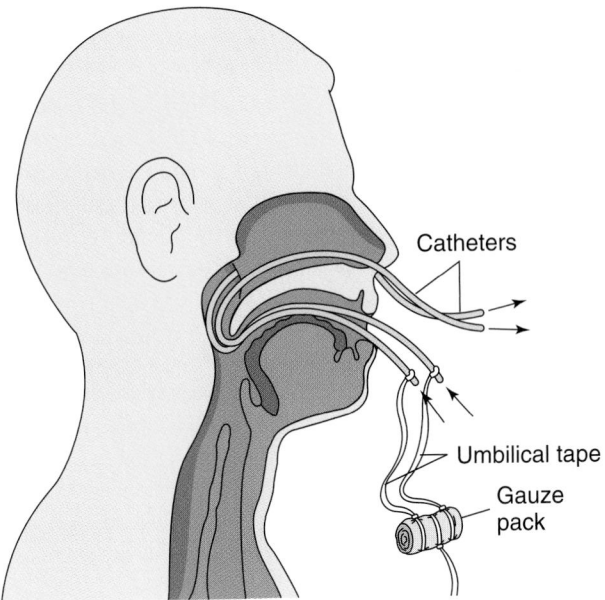

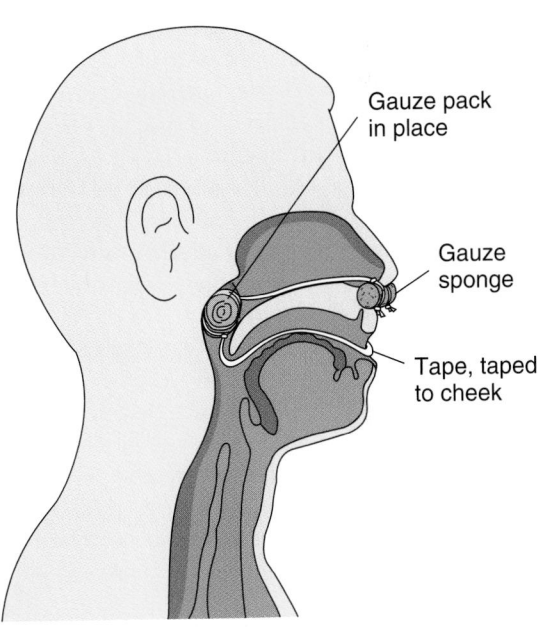

Figure 27-1 Nasal packing. *(A)* Catheters are used to pull packing into place. *(B)* Nasal packing in place.

CRITICAL THINKING

Mr. Jondahl
Mr. Jondahl is brought to the emergency room with a nosebleed. His vital signs are 140/90, 92, 20. He states that he has never had a nosebleed before. He denies any history of coagulation disorders. His current medications include captopril (Capoten), furosemide (Lasix), and ibuprofen (Motrin). What are two areas you should assess further in trying to determine a cause? (Hint: If you are not familiar with Mr. Jondahl's medications, look them up.)

Answers at end of chapter.

Nasal Polyps
Pathophysiology and Etiology

Polyps are grapelike clusters of mucosa in the nasal passages. They are usually benign, but they can obstruct the nasal passages. Though the exact cause is unknown, people with allergies are prone to developing polyps. Some patients with nasal polyps also have asthma and are allergic to aspirin. This is called a triad disease because the three components often occur together.

Treatment

Control of allergy symptoms may help control polyp development. If polyps obstruct breathing, they can be removed. This is done as an outpatient procedure under local anesthesia, using laser or endoscopic surgery. Patients are taught to avoid aspirin products following surgery because they increase the risk of postoperative bleeding and recurrence of the polyps.

Deviated Septum
Pathophysiology and Etiology

The septum dividing the nasal passages is slightly deviated in most adults. This may result from nasal trauma but often has no cause. Some may be so deviated that they block sinus drainage or interfere with breathing.

Signs and Symptoms

The patient may complain of a chronically stuffy nose or discomfort from blocked sinus drainage. Some patients experience headaches and nosebleeds.

Medical Treatment

If the deviated septum is causing chronic discomfort, a submucous resection (SMR), or **nasoseptoplasty,** is done. This surgery involves making an incision through the mucous membrane covering the septum and removing the deviated portion. Nasal packing is then placed to reduce bleeding. This is generally done as an outpatient surgical procedure, under local anesthetic.

Nursing Care

Monitor vital signs and bleeding until the patient is stable following surgery. Excessive swallowing should alert you to check for blood running down the back of the throat. The patient will have nasal packing and a "mustache dressing" of folded gauze under the nose to catch drainage.

Most patients are discharged home once they are stable, so teaching is important. The patient should maintain a semi-Fowler's position as much as possible and avoid anything that might increase pressure and cause bleeding, such as sneezing, coughing, or straining to move the bowels. Stool softeners and cough suppressants may be ordered by the physician if necessary. Aspirin and related medications are avoided because they increase the risk of bleeding. Antibiotics may be ordered if packing is in place because of the risk of infection from nasal bacteria. The physician should be contacted for specific orders if the patient is on anticoagulant therapy at home. Ice can be used to reduce swelling and bruising. Instruct the patient to contact the physician if fever, excessive pain, swelling, or bleeding occur and to return in 24 to 48 hours for removal of nasal packing. See Box 27–1 for patient teaching after nasal surgery.

Rhinoplasty

Rhinoplasty is the surgical reconstruction of the nose, usually for cosmetic purposes. It may also be done to correct deformity caused by trauma. Nursing care is similar to that for the patient after SMR, described previously and in Box 27–1.

Sinusitis
Pathophysiology and Etiology

Sinusitis is inflammation of the mucosa of one or more sinuses. It can be either acute or chronic. Chronic sinusitis is diagnosed if symptoms are present for more than 2 months and are unresponsive to treatment. The maxillary and ethmoid sinuses are the most commonly affected. The inflammation is often the result of a bacterial infection and may follow a viral upper respiratory illness. Because the mucous lining of the nose and sinuses is continuous, nasal organisms easily travel to the sinuses. When the infected mucous lining of the sinuses swells, drainage is blocked. Bacteria that normally reside in the sinuses multiply in the retained secretions. The most common infecting organisms are *Streptococcus pneumoniae* and *Haemophilus influenzae*. Other causes of sinusitis include swelling caused by allergies, fungal infection, or intubation with a nasotracheal or nasogastric tube.

BOX 27–1 **Patient Teaching after Nasal Surgery**

1. Your nose will feel stuffy and may drain. Change the moustache dressing as often as needed. *Do not* blow your nose. If you must sneeze, do so with your mouth open.
2. Drink plenty of fluids unless your physician advises otherwise.
3. Use a cool mist vaporizer to humidify air and prevent nasal drying.
4. Keep your head elevated on two pillows or sleep in a recliner chair.
5. An ice pack on your face may help reduce swelling.
6. Take pain medication as prescribed.
7. Call your physician if you have a fever higher than 101° F.
8. Return to see your physician in _____ days.

---

nasoseptoplasty: naso—nose + septo—septum + plasty—plastic surgery

rhinoplasty: rhin—nose + plasty—to mold
sinusitis: sinu—sinus + itis—inflammation

Signs and Symptoms

The patient usually has pain over the region of the affected sinuses and purulent nasal discharge. If a maxillary sinus is affected, the patient experiences pain over the cheek and upper teeth. In ethmoid sinusitis, pain occurs between and behind the eyes. Pain in the forehead typically indicates frontal sinusitis. Fever may be present in acute infection, with or without generalized fatigue and foul breath.

Complications

The patient who has received inadequate treatment, or who has not complied with treatment, is at risk for complications. Uncontrolled sinusitis may spread to surrounding areas, causing osteomyelitis, cellulitis of the orbit (infection of the soft tissues around the eye), abscess, or meningitis.

Diagnostic Tests

Uncomplicated sinusitis may be diagnosed based on symptoms alone. If repeated episodes occur, x-ray examination, a computed tomography (CT) scan, or magnetic resonance imaging (MRI) may be done to confirm the diagnosis and determine the cause. Nasal discharge may be cultured to determine appropriate antibiotic therapy.

Medical Treatment

Treatment is aimed at relieving pain and promoting sinus drainage. Adrenergic nasal sprays such as oxymetazoline (Afrin, Allerest) constrict blood vessels and therefore reduce swelling, but they should be used cautiously by patients with heart disease or hypertension because vasoconstriction increases blood pressure. Sprays may be used for up to 3 days; longer use may cause rebound congestion. Hot moist packs over the affected sinus for 1 to 2 hours twice a day may help decrease inflammation. Nasal irrigation with normal saline solution and a bulb syringe has helped some sufferers of chronic sinusitis. Acetaminophen is given for pain and fever. Codeine or meperidine may be used if pain is severe. Expectorants such as guaifenesin (Robitussin), fluids, and room humidification help liquefy secretions. Antihistamines dry and thicken secretions and are generally avoided. Antibiotics are used only if bacterial infection is suspected, as in the patient with purulent drainage and fever. If conservative treatment does not relieve symptoms, the physician may surgically drain the affected sinus and irrigate it with normal saline or an antibiotic solution.

One drainage procedure is the Caldwell-Luc procedure. The surgeon enters the maxillary sinus above the upper teeth, under the upper lip. The infected mucosa and bone are removed, and a new, larger opening is made to drain the sinus. Newer procedures use endoscopy to open and drain a chronically infected sinus.

Nursing Care

Patients with uncomplicated sinusitis are cared for at home. Instruct the patient to increase water intake to 8 to 10 glasses per day unless contraindicated. Excess water might be contraindicated in patients with fluid overload, such as those with cardiovascular compromise or kidney disease. Pressure may be relieved if the patient maintains a semi-Fowler's position, as in a reclining chair. Explain use of hot moist packs, acetaminophen, and prescribed medications. Instruct the patient to finish the antibiotic prescription even if he or she is feeling better before it is completed and to call the physician if pain becomes severe or if signs of complications such as a change in level of consciousness occur.

▶ INFECTIOUS DISORDERS
Rhinitis/Common Cold
Pathophysiology and Etiology

Rhinitis (also called coryza) is inflammation of the nasal mucous membranes. The release of histamine and other substances causes vasodilation and edema, which result in symptoms. It may occur as a reaction to allergens (sometimes called hay fever) such as pollen, dust, molds, or some foods, or it may be caused by viral or bacterial infection. Viral rhinitis is another name for the common cold.

Signs and Symptoms

Common symptoms include nasal congestion, localized itching, sneezing, and nasal discharge. Viral or bacterial rhinitis may also be accompanied by fever and malaise.

Diagnostic Tests

If allergic rhinitis is suspected, skin testing may be done to determine the offending allergens. A blood test for IgE antbodies might also be done.

Treatment

Treatment of viral rhinitis is symptomatic. Because most colds are caused by viruses, antibiotics are not effective. In one study, however, researchers found that 60 percent of patients who visited their physician for cold symptoms received a prescription for an antibiotic. This practice is not only expensive, it also increases the risk of developing strains of bacteria that are resistant to antibiotics. Explain to the patient that requesting antibiotics for a viral infection is not only ineffective but potentially dangerous. Teach the patient that rest and fluids are the most effective treatment (Nursing Care Plan Box 27–2).

Antihistamines may be used to help control symptoms by inhibiting the histamine response. Decongestants cause vasoconstriction, which reduces swelling and congestion. Any drugs that cause vasoconstriction should be used cautiously in patients with heart disease or hypertension. Severe allergies may be treated with desensitization ("allergy shots").

Pharyngitis
Pathophysiology and Etiology

Pharyngitis, or inflammation of the pharynx, is usually related to bacterial or viral infection. It may also occur as a result of trauma to the tissues. The most common bacterial

---

rhinitis: rhin—nose + itis—inflammation
pharyngitis: pharyng—pharynx + itis—inflammation

BOX 27-2   NURSING CARE PLAN FOR THE PATIENT WITH AN UPPER RESPIRATORY INFECTION

 Acute pain related to infectious process

PATIENT OUTCOMES
Patient will be comfortable as evidenced by (1) statement of increased comfort and (2) ability to sleep at night.

EVALUATION OF OUTCOMES
(1) Does patient express comfort? (2) Is patient able to sleep?

INTERVENTIONS	RATIONALE	EVALUATION
• Assess for cause of discomfort: malaise, muscle aches, fever.	Knowing cause of discomfort helps guide intervention.	Can interventions be directed toward specific symptoms?
• Offer acetaminophen or other analgesic/antipyretics as ordered.	Analgesics relieve pain. Antipyretics relieve fever, which may contribute to discomfort.	Do analgesics/antipyretics relieve discomfort?
• Offer throat lozenges and salt-water gargles as ordered for irritated throat.	Lozenges soothe irritated mucous membranes. Saltwater gargles may reduce swelling.	Do measures relieve throat irritation?
• Encourage rest.	Physical stress increases need for sleep. Rest boosts immune function.	Is patient resting comfortably?

 Hyperthermia related to infectious process

PATIENT OUTCOMES
(1) Temperature lower than 103° F. (2) No signs/symptoms of dehydration.

EVALUATION OF OUTCOMES
(1) Is fever controlled at safe level? (2) Is patient well hydrated?

INTERVENTIONS	RATIONALE	EVALUATION
• Monitor temperature daily; every 4 hours if fever present.	Screening helps detect temperature changes early.	Is patient febrile?
• If patient begins chilling, recheck temperature when chilling subsides.	Chilling indicates rising temperature.	Is chilling present? Should temperature be checked more often?
• Monitor for signs of dehydration: dry skin and mucous membranes, thirst, weakness, hypotension.	Fever causes loss of body fluids.	Are signs of dehydration present?
• Encourage oral fluids if not contraindicated.	Fluids prevent or treat dehydration.	Is patient taking fluids well?
• Administer antipyretic such as acetaminophen if fever is higher than 102.5° F or for discomfort.	Antipyretics reduce fever. Fever enhances immune function, so should only be treated if very high, if patient has a history of febrile seizures, or if patient is uncomfortable.	Is fever higher than 102.5° F? Are antipyretics indicated? Are they effective?

Risk for infection: transmission to others related to presence of infectious disease

PATIENT OUTCOMES
Risk for infection of others is reduced, as evidenced by the following: (1) Patient states measures to prevent transmission. (2) Patient takes precautions against spread.

EVALUATION OF OUTCOMES
Is transmission to others prevented?

INTERVENTIONS	RATIONALE	EVALUATION
• Assess patient's understanding of infection transmission.	Understanding of mode of transmission is essential to prevention.	Does patient understand how infection is transmitted?
• Based on patient's previous knowledge, teach patient and all caregivers the importance of good handwashing after contact with patient or patient's belongings, covering nose and mouth when coughing or sneezing, and not sharing eating or drinking utensils.	The nurse should build on patient's previous understanding and not repeat information. Handwashing prevents spread of infection. Covering nose and mouth prevents spread of infectious droplets. Many infections are transmitted via contaminated objects.	Does patient take precautions to prevent spread of infection?

infection is caused by beta-hemolytic streptococci, commonly referred to as strep throat. If strep throat is not treated with antibiotics, it can lead to rheumatic fever, glomerulonephritis, or other serious complications.

Signs and Symptoms

The most common symptom of pharyngitis is a sore throat. Some patients may also experience **dysphagia** (difficulty swallowing). The throat appears red and swollen, and **exudate** (drainage or pus) may be present. Exudate usually signifies bacterial infection and may be accompanied by fever, chills, headache, and generalized malaise.

Diagnostic Tests

A physician may order a throat culture and sensitivity test (explained in Chapter 26) to identify the causative organism and determine which antibiotic will be effective.

Treatment

If the pharyngitis is bacterial, antibiotics are ordered. Acetaminophen or throat lozenges may be used to relieve discomfort. Saltwater gargles help reduce swelling, and increased fluids (if not contraindicated) and rest are encouraged. (See Nursing Care Plan Box 27–2.)

Laryngitis
Pathophysiology and Etiology

Laryngitis is an inflammation of the mucous membrane lining the larynx (voice box). It can be caused by irritation from smoking, alcohol, or chemical exposure; or a viral, fungal, or bacterial infection. If often follows an upper respiratory infection.

Signs and Symptoms

The most common symptom is hoarseness. Cough, dysphagia, or fever may also be present.

Diagnostic Tests

A physician may use a laryngeal mirror to view the larynx. If hoarseness persists for more than 2 weeks, a laryngoscopy is done to rule out cancer of the larynx.

Treatment

Treatment includes rest, fluids, humidified air, and aspirin or acetaminophen. Antibiotics are used if bacterial infection is present. Encourage the patient to avoid speaking to rest the voice. Obtain a "magic slate" (from the speech therapy department) or paper and pen to help the patient communicate. Throat lozenges may help increase comfort. Instruct the patient to identify and avoid causative factors. (See Nursing Care Plan Box 27–2.)

Tonsillitis/Adenoiditis
Pathophysiology and Etiology

The tonsils are masses of lymphoid tissue that lie on each side of the oropharynx. They filter microorganisms, thus protecting the lungs from infection. Tonsillitis occurs when the filtering function becomes overwhelmed with a virus or bacteria, and infection results. The adenoids, a mass of lymphoid tissue located at the back of the nasopharynx, can also become involved. Tonsillitis is more common in children, but it is more serious when it occurs in adults. The most common organisms causing tonsillitis are *Streptococcus* species, *Staphylococcus aureus*, *Haemophilus influenzae*, and *Pneumococcus* species.

Signs and Symptoms

Tonsillitis usually begins suddenly with a sore throat, fever, chills, and pain on swallowing. Generalized symptoms include headache, malaise, and **myalgia.** On examination, the tonsils appear red and swollen and may have yellow or white exudate on them. The patient's voice may sound like the patient has a hot potato in his or her mouth. If the adenoids are involved, the patient may have complaints of snoring, nasal obstruction, and a nasal tone to the voice.

Diagnostic Tests

A throat culture is done to discover the causative organism and determine effective treatment. A white blood cell count helps identify whether the infection is viral or bacterial. A chest x-ray examination may be done if respiratory symptoms are present.

Treatment

Antibiotics are prescribed for bacterial infection. Acetaminophen, lozenges, and saline gargles help promote comfort. For care of the patient who is not having a tonsillectomy, see Nursing Care Plan Box 27–2.

TONSILLECTOMY. If tonsillitis becomes chronic or if an abscess occurs, a tonsillectomy may be performed, although this is not a common procedure in an adult. An adenoidectomy may be performed at the same time. After the tonsillectomy, the patient is maintained in a semi-Fowler's position to reduce swelling and promote drainage. Monitor the patient for bleeding and airway patency, and provide comfort measures. Encourage fluids for hydration; cold fluids may

⚡ CRITICAL THINKING

Mrs. Hiler
You are assessing Mrs. Hiler after a tonsillectomy. She is sleeping, but you notice that she swallows every few seconds. She has an intravenous (IV) line of normal saline solution running at 100 mL per hour. How do you respond? How many drops per minute do you set on her IV if the tubing has a drop factor of 15?

Answers at end of chapter.

---

dysphagia: dys—bad + phagia—to swallow
exudate: to sweat out
laryngitis: laryng—larynx + itis—inflammation

myalgia: myo—muscle + algia—pain

help reduce pain and bleeding. Red-colored drinks are avoided because they interfere with observation for bleeding. Suction equipment should be available for emergencies.

Influenza
Pathophysiology and Etiology

Influenza, commonly referred to as the flu, is a viral infection of the respiratory tract. Many different flu viruses have been identified, and new strains appear each year, making immunization difficult. Influenza is the cause of millions of lost work days each year. The elderly are at particular risk for complications and even death from influenza because of pre-existing chronic disease and compromised immune function.

Influenza is easily transmitted via droplets from coughs and sneezes of infected individuals, or it may be transmitted by physical contact with a person or object that harbors the virus. The incubation period from time of exposure to onset of symptoms is 1 to 3 days.

Prevention

Yearly immunization is recommended for prevention of influenza. The Centers for Disease Control and Prevention (CDC) recommends a yearly vaccine for people over age 50 and for residents of extended care facilities. People with chronic illnesses such as lung disease or asthma should also be immunized. Immunization of healthy adults can reduce the risk of influenza and decrease absenteeism from work. Immunization of health care workers helps prevent the spread to at-risk patients. Although Medicare covers the cost of a flu shot, many elders do not get one. Stress that they will not get the flu from the shot, because it does not contain any live virus. Once the shot has been administered, it takes about 2 weeks for antibodies to develop; it is then effective for about 4 months. Other preventive measures include handwashing and avoidance of individuals with influenza.

Signs and Symptoms

Symptoms of flu include abrupt onset of fever, chills, myalgia, sore throat, cough, general malaise, and headache. It can last for 2 to 5 days, with malaise lasting up to several weeks.

Complications

The most common complication of influenza is pneumonia, which may be caused by the same virus as the flu or by a secondary bacterial infection. This should be considered if the patient experiences persistent fever and shortness of breath or if the lungs develop crackles or wheezes.

Diagnostic Tests

Sputum or throat cultures may be done to determine whether there is a bacterial cause of the symptoms.

Medical Treatment

Treatment is primarily symptomatic. Acetaminophen is given for fever, headache, and myalgia. Aspirin is avoided

in children because it increases the risk for Reye's syndrome. Rest and fluids are essential. Antibiotics are used only if a secondary bacterial infection is present.

Antiviral drugs such as amantadine (Symmetrel), zanamivir (Relenza) and oseltamivir (Tamiflu) may be helpful for high-risk patients if given within 48 hours of exposure. These drugs may reduce the severity and duration of symptoms. They may also be given prophylactically to high-risk people who have not been immunized.

Nursing Care

Elderly or other high-risk patients may be hospitalized for treatment of influenza. These patients are closely monitored for signs of complications. Assess lung sounds and vital signs every 4 hours and report changes to an RN or physician. Encourage rest and fluids (if not contraindicated), and provide comfort measures. Parents are educated about avoiding aspirin to treat influenza symptoms in children under 18 to prevent Reye's syndrome. (See Nursing Care Plan Box 27–2.)

CRITICAL THINKING

Murdie

Murdie is a 97-year-old nursing home resident who develops flu symptoms. She is lethargic, confused, and feverish. Because of her mental status changes, you want to send her to the hospital, but her son asks you to please keep her where she is. She has a history of chronic obstructive pulmonary disease (COPD) and diabetes.

1. How could Murdie have caught the flu?
2. How could it have been prevented?
3. What can be done now to prevent her from developing complications that could lead to her death?
4. What other concerns do you have?

Answers at end of chapter.

▶ MALIGNANT DISORDERS
Cancer of the Larynx
Pathophysiology

Cancer of the larynx usually develops in the mucosal epithelium. It is evaluated based on the tumor-node-metastasis (TNM) staging system described in Chapter 10. It is usually a primary cancer and can spread to the lungs, liver, or lymph nodes. The prognosis for a patient with laryngeal cancer is often poor because metastasis (spread) may have occurred before the patient sought help.

Etiology

Risk factors for cancer of the larynx include a history of alcohol and tobacco use. Exposure to industrial chemicals, hardwood dust, and chronic overuse of the voice are also factors. Men are five times as likely to be affected as women.

Prevention

Prevention begins with education; you can help educate patients about the relationship between cancer of the larynx and abuse of alcohol and tobacco. It is also important to teach patients to seek help when symptoms first occur, because a delayed diagnosis may mean metastasis of the cancer and a poor prognosis. Teach them that any hoarseness that lasts longer than 2 weeks should be investigated by a physician.

Signs and Symptoms

The most common symptom is hoarseness, because the vocal cords are located in the larynx (Box 27–3). The patient may also have pain, shortness of breath, a chronic cough, and difficulty swallowing. Stridor may indicate a tumor obstructing the airway. Late signs include weight loss and halitosis (foul breath).

Diagnostic Tests

The larynx can be examined with a laryngeal mirror. Laryngoscopic examination and biopsy are used to diagnose and determine the stage of laryngeal cancer. A CT scan, MRI, or other diagnostic tests may be done to determine the presence or extent of metastasis.

Medical Treatment

If laryngeal cancer is diagnosed early in the disease, it may be treatable with radiation therapy; this treatment can preserve the patient's voice. Chemotherapy may be used with radiation or surgery, but usually it is not used alone. In more advanced cases, surgical intervention is necessary. The larynx will be either partially or completely removed (Fig. 27–2). If cancer has spread beyond the larynx, a radical neck dissection, which removes adjacent muscle, lymph nodes, and tissue, may be done. Surgery can be done using laser technology, endoscopy, or traditional methods. After a partial laryngectomy, the patient may have a permanently hoarse voice. If a total laryngectomy is done, the patient will have a permanent tracheostomy and no voice. Alternative methods of communication must be employed. A

| BOX 27-3 | Laryngeal Cancer Summary |

Symptoms

Hoarse voice
Pain
Cough
Shortness of breath
Difficulty swallowing
Weight loss
Foul breath

Diagnostic Tests

Examination with laryngeal mirror
Laryngoscopy with biopsy
Additional blood and radiographic studies to detect metastasis

Therapeutic Management

Radiation therapy
Chemotherapy (adjunct to radiation or surgery)
Endoscopic laser surgery to destroy tumor
Partial laryngectomy (preserves some voice)
Radical neck dissection with total laryngectomy (loss of voice)

person who has had a laryngectomy is sometimes referred to as a **laryngectomee.**

Nursing Process: The Patient Undergoing Laryngectomy

PREOPERATIVE CARE. In addition to routine preoperative teaching, the patient undergoing laryngectomy surgery must be prepared for the loss of ability to breathe through the mouth and nose and the loss of ability to speak. Initial instruction in communication techniques should take place before surgery to prevent the patient from feeling panicky

---

laryngectomee: larng—larynx + ectome—excision (person who has undergone laryngectomy)

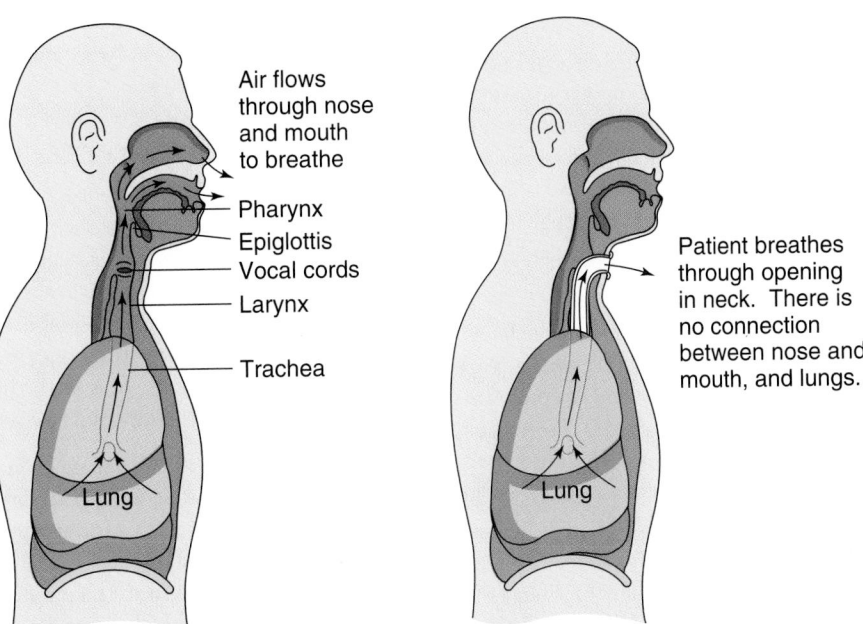

Air flows through nose and mouth to breathe

Pharynx
Epiglottis
Vocal cords
Larynx
Trachea
Lung

Patient breathes through opening in neck. There is no connection between nose and mouth, and lungs.

Lung

Figure 27-2   Laryngectomy. *(A)* Before laryngectomy. *(B)* After laryngectomy.

after surgery when he or she is unable to communicate any needs. A variety of techniques and devices are available. Consult the speech therapist before surgery to provide a picture board, magic slate, or paper and pencil. (See Chapter 46.) The patient is instructed to point to the picture that corresponds with the need or to write out his or her concern. A dietary consult is also important before surgery if the patient has been undernourished.

POSTOPERATIVE CARE

Assessment. Assessment of physical and psychosocial status, comfort, nutritional status, and ability to swallow is important both before and after surgery. After surgery, assessment of airway patency and respiratory function is vital. Monitor lung sounds, oxygen saturation, and arterial blood gases. In addition, be sure to assess the patient's understanding of the disease process and self-care needs after surgery. It is important to evaluate the patient's support systems and ability to cope with the partial or total loss of voice after surgery.

Nursing Diagnosis, Planning, Implementation, and Patient Education

Ineffective airway clearance related to excessive secretions and new tracheostomy. The patient will have a laryngectomy tube in place after surgery. A laryngectomy tube is shorter and has a larger diameter than a tracheostomy tube, but care is the same. Provide routine tracheostomy care and suctioning according to hospital policy to keep the airway clear. (See Chapter 26.) Strict sterile technique is essential. Place the patient in a semi-Fowler's position to make breathing easier, and encourage the patient to deep breathe and cough every hour. A special tracheostomy collar may be used to provide oxygen and humidification. Avoid use of powders, sprays, or other airborn materials near the patient. Monitor and record amount, color, and consistency of secretions; vital signs; oxygen saturation; lung sounds; and signs of respiratory distress. Report signs of infection or respiratory distress to the physician immediately.

Acute pain related to surgical procedure. Analgesics are given as ordered, on an around-the-clock basis or via a patient-controlled pump rather than as needed, for the first few days after surgery. If the liver has been damaged from alcohol abuse, dosages are adjusted by the physician. Narcotics are given carefully because they may reduce the cough reflex, which is vital to clearing the airway. Distraction and relaxation are also helpful.

Impaired verbal communication related to loss of vocal cords. After surgery, the nurse, speech therapist, and physician work together to provide the patient with a method of communication that best fits his or her needs (Fig. 27–3). Esophageal speech involves swallowing air and forming words as it is regurgitated back up the esophagus. Electronic devices are also available, which the patient places next to the neck or mouth. These devices use sound waves to help the patient form words. Another alternative is a surgically implanted voice prosthesis that creates a valve between the trachea and esophagus. If the patient holds a finger over the tra-

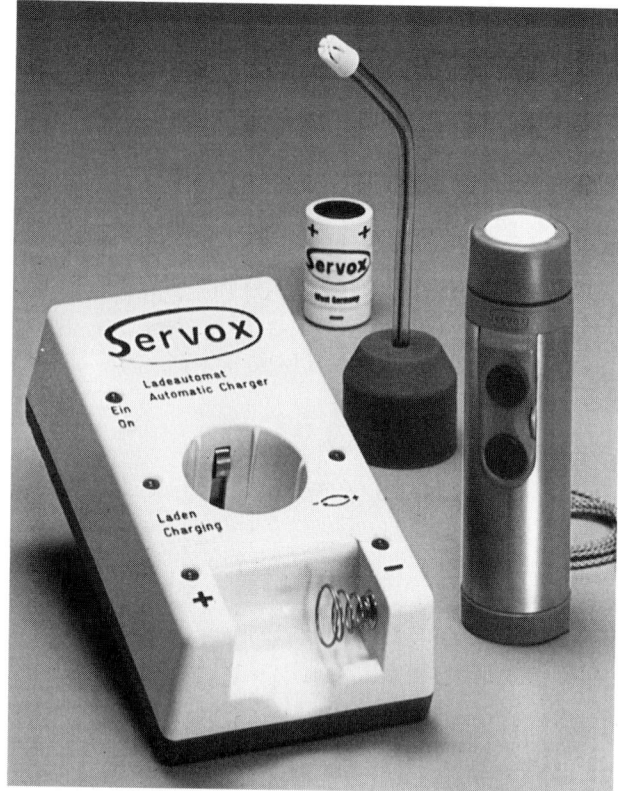

Figure 27–3 Devices to aid speech in laryngectomy patient. (Courtesy Siemens Hearing Instruments, New York.)

cheostomy, air is diverted into the esophagus and the patient forms words as the air exits via the mouth. All these devices take time to adjust to, and the patient will need support after discharge to continue to develop communication skills.

Imbalanced nutrition, less than body requirements related to absence of oral feeding. Most patients receive parenteral nutrition or tube feedings after surgery until the neck has begun to heal and swallowing can be evaluated. If the patient has a history of alcohol abuse, he or she may have been undernourished before surgery. You may need to advocate for the patient and ensure that he or she is receiving adequate calories for healing. A dietitian consult is essential.

Impaired swallowing related to edema or laryngectomy tube. The patient should be carefully assessed for the ability to swallow safely before oral feedings begin. The speech therapist may be consulted to assist with a swallowing assessment and recommendations. Assure the patient that aspiration will not occur because there is no longer a connection between the mouth and the lungs. Stay with the patient during the first attempts to eat to help alleviate anxiety.

Ineffective health maintenance related to knowledge deficit regarding self-care. The patient must be taught self-care measures for his or her laryngectomy, including how to perform tracheostomy care and suctioning. (See Chapter 26.) This is begun after an assessment of the patient's readiness to learn. Involve the significant other or family whenever possible. Referral to home nursing after discharge will provide assessment of the home environment, as well as follow-up instruction. A social service referral may be made for financial or psychosocial concerns if needed. Consult with the physician or check the local phone directory for laryngectomee support groups, and refer the patient to them if appropriate. The local branch of the American Cancer Society may also be able to provide information.

Grieving related to loss of voice. Inability to speak is a loss that cannot be overemphasized. Patients will grieve over this loss. In addition to voice loss, the patient may also be facing a career change if job-related exposure contributed to the disease or if loss of voice prevents returning to a previously held job. Assessing and involving support systems is important. The family is encouraged to communicate with the patient, even though it is often difficult, and to exercise patience as the patient learns to communicate with them. Local support groups may have names of people who have had similar experiences who are willing to visit with the patient. Of course, whenever referring to this type of visitor, it is important that the visitor has a positive attitude and has dealt effectively with a similar loss.

Disturbed body image related to change in body structure and function. In addition to changes in voice, the laryngectomee also now breaths through a hole in his or her neck. In time the patient may not need a tracheostomy tube, but the hole remains intact. This causes a change in body image. Again, a visitor who has been through this experience, and much support from staff and family, can be helpful. Portray an accepting attitude and allow the patient to share his or her feelings if the patient indicates a need to do so.

The patient must also be instructed to perform gentle range-of-motion exercises of the neck. Some patients may avoid extending the neck because of the location of the incision, causing muscle contracture and eventual inability to do so.

Many more nursing diagnoses may also be identified based on individual assessment, because of the many implications of this disease and surgery. To find more information for laryngectomees, visit the National Cancer Institute Web site www.cancer.gov. For information about ways laryngectomees can communicate, visit http://www.voice-center.com/larynx_ca.html. Many other Web sites can be found by using a search engine and to search for the term "laryngectomy."

Evaluation. When evaluating the patient's progress toward goals, ask the following questions: Does the patient verbalize an acceptable level of comfort? Is the airway clear, with-out signs of infection? Do the patient and significant others demonstrate understanding of self-care at home or have referrals to continue learning self-care at home? Does the patient indicate satisfaction with the level and quality of communication? Are nutritional needs met, as evidenced by albumin levels greater than 3.0 and stable weight? Is the patient able to swallow if taking oral nutrition? Is the patient able to grieve appropriately, and does the patient have someone to talk to if he or she wishes? Finally, does the patient show acceptance of the laryngectomy by learning to look at it and care for it? It should be noted that many of these evaluative criteria are long term and may not be seen while the patient is hospitalized, so follow-up by a home care nurse is essential.

Answers to CRITICAL THINKING

Mr. Jondahl

Consider the possibility of hypertension as a contributing factor. Mr. Jondahl's blood pressure is currently 140/90, which may be lower than normal for him because he has been bleeding. He is also on an antihypertensive drug and a diuretic. Explore the amount of ibuprofen being taken daily, because nonsteroidal anti-inflammatory drugs can prolong clotting time.

Mrs. Hiler

Mrs. Hiler may be swallowing blood. Examine the back of her throat with a flashlight. Check vital signs for evidence of impending shock. Notify a physician if bleeding is confirmed.
Use this formula to determine drops per minute: 100 mL/60 minute × 15 drops/mL = 25 drops/minute.

Murdie

1. Murdie may have contracted the flu from a visitor or a staff person at the nursing home. She is susceptible because of her age and comorbid conditions (COPD, diabetes).
2. Murdie's flu could probably have been prevented with a flu vaccination, but her son refused it because he believed it could cause her to get the flu.
3. If it is within 48 hours of symptom onset, a physician can prescribe an antiviral agent to help reduce her symptoms and shorten the course of her illness. In addition, you can provide fluids, acetaminophen, and comfort measures. You should also monitor her closely for evidence of bacterial infection or pneumonia, and report signs or symptoms immediately to a physician.
4. A major concern is that Murdie could transmit the flu to other residents or staff. Hopefully they have all been vaccinated. In addition, you must decide whether or not to send Murdie to the hospital. Check her advance directives, and talk to her son about goals for her care. If necessary, educate him about differences in nursing home and hospital care.

REVIEW QUESTIONS

1. Which of the following positions is recommended for a patient experiencing a nosebleed?
   a. Lying down with feet elevated
   b. Sitting up with neck fully extended
   c. Lying down with a small pillow under the head
   d. Sitting up leaning slightly forward

2. Which of the following is the best explanation by a nurse for why a physician did not prescribe antibiotics for influenza?
   a. "Most cases of influenza are caused by antibiotic-resistant bacteria."
   b. "Influenza is caused by viruses."
   c. "Antibiotics have too many serious side effects."
   d. "Antibiotics can interact with other medications used for influenza."

3. Which of the following responses is correct when a patient asks why her physician didn't order that new antiviral drug for her flu?
   a. "The antiviral drugs are for AIDS, not the flu."
   b. "The side effects of the antiviral drugs are worse than having the flu."
   c. "Antiviral drugs are only for children."
   d. "These drugs only work if you start them within 48 hours after flu symptoms start."

4. After a laryngectomy, which of the following assessments takes priority?
   a. Airway patency
   b. Nutritional status
   c. Lung sounds
   d. Patient acceptance of surgery

5. Which of the following communication methods will not work for the patient with a laryngectomy?
   a. Placing a finger over the stoma
   b. Providing a special valve that diverts air into the esophagus
   c. Obtaining a picture board
   d. Teaching the patient esophageal speech

6. Which of the following statements best explains why the nurse is careful when administering narcotics to a laryngectomee?
   a. Most laryngectomees have been drug addicts in the past.
   b. Even low doses of narcotics may cause respiratory arrest in the laryngectomee.
   c. Narcotics can depress the cough reflex.
   d. Laryngectomy patients have very little pain after surgery.

REFERENCES

1. Mainous, AG, et al: Antibiotics and upper respiratory infection. J Fam Pract 42:4, 1996.

28

NURSING CARE OF PATIENTS WITH LOWER RESPIRATORY TRACT DISORDERS

Paula D. Hopper

KEY TERMS

adjuvant (ad-**JOO**-vant)

anergy (**AN**-er-jee)

antitussive (**AN**-tee-TUSS-iv)

atelectasis (AT-e-**LEK**-tah-sis)

atypical (ay-**TIP**-i-kuhl)

bleb (**BLEB**)

bronchiectasis (BRONG-key-**EK**-tah-sis)

bronchitis (brong-**KIGH**-tis)

bronchodilator (BRONG-koh-**DYE**-lay-ter)

bronchospasm (**BRONG**-koh-spazm)

bulla (**BUHL**-ah)

compliance (kom-**PLIGH**-ens)

ectopic (ek-**TOP**-ik)

embolism (**EM**-boh-lizm)

emphysema (EM-fi-**SEE**-mah)

empyema (EM-pigh-**EE**-mah)

expectorant (ek-**SPEK**-tuh-rant)

exudate (**EKS**-yoo-dayt)

hemoptysis (hee-**MOP**-ti-sis)

hemothorax (HEE-moh-**THAW**-raks)

hypostatic (HIGH-poh-**STA**-tik)

immunocompromised
(IM-yoo-noh-**KAHM**-prah-mized)

induration (IN-dyoo-**RAY**-shun)

lobectomy (loh-**BEK**-tuh-mee)

mucolytic (MYOO-koh-**LIT**-ik)

paradoxical respiration
(PAR-uh-**DOK**-si-kuhl RES-pi-**RAY**-shun)

pleurodesis (PLOO-roh-**DEE**-sis)

pneumonectomy
(NEW-moh-**NEK**-tuh-mee)

pneumothorax (NEW-moh-**THAW**-raks)

polycythemia
(PAH-lee-sigh-**THEE**-mee-ah)

status asthmaticus
(**STAT**-us az-**MAT**-i-kus)

tachypnea (**TAK**-ip-NEE-uh)

thoracotomy (THAW-rah-**KAH**-tah-mee)

QUESTIONS TO GUIDE YOUR READING

1. How would you explain the pathophysiology of each of the disorders of the lower respiratory tract?

3. What are the etiologies, signs, and symptoms of each of the disorders?

4. What care would you provide for patients undergoing tests for each of the disorders?

5. What is current medical treatment for disorders of the lower respiratory tract?

6. What data should you collect when caring for patients with disorders of the lower respiratory tract?

7. What nursing care will you provide for patients with disorders of the lower respiratory tract?

8. What specific nursing care can you provide for patients experiencing impaired gas exchange, ineffective airway clearance, or ineffective breathing pattern?

9. How will you know if your nursing interventions have been effective?

Disorders of the lower respiratory tract include problems of the lower portion of the trachea, bronchi, bronchioles, and alveoli. These disorders may be related to infection, noninfectious alterations in function, neoplasm (cancer), or trauma. Any pathological condition of the lower respiratory tract can seriously impair carbon dioxide and oxygen exchange.

▶ INFECTIOUS DISORDERS

Acute Bronchitis

Bronchitis is an inflammation of the bronchial tree, which includes the right and left bronchi, secondary bronchi, and bronchioles. When the mucous membranes lining the bronchial tree become irritated and inflamed, excessive mucus is produced. The result is congested airways. Acute bronchitis is usually an isolated episode. It becomes chronic when it occurs more than 3 months out of the year for 2 consecutive years. See Chronic Bronchitis later in this chapter for more information.

Bronchiectasis
Pathophysiology

Bronchiectasis is a dilation of the bronchial airways (Fig. 28–1). The dilated areas form sacs that can remain localized or spread throughout the lungs. Secretions pool in these sacs and frequently become infected.

Etiology

Bronchiectasis usually occurs secondary to another chronic respiratory disorder, such as cystic fibrosis, asthma, tuberculosis, bronchitis, or exposure to a toxin. Infection and inflammation of the airways weakens the bronchial walls and reduces ciliary function. Airway obstruction from excessive secretions, then predisposes the patient to development of bronchiectasis.

Signs and Symptoms

The patient with bronchiectasis has recurrent lower respiratory infections. Sputum is copious and purulent and pools in the dilated airways. The accompanying cough can produce as much as 200 mL of thick, foul-smelling sputum in a single episode of coughing. Extreme airway inflammation may cause sputum to be bloody. If bronchiectasis is widespread throughout the lungs, the patient may experience dyspnea even with minimal exertion. Wheezes and crackles may be auscultated. Fever is present during active infection. Cor pulmonale (right-sided heart failure, covered in Chapter 21) and clubbing of the fingers may develop with chronic disease.

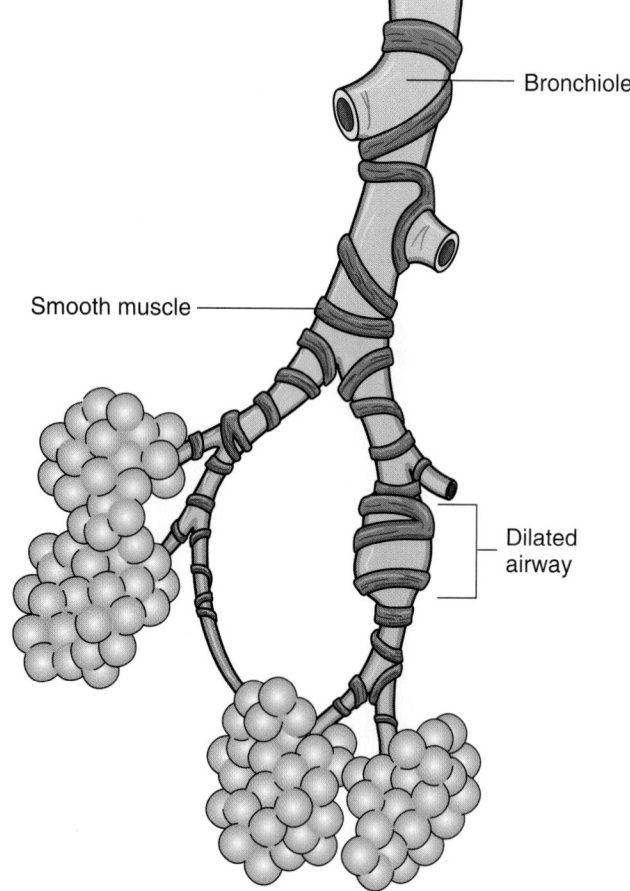

Figure 28-1   Bronchiectasis. Note dilated airway.

Diagnostic Tests

A chest x-ray examination may be done, but it may not show early disease. A computed tomography (CT) scan provides a better view of the dilated airways. Sputum cultures determine infecting organisms and guide antibiotic therapy. Additional testing is done to determine the cause of bronchiectasis.

Medical Treatment

Treatment of bronchiectasis is aimed at keeping the airway clear of secretions, controlling infection, and correcting the underlying problem. Antibiotics may be used intermittently or for prolonged periods. **Bronchodilators** improve airway obstruction. **Mucolytic** agents help thin secretions, and chest physiotherapy helps mobilize secretions so they can be more effectively expectorated. Oxygen is used if hypoxemia is present. If the affected area of the lung is localized and symptoms are severe, surgery may be considered to remove the diseased area.

---

bronchitis: bronch—airway + itis—inflammation
bronchiectasis: bronch—airway + ectasis—dilation or
   expansion

bronchodilator: broncho—airway + dilator—to expand
mucolytic: muco—mucus + lytic—break up

Pneumonia

Pneumonia is the cause of more than 10 percent of hospital admissions each year and is the most common cause of death from infection.

Pathophysiology

Pneumonia is an acute infection of the lungs, occurring when an infectious agent enters and multiplies in the lungs of a susceptible person. Infectious particles can come from the cough of an infected individual, from contaminated respiratory therapy equipment, from infections in other parts of the body, or from aspiration of bacteria from the mouth, pharynx, or stomach. Organisms from the mouth and pharynx may be related to poor oral hygiene or may be present because of a cold or influenza virus. In a healthy person, normal respiratory defense mechanisms and the immune system prevent the development of infection. In a person who is **immunocompromised,** however, even microorganisms that are normally present in the oropharynx can develop into an infection. Persons at risk for pneumonia are the very young, the elderly, and those who are immunocompromised, such as people with acquired immunodeficiency syndrome (AIDS) or another chronic illness.

When the microorganisms multiply, they release toxins that induce inflammation in the lung tissue, causing damage to mucous and alveolar membranes. This leads to the development of edema and **exudate,** which fills the alveoli and reduces the surface area available for exchange of carbon dioxide and oxygen. Some bacteria also cause necrosis of lung tissue.

Pneumonia may be confined to one lobe, or it may be scattered throughout the lungs. If it affects only one lobe, it is called lobar pneumonia. Generalized pneumonia is much more serious and is called bronchopneumonia. Bronchopneumonia occurs more often as a nosocomial (hospital-acquired) infection in hospitalized patients, the very young, or the very old.

Etiology

Pneumonia has a variety of causes, listed next.

BACTERIAL PNEUMONIA. The most common cause of bacterial pneumonia acquired in the community is *Streptococcus pneumoniae*. This is often termed pneumococcal pneumonia. This organism accounts for approximately 90 percent of all bacterial pneumonias. Other community-acquired infections are caused by *Staphylococcus aureus* and *Mycoplasma pneumonia*. Hospital-acquired pneumonias are often more serious and may be caused by *Escherichia coli*, *Haemophilus influenzae*, and *Pseudomonas aeruginosa*, among others.

VIRAL PNEUMONIA. Influenza viruses are the most common cause of viral pneumonia. The presence of viral pneumonia increases the patient's susceptibility to a secondary bacterial pneumonia. Generally, patients are less ill with viral pneumonia than with bacterial pneumonia, but they may be ill for a longer period because antibiotics are ineffective against viruses.

FUNGAL PNEUMONIA. *Candida* and *Aspergillus* are two types of fungi that can cause pneumonia. *Pneumocystis carinii* is a fungus that typically causes pneumonia in patients with AIDS.

ASPIRATION PNEUMONIA. Some pneumonias are caused by aspiration of foreign substances. This most often occurs in patients with decreased levels of consciousness or an impaired cough or gag reflex. These conditions can occur with alcohol ingestion, stroke, general anesthesia, seizures, or other serious illness. Aspiration pneumonia increases the risk for subsequent bacterial pneumonia.

HYPOSTATIC PNEUMONIA. Patients who hypoventilate because of bedrest, immobility, or shallow respirations are at risk for **hypostatic** pneumonia. Secretions pool in dependent areas of the lungs and can lead to inflammation and infection.

CRITICAL THINKING

Mr. Smith

Mr. Smith is an 86-year-old gentleman who was watching television when he couldn't sleep one night. After seeing a commercial for toilet cleaner, he decided his own toilet could use some attention. He used bleach and ammonia "to get it really clean." The combination created toxic fumes, which caused a severe chemical pneumonia. He was brought to the emergency room in acute respiratory distress.

1. As his nurse, what questions might you ask as you further assess the cause of his pneumonia?
2. What will you teach Mr. Smith related to prevention of similar episodes in the future?

Answers at end of chapter.

CHEMICAL PNEUMONIA. Inhalation of toxic chemicals can cause inflammation and tissue damage, which can lead to chemical pneumonia.

Prevention

A vaccine is available to help prevent *Streptococcus pneumoniae* pneumonias in high-risk patients and people older than age 65. It is effective about 80 to 90 percent of the time and requires only a one-time injection. Some individuals may require a repeat vaccination after 5 years. Yearly influenza vaccination is also recommended for high-risk individuals.

---

immunocompromised: immune—referring to immune system + compromised—lacks resistance
exudate: to sweat out

---

hypostatic: hypo—below + static—standing

Nursing care plays an important role in the prevention of nosocomial pneumonia. Regular coughing and deep breathing for patients on bedrest or after surgery, prevention of aspiration for patients at risk, and good handwashing practices by health care personnel can help prevent many cases (Gerontological Issues Box 28–1).

BOX 28-1   GERONTOLOGICAL ISSUES

Advanced age is a significant risk factor for serious complications from respiratory infections such as influenza and pneumococcal pneumonia. Therefore it is recommended that people over the age of 65 and individuals with chronic disease have yearly influenza vaccines and a once-in-a-lifetime pneumococcal vaccine.

Signs and Symptoms

Patients with pneumonia present with fever, shaking chills, chest pain, dyspnea, and a productive cough. Sputum is purulent or may be rust colored or blood tinged. Crackles and wheezes may be heard on lung auscultation.

Some bacterial and many viral pneumonias cause **atypical** symptoms. The patient may experience fatigue, sore throat, dry cough, or nausea and vomiting.

Elderly patients may not exhibit expected symptoms of pneumonia. New-onset confusion or lethargy in an elderly patient can indicate reduced oxygenation and should alert you to look for other symptoms or request further testing. New onset of fever or dyspnea should also cause suspicion of possible pneumonia in the elderly.

Complications

Complications from pneumonia most commonly occur in patients with other underlying chronic diseases. Pleurisy and pleural effusion (discussed later in this chapter) are two of the most common complications and generally resolve within 1 to 2 weeks. **Atelectasis** (collapsed alveoli) can occur as a result of trapped secretions and may be resolved with efforts to keep the airways clear. Other complications result from spread of infection to other parts of the body, causing meningitis, septic arthritis, pericarditis, and endocarditis. Treatment for each of these is antibiotics. Although antibiotics have greatly reduced the incidence of death related to pneumonia, it is still a common cause of death in the elderly.

Diagnostic Tests

A chest x-ray examination is done to identify the presence of pulmonary infiltrate, which is fluid leakage into the alveoli from inflammation (Fig. 28–2). In addition, sputum and blood cultures are obtained to identify the organism causing the pneumonia and determine appropriate treatment. Cultures should be obtained before antibiotics are started to avoid altering culture results. If the patient is unable to pro-

---

atypical: a—not + typical—usual
atelectasis: atel—imperfect + ectasis—expansion

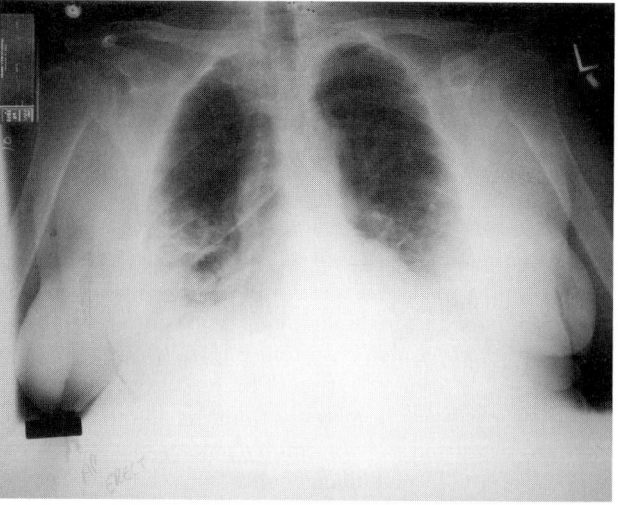

Figure 28-2   Chest x-ray examination showing infiltrates in pneumonia.

duce a sputum specimen, a nebulized mist treatment may be ordered to promote sputum expectoration. If this is unsuccessful, a bronchoscopy may be done to obtain a specimen from the very ill patient.

Medical Treatment

Broad-spectrum antibiotics are initiated before culture results are completed (be sure to obtain the specimen before starting the antibiotics). Once the culture and sensitivity report is available, specific antibiotics are ordered if the cause is bacterial. Many patients can be treated with oral antibiotics as outpatients, but hospitalization and intravenous (IV) therapy may be necessary in the elderly, chronically ill, or acutely ill individual.

Expectorants, bronchodilators, and analgesics may be given for comfort and symptom relief. Nebulized mist treatments or metered dose inhalers may be used to deliver bronchodilators. Supplemental oxygen via nasal cannula or mask is used as necessary.

Tuberculosis
Pathophysiology

Tuberculosis (TB) is an infectious disease caused by the *Mycobacterium tuberculosis* bacteria. TB primarily affects the lungs, although other areas, such as the kidneys, liver, brain, and bone, may be affected as well. M. *tuberculosis* is an acid-fast bacillus (AFB), which means that when it is stained in the laboratory and then washed with an acid, the stain remains, or stays fast. M. *tuberculosis* can live in dark places in dried sputum for months, but a few hours in direct sunlight kills it. It is spread by inhalation of the tuberculosis bacilli from respiratory droplets (droplet nuclei) of an infected person.

Once the bacilli enter the lungs, they multiply and begin to disseminate to the lymph nodes and then to other parts of the body. The patient is then infected but may or may not go on to develop clinical (active) disease. During this time the body develops immunity, which keeps the infection under control. The immune system surrounds the infected lung area with neutrophils and alveolar macrophages. This process creates a lesion called a tubercle, which seals off the bacteria and prevents spread. The bacteria within the tubercle die or become dormant, and the patient is no longer infectious. If the patient's immune system becomes compromised, however, some of the dormant bacteria can become active again, causing reinfection and active disease. Only 5 to 10 percent of infected individuals in the United States actually develop the disease, and even then it may not occur for many years (Gerontological Issues Box 28–2).

BOX 28-2   GERONTOLOGICAL ISSUES

The reduction in immune system function from aging can decrease the effectiveness of the tuberculosis antibodies in someone who previously had dormant disease. The tuberculosis bacilli can be reactivated, causing active disease.

LEARNING TIPS

If the physician orders "sputum culture for AFB," tuberculosis is suspected. Ask whether isolation precautions should be taken while waiting for culture results.

Etiology

Crowded or poorly ventilated living conditions place people at risk for becoming infected with tuberculosis. Although tuberculosis can infect any age group, the elderly are especially at risk. Elders may have contracted the disease many years before, but it reactivates as the aging process diminishes immune function. Patients with AIDS and chronic alcohol abuse have a very high risk because of their compromised immune function. In the United States, tuberculosis is also prevalent among the urban poor and minority groups.

Before 1985 the incidence of TB was steadily decreasing. Now it is again on the rise, in part because of the prevalence of AIDS, the development of antibiotic-resistant strains of the TB bacillus, and ineffective treatment programs. TB kills 2 million people each year worldwide.[1]

Prevention

Clean, well-ventilated living areas are essential to the health of all people. If a hospitalized patient is known or suspected to have tuberculosis, he or she is placed in respiratory isolation to prevent spread to staff or other patients. Special isolation rooms are ventilated to the outside. Staff should wear special high-efficiency filtration masks when in the patient's room. A regular surgical mask is *not* effective against TB. Verify with the institution's infection control department that the masks provided are effective for use with TB patients. If the patient must travel through the hallway for tests or other activities, the patient must wear a mask. Additional protective barriers, such as gowns, gloves, or goggles, are used when contact with sputum is likely.

A vaccine against tuberculosis is available and is used in areas where TB is prevalent. It is safe, but its effectiveness has been questioned. It is not used routinely in the United States.

Ultimately prevention will come from adequate treatment of patients with TB. A current concern is the development of antibiotic-resistant strains of the tuberculosis bacillus, which develop when patients are noncompliant with drug therapy. When antibiotics are taken intermittently or discontinued early, the more virulent (stronger) bacteria survive and multiply and are resistant to the drugs being used. This drug-resistant bacteria can then be passed on to someone else. It is therefore vital to teach all patients the importance of strict compliance with drug therapy. Patients who are noncompliant with drug therapy must have a visiting nurse or other health professional observe each dose of antibiotic taken. This is called directly observed therapy (DOT). The World Health Organization currently recommends a 6- to 8-month course of treatment, with the first 2 months being directly observed.

Signs and Symptoms

Active tuberculosis is characterized by a chronic productive cough, blood-tinged sputum, and drenching night sweats. A low-grade fever may be present. If effective treatment is not initiated, a downhill course occurs, with pulmonary fibrosis, **hemoptysis,** and progressive weight loss.

Complications

Spread of the tuberculosis bacilli throughout the body can result in pleurisy, pericarditis, peritonitis, meningitis, bone and joint infection, genitourinary or gastrointestinal infection, or infection of many other organs.

Diagnostic Tests

Routine screening for tuberculosis infection is usually done with a purified protein derivative (PPD) skin test. The PPD is injected intradermally; the test is considered positive if a raised area of **induration** occurs within 48 to 72 hours. If there is a red area around the induration, this is not measured. The size of induration that indicates a positive test varies based on the individual's history (Table 28–1). A red area without induration is not considered a positive result. A positive result indicates that a person has been exposed; it does not mean that active TB disease is present.

Some health care institutions use a two-step process for baseline testing of employees and residents. If an individual has a negative test, he or she is retested in 1 to 3 weeks. This is because someone who was exposed many years ago may not react to the first test. The first test acts as a "reminder" to the immune system to react. The second test will then be positive in the person with a past TB infection.

---

hemoptysis: hem—blood + ptysis—to spit
induration: in—in + durus—hard

TABLE 28-1   CLASSIFYING A TUBERCULIN SKIN TEST REACTION

Size of Induration	Considered Positive For
5 mm or more	• People with HIV infection • Close contacts • People who have had TB disease before • People who inject illicit drugs and whose HIV status is unknown
10 mm or more	• Foreign-born persons • HIV-negative persons who inject illicit drugs • Low-income groups • People who live in residential facilities • People with certain medical conditions • Children younger than age 4 • People in other groups, as identified by local public health officials
15 mm or more	• People with no risk factors for TB

Source: Retrieved August 5, 2002, from Centers for Disease Control and Prevention: Diagnosis of tuberculosis infection and disease. www.phppo.cdc.gov.
HIV = human immunodeficiency virus.

A chest x-ray examination is used as a screening tool in someone with a known positive test. Diagnosis is made based on sputum culture results.

LEARNING TIPS

Some institutions use *Candida* or mumps skin tests along with a PPD skin test. This does not mean the patient is being tested for *Candida* or mumps, because everyone generally reacts to these. Rather, the patient is being tested for **anergy,** or the inability of the immune system to react to an antigen. If the *Candida* or mumps tests produce positive results, the TB results are considered to be reliable.

The *Candida* or mumps test is administered in the same way as the PPD. If more than one test is administered, use an indelible marker to identify which test is which, and clearly document (draw a picture) of where each test was placed. Check institution policy; it may direct where the tests are administered—PPD in the right arm and *Candida* in the left, for example.

Medical Treatment

Treatment consists of specific antibiotic therapy. First-line drugs (Box 28–3) have the fewest adverse effects. However, these drugs can be toxic to the liver and nervous system, as well as having other side effects. Second-line drugs are more toxic and are reserved for cases that do not respond to first-line drug therapy. Generally two or three antibiotics are given simultaneously to allow lower doses of each individual drug and to reduce the incidence of serious side effects. Drugs must be taken for 6 to 8 months or longer. Because of the length of therapy and the incidence of side effects, you must anticipate that compliance may be a problem.

BOX 28-3   Antibiotics Used in Treatment of Tuberculosis

First-line Drugs	Second-line Drugs
Isoniazid	Ethionamide
Rifampin	Kanamycin
Streptomycin	Para-aminosalicylic acid
Ethambutol	Cycloserine
Pyrazinamide	

Nursing Care

Perform thorough respiratory and psychosocial assessments of the patient with TB. The severity of the disease determines the impact on the patient's lifestyle. It is also imperative to determine the patient's knowledge of the disease and treatment and his or her compliance with drug treatment.

Possible nursing diagnoses include impaired gas exchange, ineffective airway clearance, ineffective breathing pattern, anxiety, imbalanced nutrition, risk for infection of patient's contacts, and possible noncompliance with drug therapy or ineffective therapeutic regimen management. Diagnoses should be chosen based on individual patient data.

Nursing interventions for impaired gas exchange, airway clearance, and breathing pattern are found in Nursing Care Plan Box 28–4 for the Patient with a Lower Respiratory Tract Disorder. Anxiety may be reduced by educating the patient in self-care measures and by reassuring the patient that the disease can be controlled by careful compliance with treatment. The patient who is emaciated because of the disease will benefit from a dietitian consultation to provide specific recommendations or supplements. To prevent spread of infection to others, teach the patient to use a tissue to cover the mouth and nose when coughing or sneezing. Tissues should be flushed down the toilet or disposed of carefully in the trash. Teach all family members the importance of careful handwashing, how to manage drug therapy, and when to report side effects. Forewarn the patient that rifampin turns the urine and body fluids red.

A visiting nurse is essential to evaluate the home environment and assess the patient's ability to comply with therapy. If the patient is unable to comply with therapy, measures must be instituted to ensure that medications are taken to protect both the patient and the public. Directly observed therapy at a local health clinic or by a home health nurse may be necessary.

The patient will be followed periodically by the physician for sputum cultures and drug monitoring. Once sputum cultures are negative, the patient is no longer contagious.

The Centers for Disease Control and Prevention has an excellent Web site with lots of information about tuberculosis at www.cdc.gov. Simply type "tuberculosis" into the search window.

BOX 28-4 **NURSING CARE PLAN FOR THE PATIENT WITH A LOWER RESPIRATORY TRACT DISORDER**

Note: The three most commonly used nursing diagnoses related to respiratory disorders are presented in the following care plan. This is not a care plan for any one respiratory disorder. Rather, the student should use it as a reference for use when one of the nursing diagnoses applies to the patient, based on a thorough respiratory assessment.

 Impaired gas exchange related to decreased ventilation or perfusion

PATIENT OUTCOMES

The patient will experience improved gas exchange, as evidenced by (1) improving arterial blood gases or pulse oximetry and (2) statement of acceptable level of dyspnea.

EVALUATION OF OUTCOMES

(1) Are blood gases or SaO$_2$ improving? (2) Does patient state that dyspnea is gone or controlled at an acceptable level?

INTERVENTIONS	RATIONALE	EVALUATION
• Assess lung sounds, respiratory rate and effort, use of accessory muscles.	Respiratory rate less than 12 or more than 24 or use of accessory muscles indicates distress. Diminished lung sounds indicate possible poor air movement and impaired gas exchange.	Are lung sounds clear and audible? Is respiratory rate 12 to 20 per minute and unlabored?
• Observe skin and mucous membranes for cyanosis.	Cyanosis indicates poor oxygenation. Oral mucous membrane cyanosis indicates serious hypoxia.	Are skin and mucous membranes pink?
• Assess degree of dyspnea on a scale of 0 to 10, 0 = no dyspnea, 10 = worst dyspnea.	The patient's subjective report is the best measure of dyspnea.	Is patient's degree of dyspnea within parameters that are acceptable to patient?
• Monitor for confusion or changes in mental status.	Changes in mental status can signal impaired gas exchange.	Is patient alert and oriented? If not, could poor gas exchange be the reason?
• Monitor arterial blood gas values and pulse oximetry as ordered.	PaO$_2$ < 80 mm Hg, PaCO$_2$ > 45 mm Hg, or SaO$_2$ < 90 indicate impaired gas exchange.	Are values within patient's baseline values?
• Elevate head of bed or help patient to lean on overbed table.	Upright positioning promotes lung expansion.	Did change of position relieve some distress?
• Position with good lung dependent ("good lung down").	This position allows the healthier lung to be better perfused and increases gas exchange.	Is SaO$_2$ improved in this position?
• Administer supplemental oxygen at ≤ 2 L/min unless ordered otherwise.	Supplemental oxygen decreases hypoxia. Rates more than 2 L/min can depress hypoxic drive.	Is oxygen placed properly on patient? Does it provide relief from dyspnea?
• Teach patient relaxation exercises.	Relaxation exercises decrease perceived dyspnea.	Does patient use relaxation effectively?
• Teach patient diaphragmatic and pursed-lip breathing. (See Chapter 26.)	Breathing exercises promote relaxation and increase CO$_2$ excretion.	Does patient use breathing exercises correctly? Do they help?
• Encourage patient to stop smoking if patient is a current smoker.	Smoking is damaging to lungs and respiratory function.	Is patient receptive to smoking cessation? Are resources available?
• For severe dyspnea, ask physician about an order for intravenous morphine sulfate.	Low doses of IV morphine cause vasodilation, which helps relieve pulmonary edema and anxiety.	Does morphine provide relief from dyspnea?

 Ineffective airway clearance related to excessive secretions

PATIENT OUTCOMES

The patient will have improved airway clearance as evidenced by (1) clear breath sounds and (2) ability to cough up secretions.

EVALUATION OF OUTCOMES

(1) Are breath sounds clear? (2) Is patient able to effectively cough up and expectorate secretions?

INTERVENTIONS	RATIONALE	EVALUATION
• Assess lung sounds q4h and prn.	Crackles and wheezes may indicate excess secretions in airways.	Do lung sounds indicate retained secretions?
• Monitor amount, color, and consistency of sputum.	Thick, purulent sputum indicates infection and should be reported to the physician.	Does sputum indicate infection?
• Encourage oral fluids; use cool steam room humidifier.	Hydration decreases viscosity of secretions and aids expectoration.	Is patient able to take oral fluids? Are secretions thin and easily expectorated?

• Turn patient q2h or encourage to ambulate if able.	Movement mobilizes secretions.	Is patient mobile?
• Encourage patient to cough and deep breathe every hour and prn.	Controlled coughing following deep breaths is more effective.	Does patient cough and deep breathe effectively?
• Administer expectorants as ordered.	Expectorants help liquefy secretions and trigger the cough reflex.	Are expectorants effective?
• If patient is unable to cough up secretions, suction per institution policy.	Suctioning is necessary to remove secretions when the patient is unable to cough effectively.	Is suctioning necessary? Does it help remove secretions?
• Obtain order for chest physiotherapy or flutter valve if indicated.	Percussion and postural drainage help mobilize secretions.	Is chest physiotherapy effective and well tolerated by the patient?

▶ **Ineffective breathing pattern related to anxiety or pain**

PATIENT OUTCOMES

The patient will maintain an effective breathing pattern as evidenced by (1) respiratory rate between 12 and 20 per minute, even, and unlabored; and (2) arterial blood gas and oxygen saturation results within patient's normal range.

EVALUATION OF OUTCOMES

(1) Is patient's respiratory rate within normal limits and unlabored? (2) Does breathing pattern support normal blood gas and SaO_2 values?

INTERVENTIONS	RATIONALE	EVALUATION
• Assess respiratory rate, depth, and effort q4h and prn.	Respirations less than 12 or more than 20 may indicate an ineffective pattern.	Is respiratory pattern ineffective?
• Monitor blood gas and oxygen saturation values.	An ineffective breathing pattern will not maintain oxygenation.	Is breathing pattern adversely affecting oxygenation?
• Determine and treat the cause of ineffective breathing pattern.	Pain or anxiety can cause a patient to change the breathing pattern, and should be treated.	Is a contributing factor identifiable and correctable?
• Place patient in Fowler's or semi-Fowler's position.	This allows for maximum chest expansion.	Is the patient in a comfortable position that enables the patient to breathe effectively?
• Teach patient to use diaphragmatic breathing, with a regular 2 second in, 4 second out pattern.	Breathing exercises promote relaxation and increase CO_2 excretion.	Is the patient able to demonstrate an effective breathing pattern?

▶ ## NURSING PROCESS: THE PATIENT WITH A LOWER RESPIRATORY INFECTION

ASSESSMENT. Question the patient about a history of respiratory disorders, smoking history, and symptoms of the current illness using the WHAT'S UP? format in Chapter 26. Ask about dyspnea, activity tolerance, cough, and sputum presence, including amount, consistency, and color. Whenever possible, observe the sputum. Ask the patient to rate the degree of dyspnea on a scale of 0 to 10. Physical assessment includes vital signs, lung sounds, color of skin and mucous membranes (including presence of cyanosis), use of accessory muscles, and observation of adaptive measures such as positioning to ease respiratory distress.

NURSING DIAGNOSIS. Possible nursing diagnoses include ineffective airway clearance related to increased mucus production, impaired gas exchange related to reduced ventilation, hyperthermia, and activity intolerance related to dyspnea and malaise.

PLANNING. Nursing care is aimed at relieving symptoms and promoting comfort. The airway should be free of secretions, dyspnea and fever should be controlled, and the patient should receive assistance with activities of daily living (ADL) until activity tolerance improves.

IMPLEMENTATION. Oral or intravenous fluids, a room humidifier, expectorants, and reminders to cough and deep breathe every 1 to 2 hours help the patient to raise sputum. Chest physiotherapy may be ordered to help loosen secretions if necessary. Pain control will assist the patient in taking deep breaths and coughing. Suction is used if the patient is unable to cough effectively.

Place the patient in Fowler's or semi-Fowler's position to promote lung expansion and reduce dyspnea. Administer oxygen and antibiotics as ordered. Provide acetaminophen as ordered for fever and malaise. Offer assistance with ADLs and allow frequent rest periods until the patient is able to tolerate an increase in activity. (See Nursing Care Plan Box 28–4 for the Patient with a Lower Respiratory Tract Disorder for additional interventions.)

PATIENT EDUCATION. Instruct the patient in use of medications at home and the importance of finishing the antibiotic prescription. Explain the benefit of pacing activities to prevent fatigue and dyspnea and teach measures to prevent recurrence of infection. The patient should also be aware of signs and symptoms to report to the physician.

EVALUATION. If the plan has been effective, dyspnea will be controlled at a level that is acceptable to the patient. The patient will be able to effectively cough up secretions, which will become progressively less purulent. The patient will verbalize correct understanding of medication use at home, and activity tolerance will improve.

CRITICAL THINKING

Jim

Jim is a 36-year-old accountant with bronchiectasis secondary to cystic fibrosis. You enter his room during an episode of uncontrollable coughing and offer him support. You observe his sputum as you dispose of it—a whole Styrofoam coffee cup full of thick, bright yellow sputum; the smell makes you nauseous. Even after coughing, his lungs sound congested from retained secretions. You offer him mouth care before you leave his room.

1. What questions can you ask Jim to assess his cough?
2. What nursing diagnosis is most appropriate for Jim?
3. What nursing care can you provide to enhance secretion removal?
4. You have an order for guaifenesin 300 mg q4h prn. It is supplied as 200 mg per 5 mL. How much will you administer?
5. How would you document this episode of coughing?

Answers at end of chapter.

▶ RESTRICTIVE DISORDERS

Restrictive disorders are those problems that limit the ability of the patient to expand his or her lungs. These are caused by a decrease in the **compliance** (or elasticity) of the lungs or chest wall.

Pleurisy (Pleuritis)
Pathophysiology

Recall that the visceral and parietal pleura are the membranes that surround the lungs. Between these membranes is a serous fluid that prevents friction as the pleurae slide over each other during respiration. If the membranes become inflamed for any reason, they do not slide as easily. Instead of sliding, one membrane may "catch" on the other, causing it to stretch as the patient attempts to inspire. This causes the characteristic sharp pain on inspiration. The irritation causes an increase in the formation of pleural fluid, which in turn reduces friction and decreases pain.

Etiology

Pleurisy is usually related to another underlying respiratory disorder, such as pneumonia, tuberculosis, tumor, or trauma.

Signs and Symptoms

Pleurisy causes a sharp pain in the chest on inspiration. Pain also occurs during coughing or sneezing. Breathing may be shallow and rapid, because deep breathing increases pain. The patient may also exhibit fever, chills, and an elevated white blood cell count if the cause is infectious. A pleural friction rub is heard on auscultation.

Complications

As pleural membranes become more inflamed, serous fluid production increases, which may result in pleural effusion (see next section). If pleuritic pain is not controlled, patients have difficulty breathing deeply and coughing, which may lead to atelectasis. If infection goes untreated, empyema can result.

Diagnostic Tests

Diagnosis is based on signs and symptoms, including auscultation of a pleural friction rub. A chest x-ray examination and complete blood cell count (CBC) may be done. Additional testing is done to determine the underlying cause.

Medical Treatment

Treatment is aimed at correcting the underlying cause. Narcotics are given to control pain and facilitate deep breathing and coughing. The physician may perform a nerve block, injecting anesthetic near the intercostal nerves to block pain transmission.

Pleural Effusion
Pathophysiology

When excess fluid collects in the pleural space, it is called a pleural effusion. Fluid normally enters the pleural space from surrounding capillaries and is reabsorbed by the lymphatic system. When a pathological condition causes an increase in fluid production or inadequate reabsorption of fluid, excess fluid collects. A normal amount of pleural fluid around each lung is 1 to 15 mL. More than 25 mL of fluid is considered abnormal; as much as several liters of fluid can collect at one time. The effusion can be either transudative, forming a watery fluid from the capillaries, or exudative, with fluid containing white blood cells and protein from an inflammatory process.

Etiology

Like pleurisy, pleural effusion is generally caused by another lung disorder. It is a symptom rather than a disease. Transudative effusions may result from heart failure, liver disorders, or kidney disorders. Exudative effusions more commonly occur with lung cancer, infection, or inflammation.

Signs and Symptoms

Symptoms depend on the amount of fluid in the pleural space. The patient may or may not experience pleuritic pain. Increasing shortness of breath occurs because of the decreasing space for lung expansion. Cough and tachypnea may be present. A dull sound is heard when the affected area is percussed. Lung sounds are decreased or absent over the effusion, and a friction rub may be auscultated.

Diagnostic Tests

A chest x-ray examination is done to determine whether pleural effusion is present. If a thoracentesis is done, fluid samples are sent to the laboratory for culture and sensitivity and cytological examination. Further tests may be done to determine the cause of the effusion.

Medical Treatment

Bedrest is recommended to enhance spontaneous resolution of the effusion. If symptoms are severe, a therapeutic thoracentesis is done to remove the excess fluid from the pleural space and relieve the patient of dyspnea. See Chapter 26 for how to assist with a thoracentesis. The physician will use x-ray examinations and percussion to determine where to insert the needle to obtain the fluid. If the fluid accumulation is large or recurring, a chest tube might be placed to continuously drain the pleural space. Occasionally talc or another irritating agent will be instilled via the chest tube to cause the pleural membranes to adhere to each other, eliminating the pleural space and preventing future episodes of pleural effusion. Treatment of the underlying cause of the effusion is necessary to prevent recurrence.

Empyema

Empyema is the collection of pus in the pleural space. It is a pleural effusion that is infected. Empyema is usually a complication of pneumonia, tuberculosis, or lung abscess.

Symptoms, diagnosis, medical treatment, and nursing care are the same as the care of the patient with a pleural effusion, with an added emphasis on resolving the infection. A chest tube or surgery may be necessary to drain the area.

Atelectasis

Atelectasis is the collapse of alveoli. It most commonly occurs in postsurgical patients who do not cough and deep breathe effectively, although it can be caused by anything that causes hypoventilation. Areas of the lungs that are not well aerated become plugged with mucus, which prevents inflation of alveoli. As a result, alveoli collapse. Compression of lung tissue from effusion or a tumor can also cause atelectasis. The focus of nursing care is on prevention. Patients should be taught the importance of coughing and deep breathing whenever there is the risk for hypoventilation. Frequent position changes and ambulation are also important.

▶ NURSING PROCESS: THE PATIENT WITH A RESTRICTIVE DISORDER

ASSESSMENT. Perform a routine respiratory assessment. Monitor lung sounds for friction rub or decreasing breath sounds in any of the lobes. Assess pain level. Promptly report any increase in dyspnea, changes in vital signs or pulse oximetry, or increased white blood cell count or temperature.

NURSING DIAGNOSIS. Most patients experience fear or anxiety when they feel dyspneic. Impaired gas exchange occurs if atelectasis or hypoventilation are present. Risk for infection occurs when there is stasis of pleural fluid. Ineffective breathing pattern is identified when the patient is unable to take a deep breath due to pain on inspiration or buildup of fluid.

PLANNING AND IMPLEMENTATION. Stay with the patient during acute dyspneic episodes and encourage an effective breathing pattern. (See Chapter 26.) Analgesics reduce pain so that the patient will be better able to take deep breaths, though opioids are used cautiously because of their ability to depress respirations. Encourage coughing and deep breathing to help prevent infection and atelectasis. Administer oxygen as ordered. Bedrest is maintained during acute episodes of dyspnea. Encourage fluids unless contraindicated. Instruct the patient to report signs and symptoms of recurrence promptly to the physician. See Chapter 26 for care of the patient undergoing a thoracentesis.

EVALUATION. If interventions have been effective, the patient should report a decrease in dyspnea and anxiety. Pain will be controlled so that the patient is able to take deep breaths and cough effectively, and the patient will be free of signs and symptoms of infection.

▶ CHRONIC OBSTRUCTIVE PULMONARY DISEASE/CHRONIC AIRFLOW LIMITATION
Pathophysiology

Chronic obstructive pulmonary disease (COPD) is a group of pulmonary disorders characterized by difficulty exhaling because of narrowed or blocked airways. More effort is required to push air out through obstructed airways (Fig. 28–3). **Emphysema,** chronic bronchitis, and asthma are disorders that limit airflow. A patient with COPD often has some degree of both emphysema and chronic bronchitis. Asthma may also be present, but it differs somewhat because the airway constriction in asthma is reversible, whereas the airflow limitation in emphysema and bronchitis is not. Generally either emphysema or bronchitis predominates. COPD may also be referred to as chronic airflow limitation (CAL) or chronic obstructive lung disease (COLD). COPD develops over at least 30 years before symptoms become evident and may be advanced by the time the patient seeks treatment. Box 28–5 summarizes COPD.

Chronic Bronchitis Pathophysiology

Chronic bronchitis is similar to acute bronchitis, with symptoms occurring for at least 3 months out of the year for 2 consecutive years. The bronchial tree becomes inflamed from inhaled irritants, and impaired ciliary function reduces the ability to remove the irritants. The mucous-producing glands in the airways become hypertrophied, producing excessive thick, tenacious mucus, which obstructs airways and

emphysema: to inflate

AIR TRAPPING IN CHRONIC AIRFLOW LIMITATION

A. Air trapping from excess mucous

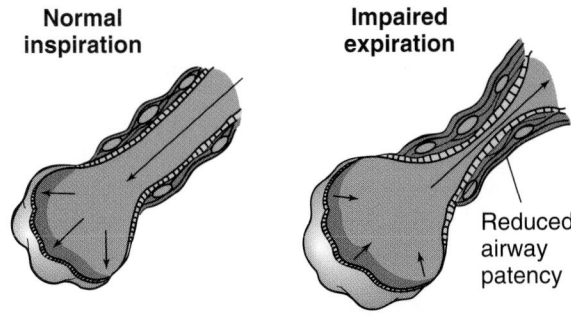

Air movement during inspiration

Air trapping during expiration

Mucous

Muscle

Alveolar wall

B. Air trapping from decreased elastic recoil

Normal inspiration

Impaired expiration

Reduced airway patency

Air trapped due to decreased elastic recoil of alveolus and collapsed airway

Figure 28-3 Air trapping in COPD.

---

traps air (Fig. 28–4). These changes lead to chronic low-grade infection.

Emphysema Pathophysiology

Emphysema affects the alveolar membranes, causing destruction of the alveolar walls and loss of elastic recoil. This also causes damage to adjacent pulmonary capillaries. Because of the loss of elastic recoil, passive expiration is impaired and air is trapped in the alveoli. Reduction in pulmonary capillaries reduces gas exchange. Emphysema can occur primarily in the respiratory bronchioles (centrilobular emphysema), with delayed alveolar damage, or in the respiratory bronchioles and alveoli (panlobular emphysema) (Fig. 28–5).

Etiology

Smoking is the single most important risk factor for COPD. Other factors include passive (secondhand) smoking, air pollution, and exposure to industrial chemicals. Some familial predisposition to chronic bronchitis has been demonstrated. A small number of individuals have an inherited deficiency of the enzyme alpha-antitrypsin (α_1AT), which causes a predisposition to the development of emphysema.

BOX 28-5 **COPD Summary**

Signs and Symptoms

Cough, may be productive
Dyspnea
Activity intolerance
Crackles, wheezes, diminished breath sounds
Barrel chest
Use of accessory muscles
Elevated PaCO$_2$, decreased PaO$_2$

Diagnostic Tests

Chest x-ray exam, CT scan
Arterial blood gas analysis (later in disease process)
CBC
Sputum analysis
Pulmonary function studies
α_1AT level if deficiency suspected

Therapeutic Management

Smoking cessation
Bronchodilators (PO, NMT, MDI)
Corticosteroids, expectorants
Flu and pneumonia vaccinations
Supplemental oxygen
Breathing exercises
Chest physiotherapy
Pulmonary rehabilitation

MDI = metered dose inhaler; NMT = nebulized mist treatment; PO = by mouth.

Patients with this inherited tendency who also smoke have a very high risk of developing the disease. Children of smoking parents are at higher risk because of secondhand smoke exposure.

Prevention

Prevention is important because no cure for COPD is currently available. Avoidance of smoking and other inhaled irritants is vital, especially in those individuals with parents or siblings with COPD.

Signs and Symptoms

The patient with COPD exhibits prolonged expiration because of obstructed air passages. Hyperinflated lungs caused by air trapping leads to the classic barrel-shaped chest. The patient with chronic bronchitis has a chronic productive cough, shortness of breath, and activity intolerance. Symptoms may initially be worse in the winter months. Crackles and wheezing are often noted on auscultation and may improve after coughing. The most characteristic symptom of emphysema is progressive shortness of breath, accompanied by activity intolerance. Use of accessory muscles is evident. Auscultation reveals diminished breath sounds. Remember that many patients exhibit symptoms of both chronic bronchitis and emphysema.

Arterial blood gases (ABGs) may be checked during an acute exacerbation of COPD and may show an increase in

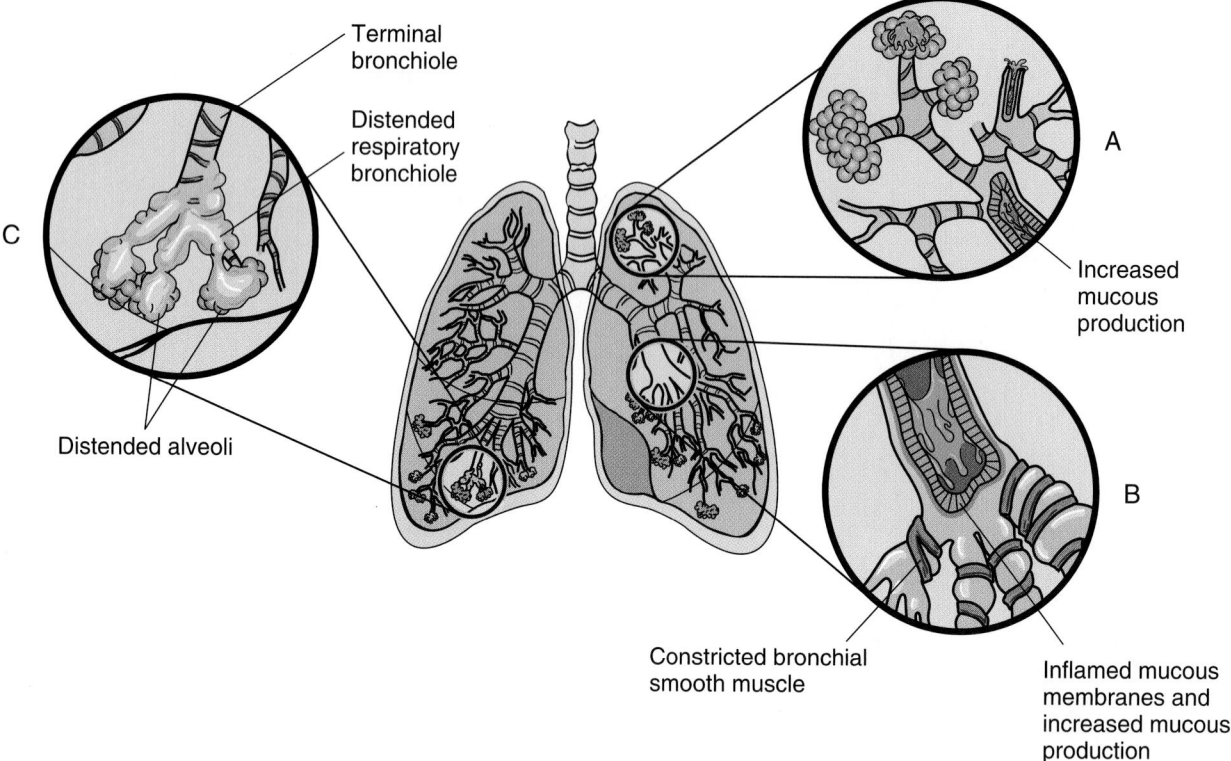

Terminal bronchiole

Distended respiratory bronchiole

C

Distended alveoli

A

Increased mucous production

B

Constricted bronchial smooth muscle

Inflamed mucous membranes and increased mucous production

Figure 28-4 *(A)* Chronic bronchitis. Note inflamed airways and excessive mucus. *(B)* Asthma. Note narrowed bronchial tubes and swollen mucous membranes. *(C)* Emphysema. Note distended respiratory bronchioles and alveoli.

PaCO$_2$ and often a low PaO$_2$. The patient develops **polycythemia** in response to reduced oxygenation, which results in a ruddy skin color. Cyanosis may also be present. Right-sided heart failure may develop because the heart has to work harder to pump to the diseased lungs. (See the section on cor pulmonale in Chapter 21.)

In late stages of COPD, patients may lose weight and become malnourished. They have difficulty eating because of severe dyspnea, and the increased work of breathing expends more calories. Chronic hypoxemia causes release of certain chemicals that may also lead to weight loss.

Complications

Some patients with emphysema develop large air spaces within the lung tissue (**bullae**) or adjacent to the pleurae (**blebs**). These are like blisters that can rupture and cause the lung to collapse. In severe COPD, cor pulmonale can develop. Death can result from respiratory infection or respiratory failure.

Diagnostic Tests

Information from a chest x-ray examination, CT scan, blood gas analysis, CBC, sputum analysis, and pulmonary function studies is correlated with the history and physical examination to diagnose COPD. An α_1AT level is checked if deficiency is suspected.

polycythemia: poly—many + cyt—cells + emia—in the blood

Medical Treatment

The most important treatment measure for COPD is the cessation of smoking. Even late in the disease process, stopping smoking can slow disease progression and prolong life. Exposure to other respiratory contaminants should also be minimized. Hair spray, body powder, and other household aerosols should be avoided.

Oxygen is administered as ordered at a flow rate of 1 to 2 L. Higher flow rates are avoided to prevent suppression of the hypoxic drive. Patients with chronic oxygen saturation levels of less than 88 percent may be placed on home oxygen. A pulmonary rehabilitation program can help the patient increase exercise tolerance. Breathing exercises help improve oxygenation and reduce anxiety. (See Chapter 26.) Bronchodilators and expectorants may be helpful. Table 28–2 lists medications that are used in the treatment of COPD and lower respiratory disorders.

A pneumococcal vaccination and yearly influenza vaccinations are recommended to reduce the risk of respiratory infection. Avoidance of crowds and exposure to people with respiratory infections is advised.

Medications commonly used include adrenergic, anticholinergic, or theophylline bronchodilators; expectorants; and, intermittently, antibiotics. Because the risk for infection is high, the patient is taught to report the onset of purulent sputum as soon as it begins, so treatment can be initiated quickly. Steroids may be used late in the disease to reduce airway inflammation. See Table 28–2 for medications used in the treatment of COPD.

Normal lungs

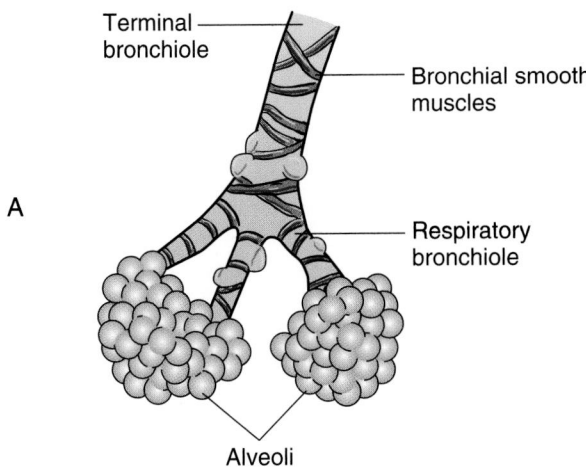

A

Centrilobular emphysema

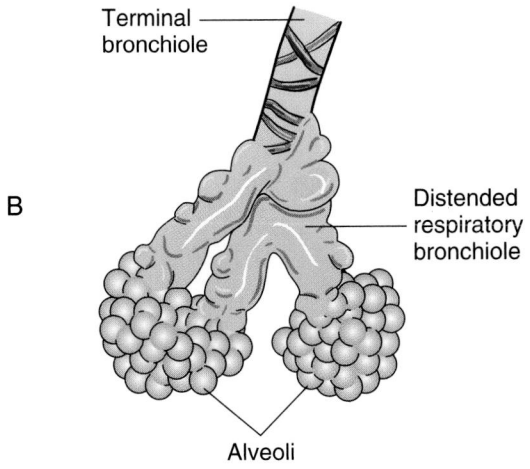

B

Panlobular emphysema

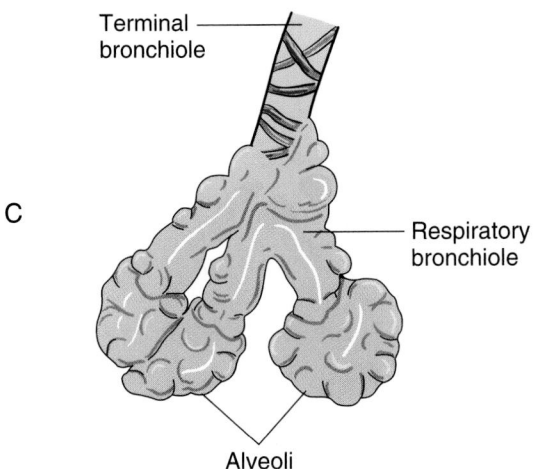

C

Figure 28-5 Types of emphysema. *(A)* Normal lungs. *(B)* Centrilobular emphysema. *(C)* Panlobular emphysema.

Good hydration and a cool mist humidifier help keep secretions loose. Chest physiotherapy may be used to help the patient remove excessive secretions. Nebulized mist treatments and metered dose inhalers are commonly used to administer bronchodilators. A dietitian consultation is helpful for the patient who is unable to maintain a desirable weight. Pulmonary rehabilitation programs can help the patient increase exercise tolerance and maintain a sense of well-being.

A newer treatment for emphysema is the surgical removal of some of the diseased lung tissue (called lung volume reduction surgery, or LVRS). This increases the space available for good lung tissue to expand, reducing dyspnea and increasing exercise tolerance. This is a high-risk procedure, but it has allowed some patients to return to a more normal activity level. Another new treatment is replacement of α_1AT. Research is ongoing to determine if either of these treatments will alter long-term outcomes of the disease. Lung transplant may be an option in some patients. Check the American Lung Association Web site for more information at www.lungusa.org.

ASTHMA

The incidence of asthma is on the rise. Nearly 12 million people in the United States have asthma. It is more prevalent in African-Americans than in whites. Nearly 5000 people die from asthma each year. Asthma deaths are more prevalent in lower socioeconomic groups, presumably because of lack of compliance with treatment regimens. With careful monitoring and treatment, however, patients with asthma can control their symptoms and lead normal lives.

Pathophysiology and Etiology

Asthma is characterized by inflammation of the mucosal lining of the bronchial tree and spasm of the bronchial smooth muscles (**bronchospasm**). This causes narrowed airways and air trapping. (See Fig. 28–4.) Symptoms are intermittent and reversible, with periods of normal airway function. About 50 percent of asthmatics develop the disorder in childhood, but contrary to popular belief, most children do not outgrow asthma. Instead, symptoms just diminish, often returning later in life.

The tendency to develop asthma is inherited. Some sources classify asthma as either allergic or idiosyncratic (unexpected). Allergic asthma is triggered by allergens such as pollen, foods, medications, animal dander, air pollution, molds, or dust mites. It is commonly seasonal. Individuals who developed asthma as children tend to have allergic asthma. Idiosyncratic asthma is generally diagnosed in adults and is related to environmental or other nonallergic factors, such as environmental irritants, smoking, and respiratory or sinus infection. Emotional upset and exercise can also trigger symptoms in some persons with asthma. Asthma frequently complicates chronic bronchitis or emphysema.

A newer finding is asthma caused by gastroesophageal reflux disease (GERD). It is believed that stomach acid may

bronchospasm: broncho—airway + spasm—convulsion, narrowing

TABLE 28-2   SELECTED MEDICATIONS USED FOR LOWER RESPIRATORY TRACT DISORDERS

Drug Class and Examples	Route	Action	Side Effects and Nursing Implications
Glucocorticoids			
Methylprednisolone (Medrol, Solumedrol)	PO IV	Potent anti-inflammatory agents, reduce inflammation in airways	Cushingoid side effects with prolonged use: moon face, sodium and water retention, buffalo hump, osteoporosis, hyperglycemia. May be given IV, PO, or inhaled. Inhaled route causes fewer side effects. Never discontinue abruptly.
Prednisone	PO		
Triamcinolone acetonide (Azmacort)	Inhaled		
Beclomethasone (Vanceril, Beclovent)	Inhaled		
Fluticasone (Flovent)	Inhaled		
Bronchodilators			
Adrenergic		Stimulate beta receptors to dilate bronchioles	Adrenergic agents cause increased heart rate, tremor, anxiety. Use with care in patients with cardiac disease. Overuse can cause rebound bronchospasm. Serevent is long acting, used bid only.
Albuterol (Ventolin, Proventil)	PO, Inhaled		
Metaproterenol (Alupent, Metaprel)	PO, Inhaled		
Pirbuterol (Maxair)	Inhaled		
Salmeterol (Serevent)	Inhaled		
Anticholinergic		Blocks parasympathetic response, causing bronchodilation	
Ipratropium (Atrovent)	Inhaled		
Methylxanthines		Relax bronchial smooth muscle to dilate airways	Most common side effects are tremor, anxiety, tachycardia, nausea, vomiting. Therapeutic theophylline level 5–15 μg/mL.
Theophylline (Theovent, Theolair, Uniphyl)	PO		
Aminophylline	PO, IV		
Combination Agents			
Albuterol and ipratropium (Combivent)	Inhaled		
Flutricasone and Salmeterol (Advair)			
Mast Cell Stabilizers			
Cromolyn sodium (Intal)	Inhaled	Stabilize mast cells to reduce histamine release	Few side effects. Effective for allergic asthma. May be used prophylactically before exercise or allergen exposure.
Nedocromyl (Tilade)			
Expectorants			
Guaifenesin (Robitussin, Humabid)	PO	Liquefy secretions and stimulate cough	Few side effects. Encourage fluids.
Antileukotrienes			
Zafirlukast (Accolate)	PO	Inhibit leukotriene synthesis or activity, a mediator of inflammation in asthma	No serious side effects. Possible elevation of liver enzymes.
Montelukast (Singulair)			
Zileuron (Zyflo)			
Antitussives			
Codeine	PO	Suppress cough reflex	Related to opioids; may be sedating at high doses. Avoid giving to patient who has secretions that need to be expectorated.
Dextromethorphan (DM suffix in cough preparations)			

NOTE: This table is an overview. A drug guide should be consulted for complete administration guidelines.
PO = by mouth.

reflux into the esophagus and then be aspirated, triggering asthma. This occurs especially at night. GERD and its treatment are discussed in Chapter 30.

Prevention

Although asthma cannot be prevented, individual episodes can be. It is important that the patient identify triggers of asthma symptoms and avoid them whenever possible. Compliance with prophylactic and maintenance therapy is also important.

Signs and Symptoms

Asthma symptoms are intermittent and are often referred to as "attacks," which may last from minutes to days. The pa-

tient complains of chest tightness, dyspnea, and difficulty moving air in and out of the lungs. Once initial symptoms are controlled, airways may remain hypersensitive and prone to asthma symptoms for many weeks.

On examination, you will note an increased respiratory rate as the patient attempts to compensate for narrowed airways. Inspiratory and expiratory wheezing is heard because of turbulent airflow through swollen airways with thick secretions and may sometimes be audible even without a stethoscope. Air is trapped in the lungs, and expiration is prolonged. A cough is common and may produce thick, clear sputum. Use of accessory muscles to breathe is a sign that the attack is severe and warrants immediate attention.

Be aware that an absence of audible wheezing may not signal improvement but rather may be an ominous sign that the patient is moving very little air. If wheezing is not heard, use of accessory muscles and peak expiratory flow rate values must be carefully evaluated. Once treatment begins to be effective and the patient is moving more air, wheezing may become audible.

Complications

Status asthmaticus occurs if bronchospasm is not controlled and symptoms are prolonged. As the patient increases the respiratory rate to compensate for narrowed airways, a lot of carbon dioxide is blown off and respiratory alkalosis occurs. If the attack is not resolved and the patient begins to tire, the patient will no longer be able to compensate and $PaCO_2$ will rise, resulting in respiratory acidosis. This can lead to respiratory failure and death if untreated.

Diagnostic Tests

Diagnosis is based on the patient's report of symptoms, physical examination, and pulmonary function studies. Peak expiratory flow rate is reduced. Arterial blood gases may initially show decreased $PaCO_2$. Late in the course of an attack, PaO_2 decreases and $PaCO_2$ increases. Allergy skin testing and increased serum IgE and eosinophil levels indicate allergic involvement and may help determine treatment.

Medical Treatment

MONITORING. Some patients monitor their peak expiratory flow rate (PEFR) at home (Fig. 28–6). This is a measure of the amount of air the patient can blow into a peak flowmeter from fully inflated lungs and is measured in liters per minute. The patient determines his or her normal PEFR during symptom-free times. If the PEFR begins to fall below the patient's personal norm, treatment that has been predetermined with the health care provider should be initiated. Readings can be charted to keep track of progress (Fig. 28–7). Often PEFR results indicate the onset of asthma before the patient experiences any symptoms (Home Health Hints).

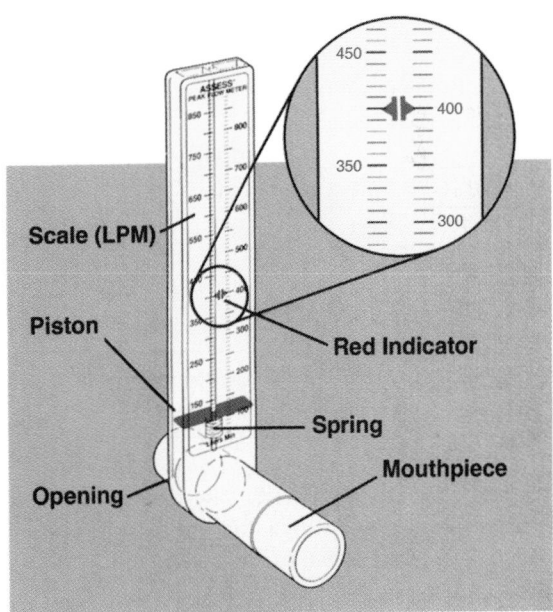

Figure 28-6 Patient with asthma using a peak flowmeter to monitor peak expiratory flow rate. (Courtesy Respironic Health-Scan Asthma & Allergy Products, Cedar Grove, NJ.)

HOME HEALTH HINTS

- When a patient is using oxygen by nasal cannula, the area around the ears can become irritated or excoriated. A small sponge-type hair roller can be placed around the tubing to protect the ears.

- When a home health patient has a metered dose inhaler (MDI), the nurse should not assume he or she is using it correctly. The patient should be observed using it. (Review procedure in Chapter 26.)

- When a patient requires more than one MDI, the canisters can be numbered in the order they are to be taken.

- To tell if an inhaler is empty, put the canister in a bowl of water. A full canister will sink and an empty canister will float.

- To help the COPD patient conserve energy, he or she can be encouraged to sit on a stool when cooking at the stove or doing dishes. Personal care activities should be spaced throughout the day (shampoo, bath, etc.).

- If the COPD patient is tempted to adjust his or her own oxygen flow rate, equipment suppliers can put on a locking flowmeter. Increasing flow rate can reduce hypoxic drive and cause hypoventilation.

- Nebulizer parts should be cleaned three times a week or every other day, using warm soapy water in a common home disinfectant solution for 30 minutes.

Name

Green zone _____   Yellow zone _____   Red zone _____

Date	AM	PM	AM	PM	AM	PM	AM	PM	AM	PM	AM	PM	AM	PM
800														
750														
700														
650														
600														
550														
500														
450														
400														
350														
300														
250														
200														
150														
100														
Notes														

Figure 28-7   Peak flow chart. The green zone is 80 to 100 percent of the patient's normal peak flow rate. The yellow zone is 50 to 80 percent of normal. The red zone is less than 50 percent of normal. The patient works with the physician to determine which actions to take when readings fall in the yellow or red zones.

AVOIDANCE OF TRIGGERS. The patient is instructed to identify and avoid asthma triggers. If triggers cannot be avoided, the patient can use bronchodilator or mast cell inhibitor metered dose inhalers (MDIs) as prescribed before exposure. MDIs can be especially useful before exercise. Animal dander and foods that cause symptoms are best avoided when possible. Eliminating carpets and curtains in bedrooms, using vinyl mattress and pillow covers, and installing a portable or central air filter can reduce dust mite exposure. Maintenance of indoor humidity between 40 and 50 percent can reduce mold growth. If cold air triggers symptoms, the patient should keep his or her nose and mouth covered when outside in cold weather. Smoking and exposure to secondary smoke are strongly discouraged.

Aspirin and nonsteroidal anti-inflammatory drugs can cause asthma symptoms in some individuals. Beta-blocking medications (propranolol, metoprolol), used commonly for hypertension, block beta receptors in the lungs, preventing the sympathetic nervous system from promoting bronchodilation. These drugs should be avoided if they make symptoms worse.

MEDICATIONS. Medications for asthma treatment may be continuous or intermittent, depending on the chronicity of symptoms. See Table 28-2 for a summary of medications used in the treatment of lower respiratory disorders. For patients with only occasional symptoms, adrenergic bronchodilators such as albuterol (Proventil, Ventolin) may be administered via MDI when symptoms occur or before exercise or other events that trigger asthma.

If the patient needs to use an adrenergic MDI more than three times a week, a mast cell inhibitor such as cromolyn sodium (Intal) or nedocromil sodium (Tilade) MDI or an inhaled steroid (Azmacort, Beclovent) may be added. Steroids delivered via metered dose inhalers have become popular and effective therapy in recent years. Because they are used topically, side effects are minimal. Instruct the patient that mast cell inhibitors and steroids must be used regularly to prevent symptoms and will not provide immediate symptom relief during an acute attack. Some patients use mast cell inhibitors 10 to 15 minutes before exposure to allergens or exercise to reduce symptoms.

LEARNING TIPS

If both bronchodilator and corticosteroid inhalers are used at the same time, instruct the patient to use the bronchodilator first. This opens the airways and allows the corticosteroid to be better distributed in the lungs.

If inhaled medications do not control symptoms, or if the patient has nocturnal symptoms, oral theophylline bronchodilators such as Theovent or Theo-Dur may be added. Alternatively, antileukotrienes may be tried; these are in a newer class of bronchodilators that have fewer side effects than theophylline. Immunotherapy (allergy shots) may be used for some patients with allergic asthma.

An acute asthma attack may be treated with an inhaled (nebulized) or subcutaneous adrenergic bronchodilator or intravenous aminophylline. Intravenous or oral corticosteroids (methylprednisolone, prednisone) are potent anti-inflammatory agents that are useful in an acute episode but are avoided for long-term therapy if possible because of their cushingoid side effects. (See the section on Cushing's syndrome in Chapter 37.) Corticosteroids must be tapered before discontinuing, to prevent withdrawal symptoms. (See the section on addisonian crisis in Chapter 37.)

Oxygen is not often necessary, because many patients hyperventilate. If the attack is prolonged and the patient becomes cyanotic or PaO_2 levels begin to fall, oxygen therapy will be used.

LEARNING TIPS

Instruct the patient to contact the health care provider if the patient is using more than two adrenergic MDI canisters per month. This has been associated with an increased risk of death.

► CYSTIC FIBROSIS

In the past, cystic fibrosis (CF) was thought to be just a childhood disease, because most affected children did not survive past puberty. However, with new treatments, patients with CF are living longer and more productive lives. Some CF patients now marry, have careers, and live well into their thirties.

Pathophysiology

CF is a disorder of the exocrine glands that affects primarily the lungs, gastrointestinal (GI) tract, and sweat glands. The disease varies in severity; some patients have no GI involvement. Abnormal sodium and chloride transport across cell membranes, causing thick, tenacious secretions, is responsible for many of the characteristic symptoms. Thick, sticky respiratory secretions that are difficult to remove cause airway obstruction, resulting in frequent respiratory infections.

Similar abnormalities in the pancreas cause blocked ducts and retained digestive enzymes. These retained enzymes digest and destroy the exocrine pancreas. The absence of digestive enzymes in the intestines causes malabsorption of essential nutrients; frequent foul-smelling, fatty stools; and excess flatus.

Patients with CF secrete sweat that is high in sodium and chloride, because these electrolytes are not reabsorbed as they pass through the sweat ducts.

Etiology

CF is a genetic disorder. Both parents must be carriers of the defective gene for CF to be present in a child. Patients who marry are counseled on the risk of potential offspring having the disease.

Signs and Symptoms

Symptoms usually first appear in infancy or childhood, although a few individuals are not diagnosed until adulthood. Respiratory symptoms are often the first visible manifestation of the disease and range from chronic sinusitis to production of thick, tenacious sputum. Patients with CF are at risk for frequent respiratory infections, manifested by an increase in cough and purulent sputum. Finger clubbing is common. Late in the disease, hemoptysis may occur related to damaged blood vessels within the lungs. Over time, bouts of infection become more frequent, with eventual loss of lung function and respiratory failure. Antibiotic-resistant bacteria are a threat to life in these individuals.

Frequent foul-smelling stools result from the lack of enzymes in the small intestine. Inability to absorb fat-soluble vitamins and poor appetite due to respiratory disease result in malnutrition. Bowel obstruction, cirrhosis, cholecystitis, and cholelithiasis are associated findings.

Chronic disease causes delayed sexual maturation in both males and females, and infertility is common.

Complications

Patients with CF are at risk for a variety of complications, including bronchiectasis, pneumothorax, cor pulmonale, and respiratory failure. Bowel obstructions can occur as a result of thick mucus binding with poorly digested fecal matter. Diabetes from pancreatic islet cell involvement may be present late in the disease. Death is usually the result of pulmonary complications.

Diagnostic Tests

Because so many different gene mutations can occur in CF, genetic testing is not helpful at this time. The standard diagnostic test is the sweat chloride test. If respiratory symptoms are accompanied by excessive amounts of sodium chloride in sweat, CF is diagnosed. The reader may recall public health campaigns that advised parents to kiss their babies and report any salty taste to their physicians.

Medical Treatment

Because there is no cure for CF, treatment is aimed at relieving symptoms. Removal of thick sputum is promoted with hydration, use of the Flutter mucous clearance device, and chest physiotherapy up to four times a day. Regular exercise also helps mobilize secretions. A hot shower may be an easy occasional alternative to loosen secretions. Nebulized mist treatments using saline or mucolytic medications may be used before chest physiotherapy. Medications to decrease the viscosity of secretions have not yet been entirely successful, but new drugs are constantly being tested. Inhaled bronchodilators are sometimes useful. High doses of ibuprofen (Motrin) may slow lung deterioration. Breathing exercises, incentive spirometry, and effective coughing techniques are also helpful. Lung transplant is a potentially promising treatment.

Prevention of infection is vital to slowing progression of lung damage. Antibiotics must be administered as soon as signs of infection occur. Prophylactic antibiotic therapy may be used. Some patients use home intravenous antibiotic therapy. Antibiotic-resistant infections are a deadly threat to the CF patient.

Pancreatic enzyme replacement (Pancrease, Viokase) helps reduce symptoms related to malabsorption and improve nutritional status. An increase in calorie requirements necessitates a high-calorie, nutrient-dense diet. For more information, visit the Cystic Fibrosis Foundation at www.cff.org.

Nursing Care

Nursing care of respiratory system problems in the patient with CF is similar to care of the patient with COPD. However, be sure to remember the special needs of the adolescent patient with this chronic, debilitating disease. Not only are normal physical growth and development delayed, but psychosocial development is also affected by repeated hospitalizations and the necessity of routine daily medication and treatments.

▶ NURSING PROCESS: THE PATIENT WITH COPD/CAL

ASSESSMENT. Perform a complete respiratory assessment as presented in Chapter 26. Frequency of assessment is dictated by the severity of the patient's condition. Note level of consciousness; poor gas exchange causes confusion and lethargy. Observe skin and mucous membranes for cyanosis. Auscultate lung sounds for adventitious sounds. Monitor cough and color and amount of sputum. Note exercise tolerance, and measure degree of dyspnea on a scale of 0 to 10. Monitor oxygen saturation or arterial blood gases. Careful documentation of findings allows you to monitor and report trends in the patient's progress.

NURSING DIAGNOSIS. A number of nursing diagnoses are appropriate for the patient with COPD. As always, choose diagnoses based on defining characteristics and the patient's individual assessment findings. Impaired gas exchange caused by poor ventilation or damaged alveoli is common and can be easily identified when dyspnea is accompanied by abnormal ABGs. Ineffective airway clearance related to copious thick secretions is also a common occurrence, especially in patients with chronic bronchitis and cystic fibrosis. Ineffective breathing pattern may occur related to anxiety. Activity intolerance related to hypoxia is often present. Related diagnoses may include imbalanced nutrition: less than requirements related to dyspnea and poor appetite, and anxiety related to dyspnea. See Nursing Care Plan Box 28–4 for the Patient with a Lower Respiratory Disorder.

PLANNING. To increase patient acceptance of the plan of care, the patient must be involved in identifying goals. The airway should remain clear of secretions. Anxiety should be manageable. Dyspnea should be controlled to allow the patient to maintain his or her desired activity level, within appropriate limitations. Nutrition status should be stable.

IMPLEMENTATION

Impaired Gas Exchange, Ineffective Airway Clearance, Ineffective Breathing Pattern. Gas exchange, airway clearance, and breathing pattern are priority problems for most respiratory patients. Interventions for these diagnoses are presented in Nursing Care Plan Box 28–4 for the Patient with a Lower Respiratory Tract Disorder.

Activity Intolerance. Encourage the patient to rest between activities. Even talking or eating can cause dyspnea in a patient with end-stage disease. A bedside commode prevents unnecessary trips to the bathroom. A shower chair and handheld showerhead conserve energy while bathing. If the patient is able to ambulate, a portable oxygen source should be used. The patient should be allowed to sleep without interruptions at night as much as possible. Most patients sleep with the head of the bed elevated; a reclining chair may be used at home. Activity level is slowly increased if the patient is able. Referral to a pulmonary rehabilitation program can help increase exercise tolerance.

Altered Nutrition. The cause of altered nutrition must first be identified. If the patient is too dyspneic to eat, scheduling rest periods and nebulized mist treatments or MDI administration before meals may be helpful. A poor appetite can be improved by creating a pleasant eating environment; providing smaller, more frequent meals of the patient's favorite foods; or encouraging family members to bring food from home for the hospitalized patient. Liquid supplements may be necessary to maintain weight. A specialized supplement such as Pulmocare provides less carbon dioxide when metabolized and may be used for patients with respiratory disease (Nutrition Notes Box 28–6).

Anxiety. Remain with the patient who is acutely anxious. A calm voice reminding the patient to breathe slowly in through the nose and out through pursed lips can be very helpful. Relaxation exercises, learned during times when anxiety is minimal, may be used at this time. Some patients learn to fold their arms just below the ribcage. They then push into their belly while exhaling to help push out trapped air and release pressure during inhalation. Antianxiety medications are given as ordered. Intravenous morphine helps acute dyspnea and anxiety, but it is usually reserved for patients with end-stage disease.

PATIENT EDUCATION. The patient must be aware of the contributing factors to the disease and eliminate them if at all possible. The patient who is a smoker should not simply be told to quit smoking; he or she should be referred to a smoking cessation program and be provided with medication, nicotine patches, or other resources and support as necessary to quit. (See Chapter 26.) Techniques for effective breathing and anxiety control should also be taught.

BOX 28-6 Nutrition Notes

Optimizing Nutrition in Patients with Respiratory Disease

Many factors affect nutrition in a patient with respiratory disease. For example, patients commonly have an inadequate food intake because of anorexia, shortness of breath, or gastrointestinal (GI) distress or a combination of these problems.

Calorie requirements are commonly increased in patients with pulmonary disease. The caloric cost of breathing ranges from 36 to 72 calories each day in normal people, but it increases to 430 to 720 calories each day in patients with chronic obstructive pulmonary disease (COPD). Both decreased food intake and increased energy requirements may contribute to the weight loss commonly seen in these patients. When caloric intake is decreased, the body begins to break down muscle stores, including the respiratory muscles.

The GI distress common in these patients may be related to malnutrition of the GI tract. Malnutrition and the resulting decrease in antibody production lower the patient's resistance to infection. Also, the malnourished patient's lungs produce less pulmonary phospholipid, a fatlike substance that assists in lubricating lung tissue and helps to protect the lungs from inhaled pathogens.

Improved nutritional status has been associated with an increased ability to wean patients from respirators or ventilators. To some extent, all the respiratory muscles atrophy from inactivity when a machine does the work for the patient. Nutrition support improves the likelihood of successful weaning in patients on artificial respiration.

Many patients with COPD have carbon dioxide retention and oxygen depletion. The medical goal for these patients is to decrease the level of carbon dioxide in their blood. Because fat calories produce less carbon dioxide when metabolized than carbohydrate calories, a diet with as much as 50 percent of the calories from fat may be prescribed. Several companies produce complete nutritional supplements with higher fat content formulated for patients with respiratory disease. Medical opinion is not unanimous on this issue because some evidence has related a high fat intake to immunosuppression in some patients.

NURSING INTERVENTIONS. Many respiratory patients are breathless and lack the energy to eat. Offer small frequent feedings of nutrient-dense foods. Select foods that require little or no chewing. To lessen abdominal pressure on the diaphragm, discourage gaseous foods.

EVALUATION. If interventions have been effective, the airway will be clear of secretions. The patient will be able to manage anxiety symptoms and complete ADLs or other desired activity without dyspnea. The patient's intake should be adequate to maintain a stable weight. If any of the patient's goals have not been met, the plan of care should be revised.

Rehabilitation of the Patient with COPD

Many institutions now have pulmonary rehabilitation departments. These programs help patients learn effective breathing techniques and how to slowly increase their exercise tolerance. You can request this referral when appropriate.

CRITICAL THINKING

Mr. Franklin

Mr. Franklin is admitted to the respiratory unit with exacerbated COPD. He has a history of emphysema and now has an acute infection complicating his disease. His lung sounds are very diminished, and he is short of breath at rest, even on 4 L of oxygen per nasal cannula. You walk into his room when he puts on his call light and find him sitting on the bedside commode with a look of panic in his eyes. He is gasping for breath, his color is gray, and his respiratory rate is 36 per minute.

1. What do you do first?
2. What can you teach Mr. Franklin to prevent an acute dyspneic episode in the future?
3. How will you document this episode?

Answers at end of chapter.

► PULMONARY VASCULAR DISORDERS

Pulmonary Embolism

Pathophysiology

An **embolism** is a foreign object that travels through the bloodstream. It may be a blood clot, air, or fat. A pulmonary embolism (PE), sometimes called pulmonary thromboembolism (PTE), is usually a blood clot that has traveled into a pulmonary artery. Resulting obstruction of blood flow causes a ventilation-perfusion mismatch, which in this case means that an area of the lung is well ventilated with air but has no blood flow, or perfusion. Because reduced or no blood supply is available to pick up the oxygen in the affected portion of the lung, it becomes pulmonary "dead space," causing seriously impaired gas exchange.

Occasionally damage occurs to a portion of the lung because of lack of oxygen. This is called lung infarction, and it is not common because oxygen is delivered to lung tissue not only from the pulmonary arteries but also via the bronchial arteries and the airways.

Etiology

Most pulmonary emboli originate in the deep veins of the lower extremities (deep vein thrombosis, or DVT). Some risk factors of DVT, and therefore PE, include surgical procedures done under general anesthesia, heart failure, fractures of the lower extremities, bedrest, obesity, and a previous history of DVT or PE. Less common causes of PE include fat emboli from compound fractures, amniotic fluid embolism during labor and delivery, and air embolism from entry of air into the bloodstream.

Prevention

Prevention of thrombi in the deep veins of the legs is the most important factor in the prevention of a pulmonary embolism. Regular ambulation is advised if the patient is able. If a patient is at risk for DVT or PE, low-dose heparin, enoxaparin, warfarin (Coumadin), or intermittent compression stockings are used to prevent thrombus formation. If a

DVT is diagnosed, prompt treatment is essential to prevent a PE.

Signs and Symptoms

The most common symptom of PE is a sudden onset of dyspnea for no apparent reason. The patient may be gasping for breath and may appear anxious. Tachycardia, tachypnea, and cough may be present. Auscultation may reveal crackles or a friction rub. If lung infarction has occurred, hemoptysis and pleuritic chest pain may also be present. Some patients have no symptoms at all. Be alert to the presence of risk factors and obtain immediate assistance if the cause of dyspnea might be PE. Death can occur if treatment is not quick and effective.

Complications

High blood pressure within the pulmonary circulation (pulmonary hypertension) may result from arterial occlusion and lead to right ventricular failure. This occurs because the right ventricle is unable to push blood into the occluded artery. As a result, the contraction becomes weak, cardiac output falls, and the patient becomes hypotensive.

Diagnostic Tests

A spiral CT scan is a new and fast type of CT scan that is noninvasive and can diagnose PE quickly. If this is not available, a lung scan (ventilation-perfusion scan) is done to assess the degree of ventilation of lung tissue and the areas of blood perfusion. If an area is well ventilated but poorly perfused (i.e., a mismatch), PE is suspected.

A pulmonary angiogram can outline the pulmonary vessels with a radiopaque dye injected via a cardiac catheter. This can show where blood flow is diminished or absent, suggesting an embolism.

Chest x-ray examination, electrocardiogram (ECG), arterial blood gas analysis, or magnetic resonance imaging (MRI) may also be done. However, many of these show changes only in the presence of a very large embolism or infarction.

Medical Treatment

The body naturally dissolves clots in 7 to 10 days. However, if the embolism is large, a thrombolytic agent might be used. These agents, such as streptokinase, urokinase, and tissue plasminogen activator (t-PA), dissolve clots and are very effective. However, they must be used within 4 to 6 hours of the clot's occurrence and are associated with a risk for hemorrhage.

If a thrombolytic agent is not used, treatment is aimed at preventing extension of the clot and the formation of additional clots. Heparin, a potent anticoagulant medication, is administered via continuous intravenous infusion. Sometimes an intermittent IV or subcutaneous route is used. Heparin is never given intramuscularly because of the risk of hematoma development. Clotting studies (partial thromboplastin time [PTT] or thrombin clotting time [TCT]) are monitored and maintained at 1.5 to 2 times the control value. Sometimes heparin therapy is initiated even before a diagnosis of PE is made. It is believed that it is safer to begin therapy and then stop if PE is not confirmed than to wait until all test results are available.

Oxygen is administered as ordered. Intubation and mechanical ventilation may be required in some cases.

Warfarin sodium, an oral anticoagulant, is used for approximately 12 weeks following the embolism to prevent recurrence. It can also be used for long-term prevention of repeated clots in patients who have risk factors that cannot be resolved. Warfarin therapy can be initiated 2 to 3 days after the initiation of heparin therapy. Because it has a slow onset of action, it may require several days for the full anticoagulant effect to occur. The patient will be on both anticoagulants for a time. Warfarin therapy is monitored regularly with prothrombin times or international normalized ratios (INRs). See Box 28-7 for education topics for patients on anticoagulant therapy.

If clots are a recurring problem, a filter may be placed into the inferior vena cava via the jugular or femoral vein. One filter that is commonly used is the Greenfield filter, which filters out clots traveling from the lower extremities toward the heart and lungs.

In patients with life-threatening symptoms, a surgical embolectomy can be performed. This is a rare procedure that is reserved for emergency situations.

Nursing Care

ASSESSMENT. The patient is assessed for respiratory distress, including respiratory rate and effort, cyanosis, confusion, and subjective feelings of dyspnea and anxiety. Lung sounds are auscultated. Sputum color and amount are noted. Arterial blood gases and oxygen saturation are mon-

---

BOX 28-7 **Education for Clients Receiving Anticoagulant Therapy**

Anticoagulants prolong the time it takes blood to clot, so it is important to prevent injury and to recognize and report signs of bleeding to the physician.

To Prevent Injury
- Wear shoes or slippers; avoid going barefoot.
- Use an electric razor to shave.
- Use a soft toothbrush.

Signs of Bleeding to Report to Physician
- Easy bruising
- Nosebleeds
- Bleeding that does not stop
- Blood in urine
- Blood in sputum
- Blood in stools or black stools

Additional Instructions
- Avoid use of aspirin because it further prolongs clotting time.
- Have your lab work done as prescribed by your physician to monitor your clotting time and medication dosage.

itored. Heart sounds and peripheral edema are monitored for signs of heart failure. Contributing factors, such as calf pain, are assessed.

NURSING DIAGNOSIS. The priority nursing diagnosis for a patient with a pulmonary embolism is impaired gas exchange. Because of the impaired perfusion of the affected area of the lung, oxygen and carbon dioxide exchange are limited. Anxiety occurs related to dyspnea. Risk for injury related to anticoagulant therapy is a concern once treatment is initiated.

PLANNING AND IMPLEMENTATION

Impaired gas exchange. Administer oxygen as ordered. Bedrest is maintained to decrease oxygen demand. Assisting the patient with turning, coughing, and deep breathing further facilitates gas exchange.

Anxiety. Stay with the patient during acute episodes of dyspnea. Answer the patient's questions, and provide information about pulmonary embolism and its treatment at a level appropriate to the patient's need and tolerance because knowledge may help decrease anxiety. Administer antianxiety agents as ordered.

Risk for injury. Monitor coagulation studies and report results to the physician. Protect the patient from injury, so that excessive bleeding does not occur. Encourage the patient to wear shoes or slippers when ambulating. A soft toothbrush and an electric razor are used. Instruct the patient to report any signs of bleeding, such as hematuria or easy bruising. (See Box 28–7.)

EVALUATION. The patient should state that dyspnea and anxiety are controlled, and verbalize understanding of anticoagulant therapy and precautions.

Pulmonary Hypertension
Pathophysiology and Etiology

Primary pulmonary hypertension occurs when the arteries that carry deoxygenated blood from the heart to the lungs become narrowed as a result of changes in the vascular smooth muscle. The result is elevated pressure in the pulmonary arteries, causing the right ventricle to work harder to push blood into them. Eventually the right ventricle fails. The reason for these vascular changes is not known. Primary pulmonary hypertension is more common in women between ages 20 and 40 and has a hereditary tendency.

Secondary pulmonary hypertension results from other disorders, such as coronary artery disease or mitral valve disease, both of which increase pressures in the left side of the heart. The pulmonary arterial pressure then rises to push blood into the left heart. Capillary destruction related to alveolar damage in COPD can also lead to secondary pulmonary hypertension. Right ventricular failure eventually occurs as the heart works to push blood against high pulmonary arterial pressures.

Signs and Symptoms

The most common symptoms include dyspnea and fatigue, which worsen over time. Crackles and decreased breath sounds are heard on auscultation. Cyanosis and **tachypnea** (rapid respiratory rate) are noted. If heart failure is present, peripheral edema and distended jugular veins are seen. Angina may result from right ventricular ischemia. Death usually occurs within 2 to 3 years of diagnosis, unless a lung or heart-lung transplant is done.

Diagnostic Tests

Arterial blood gases commonly show hypoxemia and hypocapnia. Cardiac catheterization can be done to determine high pulmonary arterial pressures. An ECG may show right ventricular hypertrophy. A chest x-ray examination, pulmonary function tests, lung scan, and pulmonary angiogram may be done to determine underlying causes in secondary pulmonary hypertension.

Treatment

No cure is available for pulmonary hypertension, except for lung or heart-lung transplant. In secondary pulmonary hypertension, the underlying disorder is treated. Supportive care includes a low-sodium diet and diuretics to reduce blood volume (and therefore pressure), oxygen, and cardiac monitoring. Vasodilators such as calcium channel blockers may be beneficial in some patients to reduce pulmonary artery pressure. Warfarin may be used to prevent clotting. A newer therapy is IV administration of prostacyclin, a vasodilator that may reverse some of the vascular changes but has many serious side effects. Nursing care is collaborative and focuses primarily on patient assessment. Fowler's or high Fowler's position may help reduce dyspnea, and bedrest and comfort measures are helpful in treating fatigue and anxiety.

▶ TRAUMA
Pneumothorax

The term **pneumothorax** literally means "air in the chest" and is used to describe conditions in which air has entered the pleural space outside the lungs. If the pneumothorax occurs without an associated injury, it is called a spontaneous pneumothorax. A secondary spontaneous pneumothorax may occur due to underlying lung disease. Traumatic pneumothorax may result from a penetrating chest injury.

Pathophysiology and Etiology

Recall that the pleural cavity has visceral and parietal pleurae. These membranes normally are separated only by a thin layer of pleural fluid. Each time a breath is taken in, the diaphragm descends, creating negative pressure in the thorax.

---

tachypnea: tachy—rapid + pnea—breathing
pneumothorax: pneumo—air + thorax—chest

This negative pressure pulls air into the lungs via the nose and mouth. If either the visceral pleura or the chest wall and parietal pleura is perforated, air enters the pleural space, negative pressure is lost, and the lung on the affected side collapses (Fig. 28–8). Each time the patient takes a breath, the temporary increase in negative pressure draws air into the pleural space via the perforation. During expiration, air may or may not be able to escape through the perforation.

SPONTANEOUS PNEUMOTHORAX. If no injury is present, the pneumothorax is considered spontaneous. This occurs mostly in tall, thin individuals and in smokers. Patients who have had one spontaneous pneumothorax are at greater risk for a recurrence. Patients with underlying lung disease (especially emphysema) may have blisterlike defects in lung tissue, called bullae or blebs, that can rupture, allowing air into the pleural space. Weakened lung tissue from lung cancer can also lead to pneumothorax.

TRAUMATIC PNEUMOTHORAX. Penetrating trauma to the chest wall and parietal pleura allows air to enter the pleural space. This can occur as a result of a knife or gunshot wound or from protruding broken ribs.

OPEN PNEUMOTHORAX. If air can enter and escape through the opening in the pleural space, it is considered an open pneumothorax.

CLOSED PNEUMOTHORAX. If air collects in the space and is unable to escape, a closed pneumothorax exists.

TENSION PNEUMOTHORAX. If a pneumothorax is closed, air, and therefore tension, builds up in the pleural space. As tension increases, pressure is placed on the heart and great vessels, pushing them away from the affected side of the chest. This is called a mediastinal shift. When the heart and vessels are compressed, venous return to the heart is impaired, resulting in reduced cardiac output and symptoms of shock. Tension pneumothorax is often related to the high pressures present with mechanical ventilation. It is a medical emergency.

HEMOTHORAX. The term **hemothorax** refers to the presence of blood in the pleural space. This can occur with or without accompanying pneumothorax (hemopneumothorax) and is often the result of traumatic injury. Other causes include lung cancer, pulmonary embolism, and anticoagulant use.

Signs and Symptoms

Sudden dyspnea, chest pain, tachypnea, restlessness, and anxiety occur with pneumothorax. On examination, asymmetrical chest expansion on inspiration may be noted. Breath sounds may be absent or diminished on the affected side. In a "sucking" chest wound, air can be heard as it enters and leaves the wound.

---

hemothorax: hem—blood + thorax—chest

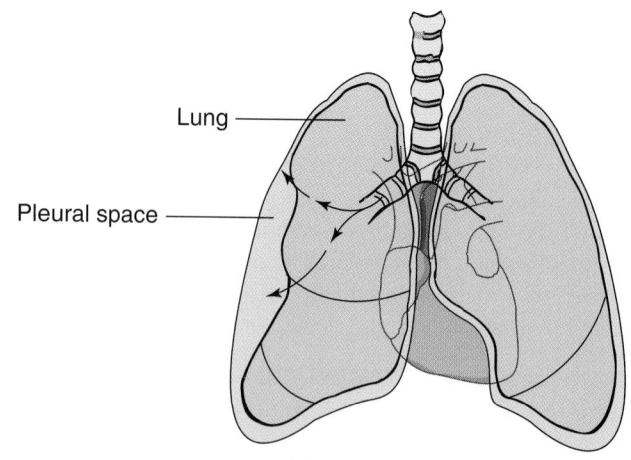

Spontaneous pneumothorax

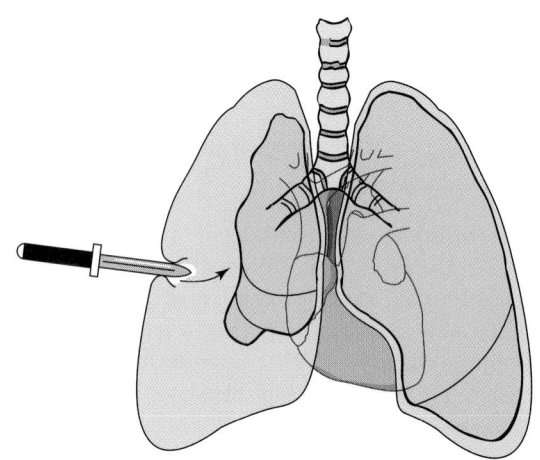

Traumatic pneumothorax

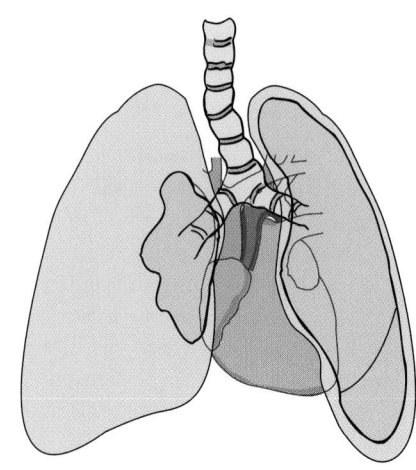

Tension pneumothorax

Figure 28-8   Types of pneumothorax. *(A)* Spontaneous pneumothorax. *(B)* Traumatic pneumothorax. *(C)* Tension pneumothorax with mediastinal shift.

If tension pneumothorax develops, the patient becomes hypoxemic and hypotensive as well. The trachea may deviate to the unaffected side. Heart sounds may be muffled. Bradycardia and shock occur if emergency intervention is not provided.

Diagnostic Tests

History, physical examination, and chest x-ray examination are used to diagnose pneumothorax. Chest x-ray examinations are repeated to monitor the resolution of the pneumothorax with treatment. Arterial blood gases and oxygen saturation are monitored throughout the course of treatment.

Medical Treatment

A small pneumothorax may absorb with no treatment other than rest, or the trapped air may be removed with a small-bore needle inserted into the pleural space. Chest tubes connected to a water seal drainage system are used to remove larger amounts of air or blood from the pleural space. See Chapter 26 for complete information about chest drainage. A Heimlich valve is another option, used for some patients who are treated at home. This is a small rubber tube that is attached to the chest tube instead of the water seal drainage system. The tube opens to allow air to escape, but then collapses during inspiration to prevent reentry of air into the pleural space. Some injuries require surgical repair before the pneumothorax can be resolved.

If the pneumothorax is recurrent, other treatments can be used to prevent additional episodes. Sterile talc or certain antibiotics (such as tetracycline) can be injected into the pleural space via thoracentesis, irritating the pleural membranes and making them stick together. This is called **pleurodesis,** or sclerosis, and prevents recurrent pneumothorax. Pleurodesis is painful; prepare the patient with an analgesic before the procedure.

Nursing Care

Nursing care of the patient with a pneumothorax involves close monitoring of the condition. A frequent and thorough assessment should be done, including level of consciousness, skin and mucous membrane color, vital signs, respiratory rate and depth, and presence of dyspnea, chest pain, restlessness, or anxiety. Regular auscultation of lung sounds provides information about reinflation of the affected lung. Any signs of increasing or tension pneumothorax are reported to the physician immediately. See Chapter 26 for care of the patient with a chest tube and water seal drainage system.

Rib Fractures
Etiology and Signs and Symptoms

Chest trauma is often accompanied by fractured ribs. Uncontrolled coughing, especially in the presence of osteoporosis or cancer can also fracture ribs. The fourth through ninth ribs are the most commonly affected. Broken ribs can be very painful, and often prevent the patient from breathing deeply or coughing effectively, which can result in atelectasis or pneumonia. Displaced ribs can also damage lung tissue, causing pneumothorax.

---

pleurodesis: pleur—pleural membrane + desis—binding

Treatment

In the past, elastic rib belts were used to stabilize the ribs while healing took place. These are no longer used because it further restricts deep breathing. Pain control is the most important treatment. Keeping the patient comfortable allows coughing and deep breathing, which in turn prevents complications such as pneumonia and atelectasis. If traditional pain control measures are ineffective, intercostal nerve blocks may be used. Ribs generally heal in about 6 weeks.

Flail Chest
Pathophysiology and Etiology

When multiple ribs are fractured, the structural support of the chest is impaired. As a result, the affected part of the chest collapses with inspiration and bulges with expiration. This is called **paradoxical respiration,** which is ineffective in ventilating the lungs and results in hypoxia.

Signs and Symptoms

The patient with a flail chest exhibits chest movement that is opposite to that usually seen with respiration. The patient is dyspneic, anxious, tachypneic, and tachycardic.

Treatment

The patient is given supplemental oxygen. Intubation and mechanical ventilation may be necessary. If lung damage has occurred, chest tubes may be necessary for reinflation.

▶ NURSING PROCESS: THE PATIENT WITH CHEST TRAUMA

The following nursing process is based on the stabilized patient. For emergency care of the patient with chest trauma, see Chapter 12.

ASSESSMENT. When caring for the patient following chest trauma, it is important to monitor respiratory status. Any sign of worsening status should be reported to the physician immediately, such as a change in vital signs or lung sounds; change in respiratory rate; increase in dyspnea, chest pain, pallor, or cyanosis; development of tracheal deviation; or new onset of anxiety. If a chest wound is present, it is cared for and closely monitored. Additional assessment may be necessary depending on the type of injury sustained.

NURSING DIAGNOSIS. Nursing diagnoses for the patient with chest trauma may include impaired gas exchange, acute pain, ineffective breathing pattern, and anxiety. Additional diagnoses may be appropriate depending on the individual patient's assessment.

PLANNING AND IMPLEMENTATION. Gas exchange is maintained with supplemental oxygen. Mechanical ventilation may be necessary. Position the patient for comfort and maximum chest expansion and encourage him or her to take deep breaths. Control pain with analgesics, so the patient is able to breathe deeply. Remember that narcotics depress respirations, so respiratory rate is carefully monitored. Splinting the chest for coughing may also be helpful to reduce pain. Additional care is necessary if the patient has chest tubes.

EVALUATION. Are pain, anxiety, and dyspnea controlled? Is the respiratory rate within normal limits? Are vital signs stable? Frequent evaluation is essential, so that failure to progress can be reported to the physician.

▶ RESPIRATORY FAILURE
Acute Respiratory Failure
Pathophysiology

Acute respiratory failure is diagnosed when the patient is unable to maintain adequate blood gas values. Hypoxemia may result from inadequate ventilation (air movement in and out of lungs) or poor oxygenation (adequate ventilation but inability to get the oxygen into the blood and therefore the cells). Some patients with chronic respiratory disease have adapted to impaired gas exchange. In these patients a drop in PaO$_2$ of 10 to 15 mm Hg is considered acute failure.

Etiology

An acute respiratory infection in a patient with chronic airway obstruction is often the precipitating factor in acute respiratory failure. Other causes include central nervous system disorders that affect breathing, such as a stroke or myasthenia gravis, inhalation of toxic substances, and aspiration.

Prevention

Avoidance of respiratory infections in patients with chronic respiratory disease is important. Patients should be instructed to notify their physician immediately if sputum becomes purulent so treatment can be initiated.

Sedatives and narcotics should be used carefully or avoided in patients with chronic respiratory disease, because these are respiratory depressants and can precipitate failure.

Signs and Symptoms

The patient with impending respiratory failure may become restless, confused, agitated, or sleepy. Arterial blood gases show decreasing PaO$_2$ and pH and increasing PaCO$_2$.

Diagnostic Tests

Arterial blood gas analysis is important in determining the presence of respiratory failure. A drop in PaO$_2$ of 10 to 15 mm Hg or a pH of less than 7.30 with associated elevated PaCO$_2$ indicates respiratory failure. Sputum cultures or chest x-ray examinations may be used to determine the cause and guide treatment. Pulse oximetry is used to continuously monitor oxygen saturation.

Medical Treatment

Carefully assess the patient and report significant findings to the physician immediately. It is easy to mistakenly treat symptoms of agitation or confusion with sedatives, which will speed the onset of respiratory failure. Oxygen therapy via nasal cannula or mask is provided at a flow rate of 1 to 2 L to prevent interference with the hypoxic drive. Antibi-

otics or other treatments are ordered to correct the underlying cause of the failure. Aminophylline and beta-adrenergic bronchodilators are used to dilate the bronchioles, promoting ventilation and secretion removal. The patient is instructed to cough and deep breathe if able. Suctioning is indicated if the patient is unable to cough effectively. Mechanical ventilation may be required.

Acute Respiratory Distress Syndrome

Acute respiratory distress syndrome (ARDS), also called adult respiratory distress syndrome, is a group of disorders that has diverse causes but similar pathophysiology, symptoms, and treatment. It is called ARDS to differentiate it from neonatal respiratory distress syndrome.

Pathophysiology and Etiology

The most common cause of ARDS is sepsis. Other causes include pneumonia, trauma, shock, narcotic overdose, inhalation of irritants, burns, pancreatitis, and aspiration, among others. Each of these causes begins a chain of events leading to alveolocapillary damage and noncardiac pulmonary edema (pulmonary edema that is not caused by heart failure). ARDS usually affects patients without a previous history of lung disease.

The alveolocapillary membranes become inflamed and damaged either by direct contact with an inhaled irritant or by chemical mediators that are released when systemic injury occurs. The membranes become leaky, so that proteins, blood cells, and fluid move from the capillaries into the interstitial space and then into the alveoli. Surfactant, a substance that reduces surface tension in the alveoli, is reduced. Alveoli collapse (atelectasis) and fibrotic changes take place. These changes cause the lungs to become stiff, or less compliant, making the patient work very hard to inspire. Blood supply to the alveoli may be adequate, but collapsed, wet alveoli are unable to oxygenate it. In other areas of the lungs, vasoconstriction reduces the ability of the vessels to pick up oxygen from functioning alveoli. Tired respiratory muscles, in combination with edema and atelectasis, reduce gas exchange and result in hypoxia. As the condition progresses, atelectasis and edema worsen and the lungs may hemorrhage. A chest x-ray examination appears white because of the excessive fluid in the lungs. These changes explain some of the older names for what is now known as ARDS: wet lung, white lung, shock lung, and stiff lung.

Prevention

Early recognition and treatment of underlying disorders is important in prevention of ARDS. Good nursing care can help reduce aspiration and some types of pneumonia.

Signs and Symptoms

Initially the patient may experience dyspnea and an increase in respiratory rate. Respiratory alkalosis results from hyperventilation. Fine inspiratory crackles may be auscultated. As the condition worsens, breathing becomes more rapid and labored and the patient becomes cyanotic. The patient is no longer able to oxygenate the blood and get rid of carbon dioxide, and respiratory acidosis occurs. Oxygen therapy does not reverse the hypoxemia. If ARDS is not reversed, eventually hypoxemia leads to decreased cardiac output and death.

Complications

Complications that can result from ARDS include heart failure, pneumothorax related to mechanical ventilation, infection, and disseminated intravascular coagulation (DIC). The death rate for ARDS in the past was 100 percent. With newer treatments, it is now between 50 and 60 percent. Most patients who survive ARDS recover completely.

Diagnostic Tests

Diagnosis is made based on history of a causative injury, physical examination, chest x-ray examination, and blood gas analysis. An ECG is done to rule out a cardiac-related cause.

Medical Treatment

The patient with ARDS is cared for in an intensive care unit. Treatment begins with oxygen therapy that is adjusted based on repeated ABG results. Intubation and mechanical ventilation is necessary in most cases, with the use of positive end-expiratory pressure (PEEP) to keep the airways open. Diuretics may be used to reduce pulmonary edema, but care must be taken to prevent fluid depletion. A pulmonary artery catheter may be used to monitor hemodynamic status. If infection is the underlying cause, antibiotics are administered. Parenteral nutrition may be given to maintain nutritional status while the patient is acutely ill. Positioning the patient with the less involved lung in the dependent position ("good lung down") allows the better lung to be well perfused with blood and may increase PaO_2. Prone positioning has also been shown to increase oxygenation in patients with ARDS.

▶ NURSING PROCESS: THE PATIENT EXPERIENCING RESPIRATORY FAILURE

ASSESSMENT. Assess the patient's degree of dyspnea on a scale of 0 to 10 if the patient is able to participate. Respiratory rate, effort, and use of accessory muscles is noted. Arterial blood gases and oxygen saturation values are monitored as ordered. The presence of cyanosis is noted.

Mental status, including restlessness, confusion, and level of consciousness, is also assessed, because reduced oxygenation can produce central nervous system (CNS) symptoms. Symptoms of the underlying cause of respiratory failure are monitored. If the cause is infectious, sputum amount and color, temperature, and white blood cell counts are monitored.

All assessment findings should be compared with earlier data. Even subtle changes in the assessment findings can be significant and should be reported.

NURSING DIAGNOSIS. Possible diagnoses include impaired gas exchange, ineffective airway clearance, and inef-

fective breathing pattern. Related diagnoses include activity intolerance, anxiety, disturbed thought processes, and self-care deficit.

PLANNING AND IMPLEMENTATION. The patient is positioned to maximize oxygenation; this might be semi-Fowler's, good lung down, prone positioning, or a combination of these. Relaxation exercises may help reduce anxiety and relieve dyspnea. Breathing exercises may be helpful if the patient is able to cooperate. Bedrest is important to reduce the demand for oxygen. Because any movement can trigger an increase in dyspnea, help position the patient in bed and anticipate needs to prevent unnecessary exertion by the patient. Meals and treatments should be spaced to allow the patient time to rest. It is often helpful to provide nebulized mist treatments (NMTs) or MDI therapy before meals, so that the patient is able to eat without excessive dyspnea. Oxygen therapy is maintained as ordered. Fluids (if not contraindicated) and coughing and deep breathing help remove secretions. Suctioning may be ordered if necessary.

L E A R N I N G T I P S

The good lung down position can help increase oxygenation in patients with lung disease. Gravity results in more blood in the dependent lung, where it can receive oxygen from the healthier lung tissue. If both lungs are diseased, the right lung down position may be beneficial, because the right lung has a larger surface area.[2]

EVALUATION. If interventions have been effective, the patient will state that dyspnea is controlled. Mental status will be normal for the patient. Airways will be kept clear at all times, and the patient's respiratory rate will be regular and within normal limits.

▶ LUNG CANCER

Lung cancer is the leading cause of cancer death in the United States. An estimated 169,500 new cases were predicted for the United States in 2001. About 28 percent of cancer deaths are from lung cancer. The 5-year survival rate for all lung cancers is 14 percent. This rate increases to 49 percent if lung cancer is diagnosed and treated early.[3] The incidence of lung cancer among men is decreasing, whereas it is on the rise in women; this is thought to be related to an increase in the number of women who smoke.

Pathophysiology

Lung cancers originate in the respiratory tract epithelium; most originate in the lining of the bronchi (Fig. 28–9). The four major types of lung cancer are identified by the type of cells that are affected. These include small cell lung cancer (SCLC), large cell carcinoma, adenocarcinoma, and squamous cell carcinoma. The latter three types are classified as non–small cell lung cancer (NSCLC).

About 20 percent of lung cancers are SCLC, sometimes called oat cell carcinoma. SCLC grows rapidly and often

Figure 28–9 Lung cancer. The black arrow marks the tumor site. (Photograph courtesy Dinesh Patel, MD. Medical Oncology, Internal Medicine, Zanesville, Ohio.)

has metastasized by the time of diagnosis. It is usually caused by smoking and is most often found centrally, near the bronchi. The patient with small cell carcinoma has a poor prognosis, with survival time averaging only 9 to 10 months.

Large cell carcinoma is a rapidly growing cancer that can occur anywhere in the lungs. It metastasizes early in the disease, so these patients also have a poor prognosis. It accounts for about 10 percent of lung cancers.

Adenocarcinoma occurs more often in women, and most often in the peripheral lung fields. It accounts for about 40 percent of lung cancers. It is slow growing but often is not diagnosed until metastasis has occurred. It is less closely linked with smoking. Prognosis may be better than other lung cancer types.

About 30 percent of lung cancers are squamous cell carcinomas. These usually originate near the bronchi and metastasize late in the disease. They are associated with a history of smoking. The prognosis for individuals with squamous cell carcinoma may also be better than for some other lung cancers.

Etiology

The most common cause of lung cancer is tobacco smoke. Cigarettes contain chemicals that cause DNA to mutate, creating changes in cells and development of tumors. Smokers have an approximately 13 times greater risk for developing lung cancer than nonsmokers. If a patient stops smoking, the risk of lung cancer decreases significantly. Unfortunately, even with all this information, 25 percent of adults in the United States continue to smoke and teen smoking is on the rise.[4]

Environmental tobacco smoke (ETS) has also been shown to cause lung cancer. A recent study by the American Cancer Society showed that women who were married to smokers but who had never smoked themselves were 20 percent more likely to die of lung cancer.[5] Other factors that contribute to increased lung cancer risk are exposure to asbestos, radon, or arsenic; air pollution; diesel exhaust; and

radiation. Genetic predisposition and a diet poor in vitamins A, C, and E may also be factors.

Prevention

The single most important way to prevent lung cancer is to reduce smoking. Many programs educate schoolchildren about the dangers of smoking. Smoking cessation programs are available for people who desire to quit. Contact your local American Cancer Society chapter for smoking cessation programs that can be recommended to patients. See Chapter 26 for more information on smoking cessation.

Signs and Symptoms

Manifestations of lung cancer depend on the location of the tumor. Commonly, patients exhibit a cough with sputum production. These symptoms may be ignored by the patient because they are also associated with smoking. Repeated respiratory infections may occur, producing thick, purulent sputum. Sputum may become bloody (hemoptysis). The patient may experience dyspnea. If the airway becomes obstructed by the tumor, wheezing or stridor may be heard. Late signs include chest pain, weight loss, anemia, and anorexia.

Complications

PLEURAL EFFUSION. Fifty percent of patients with lung cancer develop pleural effusion. Pleural fluid collects in the pleural space as a result of irritation or obstruction of lymphatic or venous drainage by the tumor.

SUPERIOR VENA CAVA SYNDROME. If the tumor obstructs the superior vena cava, blood flow is interrupted, causing distention of the jugular veins and swelling of the chest, face, and neck. Diuretics may help relieve the fluid buildup. Radiation may be used to shrink the obstruction.

ECTOPIC HORMONE PRODUCTION. Some lung cancers produce **ectopic** hormones that mimic the body's own hormones. Ectopic production of antidiuretic hormone (ADH) can produce syndrome of inappropriate ADH production (SAIDH), which is associated with fluid retention. Ectopic production of adrenocorticotropic hormone (ACTH) can cause Cushing's syndrome. These disorders are discussed in Chapter 37.

ATELECTASIS AND PNEUMONIA. Atelectasis occurs when tumor growth prevents ventilation of areas of the lung. Patients with lung cancer also have a greater risk for pneumonia.

METASTASIS. Common sites of lung cancer metastasis include the brain, bones, opposite lung, liver, adrenal gland, and lymph nodes.

Diagnostic Tests

A complete medical history and physical examination are done to look for symptoms and risk factors for lung cancer. A chest x-ray examination is done to identify a mass. How-

---

ectopic: displaced

ever, all tumors may not show up on a radiograph. A CT scan and lung scan may be done to provide more specific information about the size and location of a tumor. Sputum is analyzed for abnormal cells. Brain and bone scans are done to find metastatic lesions.

Diagnosis is confirmed with a biopsy of the lesion. A biopsy specimen may be obtained via bronchoscopy, percutaneous biopsy (a needle through the skin guided by radiograph), or mediastinoscopy (use of an endoscope into the mediastinum to look for changes in mediastinal lymph nodes).

Medical and Surgical Treatment

Tumors are staged based on the tumor-node-metastasis (TNM) staging system. (See Chapter 10.) Staging helps determine appropriate treatment (Table 28–3). If NSCLC is localized and in an early stage, it may be cured with surgical removal of the tumor. This can be accomplished with a segmental or wedge resection, which removes only the affected lung segment. A **lobectomy** (removal of a lobe) or

---

lobectomy: lobe—lobe (of lung) + ectomy—excision

TABLE 28-3		STAGES OF LUNG CANCER
Cancer Type	**Stage**	**Characteristics**
Non–small cell lung cancer	I	No metastasis to lymph nodes. Atelectasis or pneumonia may be present.
	II	Cancer has spread to local lymph nodes.
	III	Cancer has invaded chest wall and usually has spread to lymph nodes.
	IV	Tumor has metastasized to distant organs and lymph nodes.
Small cell lung cancer	Limited	Cancer is limited to one side of the chest.
	Extensive	Cancer cells are found outside one side of the chest or in pleural fluid.

removal of an entire lung may be done in more advanced cases (Fig. 28–10). Surgery is contraindicated if the cancer has spread to the other lung or has metastasized to distant areas.

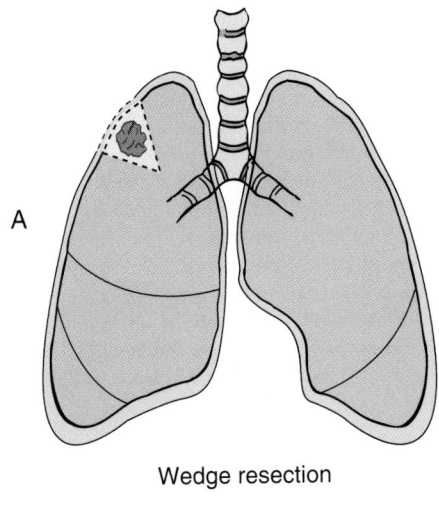

Wedge resection

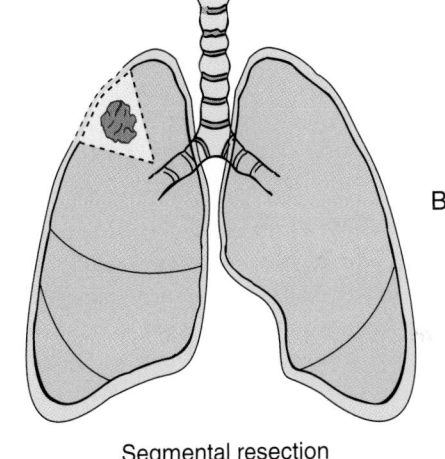

Segmental resection

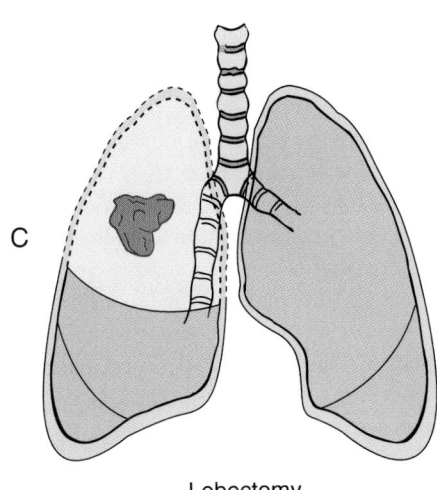

Lobectomy

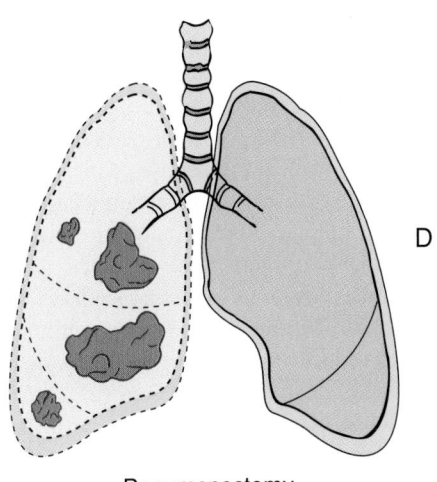

Pneumonectomy

Figure 28-10 Types of surgeries for lung cancer. *(A)* Wedge resection. *(B)* Segmental. *(C)* Lobectomy. *(D)* Pneumonectomy.

Chemotherapy is the treatment of choice in SCLC, because usually it has metastasized and may be widespread by the time of diagnosis. Radiation may be used in combination with chemotherapy. Surgery is not indicated. The goal of treatment is not cure but rather palliation of symptoms.

Radiation may be used to shrink the tumor to reduce symptoms in patients who are unable to undergo surgery. Both radiation and chemotherapy may be used before or after surgery as **adjuvant** treatments. For more information about lung cancer, visit the American Cancer Society Web site at www.cancer.org.

► NURSING PROCESS: THE PATIENT WITH LUNG CANCER

ASSESSMENT. Perform a complete biopsychosocial assessment of the patient with lung cancer. Assess and document respiratory rate and depth, skin and mucous membrane color, lung sounds, cough, and sputum amount and character. Ask the patient to rate the degree of pain and dyspnea on appropriate scales. Ask about appetite and weight loss, as well as symptoms of other complications. Note activity tolerance and fatigue.

In addition, the patient may be grieving about his or her illness and impending death. Assessment of the patient's coping strategies and support systems will help you plan care for psychosocial needs. When you recall the prognosis for patients with some types of lung cancer, the need to consider planning for terminal care becomes obvious (Ethical Consideration Box 28–8). The presence of a living will or durable power of attorney should be noted.

NURSING DIAGNOSIS. Possible diagnoses that may be experienced by the patient with lung cancer include impaired gas exchange, ineffective airway clearance, imbalanced nutrition: less than body requirements, pain, constipation related to opioid use, anticipatory grieving, and activity intolerance, among others.

PLANNING. A plan is formulated that will help the patient control episodes of dyspnea, maintain a clear airway, reduce anorexia and maintain body weight, promote comfort, and maintain regular bowel habits. Patients who have a terminal condition are assisted in coping with their impending death.

IMPLEMENTATION. The experience of dyspnea related to impaired gas exchange is frightening to patients and their families. Home oxygen therapy may be necessary. Positioning, relaxation, and breathing exercises can help reduce dyspnea and feelings of panic. Antianxiety drugs or morphine may also be helpful. Resting between activities reduces the demand for oxygen. Encourage the patient to avoid smoking and exposure to secondary smoke.

A clear airway can be promoted with a room humidifier and oral fluids to reduce viscosity of secretions, as well as regular coughing and deep breathing exercises. A nagging, nonproductive cough can be treated with an **antitussive** as

---

antitussive: anti—against + tussive—cough

BOX 28-8   ETHICAL CONSIDERATION

Mr. David Hamill, 88 years old, is admitted to a room on the surgical unit. He has been diagnosed with a metastatic tumor of the lung but does not yet know the diagnosis. Dr. Lester and the family have discussed the diagnosis. Dr. Lester decided not to tell Mr. Hammill the diagnosis because he believes that Mr. Hammill would become upset and depressed. Dr. Lester has written an order saying that the patient should not be told his diagnosis.

Two days ago, Mr. Hammill was taken to surgery for a thoracotomy. Since then, he has been asking the nurses, staff, and his family what the physician found in surgery and what the results of the pathology reports were. Dr. Lester has visited Mr. Hammill several times but has avoided talking about the diagnosis by saying that all the laboratory tests were not back yet. The family has been avoiding visiting the patient so that he will not ask them about the diagnosis. The family often asks the nurse when Mr. Hammill will be told his diagnosis. They believe the physician should tell him.

This is an unusual situation. Some elderly patients may not believe that they have a right to question a physician or be fully informed and will simply agree to whatever is suggested. Most families will express their needs, but there are some geographic areas where traditional physician-patient relationships follow a very paternalistic model. This situation provides an opportunity to examine autonomy, paternalism, and veracity.

Most nurses would be uncomfortable in a situation in which both patient and family want information. Would it be different if the patient and family were not asking for information? Would it be different if the patient was Ms. Hammill instead of Mr. Hammill? What degree of "truth" is required? What about partial truths and white lies? Can it ever be beneficial to not tell the truth? What does *paternalism* mean, and why might the physician be taking such a position with this patient? Does the hospital have an ethics committee? Could such a committee help? What options are available to the nurse or for the nurse to suggest to the family? Use the decision-making model described in Chapter 2 to determine possible alternative actions for the nurse to take.

---

ordered by the physician. Instruct the patient to notify the physician if hemoptysis is persistent. Exposure to powders, tobacco smoke, and aerosols increases airway irritation and should be eliminated. Suctioning may be necessary if the patient becomes too weak to cough effectively.

Anorexia is common in the patient with cancer. Nutrition can be maintained by eating frequent small meals. Nutritional supplements that are high in calories but easy to eat or drink may be used. A dietitian consultation is helpful. Antiemetics before meals may help control nausea. Use of spices to enhance flavor of foods may be used as the patient prefers. Mints may help reduce the metallic taste left in the mouth by some chemotherapeutic medications. Good mouth care is essential. Total parenteral nutrition may be necessary late in the disease. (See Nutrition Notes Box 28–6.)

Pain is controlled by opioids and supportive noninvasive therapies. See Chapter 9 for more information on pain control. Use of opioids for pain control necessitates attention to bowel function. Prevent constipation with the use of high-fiber foods and extra fluids if tolerated. If these conservative measures are ineffective, request an order for a bulk-forming agent, stool softener, or laxative.

Fatigue is battled with frequent rest periods and assistance with activities of daily living. Encourage the patient to identify and engage in those activities that are most important to him or her and to avoid unnecessary or undesirable activities.

The patient who is grieving should be allowed the opportunity to talk about his or her life and impending death and to express anger or sadness. You should be physically present but should not force verbalization unless the patient wishes to talk. Encourage the family to stay with the patient as much as the patient wishes. Contact a minister or spiritual counselor if the patient desires a referral.

Hospice care is available for the patient who has a terminal condition. This allows the family to have the support needed to care for the patient in his or her home or a home-like environment. Hospice nurses are knowledgeable in end-of-life care and are an invaluable resource to the family.

EVALUATION. Carefully consider the patient's individual goals when evaluating care. Is the patient comfortable and free from unnecessary dyspnea? Is the airway clear, and is nutrition maintained? Are medication side effects manageable? Have patients with terminal conditions come to terms with their impending death, and have they been able to do those things most important to them before their death?

▶ THORACIC SURGERY

A surgical incision made into the chest wall is called a **thoracotomy.** A thoracotomy may be performed for a number of reasons, including biopsy; removal of tumors, lesions, or foreign objects; to repair trauma following penetrating or crushing injuries; or to repair or revise structural problems. Open heart surgery also requires a thoracotomy and is discussed in Chapter 22.

Types of Thoracic Surgery
Pneumonectomy

A **pneumonectomy** is the surgical removal of a lung. This is usually done to treat lung cancer. It may also be used to treat severe cases of tuberculosis, bronchiectasis, or lung abscesses. Chest drainage is not usually used following a pneumonectomy, because once the lung is removed, the air in the thoracic cavity is absorbed and the cavity fills with serosanguineous fluid. At about 6 months after surgery, the fluid is coagulated and the thoracic cavity is stabilized.

Lobectomy

Lobectomy is the surgical removal of one lobe. This also may be done for lung cancer, tuberculosis, or another localized problem.

Resection

Resection refers to removal of a smaller amount of lung tissue; that is, less than one lobe. A segmental resection is the removal of one segment of a lobe; a wedge resection is removal of a small wedge of lung tissue.

Lung Transplantation

Lung transplant can benefit patients with a variety of serious pulmonary disorders, including pulmonary hypertension, emphysema, cystic fibrosis, and bronchiectasis. Either a single lung, both lungs, or heart and lungs have been successfully transplanted. Better criteria for selecting patients and donors, as well as advancements in surgical techniques, have improved outcomes for these patients.

▶ NURSING PROCESS: THE PATIENT UNDERGOING THORACIC SURGERY
Preoperative Nursing Care

Perform a thorough assessment before surgery, with a focus on the respiratory system. This gives a baseline against which to judge changes postoperatively. Routine preoperative teaching is done by the nurse and physician. The patient should understand that he or she will wake up in an intensive care environment. If at all possible, it is helpful to have the patient and family tour the intensive care unit before the surgery to decrease anxiety postoperatively. Prepare the patient for waking up after surgery with an endotracheal tube connected to a ventilator, oxygen, chest tubes, intravenous fluids, cardiac monitor, Foley catheter, and possibly an epidural catheter for pain control. Consult the surgeon for specific plans.

Advise the patient that position changes and early ambulation help prevent complications following surgery. Also instruct the patient in the use of an incentive spirometer and coughing and deep breathing techniques.

Postoperative Nursing Care

ASSESSMENT. Following thoracic surgery, patients initially are in an intensive care unit. Larger hospitals have special intensive care units specifically for surgical or thoracic patients. Here patients can be closely monitored for signs of complications. Frequent assessment of vital signs and hemodynamic stability; respiratory rate, depth, and effort; and lung sounds is performed. Remember that lung sounds are absent on the side of a pneumonectomy. An increase in pulse rate or a falling blood pressure may indicate internal bleeding and should be reported immediately. Oxygen saturation is monitored continuously. Often patients report an immediate improvement in breathing, because the pulmonary blood supply is no longer being routed to diseased lung tissue.

Assessment for tracheal deviation alerts you to the possible complication of mediastinal shift. The trachea is normally positioned straight above the sternal notch. If the trachea deviates from the midline position, the surgeon should

---

pneumonectomy: pneum—lung + ectomy—exicision

be notified immediately. Secretions are monitored and reported to the physician if they become thick, yellow or green, or foul smelling. Arterial blood gases are monitored closely. Chest tubes are usually present (except following pneumonectomy) and are monitored as in Chapter 26. Pain is assessed using a pain rating scale, and incision sites are monitored for redness, edema, or drainage. If the patient is mechanically ventilated, additional assessment of the endotracheal tube and ventilator settings will be necessary.

NURSING DIAGNOSIS, PLANNING, AND IMPLEMENTATION. After thoracic surgery, many respiratory diagnoses may be appropriate. Following are the most common.

Ineffective Airway Clearance. Regular suctioning is necessary while the patient is intubated to keep the airways free of secretions. Once extubated, the patient is reminded to cough and deep breathe regularly. Postoperative pain must be controlled for the patient to be able to cough effectively. Intravenous or oral fluids, humidified oxygen, and a room humidifier will help keep secretions thin.

Acute Pain. Pain control is important for the patient to be able to ambulate and deep breathe and cough effectively. Some institutions are successfully using epidural analgesia following thoracotomy. Because narcotics depress respirations, it is important to monitor respiratory rate before their administration. The patient is also taught to splint the incision while coughing.

Impaired Gas Exchange. Maintaining Fowler's position and use of an incentive spirometer following extubation encourage the patient to deep breathe and maximize oxygenation. Oxygen and bronchodilators are administered as ordered. Some authorities believe patients should be positioned with the operative side up, others with the operative side down. The surgeon should be consulted for specific positioning orders.

Impaired Physical Mobility. Range-of-motion exercises are performed to prevent contracture of the arm and shoulder on the affected side. This may be done passively at first, then actively when the patient is able. The patient is out of bed on the first or second postoperative day and ambulates as tolerated.

Risk for Infection. Sterile technique for suctioning and standard infection control precautions are used. The patient is extubated as soon as possible to reduce the risk of postoperative infection.

EVALUATION. The patient's airway should remain clear, and secretions should be easily coughed up. The patient should report an acceptable comfort level and be able to cough, deep breathe, and ambulate without excessive discomfort. The patient's breathing should be unlabored, with a respiratory rate of 12 to 20 per minute. The patient's affected arm and shoulder should maintain full range of motion. Signs of infection should be absent.

REVIEW QUESTIONS

1. Which of the following assessment findings in the patient with pneumonia most indicates a need to remind the patient to cough and deep breathe?
    a. The patient complains of chest pain.
    b. The patient has removed her oxygen.
    c. The nurse auscultates wheezes and crackles.
    d. The nurse notes a fever of 101° F.

2. The nurse is caring for a patient with tuberculosis who puts his light on because he needs to use the bathroom. There are green surgical masks in the isolation cabinet outside his room. What should the nurse do?
    a. Fit the mask firmly to his or her face before going into the room.
    b. Place two masks together and fit them firmly to the face.
    c. Ask the patient to put on a mask before entering the room.
    d. Ask the patient to wait while the nurse obtains a special high-efficiency mask.

3. Which of the following assessment findings does the nurse expect in the patient with emphysema?
    a. Purulent sputum
    b. Diminished breath sounds
    c. Generalized edema
    d. Dull chest pain

4. Which of the following assessment findings in the patient with pneumothorax does the nurse report immediately?
    a. Positioning of the trachea toward the unaffected side
    b. Frequent dry cough
    c. Moderate pain at the chest tube site
    d. Diminished breath sounds over the affected area

5. As the nurse enters Mr. Jones's room, he notes that the patient has become confused and combative over the past hour. Which of the following actions is appropriate first?
    a. Assess Mr. Jones; check to see if his oxygen is flowing correctly.
    b. Page the physician stat.
    c. Put up Mr. Jones's side rails and apply soft restraints.
    d. Administer an oral sedative.

6. Which of the following interventions is most appropriate for the patient with an ineffective breathing pattern?
   a. Encourage the patient to cough and deep breathe.
   b. Encourage oral fluids.
   c. Teach the patient controlled diaphragmatic breathing.
   d. Allow the patient to rest between activities.

7. Mrs. Jackson had an abdominal hysterectomy yesterday. The nurse enters her room and finds her acutely short of breath, with a look of panic in her eyes. Which of the following additional symptoms is most important as the nurse decides what to do?
   a. Mrs. Jackson complained of pain in her left leg earlier this morning.
   b. Mrs. Jackson states that she also has a headache.
   c. Mrs. Jackson has a recent history of an upper respiratory infection.
   d. Mrs. Jackson has not eaten in 24 hours.

REFERENCES

1. World Health Organization: Tuberculosis Home Page. www.who.int/gtb/, 2000.
2. Yeaw, E: How position affects oxygenation: Good lung down? Am J Nurs 92:26, 1992.
3. American Cancer Society: Cancer Facts and Figures 2001, www.cancer.org, 2001.
4. National Center for Health Statistics: Health, United States, 1998: Socioeconomic Status and Health Chartbook. Public Health Service, Hyattsville, MD, 1998.
5. Cardenas, VM, Thun, MJ, Austin H, et al: Environmental tobacco smoke and lung cancer mortality in the American Cancer Society's Cancer Prevention Study II. Cancer Causes Control 8:57, 1997.

Answers to CRITICAL THINKING

Mr. Smith

1. A complete respiratory history is taken as described in Chapter 26. An open-ended question such as "What happened to bring you to the hospital?" elicits information about the incident. In addition, questions to determine mental status and ability to make decisions and function safely on his own are appropriate. If any concerns arise, a social service consultation might be helpful for discharge planning.

2. Mr. Smith should be instructed to always read label warnings before using any cleaning products in the future and to never mix bleach and ammonia!

Jim

1. Ask questions based on the WHAT'S UP? format:
   Where (not applicable)
   How does it feel? Does the coughing cause chest pain? Are you short of breath?
   Aggravating and alleviating factors. What makes the cough worse? What seems to help? Do you use any techniques at home that are helpful?
   Timing. How often do you cough during a day? Is it interfering with sleep and rest?
   Severity. How bad is it on a scale of 0 to 10? How much sputum are you coughing up? What color is it?
   Useful other data. Are you experiencing any other symptoms with your cough (such as shortness of breath, nausea, loss of appetite)?
   Patient's perception. Is it better or worse than usual today? How can I help? (The patient with long-standing disease often knows what will help but is hesitant to ask.)

2. The most appropriate nursing diagnosis is ineffective airway clearance related to excessive secretions and ineffective cough.

3. Provide hydration with oral liquids and a room humidifier to liquefy secretions. Administer expectorants as ordered. Instruct the patient in coughing and deep breathing exercises to increase the effectiveness of his cough. Provide good oral care following expectoration of sputum to freshen the patient's mouth. Obtain an order for chest physiotherapy to help loosen and drain secretions.

4. 300 mg/200mg ×5 mL = 7.5 mL

5. "Patient coughed up 200 mL of bright yellow, foul-smelling sputum. Lungs have scattered crackles and wheezes throughout after coughing episode. **Expectorant** given; fluids encouraged. Mouth care provided."

Mr. Franklin

1. You need to do several things at once. You will begin by speaking in a calm voice and trying to help Mr. Franklin to calm himself by doing pursed lip breathing. Assure him that you will help him and won't leave. At the same time, check his oxygen to make sure it is on the ordered number of liters and that his tubing is not kinked or disconnected. Grab the bedside table for him to lean on. Call for someone to page a respiratory therapist to do an NMT if ordered. Have someone bring a pulse oximeter to check his oxygen saturation. Also call for the registered nurse (RN) to administer IV morphine if ordered. All this should take about 1 minute! Once Mr. Franklin is a bit calmer, you can find out what happened. Did the exertion of moving to the bedside commode cause his dyspnea? Check his vital signs and lung sounds, and work with the RN to determine if this represents a change in Mr. Franklin's condition that should be reported to the physician.

2. Teach Mr. Franklin that he should probably stay on bedrest until his acute exacerbation is resolved. Once he is able to start moving around, he should call for help to get up. Review his controlled breathing exercises, which he can use during movement, and encourage rest between activities.

3. "3:00: Patient up on BSC, RR 36 and labored, color gray, appeared very apprehensive. O_2 on at 4 L per min per NC, assisted to lean on overbed table. Encouraged pursed lip breathing. VS 146/64, 102, 36, SaO_2 86%. RT paged; administered PRN NMT. LS diminished, no cough. At 3:15, patient appears much more calm, RR 24 and less labored, SaO_2 90%."

expectorant: ex—out + pect—breast

UNIT SIX BIBLIOGRAPHY

Campolo, S: Spontaneous pneumothorax. American Journal of Nursing 2, 1997.

Carroll, P: Exploring chest drain options. RN 63:10, 2000.

Chang, VM: Protocol for prevention of complications of endotracheal intubation. Critical Care Nurse 15:19, 1995.

Bell, SD: use of Passy-Muir tracheostomy speaking valve in mechanically ventilated neurological patients. Critical Care Nurse 16:1, 1996.

Dunn, N: Keeping COPD patients out of the ED. RN 64:2, 2001.

Hayes, DD: Stemming the tide of pleural effusions. Nursing 2001 31:5, 2001.

Harding, M: Preparing patients for the effects of laryngectomy. Nursing Times 90(32):36, 1994.

Jablonski, RAS: If ventilator patients could talk. RN 58:32, 1995.

Lasater-Erhard, M: The effect of patient position on arterial oxygen saturation. Critical Care Nurse 15:31, 1995.

Lazzara, D: Respiratory distress: Loosening the grip. Nursing 2001 31:6, 2001.

Mainous, AG, et al: Antibiotics and upper respiratory infection. Journal of Family Practice 42:4, 1996.

Marion, BS: A turn for the better: Prone positioning of patients with ARDS. American Journal of Nursing 101:5, 2001.

Martinez, FJ, et al: Lung-volume reduction improves dyspnea, dynamic hyperinflation, and respiratory muscle function. American Journal of Respiratory and Critical Care Medicine 6:155, 1997.

Perkins, LA, and Shortall, SP: Ventilation without intubation. RN 63:1, 2000.

Petty, TL: COPD: Interventions for smoking cessation and improved ventilatory function. Geriatrics 55:12, 2000.

Ruppert, RA: The last smoke. American Journal of Nursing 99:11, 1999.

Sandhu, AK, and Mossad, SB: Influenza in the older adult: Indications for the use of vaccine and antiviral therapy. Geriatrics 56(1):43–51, 2001.

Skoner, DP: Allergic rhinitis: Definition, epidemiology, pathophysiology, detection, and diagnosis. Journal of Allergy and Clinical Immunology 108:1, 2001.

Somerson, SJ, et al: Mastering emergency airway management. American Journal of Nursing 96:24, 1996.

Swanlund, SL: Body positioning and the elderly with adult respiratory distress syndrome: Implications for nursing care. Journal of Gerontological Nursing 22:46, 1996.

Woodruff, DW: Iatrogenic injuries: Pneumothorax. RN 62:9, 1999.

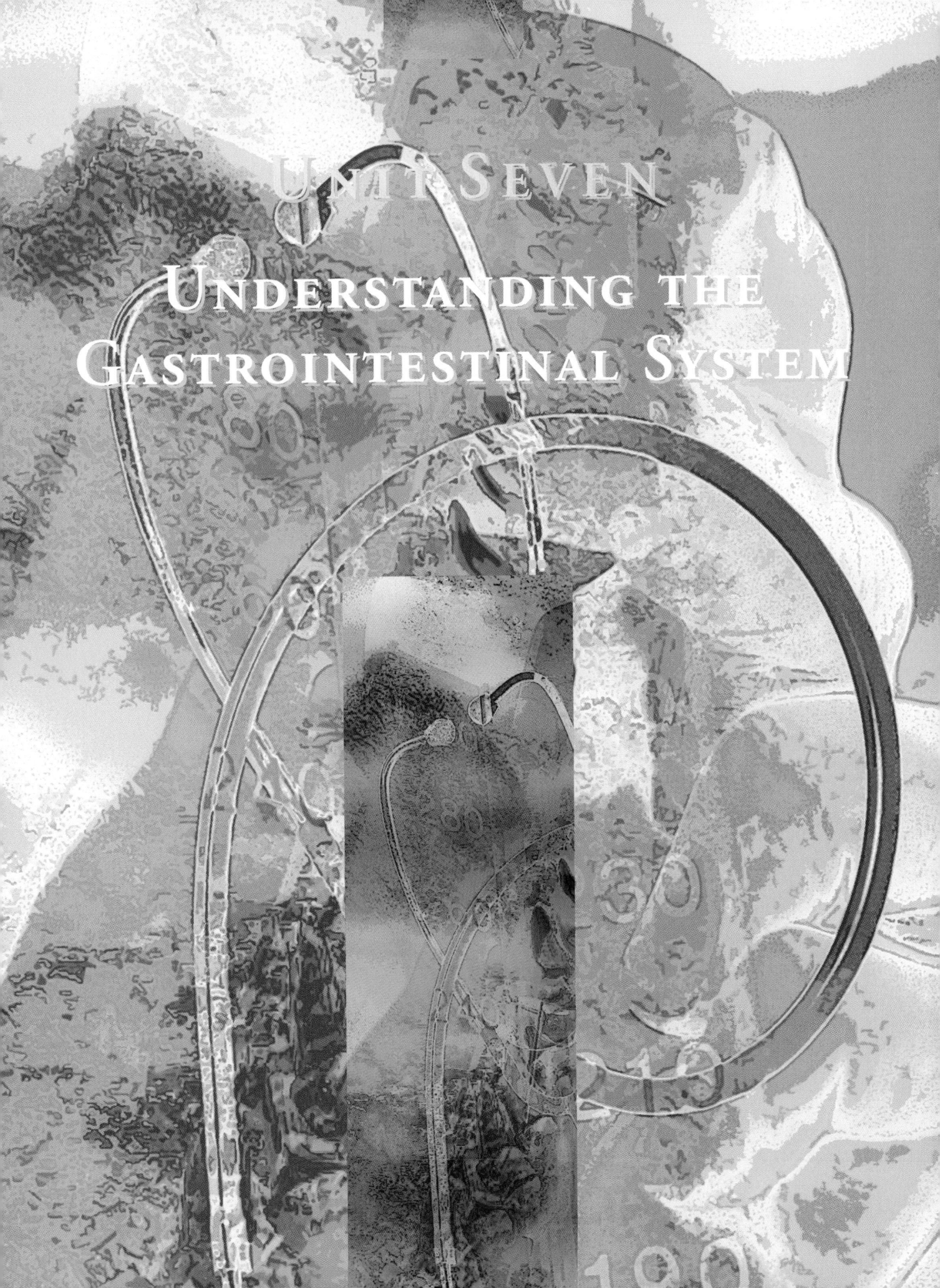

Unit Seven

Understanding the Gastrointestinal System

29

GASTROINTESTINAL SYSTEM FUNCTION, ASSESSMENT, AND THERAPEUTIC MEASURES

Sharon Gordon Dawson

KEY TERMS

basal cell secretion test (**BAY**-zuhl SELL see-**KREE**-shun TEST)

bowel sounds (BOW'L SOWNDS)

carcinoembryonic antigen (**KAR**-sin-oh-EM-bree-ah-nik **AN**-ti-jen)

colonoscopy (KOH-lun-**AHS**-kuh-pee)

endoscopy (**EN**-dohs-kuh-pee)

esophagogastroduodenoscopy (ee-**SOFF**-ah-go-GAS-troh-doo-AH-den-**AHS**-kuh-pee)

esophagoscopy (ee-soff-ah-**GAHS**-kuh-pee)

fluoroscope (**FLOOR**-o-skohp)

gastric acid stimulation test (**GAS**-trik ASS-id STIM-yoo-**LAY**-shun TEST)

gastric analysis (**GAS**-trik ah-**NAL**-i-sis)

gastroscopy (gas-**TRAHS**-kuh-pee)

gastrostomy (gas-**TRAHS**-toh-mee)

gavage (gah-**VAZH**)

impaction (im-**PAK**-shun)

lavage (lah-**VAZH**)

lower gastrointestinal series (**LOH**-er GAS-troh-in-**TES**-ti-nuhl **SEER**-ees)

occult blood test (ah-**KULT** BLUHD TEST)

peripheral parenteral nutrition (puh-**RIFF**-uh-ruhl par-**EN**-te-ruhl new-**TRISH**-un)

peristalsis (paris-**TALL**-sis)

proctosigmoidoscopy (PROK-toh-SIG-moy-**DAHS**-kuh-pee)

upper gastrointestinal series (**UH**-per GAS-troh-in-**TES**-ti-nuhl **SEER**-ees)

QUESTIONS TO GUIDE YOUR READING

1. What are the structures of the gastrointestinal system?
2. What are the functions of each organ of the gastrointestinal system?
3. How does age affect the gastrointestinal system?
4. How would you perform a gastrointestinal assessment on a patient with actual or potential problems of the mouth, esophagus, stomach, and intestines?
5. How would you prepare, teach, and provide follow-up care for patients having various diagnostic tests of the gastrointestinal tract?
6. What therapeutic interventions would you provide for patients with each type of gastrointestinal disease?

► REVIEW OF NORMAL GASTROINTESTINAL ANATOMY AND PHYSIOLOGY

The alimentary tube is part of the digestive system (Fig. 29–1). It extends from the mouth to the anus and consists of the oral cavity, pharynx, esophagus, stomach, small intestine, and large intestine (or colon). Digestion begins in the oral cavity and continues in the stomach and small intestine. Most absorption of nutrients takes place in the small intestine. The large intestine is where the majority of water is reabsorbed from digested food. Undigestible material, mainly cellulose, is then eliminated from the large intestine.

Oral Cavity and Pharynx

The boundaries of the oral cavity are the hard and soft palates superiorly, the cheeks laterally, and the floor of the mouth inferiorly. Within the oral cavity are the teeth and tongue and the openings of the ducts of the salivary glands.

The teeth begin mechanical digestion, the physical breakup of food into smaller pieces to create more surface area for the chemical digestion brought about by enzymes. The roots of the teeth are in sockets in the jawbones (the mandible and maxillae). The gums, or gingiva, cover the jawbones and surround the bases of the crowns (tops) of the teeth. The tooth sockets are lined with a periodontal membrane that produces a bonelike cement to anchor the teeth.

The tongue is made of skeletal muscle innervated by the hypoglossal nerve (twelfth cranial nerve). The papillae on the upper surface of the tongue contain taste buds, innervated by the facial and glossopharyngeal nerves (seventh and ninth cranial). The tongue is important for chewing because it keeps food between the teeth. Elevation of the tongue is the first step in swallowing.

The three pairs of salivary glands are the parotid, submandibular, and sublingual glands. Their ducts carry saliva to the oral cavity. The presence of anything in the mouth increases the rate of secretion; this is a parasympathetic response mediated by the facial and glossopharyngeal nerves. Saliva is mostly water, which is used to dissolve food for tasting and moisten the food for swallowing. The only digestive enzyme in saliva is amylase, which digests starch to maltose. Usually, however, food does not remain in the mouth long enough for amylase to have any significant effect.

The pharynx is a muscular tube that is a passageway for food exiting the oral cavity and entering the esophagus. When a mass of food is pushed backward by the tongue, the constrictor muscles of the pharynx contract as part of the swallowing reflex. This reflex is regulated by the medulla.

Esophagus

The esophagus is about 10 inches long and carries food from the pharynx to the stomach. No digestion takes place in the esophagus. **Peristalsis** (rhythmic contraction of muscles) of the muscle layer in the wall of the esophagus is one-way; food reaches the stomach even if the body is upside down. At the junction with the stomach, the lumen of the esophagus is surrounded by the lower esophageal sphincter (LES; gastroesophageal sphincter, or cardiac sphincter), a circular smooth muscle. The LES relaxes to permit food to enter the stomach and then contracts to prevent the backup of stomach contents. Incomplete closure of the LES may allow gastric juice to splash up into the esophagus.

Stomach

The stomach is in the upper left abdominal quadrant, to the left of the liver and in front of the spleen. It is a saclike organ that extends from the esophagus to the duodenum of the small intestine. Some digestion takes place in the stomach, and it also serves as a reservoir for food so that digestion may take place gradually.

The parts of the stomach are shown in Figure 29–2. The LES provides the opening from the esophagus to the stom-

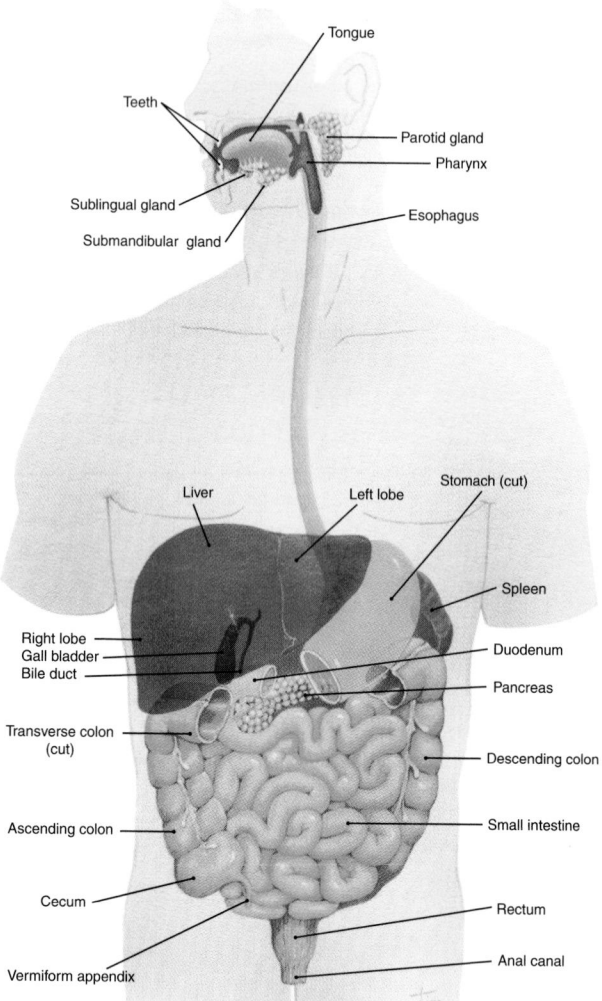

Figure 29–1 Anterior view of the digestive system. (From Scanlon, VC, and Sanders, T: Essentials of Anatomy and Physiology, ed 3. FA Davis, Philadelphia, 1999, p 353, with permission.)

The author acknowledges the contribution to this chapter by Gary Sanders Lott.

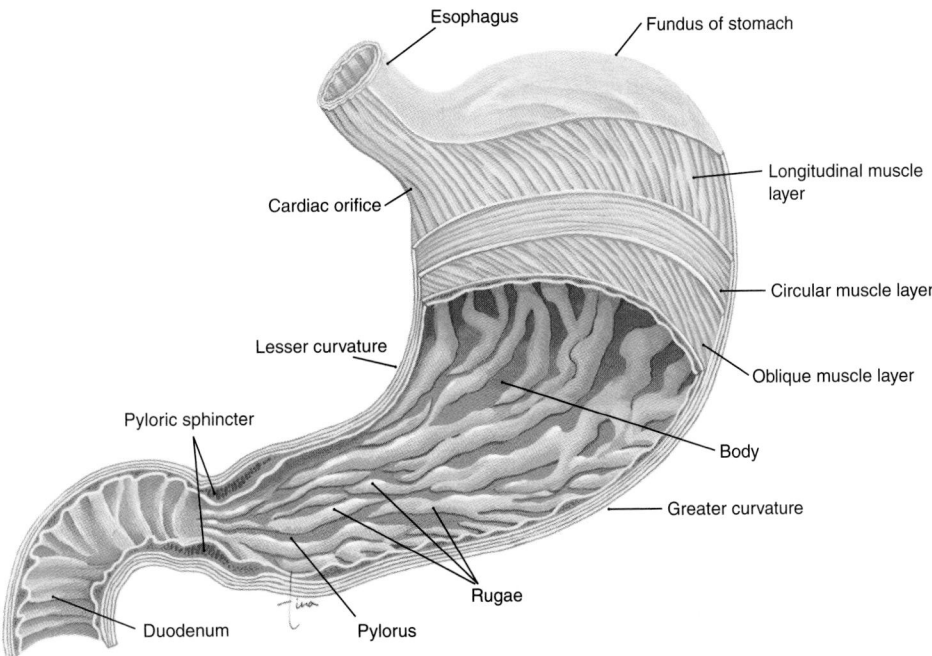

Figure 29-2   Stomach: anterior view and partial section. (From Scanlon, VC, and Sanders, T: Essentials of Anatomy and Physiology, ed 3. FA Davis, Philadelphia, 1999, p 359, with permission.)

ach. The fundus forms the upper curve of the stomach. The body of the stomach is the large, central portion, bounded laterally by the greater curvature and medially by the lesser curvature. The pylorus is adjacent to the duodenum, and the pyloric sphincter surrounds the junction of the two organs.

When the stomach is empty, the mucosa has folds called rugae. The rugae flatten out as the stomach fills and permit expansion of the lining. The mucosa contains gastric pits, the glands of the stomach that produce gastric juice. Gastric juice is mostly water and contains mucus, pepsinogen, and hydrochloric acid. Pepsinogen is an inactive enzyme that is changed to active pepsin by hydrochloric acid; pepsin begins the digestion of proteins to polypeptides. Hydrochloric acid creates the pH of 1 to 2 that is necessary for pepsin to function and to kill most microorganisms that enter the stomach.

Gastric juice is secreted at the sight or smell of food; this is a parasympathetic response. The presence of food in the stomach stimulates the secretion of the hormone gastrin by the gastric mucosa. Gastrin increases the secretion of gastric juice.

The stomach wall has three layers of smooth muscle: circular, longitudinal, and oblique. These provide for very efficient mechanical digestion to change food to a thick liquid called chyme. The pyloric sphincter contracts when the stomach is churning food and relaxes at intervals to allow small amounts of chyme to pass into the duodenum, then contracts again to prevent the backup of intestinal contents into the stomach. Carbohydrates are most readily digested by the stomach, followed by proteins and fats.

Small Intestine

The small intestine is about 1 inch in diameter and approximately 20 feet long. Within the abdominal cavity, the coils of the small intestine are encircled by the colon. The small intestine extends from the stomach to the cecum of the colon. The duodenum is the first 10 inches and contains the hepatopancreatic ampulla (ampulla of Vater), the entrance of the common bile duct. The jejunum is about 8 feet long, and the ileum is about 11 feet in length.

Digestion is completed in the small intestine, and the end products of digestion are absorbed into the blood and lymph. Bile from the liver and enzymes from the pancreas function in the small intestine (Table 29–1). When chyme enters the

TABLE 29-1	DIGESTIVE SECRETIONS		
Organ	**Enzyme or Other Secretion**	**Function**	**Site of Action**
Salivary glands	Amylase	Converts starch to maltose	Oral cavity
Stomach	Pepsin	Converts proteins to polypeptides	Stomach
	Hydrochloric acid	Changes pepsinogen to pepsin; maintains pH 1–2; destroys pathogens	Stomach
Liver	Bile salts	Emulsifies fats	Small intestine
Pancreas	Amylase	Converts starch to maltose	Small intestine
	Lipase	Converts emulsified fats to fatty acids and glycerol	Small intestine
	Trypsin	Converts polypeptides to peptides	Small intestine
Small intestine	Peptidases	Converts peptides to amino acids	Small intestine
	Sucrase, maltase, lactase	Converts disaccharides to monosaccharides	Small intestine

duodenum, the intestinal mucosa produces the enzymes sucrase, maltase, and lactase, which complete the digestion of disaccharides to monosaccharides, and the peptidases, which complete the digestion of proteins to amino acids.

The absorption of nutrients requires a large surface area, and the small intestine has extensive folds for this purpose. The circular folds (plica circulares) are macroscopic folds of the mucosa and submucosa. Villi are folds of the mucosa, and microvilli are microscopic folds of the cell membranes on the free surface of the intestinal epithelial cells. Within each villus is a capillary network and a lymph capillary called a lacteal. Water-soluble nutrients (monosaccharides, amino acids, minerals, water-soluble vitamins) are absorbed into the blood in the capillary networks. Fat-soluble vitamins and fatty acids and glycerol are absorbed into the lymph in the lacteals.

Large Intestine

The large intestine extends from the ileum of the small intestine to the anus. It is about 5 feet long and 2.5 inches in diameter. The cecum is the first part, and at its junction with the ileum is the ileocecal valve, which prevents backup of colon contents into the small intestine. Attached to the cecum is the small, dead-end appendix, apparently a vestigial organ (an incompletely developed structure that was more developed in a previous stage of the species) for people, but one that may become a site of infection.

The other parts of the colon are the ascending, transverse, and descending colon, which encircle the small intestine; the sigmoid colon, which turns medially and downward; the rectum, which is about 6 inches long; and the anal canal, the last inch that surrounds the anus (clinically, the terminal end of the colon is usually referred to as the rectum).

Although no digestion takes place in the colon, its functions are important. The colon temporarily stores and then eliminates undigestible material. The mucosa absorbs significant amounts of water and minerals, as well as the vitamins produced by the normal bacterial flora.

Elimination of feces is accomplished by the defecation reflex, a spinal cord reflex over which voluntary control may be exerted. When peristalsis propels feces into the rectum, receptors in the smooth muscle layer detect the stretching and generate impulses to the spinal cord. The returning motor impulses cause contraction of the smooth muscle of the rectum and relaxation of the internal anal sphincter, which surrounds the anus. Surrounding the internal sphincter is the external anal sphincter, which is made of skeletal muscle and may be voluntarily contracted to prevent defecation.

Aging and the Gastrointestinal System

Many changes occur in the aging gastrointestinal (GI) system (Gerontological Issues Box 29–1). The sense of taste becomes less acute, and there is greater likelihood of periodontal disease and oral cancer. There may be difficulties with chewing if teeth have been lost. Secretions throughout the GI tract are reduced, and the effectiveness of peristalsis

---

BOX 29-1 GERONTOLOGICAL ISSUES

Age-related Changes in the Gastrointestinal System

Mouth
Tooth enamel becomes harder and more brittle.
Tongue atrophy results in less acute taste sensations.
Sweet taste sensation is lost to a greater degree than sour, salt, and bitter.
Saliva production decreases by 33%.
Esophagus
Esophageal motility decreases, the esophagus dilates, and emptying is slower.
Gag reflex is weaker.
Stomach
Decreased mobility of stomach results in decreased gastric emptying.
Gastric hydrochloric acid production is decreased.
Large and Small Intestines
Fat absorption is slower in the intestine.
Large and small intestine begin to atrophy.
Mucous secretions and elasticity of rectal wall are decreased.
Weakness of intestinal wall can result in diverticulosis.
Absorption of vitamins B_1 and B_{12}, calcium, and iron is faulty.

---

diminishes because of loss of muscle elasticity and slowed motility. Indigestion may become more common, especially if the LES loses its tone, and there is greater chance of peptic ulcer. In the colon, diverticula may form. Constipation may be a problem, as may hemorrhoids. The risk of colon cancer also increases with age.

▶ NURSING ASSESSMENT

Nursing assessment of the GI system includes a patient history and physical examination (Box 29–2).

BOX 29-2 **Nursing Assessment of the Gastrointestinal System**

1. Visually inspect the abdomen for size, shape, and color of skin.
2. Does the patient have bowel sounds? Are they heard in all four quadrants? How long do bowel sounds last in each quadrant? What do they sound like?
3. Is the abdomen hard or soft? Is it distended with fluid? Does the patient have pain in the abdomen? Are there any masses?
4. Has the patient had any nausea or vomiting in the preceding 24 hours?
5. What type of diet is the patient on? Is the diet tolerated?
6. Can the patient swallow without difficulty?
7. What is the patient's weight and height? Has the patient had any recent changes in weight?
8. What are the patient's bowel patterns? How often does the patient usually have a bowel movement? What is the color? What is the consistency? Does the patient have a colostomy? If so, what does the stoma look like?
9. What drugs are being prescribed? How do the drugs relate to the GI system? Are they working? How do you know?
10. What laboratory values relate to the patient's GI system? Are they normal or abnormal? What do the results mean?

Subjective Data
Health History

Demographic data is obtained, including travel history, which may help in diagnosing the cause of GI symptoms such as diarrhea. Assessment of current signs and symptoms includes asking the WHAT'S UP? questions: where it is, how it feels, aggravating and alleviating factors, timing, severity, useful data for associated symptoms, and perception by the patient of the problem. Also documented are the patient's normal bowel pattern; changes in bowel patterns or habits; a history of any gastrointestinal diseases, such as ulcers, cancer, Crohn's disease, or colitis; or an unexplained weight loss or gain. Information about previous GI surgeries is obtained. Signs or symptoms of disease, such as bloody or tarry stools, rectal bleeding, stomach pain, or abdominal pain, are also noted.

Medications

The patient is asked about medication use such as nonsteroidal anti-inflammatory drugs (NSAIDs), aspirin, vitamins, laxatives, enemas, or antacids. Heavy use of medications that can cause irritation and bleeding in the GI tract, such as NSAIDs or aspirin, should be carefully noted. Elderly patients with arthritis often use these types of medications for pain control. The patient's knowledge of the side effects of these medications should be assessed to identify teaching needs. Elderly patients may use laxatives regularly and develop a dependence on them. Teaching may be needed on normal bowel patterns and laxative use.

CRITICAL THINKING

Mrs. Todd

Mrs. Todd, 74, has arthritis and takes eight aspirin daily for pain control. She is scheduled for an **esophagogastroduodenoscopy** (EGD) for suspected GI bleeding caused by unexplained anemia.

1. What is a likely cause of the GI bleeding?
2. What could you do to help prevent future bleeding episodes for Mrs. Todd?

Answers at end of chapter.

Nutritional Assessment

A diet history should include usual foods and fluids, allergies, appetite patterns, swallowing difficulty, and use of nutritional and herbal supplements (Cultural Consideration Box 29–3). A patient food diary can be used to provide more detailed information. Elderly patients may be on fixed incomes, which may limit their food budget and result in meal skipping or purchasing of inexpensive foods. The elderly patient's daily food intake should be explored, especially if malnutrition, financial limitations, or living alone is noted (Home Health Hints).

Also explored during a nutritional assessment are patterns of gastric acid reflux, indigestion, heartburn, nausea,

---

BOX 29-3 CULTURAL CONSIDERATION

Questions to ask when performing a cultural nutritional assessment:

1. What types of your cultural foods are available in your community?
2. What are your preferred foods over foods available and eaten?
3. Which foods do you most commonly consume?
4. How and where are your foods chosen and purchased?
5. Who prepares the food in your household?
6. Who purchases the food in your household?
7. How is your food stored for future use?
8. How is your food prepared before eaten?
9. How is any uneaten food discarded?
10. What foods do you eat to maintain your health?
11. What foods do you avoid to maintain your health?
12. What foods do you eat when you are ill?
13. What foods do you avoid when you are ill?

HOME HEALTH HINTS

- Be familiar with community nutritional support services: Women, Infant's, and Children (WIC) program, elderly nutrition sites, Meals on Wheels, school food programs, and government surplus food programs.

- Assess patient's food preparation facilities to ensure the patient's nutritional needs can be met. Some elderly patients may have outdated or spoiled food in their refrigerators or cupboards because they are unable to see dates or mold growing on foods.

- Ensure that patients are able to use appliances to heat food safely. Patients with limited vision may not see gas flames and can ignite their clothing. Confused patients might try to heat foods in the cardboard containers. If the patient is able to obtain and learn to use a microwave, it may be a safer cooking appliance.

- A feeding tube can be prevented from kinking by slipping a split straw lengthwise around the area that tends to kink and lightly taping over the split in the straw.

- Wire coat hangers make good hooks for enteral feeding solutions bags. They can be bent and hung over doorways or closet bars.

- If a 60-mL feeding syringe is not available for a bolus tube feeding or tube flushing, a measuring cup and funnel can be used.

---

vomiting, diarrhea, constipation, flatulence, and bowel incontinence, all of which may interfere with proper nutrition (Gerontological Issues Box 29–4). Acid reflux can be assessed by asking patients if they experience reflux with a bile taste or awaken with an unpleasant taste in their mouth.

Family History

Family history of close relatives with conditions that may influence the patient's GI status is assessed. Some GI problems such as colon cancer are thought to be hereditary.

BOX 29-4 GERONTOLOGICAL ISSUES

A complete bowel history should be obtained for older patients before beginning a bowel program. A bowel history includes the following:

Normal bowel evacuation pattern
Characteristics of stool
Presence of any bleeding or mucus with the stool
Use of products and medications to stimulate or slow bowel function
Report of usual diet
Amount of fluids—number and size of beverages glasses per day (beverages containing caffeine, such as coffee, tea, and sodas, do not count as fluids because of the diuretic effect of caffeine)
Exercise and physical activity
Rituals and practices related to bowel function

Cultural Influences

Many cultures have special dietary practices and restrictions (Cultural Consideration Box 29–5). Understanding these cultural influences, respecting them, and assisting the patient to maintain desired cultural practices is important for nutritional maintenance.

Objective Data
Height, Weight, and Body Mass Index

When the GI system is assessed, the patient's height and weight are obtained for planning care. The patient's ideal body weight according to height is obtained using current reference charts. Body mass index (BMI) is calculated to measure body fat and used with waist circumference meas-

---

BOX 29-5 CULTURAL CONSIDERATION

Most societies of the world use various foods and herbs for maintaining health. With increased attention to herbal therapies in the United Sates, the U.S. National Institutes of Health is studying 40 foods that are thought to fight disease. Among them are garlic, carrots, soy, cranberry juice, licorice, and green tea. Green tea has been used in Japan and China for centuries as a means of maintaining health and preventing disease.

African-Americans

Obesity is seen as positive among many African-Americans. They often view individuals who are thin as "not having enough meat on their bones." One needs to have adequate meat on his or her bones so that when an illness occurs, one can afford to lose weight. Many African-American diets are high in animal fat and fried foods and low in fiber, fruits, and vegetables.

Appalachians

The diet of some Appalachians is deficient in vitamin A, iron, and calcium. The nurse working with this population needs to do a dietary assessment and teach patients food selections that include adequate vitamin A, iron, and calcium.

Arabs

Many Arabs eat food only with their right hand because it is regarded as clean. The left hand, commonly used for toileting, is considered unclean. Thus the nurse should feed the Arab patient with the right hand regardless of the nurse's dominant handedness. Additionally, some may not drink beverages with their meals because some individuals consider it unhealthy to eat and drink at the same meal. Likewise, mixing hot and cold foods may be seen as unhealthy.

Muslim Arabs may refuse to eat meat that is not Halal (slaughtered and prepared in a ritual manner). Because Muslim Arabs are prohibited from ingesting alcohol or eating pork, they may refuse medication that includes alcohol, such as mouthwashes, toothpaste, alcohol-based syrups and elixirs, and products derived from pigs, such as insulin, gelatin-coated capsules, and skin grafts. However, if no substitute is available, Muslims are permitted to use these preparations.

The condition of the alimentary tract has priority over all other body parts in the Arab's perception of health. Gastrointestinal complaints are the most common reason Arab-Americans seek care.

Asian Indians

Among Asian Indians, nutritional deficiencies are patterned from their region of emigration. For example, beriberi (thiamine deficiency) is found in people emigrating from rice-growing areas. Pellagra (niacin deficiency) causing skin and mental disorders and diarrhea is found in people emigrating from maize-millet areas. Thiamine deficiency is common among people mostly dependent on rice. Thorough milling of rice, washing rice before cooking, and allowing the cooked rice to remain overnight before consumption the following day result in the loss of thiamine.

Commitment to the sacred cow concept has an impact on Hindus by encouraging dairy and milk use. However, lactose intolerance affects more than 10% of adults. The adequacy or inadequacy to digest lactose may be due to genetic differences among Asian Indians.

Goiter is prevalent among some Asian Indian immigrants resulting from an iodine deficiency in food and water from their homeland. Fluorosis occurs in other parts of India resulting from drinking water high in fluoride. Osteomalacia is prevalent where diets are deficient in calcium and vitamin D. Endemic dropsy is prevalent among Asian Indians emigrating from West Bengal, resulting from using mustard oil for cooking. The nurse needs to be aware of these conditions and their causes when working with Hindus and Asian Indians and teach patients prevention.

Brazilians

There is an increase in gastrointestinal distress among Brazilians when they first come to the United States, partially because many have a lactose intolerance and partially because of different methods of milk pasteurization. The nurse can assist the patient in identifying alternative food sources for Brazilian patients to obtain needed calcium in their diet.

Jewish

Among Jews, the laws regarding food are commonly referred to as the laws of Kashrut or the laws of what foods are permissible in accordance with the religious law. The term *kosher* means fit for eating; it is not a brand or form of cooking.

Foods are divided into those that are permitted (clean) and forbidden (unclean). The kosher slaughter of animals prevents undue cruelty to the animal and ensures the animal's health for its consumer. Care must be taken that all blood is drained from the animal before eating it.

BOX 29-5 CULTURAL CONSIDERATION

Among the more conservative and Orthodox Jews, milk and meat may not be mixed together, whether in cooking, serving, or eating. This involves separating the utensils used to prepare foods and the plates used to serve them. To avoid mixing foods, religious Jews have two sets of dishes, pots, and utensils: one set for milk products and one for meat.

Cheeseburgers, meat lasagna, and grated cheese on meatballs and spaghetti are not acceptable. Milk cannot be used in coffee if served with a meat meal. Nondairy creamers can be used as long as they do not contain sodium caseinate, which is derived from milk.

Fish, eggs, vegetables, and fruits are considered neutral and may be used with either dairy or meat dishes. A U with a circle around it or a K is used on food products to indicate kosher.

When working in a Jewish person's home, the nurse should not bring food into the house without knowing whether the patient is kosher. If the patient is kosher, do not use any cooking items, dishes, or silverware without knowing which are used for meat and which are used for dairy. It is important for the nurse to understand the dietary laws so as not to offend the patient. The nurse should advocate for kosher meals if they are requested and plan medication times accordingly.

Although liberal Jews decide for themselves which dietary laws, if any, they follow, many still avoid pork and pork products out of a sense of tradition and symbolism. It would be insensitive to serve pork products to Jewish patients unless they specifically request it.

Kosher meals are available in hospitals and long-term care facilities. Even though the organization may not have a kosher kitchen, frozen kosher meals can be obtained from several organizations, most of which are located in large cities with large Jewish populations. The kosher kitchen closest to your organization can be obtained by calling the Jewish synagogue nearest the organization to obtain the address and telephone number. Kosher meals arrive on paper plates with plastic utensils sealed in plastic. The nurse should not unwrap the utensils if the patient is able to do so or change the foodstuffs to another serving dish. Determining a patient's dietary preferences and practices regarding dietary laws should be done during the admission assessment.

Mexican-Americans

Good health to Mexican-Americans, which is largely a part of "God's will," can be maintained by dietary practices that keep the body in balance. To provide culturally competent care, the nurse must be aware of the hot-and-cold theory of disease when offering health teaching. Many diseases are thought to be caused by a disruption in the hot-and-cold balance theory of the body. Thus, by eating foods of the opposite variety, one may either cure or prevent specific hot-and-cold illnesses and conditions.

Examples of hot disease conditions include infection, diarrhea, sore throats, stomach ulcers, liver conditions, kidney problems, gastrointestinal upsets, and febrile conditions. Foods that are considered "cold" are therefore viewed as remedies for hot illness conditions. Cold foods include fresh fruits and vegetables, dairy products, barley water, fish, chicken, goat meat, and dried fruits. However, there are significant differences in what are considered hot and cold foods and illnesses among Mexican-American families depending on their native region in Mexico.

Examples of cold illness conditions include cancer, malaria, earaches, arthritis and related conditions, pneumonia and other pulmonary conditions, headaches, menstrual cramping, and musculoskeletal conditions. Hot foods used to treat these conditions typically include cheeses, liquor, beef, pork, spicy foods, eggs, grains other than barley, vitamins, tobacco, and onions.

---

urements to determine patient's health risk factors (Table 29–2). Excess waist circumferences (for women, more than 35 inches; for men, more than 40 inches) place people at greater risk for diabetes and cardiovascular disease.

Oral Cavity

Gastrointestinal assessment begins with the oral cavity. The lips are examined for lesions, abnormal color, and symmetry. With a penlight and tongue blade, the oral cavity is inspected for inflammation, tenderness, ulcers, swelling, bleeding, and discolorations. Any odor of the patient's breath is noted. A foul odor may indicate infection or poor oral care. The tongue should be pink with a rough texture and assessed for signs of dehydration such as dryness, cracks, or furrows. The patient's gums should be pink without swelling, redness, or irregularities. The teeth or dentures are examined for loose, broken, or absent teeth and the fit of the dentures or dental work. Ill-fitting dentures can affect the patient's nutritional intake and obstruct the airway. Loose teeth can become dislodged and aspirated into the airway. Broken teeth can be a source of pain and contribute to poor nutritional intake. The patient's knowledge of dental and oral care is assessed. The ability of the patient to perform oral care is noted and included in the plan of care if there are deficits.

Abdomen

INSPECTION. To inspect the abdomen, patients are placed in a supine position with their arms at their sides. The abdomen is visually inspected to note the condition of the skin and the contour. The contour may be rounded, flat, concave, or distended, depending on the patient's body type. Irregularities in contour may be due to distention, tumors, hernia, or previous surgeries. Scars, wounds, tubes, and ostomy device type and location are noted.

TABLE 29-2 CALCULATING BODY MASS INDEX AND WAIST MEASUREMENT

To calculate body mass index:	Step 1. Multiply body weight (in pounds) by 703.
	Step 2. Multiply height (in inches) by height.
	Step 3. Divide answer in step 1 by answer in step 2.
To obtain waist measurement:	Step 1. Place measuring tape at the level of the top of the iliac crest.
	Step 2. Pull snugly around the waist.
	Step 3. Read measurement at end expiration.

AUSCULTATION. When auscultating the patient's abdomen, the upper right quadrant is auscultated first (Fig. 29–3). Then a clockwise direction is followed to listen to the other quadrants. The stethoscope is pressed lightly on the abdomen to listen for bowel sounds, which are soft clicks and gurgles that may be heard every 5 to 15 seconds. **Bowel sounds** at this rate are considered normal. Bowel sounds are produced when peristalsis moves air and fluid through the GI tract and are categorized as normal, hyperactive, hypoactive, or absent. Hyperactive bowel sounds are usually rapid, high pitched, and loud and may occur with hunger or gastroenteritis. Hypoactive bowel sounds are bowel sounds that are infrequent and can occur in patients with a paralytic ileus or following abdominal surgery. Bowel sounds are considered absent if no sounds are auscultated after listening to all four quadrants for a minimum of 5 minutes in each quadrant. With a bowel obstruction, a high-pitched tinkling sound that is proximal to the obstruction and absent distal to the obstruction may be heard. Abnormal or absent bowel sounds are important findings and should be documented and reported to the physician.

LEARNING TIPS

In a complete bowel obstruction, air and fluid are propelled forward by peristalsis proximal to the obstruction. This produces proximal high-pitched bowel sounds when the air and fluid create turbulence as they hit the obstruction and are unable to pass. Absent bowel sounds are heard distal to the obstruction.

PALPATION. Using the same quadrant approach as mentioned before, lightly depress the abdomen with the finger pads to check for distention. Note any pain, tenderness, or rigidity. Abdominal girth is measured by placing a tape measure around the patient's abdomen at the iliac crest. A mark can be made at the measurement site so that measurements obtained by others are made at the same location for comparison. Abdominal girth is increased in patients with distention or conditions such as ascites (accumulation of fluid in the peritoneal cavity). Daily measurements should be obtained and recorded to monitor changes when abdominal girth is abnormal.

The advanced nurse practitioner or physician performs all other types of palpation. The liver is not normally palpable, but if enlarged, it may be felt below the right lower ribcage. Rebound tenderness is determined by pressing down on the abdomen a few inches and quickly releasing the pressure. If the patient feels a sharp pain during this procedure, appendicitis may be indicated.

PERCUSSION. Percussion produces a sound that identifies the density of the organs beneath and is performed by the physician or advanced nurse practitioner. Percussion is used to detect fluid, air, and masses in the abdomen and to identify size and location of abdominal organs (especially the liver and spleen). Tympanic high-pitched sounds indicate the location of air, and dull thuds indicate fluid or solid organs.

Diagnostic Tests

Commonly used tests associated with diseases of the GI tract include the following:

- Laboratory tests
- Radiographic tests
- **Endoscopy**
- **Gastric analysis**
- Cytological studies
- Magnetic resonance imaging

Observe standard precautions when obtaining specimens of body fluids, substances, or blood. Handwashing before and after the procedure, wearing gloves, and using goggles if splashing may occur are important.

Laboratory Tests

The complete blood cell count (CBC) reveals if anemia or infection are present (Table 29–3). Anemia may occur with GI bleeding or cancer. Electrolyte imbalances often occur with GI illness as a result of vomiting, diarrhea, malabsorption, or use of GI suction. **Carcinoembryonic antigen** (CEA) and carbohydrate antigen 19-9 are markers used to monitor GI cancer treatment effectiveness and detect recurrence. These markers are also found in patients with cirrhosis, hepatic disease, and alcoholic pancreatitis and in heavy smokers.

Stool Tests

Stool samples can be tested for **occult blood** (blood not seen by the naked eye). A series of three tests are usually done to increase the chances of detecting blood. False-positive occult blood results can occur with bleeding gums following a dental procedure; ingestion of red meat within 3 days before testing; ingestion of fish, turnips, or horseradish; and use of drugs, including anticoagulants, aspirin, colchicine, iron preparations in large doses, NSAIDS, and steroids.

Stool for ova (eggs) and parasites is collected to detect intestinal infections caused by parasites and their ova. The

Quadrants of the abdomen

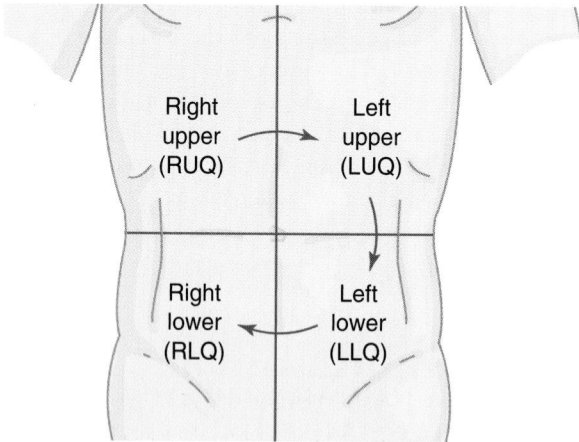

Right upper (RUQ)	Left upper (LUQ)
Right lower (RLQ)	Left lower (LLQ)

Figure 29–3 Abdominal quadrants are auscultated from the right quadrant in a clockwise manner.

TABLE 29-3	COMMON LABORATORY TESTS USED TO ASSESS GASTROINTESTINAL FUNCTION	
Test	**Normal**	**Significance of Abnormal Findings**
Carcinoembryonic Antigen		
	Less than 5 ng/mL (nonsmokers)	Increased values indicate possible colorectal cancer and inflammatory bowel disease
Complete Blood Cell Count		
Red blood cell count	4.2–5.2 million/mm³ (women) 4.5–6.2 million/mm³ (men)	Decreased values indicate possible anemia or hemorrhage
Hemoglobin	12–16 g/dL (women) 14–18 g/dL (men)	Increased values indicate possible hemoconcentration, caused by dehydration
Hematocrit	38–46% (women) 42–54% (men)	
Electrolytes		
Calcium	8.0–10.5 mg/dL	Decreased values indicate possible malabsorption
Chloride	98–107 mEq/L	Decreased values indicate possible malabsorption
Potassium	3.5–5.0 mEq/L	Decreased values indicate possible GI suction, diarrhea, vomiting, intestinal fistulas
Sodium	135–145 mEq/L	Decreased values indicate possible malabsorption and diarrhea
Fecal Analysis		
Stool for occult blood	Negative	Presence indicates possible peptic ulcer, cancer of the colon, ulcerative colitis
Stool for ova and parasites	Negative	Presence indicates infection
Stool cultures	No unusual growth	Presence of pathogens may indicate shigella, salmonella, *Staphylococcus aureus*, or *Bacillus cereus* infections
Stool for lipids	2–5 g per 24 hours (normal diet)	Increased values indicate possible malabsorption syndrome or Crohn's disease

test usually requires a series of three stool specimens collected every second or third day. The stool specimen is collected using a tongue blade, placed in a container with a preservative, and sent immediately to the laboratory. The stool must be examined within 30 minutes of collection. False-negative results can occur as a result of urine in the specimen or if the specimen is not fresh.

Stool cultures (sterile collection technique) are done to determine the presence of pathogenic organisms in the GI tract. Stool can also be examined for lipids (fat). Excessive secretion of fecal fats **(steatorrhea)** may occur in various digestive and absorptive disorders. The stools are collected for 72 hours and stored on ice if necessary before being sent to the laboratory.

Radiographic Tests
Flat Plate of the Abdomen

A flat plate of the abdomen is an x-ray examination giving an anterior-to-posterior view. Radiographs visualize abdominal organs and can detect such abnormalities as tumors, obstructions, and strictures. For an x-ray examination, the patient should be dressed in a hospital gown without any metal such as zippers, belts, or jewelry. Pregnant patients or those thought to be pregnant should avoid x-ray examinations.

Upper Gastrointestinal Series (Barium Swallow)

An **upper gastrointestinal series** (UGI series) is an x-ray examination of the esophagus, stomach, duodenum, and je-

junum using an oral liquid radiopaque contrast medium (barium) and a **fluoroscope** to outline the contours of the organs. A UGI series is used to detect such things as strictures, ulcers, tumors, polyps, hiatal hernias, and motility problems.

The patient receives nothing by mouth (nada per os in Latin, abbreviated NPO) for 6 to 8 hours before the procedure. Usually the patient eats a clear liquid supper the night before the procedure and is then NPO until the procedure is done. Because smoking can stimulate gastric motility, the patient is discouraged from smoking the morning of the procedure. Patient teaching includes information about the patient's diet before and after the procedure, the barium ingestion, and the appearance of stools afterward.

During the procedure, the patient drinks the thick, chalky barium while standing in front of a fluoroscopy tube. X-ray films are taken at specific intervals to visualize the outline of the organs and to note the passage of the barium through the GI tract.

A laxative is usually ordered after the procedure to expel the barium and prevent constipation or a barium **impaction.** The patient is instructed to drink 12 8-oz glasses of water per day for several days to prevent dehydration, which can lead to constipation. The abdomen is assessed for distention and bowel sounds. The stool is monitored to determine whether the barium has been completely eliminated. Initially the barium causes the patient's stool to be white, but it should return to its normal color within 3 days. Constipation with distention indicates a barium impaction.

Lower Gastrointestinal Series

The **lower GI series** (barium enema) is performed to visualize the position, movements, and filling of the colon. Tumors, diverticula, stenosis, obstructions, inflammation, ulcerative colitis, and polyps can be detected. The patient is placed on a low-residue or clear liquid diet for 2 days before the test to empty the bowel. Laxatives, bowel cleansing solutions (such as GoLYTELY), and enemas may be administered the evening before the test. GoLYTELY is chilled and drunk full strength with no ice, 8 oz every 10 minutes for a total of 4 L. Inform the patient that a watery diarrhea will begin in about 1 hour and continue up to 5 hours as the bowel is cleared. This is necessary for adequate visualization during the procedure. Inadequate bowel prep may result in poor test results or test cancellation. (See Fig. 32–2.) The patient either receives a clear liquid diet the morning of the test or is NPO after midnight the night before. The area around the rectum should be clean before the patient is sent for the procedure. If the patient has active inflammatory disease of the colon or suspected perforation or obstruction, a barium enema is contraindicated. Active GI bleeding may also prohibit the use of laxatives and enemas.

During the procedure, barium is instilled rectally and x-ray films are taken with or without fluoroscopy. The procedure takes about 15 minutes, and the patient is allowed to use the bathroom immediately after the procedure.

The patient's stools are monitored after the procedure to note if all the barium is passed, as with the UGI. Constipation development is monitored. The patient is encouraged to drink at least one 8-oz glass of liquid per hour for the next 24 waking hours to help remove the barium. The patient is told to report any abdominal pain, bloating, or absence of stool, all of which could indicate constipation or bowel obstruction, as well as any rectal bleeding.

Computed Tomography

Computed tomography (CT) uses a beam of radiation to allow three-dimensional visualization of abdominal structures. Diluted oral barium or other contrast media may be used to distinguish normal bowel from abnormal masses. The patient may have a clear liquid diet the morning of the test. If a contrast medium is to be used, any allergies to iodine or shellfish are assessed, a consent form is signed, and the patient is NPO for 2 to 4 hours before the procedure.

Endoscopy

Endoscopy uses a tube and a fiberoptic system (endoscope) for observing the inside of a hollow organ or cavity. A consent form is signed for any endoscopic procedure.

Esophagogastroduodenoscopy

Esophagogastroduodenoscopy (EGD) visualizes the esophagus (**esophagoscopy**), the stomach (**gastroscopy**), and the

---

esophagoscopy: oisophagos—esophagus + skopein—to examine
gastroscopy: gaster—stomach + skopein—to examine

duodenum. Abnormalities such as inflammation, cancer, bleeding, injury, and infection can be seen.

The procedure is explained to the patient. To prevent aspiration of stomach contents into the lungs if vomiting occurs, the patient is NPO for 8 to 12 hours before the procedure. Sedatives such as diazepam (Valium), meperidine (Demerol), or midazolam (Versed) may be given before the procedure to help relax the patient. A local anesthetic in spray or gargle form is administered just before the scope is inserted to inhibit the gag reflex.

The patient is placed on the left side, and the flexible, fiberoptic endoscope tube is passed orally down the GI tract (Fig. 29–4). Photos or videotapes of the procedure can be made. Biopsy or cytology specimens can be obtained.

After the procedure vital signs are checked as ordered. Patients are placed on one side to prevent aspiration while sedation and the local anesthetic wear off. Patients are NPO until the gag reflex returns (usually within 4 hours). Patients are assessed for signs of perforation, which include bleeding, fever, and dysphagia. Midesophageal perforation can cause referred substernal or epigastric pain. Blood loss secondary to perforation can lead to hematoma formation, which in turn can result in cyanosis and referred back pain. Distal esophageal perforation may result in shoulder pain, dyspnea, or symptoms similar to those of a perforated ulcer. The patient may have a sore throat for a few days.

Lower Gastrointestinal Endoscopy

PROCTOSIGMOIDOSCOPY. Proctosigmoidoscopy is the examination of the distal sigmoid colon, the rectum, and the anal canal using a rigid or flexible endoscope (sigmoidoscope). Ulcerations, punctures, lacerations, tumors, hemorrhoids, polyps, fissures, fistulas, and abscesses can be detected. Malignancies at an early stage can be detected, so an annual examination for patients 40 years old and older is recommended.

Proctosigmoidoscopy requires the lower bowel to be cleaned out. The patient usually receives a clear liquid diet 24 hours before the test and a laxative the night before the test. The morning of the procedure a warm tap-water enema or sodium biphosphate (Fleet) enema is given. Bowel prepa-

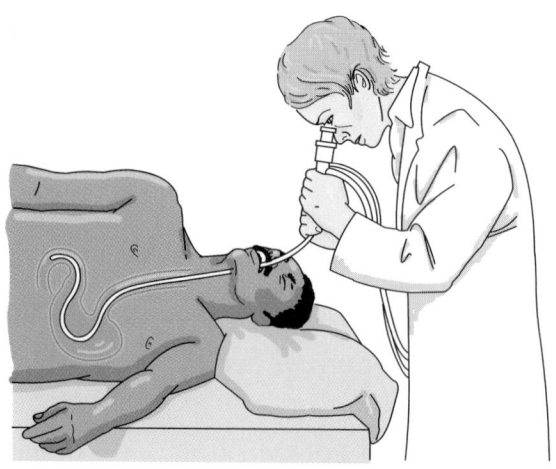

Figure 29-4 Gastroscopy.

ration may not be ordered for patients with bleeding or severe diarrhea.

The patient is positioned in a left lateral knee-to-chest position, which allows the sigmoid colon to straighten by gravity. A rigid proctoscope is used to visualize the rectum. A flexible scope is then used to permit visualization above the rectosigmoid junction. Patients are told they may feel pressure as though they are going to have a bowel movement. During the procedure, one or more small pieces of intestinal tissue may be removed (biopsy specimens). Rectal or sigmoid polyps are removed with a snare. An electrocoagulating current is used to cauterize sites to prevent or stop bleeding. Specimens are labeled and sent to the pathology laboratory immediately for examination.

After the procedure, the patient is allowed to rest for a few minutes in the supine position to avoid orthostatic hypotension when standing. Pain and flatus may occur from instilled air. The patient is observed for signs of perforation such as bleeding, pain, and fever.

COLONOSCOPY. Colonoscopy provides visualization of the lining of the large intestine to identify abnormalities through a flexible endoscope, which is inserted rectally. A biopsy specimen may be obtained or polyps be removed during the colonoscopy.

The patient receives a liquid diet 24 hours before the test and is NPO after midnight before the procedure. A bowel preparation solution such as GoLYTELY is given the night before the procedure. Drinking this solution can be unpleasant for the patient. In addition, a laxative, suppository (bisacodyl [Dulcolax]), or enema may be needed.

LEARNING TIPS

Elderly patients may experience fatigue and weakness during bowel preparation and may be unable to complete it. Monitor the patient for distress. Consult the physician if you note any patient distress during bowel preparation.

Conscious sedation (e.g., midazolam, meperidine) is used to relax and ease pain during the procedure. The patient is positioned on the left side with the knees bent. A small amount of air is instilled into the colon to help the physician visualize the bowel. The air causes pressure and may be uncomfortable for the patient. The patient is encouraged to relax and take slow deep breaths through the nose and out the mouth. Vital signs are monitored throughout the procedure to watch for a vasovagal response, which can lead to hypotension and bradycardia.

After the procedure, the patient is monitored until stable. Complications such as hemorrhage or severe pain are reported. When giving the patient discharge instructions, explain that flatus and cramping will occur for several hours after the test, that blood may be present in the stool if a

biopsy specimen was taken, and to report problems to the physician.

Gastric Analysis

Gastric analysis measures the secretions in the stomach. Diagnosis of duodenal ulcer, gastric carcinoma, pyloric or duodenal obstruction, and pernicious anemia are made. A diagnosis of pernicious anemia is ruled out by the finding of acid. A diagnosis of gastric carcinoma may be made by the presence of cancer cells in the gastric secretions. There are two tests performed in gastric analysis: the **basal cell secretion test** and the **gastric acid stimulation test.**

Before the basal cell secretion test, the patient should avoid taking any drugs that could interfere with gastric acid secretion, such as cholinergics and antacids. The patient is NPO after midnight the night before the test. For the procedure, a nasogastric (NG) tube is inserted and the contents of the stomach are suctioned out through the tube using a syringe. The NG tube is connected to wall suction, and stomach contents are collected every 15 minutes for 1 hour. The specimens are labeled according to the time they were collected and the order in which they were obtained. The gastric acid is tested for pH using indicator paper or a pH meter. The amount of gastric acid is also measured. Too much hydrochloric acid may indicate a peptic ulcer; too little could be a sign of cancer or pernicious anemia.

The gastric acid stimulation test measures the amount of gastric acid for 1 hour after subcutaneous injection of a histamine drug. If abnormal results occur, radiographic tests or endoscopy can be done to determine the cause.

Endoscopic Ultrasonography

Endoscopic ultrasonography is performed through the endoscope using sound waves. Tumors can be detected in various GI structures and organs. Pre- and postprocedure care is similar to endoscopic care. During the test the patient must lie still while a transducer with gel is moved back and forth over the abdomen to produce images.

▶ THERAPEUTIC MEASURES
Gastrointestinal Intubation

Gastrointestinal intubation is the placement of a tube within the GI tract for therapeutic or diagnostic purposes (Fig. 29–5). When the GI tube is inserted from the nares into the stomach, it is referred to as a nasogastric tube. A nasointestinal tube is a tube inserted from the nares into the intestines. A variety of tubes are available, each designed for specific purposes (Table 29–4). Nasogastric and nasointestinal tubes are inserted for a variety of reasons, but the main purposes for their use include the following:

■ To remove gas and fluids from the stomach or intestines (decompression)
■ To diagnose GI motility and to obtain gastric secretions for analysis
■ To relieve and treat obstructions or bleeding within the GI tract

---

colonoscopy: kolon—colon + skopein—to examine

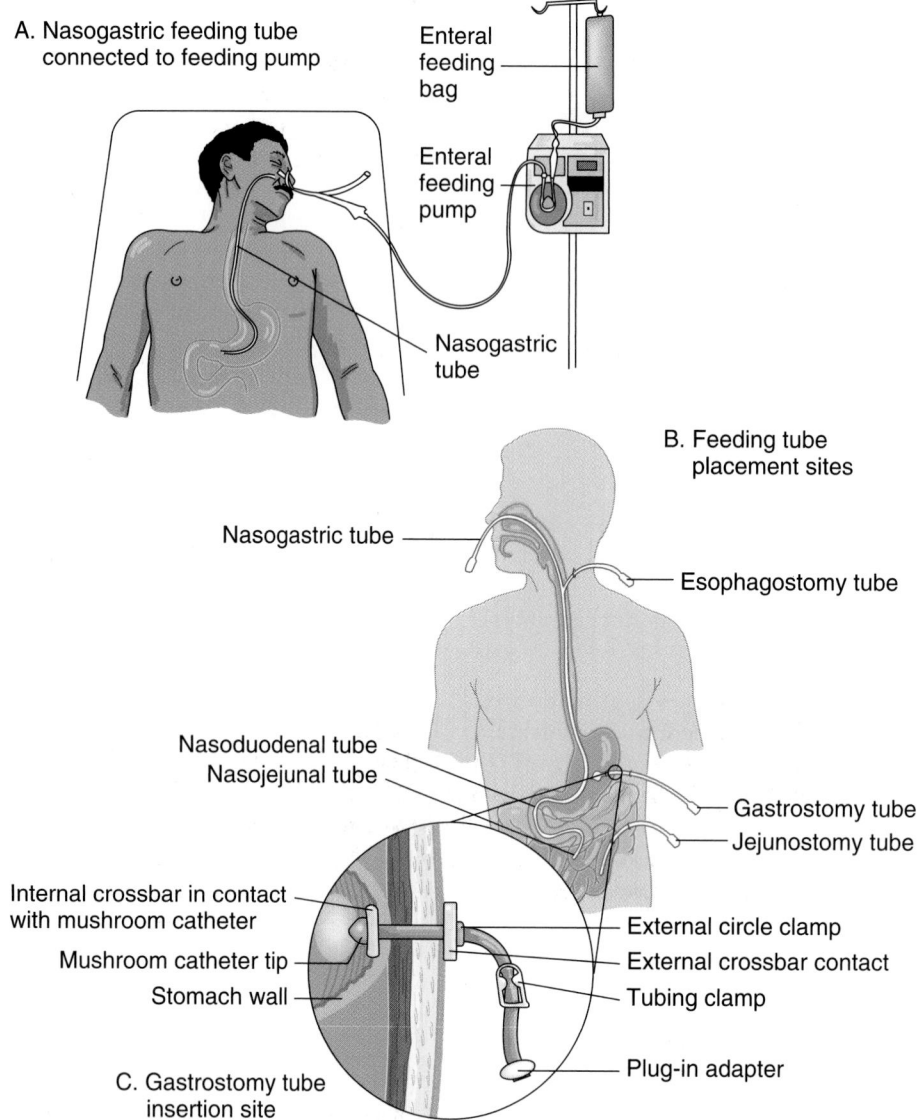

A. Nasogastric feeding tube connected to feeding pump

Enteral feeding bag

Enteral feeding pump

Nasogastric tube

B. Feeding tube placement sites

Nasogastric tube

Esophagostomy tube

Nasoduodenal tube
Nasojejunal tube

Gastrostomy tube
Jejunostomy tube

Internal crossbar in contact with mushroom catheter

Mushroom catheter tip

Stomach wall

External circle clamp
External crossbar contact
Tubing clamp

Plug-in adapter

C. Gastrostomy tube insertion site

Figure 29-5 Feeding tubes. *(A)* Nasogastric tube connected to tube feeding pump. *(B)* Feeding tube placement sites (esophagostomy, nasointestinal, gastrostomy, and jejunostomy). *(C)* Gastrostomy tube insertion site.

TABLE 29-4 GASTROINTESTINAL TUBES		
Tube	**Uses and Description**	**Nursing Considerations**
Gastric		
Levine tube	Single lumen, may be used for gastric decompression, irrigations, lavages, and feedings	Tube is not vented, so use with low intermittent suction to prevent injury to stomach lining.
Sump tube (Salem sump) *Anderson tube*	Double lumen, "pigtail" acts as an air vent and prevents excess suction, which could damage stomach lining; used for decompression, irrigations, and lavages	May be used with low continuous suction because of air vent.
Weighted, flexible feeding tubes with stylets (Nutriflex, Keofeed)	Small-bore tubes for tube feedings only; less injury, remains in place for extended periods	Suction collapses tube. Use a 10-mL syringe or greater, because smaller syringe creates too much pressure and possible rupture of tube.
Intestinal		
Miller-Abbott tube	Double-lumen tube used to drain and decompress the small intestine in cases of partial or complete obstruction; one lumen for aspiration, the other to inflate the balloon with mercury so that the tube is weighted and moves by gravity and peristalsis into the small intestine	Rarely used, this tube is inserted by the physician or a specially trained nurse. Tube is not secured with tape, but passed through gauze taped to patient's forehead to allow tube to advance into intestines. Usually the tube advances 1–2 inches every 2 hours until it reaches small intestine; turning and ambulating the patient, if possible, facilitates tube's advancement.
Cantor tube *Harris tube*	Single-lumen tubes with distal mercury-filled balloons and proximal drainage ports.	

■ To provide a means for nutrition (**gavage** feeding), hydration, and medication when the oral route is not possible or is contraindicated

■ To promote healing after esophageal, gastric, or intestinal surgery by preventing distention of the GI tract and strain on the suture lines

■ To remove toxic substances (**lavage**) that have been ingested either accidentally or intentionally and to provide for irrigation

Feeding tubes include esophagostomy, **gastrostomy,** or jejunostomy tubes. (See Fig. 29–5.) Nasogastric tubes are

usually temporary and short term. Esophagostomy, gastrostomy, or jejunostomy tubes are generally used for longer-term nutrition delivery.

General nursing care measures for insertion and maintenance of GI tubes are given in Box 29–6. Provide emotional support and explanation to the patient and significant others to facilitate the process of tube insertion and maintenance. Assessing tube placement is essential to prevent complications or death from incorrect tube placement. Nasogastric tube placement must be assessed after insertion and then intermittently to ensure that it is in the correct position and not in the lungs (most common), esophagus, pleural space, or brain.

---

gastrostomy: gaster—stomach + ostium—little opening

| BOX 29-6 | **Nursing Care for Insertion and Maintenance of Nasogastric Tubes** |

1. Explain the procedure and reason for the tube to the patient. Inform the patient how he or she can help by swallowing when instructed to do so if able.
2. Assist patient into high Fowler's position (right side lying as an alternative) as able.
3. Measure tube for correct insertion length: Hold insertion end of tube from nose tip to earlobe to xyphoid process and mark tube with tape at this point.
4. Select naris that is straightest and from which patient breaths the easiest, because tube is inserted more easily in a straight naris.
5. Lubricate tube with water-soluble lubricant before insertion.
6. Insert tube as follows:
   • As tube is inserted, aim it along the floor of the naris and laterally. Rotate the tube gently if resistance is met.
   • Encourage patient to swallow to advance the tube. Drinking water with a straw or ice chips facilitates the swallowing process if patient is able.
   • If patient is unconscious, flex patient's head to bring chin toward chest to help prevent tube from passing into the trachea.
   • Observe for coiling of the tube in patient's mouth.
   • To assist in assessing correct placement, insert tube to level of carina tracheae and listen for air at end of tube. If air is present, remove tube. If no air is heard, advance tube to stomach.
   • If at any time the patient begins to cough uncontrollably, becomes cyanotic, or begins to experience any respiratory distress, remove the tube, allow rest time, and then reattempt insertion.
7. After tube is inserted to premeasured length, confirm gastric placement per institutional policy using the following methods[1]:
   • Flexible feeding tubes should have x-ray confirmation.
   • Aspirate for gastric contents with a 30- or 60-mL catheter-tipped syringe:
     • Gastric fluid—green with sediment; colorless and clear with off-white or tan mucus; brown if digested blood is present.
     • Esophageal contents—scant fluid, aspirate is unreliable for confirmation.
     • Intestinal fluid—light to dark golden yellow to brownish green.
     • Respiratory secretions—off-white or tan mucus.
     • Pleural fluid (with stylet perforation)—watery, straw colored, may be blood-tinged from perforation.

   • Measure pH of secretions obtained to rule out respiratory placement:
     • Gastric pH range is acidic (1–5).
     • Respiratory and intestinal secretions are alkaline with pH > 6.
   • If any doubts about gastric placement exist, notify physician for x-ray order to confirm tube placement.
8. Secure tube in place with tape so that pressure is not put on the naris. Provide daily skin care to taped area to prevent skin breakdown. Slipknot a rubber band around tube and pin rubber band to patient's gown.
9. If suction is ordered, low intermittent suction is used with nonvented tubes (Levin); vented tubes (Anderson, Salem) may have continuous low suction.
10. If patient is NPO, provide frequent care to keep oral mucous membranes moist. When suction is used, prevent excessive intake of ice chips, if ordered, because electrolyte imbalances may result when the water is suctioned out along with electrolytes. Normal saline instead of water can be made into ice chips to help prevent imbalances.
11. Gastric placement is periodically confirmed, especially before instilling anything into the tube[1]:
    • Review any recent chest or abdominal x-ray report with tube status.
    • Verify tube position marking is in original position. If not, reposition as necessary.
    • Bolus feedings/medications: 4 hours after last feeding, aspirate fluid for pH and appearance.
    • Continuous feedings: Assess tolerance. If normal along with correct tube marking and any confirming x-ray reports, continue feeding.
    • For patients at high risk of dislodgement (e.g., those who are vomiting or have severe coughing) with tube movement, consider x-ray examination.
12. The tube is flushed at intervals, every 2–4 hours, to maintain patency. When tube feedings are administered, residual feeding amounts are checked at specified intervals (hourly when begun, every 4 hours thereafter) to ensure the feeding is being absorbed.
13. After tube placement, record accurate intake and output, including drainage, vomitus, and irrigation solution instilled (use only normal saline to prevent electrolyte imbalance).

Gastrostomy or jejunostomy tube placement is verified by comparing current length with documented insertion length. The tube may not be in the desired position if the current and insertion tube lengths are different, so the physician should be consulted before using the tube.

Tube Feedings

A tube feeding supplies patients with nutrition when oral intake is not possible. Feedings can be given to the patient as a supplement or to provide the patient's total nutritional needs. If the esophagus and stomach need to be bypassed, tube feedings are delivered directly into the duodenum or proximal jejunum. Reasons for administering tube feedings include inability to swallow, severe burns or trauma to the face or jaw, debilitation, mental retardation, and oropharyngeal or esophageal paralysis. Complications associated with tube feedings are presented in Table 29–5. (See Home Health Hints.)

Tube Feeding Formulas

Tube feeding formulas are chosen by the physician based on the patient's nutritional needs, the consistency of the formula, the size and location of the tube, the method of delivery, and the convenience for the patient at home. Commercially prepared formulas are composed of protein, carbohydrates, and fats. When patients receive tube feedings, their daily water needs in addition to any water supplied by the feeding should be considered. Dietitians can help calculate the patient's water needs. The water used to flush the tube or administer medications can be considered as satisfying part of the patient's daily total water needs. Dehydration can occur if the patient's water needs are not met.

CRITICAL THINKING

Mrs. Wood

Mrs. Wood is receiving a tube feeding because of dysphagia, the cause of which is being investigated. Mrs. Wood is not receiving any medications. You note that Mrs. Wood's tongue is bright red with deep furrows. She states her mouth is very dry. Her skin remains tented when skin turgor is checked.

1. What do Mrs. Wood's assessment findings indicate?
2. How would you document your assessment findings?
3. Why might Mrs. Wood be exhibiting this condition?
4. What actions could you take for this condition?

Answers at end of chapter.

Method of Delivery

Feedings are administered either by gravity or by a controlled pump that delivers continuous volume through the feeding tube. Gravity feedings are placed above the level of the stomach and dripped in by gravity slowly or given as a bolus feeding over a few minutes. Intermittent feedings are defined as either being delivered by a pump that runs continuously throughout the day and is discontinued each night or as a 4- to 6-hour volume of feeding given over 20

to 30 minutes. Intermittent feedings via a pump allow the stomach to rest at night and more closely simulate normal eating and nutrient absorption patterns. A continuous feeding administered 24 hours a day through a pump allows for small amounts to be given over a long period. Pumps are set at the specified rate to control the speed of the feeding being delivered to the patient.

When feedings are administered, patients must be positioned in a sitting or high Fowler's position to reduce the risk of aspiration. Monitor the rate carefully to avoid administering feedings too rapidly, and watch for signs that the feeding is not being absorbed. Abdominal distention, patient report of a feeling of fullness, and nausea or vomiting are indicators that the feeding is not being absorbed and should be stopped to prevent aspiration. A residual check to see how much feeding, if any, has not been absorbed is done hourly when the feeding is initiated, then every 4 hours or before giving any medications or adding more feeding for infusion. If there is more than 100 mL or the amount specified by the agency or physician, the feeding should be stopped to prevent vomiting or aspiration and the physician notified. Continuous or intermittent feedings reduce the risk of aspiration, distention, nausea, vomiting, and diarrhea.

If medications are administered during tube feedings, understand possible drug-nutrient interactions. Some medications cannot be given with certain substances. Other medications, such as enteric-coated or sustained-release medications, should not be crushed. Liquid medications should be used when possible to reduce clogging of the tube. Pharmacists and dietitians should be consulted for special considerations.

Gastrointestinal Decompression

Gastrointestinal decompression may be necessary when the stomach or small intestines become filled with air or fluid. Swallowed air and GI secretions enter the stomach and intestines and collect there if they are not propelled through the GI tract by peristalsis. Accumulating air or fluid causes distention, a feeling of fullness, and possibly pain in the abdomen. Gastric distention may occur after major abdominal surgery. Ambulating or turning the patient frequently can help prevent this. However, when GI decompression is necessary, a nasogastric tube or occasionally a nasointestinal tube may be inserted and suction applied. Nasointestinal tubes are more difficult and slower to place and may be uncomfortable, so they are not used often. The tube remains in place until full peristaltic activity (active bowel sounds and passage of flatus) has returned. The diet is then progressed as ordered and tolerated by the patient.

Total Parenteral Nutrition

Total parenteral nutrition (TPN, also known as intravenous hyperalimentation) is a method of supplying nutrients to the patient by an intravenous (IV) route. TPN solutions usually contain dextrose (sugar), amino acids (protein), vitamins, minerals, and fat (intralipid) emulsions. TPN solutions are designed to improve the patient's nutritional sta-

TABLE 29–5	COMMON MECHANICAL GASTROINTESTINAL AND METABOLIC COMPLICATIONS OF TUBE-FED PATIENTS AND PREVENTION STRATEGIES

Complication	Prevention Strategies
Mechanical	
Tube irritation	Consider using a smaller or softer tube.
	Lubricate the tube before insertion.
Tube obstruction	Flush tube after use.
	Do not mix medications with the tube feeding formula.
	Use liquid medications if available.
	Crush other medications thoroughly (if not contraindicated to crush).
	Use an infusion pump to maintain a constant flow. (See Fig. 29–5.)
Aspiration and regurgitation	Feeding should not be started until tube placement is radiographically confirmed.
	Elevate head of the patient's bed more than or equal to 30 degrees at all times.
	Discontinue feedings at least 30 to 60 minutes before treatments requiring head to be lowered (e.g., chest percussion).
	If the patient has an endotracheal tube in place, keep the cuff inflated during feedings.
	Test pH of aspirate with pH paper or meter.
	a. pH of tracheobronchial secretions is alkaline, >7.4.
	b. pH of gastric secretions is acidic, <5.0.
	c. As the tube moves from the acid stomach to the alkaline duodenum, pH will change from acid to alkaline.
	Place a black mark at the point where the tube, once properly placed, exits the nostril.
Tube displacement	Replace tube and obtain physician's order to confirm with x-ray imaging.
Gastrointestinal	
Cramping, distention, bloating, gas pains, nausea, vomiting, diarrhea*	Practice good personal hygiene when handling any feeding product.
	Bring formula to room temperature before feeding.
	Initiate and increase amount of formula gradually.
	Change to a lactose-free formula.
	Decrease fat content of formula.
	Administer drug therapy as ordered (e.g., Lactinex, kaolin-pectin, Lomotil).
	Change to formula with a lower osmolality.
	Change to formula with a different fiber content.
	Evaluate diarrhea-causing medications the patient may be receiving (e.g., antibiotics, digitalis preparations).
Metabolic	
Dehydration	Assess the patient's fluid requirements before treatment.
	Monitor hydration status.
Overhydration	Assess the patient's fluid requirements before treatment.
	Monitor hydration status.
Hyperglycemia	Initiate feedings at a low rate.
	Monitor blood glucose.
	Use hyperglycemic medication if necessary.
	Select a low-carbohydrate formula.
Hypernatremia	Assess the patient's fluid and electrolyte status before treatment.
	Provide adequate fluids.
Hyponatremia	Assess the patient's fluid and electrolyte status before treatment.
	Restrict fluids.
	Supplement feeding with rehydration solution and saline.
	Diuretic therapy may be beneficial.
Hypophosphatemia	Monitor serum levels.
	Replenish phosphorus levels before refeeding.
Hypercapnia	Select a low-carbohydrate, high-fat formula.
Hypokalemia	Monitor potassium levels.
	Supplement feeding with potassium if necessary.
Hyperkalemia	Reduce potassium intake.
	Monitor potassium levels.

*The most commonly cited complication of tube feeding is diarrhea.
From Lutz, CA, and Przytulski, KR: Nutritional and Diet Therapy, ed 2. FA Davis, Philadelphia, 1997, p 280. Reprinted with permission.

tus, achieve weight gain, and enhance the healing process. The patient with conditions such as burns, trauma, cancer, acquired immunodeficiency syndrome (AIDS), malnutrition, anorexia nervosa, or fever and those undergoing major surgery may need TPN.

Usually, registered nurses are responsible for administering TPN. A filter must be used with TPN solutions but not with lipid solutions, which are given as a separate infusion along with TPN therapy. TPN is started slowly to give the pancreas time to adjust to increasing insulin production for the high amounts of glucose in the TPN. The TPN rate is increased until the ordered rate, as tolerated by the patient, is reached. When TPN is discontinued, the patient must be gradually weaned to allow the pancreas to adjust to decreasing glucose levels. The patient, as ordered, is fed before the TPN is stopped to help prevent hypoglycemia. Signs of hypoglycemia include weakness, shakiness, sweating, and confusion.

It is important to monitor glucose levels as ordered and to look for signs of hyperglycemia in the patient receiving TPN. Refer to agency policy for obtaining glucose levels when a hyperglycemic reaction is suspected in the patient receiving TPN.

During TPN administration, the following laboratory values, as ordered, are usually monitored:

- Complete blood cell count (CBC)
- Albumin
- Glucose
- Electrolytes
- Magnesium
- Platelet count
- Prothrombin time (PT)

TPN can be irritating to the peripheral veins because it is five or six times more concentrated than blood. Therefore TPN dextrose more than 12 percent is administered through a central venous catheter into a large vein such as the subclavian or internal jugular (see Fig. 6–8). The volume in the large vein dilutes the TPN solution, so it is less irritating.

Peripheral Parenteral Nutrition

Peripheral parenteral nutrition (PPN) is a method of supplying nutrients to the patient by an IV route that is not a central vein. PPN is used for less than 10 days when the patient does not need more than 2000 calories daily. PPN solutions can contain a mixture of dextrose (of less than 12 percent), amino acids, and lipids, in addition to electrolytes or water, which can be found in routine IV solutions. The all-in-one PPN system mixes dextrose, amino acids, and lipids all in one container, which causes less irritation of veins.

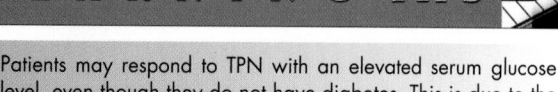

- Patients may respond to TPN with an elevated serum glucose level, even though they do not have diabetes. This is due to the high concentration of glucose used in TPN. These elevated serum glucose levels do not usually indicate that the patient who does not have diabetes has acquired the disease. After the TPN is discontinued, the serum glucose levels should return to baseline or normal levels.
- Regular insulin is given, as ordered, to control hyperglycemia during TPN therapy. The insulin is ordered either on a sliding scale (regular insulin given based on blood glucose levels measured at ordered intervals over 24 hours) or as an additive to the TPN solution.
- Administration of insulin according to a sliding scale requires a current blood glucose level. Based on the obtained glucose level, if it is elevated, specified regular insulin units may be ordered. Usually blood glucose is measured before meals, but for a patient who is not eating, as with most patients receiving TPN, there is no meal time. Instead, specified time intervals are ordered (typically every 6 hours).
- Insulin given on a sliding scale is always regular insulin. Can you figure out why? Regular insulin is rapid acting, which is what is needed to treat the current blood glucose level that was obtained to determine what insulin coverage was needed, if any.

Patients with any of the following may need to be considered for TPN or PPN:

- Any significant weight loss (10 percent or more of healthy weight)
- A decrease of oral food intake for more than 3 days
- Any significant sign of protein loss: serum albumin levels below 3.2 g/dL
- Muscle wasting
- Decreased tissue healing
- Persistent vomiting and diarrhea

Answers to CRITICAL THINKING

Mrs. Todd

1. Daily aspirin use is the most likely cause of her bleeding.
2. Medication teaching including side effects can help Mrs. Todd prevent future bleeding episodes. Assessment of pain relief needs and consultation with the physician will also help.

Mrs. Wood

1. Assessment of Mrs. Wood indicates dehydration.
2. Document as follows: "1/20/03 0800 'Mouth very dry.' Tongue bright red with deep furrows, tented turgor. Tube feeding infusing. Physician notified. K. Ohno LVN."
3. Mrs. Wood's daily water needs are not being met. She is not receiving medications that would incidentally provide water during their administration.
4. Consult a dietitian or review Mrs. Wood's daily water needs. Divide the water needs over 24 hours and ensure that water is administered. Ensure tubing is flushed per agency policy, and calculate water used toward daily water needs. Monitor intake and output. Continue assessing Mrs. Wood's signs and symptoms, and report abnormal findings.

1. The nurse is assessing a patient's bowel sounds. The nurse understands that bowel sounds heard at an irregular rate every 5 to 15 seconds should be documented as which of the following?
   - a. Abnormal
   - b. Hyperactive
   - c. Hypoactive
   - d. Normal

2. The nurse is planning care for a 78-year-old patient's elimination needs. Which of the following interventions should the nurse plan to reduce complications from the aging change of slowed motility?
   - a. Decrease ambulation.
   - b. Increase dietary fiber.
   - c. Increase dairy products.
   - d. Decrease fluid intake.

3. The nurse is caring for a patient after an EGD. Which of the following would best indicate to the nurse that it is safe for the patient to resume oral intake?
   - a. Vital signs are within normal limits.
   - b. Patient's gag reflex returns.
   - c. Patient is alert.
   - d. Patient has no nausea or vomiting.

4. The nurse is caring for a patient who is receiving a TPN infusion. Blood glucose monitoring every 6 hours is ordered to detect which of the following complications?
   - a. Hyponatremia
   - b. Hyperglycemia
   - c. Hypocalcemia
   - d. Hyperkalemia

5. Following a colonoscopy, the nurse is assisting the patient into bed and sees bright red blood on the sheets. Which of the following should be the nurse's first action?
   - a. Apply gauze dressing to perineum.
   - b. Bathe the patient.
   - c. Assess the patient's drainage.
   - d. Notify the physician immediately.

30

NURSING CARE OF PATIENTS WITH UPPER GASTROINTESTINAL DISORDERS

Sharon Gordon Dawson

KEY TERMS

anorexia (AN-oh-**REK**-see-ah)

anorexia nervosa
(AN-oh-**REK**-see-ah ner-**VOH**-sah)

aphthous stomatitis
(**AF**-thus STOH-mah-**TIGH**-tis)

bariatric (BAR-ry-**AT**-rick)

bulimia nervosa
(buh-**LEE**-mee-ah ner-**VOH**-sah)

gastrectomy (gas-**TREK**-tuh-mee)

gastritis (gas-**TRY**-tis)

gastroduodenostomy
(GAS-troh-DOO-oh-den-**AHS**-toh-mee)

gastrojejunostomy
(GAS-troh-JAY-joo-**NAHS**-toh-mee)

gastroplasty (GAS-troh-**PLAS**-tee)

Helicobacter pylori
(**HEH**-lick-co-back-tur **PIE**-lori)

hiatal hernia (high-**AY**-tuhl **HER**-nee-ah)

obesity (oh-**BEE**-si-tee)

peptic ulcer disease
(**PEP**-tick **UL**-sir di-**ZEEZ**)

Roux-en-Y (roo-ehn-**WHY**)

steatorrhea (STEE-ah-toh-**REE**-ah)

stomatitis (STOH-mah-**TIGH**-tis)

QUESTIONS TO GUIDE YOUR READING

1. What are anorexia, anorexia nervosa, and bulimia nervosa and their medical management and nursing care?

2. What is obesity, and what medical, surgical, and nursing management is used to treat it?

2. What nursing care would you give to a patient with stomatitis?

3. How would you care for patients with acute or chronic gastritis?

4. How would you explain the pathophysiology, signs and symptoms, and diagnostic testing for hiatal hernia, peptic ulcer disease, gastric bleeding, and gastric cancer?

5. What is the current pharmacological treatment for peptic ulcer disease?

6. What nursing care would you provide for patients with hiatal hernia, peptic ulcer disease, gastric bleeding, and gastric cancer?

▶ EATING DISORDERS

Anorexia

Anorexia, which is a lack of appetite, is a common symptom of many diseases and can be caused by noxious food odors, certain drugs (as an intended or side effect), emotional stress, fear, psychological problems, and infections. Prolonged anorexia with an inadequate nutritional intake can lead to serious electrolyte imbalances, which in turn can lead to cardiac dysrhythmias. Although eating is the preferred method of weight gain, other measures such as tube feedings and intravenous infusion can be used. Ask patients what causes them to lose their appetite and what improves it to plan care. Nursing actions for the patient with anorexia include documenting accurate intake and output; monitoring vital signs, electrolytes, and electrocardiograms; and monitoring the rate of the intravenous infusion and tube feeding.

Anorexia Nervosa

Anorexia nervosa is an eating disorder that is recognized by the American Psychiatric Association (Box 30–1). This disease most commonly occurs in females between the ages of 12 and 18 who are from the middle and upper classes of Western culture. Males account for less than 10 percent of the population with anorexia nervosa. Young women with low self-esteem seem to be at highest risk. Anorexia nervosa is thought to be psychological in origin. Patients may have a phobia of weight gain, are afraid of a loss of control, and are mistrusting.

Signs and Symptoms

Early signs and symptoms of anorexia nervosa include severe weight loss, low self-esteem, compulsive dieting, and an altered body image (patients imagine themselves as fat although they are within normal weight range). As the disease progresses, additional symptoms appear, including amenorrhea in females, electrolyte imbalance, cardiac dysrhythmias, constipation, dry skin, lanugo (downy hair covering body), bradycardia, hypothermia, hypotension, muscle wasting, and facial puffiness. Often patients with anorexia nervosa deny the existence of any problem. They may develop bizarre food rituals and sometimes weigh themselves several times a day. Anorexia nervosa sometimes overlaps with **bulimia nervosa** (compulsive eating with self-induced vomiting).

BOX 30-1   Diagnostic Criteria for Anorexia Nervosa

1. Refusal to maintain body weight over a minimum normal weight for age and height
2. Intense fear of gaining weight or becoming fat, even though underweight
3. Disturbance in the way in which one's body weight, shape, or size is experienced
4. In females, the absence of at least three consecutive menstrual cycles when otherwise expected to occur

Medical Management

The most important intervention for anorexia nervosa is the restoration of nutritional health; up to 18 percent of anorexia patients die as a result of the disease. During the crisis period, when severe weight loss, life-threatening electrolyte imbalances and dysrhythmias, or other symptoms occur, nutrition is supplied by intravenous infusions containing electrolytes. Tube feedings or oral feedings may also be given.

The patient's damaged self-image and self-esteem are underlying problems and must be addressed in conjunction with the nutritional aspect (Nutrition Notes Box 30–2). Both self-image and nutrition issues require treatment over a long period. Psychotherapy and behavior modification

BOX 30-2   Nutrition Notes

Supplying Nutrition in Upper Gastrointestinal Conditions

Anorexia Nervosa and Bulimia Nervosa

These eating disorders, which produce severe weight loss, require multidisciplinary treatment. Although correcting the nutritional consequences of these conditions is of major importance, to achieve a cure it is essential to treat the underlying psychological causes.

Obesity

Candidates for gastric banding or gastric bypass surgery should be carefully selected. The procedure should be viewed as one tool to assist with weight control, along with behavioral changes. It is not done to permit overeating, because in time the constructed pouch can be stretched, thus negating the surgery. Guidelines include eating three to six small balanced meals daily; chewing thoroughly and eating slowly; drinking most fluids between meals; exercising regularly; and taking a multivitamin-multimineral supplement. Potential complications of these surgeries include nausea, vomiting, bloating, heartburn, staple disruption, obstruction, dumping syndrome, and osteoporosis.

Gastroesophageal Reflux and Hiatal Hernia

Patients with gastroesophageal reflux and hiatal hernia may find symptoms alleviated by protein foods that tighten the cardiac sphincter. Substances that may be better avoided because they relax the sphincter include fat, caffeine, peppermint, spearmint, chocolate, alcohol, and nicotine. Pepper and decaffeinated coffee may be problematic because they stimulate gastric secretions. Acidic juices may be irritating as well.

Dumping Syndrome

The recommended meal pattern for patients with dumping syndrome includes six small meals per day, high in protein and low in simple sugars; fluids between rather than with meals; and reclining for half an hour after meals. Supplementation with the vitamins B_{12}, D, and folic acid and the minerals calcium and iron may be necessary to prevent deficiencies.

Gastric Cancer

If a patient has a poor prognosis following a total gastrectomy for cancer, dietary interventions should focus on symptoms the patient wishes to control. An overly restricted diet causing the patient discomfort or distress is inappropriate.

that includes participation of the patient's significant others are often used to treat anorexia nervosa.

Nursing Care

Gaining the patient's genuine cooperation by using therapeutic communication and setting realistic, mutual goals is important in establishing trust and preventing relapse. To work with patients with anorexia nervosa, a therapeutic relationship must be developed to facilitate effective interactions. Empathy, acceptance of the patient, trust, warmth, and being nonjudgmental is important. Caring for patients with anorexia nervosa is challenging. Nursing actions for these patients include obtaining vital signs, daily weights, and accurately documenting intake and output of food and fluids.

Bulimia Nervosa

Bulimia nervosa is compulsive eating with self-induced vomiting, which is commonly known as binge–purge. The bulimic patient typically eats massive amounts of food at one sitting and then purges the food by intentionally inducing vomiting so weight is not gained. Laxatives are also sometimes used by the bulimic patient to purge the body of food to avoid weight gain. Excessive exercise may also be used to control weight. As in anorexia nervosa, bulimic patients are extremely thin to the point of starvation. A high percentage of patients with bulimia are young women.

Signs and Symptoms

Patients with bulimia nervosa usually exhibit the same signs and symptoms as patients with anorexia nervosa, with a few exceptions. Bulimic patients often have enamel erosion of the front teeth caused by the acid content of the emesis. They also spend a great deal of time locked in the bathroom vomiting, especially after meals. As the electrolyte imbalance worsens, they develop metabolic alkalosis as a result of the loss of gastric acid in the stomach contents.

Medical Management

The treatment for bulimia nervosa is essentially the same as for the patient with anorexia nervosa, as discussed earlier.

Obesity

Several methods can be used to diagnose a patient as overweight or obese, although there is no one definitive measure of either. Factors such as age, body frame size, and gender can influence these measurements:

- Height-weight chart: Weight 10 to 20 percent above ideal body weight is overweight; 20 percent or more above ideal body weight is obesity.
- Waist-to-hip ratio: Waist-to-hip ration is waist measurement divided by hip measurement. If the result is more than 1.0 in men or 0.8 in women, it indicates that the patient is overweight.
- Body mass index (BMI): BMI is one of the best methods for defining **obesity.** BMI increases with age. Generally a BMI of 25 to 29.9 kg/m^2 is defined as being overweight, with more than 30 kg/m^2 being obese (Fig. 30–1). BMI can be calculated using height-to-weight ratios:

$$BMI = \frac{Weight\ (kg)}{Height\ (m^2)}$$

$$Example:\ BMI = \frac{68\ kg}{1.67\ m^2} = \frac{68\ kg}{2.7889\ m^2} = \frac{24.38\ kg}{m^2}$$

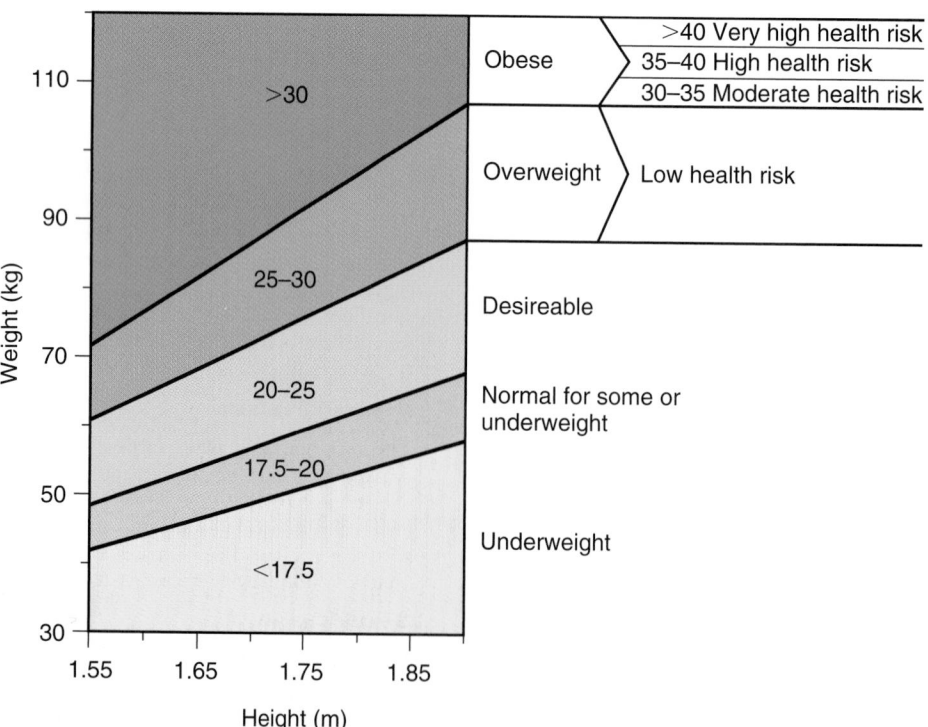

Figure 30-1 Body mass index ranges and associated level of health risk with obesity.

Obesity is caused by a caloric intake that exceeds energy expenditure. Only a small percentage of obesity is associated with a metabolic or endocrine abnormality within the body. Heath risks can result from being overweight. Diseases that are associated with obesity are called comorbidites. Comorbidites include but are not limited to atherosclerosis, heart disease, diabetes mellitus, hypertension, sleep apnea, osteoarthritis, decreased mobility, lack of self-esteem, and depression. Obesity that interferes with activities of daily living, such as breathing or walking, is known as morbid obesity. Morbid obesity refers to people whose BMI is above 40, which is about 100 pounds overweight for men and about 80 pounds overweight for women. Surgery can be an option for people whose BMI is above 40 or for people whose BMI is between 35 and 40 and who have life-threatening obesity-related diseases such as severe sleep apnea or heart disease. For more information, go to the National Heart, Lung, and Blood Institute Web site at www.nhlbisupport.com/bmi, The American Obesity Association at www.obesity.org, or The North American Association for the Study of Obesity at www.naaso.org. For information about surgery, finding a surgeon, chat rooms, and more, visit www.obesityhelp.com.

Medical Management

Initial treatment for obesity is weight loss through exercise and calorie restriction. For weight loss to occur, it is essential that the patient cooperates and has sustained motivation. Support groups such as Take Off Pounds Sensibly (TOPS) and Weight Watchers can help patients be successful. Behavior modification methods that provide rewards for successful weight loss are often included in a weight loss plan. Short-term use of medications that suppress appetite or block fat absorption may also be used.

Surgical Management

Patients who do not respond to medical methods of weight loss or whose BMI is 40 or above may have surgery to reduce their weight if they meet established criteria for the surgery. Surgical techniques produce weight loss by restriction (limiting how much the stomach can hold), or malabsorption (decreased calorie and nutrient absorption). (See Nutrition Notes Box 30–2.) The field of obesity surgery is called bariatric surgery and is designed to treat severe obesity. The word **bariatric** comes from the Greek word *baros* which means "weight." Various procedures have been used over the years, but complications have arisen with some, such as malabsorption, so they are no longer used. For a list of surgical obesity centers and surgical procedures, visit The American Society for Bariatric Surgeons at www.asbs.org.

GASTRIC BYPASS. The **Roux-en-Y** bypass is the most common gastric bypass surgery today (Fig. 30–2). In the first part of this two-step surgery, a small stomach pouch the size of a thumb is created with staples. This small pouch causes a quick satisfactory feeling of fullness during a meal, which is the key to the success of this procedure. Next, a Y-shaped section of the small intestine is attached to the pouch to al-

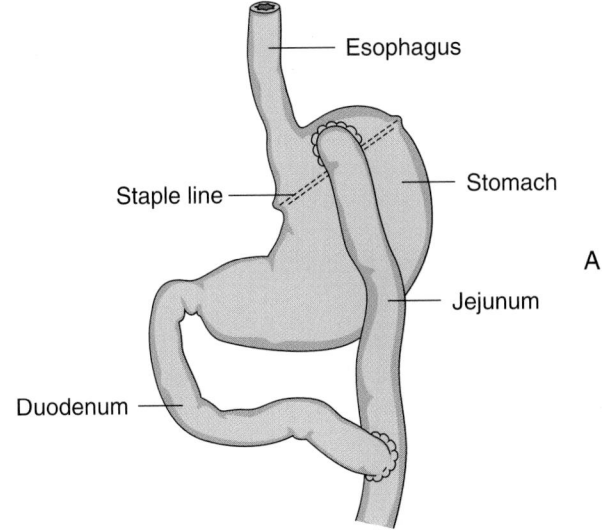

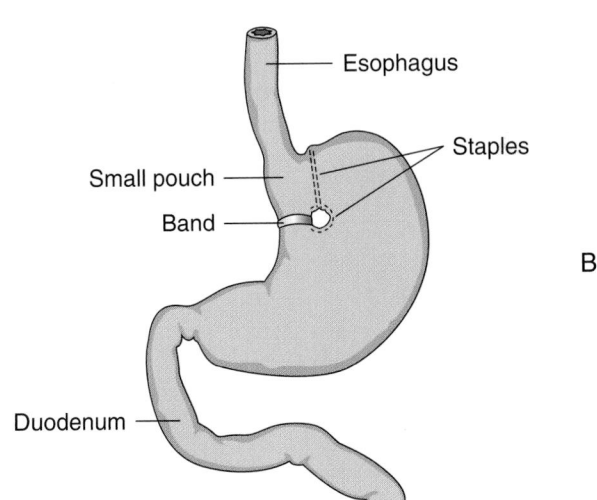

Figure 30–2 *(A)* Roux-en-Y gastric bypass. A staple line creates a small pouch at the top of the stomach, which is then attached to the jejunum. Food bypasses most of the stomach and the duodenum and goes into the jejunum. *(B)* Vertical banded gastroplasty for gastric bypass. A small stomach pouch is made with a staple line and a mesh band. A circular window made with staples allows the band to be placed around the pouch. The band restricts and slows food flow from the stomach pouch. As the small pouch fills, there is a feeling of fullness even with small meals.

low food to bypass the lower stomach and duodenum. Digestive juice flow is maintained, and food enters the jejunum within 10 minutes of eating. There is little malabsorption of food. This procedure has also been performed laparoscopically.

VERTICAL BANDED GASTROPLASTY. Vertical banded gastroplasty (VBG) is the most commonly used restrictive surgery for weight reduction and control today. (See Fig. 30–2.) There is little malabsorption of food with this procedure. About 30 percent of patients who have had VBG achieve normal weight. Other methods of gastric banding are currently under study.

COMPLICATIONS OF GASTRIC RESTRICTIVE SURGERIES.
A common side effect of restrictive surgery is vomiting caused by overeating or by not chewing food well. Severe side effects of VBG include erosion of the gastric tissue surrounding the band, breakdown of the staple line, and leaking of the stomach secretions into the abdomen. Leaking of the stomach secretions requires emergency surgery and can lead to peritonitis, a very serious infection of the peritoneum. Infection or death from any of these complications can occur.

POSTOPERATIVE CARE. Postoperative bariatric patients require care similar to that for most types of gastric surgeries. (See Nursing Management after Gastric Surgery later in this chapter.) The bariatric diet, however, is very different. Patients are started on a clear liquid diet because of the small stomach pouch that has been created. Then the diet progresses to full liquids, pureed foods, and finally, at about 6 weeks after surgery, regular foods as tolerated.

LEARNING TIPS ➡

For the patient who is, it is often necessary to have on hand special equipment for patient care. Some items you may need include the following:

Extra-large hospital bed
Extra pillows to elevate the head of the bed to ease breathing
Extra-large hospital gowns
Extra-large blood pressure cuff
Oversized wheelchair and walker
Devices for securing for Foley catheters to prevent tugging on the catheter
Abdominal binder for postoperative comfort and to reduce the occurrence of wound edge separation

---

▶ INFLAMMATORY DISORDERS
Stomatitis

Stomatitis is the general term for inflammation of the oral cavity. There are many causes of stomatitis, such as an infection or a systemic disease. The most common types of stomatitis are **aphthous stomatitis** (canker sores) and herpes simplex virus type I (also known as cold sores or fever blisters).

Aphthous Stomatitis (Canker Sores)

Aphthous stomatitis appears as small, white, painful ulcers on the inner cheeks, lips, tongue, gums, palate, or pharynx and typically lasts for several days to 2 weeks. Self-induced trauma such as biting the lips and cheeks can cause these ulcers to develop, as well as stress or exposure to irritating foods. Application of topical tetracycline several times a day usually shortens the healing time. A topical anesthetic such as benzocaine or lidocaine provides pain relief and makes it possible to eat with minimal pain.

---

gastroplasty: gastro—stomach + plasty—repair
stomatitis: stoma—mouth + itis—inflammation

Herpes Simplex Virus Type I

Herpes simplex virus type I (HSV-I) may appear as painful cold sores or fever blisters on the face, lips, perioral area, cheeks, nose, or conjunctiva. These lesions recur over time but last only for a few days each time. The onset can be provoked by fever or stress, among other things. Acyclovir ointment can be used to ease the pain, but it does not cure the lesions. Oral acyclovir may reduce recurrences. These lesions are infectious and standard precautions should be used when ointment is applied or oral care is given.

▶ ORAL CANCER

Oral cancer can occur anywhere in the mouth or throat. If detected early enough, it is curable. Oral cancer is found most commonly in patients who use alcohol or any form of tobacco. The highest incidence of oral cancer is found in the pharynx (throat), with the lowest incidence on the lips. Any oral sore that does not heal in 2 weeks should be assessed by the patient's physician. Cancerous ulcers are often painless but may become tender as the cancer progresses. In the later stages the patient may complain of difficulty in chewing, swallowing, or speaking or may have swollen cervical lymph glands. Biopsy specimens are taken to determine the presence of cancer.

Oral cancer treatment varies depending on the individualized diagnosis. Radiation, chemotherapy, and surgery are used alone or in combination to treat oral cancer. Radical or modified neck dissection is performed if the cancer has metastasized to cervical lymph nodes (Fig. 30–3). The tumor is removed along with lymph nodes, muscles, blood vessels, glands, and part of the thyroid, depending on the extent of the cancer. Drains are inserted into the incision to prevent fluid accumulation. A tracheostomy is usually performed to protect the airway and prevent obstruction. The airway must be monitored and secretions controlled to prevent aspiration. Tube feedings are usually given to meet the patient's nutritional needs because swallowing is difficult.

▶ ESOPHAGEAL CANCER

As with oral cancer, esophageal cancer is associated with the use of tobacco or alcohol. Esophageal cancer is usually detected late because of its location near many lymph nodes that allow it to metastasize. As the cancer progresses, obstruction of the esophagus can occur, with possible perforation or fistula development that may cause aspiration. The appearance of signs and symptoms usually means that the cancer is in the late stages. Signs and symptoms may include difficulty swallowing, a feeling of fullness, pain in the chest area after eating, foul breath, or regurgitation of foods if there is obstruction.

Diagnosis of esophageal cancer is usually made by esophagogastroduodenoscopy (EGD) and biopsy. Mediastinoscopy (endoscopic examination of mediastinum) is used to determine whether the cancer has spread to the lymph nodes and surrounding structures. Treatment for esophageal cancer includes radiation, chemotherapy, and surgery alone or in combination. Surgical procedures include esophageal resection (esophagogastrostomy), Dacron esophageal replacement, or use of a section of colon to replace the esophagus (esophagoenterostomy). If the tumor is inoperable, esophageal dilation or stent placement can be done to relieve dysphagia and allow food to pass through the esophagus.

▶ HIATAL HERNIA

The esophagus passes through an opening in the diaphragm called the hiatus. A **hiatal hernia** is a condition in which the lower part of the esophagus and stomach slides up through the hiatus of the diaphragm into the thorax (Fig. 30–4). Hiatal hernia occurs most commonly in women and those who are older than age 60, obese, or pregnant. A small hernia may not produce any discomfort or require treatment. However, a large hernia can cause pain, heartburn, a feeling of fullness, or reflux, which can injure the esophagus with possible ulceration and bleeding. Hiatal hernias are diagnosed by x-ray studies and fluoroscopy.

Medical Management

Treatment for hiatal hernia includes antacids; eating small meals that pass easily through the esophagus; not reclining for 1 hour after eating; elevating the head of the bed 6 to 12 inches to prevent reflux; and avoiding bedtime snacks, spicy foods, alcohol, caffeine, and smoking. (See Nutrition Notes Box 30–2).

Surgical Management

Surgical procedures can be done to prevent the herniated portion of the stomach from moving upward through the hiatus. Fundoplication is the most common surgical procedure performed, in which the stomach fundus is wrapped

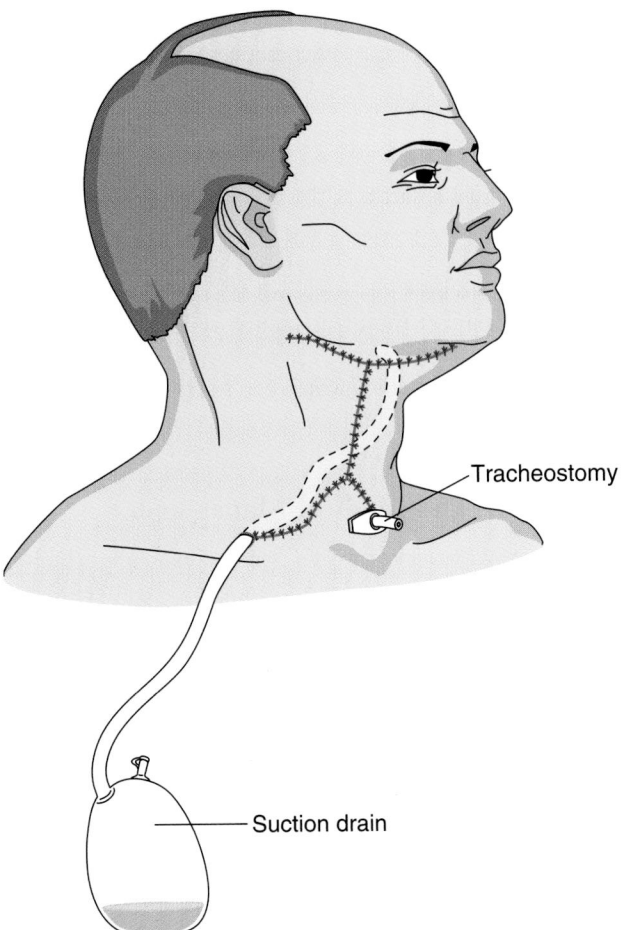

Figure 30-3 Radical neck dissection with tracheostomy tube and drains inserted.

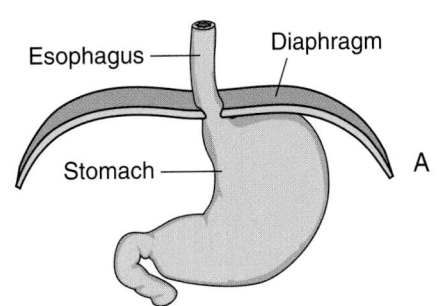

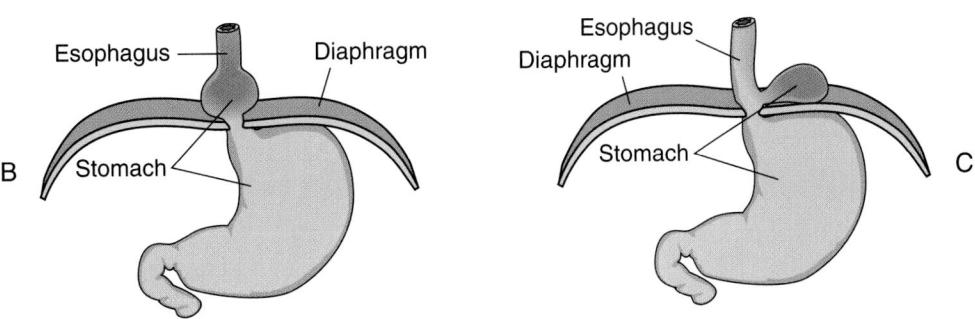

Figure 30-4 Hiatal hernia. *(A)* Normal esophagus and stomach. *(B)* Sliding hiatal hernia. *(C)* Rolling hiatal hernia.

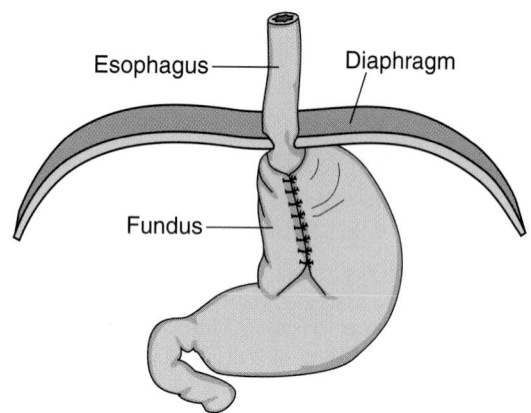

Figure 30-5   Hiatal hernia repair. Nissen fundoplication wraps the stomach fundus around the esophagus and then sutures it onto itself to hold it in place.

around the lower part of the esophagus (Fig. 30–5). Following this repair, patients are assessed for dysphagia during their first postoperative meal. If dysphagia occurs, the physician should be notified because the repair may be too tight, causing obstruction of the passage of food.

▶ GASTROESOPHAGEAL REFLUX DISEASE

Gastroesophageal reflux disease (GERD) is a condition in which gastric secretions reflux into the esophagus. The esophagus can be damaged by acidic gastric secretions and exposure to digestive enzymes. GERD is caused primarily by conditions that affect the ability of the lower esophageal sphincter to close tightly, such as hiatal hernia. Signs and symptoms of GERD include heartburn, regurgitation, dysphagia, and bleeding. Aspiration is a concern. Scar tissue can develop from the inflammation. Diagnostic tests include a barium swallow, esophagoscopy, or pH monitoring of the normally alkaline esophagus. GERD often occurs in elderly people.

Medical Management

Interventions for GERD aim to decrease the reflux of gastric secretions into the esophagus. Medications used may include antacids, histamine (H_2) receptor antagonists, cytoprotective agents (sucralfate), and cholinergic drugs that improve gastric emptying and function of the lower esophageal sphincter. Obese patients are encouraged to lose weight. A low-fat, high-protein diet is recommended because fat causes decreased functioning of the lower esophageal sphincter. Caffeine, milk products, and spicy foods should be avoided. If surgery is necessary to alleviate symptoms, a fundoplication can be done.

Nursing Management

Patients need to be educated about medications and management of their condition. They are instructed to sleep with the head of the bed elevated 4 to 6 inches, eat small meals, and avoid lying down for 2 hours after eating. Smok-

ing and alcohol intake should be avoided because they decrease functioning of the lower esophageal sphincter. Foods that cause discomfort should be identified by the patient and avoided.

▶ GASTRITIS

Gastritis is inflammation of the stomach mucosa and can be acute or chronic. Causes are listed in Box 30–3. Gastritis results when the protective mucosal barrier is broken down and allows autodigestion from hydrochloric acid and pepsin to occur. Edema of the tissue and possible hemorrhage result. With severe gastritis, the gastric mucosa can become gangrenous and perforate, which can lead to peritonitis (infection of the peritoneum). Scarring may also occur, resulting in pyloric obstruction.

Symptoms and Medical Management

The major symptom of gastritis is abdominal pain, which is often accompanied by nausea and anorexia. Treatment is removal of the irritating substance and provision of a bland diet of liquids and soft foods along with antacids. With a bland diet, the patient usually recovers in about a day.

Chronic Gastritis Type A

Chronic gastritis occurs over time and is classified as type A or type B. Type A is often referred to as autoimmune gastritis and occurs in the fundus (body of stomach). Chronic gastritis type A is diagnosed by endoscopy, upper gastrointestinal (GI) x-ray examination, and gastric aspirate analysis. (See Chapter 29.) Type A gastritis is often asymptomatic. Patients with type A gastritis usually do not secrete enough intrinsic factor from their stomach cells and as a result have difficulty absorbing vitamin B_{12}, which leads to pernicious anemia (discussed later).

Chronic Gastritis Type B

Type B gastritis affects the antrum and pylorus (lower end of the stomach near the duodenum) and is associated with *Helicobacter pylori* bacterial infection. Signs and symptoms

---

gastritis: gastr—stomach + itis—inflammation

BOX 30-3   Causes of Gastritis

Alcohol
Helicobacter pylori
Medication
Aspirin
Nonsteroidal anti-inflammatory drugs
Corticosteroids
Reflux of bile
Salmonella
Smoking
Spicy foods
Stress—physiological or psychological
Trauma

include poor appetite, heartburn after eating, belching, a sour taste in the mouth, and nausea and vomiting. Type B gastritis can also be diagnosed by endoscopy, upper gastrointestinal x-ray examination, and gastric aspirate analysis. *H. pylori* infection is treated with antibiotics.

▶ PEPTIC ULCER DISEASE

Etiology

Until 1982 the cause of peptic ulcer was poorly understood and thought to be related to stress, diet, and alcohol or caffeine ingestion. However, research results found that **peptic ulcer disease** is primarily caused by infection with the gram-negative bacterium *H. pylori*. This bacterium is responsible for 80 percent of gastric ulcers and more than 90 percent of duodenal ulcers. Two-thirds of all people are infected with *H. pylori*, and it is most common in those who are elderly, Hispanic, African-American, or in lower socioeconomic groups in the United States. The discovery of *H. pylori* has led to changes in treating and curing peptic ulcers. It is not known how *H. pylori* is transmitted, although the oral-oral or fecal-oral routes are likely. Contaminated water may also play a role. Vaccines to prevent peptic ulcers are being developed.

Peptic ulcer development is also influenced by smoking, which increases the harmful effects of *H. pylori*, alters protective mechanisms, and decreases gastric blood flow. For more information on *H. pylori*, visit www.cdc.gov or call 1-888-MY-ULCER (1-888-698-5237).

L E A R N I N G   T I P S

Most peptic ulcers are caused by an infection (*H. pylori*) that can be cured with antibiotics.

Pathophysiology

Peptic ulcer disease (PUD) is a condition in which the lining of the stomach, pylorus, duodenum, or the esophagus is eroded, usually from infection with *H. pylori*. The erosion may extend into the muscle layers or the peritoneum. Peptic ulcers occur in the portions of the gastrointestinal tract that are exposed to hydrochloric acid and pepsin. The erosion is due to an increase in the concentration or activity of hydrochloric acid and pepsin. The damaged mucosa is unable to secrete enough mucus to act as a barrier against the hydrochloric acid. Some individuals have more rapid gastric emptying, which, combined with hypersecretion of acid, creates a large amount of acid moving into the duodenum. As a result, peptic ulcers occur more often in the duodenum. Ulcers are named by their location: esophageal, gastric, or duodenal. Duodenal ulcers are more common than gastric ulcers.

Signs and Symptoms

Symptoms vary with the location of the ulcer (Table 30–1). Symptoms, including pain, may not be experienced with

TABLE 30-1	PEPTIC ULCER DISEASE SUMMARY
Signs and symptoms	Gastric: Intermittent high left epigastric or upper abdominal burning or gnawing pain, increased 1–2 hours after meals or with food
	Duodenal: Intermittent midepigastric or upper abdominal burning or cramping pain, increased 2–4 hours after meals or in the middle of the night; relieved with food or antacids
	Anorexia
	Nausea/vomiting
	Bleeding (stomach secretions or stool positive for occult blood)
Diagnostic tests *H. pylori*	Urea breath test
	IgG antibody detection test for *H. pylori*
	Biopsy
	Culture
Peptic ulcer	Upper GI series (barium swallow)
	Esophagogastroduodenoscopy
Medical treatment	Antibiotics
	Proton pump inhibitors
	H₂ antagonists
	Bismuth subsalicylate
	Sucralfate (Carafate)
	Antacids
	Bland diet
	Avoiding irritants, such as smoking, caffeine, alcohol
Complications	Bleeding
	Perforation
	Obstruction
Nursing diagnoses	Acute pain related to disruption of GI mucosa
	Risk for injury related to complications of peptic ulcer activity such as hemorrhage or perforation
	Deficient knowledge related to lack of exposure to peptic ulcer disease and its treatment

gastric or duodenal ulcers until complications such as hemorrhage, obstruction, or perforation develop. If pain does occur, patients with gastric ulcers commonly experience a burning and gnawing pain in the high left epigastric region. There may be more pain with food ingestion or 1 to 2 hours after a meal. Duodenal ulcers produce cramping or burning pain in the midepigastric or upper abdominal area. The pain occurs 2 to 4 hours after meals or in the middle of the night. This intermittent pain may be relieved by the ingestion of food or antacids. Anorexia and nausea and vomiting may also occur with either ulcer location. Bleeding may occur with massive hemorrhaging or slow oozing. Patients often have low hematocrit and hemoglobin levels, and gastric or fecal occult blood may be found, depending on where the ulcers are located.

Diagnosis

H. pylori can be diagnosed with several tests. The urea breath test is performed by having the patient drink carbon-labeled urea. The urea is metabolized rapidly if *H. pylori* is present, allowing the carbon to be absorbed and measured in exhaled carbon dioxide. An IgG antibody detection test for *H. pylori* identifies whether the patient is infected with *H. pylori*. These are both noninvasive detection tests. Biopsy specimens for the *Campylobacter*-like organism (CLO) biopsy urease test and a histological examination can be obtained during esophagogastroduodenoscopy (EGD). Biopsy is the most conclusive test for *H. pylori*. Cul-

tures of the biopsy specimen may also be done to determine antimicrobial susceptibility.

Peptic ulcers are diagnosed on the basis of symptoms, upper GI series (barium swallow), and EGD.

Medical Management

Several treatment options are used to cure *H. pylori* without recurrence (Table 30–2). The first antibiotic treatment for ulcer disease caused by *H. pylori* was approved by the Food and Drug Administration (FDA) in 1996. For better effectiveness, triple therapy with two antibiotics to decrease resistance of the bacteria and a proton pump inhibitor or H_2 antagonist is used. Treatment lasting 14 days has better eradication rates than 10-day treatments. Bismuth subsalicylate (Pepto-Bismol) may also be used for its antibacterial effects. Proton pump inhibitors are powerful agents that stop the final step of gastric acid secretion to reduce mucosa erosion and aid in healing ulcers (Table 30–3). H_2 antagonists block H_2 receptors to decrease acid secretion, although not as powerfully as gastric acid pump inhibitors. A bland diet may also be recommended, and foods known to cause discomfort to the patient, such as spicy foods, carbonated drinks, and caffeine, should be avoided until the ulcer heals.

Nursing Process
Assessment

The primary focus of nursing care for peptic ulcer disease is educating patients regarding the importance of this diagnosis, because ulcers may be caused by an infection that can be cured with antibiotics. Patients may still believe that all ulcers are caused by stress, lifestyle, or diet. Assessing the patient's knowledge aids in providing accurate information to assist the patient in managing peptic ulcer disease. Data are also collected about the patient's disease history. Identifying factors that trigger or relieve symptoms is important.

Nursing Diagnosis

Nursing Care Plan 30–4 discusses nursing diagnoses for peptic ulcer disease. The care plan should focus on the patient's understanding of the importance of taking all medication as

---

L E A R N I N G   TIPS

Basic types of medications for the stomach include the following:

- Proton pump inhibitors such as omeprazole (Prilosec) or lansoprazole (Prevacid) inhibit gastric acid from forming.
- H_2 antagonists such as cimetidine (Tagamet), famotidine (Pepcid), and ranitidine (Zantac) reduce gastric acid secretion.
- Antacids such as Maalox and Mylanta neutralize gastric acid that has already formed.
- Sucralfate (Carafate) is like a band-aid for ulcers. Carafate forms a sticky paste that adheres to the ulcer surface, providing a protective barrier to allow healing.

TABLE 30-2	MEDICATION REGIMEN OPTIONS FOR *H. PYLORI* INFECTION
Type of Therapy	**Examples of Therapy Options**
*Triple therapy**	
Two antibiotics + proton pump inhibitor	• Amoxicillin (Amoxil) + clarithromycin (Biaxin) + omeprazole (Prilosec)
	• Amoxicillin (Amoxil) + clarithromycin (Biaxin) + lansoprazole (Prevacid) (available as Prevpac, combined for convenience)
Dual therapy	
Antibiotic + proton pump inhibitor or	• Clarithromycin (Biaxin) + omeprazole (Prilosec)
	• Amoxicillin (Amoxil) + lansoprazole (Prevacid)
Antibiotic + H_2 antagonist	
Other therapy	• Clarithromycin (Biaxin) + ranitidine bismuth citrate (Tritec)
Two antibiotics + bismuth subsalicylate + H_2 antagonist	• Metronidazole (Flagyl) + tetracycline + bismuth subsalicylate (Pepto-Bismol) + H_2 antagonist

*Triple therapy has a better eradication rate.

TABLE 30-3 MEDICATIONS USED TO PROMOTE HEALING OF PEPTIC ULCERS

Medication	Action	Side Effects	Nursing Interventions
Hyposecretory Agents—H₂ Receptor Blocking Agents			
Cimetidine (Tagamet)	Inhibits gastric acid secretion by blocking H₂ receptors on gastric parietal cells.	Fever, rash, headaches, dizziness, somnolence, confusion (especially in elderly), hypotension, diarrhea, neutropenia, gynecomastia, and impotence	Monitor mental status of elderly; do not take antacids within 1 hour of Tagamet; take with meals and at bedtime; interacts with theophylline, phenytoin, warfarin, and beta blockers; continue treatment for at least 8 weeks to ensure healing.
Ranitidine (Zantac)	Inhibits gastric acid secretion by blocking H₂ receptors on gastric parietal cells.	All side effects rare, including nausea, constipation, bradycardia, increased liver enzymes, and headache	Give antacids at least 1 hour before or 2 hours after Zantac; can be given in single bedtime dose; use cautiously in patients with liver or renal disease; absorption not affected by food; interacts minimally with other drugs.
Famotidine (Pepcid)	Inhibits gastric acid secretion by blocking H₂ receptors on gastric parietal cells.	Headache, diarrhea, constipation, nausea, flatulence, increased blood urea nitrogen and creatinine, and rash	Should not be taken longer than 8 weeks without physician's order; may be given with antacids; can be given in single bedtime dose; has no significant drug interactions.
Nizatidine (Axid)	Inhibits gastric acid secretion by blocking H₂ receptors on gastric parietal cells.	Diarrhea, rash, bronchospasms, somnolence, joint pain, and sweating	Give as single bedtime dose or, if given twice a day, one dose at bedtime; assess for excessive drowsiness; monitor and record stools; do not give antacids within 1 hour of Axid; must be taken 4 to 8 weeks for ulcer healing; notify physician if somnolence or rash develops.
Proton Pump Inhibitor			
Omeprazole (Prilosec)	Binds to enzyme on gastric parietal cells to prevent final transport of hydrogen to block gastric acid secretion.	Abdominal pain, diarrhea, rash, chest pain, and weakness	Give before meal in morning; swallow capsule whole; assess for abdominal pain and bleeding; monitor complete blood cell count and liver enzymes; may give with antacids; must be taken 4 to 8 weeks for ulcer healing; notify physician if bleeding, diarrhea, headache, or abdominal pain develops.
Lansoprazole (Prevacid)	Binds to an enzyme in the presence of acidic gastric pH, preventing the final transport of hydrogen ions into the gastric lumen.	Dizziness, headache, diarrhea, abdominal pain, nausea, rash	Assess for epigastric or abdominal pain and for blood in stool, emesis, or gastric aspirate. Give before meals. Capsules may be opened and sprinkled on applesauce and taken immediately for patients with difficulty swallowing. Do not crush capsule. Tell patient not to chew capsule.
Rabeprazole (Aciphex)	Same as for lansoprazole	Headache	Tell patient to swallow tablets whole.
Antacids			
Aluminum-magnesium combinations (Riopan, Maalox, Mylanta, Gelusil)	Increases gastric pH to reduce pepsin activity; strengthens gastric mucosal barrier and esophageal sphincter tone.	Mild constipation or diarrhea	Do not give to patients with renal disease; monitor bowel movements and signs of hypermagnesemia; Riopan low in sodium; do not give within 1 to 2 hours of H₂ receptor antagonists, tetracycline, or enteric-coated tablets.
Calcium carbonate (Tums, Titralac)	Increases gastric pH to reduce pepsin activity; strengthens gastric mucosal barrier and esophageal sphincter tone.	Constipation, gastric distention, rebound hyperacidity, hypercalcemia, and hypophosphatemia	Do not give with milk; monitor for symptoms of hypercalcemia and constipation; do not give within 1 to 2 hours of H₂ receptor antagonists, tetracycline, or enteric-coated tablets.
Mucosal Barrier Fortifiers			
Sucralfate (Carafate)	In presence of mild acid condition, forms viscid and sticky gel and adheres to ulcer surface, forming a protective barrier.	Dizziness, constipation, sleepiness, nausea, and gastric discomfort	Take on an empty stomach, 1 hour before meals and at bedtime; monitor for constipation.

BOX 30-4 NURSING CARE PLAN FOR THE PATIENT WITH PEPTIC ULCER DISEASE

 Acute pain related to gastric mucosal erosion

PATIENT OUTCOMES
Patient's pain is relieved as evidenced by no report of pain and absence of nonverbal pain cues.

EVALUATION OF OUTCOMES
Is pain relieved to patient's satisfaction?

INTERVENTIONS	RATIONALE	EVALUATION
Ask patient to rate pain level on scale of 0 to 10 every 3 hours and as needed. Assess location, onset, intensity, characteristics of pain, and nonverbal pain cues.	Prompt assessment can lead to timely intervention and relief of pain.	Does patient rate pain using scale and describe pain?
Assess for factors precipitating and relieving pain.	Peptic ulcer pain may be relieved by food, antacids, or other interventions.	Is patient able to state precipitating and relieving pain factors?
Ask patient to help identify techniques for pain relief	Gaining the patient's cooperation increases compliance.	Is patient willing to participate in planning how to relieve pain?
Administer antiulcer medications as ordered.	H$_2$ receptor antagonists reduce amount of gastric acid produced, and antacids neutralize gastric acid to help relieve pain.	Do medications reduce patient's symptoms?
Provide small, frequent meals four to six times a day.	Small, frequent meals dilute and neutralize gastric acid.	Does patient report relief of gastric pain between meals?
Encourage nonacidic fluids between meals.	Nonacidic fluids decrease irritation to gastric mucosal.	Does patient identify and drink nonacidic fluids?

 Risk for injury related to complications of peptic ulcer activity such as hemorrhage and perforation

PATIENT OUTCOMES
Patient's vital signs will be maintained within normal limits and bleeding or hemorrhage will be promptly detected.

EVALUATION OF OUTCOMES
Are patient's vital signs within normal limits?

INTERVENTIONS	RATIONALE	EVALUATION
Assess for signs and symptoms of hemorrhage such as hematemesis (vomiting blood) and melena (blood in the stool).	Rapid assessment can lead to prompt intervention.	Does patient have any bleeding?
Monitor vital signs: blood pressure, pulse, respirations, and temperature.	Severe blood loss of more than 1 L per 24 hours may cause manifestations of shock such as hypotension; weak, thready pulse; chills; palpitations; and diaphoresis.	Are vital signs normal?
Maintain intravenous infusion as ordered.	Normal fluid balance prevents hypovolemia and shock due to hemorrhage.	Are intake and output balanced?
Monitor hematocrit and hemoglobin levels as ordered.	Decreased hematocrit and hemoglobin levels indicate a decrease in circulating blood volume and reduced oxygen-carrying capacity to the tissues.	Are hematocrit and hemoglobin levels normal?

directed, even if symptoms are gone. Patient noncompliance with treatment is a major cause of ulcer recurrence.

Stress Ulcers

A small number of patients who are critically ill may develop gastric or small intestinal stress ulcers from ischemia. The stress response to the illness causes reduced blood flow to the stomach and small intestine, resulting in ischemia and damage to the mucosa. The damaged mucous barrier allows acid secretions to create ulcers. Preventive treatment has dramatically reduced stress ulcers, which have a high

mortality rate because of the multiple bleeding ulcer sites. This treatment includes trauma care that quickly restores oxygen to the stomach, as well as early feeding within 24 hours of the trauma. Placement of a nasogastric (NG) tube allows testing of the gastric pH to ensure that it is higher than 5, as well as providing a means for enteral feeding. In addition, medications are given such as antacids, histamine blockers, and sucralfate (Carafate).

▶ GASTRIC BLEEDING

Gastric bleeding may be caused by ulcer perforation, tumors, gastric surgery, or other conditions. Bleeding peptic ulcers are the most common cause of blood loss into the stomach or intestine. Blood loss can be hidden (occult) blood in the stool, observable vomited blood (hematemesis), or black tarry stools (melena). When blood mixes with hydrochloric acid and enzymes in the stomach, a dark, granular material resembling coffee grounds is produced. This material can be vomited or passed through the GI system and mixed with stools. Melena occurs from slow bleeding in an upper GI area.

Signs and Symptoms

With mild bleeding, the patient may experience only slight weakness or diaphoresis. Severe blood loss (more than 1 L in 24 hours) may result in hypovolemic shock, with signs and symptoms such as hypotension; a weak, thready pulse; chills; palpitations; and diaphoresis.

Medical Management

The goal of treatment for a massive GI bleed is to prevent or treat hypovolemic shock and prevent dehydration, electrolyte imbalance, and further bleeding. The patient is kept on nothing-by-mouth (NPO) status. An intravenous (IV) line is started to replace lost fluids and administer blood if necessary. A complete blood cell count is obtained to determine the amount of blood lost. A urinary catheter is inserted to monitor output. An NG tube is inserted to assess the rate of bleeding, decompress the stomach, monitor the pH of gastric secretions, and administer saline lavage if ordered. Oxygen therapy may be required if the patient has lost a large amount of blood. To prevent aspiration with vomiting, the head of the bed is elevated. The physician may perform endoscopy to help control the bleeding. Drugs may also be instilled into the GI tract by use of an endoscope. For severe cases, surgery may be needed to remove the bleeding area or ligate bleeding vessels. Drugs such as ranitidine (Zantac) are given to decrease the secretion of gastric acid.

▶ GASTRIC CANCER

Gastric cancer refers to malignant lesions found in the stomach. It is the second most common cancer in the world and is more common in men than in women. *H. pylori* infection plays a role in gastric cancer development. Studies are being done to see whether preventing *H. pylori* infection reduces the development of gastric cancer. Other factors that may be associated with gastric cancer development include pernicious anemia; exposure to occupational substances such as lead dust, grain dust, glycol ethers, or leaded gasoline; and a diet high in smoked fish or meats. A poor prognosis is often associated with gastric cancer, because most patients have metastasis at the time of diagnosis. (See Nutrition Notes Box 30–2.)

Signs and Symptoms

Gastric cancer is rarely diagnosed in its early stages because symptoms do not appear until late in the disease. In the early stages there may not be any symptoms at all, and

metastasis to another organ, such as the liver, may have already occurred. The symptoms of gastric cancer are often mistaken for peptic ulcer disease: indigestion, anorexia, pain relieved by antacids, weight loss, and nausea and vomiting. Anemia from blood loss commonly occurs, and occult blood in the stool may be present.

Diagnosis

Diagnosis of gastric cancer is made by upper gastrointestinal x-ray examination, gastroscopy, gastric fluid analysis, and measurement of serum gastrin levels.

Medical Management

Surgical removal of the cancer is the most effective treatment for gastric cancer. Most often the cancer has already metastasized, and surgery is performed only to relieve the symptoms. Chemotherapy and radiation are sometimes used in conjunction with surgery, although they are not very effective against the cancer. Biologic therapies, new cytotoxic agents, and new delivery methods are being studied for use with gastric cancer.

▶ GASTRIC SURGERIES
Subtotal Gastrectomy

Two types of surgical interventions are used to treat upper gastrointestinal diseases. The first type is the subtotal **gastrectomy,** which is used to treat cancer in the lower two-thirds of the stomach. *Subtotal gastrectomy* is a general term used to describe any surgery that involves partial removal of the stomach. There are two types of subtotal gastrectomies: the Billroth I procedure and the Billroth II procedure.

Billroth I Procedure (Gastroduodenostomy)

In the Billroth I procedure, also known as a **gastroduodenostomy,** the surgeon removes the distal portion (75 percent) of the stomach (Fig. 30–6). The remainder of the stomach is anastomosed (surgically attached) to the duodenum. This procedure is used to treat gastric problems.

Billroth II Procedure

The Billroth II procedure, **gastrojejunostomy,** involves removal of the distal 50 percent of the stomach and reanastomosis of the proximal remnant of the stomach to the proximal jejunum. Because it results in bypassing of the duodenum, the Billroth II procedure is used to treat duodenal ulcers. Pancreatic secretions and bile are necessary for digestion and continue to be secreted into the duodenum from the common bile duct even after the partial gastrectomy.

---

gastrectomy: gastr—stomach + ectomy—to remove
gastroduodenostomy: gastro—stomach + duoden—duodenum + ostomy—mouth or opening
gastrojejunostomy: gastro—stomach + jejeun—jejunum + ostomy—mouth or opening

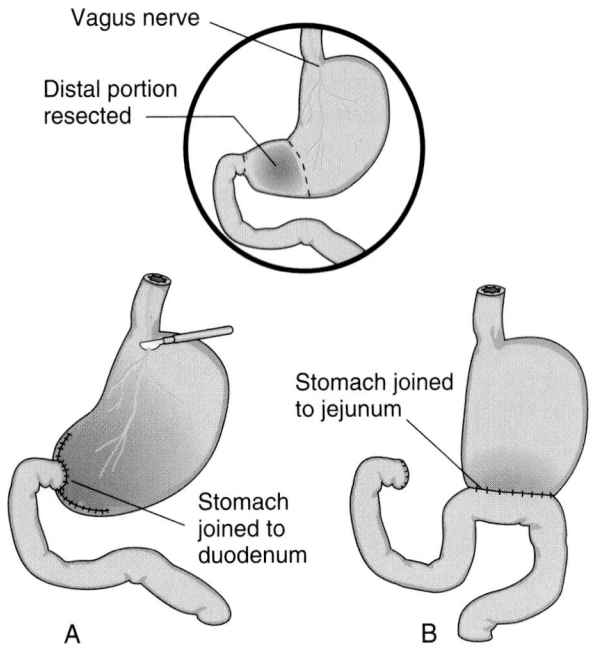

Figure 30-6 Subtotal gastrectomy involves removing the distal portion of the stomach. The remaining portion of the stomach is then sutured *(A)* to the duodenum (Billroth I procedure) or *(B)* to the proximal jejunum (Billroth II procedure). Vagotomy may also be performed.

Total Gastrectomy

Total gastrectomy, the total removal of the stomach, is the treatment for extensive gastric cancer. This surgery involves removal of the stomach, with anastomosis of the esophagus to the jejunum (Fig. 30–7).

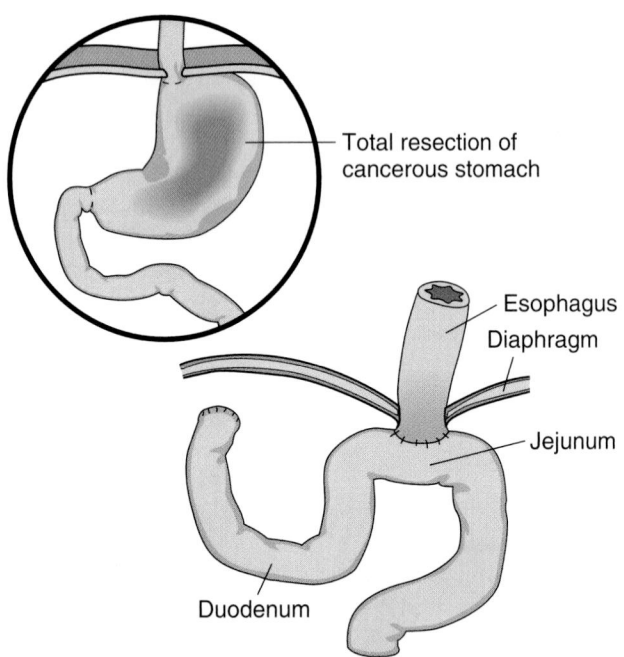

Figure 30-7 Total gastrectomy.

Vagotomy

A vagotomy, in which a section of the vagus nerve is cut, may be performed with gastric surgery. Vagotomy eliminates the vagal stimulation for hydrochloric acid and gastrin hormone secretion and slows gastric motility.

Nursing Management after Gastric Surgery

Patients' vital signs are monitored postoperatively as ordered. Respiratory status is carefully monitored because the high location of the surgical incision may cause pain, which interferes with deep breathing and coughing. Atelectasis or pneumonia can develop as a result of guarding and shallow breathing. The patient's pain is assessed and relieved, which also helps the patient's ability to deep breathe or cough without pain. The patient's IV site and infusion are monitored, and intake and output are recorded. The incisional site and dressings are observed for drainage and bleeding.

Bowel sounds are assessed. Patients may have an NG tube inserted during surgery. The nurse applies suction as ordered, which is usually low intermittent suction to prevent trauma to the gastric mucosa. The drainage from the NG tube is monitored for color and amount. If bleeding or excessive amounts of drainage are noted, they are reported to the physician. The nurse should not irrigate or reposition the NG tube following gastric surgery to prevent damaging the suture line. The patient should be assessed for abdominal distention. If distention occurs, abdominal girth should be measured and the physician notified.

It is important to teach patients how to assist in their recovery. This includes incisional care, activity or dietary restrictions, and information about prescribed medications. Patients are encouraged to ambulate early to promote a quicker recovery by improving respiratory and gastrointestinal function.

Complications of Gastric Surgery

Complications that can occur after gastric surgery include hemorrhage, acute gastric distention, nutritional problems, **steatorrhea** (fat in stools), pyloric obstruction, and dumping syndrome.

Hemorrhage

The incidence of hemorrhage after gastric surgery is very low and is most often caused by a dislodged clot at the surgical site or slippage of a suture. The patient experiencing hemorrhage exhibits restlessness, cold skin, increased pulse and respirations, and decreased temperature and blood pressure. In addition, the patient may vomit bright red blood. To prevent aspiration during vomiting, the patient is turned to one side and the head of the bed is elevated.

Following gastric surgery, patients usually have an NG tube that has been inserted in the operating room. The drainage from the tube should be assessed for color and amount. A small amount of pink or light red drainage may

steatorrhea: steatos—fat + rrhea—profuse flow

be expected for the first 12 hours, but moderate or excessive bleeding should be immediately reported to the physician. The abdominal dressing should also be assessed for any drainage or bleeding.

Gastric Distention

In the immediate postoperative period, distention of the stomach can occur if an inserted NG tube is clogged or if an NG tube has not been inserted. Symptoms of gastric distention include an enlarged abdomen, epigastric pain, tachycardia, and hypotension. The patient may complain of feeling full and may hiccup or gag repeatedly. These symptoms must be reported to the physician.

The physician usually inserts the NG tube during surgery so that the suture line is not damaged. If suction is desired, a physician's order is required. Irrigating or repositioning the NG tube is not performed by the nurse to prevent harm to the suture line. Any problems with distention or an improperly functioning NG tube are reported to the physician. The physician may need to reposition the NG tube to correct the problem. The patient's vital signs should be monitored until the patient's distention is relieved and the patient is stable.

🅼 CRITICAL THINKING

Mr. Wong
You are working the evening shift on a surgical unit. A patient, Mr. Wong, has had gastric surgery earlier that morning. He has an intravenous infusion of 1000 mL dextrose 5 percent in 0.45 normal saline running at 83 mL per hour and a nasogastric tube to low intermittent wall suction. Mr. Wong is restless and complaining of pain. His bowel sounds are absent, and his abdomen is distended. The suction canister contains no gastric output.

1. What nursing interventions in order of priority are needed to help Mr. Wong?
2. What equipment do you need to care for Mr. Wong?

Answers at end of chapter.

Nutritional Problems

Nutritional problems that commonly occur after removal of part or all of the stomach include B_{12} and folic acid deficiency and reduced absorption of calcium and vitamin D. Also, rapid entry of food into the bowel often results in inadequate absorption of food.

Following gastric surgery, patients are NPO until bowel sounds return (usually 24 to 48 hours) or the physician orders a diet. Intravenous fluid provides hydration. However, if patients are to be NPO for a longer time, they may need additional nutritional intake provided by total parenteral nutrition (TPN) through a central line. TPN is an intravenous solution given to meet caloric needs and provide fluids lost in drainage or emesis. Many patients with gastric cancer are malnourished and may require several days of TPN therapy.

After the return of bowel sounds and the removal of the nasogastric tube, clear fluids may be ordered with progression

to full liquids, then soft foods as the patient tolerates. It is important to remember that foods and fluids must be introduced into the diet gradually following gastric surgery. If the patient eats too much or too fast, regurgitation may result.

PERNICIOUS ANEMIA. Vitamin B_{12} deficiency can occur after some or all of the stomach is removed because intrinsic factor secretion is reduced or gone. Normally vitamin B_{12} combines with intrinsic factor to prevent its digestion in the stomach and promote its absorption in the intestines. Lifelong administration of vitamin B_{12} injections is required to prevent the development of pernicious anemia. The parenteral route must be used because oral vitamin B_{12} cannot be absorbed. Vitamin B_{12} is given daily initially, then monthly for life. Patients must be taught the importance of complying with this treatment for the rest of their lives to prevent pernicious anemia. Symptoms of pernicious anemia include anemia, weakness, sore tongue, numbness and tingling, and gastrointestinal upset.

Steatorrhea

Steatorrhea is the presence of excessive fat in the stools and is the result of rapid gastric emptying, which prevents adequate mixing of fat with pancreatic and biliary secretions. In most cases, steatorrhea can be controlled by reducing the intake of fat in the diet.

Pyloric Obstruction

Pyloric obstruction can occur after gastric surgery as a result of scarring, edema, or inflammation or a combination of these. The signs and symptoms are vomiting, a feeling of fullness, gastric distention, nausea after eating, loss of appetite, and weight loss. As the obstruction increases, it gradually becomes more difficult for the stomach to empty, and symptoms worsen. Conservative methods are used first, such as replacing fluids and electrolytes through intravenous fluids and decompressing the distended stomach using a nasogastric tube. Surgery may be necessary if conservative measures do not relieve the signs and symptoms. Pyloroplasty widens the exit of the pylorus to improve emptying of the stomach.

Dumping Syndrome

Dumping syndrome occurs with the rapid entry of food into the jejunum without proper mixing of the food with digestive juices. On entering the jejunum, the food draws extracellular fluid into the bowel from the circulating blood volume to dilute the high concentration of electrolytes and sugars. This rapid shift of fluids decreases the circulating

---

🅼 CRITICAL THINKING

Mrs. Lindsay
Mrs. Lindsay has had gastric surgery. You have taught her about dumping syndrome, and she is concerned about what she will eat. Create a 1-day meal plan for Mrs. Lindsay.
Answers at end of chapter.

blood volume and produces symptoms. The symptoms occur 5 to 30 minutes after eating and include dizziness, tachycardia, fainting, sweating, nausea, diarrhea, a feeling of fullness, and abdominal cramping. Additionally, the blood sugar rises, and excessive insulin is excreted in response. This release of insulin causes the patient to have symptoms of hypoglycemia about 2 hours later. Symptoms include weakness, sweating, anxiety, shakiness, confusion, and tachycardia. The patient should immediately eat some candy or drink juice containing sugar to relieve the symptoms.

The treatment for dumping syndrome includes teaching the patient to eat small, frequent meals that are high in protein and low in carbohydrates, especially refined sugars (see Nutrition Notes Box 30–2). The patient is also taught to avoid fluids 1 hour before, with, or for 2 hours after meals to prevent rapid gastric emptying. It is best for the patient to lie down after meals to delay gastric emptying. The patient is told that these symptoms may last for up to 6 months after gastric surgery but usually slowly subside over time.

REVIEW QUESTIONS

1. Which one of the following is the most important intervention for anorexia nervosa?
   a. Weigh the patient daily.
   b. Restore nutritional health.
   c. Assist with activities of daily living.
   d. Document intake and output.

2. Which of the following is the priority nursing diagnosis for a patient with peptic ulcer disease?
   a. Activity intolerance related to epigastric pain
   b. Deficient knowledge related to lack of information on ulcers
   c. Acute pain related to epigastric erosion and acid buildup
   d. Ineffective coping related to diagnosis of peptic ulcers

3. Which of the following is the purpose of H_2 antagonists?
   a. Neutralize gastric acid.
   b. Form a protective paste.
   c. Determine gastric pH levels.
   d. Inhibit secretion of gastric acid.

4. Which one of the following is primarily the cause of peptic ulcers?
   a. Eating spicy foods
   b. A stressful life
   c. A bacterial infection
   d. Excessive caffeine intake

5. Which one of the following actions should the nurse take first for a patient who has just returned from surgery after a total gastrectomy and begins to vomit bright red blood?
   a. Increase the IV rate.
   b. Administer oxygen.
   c. Place patient on side.
   d. Irrigate nasogastric tube.

Answers to CRITICAL THINKING

Mr. Wong

1. Prioritize your interventions:
   a. Take Mr. Wong's vital signs to determine whether he is stable. Pain can sometimes increase the blood pressure and pulse rate. However, gastric distention can cause pain, and once the distention is relieved, the pain caused by distention subsides.
   b. Listen to Mr. Wong's bowel sounds. Manipulation of internal organs during abdominal surgery can produce a loss of normal peristalsis for 24 to 48 hours. Expect to hear absent or hypoactive bowel sounds for the first 1 or 2 days.
   c. Next, check placement of Mr. Wong's nasogastric tube by aspirating gastric contents and verifying pH of the contents as ordered. It is important to check for abdominal placement of Mr. Wong's nasogastric tube to make sure it is not misplaced in the lungs. After abdominal placement is determined, if there is a physician's order, the nasogastric tube can be connected to suction equipment, usually set on low intermittent suction.
   d. Next check the suction equipment for ordered settings and to ensure that it is turned on. The suction setting normally is ordered to be on low. A whistling sound is heard when the tube is disconnected from the suction setup. The seals should be tight on the suction canister. When the tubing is hooked to

suction, gastric contents should be moving into the suction canister. It is important to make sure equipment is functioning properly to ensure patient safety.
   e. Check the nasogastric tube for clogs only if the physician orders aspiration or irrigation to be done. The tube is gently aspirated with a 60-mL catheter-tipped syringe. If the tube remains clogged, it is gently flushed as ordered with 10 to 20 mL of sterile normal saline.
   f. After the gastric distention has been relieved, Mr. Wong's pain level is reassessed to determine if he needs pain medication. Considering that he is less than 1 day postoperative, he probably needs it.
2. Necessary equipment includes stethoscope, 60-mL catheter-tipped syringe, gloves, goggles, and normal saline for irrigation.

Mrs. Lindsay

Although there are many variations, the following is an example of a 1-day meal plan for Mrs. Lindsay:

Breakfast: one egg, any style; ½ orange; one glass milk
Snack: one slice toast with apple butter, jelly, or jam
Lunch: 2 oz ham, ½ cup cottage cheese, four asparagus spears
Snack: ½ serving chicken salad on bed of lettuce
Dinner: 2 oz broiled fish, ½ serving corn, ½ serving broccoli
Snack: ½ cup yogurt or sherbet

31

NURSING CARE OF PATIENTS WITH LOWER GASTROINTESTINAL DISORDERS

Virginia Birnie and Deborah J. Mauffray

KEY TERMS

appendicitis (uh-PEN-di-**SIGH**-tis)

colectomy (koh-**LEK**-tuh-me)

colitis (koh-**LYE**-tis)

colostomy (koh-**LAH**-stuh-me)

constipation (KON-sti-**PAY**-shun)

diarrhea (DYE-uh-**REE**-ah)

diverticulitis (DYE-ver-tik-yoo-**LYE**-tis)

diverticulosis (DYE-ver-tik-yoo-**LOH**-sis)

enteritis (en-ter-**EYE**-tis)

fissures (**FISH**-ers)

fistulas (FIST-yoo-lahs)

hematochezia (HEM-uh-toh-**KEE**-zee-uh)

hemorrhoids (**HEM**-uh-royds)

hernia (**HER**-nee-uh)

ileostomy (ILL-ee-**AH**-stuh-me)

impaction (im-**PAK**-shun)

intussusception (IN-tuh-suh-**SEP**-shun)

megacolon (**MEG**-ah-KOH-lun)

melena (muh-**LEE**-nah)

obstipation (OB-sti-**PAY**-shun)

peristomal (PER-i-**STOH**-muhl)

peritonitis (per-i-toh-**NIGH**-tis)

stoma (**STOH**-mah)

volvulus (**VOL**-view-lus)

QUESTIONS TO GUIDE YOUR READING

1. What are the causes, signs, and symptoms of constipation and diarrhea?

2. What nursing care and teaching do patients with constipation or diarrhea require?

3. What medical treatment, nursing care, and teaching are appropriate for patients with inflammatory and infectious disorders of the lower gastrointestinal tract?

4. How would you describe irritable bowel syndrome and the nursing care for this condition?

5. What signs and symptoms of abdominal hernias should be reported to the physician?

6. What nursing care and teaching do patients with absorption disorders require?

7. What are the causes signs and symptoms of intestinal obstruction?

8. What is the medical treatment and nursing care for intestinal obstruction?

9. What is the medical treatment and nursing care for lower gastrointestinal bleeding?

10. What nursing care and teaching does a patient with an ostomy require?

The lower gastrointestinal (GI) system includes the small and large intestines, rectum, and anus. Disorders associated with this system are discussed in this chapter.

► PROBLEMS OF ELIMINATION

Constipation

Pathophysiology

Constipation occurs when the fecal mass is held in the rectal cavity for a period that is not usual for the patient. When the feces are held for a prolonged time in the rectum, the amount of water absorbed increases, making the feces drier, harder, more difficult to pass, and sometimes painful to pass.

If a patient repeatedly ignores the urge to have a bowel movement, the musculature and rectal mucous membrane become insensitive to the presence of feces. Eventually a stronger stimulus is needed to produce the peristaltic rush required for defecation. Prolonged constipation is called **obstipation.**

Etiology

There are many causes of constipation. Rectal or anal conditions such as **hemorrhoids** or **fissures** may delay defecation because of the associated pain. Metabolic or neurological conditions such as diabetes mellitus, multiple sclerosis, lupus erythematosus, or scleroderma may interfere with normal bowel innervation and function. Colon cancer may cause an obstruction that prevents normal bowel function and leads to constipation. Medications such as narcotics, tranquilizers, and antacids with aluminum decrease motility of the large intestine and may contribute to constipation. Low intake of dietary fiber and fluids decreases the bulk of the feces and causes constipation. Decreased mobility, weakness, and fatigue, especially in the elderly, reduce the strength of the muscles used for defecation, increasing the likelihood of constipation. Chronic laxative use can also contribute to constipation because the laxative overrides the bowel's ability to recognize the urge to defecate.

Prevention

Regular exercise and a diet high in fiber and fluids and are the best preventive measures for constipation. Laxatives should be used only occasionally.

Signs and Symptoms

Abdominal pain and distention, indigestion, rectal pressure, a sensation of incomplete emptying, and intestinal rumbling are indications of constipation (Table 31–1). The patient may also complain of headache, fatigue, decreased appetite, straining at stool, and elimination of hard, dry stool.

Complications

A variety of problems can result from constipation. Fecal **impaction** may result when the fecal mass is so dry it cannot be passed. Pressure on the colon mucosa from a mass of stool may cause ulcers to develop. Often, small amounts of liquid

TABLE 31–1	CONSTIPATION SUMMARY
Symptoms	Abdominal distention
	Indigestion
	Rectal pressure
	Feeling of incomplete emptying
	Straining at stool
	Hard, dry stool
	Intestinal rumbling
Diagnostic tests	History
	Physical examination
Therapeutic management	High-fiber diet
	2–3 L fluid daily
	Strengthening of abdominal muscles
	Exercise
	Bulk-forming agents
	Stool softeners
	Education

stool ooze around the fecal mass and cause incontinence of liquid stools. The incontinence may be treated with an antidiarrheal medication, which will worsen the constipation, if a thorough assessment is not performed to rule out impaction. Straining to have a bowel movement (Valsalva's maneuver) can result in cardiac, neurological, and respiratory complications. If the patient has a history of heart failure, hypertension, or recent myocardial infarction, straining can lead to cardiac rupture and death. Grossly dilated loops of the colon, known as **megacolon,** can occur proximal to the dry fecal mass and obstruct the colon. Abdominal distention occurs, and in severe cases, loops of bowel can be palpated through the abdominal wall.

Chronic laxative abuse can lead to colonic mucosal atrophy, muscle thickening, and fibrosis. These conditions can result in perforation of the colon and necessitate an emergency **colectomy.**

Diagnostic Tests

Constipation is usually self-diagnosed or diagnosed by history and physical examination. If complications are suspected, a radiographic examination, sigmoidoscopy, and stool testing for occult blood may be necessary.

Medical Treatment

Treatment of constipation depends on the cause. Fiber should be added to the diet, and exercises to strengthen abdominal muscles should be done. Behavior changes can help establish a more normal bowel pattern. These changes include setting a daily defecation time, appropriately responding to the urge to defecate, and drinking 8 oz of warm water every morning and 2 to 3 L of water every day if it is not contraindicated for other reasons. Laxative abuse should be discontinued. Bulk-forming agents such as psyllium (Metamucil) or stool softeners such as docusate sodium (Colace) should be used instead of laxatives. Ene-

---

megacolon: mega—large + colon—colon
colectomy: col—pertaining to colon + ectomy—surgical excision

mas and rectal suppositories are used only in extreme cases and are discontinued when an acute episode is resolved.

Nursing Process

ASSESSMENT. The patient may feel self-conscious or embarrassed when interviewed about bowel habits and history. Consideration should be given to the patient's feelings by postponing the discussion until rapport has been established. The nursing history should include the onset and duration of constipation, past elimination pattern, current elimination pattern, occupation, lifestyle (stress, exercise, nutrition), history of laxative or enema use, medical-surgical history, and current medications being taken. The color, consistency, and odor of the stool, as well as any intestinal symptoms, are also important.

After the interview, the patient's abdomen should be auscultated for bowel sounds. Infrequent, absent, high-pitched, or gurgling bowel sounds should be noted. The abdomen is inspected and palpated for distention and symmetry. Inspection of the perianal area may reveal fissures, external hemorrhoids, or irritation.

NURSING DIAGNOSES. Major nursing diagnoses identified may include the following:

■ Constipation
■ Anxiety related to concern about irregular elimination pattern
■ Perceived constipation for patients who abuse laxatives
■ Deficient knowledge about health maintenance practices to prevent constipation.

PLANNING. The nurse and patient collaborate on plan of care goals and interventions. The family or caregiver is included in this process. If the patient is a resident of a nursing facility or uses home care services, a representative from the facility or home care agency that is providing care should be included in this phase of the nursing process.

IMPLEMENTATION AND PATIENT EDUCATION. Patient and family or caregiver education is one of the most important aspects of treatment of constipation. Prevention of constipation may be as simple as correcting the contributing factors. The patient should be taught the factors leading to constipation and the interventions necessary to prevent it. It is important to instruct the patient and family about the physiology of defecation and the reason for obeying the urge to defecate when it occurs. Setting a specific time for defecation, such as after a meal, may facilitate the urge reflex. To ensure the most appropriate posture during defecation, patients can place their feet on a footstool to promote flexion of the hips.

What the patient considers a normal diet may not be appropriate for the correction or prevention of constipation. If constipation is caused by decreased motility and muscle tone, the patient should eat a high-fiber, high-residue diet that includes fresh fruits and vegetables and whole grains. In the older patient, adding 2 g of bran to cereal daily can significantly increase bowel movements and therefore decrease the number of laxatives, enemas, or stool softeners required (Nutrition Notes Box 31–1). The patient who be-

BOX 31-1 *Nutrition Notes*

Treating Constipation with Food Formula
Constipation may be successfully treated with 1 to 2 oz of the following mixture taken with the evening meal: 1 cup applesauce, 1 cup All-Bran cereal, and ½ cup 100-percent prune juice. Mixture may be stored in the refrigerator for 5 days and then should be discarded. In all cases of constipation, especially when increased fiber is given, adequate fluid intake is essential.

lieves that a daily bowel movement is absolutely necessary needs information and reassurance. Assure the patient that having a daily bowel movement is not always necessary.

Instructions also should include the importance of increasing activity through a daily walking program and abdominal exercises designed to improve the muscle tone. These actions improve peristalsis, promoting more spontaneous defecation.

The dangers of regular laxative use should be stressed. By increasing exercise and fluid and fiber in the diet, the patient should be able to discontinue laxative use.

EVALUATION. The plan has been effective if the patient has established a regular bowel function pattern (Table 31–2), if increased physical activity results in report of less abdominal discomfort, and if the patient verbalizes understanding of self-care measures and express satisfaction with the outcomes.

CRITICAL THINKING

Mrs. Jessie Burns
Mrs. Burns is a 93-year-old nursing home resident. You see on her chart that she has not had a bowel movement in a week. What action would you take?
Answers at end of chapter.

Diarrhea

Diarrhea occurs when fecal matter passes through the intestine rapidly, resulting in decreased absorption of water, electrolytes, and nutrients and causing frequent, watery stools. Classification and severity of diarrhea are based on the number of unformed stools in 24 hours. Large-volume

---

diarrhea: dia—through + rhea—to flow

TABLE 31-2 CRITERIA FOR REGULAR BOWEL FUNCTION

1. A regular time for defecation is routine.
2. A regular exercise program is followed.
3. Laxative use is avoided.
4. Water consumption is 2 to 3 L per day.
5. High-fiber and high-residue foods are added to the diet.
6. Consistency of stools reported are soft and formed.
7. Frequency of stools is every 1 to 3 days.

diarrhea occurs when the volume of feces is increased. Small-volume diarrhea is caused by an increase in peristalsis, without an increase in fecal volume.

Pathophysiology and Etiology

The most common cause of acute diarrhea is a bacterial or viral infection. Bacteria (normal flora) are normally found in the intestines. If these bacteria grow out of control or if bacteria or viruses are ingested in contaminated food or water, infection results. Some bacteria release toxins that irritate the intestinal mucosa, causing an inflammatory response and an increase in mucus production. Hyperperistalsis occurs, which lasts until the irritants have been excreted. The most common infectious agents are *Escherichia coli*, *Campylobacter jejuni*, *Shigella* spp., *Clostridium difficile*, *Giardia* spp., and *Salmonella* spp.

Poor tolerance or allergies to certain foods may cause diarrhea. Foods that most commonly cause diarrhea are additives (such as nutmeg or sorbitol), caffeine, milk products, meats, wheat, and potatoes. Acute diarrhea usually resolves in 7 to 14 days.

Chronic diarrhea may result from inflammatory disease, osmotic agents, excessive secretion of electrolytes, or increased intestinal motility. Inflammatory diseases such as Crohn's disease or ulcerative **colitis** (discussed later) may impair absorption, resulting in frequent, watery stools. Osmotic diarrhea results from ingestion of laxatives or other agents that prevent absorption of water or nutrients in the intestine. Additional causes of malabsorption include surgical resection or disease of certain areas of the intestinal tract, such as the terminal ileum or pylorus. Radiation therapy for cancer also may induce a malabsorption syndrome. Enteral tube feedings commonly result in diarrhea, especially when malnutrition has caused edema in the gut wall, which decreases absorption.

Increased secretion of water and electrolytes by the intestinal mucosa associated with certain hormonal disorders results in high-volume fecal output. An irritable bowel or a neurological disorder may cause increased motility problems. Also, as described earlier, diarrhea can indicate fecal impaction.

Prevention

Proper handling, storage, and refrigeration of all fresh foods minimize contact with infectious agents. Milk and milk products must be kept refrigerated and protected. Handwashing and cleaning of the kitchen and food preparation or serving items are extremely important.

It is best to start enteral feedings slowly with full-strength formula and gradually increase the rate rather than dilute the formula, thus reducing the risk of contaminating the formula by adding to it.

colitis: col—pertaining to colon + itis—inflammation

Signs and Symptoms

Initial diarrhea stools are foul smelling and may have undigested food particles and mucus. The stools may also contain blood or pus. Diarrhea resulting from food poisoning usually has an explosive onset and may be accompanied by nausea and vomiting. Abdominal cramping, distention, anorexia, intestinal rumbling, and thirst are common. Fever indicates infection. Weakness and dehydration from fluid loss may occur (Gerontological Issues Box 31–2).

BOX 31-2   GERONTOLOGICAL ISSUES

Diarrhea can cause older people to quickly become dehydrated and hypokalemic because both fluid and potassium are lost in stools. The signs and symptoms of hypokalemia include muscle weakness, hypotension, anorexia, paresthesia, and drowsiness. It can also cause cardiac dysrhythmias, such as atrial and ventricular tachycardia, premature ventricular contraction, and ventricular fibrillation, which can be fatal.

If the older peson has decreased mobility, quick access to the bathroom is important. Because of poor muscle control, older patients may be incontinent. This might embarrass patients or cause them to hurry, which increases chances of patients falling and causing other problems such as fracture, dislocation, or hematoma. Also, because older patients' skin is more sensitive resulting from poor turgor and a reduction in subcutaneous fat layers, perirectal skin excoriation can occur secondary to the acidity and digestive enzyme content of diarrheal stools.

Diagnostic Tests

The diagnosis of diarrhea is determined by the onset and progression of the disease, absence or presence of fever, laboratory examinations, and visual inspection of the stool. Evidence of bacteria, pus, and blood is checked. Diarrhea mixed with red blood cells and mucus is associated with cholera, typhoid, typhus, large-bowel cancer, or amebiasis. Diarrhea mixed with white blood cells and mucus is associated with shigellosis, intestinal tuberculosis, salmonellosis, regional enteritis, or ulcerative colitis. Bulky, frothy stool is seen in sprue and celiac disease. Pasty stools usually have a high fat content and may be associated with common bile duct obstruction, sprue, and celiac disease. "Butter stool" appearance is seen in patients with cystic fibrosis.

Medical Treatment

Replacing fluids and electrolytes is the first priority. This is done by increasing oral fluid intake, using solutions with glucose and electrolytes if ordered by the physician. Intravenous fluid replacement may be necessary for rapid hydration, especially in the very young or very old. An elimination diet can be tried to identify foods that may contribute to diarrhea. Foods known to cause diarrhea are eliminated to see if a change in bowel function occurs. Each food item is then added back into the diet, one at a time, to see which ones cause diarrhea. The patient is also encouraged to increase fiber and bulk in the diet.

If the patient has three or more watery stools per day, motility of the intestines can be decreased with the use of drugs, such as diphenoxylate (Lomotil) and loperamide (Imodium). If diarrhea is thought to be caused by antibiotics that change the normal flora of the bowel, a *Lactobacillus* granules dietary supplement (Lactinex) may be used to restore the normal flora. Antimicrobial agents are prescribed if infectious agents have been documented.

Nursing Process

ASSESSMENT. Observation of the patient's behavior and symptoms assists in identifying the cause of diarrhea. Ask the patient to describe any symptoms, when they started, and how long they have been present. Questions should include "Is there any abdominal pain, urgency, or cramping?" and "What time of the day does it happen?" Stool consistency, color, odor, and frequency are documented.

The abdomen is inspected for distention and auscultated for hyperactive bowel sounds or rumbling. The patient's usual dietary habits and any changes or recent exposure to contaminated food or water is assessed. Find out if any medications contributed to the diarrhea. If the patient has traveled recently, discover the geographic location and whether exposure to an infected person or someone with similar symptoms occurred.

Assess for symptoms of dehydration, such as tachycardia, hypotension, decreased skin turgor, weakness, thready pulse, dry mucous membranes, and oliguria. Abnormal laboratory studies that may indicate dehydration include increased serum osmolality, increased specific gravity of urine, and increased hematocrit. Decreased serum potassium may result from intestinal loss of potassium.

NURSING DIAGNOSIS. Possible nursing diagnoses include the following:

- Diarrhea related to infection or possible ingestion of irritating foods
- Risk for deficient fluid volume related to frequent passage of stools and insufficient fluid intake
- Risk for infection related to fecal contamination
- Risk for impaired skin integrity related to passage of frequent liquid stools
- Anxiety related to uncontrolled elimination
- Acute pain related to increased peristalsis

PLANNING. Patient care planning includes settings goals. The chief goals include controlling diarrhea, preventing deficient fluid volume, avoiding the of spread of infection, keeping skin intact, decreasing anxiety, and controlling pain.

IMPLEMENTATION. Measures to decrease diarrhea must be implemented. Caffeine intake must be limited because it stimulates intestinal motility. During acute diarrhea, the patient may have nothing by mouth (NPO) to promote bowel rest. Then clear liquids, such as water, juices, bouillon, and gelatin, are started, with progression to a low-residue diet (Nutrition Notes Box 31–3). Rest is also important because of the weakness associated with fluid and electrolyte loss. Intake and output (including diarrhea stools) are recorded to determine fluid balance. If output is greater than intake,

BOX 31-3 *Nutrition Notes*

Deciding When to Refer an Adult with Diarrhea for Medical Care

Most instances of diarrhea in healthy adults are self-limiting and resolve without treatment. Indications for medical consultation include the following:

- Large volumes of stool
- Severe abdominal pain
- Bloody stools
- Protracted duration
- Systemic symptoms such as fever or prostration
- Medical conditions for which fasting, dehydration, or infectious disease are hazardous

Healthy adults at minimal risk of electrolyte imbalance may institute self-treatment as follows:

- For the first 12 hours: Water or oral rehydration solutions at room temperature. Easily absorbed fluids maintain hydration. Hot or cold liquids are more likely to stimulate peristalsis.
- For the second 12 hours: Clear liquids, no caffeine or extremes of temperature. If more than 5 percent of body weight is lost, seek medical attention.
- For the third 12 hours: Full liquids. Experiment with milk in case temporary lactose intolerance has developed as a result of intestinal inflammation.
- For the fourth 12 hours: Soft diet. Include applesauce or banana for pectin and also rice, pasta, and bread without fat (digested by enzymes usually unaffected in gastroenteritis).
- By the 48th hour: Regular diet. If diarrhea has not resolved and regular diet is not tolerated, seek medical treatment.

intravenous fluid replacement may be ordered. Antidiarrheal medications are administered as directed. Control of diarrhea controls pain.

Potentially infected persons or contaminated foods must be identified to prevent the spread of infection. Thorough handwashing by patient, family, and nurse is essential. A private room to prevent infection transmission may be necessary. The perianal skin is protected from contact with liquid stools and their enzymes. This is accomplished by keeping the skin clean, dry, and protected with a moisture barrier, such as petrolatum or medicated ointment, after each bowel movement.

The patient's coping mechanisms can be increased by allowing the patient to express fear or anxiety regarding possible incontinence of liquid stools and the accompanying embarrassment. Adult disposable briefs may be used during acute episodes. Being understanding and tolerant also helps.

PATIENT EDUCATION. The patient must be taught the importance of identifying the source of diarrhea, such as foods or persons (family, neighbors) who may be infected. To prevent the spread of infection it is important to provide instruction in proper handwashing before and after handling foods and before and after toileting. The patient should also be educated about the signs and symptoms of dehydration to report.

EVALUATION. Goals have been met if fluid and electrolyte balance is achieved, frequency of diarrhea stools is decreased, skin integrity is maintained, and anxiety is controlled. Goals are also met if the patient reports an increase in comfort and verbalizes an understanding of measures to prevent reinfection.

▶ INFLAMMATORY AND INFECTIOUS DISORDERS

Many diseases of the lower GI tract are a result of inflammation in the bowel. Sometimes the inflamed areas become infected, resulting in a worsening of symptoms and necessitating antimicrobial therapy.

Appendicitis
Pathophysiology

Appendicitis is the inflammation of the appendix, the small, fingerlike appendage attached to the cecum of the large intestine. Because of the small size of the appendix, obstruction may occur, making it susceptible to infection. The resulting inflammatory process causes an increase in intraluminal pressure of the appendix.

Signs and Symptoms

Signs and symptoms of appendicitis include fever, increased white blood cells, and generalized pain in the upper abdomen. Within hours of onset the pain usually becomes localized to the right lower quadrant at McBurney's point, midway between the umbilicus and the right iliac crest (Fig. 31–1). This is one of the classic symptoms of appendicitis. Nausea, vomiting, and anorexia are also usually present.

Physical examination reveals slight abdominal muscular rigidity (guarding), normal bowel sounds, and local rebound tenderness (intensification of pain when pressure is released after palpation) in the right lower quadrant of the abdomen. Sometimes there is pain in the right lower quadrant when the left lower quadrant is palpated (Rovsing's sign). The patient may keep the right leg flexed for comfort and experience increased pain if the leg is straightened.

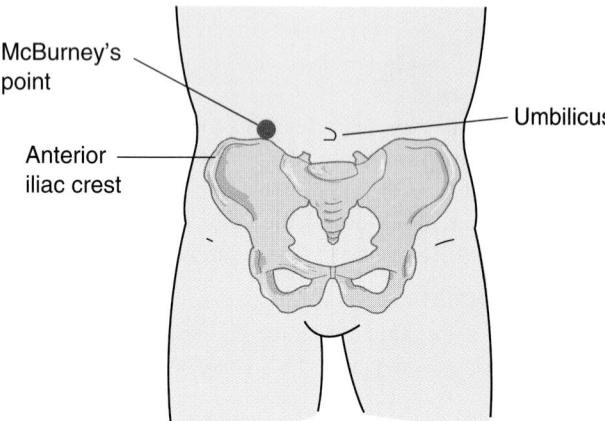

McBurney's point

Anterior iliac crest

Umbilicus

Figure 31-1 Pain at McBurney's point is a symptom of appendicitis.

appendicitis: appendic—pertaining to appendix + itis— inflammation

Complications

Perforation, abscess of the appendix, and peritonitis are major complications of appendicitis. With perforation, the pain is severe and the temperature is elevated to at least 37.7° C (100° F). An abscess is a localized collection of pus separated from the peritoneal cavity by the omentum or small bowel. This is usually treated with parenteral antibiotics and surgical drainage. An appendectomy is done about 6 weeks later. Management of **peritonitis** is discussed later in this chapter.

Diagnostic Tests

A complete blood cell count (CBC) reveals elevated leukocyte and neutrophil counts. Ultrasound or computed tomography (CT) scan reveals an enlargement in the area of the cecum.

Medical Treatment

The patient is kept NPO and surgery is done immediately unless there is evidence of perforation or peritonitis. If the appendix has ruptured, intravenous fluids and antibiotic therapy are started and surgery may be delayed for 8 hours or more. Laxatives and enemas are avoided because they may trigger or complicate a rupture. The use of a heating pad on the abdomen is avoided because the warmth may increase inflammation and risk of rupture. The patient may be discharged from the hospital on the day of surgery once the inflamed appendix is removed, pain is controlled, and the patient is afebrile. However, if there is evidence of rupture, the patient is hospitalized for 5 to 7 days to observe for signs of peritonitis or an ileus.

Nursing Care

The patient with suspected appendicitis is given nothing by mouth until a diagnosis is confirmed, in case surgery is necessary. Ice to the site of pain and maintaining semi-Fowler's position may help reduce pain while the diagnosis is being made. The patient is often readied for an appendectomy by emergency department staff, so time for preoperative teaching is limited.

After surgery the patient is NPO if no bowel sounds are present. If the appendix has ruptured, the postoperative patient may have a nasogastric tube until bowel sounds return to prevent abdominal distention and vomiting. After bowel sounds return, the diet initially consists of clear fluids and is advanced as tolerated by the patient. Vital signs and abdominal assessment are done to monitor for signs and symptoms of peritonitis. Early ambulation, coughing, deep breathing, and turning are encouraged to prevent respiratory complications. The patient may be taught to splint the incision while coughing and deep breathing to reduce pain and stress on the incision. Dressing changes are performed as ordered.

peritonitis: periton—pertaining to peritoneum + itis— inflammation

Peritonitis
Pathophysiology and Etiology

Trauma, ischemia, or tumor perforation in any abdominal organ causes leakage of the organ's contents into the peritoneal cavity. The most common cause of peritonitis is a ruptured appendix, but it may also occur after perforation of a peptic ulcer, gangrenous gallbladder, intestinal diverticula, incarcerated **hernia,** or gangrenous small bowel. It may also be a complication of peritoneal dialysis. Peritonitis results from the inflammation or infection that is caused by the leakage. The tissues become edematous and begin leaking fluid containing increasing amounts of blood, protein, cellular debris, and white blood cells. Initially the intestinal tract responds with hypermotility, but this is soon followed by paralysis (paralytic ileus).

Signs and Symptoms

Generalized abdominal pain evolves into localized pain at the site of the perforation or leakage. The area of the abdomen that is affected is extremely tender and aggravated by movement. Rebound tenderness and abdominal rigidity are present. Decreased peristalsis results in nausea and vomiting. Infection causes fever, increased white blood cells, and an elevated pulse.

Complications

Complications of peritonitis are intestinal obstruction (discussed later), hypovolemia caused by the shift of fluid into the abdomen, and septicemia from bacteria entering the bloodstream. Shock and ultimately death may result. Wound dehiscence or evisceration can occur if the patient has had abdominal surgery.

Medical Treatment

The patient is NPO because of the impaired peristalsis. Fluid and electrolyte replacement is crucial to correct hypovolemia. Abdominal distention is relieved through insertion of an nasogastric tube with low intermittent suction. Antibiotics are used to treat or prevent sepsis. Depending on the cause of the peritonitis, surgery may be performed to excise, drain, or repair the cause. An ostomy may be formed to divert feces, allowing resolution of the infection. After surgery the patient usually has a wound drain, a nasogastric (NG) tube, and a Foley catheter. Pain control is essential to overall recovery. Severely compromised patients may receive total parenteral nutrition (TPN) to meet nutritional needs for increased immune function and healing.

Nursing Process

ASSESSMENT. Pain is assessed using the WHAT'S UP? format. Abdominal distention and bowel sounds are monitored and recorded. Vital signs are monitored for fever or signs of septic shock. Intake and output are monitored and recorded accurately so that appropriate fluid replacement therapy is ordered.

NURSING DIAGNOSIS. Nursing diagnoses for the patient with peritonitis include acute pain related to inflammatory process, risk for deficient fluid volume related to fluid shifting from the circulation to the peritoneal cavity, risk for injury related to complications of peritonitis, and imbalance nutrition: less than body requirements related to nausea and NPO status.

PLANNING. Goals for the patient include pain control, stable fluid balance, early recognition and reporting of possible complications, and maintenance of adequate nutrition.

IMPLEMENTATION. Narcotics in conjunction with noninvasive measures such as position changes, diversion, and relaxation exercises are provided for pain. Semi-Fowler's position may reduce tension on the abdomen. Frequent mouth care is done if an NG tube is in place.

Reduced urinary output, dropping blood pressure, and rising pulse rate may indicate fluid volume deficit. If a fever is also present, the patient may be developing sepsis. All symptoms are reported to the physician promptly.

Respiratory complications can be prevented with coughing, deep breathing, and regular turning. Splinting the abdomen during coughing and moving may help reduce discomfort. Oxygen may be administered if the oxygen saturation is less than 95 percent. If the patient has had surgery, output from any drains is monitored. Documentation of the color, volume, odor, and consistency of all drainage is important. Care should be taken in moving the patient so that the drains are not dislodged.

If nausea occurs, the NG tube should be checked for patency. Antiemetics are administered as ordered. If the patient is to be NPO for a prolonged period (3 days or more), TPN should be considered to maintain nutritional status and prevent complications such as sepsis.

EVALUATION. A decrease in temperature and pulse rate indicate that the peritonitis is improving. The abdomen becomes less distended and softens with the return of normal bowel sounds and passing of flatus. Nursing care is effective if vital signs are stable, complications are prevented, nutritional status is maintained, and the patient reports that pain is controlled.

Diverticulosis and Diverticulitis
Pathophysiology

A diverticulum is a herniation or outpouching of the bowel mucous membrane caused by increased pressure within the colon and weakness in the bowel wall. **Diverticulosis** is a condition in which multiple diverticula are present without evidence of inflammation (Fig. 31–2). Many people have diverticulosis without knowing it because it develops gradually. When food and bacteria are trapped in a diverticulum, inflammation and infection develop. This is called **diverticulitis.**

---

diverticulosis: diverticul—blind pouch + osis—condition
diverticulitis: diverticul—blind pouch + itis—inflammation

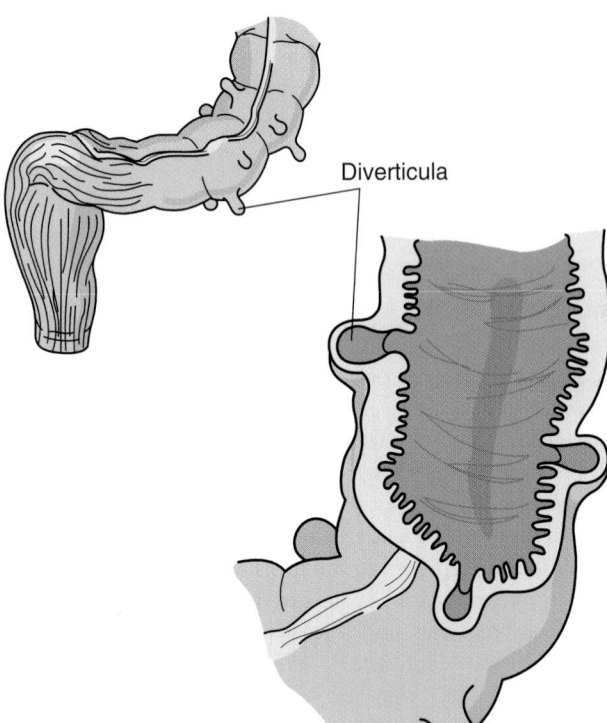

Diverticula

Figure 31-2 A diverticulum is a herniation or outpouching of the bowel mucous membrane. Multiple diverticuli are called diverticulosis. If they become inflamed or infected, the condition is called diverticulitis.

Etiology

Chronic constipation usually precedes the development of diverticulosis by many years. When the patient is chronically constipated, pressure within the bowel is increased, leading to development of diverticula. A major cause of the disease is a decreased intake of dietary fiber. Diverticulosis is most common in the sigmoid colon. A small percentage of patients with diverticulosis develop diverticulitis. People older than age 60 are the most common group to experience diverticulitis.

Prevention

Diverticulitis is prevented by increasing dietary fiber to prevent constipation and onset of diverticulosis.

Signs and Symptoms

The patient with diverticulosis is generally asymptomatic. When diverticulitis is present, the patient exhibits bowel changes, possibly alternating between constipation and diarrhea. Steady or crampy pain in the left lower quadrant of the abdomen is the most common symptom. As the condition worsens, bleeding may occur, along with weakness, fever, fatigue, and anemia. Guarding and rebound tenderness may be present. If an abscess develops, the diverticulum may rupture, leading to peritonitis (Gerontological Issues Box 31-4).

BOX 31-4 GERONTOLOGICAL ISSUES

With age-related weakness in the intestinal wall, approximately 40 percent of adults older than age 80 have diverticular disease. Clinical manifestation of this condition may include abdominal pain, rectal bleeding, nausea, and vomiting. Patients may not notice the abdominal pain until infection is present. Many times the symptoms are not reported early because patients fear it may be cancer. Due to impaired vision, the elderly may not recognize blood in the stool, which can be an indication of diverticulitis.

Diagnostic Tests

Diverticulosis is confirmed with sigmoidoscopy, colonoscopy, and barium enema. The diverticuli and specific areas of inflammation can be seen during a colonoscopy or sigmoidoscopy. If an abscess is suspected, a CT scan may be done. Barium enema may show irregular narrowing of the colon and thickened muscle walls. A stool specimen may show occult blood. An abdominal x-ray examination may be done to identify a perforated diverticulum.

Medical Treatment

Diverticulosis is managed by preventing constipation. With acute diverticulitis, the patient may be hospitalized for intravenous antibiotics and pain control. A nasogastric tube, intravenous fluids, and NPO status may be ordered until pain, nausea or vomiting, fever, and inflammation decrease. When the acute period is over, a progressive diet is started. Whether or not perforation occurs, surgical resection with anastomosis or a temporary **colostomy** (discussed later) may be done to allow the inflammation to subside and the diseased portion of the colon to rest.

Nursing Process

ASSESSMENT. The patient is assessed for signs and symptoms of diverticulitis (Table 31-3). Physical examination is done to assess for abdominal distention and tenderness. A firm mass may be palpated in the sigmoid area.

NURSING DIAGNOSIS. Constipation related to low-fiber dietary habits may be one nursing diagnosis. Others may include pain related to inflammation or infection and ineffective tissue perfusion: gastrointestinal related to infection secondary to perforation, peritonitis, or abscess.

PLANNING. Goals include normal bowel elimination (especially without straining), pain control, and prevention of risks related to infection and inflammation.

IMPLEMENTATION. Implement interventions that alleviate and prevent constipation. Unless contraindicated, fluid intake should be increased to 2 to 3 L per day. Dietary considerations for a patient with diverticulosis (without evidence of inflammation) include foods that are soft but high

colostomy: colo—pertaining to colon + stoma—mouth or opening

TABLE 31-3	SYMPTOMS ASSOCIATED WITH DIVERTICULITIS
W—Where is the pain?	Usually in the left lower quadrant
H—How does it feel? (Describe quality)	Tender, crampy
A—Aggravating and alleviating factors	Constipation and low-fiber diet may aggravate; treatment of constipation may alleviate
T—Timing (onset, duration, frequency)	Gradual onset, intermittent, gradual increase in frequency of pain events
S—Severity (0–10)	Usually 5–7
U—Useful other data/ associated symptoms	Intermittent rectal bleeding; straining at stool; constipation alternating with diarrhea; elevated white blood cells and sedimentation rate; elevated temperature and pulse rate; and pus, mucus, and blood in stool
P—Patient's perception	Fear diagnosis of cancer

in fiber, such as prunes, raisins, and peas. Unprocessed bran can be added to soups, cereals, and salads to give added bulk to the diet. Fiber should be increased in the diet slowly to prevent excess gas and cramping. Some health care providers recommend avoiding nuts or foods with small seeds that can get caught in diverticuli, such as tomatoes and raspberries, but this has not been proven to help.

If the patient is experiencing pain, administer analgesics or antispasmodic drugs as prescribed. A diet low in fiber is recommended for the patient with acute pain. Monitor the patient closely and notify the physician immediately if pain increases, especially if associated with abdominal rigidity. Increased pain may indicate that the bowel has ruptured and peritonitis is developing.

PATIENT EDUCATION. Instruct the patient on dietary changes needed to alleviate constipation and to prevent future episodes of diverticulitis. Teach the patient that laxatives or enemas may increase motility and pressure in the bowel and cause further complications. Provide the patient with information about the signs and symptoms of diverticulitis that should be reported.

EVALUATION. The goals are met when the patient reports a normal pattern of bowel elimination, pain-free bowel movements, and less abdominal pain and cramping and there is no infection.

▶ INFLAMMATORY BOWEL DISEASE
Crohn's Disease (Regional Enteritis)
Pathophysiology

Crohn's disease, also known as regional **enteritis** or granulomatous enteritis, is an inflammatory bowel disease (IBD) that can involve any part of the intestine but most commonly the terminal portion of ileum. The inflammation extends through the intestinal mucosa, which leads to the for-

enteritis: entero—intestine + itis—inflammation

mation of abscesses, **fistulas,** and fissures. As the disease progresses, obstruction occurs because the intestinal lumen narrows with inflamed mucosa and scar tissue.

Etiology

Although the exact cause of Crohn's disease has not been identified, there is a familial tendency. Other possible influences are autoimmune processes and infectious agents. Crohn's disease is most often diagnosed between the ages of 15 and 30 and occurs more often in women. There are periods of remissions and exacerbations. Physical or psychological stress may trigger exacerbations (Cultural Consideration Box 31–5).

BOX 31-5	CULTURAL CONSIDERATION

Ulcerative colitis and Crohn's disease are more common in Caucasians, persons of Jewish descent, and upper-middle-class urban populations. The incidence of Crohn's disease is increasing rapidly in Western Europe and North America. These findings support possible hereditary or environmental risk factors for IBD.

Signs and Symptoms

Crampy abdominal pains (unrelieved by defecation), weight loss, and diarrhea occur. Because the crampy pains occur after eating, the patient often does not eat in order to avoid the pain. A lack of eating and poor absorption of nutrients results in weight loss and malnutrition. Chronic diarrhea contributes to fluid deficit and electrolyte imbalance. The inflamed intestine may perforate, leading to the formation of intra-abdominal or anal fissures, abscesses, or fistulas. Some nongastrointestinal symptoms may also occur, including arthritis, skin lesions, inflammatory disorders of the eyes, and abnormalities of liver function.

Complications

In addition to malnutrition, the development of fissures, abscesses, or fistulas is the most common complication of Crohn's disease. Fistulas may include enterovaginal (small bowel to vagina), enterovesicle (small bowel to bladder), enterocutaneous (small bowel to skin), enteroentero (small bowel to small bowel), or enterocolonic (small bowel to colon) (Fig. 31–3). Fistulas communicating with organs that then drain externally can cause tremendous skin irritation, as well as increased risk of developing infections. Fistulas are corrected surgically.

Diagnostic Tests

Endoscopy (colonoscopy and sigmoidoscopy), with multiple biopsies of the diseased colon and terminal ileum, is the most conclusive diagnostic test for Crohn's disease. Endoscopy identifies areas of normal to severely inflamed mucosa. Crohn's disease is confirmed by granulomas in the biopsy specimen. A barium enema shows a classic "cobblestoning" effect and, as the disease progresses, areas of narrowing in the intestine. An elevated sedimentation rate and

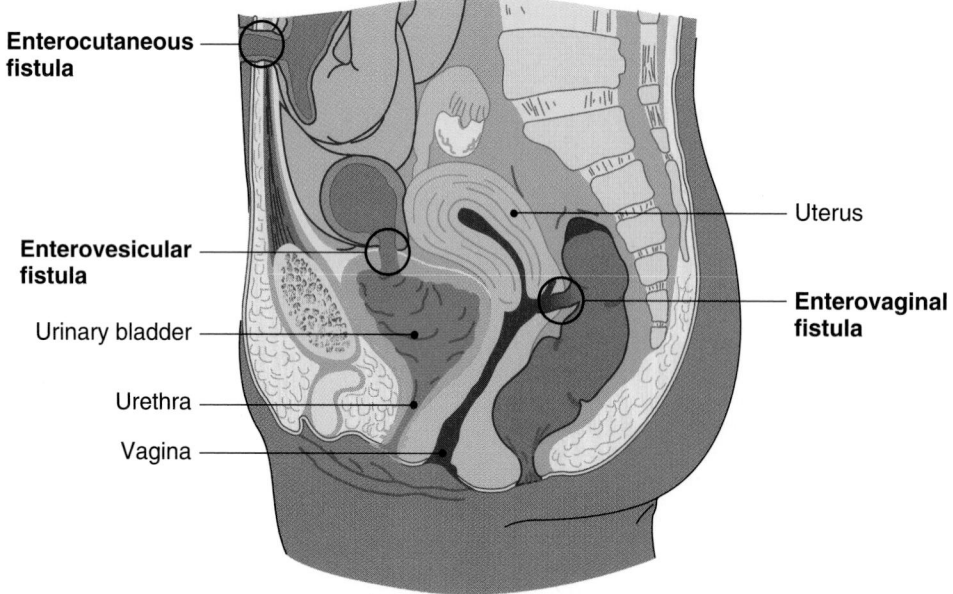

Figure 31-3 Fistulas are a common complication of Crohn's disease.

leukocytosis are present, and serum albumin levels may be low because of malnutrition or poor absorption of protein. Bowel sounds are hyperactive over the right lower quadrant, and a stool examination reveals fat content and occult blood.

Medical Treatment

Management of Crohn's disease is aimed at relieving symptoms, such as inflammation and diarrhea, and the effects of those symptoms, such as dehydration and malnutrition. Budesonide (Entocort EC), an oral steroid, is a new treatment that acts locally in the intestine to reduce inflammation. Because budesonide is not absorbed systemically, there are fewer steroid side effects (e.g., moon face and acne). Corticosteroids are used during an acute inflammation, then tapered and discontinued. Antidiarrheal medications such as loperamide (Imodium) reduce peristalsis. Sulfasalazine (Azulfidine) or mesalamine (Pentasa) is used long term to decrease inflammation. Antibiotics such as metronidazole (Flagyl) and ciprofloxacin (Cipro) are used if abscesses or peritonitis are present. In some cases, symptoms worsen. Immunosuppressants such as aszathiprine (Imuran) and cyclosporine (Sandimmune) may be used to treat patients who have not responded to other treatments or those who require steroids.

Surgery may be indicated if obstruction, stricture, fistula, or abscess is present. Surgical procedures include resection of the affected area with anastomosis, colectomy with **ileostomy,** or colectomy with ileorectal anastomosis, depending on the area of bowel involved. See details regarding intestinal ostomies later in this chapter.

Because of the similarities between Crohn's disease and ulcerative colitis, the nursing processes for both are discussed together.

Ulcerative Colitis
Pathophysiology

Ulcerative colitis is similar to Crohn's disease. Crohn's disease, however, can occur anywhere in the gastrointestinal system, whereas ulcerative colitis occurs in the large colon and rectum. Multiple ulcerations and diffuse inflammation occurs in the superficial mucosa and submucosa of the colon. The lesions spread throughout the large intestine and usually involve the rectum. The patient with ulcerative colitis has increased risk of developing colorectal cancer.

Etiology

Several theories exist to explain the cause of ulcerative colitis. These include infection, allergy, and autoimmune response. Environmental agents such as pesticides, tobacco, radiation, and food additives may precipitate an exacerbation. At one time a psychological component was suggested in the etiology of ulcerative colitis. However, anyone who experiences chronic intestinal cramps and frequent, sometimes painful bowel movements is going to be discouraged, anxious, and depressed. These behaviors may be a result of the disease, not the cause. Psychological stress may trigger or worsen an attack of symptoms. Ulcerative colitis usually begins between ages 15 and 40.

Signs and Symptoms

Abdominal pain, diarrhea, rectal bleeding, and fecal urgency are common symptoms of ulcerative colitis (Table 31-4). Anorexia, weight loss, cramping, vomiting, fever, and dehydration associated with passing 5 to 20 liquid stools a day may also occur. Along with the potential for

---

ileostomy: ileo—pertaining to ileam + stoma—mouth or opening

TABLE 31-4	INFLAMMATORY BOWEL DISEASE SUMMARY
Symptoms	Abdominal pain or cramping
	Weight loss
	Diarrhea
	Fluid and electrolyte imbalance
	Fissures, fistulas, abscesses
	Arthritis and skin lesions
	Inflammatory eye disorders
	Abnormal liver function
Diagnostic tests	Endoscopy with biopsy
	Barium enema
	Laboratory examination
	Stool examination
	Absent bowel sounds
Therapeutic management	Medications: anti-inflammatories, antidiarrheal, antibiotics, immunosuppressants, corticosteroids
	Surgery if necessary
	Avoidance of offending foods
	Elemental formula or TPN if required
	Support and education

fluid and electrolyte imbalance, there is a loss of calcium. Anemia often develops as a result of rectal bleeding. Serum albumin may be low because of malabsorption. Like Crohn's disease, arthritis, skin lesions, inflammatory disorders of the eyes, and abnormalities of liver function may also occur. Symptoms are usually intermittent, with remissions lasting from months to years.

Complications

Malnutrition and its associated problems are complications of ulcerative colitis. Other complications include the potential for hemorrhage during an acute phase, bowel obstruction, perforation, and peritonitis. The risk for colon cancer is also increased in patients with ulcerative colitis. Symptoms outside the GI tract, such as arthritis, skin lesions, and inflammatory disorders of the eyes, may also occur.

Diagnostic Tests

Examination of stool specimens is done to rule out any bacterial or amebic organisms. The stool is positive for blood. Anemia is often present because of blood loss. Leukocyte levels and erythrocyte sedimentation rate are elevated because of the chronic inflammation. Electrolytes are depleted from chronic diarrhea. There is a protein loss because of liver dysfunction and malabsorption. Endoscopy and barium enema help differentiate ulcerative colitis from other diseases of the colon with similar symptoms. Biopsy specimens taken during sigmoidoscopy and colonoscopy typically show abnormal cells.

MEDICAL TREATMENT. Foods that cause gas or diarrhea are avoided. Because the offending foods may be different for each patient, foods are tried in small amounts if they are thought to cause symptoms. In general, high-fiber foods, caffeine, spicy foods, and milk products are avoided. Total

parenteral nutrition may be needed to meet nutritional needs during acute exacerbations.

As with Crohn's disease, antidiarrheal, anti-inflammatory, and immunosuppressant agents may be given. Corticosteroids are used if needed to reduce inflammation.

If medical treatment is ineffective, surgery is considered. Because ulcerative colitis usually involves the entire large intestine, the surgery of choice is total proctocolectomy with formation of an ileostomy. Ileostomies are discussed later in this chapter.

Nursing Process for the Patient with Inflammatory Bowel Disease

ASSESSMENT. A history obtained from the patient should include identification of symptoms, including the onset, duration, frequency, and severity. Ask if there has been any correlation between exacerbations of symptoms related to dietary changes or stress. Determine the presence any food allergies or intolerances, such as milk, because some may increase diarrhea. Also, note the daily and weekly intake of caffeine, nicotine, and alcohol, because all these stimulate the bowel and can cause cramping and diarrhea.

Assess the patient for nutritional status and signs of dehydration. The amount of weight loss in 2 months can be 10 to 20 pounds. Perianal skin should be assessed for irritation and excoriation.

Assessment of emotional status, coping skills, and verbal and nonverbal behaviors is essential. The patient may withdraw from family and friends because of frequent bowel movements. Anxiety, sleep disturbances, depression, and denial can be problems. If surgery involving an ileostomy is planned, the patient is at risk for problems with body image.

NURSING DIAGNOSIS. Nursing diagnoses for the patient with IBD may include the following:

- Diarrhea related to inflammatory process
- Ineffective nutrition, less than body requirements, related to anorexia and malabsorption
- Deficient fluid volume related to anorexia, nausea, and diarrhea
- Acute pain related to increased peristalsis and cramping
- Impaired skin integrity related to frequent loose stools
- Ineffective coping related to repeated episodes of diarrhea

PLANNING. Goals include achieving normal bowel elimination, maintenance of nutrition, prevention of deficient fluid volume, relief of abdominal pain and cramping, and effective coping with the disease.

IMPLEMENTATION. Document characteristics of stools, including color, consistency, amount, frequency, and odor. Ensure the patient has quick access to the bathroom, or provide a bedside commode. Keep the environment clean and odor free to help alleviate anxiety. Administer antidiarrheal medication as prescribed. Encourage bedrest to decrease peristalsis.

Teach the patient to avoid high-fiber foods such as whole grains and raw fruits and vegetables, as well as caffeine, alcohol, and nicotine, because they stimulate the GI tract. For severe symptoms a special liquid (elemental) formula that is absorbed in the upper bowel, allowing the colon to rest, may be given. If the patient is unable to tolerate oral intake, TPN may be necessary.

Signs of deficient fluid volume are documented and reported to the physician. Daily weight and intake and output (volume of stools, fistula drainage, vomitus, and urine) are accurately recorded so the physician can order adequate intravenous (IV) fluid replacement.

Document the onset, duration, and severity of pain, and medicate as prescribed. Describe the character of the pain (dull, cramping, burning) and if the pain is associated with meals or activity.

Perianal skin should be kept clean and dry to prevent excoriation from frequent stools. Sitz baths may be helpful. Protect the skin by applying a protective moisture barrier after each stool.

Anxiety should be reduced because it aggravates the symptoms of inflammatory bowel disease. Reducing anxiety can be done by answering questions, talking in a calm, confident manner, and actively listening to the patient. If surgery involving an ostomy is discussed, obtain a consultation with a wound ostomy continence nurse (WOCN) if available. Reinforce teaching by using basic terms and providing pamphlets or showing a video of ostomy management and a picture of a stoma and allowing the patient time to handle an ostomy appliance. Allow the patient to decide what teaching information is most important, and include the patient's family or significant other. More details about ostomy care are provided later in this chapter. (See Nursing Care Plan Box 31–6 for the Patient with Inflammatory Bowel Disease.)

PATIENT EDUCATION. Information about IBD must be given in a way that promotes positive reinforcement and reassurance. Nutrition and dietary considerations need to be discussed. If the patient is on medications, instructions regarding the dose, when and how to take the drug, why it was ordered, and potential effects and side effects should be given. Patients on steroids will probably be weaned off them after surgery. The patient needs to know not to stop taking steroids suddenly to prevent adrenocortical insufficiency. Instructions in daily care are provided if the patient has an ileostomy or colostomy.

EVALUATION. After intervention, it is reasonable to expect a decrease in the frequency and amount of diarrheal stools. Nutritional status should improve, as evidenced by maintenance of ideal weight and a fluid and electrolyte balance within normal range. Other outcomes include less pain, intact skin, and the ablity to cope with the disease and treatment.

CRITICAL THINKING

Judy Moore

Judy Moore is an 18-year-old college student who has just been diagnosed with Crohn's disease. What can you do to help her adapt to this disease?

Answers at end of chapter.

BOX 31–6 NURSING CARE PLAN FOR THE PATIENT WITH INFLAMMATORY BOWEL DISEASE

 Ineffective coping related to inflammatory bowel disease

PATIENT OUTCOMES
Patient will identify strategies that will promote effective coping.

EVALUATION OF OUTCOMES
Is the patient able to state strategies for effective coping?

INTERVENTIONS	RATIONALE	EVALUATION
Assess knowledge of ulcerative colitis	Many people have little knowledge of a disease unless they know someone who has it. Inaccurate information must be corrected.	Does the patient verbalize information about ulcerative colitis and its effects on the body?
Encourage the patient to express feelings about the disease and how it may affect his or her life.	Expressing feelings about the disease and its perceived effect enables the patient to identify and talk about concerns. Once identified, these concerns can then be addressed by the health care team.	Does the patient talk about feelings regarding the potential impact of the disease on his or her life?
Determine whether the patient would like to speak with a person of similar age from the Crohn's and Colitis Foundation of America.	Speaking with someone close in age with the same disease lets the patient know that he or she is not the only person having to cope with this disorder. It can also help him or her learn some strategies for effectively coping with the disease.	Does the patient show an interest in speaking with someone with the same disease?
Identify strategies for effective coping that are acceptable to the patient.	Talking about concerns and possible solutions is a positive step. Coping strategies identified with the patient are more likely to be implemented.	Is the patient able to identify strategies for effective coping that he or she believes will work?

IRRITABLE BOWEL SYNDROME
Pathophysiology and Etiology

Irritable bowel syndrome (IBS) is a disorder of altered intestinal motility and increased sensitivity to visceral sensations. The bowel mucosa is not changed by IBS. Symptoms may be exacerbated by psychological stress or food intolerances. IBS is more common in women and in those who are young to middle aged.

Signs and Symptoms

IBS is characterized by complaints of gas, bloating, constipation, diarrhea, or alternating constipation and diarrhea. The patient also has feelings of abdominal bloating, with or without visible abdominal distention. This sensation of bloating is from increased sensitivity to visceral sensations. Other symptoms include the rectal passage of mucus, a feeling of incomplete evacuation, abdominal pain relieved by defecation, depression, anxiety, and palpitations.

Diagnostic Tests

Diagnosis of IBS is made based on history and physical examination. Stool examination, barium enema, upper GI series, and sigmoidoscopy may be done to rule out other disorders. Avoiding mild products for a time may be advised to rule out lactose intolerance.

Medical Treatment

IBS is a chronic condition, but symptoms can generally be controlled. Bran or bulk-forming laxatives such as psyllium (Metamucil) may be used to normalize stools. Foods causing gas formation are avoided. Antidepressants and relaxation exercises may help decrease stress and anxiety. Medication to decrease or increase intestinal motility (as needed) and reduce pain may also be given.

ABDOMINAL HERNIAS
Pathophysiology and Etiology

A hernia is an abnormal protrusion of an organ or structure through the wall of the cavity normally containing it, which in this case is the abdominal wall. Hernias are caused by a weakness in the abdominal wall along with increased intra-abdominal pressure, such as the pressure from coughing, straining, and heavy lifting. Obesity, pregnancy, and poor wound healing are also risk factors. The hernial sac is formed by the peritoneum protruding through the weakened muscle wall. Contents of the hernia can be small or large intestine or the omentum. Indirect hernias are caused by a defect of structural closure. Direct hernias are acquired and arise from a weakness in the abdominal wall, usually at old incisional sites.

There are many types of hernias (Fig. 31–4). Inguinal hernias are located in the groin where the spermatic cord in males or the round ligament in females emerges from the abdominal wall. This common hernia is an example of an indirect hernia and is usually seen in males.

Umbilical hernias are seen most often in obese women and in children. They are caused by a failure of the umbilical orifice to close. Ventral (incisional) hernias usually result from a weakness in the abdominal wall following abdominal surgery, especially if a drainage system was used, the patient experienced poor wound healing, the patient received inadequate nutrition, or the patient is obese.

Prevention

Improving abdominal musculature is the best way to prevent hernias. Those who do heavy lifting, tugging, or pushing should wear a support binder.

Signs and Symptoms

Unless complications occur, there are very few symptoms associated with hernias. An abnormal bulging can be seen in the affected area of the abdomen, especially when straining or coughing. It may disappear when the patient lies down. If the intestinal mass easily returns to the abdominal cavity or can be manually placed back in the abdominal cavity, it is called a reducible hernia. When adhesions or edema occur between the sac and its contents, the hernia becomes irreducible or incarcerated, meaning that the herniated bowel is trapped and cannot be returned to the abdomen.

Complications

An incarcerated hernia may become strangulated if the blood and intestinal flow are completely cut off. Incarceration leads to an intestinal obstruction and possibly gangrene and bowel perforation. Symptoms are pain at the site of the strangulation, nausea and vomiting, and colicky abdominal pain. Treatment is emergency surgery.

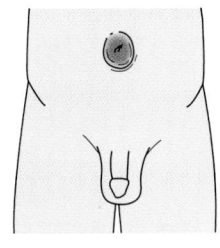 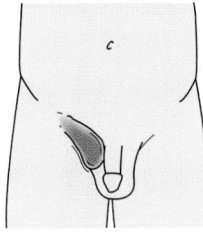 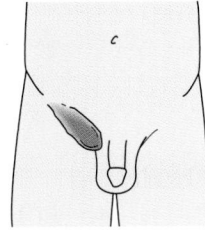 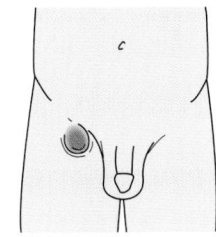

Umbilical hernia   Direct inguinal hernia   Indirect inguinal hernia   Femoral hernia

Figure 31-4 Types of hernias.

Medical Treatment

Hernias are diagnosed by physical examination. Surgery is indicated if there is strangulation or the threat of obstruction. Surgical procedures include herniorrhaphy and hernioplasty. Herniorrhaphy (surgery of choice) involves making an incision in the abdominal wall, replacing the contents of the hernial sac, and then closing the opening. Hernioplasty involves replacing the hernia into the abdomen and reinforcing the weakened muscle wall with wire, fascia, or mesh. Bowel resection or a temporary colostomy may be necessary if the hernia is strangulated. A patient who is not a good surgical candidate may be directed to use a truss or binder once the hernia is manually reduced. A truss is an external restraining device held in place with a belt used to push the hernia into the abdomen.

Nursing Care

If surgery is not planned, the patient is instructed to avoid activities that increase intra-abdominal pressure, such as lifting heavy objects. The patient is taught to recognize signs of incarceration or strangulation and the importance of notifying the physician immediately. If a truss has been ordered, the patient is taught to apply it before arising from bed each morning, while the hernia is not protruding. Special attention should be paid to maintenance of skin integrity beneath the truss.

PREOPERATIVE CARE. A simple herniorrhaphy is generally done as outpatient surgery. The patient is instructed in routine preoperative deep breathing and coughing techniques. Holding the abdomen with a splint, such as a pillow, helps support the weakened abdominal muscles during coughing and moving.

POSTOPERATIVE CARE. Care following inguinal hernia repair is similar to any postoperative care. The male patient may experience swelling of the scrotum. Ice packs and elevation of the scrotum may be ordered to reduce the swelling. Intake and output are monitored. The patient is observed for difficulty with voiding, which is not uncommon following hernia surgery, and voids before discharge. Bowel sounds are monitored, and the patient is instructed to report discomfort or distention because peristalsis may stop temporarily following surgery. Because most patients are discharged the same day as surgery, they are taught to change the dressing and report signs and symptoms of infection, such as redness, incisional drainage, or fever. The patient is also instructed to avoid lifting for 2 to 6 weeks. Specific activity limitations are provided by the surgeon. Most patients can return to work within 2 weeks.

▶ ABSORPTION DISORDERS

The process of digestion reduces nutrients to a liquid form that can be absorbed through intestinal mucosa into the portal bloodstream. Most of the more than 8000 mL of liquid with nutrients and electrolytes are absorbed proximal to the ileocecal valve.

Pathophysiology and Etiology

Malabsorption occurs when the GI system is unable to absorb one or more of the major nutrients (carbohydrates, fats, or proteins). Some causes of malabsorption are ileal dysfunction, jejunal diverticula, parasitic disease, celiac disease, enzyme deficiency, and inflammatory bowel diseases such as Crohn's disease and ulcerative colitis. The primary malabsorption disorders are tropical sprue, adult celiac disease, and lactose intolerance.

The cause of tropical sprue has not been specifically identified but is thought to be related to bacterial infection of the intestine. In adult celiac disease (sometimes called nontropical sprue), a sensitivity to gluten is thought to cause malabsorption of protein. Gluten is a protein found in wheat, barley, oats, and rye (Nutrition Notes Box 31–7).

A deficiency in lactase, an enzyme that breaks down lactose (milk sugar), causes lactose intolerance. When lactose is not digested, a high concentration of it occurs in the intestines, causing an osmotic retention of water and watery stools (Nutrition Notes Box 31–8).

Signs and Symptoms

Weight loss, weakness, and general malaise resulting in malnutrition are associated with malabsorption disorders. Signs and symptoms of sprue include frequent loose, bulky, foul stools that are gray in color and have an increased fat content. (Increased fat in stool is called steatorrhea.) Lactose intolerance causes abdominal cramping, excessive gas, and loose stools after eating milk products.

BOX 31-7 *Nutrition Notes*

Treating Celiac Disease
Celiac disease, or gluten-sensitive enteropathy, requires permanent elimination of wheat, rye, oats, and barley from the diet. Careful selection of prepared foods is mandatory because of the widespread use of these grains as thickeners. If a patient with celiac disease ingests gluten, damage to the intestine continues even in the absence of symptoms. Instruction from and follow-up by a dietitian are indicated to ensure adequate nutritional intake despite the many dietary limitations.

BOX 31-8 *Nutrition Notes*

Managing Lactose Intolerance
Patients with lactose intolerance control their conditions through trial and error, because the degree of lactose intolerance varies greatly among patients. Moderate amounts of many lactose-containing foods can be digested when taken with a mixed meal. A dietitian's services may be advised to begin dietary management after diagnosis. Ingredients to avoid include milk, milk solids, lactose, and whey. Low-lactose foods (up to 2 grams per serving) include sherbet, cheese aged longer than 90 days, processed cheese, and milk treated with lactase enzyme. In cheese-making the whey containing most of the lactose is removed, so that hard, ripened cheeses such as blue, brie, cheddar, Colby, gouda, parmesan, and Swiss are classified as low lactose. Some brands of yogurt, a high-lactose food, may be tolerated because they contain bacterial lactase.

Complications

Vitamin K deficiency and resulting hypoprothrombinemia can increase the risk of bleeding. Calcium deficiency can be severe enough to cause bone pain and neuromuscular hyperirritability, including tetany. Folic acid, vitamin B_{12}, and iron deficiencies can result in glossitis, stomatitis, anemia, and dry, rough skin.

Diagnostic Tests

See Table 31–5 for the diagnostic studies used to identify malabsorption diseases.

Medical Treatment

Folic acid, broad-spectrum antibiotics, and a high-calorie, high-protein, low-fat diet are ordered for patients with tropical sprue. Adults with celiac disease are ordered a high-calorie, high-protein, gluten-free diet to relieve the symptoms and improve nutritional status. However, because gluten is used as a filler or binder in many products, even those labeled "wheat free," diligence in identifying potentially offending foods is essential.

Lactose intolerance is treated by removing foods from the diet that contain lactose, such as milk and milk products. Some fermented milk products, such as cheese and yogurt, may be lower in lactose and better tolerated. Lactaid is an over-the-counter lactase substitute that can be taken when milk products cannot be avoided. It can be added to milk in liquid form or taken as a tablet before eating foods containing lactose. Lactaid digests about 70 percent of the lactose in foods, making them more tolerable.

Nursing Care

Nursing care involves monitoring fluid and electrolyte balance, nutritional status, and skin integrity. Daily weight and intake and output help determine if fluid loss is occurring. Intake of electrolyte-rich fluids is encouraged to replace losses. Antidiarrheal agents are given if ordered. Electrolyte levels, especially potassium, are monitored as ordered. The patient is instructed in dietary limitations. Nutritional supplements may be ordered if necessary. Perianal skin is kept clean and dry, and barrier ointments are used as needed to protect the skin from excoriation.

▶ INTESTINAL OBSTRUCTION

Intestinal obstructions occur when the flow of intestinal contents is blocked. There are two types of intestinal obstruction, mechanical and paralytic, which can be either partial or complete.

Mechanical obstruction occurs when a blockage occurs within the intestine from conditions, causing pressure on the intestinal walls such as adhesions, twisting of the bowel, or strangulated hernia. Paralytic obstruction occurs when peristalsis is impaired and the intestinal contents cannot be propelled through the bowel. Paralytic obstruction is seen following abdominal surgeries, trauma, mesenteric ischemia, or infection. The severity of the obstruction depends on the area of bowel affected, the amount of occlusion within the lumen, and the amount of disturbance in the blood flow to the bowel (Table 31–6).

Small-Bowel Obstruction
Pathophysiology

When obstruction occurs in the small bowel, a collection of intestinal contents, gas, and fluid occurs proximal to the obstruction. The distention that results stimulates gastric secretion but decreases the absorption of fluids. As distention

TABLE 31–5	DIAGNOSTIC TESTS FOR DISORDERS OF MALABSORPTION
Diagnostic Test	**Test Result and Associated Malabsorption Syndrome**
Hemoglobin and hematocrit	Decreased if anemia is present.
Mean corpuscular volume	Decreased values are found with malabsorption of vitamin B_{12}.
Upper GI series	Thickening of the intestinal mucosa, narrowed mucosa of the terminal ileum, or a change in fecal transit time are indicative of malabsorption syndrome.
D-Xylose absorption test	Decreased excretion of xylose after 5 hours is indicative of malabsorption.
Sudan stain for fecal fat	Malabsorption can be distinguished from maldigestion if this test shows abnormally large numbers of fat droplets.
72-hour stool collection for fat	Stool fat greater than 5 g per 24 hours after ingestion of 80 g of fat in 2 days implies a fat digestion disorder.

TABLE 31–6	BOWEL OBSTRUCTION SUMMARY
Symptoms	Abdominal pain
	Blood and mucus per rectum
	Feces and flatus cease
	Visible peristaltic waves in thin person
	Possible fecal vomiting
	Bowel sounds high-pitched and tinkling or absent
	Abdominal distention
	Fluid and electrolyte imbalance
Diagnostic tests	Abdominal x-ray examination
	CT scan
	CBC and electrolytes
Therapeutic management	NPO status
	Frequent mouth care
	Nasogastric tube
	Fluid and electrolyte replacement
	Medications: antibiotics, antiemetics, analgesics
	Surgery
	Education

worsens, the intraluminal pressure causes a decrease in venous and arterial capillary pressure, resulting in edema, necrosis, and eventually perforation of the intestinal wall.

Etiology

Following abdominal surgery, loops of intestine may adhere to areas in the abdomen that are not healed. This may cause a kink in the bowel that occludes the intestinal flow. These adhesions, or bands of scar tissue, are the most common cause of small bowel obstruction and are usually acquired from previous abdominal surgery or inflammation. Hernias and neoplasms are the next most common causes, followed by inflammatory bowel disease, foreign bodies, strictures, **volvulus,** and **intussusception.** A volvulus occurs when the bowel twists, occluding the lumen of the intestine. Intussusception occurs when peristalsis causes the intestine to telescope into itself (Fig. 31–5). These conditions are mechanical obstructions. Paralytic, or adynamic, ileus is a nonmechanical obstruction that occurs when the intestinal peristalsis decreases or stops because of a vascular or neuromuscular pathological condition. Box 31–9 lists causes of nonmechanical obstructions.

Signs and Symptoms

The patient initially complains of wavelike abdominal pain and vomiting. Initially, flatus and feces that is low in the

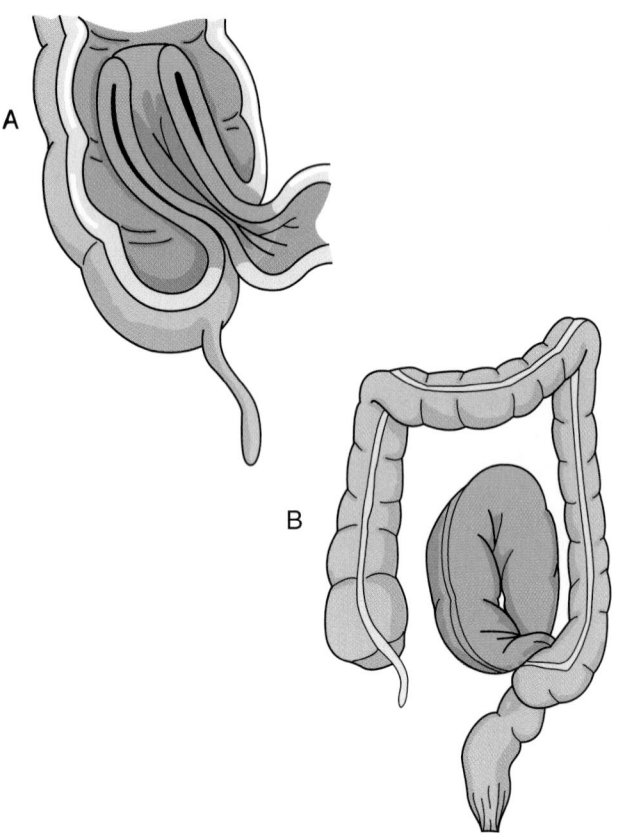

Figure 31-5   Mechanical bowel obstructions. (*A*) Intussusception. (*B*) Volvulus.

---

intussusception: intus—within + suscept—to receive

| BOX 31–9 | **Causes of Nonmechanical Obstruction** |

- Abdominal surgery and trauma
- Pneumonia
- Spinal injuries
- Hypokalemia
- Myocardial infarction
- Peritonitis
- Vascular insufficiency

bowel and blood and mucus may be passed, but this stops as the obstruction becomes worse. The symptoms progress as the obstruction worsens or becomes complete. As the obstruction becomes more extreme, peristaltic waves reverse, propelling the intestinal contents toward the mouth, eventually leading to fecal vomiting. Peristaltic waves may be visible in a thin person. Pain that is sharp and sustained may indicate perforation. In mechanical obstructions, high-pitched, tinkling bowel sounds are heard proximal to the obstruction and are absent distal to it. If the obstruction is nonmechanical, there is an absence of bowel sounds.

Loss of fluid and electrolytes leads to dehydration, with its associated symptoms of extreme thirst, drowsiness, aching, and general malaise. The lower in the gastrointestinal tract the obstruction is, the greater the abdominal distention. An uncorrected obstruction can lead to shock and possibly death.

Diagnostic Tests

Dilated loops of bowel are evident in radiographic studies and CT scans. If strangulation or perforation occurs, leukocytosis is evident. Hemoglobin and hematocrit are elevated if the patient is dehydrated, and serum electrolyte levels are decreased.

Medical Treatment

In most cases the patient is kept NPO and the bowel is decompressed using a nasogastric (or rarely an intestinal) tube, which relieves symptoms and may resolve the obstruction. An IV solution with electrolytes is initiated to correct the fluid and electrolyte imbalance. Sometimes IV antibiotics are begun. Complete mechanical obstruction requires surgical intervention, such as removal of tumors, release of adhesions, or bowel resection with anastomosis.

Large-Bowel Obstruction
Pathophysiology

Obstruction in the large bowel is less common and not usually as dramatic as small-bowel obstruction. Dehydration occurs more slowly because of the colon's ability to absorb fluid and distend well beyond its normal full capacity. If the blood supply to the colon is cut off, the patient's life is in jeopardy because of bowel strangulation and necrosis.

Etiology

Most large-bowel obstructions occur in the sigmoid colon and are caused by carcinoma, inflammatory bowel disease, diverticulitis, or benign tumors. Impaction of stool may also cause obstruction.

Signs and Symptoms

Symptoms of large-bowel obstruction develop slowly and depend on the location of the obstruction. If the obstruction is in the rectum or sigmoid, the only symptom may be constipation. As the loops of bowel distend, the patient may complain of crampy lower abdominal pain and abdominal distention. Vomiting, if it occurs, is a late sign and may be fecal. High-pitched tinkling bowel sounds may be heard. A localized tender area and mass may be felt on palpation. Large-bowel obstructions, if not diagnosed and treated, can lead to gangrene, perforation, and peritonitis (discussed earlier).

Medical Treatment

Radiological examination reveals a distended colon. If impaction is present, enemas and manual disimpaction may be effective. Other mechanical blockages may require surgical intervention.

Surgical resection of the obstructed colon may be necessary. A temporary colostomy may be indicated to allow the bowel to rest and heal. Sometimes an ileoanal anastomosis is done. A patient who is a poor surgical risk may have a cecostomy (an opening from the cecum to the abdominal wall) to allow diversion of stool.

Nursing Process

ASSESSMENT. Each quadrant of the abdomen is auscultated for 5 minutes before documenting the absence of bowel sounds. Palpation for distention, firmness, and tenderness is done. The amount and character of stool, if any, is documented. Pain is assessed using the institution's pain scale and described according to location and character, such as crampy or wavelike. Daily weight and intake and output are monitored. Skin turgor is assessed for fluid deficit. If a nasogastric tube is in place, the amount, color, and character of drainage is documented. Vital signs are monitored for signs of infection or shock.

NURSING DIAGNOSIS. Possible diagnoses include acute pain related to abdominal distention and ineffective tissue perfusion and deficient fluid volume related to collection of fluid in the intestine and vomiting.

PLANNING. Goals for the patient include pain control and prevention of dehydration and electrolyte imbalance.

IMPLEMENTATION. The nasogastric tube is maintained on low intermittent suction to relieve discomfort from distention. NPO status is maintained to rest the bowel, and frequent mouth care is provided. Medication for pain is administered as ordered. Opioids are given cautiously because they may mask symptoms of perforation and decrease intestinal motility. The patient is placed in semi-Fowler's position to reduce tension on the abdomen.

Careful monitoring of intake and output is critical. The type and amount of IV fluid replacement ordered is based on NG output. Ice chips may be given if ordered by the physician but must be used sparingly. When melted ice mixed with electrolytes and hydrochloric acid are removed from the stomach by suction, electrolyte imbalance and metabolic alkalosis occur. If surgery is anticipated, care for abdominal surgery is implemented.

EVALUATION. Goals have been met if the patient states that pain is controlled and if fluid is balanced and electrolytes are within normal limits.

CRITICAL THINKING

Mrs. Loos

Mrs. Loos is admitted for abdominal pain. You note that she has a history of diabetes mellitus. As you do your morning assessment, you find her abdomen large, firm, and tender to touch. She states that she feels nauseated. What do you do? What do you think is happening?

Answers at end of chapter.

▶ ANORECTAL PROBLEMS
Hemorrhoids

Hemorrhoids are varicose veins in the anal canal. They are caused by an increase in pressure in the veins, often from increased intra-adominal pressure. Internal hemorrhoids occur above the internal sphincter, and external hemorrhoids occur below the external sphincter. Most hemorrhoids are caused by straining during bowel movements. They are common during pregnancy. Prolonged sitting or standing, obesity, and chronic constipation also contribute to hemorrhoids. Portal hypertension related to liver disease may also be a factor.

Internal hemorrhoids are usually not painful unless they prolapse. They may bleed during bowel movements. External hemorrhoids cause itching and pain when inflamed and filled with blood (thrombosed). Inflammation and edema occur with thrombosis, causing severe pain and possibly infarction of the skin and mucosa over the hemorrhoid.

Preventing constipation, avoiding straining during defecation, and good personal hygiene relieve hemorrhoid symptoms and discomfort. Astringents, such as witch hazel, can be used for symptom relief. Sitz baths increase circulation to the area and aid in comfort and healing. Stool softeners can be used to reduce the need for straining. Other anti-inflammatory medications may be tried, such as steroid creams or suppositories. Alternating ice and heat helps relieve edema and pain with thrombosed hemorrhoids.

If hemorrhoids are prolapsed and are no longer reduced by palliative measures, more aggressive measures may be used. Sclerotherapy involves the injection of a sclerosing agent into the tissues around the hemorrhoids, causing them to shrink. Rubber band ligation uses rubber bands placed on the hemorrhoids until the tissue dies and sloughs off. Cryosurgery uses cold to freeze the hemorrhoid tissue.

Surgical hemorrhoidectomy involves surgical removal of hemorrhoids and is used in severe cases.

The patient should be instructed to consume a high-fiber diet and 2 to 3 L of fluid a day to promote regular bowel movements. The effects and side effects, proper dosage, and frequency of local or topical treatments should be explained. If the patient has surgery, analgesics should be given as needed because the many nerve endings in the anal canal can cause severe pain. Comfort measures such as a side-lying position and fresh ice packs should also be used to relieve pain. After the first postoperative day, sitz baths may be ordered. Unfortunately, a side effect of opioid analgesics is constipation, which needs to be avoided, especially in the immediate postoperative period. Because the first bowel movement can be painful and anxiety provoking, stool softeners are given and analgesics administered before the first bowel movement.

Anal Fissures

Anal fissures are cracks or ulcers in the lining of the anal canal. They are most commonly associated with constipation and stretching of the anus with passage of hard stool, although Crohn's disease or other factors may also play a role. The patient may experience bright red bleeding. Pain may be so severe that the patient delays defecation, leading to further constipation and worsening symptoms. Treatment of anal fissures involves measures to ensure soft stools, allowing fissures time to heal. Sitz baths may be used to promote circulation to the area to aid in healing. Anesthetic suppositories and nonopioid analgesics may be ordered for comfort. If conservative measures are not helpful, surgical excision may be necessary.

Anorectal Abscess

An anorectal abscess is a collection of pus in the rectal area. Common causative organisms include *E. coli*, *Proteus* spp., staphylococci, or streptococci. Symptoms include pain, redness and swelling, fever, and sometimes drainage.

Abscesses are treated with antibiotics and surgical incision and drainage of pus. The area may be left open to drain, with packing placed to assist with drainage and healing. Nursing care includes dressing or packing changes as ordered. Sitz baths are used to keep the area clean and promote healing, especially after bowel movements. The patient is instructed in the importance of keeping the area clean and dry. Other postoperative care is similar to care following hemorrhoidectomy.

▶ LOWER GASTROINTESTINAL BLEEDING

Etiology

Major causes of lower gastrointestinal bleeding are diverticulitis, polyps (growths in the colon), anal fissures, hemorrhoids, inflammatory bowel disease, and cancer. Bleeding may also occur in the upper GI tract. (See Chapter 30.)

Signs and Symptoms

Bleeding from the GI tract is evident in the stools. When blood has been in the GI tract for more than 8 hours and has come in contact with hydrochloric acid, it causes **melena,** or black and tarry stools. The presence of melena indicates bleeding above or in the small bowel. Bleeding from the colon or rectum is usually bright red **(hematochezia).**

Significant blood loss causes hypotension, light-headedness, nausea, and diaphoresis. The patient is pale and has cool skin. The onset of tachycardia and worsening hypotension indicate hypovolemic shock and should be reported to the physician immediately.

Diagnostic Tests

A thorough history is necessary to determine underlying disorders that may be causing the bleeding. Decreased hemoglobin and hematocrit result from blood loss. Blood urea nitrogen (BUN) may be elevated as a result of breakdown of proteins in the blood by the GI tract. Stool can be tested for occult blood if it is not evident on inspection. Digital examination, colonoscopy, or sigmoidoscopy may be done by the physician.

A new diagnostic test to find bleeding in the small intestine is being used by researchers. After an 8-hour fast, a pill-sized camera is swallowed to look inside previously inaccessible areas of the small intestine. The swallowed camera transmits two images per second via radiofrequencies to a recorder on a belt the patient wears. The patient can drink liquids after 2 hours and eat after 4 hours. After 8 hours the patient returns the recorder and belt to the doctor to download the images into a computer.

Medical Treatment

Treatment depends on the cause of the bleeding. Mild bleeding may resolve with conservative measures. If bleeding continues, surgery to correct diverticulosis, inflammatory bowel disease, or cancer may be considered.

Nursing Care

Stools are checked for the presence and amount of blood. Vital signs are monitored for signs of shock. Declining blood pressure and rising heart rate are reported to the physician immediately. The patient is prepared for diagnostic tests and nursing care for the underlying disorder is provided.

▶ COLON CANCER

Colon cancer is one of the most common types of internal cancer in the United States. People with a family history of colon cancer or ulcerative colitis are at higher risk of developing it themselves. The American Cancer Society estimated 135,400 new cases and 56,700 deaths from colorectal cancer in 2001.[1]

Pathophysiology and Etiology

Colon cancer originates in the epithelial lining of the colon or rectum and can occur anywhere in the large intestine. People with a personal or family history of ulcerative colitis, colon cancer, or polyps of the rectum or large intestine are at higher risk for developing cancer. Colon cancer has also been linked with previous gallbladder removal and di-

---

hematochezia: hemat—blood + chezia—in stool

etary carcinogens. A major causative agent is lack of fiber in the diet, which prolongs fecal transit time and in turn prolongs exposure to possible carcinogens. Also, bacterial flora is believed to be altered by excess fat, which converts steroids into compounds having carcinogenic properties.

Signs and Symptoms

Manifestations of colon cancer vary according to the type of tumor and the location (Fig. 31–6). A change in bowel habits is the most common symptom (Table 31–7). Blood or mucus in stools may occur. Although all tumors cause varying degrees of obstruction, those in the descending colon and rectum do not cause anemia, weight loss, nausea, or vomiting.

Complications

Complications include complete obstruction of the colon, perforation, peritonitis, and extension of the tumor to adjacent organs. Colorectal cancer can metastasize to the lymphatic system and liver.

TABLE 31-7	COLON CANCER SUMMARY
Symptoms	Change in bowel habits
	Blood or mucus in stools
	Abdominal or rectal pain
	Weight loss
	Anemia
	Obstruction
Diagnostic tests	Colonoscopy with biopsy
	Sigmoidoscopy with biopsy
	Proctosigmoidoscopy
	Barium enema
	Abdominal and rectal examination
	Fecal occult blood
Therapeutic management	Surgery, possibly colostomy
	Radiation
	Chemotherapy and/or radiation
	Medications: analgesics
	TPN as necessary
	Support and education

Diagnostic Tests

Most colorectal cancers are identified by biopsy done at the time of endoscopy (proctosigmoidoscopy, sigmoidoscopy, or colonoscopy). In addition to abdominal and rectal examination, other tests include fecal occult blood testing and barium enema. Carcinoembryonic antigen (CEA), a blood test, is used to assess response to treatment of GI cancer. CEA is present when epithelial cells rapidly divide and provides an early warning that the cancer has returned.

Medical Treatment

Small localized tumors may be excised and treated during endoscopy or laparoscopy. These procedures can also be used as palliative care for patients with advanced tumors who cannot tolerate major surgery.

Surgery is performed to either resect larger tumors and anastomose the remaining bowel or divert the feces by forming a colostomy. In rectal cancer an abdominal perineal (A & P) resection is done with the formation of a permanent end colostomy (Fig. 31–7). With an abdominal perineal resection, the anus, rectum, and part of the sigmoid colon are removed. A perineal wound, often with a drain inserted, is left where the anus was. Medical management includes radiation therapy and chemotherapy. When both are used, along with surgery, increased survival rates have been demonstrated.

LEARNING TIPS

Three incisional areas are created in an abdominal perineal resection:

1. An abdominal incision
2. A perineal incision
3. A stoma

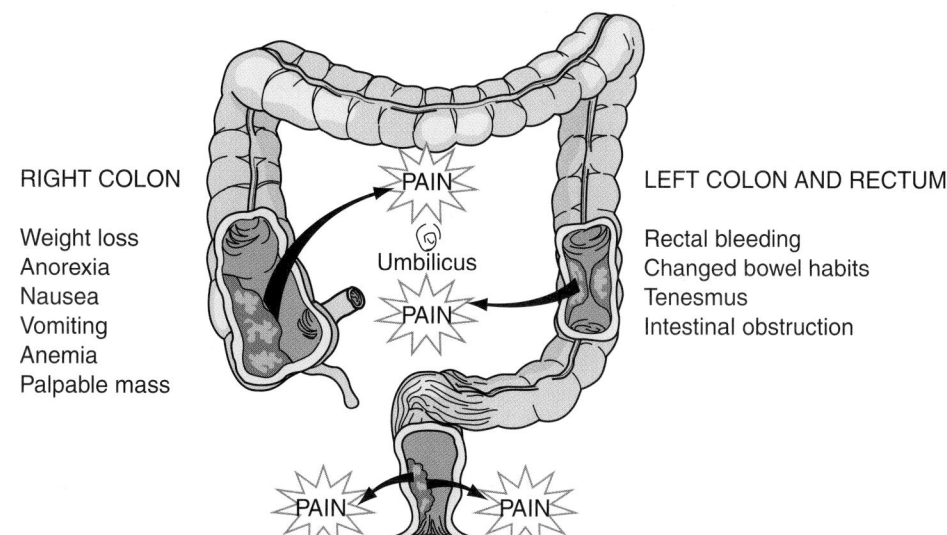

RIGHT COLON

Weight loss
Anorexia
Nausea
Vomiting
Anemia
Palpable mass

PAIN
Umbilicus
PAIN

LEFT COLON AND RECTUM

Rectal bleeding
Changed bowel habits
Tenesmus
Intestinal obstruction

PAIN   PAIN

Figure 31-6 Symptoms of carcinoma of the colon. Pain usually radiates toward the umbilicus or perianal area. (Modified from Black, JM, and Matassarin-Jacobs, E: Medical-Surgical Nursing, ed 5. WB Saunders, Philadelphia, 1997, p 1810, with permission).

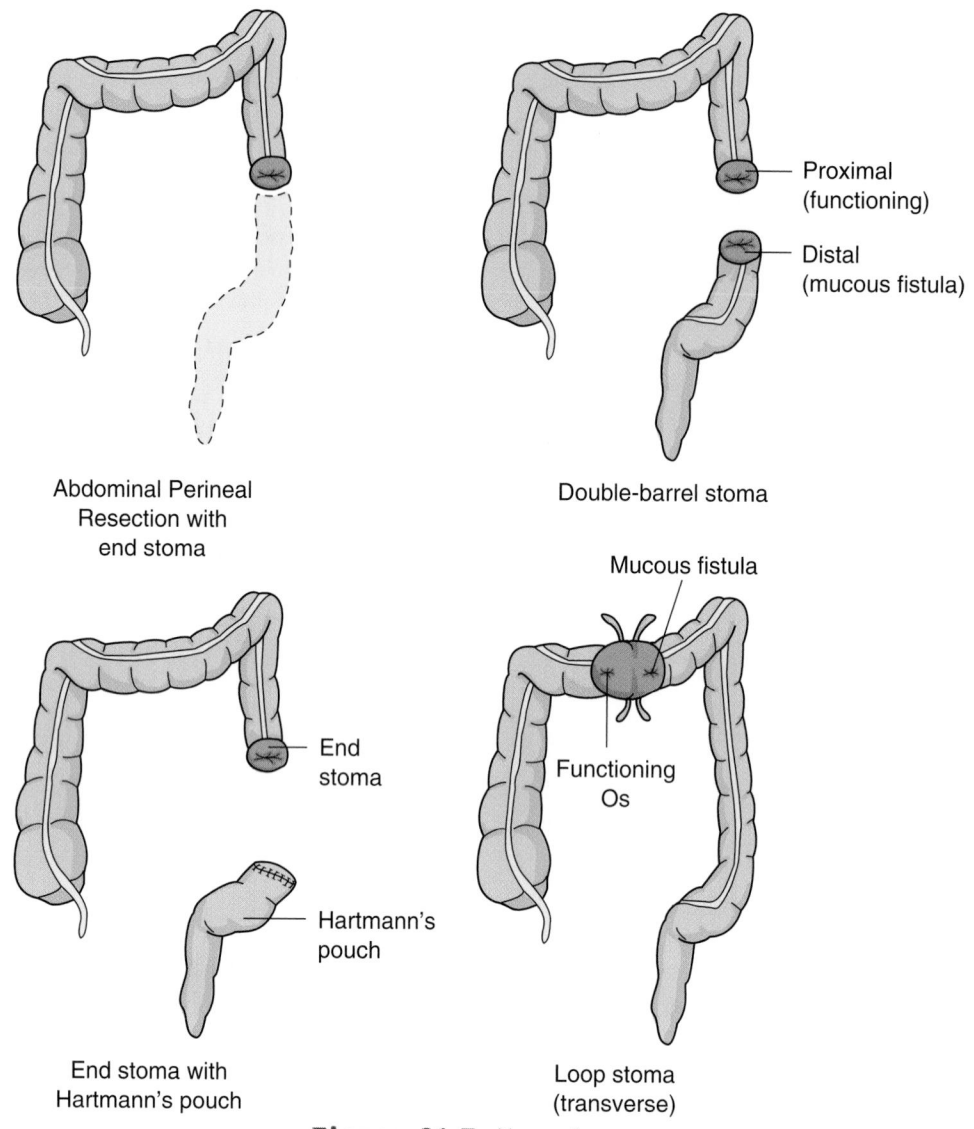

Abdominal Perineal
Resection with
end stoma

Proximal
(functioning)

Distal
(mucous fistula)

Double-barrel stoma

End
stoma

Hartmann's
pouch

End stoma with
Hartmann's pouch

Mucous fistula

Functioning
Os

Loop stoma
(transverse)

Figure 31-7   Types of stomas.

Nursing Process

ASSESSMENT. Risk factors for colon or rectal cancer are identified by asking questions about the patient's personal and family history: Is there a history of inflammatory bowel disease? What were the patient's dietary habits? What foods were usually eaten, and how much fluid was usually consumed? Prior to diagnosis, did the patient experience constipation or diarrhea? Has there been a change in bowel habits? Has mucus or blood been noted in the stools? What social habits does the patient have? Did the patient smoke, drink alcoholic beverages, exercise? Has there has been a recent weight loss? How much and in what time frame? Does the patient admit to unusual fatigue or insomnia? Stool should be checked for mucus or blood.

NURSING DIAGNOSIS. Nursing diagnoses identified, based on the assessment data, may include acute pain related to tissue compression from the tumor, anxiety related to diagnosis of cancer, imbalanced nutrition: less than body requirements related to nausea and anorexia, and deficient knowledge related to surgery and postoperative care.

PLANNING. Goals include relief from pain, alleviation of anxiety, achievement of optimum nutritional status, and understanding of self-care following surgery.

IMPLEMENTATION. Set aside time to allow the patient who so desires to talk, cry, or ask questions about the diagnosis and planned surgery. Postoperatively, administer analgesics as prescribed. Provide a quiet, relaxing atmosphere to help alleviate anxiety, and limit visitors and telephone calls if the patient prefers. Postoperative care includes monitoring vital signs, the stoma, and the return of bowel sounds and flatus, indicating peristalsis has resumed, and ambulation, coughing and deep breathing, and dressing changes as ordered. Dressings are observed for drainage. Large amounts of drainage or bleeding are reported. If a drain is inserted, often in the perineal wound, moderate amounts of serosanguineous (light pink) drainage are expected. Frequent dressing reinforcements or changes are needed to keep the area dry.

Total parenteral nutrition may be necessary to provide depleted vitamins, minerals, and nutrients if the patient has been anorexic for any length of time or has had a significant weight loss. After bowel sounds return, provide the patient with a high-protein, high-calorie diet, as ordered, that is low in residue to decrease excessive peristalsis and minimize cramping. If the patient is anemic, blood transfusions may be necessary.

EVALUATION. Expected outcomes are that the patient verbalizes less anxiety and control of pain, attains an optimum level of nutrition, and verbalizes understanding of the disease and treatment. If the patient has a colostomy, the stoma is observed by the patient and a return demonstration of the appliance change should be done before discharge.

▶ OSTOMY MANAGEMENT

An ostomy is a surgically created opening that diverts stool or urine to the outside of the body through an opening on the abdomen called a **stoma**. A stoma is the portion of bowel that is sutured onto the abdomen. The types of abdominal ostomies include ileostomy, colostomy, and urostomy. (Urinary ostomies are discussed in Chapter 35.) The stomas can be end, loop, or double barrel. See Figure 31–7 for types of ostomies.

Ileostomy

An ileostomy is an end stoma formed by bringing the terminal ileum out to the abdominal wall following a total colectomy. Two types of ileostomies can be formed: a conventional ileostomy and a continent ileostomy, such as a Kock pouch (Fig. 31–8). A conventional ileostomy is a small stoma in the right lower quadrant that requires a pouch at all times because of the continuous flow of liquid effluent.

A continent ileostomy is formed by taking a portion of the terminal ileum to construct an internal reservoir with a nipple valve. The stoma is usually flush to the skin and located near the suprapubic area. The patient is taught to insert a catheter into the stoma three or four times a day to empty the reservoir. A continent ileostomy surgery takes longer and requires additional instruction for the patient to be able to do self-care.

The patient who is opposed to having an ileostomy may have an ileoanal anastomosis done to connect the ileum to the anus, avoiding a stoma (Fig. 31–9). This is usually a two-step procedure. During the first surgery, the diseased bowel is removed. A reservoir (called a J pouch or an S pouch) is then formed from part of the ileum and connected to the anus. A temporary ileostomy is also formed to divert stool while the reservoir heals. After 2 to 3 months, the ileostomy is reversed and the patient can have bowel movements from the anus. Problems with perianal skin irritation resulting from frequent liquid stools may occur. An ileorectal anastomosis can also be performed, but this may not be a curative procedure for a patient with ulcerative colitis, because the rectum may still be diseased.

Colostomy

A colostomy is named according to where in the bowel it is formed: It may be an ascending, transverse, descending, or sigmoid colostomy. The type of effluent is dependent on the location of the bowel used (Table 31–8).

TABLE 31-8	LOCATION OF STOMA AND TYPE OF EFFLUENT	
Location of Stoma	**Type of Effluent**
Ileostomy	Liquid to mushy
Cecostomy, ascending colostomy	Liquid to mushy, foul odor
Right transverse colostomy	Mushy to semiformed
Left transverse colostomy	Semiformed, soft
Descending or sigmoid colostomy	Soft to hard formed

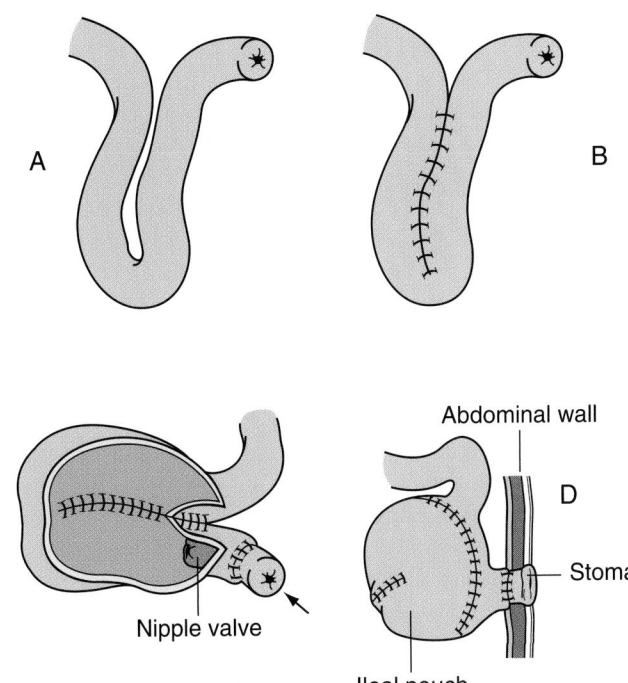

Figure 31-8 Surgical formation of continent ileostomy (Kock pouch). (*A*) Loop of terminal ileum. (*B*) Both limbs of ileum are brought together and sutured into a U shape. (*C*) Pouch created with nipple valve. (*D*) Pouch sutured to abdominal wall.

Abdominal wall

Stoma

Nipple valve

Ileal pouch

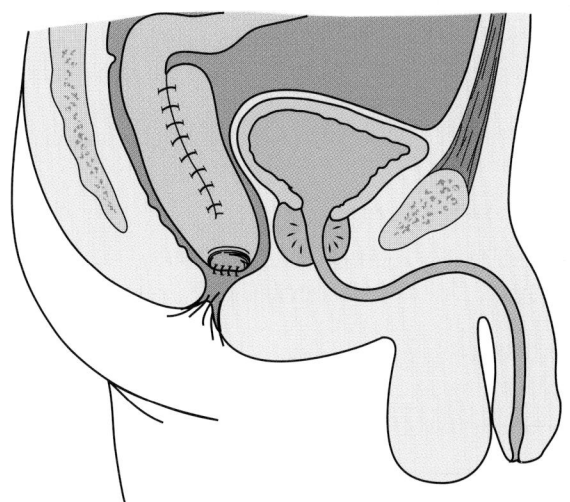

Figure 31-9 Ileal J pouch–anal anastomosis. The two-loop ileal pouch is simple to construct, provides adequate storage capacity, and is evacuated spontaneously and fully.

End Stoma

An end stoma is formed when the proximal end of the bowel is brought to the outside abdominal wall. If an abdominal perineal resection is done (usually for low rectal cancer), the rectum is removed and the proximal sigmoid or descending colon is brought out as a stoma. Another procedure that may be done involves removing the segment of diseased or injured bowel and using the proximal portion to form the stoma. The remaining limb of bowel is sutured closed and left in the peritoneal cavity so that the rectum is intact. This is called a Hartmann's pouch, or mucous fistula, and may be permanent or temporary depending on the diagnosis. Because the rectum is intact, the patient may feel the urge to defecate. This is normal because the colon continues to produce mucus. As the rectal stump fills with mucus, the sphincter is triggered and alerts the patient as if it were stool.

Loop Stoma

To create a loop stoma, a loop of bowel, usually the transverse colon, is pulled to the outside abdominal wall and a bridge is slipped under the loop to hold it in place. An incisional slit is made in the top of the exposed colon to allow stool to exit. The entire loop of bowel is not cut through.

Double-barrel Stoma

With a double-barrel stoma, the bowel is completely dissected and both ends of the colon are brought to the outside abdominal wall to form two separate stomas. The proximal stoma is the functioning stoma that expels stool. The distal stoma is called a mucous fistula because mucus produced by the bowel passes from it. A double-barrel stoma is often temporary, allowing the bowel to rest during healing after trauma or surgery.

Preoperative Care

A wound ostomy continence nurse should be consulted before surgery. The WOCN can help prepare the patient both emotionally and physically for the surgery. In addition, the WOCN has expertise in marking the stoma site for the surgeon. This involves observing the abdomen as the patient assumes different positions and noting how clothing is worn, such as where the belt rides. The site for the stoma can then be chosen so it is visible to the patient for self-care, avoids skin or fat folds, and is where clothing will not interfere with the appliance. Properly planned stoma placement can prevent agony over inability to perform self-care

and uncomfortable, leaking, or poorly fitting appliances postoperatively.

Routine preoperative instruction, including the importance of coughing and deep breathing, splinting, and early ambulation, is provided. Orders for cleansing of the bowel are performed to reduce the risk for infection following surgery. Unless the patient has chronic diarrhea related to IBD, an oral agent to cleanse the bowel (Go-LYTELY) is given. Oral and intravenous antibiotics are given as ordered.

Nursing Process

ASSESSMENT. The patient with a new ostomy has many nursing care needs. In addition to routine postoperative assessment, the stoma should be inspected at least every 8 hours. The stoma should be pink to red, moist (similar to the inside of the mouth), and well attached to the surrounding skin (Fig. 31–10). A bluish stoma indicates inadequate blood supply; a black stoma indicates necrosis. Either complication should be reported to the physician immediately for treatment, which may require that the patient return to surgery. Initially the stoma is swollen. The stoma size gradually decreases over the first few weeks following surgery. This is explained to the patient.

Skin is assessed for irritation around the pouch and under the pouch each time it is changed. Ostomy discharge (effluent) is monitored and documented. Unexpected changes, such as liquid stool from a descending ostomy, are reported.

NURSING DIAGNOSIS. A sample care plan is shown for a nursing diagnosis of disturbance in body image related to new ostomy care. (See Nursing Care Plan Box 31–10 for the Patient with an Intestinal Ostomy.)

PLANNING. The goal for the ostomy patient is to perform the self-care measures necessary for safe discharge home. The physical, emotional, psychosocial, economic, and rehabilitative aspects must be considered, and teaching must be provided to assist the patient to live with his or her ostomy. Collaborative discharge planning should include the social worker, dietitian, pharmacist, home health nurse, and WOCN. Some communities have an ostomy club that sends trained visitors with ostomies to visit with patients with new ostomies to help them adjust.

IMPLEMENTATION
Deficient Knowledge Related to New Ostomy. Recognizing patient readiness and ability to learn and perform self-care is of primary importance. The patient experiencing pain, nausea, or vomiting is not likely to be ready to look at the ostomy or learn about ostomy care. A patient who is blind, deaf, or has a language barrier requires special instructional methods. Patients with severe arthritis or other physical conditions that limit ability to perform self-care may require a specific type of ostomy appliance.

If the patient is not ready or able to learn, it is important that a family member or caregiver be included in the teaching. With short hospital stays, teaching time frames are limited and must begin soon after surgery. A home care nurse

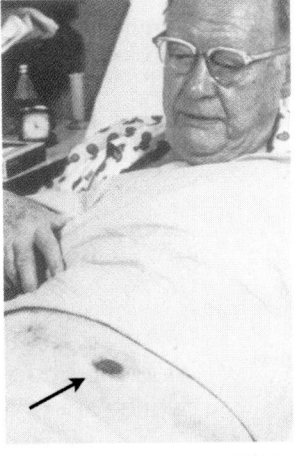 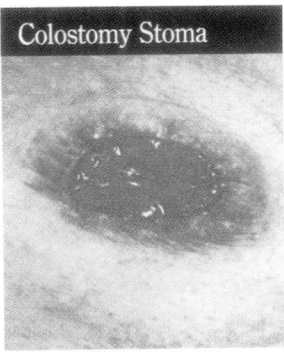

Colostomy Stoma

Man at right has a descending or "dry" colostomy.

Ileostomy Stoma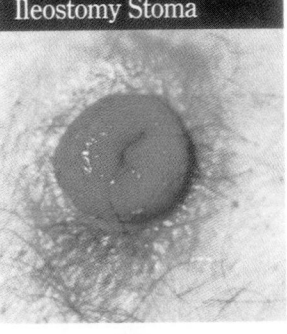

Urostomy Stoma

Loop Ostomy Stoma

Figure 31-10 Types of stomas. Note moist, pink to red appearance of healthy stoma. (Reproduced with permission of Hollister Incorporated.)

BOX 31-10 NURSING CARE PLAN FOR THE PATIENT WITH AN INTESTINAL OSTOMY

 Disturbed body image related to new ostomy

PATIENT OUTCOMES

Patient will verbalize acceptance of intestinal ostomy before discharge.

EVALUATION OF OUTCOMES

Is the patient able to verbalize acceptance of the ostomy?

INTERVENTIONS	RATIONALE	EVALUATION
Assess knowledge of self-care of ostomies.	Many people have misconceptions regarding ostomies. Identification of misconceptions and "hearsay" knowledge is important to clarify or correct.	Does the patient verbalize appropriate knowledge of ostomy care?
Encourage patient to verbalize feelings about the stoma.	Allowing the patient to express his or her feelings provides opportunity to identify and verbalize concerns, which then can be addressed by health care providers.	Does the patient discuss his or her feelings regarding the stoma to the nurse or significant other?
Explain the normal characteristics of the stoma before the patient's first look.	Helping the patient understand what to expect will help relieve anxiety. Being available to answer questions immediately also relieves anxiety.	Does the patient look at the stoma without hesitation?
Demonstrate ostomy appliance change and daily care and encourage patient participation.	When the patient observes and participates in self-care, his or her self-concept improves.	Is the patient participating in self-care? Has the patient performed return demonstration of appliance change and emptying of pouch?

often continues the teaching in the patient's home. A WOCN or equipment supplier can suggest appliances suited to individual patients' needs. Figure 31–11 shows types of appliances.

Appliance change. Depending on the type of appliance used, the appliance will need to be changed as often as every 3 days or rarely every 10 to 14 days. If leakage occurs, the appliance should be changed as soon as possible to avoid **peristomal** skin irritation. The skin barrier that is placed over the stoma on the skin should fit within one-sixteenth to one-eighth inch of the base of the stoma to prevent skin contact with stool. The stoma should be measured with each appliance change for as long as the stoma remains edematous. Because most stomas are not round, a pattern should be traced to teach the patient proper size and shape. An open-ended or drainable pouch should be used for all colostomies or ileostomies, especially during the first 8 weeks after surgery. See Figure 31–12 for the pouch application procedure.

If the patient has a left-sided (descending or sigmoid) colostomy, the bowel can be regulated either by diet or by regular irrigation of the stoma. Once bowel regulation has been achieved, the patient may use a closed-end pouch or a stoma cap.

Daily care and hygiene. The patient is instructed to empty the pouch when it is one-third to one-half full. The amount of effluent and the frequency of emptying depends on the location of the stoma in the bowel. If the pouch is allowed to get more than half full of stool, the weight of the effluent will pull on the pouch and weaken the seal of the skin bar-

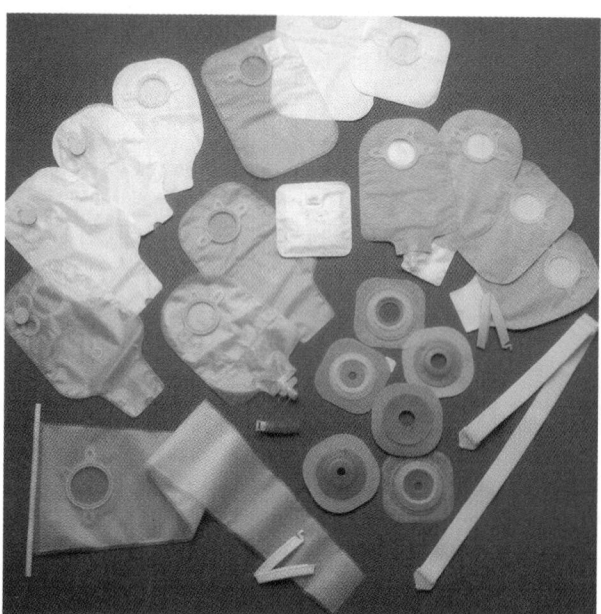

Figure 31-11 Appliances used for ostomies. The long sleeve at the lower left of the photo is used to drain the bowel following irrigation. (Reproduced with permission of Hollister Incorporated.)

peristomal: peri—surrounding + stoma—mouth or opening

Preparation of the Stomahesive Wafer with Sur-Fit Flange

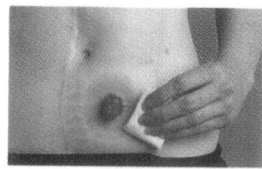

1. Cleanse the peristomal area with water and pat thoroughly dry. Measure your stoma size with the measuring guide provided and trace the proper opening on the white paper backing of the Stomahesive® disc.

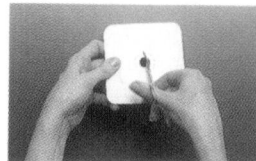

2. Leaving the white paper backing of the wafer in place, cut a hole in the wafer to the same shape and size as the base of the stoma. The best result is usually obtained by cutting from the reverse side of the wafer, using curved, short-bladed scissors.

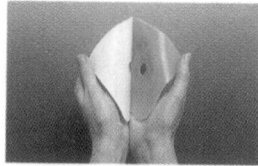

3. Peel the white paper backing from the wafer just prior to application.

4. Gaps between the wafer and the base of the stoma may be further protected by applying Stomahesive® Paste to the wafer.

Application of the Stomahesive Wafer with Sur-Fit Flange

5. Center the enlarged hole over the stoma, place on abdomen, and apply light pressure.

Figure 31-12 Preparation to apply an ostomy appliance. (Courtesy ConvaTec, a Bristol-Myers Squibb Company, Princeton, NJ, with permission.)

rier. Once the pouch is emptied, the inside of the tail of the pouch must be cleaned and dried before the clamp is replaced to help control odor. If a two-piece system is used, the pouch can be taken off, washed out, and replaced. The patient can bathe or shower with the appliance in place but needs to check the seal and retape or change it if it is loosening. Spray deodorants or chlorophyll tablets can be placed in the pouch for odor control. Chlorophyll is more effective if taken by mouth.

Dietary considerations. The patient needs to be aware of foods that contribute to odor and gas. If foods that are known to cause odor or gas are eaten, the patient should know to empty the pouch of flatus more often and to be aware that more odor is probable. It is important that the patient know the foods that contribute to and control diarrhea and what to do for constipation. A list of foods that may contribute to ileostomy blockage must be given to the ileostomy patient (Table 31–9, Nutrition Notes Box 31–11).

Colostomy irrigation. Colostomy irrigation is rarely done to regulate an ostomy anymore. However, it may be done as bowel preparation for procedures. Irrigation is perfomed in a manner similar to an enema, except that special equipment is used to instill fluid into the bowel via the stoma. Because the stoma does not have a sphincter, specially designed tubing with a cone at the end is used to irrigate the ostomy. The cone blocks the fluid that is being instilled from flowing back out the stoma.

Disturbed Body Image. Altered body image, fear of the unknown, and concerns regarding acceptance by significant

others can take control of the patient with an ostomy. Overcoming those concerns is crucial to patient adjustment. The idea of feces being expelled from the abdomen can make the patient feel dirty and abnormal. The spouse or significant other should be encouraged to view the stoma with the patient and be allowed to ask questions and participate in care. The attitude and behavior of nursing staff caring for the patient with an ostomy can significantly affect his or her ability to adapt, either positively or negatively. A trained visitor with an ostomy can help the patient see that there is hope for a normal life. (Check the local phone directory for The United Ostomy Association or ostomy club.)

One of the most common fears expressed by patients with ostomies is the fear of gas and odor. Because the stoma does not have a sphincter, flatus is expelled unexpectedly. It is helpful to inform the patient that when he or she is wearing clothes, rather than a loose hospital gown, the noise is usually muffled. The quality of appliances also allows for control of odor. The patient needs to know that if an odor is noticeable, something is wrong, and that the following questions need to be considered: (1) Was the pouch just emptied of feces or "burped" of air? (2) Is there feces on the drainage spout of the pouch? (3) Are there signs of leakage under the skin barrier? (4) If a two-piece system is used, is the pouch snapped onto the skin barrier correctly?

If leakage occurs in a public place, such as the grocery store, church, or a friend's house, the patient may be reluctant to go out again. Advise the patient to be prepared for an emergency pouch change because leakage of an appliance can occur at any time. Encourage him or her to always carry an extra appliance and to determine the cause of the leakage. Also teach the patient measures to decrease the chance of leakage and to consult a WOCN if leakage is a continuous problem.

Sexual Dysfunction. Feared change in body image can also lead to sexual dysfunction. If the male patient had an abdominal perineal resection for cancer of the rectum, there is a chance that he will be impotent. This impotence may be transient, depending on the severity of nerve damage or edema associated with the surgery. If the patient is impotent, a urologist may be consulted for interventions. Encourage patients to discuss any concerns regarding sexuality with his or her spouse or sexual partner. This may help them work through any fears or embarrassment. Attractive pouch covers can be purchased and worn to help disguise the pouch and its contents. Alternative sexual positions can also be suggested.

Ineffective Therapeutic Regimen Management. The patient may have difficulty carrying out self-care measures for a variety of reasons. The cost and availability of ostomy supplies is problematic for many patients. Most insurers, including Medicare, pay for ostomy supplies, although some limit the type of appliance and number allowed per month. Each state-funded Medicaid system is different. The type of appliance needed to eliminate leakage may not always be covered, requiring the patient to either pay the difference or wear what the insurance company will provide. If the

TABLE 31-9 FOODS THAT CAN CAUSE ILEOSTOMY BLOCKAGE

Green leafy vegetables	Mushrooms
Spinach	Nuts
Collards	Dried fruits
Mustards	Raisins
Cole slaw	Figs
Celery	Apricots
Corn, popcorn	Chinese vegetables
Foods with nondigestible	Meats with casings
peels	Sausage
Apples	Hot dogs
Grapes	Bologna
Potatoes	
Coconut	

BOX 31-11 *Nutrition Notes*

Anticipating Dietary Management of Ostomies

Ostomy patients receive a soft diet initially, progressing to a general diet as the surgeon prescribes. Stringy, high-fiber foods are avoided initially. These include celery, coconut, corn, cabbage, coleslaw, membranes on citrus fruits, peas, popcorn, spinach, dried fruit, nuts, sauerkraut, pineapple, seeds, and skins of fruits and vegetables. Some patients avoid fish, eggs, beer, and carbonated beverages because they produce excessive odor. Dietary restrictions are usually based on individual tolerance.

Patients with ostomies should be encouraged to do the following:

- Eat at regular intervals.
- Chew food well to avoid blockage at the stoma site.
- Drink adequate amounts of fluid.
- Avoid foods that produce excessive gas, loose stools, offensive odors, and undesirable bulk.
- Avoid excessive weight gain.

patient has no insurance, costs can be high. Some patients find they have to choose whether to purchase ostomy appliances or prescriptions with their limited funds. Fortunately, the pouches in most two-piece systems can be washed out and reused to save money (Home Health Hints). In some areas the availability of supplies may be limited. However, supply sources continue to increase. Local retail pharmacists and internet and mail-order companies are available.

HOME HEALTH HINTS

Ostomy bag deodorizers can increase the cost of care. Alternatives include the following: (1) Place tissue or cotton balls saturated in almond or vanilla extract inside the appliance and replace as needed. (2) Put mouthwash in the water that is used to rinse the appliance. (3) Spray a light coat of cooking spray inside the clean bag to make later cleanup easier.

Risk for Injury. Complications related to ostomies include skin and stomal problems. If there are any complications associated with the care of the ostomy, the WOCN should be consulted if available. A WOCN has had specialized instruction in caring for the stoma and peristomal skin and has a wealth of information to offer.

Peristomal skin irritation. The skin around the stoma may become irritated if the opening in the skin barrier around the stoma is cut too large and exposes skin to the GI effluent. Prolonged contact with effluent can lead to a reaction similar to a chemical burn. Tape and adhesive removal, especially if done frequently, can lead to skin shearing. Occasionally, allergic dermatitis from a sensitivity to the adhesive may develop. To prevent irritation, a skin barrier such as stomahesive should be used. Pouches are left on for several days or until leakage occurs to prevent skin shearing from frequent removal.

Peristomal hernia. Hernias may develop around the stoma as a result of weakened abdominal muscles. This can pose a problem with appliance leakage caused by the change in body contours associated with the hernia. A more flexible ostomy appliance may be helpful.

Stomal prolapse. Sometimes the weakened abdominal muscles contribute to the falling down (or out) of the intestinal mucosa. This is called a prolapse and most commonly occurs in the elderly. Pouching a prolapsed stoma can be difficult. A prolapsed stoma should be reported to the physician.

Stomal necrosis. Stomal necrosis occurs when there is circulatory compromise. This may arise as a result of vascular collapse, blockage in the mesentery of the intestines, or edema in the intestine from obstruction proximal to the stoma. The onset of necrosis should be reported to the physician immediately. Usually, necrotic tissue occurs only at the very end of the stoma and will eventually slough off, revealing viable mucosa. Odor from the necrotic tissue may be a problem.

Ileostomy blockage. It is essential that the patient with an ileostomy be taught the signs and symptoms of a blockage and how to manage one should it occur. The patient with an ileostomy may have a blockage if bowel movements are absent or a large quantity of hot liquid is passed through the stoma and is associated with abdominal cramping. With blockage, the stoma becomes edematous and the color may turn pale or dusky. Once signs and symptoms are recognized, the patient needs to consider what was eaten in the past 24 hours because certain foods are considered to cause stomal blockage. (See Table 31–9.)

Once the problem is identified, the patient should get into a tub of warm water (not too hot or cold), get into a knee-to-chest position, and sip on warm liquid, such as coffee, tea, bouillon, broth, or hot chocolate. If the blockage is partial, relief will occur fairly soon after these measures are taken. If the blockage is complete, an ileostomy lavage must be performed by a physician or WOCN.

EVALUATION. The plan of care has been effective if the patient or caregiver is able to competently care for the ostomy, the patient is able to accept the change in body image, and self-care measures to prevent or treat complications can be described by the patient or caregiver.

REHABILITATIVE NEEDS. One of the goals of care for the ostomy patient is to assist the patient to return to activities of daily living. In addition to the WOCN, a home care nurse may be able to assist the patient in becoming comfortable with self-care. Being able to perform the ostomy appliance change and daily care is vital, but ensuring that the patient returns to work or civic activities as before is also important.

The patient can generally perform any activity he or she was able to do before the ostomy. Resources available, as needed, include the United Ostomy Association, the Crohn's and Colitis Foundation of America, and the American Cancer Society (Table 31–10).

TABLE 31–10 SUPPORT GROUPS
United Ostomy Association, Inc. (UOA) 19772 MacArthur Blvd., Suite 200 Irvine, CA 92612-2405 (714) 660-8624 (voice) (714) 660-9262 (fax) (800) 826-0826 (toll free) www.uoa.org
Celiac Sprue Association PO Box 31700 Omaha, NE 68131 (402) 558-0600 www.csaceliacs.org
Crohn and Colitis Foundation of America, Inc. 386 park Ave. South 17th Floor New York, NY 10016-8804 (212) 685-3440 (voice) (212) 779-4098 (fax) (800) 932-2423 (toll free) www.ccfa.org
American Cancer Society (ACS) (800) ACS-2345 www.cancer.org

REVIEW QUESTIONS

1. A patient has ulcerative colitis. Which foods should he be taught to avoid in his diet?
    a. Fresh fruits
    b. White bread
    c. Sweet desert
    d. Meat

2. The nurse is listening to a patient's abdomen and determines that bowel sounds are absent. To make this determination, the nurse would listen for which of the following time frames?
    a. 2 minutes in each quadrant
    b. 5 minutes in each quadrant
    c. 7 minutes in each quadrant
    d. 10 minutes in each quadrant

3. The nurse is caring for a patient who has a sudden onset of diarrhea. Which of the following terms should the nurse use to document the patient's black, tarry stool?
    a. Melena
    b. Hematochesia
    c. Hematemesis
    d. Steatorrhea

4. The nurse is developing a teaching plan for a patient who is interested in lifestyle changes to help prevent colon cancer. Which of the following of the patient's dietary habits does the nurse understand may increase the risk for development of colon cancer?
    a. High fat, low fiber intake
    b. High intake of milk and milk products
    c. Low meat and protein intake
    d. Low fat, high carbohydrate intake

5. A patient with Crohn's disease is to receive sulfasalazine, 500 mg oral suspension qid. The oral suspension is available as 250 mg/5 mL. How many milliliters should the nurse give for the 0800 dose?
    a. 5 mL
    b. 10 mL
    c. 20 mL
    d. 50 mL

REFERENCES

1. 2001 Facts and Figures. American Cancer Society: http://www.cancer.org, July 20, 2001.

Answers to CRITICAL THINKING

Mrs. Jessie Burns

You need to assess the situation before intervening. First, ask Mrs. Burns or her caregivers if she had a bowel movement that was inadvertently not charted. Next, ask Mrs. Burns if she feels constipated or if she has abdominal discomfort. Assess Mrs. Burns's abdomen for distention and presence or absence of bowel sounds. A digital examination may be necessary to determine if a fecal impaction is present. If simple constipation appears to be the problem, the medical record should be checked for as-needed laxative or enema orders. Once Mrs. Burns has had a bowel movement, laxatives should be discontinued and preventive measures such as regular fluids, fiber, and exercise should be instituted.

Judy Morrow

You need to further assess how Judy perceives that Crohn's disease will affect her lifestyle. What does she know about Crohn's disease? What is she concerned about? How has Crohn's disease affected her ability to sleep, what she eats, her participation in sports, and her relationships with other people?

Convey a caring manner to Judy by being accepting of her, listening actively to her concerns, and helping her to find acceptable ways of resolving them. Provide her with information that she needs about Crohn's disease. Arrange to have a well-adapted person of approximately the same age from the Crohn's and Colitis Foundation of America meet with her to share coping strategies with her.

Mrs. Loos

1. You need to further assess Mrs. Loos before deciding what to do. Begin by asking the WHAT'S UP? questions, including exactly where the pain is occurring, how it feels, if there is anything that aggravates or alleviates the pain, when it started, how bad it is on a scale of 0 to 10, whether there are associated symptoms, and if Mrs. Loos has some insight regarding the cause of her problem. Listen for bowel sounds for 5 minutes in each quadrant. Find out when her last bowel movement was. Because of her history of diabetes, you suspect that neuropathy may be causing a nonmechanical bowel obstruction. If assessment findings confirm the possibility, the physician should be contacted. Because of the nausea and the potential obstruction, withhold food and fluids until the physician can be contacted.

2. When documenting, answer the questions what, why, when, where, who, how (either explicitly or implicitly by professional knowledge, in narrative or flow sheet format) for completeness.
    What = Patient is experiencing large, firm, tender to touch abdomen with nausea (additional assessment data should be included).
    Why = Unknown, physician notified
    When = Current date and time
    Where = Abdomen
    Who = J. Morgan, LPN

UNIT SEVEN BIBLIOGRAPHY

Bliss, DZ, et al: Supplementation with dietary fiber improves fecal incontinence. Nursing Research 50:4, 2001.

Campinha-Bacote, J: African Americans. In Purnell, L, and Paulanka, B, (eds): Transcultural health care: A culturally competent approach. FA Davis, Philadelphia, 2001.

Emery, E, et al: Banana flakes control diarrhea in enterally fed patients. Nutrition in Clinical Practice 12(2):72, 1997.

Erwin-Toth, P: Caring for a stoma is more than skin deep, Nursing 2001 31:5, 2001.

FDA approves new treatment for Crohn's disease. FDA Talk Paper, Oct. 3, T01-45, 2001.

Heitkemper, M, and Jarrett, M: It's not all in your head: Irritable bowel syndrome. American Journal of Nursing 101:1, 2001.

Heitkemper, M, et al: IBS and abuse in women. Nursing Research 50:1, 2001.

Locke, GR, et al: Risk factors for irritable bowel syndrome: Role of analgesics and food sensitivities. American Journal of Gastroenterology 95:157, 2000.

Lutz, C, and Przytulski, K: Nutrition and diet therapy. FA Davis, Philadelphia, 2001.

Martin, FL: Ulcerative colitis. American Journal of Nursing 97:38, 1997.

Metheny, N, and Clouse, R: Bedside methods for detecting aspiration in tube-fed patients. Chest 111(3):724, 1997.

O'Brien, B, et al: G-tube site care: A practical guide. RN 62:2, 1999.

Pectasides, D, et al: CEA, CA 19-9 and CA-50 in monitoring gastric carcinoma. American Journal of Clinical Oncology 20(4):348, 1997.

Purnell, L, and Paulanka, B (eds): Transcultural health care: A culturally competent approach. FA Davis, Philadelphia, 2001.

Regina-Cammon, S, and Hackshaw, M: Are we starving our patients? American Journal of Nursing 100:5, 2000.

Schiff, L: Enhanced enteral feeding formulas, RN 63:9, 2000.

Schiff, L: Ostomy products. RN 63:11, 2000.

Schmieding, N, and Walding, R: Gastric decompression in adult patients. Survey of nursing practice. Clinical Nursing Research 6(2):142, 1997.

Sercombe, J: Inflammatory bowel disease and smoking. Journal of Professional Nursing 15(7):439, 2000.

Shuster, J: OTC laxative woes. Nursing 2000 30:6, 2000.

Walsh, BA, et al: Multidisciplinary management of altered body image in the patient with an ostomy. Journal of Wound, Ostomy, and Continence Nursing 22:5, 1995.

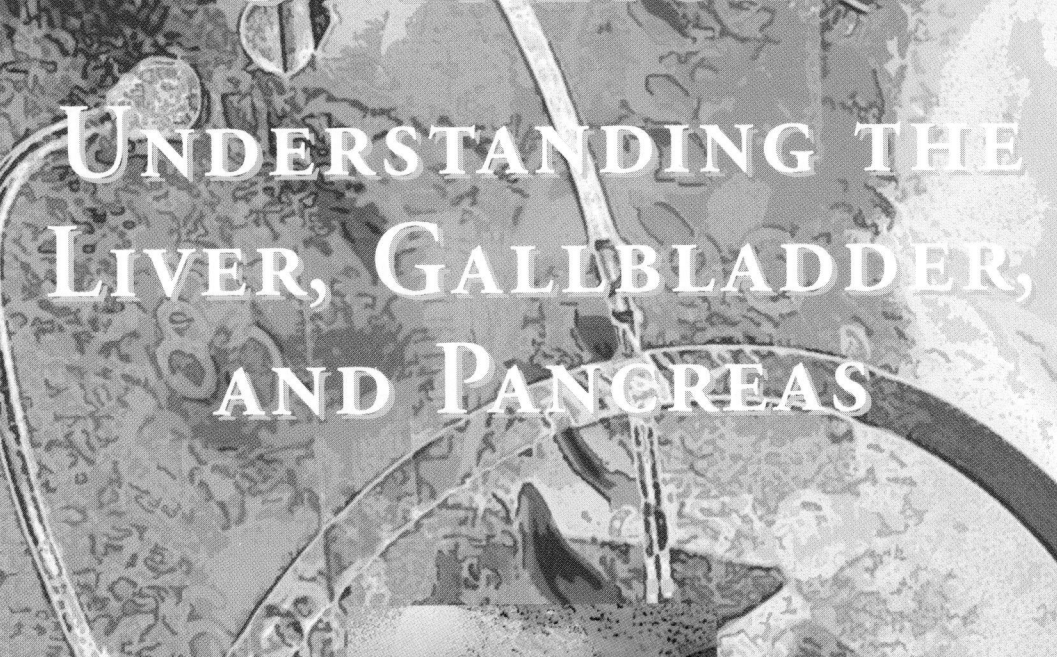

Unit Eight

Understanding the Liver, Gallbladder, and Pancreas

32

LIVER, GALLBLADDER, AND PANCREAS FUNCTIONS AND ASSESSMENT

Elaine Bishop Kennedy and Valerie Scanlon

KEY TERMS

caput medusae (**KAP**-ut mi-**DOO**-see)

esophagogastroduodenoscopy (e-SOFF-uh-go-GAS-troh-DOO-od-e-**NOS**-kuh-pee)

icterus (**ICK**-ter-us)

jaundice (**JAWN**-diss)

retrograde cholangiopancreatography (**RET**-roh-grayd koh-LAN-jee-oh-PAN-kree-ah-**TOG**-rah-fee)

spider angioma (**SPY**-der AN-jee-**OH**-mah)

striae (**STRIGH**-ee)

QUESTIONS TO GUIDE YOUR READING

1. What is the normal anatomy of the liver, gallbladder, and pancreas?
2. What are the normal functions of the liver, gallbladder, and pancreas?
3. What are the effects of aging on the liver, gallbladder, and pancreas?
4. Which data should you collect when caring for a patient with a disorder of the liver, gallbladder, or pancreas?
5. What are the techniques used in a physical examination of the abdomen conducted for a patient with possible liver, gallbladder, or pancreas disease?
6. What nursing care should you provide for patients undergoing common diagnostic tests for liver, gallbladder, or pancreatic disease?

▶ REVIEW OF NORMAL ANATOMY AND PHYSIOLOGY

The liver, gallbladder, and pancreas are called accessory organs of digestion because they produce or store digestive secretions but are not sites of the digestive process, which takes place in parts of the alimentary tube.

Liver

The liver fills the right and center of the upper abdominal cavity just below the diaphragm. The two main lobes of the liver are called the right and left lobes.

The blood supply of the liver differs from that of other organs. The liver receives oxygenated blood by way of the hepatic artery. By way of the portal vein, blood from the abdominal digestive organs and the spleen is brought to the liver before being returned to the heart. This special pathway is called hepatic portal circulation and permits the liver to regulate blood levels of nutrients or to remove potentially toxic substances such as alcohol from the blood before the blood circulates to the rest of the body.

The only digestive function of the liver is the production of bile by the hepatocytes (liver cells). Bile flows through small bile ducts, then through larger ones, and leaves the liver by way of the hepatic duct (Fig. 32–1), which joins the cystic duct of the gallbladder to form the common bile duct, which carries bile to the duodenum.

Bile is mostly water and has an excretory function in that it carries bilirubin and excess cholesterol to the intestines for elimination in feces. The digestive function of bile is accomplished by bile salts, which emulsify fats in the small intestine. Emulsification is a type of mechanical digestion in which large fat globules are broken into smaller globules but are not chemically changed. Production of bile is stimulated by the hormone secretin, which is produced by the duodenum when food enters the small intestine.

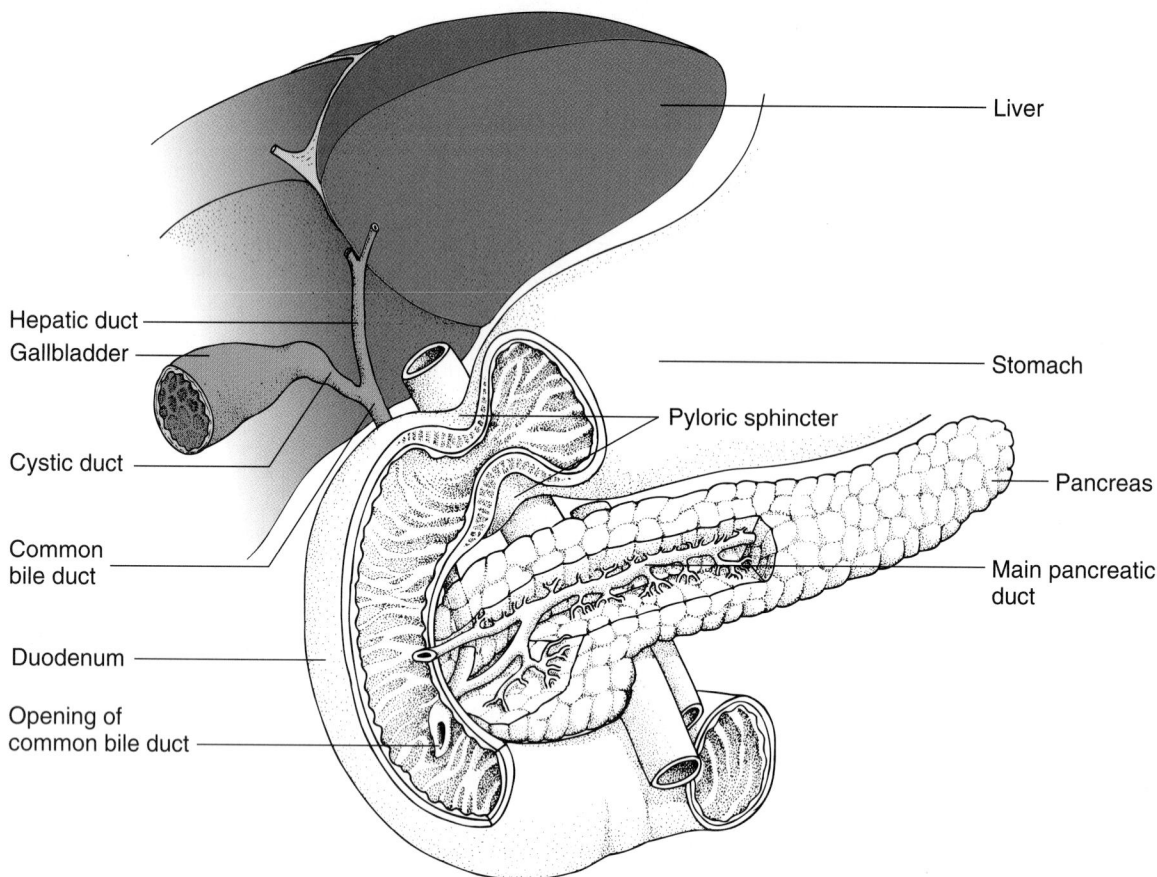

Figure 32-1   The liver, gallbladder, pancreas, and duodenum.

Functions of the Liver

The liver is involved in a great variety of metabolic functions, most of which involve the synthesis of specific enzymes. For the sake of simplicity, these functions may be grouped into categories.

CARBOHYDRATE METABOLISM. The liver regulates the blood glucose level by storing excess glucose as glycogen and changing glycogen back to glucose when the blood glucose level is low. The liver also changes other monosaccharides such as fructose and galactose to glucose, which is more readily used by cells for energy production.

AMINO ACID METABOLISM. The liver regulates the blood levels of amino acids based on tissue needs for protein synthesis. Of the 20 amino acids needed for the production of human proteins, the liver is able to synthesize 12, called the nonessential amino acids, by the process of transamination. The other eight amino acids, which the liver cannot synthesize, are called the essential amino acids. Essential amino acids are required in the diet.

Excess amino acids (those not needed for protein synthesis) undergo the process of deamination in the liver; the amino group is removed and the remaining carbon chain is converted to a simple carbohydrate that is used for energy production or converted to fat for energy storage. The amino groups are converted to urea, a nitrogenous waste product that is removed from the blood by the kidneys and excreted in urine.

LIPID METABOLISM. The liver forms lipoproteins for the transport of fats in the blood to other tissues. The liver also synthesizes cholesterol and excretes excess cholesterol into bile to be eliminated in feces.

The liver is also the main site of the process called beta oxidation, in which fatty acid molecules are split into two-carbon acetyl groups. These acetyl groups may be used by the liver to produce energy, or they may be combined to form ketones to be transported to other cells for energy production.

SYNTHESIS OF PLASMA PROTEINS. The liver synthesizes albumin, clotting factors, and globulins. Albumin, the most abundant plasma protein, helps maintain blood volume by pulling tissue fluid into capillaries. Clotting factors produced by the liver include prothrombin and fibrinogen, which circulate in the blood until needed for chemical clotting. The globulins synthesized by the liver become part of lipoproteins or act as carriers for other molecules in the blood.

PHAGOCYTOSIS BY KUPFFER CELLS. The fixed macrophages of the liver are called Kupffer cells; they phagocytize pathogens that circulate through the liver. Many of the bacteria that get to the liver come from the colon, after being absorbed as the colon absorbs water. These bacteria are normal flora of the colon but would be very harmful elsewhere in the body. Portal circulation brings this blood to the liver first, however, and the bacteria are phagocytized by Kupffer cells.

FORMATION OF BILIRUBIN. The fixed macrophages of the liver phagocytize old red blood cells (RBCs) and form bilirubin from the heme portion of their hemoglobin. The liver also removes from the blood the bilirubin formed in the spleen and red bone marrow and excretes it into bile to be eliminated in feces.

STORAGE. The liver stores the minerals iron and copper; the fat-soluble vitamins A, D, E, and K; and the water-soluble vitamin B_{12}.

DETOXIFICATION. The liver synthesizes enzymes that change harmful substances to less harmful ones. Alcohol and medications are examples of potentially toxic chemicals. The liver also converts ammonia from the colon bacteria to urea, a less toxic substance.

Gallbladder

The gallbladder is a muscular sac about 3 to 4 inches long located on the undersurface of the left lobe of the liver. Bile in the hepatic duct from the liver flows through the cystic duct (see Fig. 32–1) into the gallbladder, which stores bile until it is needed in the small intestine. The gallbladder also concentrates bile by absorbing water.

When fatty foods enter the duodenum, the duodenal mucosa secretes the hormone cholecystokinin. One function of cholecystokinin is to stimulate contraction of the smooth muscle of the wall of the gallbladder. Contraction of the gallbladder forces bile into the cystic duct, then into the common bile duct, which empties into the duodenum.

Pancreas

The pancreas is about 6 inches long, between the curve of the duodenum and the spleen in the upper right abdominal quadrant (see Fig. 32–1). The digestive secretions of the pancreas are produced by exocrine glands called acini. The small ducts of these glands unite to form larger ducts and finally the main pancreatic duct (there may be an accessory duct also), which emerges from the medial side, or head, of the pancreas and joins the common bile duct.

The pancreatic digestive enzymes are involved in the digestion of all three food types. The enzyme amylase digests starch to maltose; lipase converts emulsified fats to fatty acids and glycerol. Trypsinogen is an inactive enzyme that is changed to active trypsin in the duodenum. Trypsin digests polypeptides to shorter chains of amino acids.

The pancreas also produces a bicarbonate juice, which is alkaline because of its high sodium bicarbonate content. The function of bicarbonate juice is to neutralize the hydrochloric acid in gastric juice as it enters the duodenum from the stomach. The pH of duodenal chyme is raised to about 7.5, which prevents corrosive damage to the mucosa.

Secretion of pancreatic juice is stimulated by the hormones of the duodenal mucosa. Secretin stimulates the production of bicarbonate pancreatic juice, and cholecystokinin stimulates secretion of the pancreatic enzyme juice.

Aging and the Liver, Gallbladder, and Pancreas

The liver usually continues to function well into old age, unless damaged by pathogens such as the hepatitis viruses or by toxins such as alcohol. There is a greater tendency for gallstones to form, sometimes necessitating removal of the gallbladder. In the absence of specific pathological conditions, the pancreas usually functions well, although acute pancreatitis of unknown cause is somewhat more common in the elderly.

▶ NURSING ASSESSMENT

When assessing the patient with disease of the liver, pancreas, or gallbladder, use the same systematic approach as when assessing other organ systems. Make the patient as comfortable as possible before beginning the assessment process.

Nursing History

Obtain a patient history (Table 32–1). Elements of the patient history include a complete description of any abdominal pain or tenderness. Ask the patient which events cause or provoke pain; what actions, if any, relieve or palliate the pain; the quality of the pain, such as sharp, dull, boring, or burning; the location of the pain; the severity of the pain; and the timing, or when the pain occurs. Using the WHAT'S UP? model presented in Chapter 1 will help you remember all the elements that need to be included in a complete symptom assessment.

Ask the patient about any nausea, vomiting, or abdominal distention. Information about the timing or other common triggers of episodes of nausea or vomiting may help the practitioner identify their cause. Such information may also help determine appropriate treatment for any future nausea or vomiting. Abdominal distention in the presence of nausea and vomiting may indicate intestinal obstruction. Patients with liver, gallbladder, or pancreatic disease may also complain of feeling bloated, of having gas or belching frequently, or of right upper quadrant (RUQ) tenderness.

Question the patient about any observed changes in bowel elimination. Diarrhea may be caused by irritation of the bowel. Constipation may indicate decreased water intake or excessive water loss. Observe the patient's stool for evidence of bacteria (a foul smell), fat (stool floats on the water surface and appears greasy), pus, blood, or mucus. Patients with liver or gallbladder disease may have pale or clay-colored stools.

Patients with disease of the liver, pancreas, or gallbladder commonly have changes in appetite such as anorexia or alterations in eating preferences. Ask the patient about any abnormal weight loss or unexpected weight gain and changes in food tolerance, including the type or amount of

TABLE 32-1 QUESTIONS ASKED DURING ASSESSMENT OF THE LIVER, GALLBLADDER, AND PANCREAS

Question	Rationale
Do you have abdominal pain? (Use WHATS UP? to further assess.) Do any foods cause pain?	Pain can be associated with disease of the liver, gallbladder, or pancreas. Fatty foods can cause pain related to gallbladder disease.
Have you experienced nausea and vomiting?	These are generalized symptoms that may be associated with disease of the liver, gallbladder, or pancreas.
Does your abdomen feel distended or full?	Fluid in the abdomen, or ascites, occurs with liver disease.
What do your stools look like?	Clay-colored stools indicate liver or gallbladder disease. Black stools may indicate bleeding, which can be related to liver disease. Fatty stools occur with pancreatic disease.
Has your appetite or weight changed?	Anorexia and weight loss accompany many liver, gallbladder, and pancreatic disorders.
Does anyone in your family have liver, gallbladder, or pancreatic disease or alcoholism?	These disorders tend to run in families.
How much alcohol do you drink each day?	Excess alcohol intake is associated with liver disease and pancreatitis.
Do you bruise or bleed easily?	Bleeding is associated with liver disease, because clotting factors are made in the liver.
Do you use any over-the-counter drugs or herbal remedies?	Many drugs and herbs are toxic to the liver.
What is your occupation?	Some occupations are associated with exposure to substances that can be toxic to the liver.

offending foods. For example, patients with gallbladder disease may report that they feel nauseated or bloated after eating fried or greasy foods.

The patient's history should note whether there is a family history of liver, pancreas, or gallbladder diseases, such as diabetes mellitus, alcoholism, cancer, heart disease, or bleeding tendencies. These diseases have a high incidence within families.

Obtain a social history. Determine whether the patient ingests alcohol or uses other recreational drugs. If the patient does acknowledge using alcohol or other drugs, record the type, frequency, and amount used. You also need to ask the patient what medications are being taken with or without a physician's prescription. Many people do not consider over-the-counter preparations and herbal or natural products that they purchase without a prescription important enough to report and must be asked specifically to do so. Remember to ask specifically about the use of antacids, laxatives, vitamins, or pain relievers such as aspirin or nonsteroidal anti-inflammatory drugs (NSAIDs). See also Gerontological Issues Box 32–1.

BOX 32-1 GERONTOLOGICAL ISSUES

With increasing age, the liver decreases in volume, mass, and blood flow. These changes are significant because the liver acts to metabolize many drugs. If a patient has impaired liver function, there may be toxic levels of a drug present in the blood. It is important to assess liver function tests and perform a medication review to determine those that are metabolized by the liver. Older patients may require reduced doses of many drugs.

Ask about the patient's usual work activities and work setting. Document exposure to chemicals such as paint fumes, industrial dyes, acids, farm pesticides, or other liver-toxic substances.

Investigate the patient's activities other than work. Document reports of fatigue along with information about when the fatigue occurs. Ask the patient about stressors such as financial concerns, problems dealing with the health care environment, and any family or personal problems. Attempt to determine what coping mechanisms the patient usually employs to deal with stressors.

Physical Assessment

Be prepared to assist with a thorough physical assessment of the patient. Instead of following the usual inspect-palpate-percuss-auscultate (IPPA) format, assess the abdomen starting with inspection, then auscultation, percussion, and palpation. This is to prevent palpation from changing other assessment findings.

Inspection

Assessment begins with inspection of the patient's skin for scars, **striae** (light silver-colored or thin red lines on the abdomen), bruising, **caput medusae** (bluish-purple swollen vein pattern extending out from the navel), and **spider angiomas** (thin reddish-purple vein lines close to the skin surface). The patient's abdomen is observed for any visible masses, visible movement or peristalsis, or **jaundice** (also called **icterus,** a yellowing of the skin and the sclera of the eye).

JAUNDICE. Jaundice is a cardinal symptom of liver or gallbladder disease and red blood cell disorders. Old red blood cells are cleared from the circulatory system by phagocytes in the spleen, liver, lymph nodes, and bone marrow. In the process, the compound heme (part of hemoglobin) is split into iron and another substance that is metabolized to bilirubin. The liver is then responsible for converting bilirubin to a water-soluble compound that can be excreted in bile. If the liver is unable to convert or conjugate bilirubin to a water-soluble compound, or if bile drainage is obstructed, serum bilirubin is elevated and pigments are deposited in body tissues.

When serum bilirubin levels elevate, the patient's skin color changes to yellow. The yellow color varies from pale yellow to a striking golden orange. The color intensity is directly related to the amount of elevation of the serum bilirubin. Jaundice can be seen in nearly every body tissue and fluid where there is any amount of albumin. Pigment may occasionally be seen in cerebrospinal fluid or joint fluid. Pigment is not seen in saliva or tears. Urine becomes dark, and if bile flow to the bowel is obstructed, stools will be a light clay color (Cultural Consideration Box 32–2). Describe any abnormal finding completely in the patient's record, and report your findings promptly.

BOX 32-2	CULTURAL CONSIDERATION

The nurse should be aware of the variations of assessing for jaundice among people of color. To assess for jaundice in a dark-skinned patient, look at the sclera, conjunctiva, palms of hands, soles of feet, and in the buccal mucosal for patches of bilirubin pigment.

Auscultation

Inspection should be followed with auscultation of bowel sounds with a stethoscope. Auscultate all four quadrants for at least 1 minute in each quadrant. Normal bowel sounds are heard every 5 to 15 seconds. Bowel sounds are considered normal if they occur within a range of 5 to 30 per minute and sound like soft gurgles or clicking. If the patient has recently had surgery, bowel sounds may be absent or occur less frequently than five per minute. High-pitched, frequent sounds may indicate hypermotility of the bowel or early bowel obstruction. It may be possible to hear circulatory sounds if the stethoscope is placed over the abdominal aorta. Patients with chronic liver failure may have a humming sound over their liver. This finding usually indicates overloaded venous circulation in the liver.

Percussion

The experienced practitioner may percuss the abdomen to determine the presence of masses, fluid, or air in the abdominal cavity. The normal sounds heard during percussion of the abdomen are tympany (musical drumlike sound over hollow air-filled space) or dull (a soft thudding sound over solid organs such as the liver). If an abnormal amount of fluid has collected in the peritoneal cavity, a sloshing sound may be heard when the patient turns from side to side or shifts position.

Palpation

Light palpation of the abdomen concludes the physical assessment. Depress the abdomen not more than 1/2 to 1 inch during the palpation. Note any muscle tension, rigidity, masses, or expressions of pain. Deep palpation of the abdomen is done only by physicians and highly skilled nurses such as nurse practitioners.

▶ DIAGNOSTIC STUDIES
Laboratory Tests

Laboratory tests and normal values listed in Table 32–2 are the common tests performed to detect problems with the liver, gallbladder, and pancreas. Bilirubin is an excellent measure of liver and gallbladder functioning. In addition, certain enzymes such as alanine aminotransferase (ALT, formerly called serum glutamic pyruvic transaminase [SGPT]), aspartate aminotransferase (AST, formerly called serum glutamic oxaloacetic transaminase [SGOT]), and lactic dehydrogenase (LDH) are released by damaged liver cells. Elevations in these blood values in the absence of known trauma or heart muscle damage such as a heart attack are excellent indicators of liver damage.

Radiology

A flat-plate x-ray examination of the abdomen is usually the first diagnostic radiological procedure the physician orders. The flat-plate radiograph may show abdominal structures, abnormal masses, or blockage of normal bowel action. There is no patient preparation required.

Upper Gastrointestinal Series

An upper gastrointestinal (GI) series is usually ordered for patients complaining of nausea, vomiting, weight loss, or abdominal pain. These complaints are common for patients with liver, gallbladder, or pancreas disorders. The patient is allowed nothing by mouth (NPO) after midnight (or at least 8 hours) before the test. You should answer any patient concerns and explain that the patient will be asked to swallow a large amount of barium (a substance that is opaque on radiographs) during the procedure. The patient will be placed on an x-ray table, and radiographs will be taken in various positions and at various times during the procedure. The patient should understand that the procedure may take several hours depending on the rate at which the barium moves through the patient's gastrointestinal tract. A laxative such as Milk of Magnesia may be given after the procedure to aid in the evacuation of the barium.

Lower Gastrointestinal Series

A barium enema (BE, or lower GI series) may be ordered if the patient has reported any changes in bowel function, such as diarrhea or constipation, or blood or mucus in the stool. The patient is asked to eat a low-residue diet for 2 days before the scheduled procedure, then a clear liquid meal the night before. Fluids should be increased to prevent dehydration related to the preparation. The patient is made NPO after midnight before the test. Further, the patient is asked to use a powerful laxative such as bisacodyl (Dulcolax) or magnesium citrate (Citrate of Magnesia) to cleanse the bowel for more accurate x-ray views (Fig. 32–2). Enemas may also be ordered. Individual institutions' preparations may vary. Older patients may find the preparation routine especially taxing. Observe the patient often because defecation urgency, especially in unfamiliar surroundings, may create a fall risk.

TABLE 32-2 LABORATORY TESTS FOR LIVER, GALLBLADDER, AND PANCREAS DISORDERS

Test	Normal Range	Significance
		Blood
Alanine aminotransferase (ALT)	5–35 IU/dL	↑ in chronic liver failure and hepatitis
Albumin	3.1–4.3 g/dL	↓ in liver disease
Amylase	53–123 U/L	↑ in pancreatitis, gallstones
Ammonia	12–55 mol/L	↑ in chronic liver failure, hepatitis
Aspartate aminotransferase (AST)	8–20 units/L	↑ in chronic liver failure, viral hepatitis, acute pancreatitis
Bilirubin		
Total serum	0.1–1.0 mg/dL	↑ in liver and gallbladder disease with red blood cell destruction
Conjugated (direct)	0.0–0.4 mg/dL	↑ in gallbladder obstruction
Unconjugated (indirect)	0.1–1.0 mg/dL	↑ with red blood cell destruction or liver disease
Calcium	9–10.5 mg/dL	↓ with acute pancreatitis, liver disease, or malabsorption
Cholesterol	150–200 mg/dL	↑ in pancreatitis, gallbladder disease; ↓ may indicate severe liver disease
Lactic dehydrogenase (LDH)	110–250 IU/L	↑ in liver disease
Potassium	3.5–5.0 mEq/L	↓ with diarrhea, intestinal fistulas, vomiting, suctioning
Prothrombin time	11–12.5 s	↑ in liver disease, vitamin K deficiency
		Urine
Urine amylase	Depends on test	↑ in acute pancreatitis
Urine bilirubin	Negative	↑ in chronic liver failure, hepatitis, biliary obstruction
Urobilinogen	0.3–1.0 Ehrlich units in 2 h	↑ with destruction of red blood cells, hepatitis, chronic liver failure, obstructive jaundice
		Feces
Fecal fat	Negative	↑ in pancreatic disease
Occult blood	Negative	Positive in cancer, bleeding tendencies from ↓ vitamin K

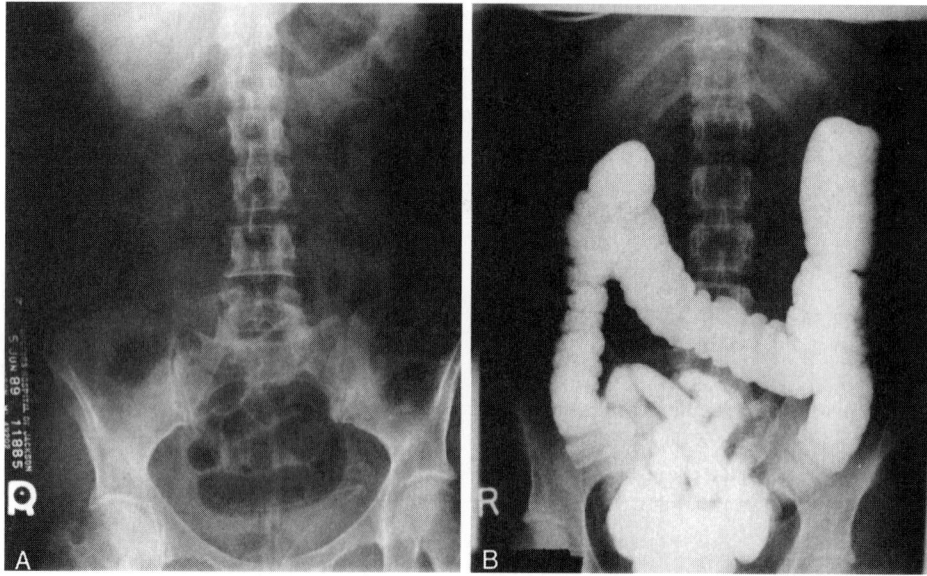

Figure 32-2 *(A)* An image of a patient who was poorly prepared for a barium enema. *(B)* An image of a patient who was adequately prepared for a barium enema. (Courtesy Dr. Russell Tobe.)

During the procedure, barium is given by enema and the patient is asked to retain the barium while radiographs are taken. The patient may experience some abdominal cramping and an urge to have a bowel movement during the procedure. The patient is told to take slow, deep breaths and to tighten the anal sphincter. The rate of flow of the barium is slowed until the cramping diminishes.

After the procedure, instruct the patient to drink plenty of fluids. Administer a mild laxative as ordered to eliminate the barium, because barium left in the bowel tends to harden and become difficult to evacuate. The stool should be observed for complete evacuation of the barium to prevent an impaction. Stools will change from clay colored or white to their normal brown color when the barium is gone.

Oral Cholecystogram

An oral cholecystogram (gallbladder series) may be ordered if the physician suspects gallstones. The patient is asked to ingest a radiopaque dye that collects in bile in the gallbladder. The dye then shows up when the patient has an x-ray examination, showing the gallbladder, ducts, and any gallstones.

In preparation, the patient is asked to eat a high-fat diet for 2 days before the test, then a low-fat diet the day before the examination. Depending on agency protocol, the patient takes radiopaque tablets (usually six), the evening before the examination. Some agencies ask the patient to take the pills for two nights before the examination. The patient takes them with water about 5 minutes apart. The patient then has nothing by mouth after midnight before the examination. The radiopaque tablets contain iodine, so be sure to ask patients about any allergies to iodine or shellfish (which contain iodine). Also advise the patient that the tablets may cause diarrhea. The procedure will take about half an hour in the x-ray department.

Computed Tomography

A computed tomography (CT) scan may be ordered for any suspected abnormalities of liver, gallbladder, or pancreas functioning. The computer-enhanced radiographs may be taken with or without contrast medium. No patient preparation is necessary. Patients are told that they will be on a narrow, hard x-ray table that is closely surrounded by a noisy metal shell. Patients with fear of closed spaces may have to make arrangements for a CT scan at an "open" scanner. There are no follow-up care requirements.

Angiography

Intravenous cholangiography may be ordered for patients with symptoms of biliary obstruction or after a cholecystectomy. This examination permits the radiologist to see the gallbladder and biliary ducts by intravenously injecting a contrast medium, which then collects in these structures. The injection of contrast medium is done about 1 hour before the examination. Radiographs are taken about every 20 minutes for 1 hour or until the structures are readily viewed. The radiopaque material is iodine based, so ask the patient about any allergies to iodine. There are no follow-up care requirements after the procedure.

Liver Scan

A liver scan involves injecting a slightly radioactive medium that is taken up by the liver. An instrument is passed over the liver that records the amount of material taken up by the liver and forms a composite "picture" of the liver. The physician may be able to determine tumors, masses, and abnormal size and patterns of blood vessel. The procedure takes a short time.

Endoscopy

The physician may order an endoscopic examination of the gastrointestinal tract for patient complaints such as bleeding or if cancer or obstruction is suspected. Endoscopic examination permits the physician to directly view the structures of portions of the gastrointestinal tract with an instrument that contains flexible optic fibers. In addition to viewing the structures, the physician can also remove polyps, take biopsy specimens, or coagulate bleeding sites that are identified.

Esophagogastroduodenoscopy

An **esophagogastroduodenoscopy** (EGD) is an endoscopic procedure that allows the physician to view the esophagus, stomach, and duodenum. The patient is asked not to take anything by mouth after about 8 P.M. the night before the examination. Because this is an invasive procedure, patients may be asked to sign an operative consent form, and a preoperative checklist may be necessary, depending on institution policy. Just before the procedure, the patient is asked to remove dentures. At the start of the procedure, the patient is given medication such as midazolam hydrochloride (Versed) or diazepam (Valium) for relaxation. Doses of the medications may be given at intervals during the procedure to keep the patient relaxed. The patient may be given atropine sulfate to dry secretions in the mouth. The physician usually sprays the back of the throat with a topical anesthetic to ease passage of the endoscope through the mouth and into the GI tract.

Follow-up care includes placing side rails up until the patient is fully alert. The patient is not given anything by mouth for up to 2 hours or until the gag reflex returns. The nurse checks the vital signs and gag reflex and observes the patient for any bleeding, complaints of pain, or signs of perforation such as abdominal pain, tenderness, or guarding.

Endoscopic Retrograde Cholangiopancreatography

Endoscopic **retrograde cholangiopancreatography** (ERCP) permits the physician to visualize the liver, gallbladder, and pancreas (Fig. 32–3). The procedure allows both direct viewing and use of contrast medium. An endoscope is passed through the esophagus to the duodenum, where dye is injected that outlines the pancreatic and bile ducts.

esophagogastroduodenoscopy: esophago—esophagus + gastro—stomach + duoden—duodenum + oscopy—to examine
retrograde cholangiopancreatography: retro—backward + grade—step + chol—bile + angio—via the vessels + pancreat—pancreas + ography—writing

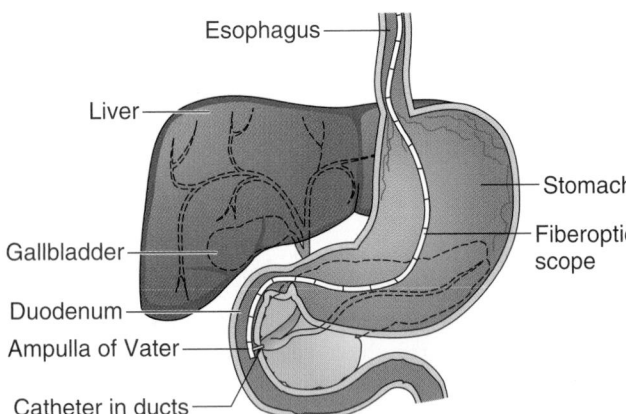

Figure 32-3   Endoscopic retrograde cholangiopancreatography. (Modified from Watson, J, and Jaffe, S: Nurse's Manual of Laboratory and Diagnostic Tests, ed 2. FA Davis, Philadelphia, 1995, p 525, with permission.)

The patient is prepared for an ERCP in the same manner as for an EGD, with nothing by mouth after 8 P.M. the night before the examination. In addition, the patient is asked about allergies to iodine. Ensure that any ordered laboratory studies, such as a prothrombin time, have been done before the procedure and that the patient has removed dentures. Follow-up care is similar to an EGD. In addition, the nurse is alert to patient complaints such as increased right upper quadrant pain, fever, or chills, which may indicate infection. Hypotension, tachycardia or rapid heart rate, increasing RUQ pain, nausea, or vomiting may indicate perforation or the onset of pancreatitis. Report such findings immediately.

Ultrasonography

The use of high-frequency sound waves through the abdomen allows the physician to view soft-tissue structures. The sound waves reflect varying images based on the density of the soft tissues in the abdomen. The patient is asked not to take anything by mouth after midnight on the day of the examination. A clear gel is applied to the abdomen and to the transducer on the sonograph. The gel improves the conduction of sound waves and thus improves the images obtained. The transducer is placed on the skin and moved over the abdomen while the technician views the sonograph screen and takes periodic pictures. The procedure takes about half an hour and requires no follow-up care.

Percutaneous Liver Biopsy

If less invasive tests do not aid in diagnosis of liver disease, a liver biopsy may be done. This may be done to identify cancer, cirrhosis, hepatitis, or other causes of liver disease. The physician generally inserts a needle through the skin and into the liver to withdraw a small sample for examination. This procedure places the patient at risk for bleeding, because the liver is highly vascular and because many patients with liver disease have faulty clotting ability.

Before the biopsy, ensure that the patient understands the procedure and that a consent has been signed if required by institution policy. You should also ensure that laboratory tests, such as a complete blood cell count and coagulation studies, have been completed and reviewed as ordered. The patient may be ordered nothing by mouth for 6 to 8 hours before the procedure. Baseline vital signs are taken, and a sedative is given if ordered.

During the procedure the nurse assists the physician to position the patient on his or her back or left side and assists the patient to hold very still while the needle is being introduced. The physician may also ask the patient to exhale and hold his or her breath during the needle insertion.

After the biopsy, the patient should remain on bedrest for 24 hours. The patient lies on the right side for the first 2 hours with a small pillow or rolled towel under the biopsy site to provide pressure and prevent bleeding. Vital signs and the site are monitored for signs of bleeding. The patient is advised to avoid coughing or straining. Analgesics are offered for comfort if ordered.

CRITICAL THINKING

Mr. Wozynski

Mr. Wozynski is admitted with chronic liver failure and jaundice. What specific laboratory value can you expect to be elevated related to his jaundice? Mr. Wozynski's physician orders a liver biopsy. Why is it important for you to check Mr. Wozynski's laboratory reports before the procedure?

Answers at end of chapter.

REVIEW QUESTIONS

1. Which of the following is a function of the liver?
   a. Synthesis of plasma proteins
   b. Elimination of carbohydrates
   c. Concentration of bile
   d. Secretion of cholecystokinin

2. Which food is most likely to stimulate the release of bile?
   a. Green beans
   b. French fried potatoes
   c. Coffee
   d. Poached egg

3. The enzymes of the pancreas are involved in the digestion of which foods?
   a. Starch and fat
   b. Starch, fat, and protein
   c. Fat and protein
   d. Starch and protein

4. Which of the following complications should the nurse monitor for after a liver biopsy?
   a. Nausea
   b. Muscle twitching
   c. Bleeding
   d. Hypoventilation

5. Which of the following nursing measures is most important after an upper or lower GI series?
   a. Offer a laxative as ordered.
   b. Place pressure on the puncture site.
   c. Check for return of a gag reflex.
   d. Keep the patient in semi-Fowler's position.

6. Mr. Sikmiller returned an hour ago from an EGD and is asking for a glass of water. Which of the following responses is best?
   a. "I'm sorry, Mr. Sikmiller, you will need to wait 4 hours before drinking anything."
   b. "I can't give you any water; would you like some ice chips?"
   c. "I will bring you some water and a laxative."
   d. "Let me check your gag reflex first to make sure it is safe to drink water."

Answers to CRITICAL THINKING

Mrs. Pearl

Mrs. Pearl is at risk for dehydration and electrolyte loss as a result of the laxative and enema preparation and NPO status. This risk is increased because of her age. Her fluid and electrolyte status should be monitored closely.

Mrs. Pearl will likely have a concern about "making it" to the bathroom during the preparation and should have a bedside commode placed within easy reach. Her call light should be answered promptly. If enemas are ordered "until clear," Mrs. Pearl will be at greater risk for fluid and electrolyte loss. If more than two or three enemas are required, the physician should be notified.

Elderly patients can become very fatigued during testing and test preparation. Mrs. Pearl should be allowed plenty of rest before and after the test. She may also have a concern about being able to hold the barium in her bowel during the test without having an "accident." She should be assured that the barium is held in with a balloon on the end of the enema catheter and that bathrooms are nearby.

Mr. Wozynski

You can expect to find that Mr. Wozynski's serum bilirubin is elevated because his liver is unable to convert or conjugate bilirubin into a water-soluble compound that can be eliminated in feces. Mr. Wozynski is at risk for bleeding because the liver is highly vascular and prone to bleed when a biopsy specimen is taken. In addition, he may not be manufacturing the necessary amount of prothrombin needed for blood clotting and is less likely to stop bleeding once the biopsy has been performed. It will be especially important to check his coagulation studies and report any elevations to the physician before the biopsy.

33

Nursing Care of Patients with Liver, Gallbladder, and Pancreatic Disorders

Elaine Bishop Kennedy

KEY TERMS

ascites (a-**SIGH**-teez)

asterixis (AS-ter-**ICK**-sis)

cholecystitis (KOH-lee-sis-**TIGH**-tis)

choledochoscopy
(koh-LED-oh-**KOS**-koh-pee)

choledocholithiasis (koh-LED-oh-koh-li-**THIGH**-ah-sis)

cholelithiasis (KOH-lee-li-**THIGH**-ah-sis)

cirrhosis (si-**ROH**-sis)

colic (**KAH**-lick)

encephalopathy
(en-SEFF-uh-**LAHP**-ah-thee)

extracorporeal shock-wave lithotripsy
(ECKS-trah-koar-**POR**-ee-uhl **SHAHK**-wayv LITH-oh-**TRIP**-see)

fetor hepaticus (**FEE**-tor he-**PAT**-i-kus)

hepatitis (HEP-uh-**TIGH**-tis)

hepatorenal syndrome
(hep-**PAT**-oh-REE-nuhl **SIN**-drohm)

laparoscopy (LAP-uh-roh-**SKOP**-ee)

pancreatectomy (partial, total) (PAN-kree-uh-**TECK**-tuh-mee)

portal hypertension
(**POR**-tuhl HIGH-per-**TEN**-shun)

T-tube (**TEE**-toob)

transjugular intrahepatic portosystemic
shunt (**TRANZ**-jug-u-lar in-tra-hep-**PAT**-ick por-to-sis-**TEM**-ick SHUNT)

varices (**VAR**-i-seez)

QUESTIONS TO GUIDE YOUR READING

1. How would you explain the causes, risk factors, and pathophysiology of the various types of viral hepatitis?

2. What is current medical treatment for patients with viral hepatitis?

3. What nursing care should you provide for the patient experiencing viral hepatitis?

4. What are the causes, risk factors, and pathophysiology of acute and chronic liver failure?

5. What is current medical treatment for acute and chronic liver failure?

6. What nursing care would you provide for patients with acute or chronic liver failure?

7. What are five common complications of chronic liver failure?

8. What is current medical treatment of patients with cancer of the liver?

9. What nursing care would you provide for patients with cancer of the liver?

10. What are the causes, risk factors, and pathophysiology of acute and chronic pancreatitis?

11. What is current medical treatment for acute and chronic pancreatitis?

12. What nursing care would you provide for patients with acute or chronic pancreatitis?

13. How would you explain the causes, prevention, medical and nursing treatment of cancer of the pancreas?

14. What are the causes, risk factors, and pathophysiology of cholecystitis and cholelithiasis?

15. What is current medical treatment for cholecystitis and cholelithiasis?

16. What nursing care would you provide for patients with cholecystitis and cholelithiasis?

17. What are the causes, prevention, complications, and medical treatment of cancer of the gallbladder?

▶ DISORDERS OF THE LIVER
Hepatitis

Hepatitis is an inflammation of the cells of the liver, usually caused by a virus. Less commonly, hepatitis may be caused by certain drugs or occasionally by bacteria. Symptoms of hepatitis range from nearly no symptoms to life-threatening symptoms from liver necrosis or death of liver tissue.

Pathophysiology

Hepatitis is usually caused by one of six viruses:

- Hepatitis A virus (HAV), sometimes called infectious hepatitis
- Hepatitis B virus (HBV), sometimes called serum hepatitis
- Hepatitis C virus (HCV), sometimes called non-A, non-B (NANB) hepatitis
- Hepatitis D virus (HDV)
- Hepatitis E virus (HEV)
- Hepatitis G virus (HGV)

The viral agents vary by mode of infection, incubation period, symptoms, diagnostic tests, and preventive vaccines (Table 33–1). The infecting organism causes inflammation of the liver, with resulting damage to liver cells and loss of liver function. If damage involves the bile canaliculi, obstructive jaundice will occur. If complications do not occur, cells regenerate and normal liver function eventually resumes.

There are approximately 60,000 new cases of hepatitis in the United States each year. The incidence of HAV increased nearly 27 percent in the mid-1980s and is the most common cause of hepatitis. HAV has a low mortality rate. However, HBV is more common among some groups, including health care workers and intravenous drug users; it has a mortality rate of about 5 percent. Hepatitis G virus is a recently discovered virus that is transmitted by blood transfusions and unprotected sex. It is not known to cause hepatitis.

Prevention

The hepatitis viruses are very resistant to a wide range of anti-infective measures, such as drying, heat, ultraviolet light exposure, freezing, and bleach and other disinfectants. At least 30 minutes in boiling water is required to guarantee their destruction. The best methods for preventing the transmission of the hepatitis viruses are careful attention to cleanliness and the use of vaccines such as immune serum globulin (ISG) or vaccines to HBV and HAV. Health care workers must use standard precautions at all times. Infection control precautions should reflect the usual mode of transmission of the particular hepatitis virus.

Immune serum globulin is a temporary, passive, nonspecific immunity to hepatitis. Permanent, active immunity is acquired from the body's own antibodies in response to actual viral infection. The active immunity is to the specific virus to which the body has developed antibodies. Vaccines to HBV are available and provide permanent, active immunity to HBV. Health care workers are strongly encouraged to be vaccinated for HBV. A vaccine for HAV has also been developed.

Public health measures such as health education programs, licensing and supervision of public facilities, screening of blood donors, and careful screening of food handlers are general measures to prevent the transmission of hepatitis viruses.

Signs and Symptoms

Hepatitis usually shows a typical pattern of loss of liver function. There are generally three stages in hepatitis:

1. The prodromal, or preicteric (prejaundice), stage lasts about 1 week. The patient complains of flulike symptoms of malaise, headache, anorexia, low-grade fever, and possibly dull right upper quadrant (RUQ) pain.
2. The icteric stage lasts 4 to 6 weeks. The patient complains of more severe fatigue, anorexia, nausea, vomiting, and malaise. The patient is also likely to have jaundice or noticeable yellowing of the skin, sclera of the eyes, and other mucous membranes. The liver is usually enlarged and tender on examination.
3. The posticteric, or convalescent, stage lasts from 2 to 4 weeks to months. The patient usually feels well during this time, but full recovery as measured by the return to normal of all liver function tests may take as long as 1 year.

Hepatitis is considered a reversible process if the patient complies with a medical regimen of adequate rest, good nutrition, and abstinence from alcohol or other liver-toxic agents. See Chapter 32 for a more complete discussion of jaundice.

Complications

Hepatitis may lead to fulminant or acute liver failure (see the following). About 5 percent of hepatitis patients progress to chronic liver failure. Some patients become asymptomatic carriers of the virus; HBV-infected carrier patients have a greater risk of developing cancer of the liver.

Diagnostic Tests

Serum liver enzymes are elevated. Serum bilirubin and urobilinogen may be elevated. The erythrocyte sedimentation rate is usually elevated from the inflammatory process. In patients with severe hepatitis, prothrombin time may be elevated (Table 33–2). Serological tests may be ordered to determine the specific virus causing the hepatitis. Each virus has specific antigen markers that serological study can reveal. The antigen markers can be further used to determine the degree of healing from the hepatitis. Abdominal x-ray examination may show an enlarged liver.

Medical Treatment

Medical treatment is aimed at providing rest and adequate nutrition for healing. There are no specific drugs or other

---

hepatitis: hepat—liver + itis—inflammation

TABLE 33-1   COMPARISON OF TYPES OF VIRAL HEPATITIS

	Hepatitis A Virus	Hepatitis B Virus	Hepatitis C Virus	Hepatitis D Virus	Hepatitis E Virus
Mode of transmission	Oral-fecal contamination of water, shellfish, eating utensils, or equipment	Blood or body fluids such as saliva, semen, and breast milk; equipment contaminated by blood	Blood transfusions, IV drug use, unprotected sex	Blood or body fluids as with HBV; strongly linked as a coinfection with HBV	Usually, contaminated water
Incubation period	3–7 weeks	2–5 months	1 week to months	Same as HBV	2–9 weeks
Symptoms	Early (prodromal): fatigue, anorexia, malaise, nausea, or vomiting. Icteric: jaundice, pale stools, amber or dark urine, RUQ pain	May have no early symptoms. Early (prodromal): 1–2 months of fatigue, malaise, anorexia, low-grade fever, nausea, headache, abdominal pain, muscle aches. Icteric: jaundice, rashes	Same as HBV, usually less severe	Similar to HAV and to HBV but more severe	Similar to HAV
Diagnostic tests	Elevated serum liver enzymes (ALT, AST), elevated serum bilirubin, HAV antigen	Elevated serum liver enzymes (ALT, AST), elevated serum bilirubin, HBV antigen	Elevated serum liver enzymes (ALT, AST), elevated serum bilirubin, HCV antigen	Elevated serum liver enzymes (ALT, AST), elevated serum bilirubin, HDV antigen	Elevated serum liver enzymes (ALT, AST), elevated serum bilirubin
Preventive vaccine	Immune globulin	Immune globulin or HBIG	None	HBIG	None
Groups at risk	Individuals in military or day care	IV drug abusers, homosexuals, healthcare workers, transplant and hemodialysis patients	Same as HBV	Same as HBV	Travelers to endemic areas

ALT = alanine aminotransferase; AST = aspartate aminotransferase; HBIG = hepatitis B immune globulin; IV = intravenous; RUQ = right upper quadrant.

TABLE 33-2 LABORATORY TESTS FOR HEPATITIS

Test	Normal Range	Significance
Alanine aminotransferase	5–35 IU/mL	Found in high concentration in liver cells; released with death of liver cells
Aspartate aminotransferase	8–20 U/L	Found in high concentrations in liver cells; released with death of liver cells
Erythrocyte sedimentation rate	Adult: Women, 1–20 mm/h Men, 1–13 mm/h	Increased with inflammation and tissue damage
Prothrombin time	8.8–11.6 s	Liver can no longer make prothrombin
Serological tests	Negative titer	Indicates exposure and probable infection with virus
Anti-HAV	Negative titer	Indicates exposure and probable infection with virus
Anti HBV	Negative titer	Indicates exposure and probable infection with virus
Anti-HCV		

medical therapies for hepatitis. Interferon therapy and antiviral medication may be used for hepatitis B or C, but they may not be effective. With proper rest and nutrition, the liver should recover.

Patients are restricted from any alcohol or drugs that are known to be toxic to the liver (Box 33–1). In addition, patients are generally placed on limited activity with bathroom privileges. Because the patient usually experiences malaise, fatigue, and anorexia, rest is advised. As the patient improves, activity may be increased if the patient does not become fatigued.

Nursing Process

ASSESSMENT. Assess the patient for subjective complaints such as malaise, fatigue, pruritus (itching), nausea, anorexia, and RUQ abdominal pain. Objective data, such as vomiting, pale stools, amber or dark-colored (tea-colored) urine, and jaundice, are recorded. The patient's vital signs are taken, and a low-grade fever or any abnormal bruising or bleeding is reported immediately.

BOX 33-1 **Common Hepatotoxic Substances**

Ethyl alcohol
Acetaminophen (Tylenol)
Acetylsalicylic acid (aspirin)
Anesthetic agents
　Halothane (Fluothane)
Diazepam (Valium)
Erythromycin estolate (Ilosone)
Isoniazid (INH)
Methyldopa (Aldomet)
Oral contraceptives
Phenobarbital (Luminal)
Phenytoin (Dilantin)
Tranquilizers
　Chlorpromazine (Thorazine)
Industrial chemicals
　Carbon tetrachloride
　Trichloroethylene
　Toluene

NURSING DIAGNOSIS. Common nursing diagnoses for the patient with viral hepatitis are as follows:

- Imbalanced nutrition, less than body requirements related to anorexia, nausea, or vomiting
- Risk for impaired skin integrity related to itching secondary to bilirubin pigment deposits in skin
- Pain related to inflammation and enlargement of the liver
- Risk for ineffective management of therapeutic regimen related to lack of knowledge of hepatitis and its treatment.

PLANNING. Major goals for the patient with viral hepatitis include that the patient will have adequate nutrition, describe pain and other discomforts as tolerable, and be able to self-manage the treatment regimen for viral hepatitis.

IMPLEMENTATION

Imbalanced Nutrition. The usual ordered diet is a high-calorie, high-protein, high-carbohydrate, low-fat diet. There are no restrictions on what the patient may eat, but the focus is to provide a well-rounded, nutritious diet. Give the patient antiemetic drugs as ordered. Provide larger meals earlier in the day because nausea tends to increase during the day. Place the patient in an upright or sitting position for meals to decrease abdominal discomfort. Serve meals in a quiet, pleasing environment without unpleasant noise or odors that may decrease appetite.

Risk for Impaired Skin Integrity. The physician may order an antihistamine to decrease the itching related to bilirubin pigment deposits in skin. Encourage the patient not to scratch, but to press firmly on the itching area. Encourage the patient to keep fingernails trimmed short so that vigorous scratching does not tear the skin. Maintain room temperature at a comfortable level to decrease perspiration, which may increase itching.

Pain. The physician may order analgesics to decrease the pain related to inflammation and enlargement of the liver. The analgesics are administered as ordered. The patient is evaluated frequently for abdominal pain and the extent of any RUQ tenderness.

Risk for Ineffective Management of Therapeutic Regimen. Teach the patient the necessity of proper home cleanliness, including handwashing after toileting and using soap and hot water to clean eating utensils, cookware, and food preparation surfaces. (See Home Health Hints.) Teach patients how hepatitis affects their body and the importance of adequate rest and proper nutrition. Teach the importance of avoiding alcohol and other liver-toxic drugs.

HOME HEALTH HINTS

- For patients with hepatitis, home health nurses are concerned with proper treatment to prevent transmission in the community and to prevent permanent liver damage to the patient.

- If possible, the patient should have a separate bedroom and bathroom. The person cleaning the bathroom should wear disposable gloves or rubber gloves and then clean the gloves with a 10-percent bleach solution. The family is advised to use liquid soap instead of bar soap.

- Contaminated linens should be washed separately from household laundry, in hot water. One cup of bleach should be added with the detergent to each load. Rubber gloves should be worn to wash the patient's laundry.

- Patients with abdominal ascites need a hospital bed at home so the patient can be positioned to aid in breathing. A physician's order must be obtained.

- The measurement of abdominal girth should be taken at each visit and recorded in the nurse's notes. The patient should weigh on the same scale, first thing in the morning, and record the weight so the nurse can document the findings.

EVALUATION. Management of the patient with hepatitis has been successful if the patient demonstrates the following:

- Maintenance of body weight to within 2 pounds of preillness weight
- No breaks, cuts, or tears on skin
- Abdominal pain or other discomfort reported as not greater than 2 on a 5-point scale
- The effects of hepatitis and the necessity of adequate rest, proper nutrition, and necessary sanitation measures can be stated

For more information, visit Hepatitis Foundation International at www.hepfi.org and the American Liver Foundation at www.liverfoundation.org.

Fulminant Liver Failure

Fulminant (acute) liver failure is an uncommon but gravely serious complication of liver disease and has a mortality rate as high as 50 percent.

Pathophysiology

Acute liver failure results from the sudden massive loss of liver tissue, or necrosis. The cause of liver damage is usually drug toxicity or HBV in the presence of HDV. The outcome

CRITICAL THINKING

Carl

Carl, a 23-year-old man, is admitted to a medical-surgical nursing unit. During the admission process, Carl talks about his trip to an African country as a missionary. He returned to this country 4 weeks ago. During his trip he sustained a serious laceration that required many sutures. Carl also mentions his fondness for seafood and that he has had several "feasts" that have included raw oysters. Carl states that since his return he has lost nearly 8 pounds, is nauseated, has frequent headaches, and tires easily. Carl also tells you that he is very irritable, which is different from his usual easygoing manner.

1. What information might lead you to suspect hepatitis A? Hepatitis B?
2. What precautions should be instituted for Carl until a diagnosis is made?
3. What nursing actions might you implement to help Carl improve his nutrition?
4. What medications should Carl avoid?
5. What information should be included in a discharge teaching plan for Carl?

Answers at end of chapter.

of the disease may be decided within 48 to 72 hours of diagnosis. Possible outcomes are reversal, need for transplantation, or death.

Etiology

Patients are usually admitted to the hospital with a diagnosis of viral hepatitis. Some patients are admitted through the emergency department with a diagnosis of drug toxicity.

Prevention

Acute liver failure may be avoided by eliminating exposure to hepatitis B or hepatotoxic, liver-damaging substances. Hepatitis B can be transmitted through body fluids, such as blood or semen from unprotected sex; intravenous drug use; and dialysis. See Box 33–1 for a list of hepatotoxic substances.

Signs and Symptoms

Signs and symptoms of the disorder include hepatic **encephalopathy,** or central nervous system dysfunction (discussed in detail in the section on chronic liver failure). The patient may suddenly lapse into extremely serious illness, starting with confusion and progressing to coma. In a matter of hours the liver shows a rapid reduction in size, a typical sign of onset of acute liver failure. In addition, there is a sudden elevation of liver enzymes, bilirubin, and prothrombin time. Marked elevation in the prothrombin time is an ominous sign.

Diagnostic Tests

Early diagnosis of acute liver failure is essential so that the process of organ procurement may begin. An otherwise

encephalopathy: encephalo—brain + pathy—disease

healthy patient may be a priority organ recipient depending on age and whether or not the patient is alcohol dependent.

Laboratory tests include the serum liver enzymes alanine aminotransferase (ALT) and aspartate aminotransferase (AST). Levels of ALT and AST may rise from 1000 mU/mL to as high as 4000 mU/mL. The serum bilirubin level is more than 2.5 mg/dL. Urobilinogen levels may be elevated. Serum potassium levels drop below 3.5 mEq/L. Blood glucose drops below 70 mg/dL. The prothrombin time is elevated above 25 seconds.

An abdominal x-ray examination may document the change in size of the liver.

Medical Treatment

Medical treatment is directed toward stopping and reversing the damage to the liver. An attempt is made to put the liver completely at rest. The patient is put on complete bedrest. All drugs are eliminated. Dialysis may be ordered if the liver damage is a result of an overdose of a hepatotoxic substance. The patient is ordered a high-calorie, low-sodium, low-protein diet. Lactulose, neomycin, magnesium citrate, or sorbitol may be given to decrease ammonia levels, but they are not always effective. (For information on how these drugs work, see the section on medical treatment for chronic liver failure.) The patient needs intensive amounts of supportive care. It may be possible to support the patient long enough to stabilize him or her for transplantation.

Complications

The patient with acute liver failure experiences metabolic alkalosis, hypokalemia, hypoglycemia, disruption of blood clotting, and possibly sepsis. Metabolic alkalosis is related to disruption of the urea production cycle and the resulting accumulation of bicarbonate. Patients with acute liver failure also experience electrolyte imbalances. The patient's kidneys excrete potassium rather than hydrogen ions in an attempt to correct the alkalosis. The patient usually has hypoglycemia from loss of glycogen stores in the damaged liver, impaired gluconeogenesis (the manufacture of glucose from other nutrients), and an elevated insulin level caused by the stress response. Further, blood clotting disorders that can lead to disseminated intravascular coagulation (DIC) develop when prothrombin time elevates.

Finally, the patient is at risk for sepsis because of poor white blood cell migration and other responses to infection. Sepsis, abscess formation, endocarditis, and meningitis account for nearly a quarter of the deaths from acute liver failure. Renal failure accounts for about 30 percent of deaths. Respiratory problems and hypotension also contribute to the deaths from acute liver failure. The patient progresses through encephalopathy to coma and death.

Nursing Process

Nursing care of the patient with acute liver failure is essentially the same as for the patient with terminal chronic liver failure, discussed next.

Chronic Liver Failure

Chronic liver failure is also called Laënnec's **cirrhosis,** or portal, nutritional, or alcoholic liver disease. Chronic liver failure is the tenth leading cause of death among the total population and is more common among men than women (Cultural Consideration Box 33–2).

BOX 33-2 **CULTURAL CONSIDERATION**

The incidence of liver disease is more common among Mexican-Americans. Risk factors include working in occupations such as mining, factories using chemicals, and farming using pesticides. Additionally, the use of alcohol is increased with the machismo of society, and cigarette smoking is common.

Egyptian-Americans may suffer from schistosomiasis, known as bilharziasis in Egypt. Schistosomiasis can lead to cirrhosis, liver failure, portal hypertension, esophageal varices, bladder cancer, and renal failure. Thus the nurse may need to screen newer Egyptian-American immigrants for this disease.

Alcohol use and abuse is common among African-American communities. For many it is a socially accepted behavior and carries little or no stigma. The mortality rate from cirrhosis of the liver among African-Americans is nearly twice that of European-Americans. The nurse needs to provide counseling, teaching the detrimental effects of alcohol, and work with African-American churches and community leaders to help prevent the detrimental effects of alcohol abuse.

Etiology

Chronic liver failure may be caused by chronic excessive alcohol ingestion, especially when excess alcohol is combined with a lack of dietary protein. Additional types of liver failure include postnecrotic, biliary, and cardiac.

■ Postnecrotic liver failure may result from massive exposure to hepatotoxins or viral hepatitis.
■ Biliary liver failure is caused by chronic inflammation and obstruction of the gallbladder and bile ducts.
■ Cardiac liver failure is caused by chronic severe congestion of the liver from heart failure. The liver congestion causes death of liver cells from lack of nutrients and oxygen.

Pathophysiology

Chronic liver failure is a progressive disease. Healthy liver cells respond to toxins such as alcohol by becoming inflamed. The liver cells are infiltrated with fat and white blood cells and are then replaced by fibrotic tissue. As the disease progresses, more and more liver cells are replaced by fatty and scar tissue. The lobes of the liver are disrupted and the liver becomes hardened and lumpy. Early in the disease, the liver is enlarged, firm, and hard from the inflammatory process. Later, the liver shrinks and is covered with gray connective tissue.

---

cirrohsis: cirrh—orange yellow + osis—condition

Prevention

Chronic liver failure may be prevented by abstinence from alcohol, eating a balanced diet with adequate amounts of protein, and avoiding exposure to infections or hepatotoxic chemicals. Patients are advised that total abstinence from alcohol should be a lifelong goal.

Signs and Symptoms

Signs and symptoms of impaired liver function include malaise, anorexia, indigestion, nausea, weight loss, diarrhea or constipation, and dull, aching RUQ pain. The liver may be enlarged, firm, and tender. Bruising of the skin, bleeding gums, anemia, and jaundice, also known as icterus, may be present. Jaundice is a common finding with hepatitis. The patient's skin may be dry or contain abnormal pigmentation. The patient may complain of severe pruritus (itching). Laboratory values reflect progressive loss of liver function. As chronic liver failure progresses, signs and symptoms of increasing loss of liver function and complications related to the increasing loss of function develop.

Complications

Complications of chronic liver failure include **hepatorenal syndrome,** blood clotting defects, **ascites, portal hypertension,** and hepatic encephalopathy. Hepatorenal syndrome occurs in about one-third of liver failure patients. Symptoms of hepatorenal syndrome include oliguria without detectable kidney damage, reduced glomerular filtration rate (GFR) with essentially no urine output or less than 200 mL per day, and nearly total sodium retention. Hepatorenal syndrome is considered an ominous sign.

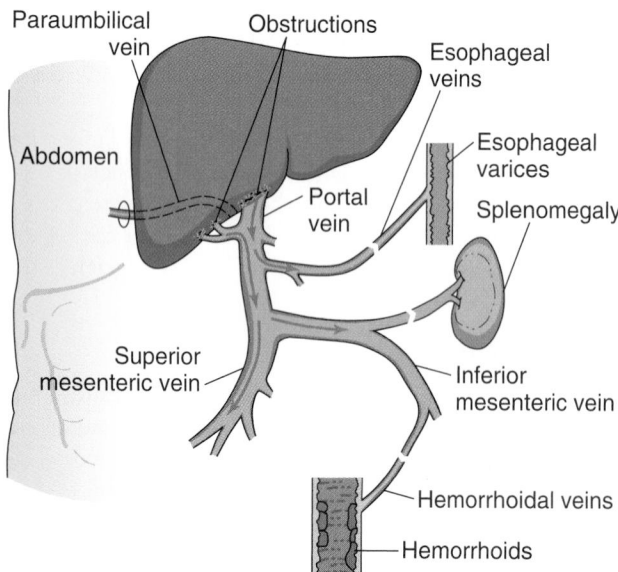

Figure 33-1 Portal hypertension. Obstruction of normal blood flow through the liver causes blood to back up into the venous system, leading to esophageal varices, splenomegaly, hemorrhoids, and caput medusae.

---

hepatorenal syndrome: hepato—liver + renal—kidneys + syndrome—group of symptoms

Blood clotting defects may develop because of impaired prothrombin and fibrinogen production in the liver. Further, the absence of bile salts prevents the absorption of fat-soluble vitamin K, which is essential for some blood clotting factors. Patients with chronic liver failure have a tendency to bruise easily and may progress to DIC or hemorrhage.

Ascites is an accumulation of serous fluid in the abdominal cavity. The fluid accumulates primarily because of low production of albumin by the failing liver. An insufficient amount of protein in the capillaries causes plasma to seep into the abdominal cavity. The accumulated fluid causes a markedly enlarged abdomen. The fluid may cause severe respiratory distress as a result of elevation of the diaphragm.

Portal hypertension is a persistent blood pressure elevation in the portal circulation of the abdomen. Liver damage causes a blockage of blood flow in the portal vein. Increased resistance from delayed drainage causes enlargement of the visible abdominal veins around the umbilicus (caput medusae), rectal hemorrhoids, enlarged spleen, and esophageal **varices** (Fig. 33–1). The most serious result of portal hypertension is bleeding esophageal varices (dilated veins). The walls of the esophageal veins are thin and tear easily. Varices usually develop from the fundus of the stomach upward and may extend into the upper esophagus. The blood-filled, thin-walled varices may tear easily from sudden excessive pressure, such as the intra-abdominal pressure that results from coughing, lifting, or straining, causing severe bleeding.

Hepatic encephalopathy is caused by the accumulation of noxious substances in the circulation. The failing liver is unable to make the toxic substances water soluble for excretion in the urine. Ammonia, a byproduct of protein metabolism, is most commonly the substance causing symptoms. Signs and symptoms of hepatic encephalopathy include progressive confusion; **asterixis,** or flapping tremors in hands caused by toxins at peripheral nerves; and **fetor hepaticus,** or foul breath caused by metabolic end products related to sulfur. Stages of hepatic encephalopathy are *early, stuporous and confused,* and *comatose.* Signs and symptoms of the stages are as follows:

- Early: The patient exhibits subtle changes in personality, fatigue, drowsiness, and changes in handwriting (the best assessment for the early stage).
- Stuporous and confused: The patient is often belligerent and irritable and develops asterixis, muscle twitching, and marked confusion.
- Comatose: The patient gradually loses consciousness and becomes comatose.

If toxic levels can be decreased and managed, the patient gradually regains consciousness. Hepatic encephalopathy represents end-stage liver failure and has a mortality rate as high as 90 percent once coma begins.

Diagnostic Tests

Liver serum enzymes, serum bilirubin, urobilinogen, serum ammonia, and prothrombin times are elevated (Table 33–3). Abdominal radiographs of patients with chronic liver failure may be expected to show ascites and enlargement of the liver. An upper gastrointestinal (UGI) series may reveal esophageal varices or evidence of gastric inflammation or ulcers. If the patient is bleeding, other arterial radiological examinations may be done to locate the specific source of bleeding.

Liver scans may be done to show abnormal liver masses or thickening. The physician may order an esophagogastroduodenoscopy (EGD) to detect any bleeding and to directly observe the esophagus, stomach, and duodenum. Small surface vessels that are bleeding can be treated by injection sclerotherapy during the EGD. The procedure uses sclerosing agents, which are chemical substances that cause the veins to inflame and scar shut.

The physician may do a liver biopsy to determine the extent and nature of the liver damage. Patients with chronic liver failure undergoing a liver biopsy need careful observation for bleeding after the procedure. (See Chapter 32.)

Medical Treatment

Medical treatment for chronic hepatic failure seeks to remove or to treat the underlying causes of the disease. In addition, medical treatment seeks to support liver regeneration and to treat the complications of liver failure.

Ascites is treated with diuretics, albumin infusions, and sodium restriction. Paracentesis is sometimes considered as an emergency measure to remove accumulated abdominal fluid. Paracentesis is not commonly done because it removes serous fluid, which contains a large amount of albumin that the liver cannot easily replace. Ascites may be treated by the nonsurgical placement of a shunt between the portal and systemic venous systems, called a **transjugular intrahepatic portosystemic shunt** (TIPS).

The purpose of the shunt is to sidetrack venous blood around the liver to the vena cava. Shunts are used for patients with severe respiratory compromise and are not as successful as originally hoped. Surgical shunts, sometimes called portacaval shunts, may be used to relieve portal hypertension (Fig. 33–2). When shunts are used, less venous blood circulates through the liver and fewer protein end

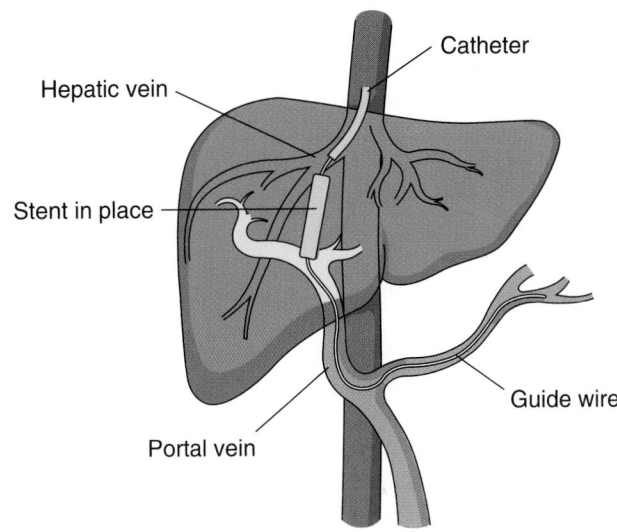

Figure 33-2 Transjugular intrahepatic portosystemic shunt. A stent is placed to shunt blood from the portal vein to the systemic circulation to divert blood flow around the diseased liver.

LEARNING TIPS

For complications of chronic liver failure, remember the pneumonic CHEAP:
C—clotting disorders
H—hepatorenal syndrome
E—encephalopathy
A—ascites
P—portal hypertension

TABLE 33-3	LABORATORY TESTS FOR CHRONIC LIVER FAILURE	
Test	**Normal Range**	**Significance**
		Blood
Alanine aminotransferase	5–35 IU/mL	Found in high concentrations in liver cells; released with death of liver cells
Albumin	3.5–5.5 g/dL	Decreased because of impaired protein synthesis; edema and ascites may result
Ammonia	15–19 µg/dL	Increased because liver cannot metabolize protein end product; contributes to hepatic encephalopathy
Bilirubin	0.3–3.0 mg/dL	Increased from increased breakdown of red blood cells
Aspartate aminotransferase	8–20 U/L	Found in high concentrations in liver cells; released with death of liver cells
Prothrombin time	Adult: Women, 1–20 mm/h Men, 1–13 mm/h	Liver can no longer make prothrombin; patient bleeds easily
		Urine
Urobilinogen	<4.0 mg/24 h	Increased urine because of filtration of excessive bilinogen in blood

asterixis: a—not + sterixis—fixed postition
fetor hepaticus: fetor—offensive odor + hepat—liver + icas—related to

transjugular intrahepatic portosystemic shunt: trans— across + jugular—jugular vein + intra—within + hepatic— liver + porto—portal (liver circulation) + systemic—systemic (circulation) + shunt—to divert

products are metabolized. For this reason, patients are put on a low-protein diet.

The medical goals for managing bleeding from esophageal varices are as follows:

■ Stop the bleeding.
■ Treat the fluid volume deficit caused by the bleeding.
■ Prevent further fluid loss.
■ Maintain fluid and electrolyte balance.

Bleeding varices are treated with vasoconstrictors such as vasopressin, with tamponade (direct pressure on the bleeding veins), or with emergency sclerotherapy to close the veins.

Tamponade is usually a temporary measure and is done with a multilumen esophagogastric tube such as the Sengstaken-Blakemore tube (Fig. 33–3). Other multilumen tubes such as the Minnesota tube may be used. The Sengstaken-Blakemore tube is inserted through the nose or mouth to the stomach. First the gastric balloon is inflated with 200 mL of air to secure the tube in its proper location. The esophageal tube is then inflated with 50 mL of air to produce tamponade. One to two pounds of traction may be applied to the tube. Occasionally the patient may wear a football helmet so that the tube can be secured to the face guard rather than putting pressure on the patient's nose. The Sengstaken-Blakemore tube is then connected to nasogastric suction.

Complications that may occur from the use of esophagogastric tamponade include (1) aspiration, (2) erosion of esophageal gastric mucosa, and (3) suffocation. With the balloons inflated, the patient cannot swallow saliva. Oral suction with a Yankauer catheter should be available. Sometimes a Salem sump tube is placed in the upper esophagus before the esophageal balloon is inflated to drain secretions.

Inflation pressure of the esophageal balloon should be maintained between 20 and 25 mm Hg. Agency procedures should clearly state how long the tube may remain in place and how often (and for how long) the esophageal balloon should be deflated. The gastric balloon remains inflated at all times until the procedure is stopped.

If the gastric tube dislodges into the esophagus, the patient can be suffocated. Keep a pair of scissors at the bedside when esophagogastric tamponade is in progress. If the gastric balloon dislodges, cut the inflation ports of the gastric and esophageal balloons to allow quick release of air.

Unfortunately, recurrence of bleeding occurs in about 20 to 60 percent of patients after successful tamponade. With each new bleeding episode, the risk of mortality increases. Sclerotherapy may be undertaken to prevent recurrence of bleeding. Sclerotherapy can be done without general anesthesia, which is an advantage for the patient with a severely damaged liver. The procedure is usually done as part of an EGD while the patient is sedated with diazepam (Valium) or midazolam hydrochloride (Versed). A topical anesthetic is used. The varices are injected with a sclerosing agent that causes thickening and closing of the dilated vessels. The procedure usually takes about an hour. After the procedure the patient may complain of chest pain for up to 72 hours. Give the prescribed analgesics and monitor the patient for pain relief. Report severe pain unrelieved by the prescribed analgesic immediately because the patient may be experiencing an esophageal perforation or ulceration, which are complications of sclerotherapy.

Hepatic encephalopathy is treated by trying to remove the toxic waste material. Saline or magnesium sulfate (MgSO$_4$) enemas may remove some of the toxic waste. The enemas may be given to cleanse the bowel of the noxious substances. Neomycin, an intestinal antibiotic, may be given by mouth, nasogastric tube, or enema. The antibiotic inhibits ammonia formation by reducing colonic bacteria that change ammonium to ammonia. Lactulose may be given by mouth to reduce the pH of the intestine and to "trap" ammonia, allowing it to be excreted in the stool. Hepatic encephalopathy is also treated by restricting or eliminating dietary protein. In severe cases, dialysis may be considered to remove ammonia (Nutrition Notes Box 33–3).

Nursing Process for Acute and Chronic Liver Failure

ASSESSMENT. A complete history and physical assessment are done. Be alert to subjective symptoms of liver slowdown such as malaise, anorexia, indigestion, nausea, severe itching, and dull, aching RUQ pain. Assess the patient for objective evidence of liver problems, such as weight loss, diarrhea or constipation, and an enlarged, firm, and tender liver. Observe the patient for dryness and bruising of the skin, bleeding gums, anemia, jaundice, and any evidence of alterations in thought processes, such as confusion, disorientation, or inability to make decisions. Figure 33–4 shows a mind map for assessing a patient with chronic liver failure.

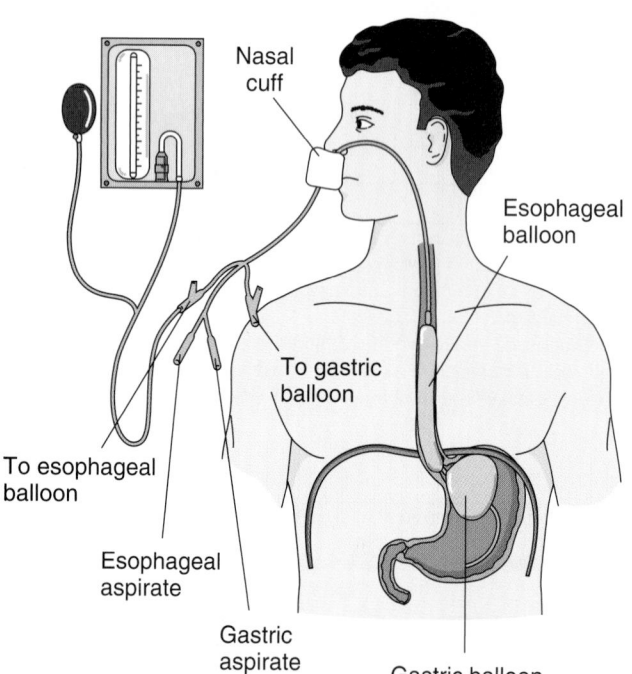

Figure 33-3 Balloon tamponade for bleeding esophageal varices. An esophageal balloon compresses bleeding vessels.

BOX 33-3  *Nutrition Notes*

Supplying Nutrients To Patients with Liver Disease

A registered dietitian may modify the diet daily for patients with liver disease. Usually these patients suffer from severe anorexia, often displaying their best appetites for breakfast. For early or mild disease, protein is encouraged to support healing; later in liver disease, protein is restricted. For esophageal varices, foods are soft, in addition to other restrictions. For liver failure, the following dietary considerations are necessary:

- Protein is restricted according to individual's ability to metabolize it.
- Complete protein foods are selected to provide all the essential amino acids.
- Homogenized milk and eggs are offered because the fat is already emulsified, requiring less bile for digestion.
- Adequate carbohydrates are given to prevent the use of tissue protein for energy.

- Fluid and sodium are likely to be restricted if ascites is present.

Hepatic encephalopathy is associated with increased serum levels of ammonia and aromatic amino acids. Foods producing high ammonia levels may be restricted (e.g., chicken, salami, ground beef, ham, bacon, gelatin, peanut butter, potatoes, onions, lima beans, egg yolk, buttermilk, blue and cheddar cheeses). Vegetable proteins and special enteral feeding formulas (e.g., Hepatic-Aid II and Travasorb-Hepatic) containing branched-chain amino acids may be given in place of foods containing aromatic amino acids that interfere with formation of dopamine and norepinephrine.

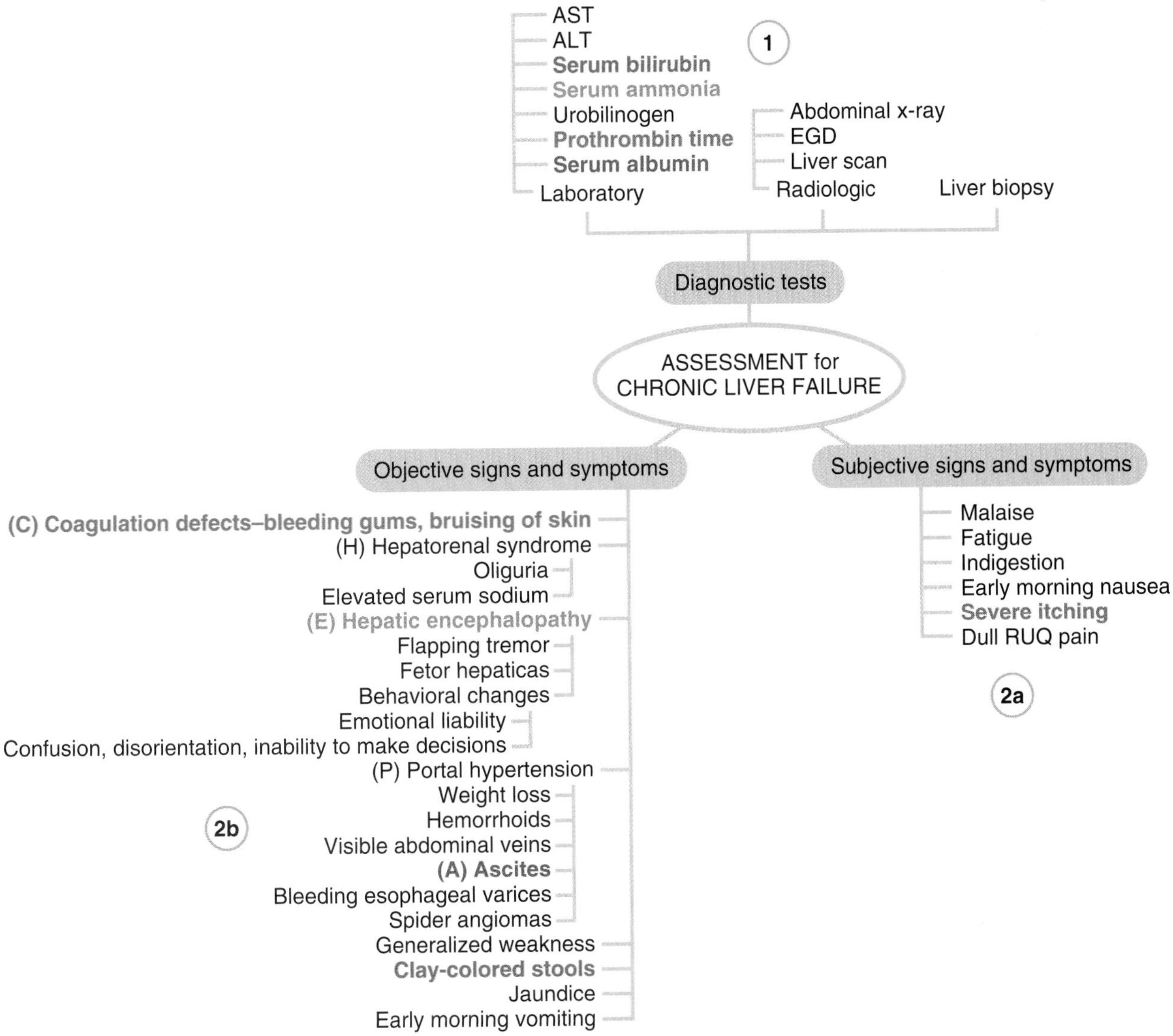

Figure 33-4 Mind map: Assessment of the patient with chronic liver failure.

NURSING DIAGNOSIS. Common nursing diagnoses for chronic liver failure include the following:

- Fluid volume excess related to portal hypertension (ascites)
- Imbalanced nutrition: less than body requirements related to disinterest in food
- Pain related to abdominal pressure
- Risk for disturbed thought processes related to elevated ammonia levels
- Risk for ineffective breathing pattern related to abnormal amounts of fluid in the abdomen
- Risk for deficient fluid volume related to bleeding (esophageal varices, clotting disorders)
- Risk for infection related to impaired immune responses

PLANNING. Major goals for the patient with liver failure include that the patient will experience no fluid excess or deficit, have adequate nutrition, be able to think clearly, breathe effectively, and be free of infection.

IMPLEMENTATION

Fluid Balance. Weigh the patient on admission to obtain a baseline weight. Measure and record the patient's weight and abdominal girth or circumference daily. Mark the place where you measured so the same site will always be measured during subsequent assessments. Report any weight gain or increase in girth promptly. Monitor the patient's low-sodium diet and maintain any ordered fluid restrictions. If intravenous fluids or albumin have been ordered, the assist in careful monitoring of the rate of infusion. Check the patient's vital signs every 4 hours, and report any evidence of difficulty breathing or changes in mental status promptly. Administer ordered diuretics as scheduled. Assist with a paracentesis as required.

Nutrition. The patient with chronic liver failure usually experiences anorexia in addition to impaired metabolism of needed nutrients. Assess the patient's bowel sounds, abdominal distention, and evidence of bleeding at least once every 8 hours. Report weight loss of more than half a pound from previous measurement. Monitor the patient's diet to ensure that any ordered protein restriction is carried out. Offer the patient frequent, small, high-calorie meals. If total parenteral nutrition (TPN) or tube feedings have been ordered, assist with the treatment and monitor the patient's response closely. Make sure that odors and other unpleasant stimuli are eliminated to prevent further worsening of appetite. Offer frequent mouth care. Administer vitamins or other medications as ordered.

Altered Thought Processes. Assess the patient's level of consciousness, speech, behavior, and neuromuscular function frequently. Neuromuscular function can be assessed by asking the patient to hold his or her arms out straight in front and steady. If asterixis, or liver flap, is present, the patient's hands will unwillingly dip and return to the horizontal position in a flapping motion. Look for changes in the patient's handwriting. Give lactulose, neomycin, magnesium citrate, or sorbitol as scheduled to decrease serum ammonia levels. Be aware that lactulose causes loose stools, and do not hold the medication when the patient develops diarrhea. Question giving medications such as sedatives, narcotics, and tranquilizers that may increase the serum ammonia levels. Reorient the patient to time and place frequently. Give simple, clear explanations of care to the patient and give the patient time to understand the explanation. Provide a safe environment for the confused or unsteady patient.

Ineffective Breathing Pattern. Assess the patient's respiratory rate, rhythm, chest movement, and skin color frequently. Assist the patient in using an incentive spirometer and in coughing gently every 2 to 4 hours. Elevate the head of the patient's bed so that the patient's lungs have maximum room for expansion. Reposition the patient at least every 2 hours. Avoid suctioning the patient if possible because bleeding may occur. Make sure that oxygen and respiratory treatments are carried out as ordered, and that the patient is responding to the treatments as desired. Assist with a thoracentesis if required. Carry out patient care in an unhurried, quiet manner. Time care to allow the patient to rest and to avoid fatigue.

Safety. The patient with chronic liver failure is at risk for hemorrhage from bleeding esophageal varices, gastrointestinal bleeding, and lack of blood clotting factors. Assess gastric secretions, stool, and urine for blood at least every 8 hours. Monitor blood clotting laboratory studies such as the prothrombin time and report any abnormal values. Caution the patient to use a soft-bristle toothbrush and an electric rather than straight razor to avoid injury. Use a small-gauge needle for injections and apply direct pressure to all puncture sites to avoid bleeding. The patient should not eat hot, spicy, or irritating foods. The patient also should not cough or blow the nose forcefully, strain, vomit, or gag if at all possible. If bleeding occurs, assist with monitoring the patient receiving blood, blood products, and fluids. You may need to assist with esophagogastric tubes if needed. Administer vasopressin and vitamin K as ordered.

Infection. Observe the patient carefully for evidence of infection. Report any rise in temperature or sudden increase in pulse and respiratory rate. Carefully evaluate laboratory studies such as the white blood cell count. The white blood cell count may not elevate or may elevate slowly because the white cell activity is impaired. The earliest warning signs of infection may be subtle changes in the patient's behavior, such as sudden restlessness, an increase in confusion, or irritability.

PATIENT EDUCATION. Teach patients how chronic liver failure is affecting their bodies. In particular, patients need to know about portal system hypertension and hepatic encephalopathy. Patients and their families need to observe for and report any confusion, tremors, or personality changes. Teach patients to get adequate rest and to avoid strenuous activity. Instruct patients about a diet high in calories, low in sodium, and high in protein if hepatic encephalopathy has not developed. Patients should know to avoid narcotics, sedatives, and tranquilizers. Patients and their families should know to promptly report any bleeding; any sign of low potassium, such as muscle cramps, nausea, or vomiting caused by diuretics; changes in mental status, such as confusion or personality changes; changes in weight; and any increase in current symptoms. The patient and the family should know that the patient must avoid alcohol. The patient should also know the importance of frequent follow-up care and laboratory studies.

EVALUATION. Successful management of the patient with chronic liver failure can be claimed if the patient demonstrates the following:

■ No weight gain or increase in abdominal girth
■ Maintenance of body weight
■ Alert and oriented demeanor
■ Respiratory rate between 16 and 20 respirations per minute with no cyanosis or changes in level of consciousness
■ No bleeding, infection, or injuries
■ Accurate knowledge of chronic liver failure and proper disease management requirements

---

CRITICAL THINKING

Mary

Mary, a 76-year-old retired business woman, has lived alone for the past 20 years, since the death of her husband. She has a long history of poor nutritional habits but does not use alcohol. During a recent visit to her physician, she was diagnosed with early chronic liver failure.

1. What risk factors does Mary have for chronic liver failure?
2. What symptoms would you expect Mary to exhibit with early chronic liver failure?
3. What values do you expect to see for serum albumin? Prothrombin time?
4. What are the two greatest concerns with portal hypertension?
5. What is the usual treatment for ascites?
6. What complication of esophageal tamponade requires that a pair of scissors be at the bedside at all times while tamponade is in progress?

Answers at end of chapter.

Transplantation

Patients with liver disease who do not respond well to medical or surgical treatment may be candidates for a liver transplant. Patients who have chronic liver failure from hepatitis or biliary disease, metabolic disorders, or hepatic vein obstruction may be evaluated for a liver transplant. In general, patients who have cancer are not considered for liver transplantation; the drugs used to suppress tissue rejection by the immune system may cause the cancer cells to grow at an increased rate.

The patient with end-stage liver failure who does not have hypertension, bleeding esophageal varices, infection, or severe cardiac disease is placed on a national list as a potential liver recipient. The patient must be as physically and emotionally stable as possible (Cultural Consideration Box 33–4).

BOX 33-4 CULTURAL CONSIDERATION

Jewish law views organ transplantation from the recipient, the living donor, the cadaver donor, and the dying donor. If the recipient's life can be prolonged without considerable risk, transplant is ordained. For a living donor to be approved, the risk to the life of the donor must be considered. One is not obligated to donate a part of himself or herself unless the risk is small. This includes kidney and bone marrow donations. The use of a cadaver for transplant is usually approved if it is saving a life. The use of skin for burns is acceptable, although there is no agreement on the use of cadaver corneas. The nurse may need to help the Jewish patient contact a rabbi when making a decision regarding organ donation or transplantation.

After the surgical implantation of a donor liver, the patient must be closely observed for evidence of donor organ rejection. The patient is placed on drugs such as cyclosporine (Cyclosporin A), azathioprine (Imuran), and prednisone (Deltasone) to suppress the immune system responses to foreign protein or tissue rejection. Observe the patient for the following signs of impending rejection:

■ Pulse greater than 100 beats per minute
■ Temperature greater than 101° F (38° C)
■ Complaints of RUQ pain
■ Increased jaundice
■ Decrease in bile from the **T-tube** or change in bile color

In addition, laboratory studies may show increased serum transaminases (ALT and AST), serum bilirubin, alkaline phosphatase, and prothrombin time. Symptoms of acute tissue rejection usually develop between the fourth and tenth postoperative days.

The patient who has received an organ transplant needs extended medical follow-up. Teach the patient to promptly report to the physician any symptoms of infection, bleeding episodes, or RUQ pain (Ethical Consideration Box 33–5).

Cancer of the Liver

Cancer of the liver is usually the result of metastasis from a primary cancer at a distant location. The liver is a likely area of involvement if cancer originated in the esophagus, lungs, breast, stomach, or colon. Patients with malignant melanoma may also have liver involvement. For some patients, liver cancer is the primary tumor site. Patients with a history of chronic hepatitis B, nutritional deficiencies, or exposure to hepatotoxins may develop cancer of the liver.

Symptoms of cancer of the liver include encephalopathy, abnormal bleeding, jaundice, and ascites. Laboratory tests show an elevated serum alkaline phosphatase. Radiological examinations may include abdominal radiographs or radioisotope scans, which show tumor growth. Liver cancer is definitively diagnosed with a positive needle biopsy combined with an ultrasonogram of the liver.

BOX 33-5 ETHICAL CONSIDERATION

Organ Transplantation

Despite widespread public and medical acceptance of organ transplantation as a highly beneficial procedure, ethical questions remain. Whenever a human organ is transplanted, a large number of people are involved, including the donor, the donor's family, and medical and nursing personnel, as well as the recipient and the recipient's family. Society in general could also be added to this mix because of the high cost of organ transplantation that is usually borne directly by tax monies or indirectly in the form of increased insurance premiums. Each one of these persons or groups has rights that may conflict with others' rights.

Most institutions that perform transplants and organizations that are involved in obtaining organs, have developed elaborate, detailed, and involved procedures to help deal with the ethical and legal issues involved in transplantation. Despite these efforts, there are still some ethical issues that should be considered whenever the issue of organ transplantation is raised.

Despite the best efforts of the medical and legal community to establish criteria for death, there are still some ethical questions about when a person is really dead. Does brain death, the most widely accepted criterion for death, really indicate that a person no longer exists as a human being? Or are there other criteria that should be examined? Such organs as hearts, lungs, and liver must come from a donor with a beating heart. Might there not be the tendency to declare brain death before it actually occurs?

One of the most difficult ethical issues involved in organ transplantation is the selection of recipients. There are many fewer organs available than there are people who need them. Many potential ethical dilemmas arise from this fact. Should someone receive an organ because he or she is rich or famous or knows the right people? The national organ recipient list attempts to list and rank all persons who need organs in a nondiscriminatory manner. Some of the important criteria include need, length of time on the list, potential for survival, prior organ transplantation, value to the community, and tissue compatibility.

Nurses can be and often are involved in some aspect of the organ donation process. Many states have passed laws that require health care workers to ask family members of potential organ donors if they have ever thought about organ donation for their dead or dying loved one. Many nurses, particularly nurses in critical care units, provide care for patients who are potential organ donors. Nurses in operating rooms may help in the surgical procedures that remove organs from a cadaver and that transplant them into a recipient's body. Many floor nurses provide postoperative care for patients who have received a transplanted organ. Home health care nurses provide follow-up care for these patients at home.

Nurses working with organ transplantation need to be sensitive to the potential for manipulation. Most people who are seeking organ transplantations are desperately ill or near death. They, and their families, can be very easily manipulated or can be very manipulating. On the other side, the families of potential organ donors are usually emotionally distraught because of the sudden and traumatic loss of a loved one. They too are very vulnerable to manipulation. As a general rule, neither the donor nor the donor's family should play any part in the selection of a recipient. Nurses must avoid making statements or giving nonverbal indications of approval or disapproval of potential recipients.

The patient with liver cancer may be a candidate for surgical removal of the affected portion of the liver. Care of the postsurgical patient is similar to other abdominal surgery patients. If surgery is not an option, the patient may receive chemotherapeutic drugs by injection directly into the affected lobe of the liver or into the hepatic artery. Intra-arterial injection of chemotherapeutic drugs has the advantage of being less toxic to the rest of the body. (See Chapter 10 for care of the patient receiving chemotherapy.)

▶ DISORDERS OF THE PANCREAS
Pancreatitis

Pancreatitis is an inflammation of the pancreas; it may be mild or severe, acute or chronic. The two forms of pancreatitis have different courses and are considered two different disorders.

Acute Pancreatitis
Pathophysiology

Inflammation of the pancreas appears to be caused by a process called autodigestion. For reasons not fully understood, pancreatic enzymes are activated while they are still within the pancreas and begin to digest the pancreas. In addition, large amounts of the enzymes are released by inflamed cells. As the pancreas digests itself, chemical cascades occur. Trypsin destroys pancreatic tissue and causes vasodilation. As capillary permeability increases, fluid is lost to the retroperitoneal space, causing shock. In addition, trypsin appears to set off another chain of events that causes the conversion of prothrombin to thrombin, so that clots form. The patient may develop DIC. (See Chapter 24.)

Etiology

Pancreatitis is most commonly associated with excessive alcohol consumption. Alcohol appears to act directly on the acinar cells of the pancreas and the pancreatic ducts to irritate and inflame the structures. Biliary disease such as **cholelithiasis** (gallstones) or cholangitis (inflammation of the bile ducts) may also trigger pancreatitis. Gallstones may plug the pancreatic duct and cause inflammation from excessive fluid pressure on sensitive ducts. The irritant effect of bile itself may cause inflammation. Blunt trauma to the abdomen or infection may trigger the process by causing ischemia, inflammation, and activation of the pancreatic enzymes. Drugs such as thiazide diuretics (HydroDiuril), estrogen, and excessive serum calcium from hyperparathyroidism are less common causes of pancreatitis.

Elderly patients and patients with a first diagnosis of pancreatitis have a higher mortality rate. In addition, patients

---

cholelithiasis: chole—bile + lith—stone + iasis—condition

who have pancreatitis associated with biliary disease have a higher mortality rate than patients with alcohol-related pancreatitis.

Prevention

Caution patients who drink alcohol to stop. Patients with biliary disease need to seek medical treatment for these conditions so that pancreatitis does not develop as a complication. Monitor patients, especially the elderly, for any abdominal complaints when they are placed on a thiazide diuretic or estrogen therapy (Cultural Consideration Box 33–6).

BOX 33-6 CULTURAL CONSIDERATION

Pancreatic disease is more common among Mexican-Americans and Chinese-Americans. Risk factors include working in occupations such as mining, factories using chemicals, and farming using pesticides. The high use of alcohol and cigarette smoking add to the risk of pancreatic disease.

Signs and Symptoms

Patients with acute pancreatitis are very ill, with dull abdominal pain, guarding, rigid abdomen, hypotension or shock, and respiratory distress from accumulation of fluid in the retroperitoneal space. The abdominal pain is generally located in the midline just below the sternum, with radiation to spine, shoulders, or low back. The location and degree of pain indicate the area of the pancreas involved and to some extent the amount of involvement. For example, if the patient complains primarily of RUQ pain, the head of the pancreas is most likely involved. Respirations are likely to be shallow as the patient attempts to splint painful areas.

The patient may have a low-grade fever, dry mucous membranes, and tachycardia. If the primary cause is biliary, the patient may complain of nausea and vomiting, and jaundice may be evident. The islets of Langerhans in the terminal one-third of the pancreas are usually not impaired.

Complications

It may be useful to think of pancreatitis as a chemical burn to the organ. As with other severe burns, death is likely to occur from secondary causes. From the onset of symptoms, cardiovascular, pulmonary (including acute respiratory distress syndrome), and renal failure are the most likely causes of death. Hemorrhage, peripheral vascular collapse, and infection are also major concerns for patients with pancreatitis. A purplish discoloration of the flanks (Turner's sign) or a purplish discoloration around the umbilicus (Cullen's sign) may occur with extensive hemorrhagic destruction of the pancreas.

Diagnostic Tests

Serum amylase (normal: 56 to 190 IU/L) and serum lipase (normal: 0 to 110 U/L) may be elevated 5 to 40 times normal. The levels usually begin to drop within 72 hours.

Urine amylase elevates and stay elevated longer. Glucose, bilirubin, alkaline phosphatase, lactic dehydrogenase, ALT, AST, cholesterol, and potassium are all elevated. Decreases are measured in serum albumin, calcium, sodium, and magnesium.

X-ray examination may show pleural effusion from local inflammatory reaction to pancreatic enzymes, pulmonary infiltrates, or a change in the size of the pancreas. Computed tomography and ultrasonography can provide more complete information about the pancreas and surrounding tissues.

Medical Treatment

Medical treatment of acute pancreatitis depends on the intensity of the symptoms. Treatment is concerned with the maintenance of life support until the inflammation resolves. The physician orders intravenous fluids, such as crystalloid, electrolyte, or colloid (such as albumin) solutions, if the patient experiences hypovolemic shock. Blood or blood products may also be ordered if the patient has massive blood loss from hemorrhage.

The patient may be given antianxiety agents to decrease oxygen demand. The patient may require supplemental oxygen if abdominal pressure, pleural effusion, or acidosis cause a severe decrease in circulating oxygen.

The physician usually orders meperidine hydrochloride (Demerol) for pain. Pain and anxiety increase pancreatic secretion by stimulating the autonomic nervous system. Morphine sulfate, long thought to be contraindicated as an analgesic because it was believed to cause smooth muscle spasm in the biliary and pancreatic ducts, may also be ordered.

The patient is usually ordered to have nothing by mouth (NPO) to rest the gastrointestinal tract. Further, the patient may need to have a nasogastric tube inserted into the stomach and attached to low suction to empty gastric contents and gas. The physician usually orders histamine (H_2) antagonists such as ranitidine hydrochloride (Zantac) to prevent stress ulcers and to decrease acid stimulation of pancreatic secretion. If the NPO therapy is prolonged or if the patient is malnourished, the patient will receive TPN (Nutrition Notes Box 33–7).

BOX 33-7 Nutrition Notes

Nourishing the Patient with Pancreatitis

For acute pancreatitis the patient may be given ice chips made with electrolyte solutions to minimize gastric secretions and subsequent loss through the nasogastric tube. Nourishment is provided intravenously via TPN. If enteral feedings are started, an elemental formula that has been "predigested" may be selected.

In chronic pancreatitis the meal pattern is six small meals per day, beginning with clear liquids and progressing to a high-carbohydrate, low-fat diet. Medium-chain triglyceride (MCT) oil may be used because of its direct absorption. Vitamin supplements, including parenteral B_{12}, are commonly administered. An increase in serum amylase levels may necessitate the return to a more restricted diet.

Additional typical drug orders include sodium bicarbonate to reverse the acidosis caused by shock, electrolytes such as calcium and magnesium to replace losses, regular insulin to combat hyperglycemia, and antibiotics to treat sepsis.

Nursing Process

ASSESSMENT. Frequently measure and record vital signs (especially blood pressure and pulse), skin color and temperature, and urinary output. Monitor nausea and vomiting. Observe the patient for evidence of respiratory distress, such as restlessness, irritability, use of accessory muscles for breathing, or dyspnea. Evaluate the patient frequently for pain.

NURSING DIAGNOSIS. The most common nursing diagnoses for the patient with acute pancreatitis are as follows:

- Pain related to edema and inflammation
- Imbalanced nutrition: less than body requirements related to vomiting, pain, NPO, gastric suction
- Risk for ineffective breathing pattern related to abdominal pain and pressure
- Risk for injury related to disturbed blood clotting mechanisms and fluid and electrolyte imbalances

PLANNING. The nurse, the patient, and other members of the health team need to work together to develop a plan for achieving pain control, adequate nutrition, and prevention of cardiorespiratory complications.

IMPLEMENTATION

Pain. Administer analgesics as ordered. Evaluate the patient for pain at least every 2 hours initially. Advise the patient to sit in an upright or slightly forward-leaning position to decrease abdominal discomfort. Space nursing care activities to allow the patient to rest. Keep the patient's immediate surroundings quiet, restful, and free from anxiety-producing stimuli.

Imbalanced Nutrition: Less than Body Requirements. Weigh the patient regularly and evaluate laboratory tests such as serum albumin (normal: 3.5 to 5.5 g/dL) and other pancreatic function tests. If the serum albumin level drops below 3.2 g/dL, the physician may order intravenous albumin as replacement. Assess the patient at least every 8 hours for bowel sounds, nausea, and vomiting.

Monitor TPN therapy, including daily blood sugar results. Patients may be ordered to have blood glucose monitoring done every 6 hours. Administer regular insulin as ordered for elevated blood sugars.

Give histamine antagonists and antacids as ordered. If the patient has a nasogastric tube, record the amount and character of the drainage every 8 hours or according to agency policy.

The patient is usually slowly progressed to solid foods, starting with carbohydrate supplements through the nasogastric tube, then tube feedings with protein and fats as they are tolerated. Finally, bland, soft foods are ordered. Report any nausea, vomiting, or increase in complaints of abdominal discomfort immediately. Patients who progress too quickly may have a relapse of the pancreatitis.

Risk for Ineffective Breathing Pattern. Carefully observe the patient for signs of respiratory distress such as tachypnea, tachycardia, dyspnea, use of accessory muscles, or presence of abnormal breath sounds or crackles. Report evidence of poor oxygenation, including changes in mental status (e.g., irritability, confusion, or increasing sleepiness) promptly. Administer supplemental oxygen as ordered. Position the patient in an upright or slightly forward-leaning position. Good pain control enables the patient to breathe more effectively.

Risk for Injury. Assess the patient for evidence of disturbed electrolyte balance. Monitor laboratory values for sodium, potassium, calcium, and magnesium daily. Evidence of hypokalemia includes muscle weakness, diminished bowel sounds, apathy, and hypotension. Evidence of hypocalcemia includes nausea, vomiting, irritability, abdominal pain, and neuromuscular irritability. Neuromuscular irritability can be determined by Chvostek's sign. (See Chapter 5.) Assess the patient for evidence of abnormal bleeding. Observe the abdomen for Turner's and Cullen's signs. Evaluate the patient's vital signs for tachycardia, tachypnea, and hypotension, which may indicate bleeding. Initially, urinary output is measured hourly using a urometer. Urine output should be greater than 30 mL/h. Report any evidence of bleeding, such as a urinary output of less than 30 mL/h or frank bleeding, promptly. Assist with monitoring the administration of volume replacement solutions or blood infusions as ordered.

PATIENT EDUCATION. Teach the patient about pancreatitis. The patient needs to know the treatment regimen and the reason for slow dietary progression from NPO to a normal diet. Teach that rest is a vital part of the treatment plan and that gradual progression to full activities is normal. The patient must exercise moderation in food, exercise, and alcohol ingestion to decrease the risk of future episodes of pancreatitis.

If the pancreatitis is the result of chronic excessive alcohol ingestion, arrange to have the patient counseled about abstinence from alcohol. Refer the patient and family to support organizations such as Alcoholics Anonymous. For more information, visit Alcoholics Anonymous at www.alcoholics-anonymous.org.

EVALUATION. The plan of care for the patient with acute pancreatitis has been successful if the patient demonstrates the following:

- Pain of 2 or less on a scale of 0 to 5
- Weight loss of less than 5 percent of total body weight
- Respirations 16 to 20, unlabored, regular
- No elevated temperature
- No abnormal bruising
- No restlessness
- Urinary output greater than 30 mL/h
- Blood pressure within 5 percent of baseline

See Nursing Care Plan Box 33–8 for the Patient with Acute Pancreatitis.

BOX 33-8 NURSING CARE PLAN FOR THE PATIENT WITH ACUTE PANCREATITIS

 Pain related to edema and inflammation

PATIENT OUTCOME
Patient experiences an increase in comfort as evidenced by statement of pain level less than 2 on a scale of 0 to 5.

EVALUATION OF OUTCOME
Does patient sate pain level is less than 2 on a pain scale of 0 to 5?

INTERVENTIONS	RATIONALE	EVALUATION
Assess the patient every 2 hours for pain by 　Asking the patient to rate pain on a scale of 0 to 5. 　Observing the patient for acute pain behaviors such as grimacing, irritability, reluctance to move, or inability to lie quietly.	Intense pain is likely to occur with acute pancreatitis. A pain scale allows for a consistent and individual evaluation of pain. Observation of acute pain behaviors, such as reluctance to move, shallow respirations, grimacing, or irritability, may be a reliable indicator of pain. Some patients with pain may deny their discomfort. However, the patient in pain may have no observable pain behaviors.	Does patient state that pain is less than 2 on a pain scale of 0 to 5, where 0 = no pain and 5 = worst possible pain? Does patient exhibit pain behaviors that differ from his or her report of pain?
Administer analgesics as ordered.	Analgesics are most effective if given before pain becomes too great.	Are analgesics effective?
Assist the patient to a position of comfort, usually high Fowler's or leaning forward slightly.	Upright position keeps abdominal organs from pressing against the inflamed pancreas.	Does positioning promote comfort?
Keep the environment free from excessive stimuli.	Quiet, restful, anxiety-free atmosphere permits the patient to relax and may decrease pain perception.	Does patient state atmosphere is relaxing?
Teach the patient alternative pain control strategies such as guided imagery and relaxation techniques.	Successful use of pain control strategies may decrease the amount of analgesics needed and give the patient a greater sense of control.	Are alternative strategies effective?

 Altered nutrition: less than body requirements related to pain, medical restrictions (NPO), and treatment (suction)

PATIENT OUTCOME
Patient will experience improved nutrition as evidenced by stable weight, albumin greater than 3.5 g/L.

EVALUATION OF OUTCOME
Is weight stable? Is albumin level greater than 3.5 g/L?

INTERVENTIONS	RATIONALE	EVALUATION
Assess the patient's nutritional status by 　Weighing the patient every other day. 　Checking serum albumin levels on laboratory studies. 　Auscultating for bowel sounds. 　Observing for nausea or vomiting. 　Monitoring blood sugar at least every 6 hours if the patient is on TPN. 　Observing for diarrhea, bloating, or steatorrhea (fatty stools).	A loss of 1 lb of body weight occurs when the body uses 3500 calories more than is taken in. Serum albumin of 3.5–5.5 g/L indicates normal protein metabolism in the absence of liver or renal disease. Nausea, vomiting, and pain are risk factors for inadequate intake. Patients on TPN are more likely to have high blood glucose. Diarrhea, bloating, or fatty stools may indicate malabsorption syndrome.	Has patient lost less than 5% of total baseline body weight? Is patient's albumin below 3.5 g/dL? Does patient have sufficient energy and strength to carry out activities of daily living?
Administer nutritional supplements, including pancreatic enzymes, as ordered.	Provides adequate nutrition.	Does patient take any supplements?
Teach the patient to avoid alcohol.	Alcohol may trigger another episode of pancreatitis.	Does patient verbalize understanding of importance of avoiding alcohol?
Teach the patient and family the signs and symptoms of diabetes mellitus.	Patients with pancreatitis are at great risk for developing diabetes mellitus, which also causes high blood glucose levels.	Does patient verbalize signs and symptoms of diabetes to report?

Continued

BOX 33-8   NURSING CARE PLAN FOR THE PATIENT WITH ACUTE PANCREATITIS—CONT'D

Teach the patient and family to self-monitor for symptoms of malabsorption syndrome, such as fatty stools, weight loss, dry skin, or bleeding.	Absence of pancreatic enzymes causes problems with digestion of fats, carbohydrates, and proteins.	Does patient verbalize understanding of symptoms of malabsorption to report?

 Risk for ineffective breathing pattern related to abdominal pressure and pain

PATIENT OUTCOME

Patient has an effective breathing pattern as evidenced by unlabored respirations, 16–20 per minute.

EVALUATION OF OUTCOME

Are respirations unlabored, 16–20 per minute?

INTERVENTIONS	RATIONALE	EVALUATION
Assess the patient's breathing patterns: Observe respirations for depth, regularity, and rate. Observe respiratory effort.	Abdominal pressure from inflammation and tissue damage under the diaphragm may cause the patient to take shallow, rapid respirations, which can tire the patient.	Are patient's respirations 16–20 per minute, unlabored, and regular? Is patient alert and oriented? Has there been a change in the level of patient's arousal?
Observe for evidence of respiratory distress, such as use of accessory muscles, use of intercostal muscles, and rapid or difficult breathing.	Abdominal pressure can force the use of additional muscles to aid in breathing.	Does patient exhibit signs of distress?
Administer oxygen as ordered.	Oxygen can decrease the amount of effort the patient must expend to breathe.	Does oxygen help patient breathe easier?
Place the patient in an upright or slightly forward-leaning position.	Relieves pressure on the diaphragm.	Is positioning effective?
Prepare the patient's food by opening cartons and lids. Cut food into bite-size portions.	Decreases the demand for oxygen.	Does patient accept assistance with food preparation?
Teach the patient to move slowly and to take frequent rests.	Helps decrease the demand for oxygen.	Does patient tolerate activity?

 Risk for injury related to disturbed blood clotting mechanisms, fluid, and electrolyte imbalances

PATIENT OUTCOME

Patient experiences no injury.

EVALUATION OF OUTCOME

Is there evidence of injury? Are signs and symptoms of impending injury recognized and reported early?

INTERVENTIONS	RATIONALE	EVALUATION
Assess the patient's fluid, electrolyte, and blood clotting mechanisms by Monitoring laboratory values of sodium, potassium, calcium, and magnesium daily. Evaluating neuromuscular status by checking Chvostek's sign.	Sodium, potassium, calcium, and magnesium need to be replaced.	Does patient have any abnormal bruising, bleeding gums, or pink urine? Is urinary output greater than 30 mL/h? Is patient's blood pressure within 5% of baseline? Do laboratory studies show that patient's electrolyte, hemoglobin, hematocrit, and blood clotting values are within acceptable ranges?
Monitoring the patient's hematocrit, hemoglobin, and blood clotting times frequently.	Patients are likely to bleed.	
Observing abdomen and flanks for Cullen's and Turner's signs. Weighing the patient daily. Measuring and recording intake and output every shift. Observing for nausea and vomiting.	Patients may lose fluids because of nausea, vomiting, diarrhea, and hemorrhage.	

BOX 33-8   NURSING CARE PLAN FOR THE PATIENT WITH ACUTE PANCREATITIS—CONT'D

Report any drop in blood pressure greater than 5% of the patient's baseline.	May indicate severe fluid loss.	Are vital signs stable?
Monitor TPN and report any difficulties.	TPN must not be stopped abruptly. Patient's blood glucose may drop sharply.	Are blood glucose levels stable?
Teach the patient to report any weakness or muscle twitching.	May indicate electrolyte imbalance.	Does patient verbalize understanding of signs and symptoms of electrolyte imbalance to report?

CRITICAL THINKING

Mrs. Samuels

Mrs. Samuels, an 85-year-old retired librarian, is admitted to the nursing unit from the emergency department with severe mid-epigastric pain that radiates to her back. On admission, she is noted to have guarding of the abdomen, and the abdomen is full and tense. Her medical record documents that she had a endoscopic retrograde cholangiopancreatograph 2 days ago for recurrent episodes of RUQ abdominal pain. She has no history of excessive alcohol intake.

1. What is the most common cause of acute pancreatitis? Does Mrs. Samuels fit the description?
2. Why do patients such as Mrs. Samuels have difficulty breathing?
3. Why is Mrs. Samuels at risk for hemorrhage?
4. What laboratory test is most likely to be abnormal in early acute pancreatitis?
5. Why are narcotics commonly ordered for acute pancreatitis?
6. Why does the physician usually order a histamine antagonist?

Answers at end of chapter.

Chronic Pancreatitis

Chronic pancreatitis is continuing pancreatic cellular damage and decreased pancreatic enzyme functioning usually following repeated occasions of acute pancreatitis.

Pathophysiology

Chronic pancreatitis is a continuous, progressive disease that replaces functioning pancreatic tissue with fibrotic tissue as a result of inflammation. Toxins from alcohol irritate the pancreatic ducts. The ducts become obstructed, dilated, and finally atrophied. The acinar or enzyme-producing cells of the pancreas ulcerate in response to inflammation. The ulceration causes further tissue damage and tissue death, and it may cause cystic sacs filled with pancreatic enzymes to form on the surface of the pancreas. The pancreas becomes smaller and hardened, and progressively smaller amounts of the enzymes are produced.

Etiology and Incidence

The major cause of chronic pancreatitis in men is excessive alcohol ingestion that causes repeated attacks of acute pancreatitis. The major cause in women is chronic obstructive biliary disease that leads to persistent inflammation of the pancreatic ducts. Other conditions known to cause chronic pancreatitis are prolonged malnutrition, cancer of the pancreas or duodenum, and prolonged use of enteral feedings, which can cause atrophy of the pancreas. The usual age for chronic pancreatitis to develop is between ages 45 and 60. The patient's mean life span is 25 years after the diagnosis of chronic pancreatitis is made. Death is often not related to pancreatic failure.

Prevention

Advise patients with an episode of acute pancreatitis from excessive alcohol ingestion that abstinence could prevent recurrence of the pancreatitis and prevent the possibility of chronic pancreatitis. Advise all patients with obstructive biliary disease to seek medical treatment for their condition to prevent the progression from acute to chronic pancreatitis. Carefully monitor patients who are unable to feed themselves for nutritionally adequate diets. Monitor routine laboratory values. Report any trend toward reduced functioning of the pancreas.

Signs and Symptoms

The signs and symptoms of chronic pancreatitis are less severe than acute pancreatitis but more long term. The patient's history will show a pattern of remissions and exacerbations over a period of years. The patient will complain of epigastric or LUQ pain, weight loss, and anorexia. Malabsorption and fat intolerance occur late in the disease. Usually the islets of Langerhans function until late stages of the disease, so diabetes mellitus is a late-occurring symptom.

Complications

Complications of chronic pancreatitis include the development of abscesses or fistulas, malabsorption syndrome, and diabetes mellitus. Abscesses and fistulas may develop when cysts filled with pancreatic enzymes burst into the abdominal cavity, causing severe inflammation and tissue necrosis. Pleural effusion may develop from inflammation just under the diaphragm. Malabsorption syndrome with fatty stools and diarrhea may develop in response to the limited amount of pancreatic enzymes produced. In addition, biliary obstruction may further complicate fat absorption. Pancreatic enzymes are essential for normal absorption of nutrients from the intestines. As the terminal third of the pancreas becomes involved and the islets of Langerhans are destroyed, the patient exhibits the classic pattern of insulin-dependent diabetes mellitus (discussed in Chapter 38).

Diagnostic Tests

Serum amylase (normal: 59 to 190 U/L) and serum lipase (normal: 0 to 110 U/L) levels are lower than normal. Fecal fat analysis shows higher than normal amounts of fat.

Both computed tomography and ultrasonography show characteristic pancreatic structural changes such as masses, calcification of ducts, cysts, and change in pancreatic size. Endoscopic retrograde cholangiopancreatography (ERCP) can locate specific obstructions and detect ductal leaks.

Medical Treatment

Medical treatment is aimed at promoting comfort and maintaining adequate nutrition. Pain is managed with analgesics. Nutrition is improved with the careful replacement of pancreatic enzymes and specially prepared nutritional supplements.

Surgery may be necessary to repair fistulas, drain cysts, or repair other damage. In some cases, ducts or sphincters may be surgically repaired. In other instances, part or all of the pancreas may be removed.

Nursing Process

ASSESSMENT. Perform frequent thorough assessments for pain. Weigh the patient frequently. Monitor laboratory values to determine the adequacy of nutrition, including serum albumin (normal: 3.5 to 5.5 g/dL) and blood glucose levels.

NURSING DIAGNOSIS. Common nursing diagnoses for the patient with chronic pancreatitis include the following:

- Pain related to edema and inflammation
- Imbalanced nutrition: less than body requirements related to inability to absorb nutrients

PLANNING AND IMPLEMENTATION
Pain. Determine with the patient what events and patterns cause or help the pain. Attempt to determine how pain is affecting the patient's eating habits and lifestyle. Administer analgesics as ordered. The patient may have chronic pain with periodic episodes of severe acute pain. The possibility exists that the patient may develop dependence on the analgesic agent. The nurse can best help by careful assessment of the patient so that medication regimens can be tailored to the patient's need.

Imbalanced Nutrition: Less Than Body Requirements. Weigh the patient daily. Administer nutritional supplements, including pancreatic enzymes, as ordered. Administer histamine antagonists or antacids as ordered. Monitor the patient's blood sugar and observe the patient carefully for evidence of diabetes mellitus. The patient with less than adequate amounts of insulin exhibits excessive thirst, frequent urination, hunger, and weight loss. (See Chapter 38 for management of the patient with diabetes mellitus.) Further, observe the patient for diarrhea, bloating, or **steatorrhea** (fatty stools). Report any evidence of continuing difficulty with digestion immediately.

PATIENT EDUCATION. Counsel the patient to avoid alcohol. Give the patient and family the warning signs and symptoms of diabetes mellitus. Teach the patient to monitor for malabsorption syndrome. The patient should check for fatty stools, report any weight loss, and report any evidence of vitamin A, D, E, or K deficiencies. Vitamin A deficiency may be evidenced by dry, scaly skin or changes in skin pigment. Vitamin D deficiency may be detected by patient complaints of bone pain. Vitamin E deficiency may be associated with anemia. Vitamin K deficiency can be detected from abnormal bruising or bleeding and prolonged clotting time. The patient should know that sudden increases in abdominal pain, increased difficulty breathing, or radiating back pain may indicate development of pancreatic cysts, abscesses, or fistulas and should be reported promptly.

EVALUATION. The plan of care for the patient with chronic pancreatitis is successful if the patient exhibits the following:

- Pain that is no more than 2 on a pain scale of 0 to 5
- No weight loss or gain of more than half a pound per week
- No diarrhea or steatorrhea
- Ability to state signs and symptoms of diabetes mellitus to report
- Ability to state signs and symptoms of pancreatic cysts, abscesses, or fistulas to report

Cancer of the Pancreas

Cancer of the pancreas is the fifth leading cause of cancer death in the United States, killing more than 20,000 people each year. More than 25,000 new cases of cancer of the pancreas are diagnosed yearly; it most often affects people between ages 65 and 79. About 70 percent of cancers of the pancreas occur in the head of the pancreas. About 30 percent of cancers are located in the body and tail of the pancreas.

Pathophysiology

Most primary tumors of the pancreas are ductal adenocarcinomas and occur in the exocrine parts of the pancreas. The tumors in the head and body of the pancreas tend to be large. Cancer of the pancreas spreads rapidly by direct extension to the stomach, gallbladder, and duodenum. Cancer located in the body of the pancreas usually spreads further and more rapidly than do masses in the head. Cancer of the pancreas may spread by the lymphatic system and through the vascular system to distant organs and lymph nodes.

Etiology

The cause of cancer of the pancreas is not known. Cancer of the pancreas has been associated with chemical carcinogens such as high-fat diets and cigarette smoking, diabetes mellitus, excessive alcohol intake, and chronic pancreatitis. It may also occur as a result of metastasis from a primary cancer of the lung, breast, kidney, or thyroid gland or malignant melanomas of the skin.

Signs and Symptoms

The patient with cancer of the pancreas usually complains of vague symptoms early in the disease process. Weight loss, pain, anorexia, nausea, vomiting, and weakness are among the vague early symptoms. Detection is often difficult because of the nonspecific complaints offered by the patient. The patient may complain of abdominal pain that is often worse at night. The pain is described as gnawing or boring, and it radiates to the back. The pain may be lessened by a side-lying position with the knees drawn up to the chest or by bending over when walking. The pain becomes increasingly severe and unrelenting as the cancer grows.

The patient may complain of a bloated feeling or fullness after eating. If the cancer obstructs the bile duct, the patient may have jaundice, dark urine, pruritus, and light-colored stools. The patient often complains of fatigue and depression. The patient's health history may include a recent diagnosis of diabetes mellitus.

Complications

Complications may occur before or after surgical treatment. Preoperative complications include malnutrition, spread of the cancer, and gastric or duodenal obstruction. Postoperative complications include infection, breakdown of the surgical site, fistula formation, diabetes mellitus, and malabsorption syndrome. If the patient has chemotherapy or radiation therapy, complications specific to those therapies may also occur.

Thrombophlebitis is a common complication of cancer of the pancreas. As the tumor grows, byproducts of the tumor growth appear to increase the levels of thromboplastic (clotting) factors in the blood, making clotting easier. The potential for thrombophlebitis increases if the patient is confined to bed or has surgery.

Diagnostic Tests

Serum amylase, lipase, alkaline phosphatase, and bilirubin levels are elevated. Blood coagulation tests, such as clotting time, may be done. Carcinoembryonic antigen (CEA) may be ordered to confirm the presence of cancer (normal: less than 5 ng/mL).

Abdominal radiographs may be ordered to determine the size of the pancreas and the presence of masses. Computed tomography and ultrasonography may be done to more precisely locate any masses in the pancreas. ERCP may be done to visualize the common ducts and to take tissue samples for microscopic analysis.

A tissue sample may be obtained by needle aspiration during ultrasonography. This procedure may cause seeding of the tumor along the needle pathway.

Medical Treatment

Medical treatment depends on the staging of the cancer. If diagnosed early, treatment may be aimed at cure. If the patient's cancer has progressed to distant involvement of other organ structures and lymph nodes, treatment is directed at easing symptoms, thus making the patient more comfortable.

Surgery may include a total or partial **pancreatectomy** or removal of all or part of the pancreas. Whipple's procedure is done to remove the head of the pancreas, parts of the stomach nearby, the lower portion of the common bile duct, and the duodenum (Fig. 33–5). Sometimes the gallbladder is also removed. Potential postoperative problems after Whipple's procedure include failure of the suture lines to hold, causing leakage of pancreatic enzymes and bile into the abdomen; pneumonia or atelectasis from shallow breathing because the incision line is directly under the diaphragm; paralytic ileus; gastric retention or ulceration; wound infection; fistula formation; unstable diabetes mellitus; and renal failure.

Other surgical procedures may remove the entire pancreas, stomach, gallbladder, duodenum, and regional lymph nodes, or merely the distal portion of the pancreas and the spleen for smaller, more localized tumors. Postoperative complications similar to those that occur after Whipple's procedure may develop.

---

pancreatectomy: pancreat—pancreas + ectomy—exicsion

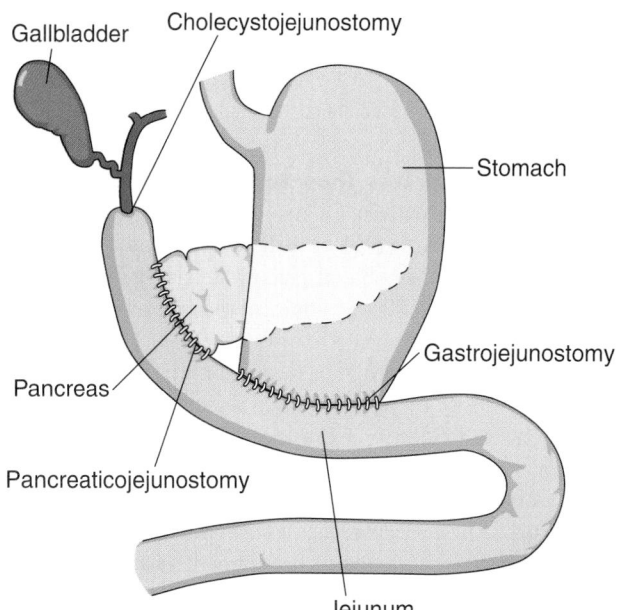

Figure 33-5 Pancreatoduodenectomy (Whipple's procedure) for cancer of the head of the pancreas.

Relief of biliary obstruction can sometimes be accomplished by implanting a stent or plastic tube in the common bile duct during an endoscopic procedure. Pain can be reduced by surgical removal of a portion of the greater splanchnic nerve.

Surgery may be followed by chemotherapy and radiation therapy. In some instances, either radiation therapy or chemotherapy may be used for relief of symptoms if the cancer has become too widespread for surgery. (See Chapter 10 for care of the patient with radiation or chemotherapy.)

Nursing Process

ASSESSMENT. Observe the patient with cancer of the pancreas for evidence of malnutrition and fluid imbalance, including weight loss, inelastic skin turgor, vomiting, fatty stools, and complaints of anorexia or nausea. Review laboratory tests, especially blood glucose, liver function studies, and clotting time. Evaluate the patient every 2 to 4 hours for pain, including what triggers and what helps relieve the pain. Observe the skin for bruising, scaling, and yellowing, and question the patient about itching. Evaluate the patient's mental status for evidence of depression.

NURSING DIAGNOSIS. The common nursing diagnoses for the patient with cancer of the pancreas include the following:

- Imbalanced nutrition: less than body requirements related to inability to digest food, anorexia, nausea, and vomiting
- Pain related to pancreatic tumor or surgical incision
- Risk for deficient fluid volume related to nasogastric tube drainage or hemorrhage
- Risk for impaired skin integrity related to malabsorption, leakage of pancreatic or bile drainage postoperatively

PLANNING. The patient and health team plan to ensure that the patient has adequate nutrition, remains free of excessive pain, and does not develop complications from treatment.

IMPLEMENTATION

Altered Nutrition: Less Than Body Requirements. Frequently assess the patient for nausea or vomiting. Auscultate bowel sounds every 4 to 8 hours. Carefully monitor TPN if ordered. Assess blood glucose levels by finger stick every 6 hours when TPN is administered. Administer regular insulin as ordered.

Administer nasogastric tube feeding or other enteral feedings (gastrostomy or jejunostomy), as scheduled. Observe the patient for evidence of intolerance to feedings, such as diarrhea, nausea, or vomiting. Give pancreatic enzyme replacements as ordered. Steatorrhea (fatty stools) may indicate that the enzyme replacement doses are not meeting the patient's needs. Report steatorrhea immediately.

Pain. The pain associated with pancreatic cancer is intense. Evaluate the patient for pain every 2 to 4 hours and give analgesics as ordered. Opioids such as meperidine hydrochloride or morphine sulfate are usually ordered. For adequate pain relief, dosages of analgesics may need to be very high. Dependence on opioid analgesics is not usually a concern because the outcome of the disease is so poor. Administer other medications, such as antidepressants, as ordered.

Place the patient in the position that provides the most comfort. The usual position is semi-Fowler's, so that the diaphragm has the most expansion room. However, if the patient is more comfortable in a side-lying position with legs drawn up, support him or her in that position. Additional methods of pain relief, such as guided imagery, distraction, or massage, may assist the patient to greater comfort.

Risk for Fluid Volume Deficit. Monitor the patient's intake and output carefully. Tachycardia, tachypnea, and low blood pressure may indicate excessive fluid loss. Monitor laboratory values, especially serum sodium, potassium, calcium, and chloride levels. If electrolyte values are low, the physician may order intravenous replacement solutions. Serum albumin should be between 6 and 8 g/dL. Report a low albumin level and assist with monitoring intravenous albumin therapy if ordered.

Coagulation studies may reveal a lowered clotting time from vitamin K deficiency. Carefully observe the patient for abnormal bruising, bleeding gums, or pink-tinged urine. Inspect the patient's abdomen and flanks for Cullen's sign (bluish discolorations around the umbilicus) and Turner's sign (bluish discolorations on the flanks), which are indications of bleeding into the retroperitoneum. If the patient has tubes or drains, carefully inspect around the incision sites and the drainage tubing for any bleeding. Teach the patient to use a soft-bristle toothbrush. Encourage use of an electric razor rather than a straight razor. The physician may order vitamin K.

High Risk for Impaired Tissue Integrity. Assess the patient for any complaints of itching. Help the patient keep fingernails short, provide frequent skin care with products free of soap or alcohol, and protect skin around drains with skin-protective barrier products and ostomy bags. Products such as calamine lotion may be ordered to decrease itching.

Exercise special care of any drains to prevent unnecessary tension that may cause the sutures to give way. Keep all drains patent, and keep drainage tubing and bags free from kinks. If the nasogastric tube must be irrigated, 10 to 20 mL of normal saline or air may be gently instilled, unless agency policy calls for a different procedure. Place the patient in semi-Fowler's position to help with gravity drainage.

PATIENT EDUCATION. Teach the patient and family self-care measures such as blood glucose monitoring, insulin administration, signs and symptoms of hyperglycemia and hypoglycemia (see Chapter 38), and the regimen for pancreatic enzyme replacement. Instruct the patient on how to manage dressing changes if he or she is to be discharged with tubes or drains. The patient and family should know the signs and symptoms of hemorrhage, gastric ulceration, infection, and fistula formation. Arrange to have the patient referred to the appropriate community health care agency for assistance with pain management. Offer a referral for hospice care.

EVALUATION. The plan of care for the patient with pancreatic cancer is successful if the patient exhibits the following:

- Maintains body weight within 5 percent of normal body weight and experiences no nausea or vomiting
- States that pain remains at 2 or less on a pain scale of 0 to 5
- Has urinary output greater than 40 mL/h, skin turgor is elastic, mucous membranes are moist, and pulse and blood pressure remain within 10 percent of patient's baseline
- Has no complaints of sudden, excessive abdominal pain or rigidity; incisions healed at the expected rate
- Can demonstrate the appropriate self-care procedures for tubes, drains, dressings, and medication administration and states the signs and symptoms of complications that are to be reported immediately

For more information, visit the National Pancreas Foundation at www.pancreasfoundation.org.

▶ DISORDERS OF THE GALLBLADDER
Cholecystitis, Cholelithiasis, and Choledocholithiasis

Gallstones and inflammations of the gallbladder and common bile duct are the most common disorders of the biliary system. These disorders are also common health problems for people in the United States: Nearly 25 million people have gallstones, and about 1 million new cases are diagnosed each year.

Pathophysiology

Cholecystitis is an inflammation of the gallbladder. It is most often a response to obstruction of the common bile duct resulting in edema and inflammation. Often bacteria invade the bile and add to the inflammation and irritation of the gallbladder. Chronic cholecystitis may be the result of repeated attacks of acute cholecystitis or chronic irritation from gallstones. The gallbladder becomes fibrotic and thickened and does not empty easily or completely.

Cholelithiasis, or stones in the gallbladder, are most often composed primarily of cholesterol. **Choledocholithiasis** refers to gallstones in the common bile duct. Although the exact cause of gallstones is unknown, one theory suggests that cholesterol may supersaturate the bile in the gallbladder. After a time, the supersaturated bile crystallizes and begins to form stones. Another type of gallstone is a pigment stone. Pigment stones appear to be composed of calcium bilirubinate, which occurs when free bilirubin combines with calcium.

---

cholecystitis: chole—bile + cyst—bladder + itis— inflammation
choledocholithiasis: chole—bile + docho—duct + lith— stone + iasis—condition

Etiology and Incidence

Pooling, or stasis, of bile within the gallbladder appears to contribute to the formation of stones. Stasis may be caused by a decreased gallbladder emptying rate or a partial obstruction in the common duct. Excessive cholesterol intake combined with a sedentary lifestyle is linked with an increased incidence of cholelithiasis.

Some low-fat diets have been linked to cholelithiasis because the diet appears to free cholesterol from body tissues; the cholesterol then crystallizes in the gallbladder before it is excreted. Further, a family history of cholelithiasis, obesity, diabetes mellitus, pregnancy, some hemolytic blood disorders, and bowel disorders such as Crohn's disease have also been linked to a higher incidence of cholelithiasis.

Cholelithiasis is responsible for about 90 percent of the cases of cholecystitis, or inflammation of the gallbladder. Women between ages 20 and 50 are about three times more likely to have gallstones than men. After age 50, the rate of gallstones is about the same for men and women (Cultural Consideration Box 33–9). Cholelithiasis is responsible for nearly 800,000 hospital admissions at a cost of more than $2 billion dollars every year. The incidence of gallstones increases with age.

BOX 33-9 **CULTURAL CONSIDERATION**

Gallbladder disease is more common among Mexican-Americans. Risk factors include working in occupations such as mining, factories using chemicals, and farming using pesticides.

Native-Americans have an increased incidence of pancreatic and gallbladder disease. It is unknown how much increased dietary risk factors may contribute to gallbladder disease. The nurse can positively affect the nutritional status of Native-Americans by teaching food preparation practices that use less fat.

Signs and Symptoms

Signs and symptoms of cholecystitis and cholelithiasis are similar. Objective symptoms include evidence of inflammation, such as an elevated temperature, pulse, and respirations; vomiting; and jaundice. Subjective symptoms include patient complaints of epigastric pain, RUQ tenderness, nausea, and indigestion. The patient may have a positive Murphy's sign, which is the inability to take a deep breath when an examiner's fingers are pressed below the liver margin.

The epigastric pain caused by cholelithiasis may also be called biliary **colic.** The pain is a steady, aching, severe pain in the epigastrium and RUQ that may radiate back to behind the right scapula or to the right shoulder. The pain usually begins suddenly after a fatty meal and lasts for 1 to 3 hours. If the pain is caused by a stone in the common bile duct (choledocholithiasis), the pain may last until the stone has passed into the duodenum. Jaundice is more commonly

---

colic: colic—spasm

present with acute choledocholithiasis because the common bile duct is blocked or inflamed.

The biliary colic caused by cholecystitis typically lasts 4 to 6 hours. The pain is made worse with movement such as breathing. The patient usually has nausea, vomiting, and a low-grade fever with the pain.

Patient complaints of heartburn, indigestion, and flatulence are more common with chronic cholecystitis. Patients often report a medical history that suggests repeated attacks of acute cholecystitis (Table 33–4).

Family history of either cholecystitis or cholelithiasis; dietary habits such as high fat intake or a recent low-fat diet; and complaints of flatulence (gas), eructation (belching), nausea, vomiting, or abdominal discomfort after a high-fat meal are common evidence of a gallbladder disorder.

Complications

Complications of cholecystitis include inflammation of the bile ducts (cholangitis), necrosis or perforation of the gallbladder, empyema (a collection of purulent drainage in the gallbladder), fistulas, and adenocarcinoma of the gallbladder. A major complication of choledocholithiasis is acute pancreatitis if the pancreatic duct is obstructed.

Diagnostic Tests

The patient may have an elevated white blood cell count (normal: 5,000 to 10,000 cells per mm³). The serum amylase levels may be elevated (normal: 59 to 190 IU/L) if the pancreas is involved or if there is a stone in the common duct.

An abdominal x-ray examination is done to determine whether the gallstone is primarily calcium. Calcium-based stones are less responsive to medical treatments. Cholesterol stones are highly responsive to medical therapy. Further radiological examinations may include an oral cholecystography. The patient is given a contrast medium either by mouth or intravenously. Radiographs then show the presence of gallstones. Cholesterol stones usually float to the top of the bile. Patients are given a high-fat meal, and the radiologist can see the gallbladder contract and empty.

Oral cholecystography is not used with patients who are jaundiced because the liver cannot transport the contrast medium to the obstructed gallbladder.

Sonograms can detect stones and may be able to determine whether the walls of the gallbladder have thickened.

An ERCP can be done to directly visualize the pancreatic ducts and bile ducts to determine the presence of stones in the common duct and occasionally to remove stones from the common duct.

Medical Treatment

Medical management for an acute episode of cholecystitis centers on pain control, prevention of infection, and maintenance of fluid and electrolyte balance. Pain control is achieved by using opioid analgesics. The analgesic agent most often ordered is meperidine hydrochloride because morphine sulfate is believed to cause spasms of the gallbladder, biliary ducts, and the sphincter of Oddi. Antispasmodics or anticholinergic drugs such as propantheline bromide (Pro-Banthine) and dicyclomine hydrochloride (Bentyl) may be ordered to decrease the biliary colic. If the patient has nausea and vomiting, an antiemetic such as prochlorperazine (Compazine) may be ordered. Patients are placed on high-protein, low-fat diets after the nausea and vomiting subside (Nutrition Notes Box 33–10).

Treatment for cholelithiasis usually involves surgical removal of the gallbladder. The surgical procedure may be cholecystectomy via **laparoscopy** or a traditional cholecystectomy. A laparoscopic cholecystectomy may be done with a laparoscope through four small puncture wounds in the abdomen. A traditional cholecystectomy is done through a long, transverse, right subcostal incision. The patient has a T-tube inserted into the common duct for several days postoperatively to ensure that bile drainage is not obstructed

---

laparoscopy: laparo—pertaining to flank + scopy—to examine

TABLE 33–4	SYMPTOMS OF GALLBLADDER DISORDERS		
	Acute Cholecystitis	**Chronic Cholecystitis**	**Cholelithiasis and Choledocholithiasis**
Biliary colic	Lasts 4–6 hours Worse with movement	Only during acute attack	Sudden onset
Jaundice	✓ (if common bile duct is inflamed or blocked)	✓	Lasts 1–3 hours Radiates to right scapula or shoulder
Low-grade fever	✓	✓	✓
Nausea, vomiting	✓	Only during acute attack	✓
Repeated attacks		✓	
Heartburn, indigestion, and flatulence		✓	
Complications	Cholangitis Necrosis or perforation Fistulas	Empyema Fistulas Adenocarcinoma	Acute pancreatitis

✓ = Usually present or commonly found on assessment.

BOX 33-10 *Nutrition Notes*

Modifying the Diet for Patients with Gallbladder Disease

During an acute attack of cholecystitis, a full liquid diet with minimal fat is usually allowed. For treatment of chronic cholecystitis, the patient is instructed to do the following:

- Correct obesity.
- Avoid troublesome and gas-forming foods.
- Decrease dietary fat each day by (1) selecting skim-milk dairy products, (2) limiting fats or oils to 3 teaspoons, and (3) consuming no more than 6 ounces of very lean meat.

Cholelithiasis has been associated with a long overnight fast that permits concentrated bile to remain in the gallbladder for an extended period.[1] Eating a light bedtime snack or drinking two glasses of water on arising if breakfast is delayed alter this risk factor.

1. Sichiere, R, Everhart, JE, and Roche, H: A prospective study of hospitalization with gallstone disease among women: Role of dietary factors, fasting period, and dieting. Am J Public Health 81:880, 1991.

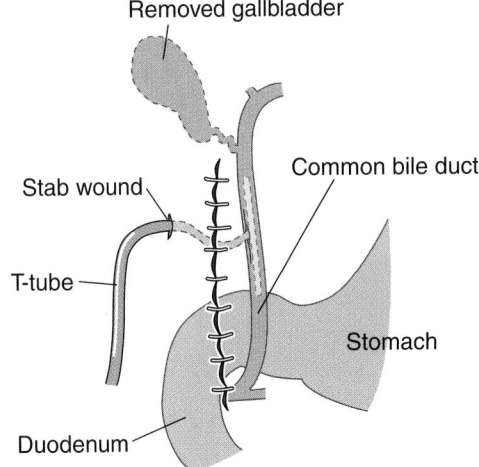

Figure 33-6 T-tube. A T-tube is used to drain bile after a cholecystectomy until swelling of the duct subsides.

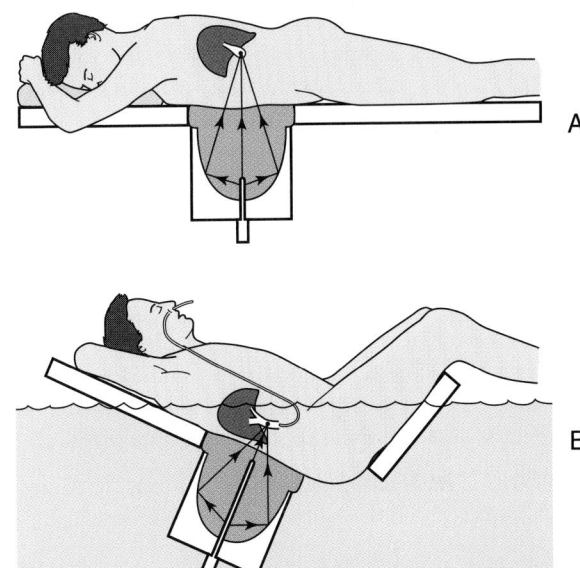

Figure 33-7 Extracorporeal shock-wave lithotripsy. Shock waves are transmitted through water to break up gallstones. *(A)* Position for stones in gallbladder. Patient is lying on a fluid-filled bag. *(B)* Position for stones in common bile duct. Patient is in a water bath.

(Fig. 33–6). The patient with a traditional cholecystectomy has incisional pain that creates difficulty with coughing and deep breathing postoperatively because deep breathing causes the diaphragm to press on the operative site. Patients are hospitalized for 2 to 3 days with a traditional cholecystectomy and 24 hours or less with a laparoscopic cholecystectomy.

Some patients are poor surgical risks and have stones in the gallbladder or biliary duct that cannot be removed easily by other methods. Such patients may have a cholecystostomy, which is an incision directly into the gallbladder to remove a stone. If the stone is in the biliary ducts, a choledocholithotomy may be done. The patient usually has a T-tube in place for several days to help remove bile from swollen structures.

Other methods of treatment include **choledochoscopy, extracorporeal shock-wave lithotripsy** (ESWL), oral drugs to dissolve stones, and direct-contact dissolving drugs. The procedure for a choledochoscopy involves the use of an endoscope to explore the common bile duct and in some instances to snare and remove any stones found.

Extracorporeal shock-wave lithotripsy uses shock waves as a noninvasive method to destroy stones in the gallbladder or biliary ducts. Patients who are considered poor surgical risks and who have few cholesterol stones that are not calcified are the most likely candidates for ESWL. The patient lies face down on a water bag over a lithotriptor (Fig. 33–7). A conductive gel is placed between the patient and the water bag. Ultrasound is used to locate the stone or

stones and to monitor the destruction of the stones. The patient requires sedation and strong analgesics during the procedure to reduce the pain and discomfort of the shock waves. After ESWL, the patient is usually put on a course of oral dissolution drugs to ensure complete removal of all stones and stone fragments.

Dissolution of stones with drugs may be attempted with chenodeoxycholic acid (chenodiol) or ursodeoxycholic acid (ursodiol). Patients who are poor surgical risks because of advanced age or severe health problems and who have cholesterol stones may be given oral dissolution drugs. The major disadvantages to the use of these drugs are that abnormal liver function studies and diarrhea are common side effects. Patients may also have an increase in serum choles-

---

choledochoscopy: chole—bile + docho—duct + scopy—to examine

extracorporeal shock-wave lithotripsy: extra—outside + corporeal—body + litho—stone + tripsy—rub or crush

terol while taking the drugs. Treatment with the dissolution drugs may take from 4 months to 2 years. Patients may need to take cholesterol-lowering drugs such as cholestyramine (Questran) in addition to the dissolution drugs to lower serum cholesterol and to decrease the probability of stones reforming.

Direct-contact dissolution drugs are administered directly into the gallbladder through a percutaneous transhepatic catheter. The physician inserts a catheter through the wall of the abdomen into the gallbladder and then injects and aspirates the dissolution agent repeatedly during the treatment. Candidates for this procedure are patients who are poor surgical risks and whose gallbladder can be seen during oral cholecystography. Disadvantages of the procedure include abnormal liver function studies, pain at the catheter site, nausea, and elevations in the white blood cell count.

Nursing Process

ASSESSMENT. Assess the patient frequently for pain, including location, intensity, and relieving and intensifying events. Take the patient's vital signs, particularly the temperature and pulse, frequently to monitor for signs of infection. Weigh the patient and inspect mucous membranes, skin turgor, and urinary output for signs of dehydration. Measure intake and output, including any emesis or drainage from nasogastric tubes or T-tubes. Observe stools and urine for color and consistency. Obstruction of bile flow may result in stools that are clay colored or have a foul, greasy appearance or urine that is dark amber or tea colored. Report either finding immediately. Evaluate laboratory studies for elevation in the white blood cell count or abnormalities in electrolytes or serum bilirubin levels.

NURSING DIAGNOSIS. Common nursing diagnoses for the patient with cholecystitis include the following:

■ Pain related to biliary colic
■ Risk for deficient fluid volume related to anorexia, nausea, vomiting, or excessive tube drainage

Additional nursing diagnoses for the patient with cholelithiasis who has a surgical procedure include the following:

■ Risk for impaired skin integrity related to surgical incision and T-tube drainage
■ Risk for ineffective breathing pattern related to abdominal incision

PLANNING. Goals for the patient with cholecystitis or cholelithiasis include pain management so that biliary colic is tolerable, management of the nausea and vomiting to prevent excessive fluid loss, management of the surgical incision to prevent infection or tissue damage from drainage, and close monitoring to detect and treat any infection.

IMPLEMENTATION
Pain. Assess the patient frequently for pain. The pain should be controlled at not more than 2 on a pain scale of

0 to 5. Give the patient meperidine hydrochloride as ordered. Give antispasmodics or anticholinergics as ordered. Place the patient is placed in a position of comfort and provide support.

Risk for Fluid Volume Deficit. Assist with administration of intravenous fluids and electrolytes as ordered while the patient is on restricted oral intake. Give antiemetics as needed if the patient experiences nausea and vomiting. After surgery, a nasogastric tube may be inserted to prevent an ileus from developing. Frequently assess for the return of bowel sounds and passage of flatus. After the nasogastric tube is removed, the patient is slowly reintroduced to a solid diet, usually high protein and low fat. Monitor T-tube drainage. About 250 to 500 mL of yellowish-green bile is common within the first 24 hours after surgery. The amount of bile diminishes over the next several days. Carefully observe the T-tube drainage unit to prevent kinking of the tubing. Pressure in the biliary drainage system from poor drainage may greatly increase the patient's pain and the risk for infection.

Risk for Impaired Skin Integrity. Change the patient's position frequently. Encourage the patient to stay in a low semi-Fowler's position as much as possible. Inspect the cholecystectomy incision frequently for excessive drainage or evidence of infection such as redness, edema, or warmth. Change dressings frequently to protect the skin around the incision site. Montgomery straps prevent skin irritation from repeated or prolonged exposure to adhesive bandages. Protect the skin with a skin barrier product or bag such as those used with colostomies if bile is leaking around the T-tube site. An enterostomal therapist can be consulted for the best choice of dressing if the facility has such a specialist. Inspect the patient's skin and the sclera of the eyes frequently for jaundice. Report patient complaints of excessive itching.

Risk for Ineffective Breathing Pattern. Deep breathing and coughing after any surgical procedure help prevent respiratory tract infections and atelectasis. Patients with high abdominal incisions are particularly reluctant to cough and deep breathe. Instruct the patient in the proper techniques before surgery and give the opportunity to practice. After surgery, encourage the patient to cough and deep breathe at every encounter. If the patient is reluctant to cough because of pain, the pain medication regimen may need to be evaluated. Assist the patient with splinting and encourage the patient to walk when permitted.

PATIENT EDUCATION. Patient education focuses on diet. Patients are put on high-protein, low-fat diets. Encourage obese patients to lose weight. After a cholecystectomy, there is a slow reintroduction of fat in the diet. Once the duodenum becomes accustomed to constant infusion of bile, the patient's individual tolerance for fat becomes the only restriction for diet.

EVALUATION. The plan of care for a patient with cholecystitis or cholelithiasis is successful if the patient exhibits the following:

■ Reports pain not greater than 2 on a pain scale of 0 to 5 or pain that is tolerable

■ No weight loss, urinary output greater than 50 mL/h, moist mucous membranes, elastic skin turgor, and no complaints of excessive thirst

■ Intact skin with no warmth, redness, swelling, or purulent drainage at the wound site; no jaundice or complaints of itching

■ Clear breath sounds and a normal white blood cell count

CRITICAL THINKING

Donna

Donna, a 23-year-old woman, is diagnosed with possible acute cholecystitis. She is 5-feet, 6-inches tall and weighs 138 pounds. She has recently stopped following an extremely low-fat diet after losing nearly 70 pounds. Her physician wishes to delay surgery until her inflammation has subsided.

1. What risk factors does Donna have for cholecystitis?
2. What diagnostic tests might be ordered to confirm Donna's diagnosis of cholecystitis?

3. What medications can you anticipate that the physician will order for Donna?
4. What type of diet will Donna need to eat after discharge?
5. When the diagnosis of cholecystitis is confirmed, what type of surgical treatment might be ordered?

Answers at end of chapter.

REVIEW QUESTIONS

1. The patient with altered thought processes related to liver failure most likely has which of the following laboratory findings?
   a. Low bilirubin
   b. Low amylase
   c. High hematocrit
   d. High ammonia

2. Jack, a patient with chronic liver failure, has an episode of bleeding. Which of the following conditions placed Jack at risk for bleeding?
   a. Portal hypertension
   b. Low vitamin K
   c. Elevated liver enzymes
   d. High-fiber diet

3. Most gallstones are composed of which of the following?
   a. Lipase
   b. Cholesterol
   c. Sodium
   d. Potassium

4. John develops jaundice and dark, amber-colored urine. Which of the following is the most likely cause?
   a. Encephalopathy
   b. Pancreatitis
   c. Bile duct obstruction
   d. Cholecystitis

5. Jeff is a 26-year-old health care worker who is diagnosed with hepatitis C virus. Which of the following questions is most important to ask Jeff?
   a. "Have you eaten any raw seafood recently?"
   b. "Have you experienced a needle stick?"
   c. "Have you made beds or handled clothing such as slippers without gloving?"
   d. "Has anyone coughed into your face when you weren't wearing a mask?"

6. Jim, age 43, is admitted to your unit with chronic pancreatitis. You recognize that an elevation in which diagnostic test indicates chronic pancreatitis?
   a. Serum bilirubin
   b. Serum calcium
   c. Serum albumin
   d. Serum amylase

7. In planning care for the newly admitted patient with acute pancreatitis, you assign the highest priority to which patient outcome?
   a. Patient expresses satisfaction with pain control.
   b. Patient verbalizes understanding of medications for home.
   c. Patient increases activity tolerance.
   d. Patient maintains normal bowel function.

REFERENCES

1. American Cancer Society: Cancer Facts and Figures. Report No. 00-300M American Cancer Society, Atlanta, 2001.

2. Purnell, L, and Paulanka, B (eds): Transcultural Health Care: A Culturally Competent Approach. FA Davis, Philadelphia, 1998.

Answers to Critical Thinking

Carl

1. Foreign travel within the past 2 months, fatigue, nausea, and irritability suggest hepatitis A virus. Recent possible exposure to materials contaminated with blood or body fluids, fatigue, headache, and nausea suggest hepatitis B virus.
2. Careful handwashing and standard precautions when handling any body fluids or feces should be instituted.
3. The nurse should plan to give an antiemetic if Carl is nauseated. Larger meals should be given early in the day, with Carl in an upright or sitting position. The nurse should also ensure that the environment is free of noxious stimulants such as unpleasant odors. The diet should be high calorie, high protein, high carbohydrate, and low fat.
4. Any medication that is known to be hepatotoxic, such as acetaminophen, aspirin, and diazepam (Valium), should be avoided.
5. Carl should be reminded that cleanliness, especially with food preparation, is essential. He should also be reminded that frequent handwashing is crucial. Carl needs to know that alcohol and other liver-toxic substances should be avoided.

Mary

1. Mary has a history of poor nutrition. Her age also puts her at risk.
2. Mary may report that she has malaise, nausea, weight loss, a change in bowel habits, and dull, aching RUQ pain.
3. Serum albumin is usually less than 3.2 g/dL. Her prothrombin time will probably be greater than 25 seconds.
4. Esophageal varices and ascites are the two greatest concerns for the patient with portal hypertension.
5. The physician will usually order diuretics, intravenous albumin infusions, and a sodium-restricted diet.
6. The possibility of suffocation from having the gastric tube dislodge into the esophagus requires that scissors be kept at the bedside at all times during tamponade.

Mrs. Samuels

1. The most common cause of acute pancreatitis is excessive alcohol intake. Mrs. Samuels denies alcohol consumption, but she does have the risk factor of having a recent ERCP, which may have dislodged a gallstone or irritated the pancreatic duct.
2. Respiratory distress may result from excess fluid accumulation in the retroperitoneal space and from shallow respirations that seek to decrease pressure from the diaphragm on the inflamed pancreas and surrounding tissues.
3. Pancreatitis is similar to a chemical burn and may cause erosion of major blood vessels in surrounding tissue.
4. Serum amylase may be elevated as much as 40 times more than normal early in acute pancreatitis.
5. Narcotics are ordered because pain is intense, and pain with anxiety stimulates the autonomic nervous system, which may stimulate greater production of pancreatic enzymes.
6. Stomach acid stimulates the production of pancreatic enzymes. Histamine antagonists decrease stomach acidity.

Donna

1. Some low-fat diets have been linked to the development of cholesterol gallstones, which then irritate the gallbladder and cause inflammation.
2. Donna's physician might order a white blood cell count, which will be elevated if she has cholecystitis. In addition, the physician may order a cholecystogram to visualize the gallbladder and its contents and the common bile duct.
3. You can anticipate that the physician will order an antibiotic, a narcotic such as meperidine hydrochloride, and possibly a antispasmodic such as the anticholinergic propantheline bromide.
4. Donna will need to eat a low-fat diet after discharge. Eventually she may be able to add more fats to her diet as her body adjusts.
5. Donna will probably have a laparoscopic cholecystectomy unless her surgeon decides that she needs a traditional cholecystectomy.

UNIT EIGHT BIBLIOGRAPHY

Brozenec, SA, and Russell, SS (eds): Core Curriculum for Medical-Surgical Nursing. Academy of Medical Surgical Nurses, Pitman, NJ, 1995.

Buckhold, KM: Who's afraid of hepatitis C? American Journal of Nursing. 100(5):26-31, 2000.

Grindel, CG, and Costello, MC: Nutritional screening: An essential assessment parameter. Medsurg Nursing 5(3):145, 1996.

Lutz, CA, and Przytulski, KR: Nutrition and Diet Therapy, ed 2. FA Davis, Philadelphia, 1997.

McCormick, ME: Endoscopic retrograde cholangiopancreatography. American Journal of Nursing 99(2):24HH-JJ, 1999.

Nursing99: Understanding gallstone formation. Author 29(4):14, 1999.

Office of Minority Health Resource Center: Liver cancer is high among American Chinese. Author, Washington, DC, 1994.

Sichiere, R, Everhart, JE, and Rothe, H: A prospective study of hospitalization with gallstone disease among women: Role of dietary factors, fasting period, and dieting. American Journal of Public Health 81:880, 1991.

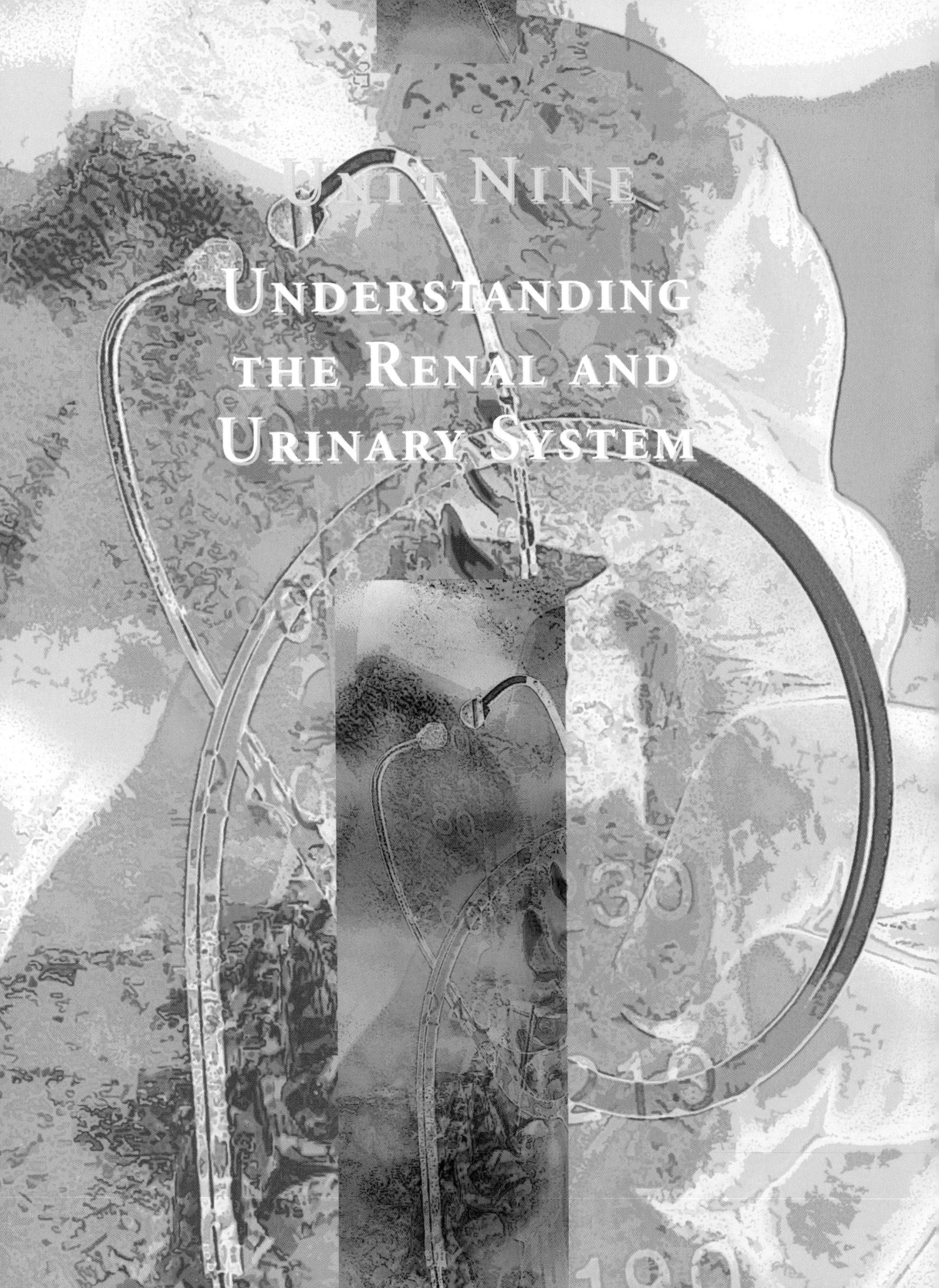

UNIT NINE

UNDERSTANDING THE RENAL AND URINARY SYSTEM

34

URINARY SYSTEM FUNCTION, ASSESSMENT, AND THERAPEUTIC MEASURES

Betty J. Ackley

KEY TERMS

cystoscopy (sis-**TAHS**-koh-pee)

dysuria (dis-**YOO**-ree-ah)

hematuria (HEM-uh-**TYOOR**-ee-ah)

percutaneous (PER-kyoo-**TAY**-nee-us)

pyelogram (**PIE**-loh-GRAM)

uremia (yoo-**REE**-mee-ah)

QUESTIONS TO GUIDE YOUR READING

1. What is the normal anatomy of the urinary system?
2. What is the normal function of the urinary system?
3. What are the effects of aging on the urinary system?
4. What is normal and oliguric 24-hour urinary output?
5. What data should you collect when caring for a patient with a disorder of the urinary system?
6. How do you collect a midstream, clean-catch urine specimen and 24-hour creatinine clearance specimen?
7. What is the meaning of an elevated serum creatinine, blood urea nitrogen, and uric acid level?
8. What is the preparation and aftercare for diagnostic tests of the urinary system?
9. Which nursing actions can be taken to decrease the risk of infection in catheterized patients?
10. What nursing care should be given to patients who are incontinent?

► REVIEW OF ANATOMY AND PHYSIOLOGY

The urinary system consists of two kidneys, two ureters, the urinary bladder, and the urethra. The kidneys form urine, and the rest of the system eliminates urine. The purpose of urine formation is the removal of potentially toxic waste products from the blood; however, the kidneys have other equally important functions as well:

■ Regulation of the blood volume by the excretion or conservation of water
■ Regulation of the electrolyte balance of the blood by the excretion or conservation of minerals
■ Regulation of the acid-base balance of the blood by the excretion or conservation of ions such as hydrogen or bicarbonate
■ Regulation of all of the above in tissue fluid

The process of urine formation thus helps maintain the normal composition, volume, and pH of blood and tissue fluid.

Kidneys

The two kidneys are located in the upper abdominal cavity behind the peritoneum on each side of the vertebral column. The upper portions of both kidneys rest on the lower surface of the diaphragm and are enclosed and protected by the lower ribcage. The kidneys are cushioned by surrounding adipose tissue, which is in turn covered by a fibrous connective membrane called the renal fascia; both help hold the kidneys in place. On the medial side of each kidney is an indentation called the hilus, where the renal artery enters and the renal vein and ureter emerge. The renal artery is a branch of the abdominal aorta, and the renal vein returns blood to the inferior vena cava. The ureter carries urine from the kidney to the urinary bladder.

Internal Structure of the Kidney

A frontal section of the kidney shows three distinct areas (Fig. 34–1). The outermost area is the renal cortex, which contains the parts of the nephrons called renal corpuscles and convoluted tubules. The middle area is the renal medulla, which contains loops of Henle and collecting tubules. The renal medulla consists of wedge-shaped pieces called renal pyramids; the apex, or papilla, of each pyramid points medially. The third area is a cavity called the renal pelvis; it is formed by the expansion of the ureter within the kidney at the hilus. Funnel-shaped extensions of the renal pelvis, called calyces, enclose the papillae of the renal pyramids. Urine flows from the pyramids into the calyces, then to the renal pelvis, and finally into the ureter.

Nephron

The nephron is the structural and functional unit of the kidney. Urine is formed in the approximately 1 million nephrons in each kidney. The two major parts of a nephron are the renal corpuscle and the renal tubule; these and their subdivisions and blood vessels are shown in Fig. 34–2.

A renal corpuscle consists of a glomerulus surrounded by a Bowman's capsule. The glomerulus is a capillary network that arises from an afferent arteriole and empties into an efferent arteriole. The diameter of the efferent arteriole is smaller than that of the afferent arteriole, which helps maintain a fairly high blood pressure in the glomerulus. Bowman's capsule is the expanded end of a renal tubule; it encloses the glomerulus. The inner layer of Bowman's capsule has pores and is highly permeable; the outer layer has no pores and is not permeable. The space between the inner and outer layers contains renal filtrate, the fluid that is formed from the blood in the glomerulus and that will eventually become urine.

The renal tubule continues from Bowman's capsule and consists of the proximal convoluted tubule, the loop of Henle, and the distal convoluted tubule. The distal convoluted tubules from several nephrons empty into a collecting tubule. Several collecting tubules then unite to form a papillary duct that empties urine into a calyx of the renal pelvis. All the parts of the renal tubule are surrounded by the peritubular capillaries, which arise from the efferent arteriole and receive the materials reabsorbed by the renal tubules.

Blood Vessels of the Kidney

The pathway of blood flow through the kidney is an essential part of the process of urine formation. Blood from the abdominal aorta enters the renal artery, which branches extensively within the kidney into smaller arteries. The smallest arteries give rise to afferent arterioles in the renal cortex. From the afferent arterioles, blood flows into the glomeruli (capillaries), to efferent arterioles, to peritubular capillaries, to veins in the kidney, to the renal vein, and finally to the inferior vena cava. In this pathway are two sets of capillaries—that is, two sites of exchanges between the blood and the surrounding tissues (in this case, the parts of the nephrons). The exchanges that take place in the capillaries of the kidneys form urine from blood plasma.

Formation of Urine

The formation of urine involves three major processes: glomerular filtration in the renal corpuscles, tubular reabsorption, and tubular secretion.

Glomerular Filtration

Filtration is the process by which blood pressure forces plasma and dissolved materials out of capillaries. In glomerular filtration, blood pressure forces plasma, dissolved substances, and small proteins out of the glomeruli and into Bowman's capsules. This fluid is then called renal filtrate.

The blood pressure in the glomeruli is relatively high, about 60 mm Hg. The pressure in Bowman's capsule is low, and its inner layer is permeable, so that approximately 20 to 25 percent of the blood that enters glomeruli becomes renal filtrate in Bowman's capsules. The larger proteins and blood cells are too large to be forced out of the glomeruli; they remain in the blood. Waste products such as urea and ammonia are dissolved in plasma, so they pass to the renal filtrate, as do dissolved nutrients and minerals. Renal filtrate is similar to blood plasma except that there is far less protein and no blood cells are present.

The glomerular filtration rate (GFR) is the amount of renal filtrate formed by the kidneys in 1 minute; it averages 100 to 125 mL. The GFR may change if the rate of blood flow through the kidney changes. If blood flow increases, the GFR increases, more filtrate is formed, and urinary output increases. If blood flow decreases, the GFR decreases, less filtrate is formed, and urinary output decreases.

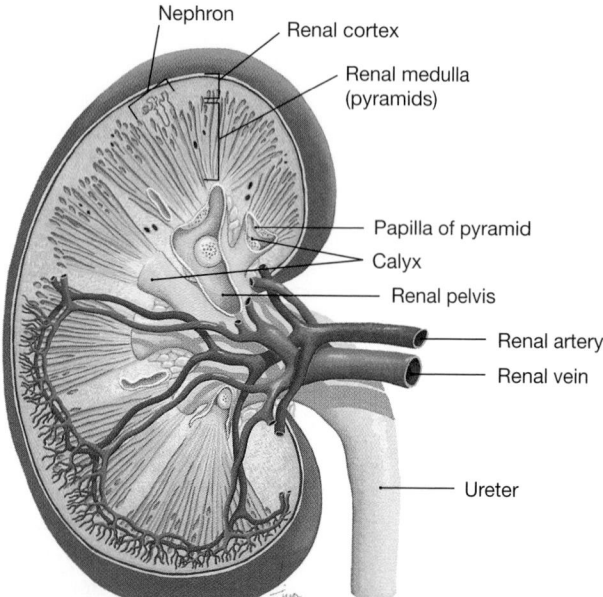

Figure 34–1   Frontal section of the left kidney. (From Scanlon, V, and Sanders, T: Essentials of Anatomy and Physiology, ed 3. FA Davis, Philadelphia, 1999, p 405, with permission.)

Labels: Nephron · Renal cortex · Renal medulla (pyramids) · Papilla of pyramid · Calyx · Renal pelvis · Renal artery · Renal vein · Ureter

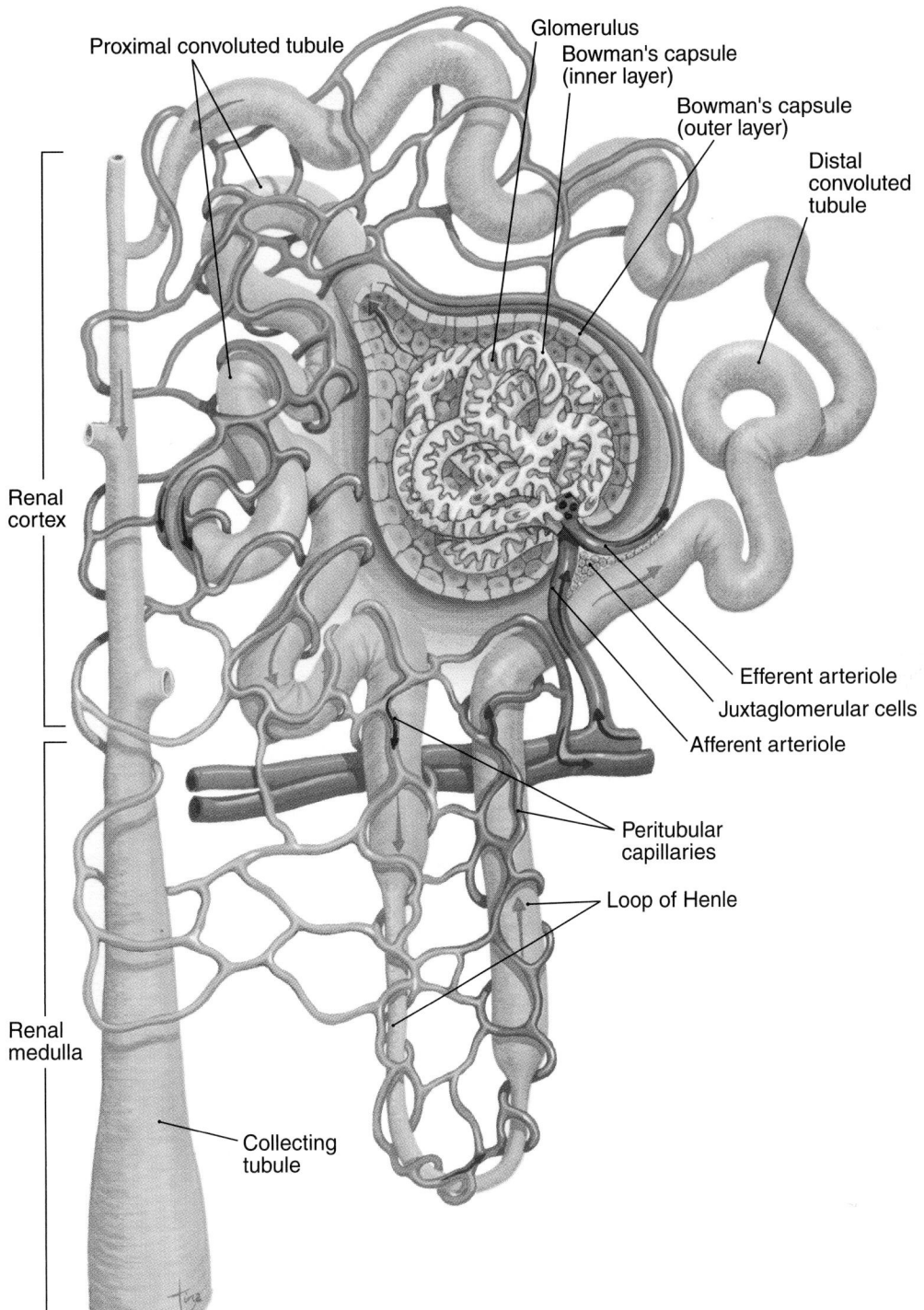

Proximal convoluted tubule

Glomerulus

Bowman's capsule
(inner layer)

Bowman's capsule
(outer layer)

Distal
convoluted
tubule

Renal
cortex

Efferent arteriole

Juxtaglomerular cells

Afferent arteriole

Peritubular
capillaries

Loop of Henle

Renal
medulla

Collecting
tubule

Figure 34-2   A nephron and its associated blood vessels. (From Scanlon, V, and Sanders, T: Essentials of Anatomy and Physiology, ed 3. FA Davis, Philadelphia, 1999, p 406, with permission.)

Tubular Reabsorption

Tubular reabsorption is the recovery of useful materials from the renal filtrate and their return to the blood in the peritubular capillaries. Approximately 99 percent of the renal filtrate formed is reabsorbed, and normal urinary output is 1000 to 2000 mL per 24 hours. Most reabsorption takes place in the proximal convoluted tubules, whose cells have microvilli that greatly increase their surface area. The distal convoluted tubules and collecting tubules are also important sites for the reabsorption of water. The mechanisms of reabsorption are active transport, passive transport, osmosis, and pinocytosis.

Active transport requires energy in the form of adenosine triphosphate (ATP); the cells of the renal tubule use energy to transport useful materials such as glucose, amino acids, vitamins, and positive ions back to the blood. For many of these substances there is a threshold level of reabsorption—that is, a limit to how much the renal tubules can remove from the filtrate. The level of a substance in the renal filtrate is directly related to its blood level. If the blood level of a substance such as glucose is normal, the filtrate level is normal, the threshold level cannot be exceeded, and no glucose appears in the urine.

Passive transport is the mechanism by which negative ions are reabsorbed. They are returned to the blood after the reabsorption of positive ions, because unlike charges attract.

The reabsorption of water is by osmosis following the reabsorption of minerals, especially sodium. The conservation of water is very important to maintain normal blood volume and blood pressure. The hormones that influence the reabsorption of water or minerals are summarized in Table 34–1.

Small proteins in the filtrate are reabsorbed by pinocytosis; the proteins become adsorbed to the membranes of the tubule cells and are engulfed and digested. Normally, all proteins in the filtrate are reabsorbed and none are found in urine.

Tubular Secretion

In tubular secretion, substances are actively secreted from the blood in the peritubular capillaries into the filtrate in the renal tubules. Waste products, such as ammonia and creatinine, and the metabolic products of medications may be secreted into the filtrate to be eliminated in urine. Hydrogen ions may be secreted by the tubule cells to help maintain the normal pH of the blood.

In summary, tubular reabsorption conserves useful materials, tubular secretion may add unwanted substances to the filtrate, and most waste products simply remain in the filtrate and are excreted in urine.

TABLE 34–1	EFFECTS OF HORMONES ON THE KIDNEYS
Hormone (Gland)	**Function**
Aldosterone (adrenal cortex)	Promotes reabsorption of sodium ions from the filtrate to the blood and excretion of potassium ions into the filtrate. Water is reabsorbed following the reabsorption of sodium.
Antidiuretic hormone (posterior pituitary)	Promotes reabsorption of water from the filtrate to the blood.
Atrial Natriuretic hormone (atria of heart)	Decreases reabsorption of sodium ions, which remain in the filtrate. More sodium and water are eliminated in the urine.
Parathyroid hormone (parathyroid glands)	Promotes reabsorption of calcium ions from the filtrate to the blood and excretion of phosphate ions into the filtrate.

Source: Scanlan, V, and Sanders, T: Essentials of Anatomy and Physiology, ed 3. FA Davis, Philadelphia, 1999, p 410, with permission.

The Kidneys and Acid-Base Balance

The kidneys are the organs most responsible for maintaining the normal pH range of blood and tissue fluid. They have the greatest ability to compensate for or correct the pH changes that are part of normal body metabolism or the result of disease.

At its simplest, this function of the kidneys may be described as follows: If body fluids are becoming too acidic, the kidneys secrete more hydrogen ions into the renal filtrate and return more bicarbonate ions back to the blood. This helps raise the pH of the blood back to normal. In the opposite situation, when the body fluids become too alkaline, the kidneys return hydrogen ions to the blood and excrete bicarbonate ions in urine. This helps lower the pH of the blood back to normal.

Other Functions of the Kidneys

Some functions of the kidneys are not related to the formation of urine. These include the secretion of renin, activation of vitamin D, and production of erythropoietin. The production of renin influences urine formation and is considered first.

When blood pressure decreases, the juxtaglomerular cells in the walls of the afferent arterioles secrete the enzyme renin. Renin then initiates the renin-angiotensin-aldosterone mechanism, which results in the formation of angiotensin II. (See Chapter 15.) Angiotensin II stimulates vasoconstriction and increases the secretion of aldosterone, both of which help raise blood pressure.

Vitamin D exists in several structural forms, which are converted to calciferol, the most active form, by the kidneys. Vitamin D is important for the efficient absorption of calcium and phosphate from food in the small intestine.

Erythropoietin is a hormone secreted by the kidneys during states of hypoxia; it stimulates the red bone marrow to increase the rate of red blood cell (RBC) production. With more RBCs in circulation, the oxygen-carrying capacity of the blood is greater and the hypoxic state may be corrected.

Elimination of Urine

The ureters, urinary bladder, and urethra do not change the composition or volume of urine but are responsible for its elimination.

Ureters

The ureters are behind the peritoneum of the dorsal abdominal cavity. Each ureter extends from the hilus of a kidney to the lower, posterior side of the urinary bladder. The smooth muscle in the wall of the ureter contracts in peristaltic waves to propel urine toward the urinary bladder. As the bladder fills, it expands and compresses the lower ends of the ureters to prevent backflow of urine.

Urinary Bladder

The urinary bladder is a muscular sac below the peritoneum and behind the pubic bones. In women the bladder is inferior to the uterus; in men the bladder is superior to the prostate gland. The functions of the bladder are the temporary storage of urine and its elimination.

Urethra

The urethra carries urine from the bladder to the exterior. Within its wall is the external urethral sphincter, which is made of skeletal muscle and is under voluntary control.

In women the urethra is 1 to 1.5 inches long and is anterior to the vagina. In men the urethra is 7 to 8 inches long and extends through the prostate gland and penis. The male urethra carries semen, as well as urine.

The Urination Reflex

Urination is a spinal cord reflex over which voluntary control may be exerted. The stimulus is the stretching of the detrusor muscle as urine accumulates in the bladder. Sensory impulses travel to the sacral spinal cord, and motor impulses return along parasympathetic nerves to the detrusor muscle, causing contraction. At the same time, the internal urethral sphincter relaxes. If the external urethral sphincter is voluntarily relaxed, urine flows into the urethra and the bladder is emptied.

Characteristics of Urine
Amount

Normal urinary output is 1000 to 2000 mL per 24 hours. Any changes in fluid intake or other fluid output (such as sweating) changes this volume.

Color

The color of urine is often referred to as straw or amber. Dilute urine is a lighter color (straw) than is concentrated urine. Freshly voided urine is clear. Cloudy urine may indicate an infection.

Specific Gravity

The usual range of specific gravity of urine is 1.010 to 1.025; this is a measure of the dissolved materials in urine. (The specific gravity of distilled water is 1.000.) The higher the specific gravity, the more dissolved material is present. Specific gravity of urine is a measure of the concentrating ability of the kidneys; the kidneys must excrete the waste products that are constantly formed in as little water as possible.

pH

The pH range of urine is 4.6 to 8.0, with an average of 6.0. Diet has the greatest influence on urinary pH. A vegetarian diet results in a more alkaline urine; a high-protein diet results in a more acidic urine.

Constituents

Urine is approximately 95 percent water, which is the solvent for waste products and salts. Nitrogenous wastes include urea, creatinine, and uric acid. Urea is formed by liver cells when excess amino acids are deaminated to be used for energy production. Creatinine comes from the metabolism

of creatine phosphate, an energy source in muscles. Uric acid comes from the metabolism of nucleic acids—that is, the breakdown of DNA and RNA.

Aging and the Urinary System

With age, the number of nephrons in the kidneys decreases, often to half the original number by age 70 or 80. The GFR also decreases; this is in part a consequence of arteriosclerosis and diminished renal blood flow. The urinary bladder decreases in size, and the tone of the detrusor muscle decreases. This may result in the need to urinate more frequently or in residual urine in the bladder after voiding. Elderly people are also more subject to infections of the urinary tract, and the changes of aging may influence medication therapy for elderly people (Gerontological Issues Box 34–1).

BOX 34-1 GERONTOLOGICAL ISSUES

Age-related Renal Changes
Certain changes typically occur in the renal system as people age. They include the following:

- The renal mass becomes smaller.
- Renal flow decreases by 50 percent, with subsequent decreased glomerular filtration rate.
- Tubular function and the exchange of substances decrease.
- Bladder muscles weaken and bladder capacity decreases, leading to increased frequency and nocturia.
- The voiding reflex is delayed.

Also, keep in mind that most drugs are excreted through the kidneys. Consequently, changes in renal function become a serious consideration for older adults who need drug therapy. Decreased renal function could slow the excretion of some drugs, keeping them in the body longer. This can increase the risk of adverse drug reactions, such as toxicity and overdose. It is important to monitor kidney function (such as creatinine and blood urea nitrogen levels) in an older person receiving drug therapy.

▶ NURSING ASSESSMENT
Health History

When recording a patient's health history, the following minimum information should be obtained:

- History of renal or urinary problems
- Family history of urinary disorders, diabetes, or hypertension (diabetes and hypertension are major contributing factors to renal failure)
- Presence of pain or burning either with voiding or over the kidney
- Voiding pattern
- New onset of symptoms of edema, shortness of breath, weight gain, abdominal fullness, or other symptoms of renal failure (see Chapter 35)
- Current medications, including over-the-counter medications and herbs
- Functional ability to take care of toileting needs

■ Fluid intake and diet
■ Age

LEARNING TIPS

To find your kidneys, put your hands on your hips with your thumbs pointing back and upward. Your thumbs are pointing to the bottom of the kidneys. This is called the flank area. Pain in this area is called flank pain.

Any symptoms should be further assessed using the WHAT'S UP? format found in Chapter 1.

If the patient has impaired kidney function or is in kidney failure, a complete head-to-toe assessment is needed, because kidney failure (renal failure) affects every system of the body. (See Chapter 35.)

Physical Assessment

The nurse first inspects the skin for color and texture. A patient with chronic renal failure may have a yellow or gray cast to the skin. The presence of crystals on the skin is called uremic frost and is a late sign of waste products building up in the blood (**uremia**). When the wastes are not filtered by the kidneys, they can come out through the skin and look like a coating of frost.

Palpation and percussion of the kidneys is done by physicians and advance practice nurses. Gentle palpation and percussion of the bladder may be done by the licensed practical nurse/licensed vocational nurse (LPN/LVN) if urine retention is suspected. If the patient has a feeling of fullness but is unable to urinate, the nurse gently palpates the suprapubic area for a full bladder. Normally the bladder is not palpable. The bladder may also be percussed. The percussion note sounds dull over a fluid-filled bladder.

Most assessment of the urinary system is done using indirect measures. Assessment of vital signs, lung sounds, edema, daily weights, and intake and output can provide valuable data related to urinary function.

Vital Signs

If renal disease is suspected, blood pressure should be assessed and documented while the patient is lying, sitting, and standing. An increase in blood pressure is commonly seen with renal disease. A drop in blood pressure accompanied by a rise in pulse rate as the patient rises to sitting or standing positions is called orthostatic or postural hypotension and may indicate fluid deficit. (See Chapter 15 for a more complete discussion of orthostatic hypotension.) A rapid respiratory rate may indicate fluid retention in the lungs.

Lung Sounds

If the patient retains more fluid than the heart can effectively pump, fluid may be retained in the lungs. This is manifested as crackles, which are popping sounds heard on inspiration and sometimes on expiration when the chest is auscultated. Wheezes may also be present. New-onset crackles and wheezes should be reported to a physician. (See Chapter 26 for assessment of the respiratory system.)

Edema

Fluid retention may be manifested as edema (excess fluid in tissues). The nurse assesses and documents the degree and location of edema. Edema may be generalized in renal failure. The nurse also looks for edema in the area around the eyes (periorbital edema). Assessment of edema is discussed in detail in Chapter 15.

Daily Weights

Weight is the single best indicator of fluid balance in the body. Patients with renal disease often have fluid imbalances. The patient should be weighed at the same time each day, in the same or similar clothing. The nurse is careful not just to document the weight, but also to look at trends in weight gain or loss. If the patient's weight is steadily increasing, fluid retention is suspected and should be reported. A patient undergoing diuresis is expected to have decreasing weights.

Intake and Output

All patients with renal disease should have careful measurement of intake and output. Individual voidings are analyzed for amount and recorded. The nurse measures and records all liquids taken in, including oral, intravenous, irrigation, tube feeding, and other fluids. Output includes urine, emesis, nasogastric effluent, wound drainage if it is copious, and any other drainage. As part of measuring output, the nurse should examine the urine for any abnormalities. Often problems show up readily in the urine, with changes such as **hematuria,** cloudiness, and foul odor seen with infection or concentrated dark-amber urine seen with dehydration.

Intake and output totals are analyzed and recorded every 8 or 12 hours or more often for unstable patients. As with daily weights, the nurse notes trends in retention or loss of fluid and reports significant changes to the physician. Ac-

CRITICAL THINKING

Mr. Nolan

It is the end of the shift. As you empty Mr. Nolan's indwelling catheter bag, you find that it has only 50 mL of concentrated urine in it. What do you do?

Answer at end of chapter.

uremia: ur—urine + emia—blood

hematuria: hemat—blood + uria—urine

curate documentation is vital, because the physician may base medication and intravenous fluid orders on intake and output results.

DIAGNOSTIC TESTS OF THE RENAL SYSTEM

Laboratory Tests

Urine Tests

URINALYSIS. A urinalysis (urine analysis) is a commonly performed diagnostic test for the renal system. The results of the urinalysis give information regarding kidney function and various body functions. Table 34–2 lists normal and abnormal findings on a urinalysis.

To collect a voided specimen for urinalysis, the nurse has the patient wash the perineum using soap and water or a special towelette from a clean-catch midstream urine collection kit. Women should be directed to wash from the front to the back of the perineum. The patient is instructed to begin to void into the toilet, and then move the collection container under the stream, and then finish voiding into the toilet. This is called a clean-catch midstream specimen. It is used to obtain the cleanest possible specimen. Female patients should be told to separate the labia with one hand and keep it separated while washing and collecting the specimen to decrease the risk of contamination of

the specimen. If the female patient is menstruating, this should be specified on the laboratory form. A tampon may be used to prevent contamination of the specimen. The uncircumcised male patient should be directed to retract the foreskin with one hand and keep it retracted while cleansing and voiding. At least 10 mL of urine should be collected.

If a urinalysis is ordered for a patient with a urinary catheter, the nurse obtains the urine specimen. This specimen is considered sterile because it is coming directly from the bladder into the urinary catheter tubing. To obtain the specimen, wear clean gloves and use an alcohol swab to clean the sample port on the catheter tubing. Insert the needle of a syringe (usually 10 mL) into the port and withdraw urine from the tubing into the syringe. Then empty the urine from the syringe into a collection container and safely dispose of the syringe.

URINE CULTURE. A urine culture is done to determine the number of bacteria present in the urine and to identify the organism causing infection in the urine. The urine should be collected before antibiotic treatment is begun to avoid affecting results. The midstream clean-catch system is used to obtain voided specimens. A physician may order a catheterized specimen if there is a risk of contamination from the vagina, if a female patient is menstruating, or if the patient is incontinent. As a general rule, a bacterial count

TABLE 34-2	URINALYSIS RESULTS	
Test	**Normal Results**	**Abnormal Results and Significance**
Color of urine	Pale yellow to amber	Dark-amber urine suggests dehydration. Yellow-brown to green urine indicates excessive bilirubin. Cloudiness of freshly voided urine indicates infection. Nearly colorless urine is seen with a large fluid intake or diabetes insipidus.
Odor of urine	Aromatic	With infection, urine becomes foul smelling. In diabetic ketoacidosis, the urine has a fruity odor. Urine that has been standing for a while develops a strong ammonia smell.
pH	4.6–8.0	The pH is greatly affected by the food eaten. pH below 4.6 is seen with metabolic and respiratory acidosis. pH above 8.0 is seen when urine has been standing or with infection because bacteria decompose urea to form ammonia.
Specific gravity	1.010–1.025	Low specific gravity indicates excessive fluid intake or diabetes insipidus. High specific gravity is seen with dehydration. A specific gravity fixed at 1.010 indicates kidney dysfunction.
Protein	0–18 mg/dL	Persistent proteinuria is seen with renal disease from damage to the glomerulus. Intermittent protein in the urine can result from strenuous exercise, dehydration, or fever. As a general rule, protein in the urine is a significant sign of renal problems.
Glucose	None	Glucose in the urine indicates diabetes mellitus, excessive glucose intake, or low renal threshold for glucose reabsorption.
Ketones	None	Ketones in the urine indicate diabetes mellitus with ketonuria or starvation from breakdown of body fats into ketones.
Bilirubin	None	Bilirubin in the urine indicates liver disorders causing jaundice. Bilirubin may appear in the urine before jaundice is visible.
Nitrite	Negative	Nitrites in the urine indicate infection in the urine. Bacteria in the urine convert nitrate to nitrite, which gives a positive reading.
Leukocyte esterase	Negative	A positive leukocyte esterase in the urine indicates infection in the urine. It determines the presence of an enzyme released by WBCs in the urine.
Red blood cells	0–4/hpf	Blood in the urine may be caused by kidney stones, infection, cancer, renal disease, or trauma.
White blood cells	0–5/hpf	WBCs in the urine indicate infection or inflammation in the urinary tract.
Casts	None to occasional hyaline cast	Casts are formed when abnormal urine contents settle into molds of the renal tubules and may be made of protein, WBCs, RBCs, or bacteria. A few hyaline casts may be found in normal urine. The presence of casts generally indicates renal damage or infection.

hpf = high-power field; WBC = white blood cells.

of 100,000 or more per milliliter of urine indicates a urinary tract infection. An amount less than that may result from contamination during specimen collection. The urine is cultured to grow and identify the kind of bacteria present. Often a sensitivity test is also ordered to determine what kind of antibiotic will be most effective in eradicating the offending bacteria.

Renal Function Tests

A number of blood tests reflect kidney function. If the kidneys are not functioning adequately, these test results will be elevated.

SERUM CREATININE. Creatinine is a waste product from muscle metabolism and is released into the bloodstream at a steady rate. Creatinine levels are a very good indicator of kidney function (normal: 0.6 to 1.5 mg/dL). A serum creatinine level above 1.5 mg/dL means there is kidney dysfunction. The higher the creatinine level, the more impaired the kidney function.

BLOOD UREA NITROGEN. Urea is a waste product of protein metabolism. The blood urea nitrogen (BUN; normal: 8 to 25 mg/dL) is not as sensitive an indicator of kidney function as the creatinine level. This is because it is readily affected by increased protein intake, dehydration, and other factors in the body. An elevated BUN level can be caused by the following factors:

- Kidney dysfunction or failure
- Decreased kidney blood supply, such as when the patient is in a state of shock or in severe heart failure
- Dehydration, because the loss of water makes the blood more concentrated (decreased BUN level is seen with overhydration)
- High-protein diet, because urea formation increases
- Gastrointestinal bleeding, because blood is absorbed as protein and converted into urea
- Steroid use, because steroids increase the rate of protein breakdown in the body

URIC ACID. Uric acid is an end product of purine metabolism and the breakdown of body proteins. The uric acid is not as diagnostic as creatinine because many factors can cause an elevated uric acid level (normal: 2 to 7 mg/dL). An elevated uric acid level can be caused by the following:

- Kidney disease
- Gout (patients with gout metabolize uric acid abnormally)
- Malnutrition
- Leukemia
- Use of thiazide diuretics (because of impairing uric acid clearance by the kidney)

CREATININE CLEARANCE. The creatinine clearance test measures the amount of creatinine cleared from the blood in a specified period by comparing the amount of creatinine in the blood with the amount of creatinine in the urine. It is an excellent indicator of renal function.

To carry out the test, urine is collected for a 24-hour period, and a sample for serum creatinine is collected some-

time during the 24 hours. The following procedure should be followed:

1. When the test is begun, the patient is directed to void and discard that urine.
2. Urine is collected for 24 hours, keeping the urine in a large container provided by the laboratory. The container is kept on ice.
3. Twenty-four hours after the test was begun, the patient is instructed to void again. This urine is added to the collection container.
4. The laboratory collects a serum creatinine during this 24-hour period.

The creatinine clearance is computed in the laboratory and is expressed in volume of blood that is cleared of creatinine in 1 minute. Normal is 85 to 125 mL. A minimum creatinine clearance of 10 mL per minute is needed to live without dialysis.

L E A R N I N G TIPS

A handy approximation to determine kidney function is to equate the creatinine clearance result to percent of renal function. For example, a creatinine clearance of 100 mL per minute = 100 percent renal function, 30 mL per minute = 30 percent renal function, and 5 mL per minute = 5 percent renal function.

Radiological Studies
Kidney-Ureter-Bladder

A kidney-ureter-bladder (KUB) is an x-ray examination of the named structures. This test is also known as a flat plate of the abdomen. It displays the outline of the renal structure and can show tumors, swollen kidneys, and calcium-based kidney stones. No special care is necessary for this test.

Intravenous Pyelogram

The intravenous **pyelogram** (IVP) is a common test. During the test, a radiopaque dye is injected into a large vein. The dye is cleared from the blood by the kidneys. Because the x-rays cannot penetrate the dye, the dye outlines the renal structures. Radiographs are taken at frequent intervals to see the dye filling the renal pelvis and going down the ureters into the bladder (Fig. 34–3).

In preparation for an IVP, the patient takes laxatives to cleanse the bowel the day before the test, following agency policy. The patient is allowed nothing by mouth (NPO) after midnight the evening before the test. As with all contrast studies, the patient should be questioned for allergies to iodine and shellfish before the test. The dye can cause allergic and anaphylactic reactions in people who are allergic to these substances, although this problem is less common since the introduction of a newer radiopaque dye. The patient should also be warned about a warm, flushing sensa-

---

pyelogram: pyelo—pelvis of the kidney + gram—radiograph

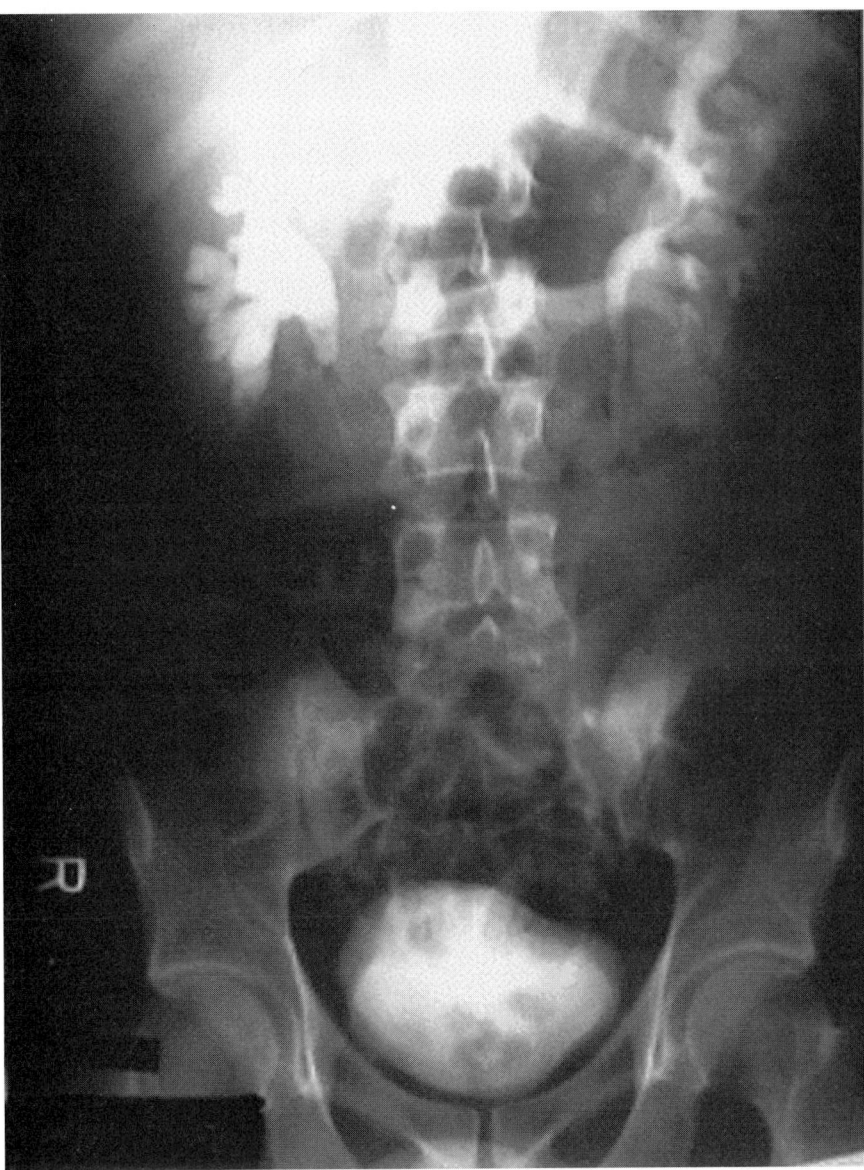

Figure 34-3 Intravenous pyelogram.

tion up the arm and sometimes all over the body when the dye is injected. A strange taste may occur as well.

Aftercare for an IVP includes having the patient drink a large amount of fluids to help clear the dye from the kidneys. On rare occasions, people can develop acute renal failure because the dye is highly concentrated and can obstruct the kidney tubules.

Renal Angiography

A renal angiogram is a test to visualize the renal arteries. The femoral artery is pierced with a needle, and a catheter is threaded up through the femoral and iliac arteries into the aorta and then the renal artery. A contrast agent is injected to make the renal arterial supply visible on x-ray examination. Iodine or shellfish allergies should be determined before the study. Angiography is done in the x-ray department with the patient awake during the procedure. The test helps the physician see blood flow to the kidneys to determine the cause and treatment of kidney disease.

Before the procedure, the patient should be NPO for 4 to 8 hours. Patient care following angiography includes bedrest for up to 12 hours to prevent bleeding at the injection site. The extremity is immobilized with a sandbag, and a pressure dressing is applied. The patient is instructed not to bend the leg, and the head of the bed is not raised more than 45 degrees. The nurse monitors vital signs, dressing, and pulses in the affected extremity frequently. Institution policy is consulted for specific care guidelines.

Endoscopic Procedures
Cystoscopy and Pyelogram

A **cystoscopy** and pyelogram (C&P) is a minor surgical procedure that involves a rigid or fiberoptic instrument (cystoscope) inserted into the bladder through the urethra. A

---

cytoscopy: cysto—bladder + scopy—to examine

light at the end of the instrument allows a physician to visualize the interior of the bladder. Commonly a pyelogram is done as well. This involves insertion of a ureteral catheter into the pelvis of the kidney. Radiopaque dye is injected through the catheter and radiographs are taken. A C&P is done for both diagnostic and therapeutic reasons. As part of the diagnosis, a physician can do the following:

■ Inspect the inside of the bladder.
■ Collect a urine specimen from either kidney.
■ Visualize the renal structure with x-rays.
■ Biopsy any suspicious growths in the bladder or ureter.

Therapeutic interventions that can be done during a C&P include the following:

■ Removal of small bladder tumors
■ Removal of stones from the bladder
■ Removal of stones from the ureters
■ Dilation of the ureters

The preparation for a C&P is the same as for any surgery. Care following a C&P includes measuring urine to make sure the patient has not developed urinary retention from swelling of the urinary meatus and encouraging fluid intake. The patient should expect some **dysuria** for 24 hours following a C&P, and the first one or two voidings may be blood tinged. Complications that can result include urinary tract infection, urinary retention, and perforation of the bladder.

Ultrasound Examination of the Kidneys
Renal Ultrasound

A renal ultrasound, also known as ultrasonography, is a noninvasive study using sound waves to examine the anatomy of the urinary tract. A transducer is passed over the skin, which has been covered with a conductive gel. Structures of the urinary system can be visualized on a screen. A renal ultrasound is used to help diagnose tumors of the kidney and to look for enlargement of the kidneys, kidney stones, and changes of renal structures with chronic infection. There is no special preparation or aftercare, and there are no known complications.

Renal Biopsy

A renal biopsy may be done to diagnose or gain more information about kidney disease. Most renal biopsies are **percutaneous** (done with a needle through the skin), though some are open (done through a surgical incision). A percutaneous biopsy may be done in the operating room, the radiology department, or the patient's room. A computed tomography (CT) scan or ultrasound is done first to locate the kidney for biopsy. The patient is in a prone position, usually with a sandbag under the abdomen, and the biopsy is taken through the flank area. The physician obtains a sample of tissue, which is examined in the laboratory.

Before the biopsy, the patient is NPO for 6 to 8 hours and a mild sedative is given. The patient should not take anticoagulants before the biopsy because of the risk of bleeding. A complete blood cell count and coagulation studies are performed. For a percutaneous biopsy, a local anesthetic is used.

Following the biopsy, the patient is observed closely for bleeding, because the kidney is highly vascular. A pressure dressing is applied to prevent bleeding, and the patient is maintained on bedrest for 24 hours. A sandbag may be used under the flank to place additional pressure on the biopsy site. Urine is inspected for blood with each voiding. Vital signs are monitored frequently for 24 hours according to agency policy. Grossly bloody urine, falling blood pressure, and rising pulse are signs of bleeding and are reported immediately. Fluids are encouraged if not contraindicated.

▶ THERAPEUTIC MEASURES
Management of Urinary Incontinence

Urinary incontinence is defined as the involuntary passing of urine and is very common. It is estimated that 10 million Americans (mostly women) suffer from incontinence. Most patients do not seek treatment. At times, urinary incontinence can be prevented by patient teaching or physician intervention. Incontinence that cannot be prevented is managed by the use of padding and absorptive products worn by the patient. With all kinds of incontinence, it is helpful for the nurse or patient to keep a urinary diary for at least several days to determine when incontinence occurs and to look for any predisposing events. The patient should be referred to a urologist specializing in the area of incontinence for a careful examination to determine the cause and identify potential medical or surgical treatment. Some areas of the country have continence clinics, which can be helpful for patients with this problem. There are several types of incontinence, discussed next.

Stress Incontinence

Stress incontinence is the involuntary loss of urine associated with increasing abdominal pressure during coughing, sneezing, laughing, or other physical activities. Stress incontinence is commonly seen in women following childbirth and after menopause.

Urge Incontinence

Urge incontinence is the involuntary loss of urine associated with an abrupt and strong desire to void. The patient typically complains of being "unable to make it to the bathroom in time." Patients with stress incontinence or urge incontinence can be taught Kegel exercises to increase perineal muscle tone. Box 34–2 explains how to teach patients to perform Kegel exercises.

Functional Incontinence

Functional incontinence is caused by chronic impairment of physical function or ability to think, leaving the patient unable to get to the toilet in time to maintain continence. This is a common cause of incontinence in the elderly.

---

dysuria: dys—difficult or painful + uria—urination
percutaneous: per—through + cutaneous—skin

BOX 34-2 Patient Education for Kegel Exercises

The purpose of Kegel exercises is to decrease the incidence of incontinence by strengthening the pubococcygeal muscle, which supports the pelvic organs. By increasing the tone of this muscle, the patient has an increased ability to tighten the muscle that encircles the urinary meatus and stop the flow of urine. This exercise can also help prevent uterine prolapse, enhance sensation during sexual intercourse, and hasten postpartum healing. It may be used by the elderly male patient to control dribbling.

1. Establish awareness of pelvic muscle function by instructing the patient to "pull in" the muscles in the perineum as if to control urination or defecation. The muscles of the buttocks, inner thigh, and abdomen are not used to do Kegel exercises.
2. To help identify the correct muscles to tighten, ask the patient to tighten the muscles that control urination. It can be helpful to use an analogy of an elevator: start squeezing at the bottom floor and then squeeze upward to the top floor.
3. Instruct the patient to tighten the pelvic muscles for 10 seconds, followed by at least 10 seconds of relaxation.
4. Advise the patient to perform these exercises 30 to 80 times per day. Help the patient determine cues to remind the patient to perform the exercises, such as stopping the stream of urine 10 times each time the patient urinates.

Source: Adapted from Urinary Incontinence in Adults: Clinical Practice Guideline. Agency for Health Care Policy and Research, Public Health Service, US Department of Health and Human Services (AHCPR Pub No 92-0038), Rockville, MD, 1992.

Total Incontinence

Total incontinence is a continuous and unpredictable loss of urine. Bladder training has been tried and proven ineffective. Often the patient with total incontinence is neurologically impaired. In these situations the nurse's priority is to keep the patient clean and dry using absorptive products. For the male patient, an external condom catheter can be effective in some situations.

NURSING PROCESS. The medical diagnoses of stress and urge incontinence are also nursing diagnoses. Many nursing interventions can be helpful to decrease these kinds of incontinence. See Nursing Care Plan Box 34–3 for the Patient with Stress Incontinence or Urge Incontinence and Nursing Care Plan Box 34–4 for the Patient with Functional Incontinence.

Management of Urine Retention

Urine retention can be caused by many factors. It can be acute, with a sudden onset of retention and no urine output, or chronic, with a slower onset of retention of urine and some urine being expelled. Acute retention often results from surgeries and is caused by anesthesia, medications, or local trauma to the urinary structures. Acute retention can be a medical emergency causing extreme pain, a large bladder, and the possibility of bladder rupture or acute renal failure. Chronic urine retention may be related to an enlarged prostate gland, a medication effect, strictures, or other causes of obstruction of the urinary tract.

BOX 34-3 NURSING CARE PLAN FOR THE PATIENT WITH STRESS INCONTINENCE OR URGE INCONTINENCE

 Stress incontinence or urge incontinence related to decreased tone of perineal muscles

PATIENT OUTCOMES

Patient will be continent of urine. Patient will state three actions that can be taken to decrease incidence of stress or urge incontinence.

EVALUATION OF OUTCOMES

Is the patient continent? Is the patient able to state three actions that can be taken to decrease the incidence of stress or urge incontinence?

INTERVENTIONS	RATIONALE	EVALUATION
	STRESS INCONTINENCE AND URGE INCONTINENCE	
Assess history of incontinence; have patient keep a voiding journal.	A journal helps identify the severity and timing of incontinence.	Does patient complete the voiding journal?
Instruct patient on how to perform Kegel exercises (see Box 34-2).	Kegel exercises increase perineal muscle tone and help prevent incontinence.	Does patient explain how to perform Kegel exercises?
Work with the patient to incorporate Kegel exercises into normal activities of daily living (e.g., do 10 pelvic muscle contractions with each voiding).	An excellent time to perform Kegel exercises is when voiding because the correct muscles are used.	Does patient perform Kegel exercises when voiding or at other cued times during the day?
Encourage patient to drink at least 2000 mL of fluid per day, preferably 3000 mL per day unless medical reason for fluid restriction.	Concentrated urine is irritating to the urinary tract and can increase the incidence of urge incontinence and dribbling.	Is urine dilute?

Encourage patient to avoid alcohol and caffeine.

Discuss use of and provide small adhesive peripads to wear in underclothing.

Refer patient to a continence clinic or to a physician specializing in incontinence.

Refer patient to supportive and educational groups such as Help for Incontinent People (HIP).

Alcohol serves as a diuretic, and caffeine is irritating to the urinary tract.

Peripads provide protection in case of incontinence.

Specialists in the area of incontinence can use medical or surgical interventions to decrease the incidence of incontinence.

Support groups can help patients deal with the embarrassment of incontinence and learn methods and resources to prevent incontinence.

Does patient explain the need to avoid alcohol and fluids containing caffeine?

Does patient have and use peripads if desired?

Does patient know resources available to further assist with treatment of incontinence?

Does patient know the names and addresses of support groups to help with incontinence?

URGE INCONTINENCE

Teach patient to void at frequent intervals (every 2 hours), then gradually increase length of time between voidings.

Teach urge inhibition techniques (distraction), such as counting back from 100 by sevens and relaxation breathing.

By emptying the bladder at frequent intervals, the incidence of urge incontinence can be decreased.

These distraction techniques can help patients reach the bathroom in time to prevent incontinence.

Does patient follow a frequent voiding schedule?

Do distraction techniques help patient prevent incontinence?

 Functional urinary incontinence related to interference with rapid voiding

PATIENT OUTCOMES

Patient will be continent of urine. Patient will state three measures to increase continence.

EVALUATION OF OUTCOMES

Is the patient continent of urine? Is the patient able to state three measures to increase continence?

INTERVENTIONS	RATIONALE	EVALUATION
Assess history of incontinence. Keep a voiding log of when patient is incontinent.	A voiding log helps demonstrate when incontinence is most likely to occur and can help determine the cause of incontinence.	Does the patient cooperate so that a voiding log can be kept?
Determine any acute causes of incontinence, including new onset of urinary tract infection, constipation or impaction, medication effect, or poor fluid intake.	These may be readily treatable causes of incontinence.	Does the patient have any easily treatable causes of incontinence?
Determine if clothing is inhibiting timely voiding. If necessary, Velcro fasteners can be appropriate, or sweat shirts and sweat pants.	Clothing can be difficult to remove for the elderly, resulting in voiding before the clothing can be removed. Clothing can be modified so that it comes off quickly.	Does the patient have easy-to-remove clothing?
Determine if there are any obstacles to reaching appropriate urine receptacle, such as poor lighting, busy bathroom, lack of assistive devices.	Obstacles can make it impossible for the patient to reach the voiding receptacle in time to prevent incontinence.	Does the patient have ready access to a voiding receptacle?
Provide appropriate urinary receptacles, such as a three-in-one commode, female or male urinal, or no-spill urinal.	Assistive devices can be helpful for the patient to increase continence.	Does the patient need and have access to an appropriate assistive device?
Initiate a voiding schedule of every 2 hours, or base schedule on voiding log. Always assist patient to the toilet when patient first awakens and before sleep. Use prompted voiding techniques: check patient regularly, provide positive reinforcement if dry, prompt patient to toilet, praise patient after toileting, return patient to toilet in a specified time.	Frequent scheduled voiding using prompting techniques can increase continence.	Does the patient receive help to do bladder training with prompted voiding?
Teach patient to set up schedule of voiding using environmental cues such as meals, bedtime, and television shows.	Environmental cues help the patient remember when it is time to void.	Can the patient indicate cues throughout the day that prompt voiding?

To assess for urinary retention, urine output is determined and the lower abdomen is palpated. An enlarged bladder can be palpated, and the dull percussion notes may extend up to and beyond the umbilicus.

A newer technique called a bladder scan is used to assess the volume of urine in the bladder (Fig 34–4). This technique uses sound waves to estimate the amount of urine in the bladder. It is painless, noninvasive, and requires no patient preparation. The nurse performs this scan at the bedside. It helps guide the need for catheterization, thereby reducing unnecessary catheterizations and associated risks. The bladder scan may be used instead of catheterization (the gold standard for determining urine retention) after the patient urinates to determine the amount of urine remaining in the bladder. Normally the bladder contains less than 50 mL after urination. A residual volume of 150 to 200 mL of urine indicates the need for treatment for urinary retention.

Urinary Catheters
Indwelling Catheters

Indwelling urinary catheters (Foley catheters) may be inserted into hospitalized patients for various justifiable reasons such as shock or urinary tract obstruction. As a general rule, catheters should be avoided if possible. Urinary incontinence is not justification for insertion of a catheter. Urinary catheters result in infection of the urinary tract in up to 44 percent of patients within 72 hours of catheterization, and up to 90 percent of patients who have indwelling catheters for 17 days develop significant bacterial infection.[1]

Bacteria enter the bladder mainly in one of two ways with an indwelling catheter: (1) through the outlet at the end of the drainage bag contaminating the urine, which is then inadvertently drained back into the bladder; or (2) around the catheter up the urethra and into the bladder. It has been demonstrated that the incidence of infection is decreased when intermittent straight catheterization is used instead of indwelling Foley catheters. Box 34–5 outlines

BOX 34-5	**Guidelines for Care of the Patient with an Indwelling Catheter**

1. Maintain a closed system. Do not separate the catheter from the tubing of the bag. Instead, collect specimens and irrigate through the sample port in the tubing.
2. Keep the catheter securely taped or fastened to the leg. This decreases traction on the catheter with back-and-forth movement of the catheter that can help bacteria enter the bladder.
3. Encourage fluids to internally irrigate the catheter, if fluids are not contraindicated because of heart or kidney disease.
4. Use good aseptic technique when emptying the drainage bag by handwashing, wearing clean gloves, and using a container designated for that patient only to collect the urine.
5. Wash the perineum with soap and water at least once a day, and again if there is any bowel incontinence.
6. Keep the tubing coiled on the bed and positioned to allow free flow of urine. *Keep the catheter bag below the level of the bladder at all times.*
7. Do not clamp catheters. Clamping a catheter results in obstruction, which increases infection. Periodic clamping has not been found to be effective in bladder retraining.
8. Remove indwelling catheters as soon as possible.

guidelines that should be followed to decrease infection in the patient with a catheter (Home Health Hints).

After an uncircumcised male is catheterized, it is essential that the foreskin is properly positioned over the glans penis and not left retracted to prevent injury. If left retracted,

HOME HEALTH HINTS

- The home health nurse should always have a sterile specimen container. This provides a quick way to get a specimen to the physician's office without the nurse having to obtain the container, saving time and money.

- When catheters plug and irrigation fails, families can be taught to take the catheter out. A syringe for removing water from the balloon should be left in the home for this purpose. The family is instructed to avoid cutting the valve stem. The family should contact the nurse to reinsert the catheter, but in the meantime, the patient's bed or chair can be padded with towels or diapers. Garbage or cleaners' bags can be used to line the mattress. The family must be instructed to notify the nurse immediately if the catheter is plugged or has been removed.

- Not all homes have adequate lighting, which is a must for inserting a catheter. If lighting is inadequate, a caregiver can be asked to hold a flashlight while the nurse inserts the catheter. The nurse should always have a flashlight with him or her. Having two catheters and catheter trays is also wise in case of a defect or contamination.

- Urinary drainage bags and leg bags should be changed a minimum of every 2 weeks in the home setting to prevent infection. The family is instructed not to clean and reuse the bags.

- To encourage fluid intake, the patient should keep a large container of water (1 to 2 quarts) next to the place the patient sits most of the day. The goal is to drink 2 quarts of water by the end of the day, unless contraindicated by other medical problems. This also simplifies measuring intake. A sport bottle helps to keep water at hand while moving about the home.

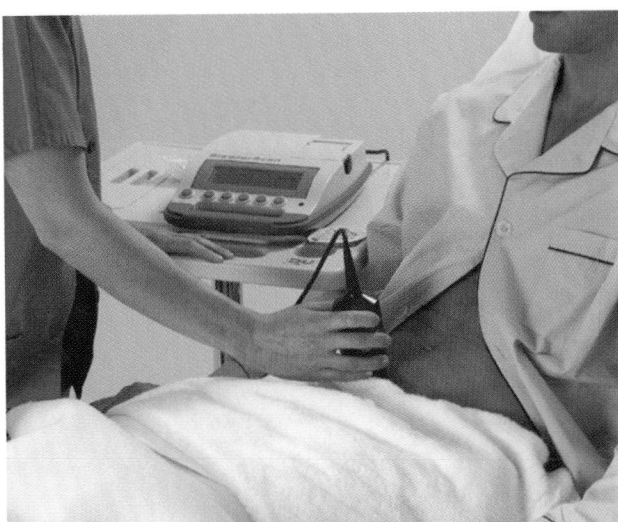

Figure 34-4   A bladder scan can be used to determine the volume of urine in a patient's bladder.

subsequent swelling may make it impossible to pull the foreskin over the glans penis later. This can then cause ischemia of the glans penis, which is an emergency. A physician must be notified immediately and may need to perform an emergency circumcision if the foreskin cannot be properly positioned. Always make sure that the foreskin is positioned properly following catheterization or perineal care.

Intermittent Catheterization

For the patient who is unable to void, the best intervention is intermittent catheterization. A postoperative patient or a patient with a neurological disorder or urine retention may benefit from intermittent catheterization. It reduces the risk of infection as long as the bladder is not allowed to overfill. A full bladder stretches the muscle fibers, which in turn reduces circulation to the bladder and increases the risk of infection.

Intermittent catheterization involves the use of a straight plastic or rubber catheter that is inserted into the urethra every 3 hours or more to empty the bladder. Once the bladder is empty, the catheter is removed. Patients may be taught to do intermittent self-catheterization (ISC) at home. Patients doing ISC may be taught to wash and reuse the same catheter repeatedly when they are in their own environment. In the hospital, however, sterile technique is used.

Suprapubic Catheter

After some surgeries of the urinary tract and in some long-term situations, a suprapubic catheter may be used. This is an indwelling catheter that is inserted through an incision in the lower abdomen directly into the bladder.

Nursing care of a suprapubic catheter involves keeping the area clean and dry, changing the dressing when the site is new, and keeping the catheter taped to prevent tension. A skin barrier such as Stomahesive may help protect the skin from urine leakage. All other care is the same as for any indwelling catheter.

Answers to CRITICAL THINKING

Mr. Nolan

You should realize that 50 mL of concentrated urine after 8 to 12 hours is not normal. Further investigation is necessary to determine the cause and seriousness of the problem. Some items to assess follow.

1. Consider Mr. Nolan's diagnosis. Is he in renal failure? Is he severely dehydrated and retaining water?
2. Ask if anyone emptied Mr. Nolan's bag earlier in the shift.
3. Look at the trends in Mr. Nolan's intake and output record. Has his output been decreasing? Is this a change?
4. Has Mr. Nolan been taking in enough fluids?
5. Look at trends in daily weights. Is Mr. Nolan's weight increasing? Is this an expected finding?
6. Listen to Mr. Nolan's lung sounds. Check for edema. Do findings indicate fluid retention?
7. Palpate Mr. Nolan's bladder. Is it distended? Maybe the catheter is blocked.
8. If a problem is identified, the physician should be contacted.

REVIEW QUESTIONS

1. A home health nurse visits a patient who is 82 years old. He uses a cane and is not incontinent. Which one of the following interventions should be included in the plan of care, based on an understanding of normal age-related changes of the urinary system, to promote his safety?
   a. Encourage fluids after 6 P.M.
   b. Limit fluids to 1000 mL per day.
   c. Provide a nightlight in the bathroom.
   d. Provide adult briefs to prevent dribbling.

2. Which of the following is the most accurate assessment of fluid balance in the patient with renal failure?
   a. Voiding pattern
   b. Daily weight
   c. Laboratory studies
   d. Skin turgor

3. A physician has ordered a midstream urine specimen for culture and sensitivity. Which of the following should be included in patient teaching for collecting this specimen?
   a. A second voided specimen is preferred.
   b. A 24-hour urine specimen is needed.
   c. As soon as the urine starts to flow, it should be collected in a sterile container.
   d. Women should keep the labia separated while voiding.

4. Which of the following is the most important nursing action that the nurse should take to prevent urinary tract infection in the catheterized patient?
   a. Force fluids to 4000 mL every 24 hours.
   b. Empty the Foley bag every 4 hours around the clock.
   c. Maintain a closed catheter system.
   d. Wash the perineum every 8 hours.

5. A patient is experiencing stress incontinence with frequent involuntary loss of urine. Which of the following directions would be most appropriate when teaching her how to perform Kegel exercises?
   a. "Tighten your rectum at frequent intervals throughout the day."
   b. "Keep your abdominal muscles tightened; do this every time you stand up."
   c. "Do at least 20 sit-ups per day."
   d. "When urinating, stop and start the stream of urine by tightening the perineal muscles."

REFERENCES

1. Crow, R, et al: Study of Patients with an Indwelling Urinary Catheter and Related Nursing Practice. Nursing Practice Research Unit, University of Surrey, Guildford, 1986.

35

Nursing Care of Patients with Disorders of the Urinary Tract

Betty J. Ackley

KEY TERMS

anuria (an-**YOO**-ree-ah)

azotemia (AY-zoh-**TEE**-me-ah)

calculi (**KAL**-kyoo-lye)

cystitis (sis-**TIGH**-tis)

glomerulonephritis
(gloh-MER-yoo-loh-ne-**FRY**-tis)

hemodialysis (HEE-moh-dye-**AL**-i-sis)

hydronephrosis
(HIGH-droh-ne-**FROH**-sis)

nephrectomy (ne-**FREK**-tuh-mee)

nephrolithotomy
(NEFF-roh-li-**THOT**-uh-mee)

nephropathy (ne-**FROP**-uh-thee)

nephrosclerosis (NEFF-roh-skle-**ROH**-sis)

nephrostomy (ne-**FRAHS**-toh-mee)

nephrotoxin (NEFF-roh-**TOK**-sin)

oliguria (AH-li-**GYOO**-ree-ah)

peritoneal dialysis
(PER-i-toh-**NEE**-uhl dye-**AL**-i-sis)

polyuria (PAH-lee-**YOOR**-ee-ah)

pyelonephritis (PYE-e-loh-ne-**FRY**-tis)

stent (STENT)

urethritis (YOO-ree-**THRIGH**-tis)

urethroplasty (yoo-**REE**-throh-PLAS-tee)

urosepsis (yoo-roh-**SEP**-sis)

QUESTIONS TO GUIDE YOUR READING

1. What are the predisposing causes, symptoms, laboratory abnormalities, and treatment of urinary tract infections?

2. What are the predisposing causes, symptoms, and treatment of kidney stones?

3. What should you teach patients with kidney stones?

4. What are symptoms of cancer of the bladder and cancer of the kidneys?

5. How do you give care to a patient with an ileal conduit or continent reservoir?

6. How would you explain the pathophysiology associated with diabetic nephropathy, nephrosclerosis, hydronephrosis, and glomerulonephritis?

7. What is the nursing care for patients with diabetic nephropathy, nephrosclerosis, hydronephrosis, and glomerulonephritis?

8. What are common nephrotoxic substances?

9. What is the difference between acute renal failure versus chronic renal failure?

10. What are common symptoms experienced by the patient in renal failure?

11. What nursing care should be given to patients in renal failure and with a hemodialysis blood access site?

12. How does hemodialysis and peritoneal dialysis work to replace the function of the kidney?

Disorders of the urinary tract include a variety of problems involving the kidneys, ureters, bladder, and urethra. These problems may arise from infection, obstructions, cancer, hereditary disorders, or chronic diseases. Some may lead to renal failure if not treated or controlled. Infection may be found in three different anatomic parts of the urinary tract: the urethra, resulting in **urethritis;** the bladder, with a diagnosis of **cystitis;** or the kidneys, with a diagnosis of **pyelonephritis.**

▶ URINARY TRACT INFECTIONS

Urinary tract infection (UTI), a general term, refers to invasion of the urinary tract by bacteria.

Predisposing Factors for Urinary Tract Infections

Urinary tract infections are almost always caused by an ascending infection, starting at the external urinary meatus and progressing toward the bladder and kidneys. Predisposing factors for UTIs include the following:

- Stasis of urine in the bladder can be caused from obstruction such as a clamped catheter or simply from not voiding frequently enough. Urine overdistends the bladder, decreasing the blood supply to the wall of the bladder, which keeps white blood cells (WBCs) from fighting contamination that may have entered the bladder. The standing urine then serves as a culture medium for bacterial growth.
- Contamination in the perineal and urethral area can be from fecal soiling, from intercourse in which bacteria are massaged into the urinary meatus, or from infection in the area, such as vaginitis, epididymitis, or prostatitis.
- Instrumentation, or having instruments or tubes inserted into the urinary meatus. The most common cause is urinary catheterization. Bacteria ascend around or within the catheter, causing infection. Many patients develop a UTI within 2 weeks of placement of an indwelling catheter.
- Reflux of urine from the urethra to the bladder or the bladder to the ureter because of faulty valves to maintain one-way flow. Reflux can be congenital, or it may be acquired as a result of previous infections.
- Previous UTIs are thought to provide a reservoir of persistent bacteria that causes reinfection.

Female patients are more likely to develop UTIs because of the shorter length of the urethra and the closer proximity of the anal orifice. The majority of UTIs are caused by the bacteria *Escherichia coli,* which is commonly found in stool. Elderly men are predisposed to infection because an enlarged prostate causes obstruction of urine flow.

---

urethritis: urethr—urethra (canal that discharges urine from bladder) + itis—inflammation
cystitis: cyst—closed sac containing fluid + itis—inflammation
pyelonephritis: pyelo—pelvis + nephr—kidney + itis—inflammation

Symptoms

UTIs are characterized by common symptoms of dysuria, urgency, frequency, and cloudy, foul-smelling urine.

Urethritis

Urethritis is inflammation of the urethra that may be due to a chemical irritant or bacterial infection or can be sexually transmitted. Bubble bath and bath salts are common irritants and should not be used by anyone with a history of UTIs. Urethritis can also be caused by spermicidal agents. Gonorrhea and chlamydia are sexually transmitted diseases that can cause urethritis in men. It is common to have some degree of urethritis in association with bladder or prostatic infections.

Symptoms of urethritis include urinary frequency, urgency, and dysuria. The male patient may have discharge from the penis. A urinalysis or urine culture is done to diagnose urethritis.

The treatment of urethritis is removal of the cause if it is caused by a chemical irritant. If urethritis is caused by a bacteria, an antibiotic is prescribed based on the results of a culture. Phenazopyridine (Pyridium), a urinary analgesic, is often used to treat dysuria. The patient should be forewarned that urine will turn orange while on phenazopyridine. If urethritis is sexually transmitted, it is important that the sexual partner also be treated.

Cystitis

Cystitis is inflammation and infection of the bladder wall. Symptoms include dysuria, frequency, urgency, and cloudy urine. Cystitis infection acquired outside the hospital is diagnosed with a routine urinalysis collected as a clean-catch midstream specimen. Changes seen in the urinalysis include cloudy urine and the presence of WBCs, bacteria, and sometimes red blood cells (RBCs) in the specimen. Nitrites are usually positive. Some laboratories also examine for leukocyte esterase, which is positive if infection is present in the urine. For the complicated UTI, such as one acquired in the hospital or a repeat infection, a urine culture and sensitivity should be done. Hospital-acquired UTIs are often caused by bacteria that are resistant to the usual antibiotics used for UTIs. A sensitivity test can identify which antibiotics will be effective against the offending organism.

The treatment of uncomplicated cystitis is most often a combination sulfa medication, such as sulfamethoxazole and trimethoprim (Bactrim, Septra). Complicated cystitis is often treated with ciprofloxacin (Cipro). Other antibiotics may be prescribed depending on the results of the urine culture and sensitivity. The patient is told to take all medications until gone, force fluids unless contraindicated, and return in a week for a follow-up urinalysis or culture to ensure that the infection is gone.

Pyelonephritis

PATHOPHYSIOLOGY. Pyelonephritis is infection of the kidneys. Pathophysiology includes formation of small abscesses throughout the kidney and gross enlargement of the kidney. The cause is usually an ascending bacterial infection. On occasion, kidney infection is caused by bacteria spreading from a distant site through the bloodstream and entering the kidney through the glomerulus.

SIGNS AND SYMPTOMS. Symptoms include urgency, frequency, dysuria, flank pain, fever, and chills. The urine is cloudy with increased WBCs, bacteria, casts, RBCs, and positive nitrites. In contrast to cystitis, the patient with pyelonephritis is much sicker and shows signs of systemic disease.

DIAGNOSTIC TESTS. Several tests are helpful to differentiate pyelonephritis from cystitis. With kidney infection, the urinalysis will show casts. Casts are microscopic particles formed in the kidney from abnormal constituents in the urine such as WBCs, RBCs, or pus. The presence of casts always indicates a problem in the kidneys. The complete blood cell count (CBC) will show an elevated WBC count. The patient will also have costovertebral (flank) tenderness.

MEDICAL MANAGEMENT. Treatment of pyelonephritis includes antibiotics based on the results of the culture and sensitivity. With severe gram-negative infections, the patient is hospitalized for intravenous (IV) antibiotics. The patient with acute pyelonephritis generally heals completely after treatment and has no lasting kidney damage.

COMPLICATIONS. Repeated kidney infections can result in scarring and loss of kidney function, leading to renal failure. Septicemia may occur from bacteria invading the bloodstream. When septicemia results from a urinary cause is called **urosepsis.** In the elderly, urosepsis can be the cause of new onset confusion. The elderly or immunocompromised patient may develop septic shock from infection in the urinary tract that has invaded the bloodstream, which may result in death.

Nursing Process: The Patient with a Urinary Tract Infection

Assessment

The patient is asked about pain on urination, flank pain, or general symptoms of infection such as fever, chills, and malaise. The urine is examined for cloudiness, blood, or foul odor. The presence of a catheter, recent instrumentation, or other predisposing factor is determined. Urinalysis and culture results are examined.

It is important to listen to the patient's concerns about the diagnosis. There can be acute embarrassment and feelings of shame associated with the diagnosis of a UTI.

Nursing Diagnosis

Possible nursing diagnoses include acute pain: dysuria and flank pain; hyperthermia; impaired urinary elimination:

dysuria, frequency, urgency, or incontinence; and risk for repeat infection.

Planning

The plan for the patient focuses on promotion of comfort, treatment of symptoms of infection, and educating the patient to prevent future infections.

Implementation

Discomfort may be treated with urinary or other analgesics. Fever is treated with acetaminophen. Fluids are encouraged to help flush bacteria from the bladder. Assistance to the bathroom is provided promptly when the patient requests help to avoid incontinence related to urgency. A bedside commode may be helpful in preventing incontinence.

Patient Education

It is very important that patients be advised to take all the prescribed antibiotic until it is gone. Commonly patients take medication for several days until they no longer have symptoms and then stop. This allows the infection to continue, and it may become chronic and resistant to antibiotics.

Patients who have one UTI commonly develop repeat infections. It is important that they receive health teaching to prevent repeated infections of the urinary tract (Box 35–1 and Nutrition Notes 35–2).

BOX 35–1 **Patient Teaching to Prevent Urinary Tract Infection**

1. Void frequently—at least every 3 hours while awake.
2. Drink up to 3000 mL of fluid a day if there are no fluid restrictions from the physician. Preferably drink water.
3. Drink one glass of cranberry juice (10 ounces) per day.
4. Take showers; avoid tub baths.
5. Wipe perineum from the front to the back after toileting.
6. Urinate after intercourse.
7. Avoid bubble bath and bath salts, perfumed feminine hygiene products, synthetic underwear, and constricting clothing such as tight jeans.
8. Take prescribed medication for UTIs until is it all gone.
9. If UTI is associated with another source of infection such as vaginitis or prostatitis, ensure that both infections are treated.

BOX 35–2  *Nutrition Notes*

Urinary Tract Infections

Instructions to increase fluid intake should specify amounts to consume or amount of urine resulting. Patients have developed electrolyte imbalances by overenthusiastically forcing fluids.

An effective intervention for urinary tract infections is increasing fluid intake (with the precaution cited above), both for its flushing effect and to excrete urinary drugs. Daily intake of cranberry juice to prevent UTI has reduced bacteriuria.

---

urosepsis: uro—urine + sepsis—infection in the blood

Evaluation

Goals have been met if the patient states that pain, urgency, and frequency are relieved and fever is controlled. It is also important to have the patient restate what was taught to ensure that correct learning occurred.

CRITICAL THINKING

Mrs. Milan

Mrs. Milan is a 25-year-old woman who recently spent a 3-day weekend getaway with her husband. On Monday she notices that she has symptoms of dysuria, frequency, and urgency. She visits her family practitioner and is diagnosed with a UTI.

1. What could have predisposed Mrs. Milan to developing a UTI?
2. What should Mrs. Milan be taught to prevent further occurrences of a UTI?
3. What urinalysis findings would you expect for Mrs. Milan?

Answers at end of chapter.

▶ UROLOGICAL OBSTRUCTIONS

Obstruction of urine flow in the urinary tract is always significant. When urine does not drain normally from the kidney, the resulting backup of urine eventually destroys the kidney by compressing the kidney structures. The causes of urological obstructions include strictures, stones, and tumors.

Urethral Strictures

A urethral stricture is a narrowing of the lumen of the urethra caused by scar tissue. A common cause of stricture is urethral injury resulting from insertion of catheters or surgical instruments. Strictures can also be caused by trauma from straddle injuries, a result of direct application of force to the perineal area, as well as untreated gonorrhea and congenital abnormalities.

The patient with a urethral stricture has a diminished urinary stream and is prone to develop UTIs because of obstruction of urine flow. Urethral strictures are often seen in elderly men. The problem becomes more apparent when attempts to insert a urinary catheter are unsuccessful because of the narrowed lumen.

Initially the treatment of a urethral stricture is mechanical dilation by a urologist, by inserting instruments to stretch open the urethra and then inserting a urinary catheter. If the stricture continues to be a problem after dilation, the area can be surgically repaired **(urethroplasty).**

The dilation process is often done at the bedside when the patient is awake. This is a painful experience for the patient, and it is helpful and caring to encourage the urologist to order pain medication before the procedure. The nursing diagnosis of acute pain is very relevant. An indwelling catheter is generally inserted after the dilation, so the nurs-

---

urethroplasty: urethro—urethra + plasty—surgical repair

ing diagnosis of risk for infection is also present. Patients need teaching on how to prevent UTIs (see Box 35–1).

Renal Calculi

Renal **calculi** (kidney stones; one stone is a calculus) are hard, generally small stones that form somewhere in the renal structures. When stones are found in the kidneys, the condition is called nephrolithiasis (Fig. 35–1).

Pathophysiology

Normally the dissolved substances in urine, including urinary salts, are diluted and readily excreted from the body. Calculi are formed when urinary salts are concentrated enough to settle out; there is often a nucleus around which the salts collect and deposit. Substances that can serve as a nucleus include pus, blood, dead tissue, a catheter, and crystals. Following are common urinary salts that make up renal calculi, arranged in order of frequency:

1. Calcium oxalate
2. Calcium phosphate
3. Magnesium ammonia
4. Uric acid
5. Cystine

The majority of renal calculi contain calcium, either in the form of calcium oxalate or calcium phosphate, but it is also possible to have combination stones.

Etiology

Causes of calculi formation include a family history of stones, chronic dehydration (causing more concentrated urinary salts), and infection, because it provides a nucleus

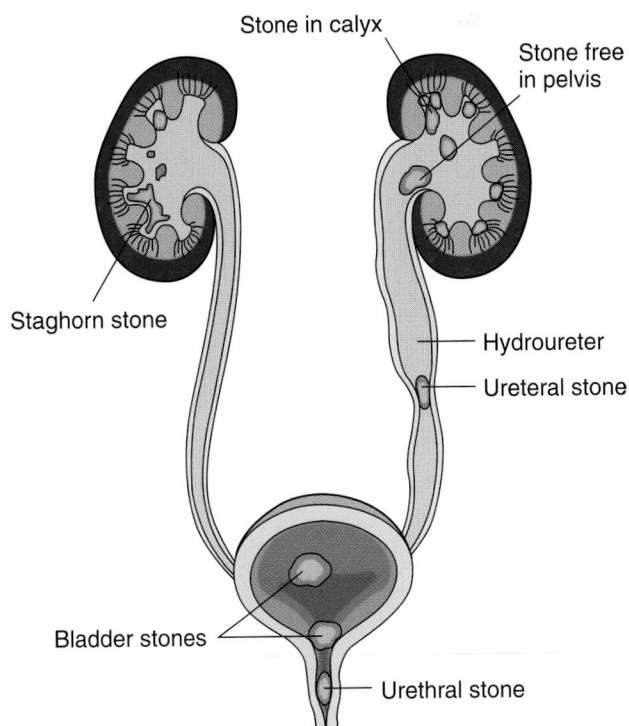

Figure 35-1 Location of calculi in the urinary tract.

for stone formation. Additional causes of calcium stones include dietary factors (see Nutrition Notes Box 35–2). Excessive amounts of calcium in the water in some geographic areas may also be a factor. Immobility causes stone formation because of the resulting urinary stasis; in addition, calcium leaves the unstressed bones during immobility, so more calcium is in the blood that is then filtered through the kidneys.

Signs and Symptoms

Symptoms of renal calculi include excruciating flank pain and renal colic; when the stone is lodged in the ureter, it is common to have pain radiate down to the genitalia. The pain results when the stone prevents urine from draining. Additional symptoms include hematuria from irritation by the stone, dysuria, frequency, urgency, and enuresis. The patient may also have costovertebral tenderness. Some patients develop nausea, vomiting, and diarrhea because of the proximity of the gastrointestinal structures.

Diagnostic Tests

The diagnosis of renal calculi may be made initially by doing a kidney-ureter-bladder (KUB; flat plate of the abdomen) examination or an intravenous pyelogram.

Medical Management

Renal calculi are treated medically if possible. Most stones are urinated out of the body. Patients can urinate stones if they are 5 mm or smaller; larger stones do not pass. If patients experience severe renal colic, they are admitted to the hospital. Intravenous fluids are administered to hydrate the patient and help flush the stone out of the body. All urine is strained to detect passage of stones, and pain medication such as morphine is given. If the patient is unable to pass the stone and infection, impaired renal function, or severe pain continues, intervention is needed.

LITHOTRIPSY. Lithotripsy therapy is used to break the stones into smaller parts that can then be removed or urinated out. Forms of lithotripsy include extracorporeal shock-wave lithotripsy (ESWL), electrohydraulic lithotripsy, laser lithotripsy, and percutaneous ultrasonic lithotripsy. With ESWL, the patient is immersed in a tub of water and ultrasonic shock waves are used to break up the stone into sand particles (Fig. 35–2), which are then urinated out. The patient is anesthetized for the procedure. After the procedure, the patient is usually discharged home after being told to increase fluid intake to help flush the sand out and to notify the urologist if there are any problems. Blood in the urine is common after lithotripsy.

SURGERY FOR RENAL CALCULI. For some patients, surgery may be necessary. The surgical procedure depends on the location of the stone. With any of these surgical procedures, postprocedure bleeding is a concern. Endoscopic procedures or open surgery can be used. Endoscopic procedures for the bladder include a cystoscopy for small stones and a cystolitholapaxy for larger stones. For cystolitholapaxy an instrument is inserted through the urethra to the bladder to crush the stone; the stone is then washed out with an irrigating solution. If the stone is lodged in a ureter, the urologist may insert an instrument into the ureter through a cystoscope to crush the stone or use an ultrasonic lithotripsy instrument to break the stone into fragments. Postoperative care following these procedures is similar to care following any cystoscopy. (See Chapter 34.) The open surgery procedure for stones in the bladder is a cystotomy and for the ureter is a ureterolithotomy.

For kidney stones, a percutaneous **nephrolithotomy** is performed, in which a scope is inserted through the skin

---

nephrolithotomy: nephro—kidney + lith—stone + otomy—incision

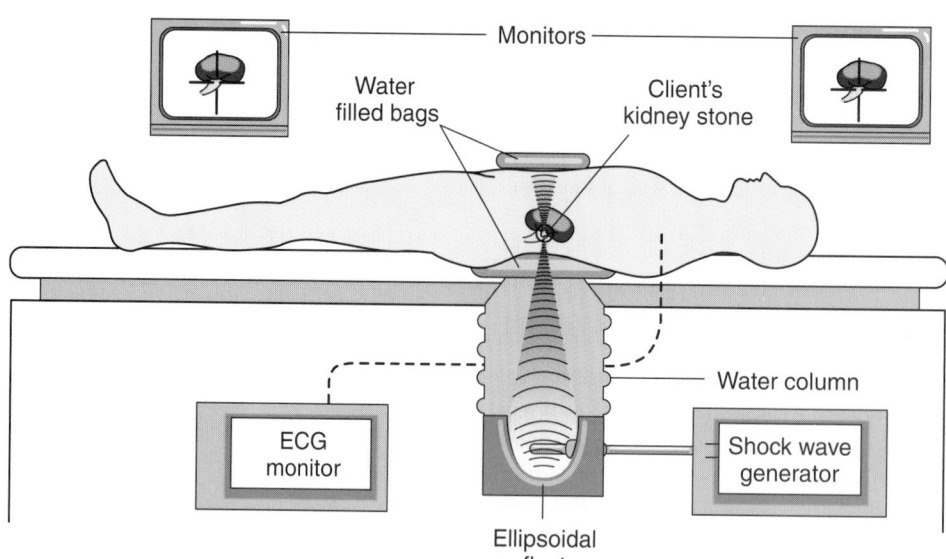

Figure 35-2 Extracorpeal shock-wave lithotripsy.

into the kidney to aid in breaking up the stone and to irrigate the renal pelvis. If the stone is very large, it may be necessary to do a nephrolithotomy, which is a surgical incision into the kidney to remove the stone. A pyelolithotomy is done to remove stones lodged in the renal pelvis.

Prevention of Renal Calculi

The patient may be advised to avoid foods that increase the risk of recurrent calculus development. Nutrition Notes Box 35–3 discusses foods that may contribute to calculi. (See also Cultural Consideration Box 35–4.) It should be noted that a low-calcium diet is generally no longer prescribed. Consult with the physician and dietitian to determine which foods should be avoided, depending on the type of stone found.

Complications of Renal Calculi

The presence of renal calculi increases the risk for UTIs because of obstruction of free flow of urine. Untreated obstruction of a stone in a ureter or the urethra can also result in retention of urine and damage to the kidney. This process is called **hydronephrosis** (discussed later).

Nursing Process

ASSESSMENT. Patients with stones are often in extreme pain and should be assessed routinely for pain. Nursing care of a patient with a renal calculus always involves careful measurement of intake and output and observation of the urine for abnormalities such as hematuria, pyuria, or passage of a stone. Temperature is monitored for onset of fever, which would indicate infection. A special strainer is used to strain all urine for stones. If a stone is found, it is saved for analysis in the laboratory. The patient is also asked about a recent history of infection, dietary or activity changes, or other risk factors for renal calculi. If the cause can be identified, teaching can be done to help prevent recurrent calculi.

NURSING DIAGNOSIS. Acute pain is a priority nursing diagnosis for any patient with renal calculi. Risk for impaired urinary elimination resulting from obstruction may be a priority issue. Risk for injury or infection from urine retention and recurrent calculi are also concerns. Deficient knowledge related to prescribed therapeutic regimen must also be addressed.

PLANNING. Planning should focus on pain relief, prevention of injury and infection, and teaching to prevent recurrent calculi.

IMPLEMENTATION. Analgesics are given to relieve pain. Fluids are encouraged to facilitate passage of the stone and to prevent dehydration if nausea and vomiting are present. Antipyretics and antibiotics may be used if infection is present. The physician is notified immediately if gross hematuria, **oliguria,** or **anuria** develops. The patient is instructed about the importance of maintaining a large fluid intake for the rest of their lives, activity increase, and diet changes if ordered.

EVALUATION. Patient outcomes are met if pain is controlled, fever and other complications are recognized and reported promptly, and the patient verbalizes understanding of self-care measures to prevent recurrent stones.

Hydronephrosis

Hydronephrosis is a condition that results from untreated obstruction in the urinary tract. It is usually treatable once the condition is detected. The obstruction of urine flow can be from a stricture in a ureter or the urethra, from kidney stones, from a tumor, or from an enlarged prostate. Because of the unrelieved obstruction, urine backs up and distends the ureters and then progresses to the kidney (Fig. 35–3). This enlargement of the kidney can be either unilateral or

---

BOX 35-3  *Nutrition Notes*

Renal Calculi

Sufficient fluid to produce 2000 mL of urine per day should be ingested to prevent concentrated urine, which enhances precipitation of crystals.

About 80 percent of kidney stones are composed of calcium oxalate, which led to early prescriptions of low-calcium diets. A large prospective study correlated higher calcium intake with fewer kidney stone occurrences, possibly because a high calcium intake inhibits gastrointestinal absorption of oxalate and consequently decreases the amount to be excreted in urine. Tables of nutritive composition of foods often are incomplete regarding oxalates; thus implementing a low-oxalate diet is problematic but recommended. Additionally, in the Curhan study, known high-oxalate foods (e.g., chocolate, nuts, tea, and spinach) were not associated with risk of kidney stones; however, intake of animal protein was directly associated and potassium and fluid intakes inversely associated with the risk of kidney stones.

Uric acid kidney stones can be a complication of gout, which is a disorder of purine metabolism. Purines are end products of digestion of certain proteins and are present in some medications. High-purine foods include organ meats, anchovies, herring, sardines in oil, meat extracts, consommé, and gravies. Low-purine foods include fruits, milk, cheese, eggs, refined grains, sugars, coffee, tea, carbonated beverages, tapioca, yeast, and vegetables (except asparagus, beans, cauliflower, mushrooms, peas, and spinach).

BOX 35-4 CULTURAL CONSIDERATION

Recurrent Calculus Development

Filipino immigrants are at high risk for developing renal stones, hyperuricemia, and gout. A shift from a traditional Filipino diet to a U.S. diet increases the occurrence of hyperuricemia, with some older Filipinos developing gout. The nurse may need to assist Filipino patients to identify food choices that will help prevent these conditions.

---

hydronephrosis: hydro—pertaining to water + nephrosis—degenerative change in kidney
oliguria: olig—small + uria—urine
anuria: an—without + uria—urine

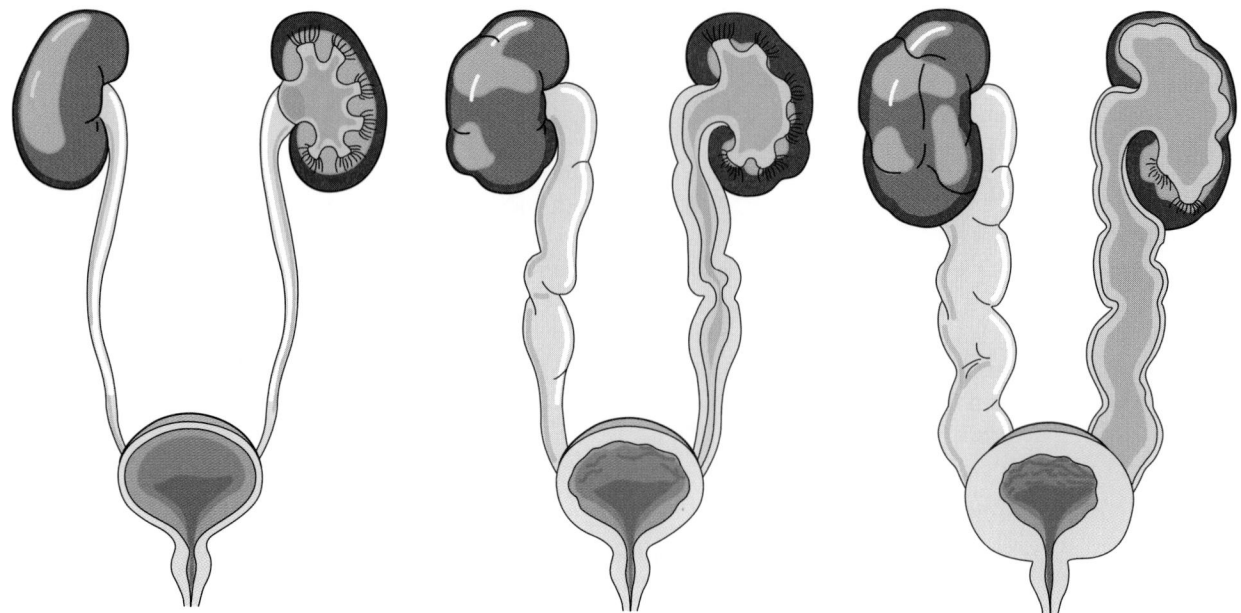

Figure 35-3 Hydronephrosis. Progressive thickening of bladder wall and dilation of ureters and kidneys results from obstruction of urine flow.

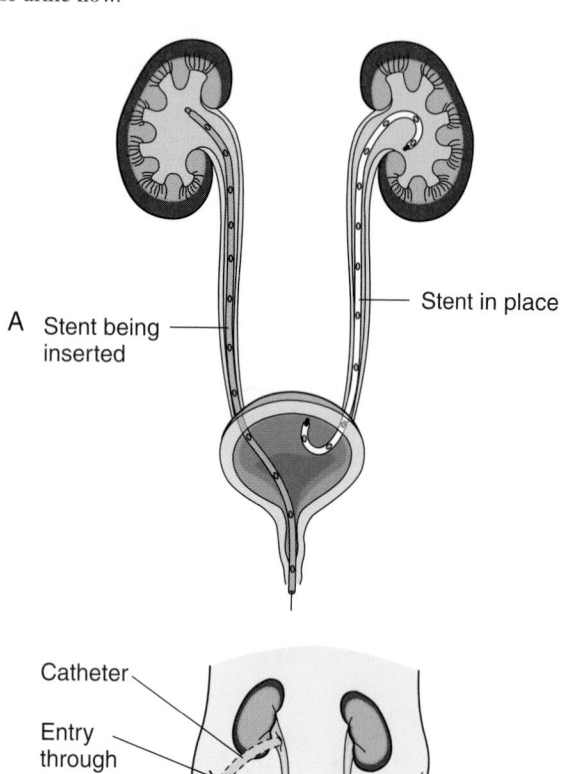

A Stent being inserted

Stent in place

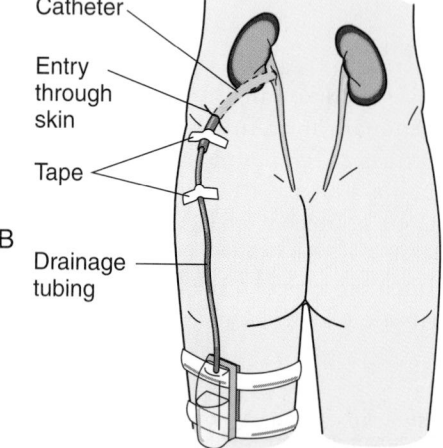

Catheter

Entry through skin

Tape

B

Drainage tubing

Posterior view

bilateral. The unrelieved pressure on the kidneys from the urine causes the kidneys to become bags filled with urine instead of functioning kidneys.

If the onset of obstruction is gradual, the patient initially may be asymptomatic. Patients often develop UTIs because of the obstruction of urine flow and may have symptoms of frequency, urgency, and dysuria. As the disease progresses, flank and back pain may occur. Eventually the patient develops symptoms of renal failure (discussed later).

The treatment of hydronephrosis always involves relieving the obstruction. Initial removal of the obstruction may be done by insertion of an indwelling urinary catheter. Long-term correction of the obstruction depends on the cause and includes treatments and surgeries to relieve obstruction from strictures, stones, tumor, or an enlarged prostate. At times the obstruction cannot be relieved because a stone is too large or removal of tumor growth would result in death of the patient. In these situations, **stents,** which are tiny tubes, may be placed inside the ureters during a cystoscopy and pyelogram (C&P) to hold them open or a **nephrostomy** tube may be inserted directly into the kidney pelvis to drain urine. A nephrostomy tube exits through an incision in the flank area and allows urine to drain into a collecting bag, so that function of the kidney can be maintained. Figure 35–4 shows a stent in place in a ureter and a nephrostomy tube.

---

nephrostomy: nephr—pertaining to the kidney + ostomy— surgically formed artificial opening to the outside

Figure 35-4 *(A)* Ureteral stents. *(B)* Nephrostomy tube inserted into renal pelvis; catheter exits through an incision on flank.

Complications associated with hydronephrosis include increased incidence of UTIs because of obstruction of urine flow and kidney failure from unrelieved pressure on the kidneys.

Intake and output is carefully measured. Urine retention can worsen the condition and must be recognized and reported promptly. If the patient has a nephrostomy tube, ensure that it is draining adequately and prevent kinking or clamping of the tube. Kinking of the tube results in continuation of the hydronephrosis, and the resulting pressure will destroy kidney function. If both a nephrostomy tube and urinary catheter are present, output from each should be measured and documented separately.

▶ TUMORS OF THE RENAL SYSTEM
Cancer of the Bladder

Cancer of the bladder is the most common kind of cancer of the urinary tract. It is more commonly seen in men ages 50 to 70.

Etiology

There is a strong correlation between smoking and bladder cancer. Exposure to industrial pollution such as aniline dyes, benzidine and naphthylamine, leather finishings, metal machinery, and petroleum processing products also increases the incidence.

Pathophysiology

Cancer of the bladder often starts as a benign growth on the bladder wall that undergoes cancerous changes. Common sites for metastasis include the liver, bones, and lungs.

Diagnostic Tests

Diagnosis of cancer of the bladder is made with a cystoscopy and biopsy, as well as from laboratory tests such as a urinalysis. An intravenous pyelogram (IVP) may also be done. The patient is also examined for metastasis to other parts of the body utilizing bone scans and x-ray examinations.

Signs and Symptoms

Cancer of the bladder usually causes painless hematuria. Blood in the urine is one of the seven warning signs of cancer from the American Cancer Society. Initially the bleeding is intermittent, which often results in people's delaying seeking treatment. As the disease progresses, the patient experiences frank hematuria, bladder irritability, urinary retention from clots obstructing the urethra, and fistula formation (an opening between the bladder and an adjoining structure such as the vagina or bowel).

Medical Management

Treatment depends on the kind of bladder cancer and the severity. For small, confined tumors, chemotherapy is instilled into the bladder through a urinary catheter, allowed to dwell, and then removed along with the catheter. Treatments are given at intervals. Systemic chemotherapy is also used and can be helpful to prolong life when other treatments are no longer indicated. The bacille Calmette-Guérin (BCG) vaccine may be instilled into the bladder to prevent recurring tumors.

Photodynamic therapy may be used, in which drugs are given that make tumors sensitive to light. When light is applied to the tumor area, cancer cells are killed.

Surgical treatment of cancer of the bladder includes a number of procedures. A cystoscopy and pyelogram with fulguration (destruction of tissue with electrical current) may be done to burn off cancerous tissue. An alternate method is use of a laser to destroy tumor tissue.

INCONTINENT URINARY DIVERSION. If it has been determined that the patient has a potentially curable disease with significant bladder involvement, complete removal of the bladder and creation of a urinary diversion may be done. A urinary diversion means that urine leaves the body in a different manner. A common incontinent surgery for urinary diversion is called an ileal conduit, an involved surgery in which a section of the ileum or colon is removed and used as a conduit for urine. The remaining portions of the bowel are stitched back together. The surgeon is careful to keep the blood and neurological supply intact to the section of bowel that has been removed. The isolated section of bowel is closed off on one end, the ureters are stitched into it, and the other end is brought out as a stoma on the abdomen that almost continuously drains urine (Fig. 35–5). The urine from an ileal conduit contains mucus because it comes through the ileum, which normally secretes mucus. The patient must wear an ostomy appliance at all times over the stoma to collect urine. Box 35–5 explains how to apply an appliance to an ileal conduit stoma.

CONTINENT URINARY DIVERSION. Several newer continent urinary diversion surgeries are being done for patient convenience. One version is the Kock pouch (continent internal ileal reservoir), which is created from a segment of ileum that has been made into a reservoir for urine (see Fig. 35–5). The ureters are implanted into the side of the reservoir. A special nipple valve is constructed and is the passageway through which the patient inserts a catheter at 4- to 6-hour intervals to drain urine. Another type of this surgery is the Indiana pouch (see Fig. 35–5). A reservoir is created using a portion of the ascending colon and terminal ileum, making a larger pouch than the Kock pouch. Additional versions of this type of surgery use other parts of the bowel and include the Mainz pouch or Florida pouch.

ORTHOTOPIC BLADDER SUBSTITUTION. The newest surgery is formation of an orthotopic bladder, using a section of the intestines to make a neobladder (neo = new) and implanting both the ureters and the urethra into the neobladder.

Different types of orthotopic bladder substitution surgery include the Studer pouch, hemi-Kock pouch and ileal W-neobladder. After this surgery the patient can void through the urethra, though incontinence may be a problem and intermittent catheterization may be needed.

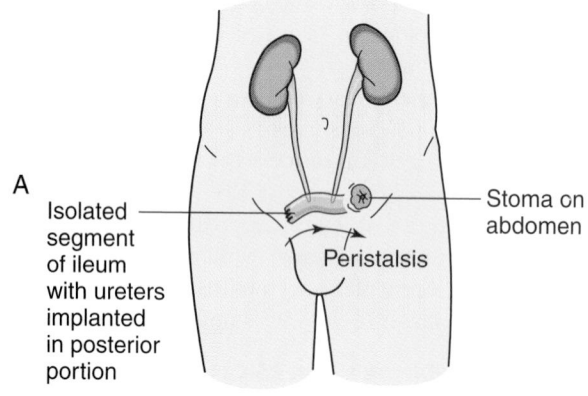

A Isolated
   segment
   of ileum
   with ureters
   implanted
   in posterior
   portion

Stoma on
abdomen

Peristalsis

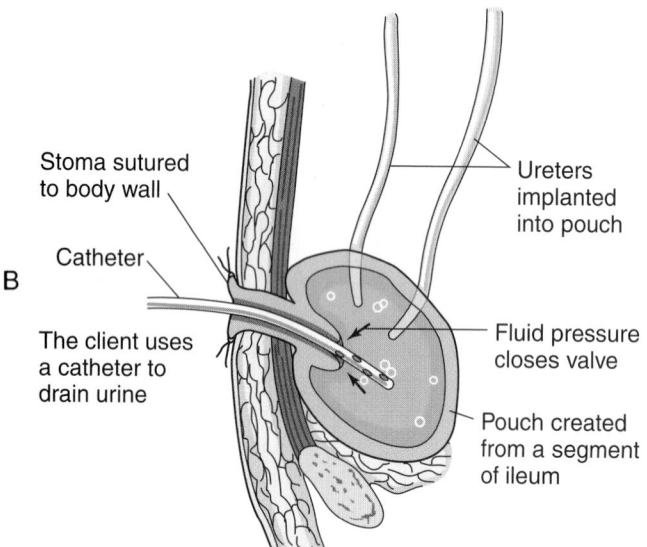

Stoma sutured
to body wall

Catheter

B

The client uses
a catheter to
drain urine

Ureters
implanted
into pouch

Fluid pressure
closes valve

Pouch created
from a segment
of ileum

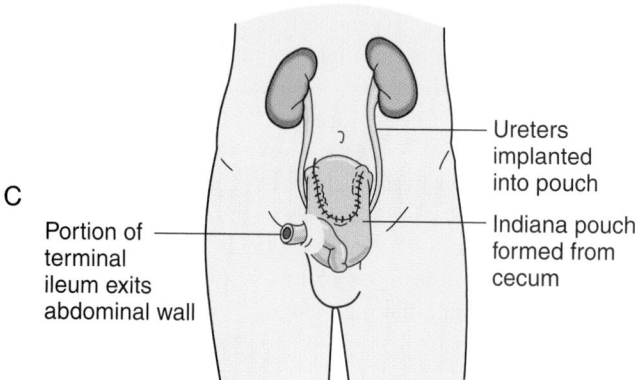

C

Portion of
terminal
ileum exits
abdominal wall

Ureters
implanted
into pouch

Indiana pouch
formed from
cecum

Figure 35-5   Urinary diversion surgery. *(A)* Ileal conduit. *(B)* Kock pouch. *(C)* Indiana pouch.

Nursing Management

Nursing care of the postoperative patient is similar to care following any major surgical procedure. (See Chapter 11.) It is important to ensure that there is adequate urinary output and to detect and report any obstruction of urine drainage early to prevent complications. The patient will

---

BOX 35-5   **Application of a Disposable Pouch to an Ileal Conduit**

1. Gather all supplies: a washcloth and towels and water, a pouch to apply with a stomahesive flange, and wicks such as Kerlix gauze to absorb urine. The nurse should wear clean gloves.
2. Empty the old pouch.
3. Gently remove soiled pouch by pushing down on skin while lifting up on the flange. Discard soiled pouch and flange.
4. Place a towel around the stoma to catch urine.
5. Cut an opening in the flange that is only $\frac{1}{16}$ to $\frac{1}{8}$ inch larger than the stoma. Once stomal shrinkage is complete, a presized pouch can be used that fits the stoma.
6. Remove paper backing from the stomahesive and set the flange to one side.
7. Clean the skin around the stoma with water. Pat dry. Immediately wrap the stoma in wicks to absorb urine. Otherwise urine will leak onto the skin, and the flange will not adhere.
8. Center the flange over the stoma, remove the wick, and immediately apply the flange. Then snap the pouch onto the flange. NOTE: The flange and pouch may be snapped together before application to the stoma.
9. Use the heat of your hand to compress the flange to ensure a good seal.
10. Ensure that the bottom of the pouch is closed off, or connect to a urinary catheter bag for nighttime or for the patient in bed most of the time.

Source: Adapted from Hampton, BG, and Bryant, RA: Ostomies and Continent Diversions: Nursing Management. Mosby, St Louis, 1992.

---

need instruction on how to care for the urinary diversion after surgery, either by frequent draining with a catheter or by wearing an appliance. Be sensitive to the patient's anxiety about caring for the urinary diversion. Body image disturbance may occur because of the change in body function. A consultation with a nurse who specializes in wound and ostomy care or an ostomy support group may be helpful both before and after surgery.

Cancer of the Kidney
Pathophysiology and Etiology

Cancer of the kidney is rare but serious. Risk factors include smoking and exposure to industrial pollution. Males have twice the incidence of females. Often the cancer has metastasized before it is diagnosed, because the kidney has such a large volume of circulating blood, which increases the risk for spread of the tumor. In addition, there are few early symptoms of the disease.

Signs and Symptoms

The three classic symptoms of kidney cancer are hematuria, dull pain in the flank area, and a mass in the area. Symptoms of metastasis may be the first manifestation of kidney cancer and include weight loss and increasing weakness.

Diagnostic Tests

A number of diagnostic tests will be done, including an IVP, cystoscopy and pyelogram, ultrasound examination of the kidneys, computed tomography (CT) scans, and magnetic resonance imaging (MRI). A definitive diagnosis is made with a renal biopsy.

Medical Management

A radical **nephrectomy** is the preferred treatment for cancer of the kidney. The kidney is removed, along with the adrenal gland and other surrounding structures, including fascia, fat, and lymph nodes, in the area. Radiation therapy, immunotherapy, or chemotherapy may be used following surgery.

Nursing Management

Nursing care of the nephrectomy patient is similar to postoperative care following any major surgery. (See Chapter 11.) Because the kidney is highly vascular, it is essential that the nurse observe for onset of bleeding and any signs of hypovolemic shock. Urine output is monitored. Changes in urine amount or color, bleeding, and signs of infection are reported.

▶ RENAL SYSTEM TRAUMA

There are many causes of trauma to the kidney, ureters, and bladder, such as motor vehicle accidents, sports injuries, falls, gunshot, and stabbing. Young males are at greatest risk for renal system trauma. Diagnostic tests include urinalysis, IVP, ultrasound, CT, and MRI. Flank pain and hematuria may be present. Treatment depends on the extent of the injury and ranges from bedrest to surgical intervention. Nursing care includes measuring intake and output, monitoring vital signs, and providing IV fluids and pain relief.

▶ POLYCYSTIC KIDNEY DISEASE

Polycystic kidney disease is a hereditary disorder that can result in renal failure. The disease is characterized by formation of multiple cysts in the kidney that can eventually replace normal kidney structures. The cysts are grapelike and contain serous fluid, blood, or urine. The patient generally first shows signs of the disease in adulthood. The initial symptoms include a dull heaviness in the flank or lumbar region and hematuria. As the disease progresses, the patient develops symptoms of renal failure (discussed later).

There is no treatment to stop the progression of polycystic kidney disease. Complications such as urinary tract infections are treated as needed. As the disease progresses, treatment for hypertension and eventual renal failure may be necessary. Because polycystic disease is hereditary, patients should be counseled about the risks of children inheriting it.

▶ CHRONIC RENAL DISEASES
Diabetic Nephropathy

Diabetic **nephropathy** is the most common cause of renal failure. It is a long-term complication of diabetes mellitus in which the effects of diabetes result in damage to the small blood vessels in the kidneys. Renal damage shows up approximately 15 to 20 years after onset of type 1 diabetes (insulin dependent), but it may also be a complication of type 2 diabetes (non–insulin dependent). Careful control of blood glucose levels reduces the risk of nephropathy in patients with diabetes.

Pathophysiology

Multiple factors contribute to diabetic nephropathy. Widespread atherosclerotic changes occur in the blood vessels of patients with diabetes, decreasing the blood supply to the kidney. Abnormal thickening of glomerular capillaries damages the glomerulus, allowing protein to leak into the urine. Patients with diabetes also commonly develop pyelonephritis and renal scarring. Another complication of diabetes, neurogenic bladder, causes incomplete bladder emptying. This results in retention of urine, which can cause infection or obstruction of urine, further damaging the kidneys.

Initially patients lose only small amounts of protein in their urine (microalbuminuria); this disease can be detected only with careful watching by the physician, utilizing frequent examinations of the urine. As the disease progresses, high-output renal failure (nonoliguria) can develop, in which a large amount of diluted urine is excreted without the usual amounts of waste product dissolved in the urine. The patient can lose large amounts of protein in the urine and may develop nephrotic syndrome, which causes massive edema because of low levels of albumin in the blood. As renal function decreases, the patient needs smaller doses of insulin because the kidney normally degrades insulin. Because the kidney is no longer able to break down insulin and excrete it, small doses of insulin circulate in the body for long periods.

Symptoms

As diabetic nephropathy progresses, urine output decreases, toxic wastes accumulate, and the patient develops chronic renal failure. Symptoms of chronic renal failure are found in Figure 35–6.

Diagnostic Tests

Diabetic nephropathy is diagnosed by careful watching of the patient with diabetes for onset of protein spillage in the urine, which is an early sign of the disease. Serum creatinine levels and 24-hour creatinine clearance tests are then done to confirm the presence and extent of diabetic nephropathy.

---

nephrectomy: nephr—kidney + ectomy—excision

nephropathy: nephro—pertaining to the kidney + pathy—disease

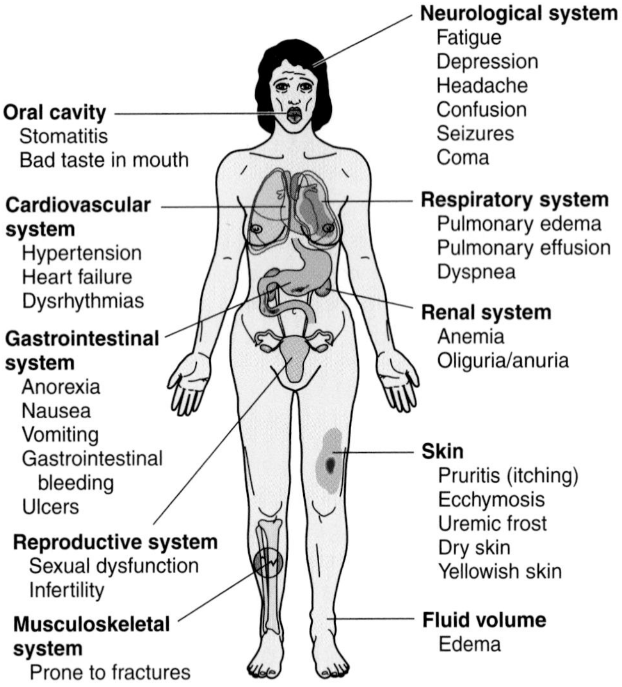

Figure 35-6 Symptoms of chronic renal failure.

Neurological system
Fatigue
Depression
Headache
Confusion
Seizures
Coma

Oral cavity
Stomatitis
Bad taste in mouth

Cardiovascular system
Hypertension
Heart failure
Dysrhythmias

Gastrointestinal system
Anorexia
Nausea
Vomiting
Gastrointestinal bleeding
Ulcers

Reproductive system
Sexual dysfunction
Infertility

Musculoskeletal system
Prone to fractures

Respiratory system
Pulmonary edema
Pulmonary effusion
Dyspnea

Renal system
Anemia
Oliguria/anuria

Skin
Pruritis (itching)
Ecchymosis
Uremic frost
Dry skin
Yellowish skin

Fluid volume
Edema

TABLE 35-1	CAUSES OF NEPHROTIC SYNDROME

- Primary nephrotic syndrome
- Glomerulonephritis
- Diabetes
- Systemic lupus erythematosus
- Infections: streptococcal, human immunodeficiency virus (HIV), hepatitis, malaria
- Cancer: leukemia, Hodgkin's disease
- Drugs: nonsteroidal anti-inflammatory drugs (NSAIDs), penicillamine, captopril

Medical Management

In the early stages of diabetic nephropathy, strict control of blood glucose levels and blood pressure and a restricted-protein diet can help slow the progress of the disease and reduce symptoms. As the disease progresses to renal failure, the patient needs dialysis to maintain life. Unfortunately, other complications related to diabetes cause patients to tolerate dialysis less well than patients with renal failure from other causes. Kidney or kidney-pancreas transplant, when available, is the treatment of choice for the patient with diabetic nephropathy and often improves the patient's chance for a healthier life.

Complications

Patients with diabetic nephropathy often have a guarded prognosis because they are vulnerable to all the complications of long-term diabetes in addition to kidney disease. (See Chapter 38.)

Nephrotic Syndrome

Nephrotic syndrome may occur as a result of other disease processes (Table 35–1). In nephrotic syndrome, large amounts of protein are lost in the urine from increased glomerular membrane permeability. As a result, serum albumin and total serum protein are decreased. Normally, albumin and other serum proteins maintain fluid within the vascular space. When levels of these proteins are low, fluid leaks from the blood vessels into tissues, resulting in edema. With very low levels of protein, ascites and massive widespread edema (anasarca) occur. In response to the low protein levels, the liver produces lipoproteins. As a result, serum cholesterol, low-density lipoproteins, and triglyceride levels are elevated.

Treatment is focused on the cause and symptoms of nephrotic syndrome. To control edema, sodium intake is restricted. A low to moderate protein intake is ordered to prevent buildup of nitrogen wastes (nitrogen wastes result from protein metabolism) from impaired kidney function. Protein intake is based on the severity of urinary protein loss. Diuretics may be used. Lipid-lowering drugs may be tried. Anticoagulants are given for thrombosis. In some cases, corticosteroids may be used.

Complications of nephrotic syndrome include impaired immune function, nutritional imbalances, and most importantly increased blood coagulation. Nursing care focuses on the edema and preventing infection. For edema, daily weights, careful intake and output measurement, and abdominal girth measurement are performed and documented. Edematous tissue must be protected from injury. Preventing malnutrition is challenging but important in maintaining normal body functions.

Nephrosclerosis

Hypertension damages the kidneys by causing sclerotic changes, such as arteriosclerosis with thickening and hardening of the renal blood vessels (**nephrosclerosis**). The arteriosclerotic changes in the kidney blood vessels result in a decreased blood supply to the kidney (ischemia of the kidney) and can eventually destroy the kidney. Symptoms of nephrosclerosis include proteinuria, hyaline casts in the urine, and, as it progresses, symptoms of renal failure.

The treatment of nephrosclerosis is treatment of hypertension. The patient is placed on antihypertensive medications or, if already on these, changed to stronger antihypertensive medications. The patient is placed on a low-sodium diet. If the patient develops renal failure, dialysis will be used to maintain life.

The prognosis is often poor because by the time the patient has developed nephrosclerosis, there is widespread arteriosclerosis throughout the body. Arteriosclerosis makes the patient prone to myocardial infarctions or cerebrovascular accidents.

The major nursing diagnosis that is relevant when the patient develops nephrosclerosis is impaired health mainte-

nephrosclerosis: nephro—pertaining to the kidney + sclerosis—hardening

nance. The priority is to help the patient learn as much about the control of hypertension as possible. The patient should also be taught the symptoms of renal failure. Once the patient has lost renal function, the nursing care plan for renal failure is appropriate.

CRITICAL THINKING

Mr. White

Mr. White is a 35-year-old African-American gentleman admitted to the intensive care unit with uncontrolled hypertension. His blood pressure is controlled by intravenous medication. His lab tests show protein and hyaline casts in the urine. He is diagnosed with nephrosclerosis.

1. What should the nurse assess as part of the morning evaluation of the patient's condition?
2. What other renal function tests are appropriate for the nurse to check?
3. What teaching does Mr. White need when his condition is more stable?

Answers at end of chapter.

► GLOMERULONEPHRITIS

Pathophysiology

Glomerulonephritis is an inflammatory disease of the glomerulus. Inflammation occurs as a result of the deposition of antigen-antibody complexes in the basement membrane of the glomerulus or from antibodies that specifically attack the basement membrane. The resulting immune reaction in the glomerulus causes inflammation, which in turn causes the glomerulus to be more porous, allowing proteins, white blood cells, and red blood cells to leak into the urine.

Etiology
Acute Poststreptococcal Glomerulonephritis

Glomerulonephritis can be caused by a variety of factors but is most commonly associated with a group A beta-hemolytic streptococcus infection following a streptococcal infection of the throat or skin. This is the most common cause in children and young adults. Antibodies form complexes with the streptococcal antigen and are deposited in the basement membrane of the glomerulus. Glomerulonephritis generally develops about 6 to 10 days after the preceding infection. Other kinds of bacteria and viruses can also be the offending infectious agent.

Goodpasture's Syndrome

Occasionally glomerulonephritis is caused by an autoimmune response, in which the person for some unknown reason forms antibodies against his or her own glomerular basement membrane. Glomerulonephritis caused by an autoimmune response usually progresses rapidly and often leads to renal failure.

Chronic Glomerulonephritis

Chronic glomerulonephritis occurs over years as a result of glomerular inflammatory disease. It is often discovered during an examination for another concern. Ultrasound, CT scan, or renal biopsy is used to diagnose the cause.

Symptoms

The symptoms of glomerulonephritis include fluid overload with oliguria, hypertension, electrolyte imbalances, and edema. Edema may begin around the eyes (periorbital edema) and face and progress to the abdomen (ascites), lungs (pleural effusion), and extremities. Flank pain may be present. Blood urea nitrogen (BUN) and creatinine levels may be elevated. Urinalysis shows red blood cells, white blood cells, albumin, and casts. The urine is dark or cola colored from old red blood cells and may be foamy because of proteinuria.

Medical Management

Most cases of acute glomerulonephritis resolve spontaneously in about a week, but some progress to renal failure. Treatment is primarily symptomatic. Sodium and fluid restrictions may be ordered along with diuretics to treat fluid retention. Medications may be given to control hypertension. If associated with a streptococcus infection, antibiotics are given to treat any remaining infection. If fluid overload is severe, dialysis may be done.

Complications

The prognosis is good for acute glomerulonephritis acquired in childhood, and the majority of children recover completely. Adults who develop glomerulonephritis may recover renal function or progress to chronic glomerulonephritis. Some patients develop rapidly progressive glomerulonephritis, which can quickly lead to renal failure. Chronic glomerulonephritis is a slow process characterized by hypertension, gradual loss of renal function, and eventual renal failure, discussed next.

Nursing Management

Nursing care for a patient with glomerulonephritis focuses on symptom relief. Vital signs are monitored because the patient may be critically ill. During the acute phase, rest is encouraged. Edema is controlled with fluid and sodium intake restrictions. Protein intake may be limited if the kidneys are not filtering protein waste products (as seen by increased serum BUN and creatinine levels). Other care is discussed in the section on chronic renal failure. Teaching the patient about preventing glomerulonephritis is important. Antibiotics for diagnosed streptococcal throat infections should be taken for prevention.

---

glomerulonephritis: glomerulo—glomerulus + nephr— kidney + itis—inflammation

▶ RENAL FAILURE

Renal failure, also called kidney failure, is diagnosed when the kidneys are no longer functioning adequately to maintain normal body processes. This results in dysfunction in almost all other parts of the body as a result of imbalances in fluid, electrolytes, and calcium levels, as well as impaired RBC formation and decreased elimination of waste products. Renal failure can be acute, with sudden onset of symptoms, or chronic, occurring gradually over time. For more information on kidneys, visit the American Kidney Fund at www.akfinc.org, the National Kidney Foundation at www.kidney.org, and the American Association of Kidney Patients at www.aakp.org.

Acute Renal Failure

Acute renal failure (ARF) occurs when loss of kidney function is sudden, with a rapid onset of hours to days.

Pathophysiology and Etiology

In acute renal failure, rapid damage to the kidney causes waste products to accumulate in the bloodstream, resulting in the symptoms of renal failure. The patient becomes oliguric, with urine output decreasing to less than 20 mL per hour. Many patients with acute renal failure recover completely with treatment directed toward correcting the cause, supporting the patient with dialysis, and prevention of complications that may lead to permanent damage. Approximately 50 percent of patients die as a result of complications of infection, pneumonia, or septicemia.

Three categories of causes can lead to acute renal failure. Each category is associated with the location of the cause in the kidney.

PRERENAL FAILURE. Prerenal (before the kidney) failure is associated with a decrease or interruption of blood supply to the kidneys. The causes may include a decrease in blood pressure as a result of dehydration, blood loss, shock, or trauma to or blockage in the arteries that carry blood to the kidneys. When the nephrons receive an inadequate blood supply, they are unable to make urine and the waste products are not adequately removed.

Prerenal failure can be diagnosed by evaluating possible causes. An arteriogram of the renal arteries is helpful to determine if the blood supply to the kidneys is decreased or blocked; angioplasty may be used to open the blockage.

INTRARENAL FAILURE. Intrarenal failure (inside the kidney) occurs when there is damage to the nephrons inside the kidney. The most common causes are from infectious processes leading to glomerulonephritis, trauma to the kidney, exposure to **nephrotoxins** (Table 35–2), allergic reactions to radiograph dyes, and severe muscle injury, which releases substances that are harmful to the kidneys.

TABLE 35–2	**COMMON NEPHROTOXINS**
Antibiotics	
Aminoglycosides	
Tetracyclines	
Cephalosporins	
Sulfonamides	
Vancomycin	
Amphotericin B	
Analgesics	
Nonsteroidal anti-inflammatory drugs	
Acetaminophen	
Phenacetin	
Salicylates	
Other Drugs	
ACE Inhibitors	
Dextran	
Mannitol	
Interleukin-2	
Cisplatin	
Amphetamines	
Heroin	
Heavy Metals	
Lead	
Mercury	
Arsenic	
Copper	
Gold	
Lithium	
Contrast Dyes	
Contrast media used for diagnostic testing such as intravenous pyelograms, cardiac catheterizations	
Organic Solvents	
Gasoline	
Glycols	
Kerosene	
Turpentine	
Tetrachloroethylene	

ACE = angiotensin-converting enzyme.

A number of substances can be toxic to the kidneys (nephrotoxic) when they enter the body. Kidney damage is most likely to occur when these substances enter the body in high concentrations or when there is preexisting kidney damage for some other reason. Environmental nephrotoxins, such as insecticides and lead paint, may be ingested by children. Many commonly administered medications can be nephrotoxic. Aminoglycosides are nephrotoxic antibiotics; when they are administered, blood levels of the drugs are carefully monitored to avoid toxic levels.

Contrast media used during tests such as intravenous pyelograms and CT scans can cause kidney damage when the patient is dehydrated or has preexisting renal damage. The medium can precipitate out in the tubules, damaging the kidney. It is important for the patient to be adequately hydrated after any diagnostic test using a contrast medium to decrease the incidence of toxicity.

---

nephrotoxins: nephro—kidney + toxin—poison

LEARNING TIPS

To protect patients' kidneys, be aware of the following:
- The patient's renal function status: Serum BUN and creatinine levels will tell you this.
- Nephrotoxic substances:
  - Diagnostic contrast media (dyes) in presence of dehydration or renal impairment
  - Medications—IV aminoglycosides (gentamicin, tobramycin, amikacin), cisplatin
  - Chemicals—arsenic, carbon tetrachloride, lead, mercuric chloride
- Preventive measures:
  - Before administering nephrotoxic dyes or medications, check serum BUN and creatinine levels.
  - Make sure peak and trough drug levels of nephrotoxic drugs are obtained on an ongoing basis per institutional policy.
  - After contrast media (dye) tests, encourage fluids to dilute and flush, flush, flush the dye away!

POSTRENAL FAILURE. Postrenal (after the kidney) failure is associated with an obstruction that blocks the flow of urine out of the body. In this case the blood supply to the kidneys and nephron function initially may be normal, but urine is unable to drain out of the kidney, resulting in backup of urine and impaired nephron function. Common causes are kidney stones, tumors of the ureters or bladder, and an enlarged prostate that blocks the flow of urine.

Diagnosis of causes of postrenal failure can be done with x-ray examination of the kidneys, ureters, and bladder. Cystoscopy will show presence of tumors, stones, or prostate enlargement. Surgical intervention may be needed to correct the problem.

Medical Management

Acute renal failure is treated by relieving the cause. The care of the patient with acute renal failure is similar to care of the patient with chronic renal failure, as explained in the next section.

CONTINUOUS RENAL REPLACEMENT THERAPY. Temporary dialysis or continuous renal replacement therapy (CRRT) may be necessary to treat renal failure and to prevent permanent renal damage (Fig. 35–7). Temporary **hemodialysis** is indicated for severe symptoms of **uremia**. He-

---

hemodialysis: hemo—blood + dialysis—passage of a solute through a membrane
uremia: ur—urea + emia—in the blood

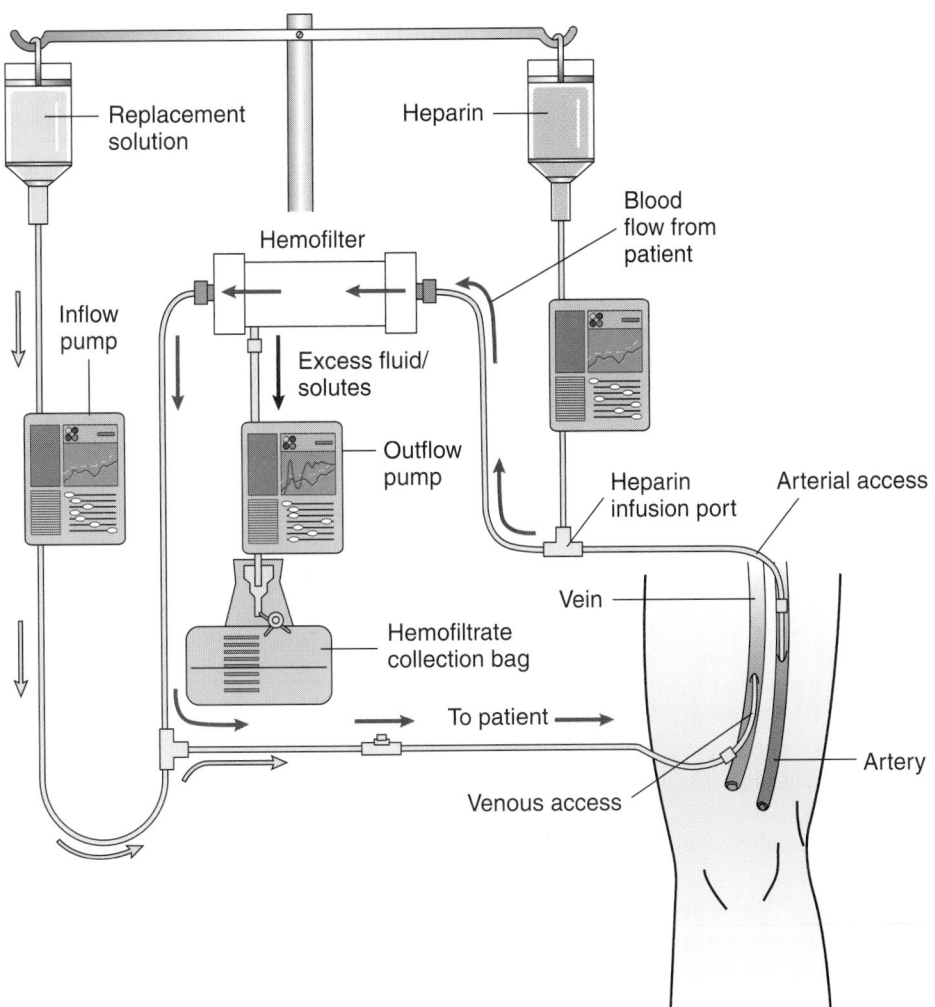

Figure 35-7 Continuous renal replacement therapy: continuous arteriovenous hemofiltration.

modialysis is intermittent, but CRRT can be used to remove fluid continuously along with the hemodialysis. CRRT is also used alone to remove fluid and solutes in a controlled, continuous manner in unstable patients with fluid overload, as in acute renal failure or pulmonary edema. CRRT is done in a variety of ways that depend on arterial or venous access of a central artery or vein. A pump is required to move the blood when venous access is used. CRRT is not as complex as hemodialysis and can be done for more than a month if needed.

During CRRT, a permeable hemofilter is attached to the vascular access. Blood flows through the hemofilter and excess fluids and solutes move into a collection bag. The remaining blood returns to the patient via the venous access. If desired, replacement fluid and electrolytes can be given through the vascular access. Monitoring intake and output, fluid and electrolytes, daily weights, hourly vital signs, and vascular access is important.

Chronic Renal Failure

Chronic renal failure (CRF) affects approximately 290,000 people in the United States. It occurs with a gradual decrease in the function of the kidneys over time. This loss of function is not reversible.

Etiology

The causes of chronic renal failure are numerous; common ones include diabetes mellitus resulting in diabetic nephropathy, chronic high blood pressure causing nephrosclerosis, glomerulonephritis, and autoimmune diseases.

Pathophysiology

When a large proportion of the nephrons are damaged or destroyed because of acute or chronic kidney disease, renal failure occurs. As the nephrons die off, the undamaged ones increase their work capacity and take over the work previously done by the dead ones, so the patient may experience significant kidney damage without showing symptoms of renal failure.

Chronic renal failure is a progressive disease process. In the early, or silent, stage the patient is usually without symptoms, even though up to 50 percent of nephron function may have been lost. This stage is often not diagnosed.

The renal insufficiency stage occurs when the patient has lost 75 percent of nephron function and some signs of mild renal failure are present. Anemia and the inability to concentrate urine may occur. The BUN and creatinine levels are slightly elevated. These patients are at risk for further damage caused by infection, dehydration, drugs, heart failure, and use of diagnostic x-ray dyes. The goal of care is to prevent further damage if possible by good control of blood sugar levels and blood pressure.

End-stage renal disease (ESRD) occurs when 90 percent of the nephrons are lost. Patients at this stage experience chronic and persistent abnormal kidney function. The BUN and creatinine levels are always elevated. These patients may make urine but not filter out the waste products, or urine production may cease. Dialysis or a kidney transplant is required to survive.

Uremia (urea in the blood) is present in chronic renal failure. Patients eventually develop problems in all body systems. Table 35–3 shows the effect of renal failure on body systems. If left untreated, the patient with uremia dies, often within weeks.

Symptoms of Renal Failure and Nursing Interventions

Patients in either acute or chronic renal failure have multiple symptoms. Some of the more common symptoms associated with renal failure are explained next (see Fig. 35–6).

Disturbance in Water Balance

Patients with renal failure experience disturbances in the removal and regulation of water balance in the body and show signs of fluid accumulation. An early symptom is edema (swelling) of the extremities, sacral area, and abdomen. Patients may complain of being short of breath. Crackles and wheezes may be present on auscultation of the lungs, which are signs of fluid accumulation in the lungs. The blood vessels in the neck may be distended, and the patient may be hypertensive. These patients may produce a large amount of dilute urine (polyuria), small amounts of urine (oliguria), or no urine (anuria).

Nursing care for disturbances in water balance involves monitoring the patient's weight at the same time each day. The patient who is retaining fluid will have an increase in weight. Intake and output are measured, and a fluid restriction will be ordered by the physician, generally 1000 mL per 24 hours. Lung sounds are monitored for crackles and wheezes. Symptoms of fluid retention are reported.

Disturbance in Electrolyte Balance

As kidney function decreases, the kidneys lose their ability to absorb and excrete electrolytes. Important electrolytes are sodium, potassium, and magnesium. When the kidneys are unable to maintain normal amounts of electrolytes in the blood, these substances accumulate at high levels and may be life threatening.

When the kidneys are unable to regulate sodium levels adequately, the patient may show signs of hypernatremia, an excessive sodium level in the blood, which causes water retention, edema, and hypertension. Hyponatremia, too little sodium, may occur when too much sodium is lost. This can occur when the patient has experienced prolonged episodes of vomiting or diarrhea or is urinating large amounts of diluted urine. Patients with hyponatremia may show signs of confusion.

Hyperkalemia (high level of potassium) presents a life-threatening situation. The patient may exhibit signs of dysrhythmia and cardiac arrest if the potassium level is too high. Patients complain of muscle weakness, abdominal cramping, and diarrhea. The nurse may identify that the pa-

---

polyuria: poly—much + uria—urine

TABLE 35-3	EFFECTS OF RENAL FAILURE ON BODY SYSTEMS
Body System	**Disease Process**
Cardiovascular	Hypertension due to fluid overload and accelerated arteriosclerosis
	Congestive heart failure/pulmonary edema due to fluid overload, increased pulmonary permeability, left ventricular failure
	Angina due to coronary artery disease, anemia
	Dysrhythmias due to electrolyte imbalance, coronary artery disease
	Edema due to fluid overload and a decrease in osmotic pressure
	Pericarditis due to presence of waste products in the pericardial sac
Pulmonary	Pleurisy/pleural effusion due to waste products in the pleural space causing inflammation with pleurisy pain and also collection of fluid resulting in effusion
Hematopoietic	Anemia due to impaired synthesis of erythropoietin, a substance needed by the bone marrow to stimulate formation of RBCs; also due to decreased life span of RBCs due to uremia and interference in folic acid action
	Bleeding tendency due to abnormal platelet function from effects of uremia
	Prone to infection due to a decrease in immune system function from uremia; renal patients can rapidly become septic and die from septic shock
Integumentary	Dry, itchy, inflamed skin due to calcium-phosphate deposits in the skin and urochrome, and a pigment of uremia, which causes the skin to be pale gray, yellow bronze; skin will have an odor of urine because skin is an organ of excretion and the body attempts to remove toxins; there is also a decrease in function of oil and sweat glands
Gastrointestinal	Stomatitis due to fluid restriction, presence of waste products in the mouth, secondary infections
	Anorexia, nausea, vomiting due to uremia
	Gastritis/gastrointestinal bleeding due to urea decomposition in gastrointestinal tract releasing ammonia that irritates and ulcerates the stomach or bowel; patient is also under stress, increasing ulcer formation, and may have platelet dysfunction
	Constipation due to electrolyte imbalances, decrease in fluid intake, decrease in activity, phosphate binders
	Diarrhea, hypermotility due to electrolyte imbalance
Neurological	Confusion due to uremic encephalopathy from an increase in urea and metabolic acids
	Peripheral neuropathy due to effects of waste products on neurological system
	Cerebrovascular accidents due to accelerated atherosclerosis
Skeletal	Bone disease due to renal osteodystrophy from hyperphosphatemia and hypocalcemia
Reproductive	Loss of libido, impotence, amenorrhea, infertility due to a decrease in hormone production

tient is confused or demonstrates disinterest in care. Hyperkalemia exists when the potassium level exceeds 5 mEq/L. These patients should be placed on a cardiac monitor and observed for cardiac dysrhythmias. A potassium level above 7 mEq/L may be life threatening. A high potassium level in the patient with renal failure may be caused by a diet high in potassium-rich foods, injuries, or blood transfusions. Nursing care should include monitoring daily laboratory values, restricting potassium intake, and reporting abnormalities. Insulin or sodium bicarbonate may be used as a temporary measure to drive excess potassium into the cells. Sodium polystyrene sulfonate (Kayexalate) may be given either orally or as a retention enema; it causes the potassium to be eliminated through the bowels. The definitive treatment for hyperkalemia is hemodialysis to remove potassium from the body. Dietary education is extremely important. The patient is instructed to avoid foods that are high in potassium (Box 35–6).

BOX 35-6	**Foods High in Potassium**

Citrus fruits and juices
Bananas
Salt substitutes
Potatoes, sweet and white
Excessive dairy products
Excessive meats
Chocolate

Calcium levels decrease because the kidneys are unable to produce the hormone that activates vitamin D, the vitamin that is necessary for the absorption of calcium. Hypocalcemia exists when the calcium level falls below 8.5 mg/dL. Also associated with a low calcium level is hyperphosphatemia, a phosphorous level above 5 mg/dL. These imbalances cause the bones to release calcium, causing patients to be prone to fractures. These patients should ambulate regularly to prevent further calcium loss from the bone.

Phosphates are supplied by some foods. The increase in phosphorus levels may result in severe itching, and patients may have open sores as a result of scratching, placing them at risk for infections. Patients complain of muscle cramps and aches. Medications to bind phosphate are given to patients with high phosphate levels. Tums and Caltrate (calcium carbonate) or PhosLo (calcium acetate) are examples of commonly ordered phosphate binders. These must be given to the patient with meals, so that the medicines can bind with the phosphates in the bowel and be eliminated.

Disturbance of Removal of Waste Products

When the kidneys are unable to remove the waste products of metabolism, **azotemia** occurs. This is an increase in the serum urea level (measured by BUN); creatinine levels are often elevated as well. These substances are the

azotemia: azo—nitrogenous waste products + temia—blood

result of protein metabolism. The patient may show signs of weakness and fatigue, confusion, seizures, twitching movements of extremities (asterixis), nausea, vomiting, and lack of appetite and may complain of a bad taste in the mouth. The nurse may smell urine on the patient's breath. The patient may have yellowish-pale skin and complain of itching.

Nursing care for patients with azotemia includes monitoring laboratory results and treating and reporting symptoms. Adequate nutrition is important; the dietitian should be consulted to recommend the appropriate low-protein diet to decrease formation of these waste products.

Good oral care and skin care is important for these patients. The nurse should offer mouth care frequently and inspect the mouth for sores. Lotion is used for itching, and the skin is observed for open areas and signs of infection. The patient is protected from injury if confusion or seizures occur. Dialysis to remove the excessive waste products in the blood is the only treatment of the underlying causes of these symptoms.

Disturbance in Maintaining Acid-Base Balance

Renal failure affects hydrogen ion excretion, causing a disturbance in acid-base balance that results in metabolic acidosis. Patients may complain of a headache, fatigue, weakness, nausea, vomiting, and lack of appetite. As the acidosis progresses, the patient shows signs of lethargy, stupor, and coma. Respirations become fast and deep as the lungs attempt to blow off carbon dioxide to correct the acidosis. See Chapter 5 for a more detailed discussion of acid-base balance.

Disturbance in Hematological Function

Failing kidneys do not produce adequate erythropoietin, the hormone that stimulates red blood cell production. Nutritional deficiencies and blood loss during dialysis also contribute to anemia. Regular injections of epoetin (Epogen, Procrit), a synthetic form of erythropoietin, helps restore RBC production and prevent anemia.

Impaired white blood cell and immune functions contribute to an increased risk for infection. The patient should be protected from potential sources of infection.

Impaired platelet function creates a risk for bleeding. The patient should be protected from injury, and signs of bleeding, such as blood in stool or emesis, should be reported.

Medical Management of Renal Failure

Renal insufficiency and early renal failure are treated symptomatically with restricted diet and fluids, medications, and careful monitoring for onset of serious problems that warrant initiation of dialysis. In later stages, dialysis is necessary to replace lost kidney function. A kidney transplant, when available, may return the patient to a nearly normal state of health.

Diet

Dietary recommendations are individualized by the dietitian and physician, based on the patient's needs. Calories are high to maintain weight and energy needs. Protein is usually restricted to limit nitrogen intake but may be increased for the patient on dialysis, because protein is lost during the dialysis process. Sodium is restricted to minimize sodium and fluid retention. Potassium is restricted, especially later in the disease when the kidneys are unable to eliminate it. Calcium is increased or supplemented because of poor absorption related to faulty vitamin D activation. Phosphorus is restricted because of high blood levels related to hypocalcemia. Saturated fat and cholesterol are restricted for patients with hyperlipidemia. Fluids are restricted to prevent overload. Most patients are given iron, vitamins, and minerals to supplement the restricted diet (Nutrition Notes Box 35–7).

BOX 35-7 *Nutrition Notes*

Understanding Dietary Changes in Renal Disease

Patients with impaired renal function require careful coordination of diet with current physiological status, which may change frequently. Six National Renal Diets have been developed for patients with renal sufficiency who are being treated with hemodialysis or peritoneal dialysis. Different diets are used for patients with and without diabetes. The following principles are offered as general guidelines.

- Caloric intake is maintained to avoid catabolism of tissue for energy. Simple carbohydrates and monounsaturated and polyunsaturated fats are given freely because their end products, carbon dioxide and water, are less likely than protein to burden the kidney. Patients with diabetes and uremia may receive higher amounts of sugar than usual because treatment of the uremia may take precedence over the diabetes. Patients with type IV hypertriglyceridemia may have to limit carbohydrates.
- Protein may be restricted when the patient's kidneys are failing but increased when the patient is treated with dialysis to compensate for losses in the dialysate. Sometimes proteins of high biologic value (e.g., eggs, meat, and dairy products) are prescribed because they are more easily converted to body protein than those of low biologic value. In other situations, vegetarian diets may be given, with the plant proteins carefully selected to manage potassium and phosphorus serum levels.
- Sodium may be restricted, depending on blood pressure, edema, and laboratory findings.
- Potassium may be restricted for patients with oliguria. Salt substitutes are often potassium compounds that are to be avoided. Potassium content in foods varies with processing and preparation methods. Patients should choose from prescribed foods only.
- Fluid restriction may be altered daily according to output. Renal insufficiency patients may receive 500 mL plus the amount of the previous day's output.

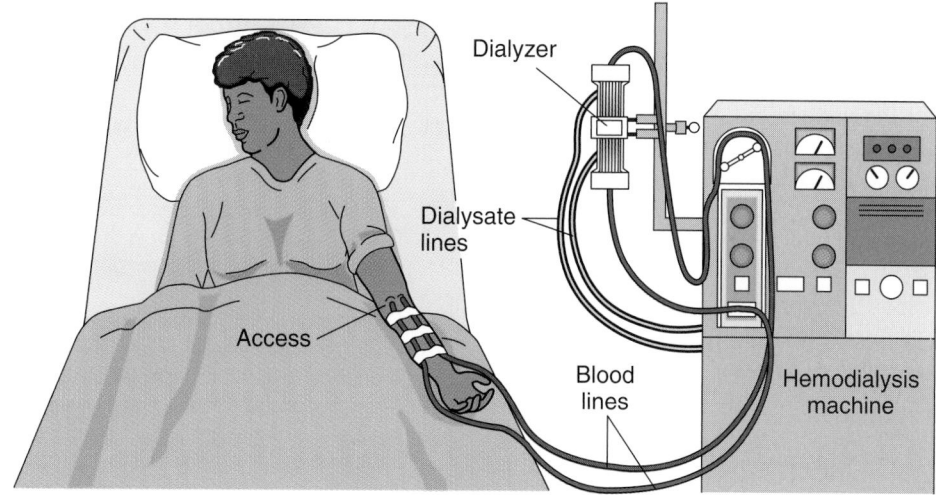

Figure 35-8   Hemodialyis.

Because restrictions are complex, the diet is a source of frustration for many patients. The nurse should assist the patient to identify foods that are palatable yet within the diet plan. The dietitian should be consulted for instruction and assistance.

Medications

Early in the disease, diuretics are given to increase output, and angiotensin-converting enzyme (ACE) inhibitors, calcium channel blockers, or beta-blocking agents may be used to control hypertension. Phosphate binders are given with meals to reduce phosphate levels. Calcium and vitamin D supplements are used to raise calcium levels. Agents to lower potassium levels are used if necessary. All drug therapy is closely monitored because diseased kidneys are unable to effectively remove medications from the body.

Dialysis

Dialysis is started when the patient develops symptoms of severe fluid overload, high potassium levels, acidosis, or symptoms of uremia that are life threatening.

HEMODIALYSIS. Hemodialysis involves the use of an artificial kidney to remove waste products and excess water from the patient's blood. The blood flows from the patient's body through tubes into an artificial kidney called a dialyzer. The artificial kidney has specialized chambers that allow the patient's blood to flow through while the waste products and excess fluids pass into a dialysate solution that circulates around the chambers. The dialysate solution carries the waste products away, and the cleansed blood is returned back into the patient's body through another tube (Fig. 35–8). A hemodialysis treatment takes 3 to 4 hours and is done three or four times a week. Hemodialysis is done at a hemodialysis center (Fig. 35–9) or in the hospital if the patient develops a complication and needs hospitalization.

Hemodialysis provides a rapid and efficient way to remove waste products from the blood. It is also an excellent

means to correct excessive fluid-overloaded states such as occur in heart failure.

Hemodialysis is not without side effects. Following a treatment, the patient normally feels weak and fatigued, sometimes even too tired to eat. Sudden drops in blood pressure may cause the patient to become weak, dizzy, and nauseated. Cardiac dysrhythmias and angina may occur. Fluid and electrolyte levels drop rapidly and cause the patient to feel lethargic and have muscle cramps. Patients are given large amounts of heparin, an anticoagulant used to keep the blood from clotting while it is in the artificial kidney; this may cause bleeding from the puncture sites, gastrointestinal tract, nose, or other sites if injury occurs.

The hemodialysis patient requires a surgically placed access site that allows the blood to be removed from the body and replaced back into the body at the time of dialysis. An access is made by joining an artery and a vein together under the skin. There are two kinds of permanent access. An arteriovenous graft (AV graft) uses a piece of special material (e.g., Gore-Tex or the new Vectra VAG, which reduces

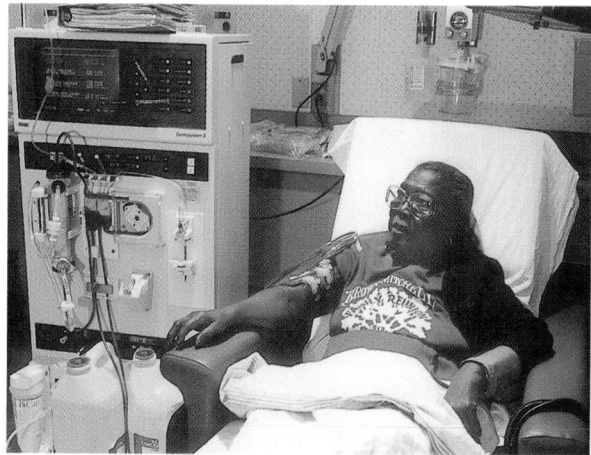

Figure 35-9   Patient undergoing hemodialysis at dialysis center.

the waiting period after surgical implantation and decreases postdialysis bleeding time) that is sewn to an artery and then attached to a vein. An AV fistula is made by sewing the vein and the artery together (Fig. 35–10). The graft and the fistula are both placed in the arm when possible. Both types of access sites take 1 to 2 weeks to mature once the surgical procedure is completed. This allows the new site to receive the high-pressure blood from the dialyzer following the treatment.

A temporary access is used for patients requiring hemodialysis before the graft or fistula is placed or is usable. This is a special hemodialysis catheter that is placed in the subclavian vein or the jugular vein in the neck or the femoral vein in the groin.

Special care of the access site must be taken because this is the patient's only way to eliminate waste products (Box 35–8). It is important that the site be carefully monitored per institution policy to detect any clotting or problems at the site. Early detection of clotting allows the surgeon an opportunity to save the access by performing a declotting procedure, rather than a total revision. Box 35–9 describes care of the patient during hemodialysis.

PERITONEAL DIALYSIS. Peritoneal dialysis provides continuous dialysis treatment and is done by the patient or family in the home. The peritoneal membrane is used as a semipermeable membrane across which excess wastes and fluids move from blood in peritoneal vessels into a dialysate solution that has been instilled into the peritoneal cavity. A peritoneal catheter is placed into the patient's peritoneal space between the two layers of the peritoneum below the waistline. This catheter is used to perform an exchange. The exchange process has three steps: filling, dwell time, and draining.

peritoneal dialysis: peritoneal—peritoneum + dialysis—passage of a solution through a membrane

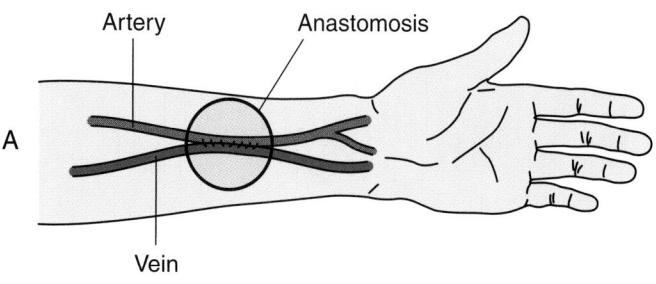
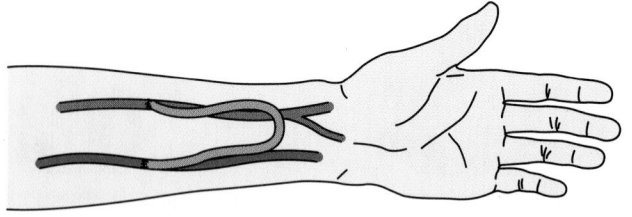

Figure 35-10 Hemodialysis access sites. *(A)* Arteriovenous fistula. *(B)* Arteriovenous vein graft.

BOX 35-8 **Care of Blood Access Graft or Fistula**

1. Watch for signs of bleeding or infection at the site.
2. Listen for a bruit at the site by placing the diaphragm of a stethoscope gently on the site. A bruit is a swishing sound made as the blood passes through the access site.
3. Gently palpate the site for a thrill, which is a buzzing or pulsing feeling that indicates good blood flow through the access site.
4. Do not take blood pressure, use a tourniquet, draw blood, or start any intravenous lines in the affected arm. Injections should be avoided if possible.
5. Many hospitals have the patient wear a red arm bracelet to signify that the arm should be protected. A sign above the bed may also be helpful.
6. Teach the patient to avoid wearing constrictive clothing or jewelry over the site.
7. Teach the patient to avoid prolonged bending or sleeping on the arm with an access.
8. Notify the physician if signs of bleeding or reduced circulation through the access site occur.

BOX 35-9 **Nursing Care of the Patient Receiving Hemodialysis**

1. Consult with the physician about medications to hold before hemodialysis. Some medications, such as antihypertensives, can be harmful when they become effective during dialysis and can reduce blood pressure to dangerously low levels. Other medications are water soluble and will be dialyzed out and thus are not effective.
2. Ensure that the patient is weighed both before dialysis in the morning and after dialysis to document weight loss as a result of fluid removal.
3. If the patient has lab tests ordered and blood needs to be drawn, coordinate this process with the dialysis nurse, who can obtain the blood samples and save the patient unnecessary needle sticks.
4. Try to get morning care done early and breakfast given before dialysis. After dialysis, patients are often exhausted and need rest.
5. When the patient returns from dialysis, weigh the patient, assess the access site for bleeding, and make sure the vital signs are stable. Administer medications that were held if not contraindicated and vital signs are stable.
6. Protect the patient's dialysis access as outlined in Box 35–4. *No dialysis access site should be utilized for any purpose other than dialysis.*

The fill step involves instilling a bag of sterile dialyzing solution (dialysate) into the patient's peritoneal cavity through the catheter. The amount of solution is usually 1500 to 2000 mL. The solution is left to dwell in the abdomen for several hours, allowing time for the waste products from the blood to pass through the peritoneal membrane into the dialysate solution (Fig. 35–11).

The solution is then drained out of the body and discarded. This process is repeated three or four times a day and is continuous for the patient. Several different treatment plans use this exchange process; the treatment plan

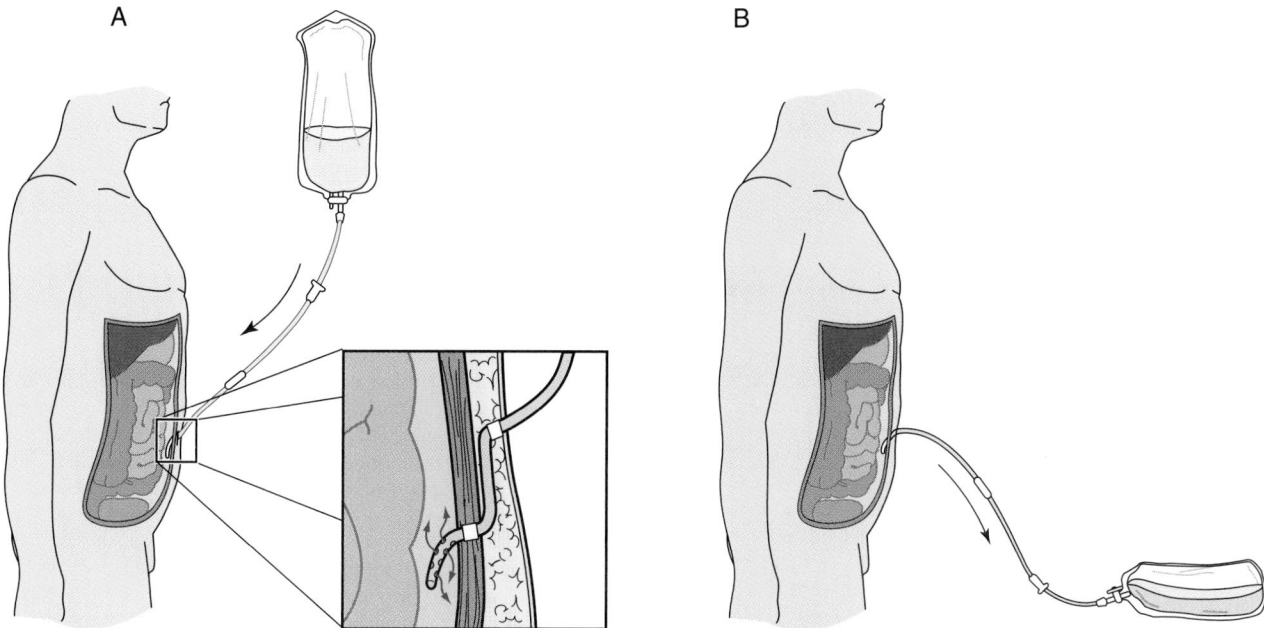

Figure 35-11   *(A)* Peritoneal dialysis works inside the body. Dialysis solution flows through a tube into the abdominal cavity, where it collects waste products from the blood. *(B)* Periodically the used dialysis solution is drained from the abdominal cavity, carrying away waste products and excess water from the blood.

that best suits the patient's needs is determined by the patient and the dialysis team.

Continuous ambulatory peritoneal dialysis (CAPD) is the most commonly used treatment plan. Usually three exchanges are done during the day and one before bedtime. Other treatment plans allow for the use of a computerized machine called a cycler to regulate the exchanges during sleeping hours. Sometimes medications are added to the dialyzing solutions, such as heparin to prevent clotting of the catheter, insulin for the patient with diabetes, or antibiotics if there is infection.

Patient and family education is extremely important for peritoneal dialysis to be successful. The patient must be taught, and be able to demonstrate that he or she is able, to do a successful exchange. Sterile technique while performing the exchanges is imperative, and the exchanges should be done in a clean environment. A major complication is peritonitis (infection of the peritoneum), which can be life threatening. The major cause of peritonitis is poor technique when connecting the bag of dialyzing solution to the peritoneal catheter. The first sign of peritonitis is usually abdominal pain. Refer to Chapter 31 for additional signs and symptoms of peritonitis. If any symptoms of peritonitis occur, the patient must contact the physician immediately, so that antibiotic treatment can begin. The patient should be taught to care of the exit site (the site where the catheter comes out of the abdomen) and the need to inspect both the site and the dialysate solution for any signs of infection.

Dietary education is also important. A dietitian can assist the patient in making appropriate choices for adequate calories, protein, and potassium intake. The peritoneal dialysis patient generally has fewer dietary and fluid restrictions than the patient on hemodialysis, because dialysis is continuous.

LEARNING TIPS

Differences between hemodialysis (HD), peritoneal dialysis (PD) and continuous renal replacement therapy include the following:

Equipment: HD requires a specialized complex dialyzer. PD and CRRT do not require the specialized dialyzer.
Training: HD requires skilled HD nurse. CRRT can be done by a non-HD nurse. PD can be done by the patient.
Timing: HD is intermittent. PD and CRRT are continuous.
Solute removal: HD and PD use the principles of osmosis and diffusion, which require a dialysate solution. CRRT uses convection, so no dialysate is needed.
Cardiovascular effects: HD may cause hypotension, which is a risk in the unstable patient. PD and CRRT have few cardiovascular effects. CRRT can be used on the unstable patient.

Kidney Transplantation

Another treatment for renal failure is kidney transplantation. A kidney transplant is a procedure in which a donor kidney is placed in the abdomen of a patient with chronic renal failure (Fig. 35–12). This healthy transplanted kidney functions as a normal functioning kidney does. The donated kidney can come from a family member; a living, nonrelated donor; or a cadaver donor. Tissue and blood types must match so that the body's immune system does not reject the donated kidney. Patients receive special drugs to help prevent rejection; these drugs must be taken for the rest of the patients' lives. Sometimes even with these drugs, the body rejects the kidney and the patient needs to go back on dialysis (Cultural Consideration Box 35–10 and Patient Perspective Box 35-11). For more information, visit the Transplant Society at www.a-s-t.org.

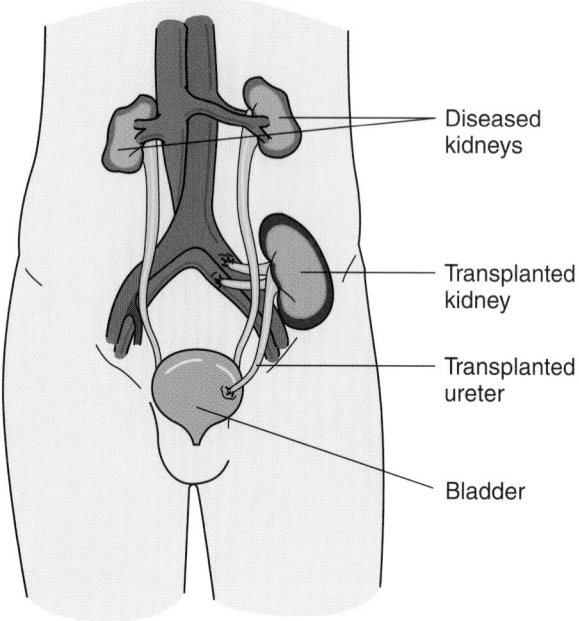

Figure 35-12 A transplanted kidney is placed in the abdomen. The patient's kidneys are usually left in place.

Kidney Transplantation: Pat

My experience with a kidney transplant spans three decades, but my overall renal illness experience also spans several years of illness and dialysis before the transplant. I think the biggest changes I have seen over the years are the involvement of patients in their own care options, as well as increased technical advances that allow transplantation to be much more successful. Some of the feelings experienced before transplantation are fear, uncertainty, and—if awaiting a donor—the guilt of knowing someone must die before you can live.

Thirty years ago there were no support groups and no one to talk to except family and doctors. Today there are many support mechanisms in place for patients and families both before and after transplantation. As a nurse, you can help patients by knowing these support resources for referral. After a transplant, it is wonderful to feel better, almost immediately. However, as wonderful as transplantation is, the side effects of the antirejection medications may be immediately felt, while other problems take years to develop. Unfortunately no patients escape the side effects of the medications. I have had breast cancer, osteoarthritis, cataracts, ulcers, skin cancer, anemia, weight gain, and other side effects of Prednisone and Imuran. However, I can assure you that this is much better than the alternative of dialysis.

Nursing Process

See Nursing Care Plan Box 35–12 for the Patient with Renal Failure for a summary of nursing care.

CRITICAL THINKING

Mrs. Jackson

Mrs. Jackson is a 56-year-old woman with a 20-year history of type 1 diabetes. She is now in chronic renal failure. She is admitted to a medical unit for beginning treatment of renal failure. She has severe edema, oliguria, and cardiac dysrhythmias.

1. What should the nurse assess as part of the morning evaluation of the patient's condition?
2. What laboratory tests are essential for the nurse to check?
3. What should the nurse assess related to Mrs. Jackson's understanding of self-care?

Answers at end of chapter.

BOX 35-10   **CULTURAL CONSIDERATION**

Renal Function and Assessment

Because many Vietnamese people believe that the body must be kept intact, even after death, they may object to removal of body parts or organ donation.

Jewish law views organ transplantation from the recipient, the living donor, the cadaver donor, and the dying donor differently. If the recipient's life can be prolonged without considerable risk, transplant is ordained. For a living donor to be approved, the risk to the life of the donor must be considered. One is not obligated to donate a part of himself or herself unless the risk is small. The use of a cadaver for transplant is usually approved if it is saving a life. The nurse may need to assist the Jewish patient contact a rabbi when making a decision regarding organ donation or transplantation.

BOX 35-12   **NURSING CARE PLAN FOR THE PATIENT WITH RENAL FAILURE**

 **Excess fluid volume related to kidneys' inability to excrete water**

PATIENT OUTCOMES

Fluid volume will be stable as evidenced by stable weight, absence of edema, lung sounds clear, and blood pressure within patient's normal parameters.

EVALUATION OF OUTCOMES

Is weight stable? Is edema absent? Are lungs clear? Is blood pressure within patient's normal parameters?

NURSING INTERVENTIONS	RATIONALE	EVALUATION
Monitor weight daily at same time; report gain of greater than 2 pounds.	Weight gain represents retention of fluid.	Is weight stable? Should physician be notified of change?

BOX 35-12 **NURSING CARE PLAN FOR THE PATIENT WITH RENAL FAILURE—CONT'D**

Monitor intake and output.	This monitors degree of fluid retention.	Is output less than intake? Is this a change?
Assess for and report shortness of breath, tachycardia, crackles in lungs, frothy sputum, heart irregularities, hypotension, cold clammy skin.	These are symptoms of heart failure that may accompany fluid overload.	Are symptoms of heart failure present?
Watch for new onset of neck vein distention with patient's head raised to 30- to 45-degree angle.	Fluid overload causes right-sided heart failure resulting in distended neck veins.	Are neck veins distended? Is this a new finding?
Monitor vital signs, including orthostatic blood pressure.	Blood pressure changes reflect fluid volume.	Is blood pressure increased?
Monitor for edema.	Edema is a symptom of fluid overload.	Is edema present? Is this a change?
Monitor activity tolerance.	Reduced activity tolerance may indicate heart failure related to fluid retention.	Is patient's tolerance of activity stable? Worsening?
Monitor serum protein and albumin levels.	Low serum protein and albumin levels contribute to edema.	Are levels within normal limits?
Maintain sodium and fluid restrictions as ordered. Develop a plan with specific allotted amounts of fluid at each meal and for medications. Teach client importance of each.	Sodium and fluids cause fluid retention in the patient with renal failure.	Does patient understand and maintain sodium and fluid restriction?

 Activity intolerance related to anemia secondary to impaired synthesis of erythropoietin by the kidneys

PATIENT OUTCOMES

Patient will be able to perform activities important to him or her.

EVALUATION OF OUTCOMES

Does patient state satisfaction with level of activity tolerance?

INTERVENTIONS	RATIONALE	EVALUATION
Assess for pale mucous membranes and skin color, dyspnea, chest pain.	These are signs and symptoms of anemia.	Does patient exhibit symptoms of anemia?
Monitor hemoglobin (Hgb), hematocrit (Hct).	Low hemoglobin and hematocrit indicate anemia.	Are Hgb and Hct within normal limits?
Watch for signs of bleeding.	Bleeding will worsen anemia.	Are signs of bleeding present?
Administer erythropoietin as ordered. Assist with blood transfusion as necessary.	Erythropoietin stimulates production of red blood cells by bone marrow.	Are Hgb and Hct rising with use of erythropoietin?
Have patient space activities with rest periods.	Rest periods decrease demand for oxygen.	Is patient able to tolerate activities with rest periods?

 Risk for injury related to bleeding tendency from platelet dysfunction and use of heparin during dialysis, and tendency for gastrointestinal bleeding

PATIENT OUTCOMES

Patient will not experience bleeding. If bleeding occurs, it will be recognized and stopped quickly.

EVALUATION OF OUTCOMES

Are signs and symptoms of bleeding absent or recognized and reported quickly?

INTERVENTIONS	RATIONALE	EVALUATION
Observe for and report blood in stool or emesis, easy bruising, bleeding from mucous membranes or puncture sites and report immediately if present.	Bleeding must be recognized quickly to prevent complications.	Does patient exhibit signs of bleeding?
Monitor Hgb, Hct, clotting studies, and platelets and report results.	Declining Hgb and Hct indicate blood loss. Declining platelet count or rising clotting times indicates increased risk for bleeding.	Are lab results stable?
Monitor vital signs.	Falling blood pressure and rising pulse may indicate volume deficit from bleeding.	Are vital signs stable?
		Can medications be given by another route?
Avoid giving injections if possible. If bleeding, apply gentle pressure to site if possible.	Injections can cause bleeding into tissue. Pressure promotes hemostasis.	Does pressure stop bleeding?

BOX 35-12   **NURSING CARE PLAN FOR THE PATIENT WITH RENAL FAILURE—CONT'D**

Teach patient to prevent injury to self and symptoms of bleeding to report.	Injury can cause bleeding. Understanding of symptoms of bleeding encourages early reporting.	Does patient verbalize understanding of instruction?

 Risk for infection related to impaired immune system function

PATIENT OUTCOMES

Patient will not develop infection as evidenced by WBCs and temperature within normal limits, no signs and symptoms of infection.

EVALUATION OF OUTCOMES

Are WBCs and temperature within normal limits?

INTERVENTIONS	RATIONALE	EVALUATION
Monitor for signs and symptoms of infection and report promptly to physician.	Early recognition of infection and prompt treatment help prevent complications.	Does patient exhibit symptoms of infection?
Protect patient from any source of infection, including infected roommates, visitors, or nursing staff.	Exposure to pathogens increases risk for infection.	Does anyone in contact with the patient have an infection?
Maintain skin integrity.	Intact skin protects against infection.	Is skin intact?
Staff and patient practice good handwashing technique.	Handwashing helps control spread of infection.	Is good handwashing being practiced?
Culture any suspected site of infection as ordered by physician.	A culture identifies pathogens and guides treatment.	Is a culture necessary?
Consult with physician about influenza and pneumonia vaccines.	Patients with impaired immune function are at risk for influenza and pneumonia.	Has the patient been vaccinated?
Teach patient and family signs and symptoms of infection to report to physician.	Early reporting of symptoms allows for prompt initiation of treatment.	Do patient and family verbalize understanding of symptoms to report?

 Impaired nutrition, less than body requirements related to restricted diet, anorexia, nausea, and vomiting, and stomatitis secondary to effect of excessive urea on the gastrointestinal system

PATIENT OUTCOMES

Patient will maintain ideal weight. Serum protein and albumin levels are within normal limits.

EVALUATION OF OUTCOMES

Are weight and lab values at desired levels?

INTERVENTIONS	RATIONALE	EVALUATION
Monitor weekly weight and serum protein and albumin levels.	Weight and lab results provide information about nutrition status.	Are weight and lab values stable?
Initiate a calorie count—consult dietitian for assistance.	A calorie count can provide information about the adequacy of the patient's diet.	Is patient receiving adequate calories?
Provide frequent oral care.	Oral care enhances appetite.	Does oral care enhance appetite?
Offer frequent small feedings and dietary supplements.	Smaller feedings are better tolerated and reduce risk of nausea.	Does patient tolerate small feedings?
Offer medications ordered for nausea before meals.	Nausea reduces appetite and must be controlled.	Are antiemetics effective?
Ensure bowel movement daily or according to patient's usual pattern.	Constipation can interfere with appetite.	Are patient's bowels functioning normally for him or her?

1. A patient is admitted to the nursing unit with a diagnosis of urinary tract infection. When teaching the patient about preventing urinary tract infections, which of the following information is most important for the nurse to include?

    a. Void frequently and after sexual intercourse.

    b. Drink large amounts of citrus juices.

    c. Eat large amounts of vegetables.

    d. Wash the perineum every 8 hours.

2. A patient has an ileal conduit for treatment of his cancer. While changing the pouch at the stoma site, the nurse notes that the stoma is constantly spilling urine. Which of the following actions should the nurse take?

    a. Notify the physician of the constant spillage.

    b. Continue changing the pouch.

    c. Remove the overflow of urine with a straight catheter.

    d. Irrigate the stoma with a sterile solution of normal saline.

3. A patient has a graft inserted into his left arm to provide a blood access for hemodialysis treatments. Which of the following interventions should the nurse do to determine patency of the graft?

    a. Observe the tubing for bright red blood.

    b. Feel for a brachial pulse on both arms.

    c. Feel for a thrill over the graft.

    d. Assess blood pressure in the affected arm.

4. A patient with renal failure who is on hemodialysis asks for a snack in the afternoon. His potassium level remains high. Which of the following foods would be contraindicated?

    a. Banana

    b. Gelatin desert

    c. Clear carbonated beverage

    d. Cranberry juice

5. The nurse is caring for a patient with glomerulonephritis. Which of the following interventions would the nurse recommend be included in the patient's plan of care?

    a. Increase fluid intake

    b. Decrease sodium intake

    c. Increase potassium intake

    d. Decrease carbohydrate intake

Answers to CRITICAL THINKING

Mrs. Milan

1. Sexual intercourse can be a predisposing factor to UTI, especially if the patient does not urinate after intercourse.
2. Mrs. Milan should be cautioned to always urinate after intercourse. See also Box 35–1.
3. The urinalysis will show WBCs, bacteria, RBCs, and positive nitrites.

Mr. White

1. Weight, intake and output, blood pressure, and lab tests should be assessed as part of the morning evaluation.
2. BUN, serum creatinine, and potassium levels should also be checked.

3. Mr. White should be taught that he needs to take antihypertensive medications, keep his follow-up visits to his physician, follow a low-sodium diet, and restrict fluids if ordered.

Mrs. Jackson

1. Assess Mrs. Jackson's weight and intake and output to monitor fluid balance. A cardiovascular system assessment should be done to see how she is tolerating the dysrhythmia.
2. BUN, serum creatinine, serum potassium, hemoglobin and hematocrit, and blood sugar are monitored.
3. Assess Mrs. Jackson's understanding of what renal failure is, how it is treated, how to follow the renal diet and fluid restrictions, and the action and importance of medications.

UNIT NINE BIBLIOGRAPHY

Ackley, B, and Ladwig, G: Nursing diagnosis handbook: A guide to planning care, ed 5. Mosby, St Louis, 2002.

Bernie, JE, Kambo, AP, and Monga M: Urinary lithiasis: Diagnostic strategies. Consultant 40(14):2313, 2000.

Cavanaugh, B: Nurse's manual of laboratory and diagnostic tests, ed 3. FA Davis, Philadelphia, 1999.

Duffin H: Intermittent self-catheterisation. Journal of Community Nursing, 14(10):29–32, 2000.

Foxman B, et al: Urinary tract infection among women aged 40 to 65: Behavior and sexual risk factors. J Clin Epidemiol 54:710, 2001.

Gray, M: Urinary retention: Management in the acute care setting, Part 1. American Journal of Nursing 100(7):40–48, 2000.

Gray, M: Urinary retention: Management in the acute care setting, Part 2. American Journal of Nursing 100(8):36–44, 2000.

Karlowicz, KA (ed): Urologic nursing: Principles and practice. WB Saunders, Philadelphia, 1995.

Lutz, CA, and Przytulski, KR: Nutrition and Diet Therapy, ed 3. FA Davis, Philadelphia, 2001.

McKinney, BC: Cut your patients' risk of nosocomial UTI. RN 58(11):20, 1995.

Moore K, et al: Bacteriuria in intermittent catheterization users: The effect of sterile versus clean reused catheters. Rehabilitation Nursing 18:306, 1993.

Mulvey, MA, Schilling, JD, and Hultgren, SJ: Establishiment of a persistent *Escherichia coli* reservoir during the acute phase of a bladder infection. Infect Immun 69(7):4572, 2001.

Resnick, B: Retraining the bladder after catheterization. American Journal of Nursing 93:46, 1993.

Retzky, SS, and Rogers, RB: Urinary incontinence in women. Clinical Symposia 47(3):1, 1995.

Saint, S: How to prevent urinary catheter-related infections in the critically ill: Recognizing when indwelling devices must be used and when they should not. Journal of Critical Illness, 15(8):419–423, 2000.

Schaeffer AJ: What do we know about the urinary tract infection-prone individual? J Infect Dis (suppl 1)183:S66, 2001.

Schmitz PG: Progressive renal insufficiency. Postgrad Med 108(1):145, 2000.

Tambyah, PA, and Maki, DG: Catheter-associated urinary tract infection is rarely symptomatic. Archives of Internal Medicine 160:678–682, 2000.

Tran, M, and Rutecki, GW: Renal disease: Tips on prevention and early recognition. Consultant 40(2):222, 2000.

Urinary Incontinence Guideline Panel: Urinary incontinence in adults: Clinical practice guideline. Agency for Health Care Policy and Research, Public Health Service, US Department of Health and Human Services (AHCPR Pub No 92-0038), Rockville, MD, 1992.

Verdell, L (ed): Help for incontinent people (HIP): Resource guide of continence products and services, ed 6. HIP, Union, SC, 1994.

Wolfe S: Meds and the dialysis patient. RN 63(7):54, 2000.

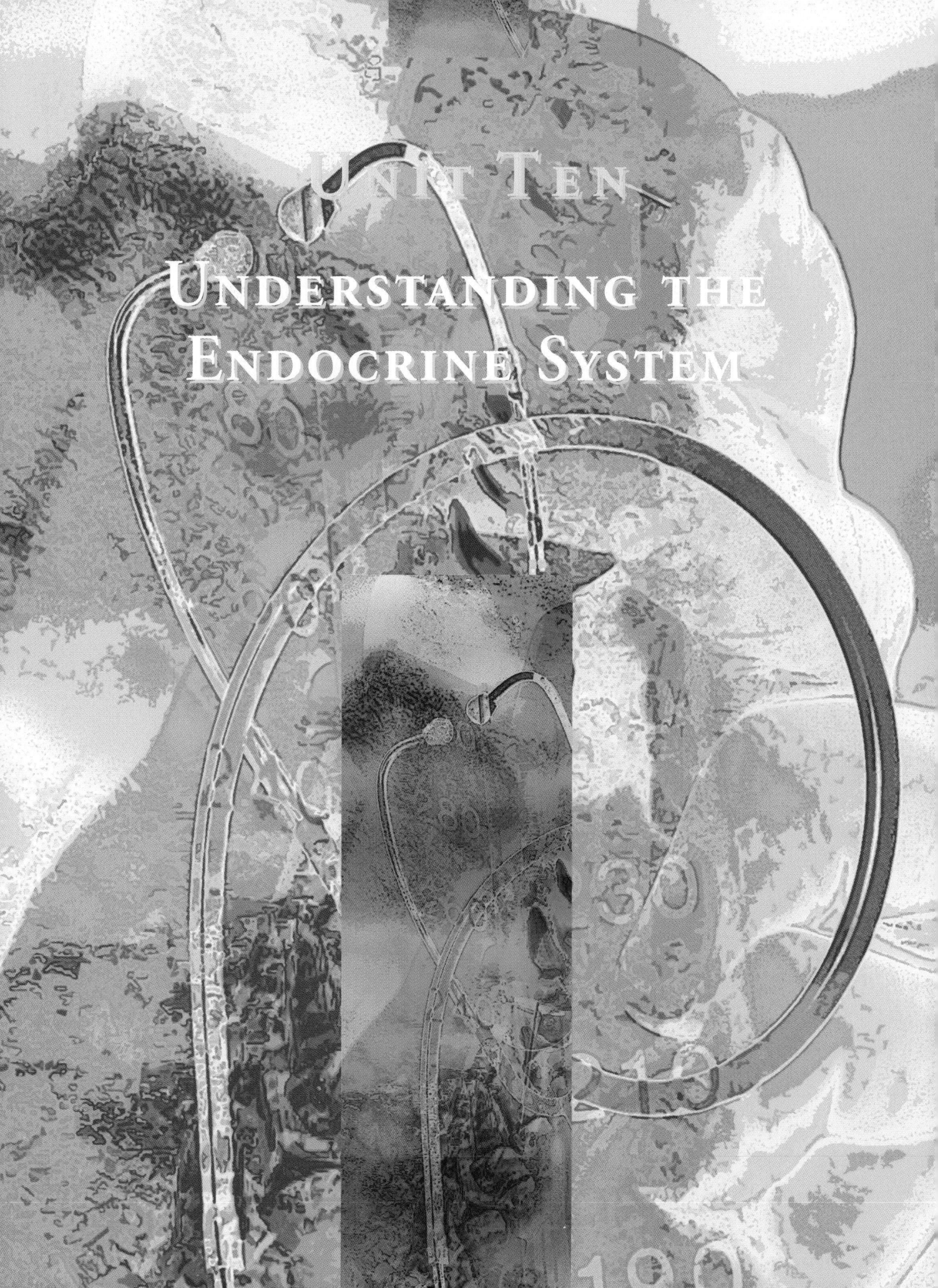

Unit Ten

Understanding the Endocrine System

36

ENDOCRINE SYSTEM FUNCTION AND ASSESSMENT

Paula D. Hopper

KEY TERMS

affect (**AF**-feckt)
exophthalmos (ECKS-off-**THAL**-mus)

QUESTIONS TO GUIDE YOUR READING

1. What are the glands of the endocrine system?
2. What is the function of each of the hormones in the endocrine system?
3. What are the effects of aging on endocrine system function?
4. What data should you collect when caring for a patient with a disorder of the endocrine system?
5. What nursing care should you provide for patients undergoing testing for an endocrine disorder?

► NORMAL ANATOMY AND PHYSIOLOGY

The endocrine system consists of the endocrine, or ductless, glands, which secrete chemicals called hormones. Unlike other organ systems, the glands of the endocrine system are anatomically separate (Fig. 36–1). Their hormones, however, are involved in all aspects of metabolism: growth, energy production, regulation of fluid and electrolyte balance and pH, resistance to stress, and reproduction. Each hormone is secreted in response to a particular and specific stimulus, is circulated by the blood throughout the body, and exerts its effects on certain target organs or tissues that have receptors for that hormone. The effects of the hormone often reverse the stimulus and ultimately lead to decreased secretion of the hormone. This is called a negative feedback mechanism; the secretion of many hormones is regulated this way. Some hormones are secreted in response to hormones from other endocrine glands.

Pituitary Gland

The pituitary gland is also called the hypophysis; it hangs by a short stalk from the hypothalamus in the brain. The two major parts are the posterior pituitary (neurohypophysis) and anterior pituitary (adenohypophysis).

Posterior Pituitary Gland

The posterior pituitary gland stores antidiuretic hormone (ADH, sometimes called vasopressin) and oxytocin, which are actually produced by the hypothalamus. Their release is stimulated by nerve impulses from the hypothalamus.

Antidiuretic hormone increases the amount of water reabsorbed by the kidney tubules, which decreases urinary output. The water is reabsorbed back into the blood, thereby maintaining normal blood volume and normal blood pressure. The stimulus for secretion of ADH is a decrease in the water content of the body—that is, dehydration. When body water is lost and not replaced, specialized cells in the hypothalamus called osmoreceptors detect the increased salt concentration of body fluids and transmit impulses to the posterior pituitary to secrete ADH to prevent the further loss of water in urine. In cases of great fluid loss, as in severe hemorrhage, the large amount of ADH secreted causes vasoconstriction, which also contributes to maintenance of a normal blood pressure.

Oxytocin causes contraction of the smooth muscle in the uterus and mammary glands. At the end of pregnancy, stretching of the cervix generates sensory impulses to the hypothalamus, which then transmits impulses to the posterior pituitary for the release of oxytocin. Oxytocin causes strong contractions of the myometrium to bring about

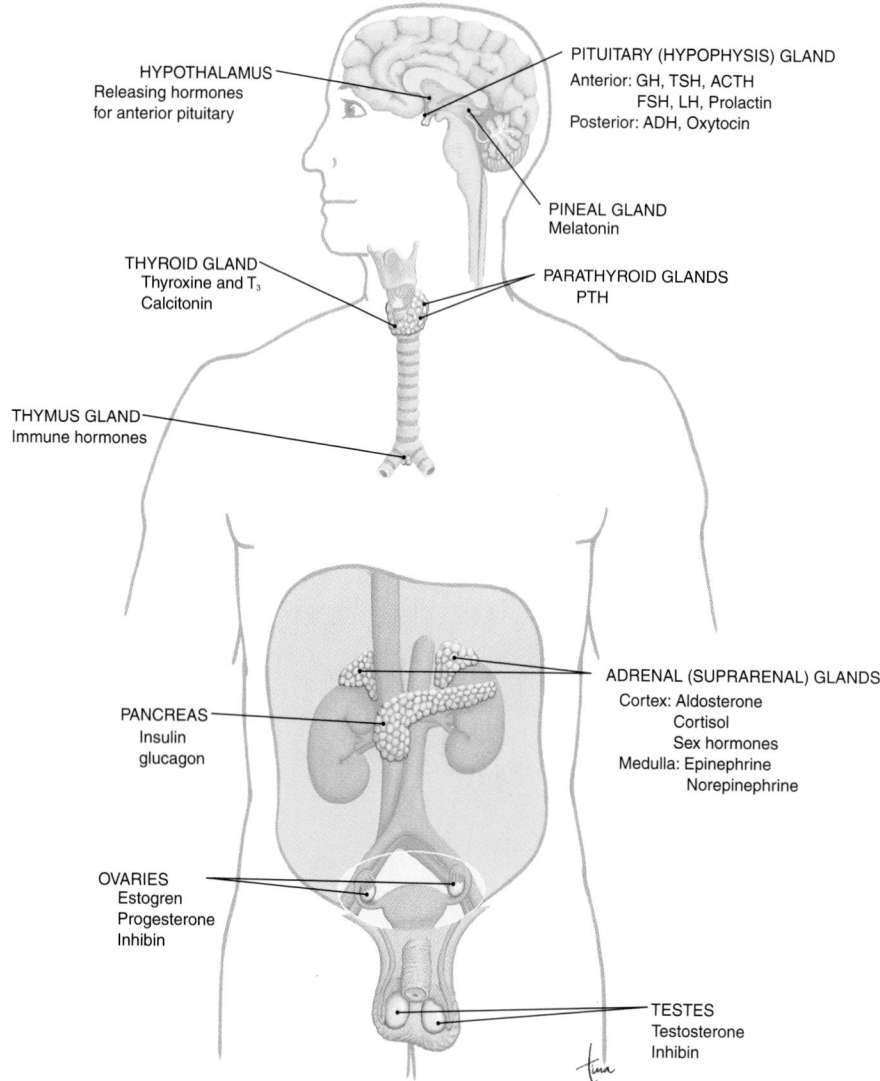

Figure 36-1 Glands of the endocrine system. (From Scanlon, V., and Sanders, T: Essentials of Anatomy and Physiology, ed 3. FA Davis, Philadelphia, 1999, p 211, with permission.)

delivery of the baby and the placenta. This is an example of a positive feedback mechanism. The placenta also produces oxytocin, but its precise role in labor and delivery has not yet been determined.

During breast-feeding, the sucking of the baby generates sensory impulses from the mother's nipple to the hypothalamus. The subsequent release of oxytocin causes contraction of the smooth muscle cells around the mammary ducts. This release of milk is sometimes called "milk ejection" or the "milk letdown" reflex.

Anterior Pituitary Gland

The anterior pituitary gland secretes its hormones in response to releasing hormones from the hypothalamus. The anterior pituitary secretes growth hormone, thyroid-stimulating hormone, adrenocorticotropic hormone, prolactin, follicle-stimulating hormone, and luteinizing hormone.

Growth hormone (GH, or somatotropin) increases cell division in those tissues capable of mitosis, which is one of

the ways it is involved in growth. It also increases the transport of amino acids into cells and their use in protein synthesis. Growth hormone also increases the release of fat from adipose tissue and the use of fats for energy production; this is important even after growth in height has ceased. The secretion of GH is regulated by growth hormone–releasing hormone (GHRH) and by growth hormone–inhibiting hormone (GHIH, or somatostatin), both from the hypothalamus. GHRH is produced during hypoglycemia or when there is a high blood level of amino acids (to be turned into protein). GHIH is secreted during hyperglycemia, when carbohydrates are available for energy production and the mobilization of fat is not necessary.

Thyroid-stimulating hormone (TSH, or thyrotropin) has only one target organ: the thyroid gland. TSH stimulates growth of the thyroid and the secretion of two of its hormones, thyroxine (T_4) and triiodothyronine (T_3). The secretion of TSH is stimulated by thyrotropin-releasing hormone (TRH) from the hypothalamus when metabolic rate decreases and there is a need for thyroxine.

Adrenocorticotropic hormone (ACTH) stimulates the secretion of cortisol and related hormones from the adrenal cortex. Corticotropin-releasing hormone (CRH) from the hypothalamus stimulates the release of ACTH. CRH is produced during any type of physiological stress situation, such as injury, disease, exercise, or hypoglycemia.

Prolactin initiates and maintains milk production by the mammary glands. The hypothalamus produces both prolactin-releasing hormone (PRH) and prolactin-inhibiting hormone (PIH); prolactin is not secreted until pregnancy is over and the levels of estrogen and progesterone (from the placenta) have dropped.

Follicle-stimulating hormone (FSH) is a gonadotropic hormone; that is, its target organs are the ovaries or testes. In women, FSH initiates growth of ova in ovarian follicles and secretion of estrogen by the cells of those follicles. In men, FSH initiates sperm production in the seminiferous tubules of the testes. FSH is secreted in response to gonadotropin-releasing hormone (GnRH) from the hypothalamus. Another hormone called inhibin (from the ovaries or testes) decreases the secretion of FSH.

Luteinizing hormone (LH) is another gonadotropic hormone whose secretion is increased by GnRH from the hypothalamus. In women, LH causes ovulation and stimulates the ruptured ovarian follicle to become the corpus luteum and begin secreting progesterone, as well as estrogen. In men, LH stimulates the secretion of testosterone by the interstitial cells of the testes.

Thyroid Gland

The thyroid gland consists of two lobes connected by a middle piece called the isthmus; the gland is located on the front and sides of the trachea just below the larynx. Three hormones are produced by the thyroid gland. T_4 and T_3 are produced in the thyroid follicles, require iodine (T_4 has four iodine atoms, T_3 has three iodine atoms), and have the same functions. Calcitonin is the third hormone; it is produced by parafollicular cells.

T_4 and T_3 regulate normal energy production and protein synthesis. They increase the rate of cell respiration of all food types (carbohydrates, fats, and excess amino acids), which increases the metabolic rate—that is, energy and heat production. They also increase the rate of protein synthesis in cells. These hormones are the most important day-to-day regulators of metabolic rate; their activity is reflected in the functioning of the heart, brain, muscles, and virtually all other organs. They are essential for normal physical growth, mental development, and reproductive maturation.

The direct stimulus for secretion of T_4 and T_3 is TSH from the anterior pituitary. The sequence of events is as follows: A decrease in metabolic rate (energy production) is detected by the hypothalamus, which secrets TRH. TRH stimulates the anterior pituitary to secrete TSH, which stimulates the thyroid to increase secretion of T_4 and T_3, which increase energy production to raise the metabolic rate. As the metabolic rate rises, negative feedback decreases the secretion of TRH from the hypothalamus until the metabolic rate drops again.

The third thyroid hormone, calcitonin, has bones as its target and seems to be of greater importance in childhood when bones are growing than in maturity when bone growth has ceased. Calcitonin decreases the reabsorption of calcium and phosphate from the bones to the blood, thereby lowering the blood levels of these minerals as they are retained in bones. This one function of calcitonin has two important results: the maintenance of normal blood levels of calcium and phosphate and the maintenance of a strong, stable bone matrix.

LEARNING TIPS

An easy way to remember the function of calcitonin is to remember *calciTONin TONes down serum calcium.*

The stimulus for secretion of calcitonin is hypercalcemia, a high blood calcium level. When the blood calcium level rises, increased calcitonin ensures that no more will be removed from bones until there is a real need for more calcium in the blood.

Parathyroid Glands

There usually are four parathyroid glands, two on the back of each lobe of the thyroid gland. The hormone they produce is called parathyroid hormone (PTH), which is an antagonist to calcitonin for the maintenance of normal blood levels of calcium and phosphate. Besides bone, the target organs of PTH are the small intestine and kidneys.

PTH increases the reabsorption of calcium and phosphate from the bones to the blood, which raises their blood levels. Absorption of calcium and phosphate from food in the small intestine is also increased by PTH through its action of activating vitamin D (calcitriol) in the kidneys. PTH also increases the reabsorption of calcium by the kidneys and the excretion of phosphate (more than is obtained from bones). Therefore the overall effect of PTH is to raise the blood calcium level and lower the blood phosphate level.

Secretion of PTH is stimulated by hypocalcemia, a low blood calcium level, and is inhibited by hypercalcemia. In adults, PTH is probably the most important regulator of the blood calcium level. Calcium ions in the blood are essential for normal excitability of neurons and muscle cells and for the process of blood clotting.

Adrenal Glands

The two adrenal (also called suprarenal) glands are located one on top of each kidney. Each adrenal gland consists of an inner adrenal medulla and an outer adrenal cortex.

Adrenal Medulla

The cells of the adrenal medulla are called chromaffin cells. They secrete epinephrine and norepinephrine, which are collectively called catecholamines and are sympathomimetic (mimicking the sympathetic nervous system). Secretion of both hormones is stimulated by sympathetic impulses from the hypothalamus in stressful situations. The functions of the catecholamines mimic and prolong those of the sympathetic nervous system, which enable the individual to respond physiologically to stress situations.

Of the two hormones, epinephrine is secreted in larger amounts (approximately four times that of norepinephrine) and has many effects. It increases the heart rate and force of contraction, stimulates vasoconstriction in skin and viscera and vasodilation in skeletal muscles, dilates the bronchioles, decreases peristalsis, stimulates the liver to convert glycogen to glucose, increases the use of fats for energy, and increases the rate of cell respiration. The most significant function of norepinephrine is to cause vasoconstriction in the skin, viscera, and skeletal muscles, thereby raising blood pressure.

Adrenal Cortex

The adrenal cortex secretes three types of steroid hormones: sex hormones, mineralocorticoids, and glucocorticoids. The sex hormones are small amounts of male androgens and even smaller amounts of female estrogens. Their function is not known with certainty, though they may contribute to the growth spurt that often occurs just before puberty and to the libido (sex drive) in adult women.

LEARNING TIPS

An easy way to remember the hormones of the adrenal cortex is to remember *salt, sugar, and sex.* Mineralocorticoids promote salt retention, glucocorticoids affect sugar (carbohydrate) metabolism, and androgens and estrogens are sex hormones.

Aldosterone is the most abundant of the mineralocorticoids, and its target organs are the kidneys. Aldosterone increases the reabsorption of sodium ions and the excretion of potassium ions by the kidney tubules. This means that sodium ions are returned to the blood and potassium ions are eliminated in urine. This function of aldosterone has important consequences. As sodium ions are reabsorbed, hydrogen ions may be excreted in exchange; this is one mechanism to prevent the accumulation of hydrogen ions that would lead to acidosis. Also as sodium ions are reabsorbed, water and negative ions such as bicarbonate follow and are thus returned to the blood. Although the reabsorption of water is an indirect effect of aldosterone, it is important for the maintenance of normal blood volume and blood pressure.

Secretion of aldosterone may be stimulated in several ways: a low blood sodium level, a high blood potassium level, or loss of blood or dehydration that lowers blood pressure. Low blood pressure activates the renin-angiotensin mechanism of the kidneys, which culminates in the formation of angiotensin II. One function of angiotensin II is to increase secretion of aldosterone. The hormone called atrial natriuretic peptide (ANP), secreted by the atria of the heart

when blood pressure or blood volume rises, seems to inhibit secretion of aldosterone (and renin) and thereby promotes elimination of sodium ions and water by the kidneys.

Cortisol is the most abundant of the glucocorticoids and has many target tissues. Cortisol stimulates the liver to change glucose to glycogen (glycogenesis) for storage. It increases the conversion of excess amino acids to carbohydrates (gluconeogenesis) for energy production and increases the use of fats for energy. By providing these secondary energy sources to most cells, cortisol ensures that whatever glucose is present will be available for the brain (the glucose-sparing effect).

Cortisol also has an anti-inflammatory effect because it blocks the effects of histamine and stabilizes the lysosomes in cells. Normal cortisol secretion seems to limit the inflammation process to what is useful for tissue repair and to prevent excessive tissue destruction. Excess cortisol has damaging effects, however: It raises blood glucose levels, decreases the immune response, and delays healing of damaged tissue.

The direct stimulus for cortisol secretion is ACTH from the anterior pituitary gland. Cortisol is also a "stress" hormone, and any type of physiological stress (injury, disease, malnutrition) stimulates the hypothalamus to secrete CRH. CRH increases the secretion of ACTH by the anterior pituitary, which increases cortisol secretion by the adrenal cortex.

Pancreas

The pancreas is located in the upper left abdominal quadrant and extends from the curve of the duodenum to the spleen. The endocrine portions of the pancreas are called islets of Langerhans (pancreatic islets); they contain alpha cells, which produce glucagon, and beta cells, which produce insulin. Delta cells in the islets secrete somatostatin, which inhibits secretion of both insulin and glucagon.

The functions of glucagon are all related to energy production. Glucagon stimulates the liver to change glycogen to glucose (glycogenolysis) and to increase the use of fats and excess amino acids for energy production. The overall effect, therefore, is to raise the blood glucose level and to make all types of food available for cell respiration.

The secretion of glucagon is stimulated by hypoglycemia, a low blood glucose level. Such a state may occur during physiological stress situations such as exercise or simply being between meals.

Insulin increases the transport of glucose from the blood into cells by increasing the permeability of cell membranes to glucose (brain and liver cells, however, are not dependent on insulin for glucose intake). Inside cells, glucose is broken down in cell respiration to release energy. The liver and muscles are also stimulated by insulin to change glucose to glycogen (glycogenesis) to be stored for later use. Insulin also enables cells to take in fatty acids and amino acids to use in the synthesis of lipids and proteins (not energy production). Insulin, therefore, decreases the blood glucose level by increasing the use of glucose for energy, promoting the storage of excess glucose, and decreasing energy production from other food sources.

Secretion of insulin is stimulated by hyperglycemia, a high blood glucose level. This state occurs after meals, especially those high in carbohydrates. It should be apparent that insulin and glucagon function as antagonists and that normal secretion of both hormones ensures a blood glucose level that fluctuates within normal limits. Table 36–1 reviews endocrine hormone function.

Aging and the Endocrine System

Most of the endocrine glands decrease their secretions with age, but normal aging usually does not lead to serious hormone deficiencies. There are decreases in adrenal cortical hormones, for example, but the levels are usually sufficient to maintain homeostasis of water, electrolytes, and nutrients. The decreased secretion of growth hormone leads to a decrease in muscle mass and an increase in fat storage. A lower basal metabolic rate is common in elderly people as the thyroid slows its secretion of thyroxine. Although insulin secretion declines somewhat, for most elderly people with a decreased glucose tolerance the cause is a decrease in the cells' sensitivity to insulin. Unless specific pathological conditions develop, the endocrine system usually continues to function adequately in old age.

TABLE 36–1 REVIEW OF ENDOCRINE FUNCTION

Hormone	Function(s)	Regulation of Secretion
Hormones of the Posterior Pituitary Gland		
Oxytocin	Promotes contraction of myometrium of uterus (labor)	Nerve impulses from hypothalamus, the result of stretching of cervix or stimulation of nipple
	Promotes release of milk from mammary glands	Secretion from placenta at the end of gestation-stimulus unknown
Antidiuretic hormone	Increases water reabsorption by the kidney tubules (water returns to the blood)	Decreased water content in the body (alcohol inhibits secretion)
Hormones of the Anterior Pituitary Gland		
Growth hormone	Increases rate of mitosis	GHRH (hypothalamus) stimulates secretion
	Increases amino acid transport into cells	
	Increases rate of protein synthesis	GHIH—somatostatin (hypothalamus) inhibits secretion
	Increases use of fats for energy	
Thyroid-stimulating hormone	Increases secretion of thyroxine and T_3 by thyroid gland	TRH (hypothalamus)
Adrenocorticotropic hormone	Increases secretion of cortisol by the adrenal cortex	CRH (hypothalamus)
Prolactin	Stimulates milk production by the mammary glands	PRH (hypothalamus) stimulates secretion
		PIH (hypothalamus) inhibits secretion
Follicle-stimulating hormone	*In women:*	GnRH (hypothalamus)
	Initiates growth of ova in ovarian follicles	
	Increases secretion of estrogen by follicle cells	
	In men:	
	Initiates sperm production in the testes	GnRH (hypothalamus)
Luteinizing hormone	*In women:*	GnRH (hypothalamus)
	Causes ovulation	
	Causes the ruptured ovarian follicle to become the corpus luteum	
	Increases secretion of progesterone by the corpus luteum	
	In men:	GnRH (hypothalamus)
	Increases secretion of testosterone by the interstitial cells of the testes	
Melanocyte-stimulating hormone (MSH)	Stimulates production of melanin by melanocytes	CRH (hypothalamus)
Hormones of the Thyroid Gland		
Thyroxine and triiodothyronine	Increase energy production from all food types	TSH (anterior pituitary)
	Increase rate of protein synthesis	
Calcitonin	Decreases the reabsorption of calcium and phosphate from bones to blood	Hypercalcemia
Hormones of the Parathyroid Glands		
Parathyroid hormone	Increases the reabsorption of calcium and phosphate from bone to blood	Hypocalcemia
	Increases absorption of calcium and phosphate by the small intestine	
	Increases the reabsorption of calcium and the excretion of phosphate by the kidneys	

Source: Adapted from Scanlon, VC, and Sanders, T: Essentials of Anatomy and Physiology, ed 3. FA Davis, Philadelphia, 1999.

TABLE 36-1   REVIEW OF ENDOCRINE FUNCTION—CONT'D

Hormone	Function(s)	Regulation of Secretion
Hormones of the Pancreas		
Glucagon	Increases conversion of glycogen to glucose in the liver	Hypoglycemia
	Increases the use of excess amino acids and of fats for energy	
Insulin	Increases glucose transport into cells and the use of glucose for energy production	Hyperglycemia
	Increases the conversion of excess glucose to glycogen in the liver and muscles	
	Increases amino acid and fatty acid transport into cells and their use in synthesis reactions	
Somatostatin	Inhibits glucagon, insulin, and growth hormone secretion	Increased levels
Hormones of the Adrenal Medulla		
Norepinephrine	Causes vasoconstriction in skin, viscera, skeletal muscles	Sympathetic impulses from the hypothalamus in stress situations
Epinephrine	Increases heart rate and force of contraction	
	Dilates bonchioles	
	Decreases peristalsis	
	Increases conversion of glycogen in glucose in the liver	
	Causes vasodilation in skeletal muscles	
	Causes vasoconstriction in skin and viscera	
	Increases use of fats for energy	
	Increases the rate of cell respiration	
Hormones of the Adrenal Cortex		
Aldosterone	Increases reabsorption of Na⁺ ions by the kidneys to the blood	Low blood Na⁺ level
		Low blood volume or blood pressure
	Increases excretion of K⁺ ions by the kidneys in urine	High blood K⁺ level
Cortisol	Increases use of fats and excess amino acids for energy	ACTH (anterior pituitary) during physiological stress
	Decreases use of glucose for energy (except for the brain)	
	Increases conversion of glucose to glycogen in the liver	
	Anti-inflammatory effect: stabilizes lysosomes and blocks the effects of histamine	
Androgens and estrogens	May be involved in growth spurt at puberty	ACTH
	Contribute to libido	
	Source of sex hormones for women after menopause	

K⁺ = potassium; Na⁺ = sodium.

▶ NURSING ASSESSMENT
Health History

When performing a health history, a number of questions can be asked to determine whether an endocrine problem exists. Often, however, you might be aware of a history of an endocrine disorder, such as diabetes or hypothyroidism. When a disorder exists or is suspected, you can do a more focused assessment. Assessment of individual disorders is provided in Chapters 37 and 38. Table 36–2 offers general questions that can help you identify new problem areas. If the assessment reveals abnormalities, they should be reported to a registered nurse or physician.

LEARNING TIPS

Some patients call diabetes mellitus "sugar." So instead of asking if anyone in their family has diabetes, you may need to ask if anyone has "sugar."

Physical Assessment of the Patient with an Endocrine Disorder

Physical assessment starts with height, weight, and vital signs. These should always be compared with the patient's baseline assessment if available. Table 36–3 includes common endocrine-related causes of physical assessment abnormalities.

Inspection

Observe the patient for mood and **affect** (emotional tone) throughout the physical assessment. Inspect the neck for thyroid enlargement. Look for eyes that bulge (**exophthalmos**). Note posture, body fat, and presence of tremor. Observe skin and hair texture and moisture. Note the presence of a moonlike face or "buffalo hump" on the upper back. Observe the lower extremities for skin and color changes that might indicate circulatory impairment. See Table 36–3 for rationales for these observations.

exopthalmos: exo—outward + ophthalmos—eye

TABLE 36-2 QUESTIONS ASKED DURING ASSESSMENT OF THE ENDOCRINE SYSTEM

Question	Rationale
Have you noticed a change in your energy level?	Lack of energy may be associated with uncontrolled diabetes, hypothyroidism, hyperthyroidism, Addison's disease, or pituitary disorders.
Have you noticed muscle spasms or twitching?	These symptoms may be associated with excessive antidiuretic hormone secretion (SIADH) or calcium depletion resulting from hypoparathyroidism.
Do you have numbness, tingling, or pain in your feet, legs, or hands?	These may be associated with neuropathy resulting from diabetes mellitus. Numbness and tingling may also indicate hypocalcemia related to hypoparathyroidism.
Have you gained or lost weight without trying?	Actual weight gain may be associated with hypothyroidism. Weight gain due to water retention may result from Cushing's syndrome, or SIADH. Weight loss may result from uncontrolled diabetes or hyperthyroidism. Weight loss due to dehydration may be related to Addison's disease.
Have you noticed excessive thirst or urination?	Excessive thirst and urination are classic symptoms of diabetes mellitus and diabetes insipidus.
Do you generally tolerate changes in environmental temperature?	Hypothyroidism can cause cold intolerance. Hyperthyroidism can cause heat intolerance.
Have you noticed a change in your mood or memory?	Mental function may be dull with hypothyroidism. Mood swings may occur with Cushing's syndrome. Agitation or confusion may result from hypoglycemia in a person with diabetes.
Does anyone in your family have a thyroid problem or diabetes?	These may be hereditary.

SIADH = syndrome of inappropriate antidiuretic hormone.

TABLE 36-3 ENDOCRINE-RELATED CAUSES OF ABNORMAL PHYSICAL ASSESSMENT FINDINGS

Assessment Finding	Possible Causes
Inappropriate mood or affect	Depressed mood or affect from hypothyroidism
	Nervousness related to hyperthyroidism
	Agitation related to low blood sugar
Weight change	Gain due to decreased metabolic rate in hypothyroidism, fluid excess
	Loss due to increased metabolic rate in hyperthyroidism; uncontrolled diabetes, dehydration
Poor skin turgor	Dehydration due to water loss in Addison's disease, diabetes mellitus, diabetes insipidus
Hyperpigmentation of skin	Addison's disease
Dry, scaly skin	Hypothyroidism
Change in pulse or temperature	Elevated due to increased metabolic rate in hyperthyroidism
	Decreased due to slowed metabolic rate in hypothyroidism
Elevated blood pressure	Increased catecholamine release in pheochromocytoma or fluid retention in Cushing's syndrome
Decreased blood pressure	Sodium and water loss in Addison's disease
Weak peripheral pulses, dusky lower extremities	Circulatory changes in diabetes mellitus
Tremor	Hyperthyroidism or pheochromocytoma
Exophthalmos (bulging eyes)	Fat deposits and edema behind the eyes in Graves' disease
Fat pads on neck and shoulders ("buffalo hump"), round face	Accumulation of fat in Cushing's syndrome
Enlarged thyroid gland	Excessive stimulation by TSH in hypothyroidism or hyperthyroidism

Palpation

The thyroid gland is the only palpable endocrine gland. The LPN/LVN may assist a physician or nurse practitioner to palpate the thyroid gland. The practitioner stands behind or in front of the seated patient and palpates the gland while the patient swallows a sip of water (Fig. 36–2). You may assist with positioning the patient, providing water, and instructing the patient to take a sip of water and hold it in his or her mouth until told to swallow. The thyroid gland should never be palpated in a patient with uncontrolled hyperthyroidism, because this may stimulate secretion of additional thyroid hormone.

Palpate all peripheral pulses. The posterior tibial and dorsalis pedis pulses may be diminished in patients with circulatory impairment. Palpate skin turgor by gently pinching a small piece of skin. If a "tent" remains, the patient may be dehydrated as a result of water loss, as in ADH deficiency.

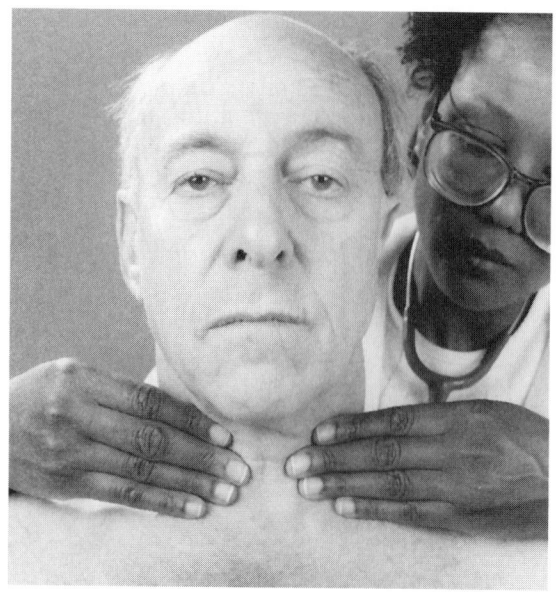

Figure 36-2   Thyroid palpation.

Auscultation and Percussion

Auscultation and percussion are not usually part of an endocrine assessment.

▶ DIAGNOSTIC TESTS

Hormone Tests

Serum Hormone Levels

Many hormones can be measured from a simple blood specimen. This is useful in diagnosing hypofunctioning or hyperfunctioning gland states. See Table 36–4 for some commonly measured hormones.

Stimulation Tests

Stimulation tests may also help determine endocrine gland function. For this type of test, a substance is injected to attempt to stimulate a gland. The hormone secreted by that gland is then measured in the blood to determine how well it responded to the stimulation. For example, in a TSH

TABLE 36-4 COMMON ENDOCRINE-RELATED LABORATORY TESTS		
Test	**Normal Values***	**Significance**
Thyroid Tests		
Thyroid-stimulating hormone	0.5–5.0 U/mL	↑ in primary hypothyroidism ↓ in primary hyperthyroidism
Triiodothyronine	75–195 ng/100 mL	↓ in hypothyroidism ↑ in hyperthyroidism
Thyroxine	4–12 µg/100 mL	↓ in hypothyroidism ↑ in hyperthyroidism
Parathyroid Tests		
Parathyroid hormone	<25 pg/mL	↑ in primary hyperparathyroidism ↓ in primary hypoparathyroidism, parathyroid trauma during thyroid surgery
Calcium	8.5–10.5 mg/100 mL	↑ in some cancers, hyperparathyroidism ↓ in hypothyroidism
Phosphorus	2.4–4.7 mg/dL	↑ in hypoparathyroidism ↓ in hyperparathyroidism
Pituitary Tests		
Growth hormone	<5 ng/mL	↑ in acromegaly ↓ in small stature
Antidiuretic hormone	2.3–3.1 pg/mL	↑ in SIADH ↓ in diabetes insipidus
Urine specific gravity	1.010–1.025	↓ in diabetes insipidus
Adrenocorticotropic hormone	<120 pg/mL at 6–8 A.M.	↑ in Addison's disease ↓ in Cushing's syndrome, long-term corticosteroid therapy
Adrenal Tests		
Cortisol	5–25 µg/100 mL	↑ in Cushing's syndrome, stress ↓ in Addison's disease, steroid withdrawal
Vanillylmandelic acid (VMA; Urine Test)	0.7–6.8 mg/24 h	↑ in pheochromocytoma
Pancreas Tests		
Fasting blood sugar (FBS)	70–110 mg/100 mL	↑ in diabetes mellitus, stress, Cushing's syndrome ↓ in hypoglycemia, Addison's disease
Oral glucose tolerance test	Blood glucose returns to normal within 3 h	Any two values over 140 diagnose diabetes mellitus
Two-hour postprandial glucose	<120 mg/dL <140 mg/dL in elderly	↑ in diabetes mellitus ↓ in hypoglycemia, gastrointestinal malabsorption
Glycosylated hemoglobin	4–7%	↑ in poor diabetes control

SIADH = syndrome of inappropriate antidiuretic hormone.
*All normal values are for a fasting test.

stimulation test, TRH is injected. If the pituitary gland responds appropriately, TSH is secreted. If the thyroid gland responds appropriately to the TSH, T_3 and T_4 levels rise. Failure of the TRH to stimulate TSH and thyroid hormone indicates a pituitary or thyroid condition. Further studies might be done to determine the cause.

Suppression Tests

Suppression tests are the opposite of stimulating tests. For this type of test, a substance is injected that is expected to suppress a hormone's release. For example, if dexamethasone (a steroid hormone) is injected, cortisol release is expected to be suppressed via the negative feedback mechanism. If the cortisol level is not suppressed, adrenal cortex dysfunction is suspected.

Urine Tests

Sometimes it is helpful to measure the amount of hormone or hormone byproduct excreted in the urine during a 24-hour period. Examples are cortisol and vanillylmandelic acid, a product of catecholamine metabolism.

To collect a 24-hour urine specimen, it is necessary to obtain a special urine container from the laboratory. It is usually an opaque container that protects the specimen from light; it may have preservative in it. Check with the laboratory to find out whether the specimen needs to be kept on ice during the test and whether the patient needs to be on a special diet before or during the test. To keep a specimen on ice, a bath basin is filled with ice and the container is placed in the basin. The ice must be refilled every few hours to keep the specimen cold. When initiating the test, ask the patient to urinate and discard the urine. The time of this first discarded voiding is considered the start of the test. Any urine collected from this time forward for 24 hours is saved. At the end of the 24-hour period, the patient is again asked to urinate, but this time the urine is saved. The entire collection is then labeled and sent to the laboratory. Instruct the patient how to do the test, place a sign on the toilet with the start and stop time of the test, and remind all staff to save the urine.

If the patient is incontinent or otherwise unable to participate in the test, a catheter may need to be inserted. If the patient already has an indwelling catheter, a new bag and tubing should be attached before the start of the test. The lab should be consulted to determine the need for a preservative or ice. Preservative can be added to the catheter bag if necessary. If ice is necessary, the bag should be kept in a basin of ice rather than hanging on the side of the bed. If the specimen must be protected from light, the bag can be covered with dark plastic or foil.

Other Laboratory Tests

Some laboratory tests may indirectly reflect the function of an endocrine gland. For example, a serum calcium level helps indicate PTH or calcitonin secretion, and a blood glucose level reflects insulin secretion.

Nuclear Scanning

A thyroid scan may be done to determine the presence of tumors or nodules. For this test, a radioactive material is injected or radioactive iodine is taken orally. The material is attracted by the thyroid gland. After a specified time, the thyroid gland is scanned with a scintillation camera. The scan will show "hot spots," which are nodules that are not malignant, or "cold spots" (areas that do not take up the radioactivity), which indicate malignancy. Cold spots may then be biopsied to confirm a diagnosis. Because such a small amount of radioactive material is used, there is no risk to the patient. The patient should be aware that the test takes approximately 30 minutes to complete.

Radiographic Tests

A computed tomography (CT) scan or magnetic resonance imaging (MRI) may be done to locate a tumor or identify hypertrophy of a gland.

Ultrasound

Ultrasound may be done of the thyroid or parathyroid glands to determine if they are enlarged or to find masses.

Biopsy

Biopsy is done to obtain tissue to examine for possible cancerous cells. The thyroid gland can be biopsied either by needle aspiration under local anesthesia or using a surgical incision.

REVIEW QUESTIONS

1. The posterior pituitary gland secretes the following hormone:
   a. Antidiuretic hormone
   b. Thyroid-stimulating hormone
   c. Growth hormone
   d. Luteinizing hormone

2. When the effects of a hormone suppress further secretion of the hormone, which of the following types of feedback has occurred?
   a. Positive feedback
   b. Negative feedback
   c. Rhythm feedback
   d. Stimulus feedback

3. The functions of T_4 and T_3 include which of the following?
   a. Retention of salt and water
   b. Maintenance of blood sugar
   c. Maintenance of blood pressure
   d. Regulation of energy production

4. Strong bones are maintained by the action of which hormone?
   a. ADH
   b. Insulin
   c. Calcitonin
   d. TRH

5. When collecting a 24-hour urine specimen for hormone measurement, the nurse should do which of the following to begin the test?
   a. Discard the first voided urine.
   b. Save the first voided urine.
   c. Discard the last voided urine.
   d. Save the first and last voided urine.

6. When explaining a thyroid scan to a patient, which of the following statements is correct?
   a. "You will take a special pill, and then an ultrasound will be taken of your neck."
   b. "You will receive an injection of radioactive material, and then a special camera will take pictures of your thyroid gland."
   c. "You will be placed into a special machine, and x-rays will be taken of your neck. It may be noisy."
   d. "You will be given a special drink, and then magnetic energy is used to visualize the thyroid area."

Answers to CRITICAL THINKING

Mrs. Trombley

1. The specimen must include all urine for a 24-hour period. Restart the test from noon.
2. To prevent this from happening again, try placing an incontinence pad or other large item directly over the toilet as a reminder to the patient. Be sure there is a sign posted above the toilet or on the toilet handle, and communicate with the nursing assistant why the urine needs to be saved.
3. "Patient voided in toilet at 1200. 24-hour urine test restarted; importance of saving urine explained to patient. RN notified." Note that it is not appropriate to "blame" the nursing assistant for making a mistake.

37

NURSING CARE OF PATIENTS WITH ENDOCRINE DISORDERS

Paula D. Hopper

QUESTIONS TO GUIDE YOUR READING

1. What are the disorders caused by variations in the hormones of the pituitary, thyroid, parathyroid, and adrenal glands?

2. How would you explain the pathophysiology of each of the endocrine disorders presented?

3. What are the etiologies, signs, and symptoms of each of the disorders?

4. What is current medical and surgical treatment of each of the selected endocrine disorders?

5. What data would you collect when caring for patients with each of the endocrine disorders discussed?

6. What nursing care will you provide for patients with each of the disorders?

7. How will you know if your nursing interventions have been effective?

A variety of disorders can be found within the endocrine system. Although the causes vary, the pathophysiology usually involves either too little or too much hormone activity. Insufficient hormone activity may be the result of hypofunction of an endocrine gland or insensitivity of the target tissue to its hormone. Excessive hormone activity may be the result of a hyperactive gland, **ectopic** hormone production, or self-administration of too much replacement hormone (Table 37–1). If you can remember the function of each hormone in the body, understanding the problems involved with an altered amount of each hormone becomes easier.

Most endocrine disorders are either primary or secondary. A primary disorder is a problem within the gland that is out of balance. Secondary disorders are caused by problems outside the gland, such as an imbalance in a tropic hormone, certain drugs, trauma, surgery, or a problem in the feedback mechanism.

---

ectopic: ec—away from normal + topic—place

LEARNING TIPS

If you can remember what each hormone does in the body, it will be easier to remember what results from imbalances of that hormone. Most symptoms of hormone hyperactivity are the opposite of symptoms of that hormone's hypoactivity.

TABLE 37–1	CAUSES OF ENDOCRINE PROBLEMS	
Insufficient Hormone Activity	**Excess Hormone Activity**
Gland hypofunction	Gland hyperfunction
Lack of tropic or stimulating hormone	Excess tropic or stimulating hormone
Target tissue insensitivity to hormone	Ectopic hormone production
	Self-administration of too much replacement hormone

▶ PITUITARY DISORDERS

Pituitary disorders often involve several hormone imbalances caused by general hypopituitarism or hyperpituitarism. Problems involving all the pituitary hormones at once are rare. For simplicity, imbalances are considered separately here.

Disorders Related to Antidiuretic Hormone Imbalance

Antidiuretic hormone (ADH, also called vasopressin) is synthesized in the hypothalamus and stored and secreted by the posterior pituitary gland. A decrease in ADH activity results in diabetes insipidus (DI). An increase in ADH activity is called syndrome of inappropriate antidiuretic hormone (SIADH). Table 37–2 compares DI and SIADH. Note how symptoms of too little ADH (water loss) are opposite from symptoms of too much ADH (water retention).

Diabetes Insipidus

PATHOPHYSIOLOGY. Diabetes insipidus is caused by a deficiency of ADH. Recall that ADH is responsible for reabsorption of water by the distal tubules and collecting ducts in the kidneys. If ADH is lacking, adequate reabsorption of water is

TABLE 37–2	COMPARISON OF ANTIDIURETIC HORMONE DISORDERS	
	Insufficient ADH	**Excess ADH**
Disorder	Diabetes insipidus	SIADH
Signs and symptoms	Polyuria, polydipsia, hypernatremia, dehydration	Fluid retention, weight gain, hyponatremia
Usual treatment	Synthetic ADH replacement	Treat cause
Priority nursing diagnoses	Risk for fluid volume deficit	Risk for fluid volume excess

prevented, leading to diuresis. In **nephrogenic** diabetes insipidus, there is enough ADH but the kidneys do not respond to it. Patients may urinate from 3 to 15 L per day. This leads to increased serum **osmolality** (concentrated blood) and dehydration. The increased osmolality and decreased blood pressure normally trigger ADH secretion, which causes water retention and dilutes the blood; in patients with DI, this does not occur. Increased osmolality also leads to extreme thirst, which usually causes the patient to drink enough fluids to maintain fluid balance. In an unconscious patient or a patient with a defective thirst mechanism, however, dehydration may quickly occur if the problem is not recognized and corrected. For more information, check out the Nephrogenic Diabetes Insipidus Foundation at www.ndif.org.

ETIOLOGY. The primary causes of diabetes insipidus are tumors or trauma to the pituitary gland. Surgery in the area of the pituitary and certain drugs, such as glucocorticoids or alcohol, may also cause DI. Occasionally the cause is **psychogenic:** The patient drinks large quantities of water in the absence of true disease. Nephrogenic DI may be inherited or acquired and is diagnosed when the kidneys do not respond to ADH. It can be triggered by certain drugs or neoplasms or by damage to the kidneys from pyelonephritis, polycystic disease, or other causes.

SIGNS AND SYMPTOMS. The patient with DI urinates frequently (**polyuria**), and nighttime urination (**nocturia**) is present. This results in high serum osmolality and low urine osmolality. Urine specific gravity is decreased (the urine is diluted and light in color).

The patient experiences extreme thirst (**polydipsia**), and large volumes of water are consumed. Often patients crave ice-cold water. If urine output exceeds fluid intake, dehydration occurs, with characteristic symptoms of hypotension, poor skin turgor, and weakness. **Hypovolemic** shock occurs if fluid balance is not restored. Dehydration and electrolyte imbalances result in a decrease in level of consciousness and death if the problem is not corrected.

The patient with DI may develop an enlarged bladder and kidney damage from constantly trying to "hold" too much urine.

DIAGNOSTIC TESTS. Diagnosis is based initially on a history of risk factors and reported symptoms. Urine specific gravity is less than 1.005 (normal: 1.010 to 1.025), and can be monitored by laboratory tests or by using reagent strips at the bedside. Plasma osmolality is measured. The actual amount of sodium in the blood may be normal, but it appears elevated in relation to the decreased amount of water. Computed tomography (CT) scan or magnetic resonance imaging (MRI) may show a pituitary tumor.

---

nephrogenic: nephro—kidney + genic—to produce
psychogenic: psycho—related to the mind + genic—to produce
polyuria: poly—much + uria—urine
nocturia: noct—night + uria—urine
polydipsia: poly—much + dipsia—thirst
hypovolemic: hypo—deficient + vol—volume + emic—blood

A water deprivation test may be done. For this test, the patient is deprived of water for up to 8 hours. Body weight and urine osmolality are tested hourly. If the urine continues to be diluted, even though the patient is not drinking and is losing weight as a result of volume depletion, DI is suspected. In the second stage of the test, the patient receives an injection of ADH, with a final urine test done 1 hour later. If the DI is nephrogenic, the kidneys do not respond to the injected ADH.

MEDICAL TREATMENT. If a pituitary tumor is involved, treatment usually involves removal of the pituitary gland (**hypophysectomy**). Medical treatment of DI involves replacement of ADH. In acute cases, vasopressin, a synthetic form of ADH, is given by the intravenous or subcutaneous route, along with intravenous fluid replacement. In patients who require long-term therapy, such as those who have had a hypophysectomy, synthetic ADH (desmopressin, or DDAVP) in the form of a nasal spray is used. Desmopressin's action lasts 18 to 24 hours. Other drugs, such as chlorpropamide (Diabinese), stimulate ADH secretion in patients with partial DI. Thiazide diuretics may be ordered to treat nephrogenic DI.

NURSING PROCESS

Assessment. When doing a nursing assessment of a patient with DI, place special attention on fluid balance. Daily weights are the most reliable method for monitoring the amount of fluid that is being lost. Accurate intake and output measurement is also helpful. Skin turgor will be poor and mucous membranes will be dry and sticky if the patient is becoming dehydrated. Monitor skin integrity because dehydration increases risk of breakdown. Monitor vital signs for signs of shock. Use a reagent strip (dipstick) or urimeter to measure urine specific gravities. Monitor serum electrolytes and osmolality as ordered, and watch for changes in level of consciousness. Assess the patient's understanding of his or her disease and treatment. Once treatment is initiated, observe the patient for fluid overload.

Nursing Diagnosis. Priority nursing diagnoses include deficient fluid volume related to failure of regulatory mechanisms and deficient knowledge related to lack of exposure to information about self-care of diabetes insipidus.

Planning and Implementation. Give the patient oral fluids if the DI is not psychogenic. If the patient's thirst mechanism is not intact, give the patient fluids every hour. Hypotonic intravenous (IV) fluids such as 0.45 percent saline may be ordered to replace intravascular volume without adding excessive sodium. IV fluids are especially important if the patient is unable to take oral fluids. Report a significant drop in blood pressure and a rising pulse to the registered nurse or physician because these may be signs of shock. If the patient is able, he or she may participate in maintaining intake and output records.

Patient Education. Teach the patient the basic pathophysiology of the disease and how to administer medications and monitor their effectiveness. Teach the patient how to measure urine specific gravity and the significance of results. Include signs and symptoms of dehydration and fluid overload. Stress the importance of daily weights; losses or gains of greater than 2 pounds in a day should be reported to the physician. Advise the patient to wear identification, such as a medical alert bracelet, that identifies the disorder.

Evaluation. If treatment has been effective, signs of dehydration will be absent and weight and vital signs will be stable. The patient should be able to explain what is happening in the disease, symptoms to report, and how to manage self-care.

Syndrome of Inappropriate Antidiuretic Hormone

PATHOPHYSIOLOGY. Syndrome of inappropriate antidiuretic hormone (SIADH) results from too much ADH in the body. This causes excess water to be reabsorbed by the kidney tubules and collecting ducts, leading to decreased urine output and fluid overload. As fluid builds up in the bloodstream, osmolality decreases and the blood becomes diluted. Normally a decreased serum osmolality inhibits release of ADH. In SIADH, however, ADH continues to be released, adding to the fluid overload.

ETIOLOGY. Bronchogenic lung cancer, duodenal cancer, or pancreatic cancer may be ectopic sites of production of an ADH-like substance. Certain drugs, such as tricyclic antidepressants and general anesthetics, may increase ADH secretion. Head trauma or surgery or a brain tumor affecting pituitary function may also cause SIADH. It may be a complication of treatment of diabetes insipidus.

SIGNS AND SYMPTOMS. Symptoms of SIADH include symptoms of fluid overload, such as weight gain (usually without edema) and dilutional hyponatremia (Box 37–1). The actual amount of sodium in the blood may be normal, but it appears low because of the diluting effect of the retained fluid. Serum osmolality is less than 275 mOsm/kg. The urine is concentrated because water is not being excreted. Muscle cramps and weakness may occur because of electrolyte imbalance. Because the osmolality of the blood is low, fluid may leak out of the vessels and cause brain

---

BOX 37–1 **Manifestations of Dilutional Hyponatremia**

Bounding pulse
Elevated or normal blood pressure
Muscle weakness
Headache
Personality changes
Nausea
Diarrhea
Convulsions
Coma

---

hypophysectomy: hypophysis—pituitary + ectomy—surgical removal

swelling. If untreated, this results in lethargy, seizures, coma, and death.

DIAGNOSTIC TESTS. Serum and urine sodium levels and osmolality are measured. A water load test may be done, which involves administering a specific amount of water, then measuring blood and urine sodium and osmolality hourly for 6 hours. The patient with SIADH retains the water instead of excreting it. Additional testing may be done to diagnose and locate an ADH-secreting tumor.

MEDICAL TREATMENT. Treatment is aimed at eliminating the cause. If a tumor is secreting ADH, surgical removal may be indicated. Symptoms may be alleviated by restricting fluids to 800 to 1000 mL per 24 hours. Normal or hypertonic saline fluids may be administered intravenously to maintain the serum sodium level. If the cause is inoperable cancer, drugs such as furosemide (Lasix) and demeclocycline (Declomycin) are used to block the action of ADH in the kidney.

NURSING PROCESS

Assessment. Fluid overload with hyponatremia is the primary concern for the patient with SIADH. To monitor fluid balance, assess vital signs, daily weights, intake and output, urine specific gravity, skin turgor, edema, and lung sounds. The patient's ability to maintain a fluid restriction is determined. Assess level of consciousness and neuromuscular function. (See Chapter 45.) Monitor laboratory tests, including serum sodium level, as ordered by the physician. Assess the patient's understanding of the disease process and treatment.

Nursing Diagnosis. Priority nursing diagnoses include risk for excess fluid volume related to compromised regulatory mechanism and knowledge deficit related lack of exposure to information about self-care of SIADH.

Planning and Implementation. Explain the importance of maintaining the fluid restriction to the patient. Hard candy may help alleviate thirst. Ice chips may also help and are counted as half the volume of fluid; that is, 100 mL of ice chips equal approximately 50 mL of water. Providing calibrated cups can help the patient maintain the restriction independently if able. To increase compliance, allow the patient to participate in planning the types and times of fluid intake. Fluids high in sodium, such as broth, cola, or tomato juice, may help correct dilutional hyponatremia. Report a change in level of consciousness immediately, and monitor the patient for seizures.

Patient Education. Instruct the patient to report any weight gain greater than 2 pounds in one day, a change in urine output, or acute thirst. Use of medical alert bracelet or other identification should be encouraged.

Evaluation. Weight should stabilize at the preillness level once treatment is begun. Serum sodium level should be within normal limits. Patients should be able to verbalize the cause of their symptoms and demonstrate self-care, including ability to maintain a fluid restriction if necessary.

▶M▶ **CRITICAL THINKING**

Mrs. Jackson

You are caring for Mrs. Jackson, a 78-year-old woman who has just returned to your unit following hip surgery. During the next 2 days, you notice that her weight increases from 118 to 124 pounds and she seems lethargic, but the nurse in report didn't seem concerned about it. You check her ankles and sacrum for edema but find none. In the afternoon, her son rushes out of the room and tells you she is becoming confused, adding that this is not like her at all.

1. What could be happening?
2. What assessment can you do to gather further data to support your suspicions?
3. What will you do?

Answers at end of chapter.

Disorders Related to Growth Hormone Imbalance

Growth hormone (GH), also called somatotropin, is responsible for normal growth of bones, cartilage, and soft tissue. GH is synthesized and secreted by the anterior pituitary gland. An excess or deficiency of GH may be related to a more generalized problem with the pituitary gland. Excess GH results in gigantism or acromegaly. A deficit of GH results in dwarfism (Fig. 37–1).

Figure 37-1   Gigantism and dwarfism.

Dwarfism

PATHOPHYSIOLOGY. Dwarfism, also called short stature, occurs when growth hormone is deficient in childhood. A deficiency of GH in adults does not affect growth but is responsible for a variety of other symptoms.

ETIOLOGY. Growth hormone may be deficient as a result of a pituitary tumor or failure of the pituitary to develop. It may be the result of infection or other trauma to the pituitary gland. It may also be deficient in some cases of neglect or severe emotional stress, causing psychosocial dwarfism. Sometimes the cause is not known (Cultural Consideration Box 37–2).

BOX 37-2 CULTURAL CONSIDERATION

Dwarfism, mostly related to a limited gene pool, often occurs among Amish communities. Ellis–van Creveld syndrome is prevalent among the Amish of Lancaster County, Pennsylvania. This syndrome is characterized by short stature and an extra digit on each hand, with some individuals having a congenital heart defect and nervous system involvement resulting in a degree of mental handicap.

SIGNS AND SYMPTOMS. Children may grow to only 3 to 4 feet in height but have normal body proportions. Sexual maturation may be slowed, related to involvement of additional pituitary hormones. Dwarfism in children is sometimes accompanied by mental retardation. In adults, symptoms include weakness, hypoglycemia, sexual dysfunction, skin changes, and increased risk for cardiovascular and cerebrovascular disease. Headaches, mental slowness, and visual disturbances may also occur.

DIAGNOSTIC TESTS. Growth hormone levels in the blood are measured by a routine laboratory test. A growth hormone stimulation test may be done by measuring GH response to induced hypoglycemia. Radiographic studies help determine the presence of a pituitary tumor and may also be used to determine bone age.

MEDICAL TREATMENT. Treatment of dwarfism in a child is administration of growth hormone. In the past, GH was derived from human pituitary glands, so treatment was expensive. Now GH can be made in a laboratory using genetic engineering, and it is more readily available to those who need it. It is administered by injection. Surgery may be indicated if a tumor is the cause.

NURSING PROCESS
Assessment. Assessment of the adult with dwarfism includes mental status, ability to cope with the effects of the disorder, and understanding of the treatment plan.

Nursing Diagnosis. Possible priority nursing diagnoses include disturbed body image, ineffective coping, and deficient knowledge. If the adult with dwarfism has accepted and coped well with the condition, diagnosis and treatment by a nurse may not be indicated.

Planning and Implementation. Approach patients with dwarfism with an attitude of acceptance and caring. Provide an opportunity to verbalize their feelings. An occupational therapist may be able to provide techniques to assist patients to adapt to an environment that is geared toward people of average height. Information about support groups may be provided to patients.

Evaluation. The goal has been achieved if the patient is able to express his or her feelings of acceptance of the disorder, demonstrate adaptive techniques, and verbalize understanding of treatment.

CRITICAL THINKING

Adoption
Three siblings were adopted to a loving home after having been in several foster homes. After a year in their new home, each child suddenly grew 6 to 8 inches. What do you think happened?
Answer at end of chapter.

Acromegaly

Acromegaly is a rare excess of growth hormone that affects adults, usually in their thirties or forties. If a GH excess occurs in children, the condition results in gigantism.

PATHOPHYSIOLOGY. Acromegaly occurs as a result of oversecretion of GH in an adult. Bones increase in size, leading to enlargement of facial features, hands, and feet. Long bones grow in width but not length, because the epiphyseal disks are closed. Subcutaneous connective tissue increases, causing a fleshy appearance. Internal organs and glands enlarge. Impaired tolerance of carbohydrates leads to elevated blood glucose.

ETIOLOGY. Excess secretion of growth hormone can be caused by pituitary **hyperplasia,** a benign pituitary tumor, or hypothalamic dysfunction.

SIGNS AND SYMPTOMS. Often the first symptom noticed is a change in hat or shoe size. The nose, jaw, brow, hands, and feet enlarge (Fig. 37–2). The teeth may be displaced, causing difficulty chewing, or dentures may no longer fit. The tongue becomes thick, causing difficulty in speaking and swallowing. Vertebral changes may lead to kyphosis. Visual disturbances may occur because of tumor pressure on the optic nerve. Headaches are a result of tumor pressure on the brain. Diabetes mellitus may develop because GH increases blood glucose and causes an increased workload for the pancreas. (See Chapter 38.) Osteoporosis and arthritis may occur. Erectile dysfunction may occur in men and **amenorrhea** in women. With treatment, soft tissues reduce in size, but bone growth is permanent.

---

hyperplasia: hyper—excessive + plasia—formation or deviation
amenorrhea: a—not + men—month + orrhea—flow

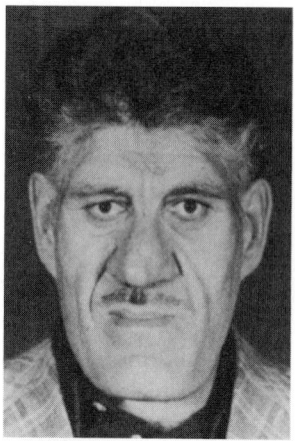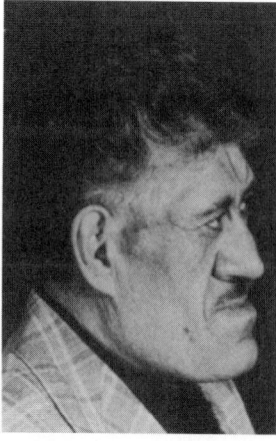

Figure 37-2   Patient with acromegaly. From Tamparo, CD, and Lewis, MA: Diseases of the Human Body, ed 2. FA Davis, Philadelphia, 1995, p. 270.

DIAGNOSTIC TESTS. Serum growth hormone levels are measured, and radiographs show abnormal bone growth.

MEDICAL TREATMENT. Treatment is aimed at the cause. Bromocriptine (Parlodel) or octreotide (Sandostatin) may decrease GH levels. Hypophysectomy or radiation may be indicated if a tumor is the cause. If the pituitary is removed, lifelong replacement of thyroid hormone, corticosteroids, and sex hormones is important to maintain homeostasis.

NURSING PROCESS

Assessment. The nurse caring for the patient with acromegaly is concerned with the patient's response to the disease. Assess safety in relation to impaired eyesight, chewing, and swallowing. Monitor serum glucose levels for onset of diabetes mellitus. Assess knowledge and acceptance of the disease. If hypophysectomy is planned, assess the patient for anxiety related to the surgery and perform a preoperative baseline neurological assessment. (See Chapter 45 and Care of the Patient Undergoing Hypophysectomy later in this chapter.)

Nursing Diagnosis. Possible nursing diagnoses include disturbed body image, risk for injury related to poor eyesight and **dysphagia,** deficient knowledge, pain, and disturbed sensory perception. Additional diagnoses are identified if diabetes mellitus or other problems exist.

Planning and Implementation. Allow the patient to verbalize feelings related to the disease. A care plan for safety is implemented if vision or swallowing is disturbed. If pain is present, comfort measures are provided.

Patient Education. Teach the patient and significant others about the disease and treatment. If the patient is having a hypophysectomy, be sure he or she understands that some symptoms will be relieved but that bone growth and visual changes may not reverse. Stress the need for lifelong hormone replacement after surgery.

---

dysphagia: dys—bad + phagia—swallowing

Evaluation. The patient who is effectively treated will have some soft tissues return to normal size. Injuries related to poor eyesight and difficulty swallowing will be avoided. The patient should be able to accurately describe self-care requirements.

Pituitary Tumors

Most tumors of the pituitary gland are benign adenomas. However, even benign tumors in the brain can cause many symptoms, including visual disturbances, symptoms of increased pressure in the brain, and symptoms related to hormone imbalances, as described earlier. Treatment for pituitary tumors is usually hypophysectomy (surgical removal of the pituitary gland). Radiation may be used alone or as an adjunct to surgery.

Care of the Patient Undergoing Hypophysectomy

Removal of the pituitary gland is called hypophysectomy. Figure 37–3 shows the transsphenoidal approach to the gland. Some tumors may necessitate removal via a transfrontal craniotomy (entry through the frontal bone of the skull).

Preoperative Care

Ensure that the patient understands the physician's explanation of surgery. Perform and document a baseline neurological assessment. Prepare the patient for what to expect following surgery. Instruct the patient to avoid any actions that increase pressure on the surgical site, such as coughing, sneezing, nose blowing, straining to move bowels, or bending from the waist. Because coughing can raise intracranial pressure and is therefore contraindicated, instruct the patient in deep breathing exercises or use of an incentive spirometer.

Postoperative Care

Perform routine neurological assessments to monitor for neurological damage. Also be sure to check urine for specific gravity, because diabetes insipidus may occur following pituitary surgery. If patients have had transsphenoidal surgery, they will have nasal packing and a "mustache dressing," which is placed under the nose to collect drips. These are left in place and not removed unless ordered by the physician. Monitor the dressing for signs of cerebrospinal fluid (CSF) leakage. CSF contains glucose, so glucose testing strips can be used to determine if drainage is actually CSF or just nasal discharge. Remind the patient to avoid any actions that increase pressure on the surgical site. Obtain orders for stool softeners and antitussives as needed. Tooth brushing is avoided until the incision line is healed. The patient may use floss and mouth rinses. The patient is placed on hormone replacement therapy following hypophysectomy. Pituitary hormones are difficult to replace, so target hormones are generally given. These may include thyroid hormone, glucocorticoids, intranasal desmopressin, and sex hormones. Instruct the patient in how to administer the hormones, as well as side effects to report.

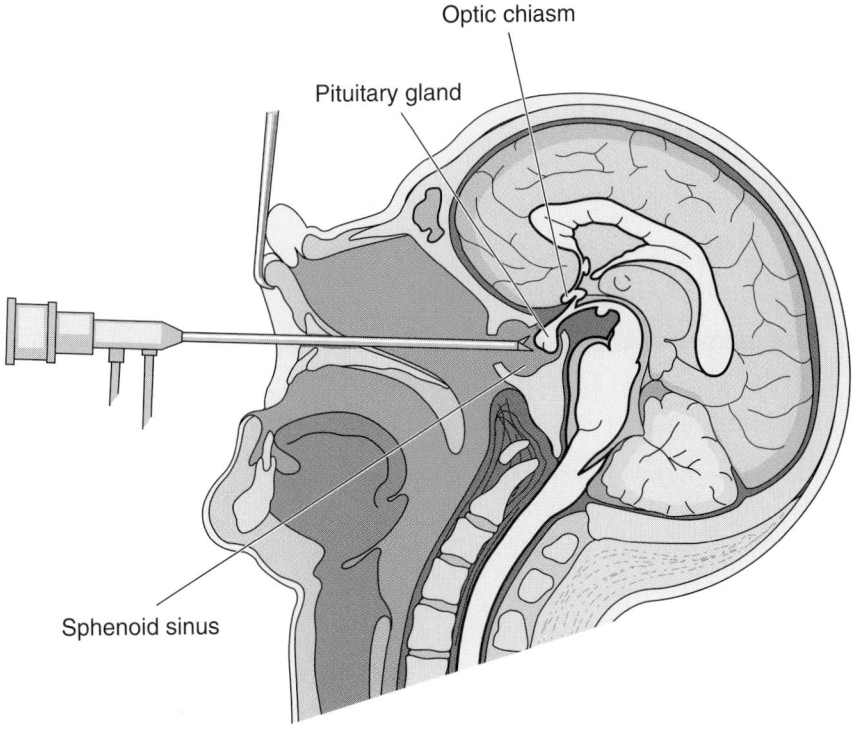

Figure 37-3 Transsphenoidal approach to pituitary gland for hypophysectomy.

DISORDERS OF THE THYROID GLAND

Triiodothyronine (T_3) and thyroxine (T_4) are secreted by the thyroid gland. These hormones may be collectively referred to as thyroid hormone (TH). Deficient secretion of these hormones results in hypothyroidism; excess TH results in hyperthyroidism. For more information on disorders of the thyroid gland, visit the American Thyroid Association at www.thyroid.org.

Hypothyroidism

Hypothyroidism occurs primarily in women 30 to 60 years old. If hypothyroidism occurs in an infant, the result is cretinism. Hypothyroidism that develops in an adult is called **myxedema.**

PATHOPHYSIOLOGY. Primary hypothyroidism occurs when the thyroid gland fails to produce enough TH even though there is enough thyroid-stimulating hormone (TSH) being secreted by the pituitary gland. The pituitary responds to the low level of TH by producing more TSH. Secondary hypothyroidism is caused by low levels of TSH or thyrotropin-releasing hormone (TRH, secreted by the hypothalamus), which fail to stimulate release of TH. Most cases of hypothyroidism are primary (Table 37–3).

Because thyroid hormones are responsible for metabolism, low levels of these hormones result in a slowed metabolic rate, which causes many of the characteristic symp-

TABLE 37-3	THYROID HORMONE ABNORMALITIES	
	Hyperthyroidism	**Hypothyroidism**
Primary	TH↑ TSH↓	TH↓ TSH↑
Secondary (pituitary cause)	TH↑ TSH↑	TH↓ TSH↓

toms of hypothyroidism. Other symptoms are related to myxedema, which refers to a nonpitting type of edema that occurs in connective tissues throughout the body.

ETIOLOGY. Primary hypothyroidism may be a result of a congenital defect, inflammation of the thyroid gland, iodine deficiency, or thyroidectomy. Hashimoto's thyroiditis is an **autoimmune** disorder that eventually destroys thyroid tissue, leading to hypothyroidism. Secondary hypothyroidism may be caused by a pituitary or hypothalamic lesion or by postpartum pituitary necrosis, a rare disorder in which the pituitary is destroyed following pregnancy and delivery. Peripheral resistance to TH may also occur.

SIGNS AND SYMPTOMS. Manifestations primarily are related to the reduced metabolic rate and include fatigue, weight gain, bradycardia, constipation, mental dullness, feeling cold, shortness of breath, and dry skin and hair (Table 37–4). Heart failure may occur because of decreased pumping strength of the heart. Altered fat metabolism causes hyperlipidemia. Myxedema causes water retention, with puffiness

---

myxedema: myx—mucus + edema—swelling

autoimmune: auto—pertaining to self + immune—exempt

TABLE 37-4 **SYMPTOMS OF THYROID DISORDERS**

	Hypothyroidism	Hyperthyroidism
Cardiovascular	Bradycardia, decreased cardiac output, cool skin, cold intolerance	Tachycardia, palpitations, increased cardiac output, warm skin, heat intolerance
Neurological	Lethargy, slowed movements, memory loss, confusion	Fatigue, restlessness, tremor, insomnia, emotional instability
Pulmonary	Dyspnea, hypoventilation	Dyspnea
Integumentary	Cool, dry skin; brittle, dry hair	Diaphoresis; warm, moist skin; fine, soft hair
Gastrointestinal	Decreased appetite, weight gain, constipation, increased serum lipid levels	Increased appetite, weight loss, frequent stools, decreased serum lipid levels
Reproductive	Decreased libido, erectile dysfunction	Decreased libido, erectile dysfunction, amenorrhea

in the face, eye area, and feet. Fluid may also accumulate around the heart, causing altered cardiac function.

COMPLICATIONS. If the metabolic rate drops so low that it becomes life threatening, the result is called myxedema coma. This can be triggered by stress, such as infection, trauma, or exposure to cold. The patient becomes hypothermic, with a temperature less than 95° F, and has a decreased respiratory rate and blood pressure. Depressed mental function and lethargy may occur. Blood glucose drops. Death may occur as a result of respiratory failure. If you note changes in mental status or vital signs, the physician should be contacted immediately. Treatment of myxedema coma involves intubation and mechanical ventilation. The patient is slowly rewarmed with blankets. Intravenous levothyroxine (Synthroid) is given.

DIAGNOSTIC TESTS. T_3 and T_4 levels are low, and TSH may be high or low, depending on the cause. If the pituitary is functioning normally, TSH is elevated in an attempt to stimulate an increase in TH. Serum cholesterol and triglycerides are elevated. Antibodies are usually present in autoimmune disease.

MEDICAL TREATMENT. Hypothyroidism is easily treated with thyroid replacement hormone. Some patients still use TH from animal thyroids. Most patients now take synthetic thyroid hormone (levothyroxine). Doses are started low and are slowly increased to prevent symptoms of hyperthyroidism or cardiac complications.

NURSING PROCESS. See Nursing Care Plan 37–3 for nursing care of the patient with hypothyroidism.

BOX 37-3 **NURSING CARE PLAN FOR THE PATIENT WITH HYPOTHYROIDISM**

 Activity intolerance related to fatigue

PATIENT OUTCOMES

(1) Patient reports lessening fatigue after treatment initiated. (2) Patient is able to carry out usual activities of daily living (ADLs).

EVALUATION OF OUTCOMES

(1) Does patient report lessening fatigue? (2) Is patient able to carry out ADLs?

INTERVENTIONS	RATIONALE	EVALUATION
Assist patient with self-care activities.	Patients with fatigue may have difficulty carrying out activities independently.	Are patient's self-care needs being met? Is assistance needed?
Allow for rest between activities.	Rest periods will enable patient to conserve energy for activities.	Does patient state rest is adequate?
Slowly increase patient's activities as medication begins to be effective.	As thyroid replacement therapy becomes effective, patient's fatigue will subside.	Does patient tolerate increases in activity?

GERIATRIC

When getting elderly patients up, watch for orthostatic hypotension.	Orthostatic hypotension is common in elderly and may cause falls.	Does patient's blood pressure drop when changing positions?

 Constipation related to slowed gastrointestinal motility

PATIENT OUTCOMES

(1) Soft, formed stool passed at preillness frequency. (2) Patient identifies measures to prevent constipation in future.

EVALUATION OF OUTCOMES

(1) Are bowels moving according to patient's preillness pattern? (2) Does patient verbalize measures to prevent constipation?

BOX 37-3	NURSING CARE PLAN FOR THE PATIENT WITH HYPOTHYROIDISM—CONT'D

INTERVENTIONS	RATIONALE	EVALUATION
Monitor and record bowel movements.	A record helps determine if a problem exists.	Does record show a problem?
Help patient follow usual preillness pattern (e.g., after morning coffee).	A schedule allows bowel movement to occur before stool becomes hard and dry.	Is patient able to identify and implement usual self-care for bowels?
Increase fluids to eight 8-ounce glasses of water daily if cardiovascular status stable.	Adequate fluid intake helps prevent hard, dry stools.	Does patient take adequate fluids?
Add fiber to diet: fresh fruit, vegetables, bran.	Fiber helps increase the number of bowel movements.	Does patient tolerate fiber? Is it effective?
Encourage regular ambulation.	Activity increases peristalsis.	Is patient able to ambulate or engage in other activity?
Use bedside commode or bathroom rather than bedpan.	The sitting position aids in evacuation.	Is sitting position effective?
Obtain physician order for stool softener if needed.	Soft stools are passed more easily.	Is stool softener needed? Effective?
If stool is impacted, break up stool digitally and gently remove.	Breaking up stool eases evacuation.	Is stool impacted? Is digital disimpaction effective?
Avoid use of enemas.	Enemas can cause fluid and electrolyte imbalances and can damage mucosa.	Does patient understand need to avoid enemas?

 Impaired skin integrity related to dry skin, inactivity

PATIENT OUTCOMES

(1) Skin is soft and moist. (2) Skin remains intact.

EVALUATION OF OUTCOMES

(1) Is skin soft and moist? (2) Is skin intact?

INTERVENTIONS	RATIONALE	EVALUATION
Assess skin daily for breakdown.	Skin lesions are more effectively treated when identified early.	Is breakdown present?
Avoid use of soap on dry areas. Try bath oil.	Soap is drying to skin.	Does use of bath oil help?
Use nondrying lotion following bath.	Lotion helps trap moisture in skin. Some lotions contain alcohol, however, which is drying.	Does patient state relief with use of lotion?
Encourage/assist with position changes at least every 2 hours.	Changing position enhances circulation to the skin, promoting healing and preventing breakdown.	Does patient change position at least every 2 hours? Are pressure areas prevented?

Altered nutrition, more than requirements, related to decreased metabolic rate

PATIENT OUTCOMES

(1) Patient will return to preillness weight. (2) Patient will verbalize understanding of dietary recommendations.

EVALUATION OF OUTCOMES

(1) Is patient approaching preillness weight? (2) Is patient able to explain dietary recommendations and how they will be implemented?

INTERVENTIONS	RATIONALE	EVALUATION
Weigh weekly and record.	Weekly weights record progress without the frustration of daily fluctuations.	Is patient losing or maintaining weight?
Consult dietitian for therapeutic diet until hypothyroidism is controlled.	The dietitian can provide food choices for gradual weight loss if necessary.	Does patient verbalize understanding of and ability to follow diet?
Encourage regular exercise within limits of fatigue.	Exercise promotes weight control.	Does patient verbalize understanding of and ability to follow exercise plan?
Counsel patient that weight should normalize once hypothyroidism is controlled.	Thyroid replacement hormone increases the metabolic rate, allowing to return to normal weight.	Does patient verbalize understanding of instruction?

GERIATRIC		
Allow patient to help determine acceptable diet modifications.	Older patients may have long-standing dietary habits that are hard to change.	Is patient satisfied with weight loss plan?

PATIENT EDUCATION. Instruct the patient in the importance of consistent use of thyroid replacement medication and regular blood tests to monitor the levels of TH. The patient needs to be aware that too much thyroid hormone will cause symptoms of hyperthyroidism. Such symptoms should be reported to the physician immediately.

CRITICAL THINKING

Mae

Mae is a 59-year-old woman who is tired all the time and has gained 16 pounds during the past year. Her physician does some blood tests and prescribes levothyroxine PO. Lab results show low T_3 and T_4 and elevated TSH.

1. Why is Mae's TSH elevated?
2. What will happen to Mae's caloric requirements as she begins treatment? Why?
3. Why should you teach Mae to check her pulse?

Answers at end of chapter.

Hyperthyroidism

Hyperthyroidism is most often diagnosed in young women. Graves' disease, which is one cause of hyperthyroidism, is more common in young women. Multinodular goiter is more common in older women.

PATHOPHYSIOLOGY. Hyperthyroidism results from excessive amounts of circulating thyroid hormone (thyrotoxicosis). Primary hyperthyroidism occurs when a problem within the thyroid gland causes excess hormone release. Secondary hyperthyroidism occurs because of excess TSH release from the pituitary or excess TRH from the hypothalamus, which overstimulate the thyroid. A high level of thyroid hormone increases the metabolic rate. It is also believed to increase the number of beta-adrenergic receptor sites in the body, which enhances the activity of norepinephrine. The resulting fight-or-flight response is the cause of many of the symptoms of hyperthyroidism.

ETIOLOGY. A variety of disorders can cause hyperthyroidism. Graves' disease is the most common cause; it is thought to be an autoimmune disorder, because thyroid-stimulating antibodies are present in the blood of these patients.

Multinodular goiter, in which thyroid nodules secrete excess TH, is sometimes associated with hyperthyroidism. A pituitary tumor may secrete excess TSH, which overstimulates the thyroid gland. A thyroid tumor may secrete TH. Patients taking thyroid hormone for hypothyroidism may take too much. Each of these problems can cause excess circulating TH and symptoms of hyperthyroidism.

SIGNS AND SYMPTOMS. Many signs and symptoms are related to the hypermetabolic state, such as heat intolerance, increased appetite with weight loss, and increased frequency of bowel movements. Nervousness, tremor, tachycardia, and palpitations are caused by the increase in sympathetic nervous system activity. Heart failure may oc-

cur because of tachycardia and the resulting inefficient pumping of the heart. See additional signs and symptoms in Table 37–4. If treatment is not begun, the patient may become manic or psychotic. Additional signs that occur only with Graves' disease include thickening of the skin on the anterior legs and exophthalmos (bulging of the eyes; Fig. 37–4) caused by swelling of the tissues behind the eyes.

Be especially alert to the elderly patient, who may not exhibit the typical signs and symptoms of hyperthyroidism. These patients may present with heart failure, atrial fibrillation, fatigue, apathy, and depression.

COMPLICATIONS

Thyrotoxic Crisis. Thyrotoxic crisis (also called thyroid storm) is a severe hyperthyroid state that can occur in hyperthyroid individuals who are untreated or who are experiencing another illness or stressor. It may also occur following thyroid surgery in patients who have been inadequately prepared with antithyroid medication. Thyrotoxic crisis can result in death in as little as 2 hours if untreated. Symptoms include tachycardia, high fever, hypertension (with eventual heart failure and hypotension), dehydration, restlessness, and delirium or coma.

If thyrotoxic crisis occurs, treatment is first directed toward relieving the life-threatening symptoms. Acetaminophen is given for the fever. Aspirin is avoided because it binds with the same serum protein as T_4, freeing additional T_4 into the circulation. Intravenous fluids and a cooling blanket may be ordered to cool the patient. A beta-adrenergic blocker, such as propranolol, is given for tachycardia. If the patient is short of breath because of cardiac dysfunction, oxygen is administered and the head of the bed is elevated. Once symptoms are controlled and the patient is safe, the underlying thyroid problem is treated.

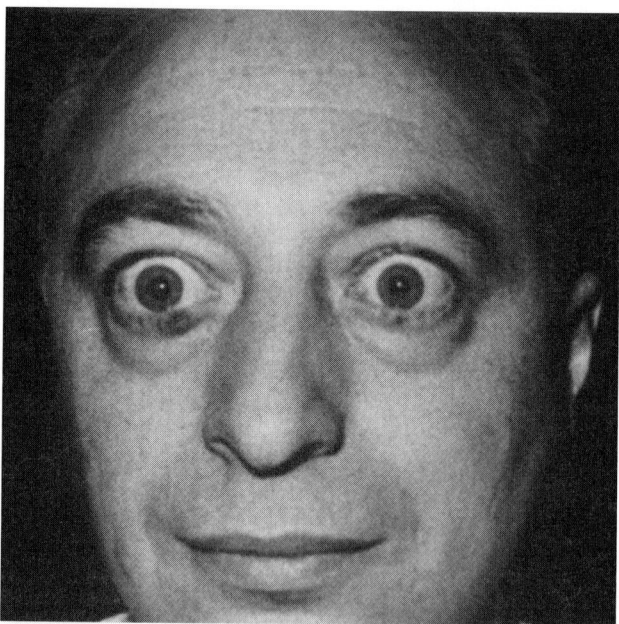

Figure 37-4 Exophthalmos caused by Graves' disease. From Tamparo, CD, and Lewis, MA: Diseases of the Human Body, ed 2. FA Davis, Philadelphia, 1995, p 275, with permission.)

Hypothyroidism. Another complication of hyperthyroidism can be hypothyroidism. This can occur as a result of long-term disease or as a result of treatment. Patients with a history of hyperthyroidism should be monitored for recurrent hyperthyroidism or the onset of hypothyroidism.

DIAGNOSTIC TESTS. Serum levels of T_3 and T_4 are elevated. TSH is low in primary hypothyroidism or high if the cause is pituitary. A TRH stimulation test may be done. A thyroid scan can be done to locate a tumor.

MEDICAL TREATMENT. Several medications can be used to treat hyperthyroidism. Propylthiouracil (PTU) and methimazole (Tapazole) inhibit the synthesis of TH. Propranolol (Inderal) is a beta-blocking medication that relieves the sympathetic nervous system symptoms. Radioactive iodine (^{131}I) may be used to destroy a portion of the thyroid gland. Oral iodine suppresses the release of thyroid hormone.

Sometimes medications alone control hyperthyroidism. If this does not occur, surgery is planned. If surgery is the treatment chosen, antithyroid medications are given to calm the thyroid before surgery. They help slow the heart rate and reduce other symptoms, making surgery safer. Iodine also reduces the vascularity of the thyroid gland, decreasing the risk of bleeding during surgery. Adequate preparation of the patient is important, because a **euthyroid** state helps prevent a postoperative thyrotoxic crisis. Nursing care of the patient undergoing a thyroidectomy is discussed later in this chapter.

NURSING PROCESS
Assessment. Monitor the patient with hyperthyroidism closely until normal thyroid activity is restored. Monitor vital signs and report any increases in pulse and blood pressure to the registered nurse or physician. Monitor lung sounds, because crackles may indicate heart failure. Assess level of anxiety and ability to cope with symptoms. Monitor weight and bowel function. Assess eyes for risk for injury caused by exophthalmos, and note degree of muscle weakness. Never palpate the thyroid gland of a patient with hyperthyroidism, because palpation may stimulate release of thyroid hormone and precipitate a thyrotoxic crisis.

Nursing Diagnosis. Priority nursing diagnoses include risk for injury related to hypermetabolic state and exophthalmos, hyperthermia, diarrhea, imbalanced nutrition, disturbed sleep pattern, and anxiety.

Planning and Implementation. Report changes in vital signs to the physician. The patient with exophthalmos may benefit from lubricating eyedrops and dark glasses. The eyes may be gently taped shut with nonallergic tape for sleeping. Elevation of the head of the bed and a low-sodium diet may decrease edema behind the eyes. If routine measures for hyperthermia are ineffective, a cooling blanket may be ordered. Diarrhea may be lessened with a low-fiber diet. A high-calorie diet with six meals a day may be necessary to meet caloric requirements. A restful environment and a mild sleeping aid may assist the patient to fall asleep. Assure the patient that proper treatment will correct symptoms.

Patient Education. Teach the patient about the disease and symptoms of hyperthyroidism or hypothyroidism to report. Also teach the patient how to take antithyroid medication if ordered.

Evaluation. If the plan of care is effective, the patient will remain free from complications or injury related to the hypermetabolic state. Eyes will be comfortable and free from injury. Body temperature will be kept within normal limits. Diarrhea will be controlled, and complications of diarrhea such as skin breakdown and dehydration avoided. The patient's weight should remain stable. The patient should report that he or she is rested on awakening and that anxiety is controlled.

CARE OF THE PATIENT RECEIVING RADIOACTIVE IODINE. If radioactive iodine is used, it is usually given orally in one dose. If the dose is high, such as for the patient with thyroid cancer, the patient is hospitalized. Patients receiving lower doses may be treated as outpatients. You should limit time spent with the patient and maintain a safe distance when providing direct care. The nurse who is pregnant should avoid caring for patients receiving radioactive iodine. Urine, vomitus, and other body secretions are contaminated and should be disposed of according to hospital policy. Many hospitals direct nurses to flush the toilet twice following disposal of contaminated material. The radiation safety officer and hospital policy should be consulted for specific precautions.

At home, the patient is instructed to avoid close contact with family members for about a week and to use careful handwashing after urinating. Oral contact with others should be avoided, and eating utensils should be washed thoroughly with soap and water. Hospital teaching protocols should be used for specific patient teaching. If the treatment is being administered for hyperthyroidism, inform the patient that symptoms will subside in about 6 to 8 weeks. In addition, the patient should be aware of symptoms of hypothyroidism to be reported, because hypothyroidism may occur up to 15 years after the treatment.

Goiter

PATHOPHYSIOLOGY AND ETIOLOGY. Enlargement of the thyroid gland is called a goiter. The thyroid gland may enlarge in response to increased TSH levels. TSH is elevated in response to low TH; iodine deficiency; pregnancy; or viral, genetic, or other conditions. When a goiter is caused by iodine deficiency or other environmental factors, it is called an endemic goiter.

A goiter may be associated with a hyperthyroid, hypothyroid, or euthyroid state. Once the cause of the goiter is removed, the gland usually returns to normal size.

Some foods and medications are **goitrogens.** These substances interfere with the body's use of iodine and include

---

euthyroid: eu—normal, healthy + thyroid

goitrogenic: goitro—goiter + genic—producing

BOX 37-4   *Nutrition Notes*

Relating Nutrition to Thyroid Disorders

Iodine Deficiency

Inland areas of all continents have iodine-poor soils. People in developing countries with only local food sources are still subject to endemic goiter; however, fortification of salt with iodine has significantly reduced the occurrence of endemic goiter in developed countries.

Most goitrogens compete with iodide for active transport into thyroid cells but are not thought to cause endemic goiter. At particular risk of iodine deficiency are strict vegetarians who consume sea salt, which contains virtually no iodine, rather than iodized salt. In a study of a group of North American vegans, 12 percent developed hypothyroidism.

Iodine Excess

Saltwater fish, shellfish, and seaweed are naturally high in iodine. Iodine toxicity has occurred in Japan. Excessive iodine can be manifested as either hyperthyroidism or hypothyroidism, sometimes producing "iodine goiter."

Remer, T, Neubert, A, and Manz, F: Increased risk of iodine deficiency with vegetarian nutrition. Br J Nutr 81:45, 1999.

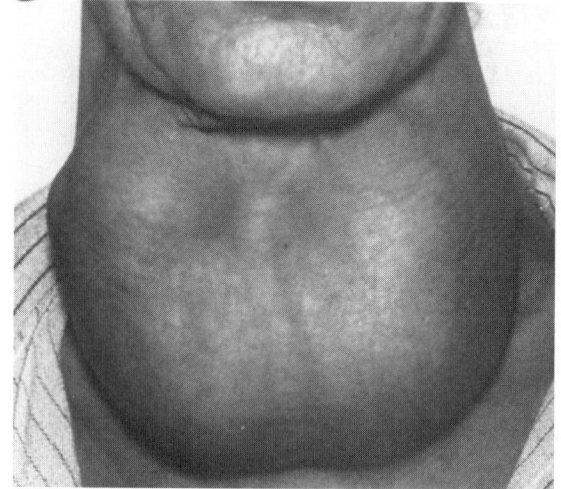

Figure 37-5   Patient with goiter.

BOX 37-5   **CULTURAL CONSIDERATION**

Because of the Chernobyl nuclear disaster in Russia in 1988, Russian immigrants are at exceptionally high risk for developing pituitary, thyroid, and parathyroid disorders and cancers. The proximity of Estonia, Latvia, Lithuania, Poland, and other Eastern European countries to Russia places immigrants and long-term visitors from these countries at risk also. The nurse needs to be alert for endocrine disorders among these populations and assist patients to arrange genetic counseling for those who desire it.

such foods as turnips, cabbage, broccoli, horseradish, cauliflower, and carrots (Nutrition Notes Box 37–4). Some goitrogenic medications include propylthiouracil, sulfonamides, lithium, and salicylates.

SIGNS AND SYMPTOMS. The thyroid gland is enlarged, with swelling apparent at the base of the neck (Fig. 37–5). The patient may have a full sensation in the neck. If the goiter is large, it may interfere with swallowing or breathing. Symptoms of hypothyroidism or hyperthyroidism may be present.

DIAGNOSTIC TESTS AND MEDICAL TREATMENT. A thyroid scan shows an enlarged thyroid gland. Serum T_3 and T_4 levels are measured to determine thyroid function. Treatment is aimed at the cause. If goitrogens are suspected, the patient is given a list of foods to be avoided. If iodine deficiency is a problem, it is added to the diet with supplements or iodized salt. A thyroidectomy may be necessary if the gland is interfering with breathing or swallowing.

NURSING CARE. Be careful to assess the effect of the goiter on breathing and swallowing. Stridor, a whistling sound, may be heard if the airway is obstructed. Stridor is an ominous sign and should be reported to the physician immediately. If the patient experiences difficulty swallowing, notify the physician and collaborate with the dietitian to provide soft foods or liquid nutrition. A swallowing study might be ordered, which can assist a speech pathologist or other expert to make specific recommendations for safe swallowing.

Cancer of the Thyroid Gland

Although thyroid cancer is rare, it is the most common cancer of the endocrine system (Cultural Consideration Box 37–5). Women are affected more often than men. Most tumors of the thyroid gland are not malignant.

ETIOLOGY. Thyroid hyperplasia may lead to thyroid cancer. Other causes include exposure to radiation, iodine deficiency, and prolonged exposure to goitrogens.

SIGNS AND SYMPTOMS. A hard, painless nodule may be palpable on the thyroid gland. Difficulty breathing or swallowing or changes in the voice may occur if the tumor is near the esophagus and trachea. Most patients with cancer of the thyroid have normal TH levels.

DIAGNOSTIC TESTS. A thyroid scan shows a "cold" nodule. This is because malignant tumors of the thyroid do not take up the radioactive iodine administered for the scan. A fine-needle aspiration biopsy confirms the diagnosis. A "hot" nodule indicates a benign tumor.

MEDICAL TREATMENT. A partial or total thyroidectomy may be done. Chemotherapy or radioactive iodine therapy may also be used, alone or following surgery.

NURSING CARE. Nursing care is determined by the symptoms the patient is experiencing. See Chapter 10 for care of the patient with cancer.

NURSING CARE OF THE PATIENT UNDERGOING THYROIDECTOMY. Patients may undergo thyroidectomy for cancer of the thyroid, hyperthyroidism, or a goiter that is causing dyspnea or dysphagia. See Chapter 11 for general care of a patient having surgery.

A total thyroidectomy is usually performed if cancer is present. After a total thyroidectomy, lifelong replacement hormone must be taken. A subtotal (partial) thyroidectomy may be done for hyperthyroidism, leaving a portion of the thyroid gland to secrete TH.

Before undergoing a thyroidectomy, the patient should be in a euthyroid state. This is accomplished with the use of antithyroid medication. Saturated solution of potassium iodide (SSKI) may also be administered to decrease the size and vascularity of the gland, reducing the risk of bleeding during surgery.

PREOPERATIVE CARE

Assessment. Assess the patient for symptoms related to an enlarged thyroid, including difficulty swallowing or breathing. Monitor TH levels. Assess nutritional status and drug history, and monitor vital signs. Assess the patient's understanding of the procedure and level of anxiety.

Nursing Diagnosis. Possible diagnoses include anxiety, deficient knowledge related to lack of exposure to information about surgery, and risk for injury.

Planning and Implementation. Reassure the patient that symptoms will be relieved after surgery. Allow the patient time to express fears and ask questions. Approach the patient slowly and calmly, because the hyperthyroid state increases the patient's feelings of anxiety. Report changes in vital signs, ensure that the patient's nutritional needs have been addressed, and verify that antithyroid drugs have been administered as ordered.

Patient Education. Explain what the patient can expect before, during, and after surgery and clarify misconceptions. (See Chapter 11.) Before surgery, teach the patient how to perform gentle range-of-motion exercises of the neck, how to support the neck during position changes, and how to use an incentive spirometer after surgery.

Evaluation. The patient will state that anxiety is controlled and that questions related to surgery have been answered. Risk for injury will be minimized by verifying that the patient is in a euthyroid state before surgery.

POSTOPERATIVE CARE

Assessment. Monitor vital signs and dressings every 15 minutes initially, progressing to every 4 hours, as ordered. Decreased blood pressure with increased pulse should alert you to the possibility of shock related to blood loss. Tachycardia and fever, along with mental status changes, may indicate thyrotoxic crisis. Check the back of the neck for pooling of blood. Because of the location of the surgery, observe for signs of respiratory distress, including an increase in respiratory rate, dyspnea, or stridor. Ask the patient to speak to detect hoarseness of the voice, which may indicate trauma to the recurrent laryngeal nerve. Monitor the patient for evidence of **tetany** (discussed later in this chapter). Report any abnormal findings to the physician immediately.

Nursing Diagnosis. Possible diagnoses include risk for injury (complications), pain related to surgical procedure, risk for imbalanced nutrition, and ineffective airway clearance due to edema at surgical site.

Planning and Implementation. Notify the physician of changes in vital signs, respiratory distress, or excessive bleeding from the surgical site. Keep a tracheostomy set at the bedside for emergency use if edema causes respiratory obstruction. Semi-Fowler's position helps reduce edema and promotes comfort. Implement routine interventions to prevent postoperative pain. Pillows or sandbags may be used to support the head. The patient should be assisted with gentle range-of-motion exercises, avoiding hyperextension of the neck, which can cause strain on the incision line. Remind the patient to do coughing and deep breathing exercises every hour. Encourage the patient to use the incentive spirometer to assist with deep breathing. When the patient's swallowing and gag reflexes are intact, clear liquids are ordered. A dietitian may be consulted to assist the patient with potential dietary changes needed following surgery. With correction of metabolic alterations, dietary needs may be significantly altered.

Patient Education. Teach the patient the importance of follow-up care, and how to administer replacement hormone if indicated. Also teach the patient or significant other how to change the dressing and to report bleeding or signs of infection at the site. Because the threat of thyrotoxic crisis may continue after discharge, the patient or significant other should be taught to immediately report unusual irritability, fever, or palpitations.

Evaluation. If the plan has been effective, complications caused by surgery will not occur or will be recognized and reported early. Pain will be prevented or controlled, and the patient will demonstrate understanding of dietary modifications and postoperative self-care.

COMPLICATIONS

Thyrotoxic Crisis. Thyrotoxic crisis may result from manipulation of the thyroid gland during surgery, with the subsequent release of large amounts of thyroid hormone. This is a rare complication because the use of antithyroid drugs before surgery has become routine. For more information on thyrotoxic crisis, see the section on hyperthyroidism earlier in this chapter.

Tetany. Tetany is caused by low calcium levels and is characterized by tingling in the fingers and perioral area (around the mouth), muscle spasms, twitching, and cardiac dysrhythmias. Muscle spasms in the larynx can lead to respiratory obstruction. Watch carefully for symptoms of tetany and report them immediately if they occur, because if the problem is not recognized quickly, death can result.

Tetany can occur if the parathyroid glands are accidentally removed during thyroid surgery. Because of the proximity of the parathyroid glands to the thyroid, it is sometimes difficult for the surgeon to avoid them. In the absence of parathyroid hormone, serum calcium levels drop and tetany results.

Intravenous calcium gluconate is given to treat acute tetany. To provide temporary relief while medications are being prepared, have the patient breathe into a paper bag. This causes mild acidosis, which increases ionization of calcium in the blood. A respiratory therapist can assist with this procedure.

▶ DISORDERS OF THE PARATHYROID GLANDS

Recall that the parathyroid glands secrete parathyroid hormone (PTH) in response to low serum calcium levels. PTH raises serum calcium levels by promoting calcium movement from bones to blood and by increasing absorption of dietary calcium. Decreased PTH activity is called hypoparathyroidism. Increased PTH activity is called hyperparathyroidism.

Hypoparathyroidism

PATHOPHYSIOLOGY. A decrease in PTH causes a decrease in bone resorption of calcium. This means that calcium stays in bones instead of being moved into the blood. The result is a decreased serum calcium level, called hypocalcemia. As calcium levels fall, phosphate levels rise.

ETIOLOGY. The most common causes of hypoparathyroidism are heredity and the accidental removal of the parathyroid glands during thyroidectomy. Because of the proximity of the glands to the thyroid, it is sometimes difficult to avoid removing them. Hypoparathyroidism also occurs following purposeful removal of the parathyroid glands for hyperparathyroidism. Another cause is hypomagnesemia, which impairs secretion of PTH. Hypomagnesemia can occur with chronic alcoholism or certain nutrition problems.

SIGNS AND SYMPTOMS. Calcium plays an important role in nerve cell stability. Hypocalcemia causes neuromuscular irritability. In acute cases, tetany may occur (see previous section on tetany), with numbness and tingling of the fingers and perioral area, muscle spasms, and twitching. (See Table 37–5.) Positive Chvostek's and Trousseau's signs are early indications of tetany. (See Figs. 5–4 and 5–5 in Chapter 5.) To check Chvostek's sign, tap on the patient's facial nerve just in front of the ear. Spasm of the face is a positive result, indicating hypocalcemia. To elicit Trousseau's sign, place a sphygmomanometer on the patient's arm and pump it to above the patient's systolic pressure. Spasm of the thumb and fingers occurs within 3 minutes if the patient has hypocalcemia.

L E A R N I N G   TIPS

To remember which test is which, remember *CH*vostek's sign is done near the *CH*eek.

In chronic hypoparathyroidism, the patient is lethargic and experiences muscle spasms. Calcifications may occur in the eyes and brain, leading to psychosis. Bone changes are evident on x-ray examination. Convulsions may occur. Death can result from laryngospasm if treatment is not provided.

DIAGNOSTIC TESTS. Chvostek's and Trousseau's signs are present. Laboratory studies show decreased serum calcium and PTH levels and increased serum phosphate. Magnesium levels may be low. Radiographs show bone changes.

MEDICAL TREATMENT. Acute cases of hypoparathyroidism are treated with intravenous calcium gluconate. Long-term treatment includes a high-calcium diet (Box 37–6), with oral calcium and vitamin D supplements. Thiazide diuretics may also be used because they reduce the amount of calcium excreted in the urine. Magnesium is given if hypomagnesemia is present.

NURSING PROCESS

Assessment. The patient at risk for hypoparathyroidism should be closely monitored for symptoms of tetany. If you suspect tetany, check for Chvostek's and Trousseau's signs. Monitor respirations closely for stridor, a sign of laryngospasm.

Nursing Diagnosis. A priority nursing diagnosis is risk for injury related to tetany. In chronic hypoparathyroidism, the patient should be assessed for knowledge deficit related to self-care of hypoparathyroidism.

Planning and Implementation. If the patient exhibits signs of tetany, you should recognize a potential emergency and notify the registered nurse and physician immediately. A tracheostomy set, endotracheal tube, and intravenous calcium are kept at the bedside for emergency use. Ensure that the patient with chronic hypoparathyroidism has knowledge of home medications and dietary recommendations. Consult the dietitian for diet instruction.

BOX 37–6	Dietary Sources of Calcium

Milk
Cheeses
Yogurt
Sardines
Oysters
Salmon
Cauliflower
Green leafy vegetables

TABLE 37-5	COMPARISON OF PARATHYROID DISORDERS	
	Insufficient PTH	**Excess PTH**
Disorder	Hypoparathyroidism	Hyperparathyroidism
Signs and symptoms	Hypocalcemia, neuromuscular irritability, tetany	Hypercalcemia, fatigue, pathological fractures
Usual treatment	Calcium replacement; high-calcium, low-phosphorus diet	Calcitonin, parathyroidectomy
Priority nursing diagnoses	Risk for injury related to tetany	Risk for injury related to bone demineralization

Evaluation. Injury is prevented through early recognition and reporting of signs and symptoms of tetany. The patient should be able to describe correct treatment for this disease.

Hyperparathyroidism

PATHOPHYSIOLOGY. Overactivity of one or more of the parathyroid glands causes an increase in PTH, with a subsequent increase in the serum calcium level (hypercalcemia). This is achieved through movement of calcium out of the bones and into the blood, absorption in the small intestine, and reabsorption by the kidneys. PTH also promotes phosphate excretion by the kidneys.

ETIOLOGY. Hyperparathyroidism is usually the result of hyperplasia or a benign tumor of the parathyroid glands, or it may be hereditary. Some cancers can also make a substance that mimics PTH and causes hypercalcemia. Secondary hyperparathyroidism occurs when the parathyroids secrete excessive PTH in response to low serum calcium levels. Serum calcium may be reduced in kidney disease because of the kidneys' failure to activate vitamin D, which is necessary for absorption of calcium in the small intestine.

SIGNS AND SYMPTOMS. Signs and symptoms of hyperparathyroidism are caused primarily by the increase in serum calcium level, although many patients are asymptomatic. Symptoms include fatigue, depression, confusion, increased urination, anorexia, nausea, vomiting, kidney stones, and cardiac dysrhythmias. The increased serum calcium level also causes gastrin secretion, resulting in abdominal pain and peptic ulcers. Because calcium is being removed from bones, joint pain and pathological fractures may occur. With severe hypercalcemia, the result may be coma and cardiac arrest.

DIAGNOSTIC TESTS. Laboratory studies include serum calcium, phosphate, and PTH levels. Radiographs may show decreased bone density.

MEDICAL TREATMENT. The patient should be monitored for bone changes and decline in renal function. Hydration with intravenous normal saline lowers the calcium level by dilution. Furosemide is given to increase renal excretion of calcium. Pamidronate (Aredia) or calcitonin may be given to prevent calcium release from bones. Mithramycin may be used to lower serum calcium levels, although its toxicity limits use to two or three doses. If hypercalcemia is severe or if the patient is at risk for bone or kidney complications, surgery to remove the diseased parathyroid glands is performed. If possible, some parathyroid tissue is left intact to continue to secrete PTH.

NURSING PROCESS

Assessment. Assess the patient for symptoms related to hypercalcemia, including muscle weakness, lethargy, bone pain, anorexia, nausea, vomiting, behavioral changes, and renal insufficiency. Monitor serum calcium levels as ordered.

Nursing Diagnosis. Nursing diagnoses depend on assessment findings. Risk for injury (fracture) related to bone demineralization, imbalanced nutrition related to nausea and vomiting, activity intolerance related to fatigue, and disturbed thought processes related to hypercalcemia are some possible diagnoses.

Planning and Implementation. The patient should be protected from injury. Treat nausea so nutritional status can be maintained while the underlying problem is being treated. Allow for adequate rest periods. Report changes in mental status immediately. If surgery is planned, preoperative and postoperative care is similar to that of the patient undergoing thyroid surgery.

Evaluation. If the plan is effective, symptoms of hypercalcemia will be recognized and reported quickly, nausea and vomiting will be controlled, and complications and injury will be prevented.

▶ DISORDERS OF THE ADRENAL GLANDS

Adrenal disorders may involve the adrenal medulla or the adrenal cortex. Hypersecretion of epinephrine from the adrenal medulla is associated with a rare tumor called a **pheochromocytoma.** Hyposecretion of epinephrine is rare and generally causes no symptoms. Hypersecretion of cortisol from the adrenal cortex results in Cushing's syndrome. Hypofunction of the adrenal cortex results in Addison's disease.

Pheochromocytoma

PATHOPHYSIOLOGY. A pheochromocytoma is an uncommon tumor that arises from the chromaffin cells of the adrenal medulla. Occasionally a pheochromocytoma occurs outside the adrenal gland. The tumor autonomously secretes catecholamines (norepinephrine and sometimes epinephrine) in excessive amounts. Ninety percent of pheochromocytomas are benign.

ETIOLOGY. The cause of pheochromocytoma is unknown. About 5 percent of cases are hereditary.

SIGNS AND SYMPTOMS. Because norepinephrine is the fight-or-flight hormone, patients with a pheochromocytoma have exaggerated fight-or-flight symptoms. These might be fairly constant, or occur in "attacks." Manifestations include hypertension, tachycardia (with heart rate greater than 100), palpitations, diaphoresis, feeling of apprehension, elevated blood glucose, and severe pounding headache. Nausea and vomiting are occasionally present. Blood glucose may increase because catecholamines inhibit insulin release from the pancreas. The most prominent characteristic, however, is intermittent unstable hypertension. Diastolic pressure may be greater than 115 mm Hg. If hypertension is uncontrolled, the patient is at risk for stroke, vision changes, and organ damage. It is estimated that about 0.1 percent of cases of hypertension are caused by a pheochromocytoma.

---

pheochromocytoma: pheo—dark + chromo—color + cyt—cell + oma—tumor

DIAGNOSTIC TESTS. Patients with a suspected pheochromocytoma have a 24-hour urine test for metanephrines and vanillylmandelic acid (VMA). These are end products of catecholamine metabolism. The patient should avoid caffeine and medications for 2 days before and during the test. Check institution policy for other dietary restrictions. If results are elevated, a CT scan or MRI is done to locate the tumor.

MEDICAL TREATMENT. Treatment for pheochromocytoma is surgical removal of one or both adrenal glands. However, the patient must be stabilized before surgery. Alpha-blocking medications such as phentolamine (Regitine) or phenoxybenzamine (Dibenzyline) dilate blood vessels to control acute hypertension. Beta-blocking medication may be added to block beta-adrenergic receptors in the heart and lungs, reducing other fight-or-flight symptoms.

NURSING CARE. Patients with pheochromocytoma are at risk for very high blood pressure and related complications. Monitor vital signs closely and report increases in blood pressure or pulse to the registered nurse or physician.

Because of the unstable hypertension and fight-or-flight state, patients are often quite anxious. Stress may precipitate a hypertensive episode. Approach the patient calmly and maintain a quiet environment. Teach the patient how the medications will reduce symptoms, and the importance of avoiding foods and beverages containing caffeine.

Adrenocortical Insufficiency

Adrenocortical insufficiency (AI) is the insufficient production of the hormones of the adrenal cortex. Primary AI is called Addison's disease.

PATHOPHYSIOLOGY. Adrenal insufficiency is associated with reduced levels of cortisol, aldosterone, or both hormones. A deficiency in androgens may exist but usually does not cause symptoms. In primary disease, ACTH levels may be elevated in an attempt to stimulate the adrenal cortex to synthesize more hormone. In secondary disease, deficient ACTH fails to stimulate adrenal steroid synthesis. In most cases the adrenal glands are atrophied, small, and misshapen and are unable to produce adequate amounts of hormone.

ETIOLOGY. Addison's disease is thought to be autoimmune; that is, the gland destroys itself in response to conditions such as tuberculosis, fungal infection, infection related to acquired immunodeficiency syndrome (AIDS), or metastatic cancer. It may also be associated with other autoimmune diseases, such as Hashimoto's thyroiditis. Adrenalectomy also results in adrenal insufficiency.

Secondary AI may be caused by dysfunction of the pituitary or hypothalamus. In addition, prolonged use of corticosteroid drugs may depress ACTH and corticotropin-releasing hormone production, which in turn reduces steroid hormone production. A patient receiving long-term corticosteroid therapy is particularly at risk for AI if the drugs are abruptly discontinued. Because the pituitary has been suppressed for a prolonged period, it may take up to a year before ACTH is produced normally again. Therefore corticosteroid therapy should be slowly tapered, never abruptly discontinued.

SIGNS AND SYMPTOMS. The most significant sign of Addison's disease is hypotension. This is related to the lack of aldosterone. Remember that aldosterone causes sodium and water retention in the kidney and potassium loss. If aldosterone is deficient, sodium and water are lost and hypotension and tachycardia result. Low cortisol levels cause hypoglycemia, weakness, fatigue, confusion, and psychosis. In primary AI, increased ACTH may produce hyperpigmentation of the skin, causing the patient to have a tanned or bronze appearance. Anorexia, nausea, and vomiting may also occur, possibly as the result of electrolyte imbalances.

COMPLICATIONS. If a patient is exposed to stress, such as infection, trauma, or psychological pressure, the body may be unable to respond normally with secretion of cortisol and an adrenal crisis can occur. Loss of large amounts of sodium and water and the resulting fluid volume deficit cause profound hypotension, dehydration, and tachycardia. Potassium retention can cause cardiac dysrhythmias. Hypoglycemia may be severe. Coma and death result if treatment is not initiated. Treatment of adrenal crisis involves rapidly restoring fluid volume and cortisol levels. Intravenous fluids, glucocorticoids, and mineralocorticoids are administered.

DIAGNOSTIC TESTS. Serum and urine steroids are measured. These include cortisol, aldosterone, and 17-ketosteroids. Blood glucose is low. Blood urea nitrogen (BUN) and hematocrit levels may appear to be elevated because of dehydration. An ACTH stimulation test may help determine whether the adrenal glands are functioning. Serum sodium and potassium levels are monitored.

MEDICAL TREATMENT. Treatment consists of replacement of glucocorticoids (hydrocortisone) and mineralocorticoids (fludrocortisone). Patients will need hormone replacement therapy for the rest of their lives. Hormones are given in divided doses, with two-thirds of the daily dose given in the morning and one-third in the evening to mimic the body's own diurnal rhythm. Remember that steroid hormones are our natural stress hormones and so are naturally elevated during times of stress. Therefore, during times of stress or illness, doses need to be increased to two to three times normal. The patient may also be placed on a high-sodium diet.

NURSING PROCESS

Assessment. The patient with Addison's disease should be assessed for understanding of and compliance with the treatment regimen. Daily weights or intake and output measurement help monitor fluid volume. Monitor serum glucose levels and symptoms of hyperkalemia and hyponatremia. Report changes in mental status. If the patient is in crisis, monitor vital signs closely and report any signs of fluid volume deficit such as orthostatic hypotension or poor skin turgor to the physician immediately.

Nursing Diagnosis. Priority diagnoses include risk for deficient fluid volume and deficient knowledge related to lack of exposure to information about self-care of Addison's disease.

Planning and Implementation. Nursing actions for the patient with adrenal insufficiency primarily involve assessment and education. Monitor fluid balance carefully, and notify the physician if symptoms of fluid volume deficit occur.

Patient Education. Teach the patient the importance of hormone replacement. Explain the need to increase medication dosage during times of stress or illness according to the physician's instructions. Help the patient identify the causes and symptoms of stress. Medical alert identification is recommended. If ordered by the physician, teach the patient and significant other how to use an emergency intramuscular hydrocortisone injection kit.

Evaluation. The patient and family should be able to describe proper self-care of Addison's disease. If the plan is successful, complications will be prevented or recognized and reported promptly.

Cushing's Syndrome

Cushing's *disease* is characterized by excess cortisol secretion resulting from secretion of too much adrenocorticotropic hormone (ACTH) by the pituitary. Cushing's *syndrome* refers to symptoms of cortisol excess caused by other factors. See Table 37–6 for a comparison of adrenal insufficiency and Cushing's syndrome.

PATHOPHYSIOLOGY. Recall that cortisol, aldosterone, and androgens are the three steroid hormones secreted by the adrenal cortex. Cortisol is essential for survival and is normally secreted in a diurnal rhythm, with levels increasing in the early morning. Secretion is increased during times of stress. In Cushing's syndrome, cortisol is hypersecreted without regard to stress or time of day. When levels of cortisol are very high, effects related to excess aldosterone and androgens are also seen.

L E A R N I N G   T I P S

Remember from Chapter 36, an easy way to remember the hormones of the adrenal cortex is to think *salt, sugar, and sex.* Aldosterone promotes salt retention, cortisol affects sugar (carbohydrate) metabolism, and androgens are sex hormones.

ETIOLOGY. Cushing's disease is caused by the hypersecretion of ACTH by the pituitary. This is most often the result of a pituitary adenoma. Sometimes ACTH is produced by a tumor in the lungs or other organs. The high levels of ACTH cause adrenal hyperplasia, which in turn increases production and release of cortisol.

The most common cause of Cushing's syndrome is prolonged use of glucocorticoid medication (e.g., prednisone) for chronic inflammatory disorders such as rheumatoid arthritis, chronic obstructive pulmonary disease, and Crohn's disease.

SIGNS AND SYMPTOMS. Most signs and symptoms of Cushing's syndrome are related to excess cortisol levels. Weight gain, truncal obesity with thin arms and legs, buffalo hump, and moon face result from deposits of adipose tissue at these sites (Fig. 37–6). Cortisol also causes insulin resistance and stimulates gluconeogenesis, which result in glucose intolerance. Some patients develop secondary diabetes mellitus. (See Chapter 38.) Muscle wasting and thin skin with purple striae occur as a result of cortisol's catabolic effect on tissues. Catabolic effects on bone lead to osteoporosis, pathological fractures, and back pain. Because cortisol has anti-inflammatory and immunosuppressive actions, the patient is at risk for infection. Hyperpigmentation of the skin may occur. Approximately 50 percent of patients experience mental status changes, from irritability to psychosis (sometimes referred to as steroid psychosis). Sodium and water retention are related to the mineralocorticoid effect. As sodium is retained, potassium is lost in the urine, causing hypokalemia. (See Chapter 5 to review these electrolyte imbalances.) Androgen effects include acne, growth of facial hair, and amenorrhea in women.

DIAGNOSTIC TESTS. Suspicion of Cushing's syndrome may initially be based on a cushingoid appearance. Plasma and urine cortisol and plasma ACTH are measured. A dexamethasone suppression test may be done.

MEDICAL TREATMENT. If a pituitary or other ACTH-secreting tumor is present, surgical removal or radiation therapy to the pituitary may be employed. If the adrenals are the primary cause of the problem, radiation or removal of the adrenal gland or glands may be performed. Drugs such as ketoconazole block production of adrenal steroids.

If the cause of Cushing's syndrome is administration of steroid medication, an every-other-day schedule or once-a-day dosing in the morning may reduce side effects. Usually steroids are prescribed as a last resort for chronic disorders that are unresponsive to other treatment. The patient and physician must weigh the risks and benefits of continuing

TABLE 37–6	**COMPARISON OF ADRENAL CORTEX HORMONE IMBALANCES**	
	Hypofunction	**Hyperfunction**
Disorder	Adrenocortical insufficiency, Addison's disease	Cushing's syndrome
Signs and symptoms	Sodium and water loss, hypotension, hypoglycemia, fatigue	Weight gain, sodium and water retention, hyperglycemia, buffalo hump, moon face
Usual treatment	Glucocorticoid and mineralocorticoid replacement	Alter steroid therapy schedule; surgery if tumor
Priority nursing diagnoses	Risk for fluid volume deficit	Risks for fluid volume excess, glucose intolerance, infection

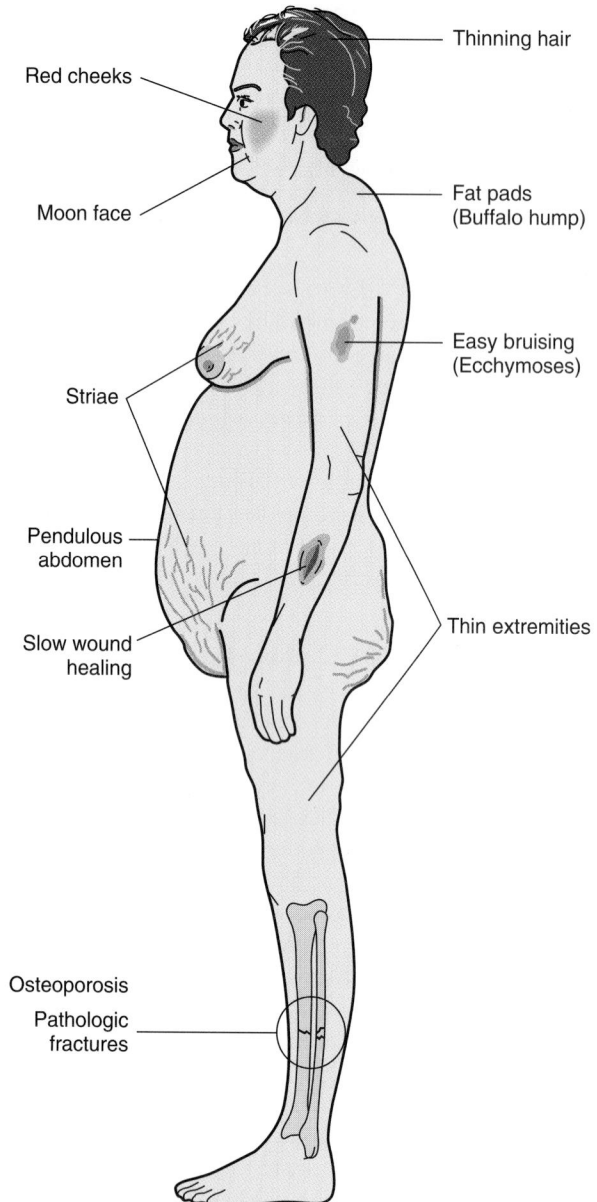

Thinning hair

Red cheeks

Moon face

Fat pads
(Buffalo hump)

Easy bruising
(Ecchymoses)

Striae

Pendulous
abdomen

Slow wound
healing

Thin extremities

Osteoporosis

Pathologic
fractures

Figure 37-6 Physical manifestations seen in Cushing's syndrome.

the medication. The physician may order a high-potassium, low-sodium, high-protein diet. Potassium supplements may be ordered. If the patient has high blood sugar, appropriate nutrition therapy for diabetes should be instituted. (See Chapter 38.)

NURSING PROCESS

Assessment. When caring for the patient with Cushing's disease or syndrome, assess the patient's drug history. Monitor vital signs and complications related to fluid and sodium excess. Auscultate the lungs for crackles, and assess extremities for edema. Assess skin integrity, and monitor capillary glucose as ordered by the physician. Watch for signs of infection.

Nursing Diagnosis. Possible diagnoses include excess fluid volume related to sodium and water retention, risk for impaired skin integrity, risk for infection, and risk for injury re-

lated to impaired glucose tolerance. Disturbed body image is common as a result of the cushingoid appearance.

Planning and Implementation. Report signs of fluid overload to the physician. Take care to protect the skin. Protect the patient from infection and instruct to avoid others who are ill. You and other health care workers should use good handwashing technique. If glucose intolerance occurs, be prepared to administer insulin, because oral hypoglycemics are not usually effective. Patients who are to receive long-term insulin therapy should be referred to diabetes education classes. Consult a dietitian for diet counseling. If surgery is planned, the patient is prepared for an adrenalectomy or hypophysectomy, depending on the cause of the disorder.

Evaluation. If care has been effective, complications of fluid overload will be recognized and treated early. The patient will have intact skin and be free from signs of infection. The patient will demonstrate skill in self-care of diabetes if indicated.

NURSING CARE OF THE PATIENT UNDERGOING ADRENALECTOMY

Preoperative Care. Monitor the patient for electrolyte imbalance and hyperglycemia. Abnormalities must be corrected before surgery. To prevent adrenal crisis, glucocorticoids are administered, because removal of the adrenals causes a sudden drop in adrenal hormones.

Postoperative Care. Following surgery, the patient receives routine postoperative care. In addition, the patient is closely monitored for changes in fluid and electrolyte balance and adrenal crisis. Patients who undergo bilateral adrenalectomy must take replacement glucocorticoid and mineralocorticoid hormones for the remainder of their life. If only one adrenal gland is removed, the remaining gland should eventually produce enough hormone to enable the patient to discontinue replacement hormone.

See Table 37-7 for a summary of endocrine disorders.

CRITICAL THINKING

Mrs. Tercini

Mrs. Tercini is a 62-year-old woman admitted to your unit in addisonian crisis. She is lethargic, with a blood pressure of 86/58, pulse 112, and respirations 18. While interviewing her daughter, you learn that Mrs. Tercini has a history of Cushing's syndrome treated with bilateral adrenalectomy 25 years ago. She has been taking 150 mcg fludrocortisone (Florinef) and 200 mg hydrocortisone daily ever since. Three days ago she developed the flu.

1. Why is an adrenalectomy done to treat Cushing's syndrome?
2. What is the most effective schedule for Mrs. Tercini's medication?
3. What precipitated this addisonian crisis?
4. Why is Mrs. Tercini's blood pressure low?
5. How could this crisis have been prevented?
6. Fludrocortisone is available as 0.1 mg tablets. How many should you administer?

Answers at end of chapter.

TABLE 37-7 SUMMARY OF ENDOCRINE PROBLEMS

Hormone	Hypofunction	Hyperfunction
Antidiuretic hormone	Diabetes insipidus—water loss	SIADH—water retention
Growth hormone	Dwarfism—short stature	Acromegaly, gigantism—bone and tissue overgrowth
Thyroid hormone	Hypothyroidism—slow metabolism	Hyperthyroidism—increased metabolism
Epinephrine	Rare	Pheochromocytoma—hypertension
Parathyroid hormone	Hypoparathyroidism—low serum calcium, osteoporosis, tetany	Hyperparathyroidism—high calcium, weakness
Cortisol	Addison's disease—sodium and water loss	Cushing's syndrome—sodium and water retention, hyperglycemia; see text

REVIEW QUESTIONS

1. When assessing a patient with diabetes insipidus, which of the following findings is expected?
    a. Edema
    b. Polyuria
    c. Heat intolerance
    d. Diarrhea

2. Which of the following expected outcomes is most appropriate for the patient with SIADH?
    a. Patient will verbalize relief from excessive thirst.
    b. Patient will verbalize understanding of importance of increasing fluid intake.
    c. Patient's daily weights will be stable.
    d. Patient will state pain is relieved.

3. Which action by the nurse is most important following hypophysectomy?
    a. Performing a routine neurological assessment
    b. Encouraging the patient to cough and deep breathe
    c. Monitoring for tracheal edema
    d. Maintaining strict intake and output

4. Which of the following statements by the patient with hypothyroidism indicates to the nurse that the plan of care has been effective?
    a. "I feel so much better now that my energy is returning."
    b. "I'm really glad the diarrhea has stopped."
    c. "I'm so glad I won't have to take medication for very long."
    d. "My fingers aren't tingling any more."

5. Which of the following nursing assessments is most important in the patient with hyperthyroidism and risk for thyrotoxic crisis?
    a. Intake and output q2h
    b. Patient's understanding of dietary restrictions
    c. Bowel sounds each shift
    d. Frequent vital signs

6. Which symptom in the patient with a goiter should the nurse report to the physician immediately?
    a. Weight gain of 2 pounds
    b. Stridor
    c. Excessive thirst
    d. Nausea

7. Following a total thyroidectomy, which of the following instructions from the physician will the nurse reinforce?
    a. "You will be taking thyroid replacement hormone for the rest of your life."
    b. "You must weigh yourself daily and report any gain or loss of more than 1 pound."
    c. "You will need to return to the physician's office for a weekly blood pressure check."
    d. "You will need to restrict your sodium and potassium intake."

8. Which of the following nursing interventions will be most helpful for the patient with acute hypertension related to pheochromocytoma?
    a. Offer the patient pain medication every 4 hours.
    b. Provide a calm, quiet environment.
    c. Assist the patient to elevate the legs.
    d. Encourage increased fluid intake.

Answers to CRITICAL THINKING

Mrs. Jackson

1. Her weight gain is most likely caused by fluid retention, which can be a result of heart failure or SIADH, among other things.
2. Assess edema, lung sounds, and vital signs. Check intake and output during the past 2 days. Monitor mental status and level of consciousness. Check recent lab work to see if her serum sodium is low. You also check a book on the unit and recall that anesthetics and morphine are possible causes of SIADH. Morphine can also cause confusion.
3. Notify the registered nurse of your findings and your suspicions. Be prepared to place Mrs. Jackson on a fluid restriction. Reassure her son that the physician is being notified of the changes he noted.

Adoption

The children's growth hormone secretion was probably suppressed because of psychosocial stress. Once they felt secure in a loving environment, growth hormone levels returned to normal.

Mae

1. Mae's TSH is elevated because her pituitary gland is working overtime to try to stimulate the underactive thyroid gland.
2. Mae's metabolism has been slow, so she has been burning fewer calories. When she starts on thyroid replacement hormone, her metabolic rate will return to normal and she will need more calories. Intake of calories should be balanced with the possible need for weight loss.
3. If Mae receives too much thyroid hormone, she will have symptoms of hyperthyroidism, including an increased pulse rate. She should know how to check her pulse and to call her physician if it is elevated.

Mrs. Tercini

1. Cushing's syndrome is caused by too much cortisol. The adrenal cortex is responsible for secreting cortisol.
2. Mrs. Tercini should take two-thirds of her daily dose of hydrocortisone and fludrocortisone in the morning and one-third in the evening. This most closely mimics the body's natural corticosteroid secretion.
3. The flu probably triggered this crisis. Illness is a stressor, and normally the body secretes steroids during stress. Because Mrs. Tercini's body is unable to produce steroids, she experiences symptoms of hypoadrenalism during stressful times.
4. Mrs. Tercini's blood pressure is low because she has insufficient circulating mineralocorticoids. Without aldosterone, sodium and water are lost and blood pressure drops.
5. Mrs. Tercini should have taken extra medication when she became ill.
6. You should give her 1.5 tablets; remember that 1 mg = 1000 mcg.

38

NURSING CARE OF PATIENTS WITH DISORDERS OF THE ENDOCRINE PANCREAS

Paula D. Hopper

KEY TERMS

diabetes mellitus
(DYE-ah-**BEE**-tis mel-**LYE**-tus)

endogenous (en-**DAH**-jen-us)

gastroparesis (GAS-troh-puh-**REE**-sus)

glycosuria (GLY-kos-**YOO**-ree-ah)

hyperglycemia
(HIGH-per-gligh-**SEE**-mee-ah)

hypoglycemia
(HIGH-poh-gligh-**SEE**-mee-ah)

ketoacidosis (KEE-toh-ass-i-**DOH**-sis)

Kussmaul's (**KOOS**-mahlz)

nephropathy (ne-**FROP**-uh-thee)

neuropathy (new-**RAH**-puh-thee)

polyphagia (PAH-lee-**FAY**-jee-ah)

retinopathy (RET-i-**NAH**-puh-thee)

QUESTIONS TO GUIDE YOUR READING

1. What is the pathophysiology of type 1 and type 2 diabetes mellitus?

2. What are the causes and risk factors for type 1 and type 2 diabetes mellitus?

3. What are the signs and symptoms of diabetes mellitus?

4. What are the causes, signs and symptoms, and treatment of high and low blood glucose levels?

5. Why are persons with diabetes prone to complications such as heart disease, blindness, and kidney failure? How can you help your patients prevent these complications?

6. What diagnostic tests are used to diagnose and monitor diabetes mellitus and its complications?

7. What types of diets are available to help patients with diabetes control their blood glucose levels?

8. How does exercise affect diabetes and blood glucose levels?

9. How do the different insulins and oral hypoglycemic agents lower blood glucose levels? What should you know when administering these medications?

10. How would you apply the nursing process to the patient with diabetes mellitus?

11. Why can surgery be especially risky for a patient with diabetes? What measures can be taken to increase the safety of the patient undergoing surgery?

12. What is reactive hypoglycemia? How is it diagnosed and treated?

▶ DIABETES MELLITUS

Diabetes mellitus (be careful not to confuse it with diabetes insipidus) is a group of metabolic diseases in which defects in insulin secretion or action result in high blood sugar (**hyperglycemia**). Approximately 16 million people in the United States have diabetes mellitus, and 5 million of those

---

diabetes: diabetes—passing through
hyperglycemia: hyper—excessive + glyc—glucose + emia— in the blood

don't know it. The direct and indirect cost (such as lost work time) of diabetes in the United States is about $98 billion per year.[1] The incidence of diabetes mellitus varies by race and ethnicity. In the United States, Native-Americans have the highest rate of diabetes of all ethnic and racial groups. African-Americans, people of Mexican heritage, Puerto Ricans, and Japanese-Americans have a higher rate of diabetes than do white ethnic groups.

Diabetes is a serious disease that can cause complications such as blindness, kidney failure, heart attacks, and strokes.

It is a leading cause of lower limb amputations in the United States. With good education and self-care, patients with diabetes can prevent or delay these complications and lead full and productive lives. A major role of the nurse is helping the patient learn to care for himself or herself effectively.

Pathophysiology

Body tissues, and the cells that compose them, use glucose for energy. Glucose is a simple sugar provided by the foods we eat. When carbohydrates are eaten they are broken down into glucose, which is then absorbed into the bloodstream. Carbohydrates provide most of the glucose used by the body; proteins and fats provide smaller amounts of glucose. Glucose is able to enter the cells only with the help of insulin, a hormone produced by the beta cells in the islets of Langerhans of the pancreas (Fig. 38–1). When insulin comes in contact with the cell membrane, it combines with a receptor that allows activation of special glucose transporters in the membrane.

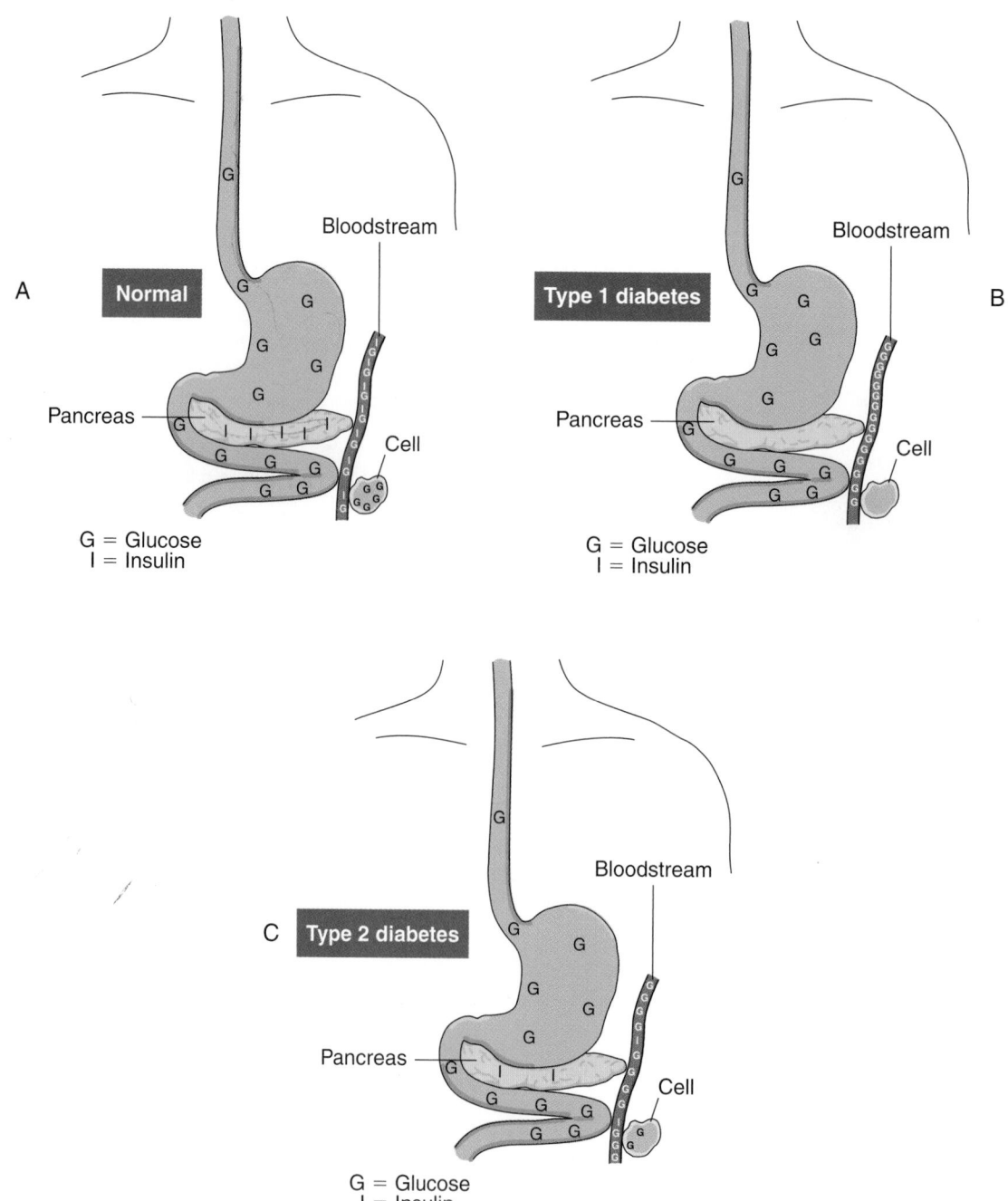

Figure 38-1   Maintenance of blood glucose levels. *(A)* Normal physiology: Foods (especially carbohydrates) are broken down into glucose, which is absorbed into the bloodstream for transport to the cells. Insulin, produced by the beta cells of the islets of Langerhans in the pancreas, is needed to "open the door" to the cells, allowing the glucose to enter. *(B)* In type 1 diabetes mellitus the pancreas does not produce insulin. Because glucose is unable to enter the cells, it builds up in the bloodstream, causing hyperglycemia. *(C)* In type 2 diabetes mellitus, insulin production is reduced. Less glucose enters the cell, and hyperglycemia results. In some cases, enough insulin is produced but cells are resistant to it.

By helping glucose enter the body's cells, insulin lowers the glucose level in the blood. Insulin also helps the body to store excess glucose in the liver in the form of glycogen. Another hormone, glucagon, is produced by the alpha cells in the islets of Langerhans. Glucagon raises the blood glucose when needed by releasing the stored glucose from the liver and muscles. Insulin and glucagon work together to keep the blood glucose at a constant level.

Diabetes results from faulty production of insulin by the beta cells in the pancreas or from inability of the body's cells to use insulin. When glucose is unable to enter body cells, it stays in the bloodstream; hyperglycemia results, and the cells are denied their energy source.

Types and Causes

Type 1 Diabetes Mellitus

Type 1 diabetes (formerly called juvenile diabetes mellitus, insulin-dependent diabetes mellitus, or IDDM) is caused by destruction of the beta cells of the pancreas. When the beta cells are destroyed, the pancreas produces no insulin at all. Insulin must be injected for the body to use food for energy. See Patient Perspectives Box 38–1 for Dave's story about having type 1 diabetes.

It is believed that the pancreas may attack itself following a viral infection (this is called an autoimmune response). Almost 85 percent of patients newly diagnosed with type 1 diabetes have islet cell antibodies in their blood. Other times, the cause of type 1 diabetes is unknown, or idiopathic. From 10 to 30 percent of type 1 diabetes cases are hereditary. The patient with type 1 diabetes is usually young and thin and is prone to develop **ketoacidosis** when blood glucose is elevated. (See Table 38–1 for a comparison of

---

ketoacidosis: keto—ketones + acid—acidic + osis—condition

TABLE 38-1 COMPARISON OF TYPE 1 AND TYPE 2 DIABETES

	Type 1 (IDDM)	Type 2 (NIDDM)
Onset	Rapid	Slow
Age at onset	Usually < 40	Usually > 40
Risk factors	Virus, autoimmune response, heredity	Heredity, obesity
Body type	Lean	Obese
High blood glucose complication	Ketoacidosis	Hyperosmolar Hyperglycemic Nonketotic syndrome
Treatment	Diet, exercise; must have insulin to survive	Diet, exercise; may need oral hypoglycemics or insulin to control blood glucose level

---

BOX 38-1 *Patient Perspective*

Dave

I was diagnosed with type 1 diabetes mellitus at age 3. I remember being left in a children's ward of the hospital with the nuns (who were nurses) in their habits and looking into the parking lot as my parents got in the car and drove away. I remember the fear and horror of being left alone. Later, I remember my doctor teaching my mother and me about the diet and monitoring my urine for glucose and ketones. This was all we had in those days (the early 1960s). I was supposed to test my urine before meals and at bedtime, just like we monitor blood glucose today. Every time I saw my doctor he would review and change my treatment based on the results. Insulin then was extracted from pig and cow cadavers.

I remember my life being pretty normal except at holidays. When my brothers and sisters were getting Halloween candy and Christmas candy canes I felt odd and left out. My mom was really tough and observant and I had to be sneaky to get away with stealing a treat or two left momentarily unobserved by those around me. I was thrilled when sugarless candy became available around age 10 or so and I could have my own candies to hoard for myself.

Hypoglycemic reactions were always a trauma-filled event at our house. They came at unexpected times and my mom sometimes blamed herself. I can honestly say that my mom deserves more credit than I have ever given her for my good fortune with my diabetes. She did not have glucometers, glycohemoglobin tests, or even diabetes specialists and educators; she just had the desire of a mother who did everything she could to make sure I had and did everything right as far as she knew.

In my late teens I took everything my doctor and my mom had done for me and trashed them. I lived with reckless abandon: I did what I wanted, ate what I wanted, and didn't even think about the disease I lived with until I was about 30. I worked hard those years in construction, and the physical activity probably delayed the complications that might otherwise have occurred.

At age 35 I began to experience the inevitable effects of long-term hyperglycemia. I developed diabetic retinopathy, the leading cause of adult-onset blindness in the United States. My vision was restored in my left eye with laser surgery. It would take another surgery (vitrectomy) to bring the vision in my right eye back to the 20/60 it is now from the 20/2000 it was before the surgery.

About that time I was referred to an endocrinologist who specialized in diabetes, and my treatment took a remarkable turn for the better. The endocrinologist began to aggressively change my medications, insulin regimen, exercise program, and diagnostic testing, all with the purpose of controlling blood glucose and preventing the onset and advancement of diabetes complications. This aggressive plan of action was taught to me by a registered nurse certified diabetes educator who worked with my doctor. I must also give enormous credit to the office staff nurses who were an integral part of the team, supporting me throughout the whole experience. The doctors and nurses made my education both understandable and pertinent. They basically put the ball in my court, and then it was up to me to take control of my own life. My wife was more than supportive throughout this time period too, and sacrificed much for the new array of medications and doctor's appointments that took a huge chunk out of our family finances.

I have mentioned briefly those involved in my life with diabetes. I take no credit for any of the results though; I believe it was God who made these choices available and miraculously reversed the effects of disease in my body. To the amazement of medical professionals, He continues to bless me with good health. It is because of all I have gone through with diabetes that I have recently completed my nursing degree and hope to become a diabetes educator myself.

type 1 and type 2 diabetes.) Diabetic ketoacidosis is discussed later in this chapter. Exciting research is currently being done on a new drug that may block the development of type 1 diabetes when administered early in the disease process.

Type 2 Diabetes Mellitus

In type 2 diabetes mellitus (formerly called adult-onset diabetes mellitus, non–insulin-dependent diabetes mellitus, or NIDDM), some insulin is still made by the pancreas, but in inadequate amounts. Sometimes the amount of insulin is normal or even high, but the tissues are resistant to it and hyperglycemia results.

Heredity is responsible for up to 90 percent of cases of type 2 diabetes. Obesity is also a major contributing factor. Many experts are concerned about the growing number of obese children in the United States because of their increased risk for diabetes, heart disease, and other obesity-related problems. Often the patient with a new diagnosis of type 2 diabetes is obese, relates a family history of diabetes, and has had a recent life stressor such as the death of a family member, illness, or loss of a job.

Gestational Diabetes

Gestational diabetes mellitus (GDM) may develop during pregnancy, especially in women with the risk factors for type 2 diabetes. The extra metabolic demands of pregnancy trigger the onset of diabetes. Blood glucose usually returns to normal after delivery, but the mother has an increased risk for type 2 diabetes in the future. If she is overweight, she should be counseled that weight loss and exercise will decrease her risk of later developing diabetes. Mothers with GDM require specialized care and should be referred to an expert in this area.

Prediabetes

Prediabetes (*impaired glucose tolerance* [IGT]) refers to blood glucose levels that are above normal but do not meet the criteria for diagnosing diabetes. An IGT diagnosis is made after an oral glucose tolerance test shows a glucose level between 140 and 200 mg/dL after 2 hours. A diagnosis of impaired fasting glucose (IFG) is made when the fasting glucose level is between 110 and 126 mg/dL.

SYNDROME X. A newer finding is the link between diabetes and a condition called metabolic syndrome, or syndrome X. This is a group of disorders commonly found together, including insulin resistance, glucose intolerance, low HDL cholesterol, high triglycerides, hypertension, abdominal obesity, a tendency to form clots, and high levels of c-reactive protein, a risk factor for heart attack. Any patient who fits this profile should be monitored closely for the onset of diabetes and should be encouraged to adopt a healthy lifestyle.

Other Causes of Diabetes

Secondary diabetes may develop as a result of another chronic illness such as pancreatitis or cystic fibrosis. Prolonged use of some drugs, such as steroids, phenytoin (Dilantin), thiazide diuretics, and thyroid hormone, may also impair insulin action and raise blood glucose. Less common causes include pancreatic trauma, certain genetic defects, and other endocrine disorders. One genetic defect is called maturity-onset diabetes of the young (MODY), which is an inherited impairment of insulin secretion that usually occurs before the age of 25.

Signs and Symptoms

Classic symptoms of diabetes mellitus include **polydipsia** (excessive thirst), **polyuria** (excessive urination), and **polyphagia** (excessive hunger). Because glucose is unable to enter the cells, the cells starve, causing hunger. The large amount of glucose in the blood causes an increase in serum concentration, or osmolality. The renal tubules are unable to reabsorb all the excess glucose that is filtered by the glomeruli, and **glycosuria** results. Large amounts of body water are required to excrete this glucose, causing polyuria, **nocturia**, and dehydration. The increased osmolality and dehydration cause polydipsia. High blood glucose may also cause fatigue, blurred vision, abdominal pain, and headaches. Ketones may build up in the blood and urine of patients with type 1 diabetes (ketoacidosis).

L E A R N I N G   TIPS

Remember the classic symptoms of diabetes by the 3 Ps: *polydipsia, polyuria,* and *polyphagia.*

Diagnostic Tests

FASTING BLOOD GLUCOSE. Diagnosis of diabetes mellitus is based on blood glucose levels measured by a laboratory. A normal blood glucose level ranges from 60 to 110 mg/dL, although different laboratories may have slightly different normal values. When the fasting blood glucose (drawn after at least 4 hours without eating) is above 126 mg/dL on two separate occasions, diabetes is diagnosed.

ORAL GLUCOSE TOLERANCE TEST. Another test to diagnose diabetes is the oral glucose tolerance test (OGTT). An OGTT measures blood glucose at intervals after the patient drinks a concentrated carbohydrate drink. Diabetes is diagnosed when the blood glucose level is 200 mg/dL or higher after 2 hours.

GLYCOHEMOGLOBIN. The glycohemoglobin test (also called glycosylated hemoglobin, or HbA$_{1c}$) is used to gather baseline data and to monitor progress of diabetes control. Glucose in the blood attaches to hemoglobin in the red

---

polydipsia: poly—many or much + dipsia—thirst
polyuria: poly—many or much + uria—urine
polyphagia: poly—many or much + phagia—to eat
glycosuria: glyc—glucose + uria—urine
nocturia: noc—by night + uria—urine

blood cells. Red blood cells live about 3 months in the body. When the glucose that is attached to the hemoglobin is measured, it reflects the average blood glucose level for the previous 3 months. Results should be approximately 4 to 7 percent, depending on the laboratory. This is a helpful measurement when blood glucose levels fluctuate and a single measurement would be misleading. It also assists in determining the degree of effectiveness of a patient's treatment plan. Newer methods allow this test to be done in a physician's office while the patient waits. Glycohemoglobin testing might be inaccurate in some people, such as those with anemia. These individuals may have a glycated serum protein test, which is a similar test that indicates glucose levels over a period of 1 to 2 weeks instead of 3 months.

ADDITIONAL TESTS. Because diabetes affects so many body systems, additional tests recommended to gather baseline data include a lipid profile, serum creatinine and urine microalbumin levels to monitor kidney function, urinalysis, and electrocardiogram.

CRITICAL THINKING

Mr. McMillan

Mr. McMillan is a 50-year-old patient brought into the emergency department with extreme fatigue and dehydration. After the physician sees him, you ask Mr. McMillan some additional questions. Based on the patient's answers, you request that the physician add a glucose level to the laboratory tests ordered. The result is 1400 mg/dL.

1. What questions would you ask Mr. McMillan if you suspected diabetes?
2. Why was Mr. McMillan fatigued?
3. Why was he dehydrated?

Answers at end of chapter.

Treatment

The only cure for diabetes is a pancreas transplant. However, diabetes can be controlled. Treatment begins with diet and exercise. Insulin is added in patients with type 1 diabetes and insulin or oral hypoglycemia medication as needed in those with type 2 diabetes. Blood glucose monitoring and education are also important to good diabetes control.

Type 2 diabetes can be prevented or delayed in some cases. In patients with a family history of diabetes, a healthy diet, exercise, and moderate weight loss are advised, along with regular blood glucose screening.

To monitor the effectiveness of treatment, patients should have regular health care follow up visits. See Figure 38–2 for American Diabetes Association recommendations for follow-up visits.

Goals of Treatment

The American Diabetes Association recommends that patients maintain a fasting plasma glucose level of less than 120 mg/dL and glycohemoglobin level of less than 7 percent

to prevent or delay complications. These goals may be adjusted in individual circumstances. For example, the patient who does not feel symptoms of **hypoglycemia** might have a higher goal of preventing undetected hypoglycemic episodes.

Medical Nutrition Therapy

The goal of medical nutrition therapy (MNT) is to achieve and maintain blood glucose and lipid levels as near to normal as possible to prevent long-term complications. For some, especially those with type 2 diabetes, weight loss and blood pressure control may be additional goals.

Because the patient with diabetes has a limited amount of insulin, either **endogenous** (from within the body) or injected, it is important to eat an amount of food that will not exceed the insulin's ability to carry it into the cells. The meal plan should include consistent amounts of carbohydrates, proteins, and fats each day (Nutrition Notes Box 38–2). Because carbohydrates contribute most to the blood glucose level, it is most important that carbohydrates are consistent from one day to the next. If a patient eats a small amount of carbohydrate one day and a large amount the next, the blood glucose fluctuates, leading to complications. It is possible to relax nutrition restrictions somewhat if the patient is willing to test blood glucose frequently at home and adjust treatment accordingly. This requires in-depth instruction by a diabetes educator.

In the past the commonly prescribed meal plan was the American Diabetes Association (ADA) diet, which was based on a basic plan with lists of amounts of foods that could be exchanged to vary the menu. The ADA no longer recommends a single meal plan, but rather advocates a complete assessment by a specially trained dietitian and individualized nutrition therapy recommendations and teaching.

A variety of meal plans are now available, as shown in Table 38–2. Because diabetes increases the risk of high serum cholesterol and triglycerides, all plans limit fat intake. Patients who use fat replacers (in foods such as fat-free baked goods or ice cream) should be aware that they still have food value and calories, so they cannot be considered "free" foods. Sodium intake is limited in individuals with hypertension. Most plans also encourage the use of complex carbohydrates such as grains, pastas, vegetables, and fruits. Simple sugars, which may cause the blood glucose to rise more rapidly, are used less but are not prohibited as they were in the past. Any meal plan should be chosen to fit the patient's lifestyle and food preferences. Preferences of patients based on their ethnic background should also be considered (Cultural Consideration Box 38–3).

The success of MNT is evaluated by monitoring glucose levels, HbA$_{1c}$, lipids, weight, blood pressure, and kidney function.

---

hypoglycemia: hypo—deficient + glyc—glucose + emia—in the blood

endogenous: endo—within + genous—to produce

The American Diabetes Association

suggests these goals for most people with diabetes. You may have different goals. You can record your goals and your results in the spaces provided below.

Every Diabetes Visit	ADA Goals	My Goals	My Results Date	Date
Review blood sugars before meals 80–120 at bedtime 100–140				
Check blood pressure below 130/80				
Review meal plan				
Review activity level				
Check weight				
Discuss questions or concerns				
At Least Every 6 Months				
Hemoglobin A$_{1c}$ below 7				
At Least Every Year				
Physical exam				
Complete foot exam				
Dilated eye exam				
Microalbumin below 30				
Cholesterol Total below 200 HDL above 45 LDL below 100				
Triglycerides below 200				
Flu shot				
Tips—Ask Your Diabetes Care Team About				
Diabetes education				
Getting a pneumonia shot				
Stopping smoking				
Taking aspirin				
Unusual symptoms				
New therapies				

I Have Diabetes

I may be having a low blood sugar reaction to insulin or a diabetes pill.

If I cannot be awakened or cannot swallow, do not try to give me anything to drink. Call 911.

If I'm awake but acting strangely, give me some regular soft drink, juice, milk, hard candy, or some sugar. If I do not get better within 15 minutes, call 911 or get me to a hospital.

My name _____

Emergency contact _____

Diabetes care provider_____

   Telephone_____

   _____

Diabetes medicines _____

_____

Allergies _____

Your diabetes care team and you work together for good diabetes care. You decide what and how much to eat, when to take your medicine, when to check your blood sugar, and how to be active. Remember, you're the leader of your diabetes care team. Review the diabetes care list on the back of this card. Ask to see your "numbers."

Call **1 800 Diabetes (342-2383)** for diabetes information.

Go to **www.diabetes.org** for online information.

Printing of this card compliments of

Order Code: 5984-01   7/2000

©2000 by American Diabetes Association

Figure 38-2 Guidelines for diabetes follow-up visits. To obtain an American Diabetes Association card in English or Spanish, call 1-800-DIABETES.

BOX 38-2  *Nutrition Notes*

Individualizing Medical Nutrition Therapy in Diabetes Mellitus

A certified diabetes educator can use many approaches in the nutritional management of diabetes. This permits a meal plan based on the patient's abilities and commitment. Several meal-planning approaches are described in Table 38–2. Overall goals and strategies differ for the two types of diabetes. In general, a patient with type 1 diabetes needs to prevent wide swings in blood glucose levels through careful timing of meals and snacks in relation to insulin therapy and activity. A patient with type 2 diabetes uses diet modifications to maintain near-normal glucose, blood pressure, and lipid levels and to lose weight as necessary.

Using ADA Exchange Lists

A food guide called the *ADA Exchange Lists*, published jointly by the American Dietetic Association and the American Diabetes Association, is used by some patients with diabetes and also by patients who are attempting to achieve a healthier body weight. This system is composed of six lists of foods plus a "free food" list. Foods in each list contain similar energy nutrients (carbohydrate, protein, and fat). For example, corn is on the starch list rather than the vegetable list because it is closer in composition to a slice of bread than to green beans. Individual food items within an exchange list are essentially equal to each other in nutrient composition and can thus be exchanged or "swapped" for each other. Exchanges were designed to be approximately equal in nutrients, not in volume; therefore, portion sizes may vary. For instance, one fruit exchange is equal to $1^1/_4$ cup of whole strawberries but to only three dates. To correctly use this method of meal planning, patients must choose the prescribed number of items from each appropriate list: starch, fruit, milk, vegetable, meat, and fat.

A specific meal plan should be given to the patient with the exchange lists. A meal plan is a food guide that shows the number of choices or exchanges the patient should eat at each meal and snack. A meal plan based on exchange lists allows patients a variety of food choices yet requires minimal calculation. Even experienced users of the exchange list system should be advised to weigh or measure their portions several times per week to avoid portion inflation.

Using Carbohydrate Counting

Carbohydrate counting is a three-level approach to teaching meal planning that focuses on only carbohydrates, but patients are told to eat about the same amount of protein each day and to choose low-fat foods. With this system the patient keeps track of the units or exchanges of carbohydrate eaten throughout the day to keep carbohydrates and insulin balanced. Carbohydrate counting is based on classifying carbohydrates together, whether from starch, fruit, or milk. For example, one slice of whole wheat bread, one orange, and 8 ounces of skim milk are each equivalent to one carbohydrate unit or exchange. This method offers more flexible food choices within a day's meal plan than the exchange list system and may achieve better control of blood glucose. However, it may entail weighing and measuring food, keeping food records, monitoring blood sugar before and after eating, and controlling body weight. Mastery of level III of carbohydrate counting permits adjusting short-acting insulin dosage using insulin-to-carbohydrate ratios and is often a prerequisite for receiving insulin pump therapy.

Using the Glycemic Index

All carbohydrates are not metabolized the same. Foods containing equal amounts of carbohydrate impact blood glucose levels differently. The glycemic index is a classification of foods according to the speed and degree of change in blood glucose levels. Compared with a standard of 100 for white bread, glucose has a glycemic index of 138, sucrose (table sugar) 83, and fructose 26. Baked russet potatoes and corn flakes have higher glycemic indices than sucrose, whereas sweet potato has a lower one. Because most meals contain a combination of nutrients, the glycemic index is rarely used in clinical practice except with highly motivated and educated patients.

TABLE 38-2 MEAL-PLANNING APPROACHES FOR DIABETES

Approach	Comments	Availability
Food Pyramid	Initial phase of teaching Provides a basic foundation in normal nutrition. Does not emphasize meal consistency.	A colorful version is available from the National Dairy Council, 10255 West Higgins Road, Suite 900, Rosemont, IL 60018-4233 1-708-803-2000
Dietary Guidelines	Initial phase of teaching Provides a basic foundation in normal nutrition, 40 pages in length. Does not emphasize meal consistency.	United States Department of Agriculture Home & Garden Bulletin #232, Local Cooperative Extension Office
The First Step in Diabetes Meal Planning	Initial phase in teaching Combines Food Pyramid and Dietary Guidelines, and provides information on meal consistency in a simplified format.	The American Dietetic Association, 216 West Jackson Boulevard, Suite 800, Chicago, IL 60606-6995 1-800-366-1655
CHO Counting	Progressive teaching tool that leads to maximum control of blood glucose and lipid levels. Decreased emphasis on balance and variety.	The American Dietetic Association, 216 West Jackson Boulevard, Suite 800, Chicago, IL 60606-6995 1-800-366-1655
Month-O-Meals	Each book contains 28 complete and interchangeable menus for breakfast, lunch, dinner, and snacks. Excellent approach for the client who "just wants to be told what and when to eat."	The American Dietetic Association, 216 West Jackson Boulevard, Suite 800, Chicago, IL 60606-6995 1-800-366-1655
Exchange Lists of the American Dietetic and the American Diabetes Associations	Allows the health-care educator to distribute all of the energy nutrients. More emphasis on the importance of eating a balanced diet than the CHO counting approach. Time consuming to learn and teach	The American Dietetic Association, 216 West Jackson Boulevard, Suite 800, Chicago, IL 60606-6995 1-800-366-1655

From Lutz, CA and Przytulski, KR, Nutrition and Diet Therapy, 3rd ed., FA Davis, 2001.

The diabetic diet for ethnic individuals may need a significant adjustment from the U.S. menu. An exchange list of foods for these patients will not be followed because their food choices are different. The patient may be labeled noncompliant, when in reality the health care worker has been culturally insensitive.

The nurse can consult the American Diabetes Association in Washington, D.C. (1-800-342-2383) to obtain meal plans for ethnic individuals such as Asians, Hispanics, African-Americans, and Native-Americans.

Exercise

Exercise is an important factor in controlling blood glucose and lipid levels. Exercise lowers blood glucose, both immediately and for approximately 24 hours after the exercise. Insulin is not needed for glucose to enter exercising muscle cells. Exercise also improves blood lipid levels and circulation, which is important for the person with diabetes, who already has an increased risk of cardiovascular disease. Patients are instructed to exercise on a regular basis, ideally 30 minutes on most days of the week, to keep blood glucose levels stable and promote health.

Some patients with complications of diabetes must be careful in their exercise choices. For example, a patient with **retinopathy** should not do anything that causes straining. (See Long-Term Complications later in this chapter.) A patient with **neuropathy** or foot problems should limit weight-bearing exercise. A physician or exercise physiologist should be consulted for an individualized exercise plan.

Persons with diabetes should always carry a quick source of sugar when exercising in case the blood glucose drops too low. Individuals on intermediate-acting insulin are taught to avoid exercising at the time of day when their blood glucose is at its lowest point (i.e., when insulin or medication is peaking) and to have a carbohydrate snack before exercising if blood glucose is less than 100 mg/dL. Exercising at similar times each day also helps prevent blood glucose fluctuations.

Patients should be cautioned to avoid exercise when their glucose level is higher than 250 mg/dL and ketones are present in the urine or if glucose is more than 300 mg/dL without ketones. This indicates that insufficient insulin is available and glycogen may be released during exercise, further increasing the serum glucose.

Medication

INSULIN. The individual with type 1 diabetes has no endogenous insulin and therefore must inject insulin daily. At this time, insulin cannot be taken by mouth, because it is a protein and is therefore digested. Researchers are trying to find ways that insulin can be given without injections—either by inhaling it intranasally or by taking it in an oral form that is not digested. For now, it is generally given subcutaneously, although regular insulin may be ordered via the intramuscular or intravenous route in urgent situations. There are several types of insulin and schedules by which it may be given. The type and schedule are determined by the physician, in collaboration with the patient, based on his or her lifestyle and willingness to spend time on injections. In general the more frequent the injections, the better the glucose control.

Insulin injections should be given in a different subcutaneous site each time to avoid injury to the tissues. A sample rotation chart is shown in Figure 38–3. Because each area absorbs insulin at a slightly different rate, it is advisable to use one area for a week, then move on to the next area. Within that area, each injection should be spaced at least 1 inch from the previous injection. Most experts recommend using primarily the torso (abdomen and buttocks) to provide more uniform absorption. Aspirating for blood before injection and rubbing the site following injection are not recommended with insulin injections.

A newer development in diabetes care is the insulin pump (Fig. 38–4). This is a small device that delivers subcutaneous insulin continuously in small (basal) amounts. The patient can then add a bolus of insulin before meals or snacks. This provides insulin administration more closely matched to a person without diabetes. It allows for tighter control of blood glucose levels and a more flexible lifestyle for the patient.

Most insulin is now synthetically produced in a laboratory and is identical to human insulin. Some patients may still use insulin from beef or pork sources. It is important that the physician include the insulin source in the prescription. Be careful to check the source when preparing insulin for injection, because insulins from different sources may act slightly differently. Some individuals may be aller-

retinopathy: retino—nervous tissue of the eye + pathy—illness
neuropathy: neuro—nervous system + pathy—illness

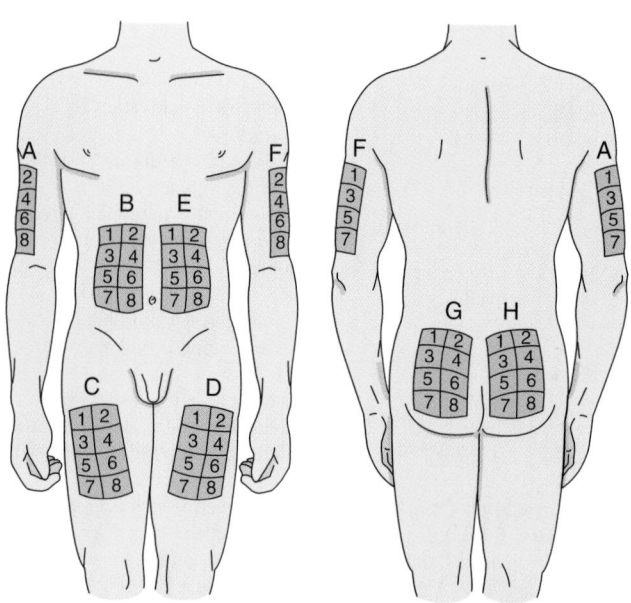

Rotation sites for injection of insulin.
Figure 38-3 Sample insulin rotation chart.

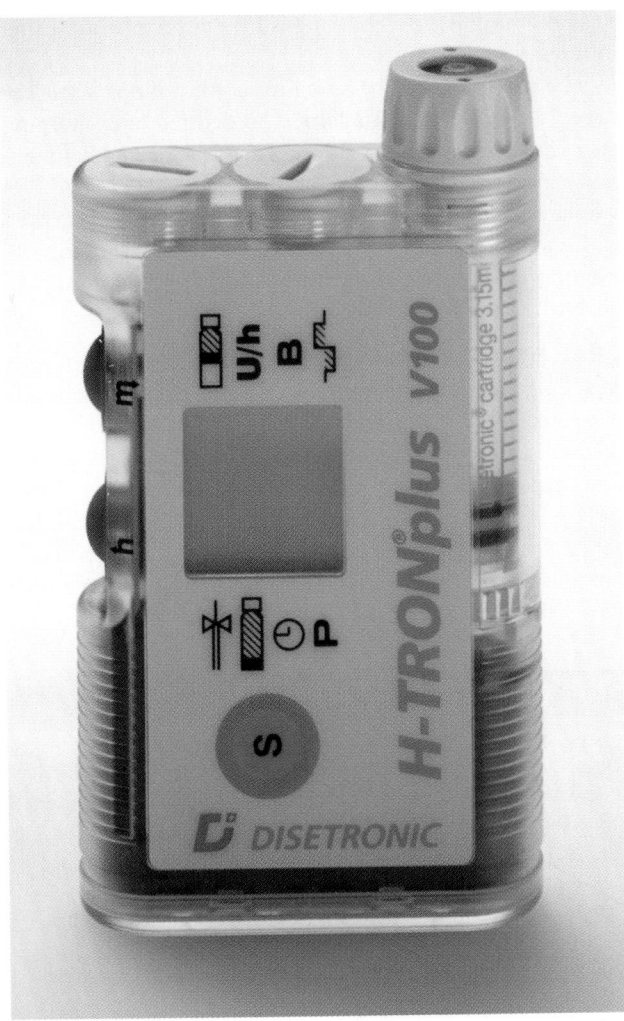

Figure 38-4   Insulin pump. (Courtesy Disetronic Medical Systems, Inc.)

gic to beef or pork preparations or may refuse them based on cultural practices (Cultural Consideration Box 38–4).

Once insulin is injected, a period elapses before it begins to lower blood glucose. This time is called the onset of action. The peak action time occurs when the insulin is working at its hardest and the blood glucose is at its lowest point. It is during this peak time that the patient is most at risk for an episode of low blood glucose. Duration is the length of time the insulin works before it is used up. Onset, peak, and duration are determined by whether the insulin is short-, intermediate-, or long-acting (Table 38–3). A newer type of

BOX 38-4   **CULTURAL CONSIDERATION**

Muslims and Jewish people do not eat pork. Therefore they should not be prescribed pork insulin.

The cow is sacred in India. Therefore the Hindu patient should not be prescribed beef insulin.

insulin, insulin glargine (Lantus AE), is now available and provides a basal insulin dose. A person who does not have diabetes is always secreting a little insulin, called a basal rate—about 1 unit per hour. When Lantus is injected at bedtime, it provides a very slow basal absorption rate for 24 hours, without a peak. The patient then injects short-acting insulin before meals during the day. This allows for more normal insulin and blood glucose levels in the body.

It is important for the individual with diabetes and the nurse to be aware of the onset, peak, and duration of any insulin given. This assists in making decisions such as when to give insulin, when to exercise, and when to be alert to low blood glucose symptoms.

LEARNING TIPS

The Evens-and-Odds Rule: To remember the onset, peak, and duration of intermediate-acting insulin, think *evens*—2, 12, and 24 hours. To remember short-acting insulin, think *odds*—1, 3, and 5 hours. These times are not exact, but they are a great memory booster when you need to think fast.

In the past, most patients with diabetes used an injection of an intermediate-acting insulin before breakfast and possibly a second injection before supper. Many patients are now choosing to take more frequent injections of short-acting insulin before meals or a combination of short- and intermediate- or long-acting insulins to achieve better, "tighter" control. These patients are often taught to adjust their insulin dose based on blood glucose level and the amount of carbohydrates eaten.

When two insulins need to be given at the same time, they can often be mixed together to prevent having to give more than one injection. See Box 38–5 for how to mix insulins in one syringe. Preset mixtures of intermediate- and short-acting insulins are available for patients who have difficulty learning to mix insulins.

TABLE 38-3 ONSET, PEAK, AND DURATION OF INSULINS					
Insulin Type	**Example**	**Brand Name**	**Onset**	**Peak**	**Duration**
Very short acting	lispro	Humalog	15 min	30–90 min	≤5 h
	aspart	NovoLog	Rapid	1–3 h	3–5 h
Short acting	regular	Humulin R, Novolin R	½–1 h	2–5 h	5–8 h
Intermediate acting	NPH	Humulin N, Novolin N	1–3 h	6–12 h	18–26 h
	Lente	Humulin L, Novolin L			
Long acting	Ultra Lente	Humulin U Ultralente	4–6 h	14–24 h	26–36 h
	Insulin glargine	Lantus AE	Unknown	No peak	24 h
Premixed insulins	70% NPH and 30% regular	Humulin 70/30	30 min	4–8 h	Up to 24 h
	or				
	50% NPH and 50% regular	Humulin 50/50			

LEARNING TIPS

When mixing insulin, remember *clear to cloudy*. Always draw up the clear insulin first. This involves injecting air into the cloudy vial first. This is because if the clear is drawn up last, the vial may be contaminated by the cloudy insulin, altering the action of the clear insulin. If cloudy insulin is unknowingly contaminated by clear insulin, the clear will become cloudy and its effect will be diminished.

BOX 38-5   **How to Mix Insulin**

1. Assemble equipment: insulins, syringe (be sure it is large enough to hold the entire insulin dose), alcohol swab, physician's order.
2. Check physician's order to confirm correct insulin types and doses of regular (clear) and an intermediate-acting (cloudy) insulin.
3. Roll the bottle of cloudy insulin to mix. Do not shake, because this will cause bubbles.
4. Wipe tops of both vials with alcohol swab.
5. Draw up and inject an amount of air equal to the dose of intermediate-acting insulin into the cloudy vial. Remove syringe from vial.
6. Draw up and inject an amount of air equal to the dose of intermediate-acting insulin into the cloudy vial. Remove syringe from vial.
7. Draw up the correct amount of clear insulin. Double-check amount with another nurse if this is the institution's policy.
8. Remove the syringe and insert into the cloudy vial. Carefully draw up the correct amount of insulin. If too much insulin is accidentally drawn into the syringe, the syringe must be discarded and the process repeated. Double-check again with another nurse.

NOTE: Some insulins cannot be mixed. Check a drug reference before mixing.

During times of stress or illness, some patients are placed on "sliding-scale" insulin. This involves determining each dose of short-acting insulin based on blood glucose results, usually before meals and at bedtime. For example, an order might read, "For blood glucose < 200, no insulin; 201–250, 2 units regular insulin; 251–300, 4 units regular insulin; 301–350, 6 units regular insulin; >350 call physician."

Problems with Insulin Therapy. Two problems that can occur with glucose control are the Somogyi effect and the dawn phenomenon. The Somogyi effect may be at fault when the patient's blood glucose seems to be rising in spite of increasing insulin doses. If insulin levels are too high, the blood glucose may drop too low, stimulating release of counterregulatory hormones (epinephrine, glucagon, corticosteroids, growth hormone) that then elevate the blood glucose. The low sugars often occur during the night, and the patient may report night sweats or morning headaches. The high morning glucose is then interpreted as hyperglycemia, and the insulin dose may be further increased, compounding the problem.

Dawn phenomenon is thought to occur because of the natural release of growth hormone and cortisol during the early morning hours. This causes hyperglycemia on arising in the morning.

The patient might be asked to monitor blood sugar between 2 and 4 A.M. in addition to bedtime and morning testing to assess whether Somogyi effect or dawn phenomenon is occurring. Correction of the Somogyi effect involves reducing the insulin dose. Dawn phenomenon is treated with careful adjustment of meals and insulin; moving the evening insulin dose to bedtime may cause the insulin to peak when the blood glucose is highest.

ORAL HYPOGLYCEMIC MEDICATION. The patient with type 2 diabetes may be able to control blood glucose levels with medical nutrition therapy and exercise alone. Oral hypoglycemic medication or insulin may also be prescribed. Oral hypoglycemics are not insulin pills. Remember that if insulin is ingested, it is digested, because it is a protein. Because most oral hypoglycemics depend on at least a partially functioning pancreas, most are not useful for patients with type 1 diabetes.

CRITICAL THINKING

Mrs. Evans

Mrs. Evans is a 68-year-old woman with type 2 diabetes who resides in an assisted living facility. She is on Lente insulin every morning.

1. If Mrs. Evans eats her breakfast at 8:00 every morning, when should she take her insulin?
2. At what time of day should she be alert for symptoms of low blood sugar?
3. What could happen if Mrs. Evans has a busy day and misses her lunch?

Answers at end of chapter.

Older oral hypoglycemic agents, called first-generation drugs, have many side effects. It is believed that they may increase the risk of death from cardiovascular disease. Second-generation drugs are more effective and have fewer side effects. Many new oral hypoglycemic agents have recently been developed (Table 38–4).

Most insulins and oral hypoglycemics should be administered before meals (check individual drugs for specific timing). Care should be taken to prevent passage of more than 30 minutes between medication administration and the meal, because this may result in a hypoglycemic episode.

If blood glucose levels are not controlled with oral hypoglycemic agents, insulin may be necessary for the person with type 2 diabetes. This does not mean the person has type 1 diabetes. Insulin may be necessary to control blood glucose, but it is not necessary to sustain life, as it is for the person with type 1 diabetes.

Self-Monitoring of Blood Glucose

The ability to test blood glucose levels at home has been a major advance in diabetes care. Blood glucose can be better controlled because of the availability of monitoring at any time, in any place. A variety of blood glucose monitors are

TABLE 38-4 ORAL HYPOGLYCEMIC AGENTS

Generic Name	Trade Names	Action	Usual Dosage	Nursing Considerations
Insulin Stimulators				
First-generation sulfonylureas				
Tolbutamide	Orinase	Stimulates insulin secretion from pancreas, increases insulin receptor sensitivity. May be given with other classes of oral agents.	500–2000 mg per day	May cause weight gain. Monitor for hypoglycemia. Teach patient to avoid alcohol. May result in cardiovascular events. Rarely used.
Chlorpropamide	Diabinese		250–500 mg per day	
Second-generation sulfonylureas				
Glipizide	Glucotrol	Stimulates insulin secretion from pancreas, increases insulin receptor sensitivity. May be given with other classes of oral agents.	2.5–40 mg per day	May cause weight gain. Monitor for hypoglycemia. Teach patient to avoid alcohol. Cheapest class. Give Glucotrol 30 minutes before meals.
Glimepiride	Amaryl		1–8 mg per day	
Glyburide	Micronase Diabeta Glynase Pres-Tabs		2.5–20 mg per day 2.5–20 mg per day 1.5–12 mg per day	
Meglitinides				
Repaglinide	Prandin	Stimulates insulin secretion from pancreas, increases insulin receptor sensitivity. May be given with other classes of oral agents.	0.5–4 mg with meals	Dosed with meals to improve postprandial hyperglycemia. Multiple dosing less convenient. Hold dose if patient skips meal.
Nateglinide	Starlix		60–240 mg with meals	
Insulin Sensitizers				
Biguanides				
Metformin	Glucophage Glucophage XR	Increases insulin-stimulated glucose transport in skeletal muscle, reduces production of glucose by liver. Reduces LDL cholesterol, triglycerides, and blood pressure. May be given with other classes of oral agents.	500 mg–2.5 g per day	Give with meals. May cause GI side effects; may enhance weight loss. Withhold if patient is having tests involving contrast dye. Contraindicated in renal and hepatic disease and CHF. Notify physician of early symptoms of lactic acidosis: hyperventilation, myalgia, malaise.
Combination agent				
Metformin/glyburide	Glucovance	Combines actions of metformin and glyburide	1.25/250–5/500 mg	See metformin and glyburide
Thiazolidinediones (glitazones)				
Pioglitazone	Actos	Reduces insulin resistance in muscles. Improves blood lipids; may lower blood pressure and improve cardiovascular risk. May be given with other classes of oral agents.	15–45 mg per day	Give with meal. Side effects include nausea, weight gain, and fluid retention. Works well with obese patients. Avoid with liver disease; monitor liver enzymes. May alter effectiveness of some birth control pills.
Rosiglitazone	Avandia		4–8 mg daily as single dose or divided	
Troglitazone	Rezulin		200–600 mg per day	Off market in 2000 because of deaths from liver failure.
Absorption Delayers				
Alpha-glucosidase inhibitors (AGIs)				
Acarbose	Precose	Lowers postprandial glucose by reducing rate of carbohydrate digestion and absorption. May be given with other classes of oral agents.	25–100 mg before each meal (tid)	Give at start of each meal. May cause diarrhea, flatulence, and abdominal pain. No weight gain or hypoglycemia risk. Multiple dosing less convenient. If used in combination with another drug and hypoglycemia occurs, treat with milk or glucose tablets, not table sugar.
Miglitol	Glyset		25–75 mg before each meal (tid)	

CHF = congestive heart failure; LDL = low-density lipoprotein.

on the market at a reasonable price (Fig. 38–5). Most of the cost involved in monitoring is in the test strips that must be used. Health insurance programs sometimes cover this cost.

Self-monitoring of blood glucose (SMBG) is generally done before meals and at bedtime by the individual on insulin who wishes to maintain tight control of blood glucose. Less frequent schedules may be prescribed for patients who are unable or unwilling to test four times a day or for patients not on insulin. Some may test before breakfast and supper, and some may vary testing times from day to day. Patients are also often recommended to periodically test 2 hours after meals. New noninvasive devices are being developed, such as a monitoring device that resembles a wrist watch. It is worn on the arm, and it checks glucose levels by extracting fluid from the arm every 20 minutes. An excellent test for long-term monitoring is the HbA_{1c}, described earlier. It is usually done every 3 to 4 months.

The physician should be consulted for desirable blood glucose ranges, because these may differ for each patient. For example, a patient whose type 2 diabetes is controlled by nutrition therapy and exercise alone may have a goal of 80 to 120 mg/dL. Patients who are prone to insulin reactions (hypoglycemia) or small children or the elderly may have higher goal ranges, such as 100 to 150 mg/dL. Lower blood glucose levels for these populations could increase the risk of hypoglycemia.

Urine Glucose and Ketone Monitoring

Urine may also be tested for glucose and for ketones. Urine glucose testing was done routinely before the development of SMBG. A variety of dipsticks and tape products are available for urine testing. If glucose appears in the urine, the patient is warned that the blood glucose is elevated, but the actual level is unknown. Most people have glucose in their urine when their blood glucose is more than 180 mg/dL. It is difficult to base treatment on urine glucose levels, and so routine urine testing for glucose is no longer recommended.

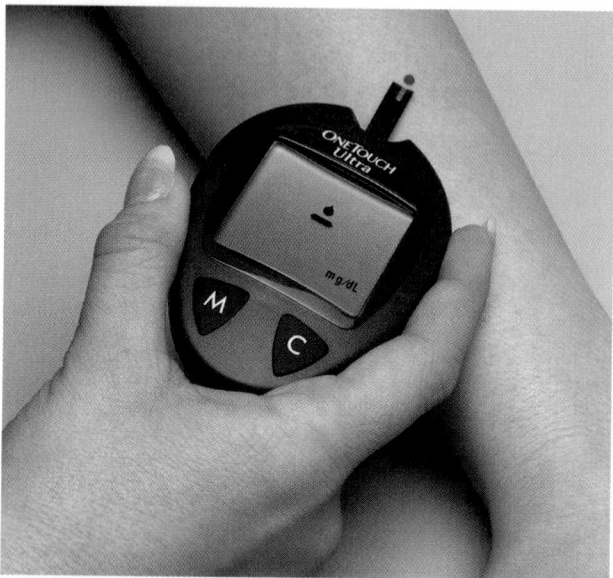

Figure 38–5   OneTouch Ultra glucose monitor. (Courtesy LifeScan.)

Urine should be tested for ketones during acute illness or stress, when blood glucose levels are consistently above 300 mg/dL, during pregnancy, or when symptoms of ketoacidosis are present. If ketones are present, the patient knows an insulin deficiency is present and should notify the physician. Patients with type 1 diabetes are most at risk for developing ketoacidosis; however, it is wise for the patient with type 2 diabetes to test for ketones if risk factors are present.

An important aspect of blood or urine monitoring is the interpretation of results. Monitoring is useless if the results are not used to improve blood glucose control. The patient should be instructed to keep a diary of blood glucose levels (Fig. 38–6). Some patients have computer software that graphs results. The patient may be taught by a diabetes educator to interpret the trends in the results, or the diary may be taken on a regular basis to the health care provider for interpretation and adjustment of the treatment plan.

See Box 38–6 for a summary of diabetes symptoms, diagnosis, and treatment.

Pancreas Transplant

If the patient is unable to control blood sugar levels with conventional treatment, a pancreas transplant may be considered. This is especially beneficial in the patient with kidney disease, who can receive both a kidney and pancreas transplant at the same time. An experimental but exciting development is the use of pancreatic islet cell transplants.

Acute Complications of Diabetes

The individual with diabetes is at risk for a variety of complications. Acute complications related to high and low blood glucose levels are treatable and can often be prevented with appropriate care.

BOX 38-6   **Diabetes Summary**

Symptoms

Polyuria
Polydipsia
Polyphagia
Fatigue
Blurred vision
Headache
Abdominal pain

Diagnostic Tests

Fasting blood glucose
HbA_{1c} (glycosylated hemoglobin)
Oral glucose tolerance test

Management

Diet
Exercise
Insulin
Oral hypoglycemic medication
Self-monitoring of blood glucose levels
Education

Day		Break-fast	Lunch	Supper	Bedtime	Urine Ketones	Notes
Sunday	Time	7:00	11:30	6:00	11:00		
	Glucose	186	108	116	142		
	Insulin	38N,6R		16N,2R			
Monday	Time	7:30	12:00	6:00	10:45	6:00-neg	Ate cake at Betty's party at 3 PM-oops!
	Glucose	171	97	302	180		
	Insulin	38N,6R		16N,2R			
Tuesday	Time						
	Glucose						
	Insulin						
Wednesday	Time						
	Glucose						
	Insulin						
Thursday	Time						
	Glucose						
	Insulin						
Friday	Time						
	Glucose						
	Insulin						
Saturday	Time						
	Glucose						
	Insulin						

Figure 38-6 Sample diary of blood glucose results and insulin use.

Hyperglycemia

When calories eaten exceed insulin available or glucose used, high blood glucose (hyperglycemia) occurs. The most common cause of hyperglycemia is eating more than the meal plan prescribes. Another major cause is stress. Stress causes the release of counter-regulatory hormones, including epinephrine, cortisol, growth hormone, and glucagon. These hormones all increase the blood glucose level. In a person without diabetes, this is an adaptive function. However, the patient with diabetes is unable to compensate for the increased blood glucose with increased insulin secretion, and hyperglycemia occurs.

Patients must be able to recognize signs and symptoms of high blood glucose levels and know what to do if they occur (Table 38–5). For many patients, these are similar to the symptoms they experienced when they were first diagnosed with diabetes. Chronic high blood glucose levels may lead to long-term complications (discussed later in this chapter).

Hypoglycemia

Low blood glucose, or hypoglycemia, occurs when there are not enough calories available in relation to circulating insulin. This is sometimes referred to as an insulin reaction. Hypoglycemia is usually defined as a blood glucose level below 50 mg/dL, although some patients have symptoms at higher glucose levels. Occasionally symptoms occur as a result of a rapid drop in blood glucose, even though the actual glucose level is normal or high. Causes of hypoglycemia may include skipping a meal, exercising more than usual, or ac-

cidentally administering too much insulin. An occasional hypoglycemic episode, treated promptly, should not lead to chronic complications. Repeated or extremely low blood glucose levels may cause neurological damage. It is therefore important to teach patients how to prevent and treat low blood sugar.

> ### CRITICAL THINKING
>
> **Jeff**
>
> Jeff is a 16-year-old who is having trouble with repeated insulin reactions. He says he has not had this trouble before, and it is interfering with his new job. What questions might you ask as you do your assessment to help him figure out how to prevent future reactions?
>
> *Answers at end of chapter.*

Symptoms of low blood glucose include hunger, sweating, pallor, tremor, palpitations, and headache. These symptoms are caused by activation of the sympathetic nervous system. As hypoglycemia progresses, the brain is deprived of glucose and neurological symptoms such as irritability, confusion, seizures, and coma may occur.

If you find a patient with symptoms of an altered blood glucose but are unable to identify whether it is high or low, do a blood glucose test. However, if the patient is exhibiting neurological symptoms, treat low blood glucose immediately. The blood glucose may then be checked and further treatment provided as indicated.

TABLE 38-5   COMPARISON OF HIGH AND LOW BLOOD GLUCOSE LEVELS

	Hyperglycemia	Hypoglycemia
Causes	Overeating Stress Illness Too little insulin or medication	Undereating, skipping a meal Too much insulin or medication Exercise
Symptoms	Polyuria Polydipsia Polyphagia Blurred vision Headache Lethargy Abdominal pain Ketonuria (if type I) Coma	Hunger Sweating Blurred vision Tremor Headache Irritability Confusion Seizures Coma
Treatment	Confirm hyperglycemia with glucose meter; if greater than 300 mg/dL, check urine for ketones and increase fluid intake. Assess cause of hyperglycemia, teach prevention. Return to prescribed treatment plan if applicable. Call physician for medication adjustment if indicated or if blood glucose is >200 mg/dL for 2 days. Call physician if patient is ill or vomiting.	Confirm hypoglycemia with glucose meter (if patient is not acutely ill). Administer 15 g fast-acting carbohydrate. Recheck glucose in 15 minutes; if still low, readminister carbohydrate. Continue checking glucose and administering fast sugar until hypoglycemia subsides; if symptoms worsen, call physician or emergency help. Glucagon subcutaneously or dextrose 50% IV may be administered if ordered. Assess cause of hypoglycemia, teach prevention.

To treat low blood glucose, administer a "fast sugar"—15 grams of carbohydrate that will enter the bloodstream quickly. This may be 4 ounces of orange juice, commercially available glucose tablets, or another quickly available source of sugar (Box 38–7). If the patient is not alert or is unable to safely swallow, contact the registered nurse (RN) for administration of intravenous (IV) glucose according to agency policy. Recheck glucose in 15 minutes. If the blood glucose does not return to normal, repeat the procedure every 15 minutes until relief occurs. Do not overtreat hypoglycemia with too much sugar, because this may cause hyperglycemia and rebound hypoglycemia. Many nurses added sugar to orange juice in the past. This is no longer recommended.

If the next meal is more than 1 hour away, follow the treatment with a protein and complex carbohydrate snack, such as crackers with cheese or peanut butter, or half of a sandwich. If symptoms worsen, the physician should be contacted. Intravenous glucose or subcutaneous glucagon may be given in emergency situations. Check hospital or agency policy for specific protocol for treating hypoglycemic episodes.

Some elderly patients with poor autonomic nervous system function or patients taking beta-adrenergic blocking medication (such as propranolol or atenolol) may not feel the symptoms of hypoglycemia. These patients should check glucose levels more frequently and keep the levels in a safe range to prevent hypoglycemic episodes.

Individuals with diabetes should be instructed to keep a fast sugar in their purse or pocket at all times. Fast sugars may also be stored in bedside tables, cars, and desks at work.

Diabetic Ketoacidosis

PATHOPHYSIOLOGY. Diabetic ketoacidosis (DKA) occurs when blood glucose levels become very high and insulin is deficient. This most commonly occurs in individuals with type 1 diabetes. DKA is often the reason a person with undiagnosed type 1 diabetes first seeks help. It may also be the result of stress or illness in a person with previously diagnosed type 1 diabetes. When there is insufficient insulin to allow glucose into cells, the cells starve. The body then breaks down fat to be used for energy. The fat breakdown releases an acid substance called ketones. As ketones build up in the blood, ketoacidosis occurs.

The body attempts to reduce acidosis by deepening respirations, thereby blowing off excess carbon dioxide. (See the section on metabolic acidosis in Chapter 5.) The deep, sighing respiratory pattern is called **Kussmaul's** respirations. The expired air has a fruity odor caused by the ketones and may be mistaken for alcohol. Some nurses have likened the odor to Juicy Fruit gum.

With such high blood glucose and the accompanying polyuria, the body becomes dehydrated very quickly. Tachycardia, hypotension, and shock can result. High blood glu-

BOX 38-7   **Fast Sugars**

4 oz orange juice
6 oz regular (not diet) soda
Miniature box of raisins
3 glucose tablets
6–8 Life Savers

cose also causes potassium to leave the cells and accumulate in the blood (hyperkalemia). The combination of dehydration, hyperkalemia, and acidosis causes the patient to develop flulike symptoms, including abdominal pain and vomiting. The patient loses consciousness and death occurs if DKA is not treated. The mortality rate for DKA is about 5 percent.

TREATMENT. Treatment includes IV fluids, IV insulin, and blood glucose monitoring. Assist with monitoring blood glucose levels closely and notify the RN or physician when the desired level is reached. Glucose should be added to the IV when the blood glucose drops to 250 mg/dL to avoid hypoglycemia. Potassium should also be monitored, because the serum potassium level drops rapidly as it reenters the cells. The cause of the DKA should be identified and treated.

Prevention of ketoacidosis involves careful monitoring of blood glucose levels at home. If the blood glucose rises above 300 mg/dL, the patient should use a urine dipstick made to detect ketones. If ketones are present, the physician should be notified. Patients should be instructed to never stop their insulin without a physician's supervision.

Hyperosmolar, Hyperglycemic, Nonketotic Syndrome

PATHOPHYSIOLOGY. Hyperosmolar, hyperglycemic, nonketotic (HHNK) syndrome occurs primarily in type 2 diabetes, when blood glucose levels are high as a result of stress or illness. Because the person with type 2 diabetes has some insulin production, cells do not starve and DKA usually does not occur. It occurs more often in the elderly.

As the blood glucose rises (hyperglycemic), polyuria causes profound dehydration, producing the hyperosmolar (concentrated) state. Blood glucose may rise as high as 1500 mg/dL. Because ketoacidosis is not present, the patient may not feel as physically ill as the patient with DKA and may delay seeking treatment. Symptoms of HHNK develop slowly and include extreme thirst, lethargy, and mental confusion. Shock, coma, and death occur if HHNK is left untreated. The mortality rate for HHNK is about 15 percent.

TREATMENT. Treatment includes IV fluids and insulin, and glucose monitoring. The cause should be identified and treated. HHNK syndrome can be prevented with careful monitoring of glucose levels at home. Patients should be instructed to drink plenty of fluids if blood glucose levels are beginning to rise, especially in times of stress and illness. They should also know when to call their physician with high blood glucose results.

Long-Term Complications

Over time, chronic hyperglycemia causes a variety of serious complications in persons with diabetes. These involve the circulatory system, eyes, kidneys, and nerves. Most of the complications involve either the large blood vessels in the body (macrovascular complications) or the tiny blood vessels, such as those in the eyes or kidneys (microvascular complications). The Diabetes Control and Complications Trial (DCCT), a large research study completed in 1993, showed that individuals with type 1 diabetes who maintain tight control of blood glucose experience fewer long-term complications than individuals who take traditional care of their diabetes. Similarly, the United Kingdom Prospective Diabetes Study (UKPDS), completed in 1998, showed that individuals with type 2 diabetes who maintain an HbA$_{1c}$ below 7 percent can significantly reduce complications. In fact, for every percent of decrease in HbA$_{1c}$, there were 25 percent fewer deaths from diabetes-related complications.[4] Unfortunately, even tight control does not guarantee the prevention of all complications.

Macrovascular Complications

CIRCULATORY SYSTEM. Individuals with diabetes develop atherosclerosis and arteriosclerosis faster than the general population. They are more likely to have hypertension and elevated low-density lipoprotein (LDL) cholesterol and triglyceride levels. High blood glucose may also affect platelet function, leading to increased clotting. These problems lead to a higher incidence of strokes, heart attacks, and poor circulation in the feet and legs. The risk of cardiovascular disease and strokes is two to four times more common in persons with diabetes than in the general population.

Blood glucose and blood pressure control is vital to help prevent these deadly complications. Patients should also avoid smoking, and maintain normal weight. Aspirin therapy to reduce platelet aggregation is recommended for patients older than 21 years of age with diabetes.

Microvascular Complications

EYES. Small blood vessels may become diseased, eventually leading to retinopathy in most patients with diabetes. Retinopathy involves damage to the tiny blood vessels that supply the eye. Small hemorrhages occur, which can cause blindness if not corrected. Diabetic retinopathy causes 12,000 to 24,000 new cases of blindness each year.[5] Newer laser surgery techniques may help improve vision after hemorrhages occur. Diabetes is also associated with a high incidence of cataracts. Patients with diabetes should have a yearly dilated eye examination.

KIDNEYS. Nephropathy is caused by damage to the tiny blood vessels within the kidneys. From 20 to 30 percent of patients with diabetes develop some degree of nephropathy. Native-Americans, Hispanics, and African-Americans have the highest risk. A primary risk factor for diabetic nephropathy is poor control of blood glucose. If nephropathy occurs, the kidneys are unable to remove waste products and excess fluid from the blood. Diabetes is the leading cause of end-stage renal (kidney) disease (ESRD). When the kidneys have lost most of their function, patients may have their blood cleansed artificially, by either hemodialysis

---

nephropathy: nephro—of the kidney + pathy—illness

or peritoneal dialysis. (See Chapter 35.) The only cure for ESRD is a kidney transplant.

Patients should be taught the importance of blood glucose control in the prevention or delay of kidney disease. Angiotensin-converting enzyme (ACE) inhibitor medications such as enalapril (Vasotec) have also been shown to slow the development of kidney problems in patients with diabetes. Patients who have both diabetes and hypertension should be placed on an ACE inhibitor. Routine urine tests are done to check for microalbuminuria (tiny amounts of protein in the urine) or microalbumin-to-creatine ratios. If microalbuminuria occurs, a low-protein diet may help delay further development of nephropathy. A trained renal dietitian should work with the patient and physician in determining the best diet for the patient.

Nerves

Another complication of diabetes is neuropathy, which is damage to nerves as a result of chronic hyperglycemia. Neuropathy can cause numbness and pain in the extremities, erectile dysfunction (impotence) in males, **gastroparesis** (delayed stomach emptying), and other problems. Unfortunately, pain caused by neuropathy is difficult to treat with traditional analgesics. Some antidepressant and anticonvulsant drugs may be helpful, and in some cases local injections of anesthetics may be used. Improved control of blood glucose levels may also help.

Infection

Persons with diabetes are prone to infection for several reasons. If injuries occur, healing may be slow because of impaired circulation. There may not be enough blood supply to heal the wound or fight an infection. For the same reason, it may be difficult for IV antibiotics to reach an infected site, and topical antibiotics may be preferable. In the presence of hyperglycemia, white blood cells become sluggish and ineffective, further reducing the body's ability to fight infection.

The incidence of periodontal (gum) disease, caused by bacteria in plaque, is also increased in individuals with diabetes. Patients must be taught to maintain good oral hygiene and make regular visits to the dentist.

Foot Complications

The combination of vascular disease, neuropathy, and risk for infection makes patients with diabetes prone to foot problems. Consider patients who have no feeling in their feet because of neuropathy. If they step on a tack, they may not feel it right away. Vascular disease prevents a good blood supply from promoting healing. If infection sets in, it is slow to resolve. Pressure points on the feet may also break down (Fig. 38-7). Neuropathy can lead to deformities of the feet, further increasing the risk for injuries.

For these reasons, diabetes is the leading cause of amputation of the lower extremities. Patients should be taught to

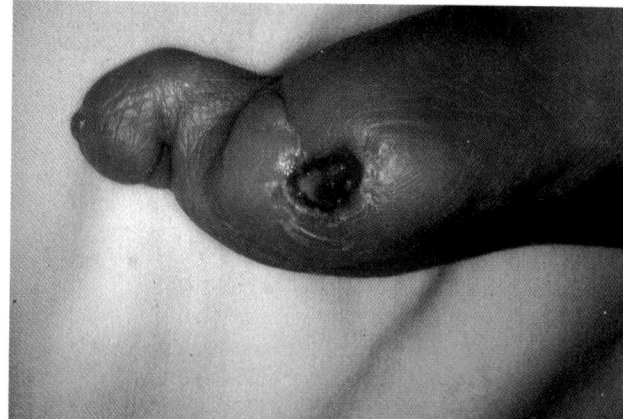

Figure 38-7 Diabetic foot ulcer. (From Goldsmith, LA, Lazarus, GS, and Tharp, MD: Adult and Pediatric Dermatology. FA Davis, Philadelphia, 1997, p 438, with permission.)

protect their feet at all times by wearing well-fitting shoes and by washing, drying, and inspecting their feet daily (Box 38-8). If any sores are noted, the patient should not delay in seeking treatment. During routine visits to the physician, the patient should be sure to remove shoes and socks so the feet can be thoroughly examined. The physician or diabetes specialist can test sensation in the feet with tiny filaments. Loss of protective sensation is an early risk factor for amputation, so any reduction in sensation is a warning sign that extra care must be taken. A podiatrist (foot doctor) can be consulted if problems occur. Specialized wound treatment centers have new healing techniques that have prevented many amputations (Ethical Consideration Box 38-9).

Special Considerations for the Patient Undergoing Surgery

Surgery is a stressor. The counter-regulatory hormones released during stress cause the blood glucose to rise, even if the patient has been fasting. High blood glucose levels interfere with healing and may promote an environment for infection. Blood glucose should be well controlled before elective surgery. Check the physician's orders for changes in

---

gastroparesis: gastro—stomach + paresis—partial paralysis

BOX 38-8	Foot Care Tips

Wash and dry feet every day. Use warm (not hot) water.
Apply lotion that does not contain alcohol, avoiding areas between toes.
Inspect feet for sores or red areas daily (have a family member help if necessary).
Report any abnormalities immediately.
Wear leather shoes and cotton socks.
Never go barefoot.
Avoid garters and tight socks.
Avoid crossing legs.
Cut toenails to natural shape of nail—not into corners.
See a podiatrist for calluses or problem toenails (avoid "bathroom surgery").
Have feet checked at least once a year, preferably 3–4 times a year, for loss of sensation.

CRITICAL THINKING

Mr. Jones

Mr. Jones is a 54-year-old banker with type 2 diabetes admitted to your unit with a tiny red area on his right heel. His admitting blood glucose is 360 mg/dL. The lesion is so small you wonder what the fuss is about. While doing his assessment, you find that he wore a new pair of shoes to work all day about a month ago and has been avoiding seeing his physician about the resulting red area. He is placed on bedrest and IV antibiotics, and within 3 days the red area has broken open and has yellow drainage. He is sent home with topical antibiotics, to be followed by a visiting nurse. The wound takes 6 months to fully heal.

1. List three risk factors for foot problems.
2. Why did the sore take so long to heal?
3. Why was bedrest important?
4. Why might topical antibiotics work better than IV antibiotics?
5. The nurse documents the following description of Mr. Jones' wound: "Small red open area on heel, with yellow drainage on dressing." What is wrong with this charting? How can you improve it?

Answers at end of chapter.

BOX 38-9 ETHICAL CONSIDERATION

Mr. Mann is 55 years old and has had diabetes for more than two decades. He has not managed his diabetes well and has subsequently suffered many complications. Mr. Mann does not regularly test his blood sugar, so he does not always get the appropriate amount of insulin. Also, he does not eat properly. He picks up fast-food during his lunch break and he binges on various snack foods before he eats his dinner, or instead of eating his dinner. He is not aware of the importance of taking care of himself to prevent long-term complications of diabetes. His present situation is the result of a neglected leg ulcer. He tried to treat it at home with over-the-counter remedies, but it did not improve.

The leg ulcer grew and became infected. Currently part of the foot is gangrenous and an amputation is recommended. The first time the surgery was mentioned, Mr. Mann became outraged and was adamant about the unacceptability of the proposed treatment. He was not going to "leave this earth" without all his parts. After further discussions with Mr. Mann, it was clear that he understood the consequences of his condition. He was more annoyed by the drainage from the ulcer than worried about the progressive gangrene.

Can Mr. Mann be treated without his permission? Should action be taken to have someone else named as Mr. Mann's guardian, despite his apparent competence? Does disagreement with the medical establishment in itself denote incompetence? Does the level of risk if treatment is refused make any difference in the health professional's response? What if Mr. Mann's refusal was based on religious beliefs or was related to some cultural tradition? How does the application of ethical theories help to sort out the critical elements when this case is examined?

insulin orders. Often patients are placed on intravenous infusions of glucose and insulin during and immediately after surgery. Monitor blood glucose levels every 2 to 4 hours or as ordered, and monitor carefully for signs and symptoms of hypoglycemia or hyperglycemia.

Patients who were not previously on insulin may be placed on insulin during surgery and postoperatively. They can generally return to their presurgical treatment plan after the stress of surgery is past.

Nursing Process

Assessment

A complete history and physical assessment should be carried out because diabetes affects every body system. Some areas on which to focus are shown in Table 38–6. It is especially important to assess each patient's knowledge of diabetes and its care, so that appropriate teaching can be done.

Nursing Diagnosis

Because diabetes affects so many different areas, nearly any nursing diagnosis may be appropriate. It is important to assess each patient as an individual and choose diagnoses

based on assessment findings. A sample care plan is shown for a diagnosis of risk for ineffective health maintenance, because patients with diabetes must learn to care for themselves to maintain their health. The actual presence of the defining characteristics should be confirmed with the patient before choosing any nursing diagnosis. (See Nursing Care Plan Box 38–10 for the Patient with Diabetes Mellitus.)

Planning and Implementation

Once diagnoses have been identified, planning takes place. This should be done with the patient and family. Diabetes affects not only the person with the disease, but the entire family as well. The desired outcome for the plan of care is that the patient is knowledgeable about and able to care for

TABLE 38-6	ASSESSMENT OF THE PATIENT WITH DIABETES MELLITUS		
Acutely Ill Patient with Newly Diagnosed Diabetes		**Patient with Previously Diagnosed Diabetes**	
Subjective Data	**Objective Data**	**Subjective Data**	**Objective Data**
• History of current problem • History of stress, illness, virus • Family history of diabetes • Current medications • Other medical or surgical conditions • Knowledge of diabetes self-care	• Vital signs • Lab values—electrolytes, blood glucose, ketones • Signs of dehydration • Fruity breath • Presence of complications if suspect diabetes was undiagnosed for period of time	• History of diabetes: type, onset, duration, degree of blood glucose control • Knowledge of self-care and degree of compliance • Support systems • History of complications	• Labs: blood glucose level, glycohemoglobin, BUN, creatinine, ketones, cholesterol, triglycerides • Condition of legs and feet; pulses, presence of circulatory or sensation impairment

BOX 38-10   NURSING CARE PLAN FOR THE PATIENT WITH DIABETES MELLITUS

 Risk for ineffective health maintenance related to knowledge deficit in the patient with newly diagnosed diabetes mellitus

PATIENT OUTCOMES

Blood glucose levels within parameters negotiated with health care provider. Patient states satisfaction with understanding of diabetes self-care.

EVALUATION OF OUTCOMES

Are blood glucose levels within parameters negotiated? Does patient state satisfaction with understanding of diabetes self-care?

INTERVENTIONS	RATIONALE	EVALUATION
Assess knowledge of diabetes self-care.	Teaching should be initiated only if a knowledge deficit exists.	Does patient exhibit knowledge of diabetes self-care?
Assist patient to collaborate with health care provider to determine appropriate blood glucose levels and action to be taken if glucose levels are too high or too low.	Appropriate blood glucose levels are different for each patient and should be determined on an individualized basis. The patient should know what blood glucose levels require notification of the health care provider.	Are blood glucose levels within parameters negotiated with health care provider? Does patient state satisfaction with knowledge of diabetes self-care? Does patient state appropriate blood glucose levels and action to take if glucose is high or low?
Teach patient to assess glucose levels before meals and at bedtime or as ordered by health care provider. Ensure that patient knows how to obtain glucose monitor and instruction for home use.	Good blood glucose control depends on knowledge of glucose levels and trends.	Does patient demonstrate correct use of glucose monitor or state how monitor and instruction will be obtained?
Teach patient to administer insulin or oral hypoglycemic agent 30 minutes before meals (<15 minutes if patient is using Lispro). Ensure that meal is begun within 30 minutes of medication. Replace any uneaten foods to prevent hypoglycemia.	The onset of most diabetes medication is about 30 minutes (see Table 38–4 for specific times). If more than 30 minutes passes before a meal, the patient might experience hypoglycemia.	Does patient state correct meal and medication schedule?
Teach technique for administering insulin if indicated.	The patient and family should be familiar with the injection procedure.	Does patient demonstrate correct injection technique?
Observe for symptoms of hypoglycemia and hyperglycemia and treat as necessary. Teach causes, prevention, recognition, and treatment of hypoglycemia and hyperglycemia.	If the patient has a good understanding of hypoglycemia and hyperglycemia, most episodes can be prevented. If hypoglycemia or hyperglycemia does occur, prompt treatment is essential to prevent complications.	Does patient state causes, prevention, symptoms, and treatment of hypoglycemia?
Consult with dietitian for nutrition therapy instruction.	The dietitian is trained to provide indepth meal plan instruction.	Is patient able to state plan for obtaining appropriate meals?
Consult with social worker as needed.	Some patients may not have the resources to carry out effective self-care.	Does patient state availability of adequate resources for self-care at home?
Provide patient with information regarding comprehensive diabetes education. Remind patient that only survival skills have been taught during initial instruction.	Instruction provided in the hospital usually is not comprehensive. Outpatient diabetes classes can provide additional self-care and health promotion information.	Does patient state plan for obtaining further diabetes education after discharge?
Assist patient to obtain medical alert card or tag that identifies diabetes.	If the patient is ever unresponsive for any reason, the health care provider would need to be aware of diabetes.	Does patient state plan to carry or wear identification at all times?
	GERIATRIC	
Assess ability to see and manipulate syringe, glucose monitor, and other equipment. Obtain assistive devices as needed.	The elderly patient may have poor eyesight or other sensory deficits.	Is patient able to manipulate equipment to safely care for self?

his or her disease. Consult the dietitian, social worker, certified diabetes educator (CDE), home care nurse, outpatient education programs, and other resources as needed. (See Home Health Hints.)

HOME HEALTH HINTS

- Some home glucose monitoring devices have a memory that the nurse can access on the visit. It gives the date, time, and blood glucose result. This is a good indication of compliance with self-monitoring performed by patient or caregiver.

- The nurse should remember to call the patient the day before performing a venipuncture for a fasting blood sugar and remind the patient not to eat after midnight.

- Older patients tend to skip meals. Assist them to identify easy but nutritious meals, such as frozen dinners. Meals-on-Wheels is another option.

- Prefilled syringes should be stored in the refrigerator flat or with needles pointing up. This prevents crystals from settling and clogging the needle.

- The patient can discard used syringes and needles in a hard plastic container such as a Clorox bottle with a screw top.

- If the patient has a visual or dexterity problem, suggest getting a syringe magnifier. A pharmacist or occupational therapist may be able to assist with obtaining one.

- The patient can use a mirror to look at the bottom of the feet or have a family member examine the patient's feet. The patient should remove shoes and socks at each physician visit for a thorough foot inspection. Catching a "red spot" early is the goal.

- Due to decreased skin sensation in some persons with diabetes, hot water heaters should be set below 120° F.

- Even if patients have had diabetes for many years, observe them preparing and injecting their insulin. This provides an opportunity to praise good technique or correct bad habits.

Unless the patient demonstrates thorough knowledge, implementation of the care plan should include teaching. As each phase of the plan is carried out, explain to the patient and family what is being done and why. Appointments should be made to sit down with the patient and family to discuss new information and allow time for questions. Provide the patient ample time to demonstrate any newly learned skills. You should be familiar with agency policy regarding licensed practical nurse/licensed vocational nurse (LPN/LVN) responsibilities for diabetes education. Many institutions have written teaching plans that are helpful. Additional information about patient education is presented later in this chapter. Two Web sites that might be helpful to both you and your patients are www.diabetes.org and www.lifeclinic.com/focus/diabetes/resources.asp.

Evaluation

The best indicator of the success of a care plan for diabetes is controlled blood glucose and glycohemoglobin levels. The patient should also be without symptoms of hypoglycemia or hyperglycemia and be able to state what to do

if they do occur. Long-term complications should be minimized. Another important indicator is the patient's statement of satisfaction and comfort with the plan and his or her ability to carry it out on a daily basis.

Patient Education

The individual with diabetes must be taught self-care if at all possible. No amount of care from a physician or nurse can replace the self-care required of the person with diabetes. The involvement of family or significant others is also important for the success and well-being of the person with diabetes.

If the patient is hospitalized at diagnosis, the initial instruction is done in the hospital. However, with hospital stays becoming shorter, you must not waste any time. Begin teaching as soon as the patient is feeling physically well enough to learn. This is usually the responsibility of the primary or registered staff nurse, although aspects of the instruction may be delegated to the LPN/LVN. Some hospitals have a certified diabetes educator who provides classroom or bedside instruction. The dietitian should be contacted to provide meal plan instruction.

Most hospitals have policies or management plans describing the instruction to be provided by the nurse. Generally this encompasses "survival skills," which include the basic information the patient needs initially to survive at home. Survival skills include medication administration, glucose monitoring, meal plan basics, and what to do if high or low blood glucose levels occur. A variety of helpful aids, such as pamphlets and videos, are available. Diabetes equipment suppliers provide kits that are full of samples and information. These are a significant help when you are teaching a patient. Also advise the patient to purchase a medical alert bracelet or necklace.

It is difficult to know how to operate and teach glucose monitoring with the variety of glucose monitors available. Many drugstores and medical supply stores not only sell the monitors but also provide training for the patient and family. You can obtain this information by calling local medical suppliers or by contacting the diabetes educator.

After discharge, the patient is referred to outpatient diabetes classes for further instruction. If classes are unavailable or if the patient is unable to leave home, a referral to a visiting nurse should be given. It is usually advisable to have a nurse present for the patient's first insulin injection at home.

Because many people with diabetes are elderly, it is important to be aware of their special needs (Gerontological Issues Box 38–11).

BOX 38–11 GERONTOLOGICAL ISSUES

Diabetes care can be a challenge for many elderly patients. Syringe magnifiers and talking glucose meters are available for those with impaired vision. Family members may be taught to draw up a week's supply of insulin for the patient. Home meal programs may help ensure an adequate diet. Elderly patients should also have an emergency call system in their home and regular contact with family members or other support people.

► REACTIVE HYPOGLYCEMIA

Reactive hypoglycemia occurs when the blood glucose drops below a normal level, usually below 50 mg/dL. Hypoglycemia is most often a complication of diabetes treatment, but at times it may occur without the presence of diabetes. It may be a warning sign of impending diabetes.

Pathophysiology

Low blood glucose may occur as an overreaction of the pancreas to eating. The pancreas senses a rising blood glucose and produces more insulin than is necessary for the use of that glucose. As a result, the blood glucose drops to below normal. Some experts believe that this is a rare condition and that many "hypoglycemic" episodes are due to activation of the sympathetic nervous system for other reasons, without true hypoglycemia.

Signs and Symptoms

Low blood glucose causes release of epinephrine, which in turn causes the blood glucose to rise. Epinephrine release causes a fight-or-flight reaction, which may produce shaking, sweating, and palpitations. Headache, chills, and confusion may also occur.

Diagnosis

Diagnosis is often based on a 5-hour glucose tolerance test, with below-normal readings between 2 and 5 hours. However, with the availability of home glucose monitors, it is now preferable for patients to monitor blood glucose levels at home. Readings should be taken in the morning on arising, 2 hours after each meal, at bedtime, and during symptoms of hypoglycemia. These results may then be taken to the physician for interpretation.

Treatment

Treatment includes frequent small meals and avoidance of fasting. Simple sugars are avoided, because they may aggravate symptoms. A high-protein, low-carbohydrate diet is stressed. See Table 38–7 for a sample diet.

TABLE 38–7	SAMPLE MEAL PLAN FOR HYPOGLYCEMIC DIET		
Exchange Group	**Sample Menu**	**Exchange Group**	**Sample Menu**
Morning	½ cup unsweetened orange juice	*Mid-afternoon*	
1 fruit	¾ cup whole-grain cereal	1 meat	1 oz low-fat cheese
1 starch	1 low-fat cheese or	1 starch	4 whole-grain crackers
1 meat	½ cup skim milk	*Evening*	
½ skim milk	Decaffeinated coffee	2–4 meat	2–4 oz lean meat
Free		1 starch	½ cup potato or pasta
Mid-morning		1 vegetable	½ cup vegetable
1 meat	1 tbsp peanut butter	1 fat	Lettuce salad with dressing
1 starch	4 whole-grain crackers	1 fruit	1 pce fresh fruit
Noon		Free	Decaffeinated coffee or tea
Chef salad		*Bedtime*	
2–4 meat	2–4 oz lean meat	1 starch and	½ sandwich (1 slice whole-grain bread and
1 vegetable	Lettuce, tomatoes, and	1 meat	1 oz lean meat)
1 fat	Dressing	1 vegetable	Fresh vegetables
1 fruit	1 small piece fresh fruit	Free	Decaffeinated beverage
1 skim milk	1 cup skim milk		
1 starch	2 breadsticks (4 × ½ in)		

REVIEW QUESTIONS

1. Which of the following is the best definition of diabetes mellitus?
   a. It is a group of metabolic diseases in which high blood sugar results from defective insulin secretion or action.
   b. It is a disease that causes polyuria and polydipsia.
   c. It is a disease characterized by macrovascular and microvascular complications.
   d. It is a complex disease of protein and fat metabolism.

2. Which of the following is a risk factor for type 2 diabetes mellitus?
   a. Cardiovascular disease
   b. Obesity
   c. Age younger than 40 years
   d. Smoking

3. Diabetes is diagnosed when the fasting blood glucose is greater than _____ on two occasions?
   a. 70 mg/dL
   b. 126 mg/dL
   c. 140 mg/dL
   d. 200 mg/dL

4. Which of the following symptoms is most commonly associated with hyperglycemia?
   a. Tremor
   b. Flank pain
   c. Sweating
   d. Polyuria

5. Protein in the urine is a sign of which long-term complication of diabetes?
   a. Nephropathy
   b. Neuropathy
   c. Retinopathy
   d. Gastroparesis

6. What is the best way for patients to avoid long-term complications of diabetes?
   a. See the doctor for a complete check up every 6 months.
   b. Check feet daily.
   c. Maintain blood sugar levels under 120 mg/dL.
   d. Follow a strict 1500 calorie diet.

7. For which of the following blood glucose results would the nurse administer a fast sugar?
   a. 48
   b. 80
   c. 126
   d. 223

8. Which meal plan is best for the patient with reactive hypoglycemia?
   a. High-carbohydrate meals
   b. Small, frequent meals
   c. Avoidance of fats and proteins
   d. Three medium to large meals daily

REFERENCES

1. American Diabetes Association: Facts and Figures: www.diabetes.org, 2001.
2. American Diabetes Association: Tests of glycemia in diabetes. Diabetes Care 24(1):81, 2001.
3. The Diabetes Control and Complications Trial Research Group: The effect of intensive treatment of diabetes on the development and progression of long-term complications in insulin-dependent diabetes mellitus. N Engl J Med 329:14, 1993.
4. U.K. Prospective Diabetes Study Group: Intensive blood-glucose control with sulphonylureas or insulin compared with conventional treatment and risk of complications in patients with type 2 diabetes. Lancet 352:837, 1998.
5. Centers for Disease Control and Prevention: Reducing the burden of diabetes. CDC Diabetes Home Page, http://www.cdc.gov, 1998.
6. American Diabetes Association: Diabetic nephropathy. Diabetes Care 24(1):69, 2001.

UNIT TEN BIBLIOGRAPHY

American Diabetes Association: Clinical Practice Recommendations 2002. Diabetes Care 25: Supplement 1, 2002.

American Diabetes Association: Facts and Figures. Retrieved 2001 from www.diabetes.org.

Bacoka, J: Action stat: Thyroid storm. Nursing 31(12):88, 2001.

Bell, N: Feet first. Diabetes Forecast 50:6, 1997.

Bianco, CM: Diabetes insipidus. American Journal of Nursing 96:8, 1996.

Cincinnati, R, and Veliko, J: Diabetes update: Oral medications. RN 64:8, 2001.

Guthrie, RA, Hinnen, D, Childs, BP: Clinical decision making. Glargine: A new basal insulin, a new opportunity. Diabetes Spectrum, 14:3, 120–123, 2001.

Isomaa, B, et al: Cardiovascular morbidity and mortality associated with the metabolic syndrome. Diabetes Care 24:4, 2001.

Jankowski, CB: Irradiating the thyroid: How to protect yourself and others. American Journal of Nursing 96:51, 1996.

Maryniuk, MD: The new shape of medical nutrition therapy. Diabetes Spectrum 13:3, 122–124, 2000.

Riddle, MC: Managing type 2 diabetes over time: Lessons from the UKPDS. Diabetes Spectrum 13:4, 194–196, 2000.

Sachse, D: Acromegaly. American Journal of Nursing, 101:11, 69–77, 2001.

Sammer, CE: How should you respond to hypoglycemia? Nursing 31:7, 2001.

The Diabetes Control and Complications Trial Research Group: The effect of intensive treatment of diabetes on the development and progression of long-term complications in insulin-dependent diabetes mellitus. New England Journal of Medicine 329:14, 1993.

UK Prospective Diabetes Study Group: Intensive blood-glucose control with sulphonylureas or insulin compared with conventional treatment and risk of complications in patients with type 2 diabetes. Lancet 352, 1998.

Watson, R: Assessing endocrine system function in older people. Nursing Older People 12:9, 27–28, 2000.

Answers to CRITICAL THINKING

Mr. McMillan

1. "Have you been eating or drinking more than usual? Have you been urinating more than usual? Do you get up at night to urinate?"
2. Fatigue occurs because the glucose is unable to enter the cells without insulin, so they are starving.
3. Mr. McMillan is dehydrated because he is losing excessive amounts of urine as his kidneys excrete extra glucose.

Mrs. Evans

1. Mrs. Evans should take her insulin at 7:30 A.M.
2. She should be alert for low blood sugar at midafternoon to just before supper. Although her insulin peaks between 1:30 and 7:30 P.M., her chances of having a hypoglycemic episode are slim once she has eaten her supper.
3. If she misses a meal, her blood sugar will drop further, increasing her risk of a hypoglycemic episode.

Jeff

What kind of new job is it? Is it more physically strenuous than his previous job? Does it interfere with his usual meal schedule?

Mr. Jones

1. Poor circulation, neuropathy, and slow wound healing place Mr. Jones at risk for problems.
2. Circulation to the foot may be poor, and white blood cells are sluggish if the blood glucose is high.
3. Any pressure on the foot while walking may further impair circulation.
4. If circulation to the area is poor, IV antibiotics may not reach the sore.
5. "Small red open area on right posterior heel, 1 cm × 1.5 cm, 2 mm deep. 2 cm area of yellow drainage on dressing." In addition, many agencies are now taking instant photos of wounds to include in the chart. If no camera is available, a drawing of the size and shape is helpful.

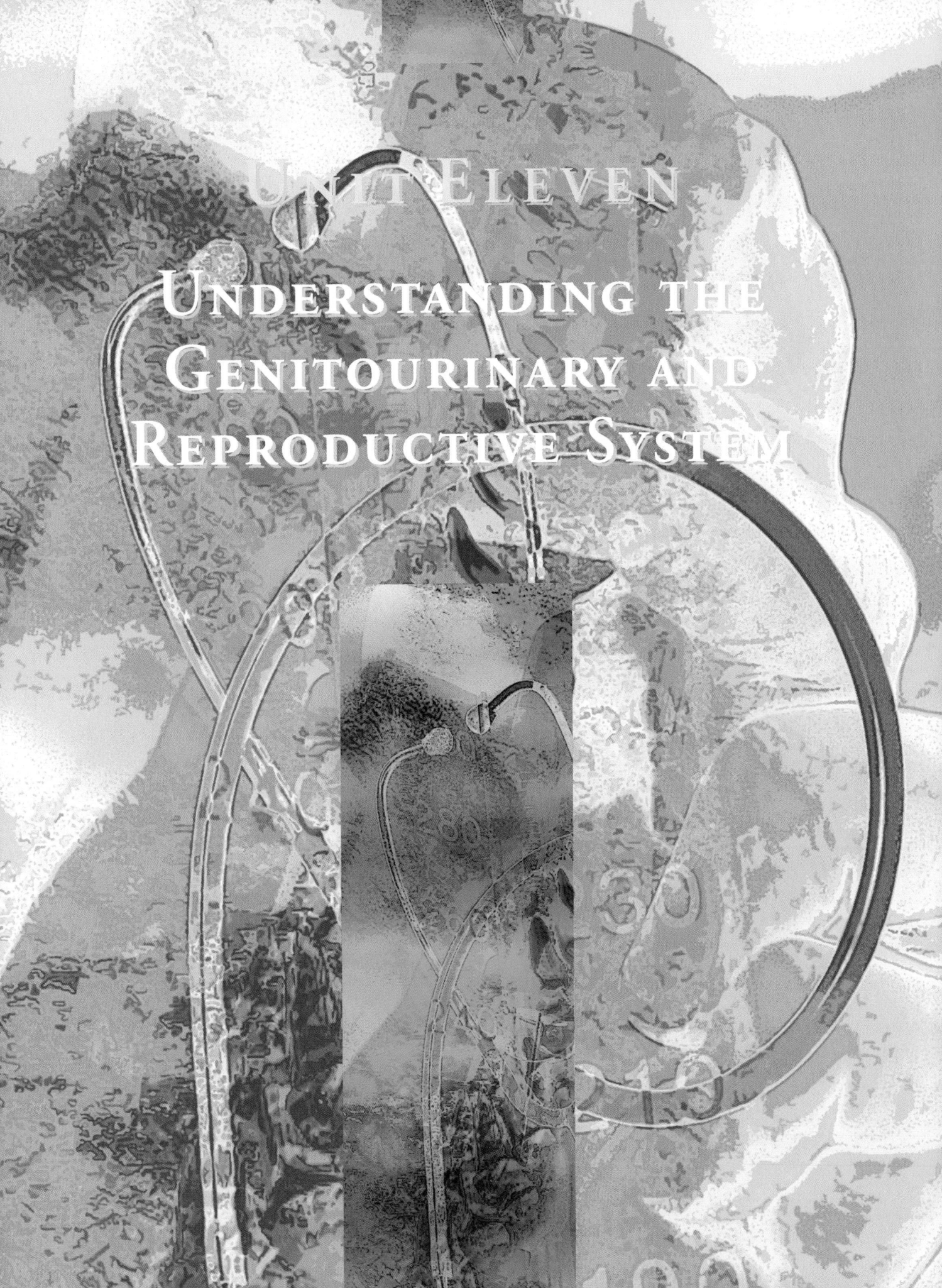

Unit Eleven

Understanding the Genitourinary and Reproductive System

39

GENITOURINARY AND REPRODUCTIVE SYSTEM FUNCTION AND ASSESSMENT

Valerie C. Scanlon, Linda Cook, and Cindy Meredith

KEY TERMS

adnexa (ad-**NECK**-sah)

bimanual (by-**MAN**-yoo-uhl)

circumcised (**SIR**-kuhm-sized)

colposcopy (kul-**POS**-koh-pee)

conization (KOH-ni-**ZAY**-shun)

culdoscopy (kul-**DOS**-koh-pee)

curet (kyoo-**RET**)

cystic (**SIS**-tik)

ejaculation (ee-JAK-yoo-**LAY**-shun)

erection (e-**REK**-shun)

gynecomastia (JIN-e-koh-**MASS**-tee-ah)

hydrocele (**HIGH**-droh-seel)

hypospadias (HIGH-poh-**SPAY**-dee-ahz)

hysterosalpingogram (HIS-tur-oh-**SAL**-pinj-oh-gram)

hysteroscopy (HIS-tur-**AHS**-koh-pee)

insufflation (in-suff-**LAY**-shun)

libido (li-**BEE**-doh)

mammography (mah-**MOG**-rah-fee)

menarche (me-**NAR**-kee)

menopause (**MEN**-oh-pawz)

orgasm (**OR**-gazm)

salpingoscopy (SAL-ping-**AHS**-koh-pee)

transillumination (TRANS-i-loo-mi-**NAY**-shun)

varicocele (**VAR**-i-koh-seel)

QUESTIONS TO GUIDE YOUR READING

1. What is the normal anatomy of the reproductive system?
2. What are the normal functions of the reproductive system?
3. Which data should you collect when caring for a patient with a disorder of the reproductive system?
4. Which diagnostic tests are commonly performed to diagnose disorders of the reproductive system?
5. What nursing care should you provide for patients undergoing each of the diagnostic tests?
6. Which therapeutic measures are commonly used for patients with disorders of the reproductive system?

▶ REVIEW OF NORMAL ANATOMY AND PHYSIOLOGY

The male and female reproductive systems produce gametes (sperm and egg cells) and ensure the union of gametes in fertilization following sexual intercourse. In women the uterus provides the site for the developing embryo/fetus until birth.

Female Reproductive System

The female reproductive system consists of the paired ovaries and fallopian tubes, the single uterus and vagina, and the external genital structures (Fig. 39–1). The mammary glands may be considered accessory organs to the system.

Ovaries

The ovaries are a pair of oval structures about 1.5 inches long on either side of the uterus in the pelvic cavity (Fig. 39–2). The ovarian ligament extends from the medial side of the ovary to the uterine wall, and the broad ligament is a fold of the peritoneum that covers the ovaries. These ligaments help keep the ovaries in place.

The ovaries produce egg cells by the process of meiosis, more specifically called oogenesis, which begins at puberty and ends at **menopause,** between the ages of 45 and 55. This process is cyclical in that usually one mature ovum is

---

menopause: men—month + pause—stop

produced every 28 days and is under hormonal control (covered in the section on the menstrual cycle). The follicles of the ovary produce the hormone estrogen and later, as the corpus luteum, secrete progesterone as well.

Fallopian Tubes

Each fallopian, or uterine, tube is about 4 inches long; the lateral end with its fringelike fimbriae encloses the ovary on its side, and the medial end opens into the uterus. The lining of a fallopian tube is ciliated epithelium; in the wall is smooth muscle. The sweeping of the cilia and the peristaltic contractions of the smooth muscle usually ensure that the ovum (or the zygote after fertilization), which has no means of self-locomotion, will reach the uterus. Fertilization usually takes place within the fallopian tube, and the zygote is swept into the uterus within 4 to 5 days.

Uterus

The uterus is about 3 inches long and 2 inches wide, superior to the urinary bladder and medial to the ovaries in the pelvic cavity. Ligaments help keep the uterus in place, tilted forward over the top of the bladder. During pregnancy the uterus increases greatly in size, contains the placenta to nourish the embryo (later called the fetus), and expels the baby at the end of gestation.

The fundus of the uterus is the upper portion above the entry of the fallopian tubes, and the body is the large central portion. The cervix is the narrow, lower end, which

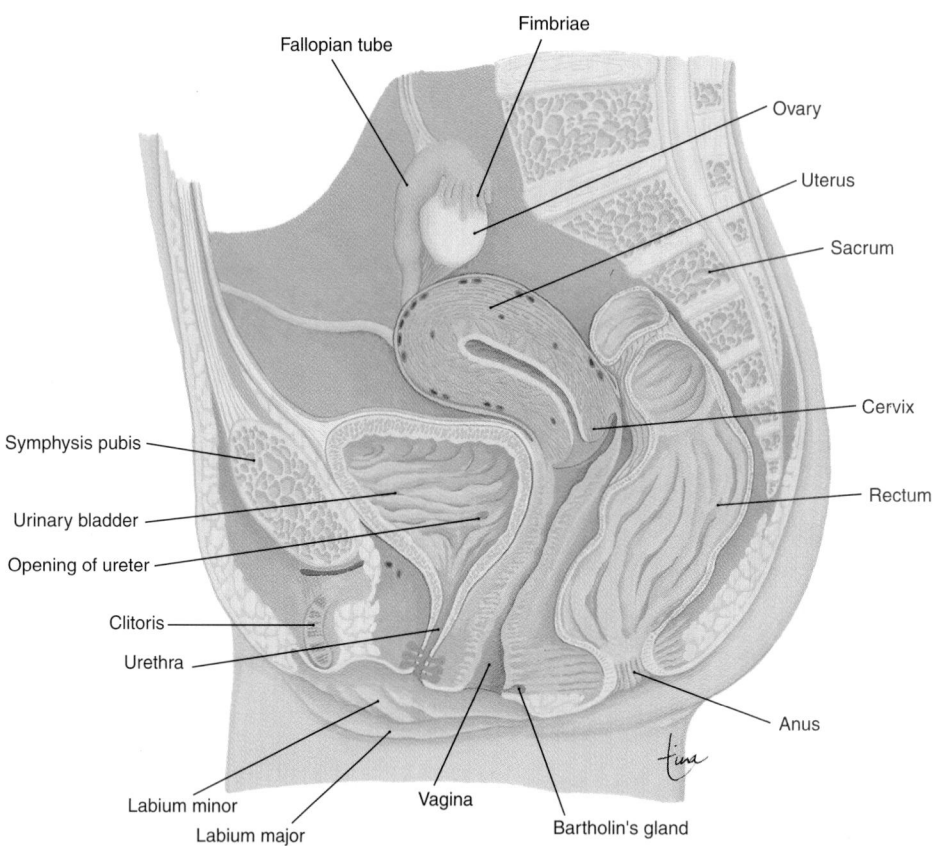

Figure 39-1 Female reproductive system in a midsagittal section. (From Scanlon, VC, and Sanders, T: Essentials of Anatomy and Physiology, ed 3. FA Davis, Philadelphia, 1999, p 446, with permission.)

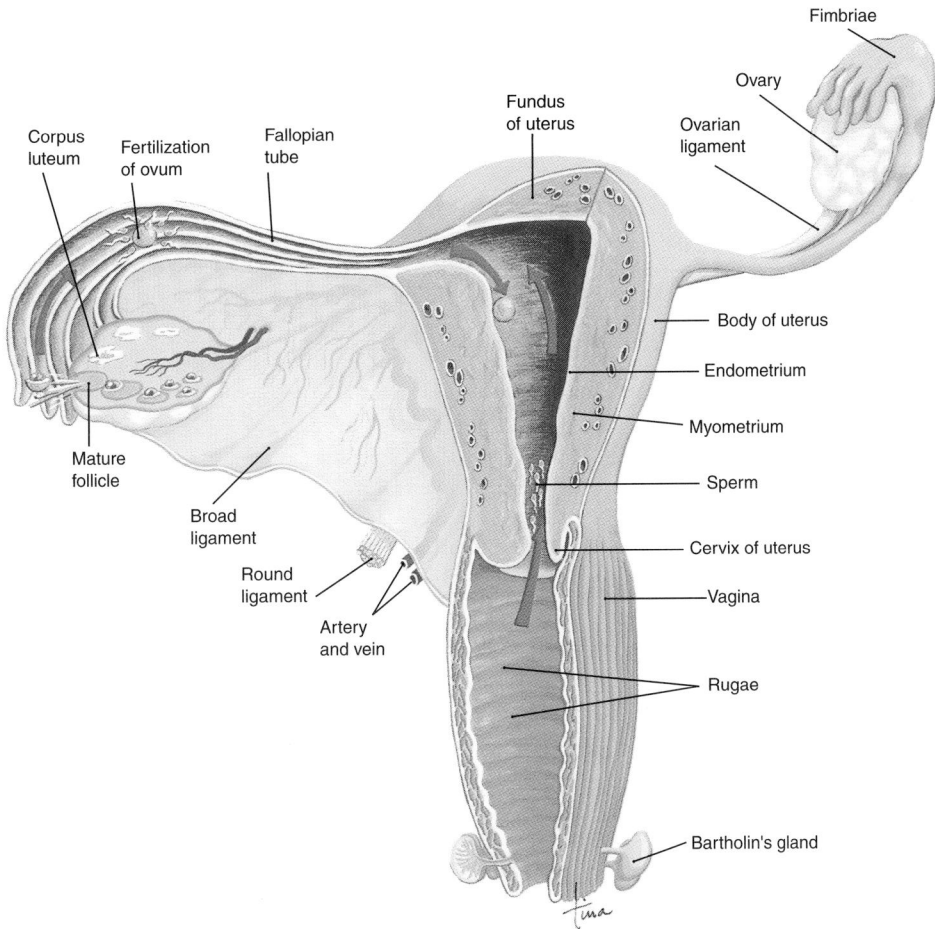

Figure 39-2 Female reproductive system in a anterior view and a longitudinal section. (From Scanlon, VC, and Sanders, T: Essentials of Anatomy and Physiology, ed 3. FA Davis, Philadelphia, 1999, p 447, with permission.)

opens into the vagina. The outermost layer of the uterine wall is the epimetrium, a fold of the peritoneum. The myometrium is the smooth muscle layer; during pregnancy these cells increase in size to accommodate the growing fetus, and they contract for labor and delivery at the end of gestation.

The lining of the uterus is the endometrium, a highly vascular mucous membrane, part of which is lost and regenerated with each menstrual cycle. During pregnancy, the endometrium forms the maternal side of the placenta.

Vagina

The vagina is a muscular tube about 4 inches long that extends from the cervix to the vaginal orifice in the perineum. It is between the urethra and the rectum. The functions of the vagina are to receive sperm from the penis during sexual intercourse, to serve as the exit for menstrual blood flow, and to serve as the birth canal at the end of pregnancy.

After puberty, the vaginal mucosa is relatively resistant to infection. The normal bacterial flora of the vagina creates an acidic pH that helps inhibit the growth of most pathogens.

External Genitals

Also called the vulva, the female external genital structures are the clitoris, labia majora and minora, and Bartholin's glands. The clitoris is a small mass of erectile tissue anterior to the urethral orifice. Its function is sensory; it responds to

sexual stimulation, and its vascular sinuses become filled with blood.

The mons pubis is a pad of fat over the pubic bones, covered with skin and pubic hair. Extending posteriorly from the mons are the lateral labia majora and the medial labia minora; these are paired folds of skin. The area between the labia minora is the vestibule; it contains the openings of the urethra and vagina. The labia cover these openings and prevent drying of their mucous membranes. Bartholin's glands, also called vestibular glands, are within the floor of the vestibule; their ducts open into the mucosa at the vaginal orifice. Their secretion keeps the mucosa moist and lubricates the vagina during sexual intercourse.

Mammary Glands

Enclosed within the breasts and surrounded by adipose tissue, the mammary glands produce milk after pregnancy. The milk enters the lactiferous ducts, which converge at the nipple. The skin around the nipple is a pigmented area called the areola. The formation of milk is under hormonal control. During pregnancy, high levels of estrogen and progesterone prepare the glands for milk production. Prolactin from the anterior pituitary causes the actual synthesis of milk after pregnancy. The sucking of the infant on the nipple stimulates the release of oxytocin from the posterior pituitary gland, which in turn stimulates the release of milk.

The Menstrual Cycle

The menstrual cycle depends on follicle-stimulating hormone (FSH) and luteinizing hormone (LH) from the anterior pituitary gland and estrogen and progesterone from the ovaries. These hormones bring about changes in the ovaries and uterus. A cycle may be described in terms of three phases: menstrual phase, follicular phase, and luteal phase.

The menstrual phase involves the loss of the endometrium in menstruation, which may last 2 to 8 days, with an average of 3 to 6 days. At this time, secretion of FSH is increasing and several ovarian follicles, each with a potential ovum, begin to develop. Table 39–1 includes a summary of these hormones.

During the follicular phase, FSH stimulates growth of ovarian follicles and secretion of estrogen by the follicle cells. The secretion of LH is also increasing, but more slowly. FSH and estrogen promote the growth and maturation of the ovum, and estrogen stimulates the growth of blood vessels to regenerate the endometrium. This phase ends with ovulation, when a sharp increase in LH causes rupture of a mature ovarian follicle.

During the luteal phase, LH causes the ruptured follicle to become the corpus luteum, which begins to secrete progesterone in addition to estrogen. Progesterone stimulates further growth of blood vessels in the endometrium and promotes the storage of nutrients such as glycogen. As progesterone secretion increases, LH secretion decreases, and if the ovum is not fertilized, the secretion of progesterone also begins to decrease. Without progesterone, the endometrium cannot be maintained and begins to slough off in menstruation. FSH secretion begins to increase (as estrogen and progesterone decrease), and the cycle begins again. Although an average cycle is 28 days, cycles of 23 to 35 days may also be considered normal.

Male Reproductive System

The male reproductive system consists of the testes and a series of ducts and glands. Sperm are produced in the testes and are transported through the reproductive ducts: epididymis, ductus deferens, ejaculatory duct, and urethra (Fig. 39–3). The reproductive glands are the seminal vesicles, prostate gland, and bulbourethral glands, all of which produce secretions that become part of semen.

Testes

The testes are located in the scrotum between the upper thighs, where the temperature is slightly lower than body temperature, which is necessary for the production of viable sperm. Each testis is about 1.5 inches long and 1 inch wide and contains the seminiferous tubules in which spermatogenesis (meiosis) takes place. In contrast to oogenesis, once started at puberty, spermatogenesis is a constant rather than cyclical process and usually continues throughout life. Also in the testes are specialized cells that produce the hormones testosterone and inhibin. Spermatogenesis is initiated by FSH from the anterior pituitary. LH from the anterior pituitary stimulates the secretion of testosterone, which contributes to the maturation of sperm. The secretion of inhibin is stimulated by testosterone; inhibin decreases the secretion of FSH, which helps keep the rate of spermatogenesis fairly constant. The functions of these hormones are summarized in Table 39–2.

A sperm cell consists of the head, which contains the 23 chromosomes; a flagellum, which provides motility; and the acrosome on the tip of the head, which contains enzymes to digest the membrane of the egg cell. Sperm from all the seminiferous tubules of a testis passes through tubules leading to the epididymis.

Epididymis, Ductus Deferens, and Ejaculatory Ducts

The epididymis is a tube about 20 feet long that is coiled on the posterior side of a testis. Smooth muscle within its wall propels sperm into the ductus deferens.

Also called the vas deferens, the ductus deferens extends from the epididymis in the scrotum through the inguinal

---

TABLE 39–1   HORMONES OF FEMALE REPRODUCTION

Hormone	Secreted By	Functions
Follicle-stimulating hormone	Anterior pituitary	Initiates development of ovarian follicles
		Stimulates secretion of estrogen by follicle cells
Luteinizing hormone	Anterior pituitary	Causes ovulation
		Converts ruptured ovarian follicle into corpus luteum
		Stimulates secretion of progesterone by corpus luteum
Estrogen	Ovary (follicle)	Promotes maturation of ovarian follicles
	Placenta	Promotes growth of blood vessels in endometrium
		Initiates development of secondary sex characteristics
		Promotes growth of duct system of mammary glands
Progesterone	Ovary (corpus luteum)	Promotes further growth of blood vessels in endometrium
	Placenta	Inhibits contractions of the myometrium during pregnancy
		Promotes growth of secretory cells of mammary glands
Inhibin	Ovary (corpus luteum)	Decreases secretion of FSH toward end of cycle
Prolactin	Anterior pituitary	Promotes production of milk after birth
Oxytocin	Posterior pituitary	Promotes release of milk

Source: Scanlon, VC, and Sanders, T: Essentials of Anatomy and Physiology, ed 3. FA Davis, Philadelphia, 1999, p 454, with permission.

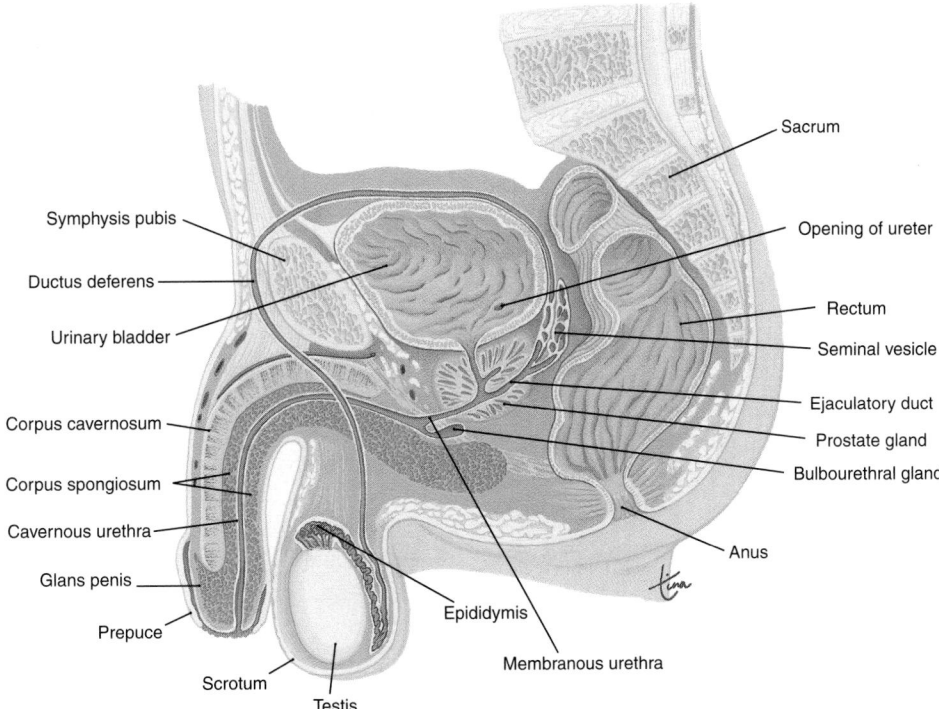

Figure 39-3   Male reproductive system in a midsagittal section. (From Scanlon, VC, and Sanders, T: Essentials of Anatomy and Physiology, ed 3. FA Davis, Philadelphia, 1999, p 442, with permission.)

TABLE 39-2	HORMONES OF MALE REPRODUCTION	
Hormone	**Secreted By**	**Functions**
Follicle-stimulating hormone	Anterior pituitary	Initiates production of sperm in the testes
Luteinizing hormone	Anterior pituitary	Stimulates secretion of testosterone by the testes
Testosterone	Testes	Promotes maturation of sperm
		Initiates development of male secondary sex characteristics
Inhibin	Testes	Decreases secretion of FSH to maintain a constant rate of spermatogenesis

Source: Scanlon, VC, and Sanders, T: Essentials of Anatomy and Physiology, ed 3. FA Davis, Philadelphia, 1999, p 443, with permission.

canal and into the abdominal cavity. The inguinal canal is an opening in the abdominal wall for the spermatic cord, a connective tissue sheath that contains the ductus deferens, testicular blood vessels, and nerves. Within the abdominal cavity, the ductus deferens extends over the top of the urinary bladder, then down the posterior side to join the ejaculatory duct on its own side.

Each of the two ejaculatory ducts receives sperm from the ductus deferens and the secretion of the seminal vesicle on its own side. Both ejaculatory ducts empty into the urethra.

Seminal Vesicles, Prostate Gland, and Bulbourethral Glands

The paired seminal vesicles are posterior to the urinary bladder. Their secretion is alkaline to enhance sperm motility and contains fructose to nourish the sperm. The duct of a seminal vesicle joins the ductus deferens on its side to form the ejaculatory duct.

The prostate is a muscular gland that surrounds the first inch of the urethra as it emerges from the urinary bladder. The secretion of the prostate is alkaline and contributes to sperm motility. The smooth muscle of the prostate contracts during **ejaculation,** causing the expulsion of semen from the urethra.

The bulbourethral glands are located below the prostate gland and empty into the urethra. Their alkaline secretion coats the interior of the urethra just before ejaculation, which neutralizes any acidic urine that might be present.

The alkaline secretions of the male reproductive glands ensure that sperm remain viable in the acidic environment of the vagina. The normal bacterial flora of the vagina create an acidic pH there, but the pH of semen is about 7.4 and permits sperm to remain motile.

Urethra and Penis

The urethra is the last of the male reproductive ducts, and its longest portion is within the penis. The penis is an external genital organ; its distal end is called the glans penis and is covered with a fold of skin called the prepuce or foreskin. Within the penis are three masses of erectile or cavernous tissue. Each consists of a framework of smooth muscle and connective tissue that contains blood sinuses, which are large, irregular vascular channels.

When blood flow through these sinuses is minimal, the penis is flaccid (soft). Sexual stimulation causes the arteries to the penis to dilate; the sinuses fill with blood, and the penis becomes erect and firm. This is brought about by parasympathetic impulses. The culmination of sexual stimulation is ejaculation, which is brought about by peristalsis of the reproductive ducts and contraction of the prostate gland.

Aging and the Reproductive System

For women there is a definite end to reproductive capability; this is called menopause and usually occurs between the ages of 45 and 55. Estrogen secretion decreases, and ovulation and menstrual cycles become irregular and finally cease. The decrease in estrogen has other effects as well. Loss of bone matrix may lead to osteoporosis and fractures; an increase in blood cholesterol makes women more likely to develop coronary artery disease; and drying of the vaginal mucosa increases susceptibility to vaginal infections.

For most men, testosterone secretion continues throughout life, as does sperm production, although both diminish with advancing age. Perhaps the most common reproductive problem for older men is enlargement of the prostate gland, called prostatic hypertrophy. As the urethra is compressed, urination may become difficult. Prostatic hypertrophy is usually benign, but cancer of the prostate is one of the more common cancers in elderly men.

▶ FEMALE ASSESSMENT

Assessment of women's reproductive health can seem challenging because of the complex relationship of physical and psychosocial factors. Hormones not only affect a multitude of body functions, they also can influence moods and mental functioning. Reproduction involves not only physical processes, but also relationships, role identifications, and self-esteem issues.

Normal Function Baselines

Knowing about expected functioning of the reproductive system is the nurse's best preparation for nursing assessment. Regular, relatively pain-free shedding of an appropriate amount of the lining of the uterus is expected from puberty through midlife or later. Intercourse is normally expected to be free of pain and infection, to occur when desired by both partners, to be satisfying, and generally to result in pregnancy over a reasonable period (unless precautions are taken). Pregnancies are expected to last approximately 40 weeks and to produce a healthy child. Physical and psychological sexual characteristics and function including **libido** are expected to be adequately maintained by hormones. Sexual functioning, desire, and fertility are expected to change throughout the process of aging. Although individuals may vary somewhat from these expected descriptions, these serve as a baseline for assessment of possible disorders. Chapter 40 further defines specific reproductive system disorders.

Much of what happens in female reproductive system disorders occurs inside the body and may not show external signs. Skill in asking appropriate questions, documenting patient statements, and describing observations is essential. Descriptions of symptoms should be thorough and follow the WHAT'S UP? format described in Chapter 1. Short quotes of the patient's own words to describe change that you have noticed may add valuable information, but be careful to use critical thinking skills when interviewing and documenting, rather than just quoting indiscriminately. Because many signs and symptoms of reproductive system disorders occur in a cyclic fashion, the patient may be asked to keep an accurate written record of occurrences, noting times and dates to identify patterns.

History

The nurse is often responsible for recording initial history and physical assessment data. Breast history questions include breast-feeding history; knowledge and practice of breast self-examination; discomforts; breast changes; and presence of any lumps, thickening, or nipple discharge.

Menstrual history questions include age of **menarche** (beginning of menstrual periods) and age of menopause (if applicable), length of cycles, length of menses, menstrual regularity, amount of bleeding, presence of clots, discomforts, and measures taken to relieve discomforts.

Obstetrical history includes number of pregnancies, pregnancy outcomes, and complications. These are generally documented using abbreviations of Latin words: G = pregnancies (from the Latin word *gravida*); P = births, whether alive or stillborn (from the Latin word *para*); A = abortions, whether spontaneous or therapeutic (from the Latin word *abortus*; a spontaneous abortion is sometimes called a miscarriage); and Roman numerals following the letter to specify the number of each. For example, three pregnancies, twins, one single birth, and one spontaneous abortion, are recorded as GIII, PIII, AI. This may also be written as $G_3P_3A_1$. Note, however, that some hospitals use different notation, further delineating history, such as, number of premature or full term births, number of living children, and number of therapeutic abortions. If uncertain of how to document, check your institutional history forms or ask the person in charge.

Medical intervention history includes any testing, operations, or treatments done on the reproductive organs and excretory system. Medications the patient is taking (for whatever reasons), height-to-weight ratio, and marked changes in weight may also provide significant data for diagnostic and care planning purposes concerning reproductive system disorders.

---

libido: sexual desire

menarche: men—month + arche—beginning

Sexual history questions include whether the patient is sexually active, sexual preference, number of partners, history of sexually transmitted disease (STD), knowledge and practice of STD risk reduction, history of participation in high-risk sexual activities, birth control knowledge and practice, and whether their level of sexual functioning is satisfactory. Many nurses feel awkward asking these questions, and patients may also feel some uneasiness with this line of questioning. A matter-of-fact attitude, an assurance of confidentiality, and an adequate explanation about why the information is needed tend to encourage patient comfort and cooperation.

Breast Assessment
Palpation

Palpation is the most important assessment technique for breast examination, because it can be used to identify alterations from normal consistency, to confirm the presence of lumps, and to locate areas of tenderness. Even mammograms are not sensitive enough to detect a small percentage of masses that can be felt by the patient or health care provider.

Breast Self-Examination

Self-palpation during breast self-examination (BSE), if done regularly and thoroughly, may be even more sensitive than physician or nurse palpation, because the patient becomes so familiar with her own breasts that she is more likely to notice subtle changes that an infrequently visited health practitioner might overlook. BSE is one of the most important health protection skills that nurses can teach to women. The few moments spent monthly on this activity may mean the difference between life and death or comfort and extreme suffering for women.

Patient Teaching

The examination procedure is simple. BSE should be done regularly once per month. One week after the beginning of the menstrual period is a good time to do BSE because some women's breasts become swollen and feel lumpier around the beginning of menses. For women who no longer have a regular menstrual period, any regular monthly schedule is fine. Although most women's breasts are not exactly the same size, marked differences between the breasts or a change in the size of one breast should be checked with a health care provider. Puckering or dimpling of skin, asymmetrical movement, and different pointing position of the nipples should also be reported to a health care provider. Whether the breasts are examined in parallel lines, a spiral formation or a wedge pattern is probably insignificant. It is important, however, to encourage that the examination be methodical and cover all areas of the breast, the tail of Spence, and the axilla (Fig. 39–4).

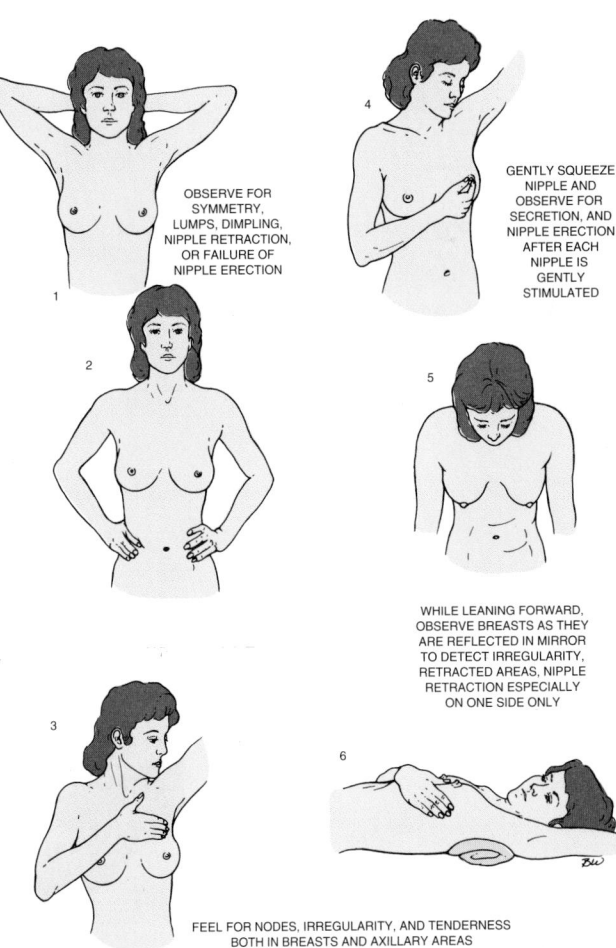

Figure 39-4 Breast self-examination. (From Venes, DJ (ed): Taber's Cyclopedic Medical Dictionary, ed 19. FA Davis, Philadelphia, 2001, with permission.)

Diagnostic Tests of the Breasts
Ultrasound and Mammography

Further assessment of the breast may be done by several methods. Ultrasound examination is done by bouncing high-frequency sound waves off the tissues within the breast to determine the density of the tissues and to map the breast structures (Fig. 39–5). This is mainly useful for distinguishing fluid-filled (cystic) lumps from solid tumors but may also be used to guide a needle for fine needle aspiration of fluid or core needle biopsies.

Mammography is a radiographic (x-ray) examination of the breast. A special machine is used that spreads and flattens the breast tissue to a thin layer to more effectively show benign and malignant growths, which might be hidden by breast structures on typical chest examination. (See Chapter 10.) Generally at least two radiographs are taken of each breast, with the machine compressing the breast top to bottom and side to tide to give comparison views of any lumps from more than one angle. If suspicious or unclear spots are seen, additional views may be taken. The American Cancer Society recommends from age 20 throughout life that women do monthly BSE, from ages 20 to 39 that they add a breast examination by a health professional every 3 years, and from age 40 that they have a yearly mammogram and breast examination by a health professional. Those who experience breast symptoms or have a strong family history of breast cancer may be advised to have more frequent examinations.

---

cystic: baglike
mammography: mammo—breast + graphy—recording

PATIENT TEACHING. Patients preparing for mammography should be advised to bathe and not to apply deodorant, powder, or any other substance to the upper body because these may cause false shadows on the test. They should be instructed that if a shadow is seen on a mammogram, further testing will be done to determine the reason for the shadow.

Thermography, Tomography, and Magnetic Resonance Imaging

There are several other less commonly used methods for diagnosis of breast disorders. Thermography is a method of mapping the breast using photographic paper, which records temperature variations throughout the tissue in different colors. Tomography takes very precise x-ray pictures of the breast, layer by layer, as it would look if it were in thin slices, and then stores these pictures in a computer. This allows for precise measurement of the position of tumors without the displacement caused by flattening the breast for a mammogram. Tomography, however, is much more expensive than mammography, so it is not a practical method to use for general screening of all women to detect possible breast cancers. Magnetic resonance imaging (MRI) uses radiofrequency radiation and magnetic fields to map the breast tissue. The equipment needed for this method is expensive, and it is unavailable in some areas.

PATIENT TEACHING. Ask patients preparing for MRI whether they have any metal inside their bodies, such as orthopedic wires, metal sutures, or artificial joint replacements, because heat is generated by the MRI and the procedure may be contraindicated.

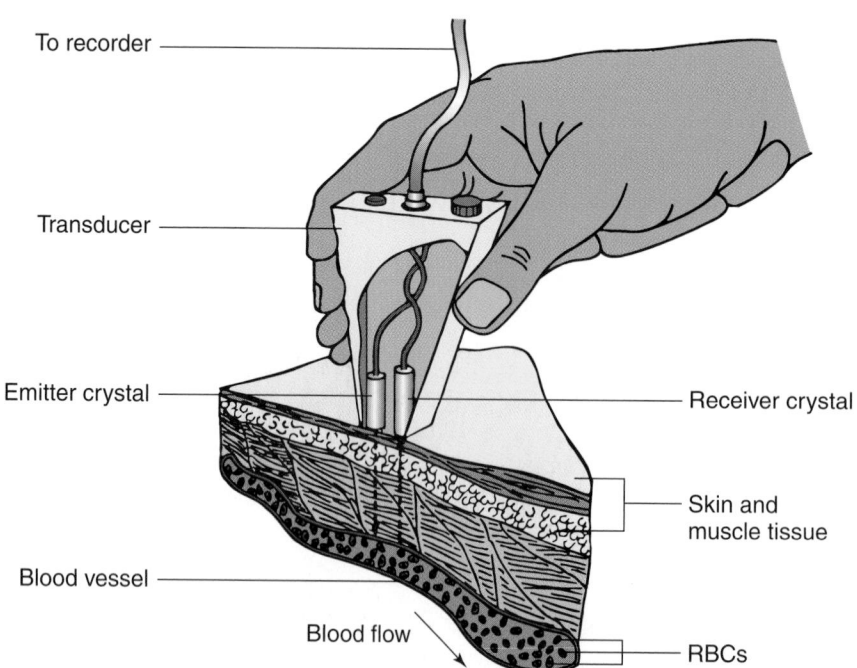

Figure 39-5 Diagnostic testing by ultrasound. (Modified from Cavanaugh, BM: Nurse's Manual of Laboratory and Diagnostic Tests, ed 3. FA Davis, Philadelphia, 1999, p 687, with permission.)

To recorder

Transducer

Emitter crystal

Receiver crystal

Skin and muscle tissue

Blood vessel

Blood flow

RBCs

Biopsy

If suspicious lumps or lesions are found in the breast by any of these methods, they may be checked using one of the other methods and then further assessed by biopsy. This procedure involves removing a small portion of tissue, fluid, or cells from the breast or lymph nodes for microscopic examination. This may be done by surgically removing a portion of tissue or by aspirating fluid or cells through a needle that is placed into the lump or lesion. Needle biopsies are often done with local anesthetic and may take place in a clinic or physician's office. More extensive biopsies may require a general anesthetic. A frozen section examination may be done in the laboratory by moistening and rapidly freezing a section of tissue, slicing it very thinly, and immediately examining it by microscope. This allows for diagnosis to be made during the course of an operation, so that the patient is spared an additional later operation for removal of cancerous tissue.

Nursing Care

As with any surgical intervention, be prepared to set out the sterile equipment and supplies needed for the procedure and ensure that a signed consent has been obtained. Following the biopsy, assess the patient for excessive bleeding and instruct about signs of impairment of healing processes. It is important that biopsy samples be clearly labeled, packaged appropriately for transport to laboratory facilities, and delivered promptly. Consult the laboratory for information about the transport container and whether a cell fixative is required in the container.

Assessment of the patient's psychological condition during breast diagnosis procedures is essential. Most women know someone who has had breast cancer. Although breast cancer screening procedures can seem routine to health care workers, they may be a cause of much anxiety for patients and their families. An understanding and calm nurse who can explain the procedures can help the assessment phase to be less traumatic.

Additional Diagnostic Tests of the Female Reproductive System

There are many tests to assess reproductive system function; this is presently an area of rapid change. The names of the individual procedures may also vary among institutions or according to particular methods used (e.g., a **salpingoscopy** is also called a fallopscopy, and a laparoscopy is the same as a peritoneoscopy). For this reason, this chapter is limited to general descriptions of categories of medical and surgical assessment procedures rather than attempting to name all tests.

Hormonal Tests

Hormonal tests are commonly used to assess functioning of the endocrine system as it relates to reproduction. They may be used to measure potential fertility, to find reasons for abnormal menses, to assess hormone-producing tumors, and to determine whether treatments to adjust hormone levels have been effective. Several hormones may be assessed at any one time. Some hormonal tests are time specific, and the samples can be useless if not gathered within a certain time range.

NURSING CARE. Consult institution policy for specific instructions for each test. Explain the procedure to the patient and provide support. Women who are undergoing hormonal tests may feel embarrassed, worried about their femininity and potential fertility, and depressed because of repeated tests, often with little positive result. Some may fear loss of their spouse's love (and perhaps relationship) if they are diagnosed as infertile.

Pelvic Examination

The pelvic examination allows visual inspection of the vagina and cervix, as well as sampling of mucus, discharge, cells, and exudates. Palpation of portions of the reproductive system and some treatments may also be done as part of the procedure.

NURSING CARE. Be prepared to assist the health care provider with the examination. Explain the procedure as you set out the supplies. Vaginal specula range from tiny virginal sizes to extra large (Fig. 39–6). The appropriate size is related to the size of the woman's (or child's) pelvis and whether she has had children or not. For a small child a nasal speculum may be used. A sterile basin of warm tap water to warm the sterile speculum and a small amount of surgical lubricant should be placed near but not on the speculum (because some tests may be affected by water or lubricant). Two clean gloves for the examiner should be placed nearby, and a light should be adjusted to illuminate the area. Other equipment may be set out according to the tests or treatments that will be carried out during the pelvic examination.

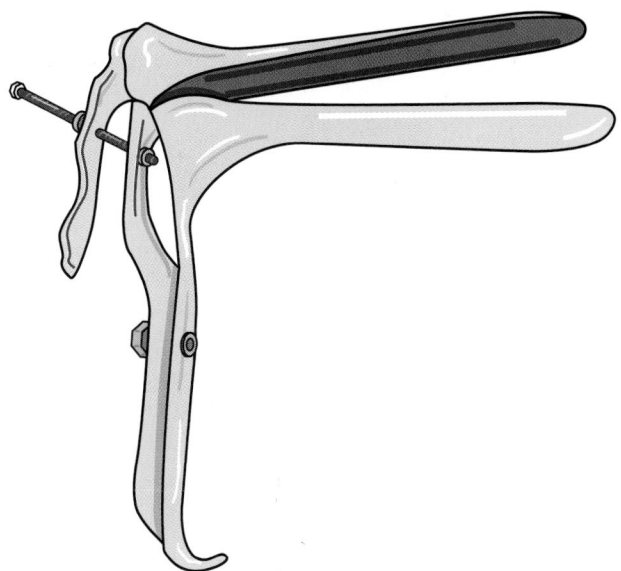

Figure 39-6 Vaginal speculum.

---

salpingoscopy: salpingo—tube + scopy—looking

Instruct the patient to empty her bladder before the examination. Place the patient in a gown (without underwear) either on her back with her arms resting down at her sides to aid relaxation of abdominal muscles or in a side-lying position, according to the health care provider's preference. Provide a sheet turned so that there is plenty of room for the legs to spread while the sheet still covers the patient on the sides.

Bimanual Palpation

Because much of the reproductive system is not visible even with a speculum, **bimanual** palpation is often done during a pelvic examination. One hand is placed on the abdomen and the other gloved hand is inserted deeply into the vagina. The uterus and **adnexa** are moved about between the two hands to feel the size, shape, and consistency of the uterus and adnexa and to check for any abnormal growths.

NURSING CARE. Explain the procedure and support the patient. Some women may be fearful, embarrassed, or tense and may find the procedure uncomfortable. Active relaxation strategies may decrease discomfort.

⌁⌁ CRITICAL THINKING

Reproductive Assessment

How might the age of the patient change your approach, plans, and teaching for patients who have disorders of the reproductive system? Consider your approach, plans, and teaching for each of the following scenarios.

1. A 2-year-old child is brought into the clinic by her mother because she has a foul smell coming from her perineal area and a slight yellowish discharge from her vagina.
2. A 19-year-old woman comes to the doctor's office where you work for renewal of her yearly birth control pill prescription. Your employer enforces regular checks for cervical changes by renewing the prescription only after a Papanicolaou (Pap) smear is done. As you start setting out the Pap smear materials, your patient expresses some reluctance to have a Pap smear today because she is so sore already.
3. Your 56-year-old patient comes in to "get things checked out" because she hurts every time that she and her husband have intercourse.

Answers at the end of chapter.

Cytology

Cytology is the study of cells taken as biopsies. There are several ways that cells from the reproductive system may be removed for microscopic examination. During a Pap smear, one or more small samples of cells are gently scraped away from the surface of the cervical canal using a small wooden spatula, tiny cylindrical brush, or long cotton-tipped applicator, and then they are smeared or rolled onto microscopic

slides for viewing. Cells may also be collected by **conization,** which involves removing a small cone-shaped sample from the cervical canal, or by punch biopsy, which removes a small core of cells. Endometrial biopsy specimens are samples of cells taken from the lining of the uterus by scraping with a small spoon-shaped tool called a **curet,** which is inserted through the cervix. Small biopsy specimens may also be taken by cutting or removing a suspicious lesion. Cells may be observed for changes indicative of hormonal secretion, cellular maturation, or abnormalities such as are seen with viral growths and cancerous or precancerous conditions.

NURSING CARE. Add appropriate sample collection materials and fixatives to the pelvic examination supplies according to the type of cytological examination being done. For a Pap smear, this usually includes one or two clear glass slides labeled with the woman's name, fixative spray, cytobrush or wooden cervical spatula, and transport box for the slides. For larger cell samples, a sterile biopsy collection container labeled with the name and site of biopsy and a bottle of fixative solution should be placed close to the pelvic examination supplies. The procedure manual or health care provider is consulted concerning the types of instruments to put out for biopsies other than Pap smears. Cells die and degrade rapidly once removed from the patient, so they must be packaged securely (generally in a preservative solution or sprayed with a fixative) for transport to laboratory facilities.

Prepare the patient by explaining the procedure and providing support. The woman may be fearful of cancer or other abnormality. Removal of the sample may cause pain, bleeding, swelling, or, later, inflammation, so the patient is monitored after the procedure and alerted to watch for and report these complications if they occur. Document the woman's status after the procedure on the chart and record that the sample was sent to the laboratory.

Swabs and Smears

Swabs and smears are done to determine which microorganisms are causing disease and which antibiotics should be used.

NURSING CARE. Add sample collection materials, including swabs, slides, and a chlamydia collection kit, to pelvic examination equipment if symptoms of vaginal infection are present. It is important to place samples of discharge into culture media that support growth. Clear media are required for some and charcoal media for others, so both types should be set out. Viral swabs require a special collection kit. Chlamydia samples are especially difficult to transport to laboratories, and special kits are available for this pathogen. Some microorganisms, such as yeasts and *Trichomonas,* can be identified well from smears on slides. Wet

---

bimanual: bi—two + manual—hands
adnexa: ad—together + nexa—to tie (usually refers to ovaries and tubes)

conization: coniz—cone-forming + ation—process
curet: scoop

mounts are smears of discharge spread onto a slide. These must be taken to the microscope immediately after they are obtained. Sodium chloride and potassium hydroxide are dropped onto individual wet mount slides before they dry to aid in identification of some microorganisms. Support the patient, who may be anxious about possible sexually transmitted diseases and effects on relationships.

Sonography

Ultrasound assessment (also called sonography) may be done to determine size, shape, development, and density of structures associated with the female reproductive system, as well as fetal measurements and some types of prenatal diagnoses. This procedure is especially useful for differentiating cysts from solid tumors, as well as locating **ectopic** pregnancies and intrauterine devices. Ultrasound may also be used to guide needles for obtaining samples of fluid or cells. Either external or vaginal transducers may be used to send and receive the signals for this procedure. Vaginal transducers are placed in a plastic sheath before insertion into the vagina. A full bladder may be required for some ultrasound tests.

NURSING CARE. Explain the procedure and support the patient. The pressure of the transducer on the skin may be painful if the underlying structures are inflamed or swollen or if the bladder is very full.

Radiographic Procedures

Several radiographic procedures may be used for diagnosis of reproductive system problems. Computed tomography (CT) scanning and MRI are used to locate tumors of the reproductive system. Structures of the female reproductive system may also be outlined by taking x-ray pictures of cavities that have been filled with a radio-opaque substance.

During a **hysterosalpingogram,** dye is injected into the uterus until it comes out the ends of the fallopian tubes. This test is useful for identifying congenital abnormalities in the shape of the uterus and blockages of the fallopian tubes.

NURSING CARE. Prepare the patient for this test according to the x-ray department manual protocol and the physician's orders (they may include a laxative, suppository, or enema). Ensure that the patient understands the procedure and that appropriate consents are signed if required. Ask about allergies to iodine or shellfish, because contrast media may contain iodine. Notify the charge nurse immediately if the patient reports an allergy. After the procedure, assess for nausea, light-headedness, and signs of allergy and promote comfort, because some cramping may occur. Discharge teaching should include signs of infection and advice that the x-ray dye may stain clothing, so a pad should be worn until any vaginal drainage stops.

Endoscopic Examinations

Several types of endoscopic examinations are done to visually inspect internal areas to diagnose (and sometimes treat) reproductive system disorders. The names of the tests vary according to the area inspected, but all these generally make use of a fiberoptic light and lens system, which is inserted through a tube called a cannula into a small incision. A laparoscopy is done to view the abdominal cavity and is useful for identifying problems such as endometriosis (Fig. 39–7). A salpingoscopy is performed to see the inside of the fallopian tubes and a **hysteroscopy** to see the inside of the

---

hysterosalpingogram: hystero—womb + salpingo—tube + gram—record
hysteroscopy: hystero—womb + scopy—looking

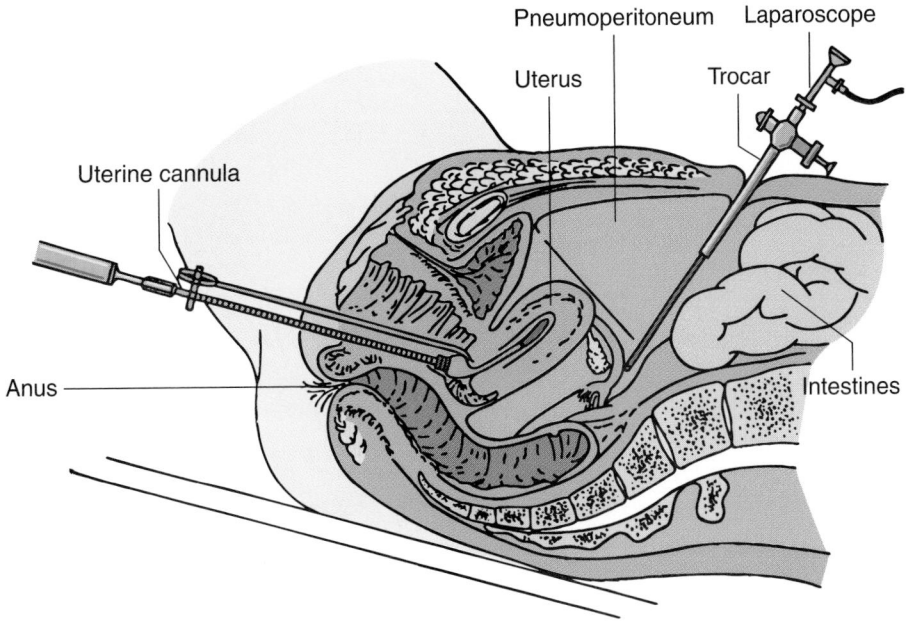

Figure 39-7   Laparoscopy. (Modified from Cavanaugh, BM: Nurse's Manual of Laboratory and Diagnostic Tests, ed 3. FA Davis, Philadelphia, 1999, p 571, with permission.)

uterus. A binocular microscope is used with an endoscope that is introduced into the vagina to closely study lesions of the cervix during a **colposcopy.** During **culdoscopy,** an endoscope is introduced into the vagina and through a small incision in the vagina into the cul-de-sac of Douglas, a cavity behind the uterus (Fig. 39–8).

NURSING CARE. Preoperatively the patient is prepared for an endoscopic examination according to institutional protocol. This generally involves asking the patient whether she has fasted as instructed, assessing vital signs, recording the time of last voiding, helping the patient into the operating room gown, and obtaining a signature on the consent form. General anesthesia may be given for some endoscopic procedures. Explain what to expect and provide support for the woman. The patient may be anxious about possible disorders.

Postoperatively, provide comfort measures. These procedures produce almost no blood loss. The woman may experience pain in the neck, shoulders, and upper back if carbon dioxide (CO_2) gas was pumped into the body compartment being examined. Called **insufflation,** this is done to increase the distance between structures, so that it is easier for the physician to see and diagnose possible disorders. The CO_2 remaining after completion of the examination travels to the highest level of the body before being absorbed, so lying flat for a few hours after the examination may decrease discomfort. If incisions were made through the abdominal wall for insertion of the endoscope and for insufflation, these are tiny and a Band-Aid or small dressing is applied.

---

colposcopy: colpo—vagina + scopy—looking
culdoscopy: culdo—cul de sac + scopy—looking
insufflation: in—in + suffl—to blow + ation—process

PATIENT TEACHING. Advise the patient to observe the sites for redness, bleeding, or any drainage, returning to the physician promptly if these occur and otherwise approximately 1 week later for suture removal (according to physician preference). If the endoscopic procedure was done transvaginally, advise the patient to wear a perineal pad until the drainage stops, to report any bright bleeding after the operative day, and to report any fever or foul-smelling discharge.

Further detail about some specific forms of the tests that have been described in this chapter are included with the disorders to which they apply in Chapters 40 and 42.

▶ **MALE ASSESSMENT**

As with the female reproductive system, the male reproductive system is a complex interaction of both physical and psychosocial factors. Unlike women, however, men may find it much more difficult in our society to talk about or admit to having problems related to reproductive health. From toilet training through adulthood, men are expected to have behaviors associated with maleness. Unfortunately, by the time some boys reach manhood, their male identity is defined by the successful functioning of their sex organs.

One of the important first steps in obtaining a male reproductive assessment is providing a comfortable, nonjudgmental, confidential atmosphere for discussion. This means you must first be knowledgeable and comfortable with sexual issues. Nurses do not hesitate to ask female patients about their menstrual history and should be equally comfortable in asking men about their **erection** or ejaculation history. Be open and straightforward with all questions and answers. It may be necessary at times to use more commonly expressed sexual words instead of medical terminology. You will discover that many men do not know the function of their prostate gland or the difference between ejaculation

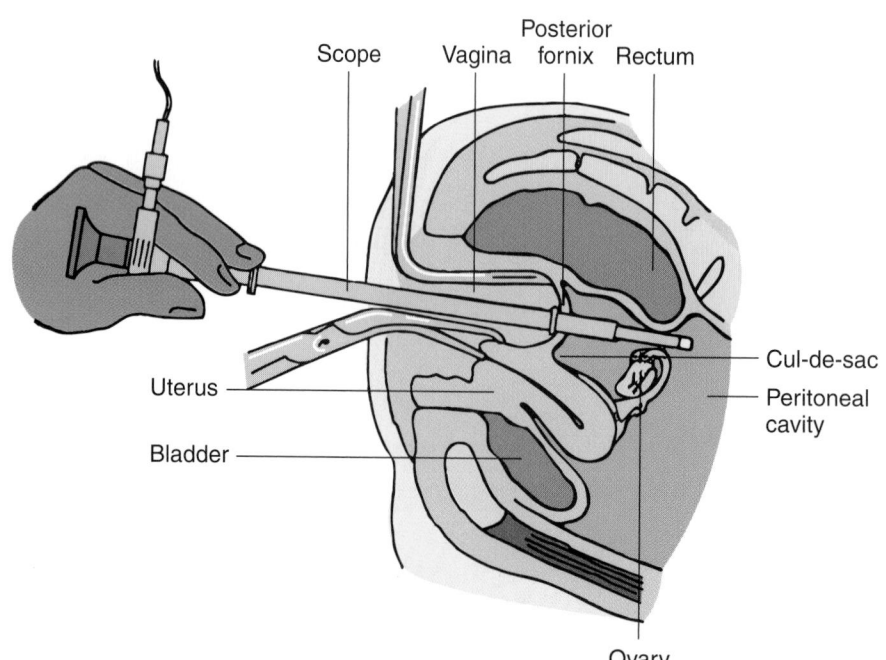

Figure 39-8   Culdoscopy. (Modified from Cavanaugh, BM: Nurse's Manual of Laboratory and Diagnostic Tests, ed 3. FA Davis, Philadelphia, 1999, p 566, with permission.)

and **orgasm.** Use the assessment as an opportunity to teach men the facts about their own sexual functioning.

History

There are some basic questions to ask a male patient during an assessment of sexual functioning. These include the following:

1. "Are you having any problems getting or keeping an acceptable erection?" Additional questions might be whether the patient can have an erection anytime (with or without a sex partner); if erection problems occur when he is under stress or taking medication; or if he is involved in substance abuse. "Do you have erections first thing in the morning or during sleep?" If the patient is able to get an erection, is it straight and firm enough for penetration? Does the erection last long enough for him to reach orgasm?

2. "Are you able to ejaculate during orgasm? Do you have any painful sensations related to the experience? Does any fluid come out during the ejaculation process, and if so, is it a small, moderate, or large amount, and is it clear, cloudy, or brown in color?" The amount and color may de-

pend on whether the patient has had recent sexual activity or genitourinary surgery or may indicate a congenital anomaly or infection.

3. For patients younger than 40 years old: "Do you practice monthly testicular self-examinations?" Men younger than age 40 should be encouraged to perform monthly testicular examinations as a cancer prevention measure.

4. For patients older than 40 years: "When was the last time you had a prostate examination?" A digital rectal examination (DRE) should be a regular part of a man's routine physical after age 40. Prostate cancer is treatable when detected early.

A more complete male sexual history not only includes information about sexual practices but also medical information. A variety of assessment data may affect a man's current and future sexual functioning (Table 39–3).

Physical Examination

The physical examination is generally performed by a physician or someone trained in physical assessments. The examination begins with the patient's general appearance. He

TABLE 39-3 MALE SEXUAL ASSESSMENT GUIDELINES		
Subject	**Data to Collect**	**Possible Impact**
Medication	Prescribed or over-the-counter medications that affect sexual desire, erection, or ejaculation (refer to Table 41-3)	Loss of sexual desire, erection, ejaculation, orgasm, or fertility
Family history	Any genetically transmitted diseases (e.g., heart problems, hypertension, diabetes), mother's use of DES during pregnancy	High risk for circulation problems that interfere with erections, congenital anomalies of reproductive organs
Personal habits	Tobacco use, caffeine intake, substance abuse/recreational drugs, steroid use, frequent use of hot tubs, long-distance driving or bide riding	Decreased blood flow to penis, loss of erection; decreased testosterone (male hormone) interferes with erection and fertility; excessive heat decreases sperm production
Personal health history	Mumps during adolescence, recent infection or fever	Decreased sperm production
Mental health	Stressful personal problems or job situation, problems with sexual partner, performance anxiety or depression	Decreased sexual desire and ability to have an erection
Circulatory/respiratory	Heart problems/surgery, high blood pressure, sickle cell disease, lung disease, sleep apnea	Decreased circulation—unable to have usable erection; decreased respiratory—activity intolerance, loss of erection, congenital anomalies of reproductive organs
Gastrointestinal	Liver infection/disease, surgery, bowel problems/surgery	Liver—decreased testosterone and increased estrogen, loss of erection; gastrointestinal/bowel—pain, loss of desire; surgery—loss of blood or nerve flow
Musculoskeletal	Painful joints, back injury/surgery, pelvic/lower back nerve damage	Pain, loss of desire; limited movement/positions, loss of erection, ejaculation, and orgasm
Neurological	Stroke, multiple sclerosis, Parkinson's disease	Limited movement/positions, loss of sensations, loss of control
Metabolic/endocrine	Diabetes, obesity, thyroid problems	Diabetes mellitus—circulation problems, retrograde ejaculation, nerve damage; obesity—decreased male hormones, excess female hormones
Genitourinary	Congenital deformity of penis/testicles, prostate problems/surgery, erection/ejaculation problems, sexually transmitted diseases	Difficulty with penetration, erection problems, retrograde ejaculation, infertility
Sexual practices	Frequency of intercourse (including positions, timing with female ovulation cycle), masturbation, use of lubricants or birth control methods, problems with erectile dysfunction or premature ejaculation	Decrease in quality and quantity of sperm that reach the female egg

DES = diethylstilbestrol.

is observed for male patterns of hair growth on the head, face, chest, arms, and legs. Normal male pubic hair pattern is "triangle up," hair growth up toward the umbilicus. The patient's height and muscle mass are noted. Men are commonly taller than 5-feet 6-inches tall, weigh more than 135 pounds, and have shoulders that are broader than their hips. The presence of excess breast tissue may indicate **gynecomastia,** an excess of female hormones. Abnormal findings in either hair patterns or muscle mass often indicate a hormone imbalance.

The penis, scrotum, and testes (testicles) are examined by observation and palpation. On observation, the penis is normally flaccid (soft) and hanging straight down. The size can vary greatly and should not be a concern unless it is unusually small (microphallus) or edematous. The left testis generally hangs slightly lower in the scrotum than the right.

The penis is examined for warts, sores (evidence of sexually transmitted diseases), swelling, curves, or lumps along the shaft. The examiner also makes sure the urethral opening is at the tip of the penis and not on the underside of the shaft **(hypospadias).** If the man is not **circumcised,** the foreskin should be pulled back carefully and the area inspected for signs of inflammation or foul-smelling discharge. The practitioner should be sure to replace the foreskin in the forward position after the examination is completed.

The scrotum and testes are carefully examined and palpated. Both testes should be present and a normal size (approximately 2 cm × 4 cm). The testes are egg shaped and should feel smooth and rubbery when lightly palpated between the thumb and fingers. The epididymis can be felt along the top edge and posterior section of each testis. The testes and scrotum are palpated for any lumps, cysts, or tumors. If a fluid-filled mass **(hydrocele)** is found, further evaluation should be done. A simple noninvasive test called **transillumination** is used to determine if the mass is fluid filled or solid. With the room lights out, a flashlight is held behind the scrotum. If the mass is fluid, a red glow appears; if it is solid, it appears opaque. The spermatic cord (made up of veins, arteries, lymphatics, nerves, and the vas deferens) is palpated and should feel firm and threadlike. If a condition called a **varicocele** is present, the area feels like a bag of worms. A varicocele, which is swelling of the veins of the spermatic cord, is one of the most common problems associated with male infertility.

The male patient is also examined for inguinal hernias by pressing up through the scrotum into each of the inguinal rings while asking him to cough or bear down. Each side is examined separately while he is in the standing position. A hernia feels like a pulsation against the examining fingertips.

A digital rectal examination may be done by an experienced practitioner. During DRE, the prostate gland is palpated by inserting a gloved, lubricated finger into the rectum

while the man is in a knee-to-chest position. The entire posterior lobe of the gland can be felt this way. The gland should feel slightly firm and without any lumps. If the prostate gland feels very hard or soft, enlarged, or contains any lumps, a rectal ultrasound with needle biopsy is often ordered. A swollen, painful prostate generally indicates that an infection is present. Remind all men older than age 40 that unless they have had a complete removal of the prostate gland they still need a DRE performed every year. Many men are under the impression that any prostate surgery means the gland has been completely removed. When simple surgery is performed, prostate tissue is left in the body and will begin to regrow over time. This prostatic tissue can become cancerous and needs to be monitored with a yearly DRE.

Testicular Self-Examination

All men after puberty should do monthly testicular self-examination (TSE) to detect any tumors or other changes in the scrotum. See Box 39–1 for instructions that can be used to teach a man how to examine his testicles. Refer also to Figure 39–9.

Breast Self-Examination

Although breast cancer in men is rare, it can occur. Men, like women, should perform monthly breast self-examination.

Diagnostic Tests of the Male Reproductive System
Ultrasound

An ultrasound may be done to diagnose or evaluate a variety of male reproductive or genitourinary problems. A transrectal ultrasound may be done to diagnose prostate cancer. For this procedure, a rectal probe transducer is inserted into the rectum and sound waves are used to evaluate the

---

BOX 39–1 **Guidelines for Monthly Testicular Self-Examination**

The examination is easiest during or right after a warm shower or bath, when the scrotum is relaxed and the testicles are hanging low. Choose one day a month to always do the examination.

1. Raise the penis up out of the way and look for any difference in size or shape of each side of the scrotum (sac). The left side usually hangs a little lower than the right.
2. Using both hands, hold the scrotum in the palms. Begin, one at a time, to gently roll each testicle between the thumb and first three fingers, feeling for any lumps or hard spots.
3. Identify the parts. The testicles should feel round, smooth, and egg shaped. The epididymis along the top and back side should feel soft and a little bit tender. The spermatic cord is a tube that runs from the epididymis and usually feels firm, smooth, and movable.
4. See your doctor immediately if you feel any lumps or unusual changes.

---

gynecomastia: gyneco—female + mastia—breast
transillumination: trans—across + illumin—light + ation—
   process

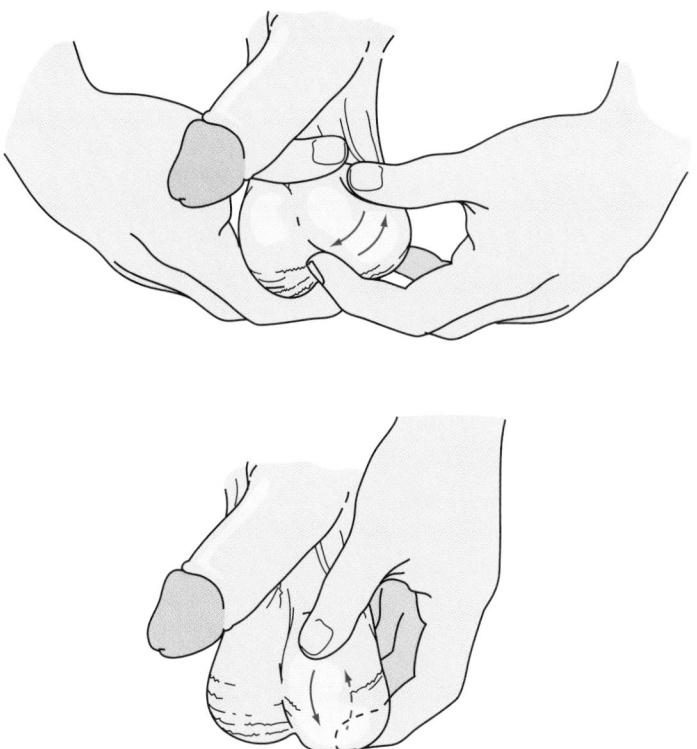

Figure 39-9 Testicular self-examination.

prostate gland. An enema may be ordered before the procedure. No special aftercare is necessary.

Pelvic or scrotal ultrasound helps evaluate and locate masses. Ultrasound may also be done to guide the needle during a fine needle biopsy.

Cystourethroscopy

Cystourethroscopy may be done to evaluate the degree of obstruction by an enlarged prostate gland. For this procedure, a Foley catheter is inserted and a dye is injected into the bladder. Radiographs are taken with the dye in the bladder and while voiding after the catheter has been removed.

NURSING CARE. The procedure is explained to the patient, and possible allergic reaction to dyes is assessed and communicated to the physician as necessary. The patient is instructed to void before the procedure. A sedative or analgesic may be ordered to help the patient relax during the procedure.

After the procedure, intake and output are measured for 24 hours and alteration from the patient's normal pattern or absence of urination are reported to the physician. Fluids are encouraged to promote excretion of the dye.

Laboratory Tests

PROSTATE-SPECIFIC ANTIGEN. Normal value of prostate-specific antigen (PSA) is less than 4 ng/L. PSA is a glycoprotein produced by prostate cells. An elevated level indicates prostatic hypertrophy or cancer.

PROSTATIC ACID PHOSPHATASE. Normal value of prostatic acid phosphatase (PAP) is less than 3 ng/mL. PAP is an enzyme that normally affects metabolism of prostate cancer cells. An elevated level indicates prostate cancer.

OTHER TESTS. If prostate cancer is suspected or diagnosed, additional tests may be done. Acid phosphatase may be elevated in metastatic prostate cancer. Alkaline phosphatase and serum calcium levels may be elevated if metastasis to the bone has occurred.

Tests for Infertility

Various hormone levels may be measured, including FSH, LH, testosterone, and adrenocorticotropic hormone (ACTH) to help determine causes of infertility in male patients.

Semen analysis may be done to provide information about infertility or to evaluate whether a vasectomy has been effective. Semen may be analyzed for sperm count, motility, and shape. Other tests determine whether the semen contains adequate nutrients to support sperm, whether antibodies to the sperm are present, and the ability of the sperm to penetrate an ovum.

NURSING CARE. The patient is instructed to refrain from ejaculation for 3 days before collecting the semen sample to avoid altering findings. Generally specimens are collected on three separate occasions over a period of 4 to 6 days. Masturbation and ejaculation directly into a sterile container are recommended to avoid loss of semen. Condoms and lubricants should be avoided. The sample should be taken to the laboratory within 1 hour of collection. Additional tests of the male reproductive system are discussed in Chapter 41.

1. Which male reproductive duct carries sperm into the abdominal cavity?
    a. Urethra
    b. Epididymis
    c. Ductus deferens
    d. Ejaculatory duct

2. Which of the following is the usual site of fertilization?
    a. Ovary
    b. Uterus
    c. Vagina
    d. Fallopian tube

3. According to the American Cancer Society, what is most effective method for detecting breast cancer?
    a. BSE
    b. Mammogram
    c. Clinical breast examination
    d. One or more of the above according to age guidelines

4. Which of the following items should be placed out in preparation for a Pap smear?
    a. Syringe
    b. Vaginal speculum in water
    c. Dry, sterile vaginal speculum
    d. Vaginal speculum with lubricant

5. Which of the following techniques is recommended for male patients younger than age 40 to detect prostate cancer early?
    a. Yearly DRE
    b. Monthly STE
    c. Yearly PSA
    d. Bimonthly bimanual examination

6. Anna Brown has just had a laparoscopy to investigate the causes of her infertility. The nurse should instruct her to lie flat in the bed for a few hours for which of the following reasons?
    a. She will rip out the stitches in her 4-inch long abdominal incision.
    b. Her blood pressure will be extremely low because of blood loss.
    c. The carbon dioxide left over from the test will travel upward and cause pain.
    d. Her uterus needs to be at the same level as her heart to prevent excessive swelling.

Answers to CRITICAL THINKING

Reproductive Assessment

1. Calm fears. Explain simply. Allow the parent to stay with the child if appropriate. Consider whether the child has possibly been abused. (If so, evidence needs to be collected and a report filed with the appropriate child protection authorities.) Place a nasal speculum out for examination. Place a forceps out for removal of a possible foreign body (not uncommon at this age). Teach that this is a normal part of the body that is to be protected and taken care of.

2. Assess knowledge and maturity. Set out supplies for a Pap smear and for swabs and smears. Teach while getting supplies ready. Explain that vaginal soreness generally needs to be treated and that the doctor must know more about the problem to do so effectively. Explain that inflammations can interfere with Pap smear results, so it may have to be repeated later after treatment. Explain culture and sensitivity testing. Teach about risk reduction and inform that oral contraceptives do not offer a barrier.

3. Try to put the woman at ease through general conversation. Set out supplies for a Pap smear (if needed) and for swabs and smears. Teach while getting supplies ready. Discuss aging and the effects of decreased estrogen in general and specifically on vaginal tissues. Inform that there are several ways to deal with the problems resulting from decreased estrogen, such as oral hormonal replacements, water-soluble vaginal lubricants, vaginal creams, estrogen patches, and estrogen receptor modulator medication.

40

Nursing Care of Women with Reproductive System Disorders

Linda Hopper Cook

KEY TERMS

agenesis (ay-**JEN**-uh-sis)

anteflexion (AN-tee-**FLECK**-shun)

anteversion (**AN**-tee-VER-zhun)

augmentation (AWG-men-**TAY**-shun)

balanitis (BAL-uh-**NIGH**-tis)

bilateral salpingo-oophorectomy (by-**LAT**-er-uhl sal-PINJ-oh-ah-fuh-**RECK**-tuh-mee)

cautery (**KAW**-ter-ee)

colporrhaphy (kohl-**POOR**-ah-fee)

contraceptive (KON-truh-**SEP**-tiv)

cryotherapy (KRY-oh-**THER**-uh-pee)

culdocentesis (KUL-doh-sen-**TEE**-sis)

culdotomy (kul-**DOT**-uh-mee)

cystocele (**SIS**-toh-seel)

cytolytic (SIGH-toh-**LIT**-ik)

dermoid (**DER**-moyd)

dilation and curettage (DIL-**AY**-shun and kyoor-e-**TAHZH**)

dysmenorrhea (DIS-men-oh-**REE**-ah)

dyspareunia (DIS-puh-**ROO**-nee-ah)

dysplasia (dis-**PLAY**-zee-ah)

ectasia (ek-**TAY**-zee-ah)

fibrocystic (FIGH-broh-**SIS**-tik)

hypertrophy (high-**PER**-truh-fee)

hypoplasia (HIGH-poh-**PLAY**-zee-ah)

hysterectomy (HISS-tuh-**RECK**-tuh-mee)

hysterotomy (HISS-tuh-**RAH**-tuh-mee)

imperforate (im-**PER**-foh-rate)

in vitro fertilization (in-**VEE**-troh FER-ti-li-**ZAY**-shun)

laparotomy (LAP-uh-**RAH**-tuh-mee)

laser ablation (LAY-zer uh-**BLAY**-shun)

leiomyoma (LYE-oh-my-**OH**-ma)

mammoplasty (MAM-oh-**PLAS**-tee)

marsupialization (mar-SOO-pee-al-i-**ZAY**-shun)

mastalgia (mass-**TAL**-jee-ah)

mastectomy (mass-**TECK**-tuh-mee)

mastitis (mass-**TIGH**-tis)

mastopexy (MAS-toh-**PEKS**-ee)

myomectomy (MY-oh-**MECK**-tuh-mee)

panhysterectomy (PAN-hiss-tuh-**RECK**-tuh-mee)

pedicle (**PED**-i-kuhl)

perimenopausal (PER-ee-MEN-oh-**PAWS**-uhl)

postcoital (post-**KOH**-i-tal)

rectocele (**RECK**-toh-seel)

retroflexion (RET-roh-**FLECK**-shun)

retrograde (**RET**-roh-grayd)

retroversion (RET-roh-**VER**-zhun)

teratoma (ter-uh-**TOH**-muh)

vaginosis (VAJ-i-**NOH**-sis)

QUESTIONS TO GUIDE YOUR READING

1. How would you explain the pathophysiology of each of the disorders of the female reproductive system?
2. What are the etiologies, signs, and symptoms of each of the disorders?
3. What care would you provide for patients undergoing tests for each of the disorders?
4. What is current medical treatment for each of the disorders?
5. What data should you collect when caring for patients with disorders of the female reproductive system?
6. What nursing care will you provide for patients with each of the covered disorders?
7. How will you know if your nursing interventions have been effective?

Reproductive system disorders can be frightening, irritating, frustrating, embarrassing, and in some cases fatal. They involve not just body parts but also roles, relationships, and sense of identity and purpose in life. Nurses can play an important role in helping women with these disorders. Women's health is an area where much research is being done. The Nurses' Health Study, conducted at Harvard Medical School, is a large ongoing study on many topics related to women's health. Learn about it at www.nurseshealthstudy.org.

▶ BREAST DISORDERS
Benign Breast Disorders

Much has been done in recent years to educate the general public concerning breast cancer. It is the leading cause of death among women ages 35 to 54; the mortality rate has not changed significantly since 1930.[1] Heightened awareness of the risks of breast cancer, however, sometimes results in excessive anxiety among women with benign breast conditions. The following section covers benign, or noncancerous, breast disorders.

Cyclic Breast Discomfort

PATHOPHYSIOLOGY, ETIOLOGIES, AND SIGNS AND SYMPTOMS. The most common breast symptoms result from cyclic variations in hormone levels. Swelling, tenderness, and sometimes pain (**mastalgia**) can be related to hormone-mediated changes within the breast tissues that prepare them for their potential role of breast-feeding.

TREATMENT. If persistent or severe, these symptoms may be treated with medications that modify hormone levels. Explaining that cyclic discomfort is temporary and not from a disease helps to allay fears.

Fibrocystic Breast Changes

PATHOPHYSIOLOGY, ETIOLOGIES, AND SIGNS AND SYMPTOMS. Overresponsiveness of cells to hormonal stimulation may cause long-term changes resulting in replacement of normal tissue with fibrous tissue, **ectasia** (overdevelopment) of cells, and blockage of ducts so that cysts form around trapped fluid. This makes the breasts feel somewhat hard and lumpy.

TREATMENT. Generally no treatment is necessary. Although **fibrocystic** changes are not cancerous, more frequent mammography or ultrasound may be advised because the fibrocystic changes may make it more difficult to feel early cancerous lumps during breast self-examination (BSE) and some types of breast cysts are associated with a higher cancer risk.[2]

Mastitis

PATHOPHYSIOLOGY, ETIOLOGIES, AND SIGNS AND SYMPTOMS. Breast infection with inflammation (**mastitis**) occurs as a result of injury and introduction of bacteria into the breast. This condition most commonly occurs while breast-feeding. The breast becomes swollen, hot, red, and painful and may form an abscess.

TREATMENT. Mastitis may be treated either with antibiotics or by incision and drainage (I & D) of the abscessed area. The location and type of infection determine whether breast-feeding should be continued.

NURSING CARE AND TEACHING. You may assist with an I & D by setting out the equipment: a wrapped sterile sharp-pointed scalpel blade, a blade handle, clean gloves, and dressing materials. A dressing may be applied over the I & D site to absorb drainage. Application of warm, moist packs may also increase comfort and promote healing for patients with mastitis.

Patient teaching should include instructions to wash hands carefully to prevent the spread of infection and to wear a supportive bra to relieve some of the discomfort.

Malignant Breast Disorders
Etiology

Research has identified factors that increase the risk of development of breast cancer: increasing age, personal or family history of breast cancer, high-fat diet, high alcohol intake, treatment with estrogens (especially when used without progestins), early menarche, late menopause, late first pregnancy, and no pregnancies.

Prevention

Exercising moderation in fat and alcohol consumption, using nonhormonal methods of birth control, and managing menopausal symptoms can reduce risk of breast cancer. However, there are many factors that cannot be controlled, so the importance of early detection cannot be overemphasized. Recent research has discovered genes (BRCA1 and BRCA2) that are linked with susceptibility to breast cancer. These findings offer the possibility of very early identification of women at the most risk of developing breast cancer (and also ovarian cancer for those with BRCA1). They can then be monitored closely for any changes and receive early treatment if cancer develops.[3]

Diagnostic Tests

Breast self-examination and clinical breast examinations are important parts of identification of cancer. Cancerous growths tend to be harder, less movable, more irregularly shaped, with less clearly defined borders than benign growths. The prognosis is good for women who have breast

---

mastalgia: mast—breast + algia—pain
ectasia: extension
fibrocystic: fibro—fibrous + cystic—saclike

mastitis: mast—breast + itis—inflammation

cancers removed in the early stages but gets worse with time. Teaching and encouraging the regular use of BSE and appropriate use of mammography can save lives. Refer to Chapter 39 for more about BSE and for explanations about diagnostic tests used to assist in determining whether tumors of the breast are malignant.

Staging

The spread (metastasis) of cancerous cells from the primary site to other areas of the body by way of the blood or lymph is denoted by staging classifications. (See Chapter 10.) The lower numbers indicate less cancer spread.

Treatment and Complications

There are five main treatment options for breast cancer: radiation therapy, chemotherapy, hormonal therapy, modification of biologic response, and surgery. These options may be used separately or in combination depending on the condition of the patient and the stage of the disease.

The possibility of metastases necessitates drastic treatment of the whole body in many cases to prevent secondary cancer growth at other sites. Both radiation and chemotherapy generally combat cancer by destroying rapidly reproducing cells. This treatment is effective against cancer but also tends to destroy normal body cells that reproduce rapidly, such as hair and the cells lining the mouth, vagina, and gastrointestinal tract. Interventions to help preserve these tissues and maintain nutrition can be helpful.

Hormonal therapy may be undertaken to deprive cancer cells of hormones that stimulate their growth. Because breast cancer cells are often estrogen sensitive, this may be accomplished by decreasing circulating estrogen levels with drugs or by blocking the use of estrogen by cancer cells as occurs with use of the drug tamoxifen citrate. Interference with estrogen levels, however, may produce menopausal symptoms and increase the risk of osteoporosis and heart disease.

Substances that modify the body's biologic responses may be given to intensify positive responses of the body (e.g., stimulate the immune system) or to decrease negative body responses. Some examples of biologic response modifiers are interferons, tumor necrosis factor, interleukins, and various experimental immunotherapy formulations. This is an area of much research and is likely to expand greatly during the next few years.

Breast surgeries to remove cancerous tissue can be disfiguring and have profound effects on the patient's self-concept. The amount of tissue removed varies depending on the size, nature, and invasiveness of the cancer. A lumpectomy removes just the tumor and a margin around it. A **mastectomy** may be partial (removing only part of the breast), simple (removing only breast tissue), or radical (removing breast tissue, underlying muscle, and surrounding lymph nodes). Recently, surgical practice has shifted from mainly radical mastectomies to more breast-conserving surgeries with radiation therapy, with similar survival rates demonstrated.[4]

Nursing Care and Teaching

See Nursing Care Plan Box 40–1 for care of the patient undergoing a mastectomy.

---

mastectomy: mast—breast + itis—inflammation

BOX 40–1   NURSING CARE PLAN FOR THE PATIENT UNDERGOING MASTECTOMY

 Risk for anxiety related to uncertainty about diagnosis, prognosis, and treatments

PATIENT OUTCOMES
Patient will verbalize a decrease in anxiety.

INTERVENTIONS	RATIONALE	EVALUATION
Teach what to expect about to the surgical experience based on patient's understanding, concerns, and willingness to learn. Support the physician's explanations, answer questions, and refer to knowledgeable sources.	Knowledge dispels unreasonable fears and helps patient to prepare to cope with stressors.	Does patient express satisfaction with amount and type of information? Does patient evidence adequate understanding of the procedure and what to expect afterward? Do patient's vital signs, verbal, and nonverbal demeanor suggest anxiety?

 Risk for ineffective breathing pattern related to pain with chest movement

PATIENT OUTCOMES
Effective breathing pattern with clearance of mucus from air passages.

INTERVENTIONS	RATIONALE	EVALUATION
Medicate to relieve pain as necessary. Encourage deep breathing and coughing each hour. Encourage use of an incentive spirometer each hour when awake.	Pain may inhibit deep breathing efforts. This helps to loosen secretions and to prevent atelectasis, pneumonia, and inadequate oxygenation of tissues.	Does patient evidence pain or guarding during chest movement? Does chest sound clear? Are skin color and oxygen saturation adequate?

BOX 40-1 NURSING CARE PLAN FOR THE PATIENT UNDERGOING MASTECTOMY—cont'd

 Risk for impaired tissue perfusion and integrity related to damage to blood and lymph vessels and tension at surgical incision site

PATIENT OUTCOMES

Incision will heal by primary intention without excessive bleeding or swelling.

INTERVENTIONS	RATIONALE	EVALUATION
Monitor vital signs, oxygen saturation, and peripheral vascular status according to hospital policy and as necessary.	Vital signs, oxygen saturation, and peripheral vascular signs reflect circulatory status.	Are vital signs stable and within normal range? Is dressing dry?
Avoid use of the affected arm for blood pressure, venipunctures, and injections.	Restrictive and invasive procedures might further compromise tissue integrity of the affected arm.	Is arm protected?
Check for bleeding, amount and color of drainage if a drain device is used, and swelling.	Excessive bleeding and swelling may further compromise tissue perfusion.	Does incisional area look swollen, smooth, or shiny? Is drainage amount and color appropriate?
Measure circumference of arms daily and compare.	Swelling causes an increase in circumference.	Is affected arm larger than unaffected arm?
Elevate affected arm if swelling occurs.	Gravity aids fluid return to the heart.	Does elevation prevent swelling?
Place items where patient may easily reach them.	Excessive movement of the arm may exert tension on incision and increase bleeding.	Can patient reach items without abducting the arm over 90 degrees?
Encourage reasonable exercise of the affected arm following postmastectomy exercises that are approved by the institution.	Reasonable exercise promotes circulation, preserves muscle and joint function, and increases self-care ability.	Is patient moving the arm appropriately and gradually increasing range of motion and self-care ability?

 Risk for ineffective coping related to cancer threat and body image disturbance

PATIENT OUTCOMES

Patient verbalizes ability to cope; seeks help and support appropriately.

INTERVENTIONS	RATIONALE	EVALUATION
Maintain an open and trusting therapeutic relationship.	Effective communication is based on trust.	Is patient talking about concerns?
Allow grieving to take place. Encourage patient to express feelings and concerns.	Loss of a breast disturbs many aspects of body image, and cancer threatens one's sense of security and reasonableness of life.	Is patient willing to look at incision area? How are family members interacting with patient? How is patient responding to family members and friends?
Encourage active problem solving. Help patient remember previous successes in coping and strategies used.	Active problem solving promotes self-efficacy and combats depression. Memory of prior success can encourage hope for future success.	Is grief being expressed? Is patient planning for the future? Is patient taking an active interest in her personal appearance?
Refer to appropriate agencies for further support as needed (e.g., American Cancer Society, Reach for Recovery, local support groups,	Social support can assist individuals to meet their needs while developing effective coping skills and strategies.	Does patient have sufficient coping skills or supports available to promote healthy living?

Alternative therapies are also available for many cancers. See Chapter 4 for information about helping patients evaluate alternative and complementary therapies. The American Cancer Society and cancer treatment centers also have people who can answer questions about experimental and alternative therapies and discuss research findings. For more information about breast cancer, visit the American Cancer Society at www.cancer.org.

Breast Modification Surgeries

Mammoplasty is surgical modification of the breast. This may be done to restore a normal shape after removal of cancerous tissues. Many women, however, undergo mammoplasty electively to reduce or increase the size or improve

---

mammoplasty: mamm(o)—breast + plasty—to mold

the shape of their breasts. A caring and nonjudgmental attitude is essential. Because nurses are very aware of the dangers involved with surgery, psychosocial issues may seem to be a trivial reason to voluntarily assume such risks to life and health. However, body image is an important component of quality of life. Patients' informed decisions should be respected if they choose this surgery.

Breast Reduction and Mastopexy

Generally in breast reduction operations the nipple is separated from the surrounding tissue except for a small section with the blood vessels and nerves that supply it (Fig. 40–1). A large wedge of tissue is removed from the bottom of the breast, the edges are sewn together, and the nipple is reimplanted in a higher position. This not only decreases the overall size of the breast, which may help with back, neck, and head pain, it also corrects excessive sagging—a common problem for women with large breasts.

A **mastopexy** involves the removal of some skin and fat with subsequent resuturing so that the breast tissues are held higher on the chest to correct sagging breasts. This procedure usually does not remove as much tissue as a breast reduction.

Augmentation and Reconstruction Mammoplasty

Augmentation is a surgery to increase the size of the breasts. An implant—either a bag containing saline solution or silicone gel or a transplanted portion of the patient's own body tissues from another area—is inserted through an incision and positioned either under or over the pectoral muscles (Fig. 40–2).

For reconstructive mammoplasty, use of the patient's own tissues is generally safer than use of artificial implants, because no foreign material is introduced into the body. For situations in which significant amounts of tissue are needed for reconstruction, a portion of tissue may be moved from one area of the body to another as a **pedicle** graft. Pedicle literally means "little foot" because the graft remains attached to a stalk (containing the blood vessels and nerves) somewhat resembling a little leg with a foot (the graft) attached. Figure 40–3 shows two options for mastectomy graft repair—using the latissimus dorsi muscle and overlying tissue on the side of the chest or using the rectus abdominis muscle of the abdomen with its overlying tissue. For both these procedures, a portion of muscle is separated from its usual attachment. Tissues overlying a part of the muscle are excised and left attached to the muscle. This segment of tissue is then pulled under the skin and superficial layers to an incision at the mastectomy site. There it is brought to the surface and attached to reconstruct a breast shape. Tissue from the buttock area or the abdomen may also be grafted onto a mastectomy site without a pedicle.

---

pedicle: ped—foot + icle—little

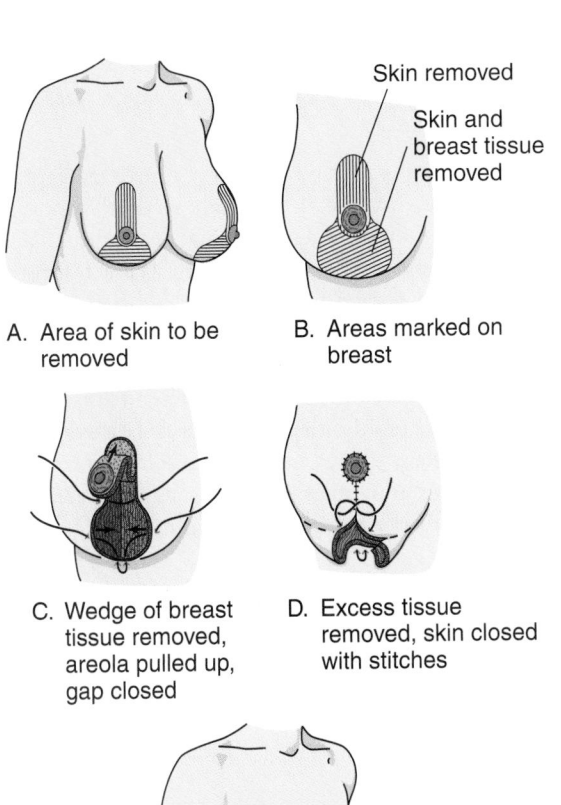

A. Area of skin to be removed

B. Areas marked on breast

C. Wedge of breast tissue removed, areola pulled up, gap closed

D. Excess tissue removed, skin closed with stitches

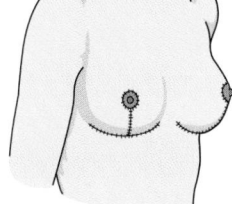

E. Post-operative appearance

Figure 40-1 Breast reduction. (Modified from Love, SM: Dr. Susan Love's Breast Book, ed 3. Perseus Publishing, Cambridge, Mass. 2000.)

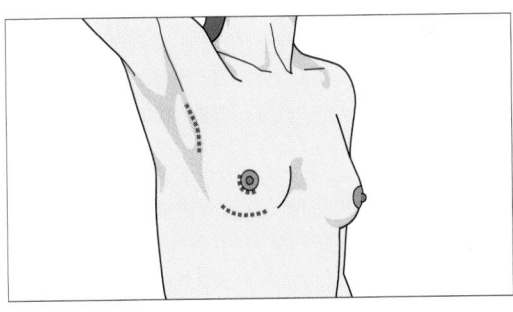

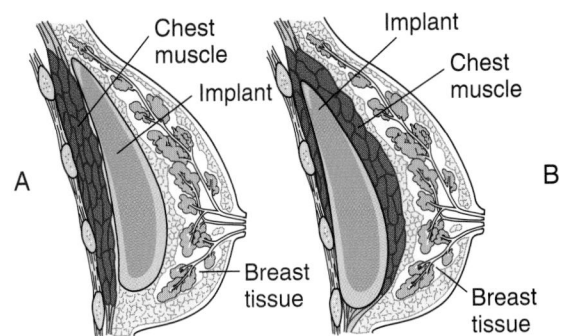

Figure 40-2 Breast implants. (A) Implant over muscle. (B) Implant under muscle. (Modified from Love, SM: Dr. Susan Love's Breast Book, ed 3. Perseus Publishing, Cambridge, Mass. 2000.)

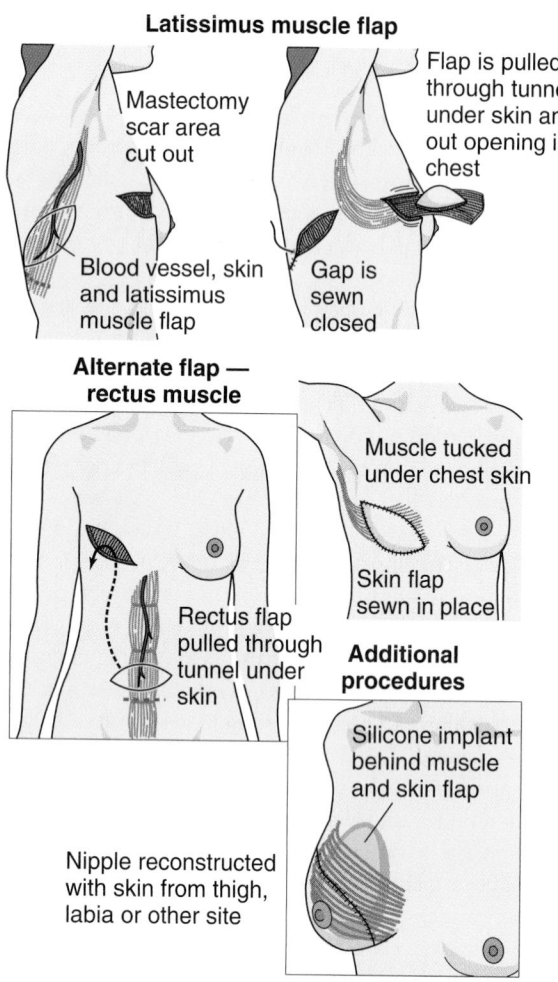

Latissimus muscle flap

Mastectomy scar area cut out

Blood vessel, skin and latissimus muscle flap

Flap is pulled through tunnel under skin and out opening in chest

Gap is sewn closed

Alternate flap — rectus muscle

Muscle tucked under chest skin

Skin flap sewn in place

Rectus flap pulled through tunnel under skin

Additional procedures

Silicone implant behind muscle and skin flap

Nipple reconstructed with skin from thigh, labia or other site

Figure 40-3   Mastectomy reconstruction. (Modified from Love, SM: Dr. Susan Love's Breast Book, ed 3. Perseus Publishing, Cambridge, Mass, 2000.)

Complications

Any of these surgeries may be complicated by infection or impaired healing. The use of silicone implants has been less than satisfactory for many women. Some women have experienced hardening of breast tissues and others have developed serious autoimmune disease problems after receiving silicone gel implants. Although actual etiology of all the problems is uncertain, many surgeries have been undertaken recently to remove silicone implants and saline implants are now used more often.

Nursing Care and Teaching

Carefully assess the healing process when changing dressings and explain to the patient how to assess healing, because not all tissues successfully attach at the new site and failure of attachment can require surgical revision. Signs of poor attachment include unnatural color of the incision, graft, or surrounding tissues; swelling; drainage; gaping of incision lines; and sloughing of the graft or edges of the site.

▶ MENSTRUAL DISORDERS
Flow and Cycle Disorders
Pathophysiology, Etiologies, and Signs and Symptoms

There are many types of menstrual abnormalities (Table 40–1). Causes can include stress, pregnancy, hormonal imbalances, metabolic imbalances (such as obesity, anorexia nervosa, and loss of too much body fat through excessive exercise), tumors (both benign and malignant), infections, organ diseases (such as liver, kidney, or thyroid disease), blood or bone marrow abnormalities, and the presence of foreign bodies in the uterus (such as intrauterine devices). Menstrual abnormalities can be distressing and can result in anemia, persistent fatigue, and sexual dysfunction. Establishment of a comfortable and open professional relationship is essential for communication about such concerns.

Diagnostic Tests

Appropriate testing to determine the cause of menstrual abnormalities can involve a thorough medical history and physical examination, swabs of vaginal discharge, a Papanicolaou (Pap) smear, pregnancy testing, urine testing, and extensive blood testing to screen for any of the disorders that may influence the menstrual cycle and flow. Generally health care providers initially test to rule out the most likely causes and then begin to test for more obscure disorders until the cause of the disorder is identified. For example, reproductive hormone levels are likely to be tested before kidney or liver function, unless the latter disorders are evident in the initial history and physical examination.

Treatment

Medical treatment of menstrual disorders generally involves manipulation of hormone levels. Surgical treatment of

TABLE 40-1 MENSTRUAL FLOW DISORDERS	
Disorder	**Description**
Amenorrhea	Menses absent for more than 6 months or three of previous cycles
	Called primary amenorrhea when menarche has not occurred by age 17
	Called secondary amenorrhea when menses are absent after menarche
Oligomenorrhea	Menstrual cycles of more than 35 days
Hypomenorrhea	Less than the expected amount of menstrual bleeding
Menorrhagia	Passing more than 80 mL of blood per menses
Hypermenorrhea	Menses lasting longer than 7 days
Polymenorrhea (also called metrorrhagia)	Menses more frequently than 21-day intervals
Menometrorrhagia (also called metromenorrhagia)	Overly long, heavy, and irregular menses

menstrual disorders can involve **dilation and curettage (D & C)**, **laser ablation** of endometrial tissue, and **hysterectomy.** During D & C the cervix is first dilated (opened wider) and then curets—sharp, spoonlike instruments—are inserted through the cervix and used to scoop out the inner lining of the uterus. Laser ablation involves targeted burning of endometrial tissue so that scar tissue that does not bleed forms. Hysterectomy (removal of the uterus), a last resort treatment, is described later in this chapter.

Nursing Care and Teaching

Estimation of the amount of blood lost during menses may be difficult because pad counts can vary widely depending on the frequency of pad changes and the portion of the pad that contains blood. The only accurate way to estimate menstrual flow is by weighing the pads (sealed in a biohazard bag) and then subtracting the weight of the original pads. A 1-g increase in pad weight equals approximately 1 mL of blood loss. You can document what the patient says for blood loss, but be sure to quote it and in parentheses place "patient estimate" after the quote.

Dysmenorrhea
Pathophysiology, Etiologies, and Signs and Symptoms

Painful menstruation, or **dysmenorrhea,** is a common problem of women, although few find it to be incapacitating. Primary dysmenorrhea is not pathological and is caused mainly by the action of endogenous prostaglandins, which stimulate uterine contractions producing cramping pain. Secondary dysmenorrhea occurs after normal menses without discomfort has been established and may be caused by some pathological condition such as endometriosis, pelvic infection, **retroversion** of the uterus, or tumors.

Diagnostic Tests

Hormonal tests may be required for primary dysmenorrhea, but laparoscopic examination, biopsies, and other various reproductive function tests (explained later in the chapter) may be required for investigation of secondary dysmenorrhea.

Treatment

Primary dysmenorrhea may be treated with drugs that inhibit prostaglandin synthesis, such as aspirin and nonsteroidal anti-inflammatory drugs (NSAIDs). Correction of secondary causes of dysmenorrhea may include such measures as hormonal adjustment, dilation and curettage, and other surgical procedures.

---

dilation and curettage: dilat(e)—to widen + ation—the
    process of + curet—scoop + tage—doing
laser ablation: laser—light amplification by stimulated
    emission of radiation + ab—away + lat—to carry + ion—
    the process
hysterectomy: hyster—womb + ec—away + tomy—cutting
dysmenorrhea: dys—painful + men(o)—month + rrhea—
    flow
retroversion: retro—back + version—turning

Nursing Care and Teaching

Several nonprescription preparations are available for treatment of dysmenorrhea, but patients should be advised to read the labels carefully, because aspirin and NSAIDs (the main component of many of the drugs) may be bought less expensively. Other added drugs in these compounds, such as diuretics, may not be necessary. If dysmenorrhea is related to uterine retroversion, assuming a knee-to-chest position may relieve the discomfort. Sudden development of dysmenorrhea in a woman with no previous menstrual discomfort should always be investigated.

Premenstrual Syndrome
Pathophysiology, Etiologies, and Signs and Symptoms

Premenstrual syndrome (PMS) is a recurrent problem for many women and may involve water retention; headaches; discomfort of joints, muscles, and breasts; changes in affect, concentration, and coordination; and sensory changes. Few women find PMS serious enough to interfere with work or relationships. The impact of ovarian hormones, aldosterone, and neurotransmitters such as monoamine oxidase and serotonin on PMS is not well understood, and further research is needed.

Medical Treatment

A variety of drugs have been given to combat PMS with varying degrees of success. Some commonly used PMS medications include drugs that affect prostaglandin production, hormonal balance, and neurotransmitter production and reuptake, as well as diuretics and supplements of calcium, magnesium, vitamin E, and vitamin B_6. Patients should be warned, however, that dosages of vitamins should not be increased without professional advice, because vitamins are medications (as well as nutrients) and high doses of some vitamins can lead to physiological damage.[5]

Nursing Care and Teaching

Being understanding and nonjudgmental is especially important, because some women who suffer from severe PMS may have been treated as if they are psychologically unbalanced because of the interaction of hormones and neurotransmitters and because of outdated ideas concerning PMS. You can help by providing educational materials on lifestyle measures, such as restriction of alcohol, caffeine, nicotine, salt, and simple sugars; participation in regular exercise; and development of stress management skills that can help women to who have a tendency to experience PMS.

Endometriosis
Pathophysiology, Etiologies, and Signs and Symptoms

Endometriosis is a condition in which functioning endometrial tissue is located outside the uterus (Fig. 40–4). Several theories have been proposed to explain development of en-

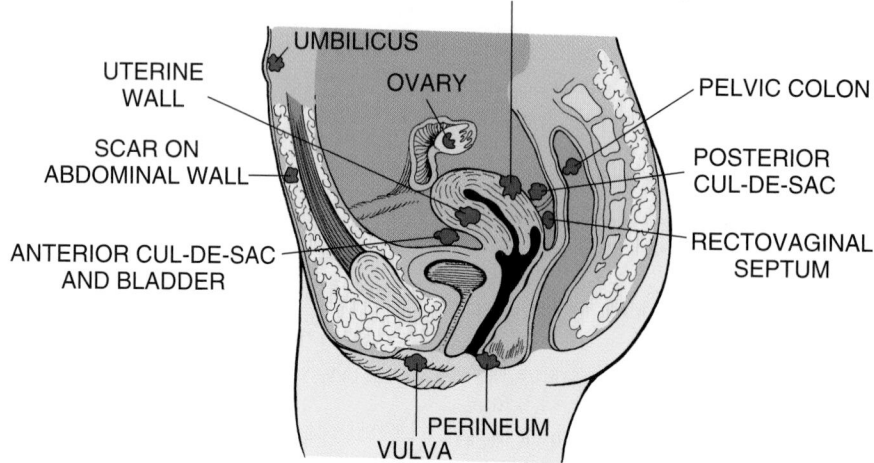

POSSIBLE SITES OF OCCURRENCE OF ENDOMETRIOSIS

Figure 40-4   Possible sites of endometriosis.

dometriosis, including faulty developmental differentiation of cells, transport of endometrial cells via blood and lymph to other parts of the body, and **retrograde** menstruation—a backward leakage of blood and tissue out through the fallopian tubes during the menstrual period.

Endometriotic cells grow in areas of sufficient blood supply—extending into other tissues such as intestinal walls, ovaries, and other abdominal structures. On a cyclic basis, mediated by ovarian hormones, these cells build up and slough just as they would in the uterus, but the sloughing and bleeding occur in the enclosed abdominal cavity or into the tissues that they have invaded. The buildup of the blood and cells can result in pain, swelling, damage to abdominal organs and structures, scar tissue development, and infertility. Ectopic endometrial tissue also produces complement component 3 and prostaglandins, increasing capillary permeability, adhesion development, and cramping.

Treatment

Surgical intervention may be required, especially if scar tissue develops into tight bands, which can cause strangulation of sections of bowel or ureters. Reduction of estrogen and prevention of ovulation either with medications or by surgical removal of the ovaries can be very effective but result in infertility and menopausal symptoms. Analgesics may be required.

Nursing Care and Teaching

The severity and persistence of the pain of endometriosis may lead to overuse of pain medication, so it is important to teach patients alternative pain relief strategies such as relaxation exercises and application of heat to the abdomen or back.

Menopause
Pathophysiology and Signs and Symptoms

Menopause is the permanent cessation of menstrual cycles resulting from decreased hormone production. This is a natural part of aging, but it is placed within this section because several related uncomfortable symptoms and conditions can occur. The climacteric (perimenopause) is the period of gradual decline in hormone production before the permanent end of menses and may last from months to several years. **Perimenopausal** physical symptoms vary widely and may include erratic menses, atrophy of urogenital tissues with a marked decrease in the amount of natural lubrication, a pH shift toward alkalinity (encouraging yeast overgrowth), and vasomotor instability (resulting in hot flashes and night sweats).

Estrogen protects women against several disease processes; the risk of heart disease and osteoporosis increase with declining estrogen production. Mental changes may also occur because of the complex interplay of reproductive hormones and neurotransmitters. It is important to acknowledge mental symptoms such as irritability, anxiety, insomnia, memory problems, and mild depression as a normal, temporary part of hormonal changes, so that perimenopausal women do not doubt their sanity.

Treatment

Hormone replacement therapy (HRT) is a controversial treatment for perimenopausal symptoms. Medications such as conjugated estrogens (Premarin), estradiol, and medroxyprogesterone acetate (Provera) may be prescribed. The National Heart, Lung, and Blood Institute of the National Institutes of Health conducted a major research project to study the risks and benefits of combined estrogen and progestin therapy for

---

retrograde: retro—backward + grade—step

perimenopausal: peri—around + men(o)—month + pausal—stopping

healthy women. However, the project was halted in July, 2002, 3 years early, because of very worrisome results. Positive findings with HRT included a one-third reduction in hip fractures, a 24% reduction in total fractures, and a 37% decrease in colorectal cancer. Disturbing findings included a 26% increase in breast cancer, a 41% increase in strokes, a 29% increase in heart attacks, doubling of venous thromboembolism rates, and an overall 22% increase in cardiovascular disease. It is uncertain what to make of this data because there was no difference in total mortality when comparing women treated with HRT and those given a placebo. Long-term results will continue to be examined, but for now, the risks of HRT have been deemed too great to continue the study.[6]

Prevention of osteoporosis begins in early adulthood. Fair-skinned, Caucasian women are at greatest risk for bone loss. Throughout life, adequate intake of calcium and vitamin D (preferably from foods) and regular weight-bearing exercise help to maximize bone mass. At menopause, some women may receive HRT or intensive treatment with bone-building medications to retard bone loss.

Complications

It is important to note that resumption of vaginal bleeding after menstruation has finally ceased can be a sign of disorder of the cells of the endometrium caused by either benign changes, such as polyps, or malignant changes and should always be investigated.

Nursing Care and Teaching

Patients who are perimenopausal often ask nurses about how to cope effectively with symptoms. Planning ahead for hot flashes by dressing in clothing that may be removed or applied in layers for comfort makes adjustment easier. Not allowing hot flashes to interrupt activities is an important strategy, as is engaging in satisfying and calming activities that contribute to a sense of serenity. Treatment of vaginal symptoms with a water-soluble moisture restorer or lubricant or with an estrogen cream (following prescription directions) can help. Eating a healthy diet that is light in caffeine, sugar, and alcohol can help women better control their bodies and minds. Looking forward to new challenges rather than toward the past may help to counteract depressive tendencies occurring with hormonal changes. It is important to remind perimenopausal women that they may still be fertile unexpectedly after several months of **amenorrhea.** To prevent conception, they need to continue to practice birth control until they receive confirmation from their physician that menopause is complete.

▶ IRRITATIONS AND INFLAMMATIONS OF THE VAGINA AND VULVA

Several causative agents can irritate the vulva and the vagina. Signs and symptoms are often similar, but there are some differences in the discharge produced in response to the disorders. See Table 40–2 for common vaginal irritations and inflammations that are not generally sexually transmitted. See Chapter 42 for information on sexually transmitted diseases.

Pathophysiology, Etiologies, and Signs and Symptoms

The normal vaginal environment is a balanced ecosystem with a pH of less than 4.2 as a result of lactic acid and hydrogen peroxide production by cells in the vagina. This acidic pH protects against the growth of many pathogenic microorganisms. A variety of normal resident microorganisms coexist relatively harmoniously unless the ecological balance is destroyed. Candidiasis, bacterial **vaginosis,** and **cytolytic** vaginitis are all instances of overgrowth of normally present, nonpathogenic microorganisms. Trichomoniasis also is included here because it can be transmitted nonsexually (on fomites, such as toilet seats), as well as sexually, and it grows well when the vaginal environment is disturbed.

Several conditions can predispose patients to an overgrowth of resident microbes: poor nutrition (especially diets high in simple sugars), inconsistent control of blood glucose levels in patients with diabetes, stress, pregnancy, marked hormonal fluctuations, pH changes, prolonged overheating of the genital area with little aeration (as happens with sitting still for long periods in overly restrictive clothing), and changes in the balance of vaginal flora types because of antibiotic treatment. Patients who have a compromised immune system may experience frequent overgrowths of resident microbes, and, conversely, vaginal infections can make women more susceptible to infection with sexually transmitted diseases, such as gonorrhea and human immunodeficiency virus (HIV). Frequent and persistent yeast infections may be one sign of HIV. Vaginosis (overgrowth) and vaginitis (inflammation) can sometimes produce irritation and inflammation in the male sexual partner as well and may lead to urethritis, **balanitis,** excoriation, and sores on the penis. If the male partner is not treated also, he may reactivate the problem for the woman. Therefore several types of medication come in "partner packs" for both partners to use. (See Table 40–2.)

Nursing Care and Teaching of the Patient Undergoing Diagnostic Testing

The patient may feel embarrassed to speak about what is bothering her. A safe question to begin with for most patients is, "Hello. What can I write on your chart as the reason for your visit today?" If embarrassment is evident, a comment that you need to know a bit about what materials to put out for examination purposes often defuses an uncomfortable situation. As you set up materials for a pelvic examination, swabs, and cultures, you can explain that some information is needed to determine how to treat the

---

amenorrhea: a—without + men(o)—month + rrhea—flow

vaginosis: vagin—vagina + osis—condition
balanitis: balan—acorn (shape of glans penis) + itis—inflammation

TABLE 40-2 COMMON VAGINAL IRRITATIONS AND INFLAMMATIONS

Disorder and Etiology	Signs and Symptoms	Discharge/ Examination	Diagnostic Tests	Usual Treatment
Candidiasis: *Candida albicans, glabrata* or *tropicalis*, overgrowth	Burning, itching, redness of vulva; burning on urination	White, cottage cheese appearance	Wet mount slides (yeasts look like tiny, budding tree branches); may be cultured	Antifungals (drugs mostly ending in–*azole*)
Bacterial vaginosis: *Gardnerella vaginalis, Mycoplasma,* or anaerobe overgrowth	None or vulvar or vaginal irritation	White or gray, homogeneous, foul-smelling discharge; pH higher than 4.5	Wet mount slides show "clue cells" or release fishy odor when potassium hydroxide is applied	Drugs such as metronidazole or clindamycin
Trichomoniasis: Trichomonas vaginalis *(may be transmitted by inanimate objects or sexually)*	Itching, irritation, foul odor, redness, dysuria	Discharge may be frothy; pH higher than 4.5; "strawberry cervix" resulting from petechiae	Wet mount slides treated with normal saline show motile cells with flagella (like tiny whips); may also be cultured	Metronidazole, 2-g single dose
Cytolytic vaginosis: Lactobacilli overgrowth, stress, some medications	Burning, irritation, pain with intercourse	Nonodorous, thick, white, pasty, or dry and flaking	Lower than normal pH as tested with pH indicator tape (or litmus strip); may be cultured	Depends on cause; alkaline douches may be prescribed
Contact vulvo-vaginitis: contact with allergens or irritating chemicals such as contraceptive creams or bubble baths	Itching, burning, redness	Generally no change from normal discharge, though may be increased	History and physical information, recent contact with chemicals	Avoidance of the offending substance; warm sitz baths or application of hydrocortisone cream
Atrophic vaginitis: estrogen levels too low to support estrogen-sensitive vaginal tissues	Vulvovaginal irritation, less lubrication, dyspareunia, increased tendency for resident microbe overgrowth	May have little or increased discharge; discharge may be watery, yellow, or green; may be blood-tinged	Maturation index may be determined during Pap test to identify atrophic cellular changes, but diagnosis is usually by history and physical information only	HRT (oral, patch, or vulvovaginal cream) or water-soluble lubricant replacing vaginal lubricants

problem. (See Chapter 39.) Often this is a good time to ask about the discharge and other signs and symptoms using the WHAT'S UP? format. (See Chapter 1.) Allow the patient privacy while she changes into a gown. Return to the room if requested as a chaperone, assistant, and support for the woman. NOTE: If any wet mount slides were made, these must be taken to the laboratory immediately while still wet. Use standard precautions to transport samples. Although swabs may be taken for culture, the health care provider may prescribe medication before the results of the swabs return because such irritations are so uncomfortable.

Nursing Care and Teaching of the Patient Undergoing Treatment

Vaginal inflammations and infections may require local application of medication either in cream, suppository, or medicated douche form. Depending on the practice standards in the area, the nurse may apply this for patients who are unable to do this for themselves or may teach patients to self-administer. Anatomically the vagina slopes backward toward the sacrum for approximately the length of an adult finger (though it can stretch longer). Application is easiest when the patient is lying down ready to sleep, because vagi-

nal medications tend to run out because of gravity when the patient stands or sits. Medicated douches may be administered with the patient sitting on a bedpan with the bed in semi-Fowler's position or, if self-administered, while sitting on a toilet. Most vaginal medications come with an applicator that either injects a dose of creamy medication or pushes a firmer, shaped dose of medication off the end of the tube when the plunger is depressed. Consult instructions supplied with the medication. Patients should be instructed to use all the medication as prescribed and to wear an absorbent pad to prevent possible staining of clothing.

Toxic Shock Syndrome
Pathophysiology, Etiologies, and Signs and Symptoms

Toxic shock syndrome (TSS), first identified in 1978, is primarily associated with superabsorbent tampon use during menstruation but can also occur with use of nasal packings and in individuals who are not menstruating and who have no packing at all. It is a severe systemic infection with strains of *Staphylococcus aureus,* which produce an epidermal toxin. The effect of the toxin on the liver, kidneys, and circulatory system makes this a life-threatening condition. A streptococcal infection can cause a similar syndrome.

Individuals with TSS may experience a sudden high fever with sore throat, headache, dizziness, confusion, redness of the palms and soles of the feet, skin rashes, blisters, and petechiae followed by peeling of the skin. Muscle weakness and pain and gastrointestinal upset also have been reported. Signs and symptoms such as these should be reported to the health care provider immediately.

Prevention

Tampon makers have removed the highly absorbent fibers that were most often associated with the syndrome from their product lines, and TSS is now rare. Women can also reduce their risk of developing TSS by substituting sanitary pads for tampons at least part of the time (nighttime use may work well), changing tampons every 4 hours, washing hands carefully before inserting anything into the vagina, not leaving female barrier **contraceptives** in place for longer than needed, and not using tampons or female barrier contraceptives in the first 12 weeks after giving birth.

Nursing Care and Teaching

All menstruating women should be taught measures to prevent TSS. They should also be taught to recognize symptoms of TSS, because early identification and treatment can save lives.

▶ DISORDERS RELATED TO THE DEVELOPMENT OF THE GENITAL ORGANS

Pathophysiology, Etiologies, and Signs and Symptoms

Several types of congenital malformations of the genital organs may affect the health of female patients. Genetic or environmental factors during pregnancy may cause these, and they may require medical or surgical treatment at some point in life. **Agenesis** of structures means that they never developed. **Hypoplasia** of reproductive tract portions means that they are underdeveloped. **Imperforate** means that expected openings do not exist. Blind pouches exist where cavities should meet but do not. The uterus can form in several different configurations, including a double uterus.

Many malformations are discovered during childhood or early adolescence, but some are identified when patients seek medical help because of dysmenorrhea, **dyspareunia** (pain with intercourse), infertility, or repeated spontaneous abortions (miscarriages).

Diagnostic Tests

Procedures such as ultrasonography, hysterosalpingography, computed tomography (CT), magnetic resonance imaging (MRI), and endoscopic examinations may be used to determine the type and extent of developmental defects.

Treatment

Some defects can be repaired surgically, but others cannot. Depending on the type and location of the defect, surgeries may be done by endoscopy or by surgical incision. Absence of hormone-producing tissue may be overcome by hormone supplements.

Nursing Care and Teaching

Patients who have these problems may struggle with self-esteem issues, such as feeling that they are somehow incomplete or have been cheated of something they desire. You can show that you are willing to listen if and when the patient wishes to talk while allowing her as much privacy as she desires.

Displacement Disorders
Pathophysiology, Etiologies, and Signs and Symptoms

The pelvic organs are suspended within the pelvis by ligaments and supported by muscles and fascia. The pubococcygeal muscle runs from the pubis to the coccyx and supplies support from below. Pregnancies (especially those producing large babies) and rapid or traumatic deliveries may result in stretching and injury of the supporting structures, which can cause displacement of the uterus, vagina, bladder, or bowel from their normal positions. The observation that some children have defective muscle support of the pelvic organs and that prolapse is more prevalent in some families seems to suggest that congenital defects and genetic inheritance may also influence displacement disorders even without pregnancy. Scarring from sexually transmitted diseases also may cause some displacement of the pelvic organs. Aging generally increases the problem because the effects of gravity over time contribute to stretching, and lower estrogen levels weaken estrogen-dependent supportive tissues. Chronic constipation, obesity, and lack of exercise also worsen these problems.

Nursing Care and Teaching

Teach patients to eat a healthy diet to avoid obesity and constipation and how to do Kegel exercises to keep the pubococcygeal muscle strong to support the organs in the pelvic cavity. There are several variations of such exercises. The important idea is that the appropriate muscles are exercised adequately to strengthen and build up ability to control muscle contractions.

1. To find the pubococcygeal muscle, tighten while urinating so that the flow of urine stops.
2. Squeeze the muscle that stopped urinary flow tightly, holding for 10 seconds, and totally relaxing the muscle afterward. Repeat 15 times per day.

---

contraceptive: contra—against + ceptive—taking in (conceiving)
agenesis: a—without + genesis—production
hypoplasia: hypo—little + plasia—shape (or form)
imperforate: im—not + perforate—pierced
dyspareunia: dys—painful or abnormal + pareunia—mating

3. Practice controlling the muscle by contracting and relaxing it to move the pelvic floor upward and downward very slowly. Thinking of an elevator helps some women. Repeat this 15 times per day.

These exercises can be done anywhere and are not apparent to anyone watching.

L E A R N I N G   TIPS

Teach patients to do Kegel exercises while waiting in lines to use otherwise wasted time to promote their health. Another suggestion is to plan specific times of day or activities that would include Kegel exercises, such as while in a car or working at a computer.

Treatment

A pessary is a supportive (usually ring-shaped) device that is placed in the proximal end of the vagina to help support the pelvic organs. These are usually removed daily at bedtime for cleaning, but some types are designed to remain in the vagina for months at a time. When pessary use is begun, it is important that the woman return to the physician for a recheck after an initial period of use to determine whether the pessary is causing pressure damage to tissues. Because the pessary is a foreign object in the vagina, increased vaginal discharge may be expected. Discharge should not be pink, bloody, or purulent.

Cystocele
Pathophysiology, Etiologies, and Signs and Symptoms

Cystocele occurs when the bladder sags into the vaginal space because of inadequate support (Fig. 40–5). A feeling of pelvic pressure and stress incontinence are common with this condition.

Treatment

Kegel exercises or the use of a pessary may help. If these measures are ineffective, anterior **colporrhaphy,** which is a surgical repair of the anterior portion of the vagina, may be necessary to correct this problem. Another possible surgical treatment involves resuspending the bladder.

Rectocele
Pathophysiology, Etiologies, and Signs and Symptoms

Rectocele occurs when a portion of the rectum sags into the vagina because of inadequate support (see Fig. 40–5). A feeling of pelvic pressure, as well as fecal incontinence, constipation, and hemorrhoids, may result.

---

cystocele: cysto—bag (bladder) + cele—hernia
colporrhaphy: colpo—vagina + rrhaphy—suture
rectocele: recto—rectum + cele—hernia

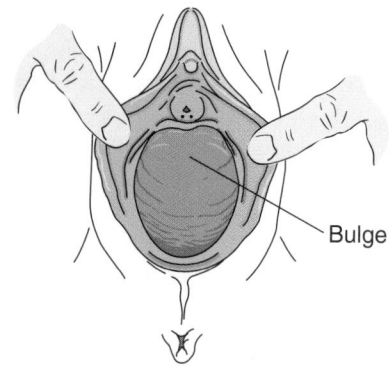

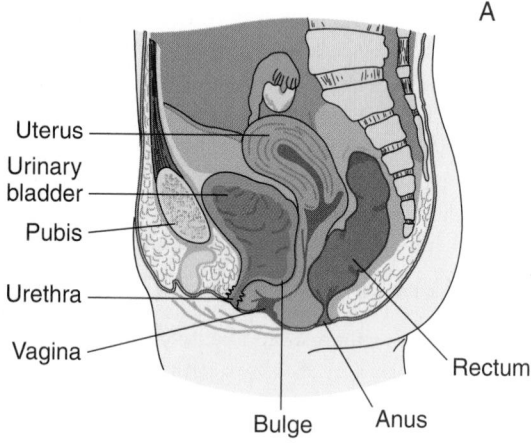

A

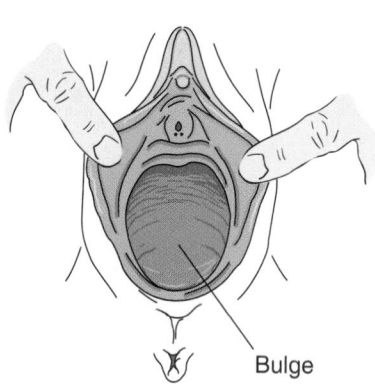

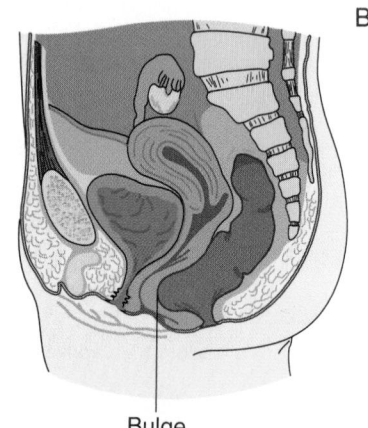

B

Figure 40-5 *(A)* Cystocele. *(B)* Rectocele.

Treatment

Kegel exercises may help strengthen the supporting muscles. The patient should maintain bowel regularity with a high-fiber diet to avoid further discomfort and sagging from bowel overdistention. Posterior colporrhaphy may be necessary to correct this problem.

Uterine Position Disorders
Pathophysiology, Etiologies, and Signs and Symptoms

The most common variations of position of the uterus are **anteversion, anteflexion,** retroversion, and **retroflexion** (Fig. 40–6). In anteversion the uterus lies too far forward, and in retroversion it lies too far backward. In anteflexion the upper portion of the uterus bends forward, and in retroflexion it bends backward.

Symptoms that may result from these uterine displacements include painful menstruation and intercourse, infertility, and repeated spontaneous abortion.

---

anteversion: ante—front + version—turning
anteflexion: ante—front + flexion—bending
retroflexion: retro—back + flexion—bending

Treatment

A pessary may correct some positional problems. If infertility or recurrent spontaneous abortion is involved or the condition is very painful, surgery to correct the condition may be undertaken.

Uterine Prolapse
Pathophysiology, Etiologies, and Signs and Symptoms

Uterine prolapse occurs when the uterus sags into the vagina (Fig. 40–7). The amount of sagging can vary and may increase over time as a result of the effects of gravity, poor pelvic support, and excessive lifting or straining. In first-degree prolapse, less than half the uterus sags into the vagina. In second-degree prolapse, the entire uterus sags into the vagina. In third-degree prolapse, the uterus sags outside the body.

Uterine prolapse can be very uncomfortable, resulting in back pain, pelvic pain, pain with intercourse (or inability to have intercourse), urinary incontinence, constipation, and the development of hemorrhoids. The pressure on the uterus also may compromise circulation, resulting in tissue necrosis. NOTE: Vaginal vault prolapse may also occur in women who have had a hysterectomy, so that the vagina turns inside out and sags downward with similar signs and

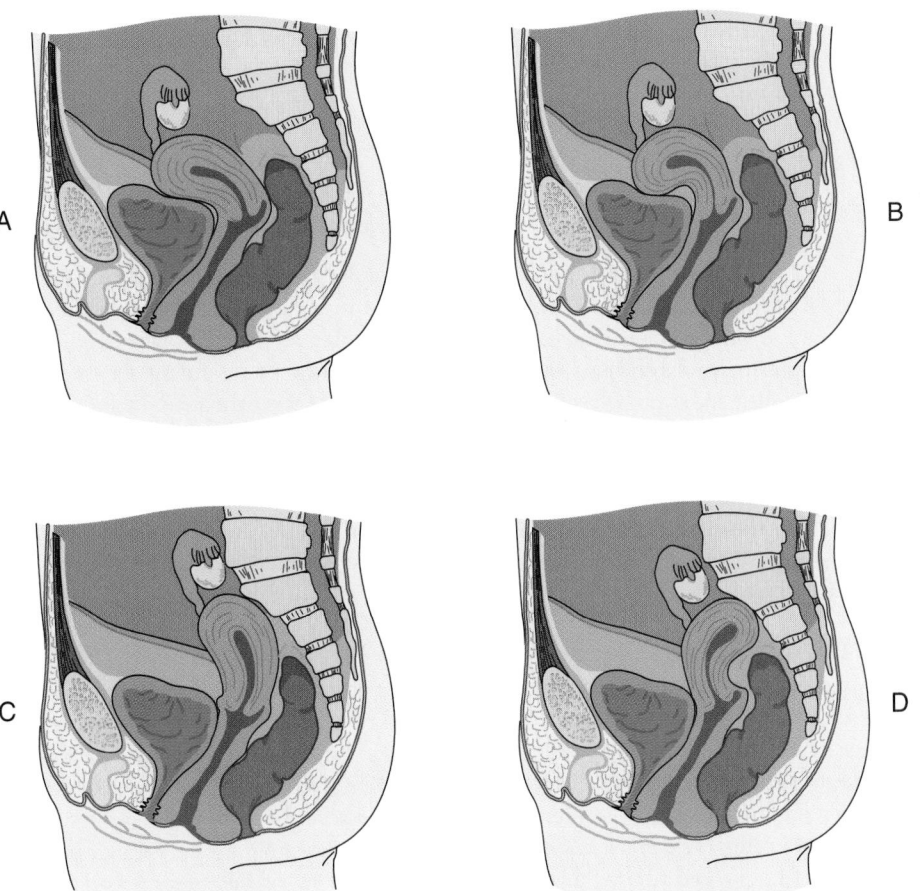

Figure 40-6 Uterine positions. *(A)* Anteversion. *(B)* Anteflexion. *(C)* Retroversion. *(D)* Retroflexion.

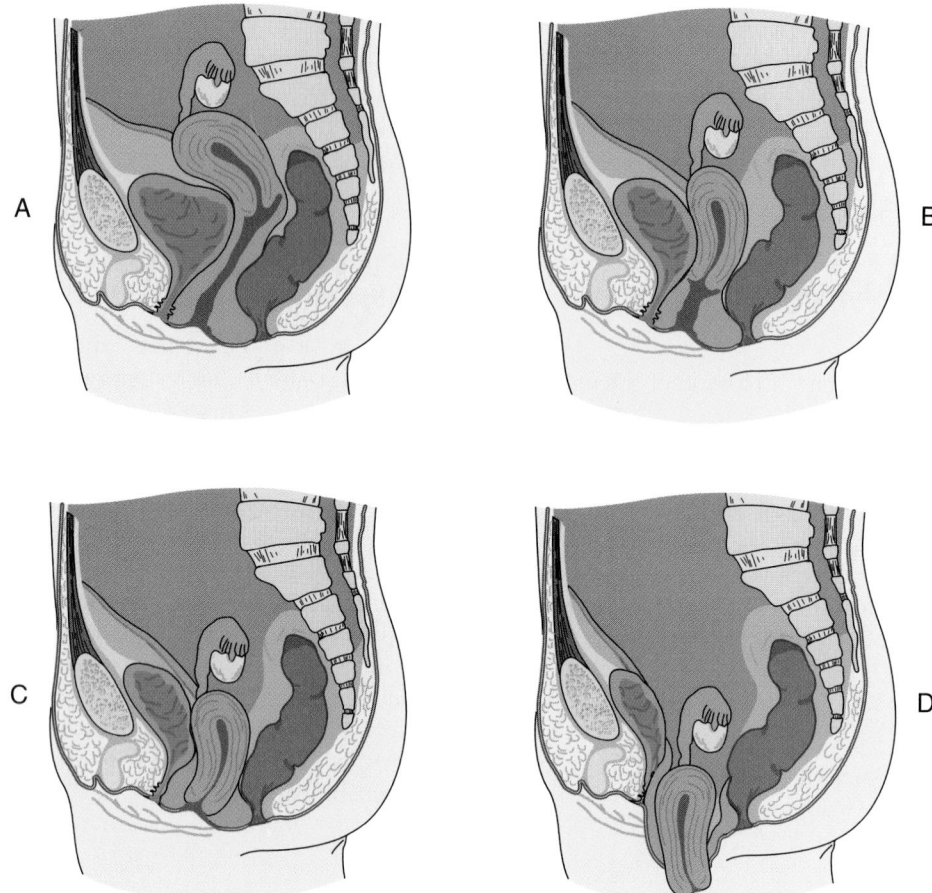

Figure 40-7 Uterine prolapse. *(A)* Normal uterus. *(B)* First-degree prolapse: Descent within the vagina. *(C)* Second-degree prolapse. *(D)* Third-degree prolapse: The vagina is completely everted.

symptoms. This condition generally requires surgical resuspension.

Treatment

Some of the more minor uterine displacements may be treated with use of a pessary. Kegel exercises may be more effective in prevention of uterine prolapse than in treatment, because once the tissues become stretched sufficiently for the uterus to sag into the vagina, the continued weight of the uterus prevents adequate contraction of the muscles. Surgery may be done to correct this problem. Although the uterus may be resuspended by shortening the muscles and fascia, hysterectomy is the more common treatment unless further childbearing is desired.

▶ FERTILITY DISORDERS

Infertility is a complicated problem that must be considered according to the causative factors. Some couples with infertility may have multiple reproductive problems. Both male and female partners should be examined. (See Chapter 41 for greater detail on male reproductive system disorders.) Often the woman sees the health care provider first and may be given a specimen container and advised to give this to her partner to provide a semen sample for analysis. See Table 40–3 for a summary of fertility disorders and diagnostic tests.

Nursing Care and Teaching of the Patient Undergoing Fertility Testing

An understanding attitude is very important, because infertility can be a cause of low self-esteem, as well as relationship problems. Patients who have been undergoing diagnostic testing or treatment for infertility can become very discouraged with the process and the expense, especially if it has been ineffective. Having to plan your sexual activity around a health care provider's directions can compromise feelings of spontaneity, enjoyment, and privacy. Extensive questioning by nurses can aggravate the situation, but avoiding conversation may convey a lack of caring. A friendly "Which test shall I help you get ready for today?" may well be enough to get the needed information. Many women undergoing infertility testing are very well informed about the test they will be having and can tell you so that you know which equipment to set out.

On the first infertility investigation visit, the nurse may teach or give a handout to the patient about keeping a precise record of her oral temperatures with a basal thermometer each morning on awakening, before any other activity. The first day of her menses is day 1 on the temperature chart. Changing levels of hormones result in slight temperature changes, which can be used to identify when ovula-

TABLE 40-3 FERTILITY DISORDERS

Disorder	Pathophysiology/Etiology	Diagnostic Tests
Male	Possible anatomic abnormalities, hormonal factors, genetic defects, inflammatory conditions, immune system disorders, difficulties with sexual function or technique, psychological factors, or exogenous influences such as drug use, radiation or chemical exposure, trauma, and excessive testicular temperatures (as may occur with prolonged hot tub use or too tight clothing)	Semen analysis of number, condition, and movement of the sperm and composition of seminal fluid using various tests
Female: Ovulation	Possible anatomic and physiological abnormalities of ovaries; hormonal balance problems related to hypothalamus, thyroid, or adrenal glands; or polycystic ovary syndrome	Basal body temperature charting, midluteal serum progesterone blood levels, luteinizing hormone levels blood or urine testing, ultrasound monitoring of a follicle for evidence of release of ovum, endometrial biopsy, observation of male hair distribution, and other hormone testing as indicated
Tubal	Possible obstruction of the fallopian tubes resulting from anatomic variations, scarring, or adhesions; prior surgeries; and inflammatory processes involving other abdominal tissues	Hysterosalpingography (described in Chapter 39), laparoscopy
Uterine	Possible abnormalities in shape or blockages within the uterus (rare cause of infertility but a potential cause of pregnancy loss before maturity), menstrual disorders involving the endometrium	Hysteroscopy (described in Chapter 39), removal of tissue samples using curet or endoscope
Other sources of infertility	Possible reproductive environmental factors such as destructive antigen-antibody responses, inappropriate pH of seminal fluid for maximal sperm motility, or substances in female partner's genital tract fluids that disable sperm	Postcoital test: couple is advised to have intercourse when luteinizing hormone and estrogen levels are high, then a specimen of cervical mucus is taken from the woman 2–12 hours later for analysis of reproductive environment

tion seems to be occurring and when particular hormone levels should be tested. Because many factors may influence temperature and cycles, explain that it may take a few months of recording to clearly identify her own pattern.

You may assist with office procedures such as endometrial biopsy, which may be done during a pelvic examination 2 or 3 days before menses is expected. A pregnancy test should be done before this procedure to avoid interfering with a pregnancy. Consult your procedure manual, laboratory manual, or health care provider for what materials to set up. The woman may receive pain medication and paracervical block anesthesia for the procedure. Assess pulse and blood pressure. A vasovagal reaction treatment kit containing epinephrine (or atropine, according to the health care provider's choice), a tourniquet, and a syringe should be kept handy for injection if a vasovagal reflex occurs during the procedure. Vasovagal reflex is a reflex stimulation of the vagal nerve that can happen when the cervix, larynx, or trachea is manipulated and results in slowing of the heart rate and decreased cardiac output, so that the blood pressure drops markedly.

Treatment

Treatment of infertility is designed to ensure that an adequate amount of sperm and an ovum can be in close proximity in the most conducive environment for fertilization. Removal of barriers such as scar tissue may require surgery.

Depending on the results of blood tests and the **postcoital** test (described in Table 40–3), adjustments of environmental factors may involve such actions as sperm washing to avoid destructive antigen-antibody responses, changing the pH of the seminal fluid to encourage sperm motility, treating the female partner to prevent substances in her genital tract fluids from disabling the sperm, or adjusting her hormone levels. The number of sperm or ova available may be increased through use of such fertility drugs as clomiphene citrate or various hormone preparations. Infertility treatments are quite complicated and expensive and change periodically as a result of ongoing research. In-depth coverage of these treatments is beyond the scope of this book.

Various methods may be used to bring the gametes into close proximity. If the problem involves inability to get the sperm close enough to the ovum (as may happen with ejaculatory problems), the physician may place a semen sample from the male partner closer to the ovum via a small catheter. **In vitro fertilization** (IVF) involves bringing ova and sperm together outside the bodies of the participants. Ova may be harvested using a long needle or an endoscope after hormonal preparation of the woman. Sperm may be obtained through masturbation; intercourse with a nonlu-

postcoital: post—after + coital—pertaining to intercourse
in vitro fertilization: in—inside + vitro—glass + fertiliz—fruitful + ation—process

bricated, nonspermicidal condom; or electrical stimulation of ejaculation for patients with spinal cord injuries.

For those whose sperm is unable to successfully penetrate the ovum, procedures involving gamete micromanipulation may be done. Under a microscope, an ovum from the female partner is partially opened by removing a portion of the outer covering to facilitate sperm penetration, or sperm may be injected into the ovum. This fertilized ovum is then reinserted into the woman's body.

When measures to improve the chances of conception using the partners' own gametes are unsuccessful, gametes from donors may be utilized. Artificial insemination by injecting another man's sperm into the genital tract of the female patient is the simplest of the donor procedures. Ova also may be harvested from a donor woman and used for in vitro fertilization using the male partner's sperm if possible. Both these procedures allow for genetic inheritance from one member of the couple. If genetic inheritance is not possible or desirable (as with familial disease carriers), both donor sperm and ova may be used for in vitro fertilization to be transferred into the female patient. Surrogacy is a situation in which an embryo from one couple is placed into a "host" mother for growth of a baby for the couple and is a topic of much ethical debate.

Nursing Care and Teaching of the Patient Undergoing Treatment

Patients who are undergoing infertility treatment may experience many upsetting and distressing feelings. Feelings of inadequacy, frustration, depression, and anger are common. If the infertility was caused by something the patient perceives as avoidable, such as sexually transmitted disease, guilt feelings may add to the psychological discomfort. Any or all of the previously described tests may be completed and some repeated many times without identification of reasons for the infertility, resulting in repeated disappointments. Some patients cling to the hope that new tests and treatments are being developed that may help them, whereas others feel that they are being used as a "guinea pig" for development of new strategies. The beginning of menses may signal a time of mourning for these couples. Depression may result after failed IVF attempts. Strained relationships may develop between marriage partners, especially if there is disagreement about the value of testing or the importance of having children.

Usually more than the desired number of embryos are implanted because it is expected that not all will survive and because this is more cost-effective with less physical risk for the mother. However, this requires heart-wrenching decisions of whether to "reduce" (abort) extra pregnancies or to risk having more than the desired number of children at once as a result of the fertility treatments.

Offer a listening ear while being careful not to give advice about treatment modalities. There are many ongoing debates among researchers and practitioners as to the value of particular procedures; consequently, strategies may vary widely from one health care provider to another. Encourage open communication among the patient, the health care provider, and the patient's significant other, and encourage decision making that is informed and based on the patient's values.

Many varieties of assistive reproductive technology are available, and the number is increasing with research. Most of the procedures are called by acronyms (names developed by using first letters of each descriptor). For example, *GIFT* means "gamete intrafallopian transfer" (putting the gametes together in the fallopian tube). Acronyms can be a useful shortcut but can be confusing. Most nurses probably do not need to know all acronyms of infertility treatments unless they work in a gynecologist's office or infertility clinic. Knowing that words in all capital letters are generally acronyms helps understanding of many procedures.

▶ REPRODUCTIVE LIFE PLANNING

Reproductive life planning is a more comprehensive term than *contraception* and implies reasoned decisions related to pregnancy timing and whether or not to have children. Nurses can contribute to the overall health and quality of life for women and families by assisting them to find the information they need to make wise choices.

There are many different types of birth control available, and several additional types are in developmental and testing stages. General categories of agents are discussed in this chapter. General knowledge of how the different types of contraceptives work can assist the nurse in answering patients' questions or helping patients find additional information. This section is not intended to be a substitute for discussion with the health care provider or for information supplied with individual products.

No numerical statements of effectiveness are included here because several different sets of statistics are currently in use. It seems likely that as products improve, hormone dosages are changed, or more data come in from users, the numbers will be adjusted. Methods are introduced in the order of general effectiveness from most to least effective (with the exception that experimental methods are discussed at the end regardless of their efficacy). Consult your clinic or health care provider for an approved, current comparison list of methods for distribution.

For some patients, the distinction of whether the birth control method actually prevents conception or only interferes with implantation or maintenance of a pregnancy is an important factor in their decision. If a patient believes life begins at conception, any action other than prevention of conception would be considered equivalent to abortion (Fig. 40–8).

Oral Contraceptives

Oral contraceptive medications are among the most widely used forms of birth control in North America. Most contain an estrogen and a progestin in combination, although some (minipills) contain only a progestin. Some work to prevent conception by inhibiting ovulation or changing the environment of the reproductive tract so that activity of the sperm is inhibited. Others do not prevent conception but

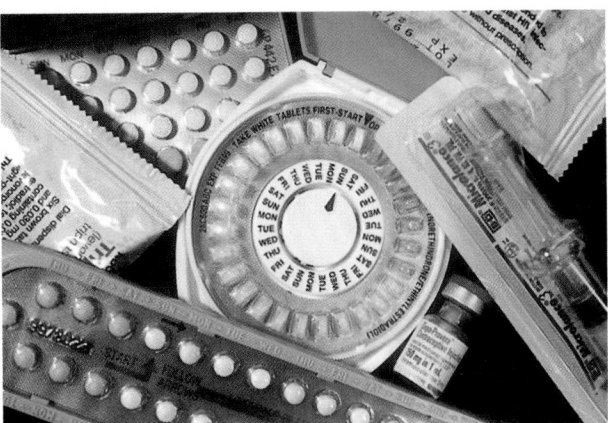

Figure 40-8 A variety of birth control methods are available.

make implantation less likely and hasten the breakdown of the corpus luteum so that pregnancy-sustaining hormones are not produced. Many of the adverse effects that occurred in the past have been overcome by adjustment of dosage levels.

Oral contraceptives may also be used in some instances to regulate irregular menses or to decrease dysmenorrhea. There is much debate about whether hormonal contraceptive agents may offer some protection against some sexually transmitted diseases based on lower statistical rates of STDs among oral contraceptive users. However, cellular changes of the cervix seen with hormonal contraceptive use actually tend to be associated with higher rates of some sexually transmitted diseases.[7] Unless some specific mechanism of prevention is demonstrated by research, it seems irresponsible to suggest that oral contraceptives alone offer protection against anything other than pregnancy. Therefore women should still be advised about the risks of contracting sexually transmitted diseases while taking an oral contraceptive.

Advantages, Disadvantages, Side Effects, and Risks

Oral contraceptives are very effective. Improvement of dysmenorrhea, increased regularity of menses, and decrease in menstrual flow may occur; however, some women experience menstrual changes such as amenorrhea, irregular or prolonged menses, and intermenstrual spotting. They require a great deal of commitment because irregular use decreases effectiveness. To encourage regular use, oral contraceptives are generally dispensed in containers labeled with the days of the week, and some companies include unmedicated pills in the package to be taken during the time of hormone cessation for menses, so that the woman only has to remember to take a pill every day.

Some women experience side effects such as acne, fluid retention, headaches, breast swelling and discomfort, midcycle bleeding, and sometimes depression. Use of an oral contraceptive also has some risks. Higher rates of blood clot formation, strokes, high blood pressure, heart attacks, and worsening of diabetes are rare occurrences with some hormonal contraceptives and are generally related to preexis-

tent disease entities. Women who smoke or have diabetes, high blood pressure, heart disease, or a history of thrombophlebitis should receive counseling about the risks of oral contraceptives and education about alternative methods of contraception.

Oral contraceptives decrease the risk of endometrial and ovarian cancer, but there is debate about risk of breast cancer, and cervical **dysplasia** (cell changes that may become cancerous) sometimes occurs among oral contraceptive users. Women should definitely be advised to have regular annual (or more frequent if abnormalities develop) Pap smears while taking oral contraceptives.

Many medications can alter the effectiveness of oral contraceptives, and women should be warned always to alert health care providers and pharmacists that they are on an oral contraceptive whenever a new medication is to be started or a regular medication is discontinued. Use of hormonal contraceptives increases the risk of vitamin B deficiencies, so a healthy diet with good sources of B vitamins is advisable.

Contraceptive Implants

Contraceptive implants involve one or more small permeable tubes that are surgically implanted through a small incision under the skin of the upper arm. These tubes slowly release medication such as levonorgestrel (Norplant) and provide rapid and continued contraceptive effects for 5 years. Three-year replacement may be recommended for larger women.

Advantages, Disadvantages, Side Effects, and Risks

Long-term contraceptive effect without having to remember to take daily medication is the main advantage of implants. Improvement of dysmenorrhea and decrease in menstrual flow may occur; however, some women experience menstrual changes such as amenorrhea, irregular or prolonged menses, and breakthrough bleeding (especially in the first year of use).

Women older than 30 years of age may experience some delay of fertility after removal of the implants. Use of these implants is contraindicated for women with history of jaundice or liver disease. If redness, swelling, or inflammation occurs at the insertion site, the patient should return to the health care provider. Change or removal is done through another small incision. Some fibrosis develops around the insertion sites.

Depot Medications

Medroxyprogesterone and norethindrone are two types of contraceptive agents available in a slow-release (depot) form that can be injected intramuscularly. Medication is continuously released for 3 months. Some other depot contraceptives require monthly injections.

---

dysplasia: dys—painful or abnormal + plasia—shape or form

Advantages, Disadvantages, Side Effects, and Risks

The main advantage of depot medications is that there is no requirement to take medication daily. Disadvantages are that the medication is not immediately effective, so another method is necessary for the first 2 weeks after the initial injection, and that fertility may not return for several months to 1 year after cessation of injections.

Alterations in menstrual flow, especially amenorrhea, are the most commonly noted side effects. Other side effects and risks are similar to those encountered with oral contraceptive use.

Estrogen-Progestogen Contraceptive Ring

An estrogen-progestogen contraceptive ring has been very effective in trials.[8] It works in much the same manner as other hormonal contraceptives by slowly releasing hormones. The user inserts it into the vagina around the cervix much in the manner of a diaphragm, but it does not provide a barrier over the cervix, so there is less risk of infection than with a diaphragm or cervical cap. It is left in place for 3 weeks and then removed for 1 week for menses to occur.

Advantages, Disadvantages, Side Effects, and Risks

Not having to remember daily medication can be an advantage to the contraceptive ring, but failing to remove it at the right time may disrupt the regularity of the menstrual cycles. Further research is necessary to determine the seriousness of any side effects and risks, but these would be likely to resemble other low-dose hormonal contraceptives.

Transdermal Contraceptive Patch

A transdermal patch has been tested in the United States and Canada. It is worn weekly for 3 weeks and is then removed for 1 week.[9]

Advantages, Disadvantages, Side Effects, and Risks

The contraceptive patch has been found to be similar to oral contraceptives in effect without having to remember to take a pill each day. Bathing, swimming and other activities can continue with the patch in place.

Barrier Methods

Barrier methods of birth control alone are less effective in preventing pregnancy than most of the previously mentioned methods. Barriers are intended to prevent sperm from reaching the ovum. There are several forms of barrier contraceptives. Effectiveness of all the barrier methods may be increased by use of a spermicidal preparation with them. Spermicidal preparations may be purchased without a prescription.

Condoms

Condoms are to be used once and then discarded into an appropriate waste receptacle. They should be stored in a cool, dry place before use and should not be stored tightly pressed, because heat and continued pressure can weaken them. Storage in a wallet or glove compartment is not advisable. Petroleum-based substances, such as Vaseline, can also weaken condoms, so use of water-soluble lubricants (preferably spermicides) should be advised.

ADVANTAGES, DISADVANTAGES, SIDE EFFECTS, AND RISKS OF MALE CONDOMS. Male condoms have long been used for contraception because they are a relatively inexpensive, totally reversible method that men can control at the time of intercourse. They provide some barrier protection against transmission of sexually transmitted disease organisms as well. An electron microscope study of a sample of nonlubricated latex condoms, however, found that the majority of those viewed had surface abnormalities, including cracking and melted areas.[10] Patients should be informed that barrier methods can reduce risk but do not absolutely prevent transmission of STDs, especially in areas of contact not covered by the barrier. The main disadvantages of condom use are interruption of foreplay for application, decreased sensation, and the possibility of slippage or breakage during intercourse. These disadvantages may be overcome by incorporating application of the condom by the female partner as a part of foreplay; using thinner, lubricated, or textured condoms to increase sensation; using the correct size condom with a reservoir or applied with approximately half an inch at the tip of the condom loose enough to serve as a reservoir for the semen (Fig. 40–9); and removal from the vagina before relaxation of the erection.

ADVANTAGES, DISADVANTAGES, SIDE EFFECTS, AND RISKS OF FEMALE CONDOMS. Female condoms are a more recent innovation that allows female initiation of contracep-

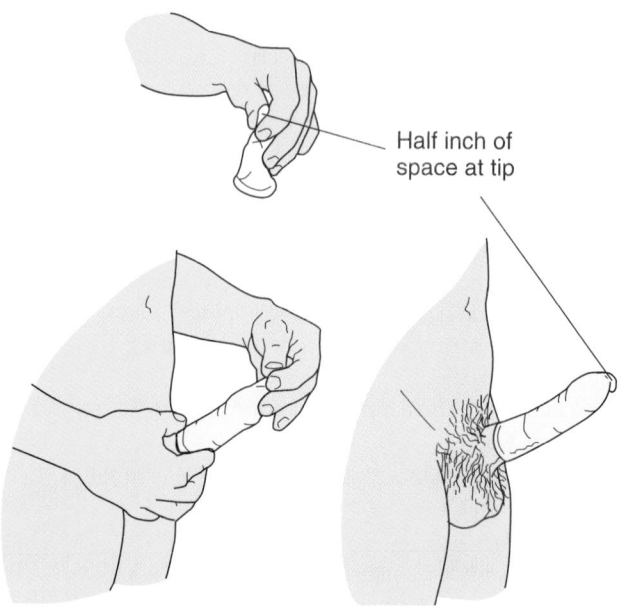

Half inch of space at tip

Figure 40-9   Correct application of a condom.

tion, as well as some barrier protection against infection with sexually transmitted diseases. Coverage of the labia by the condom may provide more of a barrier than male condoms (Fig. 40–10). Disadvantages are greater expense than male condoms, decreased sensation, the necessity to apply before

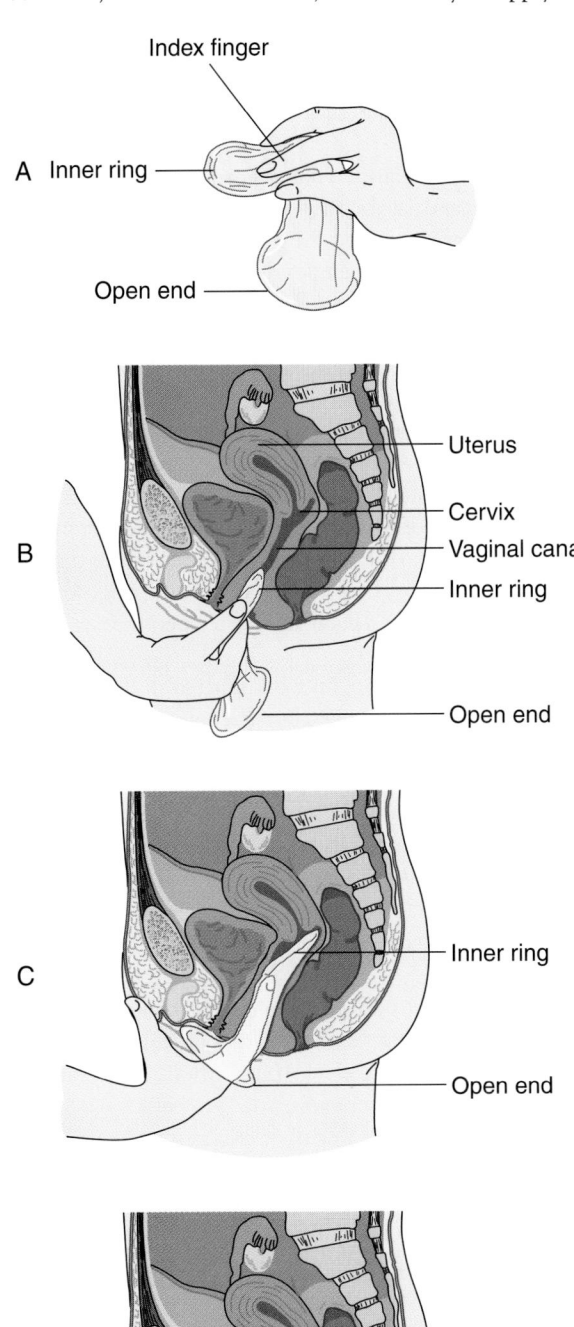

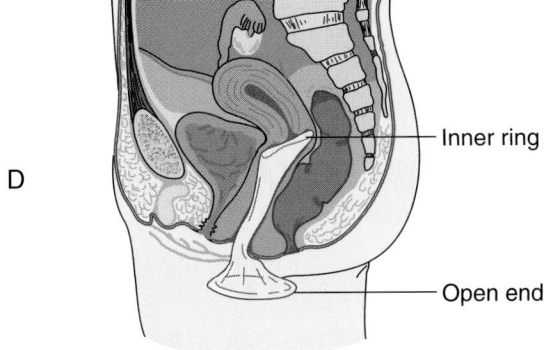

Figure 40-10 Female condom application. *(A)* Inner ring is squeezed for insertion. *(B)* Sheath is inserted similar to a tampon. *(C)* Inner ring is pushed up as far as it can go with index finger. *(D)* Condom in place.

intercourse, and the possibility of flaws in the condom material. As of the writing of this book, there had been relatively little controlled research on this type of condom.

Diaphragms, Cervical Caps, Sponges, and Lea's Shield

ADVANTAGES, DISADVANTAGES, SIDE EFFECTS, AND RISKS. Diaphragms, cervical caps, and sponges all work in the same manner by blocking the entry of sperm through the cervix. The barrier effect is enhanced by a spermicide. The sponge comes presaturated with a spermicide and generally needs to be moistened slightly with water and inserted with the indentation toward the cervix. The diaphragm and cervical cap require application of spermicide to the edges and a small amount in the cup before application. Lea's Shield is a silicone bowl-shaped barrier that may be purchased without prescription in some countries. It may offer improvements over the traditional diaphragm because it has a loop for ease of removal, fits closer to the cervix with less pressure exerted toward the urethra, and requires less spermicide than a traditional diaphragm. Little research was available on it at the time of the writing of this book.

All these methods are relatively inexpensive, are female initiated, and work without systemic medication. Diaphragms and cervical caps require initial fitting and a prescription to buy them, may need to be refitted after childbirth and the loss or gain of weight, and can last for years. These devices should be replaced periodically as the manufacturer recommends or whenever there is any evidence of hardening, cracking, or thin spots. They need to be washed with soap and water, dried, and stored in a case away from heat and sunlight between uses. Sponges do not require fitting, but they are more expensive than the other two methods and are used only once and then discarded.

Women and their partners may experience irritation or allergic reaction to the spermicide or the contraceptive device material, which would require changing birth control methods. All these methods require that the device be inserted before intercourse and left in place for several hours after intercourse. (See package insert for recommendations.) An increase in incidence of urinary tract infection has been reported with use of the diaphragm, and risk of toxic shock syndrome increases with prolonged uninterrupted use of cervical barriers. Adequate fluid intake, voiding shortly after intercourse, and removal of the device when 8 hours have passed since intercourse all help to prevent these potential problems. If urinary tract infections are recurrent using the diaphragm, changing to a cervical cap or a shield may decrease the occurrence because there is less pressure against the bladder side of the vagina.

Spermicides

Spermicidal agents also may be used alone, although use in combination with a barrier method is much more effective. They come in a variety of forms, such as creams, gels, foams, and suppositories, which kill or disable sperm so that fertilization does not occur.

Advantages, Disadvantages, Side Effects, and Risks

Spermicidal preparations are relatively inexpensive and can be male or female initiated. They do not produce systemic effects, and no hormones are involved. They are less effective alone than the previously described methods. Spermicides require application before each act of intercourse and are considered by some patients to be somewhat messy. Many contain the same ingredient—nonoxynol 9. If genital irritation or a rash occurs with spermicides, the patient should read the labels carefully to avoid contact with the same ingredient.

Intrauterine Devices

The presence of a foreign object in the uterus is thought to alter the environment, so that implantation is less likely to occur. Intrauterine devices (IUDs) are generally made of some form of plastic and may contain copper wire or a supply of a progestin that is slowly released into the system to further alter the environment, so that fertilization or implantation are hindered.

Advantages, Disadvantages, Side Effects, and Risks

The main advantage of use of an IUD is continued contraception without the necessity of remembering to take medication and without the side effects associated with medications. The disadvantages are changes in menstrual bleeding (especially increases in bleeding), cramping, and increased risk of pelvic inflammatory disease. Rarely an IUD has caused a uterine perforation. IUDs are contraindicated for women who have never been pregnant, those with uterine abnormalities, and those who have a history of anemia or heavy menstrual flow. Expulsion or displacement of the IUD can occur, so women should be taught to feel for the string before intercourse.

Insertion Procedure

Insertion of an IUD is generally done as a physician's office procedure with a nurse assisting. Usually this is done during the first 7 days of the menstrual cycle because the cervix is generally slightly dilated at this time. You will need to put out a vasovagal reaction treatment kit, an IUD insertion kit containing one or more uterine "sounds" and a tenaculum (a special long forceps) for grasping the cervix, and the IUD package. The IUD is generally inserted into the vagina through a tube that comes packaged with the IUD, which temporarily holds the IUD flat or folded so that it requires less room for insertion. When the IUD is pushed out the end of the tube, it springs into a shape that helps to keep it inside the uterus. One potential danger associated with IUD insertion is vasovagal reflex stimulation (previously described in association with endometrial biopsy). You periodically assess pulse or blood pressure during the procedure and notify the health care provider of slowing of the heart rate or a decrease in blood pressure.

Natural Family Planning

Periodic abstinence, or natural family planning, is less effective than the previously described methods. It is a method by which couples control their fertility by restricting intercourse to "safe periods" in which risk of conception is low. Many signs may be assessed to determine "safe" days, including temperature changes, cervical consistency and mucus changes, calendar timing, and awareness of symptoms of fertility.

Slight body temperature changes can indicate ovulation. During the first half of the menstrual cycle the temperature remains low, with a marked drop just before ovulation occurs. With ovulation the temperature rises and stays higher for the last half of the cycle. Women who use this assessment method should use a basal body temperature thermometer when they awaken, before doing anything else, and record it on a chart.

Cervical consistency and mucus changes may also help pinpoint ovulation. As hormone levels change, the consistency of cervical mucus changes. As ovulation approaches, there is an increase in the amount of mucus and the mucus becomes more clear, thin, slippery, and stretchable than at other times of the month. Around the time of ovulation the cervix becomes softer to touch and more open than at other times of the cycle.

Following the calendar can work fairly well if a woman's menstrual periods are regular, but becoming aware of her pattern may take time. Symptoms such as breast tenderness and midcycle discomfort (mittelschmerz) may also help identify ovulation. Users of this method should be advised to abstain for approximately 3 days before ovulation and 3 to 4 days after, because the sperm and ovum can survive for a long period in the female genital tract.

Advantages and Disadvantages

The advantages of this method are that it is inexpensive, it requires no medication, and it is the only birth control method presently approved by the Catholic church. The disadvantages are that it requires the cooperation of both partners and may interfere with spontaneity of sexual expression. It is generally not very effective as a means of birth control. It may be difficult to be accurate about ovulation times because infectious and inflammatory processes may affect temperature readings, infections and feminine hygiene products may affect cervical mucus, and irregularity of flow and symptoms may make prediction difficult.

Less Effective Methods
Coitus Interruptus

Coitus interruptus involves removal of the penis from the vagina before ejaculation occurs.

ADVANTAGES, DISADVANTAGES, AND RISKS. Although this method requires no expense or preparation, it is not very effective. Excellent control of ejaculation is required, and even the small amount of sperm that may be present in pre-ejaculatory fluid may result in pregnancy.

Postcoital Douching

The intended purpose of postcoital douching is to wash sperm out of the reproductive tract or to kill or immobilize sperm that the douche solution contacts.

ADVANTAGES, DISADVANTAGES, AND RISKS. This is relatively inexpensive and female initiated, but it is not very effective. Sperm move very rapidly once deposited, and douching may actually push the sperm upward.

Breast-Feeding

Breast-feeding is sometimes used as a method of birth control because the high blood levels of prolactin that occur with breast-feeding may suppress ovulation.

ADVANTAGES, DISADVANTAGES, AND RISKS. This method costs nothing but is not very effective. Prolactin levels can vary widely, and ovulation may resume at any time without any noticeable signs, resulting in pregnancy before even experiencing a menstrual period after the previous birth.

Ongoing Research: Future Possibilities for Contraceptive Choices

Several contraceptive agents are presently in developmental stages. A single contraceptive implant containing nomegestrol acetate (Uniplant) implanted under the skin of the upper arm or gluteal region has been tested in clinical trials and found to be effective for 1 year. This method, however, will probably be further refined and tested as a result of menstrual cycle irregularities among women using it.

Many researchers have tried to develop effective and reversible male contraceptives. A plant called Tripterygium, used in Chinese herbal medicine, when taken orally yields substances that can limit numbers and mobility of sperm, but it also has active ingredients that suppress immunity somewhat, so further investigation is necessary. Reversible injection procedures to block the seminal vas deferens are also being investigated, as are reversible injection procedures to block the fallopian tubes. Reversible birth control vaccines are also being investigated for both men and women with the goal of causing an immune response to occur at some vital point in the process of conception.

Some questions related to contraceptive vaccines include the unknown long-term repercussions of stimulating the body to respond with immunity to itself and whether governments could use vaccines as a means to control populations without their consent. Another frightening possibility is that with further removal of the threat of pregnancy even fewer people would use barrier methods and STDs would increase even more dramatically.

Sterilization
Actions

Permanent sterilization can be accomplished by either interrupting the fallopian tubes (tubal ligation) or vas deferens (vasectomy) or by removing the uterus and suturing the

---

CRITICAL THINKING

Jessica

You have just observed a patient who looks to be about 13 years old announce loudly at the clinic reception desk that she is "ready to be a responsible adult" and would like some birth control.

1. What information needs to be gathered from her?
2. What do you think she needs to know?
3. How can contraceptive teaching capitalize on her desire to be a responsible adult?

Answers at end of chapter.

---

proximal end of the vagina closed. Tubal interruption may be done by tying a suture or placing a ring or clip around each fallopian tube, by coagulating a section of the tubes, or by surgically removing a portion of the tube and suturing the ends. These procedures are usually done by laparoscope.

Advantages, Disadvantages, Side Effects, and Risks

Although this method is not absolutely certain to be permanent, the failure rate is low and has been decreasing recently with newer surgical methods. Tubal repair is sometimes requested at a later time to reestablish fertility. This requires microsurgery with anesthesia and has a poor success rate.

Patient Teaching

Patients should be advised by their surgeon about the complications of the surgery and of reversal before they sign a consent for sterilization. If any uncertainty about the surgery is evident, the physician should be notified promptly.

▶ PREGNANCY TERMINATION

Termination of pregnancy (abortion) is a difficult topic. Discussions about it are often highly charged with emotion. Both pro-life and pro-choice advocates argue on the basis of human rights—the former based on supposed rights of the fetus and the latter on supposed rights of the mother—because of the humanity of each party. There are very few people on either side of the philosophical argument, however, who would describe abortion as a healthy medical intervention. Most agree that abortion is a problematic solution to a difficult situation. There are instances in which carrying a pregnancy to term threatens the life of the mother. There are also many more instances in which a pregnancy is inconvenient or undesired. Although a discussion of the ethical issues surrounding abortion is encouraged, it is beyond the scope of this book to thoroughly address all the issues (Ethical Consideration Box 40–2).

Nursing Care, Teaching, and Patient Advocacy

Ethically an individual nurse should not be required to assist in any treatment that demands that he or she act in a way that contradicts personal moral beliefs—this would vi-

BOX 40-2 ETHICAL CONSIDERATION

The famous case of *Roe versus Wade* (1973) changed the legal status of therapeutic abortion in the United States but added a great deal of confusion to its ethical and moral status. A careful reading of the *Roe versus Wade* decision reveals that the court made no decision about the ethics or morality of elective (therapeutic) abortion. Rather, the court said that according to the U.S. Constitution, all people, including women, have a right to determine what they can do with their bodies (right to self-determination) and a right to privacy. Therefore during the first trimester a woman and her physician may decide to terminate a pregnancy without interference from the state. During the second trimester, the state may regulate the circumstances under which an abortion may occur to protect the woman's health. Once the fetus is viable, the state may also prohibit abortion. Although a woman may have a right to obtain an abortion, there is not a right to require the death of the fetus should a viable fetus be identified after the abortion procedure. Pro-choice groups put the right to self-determination and privacy central to their concerns. Pro-life groups generally believe that abortion is fundamentally killing and therefore abortion is wrong.

One of the central issues in the debate is the status of the fetus. Is the fetus considered a person with individual rights separate from those of the woman who is pregnant? Another central question is, When does human life begin? Is abortion at any stage of pregnancy morally acceptable? Each fertilized egg contains the essential genetic material for the development of a unique individual. However, many fertilized eggs never successfully implant in the uterus in the natural process of conception. Does such a factor make the timing of the procedure a critical element? Related to this problem is the status of frozen embryos created through in vitro fertilization. They too contain the essential, unique genetic material for the development of a human person. If every embryo should be respected as potential human life, what moral obligation do we have toward those embryos?

olate the nurse's rights. However, there is also an ethical duty to provide care to patients for whom the nurse is responsible. Therefore it is wise for nurses who have moral objections to abortion to carefully choose their work setting. For example, choosing to work in day surgery in a hospital that performs abortions and refusing to care for abortion patients is not a legitimate option. One way that nurses can positively influence the abortion situation is by teaching about family planning, which may lower the number of requests for abortions. Another way might be to become involved with agencies that help pregnant women to have viable alternatives to abortion.

Therapeutic Abortion for Ectopic Pregnancy

An ectopic pregnancy is the implantation of a fertilized ovum in an area other than the uterus. It may occur because of an abnormally shaped uterus or fallopian tubes that are obstructed as a result of abnormal development, scarring from STDs or other inflammatory processes, or for unknown reasons. This is a life-threatening situation for the mother, and currently abortion is the only treatment.

Therapeutic Abortion for Prenatal Abnormalities

Development of a variety of prenatal testing methods has introduced the possibility of knowing many things about a baby before birth. Prenatal testing may be done using ultrasound, samples of fluid taken from the amniotic sac or the placental villi, or blood samples from the mother. From these tests, several genetic diseases and congenital deformities can be identified. After anomalies are diagnosed, some patients choose to abort the baby. This is a very difficult decision to consider even in instances in which the baby has a fatal defect that will not allow it to live outside the uterus. Information about alternatives to abortion and possible treatments for their child are important for patients who have a serious prenatal diagnosis. No one should feel pressured to make the decision quickly to abort, but legal requirements and increasing risk for the mother may limit the time to decide. Abortion because of fetal abnormality may result in much grieving for the patient and her family.

Methods of Abortion

Several methods are available. The method is determined primarily by the length of the gestation and the goal of inflicting as little trauma to the mother's reproductive system as possible while still inducing pregnancy loss. Periods for the different abortion methods and the allowable reasons for legal abortion vary according to the laws of the state, province, or country.

Chemical Agents

The "morning after pill," or emergency contraceptive treatment, consists of postcoital administration of sufficient estrogen or an estrogen/progestin combination to cause sudden sloughing of the endometrial lining of the uterus, preventing implantation of a possibly fertilized ovum. For this to be effective, the initial dose is generally given within 72 hours of intercourse. This treatment is used in case of unexpected unprotected sexual intercourse (as with sexual assault) or unexpected risk of conception (as with condom failure). Nausea and cramping may accompany the shedding of the uterine lining, and an antiemetic may be taken to manage nausea. No advance planning before intercourse is required, so this can be abused as a casual form of birth control. These patients may need education about appropriate birth control methods.

Another example of postcoital contraception is the drug RU-486, which is a progestin antagonist. It prevents the binding of progestins at their receptors, resulting in a chemically induced abortion up to the 10th week of pregnancy. Nausea and cramping may accompany expulsion of uterine contents. There is much debate about whether RU-486 should be used at all or should be used only within specific guidelines. The United States Food and Drug Administration approved RU-486 in 2000 in an accelerated drug approval process, but legislative bills have been introduced and reintroduced to stipulate regulations for health care providers prescribing RU-486. Undoubtedly RU-486, as well as methotrexate (a cancer medication) and misoprostol (a medication to prevent stomach ulcers), all of which can stimulate abortion, will be the subject of much debate over the next few years.

Abortion Methods for Early Pregnancy

Early in a pregnancy (during approximately the first 13 weeks) there are three primary means of pregnancy termination—menstrual extraction, vacuum aspiration, and D & C. Menstrual extraction is removal of the endometrial lining by manual suction and can be done during the first 7 weeks following the last menstrual period (LMP). This can be done without anesthesia and without cervical dilation by inserting a small cannula into the cervix and aspirating with a large syringe. Vacuum aspiration is a similar process that is used from confirmation of pregnancy through the first 13 weeks. It requires cervical dilation and is generally done with local anesthesia. The patient returns home 1 to 4 hours after the procedure. D & C may also be used during the first 13 weeks. In this procedure the cervix is dilated and the uterine contents are scooped away with a curet rather than being removed by suction.

Abortion Methods for Later Pregnancy

During the second trimester the fetus is much larger, so more dilation is required. A dilation and evacuation (D & E) may be performed, in much the same manner as a D & C. Generally, dried laminaria (a type of seaweed) or some other absorbent substance is placed inside the cervical canal. This absorbs fluid and swells, thus gradually dilating the cervix. Prostaglandin may be administered either by suppository into the vagina or by injection into the amniotic sac; this usually induces uterine contractions and results in delivery a few hours later. Unfortunately, a live fetus too premature to survive may be born by this method and continue to breathe for a time until death.

An induction with either a saline or urea injection may be used for pregnancies beyond 16 weeks. A portion of amniotic fluid is removed and replaced with concentrated saline or urea solution, which kills the fetus and stimulates contractions. Sometimes saline and prostaglandins are used in combination to terminate a pregnancy.

Hysterotomy involves removal of the uterine contents through an abdominal incision in the same manner as a cesarean section. This procedure is rarely done for pregnancy termination.

Risks and Complications

Abortion involves risks. Some are the same risks inherent in childbirth, such as possible hemorrhage or introduction of infection, but there are additional risks related to inter-

ruption of natural processes and the aggressiveness with which the products of conception are removed during abortion. During an uncomplicated childbirth the uterine lining is not scraped or forcefully emptied by suction. Natural hormonal preparation for term childbirth contributes to uterine contraction after the birth, which decreases blood loss, but no such preparation occurs for abortion. Artificial dilation of the cervix may cause injury, as may introduction of the instruments used for abortion. Injured tissues are more likely to become sites for growth of microorganisms than are intact tissues. Finally, the possibility of infertility as a result of complications related to abortion, although relatively uncommon, is a risk.

Some possible physical complications following therapeutic abortion are injuries to the uterus or cervix, excessive bleeding, infection, retention of some products of conception, and possible failure of abortion. Rarely, second trimester abortions can be complicated by amniotic fluid embolism, in which amniotic fluid is absorbed into the uterine circulation because of disruption of placental attachments with instruments. Amniotic fluid in the mother's circulatory system can result in circulatory collapse and disseminated intravascular coagulation (DIC). DIC is a serious derangement of the body's blood clotting controls and, though rare, can be fatal. Clotting mechanisms are overstimulated, so that the blood begins to form clots in vessels all over the body. In response to this hypercoagulation, fibrinolytic mechanisms are overstimulated and the patient may suffer severe and widespread bleeding. Treatment of this disorder is difficult because the normal mechanisms for both sides of this clotting/clot dissolving equation are hypersensitive.

Nursing Care and Teaching

Aftercare is very important. Abortion patients rarely stay overnight, and complications may occur after they are discharged. They should be carefully assessed after the procedure for signs of bleeding. Instruct that bleeding should not exceed that of a heavy period, that the passage of clots larger than a quarter may be a sign of complications, and that the discharge should not become foul smelling. Patients should be given a phone number (available 24 hours per day, seven days per week) to call if fever, chills, excessive bleeding, or any signs of infection occur. The patient should be advised to abstain from sexual intercourse for the time specified by the health care provider (usually about 3 weeks). A grief response may occur after a pregnancy termination, even if the baby was definitely unwanted and the patient does not have strong beliefs against abortion. There is much debate about frequency of postabortion syndrome or whether such a condition exists. However, loss and trauma have occurred in any case, and reorganization of the self takes time. Availability of psychological counseling for women after abortion is very important. Women should be given a number to call if they experience psychological discomforts. The need for birth control should be assessed.

hysterotomy: hystero—womb + tomy—cutting

► TUMORS OF THE REPRODUCTIVE SYSTEM

Benign Growths

Fibroids

PATHOPHYSIOLOGY, ETIOLOGIES, AND SIGNS AND SYMPTOMS. Fibroids or **leiomyoma** (singular form is leiomyomata) are benign tumors made up of endometrial cells that have implanted on or within the walls of the uterus. These can grow very large and may cause pain or menstrual disorders, exert pressure on the bladder or bowel, cause necrosis because of pressure on the blood supply to tissues, and interfere with fertility.

TREATMENT. Because fibroids are estrogen sensitive, medical treatment may involve hormone suppression. Surgical treatment may involve **myomectomy** or hysterectomy. Myomectomy is removal of only the fibroid tumor and may be chosen to preserve fertility. Myomectomy may be done surgically through an abdominal or vaginal incision or with a laser introduced through a laparoscope. Hysterectomy may be necessary for very large fibroids or those that cause severe bleeding or discomfort.

Polyps

PATHOPHYSIOLOGY, ETIOLOGIES, AND SIGNS AND SYMPTOMS. Polyps are generally benign growths that grow inside the uterus or on the cervix and may bleed after intercourse or between menstrual cycles. These are generally teardrop shaped and are attached by a stalk. Polyps develop most often after the age of 40.

TREATMENT. Polyps are generally removed vaginally or transcervically by separating the stalk from the uterus and then stopping the bleeding by use of chemical, electrical, or laser **cautery.** Removal of polyps in the vagina may be done without anesthetic in a physician's office. Removal of polyps transcervically requires cervical dilation and is more likely to be done in a hospital with anesthesia.

Reproductive System Cysts

PATHOPHYSIOLOGY, ETIOLOGIES, AND SIGNS AND SYMPTOMS. Several types of cysts may affect women's health. Cysts of the ovaries may develop associated with incomplete ovulation, **hypertrophy** of the corpus luteum after ovulation, or inflammation of the ovary. Most ovarian cysts eventually spontaneously shrink and merely cause discomfort for a time. Chocolate cysts are formed when endometrial cells bleed into an enclosed space, as occurs with endometriosis. These are called chocolate cysts because they are filled with old blood, which has become chocolate colored. Cystoadenomas are benign growths that can sometimes undergo cellular transformation and become cancerous. Any pelvic mass in a postmenopausal woman has a high potential for malignancy and should be investigated.

TREATMENT. Most cysts are not surgically removed, but excessive size, interference with fertility, and high cancer potential may make needle drainage, biopsy, laparoscopic surgery, or **laparotomy** advisable. If cysts are painful, application of heat to the abdomen or back may help promote comfort.

Polycystic Ovary Syndrome

PATHOPHYSIOLOGY AND ETIOLOGY. Polycystic ovary syndrome (PCOS) is a complex abnormality of endocrine balance of unknown etiology. Multiple cysts on the ovaries are a sign that was discovered early and for which the disease was named, but they are not present in all cases. There seem to be strong genetic links with such family history as too much or too little hair (especially for women), severe acne, diabetes, irregular menses, and infertility. Many of the symptoms of PCOS are a result of excessive levels of insulin in the blood because of insulin resistance. Excess insulin in turn stimulates secretion of androgens.

SIGNS AND SYMPTOMS. Women with PCOS often present with infertility, obesity, and menstrual disturbances. They may also exhibit masculinization because of the excess androgen secretion. They have a higher risk for diabetes mellitus, elevated blood pressure, coronary artery disease, and endometrial cancer.

DIAGNOSTIC TESTS. Diagnostic tests may include blood tests to rule out other causes of endocrine abnormality, tests to determine whether ovulation is occurring (such as midluteal progesterone levels and basal body temperature graphing), endometrial biopsies to determine the level of proliferation and to check for endometrial cancer, and blood tests to determine lipid levels and glucose tolerance.

MEDICAL TREATMENT. Medical treatments may involve blood pressure medications, lipid control medications, and oral hypoglycemics. Diet and exercise may be recommended for weight reduction, control of lipid levels and cardiac health. Oral contraceptives may be used to normalize hormone levels and protect the endometrium for those not desiring to conceive. Ovulation-inducing medication may be used for those women who desire to conceive. If masculinization is a problem, antiandrogen medications may be prescribed. In severe cases, gonadotropin-releasing hormone (GnRH) agonists may be used to produce medical suppression of the ovaries, with results similar to removal of the ovaries. This is followed 6 months later with an estrogen/progestin combination to protect the bones from development of osteoporosis.

---

leiomyoma: leio—smooth + myom(a)—fibroid
myomectomy: myom(a)—fibroid + ec—away + tomy—cutting
cautery: branding iron
hypertrophy: hyper—too much + trophy—nourishment (growth)

---

laparotomy: laparo(o)—abdominal wall + tomy—cutting

Bartholin's Cysts

PATHOPHYSIOLOGY, ETIOLOGIES, AND SIGNS AND SYMPTOMS. Bartholin's cysts are actually infected or obstructed Bartholin's glands at either side of the vaginal opening. Excessive swelling of Bartholin's glands results in pain with sitting and with intercourse.

TREATMENT. Incision and drainage may alleviate the discomfort. If Bartholin's cyst formation occurs often, **marsupialization**—the surgical formation of a pouch around an opening made into a gland to facilitate drainage may be necessary. Sitz baths may be ordered to cleanse the area and to promote comfort and healing.

Dermoid Cysts

PATHOPHYSIOLOGY, ETIOLOGIES, SIGNS AND SYMPTOM. Rarely, for unknown reasons, a **dermoid** cyst (also called a cystic **teratoma**) may develop from a germinal cell of an ovary. This cell divides and differentiates into various tissue types such as skin, teeth, bones, hair, and even extremities in a disordered arrangement. This type of cyst may grow quite large and may occur on both ovaries at the same time.

TREATMENT. Dermoid cysts are removed by laparoscopy or laparotomy. If the cyst contained glandular tissue that was secreting hormones, adjustment of hormone levels to normal may take some time. Although most teratomas are benign, some are malignant, especially in postmenopausal women, so a biopsy is generally done on the tissue.

NURSING CARE AND TEACHING. Growth of a dermoid can be a frightening experience for a woman. Reassurance that this is merely a disordered group of cells identical to the other cells in her body, rather than a monster or deformed baby, is important.

Malignant Disorders

It is sometimes difficult to distinguish benign growths from malignant growths without biopsy results, and some benign growths can become cancerous. Malignancies can occur in all parts of the reproductive system and can occur at all ages. Although reproductive system cancers are more common in older age groups, ovarian tumors can occur even in young children. Both male and female children of women who were given diethylstilbestrol (DES) to prevent premature delivery in high-risk pregnancies have experienced a high incidence of developmental defects and cancers of the reproductive organs.

Many different types of genital cancer are possible, and discussion of every type in detail is beyond the focus of this book. This section presents a general overview of the most common cancers. Cancerous changes often can be observed or noticed, and if investigated and treated early enough, cure is often complete.

---

marsupialization: marsupial—pouch + ization—process of making
dermoid: derm—skin + oid—form
teratoma: terat—monster + oma—growth

LEARNING TIPS

Three C changes that may indicate cancer are changes in color, contour, and *consistency* of a tissue.

Vulvar Cancer

PATHOPHYSIOLOGY, ETIOLOGIES, AND SIGNS AND SYMPTOMS. Although vulvar cancer is not a common, alertness to changes of visible parts of the reproductive system such as the vulva can result in early diagnosis and requirement of less drastic treatment with more positive results. Persistent itching of the vulva or appearance of white or red patches, rough areas, skin ulcers, or wart-like growths on the vulva should not be ignored; these can be signs of precancerous or cancerous changes. Risk factors for development of vulvar cancer are an STD of any type, precancerous or cancerous changes of the anus or any of the genitalia, immune system depression, and smoking.

DIAGNOSTIC TESTS. Regular Pap smears and physical examinations can identify lesions. Biopsy of suspicious lesions is necessary to diagnose vulvar cancer.

TREATMENT. If discovered early, vulvar cancer may be treated with removal or destruction of cancerous cells. If not diagnosed early, it may require surgical removal of the entire vulva and associated lymph nodes (a radical vulvectomy) with subsequent skin grafting from other areas of the body for repair.

Cervical Cancer

PATHOPHYSIOLOGY, ETIOLOGIES, AND SIGNS AND SYMPTOMS. Changes in the cells of the cervix (called cervical intraepithelial neoplasia, or CIN) can progress to cervical cancer. Dysplastic cells (those with dysplasia) are generally less differentiated or less ordered than expected for their cell type.

Some identified risk factors for development of cervical cancer include starting sexual activity at an early age, having multiple sexual partners, having several pregnancies, smoking, and being infected with human papillomavirus or herpes simplex virus type II (HSV-II).[11] Use of oral contraceptives for several years may also increase a woman's risk of developing cervical cancer, although part of the difference in incidence may be because women using oral contraceptives may not be using barrier protection against STDs. Although some women experience slight spotting or a serosanguineous discharge with cervical cancer, many are asymptomatic until the cancer is widespread.

DIAGNOSTIC TESTS. Pap smears are the best method of screening for cervical cancer presently available, but some work is being done on self-tests that women can collect. A Pap smear determines the degree of cellular change, or dysplasia. Ranking systems vary, but Pap smear results are usually presented in categories that range from no atypical cells seen to invasive cancer evident (0 to IV). This procedure has significantly reduced the incidence of invasive cervical cancer over the years since its introduction, because cellular changes can be identified early enough for treatment before the cells become cancerous. Schiller's test may be done if a patient has an abnormal Pap smear. This involves painting the cervix with iodine. Dysplastic cells stain differently than normal ones. Biopsy is done to confirm a cancer diagnosis. Recommendations for frequency of screening vary, but most recommend that Pap smears begin at either age 18 or with the start of sexual activity and be done yearly unless abnormalities develop. After a period of normal Pap smears, some health care providers advocate longer intervals for low-risk people.

TREATMENT. Treatments for preinvasive neoplasia include **cryotherapy** (freezing), laser therapy (burning), and surgical removal of the involved area with a loop excision instrument or by conization (Fig. 40–11). All these procedures are done through the vagina, so there are no external incisions. After any of these treatments the patient is advised not to douche, use tampons, or have intercourse for approximately 2 weeks to allow for healing to take place. She should be advised to report immediately if fever or bloody or foul vaginal discharge occurs. For invasive cancers, hysterectomy, radiation implant, or chemotherapy may be done.

cryotherapy: cryo—cold + therapy—treatment

Endometrial Cancer

PATHOPHYSIOLOGY, ETIOLOGIES, AND SIGNS AND SYMPTOMS. Endometrial cancer is the most common type of uterine cancer. Most develop in response to relative estrogen excess. Abrupt changes in bleeding patterns, especially bleeding in a menopausal woman, may indicate endometrial cancer development. Estrogen excess can develop for many reasons. Estrogen levels fluctuate widely in the perimenopausal period. Obesity results in increased estrogen production that is not balanced by progestins. Estrogen replacement therapy for menopausal symptoms without the addition of progestins also has been associated with an increase in endometrial cancer, but addition of a progestin may decrease the risk of endometrial cancer to less than that of untreated women. Whether alcohol consumption increases the risk of endometrial cancer by interfering with estrogen metabolism is still a matter of debate. Some endometrial cancer, however, is unexplained by any presently known risk factors.

DIAGNOSTIC TESTS. Diagnosis is generally done by endometrial biopsy, but MRI may be used to evaluate invasiveness and involvement of lymph nodes.

TREATMENT. Depending on the stage of endometrial cancer and metastasis, treatment with hysterectomy, radiation, or chemotherapy may be used.

Ovarian Cancer

PATHOPHYSIOLOGY, ETIOLOGIES, AND SIGNS AND SYMPTOMS. Ovarian cancer is an especially insidious killer because cellular changes in the ovaries often are

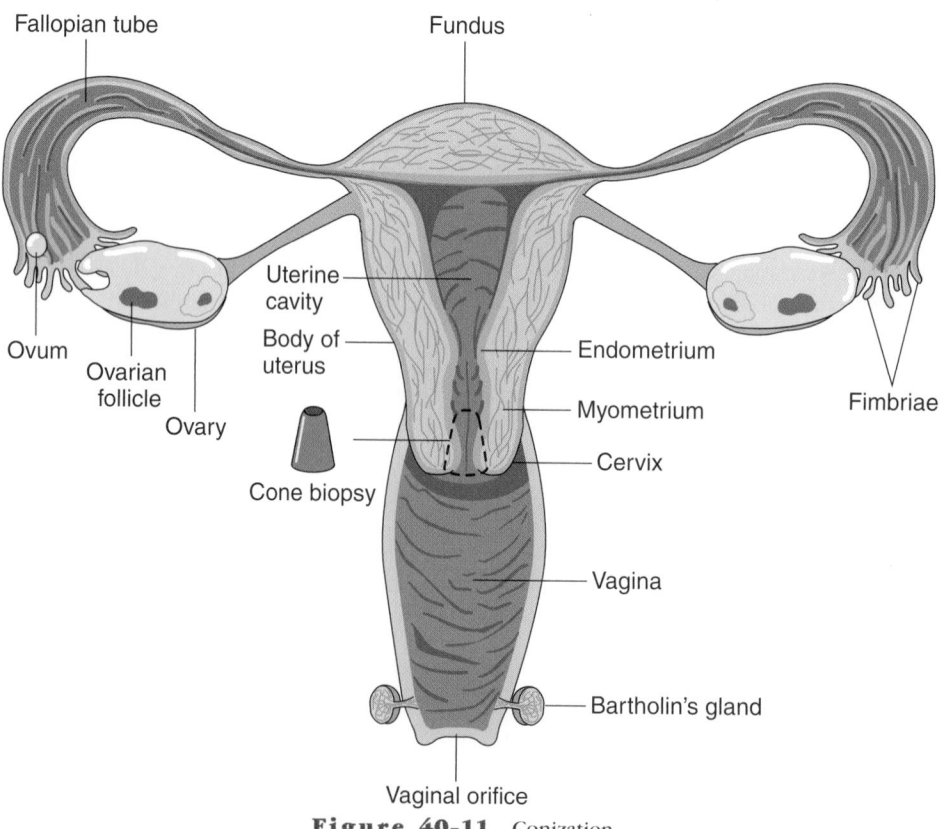

Figure 40-11 Conization.

asymptomatic until the cancer is quite advanced. Little is known about what prompts these cells to undergo malignant changes. Risk factors are not definitely identified, but some proposed factors include low fertility and number of children, late menopause, a family history of reproductive or colon cancers, and a diet rich in animal fats. Use of hormonal contraception may help prevent this, because it results in less ovulation during the woman's lifetime.

DIAGNOSTIC TESTS. Identification of abnormal growths on the ovaries may begin with bimanual examination, so it is important for women, especially in the older age groups, to continue to have regular pelvic examinations even if they are not sexually active and even if they have had a hysterectomy. Various blood tests measuring tumor marker substances, ultrasonography, CT scanning, and MRI may also be used to assist in diagnosis.

TREATMENT. Treatment may involve surgical removal of the ovaries by laparoscopy or laparotomy. Sometimes the ovaries are removed to prevent the disease in women who have a high familial risk. Radiation and combination chemotherapy may also be used.

Nursing Care and Teaching of Patients with Malignant Disorders

Radiation therapy for cancers of the reproductive system may involve the placement of radioactive implants into the patient's body for 24 to 72 hours. Avoid prolonged contact with the patient and perform care with as much of the patient's body between you and the implant as possible (e.g., for cervical implants, stand at the head of the bed rather than at the foot). Prevent inappropriate radiation of the patient's other body parts by such actions as maintaining patency of a urinary catheter to avoid unnecessary exposure of the bladder. Follow institutional guidelines for radiation precautions. Do not give care to patients with radioactive implants if you are pregnant. A foul-smelling vaginal discharge is expected after radiation by implant because of tissue destruction caused by the radiation; document the amount and character of the discharge. Chemotherapy treatments often cause severe nausea, as well as anorexia and sores of the mouth, vagina, and anus. See Chapter 10 for care of patients receiving chemotherapy.

► GYNECOLOGICAL SURGERY
Endoscopic Surgeries

Many of the surgeries performed on the reproductive system may be done using an endoscope. The scopes used contain not only magnifying lenses and a light source but may also include tiny tools for performing surgery, for removal of small areas of diseased tissue and samples, for suction, and for cauterization of bleeding vessels. Because endoscopic surgeries require tiny incisions (usually less than 1 inch long), there is less tissue disruption and very little bleeding when compared with traditional surgical techniques. Smaller incisions also present less risk of infection than traditional methods, and recuperation is generally more rapid

with fewer complications. Overall the danger to the patient is generally low for endoscopic surgeries; however, not all surgical situations may be satisfactorily handled in this way. The size of the cannula restricts the size of tissues that can be removed, unless they can be divided into smaller sections and then pulled out through the cannula. If affected areas are widespread, the scope may not be able to reach all sites. Traditional surgery still may have to be done when endoscopic surgery has been ineffective, and this can be frustrating to patients. Information gained through the prior endoscopic surgery may decrease the time required for the traditional surgery, however.

Laparoscopies are the most common type of endoscopic surgical procedure employed for women's reproductive system surgeries. This method can be used for access to the abdominal cavity and the anterior portions of the reproductive organs. Tubal ligations, tubal repairs, removal of ectopic pregnancy implantations, removal of small tumors, removal of endometriotic sections of tissue, and aspiration of fluid-filled cysts can all be done by this method.

Culdoscopies may be done to access the area at the back of the uterus. A **culdotomy,** which is an incision into the upper posterior portion of the vagina, is necessary to insert the cannula. A **culdocentesis,** which is the removal of fluid from the cul de sac of Douglas, may be done during a culdoscopy. Aftercare is much the same as for laparoscopy, although the patient should be informed that a small amount of vaginal spotting may be expected from the incision but heavy, purulent, or foul-smelling discharge could indicate infection. (See Chapter 39.)

Colposcopies are generally used to screen, diagnose, or treat problems of the cervix. The binocular microscope attached to the scope cannula, which is introduced into the vagina, allows the physician to examine dysplastic cells while they are still in their normal place and to treat cervical dysplasia as previously described. Hysteroscopy may be used to treat problems within the uterus. Removal of polyps and other growths, modification of congenital malformations such as septa (walls of tissue where there should be none), and laser ablation of endometrial tissue may all be done during hysteroscopy. The endoscope may be inserted further into the fallopian tubes to perform a salpingoscopy, allowing surgical or laser opening of blocked tubes.

Nursing Care and Teaching

Postoperative care involves careful assessment for signs of possible excessive internal bleeding, including checks of vital signs, skin color and temperature, and pain. Measures to reduce the discomfort produced by residual carbon dioxide from insufflation may include instruction to lie flat for a few hours, massaging of the back and shoulders, and administration of pain medication. Instruction about possible danger signs, any medications, and when and where to go for suture removal complete the discharge teaching.

---

culdotomy: culdo—cul de sac + tomy—cutting
culdocentesis: culdo—cul de sac+ centesis—puncturing

Hysterectomy

Removal of the uterus (hysterectomy) may be done for a variety of reasons, including abnormally heavy or painful menstruation, large fibroids or other benign tumors, severe uterine prolapse, and cancer of the uterus. It should not be done merely as a sterilization procedure, because the risks involved in hysterectomy surgery are much greater than risks associated with tubal ligation. The surgery is usually performed through an abdominal incision, but it may be done vaginally in some cases. The vagina is left intact except for suturing of the proximal end, which had been attached to the uterus, forming a blind pouch. Although less vaginal lubrication is present after hysterectomy, nerve routes are maintained and satisfactory sexual intercourse is expected to continue.

The uterus alone may be removed, and in some cases the tubes and ovaries may also be removed—a procedure called total abdominal hysterectomy with **bilateral salpingo-**oophorectomy (TAH-BSO), or **panhysterectomy.** If the ovaries are removed, the woman undergoes immediate menopause and may suffer from symptoms associated with menopause, including the increased risks of cardiovascular disease and osteoporosis. Because removal of the ovaries is usually done because of the presence of estrogen-dependent cancer, estrogen replacement is not usually feasible in these cases, and extra care, comfort, and explanation from nurses are necessary. (See Nursing Care Plan 40–3 for the Patient Undergoing Hysterectomy. In addition to the nursing diagnoses covered, also assess for anxiety related to surgery and resulting body image changes and for ineffective coping.)

---

bilateral salpingo-oophorectomy: bi—two + lateral—sided + salpingo—tubal + oophor—ovary + ec—from + tomy—cutting
panhysterectomy: pan—all + hyster—uterus + ec—from + tomy—cutting

---

BOX 40-3 **NURSING CARE PLAN FOR THE PATIENT UNDERGOING HYSTERECTOMY**

 Risk for impaired tissue integrity related to surgical incision and removal of the uterus (and possibly the ovaries)

PATIENT OUTCOMES
Incision(s) will heal by primary intention without excessive bleeding.

INTERVENTIONS	RATIONALE	EVALUATION
Monitor vital signs and oxygen saturation according to hospital policy and as necessary.	Vital signs and oxygen saturation reflect tissue perfusion status.	Are vital signs stable and within normal range?
Check for bleeding or other discharge on perineal pad and on abdominal dressing (if applicable).	Excessive bleeding may compromise tissue perfusion and slow healing. Vaginal discharge gives clues to healing of incision at the proximal end of the vagina.	Is pad or dressing dry? Is discharge foul smelling?
Assess wound healing twice a day (bid) and report any evidence of infection or inadequate healing promptly.	Early treatment of inadequate wound healing decreases postoperative complications.	Is incisional area swollen, reddened, or draining purulent material?

 Risk for impaired urinary elimination related to manipulation of the bladder and ureters during surgery, anticholinergic drugs, fluid intake changes, and fear of pain

PATIENT OUTCOMES
Patient will have urinary output 30 mL/h or more without difficulty.

INTERVENTIONS	RATIONALE	EVALUATION
Assess urinary output after surgery. Report if less than 30 mL/h or unable to void.	Inadequate urinary output can be an evidence of dehydration, low glomerular perfusion, kidney dysfunction, damage to ureter, or urinary retention.	Is output adequate? Is patient able to void without discomfort?
Assess bladder fullness using Doppler monitoring or scratch test (listening with a stethoscope, lightly scratch abdomen as you move downward from xiphoid until you hear change in sound indicating top of the bladder).	Urinary retention can cause damage to kidneys, ureters, and bladder. Scratch test and Doppler monitoring cause less discomfort and pressure to an abdominal incision area.	Does patient feel she is emptying fully when voiding? Does Doppler indicate residual urine after voiding? Where is level of sound change with scratch test?
Medicate for pain on a fixed schedule for operative day and first postoperative day (unless patient declines pain medication).	Maintenance of a consistent blood level of medication in the immediate postoperative period provides relief of pain and promotes voiding without fear of discomfort.	Does patient state that she is comfortable?

BOX 40-3 NURSING CARE PLAN FOR THE PATIENT UNDERGOING HYSTERECTOMY—cont'd

 Risk for constipation and gas related to manipulation of the bowel during surgery, opioid analgesics and anticholinergic drugs, diet changes, less exercise than usual, and fear of pain when passing stool

PATIENT OUTCOMES
Patient will pass soft formed stool without excessive gas discomfort by third postoperative day.

INTERVENTIONS	RATIONALE	EVALUATION
Assess for active bowel sounds in all four abdominal quadrants before giving anything orally. Encourage high fluid intake, and graduate diet toward a high-fiber, regular diet as soon as patient is able to tolerate if (or doctor's orders prescribe).	Manipulation of the bowel during surgery or anesthetics and other medications may interfere with bowel function. Adequate fluid and fiber in the diet softens the stool for easy passage.	Are bowel sounds sluggish, active, or hyperactive? Are bowel sounds present in all four quadrants? Is patient drinking well? Is patient ready for more normal foods, as well as liquids?
Encourage adequate exercise.	Reasonable exercise promotes peristalsis and relives gas discomfort.	Has patient dangled at bedside the day of surgery and then walked increasing amounts each day following?
Control pain with analgesics, especially before administering suppository or enema.	The presence of pain may inhibit defecation.	Is patient passing intestinal gas without difficulty? Does patient express satisfaction with pain control?
Give stool softeners, laxatives, suppositories, or enemas as ordered (check bowel protocol or standing orders).	Soft stool is easier to pass.	Has patient passed soft formed stool by the third day after surgery?

REVIEW QUESTIONS

1. Which of the following is the least effective form of contraception?
   a. Douching
   b. Condom with spermicide
   c. Diaphragm with spermicide
   d. Oral contraceptive medication

2. Following a hysterectomy, what should the nurse teach the patient to expect?
   a. Heavy bleeding for a week
   b. Symptoms of menopause

   c. Painful intercourse for approximately 6 months
   d. Monthly cramping but no menstrual flow

3. During an endometrial biopsy, the nurse observes for which of the following signs and symptoms of vasovagal response?
   a. Pain in the chest and abdomen
   b. Cramping and diaphoresis
   c. High blood pressure and tachycardia
   d. Bradycardia and falling blood pressure

4. Which of the following can be used to treat vasovagal response during gynecologic procedures?
   a. Atropine
   b. Morphine
   c. Epinephrine
   d. Norepinephrine

5. Which response by the nurse is most appropriate when a 60-year-old neighbor, who has been menopausal for several years, relates that she has begun having vaginal bleeding again?
   a. "Ignore it—it is perfectly normal."
   b. "I guess you were not really menopausal yet."
   c. "You should see a doctor to have that checked as soon as possible."
   d. "Give it time—bleeding after menopause usually goes away within a month."

6. If a mastectomy patient has a Hemovac drain in place postoperatively, nursing actions should include which of the following?
   a. Leave the drainage in the bag until the Hemovac is removed from the patient.
   b. Stop the drain by clamping it.
   c. Empty the bag as ordered and note the amount and color of drainage.
   d. No action is necessary.

REFERENCES

1. Marchant, DJ, and Falkenberry, SS: Breast cancer. In Copeland, LJ (ed): Textbook of Gynecology, ed 2. WB Saunders, Philadelphia, 2000, p 1131.
2. Ursin, G, Spicer, DV, and Bernstein, L: Breast cancer epidemiology, treatment, and prevention. In Goldman, MB, and Hatch, MC: Women and Health. Academic Press, San Diego, 2000, p 876.
3. Newman, B: Inherited genetic susceptibility and breast cancer. In Goldman, MB, and Hatch, MC: Women and Health. Academic Press, San Diego, 2000, p 886.
4. Love, SM: Dr. Susan Love's Breast Book, ed 3. Perseus Publishing, Cambridge, Mass, 2000, 368-77.
5. Mortola, JF Premenstrual syndrome. In Goldman, MB, and Hatch, MC: Women and Health. Academic Press, San Diego, 2000, p 114.
6. National Institutes of Health. National Heart, Lung, and Blood Institute: NHLBI stops trial of estrogen plus progestin due to increased breast cancer risk, lack of overall benefit. NIH News release, 2002, July 9. http://www.nhlbi.nih.gov/new/press/02-07-09.htm
7. Cates, W, Padian, NS: The interrelationship of reproductive health and sexually transmitted diseases. In Goldman, MB, and Hatch, MC: Women and Health. Academic Press, San Diego, 2000, p 381.
8. Shoupe, D: Contraception: conception control. In Wallis, L (ed): Textbook of Women's Health, Lippincott, Williams & Wilkins, Philadelphia, 1998, p 631.
9. Audet et al: Evaluation of contraceptive efficacy and cycle control of a transdermal contraceptive patch versus an oral contraceptive: A randomized controlled trial. JAMA 285:2347, 2001.
10. Rosenweig, BA, Even, A, and Budnick, LE: Observations of scanning electron microscopy detected abnormalities of non-lubricated latex condoms. Contraception 53:49, 1996.
11. Cuzick, J: Human papillomavirus testing for primary cervical cancer screening. JAMA, 283:108, 2000.

Answers to CRITICAL THINKING

Jessica

1. Some important information from Jessica would include her true age (laws vary concerning birth control for minors), her intentions, her family situation, whether she is already sexually active, and what information she wants.
2. She needs to know that being sexually active involves more risks than just pregnancy. Discussion of STDs is vital. Discussion of violence, potential for abuse, long-term effects, and psychological suffering, which may accompany early sexual activity, may also be important. The choices of birth control should be explained, including the risks, effectiveness, disadvantages, and advantages of each method.
3. Although focusing on her desire to be a responsible adult is an admirable attitude, potential scenarios can be presented for her "responsible" consideration, such as the following: What would she do if contraceptive failure resulted in a pregnancy? How would she feel if she contracted an incurable or permanently damaging sexually transmitted disease that she might pass on to someone else? How would she react now to being "dumped" by someone she cares for who is less responsible and mature? Asking about her goals and plans in life may be significant. Counseling that evidences concern for the individual at this stage may do a lot to postpone sexual activity until the patient is more mature. It is important for her to realize that choosing to delay sexual activity at this time may be the most responsible and health-promoting life decision she can make.

41

NURSING CARE OF MALE PATIENTS WITH GENITOURINARY DISORDERS

Cindy Meredith

QUESTIONS TO GUIDE YOUR READING

1. How would you explain the pathophysiologies associated with male genitourinary and reproductive disorders?

2. What are the etiologies, signs and symptoms, and treatments of prostate problems?

3. What disorders of the penis can interfere with normal function?

4. What are disorders of the testes and how do they impact sexual function?

5. What are some physical and emotional causes of erectile dysfunction?

6. What are the medical, surgical, and psychological treatment options for erectile function?

7. What is the nurse's role in helping men cope with the loss of sexual function?

8. What are male factors that interfere with fertility?

9. What are the treatment options for male infertility?

Problems affecting the male genitals and urinary system are generally difficult areas for both the patient and the nurse to deal with because of the sexual nature of the male anatomy. It is important to realize that sexuality is a natural part of each of us as human beings and should not be avoided when we provide care to patients. Very often the nurse is in an ideal position to provide important sexual health care teaching to patients. If the patient is approached in a confident, confidential manner, it can be a positive learning experience for both the patient and the nurse.

▶ PROSTATE DISORDERS

The prostate gland sits at the base of the bladder and wraps around the upper part of the male urethra like a doughnut. It is traditionally divided into lobes: anterior, posterior, median, and lateral. The primary purpose of the prostate is to provide alkaline secretions to semen and to aid in ejaculation. The prostate does not contain any hormones; however, many men fear that prostate problems and treatment will cause problems with their erections or their "nature" (sexual activities).

Prostatitis
Pathophysiology

Prostatitis, or inflammation of the prostate gland, can occur any time after puberty. The problem may be chronic or a single, acute episode. The inflammation causes the prostate gland to swell, resulting in pain, especially when standing. It eventually may lead to difficulty in passing urine as a result of an inward squeezing of the urethra that causes a mild obstruction.

Etiology

There are three basic types of prostatitis: acute bacterial, chronic bacterial, and nonbacterial. Bacterial prostatitis is most common in older men. It results in edema and inflammation of all or part of the prostate gland.

The bacteria primarily responsible for the infection are gram-negative organisms such as *Escherichia coli*; however, gram-positive and gonococcal bacteria may also play a part. The prostate gland may become infected by the following:

- Bacteria ascending the urethra
- Infected urine refluxing from the bladder into the prostatic ducts
- Bacteria in the blood or lymph supply to the gland
- Surgical instrumentation or other forms of urethral trauma.

Prevention

Ways to avert prostatitis are regular and complete emptying of the bladder to prevent urinary tract infection (UTI), avoiding excess alcohol (more than 2 to 3 ounces per day—alcohol is a bladder irritant), and avoiding certain high-risk sexual practices. Avoiding contamination of the urinary tract and factors that produce congestion of the prostate gland are the best preventive measures.

Signs and Symptoms

The most common symptoms are the same ones that occur with any UTI: complaints of urgency, frequency, hesitancy, and dysuria. Because of the location and role of the prostate gland, the patient may complain of low back, perineal, and postejaculation pain; he may also have a fever and chills.

Complications

One complication of acute bacterial prostatitis is urinary retention. If the prostate is extremely swollen, it prevents complete bladder emptying. For most men, the most troublesome complication may be a temporary problem with erections. Ascending infections, prostatic abscess, **epididymitis,** and prostatic calculi (stones) are some of the more serious and rare complications of prostatitis.

Diagnostic Tests

The first test performed is a careful, gentle, digital rectal examination (DRE) of the prostate. The prostate gland is examined by the health care provider by insertion of a gloved finger into the rectum. The examiner may find a warm, irregular, swollen, painful prostate gland. A urine culture generally is positive for bacteria. The examiner may also gently massage the prostate gland and order an expressed prostate secretion (EPS) test that reveals bacteria and a large number of white blood cells.

Medical Treatment

Acute bacterial prostatitis is usually treated medically with antibiotic therapy. The preferred treatment is trimethoprim and sulfamethoxazole (Bactrim, Septra) for 30 days. Other antibiotics may be used for chronic prostatitis.

Other forms of treatment may include anti-inflammatory agents, warm sitz baths, prostatic massage, and diet changes such as decreasing spicy foods and alcohol. In some cases, prostate surgery is necessary to remove the obstruction.

Nursing Process

ASSESSMENT. Begin the assessment by asking the patient to describe signs and symptoms that indicate evidence of a UTI, such as sudden fever, chills, and complaints of urgency, frequency, hesitancy, dysuria, and nocturia. In addition, the patient may have complaints of pain in the lower back, in the perineum, or after ejaculating. Ask if the patient has ever had a UTI or prostate infection in the past. Care must be taken to assess urinary retention resulting from obstruction. Obtain a urine culture and assist with collection of the EPS specimen, if requested, as part of the patient assessment.

NURSING DIAGNOSIS. The nursing diagnoses for patients with prostatitis may include impaired urinary elimination, and ineffective health maintenance: knowledge deficit related to causes, treatment, and prevention of prostatitis, particularly for those with their first acute episode. Because of the alteration in comfort, a diagnosis related to pain should be included. Because the prostate gland is so closely associated with sexual activity, a diagnosis of anxiety related to sexual concerns should be considered for most patients.

PLANNING. Because of the nature of prostatitis and its accompanying discomfort, the patient's sexual partner is often included in the plan of care. If the bacteria causing the problem is a sexually transmitted organism such as *Chlamydia* or *Gonococcus*, both partners should be treated with antibiotics; sexual intercourse should be avoided during treatment. Scheduled medications and frequent sitz baths may require adjustments in work or normal routines. If a urinary catheter is required, provide catheter care and monitor fluid intake and output every 8 hours.

INTERVENTIONS. Take special care to maintain privacy and sensitivity when discussing treatment with the patient. Emphasize preventive measures that will help the patient

prostatitis: prostat—prostate gland + itis—inflammation
epididymitis: epi—upon + didym—testis + itis—inflammation

avoid future infection. An area of special concern is reinfection resulting from not taking antibiotics until the medicine is completely gone. Many patients take antibiotics only until the symptoms disappear and need to be reminded to carefully follow directions for taking their medications.

EVALUATION. A clean urine culture with complete disappearance of signs and symptoms of prostatitis is the desired outcome. Prevention of chronic prostatitis can generally be achieved with patient education.

PATIENT EDUCATION. Teach the patient the causes, prevention, and treatment of prostatitis. Include risk factors such as the use of indwelling urinary catheters, poor hygiene or sexual practices, excessive intake of bladder irritants, ignoring signs and symptoms of UTIs, and poor compliance with the antibiotic treatment plan.

Encourage the patient to wash his hands and sitz bath equipment before and after each treatment. Fluids such as water and cranberry juice should be encouraged up to 2500 to 3000 mL per day unless contraindicated. Bladder irritants in the form of caffeine products (e.g., coffee, tea, cola, and chocolate), citrus juices, and alcohol should be taken in very limited amounts. Encourage the patient to empty his bladder every 2 to 3 hours even if he does not feel the urge to urinate.

Encourage the patient to discuss possible complications and questions about sexual practices with his health care provider. In some cases, sexual intercourse is encouraged as a means of relieving prostatic congestion; in other situations it may be contraindicated.

Benign Prostatic Hyperplasia

Enlargement of the prostate gland is a normal process in older men. It begins at about age 50 and happens in 75 percent of men older than age 70. Benign prostatic hyperplasia (BPH) is a nonmalignant growth of the prostate that gradually causes urinary obstruction. According to current studies, BPH does not increase a man's risk of developing cancer of the prostate.

Pathophysiology

There is a slow increase in the number of cells in the prostate gland, generally the results of aging and the male hormone dihydrotestosterone. As the size of the prostate gland increases, it begins to compress or squeeze the urethra shut. The narrowing of the urethra means the bladder must work harder to expel the urine. More effort and a longer time is required to empty the bladder. Eventually the narrowing causes an obstruction and may lead to urinary retention or eventually distention of the kidney with urine (hydronephrosis).

It is the location of the enlargement, not the amount, that causes the problem. A small growth in the prostate gland closest to the urethra may cause more problems with urination than a growth the size of an orange in the outer portion of the gland.

Etiology

There is no known cause of BPH other than normal aging. Some men think they may have caused the problem by certain sexual practices; however, there is no scientific proof at this time. Some factors that are being investigated in research studies are high-fat diet, ethnic background, and lifestyle issues.

Prevention

Because there is no known cause, there is no proven method to prevent enlargement of the prostate gland. There are many new treatments that are aimed at slowing down the enlargement process. One such treatment that is being researched is the herbal supplement saw palmetto.

Signs and Symptoms

Symptoms of BPH are usually identified in two ways (Table 41–1). A prostate symptom index score sheet has been developed, and health care providers are asking their older patients the questions as a way of assessing the seriousness of their symptoms and determining treatment options.

Complications

When BPH is untreated and obstruction is prolonged, serious complications may occur. Urine that sits in the bladder for too long can back up into the kidneys and cause hydronephrosis, renal insufficiency, or **urosepsis;** it can also damage the bladder walls, leading to bladder dysfunction, recurrent UTIs, or calculi (stones).

Diagnostic Tests

The first step is a medical history, including specific questions about the patient's symptoms. A DRE of the prostate is then conducted by the health care provider to assess for enlargement and whether the gland is hard, lumpy, or "boggy." Additional diagnostic tests are listed in Table 41–1.

Medical Treatment

If the patient has no symptoms or only mild ones, the most current medical approach is "watchful waiting." The health care provider watches for any increase in symptoms or signs that the urethra is becoming obstructed. Treatment of symptoms may include use of a catheter (indwelling or intermittent), encouraging oral fluids, and antibiotics for UTI.

Conservative medical treatment includes the use of medication to either relax the smooth muscles of the prostate and bladder neck or block the male hormone to prevent or shrink tissue growth. Alpha blockers are medications that relax the smooth muscles and include prazosin (Minipress), terazosin (Hytrin), and doxazosin (Cardura). These medications are also used to treat high blood pressure, and patients need to work closely with their health

---

urosepsis: uro—urine + sepsis—systemic infection

TABLE 41-1 BENIGN PROSTATIC HYPERPLASIA SUMMARY

Symptoms

Related to obstruction	Decrease in size and force of urinary stream
	Difficulty in starting stream or pushing to start
	Dribbling at the end of urination
	Interrupted stream, starting and stopping stream
	Urinary retention—feeling bladder is not empty
	Overflow incontinence—leaking more
Related to irritation	Nocturia
	Dysuria
	Urgency

Diagnostic Tests

Primary	Urinalysis
	BUN and serum creatinine
	Prostate-specific antigen
Secondary	**Urodynamic** flow studies
	Transrectal ultrasound of the prostate
	Cystoscopy

Therapeutic Management

Conservative	Alpha blockers
	Testosterone blockers
Nonsurgical (experimental) options	Transurethral microwave antenna (TUMA), which involves heat applied directly to the gland
	Prostatic balloon, which dilates the urethra by stretching or compressing
	Prostatic stents, which open the passageway for urine to flow
Transurethral options	Transurethral incision of the prostate (TUIP), in which surgical incisions are made into the gland to relieve obstruction
	Transurethral ultrasound-guided laser-induced prostatectomy (TULIP), in which incisions are made by laser to relieve obstruction
	Transurethral resection of the prostate (TURP), in which a resectoscope is used to remove small pieces of tissue
Open prostatectomy	Suprapubic resection, which involves making an abdominal incision and removing the gland through the bladder
	Retropubic resection, which involves making an abdominal incision and removing the gland through the prostate capsule
	Perineal resection, which involves making an incision between the scrotum and anus and removing the gland and low pelvic mass

BUN = blood urea nitrogen.

urodynamic: uro—urine + dynamic—force

care providers to avoid overdose or the negative side effects of postural hypotension. The most commonly used medication to block the action of the male hormone in the prostate gland is finasteride (Proscar). All these medications must be taken on a long-term, continuous basis to achieve results. Conservative measures are used initially unless there are recurring infections, repeated gross hematuria, bladder or kidney damage, evidence of cancer, or unsatisfactory lifestyle changes. (See Table 41–1.)

Nonsurgical invasive treatments, some of which are experimental, are available in some areas of the country in addition to surgical options. (See Table 41–1.)

Surgical Treatment

TRANSURETHRAL RESECTION OF THE PROSTATE. During the past 50 years, transurethral resection of the prostate (TURP) has been the surgical treatment used most often to relieve obstruction caused by an enlarged prostate (Fig. 41–1). The patient is anesthetized and the surgery is performed using an instrument called a resectoscope. The resectoscope is inserted into the urethra and the prostate gland is "chipped" away a piece at a time. Special surgical instruments are now being used that "vaporize" or "microwave" the pieces and cut down on the amount of bleeding during surgery. During routine TURP, the "chips" are flushed out using an irrigating solution and are sent to the laboratory to be analyzed for possible evidence of cancer. The prostate gland is not completely removed but peeled away like the rind of an orange. The prostate tissue that is left eventually grows back and can cause obstruction again at a later time. Patients need to be reminded to continue having yearly prostate examinations.

As the tissue is removed during TURP, bleeding occurs. A Foley catheter is left in place with 30 to 60 mL of sterile water inflating the balloon. The balloon is overfilled and may be secured to the leg or abdomen to **tamponade** (compress) the prostate area and stop the bleeding. Irrigation solution generally flows continuously; manual irrigation may be done for the first 24 hours to help maintain catheter patency by removing clots and chips. The health care provider removes the Foley catheter after the danger of hemorrhage has passed.

Complications associated with prostate surgery depend on the type and extent of the procedure performed. The

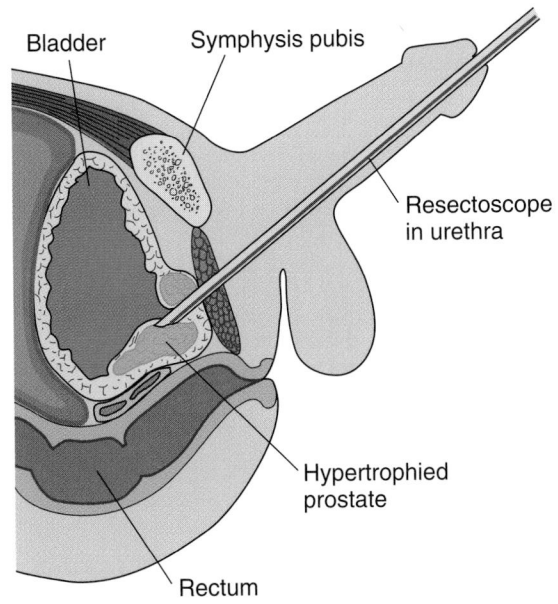

Bladder Symphysis pubis

Resectoscope in urethra

Hypertrophied prostate

Rectum

Figure 41-1 Transurethral resection of the prostate.

main medical complications include clot formation, bladder spasms, and infection. Less common complications may be urinary incontinence, hemorrhage, and erectile dysfunction (Nursing Care Plan Box 41–1).

Retrograde ejaculation is a common side effect of prostate surgery. When any of the prostate gland is removed, there is a decrease in the amount of semen produced and a part of the ejaculatory ducts may be removed. The result is that less semen is pushed outside the body, and instead it "falls back" into the bladder. This causes no harm; the semen is simply passed during the next urination.

It is important to understand that erection, ejaculation, and **orgasm** are all separate actions. Erection means the penis becomes hard, ejaculation is the release of semen, and orgasm is felt as pulsations along the urethra. Unless additional problems are present, the patient continues to have

retrograde: retro—backward + grade—step

BOX 41-1 NURSING CARE PLAN FOR THE POSTSURGICAL PATIENT HAVING TRANSURETHRAL RESECTION OF THE PROSTATE FOR BENIGN PROSTATIC HYPERPLASIA

▶ Acute pain related to bladder spasms, obstruction, or surgical process

PATIENT OUTCOMES

(1) Patient states pain has decreased. (2) Patient identifies at least two comfort measures.

EVALUATION OF OUTCOMES

(1) Does patient state that pain is decreased? (2) Is patient able to identify at least two measures that will help relieve pain?

INTERVENTIONS	RATIONALE	EVALUATION
Monitor pain every 2 to 4 hours using a pain scale for first 48 hours. Monitor for signs of bladder spasm, such as facial grimaces, irrigation solution that does not flow into bladder, urinating around catheter, multiple clots.	A pain scale is a more accurate measure of pain. Relief of spasms promotes comfort, rest, and healing.	Does patient verbalize pain as increasing or decreasing on the scale? Is number of spasms increasing or decreasing?
Give prescribed medication (analgesics, antispasmodics) and monitor response.	Medications relieve symptoms.	Does patient state relief when medications are given?
Irrigate catheter as ordered.	Removal of clots reduces spasms and pain.	Does irrigating solution go in and out easily? Are clots being removed?
Teach relaxation, deep breathing techniques.	Relaxation calms spasms and relives pain.	Is patient able to relax?

▶ Urge incontinence related to poor sphincter control

PATIENT OUTCOMES

(1) Patient identifies at least two methods to achieve dryness. (2) Patient verbalizes satisfactory control of dribbling.

EVALUATION OF OUTCOMES

(1) Does patient identify ways to prevent incontinence? Do they help? (2) Does patient verbalize satisfaction with outcome?

INTERVENTIONS	RATIONALE	EVALUATION
Teach Kegel (pelvic floor) exercises (see Chapter 40)—to be practiced every time patient urinates and throughout the day.	Strengthens muscle tone to hold urine after catheter is removed.	Is patient able to start and stop his urine stream?

| BOX 41-1 | NURSING CARE PLAN FOR THE POSTSURGICAL PATIENT HAVING TRANSURETHRAL RESECTION OF THE PROSTATE FOR BENIGN PROSTATIC HYPERPLASIA—CONT'D |

Discuss use of condom catheter or penile pads.

Instruct patient to continue drinking 2000 to 4000 mL of non-caffeinated, nonalcoholic beverages each day.

Encourage patient to discuss long-term (>6 months) incontinence problems with physician. National incontinence support groups are available.

Specific urine control devices are available for men.

Adequate nonirritating fluid intake is important for healing and preventing UTI.

Patient may need to learn self-catheterization or try medication.

Increased abdominal pressure can cause healing tissue to break apart in the prostate capsule, and excessive bleeding occurs.

Does patient indicate an informed choice of incontinence products?

Does patient drink adequate fluids even though he dribbles?

Does patient verbalize understanding of what to do if incontinence continues?

Is patient able to identify activities that would cause risk for excessive bleeding?

 Ineffective management of therapeutic regimen related to lack of knowledge of postoperative restrictions and care

PATIENT OUTCOMES

(1) Patient avoids activities that increase intra-abdominal pressure resulting in excessive bleeding. (2) Patient verbalizes understanding of how to prevent postoperative infection.

EVALUATION OF OUTCOMES

(1) Does patient verbalize understanding of how to prevent bleeding? (2) Is infection prevented?

INTERVENTIONS	RATIONALE	EVALUATION
Teach patient to avoid lifting heavy objects (>10 lb), stair climbing, driving, strenuous exercise, constipation, straining during bowel movements, and sexual activities until approved by physician (about 6 weeks). Instruct patient on proper catheter care using verbal, written, and demonstration techniques (many are sent home before catheter is removed). Include the following information: • Keep catheter bag secured to abdomen or thigh and below bladder. • Wash catheter/meatal junction with soap and water once daily. • Use clean technique to change form leg bag to night drainage bag. • Report signs and symptoms of UTI to physician immediately. • Encourage oral fluids.	Urinary tract infections are extremely dangerous and can cause death following genitourinary surgery in an elderly patient.	Can patient give a return demonstration of proper catheter care? Is patient free from signs and symptoms of infection?

 Anxiety related to concerns over loss of sexual functioning

PATIENT OUTCOME

(1) Patient verbalizes normal sexual changes that happen after prostate surgery. (2) Patient identifies support systems if needed.

INTERVENTIONS	RATIONALE	EVALUATION
Explain to patient that he will probably have retrograde ejaculation into bladder after surgery. It is not harmful and semen will come out when he urinates.	Removal of the prostate gland often results in retrograde ejaculation.	Does patient understand what will happen when he ejaculates?
Instruct patient to talk with urologist if erection problems occur.	Urologists who specialize in treatment of erectile dysfunction can be helpful.	Is patient aware of local support services?

erections and orgasmic sensations but decreased or no ejaculation.

OPEN PROSTATECTOMY. When the prostate gland is very large, is causing obstruction, or is cancerous, an open **prostatectomy** is performed (Fig. 41–2). In the **suprapubic** approach, an incision is made through the lower abdomen into the bladder. The gland is removed, and the urethra is reattached to the bladder. The retropubic approach is similar except there is no incision into the bladder. A perineal prostectomy procedure is rarely done because of the increased risk of contamination of the incision (close to the rectum), urinary incontinence, erectile dysfunction, or injury to the rectum.

An open prostatectomy means a longer hospital stay compared with other BPH surgeries. A suprapubic catheter and care for an abdominal incision increase the length of stay and the risk for complications. Follow-up home care for wound dressing changes and catheter care is an important aspect of nursing interventions for these patients.

Nursing Process

ASSESSMENT. Begin by asking the patient if he has ever had treatment or surgery for prostate trouble. Assess amount and type of fluid intake per day and whether the patient has noticed any of the symptoms of BPH.

NURSING DIAGNOSIS. Nursing diagnoses for patients with mild or no symptoms are directed at knowledge deficits related to prevention of UTIs, knowing when to report an increase in symptoms to the health care provider, and taking medication exactly as the health care provider orders. For patients scheduled for prostate surgery, the preoperative nursing diagnoses might include the following:

---

prostatectomy: prostat—prostate gland + ectomy—excision
suprapubic: supra—above + pubic—pubic bone

- Impaired urinary elimination related to obstruction of urethra
- Ineffective health maintenance related to insufficient knowledge of condition, surgery, and postsurgical activities
- Body image disturbance related to loss of normal urination
- Anxiety related to insufficient knowledge of preoperative and postoperative routines and sensations
- Pain related to urinary retention
- Disturbed sleep pattern related to frequency of urination

Nursing care after TURP follows standard guidelines for routine care (see Nursing Care Plan Box 41–1). In addition, each patient experiences individual responses and needs. Individualized care plans are often necessary, because the majority of these patients are elderly and have secondary medical problems such as cardiovascular disease.

PLANNING. Planning should include use of educational materials to discuss the expected surgical outcomes with both the patient and his family whenever possible. It is important that they know about bladder spasm sensations and that the treatment option may be a belladonna and opium (B & O) suppository. If you ask the patient if he wants a suppository for pain, the answer may be no, because he does not realize it is not for his bowels but to relieve the spasms.

Seeing the catheter bag filled with bloody drainage may also be upsetting to a patient or his family. Tell them it is normal to have this type of drainage and that what they see is actually a little blood mixed with a large amount of irrigating solution. It is important to closely monitor the urinary output in terms of amount, color, and presence of clots at least every hour for the first 24 to 48 hours postoperatively. Careful monitoring and documentation of vital signs and fluid intake and output can help prevent major complications.

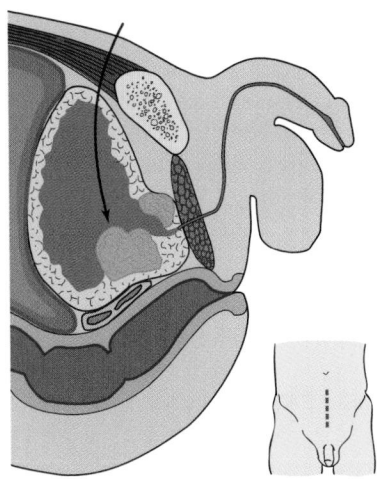
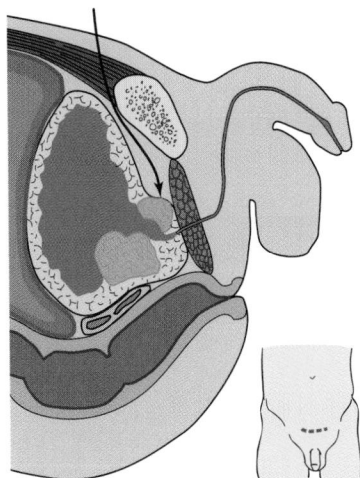
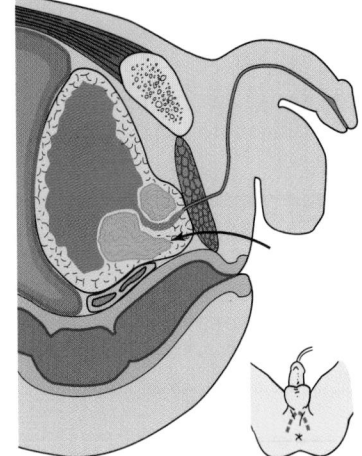

A. Suprapubic prostatectomy   B. Retropubic prostatectomy   C. Perineal prostatectomy

Figure 41-2 *(A)* Suprapubic prostatectomy. *(B)* Retropubic prostatectomy. *(C)* Perineal prostatectomy.

INTERVENTIONS. Special care should be taken to control bladder spasms, prevent excessive bleeding, and prevent infection. Encourage the patient to ask for medication before pain becomes severe. Often men are reluctant to ask for pain medication, so it is important to monitor closely for evidence of pain. Routinely offer the patient pain medication if he does not ask on his own.

Ambulation and other activity levels should be limited based on the characteristics of the drainage in the catheter bag. If the drainage is bright red or has large clots, the patient should be encouraged to lie down in bed until the color lightens and the clots are diminished. Interventions to clear the catheter require increasing either the continuous or manual irrigation process.

The risk for catheter-induced infections is very high unless sterile technique is used when the catheter is opened for irrigation. If the patient is being discharged home with a catheter, careful written instructions regarding catheter care and the need for increased fluid intake to prevent UTI are an important part of discharge planning. Patients are routinely placed on antibiotics during the operation and sometimes postoperatively to prevent infection.

EVALUATION. A patient should be discharged home with a minimum of bladder discomfort, light pink to clear urine, and no evidence of UTI. Home care nursing may be required if the patient lives alone or does not have the capacity to provide for meals, toileting, or transportation for the follow-up visit.

PATIENT EDUCATION. Prevention of complications such as postoperative bleeding and infections are critical to these patients' recovery. Encourage patients to drink up to 2500 mL per day (unless contraindicated by other medical conditions) of water, noncitrus juices (e.g., prune juice to prevent straining from constipation and cranberry juice to reduce risk of UTI), and other noncaffeinated, nonalcoholic beverages. Activities should be limited to short periods of sitting or walking followed by rest. Climbing stairs should be limited to once a day, which may mean keeping a urinal handy if a bathroom is unavailable on the level where the kitchen and television are located. The patient should not drive or engage in sexual intercourse until after he is seen for a final checkup with his health care provider (usually 4 to 6 weeks). Heavy lifting must also be avoided during this time (limited to 10 pounds or less).

CRITICAL THINKING

Mr. Atkinson

Mr. Atkinson is a 68-year-old African-American farmer with an enlarged prostate. He lives on a 75-acre farm with his wife and one son. He is scheduled for a TURP the beginning of March.

1. What postoperative instructions should he be given in light of his occupation?
2. What should you tell him if he asks about how the surgery will affect his "nature" (sexual activities)?

Answers at end of chapter.

Cancer of the Prostate

Cancer of the prostate is the second most common cause of cancer death in American men older than 60 years of age. Chances are that, unless a cause other than age is found, the numbers will continue to increase in the years to come as more men live longer.

Pathophysiology

Prostate cancer depends on testosterone to grow. The cancer cells are usually slow growing and begin in the posterior (back) or lateral (side) part of the gland. The cancer spreads by one of three routes. If it spreads by local invasion, it will move into the bladder, seminal vesicles, or peritoneum. The cancer may also spread through the lymph system to the pelvic nodes and may travel as far as the supraclavicular nodes. The third route is through the vascular system to bone, lung, and liver. Prostate cancer is staged or graded based on the growth or spread.

Etiology

Age is the primary risk factor. Prostate cancer is found most often in men older than age 65 and is rare in men younger than age 40. Other risk factors are higher levels of testosterone, high-fat diet, and immediate family history.

Prostate cancer rates are highest in African-American men and lowest in Japanese men. Occupational exposure to cadmium (e.g., welding, electroplating, alkaline battery manufacturing) has been identified as an added risk factor.

Signs and Symptoms

Symptoms are rare in the early stage (stage A) of prostate cancer. Later stages (stages B and C) include symptoms of urinary obstruction, hematuria, and urinary retention. In the advanced (metastatic) stage (stage D), symptoms may be bone pain in the back or hip, anemia, weakness, weight loss, and overall tiredness.

Complications

Early complications of prostate cancer are related to bladder problems, such as difficulty in urinating, and bladder or kidney infection. As the cancer metastasizes, the patient may develop problems such as pain, bone fractures, weight loss, and depression; eventually death may occur if treatment is not successful.

Diagnostic Tests

A routine DRE of the prostate is the first test; often the examiner finds a hard lump or hardened lobe. A blood test looking for high levels of prostate-specific antigen (PSA) or prostatic acid phosphatase (PAP) is done to detect prostate cancer. When there is a palpable tumor, the health care provider may order a transrectal ultrasound and biopsy to help confirm the diagnosis. Bone scans and other tests may be ordered to determine if the cancer has spread outside the prostate gland.

Medical Treatment

Prostate cancer in stage A may be treated with testosterone-suppressing medications, such as leuprolide (Lupron) or goserelin (Zoladex); surgery, such as TURP or open prostatectomy; or a combination of medication and radiation therapy. In stages B and C the treatment is usually a radical prostatectomy, radiation therapy, or implantation of radioactive "seeds" into the prostate (brachytherapy). In stage D, prostate cancer treatment involves relief of symptoms or blocking testosterone by bilateral **orchiectomy,** estrogen therapy (diethylstilbestrol [DES]), administration of antiandrogen (flutamide [Eulexin]), or use of agents such as leuprolide and goserelin. Sometimes chemotherapy is used to help relieve symptoms from the cancer spread.

Brachytherapy, external-beam radiation therapy, and radical prostatectomy combinations are showing favorable results in the treatment of advanced cancer. Gene therapy, vaccine, and immune-based interventions are the latest medical treatment options under investigation in the war against prostate cancer. Nontraditional prostate cancer prevention using vitamin E and selenium is currently being investigated in a 5-year nationwide study.

RADICAL PROSTATECTOMY. The radical prostatectomy procedure is generally reserved for patients with cancer of the prostate or when the gland is too large to resect using the TURP method.

The patient returns from surgery with a large indwelling catheter in the urethra and may also have a suprapubic catheter. Often there is a Penrose or sump drain in place to remove fluids from the abdominal cavity and allow the wound to heal from the inside outward. Special care must be taken to keep the incision and drain sites clean and dry. Dressings should be changed according to institution policy using sterile technique.

There are more complications associated with radical prostatectomy than any other treatment option. The major complications are hemorrhage, infection, loss of urinary control, and erectile dysfunction.

Patient Education

All men older than age 40 should be encouraged to have a yearly DRE of the prostate. Prevention and early detection are the best ways to fight prostate cancer.

▶ PENILE DISORDERS

Problems of the penis, aside from sexually transmitted diseases, are fairly rare but may cause great concern and worry for the patient. Most men have difficulty seeking help for a "private" problem. It is important to be sensitive when assessing or providing care for these patients.

Peyronie's Disease

Peyronie's disease often gives the penis a curved or crooked look when it is erect. Fibrous bands or plaques form mainly on the dorsal (top) part of the layer of tissue that surrounds one of the corpora cavernosa of the penis. The plaque may be caused by injury or inflammation of the penile tissue, or it may come and go spontaneously. If the plaque is thick enough, it can cause curvature, painful erection, difficulty in vaginal penetration, and erectile dysfunction. When conservative treatments such as vitamin E, steroids, or ultrasound do not work, surgery may be needed to remove the plaque. Patients need to be reassured that the problem is not life threatening and can be treated.

Priapism

Priapism is a painful erection that lasts too long. Anytime an erection lasts for longer than 4 to 6 hours it can become a medical emergency. The small veins in the corpora cavernosa spasm, so the blood cannot drain back out of the penis as it should. When the blood cannot drain, the tissue of the penis does not get oxygen and there can be permanent tissue damage. There may be a complete loss of erection ability after the priapism episode. Prolonged priapism can also prevent the patient from passing urine, which can lead to painful bladder and kidney problems. Some causes of priapism are prolonged sexual activity, sickle cell anemia, leukemia, widespread cancer, spinal cord injury or tumors, and use of crack cocaine or certain other drugs. Treatment in the emergency department may include ice packs, sedatives, analgesics, injection of medications directly into the penis to relax the vein spasms, needle aspiration, and irrigation of the corpora. If all else fails, surgery is done to drain the blood out of the penis.

Phimosis and Paraphimosis

Phimosis is the term used to describe a condition in which the foreskin of an uncircumcised male becomes so tight it is difficult or impossible to pull back away from the head of the penis. It may make it impossible to clean the area underneath. Smegma, a cottage cheese–like secretion made by the glands of the foreskin, becomes trapped under the foreskin and is an excellent place for the growth of bacterial and yeast infections. Treatment usually begins with antibiotics and warm soaks to the area. The physician may then cut a small slit in the foreskin to relieve the pressure and treat the infection. A full circumcision may be recommended to the patient if the problem continues or if a condom catheter is necessary for urine drainage. Phimosis is generally prevented by teaching uncircumcised males to pull the foreskin back carefully, wash with mild soap and water daily, and replace the foreskin to its normal position.

Paraphimosis occurs when the uncircumcised foreskin is pulled back, during intercourse or bathing, and not immediately replaced in a forward position. This causes constriction of the dorsal veins, which leads to edema and pain. Moderate to severe paraphimosis is a medical emergency and requires immediate care by a physician. The longer the problem continues, the greater the risk of circulation problems and possible gangrene. Again, prevention through daily cleaning and replacing the foreskin in its normal place is important.

---

orchiectomy: orchi—testes + ectomy—excision

Cancer of the Penis

Cancer of the penis has been found in men who were not circumcised as babies or have acquired the human papillomavirus (HPV). The tumor looks like a small, round, raised wart. It is one form of cancer that may be spread to the sex partner. Several research studies have found a link between cancer of the penis and cancer of the uterine cervix. Cancer of the penis may be treated with minor surgery such as a circumcision or laser removal of the growth. If the cancer has spread, the treatment may mean cutting away part or all of the penis, radiation, or chemotherapy. Finding and treating any wartlike tumor in its earliest stages is an important part of patient education.

▶ TESTICULAR DISORDERS
Cryptorchidism

Cryptorchidism (undescended testes) is a congenital condition in which a baby boy is born with one or both of his testes not in the scrotum. The testes normally drop down (descend) into the scrotum in the last 1 to 2 months before the boy is born. Many times the testes descend into the scrotum on their own by 2 years of age. If they do not descend by the age of 2, surgery should be done to correct the problem. Testes that are not brought down into the scrotum decrease a man's chances of producing a child, usually because excessive body heat damages sperm production in the testes. Studies have shown that the chances of testicular cancer are also higher if the condition is not corrected before the child reaches his teen years. If normal male sex characteristics do not develop during puberty because the testosterone level is too low, extra testosterone medication may be given. Testosterone can be administered in the form of a daily pill or long term by injection or patch.

Hydrocele

A **hydrocele** is a collection of fluid in the scrotal sac. Hydroceles are not dangerous and generally do not cause any pain. The cause is not known, and it can happen at any point during the lifetime. No treatment is necessary unless the hydrocele is so large that it causes discomfort or embarrassment or is a threat to the blood supply to the testes. If treatment is needed, the health care provider aspirates, or surgically drains, the fluid.

Varicocele

A varicocele is a condition sometimes called varicose veins of the scrotum. The main blood supply to the testes travels along the spermatic cord. The veins become dilated, and when the man is standing, the area in the scrotum begins to feel like a "bag of worms." The patient may complain of a pulling sensation, a dull ache, or scrotal pain. The sensations are most often felt when standing up. Most varicoceles occur on the left side because of the way the scrotal vein enters at a sharp angle from the left renal vein.

A varicocele is often not discovered until a couple tries to have a baby and is unable to conceive. It is believed that the varicose veins may increase the temperature of the testes and cause damage to the sperm. The most successful treatment is surgical repair of the varicose veins.

Epididymitis

The epididymis is a small tube along the back of the testes where sperm is matured for its last 10 to 12 days before it is ready to be ejaculated. Epididymitis is inflammation or infection of the epididymis that may be caused by bacteria, viruses, parasites, chemicals, or trauma. Epididymitis may be facilitated by sexual or nonsexual contact, a complication of some urological procedures, or reflux (backflow) of urine. The problem may also be associated with prostate infections and is usually painful, with the scrotal skin being tender, red, and warm to the touch.

Epididymitis is treated with antibiotics; the partner is also treated if it was sexually transmitted. Depending on the severity of the pain, the patient may be placed on bedrest with the scrotum elevated, possibly on ice packs, and also given analgesics. The pain and tenderness usually go away in about a week, although the swelling may last for several weeks. Complications may include chronic epididymitis, abscess formation, and sterility.

Orchitis

Orchitis is a rare inflammation or infection of the testes. The problem may be caused by trauma or surgical procedures; chemical substances; infection from epididymitis; UTI; or systemic diseases such as influenza, infectious mononucleosis, tuberculosis, gout, pneumonia, or mumps (after puberty). The patient has swollen, extremely tender testes; red scrotal skin; and a fever. Interventions are basically the same as for epididymitis and include bedrest, scrotal support, antibiotics, and medication to relieve pain and fever. Complications such as sterility from mumps orchitis can be prevented by giving boys the mumps vaccine at an early age.

Cancer of the Testes
Pathophysiology and Etiology

Cancer of the testes is the most common solid tumor in men 15 to 40 years of age and peaks between 20 and 34 years of age. The etiology of testicular cancer is unknown. Some of the known risk factors are cryptorchidism, family history, mother's use of DES (an estrogen preparation once used to prevent spontaneous abortion) while pregnant, white race, and high socioeconomic status. The tumors are mostly a germ cell type of cancer formed during normal embryo development.

---

cryptorchidism: crypt—hidden + orchid—testis + ism—
  condition
hydrocele: hydro—water + cele—swelling

---

orchitis: orch—testis + itis—inflammation

Prevention

The best prevention is monthly testicular self-examination (TSE). The procedure is simple and easy to learn and should be taught to males between ages 15 and 40. See Chapter 39 for instructions on TSE.

Signs and Symptoms

Early warning signs of cancer may include a small, painless lump on the side or front of the testes. The patient may also notice that the scrotum is swollen and feels heavy. Late symptoms of back pain, shortness of breath, difficulty swallowing, breast enlargement, and changes in vision or mental status indicate metastasis of the cancer.

Complications

Emotional complications can range from fear of cancer and death to feelings of loss of masculine body image and sexual function. Physical complications may involve dealing with pain and the effects of metastasis to areas such as the lungs, abdomen, or lymph nodes. Other less common areas of cancer spread are the liver, brain, and bone. Treatments such as surgery, chemotherapy, and radiation therapy can all have negative side effects on the patient and require special care.

Diagnostic Tests

When a tumor is found, several laboratory and radiographic tests are done. An ultrasound of the testes is done first. If the test shows cancer, a chest x-ray examination is done to look for spread to the lungs. A scan of the lymph nodes, liver, brain, and bones may also be ordered. Blood is drawn to look for what are called tumor markers. An example of a tumor marker for testicular cancer is beta human chorionic gonadotropin (bHCG). An exploration, biopsy, and removal of the testes is done to decide the stage of the tumor.

Testicular tumors may be staged or classified in several ways. The simplest way to stage a testicular tumor is as follows:

- Stage I—tumor only in the testes
- Stage II—tumor spread to groin lymph nodes
- Stage III—tumor spread past lymph nodes, usually to the lungs

Medical Treatment

Intervention depends on the stage of the cancer. All treatment begins with complete removal of the cancerous testes, spermatic cord, and local lymph nodes. Stage I tumor treatment then includes radiation to the groin area lymph nodes. Treatment for stage II involves chemotherapy. Stage III or metastatic cancer is treated with both radiation and chemotherapy. If the cancer is found in the beginning stages, the chances for complete recovery are about 75 percent. All patients should have regular follow-up testing.

Nursing Care

Nursing care is directed first at prevention, by teaching young men to practice monthly testicular self-examination and to see their health care provider if they notice any changes. If a diagnosis of cancer has been made, provide emotional support for the patient. If the patient wants to have children, he should be encouraged to make deposits in a sperm bank before any surgery or treatment is started. The patient and his partner may have many questions about sexual activities as they go through treatment. They should be encouraged to talk with their health care provider or a sex therapist about ways to express love and tenderness toward one another. Helping the patient deal with pain and the side effects of chemotherapy or radiation therapy are also important nursing interventions for these men.

✦ CRITICAL THINKING

Mr. Cunningham

Mr. Cunningham is a 23-year-old white college student engaged to be married next spring. While taking a shower one day he discovers a lump on his left testes.

1. What should he do?
2. What are the treatment options if Mr. Cunningham has cancer of the testes?
3. How can you help Mr. Cunningham cope with the diagnosis?

Answers at end of chapter.

▶ SEXUAL FUNCTIONING

Vasectomy

A **vasectomy** is the surgical cutting and sealing off of the vas deferens to prevent sperm from reaching the outside of the body. This 15- to 30-minute surgery is performed most often as a permanent birth control method but may also be performed on some men during prostate gland removal. The male patient should carefully discuss the surgery with his physician, so there is a clear understanding of the results following the procedure.

The testes continue to produce the male hormone testosterone and sperm. The prostate gland, along with the seminal vesicles, still ejaculate semen, but the semen does not contain sperm. There should be no major change in the way the ejaculate looks or feels following the procedure. The patient should be encouraged to continue using another birth control method for about 6 weeks after surgery to be sure there are no sperm left in the tract above the surgical site. The sperm continue to be produced in the testes but are absorbed by the body.

There are times when the man may decide he wants to have more children and asks to have the vasectomy "undone." The surgical procedure to reverse a vasectomy is called a vasovasotomy. Using microscopic instruments, the surgeon reconnects the two pieces of the vas deferens. During the surgery the physician generally tries to determine whether the testes are still producing good sperm. If the vasectomy was done less than 10 years before, the fertility

---

vasectomy: vas—vas deferens + ectomy—excision

success rate is higher. A vasovasotomy is generally not an option if the vasectomy procedure is more than 10 years old.

Erectile Dysfunction

Problems with getting or keeping an erection can happen at any age and have been a concern of men and their partners for centuries. It is a unique problem because it affects not only the man but his sex partner as well. Most men experience a temporary erection problem at some time during their life. It is often caused by stress, illness, fatigue, or an excessive use of alcohol or drugs. When the problem becomes persistent, as it does for more than 30 million men in the United States, it is time to seek medical help.

Before the 1980s, 90 percent of men who went to their physicians for help were told the problem was emotional, not physical. Through improved testing methods, it is now believed that 80 to 90 percent of erection problems have a physical cause. For more information, visit the National Library of Medicine at www.nlm.nih.gov/nih/cdc/www/91txt.html and www.medlineplus.nlm.nih.gov/medlineplus/druginfo/.

Pathophysiology

The term *impotence* means powerlessness. This term is being replaced with the more accurate term **erectile dysfunction** (ED), which describes a physical condition. Erectile dysfunction means that a man cannot get or keep a usable erection that is firm enough and long-lasting enough for satisfactory sexual intercourse. For a man to have a usable erection, several conditions must be met.

1. *Circulatory system.* The blood supply coming into the penis from the arteries must be sufficient to fill the corpora cavernosa (spongy erectile tissue inside the penis), causing the penis to become rigid. The diameter of these vessels is about the size of pencil lead. The veins in the penis must then be able to constrict (narrow down) to trap the blood in the corporal bodies to keep the penis erect. The most common cause of erectile dysfunction is failure in the circulatory system.

2. *Nervous system.* Both the sympathetic and parasympathetic nerves are involved in the erection, ejaculation, orgasm, and resting phases of the penile response cycle. There are many nerve receptors and transmitters in the spinal cord and the penis that must be intact for a usable erection. Spinal cord injuries are the most common neurological cause of erection problems. The remarkable fact about spinal cord injuries is that higher cord injuries (cervical or thoracic) have less negative impact on erections than those in the lumbar or sacral area.

3. *Hormonal system.* There are three basic male hormones involved with an erection. The most important hormone, testosterone (normal: 350 to 1000 ng/dL), affects a man's sex drive and desire. The testosterone level is highest when a man is in his late teens and begins to decline in middle age. Luteinizing hormone (LH; normal: 5 to 18 mIU/mL) stimulates the production of testosterone, and prolactin (less than 15 ng/mL) in large amounts may block testosterone.

4. *Limbic system.* This is the center in the brain that affects how we feel emotionally. It works with our five senses to stimulate the desire for sex.

All these systems can be influenced by physical, emotional, and chemical factors. A good assessment is important to determine the cause of erectile dysfunction.

Etiology

The psychological causes of erectile dysfunction usually come from stress or anxiety. Marital problems, financial worries, job frustration, or even fear of not being able to perform resulting from an isolated incidence of erection problems may lead to sexual dysfunction. Even when the cause is physical, there are psychological (emotional) side effects. Fear of failure, ridicule, or rejection by the partner lead to a lack of self-esteem or self-confidence.

The list of physical causes is long because it includes anything that may interfere with the flow of blood, the nerve supply, the balance of hormones, or the effects of emotions (Table 41–2). There are many medications and chemicals

TABLE 41-2   CAUSES OF ERECTILE DYSFUNCTION

Psychogenic	Surgical	Vascular	Cardiorespiratory	Neurological
Excessive stress in family, work, or interpersonal relationships; depression; fatigue; fear of failure to perform	Coronary artery bypass Abdominal-perineal resection Cystectomy Radical prostatectomy Pelvic lymphadenectomy Renal transplant Sympathectomy TURP	Aortic aneurysm Arteriosclerosis Aortofemoral bypass Cerebrovascular disease/accident Pelvic steal syndrome Venous leak Hypertension	Angina pectoris Myocardial infarction COPD Coronary insufficiency Congestive heart failure	Electroshock therapy Multiple sclerosis Parkinson's disease Cerebral palsy Myasthenia gravis Peripheral neuropathy Sympathectomy Tumors/transection of spinal cord Trauma to spinal cord Head injuries Spina bifida

COPD = chronic obstruction pulmonary disease.

that can interfere with desire, blood supply, or nerve transmission and cause problems with erections (Table 41–3). The most common types of medications that cause problems are high blood pressure and cardiovascular medications.

Diagnostic Tests

Assessment of erectile dysfunction has now been divided into primary and secondary diagnostic studies according to guidelines developed by the National Institutes of Health. The first step begins with a careful history that is especially designed to focus on the medical-surgical history, medications including any substance abuse, lifestyle patterns, and sexual history. The main areas to investigate are those that indicate vascular, neurological, endocrine, and psychological factors. A physical examination looks for evidence of abnormal genital disorders, hormonal imbalance (such as hair patterns or enlarged breasts), surgical interventions, decreased circulation, and lack of nerve sensations. Blood is taken to look at glucose levels; testosterone; evidence of liver, heart, or kidney disorders; signs of infection; or blood disorders. The health care provider may also order specific blood tests if other problems exist that may have an impact on erections. Some physicians are also using intracorporeal injection of vasoactive medications that can create an erection to test the blood flow in the penis. A psychological evaluation is recommended to rule out any serious issues in the sexual relationship or emotional problems that may contribute to erectile dysfunction or affect the treatment outcome.

Secondary level of testing may involve the use of sophisticated vascular flow studies to locate areas where either the blood vessels are narrowed or the veins allow the blood to drain out of the penis too rapidly. Another area of testing may be related to sleep tumescence and apnea studies. These studies are based on the physiological process of erections during rapid eye movement (REM) sleep. A "normal" man has erections every 60 to 70 minutes while he sleeps.

TABLE 41-3	**MEDICATIONS CONTRIBUTING TO ERECTILE DYSFUNCTION**
Antianxiety agents	Glucocorticoids
Antidepressants	H₂ antagonists
Antihistamines	Major tranquilizers
Antineoplastic agents	Muscle relaxants
Blood-pressure medications	Commonly abused drugs
ACE inhibitors	Alcohol
Beta-blocking agents	Amphetamines
Others	Barbiturates
Diuretics	Caffeine
Drugs for Parkinson's disease	Cocaine
Estrogens	Marijuana
	Nicotine
	Opiates

ACE = angiotensin-converting enzyme; H₂ = histamine.
NOTE: Not all drugs in a category cause erectile dysfunction.

Because of the expense of vascular flow studies and sleep studies, they are used on a limited basis.

Medical Treatment

One of the most important treatment options begins with the couple's being able to share intimate communication. No matter what may be causing the problem, if the patient and his partner are not touching, talking, and sharing feelings with one another, any treatment option is going to have limited success.

When the problem has clearly been identified as psychological, counseling therapy is the treatment of choice. If long-term therapy has been tried with only limited success, the addition of oral medication or even intracorporeal injection therapy may be added to provide a boost in confidence and self-esteem. Medical treatment for physical erection problems begins with conservative, nonsurgical treatment and then progresses to surgical options if necessary.

MEDICATION CHANGES. Sometimes all that is needed to correct the problem is a change in medication. There are many prescription, over-the-counter, and street drugs that can cause erection problems. (See Table 41–3.) It is important for the patient to talk with the health care provider before discontinuing any medication. Some men have been known to stop taking their blood pressure medication and risk a stroke or heart attack because the medication interfered with their sexual activity.

HORMONE TREATMENT. If the testosterone level is low and it is not the result of a pituitary tumor or thyroid problem, replacement hormone may be needed. The health care provider should first examine the patient carefully for any evidence of prostate cancer, because testosterone treatment can cause the cancer to grow and spread. Testosterone treatment works best when it is given by intramuscular (IM) injections. It may take several weeks to raise the hormone level high enough to see any benefits. If after about 6 to 8 weeks of treatment there is no increase in the number or quality of erections, a different treatment should be tried.

YOHIMBINE. Most men are looking for a magic pill or potion to correct the problem. Some health care providers prescribe a pill called yohimbine (Yocon, Yohimex), which is thought to be a mild aphrodisiac and vasodilator. When the cause may be psychological or mildly physical, the medication may be helpful in restoring confidence and erections.

INTRACORPOREAL MEDICATIONS. The use of vasoactive or smooth muscle relaxant medication is the latest nonsurgical treatment option. The medication (papaverine, prostaglandin E, phentolamine, or sildenafil) causes the arterioles and cavernous tissue to relax, allowing more blood to flow in and lowering peripheral resistance, trapping blood in the cavernous bodies to maintain the erection. There are now four methods of administering the medication: direct injection into the corporeal bodies, transurethral suppository pellet, transdermal patch, and pill.

Injection Method. After careful evaluation, the patient or his partner is taught how to inject the medication into the penis using a 26- or 27-gauge needle on a tuberculin syringe or a prefilled autoinjector. The injections are nearly painless and produce a natural erection in 10 to 15 minutes. The erection may last 1 to 2 hours, and patients are generally limited to a maximum of three injections per week. The dosage is regulated on an individual basis. The most serious side effect is priapism, which requires immediate reversal in a physician's office or emergency room. When the patient is taught and monitored carefully, the risk of complications is minimal.

Transurethral Suppository. The patient is instructed to urinate first to disperse the medicated pellet and help promote absorption of the medication. A tiny pellet (microsuppository) is inserted into the urethra using a specialized single-dose applicator. The medication usually begins to work in 5 to 10 minutes, and the effects last for approximately 30 to 60 minutes. The most common side effect has been pain in the penis when high dosages were used.

Transdermal Patch. A skin patch containing the vasoactive medication may be applied to the penis. Special care must be taken with some forms, so that the partner does not absorb the medication. This can be prevented by having the male wear a condom over the patch.

Oral Pill. Oral medications are now the first line of therapy used to treat ED. The most widely prescribed pill is sildenafil (Viagra). The pill is usually taken 30 to 60 minutes before sexual intercourse. It provides an easy way to administer the medication and has a success rate of between 40 and 80 percent, depending on the cause of the problem. Men using nitrate medication (antianginal agents) should avoid using the pill because it is vasoactive. It can also cause abnormal visual problems for some men.

Other forms of ED medication being evaluated are an oral form of phentolamine (Vasomax), apomorphine (Uprima, an older drug used to treat Parkinson's), and pentoxifylline (improves oxygen delivery to the penis). Investigative work using genetically designed agents and regeneration of penile nerves involving proteins called immunophilins is showing promise as a treatment option.

Other Nonsurgical Treatments

SEXUAL DEVICES AND TECHNIQUES. There are a variety of sexual aids, such as vibrators and dildos (hollow penises), that may be alternatives for patients who do not want or cannot afford expensive medical treatment. They should be encouraged to talk with a health care provider or qualified sex therapist before trying these alternatives.

Suction devices are another nonsurgical treatment option. This is an external cylinder vacuum device that fits over the penis and draws the blood up into the corporeal bodies, causing an erection. A penile ring is then slipped onto the base of the penis. Once the cylinder is removed, sexual intercourse can begin. Special care must be taken to remove the penile ring within 15 to 20 minutes to prevent tissue damage. The suction device may be used alone or with intracorporeal injection therapy for patients who have difficulty keeping an erection. The companies that manufacture the suction devices provide free videotapes and written patient instruction booklets.

Surgical Treatments

PENILE IMPLANTS (PROSTHESIS). Penile implants are a pair of solid or fluid-filled chambers that are surgically placed into the corporeal bodies in the penis to produce an erection. This treatment option has been used successfully for more than 25 years. There are two basic types of implants—noninflatable and inflatable (Fig. 41–3). The noninflatable type is economical, can be surgically implanted in less than an hour, and provides firmness for penetration. The inflatable device contains a sterile saline solution that fills the cylinders using a manual pumping action. When activated, the inflatable implant provides both firmness and an increase in diameter of the penis. Although the inflatable device is more expensive and has a

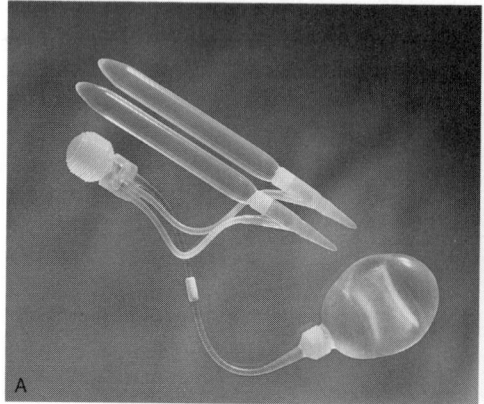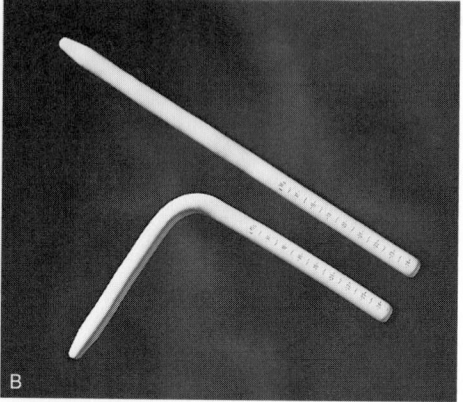

Figure 41-3 Penile implants. *(A)* Inflatable penile implant. Inflatable cylinders are implanted in the penis, the small hydraulic pump in the scrotum, and the fluid-filled reservoir in the lower abdomen. Sterile radiopaque saline from the reservoir fills the cylinders to provide an erection. (Alpha-1 implant photo courtesy Mentor Urology.) *(B)* Malleable penile implant. Malleable rods are implanted into the penis. The penis is always firm, but the rods can be bent close to the body when erection is not desired. (Acu-Form implant photo courtesy Mentor Urology.)

greater risk of mechanical failure, it provides a more natural appearing and functioning erection than the noninflatable implant.

Penile implants are considered secondary treatment options because they have a risk of complications such as mechanical failure, infection, and erosion. Patients at greatest risk for complications are those with uncontrolled diabetes and those with severe circulation problems. Patients should be taught that an implant will not restore ejaculation or orgasm if these functions have been lost before the surgery. Surgery recovery time varies from 4 to 6 weeks; the patient must receive approval from the health care provider before having sexual intercourse.

VASCULAR SURGERY. If a younger man (younger than age 35) has an erection problem caused by poor blood flow into the penis or from blood leaking out of the penis, rapidly causing the loss of the erection, corrective surgery may be performed. A bypass graft may be done to increase the blood flow into the penis or to go around a blockage (e.g., Peyronie's disease). If the blood leaves the corporeal bodies too fast, a procedure to ligate (tie off) the leaking veins may be successful. Neither surgery works well in older men because the bypass graft rapidly becomes obstructed and the natural tendency of the older body is to form collateral circulation around the veins that are tied off.

Nursing Process

ASSESSMENT. Before the assessment begins, it is important to provide privacy and ensure confidentiality. A complete sexual history, often conducted by a nurse, is taken, with special focus on factors related to the circulatory, nervous, and endocrine systems. The questions cover the following:

- Medical problems
- Surgical treatments that might interfere with blood flow or nerve supply to the groin or spine
- Genitourinary problems
- A complete list of medications used, including over-the-counter drugs and any evidence of substance abuse
- General lifestyle patterns, including stress factors, depression, and excessive use of caffeine or nicotine (causing vasoconstriction)
- Sexual patterns and practices

A physical examination is performed by the physician or health practitioner to assess for any evidence of congenital deformities, hormonal imbalance, decreased circulation, or nerve damage. Throughout the assessment process it is important to observe for any signs of psychological or emotional distress.

NURSING DIAGNOSIS. Because of the nature of the problem, it is important to recognize nursing concerns related to anxiety, fear, powerlessness, knowledge deficit, and issues related to ineffective role performance or family dynamics. The present nursing diagnosis of sexual dysfunction is very broad and does not focus on specific aspects of male erectile dysfunction.

PLANNING. Planning generally involves not only the male patient but also his sexual partner. Treatment success rates are much higher when the support person is included in the decision-making process.

Unrealistic expectations are common in the beginning period of restoration. The patient must be encouraged to communicate with his partner through both talk and touch. Restoration of an erection cannot repair a bad relationship, and often counseling is an important part of the treatment plan.

IMPLEMENTATION. Special care should be taken in selecting the treatment option that will work best for the patient and his partner. They should be encouraged to ask questions of their physician and other health care professionals involved in their care. Where available, support groups that are focused on men and their partners have provided excellent resources and information on treatment options. Conservative treatment options are tried first, with surgical interventions reserved as a last resort. Nursing care focuses both on the emotional and physical discomfort associated with the diagnosis and treatment options.

EVALUATION. The best indicator of a positive outcome is restoration of erectile function with a verbal account of satisfaction. Sometimes the physical problem is easier to correct than the emotional scars that the problem has created. It is important to evaluate both the physiological and emotional outcomes of treatment.

PATIENT EDUCATION. The nurse plays an important role in public education related to erectile dysfunction. Men need to know that they are not alone with their problem. More than 20 million men in the United States experience ongoing problems with erections. The majority of the causes are physical, and help is available through health care providers who specialize in treating erectile dysfunction.

Treatment may be as simple as a change in lifestyle or in medication. Men need to be encouraged to seek help from appropriate health care providers and recognize that erectile dysfunction is a treatable condition with positive outcomes.

Infertility

A growing number of couples in the United States are having difficulty conceiving children. Several factors can interfere with a man's ability to father a child.

Physiology

Eight fundamental male physiological factors are necessary for normal conception to occur:

- Proper endocrine function between the hypothalamus, pituitary gland, and testes
- At least one testes that produces quality sperm
- An epididymis, or storage place, to mature the sperm
- A duct system to transport sperm from the testes to the outside of the body

■ Glands that secrete the right type and amount of seminal fluid to nurture and transport the sperm
■ An intact nervous system that helps provide an erection and ejaculation
■ Semen that meets the following criteria: volume 1.5 to 5 mL, a concentration of more than 20 million sperm per milliliter, 50 to 60 percent of the sperm classified as grade 2 mobility, 60 to 80 percent of sperm with a normal shape, pH is between 7.2 and 7.8, a small amount of fructose (sugar, a food supply), and sperm (semen) that first coagulate and then liquefy
■ A basic knowledge of sexual practices, along with a willing partner.

Etiology

The factors related to infertility are divided into three general categories: pretesticular, testicular, and posttesticular.

PRETESTICULAR (ENDOCRINE) FACTORS. The first factor involves the proper functioning of the hypothalamus, the pituitary gland, and the testes. These endocrine functions are complex and are a rare (3 percent) cause of infertility. Examples of endocrine causes might be pituitary or adrenal tumors, thyroid problems, or uncontrolled diabetes.

TESTICULAR FACTORS. The two most common causes of male infertility are a varicocele (40 to 50 percent) and idiopathic causes (40 percent). It is believed that a varicocele lowers sperm count by raising the blood flow and temperature in the testes. Sperm cannot live if the temperature is too high or too low.

Congenital anomalies such as Klinefelter's syndrome (a chromosome defect) or cryptorchidism result in absent or damaged testes. Failure of a part of the male reproductive system to develop as a result of the mother's use of DES or other drugs during pregnancy may also result in fertility problems.

Certain disease or inflammatory processes may cause damage to the storage area (epididymitis) or to the testes themselves (mumps orchitis). Any high fever or viral infection can interfere with the production of sperm for up to 3 months.

Medications, radiation, substance abuse, environmental hazards, and lifestyle practices have all been identified as possible factors that can interfere with spermatogenesis (sperm production). Medications such as cimetidine (Tagamet), sulfasalazine (Azulfidine), anabolic steroids (testosterone), anticancer drugs (cyclophosphamide [Cytoxan], methotrexate), recreational drugs (cocaine, marijuana), and antihypertensives (methyldopa) have all been identified as possible causes of infertility. Radiation damage, whether as treatment for cancer or job related, tends to depend on the dosage received. A small amount of radiation causes temporary loss of sperm. Permanent sterility occurs with large doses or prolonged treatment. The relation between use of drugs such as cocaine and marijuana and infertility has been documented in several studies, but there is no clear indication of the amount or length of time it takes for these substances

to cause infertility problems. Environmental hazards such as pesticides, Agent Orange, and lead poisoning have been listed as possible causes of infertility and are still under investigation. Excessive use of hot tubs and saunas, wearing tight jeans, and long-haul truck driving have all been identified as raising the temperature level in the scrotum to the extent that the sperm production is decreased.

POSTTESTICULAR FACTORS. The most common factor in posttesticular infertility is the result of surgery or injury along the pathway from the testes to the outside of the man's body. Examples of surgical causes are vasectomy, bladder neck reconstruction, pelvic lymph node removal, or any surgery that causes retrograde ejaculation. Congenital anomalies and various types of infections may also cause infertility problems.

Prevention

Prevention involves possible lifestyle changes to avoid excessive heat to the scrotum, substance abuse, exposure to toxins, and environmental hazards. Problems related to medication or infections should be discussed with the health care provider.

Signs and Symptoms

A couple is considered infertile if they have been unsuccessful at becoming pregnant after at least 1 year of unprotected intercourse. If pregnancy has occurred during the year but there was no delivery, the problem is generally considered a female rather than a male factor.

Diagnostic Tests

Diagnosis begins with a detailed history and physical examination that looks for known male causes of infertility.

History

SEXUAL PRACTICES. Assessment includes frequency of intercourse, positions, timing (according to ovulation cycle), use of contraceptives, problems with premature ejaculation, and erection problems.

LIFESTYLE PRACTICES. Weight lifting or use of steroids, hot tubs, or saunas; tight jeans; use of nicotine, caffeine, alcohol, or marijuana; and the strength of desire for children on the part of the man are all assessed.

OCCUPATION. High stress, long periods of sitting, and exposure to environmental toxins are determined.

MEDICAL-SURGICAL HISTORY. Assessment includes any sexually transmitted viruses or diseases, endocrine problems, congenital urinary problems, serious illnesses or groin injuries, cancer, and treatment with chemotherapy or radiation.

Physical Examination

Observe for normal hair pattern and growth, muscle development, size of testes, and any evidence of a varicocele or hydrocele.

Diagnostic Tests

A semen analysis is done after collecting several specimens following a special collection technique. It is analyzed to see if it contains the right amount and type of healthy sperm needed for a pregnancy. There are several other tests that may be done, depending on the level of desire and the financial resources of the couple. Many insurance companies do not pay for the tests or treatment options for infertility.

Medical Treatment

Treatment may be as simple as making a change in sexual or lifestyle practices. If the couple is able to handle the emotional and financial strain, they may try male surgery to correct a varicocele or a variety of in vitro fertilization procedures. The success rates for in vitro fertilization range from 8 to 60 percent and generally cost several thousand dollars each time an attempt is made. Another possible option that should be presented to the couple is adoption.

You can play an important role in the emotional support a couple needs during infertility studies. It is important that the couple feel comfortable in communicating their feelings and frustrations with one another and their health care provider. You may need to be the communication link, explaining various tests and cost factors and discussing how long the couple may want to continue trying the various treatment options. It may also help them to talk with other couples or attend a support group designed for couples experiencing infertility. For more information, visit the following Websites:

www.cancer.org
www.nlm.nih.gov/medlineplus/ency/article
www.niddk.nih.gov/health/urolog/pubs
www.urologychannel.com/
www.webinfomed.com/medsearch/hpages/genitourinary.html

Answers to CRITICAL THINKING

Mr. Atkinson

1. Mr. Atkinson should be instructed not to lift anything heavier than 10 pounds for the first 6 weeks, and he will not be able to plow or drive for the first 6 weeks. It is important that his son understand his father's limitations and how important it is for him or someone else to help out with the farm chores.
2. Mr. Atkinson will notice a change in his ejaculation (either very little or none at all). If he could get an erection before surgery, the chances are very good that he will continue to be able to have intercourse; however, he will not ejaculate.

Mr. Cunningham

1. Mr. Cunningham should be encouraged to see his health care provider immediately for an evaluation to rule out cancer of the testes.
2. Depending on the stage of the cancer, he will have the cancerous testes, cord, and lymph nodes removed. He may need chemotherapy or radiation treatments as well.
3. Mr. Cunningham should be encouraged to make deposits at a certified sperm bank before any treatments. It is also important to include his future wife and his family in the decision-making process and encourage them to share their feelings and concerns with one another. Cancer support groups may also be helpful to Mr. Cunningham.

1. Which test is encouraged annually for every man older than age 40?
   a. Prostate ultrasound
   b. Serum creatinine
   c. Digital rectal examination
   d. Acid phosphatase

2. Which of the following is the greatest risk factor for prostate cancer?
   a. Smoking
   b. Aging
   c. Drug use
   d. History of STDs

3. Which of the following is the most commonly used surgical treatment for BPH?
   a. TUIP
   b. TUMA
   c. TURP
   d. TULIP

4. You are caring for Mr. Frank following transurethral resection of the prostate. He says he is having pain in his bladder, and you notice urine leakage around his catheter. Which of the following responses is best?
   a. "Bladder spasms are common after your surgery. Take some deep breaths while I get a B & O suppository."
   b. "You should not be experiencing spasms. I will notify the RN right away."
   c. "Spasms can be very painful. Would you like an injection of Demerol?"
   d. "Your catheter is leaking; we will need to replace it right away."

5. Mr. Jackson is discharged following prostate surgery. As you are doing his discharge teaching, he asks if he can return to work delivering water softener salt to homes. Which response is best?
   a. "Certainly, returning to work is a good way to get your mind off your surgery."
   b. "You should spend your time off looking for a job that involves less lifting."
   c. "It is okay to drive, but you should not lift heavy objects for 6 weeks."
   d. "You should not drive for 4 to 6 weeks or lift anything heavier than 10 pounds."

6. Which of the following is the most common cause of erectile dysfunction?
   a. Endocrine problems
   b. Circulatory problems
   c. Excessive stress
   d. Excessive alcohol use

42

NURSING CARE OF PATIENTS WITH SEXUALLY TRANSMITTED DISEASES

Linda Hopper Cook

KEY TERMS

chancre (**SHANK**-er)

condylomata acuminatum (KON-di-**LOH**-ma-tah ah-KYOOM-in-**AH**-tum)

condylomatous (KON-di-**LOH**-ma-tus)

conjunctivitis (kon-JUNK-ti-**VIGH**-tis)

cytotoxic (SIGH-toh-**TOCK**-sick)

electrocautery (ee-LECK-troh-**CAW**-tur-ee)

electrocoagulated (ee-LECK-troh-coh-**AG**-yoo-LAY-ted)

endometritis (EN-doh-me-**TRY**-tis)

epidemiological (EP-i-DEE-me-ah-**LAHJ**-i-kuhl)

gummas (**GUM**-ahs)

hepatosplenomegaly (he-PA-toh-SPLE-noh-**MEG**-ah-lee)

herpetic (her-**PET**-ick)

lymphadenopathy (lim-FAD-e-**NAH**-puh-thee)

mucopurulent cervicitis (MYOO-koh-**PYOOR**-uh-lent SIR-vi-**SIGH**-tis)

ophthalmia neonatorum (ahf-**THAL**-mee-ah NEE-oh-nuh-**TOR**-uhm)

perinatal (PAIR-ee-**NAY**-tuhl)

proctitis (prock-**TIGH**-tis)

puerperal (pyoo-**ER**-per-uhl)

sacral radiculopathy (**SAY**-krul ra-DICK-yoo-**LAH**-puh-thee)

salpingitis (SAL-pin-**JIGH**-tis)

serological (SEAR-uh-**LAJ**-ick-uhl)

urethritis (YOO-ree-**THRIGH**-tis)

verrucous (ve-**ROO**-kus)

vesicular (ve-**SICK**-yoo-ler)

vulvovaginitis (VUL-voh-VAJ-i-**NIGH**-tis)

QUESTIONS TO GUIDE YOUR READING

1. How would you explain the pathophysiology of each of the common sexually transmitted diseases (STDs)?

2. What are the etiologies, signs, and symptoms of each of the common STDs?

3. What care would you provide for patients undergoing tests for STDs?

4. What is current treatment for the common STDs?

5. What data should you collect when caring for patients with STDs?

6. What nursing care will you provide for patients with STDs?

7. How will you know if your nursing interventions have been effective?

Sexually transmitted diseases (STDs) are infections that can be transmitted through intimate contact with the genitals, mouth, or rectum of another individual. Some may also be spread by other routes such as blood or body fluids. A nurse's best protection against catching diseases from blood and body fluids from patients is the strict practice of standard precautions and maintaining your own healthy, intact skin.

Physically, STDs may cause tremendous suffering through pain, scarring of genitourinary structures, damage to other body organs, infertility, birth defects, nervous system damage, development of cancer, and even death of infected adults and sometimes their children. Psychologically and socially, these diseases also have profound effects on individuals, families, and relationships. Guilt about passing on an incurable disease to a loved one or feelings of betrayal because of being infected as a result of someone else's choices are only part of the emotional consequences of STDs.

Changing sexual mores have been associated with increasing incidence of almost all types of sexually transmitted diseases, including some previously rare diseases related to anal intercourse. Coexistence of more than one STD in an individual is also occurring more often. There are more than 50 diseases and syndromes associated with sexually transmitted diseases; the more common ones are discussed here. Human immunodeficiency virus (HIV) and acquired immunodeficiency syndrome (AIDS) are discussed separately in Chapter 54.

Symptoms, diagnostic techniques, and treatment regimens vary for different geographic areas. Availability of equipment for diagnosis, experience with patients who are atypical or antibiotic resistant, and health care provider preferences are responsible for much of the variety that is documented in the literature. As an introduction for practical/vocational nurses, general overviews are presented. Attempts have been made to present consensus from a variety of the available health care literature.

▶ DISORDERS AND SYNDROMES RELATED TO SEXUALLY TRANSMITTED DISEASES

Vulvovaginitis is an inflammation of the vulva and vagina and can be asymptomatic or involve redness, itching, burning, excoriation, pain, swelling of the vagina and labia, and discharge. A variety of sexually transmitted and nonsexually transmitted infectious agents can cause vulvovaginitis. The odor, consistency, and color of the discharge varies with the different microbes involved. Nonsexually transmitted vaginitis, vulvovaginitis, and vaginosis are described in Chapter 40. Some microorganisms may be acquired either by sexual or nonsexual routes, so they are also mentioned in this chapter. Bartholin's glands may develop ab-

scesses as a result of infection with nonsexually transmitted microbes or STDs such as gonorrhea and chlamydia.

Both STDs and nonsexually transmitted microorganisms can cause **urethritis** in men and women. In men, inflammation of the urethra, prostate, and epididymis can result in difficult, painful, and frequent urination and a urethral discharge, which may be clear, cloudy, or yellow. Partners of men with urethritis may also suffer from urethritis, but they may also develop **mucopurulent cervicitis** (MPC) and a variety of other symptoms of the particular infection. Some causative agents for urethritis include *Neisseria gonorrhoeae, Chlamydia trachomatis, Ureaplasma urealyticum, Trichomonas vaginalis, Candida albicans,* and herpes simplex. Often this disease category is divided into gonococcal urethritis caused by *N. gonorrhoeae* (GU) and nongonococcal urethritis (NGU).

MPC is an inflammation of the cervix that can produce a mucopurulent yellow exudate on the cervix or may have no noticeable symptoms. MPC during pregnancy can result in **conjunctivitis** and pneumonia in newborn babies, as well as **puerperal** fever of the mother. MPC can be caused by such organisms as *C. trachomatis, N. gonorrhoeae, T. vaginalis, C. albicans,* and herpes simplex.

MPC may spread to become pelvic inflammatory disease (PID), a chronic inflammation of the lining of the uterus **(endometritis)** and the fallopian tubes **(salpingitis),** resulting in scarring, infertility, and increased risk of ectopic pregnancy (because scar tissue may block the fallopian tubes). Two of the most common causative agents of PID are *C. trachomatis* and *N. gonorrhoeae,* often in combination, which requires treatment for both.

Proctitis is inflammation of the rectum and anus that may be due to either nonsexually transmitted microbes or to STDs. This is especially prevalent among those who practice anal intercourse, both heterosexual and homosexual. Enteritis, which is inflammation of the lining of the intestine, may occur as a result of contamination during anal intercourse. Infection with *Campylobacter* spp., *Shigella* spp., and *Giardia lamblia* can be a problem for homosexual men. Care of patients who have gastrointestinal disorders is discussed in Unit 7.

Genital ulcers are formed when papules or macules erode and leave often painful raw, pitted, or excoriated areas on or around the genitals. Not all genital ulcers are caused by STDs—injury, some non-STD viruses, some types of drug reactions, radiation, and some forms of cancer can also pro-

vulvovaginitis: vulvo—vulva + vagin—vagina + itis—inflammation

urethritis: ureth—urethra + itis—inflammation
mucopurulent cervicitis: muco—involving mucus + purulent—involving pus + cervic—cervix + itis—inflammation
conjunctivitis: conjunctiv(a)—lining of the eyelids and sclera of the eye + itis—inflammation
puerperal: childbirth
endometritis: endo—inside + metr—womb + itis—inflammation
salpingitis: salping—tube + itis—inflammation
proctitis: proc—anus + itis—inflammation

duce genital ulcers. However, several STDs can produce genital ulcers, including syphilis, herpes, chlamydia, chancroid, granuloma inguinale, and HIV. Genital ulcers from one type of disease may increase the risk of infection with other STDs during sexual activity, because the open areas present an easy portal of entry for the infecting organism.

Cellular changes can also be caused by STDs, including **condylomatous** (wartlike) growths and dysplasia or neoplasia, which may result in precancerous or cancerous conditions. Herpes viruses, HIV, and human papillomavirus (HPV) have all been linked to the development of cancer.

Chlamydia
Etiology and Signs and Symptoms

Chlamydia is the most common STD in the United States.[1] It can be transmitted sexually and by blood and body fluid contact. There are several serotypes of the bacteria *C. trachomatis*. Chlamydia is often asymptomatic in women, but urethritis, MPC, and conjunctivitis may result from this infection. Fitz-Hugh–Curtis syndrome, a surface inflammation of the liver, can also be caused by *C. trachomatis*. This inflammation may cause nausea, vomiting, and sharp pain at the base of the ribs that sometimes refers to the right shoulder and arm. A large proportion of cases of PID can be attributed to chlamydia, and it can lead to infertility and much greater risk of ectopic pregnancy. The infection can be passed from mother to baby during birth, resulting in neonatal pneumonia and conjunctivitis.

Lymphogranuloma venereum (LGV) is also caused by some serotypes of *C. trachomatis* but is more commonly seen in tropical climates or among people who emigrated from these areas. This disease also causes urethritis and proctitis, and it inflames lymph nodes that drain the pelvic area, resulting in draining sores and fistula development. Scarring from this disease can complicate vaginal deliveries.

Diagnostic Tests

There are several tests for chlamydia. Samples for culture are gathered in a special collection tube to send to a laboratory for microscopic examination. However, culturing is difficult and expensive and generally requires 2 to 6 days for results, depending on the availability of a fluorescence microscope. Research on some serological tests has shown some specificity problems, but some first-void urine samples, cervical samples, and urethral samples are being tested using RNA amplification. Because this disease is so common, it is wise to set out an unopened chlamydia collection kit within easy reach of the health care provider for each pelvic examination.

Treatment

Antibiotics ending in -*cycline* and -*mycin* are generally given to treat chlamydia in adults. The latter type is given during pregnancy because tetracyclines can deposit in de-

veloping bones and teeth. Eye drops of tetracycline or erythromycin may be given to babies shortly after birth for the prevention of conjunctivitis. Institutional policies and state regulations determine the type of eye drops to be used and whether administration of the drops requires specific consent of the parents.

Gonorrhea
Etiology and Signs and Symptoms

Almost 325,000 cases of gonorrhea were reported to the U.S. Center for Communicable Disease in 1997 (the most recently available complete statistics), but it is estimated that a true picture of the number of people who contract the disease yearly is around 800,000 and that the cost of care and treatment in the United States is approximately $1.1 billion.[2] The causal bacteria, *N. gonorrhoeae*, may be transmitted vaginally, rectally, orally, and through contact with blood and body fluids and can produce a variety of symptoms. Men may be asymptomatic or may have urethritis with a yellow urethral discharge. Women who have gonorrhea may have either no noticeable symptoms or have a sore throat, MPC, urethritis, or abnormal menstrual symptoms such as bleeding between periods. Many cases of PID are caused by gonorrhea. Intercourse with an infected partner during menstruation may be especially risky for development of PID, because removal of the cervical mucous barrier can promote the growth of the gonococcus in the higher reproductive tract. Gonorrhea can also cause Fitz-Hugh–Curtis syndrome. Fever, nausea, vomiting, and lower abdominal pain may be present. Gonorrhea may also infect the throat and the rectum and may cause disseminated gonococcal infection, resulting in inflammation of the joints, skin, meninges, and lining of the heart.

Newborns born to mothers who have gonorrhea can develop **ophthalmia neonatorum,** which involves inflammation of the conjunctiva and deeper parts of the eye and can result in blindness. The newborn may also experience disease at other sites of infection with gonorrhea at birth. Abscesses may develop where fetal scalp monitors were attached during labor, and infection of the nose, lungs, and rectum may occur.

Diagnostic Tests

Diagnosis is done by microscopic examination of smears and cultures of the discharge or identification of bacterial DNA in the urine. More than one test may be done to verify the diagnosis.

Treatment

Development of antibiotic resistance by *N. gonorrhoeae* and co-infection with other microorganisms, such as *C. trachomatis*, is making treatment more complicated. At the time of the writing of this book, ceftriaxone, cefixime,

---

condylomatous: condyl—rounded projection + oma(t)—growth + ous—like

ophthalmia neonatorum: ophthalmia—eye disease + neonatorum—of the newborn

ciprofloxacin, and ofloxacin generally are recommended (but the latter two should not be given to people younger than 18 or to pregnant women).[3] Ophthalmia neonatorum may be prevented by use of antibiotic eye drops such as erythromycin or tetracycline in the newborn at birth. Institutional policies and state regulations determine the type of eye drops to be used and whether administration of the drops requires specific consent of the parents.

Syphilis
Etiology and Signs and Symptoms

Syphilis is an ancient disease that has not disappeared, although it is overshadowed by more commonly occurring diseases such as chlamydia. The primary stage of syphilis begins with the entry of the *Treponema pallidum* spirochete through the skin or mucous membranes. Between 3 and 90 days later a papule develops at the site of entry; this sloughs off, leaving a painless, red, ulcerated area called a **chancre** (Fig. 42–1). Chancres may also develop in other areas of the body at this time. Chancre formation is generally the only symptom of this stage of syphilis. The chancre eventually heals, but the spirochete remains active in the infected individual and can be passed on to others. Secondary syphilis begins 2 to 8 weeks later and affects the body more generally, causing such problems as flulike symptoms, joint pain, hair loss, skin rashes (primarily on the soles of the hands and feet), mouth sores, and condylomatous growths in moist areas of the body.

Serious damage can occur if syphilis is untreated in the early stages. The disease may not progress to the tertiary (or late) stage for 3 to 15 years. At this stage it can involve any organ system of the body. In the tertiary stage the spirochete may form **gummas,** which are tumors of a rubbery consistency that can break down and ulcerate, leaving holes in body tissues. The gummas can damage the heart, circulatory system, and nervous system (neurosyphilis). Ulceration of gummas may destroy areas of vital tissue and lead to mental and physical disability or early death.

---

gummas: from the word meaning "rubber"—rubber tumors

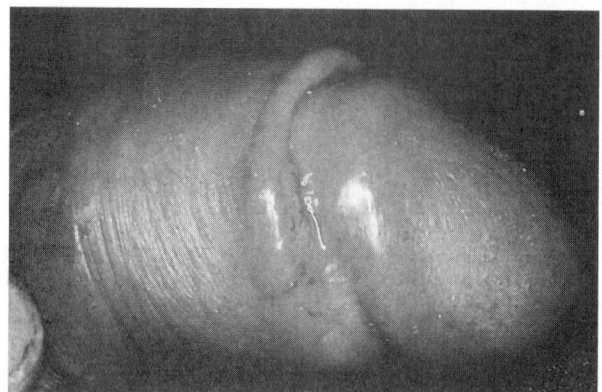

Figure 42–1 Syphilis chancre. (From Lemone, P., and Burke, KM: Medical-Surgical Nursing: Critical Thinking in Client Care. Addison-Wesley, Menlo Park, Calif, 1996, p 2069, with permission.)

Syphilis can be passed on to the unborn children of women who carry the spirochete, resulting in **hepatosplenomegaly;** increase in bilirubin; destruction of red blood cells; birth defects, especially of the face; **lymphadenopathy;** and a baby who can transmit the spirochete through nasal drainage. If left untreated, syphilis during pregnancy may cause lesions in various organs of the unborn baby and result in higher rates of spontaneous abortion, stillbirth, and premature birth.

Diagnostic Tests

Several tests for syphilis exist, and different tests may have to be used at different stages of the disease for accurate diagnosis. The organism is difficult to grow in culture and is so tiny that it cannot be seen with a light microscope. Darkfield microscopy of material from a chancre may show the organism but may also be negative in spite of definite symptoms of systemic syphilis infection. **Serological** (blood) tests are categorized as nontreponemal and treponemal. Some nontreponemal tests include the Venereal Disease Research Laboratory (VDRL) test, the rapid plasma reagin (RPR) test, and the automated reagin test (ART). These tests indirectly check for syphilis by detecting the presence of antibodies that the body forms in response to treponema and, unfortunately, in response to some other disorders, so false-positive results can occur. Treponemal tests are much more specific for syphilis but, unfortunately, can remain positive long after the disease has been effectively treated, so a combination of testing methods may be used to determine whether a patient truly has syphilis. Diagnosis of neurosyphilis is even more difficult because some testing of cerebrospinal fluid may result in false-negative results. Treponemal enzyme-linked immunosorbent assay (ELISA), fluorescent treponemal antibody absorption (FTA-ABS), and polymerase chain reaction (PCR) tests for treponemal DNA are some new methods being introduced to improve specificity.[4]

Treatment

Probenecid is added to penicillin to prolong the action to treat syphilis of nonpregnant, pregnant, and newborn patients. For those who are allergic to penicillin, doxycycline is one choice. However, when HIV and syphilis are seen in the same individual, symptoms of neurosyphilis are more likely and ceftriaxone is one recommended medication.[5]

Trichomonas
Etiology and Signs and Symptoms

Trichomoniasis is generally a sexually transmitted disease, but it may be transmitted through nonsexual contact with infected articles, because it can survive for quite a long time

---

hepatosplenomegaly: hepato—liver + spleno—spleen + megaly—enlargement
lymphadenopathy: lymph—lymph nodes + adeno—node + pathy—disorder
serological: sero—blood + logical—science

outside the body. (Therefore it was included in the chart of vaginal irritations and inflammations in Chapter 40.) Carriers of *T. vaginalis* may be asymptomatic for years until changes in vaginal or urethral conditions encourage an outbreak of the disease. A decrease in acid-producing resident bacteria, injuries to the vaginal tissues, and development of lesions from some other STDs or from some forms of cancer may activate the organism. Redness, swelling, itching, and burning of the genital area; pain with intercourse and voiding; and a frothy and foul-smelling discharge are seen with this infection.

Diagnostic Tests

The organisms can be identified by their motility and whiplike flagella when wet-mount slides of the discharge are viewed under a microscope. Trichomonas may produce abnormal Papanicolaou (Pap) smear readings, which require that more frequent Pap smears be done to provide adequate surveillance of cellular changes.

Treatment

The drug treatment of choice is metronidazole, except during the first trimester of pregnancy, when a 2-percent clindamycin vaginal cream is recommended. Some strains of *Trichomonas* may exhibit resistance but generally succumb to much higher doses of the drug. Because some people carry the organism without symptoms, sexual partners should also be treated regardless of symptoms.

Herpes
Etiology and Signs and Symptoms

Herpes infection is caused by the herpes simplex viruses types I and II (HSV-I and HSV-II). Herpesviruses have an affinity for tissues of the skin and nervous system and can lie dormant in nervous system tissues and then reactivate periodically when the body undergoes stress, fever, or immune system compromise. HSV-II can cause "fever blisters" of the mouth, as well as genital lesions. HSV-II causes a more severe genital version. After infection, vesicles develop, spontaneously rupture, and produce painful ulceration of the underlying skin tissues. However, some patients are totally asymptomatic and can still transmit the disease. Asymptomatic latent periods are generally interspersed between the **vesicular** outbreaks, but the virus may still be transmitted even during latent periods.

Initial infection with herpes may produce a flulike condition. Urethritis, cystitis, and MPC with vaginal discharge may also be evident. Infection of the spinal nerve roots by HSV may result in **sacral radiculopathy,** causing retention of urine and feces. Although rare, disseminated herpes infection can result in inflammation of the spinal cord, meninges, nerve pathways, and lymph nodes. Urethral

strictures and increased risk for development of cervical cancer in women are also consequences of herpes.

Transmission may occur from a mother to her child during pregnancy and delivery and through breast-feeding. It is estimated that one in five of all pregnant women carry herpes, although most of their babies do not develop **herpetic** disease. If infected, the baby's skin, eyes, mucous membranes, and nervous system may be involved and death from disseminated herpes infection may occur. Active lesions or asymptomatic viral shedding increase risk of transmission.

Diagnostic Tests

Testing for HSV requires special viral collection kits for swabbed or scraped specimens from lesions. Follow the directions on the viral collection kit. Blood tests are improving; the Western blot assay can determine whether the person has HSV-I or HSV-II antibodies.

Treatment

There is presently no known cure for herpes infection, although antiviral medications such as acyclovir, valacyclovir, and ganciclovir may be given to decrease the severity of symptoms. (Only acyclovir may also be given during pregnancy, a treatment that may reduce the risk of transmission to the baby).[6] Cesarean section delivery may also decrease the risk of transmission of the disease from the mother to the baby if there are active, open lesions at the time of delivery. See also Nursing Care Plan Box 42–1 for the Patient with Genital Herpes.

Genital Warts
Signs and Symptoms

Condylomata acuminatum (genital warts) are the most common sexually transmitted viral disease, and their incidence is increasing rapidly. Infection with human papillomavirus (HPV) produces the condylomata—soft, raised, **verrucous** fleshy tumors, which may also have fingerlike projections (Fig. 42–2). Some people remain asymptomatic, but they can still transmit the infection. More than 100 types of HPV have been identified, and several have been closely linked to development of cancers of the reproductive organs and anus in both males and females. There may be a long latent period of as much as 3 years' duration from the time of exposure to development of the warts.

HPV can be passed on from a pregnant woman to her fetus, resulting in the growth of genital warts on the baby, HPV infection of the baby's respiratory tract, and a possible increased risk of cancer development. HPV infection during pregnancy can cause particularly difficult problems. Genital warts tend to grow more rapidly in pregnant women

---

vesicular: vesicul—blister + ar—type
sacral radiculopathy: sacral—sacrum + radiculo—root + pathy—disorder or disease

herpetic: herpet—herpes + ic—pertaining to
condylomata acuminatum: condyl—rounded projection + oma—growth + ta—pluralizes the word (singular form is condyloma) + acuminatum—genital growths
verrucous: verruc—wart + ous—like

BOX 42-1   NURSING CARE PLAN FOR THE PATIENT WITH GENITAL HERPES

 Pain related to inflammation, skin lesions

PATIENT OUTCOMES
The patient will express pain relief and will rest and move well.

EVALUATION OF OUTCOME
Does patient state relief of pain?

INTERVENTIONS	RATIONALE	EVALUATION
Assess pain using the WHAT'S UP? format.	Assessment of the characteristic of the pain assists the nurse in providing appropriate relief measures.	Can patient describe the pain characteristics?
Recommend pain relief measures appropriate to the type and location of the pain (both alternative measures, such as heat, ice, and change of position, and medication may be offered).	Not all types of pain respond well to the same treatment.	Does patient express satisfactory relief of pain? Does patient move and rest without evidence of pain?
Document results of pain relief measures.	Documentation alerts other caregivers about what works and does not work, thus providing more consistent, effective pain relief.	Have you gained sufficient information from patient to document results?
Instruct patient about self-care for pain and STD treatment at home.	Most STDs are treated at home.	Does patient verbalize understanding of self-care measures?

 Risk for transmission of infection to others related to lack of knowledge about transmission, symptoms, and treatment

PATIENT OUTCOME
Patient will verbalize understanding of measures to prevent transmission to others.

EVALUATION OF OUTCOME
Does patient verbalize understanding of transmission prevention? Does patient practice preventive behaviors?

INTERVENTIONS	RATIONALE	EVALUATION
Assess patient's understanding of transmission, symptoms, complications, and treatment of STDs.	New instruction should be based on patient's previous knowledge.	Is patient's current understanding accurate? What teaching is necessary?
Assess whether patient is engaging in high-risk behaviors.	If patient is continuing to engage in high-risk behaviors, the risk for infection of others is high.	Is patient protecting self and others appropriately?
Use universal precautions and strict aseptic technique for *all* procedures involving blood and body fluids.	The health team, in addition to other patient contacts, must be protected.	Are universal precautions observed?
Instruct patient in appropriate strategies to reduce risk of infecting others: • Abstinence • Monogamy (if no active infection) • Use of barrier methods and spermicides • Adherence to treatment regimen	These measures may help prevent transmission of infection to others.	Does patient verbalize understanding of methods to prevent transmission and intent to practice them?
Teach patient signs and symptoms of STDs to report immediately.	Prompt treatment of patient and partners further reduces risk of transmission of infection.	Does patient verbalize understanding of signs and symptoms to report?

 Fear related to diagnosis of an incurable illness and effects on sexual relationships and reproduction

PATIENT OUTCOME
Patient will verbalize realistic and accurate information about disease process and relate control of excessive fear.

EVALUATION OF OUTCOME
Does patient relate accurate knowledge? Is fear manageable?

BOX 42-1   NURSING CARE PLAN FOR THE PATIENT WITH GENITAL HERPES—CONT'D

INTERVENTIONS	RATIONALE	EVALUATION
Assess patient's fears.	Fear is a normal response and may be appropriate.	What does patient fear?
If fear is based on misconceptions, provide factual information.	When fear is based on misconceptions, they should be corrected.	Are fears based on factual information?
Allow patient to verbalize feelings. Be empathetic, but do not offer false hope.	Sharing fears may help patient gain insight into dealing with them.	Is patient able to verbalize feelings?
Explain all procedures and treatments.	Unfamiliar procedures or treatments may contribute to fear.	Does patient understand procedures and treatments?
Help patient identify support systems and coping strategies that have worked in the past.	Methods that have worked for patient before are likely to be helpful again.	Does patient have effective coping skills and support systems?

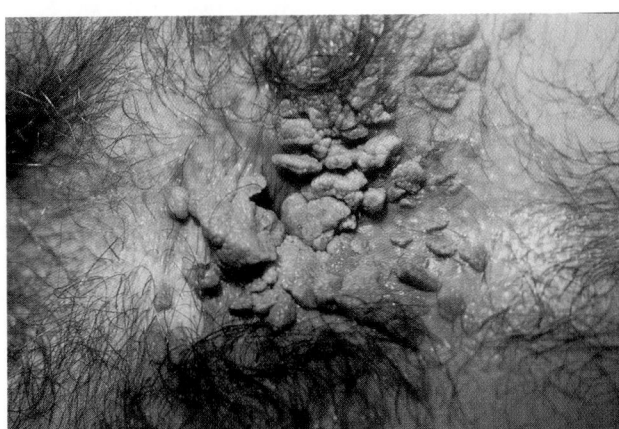

Figure 42-2 Condylomata, commonly known as genital warts. (From Lemone, P., and Burke, KM: Medical-Surgical Nursing: Critical Thinking in Client Care. Addison-Wesley, Menlo Park, Calif, 1996, p 2079, with permission.)

and to bleed more easily with injury than in nonpregnant women.

Diagnostic Tests

Diagnosis may be made by applying dilute acetic acid (vinegar) to the skin of the external genital area, vagina, cervix, and anus and then closely examining with a colposcope the areas that turn a lighter color. Biopsy specimens of the suspicious areas can be sent for further study of the cells. Other tests to diagnose HPV include an antigen test and the Southern and dot blot tests, which use radioactive probes. Cancerous changes stimulated by this virus may be identified on Pap smears.

Treatment

There is presently no known cure for papillomavirus infection. The warts may be treated by freezing, burning, or chemically destroying them or by manipulating the patient's immune system to attack the virus. Cryotherapy (freezing) of the warts may be done by touching each wart with a cryoprobe or a liquid nitrogen–soaked swab. Warts may also be burned, or **electrocoagulated** with an **electro-**

cautery or a laser. Heat causes the proteins to coagulate, resulting in death of the wart tissue. Podophyllin, trichloroacetic acid (TCA), and fluorouracil (also called 5FU) are some of the chemical agents that may be applied topically to the warts.[7] Some options are not appropriate for use during pregnancy because of their **cytotoxic** effects, which might damage the fetus, but cryosurgery and laser destruction of the wart tissue may be done during pregnancy. All treatments may require multiple applications and generally result in a great deal of discomfort as the warts degenerate, ulcerate, and slough over a long period.

Various types of immunotherapy have been used against HPV. Interferons are proteins produced by the body that can inhibit viral growth. Several types of interferon have been used to combat HPV. These substances may be applied topically, injected into the condyloma, or administered systemically. Interferons can produce side effects of flulike symptoms, a drop in the number of white blood cells, and changes in liver function. Systemic interferon treatment, however, may offer the advantage of being able to attack warts all over the body at the same time, rather than individually as with topical treatments, thus speeding the process of treatment. Research to develop vaccines against HPV strains is ongoing, but the multitude of varieties makes this difficult. Widespread destruction of tissue in the most sensitive areas of the body can cause severe discomfort. More conservative treatments, however, may not keep pace with the growth of new warts.

Home Care

Patients who have genital warts (condylomata acuminatum) burned off need to recuperate at home. If the burns (there may be multiple areas treated) are near the urethra or rectum, the patient may need a Foley catheter inserted to avoid contamination. Also, the patient is instructed to increase dietary roughage and fluids to prevent constipation. The health care provider is consulted for orders on care of burns. Sterile technique is used for dressing changes. The patient is premedicated for pain control if necessary for dressing changes.

---

electrocoagulated: electro—electrical + coagul—curdled or hardened + ated—process completed
electrocautery: electro—electrical + cautery—branding iron

cytotoxic: cyto—cell + toxic—poison

Hepatitis B
Etiology and Signs and Symptoms

There are several main hepatitis viruses, but this section deals only with hepatitis B virus (HBV), which is generally considered within the STD category because it can be transmitted through sexual contact with blood and body fluids. Early signs of hepatitis are loss of appetite, rashes, malaise, muscle and joint pain, headaches, nausea, and vomiting. As the virus affects the liver, the urine may darken and the stool color lighten (as a result of changes in bile excretion), liver enzymes may rise, and jaundice may appear. Enlargement of the spleen, enlargement and tenderness of the liver, necrosis of liver cells, cirrhosis, coma, and death may follow if the disease is severe. Chronic asymptomatic carrier status may follow hepatitis virus infection, with an increased risk of liver cancer.

During pregnancy, hepatitis B virus may be transmitted to the unborn baby, which can result in acute hepatitis and the possibility of becoming a chronic carrier of HBV.

Diagnostic Tests

Diagnosis of hepatitis is generally made using a variety of blood tests based on antigen and antibody responses.

Treatment

Supportive medical care with avoidance of drugs that require liver metabolism may help the patient through the active stage of the disease. Treatment of the disease may involve injection of serum immune globulins to confer passive immunity. It is recommended that all babies of HBV-positive mothers receive a 0.5-mL dose of HBV immune globulin less than 12 hours after birth and then be immunized with HBV vaccine 1 week, 1 month, and 6 months after birth. See Chapter 33 for more information. Interferon alpha is also available for patients with chronic hepatitis B.[8]

Prevention

Prevention is better than treatment and may be accomplished by using HBV vaccine. This is especially recommended for health care workers who come in contact with blood and body fluids. Standard precautions should be used when contact with any body fluids is expected.

Genital Parasites
Etiology and Signs and Symptoms

Genital parasites are not a true STD, but they may be transmitted during close body contact. The two most commonly seen parasites are pubic lice (*Phthirus pubis*, commonly called "crabs" because of the shape of the lice) and scabies (*Sarcoptes scabiei*). These parasites cause itching, redness, and, for scabies, tracks under the skin where the females burrow to lay their eggs.

Diagnostic Tests

Direct visual or magnified view of the parasites aids in diagnosis.

Treatment

Parasites are treated with topical insecticides. Advise the patient to refer to package inserts for application instructions and precautions to avoid reinfection.

Pediatric Sexually Transmitted Diseases

Not all STD patients are adults. Unfortunately, childhood sexual abuse does occur. You can play an important role in providing explanations in simple terms for procedures and in comforting and supporting the child. Fortunately, not all STDs grow well in immature genital tracts. However, chlamydia, syphilis, herpes, and genital warts are sometimes seen in young children. It is very important that child sexual abuse be reported and investigated. For this to occur effectively, evidence must be gathered. The health care provider may request an evidence kit from the local police for use during the examination.

Accusations of sexual abuse have sometimes been wrongly filed and have produced a great deal of trauma in children and families. Some forms of STDs may have been transmitted at or before birth and later have presented symptoms. It is probably wisest to report suspicions with documentation of evidence to a supervisor if the patient is seen in a health care setting. In the community, child welfare agencies may be contacted. Depending on the state, province, or country, the nurse may be required by law to report directly to a child protection agency any suspicion of child abuse. Children who frequently scratch their genital areas or who have a foul smell emanating from the genital area may be victims of sexual abuse or may have put a foreign object into the vagina. Examination of a small child may be done with a nasal speculum rather than a vaginal speculum. Further content on care and examination of children is addressed in pediatrics texts. Teenagers and the elderly are also at risk for STDs (Gerontological Issues Box 42–2).

BOX 42-2 GERONTOLOGICAL ISSUES

Older adults remain sexually active even though society often views them as asexual. Do not assume that because an older adult is single or widowed he or she is not sexually active. Older adults who have enjoyed active and fulfilling sex lives with a spouse or partner seek that in new relationships. Older adults who engage in high-risk sexual behaviors (multiple partners, genital-anal sex, no use of barriers during sexual intercourse) are also at risk for sexually transmitted diseases.

▶ REPORTING OF SEXUALLY TRANSMITTED DISEASES

You may also facilitate the reporting and public health follow-up of STDs by filling in patient information on the STD reporting form and placing the form in the patient's chart for completion by the health care provider. The requirements for reporting STDs may vary for different states, provinces, and countries. In some areas, laboratories are also required to submit a report form for positive reportable STD tests. Laboratory reports that are not followed by a health care provider's report may result in investigation of the situation by an STD investigator. Generally the report form has spaces for listing of sexual contacts who should be notified of possible STD exposure. Depending on the laws of the state, province, or country, the patient may notify contacts or have the health care provider do so or contacts may be notified by a public health authority that they have been listed as a sexual contact by an anonymous person who has tested positive for a particular STD.

▶ NURSING CARE AND TEACHING

STDs are usually diagnosed and treated in health care provider's offices and clinics. Hospitalization related to STDs generally occurs only in conjunction with infertility investigations, **perinatal** difficulties caused by STDs, or serious long-term complications resulting from STDs, such as AIDS. If the complication is, for instance, meningitis caused by a herpesvirus, the plan of care will be similar to the plan for any other type of meningitis, with the consideration that psychosocial concerns may differ somewhat because of the source of the infection. The nurse's role in the care of patients with STDs includes observation and documentation of signs and symptoms and providing education and patient support during times of diagnosis and treatment.

You may be expected to set out supplies for pelvic examinations and swabs and smears and to be present during the examination as a chaperone or patient support person. (See Chapter 39.) Some patient teaching may take place as the supplies are readied. Explanation of what to expect during the examination and collection of test samples may help decrease patient anxiety. Information about possible risk reduction strategies may also be needed. Pamphlets or other reading materials may be available to provide to the patient. Clarify with your employer the type and depth of information patients should be given, so that teaching is not duplicated, confused, or omitted.

Sometimes patients visit clinics or health care providers' offices for stated reasons other than STDs, yet their real concern is an STD. A female patient may state that she needs a Pap smear, when her real concern is the possibility of STDs. After the patient has been placed in an examining room, asking the date of the last Pap smear and whether she is presently experiencing any irritation, pain, or unusual growths or discharges in her private areas may clarify the request without being considered too intrusive. Explaining to the patient the need to know what examination supplies to set out for the health care provider may further help put the patient at ease. Some people misunderstand the purpose of Pap smears and believe that any problems of the reproductive organs can be diagnosed through this test. If appropriate questioning does not determine the real needs of the patient, she may believe that she has received testing for

infection that in reality was not done. If she is asymptomatic by visual inspection and does not tell the health care provider about her concerns, a Pap smear may be all that is done. Furthermore, the results for Pap smears that are done when there is an active infection may read "inflammatory changes present, repeat Pap smear after resolution of the inflammation." Although this may eventually alert the health care provider that there was an infection at the time of the Pap smear, the infective agent is not identified, another pelvic examination is probably be required, and the diagnosis is delayed needlessly (during which time the infection may have been transmitted to others).

Nursing diagnoses that may be identified for the patient with an STD may include pain related to inflammation or skin lesions, risk for transmission of infection to others, ineffective sexuality pattern related to illness and risk for transmission, and fear related to diagnosis of possible incurable illness. Your plan of care should include measures to increase patient comfort, instruction on safer sex practices and how to prevent transmission of the disease, and identification of coping behaviors that help reduce fear. Ensure that the patient understands self-administration of analgesics and other prescribed treatments, as well as signs of possible complications and how to report them, and knows that the pain will subside with effective treatment of the infection.

It has been said that the only sure prevention of sexually transmitted diseases is abstinence. Lifelong monogamy of both sexual partners in a relationship also provides excellent protection. Having a sexual relationship with someone is the **epidemiological** equivalent of engaging in sexual activity with all of that person's previous partners. Ignorance of the serious repercussions of some STDs may explain some of the apathy surrounding risk reduction.

Many myths about sexual activity are sincerely believed by some patients (Table 42–1). Assessing the patient's health beliefs and correcting misconceptions is part of the job of a nurse. It is vitally important for patients to know the sexual and lifestyle history of any potential partner before sexual activity has occurred and before emotional issues may cloud judgment. Development of a healthy relationship in which honest communication can occur generally takes time and effort.

When teaching patients, the terms *safe sex* and *STD prevention* are misnomers. You should more accurately refer to information about barrier methods as *safer sex* practices, which may decrease the risk of (but not absolutely prevent) transmission of STDs (Table 42–2).[9]

One of the most common contributing factors for infection with an STD is the consumption of alcohol or other psychoactive drugs. Patients should be advised that reduction of inhibitions and changes in judgment may result in unintended sexual encounters, which can transmit STDs. Avoiding or limiting alcohol and other drug consumption

---

perinatal: peri—around + natal—birth

epidemiological: epi—on + demio—people + logical— pertaining to

TABLE 42-1 COMMON MYTHS ABOUT SEXUALLY TRANSMITTED DISEASES

Myth	Factual Data
People who have STDs are easily identifiable.	Inspection of the potential partner's genitals before sexual activity may decrease the risk (if one does not participate in sexual activity with a person who has visible lesions), *but* . . . • Not all people who are infected have visible symptoms. • There is no standard personality or physical profile for people who can be infected with STDs—*anyone* can be infected.
Avoiding persons who have a history of casual sex, intravenous drug use, homosexual activity, bisexual activity, or a previous sexual relationship with persons who engage in these high-risk practices effectively protects one from infection with STDs.	Avoiding people with these types of history may decrease risk, *but* . . . • Not everyone is honest when responding to questions about sexual history. • Not everyone is aware of their previous partners' histories or the histories of others with whom their previous partners have had sexual relationships. • Asking these kinds of questions is difficult and may be postponed at times until emotional factors complicate such communication.
STDs never happen the first time.	Only one contact with one microorganism is necessary for infection.
Intact genital skin is impervious to the germs (and gentle sexual activity does no harm).	Intact skin is the body's first line of defense, *but* . . . • Some microorganisms can be transmitted without a noticeable tissue injury. • Minor injuries can occur during many types of sexual activity, including vaginal intercourse.
Condoms prevent the spread of all STDs.	Condoms can greatly decrease the risk of STDs, *but* . . . • Condoms can have tiny channels in the rubber (or other elastic material), which can allow microorganisms to pass through. • Condoms can break, slip off, or be applied improperly. • Petroleum-based lubricants may weaken latex condoms. • Condoms do not provide a barrier for any area other than the penis and most of the vagina (or anus). Some STDs may still be transmitted by contact of surrounding uncovered tissues.
The female condom prevents all transmissions of STDs.	It does cover more surface area, but it may have similar problems to male condoms (see previous).
Manual, oral, and anal stimulation cannot transmit STDs.	Contact of hands to genitals can allow for transmission of microorganisms through breaks in the skin. Oral sex can transmit some STD-causing microorganisms. Anal intercourse is a very high-risk activity for transmission of STDs because anal tissues are easily injured and the gastrointestinal tract can be a reservoir for many microorganisms.
Nonoxynol-9 spermicide kills all STD germs.	Nonoxynol-9 can reduce the risk of transmission of STDs, *but* . . . • Nonoxynol-9 is *not* guaranteed to kill all microorganisms.
People get AIDS only by homosexual sexual activity or by blood transfusion. A woman cannot transmit HIV to a man. A man cannot transmit HIV to a woman.	Homosexual activity may result in a higher incidence of transmission of human immunodeficiency virus, *but* . . . • HIV can be transmitted during heterosexual activity. • The gender of the individual does not protect him or her from being infected with HIV.
Sexual activity during menstruation is less likely to result in STDs.	Sexual activity during menstruation is more likely to result in transmission of some microorganisms that cause STDs because of the vulnerability of the lining of the uterus caused by sloughing of the outer layers of cells and because blood and cellular debris may serve as a nutritious medium for growth of microorganisms.
Lesbian sexual activity cannot transmit STDs.	Transmission of microorganisms can occur by contact with mouth, anal, or genital tissues or fomites (inanimate objects, such as vibrators and other sex paraphernalia) that have been contaminated with microorganisms from an infected individual—regardless of the original source of the infection.
Individuals who have not been infected after sexual activity with several people are naturally immune to STDs.	There is no known natural immunity to STDs. The individual may not yet have had contact with someone with an active STD.
Those who have had an STD and have been cured of it by taking medicine are now immune to that disease.	Infection that has been eradicated by medication does not confer immunity.
People can be certified free of all STDs by having a blood test and taking a simple medication if an infection is present.	Testing of those who suspect they may have contracted an STD and treatment (if possible) may decrease the spread of STDs, *but* . . . • No one test identifies all STDs. Some are identified by examination, and not all infected people show symptoms. • Some STDs do not show positive test results for long periods yet may be transmitted by the individual while the tests are still negative. • People may be infected with more than one causative agent at a time and each must be treated (if possible). One STD may obscure the symptoms of other concurrent STDS, so that one or more types may go unnoticed and untreated or may not be evident until other STDs have been treated. • There are no known cures for some STDs.
Oral contraceptive (OC) pills give protection against STDs.	OC preparations are *not* an antibiotic—they provide only some protection against conception. Use of a barrier method with spermicide along with the OC can decrease risk of STDs, as well as pregnancy.

TABLE 42-2 BARRIER METHODS FOR SAFER SEX

Barrier	Related Information
Male condoms	Latex condoms are less likely to break during intercourse than other types.
	Lubrication decreases the chances of breakage during use, but only water-soluble lubricants should be used, because substances such as petroleum jelly (Vaseline) may weaken the condom.
	Condoms should never be inflated to test them, because this can weaken them.
	Condoms should be applied only when the penis is erect.
	Either condoms with a reservoir tip or regular condoms that have been applied while holding approximately ½ inch of the closed end flat between the fingertips allow room for expansion by the ejaculate without creating excessive pressure, which might break the condom.
	The penis should be withdrawn after ejaculation before the erection begins to subside while holding the top of the condom securely around the penis to avoid spillage.
	Condoms should never be reused and should be discarded properly after use so that others will not come in contact with the contents.
Female condoms	Female condoms should be applied before any penetration occurs (even preejaculation fluid can contain microorganisms).
	Lubrication decreases the chances of breakage during use, but only water-soluble lubricants should be used, because substances such as petroleum jelly may weaken the condom.
	Female condoms should never be reused and should be discarded properly after use so that others will not come in contact with the contents.
Cervical caps or diaphragms	These may provide some protection for the cervix only. They are not effective barriers against STD infection.
Rubber gloves, rubber dental dams, split (opened) male condoms	These may provide some barrier protection for manual and oral sexual activity. Although some groups suggest that male condoms may be split down one side and opened or rubber dental dam material may be taped over areas that have lesions to avoid direct contact with blood and body fluid, especially during sadomasochistic sexual activity, this *very high-risk behavior* is not recommended.
Double condoms	Anal intercourse is a *very high-risk activity* for transmission of many types of STDs, as well as many intestinal organisms, and is not recommended. Homosexual networks advise wearing double condoms and using water-soluble lubricants, preferably containing nonoxynol-9, to decrease risk somewhat if engaging in this type of sexual activity.

when with potential partners may help prevent STD infection from occurring.

For those who choose to be sexually active with more than one person (or with a person who has been sexually active with others) and for those who choose to practice high-risk behaviors, barrier protection may decrease the chances of developing an STD. Information on how to use barrier methods effectively is essential. Ensure that the patient has access to these methods. For more information on STDs, visit http://www.niaid.nih.gov/factsheets/stdstats.htm and http://www.ashastd.org/.

Evaluation

Goals have been met if the patient does the following:

CRITICAL THINKING

Stephanie

As you seat a young woman in an examining room of the clinic where you work, she comments, "I am new to this area and I've heard that there are three guys in this town who have syphilis and are spreading it around. Is that true?"

1. What are some concerns this question might reflect?
2. You find out that Stephanie knows very little about syphilis. List in outline form a teaching plan that includes the information that is important for Stephanie to know about syphilis.

Answers at end of chapter.

- States that pain is controlled
- Verbalizes understanding of and intent to follow recommendations for prevention of infection transmission
- Describes safer sexual practices
- Relates accurate information and reduction in fear

▶ CONCLUSION

Neighbors, friends, or family members may seek information from you because they know nurses are educated about health issues. Such questions may be stated in indirect terms, such as, "I have a friend who is having a problem. . . ." It is often wisest to identify broad areas of information that people seem to be seeking without going into details that may be embarrassing. Suggesting that the concerned person see a health care provider or contact the local STD clinic for further investigation or information is wise. Emphasize the serious consequences of untreated STDs and the impossibility of effective self-treatment. Avoid stating that there is no cure for a particular type of STD, because research is ongoing and sometimes cures are developed and tested before the general medical community becomes aware of them.

STD diagnosis is a complicated process. Individual pathogens may affect the body in a variety of ways, and infection with more than one pathogen may further complicate the picture. Some signs and symptoms that are seen with STDs may also be caused by non–sexually transmitted infection or overgrowth of the body's resident microorganisms. General awareness of common signs and symptoms of STDs, however, is important for nurses. Some patients admitted for other diagnoses may be unaware that they have STDs as well. Nurses are often the ones who bathe and given perineal care to patients. Unusual discharges, redness, blisters, swollen areas, ulcers, and evidence of parasites in the genital area may be observed during patient care. STD awareness can sensitize you to the possible significance of patient complaints such as persistent pelvic pain, dysuria, discharges, and rectal soreness. Such problems should be accurately documented and reported, so that further investigation and possible treatment can take place.

One of the most important ways you can help those who experience STDs is by being kind, polite, nonjudgmental, and sensitive to the patient's communication. Keeping the arms crossed over the chest or other closed postures may be interpreted as evidence of a judgmental attitude regardless of your feelings. Maintaining an open posture and eye contact that is appropriate for the patient's culture relays a sense of openness and willingness to talk and preserves the possibility of continuing health promotion with these individuals in the future.

REVIEW QUESTIONS

1. Medication may be put in newborns' eyes to protect them against which of the following disorders?
   a. Hepatosplenomegaly
   b. Neonatal pneumonia
   c. Ophthalmia neonatorum
   d. Becoming a chronic carrier of HBV

2. Which of the following pathogens causes syphilis?
   a. *Treponema pallidum*
   b. *Chlamydia trachomatis*
   c. Human papillomavirus
   d. Human immunodeficiency virus

3. Which of the following has been linked to cancerous changes in cervical cells?
   a. Syphilis
   b. Gonorrhea
   c. Chlamydia
   d. Human papillomavirus

4. Which is the most accurate term to use when teaching patients about STD prevention?
   a. Safe sex strategies
   b. Safer sex strategies
   c. Risk elimination strategies

5. Lisa comes into the clinic where you work and is diagnosed with an STD. She says, "How could I have an STD? I only have sex with my boyfriend. I don't sleep around!" Which of the following responses is best?
   a. "You are right, that should have kept you safe. There just are no guarantees."
   b. "If your boyfriend is not infected, then obviously you have had sex with someone else."
   c. "Your boyfriend could be infected from his past sexual encounters. He should be tested."
   d. "Even lifelong monogamy cannot prevent many STDs."

6. What is the worst possible outcome of an STD?
   a. Pain
   b. Death
   c. Scarring
   d. Infertility

Answers to CRITICAL THINKING

Stephanie

1. Concerns might include (a) a wish to speak with a health care worker; (b) uncertainty about whether patient information will be kept confidential (give assurance that if you knew about anyone with syphilis, it would be your professional responsibility to keep it confidential); (c) fear that she might have become infected through heterosexual contact; (d) a desire to protect herself by avoiding those who have this problem; and (e) a desire for information about syphilis and its transmission routes.

2. The teaching plan might include information about (a) the spirochete that causes syphilis; (b) signs and symptoms; (c) diagnostic tests; (d) means of transmission; (e) strategies for risk reduction; (f) treatment; (g) research; and (h) rights and responsibilities of those who have the disease.

REFERENCES

1. National Institute of Allergy and Infectious Disease, National Institutes of Health: Fact sheet: STD Statistics. http://www.niaid.nih.gov/factsheets/stdstats.htm, March 2001.
2. National Institute of Allergy and Infectious Disease, National Institutes of Health: Fact sheet: Gonorrhea. http://www.niaid.nih.gov/factsheets/stdgon.htm, October 2000.
3. National Institute of Allergy and Infectious Disease, National Institutes of Health: Fact sheet: Gonorrhea. http://www.niaid.nih.gov/factsheets/stdgon.htm, October 2000.
4. Mindel, A, and Estcourt, C: Syphilis. In Stanberry, LR, and Bernstein, DI (eds): Sexually Transmitted Diseases: Vaccines, Prevention, and Control. Academic Press, San Diego, 2000, p 397.
5. Musher, DM: Early syphilis. In Holmes, KK, et al (eds): Sexually Transmitted Diseases, ed 3. McGraw-Hill, New York, 1999, p 483.
6. Corey, L, and Wald, A: Genital herpes. In Holmes, KK, et al (eds): Sexually Transmitted Diseases, ed 3. McGraw-Hill, New York, 1999, p 301.
7. National Institute of Allergy and Infectious Disease, National Institutes of Health: Fact sheet: Human Papillomavirus and Genital Warts. http://www.niaid.nih.gov/factsheets/stdhpv.htm, March 2001.
8. Zuckerman, JN, and Zuckerman, AJ: Hepatitis B virus infection. In Stanberry, LR, and Bernstein, DI (eds): Sexually Transmitted Diseases: Vaccines, Prevention, and Control. Academic Press, San Diego, 2000, p 307.
9. Rosenweig, BA, Even, A, and Budnick, LE: Observations of scanning electron microscopy detected abnormalities of non-lubricated latex condoms. Contraception 53:49, 1996.

UNIT ELEVEN BIBLIOGRAPHY

Cates, W, and Padian, NS: The interrelationship of reproductive health and sexually transmitted diseases. In Goldman, MB, and Hatch, MC: Women and Health. Academic Press, San Diego, CA, 2000, pp 381–389.

Cookson, MM: Prostate cancer: Screening and early detection. Cancer Control 8:133, 2001.

Copeland, LJ (ed.): Textbook of Gynecology, ed 2. WB Saunders, Philadelphia, 2000.

Cuzick, J: Human papillomavirus testing for primary cervical cancer screening. Journal of the American Medical Association 283:108, 2000.

Epperly, TD, and Moore, KE: Health issues in men: Part I. Common genitourinary disorders. American Family Physician. Retrieved July 15, 2000, from www.aafp.org/afp/20000615/3657.html.

Goldman, MB, and Hatch, MC: Women and Health. Academic Press, San Diego, 2000.

Hart, DM, and Norman, J (eds.): Gynecology Illustrated, ed 5. Churchill Livingstone, Edinburgh, 2000.

Holmes, KK, et al. (eds.): Sexually Transmitted Diseases, ed 3. McGraw-Hill, New York, 1999.

Lewis, RW: Epidemiology of erectile dysfunction. Urologic Clinics of North America 28:209, 2001.

Meredith, CE: Erectile dysfunction. In Meredith, CE, and Karlowicz, KA (eds.): Urologic Nursing: A Study Guide. Society of Urologic Nurses and Associates, Pitman, NJ, 1995, pp 137–141.

Meredith, CE: Male infertility. In Karlowicz, KA (ed.): Urologic Nursing: Principles and Practice. WB Saunders, Philadelphia, 1995, pp 360–372.

Mindel, A, and Estcourt, C: Syphilis. In Stanberry, LR, and Bernstein, DI (eds.): Sexually Transmitted Diseases: Vaccines, Prevention, and Control. Academic Press, San Diego, 2000, pp 387–420.

Narod, SA, et al: Tubal ligation and risk of ovarian cancer in carriers of BRCA1 or BRCA2 mutations: A case-control study. Lancet 357(9267):1467, 2001.

National Cancer Institute: NCI Fact Sheet: Oral Contraceptives and Cancer Risk - National Institute of Allergy and Infectious Disease. National Institutes of Health. Fact sheet: Chlamydia. US Department of Health and Human Services, Bethesda, MD. Retrieved October, 2000, from http://www.niaid.nih.gov/factsheets/stdclam.htm.

Newkirk, GR: Abnormal vaginal bleeding. In Sultz, JW (ed.): Textbook of Family Medicine. McGraw-Hill, 2000, pp 373–383.

Slowey, MJ: Polycystic ovary syndrome: New perspective on an old problem. Southern Medical Journal 94(2):190, 2001.

Walker, CK: Reproductive tract infections: Sexually transmitted diseases. In Copeland, LJ: Textbook of Gynecology, ed 2. WB Saunders, Philadelphia, 2000, pp 869–894.

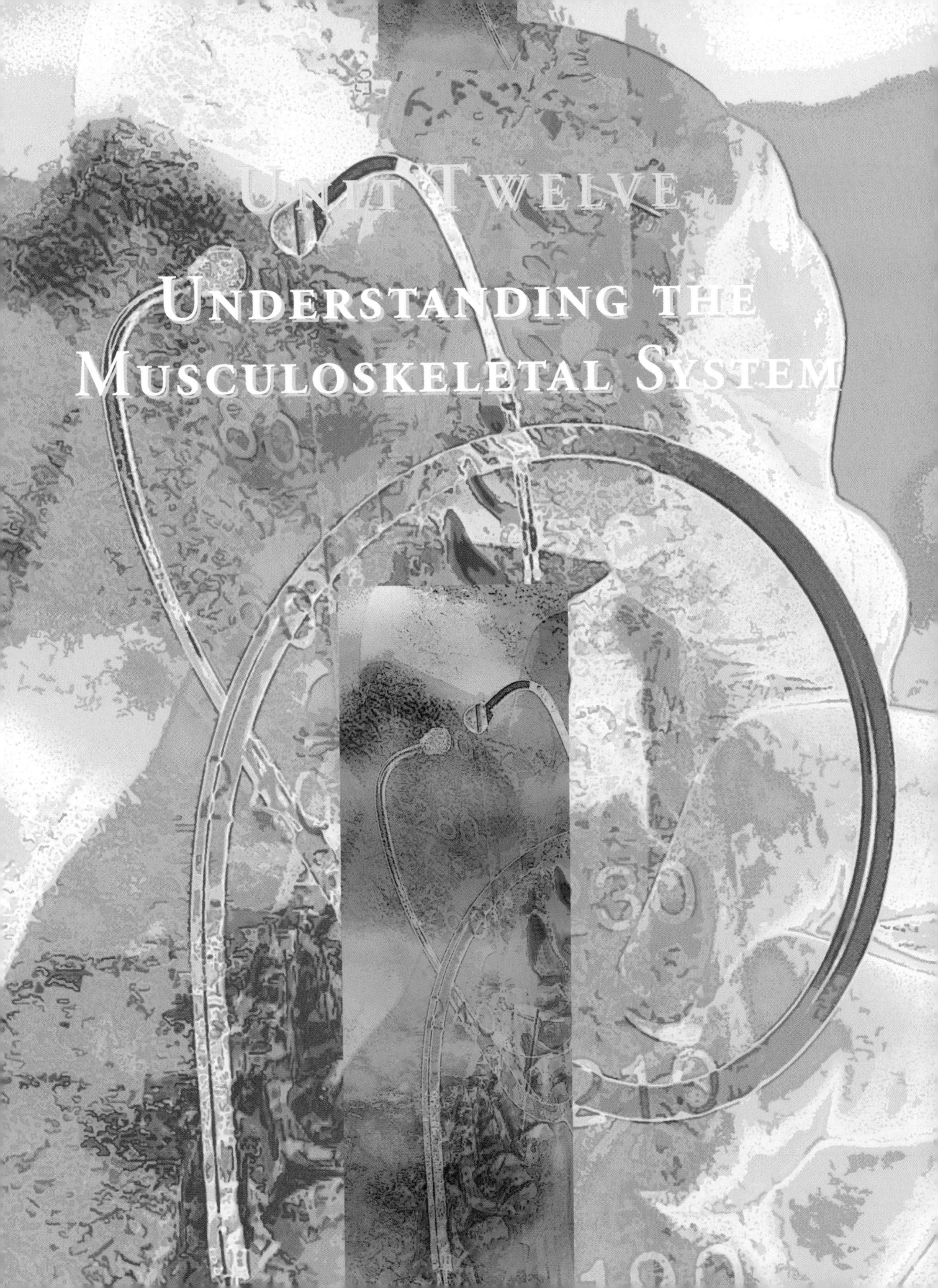

Unit Twelve

Understanding the Musculoskeletal System

43

MUSCULOSKELETAL FUNCTION AND ASSESSMENT

Rodney Kebicz and Donna D. Ignatavicius

KEY TERMS

arthroscopy (ar-**THROSS**-scop-ee)

arthrocentesis (ar-**THROW**-sen-tee-sis)

articular (ar-**TIK**-yoo-lar)

bone (BOWN)

bursae (**BURR**-sah)

crepitation (crep-i-**TAY**-shun)

hemarthrosis (heem-ar-**THROW**-sis)

joint (JOYNT)

muscle (**MUSS**-uhl)

synovitis (sin-oh-**VIGH**-tis)

vertebrae (**VER**-te-bray)

QUESTIONS TO GUIDE YOUR READING

1. What is the normal structure and function of the musculoskeletal system?
2. What major changes occur in the musculoskeletal system when a person ages?
3. Why is it important to inspect, palpate, and determine range of motion in a patient with a musculoskeletal problem?
4. What areas are reviewed when performing a neurovascular assessment?
5. What psychosocial problems do you assess for when caring for patients with musculoskeletal problems?
6. How would you describe diagnostic tests for musculoskeletal problems?
7. What nursing care would you provide for each musculoskeletal diagnostic test?

► REVIEW OF NORMAL ANATOMY AND PHYSIOLOGY

The skeletal and muscular systems may be considered one system because they work together to enable the body to move. The skeleton is the framework that supports the body and to which the voluntary **muscles** are attached. The skeletal framework includes the **joints** or articulations between **bones.** Contraction of a muscle pulls a bone and changes the angle of a joint. It is important to remember that movement would not be possible without the proper functioning of the nervous, cardiovascular, and respiratory systems. Voluntary muscles require nerve impulses to contract, a continuous supply of blood provided by the circulatory system, and oxygen provided by the respiratory system.

► SKELETAL SYSTEM TISSUES AND THEIR FUNCTIONS

The tissues that make up the skeletal system are bone tissue; cartilage, which covers most joint surfaces; and fibrous connective tissue, which forms the ligaments that connect one

bone to another and also form part of the structure of joints. The tissues of the muscular system are skeletal (also called striated or voluntary) muscle; fibrous connective tissue, which forms the tendons that connect muscle to bone; and the fasciae, the strong membranes that enclose individual muscles.

Besides its role in movement, the skeleton has other functions. It protects organs from mechanical injury. For example, the brain is protected by the skull and the heart and lungs are protected by the ribcage. Flat and irregular bones contain and protect the red bone marrow, one of the hematopoietic (blood-forming) tissues. The bones are also a storage site for excess calcium, which may be removed from bones to maintain a normal blood calcium level. Calcium in the blood is necessary for blood clotting and for the proper functioning of nerves and muscles.

Although the primary function of the muscular system is to move the skeleton, the voluntary muscles collectively contribute significantly to heat production, which maintains normal body temperature. Heat is one of the energy products of cell respiration, the process that produces

749

adenosine triphosphate (ATP), the direct energy source for muscle contraction. Another important function of the muscular system is that it aids in returning blood from the legs through muscular compression on the leg veins.

Bone Tissue and Growth of Bone

Bone tissue is composed of bone cells, called osteocytes, within a strong nonliving matrix made of calcium salts and the protein collagen. In compact bone the osteocytes and matrix are in precise arrangements called haversian systems. Compact bone is very dense and to the unaided eye appears solid. In spongy bone the arrangement of cells and matrix is less precise, giving the bone a spongy appearance. Compact bone forms the diaphyses (shafts) of the long bones of the extremities and covers the spongy bone that forms the bulk of short, flat, and irregular bones.

A living bone is covered by a fibrous connective tissue membrane called the periosteum, which is the anchor for tendons and ligaments because the collagen fibers of all these structures merge to form connections of great strength. This membrane also contains the blood vessels that enter the bone itself (most bone has a very good blood supply) and bone-producing cells called osteoblasts that are activated to initiate repair when bone is damaged.

The growth of bone from fetal life until a person attains final adult height depends on many factors. Proper nutrition provides the raw material to produce bone matrix: calcium, phosphorous, and protein. Vitamin D is essential for the efficient absorption of calcium and phosphorous from food in the small intestine. Vitamins A and C do not become part of bone but are needed for enzymes involved in the production of bone matrix (a process called calcification or ossification). Hormones directly necessary for growth include growth hormone (GH) from the anterior pituitary gland, thyroxine from the thyroid gland, and insulin from the pancreas. Growth hormone increases mitosis and protein synthesis in growing bones; thyroxine also increases protein synthesis, as well as increasing energy production from food. Insulin is essential for the efficient use of glucose to provide energy. If a child is lacking any of these hormones, growth is much slower and the child does not reach his or her genetic potential for height.

Bone is not a fixed tissue, even when growth in height has ceased. There is a constant removal and replacement of calcium and phosphate (usually the rates are equal) to maintain normal blood levels of these minerals. Parathyroid hormone secreted by the parathyroid glands increases the removal of calcium and phosphate from bones; the hormone calcitonin from the thyroid gland promotes the retention of calcium in bones, although its greatest effects may be during childhood.

Osteoblasts produce bone matrix during normal growth, to replace matrix lost during normal turnover and to repair fractures. Other cells called osteoclasts reabsorb bone matrix when more calcium is needed in the blood and during normal growth and fracture repair when excess bone must be removed as bones change shape.

The sex hormones, estrogen from the ovaries or testosterone from the testes, are important for the retention of calcium in adult bones. For women after menopause, more calcium may be removed from bones than is replaced, leading to a thinning of bone tissue and the possibility of spontaneous fractures.

Structure of the Skeleton

The 206 bones of the human skeleton are in two divisions: the axial skeleton and the appendicular skeleton. The axial skeleton consists of the skull, vertebral column, and ribcage; all are flat or irregular bones and contain red bone marrow. The appendicular skeleton consists of the bones of the arms and legs and the shoulder and pelvic girdles, by which the extremities attach to the axial skeleton (Fig. 43–1).

The long bones of the limbs are those of the arm, forearm, hand, and fingers and those of the thigh, leg, foot, and toes. All long bones have the same general structure: a central diaphysis or shaft with two ends called epiphyses. The diaphyses of long bones contain yellow bone marrow, which is mostly fat—that is, stored energy. The bones of the wrist and ankle are short bones, and those of the shoulder and pelvic girdles are considered flat bones. These bones contain red bone marrow.

Skull

The skull consists of 8 cranial bones and 14 facial bones and also contains the 3 auditory bones found in each middle ear cavity. The cranial bones that enclose and protect the brain are frontal, two parietal, two temporal, occipital, sphenoid, and ethmoid (Fig. 43–2). All the joints between cranial bones and those between most of the facial bones are immovable joints called sutures. The mandible is the only movable facial bone; the temporomandibular joint is a condyloid joint (Table 43–1). The maxillae are the upper jaw bones, which also form the front of the hard palate. The rest of the facial bones are shown in Fig. 43–2.

Vertebral Column

The vertebral column (or spinal column) is made of individual bones called **vertebrae** (see Fig. 43–1). From top to bottom there are 7 cervical, 12 thoracic, 5 lumbar, 5 sacral fused into 1 sacrum, and 4 or 5 coccygeal vertebrae fused into 1 coccyx.

The first cervical vertebra, the atlas, articulates with the occipital bone of the skull and forms a pivot joint with the axis, the second cervical vertebra. The thoracic vertebrae articulate with the posterior ends of the ribs. The lumbar vertebrae are the largest and strongest. The sacrum permits the articulation of the two hip bones, the sacroiliac joints. The coccyx is the remnant of tail vertebrae, and some muscles of the perineum are attached to it.

The vertebrae as a unit form a flexible backbone that supports the trunk and head and contains and protects the spinal cord. The joints between vertebrae are symphysis joints in which a disk of fibrous cartilage serves as a cushion and permits slight movement.

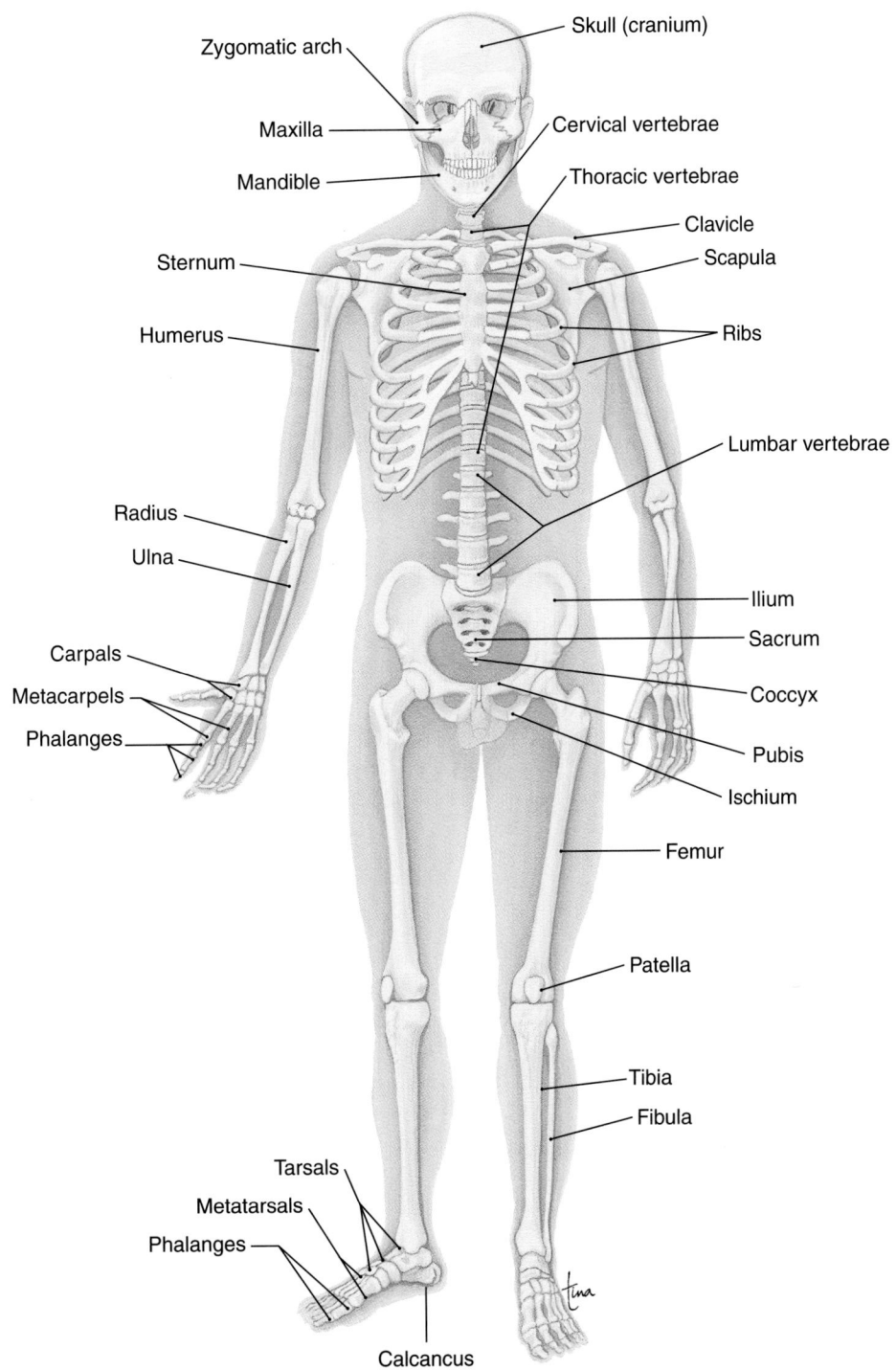

Figure 43-1 The full skeleton in anterior view. (Modified from Scanlon, VC, and Sanders, T: Anatomy and Physiology, ed 3. FA Davis, Philadelphia, 1999, p 106, with permission.)

Ribcage

The ribcage consists of the 12 pairs of ribs and the sternum, or breastbone. All the ribs connect posteriorly with the thoracic vertebrae. The seven pairs of true ribs articulate directly with the sternum by means of costal cartilages; the three pairs of false ribs join indirectly with the sternum, and the two pairs of floating ribs do not connect to the sternum at all.

The ribcage protects the heart and lungs, as well as upper abdominal organs such as the liver and spleen, from me-

chanical injury. During breathing, the flexible ribcage is pulled upward and outward by the external intercostal muscles to expand the chest cavity and bring about inhalation.

Appendicular Skeleton

The bones of the appendicular skeleton are shown in Fig. 43–1. The important joints of the appendicular skeleton are summarized in Table 43–1.

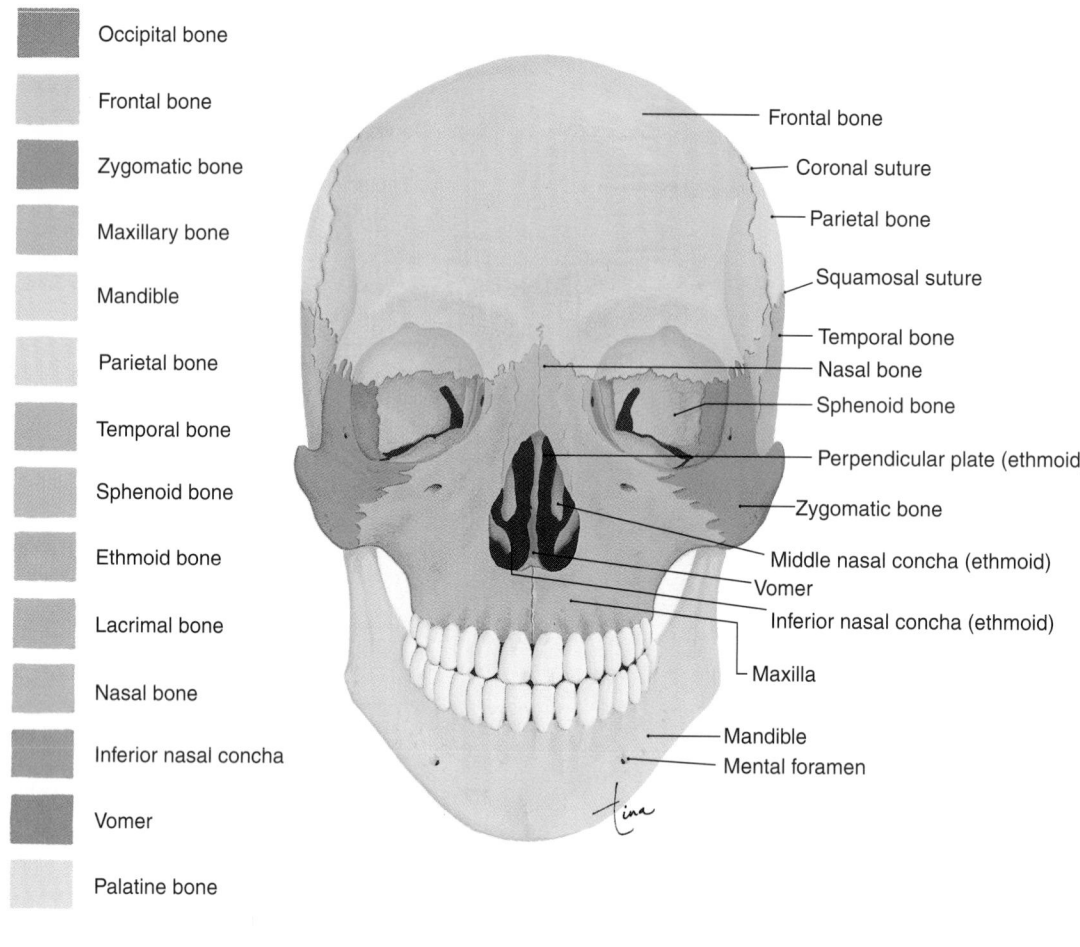

Occipital bone

Frontal bone

Zygomatic bone

Maxillary bone

Mandible

Parietal bone

Temporal bone

Sphenoid bone

Ethmoid bone

Lacrimal bone

Nasal bone

Inferior nasal concha

Vomer

Palatine bone

Frontal bone

Coronal suture

Parietal bone

Squamosal suture

Temporal bone

Nasal bone

Sphenoid bone

Perpendicular plate (ethmoid)

Zygomatic bone

Middle nasal concha (ethmoid)

Vomer

Inferior nasal concha (ethmoid)

Maxilla

Mandible

Mental foramen

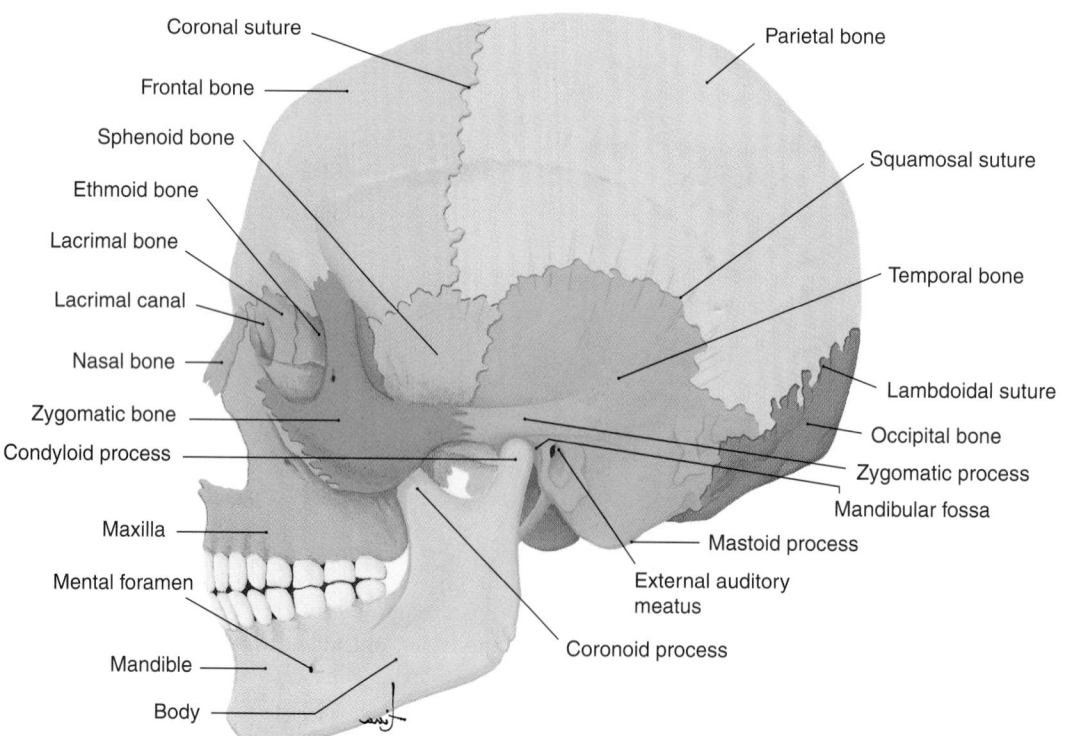

Coronal suture

Parietal bone

Frontal bone

Sphenoid bone

Squamosal suture

Ethmoid bone

Lacrimal bone

Temporal bone

Lacrimal canal

Nasal bone

Lambdoidal suture

Zygomatic bone

Occipital bone

Condyloid process

Zygomatic process

Maxilla

Mandibular fossa

Mental foramen

Mastoid process

External auditory meatus

Mandible

Coronoid process

Body

Figure 43-2   Anterior *(upper)* and lateral *(lower)* views of the skull. (Modified from Scanlon, VC, and Sanders, T: Anatomy and Physiology, ed 3. FA Davis, Philadelphia, 1999, p 107, with permission.)

TABLE 43-1 JOINTS OF THE APPENDICULAR SKELETON	
Type of Joint and Description	**Examples**
Symphysis—disk of fibrous cartilage between bones	Between vertebrae
	Between pubic bones
Ball and socket—movement in all planes	Scapula and humerus (shoulder)
	Pelvic bone and femur (hip)
Hinge—movement in one plane	Humerus and ulna (elbow)
	Femur and tibia (knee)
	Between phalanges (fingers and toes)
Condyloid—hinge with some lateral movement	Temporal bone and mandible (lower jaw)
Pivot—rotation	Atlas and axis (neck)
	Radius and ulna (distal to elbow)
Gliding—side-to-side movement	Between carpals (wrist)
Saddle—movement in several planes	Carpometacarpal of thumb

Adapted from Scanlon, VC, Sanders, T: Essentials of Anatomy and Physiology, ed 3. FA Davis, Philadelphia, 1999, p. 120, with permission.

Structure of Synovial Joints

All freely movable joints (this excludes sutures and symphyses) are synovial joints in that they share similarities of structure (Fig. 43–3). On the joint surface of each bone is the **articular** cartilage, which provides a smooth surface. The joint capsule is similar to a sleeve. It is made of fibrous connective tissue and forms a strong sheath that encloses the joint. Lining the joint capsule is the synovial membrane, which secretes synovial fluid into the joint cavity. Synovial fluid prevents friction as the bones move.

Many synovial joints also have **bursae,** which are small sacs of synovial fluid between the joint and the tendons that cross over the joint. Bursae permit the tendons to slide easily as the joint moves.

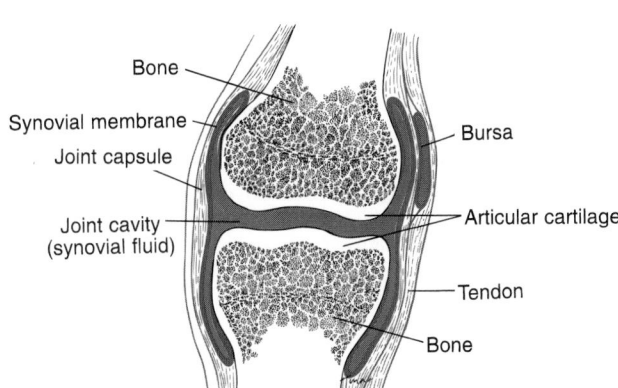

Figure 43-3 Longitudinal section through a typical synovial joint. (Modified from Scanlon, VT, and Sanders, T: Workbook for Essentials of Anatomy and Physiology, ed 3. FA Davis, Philadelphia, 1999, p 121, with permission.)

MUSCLE STRUCTURE AND ARRANGEMENTS

One muscle is made of thousands of muscle cells (fibers), which are specialized for contraction. When a muscle contracts, it shortens and pulls on a bone. Each muscle fiber receives its own motor nerve ending, and the number of fibers that contract depend on the job the muscle has to do. Muscles are anchored to bones by tendons, which are made of fibrous connective tissue. A muscle usually has at least two tendons, each attached to a different bone. The more stationary a muscle's attachment to its origin, the more movable the attachment to its insertion. The muscle itself crosses the joint formed by the two bones to which it is attached, and when the muscle contracts it pulls on the insertion and moves the bone in a specific direction.

The more than 600 muscles in the body are arranged to bring about a variety of movements (Fig. 43–4). The two general types of arrangements are the opposing antagonists and the cooperative synergists.

Antagonistic muscles have opposite functions; such arrangements are necessary because muscles can only pull, not push. If the biceps brachii, for example, flexes the forearm, an antagonist, the triceps brachii, is needed to extend the forearm. Other examples of antagonists are the quadriceps femoris and hamstring groups, the pectoralis major and the latissimus dorsi, and the tibialis anterior and gastrocnemius.

Synergistic muscles have similar functions or work together to perform a particular function. The brachioradialis is a synergist to the biceps brachii for flexion of the forearm; the sartorius is a synergist to the quadriceps group for flexion of the thigh. Synergists are necessary to provide slight differences in angles when joints are moved.

ROLE OF THE NERVOUS SYSTEM

Skeletal muscles are voluntary muscles in that nerve impulses are essential to cause contraction. Such nerve impulses originate in the motor areas of the frontal lobes of the cerebral cortex. The coordination of voluntary movement is a function of the cerebellum. The cerebellum also regulates muscle tone, the state of slight contraction usually present in muscles. Good muscle tone is important for posture and for good coordination.

Neuromuscular Junction

Each of the thousands of fibers in a muscle has its own motor nerve ending; the neuromuscular junction is the termination of the motor neuron on the muscle fiber. (See Fig. 47–2). The axon terminal is the enlarged tip of the motor neuron. It contains sacs of the neurotransmitter acetylcholine. The membrane of the muscle fiber, called the sarcolemma, contains receptor sites for acetylcholine and also the inactivator cholinesterase. The synaptic cleft is the small space between the axon terminal and the sarcolemma.

When a nerve impulse arrives at the axon terminal, it causes the release of acetylcholine, which diffuses across the

A

B

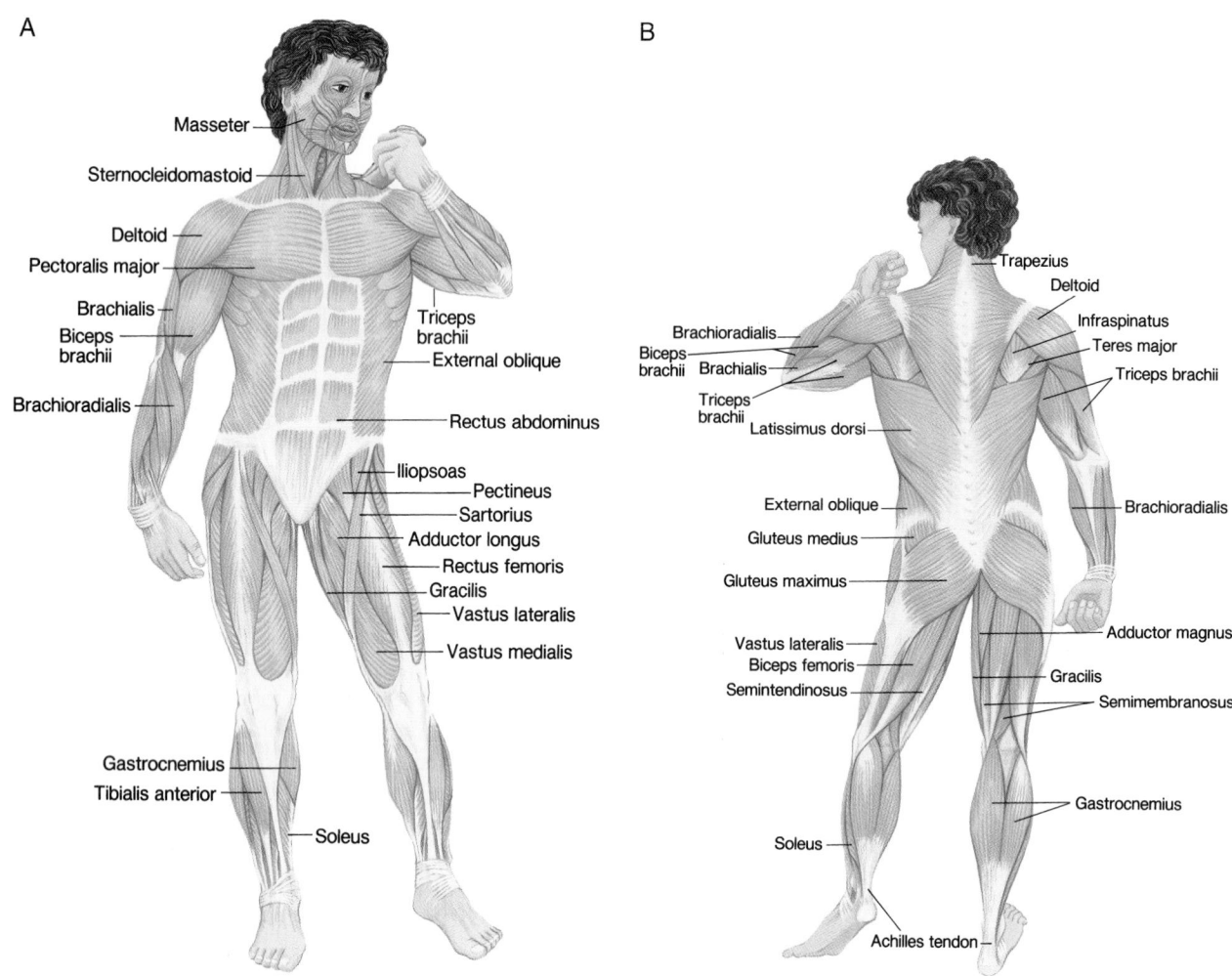

Figure 43-4 Major muscles. *(A)* Anterior view. *(B)* Posterior view. (Modified from Scanlon, VC, and Sanders, T: Essentials of Anatomy and Physiology, ed 3. FA Davis, Philadelphia, 1999, pp 140–141, with permission.)

synaptic cleft and bonds to the acetylcholine receptors on the sarcolemma. This makes the sarcolemma permeable to sodium ions, which rush into the cell and generate an electrical impulse (an action potential) along the entire sarcolemma. This electrical change triggers a series of reactions in the internal units of contraction called sarcomeres. Put simply, filaments of the protein actin slide over filaments of another protein called myosin, and the sarcomere shortens. All the thousands of sarcomeres in a muscle fiber shorten, and the entire cell contracts. If a muscle has little work to do, few of its many muscle fibers contract, but if the muscle has more work to do, more of its muscle fibers contract.

▶ AGING AND THE MUSCULOSKELETAL SYSTEM

The amount of calcium in bones depends on several factors. Good nutrition is certainly one factor, but age is another, especially for women. One function of estrogen or testosterone is the maintenance of a strong bone matrix. For women after menopause, bone matrix loses more calcium than is replaced. Weight-bearing joints are also subject to damage after many years. Often the articular carti-

lage wears down and becomes rough, leading to pain and stiffness.

Muscle strength declines with age as the process of protein synthesis decreases. Such loss of strength need not be drastic because aging muscles benefit from regular exercise, which has been shown to increase strength and reduce falls and accidents (Gerontological Issues Box 43–1).

BOX 43-1 GERONTOLOGICAL ISSUES

Age-related Changes in the Older Adult
Age-related changes can lead to impaired mobility, an increased risk for falls, and pain. Common age-related musculoskeletal changes include the following:

- Muscle mass and strength decline.
- Number of muscle cells decrease and are replaced by fibrous connective tissue.
- Elasticity of ligaments, tendons, and cartilage decrease, resulting in weaker bones.
- Intervertebral spaces decrease from loss of water.
- Posture and gait change. (Men develop a wider stance and take smaller steps; women have a narrow stance and walk with a waddling gait.)

NURSING ASSESSMENT OF THE MUSCULOSKELETAL SYSTEM

The initial assessment begins with a history that includes the effects the condition is having on the patient's life. It then proceeds to a physical and psychosocial assessment (Table 43–2). Frequent neurovascular assessments may be needed if there is a risk of circulation impairment, such as if the patient has a fracture or musculoskeletal surgery (Table 43–3).

Subjective Data
History

The patient's history should include the following:

- If there was an injury, how it happened and when it happened
- Occupation and activities, including sports and other physical activities
- Risk factors for musculoskeletal problems and family history of musculoskeletal problems (to detect hereditary problems)
- Current health status, including ongoing or chronic medical conditions (such as heart disease, diabetes, lung conditions)
- Diet history (including whether calcium and vitamin D intake are adequate to ensure proper bone and muscle maintenance and repair)
- Information specific to the patient's musculoskeletal problems

Patients with musculoskeletal problems frequently report pain or related stiffness and tenderness as a major concern. The pain may be acute or chronic and may limit the patient in everyday life. Assessment includes previous diagnoses, pain severity, medications, treatments, and procedures the patient uses to alleviate the pain. The WHAT'S UP? model can be used to assess the patient's pain. (See Chapter 1.)

Physical Assessment

Three areas of musculoskeletal assessment are important: inspection, palpation, and range of motion (ROM). If the patient is able to walk, inspect the patient's posture and gait, noting poor posture or alterations in movement, such as limping. Note the use of mobility aids, such as a cane or walker. Document other gross deformities, such as unequal limbs, malalignment, or contractures. Spinal deformities are especially significant because they can compromise breathing and balance. Inspect the joints and muscles of the arms, hands, legs, and feet for deformity, redness, swelling, increased temperature, or **crepitation** (grating sound as joint or bone moves). Also note the patient's general nutritional status (e.g., normal, obesity, emaciated).

After inspection, gently palpate for warmth, swelling, and tenderness in the areas of swelling; redness; and areas where the patient reported pain (being careful to minimize the pain this may cause). For example, reddened joints should be palpated for **synovitis** (swollen synovial tissue

---

synovitis: synovia—joint + itis—inflammation

TABLE 43-2	NURSING ASSESSMENT OF MUSCULOSKELETAL SYSTEM
History	**Significance**
Age, gender, socioeconomic status	Increased age, being female, and lower socioeconomic status increases risk of musculoskeletal injury/problems.
History of the injury (if there was one)	Provides information that helps in the diagnosis of the problem, as well as making you aware of possible complications of the injury.
Occupation	Enables you to begin planning for discharge teaching if the patient has to alter his or her employment.
Activities patient participates in	Provides information regarding the level of activity the patient had before the concern.
Risk factors for musculoskeletal problems	Smoking and a sedentary lifestyle are risk factors for musculoskeletal problems.
Family history	Some musculoskeletal conditions have genetic and familial tendencies.
Current health status	Coexisting illnesses or conditions may affect the patient's care.
Diet history	Dietary intake such as calcium and vitamin D influence some musculoskeletal disorders.
Allergies	Prevents exposure to medication or compounds used in diagnostic tests, treatments, and therapies.
Pain	Provides information about severity of the condition and effectiveness of the treatment and therapy.
Physical Assessment	**Possible Abnormal Findings**
Inspect, palpate, and observe range of motion of affected areas	Altered gait, tone, size, shape, posture, contractures, deformities, range of motion, pain, activities of daily living, nerve function, sensation, movement, and weakness can be determined.
Psychosocial Assessment	**Significance**
Assess for deformities, changes in body image, self-concept, socialization, employment.	The patient may need assistance with strategies to cope with the stress of a possible chronic musculoskeletal condition. Some musculoskeletal conditions require lifestyle alterations that can cause increased stress and difficulties in coping.

TABLE 43-3	NEUROVASCULAR ASSESSMENT
Assess For	**Note and Report**
Color	Pallor, cyanosis, redness, or discoloration
Temperature	Unusual coolness or warmth
Pain	Pain that is worse on passive motion, pain that no longer responds to analgesics
Movement	Alterations in movement
Sensation	Alterations in feeling, tingling or paresthesias
Pulses	Diminished or absent distal pulses
Capillary refill	Nailbed that does not blanch in 3–5 seconds

CRITICAL THINKING

Mr. Smith

Mr. Smith, age 80, is brought to the emergency department with a fractured left hip. He is positioned for comfort while you collect data.

1. What information should you obtain in Mr. Smith's history?
2. What should be assessed in Mr. Smith's physical examination?

Answers at end of chapter.

within the joint) or the presence of bony nodes. In some cases, joints and muscles may seem healthy but are tender when palpated.

Next assess joint mobility. Stabilize the body area proximal to the joint being moved. Observe the patient's ROM for performing independent activities of daily living. Pay particular attention to the hands and observe movement in finger joints. For a quick and easy assessment of range of motion in the hands, ask the patient to touch each finger, one by one, to the thumb (known as opposition) and then to make a fist.

Also assess the size, shape, strength, and tone of muscles. Evaluate bilateral muscle strength by asking the patient to grip your hands. This enables you to feel the strength and equality. Pushing an extremity against your hand provides a general indication of muscle strength. More specific evaluation is performed by a physical therapist (PT) or an occupational therapist (OT). Using a scale of 0 to 5 (0 = paralysis and 5 = moving a muscle against resistance), the PT or OT measures the strength of each muscle group and rates it as a fraction. For example, 5/5 means that the patient reached a 5 out of 5 possible on the muscle strength scale.

Psychosocial Assessment

Deformities resulting from arthritis or other musculoskeletal disorders can affect a patient's body image and self-concept. (See Chapter 44.) Chronic pain may keep the patient from socializing or from working. Many work days are lost as a result of both acute and chronic musculoskeletal problems. Patients often avoid social events and tend to withdraw from people. Data collection should include questions related to the psychological effects of the musculoskeletal disorder.

Patients may experience a tremendous amount of psychological stress resulting from the pain, loss of income, and withdrawal from friends and family. The nurse assesses the patient's ability to cope, asking what coping strategies have been used in the past for other life stressors. Support systems for the patient need to be identified, especially spiritual and social systems. As needed, consult the appropriate member of the health care team (social work, clergy, support groups) to ensure that the patient's psychosocial needs are being met.

DIAGNOSTIC TESTS

Diagnosis of musculoskeletal problems is assisted by laboratory tests and diagnostic imaging (including x-ray examinations and nonradiological tests). Specific tests for patients with connective tissue diseases are described in Chapter 44.

Laboratory Tests
Serum Calcium and Phosphorus

Bone disorders commonly cause changes in calcium and phosphorus (or phosphate) levels. When a person is healthy, calcium and phosphorus have an inverse relationship. This means that when serum calcium increases, serum phosphorus decreases, and vice versa. Some disorders, however, cause an increase in both values or a decrease in both values. Calcium and phosphorus levels are regulated by calcitonin from the thyroid gland and parathyroid hormone from the parathyroid glands. When these glands are not functioning properly, alterations in calcium and phosphorus levels can occur.

Serum calcium tends to decrease in patients with osteoporosis or in people who consume inadequate amounts of calcium in their diets. Serum calcium levels increase in patients with bone cancer, particularly those with metastatic disease.

Alkaline Phosphatase

Alkaline phosphatase (ALP) is an enzyme that increases when bone or liver tissue is damaged. In metabolic bone diseases and bone cancer, ALP increases to reflect osteoblast (bone-forming cell) activity.

Serum Muscle Enzymes

When muscle tissue is damaged, a number of serum enzymes are released into the bloodstream, including skeletal muscle creatine kinase (CK-MM [CK3]), aldolase (ALD), aspartate aminotransferase (AST), and lactate dehydrogenase (LDH). These enzymes increase in certain muscle diseases such as muscular dystrophy, polymyositis, and dermatomyositis.

Radiographic Tests
Standard X-rays

The skeleton and dense or inflamed tissues surrounding joints can be visualized on standard x-ray examination. Radiographs can readily detect bone abnormalities and deformities. An x-ray examination can determine bone density, texture, changes in alignment and bone relationship, erosion, swelling, and intactness. In addition, x-ray examinations can be useful in identifying certain soft tissue damage (e.g., ligaments and tendons) because of alterations in bone position and spacing.

Although there is no special nursing care associated with x-ray examinations, you should inform patients that they will have to lie still during the examination and that the x-ray table will be cold and hard.

Computed Tomography

Tomograms radiographs that produce a clearer image by focusing on a particular slice of bone or soft tissue, such as ligaments and tendons. Computed tomography (CT) is especially helpful for diagnosing problems of the joints or vertebral column (Fig. 43–5). It may be used with or without a contrast medium (similar to a dye), which is given orally or intravenously. If a contrast medium is used, check that the patient is not allergic to iodine or shellfish (which contains iodine) because some contrast mediums have an iodine base. Because of the possibility of allergic responses or nausea from the contrast medium, the patient should be given nothing by mouth (NPO) for at least 4 hours before the test.

Inform patients that they must lie completely still during the test and that they will be surrounded by the scanner during the test. Headphones are worn for communication with the technician and to listen to soothing music of the patient's choice. Reports of claustrophobia and annoying clicking sounds made by the scanner while it is rotating are common.

Arthrography

An x-ray examination of any synovial joint can be performed for patients with suspected joint trauma. The most common joints tested are the knee and shoulder. For this test, a contrast medium or air is injected into the joint to help visualize the soft tissues. Inform the patient that the test may be temporarily uncomfortable while the contrast medium is injected. After the procedure, the joint is usually swollen for several days. Ice and elevation help diminish swelling. Strenuous physical activity should be avoided for 12 to 24 hours.

Myelogram

During a myelogram, a contrast medium is injected into the subarachnoid space so that the spine and spinal cord can be visualized for disk herniation or other abnormalities. Inform patients that they may be positioned head down for a short period to allow the contrast medium to flow up to the level of the neck. After the procedure, assess the patient for headache (a possible complication of the procedure) or nausea. The patient must lie in bed with the head raised to a maximum 45-degree angle for at least 3 hours or as ordered by the physician.

Other Diagnostic Tests
Magnetic Resonance Imaging

Magnetic resonance imaging (MRI), with or without contrast media, is a commonly performed test to diagnose musculoskeletal problems, especially those involving soft tissue (Fig. 43–6). MRI uses electromagnets and is more accurate than CT for diagnosing many problems of the vertebral column. If the patient has had previous spinal surgery, a contrast medium is used.

The image is produced by interaction of magnetic fields and radio waves. For very large patients or those who are claustrophobic, the open MRI offers a comfortable

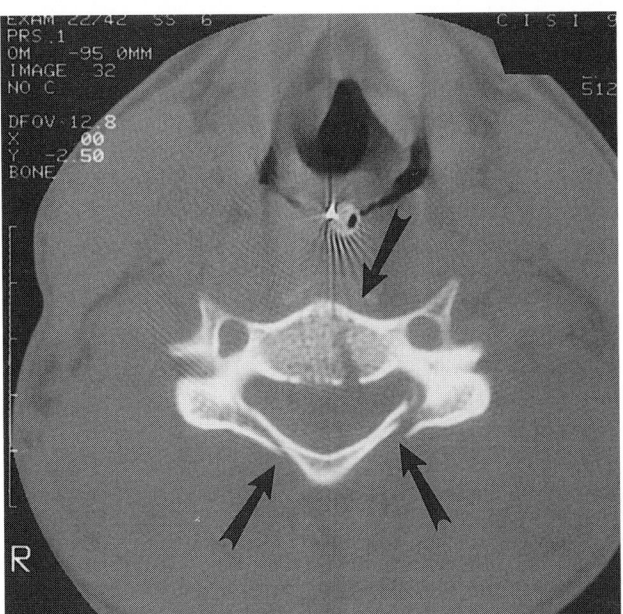

Figure 43-5 Computed tomography scan of fifth cervical vertebra showing a burst fracture of the vertebral body (top arrow) and both laminae (bottom arrows). (From McKinnis, LN: Fundamentals of Orthopedic Radiology. FA Davis, Philadelphia, 1997, p 23, with permission.)

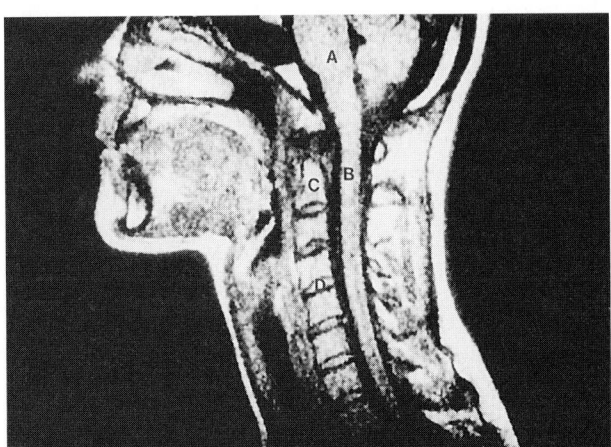

Figure 43-6 Magnetic resonance image of a normal cervical spine. (A) Cerebellum. (B) Spinal cord. (C) Marrow of C2 vertebral body. (D) C45 intervertebral disk. (From McKinis, LG: Fundamentals of Orthopedic Radiology. FA Davis, Philadelphia, 1997, p 26, with permission.)

alternative to the traditional machine (Patient Perspective Box 43–2).

The use of an electromagnet in the machine necessitates the removal of anything metal or with metal components from the patient's body. Pacemakers, surgical clips, and any other internally implanted metal device or apparatus are contraindications for MRI (Box 43–3).

Nuclear Medicine Scans

Several tests are performed using radioactive material to help visualize bone and other tissues. A bone scan allows visualization of the entire skeleton. It is used primarily for detecting metastatic bone disease, osteomyelitis (bone infection), and unexplained bone pain. The patient is injected with a radioisotope 2 to 3 hours before the scan. The radioisotope is attracted to bone and therefore travels to bone tissue. Reassure patients that this substance is not harmful.

For an accurate test, the patient must be able to lie still for 30 to 60 minutes during scanning. Patients who are elderly, restless, agitated, or in pain may therefore find this test

---

BOX 43–3   **Preparing the Patient for Magnetic Resonance Imaging**

- Is the patient pregnant?
- Does the patient have magnetic metal fragments or implants, such as an aneurysm clip?
- If the patient has an IV catheter, can it be converted to a saline lock temporarily?
- Is the patient claustrophobic?
- Does the patient have a pacemaker or electronic implant?
- Can the patient be without supplemental oxygen for an hour?
- Can the patient tolerate the supine position for 20 to 30 minutes?
- Can the patient lie still for 20 to 30 minutes?
- Does the patient need life support equipment?
- Can the patient communicate clearly and understand verbal communication?

Source: Ignatavicius, DD, Workman, ML, and Mishler, MA: Medical-Surgical Nursing: A Nursing Process Approach, ed 2, WB Saunders, Philadelphia, 1995, Chart 49–3, with permission.

---

uncomfortable. Sedatives or analgesics may have to be administered before or during the procedure. The physician looks for "hot spots," indicating areas where the radioactive substance is concentrated. These hot spots indicate abnormal bone metabolism, a sign of bone disease.

LEARNING TIPS

Hot spots are created because increased circulation occurs in abnormal bone areas, resulting in increased amounts of the radioactive substance being transported to the abnormal area.

Gallium/Thallium Scans

A gallium or thallium scan is similar to a bone scan but is more specific and sensitive as a diagnostic test. Gallium not only migrates to bone but also to brain and breast tissue and is therefore used to diagnose problems in these tissues as well.

Traditionally used for heart problems, thallium is now used for evaluation of bone cancers. Thallium is best for detecting osteosarcoma. Like the bone scan, these scans are not harmful to the patient.

Arthroscopy

An arthroscope is a tubular instrument that allows the surgeon to directly visualize a joint. The knee and shoulder are the joints most often evaluated. Because **arthroscopy** is an invasive procedure performed under local or light general anesthesia, the patient is treated as a surgical candidate.

BOX 43–2   *Patient Perspective*

Undergoing an MRI

I was told that the MRI scanner was small and could cause feelings of claustrophobia. When asked if I was claustrophobic, I said no and so was not offered an antianxiety medication before the procedure. I really am not claustrophobic, but I decided to keep my eyes closed during the test to make sure to prevent these feelings. As the table I was lying on moved into the scanner, my heart began beating fast. I shut my eyes and imagined myself walking on the beach, which is a favorite place for me. The cool air that was blowing in the machine I imagined to be the wind blowing. That air really is essential in keeping you cool and the claustrophobic feeling away. I had on headphones through which the music I had brought with me was playing. I focused on the music and sang along to myself as I "walked on the beach." I felt calmer as I did this. Through the headphones the technician kept me informed of how much longer the test would be and when the loud pounding noise would start. This really helped and I did not mind the noise. I knew that I had to be very still for the test, so I gave myself pep talks. "Okay, only 10 more minutes. Just lie still and this will be over soon."

Then, halfway through the MRI, I accidentally opened my eyes for an instant. "Uh, oh." I quickly shut them again with my heart racing and a feeling of panic rising. The wall of the MRI machine was only inches from my face. I quickly focused on the music again, told myself I could get through this, and focused on calming myself down by "going back to the beach." Before I knew it, the test was over and I was told that because I held so still it was a great test that went more quickly than usual. I sure was glad to hear that! My personal coping techniques really helped me through the MRI. Without them I would have panicked and been unable to complete the test. The information the nurse gave me helped me know what to expect during the procedure so I could prepare myself to cope with the test. Providing information to your patients on what to expect during the test, as well as coping methods to use, can help them successfully complete an MRI.

---

arthroscopy: arthro—joint + scopy—to examine

Arthroscopy is done in same-day surgery settings. The surgeon makes several small incisions and distends the joint with injected saline. The scope is inserted and the joint is visualized from different angles. The joint is moved through range of motion, so tears, defects, or other soft tissue damage can be repaired through the scope using special instrumentation. Depending on the extent of the surgery, a bulky or small dressing wrapped with an elastic bandage may be applied.

The nurse in the postanesthesia care unit (PACU) assesses the neurovascular status of the surgical foot and leg frequently. (See Table 43–3.) If the patient had a diagnostic arthroscopy and no surgical repair, the PACU nurse encourages the patient to exercise the leg, including straight-leg raises. A mild analgesic usually relieves pain, and the patient returns to regular activities in 24 to 48 hours. If a surgical repair was performed, the patient may have activity restriction and need a stronger analgesic, such as oxycodone with acetaminophen (Tylox, Percocet).

Although complications are not common, the nurse monitors and teaches the patient to watch for and report to the physician the following:

- Thrombophlebitis (blood clot and vein inflammation)
- Infection (fever or warmth, pain, redness, swelling at surgical site)
- Increased joint pain

If a repair was done during the surgery, the patient is seen by the physician in 1 week to check for complications and progress. The patient may need crutches for the first week to limit weight bearing, depending on the surgical procedure performed. Physical or occupational therapy may be ordered (Home Health Hints).

HOME HEALTH HINTS

- Patients are considered homebound if (1) they are bedbound or require the maximum assistance to ambulate while using a walker or to transfer; (2) they can ambulate with only moderate assistance while using a cane to negotiate uneven surfaces; or (3) they can leave home only for periods of relatively short duration or for need of medical treatment.

- When testing strength, extend two or three fingers and ask the patient to squeeze the fingers.

- Observe patients moving around a room or bed. If they are clumsy or have involuntary movement, make efforts during that visit and subsequent visits to protect them from potential injury. Research has shown that pain or fear of falling may prevent a patient from moving and functioning to maximum potential.

- Encourage patients to wear flat, sturdy, rubber-soled shoes to prevent slipping, tripping, or spraining an ankle.

- Use sand or cat box filler on icy steps to increase traction, preventing slips and falls.

- Patients who use walkers can get pressure ulcers on their palms. One way to relieve the pressure is to wear padded cycling gloves that leave the fingers free.

CRITICAL THINKING

Mrs. Jones

Mrs. Jones was walking down the street when, without warning, she suddenly fell to the ground with extreme pain in her left leg. She was taken to the hospital, where it was determined that the greater trochanter of her left femur was fractured.

1. What information should you obtain from Mrs. Jones?
2. What possible condition may be the cause of her fracture?
3. What tests may be performed to identify the condition creating her problem?

Answers at end of chapter.

CRITICAL THINKING

Mr. Allan

Mr. Allan, age 45, comes to emergency with extreme pain in his lower back. The pain radiates down his right buttock and down the back of his leg to his knee. He tells you that he hurt his back picking up a box in the warehouse where he works.

1. What other information should you obtain from Mr. Allan?
2. What is a probable cause of Mr. Allan's pain?
3. What tests, procedures, and treatments may be done for Mr. Allan's condition?
4. How might this injury impact Mr. Allan's life?
5. Mr. Allan is to receive morphine 10 mg by intermuscular injection. You have available morphine 15 mg/mL. How many milliliters will you give?

Answers at end of chapter.

Bone or Muscle Biopsy

Bone or muscle tissue can be surgically extracted for microscopic examination to confirm cancer, infection (bone biopsy), inflammation, or damage (muscle biopsy). Muscle can also be biopsied to diagnose malignant hyperthermia, a genetic disorder. (See Chapter 11). Two techniques are used to retrieve muscle tissue: a needle (closed) biopsy or an incisional (open) biopsy.

A closed biopsy can be performed in the patient's room or special procedures area. After local or general anesthesia, the physician inserts a long needle into the tissue for extraction of a sample.

The open biopsy is performed in the operating suite under general anesthesia. A small incision is made and a section of bone or muscle is removed. A sterile pressure dressing is applied because bone is highly vascular.

The nurse inspects the biopsy site for bleeding, swelling, and hematoma formation. Increased pain that is unresponsive to analgesic medication may indicate bleeding in the soft tissue. The area is not moved for 8 to 12 hours to prevent bleeding. Vital signs and neurovascular assessments are monitored. (See Table 43–3.)

Ultrasonography

Sound waves are used to detect osteomyelitis (bone infection), soft tissue disorders, traumatic joint injuries, and surgical hardware placement. For this noninvasive procedure, the technologist first applies a jellylike conducting substance over the area to be tested. A transducer is moved over the area while the ultrasound machine records the images. No special patient instructions are needed for this test.

Arthrocentesis

Arthrocentesis is a diagnostic or therapeutic procedure where synovial fluid is aspirated from a joint for analysis or to relieve pressure (pain) from effusion. Analysis of the synovial fluid aids in the diagnosis of noninflammatory conditions, septic arthritis, crystal detection, and **hemarthrosis** (blood in the joint cavity). Using aseptic technique, the physician provides local anesthetic and then uses a needle to aspirate the contents of the joint space. After the procedure

is complete, the physician can instill anti-inflammatory medications such as a corticosteroid. The site is covered with a sterile dressing. The main nursing intervention is to monitor the patient and teach the importance of observing for infection or hemarthrosis.

Nerve Conduction Studies

Electromyography (EMG) measures a muscle's electrical impulses. This aids in the diagnosis of muscle diseases or nerve damage. There are no special nursing actions related to this procedure other than informing the patient what will occur. Occasionally, slight discomfort occurs at the site where the study occurred. Warm compresses or mild analgesics can be offered for pain relief.

REVIEW QUESTIONS

1. Which of the following is the function of synovial fluid in joints?
   a. Exchange nutrients
   b. Prevent friction
   c. Absorb water
   d. Wear away rough surfaces

2. Which of the following assessments is included in neurovascular checks of the lower extremities?
   a. Radial pulses
   b. Checking for clubbing
   c. Femoral pulses
   d. Biceps reflex

3. A patient has undergone an arthroscopy. Two hours after the procedure his pedal pulses are diminished from the previous assessment. What should the nurse do?
   a. Take his vital signs.
   b. Notify the physician.
   c. Perform a neurovascular assessment.
   d. Change the dressing and rewrap the elastic wrap.

4. A patient has been diagnosed with a musculoskeletal disease that causes decreased bone density. Which assessment question is most appropriate by the nurse?
   a. "Do you have any broken bones?"
   b. "Has your doctor informed you to not exercise in case you break a bone?"
   c. "What forms of physical activity are you able to participate in?"
   d. "Do any of your spouse's relatives have problems with their bones?"

5. A patient is scheduled for an MRI of his pelvis. Which of the following would the nurse do if during data collection the nurse found out that he had had previous heart surgery?
   a. Ask the patient if he has any metal in his body.
   b. Order a chest x-ray examination to identify if he has any metal objects in his body.
   c. Cancel the MRI.
   d. Inform the physician.

Answers to CRITICAL THINKING

Mr. Smith

1. You should determine if Mr. Smith has any allergies, how and when the injury occurred, if he has had any previous surgeries, what medications he takes, his medical history, and any past problems with anesthesia (in Mr. Smith or his family).
2. You should assess his left leg compared with his right leg, including limb length, deformity, pain, loss of range of motion, edema, and ecchymosis, and perform neurovascular checks, including movement, sensation (numbness/tingling), presence of pulses, skin temperature, color, and capillary refill.

Mrs. Jones

1. You should assess Mrs. Smith's age, her diet (does she have a low-calcium or vitamin D–deficient diet?), what she was doing at the time of the break, whether anything like this has happened before, whether anything similar has happened to any of her relatives, her pain level, when she ate last, her medications, her medical history, whether she smokes, and whether she has any allergies.
2. There is a possibility that Mrs. Jones has osteoporosis and that she has experienced a pathological fracture from decreased bone density. This is a common occurrence in postmenopausal women.

3. X-ray examinations, bone scans, bone density tests, and laboratory tests such as serum calcium, phosphorus, acid phosphatase, thyroid, and vitamin D levels are tests that might be performed.

Mr. Allan

1. You should determine what occurred, when it occurred, how it happened, if the pain was immediate, where exactly the pain is, whether anything makes it better or worse, if any kind of treatment was started, what kind of job Mr. Allen has, whether he has had any back problems in the past, what medications he is on, and whether he has any allergies.
2. Mr. Allan may have ruptured a disk in his vertebral column, which is a common occurrence with improper lifting or trying to lift something too heavy.
3. Test might include x-ray examination, myelogram, MRI, bone scan, and CT scan. Conservative treatment will be attempted first but ultimately a diskectomy may be necessary.
4. Many back injuries result in lifelong chronic pain. Depending on the severity of the injury and the effectiveness of therapy, Mr. Allan may have to limit his physical and social activities. He may also have to find another type of employment.
5. $\frac{\text{Desired dose}}{\text{Dose on hand}} \times \text{vehicle} = \frac{10 \text{ mg}}{15 \text{ mg}} \times 1 \text{ mL} = 0.67 \text{ mL}$

44

NURSING CARE OF PATIENTS WITH MUSCULOSKELETAL AND CONNECTIVE TISSUE DISORDERS

Rod Kebicz and Donna D. Ignatavicius

KEY TERMS

arthritis (ar-**THRYE**-tis)

arthroplasty (AR-throw-**PLAS**-te)

avascular necrosis
(a-**VAS**-kue-lar ne-**KROW**-sis)

fasciotomy (fash-e-**OTT**-oh-me)

hemipelvectomy (hem-e-pell-**VEC**-toe-me)

hyperuricemia
(HIGH-per-yoor-a-**SEE**-me-ah)

osteomyelitis (AHS-tee-oh-my-**LIGHT**-tis)

osteosarcoma
(AHS-tee-oh-sar-**KOH**-mah)

polymyositis (PAH-lee-my-oh-**SIGH**-tis)

replantation (re-plan-**TAY**-shun)

scleroderma (SKLER-ah-**DER**-ma)

vasculitis (VAS-kue-**LIGH**-tis)

QUESTIONS TO GUIDE YOUR READING

1. How would you differentiate between the care for osteoarthritis and rheumatoid arthritis?

2. What are the pathophysiology of and treatment for gout?

3. Which nursing interventions are appropriate when caring for patients with systemic lupus erythematosus, progressive systemic sclerosis, and polymyositis?

4. What would you include when preparing a plan of care for the patient undergoing a total joint replacement?

5. What patient education would be included for a patient with a lower extremity amputation and prosthesis?

6. What are the signs and symptoms of fractures?

7. Which nursing interventions are appropriate when caring for a patient in a cast or traction?

8. What would you include in a plan of care for a patient with a fractured hip?

9. What are the common causes of osteomyelitis?

10. What are the risk factors for the development of osteoporosis?

11. What signs and symptoms may be seen in patients with Paget's disease?

12. What are the characteristics of and treatments for patients with primary and metastatic bone tumors?

► CONNECTIVE TISSUE DISORDERS

Connective tissue disorders comprise a group of more than 100 diseases in which the major signs and symptoms result from joint involvement. Some connective tissue diseases affect only one part of the body; others affect many body organs and systems. Several of the most common disorders are discussed here, including osteoarthritis, rheumatoid **arthritis,** gout, systemic lupus erythematosus, progressive systemic sclerosis, and **polymyositis.**

---

arthritis: arthr(on)—joint + itis—inflammation
polymyositis: poly—many + myo—muscle + itis—inflammation

Osteoarthritis

Osteoarthritis (OA) is the most common type of connective tissue disorder, affecting more than 20 million people in the United States. The term *arthritis* means inflammation of the joint, but OA is not a primary inflammatory process. Therefore some health care providers may refer to this disorder as *degenerative joint disease*. This term better reflects its pathophysiology.

Pathophysiology

Osteoarthritis occurs when the articular cartilage and bone ends of joints slowly deteriorate (Table 44–1). The joint space narrows, bone spurs develop, and the joint may become somewhat inflamed. The repair process is not able to overcome the rapid loss of cartilage and bone, eventually resulting in joint deformities, pain, and immobility, leading to the patient's functional decline. Weight-bearing joints (hips and knees), hands, and the vertebral column are most often affected (Fig. 44–1).

Causes and Types

The most common type of OA is primary (idiopathic) osteoarthritis. The cause of OA is unknown, but several risk factors have been identified. Aging, obesity, and physical activities that create mechanical stress on joints are major risks. Each of these factors cause prolonged or excessive "wear and tear" on synovial joints. The majority of people older than 60 years of age have some degree of symptomatic joint degeneration. Native-Americans are affected more often than other groups, but the reason for this is unknown.

Patients with secondary osteoarthritis develop their joint degeneration as a result of trauma, sepsis, congenital anomalies, certain metabolic diseases (such as Paget's disease), or systemic inflammatory connective tissue disorders such as rheumatoid arthritis.

Signs and Symptoms

The patient usually seeks medical attention when joint pain and stiffness become severe or the patient has problems

TABLE 44–1	COMPARISON OF OSTEOARTHRITIS AND RHEUMATOID ARTHRITIS	
	Osteoarthritis	**Rheumatoid Arthritis**
Pathophysiology	Articular cartilage and bone ends deteriorate	Inflammatory cells cause synovitis
	Joint is inflamed	Synovium become thick and fluid accumulates, causing swelling and pain
		Joint becomes deformed
Etiology	Primary (idiopathic):	Autoimmune disease
	–Cause unknown	Can occur at any age (including
	–Risk factors include age, obesity, activities causing joint stress	juvenile rheumatoid arthritis)
	Secondary:	Cause unknown
	–Causes include trauma, sepsis, congenital abnormalities, metabolic disorders (Paget's disease), rheumatoid arthritis	Familial history possible
	Joint pain and stiffness	Symptoms vary according to disease process
	Alterations in ADLs	Early symptoms:
	Pain increases with activity and decreases with rest	–Bilateral and symmetrical joint inflammation
	Nodes on joints of fingers (Heberden's nodes, Bouchard's nodes)	–Redness, warmth, swelling, stiffness, pain
		–Stiffness after resting (morning stiffness)
		–Activity decreases pain and stiffness
		–Low-grade fever, weakness, fatigue, anorexia (mild weight loss)
		–Organ system involvement
		Late symptoms:
		–Joint deformity
		–Secondary osteoporosis
Treatment	Medication:	Medication:
	–NSAIDs	–Salicylates
	–Acetaminophen	–NSAIDs
	–Muscle relaxants	–Gold treatment
	–Cox-2 inhibitors	–Methotrexate
	Balanced rest and exercise	–Prednisone
	Splinting of joint to promote rest	Heat and cold
	Heat and cold	Balanced rest and activity
	Diet for weight loss	Surgery for total joint replacement
	Complementary therapies	
	Surgery for total joint replacement	

ADLs = activities of daily living.

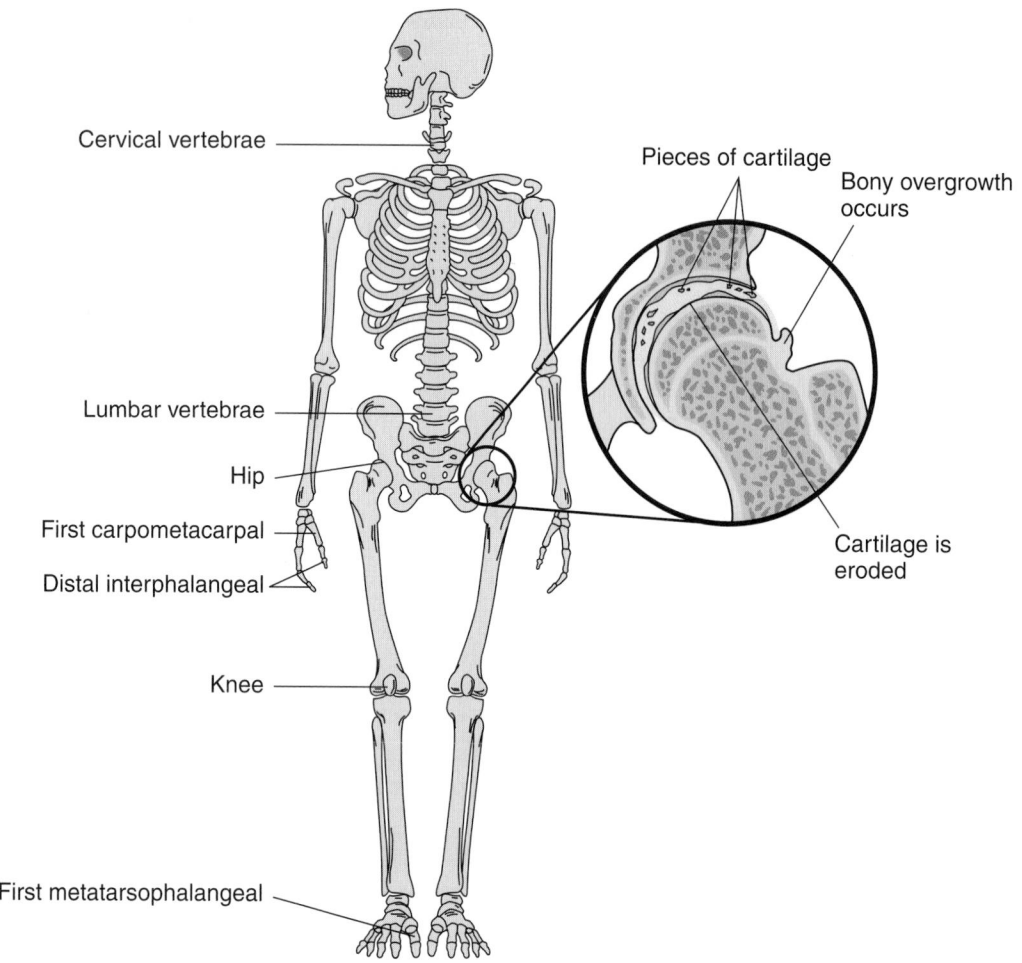

Cervical vertebrae

Pieces of cartilage

Bony overgrowth occurs

Lumbar vertebrae

Hip

First carpometacarpal

Distal interphalangeal

Knee

Cartilage is eroded

First metatarsophalangeal

Figure 44-1 Common joints affected by osteoarthritis and the changes that result in the joint.

with everyday activities. One or more joints may be affected, most commonly in the hands, hips, knees, spine, and feet. Joint pain intensifies after physical activity but lessens following rest. If the vertebral column is involved, the patient complains of radiating pain and muscle spasms in the extremity innervated by the area affected.

About half of patients with OA have bony nodes on the joints of their fingers, called Heberden's and Bouchard's nodes. Women tend to have them more often than men, and they may or may not be painful. The nodes have a familial tendency and are often a cosmetic concern to female patients.

Diagnostic Tests

X-ray examinations are useful in outlining joint structure and detecting bone changes. A computed tomography (CT) scan or magnetic resonance imaging (MRI) may be used to diagnose various joint involvement. Analysis of synovial fluid can aid in the diagnoses of OA while ruling out other pathological conditions of the joint.

Treatment

There is no curative therapy currently available for OA. Management of patients with OA centers on pain control, which is accomplished by drug therapy, other pain relief

CRITICAL THINKING

Mr. Dennis

Mr. Dennis is a 59-year-old overweight carpenter who visits his physician with complaints of knee and wrist pain. He has noticed that it is becoming increasingly difficult to climb a ladder or use a hammer. The physician suspects osteoarthritis.

1. What questions will the physician most likely ask Mr. Dennis?
2. What risk factors does he have?
3. What other signs and symptoms might he have?

Answers at end of chapter.

measures, or ultimately surgery. An interdisciplinary approach is needed to prevent decreased mobility and preserve joint function.

Synvisc is a newer therapy that is injected directly into osteoarthritic knees and acts like healthy, cushioning synovial fluid. Pain is relieved and flexibility restored when the knee joint is again lubricated and cushioned. For more information about this therapy, visit www.synvisc.com.

MEDICATION. Drug therapy, often used in combination with other therapies to reduce pain, is commonly used for patients with OA. The most typically used drugs are nonsteroidal anti-inflammatory drugs (NSAIDs) (Table 44–2).

TABLE 44-2 COMMON DRUGS USED TO TREAT CONNECTIVE TISSUE DISEASES: OSTEOARTHRITIS, RHEUMATOID ARTHRITIS, AND OTHERS

	Type	Drug: Generic (Trade)
NSAIDs	Nonselective cyclooxygenase (COX-1 and COX-2) inhibitors	Acetylsalicylic acid (aspirin)
		Diclofenac sodium (Voltaren)
		Diflunisal (Dolobid)
		Etodolac (Lodine; osteoarthritis only)
		Fenoprofen (Nalfon)
		Flurbiprofen (Ansaid)
		Ibuprofen (Motrin)
		Indomethacin (Indocin)
		Ketoprofen (Orudis)
		Naproxen (Naprosyn)
		Oxaprozin (Daypro)
		Piroxicam (Feldene)
		Sulindac (Clinoril)
		Tolmetin (Tolectin)
	Selective cyclooxygenase (COX-2) inhibitors	Celecoxib (Celebrex)
		Rofecoxib (Vioxx)
Corticosteroids	Intermediate-acting	Prednisone (Deltasone)
Disease-modifying antirheumatic drugs (DMARDs)	Gold Preparations	Auranofin (Ridaura)
		Aurothioglucose (Solganal)
	Immunosuppressives	Azathioprine (Imuran)
		Cyclophosphamide (Cytoxan)
		Etanercept (Enbrel)
		Leflunomide (Arava; rheumatoid arthritis only)
		Methotrexate (Mexate)
	Antimalarials	Chloroquine (Aralen)

These drugs have analgesic and anti-inflammatory effects but may cause side effects if not carefully monitored. Common side effects include gastrointestinal (GI) distress; bleeding tendencies, which can be severe; and sodium and fluid retention. Newer NSAIDs called COX-2 inhibitors are effective for short-term treatment of the pain of OA. The benefit of COX-2 inhibitors (e.g., celecoxib [Celebrex] and rofecoxib [Vioxx]) is minimal GI effects and decreased likelihood of bleeding problems. COX-2 inhibitors are indicated for those at high risk of developing GI problems. Elderly patients receiving NSAIDs on a routine basis should be carefully monitored for congestive heart failure or high blood pressure as a result of fluid retention.

REST AND EXERCISE. Joint pain from OA tends to decrease with rest; therefore pain is less severe in the morning. Activities should be scheduled at this time. A severely inflamed joint may be splinted by the occupational or physical therapist to promote rest to a selected joint. However, rest must be balanced with exercise to prevent muscle atrophy from disuse. Exercise has been identified as a means to maintain general health, range of motion, and muscle strength, while decreasing anxiety and depression. To minimize muscle atrophy, patients should be encouraged to perform exercises to strengthen their quadriceps if they have OA of the knee.

Joints should always be placed in their functional position—that is, a position that does not lead to contractures. For example, only a small pillow should be placed under the head when sleeping to prevent excessive neck flexion.

HEAT AND COLD. The patient with OA usually prefers heat therapy unless the joint is acutely inflamed. Hot packs, warm compresses, warm showers, moist heating pads, and paraffin dips provide sources of heat for the patient. Cold therapy minimizes inflammation while altering cutaneous pain receptors, thereby decreasing pain. Cold packs should be applied for no longer than 20 minutes at a time.

DIET. The obese or overweight patient benefits from losing weight to decrease joint stress, on weight-bearing joints, thereby reducing pain. If the patient is on medications that can alter fluid volumes (corticosteroids), a diet low in sodium may be appropriate.

COMPLEMENTARY THERAPIES. The popularity of complementary therapies to reduce pain and stress has grown tremendously. Imagery, music therapy, acupressure, acupuncture, and other holistic modalities that foster the mind–body–spirit connection work well for many people. Homeopathic therapies such as glucosamine and chondroitin have been suggested to improve OA. Recent studies have demonstrated that these two therapies are effective in OA therapy; however, further research must be conducted before they become an accepted and recommended therapy for OA.

SURGERY. If the patient's pain is not successfully managed, a total joint replacement (TJR) may be indicated. A TJR is the most common type of **arthroplasty.** (See Musculoskeletal Surgery.)

arthroplasty: arthro—joint + plasty—creation of

Nursing Process

ASSESSMENT. The patient's complaint of pain is assessed and the joints observed for signs of inflammation or deformity. Also assessed are function, alterations in activities of daily living (ADLs), and mobility. (See Chapter 43.)

NURSING DIAGNOSIS. The primary nursing diagnoses for the patient with OA include but are not limited to the following:

- Activity intolerance
- Chronic pain
- Chronic sorrow
- Disturbed body image
- Impaired physical mobility
- Ineffective health maintenance
- Risk for spiritual distress
- Deficient self-care

PLANNING. The goal of all the disciplines for all therapies related to joint conditions is to decrease suffering, increase quality of life, and decrease morbidity, all in a cost-effective manner. Collaboration with the interdisciplinary health team, especially the physician and physical or occupational therapist, is done to meet the goals of pain management, to improve mobility, and to minimize further joint destruction. For patients who need weight reduction, the dietitian is an important resource. The social worker, discharge planner, or case manager can help make arrangements for home care if needed. A home care aide may be necessary to help the patient with ADLs and mobility.

IMPLEMENTATION. Interventions focus on pain management, joint preservation, enhanced quality of life, and education.

EVALUATION. After pain management modalities are implemented, it is especially important to note the patient's response to these interventions. The best indicator of whether the plan of care was successful for the patient with OA is the patient's report of pain. If pain subsides or decreases, the plan was successful. If the pain worsens, surgery may be needed. Another important outcome is the patient's satisfaction with his or her ability to perform ADLs.

Patient Education

A vital function of each member of the health care team is health teaching. The patient with OA is seldom admitted to the hospital for treatment of OA unless surgery is scheduled. However, many patients with OA are admitted for other reasons, and their arthritis needs must also be considered in the comprehensive plan of care. Most patients residing in nursing homes also have OA, which can affect their participation in recreational activities, as well as their ADLs.

In any setting, including the home, patients can be taught ways to protect their joints and conserve energy. Nurses need to teach patients and their families how to promote health. Box 44–1 lists tips for joint protection and en-

---

> **BOX 44-1** **Energy Conservation for the Patient with Arthritis**
>
> - Balance activity with rest. Take one or two naps each day.
> - Pace yourself; do not plan too much for one day.
> - Set priorities. Determine which activities are most important, and do them first.
> - Delegate responsibility and tasks to your family and friends.
> - Plan ahead to prevent last-minute rushing and stress.
> - Learn your own activity tolerance and do not exceed it.
>
> *Source:* Ignatavicius, DD, Workman, ML, and Mishler, MA: Medical-Surgical Nursing: A Nursing Process Approach, ed 3. WB Saunders, Philadelphia, 1998, with permission.

ergy conservation. For information on education al materials and self-help courses, visit the Arthritis Foundation at www.arthritis.org.

Rheumatoid Arthritis

Rheumatoid arthritis (RA) is a chronic, progressive, systemic inflammatory disease that destroys synovial joints and other connective tissues, including major organs. It affects women three times more often than men and Native-Americans more often than other ethnic groups. Rheumatoid arthritis can occur at any age; when it occurs in children it is called juvenile RA (JRA). The peak onset of RA is 30 to 60 years of age, and it affects 1 to 3 percent of the population in the United States. The etiology of RA is still unknown; however, there are indications that genetic predisposition and the environment play a role in triggering its development.

Pathophysiology

Inflammatory cells and chemicals cause **synovitis,** an inflammation of the synovium (the lining of the joint capsule). As the inflammation progresses, the synovium becomes thick and fluid accumulation causes joint swelling and pain. A destructive pannus (new synovial tissue growth infiltrated with inflammatory cells) erodes the joint cartilage and eventually destroys the bone within the joint (Fig. 44–2). Ultimately the pannus is converted to bony tissue, resulting in loss of mobility. Joint deformity and bone loss are common in late RA. (See Table 44–1.)

Synovial joints are not the only connective tissues involved in RA. Any connective tissue may be affected, including blood vessels, nerves, kidneys, pericardium, lungs, and subcutaneous tissue. The result of body system involvement is malfunction or failure of the organ or system. Death can occur if the disease does not respond to treatment.

Many patients experience spontaneous remissions and exacerbations (flare-ups) of RA. The symptoms of the disease may disappear without treatment for months or years. Then the disease may exacerbate just as unpredictably. Exacerbations usually occur when the patient experiences physical or emotional stress, such as surgery or infection.

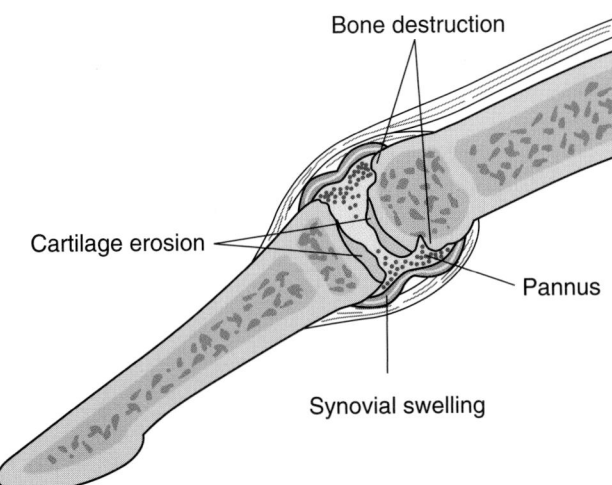

Figure 44-2 Rheumatoid arthritis.

Etiology

The exact cause of RA is unknown. An autoimmune response occurs that affects the synovial membrane of the joints; it is unknown what triggers the initial response. Antibodies (called rheumatoid factor) are often found in patients with RA. It is suggested that these antibodies join with other antibodies and form antibody complexes. These complexes lodge in synovium and other connective tissues, causing local and systemic inflammation, and may be responsible for the destructive changes of RA in body tissues.

The origin of the rheumatoid factor is not clear, but a genetic predisposition is likely. RA affects people with a family history of the disease two to three times more often than the rest of the population.

Signs and Symptoms

Signs and symptoms vary because the disease progresses in different patterns and rates from person to person. In general the signs and symptoms can be divided into early and late manifestations (Table 44–3).

The typical pattern of joint inflammation is bilateral and symmetrical. The disease usually begins in the upper extremities and progresses to other joints over many years (Fig. 44–3). Affected joints are slightly reddened, warm, swollen, stiff, and painful. The patient with RA often has morning stiffness lasting for up to an hour, and those with severe disease may complain of stiffness all day. Generally activity decreases pain and stiffness.

Because of the systemic nature of RA, the patient may have a low-grade fever, malaise, depression, lymphadenopathy, weakness, fatigue, anorexia, and weight loss. As the disease worsens, major organs or body systems are affected. Joint deformities occur as a late symptom, and secondary osteoporosis (bone loss) can lead to fractures.

Several associated syndromes are seen in some patients with rheumatoid arthritis. For example, Sjögren's syndrome is an inflammation of tear ducts (causing dry eyes) and salivary glands (causing dry mouth). Felty's syndrome is less common and is characterized by an enlarged liver and spleen and leukopenia (decreased white blood cell count).

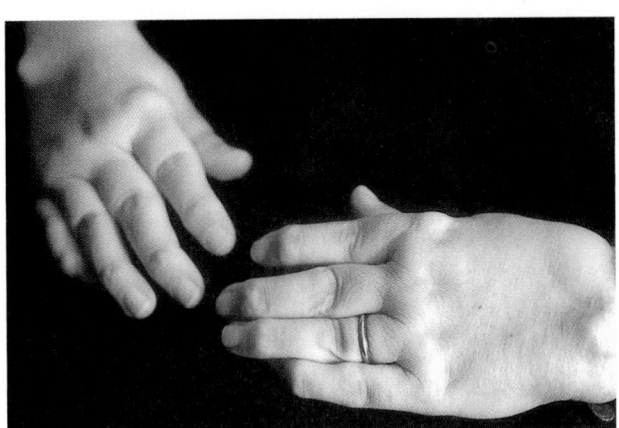

Figure 44-3 Joint abnormalities in hands of patient with rheumatoid arthritis.

TABLE 44-3	**KEY FEATURES OF RHEUMATOID ARTHRITIS**	
	Early Manifestations	**Late Manifestations**
Joint	Inflammation	Deformities (such as swan neck or ulnar deviation)
		Moderate to severe pain and morning stiffness
Systemic	Low-grade fever	Osteoporosis
	Fatigue	Severe fatigue
	Weakness	Anemia
	Anorexia	Weight loss
	Paresthesias	Subcutaneous nodules
		Peripheral neuropathy
		Vasculitis
		Pericarditis
		Fibrotic lung disease
		Sjögren's syndrome
		Renal disease

Source: Ignatavicius, DD, Workman, ML, and Mishler, MA: Medical-Surgical Nursing: A Nursing Process Approach, ed 3. WB Saunders, Philadelphia, 1998, with permission.

Diagnosis

No specific diagnostic test confirms RA, but several laboratory tests help support the diagnosis. An increase in white blood cells and platelets is typical, unless the patient has Felty's syndrome. A group of immunologic tests are usually performed, and typical findings for patients with RA include the following:

- Presence of rheumatoid factor (RF) in serum
- Decreased red blood cell (RBC) count
- Decreased C4 complement
- Increased erythrocyte sedimentation rate (ESR)
- Positive antinuclear antibody (ANA) test
- Positive C-reactive protein (CRP) test

RF can indicate the aggressiveness of the disease. However, it is not specific to RA and can also be found in systemic lupus erythematosus, connective tissue disease, and myositis. The ESR is also obtained to evaluate the effectiveness of treatment. If the disease responds to treatment, the ESR decreases. The higher the ESR, the more active the disease process.

L E A R N I N G   TIPS

The ESR is a general screening test for inflammation. It measures the amount of time it takes for RBCs to settle to the bottom of a test tube. In the presence of inflammation, RBCs settle faster in the tube. Therefore the ESR increases with the presence of inflammation.

X-ray examination and MRI detect joint damage and bone loss, especially in the vertebral column. A bone or joint scan assesses the extent of joint involvement throughout the body. For some patients, an **arthrocentesis** may be performed; the synovial fluid is cloudy, milky, or dark yellow with inflammatory cells present.

Medical Treatment

Like patients with osteoarthritis, patients with RA experience chronic joint pain. Pain can interfere with mobility or the ability to perform ADLs. Drug therapy is often needed to relieve or reduce pain as well as to slow the progression of the disease.

MEDICATION. Treatment for RA includes disease-modifying antirheumatic drugs (DMARDs), which can prevent joint destruction, deformity, and disability with early single or combination drug use; NSAIDs; and corticosteroids. (See Table 44–2.) Newer DMARDs such as leflunomide (Arava) and etanercept (Enbrel) are used to slow the progression of RA. Leflunomide taken orally has antiproliferative and anti-inflammatory properties. Etanercept inhibits tumor necrosis factor, which is involved in the inflammatory process, and is given subcutaneously twice a week. Low-dose methotrexate (MTX) or gold therapy is given to induce disease remission. NSAIDs such as aspirin, ibupro-

fen, and COX-2 inhibitors are prescribed for pain and stiffness, although they do not slow the disease process. Prednisone is a corticosteroid used to induce disease remission. Many of these medications have potentially serious side effects that must be monitored carefully.

Complementary therapies that may help decrease inflammation or pain include capsaicin cream, fish oil, magnetic therapy, and antioxidants such as vitamin C, vitamin E, and betacarotene. (See Chapter 4.)

HEAT AND COLD. Heat applications or hot showers help decrease joint stiffness and make exercise easier for the patient. For acutely inflamed, or "hot," joints, cold applications are preferred. As for patients with osteoarthritis, a program that balances rest and exercise later in the day is most beneficial for the patient.

SURGERY. If nonsurgical approaches are not effective in relieving arthritic pain, the patient may have a total joint replacement (discussed later). In general, patients with RA who have surgery are not as successful when compared with patients with osteoarthritis. The presence of a systemic disease predisposes patients with RA to more postoperative complications.

Nursing Process

ASSESSMENT. A thorough history and physical assessment are needed for the patient with RA because the disease can involve every system of the body. In addition to assessing physical signs and symptoms, assess the patient for psychosocial, functional, and vocational needs.

Body changes resulting from joint deformity may lead to poor self-esteem and body image. Patients may feel helpless as they lose control over a disease process that is affecting many parts of their body. Chronic pain and suffering negatively affect the quality of one's life, and the patient may experience depression.

After having the disease for approximately 15 years, fewer than half of RA patients are totally independent in their ADLs. These limitations may place a burden on family members, who must be included in the care of the patient with RA.

Many patients with the disease are young or middle aged. RA can impair their ability to work, depending on the type of job they have. Chronic fatigue requires frequent rest periods that may not fit into a day's work schedule. The health care team assesses the patient's work skills to determine the need for changes in the workplace or a need to train for a new type of work.

NURSING DIAGNOSIS. The nursing diagnoses depend on the severity of the disorder and whether or not organs or systems are involved. All patients with RA typically experience but are not limited to the following:

- Acute pain
- Disturbed body image
- Fatigue
- Deficient self-care
- Impaired physical mobility

PLANNING. The plan of care for the patient with RA must be interdisciplinary, involving many members of the health team. For most patients, the physical therapist, occupational therapist, and social worker or case manager are vital in successfully meeting patient outcomes and improving quality of life. Family and significant others should be an integral part of the planning process.

IMPLEMENTATION. Each patient's plan of care is individualized. A major intervention is patient teaching. In addition, the interdisciplinary plan of care is carried out.

EVALUATION. The expected outcomes for the patient are relief of pain, disease remission, and improved ability to carry out ADLs. Medications are adjusted to control pain as well as to decrease the inflammatory process. ESR is used as an indicator of how well the treatment plan worked because its value decreases as inflammation subsides.

CRITICAL THINKING

Mrs. Summers

Mrs. Summers is a 48-year-old nurse who has had upper extremity joint pain and swelling for about 4 years. She was recently diagnosed with Rheumatoid arthritis (RA) but has no systemic involvement other than extreme fatigue at this time. She is concerned that she will have to give up providing direct patient care on a busy medical unit in the local hospital.

1. What questions might you ask her at this time about her illness?
2. What should you teach her about pain management?

Answers at end of chapter.

Patient Education

The patient with RA needs extensive patient education regarding the disease process, medication management, and the comprehensive plan of care. Many fads and myths published in popular tabloids are available, and some publicized "cures" can actually be harmful to the patient.

In collaboration with health team members, help the patient plan a daily schedule that balances rest and exercise. (See Box 44–1.) Child care responsibilities and other day-to-day activities need to be scheduled. A vocational counselor may be necessary for job training if the patient needs to pursue a different occupation. Patients who are unable to work may be able to qualify for disability benefits through the Social Security program.

Inform the patient about community resources. For example, the local chapter of the Arthritis Foundation provides support groups, information, and other resources for patients with RA and other types of connective tissue disorders. (See the Web Link in the Osteoarthritis section of this chapter.)

Gout

Gout (sometimes referred to as gouty arthritis) is an easily treated systemic connective tissue disorder. Men, especially those middle aged and older, are affected more than women. Patients with gout are seldom hospitalized for their disease.

Pathophysiology

Uric acid is a waste product resulting from the breakdown of proteins (purines) in the body. Urate crystals, formed because of excessive uric acid (**hyperuricemia**) build up and are deposited in joints and other connective tissues, causing severe inflammation. When an "attack" of gout occurs, the patient unexpectedly has severe pain and inflammation in one or more small joints, usually the great toe. The inflammation may resolve in several days with or without treatment. Months or years may pass between attacks, and the patient may have no signs or symptoms of joint inflammation between episodes. Urate deposits may appear under the skin (tophi) (Fig. 44–4) or in the kidneys or urinary system, causing stone (calculi) formation. (See Chapter 35.)

Causes and Types

The causes and types of gout are well known. Primary gout is the most common and is caused by an inherited problem with purine metabolism. Uric acid production is greater than the kidneys' ability to excrete it. Therefore the amount of uric acid in the blood increases. About 25 percent of patients have a family history of primary gout. Acute attacks of gout may be triggered by stress, alcohol, illness, trauma, dieting, or certain medications.

Patients with secondary gout also experience hyperuricemia, but the increase is the result of another health problem, such as renal insufficiency, or medications, such as diuretic therapy and certain chemotherapeutic agents.

Signs and Symptoms

ACUTE GOUT. Patients with acute gout have one or more severely inflamed joints, usually small joints, often in the joint of the great toe. The joint is swollen, red, hot, and usually too painful to be touched.

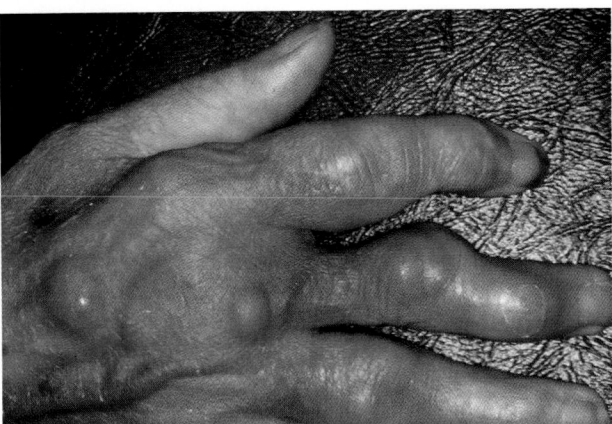

Figure 44-4 Gout: subcutaneous nontender lesions near joints. (From Goldsmith, LA, et al: Adult and Pediatric Dermatology. FA Davis, Philadelphia, 1997, p 405, with permission.)

---

hyperuricemia: hyper—excessive + uric—uric acid + emia—in blood

CHRONIC GOUT. Patients with chronic gout may not have obvious signs and symptoms. Tophi, small urate deposits under the skin, are not commonly seen today because management of patients with gout has improved. If they are present, they tend to appear most often on the outer ear. Renal stones develop in about 20 percent of patients with gout. Various diagnostic tests may be needed to determine stone formation.

Diagnostic Tests

Diagnosis of gout is based on the serum uric acid level. The level at which uric acid levels become pathological are controversial because of normal variations within a patient, as well as different normal range values for men, women, and children. Any result above the maximum normal level of 7.5 mg/dL should be further evaluated. Joint fluid aspiration analysis can identify uric acid crystals in the synovial fluid, further confirming the diagnosis of gout. Urinary uric acid levels may also be assessed. To determine if kidney damage has occurred, renal function tests such as blood urea nitrogen (BUN) and serum creatinine are evaluated. Radiographs of the kidney determine the presence of calculi.

Treatment

MEDICATION. The treatment of secondary gout is management or removal of the underlying cause. Drug therapy is the first-line treatment for primary gout. When the patient has an acute gout episode, the physician usually prescribes either colchicine or an NSAID, which interferes with the white blood cell (WBC) inflammatory response to urate crystals. The patient usual takes these medications until joint inflammation subsides.

Uricosuric agents (medications used to decrease uric acid) are the drug of choice when trying to decrease serum levels. Allopurinol (Zyloprim) is the preferred drug for chronic gout. Allopurinol decreases uric acid production, necessitating several weeks of therapy before the medication becomes effective. The patient must take it every day to keep the uric acid level within the normal range. Probenecid (Benemid) may also be used temporarily to increase renal excretion of uric acid. The patient's serum uric acid level is monitored periodically.

DIET. There is no special therapeutic diet for patients with gout; however, certain foods should be avoided or consumed in moderation (Box 44–2). The patient should be instructed to avoid all forms of aspirin and diuretics because

---

BOX 44-2   **Health Promotion for Patients with Gout**

- Avoid high-purine (protein) foods, such as organ meats, shellfish, and oily fish (e.g., sardines).
- Avoid alcohol.
- Drink plenty of fluids, especially water.
- Avoid all forms of aspirin and drugs containing aspirin.
- Avoid diuretics.
- Avoid excessive physical or emotional stress.

---

they can trigger an attack. Increasing daily fluid intake is also important to help prevent kidney stones.

Systemic Lupus Erythematosus

The word *lupus* comes from the Latin word for wolf and refers to the masklike facial rash that patients with lupus may develop. The rash is red, and thus the word *erythematosus*, meaning reddened, was added to describe the disease.

Most patients with lupus have the systemic type, but a small percentage have the type that affects only the skin, a condition called discoid lupus erythematosus. Discoid lupus is not life threatening; systemic lupus erythematosus (SLE) can be life threatening because it is a progressive, systemic inflammatory disease that can cause major body organ and system failure. Although this definition seems similar to the definition of RA, one distinct difference exists. Patients with SLE typically have more body organ development earlier in their disease than do patients with RA.

Pathophysiology

SLE is an autoimmune disease characterized by spontaneous remissions and exacerbations. Production of abnormal antibodies (antinuclear antibodies [ANA]), immune complex formation, and complement system activation results in autoimmune effects on the patient's healthy connective tissue. Many of the manifestations result from recurring injuries to the patient's vascular system. The immune complexes that result lodge in the blood and organs, leading to inflammation, damage, and possibly death. The cause of SLE is unknown, but the disorder tends to occur in families. Identified chromosomal markers indicate a genetic link.

The mortality rate for patients with the disease has improved greatly during the past 30 years. The leading causes of death are kidney failure, heart failure, and central nervous system involvement. Lupus most often affects women between ages 15 and 40, at a rate 8 to 10 times more often than for men. African-American women are at a greater risk for the disease than women of other racial or ethnic backgrounds; however, there is also an increased risk in Asian and Chinese populations.

Signs and Symptoms

Unfortunately there is no classic textbook description of patients with SLE. Some patients have a very mild form of the disease in which the skin and joints are affected. Others have devastating signs and symptoms when the disease affects multiple body systems at the same time.

The classic feature of lupus is the characteristic butterfly rash, although only half of patients develop it. It is a raised, reddened rash over the bridge of the nose that extends to both cheeks (Fig. 44–5). The rash is usually dry and may itch. It commonly is photosensitive, tends to worsen during an exacerbation, and can be triggered by exposure to sun or ultraviolet light or by physical stressors, such as pregnancy or infection. Instead of the butterfly rash, some patients have discoid (coinlike) skin lesions on other parts of the

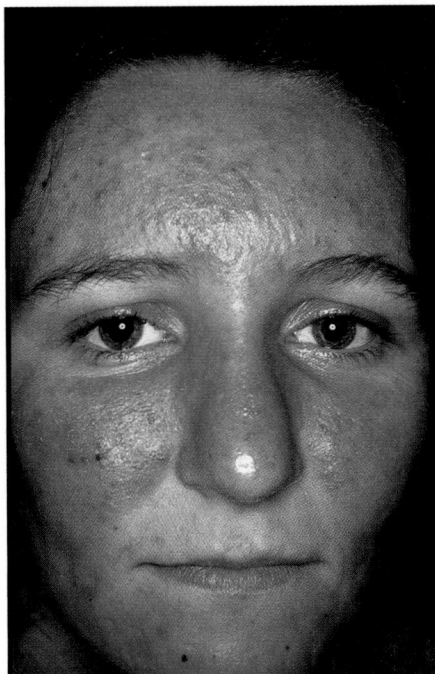

Figure 44-5 Lupus erythematosus: red papules and plaques in butterfly pattern on face. (From Goldsmith, LA, et al: Adult and Pediatric Dermatology. FA Davis, Philadelphia, 1997, p 230, with permission.)

TABLE 44-4	KEY FEATURES OF SYSTEMIC LUPUS ERYTHEMATOSUS (SLE) AND PROGRESSIVE SYSTEMIC SCLEROSIS (PSS)
SLE	**PSS**
Skin Manifestations	
• Inflamed, red rash	• Inflammation
• Discoid lesions	• Fibrosis
	• Sclerosis
	• Edema
Renal Manifestations	
• Nephritis	• Renal failure
Cardiovascular Manifestations	
• Pericarditis	• Myocardial fibrosis
• Raynaud's phenomenon	• Raynaud's phenomenon
Pulmonary Manifestations	
• Pleural effusions	• Interstitial fibrosis
Neurological Manifestations	
• CNS lupus	• Not common
Gastrointestinal Manifestations	
• Abdominal pain	• Esophagitis
	• Ulcers
Musculoskeletal Manifestations	
• Joint inflammation	• Joint inflammation
• Myositis	• Myositis
Other Manifestations	
• Fever	• Fever
• Fatigue	• Fatigue
• Anorexia	• Anorexia
• Vasculitis	• Vasculitis

Source: Ignatavicius, DD, Workman, ML, and Mishler, MA: Medical-Surgical Nursing: A Nursing Process Approach, ed 3. WB Saunders, Philadelphia, 1998, with permission.
CNS = central nervous system.

body. During flare-ups, a fever develops. Fatigue, arthralgia, myalgia, malaise, weight loss, mucosal ulcers, and alopecia are other possible signs and symptoms of SLE. Table 44–4 lists other possible signs and symptoms of SLE.

Diagnostic Tests

Skin lesions can be biopsied and examined microscopically for signs of inflammation. Patients with suspected SLE are evaluated using the same immunologically based laboratory tests that are used to assess patients with rheumatoid arthritis. These tests include ESR (to detect systemic inflammation) and ANA titers (to detect the presence of abnormal antibodies, sometimes called LE cells). A new blood test has been developed that aids in the diagnosis of SLE. Systemic lupus erythematosus patients make antibodies against serine/arginine rich (SR) proteins (which are important in cell division). Seventy percent of SLE patients react positively to the SR proteins. This new diagnostic tool should help in identifying those SLE patients who do not produce some of the other antibodies looked for in the diagnosis of SLE. Although no laboratory test confirms a diagnosis of lupus, the results of the immunologic tests may support the diagnosis.

Treatment

Treatment of SLE focuses on decreasing inflammation and preventing life-threatening organ damage. At present, the therapy of choice includes medications to treat the symptoms or the body systems affected. Research is ongoing regarding the possible cause of SLE. Researchers have found the general location of a gene that is believed to predispose a person to lupus. Identifying the genetic cause of lupus will enable researchers to develop new methods of therapy, including gene therapy.

MEDICATION. Depending on the area of the body affected, medications are prescribed according to the patient's needs. NSAIDs, acetaminophen, corticosteroids, antimalarials, and immunomodulating drugs may all be part of the medication regimen. Topical cortisone preparations may help reduce skin inflammation and promote fading of skin lesions. For some patients with discoid or systemic lupus, the antimalarial drug chloroquine (Aralen) may be prescribed.

Patients who experience joint inflammation are usually placed on a NSAID. Patients with organ or major body system involvement are given more potent drugs that suppress the immune process, including oral steroids such as prednisone or chemotherapeutic agents such as azathioprine (Imuran). These drugs have serious side effects, and patients receiving them are monitored very carefully. In addition to monitoring for a variety of side effects, patients must be taught to avoid people with infections because they are immunocompromised while taking any of these medications.

Nursing Process

ASSESSMENT. Assess the patient to determine the extent and severity of signs and symptoms, such as pain, fatigue, skin lesions, and fever. The plan of care for the patient is individualized because every patient who has SLE is unique.

NURSING DIAGNOSIS. The major nursing diagnoses related to SLE include but are not limited to the following:

- Acute pain
- Chronic sorrow
- Fatigue
- Disturbed body image
- Ineffective coping
- Ineffective health maintenance
- Powerlessness
- Risk for impaired skin integrity
- Deficient self-care

PLANNING. Collaborate with other health care team members as needed to promote or maintain function. For joint involvement, physical therapy is beneficial for exercise. Identify coping strategies that the patient has used in the past and wishes to use again. Young patients are often concerned about issues related to sexuality, pregnancy, and child rearing. These lifestyle concerns must be addressed with the appropriate health care professional.

IMPLEMENTATION. One of the most important nursing roles is coordination of care and patient education. The health of the patient with SLE can vary from seemingly healthy to severely ill in a critical care unit. These extremes can cause anxiety and concern about whether the patient may die from disease complications. Emotional support is offered to both the patient and family. Ensuring that the patient's medication regimen is administered appropriately and monitoring for side effects and potential complications of therapy are important nursing interventions.

EVALUATION. The expected outcome of care is that the patient can function daily without severe pain or fatigue and can avoid exacerbations of the disease.

Patient Education

Teach the patient about skin care and ways to prevent disease exacerbations. The patient should avoid prolonged exposure to sunlight or other forms of ultraviolet light, which can precipitate a flare-up of the illness. In addition, teach the patient how to care for the skin using mild soap and nondrying substances. Box 44–3 lists guidelines for skin protection and care for the patient with lupus.

It is also important to teach preventative measures to reduce exacerbations. Exercise can prevent muscle weakness and fatigue. The patient should be encouraged to be immunized against specific infections. Methods of stress reduction should be identified and utilized.

The Arthritis Foundation and the Lupus Foundation are national organizations that can provide information, assistance, and community support groups for patients diagnosed with lupus. For more information visit the Lupus Foundation of America at www.lupus.org.

| BOX 44–3 | **Skin Protection and Care for Patients with Lupus Erythematosus** |

- Cleanse skin with a mild soap.
- Dry skin thoroughly by patting rather than rubbing.
- Apply lotion liberally to dry skin areas.
- Avoid powder and other drying agents, such as rubbing alcohol.
- Use cosmetics that contain moisturizers.
- Avoid direct sunlight and any other type of ultraviolet lighting (including tanning beds).
- Wear a large-brimmed hat, long sleeves, and long pants when in the sun.
- Use a sun-blocking agent with at least a sunburn protection factor (SPF) of 30.
- Inspect skin daily for open areas and rashes.

Source: Ignatavicius, DD, Workman, ML, and Mishler, MA: Medical-Surgical Nursing: A Nursing Process Approach, ed 3. WB Saunders, Philadelphia, 1998, with permission.

Scleroderma

Scleroderma is similar to SLE in that it can affect multiple organs and other connective tissues. Scleroderma is not as common as SLE but has a higher mortality rate. There are two types of scleroderma: localized and systemic. As the name implies, systemic scleroderma can affect any part of the body. Among other names for systemic scleroderma are diffuse and progressive systemic scleroderma (PSS).

Pathophysiology

Scleroderma is characterized by inflammation that ultimately develops into fibrosis (scarring) and then sclerosis (hardening) of tissues. The disease is an autoimmune response to the body's normal tissues. Like some of the other systemic connective tissue diseases, abnormal antibodies damage healthy tissue, resulting in inflammation, which then triggers overproduction of collagen, which is deposited in the skin.

PSS affects women three to four times more often than men, usually between ages 30 and 50. The disease tends to progress rapidly and does not respond well to treatment. Spontaneous remissions and exacerbations can occur.

Signs and Symptoms

When discussing the signs and symptoms of scleroderma, it is appropriate to discuss PSS because PSS can affect any body system. Although arthritis and fatigue are commonly seen, the most obvious sign of PSS is scleroderma, which is manifested at first by pitting edema, starting in the upper extremities. The skin is taut, shiny, and without wrinkles. If occurring as part of PSS, the swelling is replaced by tightening, hardening, and thickening of skin tissue. The skin then loses its elasticity, range of motion is decreased, and skin ulcers may appear. As the disease progresses, the patient loses range of motion and becomes contracted.

---

scleroderma: sclero—hardening + derma—skin

The same pathophysiological process affects certain body systems, especially the kidneys, lungs, heart, and gastrointestinal tract. If any of these systems is affected, the corresponding signs and symptoms are present. For example, gastrointestinal tract involvement usually manifests as esophagitis, dysphagia (difficulty swallowing), and decreased intestinal peristalsis caused by decreased smooth muscle elasticity.

The prognosis is thought to be worse when the patient has CREST syndrome, a group of signs and symptoms occurring at the same time:

- **C**alcinosis (calcium deposits)
- **R**aynaud's phenomenon (severe vasospasms of the small vessels in the hands and feet)
- **E**sophageal dysmotility (decreased activity)
- **S**clerodactyly (scleroderma of the finger digits)
- **T**elangiectasia (spiderlike skin lesions)

Diagnosis

The patient's clinical history and physical manifestations aid in the diagnosis of scleroderma. Laboratory tests, pulmonary function tests, or x-ray examinations are used to determine the severity of organ involvement or if other diagnostic testing is not helpful in diagnosing this condition.

Treatment

The goal of medical management is to slow the progression of the disease. Systemic steroids, such as prednisone, and immunosuppressant drugs are used in large doses and in combination during a flare-up of PSS.

Other care approaches are directed toward symptom management. Skin protective measures can help minimize the chance of ulcerations or irritation. Box 44–3 is also appropriate for PSS. For example, teach the patient to use mild soaps and lotions to moisturize the skin.

If the patient has esophageal involvement, small, frequent, bland meals are better tolerated than large, spicy ones. Difficulty swallowing may necessitate cutting the food into smaller, more manageable portions or by providing food that is pureed (thicker liquids are easier to swallow than thin liquids). Medications to treat esophageal reflux, such as antacids and histamine blockers, may be prescribed.

Patients who have Raynaud's phenomenon or other types of **vasculitis** usually experience severe pain when small blood vessels constrict. Joints may also be painful. Pain management is a priority in the care of patients with PSS. A bed cradle or footboard keeps bed covers away from skin. Socks and gloves may keep the fingers and toes warm, thus diminishing pain. Minimizing exposure to cold and avoiding stressful situations, stopping smoking, and taking certain medications, such as calcium channel blockers, antiadrenergic agents, and angiotensin-converting enzyme (ACE) inhibitors, can all help promote circulation or minimize the likelihood of an attack. Research has suggested

that antioxidant therapy may provide another approach to treating PSS.

Rehabilitative therapy may be needed to help the patient be as independent as possible with activities of daily living and mobility. Collaborate with other members of the interdisciplinary team to individualize care.

Polymyositis

Polymyositis is a disease with an unknown cause that results in diffuse inflammation of skeletal muscle, leading to weakness, atrophy, and degeneration. When a rash is present with muscle inflammation, the disease is called dermatomyositis. The disease is progressive; however, remissions and exacerbations are common. Women are affected more than men, especially in their middle-aged years.

The shoulder and pelvic girdle muscles (proximal muscles) are most commonly affected. The patient may have associated conditions such as arthritis, fatigue, and possibly Raynaud's phenomenon (spasms and constriction of small vessels in the hands and feet). Patients with dermatomyositis also have the classic heliotrope (lilac) rash and periorbital (around the eyes) swelling. Malignant tumors occur in patients with these diseases more often than in the rest of the population.

Patients are treated symptomatically, using an interdisciplinary approach, to maintain optimum function. The drug of choice is high doses of prednisone. Side effects, such as immunosuppression, can occur with prednisone.

Muscular Dystrophy

Muscular dystrophy (MD) is a group of disorders resulting in loss of muscle tissue and progressive muscle weakness. A number of the disorders are diagnosed in childhood (e.g., Duchenne's MD is most common in children); however, other forms of MD, such as myotonic MD, are most common in adults. In addition, individuals with MD are now living longer into adulthood as a result of advances in treatment.

Pathophysiology and Etiology

Muscular dystrophy has a genetic origin. However, the exact cause is unknown. Skeletal (voluntary) muscle fibers degenerate and atrophy. This loss of muscle tissue results in muscle weakness and wasting. Muscle tissue is replaced by connective tissue. These changes in muscle tissue result in increasing disability and deformity. Life expectancy after diagnosis depends on the type of MD.

Signs and Symptoms

Signs and symptoms usually become apparent in childhood. Difficulty walking and muscle weakness in the arms, legs, and trunk are indicators of MD. Individuals with MD may have difficulty raising their arms above their heads or climbing stairs. Other signs and symptoms include frequent falls, developmental delays involving muscle skills, drooping eyelids (ptosis), drooling, intellectual retardation (only in some types of MD), contractures, and skeletal deformities.

---

vasculitis: vascul—blood vessel + itis—inflammation

Diagnosis

An increase in serum creatinine phosphokinase (CPK) caused by muscle atrophy is present in MD. Electromyography (EMG) and muscle biopsy can be used for diagnosis. Lactic dehydrogenase (LDH) and the isoenzymes, myoglobin (urine or serum), creatinine (urine or serum), CPK isoenzymes, and aspartate aminotransferase (AST) levels may also be altered in patients with MD.

Medical Treatment

Goals include supportive care and prevention of complications. Treatment regimens focus on controlling symptoms and maximizing quality of life. Keeping the patient as active as possible is a priority in the planning of care. Exercise programs (e.g., range of motion, physical therapy) help prevent muscle tightness, contractures, and atrophy. Splints and braces provide support during ADL. Surgery may be done to correct deformities. The potential benefit of gene therapy is currently being investigated. Some of the current research indicates that gene therapy could prove effective with some types of MD.

Nursing Process

ASSESSMENT. Assess for muscle weakness, noting areas of the body are affected and the severity of the weakness. Asking the patient and family what activities can be done with and without assistance helps determine the plan of care.

PLANNING. The patient and family should be involved in all aspects of care. The goal for impaired physical mobility is to maintain activity or to use adaptive equipment. The goal for ineffective breathing pattern is to maintain arterial blood gases within normal limits.

NURSING DIAGNOSIS. Priority diagnoses are impaired physical mobility related to muscle weakness and ineffective breathing pattern related to muscle weakness.

IMPLEMENTATION. Providing assistive devices (e.g., braces, splints, wheelchair) increases mobility. Range-of-motion exercises and other physical therapy prevent contractures and improve muscle strength. Encouraging the patient to do as much as possible increases independence and helps maintain muscle function. Respiratory rate and effort should be monitored every 4 hours. Measures to prevent skin breakdown need to be instituted, such as frequent turning or repositioning and teaching patients to shift their weight if able every 15 minutes while sitting or lying.

PATIENT EDUCATION. The patient and family need to understand the importance of physical therapy in maintaining function and preventing complications. National organizations and support groups provide information, resources, and emotional support. Family members need to encourage the patient to have activity and rest periods. As with any neuromuscular condition, the patient needs to avoid exposure to the cold and persons with infections. For more information on muscular dystrophy, visit www.mdausa.org or www.mdac.ca.

EVALUATION. The patient's goals are met if the patient maintains muscle function and desired level of activity; no evidence of skin breakdown, contractures, or injury is present; and the patient maintains or improves pulmonary status.

▶ MUSCULOSKELETAL SURGERY

Some health problems cannot be managed conservatively and require surgery. Other disorders are initially treated medically but may need surgery if treatment is unsuccessful. The most common surgeries are discussed here.

Total Joint Replacement

Total joint replacement is most often performed for patients who have some type of connective tissue disease in which their joints become severely deteriorated. TJR may also be done for patients on long-term steroid therapy, such as patients with SLE or asthma. Long-term use of steroids and complications of joint replacement can cause **avascular necrosis** (AVN), a condition in which bone tissue dies (usually the femoral head) as a result of impaired blood supply. Advanced AVN is very painful and usually does not respond to conservative pain relief measures. The primary goal of total joint replacement is to relieve severe chronic pain and improve ability to carry out ADLs when no other treatment is successful.

The most common surgeries are the total hip replacement (THR) and total knee replacement (TKR), although any synovial joint can be replaced. Another term used for joint replacement is *arthroplasty*. The replacement devices, sometimes referred to as prostheses, are made of metal, ceramic, plastic, or a combination of these materials. Some prostheses are held in place by cement. Others are secured by the patient's bone as it grafts and connects to the prosthesis. Bone substitutes, also called biologics, are being used more often when the amount of available bone is insufficient to provide a good base of support for the replacement devices. Bone glues and fillers such as Osteoset or Proosteon and bone stimulants such as Allomatrix help in providing better support for the prosthetics used.

Total Hip Replacement

A total hip replacement uses a two-piece device consisting of an acetabular cup that is inserted into the pelvic acetabulum and a femoral component that is inserted into the femur to replace the femoral head and neck (Fig. 44–6). The average life span of a cemented THR is about 10 years. Noncemented prostheses used in younger patients may last longer.

PREOPERATIVE CARE. Total joint surgery is an elective procedure and scheduled far enough in advance to allow ample time for preoperative teaching and screening. A case manager (registered nurse or social worker) may be assigned to assess the patient's needs and the support systems that are

---

avascular necrosis: a—without + vascular—blood + necrosis—death

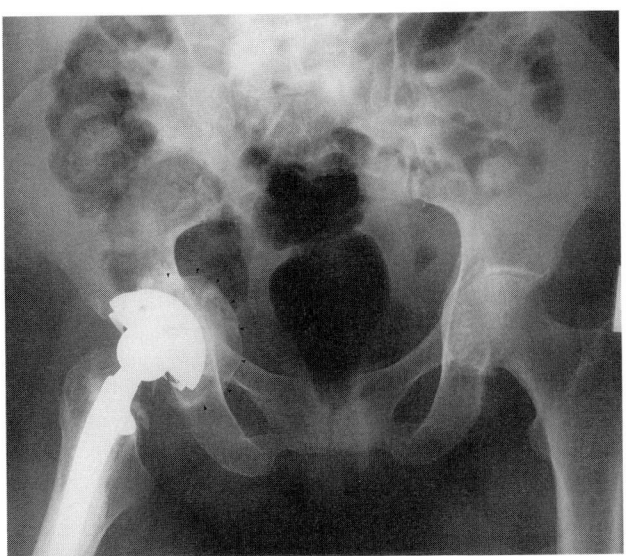

Figure 44-6 Total hip arthroplasty of arthritic right hip. (From McKinnis, LN: Fundamentals of Orthopedic Radiology. FA Davis, Philadelphia, 1997, with permission.)

available postoperatively. It is important for the patient to have a caregiver who can assist the patient after surgery.

The patient is taught about the surgery and what to expect postoperatively. Some patients are scheduled to meet with the physical therapist to learn postoperative exercises and how to ambulate with a walker or crutches. Some institutions have total joint education programs, which are a series of educational sessions designed to make the recovery process smoother and more effective for the patient.

Depending on the amount of blood loss during surgery, some patients receive postoperative blood transfusions. Because total joint surgery is an elective procedure, the physician may order autologous blood donation by the patient. The patient donates blood before surgery per guidelines (e.g., time frames specified, hemoglobin levels normal), which is then available for reinfusion postoperatively as needed. This predeposited blood donation is cost effective and reassures patients who are concerned about receiving blood from other donors.

Patients are admitted to the hospital the morning of surgery (or occasionally, if there is a coexisting medical condition, the day before surgery) and are usually transferred to a surgical or medical-surgical unit after recovery in the postanesthesia care unit (PACU). The patient's length of stay on this unit varies from 3 to 5 days, depending on the patient's age and progress.

POSTOPERATIVE CARE. In addition to providing the general postoperative care that all patients undergoing general or epidural anesthesia require, plan and implement interventions to help prevent the following common complications of THR. (See Chapter 11.)

Hip Dislocation. The most common postoperative complication for the patient having a THR is subluxation (partial dislocation) or total dislocation. Dislocation occurs when the femoral component becomes dislodged from the acetab-

ular cup. The patient experiences increased hip pain, shortening of the surgical leg, and possibly rotation of the surgical leg. If any of these signs and symptoms occur, notify the surgeon immediately and keep the patient in bed. Additional analgesics may be ordered until the patient can be taken to the operating room. Under anesthesia, the surgeon manipulates the hip back into alignment and immobilizes the leg until healing occurs.

Prevention of dislocation is a major nursing responsibility. Correct positioning of the surgical leg is critical. The primary goals are to prevent hip adduction (across the body's midline) and hyperflexion (bending forward more than 90 degrees). To accomplish these goals, place the patient returning from PACU in a supine position with the head slightly elevated. A trapezoid-shaped abduction pillow (sometimes called a triangular pillow), splint, wedge, or regular bed pillows may be used between the legs to prevent adduction (Fig. 44–7). The patient is turned to the side specified by the physician, with hip adduction avoided. The patient is turned with adductor pillow or three regular pillows (one proximal and two distal) in place between the legs. When turning, it is important to turn the hip and legs simultaneously to minimize the chance of dislocation.

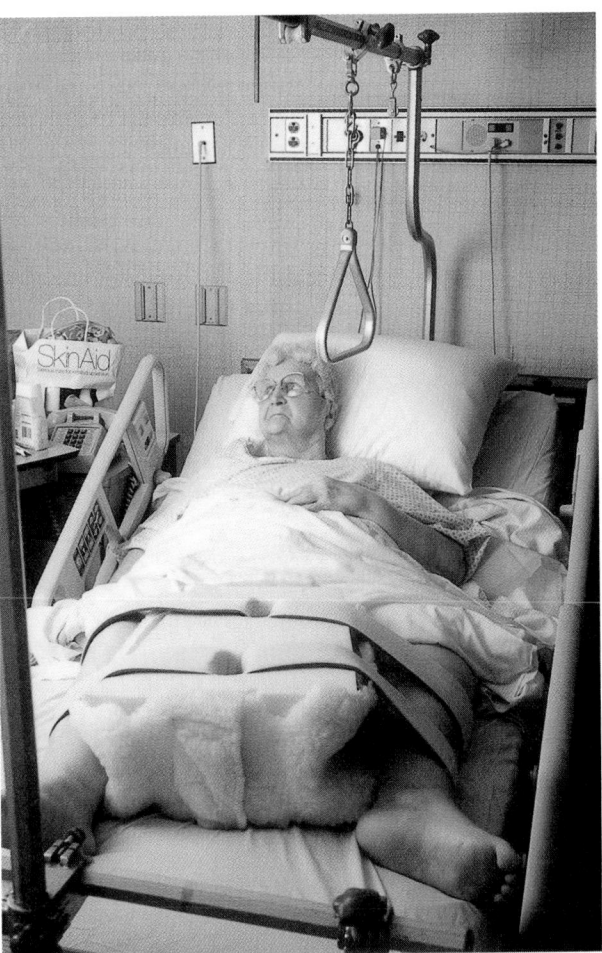

Figure 44-7 Abductor pillow is used to prevent adduction and hip dislocation.

To prevent hyperflexion, some surgeons initially allow the patient to sit at no more than a 60-degree angle in a reclining chair. The patient's position is progressed to 90 degrees, the maximum allowed to prevent hyperflexion (Fig. 44–8). While the patient is on bedrest, the use of a fracture (also called a slipper) pan when toileting the patient is recommended to minimize discomfort and to prevent the possibility of dislocation.

Skin Breakdown. Because most patients having total joint replacements are elderly, skin breakdown is a major concern as part of postoperative care. Turning the patient at least every 2 hours and keeping the heels off the bed are the key nursing interventions to prevent pressure ulcers. Heels and the sacrum are vulnerable and can break down in 24 hours. A reddened area that does not blanch is a stage 1 pressure ulcer and must be treated aggressively to prevent progression to other stages. Prophylactic application of DuoDerm dressing pads, as well as heel Poseys, help to decrease the chance of skin breakdown of the heels.

Patients who are incontinent must be kept clean and dry. Toileting the patient every 2 hours, using a protective barrier cream, and avoiding the use of diapers also help prevent skin problems related to incontinence. Box 44–4 describes additional nursing interventions that meet the needs of postoperative patients recovering from THR.

Infection. Orthopedic surgery patients are at an increased risk for infection because of the nature of the surgery and because the patients are often elderly, with an already increased risk for postoperative complications. The surgeon may order a prophylactic intravenous antibiotic preoperatively, often administers antibiotics intraoperatively, and may continue antibiotics for 24 hours postoperatively.

Once the initial surgical dressing is removed, observe the incision routinely for signs and symptoms of infection. Monitor temperature carefully. An elderly patient may not experience a fever but may appear confused instead.

BOX 44–4   Total Hip Replacement

- Use an abduction pillow or splint to prevent adduction after surgery.
- Protect the patient's heels while on bedrest to prevent pressure sores.
- Do not rely on fever as a sign of infection; elderly patients often have infection without fever. Decreasing mental status typically occurs when the patient has an infection.
- When assisting the patient out of bed, move the patient slowly to prevent orthostatic (postural) hypotension.
- Encourage the patient to deep breathe and cough and use the incentive spirometer every 2 hours to prevent atelectasis and pneumonia.
- As soon as permitted, get the patient out of bed to prevent complications of immobility.
- Anticipate the patient's need for pain medication, especially if the patient is unable to verbalize the need for pain control.
- Expect a temporary change in mental state immediately after surgery as a result of the anesthesia and unfamiliar sensory stimuli. Reorient the patient frequently.

Source: Ignatavicius, DD, Workman, ML, and Mishler, MA: Medical-Surgical Nursing: A Nursing Process Approach, ed 3, WB Saunders, Philadelphia, 1998, with permission.

Infection may not occur during the patient's hospital stay but can occur 1 or more years later. If this late infection does not respond to antibiotics, the prosthesis may be removed and replaced. To prevent infection, antibiotics are often instilled directly into the wound during surgery as beads, as part of the cement mixture, or as an irrigating solution.

Depending on the institution's policies, the surgeon may or may not remove the initial dressing. Regardless of who removes the dressing, meticulous aseptic care of the surgical wound is important to minimize the chance of infection. Care of the incision, as well as exit sites for drains, needs to be performed aseptically.

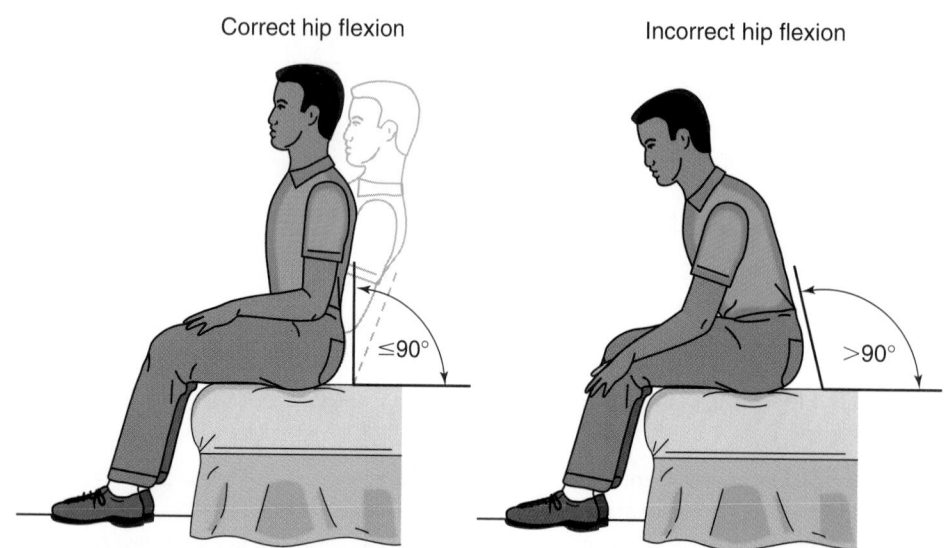

Figure 44-8   Hip flexion after total hip replacement should be 90 degrees or less to prevent dislocation.

Bleeding. Like any surgical wound, some bleeding is expected. In joint replacement surgery, up to two-thirds of the blood loss can occur postoperatively. The patient has at least one surgical drain (e.g., Hemovac or Jackson-Pratt) that is emptied every 8 to 12 hours and as required for the first day or two. On the second or third postoperative day, the patient's hemoglobin and hematocrit may decrease to the point that blood transfusion is needed. The patient may receive the preoperatively donated autologous blood or may receive salvaged operative or postoperative blood. Using a cell saver (sometimes called an orthopat, which stands for orthopedic patient autotransfusion) during surgery, about 50 percent of blood that is lost can be recovered and saved for reinfusion into the same patient. Postoperatively, blood can be replaced by collecting shed blood via suction into a reservoir, then filtering and reinfusing it within 6 hours of collection using the orthopat. Monitoring for blood loss and signs of shock are important nursing considerations.

Neurovascular Compromise. For any musculoskeletal surgery or injury, frequent neurovascular assessments for color, warmth, circulation, and movement are performed when vital signs are checked. The procedure and significance of these assessments are described in Chapter 43.

Pain. Because patients undergoing THR are in chronic pain preoperatively, some patients report that they have less pain postoperatively than they had before surgery. Initially pain is typically managed by epidural analgesia, patient-controlled analgesia (PCA), or injections with analgesics. After the first postoperative day, the patient usually progresses to oral opioid analgesia with a drug such as Percocet or Tylenol with codeine.

Ambulation. Care for the patient having a THR is interdisciplinary. The patient usually gets out of bed and into a chair the night of surgery or early the next day. Ensure that the patient does not adduct or hyperflex the surgical hip during transfer to the chair. The chair should have a straight back and be high enough to prevent excessive flexion. The toilet seat should also be raised for the same purpose. Permitted amounts of weight bearing depend on the type of prosthesis that is used. In general, weight bearing to tolerance or full weight bearing is used for cemented prostheses. If an uncemented device is used, the patient may be restricted to toe-touch, or partial, weight bearing or featherweight bearing.

Early ambulation helps prevent postoperative complications such as atelectasis and deep vein thrombosis (DVT). The physical therapist works with the patient for ambulation with a walker or crutches. Crutches are reserved for young patients. After 4 to 6 weeks, the patient is progressed to a cane. The patient does not need an ambulatory device if there is no limping.

Thromboembolic Complications. Patients having hip surgery are at greatest risk for DVT or pulmonary embolus (PE). Elderly patients are especially at risk because of compromised circulation. Obese patients and those with a history of thromboembolic (TE) problems are also at an exceptionally high risk for potentially fatal problems.

Thigh-high elastic stockings and sequential compression devices (SCDs) may be used while the patient is hospitalized. (See Chapter 11.) The surgeon orders an anticoagulant medication to help prevent clot formation, including subcutaneous low-molecular-weight heparin (enoxaparin [Lovenox]) or warfarin (Coumadin). Occasionally, heparin is still used, and if so, it is important to monitor for heparin-induced thrombocytopenia, which can occur as early as 3 days after the start of heparin therapy. The ordered daily dosage of these drugs is determined by coagulation studies. Partial thromboplastin times are monitored for patients on heparin. Prothrombin time and international normalized ratio (INR) are monitored when giving warfarin.

LEARNING TIPS

When giving enoxaparin, follow manufacturer's instructions for administration. The air bubble should not be removed from the prefilled syringe before administration to ensure the whole dose is given.

Because most DVT occur in the lower extremities, leg exercises are started in the immediate postoperative period and continued until the patient is fully ambulatory. The physical therapist teaches the patient how to perform foot and ankle exercises such as heel pumping, foot circles, and straight-leg raises (SLRs). The patient also performs quadriceps-setting exercises (quad sets) by straightening the legs and pushing the back of the knees toward the bed. Remind the patient to do several sets of these exercises each day to improve muscle tone and to help prevent blood clots in the leg.

Self-care. Because of restrictions in hip flexion, patients are instructed not to bend forward to tie shoes or put on pants. The occupational therapist provides adaptive or assistive devices, such as dressing sticks and long-handled shoe horns, to assist the patient in being independent in activities of daily living.

If the patient is medically stable, he or she is discharged home for rehabilitation or to a subacute care unit, rehabilitation unit, or nursing home for short-term rehabilitation, lasting a week or less. The rehabilitation program that began in the hospital continues after discharge until the patient is independent in ambulation and self-care.

Before hospital discharge, the interdisciplinary team provides patient education for home care, including hip precautions that need to be used until the surgeon reevaluates the patient at the 6- to 8-week follow-up visit (Box 44–5 and Home Health Hints).

Total Knee Replacement

The knee is the second most commonly replaced joint. Compared with the hip, it is a much more complicated joint and requires three components for replacement: a femoral component, a tibial component, and a patellar button (Fig. 44–9).

BOX 44-5 **Patient Education for Total Hip Replacement**

Hip Precautions

- Do not sit or stand for prolonged periods.
- Do not cross legs beyond midline of body.
- Do not bend hips more than 90 degrees.
- Use an ambulatory aid, such as a walker, when walking.
- Use assistive/adaptive devices for dressing, such as for putting on shoes and socks.
- Resume sexual intercourse as usual, but when doing so, use the hip precautions learned in the hospital.

Source: Ignatavicius, DD, Workman, ML, and Mishler, MA: Medical-Surgical Nursing: A Nursing Process Approach, ed 3. WB Saunders, Philadelphia, 1998, with permission.

HOME HEALTH HINTS

If equipment or modifications to the home are needed following hospitalization for an orthopedic problem, it is best if they can be arranged or obtained before discharge.

Peak incidents of DVT after hip or knee surgery is highest by the fifth postoperative day and that risk persists for up to 12 weeks. Be alert for signs of DVT: warmth, redness, edema, Homans' sign, and protective behavior of the affected leg.

Research has shown that pain or fear of falling may prevent a patient from moving and functioning to maximum potential. Encourage patients to wear flat, sturdy, rubber-soled shoes to prevent slipping, tripping. or turning an ankle.

Patients who use walkers can get pressure ulcers on their palms. One way to relieve the pressure is to wear padded cycling gloves that leave the fingers free.

A patient on crutches can use the crutch to prop a casted leg or foot.

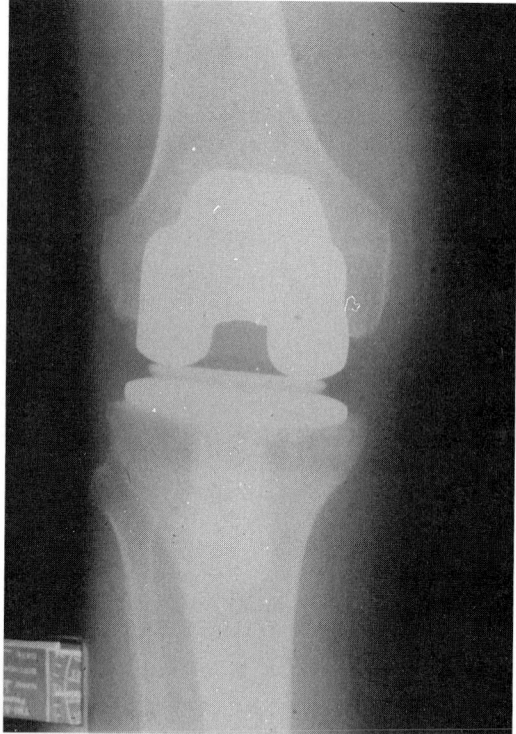

Figure 44-9 Knee joint replacement. (From Richardson, JK, and Iglarsh, ZA: Clinical Orthopaedic Physical Therapy. WB Saunders, Philadelphia, 1994, p 651, with permission.)

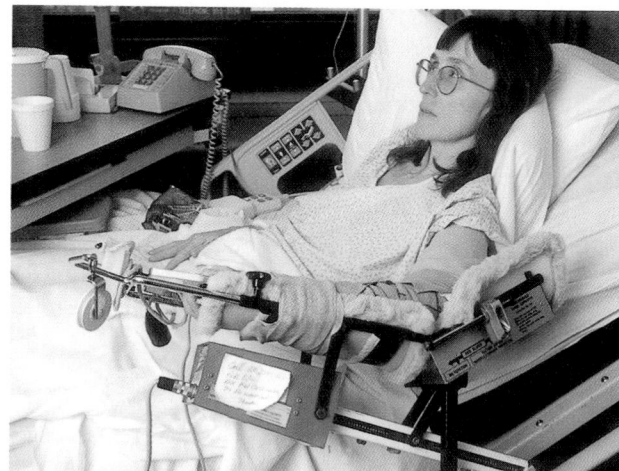

Figure 44-10 A continuous passive motion machine is used following knee or elbow (as shown here) joint replacement to increase joint mobility and enhance recovery. The CPM machine slowly moves along the track at the set degree of flexion and speed.

Care for the patient with a TKR is similar to that required for a patient with a hip replacement. Although precautions to prevent dislocation are not applicable for the patient with a knee replacement, other medical complications described for THR, such as deep vein thrombosis, may be seen in the patient undergoing knee replacement. (See Box 44–4.)

Most surgeons order a continuous passive motion machine (CPM) for the leg that is to be operated on. This motorized machine has a flexible extremity rest (for either the leg or arm) that glides back and forth on a track (Fig. 44–10). The CPM is set at the degree of flexion and speed ordered by the physician and is usually begun in the PACU. The CPM can be applied by a nurse, physical therapist, or technician and is used either intermittently, up to 8 to 12 hours a day, or continuously while the patient is in bed. The purpose of CPM is to keep the knee joint mobile. Nursing care associated with the use of the machine is summarized in Box 44–6.

Amputation

Simply defined, an amputation is the removal of a body part, which can be as limited as removing part of a finger or as devastating as removing nearly half the body. Amputations may be surgical as a result of disease or traumatic as a result of an accident. Surgical amputations are the most common type and are most often scheduled as elective surgery.

BOX 44-6 **Using a Continuous Passive Motion Machine**

- Ensure that the joint being moved is properly positioned on the machine.
- Ensure that the machine is well padded with sheepskin or other similar material.
- Check the cycle and range-of-motion settings at least once per shift (every 8 hours).
- If the patient is confused, place the controls to the machine out of the patient's reach.
- Assess the patient's response to the machine.
- Turn off the machine while the patient is having a meal in bed.
- When the machine is not in use, do not store it on the floor.

Source: Ignatavicius, DD, Workman, ML, and Mishler, MA: Medical-Surgical Nursing: A Nursing Process Approach, ed 3, WB Saunders, Philadelphia, 1998, with permission.

SURGICAL AMPUTATIONS. The main indication for surgical amputations is ischemia from peripheral vascular disease in the elderly. The rate of lower extremity amputation is much greater in the diabetic patient than in the nondiabetic patient. (See discussion of diabetes in Chapter 38.) Surgical amputations may also be done for bone tumors, thermal injuries (frostbite), congenital problems, or infections.

TRAUMATIC AMPUTATIONS. Traumatic amputations occur from accidents, often in young and middle-aged adults. Industrial machinery, motor vehicles, lawn mowers, chain saws, and snow blowers are common causes of accidental amputation.

Because in these patients the amputated part is usually healthy, attempts at **replantation** may occur. One of the most common replantations is one or more fingers. The current recommendation for prehospital care of the severed body part is to wrap it in a cool, slightly moist cloth and place it in a sealed plastic bag. The bag may be submerged in cold water until the body part is transported to the hospital.

The surgical procedure is performed by specialists who operate using a microscope. Nerves, vessels, and muscle must be reattached. These procedures are generally performed at large tertiary care centers that have specialty practitioners and equipment for replantation.

LEVELS OF AMPUTATION. The most common surgical amputation is part of the lower extremity. The loss of any or all of the small toes presents little problem. However, the loss of the great toe is more important because balance and gait are affected. Midfoot amputations are preferred over below-the-knee amputations (BKAs) for peripheral vascular disease. For the Syme amputation, the surgeon removes most of the foot but leaves the ankle intact for ambulation and weight bearing.

If the lower leg is amputated, a BKA is preferred over an above-the-knee amputation (AKA) to preserve joint function. The higher the level of amputation, the more energy is required for ambulation. Hip disarticulation (removal through the hip joint) and **hemipelvectomy** (removal through part of the pelvis) are reserved for young patients who have cancer or severe trauma. Rarely, a hemicorporectomy (hemipelvectomy plus a translumbar amputation) is performed as a last resort for young patients with cancer. This radical surgery removes nearly half of the body and requires both bowel and urinary diversion surgeries (ostomies) as well.

Upper extremity amputations are usually more significant than lower extremity amputations and more often result from trauma. The arms and hands are necessary for performing activities of daily living. Early replacement with a prosthesis is crucial for the patient with an upper extremity amputation.

PREOPERATIVE CARE. Patients who are scheduled for elective amputations have the advantage of time for preoperative teaching, prosthesis fitting, and adjustment to the loss of part of their bodies. Preoperative teaching is started in the surgeon's office. Postoperative and rehabilitative care are reviewed with the patient and family or significant other.

Preoperatively the patient should be referred to a certified prosthetist-orthotist (CPO) to begin plans for replacing the removed body part with a prosthesis.

Disturbed body image is a common nursing diagnosis for the patient having an amputation. If possible, it is helpful for the preoperative patient to meet with a rehabilitated amputee. Assess the patient's reaction to having an amputation with the expectation that the patient will experience many of the stages of loss and grieving. Support systems and coping mechanisms are identified that can help the patient through the surgery and postoperative period.

POSTOPERATIVE CARE. In addition to the general postoperative care, plan and implement interventions to help prevent postoperative complications. (See Chapter 11.)

Hemorrhage. When a patient loses part of the body, either by surgery or trauma, blood vessels are severed or damaged. The patient returns from surgery with a large pressure dressing that is secured with an elastic wrap. Assess the closest proximal pulse between the heart and the amputated body part for strength and compare findings with the nonsurgical extremity. Assess the bulky dressing for bloody drainage. If blood is on the dressing when the patient is admitted to the PACU or the surgical unit, circle, date, and time the drainage. The area is closely monitored for enlargement. If bleeding continues, the surgeon is notified immediately. A tourniquet should be readily available in the event that severe hemorrhage occurs.

---

replantation: re—again + plant—to plant + ation—process

hemipelvectomy: hemi—half + pelv—pelvis + ectomy—removal of

After the dressing is removed, observe for adequate perfusion to the skin flap at the end of the residual limb, referred to as the stump. The skin should be pink in a light-skinned patient and not discolored (lighter or darker than other skin pigmentation) in a dark-skinned patient. The residual limb should be warm but not hot.

Infection. Infection of the wound can be problematic, especially if the infection enters the bone (**osteomyelitis**). Inspect the wound for intense redness or drainage. If temperature is elevated, it could indicate a wound or other type of infection.

Pain. In addition to the usual incisional pain that is expected following a surgical procedure, phantom limb pain occurs in as many as 80 percent of all amputees. The patient complains of severe pain where the removed body part was located. The pain may be described as either intense burning, a crushing sensation, or cramping. This type of pain is not common in patients who have traumatic amputations.

Phantom limb pain can be triggered by touching the residual limb, feeling fatigued, or experiencing emotional stress. Although it occurs most often in the immediate postoperative period, phantom limb pain may occur at any time during the first postoperative year. The cause is not clear.

Never doubt that the patient is experiencing phantom limb pain. Treat the pain aggressively with medications and complementary therapies. The surgeon prescribes medication based on the type of pain sensation the patient experiences. For example, anticonvulsants, such as phenytoin (Dilantin), are used for knifelike pain. Beta-blocking agents, such as propranolol (Inderal), are appropriate for burning sensations, and gabapentin (Neurontin) or amitriptyline (Elavil) can be used for nerve pain. To complement traditional therapy, a number of therapies may be useful, including biofeedback, massage, imagery, hypnosis, acupuncture, acupressure, and distraction.

Mobility and Ambulation. To reduce surgical swelling, cold application may be ordered. Alternately, the residual limb may be elevated on a pillow for 24 hours or less. Continued use of a pillow for elevation can lead to flexion contractures, especially for patients with a BKA or an AKA. If the hip becomes contracted, using a prosthesis will not be possible because the patient will not be able to walk. Check the limb periodically to ensure that it lies completely flat on the bed. The patient should avoid positions of flexion such as sitting for long periods. If the patient is able, lying prone (on stomach) for 30 minutes four times daily helps prevent contracture.

Postoperative care of the patient experiencing an amputation is interdisciplinary, often requiring an extensive rehabilitation program in a subacute unit, nursing home, or on an ambulatory basis. The physical therapist teaches the patient muscle-strengthening exercises that help with ambulation and transfers and prevent flexion contractures.

A trapeze and overhead bed frame aid in strengthening the upper extremities and help the patient move around in the bed.

Prosthesis. The residual limb must be prepared for wearing the prosthesis. A temporary prosthesis may be worn until the swelling subsides.

The residual limb is wrapped at least every 8 hours using an elastic wrap (Ace or tensor) in a figure-of-eight fashion (Fig. 44–11). It is important to perform neurovascular checks and assess the residual limb for infection and alterations in tissue integrity at each rewrapping. Begin with the most distal portion and proceed proximally until the bandage is secured to the most proximal joint. The bandage should be tighter at the distal end.

The prosthesis requires special care and attention. Box 44–7 summarizes discharge teaching for the amputee with a prosthesis.

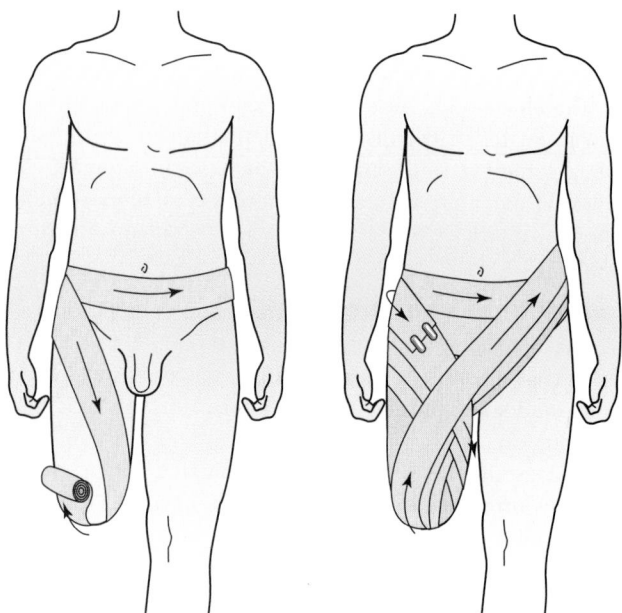

Figure 44-11 Application of elastic wraps on an above-the-knee amputation helps mold the stump for a prosthesis.

BOX 44-7 **Patient Education for Prosthesis Care**

- Have a wooden prosthesis refinished at least every 6 months.
- Clean the prosthesis socket with mild soap and water and dry it completely.
- Replace worn inserts and liners when they become too soiled to clean adequately.
- Check all mechanical parts such as bolts periodically for unusual sounds or movement.
- Grease the mechanical parts as instructed by prosthetist.
- Use garters to keep socks or stockings in place.
- Replace shoes when they wear out with new ones of the same height and type.

Source: Ignatavicius, DD, Workman, ML, and Mishler, MA: Medical-Surgical Nursing: A Nursing Process Approach, ed 3. WB Saunders, Philadelphia, 1998, with permission.

---

osteomyelitis: osteo—bone + myel—bone marrow + itis—inflammation

Lifestyle Adaptation. The patient may feel that life will be markedly changed as a result of the amputation. With the technological advances in prostheses, most patients who worked before surgery are able to return to their jobs after surgery. If the discharge planner or case manager thinks it is needed, a job analysis may be conducted by a vocational analyst or specialized case manager. Many patients with amputations are able to bowl, ski, hike, and experience all the recreational hobbies that they were able to do before surgery.

A supportive family or significant other is vital to help the patient adjust to body image change. Consider the need for a sexual counselor or psychologist if indicated. For any patient with an amputation, help the patient set realistic expectations.

For the patient who is not a candidate for a prosthesis, home adaptations for a wheelchair may be needed. The patient must have access to toileting facilities and areas necessary for self-care. Structural changes in the living environment may be necessary before the patient can be discharged from rehabilitation.

A small percentage of amputees return to their nursing home environment without prostheses. These patients need rehabilitation to ensure that they can be as independent as possible.

▶ BONE AND SOFT TISSUE DISORDERS

The musculoskeletal system is the second largest system in the body. A variety of injuries and diseases can affect bone, soft tissue, or both. The most common problems are discussed in this section.

Strains

A strain is a soft tissue injury that occurs when a muscle or tendon is excessively stretched. Causes of strains include falls, excessive exercise, and lifting heavy items without using proper body mechanics. Back and ankle injuries are common. Strains can be mild, moderate, or severe. A mild strain causes minimal inflammation; swelling and tenderness are present. A moderate strain involves partial tearing of the muscle or tendon fibers. Pain and inability to move the affected body part result. The most severe strain occurs when a muscle or tendon is ruptured, with separation of muscle from muscle, tendon from muscle, or tendon from bone. Severe pain and disability result from this injury.

RICE is an acronym for rest, ice, compression, and elevation. These four components are the basis of therapy for strain injuries. Immediately after a strain, ice should be applied to decrease pain, swelling, and inflammation. Applying an elastic bandage (compression) and elevating the affected area (if appropriate) provide support and minimize swelling. Once inflammation subsides, heat application (15 to 30 minutes four times a day) brings increased blood flow to the injured area for healing. Activity is limited (depending on the severity of the injury, casting may even be required for immobilization) until the soft tissue heals, and anti-inflammatory drugs are prescribed. Muscle relaxants may be used for muscle pulls as well. Exercise may begin as early as 2 to 5 days after the injury (depending on the severity of the injury), but it may take 1 to 3 weeks of immobility before exercise can begin. For more severe strains, surgery to repair the tear or rupture may be needed. These procedures are done on an ambulatory, same-day-surgery basis.

Sprains

A sprain is excessive stretching of one or more ligaments that usually results from twisting movements during a sports activity, exercise, or fall. Like strains, sprains also vary in severity. A mild sprain involves tearing of just a few ligament fibers and causes tenderness. In a moderate sprain, more fibers are torn but the stability of the joint is not affected. A moderate sprain is uncomfortable, especially with activity. A severe sprain causes instability of the joint and usually requires surgical intervention for tissue repair or grafting. Pain and inflammation prevent mobility.

For mild sprains, RICE is again used as a therapy method. Rest, ice applications, and a compression (elastic) bandage and elevation are used for several days until swelling and pain diminish. Anti-inflammatory drugs are also used to decrease inflammation and control pain. Moderate sprains may need immobilization with a brace or cast until healing occurs.

Carpal Tunnel Syndrome
Pathophysiology

Carpal tunnel syndrome results in the compression of the median nerve within the carpal tunnel when swelling in the tunnel occurs. This swelling can result from edema, trauma, rheumatoid arthritis, or repetitive hand movements (repetitive motion injury) as used in some occupations such as typing or cash register operation.

Signs and Symptoms

Carpal tunnel syndrome usually results in slow-onset finger, hand and arm pain, and numbness. Painful tingling and paresthesias may also be present. Eventually, fine motor deficits and then muscle weakness may develop.

Diagnosis

Diagnosis is based on signs and symptoms, along with the patient's history. A positive Phalen's test (numbness with wrist flexion) is indicative of carpal tunnel syndrome. EMG can also be used to detect nerve abnormalities.

Medical Management

Medical treatment focuses on relieving the inflammation and resting the wrist. A splint is often ordered for the patient to wear. Medications to reduce pain and inflammation are ordered, such as aspirin and NSAIDs. Cortisone may be injected into the carpal tunnel to decrease pain and inflammation.

For some patients, surgery may be necessary. The surgeon may use an open incision or may perform an endoscopy. The median nerve is released from compression during the surgery, thus correcting the problem of the nerve and the surrounding area becoming inflamed. Physiotherapy helps in the recovery of function.

Nursing Management

Educate the patient on methods to prevent carpal tunnel syndrome, such as frequent short breaks during the work day, interspersing ongoing tasks with repetitive movements throughout the day, and using ergonomically appropriate devices to minimize the pressure placed in the area of the wrist.

Provide pain relief as ordered, and if surgery is performed, provide routine preoperative and postoperative care. Postoperatively elevate the patient's hand and use a splint as ordered for up to 2 weeks. Lifting is restricted for several weeks. The patient is taught to report signs and symptoms of neurovascular compromise, such as numbness and tingling, coolness, lack of pulse, pale skin or nailbeds, or limited movement. The patient may need family assistance with ADLs.

Fractures

A fracture is a break in a bone and can occur at any age and in any bone. Some fractures are minor and are treated on an ambulatory basis; others are more complex and require surgical intervention with hospitalization and rehabilitation.

Pathophysiology

Bone is a dynamic, changing tissue. When it is broken, the body immediately begins to repair the injury (Fig. 44–12). For an adult, within 48 to 72 hours after the injury a hematoma (blood clot) forms at the fracture site, because bone has a rich blood supply. Various cells that begin the healing process are attracted to the damaged bone. In about a week or so, a nonbony union called a callus develops and can be seen on x-ray examination. As healing continues, osteoclasts (bone-destroying cells) resorb any necrotic bone and osteoblasts (bone-building cells) make new bone as a replacement. This process is sometimes referred to as bone remodeling. Young, healthy adult bone completely heals in about 6 weeks; however, it can take up to a year before the whole process of remodeling is complete. An older person takes longer to heal, and children tend to heal more quickly.

Causes and Types

The major reason for a fracture is trauma, usually from either a fall or a motor vehicle accident. Bone disease, such as osteoporosis and metastatic bone cancer, malnutrition, and

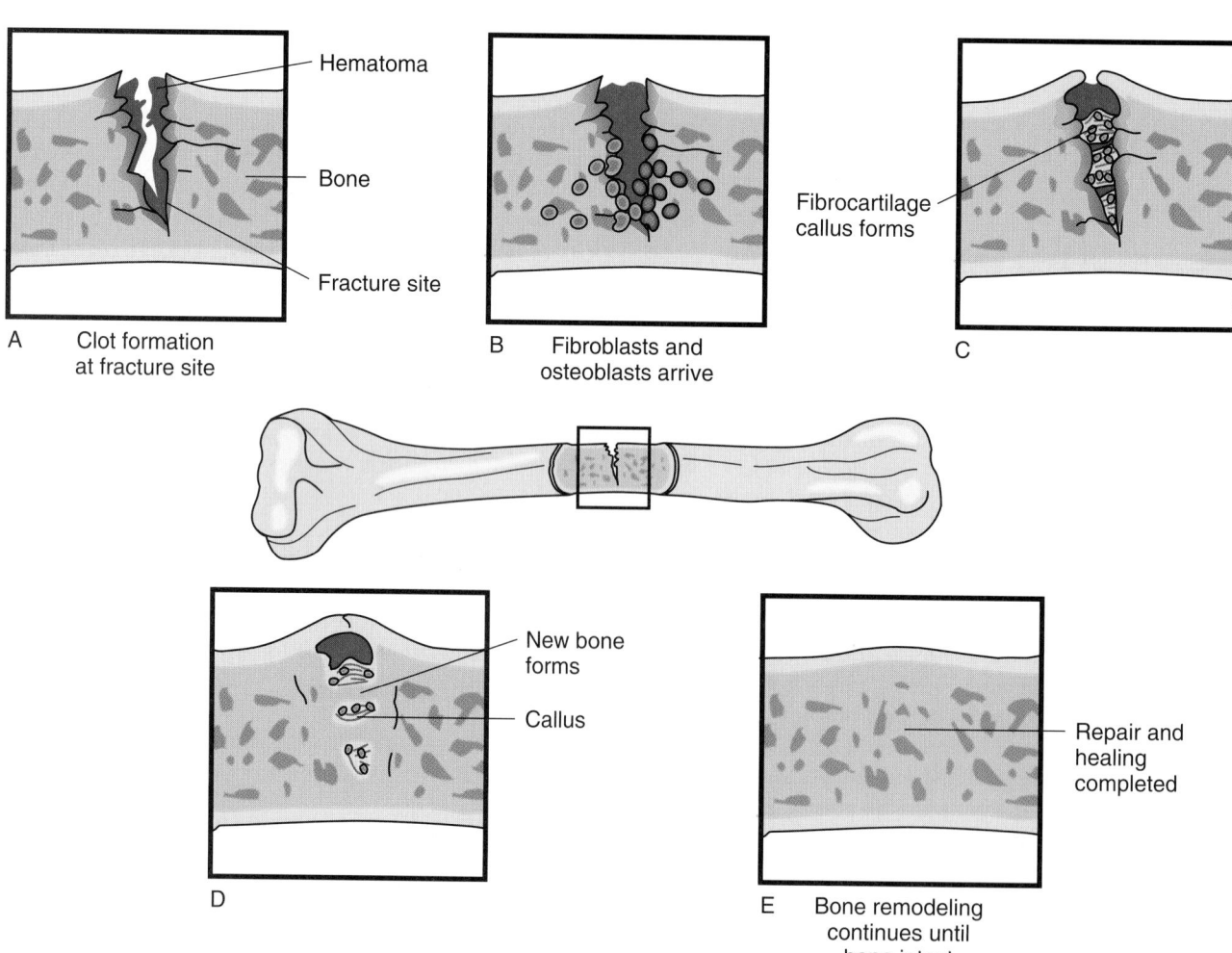

Figure 44-12 Fracture healing phases.

regular drinking of soda pop can lead to fractures as well. Fractures resulting from any of these diseases are referred to as pathological fractures. One of the most common types of fractures is the hip fracture, which occurs in middle-aged and elderly women who have osteoporosis (irreversible bone loss), a condition discussed later in this chapter.

Fractures can be classified in several ways: by the extent of the fracture, the extent of the associated soft tissue damage, or the configuration of the bone after it breaks. A fracture that is complete, breaking the bone into two separate pieces, is called a displaced fracture. An incomplete fracture does not divide the bone into two pieces; it may also be referred to as a nondisplaced fracture. Complete fractures have the potential to be life threatening because sharp bone fragments can sever blood vessels and nerves.

L E A R N I N G TIPS

The risk of developing a hip fracture in men or women can be dramatically reduced with the use of hip protectors. Hip protectors are pads sewn into underwear or worn on top of clothing with a belt. The pad directs a fall's energy away from the hip area to prevent a fracture. For more information on some brands of hip protectors that are available or are being studied, visit www.safehip.com, www.hipsaver.com, and www.hipguard.net.

Inform patients and families of the benefits of purchasing a hip protector if your institution does not provide them. (Present research findings supporting increased patient safety and decreased legal fracture risk for the institution to encourage your administrators to provide hip protectors.) Hip protector patient teaching should include the following:

- With time the patient will become used to wearing the protectors.
- The patient should wear the protectors at night for trips to the bathroom (a high-risk time for falls).
- The protector will not show under loose clothing.
- The protector can be laundered.
- Styles are available for incontinence care and easy changing.

The cost in hip fracture care ($33,000) and quality-of-life savings make this a valuable tool in fracture risk reduction. (Up to 24 percent of persons older than age 50 who suffer hip fracture die within 1 year; up to 40 percent need extended care.)

A fracture may also be classified as open or closed. In an open (or compound) fracture, the bone breaks the skin. A closed fracture does not disrupt the skin. Open fractures are more likely to become infected than closed fractures.

Another way to describe fractures is by the way that the bone breaks, such as in a spiral or oblique fashion (Fig. 44–13). These fractures may be open or closed, complete or incomplete. Table 44–5 describes the types of fractures.

Signs and Symptoms

This section focuses on fractures of upper and lower extremities. If the patient sustains a hairline (microscopic) fracture, the signs and symptoms are not readily observable. The patient may complain of tenderness over the site of the injury or more severe pain when moving the affected part of the body. The patient with a hip fracture usually complains of pain either in the groin area (the hip is a deep joint) or at the back of the knee (referred pain). If the fracture is displaced, the limb is often shortened because of contraction of the muscles pulling on the bone sections.

In addition to pain, patients with more complex fractures experience limb rotation or deformity and shortening of the limb (if a limb bone is broken). Range of motion is decreased. If the affected part is moved, a continuous grating sound (crepitation) caused by bone fragments rubbing on each other may be heard. The extremity should not be moved (to try and reposition the bone alignment) if crepitation is present.

Inspect the skin for intactness. A patient with a closed fracture may have ecchymosis (bruising) over the fractured bone from bleeding into the soft underlying tissue. Ecchymosis may not develop for several days after the injury. Swelling may also be present and can impair blood flow, causing marked neurovascular compromise. In an open fracture, one or more bone ends pierce the skin, causing a wound.

Diagnostic Tests

An x-ray examination usually visualizes bone fractures, showing bone malalignment or disruption. Computed tomography may be needed to help detect fractures of complex areas, such as the hip and pelvis. Magnetic resonance imaging is useful in determining the extent of associated soft tissue damage.

For patients experiencing moderate to severe bleeding, a hemoglobin and hematocrit level is obtained. If extensive soft tissue damage is present, the ESR is usually elevated, indicating the expected inflammatory response. The physician may order a serum calcium level to determine baseline values because bone repair requires a sufficient amount of calcium and other minerals.

Emergency Treatment

A patient with a suspected fracture often has injuries elsewhere in the body. Assess the patient for respiratory distress, bleeding, and head or spine injury. If any of these problems occur, emergency treatment is provided before concern is given to extremity or other fractures.

The treatment of fractures depends on the type and extent of the injury. Emergency treatment is essential to prevent possible life-threatening complications. Box 44–8 describes the emergency interventions for the patient with an extremity fracture.

L E A R N I N G TIPS

For emergency care of a suspected fracture, do not try to reposition the limb. Splint it as it lies.

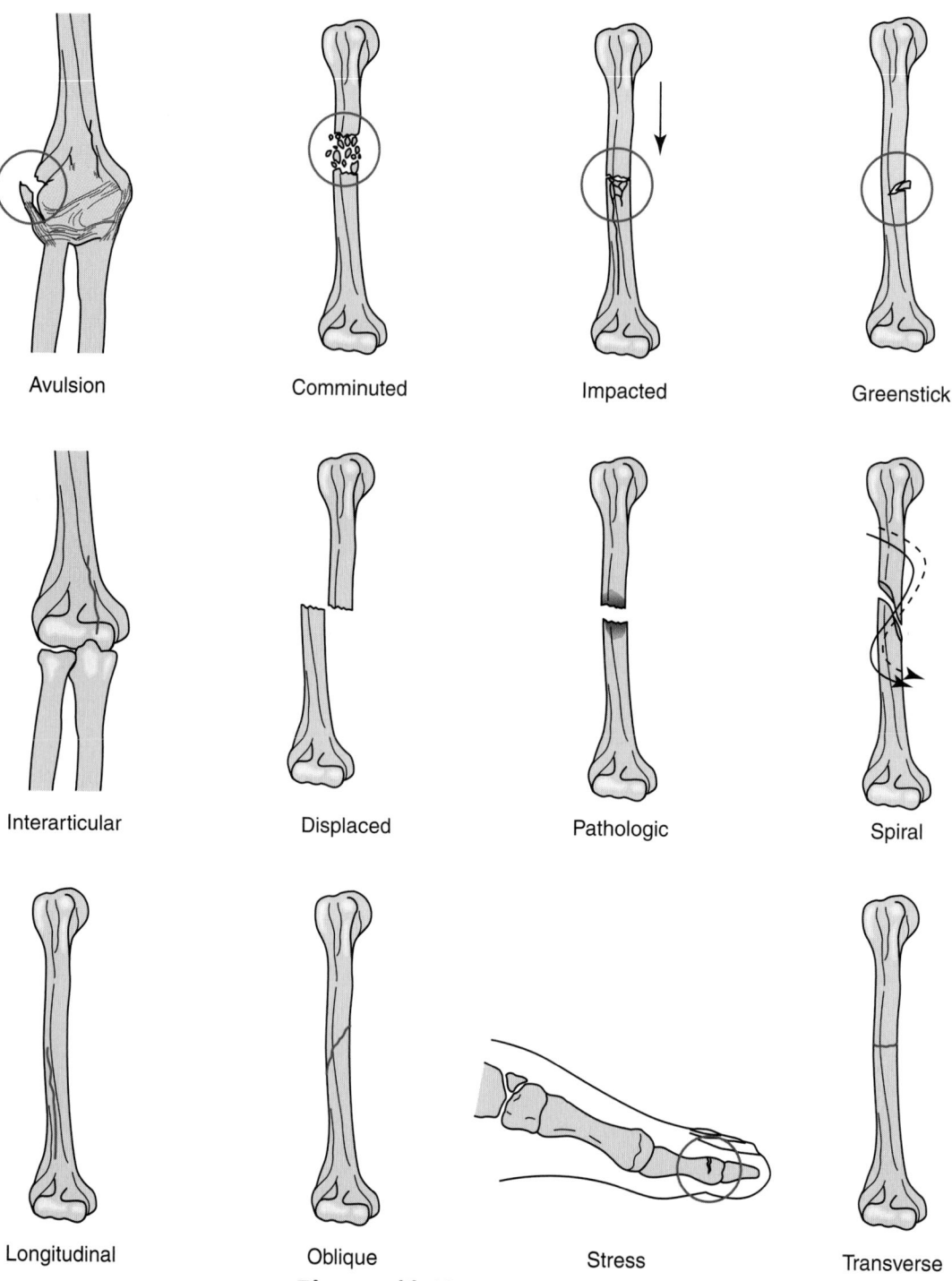

Figure 44-13 Types of fractures.

TABLE 44-5	TYPES OF FRACTURES
Fracture Type	**Description**
Avulsion	Piece of bone is torn away from the main bone while still attached to a ligament or tendon.
Comminuted	Bone splintered or shattered into numerous fragments. Often occurs in crushing injuries.
Impacted	Bone is forcibly pushed together, resulting in bone being pushed into bone.
Greenstick	Bone is bent and fractures on the outer arc of the bend. Often seen in children.
Interarticular	Fracture involves bones within a joint
Displaced	Bone pieces are out of normal alignment. One or both pieces may be out of alignment.
Pathological (also called neoplastic)	Caused by bone's being weakened either by pressure from a tumor or an actual tumor within the bone.
Spiral	Fracture curves around the shaft of the bone.
Longitudinal	Fracture occurs along the length of the bone.
Oblique	Fracture occurs diagonally or at an oblique angle across the bone
Stress	Results in the bone being fractured across one cortex. This is an incomplete fracture.
Transverse	Bone fractured horizontally.
Depressed	Bone pushed inward. Often seen with skull and facial fractures.

BOX 44-8 **Emergency Care of the Patient with an Extremity Fracture**

1. Remove the patient's clothing (cut if necessary) to inspect the affected area.
2. Apply direct pressure on the area if there is bleeding and pressure over the proximal artery nearest the fracture.
3. Keep the patient warm and in a supine position.
4. Check the neurovascular status of the area distal to the extremity: temperature, color, sensation, movement, and capillary refill. Compare affected and unaffected limbs.
5. Immobilize the extremity by splinting; include joints above and below the fracture site.
6. Cover the affected area with a clean cloth (e.g., a handkerchief).

Source: Ignatavicius, DD, Workman, ML, and Mishler, MA: Medical-Surgical Nursing: A Nursing Process Approach, ed 3. WB Saunders, Philadelphia, 1998, with permission.

Fracture Management

The two goals of fracture management are reduction, or realignment of bone ends, and immobilization of the fractured bone with bandages, casts, traction, or a fixation device. These interventions prevent further injury, reduce pain, and promote healing.

CLOSED REDUCTION. Closed reduction is the most common treatment for simple fractures. While manually pulling on the bone (limb), the physician manipulates the bone ends into realignment. Analgesia is typically used before the procedure. An x-ray examination is done to confirm that the bone ends are aligned before the area is immobilized.

BANDAGES AND SPLINTS. For some areas of the body, such as the clavicle or wrist, an elastic or muslin bandage or a splint may be used to immobilize the bone during the healing phase. Splints can be used when the fracture has some associated soft tissue damage that needs care or if there is an expectation of swelling. It is important that the splint be well padded, thereby preventing skin breakdown or unnecessary pressure. Perform neurovascular assessments to ensure adequate blood flow to the area. (See Chapter 43.)

CASTS. Casts provide a strong support for fractured bones, thereby aiding in early mobility and decreased pain. They are also used to correct deformities and to support weak joints while restricting movement. The type of cast used depends on the reason the cast applied. For more extensive fractures or for weight-bearing areas, a more rigid and durable cast is used for immobilization. Once the need for the cast is resolved (e.g., when bone healing is complete), the cast is removed.

Several types of materials are used for casts, including the traditional plaster of paris (anhydrous calcium sulfate) and a variety of synthetic products such as fiberglass. The plaster cast is used for large casts and for weight-bearing areas. Because of a chemical reaction that occurs when the plaster is wet, the cast feels hot when applied for about 30 minutes and then feels cool, taking anywhere from 24 to 72 hours to completely dry. The cast is dry when it feels hard and firm, is odorless, and is shiny white. Keep the wet cast open to air to aid in drying. A wet cast should be handled with the palms of the hand ("palming the cast") to prevent indentations or a change in the shape of the cast (Fig. 44–14). This prevents the possibility of pressure points forming inside of the cast. Unlike plaster of paris, synthetic material casts such as fiberglass harden quickly and dry in less than 2 hours.

A casted limb is elevated for 24 to 48 hours, and ice can be applied above and below the cast to reduce swelling. Assess the cast for dryness, tightness, drainage, and odor. A serious complication of a cast being too tight is compartment syndrome (discussed later). If the cast becomes too tight, the physician orders it to be cut (bivalved) with a cast cutter to relieve pressure and prevent pressure necrosis of the underlying skin (Fig. 44–15). If a wound is present or an odor is detected, the physician cuts a window opening into the cast to treat the underlying skin problem, often an infected area. The cast window should always be taped in

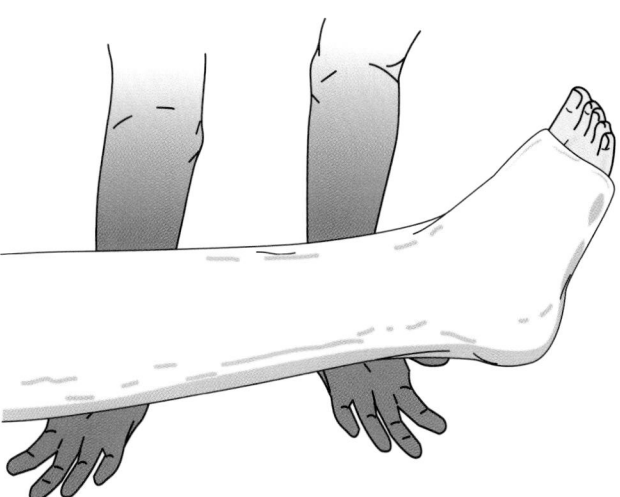

Figure 44-14 A wet plaster cast is moved with the palms of the hand to prevent making indentations in the plaster that could become pressure points.

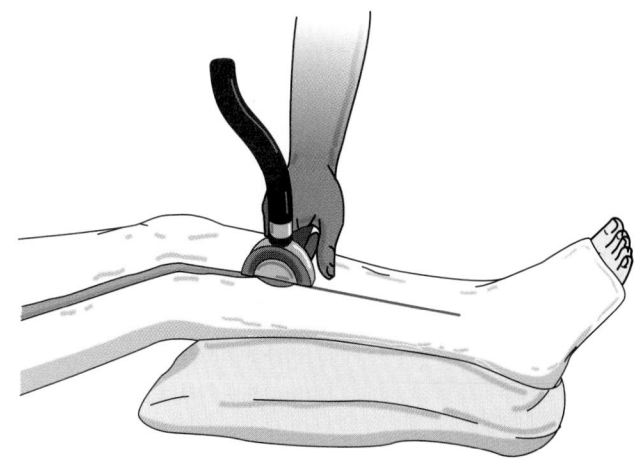

Figure 44-15 Bivalving a cast with a cast saw.

place when wound care is not being provided to prevent the skin from "popping up" through the window and developing pressure points and ischemia. Box 44–9 lists important nursing interventions when caring for a patient who has a cast.

TRACTION. Casts can be worn in or out of a hospital setting, but traction for fracture treatment usually requires that the patient be hospitalized. As a general definition, traction is the application of a pulling force to a part of the body to provide fracture reduction, alignment, or pain relief. Although still used in certain situations, improvements in surgical techniques and orthopedic devices have decreased the use of traction.

Traction is classified as either continuous or intermittent. Continuous traction is required for fracture management; intermittent traction, although not commonly used,

---

BOX 44–9 The Patient with a Cast

1. Monitor the neurovascular status of the casted extremity every 1–2 hours for the first 24 hours and every 4 hours thereafter.
   a. Perform a circulation check, as described in Ch. 43.
   b. Ask the patient if the cast feels too tight.
   c. Have the cast cutter available.
2. Maintain integrity of the cast.
   a. Turn the patient every 1–2 hours.
   b. Use the palms of the hands when handling a wet cast.
   c. Do not turn the patient by holding on to the abductor bar (e.g., hip spica).
   d. Do not cover a wet cast or place it on a plastic-coated pillow.
   e. Protect other parts from irritation caused by the rough surface of a cast made from synthetics.
   f. Keep set plaster cast dry during bathing by covering it completely with plastic (also, tuck plastic into the ends to prevent water seepage under the cast).
   g. Immerse a synthetic cast in water during bathing, if permitted.
   h. Clean a soiled plaster cast with mild detergent and a damp cloth as necessary.
   i. Inspect the cast when performing circulation checks for crumbling and cracking.
3. Maintain skin integrity.
   a. Examine the skin around the cast edges for redness and irritation.
   b. Trim the edges of the cast to prevent roughness.
   c. Petal the edge with 1- to 2-inch adhesive strips if stockinette edging is not used.
   d. Do not use lotions or powder on the skin around the cast.
   e. Teach the patient not to place foreign objects inside the cast (e.g., wire hanger to scratch under the cast).
   f. Smell the cast for foul odor and palpate for hot areas every shift.
   g. Inspect the cast for an increase in drainage every shift.

Source: Ignatavicius, DD, Workman, ML, and Mishler, MA: Medical-Surgical Nursing: A Nursing Process Approach, ed 3. WB Saunders, Philadelphia, 1998, with permission.

---

may be applied for patients experiencing muscle spasm. Most traction is either the "skin" or skeletal type. Skin traction typically involves the use of a Velcro boot (Buck's traction), sling (Russell's traction, or knee sling), belt (pelvic), or halter, which is secured around a part of the body (Fig. 44–16). This type of traction does not promote bone alignment or healing but is used instead for relief of painful muscle spasms that often accompany fractures. Buck's traction is indicated for patients with hip fractures. The weight applied is between 5 and 10 pounds (2.2 to 4.5 kg). Skeletal traction, also called balanced suspension, involves the use of pins (Steinmann), screws, wires (Kirschner), or tongs (Gardner-Wells, Crutchfield), which are surgically inserted into the bone for the purpose of alignment while the fracture heals (Fig. 44–17). From 20 to 40 pounds (9 to 18 kg) of weight are usually applied for skeletal traction, depending on the physician's order.

Balanced suspension maintains the traction while allowing the patient some mobility in bed. A Thomas (or T) splint with Pearson's attachment can be used to provide balanced suspension for the lower extremity. The patient's leg rests on a suspended sheepskin-covered splint.

Caring for the patient in traction includes frequently monitoring neurovascular status for impaired blood flow, checking the equipment to ensure proper functioning, and monitoring skin condition for pressure points or irritation from equipment. Traction must be maintained at all times for fractures. All knots, ropes, weights, and pulleys are inspected every 8 to 12 hours for any loosening and intactness. Weights are to hang unobstructed and should never touch the floor or be removed or lifted. The patient's feet should not rest against the end of the bed. Assistance should be obtained to reposition the patient in bed to prevent lifting injuries, especially with heavy weights in use.

For patients in skeletal traction, pin sites are observed for redness and drainage. Clear, odorless drainage is expected. Some agencies or physicians advocate special solutions or ointments for the skin around the pins (pin care). Others recommend no cleaning to maintain skin integrity. Follow agency policy or physician's order for pin care. Depending on the type of pin used, the pin ends may be covered to protect the patient (and health care workers) from being injured.

Patients who have traction are immobilized for an extended period and often experience problems associated with immobility. For example, pressure ulcers on heels are common among elderly patients in traction. Another common concern is the person's psychosocial health. Ensure that the patient does not become socially isolated because of the need for extended bedrest.

OPEN REDUCTION WITH INTERNAL FIXATION. An open reduction with internal fixation (ORIF) is a treatment reserved for patients who cannot be managed by casts or traction. One of the most common indications for this surgical procedure is fractured hips. Fractures of the hip involve the proximal femur and affect the elderly more than any other

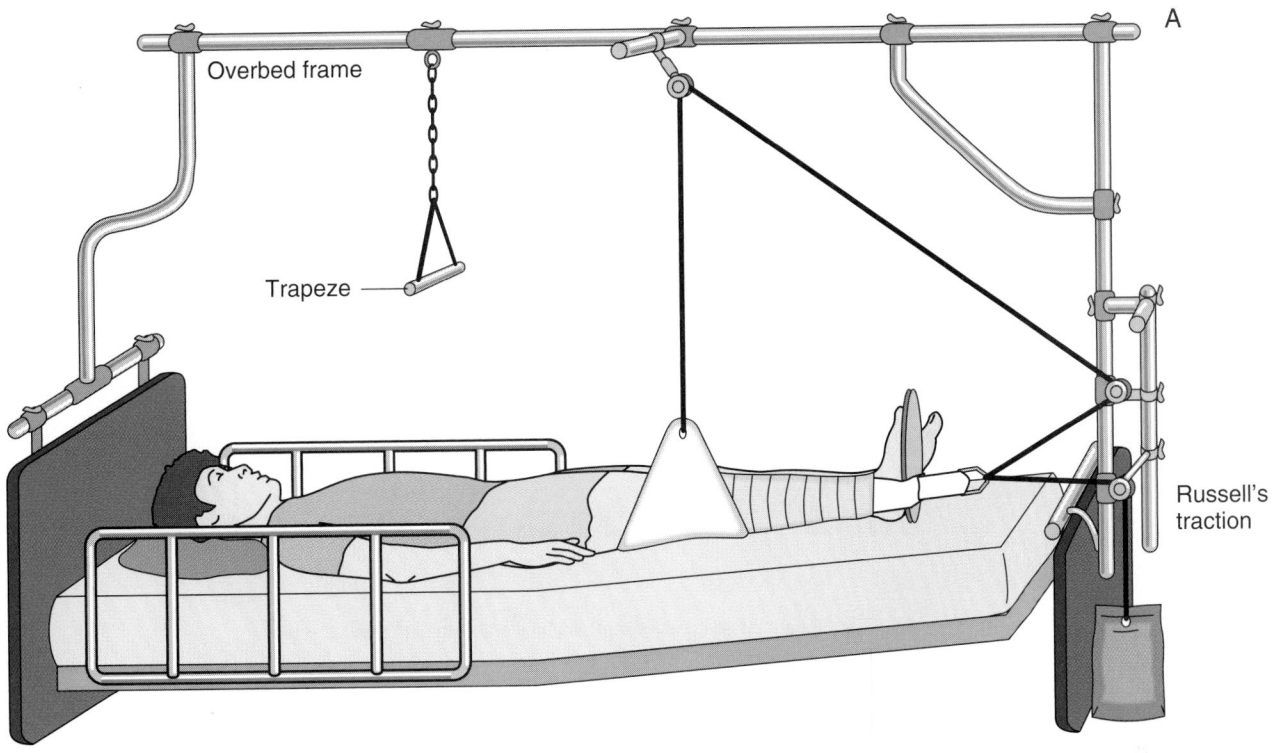

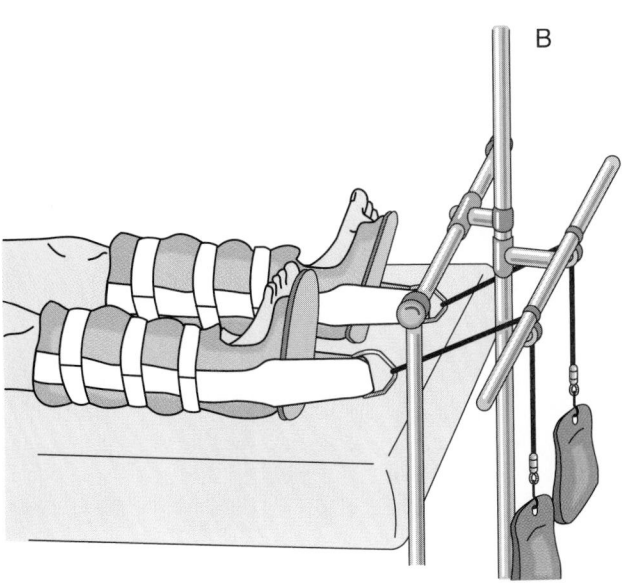

Figure 44-16 Types of skin traction. *(A)* Russell's traction. *(B)* Buck's (boot) traction.

age group. ORIF of the hip allows early mobilization while the bone is healing.

As the name implies, the bone ends are realigned (reduced) by direct visualization through a surgical incision. The bone ends are held in place by internal fixation (IF) devices such as metal plates and screws or by a prosthesis with a femoral component similar to that for total joint replacement (Fig. 44–18). For hip surgery, the IF device is not removed after the fracture heals. For ankle or long bone surgery, the hardware may need to be removed after healing to prevent it from loosening over time (Nursing Care Plan Box 44–10).

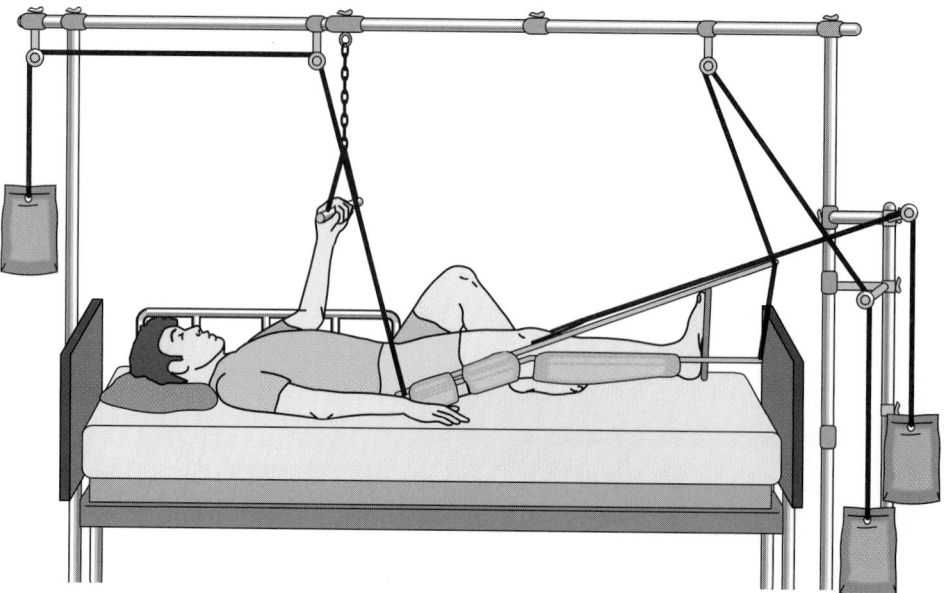

Figure 44-17 Balanced suspension and skeletal traction for femur fracture.

CRITICAL THINKING

Mrs. Brown

Mrs. Brown, a long-time resident of Happy Hills Care Center, was found lying in the dayroom on her left side, moaning and holding her leg at 10 A.M. (1/4/03). She cried out with any movement and said she fell and broke her leg. The supervisor notified the paramedics and Dr. Jones. Her vital signs are blood pressure 150/84, pulse 100, respirations 20. The licensed practical nurse (LPN) remained with Mrs. Brown and instructed her not to move until help arrived. The LPN got blankets and a pillow for her head. The paramedics arrived quickly and took Mrs. Brown to nearby Grace Hospital by ambulance, where she was diagnosed as having a nondisplaced femoral neck (hip) fracture. Dr. Jones ordered 5 pounds of Buck's traction. Mrs. Brown is restless and picking at her bedcovers when you assess her at the beginning of your shift.

1. How would you document the incident of Mrs. Brown's fall?
2. What is the purpose of Buck's traction for Mrs. Brown?
3. What are your nursing responsibilities while caring for Mrs. Brown?
4. What might explain her restlessness?

Answers at end of chapter.

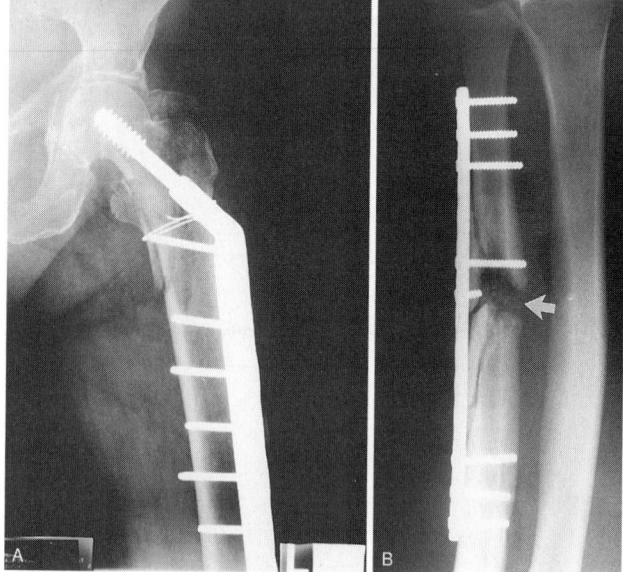

Figure 44-18 Internal fixation. *(A)* Intertrochanteric fracture of the hip with fracture fixation via a side plate and screw combination device. *(B)* Side plate and screw fixation of radial fracture. (From McKinnis, LN: Fundamentals of Orthopedic Radiology. FA Davis, Philadelphia, 1997, with permission.)

CRITICAL THINKING

Tommy Martin

Tommy, 18, was in a motor vehicle accident, resulting in a fractured pelvis. He is to be in pelvic traction for about 8 to 10 weeks.

1. Identify three nursing diagnoses related to Tommy's physical or emotional well-being.
2. What are some of the nursing interventions related to these diagnoses?

Answers at end of chapter.

EXTERNAL FIXATION. An alternative treatment for some fractures is external fixation. External fixation is used when there has been severe bone damage, such as in crushed or splintered fractures, or if there have been numerous fractures along the bone. After the fracture is reduced, the physician surgically inserts pins into the bone; the pins are held in place by an external metal frame to prevent bone movement (Fig. 44–19). External fixation is ideal for the patient who has an open fracture with soft tissue damage

BOX 44-10 NURSING CARE PLAN FOR THE PATIENT AFTER OPEN REDUCTION WITH INTERNAL FIXATION (ORIF) OF THE HIP

▶ Pain related to surgical wound

PATIENT OUTCOME
Patient states that pain relief is satisfactory.

EVALUATION OF OUTCOME
Does patient state that pain is absent or at tolerable level (pain rated 0 to 2 on pain assessment scale)?

INTERVENTION	RATIONALE	EVALUATION
Give pain medication as needed; anticipate need for pain medication.	Pain medication relieves pain, especially if given before pain is severe.	Does patient state pain is relieved?
Give pain medication before activity (e.g., session with physical therapist).	Increased activity can cause pain.	Is patient restless or agitated during activity?
Use nondrug pain relief measures, such as distraction, guided imagery, other relaxation techniques.	Analgesic therapy is enhanced with complementary pain relief measures.	Does patient report pain relief is enhanced with music or relaxation?
Use fracture bedpan.	Fracture bedpans are more comfortable and easier to position for patients.	Is patient able to use fracture pan with comfort?

▶ Impaired physical mobility related to hip precautions and surgical pain

PATIENT OUTCOME
Patient will maintain desired level of activity.

EVALUATION OF OUTCOME
Does patient maintain activity desired?

INTERVENTION	RATIONALE	EVALUATION
Reinforce transfer and ambulation techniques.	Activity is restricted due to hip precautions and weight-bearing limitations.	Does patient transfer and ambulate as instructed by physical therapist?
Place overhead frame and trapeze on bed; teach patient how to use it.	Patient mobility is increased and pain decreased with use of trapeze for movement.	Does patient use overhead frame and trapeze for movement in bed with less pain?
Assess patient for and take measures to prevent complications of immobility: turn patient every 2 hours and check skin; keep heels off bed; teach patient to deep breathe and cough q2h; teach use of incentive spirometer.	Immobility complications can occur if preventive measures are not used.	Does patient experience complications of immobility?
• Apply thigh-high elastic stockings or sequential compression device to unaffected limb as ordered.		
• Give anticoagulants as ordered.		
• Get patient out of bed as soon as ordered.		
• Ambulate patient as early as possible.		
• Remind patient to practice leg exercises.		

that needs to be treated at the same time. Like skeletal traction, the patient with this device is at risk for pin site infection. Pin care may or may not be required, but pin sites are observed frequently for signs and symptoms of infection. See Nursing Care Plan Box 44–11.

NONUNION MODALITIES. Although most bones heal properly with the correct treatment, some patients experience malunion (malalignment of healed bone) or nonunion

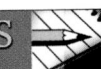 **L E A R N I N G   TIPS**

If you have to move an extremity that has an external fixation device, grasp the device and lift, raise, or move the limb as needed. By grasping the device, there is less movement of the healing bone and therefore less trauma to the site of healing and less pain with movement. Care must be taken not to loosen any fasteners holding the pins in place.

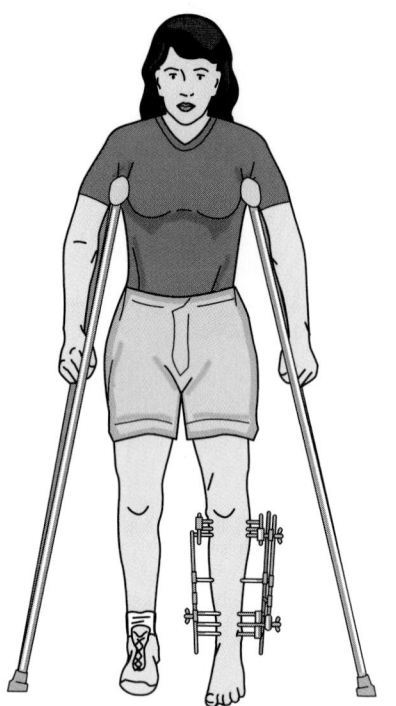

Figure 44-19   External fixation for complex fractures and wound care.

(delayed or no healing). A number of variables influence how a bone heals, including age, nutritional status, and the presence of other diseases that alter the healing process, such as diabetes mellitus.

Several methods for treating nonunion are available, including electrical bone stimulation and bone grafting. For selected patients, bone stimulation may be effective in promoting healing; the exact mechanism of action is not known. Bone grafting involves adding packed bone to the fracture site in an attempt to facilitate healing. Bone stimulating compounds such as Osteoset, Proosteon or Allomatrix are being used to promote bone growth in patients. These compounds are used during surgical procedures as glue, cement, or filler.

Another fracture healing method is low-intensity pulsed ultrasound (also called Exogen therapy). Ultrasound treatment has provided excellent results for slow-healing fractures, as well as for new fractures. The patient applies the treatment for about 20 minutes each day.

Complications of Fractures

Monitor for possible complications and implement interventions to prevent them. The most common complications include hemorrhage, infection, and thromboembolitic

---

BOX 44-11	NURSING CARE PLAN FOR THE PATIENT WITH EXTERNAL FIXATION OF THE LOWER EXTREMITY

 Risk for infection related to skin integrity impairment

PATIENT OUTCOMES
Patient does not develop an infection.

EVALUATION OF OUTCOME
Does patient remain free from infection?

INTERVENTIONS	RATIONALE	EVALUATION
Inspect dressings, wounds, pin sites for signs and symptoms of infection.	Signs and symptoms of infection could include warmth, redness, heat, swelling, drainage, pain.	Are any wounds infected?
Monitor color of and measure wound drainage.	Wound drainage color and amount can indicate severity of infection.	Does wound have large amount of purulent drainage?
Change dressings or provide wound and pin care per facility policy using aseptic technique.	Use of aseptic technique minimizes chance of infection. Pin wound sites should be free of crusting, which promotes infections because of decreased skin integrity.	Are pin sites clean with no crusting?
Monitor vital signs frequently.	Alterations in vital signs can indicate infection.	Are vital signs within baseline findings?

 Impaired physical mobility related to the external fixation (EF) device

PATIENT OUTCOMES
Patient will maintain desired level of mobility/activity.

EVALUATION OF OUTCOME
Has patient maintained desired level of mobility and activity?

INTERVENTIONS	RATIONALE	EVALUATION
Reinforce transfer and ambulation techniques.	Depending on severity of fracture and size of EF device, there may be special needs to transfer and ambulate.	Does patient transfer and ambulate as instructed?

BOX 44-11 NURSING CARE PLAN FOR THE PATIENT WITH EXTERNAL FIXATION OF THE LOWER EXTREMITY—cont'd

Place overhead frame and trapeze on bed; teach patient how to use them.	Patient mobility is increased and pain decreased with use of trapeze for movement.	Does patient use overhead frame and trapeze for movement with less pain?
Teach patient how to move limb using EF device.	Providing patient with instruction on moving the extremity promotes independence and minimizes pain.	Does patient move the extremity using EF device?
Assess patient for and take measures to prevent complications of immobility. Promote early ambulation to minimize complications. Include other disciplines such as the physiotherapist in promoting and teaching about ambulation.	Immobility complications can occur if preventative measures are not used. EF devices allow for earlier ambulation. Physiotherapy can provide initial or reinforce the education needed to promote ambulation (e.g., with crutch walking).	Does patient have any complications of immobility? Has patient used information learned from other disciplines to aid ambulation?

 Disturbed body image related to external fixation device

PATIENT OUTCOMES

Patient will not experience disturbed body image while EF device is in place.

EVALUATION OF OUTCOME

Does patient experience disturbed body image resulting from EF device?

INTERVENTIONS	RATIONALE	EVALUATION
If possible, explain to patient preoperatively what EF device will look like.	Preparing patient for what to expect postoperatively increases likelihood of acceptance and minimizes the unknown.	Was patient able to verbalize why device is to be used and what device will look like?
Reinforce the idea that EF device will decrease discomfort and allow for earlier ambulation.	Promoting early ambulation and increased comfort enhance acceptance.	Did patient understand benefit of EF device allowing for early ambulation and increased comfort?
Provide psychological support and an environment of acceptance.	Accepting your patient and allowing for discussion of concerns promotes a sense of well-being and acceptance of EF device.	Did patient feel comfortable in expressing concerns related to body image?

complications. These problems are discussed in more detail in other parts of this text. Although they do not occur often, acute compartment syndrome and fat embolism syndrome (more common with fractures of long bones) can be life-threatening complications seen with fractures.

HEMORRHAGE. Bone is highly vascular, and damage or surgery to bone can cause bleeding. Assess for bleeding and monitor vital signs carefully. Hypovolemic shock may result from severe hemorrhage. (See Chapter 8.)

INFECTION. Trauma predisposes the body to infection, especially when the skin, the body's first line of defense, is disrupted. Wound infections, pin site infections, and osteomyelitis (bone infection) are common. Hospital-acquired infections, such as pneumonia or urinary tract infection, can occur in elderly patients who are immobilized for extensive periods while their fractures heal.

THROMBOEMBOLITIC COMPLICATIONS. Deep vein thrombosis or pulmonary embolus (PE) also develop in patients who are immobile because of trauma or surgery. Thromboembolitic complications are the most common problems of lower extremity surgery or trauma and the most fatal complication of musculoskeletal surgery, particularly in the elderly. Leg exercises, early ambulation, and anticoagulant therapy, usually using low-molecular-weight heparin

such as dalteparin (Fragmin) or enoxaparin (Lovenox), help prevent these problems.

ACUTE COMPARTMENT SYNDROME. Compartments are sheaths of fibrous tissue that support and partition nerves, muscles, and blood vessels, primarily in the extremities (Fig. 44-20). There are several compartments within each extremity. Acute compartment syndrome (ACS) is a serious problem in which the pressure within one or more extremity compartments increases, causing massive circulation impairment to the area. An external device such as a cast or bulky dressing can increase pressure when there is tissue swelling or compression in the area. The early symptom of ACS is the patient's report of severe, increasing pain that is not relieved with narcotics and occurs more on active movement than passive movement. Decreased sensation follows before ischemia becomes severe. In severe ACS, the patient has the six *P*s:

- Pain (severe, unrelenting, and increased with passive stretching)
- Paresthesia (painful tingling or burning)
- Paralysis (late symptom)
- Pallor (but there may be warmth or redness over the area)
- Pulselessness (late and ominous sign)
- Poikilothermia (temperature matches environment; i.e., the extremity is cool to touch)

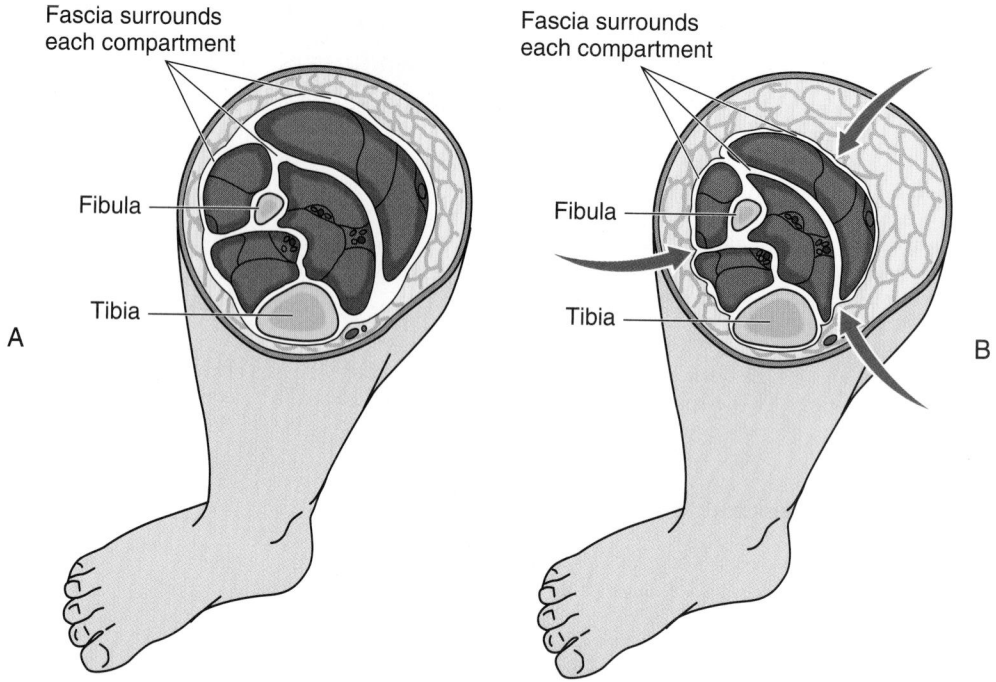

Figure 44-20 *(A)* Lower leg compartments. Each compartment contains muscles, an artery, a vein, and a nerve. *(B)* Compartment syndrome. Increased pressure in a compartment compresses structures within the compartment.

Relief of pressure is the goal. It may be accomplished by removing the source of pressure, such as bivalving a cast, or by performing a **fasciotomy,** which is an incision into the fascia that encloses the compartment. This incision allows the compartment tissue room to expand and relieves the pressure. If more than one compartment has increased pressure, multiple fasciotomies are required. These surgical wounds remain open until the pressure decreases. Then they are closed and may require skin grafting. If this condition continues without pressure relief, tissue necrosis, infection, extremity contracture, or renal failure may result. Renal failure is a potentially fatal complication of ACS.

🔊 CRITICAL THINKING

Mr. Andrews

Mr. Andrews has suffered a nondisplaced fracture of his right femur. He has a cast on from his groin to the middle of his foot. An hour ago he received 10 mg of morphine intravenously (IV) and he is complaining of continuing and increasing pain.

1. What nursing assessment should now be performed?
2. What might be happening with Mr. Andrews?
3. What interventions may be necessary?

Answers at end of chapter.

FAT EMBOLISM SYNDROME. Fat embolism syndrome (FES) is another serious complication in which small fat globules are released from yellow bone marrow into the bloodstream. The globules then travel to the lung fields,

causing respiratory distress. This process most often occurs when long bones (especially the femoral shaft) are fractured or when the patient has multiple fractures. The elderly patient with a fractured hip is also at a high risk for FES. This condition can occur up to 72 hours after the initial injury or procedure.

The earliest manifestation of FES is altered mental status resulting from a low arterial oxygen level. The patient then experiences tachycardia, tachypnea, fever, high blood pressure, and severe respiratory distress (shortness of breath). Most patients also have a measleslike rash, called petechiae, over the upper body. Even when aggressively treated, patients with FES often die from the pulmonary edema that typically develops. Note early FES signs and symptoms and report them to the physician immediately. If a fat embolism is suspected, the following actions should be taken:

- Promote oxygenation by administering oxygen at 2 L per minute via nasal cannula.
- Maintain bedrest and keep movement of extremity to a minimum.
- Prepare patient for a chest x-ray examination or lung scan.
- Administer intravenous fluids as ordered.
- Administer corticosteroids as ordered.
- Provide emotional support and calm environment.

Nursing Process

Caring for the patient with a fracture may require an interdisciplinary approach, especially for complex or multiple fractures. Coordinate care with other health team members.

ASSESSMENT. The most important aspect of assessment for the patient with a fracture is frequent monitoring of neu-

---

fasciotomy: fascia—fibrous tissue + otomy—opening into

rovascular status distal to the fracture site. As mentioned earlier, acute compartment syndrome is a potentially limb- or life-threatening complication that results when blood flow is impaired. Chapter 43 describes the procedure for this assessment and its significance.

Pain is managed by both medications and complementary therapies. Bone pain can be excruciating and must be treated aggressively. For the patient who cannot report pain, particularly a confused or comatose patient, ensure that pain relief is maintained.

LEARNING TIPS

A confused or comatose patient may not be able to report pain, the most reliable indicator of pain. Nonverbal indicators (e.g., grimacing, restlessness, elevated blood pressure and heart rate) are not reliable for pain assessment and should not be used to determine pain absence. You owe it to your patients to prevent their pain by anticipating it and treating it in advance. You can do this by recognizing causes of pain and understanding that the effects of mild but repetitive pain (as in turning several times a day) can adversely affect the patient (such as by leading to exhaustion). Causes of pain include conditions or diseases (such as fractures, surgery, trauma, or cancer), procedures (such as turning or wound care), and biomedical devices (such as orthopedic fixation devices, wound drains, Foley catheters, nasogastric tubes, and chest tubes).

With few patients medicated before painful procedures, some of which may be done several times a day (turning), confused or comatose patients are at greater risk for lack of pain relief. Provide analgesics before painful procedures and on a regular basis when pain is assumed to be present to keep your patients comfortable. For anticipated pain, the acronym APP (assume pain present) is used instead of a pain rating scale. Use accepted clinical practice guidelines as a rationale for providing your patients the best possible pain relief. Share pain research findings with administrators in your institution to establish policies that support proactive pain management for all patients.

NURSING DIAGNOSIS. The common nursing diagnoses for the patient with a fracture may include the following:

■ Acute pain
■ Impaired physical mobility
■ Impaired walking
■ Ineffective health maintenance
■ Risk for peripheral neurovascular dysfunction
■ Risk for ineffective tissue perfusion
■ Risk for ineffective skin integrity

PLANNING. The plan of care is derived from the nursing diagnoses and risk for medical complications. Patients with uncomplicated fractures are treated on an outpatient basis. Patients with multiple or more complex fractures, especially involving soft tissue injury, are admitted to the hospital for care. To ensure continuity of care, the case manager involves the various disciplines involved in the patient's care from the time of admission through the continuing care phase.

IMPLEMENTATION. The most common fracture that leads to hospitalization is the hip fracture. Nursing Care Plan 44–10 highlights the most important aspects of care for the patient with a fractured hip.

EVALUATION. The expected outcomes for the patient are relief of pain, healing without complications, and return to or improvement in their previous level of functioning. The interdisciplinary team evaluates care to ensure that it is directed toward meeting these outcomes.

Patient Education

If the patient has a cast, review the appropriate instructions for cast care. Teach the patient not to insert objects (e.g., pencils, sharp objects) into the cast to scratch an itchy area. Health teaching is also important for care of the extremity after cast removal (Box 44–12). The patient may also have a wound including pin site care that will be managed at home. Teach the patient and caregiver how to assess and dress the wound and when to report changes such as signs and symptoms of infection.

Nutritional education is also essential. The body needs adequate protein, calories, vitamins, and minerals for healing to occur. Unless otherwise contraindicated, milkshakes and instant breakfast preparations are good sources of additional protein and calories, as well as a source of calcium.

Osteomyelitis

Osteomyelitis is an infection of bone that can be either acute or chronic. A bone infection lasting less than 4 weeks is considered acute; one that lasts more than 4 weeks is chronic.

Pathophysiology

Regardless of type, osteomyelitis results from invasion of bacteria into bone and surrounding soft tissues. Inflammation occurs, followed by ischemia (decreased blood flow) (Fig. 44–21). Bone tissue then becomes necrotic (dies), which retards healing and causes more infection, often as a bone abscess.

BOX 44-12 **Care of the Extremity After Cast Removal**

- Remove scaly, dead skin carefully by soaking—do not scrub.
- Move the extremity carefully. Expect discomfort, weakness, and decreased range of motion.
- Support the extremity with pillows or your orthotic device until strength and movement return.
- Exercise slowly as instructed by your physical therapist.
- Wear support stockings or elastic bandages to prevent swelling (for lower extremity).

Source: Ignatavicius, DD, Workman, ML, and Mishler, MA: Medical-Surgical Nursing: A Nursing Process Approach, ed 3. WB Saunders, Philadelphia, 1998, with permission.

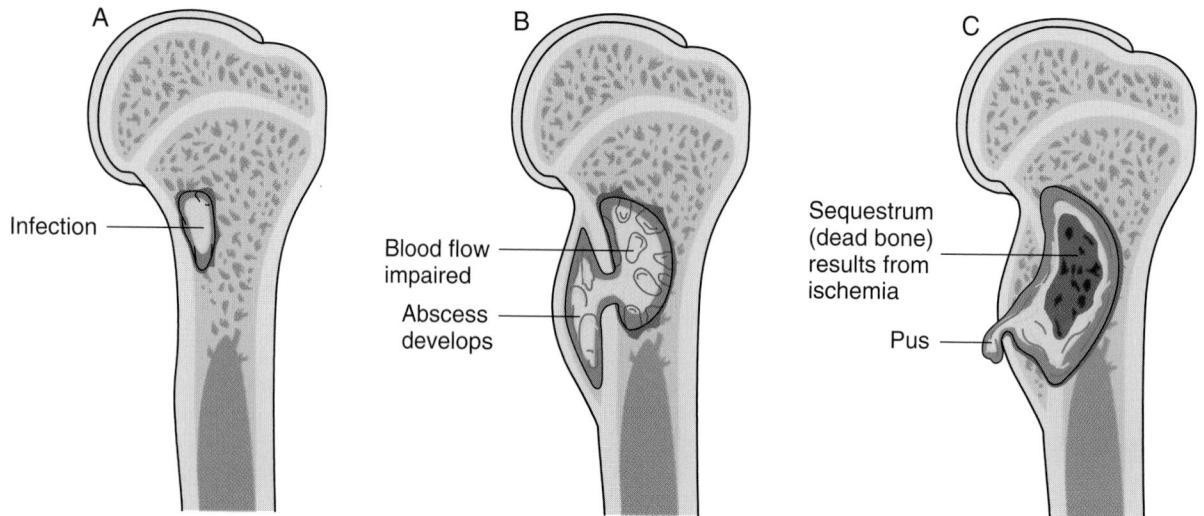

Figure 44-21 Sequence of osteomyelitis development.*(A)* Infection begins. *(B)* Blood flow is blocked in the area of infection. An abscess with pus forms. *(C)* Bone dies within the infection site, and pus formation continues.

Pathogens enter bone in several ways. Direct inoculation means that an injury to the body allows the offending microbes direct access to bone tissue. An open fracture is an example of that process. Contiguous spread occurs when surrounding soft tissue becomes infected. An example is the patient with cellulitis whose infection then spreads to underlying bone. In hematogenous spread, an infection beginning in another part of the body migrates to bone. For instance, a patient with a total hip replacement may get osteomyelitis from a urinary tract infection.

Causes and Types

Penetrating trauma leads to acute osteomyelitis by direct inoculation. The most common pathogen is *Pseudomonas aeruginosa*. The leading cause of contiguous spread is a slow-healing foot ulcer in the patient who has diabetes mellitus or peripheral vascular disease. Multiple organisms may be present in the wound and subsequently the bone. Hematogenous spread results from bacteremia (infection of the blood), underlying disease, or nonpenetrating trauma. Long-term intravenous catheters are primary sources of infection.

Signs and Symptoms

The patient with acute osteomyelitis has fever, as well as local signs of inflammation, such as tenderness, redness, heat, pain, and swelling. Pain may be the only apparent complaint. Ulceration, drainage, and localized pain are typical signs and symptoms of chronic osteomyelitis.

Diagnostic Tests

The patient with osteomyelitis typically has an elevated leukocyte (white blood cell) count, an elevated erythrocyte sedimentation rate, and positive bone biopsy for infection. Some patients also have a positive blood culture. MRIs, x-ray examinations, and CT scans can show areas of infection.

Treatment

Long-term antibiotic therapy is the treatment of choice for patients with bone infection. Infection in bone tissue is difficult to resolve and may require weeks to months of medication. Patients often administer their intravenous antibiotics at home rather than have a costly stay in a hospital. Central venous access devices, such as Hickman, Broviac, or Groshong catheters, as well as peripherally inserted central catheters (PICC), are used for intravenous drug administration. Teach the patient and caregiver about the side effects, toxicity, interactions, and precautions for antibiotic therapy. A home care nurse may be needed to assist the patient.

If a soft tissue wound is present, irrigations and dressing changes are performed by the home health nurse or taught to the patient and family. Teach the patient about the importance of handwashing and how to avoid the spread of pathogens. Some patients are placed on standard precautions or other isolation precautions, depending on the offending organism.

Antibiotic therapy alone may not resolve the infection. Patients with chronic osteomyelitis may require surgery to remove necrotic bone tissue or replace it with healthy bone tissue. Amputations are reserved for patients who have massive infections that have not responded to one or more of the conventional treatments.

Osteoporosis

Osteoporosis is a common metabolic disorder in which the bone loses its density, resulting in fragile bones and possibly fractures. The wrist, hip, and vertebral column are most commonly involved.

Pathophysiology

Bone is dynamic, constantly building new tissue and resorbing (breaking down) old tissue. Bone density (mass) peaks between 30 and 35 years of age. After these peak years, the rate of bone breakdown exceeds the rate of bone building. The result is irreversible bone loss that worsens with aging. Trabecular (cancellous) bone is lost first, followed by a loss of cortical (compact) bone. As a result, more than 1.5 million fractures occur each year in people older than age 45. Hip fractures are among the most common, accounting for more than 250,000 per year. The mortality rate for hip fractures is about 50 percent during the first year after the fracture. For postmenopausal women, decreased estrogen appears to slow down the absorption of calcium, resulting in an increased bone loss.

Cause and Types

Osteoporosis is either primary or secondary. Primary osteoporosis is the most common and is not associated with another disease or health problem. Risk factors for primary osteoporosis include the following:

- Caucasian or Asian heritage, postmenopausal, female (less estrogen available to protect bone)
- Sedentary lifestyle
- Decreased calcium intake
- Lack of vitamin D (to absorb calcium)
- Excessive alcohol consumption
- Cigarette smoking
- Excessive caffeine intake
- Petite body build

Secondary osteoporosis results from an associated medical condition, such as hyperparathyroidism; long-term drug therapy, especially steroids; and prolonged immobility, such as that seen with patients who have a spinal cord injury.

Signs and Symptoms

Most women do not realize they have osteoporosis until they fracture a bone. During the late middle years, the classic "dowager's hump," or kyphosis of the spine, is usually present. The patient's height decreases and back pain may be present. The patient may be embarrassed by the change in body image and may have curtailed social activities. Some patients have difficulty finding clothes that fit comfortably.

Diagnostic Tests

X-ray examination of the bone is not helpful in diagnosing bone loss in its early stages. Computed tomography and quantitative CT scans detect early spine changes and measure bone density. Ultrasound can also be used to screen bone mass. Dual-energy x-ray absorptiometry (DXA) is used as a screening tool to measure bone mineral content. This test is noninvasive and is currently the most widely used technique to measure bone density.

Serum calcium and vitamin D levels may be decreased, and serum phosphorus may be increased. With severe bone loss, alkaline phosphatase levels may be elevated, confirming bone damage.

Treatment

The cornerstone of treatment for osteoporosis is medication and avoidance of modifiable risk factors to prevent bone loss.

MEDICATION. Medication may be used for prevention or treatment purposes. The current drugs of choice calcium supplements, vitamin D, and biophosphonates, such as alendronate (Fosamax) and risedronate (Actonel).

Calcium is also important to prevent bone loss. If serum calcium falls below normal levels, the parathyroid glands stimulate the bone to release calcium into the bloodstream. The result is demineralized bone. Therefore calcium supplements are an important aspect of treatment. Teach the patient to drink plenty of fluids to prevent calcium-based urinary stones. Vitamin D supplementation, to aid calcium absorption, may also be necessary for patients who have inadequate sunlight exposure (institutionalized people) or who cannot metabolize vitamin D.

Alendronate and risedronate are used to prevent or slow the progress of osteoporosis. They suppress osteoclast activity to prevent the breakdown of bone. Although side effects are not common, serious cases of esophagitis and esophageal ulcers have been reported. Therefore teach the patient to take the drug early in the morning and follow it with a full glass of water. The patient should not lie down for at least 1 hour after taking the drug.

The newest drug class for osteoporosis is the selective estrogen receptor modulator (SERM). Recently approved for use, raloxifene (Evista) increases bone mass 2 to 3 percent each year. SERM drugs are designed to mimic estrogen in some parts of the body while blocking its effects elsewhere.

Other drugs that may be used include testosterone (the male hormone that helps build bone), calcitonin (nasal spray, injection), and sodium fluoride. All these medications have major disadvantages and are consequently not commonly given. Any drug used to prevent or control osteoporosis must be administered under the supervision of a physician including supplements.

DIET. Increasing calcium and fluids are the main dietary considerations for women. Teach patients what foods are high in calcium, such as dairy products and dark green, leafy vegetables. If the patient consumes excessive caffeine or alcohol, teach about the need to avoid these substances. For more information, visit the National Osteoporosis Foundation at www.nof.org.

EXERCISE. Weight-bearing exercise, especially walking, stimulates bone building. The patient should wear well-supporting, nonskid shoes at all times and avoid uneven surfaces that could contribute to falls. Resistance exercise such as weight training or the use of some of the equipment available at fitness centers is also beneficial.

FALL PREVENTION. Osteoporotic bone may cause a pathological fracture in which the hip breaks before the fall. For other patients, a fall can cause a hip or other fracture. Therefore fall prevention programs in hospitals and nursing homes are important.

In collaboration with the physical or occupational therapist, case manager, or discharge planner, assess the patient's home environment. The patient and family are taught how to create a hazard-free environment, such as avoiding scatter rugs and slippery floors. Walking paths in the home must be kept free of clutter to prevent falls. If needed, a walker or cane provides additional support.

Paget's Disease

Paget's disease, also called osteitis deformans, is a metabolic bone disease in which increased bone loss results in large, disorganized bone deposits throughout the body. It is primarily a disease of the elderly.

Pathophysiology

Three phases of the disorder have been described: active, mixed, and inactive. A prolific increase in osteoclasts (cells that break down bone) causes massive bone deformity and destruction. Osteoblasts (bone-building cells) then react to form new bone. However, the result is disorganized in structure. Finally, when osteoblastic activity exceeds the osteoclastic activity, the inactive phase occurs. The newly formed bone becomes sclerotic with increased vascularity.

Paget's disease can affect one or multiple bones. The most common areas involved are the femur, skull, vertebrae, and pelvis.

Causes and Types

The exact cause of this disease is not known, but it tends to run in families. Paget's disease may be the result of a latent viral infection contracted in young adulthood. It is more common in Europe than in the United States.

Signs and Symptoms

Most patients with Paget's disease have no obvious symptoms, particularly when the disorder is confined to one bone. For patients with more severe disease, signs and symptoms are varied and potentially fatal. Table 44–6 lists the key features of this disease.

Diagnostic Tests

Diagnosis may be made solely on x-ray findings. Radiographs of pagetic bone show punched-out areas indicating increased bone resorption. The overall mass of bone may be enlarged, depending on the phase of the disorder. Deformities, fractures, and arthritic changes are not uncommon. Bone scans can also be used to help in the diagnosis.

The primary laboratory findings are an increased alkaline phosphatase (ALP) and an increase in urinary hydroxyproline. Pyrilinks and Osteomark are urine tests that can be used in place of the urinary hydroxyproline test. ALP reflects bone damage. Urinary hydroxyproline indicates an increase in bone turnover. The higher the level, the more severe the disease. Calcium levels in both blood and urine are elevated as damaged bone releases calcium into the bloodstream.

Treatment

Nonsurgical management is employed to relieve pain and promote a reasonable quality of life for the patient. For mild disease, NSAIDs are given.

MEDICATION. The purpose of drug therapy for the patient with Paget's disease is to relieve pain and decrease bone loss. Calcitonin (Calcimar) is a thyroid hormone that is often effective in initiating a remission of the disease. It appears to decrease bone loss while also decreasing pain. If effective, the ALP level decreases. The usual duration of therapy is 6 months, followed by 6 months of etidronate disodium (Didronel) or another biphosphate drug. Its action is similar to that of calcitonin, and it must be taken on an empty stomach.

Plicamycin (Mithramycin, Mithracin) is a potent anticancer drug and antibiotic that is reserved for patients with severe hypercalcemia or severe disease with neurological involvement. This drug suppresses both osteoclastic and osteoblastic activity within days, but it has serious adverse effects. As with all drugs, observe for toxic effects such as liver and kidney failure. Platelet count is monitored because the drug can decrease platelet production. When liver enzymes become too high, the drug is temporarily discontinued until they return to baseline.

Alendronate is a bone resorption inhibitor and calcium regulator. Intravenous dosing for 5 days may initiate a disease remission.

TABLE 44–6	KEY FEATURES OF PAGET'S DISEASE OF THE BONE
Musculoskeletal manifestations	Bone and joint pain (may be in a single bone) that is aching, poorly described, and aggravated by walking
	Low back and sciatic nerve pain
	Bowing of long bones
	Loss of normal spinal curvature
	Enlarged, thick skull
	Pathological fractures
	Osteogenic sarcoma
Skin manifestations	Flushed, warm skin
Other manifestations	Apathy, lethargy, fatigue
	Hyperparathyroidism
	Gout
	Urinary or renal stones
	Heart failure from fluid overload

Source: Ignatavicius, DD, Workman, ML, and Mishler, MA: Medical-Surgical Nursing: A Nursing Process Approach, ed 3. WB Saunders, Philadelphia, 1998, with permission.

OTHER TREATMENT. Additional management for Paget's disease depends on the clinical manifestations. For example, if the patient has osteogenic sarcoma, the appropriate treatment for bone tumors is initiated.

Bone Cancer

Bone tumors may be benign or malignant. Malignant tumors may be either primary (originating in the bone) or metastatic, originating from another location and migrating to bone. Primary bone tumors tend to develop in people under 30 years of age and account for only a small percentage of bone cancers. Metastatic lesions are much more common and most often affect the elderly.

Pathophysiology

The pathophysiology depends on the type of bone cancer. The cause of bone cancer is not known.

Primary Malignant Tumors

Osteosarcoma, or osteogenic sarcoma, is the most common primary malignant bone tumor. It is a fairly large tumor that typically metastasizes to the lung within 2 years of diagnosis and treatment. More than 50 percent of osteosarcomas occur in the distal femur in young men. Older patients with Paget's disease may also develop these lesions.

Ewing's sarcoma is the most malignant bone tumor. In addition to local pain and swelling, systemic signs and symptoms, including low-grade fever, leukocytosis, and anemia, are common. The pelvis and lower extremity are most often affected in children and young men.

Patients with a chondrosarcoma (cancer of cartilaginous cells) have a better prognosis than those with the previously described types of bone cancer. This type of cancer occurs in middle-aged and older people.

METASTATIC BONE DISEASE. Primary malignant tumors that occur in the prostate, breast, lung, and thyroid gland are called bone-seeking cancers because they migrate to bone more than any other primary cancer. Once cancer has metastasized, multiple bone sites are typically seen. Patho-

logical fractures and severe pain are major concerns in managing metastatic disease. See Chapter 10 regarding caring for patients with cancer.

Signs and Symptoms

Primary tumors cause local swelling and pain at the site. A tender, palpable mass is often present. Metastatic disease is not as visible, but the patient complains of diffuse severe pain, eventually leading to marked disability.

Diagnostic Tests

Diagnosis of bone cancer is made by x-ray examination, computed tomography, bone scan, bone biopsy, or MRI. Chapter 43 discusses these tests in detail.

The patient with metastatic disease has an elevated ALP level and possibly an elevated erythrocyte sedimentation rate, indicating secondary tissue inflammation.

Treatment

Management of bone cancer depends on the type and extent of the tumor. The treatment of primary bone tumors is usually surgery, often combined with chemotherapy or radiation. The surgeon attempts to salvage the limb and performs a resection of the tumor. For patients with Ewing's sarcoma or early osteosarcoma, external radiation may be the treatment of choice to reduce tumor size and pain.

Care of the postoperative patient is similar to that for any patient undergoing musculoskeletal surgery. Monitoring neurovascular status of the limb to be operated on is a vital nursing intervention. (See Chapter 43.) Other general postoperative care is discussed in Chapter 11.

For metastatic bone disease, surgery is not appropriate. External radiation is given primarily for palliation. The radiation is directed toward the most painful sites in an attempt to shrink them and provide more comfort for the patient.

Nursing care for the patient with bone cancer is not unlike that for patients with any other type of cancer. Help the patient adjust to the diagnosis and refer the patient to resources such as the American Cancer Society and its various support groups. Chapter 10 describes the nursing care associated with chemotherapy and radiation therapy. For more information, visit the American Cancer Society at www.cancer.org.

---

osteosarcoma: osteo—bone + sarc—flesh + oma—tumor

1. A patient has osteoporosis and is scheduled for a right total hip replacement. The nurse should include which of the following postoperative leg positions in the preoperative teaching plan?
   a. Maintain legs in adduction.
   b. Maintain legs in abduction.
   c. Maintain internal leg rotation.
   d. Maintain more than 90-degree hip flexion.

2. A patient has a 36-hour-old fractured femur. He had morphine 5 mg intramuscularly 1 hour ago and is reporting severe unrelieved pain. Which nursing action is most appropriate?
   a. Give pain medication.
   b. Adjust the traction.
   c. Bivalve the cast.
   d. Notify the patient's physician.

3. The nurse is caring for a patient who just had a plaster cast applied. Which action should the nurse take to facilitate cast drying?
   a. Cover the cast with blankets to provide extra warmth.
   b. Turn the patient every 2 hours.

   c. Increase the room temperature.
   d. Apply a heating pad.

4. A patient has Paget's disease. A priority nursing diagnosis for the patient includes which of the following?
   a. Pain
   b. Deficient knowledge
   c. Excess fluid volume
   d. Deficient fluid volume

5. A patient who had a total knee replacement is to receive Toradol 15 mg intramuscularly q6h as needed for pain. The Toradol comes as 30 mg/mL. How many milliliters should the nurse give?
   a. 0.05 mL
   b. 0.15 mL
   c. 0.5 mL
   d. 1.5 mL

REFERENCES

1. National Osteoporosis Foundation: Disease Statistics. http://www.nof.org/osteoporosis/stats.htm, 2001.
2. American Academy of Orthopaedic Surgeons: Falls and Hip Fractures. http://orthoinfo.aaos.org, 2001.

3. Puntillo, KA, et al: Patients' perceptions and responses to procedural pain: Results from Thunder Project II. Am J Crit Care 10(4):238, 2001.

Answers to Critical Thinking

Mr. Dennis

1. The physician would ask Mr. Dennis the following questions:

   - "What is your typical day on the job like?"
   - "Do certain activities increase joint pain?"
   - "When is your pain worse—after activity or after rest?"
   - "How long have you experienced joint pain?"
   - "What relieves the joint pain?"

2. Risk factors include that he is overweight, is in late middle age, and has a physically demanding job.
3. Other signs and symptoms may include bony nodules on his fingers (such as Heberden's nodes) and secondary inflammation causing joint swelling.

Mrs. Summers

1. Ask Mrs. Summers about the following:

   - The nature of her pain
   - If it is worse after activity or rest
   - If she experiences joint stiffness and, if so, when

   Follow the WHAT'S UP? method of pain assessment.
2. Teach her to do the following:

   - Balance rest with exercise.
   - Use ice for very hot, swollen joints.
   - Use heat to decrease stiffness.

Mrs. Brown

1. When documenting, answer (either explicitly or implicitly by professional knowledge, in narrative or flow sheet format) what, why, when, where, who, and how, for completeness.
   What = Mrs. Brown was found on the floor lying on her left side, moaning and holding her leg, crying out with any movement.
   Why = Fell
   When = 10 A.M. on January 4, 2003
   Where = Dayroom
   Who = I. Smith, LPN
   1/4/03 1000 Found patient on floor in dayroom lying on her left side, moaning and holding her leg, crying out with any movement. Stated, "I fell. I think my leg is broken." Supervisor immediately notified, and paramedics and Dr. Jones called. Vital signs 150/84, 100, 20. Remained with patient and instructed not to move until paramedics arrive. Blankets applied and pillow placed under head for comfort. 10:30 Paramedics transported by ambulance to Grace Hospital.
   I. Smith, LPN
2. The purpose of the traction is to reduce the muscle spasms that often accompany fractures and to increase comfort.
3. Nursing responsibilities include the following:

   - Check neurovascular status frequently.
   - Check equipment, including rope, pulleys, knots, and weights.

   - Do not allow the weights to rest on the floor.
   - Assess the patient's skin often for areas of potential breakdown.
   - Remove and rewrap the elastic bandages, maintaining the traction, at least every shift. Provide skin care during this time.

4. Her restlessness is most likely the result of pain. She may be unable to state that she is in pain. Evaluate behaviors such as restlessness and other nonverbal cues to evaluate pain management needs. She may also be experiencing shock as a result of blood loss from the fracture. Assess vital signs, and check the area of the fracture for increased signs of bruising or swelling.

Tommy Martin

1. Possible nursing diagnoses include but are not limited to the following:

   - Potential for social isolation related to extended need for immobilization
   - Potential for complications of inactivity related to extended need for traction
   - Potential for boredom related to extended need for bedrest

2. Nursing interventions related to the diagnoses include the following:

   - To address social isolation, encourage Tommy's friends to come visit; have an occupational therapist assess Tommy's needs; and alternate family visitors.
   - To avoid complications of inactivity, ensure Tommy does the exercises recommended by occupational and physical therapists; reposition him every 2 to 3 hours; ensure Tommy's diet is appropriate for good bone healing; have trapeze set up for Tommy to use; assess for signs and symptoms of skin breakdown; and assess for signs and symptoms of pulmonary embolism.
   - To alleviate boredom, encourage Tommy to listen to music; encourage visitors; and ensure access to hobbies, videos, books, magazines, and comics.

Mr. Andrews

1. Nursing assessments should include the following:

   - Perform a neurovascular check.
   - Perform a further pain assessment.
   - Ask Mr. Andrews to move his limb and see if the pain worsens.
   - Take his vital signs.
   - Assess for the 6 *P*s (pulselessness, paresthesia, paralysis, pallor, pain, poikilothermia).

2. He might be experiencing compartment syndrome.
3. Interventions may include the following:

   - Bivalving cast
   - Possible fasciotomy

UNIT TWELVE BIBLIOGRAPHY

Curry, L, and Hogstel, M: Osteoporosis. American Journal of Nursing 102(1):26, 2002.

Day, MW: Actionstat: Compartment syndrome. Nursing 1999 29(6):33, 1999.

Kannus, P, et al: Prevention of hip fracture in elderly people with use of a hip protector. New England Journal of Medicine 343(2):1506, 2000.

MacDonald, HB: Nutrition bites. Soda pop and fractures—A wake-up call. Info Nursing 2000 31(30):2, 2000.

Noble, SL, King, DS, and Olutade, JI: COX-2 inhibitors' place in therapy. American Family Physician 61:3669, 2000.

Pasero, C, and McCaffery, M: Pain in the critically ill. American Journal of Nursing 102(1):59, 2002.

Ramsburg, KL: Rheumatoid arthritis. American Journal of Nursing 100(11):40, 2000.

Wiger, P, and Styf, JR: Effects of limb elevation on abnormally increased intramuscular pressure, blood perfusion pressure and foot sensation: An experimental study in humans. Journal of Orthopaedic Trauma 12(5): 343, 1998.

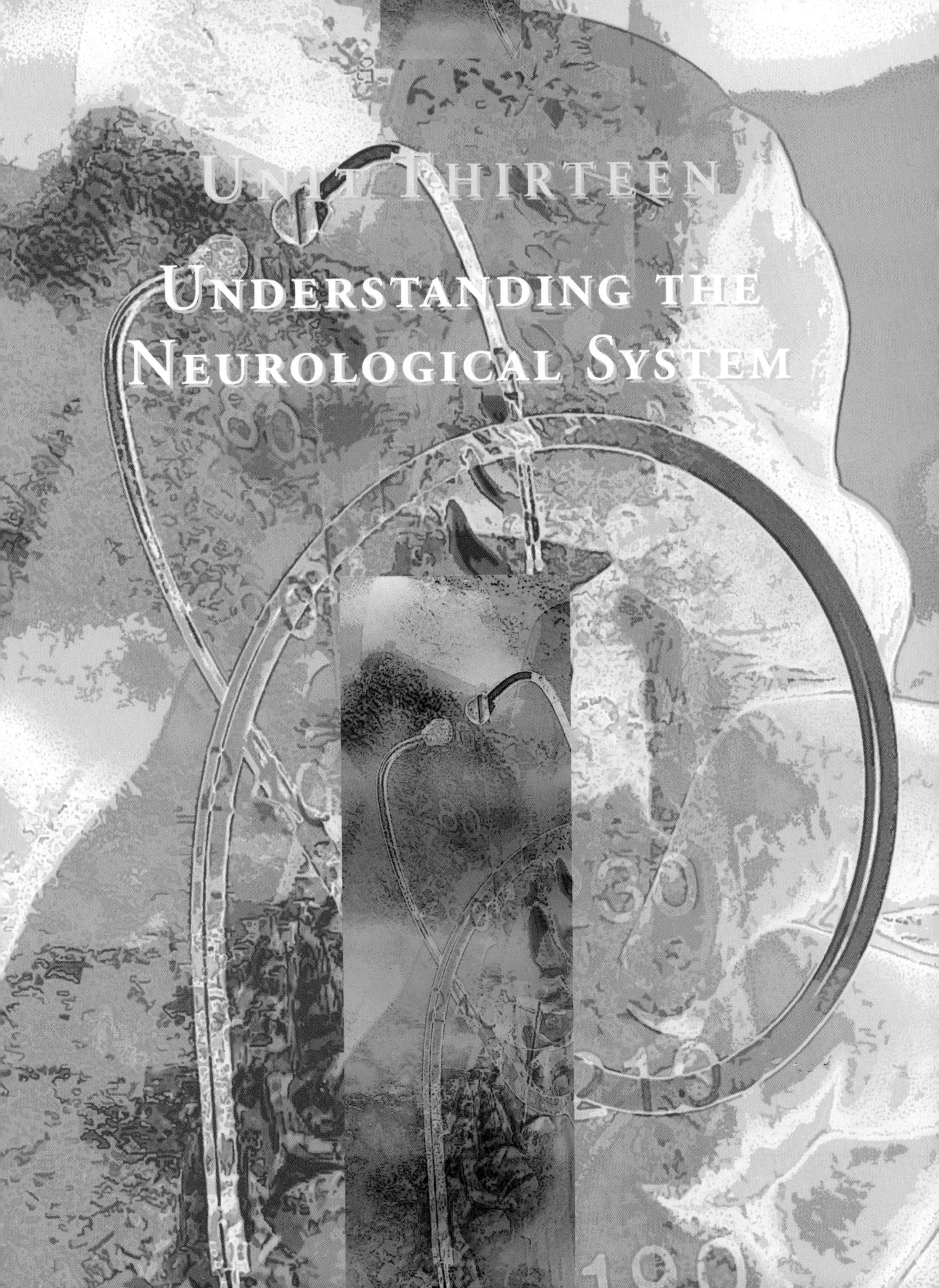

UNIT THIRTEEN

UNDERSTANDING THE NEUROLOGICAL SYSTEM

45

NEUROLOGICAL FUNCTION, ASSESSMENT, AND THERAPEUTIC MEASURES

George Byron Smith, Valerie C. Scanlon, and Sally Schnell

KEY TERMS

anisocoria (an-i-soh-**KOH**-ree-ah)

aphasia (ah-**FAY**-zee-ah)

cerebrovascular
(SER-ee-broh-**VAS**-kyoo-lur)

contractures (kon-**TRAK**-churs)

decerebrate (dee-**SER**-e-brayt)

decorticate (dee-**KOR**-ti-kayt)

dysarthria (dis-**AR**-three-ah)

dysphagia (dis-**FAYJ**-ee-ah)

electroencephalogram
(ee-LEK-troh-en-**SEFF**-uh-loh-gram)

hemiparesis (hem-ee-puh-**REE**-sis)

myelogram (**MY**-e-loh-gram)

nystagmus (nis-**TAG**-muss)

paresis (puh-**REE**-sis)

paresthesia (PAR-es-**THEE**-zee-ah)

subarachnoid (SUB-uh-**RAK**-noyd)

QUESTIONS TO GUIDE YOUR READING

1. What is the normal anatomy of the nervous system?
2. What is the normal function of the nervous system?
3. What data should you collect when caring for a patient with a disorder of the nervous system?
4. What are the effects of aging on the nervous system?
5. What are the diagnostic tests commonly performed to diagnose disorders of the nervous system?
6. What nursing care should you provide for patients undergoing each of the diagnostic tests for disorders of the nervous system?
7. What are common therapeutic measures used for patients with disorders of the nervous system?

▶ REVIEW OF NORMAL ANATOMY AND PHYSIOLOGY

The nervous system is one of the body's control systems; by means of electrochemical impulses we are able to detect changes and feel sensations, initiate appropriate responses to changes, and organize and store information for future use. Some of this is conscious activity, but much of it is reflexive in nature and happens without our awareness.

The nervous system has two divisions. The central nervous system (CNS) consists of the brain and spinal cord. The peripheral nervous system (PNS) consists of cranial nerves and spinal nerves, which include the nerves of the autonomic nervous system (ANS).

Nerve Tissue

Nerve tissue consists of neurons and specialized supporting cells. There are many kinds of neurons (nerve cells or nerve fibers), but they all have the same general structure (Fig. 45–1). The cell body contains the nucleus and is essential for the continued life of the neuron. All neuron cell bodies are found in the brain or spinal cord or within the trunk of the body; in these locations they are protected by bone. A

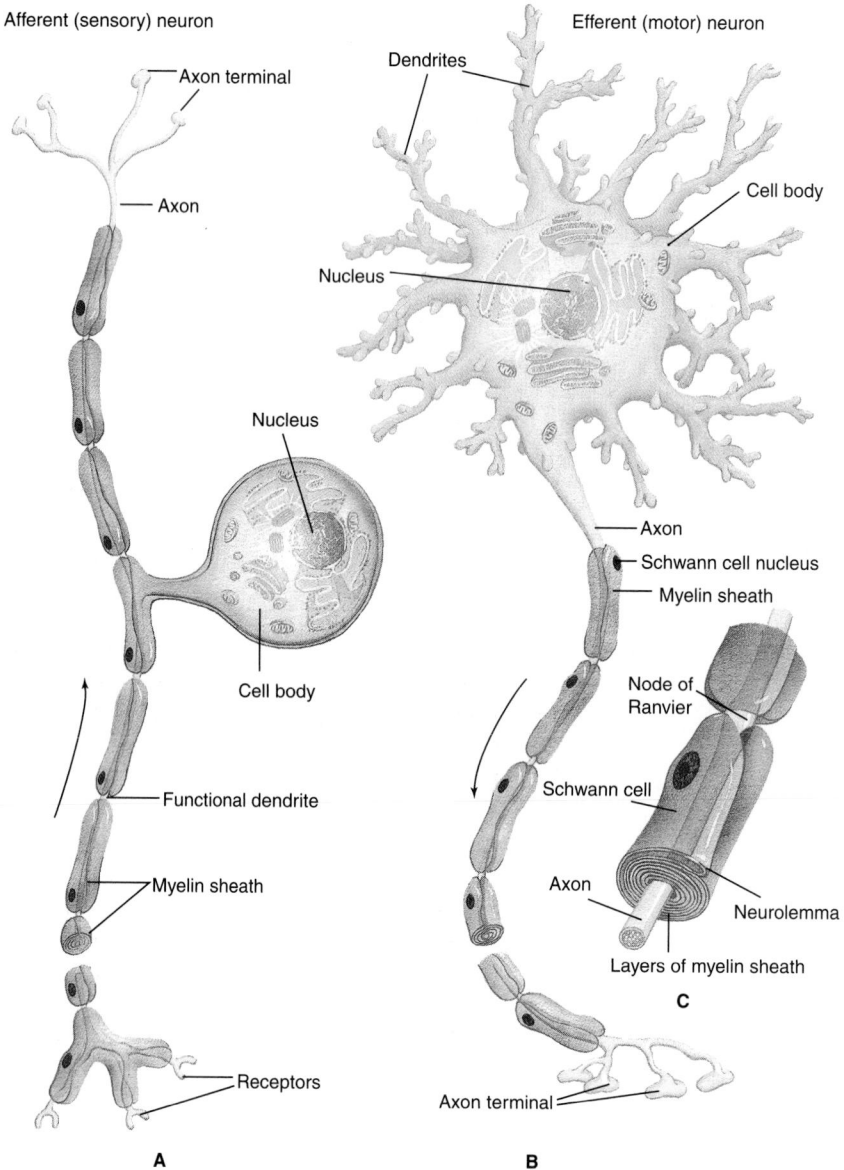

Figure 45-1 Structure of sensory and motor neurons. (From Scanlon, VC, and Sanders, T: Essentials of Anatomy and Physiology, ed 3. FA Davis, Philadelphia, 1999, p 155, with permission.)

neuron may have one or many dendrites, which are extensions that carry impulses toward the cell body. A neuron has one axon that transmits impulses away from the cell body. It is the cell membrane of the dendrites, cell body, and axon that carries the electrical nerve impulse.

In the peripheral nervous system, axons and dendrites are wrapped in specialized cells called Schwann cells. The concentric layers of cell membrane of a Schwann cell form the myelin sheath. Myelin is a phospholipid that electrically insulates neurons from one another. The spaces between adjacent Schwann cells are called nodes of Ranvier (neurofibril nodes); only these parts of the neuron cell membrane depolarize when an electrical impulse is transmitted, which makes impulse conduction rapid. The nuclei and cytoplasm of Schwann cells are outside the myelin sheath and form the neurolemma. If a peripheral nerve is severed and reattached, the individual axons and dendrites

may regrow through the tunnels provided by the neurolemma. Growth factors produced by the Schwann cells are also believed to contribute to such regeneration.

In the central nervous system, the myelin sheaths (but not a neurolemma) are formed by oligodendrocytes, one of the neuroglia, the specialized cells found only in the brain and spinal cord. See Table 45–1 for the names and functions of the other neuroglial cells.

Synapses

When the axon of a neuron must transmit an impulse to the dendrite or cell body of another neuron, the impulse must cross a small gap called a synapse. An electrical impulse is incapable of crossing this microscopic space, and at synapses impulse transmission becomes chemical. The end of the axon (the presynaptic neuron) is called the synaptic knob and contains a chemical neurotransmitter that is released

TABLE 45-1 NEUROGLIA

Name	Function
Oligodendrocytes	Produce the myelin sheath to electrically insulate neurons of the CNS
Microglia	Capable of movement and phagocytosis of pathogens and damaged tissue
Astrocytes	Contribute to the blood-brain barrier, which prevents potentially toxic waste products in the blood from diffusing out into brain tissue; disadvantage: some useful medications cannot cross it, which becomes important during brain infection, inflammation, or other disease
Ependyma	Line the ventricles of the brain; many of the cells are ciliated; involved in the circulation of cerebrospinal fluid

into the synapse by the arrival of the electrical impulse. The neurotransmitter diffuses across the synapse and combines with specific receptor sites on the postsynaptic membrane (Fig. 45–2). At excitatory synapses, the neurotransmitter makes the postsynaptic membrane more permeable to sodium ions, which rush into the cell, initiating an electrical impulse on the membrane of the postsynaptic neuron. The neurotransmitter is then inactivated to prevent continuous impulses. For example, the neurotransmitter acetylcholine is inactivated by the chemical called cholinesterase; each transmitter has its own specific inactivator.

Some synapses are inhibitory synapses in that the neurotransmitter makes the postsynaptic membrane more permeable to potassium ions, which leave the cell and make the membrane resistant to the electrical change required for an impulse. Thus the electrical impulse is stopped. Inhibitory synapses are important for things such as slowing the heart rate or balancing the excitatory impulses transmitted to skeletal muscles, which prevents excessive contraction and is important for coordination.

At synapses, impulse transmission is one-way only because the neurotransmitter is released only by the presynaptic neuron; the impulse cannot go backward. This is important for the normal activity of the functional types of neurons. The relative complexity of synapses also makes them a potential target for the actions of medications.

Types of Neurons

A useful classification of neurons is a functional one; a neuron is either a sensory neuron, a motor neuron, or an interneuron. Sensory (afferent) neurons transmit impulses from receptors to the central nervous system. Receptors are specialized to detect external or internal changes and then generate electrical impulses. Sensory neurons from receptors in the skin, skeletal muscles, and joints are called somatic; those from receptors in internal organs are called visceral sensory neurons.

Motor (efferent) neurons transmit impulses from the central nervous system to effectors—that is, muscles and glands. Motor neurons to skeletal muscle are called somatic; those to smooth muscle, cardiac muscle, and glands are called visceral. Sensory and motor neurons make up the peripheral nervous system. Visceral motor neurons form the autonomic nervous system, a specialized part of the PNS.

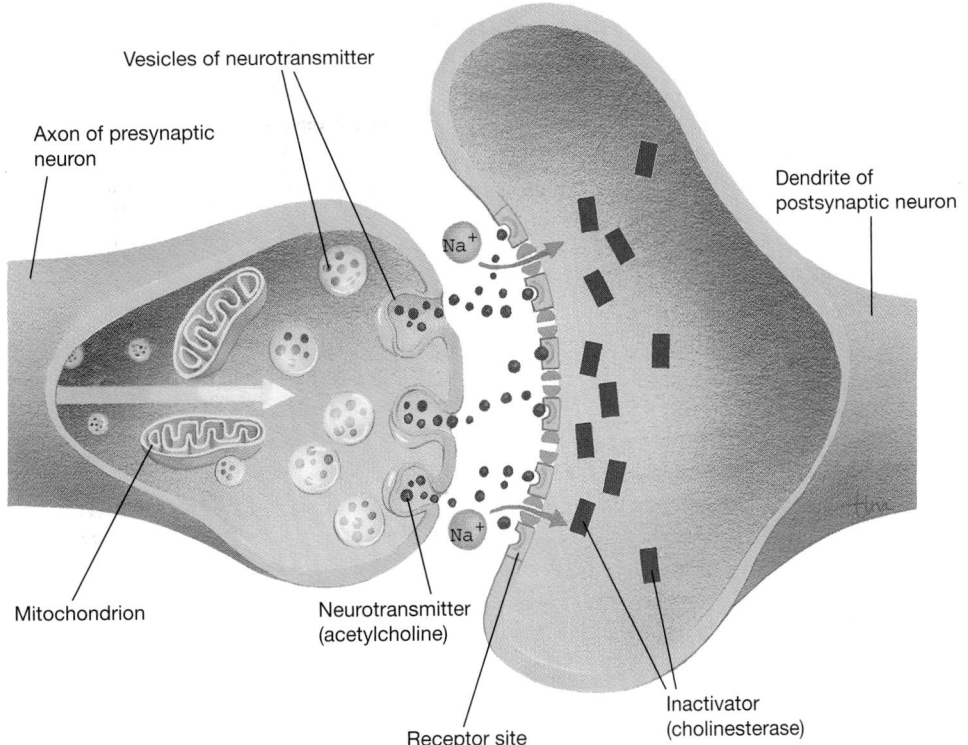

Figure 45-2 Structure of a synapse, and the effect of a neurotransmitter such as acetylcholine. (From Scanlon, VC, and Sanders, T: *Essentials of Anatomy and Physiology,* ed 3. FA Davis, Philadelphia, 1999, p 157, with permission.)

LEARNING TIPS

To remember the difference between afferent and efferent, try these clues:
Afferent: *A* is for affect or sense.
Efferent: *E* is for effect or action.
or
A before E: You have to feel or sense (afferent) a stimulus before you can take action (efferent).

Interneurons (or association neurons) are found entirely within the central nervous system. Each is specialized to transmit sensory or motor impulses or to integrate these functions. Such integration is involved in thinking and learning.

Nerves and Nerve Tracts

A nerve is a group of peripheral axons, dendrites, or both, with blood vessels and connective tissue. Most peripheral nerves are mixed; that is, they contain both sensory and motor neurons. An example of a purely sensory nerve is the optic nerve for vision; the autonomic nerves are purely motor nerves.

A nerve tract is a group of neurons within the central nervous system; such tracts are often called white matter because the myelin sheaths of the individual neurons are white. A nerve tract within the spinal cord carries either sensory or motor impulses; those within the brain may have sensory, motor, or integrative functions.

Nerve Impulse

A nerve impulse, which may also be called an action potential, is an electrical change brought about by the movement of ions across the neuron cell membrane. When a neuron is not carrying an impulse, it is in a state of polarization with a positive charge outside the membrane and a relatively negative charge inside the membrane. Sodium ions are more abundant outside the cell, and potassium and negative ions are more abundant inside the cell. A stimulus makes the membrane very permeable to sodium ions, which rush into the cell, making the inside positive and the outside relatively negative. This reversal of charges is called depolarization and spreads from the point of the stimulus along the entire neuron membrane.

Immediately following depolarization, the membrane becomes very permeable to potassium ions, which rush out of the cell. This is called repolarization and restores the positive charge outside and the negative charge inside. The sodium and potassium pumps return the sodium ions back outside and the potassium ions inside, and the neuron is polarized again and ready to respond to another stimulus. A neuron is capable of transmitting hundreds of impulses per second, and at great speed, many meters per second.

Spinal Cord

The spinal cord transmits impulses to and from the brain and is the integrating center for the spinal cord reflexes.

The spinal cord is within the vertebral canal formed by the vertebrae and extends from the foramen magnum of the occipital bone to the disk between the first and second lumbar vertebrae. The spinal nerves emerge from the intervertebral foramina.

In cross section the spinal cord is round or oval; internally it has an H-shaped mass of gray matter surrounded by white matter (Fig. 45–3). The gray matter is the cell bodies of motor neurons and interneurons. The white matter is the myelinated axons and dendrites of the interneurons. These nerve fibers are arranged in tracts based on their functions; ascending tracts transmit sensory impulses to the brain, and descending tracts transmit motor impulses from the brain to motor neurons. The central canal of the spinal cord is a small tunnel that is continuous with the ventricles of the brain; it contains cerebrospinal fluid (CSF).

Spinal Nerves

There are 31 pairs of spinal nerves, named according to their respective vertebrae: 8 cervical pairs, 12 thoracic pairs, 5 lumbar pairs, 5 sacral pairs, and 1 very small coccygeal pair. These nerves are often referred to by letter and number: the second cervical nerve is C2, the tenth thoracic is T10, and so on.

In general the cervical nerves supply the back of the head; the neck, shoulders, and arms; and the diaphragm (the phrenic nerves). The first thoracic nerve also contributes to nerves in the arms. The remaining thoracic nerves supply the trunk of the body. The lumbar and sacral nerves supply the hips, pelvic cavity, and legs. The small coccygeal pair (Co1) supplies the area around the coccyx.

Each spinal nerve has two roots, which are neurons entering or leaving the spinal cord. The dorsal root is made of sensory neurons that carry impulses into the spinal cord. The dorsal root ganglion is an enlargement of this root that contains the cell bodies of these sensory neurons. The ventral root is the motor root; it is made of motor neurons that carry impulses from the spinal cord to muscles or glands (their cell bodies are in the gray matter of the spinal cord). When the two roots merge, the nerve thus formed is a mixed nerve.

Spinal Cord Reflexes

A reflex is an involuntary response to a stimulus, an automatic reaction triggered by a specific change. Spinal cord reflexes are those that do not depend directly on the brain, although the brain may inhibit or enhance them.

Reflex Arc

A reflex arc is the pathway nerve impulses travel when a reflex is elicited. There are five parts:

1. **Receptors** detect a change (the stimulus) and generate impulses.
2. **Sensory neurons** transmit impulses from receptors to the CNS.
3. **The central nervous system** contains one or more synapses and the interneurons that may be part of the pathway.

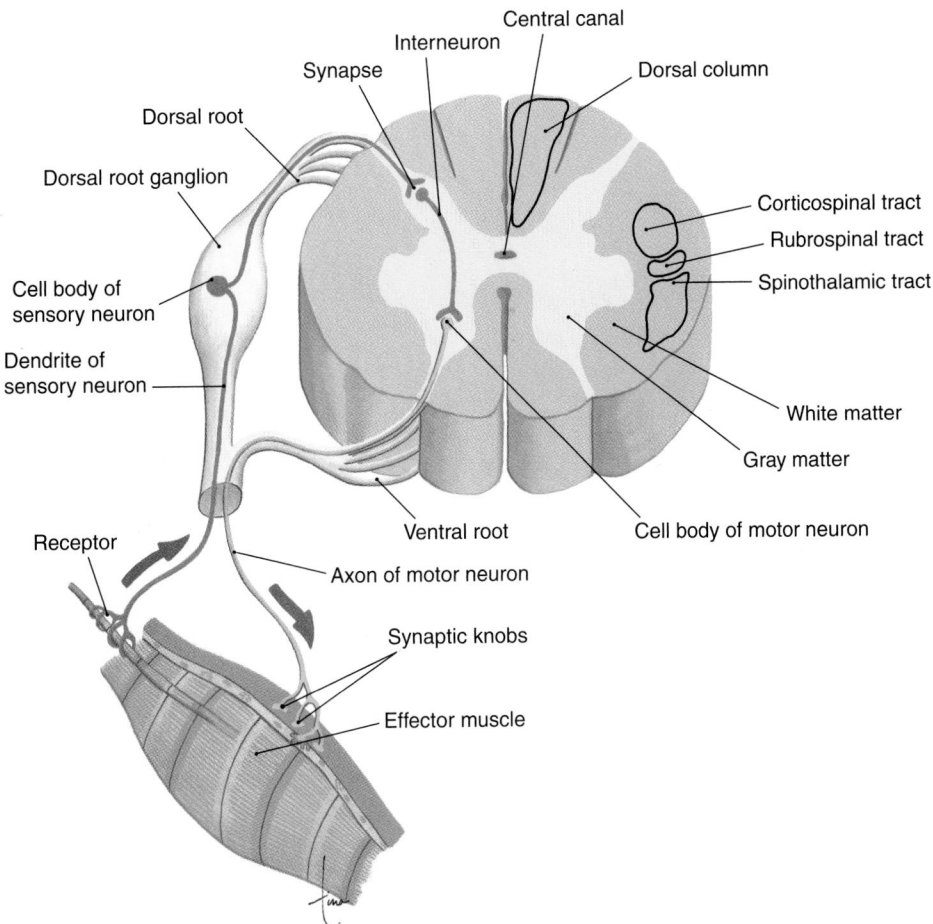

Figure 45-3 Spinal cord in cross section, with nerve roots and meninges. (From Scanlon, VC, and Sanders, T: Essentials of Anatomy and Physiology, ed 3. FA Davis, Philadelphia, 1999, p 158, with permission.

4. **Motor neurons** transmit impulses from the CNS to an effector.
5. **Effector** performs its characteristic action.

The spinal cord reflexes include stretch reflexes and flexor reflexes. In a stretch reflex, a muscle that is stretched automatically contracts; an example is the familiar patellar, or knee-jerk, reflex, but all skeletal muscles have such a reflex. The purpose of these reflexes is to keep us upright (because gravity exerts a constant pull on the body) without our having to think about it. Flexor reflexes may also be called withdrawal reflexes; the stimulus is something painful and the response is to pull away from it. Again, this occurs without the need for conscious thought; the brain is not directly involved.

The clinical testing of certain spinal cord reflexes provides a way to assess the functioning of their reflex arcs. If the patellar reflex, for example, were absent, the problem might be in the quadriceps femoris muscle, the femoral nerve, or the spinal cord itself. If the reflex is present, however, it indicates that all parts of the reflex arc are functioning normally.

Brain

The brain consists of many parts, which function as an integrated whole. The major parts are the medulla, pons, and midbrain (the brainstem); the cerebellum; the hypothalamus and thalamus; and the cerebrum (Fig. 45–4).

Ventricles

The ventricles are four cavities within the brain: two lateral ventricles within the cerebral hemispheres, the third ventricle within the thalamus and hypothalamus, and the fourth ventricle between the medulla and cerebellum. Each ventricle contains a capillary network called a choroid plexus, which forms cerebrospinal fluid (the tissue fluid of the CNS) from blood plasma.

Medulla

The medulla is anterior to the cerebellum, extends from the spinal cord to the pons, and regulates our most vital functions. Within the medulla are cardiac centers that regulate heart rate, respiratory centers that regulate breathing, and vasomotor centers that regulate the diameter of blood vessels and therefore blood pressure. Also in the medulla are reflex centers for coughing, sneezing, swallowing, and vomiting.

Pons

The pons is anterior to the upper portion of the medulla. Within the pons are two respiratory centers that work with those in the medulla to produce a normal breathing rhythm.

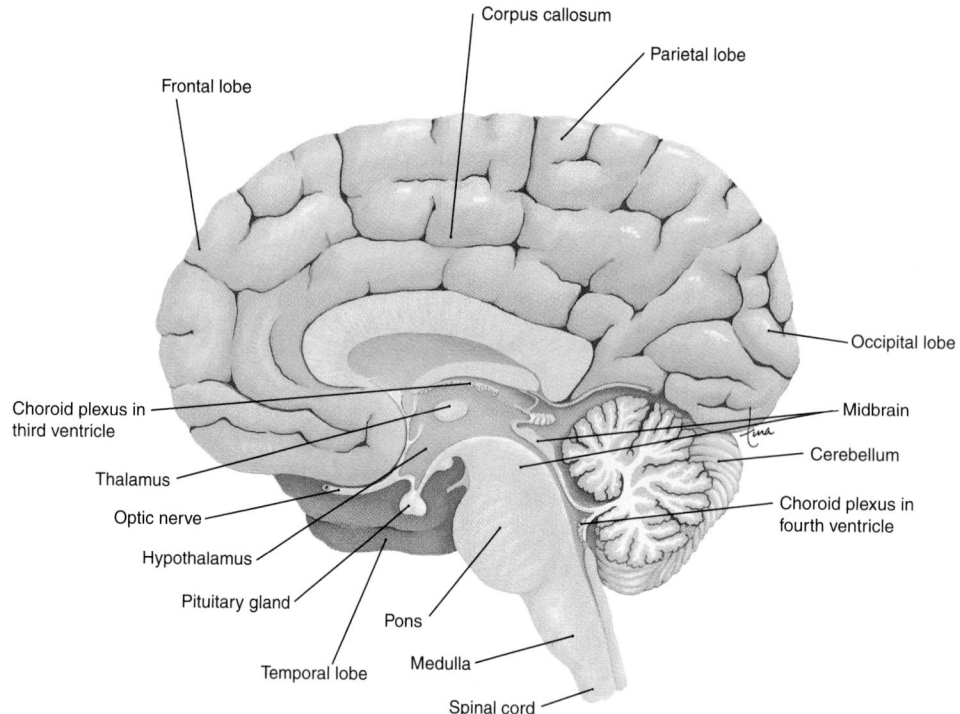

Figure 45-4 The brain's midsagittal section; medial surface of the right cerebral hemisphere. (From Scanlon, VC, and Sanders, T: Essentials of Anatomy and Physiology, ed 3. FA Davis, Philadelphia, 1999, p 165, with permission.)

Midbrain

The midbrain extends from the pons to the hypothalamus and encloses the cerebral aqueduct, a tunnel that connects the third and fourth ventricles. Primarily a reflex center, the midbrain regulates visual reflexes (coordinated movement of the eyes), auditory reflexes (turning the ear to a sound), and righting reflexes that keep the head upright and contribute to balance.

Cerebellum

The cerebellum is posterior to the medulla and pons, separated from them by the fourth ventricle; it is overlapped by the occipital lobes of the cerebrum. The functions of the cerebellum are concerned with the involuntary aspects of voluntary movement: coordination, regulation of muscle tone, the appropriate trajectory and endpoint of movements, and the maintenance of posture and balance or equilibrium. For the maintenance of balance, the cerebellum (and midbrain) uses sensory information provided by the receptors in the inner ear that detect movement and changes in position of the head.

Hypothalamus

The hypothalamus is located above the pituitary gland and below the thalamus. It has many diverse functions:

1. Production of antidiuretic hormone (ADH) and oxytocin; these hormones are then stored in the posterior pituitary gland. ADH increases the reabsorption of water by the kidneys and thus helps maintain blood volume. Oxytocin causes contractions of the myometrium of the uterus to bring about labor and delivery.
2. Production of releasing hormones that stimulate secretion of the hormones of the anterior pituitary gland. An example is growth hormone–releasing hormone (GHRH), which stimulates the anterior pituitary to secrete growth hormone.
3. Regulation of body temperature by promoting responses such as shivering in a cold environment or sweating in a warm environment.
4. Regulation of food intake; the hypothalamus is believed to respond to changes in blood nutrient levels or chemicals secreted by adipose tissue and bring about feelings of hunger or fullness.
5. Integration of the functioning of the autonomic nervous system, which is covered in a later section.
6. Stimulation of visceral responses in emotional situations, such as increased heart rate when angry or afraid. The neurological basis of emotions is not well understood, but the hypothalamus brings about bodily changes by way of the autonomic nervous system.

Thalamus

The thalamus is above the hypothalamus and below the cerebrum; its functions are concerned with sensation. Sensory pathways (except olfactory ones) to the brain converge in the thalamus, which begins to integrate sensations, which in turn permits more rapid interpretation by the cerebrum. The thalamus is also capable of suppressing unimportant sensations, permitting the cerebrum to concentrate without the distraction of minor sensations.

Cerebrum

The two cerebral hemispheres form the largest part of the human brain. The right and left hemispheres are connected by the corpus callosum, a band of about 200 million nerve fibers that allows each hemisphere to know what is going on in the other.

The cerebral cortex is the surface of the cerebrum; it is gray matter that consists of the cell bodies of neurons. The cerebral cortex is folded extensively into convolutions (or gyri) that permit more space for neurons. The grooves between the folds are called fissures or sulci. Interior to the gray matter is white matter, myelinated axons and dendrites that connect the parts of the cerebral cortex to one another and the cerebrum to other parts of the brain. The cerebral cortex is divided into lobes, whose functions have been extensively mapped.

The frontal lobes contain the motor areas that generate the impulses that bring about voluntary movement. Each motor area controls movement on the opposite side of the body. Also in the frontal lobe, usually only the left lobe, is Broca's motor speech area, which controls the movements involved in speaking.

The parietal lobes contain the general sensory areas for the cutaneous senses and conscious muscle sense. This is where these sensations are felt and interpreted.

The temporal lobes contain sensory areas for hearing, smell, and taste. Also in the temporal and parietal lobes, usually only on the left side, are speech areas involved in the thought that precedes speech.

The occipital lobes contain the visual areas that receive impulses from the retinas of the eyes; they "see" and interpret what is being seen.

In all lobes of the cerebral cortex are association areas that enable us to learn, remember, and think and probably give us our individual personalities. These have not yet been as precisely mapped as the sensory and motor areas.

Deep within the white matter of the cerebral hemispheres are masses of gray matter called the basal ganglia. Their functions are concerned with certain subconscious aspects of voluntary movement: regulation of muscle tone, inhibiting tremor, and use of accessory movements such as gestures when speaking.

Meninges and Cerebrospinal Fluid

The meninges are the three layers of connective tissue that cover the central nervous system. The outermost is the dura mater, made of thick, fibrous connective tissue. The middle layer is called the arachnoid membrane, which has a web-like appearance, and the inner layer is the pia mater, very thin connective tissue on the surface of the brain and spinal cord. Between the arachnoid membrane and the pia mater is the **subarachnoid** space, which contains cerebrospinal fluid.

Each of the four ventricles of the brain contains a choroid plexus, a capillary network that forms cerebrospinal fluid from blood plasma. This is a continuous process, and the cerebrospinal fluid then circulates from the ventricles to the central canal of the spinal cord and to the subarachnoid spaces around the brain and spinal cord. From the cranial subarachnoid space, cerebrospinal fluid is reabsorbed back to the blood through arachnoid villi that project into the cranial venous sinuses, large veins between the two layers of the cranial dura mater. The rate of reabsorption usually equals the rate of production.

As the tissue fluid of the CNS, cerebrospinal fluid permits the exchanges of nutrients and wastes between the blood and CNS neurons. It also acts as a cushion or shock absorber for the CNS. The pressure and constituents of cerebrospinal fluid may be determined by means of a lumbar puncture (spinal tap) and may be helpful in the diagnosis of diseases such as meningitis.

Cranial Nerves

The 12 pairs of cranial nerves emerge from the brainstem or other parts of the brain; some are purely sensory nerves, whereas others are mixed nerves. The impulses for sight, smell, hearing, taste, and equilibrium are all carried by cranial nerves to their respective sensory areas in the brain. Other cranial nerves carry motor impulses to muscles of the face or to glands. The functions of all the cranial nerves are summarized in Table 45–2.

Autonomic Nervous System

The ANS is part of the peripheral nervous system in that it consists of the motor portions of some cranial and spinal nerves. These are the visceral motor neurons to visceral effectors—that is, smooth muscle, cardiac muscle, and glands. The ANS has two divisions, sympathetic and parasympathetic; often they function in opposition to each other, and their activity is integrated by the hypothalamus.

An autonomic nerve pathway from the CNS to a visceral effector consists of two motor neurons that synapse in a ganglion outside the CNS (Fig. 45–5). The first neuron is called the preganglionic neuron, from the CNS to the ganglion. The second neuron is called the postganglionic neuron, from the ganglion to the visceral effector. The ganglia are actually the cell bodies of the postganglionic neurons.

Sympathetic Division

The cell bodies of the sympathetic preganglionic neurons are in the thoracic and some of the lumbar segments of the spinal cord. The axons of these neurons extend to the sympathetic ganglia, most of which are in two chains just outside the spinal column. Within the ganglia are the synapses between the preganglionic and postganglionic neurons; the axons of the postganglionic neurons then go to the visceral effectors. One preganglionic neuron often synapses with many postganglionic neurons to many effectors; this permits widespread responses in many organs.

---

subarachnoid: sub—below + arachnoid—middle layer of the meninges

TABLE 45-2	CRANIAL NERVES	
Number	**Name**	**Function**
I	Olfactory	Sense of smell
II	Optic	Sense of sight
III	Oculomotor	Movement of eyeball; constriction of pupil for bright light or near vision
IV	Trochlear	Movement of eyeball
V	Trigeminal	Sensation in face, scalp, and teeth; contraction of chewing muscles
VI	Abducens	Movement of eyeball
VII	Facial	Sense of taste; contraction of facial muscles; secretion of saliva
VIII	Vestibulocochlear	Sense of hearing; sense of equilibrium
IX	Glossopharyngeal	Sense of taste; secretion of saliva; sensory for cardiac, respiratory, and blood pressure reflexes; contraction of pharynx
X	Vagus	Sensory in cardiac, respiratory, and blood pressure reflexes; sensory and motor to larynx (speaking); decreases heart rate; contraction of alimentary tube (peristalsis); increases digestive secretions
XI	Accessory	Contraction of neck and shoulder muscles; motor to larynx (speaking)
XII	Hypoglossal	Movement of the tongue

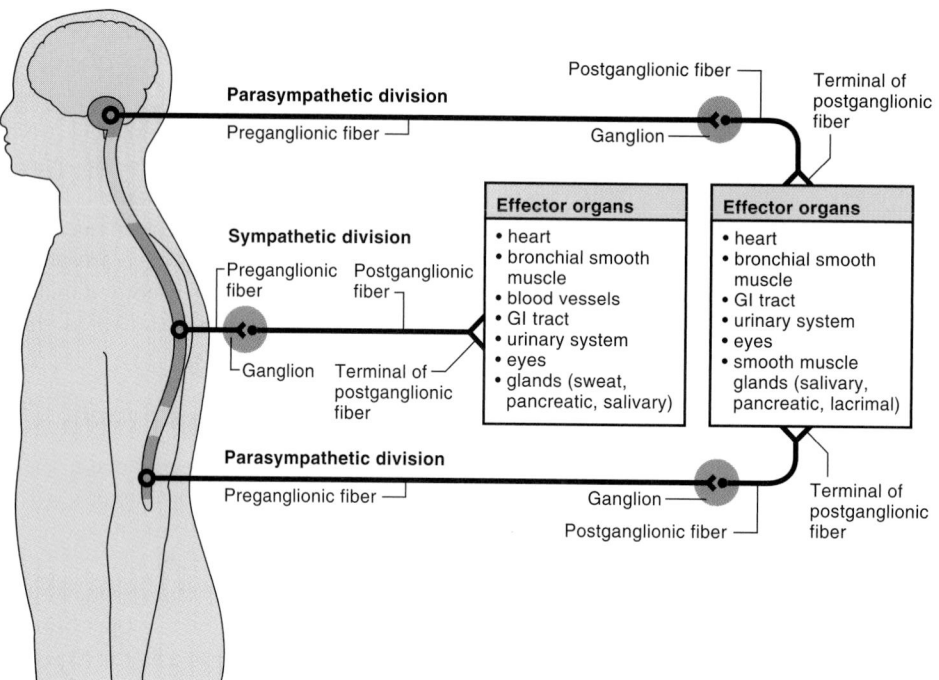

Figure 45-5 Autonomic nervous system. (Modified from Morton, PG: Health Assessment on Nursing, ed 2. FA Davis, Philadelphia, p 467, with permission.)

The sympathetic division is dominant in stressful situations such as fear, anger, anxiety, and exercise, and the responses it brings about involve preparedness for physical activity, whether or not it is actually needed. (Table 45–3 summarizes both ANS divisions.) The heart rate increases, vasodilation in skeletal muscles supplies them with more oxygen, the bronchioles dilate to take in more air, and the liver changes glycogen to glucose to provide energy. Relatively unimportant activities such as digestion are slowed, and vasoconstriction in the skin and viscera permits greater blood flow to more vital organs such as the brain, heart, and muscles.

The neurotransmitters of the sympathetic division are acetylcholine and norepinephrine. Acetylcholine is released by sympathetic preganglionic neurons; its inactivator is cholinesterase. Norepinephrine is released by most sym-

pathetic postganglionic neurons at the synapses with the effector cells; its inactivator is catechol O-methyltransferase.

Parasympathetic Division

The cell bodies of the parasympathetic preganglionic neurons are in the brainstem and the sacral segments of the spinal cord. The axons of these neurons are in cranial nerve pairs 3, 7, 9, and 10 and in some sacral nerves, and they extend to the parasympathetic ganglia. These ganglia are close to or actually in the visceral effector and contain the postganglionic cell bodies, with very short axons to the cells of the visceral effector. One preganglionic neuron synapses with just a few postganglionic neurons to only one effector. This permits localized responses.

The parasympathetic division dominates during relaxed, nonstressful situations to promote normal functioning of

TABLE 45-3 FUNCTIONS OF THE AUTONOMIC NERVOUS SYSTEM		
Organ	**Sympathetic Response**	**Parasympathetic Response**
Heart (cardiac muscle)	Increase rate	Decrease rate (to normal)
Bronchioles (smooth muscle)	Dilate	Constrict (to normal)
Iris (smooth muscle)	Pupil dilates	Pupil constricts (to normal)
Salivary glands	Decrease secretion	Increase secretion (to normal)
Stomach and intestines (smooth muscle)	Decrease peristalsis	Increase peristalsis for normal digestion
Stomach and intestines (glands)	Decrease secretion	Increase secretion for normal digestion
Internal anal sphincter	Contract to prevent defecation	Relax to permit defecation
Urinary bladder (smooth muscle)	Relax to prevent urination	Contract for normal urination
Internal urethral sphincter	Contract to prevent urination	Relax to permit urination
Liver	Change glycogen to glucose	None
Sweat glands	Increase secretion	None
Blood vessels in skin and viscera (smooth muscle)	Constrict	None
Blood vessels in skeletal muscle (smooth muscle)	Dilate	None
Adrenal glands	Increase secretion of epinephrine and norepinephrine	None

several organ systems. Digestion proceeds normally, with increased secretions and peristalsis; defecation and urination may occur, and the heart beats at a normal resting rate. (See Table 45–3.)

L E A R N I N G   TIPS

Sympathetic—*S* for STRESS RESPONSE: The autonomic nervous system is a primitive response that affected early humans automatically and involuntarily, thus increasing the likelihood of surviving in the wilderness. It is often referred to as the fight-or-flight response. When you think of the sympathetic nervous system, think about getting away from a man-eating lion: You need dilated pupils to see the path better; copious production of sweat to provide the body lubrication in case you need to run between two tight trees or stones; increased rate and force of heartbeat to ensure that enough blood gets to the extremities so you can run faster; dilated bronchioles to get more oxygen to your muscles; decreased digestion so you won't get hungry while you are trying to get away from the lion; decreased urine output so you won't have to stop for the restroom; and increased mental alertness so you are always aware of where the lion is.

Parasympathetic—*P* for PARACHUTE or PEACEFUL: The parasympathetic nervous systems brings the body back to balance and rest. It is sometimes referred to as the rest-and-digest response. Think, *I just got away from a man-eating lion, now my body can go back to normal and start digesting and urinating again!*

This is a great way to remember the responses, rather than an accurate description of the physiology involved.

Acetylcholine is the neurotransmitter at all parasympathetic synapses, both preganglionic and postganglionic; it is inactivated by cholinesterase.

Aging and the Nervous System

With age the brain loses neurons, but this is only a small percentage of the total and is not the usual cause of mental impairment in the elderly; far more common causes are depression, malnutrition, hypotension, and the side effects of medications. Some forgetfulness is to be expected, however, as is a decreased ability for problem solving. Voluntary movements become slower, as do reflexes and reaction time (Table 45–4).

► NURSING ASSESSMENT OF THE NEUROLOGICAL SYSTEM

The focus of the nursing neurological assessment is to establish the present function of the patient's neurological system and to detect changes from previous assessments. A complete neurological assessment, intended to determine the existence of neurological disease, is typically performed by a physician or nurse practitioner. A baseline neurological assessment should be performed on every patient admission (Table 45–5). In addition to giving valuable information about the current functioning of the patient's neurological system, the assessment provides a basis for comparison. This is especially important if the patient has

TABLE 45-4 AGE-RELATED CHANGES AND CLINICAL IMPLICATIONS	
Age-Related Change	**Clinical Implications**
Decreased blood flow to the brain	Increased risk of syncope, changes in some mental functions
Deposition of the aging pigment lipofuscin in nerve cells and amyloid in blood vessels	Impairment in cognition, reasoning, judgment, and orientation
Altered sleep patterns	Less sleep required and sleeping for shorter periods of time; increased sleep disturbances
Decreased vibratory sense	Equilibrium altered, leading to impaired gait
Decreased postural stability	Accidents and falls
Decreased reaction time	Greater time needed for patient responses and task performance
Progressive loss of dendrites with progression to fragmentation and cell death	Failing short-term memory

TABLE 45-5	BASIC NEUROLOGICAL ASSESSMENT

1. Assess level of consciousness (patient's response to verbal or tactile stimulation) and orientation.
2. Obtain vital signs (specifically blood pressure, pulse, and respirations).
3. Check pupillary response to light.
4. Assess strength and equality of hand grip and movement of extremities.
5. Determine ability to sense touch or pain in extremities.

chronic neurological deficits on admission. One example is a patient, admitted for placement of a prosthetic hip, who has had a previous **cerebrovascular** accident resulting in **paresis** of the right arm. A complete neurological assessment would document that the right arm is weaker than the left. If during the postoperative course the assessment demonstrates that both arms are equal in strength, the patient would be evaluated for possible causes of weakening of the left arm.

The results of the baseline assessment are invaluable in planning and implementing safe care. For example, patients who give a history of seizure activity need careful monitoring, and all staff members who interact with such patients should be aware of how to respond to a seizure. Patients with **dysphagia** (difficulty swallowing) may need to have restrictions placed on the types of food or fluids they can have. This information must be consistently communicated to all staff who are involved in the patient's care.

The patient's admitting diagnosis, the presence of any chronic neurological disorders, and the current functioning of the patient's neurological system all influence how often neurological assessments should be done. Physician or nurse

---

cerebrovascular: cerebro—brain + vascular—vessels
paresis: partial paralysis
dysphagia: dys—difficult + phagia—eating

practitioner orders for neurological assessments may vary from every 15 minutes for an acutely ill or injured patient, to every 8 hours for a patient who is close to being discharged, to every 24 hours for a resident living in long-term care. It is always appropriate to assess a patient more often than ordered, based on observed changes in the patient's condition, and to communicate the findings of those assessments to the physician or nurse practitioner. The changes noted while assessing the patient may indicate changes in the central nervous system. Rapid detection and intervention may mean the difference between chronic dysfunction or recovery or even between life and death for the patient.

Subjective Data

To understand the patient's neurological problems, ask about past and current symptoms; use of prescribed and over-the-counter medications; use of recreational drugs; surgeries; treatments; and risk factors such as family history, diet, exercise, sedentary lifestyle, caffeine intake, and recent stressors. Assessment of symptoms, as with other body systems, includes asking the WHAT'S UP? questions: *w*here it is, *h*ow it is, *a*ggravating and alleviating factors, *t*iming, *s*everity, *u*seful data related to associated symptoms, and *p*erception of the problem by the patient.

Health History

The nurse obtains a history of the patient's general health and then focuses on the neurological symptoms. Symptoms of neurological disorders vary in type, location, and intensity. It is important to remember that some neurological disorders can affect the patient's ability to think, remember, speak, or interpret stimuli. It may be necessary to question significant others about duration and severity of symptoms. Some patients may not be able to recognize their own neurological deficits. In these cases the significant other usually initiates contact with the health care system and provides the medical and social history. See Table 45–6 for sample questions to ask if the patient has a neurological problem.

TABLE 45-6	NEUROLOGICAL HEALTH HISTORY	
Questions		**Rationale**
Assessment of a Symptom: WHAT'S UP? Format		
Where is the symptom, sensation, or lack of sensation? Does it radiate? Is it unilateral or bilateral?		Location of the symptom helps determine its cause.
How does it feel? Painful, achy, tingly, or numb? Some patients may use phrases such as "pins and needles in my arm" or "it feels like my leg is asleep" to describe paresthesia. Nerve pain may be described as "shooting" or "burning."		These symptoms may be assciated with strokes, spinal cord injury, neuromuscular disorders, or other neurological problems.
Aggravating/alleviating factors: Does exercise or or fatigue make it worse? Does rest help? Do applications of ice or heat help relieve symptoms? Do changes in the weather have an impact?		Paresis or tremors are more pronouned if the patient is fatigued. Patients with multiple sclerosis may have exacerbations of symptoms in extreme weather.
Timing: How long has the patient had symptoms? Have the symptoms changed? Have they moved or spread to another part of the body? Are particular times of the day better or worse?		Patients with multiple sclerosis may experience symptoms that change in location or intensity. Patients with back or neck pain may complain of stiffness in the morning or increased pain after exertion.
Severity: How severe are the symptoms on a scale of 0 to 10? How do the symptoms affect the patient's lifestyle, job, community service, and ability to carry out ADLs?		Elderly patients may be severely compromised by relatively mild deficits. Patients may be unable to stand up and walk long enough to shop for groceries or clean their house.

TABLE 45-6	NEUROLOGICAL HEALTH HISTORY—CONT'D
Questions	**Rationale**
Assessment of a Symptom: WHAT'S UP? Format	
Useful other data: Is the patient on medication? Does the patient have any other illnesses?	Some medications may cause neurological symptoms. Some neurological disorders can be complications of other illnesses, such as diabetic neuropathy.
Patient's perception: How do symptoms affect personal relationships and family obligations? What is the impact of the disorder on the patient's significant others, especially if they are the patient's caregivers?	Patients may need encouragement to discuss the impact of **hemiparesis** on sexual function or the effect of memory impairment on family relationships and roles. Patients with cognitive deficits may fear that they have Alzheimer's disease or are "going crazy."
Level of Consciousness (LOC)	
What is the patient's LOC based on the Glasgow Coma Scale?	The GCS determines the patient's level of responsiveness to stimuli. Changes in LOC can result from a variety of central nervous system problems.
Mental Status	
Ask: What is your name? What is the month? Year? Where are you now? Why are you here?	Disorientation is often an initial sign of a neurological disorder.
Are the patient's appearance and behavior, including dress, grooming, posture, gestures, facial expression, and motor activity, appropriate?	Appearance and behavior may be inappropriate to the situation in some neurological disorders.
Intellectual Function	
Can the patient count backward from 100 or subtract 7 from 100, then 7 from that answer, and so on (serial 7s)?	Most people with intact neurological function can complete serial 7s in about 1½ minutes.
Can the patient to interpret a proverb, such as "People who live in glass houses shouldn't throw stones"?	Ability to interpret a proverb indicates abstract reasoning ability. An answer such as "You might break a window" indicates concrete thinking.
Thought Content	
Are the patient's thoughts spontaneous, clear, relevant and coherent? Does the patient have any fixed ideas, illusions, or preoccupations? What are the patient's insights into these thoughts?	Preoccupation with death or morbid events, evidence of hallucinations, and paranoid ideation are all important and require further evaluation.
Emotional Status	
Are mood and affect natural and even, or irritable, angry, anxious, apathetic, or euphoric? Are they appropriate to the situation? Does the patient's mood fluctuate normally, or does it swing from joy to sadness during the interview? Is affect appropriate to words and thought content? Is verbal communication consistent with nonverbal cues?	Emotional lability may be indicative of stroke, decreased LOC, brain tumor, or other neurological disorders.
Perception	
Can the patient identify objects correctly or identify objects functions? For example, have patient identify a pencil and what it is used for.	Agnosia (inability to interpret or recognize familiar objects) can occur in cerebral vascular accidents and brain lesions.
Motor Ability	
Can the patient throw a ball, move a chair, walk in a straight line? Is the patient's motor strength equal bilaterally and appropriate? Can the patient move all extremities?	Failure to perform these tasks suggests cerebral dysfunction. Paralysis (inability to move extremity) may suggest stroke or spinal cord injury.
Language Ability	
Does the patient answer questions appropriately? Can the patient read a sentence from a newspaper and explain its meaning? Can the patient write his or her name?	Different types of aphasia can result from injury to different parts of the brain.

hemiparesis: hemi—divided in half on a vertical plane + paresis—weakness

In addition to using these questions, the nurse observes the patient during the health history. Is he or she shifting positions and exhibiting signs of discomfort? Is the patient able to change position and move about easily? Is he or she able to carry on a coherent conversation?

Physical Assessment

The physical assessment of the patient begins when you first meet the patient and make an overall evaluation of the patient's mental and physical status. The neurological system is assessed using inspection, palpation, and percussion (with a reflex hammer). When conducting the mental status and cognitive portions of the assessment, be aware that fatigue or illness may alter findings. When interpreting findings consider the patient's age, educational background, and cultural orientation.

Level of Consciousness

The level of consciousness exists along a continuum from full wakefulness, alertness, and cooperation, to unresponsiveness, to any form of external stimuli. A fully conscious patient responds to questions spontaneously. As consciousness becomes impaired, a patient may show irritability, a shortened attention span, or an unwillingness to cooperate. The level of consciousness should be the first thing assessed during a neurological examination because the information obtained can be used to modify the remainder of the examination if necessary. Keep in mind that a decrease in the level of consciousness can be caused by hypoxia, hypoglycemia, intoxication, or other problems, as well as dysfunction of the neurological system.

Many health care institutions use the Glasgow Coma Scale (GCS), which is an international scale used to assess level of consciousness (LOC) and document findings (Table 45–7). The GCS is based on simple and clearly defined parameters of patient responses that provide for consistent assessment data. It is used to evaluate patients who have a potential for rapid deterioration in consciousness. The GCS assesses three parameters of consciousness: eye opening, verbal response, and motor response.

The first section of the GCS evaluates the stimulus needed to get patients to open their eyes. Individuals with normal neurological function open their eyes spontaneously (score of 4). People with mildly decreased LOC require verbal stimuli, such as calling out the patient's name or asking the patient to open his or her eyes (score of 3). The nurse should not confuse this response with patients who open their eyes to any loud noise. That is a reflex and indicates a lower level of consciousness. A lower level of consciousness is indicated by eye opening in response to painful stimuli (score of 2). Pressure on the nailbed or trapezius muscle are two common ways of providing a painful stimulus. Care must be taken not to damage or bruise the skin or underlying tissue. Only enough pressure to elicit a response should be used. If no response is noted, the patient receives a score of 1 for this section of the GCS.

Some patients may be unable to open their eyes because of facial trauma or swelling. This should be recorded as "C," indicating eyes closed, rather than "1" for no response. Physical inability to open the eyes does not indicate a decreased level of consciousness.

The second section of the GCS addresses verbal response to the examiner. Orientation refers to the patient's cognitive awareness. Patients are often referred to as being oriented to person, place, and time, or "oriented times 3." Some clinicians include a fourth category, situation, indicating whether the patient understands what is happening to her or him. A patient who is fully oriented receives a score of 4 on the GCS. Assessment of orientation should not be superficial. Typical questions include "What is your name? Where are you? What day is it?" (Keep in mind that we all forget the date from time to time!) Patients may be able to recall the name of the hospital and the day but have no idea why they are in the hospital. Remember, a resident of a nursing home may consider the facility his or her home and is not necessarily considered disoriented. Be sure your question is appropriate to the patient's living conditions and lifestyle. If the patient is unable to speak because of a stroke (expressive **aphasia**), do not rule out the possibility that the patient is oriented. Give expressively aphasic patients yes-or-no questions such as "Are you in a grocery store? Are you in a bowling alley? Are you in a hospital?" and "Is the month June? Is it September?" Patients can answer with a shake of the head or blinks or hand squeezes, as instructed. Patients who are unable to understand what is being said to them are referred to as receptively aphasic. These patients may be able to match a spoken name with a picture of a family member.

Confusion and *disorientation* are terms that are sometimes used interchangeably. If the patient is confused, it is helpful to determine the extent of the confusion. Patients who have been in a hospital or long-term care facility for an extended period may lose track of the date but still be oriented to person and place. A patient who can tell the nurse the year and who is president but not the precise date is less confused than someone who thinks it is 1945 and Franklin Roosevelt is president. Patients who are acutely disoriented should have the correct information reinforced for them frequently. A calendar in an easily visible place, telling patients about current events in the family, and telling patients the date and why they are in the hospital or long-term care facility all reinforce orientation. Patients with short-term memory deficits may understand the information they are told but be unable to retain it long enough to use it.

TABLE 45–7	GLASGOW COMA SCALE	
Eye Opening	Spontaneous	4
	To verbal stimulus	3
	To painful stimulus	2
	No response	1
Verbal Response	Spontaneous	5
	Confused conversation	4
	Inappropriate words	3
	Incomprehensible sounds	2
	No response	1
Motor Response	Obeys commands	6
	Localizes pain	5
	Withdraws from pain	4
	Abnormal flexion	3
	Abnormal extension	2
	No response	1

aphasia: a—absence + phasia—speech

A GCS score of 3 is given to patients who are not able to carry on a conversation and are limited to using words inappropriately. This includes those patients who spontaneously call out words that do not have meaning in that situation or who merely call out profanities at random. Incomprehensible sounds are those vocalizations that are not recognizable as words, and these patients are given a score of 2. Again, a score of 1 is given for no response. Patients who are intubated or have a tracheostomy should not automatically be scored as having no response. Head nodding, mouthing words, or using a picture board are all ways for intubated patients to respond to questions and have their orientation status assessed.

Typically, motor response is scored based on the best arm function that can be elicited. For some patients this part of the examination must be modified. A patient who is paralyzed may be able only to blink. If the patient can blink on command, that is scored as "6, obeys command." Patients may be unable to use their arms but have voluntary movement of their legs. If these patients follow commands with their legs, they receive a score of 6.

Localizing pain indicates that the patient has an adequate level of consciousness to recognize where the pain is coming from and to attempt to push it away. An example is the patient who pushes against the your hand when pressure is applied to the patient's nailbed. The ability to localize pain is scored as a 5. Withdrawal from pain indicates less function of the cerebral cortex. In this case the patient is able to recognize the source of the painful stimuli but withdraws from it instead of trying to push the source away. This withdrawal is rated as a 4. Abnormal flexion posturing, characterized by flexion of the arms at the elbow, bringing the hands up toward the chest with the legs extended (**decorticate** posturing; Fig. 45–6A), indicates significant impairment of the cerebral functioning. A score of 3 is given to patients who exhibit abnormal flexion posturing. Abnormal extension posturing, or **decerebrate** posturing, indicates damage in the area of the brainstem (Fig. 45–6B). In this case, both the upper and lower extremities are extended and the arms are internally rotated. Abnormal extension posturing is scored as a 2. As with the other categories, no response is rated as 1.

The total possible score on the GCS ranges from 3 to 15. A score of less than 7 indicates a comatose patient and a score of 15 indicates the patient is fully alert and oriented. When used to score the effects of a head injury, a score of 13 or 14 indicates mild head injury, 9 to 12 indicates moderate injury, and any score of 8 or below indicates severe head injury. For all the categories of the GCS, the type of painful stimuli required to elicit a response should be documented. Deterioration in the patient's condition (i.e., a lowering of the GCS score) should be reported to the physician promptly. Changes in GCS are often used as an indication of the need for intracranial pressure monitoring.

A. Decorticate posturing

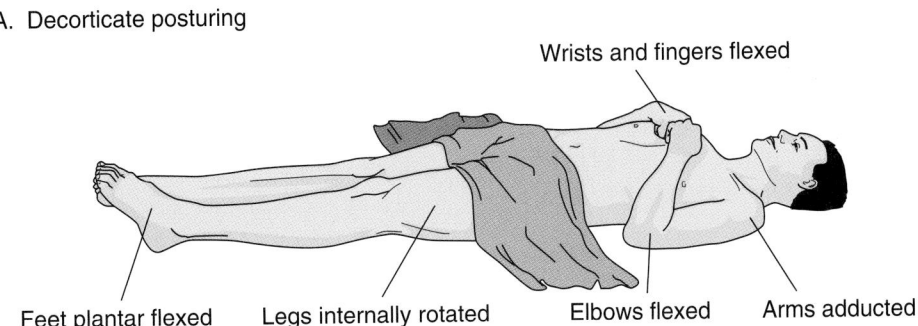

Wrists and fingers flexed

Feet plantar flexed   Legs internally rotated   Elbows flexed   Arms adducted

B. Decerebrate posturing

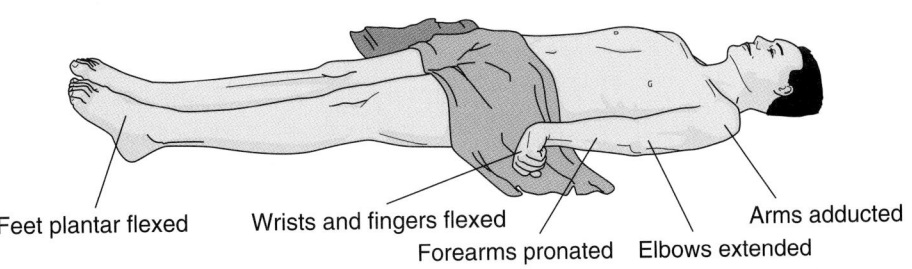

Feet plantar flexed   Wrists and fingers flexed   Arms adducted
Forearms pronated   Elbows extended

Figure 45-6   Abnormal posturing.

Mental Status

You can learn a great deal about a patient's mental capacities and emotional state by simply interacting with the patient. The mental status examination consists of observing the patient's verbal and nonverbal responses to questions and specific requests. Behavior, moods, hygiene, grooming, and choice of dress reveal pertinent informtion about mental status. Question the patient to determine the patient's cognitive functioning, thought processes, and perceptions.

Assessment of the Eyes

Assessment of pupillary response is an important part of the recurrent neurological assessment, as well as the cranial nerve evaluation. The size of the pupils at rest is documented. Many institutions use a millimeter gauge for measuring pupils. This allows an objective description of the size. If the patient's pupils are unusually large or small, you should determine if the patient has had any eyedrops instilled or has taken any medications that might affect the size of the pupils. Approximately 17 percent of people have unequal pupils (**anisocoria**) without any underlying pathological condition. If the patient's pupils are unequal in size, without a correlating diagnosis or symptoms, ask the patient or his or her significant others if this is the patient's normal state. Development of unequal pupils in a patient who previously had equal pupils is an emergency and should be reported to the physician immediately. Any deviation from the normal round shape of the pupils is documented.

Once the resting size of the pupils has been noted, the next step is to assess their response to light. In a darkened room, a light source (such as a flashlight) is directed at the pupil from the lateral aspect of the eye. This allows the examiner to see the direct and the consensual response. A consensual response means that when one pupil is exposed to direct light the other pupil also constricts. Absence of a consensual response may indicate a pathological condition in the area of the optic chiasm. Typically the speed of light reaction is rated as brisk, sluggish, or absent. Differences in the speed or size of constriction between the two pupils should be reported to the practitioner.

Accommodation refers to the patient's ability to focus on an object as it moves closer. Ask the patient to watch the your finger as you hold it about 18 inches in front of the patient's face. As you move the finger toward the patient's face, the patient's eyes should turn toward the midline and the pupils should constrict. It is important to remember that only patients who can follow commands can be assessed for accommodation.

The eyes are evaluated for limitations in range of motion and for smoothness and coordination of movements. Eyes that move in the same direction in a coordinated manner are said to have a conjugate gaze. Conversely, a dysconjugate gaze is movement of the eyes in different directions. Some patients may be unable to move one or both eyes in a specific direction; this is called opthalmoplegia. It is often documented as "limited extraocular movements." Always document what the limitation is (e.g., "Patient is unable to look laterally with left eye"). This allows colleagues to compare their findings with yours and detect any changes.

Nystagmus is involuntary movement of the eyes. Nystagmus varies in the speed of the movement and the direction. Horizontal nystagmus is the most common. Common causes of nystagmus are phenytoin (Dilantin) toxicity and injury to the brainstem.

Assessment of Muscle Function

After assessment of level of consciousness, mental status, and pupillary response, prepare to evaluate muscle function. Muscle groups are compared for symmetry of size and strength. Compression of a specific nerve root may cause atrophy and weakness of the corresponding muscle (e.g., C5–6 compression leading to atrophy and weakness of the biceps).

Assess muscle groups systematically in the upper extremities and then the lower extremities, comparing right to left. The patient's age and general physical condition should be kept in mind when evaluating muscle strength. One does not expect the same amount of strength from a 75-year-old woman as from a 20-year-old man. If the patient has chronic neurological deficits, ask if the results of the assessment are different from his or her usual functioning.

Many health care providers use a 5-point scale to document muscle strength. A score of 5 indicates a patient who is able to move the extremity against gravity and the resistance of the examiner, displaying normal muscle strength. If the examiner is able to provide more resistance than the patient can overcome with active movement, the score is 4. If the patient is able to move the extremity only against gravity, but not resistance, the score is 3. If gravity must be eliminated by having the examiner support the extremity to allow the patient to move the extremity, the score is 2. A score of 1 is given if there is no active movement of the extremity, but a minimum muscular contraction can be palpated. If the examiner is unable to detect any muscular function, a score of 0 is given.

To test the deltoids, ask the patient to raise his or her arms at the shoulder. Have the patient resist as you push down on the upper arms. The biceps are tested by having the patient flex the arm at the elbow and bring the palm toward the face, then resist as you attempt to straighten the arm by pulling on the forearm. Tell the patient to "make a muscle." With the arm similarly flexed, ask the patient to straighten the arm while you resist the movement.

Hand grasps are tested by having the patient squeeze your fingers. Remember to cross your index and middle fingers to prevent the patient from hurting your fingers. If the patient does not release the grasp when told to, it is a reflex grasp, not a response to command. A reflex palmar grasp may indicate a pathological condition of the frontal lobe.

Assess the patient for arm drift by asking the patient to hold both arms straight in front with the palms upward while keeping the eyes closed. A downward drift of the arm,

---

anisocoria: aniso—unequal + coria—pupil

or rotation so that the palm is down, indicates impairment of the opposite side of the brain. If a pathological condition is present, arm drift may be apparent before differences in muscle strength can be detected.

Assessment of leg muscle strength begins with the iliopsoas muscle. Place your hand on the patient's thigh and ask the patient to raise the leg, flexing at the hip. Hip adductors are tested by having the patient bring his or her legs together against your hands. The hip abductors and gluteus medius and minimus are tested by having the patient move the legs apart against resistance. Hip extension by the gluteus maximus is tested by placing the hand under the thigh and having the patient push down with the leg. The quadriceps femoris extends the knee and is tested by having the patient attempt to straighten the leg at the knee. The hamstrings are responsible for knee flexion and are evaluated by having the patient attempt to keep the heel of the foot against the bed or chair rung. Dorsiflexion is tested by having the patient pull the toes toward the head. Plantar flexion is tested by having the patient push against the examiner's hand with the ball of the foot.

Babinski's reflex is tested by stroking the sole of the foot. Normal response is flexion of the great toe. If the great toe extends and the other toes fan out, neurological dysfunction should be suspected if the patient is more than 6 months old. Deep tendon reflexes are not usually part of the nursing neurological assessment. The patient's gait should be assessed to detect any neurological dysfunction and also to assess ability to ambulate safely. Patients who stagger, weave, or bump into objects may need assistance when out of bed.

Romberg's test is performed by having the patient stand with feet together and eyes closed. Be sure to stand close to the patient, especially if he or she is an older adult, to prevent falling. A negative Romberg's test means that the patient experiences minimal swaying for up to 20 seconds. A patient who experiences swaying or who leans to one side is said to have a positive Romberg's test. A positive Romberg's test may be seen in cerebellar dysfunction.

LEARNING TIPS

A positive Romberg's test in an older adult is expected as a result of the normal aging changes of the cerebellum.

Assessment of Cranial Nerves

The cranial nerves are usually not examined in depth during routine bedside neurological assessment. The following recommendations assume a patient who is able to cooperate with the examiner and are intended to give a superficial assessment of cranial nerve function. The olfactory nerve can be tested by asking the patient to identify common scents such as coffee or cinnamon. The optic nerve can be assessed by asking the patient to read something, identify a picture, or tell you how many fingers you are holding up. Check the pupils for consensual reaction to light, indicating function

of the oculomotor nerve. Evaluate extraocular movements by asking the patient to follow your finger while it is moved in front of the patient's eyes. Extraocular movements are controlled by the oculomotor, trochlear, and abducen nerves. Lightly touching different parts of the face with a tissue tests the trigeminal nerve for decreased sensation. Evaluate the face for symmetry at rest and during movement. Asking the patient to frown, smile broadly, and wrinkle the forehead tests the facial nerve. Conversing with the patient gives a basic evaluation of the auditory portion of the vestibulocochlear, or acoustic, nerve. Dysfunction of the acoustic nerve may also cause difficulties with balance, which may be observed as the patient ambulates. Typically the glossopharyngeal and vagus nerves are tested together by assessing the gag reflex and by asking the patient to say "Ah." The spinal accessory nerve is assessed by asking the patient to shrug the shoulders against resistance. Ask the patient to stick out the tongue and move it from side to side. Evaluate for symmetry of size and movement as an indication of function of the hypoglossal nerve.

LEARNING TIPS

The cranial nerves are easier to remember when a mnemonic device is used:

On	Olfactory
Old	Optic
Olympus	Oculomotor
Towering	Trochlear
Top	Trigeminal
A	Abducens
Finn	Facial
And	Acoustic
German	Glossopharyngeal
Viewed	Vagal
Some	Spinal accessory
Hops	Hypoglossal

In all cases the findings of the neurological examination should be correlated with the remainder of the assessment findings. A decreased level of consciousness, coupled with a decreased oxygen saturation on pulse oximetry, point to hypoxia as a cause. Correlation of vital signs with neurological

CRITICAL THINKING

Tim Thompson

You are caring for Tim, a 78-year-old man admitted with heart problems. As you enter his room with his afternoon medications, you find him confused. He thinks he is at home, that the year is 1968, and does not understand who you are or why you are there. He recognizes his wife, who is at his bedside, and knows his own name.

1. How would you describe and document his mental status?
2. What additional data do you need to decide how to proceed?
3. What may have contributed to his confusion?

Answers at end of chapter.

signs is particularly important. Bradycardia, increasing systolic blood pressure, and widening pulse pressure, commonly referred to as Cushing's response, are late indications of increasing intracranial pressure. These findings, in conjunction with a unilateral dilated pupil, may indicate impending herniation of the brain (discussed further in Chapter 46).

▶ DIAGNOSTIC TESTS
Laboratory Tests

Specific diagnostic blood tests do not exist for neurological disorders. Measurement of erythrocyte sedimentation rate (ESR) and white blood cell (WBC) count may indicate an infection, such as meningitis. Hormone levels, such as prolactin or cortisol, may indicate dysfunction of the pituitary gland related to a brain tumor.

Lumbar Puncture

Cerebrospinal fluid may be obtained via lumbar puncture and evaluated for glucose and protein levels, presence of bacteria and white blood cells, levels of immunoglobulin, and culture and sensitivity. Cerebrospinal fluid samples should be sent to the laboratory immediately following the procedure.

LEARNING TIPS

The idea of a needle being introduced into the spinal canal is frightening to many people. Give simple, clear directions to the patient; help the patient maintain his or her position; and provide emotional support throughout the procedure.

Typically the lumbar puncture needle is placed at the level of L3–4 or L4–5 in an adult. Because the spinal cord ends at the L1 level, this placement prevents damage to the cord by the needle. If there is any difficulty inserting the needle, this may be done under fluroscopy to help guide needle placement.

Nursing Care

Informed consent is obtained prior to the procedure. Assist the patient into a side-lying position with his or her back as close to the edge of the bed nearest the practitioner as possible. Depending on the patient's condition, you may need to help the patient flex his or her knees up to the chest (Fig. 45–7). This position maximizes the space between the vertebrae, which makes it easier for the physician or nurse practitioner to insert the needle. An alternative position is to have the patient sit with the back perpendicular to the edge of the bed. Leaning over a bedside table may help the patient maintain the position.

After the lumbar puncture is completed, instruct the patient to remain on bedrest with the head of the bed flat for 6 to 8 hours, as ordered by the physician, and to increase oral intake of fluids. Keeping the head flat decreases the likelihood of leakage of cerebrospinal fluid from the puncture site, which can result in a severe headache. Increasing fluid intake promotes replacement of the fluid that was removed. Check the puncture site for leakage of cerebrospinal fluid and report any leakage to the physician or nurse practitioner. Assess the patient for headache and if necessary obtain an order for analgesia. Label and send the specimens to the laboratory as ordered.

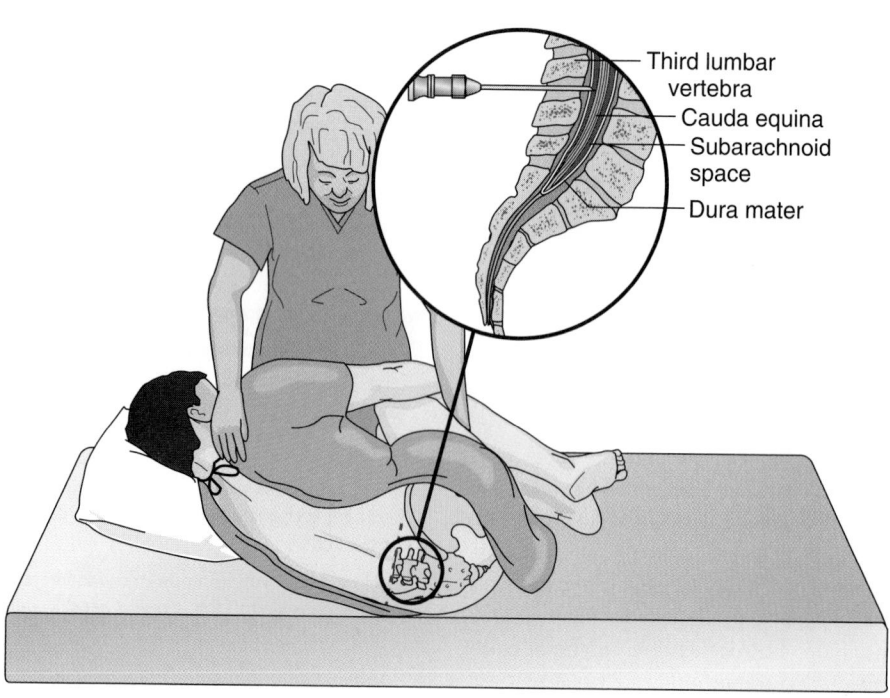

Figure 45-7 Position for lumbar puncture.

X-ray Examination

Spinal x-ray examinations are done to determine the status of individual vertebrae and their relationship to one another. If the patient experiences pain with certain movements, he or she may be asked to flex and extend the area of the spine being examined while the radiographs are taken. This allows detection of abnormal movement of the vertebrae. If the patient has possibly sustained trauma to the spine, particularly the cervical spine, radiographs are taken before immobilizing devices are removed. Skull radiographs may be taken to detect skull fractures or foreign bodies. No special nursing care is required.

Computed Tomography

A computed tomography (CT) scan is commonly the test ordered if the patient initially consults a general practitioner. CT scans for the purpose of diagnosing neurological disorders can be performed on the brain or the spine. Some of the disorders that can be detected by CT are tumors, skull fractures, and abscesses. The scan may be performed with or without radiopaque contrast material to enhance the clarity of the images that are recorded. If contrast material is used, a series of images is filmed and then the contrast material is given intravenously and another series of images is filmed. The patient should be questioned about any allergies to contrast material, iodine, or shellfish. The blood urea nitrogen (BUN) and creatinine levels should be checked before administration of contrast material because it is excreted through the kidneys. Patients with elevated BUN and creatinine or known renal disease may be unable to tolerate the contrast material. Contrast material is most commonly used if a tumor is suspected or following surgery in the area to be scanned. CT scans are commonly used in emergency evaluations because they can be done quickly, an important consideration if the patient is ventilated or unstable.

Nursing Care

During the CT scan, the patient must lie still on a moveable table. Noncontrast scans take approximately 10 minutes; contrast scans take between 20 and 30 minutes. Patients who are receiving dye should be warned that they may feel a sensation of warmth; warmth in the groin area may make them feel as though they have been incontinent of urine. Nausea, diaphoresis, itching, and difficulty breathing may indicate allergy to the dye and should be reported immediately to the physician or nurse practitioner. Sedation may be required for patients who are agitated or disoriented. Patients who are in pain may require pain medication before the examination.

Magnetic Resonance Imaging

Magnetic resonance imaging (MRI) gives a more detailed picture of soft tissue than a CT scan. It is not as useful when looking for bony abnormalities. MRI is a longer procedure and commonly requires transportation to distant areas of the hospital, which may be problematic for unstable, disoriented, or ventilated patients. As with a CT scan, the MRI can be done with or without contrast material. Some facilities have the capability to perform magnetic resonance angiograms (MRA). This test allows visualization of blood vessels and assessment of blood flow without being as invasive as a traditional angiogram.

Nursing Care

Because of the magnetic fields being used, there are restrictions placed on patients undergoing an MRI and the health care personnel who work within the MRI facility. Individuals with pacemakers or any type of metallic prosthesis are not able to undergo MRI or be in the room when one is performed. This is because the magnetic field is so strong that it could dislodge the prosthesis or pacemaker. Patients are asked to remove all metal objects, such as jewelry or hair clips, before the procedure. Individuals who may have accidentally acquired metallic foreign bodies (e.g., metal slivers in the eye or shrapnel that was not removed) may need an x-ray examination to determine the presence or absence of such objects. Carefully question patients or their significant others regarding any possible contraindications. MRI is commonly used when herniated intravertebral disks are suspected. It may be difficult for the patient to lie in one position for a prolonged period. The patient's need for pain medication should be assessed before the procedure. Use of pillows for positioning may improve comfort. The narrow, tunnel-like structure of the MRI unit causes claustrophobia in some patients; some patients may require use of sedatives or open MRI units. Encourage the use of deep breathing, guided imagery, and other relaxation techniques.

Angiogram

Angiogram is an x-ray study of blood vessels that is used when an abnormality of cerebral or spinal blood vessels is suspected or to obtain information about blood supply to a tumor. Following injection of a local anesthetic, a catheter is inserted through the femoral artery and advanced until contrast material can be injected into the cerebral vessels. The dye then shows the vessels on the radiograph and provides information about the structure of specific vessels, as well as overall circulation to the area.

Nursing Care

Before an angiogram, the patient receives a clear liquid diet and has an intravenous needle in place. Informed consent must be obtained. BUN and creatinine levels are evaluated because the contrast material is excreted through the kidneys. Potential for bleeding is assessed by prothrombin time and partial thromboplastin time tests because a puncture is being made in a large artery. Typically the patient receives some type of sedation before being transported to the angiography suite. During the injection of the contrast material, the patient may complain of severe heat sensations and a metallic taste in the mouth. The patient must lie still while the radiographs are being taken and so should be told about the sensations he or she may experience. Patients who are disoriented or agitated may require sedation to complete the test.

Following the procedure, pressure is maintained on the catheter insertion site and the patient is kept flat in bed for 6 to 8 hours to prevent bleeding from the insertion site. The patient may turn from side to side but must keep the affected leg straight. In addition to assessing vital signs, the nurse evaluates the catheter insertion site and the presence and quality of the popliteal and pedal pulses in the affected leg. Decrease or loss of the pedal pulse may indicate a clot in the femoral artery and should be reported to the physician immediately. Patients should be encouraged to increase oral intake in addition to the intravenous fluids that are administered to aid in the excretion of the contrast material.

Myelogram

A **myelogram** is an x-ray examination of the spinal canal and its contents. Following a lumbar puncture, cerebrospinal fluid is removed and sent for laboratory analysis. Contrast material is then injected into the subarachnoid space. The patient is moved into various positions and radiographs are taken. Compression of nerve roots, herniation of intravertebral disks, and blockage of cerebrospinal fluid circulation may all be detected by myelogram.

Nursing Care

Following the procedure the patient is kept on bedrest with the head elevated. This lessens the possibility of the contrast material getting into the cerebral cerebrospinal fluid circulation. The contrast material used for myelograms can lower the seizure threshold in some patients. Any patient with a known seizure disorder should have serum levels of anticonvulsants evaluated and be carefully observed for signs of seizures.

Because of their invasive nature, a separate informed consent form is normally required for angiogram, lumbar puncture, and myelogram. The physician performing the test explains the risks, benefits, and possible complications of the examination. Patients who need these diagnostic procedures, particularly lumbar puncture and angiogram, may have cognitive deficits; therefore it may be necessary to obtain consent from the legal next of kin.

Electroencephalogram

Evaluation of the electrical activity of the brain is obtained through an **electroencephalogram** (EEG). Electrodes are attached to the scalp with an adhesive. Electrical activity is transmitted through the electrodes to a tracing. Analysis of the tracing can identify areas of abnormality, such as seizure focus or areas of slowed activity.

---

myelogram: myelo—referring to the spinal cord + gram—picture
electroencephalogram: electro—electrical activity + encephalo—referring to the brain + gram—picture

Nursing Care

Before the test, make sure that the patient's hair is clean and dry. The physician may write orders to withhold any sedatives to prevent interference with the EEG, and patients may be weaned from their anticonvulsants if the goal of the test is to identify the seizure focus during a seizure. These patients must be very carefully monitored and protected from harm. Typically they undergo videotaping while the EEG is performed. Following the procedure, the adhesive must be washed from the hair. Assist the hospitalized patient to do this as soon as possible, before it becomes hardened and difficult to remove.

► THERAPEUTIC MEASURES
Moving and Positioning

Patients who have pain may need help in changing positions and ambulating. Use of heat, cold, or analgesics may allow the patient to be more independent in mobility. Patients with paresis, paralysis, or **paresthesia**s may be partially or completely dependent in moving and positioning. Care should be taken to maintain the body in functional positions when routine position changes are made. If the patient experiences sensory loss, the person assisting with position changes must ensure that no part of the body is inadvertently compressed (e.g., a hand caught under a hip or the scrotum compressed between the legs). Collaboration with the physical therapist may yield positioning techniques that maximize the chance of useful recovery.

Contractures and footdrop are complications that are often associated with neurological disorders. Contractures are permanent muscle contractions with fibrosis of connective tissue that occur from lack of use of a muscle or muscle group. They cause permanent deformities and prevent normal functioning of the affected part. Footdrop occurs when the feet are not supported in a functional position and become contracted in a position of plantar flexion (Fig. 45–8). Some of the interventions used to prevent footdrop are foot boards, high-top tennis shoes, and splints. Splints are commonly used to prevent contractures of the upper and lower extremities and to keep the affected part in a functional position. If splints are used, the patient must be evaluated for discomfort and skin breakdown at the splint site.

Mobilization should be begun as soon as the patient is medically stable. Initially this may involve the use of a cardiac chair if the patient is unable to bear weight. Transfer of the patient to a bedside chair or use of ambulation aids may require a multidisciplinary approach. The nurse must recognize any physical or cognitive deficits that may affect safety and adjust the environment to protect the patient. This includes communicating any safety concerns to unlicensed personnel who interact with the patient.

Activities of Daily Living

The effects of neurological disorders on activities of daily living (ADLs) may range from an inconvenience to com-

---

paresthesia: para—beside + asthesia—sensations

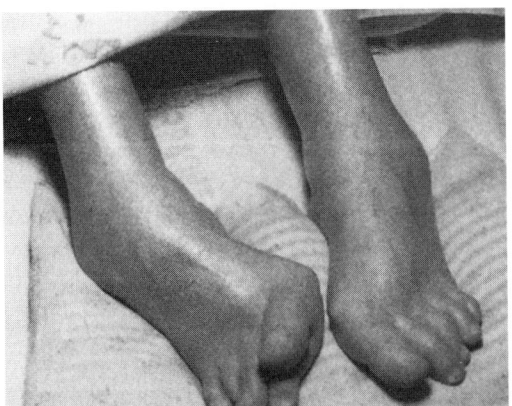

Figure 45-8   Contractures, footdrop. (From Hegner, B:Assisting in Long Term Care, ed 3. Delmar Publishers, Albany, NY, 1998, with permission.)

plete dependence. Patients may have trouble bending over to put on their shoes and socks, lifting a full cooking pot, or caring for an infant. A high-level quadriplegic may be completely unable to perform ADLs but can be taught to direct his or her own personal care. Patients should be encouraged to use strategies learned in occupational or physical therapy.

Assessment should include a discussion of the strategies the patient normally uses to accomplish ADLs. Every attempt should be made to continue to use these strategies. This is particularly true if the patient is admitted to a long-term care facility. Patients who have normal cognitive function should be included in care planning and encouraged to work collaboratively with caregivers. If the strategies the patient uses during ADLs must be changed, the rationale for the changes should be explained to the patient and significant others. An example is if the patient is using a transfer technique that is unsafe for the patient or caregivers. If patients have impaired cognitive function, try to maintain a specific routine that is as close to their normal environment as possible. Normalizing routines may help patients adapt to a change in environment and maximize their ability to function.

Communication

The communication problems associated with neurological disorders have a variety of etiologies. Some neurological disorders cause difficulty speaking **(dysarthria).** Dysfunction of the lips, tongue, or jaw makes speech difficult or impossible to understand. When dysarthric individuals know what they want to say but cannot be understood, the frustration level is high. This frustration is compounded if the patients are treated as if they have cognitive deficits merely because they have difficulty communicating.

Patients who have had a stroke can experience different types of aphasia. Expressive aphasia is difficulty or inability to verbally communicate with others. The patient may be able to speak in sentences but inappropriately substitute

words, such as "The sky is dish." Word-finding difficulty is another type of expressive aphasia. These patients may tell you "I want a . . ." and then be unable to complete the sentence. In severe cases of aphasia the patient may make sounds that resemble words or may only utter sounds. For individuals with no intelligible speech or with word-finding difficulty, a picture board with commonly used items may facilitate communication. (See an example of a picture board in Chapter 46.) Keep in mind that patients with expressive aphasia may answer yes to all questions, rather than just those for which yes is correct. The same is true of answering no. This is one reason why a nurse should never ask a patient, "Are you Mrs. Gonzalez?" An aphasic patient may say yes even if that is not her name. Instead, ask the patient to state her name. If she cannot state her name, check her identification band.

For patients who substitute words, simply correct the substitution and continue the conversation. Patients with expressive aphasia are often very aware of and frustrated by their difficulty communicating. Give them time to try to express themselves. If you cannot understand them, offer possibilities based on the situation. If the patient is sitting in the chair, ask if he or she wants to go back to bed or wants to use the bathroom. If the patient is restless, ask if he or she is in pain.

Some patients use the same word in response to all questions, and for some patients that word is a profanity. This is very difficult for significant others to deal with, particularly if swearing is not something the patient normally did. Make it clear to the family that you understand that this behavior is part of the patient's illness.

Receptive aphasia affects the patient's ability to understand spoken language. Again, the severity of the aphasia varies. Some patients may understand simple directions such as "sit down" or "squeeze my fingers." In other cases the nurse may need to pantomime the action the nurse wants the patient to perform, such as showing the patient pills and then mimicking taking the pills and drinking water.

LEARNING TIPS

If the patient has receptive aphasia, assume that he or she cannot understand or follow safety instructions, such as "Do not stand up until I get back." Even going to the bathroom to get a towel can give a patient enough time to try to stand up and subsequently fall.

Nutrition

Alterations in ability to maintain an adequate nutritional intake can have many causes. The level of consciousness may be depressed enough that the patient does not recognize that she or he is hungry or thirsty. Decreased level of consciousness or cranial nerve dysfunction may impair the patient's ability to swallow safely. Severe weakness may limit the patient's ability to take in enough food to meet the body's requirements. These conditions are often compounded by the increased metabolic rate that accompanies neurological injury or illness.

---

dysarthria: dys—dysfunctional + arthria—movement of the joints used in speech

If there is any question of the patient's ability to swallow, a swallowing evaluation should be performed by a speech therapist. Food coloring can be added to food or fluids and the patient watched until the food is swallowed. If respiratory secretions are the same color as the food coloring, the patient has aspirated and is not able to safely swallow the food or fluid. Some institutions use a radiological examination to evaluate the ability to swallow. A small amount of barium is added to the food or fluid, and fluoroscopy is used while the patient swallows. This allows visualization of the path of the food or fluid. Patients with swallowing difficulty (dysphagia) may have better success with foods or thick liquids rather than thin fluids. Liquids may be thickened with special thickening agents to allow easier swallowing. All patients should be positioned as upright as possible while eating or drinking, and patients who have difficulty swallowing should be monitored during eating.

If weakness or fatigue is the cause of decreased nutritional intake, several modifications are possible. Serving small portions of food frequently can increase intake. Using high-protein, high-calorie foods and supplements increases the nutritional content of small amounts of foods.

For patients who are unable to swallow or who cannot swallow enough food, enteral tube feedings may be required. If enteral feedings are anticipated to be for a short duration, a nasogastric tube may be used. The disadvantages of nasogastric tubes include impairment of the integrity of nasal skin and the risk of aspiration. The risk of aspiration in neurologically impaired patients who have cognitive impairments is increased because these patients may pull out the nasogastric tube because they do not understand its purpose. If long-term enteral feedings are anticipated, a gastrostomy tube may be placed directly through the abdominal wall into the stomach. This feeding method has the advantage of eliminating the risks of aspiration and nasal skin breakdown.

Answers to CRITICAL THINKING

Tim Thompson

1. He is alert but confused, oriented to person only.
2. The nurse should ask his wife if this has ever happened before; check his medical history for any disorders that may contribute to neurological dysfunction; do a quick neurological assessment to determine if any additional deficits exist; check vital signs and pulse oximetry if available; and notify the physician immediately if the symptoms are a new finding.
3. Some possible explanations to explore include hypoxemia, stroke, worsening heart problems causing inadequate flow of blood to the brain, hypoglycemia, or even confusion related to a sudden transition from home to an unfamiliar environment.

REVIEW QUESTIONS

1. Which of the following are neurons that carry impulses from the CNS to effectors?
   a. Mixed
   b. Motor
   c. Afferent
   d. Sensory

2. Which of the following is a symptom of increasing intracranial pressure?
   a. Constricted pupils
   b. Decreasing level of consciousness
   c. Narrowing pulse pressure
   d. Bradypnea

3. Which of the following actions by the nurse is the best way to determine if a patient with expressive aphasia is oriented?
   a. Ask yes-or-no questions.
   b. Ask the patient to name the family member in the room.
   c. Ask the patient who the current president is.
   d. Have the patient count backward from 10.

4. Asking the patient to stick out his or her tongue evaluates the function of which cranial nerve?
   a. IV—trochlear
   b. Vtrigeminal
   c. IX—glossopharyngeal
   d. XIIhypoglossal

5. Which of the following nursing interventions can help prevent footdrop?
   a. Positioning the patient in the left lateral position
   b. Providing daily foot massage
   c. Using high-top tennis shoes
   d. Maintaining an upright position as much as possible

6. Which of the following activities should be encouraged when a patient returns from a CT scan using a contrast medium?
   a. Ambulation
   b. Drinking fluids
   c. Turning side to side
   d. Coughing and deep breathing

7. When observing a patient's gait, the nurse is checking the integrity of what?
   a. Cerebellum
   b. Brainstem
   c. Diencephalon
   d. Cerebrum

8. Which of the following areas of the brain controls speech?
   a. Left occipital lobe
   b. Broca's area
   c. Midbrain
   d. Brainstem

46

NURSING CARE OF PATIENTS WITH CENTRAL NERVOUS SYSTEM DISORDERS

George Byron Smith and Sally Schnell

KEY TERMS

akinesia (A-ki-**NEE**-zee-uh)

ataxia (ah-**TAK**-see-ah)

bradykinesia (BRAY-dee-kin-**EE**-zee-ah)

contracture (kon-**TRAK**-chur)

contralateral (KON-truh-**LAT**-er-uhl)

craniectomy (KRAY-nee-**EK**-tuh-mee)

cranioplasty (**KRAY**-nee-oh-plas-tee)

craniotomy (KRAY-nee-**AHT**-oh-mee)

dementia (dee-**MEN**-cha)

dysreflexia (DIS-re-**FLEK**-see-ah)

encephalitis (EN-seff-uh-**LYE**-tis)

encephalopathy (en-SEFF-uh-**LAHP**-ah-thee)

endarterectomy (end-AR-tur-**ECK**-tuh-mee)

epidural (EP-i-**DUHR**-uhl)

flaccid (**FLA**-sid)

hemiparesis (hem-ee-puh-**REE**-sis)

hemiplegia (hem-ee-**PLEE**-jee-ah)

hydrocephalus (HIGH-droh-**SEF**-uh-luhs)

ipsilateral (IP-si-**LAT**-er-uhl)

laminectomy (LAM-i-**NEK**-toh-mee)

meningitis (MEN-in-**JIGH**-tis)

nuchal rigidity (**NEW**-kuhl re-**JID**-i-tee)

paraparesis (PAR-ah-puh-**REE**-sis)

paraplegia (PAR-ah-**PLEE**-jee-ah)

photophobia (FOH-tuh-**FOH**-bee-ah)

postictal (pohst-**IK**-tuhl)

prodromal (proh-**DROH**-muhl)

quadriparesis (KWA-dri-puh-**REE**-sis)

quadriplegia (KWA-dri-**PLEE**-jee-ah)

subdural (sub-**DUHR**-uhl)

thrombolytic (throm-boh-**LIT**-ik)

turbid (**TER**-bid)

QUESTIONS TO GUIDE YOUR READING

1. What are the causes, risk factors, and pathophysiology of major disorders of the central nervous system?

2. What is the appropriate plan of care for a patient with a major disorder of the central nervous system?

3. What are the primary treatments for the major disorders of the central nervous system, including surgery, intracranial pressure monitoring, medication, and rehabilitation?

4. What are the causes, risk factors, and pathophysiology of traumatic injuries to the brain and spinal cord?

5. What is the appropriate plan of care for a patient with a traumatic injury to the brain or spinal cord?

6. What are the causes, risk factors, and pathophysiology associated with degenerative neuromuscular disorders?

7. What is the appropriate plan of care for a patient with degenerative neuromuscular disorders?

8. What are the causes, risk factors, pathophysiology, and progression of symptoms seen in patients with Alzheimer's disease?

9. What are the steps in the nursing process in meeting the diverse needs of patients with Alzheimer's disease and their families?

Disorders of the central nervous system (CNS) include problems originating in the brain and spinal cord. Because the CNS is the control center for the entire body, disorders in this system can cause symptoms in any part of the body, ranging from pain to paralysis, confusion, and coma. This chapter presents nursing care of patients with these disorders.

▶ CENTRAL NERVOUS SYSTEM INFECTIONS

Infectious agents may enter the central nervous system via a variety of routes (Table 46–1). Anything that depresses the patient's immune system, such as steroid administration, chemotherapy, radiation therapy, and malnutrition, can make the patient more vulnerable to infection.

Meningitis

Meningitis is a purulent infection caused by bacterial or viral invasion of the pia mater, arachnoid, and subarachnoid space surrounding the brain and spinal cord. The membranes react to the infection with an inflammatory response. Bacterial meningitis is more common than viral. Occasionally meningitis is caused by a noninfectious irritant, such as dye from a myelogram.

Pathophysiology and Etiology

The most common organisms causing meningitis include meningococcus, pneumococcus, and *Haemophilus influenzae*. In the United States, as of the late 1990s—more than 10 years after the introduction of the *H. influenzae* serotype b vaccine (HIB)—the most common agents are *Neisseria meningitidis* and *Streptococcus pneumoniae*. During the early 1990s, about 25,000 cases occurred annually in the United States, with 70 percent identified among children age 5 or younger. Each year more than 2000 deaths were attributed to bacterial meningitis. The infection generally begins in another area, such as the upper respiratory tract, enters the blood, and invades the CNS, especially if the immune system is not functioning effectively. Bacterial invasion leads

to a rapidly increased blood supply to the meninges with massive neutrophil migration. The neutrophils then engulf the bacteria and disintegrate, causing purulent material to form. Exudate from tissue destruction also contributes to this purulent material. The purulent material causes the meninges to become inflamed, and intracranial pressure increases. Cranial nerve function may be transiently or permanently affected by meningitis. Some of the effects are listed in Table 46–2.

A common endocrine disorder associated with meningitis is a cycle of increased intracranial pressure leading to excessive release of antidiuretic hormone (ADH). ADH acts by inhibiting urination. This in turn leads to water retention, oliguria (reduced urination), hypervolemia (excess blood volume), hyponatremia (low serum sodium), and further increases in intracranial pressure. (See Increased Intracranial Pressure later in this chapter.)

Prevention

Vaccines are available against *H. influenzae* and *S. pneumoniae*. These are especially recommended for immunocompromised patients. Other vaccines are being developed.

Signs and Symptoms

The initial symptom of meningitis is a severe headache, caused by tension on blood vessels and irritation of the pain-sensitive dura mater. Fever is commonly 101° to 103° F or higher. If the meningitis is not resolved, pressure on the brainstem may elevate the temperature to 105° F or higher during the terminal phase. **Photophobia** may be present. The patient with meningococcal meningitis has a petechial rash on the skin and mucous membranes.

Nuchal rigidity (pain and stiffness when the neck is moved) is caused by spasm of the extensor muscles of the neck. Kernig's sign and Brudzinski's sign are often seen in patients suffering from meningitis. Both signs are caused by inflammation of the meninges and spinal nerve roots. Brudzinski's sign is positive when flexion of the patient's neck causes the hips and knees to flex. To elicit Kernig's sign the examiner flexes the patient's hip to 90 degrees and tries to extend the patient's knee. The sign is positive if the patient experiences pain and spasm of the hamstring (Fig. 46–1). The nausea and vomiting associated with meningitis are caused by direct irritation of brain tissue and by increased intracranial pressure (ICP).

TABLE 46–1	ROUTES OF ENTRY FOR CENTRAL NERVOUS SYSTEM INFECTIONS
Route of Entry	**Examples**
Bloodstream	Insect bite
	Otitis media
Direct extension	Fracture of frontal or facial bones
Cerebrospinal fluid	Dural tear
	Poor sterile technique
Nose or mouth	Meningococcus meningitis
In utero	Contamination of amniotic fluid
	Rubella
	Vaginal infection

TABLE 46–2	CRANIAL NERVES AFFECTED BY MENINGITIS
Cranial Nerve Affected	**Manifestation**
III, IV, VI	Ocular palsies
	Unequal and sluggishly reactive pupils
VII	Facial weakness
VIII	Deafness and vertigo

meningitis: mening—membranous covering of the brain + itis—inflammation

photophobia: photo—light + phobia—fear or intolerance

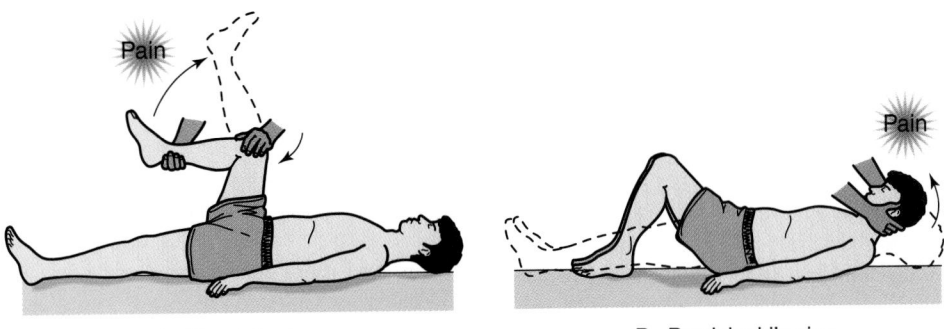

A. Kernig's sign   B. Brudzinski's sign

Figure 46-1 *(A)* Kernig's sign. *(B)* Brudzinski's sign.

Encephalopathy refers to the mental status changes seen in patients with meningitis. These are manifested as short attention span, poor memory, disorientation, difficulty following commands, and a tendency to misinterpret environmental stimuli. Late signs of meningitis include lethargy and seizures.

Acute Complications of Meningitis

Hydrocephalus can occur when purulent material interferes with circulation and reabsorption of cerebrospinal fluid. The patient may begin to improve following the initiation of treatment and then experience neurological deterioration as hydrocephalus develops. Careful and frequent neurological assessments allow for the detection of subtle changes.

Irritation of the cerebral cortex makes patients prone to seizures. Seizure precautions should be in place for all patients with meningitis (Table 46–3). An oral airway should be used only if the patient's mouth is open. Never attempt to insert something into the mouth of a seizing patient if the jaws are clenched.

As with all patients with neurological disorders, frequent assessment of respiratory function is necessary. Inflammation and edema may place pressure on the respiratory center and result in hypoxia. Hypoxia causes vasodilation, which increases intracranial pressure. If the patient experiences significant neurological and respiratory impairment, intubation and mechanical ventilation may be necessary.

Long-Term Complications

Resolution of meningitis depends on how quickly and effectively the disease is treated. Some individuals experience no lasting effects. Other patients have permanent neurological deficits. Cranial nerve damage may leave the patient blind or deaf. Seizures may continue to occur even after the acute phase of the illness has passed. Cognitive deficits ranging from memory impairment to profound learning disabilities may occur.

Diagnostic Tests

A lumbar puncture is the most informative diagnostic test for a patient with suspected meningitis. (See Chapter 45.) Viral meningitis is characterized by clear cerebrospinal fluid with normal glucose level and normal or slightly increased protein level. No bacteria are seen, but the white blood cell count is usually increased. In contrast, the cerebrospinal fluid of an individual with bacterial meningitis is **turbid,** or cloudy, because of the massive number of white blood cells. Bacteria are identified by Gram's stain and culture. The bacteria utilize the glucose normally found in cerebrospinal fluid (CSF), thereby lowering the glucose level. The amount of protein in the cerebrospinal fluid is elevated.

Medical Treatment

Meningitis can be fatal if not promptly treated. Broad-spectrum antibiotics are administered intravenously. After a culture and sensitivity test is done on CSF, the antibiotic may be changed. Antibiotics are not effective in the treatment of viral meningitis.

Symptom management is the same for viral or bacterial meningitis. Antipyretics such as acetaminophen are used to control the fever. A cooling blanket may also be used. Care

TABLE 46-3	**INTERVENTIONS FOR SEIZURES**

Seizure Precautions
- Pad side rails of hospital bed with commercial pads or bath blankets folded over and pinned in place.
- Keep call light within reach.
- Assist patient when ambulating.
- Keep suction and oral airway at bedside.

Nursing Care during a Seizure
- Stay with patient.
- Do not restrain patient.
- Protect from injury (move nearby objects).
- Loosen tight clothing.
- Turn to side when able to prevent occlusion of airway or aspiration.
- Suction if needed.
- Monitor vital signs when able.
- Be prepared to assist with breathing if necessary.
- Observe and document progression of symptoms.

---

encephalopathy: encephalo—brain + pathy—illness
hydrocephalus: hydro—water + cephalus—head

should be taken to avoid cooling the patient too much and causing shivering, because this increases the metabolic demand for oxygen and glucose. A quiet, dark environment lessens the stimulation to a patient who has a headache or photophobia (sensitivity to light) and who may be agitated, disoriented, or at risk for seizures.

Pain medications are given to lessen head and neck pain. Opioids are rarely used because of the risk of masking neurological changes. Codeine products are preferred because they are less sedating than other opioids and do not affect pupil response. Nausea and vomiting are controlled by administering antiemetic medications. The patient with meningococcal meningitis should be placed in isolation, because it can be transmitted to others.

Patients may become agitated and attempt to leave the hospital because they do not comprehend or remember how ill they are. An important aspect of nursing care focuses on keeping patients from harming themselves. It is very upsetting to families to see a loved one acting agitated or disoriented. It is important to teach the family about symptoms and treatment goals for the patient (Box 46–1).

Encephalitis
Pathophysiology

Encephalitis is inflammation of brain tissue. Nerve cell damage, edema, and necrosis cause neurological findings localized to the specific areas of the brain affected. Hemorrhage may occur in some types of encephalitis. Increased intracranial pressure may lead to herniation of the brain.

BOX 46–1 **Meningitis Summary**

Symptoms
Nuchal rigidity
Positive Kernig's sign
Positive Brudzinski's sign
Fever
Photophobia
Petechial rash on skin and mucous membranes
Encephalopathy

Diagnostic Tests
Lumbar puncture with CSF analysis

Therapeutic Management
Antimicrobials (if bacterial)
Seizure precautions
Symptom management
Antipyretics
Pain management
Reduction of environmental stimuli
Education

Etiology

Viruses are the most common cause of encephalitis. They may be specifically related to a particular time of year or geographic location. Some viruses, such as West Nile virus, are carried by ticks or mosquitoes. Others are systemic viral infections, such as infectious mononucleosis or mumps, that spread to the brain. Parasites, toxic substances, bacteria, vaccines, and fungi are other potential causes of encephalitis.

Herpes simplex is the most common non–insect-borne virus to cause encephalitis. The majority of individuals harbor herpes simplex virus type 1 in a dormant state. This is the virus responsible for cold sores on the oral mucous membranes. Infectious diseases, fever, and emotional stress are possible reasons for the virus becoming active, but the exact mechanism is not known.

Signs and Symptoms

As with meningitis, headache and fever are common presenting symptoms. The patient may also complain of nausea and vomiting and general malaise. These symptoms usually develop over a period of several days.

The viruses cause nuchal rigidity, confusion, decreased level of consciousness, seizures, sensitivity to light, **ataxia,** and tremors. The patient may have **hemiparesis** and exhibit abnormal deep tendon reflexes.

The patient with herpes encephalitis develops edema and necrosis (sometimes associated with hemorrhage), most commonly in the temporal lobes. This significant cerebral edema causes increased ICP and can lead to herniation of the brain. If the patient becomes comatose before treatment is begun, the mortality rate may be as high as 70 to 80 percent. The first 72 hours, when cerebral edema is worst, is the most likely time for death to occur.

Long-Term Complications of Encephalitis

Patients who have had encephalitis are often left with cognitive disabilities and personality changes. Ongoing seizures, motor deficits, and blindness may also occur. Deterioration in cognition and personality control are particularly stressful for significant others. The patient's behavioral control is a major factor in determining discharge plans. Assist significant others to realistically assess the patient's functional level and the family's ability to care for the patient. In-home care, outpatient therapy, and adult day care are options to explore. For some severely impaired individuals, custodial care may be the only feasible and safe discharge option.

Diagnostic Tests

Computed tomography (CT) scan, lumbar puncture to obtain cerebrospinal fluid, and electroencephalogram (EEG) are used to diagnose encephalitis. Cerebrospinal fluid analysis typically reveals increased white blood cell count and protein level and normal glucose levels. Breakdown of blood after cerebral hemorrhage results in yellow CSF.

---

encephalitis: encephalo—brain + itis—inflammation

hemiparesis: hemi—one side + paresis—partial paralysis

Treatment

No specific treatment is currently available for insect-borne encephalitis. Careful neurological assessment and a symptomatic approach to care help prevent complications and improve survival. Medications to reduce pain and fever and anticonvulsant medication are often administered. Significant others need emotional support and ongoing teaching.

Acyclovir (Zovirax) is given intravenously to treat herpes simplex encephalitis. Therapy should be instituted as soon as possible after diagnosis. The most favorable outcomes are noted when acyclovir is administered before the Glasgow Coma Scale (GCS) score is less than 10. Even with prompt treatment, only 38 percent of patients with herpes simplex encephalitis regain normal function.

▶ INCREASED INTRACRANIAL PRESSURE
Pathophysiology and Monitoring

Any patient with an intracranial pathological condition is potentially at risk for increased intracranial pressure. ICP is the pressure exerted within the cranial cavity by its components (blood, brain, and cerebrospinal fluid). The normal ICP is 0 to 15 mm Hg. This pressure fluctuates with normal physiological changes, such as arterial pulsations, changes in position, and increases in intrathoracic pressure (e.g., coughing or sneezing). Common causes of increased ICP include brain trauma, intracranial hemorrhage, and brain tumors. Prompt detection of changes in neurological status indicating increased intracranial pressure allows intervention aimed at preventing permanent brain damage.

The Monro-Kellie doctrine states that the skull is a rigid compartment containing three components: brain, blood, and cerebrospinal fluid. If an increase in one component is not accompanied by a decrease in one or both of the other components, the result is increased intracranial pressure

(ICP). The consequences of increased ICP depend on the degree of elevation and the speed with which the ICP increases. Patients with slow-growing tumors may have significantly increased intracranial pressure before they develop symptoms. Conversely, patients with a subarachnoid hemorrhage may sustain a sudden sharp increase in intracranial pressure.

The normally functioning body has several methods of compensating for increased intracranial pressure. Cerebrospinal fluid can be shunted into the spinal subarachnoid space. Hyperventilation may trigger constriction of cerebral blood vessels, decreasing the amount of blood within the cranial vault. These compensatory mechanisms are temporary and not particularly effective if the increase in ICP is sudden and severe.

Signs and Symptoms

Increased ICP may first be manifested by restlessness, irritability, and decreased level of consciousness, because cerebral cortex function is impaired. If not intubated, the patient may hyperventilate, causing vasoconstriction as the body attempts to compensate. As the pressure increases, the oculomotor nerve may be compressed on the side of the impairment. Compression of the outermost fibers of the oculomotor nerve results in diminished reactivity and dilation of the pupil. As the fibers become increasingly compressed, the pupil stops reacting to light. If the compression continues, both pupils become fixed and dilated.

Vital sign changes are a late indication of increasing intracranial pressure. Cushing's response is a classic late sign of increased ICP. Cushing's response is characterized by bradycardia and increasing systolic blood pressure while diastolic blood pressure remains the same, resulting in widening pulse pressure (Fig. 46–2). By the time these symptoms appear, the intracranial pressure is significantly increased and interventions may not be successful.

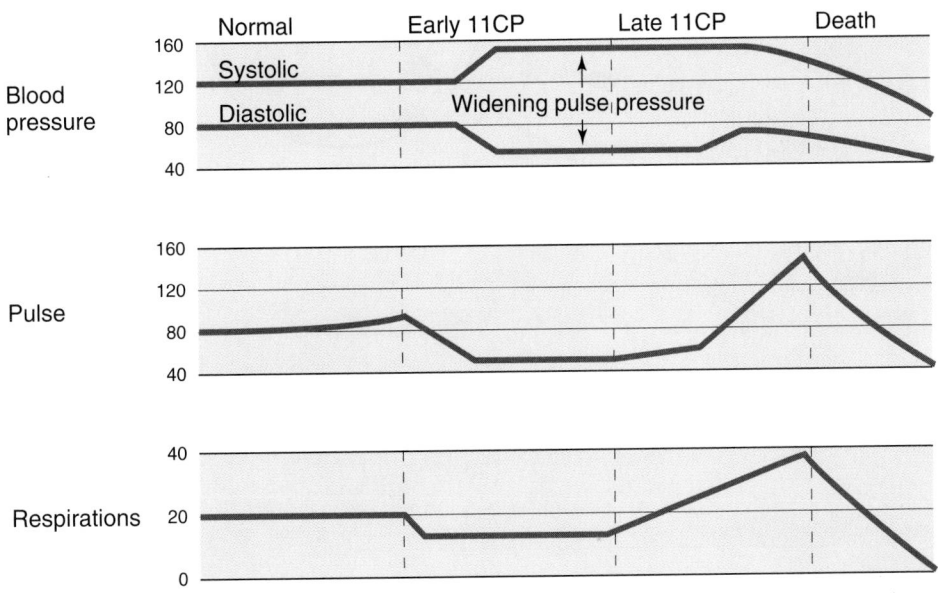

Figure 46-2 Cushing's response. Changes in vital signs that accompany increased intracranial pressure.

Monitoring

ICP monitoring has been the standard of care for patients with severe head injury for several years. ICP monitoring allows for early detection of the pressure on the brain before changes in symptoms can be seen. Intracranial monitoring is indicated when the patient demonstrates development of unequal pupils. Other reasons for monitoring include a CT scan demonstrating midline shift and the decision to use neuromuscular blocking agents and sedatives. The most common methods of monitoring ICP in adults are external ventricular drain, subarachnoid bolt, and intraparenchymal monitor. All three methods involve anesthetizing the scalp and drilling a burr hole in the skull. Patients who are rest-

less or agitated may require sedation for the procedure to be completed safely. Informed consent must be obtained, but in some emergent cases the monitor is inserted without consent.

Placement of a catheter into one of the lateral ventricles is referred to as external ventricular drainage (Fig. 46–3). Drainage of cerebrospinal fluid reduces intracranial pressure, which can be therapeutic, as well as a way of monitoring pressure. Disadvantages to this method include difficulty in locating the ventricle for insertion of the catheter and clotting of the catheter by blood in cerebrospinal fluid.

To allow communication with the subarachnoid space, a subarachnoid bolt is tightly screwed into the burr hole after the dura has been punctured (Fig. 46–4). The advantage to

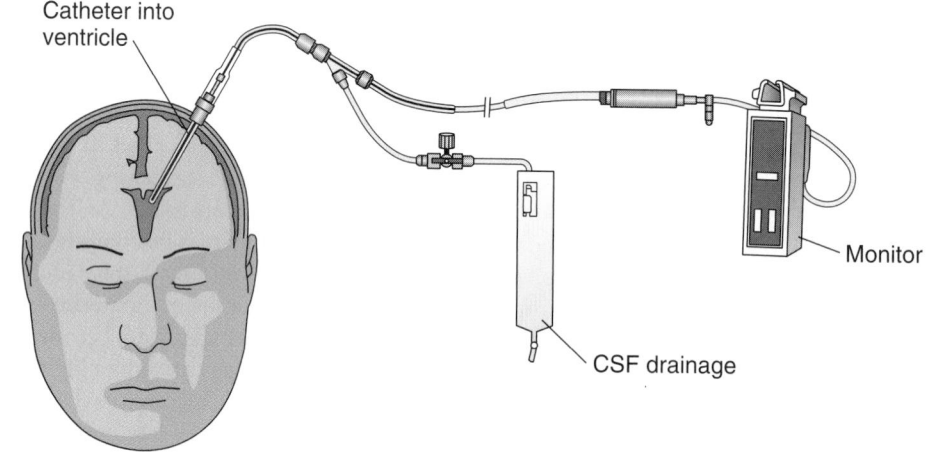

Figure 46-3 Ventricular drain. A catheter into the ventricle allows ICP monitoring and CSF drainage.

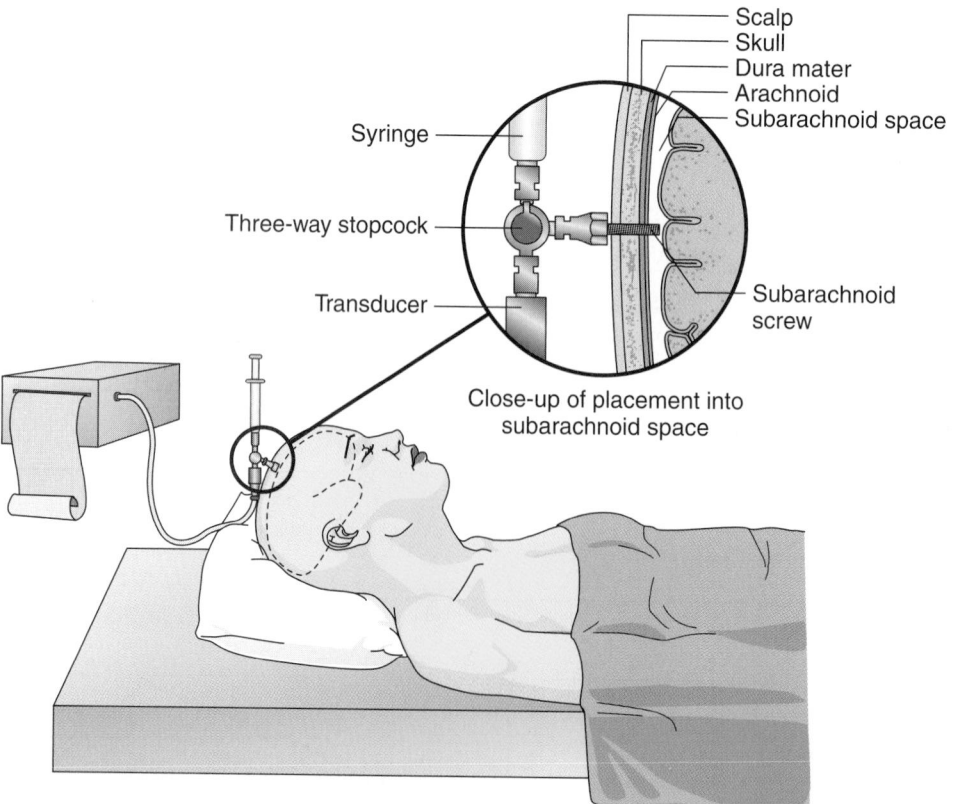

Figure 46-4 Subarachnoid bolt monitor.

a subarachnoid bolt is ease of placement. Disadvantages include occlusion of the sensor portion of the bolt with brain tissue and inability to drain CSF. An intraparenchymal monitor is placed directly into brain tissue. Some physicians believe that this most accurately reflects the actual situation within the skull. These monitors cannot be used to drain CSF and may become occluded by brain tissue.

Patients with ICP monitors are cared for in an intensive care unit (ICU) and require aggressive nursing care to prevent complications. These patients are often mechanically ventilated and may be pharmacologically paralyzed and sedated. They may be completely dependent for all care. In addition to meeting the patient's physiological needs and preventing complications, provide education and emotional support to the significant others. Patients who require intracranial pressure monitoring are seriously ill, and their significant others need frequent honest communication regarding the clinical situation and prognosis.

Nursing Process: The Patient with an Infectious or Inflammatory Neurological Disorder

ASSESSMENT. Collaborate with the registered nurse (RN) to obtain a complete history from the patient, if feasible, and from significant others. Particular attention is paid to exposure to risk factors and symptoms of generalized malaise. The physical assessment must include all body systems because neurological impairment affects the entire person. Following the initial assessment, serial neurological assessments continue to be important to detect and report changes promptly. Pupil response, level of consciousness (LOC), and vital signs are monitored for signs of increased intracranial pressure (Box 46–2). Headache is monitored on a pain scale. The Glasgow Coma Scale, presented in Chapter 45, is a valuable tool to monitor level of consciousness.

NURSING DIAGNOSIS, PLANNING, AND IMPLEMENTATION. Plan to meet the physical needs of the patient, as well as the emotional needs of the patient and significant others, during the acute phase of illness. Whenever possible, the goal is to support the patient and prevent long-term complications. This requires careful monitoring and timely reporting of changes in neurological status. Possible nursing diagnoses and interventions for the patient with a central nervous system infection include the following:

| BOX 46-2 | **Signs and Symptoms of Increased Intracranial Pressure** |

Vomiting
Headache
Dilated pupil on affected side
Hemiparesis or hemiplegia
Decorticate then decerebrate posturing
Decreasing level of consciousness
Increasing systolic blood pressure
Increasing then decreasing pulse rate
Rising temperature

Decreased Intracranial Adaptive Capacity Related to Infectious or Inflammatory Process. Decreased adaptive capacity refers to the patient's inability to maintain normal intracranial pressure. Measures such as avoiding Valsalva's maneuver during bowel movements, avoiding rectal temperatures, and avoiding flexion of the neck when positioning the patient help prevent increases in ICP. The patient with increased ICP is usually cared for in an intensive care setting, and the licensed practical nurse/licensed vocational nurse (LPN/LVN) collaborates with the registered nurse in implementing care. See Table 46–4 for additional measures to prevent increased ICP.

Pain Related to Headache and Nuchal Rigidity. Headache and other alterations in comfort are difficult to treat. Opioid analgesics (with the possible exception of codeine) are usually avoided because they mask neurological symptoms and make detection of changes difficult. If they are used, monitor the patient carefully for changes in level of consciousness or other neurological changes. Assist patient to whatever position is most comfortable. A dark, quiet room with few distractions may also reduce headache.

Hyperthermia Related to Infectious Process. Control fever with acetaminophen or aspirin because a high temperature can increase the risk for seizures. A cooling mattress and tepid sponge baths may be necessary but are uncomfortable for the patient. Comfort can be increased and shivering reduced by cooling the patient gradually and wrapping extremities in bath blankets during cooling mattress therapy.

TABLE 46-4	**INTERVENTIONS TO PREVENT INCREASED INTRACRANIAL PRESSURE**
Action	**Rationale**
Keep head of bed elevated 30 degrees unless contraindicated.	Head elevation reduces ICP in some patients.
Avoid flexing the neck; keep head and neck in midline position.	Neck flexion may obstruct venous outflow.
Administer antiemetics or antitussives as necessary to prevent vomiting and cough.	Coughing and vomiting can increase ICP.
Administer stool softeners.	Straining for bowel movement can increase ICP.
Minimize suctioning. If absolutely necessary, oxygenate first and limit suction passes to one or two.	Suctioning can increase ICP.
Avoid hip flexion.	Hip flexion can increase intra-abdominal and thoracic pressure, which can increase ICP.
Prevent unnecessary noise and startling the patient.	Noxious stimuli can increase ICP in some patients.
Space care activities to provide rest between each disturbance.	Clustering care activities may increase ICP.

Risk for Injury Related to Seizures Or Falls. Maintain patient safety with seizure precautions (see the section on seizures later in this chapter), use of side rails, and reminders not to get up without help if indicated. Having a family member stay with the patient can help the patient feel more secure and prevent falls if the patient's ability to remember instructions is impaired.

Disturbed Sensory Perception Related to Cranial Nerve Involvement. Sensory problems can increase risk of injury. Monitor the patient closely for changes in level of consciousness. See Chapters 48 and 49 for specific interventions for patients with sensory problems.

Impaired Physical Mobility Related to Long-Term Complications of the Disease. Physical mobility should be maintained as much as possible. Encourage the patient to move and turn in bed and ambulate when able. Long-term mobility problems such as hemiparesis may be reduced with the help of a physical therapy consultation. Measures to prevent impaired skin integrity should be implemented.

EVALUATION. Successful nursing management of a patient with an infectious or inflammatory neurological disorder is evidenced by a patient who is comfortable and with no preventable complications such as pressure ulcers or **contractures.** This increases the possibility that the patient with a neurological deficit will benefit from rehabilitation and improve his or her level of functioning.

Patient Education

The nature and focus of teaching depend on the patient's level of consciousness and cognitive status. When appropriate, both the patient and significant others should be included in the education process. If the patient is not able to participate, the significant others become the focus of teaching.

Describing the brain as in control of body functions may help significant others to understand some of the symptoms of neurological disorders. The spinal cord can be compared to a telephone cord, with hundreds of tiny individual wires (nerves) making up the cord. The specific wires affected by disease determine the symptoms the patient experiences.

CRITICAL THINKING

Mr. Chung

Mr. Chung is an 18-year-old Asian college student. He comes to the emergency department complaining of headache, stiff neck, and fever. On physical assessment you notice a petechial rash on his legs and torso. The physician diagnoses meningococcal meningitis.

1. What tests are likely to be performed?
2. How should patient education be planned for Mr. Chung?
3. What infection control practices should be instituted?
4. What comfort measures might you offer to Mr. Chung?
5. What concerns do you have about how Mr. Chung contracted his illness?

Answers at end of chapter.

▶ HEADACHES

As mentioned throughout this chapter, headache is a common symptom of neurological disorders. However, most headaches are transient events and do not indicate a serious pathological condition. If headaches are recurrent, persistent, or increasing in severity, the patient should undergo a neurological evaluation.

Because the causes, signs and symptoms, pathophysiology, and treatment of headaches vary based on the type of headache experienced, these subjects are discussed separately for each type of headache.

Types of Headaches
Tension or Muscle Contraction Headaches

Persistent contraction of the scalp, facial, cervical, and upper thoracic muscles can cause tension headaches. A cycle of muscle tension, muscle tenderness, and further muscle tension is established. This cycle may or may not be associated with vasodilation of cerebral arteries. Headaches of this type may be associated with premenstrual syndrome or psychosocial stressors such as anxiety, emotional distress, or depression. Symptoms typically develop gradually. Radiation of pain to the crown of the head and base of the skull, with variations in location and intensity, is common. *Pressure, aching, steady,* and *tight* are some of the words patients use to describe the pain of tension headaches.

Care must be taken to thoroughly rule out physical causes before attributing the headache to psychosocial origins. Symptom management may include the use of relaxation techniques, massage of the affected muscles, rest, localized heat application, nonnarcotic analgesics, and appropriate counseling.

Migraine Headaches

A number of theories have been proposed to explain migraine headaches. It is believed that cerebral vasoconstriction followed by vasodilation is a major factor. This response may be triggered by the trigeminal nerve, which in turn stimulates release of substance P, a pain transmitter, into the vessels. Serotonin is another neurotransmitter that may play a role in migraine pain. The tendency to develop migraine headaches is often hereditary. Migraine episodes can be triggered by a variety of factors, including noise, bright light, alcohol, and stress. Some patients can identify specific foods that trigger the headache.

Symptoms include a **prodromal** period, in which the patient may experience several hours to days of changes in mood and appetite, drowsiness, or frequent yawning. Some patients experience additional neurological symptoms, such as visual changes (e.g., an aura), immediately before the pain begins. Many patients report seeing flashing lights. The headache that follows is often accompanied by nausea and sometimes vomiting and may last for hours to days. Commonly used descriptors of migraine pain include *throbbing, boring, viselike,* and *pounding.* It is usually on one side of the head. Noise and light tend to exacerbate the headache, leading the patient to rest in a dark, quiet environment.

Treatment of migraine may be prophylactic or directed at an acute episode. Prophylactic treatment is usually reserved for those patients experiencing one or more headaches per week. Dietary restrictions may be helpful if precipitating foods or beverages can be identified. Nifedipine (a calcium channel blocker), amitriptyline (a tricyclic antidepressant), and propranolol (a beta-adrenergic blocker) are prophylactic treatments that may be effective. Propranolol and nifedipine should be used cautiously because of the potential for lowering blood pressure (BP). This is particularly true of young, slender females, who may normally have a low blood pressure. Amitriptyline may cause drowsiness, dry mouth, and weight gain. None of these medications should be stopped abruptly after long-term use.

Several types of medications are available to treat the acute migraine headache. Ergot (Cafergot), a vasoconstrictor, is effective only if taken before the vessel walls become edematous, usually within 30 to 60 minutes of headache onset. Sumatriptan (Imitrex) and zolmitriptan (Zomig) are newer medications available for migraine relief. These drugs work at the serotonin receptor sites and have a vasoconstricting action. Sumatriptan is available in both intramuscular and oral forms. The potentially additive nature of multidrug medications and opioids requires careful monitoring.

Cluster Headaches

Vascular disturbance, stress, anxiety, and emotional distress are all proposed causes of cluster headaches. As indicated by the name, these headaches tend to occur in clusters during a time span of several days to weeks. Months or even years may pass between episodes. Alcohol consumption may worsen the episodes.

The patient may state that the headache begins suddenly, typically at the same time of night. *Throbbing* and *excruciating* are often the adjectives used by the patient. The headache tends to be unilateral, affecting the nose, eye, and forehead. A bloodshot, teary appearance of the affected eye is common.

Because of the brief nature of cluster headaches, treatment is difficult. A quiet, dark environment and cold compresses may lessen the intensity of the pain. Nonsteroidal anti-inflammatory drugs (NSAIDs) or tricyclic antidepressants may be prescribed.

Diagnosis

Most headaches are diagnosed based on the patient's history and symptoms, after other causes have been ruled out. Magnetic resonance imaging (MRI), CT, or other testing may be done to make sure that a brain tumor or other structural problem is not causing the headaches.

Nursing Care
Assessment

The WHAT'S UP? mnemonic is particularly useful in helping the patient provide useful information regarding the headache.

W—Where is the pain? Does it remain in one place or radiate to other areas of the head? Does the headache consistently start in one place?

H—How does the headache feel? Is it throbbing, steady, dull, bandlike, or does it have other qualities?

A—Aggravating or alleviating factors should be assessed. Some aggravating factors include red wine, caffeine, chocolate, and foods containing nitrates. Other factors include particular stages of the menstrual cycle, emotional stress, and tension. Alleviating factors might include lying down in a dark room, cold compresses, and over-the-counter medications.

T—Timing may be a factor for a patient who experiences headaches just before or during her menstrual period. For other patients, there may be no predictive timing. Also ask how long the headache lasted.

S—Ask the patient to rate the severity on a scale of 0 to 10. Is the severity consistent or does it vary from headache to headache?

U—Are there associated symptoms, such as nausea, vomiting, or bloodshot eyes?

P—Determine the patient's perception of the headache. Does it interfere with the patient's life? If so, how? Has the patient had a previous evaluation of headaches?

Patient Education

The first step in patient education is to help the patient identify and reduce or eliminate aggravating factors. This can be accomplished by keeping a headache diary for a time, recording the time of day the headache occurs, foods eaten or other aggravating factors, description of the pain, identification of associated symptoms such as nausea or visual disturbances, and other factors related to headache symptoms. This can help the patient lessen the frequency and intensity of attacks and provides a sense of control over his or her illness. Similarly, encouraging the patient to use alleviating techniques such as biofeedback or stress reduction helps the patient participate in the treatment of the headache. Relaxation exercises or warm, moist compresses may be helpful for tension headaches. A dark room and rest are essential during a migraine headache.

Education regarding medications, appropriate dosage, expected action, side effects, and consequences of misuse is essential. Depending on the patient's learning ability and interest, information may be given orally or in written format. Drawings, diagrams, and commercially prepared drug information are potential teaching aids.

► CEREBROVASCULAR DISORDERS
Transient Ischemic Attack

Transient ischemic attack (TIA) is a temporary impairment of the cerebral circulation causing neurological impairment. It is characterized by focal neurological deficits, typically minutes to hours in duration. Symptoms resolve completely within 24 hours. Symptoms that last longer than 24 hours but do not cause permanent neurological changes are called reversible ischemic neurological deficits (RIND).

Pathophysiology

Cerebral function is dependent on oxygen and glucose delivery to neurons. Interruption of blood flow to the brain deprives neurons of needed glucose and oxygen. The particular vessel or vessels involved determine the area of the brain affected and therefore the symptoms observed. The duration of ischemia determines whether the symptoms are transient or permanent. A transient ischemic attack may be a warning of an impending cerebrovascular accident.

Etiology

Atherosclerosis resulting in narrowing of arterial diameter is the most common cause of a transient ischemic attack. Although any vessel may be involved, the bifurcation of the common carotid artery into the internal and external branches is the most common location for cerebral occlusion. Emboli may lodge in cerebral vessels, resulting in occlusion, ischemia, and infarct. Emboli may break off of arterial plaque or be released into the circulation during atrial fibrillation.

Signs and Symptoms

Symptoms depend on the area of the brain affected. Common symptoms include visual disturbances, difficulty with speech, weakness or paralysis on one side of the body, and transient confusion. They can last a few minutes to a few hours and subside within 24 hours. See Table 46–5 for symptoms associated with occlusion of specific arteries.

Diagnostic Tests

Carotid Doppler testing can determine if stenosis of the carotid arteries exists. This noninvasive test involves bouncing sound waves off the carotid arteries to determine the velocity and turbulence of blood flow.

Patients may undergo an echocardiogram to determine the presence of heart disease that may increase risk of thrombus formation. CT scan or MRI may be used to assess for previous, possibly asymptomatic, infarctions. A cerebral angiogram may be done to determine the patency of cerebral vessels and the status of any collateral circulation. Refer to Chapter 45 for a description of this examination.

Treatment

MEDICAL MANAGEMENT. Medical management focuses on controlling the cause of the transient ischemic attack. Medications are used to control atrial fibrillation or hypertension. Warfarin (Coumadin) may be prescribed for patients prone to clot development. Education regarding safety precautions is essential for these patients. They should be instructed to use electric razors to minimize nicks and to be careful to avoid any injury that might cause bleeding. The possible development of a cerebral hematoma following a fall or blow to the head must be stressed to the patient and significant others. Additional precautions are found in Chapter 24.

Antiplatelet drugs such as ticlopidine (Ticlid) or aspirin are often prescribed for patients experiencing transient ischemic attacks. Decreasing platelet aggregation lessens the likelihood of thrombus formation. If aspirin is prescribed for its antiplatelet properties, the patient should be instructed not to use aspirin or products containing aspirin for pain relief. Because ticlopidine is metabolized by the liver, it may elevate liver enzymes and regular blood tests are required.

SURGICAL MANAGEMENT. If carotid stenosis of greater than 70 percent is detected, a carotid **endarterectomy** may be performed. During this surgical procedure, the carotid artery is opened and the plaque removed. Nursing care focuses on careful neurological assessment for signs of deterioration related to ischemia. The incision is monitored for hematoma development and bleeding. Development of a hematoma can compromise the patient's airway. Bleeding at the suture line, particularly of bright red blood, may indicate failure of the sutures. This emergency situation requires prompt response to prevent massive blood loss. Balloon angioplasty for carotid stenosis is being investigated as a potential treatment in large facilities.

Acute Complications

By definition, a transient ischemic attack is an event that results in no permanent neurological deficit. However, patients can experience complications from events that occur during a transient ischemic attack. Falls may result in broken bones, abrasions and lacerations, or cerebral hematomas, particularly if the patient is on anticoagulants. Motor vehicle accidents may occur if the patient experiences visual or cognitive impairments. Hemiparesis can cause patients to burn or cut themselves if cooking or operating power tools. There are no long-term complications of transient ischemic attacks.

---

endarterectomy: endo—inside + arter—artery + ectomy—surgical removal of

TABLE 46–5	SYMPTOMS OF CEREBROVASCULAR ACCIDENT ACCORDING TO ARTERY AFFECTED				
	Hemiparesis	**Dysphasia**	**Visual Changes**	**Altered Level of Consciousness**	**Ataxia**
Carotid	X	X	X	X	
Middle cerebral	X	X	X	X	
Vertebrobasilar			X		X

Cerebrovascular Accident

Cerebrovascular accident (CVA, also known as a stroke) is the infarction (death) of brain tissue caused by the disruption of blood flow to the brain. It is characterized by focal neurological deficits specific to the area of the brain involved. Although CVAs are most commonly associated with elderly patients, they affect approximately 500,000 people of all ages each year. A newer term for CVA is *brain attack*. This reminds us that, like a heart attack, CVA is an urgent condition that can be treated if medical care is sought immediately. Exciting new treatment developments may now allow patients to be deficit free if help is sought in time.

Pathophysiology

Infarction of brain tissue happens when there is inadequate blood flow to an area of the brain. When blood flow is severely compromised or absent, the oxygen and glucose needed to meet the brain's metabolic needs are not available. The brain has no capability to store oxygen or glucose, so it relies on a constant supply of these nutrients. If the supply of oxygen and glucose is stopped, the brain tissue dies. In contrast to TIA, a brain attack can cause permanent damage if it is not reversed with timely treatment.

Causes and Types

A brain attack can either be ischemic (from deficient blood supply to the brain) or hemorrhagic (from bleeding into the brain). An ischemic brain attack is either thrombotic or embolic. Thrombosis or atherosclerosis can narrow or completely occlude a vessel. People with diabetes and hypertension have an increased risk for atherosclerosis and stroke. Emboli can lodge in a cerebral vessel, stopping the blood flow to that area of the brain. Emboli can occur as a result of endocarditis, atrial fibrillation, or valvular disease but can be dissolved with **thrombolytic** medication if treatment is sought quickly.

Rupture of a cerebral blood vessel can result in a hemorrhagic brain attack. The most common cause of an intracerebral hemorrhage is poorly controlled hypertension. Another cause is a ruptured aneurysm. These hemorrhages tend to occur deep within the brain tissue. This type of infarct has the slowest rate of recovery and the highest probability of leaving the patient with extensive neurological deficits. There is no medication to reverse the effects of this type of brain attack.

The most common etiology of brain attack in younger patients is illicit drug usage. PCP, crack, cocaine, amphetamines, and heroin have all been associated with cerebrovascular accident from subarachnoid or intracerebral hemorrhage, because these drugs raise the blood pressure and increase pressure within the cerebral vessels.

Prevention

Incidence of CVA can be lessened by reduction of risk factors. Risk factors include hypertension, smoking, atherosclerosis, diabetes, and cardiac problems that cause emboli to form. Keeping hypertension and diabetes controlled can go a long way in preventing strokes. Emboli can be prevented with warfarin in individuals with high risk. Aspirin or ticlopidine may also be used. If atherosclerosis is a concern, a carotid endarterectomy may be done.

It is important to educate all patients about new treatments for stroke and the potential for reversal of symptoms with the use of thrombolytic agents. Patients must be aware of symptoms of stroke and the importance of emergency treatment to maximize the potential for prevention of neurological deficits. *Too often patients ignore early symptoms or delay calling for help. This delay can mean the difference between leading a normal life and permanent disability.*

Signs and Symptoms

As with transient ischemic attacks, the signs and symptoms of a cerebrovascular accident relate to the specific vessel involved. (See Table 46–5.) There are some symptoms common to cerebrovascular accidents of different etiologies. These may include changes in level of consciousness, fever, headache, vomiting, and seizures. Other common symptoms include numbness, weakness, or paralysis of one side of the face and one arm or leg; trouble understanding language or speaking coherently; vision changes; and impaired coordination or balance. Additional short-term and long-term manifestations are discussed next.

SHORT-TERM EFFECTS
Neurological Deterioration. Patients suffering from a CVA develop increased intracranial pressure, which further adds to brain damage. Patients are also vulnerable to repeated cerebrovascular accidents. Careful serial neurological assessments are needed to promptly detect and report changes.

Respiratory Compromise. Respiratory compromise may occur related to an increase in intracranial pressure. Patients with CVA are prone to aspiration because of decreased level of consciousness or impaired swallowing ability. Patients should be suctioned as needed to keep the airway clear. If the patient vomits, he or she should be turned to the side to reduce the risk of aspiration. Oral feedings should be begun carefully and progressed slowly only after the patient is alert and the ability to swallow safely has been determined by an appropriate swallowing evaluation.

LONG-TERM EFFECTS
Motor Function. The side of the body opposite the side of the cerebral infarct is affected, because nerve fibers cross over as they pass from the brain to the spinal cord (Fig. 46–5.) The affected extremities may be weak or totally paralyzed **(hemiplegia).** An extremity that has no muscle tone

---

thrombolytic: thrombo—clot + lytic—causing breakdown

hemiplegia: hemi—one side + plegia—paralysis

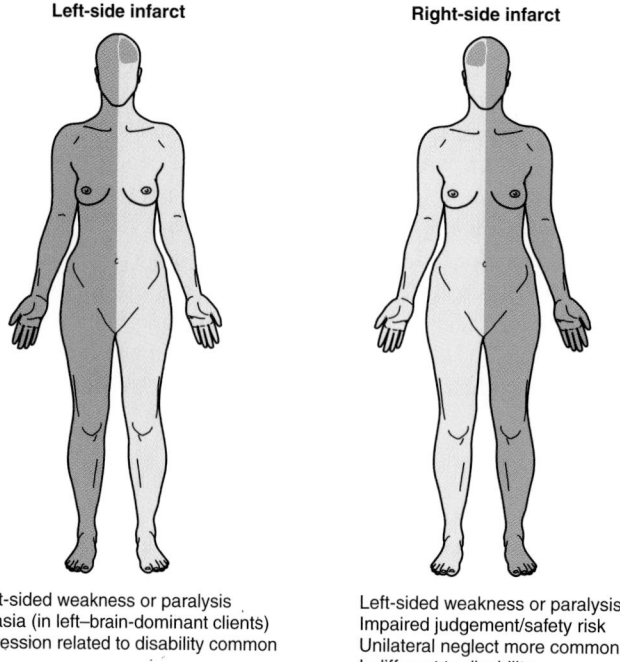

Figure 46-5 The side opposite the infarct is affected in a brain attack.

or movement at all is called **flaccid.** Depending on the artery affected, the arm may be weaker than the leg or vice versa. These patients are particularly prone to contractures, which cause permanent immobility of a muscle or joint from fibrosis of connective tissue. Adaptation or assistance with activities of daily living (ADLs) is required. Sensory changes may prevent the patient from being aware of injuries to the affected side. Patients should be mobilized within 24 to 48 hours if possible to prevent complications of immobility. Physical and occupational therapy are provided to maximize functioning, and the patient must be taught to be aware of and protect the involved limbs.

Motor involvement often affects swallowing, control of urination, and bowel function. Difficulty swallowing (dysphagia) is a major problem for patients after a stroke and can cause aspiration pneumonia, which can lead to death.

Aphasia. If the infarct is on the dominant side of the brain, the speech center will probably be affected. Aphasia may be expressive, receptive, or global. Persons with expressive aphasia know what they want to say but are unable to speak in a way that can be understood. They may be able to say words but are not able to form coherent speech, such as the patient who picks up a fork but calls it a comb. A patient with receptive aphasia does not understand what is said to him or her. In this situation it is easy to attempt to speak louder to try to help the patient understand. Remember that it is not the patient's hearing that is affected. Global aphasia is a combination of expressive and receptive disturbance.

Emotional Lability. Emotional lability, or instability, is a common consequence of cerebrovascular accident. Patients may move rapidly from profound sadness to an almost euphoric state and back again. Laughing or crying may have

no relationship to the patient's situation at any given moment. Families can be upset by this behavior, because they do not understand why a once happy person is now crying all the time or why the patient laughs inappropriately. You can help by explaining that these responses probably do not reflect how the patient is feeling, but rather are a manifestation of the stroke damage.

Impaired Judgment. Patients who have had a CVA, particularly those with right-sided lesions, present safety risks. Patients may have poor understanding of their own limitations and believe that they are capable of performing tasks they did before the CVA. Precautions must be taken to protect the patient from injury.

If the frontal lobes are involved, learned social behaviors may be lost. The patient may undress in public, use profanity, or make inappropriate sexual advances. These behaviors are extremely difficult for significant others to cope with. Education and emotional support of the significant others are essential. Allowing them to vent their frustration and anger may facilitate coping. Distracting the patient from inappropriate behavior may help. The patient should never be reprimanded or punished, because he or she no longer has the cognitive ability to control the behaviors.

Unilateral Neglect. The phenomenon of unilateral neglect is seen predominantly in patients who have right hemisphere infarcts. These individuals do not acknowledge the left side of their environment. In severe cases the patient may forget to dress the left side of the body. Initially these patients should be approached from the right side. Essential items such as the call light and telephone are placed on the patient's right side. Position the bed so that the patient's right side is toward the door. Gradually the health team can begin teaching the patient to focus on the left side. This involves teaching the patient to purposefully check where the left limbs are positioned and to look for safety risks. The patient can learn to turn his or her head and scan the environment. Patients may need reminders to turn their plates during meals to recognize the food on the left side of the plate.

Homonymous Hemianopsia. Some patients experience a loss of visual field in the same side of each eye. This is called homonymous hemianopsia, and it can cause the patient to ignore one side of a dinner tray or neglect to care for one side of the body.

When preparing a room for a patient with CVA, the caretaker should choose a room in which the patient can have the unaffected side toward the door if at all possible.

Diagnostic Tests

A CT scan can help determine the size and location of the infarct and whether the cause is hemorrhagic or occlusion of the artery. An EEG, arteriogram, or MRI may be done. If an embolism is suspected, cardiac tests may be done to determine the source.

Treatment

MEDICAL MANAGEMENT. Thrombolytic therapy is a recent development in the treatment of ischemic brain attack. Intracerebral hemorrhage must be ruled out before thrombolytic therapy is instituted. The goal of thrombolytic agents is to actually break down the thrombus causing the occlusion, which can potentially prevent or completely reverse the symptoms of stroke. Plasmin is the enzyme that causes thrombi to break down. Thrombolytic agents accomplish thrombus lysis by causing the conversion of plasminogen to plasmin. Commonly used thrombolytic agents are streptokinase, urokinase, and tissue-type plasminogen activator (TPA). Newer thrombolytics such as tenecteplase are faster and easier to administer. Cerebral hemorrhage is a major complication of thrombolytic therapy. Patients treated effectively with thrombolytic therapy may be able to leave the hospital within 1 or 2 days with no residual effects.

To be effective, thrombolytic therapy must be administered within 3 hours of the onset of symptoms. This time frame has clear implications for nurses. Patients with neurological symptoms who are seen in the emergency department must be assessed and stabilized promptly. A CT scan and, if an ischemic cause is suspected, an angiogram are performed quickly. Some people may be surprised to learn that there are treatments available for brain attack. Because of time constraints, individuals may be asked to make treatment decisions before other family members are able to arrive. This places a significant burden on people who are already experiencing stress. You can help ease this burden by explaining the time factor, repeating information as needed, and ensuring that the individuals involved have the opportunity to ask questions.

Time is brain. This means the faster the patient with CVA receives treatment, the more brain (and brain function) may be saved.

Management of the airway and control of hypertension if present are vital for the patient. Care is taken not to lower the blood pressure too quickly or too far. If the patient has long-standing hypertension, lowering the blood pressure to a "normal" level may actually cause further ischemia. Current recommendations are to not lower blood pressure more than 10 percent from baseline at one time. Vasoactive drugs must be used carefully and blood pressure monitored frequently. If the patient experiences neurological deterioration, the physician may adjust treatment to increase the blood pressure. Some physicians believe that the blood pressure should not be lowered because chronically hypertensive patients require a higher BP to maintain adequate cerebral perfusion. As soon as the patient is stabilized, a CT scan is performed to identify any cerebral hemorrhage. Further treatment is dependent on the presence or absence of hemorrhage.

If a CVA is caused by an embolism, anticoagulants may be prescribed. Treatment begins with intravenous heparin sodium. When the clotting studies reach a therapeutic level, oral warfarin is begun. The heparin can be stopped when the warfarin level is therapeutic. Patients need monitoring of prothrombin time (PT) and international normalized ratio (INR) and adjustment of warfarin dosage as long as they are on warfarin. These patients must be instructed to inform all health care providers, including dentists, that they are taking warfarin. The warfarin dosage needs to be adjusted before any invasive procedures are performed. Antiplatelet drugs may be used to lessen platelet aggregation.

SURGICAL MANAGEMENT. Surgical treatment to prevent a cerebrovascular accident was discussed earlier in the section on transient ischemic attacks. Occasionally a **craniotomy** is performed to remove an embolus or thrombus or to repair an aneurysm. Endarterectomy or balloon angioplasty may be performed if the attack is due to carotid blockage. Surgery must be performed within a few hours of the ischemic event if there is to be any possibility of preventing permanent neurological deficits (Box 46–3). For more information, visit the National Stroke Association at www.stroke.org.

---

craniotomy: crani—skull + otomy

BOX 46-3	Cerebrovascular Accident Summary

Symptoms

Numbness, weakness, or paralysis of one side of face or body
Difficulty understanding or speaking language
Vision disturbance
Impaired coordination or balance
Changes in LOC
Headache
Seizures

Diagnostic Tests

CT scan
MRI
EEG
Angiogram

Therapeutic Management

Thrombolytic, anticoagulant, or antiplatelet therapy (for ischemic strokes)
Management of airway
Control of hypertension
Maintenance of safety
Rehabilitation
Physical therapy
Occupational therapy
Speech therapy

Nursing Care

See Nursing Process: The Patient with a Cerebrovascular Disorder. Also see Nursing Care Plan Box 46–4.

Cerebral Aneurysm and Subarachnoid Hemorrhage

A cerebral aneurysm is a weakness in the wall of an artery. It may be congenital, traumatic, or the result of disease. If the aneurysm ruptures, a subarachnoid hemorrhage results. It is unknown what causes the formation of congenital aneurysms or what causes them to rupture. Unruptured aneurysms are typically asymptomatic. The exception to this is very large aneurysms, which can cause symptoms similar to brain tumors. Aneurysms often affect young, otherwise healthy adults.

Pathophysiology and Etiology

Aneurysms can occur in any of the cerebral arteries. Eighty percent of cerebral aneurysms occur in the circle of Willis. The most common site is at the bifurcation of an artery. It is theorized that increased turbulence at the bifurcation causes an outpouching of a congenitally weak arterial wall.

Subarachnoid hemorrhage is the collection of blood beneath the arachnoid mater following aneurysm rupture. Rupture of an arteriovenous malformation (Fig. 46–6) or

---

BOX 46-4	NURSING CARE PLAN FOR THE PATIENT WITH CEREBROVASCULAR ACCIDENT (STROKE)

 Self-care deficit related to immobility and muscle weakness

PATIENT OUTCOMES

Patient will increase tolerance and endurance for therapies, avoid complications of immobility, and become as independent as possible in activities of daily living.

EVALUATION OF OUTCOMES

Does patient demonstrate techniques and behaviors that enable resumption of activities? Is there absence of contractures and footdrop? Does patient perform self-care activities within level of own ability?

INTERVENTIONS	RATIONALE	EVALUATION
Assess ability to perform ADLs.	Assessment is essential to plan appropriate care.	What can patient do? What will patient need help with?
Encourage early mobilization (bedside commode, chair).	Early mobilization prevents complications of bedrest.	Is patient able to move with increasing independence? Are complications prevented?
Avoid doing things patient can do for self, but provide assistance as necessary.	Independence increases self-esteem and motivation.	Is patient becoming more independent?
Allow patient sufficient time to accomplish tasks. Alternate rest with activity.	Too much activity without rest tires patient and may cause frustration.	Are tiring and frustration avoided?
Teach family techniques to care for patient. Encourage family to attend therapy sessions.	The family will be caring for patient at home unless patient is placed in a nursing home.	Are family members able to demonstrate care activities?

 Impaired verbal communication related to aphasia

PATIENT OUTCOMES

Patient will use strategies to effectively communicate needs.

EVALUATION OF OUTCOMES

Is patient able to communicate needs?

BOX 46-4   NURSING CARE PLAN FOR THE PATIENT WITH CEREBROVASCULAR ACCIDENT (STROKE)

INTERVENTIONS	RATIONALE	EVALUATION
Assess type and degree of dysfunction.	Assessment helps determine strategies that will best help patient.	Does patient understand what is being said?
Request consult with speech therapist.	A speech therapist will assist with assessment and recommendations for communication.	Does patient make self understood? Are recommendations helpful?
Maintain a calm, quiet, unhurried atmosphere.	Distractions can be frustrating to the patient and make communication more difficult.	Does patient become frustrated when trying to communicate needs?
Use alternative methods of communication as necessary (e.g., writing, communication board, gestures).	Alternative methods may help patient communicate needs.	Is patient able to use alternative methods to effectively communicate needs?

 Disturbed sensory perception related to paresthesias, visual changes

PATIENT OUTCOMES

The patient will adapt to sensory perceptual changes in vision, sensation, and proprioception and be free from related injury.

EVALUATION OF OUTCOMES

Is patient able to compensate for changes in ability and presence of residual deficits? Is patient free from injury?

INTERVENTIONS	RATIONALE	EVALUATION
Assess patient's sensory awareness (sensation, vision).	Diminished sensory awareness can place patient at risk for injury.	Can patient differentiate between hot and cold, dull and sharp? Differentiate position of body parts? Feel pressure?
Arrange environment in a consistent manner.	A familiar environment enhances safety.	Can patient locate needed items?
Teach patient to scan the environment.	Scanning compensates for a visual deficit.	Does scanning help patient to locate items?
Assist patient to reposition at least every 2 hours, making sure to prevent areas of pressure.	Patient may not be able to reposition self and may not feel excess pressure that can cause skin breakdown.	Is skin free from breakdown?
Keep call light within reach on patient's good side.	Patient may not be able to find call light on affected side.	Is patient able to use call light when needed?

 Impaired nutrition: less than body requirements related to motor and swallowing dysfunction

PATIENT OUTCOMES

Patient will eat and drink foods and fluids without aspiration and maintain usual or attain ideal body weight.

EVALUATION OF OUTCOMES

Is patient able to consume adequate foods and fluids? Is patient at ideal or usual body weight?

INTERVENTIONS	RATIONALE	EVALUATION
Assess patient's ability to chew and swallow. Assess height and weight.	Patient should not be fed until ability to swallow safely is determined. Aspiration can cause pneumonia and death.	How well can patient swallow? Are height and weight within normal limits?
Request swallowing study/dietitian consult if indicated.	A swallowing study can detect risk for aspiration. A dietitian can provide foods that are easily swallowed.	Does study show safe swallowing? Are appropriate foods provided?
Institute swallowing safety measures. (See Nutrition Notes Box 46-5.)	These interventions help prevent aspiration.	Does patient swallow without aspiration?
Teach patient to chew on unaffected side. Check affected cheek for pocketing of food.	Patient may not be aware of foods on affected side.	Does patient pocket food?

head trauma may also result in subarachnoid hemorrhage. The presence of blood outside the blood vessels is very irritating to brain tissue. It is believed that irritation from blood breakdown is the major cause of vasospasm, a common complication of subarachnoid hemorrhage.

It is unclear what causes an aneurysm to rupture at a given time. Some individuals experience a subarachnoid hemorrhage while performing Valsalva's maneuver, engaging in sexual activity, or physically exerting themselves. For other patients the aneurysm ruptures during a quiet, nonactive pe-

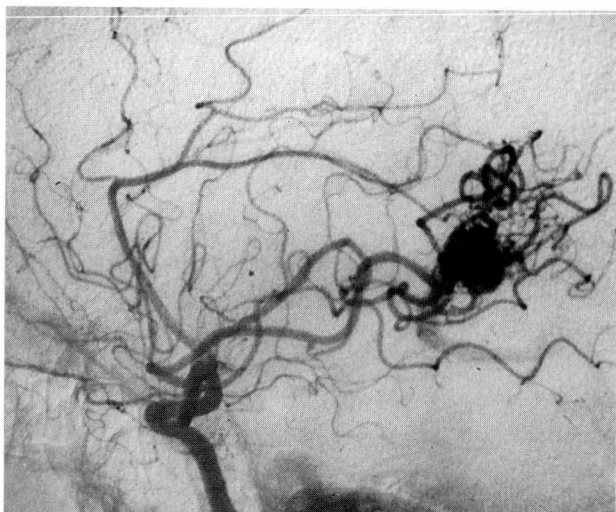

Figure 46-6 Arteriovenous malformation. Note tangled vessels.

riod. If the aneurysm rupture is associated with a particular activity, the patient may be very frightened of engaging in that activity again. This may have a negative effect on the patient's interpersonal relationships if the associated activity was sexual in nature. The patient's partner may feel guilty or responsible for the hemorrhage. Emotional support and confidentiality regarding associated events help both the patient and significant other.

Signs and Symptoms

Some patients experience a small hemorrhage before diagnosis of subarachnoid hemorrhage. This leakage of blood may cause a mild headache, vomiting, or disorientation. The symptoms may be attributed to a flulike syndrome. Patients may dismiss the symptoms and not seek medical care.

The most common presentation of rupture of an aneurysm is sudden onset of severe headaches. Typically patients state, "I have never had a headache this bad in my life." Patients may hold their heads and moan or cry in pain. Sensitivity to light is a common finding. This may make patients reluctant to cooperate with pupil checks.

Level of consciousness varies based on the severity of the hemorrhage. Patients may be alert and coherent, may lose consciousness immediately, or may gradually become less responsive. The decreased level of consciousness is caused by increased ICP and impairment of cerebral blood flow. Patients may experience generalized seizures.

Blood in the subarachnoid space causes meningeal irritation. The patient may complain of nuchal rigidity. The most commonly affected cranial nerves are III and VI. This is manifested as an enlarged pupil or disconjugate gaze. Motor dysfunction may involve one or both limbs on the side opposite the hemorrhage.

Diagnostic Tests

Because of the severe nature of the symptoms, patients with subarachnoid hemorrhage almost always come to the emergency department rather than seeking care from routine health care providers. A CT scan is done to identify the presence and location of a hemorrhage. Precise diagnosis of an aneurysm requires a cerebral angiogram. The contrast material will fill the aneurysm if one exists. For a patient with a severe headache and facing a life-threatening illness, this test can be very frightening. If the patient's neurological status does not allow him or her to cooperate, sedation may be required before and during the examination.

Treatment

SURGICAL MANAGEMENT. There is no cure for subarachnoid hemorrhage. Treatment consists of correcting the cause of the hemorrhage if possible. Preventing or managing complications and providing supportive care are important aspects of nursing care. Definitive treatment of the aneurysm involves performing a craniotomy and exposing the aneurysm. If the aneurysm has a neck (berry aneurysm), it is identified and clamped with a metal clip (Fig. 46–7). An aneurysm without a neck may be wrapped with very fine sterile muslin. This provides stability to the aneurysm walls, lessening the chance of rupture. In some situations it is possible to clamp the artery on either side of the aneurysm, removing that portion of the vessel, and the aneurysm, from the circulation.

MEDICAL MANAGEMENT. Nonsurgical intervention may be provided for aneurysms that are inoperable because of size, configuration, or the patient's medical status. A foreign material such as a tiny metallic coil or fibrin glue may be introduced into the aneurysm. A thrombus develops around the foreign body and, if the treatment is successful, occludes the aneurysm. The goal is to fill the aneurysm enough to prevent blood flowing into it without causing rupture.

Patients experiencing a subarachnoid hemorrhage are cared for in an intensive care unit setting. They typically have an arterial line and a central venous pressure monitoring catheter. The blood pressure is carefully monitored, because high pressures increase the risk of rerupture of the aneurysm and low pressures may be associated with ischemia. Values outside parameters identified by the physician

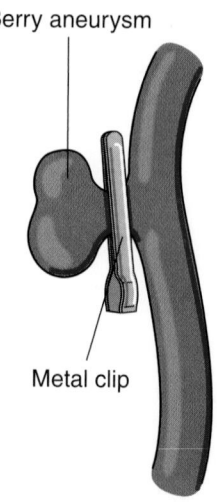

Berry aneurysm

Metal clip

Figure 46-7 Surgical management of aneurysms.

are reported. Typically the systolic blood pressure is kept between 120 and 160. Vasoactive drugs may be required to maintain blood pressure within the prescribed parameters.

Acute Complications

REBLEEDING. Recurrent rupture of cerebral aneurysm carries significant morbidity and mortality rates. Patients are at risk for rebleeding until the aneurysm is surgically repaired. If the aneurysm is wrapped or embolized, there is a risk of rebleeding, but it is much less than if the aneurysm is left untreated.

HYDROCEPHALUS. Blood within the ventricular system interferes with the circulation, and reabsorption of CSF and hydrocephalus may develop. Early in the course of subarachnoid hemorrhage an external ventricular drain may be used to treat hydrocephalus (discussed earlier).

Approximately 25 percent of patients with subarachnoid hemorrhage require placement of a ventriculoperitoneal shunt to treat their hydrocephalus (Fig. 46–8). This surgical procedure involves placement of a ventricular catheter. This catheter is then connected to a valve, which regulates the rate of cerebrospinal fluid drainage. Another catheter connects to the valve and is passed down to the peritoneal cavity. The cerebrospinal fluid drains out of the peritoneal catheter and is absorbed into the peritoneal cavity.

VASOSPASM. Vasospasm is responsible for the majority of long-term complications of subarachnoid hemorrhage. Vasospasm is the narrowing of a blood vessel diameter. Although it typically begins in the vessel giving rise to the aneurysm, vasospasm may spread to other vessels. This ex-

plains why the ischemia or infarct caused by vasospasm can be so widespread and devastating.

The long-term complications of subarachnoid hemorrhage are similar to those of cerebrovascular accident.

Nursing Process: The Patient with a Cerebrovascular Disorder

ASSESSMENT. Patients with cerebrovascular disorders require careful serial assessments of their neurological status, as described in Chapter 45. Subtle changes in orientation, level of consciousness, or motor strength may indicate ischemia or increasing intracranial pressure. By detecting and reporting such changes quickly, you can improve the likelihood of a successful outcome for the patient. In addition to assessing for changes in status, determine the patient's ability to use the call light. Assess swallowing ability before offering food or drink to prevent aspiration. This may involve radiological swallowing studies. The patient should not be fed if swallowing ability is in question. Determine the patient's ability to move and need for assistance with turning in bed. The patient and family can also provide information about the patient's previous level of functioning.

NURSING DIAGNOSIS, PLANNING, AND IMPLEMENTATION. Goals during the acute phase of the CVA include prevention of complications and recurrent CVA. Discharge planning should be initiated as soon as possible after admission, though it may be difficult to anticipate needs until the extent of the patient's deficits are known. Most patients require some rehabilitation to reach an optimum level of function. Consultation with the patient, significant others, and social worker or discharge planner should be ongoing. Possible nursing diagnoses and interventions during the acute phase of illness include the following.

Ineffective Tissue Perfusion Related to Increased ICP. Patients are at risk for increased intracranial pressure following a CVA. The patient in intensive care may have an intracranial pressure monitor. Any activities or interventions that increase ICP should be avoided, such as repeated suctioning, coughing, and vigorous turning. (See Table 46–4.)

Risk for Injury Related to Seizure or Repeat CVA. Precautions should be taken to prevent injury in the event of a seizure. (See Table 46–3.) If the patient has had a seizure, anticonvulsant medications may be ordered. Patients who have had a CVA are also at risk for a repeat CVA. Changes in neurological status should be reported immediately to the physician. Patients are also at risk for falls because of motor and sensory deficits and impaired judgment. Patients should be assisted with transfers and ambulation. Because many falls occur as patients are attempting to get up to use the bathroom, assisting the patient to the bathroom or bedside commode on a routine basis can help. Care should be taken to keep the patient's path unobstructed. Family members should be instructed in fall prevention if the patient is going home.

Imbalanced Nutrition Related to Impaired Swallowing and Motor Deficits. If the patient has difficulty swallowing

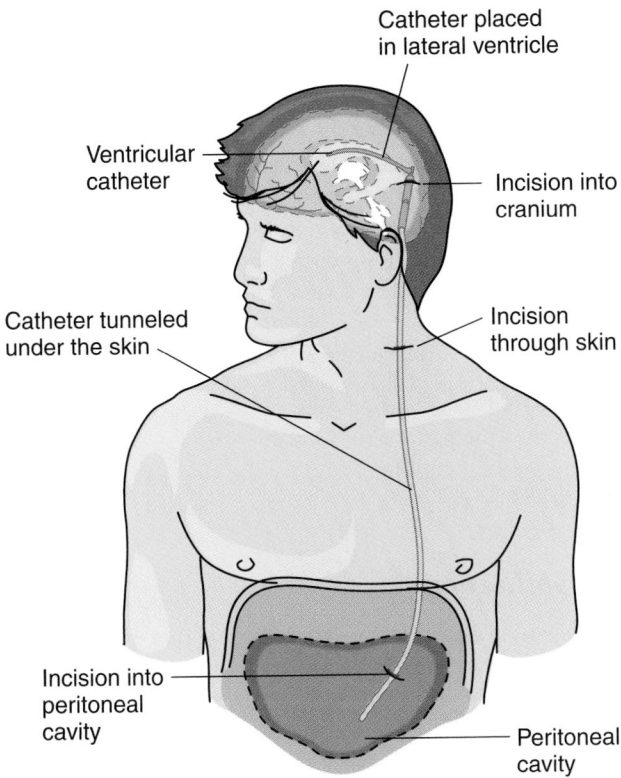

Catheter placed
in lateral ventricle

Ventricular
catheter

Incision into
cranium

Catheter tunneled
under the skin

Incision
through skin

Incision into
peritoneal
cavity

Peritoneal
cavity

Figure 46-8 A ventriculoperitoneal shunt drains cerebrospinal fluid into the peritoneal cavity.

or is unable to self-feed, altered nutrition is a concern. The physician may order a swallowing study to determine the extent of the problem. A speech-language pathologist assesses the patient and makes recommendations for safe swallowing techniques. Measures to prevent aspiration generally include staying with the patient during meals, having the patient in a chair or high Fowler's position for meals, avoiding straws, using a thickening agent for thin liquids, and having the patient swallow twice after each bite. Check the patient's mouth after each bite, because patients may pocket food on the affected side of the mouth. See Nutrition Notes Box 46–5 for additional interventions. Some patients may require a feeding tube. Advance directives should be consulted before a feeding tube is placed.

Impaired Mobility Related to Motor Deficits. The patient with hemiplegia has difficulty turning or repositioning. If the patient has sensory deficits, pain or pressure are not noticed and injury may occur. Careful and frequent repositioning is essential to prevent skin, respiratory, and musculoskeletal complications. Pillows can be used to maintain the body in good alignment and promote comfort. Skin should be inspected each time the patient is repositioned. Range-of-motion exercises should be begun within 24 hours of admission to help prevent contractures. Splints may be used for some patients to maintain a functional position of extremities. Some patients experience injury to the shoulder of an affected arm. Care should be taken to support the arm on pillows when it is in a dependent position. Use a lift sheet to reposition the patient in bed, rather than pulling on the arms. A physical therapist can be consulted to assess the patient and make specific recommendations related to mobility.

BOX 46-5   *Nutrition Notes*

Feeding Patients with Swallowing Disorders

Suggestions for persons who have difficulty swallowing include the following:

- Eat slowly.
- Avoid distractions while eating.
- Do not talk while eating.
- Sit up straight while eating.
- Use a regular teaspoon, taking only half a teaspoonful at a time.
- Swallow completely between bites or sips.

Possible nursing interventions to aid a person with a swallowing disorder include the following:

- Remove loose dentures.
- Position the head correctly. A speech therapist can help determine the optimum head position. A hemiplegic patient often benefits from turning the head toward the weak side.
- When spoon feeding a patient with hemiplegia, place the food on the unaffected side of the tongue.
- Consult a dietitian concerning appropriate textures. Often thicker substances are better managed than liquids. Although infant cereal can be an effective thickener, it may be rejected on a psychological basis in favor of instant potato flakes, unflavored gelatin, or a commercial thickener.

Impaired Verbal Communication Related to Aphasia. Aphasia can cause great frustration to the patient, family, and caregivers. Care depends on whether the aphasia is receptive or expressive. If you have determined that the patient's responses are valid, asking yes/no questions may be helpful. Gestures and visual aids may be tried. Contact a speech or occupational therapist to acquire a picture board for the patient. This board has pictures on it that the patient can point to, such as a glass of water, tissue, toilet, hungry, too hot, and others (Fig. 46–9). Some patients can relearn language skills with the help of a speech therapist. It is important that staff and family continue to speak to the patient, because he or she may understand what is being said but may be unable to respond.

Deficient Knowledge Related to Diagnosis and Treatment. The patient and significant other are likely to be very frightened of what is happening. Correct information about what a CVA is, tests and procedures, and rationale for care activities help reduce anxiety. Information should be presented in small amounts and as simply as possible. Orientation to the ICU or other setting and the constant monitoring provided helps reassure the patient and significant others that he or she is receiving competent care. If the patient is unable to communicate, do not assume that he or she cannot hear and understand. Every effort should be made to speak to the patient and to keep conversation appropriate when it is within the patient's range of hearing.

Risk for Caregiver Role Strain Related to New Responsibilities. Cerebrovascular accidents, even those that leave relatively mild residual effects, have a significant impact on psychosocial functioning. Patients and their significant others may experience changes in roles, responsibilities, finances, and intimacy. Significant others should be encouraged to assess how the patient's functional level will affect their lives. Encourage them to identify support systems and make use of community resources. Assumption of roles or responsibilities previously fulfilled by the patient may be very stressful to significant others. The nurse and social worker can help them identify priorities and plan ways of adapting to change. (See Nursing Care Plan 46–4 for the Patient with a Cerebrovascular Accident.)

EVALUATION. As with all neurological disorders, the goal is to return the patient to his or her previous level of functioning. If this is not feasible, success is measured by the patient's lack of preventable complications and readiness for rehabilitation.

Rehabilitation

If the patient is able to tolerate intensive therapy, discharge from the hospital may be to a rehabilitation center. With rehabilitation, most patients can learn to walk, some with a walker or cane. Speech therapy can help the patient learn to communicate. Many patients can learn to take care of themselves. Some may be able to drive again and even have a job.

Figure 46-9 Picture board. (Courtesy Visiboard, © 1987, Interactive Therapeutics, Inc., Stow, Ohio. Reprinted with permission.)

Continued

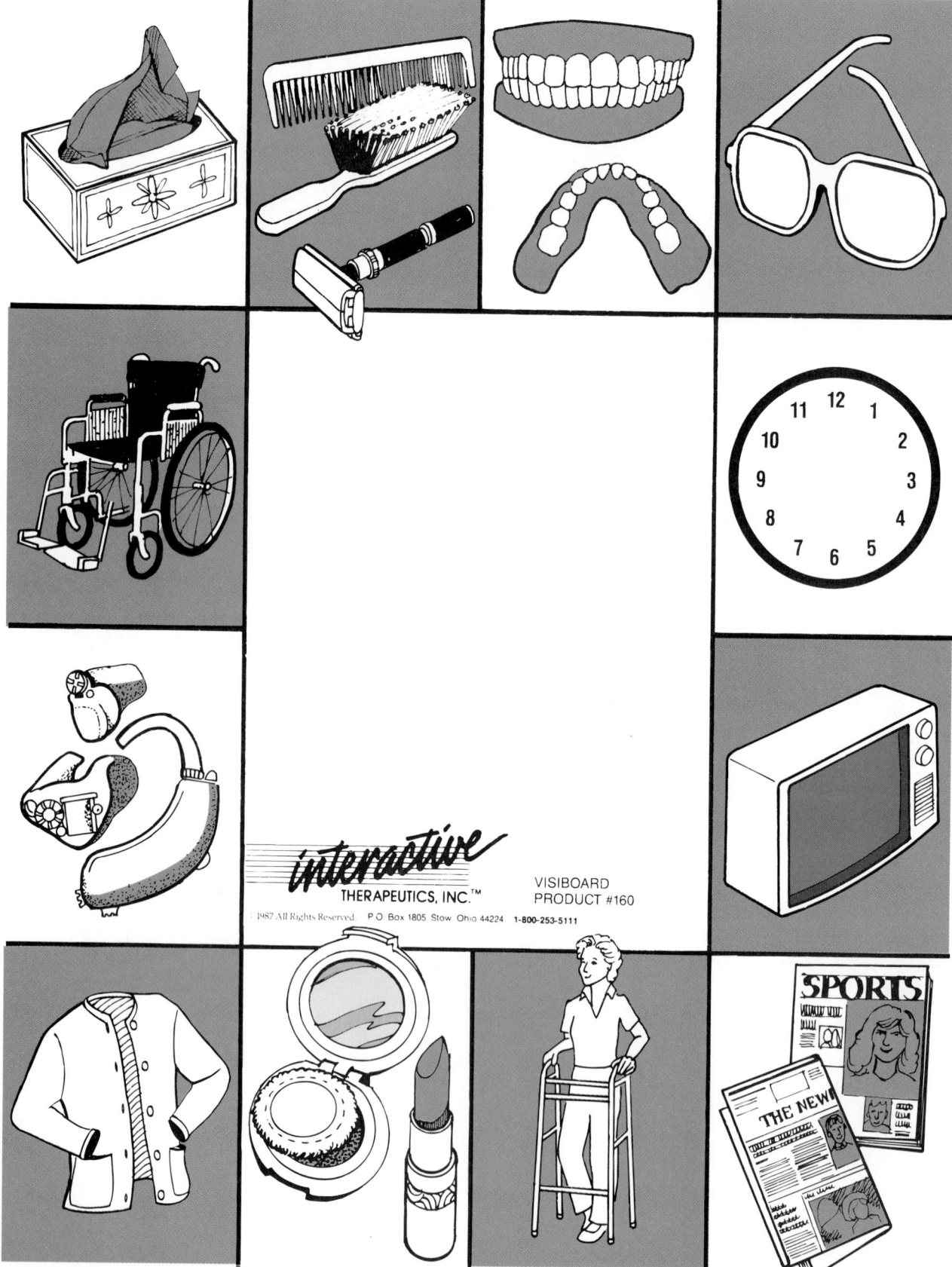

Figure 46-9, cont'd   Picture board. (Courtesy Visiboard, © 1987, Interactive Therapeutics, Inc., Stow, Ohio. Reprinted with permission.)

CRITICAL THINKING

Mrs. Washington

Mrs. Washington is a 68-year-old African-American retired office worker. She was admitted to your unit following a right-sided intracerebral hemorrhage. Her daughter states that Mrs. Washington has taken antihypertensive medication for the last 20 years. However, she states that her mother has been "forgetful" lately and that there are five more pills in the medicine bottle than expected. On admission Mrs. Washington is oriented only to person and has hemiparesis.

1. What may have precipitated Mrs. Washington's CVA?
2. On which side are Mrs. Washington's extremities affected?
3. List two safety concerns and strategies to promote patient safety.
4. List at least two educational needs for Mrs. Washington and her daughter.

Answers at end of chapter.

► SEIZURE DISORDERS

Seizures

A seizure is defined as an abnormal electrical discharge within the neuronal structure of the brain. A seizure may be a symptom of epilepsy or of other neurological disorders such as a brain tumor or brain attack. Epilepsy is a chronic neurological disorder characterized by recurrent seizure activity.

Pathophysiology

The normal stability of the neuron cell membrane is impaired in individuals with epilepsy. This instability allows for abnormal electrical discharges to occur. These discharges cause the characteristic symptoms seen during a seizure.

Seizures can be classified as partial or generalized. Partial seizures begin on one side of the cerebral cortex. In some cases the electrical discharge spreads to the other hemisphere and the seizure becomes generalized. Generalized seizures are characterized by involvement of both cerebral hemispheres.

Etiology

Epilepsy may be acquired or idiopathic. Causes of acquired epilepsy include traumatic brain injury and anoxic events. No cause has been identified for idiopathic epilepsy. The most common time for idiopathic epilepsy to begin is before age 20. New-onset seizures after this age are most commonly caused by an underlying neurological disorder.

Signs and Symptoms

Symptoms of seizure activity correlate with the area of the brain where the seizure begins. Some patients experience an aura or sensation that warns the patient that a seizure is about to occur. An aura may be a visual distortion, a noxious odor, or an unusual sound. Patients who experience an aura may have enough time to sit or lie down before the seizure starts, thereby minimizing the chance of injury.

PARTIAL SEIZURES. Repetitive, purposeless behaviors, called automatisms, are the classic symptom of partial seizures. The patient appears to be in a dreamlike state while picking at his or her clothing, chewing, or smacking his or her lips. Patients may be labeled as mentally ill, particularly if automatisms include unacceptable social behaviors such as spitting or fondling themselves. Patients are not aware of their behavior or that it is inappropriate. If the patient does not lose consciousness, the seizure is labeled as simple partial and usually lasts less than 1 minute. Older terms for simple partial seizures include *jacksonian* and *focal motor*. If consciousness is lost, it is called a complex partial seizure or psychomotor seizure. It may last from 2 to 15 minutes.

Partial seizures arising from the parietal lobe may cause paresthesias on the side of the body opposite the seizure focus. Visual disturbances are seen if the occipital lobe is the originating site. Involvement of the motor cortex results in involuntary movements of the opposite side of the body. Typically movements begin in the arm and hand and may spread to the leg and face.

GENERALIZED SEIZURES. Generalized seizures affect the entire brain. Two types of generalized seizures are absence seizures and tonic-clonic seizures. Absence seizures, sometimes referred to as petit mal seizures, occur most often in children and are manifested by a period of staring that lasts several seconds.

Tonic-clonic seizures are what most laypeople envision when they think of seizures. They are sometimes called grand mal seizures or convulsions. Tonic-clonic seizures follow a typical progression. Aura and loss of consciousness may or may not occur. The tonic phase, lasting 30 to 60 seconds, is characterized by rigidity, causing the patient to fall if not lying down. The pupils are fixed and dilated, the hands and jaws are clenched, and the patient may temporarily stop breathing. The clonic phase is signaled by contraction and relaxation of all muscles in a jerky, rhythmic fashion. The extremities may move forcefully, causing injury if the patient strikes furniture or walls. The patient is often incontinent. Biting the lips or tongue may cause bleeding. An oral airway can be used to prevent self-injury if the patient's mouth is open. Never attempt to force an airway or anything else into the patient's mouth if the jaws are clenched.

The **postictal** period is the recovery period after a seizure. Following a partial seizure the postictal phase may be no more than a few minutes of disorientation. Patients who experience a generalized seizure may sleep deeply for 30 minutes to several hours. Following this deep sleep, patients may complain of headache, confusion, and fatigue. Patients may realize that they had a seizure but not remember the event itself.

Diagnostic Tests

An EEG is the most useful test for evaluating seizures. An EEG can determine where in the brain the seizures start, the

---

postictal: post—after + ictal—seizure

frequency and duration of seizures, and the presence of subclinical (asymptomatic) seizures. Sleep deprivation and flashing light stimulation may be used to evaluate the seizure threshold. See Chapter 45 for more information on EEG.

Medical Management

If an underlying cause for the seizure is identified, treatment focuses on correcting the cause. If no cause is found or if the seizures continue despite treatment of concurrent disorders, treatment focuses on the seizure activity.

Numerous anticonvulsant medications are available, each with specific actions, therapeutic ranges, and potential side effects (Table 46–6). Typically the patient is started on one drug and the dosage is increased until therapeutic levels are attained or side effects become troublesome. If seizures are not controlled on a single drug, another medication is added. Most anticonvulsants require periodic blood tests to monitor serum levels and kidney and liver function. If seizures continue despite anticonvulsant therapy, surgical intervention may be considered.

Surgical Management

The success of surgical intervention for epilepsy depends on identification of an epileptic focus within nonvital brain tissue. If no focus is identified or if it is in a vital area such as the motor cortex or speech center, surgery is not feasible. The surgeon attempts to resect the area affected to prevent spread of seizure activity. In some cases, seizures may be cured, but in others the goal is to reduce the frequency or severity of the seizures.

The preoperative assessment for epilepsy surgery is an extensive multistage process. Thorough assessment and teaching are essential. To adequately identify seizure foci, the patient is weaned off anticonvulsant therapy. Increasing the frequency of seizures is anxiety provoking to patients and significant others.

Emergency Care

The prime objective in caring for a patient experiencing a seizure is to prevent injury. Side rails should be padded to prevent injury if the patient strikes his or her extremities

against them. If the patient falls to the floor, move furniture out of the way. A small pillow should be placed under the patient's head to prevent striking it on the floor, taking care that the airway does not become occluded. If possible, turn the patient on his or her side to prevent aspiration if vomiting occurs. An oral airway and suction should be readily available. Do not force an airway in once the seizure has begun. The individual should not be restrained, because this may increase the risk of injury. Observe and document eye deviation, incontinence, which part of the body was first involved, and progression of the seizure (Patient Perspective Box 46–6).

Status Epilepticus

Status epilepticus is characterized by at least 30 minutes of repetitive seizure activity without a return to consciousness. This is a medical emergency and requires prompt intervention to prevent irreversible neurological damage. Abrupt cessation of anticonvulsant therapy is the usual cause of status epilepticus.

Seizure activity precipitates a significant increase in the brain's need for glucose and oxygen. This metabolic demand is even greater during status epilepticus. Irreversible neuronal damage may occur if cerebral metabolic needs cannot be fulfilled. Adequate oxygenation must be maintained, if neces-

BOX 46-6 Patient Perspective

I have had seizures for 35 years and as a result of falling during seizures have experienced cuts, bruises, and a broken bone. I usually have an aura that lets me know a seizure is about to occur. This is helpful if I can get myself to a safe place to prevent falling or being injured. When a patient is having a seizure you can best help by using padding such as pillows or blankets for protection, talking calmly, and using gentle touch to prevent injury. You should not sit on or hold down someone during a seizure. I have had the frightening experience of waking up with a nurse sitting on me and holding down my arms. After you have protected the patient, let them come out of the seizure on their own. When the seizure is over, I usually want to sleep because seizures are exhausting and uncomfortable. I will rest better knowing you are watching over me to keep me safe.

TABLE 46-6 SELECTED ANTICONVULSANT MEDICATIONS

Medication	Action	Side Effects/Comments
Phenytoin (Dilantin)	Limits seizure propagation; may also be used for some cardiac dysrhythmias and some types of nerve pain	May cause gingival hyperplasia, nausea, ataxia, rash; regular dental care essential; therapeutic level is 10–20 μg/mL
Phenobarbital (Luminal)	CNS depressant; raises seizure threshold	Causes drowsiness; often given with phenytoin
Carbamazepine (Tegretol)	Decreases synaptic transmission in CNS; may be used for some types of neuralgias	Common side effects: drowsiness and ataxia; used if other drugs ineffective; therapeutic level 6–12 μg/mL
Valproic acid (Depakote)	Increases GABA, an inhibitory neurotransmitter in CNS	Causes GI upset, nausea, vomiting; therapeutic level 50–100 μg/mL
Clonazepam (Klonapin)	CNS sedative; may also be used for neuralgia or restless leg syndrome	Side effects: drowsiness, ataxia, changes in behavior; therapeutic level 20–80 ng/ml
Gabapentin (Neurantin)	Unknown	Causes fatigue, dizziness. May cause leukopenia.

GABA = gamma-aminobutyric acid.

sary by intubating and mechanically ventilating the patient. These patients are also at significant risk for aspiration.

Intravenous diazepam (Valium) or lorazepam (Ativan) is given to stop the seizures. Because both these drugs may cause respiratory depression, careful airway management is required. After obtaining serum drug levels, anticonvulsant therapy is adjusted to achieve therapeutic levels.

If seizures remain resistant to treatment, a barbiturate coma may be induced with intravenous pentobarbital. The last line of treatment for status epilepticus is general anesthesia or pharmacological paralysis. Both these therapies require intubation, mechanical ventilation, and management in an ICU setting. Continuous EEG monitoring is used to verify that the seizures have actually stopped. A patient treated with neuromuscular blockade drugs may still be seizing but have no visible manifestations. For more information, visit the Epilepsy Foundation of America at www.efa.org.

Nursing Management of Seizures
Assessment

Perform a general neurological assessment of the patient with a history of seizures. Determine the type of seizure manifestations and type of aura if any. Assess the patient's knowledge of the disease and its treatment. It is important to assess whether the patient has the resources to purchase prescribed anticonvulsant medications and whether the medication regimen is adhered to. Drug levels may help determine degree of compliance with therapy.

Nursing Diagnosis, Planning, and Implementation

The goal is for the patient and significant others to be able to manage the treatment effectively to prevent seizures. If seizures occur, the goal is to prevent injury. Priority nursing diagnoses for the patient at risk for seizures include the following.

RISK FOR INJURY RELATED TO SEIZURE ACTIVITY. Instruct the patient to recognize the aura and to get to safety if it occurs. This may mean lying down away from furniture or other objects that may cause harm. For the patient admitted to a health care institution, seizure precautions are instituted. If a seizure does occur, the patient's safety is maintained. See Table 46–3 for precautions and interventions for seizures. Encourage all patients to wear medical alert jewelry or other identification to alert others to the presence of seizure disorder.

Some patients can identify conditions that trigger seizures. Hypoglycemia, hypoxia, and hyponatremia are all potential triggers of hypersensitive neurons. Teach the patient the importance of a consistent schedule of eating and sleeping.

RISK FOR INEFFECTIVE MANAGEMENT OF THERAPEUTIC REGIMEN RELATED TO COMPLEX REGIMEN AND POSSIBLE LACK OF RESOURCES. Patients with seizures may have several medications to take several times each day. This makes compliance difficult at best. Medication teaching is vital. Patients need to understand dosing, potential side effects, possible interactions with alcohol and other medications, and the importance of regular blood tests. If finances are a concern, the patient may be unable to obtain prescriptions. Patients must understand the risk for seizures and status epilepticus if medications are stopped abruptly. A nurse or social worker can help the patient apply for assistance to pay for medications if necessary.

ANXIETY RELATED TO RISK FOR SEIZURES. Epilepsy is a chronic disorder with significant impact on lifestyle. Patients may be ashamed to acknowledge their condition and may try to hide it from others. Provide the patient with current, accurate information and help him or her identify coping strategies to maximize independence. Support groups may also be helpful.

INEFFECTIVE ROLE PERFORMANCE RELATED TO POSSIBLE DISABLING DISORDER. Finances can be a major concern to these patients. Some patients with epilepsy experience hiring discrimination, or they may not qualify for some jobs in which safety is a concern. Remind patients that falsifying information on job applications may be grounds for dismissal. Refusal of health insurance coverage can create financial hardships for patients on long-term medications. Most patients whose seizures are controlled can work and lead productive lives. You can help patients explore options for financial assistance if necessary.

Patients with poorly controlled seizures should not operate motor vehicles. In our society a driver's license is a sign of adulthood and independence, and patients who cannot drive may experience lowered self-esteem. Job opportunities may be limited for patients who depend on public transportation. Encourage the patient to obtain a state identification card. This can be used in place of a driver's license for identification.

Patients may limit interpersonal relationships out of fear of having a seizure. The involuntary movements, sounds, and possible incontinence that occur with seizures are embarrassing to patients and can be frightening to laypeople. Role playing may help the patient determine when and how to confide in others.

Evaluation

Successful care of a patient with epilepsy is manifested by a decrease in seizures to the lowest possible frequency. Patient verbalization of understanding of needed lifestyle changes is another indication of success. Patients should be able to state measures to prevent injury if a seizure should occur and should verbalize understanding of all medications and their administration schedules. Therapeutic drug levels may be measured to evaluate compliance with the medication regimen.

▶ TRAUMATIC BRAIN INJURY

Traumatic brain injury is a major cause of death and disability in adults. Young males make up a large proportion of brain injury victims. The use of alcohol and illicit drugs is often associated with brain injury.

Pathophysiology

Traumatic brain injury is a complex phenomenon with results ranging from no detectable effect to persistent vegetative state. Trauma can result in hemorrhage, contusion or laceration of the brain, and damage at the cellular level. In addition to the primary insult, the brain injury may be compounded by cerebral edema, hyperemia, or hydrocephalus.

Etiology

Motor vehicle accidents account for the largest percentage of traumatic brain injuries. Violent assaults are increasing in frequency and may be accompanied by penetrating brain injuries. Falls and sports-related injuries are also common causes of traumatic brain injury.

The brain is susceptible to several types of injury. *Acceleration injury* is the term used to describe a moving object hitting a stationary head. An example of this type of injury is a patient who is hit in the head with a baseball bat. A deceleration injury occurs when the head is in motion and strikes a stationary surface. This type of injury is seen in patients who trip and fall, hitting their head on furniture or the floor.

A combination of acceleration-deceleration injury occurs when the stationary head is hit by a mobile object and the head then strikes a stationary surface. A soccer player who sustains a blow to the head and then hits the ground with his or her head may sustain an acceleration-deceleration injury.

Rotational injuries have the potential to cause shearing damage to the brain, as well as laceration and contusions. Rotational injuries may be caused by a direct blow to the head or may occur during a motor vehicle accident in which the vehicle is struck from the side. Twisting of the brainstem can damage the reticular activating system, causing loss of consciousness. Movement of the brain within the skull may result in bruising or tearing of brain tissue where it comes in contact with the inside of the skull.

Types of Brain Injury and Signs and Symptoms

CONCUSSION. Cerebral concussion is considered a mild brain injury. If there is a loss of consciousness, it is for 5 minutes or less. Concussion is characterized by headache, dizziness, or nausea and vomiting. The patient may complain of amnesia of events before or after the trauma. On clinical examination there is no skull or dura injury and no abnormality detected on CT or MRI.

CONTUSION. Cerebral contusion is characterized by bruising of brain tissue, possibly accompanied by hemorrhage. There may be multiple areas of contusion, depending on the causative mechanism. Severe contusions can result in diffuse axonal injury. The symptoms of a cerebral contusion depend on the area of the brain involved.

Brainstem contusions affect level of consciousness. The decreased level of consciousness may be transient or permanent. Respirations, pupil reaction, eye movement, and motor response to stimuli may also be affected. The autonomic nervous system may be affected by edema or by hypothalamic injury, causing rapid heart rate and respiratory rate, fever, and diaphoresis.

HEMATOMA. **Subdural** hematomas are classified as acute or chronic based on the time interval between injury and onset of symptoms. Acute subdural hematoma is characterized by appearance of symptoms within 24 hours following injury. The bleeding is typically venous in nature and accumulates between the dura and arachnoid membranes (Fig. 46–10). Approximately 24 percent of patients who sustain a severe brain injury develop an acute subdural hematoma. Damage to the brain tissue itself may cause an altered level of consciousness. Therefore it can be difficult to recognize a subdural hematoma based only on clinical examination. As the subdural hematoma increases in size, the patient may exhibit one-sided paralysis of extraocular movement, extremity weakness, or dilation of the pupil. Level of consciousness may deteriorate further as ICP increases.

Elderly and alcoholic individuals are particularly prone to chronic subdural hematomas. Atrophy of the brain, common in these populations, stretches the veins between the brain and the dura. A seemingly minor fall or blow to the head can cause these stretched veins to rupture and bleed. Often there are no other injuries associated with the trauma. Because a chronic subdural hematoma can develop weeks to months after the injury, the patient may not remember an injury occurring.

The patient with a chronic subdural hematoma may be forgetful, lethargic, or irritable or may complain of a headache. If the hematoma persists or increases in size, the

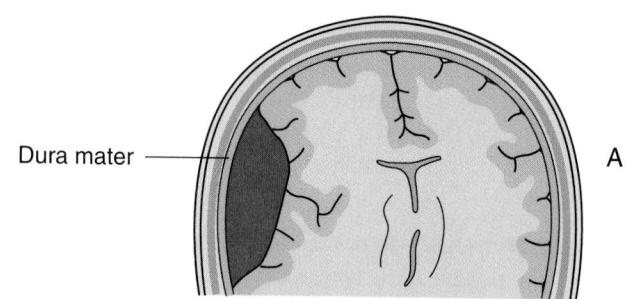

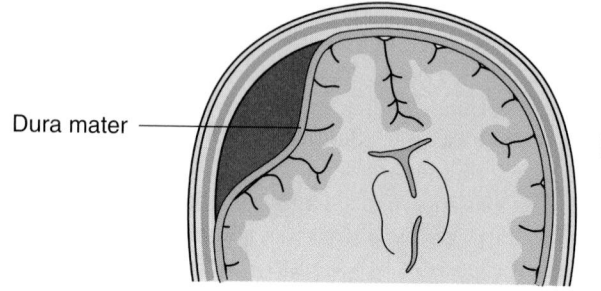

Figure 46-10 *(A)* Subdural hematoma is usually venous and forms between the dura and the arachnoid membranes. *(B)* An epidural hematoma is usually from an arterial bleed and forms between the dura mater and the skull.

---

subdural: sub—below + dural—pertaining to the dura mater

patient may develop hemiparesis and pupillary changes. The patient or significant other may not associate the symptoms with a previous injury and therefore may delay seeking medical care.

Approximately 10 percent of patients with severe brain injuries develop **epidural** hematomas. This collection of blood between the dura mater and skull is usually arterial in nature and is often associated with skull fracture. (See Fig. 46–10.) Arterial bleeding can cause the hematoma to become large very quickly. Patients with epidural hematoma typically exhibit a progressive course of symptoms. The patient loses consciousness directly after the injury; he or she then regains consciousness and is coherent for a brief period. The patient then develops a dilated pupil and paralyzed extraocular muscles on the side of the hematoma and becomes less responsive. If there is no intervention, the patient becomes unresponsive. Seizures or hemiparesis may occur. Once the patient exhibits symptoms, the deterioration may be rapid. Airway management and control of ICP must be instituted immediately. If ICP is not controlled, the patient will die.

Diagnostic Tests

CT scan is usually the first imaging test performed on the brain-injured patient. It is faster and more accessible than MRI. This is particularly important for unstable patients or those with multiple injuries. It is easier to identify skull fractures on CT than on MRI. MRI may be used later to identify damage to the brain tissue.

Neuropsychological testing can be useful in assessing the patient's cognitive function. This information helps direct rehabilitation placement, discharge planning, and return to work or school. Neuropsychological testing identifies problems with memory, judgment, learning, and comprehension. Compensation strategies can be suggested to the patient and significant others based on the results.

Treatment

SURGICAL MANAGEMENT. Surgical treatment of hematomas is discussed under intracranial surgery later in this chapter.

MEDICAL MANAGEMENT. Medical management of traumatic brain injury involves control of ICP and support of body functions. Brain-injured patients may be partially or completely dependent for maintenance of respiration, nutrition, elimination, movement, and skin integrity.

A variety of techniques are used to control intracranial pressure in the patient with moderate or severe brain injury. The first step is to insert an ICP monitor to allow determination of the ICP. Refer to the section on increased ICP earlier in this chapter for further information.

If ICP remains elevated despite drainage of cerebrospinal fluid, the next step is use of an osmotic diuretic. The most commonly used drug is intravenous mannitol (Osmitrol). Mannitol utilizes osmosis to pull fluid into the intravascular

space and eliminate it via the renal system. Serum osmolarity and electrolytes must be carefully monitored when mannitol is being administered. Some patients experience a rebound increase in ICP after the mannitol wears off.

Mechanical hyperventilation is the next step if the patient is still experiencing increased ICP. Hyperventilation is effective in lowering ICP because it causes vasoconstriction. This allows less blood into the cranium, thereby lowering ICP. Research has demonstrated, however, that aggressive hyperventilation, particularly within the first 24 hours after injury, may induce ischemia in the already compromised brain. Therefore hyperventilation is now reserved for increased ICP that does not respond to other treatments.

High-dose barbiturate therapy may be used to induce a therapeutic coma, which reduces the metabolic needs of the brain during the acute phase following injury. These patients are completely dependent for all their needs and care. They will be mechanically ventilated and cared for in an ICU setting. Vasopressors may be required to maintain blood pressure, and the patient's temperature should be kept as normal as possible.

If none of these interventions is successful, the patient may experience uncontrolled edema or herniation of brain tissue (Fig. 46–11). Herniation is displacement of brain

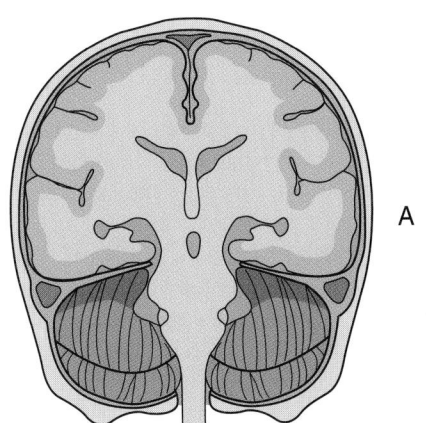

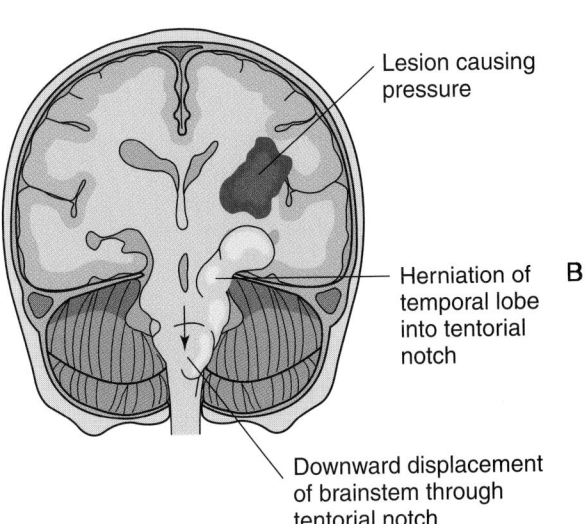

Figure 46–11 Herniation of the brain. *(A)* Normal brain. *(B)* Herniation of brain tissue into tentorial notch.

---

epidural: epi—above + dural—pertaining to the dura mater

tissue out of its normal anatomic location. This displacement prevents function of the herniated tissue and places pressure on other vital structures, most commonly the brainstem. Herniation usually results in brain death.

Patients who experience brain death may be suitable organ donor candidates. For some significant others, the opportunity to donate their loved one's organs provides some sense of purpose in the death (Box 46–7).

Acute Complications of Traumatic Brain Injury

DIABETES INSIPIDUS. Edema or direct injury affects the posterior portion of the pituitary gland or hypothalamus. Inadequate release of antidiuretic hormone results in polyuria and, if the patient is awake, polydipsia. Fluid replacement and intravenous vasopressin are used to maintain fluid and electrolyte balance.

ACUTE HYDROCEPHALUS. Cerebral edema can interfere with cerebrospinal fluid circulation, causing hydrocephalus. Initial treatment is with an external ventricular drain, followed by a ventriculoperitoneal shunt if necessary.

LABILE VITAL SIGNS. Direct trauma to or pressure on the brainstem can cause fluctuations in blood pressure, cardiac rhythm, or respiratory pattern. Treatment is aimed at control of intracranial pressure.

Long-Term Complications

POSTTRAUMATIC SYNDROME. Patients who sustain a concussion may experience ongoing, somewhat vague symptoms. They complain of headache, fatigue, difficulty concentrating, depression, or memory impairment. Symptoms may be severe enough to interfere with work, school, and interpersonal relationships. Neuropsychological testing may provide objective evidence of cognitive dysfunction and establish the need for cognitive rehabilitation. Symptoms may take 3 to 12 months to resolve.

COGNITIVE AND PERSONALITY CHANGES. Alterations in personality and cognition may be the most difficult

---

BOX 46–7 **Traumatic Brain Injury Summary**

Signs and Symptoms
Loss of consciousness (note total time)
Loss of memory before or after the injury
Increased ICP

Diagnostic Tests
CT scan
Neuropsychological testing

Therapeutic Management
Control ICP
Maintain respiratory function
Maintain diet/nutrition
Maintain skin integrity
Prevent complications
Education

---

long-term complication for patients and significant others to adjust to. The patient may have significant short-term memory impairment. This limits his or her ability to learn new information and may interfere with ability to function at work or school. Impaired judgment can make the patient a safety risk to self or others. It also affects social functioning.

Emotional lability, loss of social inhibitions, and personality changes may occur. These consequences of traumatic brain injury have a profound effect on the patient and significant others. Spouses may state, "This is not the person I married." If behavior is violent, bizarre, or profane, children may be unwilling to bring their friends home and may become socially isolated. Young children, in particular, have difficulty understanding why the parent is behaving so differently. Disintegration of relationships is not uncommon following traumatic brain injury.

Neuropsychological testing objectively identifies problems. These deficits can then be addressed with cognitive rehabilitation. Individual and family counseling may be of benefit. Support groups for patients and significant others are often helpful.

Motor and speech impairment are additional possible long-term complications of traumatic brain injury. Intensive rehabilitation provides the best opportunity for maximizing recovery. For more information, visit the Brain Injury Association at www.biausa.org.

Nursing Process: The Patient with Traumatic Brain Injury
Assessment

After stabilization in the emergency department, care of the patient with a traumatic brain injury is in the intensive care setting, where ICP can be carefully monitored. Neurological status is assessed frequently, including Glasgow Coma Scale score, pupil responses, muscle strength, and vital signs. Review Box 46–2 for signs of increased ICP. Once the patient is stabilized, neurological damage is assessed. Identification of deficits guides nursing care. Assessment of discharge needs should also begin as soon as possible. The patient may require extensive rehabilitation, and early referral may speed transfer to an appropriate facility.

Nursing Diagnosis, Planning, and Implementation

The patient with a traumatic brain injury may have many nursing problems. The goals of care are to prevent further injury or complications while the patient stabilizes and then to provide rehabilitation to maximize functioning. Assessment and rapid reporting of changes is an important intervention in maximizing recovery potential. Common diagnoses include the following.

INEFFECTIVE CEREBRAL TISSUE PERFUSION RELATED TO INCREASED ICP. If assessment reveals signs of increased ICP, the physician is notified. See Table 46–4 for interventions to prevent increased ICP.

INEFFECTIVE BREATHING PATTERN RELATED TO PRESSURE ON RESPIRATORY CENTER. Respiratory rate and depth are closely monitored. Arterial blood gases help determine effectiveness of respiration. Mechanical ventilation may be necessary.

INEFFECTIVE AIRWAY CLEARANCE RELATED TO REDUCED COUGH REFLEX AND DECREASED LEVEL OF CONSCIOUSNESS. Monitor airway and breath sounds. If the patient is unable to cough effectively, suctioning may be necessary. Suction passes should be limited to one or two at a time to prevent increased ICP.

PAIN RELATED TO TISSUE DAMAGE. The only sign of pain in the unconscious patient may be restlessness or a change in vital signs. Opioids are avoided because they further depress LOC and may mask changes in neurological status. Codeine may be ordered because it has less effect on LOC and pupils.

IMPAIRED PHYSICAL MOBILITY RELATED TO DECREASED LOC. Mobility is maintained with range-of-motion exercises at least three times a day and position changes every 1 to 2 hours (if ICP is not elevated). Monitor skin closely for breakdown. Position the patient in functional body alignment to prevent contractures that can interfere with function after recovery. Physical and occupational therapy can be instituted once the patient is stable.

SELF-CARE DEFICIT RELATED TO DECREASED LOC. Initially the patient may be totally dependent for routine care. Activities should be spaced to allow for rest because clustered activities may increase ICP. Monitor bowel and bladder function carefully. A bowel program may be necessary to maintain regular bowel movements. An indwelling catheter may be used immediately following the injury to monitor fluid balance. Long-term catheterization is avoided, however, because of risk of infection.

SENSORY-PERCEPTUAL DISTURBANCE RELATED TO CRANIAL NERVE DAMAGE. The patient may not have a normal corneal reflex and may not be able to protect his or her eyes. Lubricating eye drops and taping the eyes shut help protect the eyes. See Chapters 48 and 49 for interventions for specific sensory problems.

RISK FOR INJURY RELATED TO DECREASED LOC, RISK FOR SEIZURES. If the patient is at risk for injury, routine safety precautions are instituted. Seizure precautions are instituted. (See Table 46–3.)

DISTURBED THOUGHT PROCESSES RELATED TO DECREASED LOC AND BRAIN DAMAGE. Confusion increases risk for injury. The confused patient is reoriented as needed. A calendar, clock, family photos, and other familiar items in the patient's room may help promote orientation. You may wish to ask a family member or significant other to stay with the patient to prevent the need for restraints to maintain safety.

INEFFECTIVE ROLE PERFORMANCE RELATED TO LONG-TERM EFFECTS OF INJURY. The effects of a head injury can have a profound impact on the patient and family; provide emotional support as needed. The entire health team is needed to coordinate discharge to a safe environment. A social worker consult can help assess specific needs and identify resources for physical care and financial assistance if necessary.

Evaluation

The plan of care has been successful if the patient shows no unexpected worsening of neurological function and injuries and complications are prevented. The airway is clear. The patient is kept comfortable, and self-care needs are met.

Rehabilitation

Once the patient is stabilized, evaluation for discharge to a rehabilitation facility is done. The patient must be able to physically tolerate the rehabilitation program. The patient is taught to function as independently as possible. The family must be prepared for changes in the patient's ability to function and possible changes in personality. It may take months to years before the patient reaches his or her maximum potential. In some cases of severe brain damage or continued comatose state, rehabilitation is not feasible and the patient is discharged to home or a long-term facility for custodial care.

▶ BRAIN TUMORS

Brain tumors are neoplastic growths of the brain or meninges. They may be characterized by vague symptoms such as headache or visual changes or by focal neurological deficits such as hemiparesis or seizures.

Pathophysiology and Etiology

Brain tumors cause symptoms by either compressing or infiltrating brain tissue. Tumors may arise from central nervous system cells or may metastasize from other locations in the body. Primary brain tumors rarely metastasize. If they do metastasize, it is to the spine.

There is no established cause for primary brain tumors. It is unclear what causes the cells to begin reproducing in an uncontrolled fashion. Brain tumors can be classified in several different ways. The traditional distinction of benign and malignant is less applicable when discussing brain tumors. A benign tumor in the brainstem may be fatal, whereas a malignant tumor in the frontal lobe may not. Location of the tumor can be just as important a factor in outcome as the cell type.

Primary tumors are those arising from cells of the central nervous system. From 80 to 90 percent of brain tumors are primary in nature. Secondary tumors are those that metastasize from primary malignancies elsewhere in the body (Fig. 46–12). Intra-axial tumors are those that arise from the glial cells within the cerebrum, cerebellum, or brainstem. These tumors infiltrate and invade brain tissue itself. Extra-axial tumors arise from the skull, meninges, pituitary gland, or cranial nerves. These tumors have a compressive effect on the brain.

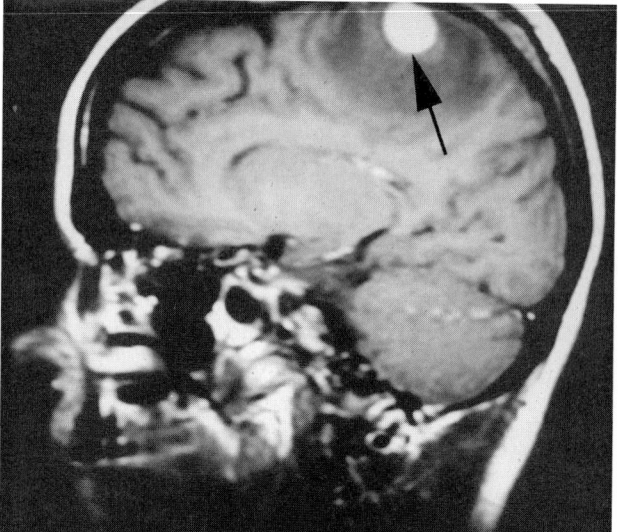

Figure 46-12   Metastatic brain tumor. This patient's primary cancer was in the lung.

Approximately 10 to 20 percent of brain tumors are metastatic from a primary malignancy elsewhere in the body. These tumors commonly spread via the arterial system. If untreated, they cause increased ICP. This may be the cause of the patient's death, rather than the primary malignancy.

Signs and Symptoms

The symptoms of a brain tumor are directly related to the location of the tumor in the brain and to the rate of growth. Slow-growing types of tumors such as meningiomas (a tumor arising from the meninges; Fig. 46–13) can get to be

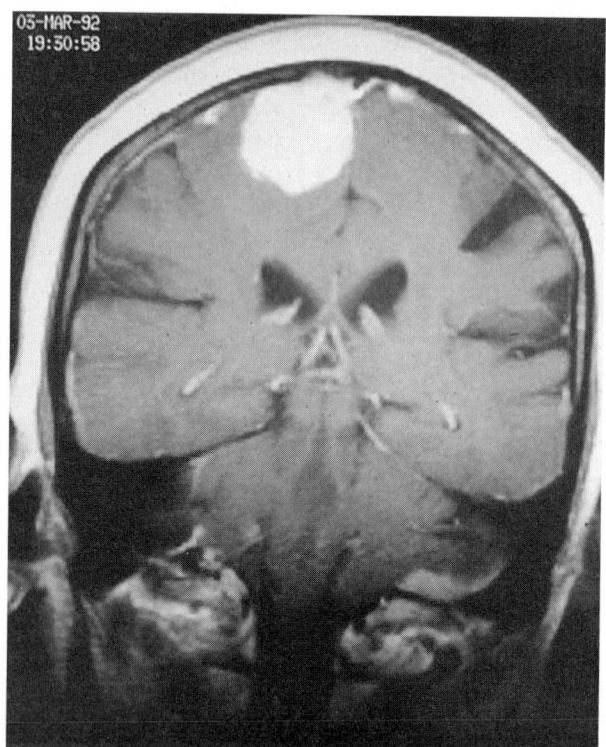

Figure 46-13   Meningioma.

quite large before becoming symptomatic. Conversely, glioblastoma multiforme or metastatic tumors may abruptly cause seizures or hemiparesis. Other types of tumors include oligodendroglioma, astrocytoma, and acoustic neuroma. The suffix *-oma* refers to tumor. The prefix denotes the type of cell the tumor arose from.

Symptoms can include seizures, motor and sensory deficits, headaches, and visual disturbances. If the pituitary gland is involved, additional symptoms such as abnormal growth or fluid volume changes are related to changes in hormone secretion.

LEARNING TIPS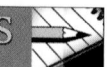

The phrase *P.T. Barnum loves kids* can be used to help remember the locations of tumors that most commonly metastasize to the brain:

P	Prostate
T	Thyroid
Barnum	Breast
Loves	Liver
Kids	Kidney

Diagnostic Tests

MRI gives the clearest images of a brain tumor. Many health care providers order a CT scan first because it is cheaper. If the tumor appears to be highly vascular or in close proximity to major blood vessels, an angiogram may be performed. It is now possible to do magnetic resonance angiograms. This involves the intravenous administration of contrast material and is much less invasive than a traditional angiogram. If the tumor is in the region of the pituitary gland, serum hormone levels are evaluated.

Treatment

Surgical treatment involves removal of the tumor or as much of the tumor as possible. Care of the patient undergoing intracranial surgery is discussed later in this chapter.

Medical Management

Medical treatment consists of controlling symptoms and administering additional therapies. Patients who have a seizure are placed on anticonvulsants. If significant cerebral edema is noted on the CT or MRI or if the patient is suffering from headaches, dexamethasone (Decadron) may be prescribed to lessen the edema. Typically these patients do not require narcotics for pain relief.

Radiation Therapy

External beam radiation therapy is standard treatment for many patients with a brain tumor. The therapy is typically given 5 days a week for 6 weeks. Some clinicians use a hyperfractionated schedule, in which the patient has therapy twice a day for less time. Brachytherapy is a means of delivering radiation therapy directly to the tumor. Small catheters are implanted in the tumor and then tiny radioactive particles are inserted into the catheters. The treatment typically takes 3 to 5 days. During this time the patient is confined to a private room and interaction with visitors and staff is kept to a minimum because of radioactivity. This therapy is not appropriate for confused individuals because they may not be able to cooperate with restrictions.

Stereotactic radiosurgery is a technique that utilizes small amounts of radiation directed at the tumor from different angles. A metal frame is affixed to the patient's skull, and the tumor is visualized within the framework on a CT or MRI. A computer plan is generated to direct the radiation. Because multiple small sources are used, the normal brain tissue receives very little radiation while the majority of the radiation accumulates in the tumor.

Chemotherapy

The blood-brain barrier is a protective mechanism that prevents many injurious substances from reaching brain tissue. Unfortunately, it is also effective in preventing chemotherapeutic agents from reaching the brain. To penetrate the blood-brain barrier, very large doses of chemotherapy may be required. These doses may not be well tolerated by other body systems. New treatments are currently being investigated. Some clinicians place chemotherapeutic substances in the cavity left by surgical resection. Others disrupt the blood-brain barrier with mannitol and then deliver intra-arterial chemotherapy under general anesthesia. Gene therapy is also being used in an effort to kill malignant cells.

Complementary Therapies

The rate of success for treatment of brain tumors is not as high as treatment of other neoplasms. Patients may be drawn to nontraditional therapies both as cures and for treatment of symptoms. Encourage patients to look at each option in a rational manner. Some questions they should ask themselves include the following:

■ Will this interfere with any of my other treatments or medications?
■ What is the cost?
■ What are the side effects?
■ Is there any objective information (research) available?
■ What does my physician think of this?

Additional information on evaluation of complementary therapies is found in Chapter 4.

Acute and Long-Term Complications

It is difficult to distinguish between symptoms of a brain tumor and complications of treatment. Seizures, headaches, memory impairment, cognitive changes, and ataxia may be symptoms of the tumor or the result of surgery or radiation therapy. Patients may experience hemiparesis or aphasia following surgery. If the tumor continues to grow despite treatment, the patient experiences further decline in function. Gradually the patient becomes more lethargic and unre-

sponsive. Once the patient becomes comatose, death occurs within a matter of days, particularly if artificial nutrition and hydration are not administered.

Nursing Process: The Patient with a Brain Tumor

Assessment

Perform routine neurological assessments to determine level of functioning and presence of neurological deficits such as vision changes, movement problems, altered thought processes, or changes in level of consciousness. A pain assessment is done using the WHAT'S UP? format presented in Chapter 1. Asking the patient open-ended questions about how he or she is coping with the diagnosis may help open up communication about fears and grieving.

Nursing Diagnosis, Planning, and Implementation

The patient who is not anticipating surgery or curative treatment is assisted to function as effectively as possible for as long as possible. Nursing diagnoses are based on actual problems the patient is experiencing. Possible diagnoses and interventions are below.

DISTURBED THOUGHT PROCESSES RELATED TO INVOLVEMENT OF BRAIN TISSUE. If the patient is disoriented, implement routine precautions to keep the patient safe. Reorient the patient as needed. The use of calendars, clocks, and familiar furnishings may help reorient the patient. A sudden worsening in thought processes should be reported to the RN or physician immediately. The cause may be reversible.

SELF-CARE DEFICIT RELATED TO IMPAIRED MOBILITY AND THOUGHT PROCESSES. The patient may be unable to care for himself or herself because of altered thought processes or mobility problems. Provide assistance as needed, while encouraging the patient to maintain as much independence as possible. If the patient becomes totally dependent, a long-term care facility may become necessary.

PAIN RELATED TO CEREBRAL EDEMA. If the patient experiences headaches or other discomforts, pain medication is provided. Nonnarcotic medications are preferred because they do not alter the level of consciousness. If these are not effective, codeine preparations, which have a minimal effect on LOC, may be prescribed.

SENSORY-PERCEPTUAL DISTURBANCE RELATED TO CRANIAL NERVE INVOLVEMENT. Alterations in sensations can take on a variety of forms, including changes in vision, hearing, and paresthesia. See Chapters 48 and 49 for interventions for patients with sensory deficits. Care must be taken to protect the patient's skin if tactile sensation is reduced. Assist the patient out of bed and into a different environment to help prevent sensory deprivation. Remember to provide sensory stimulation such as conversation, radio, and television if tolerated and enjoyed by the patient.

RISK FOR INJURY RELATED TO SEIZURES OR SENSORY DEFICITS. In addition to routine safety precautions, the patient is protected from injury that may occur as a result of seizures. Seizure precautions are listed in Table 46–3.

ANTICIPATORY GRIEVING RELATED TO POTENTIAL LOSS OF FUNCTION AND DEATH. Both the family and patient may be grieving loss of function, as well as a terminal prognosis. Patients and their significant others require emotional support to face the challenges presented by the diagnosis of brain tumor. They must make decisions regarding treatment at the same time that they may be facing a deterioration in function. How aggressively to pursue treatment and when to change the focus from treatment to palliation are questions that confront these individuals. Patients should be encouraged to organize their personal affairs and make their decisions known while they are still able. Referral to a support group, social worker, or pastoral care person may be beneficial.

Evaluation

An effective plan will result in a patient who is safe and comfortable. The patient and significant others will be able to express feelings and have the support necessary to plan and carry out necessary care.

▶ INTRACRANIAL SURGERY

The primary purpose of intracranial surgery is to remove a mass lesion. These types of lesions include hematomas, tumors, arteriovenous malformations, and, occasionally, contused brain tissue. Other indications for surgery include elevation of a depressed skull fracture, removal of a foreign body, débridement of a wound, or resection of a seizure focus. *Craniotomy* refers to any surgical opening in the skull. A burr hole is an opening into the cranium made with a drill. **Craniectomy** is the term used to describe removal of part of the cranial bone. **Cranioplasty** refers to repair of bone or use of a prosthesis to replace bone following surgery.

The goal of intracranial tumor surgery for a tumor is gross total resection of the tumor. This involves removal of all visible tumor, called debulking. Even with the use of an operative microscope, there may be viable tumor cells left behind. It is these cells that give rise to recurrence. If all the tumor cannot be removed, the surgeon debulks as much as possible. By debulking the tumor, the surgeon reduces the amount of neoplasm, thereby giving radiation therapy or chemotherapy less of a burden to combat. In some cases it is not feasible to attempt more than a biopsy of the tumor. Location of the tumor or the patient's age or medical condition may not allow the patient to tolerate a full craniotomy. The biopsy may be done under local or general anesthesia, depending on the patient's condition. The goal of a biopsy is to obtain tissue that allows pathological diagnosis of the tumor. The diagnosis then guides any further treatment.

Intracranial surgery is usually performed under general anesthesia. Occasionally a procedure requires that the patient be awake and cooperative.

Preoperative Care

Preoperative care of the patient undergoing intracranial surgery is similar to that of patients having other surgeries. (See Chapter 11.) The patient undergoes a laboratory workup and anesthesia evaluation. If the patient has cognitive impairments, it is important that a significant other be available to provide information. A thorough baseline neurological assessment should be documented.

Patient education is important preoperatively. The extent of education depends on the patient's ability to absorb new information. This is influenced by the disease process, cognitive functioning, anxiety, and educational level. Significant others are involved as needed. Information about the disease process and surgery are provided by the surgeon. You can play an important role in reinforcing and clarifying the information presented.

Anxiety is also a significant concern before surgery. The patient is anticipating serious surgery, as well as an unknown outcome. Allow time for the patient and significant others to express their fears and ask questions. Honest and accurate information should be provided.

Significant others should be prepared for how the patient will look after surgery. A preoperative visit to the intensive care unit may help prevent some anxiety postoperatively. Significant others should be accompanied on this visit by a knowledgeable nurse who can explain what they are seeing.

Surgery may last 2 hours for a biopsy to 12 hours or longer for more intricate procedures. Patients and significant others should be prepared for the idea that some or all of the patient's hair will be shaved off. Some people prefer to have all their hair shaved rather than just part. The patient should be prepared to see his or her face swollen after surgery, particularly around the eyes. The periorbital region may be bruised. Many patients wish to wear a scarf or scrub cap after the dressing is removed.

Postoperative Care
Assessment

After intracranial surgery, plan to assist the RN with frequent neurological assessments in addition to routine postoperative monitoring. Patients should have their neurological status assessed every hour for the first 24 hours or as ordered by the physician. Any deterioration in status should be immediately reported to the physician. Many patients undergo a CT scan within the first 24 hours following surgery to assess cerebral edema.

Nursing Diagnosis, Planning, and Implementation

The primary goal following intracranial surgery is prevention of complications. Once the patient is stabilized, goals can change to longer-term outcomes such as acceptance of changes in body image and understanding of self-care following discharge. If the patient has severe deficits following surgery, rehabilitation or long-term care may become necessary. A consultation with a social worker can help with planning for this transition. Priority nursing diagnoses include the following.

RISK FOR INEFFECTIVE CEREBRAL TISSUE PERFUSION RELATED TO EDEMA OF THE OPERATIVE SITE. Patients who have undergone intracranial surgery are placed in an intensive care unit postoperatively. They are typically positioned with the head of the bed at 30 degrees or higher, unless ordered otherwise, to promote venous drainage and minimize increases in intracranial pressure. The exception to this is patients who have had a chronic subdural hematoma removed, who must remain flat. Patients may turn from side to side or lie on their back. The patient should be encouraged not to lie on the operative side. Seizure precautions are instituted because the patient is at risk for seizures.

The patient may have an intracranial monitor in place following surgery to monitor intracranial pressure. Some patients may also have central venous pressure catheters or pulmonary artery catheters. Urinary catheters are used during the immediate postoperative period to accurately monitor fluid balance.

Dressings should be monitored for drainage. Drainage that is blood tinged in the center with a yellowish ring around it may be CSF leakage. A suspected CSF leak should be reported to the RN or physician immediately.

PAIN RELATED TO SURGICAL PROCEDURE. Intramuscular codeine and oral acetaminophen with codeine (Tylenol with Codeine Number 3) are the pain medications of choice for patients after intracranial surgery. These medications provide pain relief without causing sedation, which can interfere with the neurological assessment. Opioids are not used because they depress respirations and can mask signs of increasing intracranial pressure. Most patients complain of a headache postoperatively, but not severe pain.

RISK FOR INFECTION RELATED TO SURGICAL PROCEDURE. As with any surgery, a break in the integrity of the skin (and in this case skull) creates a risk for entry of pathogenic organisms. Infection following craniotomy can be deadly. All care of the incision, dressing, and monitoring equipment sites should be done using strict aseptic technique. Any signs of infection should be reported immediately.

IMPAIRED PHYSICAL MOBILITY RELATED TO MOTOR DEFICITS. Some patients have residual effects from the tumor or other pathological condition or from the surgery. These can range from weakness and paresthesias to paralysis. To prevent contractures and skin breakdown, the patient should be turned every 1 to 2 hours after surgery and avoid lying on the operative site unless specifically permitted by the surgeon. Careful positioning in correct body alignment also helps prevent contractures or other injuries. High-top tennis shoes, trochanter rolls, and slings can be used to keep the body in alignment. Range-of-motion exercises and physical therapy are initiated when permitted by the surgeon. An occupational therapist may also be able to assist the patient in learning to perform ADLs with a new disability.

BODY IMAGE DISTURBANCE RELATED TO CHANGES IN APPEARANCE OR FUNCTION. The patient will have a shaved or partially shaved head after surgery. In addition, depending on the outcome of surgery, residual motor or sensory deficits may be present. Allow the patient to express his or her feelings if desired. Portray an accepting attitude. A turban, scarf, or hat may help conceal a shaved head.

DEFICIENT KNOWLEDGE RELATED TO CHANGE IN TREATMENT REGIMEN FOLLOWING SURGERY. If the patient will be discharged home, discuss any activity restrictions, particularly driving. Evaluate the patient's understanding of the medication regimen and wound care. Have the patient and significant others verbalize the signs of infection or other possible complications to report.

Evaluation

Interventions have been effective if infection and other complications have been prevented and the patient states that pain is controlled. The patient might be able to look in the mirror and begin to show evidence of acceptance of changes in body image, though this may not happen until after discharge from the hospital. The patient and significant others should be able to describe appropriate follow-up care.

CRITICAL THINKING

Mr. Esposito

Mr. Esposito is a 24-year-old white male who was involved in a motor vehicle accident. His blood alcohol level was 0.24. Mr. Esposito has no preexisting medical problems. Emergency medical services personnel report that Mr. Esposito was unconscious on their arrival at the scene and then became alert and combative. His CT scan shows a left-sided epidural hematoma. Mr. Esposito is admitted to your unit for observation.

1. What symptoms would you expect to see if Mr. Esposito's hematoma increases in size?
2. What emergency preparations should you have ready?
3. What psychosocial assessments should you perform?

Answers at end of chapter.

▶ SPINAL DISORDERS
Herniated Disks

Herniated intravertebral disks are a common health problem. They are characterized by pain and paresthesias that follow a radicular (nerve path) pattern. It is not uncommon for patients to have more than one herniated disk or to have herniated disks in different areas of the spine.

Pathophysiology

When the disk between two vertebrae herniates, it moves out of its normal anatomic position. In most cases the annulus fibrosus, the tough outer ring of the disk, tears. This allows escape of the nucleus pulposus, the soft inner portion of the disk. Displacement of the disk compresses one or more nerve roots, causing the characteristic symptoms (Fig. 46–14).

Etiology

In some cases a specific event can be correlated with a herniated disk. The patient may describe a fall, lifting a heavy object, or a motor vehicle crash. In other instances the patient cannot identify a triggering incident.

Signs and Symptoms

Cervical disk herniation causes pain and muscle spasm in the neck. The patient may exhibit decreased range of motion secondary to pain. Hand and arm pain is unilateral (one sided) and follows the distribution of the spinal nerve root. Patients often complain of numbness or tingling in the extremity. Asymmetric weakness and atrophy of specific muscle groups may be detected. If weakness involves the entire extremity, it is unlikely that disk herniation is the etiology. The severity of the pain or paresthesia does not correlate directly with the severity of the nerve compression. However, weakness and atrophy are indicators of significant nerve compression.

Thoracic herniated disks are not common. This portion of the spine is the least mobile, therefore less stress is exerted on the disk. Patients with herniated thoracic disks may complain of pain in the back. It is uncommon to detect muscular weakness.

A herniated lumbar disk is typically characterized by low back pain, pain radiating down one leg, paresthesias, and weakness. The patient may limp on the affected leg or may have difficulty walking on his or her heels or toes. Muscle spasm is often present. Pain and muscle spasm may limit the patient's range of motion. Depending on the disk affected, the knee or ankle deep tendon reflex may be decreased or absent. A severely herniated L5–S1 disk may affect bowel or bladder continence. This is an emergency situation and should be reported immediately.

The WHAT'S UP? mnemonic can be used to assess symptoms of herniated disks at any level:

W—Where is the pain? Does it radiate into an extremity? In what distribution?
H—How does it feel? Sharp, stabbing, burning?
A—Do certain positions or activities alleviate or aggravate the pain? Holding the affected arm above the head may alleviate cervical pain. Sitting places pressure on disks and aggravates lumbar pain. Lying down may relieve it.
T—Is there a correlation between time and pain? Some patients have more pain at the end of the day. Is the pain constant or intermittent?
S—Ask the patient to rate the severity of the pain on a scale of 0 to 10. Which is the most painful, the spine or the extremity?
U—Ask the patient to identify associated symptoms such as numbness, tingling, or weakness.
P—What is the patient's perception of the pain? Is it interfering with work or other aspects of the patient's life?

Diagnostic Tests

An MRI will detect herniation of a disk and compression or abnormality of the spinal cord. If the patient has previously had surgery in the area of the suspected herniation, the MRI is done with and without contrast to differentiate between scar tissue and herniated disk.

If the patient cannot tolerate MRI or if the MRI does not provide enough information, a myelogram is done. Refer to Chapter 45 for a description of both tests.

Treatment

Most clinicians and patients prefer to try conservative medical therapy before performing surgery for a herniated disk.

MEDICAL MANAGEMENT

Rest. In the past, bedrest was advised as part of conservative management. The current recommendation is 1 or 2 days of bedrest, followed by a careful, gradual increase in activity.

Physical Therapy. Physical therapy can be very useful for some patients. A gradually progressive course of exercise strengthens the muscles. This is particularly important in the lumbar spine, where the muscles help stabilize the spine. Techniques such as ultrasound, heat, ice, and deep massage can decrease muscle spasm and allow for increased range of motion. Instructions in proper body mechanics and strategies for avoiding reinjury are an important component of physical therapy.

A transcutaneous electrical nerve stimulator (commonly called a TENS unit) is a noninvasive pain-relief technique. Small electrodes are placed on the skin around the area of the pain. The device then transmits a low-voltage electrical current through the skin. The patient feels a tingling or buzzing sensation, which may block the pain impulses. A physical therapist or pain specialist teaches the patient where to place the electrodes and how to operate the unit. The patient decides when to use it and at what settings. This allows the patient to actively par-

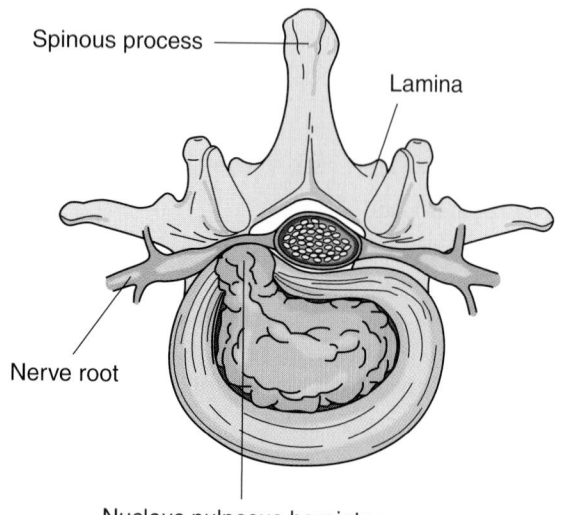

Spinous process

Lamina

Nerve root

Nucleus pulposus herniates and compresses nerve root

Figure 46–14 A herniated disk places pressure on a spinal nerve root.

ticipate in his or her care and have some control over the pain level.

Traction. Cervical traction is a noninvasive technique sometimes used by physical therapists for patients with herniated cervical disks. The patient's head is placed in a halterlike device. A series of ropes and pulleys connect the halter to a weight. This gently pulls the head away from the shoulders. The rationale is that this traction slightly separates the vertebral bodies and may allow the disk to return to its proper position. If it is effective in relieving the patient's pain, cervical traction may be done at home on an as-needed basis. Traction is discontinued immediately if it increases the patient's pain. Lumbar traction is not particularly effective because the lumbar paraspinal muscles are very large and strong. The amount of traction needed to overcome the muscular resistance can cause injury.

Medication. Muscle relaxants are often prescribed for patients who are experiencing spasm. These medications decrease pain by decreasing the spasm, helping the patient increase range of motion and activity. Muscle spasm is actually a protective mechanism. Muscles tighten and become painful, causing the patient to limit movement. This lessens the chance that the disk will be further injured. However, chronic spasm can cause tearing and scarring of the muscles. It is hard to predict which muscle relaxant will be most effective for a given patient. Patients should be warned that drowsiness is a common side effect of many muscle relaxants. They should be cautioned against driving or operating machinery until they determine how well they tolerate the medication. Diazepam is an effective muscle relaxant; however, it has a strong potential for addiction. Therefore it is usually used only if muscle spasm cannot be adequately treated with other medications.

Inflammation of the nerve root is caused by compression and irritation from the herniated disk. NSAIDs can be effective in reducing this inflammation, but there is no way of predicting response to a given drug. It may be necessary for the patient to try several nonsteroidal drugs before an effective one is found. Because several of these drugs are now available without prescription, the patient should be cautioned not to use a nonprescription NSAID at the same time as a prescription NSAID. Patients should be instructed to report any stomach upset to the clinician, because NSAIDs have the potential to cause gastric bleeding. Occasionally, oral steroids are used on a short-term basis for patients with severe inflammation that does not respond to other treatments. A rapidly tapering dose of steroid over 1 week is often prescribed. Steroids may also cause gastric upset, in addition to elevated serum glucose levels. Instruct patients with diabetes to monitor glucose levels closely and consult a physician if the levels are outside their normal parameters.

Epidural injections are an option for patients who are unable or do not wish to have surgery. A mixture of medications, typically a steroid, long-acting anesthetic, and long-acting pain reliever, is injected into the epidural space. The anesthetic provides immediate relief while the steroid reduces swelling for a longer-lasting effect. If relief is obtained, the injection can be repeated every 3 to 4 months.

The use of pain medication is a subject of concern in the treatment of patients with herniated disks. Opioids may be appropriate for short-term use. This includes patients who are trying conservative therapy or those who are not able to have surgery immediately. If surgery is not an option or is not successful, the condition may be a chronic source of pain. In that circumstance, the physician is unlikely to prescribe ongoing opioids. The patient, on the other hand, may wish to continue using opioids because they provide substantial pain relief. The physician and patient must discuss the potential complications of long-term opioid use, such as constipation, tolerance, and dependence. Patients taking opioids should not consume alcohol.

SURGICAL MANAGEMENT. Several types of surgery can be done. A **laminectomy** removes one of the laminae, the flat pieces of bone on each side of a vertebra. This may be done to relieve pressure or to gain access for removal of a herniated disk. A diskectomy removes the entire disk. A spinal fusion uses a bone graft to fuse two vertebrae together if the area is unstable. Surgery may be done through a microscope for less scarring and better recovery. Most patients are discharged within 24 hours of surgery. This is in contrast to several weeks in the hospital on bedrest, which was necessary in the recent past.

A diskectomy is generally done for a herniated cervical disk. This can be accomplished via an anterior or posterior approach. Most surgeons use the anterior approach for cervical herniations, because the muscles in the front of the neck are much smaller and more moveable than those in the back of the neck. Therefore there is less pain and muscle spasm following surgery. It is also safer than the posterior approach, which involves more maneuvering around the spinal cord.

Most surgeons replace the disk with bone or another material. This prevents collapse of the disk space and creates a spinal fusion. If bone is used, it may be harvested from the patient's iliac crest or donated from a cadaver. Mobility of the spine is lost in the area of a fusion. Spinal fusions may also be done to correct instability of the spine from other causes, such as scoliosis or degenerative disorders.

A posterior approach is used for a herniated lumbar disk. Typically the vertical incision is 1 to 2 inches long. It is necessary to pull some of the muscle away from the bone, which accounts for some of the postoperative pain that patients experience. A laminectomy is done, and the herniated portion of the disk is resected. The remainder of the disk continues to provide a cushion between the intravertebral bodies. The surgeon removes any free fragments and any disk material that appears unstable.

Percutaneous diskectomy involves insertion of a large needle into the disk under local anesthesia to aspirate herniated disk material. This technique is not used for severely

---

laminectomy: lamin—posterior portion of the vertebra + ectomy—surgical removal of

herniated disks. Laser disk surgery may be used to disintegrate the herniated tissue. Laparoscopic techniques may also be used.

Complications after Surgery

HEMORRHAGE. As with any surgery, intraoperative hemorrhage is possible. It is not common in disk surgery. If a postoperative hemorrhage occurs in a patient who has had an anterior cervical diskectomy, the airway may become occluded. The patient is monitored for bleeding from the incision and respiratory distress.

NERVE ROOT DAMAGE. If the nerve root is severed during surgery, the patient has loss of motor and sensory function in that distribution. This may result in decreased use of the extremity. If the nerve root is damaged or excessive scarring occurs, the patient may experience pain, weakness, or paresthesias. In some cases, physical therapy and NSAIDs may be effective in improving function and lessening pain.

REHERNIATION. Lumbar disks may reherniate. This can occur anywhere from 1 week to several years after the initial surgery. If the reherniation occurs within a few weeks to months after the first surgery, the patient usually undergoes another microdiskectomy. Reherniation of the cervical disk does not occur because the entire disk is removed.

HERNIATION OF ANOTHER DISK. Fusion of the cervical spine results in loss of movement at that motion segment. This can place increased stress on the disks above and below the fusion. This may increase the risk of another herniated disk, especially if the patient already has degeneration of other disks. The patient should be instructed to maintain an exercise program and to frequently move the spine through range-of-motion exercises.

Spinal Stenosis

Spinal stenosis is a condition in which the spinal canal compresses the spinal cord (Fig. 46–15). Arthritis is a major cause of spinal stenosis. The facet joints of the spine become inflamed and enlarged, narrowing the diameter of the spinal canal and compressing the spinal cord. Patients may complain of pain and weakness. Compression of the cervical portion of the spinal cord may result in hyperreflexia and weakness of the legs and arms.

A laminectomy may be done to relieve pressure on the spinal cord. The size of the incision depends on the number of vertebrae involved. Patients are typically in the hospital for 4 or 5 days. These patients are often elderly and may have concurrent illnesses. They may require inpatient rehabilitation before returning to their homes.

Nursing Process: The Patient Having Spinal Surgery
Preoperative Care

Routine preoperative care is appropriate for the patient undergoing spinal surgery. In addition to routine teaching, the patient is instructed in how to logroll following surgery.

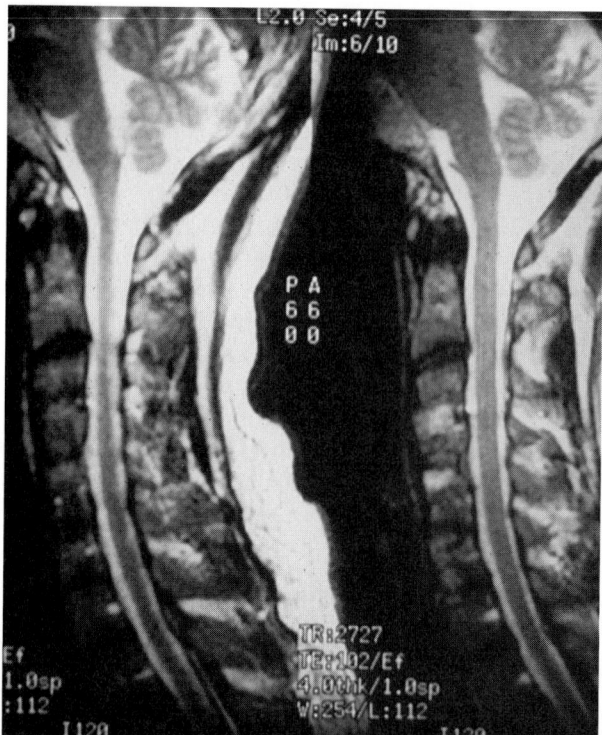

Figure 46-15 Stenosis of the cervical spine *(left)*. Compare with normal spinal column *(right)*.

This procedure involves keeping the body in alignment and rolling as a unit, without twisting the spine, to prevent injury to the operative site.

Postoperative Care

ASSESSMENT. In addition to routine postoperative assessment, the extremities are monitored for changes in circulation, movement, and sensation. Circulation is checked by the color, warmth, and presence of pulses in the extremity. Movement is assessed by asking the patient to move the extremity. Sensation is assessed by gently touching the patient's extremity and asking if feeling is present. Any changes are reported immediately to the physician because this may indicate nerve or circulatory damage. Pain is assessed. The pain that necessitated surgery should be relieved, but the patient may still have muscle and incisional pain. The patient should be reassured that it will gradually subside. The surgical dressing and drain if present are monitored for CSF drainage or bleeding. If bone was taken from a separate donor site, this site must also be monitored. Intake and output are measured to ensure that the patient is able to void. The physician is notified if the patient has difficulty voiding.

NURSING DIAGNOSIS, PLANNING, AND IMPLEMENTATION. Goals of nursing are to keep the patient safe and free from injury or complications and free from pain. Gradual return to normal physical activity is expected. Possible postoperative diagnoses include the following.

Pain Related to Surgical Procedure. If a local anesthetic was injected into the surgical site during surgery, the patient may not have pain immediately postoperatively. Muscle relaxants, analgesics, and NSAIDs are given as ordered. The patient is positioned in bed in correct body alignment and may have orders to be kept flat for 6 to 8 hours. A pillow between the legs when lying on the side promotes alignment and comfort.

Risk for Impaired Physical Mobility Related to Neuromuscular Impairment. The patient is assisted to logroll to get out of bed and ambulate on the first postoperative day. If spinal fusion has been done, the fused area of the spine will be immobile. The patient with a cervical laminectomy may have a soft cervical collar for neck support.

Risk for Impaired Urinary Elimination Related to Effects of Surgery. Patients may have difficulty voiding following lumbar surgery because of anesthesia, immobility, or occasionally because of nerve damage related to surgery. Getting up to go to the bathroom (or standing for men) and running water should promote bladder emptying. If difficulty urinating occurs, the physician should be contacted for an order for intermittent catheterization until the problem resolves.

EVALUATION. The patient is expected to be free of complications and pain, be able to urinate, be able to move all extremities, and return gradually to preillness activity.

▶ SPINAL CORD INJURIES

Injuries to the spinal cord affect people of all ages but take their greatest toll on young people. These injuries are characterized by decrease or loss of sensory and motor function below the level of the injury.

Pathophysiology

The spinal cord is made up of nerve fibers that allow communication between the brain and the rest of the body. Damage to the spinal cord results in interference with this communication process. Damage may be caused by bruising, tearing, cutting, edema, or bleeding into the cord. The damage may be caused by external forces or by fragments of fractured bone.

Causes and Types

The causes of spinal cord injury are similar to those of traumatic brain injury. It is not uncommon for a patient to have both a spinal cord injury and traumatic brain injury. Motor vehicle crashes, falls, and sports-related injuries are common causes. Diving into shallow water is often the cause of cervical cord injury. Assaults may cause cord injury if a knife or bullet penetrates the spinal cord.

Spinal cord injuries may be classified by location or by degree of damage to the cord. A complete spinal cord injury means that there is no motor or sensory function below the level of the injury. An incomplete lesion means that there is some function remaining. This does not necessarily mean that the remaining function will be useful to the patient. Some patients find that having areas where sensation is intact may be more painful than useful.

The cervical and lumbar portions of the spine are injured more often than the thoracic or sacral segments. This is because the cervical and lumbar areas are the most mobile portions of the spine.

Signs and Symptoms
Cervical Injuries

Cervical cord injuries can affect all four extremities, causing paralysis and paresthesias, impaired respiration, and loss of bowel and bladder control. Paralysis of all four extremities is called **quadriplegia;** weakness of all extremities is called **quadriparesis.** If the injury is at C3 or above, the injury is usually fatal because muscles used for breathing are paralyzed. An injury at the fourth or fifth cervical vertebrae affects breathing and may necessitate some type of ventilatory support. These patients typically need long-term assistance with activities of daily living (Fig. 46–16).

Thoracic and Lumbar Injuries

Thoracic and lumbar injuries affect the legs, bowel, and bladder. Paralysis of the legs is called **paraplegia;** weakness of the legs is called **paraparesis.** Sacral injuries affect bowel and bladder continence and may affect foot function. Individuals with thoracic, lumbar, and sacral injuries can usually learn to perform activities of daily living independently.

Spinal Shock

Spinal cord injury has a profound effect on the autonomic nervous system. Immediately following injury the cord below the injury stops functioning completely. This causes a disruption of the sympathetic nervous system, resulting in vasodilation, hypotension, and bradycardia (neurogenic shock or spinal shock). Dilation of the blood vessels allows more blood flow just beneath the skin. This blood cools and is circulated throughout the body, causing hypothermia. The patient is unable to maintain control of body temperature. Keep the patient covered as much as possible but avoid overheating. In addition, all reflexes below the level of the injury are lost, and retention of urine and feces occurs. Spinal shock can last from a week to many weeks in some patients.

Complications
Infection

Impaired respiratory effort, decreased cough, mechanical ventilation, and immobility all predispose the cervical cord–injured patient to pneumonia. Catheterization, whether indwelling or intermittent, places patients at risk for urinary tract infection.

---

quadriplegia: quad—four + plegia—paralysis
quadriparesis: quad—four + paresis—partial paralysis
paraplegia: para—beside + plegia—paralysis
paraparesis: para—beside + paresis—partial paralysis

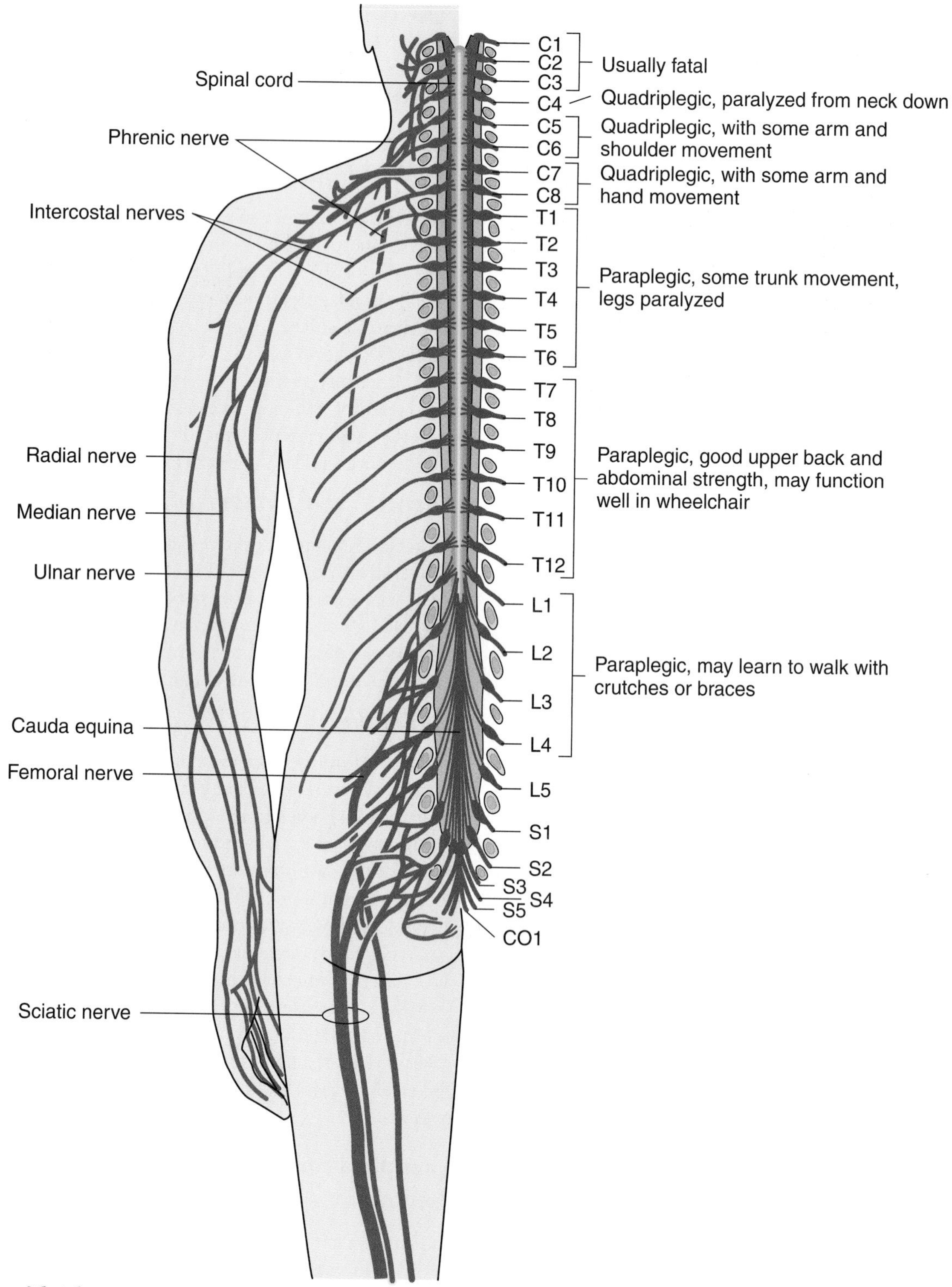

Figure 46-16   Spinal cord injury—quadriplegia versus paraplegia. (Modified from Scanlon, VC, and Sanders, T: Workbook for Essentials of Anatomy and Physiology, ed 2. FA Davis, Philadelphia, 1995, p 167, with permission.)

Deep Vein Thrombosis

Lack of movement in the legs inhibits normal blood circulation. Compression stockings, sequential compression devices, and subcutaneous heparin may be used separately or together to reduce the risk of deep vein thrombosis.

Orthostatic Hypotension

Most spinal cord–injured patients no longer have muscular function in their legs to promote venous return to the heart. They also have impaired vasoconstriction. This leads to pooling of the blood in the legs when the patient moves from a supine to a sitting position. If the movement is sudden, the patient may faint. Gradual elevation of the head, use of elastic stockings, and a reclining wheelchair help lessen this response.

Skin Breakdown

Patients or their caregivers must be diligent about relieving pressure on the skin by position changes and cushioning of bony prominences. Development of pressure ulcers can lead to infection and loss of skin, muscle, or bone. Treatment of pressure ulcers is time consuming and expensive and may interfere with work or school.

Renal Complications

Urinary tract infections are an ongoing concern to spinal cord–injured patients. Both urinary reflux and untreated urinary tract infections can cause permanent damage to the kidneys.

Depression and Substance Abuse

Patients with spinal cord injury have a higher than average incidence of depression and substance abuse. Both these factors can interfere with the patient's ability to care for himself or herself. Individual or family counseling may be helpful. Some rehabilitation centers have support groups for spinal cord–injured patients.

Autonomic Dysreflexia

This life-threatening complication occurs in patients with injuries above the T6 level. The spinal cord injury impairs the normal equilibrium between the sympathetic and parasympathetic autonomic nervous system. Some type of noxious stimuli below the spinal cord injury causes activation of the sympathetic system. This response continues unchecked because the parasympathetic responses cannot descend past the spinal cord injury.

The most common cause of autonomic **dysreflexia** is bladder distention. Other causes include bowel impaction, urinary tract infection, ingrown toenails, pressure ulcers, pain, and labor. Stimulation of the sympathetic nervous system results in cool, pale skin; gooseflesh; and vasoconstriction below the level of the injury. Blood pressure may rise as high as 300 mm Hg systolic. The parasympathetic response results in vasodilation, causing flushing and diaphoresis above the lesion, and bradycardia as low as 30 beats per minute. The patient complains of a pounding headache and nasal congestion secondary to the dilated blood vessels.

Diagnostic Tests

Plain radiographs are done to identify fractures or displacement of vertebrae. A CT scan is also useful for identifying fractures. MRI may demonstrate lesions within the cord.

---

dysreflexia: dys—abnormal + reflexia—reflex activity

Treatment

Patients with spinal cord injuries typically are brought to the emergency department. They should be kept immobilized until they are assessed by the physician. If injury to the spinal cord is detected, the patient needs to remain immobilized.

Emergency Medical Management

Emergency management involves careful monitoring of vital signs and airway and keeping the patient immobilized. Intubation and mechanical ventilation may be necessary. Intravenous normal saline may be used for fluid replacement. The physician does not rely on fluid administration alone to correct hypotension. It is possible to administer enough fluid to cause pulmonary edema and not correct the hypotension. Vasoactive drugs may be required. Various medications to reduce the extent of injury, including intravenous methylprednisolone (a steroid), are currently being researched.

RESPIRATORY MANAGEMENT. Patients with injuries above C4–5 have some degree of respiratory impairment. The patient may require a tracheostomy and continuous mechanical ventilation or require a ventilator only at night or when fatigued. Some patients are able to breathe by using a phrenic nerve stimulator. This device, similar to a pacemaker, artificially stimulates the phrenic nerve, causing the diaphragm to move. These patients use a mechanical ventilator at night. This lessens the stress on the phrenic nerve and removes the risk of the system failing while the patient is asleep.

Patients may be breathing independently when they first arrive in the emergency department and then experience respiratory compromise as the spinal cord becomes edematous. Edema can compress the spinal cord above the lesion, leading to symptoms at a higher level. This deterioration is usually temporary. Fatigue of the accessory muscles may also cause respiratory compromise. The intercostal muscles are not normally of major importance in respiration. However, if the diaphragm is paralyzed, the intercostal muscles become very important. As these muscles fatigue, the patient's breathing becomes shallow and rapid. Elective intubation and mechanical ventilation protect the patient from expending huge amounts of energy trying to breathe. Feeling their breathing becoming more labored is terrifying to these patients, and they need to be reassured that it is probably a temporary setback. As the edema recedes and the accessory muscles become stronger, the patient is weaned from the ventilator.

GASTROINTESTINAL MANAGEMENT. Absence of bowel sounds is a common finding on examination. Oral or enteral feedings are not started until bowel function resumes. The metabolic needs of the patients are influenced by the work of breathing and the extent of other injuries. If positioning or paralytic ileus precludes oral or enteral feedings, hyperalimentation is begun.

GENITOURINARY MANAGEMENT. An indwelling urinary catheter is placed to prevent bladder distention and protect skin integrity until spinal shock resolves. Once it is determined what degree of hand function the patient will have, a bladder management program is devised.

IMMOBILIZATION. The cervical spine may be immobilized with skeletal traction such as Crutchfield or Gardner-Wells tongs (Fig. 46–17). Some patients have a halo brace, a device that attaches to the skull with four small pins. The skull ring attaches to a rigid plastic vest by four poles (Fig. 46–18). This device keeps the head and neck immobile while fusion and healing take place. The advantage over traction is that the patient is not confined to bed.

Surgical Management

The goal of surgery following spinal cord injury is to stabilize the bony elements of the spine and relieve pressure on the spinal cord. Surgery may or may not improve functional outcome.

Stabilization of the spine allows for earlier mobilization of the patient. This decreases the risk of complications from immobility and quickens the transition to rehabilitation. Patients who have been in cervical traction before surgery may be placed in a halo brace postoperatively.

Unstable thoracic and lumbar fractures may also be treated with surgical implantation of rods to stabilize the spine. It is more difficult to stabilize these areas in the postoperative recovery period. Patients may wear a supportive corset, a rigid brace, or occasionally a body cast to supplement the support provided by the internal fixation devices (Box 46–8). For more information, visit the Spinal Cord Injury Information Network at http://spinalcord.uab.edu.

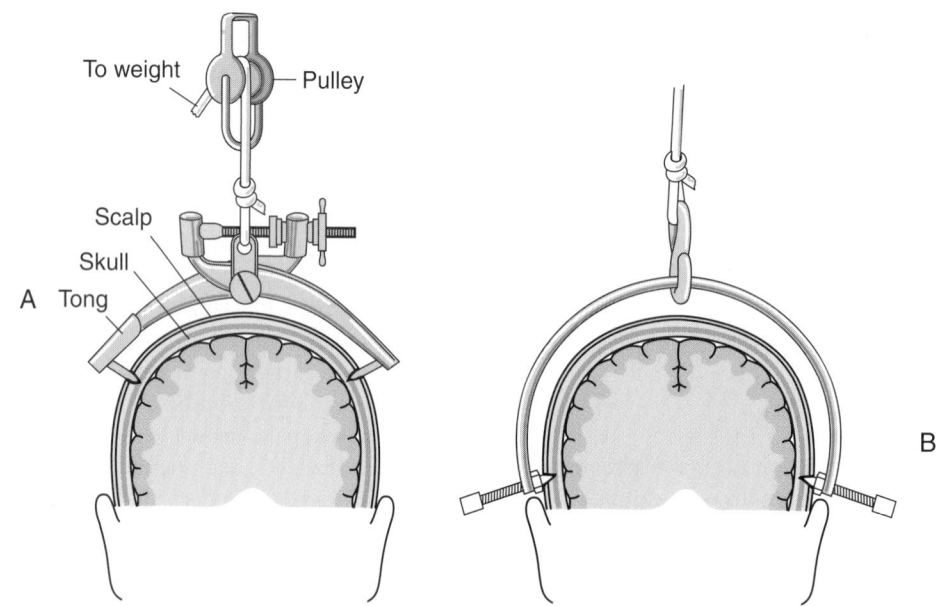

Figure 46-17 Skeletal traction for cervical injuries. *(A)* Crutchfield tongs. *(B)* Gardner-Wells tongs.

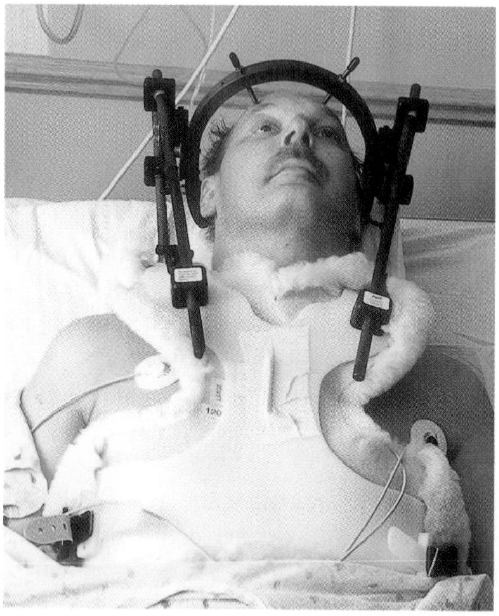

Figure 46-18 Halo brace.

BOX 46-8 Spinal Cord Injury Summary

Symptoms

Flaccid paralysis and paresthesia (dependent on level of the lesion)
Loss of reflex activity below the level of the lesion
Spinal shock initially (bradycardia, hypotension)
Risk for autonomic dysreflexia (injuries above sixth thoracic vertebra)

Diagnostic Tests

Radiograph
CT scan
MRI

Therapeutic Management

Immobilization
Maintenance of airway and respiratory status
Bowel and bladder training
Nutrition/diet
Activity/rehabilitation
Prevention of dysreflexia
Prevention of skin breakdown
Sexual counseling
Education

Nursing Process

Assessment

Patients with spinal cord injury need ongoing evaluation of all body systems. Frequent neurological and respiratory assessments are done. Early assessment of the patient's support system can help with discharge and rehabilitation planning.

Nursing Diagnosis, Planning, and Implementation

Initial goals for the patient include maintenance of safety and prevention of complications. Long-term goals include rehabilitation and maximizing remaining function.

IMPAIRED PHYSICAL MOBILITY RELATED TO INJURY. Range-of-motion exercises help prevent contractures. Regular turning and repositioning help prevent respiratory and other complications. Some patients are placed on special beds (such as the Roto-Rest bed) that move the patient frequently without the need for turning. Splints may be used to maintain functional positioning of extremities. This may include the use of high-top tennis shoes to prevent footdrop. If the patient is ambulatory, he or she should not walk alone initially. Special care should be provided when ambulating the patient with a halo brace. The halo brace alters the patient's center of gravity and requires an adjustment of the patient's sense of balance. Falling while in the halo brace can cause further injury to the spinal cord.

IMPAIRED URINARY ELIMINATION RELATED TO NERVE DAMAGE. Male patients may use an external, condom type of catheter if the bladder empties reflexively when full. If the bladder does not empty reflexively or if the patient does not wish to wear a leg bag to collect urine, the physician may order intermittent catheterization. Clean technique is used to intermittently insert a catheter and empty the bladder. (See Chapter 34.) Male patients with good hand function can be taught to catheterize themselves. Patients with poor hand function may need a caregiver to perform intermittent catheterization or use an indwelling catheter.

Female patients with good hand function may still find self-catheterization difficult because of positioning problems. They may choose to use indwelling catheters. External urine collection devices for female patients are available but may be cumbersome.

Reliance on incontinence pads is discouraged for all patients because of the threat to skin integrity. The importance of adequate fluid intake must be stressed to patients. Some individuals limit fluid intake in an effort to lessen the chances of incontinence. This predisposes the patient to constipation, urinary tract infection, and renal calculi.

CONSTIPATION RELATED TO IMMOBILITY AND NERVE DAMAGE. Once oral feedings are begun, a bowel management program should be instituted. The patient has decreased or absent sphincter tone. This, combined with an inability to detect the need to defecate, puts the patient at risk for incontinence. Slowed bowel motility, as well as generalized immobility, contributes to constipation. A high-fiber diet with adequate fluid intake is important. Use of a suppository on a scheduled daily or every-other-day basis enables most patients to maintain bowel continence.

SELF-CARE DEFICIT RELATED TO PARALYSIS. Explaining the rationale for nursing activities prepares the patient and significant others to assume responsibility for care. Encourage the patient and significant others to participate in hands-on care as much as possible. If the patient will not be able to perform self-care, assist him or her to learn to direct care. Physical and occupational therapists can help the patient adapt to a wheelchair or other mobility aids. Most patients spend some time in a rehabilitation facility to learn to function independently. Some patients may require long-term care.

RISK FOR INEFFECTIVE COPING RELATED TO LIFE CRISIS. The psychosocial impact of a spinal cord injury is devastating. Patients may be afraid of dying yet express the feeling that they would be better off dead than being paralyzed. These types of statements are difficult for significant others to cope with. Encourage the patient to focus on reasons for living (e.g., seeing children or grandchildren grow up or having a loving partner). Consult a social worker or pastoral care worker who is familiar with spinal cord–injured patients. Support groups may also be helpful.

RISK FOR AUTONOMIC DYSREFLEXIA RELATED TO STIMULI BELOW THE LEVEL OF INJURY. If you suspect autonomic dysreflexia, immediately take the patient's blood pressure and continue to monitor it every 5 minutes. Remember that patients with spinal cord injury are typically hypotensive, so a finding of even mild hypertension may represent a dramatic increase from their baseline blood pressure.

Uncontrolled blood pressure may cause seizures, intracerebral hemorrhage, or death. The goal of treatment is to identify the cause and relieve it without increasing the sympathetic nervous system response. Place the patient in a Fowler's position to utilize the effect of orthostasis to control blood pressure. Evaluate the indwelling catheter for patency. If it is not patent or a catheter is not in place, obtain an order to insert one immediately. Monitor blood pressure during catheterization.

A rectal examination is performed to determine if an impaction is present. Apply anesthetic ointment to the rectum before disimpaction, because further rectal stimulation may exacerbate symptoms. Simultaneously monitor blood pressure and stop disimpaction if the blood pressure increases.

If bowel or bladder distention is not present, examine the patient for other causative mechanisms. If a cause cannot be identified, or removal of the cause does not relieve hypertension, notify the physician immediately; an antihypertensive agent may be ordered. If hypertension is treated with medication, remember that the blood pressure may decrease rapidly once the cause of the autonomic dysreflexia is corrected. Continue to carefully monitor blood pressure.

Once the acute episode is past, work with significant others to devise a plan to prevent reoccurrence. The patient should be taught how to direct caregivers in treating autonomic dysreflexia.

INEFFECTIVE AIRWAY CLEARANCE RELATED TO INEFFECTIVE COUGH AND DECREASED MUSCLE CONTROL. Patients with cervical injuries have difficulty clearing secretions. They no longer have adequate muscle strength to cough effectively. Initially, suctioning is required to keep the airway clear. Once the patient is stable, an assisted cough may be used to clear secretions. This involves the nurse or other caregiver gently pushing upward and inward on the patient's chest while the patient coughs as strongly as possible. It is similar to the Heimlich maneuver but not as forceful. Humidified air and oral or enteral fluids help keep secretions thin and mobile.

RISK FOR IMPAIRED SKIN INTEGRITY RELATED TO IMMOBILITY AND POSSIBLE PARESTHESIAS. Extreme care must be taken to protect the skin of the injured patient, beginning with the initial emergency treatment. In the emergency department be sure to remove anything between the patient and the backboard. Patients have developed pressure ulcers from lying on keys or other objects in their pockets. If on a Roto-Rest bed, make sure the patient is not sliding as the bed turns. This could cause shearing of the skin. When permitted by the physician, turn the patient frequently and assess bony prominences for redness. Ensure that the patient's extremities do not get caught in side rails or wheelchair spokes. If a patient is in traction or a halo brace, assess pin sites frequently. Keep the sites clean and dry and report any sign of infection.

RISK FOR SEXUAL DYSFUNCTION RELATED TO AUTONOMIC NERVOUS SYSTEM DYSFUNCTION. Male patients with paraplegia usually have difficulty achieving and maintaining an erection. Male patients with quadriplegia may develop an erection during any penile stimulation. This includes catheterization and can be embarrassing to the patient. If this occurs, discontinue the procedure and continue at a later time if possible. It should be treated with a matter-of-fact attitude. Males with spinal cord injuries do not ejaculate in the normal manner. Consultation with a fertility specialist or urologist may provide some help for conception if desired.

Spinal cord injury does not impair female fertility. Patients who wish to become pregnant should seek an obstetrician familiar with spinal cord injuries. See Table 46–7 for contraception for females with spinal cord injuries.

RISK FOR INEFFECTIVE ROLE PERFORMANCE RELATED TO EFFECTS OF INJURY. Interpersonal relationships are significantly stressed by spinal cord injury. Both short-term and long-term relationships may become stronger or disintegrate. Patients and significant others should be encouraged to draw support from all available sources. Friends, family, and members of the patient's religious affiliation can provide emotional and physical help.

Loss of income may be temporary or permanent. Financial concerns add to the burden of spinal cord injury. Not all insurance policies cover the extensive inpatient rehabilita-

TABLE 46-7	BIRTH CONTROL ISSUES FOR PATIENTS WITH SPINAL CORD INJURY
Method	**Comments**
Oral contraceptives	Contraindicated because of the risk of deep vein thrombosis
Diaphragm	May be difficult for a patient with poor hand function to insert
Intrauterine device	Patient may not feel IUD move out of position; patient may not feel perforation of uterus
Norplant	No contraindications
Condom	No contraindications

tion needed by spinal cord–injured patients. Uninsured patients may have a difficult time finding a rehabilitation program that will accept them. Adaptive equipment is expensive and may not be covered by insurance. A social worker can help the patient gain access to appropriate assistance.

Patient Education

Education of the spinal cord–injured patient is an ongoing process. It begins with a basic explanation of the anatomy and physiology of the spine. Treatment of spinal cord injury and prognosis are early learning needs.

As the patient's condition stabilizes, teaching focuses on caring for the patient. This can seem overwhelming to the patient and caregiver. Break tasks down into simple steps. Focus on one body system at a time. For example, do not try to teach suctioning and bowel care at the same time. Begin with the basics; extensive teaching will be done in the rehabilitation setting.

Be aware that performing tasks such as catheterization or bowel care may interfere with feelings of intimacy between partners. Encourage partners to verbalize these feelings and make alternative arrangements for care if possible. Some patients with the financial resources to do so may choose to hire an attendant rather than rely on significant others for personal care.

Both patients and caregivers should be involved in determining contingency plans. These include what to do in the event of a power failure, fire, or illness of the caregiver.

Spinal cord–injured patients experience the same basic health care needs as noninjured individuals. Encourage the patient to establish a relationship with a primary practitioner who is familiar with spinal cord injury. This facilitates the patient's adaptation to changes in his or her condition. Female patients may find that they do not fit in the wheelchair as pregnancy advances. A patient who requires cardiac bypass may be unable to transfer or propel the wheelchair because use of the arms is impaired by the sternal incision. These and many other situations require creativity and flexibility on the part of the patient, caregiver, and health professional.

Evaluation

Avoidance of preventable complications is one sign of a successful care plan. The patient should have regular bowel and bladder function. Mobility is maximized based on the patient's rehabilitation potential. The patient and significant others should verbalize basic understanding of spinal cord injury and self-care.

▶ DEGENERATIVE NEUROMUSCULAR DISORDERS

Degenerative nervous system disorders involve the extrapyramidal tracts and basal ganglia in the brain. The extrapyramidal tracts are the efferent pathways outside the pyramidal system that connect the cerebral cortex with the spinal nerve pathways. Because these are motor pathways, movement rather than sensation is affected. The resulting symptoms can include rigidity, tremor, and abnormal movements.

Parkinson's Disease

Parkinson's disease is a chronic degenerative movement disorder that arises in the basal ganglia in the cerebrum. It usually begins in the fourth or fifth decade of life, with symptoms becoming progressively worse as the patient ages. The disease is characterized by tremors, changes in posture and gait, rigidity, and slowness of movements. Approximately 1 million people in the United States are currently living with Parkinson's disease. It occurs most often in people older than age 50.

Pathophysiology

The substantia nigra is a group of cells located within the basal ganglia, which is situated deep within the brain. These cells are responsible for the production of dopamine, an inhibitory neurotransmitter. Dopamine facilitates the transmission of impulses from one neuron to another. Parkinson's disease is caused by widespread destruction of the cells of the substantia nigra, resulting in decreased dopamine production. Loss of dopamine function results in impairment of semiautomatic movements. Dopamine also plays a part in the fight-or-flight mechanism by activating epinephrine and norepinephrine.

Acetylcholine, an excitatory neurotransmitter, is secreted normally in individuals with Parkinson's disease. The normal counterbalance of acetylcholine and dopamine is interrupted in these patients, causing a relative excess of acetylcholine, which results in the tremor, muscle rigidity, and **akinesia** (loss of muscle movement) characteristic of Parkinson's disease.

Etiology

The etiology of Parkinson's disease is unknown. It was first described in 1817 by London surgeon James Parkinson. Although scientists now know that the symptoms are caused by death of dopamine-producing cells in the substantia nigra, they do not know what causes the cells to die. Parkinson-like symptoms, referred to as parkinsonism, may be associated with use of certain drugs, such as phenothiazines.

Parkinsonism was also linked to an outbreak of encephalitis in the 1920s.

Signs and Symptoms

The onset of symptoms in patients with Parkinson's disease is usually gradual and subtle. A substantial percentage of the dopamine-producing cells are nonfunctional before the patient becomes symptomatic. Symptoms may be mistakenly attributed to aging or fatigue. In retrospect, patients and their significant others often identify a long period in which symptoms were present but not identified as symptoms of Parkinson's disease.

The primary symptoms of Parkinson's disease are muscular rigidity, **bradykinesia** (slow movement) or akinesia, changes in posture, and tremors. The brain is no longer able to direct the muscles to perform in the usual manner. This lack of communication between the brain and the muscles can have a profound impact on the patient's ability to ambulate safely, perform ADLs and job functions, or enjoy leisure activities. The symptoms may also have a significant negative impact on the patient's self-esteem.

The patient may have difficulty in initiating movement; this may be particularly apparent when the patient attempts to start walking, rise from a sitting position, or begin dressing. Because considerable effort is required to move the rigid muscles, the patient performs voluntary movements very slowly.

The extensor muscles are more affected by Parkinson's disease than the flexor muscles. This impaired function of the extensor muscles results in the stooped posture typical of patients with Parkinson's disease (Fig. 46–19). Flexion of the hips, knees, and neck shifts the center of gravity forward. The gait is characterized by shuffling, short steps. This

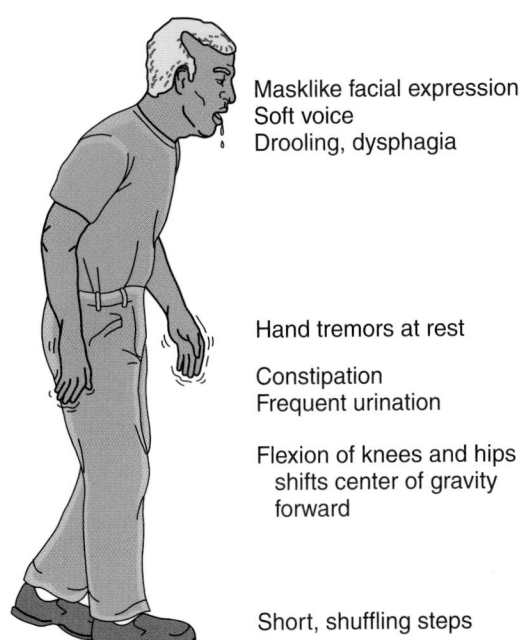

Masklike facial expression
Soft voice
Drooling, dysphagia

Hand tremors at rest

Constipation
Frequent urination

Flexion of knees and hips shifts center of gravity forward

Short, shuffling steps

Figure 46-19 Manifestations of Parkinson's disease.

---

akinesia: a—not + kinesia—movement

bradykinesia: brady—slow + kinesia—movement

shuffling gait may increase in speed once the patient finally gets walking, and the patient may have difficulty stopping. The patient maintains a broad base when making turns to try to compensate for imbalance. These changes place patients at high risk for falls. Slowness of movement and stiff muscles make it much harder for patients to catch themselves if they start to fall or to relax the muscles to minimize injury.

Tremors typically begin in the hand and then progress to the **ipsilateral** foot. In most patients the tremor then moves to the **contralateral** side. Many patients identify one side of the body as being more affected by the tremor than the other. Tremor of the hand has been described as a pill-rolling tremor; the thumb typically moves back and forth across the fingers and looks like the patient is rolling a pill. Tremors typically lessen or disappear during movement and are more noticeable when the extremity is at rest or when trying to hold an object still (this is called a resting tremor). The tremors disappear when the patient is asleep. The inability to hold an object still can make simple acts such as drinking a glass of water or reading a book nearly impossible. The signs and symptoms of Parkinson's disease tend to increase in severity when the patient becomes fatigued.

Another type of tremor, a benign familial (or essential) tremor, may sometimes be mistaken for Parkinson's disease. Treatment is different for each. See Table 46–8 for differentiation of these tremors. The secondary symptoms of Parkinson's disease include generalized weakness, muscle fatigue and cramping, and difficulty with fine motor activities. This fine motor dysfunction may make it difficult for the patient to button a shirt or tie shoes. Handwriting typically deteriorates as the disease progresses. A soft, monotone voice and masklike facial expression may make the patient appear to be lacking in emotional responses. It may be necessary to ask patients about their emotional status and help them develop ways of expressing their emotions. The normal blink response is diminished, so the patient and significant others must be educated about eye care to prevent corneal abrasions.

TABLE 46–8	Symptoms of Parkinson's Disease versus Essential Tremor	
Disease	**Parkinson's Tremor**	**Benign Familial**
Resting tremor	Yes	No
Intention tremor (with movement)	No	Yes
Pill-rolling tremor	Yes	No
Head/voice tremor	No	Yes
Relieved with beta-blocking medication (propranolol)	No	Yes
Relieved with anti-Parkinson's medications	Yes	No

ipsilateral: ipsi—same + lateral—side
contralateral: contra—opposite + lateral—side

Dysfunction of the autonomic system may be manifested by diaphoresis, constipation, orthostatic hypotension, drooling, dysphagia, seborrhea, and frequent urination. Patients who experience seborrhea and diaphoresis need frequent attention to personal hygiene. Drooling and dysphagia may make the patient reluctant to appear in public. Slowness in initiating walking, balance problems, and frequent urination place the patient at risk for urinary incontinence, which may also increase the patient's reluctance to leave home.

Late in the disease, mental function may become slowed and the patient may become demented. This is compounded by the side effects of many anti-Parkinson's drugs. Death is usually from complications of immobility.

Complications

The most typical acute complications of Parkinson's disease are related to the patient's difficulties with mobility and balance. These patients are very prone to falls, which may result in injuries ranging from bruises or fractures to head or spinal cord injuries. Constipation is common because of decreased activity, diminished ability to take in food and fluids, and side effects of anticholinergic medications. Patients are encouraged to increase the fiber and roughage in their diet. If constipation is not alleviated by dietary modifications, the patient may need to use stool softeners. The patient should be counseled not to rely on laxatives or enemas.

Muscular rigidity and bradykinesia contribute to joint immobility, which decreases patients' ability to ambulate and care for themselves. Position changes may be painful for patients. A turning sheet and adequate personnel are necessary when turning a patient in bed to prevent stress on the joints. Tremors interfere with ADLs, consume immense amounts of energy, and may prevent the patient from working or performing leisure activities. Swallowing may become so impaired that enteral (tube) feeding is required. Depression is a common complication at any stage of Parkinson's disease and may compromise communication, ability to learn, and performance of ADLs. Patients may require counseling or antidepressants.

Diagnostic Tests

No specific tests are used to diagnose Parkinson's disease. The diagnosis is based on the history given by the patient and a thorough physical examination.

Medical Treatment

There is no cure for Parkinson's disease. Treatment is aimed at controlling symptoms and maximizing the patient's functional level. Drugs used to control symptoms are listed in Table 46–9.

Many patients with Parkinson's disease experience fluctuations in motor function related to their drug therapy. This is referred to as the on-off phenomenon. Patients may experience a decreased response to levodopa, or off period, particularly as the dose is wearing off. As the disease progresses, patients may notice that the off periods become less

TABLE 46-9	Medications Used to Treat Parkinson's Disease		
Class	**Medication**	**Action**	**Side Effects/Comments**
Anticholinergic	Trihexyphenidyl (Artane)	Blocks the action of acetylcholine to control tremor and salivation.	Urine retention, dry mouth, constipation, blurring of memory, dizziness, confusion
Dopamine agonists	Amantadine (Symmetrel)	Facilitates production and secretion of dopamine.	Leg edema, hypotension, dizziness, confusion
	Levodopa (L-Dopa)	Levodopa is converted into dopamine in the brain by the amino acid decarboxylase. Reduces tremor, rigidity, and bradykinesia.	May cause nausea, vomiting, dyskinesias Breakdown products of protein metabolism compete with levodopa for transport from intestine to brain Take 15–30 minutes before meals and minimize protein intake during active times
	Levodopa/carbidopa combination (Sinemet)	Carbidopa is a peripheral decarboxylase inhibitor that prevents peripheral breakdown of levodopa so more is available in the CNS.	
	Pramipexole (Mirapex)	Improves motor function; may protect dopamine neurons.	Nausea, dizziness, weakness; may cause sudden excessive sleepiness
Monamine oxidase B inhibitor	Selegiline (Eldepryl)	Blocks the metabolism of central dopamine, increasing dopamine in CNS.	Nausea, dizziness, confusion; may have potential to slow progression of Parkinson's disease
COMT inhibitor	Entacapone (Comtan)	Blocks the enzyme COMT to prevent breakdown of levodopa, prolonging levodopa action. For use with Sinemet.	Dyskinesias, orthostatic hypotension, hallucinations, nausea, yellow-orange urine

predictable and occur more rapidly. The patient may have a delayed or absent response to the next dose of levodopa, resulting in the patient being stuck in the off stage and being significantly disabled for that period. Fluctuations in motor function may be accompanied by other symptoms such as pain, diaphoresis, anxiety attacks, hallucinations, or mood swings. These symptoms significantly increase the disability associated with the episodes.

Patients who are taking maximum doses of medication for Parkinson's symptoms may benefit from a "drug holiday." During a drug holiday, patients are taken off all drugs for a time, then restarted on lower doses. Hospitalization may be necessary during this time to maintain patient safety.

Surgical Treatment

Pallidotomy is an option for patients whose rigidity, tremor, and bradykinesia are uncontrollable by medical management. During this stereotactic procedure, a destructive lesion is placed in the basal ganglia. The surgery is only performed on one side at a time. The patient remains awake during the surgery to make sure that the lesion is being placed in the appropriate location. These patients need a great deal of education and support before and during the surgery. Some centers are experimenting with implanting fetal tissue or adrenal gland tissue in the brain to produce dopamine. For more information, visit the National Parkinson Foundation at www.parkinson.org.

Nursing Process: The Patient with Parkinson's Disease

ASSESSMENT. Assess the patient for symptoms of Parkinson's disease and their effect on level of functioning. Observe ability to move, walk, and perform ADLs. Determine risk for injury related to immobility or falls. Assess nutritional status and condition of skin. Identify presence of confusion and side effects of medications. Psychosocial assessment includes the patient's and caregiver's response to the disease, coping strategies, and support systems.

NURSING DIAGNOSIS, PLANNING, AND IMPLEMENTATION. The patient with Parkinson's disease is at risk for many problems. Typical diagnoses include the following.

Impaired Physical Mobility Related to Muscle Stiffness and Tremor. Patients often plan their daily activities based on anticipated response to their medications. This allows them to be as active as possible within the restrictions of the medication schedule. Encourage patients to determine their own best schedule for activities that require mobility. For patients in skilled nursing facilities, plan leisure activities around the patient's most active times.

Physical and occupational therapy can help maintain mobility, provide assistive devices, and provide diversional activities. Provide assistance with range-of-motion exercises. Teach patients who have difficulty initiating walking to pick up their foot as though attempting to step over something to take the first step. It may also help to take several steps in place before starting to walk.

Self-Care Deficit Related to Reduced Mobility. The most common setting for treatment of the patient with Parkinson's disease is the home. Encourage the patient to participate in ADLs as much as possible. The occupational therapist can assist with devices and strategies for maintaining independence. Patients have usually developed their own coping strategies, such as wearing clothing without buttons or shoes with adherent fasteners. As self-care abilities further decline, caregivers provide more assistance with ADLs.

As the patient ages, so do the significant others who are providing care. The point may be reached at which the caregiver is no longer able to meet the increasing needs of the patient. The decision to place the patient in a skilled nursing facility is extremely difficult and emotional. Every effort should be made to retain as much of the patient's usual routine as possible during hospitalization or admission to a skilled nursing facility.

Imbalanced Nutrition Related to Dysphagia and Reduced Mobility. Assist patients to open packages and prepare meals, so they can feed themselves if at all possible. If the patient has a severe tremor, a spoon may be safer than a fork for self-feeding. Finger foods may also be helpful. A cup with a lid and spout can help minimize spilling. Patients in the advanced stages of Parkinson's disease are at high risk for aspiration because they have difficulty swallowing. Adding thickening agents to liquids and assisting the patient to a chair or high Fowler's position for meals may help prevent aspiration. Meals that are high in fiber help prevent constipation. Small, frequent meals may be less overwhelming to the patient who must eat slowly.

Disturbed Thought Processes Related to Effects of Disease and Medication. Memory impairment is one of the most distressing symptoms that patients with Parkinson's disease experience. Many patients state that memory impairment and loss of social outlets are more troublesome than the physical impairments. Assist the patient to devise coping methods, such as written daily schedules, calendars, and reminders to take medications. If the patient becomes confused, frequent reorientation may be helpful.

Caregiver Role Strain Related to Demands of Caring for Patient. The significant others and caregivers of the patient with Parkinson's disease should be included in the plan of care. Encourage caregivers to utilize all community, personal, and governmental support systems available. Caregivers may need to be reminded that if they neglect their own health, both physical and mental, it will have a negative impact on the patient as well. Some caregivers may only need an hour or two away on an occasional basis. Others may require a more extended break. Options for relief from caregiving range from having a friend or neighbor visit to employing a home health aide or utilizing adult day care on a part-time or full-time basis. Some skilled care facilities offer respite care, in which the patient is admitted for a short time. This may be a viable option for caregivers

who must be hospitalized for their own health care. A social worker may be able to assist with identification of resources.

Risk for Injury Related to Reduced Mobility and Balance. The patient is at risk for injury from falls related to problems with mobility. If the patient is in the hospital or extended care facility, keep the call light within reach at all times. The bed should be kept in the low position, with side rails raised. Restraints should be avoided, but alarm systems are available that alert the staff that the patient is getting up. The environment should be kept free from clutter, throw rugs, or other items that may cause a patient to trip. Remind the patient to request assistance with ambulation. Walkers and other assistive devices may be helpful.

CRITICAL THINKING

Ms. Simpson

Ms. Simpson is a 47-year-old Caucasian female. She has had Parkinson's disease for the last 5 years, and the symptoms are becoming progressively worse. She is now admitted for a urinary tract infection.

1. What problems do you foresee when caring for Ms. Simpson?
2. What safety measures should you implement?

Answers at end of chapter.

Huntington's Disease

Huntington's disease is a progressive, hereditary, degenerative, incurable neurological disorder. It was first described in 1872 by George Huntington, a general practitioner from New York. The uncontrolled movements associated with Huntington's disease caused some sufferers in the seventeenth century to be accused of and executed for witchcraft. Many of the cases around the world can be traced back to specific individuals.

Pathophysiology and Etiology

Huntington's disease is inherited in the autosomal dominant manner, meaning that each offspring of an affected parent has a 50-percent chance of inheriting the disorder. It is uncertain what caused the mutation of the gene responsible for Huntington's disease. Structurally the disease is characterized by degeneration of the corpus striatum, caudate nucleus, and other deep nuclei of the brain and portions of the cerebral cortex.

Signs and Symptoms

Signs and symptoms develop slowly and become progressively more apparent. Cognitive signs may be noticed before movement problems. Patients who are not aware of their hereditary risk for Huntington's disease may be incorrectly diagnosed as being mentally ill or alcoholic.

The patient may display personality changes and inappropriate behavior. The patient may be euphoric or irritable and may rapidly alternate between moods. Paranoia is common, and behavior may become violent as **dementia** (mental impairment) worsens. The patient eventually becomes so demented that he or she is incontinent and totally dependent on others for care. These symptoms are difficult for caregivers, whether family members or professionals, to cope with. They are particularly devastating for offspring, who may or may not know whether they have inherited the disease.

Physical symptoms also develop slowly. Huntington's disease is characterized by involuntary, irregular, jerky, dancelike (choreiform) movements. Initially these symptoms may take the form of mild fidgeting and facial grimacing. In the early stages of the disease the patient may try to cover the movements by incorporating them into a voluntary movement such as crossing the arms or scratching. The involuntary movements usually start in the arms, face, and neck and progressively involve the remainder of the body. Patients display hesitant speech, eye blinking, irregular trunk movements, abnormal tilt of the head, and constant motion (Fig. 46–20). The gait is wide, and the patient may appear to be dancing. Emotional upset, stress, or trying to perform a voluntary task can significantly increase the severity and rate of the abnormal movements. The movements typically diminish or disappear during sleep. Dysphagia may significantly impair the patient's nutritional status.

Depression and suicide are common in the earlier stages of the disease. As the disease progresses, the patient becomes more and more dependent. Aspiration resulting in respiratory failure is the primary cause of death. Life span following diagnosis is about 10 to 20 years.

Diagnostic Tests

Huntington's disease has typically been diagnosed based on the clinical examination and a family history of the disease. MRI or CT may be helpful. Genetic testing is available for prenatal use and to determine if an individual has Huntington's disease before he or she becomes symptomatic. This is a significant breakthrough because Huntington's disease does not become symptomatic until patients are in their thirties or forties, when they may already have children who may be affected.

Medical Treatment

Because there is no cure, treatment of Huntington's disease focuses on minimizing symptoms and preventing complications. Antipsychotic, antidepressant, and antichorea drugs may be used to treat both the involuntary movements and behavioral outbursts. Some medical centers are transplanting fetal nerve tissue in patients with Huntington's disease. For more information, visit the Huntington's Disease Society of America at www.kumc.edu/hospital/huntingtons.

Nursing Management

Patients with Huntington's disease are typically cared for on an outpatient basis. When a patient with Huntington's disease is admitted to an inpatient facility, it is important to obtain as much information as possible about that person's response to medication, daily routine, and emotional and cognitive functioning from the caregivers. For example, knowing that a certain patient is intensely afraid of bathtubs but willingly takes showers can prevent unnecessary struggles and outbursts. Providing some objects from home may make the new environment seem less threatening. The caregivers may relate that the patient has better cognitive functioning at a particular time of day. As the dementia progresses, the patient responds less to attempts at reasoning. Giving directions in a calm but firm tone may help the patient cooperate with activities. The environment should be modified to keep the patient safe. Keep in mind that forceful, involuntary movements of the patient's extremities can happen at any time. These movements should never be misinterpreted as an attempt to harm caregivers.

Difficulty swallowing typically begins toward the middle of the disease course. Patients exhibit trouble swallowing liquids in particular. At this stage it may still be possible to teach the patient to hold the chin down to the chest while swallowing, which lessens the chance of aspiration. Patients

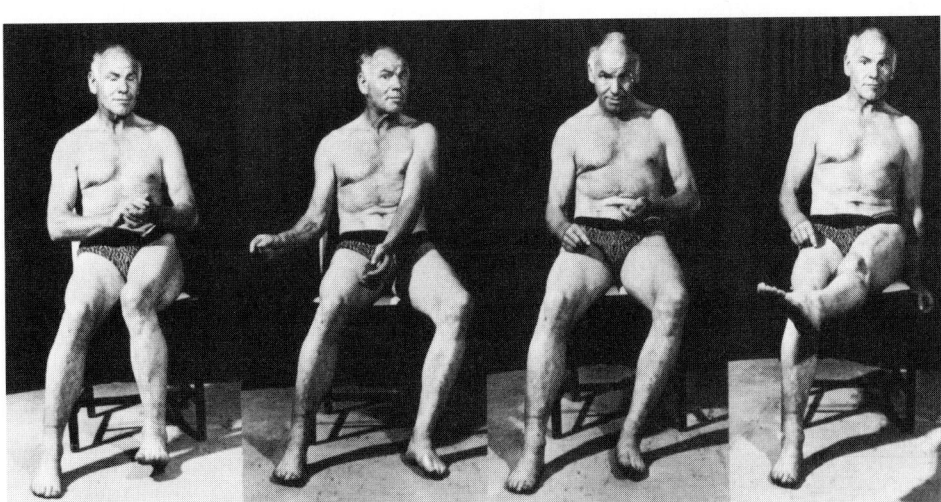

Figure 46-20 A 47-year old patient with Huntington's disease. Note constant fidgety movement. (From Spillane, JD: An Atlas of Clinical Neurology. Oxford University Press, New York, 1968, p 219, by permission.)

should sit straight upright while eating. Thickening agents may be added to thin liquids to help prevent aspiration. Adaptive devices may prolong the patient's ability to eat independently. Soft foods that are easily manipulated in the mouth are most suitable. These patients may have difficulty taking in adequate calories to maintain a normal body weight, even if a caregiver assists with feeding them. One of the many ethical issues faced by these patients and their significant others is whether artificial feeding should be used, and if so, for how long. Patients and their significant others should be encouraged to discuss end-of-life decisions early in the course of the disease.

Alzheimer's Disease

Alzheimer's disease (also called dementia of the Alzheimer's type, or DAT) is the most common of several types of dementia. Dementia is "a loss of intellectual function (thinking, remembering, and reasoning) so severe that it interferes with an individual's daily functioning and eventually results in death."[1]

Alois Alzheimer, a German neurologist, first described the disease in 1907. He described pathological changes, now referred to as neurofibrillary tangles and neuritic plaques, that he discovered while performing an autopsy. Alzheimer's disease is a progressively degenerative disease that is inevitably fatal. The incidence of Alzheimer's disease is more common in women than men and doubles for every 5 years the person lives beyond age 65.

Etiology and Pathophysiology

Many etiologies have been theorized for Alzheimer's disease, including a genetic cause. Chromosome 21 is the location for the gene sequence that is associated with Alzheimer's disease. Chromosome 21 is also the location of the genetic abnormality responsible for Down syndrome. Patients older than age 40 who have Down syndrome usually develop Alzheimer's disease. The exact correlation between the two disorders is still being studied.

Although the exact cause of Alzheimer's disease is unknown, the structural changes associated with it have been well documented. An abnormality exists within the protein of the cell membrane of a neuron. As the axon terminals and dendrite branches disintegrate, they collect in neuritic plaques. Within the normal brain is a precise arrangement of filaments and tubules, responsible for cell integrity. Individuals with Alzheimer's disease develop neurofibrillary tangles instead of the normal orderly arrangement. Instead of remaining a small area of abnormality, these neuritic plaques and neurofibrillary tangles spread via axons to other areas of the brain.

Advancement of neurofibrillary tangles and neuritic plaques typically affect the hippocampus first, resulting in short-term memory dysfunction. As the tangles and plaques spread to the temporal lobe, the memory impairment becomes more severe. It may be at this point that the patient accesses the health care system. Personality changes and incontinence are inevitable results of Alzheimer's disease.

These symptoms can be attributed to the spread of plaques and tangles to the frontal lobes of the brain.

It is believed that the younger the patient is at the time of onset, the faster the neurofibrillary tangles and neuritic plaques spread. Therefore these patients tend to deteriorate faster, require complete care earlier, and have a shorter life span.

One area of the brain that is left relatively untouched by tangles and plaques is the subcortical area. This structure is responsible for our subconscious urge to survive. The needs for basic requirements such as shelter, food and water, security, and reproduction are controlled by the subcortical area, as are emotional responses to situations. The patient with Alzheimer's disease may experience hunger but no longer know how to meet that basic need. Left to their own devices, these individuals would starve.

Signs and Symptoms

The signs and symptoms of Alzheimer's disease are typically broken down into three stages. The early stage lasts from 2 to 4 years and is characterized by increasing forgetfulness. At this stage the patient may attempt to cope by using lists and reminders. Interest in day-to-day activities, acquaintances, and surroundings tends to diminish. The patient is reluctant to take on tasks because of uncertainty in how to perform them. If the patient is still working, his or her performance deteriorates and may result in being terminated from the job.

The middle stage is the longest in duration, lasting 2 to 12 years. Progressive memory loss is demonstrated by difficulty doing simple calculations or answering questions. Patients may become irritable, particularly when asked to perform a task that they know they should be able to perform but cannot. It may help the patient to break down the task into manageable steps. Depression is common. Aphasia and the resulting inability to make themselves understood may exacerbate patients' irritability. It is during the middle stage, as cognitive function significantly deteriorates, that the patient becomes more physically active. The normal sleep-wake cycle is disrupted, and the patient tends to wander aimlessly, particularly at night. The patient may become lost in familiar surroundings, which compounds the anxiety that typically develops during this stage. Hallucinations and seizures may occur. Management of day-to-day activities such as feeding a pet or paying bills becomes overwhelming. Personal hygiene deteriorates, as does appropriate social behavior. Patients may make up stories to cover for deficits, saying that possessions they misplaced were stolen. Some patients hoard food or money.

The third stage of Alzheimer's disease is characterized by progression to complete dependency. The patient loses the ability to converse or control bowel or bladder function. Constant supervision is required, if the patient is still mobile, to protect from wandering and avoid injury. Emotional control and ability to recognize significant others are lost. This lack of recognition is particularly devastating for family members. Eventually the patient is unable to move independently, swallow, or express needs. Death occurs from complications of immobility.

The duration of the final stage of Alzheimer's disease, characterized by complete dependence, depends in part on the physical stamina and general health of the individual. The healthier the patient, the longer the body will continue to function. Another factor is the decisions that have been made regarding artificial feeding and respiratory support. Few significant others or health care practitioners advocate intubation and mechanical ventilation for patients with Alzheimer's disease. The issue of enteral feedings, however, is an emotional one with few easy answers. The use of enteral feedings can prolong the patient's life, despite the absence of cognitive functioning. As with patients suffering from Huntington's disease, every effort should be made to determine the patient's wishes before cognitive impairment makes that impossible. See Table 46–10 for a comparison of the symptoms of Parkinson's, Huntington's, and Alzheimer's diseases.

Diagnostic Tests

The only absolute method of confirming a diagnosis of Alzheimer's is by pathological examination at autopsy. In actuality, the disease is diagnosed on the basis of clinical examination, history, and elimination of other possible causes of the symptoms. MRI may reveal the presence of the classic neurofibrillary tangles and neuritic plaques. Positron emission tomography (PET) and single photon emission computed tomography (SPECT) scans show areas of neuronal inactivity. Newer tests are being evaluated that hold promise for better diagnosis.

Medical Treatment

There is no known cure for Alzheimer's disease. Treatment has traditionally focused on minimizing the effects of the disease and maintaining independence as long as possible. Tacrine (Cognex) was the first drug released expressly for the treatment of Alzheimer's disease. Tacrine is thought to inhibit the breakdown of the neurotransmitter acetylcholine. Increased levels of acetylcholine in the brain allow better functioning of the remaining neurons. Tacrine appears most effective for those patients who exhibit mild to moderate symptoms of Alzheimer's disease. A minimum of 19 weeks may be required to notice any effects of the drug. Use of tacrine diminishes the amount of medical care and social service interventions required and delays admission to skilled nursing facilities. This delay in institutionaliza-

tion can result in significant positive impact on quality of life, as well as thousands of dollars in savings. Newer drugs such as donepezil (Aricept) also inhibit the breakdown of acetylcholine in the brain but have fewer side effects than tacrine.

Antidepressants, antipsychotics, and antianxiety drugs may be used as a last resort to control symptoms of depression and behavioral disturbances, but they do not treat the dementia. Patients should be carefully monitored for drug interactions and side effects. For more information, visit the Alzheimer's Association at www.alz.org.

Nursing Process: The Patient with Dementia

ASSESSMENT. Assess mental status, including memory, orientation, and judgment. (See Chapter 45.) Assess patients' functional level to determine the level of self-care they are able to engage in. The abilities of the family or caregiver to provide the care needed are also assessed, so that appropriate referrals can be made. Availability and use of resources are determined. Nutritional status and usual food preferences are assessed to aid in planning for adequate nutritional intake.

NURSING DIAGNOSIS, PLANNING, AND IMPLEMENTATION. Patients with dementia are at risk for numerous problems. Goals include helping patients maintain the ability to care for themselves for as long as possible. Maintenance of safety and nutrition are priorities (Box 46–9). Incontinence should be minimized and skin protected. Goals should include not only the patient but the caregiver as well. If the caregiver does not have adequate support, the burden of caring for the patient may become overwhelming.

Risk for Injury. Many individuals who have a loved one with dementia wish to keep that individual at home as long as possible. Caregivers need a great deal of education and support. Although every effort should be made to retain the patient's dignity, devices designed to protect small children may be effectively used in the home. (See Home Health Hints.) Baby gates at stairways and baby latches on doors and outlets provide relatively unobtrusive protection. Because they may misinterpret the environment, patients may find mirrors frightening. Forgetfulness and impaired judgment make safety a major concern for Alzheimer's patients who live at home. Patients may strongly resist efforts to keep them from cooking or driving. It may be necessary to

TABLE 46-10 Symptoms of Parkinson's Disease, Huntington's Disease, and Alzheimer's Disease

	Parkinson's Disease	Huntington's Disease	Alzheimer's Disease
Tremors	Present	Absent	Absent
Bradykinesia/akinesia	Present	Absent	Absent
Muscle rigidity	Present	Absent	Absent
Memory dysfunction	Late	Late	Early
Cognitive dysfunction	Late	Present	Early
Inability to perform ADLs	Progressive	Progressive	Progressive
Involuntary movements	Absent	Present	Absent
Depression	Present	Present	Present

BOX 46-9 **Managing Mealtimes for Patients with Dementia**

Dining rooms should be quiet and have adequate lighting. Occupying the same chair for every meal lends familiarity. Serving one course at a time and providing necessary but not extraneous flatware, large handled if necessary, limits distractions. Dishes with high sides enable the patient to scoop the food onto a spoon or fork. Finger foods that the patient can manage may increase intake with minimal staff assistance.

Patients with dementia must be reminded of the steps involved in self-feeding: putting the food on the spoon, directing it to the mouth, swallowing. Use verbal cues or guide the patient's hand to start the necessary movement. Despite the surroundings, common courtesies model social expectations. Introducing the patient to the other persons at the table, providing a cup rather than a carton for milk, and offering foods separately rather than mixing them all together maintain a person's dignity.[1,2]

Quiet music with a slow tempo—at or below the human heart rate—has been used to dampen environmental noises that might otherwise startle patients. Fewer incidents of agitated behaviors occurred during the weeks that music was played compared with weeks without music.[3] Because staff members listened along with the patients, perhaps some of the effect was obtained by relaxing them also.

1. Kayser-Jones, J, and Schell, E: The mealtime experience of a cognitively impaired elder: ineffective and effective strategies. J Gerontol Nurs 23:33, 1997.
2. Tully, MW, et al: The eating behavior scale: a simple method of assessing functional ability in patients with Alzheimer's disease. J Gerontol Nurs 23:9, 1997.
3. Denney, A: Quiet music: an intervention for mealtime agitation? J Gerontol Nurs 23:16, 1997.

HOME HEALTH HINTS

To assess a patient's neurological status at home:

- Note whether the patient's clothes are matched and properly fastened. Is the patient clean and well groomed?
- Observe the patient during bathing, grooming, or dressing to assess motor function and coordination.
- Assess energy level by noting if the patient makes frequent requests to sit or lie down.
- Observe the patient's gait for steadiness.

To help the patient perform ADLs easier at home:

- Ask permission to move furniture and small rugs in order to provide a clear path for ambulation.
- Position frequently used items such as a comb, glass of water, eyeglasses, books, tissues, and phone where they are easily accessible.
- Recommend shoes with Velcro closures.
- Use chairs with armrests—the patient can use the armrests to push against to stand.
- Keep the patient cleaner at meals with a clip-on bib such as those used at the dentist's office. Attach clips from suspenders to a piece of elastic and place around the back of the patient's neck. Attach a clean napkin or washcloth for each meal.

To help the patient with Alzheimer's disease who has perceptual deficits:

- Have things used together the same color (e.g., toothbrush and toothpaste).
- Contrast colors in the environment to help patients function independently—slipper color should be different than the floor, a dark-colored placemat can be used under light dishes, and the first and last steps of a stairway can be painted a contrasting color.
- Use a bath or shower seat, handheld shower head, and soothing music to help the patient feel safe and oriented to the task while bathing.
- Cover doorknobs with a piece of cloth to keep the patient from wandering away.

remove car keys from the patient's access. If patients live alone, they may resist changes in living arrangements.

The patient with dementia is typically admitted to a skilled nursing facility when the significant other can no longer handle care at home. Patients may wander, making them prone to injury. Doors of units should be equipped with alarms indicating when they have been opened. A medical alert bracelet can be worn to identify patients in case they wander and become lost. Special Alzheimer's units are available at many long-term care facilities.

Imbalanced Nutrition. Maintaining adequate nutrition intake for a patient with limited attention span is a nursing challenge. Because these patients have difficulty making choices, one food at a time should be offered. Frequent high-calorie meals and snacks that can be eaten with the fingers may help increase intake.

Disturbed Thought Processes. To facilitate communication, gently touch patients to get their attention before speaking. Gestures, simple phrases, and a calm quiet setting may also help. Patients must be allowed adequate time for

the information to be processed before a response is expected.

Difficult behavior may be caused by inability to express needs or fears. The patient may feel the urge to urinate but can no longer communicate this need or accomplish the process independently. Because understanding the environment is difficult, help the patient to maintain a predictable routine. Scheduled voiding may minimize incontinence. Patients feel safer and may cope better in a stable environment.

Patients often rummage through drawers, closets, or boxes. Unfortunately, these patients do not recognize the difference between their own possessions and those of others, which becomes more problematic if they are in a nursing facility. They may not know what they are looking for, but they do feel that they need to find something. Giving the patient a box of safe, familiar items, such as empty thread spools, may occupy the patient repeatedly.

Incontinence. Regular toileting for both bowel and bladder can help prevent episodes of incontinence. As incontinence becomes more frequent, adult briefs should be used.

An indwelling catheter should be avoided because of the risk for injury and infection. Briefs should be changed regularly to prevent skin breakdown.

Caregiver Role Strain. Dementia often takes at least as much of a toll on caregivers as it does on the sufferer. Fear of the diagnosis may keep people from seeking medical care. Some patients consider suicide because they fear losing their dignity and becoming a burden to their significant others. As the disease progresses, the patient gradually loses awareness of the neurological deterioration. Occasional lucid moments can be very difficult for patient and caregiver as they realize what has been lost.

Caregivers should be encouraged to share their perspectives and coping strategies with others in similar situations.

Support, both informal and formal, is vital to caregiver coping. Caregivers of patients with Alzheimer's disease have unique concerns and needs. You can help caregivers find Alzheimer's support groups and resources.

EVALUATION. If the plan of care is successful, the patient with dementia will remain safe and without injury in an environment that is as comfortable and nonrestrictive as possible. If nutrition is adequate, the patient's weight will remain stable. The patient's needs will be anticipated so that communication frustrations will be kept to a minimum. Incontinence will be managed without skin breakdown. Patients and caregivers should be able to identify resources and obtain relief from caregiver activities when needed.

REVIEW QUESTIONS

1. Which of the following problems predisposed Jennie to develop meningitis?
    a. A muscle injury in her back
    b. A migraine headache
    c. A sore throat for 3 days
    d. Vision changes

2. Mr. Delmar has receptive aphasia. This means that he has difficulty with which of the following?
    a. Swallowing
    b. Speaking
    c. Hearing
    d. Understanding language

3. David complains of seeing flashing lights. You know he has a history of seizures. Which of the following actions do you take first?
    a. Help him lie down in a safe place.
    b. Record the events of the seizure.
    c. Take him to the emergency department.
    d. Assess his eyes.

4. Teresa is admitted following a traumatic brain injury. Which of the following actions do you take to help prevent increased ICP?
    a. Cluster care so she can have long periods of rest.
    b. Keep the head of her bed elevated at 30 degrees.
    c. Suction frequently to keep her airway clear.
    d. Do not give her anything by mouth.

5. Intracranial hemorrhage can be caused by which of the following?
    a. Hypertension
    b. Trauma
    c. Ruptured aneurysm
    d. All of the above

6. Jason is admitted following a T4 spinal injury. When taking his morning vital signs, you note that he appears restless and his blood pressure is elevated. Which of the following actions is appropriate?
    a. Recheck his blood pressure in an hour.
    b. No action is necessary.
    c. Check for a full bladder.
    d. Encourage him to express his anxiety.

7. The symptoms of Parkinson's disease are caused by depletion of which neurotransmitter?
    a. Dopamine
    b. Acetylcholine
    c. Serotonin
    d. Norepinephrine

8. The symptoms of Alzheimer's disease are associated with depletion of which neurotransmitter?
    a. Dopamine
    b. Acetylcholine
    c. Serotonin
    d. Norepinephrine

REFERENCES

1. Alzheimer's Association: Fact Sheet. http://www.alz.org/dinfo/factsheet/ADFS.html, 1998.

Answers to CRITICAL THINKING

Mr. Chung

1. Likely tests include CT scan and lumbar puncture.
2. You should use short, simple sentences, because he may be very anxious or disoriented. Involve his family. Further education can be provided when he is feeling better.
3. Because meningococcal meningitis is contagious, he should be placed in isolation. Gloves, gowns, and masks should be used. Explain the need for these practices to Mr. Chung and his visitors.
4. Comfort measures include tepid baths; a quiet, dark environment; and minimal stimulation. Administer acetaminophen and analgesics as ordered.
5. The health service at his college should be notified of his diagnosis. Close contacts may require prophylactic treatment. If Mr. Chung lives at home rather than at college, his family should be advised to see their family practitioner and begin prophylactic treatment.

Mrs. Washington

1. Uncontrolled hypertension might have precipitated Mrs. Washington's CVA.
2. Her left extremities are affected.
3. Mrs. Washington is disoriented. Her room should be as close to the nurse's station as possible. Reorient her to her surroundings and condition frequently. Keep side rails up when Mrs. Washington is alone.
   Mrs. Washington is also hemiparetic. Obtain a commode because Mrs. Washington will probably not be able to walk to the bathroom. Place the call light and telephone on her right side. Assist Mrs. Washington with positioning to prevent injury to hemiparetic limbs.

4. You should teach Mrs. Washington and her daughter about the relationship of uncontrolled hypertension to intracranial hemorrhage; options for inpatient, outpatient, and in-home therapy; and memory strategies to prevent missed medication doses (e.g., weekly pill box, keeping medications with breakfast food or an alarm clock or watch).

Mr. Esposito

1. You might expect to see impaired speech, right-sided weakness, and a rapid decrease in consciousness if Mr. Esposito's hematoma is enlarged.
2. Intubation equipment, mannitol, and intravenous access should be ready. He should be given nothing by mouth (NPO) and the results of laboratory tests should be ready in the event of emergency surgery. The location of Mr. Esposito's next of kin must be known.
3. Who are Mr. Esposito's support people? Was this drinking episode an isolated incident or a chronic problem that should be addressed?

Ms. Simpson

1. Urinary tract infection is often accompanied by urinary urgency. Ms. Simpson may have difficulty getting to the bathroom quickly and safely.
2. Keep a bedside commode nearby if the bathroom is not close. Assist Ms. Simpson to the bathroom or commode at regular intervals to prevent urgency. Remind her to ask for help if she needs to get up. Make sure that her call light is within reach.

47

NURSING CARE OF PATIENTS WITH PERIPHERAL NERVOUS SYSTEM DISORDERS

George Byron Smith, Marsha A. Miles, and Deborah L. Roush

QUESTIONS TO GUIDE YOUR READING

1. What disorders are caused by disruption of the peripheral nervous system?

2. What are the pathophysiology, major signs and symptoms, and complications of selected peripheral nervous system disorders?

3. What are the medical and surgical treatments of selected peripheral nervous system disorders?

4. What are the steps in the nursing process for the patient with a peripheral nervous system disorder?

▶ PERIPHERAL NERVOUS SYSTEM DISORDERS

The peripheral nervous system (PNS) consists of all nervous system structures outside the central nervous system (CNS). A variety of disorders affect the PNS. Some of these disorders become chronic and cause **degeneration** of body systems. Some other disorders are of a temporary nature. Two common types of PNS disorders are discussed in this chapter. Neuromuscular disorders compose one group. The second group includes cranial nerve disorders. Both types of disorders present a challenge to the nurse caring for the patient and family.

Neuromuscular Disorders

This group of neurological conditions is chronic and degenerative in nature. Neuromuscular disorders involve a disruption of the transmission of impulses between neurons and the muscles that they stimulate (Fig. 47–1). This breakdown in transmission results in muscle weakness. If the muscles of the respiratory system are affected, deadly complications can develop, including pneumonia and respiratory failure. Common neuromuscular disorders include multiple **sclerosis** (MS), myasthenia gravis (MG), amyotrophic lateral sclerosis (ALS), and Guillain-Barré syndrome

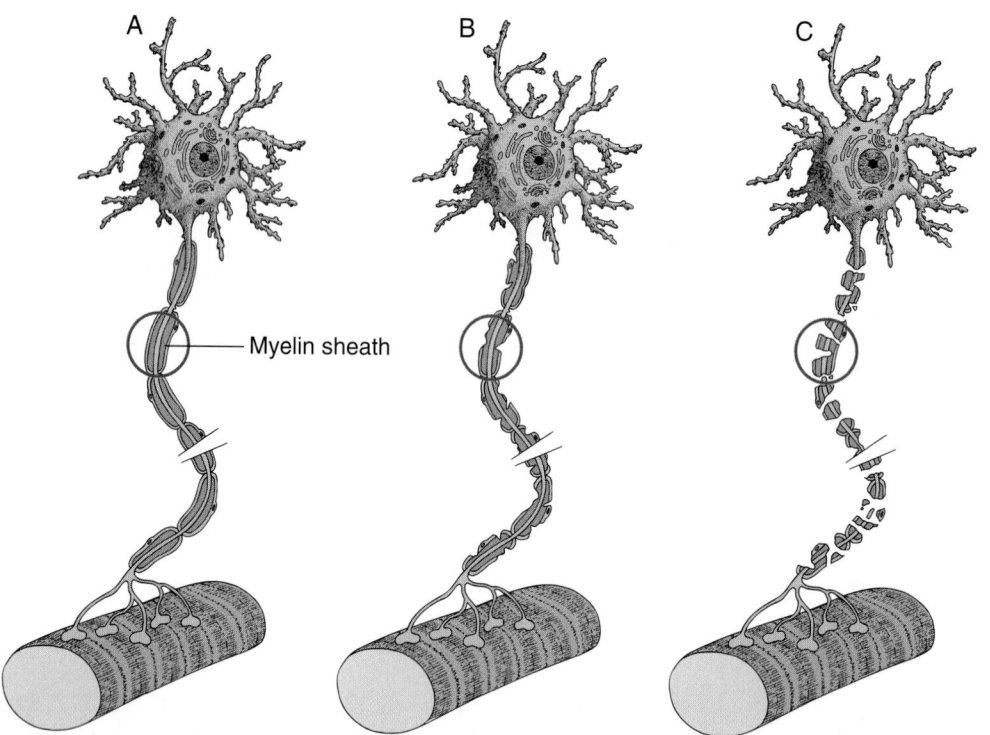

Figure 47-1   The myelin sheath breaks down in multiple sclerosis, interrupting transmission of nerve impulses. *(A)* Normal myelin sheath. *(B)* Myelin beginning to break down. *(C)* Total myelin disruption. (Modified from Scanlon, VC, and Sanders, T: Workbook for Essentials of Anatomy and Physiology, ed 2. FA Davis, Philadelphia, 1995, p 109, with permission.)

(GBS). An additional neuropathic disorder is discussed in Cultural Consideration Box 47–1.

Multiple Sclerosis

ETIOLOGY. Multiple sclerosis is a chronic progressive degenerative disease that affects the myelin sheath on the neurons in the central nervous system. It results from an autoimmune process and may be affected by viral infections, heredity, and other factors. However, the cause of MS is not known. This disease affects 36 to 80 per 100,000 persons in the United States. Onset of the disease usually occurs between ages 15 and 50. Women are affected more than men.

PATHOPHYSIOLOGY. Myelin is responsible for the smooth transmission of nerve impulses. Muscles contract when nerve impulses stimulate the muscle tissue. If the myelin

BOX 47-1   CULTURAL CONSIDERATION

Navajo neuropathy is unique to the Navajo Indian population. Characteristics include poor weight gain, short stature, sexual infantilism, serious systemic infections, and liver derangement. Manifestations include weakness, hypotonia, areflexia, loss of sensation in the extremities, corneal ulcerations, acral mutilation, and painless fractures.[1] Nerve biopsies show a nearly complete absence of myelinated fibers, which is different from other neuropathies, which present as a gradual demyelination process. Individuals who survive have many complications and are generally ventilator dependent. None have been known to survive past the age of 24.

1. Singleton, R, et al: Neuropathy in Navajo children: clinical and epidemiologic features. Neurology 40(2):363, 1990.

sheath is damaged, nerve impulses cannot be transmitted to the muscle and contraction of the muscle does not occur. In multiple sclerosis, the myelin sheath begins to break down (degenerate) as a result of the activation of the body's immune system. The nerve becomes inflamed and edematous. Nerve impulses to the muscles slow down. As the disease progresses, sclerosis from scar tissue damages the nerve. Nerve impulses become completely blocked, causing permanent loss of muscle function in that area of the body.

SIGNS AND SYMPTOMS. The patient with MS presents with muscle weakness, tingling sensations, and numbness. These symptoms may begin slowly over weeks to months or start suddenly and dramatically. MS affects many systems of the body. Box 47–2 lists problems experienced by patients with MS. A variety of factors can trigger the onset or aggravate the condition. These factors include extreme heat and cold, fatigue, infection, and physical and emotional stress. Periods of **exacerbation** and remission of symptoms lead patients with MS to be uncertain about when the disease will flare up and what system of the body will be affected. Intense fatigue is a common complaint among patients. Therefore immobility can become a problem. Accidents and falls are common because of muscular weakness of the trunk and extremities. Pneumonia can occur from immobility and from weakness of the diaphragm and intercostal muscles. Death, usually resulting from respiratory infection, typically occurs 20 to 35 years after diagnosis.

DIAGNOSTIC TESTS. Diagnosis is based on the history and signs and symptoms experienced by the patient. MS cannot be diagnosed by a specific test. Analysis of cerebrospinal fluid

BOX 47-2 Problems Associated with Multiple Sclerosis

Weakness/paralysis of limbs, trunk or head	Impaired hearing
	Nystagmus
Diplopia (double vision)	Ataxia
Slurred speech	Dysarthria
Spasticity of muscles	Dysphagia
Numbness and tingling	Constipation
Patchy blindness (scotomas)	Spastic (uninhibited) bladder
Blurred vision	Flaccid (hypotonic) bladder
Vertigo	Sexual dysfunction
Tinnitus	Anger, depression, euphoria

(CSF) may show an increase in oligoclonal immunoglobulin G (IgG). Magnetic resonance imaging (MRI) may be helpful in diagnosis because sclerotic plaques can be detected.

MEDICAL TREATMENT. MS has no cure. Interferon therapy may reduce exacerbations and delay disability. Other treatment is supportive and symptomatic. Steroids are given to de-crease inflammation and edema at the neuron, which may relieve some symptoms. Immunosuppressant drugs such as azathioprine (Imuran) and cyclophosphamide (Cytoxan) may be given to depress the immune system. Anticonvulsants such as phenytoin (Dilantin) and carbamazepine (Tegretol) help relieve neuropathic pain. Valium (Diazepam), baclofen (Lioresal), and physical therapy assist in controlling muscle spasms. Bladder problems are treated with several different medications such as bethanechol (Urecholine), neostigmine (Prostigmin), propantheline (Pro-Banthine), and oxybutynin (Ditropan). Rehabilitation after an acute episode includes physical, speech, and occupational therapies. Rehabilitation therapy assists the patient and family in adapting the home environment. Instruction in the use of assistive devices (e.g., braces, canes, wheelchairs, splints) by physical and occupational therapists allows the patient increased mobility and independence. Patients who develop speech difficulties benefit from speech therapy.

NURSING CARE. See Nursing Care Plan Box 47–3 for Care of the Patient with a Progressive Neuromuscular Disorder.

BOX 47-3 NURSING CARE PLAN FOR THE PATIENT WITH A PROGRESSIVE NEUROMUSCULAR DISORDER

 Impaired Physical Mobility Related to Muscle Weakness

PATIENT OUTCOMES

Patient will identify measures to help maintain mobility. Patient will perform exercises that help maintain current mobility. Patient will maintain optimum activity level.

EVALUATION OF OUTCOMES

Can patient identify measures that will help maintain mobility? Does patient perform exercises that help maintain mobility? Is optimum activity level maintained?

INTERVENTIONS	RATIONALE	EVALUATION OF INTERVENTIONS
Determine current level of mobility.	Provides information to formulate plan of care.	What is patient's present level of mobility?
Identify factors that affect ability to be mobile and active.	Provides opportunity to seek answers for problems.	Is patient able to identify factors that help or hinder mobility?
Encourage patient to perform self-care to maximum ability.	Promotes sense of control and independence for patient.	Does patient perform self-care activities? Is assistance required?
Consult physical therapist (PT) or occupational therapist (OT) to provide assistive devices for walking (canes, braces, walker, wheelchair) and other activities.	Assistive devices decrease fatigue, promote independence, comfort, and safety.	Does patient use assistive devices safely during activities? Do they help keep patient active?
Reposition frequently when patient is immobile.	Prevents skin breakdown and stasis of pulmonary secretions.	Is patient free from complications of immobility?
Provide active/passive range-of-motion (ROM) exercises on a regular basis.	Prevents contractures and disuse atrophy.	Does patient have any contractures or atrophy?
Plan activities with a balance of frequent rest periods.	Rest decreases fatigue.	Is fatigue controlled?
Administer medications as ordered.	Medications may slow progress of disease and reduce symptoms.	Are symptoms controlled?

 Ineffective airway clearance related to muscle weakness, impaired cough and gag reflex

PATIENT OUTCOMES

Patient will maintain a patent airway. Patient will be free of signs and symptoms of respiratory distress.

EVALUATION OF OUTCOMES

Is patient's airway patent? Is patient free of signs and symptoms of respiratory distress?

BOX 47-3	NURSING CARE PLAN FOR THE PATIENT WITH A PROGRESSIVE NEUROMUSCULAR DISORDER—CONT'D

INTERVENTIONS	RATIONALE	EVALUATION OF INTERVENTIONS
Monitor respiratory rate and depth, oxygen saturation, and arterial blood gasses (as ordered).	Increasing respiratory distress indicates progressing muscle weakness that may require mechanical ventilation or end-of-life decisions.	Is patient's respiratory rate status stable or is intervention indicated?
Encourage patient to cough and deep breathe every 2 hours.	Effective coughing helps keep airway clear.	Does patient have the strength to cough effectively?
Observe patient for breathlessness while speaking.	Inability to speak without breathlessness indicates declining respiratory function.	Is patient able to finish sentences without needing to take a breath?
Elevate head of bed.	Fowler's position improves lung expansion, decreases work of breathing, improves cough efforts, and decreases risk for aspiration.	Does elevation of head of bed help relieve dyspnea and prevent aspiration?
Evaluate cough, swallow, and gag reflexes frequently. Notify physician if absent.	Frequent evaluation of reflexes is needed to prevent aspiration, respiratory infections, and respiratory failure.	Is patient able to cough effectively? Is gag reflex intact?
Suction secretions as needed, noting color and amount of secretions.	Muscle weakness may result in inability to clear airway.	Does patient require suctioning to clear airway? What color are secretions?

 Risk for imbalanced nutrition related to weakness or incoordination of muscles of chewing and swallowing.

PATIENT OUTCOMES
Patient will maintain body weight within normal limits for height and frame.

EVALUATION OF OUTCOMES
Is patient's weight stable and within normal limits (WNL)?

INTERVENTIONS	RATIONALE	EVALUATION OF INTERVENTIONS
Evaluate cough, swallow, and gag reflexes frequently. Notify physician if absent.	If patient is unable to swallow, a feeding tube may be indicated, depending on patient's wishes.	Does patient eat and drink without aspirating?
Offer soft, easy to chew and swallow foods.	Soft foods require less effort to chew and are less fatiguing.	Is patient able to chew and swallow without excessive fatigue?
Institute swallowing precautions as needed. (See Nutrition Notes Box 46-5.)	Swallowing precautions help prevent aspiration and allow patient to maintain oral intake as long as possible.	Do precautions prevent aspiration?
Request speech therapy and dietitian consultations as indicated.	Speech therapist can help evaluate swallowing and make recommendations. Dietitian can provide appropriate foods.	Are consults indicated? Are recommendations implemented?

 Impaired verbal communication related to impaired respiratory and muscle function.

PATIENT OUTCOMES
Patient will be able to communicate needs.

EVALUATION OF OUTCOMES
Does patient indicate that needs are met with a minimum of frustration?

INTERVENTIONS	RATIONALE	EVALUATION OF INTERVENTIONS
Assess ability to speak and communicate.	Assessment is essential to planning appropriate communication interventions.	Can patient speak or communicate needs?
Request referral to speech therapist for assistance if indicated.	Speech therapist can recommend appropriate alternative communication techniques.	Is speech therapy referral indicated? Is referral completed?
Assess for nonverbal signs of pain or distress, such as restlessness, agitation, grimacing.	Patient may not be able to tell you if he or she is in pain or distress.	Are signs of pain or distress present? Are they attended to?
Use picture board or paper and pencil. Ask questions that require yes or no answer.	These do not require the patient to speak to communicate.	Do alternative methods help patient communicate needs?
Use nonhurried, calm, and caring approach while providing care.	This will help decrease anxiety and provide emotional support to patient and family	Do patient and family appear anxious? Does calm approach help?
Explain all procedures.	Patient can still hear and needs to know what is happening.	Does patient indicate understanding?

In addition to routine care, instruct the patient to avoid factors that can exacerbate symptoms. This includes avoiding stressful situations as much as possible, as well as avoiding infection and illness. Any infection, especially respiratory, should be reported immediately to a physician. Two sources of information on MS are the National Multiple Sclerosis Society at www.nmss.org and the Multiple Sclerosis Foundation at www.msfacts.org.

LEARNING TIPS

Myelin facilitates impulse transmission to the muscle. If myelin is interrupted, the impulse cannot get to the muscle efficiently.

Myasthenia Gravis

ETIOLOGY. Myasthenia gravis (MG) means "grave muscle weakness" or weakness of the voluntary or striated muscles of the body. Myasthenia gravis is an autoimmune process. No specific cause has been found for MG. However, current thought is that a virus may initiate the disease. Prevention of MG is not possible at this time. Incidence of the disease is estimated from 43 to 84 persons per million. Peak age of onset in women is 20 to 30 years. MG occurs slightly more often in women than men.

PATHOPHYSIOLOGY. MG is a disease of the neuromuscular junction (Fig. 47–2). At the neuromuscular junction, the neuron releases the chemical neurotransmitter acetylcholine (ACh), which crosses the synaptic cleft. Receptors on the muscle tissue take up ACh, and contraction of the muscle results. In MG the body's immune system is activated, producing antibodies that attack and destroy ACh receptors at the neuromuscular junction. ACh cannot stimulate muscle contraction because the number of ACh receptors has been reduced, resulting in loss of voluntary muscle strength.

SIGNS AND SYMPTOMS. MG results in progressive extreme muscle weakness. Muscles are strongest in the morning, when the person is rested. Activity causes the muscles to fatigue easily, but rest allows the muscles to regain strength. Activities affected by MG include eye and eyelid movements, chewing, swallowing, speaking, and breathing, as well as skeletal muscle function. Patients often present with drooping of eyelids **(ptosis).** Facial expressions are masklike. After long conversations, the voice fades. Falls occur because of weakness of the arm and leg muscles. Patients with MG experience periods of exacerbation and remission of symptoms, as do patients with multiple sclerosis. Exacerbations can be caused by emotional or physical stress, such as pregnancy, menses, illness, trauma, extremes in temperature, electrolyte imbalance, surgery, and drugs that block actions at the neuromuscular junction.

COMPLICATIONS. Major complications associated with MG result from weakness of muscles that assist with swallowing and breathing. Aspiration, respiratory infections, and respiratory failure are the leading causes of death. Sudden onset of muscle weakness in patients with MG resulting from not enough medication is called a myasthenic crisis. Overmedication with **anticholinesterase** drugs causes a cholinergic crisis (Table 47–1). Both crises require immediate medical attention.

LEARNING TIPS

Symptoms of cholinergic crisis can be remembered with the acronym SLUDGE: salivation, lacrimation, urination, diarrhea, gastrointestinal cramping, and emesis. A severe crisis has been described as "liquid pouring out of every body orifice."

anticholinesterase: anti—against + cholinesterase—chemical that breaks down acetylcholine

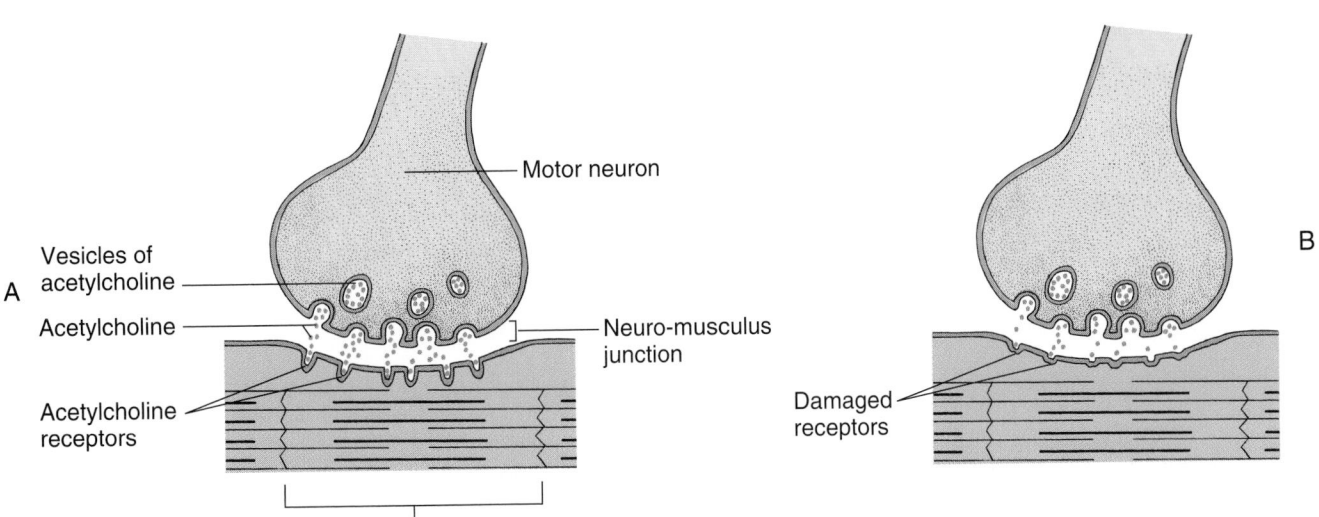

Vesicles of acetylcholine — Motor neuron — Neuro-musculus junction — Acetylcholine — Acetylcholine receptors — Sarcomere — Damaged receptors — A — B

Figure 47-2 Myasthenia gravis. (A) Normal neuromuscular junction. (B) Note damaged acetylcholine receptor sites in myasthenia gravis. (Modified from Scanlon, VC, and Sanders, T: Workbook for Essentials of Anatomy and Physiology, ed 2. FA Davis, Philadelphia, 1995, p 97, with permission.)

TABLE 47-1	COMPARISON OF MYASTHENIC CRISIS AND CHOLINERGIC CRISIS
Myasthenic Crisis	**Cholinergic Crisis**
Cause	*Cause*
Too little medication	Too much medication
Signs and Symptoms	*Signs and Symptoms*
Ptosis	Increasing muscle weakness
Difficulty swallowing	Dyspnea
Difficulty speaking	Salivation
Dyspnea	Nausea or vomiting
Weakness	Abdominal cramping
	Sweating
	Increased bronchial secretions
	Miosis (contraction of pupils)

DIAGNOSTIC TESTS. Diagnosis of myasthenia gravis is based on history of symptoms and physical examination of the patient. A simple test involves the patient looking upward for 2 to 3 minutes. Increased droop of the eyelids (ptosis) occurs if MG is present. After a brief rest, the eyelids can be opened without difficulty. Another test involves an intravenous injection of edrophonium (Tensilon, an anticholinesterase drug). If muscle strength improves dramatically (e.g., the patient can suddenly open the eyes wide), MG is diagnosed. However, improvement is only temporary. An increased number of anti-ACh receptor antibodies in the blood are present in 90 percent of patients with MG. Electromyography (EMG) may be done to rule out other conditions.

MEDICAL TREATMENT. As with multiple sclerosis, no cure has been found for MG. Treatment options include the use of drugs and **plasmapheresis** (Table 47–2). Removal of the thymus gland (thymectomy) can decrease production of

plasmapheresis: plasma—liquid of blood + pheresis—removal

ACh receptor antibodies and decrease symptoms in most patients. Medications used to treat MG include the anticholinesterase drugs neostigmine (Prostigmin) and pyridostigmine (Mestinon). These drugs improve symptoms of MG by destroying the acetylcholinesterase that breaks down ACh. Remember that ACh causes muscles to contract. If ACh is allowed more time to attach to muscle tissue receptors, the muscle contracts and strength is increased. Steroids such as prednisone and immunosuppressants are used to suppress the body's immune response. Plasmapheresis (similar to dialysis) can be used to remove antibodies from the patient's blood.

NURSING CARE. See Nursing Care Plan 47–3 for Care of the Patient with a Progressive Neuromuscular Disorder. In addition, anticholinesterase drugs should be scheduled so that peak action occurs at times when increased muscle strength is needed for activities such as meals and physical therapy. Be aware of symptoms and treatment of myasthenic and cholinergic crises, so you can determine what immediate actions need to be taken.

Patient Education. Instruct the patient that nutritious, well-balanced meals provide caloric intake to maintain strength and resistance to infections. Teach the patient to schedule activities such as grocery shopping or errands at times when medication is at peak action, so that muscle strength is increased. Teach methods to conserve energy, such as sitting down to do grooming and housekeeping activities whenever possible. Teach the importance of avoidance of persons with infections and exposure to cold to minimize risk for respiratory infections, which can exacerbate symptoms and increase risk for ineffective airway clearance. Teach signs and symptoms of crisis conditions because both crises constitute medical emergencies and require immediate medical attention. (See Table 47–1.) Provide information about support groups that can provide encouragement and assistance to patients and their families. Visit the Myasthenia Gravis Foundation of America at www.myasthenia.org.

TABLE 47-2	PLASMAPHERESIS

Plasmapheresis, also known as plasma exchange therapy, is a procedure that removes the plasma component from whole blood. The goal of this therapy is to remove inflammatory agents through exchanging plasma to suppress the immune response and inflammation.

Preprocedure Nursing Care	Postprocedure Nursing Care
Teach patient about the procedure and what to expect, including what the machine looks like (similar, but smaller than a dialysis machine), the need for arterial and venous access sites, and the length of the procedure (2 to 5 hours).	Observe the patient for signs of dehydration (hypovolemia), such as dizziness and hypotension.
The physician may order medications held until after the procedure. Assess baseline vital signs and weight.	Apply pressure dressings to access sites. Monitor patient for infection and bruits at the intravenous port site.
Assess complete blood cell count (CBC), platelet count, and clotting studies.	Monitor electrolytes and signs of electrolyte loss. Report imbalances, and administer replacement electrolytes as ordered.
Check blood type and crossmatch for replacement blood products.	Reevaluate preprocedure laboratory data, such as CBC, platelet count, and clotting times.

LEARNING TIPS

Rest promotes an increase in ACh, which results in an increase in muscle strength.

CRITICAL THINKING

Jamie
Jamie is referred to a neurologist with complaints of muscle weakness.

1. What history can help differentiate between MS and MG?
2. What assessment can be done to differentiate between MS and MG?

Answers at end of chapter.

Amyotrophic Lateral Sclerosis

PATHOPHYSIOLOGY AND ETIOLOGY. Amyotrophic lateral sclerosis (ALS; also called Lou Gehrig's disease) is a progressive, degenerative condition that affects motor neurons. Within the brain and spinal cord, motor neurons begin to degenerate and form scar tissue. Transmission of nerve impulses is blocked. Without stimulation, **atrophy** of muscle tissue occurs. Muscle strength and coordination decrease. As the disease progresses, more muscle groups, including muscles controlling breathing and swallowing, become involved. However, the ability to think and reason is not affected. A specific cause has not been discovered, though ALS may have a genetic predisposition in some cases.

ALS can occur at any age, but usually does not appear until adulthood. The incidence in the United States is 1.5 per 100,000. The usual age of onset is between 40 and 70. ALS is more prevalent in men than women.

SIGNS AND SYMPTOMS. Symptoms are vague early in the course of ALS. Primary symptoms include progressive muscle weakness and decreased coordination of arms, legs, and trunk. Atrophy of muscles and twitching (**fasciculations**) also occur. Muscle spasms can cause pain. Difficulty with chewing and swallowing place the patient at a risk for choking and aspiration as the disease progresses. Inappropriate emotional outbursts of laughing and crying may occur. Speech becomes increasingly difficult. Bladder and bowel functions remain intact, yet problems such as constipation, urinary urgency, hesitancy, or frequency may occur. Late in the disease, communication becomes limited to moving and blinking of eyes in response to questions. Pulmonary function becomes severely compromised to the point of requiring mechanical assistance (ventilator). Other complications that may occur include extreme malnutrition, falls, pulmonary emboli, and congestive heart failure. ALS eventually leads to death from respiratory complications (atelec-

tasis, respiratory failure, and pneumonia). Death usually occurs 3 to 5 years after diagnosis.

DIAGNOSTIC TESTS. Diagnosis is made from clinical symptoms. Other tests may be done to rule out other conditions (CSF analysis, electroencephalogram [EEG], nerve biopsy, EMG). Blood enzymes may be increased as a result of muscle atrophy.

MEDICAL TREATMENT. Goals of treatment are aimed at improving function as long as possible and emotionally supporting the patient and family through the process. Baclofen and diazepam may be given to relieve muscle spasticity. Quinine is used for muscle cramps. Nonpharmacological measures such as physical therapy, massage, position changes, and diversional activities may be instituted for pain control. Tube feedings via a surgically placed gastrostomy tube help provide adequate nutrition. Prevention of infections, such as pneumonia and urinary tract infection (UTI), is vital. Meticulous skin care minimizes the incidence of pressure ulcers. Rehabilitation therapy, including physical, occupational, and speech therapy, allows the patient to maximize function and control. Therapy also decreases the occurrence of complications such as aspiration, falls, and contractures. Support groups and counseling provide emotional support for the patient and family.

NURSING CARE. See Nursing Care Plan 47–1 for Care of the Patient with a Progressive Neuromuscular Disorder.

Patient Education. Reinforce information given by the physician to the patient and family about ALS and its prognosis. Referral to support groups can provide emotional support as the patient and family deal with the reality of eventual death. Rehabilitation using assistive devices and exercises helps prevent complications. Teaching family members how to perform physical therapy and other health care activities allows the patient to spend as much time as possible at home. Teach the patient to avoid exposure to persons with infections, because an infection can be deadly to the patient with a debilitating disease.

LEARNING TIPS

A person with ALS has an intact mind—it is the body that is deteriorating.

CRITICAL THINKING

Mr. Miller
Mr. Miller has been having difficulty swallowing. He is diagnosed with ALS.

1. What are the priority nursing diagnoses for him?
2. How can the patient and his family be supported in coping with this disease?

Answers at end of chapter.

---

atrophy: a—without + trophy—nourishment

Guillain-Barré Syndrome

ETIOLOGY. Guillain-Barré syndrome (GBS) is also called acute inflammatory **polyneuropathy.** This term is more descriptive of the actual disease process. GBS is an inflammatory disorder characterized by abrupt onset of symmetrical paresis (weakness) that progresses to paralysis. It is believed to be caused by an autoimmune response to some type of viral infection or vaccination, though the exact cause is not known. Usually the viral illness occurs within 2 weeks of the onset of symptoms. Incidence is 1 to 2 cases per 100,000 people, with men and women being equally affected. Average age at onset is 30 to 50.

GBS has worldwide distribution, is not seasonal, and affects individuals of all races and ages. Higher rates, however, have been found in people 45 years and older. The incidence of the disease is 50 to 60 percent higher in Caucasians than in African-Americans.

PATHOPHYSIOLOGY. The myelin sheath of the spinal and cranial nerves is destroyed by a diffuse inflammatory reaction. The peripheral nerves are infiltrated by lymphocytes, which leads to edema and inflammation. Segmental **demyelination** causes axonal atrophy, resulting in slowed or blocked nerve conduction. Typically the demyelination begins in the most distal nerves and ascends in a symmetrical fashion. **Remyelination,** which is a much slower process, occurs in a descending pattern and is accompanied by a resolution of symptoms.

There are four recognized variants of GBS. The most common form is ascending Guillain-Barré. It is characterized by progressive weakness and numbness that begins in the legs and ascends up the body. The numbness tends to be mild, but the muscle weakness usually progresses to paralysis. The paralysis may ascend all the way to the cranial nerves or stop anywhere between the legs and head. Deep tendon reflexes are either depressed or absent. In approximately 50 percent of patients with ascending Guillain-Barré, respiratory function becomes compromised.

Descending Guillain-Barré is less common. It affects the cranial nerves that originate in the brainstem first. These patients present with difficulty swallowing and speaking. The weakness progresses downward toward the legs. Respiratory compromise is rapid. Numbness is more problematic in the hands than in the feet, and the reflexes are diminished or absent.

Miller Fisher syndrome, a variant of GBS, is rare. Typically there is no respiratory compromise or sensory loss. The classic symptoms are profound ataxia, absence of reflexes, and paralysis of the extraocular muscles. Some people believe that the fourth form, pure motor Guillain-Barré, is actually a milder version of ascending Guillain-Barré. The symptoms are the same, except for the lack of numbness or paresthesias.

SIGNS AND SYMPTOMS. GBS is divided into three stages. The first stage starts with the onset of symptoms and lasts until the progression of symptoms stops. This stage can last from 24 hours to 3 weeks and is characterized by abrupt and rapid onset of muscle weakness and paralysis, with little or no muscle atrophy. Many patients give a history of a recent viral illness or vaccination, supporting the theory that the cause is autoimmune in nature. The degree of respiratory involvement correlates to the type of GBS and the level of paralysis. Patients with ascending Guillain-Barré may gradually notice a reduced ability to take deep breaths or carry on conversations and may feel short of breath. These patients are terrified that they will not be able to breathe. Patients with descending Guillain-Barré may require intubation on an emergent basis.

The autonomic nervous system is often affected by GBS. Patients may experience labile blood pressure, cardiac dysrhythmias, urinary retention, paralytic ileus, or syndrome of inappropriate antidiuretic hormone (SIADH). Patient complaints of discomfort range from annoying numbness and cramping to severe pain. The discomfort is exacerbated by the patient's inability to move voluntarily.

The second stage is the plateau stage, when symptoms are most severe but progression has stopped. It may last from 2 to 14 days. Patients may become discouraged if no improvement is evident.

Axonal regeneration and remyelination occur during the recovery phase. This stage lasts from 6 to 24 months. Symptoms slowly improve. Most patients with GBS recover completely within a few months to a year. A few patients experience chronic disability.

COMPLICATIONS. Complications that can occur include respiratory failure, infection, and depression. Fatigue and paralysis of the respiratory muscles lead to insufficient respiratory effort. Some patients with impending respiratory failure attempt to convince the staff that they are not in distress and do not need to be intubated. Discussion of the possible need for intubation early in the patient's illness is important. Constant monitoring of respiratory parameters and continuous pulse oximetry provide information indicating the need for immediate intervention.

Patients with GBS are prone to pneumonia and UTIs. Maintaining infection control practices and maximizing the patient's nutritional status help decrease the likelihood of infections occurring. Immobility leads to such problems as skin breakdown, pulmonary embolus, deep vein thrombosis, and muscle atrophy. Patients with Guillain-Barré have little time to adjust to their illness and deterioration. They fear that they will not recover function. Calm, supportive reassurance is important.

DIAGNOSTIC TESTS. A lumbar puncture is performed to obtain CSF. The CSF analysis shows a normal cell count with an elevated protein level. Electromyogram and nerve conduction velocity tests are done to evaluate nerve function.

MEDICAL TREATMENT. During the first stage, patients are partially or completely dependent for all needs. They are often frightened and anxious. In an effort to reduce inflammation, steroids are often administered. Plasmapheresis is used to remove the patient's plasma and replace it with albumin. This procedure is thought to lessen the body's immune response. To be most effective, plasmapheresis should begin 7 to 14 days from the onset of symptoms.

During the plateau phase, patients may become discouraged because they are not getting any better. Emotional support is important during this phase.

Axonal regeneration and remyelination occur during the recovery phase. Intensive rehabilitation helps the patient regain function during this phase.

NURSING CARE. See Nursing Care Plan 47–3 for Care of the Patient with a Progressive Neuromuscular Disorder. The goal of therapy is to support body systems until the patient recovers. Serial assessments of vital capacity and ABGs reveal deterioration in respiratory function. Monitor gag, corneal, and swallowing reflexes, so protective interventions can be implemented if necessary. Manage pain with administration of narcotics and nonpharmacological methods such as position changes, massage, and diversional activities. Nutritional needs may be met via tube feedings or parenteral nutrition if the patient is unable to swallow. Communication boards provide a means for the patient to indicate needs to staff. Because recovery can be prolonged, diversional activities such as visits from family and friends, listening to music or relaxation tapes, and watching television or videos can help alleviate boredom, loneliness, and depression. As the patient begins to regain function, encourage participation in therapy and point out any returning function to the patient and family.

Patient Education. All procedures should be explained to the patient and family. The patient and family need to understand the reasons for continuous respiratory monitoring. Patients may deny any respiratory difficulty because of a fear of intubation and mechanical ventilation. Informing the patient about the possibility of respiratory support and the measures taken to alleviate discomfort help decrease anxiety and encourage patient cooperation. Information about the disease, treatments, and recovery should be given, because recovery may take months or years. Educating family members about how to perform specific patient care activities encourages participation and prepare the patient and family for discharge.

Cranial Nerve Disorders

Cranial nerves are the peripheral nerves of the brain. There are 12 pairs of cranial nerves. Areas that the cranial nerves innervate include the head, neck, and special sensory structures. Cranial nerve problems are classified as peripheral **neuropathies.** Disorders may affect the sensory, motor, or both branches of a single nerve. Causes of cranial nerve disorders include tumors, infections, inflammation, trauma, and unknown causes. Two common cranial nerve problems are trigeminal **neuralgia** (tic douloureux) and Bell's palsy.

---

neuropathies: neuro—nerve + pathies—disease
neuralgia: neur—nerve + algia—pain

Trigeminal Neuralgia

PATHOPHYSIOLOGY AND ETIOLOGY. Trigeminal neuralgia involves the fifth cranial (trigeminal) nerve. This cranial nerve has three branches that include both sensory and motor functions. The branches innervate areas of the face, including the forehead, nose, cheek, and jaw. Trigeminal neuralgia affects only the sensory portion of the nerve. Irritation or chronic compression of the nerve is suspected to initiate onset of symptoms. The incident rate per year is 4.3 per 100,000 persons. This condition is seen more often in women and usually begins around age 50 to 60.

SIGNS AND SYMPTOMS. Intense recurring episodes of pain, described as sudden, jabbing, burning, or knifelike, characterize this condition. Episodes of pain begin and end suddenly, lasting a few seconds to minutes. Attacks can occur in clusters up to hundreds of times daily. However, some patients experience only a few attacks per year. Pain is felt in the skin on one side of the face. Slight touching, cold breezes, talking, or chewing can trigger attacks of pain. The areas of the face where pain is triggered are referred to as trigger zones. Areas affected include the lips, upper or lower gums, cheeks, forehead, or side of nose (Fig. 47–3). Sleep provides a period of relief from the pain. Therefore persons with trigeminal neuralgia may sleep most of the time to avoid painful attacks. They may also refrain from activities such as talking, face washing, teeth brushing, shaving, and eating to prevent pain. Frequent blinking and tearing of the eye on the affected side also occurs.

DIAGNOSTIC TESTS. History of symptoms and direct observation of an attack confirm diagnosis. Radiological studies, including computed tomography (CT) scan and MRI, may be used to rule out other causes of the pain.

MEDICAL TREATMENT. Initial management includes the use of the anticonvulsants phenytoin and carbamazepine to reduce transmission of nerve impulses. Most persons experience relief with medications. These drugs cause bone marrow suppression, so routine complete blood counts are necessary. However, medications do not offer a permanent

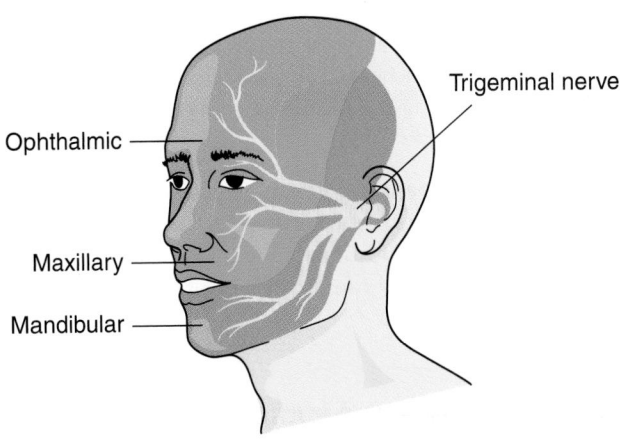

Figure 47-3 Areas innervated by the three main branches of the trigeminal nerve (cranial nerve V) are affected in trigeminal neuralgia.

solution because they lose their effectiveness after a period. Another treatment option is nerve blocks using local anesthetics. This option offers 8 to 16 months of relief. If medications and nerve blocks do not provide relief, surgical options are available. Surgery is done to identify and remove the cause of irritation and inflammation of the nerve. Radiofrequency ablation is used to destroy some of the nerve branches, resulting in anesthesia of the area.

NURSING PROCESS

Assessment. Assess attacks using the WHAT'S UP? format, being sure to include factors that trigger pain. Assess the effect of the pain on the patient's life, including nutritional status, general and oral hygiene, behavior, and emotional state.

Nursing Diagnosis. The priority diagnosis is pain related to inflammation or compression of the trigeminal nerve. Other important diagnoses include imbalanced nutrition: less than body requirements, and anxiety.

Planning and Implementation. Administer analgesics as needed for pain. Alternative pain relief measures such as biofeedback and use of diversional activities may be used in addition to drugs or if drugs do not work. Evaluation of effectiveness of pain relief measures is important. The patient's environment needs to be monitored and kept free of potential triggers. The room should be free of drafts and kept at an even, moderate temperature. Care must be taken to avoid touching the patient's face. Hygiene measures should include the use of lukewarm water, soft cloths, and solutions not requiring rinsing when cleansing the face. A soft-bristled toothbrush or a warm mouthwash can provide adequate oral hygiene. The patient must be given control over when and how care is provided. Hygiene may be avoided when an attack occurs. Small, frequent meals that are high in protein and calories and easy to chew help meet nutritional needs.

Patient Education. Instruct the patient to chew on the unaffected side and to avoid very hot or very cold foods and drinks. The importance of meticulous oral hygiene should be emphasized. Electric razors may be more comfortable for men who shave. If corneal sensation is lost, goggles and sunglasses should be used as needed to protect the affected eye. An eye patch may be needed at night to prevent injury during sleep. Protecting the face from cold or windy weather helps prevent attacks.

Evaluation. The patient will express adequate pain relief with a decrease in or absence of attacks. Caloric intake will be adequate to meet nutritional needs. The patient will verbalize understanding of the use of measures to prevent recurrent attacks.

Bell's Palsy

PATHOPHYSIOLOGY AND ETIOLOGY. In Bell's palsy, cranial nerve VII (facial) becomes inflamed and edematous or compression of the blood vessel feeding the nerve occurs causing interruption of nerve impulses. Loss of motor control occurs on one side of the face. Contracture of facial muscles may occur if recovery is slow. The etiology is unknown; it may be the result of an inflammatory process. The incidence is 23 cases per 100,000, with men and women affected equally. Although the incidence may be slightly higher among diabetics, the condition occurs in all ages and at all times of the year.

SIGNS AND SYMPTOMS. Onset of symptoms occurs over a 2- to 5-day period. Pain behind the ear may precede the onset of facial paralysis. The patient may be unable to close the eyelid, wrinkle the forehead, smile, raise the eyebrow, or close the lips effectively. The mouth is pulled toward the unaffected side (Fig. 47–4). Drooling of saliva occurs, and the affected eye has constant tearing. Sense of taste is lost over the anterior two-thirds of the tongue. Speech difficulties are present.

DIAGNOSTIC TESTS. History of the onset of symptoms is used to diagnose Bell's palsy. Observation of the patient confirms the diagnosis. An EMG may be done. The possibility of a stroke is ruled out.

MEDICAL TREATMENT. Prevention of complications is the goal of treatment, because 80 percent of patients should have complete recovery of function within weeks to months. However, recovery may take up to a year. Prednisone may be given over 7 to 10 days to decrease edema. Analgesics are given for pain control. Moist heat with gentle massage to the face and ear also eases pain. Use of a facial sling aids in eating and supports facial muscles.

NURSING PROCESS

Assessment. Assess facial muscles for signs of weakness. Evaluate for other signs and symptoms of Bell's palsy and carefully document findings.

Nursing Diagnosis. Priority diagnoses include pain related to inflammation of cranial nerve VII (facial), imbalanced nutrition: less than body requirements related to inability to chew, risk for trauma to eye related to inability to blink, and body image disturbance.

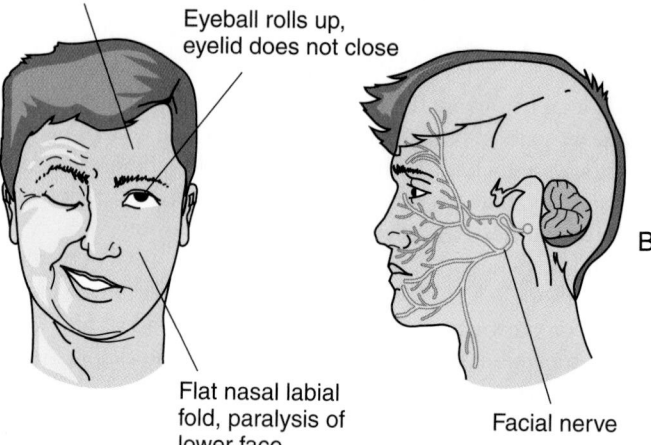

Figure 47-4 Bell's palsy. *(A)* Note weakness of affected side of face. *(B)* Distribution of facial nerve.

Planning and Implementation. Administration of analgesics, moist heat, and massage can help relieve pain. Eye drops or eye ointment as ordered by the physician and a patch are used to protect the eye. Facial exercises several times a day prevent muscle atrophy. Use of a facial sling aids in eating meals and prevents atrophy. A nutritious, well-balanced diet in a consistency the patient can tolerate is important to maintain nutritional status.

Patient Education. Instruct the patient to chew on the unaffected side. Emphasize meticulous oral and eye care. Have the patient demonstrate facial exercises and the use of a facial sling before discharge. Support from family and friends should be encouraged (Home Health Hints).

Evaluation. The patient will express adequate pain control. Caloric intake will meet nutritional needs as evidenced by maintenance of weight and normal laboratory values. The affected eye will remain free of injury.

HOME HEALTH HINTS

- If the patient has difficulty speaking (dysarthria), try using magnetic alphabet letters. Ask questions that require a yes or no answer. The patient can respond with the letter *y* or *n*.

- Bedside commode lids can be kept up with patches of Velcro attached to the seat and the frame.

LEARNING TIPS

Trigeminal neuralgia (cranial nerve V) is a sensory disorder; Bell's palsy (cranial nerve VII) is a motor disorder.

Answers to CRITICAL THINKING

Jamie

1. Muscle weakness caused by myasthenia gravis improves with rest.
2. Have Jamie look up for 2 to 3 minutes. If ptosis occurs, let the patient close her eyes for several minutes. If she can open her eyelids and look up, myasthenia gravis is confirmed.

Mr. Miller

1. Nursing diagnoses include ineffective airway clearance related to muscle weakness and risk for aspiration related to muscle weakness. If a patient's respiratory system is compromised by a disease, nursing care should be focused on maintaining pulmonary function to preserve life.
2. Compassionate care for the patient and providing information about the disease and its prognosis to the patient and family establish an honest and supportive environment. Support groups provide resources and emotional support.

REVIEW QUESTIONS

1. Tegretol may be given to a patient with MS for what purpose?
   a. Decrease inflammation
   b. Depress the immune system
   c. Help relieve pain
   d. Control bladder spasms

2. What medication is used to help diagnose myasthenia gravis?
   a. Prostigmin
   b. Tensilon
   c. Acetylcholine
   d. Prednisone

3. Karesa comes to the clinic complaining of a burning pain on her right cheek. Which of the following findings further supports a diagnosis of trigeminal neuralgia?
   a. Tearing of the eye
   b. Inability to close the affected eye
   c. Asymmetry of facial expressions
   d. Lack of a pupillary response

4. Which of the following systems should take priority when doing a nursing assessment for a patient with ALS?
   a. Skin assessment
   b. Bowel function assessment
   c. Bladder function assessment
   d. Respiratory assessment

5. In caring for a patient admitted with a diagnosis of Guillain-Barré syndrome, the nurse should continuously monitor the patient for which of the following?
   a. Increasing pain
   b. Urinary retention
   c. Respiratory distress
   d. Blurred vision

6. Bell's palsy is a disorder of which cranial nerve?
   a. Third
   b. Fifth
   c. Seventh
   d. Ninth

UNIT THIRTEEN BIBLIOGRAPHY

Abudi, S, Bar-Tal, Y, Ziv, L, and Fish, M: Parkinson's disease symptoms—Patients' perceptions. Journal of Advanced Nursing 25(1):54, 1997.

Adams, RD, Victor, M, and Ropper, AH: Principles of Neurology, ed 7. McGraw-Hill, New York, 2000.

Addison, R, et al: Stroke, catheters and constipation: Action plans. Nursing Times 97(30):54–55, 2001.

Boyne, L: Meningococcal infection. Nursing Standard, 16(7):47–55, 2001.

Cunning, S: When the Dx is myasthenia gravis. RN 63(4):26–31, 2000.

Greenwood, D, Loewenthal D, and Rose, T: A relational approach to providing care for a person suffering from dementia. Journal of Advanced Nursing 36(4):583–590, 2001.

Gresham, GE, et al: Post-stroke rehabilitation: Assessment, referral, and patient management. Clinical Practice Guideline, No. 16. AHCPR Pub. No. 95-0662. US Department of Health and Human Services, Public Health Service, Agency for Health Care Policy and Research, 1995.

Guin, PR: Advances in spinal cord injury care. Critical Care Nursing Clinics of North America 13(3):399–409, 2001.

Habel, M: Continuing education. Brain attack: New stroke treatments, education can limit disabilities. NurseWeek 2001 6(6):27–29, 2001.

Hickey, JV: The Clinical Practice of Neurological and Neurosurgical Nursing, ed 4. JB Lippincott, Philadelphia, 1997.

Hilton, G: Emergency. Acute head injury: Distinguishing subdural from epidural hematoma. American Journal of Nursing 101(9):51–52, 2001.

Hoeman, SP: Rehabilitation in nursing: Process and application, ed 2. Mosby, St Louis, 1996.

Miller, CM: The lived experience of relapsing multiple sclerosis: A phenomenological study. Journal of Neuroscience Nursing 29:5, 1997.

Mistretta, EF, and Kee, CC: Caring for Alzheimer's residents in dedicated units. Developing and using expertise. Journal of Gerontological Nursing 23(2):41, 1997.

Mittelman, MS, et al: Family intervention to delay nursing home placement of patients with Alzheimer's disease. Journal of the American Medical Association 276(21):1725, 1996.

Sommers, MS: Disease and disorders: A nursing therapeutics manual, ed 2. FA Davis, Philadelphia, 2002.

Thames, D: NGNA. Alzheimer's disease and alternative approaches to care: A clinical snapshot. Geriatric Nursing 2001 22(5):270, 2001.

Wasson, K, Tate, H, and Hayes, C: Food refusal and dysphagia in older people with dementia: Ethical and practical issues. International Journal of Palliative Nursing. 7(10):465–471, 2001.

Yee, CA: Getting a grip on myasthenia gravis. Nursing2002, 32(1), 2002.

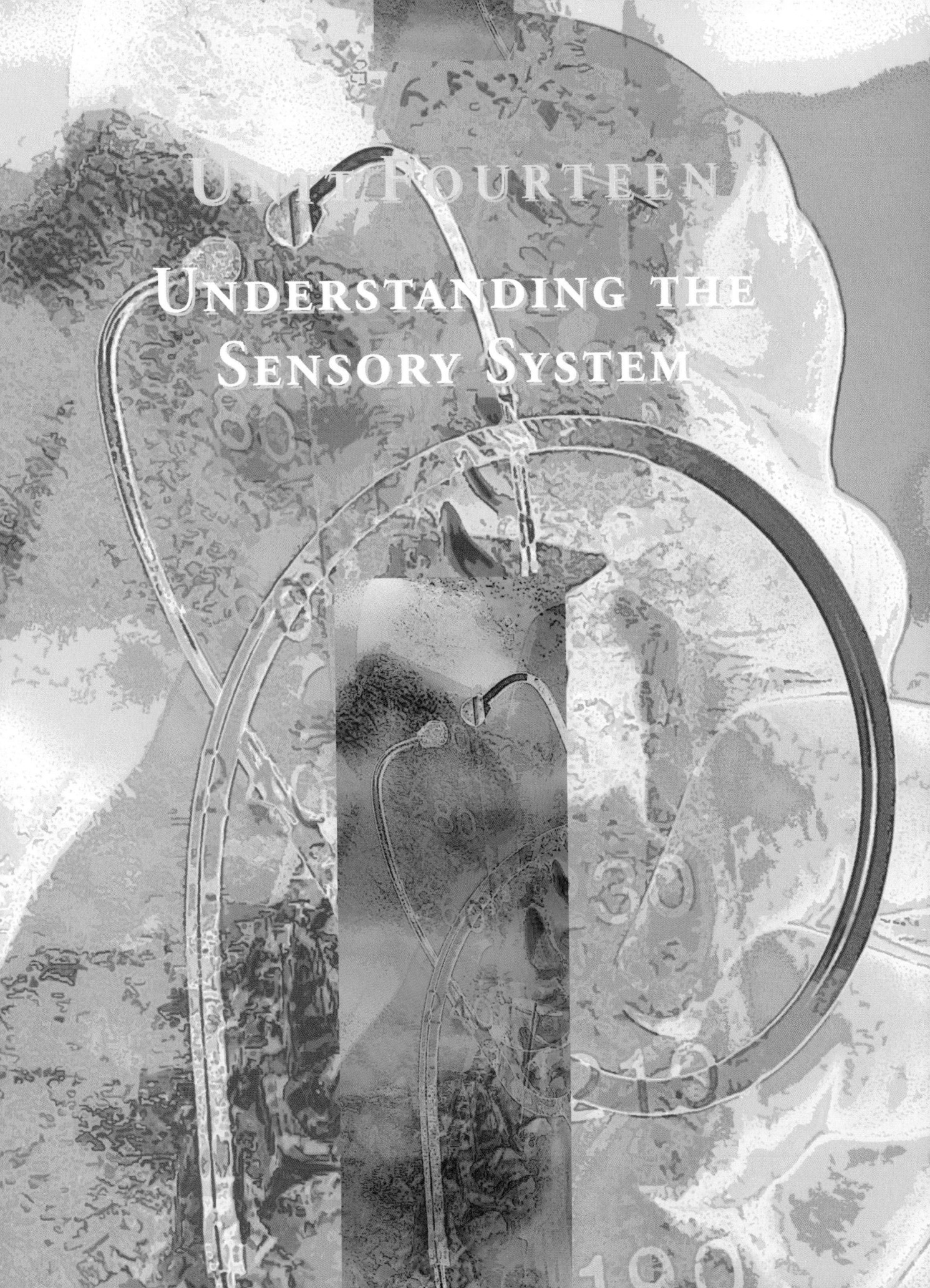

UNIT FOURTEEN

UNDERSTANDING THE
SENSORY SYSTEM

48

Sensory System Function, Assessment, and Therapeutic Measures: Vision and Hearing

Debra Aucoin-Ratcliff and Lazette Nowicki

KEY TERMS

accommodation (uh-KOM-uh-**DAY**-shun)

arcus senilus (ARkus se-**NILL**-us)

cochlear implant (**KOK**-lee-er **IM**-plant)

consensual response
(kon-**SEN**-shoo-uhl ree-**SPONS**)

electroretinography
(ee-LEK-troh-RET-in-**AHG**-ruh-fee)

esotropia (ESS-oh-**TROH**-pee-ah)

exotropia (EKS-oh-**TROH**-pee-ah)

hearing aid (**HEER**-ing **AYD**)

ophthalmologist (AHF-thal-**MAH**-luh-jist)

ophthalmoscope (ahf-**THAL**-muh-skohp)

optician (ahp-**TISH**-uhn)

optometrist (ahp-**TOM**-uh-trist)

otalgia (oh-**TAL**-jee-ah)

otorrhea (OH-toh-**REE**-ah)

ototoxic (OH-toh-**TOK**-sik)

Rinne test (**RIN**-nee **TEST**)

Romberg's test (**RAHM**-bergs **TEST**)

Snellen's chart (**SNEL**-en **CHART**)

tropia (**TROH**-pee-ah)

Weber test (**VAY**-ber **TEST**)

QUESTIONS TO GUIDE YOUR READING

1. What is the normal anatomy of the sensory systems?
2. What is the normal function of the sensory system?
3. What data should you collect when caring for a patient with a disorder of the sensory system?
4. What are the diagnostic tests commonly performed to diagnose disorders of the sensory system?
5. What nursing care should you provide for patients undergoing each of the diagnostic tests?
6. What are the common therapeutic measures used for patients with disorders of the sensory system?

Our eyes and ears provide us with a great deal of sensory information. It is difficult to imagine what it would be like not to see or hear the world around us. Nurses have an important role in assessing vision and hearing. Patients depend on health care personnel to assist them in maintaining these primary senses. To learn more about ways to promote vision and hearing health, visit http://web.health.gov/healthypeople.

► VISION
Normal Anatomy and Physiology of the Eye

The eye contains the receptors for vision and a refracting system that focuses light rays on these receptors in the retina.

External Structures

The eyelids are the protective covers for the front of the eyeball; on the border of each lid are eyelashes that help keep dust out of the eyes. The eyelids are lined with a thin transparent membrane called the conjunctiva, which is also folded over the white of the eye.

Associated with each eyeball is a lacrimal gland located within the bony socket at the upper, outer corner of the eye-ball. Small ducts take tears to the front of the eyeball, and blinking helps spread the tears over the surface. Tears contain lysozyme, an enzyme that inhibits the growth of most bacteria on the surface of the eye. The lacrimal canals at the medial corner of each eye collect tears, which then drain into the lacrimal sac to the nasolacrimal duct to the nasal cavities.

Structure of the Eyeball

Most of the eyeball is within the orbit, the bony socket that provides protection from trauma. The six extrinsic muscles that move the eyeball are attached to the orbit and to the outer surface of the eyeball. There are four rectus muscles that move the eyeball side to side or up and down and two oblique muscles that rotate the eye. The cranial nerves that innervate these muscles are the oculomotor, trochlear, and abducens (third, fourth, and sixth cranial).

The wall of the eyeball has three layers: the outer sclera, the middle choroid (uvea), and the inner retina. The sclera is made of fibrous connective tissue that is visible as the white of the eye. The most anterior portion is the transparent cornea (Fig. 48–1), which has no capillaries and is the first part of the eye that refracts light rays.

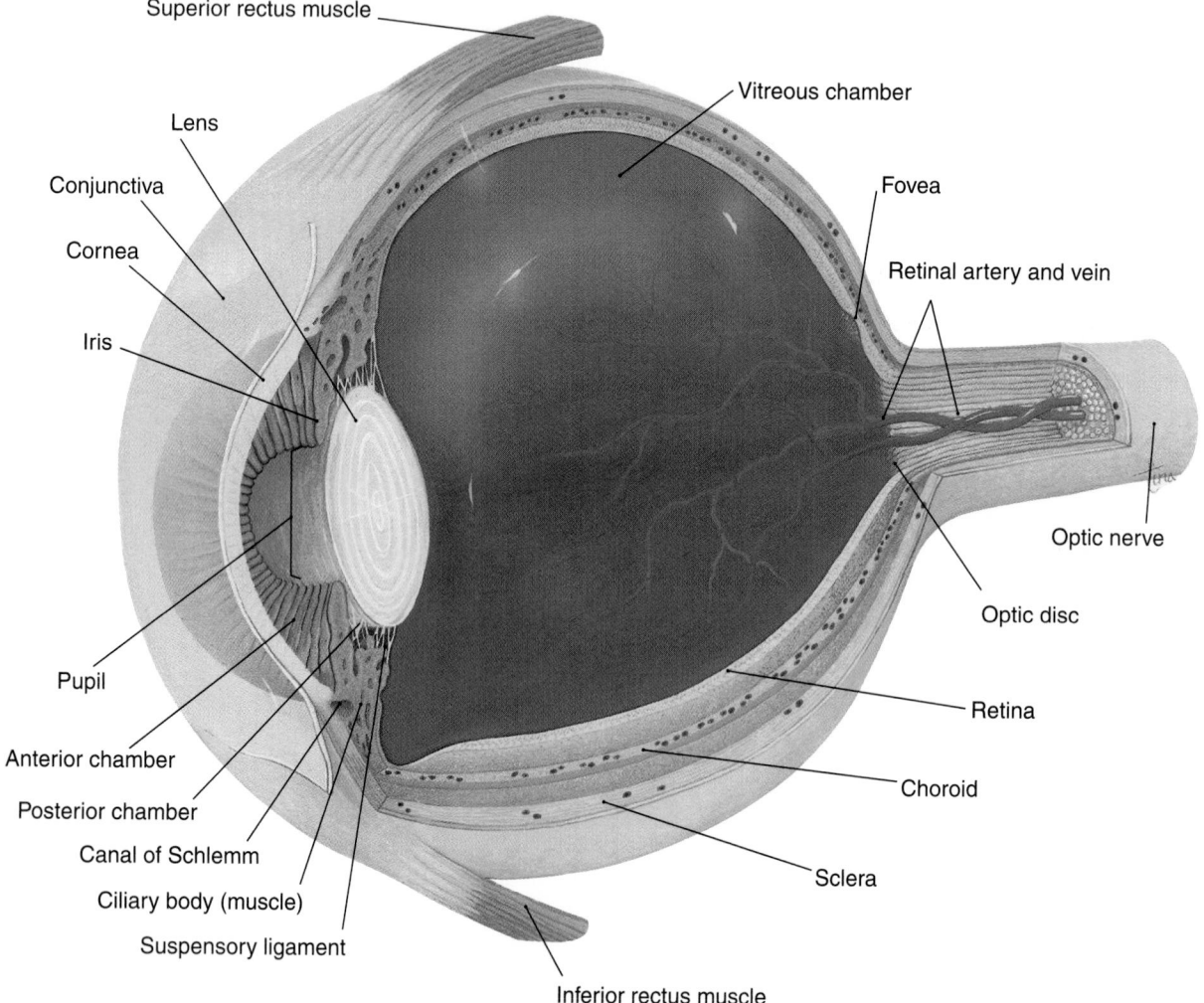

Figure 48-1   Internal anatomy of the eye. (From Scanlon, VC, and Sanders, T: Essentials of Anatomy and Physiology, ed 3. FA Davis, Philadelphia, 1999, p 192, with permission.)

The choroid layer contains blood vessels and a dark-blue pigment that prevents glare within the eyeball by absorbing light. The anterior of the choroid is modified into the ciliary body and the iris. The ciliary body (or muscle) is a circular muscle that surrounds the edge of the lens and is connected to the lens by suspensory ligaments. The lens is made of a transparent, elastic protein and, like the cornea, has no capillaries. The shape of the lens is changed by the ciliary muscle, which permits the focusing of light from objects at varying distances.

In front of the lens is the circular iris, which is made of two sets of smooth muscle fibers that change the diameter of the pupil, the central opening. Contraction of the radial fibers is a sympathetic response and dilates the pupil. Contraction of the circular fibers is a parasympathetic response (mediated by the oculomotor nerves) and constricts the pupil. Pupillary constriction is a reflex that protects the retina from intense light or that permits more acute near vision.

The retina lines the posterior two-thirds of the eyeball and contains the rods and cones, the receptors for vision. Rods detect only the presence of light, whereas cones detect the different wavelengths of light as colors. The fovea centralis is a small depression in the macula lutea of the retina, directly behind the center of the lens, and contains only cones. It is therefore the area of most acute color vision. Rods are proportionally more abundant toward the periphery of the retina, and for this reason night vision is best at the sides of the visual field.

Neurons called ganglion neurons transmit the impulses generated by the rods and cones. These neurons all converge at the optic disc and pass through the wall of the eyeball as the optic nerve. The optic disc may also be called the blind spot because no rods or cones are present.

Cavities of the Eyeball

There are two cavities within the eye, posterior and anterior. The larger posterior cavity is between the lens and retina and contains vitreous humor. This semisolid substance helps keep the retina in place.

The anterior cavity is between the cornea and the front of the lens and contains aqueous humor, the tissue fluid of the eyeball that nourishes the lens and cornea. Aqueous humor is formed by capillaries in the ciliary body, flows anteriorly through the pupil, and is reabsorbed by the canal of Schlemm (scleral venous sinus) at the junction of the iris and the cornea. The rate of reabsorption normally equals the rate of production.

Physiology of Vision

Vision involves the focusing of light rays on the retina and the transmission of the subsequent nerve impulses to the visual areas of the cerebral cortex.

The refractive structures of the eye are, in order, the cornea, aqueous humor, lens, and vitreous humor. The lens is the only adjustable part of this focusing system. When the eye is focused on a distant object, the ciliary muscle is relaxed and the lens is elongated and thin. When the eye is focused on a near object, the ciliary muscle contracts and

forms a smaller circle and the elastic lens recoils and bulges in the middle and has greater refractive power.

When light rays strike the retina, they stimulate chemical reactions in the rods and cones. Receptors contain a light-absorbing molecule called retinal (a derivative of vitamin A) bonded to a protein called an opsin. In the rods, light stimulates the breakdown of rhodopsin into scotopsin and retinal; this generates a nerve impulse for transmission. The cones also contain retinal, and similar reactions take place. The opsins of the cones are specialized to respond to a portion of the visible light spectrum; there are red-absorbing, blue-absorbing, and green-absorbing cones. The chemical reactions within the cones also generate electrical nerve impulses.

The impulses from the rods and cones are transmitted to the ganglion neurons, which converge at the optic disc and become the optic nerve. The optic nerves from both eyes converge at the optic chiasma, just in front of the pituitary gland. Here, the medial fibers of each optic nerve cross to the other side. This crossing permits each visual area to receive impulses from both eyes, which is important for binocular vision.

The visual areas are in the occipital lobes of the cerebral cortex. It is here that the upside-down retinal images are righted and the slightly different pictures from the two eyes are integrated into one image; this is binocular vision, which also provides depth perception.

Aging and the Eye

The most common changes in the aging eye are those in the lens. Over a long period the lens may become partially or totally opaque. The lens also loses its elasticity with age; most people become farsighted as they get older and by age 40 begin to need correction with glasses. Peripheral vision losses may occur. Depth perception decreases and glare is more difficult to adjust to, which can affect safety. Color vision fades with poorer discrimination of blue, green, and violet colors. Red, yellow, and orange colors are seen best.

Nursing Assessment of the Eye and Visual Status

As with most examinations, nursing assessment of the eye begins with the collection of subjective data, then moves to observation and testing, and finally a more invasive physical examination is performed. Licensed practical nurses/licensed vocational nurses (LPN/LVNs) generally do not conduct invasive examinations on the eye, but rather assist the advanced practitioner in conducting this portion of data collection.

Subjective Data

The nurse interviews patients and collects data about family history that may affect vision, particularly glaucoma, diabetes, blindness, and cataracts. Because many eye disorders are genetically transmitted, this information alerts the nurse to possible alterations in eye health. Patients should be asked about their general health and the presence of diseases such as diabetes and hypertension. The nurse determines the types of medication the patient is taking to assess for any ocular (eye) effects. Last, the nurse asks the patient about any changes in visual acuity or symptoms of abnormality (Table 48–1).

Objective Data

VISUAL ACUITY. Objective data collection begins by assessing the patient's visual acuity. Visual acuity is measured in a variety of ways but usually starts with the use of the **Snellen's chart,** E chart, or handheld visual acuity chart (Rosenbaum card) to test near and far vision. Snellen's chart is imprinted with alphabetical letters graduating in size from the smallest on the bottom to the largest on the top (Fig. 48–2). The examiner measures 20 feet and marks the distance on the floor. The examiner then asks the patient to cover one eye with a 3 × 5 card or eye cover and then read out loud an indicated line of letters. The lowest line on the chart that the patient is able to read accurately is used to indicate visual acuity for that eye. Normal vision is 20/20, which means the patient can read at 20 feet what the normal eye can read at 20 feet. Visual impairment occurs at 20/70 and legal blindness at 20/200 or more with correction. An example of findings is the patient who identifies all the letters correctly on the line marked 30; this patient has a visual acuity of 20/30. This means that the patient can see at 20 feet what the average individual can see at 30 feet. The examination is conducted on both eyes separately, then together, and documented as follows: "oculus dexter (OD) 20/30, oculus sinister (OS) 20/20, oculus uterque (OU) 20/20." In addition to identifying the eye tested, the examiner conducts the examination with and without the patient's corrective lenses, if applicable. When corrective lenses are used, documentation reflects this as

"OD 20/100 without correction, OD 20/20 with correction." The E chart is used for patients who are illiterate. The patient is asked to indicate the direction of the E-shaped figure. The handheld visual acuity chart is used to indicate visual acuity by having the patient hold the card approximately 14 inches from the eyes. The test is conducted and documented in the same way as the Snellen's and E chart examinations. For resources on helping those with blindness, visit the American Foundation for the Blind at www.afb.org or the National Federation of the Blind at www.nfb.org.

Visual Fields by Confrontation. The examiner also tests peripheral vision, which is the ability of the eye to see objects peripherally while the eye is fixed or kept in one position. This is also known as testing visual fields by confrontation. To do this, the examiner compares his or her own ability to see peripheral objects with that of the patient. This test should be done with an examiner who has normal peripheral vision. The examiner stands 2 feet in front of the patient and instructs the patient to cover one eye. The examiner covers his or her own corresponding eye (e.g., if the patient's right eye is covered, the examiner's left eye is covered). The examiner uses the arm opposite the covered eye, extends it to the space midway between the patient and the examiner, and brings it toward the eye from three directions: superior, inferior, and temporal (middle). The examiner wiggles the finger while moving the arm. The examiner asks the patient to look straight ahead and indicate at what point he or she is able to see the examiner's finger. One eye is tested and then the other. The patient has full visual fields if the point at which the patient sees the finger matches that of the examiner. The examiner documents the results as "visual fields equal to examiner," "full

TABLE 48-1	SUBJECTIVE NURSING ASSESSMENT DATA QUESTIONS
Obtain information about the family history.	Do you have any family members with a history of diabetes? Hypertension? Cataracts? Glaucoma? Blindness? Diabetes mellitus?
	Do any family members wear glasses or contact lenses? Is their vision corrected with the lens?
Find out about the patient's general health.	How would you describe your general health?
	What health problems do you currently have? How are they treated?
	What health problems have you had in the past?
	Have you ever had trauma to your eyes?
	What medications do you take?
	How often do you have eye examinations?
	When was the last time you had an eye examination?
Obtain information about visual acuity.	Do you wear glasses or contact lenses?
	Have you had any changes in vision such as difficulty seeing distances, difficulty seeing close up, difficulty seeing at night?
	Do you see things double?
	Do you have clouded vision?
	Do you see halos around lights?
	Does it look like you are looking through a veil or web?
	Is there sensitivity to light?
	Is there pain? Itching? Tearing? Burning?
	Do you have headaches? If so, what are the precipitating events?

Figure 48-2 Using Snellen's chart to assess visual acuity.

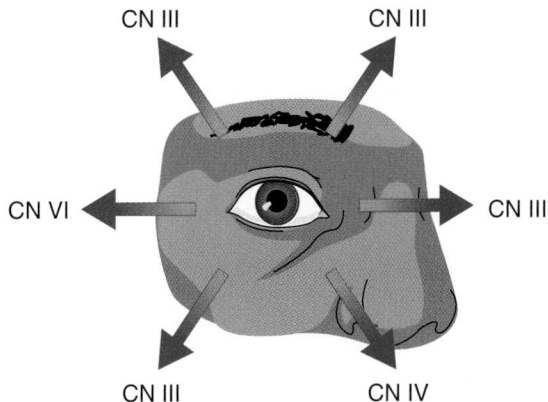

Figure 48-3 Six cardinal fields of gaze.

visual fields," or, if abnormal, "visual fields unequal to examiner in. . ." (identify position, e.g., left superior).

MUSCLE BALANCE AND EYE MOVEMENT. The examiner tests extraocular muscle balance and cranial nerve function by instructing the patient to look straight ahead and follow the examiner's finger movement without moving his or her head. As with the confrontation test, the patient and examiner face each other either standing or sitting. The examiner moves his or her finger in the six cardinal fields of gaze, coming back to the point of origin between each field of gaze (Fig. 48–3). If the patient's eyes are able to follow the examiner's finger in all fields of gaze without nystagmus, the patient is assessed to have adequate extraocular muscle strength and innervation. Nystagmus is an involuntary, cyclical, rapid movement of the eyes in response to vertical, horizontal, or rotary movement.

The corneal light reflex test is used to assess muscle balance. This test is conducted by shining a penlight toward the cornea while the patient is staring at an object straight ahead. The light reflection should be at exactly the same place on both pupils. If the eyes lack symmetry, muscle weakness could be present.

The cover test is used in conjunction with an abnormal corneal light reflex test to evaluate muscle balance. The patient is asked to look straight ahead at a far object. The examiner covers one of the patient's eyes with a 3 × 5 card. The uncovered eye should have a steady gaze; if it moves, there may be muscle weakness. Next, the cover is quickly removed and the action of this eye is observed. If this eye moves to fixate on the light instead of staring straight ahead, it indicates a drifting of the eye when it was covered, which is a sign of muscle weakness. This deviation of the eye away from the visual axis is known as **tropia.** Deviation

of the eye toward the nose is known as **esotropia,** movement laterally is known as **exotropia,** and downward deviation is hypotropia.

PUPILLARY REFLEXES. The pupils are observed. They should be round, symmetrical, and reactive to light. To test pupillary response to light, both consensual and direct examinations should be completed. A slightly darkened room works best. The patient is asked to look straight ahead, and the size of the pupil is noted. A penlight is shone toward the pupil from a lateral position, and the movement of the pupil is observed. The pupil should quickly constrict. The size of the pupil is noted when it constricts. This is known as direct response.

To conduct a consensual pupil examination, observe the eye just tested for reaction while shining the penlight into the other eye. The observed pupil should constrict. This is known as **consensual response.** Repeat the procedure for the opposite eye.

The examiner now proceeds to test for **accommodation.** Accommodation is the ability of the pupil to respond to near and far distances. The patient is told to focus on an object far away. The size and shape of the pupils are observed. The examiner continues to observe the pupils as the patient focuses on a near object (the examiner's penlight or finger) held approximately 5 inches from the patient's face. Normally the patient's eyes turn inward and the pupils constrict. These responses, convergence and constriction, are called accommodation (Gerontological Issues Box 48–1). Examiners use the acronym PERRLA to indicate pupils equal, round, reactive to light, accommodation. If accommodation is not tested along with the other tests, the examiner may use the acronym PERRL.

INSPECTION AND PALPATION OF EXTERNAL STRUCTURES. The extraocular structures are inspected beginning with the eyebrows. The presence of eyebrows, symmetry, hair texture, size, and extension of the brow are noted. The examiner inspects and palpates the orbital area for edema,

---

esotropia: eso—inward + tropia—movement of the eye
exotropia: exo—out + tropia—movement of the eye

Age-related Changes in Vision and Hearing

Vision

Elderly people commonly have the following changes in their vision:

- Presbyopia, an inability to focus up close because of decreased elasticity in the ocular lens
- Narrowing of the visual field and more difficulty with peripheral vision
- Decreased pupil size and responsiveness to light
- Difficulty with vision in dim areas or at night (requires more light to see adequately)
- Increased opacity of the lens, which causes sensitivity to glare, blurred vision, and interference with night vision
- Yellowing of the lens, which makes the person less able to differentiate low-tone colors of blues, greens, and violets (yellow, orange, and red hues are more clearly visible)
- Distorted depth perception and difficulty correctly judging the height of curbs and steps
- Decreased lacrimal secretions

Because visual accommodation is decreased with aging, elderly people have an increased risk of falling. An older person has difficulty making a visual adjustment when moving from a well-lit room into the evening darkness, for example, or when stepping out of a dark area into the sunlight.

The increased time needed to accommodate to near and far and dark and light is often the reason that older adults do not drive at night. Usually they complain that the light from oncoming traffic blinds them or that their eyes do not focus properly.

Hearing

Elderly people commonly have presbycusis (progressive hearing loss caused by loss of hair cells and decreased blood supply) and a decreased ability to hear high-frequency sounds. Deafness or decreased hearing acuity is one of the main reasons that older adults withdraw from social activities. The loss of high-pitched hearing causes the older adult to hear distracting background noises more clearly than conversation.

Older adults who are deaf may require adaptive equipment in their home for safety. The use of a hearing aid can increase hearing for elderly people who do not have nerve damage deafness. The use of flashing lights instead of buzzers or alarms increases the safety of an older adult who is not able to hear a smoke detector or fire alarm.

---

lesions, puffiness, and tenderness. Then the eyelids are inspected for symmetry, presence of eyelashes, eyelash position, tremors, flakiness, redness, and swelling. The patient is asked to open and close the eyelids. When open, the eyelid should cover the iris margin but not the pupil. The distance between the upper and lower eyelid, known as the palpebral fissure, is inspected; it should be equal in both eyes. If the palpebral fissure is nonsymmetrical, observe for ptosis, a drooping of the eyelid, which is commonly seen in stroke patients. Next the medial canthus of the lower lid is gently palpated and observed for exudate. The eyelids are palpated for nodules while the eye is palpated for firmness over the closed eyelid.

The lower eyelid is pulled down, and the patient is asked to look upward. The conjunctiva and sclera are inspected for color, discharge, and pterygium (thickening of the con-

junctiva). To inspect the upper eyelid, the upper lid is everted over a cotton-tipped applicator. The patient blinks to return the eyelid to its resting position when the inspection is complete.

The external eyes are inspected for color and symmetry of the irises, clarity of the cornea, and depth and clarity of the anterior chamber. Shining a light obliquely across the cornea assesses the clearness of the cornea. The cornea should be transparent without cloudiness. In individuals older than 40 years of age, there may be bilateral opaque whitening of the outer rim of the cornea known as **arcus senilus.** It is caused from lipid deposits and is considered normal. It does not affect vision. The anterior chamber (the area between the cornea and the iris) of the eye is inspected using oblique light. The anterior chamber should be clear when the light shines on it.

INTERNAL EYE EXAMINATION. Examination of the internal eye is done by the advanced practitioner. The LPN/LVN may be required to explain the procedure to the patient and to assist the practitioner in the examination. To perform the internal eye examination, specialized equipment must be used. It is useful, but not always necessary, to have the pupil dilated for the internal eye examination. Having a dark room allows the pupil to dilate, as does the application of anticholinergic mydriatic eye drops.

The instrument used to examine the internal eye is called an **opthalmoscope.** The opthalmoscope magnifies the structures of the eye, so the examiner can visualize the retina, optic nerve, blood vessels, and macula. The device is handheld and has a light source that is directed into the patient's internal eye. The patient should be instructed to hold the head still with the eyes focused on a distant object. The patient should be notified that the bright light might be uncomfortable. The **ophthalmologist** may examine the eye using a stationary device called a slit-lamp microscope rather than the handheld opthalmoscope. The patient is seated and rests the chin on a support. This examination allows the examiner to visualize the internal eye by use of a microscope and light source directed into the eye.

Intraocular Pressure. Estimation of intraocular pressure is measured by using one of several types of tonometer. Generally the procedure is performed with the patient lying down, and anesthetic drops may be instilled. One type of tonometer testing uses a puff of air to make an indentation in the cornea to measure intraocular pressure. Readings above the normal range may indicate glaucoma.

Diagnostic Tests

CULTURE. If there is exudate from any portion of the eye or surrounding structure, a culture may be ordered. Results of the culture determine if antiinfective treatment is necessary.

FLUORESCEIN ANGIOGRAPHY. Fluorescein angiography is a procedure used to monitor, diagnose, and treat eye diseases. The patient is assessed for dye allergies before the procedure. Then the pupil is dilated and fluorescence dye is injected into the patient's venous system. The dye travels to the retinal arteriovenous circulation, and the eye is examined via a slit-lamp microscope. The blood vessels in the eye are extremely visible with the addition of the dye.

ELECTRORETINOGRAPHY. Electoretinography is useful in diagnosing diseases of the rods and cones of the eye. The procedure evaluates differences in the electrical potential between the cornea and retina in response to light wavelengths and intensity. The test is conducted by placing contact lenses with electrodes directly on the eye.

ULTRASONOGRAPHY. Ultrasound is useful as an examination tool when the internal eye cannot be visualized directly because of obstructions such as corneal opacities or bloody vitreous. The eye is anesthetized with instillation of anesthetic drops, and a transducer probe is placed on the eye to perform the ultrasound. Patients should be instructed to keep the eye and head still during the procedure.

RADIOLOGICAL TESTS. Several radiological examinations are used to assess eye health. X-ray films are used to view bone structure and tumors. Computed tomography (CT) and magnetic resonance imaging (MRI) are used to visualize ocular structures and abnormalities of the eye and surrounding tissues.

Therapeutic Measures

Nurses have an important role in educating individuals, families, and the community about the care of healthy eyes. Nurses often have the opportunity to screen and educate people about the prevention of disease and impairment. To learn more about ways to promote vision and hearing health, visit www.lighthouse.org.

Regular Eye Examinations

Regular eye examinations should be encouraged. Individuals who are not known to have visual deficits and do not have diseases associated with visual loss such as diabetes should have their eyes examined at regular intervals throughout their life. Screening tests are usually done during an annual physical examination to detect gross visual deficits. Patients who wear corrective lenses or have disease processes that place them at risk for visual loss should have their eyes examined by an eye care provider at least yearly.

Eye care providers include the ophthalmologist and **optometrist.** An ophthalmologist is a physician who specializes in the comprehensive care of the eyes and visual system, including diagnosing and treating eye diseases. An optometrist is a health care provider who specializes in visual examinations, diagnosis, and treatment of visual problems, such as prescribing lenses. The optometrist is not a physician but is identified as a doctor of optometry. An **optician** is a person trained to grind and fit lenses according to prescriptions written by the ophthalmologist or optometrist.

Eye Hygiene

Individuals should be careful to keep debris out of their eyes to prevent scratching of the eye's delicate surfaces. When a foreign object gets into the eye, such as dirt or an eyelash, the individual should be taught not to rub the eye but to allow tears to wash out the object. This can be done by pulling the eyelid down over the eye for a brief time. When wiping the eyes, the nurse should wipe from the inner canthus to the outer canthus.

Nutrition for Eye Health

Adequate nutrition is important not only for the whole body but the eye as well (Nutrition Notes Box 48–2). Eye disorders related to inadequate vitamin intake include corneal damage and night blindness from lack of vitamin A and optic neuritis as a result of vitamin B deficiency.

BOX 48-2 *Nutrition Notes*

Interpreting the Role of Antioxidants in Eye Disease

In developing countries, vitamin A deficiency is a leading cause of blindness. In developed countries, antioxidants have been investigated in relation to macular degeneration and cataracts, two conditions that also lead to blindness. The strongest evidence links consumption of fruits and vegetables to a reduced risk of macular degeneration, cataract, and other diseases. The role of specific nutrients in the fruits and vegetables is less clear. Therefore eating fruits and vegetables (five servings a day, according to the Food Guide Pyramid) is recommended. If a person chooses to take supplements, a multivitamin and multimineral product at no more than recommended daily levels is advocated rather than separate preparations of individual nutrients.

Eye Safety and Prevention

Many people in the United States suffer eye injuries each year. Common household activities are responsible for the majority of injuries. Activities such as microwave cooking, lawn care, and shooting rubber bands and BB guns all contribute to eye injury. Many of these injuries could be prevented with education and implementation of safety measures (Table 48–2).

Eye Irrigation

It is sometimes necessary to irrigate foreign bodies or chemical substances out of the eye. The nurse prepares the patient by explaining the procedure. Usually an isotonic solution is used to irrigate the eye. Refer to Box 48–3 and Figure 48–4.

Medication Administration

A variety of drugs are available for eye application. Most of the drugs are applied as drops, ointments, or irrigations. The nurse must know the usual dosage and strength, desired action, side effects, and contraindications of the medication being administered to prevent harm to the patient. Systemic adverse reactions can occur and medical diseases can be exacerbated from the administration of eye medications. The elderly are especially susceptible to this because they have more chronic diseases, as well as long-term use of oph-

TABLE 48-2 EYE SAFETY AND INJURY PREVENTION

To Protect From:	Use These Eye Safety Measures:
Foreign objects	Wear safety goggles.
	Avoid mowing over rocks or sticks.
	Always wear safety goggles when using lawn edging yard devices.
Chemical splashes	Use splash shields when working with chemicals such as cleaning solution or body fluids.
	Close eyes to avoid getting hair spray in them.
Contact lens abrasions/infections	Follow manufacturer's or eye care professional's directions for length of use and cleaning procedures. Do not overwear lenses.
Ultraviolet light (UV)	Wear UV-protected sunglasses when outdoors.
	Instruct patients to wear sunglasses with side shields after administration of mydriatics.
	Wear hats to shield sun.
Visual deficits in adult with corrective lenses	Update prescription of glasses yearly. Glasses should fit properly, be clean, and be free of scratches.
Eye strain from computer usage	The position of the bottom of the monitor should be 20 degrees below the line of sight and should be positioned 13 to 18 inches from the eyes.
	The light in the room should prevent glare.
	Increase the font size on the screen if letters appear too small.
	If dry eyes are a problem while using a computer, adjust the monitor to a lower level so the eyes do not have to open as wide, which increases evaporation.
Eye injury from sports	Wear protective eyewear with polycarbonate lenses.
	Wear facemasks or helmets while participating in any high-contact or high-impact sports.

BOX 48-3 Eye Irrigation

1. Explain the procedure to the patient.
2. Wash hands.
3. Gather equipment. For low-volume irrigation, a prefilled squeezable bottle is used. For large-volume irrigation, an intravenous (IV) bag of isotonic solution such as normal saline or lactated Ringer's is used. Attach IV tubing to the bag and flush the line.
4. Apply anesthetic drops, if ordered.
5. Place a basin by the side of the patient's head and pad the area with towels to absorb irrigant.
6. Apply gloves.
7. The eye may be irrigated by holding the distal end of the IV tubing at the inner canthus of the eye, or a Morgan lens may be attached. (See Fig. 48–4.) The lens is placed directly on the anesthetized eye, and the tubing is connected to the IV bag tubing. Proceed with irrigation using a slow, steady stream of irrigant. Generally, use of the lens is more comfortable for patients because the eyelids do not need to be held open.
8. Assess patient's tolerance to the procedure.
9. Remove Morgan lens.
10. Remove gloves. Wash hands.
11. Document assessment, type and amount of irrigant, and patient's tolerance to the procedure.

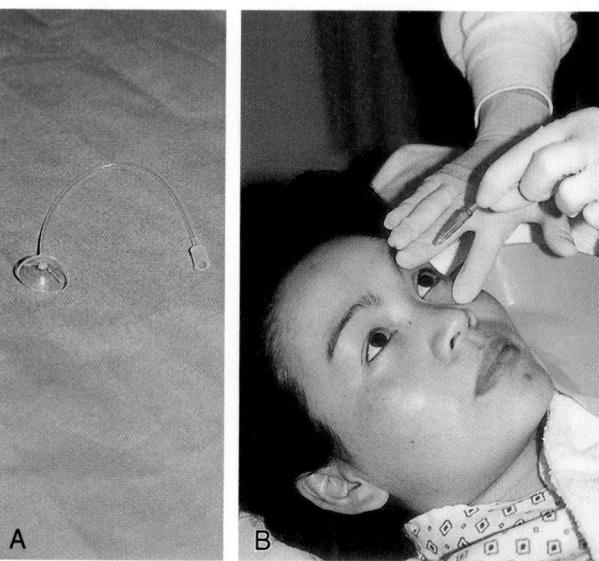

Figure 48-4 *(A)* Morgan lens. *(B)* Irrigation of eye.

thalmic agents. These agents can interact with other medications the patient is taking. The nurse needs to observe patients for possible reactions.

Chapter 49 discusses specific ophthalmic medications and their uses. Boxes 48–4 and 48–5 identify the steps in the application of eyedrops and eye ointments, respectively. Whenever eye medications, especially eye drops, are administered, the punctum (tear duct) of the eye should have pressure applied to it by either the nurse wearing gloves or the patient if able for at least 1 minute. This reduces systemic absorption of the medication via the punctum. Some eye medications can have serious cardiac or respiratory effects, and patients have had life-threatening reactions to them. The nurse should educate the patient on the proper instillation of eye medications to reduce these reactions.

LEARNING TIPS

Elderly patients, when instilling their own eye drops, may not feel the drops go in. Teaching patients to refrigerate the drops, if not contraindicated, for 15 to 30 minutes before instillation helps them feel if the drops go into the eye or on the face.

BOX 48-4 **Administration of Eye Drops**

1. Explain procedure to the patient.
2. Check medication for dosage, strength, side effects, contraindications, and expiration date.
3. Wash hands and apply gloves.
4. Instruct patient to tilt head backward and look up toward the ceiling.
5. Gently pull the lower lid down and out. This forms a pocket to catch the eye drop.
6. Approach the patient's eye from the side and instill the prescribed amount of medication into the pocket. Be careful to avoid touching the patient's eye or surrounding structure with the tip of the dropper. It is helpful for the nurse, and the patient who is self-administering eye drops, to use the forehead as a stabilizing area for the hand administering the drop.
7. Release the lower eyelid.
8. Gently apply pressure with a tissue to the punctum (over the tear duct) for at least 1 minute to prevent the medication from being systemically absorbed. The nurse or patient can do this.
9. Wipe any excess medication off of the eyelids or cheek.
10. Remove gloves. Wash hands.
11. Document medication administration and the patient's tolerance to the procedure.

BOX 48-5 **Administration of Eye Ointments**

1. Explain procedure to the patient.
2. Check medication for dosage, strength, side effects, contraindications, and expiration date.
3. Wash hands and apply gloves.
4. Instruct patient to tilt head backward and look up toward the ceiling.
5. Gently pull the lower lid down. This forms a pocket into which the ointment is placed.
6. Express the ointment directly into the exposed palpebral conjunctiva in the direction of inner to outer canthus. Be careful to avoid touching the patient's eye or surrounding structure with the tip of the ointment tube.
7. Release the lower eyelid over the ointment.
8. Instruct the patient to gently close the eyes.
9. Remove gloves. Wash hands.
10. Instruct the patient that vision may be blurred while the ointment is in the eye.
11. Document medication administration and the patient's tolerance to procedure.

EYE PATCHING. After treating an injured or infected eye, the physician may order the eye to be patched. The nurse applies ointment or drops if ordered, requests that the patient keep the eyelid shut, and then places a disposable, cotton gauze eye patch over the depression of the eye socket. If the patient has a deep eye socket, the nurse may need to place two pads over the socket to assist the eyelids in remaining closed. The purpose of eye patching is to protect the eye from further damage by keeping the lids closed. Sometimes an additional metal shield is placed over the soft pads to protect the eye from external injury. The patch is taped in place and the patient instructed to rest the eyes. The nurse should suggest quiet activities to the patient such as listening to music or an audiotaped book or sleeping. Watching television or reading is not recommended because the patched eye follows the movement of the unpatched eye.

► HEARING
Normal Anatomy and Physiology of the Ear

The ear consists of three areas: the outer ear, the middle ear, and the inner ear (Fig. 48–5). The inner ear contains the receptors for the senses of hearing and equilibrium.

Outer Ear

The outer ear consists of the auricle (or pinna) and the ear canal. The auricle is made of cartilage covered with skin. The ear canal is a tunnel into the temporal bone that curves slightly forward and downward. The canal is lined with skin that contains ceruminous glands. Cerumen, or earwax, is the secretion that keeps the eardrum pliable and, because it is sticky, traps dust.

Middle Ear

The middle ear is an air-filled cavity in the temporal bone. The eardrum (or tympanic membrane) is stretched across the end of the ear canal and vibrates when sound waves strike it. These vibrations are transmitted to the three auditory bones—the malleus, incus, and stapes. The stapes then transmits vibrations to the fluid-filled inner ear at the oval window.

The eustachian tube (or auditory tube) extends from the middle ear to the nasopharynx and permits air to enter or leave the middle ear cavity. The air pressure in the middle ear must be the same as the external atmospheric pressure for the eardrum to vibrate properly. Swallowing or yawning opens the eustachian tubes and permits equalization of these pressures.

Inner Ear

The inner ear is a cavity in the temporal bone called the bony labyrinth, lined with membrane called the membranous labyrinth. The fluid between bone and membrane is called perilymph, and that within the membrane is called endolymph. These membranous structures are the cochlea, concerned with hearing, and the utricle, saccule, and semicircular canals, all concerned with equilibrium.

The cochlea is shaped like a snail shell and partitioned internally into three fluid-filled canals. The medial canal is the cochlear duct, which contains the receptors for hearing in the organ of Corti (spiral organ). The receptors are called hair cells (their projections are stereocilia), which contain endings of the cochlear branch of the eighth cranial nerve. A membrane called the tectorial membrane hangs over the hair cells.

The process of hearing involves the transmission of vibrations and the generation of nerve impulses. When sound waves enter the ear canal, vibrations are transmitted by the

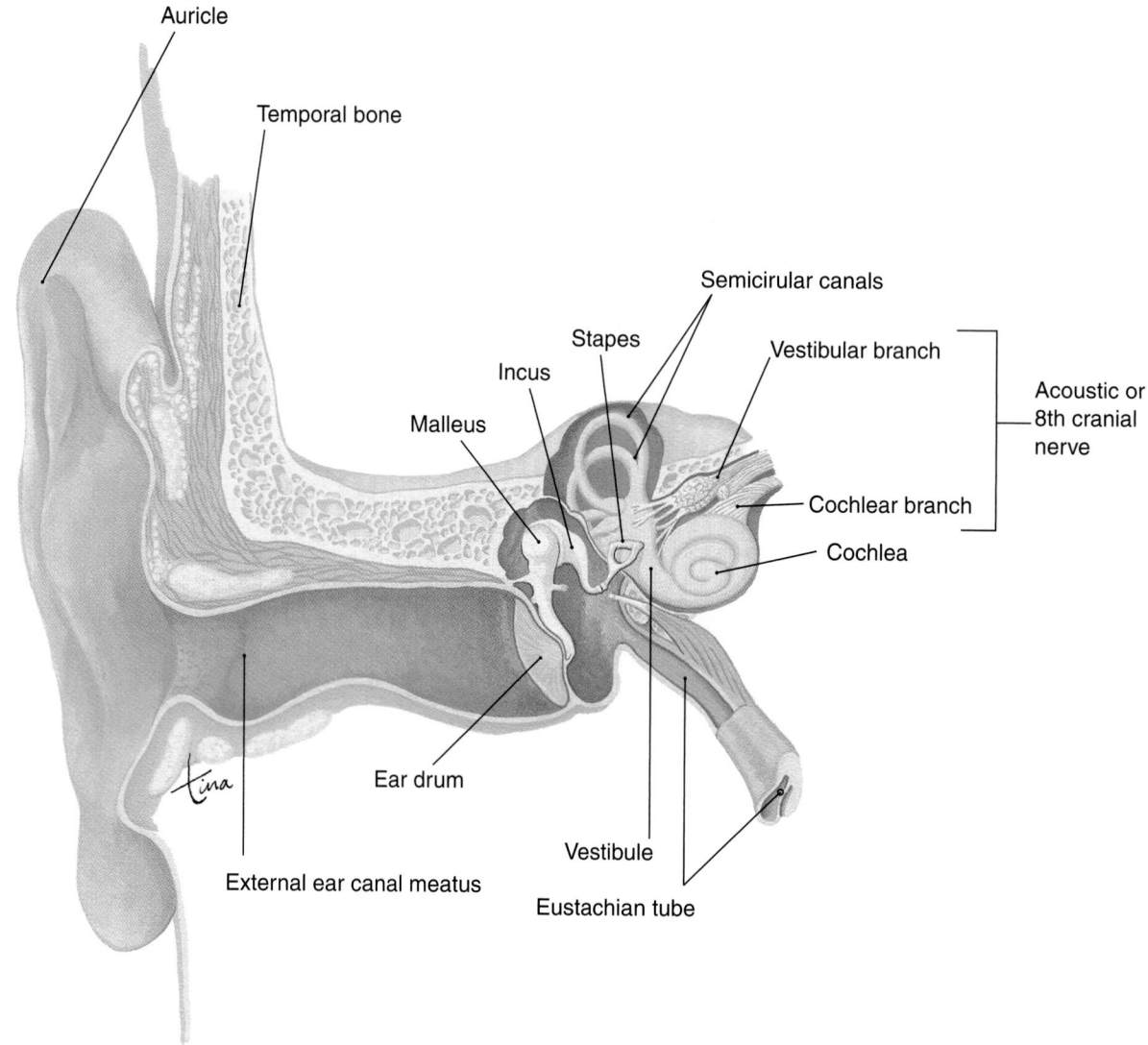

Figure 48-5 The ear in frontal section through the right temporal bone. (From Scanlon, VC, and Sanders, T: Essentials of Anatomy and Physiology, ed 3. FA Davis, Philadelphia, 1999, p 198, with permission.)

following structures: eardrum, malleus, incus, stapes, oval window of the inner ear, perilymph and endolymph within the cochlea, and hair cells of the organ of Corti. When the hair cells bend, they generate impulses that are carried by the eighth cranial nerve to the brain. The auditory areas, for both hearing and interpretation, are in the temporal lobes of the cerebral cortex.

The utricle and saccule are membranous sacs between the cochlea and semicircular canals. Each contains a patch of hair cells embedded in a gelatinous structure that contains otoliths, small crystals of calcium carbonate. The hair cells bend in response to the pull of gravity on the otoliths as the position of the head changes. The impulses generated are carried by the vestibular branch of the eighth cranial nerve to the cerebellum, midbrain, and temporal lobes of the cerebrum. The cerebellum and midbrain use this information to maintain equilibrium at a subconscious level; the cerebrum provides a conscious awareness of the position of the head.

The three semicircular canals are fluid-filled membranous ovals oriented in three planes. At the base of each is an enlarged portion called the ampulla, which contains hair cells (the crista) that are affected by movement. As the body moves forward, for example, the hair cells at first bend backward. The bending of the hair cells generates impulses carried by the vestibular branch of the eighth cranial nerve to the cerebellum, midbrain, and temporal lobes of the cerebrum. These impulses are interpreted as starting or stopping, turning, or changing speeds, and this information is used to maintain equilibrium while a person is moving.

Aging and the Ear

In the ear, cumulative damage to the hair cells in the organ of Corti usually becomes apparent sometime after the age of 60. Hair cells that have been damaged by a lifetime of noise cannot be replaced. Sounds in high-pitched ranges are usually those lost first (presbycusis), whereas hearing may still be adequate for lower-pitched ranges. The high-pitched sounds *f*, *s*, *k*, and *sh* are usually lost first. It becomes more difficult to filter out background noises, so noisy environments make it difficult to hear conversations.

L E A R N I N G   TIPS

Presbycusis is the loss of hearing high-pitched sounds (pitch = cycles per second; loudness = decibels). Because the ability to hear pitch is lost rather than loudness, it is not helpful to talk louder to a patient with this type of hearing loss. In fact, talking louder can make it more difficult to discriminate sounds. It is important to know the type of hearing loss a patient has.

Nursing Assessment of the Ear and Hearing Status

A nursing assessment of the ear includes obtaining the patient's health history and performing a physical examination. A complete nursing assessment is conducted on admission. Provide privacy and make the patient as comfortable as possible before beginning the nursing assessment. A quiet environment is helpful for an accurate assessment of hearing. During the initial assessment, observe the patient's behavior and note any information the patient shares.

Subjective Data

To understand the patient's ear disorder, perform a focused health history. Use knowledge of pathophysiology to guide questions in an appropriate and complete manner. Assessment of symptoms includes asking the WHAT'S UP? questions: where it is, how it feels, aggravating and alleviating factors, timing, severity, useful data for associated symptoms, and perception of the problem by the patient.

HEALTH HISTORY. Obtaining the patient's self-appraisal of his or her hearing or related symptoms is completed during the health history. You should also gather information about medications, surgeries, treatments, allergies, and habits. The health history helps in formulating the nursing care plan.

Symptoms and complaints related to the ear include decreased or loss of hearing, **otorrhea** (discharge), **otalgia** (ear pain), itching, fullness, tinnitus (ringing, buzzing, or roaring in the ears), or vertigo (dizziness). If any of these symptoms or complaints is positive, explore the symptoms in more detail using the WHAT'S UP? format. Record all the information accurately and completely.

Other information about the patient's medical history, including previous ear problems and use of **hearing aid**s or assistive hearing devices, is obtained. You should also ask about any surgeries, allergies, recent upper respiratory infections, history of infections, injury to the ear, hospitaliza-

tions, swimming habits, exposure to pressure changes (flying or diving), medical diseases, and exposure to any loud noises. Positive findings should be assessed using the WHAT'S UP? format and results recorded.

Information about current and past medications should be obtained. Many medications are potentially toxic to the ear and can cause hearing loss or decreased hearing. Pay particular attention to any exposure to medications that are potentially **ototoxic**, such as certain antibiotics or diuretics. (See Chapter 49.)

Family history related to ear disorders includes any hearing problems or hearing loss and family members with Ménière's disease. Significant findings are recorded, including the relationship of the family member with the problem to the patient.

Information about the patient's care of the ears is also gathered. It is important to assess what preventive measures the patient practices and what the patient's learning needs are concerning care and protection of the ears. Determine how the patient cleans the ears, any exposure to loud noises during recreational activities or during work activities, any changes in ability to hear, and any exposure to ototoxic medications. You should determine if the patient has had a hearing evaluation and if there is a history of ear problems. Instruct the patient in ways to care for the ears and maintain ear health (Table 48–3).

Objective Data

Physical assessment of the ear begins by observing the behaviors of the patient. Note how the patient communicates. Observe how the patient talks, noting any slurred speech or words. Certain behaviors can be early indicators of hearing loss, as listed in Box 48–6. Examination of the ear includes inspection, palpation, testing auditory acuity, and, for the advanced practitioner, otoscopic examination.

INSPECTION AND PALPATION OF THE EXTERNAL EAR. Inspection of the external ear begins with examining the auricle. A penlight or otoscope may be used to improve visualization of the external ear. The external ear should be inspected for size, symmetry, configuration, and angle of attachment. Note any obvious deformities or scars. The skin should be smooth and without breaks, particularly behind the ear in the crevice. The color should be uniform, without signs of inflammation. To inspect the external ear canal, tip the adult patient's head to the side and uses a penlight or otoscope to inspect the canal. Note any drainage or cerumen (wax), including the color, odor, and clarity of the drainage. The skin should be smooth and without inflammation, edema, or breaks. There should be no lesions, foreign bodies, erythema, or edema observed within the external ear canal. Inspection of the external ear canal should be completed before obtaining an infrared ear temperature, because the presence of cerumen can alter the accuracy of the reading.

---

otorrhea: oto—related to the ear + rrhea—to flow
otalgia: ot—related to the ear + algia—signifying pain

ototoxic: oto—related to the ear + toxic—poison

TABLE 48-3 PREVENTION OF EAR PROBLEMS		
Activity	**Patient Education**	**Rationale**
Care of external ear	Wash external ear with soap and water only.	Keeps external ear clean.
	Do not routinely remove wax from the ear canal.	The ear is generally self-cleaning. Wax is normally removed during showering. Wax serves as a protective mechanism to lubricate and trap foreign material.
Preventing ear trauma	Avoid inserting any objects or solutions into the ear. Avoid swimming in polluted areas.	Prevents traumatizing the ear and tympanic membrane or exposing the ear to infection.
	Avoid flying when the ear or upper respiratory system is congested.	Prevents barotrauma due to pressure changes.
Preventing damage from noise pollution	Avoid exposure to excessive occupational noise levels.	Normal speech is 60 decibels; heavy traffic is 70 decibels. Above 80 decibels is uncomfortable. If there is ringing in the ear, damage may be occurring. Occupational noise is the primary cause of hearing loss.
	Avoid other causes of excessive noise such as use of firearms and high-intensity music.	Hearing loss can occur due to exposure to loud noises.
	Use protective earplugs or earmuffs if exposure to noise cannot be avoided.	Protects ears from hearing loss by decreasing exposure to loud noises.
Early detection of hearing loss	Instruct adults to have hearing checked every 2 to 3 years.	Degenerative changes occur in the ear with aging.
	Monitor for side effects of ototoxic drugs. Instruct patient to report any dizziness, decreased hearing acuity, or tinnitus when taking ototoxic medications.	Prevents side effects of medications from causing hearing loss.
	Caution elderly patients who use aspirin that it is ototoxic.	Elderly patients may have hearing loss and not be able to hear the tinnitus.
	Instruct patient to report to physician any prolonged symptoms of ear pain, swelling, drainage, or plugged feeling.	Many medical problems can be prevented with prompt treatment.
	Instruct patient to blow nose with both nostrils open during upper respiratory infections (colds).	Prevents infected secretions from moving up the eustachian tubes into the middle ear.

BOX 48-6 **Behaviors Indicating Hearing Loss**

Adults with hearing loss may display any or all of the following behaviors:

- Turns up the volume on the television or radio
- Frequently asks, "What did you say?"
- Leans forward or turns head to one side during conversations to hear better
- Cups hand around ear during conversation
- Complains of people talking softly or mumbling
- Speaks in an unusually quiet or loud voice
- Answers questions inappropriately or not at all
- Has difficulty hearing high-frequency consonants
- Avoids group activities
- Shows loss of sense of humor
- Face looks strained or serious during conversations
- Appears to ignore people or aloof, does not participate
- Is irritable or sensitive in interpersonal relations
- Complains of ringing, buzzing, or roaring noise in the ears

Next the auricles are palpated and any tophi, lesions, or masses are noted. Tophi are deposits of uric acid crystals that appear as small, hard nodules in the helix (external ear margin); they may also occur in gout. The auricle should be nontender when it is palpated; tenderness can indicate an external ear infection. A downward protrusion of the helix, called Darwin's tubercle, is a normal finding. The mastoid process should be smooth and hard when palpated. The mastoid process can be of different sizes but should not be tender or swollen.

AUDITORY ACUITY TESTING. Auditory function can be grossly evaluated using three different assessment tests. The whisper voice test is one test to check hearing function in each ear. The patient occludes one ear with a finger, and the nurse stands 1 to 2 feet away on the opposite side. The nurse whispers two-syllable words toward the unoccluded ear. The patient restates the whispered words. The nurse should be by the patient's side to prevent the patient from lip reading. The nurse's voice can be increased from a soft, medium, or loud whisper to a soft, medium, or loud voice. The process is repeated on the other ear. The patient is asked if hearing is better in one ear than in the other ear. The patient should be able to hear a soft whisper equally well in both ears. Findings of one ear hearing better than the other or inability to hear a soft whisper can be indicative of hearing impairment. Results of the test are documented.

A second acuity test is the **Rinne test.** This test is performed with a tuning fork and is useful for differentiating between conductive and sensorineural hearing loss. To perform the test, strike the tuning fork and place it on the patient's mastoid process (Fig. 48–6). Verify that the patient is able to hear the tuning fork and then instruct the patient to say immediately when the sound is no longer heard. When

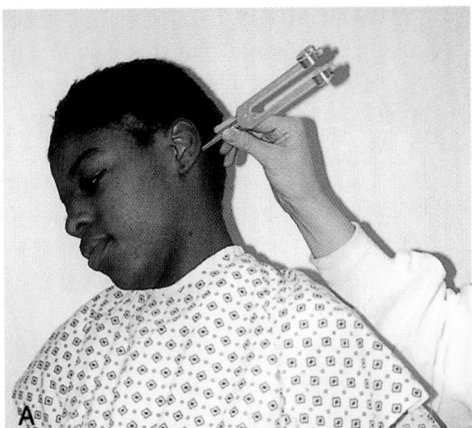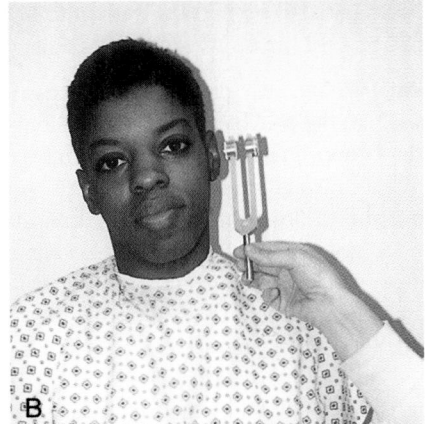

Figure 48-6 Rinne test. *(A)* Bone conduction. *(B)* Air conduction.

the patient indicates that the sound is not heard, place the vibrating tuning fork 2 inches in front of the ear. (See Fig. 48–6.) Ask the patient if he or she hears the tuning fork and then to indicate when the sound is no longer heard. Normally, air conduction (AC) is heard twice as long as bone conduction (BC). The patient reports this by hearing the tuning fork when placed in front of the ear (AC) after no longer hearing the tuning fork placed on the mastoid process (BC). Normal results are recorded as "AC > BC" (air conduction is greater than bone conduction). The test is repeated on the other ear and findings recorded. Abnormal findings can indicate conduction or sensorineural problems. See Table 48–4 for a summary of possible findings.

The **Weber test** is a third test to assess hearing acuity. The Weber test is also performed using a tuning fork. Place the vibrating tuning fork on the center of the patient's forehead or head (Fig. 48–7). Verify that the patient can hear the tuning fork. Then, with a positive answer, asks the patient if he or she hears the sound better in the left ear, better in the right ear, or the same in both ears. It is important to give the patient three choices from which to choose. Normally the patient hears the sound the same in both ears. Table 48–4 identifies abnormal findings.

BALANCE TESTING. When the patient complains of dizziness, nystagmus, or problems with equilibrium, simple tests can be performed to assess vestibular function. The first test is simply to observe the patient's gait by having the patient walk away from the examiner and then walk back to the examiner. Note the patient's balance, posture, and movement of arms and legs. The patient should be able to walk with an upright position with no difficulties in balance or movement.

Romberg's test, or falling test, is another simple test to assess vestibular function. Instruct the patient to stand with feet together, first with eyes open and then with eyes closed. Normally the patient has no difficulty maintaining a standing position with only minimal swaying. If the patient has difficulty maintaining balance or loses balance (a positive Romberg's test), it can indicate an inner ear problem. If a

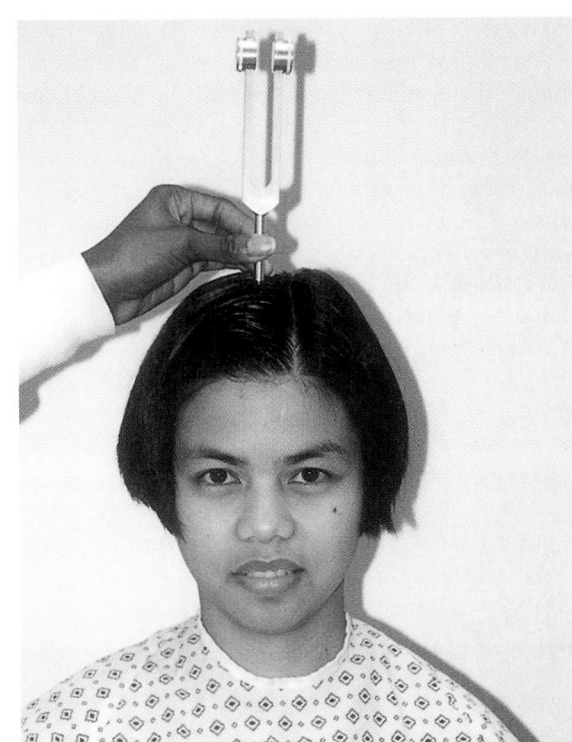

Figure 48-7 Weber test.

TABLE 48-4	AUDITORY ACUITY TUNING FORK TESTS		
Test	**Expected Results**	**Conductive Hearing Loss**	**Sensorineural Hearing Loss**
Rinne Test	Air conduction heard longer than bone conduction	Bone conduction heard longer than air conduction in affected ear	Air conduction heard longer than bone conduction in affected ear (may be less than 2:1 ratio).
Weber Test	Tone heard in center of the head; no lateralization	Sound heard louder in affected ear	Sound heard louder in better ear

fall appears likely, be prepared to support the patient to prevent injury.

OTOSCOPIC EXAMINATION. An otoscope is an instrument consisting of a handle, a light source, a magnifying lens, and an optional speculum for inserting in the ear. Some otoscopes have a pneumatic device for injecting air into the canal to test the eardrum's mobility and integrity. The otoscope is used to visualize the external ear, ear canal, and tympanic membrane. Otoscopic examination is completed to identify specific disorders or infections, remove wax, or remove foreign bodies. Examination of the ear canal should be completed during insertion and removal of the speculum. The ear canal should be smooth and empty. There should be no redness, scaliness, swelling, drainage, nodules, foreign objects, or excessive wax. The internal otoscopic examination is conducted to examine the eardrum and is done by the experienced practitioner. The eardrum should appear slightly conical, shiny, and smooth and be a pearly gray color.

Diagnostic Tests

AUDIOMETRIC TESTING. Audiometric testing is used as a screening tool to determine the type and degree of hearing loss. An audiologist conducts the hearing tests in a soundproof booth. The audiometer produces a stimulus that consists of a musical tone, pure tone, or speech. To test air conduction, the patient is placed in the booth, wears earphones, and signals the audiologist when and if the tone is heard. Each ear is tested separately as the patient is exposed to sounds of varying frequency or pitch (hertz) and intensity (decibels). By varying the levels of the sound, a hearing level is established (Table 48–5). The use of earphones measures air conduction, level of speech hearing, and understanding of speech. During bone conduction testing, a vibrator is placed on the mastoid process and the earphones are removed. Testing proceeds as with air conduction.

A patient with normal hearing should have the same air conduction as bone conduction hearing levels. Alterations in testing air and bone conduction hearing can provide information about the location and type of hearing loss.

TABLE 48-5	COMMON NOISE LEVELS
Human hearing threshold	0–25 dB
Quiet room	30–40 dB
Conversational speech	60 dB
Heavy traffic	70 dB
Telephone	70 dB
Alarm clock	80 dB
Vacuum cleaner	80 dB
Unsafe noise levels begin	90 dB
Circular saw	100 dB
Rock music	120 dB
Jet planes	120–130 dB
Pain threshold	130 dB
Firearms	140 dB

TYMPANOMETRY. Tympanometry is a test used to measure compliance of the tympanic membrane and differentiate problems in the middle ear. Varying amounts of pressure are applied to the tympanic membrane, and the results create a distinctive response recorded on a graph called a tympanogram. The test is useful in determining the amount of negative pressure within the middle ear. The patient is informed that the tympanometry may cause transient vertigo. The patient should report any nausea or dizziness experienced during the test.

CALORIC TEST. The caloric test is used to test the function of the eighth cranial nerve and to assess vestibular reflexes of the inner ear that control balance. The test is performed first on one ear and then the other. Warm (44.5° C or 112° F) or cold (30° C or 86° F) water is instilled into the ear canal. This stimulates the endolymph of the semicircular canals, which stimulates movement of the head. Nystagmus is a normal response. The patient may also experience dizziness. No nystagmus is seen if the patient has a disease of the labyrinth such as Ménière's disease. The test is contraindicated if the patient has a perforated tympanic membrane. Otoscopic examination should be completed before this test to assess for excessive cerumen or perforated tympanic membrane.

ELECTRONYSTAGMOGRAM. The electronystagmogram is used to diagnose the causes of unilateral hearing loss of unknown origin, vertigo, or ringing in the ears. It is similar to the caloric test. The test is usually completed in a darkened room. Five electrodes are taped to the patient's clean face at certain positions around the eye. The electrodes measure nystagmus in response to vestibular stimulation. Measurements are taken at rest, looking at different objects, with eyes open and closed, in different positions, with water of different temperatures, and with air. Usually tranquilizers, alcohol, stimulants, and antivertigo agents are held for 1 to 5 days before the test. The patient should also avoid tobacco and caffeine on the day of the test. The test is contraindicated in patients who have pacemakers. The patient may experience nausea, vertigo, or weakness following the test.

COMPUTED TOMOGRAPHY. A CT scan produces radiographs similar to those used in conventional radiography. A special scanner system produces cross-sectional images of anatomic structures without superimposing tissues on each other. CT is useful for visualizing the temporal and mastoid bones, the middle and inner ears, and the eustachian tube. The patient should remove hairpins and jewelry from the area of visualization.

MAGNETIC RESONANCE IMAGING. MRI produces cross-sectional images of the human anatomy through exposure to magnetic energy sources without using radiation. It is useful in differentiating between healthy and diseased tissue. The MRI allows the membranous organs, nerve, and blood vessels of the temporal bone to be examined. The test is contraindicated in patients with implanted heart valves, surgical and aneurysm clips, and internal orthopedic screws and rods. The patient should remove dental bridges and appliances, credit cards, keys, hairclips, shoes, belts, jewelry or clothing with metal fasteners, wigs, and hairpieces before entering the magnetic resonance room.

LABORATORY TESTS

Culture. Culture of drainage from the ear canal or surgical incision is important in diagnosis and treatment of acute infections. Identifying the organism responsible for the infection allows the appropriate antibiotic to be used. Often with chronic infections, the culture is less helpful because gram-negative bacilli cover up the original pathogen. Drainage from the external ear is collected using a sterile cotton-tipped or polyester-tipped swab. Samples should be taken to the laboratory immediately.

Pathological Examination. Pathological examination of tissue obtained during surgery is completed to rule out a malignancy and identify any unusual problems. A cholesteatoma (cyst of epithelial cells and cholesterol found in the middle ear) is usually documented by a pathological examination.

▄▀ CRITICAL THINKING

Mr. Frank

Mr. Frank's wife expressed concern about his changing behavior during the last 6 months. She reports that Mr. Frank no longer enjoys talking to neighbors or visiting with friends in their church group, is irritable, has lost his sense of humor, and does not always answer her questions appropriately.

1. What do you suspect is wrong with Mr. Frank?
2. What examination techniques or tests might you use to assess Mr. Frank's signs and symptoms?
3. What are the expected findings?

Answers at end of chapter.

Therapeutic Measures
Medications

The medications most often used to treat ear disorders include anti-infectives, anti-inflammatories, antihistamines, decongestants, cerumenolytics, and diuretics. Anti-infectives can be administered systemically or as a topical solution. Ear medications are generally in a liquid form for ease in administration as drops. Box 48–7 and Figure 48–8 guide the nurse in administering ear drops. Anti-inflammatories, antihistamines, and decongestants are used with acute infections to reduce nasal and middle ear congestion. Cerumenolytics are used to soften cerumen and remove it from the ear canal. Diuretics are used with some inner ear disorders to reduce pressure caused by fluids.

Ear Health Maintenance

Routine cleaning and care of the ears should be taught to all patients. Patient education should include prevention of trauma, prevention of hearing loss, and early detection of

---

| BOX 48-7 | Administration of Ear Drops |

1. Wash hands
2. Ensure medication is at room temperature.
3. Position patient sitting up with head tilted toward unaffected side or side lying on unaffected side.
4. Pull auricle down and back on children and pull auricle up and back for adults.
5. Instill prescribed number of drops, being careful not to touch the tip of the dropper to prevent contamination.
6. Have patient remain in position for 2 to 3 minutes.
7. A small cotton plug may be inserted to prevent medication from running out of ear.
8. Wash hands.

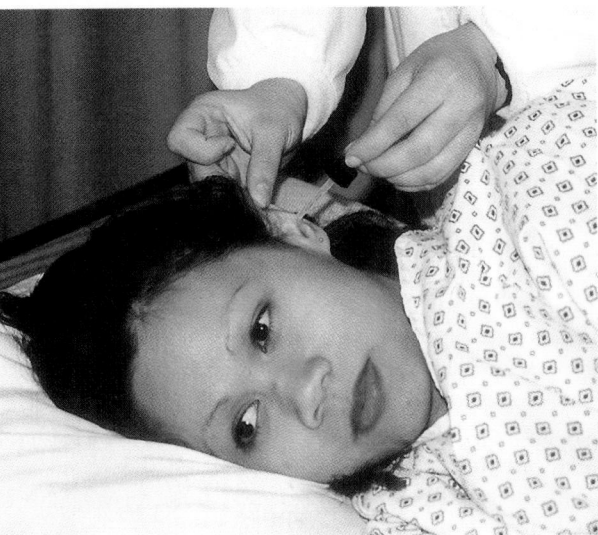

Figure 48-8 Ear drop administration.

hearing loss. All patients can benefit from this type of education, as found in Table 48–3.

Assistive Devices

Hearing aids are instruments that amplify sounds. (See Chapter 49.) Certain hearing aids may be designed to amplify sounds and attenuate certain portions of the sound signal. A microphone receives the sounds and converts them to electrical signals. These signals are amplified, and a receiver then converts the signal to sound. A small battery serves as the energy source. Digital hearing aids contain computers that provide clearer and crisper sound that is tailored to the person's hearing loss. Digital hearing aids are more expensive than analog hearing aids. There are four types of commonly used hearing aids:

1. The in-the-ear aid fits into the ear. It is small and unobtrusive to the wearer and others.
2. The behind-the-ear, or postauricular, aid is the most common type. This fits behind the ear and is comfortable to wear.
3. The all-in-one eyeglass aid combines eyeglasses with a hearing aid and is the least commonly used type.
4. The body-worn aid has a fitted ear mold inserted into the external ear and is connected to a receiver. The receiver is

wired to a transmitter, which is worn around the neck. The wearer is not able to hide the receiver and wires.

To care for a hearing aid, ensure that it is turned off and the battery is removed when it is not in use. This reduces battery expense for the patient, who may be on a fixed income. When turning the hearing aid on, the volume should be turned up just until it squeals and then turned down until the patient indicates it is at the appropriate level for hearing. At least weekly, clean the hearing aid by washing the ear mold portion with either a dry cloth or a damp, soapy cloth and then rinsing with a damp cloth. A brush may come with the hearing aid for cleaning, or a cotton-tipped swab can be used to clean the small tip that fits into the ear.

Another type of hearing aid is the implantable middle ear hearing device, called the Vibrant Soundbridge, for those with a sensorineural hearing loss. It provides sound perception by enhancing the normal middle ear hearing function. An audio processor picks up environmental sound and transmits it to the receiver implanted under the skin. The sound is then sent to a tiny floating transducer that directly vibrates the ossicles, which sends the message to the cochlea, as in normal hearing. The hair cells in the cochlea stimulate the auditory nerve, and the brain interprets the message as sound. For more information or to see the Vibrant Soundbridge, visit www.symphonix.com.

A person who is profoundly deaf and has lost all hearing may use a **cochlear implant.** All cochlear implants feature a microelectronic processor for converting the sound into electrical signals, a transmission system to relay signals to the implanted parts, and a long, slender electrode placed in the cochlea to deliver the electrical stimuli directly to the fiber of the auditory nerve. The electrode is surgically placed. Patients commonly have difficulty understanding and learning speech, even with the cochlear implant.

Diet

The patient with an ear problem usually does not have any diet modifications. However, the patient with Ménière's disease may benefit from a lower-sodium diet to prevent retention of fluid. Increased fluid may contribute to Ménière's disease symptoms.

Hearing Ear Dogs

Hearing ear dogs are now available to assist those with hearing problems. The dogs are trained to respond to sounds that the person who is hard of hearing cannot hear. Examples include a crying baby, oven timer, and smoke alarm. These dogs provide a valuable service that enriches the lives of those with hearing problems.

Answers to CRITICAL THINKING

Mr. Frank

1. He is exhibiting behaviors of hearing loss.
2. Ear inspection, a whisper voice test, a Rinne test, and a Weber test might be performed.
3. For inspection of ear, cerumen impaction may be found. For a whisper voice test, the whisper is not heard in affected ear. For a Rinne test, bone conduction is heard longer than air conduction in affected ear. For a Weber test, sound is heard louder in affected ear.

REVIEW QUESTIONS

1. Which of the following methods can be used to assess visual fields?
   a. Inspection with an ophthalmoscope
   b. Fluorescein angiography
   c. Testing vision with Snellen's chart
   d. Comparing the patient's visual fields with your own

2. Which of the following should the nurse do to reduce systemic absorption of eye drops?
   a. Instruct the patient to keep eyes closed for 30 seconds after administration
   b. Apply pressure to punctum of the eye for 1 minute after application
   c. Keep eye drops in the refrigerator before applying
   d. Have the patient look up for 1 minute after application

3. Which of the following patient behaviors is indicative of hearing loss?
   a. Turns volume low on the television
   b. Is irritable or sensitive in interpersonal relations
   c. Answers questions appropriately
   d. Complains of people talking too loudly

4. Which of the following is the most important nursing intervention during Romberg's test?
   a. Ensure patient safety
   b. Whisper softly into each ear
   c. Ensure a quiet environment
   d. Remove all cerumen from ear canal

5. Which of the following patient statements indicates that the patient understands ear care teaching?
   a. "I should insert a cotton swab into my ear canal for cleaning."
   b. "I should not get my external ear wet during bathing."
   c. "Ear wax should be routinely removed."
   d. "Aspirin can be toxic to the ears."

49

NURSING CARE OF PATIENTS WITH SENSORY DISORDERS: VISION AND HEARING

Lazette Nowicki and Debra Aucoin-Ratcliff

KEY TERMS

astigmatism (uh-**STIG**-mah-TIZM)

blepharitis (BLEF-uh-**RIGH**-tis)

blindness (**BLYND**-ness)

carbuncle (**KAR**-bung-kull)

cataract (**KAT**-uh-rakt)

chalazion (kah-**LAY**-zee-on)

conductive hearing loss (kon-**DUK**-tiv **HEER**-ing **LOSS**)

conjunctivitis (kon-JUNK-ti-**VIGH**-tis)

enucleation (ee-NEW-klee-**AY**-shun)

external otitis (eks-**TER**-nuhl oh-**TIGH**-tis)

furuncle (**FYOOR**-ung-kull)

glaucoma (glaw-**KOH**-mah)

hordeolum (hor-**DEE**-oh-lum)

hyperopia (HIGH-per-**OH**-pee-ah)

macular (**MAK**-yoo-lar)

Ménière's disease (MAY-nee-**AIRZ** di-**ZEEZ**)

miotics (my-**AH**-tiks)

myopia (my-**OH**-pee-ah)

myringoplasty (mir-**IN**-goh-PLASS-tee)

myringotomy (MIR-in-**GOT**-uh-mee)

otosclerosis (OH-toh-skle-**ROH**-sis)

photophobia (FOH-toh-**FOH**-bee-ah)

presbycusis (PRESS-bee-**KYOO**-sis)

presbyopia (PREZ-bee-**OH**-pee-ah)

retinopathy (ret-i-**NAH**-puh-thee)

sensorineural (SEN-suh-ree-**NEW**-ruhl)

stapedectomy (stuh-puh-**DEK**-tuh-mee)

QUESTIONS TO GUIDE YOUR READING

1. How would you explain the pathophysiology of each of the disorders of the sensory system?
2. How would you define blindness and the refractive errors of vision?
3. What are the etiologies, signs, and symptoms of sensory disorders?
4. What care would you provide for patients undergoing tests for sensory disorders?
5. What is the medical treatment for each sensory disorder?
6. What medications are contraindicated for patients with acute angle-closure glaucoma?
7. What are three ototoxic drugs?
8. What data should you collect when caring for patients with disorders of the sensory system?
9. What nursing care will you provide for patients with disorders of the eye or ear?
10. What nursing care interventions would you use for the patient with a hearing impairment?
11. How will you know if your nursing interventions have been effective?

Early detection and treatment of sensory injuries or diseases can reduce their impact. Any disturbance in vision or hearing disrupts a person's role performance, safety, and activities of daily living (ADLs). Treatment for sensory disturbances, such as glasses for refractive errors or hearing aids, can interfere with an individual's self-concept and body image. Nurses play an important role in recognizing symptoms of visual and hearing disorders and in assisting the individual to follow treatment, prevent recurrence, and learn new adaptive skills.

► VISION
Infections and Inflammation

Infections and inflammation of the eye and surrounding structures can be bacterial or viral in origin. The eye may become aggravated by allergens, chemical substances, or mechanical irritation, leading to infection by microorganisms. Mechanical irritation may be caused by sunburn or bacterial infection. Inflammation results from allergies to environmental substances or by irritation of chemical irritants found in perfumes, makeup, sprays, or plants. Viral agents that cause infection include herpes simplex virus, cytomegalovirus, and human adenovirus. Bacterial agents that infect the eye include *Staphylococcus* and *Streptococcus*. The most common type of acute infection is **conjunctivitis** (Cultural Consideration Box 49–1).

BOX 49-1 **CULTURAL CONSIDERATION**

Vision

Onchocerciasis, commonly called river blindness, is a filarial (worm) infection in which larvae infect almost all ocular tissues. Blindness typically follows infestation of the choroids, retina, and optic nerve. The disease is spread by blood-sucking insects such as flies, gnats, or mosquitoes that ingest the larvae from blood and then inject another host. This infection affects more than 18 million people and is predominant in Africa, South America, and Central America. All of the 18 million infected, 1 to 2 million are either blind or visually impaired. The use of the medication ivermectin has revolutionized the treatment of this disease. Ivermectin is administered annually to the infected individual for 10 years or longer.

Trachoma, a form of conjunctivitis, is a common, chronic disease that affects approximately 600 million people worldwide. It is primarily seen among low-income persons in the Mediterranean, Africa, Brazil, and the Far East. Trachoma is caused by a viral strain of *Chlamydia trachomatis* that is highly contagious. Following the acute conjunctivitis phase, the eyelids shrink as a result of scarring. The shrinking tends to pull the eyelashes inward (entropion), which may scratch the cornea. In addition, granulations form on the inner eyelids. This painful condition may eventually lead to corneal ulceration and blindness. Trachoma is medically treated with topical and oral erythromycin or tetracycline.

---

conjunctivitis: conjunctive—joining membrane + itis—inflammation

Conjunctivitis

Conjunctivitis is inflammation of the conjunctiva caused by either a virus or bacteria. Viral conjunctivitis occurs more commonly than bacterial conjunctivitis and is highly contagious. The virus is usually transmitted via contaminated eye secretions on the hand that then touches or rubs an eye, which infects the eye. The virus is hardy and may live on dry surfaces for 2 weeks or more. Viral conjunctivitis lasts 2 to 4 weeks. Bacterial conjunctivitis (commonly called pinkeye) usually is due to staphylococcal or streptococcal bacteria and is also highly contagious. Conjunctivitis can also be caused from the organisms *Haemophilus influenzae*, *Chlamydia trachomatis*, and *Neisseria gonorrhoeae*. Conjunctivitis is commonly transmitted among children and then among family members. Interestingly, conjunctivitis may also be caused from the use of nonprescription decongestant eye drops containing vasoconstrictors. When the eye drop is discontinued, a pharmacologically induced rebound phenomenon may occur. The eye vessels dilate and may become ischemic. This condition eventually subsides.

The symptoms of conjunctivitis include conjunctival redness and crusting exudate on the lids and corners of the eyes. Individuals may complain that their eyes itch and are painful. The eyes may tear excessively in response to the irritation.

Viral conjunctivitis is treated by supportive measures, which seek to keep the patient comfortable until the infection resolves on its own. Treatment includes eyewashes or eye irrigations, which cleanse the conjunctiva and relieve the inflammation and pain. Bacterial conjunctivitis is treated with antibiotic eye drops or ointments (Table 49–1). Eye drops are generally preferred by adults because they do not impair vision. Ointments are commonly used when the eye is resting (at night) or with children, who may squeeze their eyes shut and cry when ocular medications are applied, thus expelling the medication. With either type of conjunctivitis, handwashing is the best means of preventing the spread of the disease.

Blepharitis

Blepharitis, an inflammation of the eyelid margins, is a chronic inflammatory process. There are two types: seborrheic blepharitis and ulcerative blepharitis. Cause may include staphylococcal infection, seborrhea (dandruff), rosacea (a chronic disease of the skin usually affecting middle-aged and older adults), dry eye, or abnormalities of the meibomian glands and their lipid secretions. Seborrheic blepharitis is characterized by reddened eyelids with scales and flaking at base of the lashes. Ulcerative blepharitis produces crusts at eyelashes, reddened eyes, and inflamed corneas. Eyelids chronically infected with *Staphylococcus* may become thickened and eyelashes may be lost.

Treatment requires a commitment to long-term daily cleansing with cotton-tipped swabs dipped in diluted baby shampoo or sterile eyelid cleanser solutions to prevent infection. If infection occurs, antistaphylococcal antibiotic ointment (bacitracin, erythromycin) is applied to the lid

TABLE 49–1 Ophthalmic Medications

Drug Type	Uses	Nursing Considerations
Diagnostic Aids Fluorescein sodium	Staining of eye. Lesions of foreign objects pick up bright yellow-orange stain so abnormality can be detected.	Stain needs to be irrigated out of eye when examination is complete. Stain is colorfast, so caution should be used when irrigating.
Topical anesthetics	Provides local anesthesia to area, making examination painless. Also used to reduce pain of injury.	Corneal anesthesia is achieved within 1 minute and lasts about 15 minutes. The eye must be protected because the blink reflex is temporarily lost. The lid should be kept closed to keep eye moist when examination and treatment are completed.
Anti-infectives Antibiotics	Combats eye infections of bacterial origin.	Patients must be asked about previous allergic reaction to any ophthalmic or systemic medications.
Antivirals	Combats eye infections of viral origin.	To minimize systemic absorption of anti-infectives, apply pressure on tear duct up to 5 minutes after medication is applied.
Antifungals	Combats eye infections of fungal origin.	
Anti-inflammatories Steroidal Nonsteroidal	Used to reduce inflammation of the conjunctiva, cornea, or eyelids due to infection, edema, allergic reaction, or burns.	To minimize systemic absorption of anti-inflammatories, apply pressure on tear duct up to 5 minutes after medication is applied. Long-term use of corticosteroids can contribute to cataract formation.
Lubricants	Used to moisten the eyes in healthy and ill persons. Lubricants maintain the moisture on the eyeball, which contributes to maintenance of the epithelial surface.	Lubricants come in liquid and ointment forms. For patients who have ointments placed in the eye during surgery to prevent eye dryness, patient teaching should be done, informing patients that vision will be distorted in the presence of ophthalmic ointments.
Miotics	Used to lower the intraocular pressure by stimulating papillary and ciliary sphincter muscles. This assists in improving blood flow to the retina and flow of aqueous humor.	There are two types of miotics, which work differently to reduce ocular pressure—cholinergics and cholinesterase inhibitors. Miotic side effects include headache, eye pain, and brow pain. Systemic absorption can cause nausea, vomiting, diarrhea, respiratory attacks in patients with asthma, and respiratory difficulty. Pilocarpine, a miotic, causes miosis (contraction of the pupil). Expect to see a smaller than normal pupil with little if any reaction to light.
Carbonic Anhydrase Inhibitors	Used to decrease aqueous humor formation and decrease intraocular pressure. Used primarily for treatment of glaucoma when other miotics have not been successful.	Side effects include lethargy, anorexia, depression, nausea, and vomiting. Do not administer to persons allergic to sulfonamides. Carbonic anhydrase inhibitors may also cause photosensitivity. Use of the medication may cause dry eyes and dry oral membranes. Encourage patient to maintain eye and oral hygiene.
Osmotics	Used to reduce intraocular pressure in emergency situations such as acute open-angle glaucoma or used preoperatively and postoperatively to decrease vitreous humor volume, thereby reducing intraocular pressure.	Disorientation, especially in elderly, may be caused by change in electrolytes secondary to use of osmotics. Monitor for headache, nausea, vomiting, and confusion.
Anticholinergic Mydriatics	Used to dilate the pupils for examination or surgical procedures.	If pupils are dilated, they can no longer protect the eye from bright light. Instruct patient to wear dark glasses until the effects of the drug have worn off.
Cycloplegics	Used to paralyze the muscles of accommodation for examination or surgical procedures.	Contraindicated in patients with glaucoma because of increase in intraocular pressure with use. Side effects include tachycardia, dry mouth, and symptoms of atropine toxicity.

margins one to four times a day after the eyelids have been cleansed. Warm compresses may also be used.

Hordeolum and Chalazion

Another type of eyelid infection is a **hordeolum.** An external hordeolum (sty) is a small staphylococcal abscess in the sebaceous gland at the base of the eyelash (either Zeis' glands or the glands of Moll). Use of cosmetics on the eyes may contribute to hordeolum formation. Styes are small, raised, reddened areas. A second type of abscess, **chalazion** (internal hordeolum), may form in the connective tissue of the eyelids, specifically in the meibomian glands. A chalazion is larger than an external hordeolum. Styes do not cause discomfort; however, a chalazion often puts pressure on the cornea, causing more discomfort.

Hordeolum usually form and heal spontaneously within a few days and require no treatment. Chalazions may require surgical incision and drainage (I & D) if they do not drain spontaneously. If either type of abscess persists, administration of oral antibiotics may be prescribed along with application of warm compresses to aid healing.

Keratitis

PATHOPHYSIOLOGY AND ETIOLOGY. Keratitis is inflammation of the cornea and may be acute or chronic and superficial or deep. The depth of keratitis is determined by the layers of the cornea that may be affected. Keratitis may be associated with bacterial conjunctivitis, a viral infection such as herpes simplex, a corneal ulcer, or diseases such as tuberculosis and syphilis. Children may develop keratitis from vitamin A deficiency, allergic reactions, or viral diseases such as mumps or measles. Herpes simplex keratitis is the most common corneal infection in developed countries, with bacterial and fungal infections more prevalent throughout the rest of the world. People who wear contact lenses or have dry eyes, practice poor contact lens hygiene, have decreased corneal sensation, or are immunosuppressed are at increased risk of keratitis. Overnight wearing of soft contact lenses increase the risk even more. *Pseudomonas aeruginosa* is the pathogen most commonly associated with infection following overnight wear of soft contact lenses. If this infection occurs, the patient may be advised to dispose of the contaminated lenses and be treated with antibiotics.

SIGNS AND SYMPTOMS. The cornea has many pain receptors, so any inflammation of the cornea is painful. This pain increases with movement of the lid over the cornea. Other symptoms of keratitis include decreased vision, **photophobia** (sensitivity to light), tearing, and blepharospasm (spasm of the eyelids). The conjunctiva often appears reddened. In advanced cases the cornea may appear opaque (cloudy).

---

photophobia: photo—light + phobia—fear of

DIAGNOSTIC TESTS. Assessment of keratitis or corneal ulcer is made by use of a slit lamp or a handheld light. The cornea is examined by shining the light source obliquely (diagonally) across the cornea to show opacity in the cornea. Fluorescein stain may also be used to outline the area of involvement. When the stained area is viewed with a blue light, the disruption in the corneal surface shows up clear. If the patient is having pain from blepharospasm (contraction of the orbicularis oculi muscle), the examiner may instill a topical ophthalmic anesthetic such as proparacaine.

MEDICAL TREATMENT. Medical treatment may include topical antibiotics, antiviral medications for herpes simplex, cycloplegic agents (to keep iris and ciliary body at rest), and warm compresses. If the cornea is severely damaged, corneal transplant may be required. The eye may be patched to decrease the amount of eyelid movement over the cornea during healing.

COMPLICATIONS. Corneal infections are usually serious and are often sight threatening. The corneal tissue may become thin and susceptible to perforation. Untreated, keratitis can cause permanent scarring of the cornea, resulting in permanent loss of vision

Nursing Process: The Patient with Inflammation and Infection of the Eye

ASSESSMENT. Assessment of symptoms for any eye problem includes gathering subjective and objective data (Table 49–2). Objective assessment data includes the condition of the conjunctiva, eyelids and eyelashes; presence of exudate, tearing, any visible abscess on palpebral border, or a palpable abscess in eyelid; opacity of the cornea; and visual acuity testing comparing unaffected and affected eyes.

NURSING DIAGNOSIS. The major nursing diagnoses for inflammation and infection of the eye include but are not limited to the following:

■ Acute pain related to inflammation or infection of the eye or surrounding tissues
■ Disturbed sensory perception: visual related to altered sensory reception

TABLE 49-2	SUBJECTIVE ASSESSMENT OF EYE INFLAMMATION AND INFECTION CONDITIONS
W	Where is it? What part of the eye is affected? Eyelid, conjunctiva, cornea?
H	How does it feel? Pressure? Itchy? Painful? No pain? Irritated? Spasm?
A	Aggravating and alleviating factors. Is it worse when rubbing eyes or blinking? Is there photosensitivity?
T	Timing. Was there exposure to a pathogen? Previous infection or irritation? How long have symptoms persisted?
S	Severity. Is there visual impairment? Does pain affect ADLs?
U	Useful data for associated symptoms. Is patient infected with lice? Immunosuppression? Do other members of the family or peer group have symptoms? Are decongestant eye drops used? Is there exudate? Are the eyelids stuck together on awakening? Does patient wear contact lenses, soft contact lenses overnight, disposable contact lenses? Does patient have dry eyes? Is patient infected with tuberculosis, syphilis, HIV? What is typical eye hygiene?
P	Perception by the patient of the problem. What does patient think is wrong?

HIV = human immunodeficiency virus.

■ Risk for injury related to visual impairment

■ Risk for infection related to poor eye hygiene, use of contact lenses

■ Deficient knowledge related to eye disease process, prevention, and treatment from lack of previous experience

PLANNING. The patient and family are included in the planning phase to help patients return to their preillness state and prevent further eye disorders. Patient goals may include the following:

■ Indicates pain is decreased or absent as evidenced by a lower rating on a pain scale

■ Vision returns to preillness state

■ Injury does not occur as a result of impairment

■ Does not develop eye infection

■ Explains disease process, prevention, and treatment measures

■ Demonstrates treatment regimen correctly, such as administration of eye drops

NURSING INTERVENTIONS. Nursing care focuses on relieving the patient's pain, promoting safety, maintaining eye function, educating the patient about the disorder, applying medication if ordered, performing eye hygiene, and providing preventive eye care.

Assess the patient for pain. Use of dark glasses, rubbing the eye, squinting, and avoiding light may be indicators of pain that should be assessed. Eye pain is generally treated with topical anesthetic drops or ointments, antibiotics, and antiinflammatory agents. Warm or cool packs may also assist in soothing the eye. Patching of the affected eye also helps reduce pain by decreasing the movement of the eye across the eyelid. For severe pain, analgesics may also be prescribed. Other methods of pain reduction, such as guided imagery, relaxation techniques, music, or distraction, are explored.

Visual impairment of any type raises safety concerns for the patient. Safety is promoted by assessing and planning for visual impairments that may be present. Inflamed eyes often do not focus well and may have exudate, tearing, or ointment present, which interfere with vision. Patients with one eye patched should be advised that depth perception is altered and they should not drive. They must be taught to be cautious and careful when ambulating and reaching for things.

Interventions to maintain eye function must be implemented. If the patient is to rest the eye, reading and television should be discouraged because they require use of the eyes. Quiet activity, which can be carried out with the eyes closed, is best. Listening to music, radio, or a recorded book may provide distraction and rest for the eye. Contact lenses should be avoided when the eye or surrounding structure is inflamed. When the eye has healed and infection is gone, contact wear usually can be resumed. Contact lenses must be sterilized before use to prevent reinfection of the eye. Soft contact lenses that cannot be sterilized need to be discarded.

Patients require education in prevention, care of the affected eye, medication administration, safety issues, and

outcomes. Patients should demonstrate the administration of ointments or drops after teaching has occurred. The patient and family are taught how to prevent spreading the infection if it is contagious. The patient is also taught good eye hygiene to prevent further complications.

EVALUATION. The goals for the patient are met if the following occur:

■ Pain is reduced to an acceptable rating

■ Vision improves or returns to preillness level

■ Injury does not occur as a result of visual impairment

■ Infection does not occur as a result of poor eye hygiene or contact lens wear

■ Patient explains disease process, prevention, or treatment regimen accurately

■ Prescribed treatment is stated or demonstrated correctly (e.g., administering eye drops or ointments)

Refractive Errors
Pathophysiology and Etiology

Refraction refers to the bending of light rays as they enter the eye. Emmetropia, or normal vision, means that light rays are bent to focus images precisely on the macula of the retina. *Ametropia* is a term used to describe any refractive error. When an image is not clearly focused on the retina, refractive error is present. Refractive errors account for the largest number of impairments in vision. Ametropia occurs when parallel light rays entering the eye are not refracted to focus on the retina. There are four common ametropic disorders: **myopia, hyperopia, astigmatism,** and **presbyopia.**

HYPEROPIA. Hyperopia, also known as farsightedness, is caused by light rays focusing behind the retina (Fig. 49–1). People who are hyperopic see images that are far away more clearly than images that are close. Physiologically the globe or eyeball is too short from the front to the back, causing the light rays to focus beyond the retina. Hyperopia is corrected with convex lenses.

MYOPIA. Myopia, commonly referred to as nearsightedness, is caused by light rays focusing in front of the retina. The eyeball is elongated and thus the light rays do not reach the retina. Persons with myopia hold things close to their eyes to see them better. Distance vision is blurred. Myopia is corrected with concave lenses. (See Fig. 49-1.)

ASTIGMATISM. Astigmatism results from unequal curvatures in the shape of the cornea. When parallel light rays enter the eye, the irregular cornea causes the light rays to be refracted to focus on two different points. This can result in either myopic or hyperopic astigmatism. The person with astigmatism has blurred vision with distortion. The corneal irregularities can be caused by injury, inflammation, corneal surgery, or an inherited autosomal dominant trait.

PRESBYOPIA. Presbyopia is a condition in which the crystalline lenses lose their elasticity, resulting in a decrease in

---

presbyopia: presby—old age + opia—concerning vision

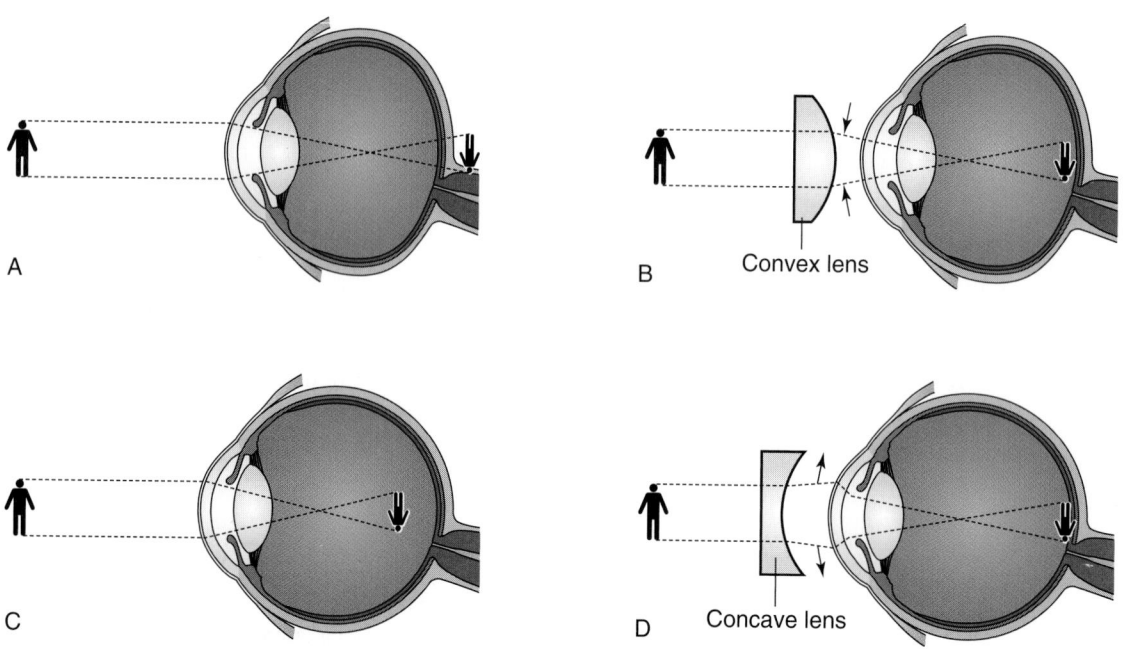

Figure 49-1 Refractive disorders. *(A)* Hyperopia (farsighted). The eyeball is too short, causing the image to focus beyond the retina. *(B)* Corrected hyperopia. *(C)* Myopia (nearsighted). A long eyeball causes the image to focus in front of the retina. *(D)* Corrected myopia.

ability to focus on close objects. The loss of elasticity causes light rays to focus beyond the retina, resulting in hyperopia. This condition usually is associated with aging and generally occurs after age 40. If an individual has preexisting hyperopia, the onset of presbyopia may occur earlier than 40 years. Likewise, if an individual has myopia, presbyopia may correct the myopia by projecting the light rays directly on the retina. Because accommodation for close vision is accomplished by lens contraction, people with presbyopia exhibit the inability to see objects at close range. They often compensate for blurred close vision by holding objects to be viewed farther away. Complaints of eyestrain and mild frontal headache are common. These symptoms are relieved with eye rest and corrective lenses.

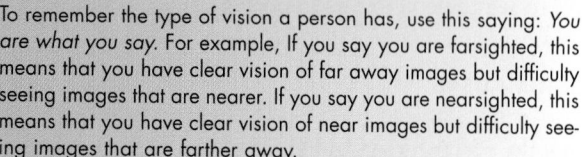

To remember the type of vision a person has, use this saying: *You are what you say.* For example, If you say you are farsighted, this means that you have clear vision of far away images but difficulty seeing images that are nearer. If you say you are nearsighted, this means that you have clear vision of near images but difficulty seeing images that are farther away.

Signs and Symptoms

Individuals with refractive errors often complain of difficulty reading or seeing objects. Often the eyestrain that occurs as one attempts to improve visual acuity causes headache. Myopic individuals may hold reading materials close to the eyes. Hyperopic individuals hold reading materials farther away from their eyes.

Diagnostic Tests

A refractive error may be roughly estimated by use of Snellen's chart or by determining the individual's vision at different distances and comparing it with that of the examiner. To obtain a more definitive refractive error measurement, a retinoscopic examination is necessary. Before this examination, a cycloplegic drug is often instilled. (See Table 49–1.) A cycloplegic drug dilates the pupil and temporarily paralyzes the ciliary muscle, thus preventing accommodation. During the examination, an ophthalmologist or optometrist examines the internal and external eye and uses trial lenses via a retinoscope to assess the type of lens best suited to correct the refractive error. If convex-shaped trial lenses correct the focusing power of the eyes, the patient is determined to be hyperopic. If concave-shaped trial lenses correct the focusing power, the patient is said to be myopic. The amount of focusing power needed in the trial lens to correct the visual defect indicates the degree of refractive error. Left and right eyes of the same person may not have the same degree of refractive error. If a cycloplegic agent has been used, patients need to be told that blurred vision will be present and sunglasses need to be worn until the agent wears off. In addition, the patient should be instructed that driving and reading are not possible until the effect of the cycloplegic drug is gone.

Medical and Surgical Treatment

Refractive errors are commonly treated with corrective lenses, either eyeglasses or contact lenses. The lenses bend the parallel light rays so that they converge on the macula portion of the retina. Incisional radial keratotomy and photorefractive keratectomy (PRK) are surgical procedures used to correct refractive error. With incisional radial keratotomy, surgical incisions are made on the cornea to reshape it. PRK utilizes laser technology to accomplish the same goal of reshaping the cornea (Box 49–2). The cornea is made flatter for individuals with myopia and more cone shaped for those with hyperopia.

Complications

Complications of corrective lens use are primarily related to safety. Eyeglasses can be broken. Eyeglass lenses can be made with special polymers that do not break as easily as traditional glass. It is a myth that if corrective lenses are not worn, vision becomes worse. Complications of contact lens use include corneal abrasions, infections, and keratitis. Incisional radial keratotomy and PRK both have surgical risks and may not always be successful.

Blindness

Blindness is described in many terms that are often reflective of the degree of visual impairment an individual has. Generally, blindness is the complete or almost complete absence of the sense of sight. Terms such as *profound blindness*, *partially sighted*, and *blind* may all have different meanings. Some people consider the terms *blind* and *partially sighted* to be negative and prefer the term *visually impaired* to describe their condition. For information or resources, visit the following sites:

■ American Foundation for the Blind at http://afb.org
■ Canadian National Institute for the Blind at www.cnib.ca
■ National Federation of the Blind at www.nfb.org
■ Prevent Blindness America at www.prevent-blindness.org

Pathophysiology and Etiology

Few people are born blind. Blindness is caused by a variety of factors, including trauma, complications from various diseases such as hypertension and diabetes, conditions such as **cataracts** and **glaucoma,** and, in children, malnutrition, infectious diseases, and parasitic infestations. Blindness is produced when there is an obstacle to the rays of light on their way to the optic nerve or by disease of the optic nerve or tract of the part of the brain connected with vision. Blindness may be permanent or transient or complete or partial or may occur only in darkness (night blindness). There are blind people in every age group, but about half the blind people in the United States are older adults.

BOX 49–2   Laser Treatment

Laser is an acronym for light amplification by stimulated emission of radiation. Lasers are devices that amplify light and produce synchronized light waves. Lasers are based on the principle that atoms, molecules, and ions can be excited by absorption of thermal, electrical, or light energy. After this energy is absorbed, the atoms, molecules, or ions give off a beam of synchronized light waves. By using this extremely intense, highly directional, pure-colored light, lasers can be used for a variety of purposes, such as making incisions, removing tissue, or stopping bleeding.

Signs and Symptoms

Aside from a general loss of vision, patients may describe their visual image as blurred, distorted, or absent in specific areas of the visual field. Objects may appear dark or absent around the peripheral field in glaucoma or retinitis pigmentosa. Retinitis pigmentosa causes this visual disturbance because the pigmented layer of the retina has degenerated. The center of the visual field may appear dark for individuals with diabetic **retinopathy** or **macular** degeneration. Half the visual field may be impaired in patients with hemianopia. This results from a defect in the optic pathways in the brain and is often seen with stroke. Patients may report that the visual field appears blurry or hazy in corneal visual problems, cataracts, diabetic retinopathy, or refractive errors (Fig. 49–2).

Diagnostic Tests

Diagnostic tests are usually done to determine the exact cause of the blindness and may include a visual field examination, tonometry, and slit-lamp microscope examination. Retinal angiography is used to follow blood flow through the retinal vessels and to detect vascular changes. Ultrasonography may be used to visualize changes in the posterior eye that cannot be directly examined because of other pathological conditions, such as a cloudy cornea, a bloody vitreous, or an opaque lens.

Medical Treatment

Medical treatment for blindness centers on treating the underlying condition and preventing further impairment. Depending on the cause of the blindness, treatment may include medication prescription, surgical intervention, corrective eyewear prescription, and referral to supportive services.

Nursing Process: The Patient with Visual Impairment

ASSESSMENT. Collection of subjective data for assessment of visual impairment can be made using the WHAT'S UP? format (Table 49–3). Collection of objective data includes observations of the patient. Is there squinting? Rubbing of eyes? Is the patient using compensatory measures—magnifying glass, sitting close to television, using large-print reading materials, avoiding reading, using eyeglasses? Psychosocial data are important because a blind person may be withdrawn or socially isolated, have low self-esteem or poor coping mechanisms, or have poor interpersonal skills as a result of the visual impairment.

NURSING DIAGNOSIS. The major nursing diagnoses for visual impairment include but are not limited to the following:

■ Disturbed sensory perception: visual related to altered sensory reception

---

retinopathy: retino—having to do with the retina + pathy—illness, disease, or suffering

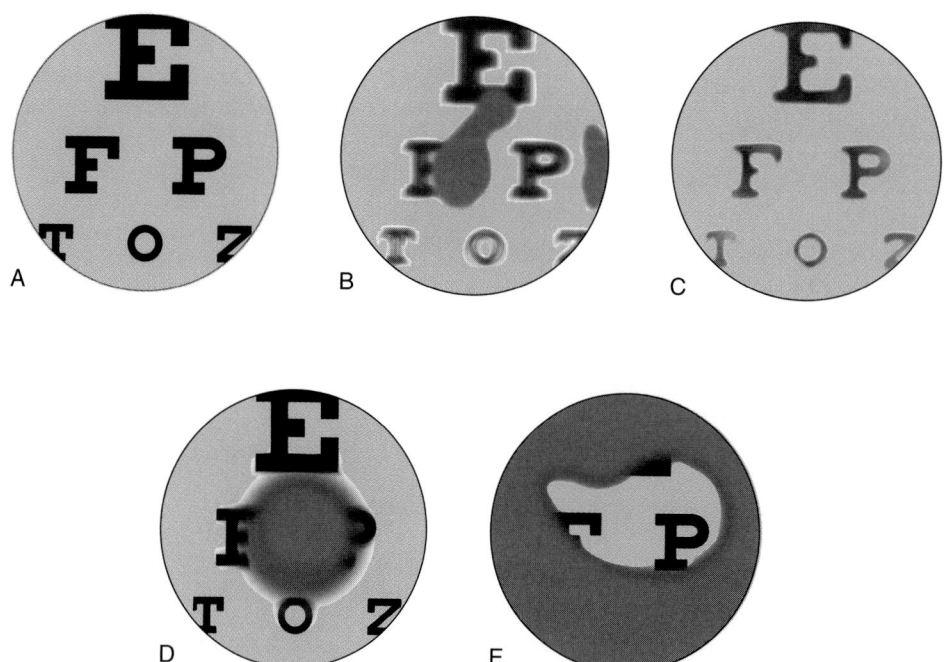

Figure 49-2   Visual field abnormalities. *(A)* Normal vision. *(B)* Diabetic retinopathy. *(C)* Cataracts. *(D)* Macular degeneration. *(E)* Advanced glaucoma.

TABLE 49-3	SUBJECTIVE ASSESSMENT OF VISUAL IMPAIRMENTS
W	Where is it? What part of the visual field is affected? If there is vision, what are the characteristics of what can be seen? Blurry? Hazy? Dark? Halos around lights?
H	How does it feel? Is there associated pain with the visual impairment? Headaches? How does it make the patient feel? Fearful? Anxious? Depressed? Helpless? Hopeless? Accepting?
A	Aggravating and alleviating factors. Is it worse when reading? Is it worse when watching TV? Does it affect the patient only at night? Is vision better at distances or close up?
T	Timing. When did the symptoms start? Do they come and go? Is the impairment progressively getting worse? Was onset sudden?
S	Severity. Does the impairment affect the patient's ADLs? If so, how severely? Does the patient need assistance to cook, dress, bathe, read mail, pay bills, access health care, obtain transportation, maintain household, shop?
U	Useful data for associated symptoms. Does the patient have diabetes, hypertension, a family history of retinitis pigmentosa, a history of eye infection, or eye trauma? Has the patient recently traveled out of the country?
P	Perception of the problem by the patient. What does the patient think is wrong? How severe does the patient perceive the impairment to be?

- Self-care deficit (specify area) related to visual impairment
- Risk for injury related to visual impairment
- Risk for impaired home maintenance related to lack of assistance, lack of rehabilitation, or other factors
- Interrupted family processes related to change in role secondary to visual impairment
- Ineffective role performance related to visual impairment, lack of rehabilitation
- Deficient knowledge related to disease process, prevention, and treatment due to lack of prior exposure
- Deficient diversional activity related to transition from sighted to visually impaired
- Fear related to blindness
- Anxiety related to sensory perception changes (visual)

See also Nursing Care Plan Box 49–3 for the Patient with Visual Loss.

PLANNING. A patient's level of independence must be included in the planning phase. If patients have minimal visual impairment or have attended rehabilitation, they may be able to function independently. If a patient has recently become visually impaired, he or she may be completely dependent until learning alternative ways of coping with this impairment. Planning focuses on meeting self-care needs, keeping the patient safe from injury, supporting the grieving process, and helping the patient acquire knowledge of agencies, services, and devices that allow maintenance of independence. Families must be included in the planning phase because they need to understand and be supportive of the self-image and role performance changes that may occur. Patient goals include the following:

- Maintains ability to complete activities of daily living with minimal assistance

BOX 49-3 **NURSING CARE PLAN FOR THE PATIENT WITH VISUAL LOSS**

▶ **Disturbed sensory perception: visual related to altered sensory reception**

PATIENT OUTCOMES

Patient will attain optimum level of sensory stimulation. Patient will become aware of visual impairment and ways to compensate. Patient will demonstrate ability to perform activities of daily living, with assistance if necessary.

EVALUATION OF OUTCOMES

Patient perceives maximum visual sensory input. Patient is able to compensate for sensory impairment by using other senses and resources. Patient is able to perform activities of daily living as independently as possible.

INTERVENTIONS	RATIONALE	EVALUATION
Assess visual acuity using a standard Snellen's vision chart. If the individual is unable to read letters, use directional arrows or pictures.	Determines patient's ability to see.	Does patient have 20/20 vision? Is there an impairment? If so, how severe is it?
Assess visual fields using the cover test or confrontation test.	Identifies deficits in visual fields.	Are visual fields of patient equal to examiner's? Is there a deficit? Is it bilateral or unilateral?
Structure environment to compensate for visual loss by adding color and contrast (e.g., chairs and carpeting should be in contrasting colors, bright tape or paint on stairs, medicine bottles color coded with colored dot stickers).	Makes the environment easier to visualize and interpret and assists in depth perception and identifying medications.	Does the environment have clearly delineated walkways, sitting areas, and doorways? Are areas with changes in elevation clearly identified using contrasting tape or paint? Is there a way for patient to safely self-administer medications?
Structure environment to compensate for visual loss by use of large-print directional signs and arrows, well-lit areas, nonglare surfaces, consistent placement of objects, traffic areas free of clutter.	Large directional signs assist the individual in maintaining orientation. Shiny floors or areas with bright window glass can impair vision. Traffic areas free of clutter assist in preventing injury.	Can the visually impaired individual identify locations such as bathroom, dining room, and office areas? Is the individual able to ambulate freely without safety hazards?
Provide for optimum care of assistive appliances such as eyeglasses, including maintenance of proper prescription, fit, and cleaning.	Improperly fitting or dirty eyeglasses may impair vision even further. Older adults should have their eyeglass prescription checked yearly.	Do eyeglasses fit properly? Are lenses clean? Is prescription current?
Introduce other assistive devices such as handheld magnifying glasses, tableside magnifiers, television magnifiers, large-print items, and phone dial covers with large numbers, talking watches, alarm clocks, and calculators.	Patients may not be aware of assistive devices, which may help them adapt to visual loss and continue previous activities such as watching TV or reading letters and magazines. Allows people to rely on hearing rather than vision.	Is patient aware of assistive devices that allow participation in previously enjoyed activities such as TV or reading? Is the visually impaired patient able to pay bills? Read mail? Communicate on telephone?
Allow patient to verbalize feelings and grieving about visual loss.	Losing a primary sense such as vision can be devastating for patients. Opportunity to ventilate feelings assists in processing the loss.	Is patient able to verbalize feelings about visual impairment and its loss?
Identify coping strategies that have been successful for patient in the past.	By identifying successful coping strategies, the nurse assists patient in dealing with stress or visual loss. A positive approach for nurses to use focuses on individual's capabilities rather than deficits.	Is patient able to identify successful coping strategies and use them to deal with the stress of visual impairment?
Refer to specialized clinician such as ophthalmologist or occupational therapist or to specialized resources such as American Federation for the Blind or Prevent Blindness America.	Specialized clinicians can provide detailed examination and treatment for the disorder. Specialized resource groups have networks in place to assist individuals in coping with loss and assisting with maximizing abilities.	Does patient know who to call for detailed examination and treatment of problems? Does patient know that there are specialized clinicians and resource groups to help with the visual impairment? Does patient know how to access these specialists?

- Remains free of injury
- Demonstrates ability to access agencies and services for visually impaired
- Maintains healthy relationships among family members
- Maintains positive self-esteem

- Verbalizes understanding of disease process, prevention, and treatment
- Participates in diversional activities
- Verbalizes lessened fear or anxiety about perceptual changes

NURSING INTERVENTIONS. Nursing care begins by understanding how to interact with the visually impaired patient (Box 49-4). Patient teaching is included in the plan of care. The goal of the teaching is to promote independence and safety for the patient while in the hospital and in the home.

Organizations exist whose mission is to enhance the independence of visually impaired persons. Referring patients to these resources enhances their ability to maintain independence. Visit the following organizations:

- American Academy of Ophthalmology at www.eyenet.org
- Americans with Disabilities Act Document Center at www.janweb.icdi.wvu.edu/kinder
- Guide Dogs for the Blind at www.guidedogs.com
- National Association for Visually Handicapped at www.navh.org
- National Eye Institute at www.nei.nih.gov
- National Eye Care Project (for qualified seniors) at www.aas.org

---

BOX 49-4 **Interacting with the Visually Impaired Patient**

- People entering a room and at each contact with the patient should identify themselves.
- Post a sign on the door or over the bed that identifies the patient's visual status so that others can interact appropriately.
- Remember that the individual is not having hearing problems, so use a normal tone of voice and do not yell.
- Ask visually impaired patients what their needs are; do not assume they need help with everything.
- Do not hesitate to use the words *blind* and *see*.
- Talk directly to the impaired patient, not through a companion.
- At mealtime, explain the location of items on the tray by comparing their position to the numbers on a clock (e.g., milk is at 2 o'clock, peas are at 7 o'clock).
- Explain any activity going on in the room or within the patient's auditory range.
- Explain procedures before beginning them. Speak to the patient before touching.
- When walking, allow the patient to grasp an assistant's arm and walk a half step behind. Be aware of obstacles on either side when walking.
- When seating a patient, place the patient's hand on the arm of the chair.
- Tell the patient when leaving the room or area so the patient does not continue conversation in an empty room, which may cause embarrassment.
- When orienting the patient to the hospital room, explain the location of items the patient may need, such as the water pitcher, call light, bed controls, urinal, tissues. Attempt to keep these items in the same place at all times.
- If the patient has a Seeing Eye dog, do not play with the dog, pet it, or feed it without consulting the patient—the dog is working! Make sure the patient's dog is near the bed, on a mat provided especially for the dog, preferably on the side of the bed that is less likely to be used by staff. Instruct staff and visitors about the Seeing Eye dog.

- Talking books: National Library Service for the Blind and Visually Handicapped at http://lcweb.loc.gov/nls/nls.html

EVALUATION. Evaluation of the nursing care plan is based on patient attainment of goals. The goals for the visually impaired patient are met if the following occur:

- Demonstrates ability to complete activities of daily living with increasing independence
- Remains free of injury
- Demonstrates ability to assess agencies and services for visually impaired
- Maintains healthy relationships among family members
- Maintains positive self-esteem
- Verbalizes understanding of disease process, prevention, and treatment
- Participates in diversional activities
- Verbalizes lessened fear or anxiety about perceptual changes

Diabetic Retinopathy
Pathophysiology and Etiology

Retinopathy is a disorder in which there are vascular changes in the retinal blood vessels. The most common incidence of retinopathy is found in persons with diabetes. It is estimated that half of the people with diabetes in the United States have at least early signs of retinopathy. The pathological changes that occur with diabetic retinopathy are related to excess glucose, changes in the retinal capillary walls, formation of microaneurysms, and constriction of retinal blood vessels. Three stages of diabetic retinopathy have been identified: background retinopathy, preproliferative retinopathy, and proliferative retinopathy.

Background retinopathy is the earliest stage, in which microaneurysms form on the retinal capillary walls. These microaneurysms may leak blood into the central retina or macula. If the leakage causes edema, the patient may notice a decrease in color discrimination and visual acuity.

The second stage, preproliferative retinopathy, is characterized by swollen and irregularly dilated veins, which results in sluggish or blocked blood flow. Patients generally are not aware of this stage because there are no symptoms.

Proliferative retinopathy, the third stage, is characterized by the formation of new blood vessels growing into the retinal and optic disc area as an attempt to increase the blood supply to the retina. The newly formed blood vessels are fragile and often leak blood into the vitreous and retina. In addition to leaking, the newer vessels may grow into the vitreous, which causes a traction effect, pulling the vitreous away from the retina and subsequently pulling the retina away from the choroid. This condition is called retinal detachment (discussed later).

Signs and Symptoms

Individuals may experience a reduction in central visual acuity or color vision as a result of macular edema. (See Fig. 49–2.) Many patients with diabetic retinopathy do not have any symptoms until the proliferative stage, at which point vision is lost. Visual loss at the last stage usually cannot be restored.

Diagnostic Tests

Diabetic retinopathy, as well as the other retinopathies, can be diagnosed only on examination of the internal eye. The examination is conducted with an ophthalmoscope following dilation of the pupil using a cycloplegic agent. The examination may be enhanced by use of retinoangiography. In the initial stages, vessels may appear swollen and tortuous (twisted).

Medical Treatment

Treatment of diabetic retinopathy focuses on stopping the leakage of blood and fluid into the vitreous and retina. The leaking microaneurysms are sealed by use of laser photocoagulation. (See Box 49–2.) If blood has already leaked into the vitreous, a vitrectomy is performed. During a vitrectomy, the vitreous humor is drained out of the eye chamber and replaced with saline or silicon oil. The replacement fluid is necessary to support the structures of the eyeball until healing can occur. Further treatment may be needed if the patient has sustained retinal detachment.

Complications

Early treatment for diabetic retinopathy is highly successful in preventing further visual loss; however, visual loss cannot be reversed. For this reason it is very important for patients with diabetes to have a comprehensive eye examination through dilated pupils at least once each year or as directed by their physician. Careful control of diabetes during the first 5 years following diagnosis reduces the occurrence and delays the onset of diabetic retinopathy.

Nursing Process: The Patient with Retinopathy

ASSESSMENT. Nursing assessment for diabetic retinopathy focuses on risk factors associated with the incidence of the disease. The patient may not have any symptoms. If patients with diabetes do have changes in perceptions of visual acuity or color discrimination, they should immediately contact their physician.

NURSING DIAGNOSIS. Nursing diagnoses for diabetic retinopathy include but are not limited to the following:

- Risk for (or actual) impaired home maintenance
- Risk for (or actual) ineffective therapeutic regimen management
- Disturbed sensory perception: visual related to altered sensory reception and transmission

PLANNING. The planning phase of the nursing process focuses on prevention of visual loss by early detection and treatment. If the patient has entered phase three and is already visually impaired, the nursing care plan for the visually impaired patient is used. Additionally, planning includes a method for monitoring blood glucose and for drawing up and administering the correct amount of insulin

for the visually impaired patient with diabetes. Specialty devices are available that can be preset to draw up amounts of insulin for the visually impaired patient with diabetes. Family members may have to assist the patient.

NURSING INTERVENTIONS. Nursing interventions for diabetic retinopathy focus on preventive eye care. Patients are taught the importance of yearly comprehensive eye examinations. Assisting patients to keep their diabetes under control also helps reduce the onset of this condition. If the patient with diabetes has experienced visual loss, nursing interventions focus on assisting the individual with home and health maintenance.

EVALUATION. Patient goals are met if the patient is able to do the following:

- Manage home maintenance
- Manage therapeutic regimen
- Prevent further visual impairment via preventive care (e.g., manage blood sugars within normal limits)

Retinal Detachment
Pathophysiology and Etiology

Retinal detachment is a separation of the retina from the choroid layer of the eye that allows fluid to enter the space between the layers. There are three types of retinal detachment:

1. Rhegmatogenous retinal detachment is caused by a hole or tear in the retina that allows fluid to flow between the two layers. The tears are related to degenerative changes in the retina or vitreous. This type of retinal detachment can also be precipitated by moderate trauma, such as stooping or lifting weights, or by direct trauma to the eye. The incidence of rhegmatogenous detachment increases with age.
2. Nonrhegmatogenous tractional detachment occurs when fibrous tissue in the vitreous humor attaches to the sensory retina and, as it contracts, pulls the retina away from its normal position. It occurs in patients with sickle cell disease or diabetes mellitus.
3. Exudative detachment occurs when fluid or exudate accumulates in the subretinal space and separates the layers. Exudative detachment occurs most often in conditions such as advanced hypertension, pre-eclampsia, or eclampsia and from intraocular tumors.

Signs and Symptoms

Patients experiencing a retinal detachment report a sudden change in vision. Initially, as the retina is pulled, "flashing lights" are reported, and then "floaters" are seen. The flashing lights are caused by vitreous traction on the retina, and the floaters are caused by the hemorrhage of vitreous fluid or blood. When the retina detaches, the patient describes it as "looking through a veil" or "cobwebs" and finally "like a curtain being lowered over the field of vision," with darkness resulting. There is no pain because the retina does not contain sensory nerves. On visual examination the patient generally has a loss of peripheral vision when the visual fields are tested and a loss of acuity in the affected eye.

Diagnostic Tests

Indirect ophthalmoscopy is used by the physician to examine the interior of the eye. This examination allows the examiner to visualize the retina, which may be pale, opaque, and in folds with retinal detachment. The examiner is able to diagnose the type of detachment based on this examination. If there are lesions in the eye, the slit-lamp examination allows the examiner to magnify the lesions.

Medical Treatment

Prompt medical treatment must be sought to prevent loss of vision. One of several procedures may be performed to reattach the retina to prevent blindness.

- Laser reattachment involves focusing a laser beam on the detached area of the retina and causing a controlled burn, which reattaches the layers together by forming an adhesion. (See Box 49–2.) This procedure is used when only a small area of the retina is involved.
- Cryosurgery involves the placement of a supercooled probe on the sclera. The probe causes injury to the tissue, forming an adhesion, a principle similar to the laser procedure.
- Electrodiathermy, the least used procedure, involves placement of an electrode needle into the sclera to allow fluid that has accumulated to drain. The retina later adheres to the choroid layer.
- Scleral buckling is a surgical procedure that involves placing a silicon implant in conjunction with a beltlike device around the sclera to bring the choroid in contact with the retina. Cryosurgery or laser is used before the buckling procedure to seal the tear and form a scar that helps adhere the retina and choroid layers together.
- Pneumatic retinopexy is a procedure that can be conducted in the physician's office and is time consuming for the patient. This procedure involves injecting air or gas into the chamber to hold the retina in place. The patient must be extremely compliant with the treatment regimen, reclining for about 16 hours before the procedure to allow the retina to fall back toward the choroid. Because air rises, the patient must maintain a position that keeps the air bubble against the detached area for up to 8 hours a day for 3 weeks.

Complications

With any of the retinal reattachment procedures there is risk of increased intraocular pressure (IOP) and recurrent detachment. The patient is also at risk for future breaks in the retina.

Nursing Process: The Patient with Retinal Detachment

Nursing process for patients with retinal detachment can be found in the section on nursing process for patients undergoing eye surgery. Assessment data specific to retinal detachment follows.

ASSESSMENT. Subjective data collected includes patient observation of the loss of peripheral vision, any change in visual acuity, and the presence of floaters, flashing lights, cobwebs, or veil-like visual impairments. There should be an absence of pain. Objective data collected includes the patient's visual acuity, visual fields, ability to perform ADLs, and level of anxiety.

CRITICAL THINKING

Mr. Samuel

Mr. Samuel, age 65, is working in the yard when a branch strikes his right eye. He sees flashes of light and then a short time later a dark shadow out of the right eye.

1. What should Mr. Samuel do?
2. After having a scleral buckling procedure, Mr. Samuel reports nausea. What action should you take?

Answers at end of chapter.

Glaucoma

Glaucoma is a group of diseases characterized by abnormal pressure within the eyeball. This pressure causes damage to the cells of the optic nerve, the structure responsible for transmitting visual information from the eye to the brain. The damage is silent, progressive, and irreversible until the end stages, when loss of peripheral vision occurs, followed by reductions in central vision and eventually blindness. (See Fig. 49–2.) Once glaucoma occurs, the patient will always have it and must follow treatment to maintain stable intraocular eye pressures.

Pathophysiology

The most common form of glaucoma, called primary, consists of two types: primary open-angle glaucoma (POAG) and acute angle-closure glaucoma (AACG). Secondary glaucoma may be caused by infections, tumors, or injuries. A third form, congenital glaucoma, primarily is due to developmental abnormalities.

AACG occurs in people who have an anatomically narrowed angle at the junction where the iris meets the cornea. When nearby eye structures such as the iris protrude into the anterior chamber, the angle is occluded, which blocks the flow of aqueous fluid. This is considered a medical emergency and results in partial or total blindness if not treated. POAG occurs when the drainage system of the eye, the trabecular meshwork and Schlemm's canal, degenerate and subsequently block the flow of aqueous humor.

Etiology and Prevention

The incidence of AACG is highest among Asians, women older than age 45, and nearsighted individuals; the incidence of POAG increases in those older than 40 years of age (older than age 50 for European-Americans, older than age 35 for African-Americans), in persons with diabetes, and in those with a family history of glaucoma and is four to five times more prevalent in African-Americans than European-Americans. Those in high-risk groups should have yearly eye examinations for glaucoma detection.

Signs and Symptoms

An ophthalmic emergency, AACG typically has a unilateral, rapid onset. The patient may complain of severe pain over the affected eye, blurred vision, rainbows around lights, eye redness, a steamy-appearing cornea, photophobia, and tearing. Nausea and vomiting may occur from the increased IOP.

POAG develops bilaterally. The onset is usually gradual and painless, so the patient may not experience noticeable symptoms or after time may experience mild aching in the eyes, headache, halos around lights, or frequent visual changes that are not corrected with eyeglasses.

Diagnostic Tests

Traditionally, tonometry is utilized to detect increased IOP (normal IOP: 12 to 20 mm Hg). Applanation tonometry uses a tiny instrument to apply pressure to the anesthetized cornea. Noncontact tonometry is performed with a tonometer mounted on a slit-lamp microscope using a warm puff of air that flattens an anesthetized area of the cornea to obtain a pressure reading. In contact tonometry the instrument is placed directly on the anesthetized cornea to measure eye pressure. Tonometry is not adequate to detect glaucoma alone, so three other methods are used. The optic nerve is examined with an ophthalmoscope through dilated pupils, visual field examination looks for loss of peripheral vision, and the angle where the iris meets the cornea is checked. In AACG, IOP may exceed 50 mm Hg. A new glaucoma screening device, the GDx Access, uses infrared laser technology to measure the thickness of the retinal nerve fiber layers to identify damage to the fibers. The advantages of this diagnostic tool are that it catches glaucoma earlier than other tests, which allows greater vision to be saved with treatment, and it is painless.

Medical Treatment

The first-line treatment for glaucoma focuses on opening the aqueous flow by administering cholinergic agents (**miotics**) such as carbachol (Isopto) or pilocarpine (Pilocar) to constrict the pupil. When the pupil is constricted, the iris pulls away from the drainage canal so that the aqueous fluid can flow freely. A second type of medication may be given to slow the production of aqueous fluid. These included carbonic anhydrase inhibitors such as acetazolamide (Diamox), adrenergic agonists such as dipivefrin (Propine), or beta blockers such as timolol (Timoptic). Slowing the aqueous fluid production helps decrease IOP. Additionally, the physician may order steroid eye drops to reduce inflammation. The patient experiencing an acute attack of AACG is given these medications and mannitol, a hyperosmolar agent, to rapidly reduce IOP, as well as analgesics and ordered to stay maintain complete bedrest.

Patients with glaucoma are required to administer lifelong eye-drop medications twice or more daily. In the absence of symptoms, compliance is often an issue. Other fac-

tors that contribute to noncompliance include age of the patient, inability to afford the medication, and lack of understanding of the disease process. Patients need to carry medical alert identification indicating they have glaucoma and what their medications are. This can help prevent administration of contraindicated medications in emergency situations.

Certain medications, regardless of their route, are contraindicated in AACG and can result in blindness if given to a patient with AACG. These medications include any anticholinergics such as atropine and antihistamines such as diphenhydramine (Benadryl) or hydroxyzine (Vistaril) because they are mydriatics. Before a medication is given, it should be determined that it is not contraindicated in glaucoma to prevent blindness from occurring.

LEARNING TIPS

- Mydriatic medications are contraindicated in acute angle-closure glaucoma because they can cause an acute episode of increased IOP by dilating the pupil and pushing the iris back, blocking the outflow of aqueous humor.
- Miotic medications constrict the pupil and so may be given to patients with acute angle-closure glaucoma.
- To remember what miotic medications and mydriatic medications do, so that the appropriate medication is given and contraindicated ones are never given, remember the following:

D = dilate = my**d**riatic = do not give
No D = constricts = miotic = okay to give

Surgical Management

When medication is no longer able to control the aqueous humor flow, surgical intervention may become necessary. Surgery focuses on creating an area for the aqueous humor to flow freely, thus preventing increased IOP. For AACG, laser peripheral iridotomy or surgical iridectomy is performed. Laser iridotomy is a noninvasive procedure utilizing a laser to remove a portion of the iris, thus allowing aqueous fluid to flow through the area. Prophylactic iridotomy may be performed on the other eye to prevent AACG. POAG is treated with argon laser trabeculoplasty (noninvasive laser beam creates openings in trabecular meshwork), trabeculectomy (part of iris and trabecular meshwork removed), or cyclocryotherapy (cryoprobe destroys part of ciliary body) (Fig. 49–3).

Nursing Process: The Patient with Glaucoma

ASSESSMENT. The patient should be assessed for loss of both central and peripheral vision, discomfort, understanding of disease and compliance with treatment regimen, and ability to conduct activities of daily living.

NURSING DIAGNOSIS. Nursing diagnoses may include the following:

- Pain related to increased intraocular pressure
- Disturbed sensory perception: visual related to altered sensory reception

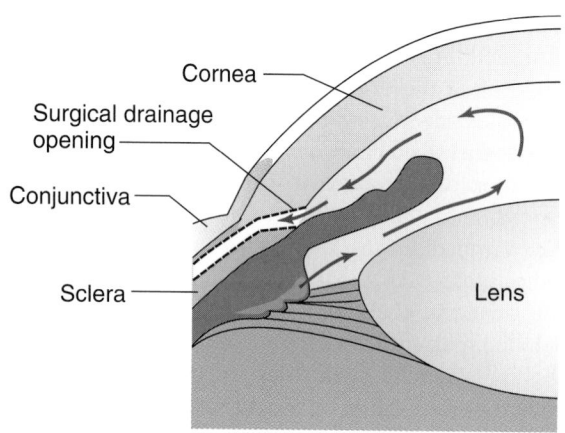

Figure 49-3   Flow of aqueous humor after trabeculoplasty (arrows).

- Self-care deficit related to decreased vision
- Anxiety related to partial or total visual loss
- Risk for injury related to decreased vision
- Impaired home maintenance related to decreased vision
- Deficient knowledge related to medical regimen, disease process due to no prior experience

PLANNING. Planning for nursing interventions needs to take into account the patient's level of understanding of disease process and medical regimen and ability to comply with the time-consuming medication regimen. The goal of nursing care for the glaucoma patient is to prevent further visual loss and to promote comfort if the patient is experiencing pain as in acute glaucoma. The patient who needs surgical intervention has additional goals (see Nursing Process: The Patient Having Eye Surgery).

NURSING INTERVENTIONS. The patient is taught how to administer medications and performs a return demonstration to ensure that eye drops are administered properly. If the patient has trouble with a steady hand when administering eye drops, teach the patient to rest his or her hand on the forehead to steady the hand. If the patient is unable to see the label on the eye drop bottle, consider large-print labels or audiotaped directions. For patients with multiple medications, consider using large, multicolored dot stickers placed on medication bottle with a corresponding direction card with a matching colored dot.

Patients are taught the need for having regular eye examinations through dilated pupils. Family members should also be advised that they are at increased risk of developing glaucoma and should have regular eye examinations.

Analgesics are given as needed for acute glaucoma. The patient is also assisted with self-care as needed. Patients are allowed to verbalize their concerns about losing their sight. Other interventions that may be implemented if the patient is experiencing severe visual loss or having surgery are in the nursing process sections for impaired vision and the patient having eye surgery, respectively.

EVALUATION. Patient goals are met if the patient does the following:

- Maintains an acceptable level of comfort
- Has no further loss of vision
- Is able to care for self with assistance if needed
- Expresses concerns and anxieties
- Does not suffer injury as a result of the visual impairment
- Is able to manage home maintenance with assistance if needed
- Demonstrates correct instillation of eye medications
- Is able to verbalize understanding of condition and treatment

Cataracts
Pathophysiology and Etiology

A cataract is an opacity in the lens of the eye that may cause a loss of visual acuity. (See Fig. 49-2.) Vision is diminished because the light rays are unable to get to the retina through the clouded lens.

Factors that contribute to cataract development may include age, ultraviolet radiation (sunlight), diabetes, smoking, steroids, nutritional deficiencies, alcohol consumption, intraocular infections, trauma, and congenital defects.

Signs and Symptoms

Cataracts are painless. Symptoms of cataract formation may include halos around lights, difficulty reading fine print or seeing in bright light, increased sensitivity to glare such as when driving at night, double or hazy vision, and decreased color vision.

Diagnostic Tests

Cataract formation is diagnosed through an eye examination. Visual acuity is tested for near and far vision. The direct ophthalmoscope and slit-lamp microscope are used to examine the lens and other internal structures.

Surgical Treatment

The only treatment for cataract formation is surgical removal of the cloudy lens. With the no-stitch cataract operation, there are no postoperative activity restrictions and vision is fine in about 2 days. After the lens is removed, there are several treatment options to correct the visual deficit that occurs when the eye is aphakic (absence of lens) and cannot accommodate or refract light properly. One treatment option is to provide the patient with eyeglasses or contact lenses that help correct the visual deficit. Another option is to replace the lens with a synthetic intraocular lens. For more information, visit the American Society of Cataract and Refractive Surgery at www.ascrs.org.

Complications

Complications of cataract surgery are rare but include inflammation, increased IOP, macular edema, retinal detachment, vitreous loss, hyphema, endophthalmitis, and expulsive hemorrhage.

Nursing Process: The Patient with Cataracts

Preoperative and postoperative nursing care is the primary nursing responsibility for the patient with cataracts. The patient is assessed for visual deficits to assist care planning, as well as knowledge needs about the disease process, surgical intervention, postoperative care, and medical regimen. The majority of patients undergoing cataract surgery have same-day surgery and then go home. The home situation, the ability of the patient or family member to follow the medical regimen, and transportation to and from the hospital for the patient are evaluated.

Nursing Process: The Patient Having Eye Surgery
Assessment

Subjective assessment data for patients having eye surgery can be collected using the WHAT'S UP? format (Table 49–4). Objective data may include visual acuity and peripheral field measurements. Visual acuity should be tested with and without corrective lenses. Eye tearing, redness, or swelling is noted.

Nursing Diagnosis

Nursing diagnoses for the patient undergoing eye surgery may include but are not limited to the following:

- Disturbed sensory perception: visual related to altered sensory reception
- Deficient knowledge related to preoperative and postoperative eye care
- Risk for infection related to surgical procedure
- Risk for injury related to altered visual acuity
- Anxiety related to visual alteration
- Fear related to surgery

Planning

The patient goals are to remain free of injury, to prevent further visual impairment by rapid diagnosis and treatment, and to have minimal anxiety surrounding the visual alteration, treatment, and recovery.

Nursing Interventions

A key nursing intervention involves patient teaching about the disease process, surgical intervention, preoperative and postoperative activity restrictions, use of dark glasses to decrease the discomfort of photophobia, use of correct technique for administration of eye medications, reporting for medical follow-up as instructed, and protecting the eye from

further injury. In some types of cataract surgery, patients may be advised to avoid activities that might increase intraocular pressure, such as vomiting, coughing, sneezing, straining, or bending over. They should not drive a car. They are told to return to the hospital if they experience sudden, worsening pain, increase in watery or bloody discharge, or sudden loss of vision, because these are signs of hemorrhage.

Anxiety is reduced by allowing patients the opportunity to discuss their feelings about the visual loss, by answering questions honestly, and by explaining any restrictions in activity. To help prevent injury, patients should be aware that their depth perception may be affected, which can result in falls. Patient should walk carefully and use clearly marked stairs. To prevent spills and slippery floors, beverages can be poured ahead of the surgery and stored in the refrigerator in single-serving glasses.

Evaluation

The patient goals have been met if the following occur:

- The patient is able to regain visual loss as a result of corrective treatment
- Injury is prevented
- Anxiety is lessened

Macular Degeneration
Pathophysiology and Etiology

Age-related macular degeneration (ARMD) is the leading cause of visual impairment in U.S. residents older than age 50. It involves a deterioration in the macula, the area on the retina where light rays converge for the sharp, central vision needed for reading and seeing small objects. The macula is also responsible for color vision (Fig. 49–4). There are two types of ARMD: dry (atrophic) and wet (exudative). In the dry form, photoreceptors in the macula fail to function and are not replaced because of advancing age. This accounts for 70 to 90 percent of the cases. In the wet form, retinal tissue degenerates, allowing vitreous fluid or blood into the subretinal space. New blood vessels are formed and compromise the macular tissue, causing subretinal edema. Eventually, fibrous scar tissue is formed, severely limiting central vision.

People at risk of developing macular degeneration include those older than age 60, those with a family history of macular degeneration, persons with diabetes, people who

TABLE 49-4	**SUBJECTIVE ASSESSMENT FOR PATIENTS HAVING EYE SURGERY**
W	Where is the visual disturbance? Is it centrally located? Peripherally? Throughout the entire visual field? Unilateral? Bilateral?
H	How does it feel? Painful? Is there an absence of pain?
A	Aggravating or alleviating factors. Is it worse in bright light or at night? Better when resting eyes or with head of bed elevated?
T	Timing. Was there a sudden onset? Gradual onset?
S	Severity. Does it affect ADLs? Does it affect close-up work?
U	Useful data for associated symptoms. Does the patient suffer from hypertension? Diabetes? Has there been trauma? Vascular disease? What is the level of anxiety? Is the patient older than age 50?
P	Perception of the problem by the patient. Will the visual disturbance impair ability to carry out ADLs? Ability to comply with medical regimen? Ability to manage home maintenance?

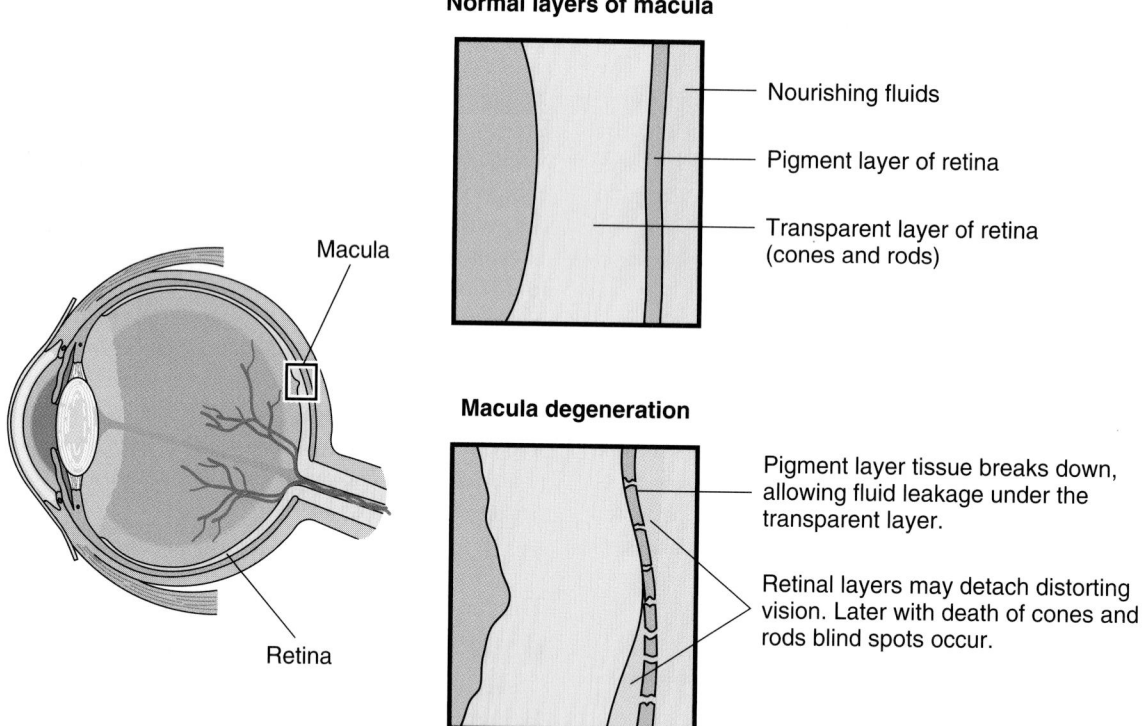

Figure 49-4 Macular degeneration. The macula is a small area of the retina responsible for central and color vision.

smoke, those frequently exposed to ultraviolet light, and Caucasian people.

Signs and Symptoms

Macular degeneration of the dry type is characterized by slow, progressive loss of central and near vision. (See Fig. 49–2.) Although individuals usually have the condition in both eyes, each eye may be affected in different degrees. Macular degeneration of the wet type has the same loss of central and near vision, but the onset is sudden. The loss can occur in one or both eyes. This visual loss is described as blurred vision, distortion of straight lines, and a dark or empty spot in the central area of vision. For some patients there may be a decreased ability to distinguish colors.

Diagnostic Tests

Examination of the patient begins with visual acuity for near and far vision and an examination of the internal eye structures with an ophthalmoscope. The examiner uses an Amsler grid (Fig. 49–5) to detect central vision distortion and a color vision test to evaluate color differences. Patients are given an Amsler grid to take home and look at on a regular basis to monitor their vision changes. If any of the grid lines look crooked or disappear, the patient should contact the physician. Intravenous fluorescein (dye) angiography to look at the retina may also be utilized to evaluate blood vessel leakage or abnormalities in the eye.

Medical Treatment

Unfortunately, there is no treatment for the dry type of ARMD. Most patients with the dry type do not lose pe-

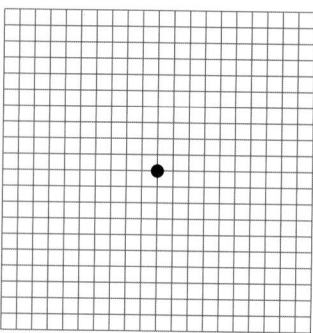

Figure 49-5 Amsler grid is used to identify central vision blind spots or distortions.

ripheral vision or become totally blind, but most are classified as legally blind (less than 20/200 vision with correction). Special low-vision lenses can enhance remaining vision. If the wet type of ARMD is diagnosed early, argon laser photocoagulation can seal the leaking blood vessels, slowing the rate of vision loss. If the patient receives argon laser photocoagulation, there is a small, permanent blind spot at the point of laser contact with the macula. With either type of ARMD, patients have significant visual loss and need to adapt their patterns of daily living.

Research is exploring cell transplant in an attempt to provide a cure in the future. Studies are focusing on transplanting retinal pigmented epithelial cells that grow and function normally.

Nursing Process

See the nursing process section for the visually impaired.

Trauma

Emergencies and trauma of the eye must be assessed immediately so that proper treatment can be initiated. Injuries to the eye include foreign bodies, burns, abrasions, lacerations, and penetrating wounds. Treatment for chemical burns and sudden, painless loss of vision should be initiated within minutes to preserve vision.

Pathophysiology and Etiology

Foreign bodies are the most common cause of corneal injury. Dust particles or propellants may lodge in the conjunctiva or cornea. Patients naturally rub their eyes to dislodge the object, which further irritates the cornea. Burns may occur from chemical, ultraviolet, and direct heat sources. Depending on the agent causing the burn, it may be superficial or deep. Abrasions and lacerations usually occur as a result of something dragging across the eye, such as a fingernail or clothing.

Penetrating wounds are the most serious eye injury. Eye structures may be damaged permanently with complete blindness resulting. A penetrating wound also puts the patient at great risk of infection.

Signs and Symptoms

Foreign bodies produce pain when the eyeball or eyelid moves, causing the foreign body to drag over the opposing surface. Usually the eye tears excessively in an attempt to irrigate the noxious substance out of the eye.

Injuries that irritate or penetrate layers of the cornea range from mild to severe pain. With corneal abrasions, the pain sensation may be delayed for several hours. Other symptoms that may be seen with abrasions, lacerations, and foreign bodies include conjunctival redness, photosensitivity, decreased visual acuity, erythema, and pruritus.

Acute pain and burning are characteristic symptoms of a burn to the eye. Chemical burns must be treated immediately with an eyewash or irrigation to remove the caustic substance from the eye.

Penetrating wounds may result in a variety of symptoms depending on the area of the eye involved and the extent of the damage. If the nerve has been damaged, the patient may have no pain.

Diagnostic Tests

With any eye trauma or injury, visual acuity must be tested. It is important to establish baseline acuity to evaluate effectiveness of treatment, although many patients resist acuity testing because of the discomfort. Testing includes examination by slit-lamp microscope and direct ophthalmoscope. Fluorescein staining is used to evaluate abrasions.

Medical Treatment

Foreign bodies are treated with a normal saline flush to irrigate the object out of the eye or to a point where it can be removed with a swab. Topical antibiotic ointment is prescribed to prevent infection.

Most chemical burns are treated with a 15- to 20-minute irrigation of either tap water at the work site or sterile solution in the health care facility. Topical antibiotic ointments are usually prescribed. Burns from heat or ultraviolet (UV) radiation are not irrigated.

Abrasions and lacerations are generally treated with anti-infective ointments or drops after cleansing the eye with a normal saline solution.

An eye specialist treats penetrating wounds. At initial injury, both eyes should be covered to prevent ocular movement. If there is a protruding object, it should be stabilized but not removed until the physician can assess the patient. The nature and extent of the penetrating wound determine treatment.

Complications

If the eye cannot be saved via medical treatment, it may be necessary to surgically remove the eye. This procedure is called **enucleation** (entire eyeball removal).

Nursing Process: The Patient with Eye Trauma

ASSESSMENT AND INTERVENTION

Foreign bodies. The eye is inspected for foreign bodies, which may be visible on the eyeball. The lids should be everted to examine the surface. Then the eye is irrigated.

Burns. Assessment of the type of burn is done because treatment options vary. Immediate irrigation of the eyes is performed once it has been established that a chemical burn has taken place. Medication and eye patching are done as indicated.

Abrasions and Lacerations. The eye is assessed for visible lacerations and then cleansed with medication and patching of the eye as indicated.

Penetrating Wounds. The patient is kept calm and relaxed to minimize eye movement and increased IOP. If a protruding object is present, the object is stabilized with tape or other supports.

NURSING DIAGNOSIS. Nursing diagnoses for the patient with eye trauma include the following:

■ Acute pain related to inflammatory process and injury
■ Disturbed sensory perception related to trauma or disease of the eye
■ Risk for infection
■ Anxiety related to visual-sensory deficit
■ Deficient knowledge related to medical regimen due to lack of prior experience

PLANNING. Planning for the eye trauma patient must take into account the ability of the patient to carry out the medical regimen, to verbalize concerns about the visual alteration, and to verbalize comfort levels.

---

enucleation: e—removed from + nuclear—center

EVALUATION. Patient goals have been met if the following occur:

■ Pain level is within an acceptable range for the patient
■ Vision is retained
■ Patient remains free of infection
■ Patient is able to verbalize a reduction in anxiety
■ Patient is able to verbalize care of the eye

➤ HEARING

Hearing Loss

Hearing loss is the most common disability in the United States and can be acquired or congenital. Hearing impairment ranges from difficulty understanding words or hearing certain sounds to total deafness. Hearing impairment can affect communication, social activities, and work activities. Hearing impairment can diminish the individual's quality of life. Nurses have a responsibility to communicate with the hearing impaired and provide necessary information regarding health care. For more information on hearing impairment, visit the following sites:

■ American Academy of Ear, Nose and Throat athttp://entnet.org
■ Canadian Hard of Hearing Association at www.cyberus. ca/~chhanational/english.html
■ National Information Center on Deafness at www.galludet.edu/~nicd
■ National Institute on Deafness and Other Communication Disorders at www.nidcd.nih.gov
■ National Organization for the Deaf at www.nad.org

Conductive Hearing Loss

Conductive hearing loss is any interference with the conduction of sound impulses through the external auditory canal, the eardrum, or the middle ear. The inner ear is not involved in a pure conductive hearing loss. Conductive hearing loss can be caused by anything that interferes with the ability of the sound wave to reach the inner ear. Conductive hearing loss is a mechanical problem. Causes of conductive hearing loss include cerumen, foreign bodies, infection, perforation of the tympanic membrane, trauma, fluid in the middle ear, cysts, tumor, and **otosclerosis.** Many causes of conductive hearing loss, such as infection, foreign bodies, and impacted cerumen, can be corrected. Hearing devices may improve hearing for conditions that cannot be corrected, such as scarred tympanic membrane or otosclerosis. Hearing devices are most effective with conductive hearing loss when no inner ear and nerve damage are present.

Sensorineural Hearing Loss

Sensory hearing loss originates in the cochlea and involves the hair cells and nerve endings. Neural hearing loss originates in the nerve or brainstem. **Sensorineural** hearing loss results from disease or trauma to the sensory or neural components of the inner ear. Some of the causes of nerve deaf-

otosclerosis: oto—ear + sclerosis—hardening

ness are complications of infections (such as measles, mumps, and meningitis), ototoxic drugs (Table 49–5), trauma, noise, neuromas, arteriosclerosis, and the aging process.

Presbycusis is hearing loss caused by the aging process that results from degeneration of the organ of Corti. This degenerative process usually begins in the fifth decade of life. The individual develops an inability to decipher high-frequency sounds (consonants s, z, t, f, g). This interferes with the individual's ability to understand what has been said, especially in noisy environments. The aging individual commonly has more difficulty understanding higher-pitched female voices than lower-pitched male voices.

Other Types of Hearing Loss

Mixed hearing loss occurs when an individual has both conductive and sensorineural hearing loss. This can be caused by a combination of any of the disorders previously mentioned. Central hearing loss occurs when the central nervous system cannot interpret normal auditory signals. This condition occurs with such disorders as cerebrovascular accidents and tumors. Functional hearing loss is a hearing loss for which no organic cause or lesion can be found. It is also called psychogenic hearing loss and is precipitated by emotional stress.

Medical and Surgical Management

Medical management consists of improving the patient's hearing. The majority of persons with ear disorders have some degree of hearing loss. With any permanent hearing loss, the use of a hearing aid should always be considered. A hearing aid is designed to amplify sound or attenuate certain portions of the sound signal and amplify other sounds. Various types of hearing aids are available (Fig. 49–6). The in-the-ear aid is a small device that fits in the ear canal. The in-the-ear aid is unobtrusive and may be preferred by the individual. The behind-the-ear aid is worn postauricular. The all-in-one eyeglass aid is attached to glasses and is positioned behind the ear. An older model is the body-worn hearing aid worn around the neck or connected to clothing.

TABLE 49–5	OTOTOXIC DRUGS
Aminoglycosides (antibiotic)	
Streptomycin	Neomycin
Gentamycin	Netilmicin
Amikacin	Kanamycin
Tobramycin	
Other antibiotics	
Vancomycin	Minocycline
Erythromycin	
Diuretics	
Furosemide	Hydrochlorothiazide
Bumetanide	
Other drugs	
Salicylates	Cisplatin
Indomethacin	Methotrexate
Quinidine	

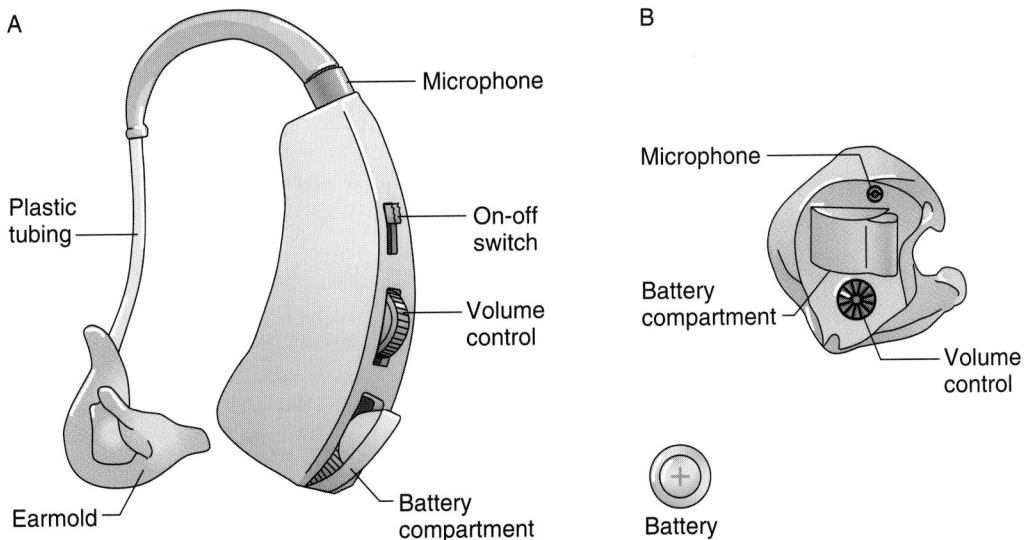

Figure 49-6 Hearing aids. *(A)* Behind-the-ear hearing aid. *(B)* In-the-ear hearing aid.

Surgical intervention may be available for patients whose hearing is not improved with hearing aids. Implantable middle ear hearing aids can improve sound perception for patients with moderate-to-severe sensorineural hearing loss. Cochlear implants are surgically placed electrical devices that receive sound and transmit the resulting electrical signal to electrodes implanted in the cochlear of the ear. The signal stimulates the cochlea, allowing the patient to hear. The cochlear implants are able to restore up to half of the patient's hearing.

Nursing Management

Nursing management includes identifying those patients at risk for hearing impairment. Patients with renal or hepatic disease, using two or more ototoxic drugs, or previously having used ototoxic drugs are at risk for developing hearing impairment. Monitor for signs of vertigo, horizontal nystagmus, nausea, vomiting, and spinning or rocking sensation while sitting still.

Nursing management for the patient with hearing impairment focuses on enhancing communication and quality of life (Nursing Care Plan Box 49–5 for the Patient with Hearing Loss). Social isolation may be more pronounced for the hearing impaired than it is for the visually impaired. Families should be included in discussions about therapeutic hearing devices and their care, enhancing communication, and limiting isolation.

Nursing Process: The Patient Who Is Hearing Impaired

ASSESSMENT. To collect assessment data for the hearing impaired patient, ask family members, as well as the patient, questions related to the patient's hearing status. Assessment of symptoms for the patient with hearing impairment includes gathering subjective data using the WHAT'S UP? format (Table 49–6).

Objective data focus on obtaining a gross screening of hearing function. Assessment should start with engaging in

TABLE 49-6	SUBJECTIVE ASSESSMENT OF HEARING IMPAIRMENT
W	Where is it? Are both ears affected? Is one side worse than the other?
H	How does it feel? Are certain words unclear, or entire conversations? Are high-frequency sounds (consonants s, t, z, f, g, and female voices) unclear or difficult to understand? Is there any pain associated with the hearing loss? Any tinnitus or vertigo?
A	Aggravating and alleviating factors. Is hearing worse in large groups or when there is a lot of background noise? Is hearing improved in a quiet environment or when speaking only to an individual? Is it easier to understand someone when seeing the person's lips move? Does the patient own or use any assistive hearing devices? Are they effective? What type is used?
T	Timing. When did the hearing loss start? Was it gradual or sudden? Is the hearing loss associated with any illness or traumatic event? Is it associated with any recent flying? Any history of ototoxic drug use?
S	Severity. Does it cause communication impairment? How much? Does it affect ADLs? Does it affect or limit usual social activities? Have family or friends commented on decreased hearing? Does patient avoid communication or social activities because of difficulty hearing? Is patient having difficulties hearing telephone voices, radio, television, or movies?
U	Useful data for associated symptoms. Is there any fever, nausea, vomiting, or dizziness? Is there any history of occupational or environmental exposure to loud noises? What are the usual ear self-care habits? Any history of impacted cerumen? Has patient ever had cerumen removed from ears?
P	Perception of problem by the patient. What does the patient feel is wrong? Does the patient think that he or she has a hearing problem? How does the patient feel about hearing assistive devices? How does the patient perceive the hearing loss, and how is it influencing the patient's life?

BOX 49-5   NURSING CARE PLAN FOR THE PATIENT WITH HEARING LOSS

 Disturbed sensory perception: hearing related to altered sensory reception and transmission

PATIENT OUTCOMES

Patient will attain optimum level of sensory stimulation. Patient will become aware of auditory impairment and ways to compensate. Patient will demonstrate ability to perform activities of daily living, with assistance if necessary.

EVALUATION OF OUTCOMES

Patient perceives maximum auditory sensory input. Patient is able to compensate for sensory impairment by using other senses and resources. Patient is able to perform activities of daily living as independently as possible.

INTERVENTIONS	RATIONALE	EVALUATION
Begin assessment of hearing by inspecting ear canals for mechanical obstruction. If cerumen is found, the use of a softening product is recommended to assist in wax removal. If canal is clear, continue assessment of hearing by use of a tuning fork, loud ticking clock, or verbal cues to determine auditory ability at various distances.	Hearing loss may occur due to the buildup of cerumen in the auditory canal. Determination of hearing ability assists nurse in developing interventions appropriate to the hearing level of patient.	Is ear canal free of mechanical obstruction? Is patient able to hear verbal input? If not, how severe is the impairment?
Enhance hearing by giving auditory cues in quiet surroundings.	Background noise such as television, radio, or large numbers of people make hearing more difficult.	Are auditory cues being delivered in an environment free of extraneous background noises? Are auditory cues being understood by patient?
Enhance understanding of auditory cues by getting patient's attention before speaking, speak slowly with careful enunciation of words, add hand gestures, speak face to face with impaired person, and adjust pitch downward without increasing volume.	Hearing is enhanced when additional cues assist the impaired individual in understanding the message. Use of hand gestures to point, lip-reading, facial expression, and lower pitch all assist communication.	Are instructions given in step-by-step format with written cues?
Structure environment to compensate for hearing loss by adding visual indicators to telephone ringer, doorbell, smoke detectors, and other emergency sounds.	Assists in communication.	Is patient able to receive input in ways other than auditory?
Provide for optimum care of assistive appliances such as hearing aids by making sure that cerumen has been cleaned from the device, that batteries are charged, and that appliance is placed correctly in ear.	Appliances that are not functioning properly will not assist patient in hearing.	Is patient's hearing aid in place correctly? Is there cerumen blocking sound conduction? Do batteries work?
Introduce other assistive devices such as hearing amplifiers, telephone amplifiers, telephones with extra-loud bells, written communication, and sign language.	Patients may not be aware of assistive devices, which may help them adapt to hearing loss and continue previous activities such as talking on telephone or listening to television.	Is patient aware of assistive devices that will allow him or her to continue to verbally communicate with others? Is patient able to use the devices to compensate for auditory impairment?
Allow patient to verbalize feelings and grieving about hearing loss.	Losing a primary sense such as hearing can be devastating for patients. Opportunity to ventilate feelings assists in processing the loss.	Is patient able to verbalize feelings about the auditory impairment and its loss?
Identify coping strategies that have been successful for patient in the past.	By identifying successful coping strategies, the nurse assists patient in dealing with stress of hearing loss. A positive approach for nurses to use focuses on the individual's capabilities rather than deficits.	Is patient able to identify successful coping strategies and use them to deal with stress of hearing impairment?
Refer to specialized clinician such as audiologist or occupational therapist or to specialized resources such as National Association of the Deaf or American Speech, Language and Hearing Association.	Specialized clinicians can provide detailed examination and treatment for the disorder. Specialized resource groups have networks in place to assist individuals in coping with the loss and assisting them with maximizing abilities.	Does patient know who to call for detailed examination and treatment of problems? Does patient know that there are specialized clinicians and resource groups to help with hearing impairment? Does patient know how to access these specialists?

normal conversation with the patient. Observe the patient for any difficulty understanding conversation or interview questions. Clarity of the patient's speech is also determined during the interview. Physical assessment includes the whisper voice, Rinne, and Weber tests. Test results provide an estimate of conductive or sensorineural hearing loss. The patient should be assessed for the underlying cause of the problem to determine if it is an external, middle, or inner ear problem. Examination of the external ear may reveal an external ear problem. The experienced practitioner may examine the ear canal for impacted cerumen or a tympanic membrane problem. Any assistive hearing devices should be noted and inspected for proper functioning. The results of the examination are documented and communicated with other health team members.

NURSING DIAGNOSIS. The major nursing diagnoses for the patient with hearing impairment may include but are not limited to the following:

- Disturbed sensory perception: auditory related to noise exposure, age, trauma, or ear disorder
- Impaired verbal communication related to impaired hearing
- Impaired social interaction related to impaired hearing and decreased communication skills
- Disturbed body image related to impaired hearing and use of assistive hearing devices
- Ineffective coping related to difficult communication
- Deficient knowledge related to care of hearing aid due to lack of prior experience

PLANNING. Planning focuses on helping the patient optimize hearing, promoting communication, and promoting adjustment to impaired hearing. The patient's goals include the following:

- Establish effective method of communication
- Maintain usual social activities
- Verbalize acceptance of altered appearance when using assistive device
- Develop coping responses to the emotional reaction of the hearing impairment
- Demonstrate care of hearing aid

NURSING INTERVENTIONS. Nursing interventions focus on establishing and maintaining effective communication. Communicating with the hearing impaired requires sensitivity to the patient's needs. Reducing or eliminating background noise provides a quiet environment. Allow extra time for patient interactions and maintain eye contact. The patient should be faced while speaking, because many hearing-impaired individuals supplement hearing with lip reading. Speak at a normal rate and volume, avoiding overarticulating or shouting and keeping hands away from the mouth when talking. If the patient hears better out of one ear, speak toward that ear. A paper and pencil should be available so the patient may write responses or questions. Using gestures enhances communication. The patient is in-

formed of topics to be discussed and when a change of topic occurs. Monitor the patient's understanding by having the patient repeat important information. During communications, avoid the appearance of frustration. Box 49–6 has additional information on communicating with the hearing impaired.

The patient is encouraged to maintain usual social activities. Assess the patient's normal activities and assist with participation in the activities by arranging transportation, scheduling activities, and involving other persons.

The patient's feelings are discussed regarding the use of assistive devices. If the patient has a hearing aid or is getting one for the first time, ensure that the patient is able to operate and care for it (Box 49–7). Check to see that hearing aid batteries are working and the volume is set at a minimal level to reduce humming. Ways to minimize the appearance of the assistive device are explored. Women may use a different hairstyle to cover the device. Stress the positive results of using the assistive device.

The patient is asked to identify prior coping methods and discuss strategies to improve hearing. Help the patient identify additional coping strategies to deal with decreased hearing.

BOX 49-6 Communicating with the Hearing Impaired

1. Get the person's attention before beginning to speak.
2. Face and stand close to the person being spoken to and maintain eye contact.
3. Avoid standing in the glare of bright sunlight or other bright lights.
4. Speak clearly, at a normal rate and volume. Do not shout or overarticulate.
5. Inform the listener of topics to be discussed and when a change of topic occurs. Stick to a topic for a while and avoid quick shifts.
6. Use short sentences and assess for understanding. If the listener does not understand after the message is repeated, rephrase the message. If the listener has difficulty with high-pitched sounds, lower the voice pitch.
7. Allow extra time for the listener to respond and do not rush the listener.
8. Ensure an optimum environment by reducing background noises by turning off television and radio, closing the door, or moving to a quieter area.
9. Encourage nonverbal communication such as touch or gestures as appropriate.
10. If the listener uses a hearing device, ensure that it is operational and in place before beginning to communicate. Give the person time to adjust the hearing device before speaking.
11. Do not smile, chew gum, or cover the mouth when talking.
12. Use active listening with attentive body posture, pleasant facial expressions, and a calm, unhurried manner.
13. Do not avoid conversation with a person who has hearing loss.
14. Use written communication if unable to communicate verbally.

BOX 49-7 **Care of Hearing Aids**

1. Insert hearing aid over a soft surface such as a pillow to prevent damage if the hearing aid is dropped during insertion.
2. Remove hearing aid before showering or bathing. Do not immerse in water.
3. Turn the hearing aid off when not in use to conserve battery.
4. Do not expose the hearing aid to extreme heat or cold.
5. Clean the hearing aid daily with a dry, soft cloth. Clean ear mold with small brush or toothpick to keep free of earwax.
6. Turn off the hearing aid and turn the volume down before inserting. Turn hearing aid on and increase volume once it is inserted.
7. Minimize whistling noise by ensuring that the volume is not too high, the aid fits securely, and the aid is free from earwax.
8. Check battery or lower the volume if sound is not clear or is intermittent. Buzzing noise may indicate that the battery door is not completely closed.
9. Do not expose the hearing aid to medicinal or hair sprays. Apply sprays before inserting hearing aid.

EVALUATION. The patient's goals are met if the following occur:

■ Patient communicates effectively
■ Patient engages in usual social activities
■ Patient verbalizes acceptance of assistive hearing device
■ Patient copes with emotional reaction of hearing impairment
■ Patient demonstrates care of hearing aid

External Ear

Infections

PATHOPHYSIOLOGY AND ETIOLOGY. Infections are the most common disorder of the external ear, with **external otitis** being the most common infection. Exposure to moisture, contamination, or local trauma provides an ideal environment for pathological growth in the external ear, which results in external otitis. It may be caused by bacterial or fungal pathogens. Staphylococci are the most common causative organisms, but other gram-negative or gram-positive bacteria can cause problems. *Pneumocystis* infections have been seen in patients who have human immunodeficiency virus (HIV). A diffuse bacterial external otitis that occurs when water is left in the ear after swimming is known as swimmer's ear. External otitis occurs more often in the summer months than in the winter months. However, swimmer's ear can be seen year round in patients who swim indoors.

A localized infection called ear canal **furuncle** or abscess results when a hair follicle becomes infected. A **carbuncle** forms when several hair follicles are involved forming the abscess. Most furuncles and carbuncles erupt and drain spontaneously. Otomycosis is an infection caused by fungal growth and is typically seen after topical corticosteroid or antibiotic use. Otomycosis occurs more often in hot weather. An infection of the auricle is called perichondritis, which can result in necrosis of cartilage.

SIGNS AND SYMPTOMS. The most common sign of infection of the external ear is pain. An early indication of infection is pain with gentle pulling on the pinna. The patient may also experience pain when moving the jaw or when the otoscope is inserted into the ear canal. Pruritus (itching) is also a common symptom and can be an early sign of infection. Signs of inflammation are present on the external ear. The ear canal may become swollen or occluded, and as a result hearing may be diminished. Redness, swelling, and drainage can be observed during otoscopic examination. If drainage is present, it usually starts out clear and becomes purulent as the disease progresses. The patient may also be febrile.

DIAGNOSTIC TESTS. Laboratory tests such as a complete blood cell count (CBC) and cultures of discharge may be completed to diagnose infections. The white blood cell (WBC) count may be elevated. Culture and sensitivity tests isolate the specific infective organism, as well as antibiotics to treat the infection. Rinne and Weber tests can indicate conductive hearing impairment.

Impacted Cerumen

PATHOPHYSIOLOGY AND ETIOLOGY. Normally the ear is self-cleaning. However, cerumen may become impacted, blocking the ear canal. People with large amounts of hair in the ear canal or who work in dusty or dirty areas are prone to cerumen impaction. Improper cleaning can also result in cerumen impaction. The older adult is at risk to develop impacted cerumen. This occurs because the amount of cerumen secreted is decreased and because of increased amounts of keratin. These two factors cause the cerumen to be drier, harder, and more easily impacted. Patients with hearing aids tend to have problems with impacted cerumen. Patients with bony growths secondary to osteophyte or osteoma are at risk for cerumen impaction.

SIGNS AND SYMPTOMS. The patient may experience hearing loss, a feeling of fullness, or blocked ear if cerumen has become impacted. Otoscopic examination reveals cerumen blocking the ear canal.

DIAGNOSTIC STUDIES. Audiometric testing reveals conductive hearing loss in the affected ear. Hearing acuity can be decreased by 45 decibels because of an impacted cerumen. Whisper voice, Rinne, and Weber tests also indicate conductive hearing loss. (See Chapter 48.)

Masses

PATHOPHYSIOLOGY AND ETIOLOGY. Benign masses of the external ear are usually cysts resulting from sebaceous glands. Other benign masses are lipomas, warts, keloids, and infectious polyps. Infectious polyps usually arise from the middle ear and enter the external ear through a hole in the tympanic membrane. Actinic keratosis is a precancerous lesion that can be found on the auricle and may be seen in the elderly. Malignant tumors such as basal cell carcinoma on the pinna and squamous cell in the ear canal may develop. These tumors can spread to surrounding tissue and bones if not treated.

SIGNS AND SYMPTOMS. Changes in the appearance of the skin can occur with benign or malignant masses. Usually impaired conductive or sensorineural hearing occurs with masses. Pain is another symptom and is usually described as deep pain radiating inward on the affected side. Ear drainage may be present. As the condition progresses, facial paralysis occurs. Visualization of the mass may be observed during otoscopic examination.

DIAGNOSTIC STUDIES. A biopsy may be obtained to determine if the mass is benign or malignant. Imaging studies are also used to diagnose tumors. Audiometric studies reveal any hearing impairment.

Trauma

PATHOPHYSIOLOGY AND ETIOLOGY. Injuries to the external ear are commonly caused by a blow to the head, automobile accidents, burns, foreign bodies lodged in the ear canal, and cold temperatures. Foreign bodies in the ear canal are common among children, with small toys being the most common object. Cotton ball pieces and insects are the most common foreign bodies found in adults.

SIGNS AND SYMPTOMS. Lacerations, contusions, hematomas, abrasions, erythema, and blistering are signs seen with thermal or physical trauma. Repeated trauma to the ear can cause hypertrophy, also known as cauliflower ear. This is common among boxers. Conductive hearing loss can occur if the ear canal is partially or totally blocked. Patients who have contusions or hematomas commonly complain of numbness, pain, and paresthesia of the auricle. The patient may or may not have symptoms associated with foreign bodies. These symptoms include decreased hearing, itching, pain, and infection. Examination of the ear canal with a penlight usually reveals the foreign body. Care should be taken during otoscopic examination not to push the foreign body further into the ear canal.

DIAGNOSTIC STUDIES. Imaging studies may be needed to determine the extent of the trauma. Audiometric, whisper voice, Rinne, and Weber testing may demonstrate conductive hearing loss.

Complications of External Ear Disorders

Complications can result from delayed treatment, no treatment, or spreading of the external ear disorder. If not treated, infections can spread, causing cellulitis, abscesses, middle ear infection, and septicemia. Metastasis can occur if malignant tumors are not treated. Infection, trauma, and malignant tumors may cause temporary or permanent hearing loss, disfigurement, discoloration, and scarring. Prompt identification and treatment of external ear disorders can prevent many complications.

Medical and Surgical Management of External Ear Disorders

External ear infections are treated with topical antibiotics in the form of drops or ointment. Systemic antibiotics are used for severe infections that are localized or have spread to surrounding tissues. Topical or systemic steroids may be used to treat inflammation. The ear is thoroughly cleaned before starting any topical treatment. If the external ear canal has drainage or is swollen shut, a wick may be inserted. The wick serves to aid in removing drainage or to aid in administering medication into the ear canal. Cerumen may be removed with installations, a blunt ear curette, or a wire ear curette. Installation is not used with history of perforated tympanic membrane (eardrum). External ear disorders are usually painful, and analgesics are used to control pain.

Débridement, surgical repair, or application of a protective covering may be done when trauma occurs to the external ear. Surgical management consists of incision and drainage of abscesses. Excision of cysts or cutaneous carcinomas may also be required.

Nursing Process: The Patient with External Ear Disorders

ASSESSMENT. Subjective data are obtained in a patient history. Data include any reports of pain, fullness, previous cerumen impaction, itching, or hearing loss, as well as onset, duration, and severity of symptoms. Additional data include patient's occupation, previous ear problems, use of a hearing aid, and typical ear hygiene. Inspection and palpation are primarily used to obtain objective data. Observe for redness, swelling, drainage, furuncles, carbuncles, lesions, abrasions, lacerations, growths, cerumen, scaliness, or crusting. The patient may report pain when the ear is palpated. Basic hearing acuity tests are conducted to evaluate hearing loss. (See Chapter 48.)

NURSING DIAGNOSIS. The major nursing diagnoses for external ear disorders may include but are not limited to the following:

■ Acute pain related to inflammation or trauma
■ Disturbed sensory perception: auditory related to altered sensory reception
■ Risk for injury related to self-cleaning of external ear
■ Deficient knowledge related to lack of information on preventive ear care

PLANNING. Planning focuses on helping the patient return to his or her preillness state and preventing any further ear disorders. The patient's goals are as follows:

■ Indicates pain is decreased or absent as evidenced by a lower rating on a pain scale
■ Hearing returns to preillness state
■ Explains or demonstrates prescribed treatment
■ Explains or demonstrates procedures to maintain wellness of the external ear

NURSING INTERVENTIONS. Nursing care focuses on relieving the patient's pain, promoting hearing, maintaining ear function, and educating the patient about the disease and preventive ear care. The patient should be assessed for nonverbal signs of ear pain. External ear pain is relieved by pharmacological and nonpharmacological interventions.

Identify with the patient an optimum analgesic schedule to promote comfort. Nonpharmacological methods such as relaxation, massage, music, guided imagery, or distraction techniques can also be used to relieve pain. Heat may be applied to the area to promote comfort. Liquid or soft foods may be offered if the patient experiences pain when chewing.

Interventions to restore hearing and relieve blockage of the ear canal are important. Topical antibiotics and antiinflammatory medications should be administered using asep-

tic technique. If the patient has a wick inserted into the ear canal, monitor for drainage. Experienced practitioners or physicians remove cerumen (Box 49–8; Fig. 49–7).

Instruct the patient how to complete the prescribed treatment and maintain ear health (Box 49–9). Patient teaching includes how to administer ear drops or ointments. The patient should keep the ear clean and dry. During an infection, cotton with petroleum jelly or earplugs should be used to avoid getting water in the ears.

EVALUATION. The goals for the patient are met if the following occur:

- Indicates pain is decreased or absent as evidenced by a lower rating on a pain scale
- Hearing improves or returns to preillness level
- States or demonstrates prescribed treatment (e.g., administering ear drops or ointments)
- Explains or demonstrates measures to maintain wellness of the external ear

---

BOX 49–8   Removing Cerumen

Instillations and irrigations should not be used on any person with a history of perforated tympanic membrane. Commercial ceruminolytics or common products such as baby oil, mineral oil, and virgin olive oil can be used to soften impacted cerumen and aid in the removal of the impacted wax. The patient should instill several drops of the solution at bedtime and then place a cotton plug in the ear to hold the solution in place. Excess oil and drainage are removed in the morning. Earwax is usually softened for 3 to 4 days before an irrigation is attempted. Patients prone to cerumen buildup should be taught how to safety remove earwax. A few drops of half-strength peroxide may be instilled into the ear canal during the day and three drops of glycerin instilled at bedtime. This can be repeated each week to minimize wax buildup.

The ear can be irrigated with an ear irrigation syringe or a Water Pik. The irrigation solution, usually water, should be warmed to body temperature. The patient is draped with a protective plastic drape, and a basin is placed below the ear to catch the irrigating solution. The patient sits with the ear toward the nurse and the head tilted toward the opposite ear. (See Fig. 49–12.) The external ear is pulled upward and backward for the adult. A low-pressure stream of water is directed toward the top of the ear canal. Care is taken not to obstruct the canal with the syringe so that the irrigation solution can backflow out of the canal. Ensure that only the tip of the syringe is in the ear canal to prevent perforation of the eardrum.

BOX 49–9   Ear Care

1. Cleanse the external ear with a wet washcloth. Gently cleanse the helix.
2. Never insert anything into the ear canal, including hairpins, cotton-tipped applicators, matchsticks, safety pins, toothpicks, paper clips, and fingers.
3. An individual with a history of ear infections, perforated tympanic membrane, or swimmer's ear should prevent moisture from entering the ear canal. Avoid swimming in contaminated water. Moisture or water in the ear canal can be prevented by using special earplugs or by using a piece of cotton rolled into a cylinder and covered with petroleum jelly.
4. Avoid home remedies for ear care without consulting a physician.
5. An individual with an upper respiratory infection should gently blow the nose with both nares open to prevent infection being forced up the eustachian tubes.

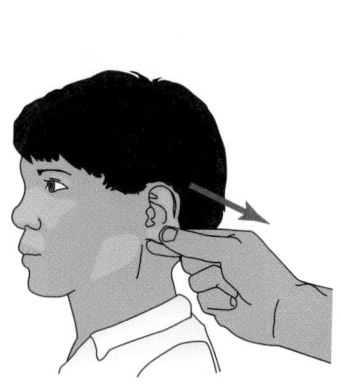

A   Pull ear back and down to straighten ear canal in a child

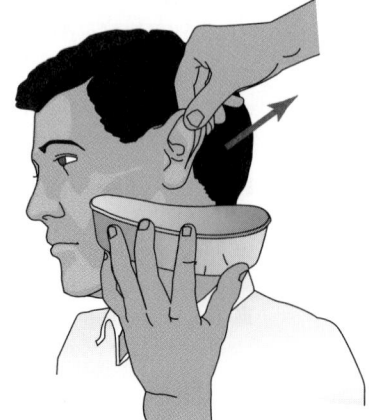

B   Pull ear up and back to straighten ear canal in an adult

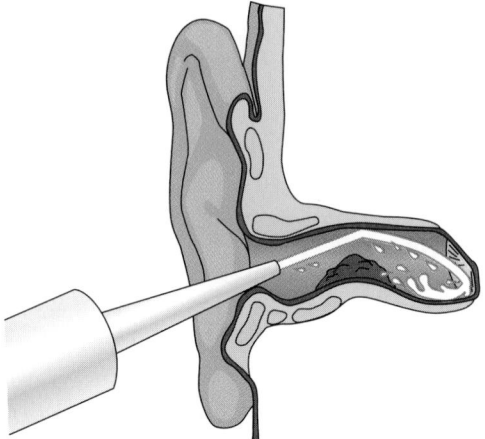

C   Irrigation – Fluid is aimed off top of ear canal wall behind impacted cerumen

Figure 49-7   Ear irrigation. *(A)* Child. *(B)* Adult.

Middle Ear, Tympanic Membrane, and Mastoid Disorders

Infections

PATHOPHYSIOLOGY AND ETIOLOGY. Otitis media is the most common disease of the middle ear. *Otitis media* is a general term for inflammation of the middle ear, mastoid, and eustachian tube. Inflammation of the nasopharynx causes most cases of otitis media. As inflammation occurs the nasopharynx mucosa becomes edematous and discharge is produced. When fluid, pus, or air builds up in the middle ear, the eustachian tube becomes blocked, and this impairs middle ear ventilation.

There are several types of otitis media in which inflammation can occur alone, with infective drainage, or with noninfective drainage. The first type of otitis media is otitis media without effusion. This is an inflammation of the middle ear mucosa without drainage. The second type of otitis media occurs when there is a bacterial infection of the middle ear mucosa. This is called acute otitis media, suppurative otitis media, or purulent otitis media. The infected fluid becomes trapped in the middle ear. If the infection continues longer than 3 months, chronic otitis media results. The third type of otitis media is otitis media with effusion. Other names include serous otitis media, nonsuppurative otitis media, and glue ear. With this type of otitis media, noninfective fluid accumulates within the middle ear.

SIGNS AND SYMPTOMS. Acute otitis media commonly follows an upper respiratory infection. A fever, earache, and feeling of fullness in the affected ear are common symptoms. As purulent drainage forms, there is pain and conductive hearing loss. Nausea and vomiting may also be present. Purulent drainage may be evident in the external ear canal if the tympanic membrane ruptures. Mastoid tenderness indicates that the infection may have spread to the mastoid area. Otoscopic examination reveals a reddened, bulging tympanic membrane.

Symptoms of otitis media with effusion may go undetected in adults because there are no signs of infection. The patient may complain of fullness, bubbling, or crackling in the ear. The patient may have a slight conductive hearing loss or allergies or be a mouth breather. Otoscopic examination can reveal a bulging tympanic membrane with a fluid level, but the eardrum is not reddened.

DIAGNOSTIC STUDIES. Laboratory studies may indicate an elevated WBC count. Ear cultures may be obtained on any drainage to identify the specific infective organism. Conductive hearing loss is usually present on audiometric studies and Rinne, Weber, and whisper voice tests. Imaging studies may be done to diagnose infection.

COMPLICATIONS. A perforation may occur with an acute or chronic infection. Buildup of fluid and pressure in the middle ear can cause a spontaneous perforation of the tympanic membrane. The patient usually experiences pain before the rupture and relief of pain after the rupture. The fluid in the middle ear moves through the perforation into the ear canal, relieving the pressure and pain. A tympanic membrane perforation causes hearing loss. The location and size of the perforation determine the extent of hearing loss. Damage to the ossicles can also occur with perforation.

Repeated infections in the middle ear or mastoid can cause a cholesteatoma, which is an epithelial cystlike sac that fills with debris such as degenerated skin and sebaceous material. The cholesteatoma starts in the external ear canal and spreads to the middle ear through a perforation in the tympanic membrane. Damage occurs in the middle ear structures as a result of pressure necrosis. The cholesteatoma causes conductive hearing loss. As the disease progresses, facial paralysis and vertigo may occur.

Tympanosclerosis is another complication of repeated middle ear infections. Tympanosclerosis consists of deposits of collagen and calcium on the tympanic membrane. The condition can slowly progress over time to the area around the middle ear ossicles. These deposits appear as chalky white plaques on the tympanic membrane and contribute to conductive hearing loss.

Mastoiditis can occur if acute otitis media is not treated. The infection spreads to the mastoid area, causing pain. Since the use of antibiotics, acute mastoiditis is relatively uncommon. Chronic mastoiditis is still seen with repeated middle ear infections.

MEDICAL AND SURGICAL TREATMENT. Medical treatment consists of treating the infections with antibiotics. Amoxicillin, penicillin V, erythromycin, cefaclor, and cotrimoxazole are commonly used. Analgesics such as aspirin, acetaminophen, or codeine or ear drops control the pain.

Surgical intervention includes several techniques. Paracentesis may be performed with a needle and syringe. The tympanic membrane is punctured with the needle, and the fluid is drained from the middle ear. A **myringotomy** may also be performed. During this procedure, an incision is made on the tympanic membrane and fluid is allowed to drain out or is suctioned out of the middle ear. An other technique is laser-assisted myringotomy, which vaporizes the tympanic membrane. Various types of transtympanic tubes may be inserted to keep the incision open. The transtympanic tube keeps the incision in the tympanic membrane open, equalizes pressure, and prevents further fluid formation and buildup. The transtympanic tubes are left in place until the infection is cured. Most tubes spontaneously extrude in 3 to 12 months and rarely have to be removed.

Reconstructive repair of a perforated tympanic membrane is called a **myringoplasty.** One technique involves placing Gelfoam over the perforation. A graft from the temporal muscle behind the ear or tissue from the external ear is then placed over the perforation and Gelfoam. The Gelfoam is absorbed, and the graft repairs the perforation.

A mastoidectomy involves incision, drainage, and surgical removal of the mastoid process if the infection has spread to the mastoid area.

---

myringoplasty: myringo—tympanic membrane + plasty—surgical repair

Otosclerosis

PATHOPHYSIOLOGY AND ETIOLOGY. Otosclerosis, or hardening of the ear, results from the formation of new bone along the stapes. With the new bone growth, the stapes becomes immobile and causes conductive hearing loss. The formation of the new bone growth begins in adolescence or early adulthood and progresses slowly. Hearing loss is most apparent after the fourth decade. Otosclerosis is more common in women than in men. The disease usually affects both ears. Although the exact cause of otosclerosis is not known, most patients have a family history of the disease. It is therefore thought to be a hereditary disease.

SIGNS AND SYMPTOMS. The primary symptom of otosclerosis is progressive hearing loss. The patient usually experiences bilateral conductive hearing loss, particularly with soft, low tones. Usually medical assistance is sought when the hearing loss interferes with the patient's ability to hear conversations. The patient may also experience tinnitus. Otoscopic examination reveals a pinkish-orange tympanic membrane because of vascular and bony changes in the middle ear.

DIAGNOSTIC STUDIES. Audiometric testing indicates the type and extent of the hearing loss. Imaging studies indicate the location and the extent of the excessive bone growth. Whisper voice test and normal conversation show decreased hearing. The patient hears best with bone conduction in the Rinne test, whereas lateralization to the most affected ear occurs with the Weber test.

MEDICAL AND SURGICAL MANAGEMENT. There is no cure for otosclerosis, but hearing aids may be used to improve hearing for the patient. The hearing aid is most effective for conductive hearing loss when there is no sensorineural involvement.

Although total restoration of hearing is not possible, reconstruction of necrotic ossicles is done to restore some of the patient's hearing. Various methods are used to reposition and replace some or all of the ossicles. Unfortunately, the surgeries are not always successful over the long term. Ossiculoplasty is the reconstruction of the ossicles. Prostheses made of plastic, ceramic, or human bone are used to replace the necrotic ossicles. Total or partial ossicular replacement prosthesis may be used. The **stapedectomy** is the treatment of choice for otosclerosis. Either part or all of the stapes is removed and replaced with a prosthesis. The prosthesis is placed between the incus and the oval window. Advances in surgical treatment include the use of lasers for improved visualization, less trauma, and greater precision during surgery. The goal is to restore vibration from the tympanic membrane to the oval window and allow sound transmission. Many patients experience improved hearing immediately, others not until swelling subsides. Complications of ossiculoplasty and stapedectomy include extrusion of the prosthesis, infection, hearing loss, dizziness, and facial nerve damage. Some patients may have the surgery repeated if complications develop.

Nursing Management

The operative ear is placed upward when lying in bed. An earplug may be used to help keep the area aseptic; the proximity of the brain makes this necessary to prevent brain infection. Activity orders may vary. The patient may be dizzy and experience nausea. Antiemetics should be given promptly to prevent vomiting. The patient's safety should be ensured if dizziness occurs. To prevent dislodgment or damage to the prosthesis, patient are instructed not to cough, sneeze, blow their nose, vomit, fly in an airplane, lift heavy objects, or shower. If the patient develops a cold, the physician should be contacted.

⏚▔ CRITICAL THINKING

Mrs. Smith

Mrs. Smith is an 83-year-old woman who is scheduled to be discharged from the hospital following a stapedectomy. She lives alone at home and is able to care for herself.

1. How would you communicate with Mrs. Smith to ensure that she understands the discharge instructions?
2. What teaching methods would you use to enhance communication?
3. What ear care instructions would you give her?

Answers at end of chapter.

Trauma

ETIOLOGY AND PHYSIOLOGY. Trauma such as a blasting force, a blunt injury to the side of the head, or sudden changes in atmospheric pressure can cause the tympanic membrane to perforate and middle ear ossicles to fracture. Blast injuries cause injury from the direct pressure on the ear. Blunt injury to the head can cause temporal skull fractures and trauma to both the middle and inner ear. Barotrauma caused by sudden changes in atmospheric pressure in the ears can occur during scuba diving and airplane take-offs and landings. Pressure changes can occur during normal atmospheric conditions such as nose blowing, heavy lifting, and sneezing. During these rapid changes of pressure, the eustachian tube does not ventilate because of occlusion or dysfunction and a negative pressure develops in the middle ear. The resulting pressure can cause the tympanic membrane to rupture or cause damage to the middle and inner ear.

SIGNS AND SYMPTOMS. Pain and hearing loss are the most common symptoms associated with trauma. Other signs and symptoms of barotrauma include fullness of the ears, vertigo, nausea, disorientation, edema of the affected area, and hemorrhage in the external or middle ear. In severe cases of barotrauma when scuba diving, these symptoms can cause drowning or cerebral air embolism from an overly rapid ascent. Otoscopic examination may reveal a retracted, reddened, and edematous tympanic membrane.

---

stapedectomy: stape(s)—stirrup + ectomy—excision of

DIAGNOSTIC STUDIES. Audiometric studies are completed to determine the hearing loss. Imaging studies may be done to determine the extent of middle and inner ear damage. Conductive or sensorineural hearing loss may be evident, depending on the extent and location of the damage.

Nursing Process: The Patient with Middle Ear, Tympanic Membrane, and Mastoid Disorders

ASSESSMENT. Assessment of symptoms includes asking the patient for subjective data using the WHAT'S UP? format (Table 49–7). The external ear should be inspected and palpated to obtain objective data. Pain with palpation is indicative of external ear problems, not middle ear problems. Pain over the mastoid area can indicate a mastoid problem. The middle ear and mastoid cavity cannot be visualized directly. The tympanic membrane is the only middle ear structure that can be directly visualized by the experienced practitioner with an otoscope. Objective assessment should also include vital signs, noting any elevation in temperature. Hearing acuity should be screened by the experienced practitioner using the whisper voice, Rinne, and Weber tests. Any drainage from the ear should be noted and described.

NURSING DIAGNOSIS. Nursing diagnoses for middle ear disorders may include but are not limited to the following:

- Risk for infection related to broken skin, pressure necrosis, chronic disease, or surgical procedure
- Disturbed sensory perception: auditory related to altered sensory reception
- Acute pain related to fluid accumulation, inflammation, or infection
- Fear related to hearing loss and lack of information
- Deficient knowledge related to lack of exposure to information due to no prior experience

PLANNING. The patient and family are included in planning care. Planning focuses on improving hearing, controlling infection, and improving the quality of life. The patient's goals include but are not limited to the following:

- Exhibits no signs of infection (no drainage from ear, no tenderness over mastoid, negative culture, afebrile)

- Hearing improves or stabilizes
- Indicates pain is decreased or absent as evidenced by a lower rating on a pain scale
- States methods for preventing problems in the middle ear, tympanic membrane, and mastoid process
- States rationale and desired outcome of any impending surgery

NURSING INTERVENTIONS. The family should be included in teaching sessions. The patient is instructed about medications. The patient is instructed to complete all therapy as directed by the physician. Ensure that the patient knows how to administer ear drops and ear ointment. Care of the ears to prevent infection is discussed. The patient is also taught to avoid getting water in the ears.

If medical or surgical management is necessary to improve the patient's hearing, the patient is instructed on this treatment. Methods to improve communication with the hearing impaired are used. (See Nursing Care Plan Box 49–3.)

Monitor pain using a pain scale, and provide measures to promote comfort. Nonpharmacological measures for pain reduction may include heat, distraction, and relaxation techniques. If analgesics are used, the optimum schedule for pain control is identified.

The patient is instructed on methods to maintain ear health and prevent ear damage from trauma, noise exposure, and environmental or occupational conditions. To equalize ear pressure, the patient should yawn or perform the jaw-thrust maneuver (opening mouth wide and moving jaw). Preventing the spread of upper respiratory infections up the eustachian tube is accomplished by telling the patient not to blow the nose by pinching off a nares. If the patient's pain worsens, hearing decreases, or drainage from ear is present, the patient should seek further medical attention. Prior to surgical procedures, the patient's knowledge is assessed and instructions are given as needed (Table 49–8).

EVALUATION. The goals for the patient are met if the following occur:

- Does not have ear drainage or pain over mastoid; has negative culture and remains afebrile
- Responds appropriately to auditory cues indicating improvement of hearing

TABLE 49–7	SUBJECTIVE ASSESSMENT OF MIDDLE EAR, TYMPANIC MEMBRANE, AND MASTOID DISORDERS
W	Where is it? Are both ears affected? Is it deep within the head?
H	How does it feel? Is there pressure? Fullness? Is it painful—sharp, dull, continuous, intermittent, throbbing, localized? No pain?
A	Aggravating and alleviating factors. Is it worse with change of position? Worse with movement? Is there relief after drainage? Relief with change of position? Relief with heat or analgesics?
T	Timing. When did it start? Has there been any recent upper respiratory infection, airline travel, scuba diving, trauma, or weight lifting? Was it a gradual or sudden onset? How long have symptoms persisted? Has there been a change in symptoms?
S	Severity. Does it cause hearing impairment? How much? Does it affect ADLs?
U	Useful data for associated symptoms. Is there any fever, drainage from the ear canal, nausea, vomiting, dizziness? Is there a family history of otosclerosis? Any previous ear problems or ear surgeries? Any occupational or recreational risk factors, such as scuba diving, weight lifting, or frequent airline travel?
P	Perception by the patient of the problem. What does the patient think is wrong? Has problem occurred before? If so, how was it the same and what was different?

TABLE 49-8	PREOPERATIVE AND POSTOPERATIVE NURSING INTERVENTIONS FOR THE PATIENT HAVING EAR SURGERY

Preoperative Care

Nursing care for the patient undergoing ear surgery begins as soon as the decision to have surgery is made. The nurse collects data, determines if the patient understands the events, assesses the patient's mental readiness, and obtains baseline physiological data.

1. Assess understanding of the surgery and explain whether local or general anesthesia will be used.
2. Help alleviate the patient's fear by encouraging the patient to ask questions. Ensure that all questions are answered before the surgery.
3. Explain the type of pain, any packing or dressings that may be in place postoperatively, and any other postoperative restrictions that may be needed.
4. Establish baseline vital signs and document findings.
5. Ensure that the operative permit is signed.
6. Determine current medications the patient is taking and document in the patient's record.
7. Assess if the patient understands that surgery does not always correct impaired hearing.
8. Leave any hearing devices in place as long as possible before the surgery.

Postoperative Care

Postoperatively the nurse is responsible for assessing the patient's physiological status. The nurse is also responsible for ensuring that the patient and family members understand discharge instructions.

1. Some degree of pain may be expected, even with minor procedures. Explain how and when to take pain medication when the patient is discharged.
2. Monitor postoperative vital signs and return to presurgical baseline.
3. Tell patients that if an occlusive dressing is in place, hearing may be decreased until the dressing is removed.
4. Instruct patients with tubes to avoid getting water in the ear. A shower cap or earplugs may be used.
5. Instruct the patient to seek medical attention if excessive bleeding or drainage occurs. If a cotton plug is to be left in place, instruct the patient to change it daily.
6. Teach the patient, unless contraindicated, to blow the nose very gently one side at a time for the first week after surgery. Instruct the patient to sneeze or cough with the mouth open for 1 week after surgery.
7. Avoid airplane flights for 1 week after surgery. For sensations of ear pressure, hold nose, close mouth, and swallow to equalize pressure.
8. The patient should avoid strenuous work for several weeks. The patient may return to work in a few days, depending on the type of surgery and the type of work the patient does.
9. Tell the patient to take prescribed medication and antibiotics as ordered.
10. Have the patient arrange for follow-up appointment by calling physician's office.

■ States that no pain is present or pain is decreased
■ Verbalizes care of ears, methods to prevent further infection; describes signs requiring medical attention
■ Verbalizes rationale and outcome for any upcoming surgery

Inner Ear
Labyrinthitis

PATHOPHYSIOLOGY AND ETIOLOGY. Labyrinthitis is an inflammation or infection of the inner ear and can be caused by either viral or bacterial pathogens. The bacteria or virus enters the inner ear from the middle ear, meninges, or bloodstream. Serous labyrinthitis is a type of acute labyrinthitis that sometimes follows drug intoxication or overindulgence in alcohol. It can also be caused by an allergy. Diffuse suppurative labyrinthitis occurs when acute or chronic otitis media spreads into the inner ear or after middle ear or mastoid surgery. Destruction of soft tissue structures from the infection can cause permanent hearing loss.

SIGNS AND SYMPTOMS. Vertigo, tinnitus, and sensorineural hearing loss are the most common symptoms. Vertigo, or dizziness, occurs when the vestibular structures are involved. Tinnitus, or ringing in the ear, occurs when the infection is located in the cochlea. Sensorineural hearing loss can be caused by infections in the cochlea or vestibular structures. Nystagmus on the affected side may occur. Other signs and symptoms include pain, fever, ataxia, nausea, vomiting, and beginning nerve deafness.

DIAGNOSTIC TESTS. Laboratory tests such as a CBC may be completed to diagnose infection. Thorough hearing evaluation by an audiologist may reveal mild to complete hearing loss. Rinne and Weber tests can indicate conductive or sensorineural hearing loss.

MEDICAL MANAGEMENT. Antibiotics are used to treat bacterial inner ear infections. Viral infections usually run their course in about 1 week. Mild sedation may help the patient relax. Although there is no specific medicine to relieve dizziness, antihistamines can be used if they prove helpful on an individual basis. Patients may be placed on bedrest.

NURSING MANAGEMENT. Nursing management includes helping the patient manage symptoms, self-care, and educating the patient about safety issues while on bedrest and sedatives. The patient should avoid turning the head quickly to help alleviate the vertigo. The patient is assisted to cope with anxiety that may be present because of the frustration surrounding hearing loss or loss of work.

Neoplastic Disorders

PATHOPHYSIOLOGY AND ETIOLOGY. Inner ear tumors can be benign or malignant. Acoustic neuroma, a tumor of the eighth cranial nerve, is the most common benign tumor. It is slow growing, occurs at any age, and usually occurs unilaterally. As it spreads, it compresses the nerve and adjacent structures. Malignant tumors arising from the inner ear are rare. Squamous and basal carcinomas arise from the epidermal lining of the inner ear.

SIGNS AND SYMPTOMS. Early symptoms of an acoustic neuroma include progressive unilateral sensorineural hearing loss of high-pitched sounds, unilateral tinnitus, and intermittent vertigo. Headache, pain, and balance disorders may also be present. Symptoms progress as the tumor spreads to other structures. Most malignant tumors grow quickly. The symptoms vary depending on the area of the ear that is involved.

DIAGNOSTIC TESTS. Neurological, audiometric, and vestibular testing are used to diagnose the neuroma. Auditory brainstem evoked response (ABR) and electronystagmography (ENG) are completed. Examination of the cerebrospinal fluid shows increased protein. Computed tomography (CT) and magnetic resonance imaging (MRI) are used to determine size and location of the tumor.

MEDICAL AND SURGICAL TREATMENT. The preferred method of treatment involves surgical removal of the tumor. The labyrinth is destroyed, with a resulting permanent hearing loss. Steroids and radiation may be used to decrease the size of the tumor or for inoperable tumors.

NURSING MANAGEMENT. Nursing management focuses on preparing the patient for surgery and adjusting to the diagnosis and the resulting hearing loss.

Ménière's Disease

PATHOPHYSIOLOGY AND ETIOLOGY. Ménière's disease is a balance disorder. Its cause is unknown. With the disease, there is a dilation of the membranous labyrinth resulting from a disturbance in the fluid physiology of the endolymphatic system. The exact etiology is unknown but is thought to stem from hypersecretion, hypoabsorption, deficit membrane permeability, allergy, virus, hormonal imbalance, or mental stress. The disease usually develops between 40 and 60 years of age. The symptoms range from vague to severe and debilitating.

SIGNS AND SYMPTOMS. A triad of symptoms of vertigo, hearing loss, and tinnitus characterizes Ménière's disease. Recurring episodic bouts of the incapacitating triad of symptoms and nausea and vomiting occur with Ménière's disease. The attacks may occur suddenly, or the patient may experience warning signs such as headache or fullness in the ears. During an acute episode, the patient experiences vertigo that lasts 2 to 4 hours. The vertigo is usually accompanied by nausea and vomiting, followed by dizziness and unsteadiness. The patient is uncoordinated and has gait changes when walking. Hearing loss is often described as a fluctuating fullness in the ears. Tinnitus is present. Irritability, depression, and withdrawal are common behavioral changes. The vital signs usually remain normal. It takes several weeks for symptoms to resolve, and hearing loss in the affected ear remains. The patient then enters a stage of remission until the next attack. The acute episodes occur two to three times per year. Eventually the patient

has complete remission with some degree of permanent hearing loss.

DIAGNOSTIC TESTS. Diagnostic tests include audiometric studies, neurological testing, and radiographs of the internal ear. Audiometric studies identify the type and magnitude of the hearing loss. Neurological testing and radiographic studies are done to rule out other pathological conditions. A caloric stimulation test may demonstrate a difference in eye movement.

MEDICAL AND SURGICAL TREATMENT. Medical treatment consists of symptomatic treatment for acute attacks and prophylactic treatment between attacks. Tranquilizers and vagal blockers may be needed during acute attacks. Salt-restricted diet, diuretics, antihistamines, and vasodilators are used during prophylactic treatment. The patient should avoid alcohol, caffeine, and tobacco use. The patient may be placed on bedrest during acute attacks. Most patients respond to medical protocol but continue to have acute attacks. Some patients who do not respond to treatment may be placed on low doses of methotrexate. The goals of medical treatment are to preserve hearing and reduce symptoms.

Surgical treatment is used only when medical management has failed. When involvement is unilateral, a labyrinthectomy is performed. This causes complete loss of hearing in that ear. Another surgical intervention establishes a shunt from the inner ear to the subarachnoid space. This procedure helps drain the fluid and prevent future hearing loss. Another surgical treatment is intratympanic gentamycin injection, which is usually done in the physician's office.

NURSING MANAGEMENT. Nursing management focuses on managing the patient's symptoms and providing safety during the acute attacks. Administer medication, monitor fluid and nutritional status, and ensure safety. Because of the unpredictability of Ménière's disease, the nursing care focuses on emotional support for the patient during periods of remission. Provide emotional support and resources to help the patient cope with the unpredictable nature of the disease and the physical impairments associated with the disease.

Nursing Process: The Patient with Inner Ear Disorders

ASSESSMENT. Assessment of symptoms for the patient with inner ear disorders includes asking the patient for subjective data using the WHAT'S UP? format (Table 49–9). Objective data include assessment of gross hearing. The whisper voice, Rinne, and Weber tests and a physical examination can be performed. The patient should be assessed for any nutritional deficiencies, including dehydration, weight loss, or weight gain. Any musculoskeletal abnormalities such as unsteady gait are also noted. The patient's vital signs are taken to determine if symptoms are associated with an infection. Laboratory data and diagnostic data are examined for abnormal findings.

TABLE 49-9	SUBJECTIVE ASSESSMENT OF INNER EAR DISORDERS
W	Where is it? Are both ears affected?
H	How does it feel? Is there pressure? Fullness? Vertigo? Tinnitus? Is it painful—sharp, dull continuous, intermittent, throbbing, localized? No pain?
A	Aggravating and alleviating factors. Is it worse with change of position? Worse with movement? Is there relief with medications? Is the patient taking current medications? Are there any allergies?
T	Timing. When did it start? Was it a gradual or sudden onset? How long have symptoms persisted? Do symptoms progress in a set timing pattern? Do symptoms occur together or separately? Has there been a change in symptoms?
S	Severity. Does it cause hearing impairment? How much? Does it affect ADLs, nutritional intake, work, or leisure?
U	Useful data for associated symptoms. Any fever, nausea, vomiting, or dizziness? Any previous ear problems or ear surgeries? Headache?
P	Perception by the patient of the problem. What does the patient think is wrong? Has patient had this problem before? If so, what was the same and what was different?

NURSING DIAGNOSIS. The major nursing diagnoses for internal ear disorders may include but are not limited to the following:

■ Anxiety related to unpredictability of sudden and severe acute attacks
■ Fear related to potential permanent hearing loss
■ Ineffective role performance related to impaired equilibrium
■ Impaired social interaction related to communication barriers
■ Grieving related to hearing loss
■ Risk for injury related to impaired equilibrium
■ Deficient fluid volume related to nausea and vomiting
■ Nutrition: less than body requirements related to nausea and vomiting
■ Deficient knowledge related to lack of information of diagnosis and treatment due to no prior experience

PLANNING. Planning focuses on helping the patient maintain a normal lifestyle, remain free of injuries, cope with the illness or hearing loss, and maintain adequate nutrition and hydration. The patient's goals are as follows:

■ Signs of anxiety are decreased
■ Is not injured from falling
■ Identify ways of increasing meaningful relationships and diversional activities
■ Experiences adequate nutrition and hydration

NURSING INTERVENTIONS. Nursing care focuses on relieving anxiety, protecting the patient from injuries, maintaining and developing relationships, and providing adequate nutrition. Encourage the patient to explore concerns about hearing loss and the unpredictability of acute attacks. The patient's understanding about the disorder should be explored and additional information provided as needed. A medication regimen to help control symptoms should be explained. Explain methods to minimize symptoms during acute episodes, such as decreasing movement.

Nursing measures to prevent injury include assessing the patient for any pattern of dizziness and environmental hazards. Instruct the patient to seek assistance before ambulating. Antivertiginous drugs are administered, and the patient is taught the medication regimen. Environmental hazards such as throw rugs, electrical cords in walkways, and poor lighting should be removed by the family.

Assist the patient in maintaining relationships and activities that were present before the disorder. Provide information about disease processes to ensure that the patient is informed about the disease. Diversional activities are planned.

EVALUATION. The goals for the patient are met if the following occur:

■ Signs of anxiety are decreased
■ Patient remains free from injury
■ Patient verbalizes satisfaction with relationships
■ Patient maintains weight within normal range

Answers to CRITICAL THINKING

Mr. Samuel

1. Mr. Samuel should seek assistance, patch both eyes, and have someone take him to receive medical treatment immediately.
2. Ensure that an antiemetic is ordered postoperatively on the patient's return to the unit. When Mr. Samuel reports nausea, the antiemetic should be given promptly.

Mrs. Smith

1. Gain her attention, face and stand in her visual field, avoid glare, speak clearly, inform her of topics to be discussed, assess for understanding, allow extra time, reduce background noises, use nonverbal communication, and do not cover your mouth when talking.
2. Use active listening. Use written communication to enhance spoken words. Use demonstration and return demonstration. Allow questions. Do not hurry. Provide information in short segments. Reassess understanding at each session.
3. Place the operative ear upward when lying in bed. Use ear plug as ordered. Do not cough, sneeze, blow nose, vomit, fly in an airplane, lift heavy objects, or shower. If a cold develops, call the physician. If dizzy, be careful when up.

REVIEW QUESTIONS

1. A patient comes to the physician's office for an eye examination. The patient has a family history of macular degeneration. Which of the following symptoms is expected with macular degeneration?
   a. Loss of peripheral vision
   b. Sudden darkness
   c. Dull ache in the eyes
   d. Loss of central vision

2. A nurse is working on a postoperative unit for eye surgery. In planning interventions for an eye surgery patient, the nurse understands that which of the following patients needs specific positioning orders postoperatively to prevent complications?
   a. 4-year-old child after removal of congenital cataract
   b. 30-year-old woman after scleral buckling
   c. 52-year-old man after trabeculectomy
   d. 82-year-old man after corneal transplant

3. A nurse is caring for a patient who has a history of acute angle-closure glaucoma and is scheduled for surgery. Which of the following medication orders should the nurse clarify to prevent serious eye complications?
   a. morphine
   b. cefazolin (Kefzol)
   c. atropine
   d. ranitidine (Zantac)

4. A patient comes to the health clinic for a suspected ear infection. Which of the following assessment findings does the nurse expect with an external ear infection?
   a. Pain
   b. Fullness in ears
   c. Fever
   d. Dizziness

5. The nurse is contributing to the plan of care for a patient with Ménière's disease. Which one of the following is the primary goal for a patient with Ménière's disease that the nurse should recommend be included in the plan of care?
   a. Prevent dehydration
   b. Decrease pain
   c. Prevent injury
   d. Preserve hearing

Age-Related Eye Disease Research Group: A randomized, placebo-controlled, clinical trial of high-dose supplementation with vitamins C and E, beta carotene, and zinc for age-related cataract and vision loss. AREDS Report No. 9. Archives of Ophthalmology 119:1439–1452, 2001.

Eichenbaum, J: Vitamins for cataracts and macular degeneration. Journal of Ophthalmic Nursing & Technology 15:2, 1996.

Epstein, S: What you should know about ototoxic medications. Self Help for Hard of Hearing People Journal available online at http://www.pub.utdallas.edu/dybala/theaudpa/blah/shhh/ototoxic.htm.

Harvard Health Letter: Age, hearing loss and hearing aids. Author 26(1), 2000.

Henney, JE: New hearing implant approved. Journal of the American Medical Association 284(13):1640, 2000.

Kilpatrick, JK, et al: Low-dose methotrexate management of patients with bilateral Meniere's disease. Ear, Nose and Throat Journal 79(2):82, 2000.

Lyle, BJ, et al: Antioxidant intake and risk of incident age-related nuclear cataracts in the Beaver Dam Eye Study. American Journal of Epidemiology 149:801, 1999a.

McConnell, E: Communicating with a hearing-impaired patient. Nursing 1998 28(1):32, 1998.

McCord, H, and McVeigh, G: Eat your spinach, save your eyesight. Prevention 52(9):60, 2000.

McHugh, M, and Schaller, P: Ergonomic nursing workstation design to prevent cumulative trauma disorders. Computers in Nursing 15:5, 1997.

Reiley, JS, Deutsch, ES, and Cook, S: Laser-assisted myringotomy for otitis media: A feasibility study with short-term follow-up. Ear, Nose and Throat Journal 79(8):650, 2000.

Smith, W, et al: Dietary antioxidants and age-related maculopathy: The Blue Mountains Eye Study. Ophthalmology 106:761, 1999.

Tuffs University Health & Nutrition Letter: New treatment may help stem vision loss from macular degeneration. Author 18(7):3, 2000.

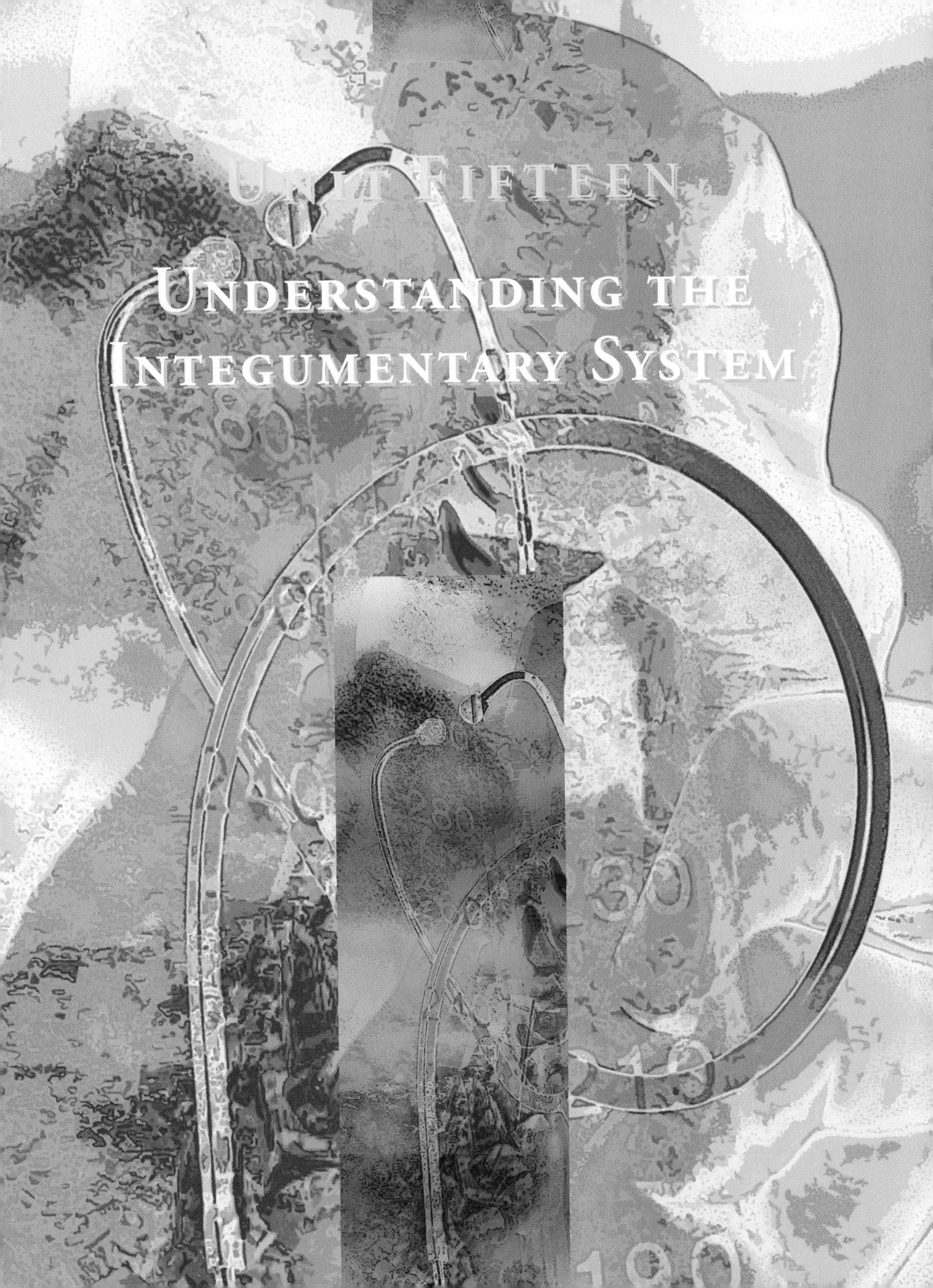

Unit Fifteen

Understanding the Integumentary System

50

INTEGUMENTARY FUNCTION, ASSESSMENT, AND THERAPEUTIC MEASURES

Rita Bolek Trofino and Valerie C. Scanlon

KEY TERMS

alopecia (AL-oh-**PEE**-she-ah)

ecchymosis (eck-uh-**MOH**-sis)

erythema (ER-i-**THEE**-mah)

petechiae (pe-**TEE**-kee-eye)

turgor (**TER**-ger)

QUESTIONS TO GUIDE YOUR READING

1. What is the normal anatomy of the integumentary system?
2. What is the normal function of the integumentary system.?
3. What are the effects of aging on the integumentary system?
4. What data should you collect when caring for a patient with a disorder of the integumentary system?
5. What laboratory and diagnostic tests are commonly performed to diagnosis integumentary disorders?
6. What common therapeutic measures are used for patients with integumentary disorders?
7. What nursing care should you provide for patients with disorders of the integumentary system?

▶ REVIEW OF NORMAL ANATOMY AND PHYSIOLOGY

The skin, its accessory structures, and the subcutaneous tissue make up the integumentary system, the covering of the body that separates the living internal environment from the external environment. The skin itself is considered an organ and consists of two layers, the outer epidermis and the inner dermis (Fig. 50–1).

Epidermis

The epidermis is made of stratified squamous epithelial tissue and is avascular, meaning that it has no capillaries within it. Its nourishment comes from the dermis beneath it. The epidermis is thickest on the palms of the hands and soles of the feet. The innermost layer of the epidermis is called the stratum germinativum, and it is here that mitosis takes place to produce new epidermal cells. The rate of mitosis is fairly constant, but it may be increased by chronic pressure on the skin (as in callus formation). The new cells produce the protein keratin. As they are pushed to the surface of the skin, they die and become the stratum corneum, the outermost of the epidermal layers.

The stratum corneum is many layers of dead cells; all that remains is their keratin. An unbroken stratum corneum is an effective barrier against pathogens and most chemicals, although even microscopic breaks are sufficient to permit their entry. Keratin is relatively waterproof, so it prevents the loss of water—that is, dehydration—and also prevents the entry of excess water by way of the body surface. As dead cells are worn off the surface of the skin (which also contributes to the removal of pathogens), they are continuously replaced by cells from within. Loss of large portions of the stratum corneum, as with extensive third-degree burns, greatly increases the risk for infection and dehydration.

Melanocytes are cells in the lower epidermis that produce the protein melanin; the amount of melanin is a genetic characteristic and gives color to skin and hair. When the skin is exposed to ultraviolet (UV) rays (part of sunlight), production of melanin increases and it is incorporated into the epidermal cells before they die, making the cells darker. Melanin is a pigment barrier to prevent further

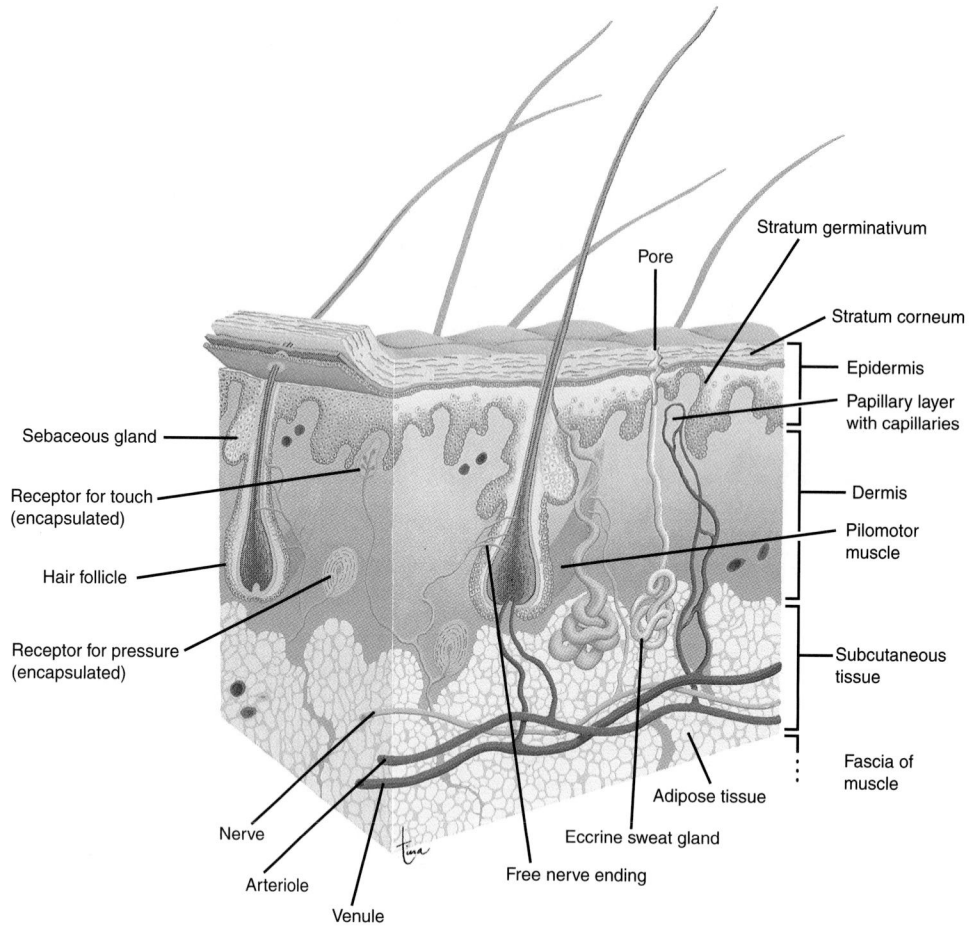

Figure 50-1 Structure of the skin and subcutaneous tissue. (From Scanlon, VC, and Sanders, T: Essentials of Anatomy and Physiology, ed 3. FA Davis, Philadelphia, 1999, p 85, with permission.)

exposure of living skin (the stratum germinativum within) to ultraviolet rays. Ultraviolet rays are considered mutagenic—that is, capable of damaging the DNA within cells and causing mutations that may result in malignant cells.

Also in the epidermis (and dermis) are Langerhans cells, a type of macrophage that presents foreign antigens to helper T cells. This is the first step in the destruction of pathogens that have penetrated the epidermis.

Dermis

The dermis is made of fibrous connective tissue; the cells present are called fibroblasts. Fibroblasts produce the protein fibers collagen and elastin. Collagen fibers are very strong and form the bulk of the dermis; elastin fibers are capable of recoil and make the dermis somewhat stretchable. Within the dermis are the hair and nail follicles, glands, nerve endings, and blood vessels. The capillaries in the papillary layer of the dermis are important to nourish the stratum germinativum, which has no capillaries of its own.

Hair

Hair develops in epidermal structures called follicles. At the base of a follicle is the hair root, a group of cells that undergoes mitosis to produce the hair shaft. The cells quickly die after producing keratin and incorporating melanin. Human hair with significant functions includes the eyelashes and eyebrows, which keep dust and sweat out of the eyes, and nostril hair, which filters air entering the nasal cavities. Hair on the head provides thermal insulation (sparse body hair does not).

Nails

Nail follicles are found at the ends of the fingers and toes, and growth of nails is similar to growth of hair. Mitosis in the nail root is a continuous process to produce new cells, which contain keratin. As these cells die, they form the visible nail. Nails protect the ends of the digits from mechanical injury and are useful for picking up small objects.

Receptors

The sensory receptors in the dermis are those for the cutaneous senses. Free nerve endings are the receptors for heat, cold, and pain; encapsulated nerve endings are specific for touch, and pressure. The sensitivity of an area of skin is determined by the number of receptors present.

Sebaceous Glands

The ducts of sebaceous glands open into hair follicles or directly to the skin surface. Their secretion is sebum, a lipid substance that inhibits the growth of some bacteria and prevents drying of skin and hair. Skin that is dry tends to crack or fissure more easily, and even these small breaks in the epidermis are potential portals of entry for pathogens.

Sudoriferous (Sweat) Glands

Sudoriferous glands are also known as sweat glands. There are two kinds of sudoriferous glands: apocrine and eccrine. Apocrine glands are really modified scent glands and are most numerous in the axillae and genital area; they are activated by stress and emotions.

Eccrine glands are found throughout the dermis but are most numerous on the face, palms, and soles. They are activated by high temperatures or by exercise and secrete sweat onto the skin surface. The sweat is evaporated by excess body heat, which is a very effective cooling mechanism, although it does have the potential to lead to dehydration if water is not replaced by drinking.

Modified sweat glands called ceruminous glands are found in the dermis of the ear canals. Their secretion is called cerumen or ear wax. Cerumen prevents drying of the outer surface of the eardrum. Excess cerumen, however, may become impacted against the eardrum, prevent it from vibrating properly, and diminish the acuity of hearing.

Blood Vessels

The blood vessels in the dermis serve the usual function of tissue nourishment, but the arterioles are also involved in the maintenance of body temperature. Blood carries the heat produced by active organs and distributes it throughout the body. In a warm environment, dilation of blood vessels in the dermis increases blood flow and loss of heat to air or clothing (if cooler than the body). Constriction of blood vessels in a cold environment decreases blood flow to the skin and conserves body heat.

Stressful situations also bring about vasoconstriction in the dermis, which allows blood to circulate to more vital organs, such as the heart, liver, brain, or muscles.

Other functions of the skin are the formation of vitamin D from cholesterol when the skin is exposed to the UV rays of the sun and the excretion of small amounts of urea and sodium chloride in sweat.

Subcutaneous Tissue

The subcutaneous tissue, between the dermis and the muscles, is made of areolar connective tissue and adipose tissue. Although an unbroken stratum corneum is an excellent barrier to pathogens, even small breaks provide portals of entry. In the subcutaneous tissue are numerous white blood cells that destroy any pathogens that have entered by way of broken skin. Subcutaneous adipose tissue cushions some bones and provides some insulation from cold, but its most important function is energy storage. Excess nutrients are changed to fat and stored as potential energy for times when food intake may decrease.

Aging and the Integumentary System

The effects of age on the integumentary system are often quite visible. Cell division in the epidermis slows, fibroblasts in the dermis die and are not replaced, and both skin layers become thinner and more fragile. Both the collagen and elastin fibers in the dermis deteriorate, causing wrinkles. Sebaceous glands and sweat glands decrease their activity. The skin becomes dry, and temperature regulation in hot weather becomes more difficult. There is often less fat in the subcutaneous layer, which may make an elderly person more sensitive to cold temperatures. Hair follicles become inactive, and the hair thins. Melanocytes die, and the hair that remains becomes gray.

▶ NURSING ASSESSMENT
Health History

Skin problems are a fairly common complaint for the patient entering the health system. Many factors can influence the integumentary system. A skin problem may be the only complaint the patient has, or it may be a manifestation of an underlying systemic condition or psychological stress. Most important, the skin visibly communicates the patient's health. Therefore the questions that are posed to the patient are important in determining if the skin problem is a disease entity of its own or a sign of a more systemic disorder. Table 50–1 provides examples of general questions that can be asked of the patient to elicit information.

If further assessment of a particular problem area is necessary, the WHAT'S UP? line of questioning may be used. For example, if the patient has a rash, you can respond with the following questions:

Where is it? Is that the only area where you have a rash?
How does it feel? Does it itch? Burn?
Aggravating and alleviating factors. Does scratching aggravate it? Does anything else aggravate it, such as soaps and detergents? What relieves it? How have you treated it in the past?
Timing. How long have you had this problem? Does it recur?
Severity. How bad is the discomfort on a scale of 0 to 10, with 0 being comfortable and 10 being unable to touch the area?
Useful other data. Do you have other symptoms besides the rash, such as itching, discharge, tingling, or loss of sensation?
Patient's perception. What do you think is causing your rash?

Physical Assessment

Assessment of the skin involves not only the entire skin area, but also the hair, nails, scalp, and mucous membranes. The main techniques utilized in physical assessment of the skin are inspection and palpation. Ensure that the patient is disrobed but adequately draped in a well-lighted and warm environment. A handheld magnifying glass or penlight may be utilized to see small details and further illuminate the area.

Normally the skin is intact, with no abrasions, and is smooth, dry, well hydrated, and warm. Skin **turgor** is firm and

TABLE 50-1	QUESTIONS ASKED DURING NURSING ASSESSMENT OF THE INTEGUMENTARY SYSTEM	
Question	**Rationale**
Do you (or does anyone in your family) have a history of dryness, rashes, itching, skin diseases, psoriasis, eczema, dermatitis, asthma, hay fever, hives, or allergies?	These conditions may be hereditary.
Have you noticed any changes in your skin, such as a sore that does not heal, rashes, lumps, or a change in an existing mole?	A sore that does not heal, a lump, or change in a mole may indicate cancer. Slow healing may also occur with vascular disease or diabetes mellitus. Rashes are associated with a variety of conditions.
Have you had any recent trauma to your skin?	A break in skin integrity can lead to infection.
Do you have a tendency to sunburn easily? Do you use sunblock?	Repeat sunburns are a risk factor for skin cancer.
Do you go to tanning salons or utilize sun lamps or tanning pills?	These are risk factors for skin cancer.
How often to you bathe or shower? What kind of soap do you use?	Some soaps may cause allergic reactions.
Do you wear a wig or hairpiece?	Adequate assessment of the scalp requires permission for removal of a wig or hairpiece.
Have you noticed a change in the growth or loss of your hair?	Hair loss can result from systemic illness or treatment or sometimes from infections or hair care products.
Have you experienced recent trauma or changes in your nails? Do you wear artificial nails?	Nail changes may be caused by circulatory problems. Artificial nails may mask changes.
Do you or any members of your immediate family or your coworkers have recent skin complaints?	Some skin disorders are contagious.
What medications do you take every day (prescription or nonprescription)? What is the dosage and frequency?	The patient may be taking medication for a skin disorder. Many medications cause skin reactions, from hives and photosensitivity to serious inflammatory conditions.
What medications did you take most recently? When did you take your last dose?	This might help pinpoint the cause of a new reaction.
What is your occupation?	Occupational exposures can lead to skin problems.
What recreational activities to you participate in?	Skin disorders can be caused by gym equipment that was not cleaned properly. Poison ivy may result from jogging in wooded areas.
Have you traveled recently?	This could help to pinpoint causes of suspicious skin changes.
Is there anything in your current environment, at home or work, that may be causing any skin problems (e.g., animals, plants, chemicals, infections, new carpeting, or new soaps or detergents)?	Various environmental factors can be causes of contact dermatitis; release of some chemicals can cause skin disorders.
Is there anything that touches your skin that causes a rash?	This may help pinpoint causes of contact dermatitis.

elastic. The skin surface is flexible and soft. Skin color ranges from light to ruddy pink or olive in white-skinned patients and light brown to deep brown in dark-skinned patients.

You need to be aware of normal developmental changes when performing an assessment. The skin of the neonate is very thin and friable (easily broken). During adolescence, the skin becomes thicker, with active sebaceous, eccrine, and apocrine glands. Body hair also changes during adolescence as a result of hormonal influences. In older patients the skin loses some of its elasticity and moisture. There is decreased activity of sebaceous and sweat glands. The older patient's skin is thinner, more fragile, and more wrinkled.

Inspection

Inspect each area of the skin, including nails, hair, scalp, and mucous membranes, for color, moisture, lesions, edema, intactness, vascular markings, turgor, and cleanliness. This examination should be done in an orderly sequence, such as hair, scalp, nails, buccal mucosa, and then the general skin surface from head to toe (Gerontological Issues Box 50–1).

COLOR. Skin color can be influenced by many factors, including the temperature of the patient, oxygenation, blood flow, exposure to UV rays, and positioning. Because skin color can differ genetically from very light to very dark, skin assessment can be difficult for the novice practitioner.

Commonly noted alterations can include pallor, **erythema** (redness), jaundice, cyanosis, and brown color. Pallor is a paleness or decrease in color and can be caused by vasoconstriction, decreased blood flow, or decreased hemoglobin levels from anemia. Pallor is best assessed on the face, conjunctivae, nailbeds, and lips. Erythema, or red discoloration, may indicate circulatory changes and can be caused by vasodilation or increased blood flow to the skin from fever or inflammation. Erythema is best assessed on the face or area of trauma.

Jaundice, a yellow-orange discoloration, may occur as a result of liver disease. The best place to inspect for jaundice is in the sclera of the eye. Cyanosis, or bluish discoloration, may indicate a cardiac, pulmonary, or perfusion problem. The best places to inspect for cyanosis are the lips, nailbeds, conjunctivae, and palms. People of Mediterranean descent normally have a bluish tone to their lips; this is not cyanosis.

A brown color may be caused by increased melanin production and can indicate chronic exposure to sunlight or pregnancy. This is best assessed on the face, areola, nipples,

BOX 50–1 GERONTOLOGICAL ISSUES

In acute care settings, priorities are determined by medical diagnoses and often center around cardiovascular, respiratory, nutrition, comfort, or other immediate concerns. The feet may be forgotten in the rush to care for the patient and plan a timely discharge.

Feet are also viewed by some as dirty; washing the feet may be seen as a lowly job. It may be assumed that people take care of their own feet. However, many older people are unable to bend down or bring the feet up high enough to see or care for them.

For these reasons it is especially important for the nurse to assess and care for the feet, both in institutional settings and at home. General guidelines for assessment include the following:

- Inspect feet for redness or pressure ulcers over bony prominences
- Inspect feet for dryness or cracking
- Inspect between toes for cracking, wounds, or excess moisture
- Inspect and palpate for callouses
- Palpate dorsalis pedis and posterior tibial pulses for circulatory status
- Assess patient's sensation using a wisp of cotton or light touch

Hints to promote healthy feet:

- Soak the patient's feet briefly in warm water and wash using a gentle soap. Test the water to be sure it is not too warm, especially for the patient with reduced sensation.
- Thoroughly dry the feet, including between the toes. Water left to evaporate can cause drying and cracking.
- Use a pumice stone to help remove dry dead skin over heels or calluses. Work gently, rubbing the stone in one direction only and removing only a small amount of dead skin at any one time.
- Use a cream or lotion that does not contain alcohol to moisturize the feet. Apply it with a gentle massage while moving the patient's feet through range-of-motion exercises. To prevent falls, never apply lotion before the patient steps into the tub or shower.
- Use gauze or a commercially made pad to decrease pressure and friction in areas between toes that cross or other areas where breakdown is likely.
- Encourage the patient to wear cotton socks and shoes or slippers to avoid injury to the feet and prevent falls.

and areas exposed to the sun. A brownish color may also be the result of chronic peripheral vascular disease, especially noted on the lower extremities.

LESIONS. A lesion is any change or injury to tissue. Assessment of skin lesions helps determine the cause of a skin disorder. Lesions are described as primary or secondary. Primary lesions are the initial reaction to a disease process. Secondary lesions are the changes that take place in the primary lesion because of trauma, scratching, infection, or various stages of a disease. Lesions are further described according to type and appearance (Fig. 50–2).

When assessing and documenting skin lesions, note the color or colors of the lesion, the size (usually in centimeters), location, distribution, and configuration. *Configuration* refers to the pattern of the lesions, shown in Figure 50–3. Also note any exudate, including amount, color, and odor, and any accompanying symptoms. Gently stretching the skin over the rash area makes it stand out more for further assessment.

In general, healthy patients with naturally dark skin have a reddish undertone, with pinkish buccal mucosa, tongue, nails, and lips. If a dark-skinned patient is pale, the mucous membranes has an ash-gray color, lips and nailbeds appear paler than usual, and the skin appears yellow-brown to ash gray. Erythema presents as a purplish-gray color. Cyanosis presents as a gray cast to the skin. The nailbeds, palms, and soles may have a bluish cast. Jaundice can be noted in the oral mucosa, particularly the hard palate and the sclera closest to the cornea.

MOISTURE/DRYNESS. The assessment of moisture provides clues to the patient's level of hydration. Observe the skin for dryness, moisture, scales, and flakes. Moisture may be found in skin fold areas. The skin should normally be smooth and dry. Flaking and scaling of the skin indicate dry skin.

EDEMA. Edema occurs because of a buildup of fluid in the tissues. Edema can cause the skin to become stretched, dry, and shiny. The location, distribution, and color of edematous areas are determined and documented. If edema is unilateral, compare it with the opposite side of the body. Edematous extremities can be measured to track improvement or worsening of the condition. Dependent edema is edema that occurs in the part of the body that is at the lowest point, typically noted in the feet and ankles or in the sacrum if the patient is lying down.

VASCULAR MARKINGS. Vascular markings may be classified as normal and abnormal. Two common abnormal vascular changes are **petechiae** and **ecchymosis**. Petechiae are small purplish hemorrhagic spots, smaller than 0.5 mm in diameter. In the darker-skinned patient, petechiae are usually not visible on the skin but can be visualized in the conjunctivae and oral mucosa. Ecchymosis is a bruise in which the color changes from blue-black to greenish-brown or yellow over time.

GENERAL INTEGRITY AND CLEANLINESS. Assess the integrity of the skin. Elderly patients have thin, fragile skin that may have breaks or tears. Be sure to check between toes and skin folds and under a pendulous abdomen or breasts. Check over bony prominences for signs of pressure. Note general cleanliness and odors.

Palpation

Palpation is utilized in conjunction with inspection. Use the dorsum (back) of the hand to palpate temperature, because this part of the hand is most sensitive to changes in temperature. Use the fingertips to gently palpate over the skin to

---

ecchymosis: ec—out + cchymos—juice + is—condition

PRIMARY LESIONS

Macule:
Flat, nonpalpable change in skin color, with different sizes, shapes, color; usually smaller than 1 cm. (e.g. rubella, scarlet fever freckles)

Papule:
Palpable solid raised lesion that is less than 1 cm. in diameter due to superficial thickening in the epidermis. (e.g. ringworm, rosea, wart, mole)
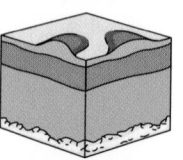

Nodule:
Solid elevated lesion that is larger and deeper than a papule (e.g. fibroma, intradermal nevi)
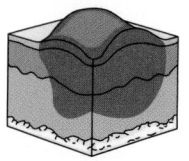

Vesicle:
A small, blister-like raised area of the skin that contains serous fluid, up to 1 cm. in diameter (e.g. poison ivy, shingles, chicken pox)
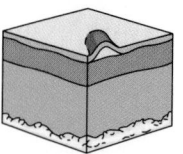

Bulla:
A fluid-filled vesicle or blister larger than 1 cm. (e.g. burns, contact dermatitis)
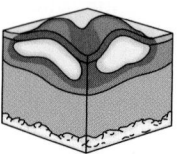

Pustule:
Small elevation of skin or vesicle or bulla that contains lymph or pus (e.g. impetigo, scabies, acne)
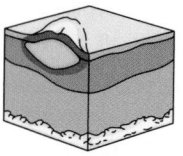

Wheal:
Round, transient elevation of the skin caused by dermal edema and surrounding capillary dilatation; white in center and red in periphery. (e.g. hives, insect bites)
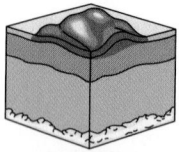

Plaque:
A patch or solid, raised lesion on the skin or mucous membrane that is greater than 1 cm. in diameter (e.g. psoriasis)
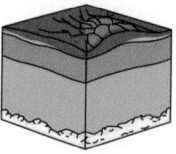

Cyst:
A closed sac or pouch which consists of semisolid, solid, or liquid material (e.g. sebaceous cyst)
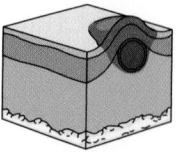

SECONDARY LESIONS

Scales:
Dry exfoliation of dead epidermis that may develop as a result of inflammatory changes (e.g. very dry skin, cradle cap, psoriasis)
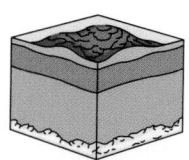

Crusts:
A scab formed by dry serum, pus, or blood (e.g. infected dermatitis, impetigo)
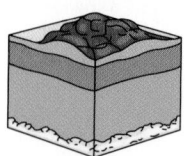

Excoriations:
Traumatized abrasions of the epidermis or linear scratch marks (e.g. scabies, dermatitis, burns)

Fissure:
A slit or cracklike sore that extends into dermis usually due to continuous inflammation and drying (e.g. athlete's foot, anal fissure)

Ulcer:
An open sore or lesion that extends to the dermis (e.g. pressure sores)
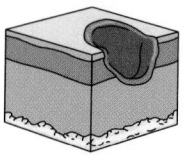

Lichenification:
Thickening and hardening of skin from continued irritation such as intense scratching looks like surface of mass.
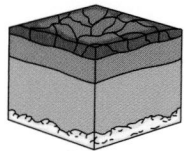

Scar:
A mark left in the skin due to fibrotic changes following healing of a wound or sore or surgical incision.
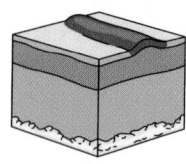

Figure 50-2   Description of skin lesions.

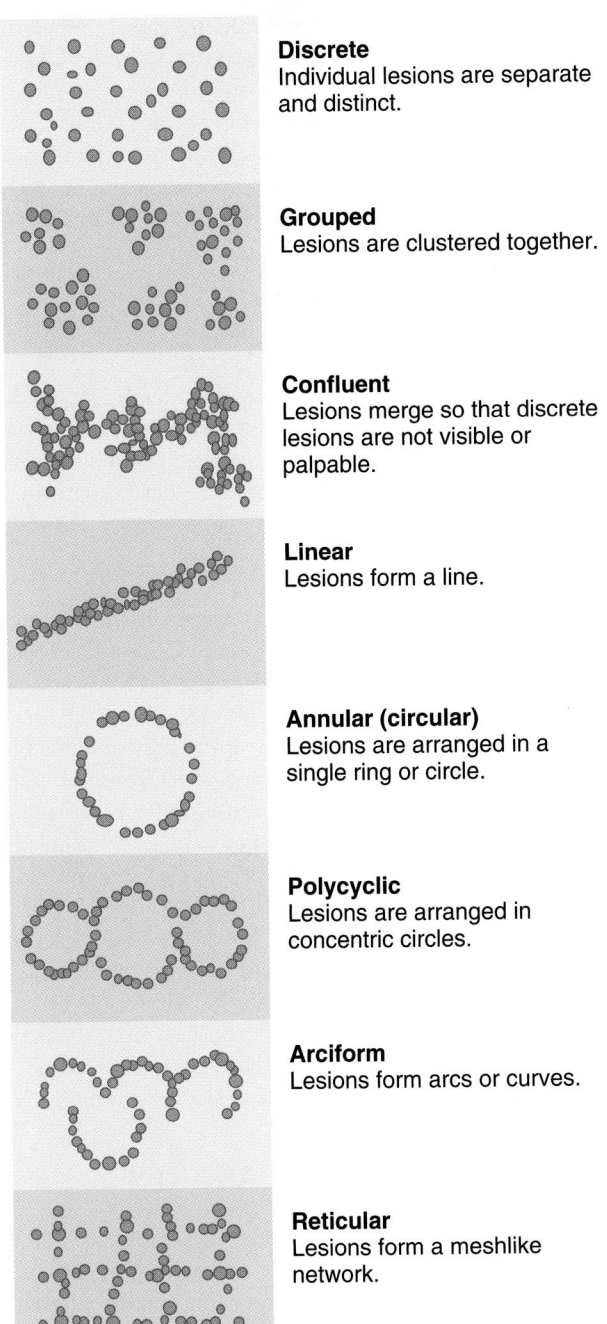

Discrete
Individual lesions are separate and distinct.

Grouped
Lesions are clustered together.

Confluent
Lesions merge so that discrete lesions are not visible or palpable.

Linear
Lesions form a line.

Annular (circular)
Lesions are arranged in a single ring or circle.

Polycyclic
Lesions are arranged in concentric circles.

Arciform
Lesions form arcs or curves.

Reticular
Lesions form a meshlike network.

Figure 50-3 To assess configuration, observe the relationship of the lesions to each other. Then characterize the configuration by one of the patterns illustrated in the chart.

determine size, contour (flat, raised, depressed), and consistency (soft or indurated) of lesions. If the lesion is moist or draining, wear gloves to protect against the spread of infectious organisms. Note the degree of pain or discomfort associated with light palpation of lesions.

Assess turgor and texture of the skin. Skin turgor is a measure of the amount of skin elasticity. To assess for turgor, the skin on the back of the forearm or over the sternum is pinched between the thumb and forefinger and then released. Normally the skin lifts easily and then quickly returns to its normal state. Poor skin turgor is indicated by

"tenting" of the skin, with more gradual return to its normal state. Poor skin turgor may indicate dehydration. Normal aging of skin produces some loss of skin elasticity; the preferred place to check skin turgor in the elderly is over the sternum.

If edema is suspected, palpate those areas to assess for tenderness, mobility, and consistency. When pressure from your fingers leaves an indentation, this is called pitting edema. Pitting edema is classified by its depth. Press the edematous area (against bone, if possible) with your thumb for 5 seconds and then release. One way to measure edema is to measure depth of the pitting in millimeters. For example:

1+ edema = 1 mm depth, or trace edema
2+ edema = 2 mm depth of indentation, or a small amount
3+ = moderate edema
4+ = a large amount of edema

Hair distribution over the entire body is palpated. Assess for hair color, quantity, thickness, and texture. Note any areas of **alopecia** (hair loss). Determine any recent changes in color and growth pattern. Note cleanliness, itching, redness, scaling, flakes, and tenderness. If lesions or lice are suspected, utilize disposable gloves to avoid spread of infection.

Terminal hair is the hair on the scalp, eyebrows, axillae, and pubic areas and on the face and chest of males. Vellus hairs are the soft, tiny hairs covering the body. Normally body hair has uniform distribution. Male and female pubic hair distribution may be noted. Scalp hair can normally be thick, thin, coarse, smooth, shiny, curly, or straight. Describe scalp hair distribution and cleanliness.

Nails can reflect the patient's general health. Assess fingers and nails for color, shape, texture, thickness, and abnormalities. Normally for all patients the nail appears pink, smooth, hard, and slightly convex (160-degree angle), with a firm base. The nails of elderly patients may have a yellowish-gray color, thickening, and ridges. Brown or black pigmentation between the nail and nail base is normal in dark-skinned patients. Abnormal findings include clubbing, which may indicate hypoxia, and spoon nails (concave nails, also called koilonychia), which may be associated with anemia. (See Chapter 26.) Thick nails may indicate fungal infection. Palpate for nail consistency and observes for redness, swelling, or tenderness around the nail area. See Table 50–2 for other nail abnormalities.

TABLE 50-2	ABNORMALITIES OF THE NAILS	
Nail Abnormality	**Description**	**Causes**
Beau's lines	Transverse depressions in the nails	Systemic illnesses or nail injury
Splinter hemorrhages	Red or brown streaks in the nailbed	Minor trauma, subacute bacterial endocarditis, or trichinosis
Paronychia	Inflammation of the skin at the base of the nail	Local infection or trauma

Describe any abnormal skin conditions in detail. Include findings such as color of lesion, pain, swelling, redness, location, size, drainage (including amount, color, and odor), and eruption patterns.

DIAGNOSTIC TESTS
Laboratory Tests
Cultures

Skin cultures are done to determine the presence of fungi, bacteria, and viruses. When a fungal infection is suspected, gently scrape scales from the lesion into a Petri dish or other indicated container. The specimen is then treated with a 10-percent potassium hydroxide (KOH) solution to make fungi more prominent. The specimen can remain at room temperature until sent to the laboratory.

LEARNING TIPS

When scraping scales, position the patient so that the skin lesion is vertical. Place the slide against the skin below the lesion. Be sure to wash your hands before and after, and wear gloves when collecting specimens.

If a viral culture is ordered, the fluid is expressed (gently squeezed) from an intact vesicle, collected with a sterile cotton swab, and placed in a special viral culture tube. If the lesion has crusts, they are removed or punctured before swabbing. The viral culture tube must be kept in ice and sent to the laboratory as soon as possible.

Wound cultures may also be collected with a sterile swab or wound culture kit. See Box 50–2 for specific instructions.

Skin Biopsy

A skin biopsy is indicated for deeper infections, to establish an accurate diagnosis, or for the evaluation of current treatment. A biopsy is an excision of a small piece of tissue for microscopic assessment. Three common types of skin biopsies are punch, shave, and incisional.

A punch biopsy uses a small round cutting instrument, called a punch, to cut a cylinder-shaped plug of tissue for a full-thickness specimen. An incisional biopsy is performed with a scalpel to make a deep incision and almost always requires a suture closure. A shave biopsy removes just the area that has risen above the rest of the skin.

BOX 50–2 **Steps in Culturing a Wound**

1. Use sterile saline to remove excess drainage and debris from the wound. Purulent material may have different bacteria than those actually causing the infection.
2. Using a sterile calcium aginate swab in a rotating motion, swab wound and wound edges 10 times in a diagonal pattern across the entire surface of the wound.
3. Do not swab over eschar.

For all biopsies, explain the procedure, assist in preparing a sterile field, calm and comfort the patient during the procedure, and assist in dressing the site following the procedure. The most uncomfortable part of the procedure is usually the injection of the local anesthetic agent. Explaining the procedure and providing appropriate calming techniques can make the procedure less traumatic to the patient.

Other Diagnostic Tests
Wood's Light Examination

Wood's light examination is the use of ultraviolet rays to detect fluorescent materials in the skin and hair present in certain diseases such as tinea capitis (ringworm). This examination is performed with a handheld black light in a darkened room.

Skin Testing

Patch and scratch tests are performed when allergic contact dermatitis is suspected. These are usually done by a dermatologist on uninvolved skin, such as the upper back or arms. Any hair in the area must first be shaved.

For the scratch test, the skin is superficially scratched or pricked with an allergen for an immediate reaction. If a reaction such as a wheal occurs, the test is positive for that allergen. Resuscitation equipment should be in the immediate vicinity in the event of a severe allergic (anaphylactic) reaction.

With the patch test, a delayed hypersensitivity reaction develops in 48 to 96 hours. The allergens are applied under occlusive tape patches. For this test, the skin should be free of oils. Cleanse the skin first with alcohol to promote patch adhesion. The test site must remain dry and free from moisture. The patch is removed in 2 days. Any reaction is noted, with a final reading in 2 to 5 days.

▶ THERAPEUTIC MEASURES
Open Wet Dressings

Wet compresses may be ordered for acute, weeping, crusted, inflammatory, or ulcerative lesions. The purpose of these dressings is to decrease inflammation, cleanse and dry the wound, and continue drainage of infected areas. They may be ordered either sterile or clean, depending on the risk for infection. The solutions commonly consist of room temperature to cool tap water or normal saline, aluminum acetate solution (Burrow's solution), or magnesium sulfate. The dressing is saturated with the solution before it is applied. Wet dressings are usually applied every 3 to 4 hours for 15 to 30 minutes.

LEARNING TIPS

To prevent chilling, no more than one-third of the body should be treated at one time. Keep the patient warm during this treatment.

Wet dressings should not be prescribed for more than 72 hours, because the skin may become too dry or macerated. If cool compresses are used, they should be reapplied every 5 to 10 minutes, because they become too warm from body heat. If warm compresses are utilized, monitor the skin closely to prevent burns.

Balneotherapy

Balneotherapy (therapeutic bath) is useful in applying medications to large areas of the skin, as well as for débridement, or removing old crusts; for removing old medications; and to relieve itching and inflammation. The temperature of the water should be kept at a comfortable level, avoiding hot baths. The bath should last for 15 to 30 minutes, while maintaining its warmth. Fill the tub half full. Keep the room warm to minimize changes in temperature. Advise the patient to wear loose clothing after the bath.

LEARNING TIPS

A bath mat should be used for treatment baths because some may make the tub slippery.

Water and saline are utilized for weeping, oozing, and erythematous lesions. Colloidal baths (such as oatmeal or Aveeno) are utilized for widely distributed skin lesions, for drying, and for relief of itching. Medicated tar baths, such as Almar-Tar or Bainetar, are used for chronic eczema problems and psoriasis. Any loose skin crusts can be removed after the bath. The room should be well ventilated, because tars are volatile.

To increase hydration after the bath, a lubricating agent is applied to damp skin, if emollient action is prescribed. Bath oils, such as Alpha-Keri, Avenol, and Lubath, are used for lubrication and itching.

Topical Medications

Many types of topical medications are used to treat skin conditions. These include lotions, ointments and creams, powders, gels, pastes, and intralesional therapy. Systemic medications may also be given for more serious conditions.

Lotions tend to cool the skin through water evaporation. They may also have a protective effect and may be antipruritic (treat itching). Lotions are usually applied with cotton gauze, gloves, or a soft brush.

Ointments and creams have a varied base (greasy, non-greasy, or penetrating), depending on the drug applied. These medications can protect the skin, provide lubrication, and prevent water loss. They are used for localized or chronic skin conditions. Ointments and creams can cause some reduction in blood flow to the skin. They are applied with a gloved hand or wooden tongue depressor.

Powders usually have a zinc oxide, talc, or cornstarch base. They act as a hygroscopic agent, to absorb moisture and reduce friction. Powders are usually applied with a shaker top. Powders are avoided around patients with respiratory disease or tracheostomies.

Gels, or semisolid emulsions, become liquid with topical application. They are usually greaseless and do not stain. Many topical steroids are prescribed in this manner.

Pastes are semisolid substances comprised of ointments and powders. They are used for inflammatory disorders. Mineral oil can facilitate removal of these agents.

Topical corticosteroids are used to reduce or relieve pain and itching by decreasing inflammation. Steroids should be used sparingly and according to package directions. Overuse of topical steroids can cause thinning of the skin. Caution is needed when used on the face to prevent glaucoma, cataracts, and perioral dermatitis.

Intralesional therapy has an anti-inflammatory action. This procedure utilizes a tuberculin syringe, most often of a sterile suspension of a corticosteroid, injected just below the lesion. Local atrophy may occur if the injection is made into subcutaneous tissue. Common conditions that are treated with this therapy include psoriasis and keloids.

CRITICAL THINKING

Mr. Evans

Mr. Evans comes to the doctor's office with atrophic skin (thin, shiny, pink, with visible vessels) at the area of psoriasis where he is applying his corticosteroid ointment. He states that he has been generously applying a thick layer four times a day.

1. What might be the cause of this condition?
2. What should you include when you document his skin condition?

Answer at end of chapter.

Dressings

Dressings may be used to enhance absorption of topical medications, promote retention of moisture, prevent evaporation of medication, and reduce pain and itching. Occlusive dressings (for sealing the wound) are commonly used for skin disorders. An airtight plastic film is applied directly over the topical agent. Corticosteroids are also available as a special plastic surgical tape and can be cut to size. See Nursing Care Plan Box 50–3 for care of the patient with an occlusive dressing. Proper application of a plastic wrap dressing includes washing the area, lightly patting dry, applying the medication to moist skin, covering the medicated area with plastic wrap, and covering with a dressing to seal the edges. Wet dressings and ointments should only be applied to affected areas, *not* to healthy intact skin, because this can cause maceration of good skin. Plastic wrap dressings should be used for no more than 10 to 12 hours a day.

BOX 50-3   NURSING CARE PLAN FOR THE PATIENT WITH AN OCCLUSIVE DRESSING

 Impaired skin integrity related to open lesions

PATIENT OUTCOME

Improved skin integrity as evidenced by reduction in size of lesion.

EVALUATION OF OUTCOME

Is there a decrease in wound size?

INTERVENTIONS	RATIONALE	EVALUATION
Assess areas of lesions for changes three times a day or as ordered.	Areas of redness, swelling, pain, and drainage may indicate infection.	Are lesions free of redness, swelling, pain, and drainage?
Assess lesions for presence or absence of dead tissue and exudates.	Indicates areas of healing and infection.	Are lesions free of exudates and dead tissue?
Cleanse wound as prescribed (see text for specific bathing instructions). Lightly pat dry.	Helps provide a healthy granulation area for healing.	Is wound clean and free of debris, crusts, and exudate?
Apply prescribed topical agent (see text for specifics) to moist skin. Apply sparingly or as directed.	Various agents have specific properties (control bacterial growth, prevent itching, have a protective effect, provide lubrication, relieve pain, or decrease inflammation).	Does area exhibit signs of healing (e.g., decrease in size and numbers of lesions, free from infection, less itching)?
Apply plastic film, cut to size. Cover with an appropriate dressing to seal edges.	Enhances absorption of medication and helps retain moisture.	Is the topical agent adherent to the skin?
Remove dressing for 12 out of 24 hours.	Continued use may cause skin atrophy, folliculitis, erythema, and systemic absorption of medication.	Are there signs of healthy granulation tissue? Is skin pink? Are there less open areas? Is dressing removed for at least 12 hours?

 Body image disturbance related to presence of lesion or wound

PATIENT OUTCOMES

Patient verbalizes acceptance of condition. Patient is willing to participate in care of lesion or wound.

EVALUATION OF OUTCOMES

Does patient verbalize acceptance of condition? Does patient participate in care of lesions?

INTERVENTIONS	RATIONALE	EVALUATION
Assess patient's feelings regarding condition.	Provides a baseline for care. If patient denies condition, he or she may not comply with care.	Does patient state willingness to follow care instructions?
Care for patient with an accepting attitude.	Patient will be aware of nuances in nurse's behavior.	Does patient allow nurse to partake in care of lesion or wound?
Allow opportunities for patient to verbalize concerns about condition.	Verbalizations allow patient to begin to accept and problem solve.	Does patient verbalize feelings appropriately?
Provide referrals to support groups and counselors as appropriate.	Patient may benefit from talking to others with similar condition or to another professional for objective evaluation.	Is patient receptive to appropriate referrals?
Assist patient in concealing lesion or wound in a safe and appropriate manner.	Long sleeves and long pants may help conceal lesions, protect lesions, and prevent further skin damage.	Is patient accepting of appearance of lesions? Are lesions or wounds visible?

 Bathing/hygiene self-care deficit, related to presence of lesions or wound and discomfort

PATIENT OUTCOMES

Patient verbalizes importance of good hygiene. Patient is willing to participate in bathing/hygiene.

EVALUATION OF OUTCOMES

Does patient verbalize importance of good hygiene? Is patient clean?

INTERVENTIONS	RATIONALE	EVALUATION
Assess patient's level of hygiene.	Provides a baseline for care.	Is patient's level of hygiene at an acceptable level?
Instruct patient in appropriate bathing/hygiene:	Patient needs to be able to properly cleanse lesions to prevent infection. Avoidance of fric-	Is patient able to verbalize understanding, as well as demonstrate good bathing techniques?

BOX 50-3 **NURSING CARE PLAN FOR THE PATIENT WITH AN OCCLUSIVE DRESSING—CONT'D**

- Avoid strong detergents and soaps; utilize gentle emollient soaps or prescribed soaps
- Gently stroke areas of lesions
- Pat dry; no friction
- Maintain a little moisture on skin
- Maintain comfortable environmental temperature
- Have temperature of bath at a comfortable level to patient, but not too hot

tion and strong soaps prevents further trauma to skin. Patient will not shiver in comfortable temperatures.

Are lesions free of infection?

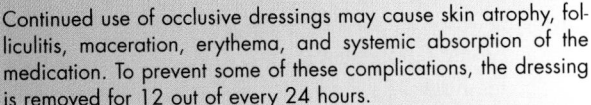

Continued use of occlusive dressings may cause skin atrophy, folliculitis, maceration, erythema, and systemic absorption of the medication. To prevent some of these complications, the dressing is removed for 12 out of every 24 hours.

Other dressings commonly used with topical treatments for skin conditions include gauze or cotton cloth held in place with small, stretchable tubular material (e.g., Surgitube, Tube gauze) for fingers, toes, and extremities; disposable polyethylene gloves sealed at the wrist; cotton socks or plastic bags for the feet; cotton cloth held in place with tubular material for the extremities; disposable diapers or cotton diapers for the groin and perineal areas; cotton cloth held in place with dress shields for the axillae; cotton or light flannel pajamas for the trunk; a shower cap for the scalp; and a face mask made from gauze and stretchable dressings with holes cut out for eyes, nose, mouth, and ears. The patient's primary care provider should specify the type of dressing and particular materials needed for this dressing.

Other Dressings

A variety of other types of dressing materials are available for wound and skin care. Transparent dressings (e.g., Op-Site, Tegaderm) can be used over skin tears or intravenous insertion sites. Hydrocolloid dressings can help protect areas exposed to pressure and treat pressure ulcers in early stages. Gels, pastes, and granules can be used to fill in deep wounds to promote granulation and aid healing. See Chapter 51 for additional dressings used specifically for pressure ulcers.

▶ WOUND HEALING

Wounds can heal by first intention, second intention, and third intention. The edges of the wound are approximated with staples or sutures with first-intention healing. This usually results in minimal scarring. With healing by second intention, the wound is usually left open and allowed to heal by granulation. Scarring is usually extensive with prolonged healing. With healing by third intention, an infected wound is left open until there is no evidence of infection and the wound is then surgically closed. Additional information about wound healing is found in Chapter 11.

Answers to CRITICAL THINKING

Mr. Evans

1. He may be sensitive or allergic to the medication. Most likely, he is applying too much too often. This ointment is applied as a thin layer, and usually only twice daily.
2. Note the size (usually in centimeters), location, color, distribution, and configuration. Describe exactly what you see, avoiding judgments about what you think it is.

REVIEW QUESTIONS

1. Which of the following is the protein in epidermal cells that makes the skin relatively waterproof?
    a. Collagen
    b. Keratin
    c. Melanin
    d. Elastin

2. Which of the following skin lesions is typically seen in chickenpox?
    a. Macule
    b. Papule
    c. Vesicle
    d. Wheal

3. Why should wet dressings be applied only to one-third of the body at one time?
    a. So the rest of the body can be observed for reaction to the dressing
    b. To prevent chilling the patient
    c. To prevent absorption of too much water, causing fluid overload
    d. To enable the patient to be more mobile

4. Which of the following terms describes a small, flat, red lesion?
    a. Macule
    b. Papule
    c. Bulla
    d. Crust

5. Where can pallor best be assessed in a dark-skinned individual?
    a. Upper back
    b. Ear lobes
    c. Antecubital fossa
    d. Mucous membranes

6. What equipment is most important to have readily available when a patient is undergoing testing for allergies?
    a. Resuscitation equipment
    b. Flashlight
    c. Measuring device
    d. Alcohol and cotton swabs

51

NURSING CARE OF PATIENTS WITH SKIN DISORDERS AND BURNS

Rita Bolek Trofino

QUESTIONS TO GUIDE YOUR READING

1. How would you explain the pathophysiology of each of the skin disorders listed in this chapter?

2. What are the etiologies, signs, and symptoms of each of the skin disorders listed in this chapter?

3. What is the current medical treatment for each of the skin disorders listed?

4. What data should you collect when caring for patients with disorders of the integumentary system?

5. What nursing care will you provide for patients with each of the covered skin disorders?

6. How will you know if your nursing interventions have been effective?

Skin disorders cover a wide array of diseases and conditions. Disorders can be generalized or localized, acute, chronic, or traumatic. This chapter discusses common skin disorders encountered by nurses. An excellent resource on skin disorders, with pictures, is at www.nsc.gov.sg/commskin/skin.html. The American Academy of Dermatology can be accessed at www.aad.org.

▶ PRESSURE ULCERS
Pathophysiology and Etiology

Pressure ulcers are often referred to by patients with old terms such as bedsores, decubitus ulcers, or pressure sores. Essentially a pressure ulcer is a sore caused by prolonged pressure against the skin. This may occur from spending a prolonged period in one position, causing the weight of the body to compress the capillaries against a bed or chair, especially over bony prominences. Pressure ulcers are the result of tissue anoxia and begin to develop within 20 to 40 minutes of unrelieved pressure on the skin. Other causes include pressure from a tight splint or cast, traction, or other device. Those at risk are immobile patients, those with decreased circulation, and those with impaired sensory perception or neurological function.

Mechanical forces (pressure, friction, and shear) lead to the formation of pressure ulcers. The pressure level that closes capillaries in healthy people is 25 to 32 mm Hg. When pressure applied to the skin is greater than the pressure in the capillary bed, it can impair cellular metabolism. It decreases the blood supply to the tissues and eventually causes tissue ischemia. This reduction in blood flow causes **blanching** of the skin. The longer the pressure occurs, the greater the risk of skin breakdown and the development of a pressure ulcer.

Friction is the rubbing of the skin surface with an external mechanical force. Also referred to as "sheet burns," this can happen when the patient is dragged or pulled across bed linens instead of being lifted.

Shearing occurs when the patient slides down in bed when the head of the bed is raised, or when being pulled or repositioned without being lifted off the sheets. With shearing, the skin and subcutaneous tissue remain stationary and the fat, muscle, and bone shift in the direction of body movement. As a result, there is damage deep within the tissues.

Any patient experiencing prolonged pressure is at risk for a pressure ulcer. Elderly patients have increased risk because of normal aging changes of the skin. Because thin patients have little padding when pressure is present, they have the greatest pressure applied to their capillaries. Obesity also is a contributing factor, because adipose tissue is poorly vascularized and is therefore more likely to develop ischemic changes. Impaired peripheral circulation also makes the skin more susceptible to ischemic damage.

Prevention

There are many interventions for the prevention of pressure ulcers. Assess and document the condition of the skin daily, so all are aware of developing problems. Gently cleanse the skin daily with tepid water and mild soap to prevent drying. To reduce friction, pat the skin dry rather than rubbing it dry. After bathing, daily lifelong lubrication of the skin with moisturizers is important to prevent dryness. Thoroughly dry skin-to-skin surfaces, such as under the breasts, skin folds (especially in the groin and abdominal folds), and between the toes, to prevent prolonged exposure to moisture. If incontinence is a problem, clean the skin promptly with tepid water and mild soap, pat dry, and apply a moisture barrier to prevent breakdown. Bony prominences or reddened skin areas should never be massaged; research has shown that blood vessels are damaged by massage when ischemia is present or when they lie over a bone.

Pressure on the skin should be avoided. Teach patients to shift their weight every 15 minutes if possible when lying or sitting. When the patient is immobile, the highest level of mobility should be maintained; frequent active or passive range-of-motion exercises should be performed, as well as turning according to a written repositioning schedule. If patients are on bedrest, turn and reposition them at least every 2 hours, but preferably more often because ischemia development begins after 20 to 40 minutes of pressure. When positioning patients on their side, place them at a 30-degree angle or less and not directly on their trochanter, because this area is especially sensitive to pressure and can quickly break down. If patients are placed on the trochanter, they usually become restless and squirm around to get off the trochanter. If the patient is seated in a chair, repositioning every hour is important. A mobility program specific to the patient must be developed.

The patient's heels should not rest on the bed surface. They should be elevated off the bed with pillows placed lengthwise under the calf or with heel elevators. Take care so pressure is not applied on the calf from the pillows. Be sure to also protect the patient's elbows, sacrum, scapula, ears, and occipital area from pressure.

Donut-shaped cushions should never be used. They create a circle of pressure that cuts off the circulation to the surrounding tissue, promoting ischemia rather than preventing it. Pad skin contact surfaces, especially bony prominences, so they do not press against each other. (For example, place a small pillow between the knees when the patient is in a side-lying position.) Provide an appropriate pressure-relieving or pressure-reducing mattress and chair cushion for immobile patients. To avoid friction, use a sheet to lift and move patients; provide a trapeze to assist patients to move themselves. Prevent malnutrition and dehydration by ensuring an adequate intake of protein, calories, and fluid; provide 2500 mL of fluid each day if not contraindicated by other medical problems. See Gerontological Issues Box 51–1 for additional preventive measures.

Signs and Symptoms

The most common sites for pressure ulcers are the sacrum, heels, elbows, lateral malleoli, greater trochanters, and ischial tuberosities. Most patients experience pain at the ulcer site. A report of pain requires continual assessment, documentation, and treatment.

L E A R N I N G TIPS

Pressure ulcers may be described according to a three-color system.

- *Black* wounds indicate necrosis.
- *Yellow* wounds have exudate and are infected.
- *Red* wounds are pink or red and are in the healing stage.

A wound may contain a mixture of black, yellow, and red colors. Necrotic wounds are the worst, because they have dead tissue. Beefy red wounds are desired because they are healing wounds. It is important to consider treating the worst color present first or healing will be delayed. For example, if a wound is both yellow and black, the dead tissue must be removed first before the infection can be effectively treated. This color system is a helpful system for patients and families to use to describe wounds to the home care nurse, because colors are easily recognized and understood by most people.

Complications

Wound infection is a common complication. New ulcers can also appear, and the present ulcer can progress to a deeper wound. Wounds that do not heal or take a prolonged time to heal also occur and may cause pain.

Diagnostic Tests

All pressure ulcers are considered to be colonized with bacteria. This means that bacteria are present, but the wound is not necessarily infected. In most cases, adequate cleansing and debridement can prevent bacterial colonization from advancing to clinical infection. Swab cultures and culture and sensitivity tests may be done to identify the causative organism in suspected infection sites. (See Chapter 50 for instructions for obtaining a culture.) Results need to be interpreted to distinguish between true wound infection and bacterial colonization. If the wound is healing by second intention, it becomes colonized by bacterial flora on the skin and from the environment. If, however, the wound is extensive, bacterial growth may exceed the local tissue defenses and a true wound infection results.

If the wound does not demonstrate any healing or if an ischemic ulcer is suspected, noninvasive and invasive arterial blood supply studies are recommended. Also, quantita-

tive wound biopsies may be performed for large, extensive wounds.

Medical Treatment

Treatment varies according to the size, depth, and stage of the pressure ulcer, as well as special needs of the patient and health care provider preference. All pressure must be removed from the affected area for healing to occur. Cleanliness must be maintained. Basic treatment includes debridement, cleansing, and dressing of the wound to provide a moist and healing environment.

LEARNING TIPS

The epidermis skates on moisture, so the wound must be kept moist to heal.

DEBRIDEMENT. Debridement is the removal of dead or nonviable tissue from a wound. It may be done surgically or nonsurgically. Nonsurgical debridement methods include mechanical, enzymatic, and autolytic. Surgical debridement is used only if the patient has sepsis or **cellulitis** or to remove extensive **eschar.** Eschar is a black, hard scab or dry crust that forms from necrotic tissue. It may hide the true depth of the wound and must be removed for the wound to heal.

Scissors and forceps can be used for mechanical debridement to selectively debride nonviable tissue. Dextranomer beads, another method of mechanical debridement, may also be sprinkled over the wound to absorb exudate and all other products of tissue breakdown, as well as surface bacteria. Whirlpool baths and wet-to-dry saline gauze dressings may also be used for mechanical debridement. The wet gauze is placed directly on the wound (avoiding surrounding healthy tissue) and allowed to dry completely. The drying process causes the gauze to adhere to the wound; when it is pulled off, tissue is pulled off with it. This results in nonselective debridement, because viable tissue may also be removed in this process. These methods are painful, so the patient should be premedicated for pain and assessed frequently.

Enzymatic debridement involves the application of a topical debriding agent. These agents vary as to application methods, so careful reading of instructions is necessary. Most of these debriding agents are proteolytic enzymes that selectively digest necrotic tissue.

Autolytic debridement is the use of a synthetic dressing or moisture-retentive dressing over the ulcer. The eschar is then self-digested via the action of the enzymes that are present in the fluid environment of the wound. This method is not used for infected wounds.

---

cellulitis: cellu—cell + itis—inflammation
eschar: eschara—scab

Surgical debridement is the removal of devitalized tissue, or thick, adherent eschar, utilizing a scalpel, scissors, or other sharp instrument. Depending on the amount of debridement to be done, this may be performed in the operating room, the treatment room, or the patient's room. Following surgical debridement, grafting may be required to close the wound. This becomes necessary if it is a full-thickness ulcer, if there is loss of joint function, or for cosmetic purposes. The patient is assessed continually for pain during the procedure, especially if there is a donor site for grafting.

WOUND CLEANSING. The ulcer should be thoroughly cleansed via whirlpool, handheld shower head, or irrigating system with a pressure between 4 and 15 pounds per square inch (psi), such as a 30-mL syringe with an 18-gauge needle. Pressure less than 4 psi does not adequately cleanse the wound, and greater than 15 psi may damage tissue. If an irrigating system is used, 250 mL of normal saline (or sometimes tap water for home wound care) should be used to thoroughly cleanse the wound. If the wound is red, gentle irrigation with a needleless 30- to 60-mL syringe should be used to prevent trauma and bleeding. When bleeding occurs, wound healing has been impaired. However, if the wound has been diagnosed as being infected, pressure flushing with a 30- to 60-mL syringe and an 18-gauge needle is needed to help remove bacteria.

LEARNING TIPS

Dilution is the solution to wound pollution!

Once the wound is cleansed and debrided, apply a dressing. The wound heals more rapidly in a moist environment, with minimal bacterial colonization and a healing temperature. This takes 12 hours to occur after the wound is covered with an occlusive dressing. If a dressing is frequently removed, the wound may not reach its healing temperature and healing may be impaired. When possible, the dressing should be left in place for extended periods. Infected wounds are not covered with occlusive dressings; draining wounds may require frequent dressing changes.

WOUND DRESSINGS. Dressings vary according to size, location, depth, stage of ulcer, and preference of the ordering practitioner. Commonly used dressing materials include hydrogel dressings, polyurethane films, hydrocolloid wafers, biologic dressings, alginates, and cotton gauze. These materials promote an optimum healing environment. Hypoallergenic tape should be used to secure dressings if tape is necessary. Protective paste may be applied to protect nonaffected tissue from topical agents. In all cases, pressure should be kept off the wound. No treatment will be effective if pressure continues to damage the tissue.

Nursing Process

ASSESSMENT. Provide an ongoing assessment of the status of the pressure ulcer, as well as underlying causes and impediments to healing. Monitor for risk factors such as prolonged immobility, incontinence, and inadequate hydration and nutrition.

Use transparency film or a disposable ruler to measure the diameter of the ulcer in centimeters. Depth can be measured with a cotton-tipped applicator. Also, gently probe a cotton-tipped applicator under the skin edges to detect tunneling and measure lateral tissue destruction.

There are several different staging systems for pressure sores based on the depth of tissue destroyed. In general, the staging systems are categorized from I to IV. See Figure 51–1 for photos of ulcers at each stage.

- Stage I: The skin is still intact, but the area is red and does not blanch. There may also be warmth, hardness, and discoloration of the skin. Be aware that even though the skin is intact, there may be deeper tissue damage that is difficult to observe.
- Stage II: There is a break in the skin, with partial-thickness skin loss of epidermis, dermis, or both. The ulcer may appear as an abrasion, a shallow crater, or a blister.
- Stage III: There is full-thickness skin loss, which extends to the subcutaneous tissue, but not fascia. There may be undermining of adjacent tissue. The ulcer looks like a deep crater and may have eschar.
- Stage IV: There is full-thickness skin loss with damage to the muscle, bone, or support structures such as tendons. There may be undermining and sinus tracts (tunneling).

Assess wound exudate. Two common types of wound exudate are serosanguineous and **purulent.** Serosanguineous exudate is fluid consisting of serum and blood. It is blood-tinged, amber-colored fluid. Purulent fluid is a fluid that contains pus. It can vary in color and have different odors, which are suggestive of different wound colonizations. Creamy yellow pus may indicate *Staphylococcus.* Beige pus that has a fishy odor may suggest *Proteus.* Green-blue pus with a fruity odor may indicate *Pseudomonas.* Brown pus with a fecal odor may suggest *Bacteroides.*

Gently palpate the wound with a gloved hand to assess the texture of granulations. If the granulations are healthy, they have a slightly spongy texture.

Document all findings carefully in the medical record, so all health team members can monitor progress of healing. Many institutions have specific forms for drawing pictures of the locations and sizes of wounds. There may be a special instant camera to document progress. Follow policy at the institution where you work.

---

purulent: purulentus—pus

DuoDERM® "S.O.S." Wound Care

"S.O.S."—"SELECTION ON SIGHT"

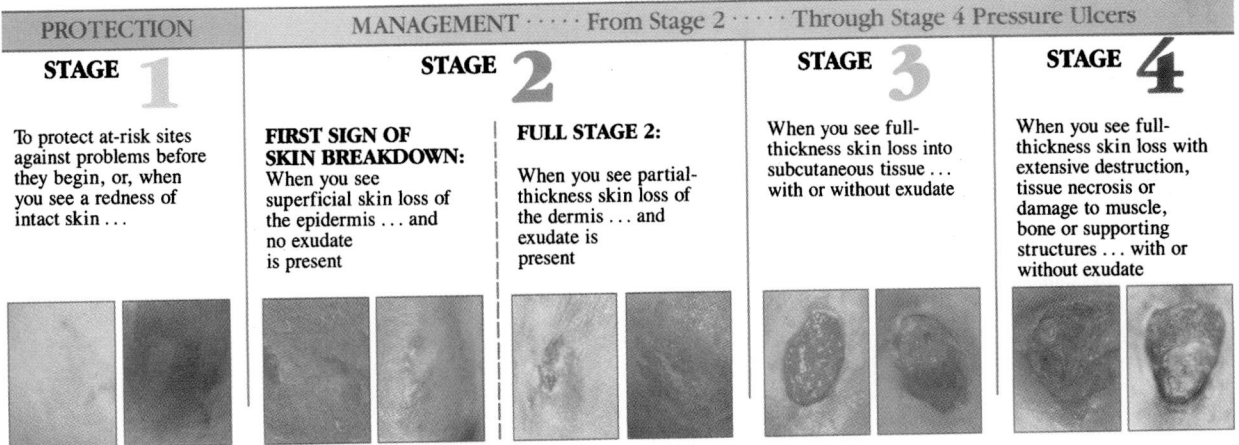

PROTECTION	MANAGEMENT · · · · · From Stage 2 · · · · · Through Stage 4 Pressure Ulcers		

STAGE 1

To protect at-risk sites against problems before they begin, or, when you see a redness of intact skin . . .

STAGE 2

FIRST SIGN OF SKIN BREAKDOWN: When you see superficial skin loss of the epidermis . . . and no exudate is present

FULL STAGE 2: When you see partial-thickness skin loss of the dermis . . . and exudate is present

STAGE 3

When you see full-thickness skin loss into subcutaneous tissue . . . with or without exudate

STAGE 4

When you see full-thickness skin loss with extensive destruction, tissue necrosis or damage to muscle, bone or supporting structures . . . with or without exudate

Figure 51-1 Pressure ulcers. (Courtesy ConvaTec, a Bristol-Myers Squibb Company, Princeton, NJ, with permission.)

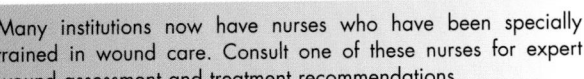

Many institutions now have nurses who have been specially trained in wound care. Consult one of these nurses for expert wound assessment and treatment recommendations.

NURSING DIAGNOSIS. Possible nursing diagnoses include impaired skin integrity related to immobility or pressure on skin surface; risk for infection related to open wound; pain related to ulcer and treatments; and ineffective coping related to chronic condition of ulcer.

NURSING CARE. See Nursing Care Plan Box 51–2 for the Patient with a Pressure Ulcer.

BOX 51-2 NURSING CARE PLAN FOR THE PATIENT WITH A PRESSURE ULCER

 Impaired skin integrity related to pressure on skin surface

PATIENT OUTCOMES

Skin integrity is improved as evidenced by decrease in wound size, no further development of other pressure ulcers.

EVALUATION OF OUTCOMES

Is there a decrease in wound size? Are there any new pressure ulcers?

INTERVENTIONS	RATIONALE	EVALUATION
Assess status of pressure ulcer according to stage, color, exudates, texture, size, and depth.	Provides baseline data on which care is based.	What stage is ulcer? Are there any other outstanding characteristics?
Assess cause of pressure (e.g., immobility, friction, shearing).	Allows for correction and also prevents further trauma.	What is the cause of this ulcer?
Cleanse wound gently with warm water; rinse; pat dry gently with gauze.	Reduces number of bacteria. Drying prevents maceration of skin. Gentle handling prevents further trauma.	Is wound clean and dry?
Debride wound as prescribed (method depends on patient's condition and goals of care).	Debridement removes drainage and wound debris. Permits granulation of tissue.	Does wound look clean and free of debris?
Dress wound appropriately for prescribed topical agent.	Protects underlying wound and helps promote healing.	Is dressing applied appropriately?
Position patient off the ulcer.	Prevents further pressure and trauma on ulcer.	Is patient positioned off the ulcer?
If a leg ulcer, provide for frequent rest periods with leg elevation; if immobile, reposition every 2 hours.	Prevents further tissue breakdown.	Is leg elevated? Is patient repositioned every 2 hours?

BOX 51-2 NURSING CARE PLAN FOR THE PATIENT WITH A PRESSURE ULCER—CONT'D

Risk for infection related to open wound

PATIENT OUTCOMES

Patient will not experience further wound infection or systemic sepsis. (Total elimination of bacteria is impossible due to nature of the condition.)

EVALUATION OF OUTCOMES

Is patient free from signs and symptoms of further infection? Is patient free from systemic infection?

INTERVENTIONS	RATIONALE	EVALUATION
Assess ulcer at every dressing change or at least every 24 hours. Note color and appearance of skin; diameter and depth of ulcer; areas of tenderness, swelling, redness, and heat; and drainage.	Allows for early recognition of infection and response to treatment.	Are signs of infection present?
Monitor temperature at least every 12 hours.	Elevated body temperature is one sign of infection.	Is patient afebrile?
Provide meticulous wound care (see under Impaired Skin Integrity).	Helps decrease the level of contamination. Prevents further infection.	Is wound showing signs of healing?
Use thorough handwashing techniques. Use sterile technique for dressing changes.	Prevents cross-contamination.	Does nurse take proper wound precautions?

Pain related to ulcer and treatments

PATIENT OUTCOMES

Patient will be as comfortable and as pain free as possible as evidenced by statement of increased comfort, statement of decreased pain, and ability to sleep at night.

EVALUATION OF OUTCOMES

Does patient express comfort? Does patient express a decrease in pain? Is patient able to sleep?

INTERVENTIONS	RATIONALE	EVALUATION
Assess level of pain through verbalizations, facial expressions, and positioning of body.	Monitors level of pain and response to therapy.	At what level is pain? Is it better or worse with treatment?
Offer analgesics as prescribed. Decrease anxiety with relaxation techniques (e.g., distraction, music).	Analgesics help relieve pain. Relaxation can lessen pain intensity.	Do analgesics relieve pain? Is patient less anxious? Does patient verbalize less pain?
Maintain a comfortable environment: cover; provide for privacy; position in good alignment and comfortably; and maintain a comfortable room temperature.	Relaxes patient and lessens intensity of discomfort.	Does patient express an increase in comfort?

INFLAMMATORY SKIN DISORDERS

Dermatitis

Pathophysiology and Etiology

Dermatitis is the inflammation of the skin and is characterized by itching, redness, and skin lesions, with varying borders and distribution patterns. Dermatitis can be caused by exposure to allergens or irritants, by heredity, or by emotional stress. Many times the cause is not known; in this case the terms *eczema* or *nonspecific eczematous dermatitis* may be used. The terms *eczema* and *dermatitis* are sometimes used interchangeably. There are three common types of dermatitis: contact dermatitis, atopic dermatitis, and seborrheic dermatitis. All types tend to be chronic and respond well to treatment, but are prone to recur. See Table 51–1 for common types of dermatitis.

Prevention

The patient should prevent irritation to the skin by avoiding irritants, allergens, excessive heat, and dryness and by controlling perspiration. Baths should be short, and water should be tepid. Deodorant soaps should be avoided; mild

---

dermatitis: derma—skin + itis—inflammation

TABLE 51-1	COMMON TYPES OF DERMATITIS
Type	**Description**
Contact	Acute or chronic condition; caused by contact with irritant or allergen
Irritant contact	Caused by direct contact with an irritating substance, such as soap, detergent, strong medication, astringent, cosmetic, or industrial chemical
Allergic contact	From contact with an allergen, such as perfume, tanning lotion, medication, hair dye, poison ivy, poison oak; contact results in cell-mediated immune response
Atopic	Chronic inherited condition; may be associated with respiratory allergies or asthma; can vary between bright red maculas, papules, oozing, **lichenified,** and hyperpigmented
Seborrheic	Chronic, inflammatory disease usually accompanied by scaling, itching, and inflammation; **seborrhea** is excessive production of sebaceous secretions; found in areas with abundant sebaceous glands (scalp, face, axilla, genitocrural areas) and where there are folds of skin; can appear as dry, moist, or greasy scales, yellow or pink-yellow crusts, redness, and dry flakiness; can be associated with emotional stress; may be genetic predisposition

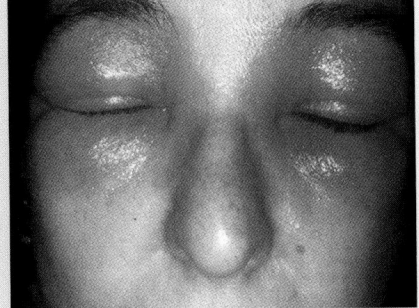

Contact dermatitis caused by nail polish

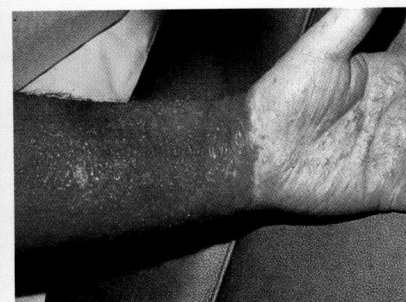

Contact dermatitis caused by topical anesthetic.

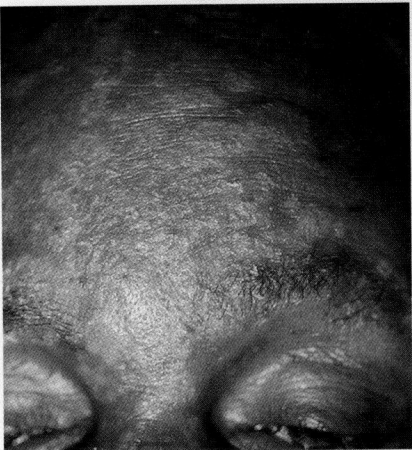

Seborreic dermatitis.

superfatted soaps are recommended instead. Dry skin is lubricated with creams, oils, or ointments as appropriate. Itching and scratching are prevented as much as possible.

Signs and Symptoms

Itching and rashes or lesions are the main clinical manifestations of dermatitis. The lesions vary depending on the type and location of dermatitis. Rashes and lesions can be dry, flaky scales; yellow crusts; red; fissures; maculas; papules; and vesicles. (These are described in Chapter 50, Fig. 50–2). Scratching can make any of these lesions worse.

LEARNING TIPS

Itching and scratching can occur during sleep, causing the rash to worsen. Place mitts on baby's hands or have the adult or older child wear cotton gloves at night.

Complications

The lesion or rash worsens with continued irritation, exposure to offending agents, or scratching. Infections of the skin are common and may be due to the many open areas and breaks in the skin, as well as the patient's reluctance to properly wash the affected area because of pain from the lesions. Some infections can also become systemic.

Diagnostic Tests

Diagnosis is usually based on history, symptoms, and clinical findings. If infection is suspected, cultures of the lesions may be ordered to identify the infecting agent.

Medical Treatment

Treatment varies according to symptoms. Basic treatment objectives are to control itching; alleviate discomfort and pain; decrease inflammation; control or prevent crust formation and oozing; prevent infection; prevent further damage to the skin; and heal the skin as much as possible.

Itching and discomfort can be somewhat relieved by antihistamines, analgesics, and antipruritic medications as ordered. Colloidal oatmeal preparations added to baths can also help relieve some of the **pruritus.**

Steroids may be used to suppress inflammation. They may be administered as a topical, intralesional, or systemic agent. The specific type and vehicle used depends on the type of lesion, the body area involved, and the extent of the lesion. Topical administration is preferred if possible, because systemic steroids can cause systemic side effects, including adrenal suppression.

pruritus: prur—itch + itis—condition

Tub baths and wet dressings help control oozing and prevent further crust formation. These interventions serve to loosen exudates, scales, and other wound debris, providing a clean area for topical application of medication. Skin is protected by lightly patting dry, avoiding friction, avoiding hot water, and using a sunscreen agent when outdoors.

Nursing Process

ASSESSMENT. You can use the WHAT'S UP? format to assess the rash, as described in Chapter 50.

Also refer to Chapter 50, Table 50–1 for specific questions to ask. Observe the rash or lesions for character, distribution, description, skin tenderness, signs of scratching, and other associated problems. (See Fig. 50–2.)

NURSING DIAGNOSIS. Possible nursing diagnoses include impaired skin integrity related to rash, lesions, and scratching; disturbed body image related to visible rash or lesions; and deficient knowledge related to disease and treatment.

PLANNING. The goals of treatment are to keep the skin intact or improve integrity, prevent infection, and maintain comfort. The patient will have improved body image as evidenced by a statement of acceptance of the condition and ability to socialize with others. The patient will verbalize understanding of the condition and demonstrate ability to perform self-care measures.

IMPLEMENTATION

Impaired Skin Integrity. Cleanse the area as ordered by the physician, taking care not to further irritate the skin. Cool moist compresses, dressings, or tepid tub baths may be ordered to help relieve inflammation and itching, debride lesions, and soften crusts and scales. Pat the skin dry rather than rubbing to prevent further trauma.

Apply topical agents as ordered to help suppress inflammation. Administration at bedtime helps promote comfortable sleep. Many antihistamines also have a sedative effect. If lesions are generalized, protein can be lost through oozing of serum. Patients are encouraged to eat a high-protein diet to promote healing and replace lost protein.

Gloves or mitts, especially at night, help prevent scratching. The patient is also encouraged to keep fingernails short to prevent scratching. Application of slight pressure with a clean cloth may help relieve itching. Relaxation exercises or referral to a support group may help the patient cope with distressing symptoms.

Disturbed Body Image. Allow patients to verbalize concerns if they wish. Refer to a support group, if available, to get help from others in similar circumstances. Display an accepting attitude while caring for skin lesions. The patient will be quick to pick up your reaction to the lesions, especially if it is negative. Encourage the patient to participate in skin care to allow more control over the situation. Long sleeves or other appropriate covering may make the lesions less noticeable and the patient more comfortable.

Deficient Knowledge. Instruct the patient in application of topical agents and dressings. Overuse of medications can

further traumatize skin, so application of a very thin layer is advised. Instruct the patient in how to recognize changes, improvement, or flare-ups of the disorder and what symptoms to report to the health care provider. Because most skin conditions are cared for at home, it is important for the patient to have the skills needed to monitor the condition and carry out treatment appropriately. Advise the patient to avoid overexposure to sun to prevent skin damage. Sunscreen agents should be used when outdoors. A humidifier in the home helps maintain hydration of skin and control itching during dry weather, especially in winter. It is important for the patient to know if the condition is chronic and how to prevent flare-ups if possible.

L E A R N I N G   TIPS

When applying topical medications, *more is not better!*

EVALUATION. If medical and nursing care have been effective, the lesions will be controlled or in remission, the patient will state that itching and other discomforts are controlled, the patient will be able to socialize without undue difficulty, and the patient will be able to describe and demonstrate self-care measures.

Psoriasis

PATHOPHYSIOLOGY AND ETIOLOGY. Psoriasis is a chronic inflammatory skin disorder in which the epidermal cells proliferate abnormally fast. Usually, epidermal cells take about 27 days to shed. With psoriasis, the cells shed every 4 to 5 days. The abnormal keratin forms loosely adherent scales with dermal inflammation.

Psoriasis is characterized by exacerbations and remissions. The cause is not known; however, often there may be a family history. The onset can be at any age, with 27 years the average age. The condition can be severe if the onset is in childhood. Many factors can influence the suppression and outbreak of lesions, but this varies from individual to individual. Sun and humidity may suppress lesions. Aggravating factors include streptococcal pharyngitis, emotional upset, stress, hormonal changes, cold weather, skin trauma, and certain drugs (e.g., antimalarials, lithium, beta blockers).

PREVENTION. Because the exact etiology is not known, measures to prevent exacerbation of symptoms are specific to the patient's circumstances. General preventive measures include avoidance of upper respiratory infections, especially streptococcal infections; avoidance of or coping with emotional stress; avoidance of skin trauma, including sunburns; and avoidance of medications that may precipitate a flare-up.

SIGNS AND SYMPTOMS. Signs and symptoms vary according to the patient and the particular type of psoriasis. Lesions are red papules that join to form plaques with distinct borders (Fig. 51–2). Silvery scales develop on un-

psoriasis: psor—itch + iasis—inflammation

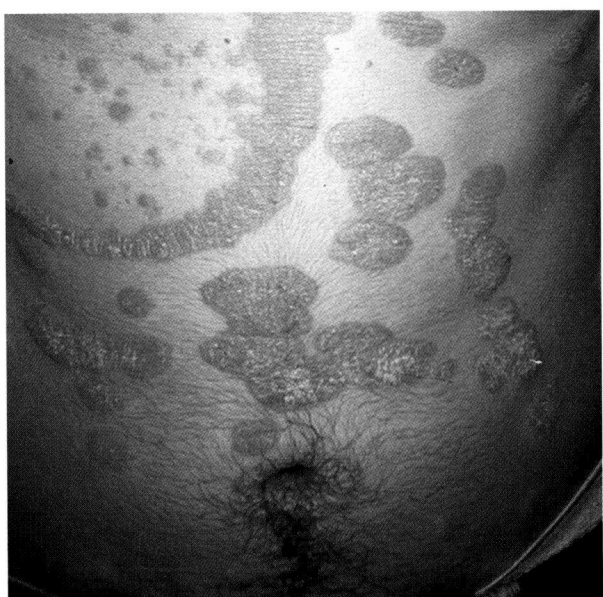

Figure 51-2 Psoriasis. Note bright red scaly plaque with silvery scale. (From Goldsmith, LA, Lazarus, GS, and Tharp, MD: Adult and Pediatric Dermatology. FA Davis, Philadelphia, 1997, p 258, with permission.)

treated lesions. Areas most often affected are the elbows and knees, scalp, umbilicus, and genitals. Other signs and symptoms include nail involvement, intergluteal pinking, itching, and dry or brittle hair.

COMPLICATIONS. Because of the nature of the disease, with its lesions and itching, secondary infections can occur. Psoriatic arthritis may develop after the psoriasis has developed, with nail changes and destructive arthritis of large joints, the spine, and interphalangeal joints. If the psoriasis becomes severe and widespread, fever, chills, increased cardiac output, and benign lymphadenopathy can result.

DIAGNOSTIC TESTS. Testing depends on the severity of the psoriasis. Normally this disease is diagnosed by physical assessment alone. Diagnostic tests may be performed to rule out concurrent disease or secondary infections.

MEDICAL TREATMENT. Treatment varies according to the type and extent of the disease, as well as physical preference. Psoriasis is a chronic disease with remissions and exacerbations. Basic treatment objectives are to decrease the rapid epidermal proliferation, inflammation, and itching and scaling. Usually the patient is instructed to bathe daily in a tub bath, using a soft brush to assist in the removal of scales.

Topical therapy includes steroids, salicylic acid, keratolytics, coal tar, anthralin, ultraviolet (UV) light, and, in severe cases, antimetabolite chemotherapy agents. Topical corticosteroids may be used for their anti-inflammatory effect. Occlusive dressings are commonly used to enhance penetration of medications. (See Chapter 50.) Keratolytic ointments or gels enhance the effects of salicylic acid to loosen or remove scales.

Tar preparations are usually prescribed along with the corticosteroids for conditions that warrant it. The tar acts as an antimitotic, slowing the epidermal cell division. Occlusive dressings are not used with tars. Anthralin is a substance extracted from coal tar. It also suppresses mitotic activity. The anthralin may be mixed with salicylic acid in a stiff paste. The patient must be closely observed, because the anthralin is a strong irritant and can cause chemical burns. It is usually applied for no longer than 2 hours. Both coal tar and anthralin are commonly used in combination with UV light and are usually administered in inpatient settings or specialized outpatient clinics.

Ultraviolet light may be designated as UVB, shorter wavelength, or UVA, longer wavelengths. UVA is from an artificial source, such as special mercury vapor lamps or special cabinets. The amount of exposure depends on the patient's condition, pigmentation, and susceptibility to burning. The patient must wear eye guards during treatments. Oral psoralen tablets (a photosensitizing agent) followed by exposure to UVA is called PUVA therapy. PUVA therapy temporarily inhibits DNA synthesis, which is antimitotic. Because psoralen is a photosensitizing agent, the patient must not only wear dark glasses during the treatment period, but also for the entire day after treatments. The long-term safety of PUVA therapy is still unknown. Possible side effects include increased skin carcinomas, premature skin aging, and actinic keratosis (premalignant lesions of the skin). The patient should be observed closely for redness, tenderness, edema, and eye changes. Therefore initial and follow-up eye examinations, skin biopsies, urinalysis, and blood tests may be ordered.

Antimebolites are reserved for the most severe cases, as a last resort. Methotrexate is the most common agent given. Because of its hepatotoxicity, it is contraindicated in patients with liver disease, alcoholism, renal disease, and bone marrow suppression. Before therapy, a liver biopsy and routine laboratory tests are completed.

⚡ CRITICAL THINKING

Mrs. Long

Mrs. Long arrives at the health clinic to complain that the prescribed shampoo she is using for her scalp is not working. She states that she washes her hair thoroughly with the medicated shampoo and immediately rinses completely. She wants to know why her scalp shows no signs of improvement. What should you tell her?

Answers at end of chapter.

Nursing Care

Nursing care for the patient with psoriasis is the same as nursing care for the patient with dermatitis. The only addition is to encourage frequent periods of rest to enhance the antimitotic effects of the therapeutic agents.

▶ INFECTIOUS SKIN DISORDERS

A variety of infections can affect the skin. The most common disorders are discussed in this section. See Table 51–2 for a summary of additional skin infections.

TABLE 51-2	INFECTIOUS SKIN DISORDERS		
Type	**Description**	**Complications**	**Treatment/Nursing Care**
Impetigo contagiosa	Common contagious, infectious, inflammatory skin disorders usually caused by streptococcus or *Staphylococcus aureus*; sources of infection include swimming pools, pets, dirty fingernails, beauty and barber shops, and contaminated clothing, towels, sheets; may occur secondary to scrapes, cuts, insect bites, burns, dermatitis, poison ivy. Primary skin infection can appear on exposed areas of the body (extremities, hands, face, neck) or skin-fold areas (axillae). Rash appears as oozing, thin-roofed vesicle that rapidly grows and develops a honey-colored crust; crusts are easily removed, and new crusts appear; lesions heal in 1 to 2 weeks if allowed to dry.	Glomerulonephritis resulting from a particular strain of streptococcus infection. Lesions may spread from one skin area to another. Lesions may persist if not permitted to dry. Secondary **pyoderma,** or acute inflammatory purulent dermatitis, may occur if lesions are unresponsive to treatment.	Systemic antibiotics are administered as prescribed. Topical antibiotics are used after crust removal. Gentle washing with a mild soap, or soaking with warm, moist compresses, aids in crust removal, removes debris, and provides a clean bed for topical therapy. Appropriate antipyretic are prescribed as necessary. Keep fingernails short and clean. Glove or mitt hands as necessary to prevent scratching. Patient must remain home until all lesions are healed. Teach proper disposal or washing of any material that comes in contact with lesions. Appropriate hygienic practices must be adhered to in preventing skin-to-skin or person-to-person spread. Observe client for 6 to 7 weeks for signs/symptoms of glomerulonephritis.

Impetigo on the face.

Herpes Simplex
Pathophysiology and Etiology

Herpes simplex virus (HSV) is a common viral infection that tends to recur repeatedly. There are two types: type I virus (HSV-I), which occurs above the waist and causes the fever blister or cold sore (Fig. 51–3); and type II virus (HSV-II), which occurs below the waist and causes genital herpes.

The primary infection occurs through direct contact, respiratory droplet, or fluid exposure from another infected person. Following this, the virus lies dormant in nerve ganglia near the spinal column, where the immune system cannot destroy it. The patient is asymptomatic at this time.

Recurrence of the infection can happen spontaneously or be triggered by fever, sunburn, stress, illness, menses, fatigue, or injury. The secondary lesion may appear isolated or as groups of small vesicles or pustules on an erythematous base. Crusts eventually form, and the lesions heal in about 1 week. The lesions are contagious for 2 to 4 days before dry crusts form.

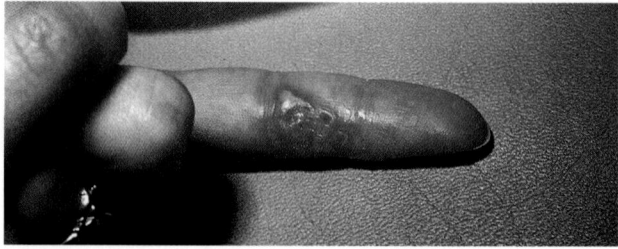

Figure 51-3 Herpes simplex. (From Goldsmith, LA, Lazarus, GS, and Tharp, MD: Adult and Pediatric Dermatology. FA Davis, Philadelphia, 1997, p 306, with permission.)

Prevention

Avoidance of contact with a known infected lesion during the blistering phase can prevent the primary lesions. Patients should also be taught to avoid sharing contaminated items such as toothbrushes, lipsticks, and drinking glasses. This disease can recur spontaneously. Avoidance of certain stressors, such as sunburn, injury, and fatigue, may delay a recurrence. The use of sunscreens, especially on the lips, may be helpful.

TABLE 51-2	INFECTIOUS SKIN DISORDERS—CONT'D		
Type	**Description**	**Complications**	**Treatment/Nursing Care**
Furuncles and carbuncles	A furuncle is a small, tender boil that occurs deep in one or more hair follicles and spreads to surrounding dermis; may be single or multiple; usually caused by *Staphylococcus*; usually occurs on body areas prone to excessive perspiration, friction, and irritation (e.g., buttocks, axillae); can recur; the boil eventually comes to a soft yellow, black, or white head; there is localized pain, tenderness, and surrounding cellulitis; lymphadenopathy may be present. A carbuncle is an extension of a furuncle; an abscess of skin and subcutaneous tissue; deeper than furuncle; caused by *Staphylococcus*; usually appears where skin is thick, fibrous, and inelastic (e.g., back of neck, upper back, and buttocks); associated symptoms may include fevers, pain, leukocytosis, prostration. Both tend to occur in debilitated clients, and more often in diabetics.	Furuncles may progress to carbuncles. Carbuncles may progress to infection of bloodstream. Further spread of infection can occur to self and others. Occasionally, scarring may occur.	Prevent trauma; avoid squeezing or irritation. Cleanse surrounding skin with antibacterial soap, followed by application of antibacterial ointment. Surgical incision and drainage may be performed. Cover draining lesion with dressing. Follow standard precautions. Double bag all soiled dressings and dispose of properly. Systemic antibiotic therapy (based on sensitivity studies) is instituted for carbuncles or spreading furuncles. Analgesia and antipyretics are ordered as necessary. Bed rest is advised with carbuncles or furuncles on perineal or anal regions. Cover mattress and pillows with plastic and wipe daily with a disinfectant. Wash all linens, towels, and clothing after each use. Properly discard razor blades after each use. Strict hand washing is maintained to prevent cross-contamination.

Signs and Symptoms

Some patients may have a prodromal phase of burning or tingling at the site for a few hours before eruption. The area becomes erythematous and swollen. Vesicles and pustules erupt in 1 to 2 days. There may also be redness with no blistering. Lesions can burn, itch, and be painful. The attacks vary in frequency but diminish with age. The patient is contagious until scabs are formed.

Complications

If herpes simplex is present in the vagina at childbirth, the newborn may be infected (meningoencephalitis or a panvisceral infection may occur). If the person touches the affected area and then rubs the eyes, the eyes can become severely infected.

Diagnostic Tests

Cultures of the lesions provide a definitive diagnosis. Most lesions are diagnosed on the basis of history, signs, and symptoms.

Medical Treatment

There is no complete cure. Recurrences will happen. Topical acyclovir (Zovirax) ointment is the drug of choice for primary lesions, to suppress the multiplication of vesicles. It does not benefit secondary lesions. Oral acyclovir may be recommended for severe or frequent attacks (six or more attacks per year) or for patients who are immunosuppressed. Various lotions, creams, and ointments may be prescribed to accelerate drying and healing of lesions (e.g., camphor, phenol, alcohol). Antibiotics may be indicated for secondary infections.

Nursing Care

Educating the patient about the disease and recurrences is very important. Instruct the patient about when the infection is contagious and how to prevent spreading the virus from one part of the body to another or to other individuals.

Herpes Zoster (Shingles)
Pathophysiology and Etiology

Herpes zoster, or shingles, is an acute inflammatory and infectious disorder that produces a painful vesicular eruption on bright red edematous plaques along the distribution of nerves from one or more posterior ganglia. This eruption follows the course of the cutaneous sensory nerve and is almost always unilateral (one sided) (Fig. 51-4).

Herpes zoster is caused by the varicella zoster virus. This virus appears identical to the one that causes chickenpox. It is thought that herpes zoster is a reactivation of this latent varicella virus. The incubation period is 7 to 21 days. The vesicles appear in 3 to 4 days. Eruption usually occurs posteriorly and progresses anteriorly and peripherally along the dermatome. The total duration of the disease can vary from 10 days to 5 weeks.

This disease occurs most commonly in the elderly or in those who have a diminished resistance, such as the patient with acquired immunodeficiency syndrome (AIDS), the patient on immunosuppressant agents, or the patient with a malignancy or injury to the spine or a cranial nerve.

Prevention

Avoidance of the person with this disease during the contagious phase (a few days before eruption until vesicles dry or scab) is the best prevention.

Signs and Symptoms

In addition to the vesicles and plaques, there may be irritation, itching, fever, malaise, and, depending on the location of lesions, visceral involvement. Lesions may be very painful; the likelihood of pain increases with age.

Complications

Postherpetic neuralgia, persistent dermatomal pain, and hyperesthesia are common in the elderly and can last for weeks to months after the lesions have healed. The incidence and severity of these complications increases with age.

Ophthalmic herpes zoster affects the fifth cranial nerve and can be a serious complication. Consultation with an ophthalmologist is imperative because this complication can affect eyesight. Other complications can occur with facial and acoustic nerve involvement, including hearing loss, tinnitus, facial paralysis, and vertigo. Full-thickness skin necrosis and scarring can occur if lesions do not heal properly; systemic infection can occur from scratching, causing the virus to enter the bloodstream.

Diagnostic Tests

Diagnosis is usually confirmed by the clinical picture of the patient and associated signs and symptoms. Cultures may be ordered if secondary bacterial infections are suspected.

Medical Treatment

Treatment is aimed at controlling the outbreak, reducing pain and discomfort, and preventing complications. Acyclovir, either intravenous (IV), oral, or topical, may be prescribed in the early stages of the initial infection, for a severe outbreak, and if the patient is immunosuppressed or debilitated. Acyclovir does not cure, but it may help control the initial outbreak. Analgesics are prescribed for pain and discomfort. Cor-

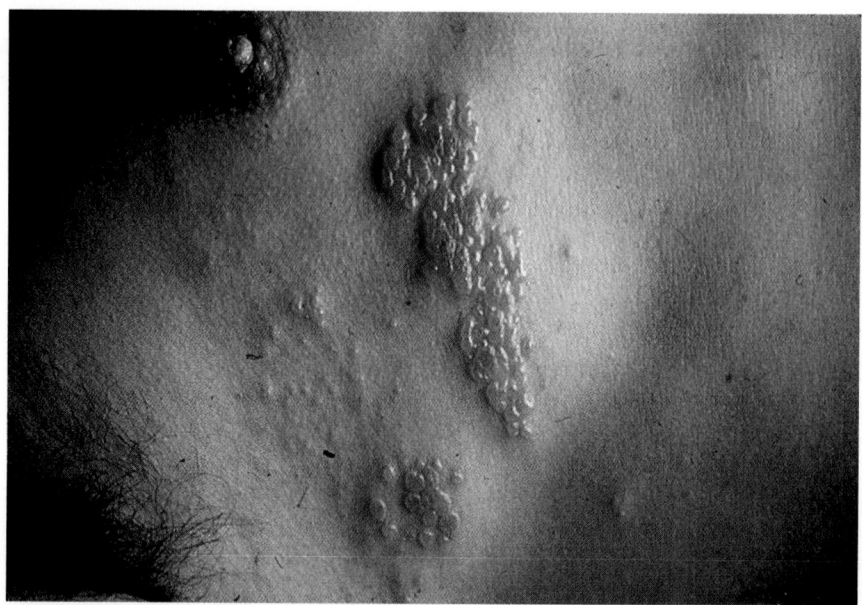

Figure 51-4 Herpes zoster (shingles). (From Goldsmith, LA, Lazarus, GS, and Tharp, MD: Adult and Pediatric Dermatology. FA Davis, Philadelphia, 1997, p 307, with permission.)

ticosteroids may be administered to prevent postherpetic neuralgia and reduce pain but are not used for ocular involvement. Topical steroids should not be applied if a secondary infection is present, because they suppress the immune system. Antihistamines are administered to control itching. Antibiotics are prescribed for secondary bacterial infections.

Nursing Care

Appropriate precautions during the contagious stage are necessary to prevent spread. Cool compresses two or three times a day help cleanse and dry lesions, as well as lessen itching. Hyperesthesia may be somewhat relieved with stockings, wraps, or a tight T-shirt that provides continuous firm pressure. Any appropriate measures to increase comfort should be initiated.

Fungal Infections
Pathophysiology and Etiology

Dermatophytosis, or a fungal infection of the skin, occurs when there is an impairment of the skin integrity in a warm, moist environment. This infection occurs through direct contact with infected humans, animals, or objects. *Tinea* is the term used to describe fungal skin infections; the name used after tinea indicates the body area affected. For example, tinea capitis is a fungal infection of the scalp. Common fungal infections and treatments are described in Table 51–3.

---

dermatophytosis: derma—skin + phyton—plant + osis—condition

TABLE 51–3	FUNGAL INFECTIONS	
Type	**Description**	**Treatment/Nursing Care**
Tinea pedis (athlete's foot)	Common fungal infection, most frequently seen in those with warm, moist, sweaty feet; occlusive shoes; or friction/trauma to the feet. Three types: chronic plantar scaling, acute vesicular, and interdigital. Chronic plantar scaling will have slight redness and mild to severe scaling; fold lines on sole appear to have white powder because of scaling; may be toenail involvement; itching is usually not present. Acute vesicular appears as a sudden eruption of small, painful, itchy vesicles; may also accompany chronic plantar scaling. Interdigital is more common; there is erosion, scaling, and fissuring in toe webs; area is painful, burning, and itchy, and there is usually an offensive odor.	Chronic plantar scaling may be treated with kerolytics and topical antifungal agents; these agents help in relieving symptoms and improve appearance; they are not curative. Acute vesicular is treated with soaks or baths two or three times a day for 2 to 3 days to dry up blisters; astringent paint is applied to debrided areas; topical corticosteroids help relieve itching. Interdigital may be treated with combined antifungal and antibacterial therapies or antifungals alone; soak feet twice daily and dry well. Teach patient prevention measures: keep feet dry; dry carefully between toes; apply foot powder to absorb perspiration; wear cotton socks to absorb perspiration; if weather permits, use perforated shoes or sandals; avoid plastic or rubber-soled shoes; wear water shoes in public showers and near swimming pools. Apply topical agents properly: apply antifungals thinly; treat for time specified, even after apparent clearing.
Tinea capitas (ringworm of scalp)	Contagious; commonly causes hair loss in children. Appears as scattered round, red, scaly patches; small papules or pustules may be evident at edges of patches; hair is brittle at site, breaks off, and temporary areas of baldness result; may be mild itching; kerion inflammation may occur after weeks.	Teach prevention measures: never share combs, brushes, pillowcases, or headgear. Systemic antifungals are prescribed because of high relapse rate with topical agents; review side effects with client. Oral corticosteroids are indicated for kerion inflammation to help prevent alopecia; review side effects with client. Instruct family on contagious aspect of disease; assess other family members and pets for organism.

Tinea capitis.

TABLE 51-3	**FUNGAL INFECTIONS — CONT'D**	
Type	**Description**	**Treatment/Nursing Care**
Tinea corporis (tinea circinata; ringworm of body)	Erythematous macule that progresses to rings of vesicles or scale with a clear center that appears alone or in clusters; usually occurs on exposed areas of body; can be moderately to intensely itchy. Infected pet is common source of infection.	Teach prevention measures: keep skin areas, especially folds, dry; use clean towel and washcloth daily; wear cotton clothing, especially on hot, humid days. Topical antifungals are prescribed for small, localized lesions. Oral antifungals are indicated for severe, widespread, resistant, or follicular cases. Topical corticosteroids are prescribed for itching.
Tinea cruris (jock itch)	Ringworm of groin that may extend to inner thighs and buttocks area; may occur with tinea pedis; often in obese people who are athletic. Lesion first appears as a small red scaly patch and then progresses to a sharply demarcated plaque with elevated scaly or vesicular borders; itching can range from absent to severe.	Teach patient prevention measures: avoid heat, moisture, and friction. Topical antifungals are prescribed; apply in a thin layer to rash and a few centimeters beyond border. Oral antifungals may be indicated for widespread cases or those resistant to topical therapy. Topical corticosteroids may be prescribed for itching.
Tinea unguium	Ringworm of the nails; also referred to as **onychomycosis.** Chronic fungal infection of nails, usually the toenails; a lifelong disease. There is yellow thickening of nail plate; it is friable and lusterless; eventually crumbly debris accumulates under free edge of the nail and causes nail plate to become separated; over time, the nail may become thickened, painful, and destroyed.	Systemic antifungals are rarely given for toenail involvement, but may be prescribed for fingernail involvement (review side effects). Topical antifungals are usually ineffective. Nail may have to be surgically removed (nail avulsion). Explain high relapse rate to patient. Keep nails neatly trimmed and buffed flat; gently scrape out any nail debris.

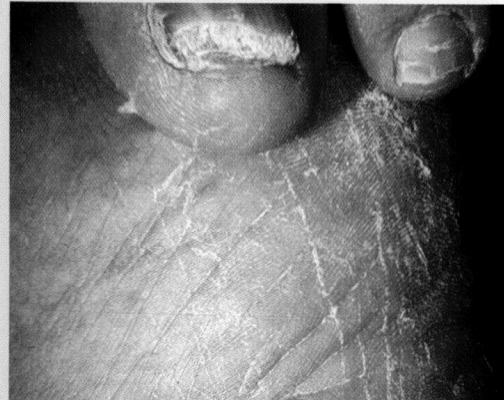

Tinea cruris. The inner thigh is the typical location for tinea cruris, or "jock itch." The border is pronounced and scaly.

Onychomycosis (tinea unguium).

Cellulitis

Pathophysiology and Etiology

Cellulitis is the inflammation of skin cells or cellular or connective tissue resulting from a generalized infection, usually with *Staphylococcus* or *Streptococcus* spp. It can occur as a result of skin trauma or a secondary bacterial infection of an open wound, such as a pressure sore, or it may be unrelated to skin trauma. It most often occurs in the extremities, usually the lower legs.

Prevention

Good hygiene and prevention of cross-contamination are important. If there is an open wound, preventing infection and promoting healing are critical.

Signs and Symptoms

The initial sign of cellulitis is a localized area of inflammation that may become more generalized if not treated properly. Common clinical manifestations include warmth, redness, localized edema, pain, tenderness, fever, and lymphadenopathy. It may be seen in any areas of an open wound, with skin trauma, and in the lower legs. The infection can worsen rapidly if not treated properly.

Diagnostic Tests

Culture and sensitivity testing of any pustules or drainage is necessary to identify the infecting organism. Blood cultures may also be indicated to rule out bacteremia.

Medical Treatment

Topical and systemic antibiotics are prescribed according to culture and sensitivity test results. Debridement of nonviable tissue is necessary when there is an open wound. Systemic antibiotics are indicated if fever and lymphadenopathy are present.

Nursing Care

Assess for a history of recent skin trauma. The affected extremity can be measured to determine degree of edema and to monitor for improvement. Temperature and other signs of worsening or systemic infection are monitored for and reported. The affected extremity should be elevated to decrease edema. Analgesics and application of warm compresses as ordered may increase comfort. Antibiotics are administered as ordered. Use of standard precautions, including frequent, thorough handwashing, is a must to prevent spread of infection. The patient should be taught infection control measures, especially if home dressing changes are ordered.

Acne Vulgaris

Pathophysiology and Etiology

Acne vulgaris is a common skin disorder of the sebaceous glands and their hair follicles that usually occurs on the face, chest, upper back, and shoulders. The etiology is multifocal. The most common cause is hormonal changes.

The sebaceous glands are under endocrine control, especially the androgens. Stimulation of androgens (e.g., during adolescence or the menstrual cycle) in turn stimulates the sebaceous glands to increase sebum production. This, along with gradual obstruction of the pilosebaceous ducts with accumulated debris, ruptures the sebaceous glands, which causes an inflammatory reaction that may lead to papules, pustules, nodules, and cysts. Acne occurs when the ducts through which this sebum flows become plugged.

Other factors that influence occurrence and severity of acne include a hereditary tendency, stress, and external irritants such as strong soaps or cosmetics. It is not related to diet, chocolate, sexual activity, or uncleanliness.

Prevention

Acne vulgaris occurs regardless of interventions; however, certain interventions can lessen the severity or prevent complications. Avoidance of "picking" pimples prevents further inflammation and scarring. The patient should avoid excessive washing, irritants, and abrasives.

Signs and Symptoms

The initial lesions are comedones. Closed comedones, or whiteheads, are small white papules with tiny follicular openings. These may eventually become open **comedo,** or blackheads. The color is not caused by dirt but to lipids and melanin pigment. Scarring occurs as a result of significant skin inflammation; picking can worsen inflammation and lead to further scarring. The resulting inflammation can lead to papules, pustules, nodules (Fig. 51–5), cysts, or abscesses.

Medical Treatment

Medical treatment prevents new lesions and helps control current lesions. Effective topical agents include benzoyl peroxide (Desquam-X; Benzagel), an antibacterial that may help prevent pore plugging; antibiotics (erythromycin,

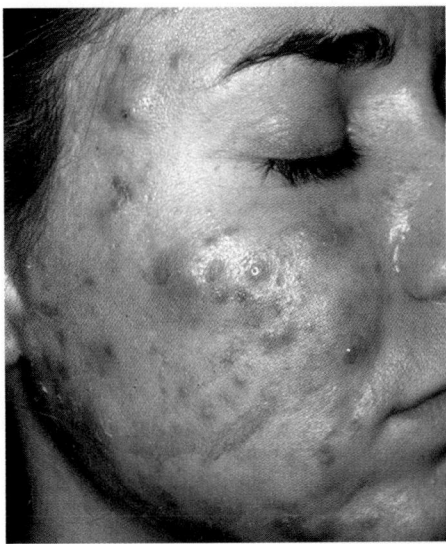

Figure 51-5 Acne vulgaris. (From Goldsmith, LA, Lazarus, and GS, Tharp, MD: Adult and Pediatric Dermatology. FA Davis, Philadelphia, 1997, p 351, with permission.)

tetracycline) to kill bacteria in follicles; and vitamin A acid (Retin-A, tretinoin) to loosen pore plugs and prevent occurrence of new comedones. Topical agents may be used alone or in combination. It may take 3 to 6 weeks before improvement is seen.

Systemic antibiotics (long term, low dose) and isotretinoin (Retin-A) are usually reserved for severe cases; the patient must be closely monitored for side effects. Estrogen therapy (oral contraceptives) may also be prescribed for young women; however, the risks often outweigh the benefits. Women should be aware that some antibiotics reduce the effectiveness of oral contraceptives. Systemic corticosteroids may occasionally be prescribed for severe nodular acne, but they are associated with severe side effects.

Other medical treatments include comedo extraction, intralesional injections of corticosteroids, cryosurgery (freezing with liquid nitrogen), mild peeling (UV light, carbon dioxide, liquid nitrogen, mild acid), dermabrasion (deep chemical peel), excision of scars, and injection of fibrin or collagen below the scars. These treatments depend on the severity, age, condition, and physician and patient preference.

Nursing Care

Review all medication instructions with the patient. Antibiotics should be used as directed to avoid development of antibiotic-resistant strains of bacteria. Encourage the patient to gently wash the face twice daily, especially before topical medication application, using a mild soap. All topical agents should be applied with clean hands to acne-prone areas, not just where the acne is. These agents must be applied to dry skin. Medications should not be applied near eyes, nasolabial folds, and corners of the mouth because of the potential for irritation.

L E A R N I N G   T I P S

Topical benzoyl peroxide may bleach colored fabrics. Have the patient wear a white cotton T-shirt under clothing if benzoyl peroxide is used on the back, and use an old or white pillowcase at night.

If the patient is ordered a combination of topical agents, unless contraindicated, the tretinoin is used at night and the others in the morning or afternoon. Tretinoin could be neutralized if mixed directly with other agents. The patient must be careful with sun or sunlamp exposure while using tretinoin. Also, remind the patient that it may be necessary to continue treatment even when the skin clears.

Above all, try to dispel misconceptions regarding the cause of the acne and give appropriate health education. Advise the patient to keep hands away from the face and especially not to touch or squeeze pimples. Keep hair clean and off the face. Avoid cosmetics, lotions, and shaving creams on the face.

▶ PARASITIC SKIN DISORDERS
Pediculosis
Pathophysiology and Etiology

Pediculosis is an infestation by lice. There are three basic types: pediculosis capitis (head lice), pediculosis corporis (body lice), and pediculosis pubis (pubic or crab lice). Generally the lice bite the skin and feed on human blood, leaving their eggs and excrement, which can cause intense itching. The lice are oval and are approximately 2 mm in length.

In pediculosis capitis the female louse lays eggs (nits) close to the scalp, where the nits become firmly attached to hair shaft. The most common areas of infestation are the back of the scalp and behind the ears. The nits are about 1 to 3 mm in length and appear silvery white and glistening. Transmission is by direct contact or contact with infested objects, such as combs, brushes, wigs, hats, and bedding. It is most common in children and people with long hair.

Pediculosis corporis is caused by body lice that lay eggs in the seams of clothing and then pierce the skin. Areas of the skin usually involved are the neck, trunk, and thighs.

Pediculosis pubis is caused by crab lice. It is generally localized in the genital region, but it can also be seen on hairs of the chest, axillae, eyelashes, and beard. The lice are about 2 mm in length and appear crablike. It is chiefly transmitted through sexual contact or to a lesser degree by infested bed linen.

Prevention

Obvious prevention is avoidance of contact with an infected person or object. Brushes, combs, hats, and other personal items should not be shared. Good personal hygiene and routine clothes washing are other preventive measures; however, even someone with meticulous hygiene can develop this infection if there is contact with the organism.

Signs and Symptoms

Pediculosis capitis can result in no itching or intense itching and scratching, especially at the back of the head. Nits may be noticeably attached to hair. A papular rash may be seen.

Pediculosis corporis may appear as minute hemorrhagic points. Excoriations may be noted on the back, shoulders, abdomen, and extremities. It may also cause intense itching.

Pediculosis pubis results in mild to severe itching, especially at night. Black or reddish-brown dots (lice excreta) may be noted at the base of hairs or in underclothing. Gray-blue macules may also be noted on the trunk, thighs, and axillae; this is the result of the insects' saliva mixing with bilirubin.

Complications

Secondary bacterial infections can occur with pediculosis capitis, resulting in impetigo, furuncles, pustules, crusts, and matted hair. Secondary lesions that can occur with pediculosis corporis include parallel linear scratches, hyperemia, eczema, and hyperpigmentation. Most important, body lice

pediculosis: pedicul(us)—louse + osis—condition

may be vectors for rickettsiae disease. Complications with pediculosis pubis include dermatitis and the coexistence of other sexually transmitted diseases.

Diagnostic Tests

Diagnosis is through history and physical assessment. The patient may also be tested for sexually transmitted diseases if pediculosis pubis is present.

Medical Treatment

Medical treatment is aimed at killing the parasites and mechanically removing nits. Pediculicides containing pyrethrum or permethrin are the most commonly recommended compounds. These agents should kill the lice and nits, although some lice develop pesticide resistance, making mechanical removal necessary. Permethrin (Nix) remains active for about a week, killing the adult lice immediately and the nits when they hatch days later. Pyrethrins (RID, A-200 Pyrinate) must be reapplied in 1 week to kill newly hatched lice.

Complications are treated, as appropriate, with antipruritics, topical corticosteroids, and systemic antibiotics. Physostigmine ophthalmic ointment is applied to affected eyebrows and eyelashes. Other medications should not be applied to eyebrows or eyelashes.

Nursing Care

Reassure the patient and family that head lice can happen to anyone, and this is not a sign of uncleanliness. These infections are treated on an outpatient basis, so patient education is important. Package instructions should be followed for correct usage of all medications.

Instruct the patient to bathe with soap and water and to disinfect combs and brushes in hot, medicated soapy water. A fine-toothed comb dipped in vinegar can be used to remove nits from hairy areas. Nits can be removed from eyebrows and eyelashes with a cotton-tipped applicator after treatment. Clothing, linens, and towels should be laundered in hot water and detergent; unwashable clothing should be dry-cleaned or sealed in a plastic bag for 10 days. Treatment should be started immediately to prevent rapid spread. Family members and close contacts (sexual contact with pediculosis pubis) should be examined for infestation and should put on clean clothing.

Shampoos and lotions kill nits, but they do not remove them. To loosen nits from the scalp, the hair may be soaked in a solution of equal parts vinegar and water and a shower cap worn for 15 minutes. The hair is then combed with a fine-toothed comb and thoroughly rinsed or shampooed to mechanically remove the nits. Children may return to school after adequate medical treatment, even if dead nits are still present.

LEARNING TIPS

It is not possible to dry-clean or wash all infected items, such as mattresses and upholstered furniture. Adult lice can live away from humans for only 3 to 4 days. Therefore simply vacuum the upholstered furniture. The lice die in 3 to 4 days without human contact.

Scabies
Pathophysiology and Etiology

Scabies is a contagious skin disease caused by the mite *Sarcoptes scabiei*. It is results from intimate or prolonged skin contact or prolonged contact with infected clothing, bedding, or animals (e.g., dogs, cats, other small animals). The parasite burrows into the superficial layer of the skin (Fig. 51–6). These burrows appear as short, wavy, brownish-black

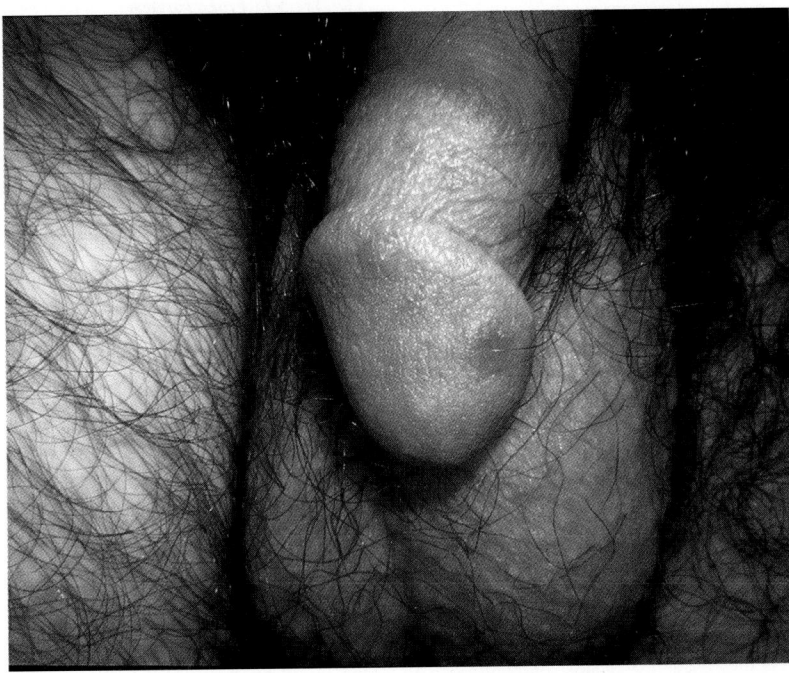

Figure 51-6 Scabies. (From Goldsmith, LA, Lazarus, GS, and Tharp, MD: Adult and Pediatric Dermatology, FA Davis, Philadelphia, 1997, p 295, with permission.)

lines. The patient is asymptomatic while the organism multiplies, but it is most contagious at this time. Symptoms do not occur for almost 4 weeks from time of contact.

Prevention

All persons (and animals) in intimate contact with an infected patient should be treated at the same time to eliminate the mites. The mites survive less than 24 hours without human contact. Therefore bed linen, clothes, and towels should be washed, but furnishings need not be cleaned. Clean clothing and linens should be applied.

Signs and Symptoms

The major complaints are itching and rash. Itching can be intense, especially at night. The itching occurs 1 month after infestation and may persist for days to weeks after treatment. The rash may appear as small, scattered erythematous papules, concentrated in finger webs, axillae, wrist folds, umbilicus, groin, and genitals. Male patients may exhibit excoriated papules on the penis and groin area.

Complications

Hypersensitivity reactions to the mite can result in crusted lesions, vesicles, pustules, excoriations, and bacterial superinfections.

Diagnostic Tests

Diagnosis is confirmed by a superficial shaving of a lesion and microscopic evaluation for adult mite, eggs, or feces.

Medical Treatment

Topical scabicides are used for chemical disinfection. Usually the cream or lotion is applied in a thin layer to the entire body from neck to feet (including genitals, umbilicus, and skin-fold areas), is left on overnight (8 to 12 hours), and is washed off in the morning; however, package instructions should be referred to for each medication. One or two applications is usually curative, depending on the agent prescribed. Antipruritics and corticosteroids may be prescribed for itching.

Nursing Care

A warm soapy bath or shower removes scales and skin debris. Advise the patient to apply the topical medication as ordered; not to repeatedly use scabicides, because it can increase itching and cause further skin irritation; to follow medication directions; to treat family members and close contacts simultaneously to eliminate mites; to wear clean clothing; and to use clean linens. Remind the patient that itching may continue for up to 2 weeks after treatment, until the allergic reaction subsides. (Dead mites remain in the epidermis until exfoliated.)

► PEMPHIGUS
Pathophysiology and Etiology

Pemphigus is an acute or chronic serious skin disease characterized by the appearance of bullae (large fluid-filled blisters) of various sizes on otherwise normal skin and mucous membranes. The etiology is unknown, but it is probably due to an autoimmune disorder. It usually occurs in patients from middle to older age.

Successive crops of bullae suddenly appear on skin or mucous membranes. The bullae are fragile and flaccid. They enlarge, rupture, and form painful, raw, eroded, partial-thickness wounds that bleed, ooze, and form crusts. Pemphigus usually originates in the oral mucosa and then spreads to the trunk. Large areas of the body become involved.

Signs and Symptoms

Besides the appearance of the blisters, the patient experiences pain, burning, and itching. The lesions have a foul smell. Involvement of the oral mucosa can interfere with chewing, swallowing, and talking. The patient is in constant misery.

Complications

The major complication is a bacterial secondary infection. There is a high morbidity and mortality rate associated with this disease.

Diagnostic Tests

A positive Nikolsky's sign is a characteristic finding. This occurs when there is sloughing or blistering of normal skin when minimal pressure is applied. A biopsy of a blister reveals acantholysis, or separation of epidermal cells from one another.

Medical Treatment

Treatment is aimed at controlling the disease, healing the skin, and preventing complications. Corticosteroids in large doses and cytotoxic agents are prescribed to control the disease and bring about remission. Medicated mouthwashes may be prescribed for mouth lesions. Analgesics and antipruritics are prescribed according to the patient's specific signs and symptoms. Because of fluid, blood, and protein losses through the partial-thickness injury, a high-protein, high-calorie diet is recommended along with appropriate fluid replacement therapy.

Nursing Care

The patient is educated on the effects and side effects of medications. Fluid balance is monitored with regular intake and output, body weight, and blood pressure measurement.

LEARNING TIPS

Animals infected with scabies should be treated by a veterinarian.

---

pemphigus: pemphix—blister

The patient is encouraged to maintain adequate fluid intake. Administer tepid wet dressings or baths to lessen secondary infection, cleanse the area, decrease odor, and increase the comfort level of the patient. Potassium permanganate baths may decrease infection and clean and deodorize the area. Always thoroughly dissolve potassium permanganate crystals in a small container before adding to tub water. Undissolved crystals may further damage and burn the skin. Dry the patient thoroughly after the bath. Do not use tape on the patient because this may cause further blistering. Talcum powder may be indicated to keep the skin from sticking to linens and bedclothes. Maintain meticulous oral hygiene. Offer cool drinks often to lessen discomfort. Provide appropriate psychosocial support necessary because of the length of illness, the chronic nature of the condition, and the physical appearance of lesions.

▶ BURNS

Many people are hospitalized each year for burns. Burns affect not only the skin but every major body system. Smoke inhalation and wound infections complicate care of the patient who has been burned.

Pathophysiology and Signs and Symptoms

Burns are wounds caused by an energy transfer from a heat source to the body, heating the tissue enough to cause damage. Locally, the heat denatures cellular protein and interrupts the blood supply. The three zones of tissue damage are described in Figure 51–7.

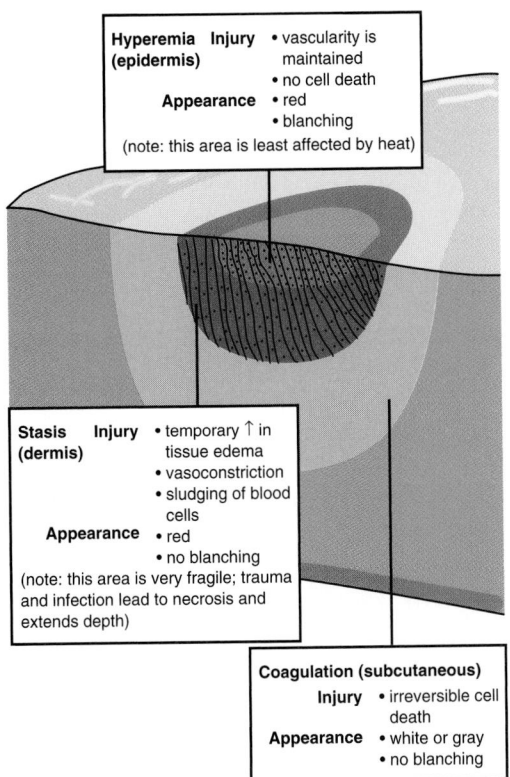

Hyperemia Injury (epidermis)
- vascularity is maintained
- no cell death

Appearance
- red
- blanching

(note: this area is least affected by heat)

Stasis Injury (dermis)
- temporary ↑ in tissue edema
- vasoconstriction
- sludging of blood cells

Appearance
- red
- no blanching

(note: this area is very fragile; trauma and infection lead to necrosis and extends depth)

Coagulation (subcutaneous)

Injury
- irreversible cell death

Appearance
- white or gray
- no blanching

Figure 51-7 Three zones of tissue damage. (Modified from Ruppert, SD, Kernick, JG, and Dolan, JT: Dolan's Critical Care Nursing. FA Davis, Philadelphia, 1996, p 942, with permission.)

The amount of skin damage is related to (1) the temperature of the burning agent; (2) the burning agent itself; (3) the duration of exposure; (4) the conductivity of tissue; and (5) the thickness of the involved dermal structures. Alterations in normal skin functioning resulting from a major burn injury include loss of protective functions, impaired temperature regulation, increased risk for infection, changes in sensory function, loss of fluids, impaired skin regeneration, and impaired secretory and excretory functions.

SYSTEMIC RESPONSES. The alterations in the functional capacity of the skin affect virtually all major body systems. Following a major burn, increased capillary permeability leads to the leakage of plasma and proteins into the tissue, resulting in the formation of edema and loss of intravascular volume. There is also an evaporative water loss through burn tissue that can be 4 to 15 times normal. The increase in metabolism leads to further water loss through the respiratory system.

Cardiac function is affected. There is an initial decrease in cardiac output, which is further compromised by the circulating plasma volume loss. Severe hematological changes resulting from tissue damage and vascular changes occur in patients with major burns. Plasma moves into the interstitial space because of increased capillary permeability. In the first 48 hours after a burn, fluid shifts lead to hypovolemia and, if untreated, hypovolemic shock. Loss of intravascular fluid causes an increase in hematocrit, and there is red blood cell destruction. The intense heat decreases platelet function and half-life. Leukocyte and platelet aggregation may progress to thrombosis.

Metabolic demands are very high in patients with burns. A high metabolic rate proportional to the severity of the burn is usually maintained until wound closure. This hypermetabolism is further compromised by associated injuries, surgical interventions, and the stress response. Severe catabolism also begins early and is associated with a negative nitrogen balance, weight loss, and decreased wound healing. Elevated catecholamine levels are triggered by the stress response. This, along with elevated glucagon levels, can stimulate hyperglycemia.

Gastrointestinal problems that can develop with a major burn include gastric dilation, Curling's ulcer, paralytic ileus, and superior mesenteric artery syndrome. Most of these problems occur in response to fluid shifting, dehydration, opioid analgesics, immobility, depressed gastric motility, and the stress response.

Acute renal insufficiency can occur as a result of hypovolemia and decreased cardiac output. Fluid loss and inadequate fluid replacement can lead to decreased renal plasma flow and glomerular filtration rate. With an electrical burn injury, renal damage can occur from direct electrical current or the formation of myoglobin casts (because of the muscle destruction), which can cause acute tubular necrosis.

Pulmonary effects are mostly related to smoke inhalation. However, hyperventilation is usually proportional to the severity of the burn. There is increased oxygen consumption resulting from the hypermetabolic state, fear, anxiety, and pain.

Immunologically, with the skin destroyed, the body loses its first line of defense against infection. Major burns also cause a depression of the immunoglobulins IgA, IgG, and IgM.

CLASSIFICATION OF BURN INJURIES. The severity of a burn injury is influenced by the depth of destruction, percentage of injury, cause of the burn, age of the patient, concomitant injuries, medical history (e.g., heart disease, diabetes), and location of the burn wound. Table 51–4 describes classification of burn depth (Figs. 51–8, 51–9, and 51–10).

The size of the burn wound is determined by an estimation of the extent of the burn injury. A common method is the rule of nines. This method divides the body into segments whose areas are either 9 percent or multiples of 9 percent of the total body surface, with the perineum counted as 1 percent. This formula is easy, but it is not accurate in assessing children. A more accurate method uses a table with a relative anatomic scale or diagram that estimates total burned area by ages and by smaller anatomic areas of the body. Figure 51–11 provides an example of each method.

Etiology

Burn injuries have many causes. The most common causes include flame, contact, chemical, electrical, and radiation (Table 51–5).

Complications

A major complication that can occur with a flame burn in an enclosed space is inhalation injury. Infection is another common complication with a major burn. The incidence of infection increases with the size of the burn wound, because the first line of defense against microorganisms is the skin.

Neurovascular compromise can also occur with a major burn. Eschar formation creates pressure and contributes to decreasing blood flow to areas distal to the burned area. Other systemic complications are reviewed under Systemic Responses earlier in this chapter.

Diagnostic Tests

Burns are diagnosed through clinical manifestations. Various diagnostic tests are performed for systemic reactions, infection, and other complications. Common tests include complete blood cell count (CBC) and differential, blood urea nitrogen (BUN), serum glucose and electrolytes, arte-

TABLE 51-4		**CLASSIFICATION OF BURN DEPTH**			
Classification	**Formerly**	**Areas Involved**	**Appearance**	**Sensitivity**	**Healing Time**
Partial-thickness (superficial)	1°–2°	Epidermis Papillae of dermis	Bright red to pink. Blanches to touch. Serum-filled blisters. Glistening, moist.	Sensitive to air, temperature, and touch.	7–10 days
Partial-thickness (deep)	2°	Epidermis, ½ to ⅞ of dermis Appendage usually present	Blisters may be present. Pink to light red to white. Soft and pliable. Blanching present.	Pressure may be painful from exposed nerve endings.	14–21 days. May need grafting to decrease scarring.
Full-thickness	3°–4°	Epidermis Dermis Tissue Muscle Bone	Snowy white, gray, or brown. Texture is firm and leathery. Inelastic.	No pain as nerve endings are destroyed, unless surrounded by areas of partial-thickness burns.	Needs grafting to complete healing.

Source: Trofino, RB: Nursing management of the patient with burns. In Ruppert, SD, Kernick, JG, and Dolan, JT (eds): Dolan's Critical Care Nursing. FA Davis, Philadelphia, 1996, p 943.

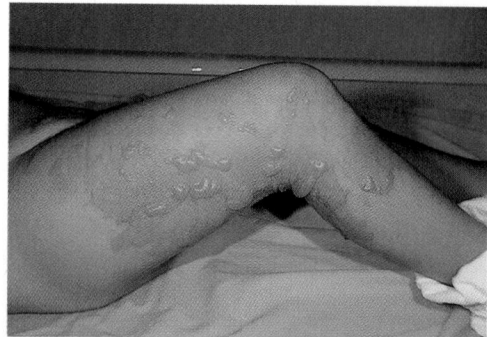

Figure 51-8   Partial-thickness burn. (From Trofino, RB: Nursing Care of the Burn-Injured Patient. FA Davis, Philadelphia, 1991, plate 1, with permission.)

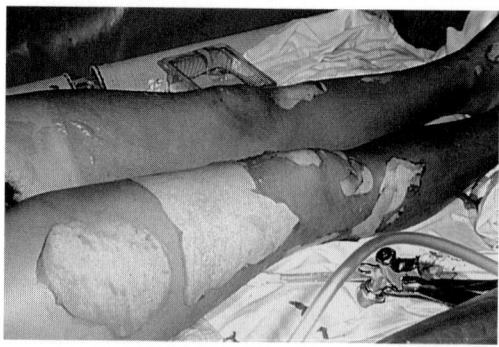

Figure 51-9   Partial-thickness burn. (From Trofino, RB: Nursing Care of the Burn-Injured Patient. FA Davis, Philadelphia, 1991, plate 2, with permission.)

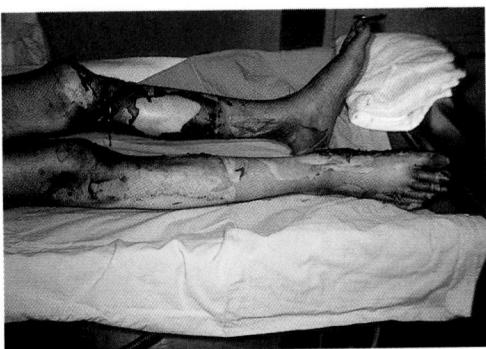

Figure 51-10 Full-thickness burn. (From Trofino, RB: Nursing Care of the Burn-Injured Patient. FA Davis, Philadelphia, 1991, plate 3, with permission.)

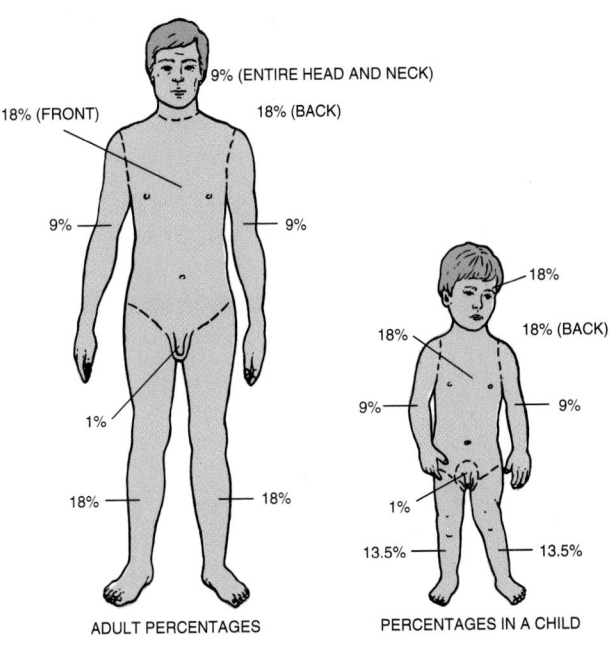

ADULT PERCENTAGES

PERCENTAGES IN A CHILD

RULE OF NINES

Figure 51-11 Estimation of extent of burn injury. (From Venes, D (ed): Taber's Cyclopedic Medical Dictionary, ed 19. FA Davis, Philadelphia, 2001. Beth Anne Willert, M.S., Dictionary Illustrator.)

rial blood gases, blood protein and albumin, urine cultures, urinalysis, clotting studies, cervical spine series, electrocardiogram, wound cultures, and, if there is a suspected inhalation injury, arterial blood gasses, bronchoscopy, and carboxyhemoglobin levels.

Medical Treatment

Medical treatment varies according to the severity of the burn and the stage the patient is in. The treatment of the patient is managed over three overlapping stages (Table 51–6).

EMERGENT STAGE. At the time of injury, the burning process must be stopped. The clothes are removed, and the wound is cooled with tepid water and covered with clean sheets to decrease shivering and contamination. The burn wound itself takes a lower priority to the ABCs (airway, breathing, circulation) of trauma resuscitation. The patient should be stabilized in terms of fractures, hemorrhage, spine immobilization, and other injuries. Inhalation injury is suspected if the patient sustained a burn from a fire in an enclosed space or was exposed to smoldering materials, if the face and neck were burned, if there are vocal changes, and

if the patient is coughing up carbon particles. Intravenous fluids are given to prevent and treat hypovolemic shock. The patient is treated for pain with appropriate IV opioid analgesics. Patient-controlled analgesia (PCA) is very effective.

An accurate history of the injury is obtained to determine severity, probable complications, and any associated trauma. The patient's medical history is also obtained. Admission to the facility and burn care treatment are explained to the patient and family.

ACUTE STAGE. If the patient is in a facility with a special burn unit, multidisciplinary care from a burn team is pro-

TABLE 51-5	COMMON CAUSES OF BURNS
Flame	House fire is a common cause.
	Usually associated with an inhalation injury.
	Flash injury occurs from a sudden ignition or explosion.
Contact	Hot tar, hot metals, hot grease produce a full-thickness injury on contact.
Scald	A burn from hot liquid.
	Common in children less than 5 years and adults older than 65 years.
	With an immersion scald, there are usually no splash marks; usually involves lower regions of body.
Chemical	Usually occurs in an industrial setting.
	Extent and depth of injury are directly proportional to concentration and quantity of agent, duration of contact, and chemical activity and penetrability of agent.
Electrical	One of the most serious types of burn injury; can be full thickness with possible loss of limbs, as well as cause internal injuries.
	Entry wound is usually ischemic, charred, and depressed.
	Exit wound may have an explosive appearance.
	Extent of injury depends on voltage, resistance of body, type of current, amperage, pathway of current, and duration of contact.
	Bones offer greatest resistance to the current; can have much damage.
	Tissue fluid, blood, and nerves offer least resistance; therefore the current travels this path.
Radiation	Can occur in an industrial setting, due to treatment of diseases, or from ultraviolet light (sun or tanning salons).
	Severity depends on type of radiation, duration of exposure, depth of penetration, distance from source, and absorbed dose.

TABLE 51-6	STAGES OF BURN CARE
Stage	**Duration**
I (emergent)	From onset of injury to completion of fluid resuscitation
II (acute)	From start of diuresis to near completion of wound closure
III (rehabilitation)	From wound closure to return of optimal level of physical and psychosocial function

Source: Trofino, RB: Nursing management of the patient with burns. In Ruppert, SD, Kernick, JG, and Dolan, JT (eds): Dolan's Critical Care Nursing. FA Davis, Philadelphia, 1996, p 948.

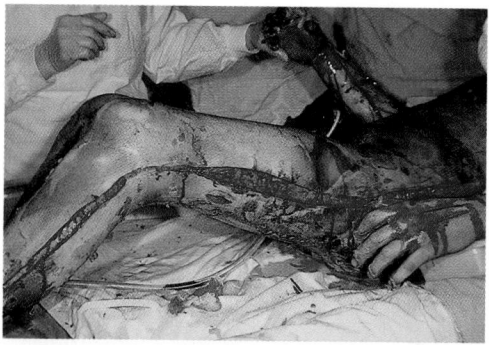

Figure 51-12 Escharotomy. (From Trofino, RB: Nursing Care of the Burn-Injured Patient. FA Davis, Philadelphia, 1991, plate 16, with permission.)

vided during the acute stage. Management goals include wound closure with no infection; minimum scarring; maximum function; maintenance of comfort as much as possible; adequate nutritional support; and maintenance of fluid, electrolyte, and acid-base balance. The patient continues to be medicated for pain as needed, especially before painful treatments. Nutritional support is maintained via a small Silastic nasogastric feeding tube.

The wound is cleansed and debrided daily to promote healing, prevent infection, and provide a clean bed for grafting. Wound cleansing is achieved by tubing with a Hubbard tank, showering using a shower trolley or shower chair, and bedside care.

Debridement, or the removal of nonviable tissue (eschar), can be mechanical, chemical, surgical, or a combination of these methods. Mechanical debridement can involve the use of scissors and forceps to manually excise loose nonviable tissue or the use of wet-to-moist or wet-to-dry fine mesh gauze. Chemical debridement involves the use of a proteolytic enzymatic debriding agent that digests necrotic tissue. Surgical debridement is the excision of full-thickness and deep partial-thickness burns. This method is followed by an application of a skin graft.

If the patient has a circumferential burn (one that surrounds an extremity or area), there is an increase in tissue pressure secondary to tissue edema. The burn then acts like a tourniquet, impeding arterial and venous flow. Common sites for these burns are the extremities, trunk, and chest. If this occurs on the chest and trunk, respiratory insufficiency can occur as a result of restricted chest expansion. An **escharotomy** is immediately necessary to relieve this pressure. An escharotomy is a linear excision through the eschar to the superficial fat that allows for expansion of the skin and return of blood flow (Fig. 51-12).

Once the area is cleaned, the burn dressing and topical treatment are prescribed. The type of dressing and topical agent chosen are dependent on the area involved, extent and depth of injury, and physician preference. Several common topical agents are listed in Table 51-7.

Dressings may be open, closed, biologic, synthetic, or a combination. The open method is the use of a topical agent without any dressing. The closed method involves the use of an occlusive dressing over the wound. General principles for dressings include the following:

1. Limit the bulk of the dressing to facilitate range of motion.
2. Never wrap skin-to-skin surfaces (e.g., wrap fingers or toes separately; place a donut gauze dressing around the ear).
3. Base dressings on the size of wounds, absorption, protection, and type of debridement.
4. Wrap extremities distal to proximal to promote venous return.
5. Elevate affected extremities.

Biologic dressing refers to tissue from living or deceased humans (cadaver skin) or deceased animals (e.g., pigskin). These dressings may be used as donor site dressings, to manage a partial-thickness burn, and to cover the clean, excised wound before autografting. Biologic dressings assist with wound healing and stimulate **epithelialization.**

Synthetic dressings are used in the management of partial-thickness burns and donor sites. These dressings are more readily available, less costly, and easier to store than biologic dressings. They are made from a variety of materials and come in many different sizes and shapes. Most of these dressings contain no antimicrobial agents.

Biologic and synthetic dressings are used as temporary wound coverings over clean partial- and full-thickness injuries. These skin substitutes help maintain the wound surface until healing occurs, a donor site becomes available, or the wound is ready for autografting.

Skin Grafts. An autograft is a skin graft from the patient's unburned skin to be placed on the clean excised burn. The two common types of autografts are the split-thickness skin graft (STSG), which includes the epidermis and part of the dermis, and the full-thickness skin graft (FTSG), which includes the epidermis and entire dermal layer.

An STSG (0.006 to 0.016 inch) may be applied as a sheet graft or a meshed graft. A sheet graft is used for cos-

escharotomy: eschara—scab + otomy—incision

epithelialization: epi—over + thele—nipple + ization—condition

TABLE 51-7	COMMON TOPICAL ANTIBIOTIC AGENTS		
Agent	**Dressings**	**Advantages**	**Disadvantages**
Silver sulfadiazine 1% cream	Buttered on. With light dressings once or twice a day.	Broad spectrum. Low toxicity. Painless, easy to apply and remove. Can be used with or without dressings.	Intermediate penetration of eschar; leukopenia.
Mafenide acetate (Sulfamylon)	Buttered on. Open exposure method. Applied 3–4 times daily.	Broad spectrum. Rapid, deep penetration of eschar. Rapid excretion.	Pain on application. Pulmonary toxicity. Metabolic acidosis. Inhibits wound healing. Hypersensitivity.
Silver nitrate solution 0.5%	Wet dressings. Change bid. Resoak every 2 hours.	Broad spectrum. Nonallergenic. Low toxicity. Inexpensive. Does not interfere with healing.	Poor penetration of eschar. Ineffective on established wound infections. Can cause an electrolyte imbalance. Discoloration of wound and environment makes assessment difficult.
Bacitracin	Buttered on. Reapply every 4–6 hours.	No pain. Clear, odorless. Useful for face burns. Softens eschar.	Poor penetration of eschar. Not effective in reducing sepsis in large burns. Occasional allergic sensitivity.
Gentamicin	Apply gently 3–4 times daily.	Broad spectrum. May be covered or left open to air.	Ototoxic. Nephrotoxic. Pain on application.
Nitrofurazone	Apply thin layer directly to wounds or impregnate into gauze. Change dressings twice daily.	Bactericidal. Broad spectrum.	Painful application. May lead to overgrowth of fungus and *Pseudomonas*.

Source: Modified from Trofino, RB: Nursing management of the patient with burns. In Ruppert, SD, Kernick, JG, and Dolan, JT (eds): Dolan's Critical Care Nursing. FA Davis, Philadelphia, 1996, p 951.

metic effect, such as for a face, neck, upper chest, breast, or hand burn. It is placed on the area as a full sheet. A meshed graft is passed through a mesher that produces tiny splits in the skin, similar to a fishnet, with openings in the shape of diamonds (Fig. 51–13), to permit the skin to expand 1.5 to 9 times its original size. The meshing allows for coverage of a large burn area with a small piece of skin by stretching it and securing it with sutures or staples. A mesh graft is especially useful when there are extensive burns resulting in few available donor sites. Graft "take," or vascularization, is complete in about 3 to 5 days.

Full-thickness skin grafts (0.035 to 0.040 inch) can be sheet grafts or pedicle flaps. FTSGs are used over areas of muscle mass, soft tissue loss, hands, feet, and eyelids. They are not used for extensive wounds, because the donor sites usually require an STSG for closure, or closure from the wound edges. A pedicle graft or flap includes the skin flap and subcutaneous tissue that is attached by its pedicle to a blood supply (artery and vein); it is then attached to the area in need of grafting. Once the distal part of the graft takes, it remains in place and the flap is divided, with the remainder returning to the original site. Pedicle flaps are not as popular as free skin flaps, because they require more than one surgery and take longer for the graft site and donor site to heal. Table 51–8 provides a comparison of split-thickness and full-thickness grafts.

Donor sites are considered a partial-thickness wound. Donor sites usually heal in 10 to 14 days, but this is dependent on thickness and method of grafting and the general health of the patient. Treatment for the donor site varies with the individual patient, the area of the body, and physician preference. Considerations for care include promoting comfort and preventing trauma and infection. An

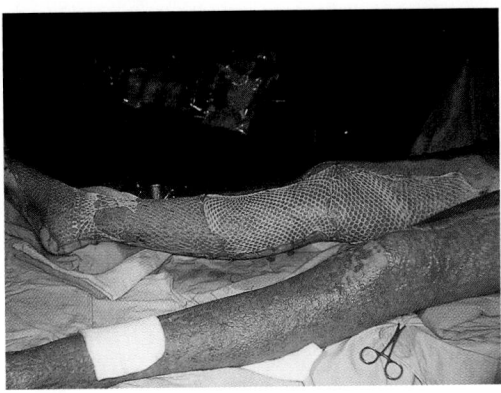

Figure 51-13   Meshed graft. (From Trofino, RB: Nursing Care of the Burn-Injured Patient. FA Davis, Philadelphia, 1991, plate 17, with permission.)

outer dressing, used to apply pressure for hemostasis, remains in place for 1 to 2 days. For dry exposure healing, a bed cradle may be used to promote circulation and avoid pressure on the site. A heat lamp (60 to 100 watts) may be employed to assist in drying, but it must be kept at least 2 feet away from the site. Loose, separating gauze is trimmed to avoid accidental trauma. The donor site is very painful. Appropriate pain medications are provided, along with nonpharmacological measures (e.g., back rub, distraction).

With any type of graft, the patient must keep the graft site immobilized until the graft takes, to prevent movement or slippage of the grafted skin. Dressings may be bulky to assist in immobilization. These dressings must not be disturbed. The involved area requires frequent circulatory checks (color, warmth, sensation, pulses, and capillary refill). Any involved extremities must be elevated to maintain circulation. Table 51–9 describes factors affecting graft viability. A

TABLE 51-8	COMPARISON OF SPLIT-THICKNESS AND FULL-THICKNESS SKIN GRAFTS	
	Split-Thickness	**Full-Thickness**
Layers	Epidermis	Epidermis
	Partial layer of dermis	Entire dermal layer
Advantages	Donor site may be reused	Allows more elasticity over joints
	Healing of donor site is more rapid, results in good "take"	Can reconstruct cosmetic defects
		Soft, pliable
		Gives full appearance
		Provides good color-match
		Less hyperpigmentation
		May allow hair growth
Disadvantages	Prone to chronic breakdown	Donor site takes longer to heal
	Likely to hypertrophy	Requires split-thickness graft to heal or closure from wound edges
	More likely to contract	

Source: Konop, D: General local treatment. In Trofino, RB (ed): Nursing Care of the Burn-Injured Patient. FA Davis, Philadelphia, 1991, p 61.

TABLE 51-9	FACTORS AFFECTING GRAFT VIABILITY
Factors Inhibiting Graft "Take"	**Factors Promoting Graft "Take"**
Infection	Adequate hemostasis
Necrotic skin (tissue)	Anatomic location of graft
Anatomic location of graft	Smooth contour
Perineum	Nonjoints
Axilla	Graft secured well
Buttocks	Immobilization of graft area
Poor-quality donor skin	Good nutritional status
Poor nutritional status	
Bleeding	
Mechanical trauma	
Shock	

Source: Konop, D: General local treatment. In Trofino, RB (ed): Nursing Care of the Burn-Injured Patient. FA Davis, Philadelphia, 1991, p 62.

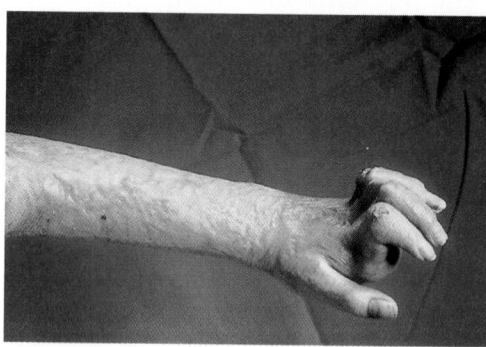

Figure 51-14 Burn deformity: contracture. (From Trofino, RB: Nursing Care of the Burn-Injured Patient. FA Davis, Philadelphia, 1991, plate 36, with permission.)

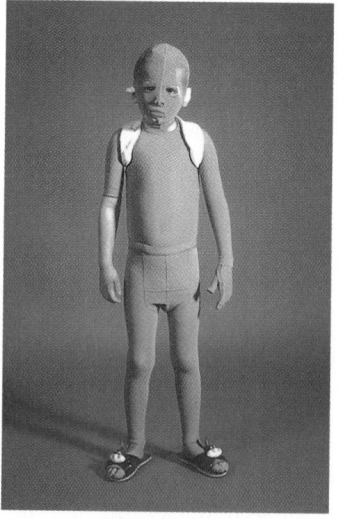

Figure 51-15 Full-body pressure garment (From Trofino, RB: Nursing Care of the Burn-Injured Patient. FA Davis, Philadelphia, 1991, plate 39, with permission.)

graft has been successful if there is good adherence of the graft to the wound with no evidence of necrosis or infection.

REHABILITATION PHASE. The therapy started during the acute phase continues in the rehabilitation phase. There is wound closure, and the goal is to return the patient to an optimum level of physical and psychosocial function. This may take months to years to accomplish. Reconstructive surgery can be ongoing for many years.

Two things to keep in mind when caring for the patient with a major burn are that (1) the most comfortable position (flexion) is the position of contracture, and (2) the burn wound will shorten until it meets an opposing force. To avoid contractures (Fig. 51–14), a specific exercise program is begun 24 to 48 hours after injury, along with the use of splinting devices to maintain proper positioning and stretching. Hypertrophic scarring, or a proliferation of scar tissue, can be minimized or prevented through the use of a pressure garment (Fig. 51–15).

The burn affects the patient's psychosocial status in many ways. The magnitude of these effects are related to the age of the patient, location of the burn (e.g., face, hands), recovery from injury, cause of the injury (especially

if related to negligence or a deliberate act), and ability to continue at preburn level of normal daily activities, disruption of role, family involvement, and general health and coping ability. Treatment involves the patient and significant others. Support groups, counselors, and psychiatrists should be utilized appropriately.

Nursing Process

ASSESSMENT. A major burn is painful and frightening to the patient and frightening to the family. Elicit information from the patient, family, and rescuers. If the injury occurred in an enclosed space with flames or smoldering materials, inhalation injury is suspected. If an electrical injury, ask about voltage, duration of contact, host susceptibility (wet or dry skin), entry and exit sites, and associated falls. With chemical burns, determine type of agent and duration of exposure.

General information to assess in all burns (in addition to normally assessed data, such as medical history, known allergies, and current medications) include extent, depth, type and location of the burn; duration of contact with the burning agent; amount and location of pain; and associated injuries. Determine the immediate first aid treatment provided at the scene. Elicit psychosocial information: other people injured; additional losses (home, pets); whether the patient was at fault; and how this injury affects the patient's role function.

NURSING DIAGNOSIS. Priority nursing diagnoses for a patient with a major burn include impaired skin integrity; impaired gas exchange; deficient fluid volume; pain; ineffective peripheral tissue perfusion; and risk for sepsis. These are covered in Nursing Care Plan Box 51–3. Additional possible diagnoses are imbalanced nutrition, activity intolerance, self-care deficit, disturbed body image, and ineffective coping.

BOX 51-3 NURSING CARE PLAN FOR THE PATIENT WITH A BURN INJURY

▶ Impaired skin integrity related to thermal injury

PATIENT OUTCOMES

Skin integrity is improved as evidenced by healing of burned areas with no infection present, healing of burning process.

EVALUATION OF OUTCOMES

Is burned area healed? Is it free from infection? Did burning process stop?

INTERVENTIONS	RATIONALE	EVALUATION
Assess burning process. If heat is felt on wound, cool with tepid tap water or sterile water.	Depth of injury increases with length of exposure to burning agent.	Is heat felt over wounds?
Assist registered nurse (RN) or physician to assess the burn area for extent (percentage) and depth (partial thickness, full thickness) of injury.	Provides basis for triage of care. Important also for calculating resuscitation fluid therapy.	What is the estimation of percentage of burn injury? What is depth of injury?
Remove clothing and jewelry.	These items can retain heat and thermal agent, therefore increasing depth of injury. Jewelry can be constrictive when edema develops.	Are clothing and jewelry removed?
Do not apply ice.	Ice causes vasoconstriction, further increasing wound damage. Ice also causes a decrease in core body temperature, which may promote shock.	Is water tepid? Has use of ice been avoided?
Cover patient with clean sheet or blanket.	Prevents excessive heat loss. Decreases pain from air exposure. Protects patient from environmental contamination.	Is patient covered?
Obtain history of burning agent.	Provides information related to depth, duration of contact, and resistance of tissues. If fire scenario, provides clues to possible inhalation injury.	What caused this thermal injury? How long was patient in contact with agent?
Initiate immediate copious tepid water lavage for 20 minutes for all chemical burns, along with simultaneous removal of contaminated clothing. (Do not neutralize chemical, because this takes too much time and resulting reaction may generate heat and cause further skin injury.) Brush off dry chemicals before lavage. Use heavy rubber gloves or thick gauze for removal of clothing.	Dilution and removal of chemical agent halts burning process. Lavage dissipates heat. Gloves and gauze are necessary to protect health care workers from injury.	Has lavage been initiated? Were there any injuries to health care workers?
Cleanse wound via tubbing or showers.	Promotes healing and helps decrease infection.	Is burn wound clean and free of wound debris?

BOX 51-3   NURSING CARE PLAN FOR THE PATIENT WITH A BURN INJURY—CONT'D

Assist RN or physician with debriding wound via surgical, chemical, or mechanical means. Apply topical agent as prescribed.
Apply dressing as prescribed. Use common practices:
1. Do not wrap skin surface to skin surface (e.g., wrap fingers and toes separately; donut bandage around ears).
2. Limit bulk of dressings.
3. Wrap extremities distal to proximal.

Promotes healing and healthy granulation bed. Most agents prevent infection and promote healing.

Dressing types vary and are influenced by area, extent, and depth of injury, as well as by topical agent used. Dressing protects burn area and promotes healing.
Wrapping separately prevents webbing and contractures.
Mobility is enhanced with less bulky dressing. Circulation is increased when extremities are wrapped distal to proximal.

Is there any eschar? Is wound free of wound debris?
Is agent applied as directed?

Is dressing applied appropriately?

 Impaired gas exchange related to upper airway edema, carbon monoxide poisoning, edema of alveolar capillary membranes

PATIENT OUTCOMES

Gas exchange will be improved as evidenced by patent airway, CO level less than 10 percent, clear lung sounds, PaO_2 80–100 mm Hg, $PaCO_2$ 35–45 mm Hg, responsive and aware.

EVALUATION OF OUTCOMES

Are blood levels improved: CO, PaO_2, $PaCO_2$? Do the lungs sound clear on auscultation? Is patient aware of surroundings? Are there no signs of respiratory distress (e.g., retractions, nasal flaring, use of accessory muscles)?

INTERVENTIONS	RATIONALE	EVALUATION
Assess respiratory status: auscultate breath sounds every 15 minutes or as necessary; note any adventitious breath sounds; observe for chest excursion: monitor ability to cough.	Detects changes in pulmonary function to alter therapy.	What is patient's respiratory status? Are any adventitious lung sounds noted?
Monitor arterial blood gases and CO level.	Assesses level of oxygenation. Helps guide oxygen therapy.	What are the patient's blood gas levels? Are they abnormal?
Monitor for nasal flaring, retractions, wheezing, and stridor.	Stridor may signal upper airway involvement; nasal flaring, retractions, and wheezing may indicate lower airway involvement.	Does patient exhibit nasal flaring, retractions, wheezing, and stridor?
Administer humidified 100-percent oxygen by tight-fitting face mask for the breathing patient.	Provides oxygen for adequate gas exchange.	Is oxygen administered appropriately? Are blood gases improving?
Elevate head of bed (if no cervical spine injuries or no history of multiple trauma).	Decreases swelling of face and neck.	Is head of bed elevated? Is there any change in facial or neck swelling?
Provide appropriate pulmonary care: turn, cough, deep breathe every 2–4 hours. Incentive spirometer every 2–4 hours, suction frequently as needed.	Mobilizes secretions and promotes lung expansion.	Is patient on scheduled activities for vigorous pulmonary care?
Obtain sputum cultures. Note amount, color, and consistency of pulmonary secretions.	Carbonaceous sputum is diagnostic for smoke inhalation injury. Infection changes color, amount, and consistency of sputum. Assists in selection of appropriate antibiotic.	Is patient coughing up any sputum? Has character of sputum been documented?
Administer bronchodilators and antibiotics as prescribed.	Bronchodilators decrease bronchospasms and edema; antibiotics fight infection.	Are medications given appropriately?

 Deficient fluid volume related to evaporative losses from wound, capillary leak, and decreased fluid intake

PATIENT OUTCOMES

Patient will maintain adequate circulating volume as evidenced by urine output of 50 mL/h (in the adult), blood pressure within normal limits, heart rate at about 100 beats per minute (adult), and stabilized body weight.

EVALUATION OF OUTCOMES

Is urine output maintained at least at 50 mL/h? Are the blood pressure and heart rate within normal limits? Is patient's weight stable?

BOX 51-3   NURSING CARE PLAN FOR THE PATIENT WITH A BURN INJURY—CONT'D

INTERVENTIONS	RATIONALE	EVALUATION
Obtain admission weight and monitor weight daily.	Helps measure fluid loss or gain.	Is patient's weight documented? Is it stable?
Record intake and output (I & O) hourly.	Serves as guide for fluid loss and replacement.	What is patient's I & O?
Assess for signs and symptoms of hypovolemia (hypotension, tachycardia, tachypnea, extreme thirst, restlessness, disorientation).	Fluid volume loss is multifocal (e.g., through increased capillary permeability, insensible loss).	Does patient exhibit any signs or symptoms of hypovolemia?
Monitor electrolytes, CBC.	Serves as guide for electrolyte replacement and blood product replacement.	What are patient's lab values? Are they within normal limits?
Administer IV fluids as ordered. Insert large-bore catheter.	Fluid replacement begins immediately. Large vessels are needed for rapid delivery of fluids.	Is patient's fluid replacement adequate? Is catheter patent?
Insert indwelling urinary catheter.	Fluid replacement is titrated based on urine output.	Is catheter patent?
Monitor urine for amount, specific gravity, and hemochromogens.	Specific gravity can predict volume replacement; hemochromogens can cause renal-tubular damage.	What are patient's urine values?
Administer osmotic diuretics as ordered; monitor response to therapy.	Decreased urinary output can be caused by decreased renal flow (due to myoglobin in urine).	What is urinary output? Has it changed due to therapy?
Assess gastrointestinal function for absence of bowel sounds. Maintain nasogastric tube.	Splanchnic constriction due to hypovolemia can cause a paralytic ileus.	Are patient's bowel sounds normal? Is nasogastric tube patent?

 Pain related to burns or graft donor sites

PATIENT OUTCOMES

Patient will experience pain control as evidenced by verbalizations of pain tolerance, nonverbal cues: less thrashing; better rest or sleep; body positioning.

EVALUATION OF OUTCOMES

Does patient verbalize better pain control? How many hours of rest/sleep does patient have in 24 hours? Does patient state she or he feels rested?

INTERVENTIONS	RATIONALE	EVALUATION
Assess level of pain: nature, location, intensity, and duration at various times (during procedures, at rest). Rate pain on visual analog scale.	Provides baseline to monitor response to therapy.	Is patient's individual response to pain documented?
Observe for varied responses to pain: increase in blood pressure, pulse, respiration; increased restlessness and irritability; increased muscle tension; facial grimaces; guarding.	Responses to pain are variable. These parameters change in response to pain.	What are patient's responses to pain? Do responses change with treatment?
Acknowledge presence of pain. Explain causes of pain.	Encourages understanding.	Is patient more trusting of the treatments?
Administer narcotics IV. Utilize PCA as appropriate.	IV administration is necessary due to edema and poor tissue perfusion. Narcotics are necessary for severe burn pain. PCA allows patient more control.	Is patient being medicated for pain appropriately?
Offer diversional activities (e.g., music, TV, books, games, relaxation techniques).	Helps patient focus on something other than pain.	Does patient utilize diversional activities? Do they help?
Properly position patient. Elevate burned extremities.	Increases comfort. Elevation decreases edema and pain.	Is patient positioned as comfortably as possible? Are extremities elevated?
Maintain comfortable environment (e.g., bed cradle; comfortable environmental temperature, 86–91.4° F; quiet environment).	Pressure from bed linens may cause discomfort; with loss of integument, body cannot self-regulate temperature.	Does patient verbalize comfort of environment?

BOX 51-3 NURSING CARE PLAN FOR THE PATIENT WITH A BURN INJURY—CONT'D

 Ineffective peripheral tissue perfusion related to circumferential burns, blood loss, decreased cardiac output

PATIENT OUTCOMES

Patient will maintain adequate tissue perfusion as evidenced by presence of peripheral pulses; minimal edema; adequate circulation, sensation, and motion; and warm extremities.

EVALUATION OF OUTCOMES

Are peripheral pulses present? Are extremities warm, with adequate sensation, motion, and circulation? Is edema decreased?

INTERVENTIONS	RATIONALE	EVALUATION
Elevate burned extremity above level of the heart.	Enhances venous return and decreases edema formation.	Are all burned extremities elevated above heart level? Is edema decreasing?
Assess pulses on burned extremities every 15 minutes.	Assesses need for escharotomy.	Are pulses present and documented?
Use Doppler as necessary. Assess capillary refill, sensation, color, swelling, and motion.	Assesses peripheral perfusion.	Is the extremity warm, with adequate color, sensation, motion, and capillary refill?
Assess for numbness, tingling, and increased pain in burned extremity.	Can be indicative of increased pressure from edema.	Does patient complain of numbness, tingling, or pain?
Measure circumference of burned extremities.	Monitors edema formation.	Is there change in measurements of circumference?
Apply burn dressing loosely.	Prevents constriction and allows for expansion as edema forms.	Is dressing too tight?
Assist with muscle compartment pressures.	Helps determine need for escharotomy (if pressure greater than 25 mm Hg).	What is patient's pressure?
Assist with escharotomy as necessary.	If indicated, removal of eschar allows for edema expansion and permits peripheral perfusion.	Does patient require an escharotomy? Is edema relieved?

 Risk for sepsis related to wound infection

PATIENT OUTCOMES

Patient will not develop a wound infection.

EVALUATION OF OUTCOMES

Is there healthy granulation tissue on unhealed areas with less than 10^5 colonies of bacteria (wound culture)? Are donor sites free of infection? Have skin grafts taken? Is there absence of clinical manifestation of infection (temperature 98.6° F; normal WBC count)?

INTERVENTIONS	RATIONALE	EVALUATION
Use sterile technique with wound care.	The unhealed burn wound is a culture medium for bacterial growth.	Is sterile technique used for all wound care?
Maintain protective isolation with good handwashing technique.	Prevents spread of bacteria from patient to patient or nurse to patient.	Do all persons in contact with patient maintain proper precautions?
Administer immunosupportive medications as prescribed: tetanus and gamma globulin.	Immunoglobulins are depressed at time of severe burn injury.	Does patient require these medications?
Perform wound care as prescribed, which may include the following: inspect and debride wounds daily; culture wound three times a week or at sign of infection; shave hair at least 1 inch around burn areas (excluding eyebrows); inspect invasive line sites for inflammation (especially if line is through a burn area).	Provides quick identification of bacterial wound invasion and decreases incidence of infection. Presence of hair increases medium for bacterial growth.	What does wound look like? Is it debrided? What are culture results? Is there any hair near burn or line sites?
Continually assess for signs and symptoms of sepsis: temperature elevation; change in sensorium; changes in vital signs and bowel sounds; decreased output; positive blood/wound cultures.	The burn patient may experience several septic episodes until wound is healed.	Does patient exhibit any signs or symptoms of sepsis?
Administer systemic antibiotics and topical agents as prescribed.	Useful in eliminating or controlling infection. Systemic antibiotics are prescribed based on results of wound cultures.	Does patient require systemic antibiotics? Are they working? Are topical agents applied appropriately? Is wound healing?

PLANNING, IMPLEMENTATION, AND EVALUATION. See Nursing Care Plan 51–3 for the Patient with a Burn Injury. For more information on burns, go to the American Burn Association Website at www.ameriburn.org.

▶ SKIN LESIONS

Skin lesions can be either benign (noncancerous) or malignant. Benign lesions are described in Table 51–10. Malignant lesions are discussed in the next section. See also Cultural Consideration Box 51–4.

Malignant Skin Lesions
Pathophysiology and Etiology

The most common skin malignancies include basal cell carcinoma, squamous cell carcinoma, and malignant melanoma. The major cause of skin malignancies is overexposure to ultraviolet rays, most commonly sunlight. Other factors include being fair skinned and blue eyed; genetic tendencies; history of x-ray therapy; exposure to certain chemical agents (e.g., arsenic, paraffin, coal tar); burn scars; chronic osteomyelitis; and immunosuppressive therapy.

BOX 51–4	CULTURAL CONSIDERATION

Some African-American men have facial hair that is kinky, curls back on itself, and penetrates the skin, which can result in pustules and small keloids. Many use depilatories or electric razors to prevent nicking the skin, which can also cause keloids.

Darker-skinned people have an increased incidence of birthmarks and mongolian spots compared with lighter-skinned people. Mongolian spots disappear over time. The nurse must be cautious to not mistake these spots for bruising indicating injury or abuse.

Darker-skinned people have a tendency toward an overgrowth of connective tissue components concerned with the protection against infection and repair after injury. Keloid formation is one example of this tendency toward overgrowth of connective tissue. Lymphoma and systemic lupus erythematosus may occur due to this overgrowth of connective tissue.

For people with light skin, such as Germans, Polish, and Irish, prolonged exposure to the sun may increase the incidence of skin cancer. The nurse needs to teach patients to protect themselves from sun exposure to reduce their risk of skin cancer. Nevi (freckles and skin discolorations) occur more often in lighter-skinned individuals. They are most common in European-Americans, followed by Asians, and then darker-skinned African-Americans.

TABLE 51–10	BENIGN SKIN LESIONS	
Type	**Description**	**Treatment**
Cyst	A saclike growth with a definite wall that may contain liquid, semifluid, or solid material. An epidermal cyst is a saclike growth of the upper portion of a hair follicle. It is due to blockage of the pilosebaceous follicle. It is a soft hemispherical module, usually with an overlying comedo, that is usually seen on the face, neck, or upper trunk. It is usually asymptomatic. A pilar cyst, or sebaceous cyst, is a saclike growth of the middle portion of the hair follicle that contains hair and cuticle-like material. It is a hard, hemispherical nodule without a surmounted pore that is usually seen on the scalp.	If bothersome, it may be surgically excised. If excision is done, the entire cyst wall is removed to prevent recurrence.

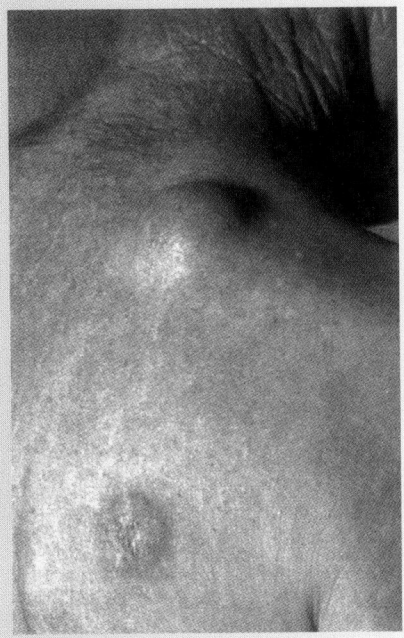

Epidermoid cyst.

TABLE 51-10 BENIGN SKIN LESIONS—CONT'D

Type	Description	Treatment
Seborrheic keratosis	A benign skin lesion that is pigmented light tan to dark brown patches. The plaques or papules have a "stuck on" appearance caused by the proliferation of epidermal cells and keratin piled on the skin surface. Cause is unknown, but it tends to occur in middle-aged to older patients, most commonly on the trunk, scalp, face, and extremities.	Treatment is cosmetic only, or if lesion becomes irritated from friction. Liquid nitrogen cryotherapy or light curettage is performed if necessary for removal.

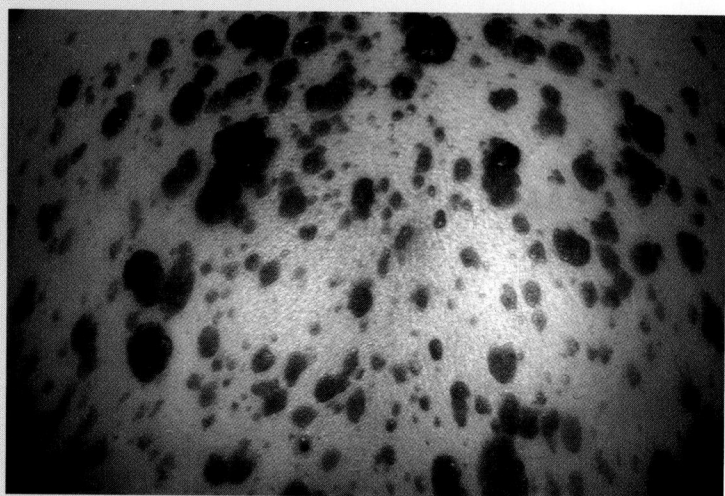

Seborrheic keratosis.

Type	Description	Treatment
Keloid	A benign growth of fibrous tissue (scar formation) at the site of trauma or surgical incision; occurs in various sizes. Growth of tissue is out of proportion to what is needed for normal healing. The benign wartlike lesion or nodule extends beyond the original injury and occurs mainly in middle-aged and elderly clients and darker-skinned patients.	Treatment varies, is not always successful, and is difficult; a larger scar may ensue. Treatments include surgical excision, intralesional steroid therapy, low-dose radiation, and pressure garments worn over the area, or a combination of these therapies.
Pigmented nevi	A benign, flesh-colored to dark brown macule or papule located randomly over the entire skin surface of the body. Can be inherited or acquired and occurs mostly in light-skinned patients. Usually begin to appear between 1 to 4 years of age, increasing in number into adulthood. Some contain a few hairs. There are many variations. Rate of transformation to a malignant melanoma is higher in congenital moles and larger lesions. Clinical signs to observe for in differentiating between a mole and a melanoma include change in color or size; inflammation of surrounding skin; irregular borders; spreading borders; variegated colors, especially a bluish pigmentation; bleeding; and oozing, crusting, and itching. Usually nevi larger than 1 cm should be carefully examined.	Treatment is indicated for any of the previously listed indications of melanoma; unsightly nevi (cosmetic); repeated irritation (rubbing from belt, bra); trauma; large moles; and client conviction of a change in the mole. Surgical removal can include excision (preferred) or surgical shave. All excised moles should be examined histologically.

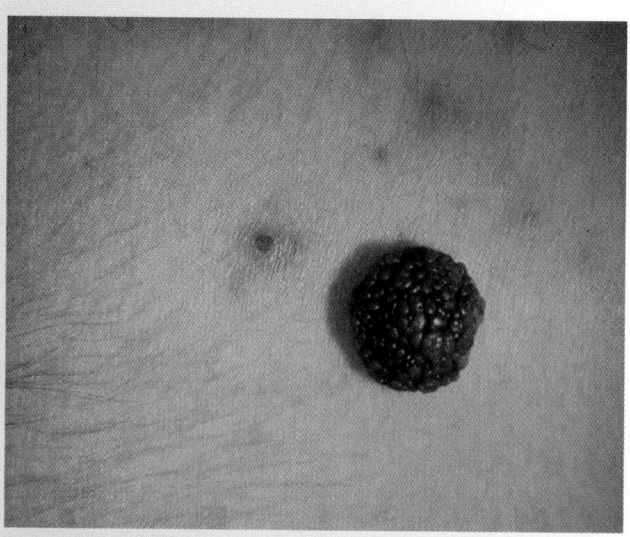

Dermal mole.

TABLE 51-10 BENIGN SKIN LESIONS — CONT'D

Type	Description	Treatment
Warts	Small, common, benign growths of the skin resulting from the hypertrophy of the papillae and epidermis. Caused by a virus. Common warts, often seen on hands and fingers, appear as raised, flesh-colored papules that have a rough surface. These warts may crack, fissure, bleed, and be painful to lateral pinching and direct, firm pressure. Plantar warts occur on the sole of the foot. They may appear granular, pitted, or protuberant, with a callous of surrounding normal skin. Incubation period can be several weeks to months. Virus is spread by direct contact into areas of broken skin or to other nails by nail cuticle biting.	If no pain or discomfort, no treatment may be indicated. Patient should be cautioned not to spread lesions by picking or biting them. Treatment is indicated for symptomatic warts and for cosmetic purposes. General treatments include kerolytic agents (e.g., salicylic acid plasters) to soften and reduce keratin; cryotherapy (liquid nitrogen); and light electrodesiccation and curettage (requires local anesthesia). Treatment of choice is usually cryotherapy, because local anesthesia is not necessary and it leaves little scarring.

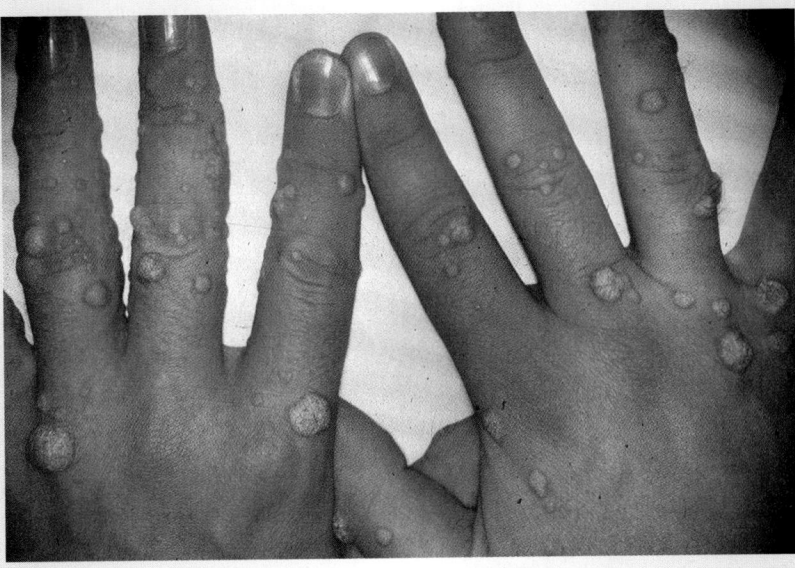

Warts.

Type	Description	Treatment
Hemangiomas (angiomas)	Benign vascular tumors of dilated blood vessels that can have varied clinical manifestations. Nevus flammeus involves mature capillaries on the face and neck. It is a congenital neoplasm that appears as a pink-red to bluish-purple macular patch. Port-wine stains or port-wine angiomas appear as violet-red macular patches, usually singular lesions, growing proportionately as the child grows. These lesions can persist indefinitely. Cherry hemangiomas are commonly seen in the elderly patient. They appear as small round papules that can vary in color from red to purple.	Nevus flammeus is usually treated for cosmetic reasons. Port-wine stains, if large enough, may require surgical excision with skin grafting. Laser therapy may also be used. Noninvasive treatment is the use of cosmetics to camouflage the affected area. Treatment for cherry hemangiomas is usually not prescribed, except for cosmetic purposes.

Basal cell carcinoma (Fig. 51–16) arises from the basal cell layer of the epidermis. It is the most common type of skin cancer. This tumor is mainly seen on sun-exposed areas of the body. The lesion appears as a small pearly or translucent papule with a rolled, waxy edge; depressed center; telangiectasia (lesion formed by dilation of vessels); crusting; and ulceration. Metastasis is rare, though it may be locally invasive.

Squamous cell carcinoma arises from the epidermis. This tumor can occur on sun-exposed areas of the skin and mucous membranes and is mainly seen on the lower lip, neck, tongue, head, and dorsa of the hands. It can occur on normal skin or a preexisting lesion (actinic keratosis). The lesion appears as a single, crusted, scaled, eroded papule, nodule, or plaque (Fig. 51–17). A neglected lesion appears more rough, scaly, and darker colored. The lesion is fragile and prone to oozing and bleeding. This is a truly invasive carcinoma. Metastasis is related to histological type, depth of invasion, and size of the lesion.

Malignant melanoma (Fig. 51-18), as the name implies, is a malignant growth of pigment cells (melanocytes). It is highly metastatic, with a higher mortality rate than basal or

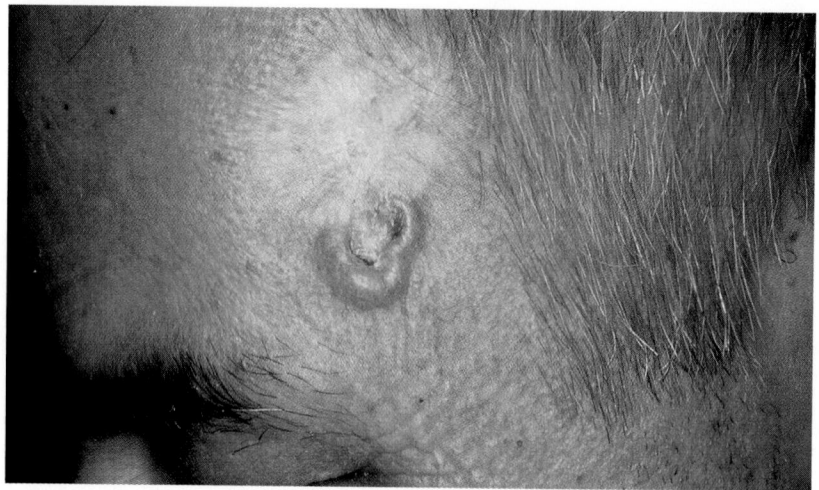

Figure 51-16 Basal cell carcinoma. Note pearly, flesh-colored papule with depressed center and rolled edge. (From Goldsmith, LA, Lazarus, and GS, Tharp, MD: Adult and Pediatric Dermatology. FA Davis, Philadelphia, 1997, p 158, with permission.)

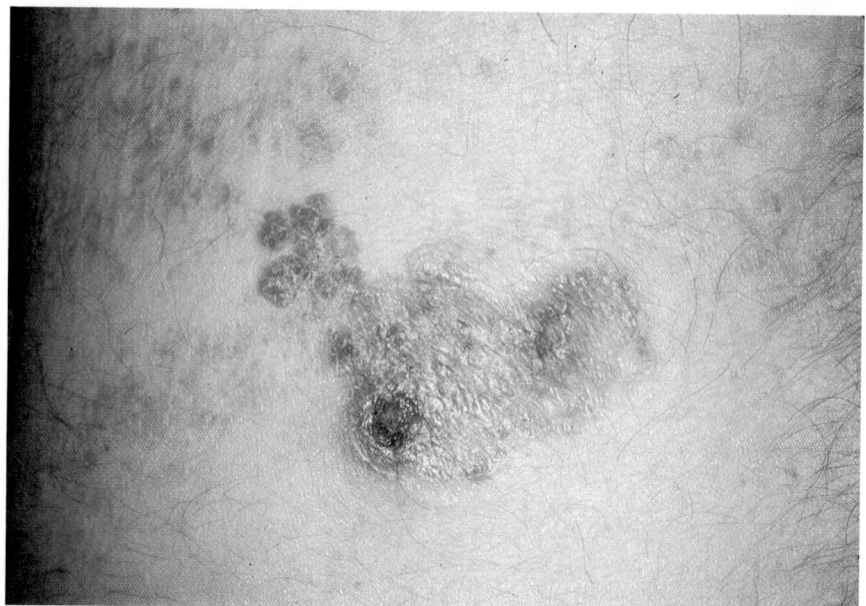

Figure 51-17 Squamous cell carcinoma. Surface is fragile and bleeds easily. (From Goldsmith, LA, Lazarus, GS, and Tharp, MD: Adult and Pediatric Dermatology, FA Davis, Philadelphia. 1997, p 237, with permission.)

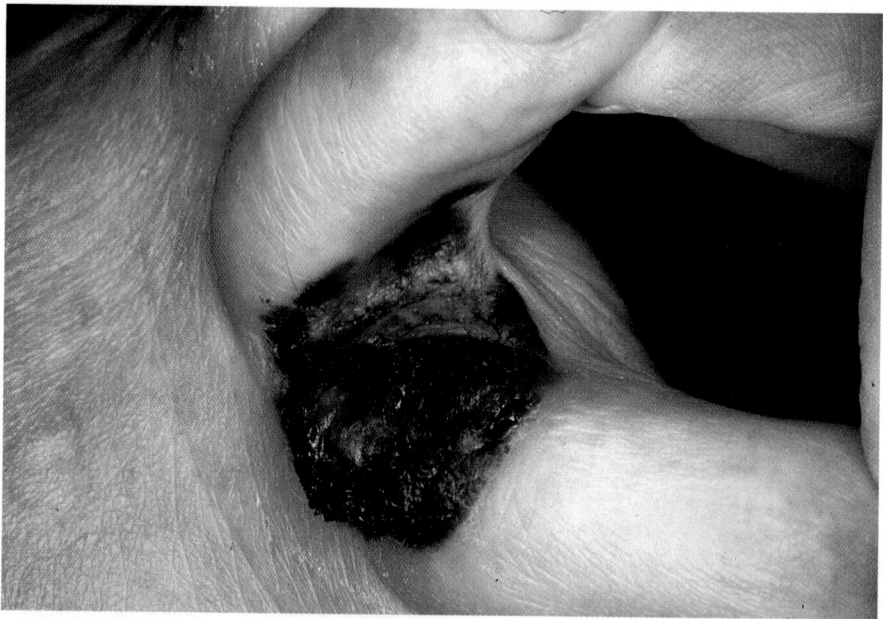

Figure 51-18 Malignant melanoma. (From Goldsmith, LA, Lazarus, GS, Tharp, MD: Adult and Pediatric Dermatology. FA Davis, Philadelphia, 1997, p 137, with permission.)

squamous cell carcinoma. This tumor can occur anywhere on the body, and about half arise from pre-existing nevi or moles. There are three general types: lentigo maligna, superficial spreading, and nodular.

Lentigo maligna melanoma (Fig. 51–19) appears as a slow-growing dark macule on exposed skin surfaces (especially the face) of elderly patients. The lesion has irregular borders and brown, tan, and black coloring. Prognosis is good if treated in the early stage.

Superficial spreading melanoma is the most common melanoma. It can occur anywhere on the body and is usually seen in middle-aged persons. The lesion appears as a slightly elevated plaque with an irregular border. The coloring of the lesion varies in combinations of black, brown, and pink. The fragile surface may bleed or ooze. Eventually the plaque develops into a nodule. The cure rate is excellent when it is in the plaque phase; prognosis is poor with the nodular phase.

Nodular melanoma occurs suddenly as a spherical papule or nodule on the skin or in a mole. Coloration in blue-black, blue-gray, or reddish-blue color that may have a rim of inflammation. The lesion is fragile and bleeds easily. Metastasis occurs rapidly. This type of melanoma has the least favorable prognosis. Early diagnosis and treatment is imperative.

Prevention

Most types of skin cancer can be prevented by limiting or avoiding direct exposure to ultraviolet rays (sun, tanning booths). If exposure to the sun is necessary, exposure should be avoided during its highest intensity (10 A.M. to 2 P.M.). The patient should use a protective sunscreen with sun protection factor (SPF) of 15 or more. The patient should also wear sun-protective clothing, such as hats and long sleeves. The patient should seek medical advice if there is a change in color, size, shape, sensation, or character of a lesion or mole.

Diagnostic Tests

A preliminary diagnosis can be based on the appearance of the lesion. A definitive diagnosis is made by biopsy. Other tests are performed based on the results of the pathological examination.

Medical Treatment

Medical treatment depends on the type, thickness, and location of the lesion; the stage of the disease; and the age and general health of the patient. Generally lesions are surgically excised with a 1- to 2-cm margin. Regional node dissection varies; it may be advised if the nodes in the area drain to one group. Grafting may be necessary for closure or repair. Chemotherapy may be used for metastasis. Radiation therapy may be used as adjunct treatment, or may be recommended for patients with a deeply invasive tumor or those who are poor surgical risks. Other therapies that also may be used include cryosurgery and curettage and electrodesiccation.

Nursing Care

Perform a complete skin assessment. Lesions are palpated to determine texture, size, and firmness. All lesions should be described as to size, location, color, surface characteristics, pain, discomfort, itching, and bleeding. Note when the patient first discovered the lesion.

Nursing care of the patient with cancer is documented in Chapter 10. Specific nursing care related to cryosurgery includes preparing the patient for the procedure. Minor discomfort can be expected with little or no local anesthesia. Expect swelling, local tenderness, and hemorrhagic blister formation 1 to 2 days after the procedure. After the procedure, the area is cleansed as ordered and prescribed ointments are applied.

Specific nursing care for curettage and electrodesiccation include preparing the patient for the procedure. After local

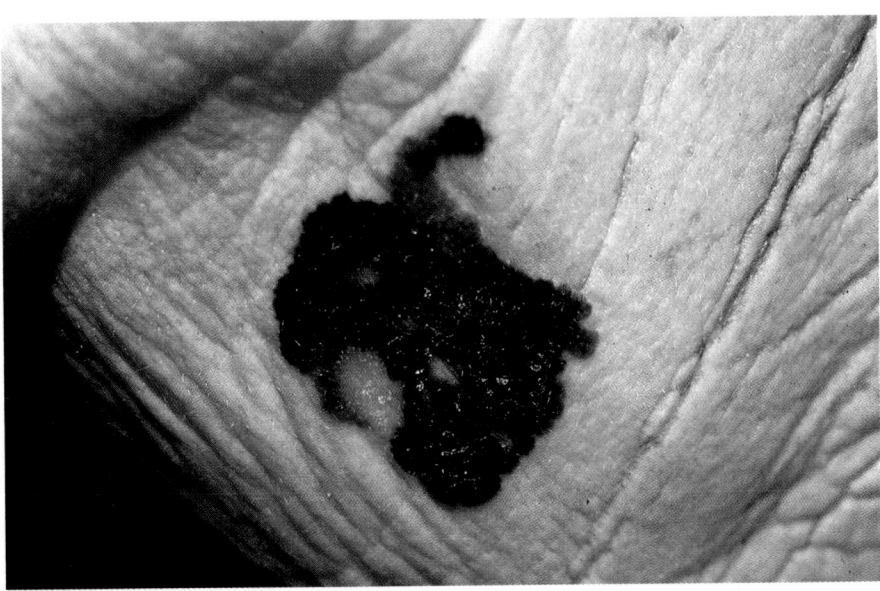

Figure 51-19 Lentigo maligna. (From Goldsmith, LA, Lazarus, and GS, Tharp, MD: Adult and Pediatric Dermatology. FA Davis, Philadelphia, 1997, p 55, with permission.)

anesthesia, a dermal curette is used to scrape away the lesion, followed by electrodesiccation of the remaining wound; the wound heals by secondary intention, usually with minimal scarring. After the procedure the wound is cleansed and dressed as prescribed.

▶ DERMATOLOGIC SURGERY

Plastic or reconstructive surgery is performed to correct certain defects, scars, and malformations, as well as to restore function or prevent further loss of function. This type of surgery is usually an elective procedure; it may be prescribed by the physician or it may be the wish of the patient in hopes of improving his or her body image. Common types of plastic surgical procedures are listed in Table 51–11.

Answers to CRITICAL THINKING

Mrs. Long

You should ask Mrs. Long if she read the package instructions. She would find that for medicated scalp shampoos to work properly, they must remain on the scalp for several minutes. Package instruction should be carefully checked for each product, because they vary from product to product.

HOME HEALTH HINTS

- A wound-measuring device that will not be misplaced is your hand. Measure your hand, such as the nailbed of the particular finger or a joint of a finger. Use these as a guide to determine wound measurements.

- Sanitary pads make great cushions for bony prominences. You can also place them in a cotton sock for better molding purposes.

- A handheld shower head is useful in debriding some leg ulcers. Do not use it if it is too painful.

- To relieve itchy skin (pruritus), oatmeal baths are sometimes prescribed. An inexpensive way to do this is to place a half cup of quick-cooking oatmeal in a cotton sock. Put it under the faucet as you fill the tub and ring out the sock.

- Instruct patients to prevent red, dried, cracked skin on hands by doing the following: wear gloves outside in the cold or windy weather to prevent chapping; avoid overheating the house; use a humidifier to keep the air moist; apply hand lotion two or three times a day and after each hand washing; use soaps with added oil and avoid those with deodorants; use sunscreen with an SPF factor of at least 15; and stop smoking (smoking reduces blood flow to the skin).

TABLE 51-11 COMMON PLASTIC SURGICAL PROCEDURES

Operation	Description	Purpose	Possible Complications	Postoperative Nursing Treatment Considerations
Rhinoplasty (nose)	Removal of excessive nasal cartilage, tissue, or bone; reshaping of nose.	Correct congenital or acquired septal defects; improve cosmetic shape of nose.	Hemorrhage, hematoma; temporary ecchymosis and edema; infection, septal perforation.	Monitor dressing and packings for bright-red bleeding; monitor vital signs and level of consciousness; maintain semi-Fowler's position to minimize edema.
Blepharoplasty (eyelid)	Incisions on upper and lower lids with excision of fat and skin and primary closure.	Removal of bags under eyes and wrinkles and bulges.	Corneal injury; hematoma; ectropion; rarely visual loss and wound infection.	Eye dressings; antibiotic ointment around eyes and lids; discoloration and swelling usually subsides in about 10 days; maintain semi-Fowler's position to minimize edema.
Rhytidoplasty (facelift)	Incision anterior to ear with removal of excessive skin and tissue; the subcutaneous tissue and fascia are folded and stretched.	Removal of excessive wrinkling or sagging skin.	Hemorrhage; hematoma, ecchymosis, and edema (temporary); wound infection, facial nerve damage.	Surgical improvement lasts from 5–10 years; antibiotic ointment to suture line; maintain semi-Fowler's position to minimize edema.
Otoplasty (ear)	Incision of ear for correction of defect.	Correct congenital defects; correct deformities; improve cosmetic shape of ear.	Hemorrhage; hematoma; edema; wound infection.	Ear dressing for about 1 week; protect ear at times of sleep for about 3 weeks.
Breast augmentation	Skin incision to insert breast implant.	Improve cosmetic shape and size of breasts.	Hemorrhage; hematoma, wound infection.	Dressing to site; antibiotic ointment to suture line. Drains may be in place— note color and amount of drainage.
Breast reduction	Skin incision to excise excess breast tissue and skin.	Improve cosmetic shape and size of breasts; comfort measure.	Hemorrhage; hematoma, wound infection; wound dehiscence; necrosis of areola and nipple area.	Dressing to site; antibiotic ointment to suture line. Drains may be in place— note color and amount of drainage.

REVIEW QUESTIONS

1. Which of the following mechanical forces can lead to formation of a skin ulcer?
    a. Pressure
    b. Friction
    c. Shear
    d. All of the above

2. Which of the following actions by the nurse is appropriate when caring for the patient with dermatitis?
    a. Bathe in hot oatmeal baths.
    b. Dry vigorously to prevent moisture buildup.
    c. Apply gloves to hands at night.
    d. Apply a thick layer of the prescribed topical agent.

3. Psoriasis is an inflammatory skin disorder that is characterized by which underlying condition?
    a. Epidermal proliferation
    b. Excessive subcutaneous fat
    c. Herpes infection
    d. Excessive melanin production

4. Which of the following is appropriate instruction for the parent of a child with impetigo contagiosa, to help prevent spread of infection? Send back to school:
    a. One week after treatment is started
    b. After the lesions scab
    c. When all lesions are healed
    d. After spread of lesions has stopped

5. Which of the following is appropriate patient education about treating scabies?
    a. Dry-clean all linens, towels, clothes.
    b. Throw away infested mattresses.
    c. Wash linens, towels, clothes.
    d. Remove infested pets.

6. Jan Smith, 42 years old, is admitted to the emergency department with flame burns to her entire chest, back, and upper extremities. Using the rule of nines, you estimate the percentage of burns to be which of the following:
    a. 36 percent
    b. 45 percent
    c. 54 percent
    d. 64 percent

7. Which of the following actions is appropriate initial treatment of a chemical burn?
    a. Neutralize the chemical.
    b. Lavage with water.
    c. Apply the prescribed topical agent.
    d. Wrap the patient in sterile sheets.

8. Which of the following is the most common cause of malignant skin lesions?
    a. Overexposure to ultraviolet rays
    b. Genetic predisposition
    c. Fair skin, blue eyes, red hair
    d. Numerous moles on body

UNIT FIFTEEN BIBLIOGRAPHY

Altman, G, Buchsel, P, and Coxon, V: Fundamental and Advanced Nursing Skills. Delmar/Thomson Learning, Albany, NY, 2000.

Bergstrom, N, et al: Treatment of pressure ulcers. Clinical Practice Guideline, No. 15. AHCPR Publication No. 95-0652. US Department of Health and Human Services, Public Health Service, Agency for Health Care Policy and Research, Rockville, MD, December 1994.

De Witt, S: Nursing assessment of the skin and dermatologic lesions. Nursing Clinics of North America 25(1):235, 1990.

Doctor, JN, et al: Health outcome for burn survivors. Journal of Burn Care & Rehabilitation 18:490, 1997.

Gaskin, FC: Detection of cyanosis in the person with dark skin. Journal of National Black Nurses' Association 1(1):52, 1986.

Goldsmith, LA, Lazarus, GS, and Tharp, MD: Adult and Pediatric Dermatology: A Color Guide to Diagnosis and Treatment. FA Davis, Philadelphia, 1997.

Hogstel, M: Nursing Care of the Older Adult. Delmar/Thomson Learning, Albany, NY, 2001.

Lawrence, JC: Dressings and wound infection. American Journal of Surgery 167:215, 1994.

Lazarus, GS, et al: Definitions and guidelines for assessment of wounds and evaluation of healing. Archives of Dermatology 130:489, 1994.

McConnel, E: Clinical do's and don'ts: Assessing the skin. Nursing 1992 22(4):86, 1992.

Murray, M, and Blaylock, B: Maintaining effective pressure ulcer prevention programs. Medsurg Nursing 3(2):85, 1994.

Reeves, JR, and Maibach, H: Clinical Dermatology Illustrated, ed 3. FA Davis, Philadelphia, 1998.

Ruppert, SD, et al: Dolan's Critical Care Nursing, ed 2. FA Davis, Philadelphia, 1998.

Scanlon, VC, and Sanders, T: Essentials of Anatomy and Physiology, ed. 3. FA Davis, Philadelphia, 1999.

Torte, SJ, and Hanifin, JM: Current management and therapy of atopic dermatitis. Journal of the American Academy of Dermatology 44(4): 13–15, 2001.

Trofino, RBT: Nursing Care of the Burn-injured Patient. FA Davis, Philadelphia, 1991.

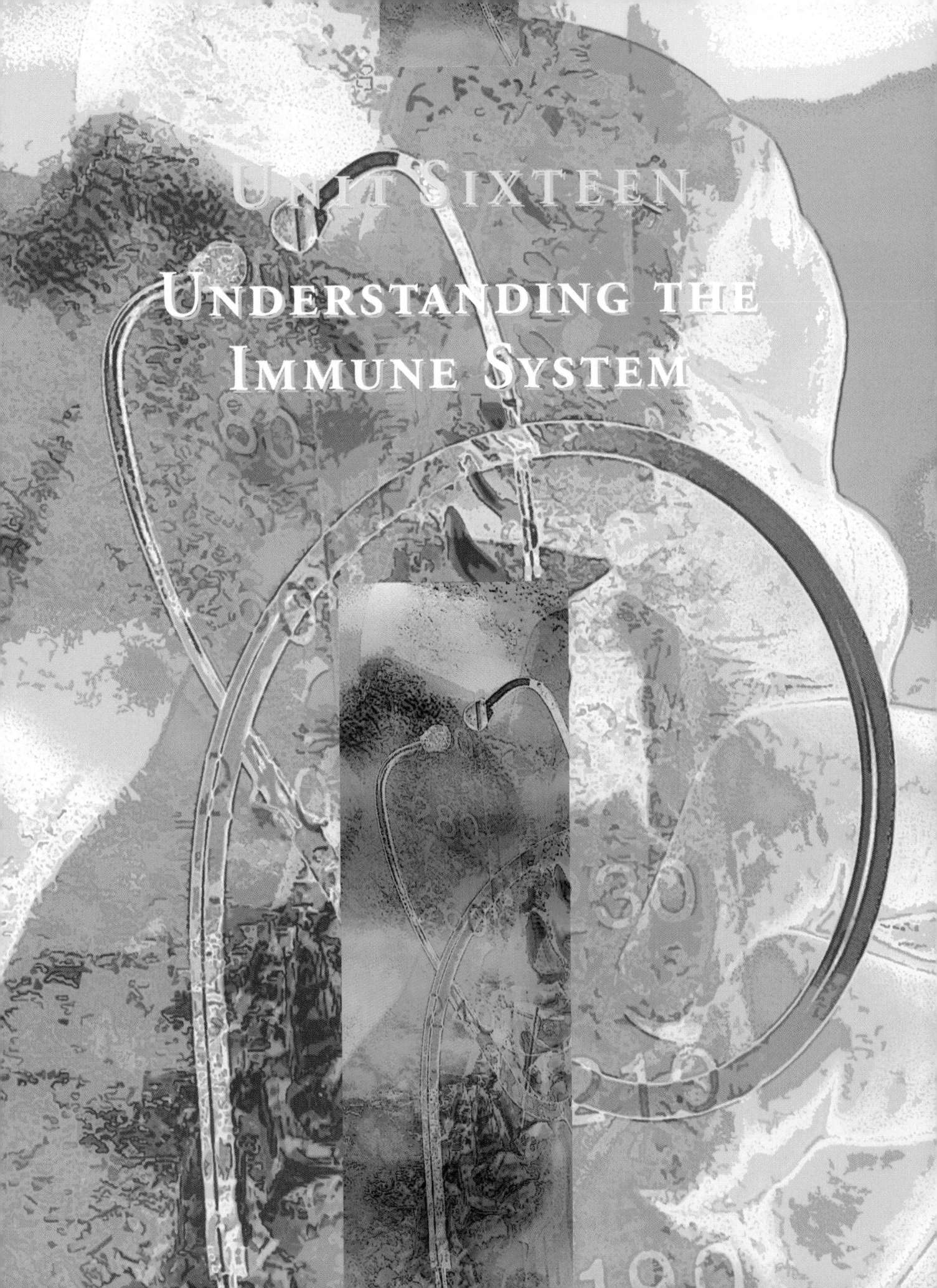

UNIT SIXTEEN

UNDERSTANDING THE
IMMUNE SYSTEM

52

Immune System Function, Assessment, and Therapeutic Measures

Sharon M. Nowak

QUESTIONS TO GUIDE YOUR READING

1. What is the normal function of the immune system?

2. What is the function of antigens, lymphocytes, T cells, and B cells?

3. How do antibodies function in an immune response?

4. What occurs in a cell-mediated immune response?

5. What occurs in a humoral immune response?

6. What occurs in an antibody response?

7. How does aging affect the immune system?

8. What data is collected when caring for a patient with a disorder of the immune system?

9. What nursing care is provided for patients undergoing diagnostic tests for the immune system?

10. What are common therapeutic measures used for a patient with disorders of the immune system?

▶ NORMAL IMMUNE ANATOMY AND PHYSIOLOGY

Immunity is defined as the ability to destroy pathogens or other foreign material and to prevent further cases of certain infectious diseases. Immunity is most often thought of in terms of the body's response to microorganisms such as bacteria, viruses, and fungi, all of which are foreign to the body. However, immunity also involves processes directed toward other cells or substances that are identified by the body, correctly or incorrectly, as foreign. Malignant cells are also foreign in that they have mutated from the normal state and are usually destroyed by the immune system before they become cancer. Transplanted organs unfortunately are also perceived as foreign; rejection of a transplanted organ is an immune response. Occasionally the immune system mistakenly reacts to part of the body itself (autoimmune disease).

The immune system consists of lymphoid organs, **lymphocyte**s and other white blood cells, and the many chemicals produced that are involved in activation of our own cells for the destruction of foreign **antigen**s (Fig. 52–1). The lymphatic system consists of lymphatic vessels that help return tissue fluid to the circulatory system; lymph nodes and nodules, which are masses of lymphatic tissue that differ in size and location; the spleen, which phagocytizes pathogens and produces **white blood cells** and antibodies; and the thymus, which functions primarily in childhood and shrinks before adulthood. Lymph nodes are grouped along lymph vessels to destroy foreign material. Three major groups are the cervical, axillary, and inguinal nodes. Lymph nodules

---

lymphocyte: lympho—lymph + kytos—cell
antigen: anti—against + gennan—to produce

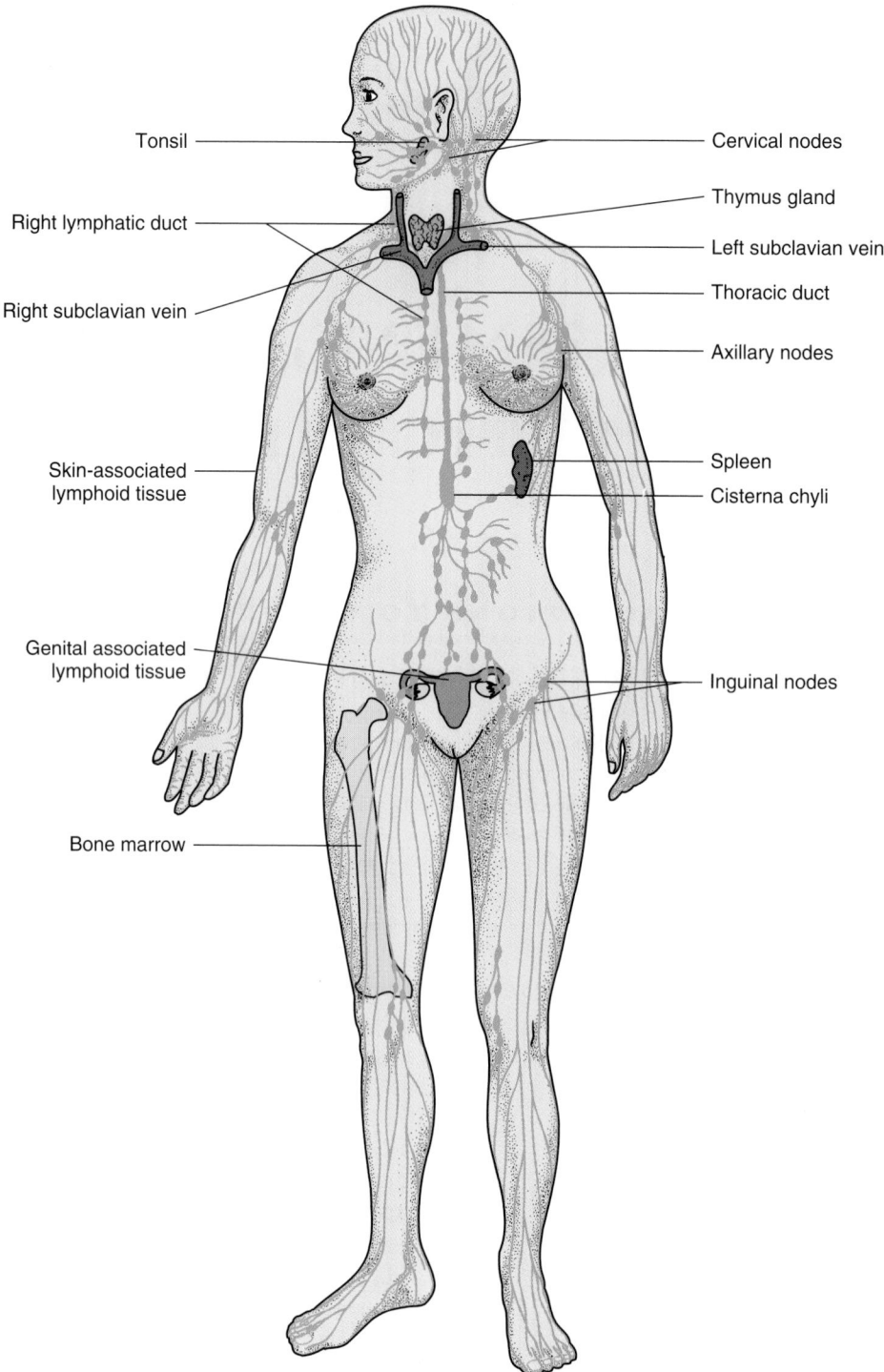

Figure 52-1   Immune system organs, lymph vessels, and major lymph nodes.

are smaller than nodes and are found under the surface of mucous membranes (e.g., tonsils).

Antigens

Antigens are chemical markers that identify cells or molecules. Examples of molecular antigens include bacterial toxins, plant pollens or proteins that trigger allergies, and the protein products of viral activity within cells. Human cells (except red blood cells [RBCs]) have their own self-antigens,

thousands of markers that identify the cell as belonging in the body. These are the major histocompatibility complex (MHC) antigens, also called human leukocyte antigens (HLAs), which are genetically determined. The MHC antigens of identical twins are identical. These MHC antigens serve as a comparison for cells of the immune system; the antigens of foreign cells do not "match" MHC antigens and may therefore be recognized as foreign and destroyed in one of several ways.

Lymphocytes

There are three types of lymphocytes: natural killer (NK) cells, T cells, and B cells, each with very different functions.

Natural Killer Cells

Natural killer cells are found in the blood, bone marrow, lymph nodes, and spleen and are able to destroy many kinds of pathogens and tumor cells. How NK cells recognize foreign antigens is not known with certainty, but they may respond to the absence of MHC antigens on foreign cells. It is believed that NK cells destroy foreign cells by rupturing their membranes or by some other form of direct contact. The action of NK cells is considered a nonspecific resistance mechanism because it is effective against a variety of foreign antigens.

T Cells and B Cells

The lymphocytes called T cells and B cells are involved in specific immune responses; that is, each cell is genetically programmed to respond to one kind of foreign antigen. It is estimated that the human immune system can respond to hundreds of millions of different foreign antigens.

In the fetus, both T cells and B cells develop in the bone marrow. T cells then migrate to the thymus, where the thymic hormones bring about their maturation. From the thymus, T cells migrate to the lymph nodes and nodules and to the spleen. B cells mature in the bone marrow and migrate directly to lymphatic tissue. When activated during an immune response, some B cells become plasma cells that produce antibodies to a specific foreign antigen. Meanwhile, other B cells become memory cells and remember the specific antigen for future encounters. Upon subsequent exposures to the antigen, the specific memory cells turn on the immune response immediately, creating a quick response and usually a stronger response.

Antibodies

Antibodies are also called immunoglobulins (Ig) or gamma globulins and are proteins produced by plasma cells in response to foreign antigens. Antibodies do not themselves destroy foreign antigens, but rather become attached to such antigens to "label" them for destruction. Each **antibody** is specific for only one antigen, and the B cells (those that become plasma cells) of an individual are capable of producing millions of different antibodies. There are five classes of human antibodies, designated by letter names: IgG, IgM, IgA, IgE, and IgD. Their functions are summarized in Table 52–1.

Mechanisms of Immunity

The two mechanisms of immunity are **cell-mediated immunity,** which involves T cells, and humoral immunity, which involves both T cells and B cells. Although the mechanisms are different, invasion by a pathogen often triggers both.

The first step in the destruction of a foreign antigen is the recognition of it as foreign. B cells in lymphatic tissue are able to recognize the foreign antigens for which they are genetically programmed and become activated B cells. Their activation is greatly enhanced if the foreign antigen is presented to them by antigen-processing cells called dendritic cells. A specific type of T cell called helper T cells (CD4) also recognize the foreign antigen and provide a further stimulation to B cells, causing them to divide (proliferate) and become more specialized (differentiate). Helper T cells are assisted in their recognition of foreign antigens by macrophages or other antigen-processing cells, which present the foreign antigen, as well as their own MHC antigens to the T cell, providing a comparison. The macrophages also provide chemical stimulation to the T cells, which then begin to divide and become more specialized.

Cell-Mediated Immunity

This mechanism of immunity does not involve the production of antibodies, but it is effective against intracellular pathogens (such as viruses), fungi, malignant cells, and grafts of foreign tissue. The first step is the recognition of the foreign antigen by helper T cells, assisted by macrophages. The activated T cells divide many times and become specialized in one of several ways. Cytotoxic, or killer, T cells (CD8) are able to lyse cells such as cancer cells or those infected by viruses or other intracellular parasites. They also release chemicals that activate phagocytes such as macrophages and **neutrophil**s.

---

neutrophil: neutro—neuter + philein—to love

TABLE 52-1	**CLASSES OF ANTIBODIES**	
Name	**Location**	**Function**
IgG	Blood, extracellular fluid	Crosses the placenta to provide passive immunity for newborns
		Provides long-term immunity following a vaccine or illness recovery
IgA	External secretions: tears, saliva, etc.	Provides passive immunity for breast-fed infants
		Found in secretions of all mucous membranes
IgM	Blood	Produced first by the maturing immune system of the infant
IgD	B cells	Produced first during an infection (IgG production follows)
IgE	Mast cells or basophils	Are antigen-specific receptors on B lymphocytes
		Important in allergic reactions; mast cells release histamine

Source: From Scanlon, V, and Sanders, T: Understanding Human Structure and Function. FA Davis, Philadelphia, 1999, p 311.)

Memory T cells remember the specific foreign antigen and quickly activate an immune response should the antigen reappear. Suppressor T cells are believed to inhibit the proliferation of both T cells and B cells, which limits the immune response to just what is needed and no more.

Humoral Immunity

Humoral immunity is also called antibody-mediated immunity and does indeed involve antibody production. Again the first step is the recognition of the antigen as foreign, this time by B cells. The helper T cells that recognize the antigen further stimulate the B cells to proliferate and differentiate. Some B cells become plasma cells that produce antibodies specific for this particular antigen. Other B cells become memory B cells that will remember this antigen and initiate a rapid response should it return.

Although B cells are stationary, the antibodies produced by plasma cells circulate throughout the body. The antibodies bond to the antigen, forming an antigen-antibody complex. This is termed opsonization, which means that the antigen is now "labeled" for phagocytosis by macrophages or neutrophils. The antigen-antibody complex also stimulates the process of complement fixation.

Complement is a group of about 20 plasma proteins that circulate in the blood until activated by an antigen-antibody complex. The activation of complement may result in the formation of an enzyme complex that lyses the cell and brings about its death. Other complement proteins bind to foreign antigens and serve as further labels to attract macrophages.

▶ ANTIBODY RESPONSES

The first exposure to a foreign antigen stimulates antibody production, but the antibodies are produced so slowly and in such small amounts that this production may be too late to prevent the disease. However, with time, the person has antibodies and memory cells that are specific for that pathogen. On a second exposure to the antigen, the memory cells initiate rapid production of large amounts of antibody, often enough to prevent a second case of the illness

(Fig. 52–2). This is the basis for the protection given by vaccines. A vaccine contains an antigen that is not pathogenic (e.g., bacterial capsules, in the case of the pneumococcal vaccine). The vaccine stimulates the formation of antibodies and memory cells.

Antibodies may also neutralize viruses; that is, they attach to a virus and render it unable to enter a cell. Outside living cells, viruses cannot reproduce, and those coated with antibodies are phagocytized by macrophages. Another aspect of our defenses against viruses is interferon, a chemical produced by cells infected with viruses. Although it does not help the infected cell, interferon protects surrounding cells by enabling them to resist viral reproduction by limiting or slowing growth of the virus.

Antibodies are also involved in allergic responses, in which the immune system responds to foreign but harmless antigens (an allergen), such as plant pollen. IgE antibodies bond to mast cells, which break down and release histamine and other chemicals that contribute to inflammation. **Anaphylactic** shock is this very same allergic reaction, only massive in nature. It is characterized by loss of plasma from capillaries (an effect of histamine) and a sudden drop in intravascular blood volume and blood pressure.

▶ TYPES OF IMMUNITY

Two categories of immunity are **passive immunity** and **active immunity.** In passive immunity, antibodies are not produced by the individual but are obtained from another source. One form of naturally acquired passive immunity includes placental transmission of antibodies from mother to fetus and transmission of antibodies in breast milk. Artificially acquired passive immunity involves injection of preformed antibodies; this may help prevent disease after exposure to a pathogen such as the hepatitis A virus. Passive immunity is always temporary, in that antibodies from another source eventually break down.

---

anaphylactic: ana—up + phylaxis—protection

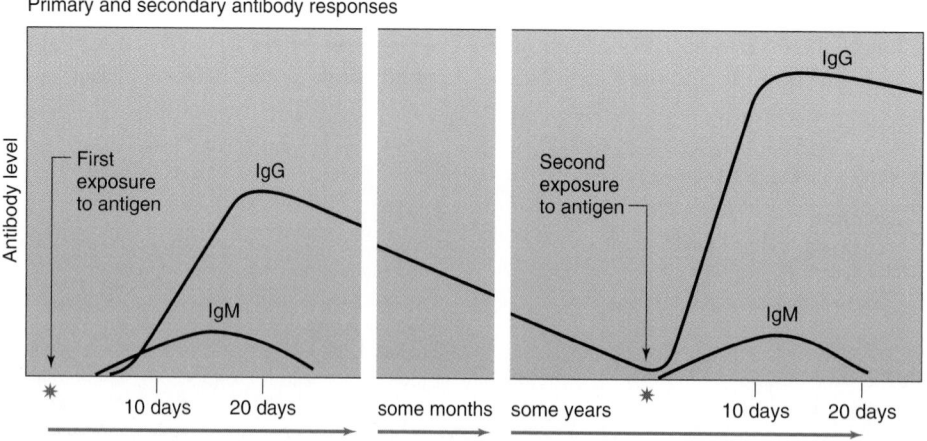

Primary and secondary antibody responses

Figure 52-2 Antibody responses to a first and then subsequent exposure to a pathogen. (From Scanlon, V, and Sanders, T: Understanding Human Structure and Function. FA Davis, Philadelphia, 1999, p 317.)

Active immunity means that the individual produces his or her own antibodies. An example of naturally acquired active immunity occurs when a person recovers from a disease and then has antibodies and memory cells specific for that pathogen. Artificially acquired active immunity occurs as the result of a vaccine that stimulates production of antibodies and memory cells. The duration of active immunity depends on the particular disease or vaccine; some confer lifelong immunity, but others do not.

► AGING AND THE IMMUNE SYSTEM

The efficiency of the immune system decreases with age. As such, elderly people are more susceptible to infections, especially secondary infections such as the development of pneumonia following the flu (Gerontological Issues Box 52–1). Autoimmune disorders are also more common among older people; the immune system mistakenly perceives the body's own tissues as foreign and initiates their destruction. The incidence of cancer is also higher; malignant cells that might once have been quickly destroyed by the immune system remain alive and proliferate.

BOX 52-1	GERONTOLOGICAL ISSUES

Significant changes occur in the immune system of the older adult. These changes are known as immune senescence, which refers to a decline in immune system function. Some specific changes include the following:

- Thymus gland decreases in size, increases production of immature T cells, and has a subsequent decline in response to antigens.
- Antibody response to pneumococcus, influenza, and tetanus decreases, creating need for vaccine.

Immunizations to support the immune responses of older adults include the following:
- Diphtheria tetanus booster every 10 years
- Pneumovax once in a lifetime (reimmunization or boosters not necessary)
- Influenza vaccine yearly (before influenza season)
- Hepatitis B vaccine if medium to high risk for exposure to hepatitis B.

► IMMUNE SYSTEM ASSESSMENT
Nursing Assessment

A thorough assessment of the immunologically compromised patient is the first step in the nursing process. It gives substance to the nursing care plan, which guides nursing practice. Because disorders of the immune system can affect every system in the body, it is vitally important that a thorough head-to-toe assessment and history is performed.

Subjective Data

DEMOGRAPHIC DATA. The patient's age, gender, race, and ethnic background data are important because some disease processes tend to be associated with a particular gender, race, or age group. For instance, systemic lupus erythematosus

(SLE), an autoimmune disorder, tends to affect women eight times more than men. Inquiring about the patient's place of birth may give insight to ethnic ties that are not initially evident in the interview. Also, where the patient lives and has lived may shed light on the patient's current problem.

HISTORY. Inquiry into medication, food, and environmental allergies such as dust, pollen, and insects should include the patient's allergies, as well as those of family members. There may be a familial tendency to react to substances, so a previous exposure and sensitization to a substance may not be necessary before a severe reaction occurs. Many times if there is a family history of a severe reaction (anaphylaxis) to penicillin, a physician does not order penicillin for the patient. Some reactions can be fatal within minutes without immediate medical treatment.

As with allergies, inquiring about past and present medical conditions or disease processes should also include a family history, as well as the patient's history. Many atopic (allergic) disorders such as allergic rhinitis and asthma or autoimmune disorders such as SLE and ankylosing spondylitis are thought to be either familial or have a genetic predisposition in certain races or cultures (Cultural Consideration Box 52–2).

BOX 52-2	CULTURAL CONSIDERATION

The Navajo people have a high incidence of severe combined immunodeficiency syndrome (SCIDS), and immunodeficiency syndrome unrelated to acquired immunodeficiency syndrome (AIDS). SCIDS results in a failure of the antibody response and cell-mediated immunity. Infants who survive initially are sent to tertiary care facilities. They must receive gamma globulin on a regular basis until a bone marrow transplant can be performed. Thus far, studies indicate that SCIDS is unique to this Navajo population.

The patient's previous surgeries may play an important role in the patient's current condition or give clues about prior health. If the patient has had his or her thymus gland removed (thymectomy), T-cell production may be altered, which affects the cell-mediated immune response. If the patient has had his or her spleen removed (splenectomy), lymphocyte and plasma cell production may be altered, which affects humoral immune response.

Further questioning about previous blood transfusions, radiation exposure (therapeutic or accidental), and current medications (prescription and over-the-counter drugs) provides data that may be associated with the current problem. For example, corticosteroids and immunosuppressants alter the immune response. Other medications, such as some anti-infectives or antineoplastics, depress the bone marrow, resulting in decreased production of the cells made in the bone marrow. Bone marrow depression of white blood cells can alter cell-mediated and humoral immune responses.

A patient's lifestyle may influence immune system function and should be assessed. Exposures to caustic chemicals or fumes at work may lead to systemic or topical immune reactions or bone marrow depression. Anaphylactic reactions can be caused by exposure to latex, which is found in gloves and other medical products that health care workers and their patients touch. Be aware of this potentially life-threatening reaction and know the agency's latex allergy protocol. These patients should wear a medical alert bracelet and carry an anaphylactic epinephrine kit. It is also important to know if the patient uses illicit drugs or has multiple sexual partners or partners of the same gender. These behaviors place the patient at risk for contracting the human immunodeficiency virus (HIV). Knowing the patient's diet habits and supplemental vitamins gives insight into the potential reserve of the immune system for fighting infection.

The patient's life stressors, coping behaviors, and support systems should be explored. Stress (environmental, physical, and psychological) can depress the immune system's function. Coping behaviors are essential in keeping stress within manageable limits to maintain an optimum-functioning immune system. Support systems play an important role in coping with stress and should be encouraged and nurtured by nurses.

Current Problem

Use the WHAT'S UP? format to collect data about the current immune system problem. For immune disorders, the patient is asked the following questions:

- **W**here is it? What part of the body is affected?
- **H**ow does it feel? Painful? Itching?
- **A**ggravating and alleviating factors?
- **T**iming. Was there exposure to a pathogen? Did you have a previous infection? Does it occur only in certain settings? Did you have chemotherapy or radiation therapy? How long have symptoms persisted?
- **S**everity. Does it affect activities of daily living (ADLs)? Work? Roles?
- **U**seful data for associated symptoms. Immunosuppression? Family history? Allergies?
- **P**erception of the patient of the problem. What do you think is wrong?

When investigating the patient's current problem, the following are examples of common signs and symptoms that may be found with immune disorders: fever, fatigue, joint pain, swollen glands, weight loss, and skin rash.

CRITICAL THINKING

Laura

Laura is scheduled for a lymph node biopsy and is seen in preadmission testing before surgery. As the licensed practical nurse/licensed vocational nurse (LPN/LVN) prepares to draw blood specimens, he learns that Laura is allergic to latex.

1. Why is this patient information important?
2. What should the LPN/LVN do next?
3. What precautions should the LPN/LVN use to draw the blood specimen?

Answers at end of chapter.

Objective Data

PHYSICAL ASSESSMENT. The physical assessment begins by observing the patient's general appearance, color, posture, gait, and facial expressions. While talking with the patient, alertness and orientation to person, place, and time are assessed. Patients with various immune disorders, such as SLE or acquired immunodeficiency syndrome (AIDS), may exhibit mentation changes, especially if they develop central nervous system complications in advanced stages.

The patient's skin, bone joints, and nailbeds should be carefully examined. Any cyanosis or erythema (redness) is noted. Rashes should be examined for size, shape, location, texture, drainage, and pruritus (itching). Painless, small purple lesions are Kaposi's sarcoma (Fig. 52–3). This sarcoma is often associated with AIDS. Bone joints, especially of the hands, should be inspected for swollen, painful nodules that are associated with rheumatoid arthritis. Any swelling, tenderness, or limited range of motion is noted. Nails are assessed; if they separate from the nailbed, it is called onycholysis and can be associated with thyroiditis.

Vision and hearing changes can be associated with an immunologic disorder. Eye movements are assessed by testing the six cardinal positions of gaze for muscle weakness. (See Chapter 48.) The conjunctiva should be deep pink and moist. If the patient is anemic, the conjunctiva is pale. Edema around the eyes (periorbital edema) may indicate hypothyroidism or renal disease. Darkened areas under the eyes (shiners) are common in patients experiencing allergic rhinitis.

Adventitious lung sounds, such as wheezing, may indicate asthma or an allergic response. Crackles are often associated with an upper respiratory infection. Crackles heard with a dry cough and labored tachypnea are a sign of *Pneumocystis carinii* pneumonia, a common opportunistic lung infection of patients with AIDS. A pleural or pericardial friction rub may be a sign of rheumatoid arthritis or SLE.

Lymph nodes should be inspected and then gently palpated. (See Fig. 52–1.) Normally, lymph nodes are not palpable in the adult (usually they can be palpated in young children). If on palpation a lymph node is enlarged, the following characteristics need to be noted: location, size, shape, tenderness, temperature, consistency, mobility, symmetry, pulsation, and if red streaks, redness, or edema is

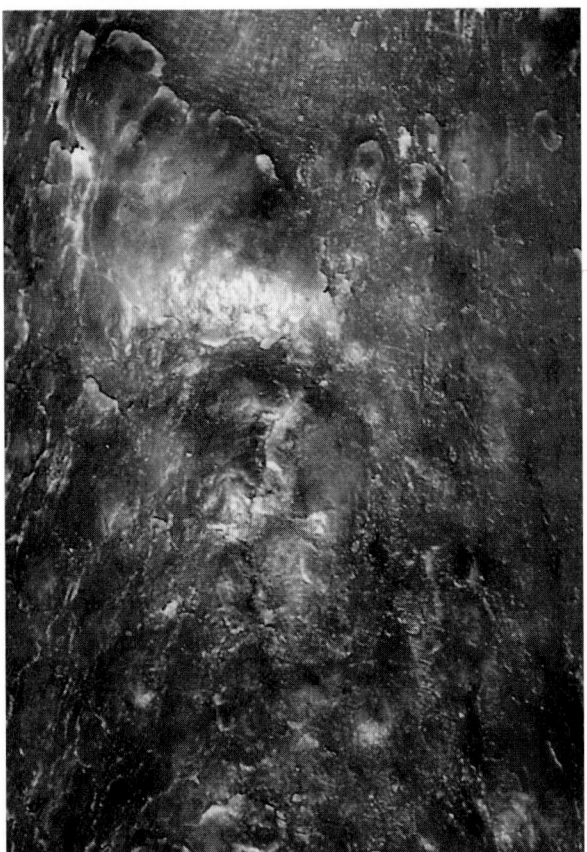

Figure 52-3 Kaposi's sarcoma of the skin. (From Wells, J: Clinical Immunology Illustrated. Adis International, Philadelphia, 1986, p 96, with permission.)

present. Generally when lymph nodes are tender and enlarged, inflammation is present. Lymph nodes that are nontender, hard, fixed, and enlarged are often associated with cancer.

If enlarged, the spleen may be palpable (by advanced practitioner) in the left upper quadrant of the abdomen with disorders in which there is an overproduction or excessive destruction of red blood cells.

Gastrointestinal signs and symptoms such as nausea, vomiting, and diarrhea should be noted because they may have an immunologic basis. The renal system may also show impairment because of an immunologic disorder, exhibited through a change in urinary output, flank pain, edema, weight gain, or elevated renal function studies.

A general neurological assessment of muscle strength and coordination, changes, or abnormalities is noted. Changes may be an indication of an immunologically based disorder such as multiple sclerosis or myasthenia gravis.

Diagnostic Tests
Blood Tests

Initially, screening tests are performed. Table 52–2 describes the most common screening blood tests for patients with allergic, autoimmune, or immune disorders. If the screening tests detect an abnormality, further specific tests may be ordered to precisely identify the disorders. Table 52–3 examines the more common of these specific immune tests.

Radiographic Tests

A chest x-ray examination, magnetic resonance imaging (MRI) or computed tomography (CT) scan may also be performed to identify size and density of structures, as well as new abnormal growths.

Biopsies

Biopsies of organs or structures possibly affected by an immune disorder aid in confirming a diagnosis, determining a prognosis, or evaluating a treatment. Biopsies may include lymph node biopsies, organ biopsies, or bone marrow aspirations. These procedures require a consent to be signed by the patient. Biopsies may be performed as inpatient, outpatient, or physician's office procedures. The specific biopsy being performed determines any required patient preparation or postprocedure care.

Skin Tests

When the immune system is intact, skin tests can be used to test for infectious diseases such as *Candida*, tetanus, or tuberculosis (purified protein derivative [PPD] test) or to identify the allergen responsible for symptoms of allergic rhinitis or drug and food allergies. Patients may also have skin testing performed to determine the specific allergen they are sensitive to. This information allows the patient to either avoid the allergen or have immunotherapy (allergy shots) to desensitize the immune system to the specific offending allergen.

Gene Testing

Mapping of the human genome has made current and future gene testing and manipulation possible. Scientists are now able to test for numerous diseases, predisposition to diseases, and enzyme deficiencies that can alter the immune response. These tests are possible but are still new and expensive.

▶ THERAPEUTIC MEASURES
Allergies

A medical alert bracelet or some sort of readily available identification of an allergy should be worn by the patient. It is important to always be aware of patient allergies. Allergies should be noted before giving any medications or foods. All allergies, including those to food, must be taken seriously.

Food allergies create serious management problems and have caused numerous deaths from anaphylactic shock (Nutrition Notes Box 52-3). The offending allergen may be present in extremely tiny amounts. Sometimes a food product is contaminated by a previous batch of food made with the same equipment. Occasionally the allergen may enter the body not by ingestion but by inhalation or contact with skin or mucous membranes.

If an antigen is environmental, such as bee or wasp stings, carrying an epinephrine pen may be vital because the allergic person may not be near medical treatment when it is needed. Epinephrine is the drug of choice for an

TABLE 52-2	COMMON BLOOD SCREENING TESTS FOR ALLERGIC, IMMUNE, AND AUTOIMMUNE DISORDERS	
Test	**Explanation of Normal Values**	**Explanation of Abnormal Values**
Red blood cell count	Number of red blood cells per mm of blood	Decreased in anemias, multiple myeloma; may be low in rheumatoid arthritis
Total white blood cell count (WBC)	Number of white blood cells per millimeter of blood	Increases in acute infection, inflammation, and leukemias Decreased in chronic infections, cancer, bone marrow failure and immunosuppression (AIDS), chronic cortisol therapy, antibiotic therapy, chemotherapy, myelosuppression
WBC differential	Percentage of type of white blood cells in 1 mm of blood	
Neutrophils		Increased in acute bacterial infections Decreased in acute viral infections, chemotherapy, radiation therapy, aplastic anemia, and neoplastic invasion of bone marrow
Lymphocytes		Increased in leukemias, pertussis, tuberculosis, and viral infections Decreased in acute bacterial infections, AIDS
Eosinophils		Increased in allergic reactions and parasitic infestations Decreased in corticosteroid therapy, bone marrow failure, and immunosuppression
Basophils		Increased in allergic reactions Decreased in bone marrow failure and immunosuppression
Serum IgG levels	Amount of IgG in milligrams per deciliter of blood and percentage of IgG of total immunoglobulins Normal: 70–75%	Increased in infection, AIDS, and autoimmune disorders Decreased in lymphocytic leukemia and agammaglobulinemia
Serum IgA levels	Amount of IgA in milligrams per deciliter of blood and percentage of IgA of total immunoglobulins Normal: 10–15%	Increased in chronic infections, autoimmune disorders Decreased in immunosuppression, lymphocytic leukemias, and agammaglobulinemia
Serum IgM levels	Amount of IgM in milligrams per deciliter of blood and percentage of IgM of total immunoglobulins Normal: 10%	Increased in autoimmune disorders, systemic lupus erythematosus Decreased in immunosuppression and chronic lymphocytic leukemia
Serum IgE levels	Amount of IgE in milligrams per deciliter of blood and percentage of IgE of total immunoglobulins Normal: 0.002%	Increased in asthma and allergic conditions Decreased in immunosuppression
Complement assay	Measures the amount of each of the components in the complement system	Increased in thyroiditis, rheumatoid arthritis, ulcerative colitis Decreased in systemic lupus erythematosus, acute poststreptococcal glomerulonephritis, acute serum sickness, multiple myeloma, and hereditary angioedema
Sedimentation rate	Measures the red blood cell descent (in millimeters) in test tube after being in normal saline for 1 hour	Increased in inflammatory, infectious, necrotic, and cancerous conditions due to the increased protein content in plasma
C-reactive protein test	An abnormal protein found in plasma during acute inflammatory processes; more sensitive than sedimentation rate	Increased in rheumatoid arthritis, cancer, systemic lupus erythematosus Suppressed by aspirin and steroids

anaphylactic reaction. Time plays a crucial role when this type of reaction occurs. The patient must seek immediate medical care and begin treatment by administering epinephrine until help is available.

An anaphylaxis kit is often prescribed for patients with allergies to food or insect stings. The kit contains injectable epinephrine, oral chewable antihistamine tablets, a tourniquet, and instructions for use. The patient must carry the kit at all times when insect stings are possible. The patient must be taught how to use the kit and give the subcutaneous epinephrine injection. The tourniquet, applied to the extremity that was stung, slows blood flow carrying the allergen to reduce its spread in the body. The tourniquet is released every 15 minutes to allow some blood flow in the extremity.

Immunotherapy

To help desensitize a patient with anaphylactic reactions to allergens or with chronic allergic symptoms, immunotherapy involves injecting small amounts of an extract of the allergen. Over time the strength of the allergen injected is increased, until the desired hyposensitivity is reached. The subcutaneous injections are given once or twice a week initially, then every few weeks indefinitely for years. It is important that the patient does not miss a dose. If this happens, the allergen strength may need to be reduced. This therapy is helpful for insect sting allergies.

When administering the injection, it is important to understand that an anaphylactic reaction can occur. A physi-

TABLE 52-3 COMMON TESTS PERFORMED FOR ALLERGIC, IMMUNE, AND AUTOIMMUNE DISORDERS

Test	Explanation of Test	Explanation of Abnormal Values
Rheumatoid factor (RF)	An abnormal protein found in serum when IgM reacts with an abnormal IgG; found in 80% of patients with rheumatoid arthritis, as well as other autoimmune disorders	Increased in rheumatoid arthritis, SLE, leukemia, tuberculosis, older age, scleroderma, infectious mononucleosis
Antinuclear antibodies (ANA)	Measures antibodies that attack the cell's nuclei	Most commonly present in SLE, leukemia, scleroderma, rheumatoid arthritis, and myasthenia gravis; many medications influence these levels
Lupus erythematosus test (LE Prep)	Nonspecific test for lupus ANA is a more sensitive test for lupus	Increased in SLE, rheumatoid arthritis, scleroderma; many medications influence levels
Western blot	Tests for antibodies to HIV; technical and labor-intensive test, therefore not used as a screening test; used as a confirmation test; positives are difficult to determine	Positive when antibodies for HIV are present
Murex SUDS	Tests for antibodies to HIV-1; manual, qualitative test; visually read in 10 minutes; used as a screening test	Positive when antibodies for HIV are present
Enzyme-linked immuno-sorbent assay (ELISA)	Tests for antibodies to HIV; inexpensive and easily performed, therefore a good screening test for HIV; high incidence of false-positive results due to the test's nonspecificity	Positive when antibodies for HIV are present, as well as numerous other viruses
p24 Antigen	Tests for the formation of the p24 antigen, which occurs before the production of antibodies, therefore testing can occur during the window between infection and seroconversion (antibody production); p24 antigen is present for only a short time and is replaced with the antibodies	Positive when p24 antigen produced by HIV is present
Polymerase chain reaction	Detects nonreplicating viral genomes; requires only 1–2 mL of blood; costly and time consuming, therefore not a screening test	Positive when HIV is present
CD4+ T cells	Measures the actual number and percentage of CD4+ T cells in a specimen	Increased in allergy-proven patients Decreased in cancer, AIDS, and immunosuppression
CD8+ T cells	Measures the actual number and percentage of CD8+ T cells in a specimen	Increased in viral infections Decreased in SLE

cian and emergency equipment should be readily available. The patient should be observed following the injection for about 20 to 30 minutes to detect a reaction. The patient should be taught that a reaction could occur up to 24 hours after the injection and how to respond.

Medications

Medications are one of the primary treatment options for immune disorders. General categories of these medications include epinephrine, corticosteroids, antihistamines, histamine (H_2) blockers, decongestants, mast-cell stabilizing drugs, antivirals, antibiotics, immunosuppressants, interferon, leukotriene antagonists, and hormone therapy. (See Chapter 53.)

Surgical Management

In some cases, splenectomy is necessary to control symptoms of an immunologic disorder.

New Therapies

Monoclonal antibodies can be produced against a variety of antigens. A monoclonal antibody is made by cloning one specific antibody and then growing unlimited amounts of it in tissue cultures. Many uses are being found for these antibodies, such as in dealing with transplant rejections. Further study will lead to expanded knowledge on these antibodies.

Recombinant DNA technology combines genes from one organism with genes from another. This therapy is used for replacing an abnormal or missing gene to produce a normal gene. The normal gene can then be injected into the patient in an attempt to cure the disorder as the patient's body reproduces the normal genes. T lymphocytedirected gene transfer for severe combined immune deficiency has been performed successfully. Along these same lines is the research regarding injection of the precursors to all cells, the stem cell, into abnormal areas to produce normal cells. Studies continue in this area for possible uses of gene therapy, and since the completion of the mapping portion of the Human Genome Project, new discoveries in this area occur daily. For additional up-to-date information on these topics, visit the National Institutes of Health at www.nhgri.nih.gov or www.ncbi.nlm.nih.gov.

BOX 52-3 · Nutrition Notes

Respecting Food Allergies

Allergy to food can be fatal. An analysis of 13 anaphylactic reactions to food identifies points along the critical path that had the potential to alter the outcome. Of 13 cases identified, 6 were fatal. In all cases the patient was known to be have asthma and to be allergic to some food. None of the patients were aware that the allergen was present in the foods consumed. The items were candy, cookies, and pastry. Symptoms began soon after ingestion but in some cases abated before becoming severe. Of particular significance is the fact that fewer than half the children had self-injectable epinephrine prescribed and only one of the six children used a dose. The average delay between ingestion of the allergenic food and receipt of a dose of epinephrine was two and one-half times as long in the fatal cases as in the nonfatal cases. Also, more of the fatal cases occurred in public places rather than a private home. These and other similarities and differences are tabulated below, followed by recommendations for the management of potential for anaphylaxis.

Situational Factor	Fatal Cases		Near-Fatal Cases	
Age of patient (average and range)	11.5 years		12.4 years	
	2–16 years		9–17 years	
Gender	1 male		2 males	
	5 females		5 females	
Known to have asthma	6		7	
Cause of anaphylactic reaction	Peanuts	3	Filberts	2
	Cashews	2	Milk	2
	Eggs	1	Peanuts	1
			Brazil nuts	1
			Walnuts	1
Average number of foods identified as allergenic for each patient	1.7		2	
Onset of symptoms	3–30 minutes		1–5 minutes	
Location	School	4	Home	2
	Fair	1	Relative's home	2
	Home	1	Friend's home	2
			Vacation home	1
Parent present at site	3		4	
Had a prescription for self-injectable epinephrine	3		3	
Used the self-injectable dose	0		1	
Number of minutes after ingestion that any epinephrine was given (average and range)	93.3		36.4	
	15–180		10–130	

Several recommendations came from this study:

- Epinephrine should be prescribed, kept available, and used for patients with IgE-mediated food allergies.
- Children and adolescents who have an allergic reaction to food should be observed for 3 to 4 hours after the reaction at a center capable of dealing with anaphylaxis.
- Parents of such children should be taught to ensure a rapid response by schools and other institutions.

Summarized from Sampson AA, Mendelson I, Rosen JP: Fatal and near-fatal anaphylactic reactions to food in children and adolescents. New England Journal of Medicine 327:380, 1992.

Answers to CRITICAL THINKING

Laura

1. The patient can have an anaphylactic reaction if exposed to latex, which may result in death for some patients.
2. The LPN/LVN should follow the agency's latex allergy protocol, enter this information into the patient's medical record, notify surgery scheduling so latex precaution protocols can be planned for surgery, and have the patient's physician informed.
3. Follow the agency's protocol, the nurse should wear nonlatex gloves and use nonlatex equipment to draw the specimens.

REVIEW QUESTIONS

1. The nurse who is teaching a patient about vaccines would be correct in teaching that a vaccine provides which of the following types of immunity?
   a. Naturally acquired passive immunity
   b. Artificially acquired passive immunity
   c. Naturally acquired active immunity
   d. Artificially acquired active immunity

2. The nurse is assisting with data collection. Which one of the following past surgeries may be useful in assessing potential immune system dysfunction?
   a. Splenectomy
   b. Thyroidectomy
   c. Pneumonectomy
   d. Parathyroidectomy

3. The nurse is assisting with data collection and the patient reports tenderness in the cervical lymph nodes. The nurse recognizes that lymph nodes that are enlarged and tender usually indicate which of the following problems?
   a. Cancer
   b. Degeneration
   c. Inflammation
   d. Arthritis

4. The nurse is caring for a patient with suspected HIV. The nurse anticipates that which of the following is a confirmation test that will be ordered to test for HIV antibodies?
   a. Murex SUDS
   b. Enzyme-linked immunosorbent assay
   c. Western blot
   d. p24 antigen testing

5. The nurse is giving immunizations. Which of the following vaccines are recommended annually for the elderly patient?
   a. Influenza
   b. Pneumovax
   c. Diphtheria tetanus
   d. Polio

53

NURSING CARE OF PATIENTS WITH IMMUNE DISORDERS

Sharon M. Nowak

QUESTIONS TO GUIDE YOUR READING

1. How would you explain the immunologic mechanism for the four types of hypersensitivities?
2. How would you explain the pathophysiology of disorders of the immune system?
3. What are the etiologies, signs, and symptoms of immune system disorders?
4. What care would you provide for patients undergoing tests for immune system disorders?
5. What is the current medical treatment for immune system disorders?
6. What data are collected when caring for patients with disorders of the immune system?
7. What factors alter or influence the self-recognition portion of the immune system?
8. What nursing care will you provide for patients with disorders of the immune system?
9. How will you know if your nursing interventions have been effective?

Disorders of the immune system can be divided into three categories. The first category is hypersensitivity reactions, which include conditions such as **anaphylaxis,** hemolytic transfusion reactions, measles, and transplant rejections. Autoimmune disorders (e.g., rheumatoid arthritis, ulcerative colitis, and multiple sclerosis) are the second category. The third category includes the immune deficiencies, such as hypogammaglobulinemia and acquired immunodeficiency syndrome (AIDS). (See Chapter 54.)

▶ HYPERSENSITIVITY REACTIONS

The immune system is an adaptive, protective system of the body. However, there are times when this system causes injury to the body because of its exaggerated response. One of these occasions is when a hypersensitivity reaction occurs.

In the past these reactions were classified as either immediate or delayed hypersensitivity. Gell and Coombs have developed a more precise classification system, which is used today. This four-division system (Types I, II, III, and IV) classifies hypersensitivity reactions according to the way the tissue is injured.

Type I

A type I reaction, an anaphylactic reaction, is an immediately occurring reaction when exposure to a specific antigen occurs. The reaction can be mild to severe and life threatening. The patient must have had previous exposure (sensitization) to the antigen. During this exposure, immunoglobulin E (IgE) antibodies are made and attach to mast cells throughout the body. When a subsequent exposure occurs, the antigen causes IgE to trigger mast cells to release their contents. One of the substances released is **histamine,** which causes vasodilation, changes in vascular permeability, an increase in mucus production, and contraction of various smooth muscles.

If the second antigen exposure is localized, the reaction is small and remains local. However, if the exposure is

---

anaphylaxis: ana—up + phylaxis—protection

systemic, the reaction is massive and widespread. Respiratory allergies, such as allergic rhinitis and allergic asthma, with associated disorders of atopic dermatitis, tend to be reactions of a larger scale. Anaphylaxis, **urticaria,** and **angioedema** are the severest forms of type I reactions.

A type I reaction occurs when the patient has a positive reaction to a scratch test. A scratch test is done to identify specific allergens to which a patient is reactive. Tiny amounts of a variety of common allergens are scratched onto the skin, which is then observed for indications of a reaction: redness, edema, and pruritus. If these indicators occur, it is considered to be a local reaction.

Allergic Rhinitis

Allergic rhinitis is the most common form of allergy. When symptoms occur throughout the year, it is called perennial allergic rhinitis. If the symptoms occur seasonally, it is called hay fever. The causative antigens are environmental and airborne.

PATHOPHYSIOLOGY. Allergic rhinitis is the result of an antigen-antibody reaction. Ciliary action decreases and mucus secretions increase. Vasodilation and local tissue edema occur.

SIGNS AND SYMPTOMS. Signs and symptoms vary in intensity and include sneezing, nasal itching, profuse watery rhinorrhea (runny nose), and itchy red eyes. The nasal mucosa is pale, cyanotic, and edematous. Frequently there are dark circles under the eyes, called allergic shiners, caused by venous congestion in the maxillary sinuses.

COMPLICATIONS. Allergic rhinitis is minor and self-limiting for most patients. Repeated occurrences of allergic rhinitis can lead to infection of the sinuses (sinusitis), nasal polyps, asthma, or chronic bronchitis.

DIAGNOSTIC TESTS. Diagnosis is generally made from a detailed history and visual inspection of the nasal passages. Sometimes, skin testing is performed to identify the specific offending allergens to allow avoidance of the allergen. However, skin testing is expensive, does not always identify the allergen, and has limited usefulness for allergens that cannot be easily avoided once identified.

MEDICAL TREATMENT. Initial treatment involves eliminating the offending environmental stimuli. Antihistamines and nasal decongestants may be prescribed for symptomatic relief. If the symptoms are severe, corticosteroids via inhalation or nasal spray may also be given. Immunotherapy, referred to as allergy shots, is reserved for patients with severe or debilitating symptoms. Immunotherapy involves receiving weekly subcutaneous injections of tiny amounts of the offending antigen. Once a tolerance to a particular dose is reached, the amount of antigen is slightly increased. This therapy continues until the patient no longer exhibits symptoms when exposed to the environmental antigen.

NURSING MANAGEMENT. Education of the patient and significant other is an important nursing function. Monitoring the patient's symptoms and compliance with the treatment plan is also a nursing responsibility. Skin testing and immunotherapy injections, given in the allergist's office, maybe performed. Being alert to signs of anaphylaxis during these procedures is essential.

Education. The teaching plan focuses on the patient avoiding allergens and following the therapeutic regimen. Exploring creative ways to prevent allergen exposure is helpful. Allergen avoidance might involve wearing a mask when mowing the lawn or working outdoors or having heating ducts cleaned or heating registers covered with filters. Frequent home vacuuming and dusting is recommended. Ensuring that patients understand prescribed medications and their correct usage is important.

Atopic Dermatitis

PATHOPHYSIOLOGY. Atopic dermatitis, often called eczema, is an inflammatory skin response. The skin lesions are not typical for a type I hypersensitivity reaction, and there usually is not a specific antigen identified as the cause. However, it is believed that the pathophysiology of atopic dermatitis is a type I hypersensitivity reaction, mediated by IgE antibodies, because it is commonly found in patients with allergic rhinitis or allergic asthma.

SIGNS AND SYMPTOMS. Initially there is pruritus, edema, and extremely dry skin, which is followed by eruptions of tiny vesicles (blisters); these eventually break open, crust over, and scale off. There is decreased sweating in these areas with the skin eventually thickening in the areas of dermatitis.

COMPLICATIONS. With open skin lesions, there is always a risk for infection. Staphylococci are found on the dermal layer of the skin, so these bacteria tend to be the most common source of secondary infections of the affected area of skin.

DIAGNOSTIC TESTS. There are no tests to confirm this diagnosis. A detailed history and physical examination are used to diagnose it. Of course, if there is an infection, culture and sensitivity tests may be ordered to determine the infecting organism and appropriate treatment.

MEDICAL TREATMENT. Treatment is focuses on symptoms. Oil-in-water lubricants such as Alpha-Keri oil tend to be the most effective for dryness. Topical corticosteroids may be ordered for their anti-inflammatory properties. If skin lesions become infected, topical or systemic antibiotics are prescribed.

NURSING MANAGEMENT. Assess and document the lesions for signs of infection: redness, warmth, swelling, and purulent drainage. Administer medications and lubricants as ordered.

Education. Instruct the patient on the signs and symptoms of infection. Assessment of the patient's knowledge about

angiodema: angeion—vessel + oidema—swelling

prescribed medications and over-the-counter preparations is needed for teaching and reinforcement of knowledge. Teach the patient that humidification during the winter months is a necessity, cotton clothing may minimize irritation, and cool soaks decrease pruritus.

Anaphylaxis

PATHOPHYSIOLOGY. Anaphylaxis is a severe systemic type I hypersensitivity reaction. IgE antibodies produced from previous antigen sensitization are attached to mast cells throughout the body. In this reaction the antigen is introduced at a systemic level, which causes widespread release of histamine and other chemical mediators contained within the mast cells.

ETIOLOGY. There are numerous causes of anaphylaxis. See Table 53–1 for potential causes.

SIGNS AND SYMPTOMS. Anaphylaxis produces sudden and life threatening signs and symptoms. Generalized smooth muscle spasms occur, causing bronchial narrowing and creating stridor, wheezing, dyspnea, and laryngeal edema, which can lead to respiratory arrest. Cramping, diarrhea, nausea, and vomiting also result from these spasms. Capillary permeability increases, allowing fluid to shift from the vessels to the interstitium. This causes hypotension, tachycardia, and an increase in respiratory symptoms. The blood volume within the vessels decreases while the blood vessels dilate, resulting in a further decrease in circulating blood volume. The dilation also causes diffuse erythema (redness) and warmth of the skin. Neurological changes include apprehension, drowsiness, profound restlessness, headache, and possible seizures.

TABLE 53–1	SUBSTANCES THAT COMMONLY TRIGGER ANAPHYLACTIC REACTIONS	
Antibiotics		*Foods*
Penicillins		Beans
Sulfonamides		Chocolate
Tetracyclines		Eggs
Cephalosporins		Fruits (e.g., strawberries)
Amphotericin B		Grains (e.g., wheat)
Aminoglycosides		Nuts
Medical Products		Shellfish
Latex rubber		*Pollens*
Diagnostic Agents		Grass
Contrast dyes		Ragweed
Anesthetics/Antiarrhythmics		*Proteins*
Lidocaine		Horse serum
Procaine		Rabbit serum
Other Medications		*Venoms*
Barbiturates		Bees, wasps, hornets
Phenytoin (Dilantin)		Fire ants
Protamine		Snakes
Salicylates		*Hormones*
Diazepam (Valium)		Insulin
Food Additives		Vasopressin
Bisulfites		Estradiol
Monosodium glutamate (MSG)		Adrenocorticotropic hormone

COMPLICATIONS. The most profound complications of an anaphylactic reaction are respiratory and cardiac arrest. Immediate treatment is necessary to prevent death.

DIAGNOSTIC TESTS. There is no time for tests to be performed during an anaphylactic reaction other than those needed to guide symptom treatment, such as arterial blood gases or electrocardiogram (ECG) monitoring. Anaphylaxis is diagnosed based on physical assessment and history from the patient or significant other. After the patient's recovery, allergen testing may be considered for future prevention.

MEDICAL TREATMENT. Intravenous access is a priority for administration of intravenous epinephrine and vasopressor drugs (dopamine) to increase blood pressure. Oxygen therapy is started. If respiratory symptoms are severe, a tracheostomy or endotracheal intubation may be necessary, with mechanical ventilation. Antihistamines and corticosteroids may also be given orally, by injection, or intravenously.

NURSING MANAGEMENT. Early recognition of anaphylaxis is important. Staying with the patient while calling for immediate help minimizes the patient's anxiety and allows close monitoring of vital signs and symptoms. Maintaining a patent airway is vital. Anticipation and preparation of emergency medication and equipment facilitates rapid treatment.

Education. The patient who is prone to anaphylactic reactions should be educated on the importance of avoiding the offending antigen and wearing medical alert identification. If the antigen is environmental, such as an insect sting or food, obtaining a prescription and providing instruction on the use of an anaphylaxis kit or an epinephrine pen is crucial. (See Chapter 52).

Urticaria

PATHOPHYSIOLOGY. Urticaria (hives) is a type I hypersensitivity reaction. The antigen-stimulated reaction of IgE antibodies causing the release of mast cell contents, especially histamine, triggers urticaria.

ETIOLOGY. There are numerous causes of urticaria. In addition to medications and foods, cold, local heat, pressure, and stress can also cause urticaria. Many patients with underlying chronic conditions, such as systemic lupus erythematosus, lymphoma, hyperthyroidism, or cancer, are susceptible to urticaria.

SIGNS AND SYMPTOMS. The lesions of urticaria are raised, pruritic, nontender, and erythematous wheals on the skin. They tend to be concentrated on the trunk and proximal extremities.

DIAGNOSTIC TESTS. Diagnosis is made on the basis of physical examination and patient history.

MEDICAL TREATMENT. Treatment depends on the degree of symptoms. In the most severe cases, epinephrine may be given to quickly resolve the urticaria. Corticosteroids may be prescribed orally, topically, or intravenously. Antihistamines and histamine (H₂) blockers, such as cimetidine and ranitidine, may aid in resolution by blocking the release of histamine.

NURSING MANAGEMENT. Nursing management includes monitoring the patient's symptoms, administering prescribed medications, and evaluating the patient's response to medications. In severe cases, maintaining a patent airway is a priority. Cool soaks and diversionary activities may help with pruritus.

Education. Instruct the patient on the use and side effects of prescribed medications. Be creative in investigating with the patient ways to avoid the causative agent of the urticaria. Some patients benefit from instruction on stress management and relaxation techniques.

Angioedema

PATHOPHYSIOLOGY AND ETIOLOGY. Angioedema is a form of urticaria. The pathophysiology and etiology of angioedema are the same as for urticaria. However, angioedema affects submucosal and subcutaneous tissue rather than the skin.

SIGNS AND SYMPTOMS. Angioedema is painless and minimally pruritic, with dermal erythematous and subcutaneous eruptions. There is also skin and mucous membrane edema. The eruptions may last longer than with urticaria.

DIAGNOSTIC TESTS. A comprehensive history and physical examination confirm the diagnosis. Skin testing may be performed to determine the specific antigen.

MEDICAL TREATMENT. The most basic treatment involves avoidance of the antigen. Symptomatic relief may be obtained through the use of antihistamines and corticosteroids. For long-term treatment, immunotherapy for allergen desensitization may be indicated.

NURSING MANAGEMENT. Assessment of the patient's symptoms is needed, as well as the administration and then evaluation of the patient's response to medications. If the reaction is severe, maintenance of a patent airway is a priority.

Education. Assessing the patient's baseline knowledge of the condition and medications is necessary. Further education may be indicated. The patient and nurse may need to find creative ways for the patient to either avoid or minimize future exposure to the causative antigen.

Type II

A type II hypersensitivity reaction involves the destruction of a cell or substance that has an antigen attached to its cell membrane, which is sensed by either immunoglobulin G (IgG) or immunoglobulin M (IgM) as being a foreign antigen. When an antigen marker is sensed as foreign, an antibody attaches to the antigen on the cell membrane, causing lysis of the cell or accelerated phagocytosis (engulfing and ingestion). When a cell is foreign, such as a bacterium, this process is beneficial. However, sometimes antigens on a red blood cell (RBC) can be sensed as foreign for the different blood types, which results in the RBC being destroyed.

Hemolytic Transfusion Reaction

PATHOPHYSIOLOGY. A hemolytic transfusion reaction is a type II hypersensitivity reaction in which RBCs with antigens foreign to the individual are rapidly lysed. The rapid RBC lysis results in a massive amount of cellular debris that occludes the blood vessels throughout the body. This leads to ischemia and necrosis of tissue and organs and can be life threatening.

ETIOLOGY. Occasionally, antibodies form after a bacterial or viral infection. However, prior sensitization is usually from a previous blood transfusion or past pregnancy. If maternal and fetal blood Rh factors (RBC surface antigens) are different, the mother becomes sensitized by the fetal Rh type, which can affect future fetuses. For example, a Rhₒ(D)-negative pregnant woman becomes sensitized by a Rhₒ(D)-positive fetus. As a result, the blood cells of future Rhₒ(D)-positive fetuses can be destroyed by maternal antibodies crossing the placenta.

SIGNS AND SYMPTOMS. With a hemolytic transfusion reaction, there is usually a rather sudden onset of low back or chest pain, hypotension, fever rising more than 1.8° F (1° C), chills, tachycardia, tachypnea, wheezing, dyspnea, urticaria, and anxiety. The patient may also complain of a headache and nausea.

COMPLICATIONS. Patients who survive severe reactions are at risk for developing shock and acute renal failure because of the massive occlusion of blood vessels.

DIAGNOSTIC TESTS. The direct Coombs' test confirms this diagnosis. In the laboratory a small amount of the patient's RBCs are washed to remove any unattached antibodies. Antihuman globulin is added to see if agglutination (clumping) of the RBCs results. If agglutination occurs, an immune reaction such as a hemolytic transfusion reaction is taking place.

MEDICAL TREATMENT. To prevent production of anti-Rho(D) antibodies, a Rho(D) immune globulin (RhoGAM) injection is given to Rho(D)-negative patients accidentally given Rho(D)-positive blood or exposed to Rho(D)-positive fetal blood by delivery, miscarriage, abortion, amniocentesis or intra-abdominal trauma. When antibodies do not form, then a hemolytic reaction can be prevented occur.

If a hemolytic reaction does occur, the medical treatment depends on the severity of the reaction and what tissue and organs are specifically affected by the reaction. Antihistamines, corticosteroids, or even epinephrine may be given. Diuretics may be given to assist the kidneys with excretion if they become occluded with cellular debris.

NURSING MANAGEMENT. Prevention of hemolytic reactions is crucial. Following strict institutional guidelines for blood transfusion administration helps ensure the patient's safety. After blood is released from the hospital blood bank, two nurses, designated per institutional policy, double-check specified data. At the bedside, transfusion guidelines include double-checking the patient's name and identification number on the chart, unit of blood, and patient's identification bracelet, as well as checking the patient's blood type in the chart, on the unit of blood, and paperwork with the unit of blood.

Agency policy is followed for taking vital signs during a blood transfusion. Usually, baseline vital signs are taken before the beginning of the blood transfusion, then every 15 minutes for 30 minutes, then every half hour or hourly, and again when the transfusion is completed. It takes only a small amount of blood to trigger a hemolytic transfusion reaction, so it is critical to stay with the patient at the bedside during the first 15 minutes of any blood transfusion. This enables detection of a blood transfusion reaction early for quick action to minimize cell destruction and complications, including death.

LEARNING TIPS

Every unit of blood, even of the same blood type, is unique and can trigger a blood transfusion reaction. Careful monitoring with every transfusion is necessary.

If symptoms of a reaction are noted, the blood is immediately stopped and agency policy for a suspected transfusion reaction is followed. A normal saline infusion is started to keep the vein patent. The physician and blood bank are immediately notified. A nurse remains with the patient for reassurance and monitoring of symptoms and vital signs. If a blood incompatibility is suspected, the unused blood and blood tubing is returned to the blood bank for testing. A series of blood and urine specimens are collected and sent to the laboratory for analysis. The physician's orders are followed to treat the patient's symptoms.

Education. Autologous (self) blood donation, as ordered by the physician, may be an option for patients having elective surgery, to avoid a transfusion reaction. Once patients have had a hemolytic transfusion reaction, it is important to inform them and their significant others of the reaction. Patients should inform their health care providers of this reaction, so that specific blood tests are performed for less common antibodies if the patient is ever typed for a blood transfusion again.

Type III

A type III hypersensitivity reaction involves immune complexes formed by antigens and antibodies, usually of the IgG type. The patient is sensitized with an initial exposure to the antigen, and on a subsequent exposure the reaction occurs. The reaction is localized and evolves over several hours, with symptoms ranging from a red, edematous skin lesion to hemorrhage and necrosis. The process involves the formation of antigen-antibody complexes within the blood vessels, as the antigen is absorbed through the vessel wall. Neutrophils are attracted to the area and release enzymes that ultimately lead to blood vessel damage.

Serum Sickness

PATHOPHYSIOLOGY. Serum sickness is a type III hypersensitivity immune reaction in which antigen-antibody complexes are formed and cause symptoms of inflammation.

ETIOLOGY. In the past, serum sickness occurred after inoculation of equine (horse) antiserum for diseases such as tetanus and diphtheria. With the refinement of these vaccines, serum sickness does not usually occur from them. Today serum sickness is seen occasionally after administration of penicillin and sulfonamide.

SIGNS AND SYMPTOMS. The signs and symptoms usually occur 7 to 10 days after the exposure. The most predominant manifestation is severe urticaria and angioedema. The patient may experience a fever, malaise, muscle soreness, arthralgia, splenomegaly, and occasionally nausea, vomiting, and diarrhea. Lymphadenopathy may occur, especially at the lymph nodes closest to the antigen entry site.

COMPLICATIONS. Serum sickness is usually a brief and self-limiting condition, although at times it can become chronic. Systemic complications may arise, such as renal failure.

DIAGNOSTIC TESTS. With serum sickness there is often a slight elevation in the white blood cell count and the sedimentation rate, and a complement assay decreases.

MEDICAL TREATMENT. Antipyretics may be given for the fever and analgesics for the arthralgia. Antihistamines and epinephrine may be prescribed for the urticaria and angioedema. If the symptoms continue to persist, corticosteroids may be ordered.

NURSING MANAGEMENT. Nursing care focuses on assessing symptoms, evaluating the effectiveness of prescribed medications, and educating the patient regarding the therapeutic regimen and measures to prevent further exposure to the antigen. Identification of the causative agent may be done through the history-taking process.

Education. Knowledge of the disease and offending agent is important for the patient to prevent a recurrence of the condition.

Type IV

A type IV hypersensitivity reaction, also called a delayed reaction, occurs when a sensitized T lymphocyte comes in contact with the particular antigen to which it is sensitized. The resulting necrosis is caused by the actions of the macrophages and the various T lymphocytes involved in the cell-mediated immune response.

Contact Dermatitis

PATHOPHYSIOLOGY. When a substance or chemical comes in contact with the skin, it is absorbed into the skin and binds with special skin proteins called haptens. With the first contact, there is no reaction or symptoms, but within 7 to 10 days T memory cells are formed. Therefore on subsequent exposures the T memory cells quickly become activated T cells, which secrete the chemicals that cause symptoms.

ETIOLOGY. Poison ivy and poison oak are the most common irritants causing this reaction. Latex rubber has also been found to be a cause for contact dermatitis and can trigger type I anaphylactic reactions.

Latex Allergy. Latex allergy is a serious problem for health care workers. Anaphylactic reactions to latex can be fatal. Exposure to latex for health care workers has increased dramatically since the implementation of universal precautions and the use of latex gloves began in 1987. Many times latex gloves are worn unnecessarily, which increases exposure to the latex protein. For patients who are allergic to latex, special protocols are followed using latex-free equipment. For information about latex allergy, visit the American Academy of Allergy, Asthma, and Immunology at www.aaaai.org. Also visit the U.S. Food and Drug Administration at www.fda.gov.

SIGNS AND SYMPTOMS. Within a number of hours from the exposure, the area of contact becomes reddened and pruritic, with fragile vesicles. Secondary infections may develop. (See earlier discussion of atopic dermatitis.)

DIAGNOSTIC TESTS. Diagnosis is usually made by assessment of the skin and lesions, as well as a detailed patient history.

MEDICAL TREATMENT. Treatment consists of controlling symptoms. Oral or topical antihistamines and topical drying agents may be used. Topical corticosteroids may be used, and if the symptoms are severe, systemic corticosteroids may be prescribed.

NURSING MANAGEMENT. To aid in the relief of symptoms, tepid baking soda baths or colloidal oatmeal baths (e.g., Aveeno) may help dry the vesicle and minimize the pruritus (Nursing Care Plan Box 53–1). Administer and evaluate the effectiveness of medications.

Education. The patient is instructed to wash with a brown soap (e.g., Fels-Naptha) or, if unavailable, any soap when contact with the offending agent is suspected. The best advice is to avoid contact with the agent. The patient should also be instructed not to scratch the skin to prevent the spread of the dermatitis, as well as infection development.

Transplant Rejection

ETIOLOGY AND PATHOPHYSIOLOGY. Any form of transplanted living tissue is sensed as foreign material by the immune system. Lymphocytes become sensitized during an induction phase immediately after the tissue is transplanted. The sensitized lymphocytes invade the transplanted tissue and destroy it via the release of chemicals and macrophage activity.

SIGNS AND SYMPTOMS. Various signs and symptoms occur depending on the transplanted tissue or organ that is involved and the severity of the reaction. Signs and symptoms reflect failure of the organ or tissue, such as renal failure for a rejected kidney.

COMPLICATIONS. A total failure and loss of the transplanted tissue or organ can occur, or the tissue or organ can be damaged from immunologic reactions and not function at full capacity. The greatest cause of death following a transplant is from infection. Immunosuppression therapy, which is necessary to prevent tissue rejection following the transplant, is a major contributory factor for severe infection development. Because the immune system is suppressed, it is unable to effectively fight infections.

DIAGNOSTIC TESTS. Biopsy, scans, blood tests, arteriogram, and ultrasound are some tests that may be performed to aid in diagnosing a transplant rejection.

MEDICAL TREATMENT. Depending on the type of transplant, the body's immunologic system is prepared before surgery with medications, transfusions, or radiation to minimize the risk of rejection.

NURSING MANAGEMENT. Nursing care depends greatly on the type of transplant performed. Initially the patient is in an intensive care unit under close observation and support. Observing for signs of rejection is a priority throughout the patient's hospitalization. Another consideration for nursing care is the psychological support of the patient and family. Many patients wait on a transplant list a long time before a donor match is found. Once a matching donor is found there is usually great elation. Yet if a donor's death made the transplant possible, the patient and family may be simultaneously feeling a profound sadness for the donor's family. Patients need time to verbalize feelings and understand that these feelings are normal and diminish with time. Also the fear of transplant rejection is always present and must be discussed.

Education. Rejection can take place weeks, months, or years following a transplant (with decreasing risk). The patient and family need to be educated about specific signs and symptoms of rejection. Also, because infection is a major complication resulting from long-term immunosuppressive medications, the patient and family need to know signs and symptoms of infection and when to notify the physician of problems. Steroid use may mask the symptoms of infection, so small indicators such as a low-grade fever should be promptly reported. Education regarding prescribed medications is a must, because long-term success of a transplant is dependent on compliance with immunosuppressive medication therapy. Avoidance of people with colds or infections is also important to reduce the patient's infection risk.

BOX 53-1 NURSING CARE PLAN FOR THE PATIENT WITH CONTACT DERMATITIS

 Risk for impaired skin integrity related to effects of allergic reaction

PATIENT OUTCOMES

Patient's skin will remain intact.

EVALUATION OF OUTCOME

Is patient's skin intact? If not intact, is skin healing? Does patient express a plan for preventing impaired skin integrity?

INTERVENTIONS	RATIONALE	EVALUATION
Assess and document skin and lesions.	Assessment provides a basis for intervention planning and evaluation of healing.	Are lesions present? Are lesions healing?
Teach patient to keep fingernails short and clean.	Short, clean nails cause less damage or infection if scratching occurs.	Does skin remain intact in spite of scratching?
Teach patient to apply clean, white cotton clothing over affected area (soaks, gloves/mittens, undershirt), especially at bedtime.	Cotton allows air movement. White cloth is less irritating than those with dyes. Scratching is decreased during sleep with the use of gloves/mittens or covering affected area.	Are symptoms of skin irritation reduced?
Teach patient to use gentle rubbing or pressure instead of scratching.	Use of gentle rubbing or pressure instead of scratching cause less skin trauma.	Does skin remain intact in spite of itchy sensation?

 Ineffective health maintenance related to lack of knowledge of methods to decrease inflammation and pruritus and reduce episodes of inflammation.

PATIENT OUTCOMES

Patient or caregiver will follow the mutually agreed upon plan of care.

EVALUATION OF OUTCOME

Can patient express knowledge of etiology, signs and symptoms, and treatment plan? Does patient discuss any emotional, social, financial, or material blocks to attaining treatment goals?

INTERVENTIONS	RATIONALE	EVALUATION
Assess patient's knowledge of disease and causes.	Assessment provides a basis for the teaching plan.	Does patient state baseline knowledge?
Assess patient's values and beliefs regarding plan of care.	Patients are more compliant if their belief system fits into plan of care.	Does patient's belief system work with plan of care?
Assess barrier's to patient's abliity to carry out plan of care and plan interventions to decrease barriers.	Barriers can prevent patient from carrying out plan of care.	Are barrier's identified? Are solutions to barriers planned?
Teach patient to wear medical alert identification for allergen.	With allergen identification, prompt medical care can be given in case patient is unable to give information.	Does patient agree to use allergen identification?
Discuss methods of avoiding allergen with patient.	Understanding prevention methods can help prevent allergen exposure.	Can patient state methods can help prevent allergen exposure?

▶ AUTOIMMUNE DISORDERS

In autoimmune disorders the immune system no longer recognizes the body's normal cells as self and not foreign. Instead the antigens on these normal body cells are recognized as foreign material, and an immune response to destroy them is launched.

A number of factors either cause or influence this breakdown of self-recognition, including viral infections, drugs, and cross-reactive antibodies. Some microbes stimulate the production of antibodies but are so closely related to normal cell antigens that the antibodies also attack some normal cells. Hormones have also been found to influence this breakdown of self-recognition.

Table 53–2 lists additional autoimmune disorders and the chapters in which they are discussed.

TABLE 53-2 AUTOIMMUNE DISORDERS

Idiopathic thrombocytopenic purpura	Refer to Chapter 24.
Multiple sclerosis	Refer to Chapter 47.
Myasthenia gravis	Refer to Chapter 47.
Rheumatoid arthritis	Refer to Chapter 44.
Systemic lupus erythematosus	Refer to Chapter 44.
Ulcerative colitis	Refer to Chapter 31.

Pernicious Anemia

PATHOPHYSIOLOGY. Antibodies against the gastric parietal cells and intrinsic factor lead to destruction of these cells and decrease secretion and function of intrinsic factor.

Because intrinsic factor is needed for vitamin B_{12} to be absorbed in the small bowel, a vitamin B_{12} deficiency ensues and production of RBCs is decreased.

ETIOLOGY. There tends to be a familial tendency toward pernicious anemia in relation to immune causes. Non–immune-related causes of pernicious anemia include any type of gastric or small bowel resections coupled with no or inadequate vitamin B_{12} or intrinsic factor replacement.

SIGNS AND SYMPTOMS. The patient experiences increasing weakness, loss of appetite, glossitis, and pallor. Irritability, confusion, and numbness or tingling in the extremities (peripheral neuropathy) occur because the nervous system is affected.

COMPLICATIONS. Patients with pernicious anemia have a greater risk of developing gastric carcinoma.

DIAGNOSTIC TESTS. On microscopic examination of the patient's RBCs, macrocytic (enlarged cells) anemia is diagnosed. Macrocytic anemia and low vitamin B_{12} levels are indicators of pernicious anemia and folic acid deficiency. To further determine if the diagnosis is pernicious anemia or folic acid deficiency, the Schilling test can be performed. For the Schilling test, radioactive vitamin B_{12} is administered to the patient. The patient's urine is then collected for 24 hours (48 hours for patients with renal disease), and the amount of radioactive vitamin B_{12} excreted in the urine is measured. If intrinsic factor is decreased, gastric absorption of vitamin B_{12} is also decreased, so that more vitamin B_{12} is excreted in the urine. Gastric secretion analysis is done to measure levels of hydrochloric acid (HCl) because low or absent HCl may be indicative of pernicious anemia.

MEDICAL TREATMENT. Treatment with corticosteroids may rectify the problem if it is immunologically based. Otherwise, weekly and then monthly intramuscular injections of vitamin B_{12} must be given for life to prevent pernicious anemia.

NURSING MANAGEMENT. Vitamin B_{12} is administered as ordered. Care related to fatigue and safety are important. Ambulation, frequent rest periods, and providing assistance with activities of daily living (ADLs) as indicated by the patient's activity tolerance are helpful for the patient with anemia.

Education. The patient and family need education regarding medication therapy. If vitamin B_{12} injections are prescribed, they must understand that this is a lifelong need to prevent the return of symptoms. Patients should not miss injections, periodic B_{12} level testing, or follow-up appointments.

Idiopathic Autoimmune Hemolytic Anemia

PATHOPHYSIOLOGY. Autoantibodies, for no known reason, are produced that attach to RBCs and cause them to either lyse or agglutinate (clump). When lysis occurs, fragments of the destroyed RBCs circulate in the blood. If agglutination occurs, occlusions in the small blood vessels result from the clumping and then tissue ischemia follows.

SIGNS AND SYMPTOMS. Clinical manifestations vary from mild fatigue and pallor to severe hypotension, dyspnea, palpitations, and jaundice.

DIAGNOSTIC TESTS. The RBC count, hemoglobin (Hgb), and hematocrit (Hct) are low, and microscopic examination reveals fragmented RBCs. Lactate dehydrogenase (LDH) is elevated because of RBC destruction and tissue ischemia.

MEDICAL TREATMENT. Supportive measures such as supplemental oxygen may be initiated. Folic acid may be prescribed to increase production of RBCs. The use of immunosuppressive medications and corticosteroids may be useful in obtaining remission. In more severe cases, blood transfusions and erythrocytapheresis (a process whereby abnormal RBCs are removed and replaced with normal RBCs) may be instituted. A splenectomy may be performed in an attempt to stop the destruction of RBCs for severe cases.

NURSING MANAGEMENT. The patient's signs and symptoms should be monitored and reported as necessary. Frequent rest periods should be planned into the patient's daily routine to prevent fatigue. Blood products are administered as ordered to replace RBCs. The patient and family are instructed on the medical regimen, and their understanding is verified.

Hashimoto's Thyroiditis

PATHOPHYSIOLOGY. Autoantibodies for thyroid-stimulating hormone (TSH) form in Hashimoto's thyroiditis. However, instead of inactivating TSH, the autoantibodies bind with the hormone receptors on the thyroid gland and stimulate the thyroid gland to secrete thyroid hormones. The thyroid gland enlarges as a result of this overstimulation (hyperthyroidism). It becomes infiltrated with lymphocytes and phagocytes, causing inflammation and further enlargement. Then different autoantibodies appear that destroy thyroid cells, which slows secretion activity, causing hypothyroidism.

ETIOLOGY. The exact cause is unknown, although it occurs in females eight times more often than in males. It is also more common in people 30 to 50 years old and patients with Down syndrome and Turner's syndrome.

SIGNS AND SYMPTOMS. Initially the manifestations are those of hyperthyroidism, such as restlessness, tremors, chest pain, increased appetite, diarrhea, moist skin, heat intolerance, and weight loss. These manifestations may go unrecognized and progress quickly into hypothyroidism. At this point, an enlarged thyroid gland (goiter) may be seen. Clinical manifestations may include fatigue, bradycardia, hypotension, dyspnea, anorexia, constipation, dry skin, weight gain, sensitivity to cold, facial puffiness, and a slowing of mental processes.

DIAGNOSTIC TESTS. Immunofluorescent assay, a test that detects antigens on cells using an antibody with a fluorescent tag, detects antithyroid antibodies. Serum thyroid-stimulating hormone levels are elevated, while triiodothyronine (T_3) and thyroxine (T_4) levels are low. A thyroid scan is also done.

MEDICAL TREATMENT. Thyroid hormone replacement therapy of thyroxine is the primary means of treatment. Lifelong thyroid hormone therapy is needed.

NURSING MANAGEMENT. If the patient has a goiter, a soft diet may be necessary for comfort. Frequent rest periods may be necessary, as well as slowly increasing patient activity. Antiembolic stockings may help prevent venous stasis during the low-energy, decreased-activity phase. Daily weights and monitoring intake and output when cardiac status is compromised are important to detect abnormalities such as fluid retention. Because weight gain and facial puffiness alter patients' self-image, they need an opportunity to verbalize their feelings to help them adjust to this disease process.

Education. Patients taking thyroid hormone replacement therapy should avoid foods high in iodine. The diet should also consist of large amounts of fiber to combat constipation. During the hyperthyroidism phase, a diet high in protein and carbohydrates encourages weight gain. Education regarding prescribed medications is also needed.

Ankylosing Spondylitis

PATHOPHYSIOLOGY. Ankylosing spondylitis, also called rheumatoid spondylitis, is a chronic progressive inflammatory disease of the sacroiliac, costovertebral, and large peripheral joints. The inflammatory process begins in the lower region of the back and progresses upward. A specific histocompatibility antigen (antigen that identifies self), human leukocyte antigen (HLA) B27, is formed that stimulates an immune response.

ETIOLOGY. There is strong evidence of a familial tendency, but no other specific causes are known. Ankylosing spondylitis tends to afflict men more than women.

SIGNS AND SYMPTOMS. There is an insidious onset of lower back stiffness and pain, which is worse in the morning. As the disease progresses, the pain worsens and there are spasms of the back muscles. The normal curvature of the lower back (lordosis) flattens, and the curvature of the upper back increases (kyphosis). Patients may also experience fatigue, anorexia, and weight loss.

COMPLICATIONS. Because of thoracic structural changes, carditis, pericarditis, and pulmonary fibrosis may develop occasionally.

DIAGNOSTIC TESTS. A culmination of findings, such as a positive family history, radiographs of the joints, a positive

---

ankylosing spondylitis: ankyle—stiff joint + osing— condition + spondyl—vertebrae + itis—inflammation

HLA-B27 blood test, and negative Rh, confirms a diagnosis of ankylosing spondylitis. There are no specific immunologic tests to diagnose ankylosing spondylitis.

MEDICAL TREATMENT. Because there is no cure for ankylosing spondylitis, treatment consists of measures to minimize the symptoms. Analgesics for pain relief, anti-inflammatory agents to decrease joint inflammation, and physical therapy to maintain muscle strength and joint range of motion are used. Surgery can be done to replace fused joints. For kyphosis, cervical or lumbar osteotomy can be performed.

NURSING MANAGEMENT. Nursing care focuses on patient education and administration and evaluation of prescribed medications.

Education. The patient and family need information about the disease process. Proper posture and range-of-motion exercises are taught and reinforced by nursing staff. The patient should also be instructed not to stay in any one position for any length of time. The patient should sleep on a mattress that is firm without a pillow to help reduce pain and stiffness.

⊶⊶ CRITICAL THINKING

Mr. Beck

Mr. Beck, a 55-year-old truck driver who was recently diagnosed with ankylosing spondylitis, verbalizes concern about how this diagnosis will affect his ability to work.

1. How would you answer his questions?
   a. "What is happening to me?"
   b. "Will I have to quit my job driving an interstate truck?"
   c. "Am I eventually going to be in really bad pain?"
   d. "Am I eventually going to be dependent on someone?"
2. Mr. Beck plans to continue driving his truck and therefore has a need upon discharge for specific interventions that will help him maintain his independence. Why are each of the following instructions given to Mr. Beck upon discharge?
   a. Perform range-of-motion exercises daily.
   b. Do not stay in one position too long. You may need to stop and walk around often.
   c. Sleep on a firm mattress without a pillow.
   d. Maintain good posture, even when driving the truck.

Answers at end of chapter.

▶ IMMUNE DEFICIENCIES

Immune deficiencies occur when one or more components of the immune system are either completely absent or deficient in quantities sufficient to elicit or sustain an adequate immune response to combat an infectious agent.

Hypogammaglobulinemia

PATHOPHYSIOLOGY AND ETIOLOGY. This condition is either a hereditary congenital disorder or acquired after childhood from unknown causes. It is characterized by the absence or deficiency of one or more of the five classes of immunoglobulins (IgG, IgM, IgA, IgD, and IgE) from defective B-cell function. The lack of normal function of these antibodies makes the patient prone to infections. The congenital form of this disorder affects males. Patients usually have a normal life span.

SIGNS AND SYMPTOMS. The infant is usually asymptomatic until 6 months of age, when the maternal immunoglobulins are gone. At this time the infant begins having many recurrent infections, especially from *Staphylococcus* and *Streptococcus* organisms.

DIAGNOSIS. Until the infant is 9 months old, diagnosis is extremely difficult. At 9 months of age, immunoelectrophoresis, which measures the level of each immunoglobulin, can be performed.

MEDICAL TREATMENT. Treatment is aimed at minimizing infections while increasing the immune system through injections of immunoglobulin. These injections mainly contain IgG, so fresh frozen plasma is given to replace IgM. IgA cannot be replaced, increasing the risk for frequent pulmonary infections.

NURSING MANAGEMENT. The infant is monitored for infections. The family is educated on signs and symptoms of a variety of infections and the importance in seeking medical help immediately. The infant should not be around crowds. Good nutrition, hydration, and hygiene are important in preventing infections. Any break in the skin must be cleansed immediately and monitored for infection development. Genetic counseling may be recommended for parents.

Answers to CRITICAL THINKING

Mr. Beck

1. a. "Human leukocyte antigen B27 is formed, stimulating a chronic immune (inflammatory) response specifically in the sacroiliac, costovertebral, and large peripheral joints. This leads to thickening of the joints, joint pain, and stiffness."
   b. "No, you shouldn't have to quite your job, but you may need to alter the way you go about driving your truck."
   c. "No, you may not eventually be in severe pain, with use of medications and exercise."
   d. "No, this disease may not affect your independence, with proper treatment and rehabilitation."
2. a. "Range-of-motion exercises will help maintain joint mobility and a full range of motion and prevent contractures from forming."
   b "Again, this frequent movement prevents stiffness and joint pain and contractures of joints."
   c. "Sleeping on a firm mattress without a pillow keeps the spine in correct alignment, which in turn helps prevent progressive changes in spine alignment (kyphosis, scoliosis), which affect various major body systems (respiratory, etc.)."
   d. "Again, good posture will aid in preventing bone deformities."

REVIEW QUESTIONS

1. The nurse is contributing to the plan of care for a patient with allergic rhinitis. Which of the following interventions should the nurse anticipate will be included in the treatment plan?
   a. Epinephrine
   b. Gold salts
   c. Anticholinergic medication
   d. Avoiding environmental stimuli

2. A patient who has an allergy to penicillin is receiving preoperative medications, which include ranitidine (Zantac), metoclopramide (Reglan), and cefazolin (Ancef) intravenously. Fifteen minutes after the cefazolin is started, the patient reports an uneasy feeling, as well as feeling very warm. The nurse would appropriately recognize that the patient is experiencing which of the following?
   a. Urticaria
   b. Anaphylaxis
   c. Angioedema
   d. Contact dermatitis

3. The nurse is contributing to the plan of care for a patient with ankylosing spondylitis. Which of the following interventions would be appropriate for the nurse to suggest?
   a. Avoid massage.
   b. Use interferon injections.
   c. Avoid one position for prolonged periods.
   d. Use topical steroids.

4. The nurse is contributing to the plan of care for a patient with an immune system deficiency. Which of the following complications should the nurse consider is most likely to occur when recommending prevention interventions?
   a. Overwhelming infection
   b. Fatal allergic reaction
   c. Asphyxiation
   d. Delayed anaphylactic reaction

5. A patient is admitted with an autoimmune disease and asks the nurse what *autoimmune* means. Which of the following would be the appropriate response by the nurse?
   a. "Immune cells produce too many antibodies."
   b. "Immune cells grow and multiply too rapidly."
   c. "Immune cells are not produced in sufficient amounts."
   d. "Immune cells are unable to distinguish between 'self' and 'not self.'"

54

NURSING CARE OF PATIENTS WITH HIV DISEASE AND AIDS

George Byron Smith

KEY TERMS

acquired immunodeficiency syndrome (uh-**KWHY**-erd im-YOO-noh-de-**FISH**-en-see **SIN**-drohm)

cytomegalovirus (sigh-**TOW**-meg-ul-low-vigh-rus)

human immunodeficiency virus (**HYOO**-man im-YOO-noh-dee-**FISH**-en-see **VIGH**-rus)

Kaposi's sarcoma (Ka-**POE**-sees sar-**CO**-mah)

Pneumocystis carinii pneumonia (new-moh-**SIS**-tis ca-**RIN**-ee-eye new-**MOH**-nee-ah)

QUESTIONS TO GUIDE YOUR READING

1. What is human immunodeficiency virus (HIV) and how is it transmitted?
2. What is the definition of acquired immunodeficiency syndrome (AIDS)?
3. What would you include in a teaching plan to prevent HIV infection?
4. What are prevention measures to decrease infection and opportunistic diseases for patients with HIV?
5. What would you include in a teaching plan for a patient with HIV receiving highly active antiretroviral therapy?
6. What nursing care will you provide for patients with HIV/AIDS related to infection prevention and maintaining nutritional status?
7. What teaching will you reinforce for the patient with HIV/AIDS to prevent infection and maintain health?

A cquired **immunodeficiency syndrome** (AIDS) is the final phase of a chronic, progressive immune function disorder. AIDS is caused by the **human immunodeficiency virus** (HIV). Not all HIV-infected people have AIDS. The Centers for Disease Control and Prevention (CDC) defines the diagnosis of AIDS as being when an HIV-infected person has the following:

■ Less than 200 CD4+ T lymphocytes cells per microliter
  or
■ CD4+ T lymphocyte percentage of total lymphocytes under 14
  and
■ Opportunistic clinical disease (Box 54–1)

LEARNING TIPS

HIV disease is no longer characterized as a life-ending illness. With highly active antiretroviral therapy (HAART), HIV disease is a chronic, progressive immune disorder.

There is no cure for AIDS, which develops after a long period of HIV infection and may eventually be fatal.

AIDS, more than any other chronic disease in recent history, is a multidimensional disease that challenges nurses to call into play their physical, emotional, social, and spiritual care skills. As you care for patients who are HIV positive or

BOX 54-1 **CDC Conditions in the AIDS Surveillance Case Definition**

CD4+ T-lymphocyte count below 200/μL, or a CD4+ T-lymphocyte percentage under 14 of total lymphocytes, or the presence of one of the following specified clinical conditions:

- Bacterial infection, multiple or recurrent
- Candidiasis of bronchi, trachea, or lung
- Candidiasis, esophageal
- Cervical cancer, invasive
- Coccidioidomycosis, disseminated or extrapulmonary
- Cryptococcosis, extrapulmonary
- Cytomegalovirus (CMV) disease (not including liver, spleen, or nodes)
- CMV retinitis and loss of vision
- Encephalopathy
- Herpes simplex, chronic ulcers; or bronchitis, pneumonitis, or esophagitis
- Histoplasmosis, disseminated or extrapulmonary
- Isosporiasis, chronic intestinal
- Kaposi's sarcoma
- Lymphoid interstitial pneumonia or pulmonary lymphoid hyperplasia
- Lymphoma, Burkitt's
- Lymphoma of the brain, primary
- *Mycobacterium avium intracellulare* complex or *Mycobacterium kansasii,* disseminated or extrapulmonary
- *Pneumocystis carinii* pneumonia
- Pneumonia, recurrent
- Progressive multifocal leukoencephalopathy
- *Salmonella* septicemia, recurrent
- Toxoplasmosis of brain
- Wasting syndrome

Modified from Centers for Disease Control and Prevention: Revised classification system of HIV infection and expanded surveillance case definition for AIDS among adolescents and adults, MMWR 41(RR17):2, 1992.

have AIDS, it is important for you to understand current information. Being informed helps you to provide caring, competent, nonjudgmental care without fear. Knowledge and treatment related to HIV and AIDS rapidly changes with new discoveries. Current information and guidelines can be found at HIV/AIDS Treatment Information Service, www.hivatis.org or 1-800-HIV-0440 (1-800-448-0440). For other resources, see Table 54-1.

▶ INCIDENCE

The HIV and AIDS epidemic was first reported by the CDC in June 1981 (Table 54-2). A rapid increase in AIDS was seen through the 1980s, followed by a decrease in the later 1990s. As we move into the third decade of the epidemic, an estimated 40,000 new cases of HIV infection continue to be reported each year in the United States. Increases in HIV infection are occurring the fastest in women and men who have sex with men, especially in racial and ethnic minority groups. In the United States from 1981 though 2000, 774,467 persons developed AIDS; of these, 448,060 died. The number of persons living today with AIDS is at its highest level because of treatment with Highly Active Antiretroviral Therapy (HAART), which was introduced in 1996. Although any age group can be affected by HIV and AIDS, 85 percent of those diagnosed with AIDS have been 20 to 49 years old. Given the long latency period between infection with HIV and an AIDS-defining illness, most people diagnosed with HIV disease younger than the age of 30 years were probably infected as adolescents. The age group older than 50 is seeing an increase in HIV infection. Of those with AIDS, male-to-male sex has caused the greatest exposure, followed by the use of injected drugs and by heterosexual contact. However, since 1996 heterosexual contact has resulted in more AIDS cases than male-to-male sex

TABLE 54-1	STAYING CURRENT: HIV/AIDS INFORMATION RESOURCES	
AIDS Clinical Trials Information Service	www.actis.org (800) 874-2572 Information available in English and Spanish.
Association of Nurses in AIDS Care	www.anacnet.org (800) 260-6780
CDC National AIDS Hotline in English	(800) 342-2437 ([800]-342-AIDS)
CDC National AIDS Hotline in Spanish	(800) 344-7432
CDC National AIDS Hotline TTY service for the deaf	(800) 243-7889
CDC National AIDS Hotline Website	www.cdc.gov/nchstp/hiv_aids/hivinfo/nah.htm
CDC National Center for HIV, STD, and TB Prevention	www.cdc.gov/nchstp/od/nchstp.html
CDC National Prevention Information Network	www.cdcnpin.org (800) 458-5231 Information available in English and Spanish.
HIV/AIDS Treatment Information Service	www.hivatis.org (800) 448-0440 Information available in English and Spanish.
Institute of Medicine, Committee on HIV Prevention Strategies in the U.S.	www.nap.edu/html/hiv_prevention
Journal of the American Medical Association HIV/AIDS Resource Center	www.ama-assn.org/special/hiv (312) 464-2405
National Clinicians' Post-Exposure Prophylaxis Hotline	http://pepline.ucsf.edu/PEPline (888) 448-4911
National Library of Medicine	www.healthgate.com/ama/search.html (888) 346-3656

TABLE 54-2	Data for U.S. Adults/Adolescents with AIDS, 1981–2000	
Data Category	**Trends for Totals from 1981–1987 vs. Totals for 1996–2000**	**Total Cases**
Gender		
Male	Declined from 92% to 77.4%	640,022
Female	Increased from 8% to 22.6%	134,441
Years of Age		
≤12	Declined from 1.5% to 0.7%	8,908
13–19	Increased from 0.4% to 0.7%	4,061
20–29	Declined from 20.9% to 13.7%	128,727
30–39	Declined from 46.2% to 43.1%	345,824
40–49	Increased from 20.8% to 29.5%	202,901
50–59	Increased from 7.3% to 9.1%	61,311
>59	Increased from 2.8% to 3.2%	22,734
Ethnicity/Race		
White	Declined from 59.7% to 34%	778,220
Black	Declined from 25.5% to 44.9%	292,522
Hispanic	Increased from 14% to 19.7%	141,694
Asian/Pacific Islander	Increased from 0.6% to 0.8%	5,728
American Indian/Alaskan Native	Increased from 0.1% to 0.4%	2,337
Deceased	Declined from 95.5% to 22.6%	448,060

Source: Data from Centers for Disease Control and Prevention: HIV and AIDS—United States, 1981-2000. MMWR 50(21), 2001.

or injection drug use. Current statistics can be found at www.cdc.gov.

▶ PATHOPHYSIOLOGY

Infection with the HIV virus causes destruction of immune cells. There are two identified subtypes of the HIV virus: HIV-1 and HIV-2. Both of these subtypes can cause AIDS. HIV-1 is found in Asia, Europe, and the Western Hemisphere. HIV-2 is found in West Africa. Without a normally functioning immune system, infections and cancers may take over. AIDS is the result of this immunodeficiency.

HIV is a retrovirus (which only has ribonucleic acid [RNA] for genetic material). After the HIV viral particle is taken into a human cell, its covering is destroyed to expose its viral RNA. The retrovirus then uses an enzyme called reverse transcriptase to force the human cell to produce a new piece of deoxyribonucleic acid (DNA) from the viral RNA. The new DNA is integrated into the person's cellular DNA. As a result, the human cell creates more viral particles, which spread through the lymphoid system (Fig. 54–1).

HIV is attracted to immune cells with a surface-attaching site referred to as a CD4 receptor. Cells with CD4 receptors include lymphocytes (called CD4+ T lymphocytes, T4 lymphocytes, or helper T lymphocytes) and macrophages (in which HIV hides). The CD4+ T lymphocytes are the primary targets for HIV infection. Because CD4+ T lymphocytes orchestrate all immune functions, HIV's attack of these cells results in progressive impairment of the body's immune response. The CD4+ T lymphocytes do not function normally and are too busy replicating more HIV to perform their own immune function.

After exposure to HIV, the person infected with HIV may develop a mononucleosis-like syndrome (called an acute retroviral syndrome) (Fig. 54–2). The mononucleosis-like symptoms (e.g., fever, rash, joint pain, lymphadenopathy, or malaise) are mild and are usually not attributed to an HIV infection. After the person has been infected with HIV, other immune system components form antibodies to fight the HIV. Detection of these antibodies via laboratory testing in most cases shows that the person was exposed to HIV. HIV antibodies typically become present within 3 weeks to 3 months. When antibodies are present, the person is said to be HIV positive.

LEARNING TIPS

Being HIV positive means that the person has been infected with the HIV virus. It does not mean that the person has AIDS.

Progression

The initial infection is followed by a relatively symptom-free period called the clinical latency stage. The virus remains in the lymph nodes, liver, and spleen and reproduces. A decrease in CD4+ T lymphocytes continues. This period, from infection to the beginning of the symptomatic stage, varies for each person and averages 8 to 12 years. During this stage, the person is considered to be HIV-infected.

During the early symptomatic stage of HIV disease, symptoms of the weakening immune system are seen. When the immune system is severely weakened, opportunistic infections and cancers occur. At this point, the person is diagnosed with AIDS.

LEARNING TIPS

Opportunistic diseases are referred to as opportunistic because a normal immune system would prevent them from occurring. With an impaired immune system, invading organisms have the "opportunity" to survive.

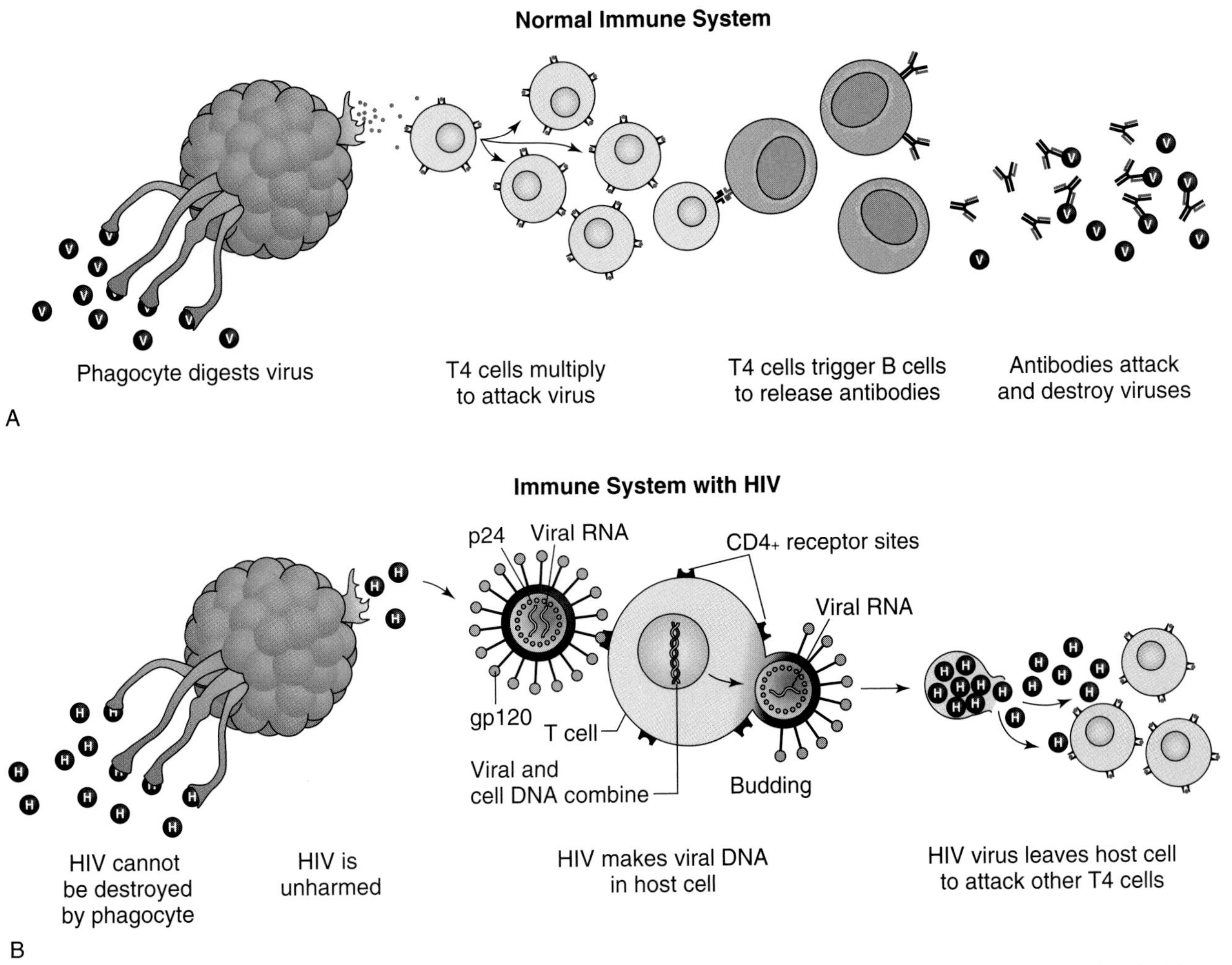

Normal Immune System

Phagocyte digests virus

T4 cells multiply
to attack virus

T4 cells trigger B cells
to release antibodies

Antibodies attack
and destroy viruses

A

Immune System with HIV

p24   Viral RNA

CD4+ receptor sites

Viral RNA

gp120   T cell

Viral and
cell DNA combine

Budding

HIV cannot
be destroyed
by phagocyte

HIV is
unharmed

HIV makes viral DNA
in host cell

HIV virus leaves host cell
to attack other T4 cells

B

Figure 54-1   *(A)* Normal immune system. *(B)* HIV contains several proteins: gp 120 protein around it and viral RNA and p24 protein inside. The gp 120 proteins attach to CD4+ receptors of T lymphocytes; HIV enters the cell and makes viral DNA; the enslaved host cell produces new viruses that bud, which destroys the host cell's membrane, causing cellular death and allowing the virus to leave to attack other CD4+ T-lymphocyte cells.

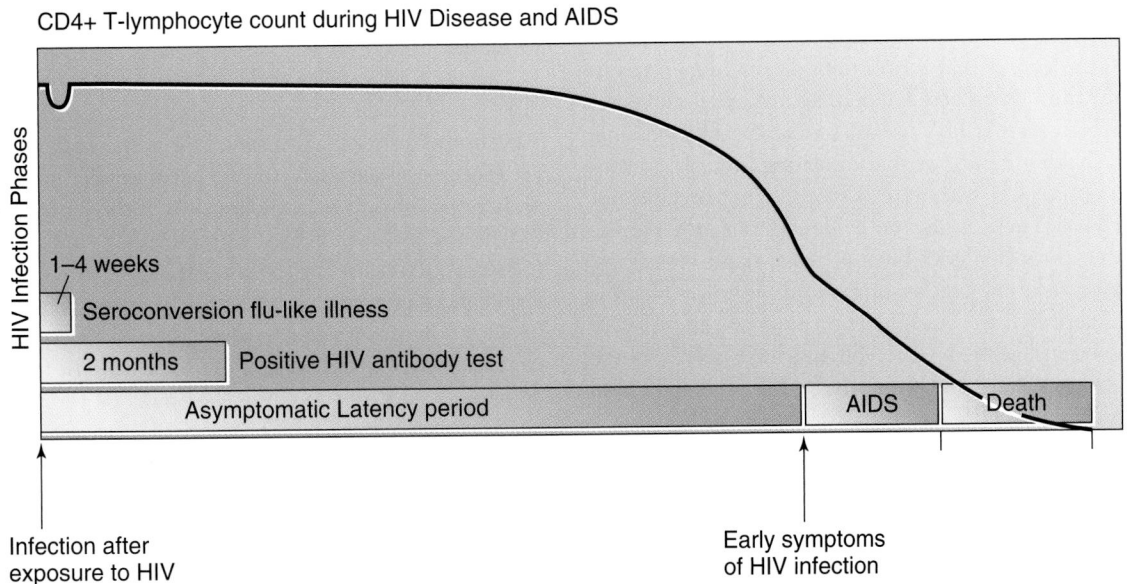

CD4+ T-lymphocyte count during HIV Disease and AIDS

HIV Infection Phases

1–4 weeks

Seroconversion flu-like illness

2 months   Positive HIV antibody test

Asymptomatic Latency period   AIDS   Death

Infection after
exposure to HIV

Early symptoms
of HIV infection

Figure 54-2   Typical phases of HIV infection, AIDS development, and CD4+ T-lymphocyte counts. Length of latency period varies but is usually many years. As CD4+ T-lymphocyte counts drop symptoms, AIDS and then death result.

▶ PREVENTION

Prevention and education are the best ways to manage the HIV/AIDS epidemic. Education regarding the disease and its transmission should begin with the older school-age child and include the general population, as well as older adults (Gerontological Issues Box 54–2).

BOX 54-2   GERONTOLOGICAL ISSUES

Nearly 10 percent of diagnosed AIDS cases occur in people age 50 and older. Numbers will continue to rise as the older population increases, yet little attention has been given to this age group. Lack of attention to this group's needs may be due an ageist view of the elderly and sexual activity.

It is important to understand that HIV infection can occur at any age. Older adults should be asked about their sexual and drug use history and be given preventive education and information about products that reduce transmission of HIV and sexually transmitted diseases. At-risk people older than age 50 are less likely than younger at-risk adults to use condoms during sex or to be tested for HIV. In this age group, condoms are often thought of as a birth control measure.

HIV/AIDS is thought of as being a disease of young, sexually active people. Older people remain sexually active, and previously monogamous people find themselves dating again because they are divorced or widowed and engaging in sexual activity, often without using condoms. Older people are also contracting the virus through homosexual contact.

A decline in the older adult's immune system increases the risk for infection with HIV. Because older adults are not usually taught HIV prevention, the rise in HIV infection among the elderly is expected to continue. With AIDS death rates dropping as a result of more effective treatments, the number of older adults living with HIV will increase.

Symptoms of HIV in older adults may be confused with commonly perceived problems of aging, such as fatigue, decreased endurance, and altered cognitive status. The effect of HIV on the brain is often misdiagnosed as Alzheimer's disease. The misdiagnosis keeps the person from obtaining treatment for HIV/AIDS.

Mode of Transmission

HIV is a fragile virus that is transmitted from human to human only through infected blood, semen, vaginal secretions, or breast milk. HIV is not spread casually. Kissing, hugging, shaking hands, or sharing eating utensils, towels, or bathroom fixtures with an HIV-positive person does not transmit HIV. There is also no evidence that HIV can be transmitted through insect bites or tears, nasal secretions, saliva, sweat, sputum, emesis, urine, or feces unless blood is present. Even then the transmission risk is considered very low. Blood transfusions before 1985 may have been a source for HIV transmission. However, since 1985, donated blood has been tested for HIV.

HIV needs a portal of entry into the body, such as a tear in a mucous membrane or nonintact skin, or access to the bloodstream or lymphatic tissue. Routes of transmission for HIV are as follows:

■ Sexual—anal, vaginal, or oral sexual contact of mucous membranes to infected body fluids

■ Parenteral—injection by needles contaminated with infected blood; sharing needles; transfusions of contaminated blood products

■ Perinatal—infected mother to fetus/infant during pregnancy, childbirth, or breast-feeding

---

LEARNING TIPS

An HIV-infected person can transmit HIV to others within a few days after initial infection and then throughout all stages of the disease. This is thought to be true even when HAART treatment has driven the viral load below detectable levels.

Counseling

Early knowledge of HIV status aids in reducing the spread of HIV infection. The CDC has recommended guidelines for counseling and testing people at risk for HIV. Counseling should be done by trained workers to help the patient make an informed decision about testing. Education is also provided that fits the patient's personal risk situation. Posttest counseling is provided to help the patient understand the test results, how to inform sexual partners and drug needle sharers, risk factor reduction, and care options if needed.

Sexual Transmission

HIV is transmitted more easily to women than men. This is due to the vagina having a greater amount of mucous membrane than the penis, as well as the increased amount of virus found in semen as compared with vaginal secretions. Sexual acts that are the riskiest for transmission of HIV are those that promote contact between infected body fluids and mucous membranes or nonintact skin. These risky acts include oral sex or anal sex. Anal sex is the riskiest type of sexual act, for either gender, because it often results in tearing of the mucous membrane that is then exposed to infected semen.

Safer Sex Practices

Abstaining from sexual intercourse is the only safe way to prevent sexual exposure to HIV. A mutually monogamous sexual relationship is considered safe if the partners are not or will not become infected with HIV. Alternative methods of sexual activities can be used that prevent exposure to blood and secretions, such as massage or masturbation. The benefits of limiting sexual partners should be explained. For vaginal or anal intercourse, the use of male or female condoms and safer sex techniques should be discussed (Box 54–3). Latex gloves protect hands during genital or anal contact. Dental dams (latex sheets) should be used as a barrier between the mouth and genitals or anus.

Parenteral Transmission

The best way to prevent parenteral transmission of HIV is to avoid injection drug use. Drug injection equipment

Patient Education: Condom Use to Prevent HIV Transmission

Condoms should be:

- New for each intercourse act.
- Latex (or polyurethane if allergic to latex), because other materials have large pores that allow HIV to pass.
- Nondamaged, current (not past expiration date).
- Applied before partner is touched. (Tip of condom is held while unrolled over erect penis, allowing room at tip for semen collection.)
- Used with adequate amounts of only water-soluble lubricants. (Petroleum or oil-based lubricants such as petroleum jelly, cooking oil, shortening, or lotions can damage latex condoms.)
- Replaced if broken. (If ejaculation occurs before replacement, immediate use of a spermicide may give some protection.)
- Withdrawn from partner by holding condom against base of erect penis to avoid semen leakage.

should not be shared. If a person who uses injected drugs is unable or unwilling to stop sharing needles, he or she should be carefully taught how to clean the needle and syringe: The needle and syringe are filled and flushed with water. The syringe is then filled with bleach and shaken for 1 minute. The bleach is flushed through the needle and syringe and then they are rinsed with water. Additionally, sexual activity should not occur when judgment is impaired from drug use, because protective measures may not be used.

Autologous (one's own) blood transfusion, when possible, is the safest type of transfusion to prevent HIV infection. Screening of blood for HIV antibodies is done on all donated blood in the United States. There is a very small chance of HIV transmission from blood that is infected but has not yet had time to develop antibodies.

Perinatal Transmission

The U.S. Public Health Service guidelines for HIV screening of pregnant women recommend that voluntary HIV pretest counseling and testing are offered during prenatal care for all pregnant women.[1] Pregnant women who are HIV positive can reduce the risk of perinatal HIV transmission to 2 percent by taking zidovudine (or other antiretroviral therapy) during pregnancy, labor, and delivery. At the time of delivery, pregnant women who have not been tested for HIV or received prophylactic treatment should be offered testing for HIV. After delivery, the infant is given zidovudine for 6 weeks.

Health Care Workers and HIV Prevention

Needle-stick injury is a source of HIV transmission to health care workers. The Needlestick Safety and Prevention Act is designed to aid health care workers in preventing needle-stick injuries. Careful use of needles—including not recapping needles, use of needless systems, and needle safety devices—can reduce needle-stick injury. To prevent HIV after a needle stick or other forms of exposure to patient body fluids, the CDC has developed postexposure prophylaxis guidelines (Table 54–3).[2]

Standard Precautions

All those at risk of contacting blood and body fluids and substances must follow CDC standard precautions to reduce their risk of HIV exposure. (See Chapter 7.) Health care workers should always use these precautions for all patients to protect themselves and other patients. Additionally, to protect patients, HIV-infected health care workers with open lesions or weeping dermatitis should avoid direct patient care.

TABLE 54-3	OCCUPATIONAL HIV EXPOSURE: POSTEXPOSURE PROPHYLAXIS
Consider occupational exposure an urgent medical condition.	
Immediate care	Wash exposure site with soap and water.
	For mucous membrane exposure, flush with water.
Consider exposure risk	What is the type of fluid person was exposed to?
	What is the type of exposure?
Test exposure source	Assess risk of infection and test known sources for HIV antibody
Postexposure prophylaxis	Start treatment within hours of exposure for 4 weeks.
	Consider pregnancy test if of childbearing age.
	Basic prophylaxis regimen or alternate basic prophylaxis regimen available: zidovudine (Retrovir, ZDV, AZT) 300 bid or 200 tid and lamivudine (Epivir, 3TC) 150 mg bid; or as combination tablet of both drugs: Combivir one tablet bid.
	Expanded prophylaxis regimen (basic prophylaxis regimen and one of the following): indinavir (Crixivan, IDV) 800 mg q8h; nelfinavir (Viracept, NFV) 750 mg tid or 1250 mg bid; efavirenz (Sustiva, EFV) 600 mg qh; or abacavir (Ziagen, ABC) 300 mg bid. Trizivir, a combination of ZDV, 3TC, and ABC, may be used.
Follow-up testing and counseling	HIV-antibody testing at baseline, 6 weeks, 3 months, and 6 months after exposure or if a retroviral syndrome type illness occurs.
	During postexposure care, exposed person should use precautions to prevent secondary transmission of HIV, including abstaining from sex or using condoms; no donation of blood, plasma, organs, or semen; if breast-feeding, consider stopping.

Source: CDC: Basic and Expanded HIV Postexposure Prophylaxis Regimens. MMWR 50(RR11):47, 2001.

► SIGNS AND SYMPTOMS

Initially after HIV infection, the patient may have no symptoms or may have mononucleosis-like symptoms, such as extreme fatigue, headache, fever, lymphadenopathy (enlarged lymph nodes in two different sites other than inguinal nodes), diarrhea, or a sore throat. These symptoms generally develop 6 to 12 weeks after HIV exposure and may last a few days to weeks.

Each patient's response to HIV is unique. After an extended asymptomatic phase, HIV infection progresses to a symptomatic stage when the virus has greatly impaired the immune system. The patient may exhibit shortness of breath, fever, weight loss, fatigue, night sweats, persistent diarrhea, oral or vaginal candidiasis ulcers, dry skin, skin lesions, peripheral neuropathy, shingles (varicella zoster virus reactivation), seizures, or dementia. In the final stage of HIV infection, AIDS is diagnosed as opportunistic infections and diseases, with their specific signs and symptoms, occur. (See Box 54–1.)

► COMPLICATIONS

Many complications are seen with HIV/AIDS that vary from patient to patient. Some of the common complications are discussed next.

AIDS Wasting Syndrome

AIDS wasting syndrome occurs in most patients with AIDS. The syndrome is defined by the occurrence of the following:

- Involuntary baseline body weight loss of more than 10 percent
- Chronic weakness or fever for more than 30 days or chronic diarrhea of two loose stools daily for more than 30 days

Several factors contribute to this syndrome: decreased appetite, oral lesions, altered metabolism, malabsorption, gastrointestinal (GI) infections, diarrhea, medication side effects, and cognitive impairment. The progressive weight loss impairs function of all body systems from malnourishment. Careful intervention, planning, and education of the patient when HIV is first diagnosed can help maintain body weight.

AIDS wasting syndrome is challenging. Do not give up! Be creative in developing interventions to help increase the patient's appetite.

Opportunistic Infection and Cancer

Opportunistic infections are a primary complication of HIV/AIDS because the person with HIV/AIDS has a very impaired immune system. Other types of infections that occur even with a healthy immune system are seen as well, such as tuberculosis. Opportunistic infections are not spread to those with healthy immune systems, but other types of infections can be. Cancer incidence also rises when the immune system is impaired.

Candida Albicans

Candida albicans is a fungus normally found in the GI tract that does not infect a person with a healthy immune system. In AIDS, overgrowth of this fungus occurs from the impaired immune system. Candidiasis of the mouth or esophagus occurs often with AIDS. Signs and symptoms of candidiasis include mouth or esophageal pain, dysphagia, and yellow-white plaques that look like cottage cheese in the mouth and throat. Nutrition can be affected by oral or esophageal candidiasis. Vaginal candidiasis can also occur in women with AIDS. Severe itching and a white discharge occur.

Cytomegalovirus

Cytomegalovirus (CMV) is a common viral infection that can affect many areas of the body. The eye is a common site (cytomegalovirus retinitis). Vision impairment ranges from little impairment to total blindness. A variety of symptoms are seen when other areas are involved, such as fever, fatigue, diarrhea, GI upset, and hepatitis, to name a few.

Pneumocystis carinii pneumonia

Pneumocystis carinii pneumonia (PCP) is caused by a fungus (formerly thought to be a protozoa) that produces shortness of breath, fever, nonproductive cough, and fatigue. PCP is the most common opportunistic infection in HIV/AIDS. To prevent PCP, oral trimethoprim-sulfamethoxazole (Septra) is given prophylactically. Pentamidine isethionate (Pentam) inhalation therapy is also used to prevent PCP when CD4+ T-lymphocyte counts fall below $200/\mu L$. PCP may be treated with oxygen, oral or intravenous trimethoprim-sulfamethoxazole, or parenteral pentamidine isethionate. Steroids may also be given to reduce lung inflammation.

Tuberculosis

Tuberculosis may occur in up to 10 percent of those with AIDS. Symptoms include dyspnea, cough, chest pain, fever, night sweats, and weight loss. A positive purified protein derivative (PPD) skin test should be performed at least annually in patients with HIV infection. Induration of 5 mm or more is defined as a positive result in patients with HIV infection. (See Chapter 28.)

When assessing the results of a PPD test, only the raised area (if present) is measured and recorded (in millimeters).

Viral Infections

Herpes simplex virus infections are found in the oral, genital, or rectal area of those with AIDS. Symptoms include blisterlike lesions that rupture and leave ulcerations, fever, pain, or bleeding. Varicella zoster virus infection is usually the result of the virus being present from a prior episode of the chickenpox. The virus remains present in the nerve ganglia. With depressed immune function, the virus can cause shingles.

Kaposi's Sarcoma

Kaposi's sarcoma (KS) is the most common cancer associated with AIDS. Painless, small purple-blue lesions on the skin occur. (See Fig. 52–3.) The lesions may scale, ulcerate, and bleed. KS may occur anywhere on the body in patients with HIV. Treatment includes bleomycin, cryotherapy, or radiation.

AIDS Dementia Complex

HIV infection of the brain or other parts of the central nervous system results in AIDS dementia complex (ADC). Symptoms range from mild to severe and may include memory impairment, personality changes, hallucinations, leg weakness, loss of balance, and slower responses. ADC is a common complication of HIV/AIDS. With ADC, safety is an important consideration for the patient and caregiver.

▶ DIAGNOSIS

Finger-stick blood, oral fluid, and urine and serum specimens are used for HIV testing. Rapid HIV testing provides results the same day. Home sample collection devices can be purchased; the sample is sent in for HIV testing, with the person calling for results, counseling, and referral if needed.

HIV Antibody Tests

After HIV infection, antibodies may not be formed for 3 weeks to 3 months or longer in some cases. The typical HIV antibody testing pattern is as follows:

- Enzyme-linked immunosorbent assay (ELISA) test is done to detect antibodies in the patient's blood to HIV antigen on test plates.
- If positive, the ELISA test is repeated, because false positives may occur (0.1 percent).
- If the ELISA test is again positive, the Western blot test is done to detect the presence of antibodies to four major HIV antigens. The test is positive if two antibodies are present.
- If all test results are positive, the patient is HIV-antibody positive.
- Other tests can be used, especially if initial test results are not conclusive.

Complete Blood Cell Count/ Lymphocyte Count

Because patients with HIV are susceptible to leukopenia, lymphopenia, anemia, and thrombocytopenia related to HIV infection and as a complication of antiretroviral therapy, a complete blood cell count (CBC) including a lymphocyte count should be obtained. The CBC should be repeated at 3- to 4-month intervals or more often if the patient's clinical course is unstable.

CD4+/CD8+ T-Lymphocyte Count

The count of CD4+ and CD8+ (cytotoxic cells) T lymphocytes is essential for evaluating the status of the immune system. In healthy adults, CD4+ levels average 500 to 1600 cells per mm^3. In HIV/AIDS, CD4+ cell levels drop but CD8+ cell levels do not. A low ratio of CD4+ cells to CD8+ cells is seen as HIV/AIDS progresses. It is recommended that CD4+/CD8+ T-lymphocyte counts be performed at 4-month intervals for most patients.

Viral Load Testing

Viral load testing measures the amount of HIV RNA in plasma and is extremely important for determining prognosis and monitoring the response to antiretroviral therapy. Combination antiretroviral regimens usually produce a 50-percent decrease in total-body HIV levels within just a few days. Viral loads should be performed 1 month after initiation of new treatments and at 4-month intervals thereafter.

General Tests

Standard serological testing for syphilis is recommended annually in patients who are sexually active. Hepatitis serology and liver chemistry panels are indicated in the early evaluation of most patients because of the high incidence of concurrent hepatitis for HIV-positive patients.

▶ MEDICAL TREATMENT

Because there is no cure for HIV/AIDS, the goal of therapy is to prevent or delay development of opportunistic diseases. Treatment with antiretroviral drugs is recommended when the CD4+ T-lymphocyte cell count falls below 350 cells per milliliter (Table 54–4). To increase life expectancy and treatment cost effectiveness, it may be recommended that patients be prophylactically treated for opportunistic infections, especially hepatitis A and hepatitis B virus, herpes simplex virus, and *Pneumocystis carinii* pneumonia (Table 54–5). Other opportunistic infections are treated with appropriate medications.

Antiretroviral drugs that inhibit reproduction of the virus (but do not kill it) are used to treat HIV infection. These antiretroviral agents have been developed to act predominately on processes specific to the virus particle to keep the integrity of the host cell. Several potential strategies specifically aimed at interruption of the viral life cycle have been defined, including the following:

- Preventing the virus from attaching to the CD4+ receptor of the T4 lymphocyte
- Interfering with "uncoating" of the virus within the cell, the first essential step in allowing the virus to integrate into the cell's DNA
- Inhibiting reverse transcriptase (RT), a viral enzyme specific to retroviruses, which enables the virus to make a DNA copy from single-stranded viral RNA prior to integration into cellular DNA
- Blocking viral regulatory and transactivating proteins, which are involved in the transcription and translation

TABLE 54–4 Medications for HIV infection

Drug	Adult Dose	Nursing Considerations
Non-Nucleoside Analog Reverse Transcriptase Inhibitors		
delavirdine (Rescriptor)	PO: 400 mg tid	Monitor WBC count, liver tests, especially with history of hepatitis B or C. Do not give within 1 hour of antacids or didanosine (ddl, Videx).
efavirenz (Sustiva)	PO: 600 mg qd	Monitor for rash (especially during first month); may be severe and life threatening. Teach patient to report rash. Stevens-Johnson syndrome (systemic erythema with rash) may occur. Nausea common. Avoid giving with high-fat meal.
nevirapine (Viramune)	PO: 200 mg qd for 2 weeks, then 200 mg bid	Monitor for rash (especially 1st month); may be life threatening and require stopping drug immediately. Teach patient to report rash immediately. Stevens-Johnson syndrome may occur. Monitor liver tests. For contraception, advise a nonhormonal contraceptive during therapy.
Nucleoside Analog Reverse Transcriptase Inhibitors		
abacavir (Ziagen)	PO: 300 mg q12h	Flulike symptoms indicate need to discontinue drug or life-threatening condition may develop.
didanosine (ddl, Videx)	PO: 125–300 mg q12h	Do not give with food. Monitor CBC for bone marrow suppression.
lamivudine) (Epivir, 3TC+	PO: 150 mg q8h	Monitor for peripheral neuropathy—extremity numbness, tingling, burning, pain—and report. Advise patient to avoid fatty foods.
stavudine (d4T, Zerit)	PO: 40 mg q12h	Monitor for peripheral neuropathy—extremity numbness, tingling, burning, pain—report. Monitor liver function tests and for lactic acidosis, which can be fatal.
zalcitabine (ddC, HIVID)	PO: 0.75 mg q8h	Monitor for peripheral neuropathy—extremity numbness, tingling, burning, and pain—and report, so drug may be promptly stopped to prevent irreversible severe pain. Monitor for pancreatitis, which can be fatal: nausea; vomiting; abdominal pain; rising levels of serum amylase, lipase, triglycerides; decreasing calcium. Monitor liver function tests and for lactic acidosis, which can be fatal.
zidovudine (Retrovir)	PO: 100 mg q4h or 200 mg tid or 300 mg q12h	Monitor for dizziness, seizures. Reduces bone marrow function and is toxic to liver and kidneys.
Nucleotide Analog Reverse Transcriptase Inhibitors		
tenofovir disoproxil fumarate (Viread)	PO: 300 mg qd	Monitor for hepatomegaly with steatosis for lactic acidosis, which can be fatal, especially in women. Use cautiously with history of liver disease, obesity, prolonged nucleoside use. Nausea, vomiting, diarrhea common. Take with food. Take 2 hours before or 1 hour after didanosine (ddl, Videx).
Protease Inhibitors		
amprenavir (Agenerase)	PO: 1200 mg q12h	Assess for sulfonamide allergy. Monitor for rash, may be life threatening and require stopping of drug. Teach patient to report rash. Monitor blood glucose. Avoid vitamin E. For contraception, advise a nonhormonal contraceptive during therapy.
indinavir (Crixivan)	PO: 800 mg q8h	Avoid giving with high-fat, high-protein meals. Do not give within 1 hour of didanosine (ddl, Videx). May cause hyperglycemia. Increase fluids because kidney stones may develop.
lopinavir (Kaletra, ABT-378/r)	PO: 400 mg q12h	Assess for sulfonamide allergy. Give with food. Give with 100 mg ritonavir.
nelfinavir (Viracept)	PO: 750 mg q8h or 1250 mg bid	Monitor for seizures. May cause hyperglycemia. Give with food. Diarrhea may occur. For contraception, advise a nonhormonal contraceptive during therapy.
ritonavir (Norvir)	PO: 300 mg bid once, then 400 mg bid 3 days, then 500 mg bid once, then 600 mg bid	Do not confuse with zidovudine (Retrovir). Monitor for seizures, hyperglycemia. Nausea, vomiting, diarrhea common. For contraception, advise a nonhormonal contraceptive during therapy.

WBC = white blood cells.

TABLE 54-4 MEDICATIONS FOR HIV INFECTION—CONT'D

saquinavir (Fortovase, Invirase)	PO: 600 mg q8h	Must be taken within 2 hours of meal for increased blood concentration. Monitor for seizures, hyperglycemia, Stevens-Johnson syndrome. Nausea, vomiting, diarrhea common. Advise sunscreen because photosensitivity occurs.
Ribonucleotide Reductase Inhibitors		
hydroxyurea (Hydrea)	PO: 500–1000 mg daily	Monitor CBC for bone marrow suppression.

TABLE 54-5 TREATMENT FOR AIDS-RELATED CONDITIONS

Opportunistic Infection/Complication	Treatment
Candidiasis	ketoconazole (Nizoral); fluconazole (Diflucan); amphotericin B (Fungizone)
Cytomegalovirus retinitis	ganciclovir (Cytovene)
Hepatitis B virus	hepatitis B virus vaccine when HIV infection diagnosed
Herpes simplex, herpes zoster, varicella zoster	acyclovir (Zovirax) therapy
Influenza	annual influenza vaccine
Pneumococcal pneumonia	pneumococcal vaccine when HIV infection diagnosed
Pneumocystis carinii pneumonia	trimethoprim-sulfamethoxazole (Bactrim, Septra); pentamidine isethionate
Tuberculosis	PPD testing; drug therapy per CDC guidelines: pyrazinamide, isoniazid (Laniazid, Isotamine), ethambutol (Myambutol)
HIV wasting	Patient education: Eat three high-calorie and high-protein meals with snacks daily. Drink liquids before meals. Eat low-residue diet for diarrhea control. Control odors. Develop easy meal plan: favorite foods, meal programs, frozen dinners, cold food to control nausea. Use antiemetics. Rest, listen to music. Numb painful oral sores with ice, Popsicles, or topical analgesic; avoid spicy foods. Use artificial saliva for dry mouth. Use nutritional supplements. Use food stamps or free meal programs as needed. Exercise to increase muscle mass. Take medications prescribed to treat HIV wasting.

of viral RNA proteins from proviral DNA as the virus goes from the quiet, integrated state to active replication

■ Inhibiting protease, a viral enzyme responsible for the adherence of viral proteins both before proviral integration and as the viral particles recombine into functional proteins needed for viral maturation

■ Preventing viral assembly and budding out of the cell

For more information, visit the Medscape quick reference guide to antiretrovirals at www.medscape.com.

Highly Active Antiretroviral Therapy

Early, aggressive treatment with multiple-drug therapy, aimed at reducing the viral load to an undetectable amount, is the most effective line of treatment for HIV. "Cocktails" of multiple antiretroviral drugs (i.e., HAART) have reduced virus loads in the bloodstream and increased CD4+ T-lymphocyte counts, resulting in prolonged survival. HAART, in use since 1996, has made monotherapy (using one drug) an outdated concept, except for some cases involving pregnant women. Several combination drug regimens can be used. A current common cocktail consists of two nucleoside analog and a nonnucleoside analog. The use of HAART is important to help reduce drug resistance, which is a common cause of treatment failure.

Many HIV drugs have side effects. If they occur, the drug regimen may be changed or interventions may be used to help control side effects. The patient should be taught to always report side effects, especially rashes (such as with trimethoprim-sulfamethoxazole) and abdominal pain (such as with zidovudine [AZT]); they could be serious.

Nucleoside Analog Reverse Transcriptase Inhibitors

Nucleoside analog reverse transcriptase inhibitors (NARTI) inhibit production of reverse transcriptase and viral replication.

Non-Nucleoside Reverse Transcriptase Inhibitors

Non-nucleoside reverse transcriptase inhibitors (NNRTI) have a high affinity for the active site of HIV reverse transcriptase and block it. These agents are best employed in combination regimens for patients who have not received prior antiretroviral therapy.

Protease Inhibitors

Protease inhibitors (PI) bind to the active site of the HIV protease enzyme whose function is to cut reproduced HIV strands. Thus PIs interrupt the formation of mature viral particles and reduce viral replication by as much as 99 percent. PIs have rapid resistance development. However, in combination with nucleoside analogs, they can be used as part of a cocktail.

Nucleotide Reverse Transcriptase Inhibitors

Tenofovir disoproxil fumarate (Viread) is the newest drug approved for HIV combination treatment in patients on prior antiretroviral therapy. Tenofovir is a nucleotide reverse transcriptase inhibitor that also blocks reverse transcription. The benefit of tenofovir is its longer action, so it can be taken once a day as a 300-mg dose. Tenofovir should be taken at the same time each day with a full meal.

LEARNING TIPS

Key points to remember:

- HIV and AIDS are disease labels, not people labels.
- Each person reacts to an HIV or AIDS diagnosis differently.
- It is not HIV that ultimately causes death, it is compromised immunity and the invasion of an opportunistic infection or disease that the patient's body is unable to successfully fight off, even with medical intervention.
- With today's current successful antiretroviral treatments, HIV has changed from being a life-ending infection to more of a chronic disease requiring constant management.

► NURSING MANAGEMENT

Assessment

Ongoing assessment is important for the patient with HIV/AIDS to detect problems early. Health history information is obtained (Box 54-4). Determining patient understanding of HIV/AIDS information is useful for planning and teaching. A physical assessment provides data on the effects of HIV/AIDS. Signs and symptoms of opportunistic diseases are noted.

Nursing Diagnosis

Nursing care is individualized to the patient's presenting symptoms. Primary nursing diagnoses for HIV/AIDS include but are not limited to the following:

- Risk for infection related to decreased immune function
- Acute or chronic pain related to neuropathy, cancer, infection, or dyspnea

---

BOX 54-4 **Assessment: Health History Information for HIV/AIDS**

Demographic data: Gender, age, marital status, occupation, residence
Date of diagnosis of HIV/AIDS
Current health status and concerns
Allergies
Medications
Height/weight: weight loss
Blood transfusions/treatment for hemophilia
Infections/cancers (see Box 54-1)
Sexually transmitted diseases
Safer sex practices
Injection drug use

- Fatigue related to HIV infection
- Imbalanced nutrition: less than body requirements related to anorexia, nausea and vomiting, increased caloric need, diarrhea, dysphagia, painful oral lesions
- Impaired oral mucous membranes related to decreased immune function
- Diarrhea related to infection, medications
- Impaired skin integrity related to infection, cancer, immobility, incontinence
- Social isolation related to fear, infection control, transmission of virus
- Risk for situational low self-esteem related to decreased self-esteem and body image changes
- Deficient knowledge related to lack of prior experience with HIV/AIDS and treatment

Additional diagnoses may include the following:

- Anticipatory grieving related to loss of function or death
- Disturbed thought processes related to AIDS dementia
- Disabled family coping related to chronic, progressive disease
- Ineffective coping related to chronic progressive disease
- Ineffective sexuality pattern related to disease transmission
- Risk for injury related to weakness, fatigue, sedation, neurological impairment

See Nursing Care Plan Box 54-5.

Implementation

HIV/AIDS affects every system of the body and every aspect of a person's life. Nurses have the opportunity to positively influence the patient's experience with HIV/AIDS with a nonjudgmental approach, empathy, and psychological support.

Risk for Infection

To reduce infection risk the patient should be taught to avoid large crowds; wash hands frequently, especially before eating; bathe daily; avoid sharing personal grooming items (e.g., toothbrush, toothpaste, deodorant); cleanse

BOX 54-5 | NURSING CARE PLAN FOR THE PATIENT WITH ACQUIRED IMMUNODEFICIENCY SYNDROME (AIDS)*

 Risk for infection, nosocomial related to weakened immune system, skin breakdown, intravenous therapy, and possible invasive procedures

PATIENT OUTCOMES
Patient will remain free of nosocomial infections. Patient will describe measures to maintain skin integrity and avoid infections.

EVALUATION OF OUTCOMES
Is patient free of nosocomial infections? Can patient explain and demonstrate skin maintenance techniques?

INTERVENTIONS	RATIONALE	EVALUATION
Assess patient's risk factors, such as skin condition, laboratory results, portals of entry for infections, and presence of any infections.	Status of these assessment factors determines plan for care.	Does patient have intact skin or nonreddened, nonpurulent sites of interrupted integrity?
Caregivers should use standard precautions and strict aseptic technique for *all* patients and procedures.	Transmission of microorganisms can occur in both directions.	Do all caregivers use standard precautions?
Instruct visitors about techniques to avoid transmission of infection, such as handwashing and not visiting when they have an infection. The nurse with an infection, especially respiratory infection, should not care for patient with AIDS.	The immune system is damaged by HIV. Ability to combat infections is severely compromised. A minor infection for most people may kill a person who has AIDS.	Do those with infections avoid contact with patient until their infection is resolved? Do laboratory tests indicate that patient is so immunocompromised that reverse isolation may be necessary?
Promote skin integrity by frequent turning, optimum mobilization, protective mattress and chair pads, application of emollient to dry areas, and prompt treatment of any injuries.	Skin is the body's first line of defense.	Does patient's skin remain intact and infection free?
Teach strategies for skin care and avoidance of infection to patient.	Self-care offers a measure of control in frequently uncontrollable situation.	Does patient satisfactorily explain or demonstrate good skin care and knowledge of how to avoid infection?

 Risk for injury related to impaired mobility, weakness, fatigue, possible electrolyte imbalances, neurological impairment, and sedative effects of pain medications

PATIENT OUTCOMES
Care and mobility needs will be met without injury.

EVALUATION OF OUTCOMES
Does patient remain free from injury?

INTERVENTIONS	RATIONALE	EVALUATION
Assess patient's abilities and disabilities.	Particular disabilities may increase danger to patient.	Does patient have deficits?
Look for potential hazards in environment (hospital or home) and eliminate as many hazards as possible.	Awareness of hazards is necessary to decrease occurrence of accidents and injuries.	Are there any hazards in the environment presently?
Instruct patient about how to avoid hazards (if cognitively and physically able to comply).	Patients can help avoidance of injury if they understand hazards.	Is patient effectively avoiding hazards, or is patient a danger to self?
Encourage self-care as much as is feasible without tiring patient.	Self-care promotes feelings of self-efficacy and can combat depression.	Does patient evidence satisfaction with self-care efforts?
Assist with care activities as needed.	Varying levels of assistance with care are necessary due to the debilitating nature of disease.	Are patient's care and mobility needs being met satisfactorily without injury?

*Written by Linda Hopper Cook.

BOX 54-5 **NURSING CARE PLAN FOR THE PATIENT WITH ACQUIRED IMMUNODEFICIENCY SYNDROME (AIDS)—CONT'D**

Institute safety measures as required, such as close observation, frequent reorientation, two staff members for ambulation, use of side rails, bed motion alarm, or room near nurse's station.	Protection of patient against inadvertent removal of tubes or equipment, falls, and other injuries may require extraordinary measures due to neurological damage.	Are safety measures effective for patient? Does patient respond negatively to the protective measures? Can these be modified to be less offensive?

 Ineffective individual coping related to terminal disease and progressive debility

PATIENT OUTCOME

Patient will show use of effective coping skills.

EVALUATION OF OUTCOME

Does patient show use of effective coping skills?

INTERVENTIONS	**RATIONALE**	**EVALUATION**
Establish and maintain open and trusting therapeutic relationship. Allow grieving to take place (keeping a journal has been effective for some patients who have AIDS). Encourage patient to express feelings and concerns. Contact counselor, chaplain, or AIDS support worker if patient so desires. Provide patient with desired information or refer to others who can supply information. Ask if patient would like information about support group and arrange such.	Effective communication is based on trust—assurance of confidentiality is essential. AIDS brings about losses of health, strength, employment, and in many cases friends and threatens one's sense of security and reasonableness of life. Healthy grieving is a natural coping response. Talking about feelings and concerns helps defuse anger, clarify needs, and relieve tension. Knowledge dispels unreasonable fears and helps patient prepare adequately to cope with stressors. Social support can help patient cope.	Does patient talk about concerns with nurse? Is grief being expressed? Is finishing things that matter to him or her? How are family members, friends, and support persons interacting with patient? How is patient responding to family members, friends, and support persons? Does patient evidence enough understanding of disease to be able to cope effectively? Does patient show satisfaction with coping resources?

toothbrush; and wash all dishes between uses (Table 54–6). In addition, the patient is taught signs of infection to report to health care provider (Box 54–6). Treatment of opportunistic infections is most effective when begun early.

Impaired Gas Exchange

If a patient develops a condition that interferes with respiration, the goal is to maintain oxygenation within normal limits and reduce dyspnea. PCP is a common respiratory infection with HIV infection. (See Table 54–5.) Monitoring the patient's vital signs, including respiratory rate, depth, and rhythm and oxygen saturation, is important. Medications and oxygen therapy are given as ordered. Positioning the patient for comfort is often done by raising the head of the bed. Assisting with activities helps reduce fatigue.

Pain

Pain may occur from a variety of causes. Treatment is focused on the cause to achieve pain control and relief. Medications may be given as ordered. Complementary therapy may be used by patients. (See Chapter 4.) Measures such as heat or cold, massage, and frequent position changes may be helpful.

BOX 54-6 **Patient Teaching: Signs and Symptoms of Opportunistic Infections to Report**

Instruct patient to monitor temperature daily and to report the following symptoms to a health care provider immediately:
New fever higher than 100° F (38.5° C) or a change in fever pattern if low-grade fevers are commonly present
Cough, shortness of breath, fever, or chest tightness, which may be signs of early pneumonia
Signs of central nervous system infection, such as severe headache; stiff neck; visual changes; problems with balance, walking, or speech; weakness of an arm or leg; and changes in moods or memory
Foul-smelling drainage or pus
Cloudy or foul-smelling urine
Signs of dehydration, such as a dry mouth, dark concentrated urine, or dizziness when standing
Diarrhea lasting longer than 48 hours; more than six stools a day; watery, mucous, or bloody stools
Rashes (possible side effect of medication)
Sore mouth or tongue, difficulty swallowing, white patches on tongue or back of mouth
Worsening fatigue

TABLE 54-6	PATIENT TEACHING: PREVENTING OPPORTUNISTIC INFECTIONS	
Precaution Category	**Reducing Exposure Risk**	**To Protect From**
Environmental/Occupational	Consider risk for exposure to infectious agents from the following: health care settings, correctional facilities, homeless shelters.	Tuberculosis
	Child care settings: Wash hands after diaper changing/body fluid contact.	CMV, cryptosporidiosis, hepatitis A, giardiasis
	Animal contact: Exposure possible from veterinary work, pet stores, farms.	Cryptosporidiosis, toxoplasmosis salmonellosis, campylobacteriosis
	Gardening/soil contact: Avoid gardening/houseplant care/bird-roosting site soil, cleaning chicken coops. Wash hands after soil contact.	Cryptosporidiosis, toxoplasmosis, histoplasmosis, coccidioidomycosis
Food/water	General measures for home or restaurants include the following:	Foodborne and waterborne infections caused by bacterial, viral, protozoal, or parasitic pathogens
	Food handlers should practice good handwashing and hygiene.	
	Discard food past expiration date, dented or swollen cans.	
	Maintain adequate refrigeration and cooking temperatures.	
	Control insects and rodents to prevent food contamination.	
	Foods to avoid:	
	Raw/undercooked eggs and foods with raw eggs, such as hollandaise sauce, Caesar dressing, mayonnaise, uncooked batters, ice cream, eggnog.	
	Raw/undercooked poultry, meat, seafood.	
	Unpasteurized milk/dairy products and fruit juice, raw seed sprouts.	
	Soft cheeses, which may harbor bacteria: feta, Brie, Camembert, blue veined, queso fresco.	
	Wash produce and avoid unwashed produce on salad bars.	
	Avoid cross-contamination of foods with uncooked meat.	
	Cook meat until internal temperature is 180° F for poultry, 165° F for red meats, with no trace of pink.	
	Foods to avoid or cook until steaming hot: leftovers, ready-to-eat, delicatessen foods, refrigerated pates, and meat spreads.	Cryptosporidiosis, giardiasis
	Water safety:	
	Avoid public drinking fountains.	
	Avoid drinking or swallowing water directly from lakes or rivers.	
	Use safe water supply or boil water 1 minute when unsure.	
	Use safe, solid-block activated carbon home water filter (absolute 1-μm filter; reverse osmosis; certified for cyst removal); have someone change filter or wear gloves when doing so, because organisms collect on outside of filter.	
	Avoid beverages made from tap water in public places.	
	Drink bottled water purified from reverse osmosis, filtration through absolute 1-μm filter, or distillation (only safe methods). For information, contact www.bottledwater.org.	
	Bottled or canned carbonated soft drinks, commercially packaged nonrefrigerated beverages, pasteurized beverages, and beers are safe.	
Sexual	Always use latex condom for every sex act.	STDs, herpes simplex virus, cytomegalovirus, human papillomavirus, resistant HIV strain
	Avoid oral-anal contact or use dental dams; use latex gloves for hand-anal contact; wash hands and genitals with warm soapy water after contact.	Intestinal infections: amebiasis, hepatitis A, cryptosporidiosis, shigellosis, campylobacteriosis, giardiasis.
	Get hepatitis A vaccine.	
	Get hepatitis B vaccine	Hepatitis B
Injection drug use	Get hepatitis A and B vaccines.	Hepatitis A, hepatitis B, hepatitis C, resistant HIV strain
	Stop using injection drugs and enter substance abuse treatment.	
	If unable to stop, never reuse or share syringes, needles, water, or drug preparation equipment.	
	If shared, use bleach and water to clean equipment.	
	Use sterile syringes from pharmacies or community syringe exchange programs and dispose of safely.	
	Use clean water and equipment and new alcohol swab.	

TABLE 54-6 PATIENT TEACHING: PREVENTING OPPORTUNISTIC INFECTIONS—CONT'D		
Precaution Category	**Reducing Exposure Risk**	**To Protect From**
Pet-related Counsel on pet contact risks but recognize emotional benefits of pets.	Avoid pet feces/diarrhea; seek veterinary treatment for pet's illness. For new pet, avoid those younger than 6 months old (and cats younger than 1 year old); obtain pets from known sanitary source; avoid strays; wash hands after handling pets. Avoid exotic pets. Cat ownership increases risk from litter box cleaning, scratches, bites, licking, fleas. Unhealthy birds may transmit infectious organisms. Avoid reptiles, turtles, chicks, and ducklings. Wear gloves for cleaning aquariums.	*Cryptosporidium, Salmonella, Campylobacter* spp. infection Toxoplasmosis, *Bartonella* spp. infection, salmonellosis, campylobacteriosis *Cryptococcus neoformans, Mycobacterium avium, Histoplasma capsulatum* infection Salmonellosis *Mycobacterium marinum* infection
Travel	Consult health care providers on travel to developing countries. Traveler's diarrhea prophylaxis is not recommended. Carry supply of antimicrobial agent to take for diarrhea. Consider prophylaxis for other types of exposures. Avoid raw fruits, vegetables, raw/ undercooked seafood or meat, tap water, ice from tap water, unpasteurized milk/dairy products, items from street vendors. Safe items include steaming-hot foods, self-peeled fruits, bottled (especially carbonated) beverages, hot coffee/tea, beer, wine, and water boiled 1 minute. Avoid soil/sand contact by wearing shoes, using beach towels.	Opportunistic pathogens, foodborne and waterborne infections

Source: Data from U.S. Department of Health and Human Services: 2001 USPHS/IDSA Guidelines for the Prevention of Opportunistic Infections in Persons Infected with Human Immunodeficiency Virus. http://hivatis.org, November 28, 2001.
STD = sexually transmitted disease.

Fatigue

Many patients with HIV experience fatigue. Other causes of fatigue include infections, medications, anemia, dehydration, depression, and poor nutrition. The patient can help manage fatigue by alternating periods of activity and rest. Tasks that use more energy should be planned at times when the patient is most energetic.

Imbalanced Nutrition: Less than Body Requirements

Maintaining general health and nutrition is important for a healthy immune system. For patients with HIV, maintaining health is vital. Patients with AIDS often have difficulty maintaining adequate nutrition and preventing weight loss. Many factors interfere with nutrition in HIV/AIDS (e.g., anorexia, oral lesions, nausea and vomiting, diarrhea, or wasting syndrome). The cause needs to be identified for appropriate intervention planning.

The patient's baseline weight is obtained. Then ongoing monitoring of the patient's weight, calorie intake, and intake and output is done. A dietitian is consulted to help plan nutritious and affordable meals. The patient should be on a low-microbial (reduces infection), high-calorie, high-protein diet. Small, frequent meals may be helpful. Vitamins, nutritional supplements, tube feedings, or parenteral nutrition (partial parenteral nutrition [PPN] or total parenteral nutrition [TPN]) may be needed to maintain adequate nutrition. Antiemetics are used to control nausea and

vomiting. Creative interventions and resources may be needed to ensure the patient receives adequate nutrition. (See Table 54-5.)

Along with adequate nutrition, exercise helps maintain muscle mass, promotes relaxation, aids sleep, and gives an individual a sense of control and well-being. An effective exercise program includes exercises that increase strength, flexibility, and endurance. Becoming physically fit is a lifestyle change that requires dedication.

Impaired Oral Mucous Membranes

Oral or esophageal candidiasis is common. The painful lesions interfere with swallowing and nutrition. In patients with AIDS who smoke there is an increased incidence of oral thrush (candidiasis). Therefore patients should be encouraged to quit smoking. Antifungal medication is given, such as ketoconazole (Nizoral) or fluconazole (Diflucan). Mouth care is very important. A soft toothbrush promotes comfort. Viscous lidocaine can be given to decrease pain during eating.

Diarrhea

Diarrhea often occurs in patient with AIDS. For diarrhea caused by opportunistic causes (except *Salmonella*), an antimotility agent (diphenoxylate/atropine [Lomotil]) may be prescribed. Consulting the dietitian may be helpful in making diet changes that reduce diarrhea (e.g., low-residue diet, no dairy products, no spicy foods, no caffeine or alcohol). Sitz baths may be soothing. Thorough cleansing of the rectal area after each stool is a must. Ointments may be applied to protect and soothe the anal area from excoriation.

Impaired Skin Integrity

Skin infections or malignancies can occur. Medications can cause skin infections that can be life threatening and must be reported immediately. Kaposi's sarcoma is the most common skin condition (discussed earlier). To help maintain a positive body image, patients can cover the lesions by using hats, clothing, or makeup (with closed lesions).

Social Isolation

Many patients with HIV/AIDS face discrimination, rejection, and isolation. The Americans with Disabilities Act (ADA) makes discrimination toward patients with AIDS illegal. Relatives, friends, and others sometimes avoid or refuse to have anything to do with the person with HIV (Ethical Consideration Box 54–7). Misunderstanding and fear lead to misuse of infection control procedures and increase a patient's isolation. Being knowledgeable about the transmission of HIV allows interaction with the patient to reduce feelings of isolation. Providing patient education to reduce fear of HIV transmission also decreases isolation.

BOX 54-7 ETHICAL CONSIDERATION

Caring for a Patient with AIDS

Edward, 28, is hospitalized in a large research hospital in California. He is in the terminal stages of AIDS, with widespread metastatic cancer and lung infections. He wants to see his parents before he dies. However, when he left his middle-class home on a ranch in Oklahoma in the middle of the Bible Belt to attend college, his parents had no knowledge of his homosexuality. They have not seen him since that time and still have no knowledge that he is homosexual.

Edward does not want his parents to know about his homosexuality or that he has AIDS. He feels that their rigid religious beliefs may make them disown him and cause a scene in the hospital. Before they come to visit him at the hospital, he asks the licensed practical nurse/licensed vocational nurse (LPN/LVN) who is assigned to care for him to answer any of his parents' questions about his illness by telling them he has leukemia, pneumonia, or another rare disease that they would know nothing about. After the third day of visiting their severely ill son, Edward's parents directly ask the LPN/LVN, "Does our son have AIDS?"

How should the LPN/LVN respond? To whom does the LPN/LVN owe the greatest obligation, the patient or his family? What principles are in conflict? What virtues play a role in this situation? Use the decision-making model and determine several possible solutions for this situation.

Risk for Situational Low Self-esteem

Changes in self-esteem and self-concept occur from several of the effects of HIV infection. Patients often experience changes in their relationships with others and in day-to-day activities such as work. Dramatic changes in appearance may occur that alter body image and reduce self-esteem.

Nurses can assist patients in maintaining self-esteem and self-concept by ensuring a climate of acceptance and promoting a trusting relationship. Patients should be encouraged to express feelings if ready and to identify positive aspects of self. Emotional and spiritual support of the patient can help improve self-esteem.

Stress impacts health. Having others to talk with and provide support is essential to stress management. Identifying and maintaining social networks is important. Patients should be taught a variety of relaxation strategies and techniques to reduce stress. Relaxation strategies can range from working on a favorite hobby to talking with friends. Relaxation techniques include progressive muscle relaxation and imagery to aid in relaxation. Stress management techniques are most effective when used every day, not just during times of stress.

Deficient Knowledge

Extensive teaching is needed for patients to understand this chronic, life-threatening disease that alters everything in their lives. (See also the Prevention section.) Adherence to treatment and medication regimens is necessary to prevent infection and prolong the time before AIDS occurs.

MEDICATIONS. It is vital that medication doses are not missed; this must be stressed during patient teaching. Tell the patient to take medications exactly as instructed. If a dose is missed, it should be taken as soon as possible unless it is very close to the time of the next dose. Doses should not be doubled. Missing doses of medication could cause therapy failure because viral loads can increase. Tell the patient to consult the physician with any questions or side effects promptly. Many of these medications can cause severe reactions.

FOOD AND WATER SAFETY. Food and water safety is vital to an immunocompromised patient. Bacterial, viral, protozoal, or parasitic pathogens can cause foodborne and waterborne infections. The patient must be taught methods to prevent foodborne and waterborne infections for all phases of handling. (See Table 54–6.) Kitchen counters and food preparation appliances (e.g., cutting boards, can openers) should be disinfected with a bleach solution. Freezing does not kill bacteria in foods. Foods should not be thawed at room temperature. Dating and using the oldest foods first is helpful. If symptoms of foodborne or waterborne infection develop (e.g., diarrhea, nausea, vomiting, abdominal cramps, headache, fever), teach patient to report them immediately

RESOURCES. Financial resources may need to be addressed so that food and medications can be obtained. Treatment can be expensive, and the patient be unable to work. However, with the new combination antiretroviral therapies, individuals are now able to continue working longer. Referrals to financial resources and support groups can be helpful.

Community and Home Health Care

During the course of HIV infection, the patient will be primarily at home. Hospitalization may be needed intermittently for acute illness. As the disease progresses, the patient may require more care from caregivers and home health nurses. The home health nurse provides physical care, establishes a therapeutic relationship with patients with AIDS and their significant others, and coordinates care with other health care team members (Home Health Hints). Caregivers should be assessed for caregiver role strain. Support services should be identified, such as community AIDS organizations, Meals on Wheels, respite care services, community mental health, and Internet support groups. Respite care provides the caregiver time away from the caregiver role to reduce stress. When a patient is terminal, comfort care and emotional support for the family are essential. Hospice care may be used at this time.

Evaluation

Patient goals are met if the patient remains free from infection and maintains desired quality of life and activities as long as possible. As the disease progresses, goals are met if the patient's needs are met and the patient's dignity is maintained.

CRITICAL THINKING

Zoe Sampson

Zoe, 22, is diagnosed as HIV positive. She is tearful and asks many questions.
1. How would you answer her questions?
   a. "Am I going to die?"
   b. "How is AIDS diagnosed?"
   c. "Can my boyfriend get it?"
2. What food and water safety methods would you teach her?
3. Years later Zoe loses weight and becomes malnourished. What interventions can you use to promote adequate nutrition?

Answers at end of chapter.

HOME HEALTH HINTS

- When providing patient care, wash hands frequently, follow standard precautions, and use plastic bags to contain soiled items or clothing.
- Teach patients with AIDS and their families how to properly clean and disinfect the home to prevent infections. The recommended disinfectant is a diluted bleach mixture that contains 1 part household bleach to 10 parts water.
- Wearing gloves, clean body fluid spills with soap and water.
- Flush body fluids, solid waste, and contaminated solutions down toilet.
- Use disinfectant to (a) disinfect spill areas; (b) clean toilet seats and bathroom fixtures; (c) clean inside the refrigerator to avoid mold growth.
- Rinse clothing and then wash separately from other clothes with 1 cup of bleach if soiled with blood, urine, feces, or semen.
- Patients with AIDS do not require separate sets of dishes or silverware. Dishes and silverware are washed in hot, soapy water and rinsed thoroughly or placed in dishwasher.
- Dispose of sharps (e.g., needles, razors) in a rigid labeled container (such as a tin can with a sealable lid). Add 1:10 bleach solution to disinfect. Tape the lid. Place in a bag and dispose of in the trash.
- Contaminated articles are disposed of by bagging in plastic and placing in the trash.
- Teach the family of a patient with AIDS which signs and symptoms to report to the physician or nurse immediately: fever; increased dyspnea; pain; change in sputum production; upper respiratory tract infection; pneumonia; respiratory distress syndrome; diarrhea five times a day or more for 5 days; uncontrolled weight loss greater than 10 pounds in the last month; persistent headaches; falling; seizures; mental status changes, including memory loss and personality changes; rashes and skin changes; difficulty swallowing; and problems with urination.

REVIEW QUESTIONS

1. The nurse is caring for a patient who is newly diagnosed with HIV infection. The patient asks what his future health status will be. The best response for the nurse to give is based on the understanding that HIV disease and AIDS are now characterized as which of the following?
   a. An acute disease
   b. A life-ending disease
   c. A chronically managed disease
   d. A disease with remissions and exacerbations

2. The nurse would evaluate the patient as understanding modes of HIV transmission if the patient stated that the modes of HIV transmission include which of the following?
   a. Saliva, tears, fecal-oral contamination

   b. Close physical contact involving skin surfaces, mosquito bites
   c. Sharing towels, sharing eating utensils, skin contact
   d. Unprotected sex with HIV infected partner, contact with infected blood/products

3. The nurse would recognize the patient needs further reinforcement of knowledge if the patient stated that one of the goals of highly active antiretroviral therapy is which of the following?
   a. Reduce the viral load
   b. Improve survival rates
   c. Decrease CD4+ T lymphocytes
   d. Delay the progression of HIV disease

4. The nurse would recognize that the patient is having a reaction to zidovudine (AZT) if which of the following occurred?
    a. Rash
    b. Edema
    c. Abdominal pain
    d. Blurred vision

5. The nurse is to give lamivudine (3TC) 150 mg PO now and has available a 10 mg/mL solution. How many milliliters should the nurse give?
    a. 0.5 mL
    b. 1 mL
    c. 10 mL
    d. 15 mL

REFERENCES

1. Centers for Disease Control and Prevention: Revised U.S. Public Health Service Recommendations for HIV Screening of Pregnant Women. MMWR, 50(RR19), 2001.

2. Centers for Disease Control and Prevention: Basic and Expanded HIV Postexposure Prophylaxis Regimens. MMWR 50(RR11), 2001.

Answers to CRITICAL THINKING

Zoe Sampson

1. a. There currently is no cure for HIV/AIDS; however, medications are available that slow the disease's progression and make HIV a manageable chronic illness. Research continues on finding improved treatments and a cure.
    b. AIDS is diagnosed when CD4+ T-lymphocyte counts are below 200 cells per microliter or CD4+ T-lymphocyte percentage is under 14 of total lymphocytes and an opportunistic clinical disease, as defined by the CDC, is present in an HIV-infected person.
    c. Yes, your boyfriend could become infected through exposure to your blood or vaginal secretions. You need to learn about preventive measures and discuss them with him.

2. Food handlers must maintain good handwashing and hygiene practices. Discard food past the expiration date and dented or swollen cans. Ensure adequate refrigeration and cooking. Control insects and rodents to prevent food contamination Avoid public drinking fountains. Drink purified bottled water. Use a safe water supply or boil water 1 minute when unsure. Avoid unpasteurized milk, other dairy products, and fruit juice and raw seed sprouts. Avoid raw and undercooked eggs, meats, and seafood.

3. Eat three high-calorie, high-protein meals and snacks daily. Drink liquids before meals. Eat a low-residue diet for diarrhea control. Develop easy meal plan use antiemetics. Numb painful oral sores. Avoid spicy foods. Refer for food stamps/free meal programs if necessary. Engage in regular exercise.

55

NURSING CARE OF PATIENTS WITH MENTAL HEALTH DISORDERS

Kathy Neeb

KEY TERMS

abuse (uh-**BYOOS**)

adaptation (ad-dap-**TAY**-shun)

addiction (uh-**DIK**-shun)

affect (**AF**-feckt)

anxiety (ang-**ZIGH**-uh-tee)

behavior management
(be-**HAYV**-yer **MAN**-ij-ment)

biofeedback (BYE-oh-**FEED**-bak)

bipolar (bye-**POH**-ler)

codependence (KO-de-**PEN**-dense)

cognitive (**KAHG**-ni-tiv)

compulsion (kum-**PUHL**-shun)

conversion disorder
(kon-VER-zhun dis-**OR**-der)

coping (**KOH**-ping)

delirium tremens
(dee-LIR-ee-uhm **TREE**-menz)

delusions (dee-**LOO**-zhuns)

dependence (di-**PEN**-dens)

displacement (dis-**PLAYSS**-ment)

dysfunctional (dis-**FUNCK**-shun-uhl)

electroconvulsive therapy
(ee-**LEK**-troh kun-**VUL**-siv **THER**-uh-pee)

eustress (**YOO**-stress)

hallucinations (huh-LOO-si-**NAY**-shuns)

illusions (i-**LOO**-zhuns)

imagery (**IM**-ij-ree)

mania (**MAY**-nee-ah)

mental health (**MEN**-tuhl HELLTH)

mental illness (**MEN**-tuhl **ILL**-ness)

milieu (me-**LYU**)

obsession (ub-**SESH**-un)

orientation (OR-ee-en-**TAY**-shun)

paranoia (PAR-uh-**NOY**-uh)

phobia (**FOH**-bee-ah)

psychoanalysis (SIGH-koh-uh-**NAL**-i-sis)

psychopharmacology
(SIGH-koh-FAR-meh-**KAHL**-uh-jee)

psychosomatic (SIGH-koh-soh-**MAT**-ik)

psychotherapy (SIGH-koh-**THER**-uh-pee)

somatoform (soh-**MAT**-uh-form)

stress (STRESS)

stressor (**STRESS**-er)

tolerance (**TALL**-ler-ens)

withdrawal (with-**DRAW**-ul)

QUESTIONS TO GUIDE YOUR READING

1. How do you define mental health and mental illness?

2. What are the components of a mental health status assessment?

3. What is the DSM-IV, and what are other methods for diagnosing mental illness?

4. How do you identify common defense mechanisms?

5. What is a therapeutic milieu?

6. How do you explain classifications, uses, actions, side effects, and nursing considerations for selected classifications of psychoactive medications?

7. What are the methods for psychoanalysis, behavior management, rational-emotive therapy, humanistic/person-centered therapy, counseling, electroconvulsive therapy, group therapy, and relaxation therapy?

8. What is the LPN/LVN's role in therapy?

9. What are the main symptoms, treatment modalities, and nursing care for anxiety disorders, mood disorders, somatoform disorders, schizophrenia, substance abuse disorders, and codependency?

▶ NORMAL FUNCTION

The study of **mental illness** is fairly new to the health care field. There are differing opinions within the mental health community as to what **mental health** and mental illness are. Mental health has been defined in many ways. These definitions include the ability to do the following:

- Be flexible
- Be successful
- Form close relationships
- Make appropriate judgments
- Solve problems
- Cope with daily **stress**
- Have a positive sense of self

Mental illness is defined as experiencing the following:

- Impaired ability to think
- Impaired ability to feel
- Impaired ability to make sound judgments
- Difficulty or inability to cope with reality
- Difficulty or inability to form strong personal relationships.

It is important to remember that mental health and mental illness exist on a continuum. It is natural for emotions to ebb and flow from day to day in response to the degree of stress that is experienced. People who remain mentally healthy are able to keep their stress in perspective. Others are not able to do so, and over time they may develop physical or emotional illnesses as a result of the constant stress in their life.

Visualize the seesaw that children play on. Mental health and mental illness are like a seesaw. When children of approximately equal weight get on each end of the seesaw, they can balance each other and keep the seesaw even. Mentally healthy people keep themselves in a state of emotional balance. Sometimes one child weighs just a little more than the other, and the seesaw tips just a little to one side. Mentally healthy people can cope with this fluctuation. Sometimes another child gets on or one child greatly outweighs the other, and the seesaw gets out of balance completely; one end goes way up while the other goes way down, and it stays there until someone alters the balance. Ultimately it must be the patient who finds his or her own balance. When people's moods are way down or way up, they are not in emotional homeostasis.

The discussion surrounding the etiology (causes) of mental illness continues to revolve around the "nature versus nurture" or "organic versus inorganic or functional" arguments. The connections between physical and emotional health are so closely intertwined that it is sometimes hard to decide if emotional causes trigger physical responses or vice versa. Nurses must have a basic understanding of both the nature and nurture schools of thought on the causes of alterations in mental health.

Explanations of mental illness include concepts from the psychoanalytic (or psychological) theory and the psychobiologic (or biologic) theory. When pertinent, other theories (e.g., behavioral, environmental) are also presented. Most mental illnesses have no absolute cause. Some etiologic theories have stronger positive correlations to illnesses than others. When it is appropriate, this chapter gives the most current or most widely accepted view of an etiology.

Many professionals in the field of psychology believe that social and cultural environments have a great influence on the way people develop and process life experiences. Some psychoanalysts believe that some cultural traditions and beliefs cause disturbances in personal relationships, which can lead to forms of emotional disturbances. It is part of the nurse's role to take time to learn about the traits that are common among people and those traits that are different. It is important to have an understanding of people's customs and beliefs to avoid unrealistic expectations of patients. See Cultural Consideration Box 55–1 for more detailed information about culture and mental health issues.

Spirituality and religion are extremely important to some patients and unimportant to others. A person's success in recuperating from physical or emotional illness may be deeply tied to his or her spirituality. It is necessary to be comfortable talking to the patient about spiritual needs while being careful not to impose personal values on the patient. If you are not comfortable in these situations, you should offer to call the chaplain in the facility or the spiritual leader of the patient's choice. It is important to keep the lines of communication open. People learn by sharing with each other. It is much better to ask a person about something than to make an assumption about it. For more information about mental health and illness, visit the National Institute of Mental Health at www.nimh.nih.gov or the American Psychiatric Association at www.psych.org. For a list of other Websites, visit http://mentalhealth.about.com.

▶ NURSING ASSESSMENT

During the data collection/assessment part of the nursing process, the mental status examination is performed. This is a series of questions and activities that evaluate eight areas: appearance and behavior, level of awareness and reality **orientation,** thinking/content of thought, memory, speech and ability to communicate, mood and **affect,** judgment, and perception. There are a number of different tools of varying names, lengths, and formats used to assess the patient's mental capabilities. See Table 55–1 for a sample mental status examination.

After data have been collected, the licensed practical nurse/licensed vocational nurse (LPN/LVN) collaborates with the registered nurse (RN) to develop nursing diagnoses. Box 55–2 lists nursing diagnoses commonly used when caring for patients experiencing mental health problems.

▶ DIAGNOSTIC TESTS

The diagnostic tool that is used most widely by psychiatrists and psychologists is the *Diagnostic and Statistical Manual of Mental Disorders IV,* or DSM-IV. The DSM-IV uses a system for grouping illnesses into categories of clinical disorders. This is a complex diagnostic tool. Although the LPN is not responsible to complete the assessment, he or she may contribute valuable information.

BOX 55-1 CULTURAL CONSIDERATION

Among many Haitians, some African-Americans, and some other groups, conjure (practicing magic) and root doctors are believed to know more about mental illness than Western-educated physicians. Some depressive and obsessive behaviors are viewed as culture-bound syndromes. These behaviors are expected of some Haitians, and those affected fulfill some expected roles in the society. Some of these illnesses are viewed as having no cure. Thus the nurse may need to include folk healers when working with African-Americans and patients of Haitian descent.

Smoking, alcoholism, and deaths from suicide or violence are prevalent problems in the American culture. Violent deaths account for high mortality rates among adolescents and young adults of African-American, Cuban, Mexican-American, and Puerto Rican origin. Programs targeting these populations should be personalized to include adolescents and their families. Given their strong family values, an important approach is to begin early in church groups or family settings.

Hispanics require lower doses of antidepressants and experience greater side effects than Caucasians. Thus careful observations need to be undertaken when observing for side effects of medications in patients of Mexican heritage.

One of the few drugs found to have a higher rate of side effects in individuals of Ashkenazi ancestry is clozopine, used to treat schizophrenia. Twenty percent of Jewish patients taking this drug developed agranulocytosis compared with about 1 percent of non-Jewish patients. Thus the nurse may need to suggest testing for agranulocytosis when Jewish patients are prescribed clozopine.

Asian patients require lower dosages and have side effects at lower dosages than Caucasians for a variety of psychoactive drugs (e.g., lithium, haloperidol) even when matched with body weight. Asians commonly believe that Western drugs are too strong for them and take less than prescribed. Most Asians are more sensitive to alcohol, resulting in facial flushing, palpitations, and tachycardia.

Many Vietnamese people believe that mental illness results from offending a deity and that such a condition brings disgrace to the family and must be concealed. A shaman may be enlisted to help. additional therapy is sought only with the greatest discretion and often after a dangerous delay. Unmistakable emotional disturbance is usually attributed to possession by malicious spirits, the bad luck of familial inheritance, or, for Buddhists, bad karma accumulated by misdeeds in past lives. Additionally, the term *psychiatrist* has no direct translation in Vietnamese and may be interpreted to mean nerve physician or specialist who treats crazy people. The nervous system sometimes is seen as the source of mental problems, neurosis being thought of as "weakness of the nerves" and psychosis as "turmoil of the nerves."

Poles and other groups from Eastern Europe usually look for a physical basis of disease before considering a mental disorder. If mental health problems exist, home visits are preferred to clinic visits and talk therapies without suitable psychosocial strategies are not maintained unless interventions are action oriented. In addition, Poles look to other family members and the community to assess appropriateness of treatments. Polish-Americans often seek self-help groups such as Alcoholics Anonymous before seeing a health care provider. A family physician is preferred over a specialist.

Mental illness among the Navajo is perceived as resulting from placing a curse on an individual. In these instances a healer who deals with dreams or a crystal gazer is consulted. Individuals may wear turquoise to ward off evil; however, an individual who wears too much turquoise is sometimes thought to be an evil person and someone to avoid. In some tribes, mental illness may mean that the affected person has special powers. Additionally, many Native Americans metabolize alcohol at a faster rate than European-Americans, resulting in them having a lowered tolerance for alcohol.

Mental illness in the Korean culture is viewed as a stigma. Hwabyung, a traditional Korean illness, occurs from the suppression of sadness, depression, worry, anger, fright, and fear. These emotions are commonly related to conflicts with close relatives or significant others. Symptoms include physical complaints such as headaches, poor appetite, insomnia, and lack of energy. Most people accept the symptoms as inevitable, although currently programs are underway to change the stigma.

Newer immigrants from Ireland have a higher incidence of mental illness than the rest of the population. Undocumented Irish immigrants in the United States report more stressors and mental health problems than their legal counterparts. Because many Irish people have difficulty expressing emotions, health care providers may need to encourage Irish-Americans to express their concerns.

Mental illness is strongly stigmatized among Iranians and is thought to be genetic. Should a family member have mental illness, it is likely to be called a "neurological disorder" so as not to stigmatize the family, which may result in daughters having a lesser chance of marrying. Psychotherapeutic help may be avoided because of stigma or because it is perceived as irrelevant. People prefer a medicine that might cure them. There is a tendency to pay more attention to somatic symptoms when under emotional stress; Iranians consider psychopharmacology most effective, and have a high rate of compliance.

Because of the stigma attached to seeking professional psychiatric help, many Hindus do not access the health care system. Instead, family and friends seem to be the best help, and a general belief is that time is the best healer. Physical and mental illnesses are considered God's will, or karma, and are associated with a fatalistic attitude.

Among Greeks, mental illness is accompanied by social stigma for the afflicted individual, as well as the relatives. The shame originates in the notion that mental illness is hereditary, and afflicted individuals are viewed as having lifelong conditions that pollute the bloodline.

Because of social stigma attached to mental illness and retardation, Arabs may keep their family from public view. However, when Arab patients suffer from mental distress, they seek medical care. They are likely to have complaints such as abdominal pain, lassitude, anorexia, or shortness of breath. Arabs have an increased tendency to experience elevated blood levels and adverse affects when customary dosages of antidepressants are prescribed. Patients often expect and insist on somatic treatment, at least vitamins and tonics.

African-Americans may be at a greater risk for being misdiagnosed with a psychiatric disorder than Caucasians. African-Americans with psychiatric disorders are more likely to have hallucinations, delusions, somatization, and hostility, even when controlled for socioeconomic class. Maintaining direct eye contact with some African-Americans may be misinterpreted as aggressive behavior. Thus nurses must take these nonverbal behaviors into consideration when working with the patient with an emotional or mental concern. Additionally, African-Americans are more susceptible to tricyclic antidepressants and thus experience more toxic side effects.[1]

TABLE 55-1	SAMPLE MENTAL STATUS EXAMINATION		
Area of Assessment	**Type of Assessment**	**Normal Parameters**	**Alterations from Normal**
Appearance and behavior	Objective and subjective observations about dress, hygiene, posture, and appearance and about the patient's actions and reactions to health care personnel.	Clean, combed hair. Clothing intact and appropriate to weather or situation. Teeth/dentures in good repair. Posture erect. Cooperates with health care personnel.	Displays either unusual apathy or concern about appearance. Displays uncooperative, hostile, or suspicious behaviors toward health care personnel.
Level of awareness and orientation	Subjective and objective assessment of the patient's degree of alertness (wakefulness) and the degree of the patient's knowledge of self.	Awareness is measured on a continuum that ranges from unconsciousness to mania. "Normal alertness" is the desired behavior. Facilities may provide a standard guideline for helping with this assessment, but subjective observations can be documented as well if the patient is not able to stay awake for even short intervals or if the patient is overly active and has difficulty staying in one place.	Outcome is not considered within accepted normal limits if the patient is difficult to arouse and keep awake or if the patient has difficulty feeling calm. Abnormal results of orientation are the patient's inability to correctly answer questions pertaining to the self or to commonly known social information, such as who is the president.
Thinking/content of thought	Subjective assessment of what the patient is thinking and the process the patient uses in his or her thinking.	Formal testing may be done by the psychologist or psychiatrist to determine the patient's general thought content and pattern. Nurses may contribute to the assessment of thought by documenting statements the patient makes regarding daily care and routines.	Behaviors including flight of ideas, loose associations, phobias, delusions, and obsessions may become apparent.
Memory	Subjective assessment of the mind's ability to recall recent and remote (long-term) information.	Recent memory: Recall of events that are immediately past or within 2 weeks before the assessment, such as a recent news event. One measurement technique is to verbally list five items. After 1 minute, patient can recall four or five of those items. Continue with assessment, and at 5 minutes patient should be able to recall three or four of the items. Remote memory: Recall of events of the past beyond 2 weeks before assessment. Patients may be asked where they were born, where they went to grade school, etc.	Inability to accurately perform recent or remote (long-term) recall exercises within parameters. May indicate symptom of delirium or dementia.
Speech and ability to communicate	Objective and subjective assessment of aspects of how patient uses verbal and nonverbal communication.	Patient can coherently produce words appropriate to age, education, and life experience. Rate of speech reflects other psychomotor activity (e.g., faster if the patient is agitated). Volume is not too soft or too loud. Stuttering, repetition of words, and words that the patient makes up (neologisms) are also assessed.	Limited speech production. Rate of speech is inconsistent with other psychomotor activity. Volume is not appropriate to situation (speaks in very loud volume even when asked to speak more quietly). Presence of stuttering, word repetition, or neologisms may indicate physical or psychological illness.
Mood and affect	Subjective and objective assessment of the patient's stated feelings and emotions. Affect measures the outward expression of those feelings.	Mood is the stated emotional condition of the patient and should fluctuate to reflect situations as they occur. Facial expression, body language (affect) should match (be congruent with) the stated mood. Affect should change to fluctuate with the changes in mood.	Mood and affect do not match (e.g., facial expression does not appear sad while the patient is expressing sad feelings).

TABLE 55-1   SAMPLE MENTAL STATUS EXAMINATION — CONT'D

Area of Assessment	Type of Assessment	Normal Parameters	Alterations from Normal
Judgment	Subjective assessment of a patient's ability to make appropriate decisions about his or her situation or to understand concepts.	Give the patient a proverb or situation to solve, such as "You can't teach an old dog new tricks." The patient should be able to give some sort of acceptable interpretation, such as "Old habits are hard to break" or "It is hard to learn something new." Another example is to ask the patient what he or she would do if a small child was lost in the store. An appropriate response might be "Call the manager" or "Try to calm the child."	Patient cannot interpret the sayings in some acceptable manner. Patient cannot complete problem-solving questions appropriately. The patient might answer very literally, "Dogs can't learn anything when they get old" or might say, "I would go through the child's pockets looking for phone numbers."
Perception	Assesses the way a person experiences reality. Assessment is based on the patient's statements about his or her environment and the behaviors expressed in association with those statements. Nurses and health team members must document this often subjective information in very objective terms.	All five senses are monitored for interaction with the patient's reality. The patient's insight into his or her condition is also assessed.	Presence of hallucinations and illusions (schizophrenia). Individuals who are not within normal boundaries of judgment or insight will not be able to state understanding of the origin of the illness and the behaviors that are associated with it.

BOX 55-2   Nursing Diagnoses Commonly Used for Patients with Mental Health Problems

Disturbed thought processes
Ineffective role performance
Anxiety (mild to panic)
Disturbed body image
Dysfunctional grieving
Impaired social interaction
Ineffective coping
Disturbed personal identity
Powerlessness
Risk for violence (toward self or others)
Self-care deficit
Low self-esteem (chronic or situational)
Disturbed sensory-perception
Sexual dysfunction
Disturbed sleep pattern
Social isolation

Physicians also use other diagnostic criteria to diagnose mental illness. It is important to rule out physical illness as a cause of symptoms. The physician may choose to refer the patient to a psychiatrist or psychologist for further testing and diagnosis.

There are also batteries of psychological tests that can be administered and interpreted by psychiatrists or psychologists. Age, hand tremors, vision, language barriers, educational background, and the interpretation of the psychiatrist or psychologist are some factors that can influence the results of these tests.

Tests that may be performed to either confirm or rule out a diagnosis of a mental illness include the following:

■ Laboratory tests (e.g., to rule out electrolyte imbalances, dehydration, drug toxicity)
■ Computed tomography (CT) scans (to rule out tumors, lesions, or other physical problems)
■ Positron emission tomography (PET) scans (to identify how the parts of the brain are functioning by showing chemical activity or metabolism)

► COPING AND DEFENSE MECHANISMS

"Oh, just learn to cope with it." "Get a grip." "Don't make a mountain out of a molehill." These are pieces of advice that people may have heard or given at some point. But what do they mean? What is **coping**? Coping is the way one adapts psychologically, physically, and behaviorally to a **stressor.** Individuals have different methods of coping or dealing with their stressors. Culture, religion, individual belief systems, experience, and personal choice influence an individual's responses to stress. It is not the value of a be-

---

CRITICAL THINKING

Joe

Joe is noted wailing loudly and continuously after the death of his wife. It is disturbing the other patients on the wing, and one of the nurses comments, "He is a real nut case. Get him out of here." What is an appropriate response to this nurse and patient? How would you document his behavior?

Answers at end of chapter.

havior that we assess as nurses; it is the desired outcome that is important. What is an effective coping skill? Is it healthy? Does it work? How do we as nurses observe and measure it?

Effective coping skills offer healthy choices for dealing with stressors. Hospitalization is stressful for patients and families. Many things are unknown and unfamiliar. The patient may not understand the illness or the implications of the treatment plan. It is common for patients to use mechanisms to help them cope during hospitalization. The process of effective coping is sometimes called **adaptation.** Allowing the patient to practice new coping techniques will give him or her confidence and will decrease the stress that can accompany change.

Often the dividing line between effective and ineffective coping is the frequency of its use. For instance, a little worry or **anxiety** can be positive. Generally, when there is a little tension, people are more alert and ready to respond. The fight-or-flight mechanism can actually help one adapt to a new situation. However, too much worry begins to cloud the consciousness and interfere with the ability to make appropriate choices and to recall the new adaptive tools one has learned. One of the most helpful roles you can perform is to listen to the patient's thoughts and feelings about the stressor and then provide information that will reinforce the patient's positive feelings.

Sometimes coping is ineffective. When conscious techniques are not successful, humans often unconsciously fall into habits that give the illusion of coping. These habits are called defense mechanisms (or coping or mental mechanisms). Defense mechanisms are mental pressure valves. The purpose of defense mechanisms is to reduce or eliminate anxiety. They give the impression that they are helping alleviate the stress level. When used in very small doses, defense mechanisms can be helpful. When they are overused, they can become ineffective and unhealthy. People are not born with these behaviors; they are learned as responses to stress. Many times they develop by the age of 10 years old. They appear to be conscious, but they are, for the most part, unconscious mechanisms. Some commonly used defense mechanisms are listed in Table 55–2.

♥▬ CRITICAL THINKING

Mrs. Beison

Mrs. Beison, a 44-year-old mother of three teenagers, is diagnosed with breast cancer. She is refusing treatment because she does not believe she has cancer. She says if anything is really wrong, her vitamins will take care of it.

What coping mechanism is Mrs. Beison using? Is it effective or ineffective? Why? How can you help?

Answers at end of chapter.

▶ THERAPEUTIC MEASURES

People who experience alterations in their mental health have special treatment needs. When emotional health is threatened, many other daily activities can be altered as well. **Cognitive** ability (the ability to think rationally and to

process those thoughts) can be impaired. Emotional responses can be decreased or even absent in some conditions. This can be extremely frightening and can lead to a worsening of the mental disorder or even the development of another disorder. This section will provide an overview of selected therapies.

Milieu

One area over which you can have some control is the therapeutic environment. In the mental health setting, this therapeutic environment is called the **milieu** or therapeutic milieu. It is believed that environment has an effect on behavior.

A therapeutic milieu is an environment that provides safety and help during the patient's stay. Furnishings may be arranged to look like a home. Patients may help govern the running of the unit, with regular meetings to set rules and assign tasks. Patients may help with dishes or other small jobs. Structured activities are provided to assist the patient to learn coping and social skills. Some things you can do include keeping the area calm and quiet; arranging for roommate changes if needed; and assisting with the running of the unit. As the patient progresses, the milieu is changed to allow the patient to take on more responsibility.

Psychopharmacology

Psychopharmacology is the use of medications to treat psychological disorders. Since the introduction of the phenothiazine class of drugs in the 1950s, the number of medications available for treating mental health disorders has increased greatly. The reason for using medications is twofold: First, the medications control symptoms, helping the patient feel more comfortable emotionally. Second, the patient is generally more receptive and able to focus on other types of therapy if medications are effective. Classifications of psychoactive drugs are provided in Table 55–3. See Chapter 4 for some commonly used herbal therapies. See also Nutrition Notes Box 55–3 for the impact of nutrition on the effects of some drugs.

Psychotherapies

Psychotherapy is the term used to describe the form of treatment chosen by the psychologist or psychiatrist. The goals of psychotherapy include the following:

■ Decrease the patient's emotional discomfort
■ Increase the patient's social functioning
■ Increase the ability of the patient to behave or perform in a manner appropriate to the situation

Several specific types of therapy that are typically used are described next.

---

psychopharmacology: psycho—soul or mind +
pharmaco—drug or medicine + ology—study of

TABLE 55-2   DEFENSE MECHANISMS

Mechanism	Description	Examples
Denial	Usually the first defense learned and used. Unconscious refusal to see reality. Is *not* conscious lying.	The alcoholic states, "I can quit any time I want to."
Repression (stuffing)	An unconscious "burying" or "forgetting" mechanism. Excludes or withholds from our consciousness events or situations that are unbearable.	A step deeper than "denial." A patient may "forget" about an appointment he or she does not want to keep.
Rationalizing	Using a logical-sounding excuse to cover up true thoughts and feelings. Is the most frequently used defense mechanism.	1. "I did not make a medication error; I followed the doctor's order." 2. "I failed the test because the teacher wrote bad questions."
Compensation	Making up for something we perceive as an inadequacy by developing some other desirable trait.	1. The small boy who wants to be a basketball center instead becomes an honor roll student. 2. The physically unattractive person who wants to model instead becomes a famous designer.
Reaction formation (overcompensation)	Similar to compensation, except the person usually develops the exact opposite trait.	1. The small boy who wants to be a basketball center becomes a political voice to decrease the emphasis of sports in the elementary grades. 2. The physically unattractive person who wants to be a model speaks out for eliminating beauty pageants.
Regression	Emotionally returning to an earlier time in life when patient experienced far less stress. Commonly seen in patients while hospitalized. NOTE: Everyone will not go back to the same developmental age. This is highly individualized.	1. Children who are toilet trained beginning to wet themselves. 2. Adults who may start crying and have a "temper tantrum."
Projection (scapegoating)	Blaming others. A mental or verbal "finger-pointing" at another for patient's own problem. **Memory tool:** Think of a projector at the movie theater. It "points" the images of the film onto the screen just as a person using this defense mechanism "points" blame on another person or situation.	1. "I didn't get the promotion because you don't like me." 2. "I'm overweight because you make me nervous."
Displacement (transference)	The "kick-the-dog syndrome." Transferring anger and hostility to another person or object that is perceived to be less powerful than yourself.	Parent loses job without notice; goes home and verbally abuses spouse, who unjustly punishes child, who slaps the dog.
Restitution (undoing)	Make amends for a behavior one thinks is unacceptable. Makes an attempt at reducing guilt.	1. Giving a treat to a child who is being punished for a wrongdoing. 2. The person who sees someone lose a wallet with a large amount of cash does not return the wallet, but puts extra in the collection plate at the next church service.
Conversion reaction	Anxiety is channeled into physical symptoms. NOTE: Often the symptoms disappear soon after the threat is over.	Nausea develops the night before a major exam, causing the person to miss the exam. Nausea may disappear soon after the scheduled test is finished.
Avoidance	Unconsciously staying away from events or situations that might open feelings of aggression or anxiety.	"I can't go to the class reunion tonight. I'm just so tired, I have to sleep."

Psychoanalysis

Psychoanalysis is the form of therapy that was developed from the theory of Sigmund Freud. In psychoanalysis the focus is on the cause of the problem, which is believed to be buried somewhere in the patient's unconscious. The therapist uses questioning and memory probing techniques to investigate the patient's past in an effort to determine where the problem began.

psychoanalysis: psycho—soul or mind + analysis—dissolving

Behavior Management

Behavior management (also called behavior modification) is a treatment method that is a result of the studies of behavioral theorists such as Skinner and Pavlov. It is a common treatment modality used in long-term care facilities and facilities that treat patients with alterations in mental health.

According to behavior management theory, behavior can be changed by either positive or negative reinforcement. Positive reinforcement is the act of rewarding the patient with something pleasant when the desired behavior

TABLE 55–3 MEDICATIONS USED FOR ALTERATIONS IN MENTAL HEALTH

Classification and Common Trade Names	Uses	Side Effects	Nursing Considerations	Patient Teaching
Antipsychotics chlorpromazine (Thorazine), haloperidol (Haldol), trifluoperazine (Stelazine), fluphenazine (Prolixin), clozapine (Clozaril), risperidone (Risperdal)	Treatment of schizophrenia and other psychotic behavior that is violent or potentially violent	Blood dyscrasias, photosensitivity Darkening of the skin Extrapyramidal side effects (EPSEs): parkinsonism, akathisia, dystonia, tardive dyskinesia	Observe for any signs of EPSEs. Monitor blood work for any kind of abnormality. Discontinue slowly. Antacids decrease absorption of antipsychotics.	These are very strong medications. Wear a large-rimmed hat, cover all exposed skin, and use a sunscreen when in the sun, especially if using chlorpromazine. Avoid alcohol, sleeping pills, and other medications. Do not alter dose before discussing it with the doctor. Take medication 1–2 hours before going to bed. Take antacids 1–2 hours after oral doses of antipsychotics.
Antianxiety Drugs alprazolam (Xanax), buspirone (BuSpar), diazepam (Valium), lorazepam (Ativan)	Decrease the effects of stress or mild depression without causing sedation	Can cause physical and psychological dependence, drowsiness, lethargy, fainting, transient hypotension, nausea, and vomiting; if discontinued abruptly, severe side effects, including nausea, hypotension, and fatal grand mal seizures	Teach the patient and family that it is not safe to drive or use alcohol while using this classification of medication.	
Antidepressants amitriptyline (Elavil), perphenazine/amitriptyline (Triavil), imipramine (Tofranil), nortriptyline (Pamelor), paroxetine (Paxil), fluoxetine (Prozac), sertraline (Zoloft), amoxapine (Asendin), maprotiline (Ludiomil), tranylcypromine (Parnate), phenelzine (Nardil), isocarboxazid (Marplan)	Treatment of depression	Dependence, sedation, dry mouth, agitation, hypotension, vertigo, blood dyscrasias	Encourage patients to continue taking the medication during this time, because they may not feel any change in their mood for up to 3 weeks after starting medication. Administer intramuscular (IM) dosages deeply, slowly, and into large muscle masses. Z-track method of IM administration is preferred.	Reinforce the teaching that these medications may take 2–3 weeks to reach a therapeutic level and provide the desired effect for the patient. See Nutrition 1 Notes Box 55–2.
Stimulants phenylpropanolamine (Acutrim, Dexatrim), dextroamphetamine (Dexedrine), methylphenidate (Ritalin)	Promotes alertness, diminishes appetite, combats narcolepsy, local anesthesia	Rapid or irregular heart rates, hypertension, hyperactivity, hand tremor, rapid speech, confusion, depression, seizures, suicidal thoughts, psychological dependence	Physical or psychological dependence, especially with long-term use. Patients with diabetes should be informed that amphetamines can cause changes in their insulin requirements. Can also cause changes in judgment; people should use extreme caution when driving or operating equipment, and should avoid these activities if possible. They should be avoided in children under 12 years old. Monitor blood pressure carefully (at least every 4 hours when beginning the course of treatment).	Patients with diabetes should monitor insulin carefully and inform the doctor of any changes. Use extreme caution when driving or operating equipment, and avoid these activities if possible. Do not stop medication without consulting doctor. Use plain water to rinse mouth or use hard, sugarless candy to relieve dry mouth.
Antiparkinsonism Agents benztropine (Cogentin), trihexyphenidyl (Artane), biperiden (Akineton)	Decrease the effects of drug-induced and non-drug-induced symptoms of parkinsonism	Blurred vision, dry mouth, dizziness, drowsiness, confusion, tachycardia, urinary retention, constipation, changes in blood pressure, and melanoma		Use hard, sugarless candy to combat the effects of dry mouth. Increase dietary roughage to maintain bowel functioning. May cause drowsiness, so should not drive or operate equipment until the response to medication is established.

BOX 55-3 *Nutrition Notes*

Anticipating Food-Drug Interactions in Patients with Mental Health Disorders

Monoamine Oxidase Inhibitors. Some antidepressant drugs (e.g., phenelzine, tranylcypromine) are monoamine oxidase inhibitors (MAO inhibitors). These drugs counteract depression by preventing the breakdown of dopamine and tyramine, an intermediate product in the conversion of tyrosine to epinephrine. The increased concentration of these chemicals in the central nervous system elevates the patient's mood.

When a patient taking an MAO inhibitor consumes foods high in tyramine, the drug prevents the normal breakdown of tyramine, leading to excessive epinephrine. Hypertension results, sometimes severe enough to cause intracranial hemorrhage. Some other drugs (e.g., furazolidone, isoniazid, procarbazine, selegiline) produce similar reactions with tyramine-containing foods.

All food groups except breads and cereals have some items with sufficient tyramine to be create problems for patients taking MAO inhibitors. Examples of common foods to be avoided are bananas, aged cheese, yogurt, bologna, salami, pepperoni, summer sausage, chocolate, beer, and wine.

Lithium Carbonate. Lithium carbonate, used to treat bipolar disorder, is absorbed, distributed, and excreted alongside sodium. Fluctuations in sodium intake affect the metabolism of lithium. Thus decreased sodium intake with decreased fluid intake may lead to retention of lithium and overmedication. Increased sodium intake from food or medications and increased fluid intake may hasten excretion of lithium, resulting in worsening signs and symptoms of mania.

has been performed. For instance, if Mrs. Powell has the habit of using foul language in an attempt to have a need met, the desired behavior change might be to come to a staff member and ask quietly for what she needs. Mrs. Powell loves to be outside but is not allowed out except at supervised times. A suitable positive reinforcement for her might be to allow 15 more minutes outdoors when she remembers to come ask for her needs quietly.

Negative reinforcement is the act of responding to the undesired behavior by taking away a privilege or adding a responsibility. Negative reinforcement can be misinterpreted as punishment. Parents who "ground" their children for unacceptable behavior are using negative reinforcement; requiring the child to perform extra household tasks for a stated period is reinforcing the fact that the behavior has consequences. The child may not repeat the undesired behavior after negative reinforcement has been used. It is necessary to be very careful when performing behavior management with patients to avoid an infraction of the Patient's Bill of Rights. A signed consent from the patient is advised when using this form of therapy.

The patient must understand the consequences of the behavior to be changed and the purpose for the type of consequence that is chosen. If the person is not capable of understanding the situation or is not able to remember the consequences because of some other problem, behavior management could be considered a questionable alternative to other kinds of treatment.

Cognitive Therapies

RATIONAL-EMOTIVE THERAPY. Dr. Albert Ellis and other cognitive therapists believe that people teach themselves to be ill because of the way they think about their situations. Cognitive (or cognitive/behavioral) therapy stresses ways of rethinking situations. The therapist confronts the patient with certain behaviors and then works out ways of thinking about them differently.

Feeling sad about an unpleasant experience (such as the death of a loved one) is acceptable and normal, but long-term depression about the death is an extreme emotion and therefore considered to be unhealthy. In this situation the patient might be helped to see the death as a sad loss, but extreme, long-term depression would be viewed as an inappropriate response.

Cognitive therapies are gaining in popularity because they are usually significantly more short term than psychoanalysis and therefore less costly to the patient. It is commonly performed in groups. The patients are given "homework" that is specific to their needs. Patients practice their assignments between sessions.

Person-Centered/Humanistic Therapy

Abraham Maslow and Carl Rogers are two theorists who are often credited with the concept of person-centered or humanistic therapy. In this form of treatment, caregivers focus on the whole person and work in the "present." It is not important in humanistic treatment to understand the cause of the problem or what happened in the person's past; what is important is the here and now. With this therapy the patient learns to see himself or herself as a person who has value and who is respected by others.

Nursing is very strongly centered in person-centered principles. Three qualities are essential for caregivers: empathy, which is the ability to identify with the patient's feelings without actually experiencing them with the patient; unconditional positive regard (respect); and genuineness or honesty. Although you may not be an active participant in the actual therapy sessions with patients, it is important to maintain these three qualities in all therapeutic relationships.

Counseling

Counseling is the provision of help or guidance by a health care professional. The area of counseling is licensed and regulated differently not only state by state but sometimes municipality by municipality. Nurses prepared at an LPN/LVN level or at an RN level can, in some areas and with special advanced education, practice some forms of counseling.

You may be asked or expected to accompany patients to counseling sessions or even to facilitate a group discussion. Remember that these are confidential sessions, even if they are group oriented. Patients are there to work; others are there by invitation for special reasons.

Group Therapy

Groups are formed for many reasons; they can be ongoing or short term, depending on the needs of the patients or the type of disorder. For example, Alcoholics Anonymous (AA) and similar 12-step self-help groups are well-established, ongoing groups formed around treatment of a specific problem. Family counseling sessions may occur with individual therapists with a specialty in the problem area for that family. Marriage counseling may be done in a group with other couples. Many times, peer counselors are used.

Therapists and counselors are tools, or facilitators, in the therapeutic process. They do not heal the patient; the patient heals himself or herself. Patients must take the suggestions given by the therapist, try them, and see what works for them. You can help by reinforcing the good work patients do in learning to stay mentally healthy and develop more effective life skills.

Electroconvulsive Therapy

Electroconvulsive therapy (ECT), sometimes called electroshock therapy (EST), is another form of treatment that is used in special situations. Some experts believe ECT works because it breaks up thought patterns in the brain, helping people to forget painful past experiences. Others believe it stimulates increases in the neurotransmitters serotonin, neoorpinephrine, and dopamine. ECT is very frightening to some people. Many changes have been made in this form of therapy since the 1940s to make it safer for the patient.

ECT often takes place in the recovery room of an operating suite, where there is ready access to emergency equipment. Written consent is required. Patients are given a short-acting anesthetic before the treatment and a smooth muscle relaxant to minimize injury. Patients are secured with restraints during the treatment, so movement is minimal. Carefully monitor blood pressure and pulse before and after the treatment. During treatment, an electric stimulus is delivered to the brain via electrodes. The amount of electrical energy used is individualized to the patient. A treatment usually lasts only a few seconds, and if you are slow to look, you might miss seeing any kind of seizure activity (convulsion). Often, because of the medications given, only a toe or a finger twitches slightly.

Side effects of ECT can be unpleasant. The patient may feel confused and forgetful immediately after the treatment. This can be from a combination of the ECT itself and the medication that was used before the treatment. If there has been a strong seizure, the patient may have some muscle soreness.

Electroconvulsive therapy is not used indiscriminately. It is used when other therapies have not been helpful, and it is usually reserved for severe or long-term depression and certain types of schizophrenia (Patient Perspective Box 55-4).

The nurse's responsibilities include careful monitoring of vital signs and accurate documentation relating to the patient's subjective and objective responses to the treatment. The patient should receive nothing by mouth (NPO) for at least 4 hours before a treatment. Remind the patient to empty his or her bladder and to remove dentures, contact lenses, hair pins, and so on. Because of the possibility of

---

BOX 55-4 Patient Perspective

My mom is now 83 and has had a total of 35 electroconvulsive therapy (ECT) treatments in her lifetime. If you met her, you'd never know; you'd find her delightful. She's a sweet little plump German lady with a big heart. I am a nurse, and when I tell fellow nurses about my mom, they often ask me why I didn't get her on antidepressants. I want to scream, "How dumb do you think I am?!" Of course Mom is on antidepressants. But at intervals they don't work and she sinks into severe depression. My choice then is to help her have ECT or let her stay depressed and miserable and put her in a nursing home. And she would soon die, because when she is depressed she refuses to move, doesn't sleep, and is horribly miserable.

The first time my mom was scheduled for ECT, one of the nurses in our local community hospital told her to refuse it, that no one should have to go through that. It was a cruel thing to do. My mom doesn't do well in counseling—she doesn't believe in it. In her mind you don't talk about your "dirty linen." My mom was in an abusive relationship with my father for 47 years, and she hid all the problems away and doesn't talk about them to this day. I'm so grateful to have my mom doing okay and grateful that ECT treatments exist. With ECT my mother is doing well and enjoying life. Without treatment she would be gone. Please understand that there are times when ECT treatments are the best thing for severely depressed people, when other treatments have been ineffective.

---

confusion and forgetfulness, it is common to restrict activity for 24 hours after a treatment. Stay with the patient until he or she is oriented and able to care for himself or herself. Ensuring that the person is kept safe after therapy is a major concern.

Relaxation Therapy

A variety of relaxation techniques can be taught to help patients manage their responses to stress. Relaxation exercises such as deep, rhythmic breathing can increase oxygenation and provide distraction from stressors. Breathing exercises may be coupled with progressive muscle relaxation exercises. For this technique, patients are taught to start at the head and neck and systematically tense and then relax muscle groups as they progress toward the lower extremities. Soft music may enhance the patient's ability to fully relax.

Imagery is the use of the imagination to promote relaxation. For this technique, the patient is taught to imagine a pleasurable experience from his or her past, such as lying on a beach or soaking in a warm bath. Use of all senses is encouraged—for a beach image, the patient might see the beach, feel the warm sun, smell the salt air, and hear the waves crashing against the shore. The patient might also be taught to visualize being successful in a problem situation.

Biofeedback uses computerized or other instruments to provide feedback about the patient's degree of relaxation. Heart rate, skin temperature, and muscle tension might be measured; the patient is taught by a specialist to control these variables with a variety of relaxation techniques.

Relaxation techniques may be used individually, but they are often used in combination with each other or with other therapies for maximum effect.

▶ MENTAL HEALTH DISORDERS
Anxiety Disorders

Stress is everywhere in our society. Stress produces anxiety. Most often, stress is associated with negative situations, but the good things that happen to us, such as weddings and job promotions, also produce stress. The stress from positive experiences is called **eustress.** Eustress can produce just as much anxiety as the negative stressors. A stressor is any person or situation that produces an anxiety response. Stress and stressors are different for each person; therefore it is important to ask patients what their personal stress producers are.

Anxiety is the uncomfortable feeling of dread that occurs in response to extreme or prolonged periods of stress. It is commonly ranked as mild, moderate, severe, or panic. It is believed that a mild amount of anxiety is a normal part of being human and that mild anxiety is necessary to change and develop new ways of coping with stress.

Anxiety may also be influenced by one's culture. It may be acceptable for some people to acknowledge and discuss stress, but others may believe that one does not discuss personal problems with others. This cultural behavior can be a challenge for the nurse during an assessment.

Anxiety is usually referred to as either free-floating anxiety or signal anxiety. Free-floating anxiety is described as a general feeling of impending doom. The person cannot pinpoint the cause but might say something like "I just know something bad is going to happen if I go on vacation." Sig-

nal anxiety, on the other hand, is an uncomfortable response to a known stressor ("Finals are only a week away and I've got that nausea again.") Both types of anxiety are involved in the various anxiety disorders.

Etiologic Theories

Psychoanalytical theory says that anxiety is a conflict between the id (the "all for me" part of the personality) and the superego (the conscience), which was repressed in early development but emerges again in adulthood.

Biologic theory looks at this situation differently. Biologic theories consider the sympathoadrenal (fight-or-flight) responses to stress and observe that the blood vessels constrict because epinephrine and norepinephrine have been released, causing blood pressure to rise. If the body adapts to the stress, hormone levels adjust to compensate for the epinephrine-norepinephrine release and body functions return to homeostasis. If the body does not adapt to the stress, the immune system is challenged, lymph nodes swell, and risk for physical illness increases.

The LPN/LVN usually observes mental illness in conjunction with medical-surgical illness. It is important to recognize the relationship between physical and emotional responses to stress. Some examples of medical conditions and the effects of the body's adaptation response to stress are shown in Table 55–4.

Differential Diagnosis

Because there are so many symptoms associated with anxiety disorders, it is important for people to have a complete physical examination before diagnosing an anxiety disorder. More than one condition may occur at the same time.

---

eustress: eu—normal or good + stress

TABLE 55–4	ADAPTATION RESPONSES TO STRESS	
Stress-related Medical Condition	**Body's Adaptation to the Stress**	**Outcome of Stress on the Body**
Lowered immunity	Interferes with effectiveness of the body's antibodies; possibly related to interactions among the hypothalamus, pituitary gland, adrenal glands, and immune system	Increased susceptibility to colds and other viruses and illnesses
Burnout	Associated with stress-related depression	Emotional detachment
Migraine, cluster, and tension headaches	Tightening skeletal muscles, dilating of cranial arteries	Nausea, vomiting, tight feeling in or around head and shoulders, tinnitus, inability to tolerate light, weakness of a limb
Stress (peptic) ulcers	Stress contributes to the formation of ulcers by stimulating the vagus nerve and ultimately leading to hypersecretion of hydrochloric acid	Nausea, vomiting, gastrointestinal bleeding, perforation of intestinal walls
Hypertension	Role of stress not positively known; thought to contribute to hypertension by negatively interacting with the kidneys, autonomic nervous system, and endocrine system	Resistance to blood flow through the cardiovascular system, causing pressure on the arteries; can lead to stroke, heart attack, and kidney failure
Coronary artery disease	Stressor increases the amount of epinephrine and norepinephrine	Coronary vessels dilate, pulse and respirations increase
Cancer	Stress lowers immune response	Lowered immunity may allow for overcolonization of opportunistic cancer cells
Asthma	Autonomic nervous system stimulates mucus, increases blood flow, and constricts bronchial tubes; may be associated with other stress-related conditions such as allergy and viral infection	Wheezing, coughing, dyspnea, apprehension; may lead to respiratory infections, respiratory failure, or pneumothorax

Types of Anxiety Disorders

PHOBIA. Phobia is the most common of the anxiety disorders. Phobia is defined as an irrational fear of a specific object or situation. The person is very aware of the fear and even the fact that it is irrational, but the person is unable to gain control over the stressor and the fear continues.

The psychoanalytical view implies that it really is not the object that is the source of the fear, but rather the fear is a result of a defense mechanism called **displacement.** For example, the person with a phobia of snakes may have seen a frightening movie in which someone died from a snakebite. The stated object of the phobia would be interpreted as a symbol for the underlying cause of the fear.

PANIC DISORDER. Panic is a state of extreme fear that cannot be controlled. It is also referred to as panic attack, and lay people may not consider it to be a serious disorder.

Panic episodes present quickly. Patients must exhibit several episodes within a specified time frame to be given the diagnosis of panic disorder. Some of the symptoms associated with panic disorder include the following:

- Fear (usually of dying, losing control of self, or "going crazy")
- Dissociation (feeling that it is happening to someone else or not happening at all)
- Nausea
- Diaphoresis
- Chest pain
- Increased pulse
- Shaking

GENERALIZED ANXIETY DISORDER. In generalized anxiety disorder (GAD) the anxiety itself (also referred to as excessive worry or severe stress) is the expressed symptom. Symptoms that may be present in GAD include the following:

- Restlessness or feeling "on edge"
- Shaking
- Palpitations
- Dry mouth
- Nausea, vomiting
- Easy frightening
- Hot flashes
- Chills
- Polyuria
- Difficulty swallowing

OBSESSIVE-COMPULSIVE DISORDER. Obsessive-compulsive disorder (OCD) is a different type of anxiety disorder. It consists of two parts: **obsession** (repetitive thought, urge, or emotion) and **compulsion** (repetitive act that may appear purposeful). An example of obsessive-compulsive disorder is the need to check that the doors are locked numerous times before one is able to sleep or leave the house. In reality, this need to repetitively check the locks may prevent the person from sleeping or leaving at all. Behaviors become very ritualistic. The person with OCD is unable to stop the thought or the action. Performing the thought or the action is the mechanism that reduces the anxiety.

POSTTRAUMATIC STRESS DISORDER. Posttraumatic stress disorder (PTSD) develops in response to some unexpected emotional or physical trauma that could not be controlled. People who have fought in wars, who have been raped, or who have survived violent storms or violent acts (such as the Oklahoma Federal Building bombing in 1995) are examples of those who are susceptible to suffering from this disorder.

A condition that is associated with PTSD is survivor guilt. This is the feeling of guilt expressed by those who survived. A survivor of an airline crash may say, "Why me? Why did I make it? I should have died too!"

Symptoms may appear immediately or may be repressed until years later. Symptoms include the following:

- Flashbacks in which the person may relive and act out the traumatic event
- Social withdrawal
- Feelings of low self-esteem
- Changes in relationships with significant others
- Difficulty forming new relationships
- Irritability and outbursts of anger seemingly for no obvious reason
- Depression
- Chemical dependency

Medical Treatment for Patients with Anxiety Disorders

Treatment is individualized for the patient and may include one or more of the following: psychopharmacology, individual psychotherapy, group therapy, systematic desensitization, hypnosis, imagery, relaxation exercises, and biofeedback.

Psychopharmacology usually involves the antianxiety classification of medications. The benzodiazepines, such as diazepam (Valium) and alprazolam (Xanax), are commonly used and are effective in most cases. Use of antianxiety drugs is short term whenever possible because of the strong potential for chemical dependency. Individuals who have anxiety disorders and who are chemically dependent on other substances are managed with other medications that have calming qualities but do not have the same high potential for **addiction** as the antianxiety drugs. Hydroxyzine hydrochloride (Atarax), clonidine (Catapres), and sertraline (Zoloft) are common alternatives in this situation.[2]

In systematic desensitization, the patient is gradually exposed to the object that causes the anxiety. Hypnosis places the patient in a subconscious state, then helps the patient recall events that may be producing anxiety so they can be dealt with. Other therapies were discussed earlier.

Nursing Interventions for Patients with Anxiety Disorders

1. Maintain a calm milieu.
2. Maintain open communication. Encourage the patient to verbalize thoughts and feelings. Honesty in dealing with patients helps them learn to trust others and enhances their self-esteem. Observe nonverbal communication.

3. Observe for signs of suicidal thoughts.
4. Report and document any changes in behavior. Any change can be significant to that patient's care. Positive or negative alterations in the way a patient responds to the nursing staff, to the treatment plan, or to other people and situations should be reported.
5. Encourage activities. Activities that are enjoyable and nonstressful provide diversion and give staff an opportunity to provide positive feedback about the progress the patient is making. The patient should not be put in a competitive situation.

CRITICAL THINKING

Tommy

Tommy has come to your clinic with numerous cracks on his hands. They are bleeding and very sore. Tommy tells you that he has to wash his hands all the time. His mother says he washes for 2 to 3 hours at a time and he will not stop when she tells him to. The doctor has diagnosed Tommy with obsessive-compulsive disorder and has explained the illness to Tommy and his mother. When the doctor leaves the room, Tommy's mother begins to cry. "What did he just say? What am I supposed to do? What did I do wrong that Tommy got this illness?" How do you respond?

Answers at end of chapter.

Mood Disorders

Mood disorders (also called affective disorders) are disorders in which the major symptom is extreme changes in mood (emotions) and affect (the outward expression of the mood). Sadness becomes depression when it lasts a very long time (generally 2 years or more) or when it begins to interfere with normal day-to-day functioning. People of all age groups and all ethnic and socioeconomic groups can develop mood disorders.

Etiologic Theories

Psychoanalytical theory indicates that people who have suffered loss in their lives are at risk for developing depression. Depression is also associated with unresolved anger and has been explained as "anger turned inward." In other words, people who cannot or do not deal appropriately with situations that anger them tend to repress the anger (turn it inside) and become depressed.

Cognitive theorists believe that the way people perceive events and situations may lead to depression. Instead of thinking about failing an exam as being unfortunate and disappointing, some people with tendencies toward depression may exaggerate the emotion and turn the situation into something much deeper, such as thoughts of "I'm stupid" or "I'll never get anywhere."

Biologic theories offer genetic links and neurochemical interactions as two etiologies. Serotonin and norepinephrine have an effect on mood; if these neurochemicals are el-

evated, mood is elevated, and if they are low, mood is low. Biologic theorists also believe that there is a connection between these neurochemicals and female hormones.

Differential Diagnosis

Symptoms of depression may occur as a result of other disorders, such as schizophrenia or drug side effects or overuse. Symptoms can mimic heart failure, nutritional deficiencies, fluid and electrolyte imbalances, infections, or diabetes.

Types of Mood Disorders

MAJOR (UNIPOLAR) DEPRESSION. People who develop major depression exhibit a vast array of symptoms. Some behavioral and physical symptoms of major depression include the following:

- Sad mood
- Loss of pleasure in things that are usually pleasurable
- **Hallucinations** or **delusions** (possible but uncommon)
- Weight loss or gain resulting from changes in appetite
- Sleep pattern disturbances
- Increased fatigue
- Increased agitation
- Increase or decrease in normal activity
- Lowered pleasure for life, including sexual activity
- Social withdrawal
- Decreased ability to think, remember, or concentrate
- Suicidal thoughts.

BIPOLAR DEPRESSION. Approximately 2 million people in the United States suffer from **bipolar** depression. Bipolar depression (also called manic-depressive illness or bipolar disorder) is a type of depression with both **mania** (extreme elation or agitation) and extreme depression. The mania is part of the depression. Think of a color you like. Now think of all the shades of that color from the very palest shade to the brightest shade (like the color cards that you can get in paint stores). Both extremes of color are very different, but both are still part of the same color. This is one way to remember that mania and depression are opposite poles of the same illness. It takes only one episode of mania to be diagnosed with bipolar disorder, but it is more common to see the patient alternating, or cycling, from one "pole" to the other. Individuals can cycle slowly (over weeks or months or even years), or they can be "rapid cyclers" who can change moods several times in an hour.

Common signs of depression were covered in the preceding section. Common signs of mania include the following:

- Excessive high (euphoric) moods
- Sustained period of behavior that is different from usual
- Increased energy, activity, restlessness, racing thoughts, and rapid talking
- Decreased need for sleep
- Unrealistic beliefs in one's abilities and powers
- Extreme irritability and distractibility
- Uncharacteristically poor judgment
- Increased sexual drive

- **Abuse** of drugs, particularly cocaine, alcohol, and sleeping medications
- Obnoxious, provocative, or intrusive behavior
- Denial that anything is wrong.

Medical Treatment for Patients with Mood Disorders

With treatment, it is estimated that approximately 80 percent of people with serious depression can be helped in a matter of a few weeks. Some common medical treatments for depression include the following:

- Lithium
- Antidepressants
- Psychotherapy
- Electroconvulsive therapy.

Lithium is the drug of choice in most instances for the treatment of bipolar depression. Antidepressants such as amitriptyline (Elavil), fluoxetine hydrochloride (Prozac), and trazodone hydrochloride (Desyrel) and antianxiety drugs such as lorazepam (Ativan) may also be used.

Psychotherapy for the patient and family may be helpful in understanding the illness and learning problem solving and other new adaptive coping behaviors. For young children, play therapy is the most common and effective form of therapy. Electroconvulsive therapy is an option when other therapy is not effective.

Nursing Interventions for Patients with Mood Disorders

- Be patient! The patient moves and thinks more slowly when depressed.
- Monitor lithium levels. Normal range of lithium is 1.0 to 1.5 mEq/L when receiving loading doses and 0.6 to 1.2 mEq/L when on maintenance doses.
- Be honest in communicating thoughts and feelings with the patient.
- Be consistent in communication and in all areas of implementation of the care plan.
- Provide appropriate activities. Activity needs to be planned carefully and should be age appropriate. Activity that includes small groups can be helpful for building esteem. Activities that can be done alone are helpful when it is necessary to decrease stimulation.
- Maintain nutrition. Physical health is important for mental health. The dietitian may be consulted for assistance.
- Use therapeutic communication and active listening. This is not a time for false reassurance. Do not use phrases such as "Cheer up!" or "It could be worse!" Making an observation such as "I like the way you look in that blue outfit. Is it new?" may go further in helping the person improve self-esteem than "Cheer up!" (See Gerontological Issues Box 55–5 for interventions for older adults who are suicidal.)
- Ensure that a referral to a community support agency is made for long-term follow-up.

BOX 55-5 GERONTOLOGICAL ISSUES

Suicide and the Older Adult

Older adults are not immune to suicidal thoughts. White males over the age of 75 have an especially high suicide rate.

Comments by any older adult referring to hopelessness or desire to die must be explored to assess suicide risk. The following comments could be a reflection of suicide potential in an older adult who is depressed:

"Living is harder than dying could ever be."
"I am a used-up old man who is a burden for everyone."
"I am useless. I can't do anything anymore."
"I don't know why God won't take me."

To adequately assess suicide potential, the nurse must ask questions that establish whether the older adult has done the following:

- Thought about ending his or her life
- Attempted to end his or her life in the past
- Developed a plan to end his or her life
- Set the plan into action (i.e., bought a gun, has a full bottle of pills in the bed stand)

Any older adult who has a plan to end his or her life and has the ability or resources to do so must be immediately referred for psychological evaluation. Never leave a person with suicidal thoughts alone.

Crisis Intervention for an Older Adult Who Is Suicidal

- Remove any items that the older adult could use to inflict an injury or end his or her life, such as razors, jewelry with pins or sharp points, and mirrors.
- Make arrangements for direct supervision and observation that are reliable, considering personnel and family resources. Often, hospital admission is the most appropriate intervention for a person at a very high risk for suicide.
- Help the older adult talk about the crisis or life event that has devastated his or her desire to live. For example, encourage reminiscence about the patient's spouse, or allow the older person to express the frustration of being unable to physically meet the daily demands of life.
- Develop a "do no harm" or suicide contract with the older adult. Outline a short-term, structured plan to keep the older adult safe. Focus on decreasing social isolation by requiring personal social contacts (e.g., stay at daughter's home for a weekend; go to the senior center for lunch; call a specific person who is willing and wants to listen to feelings and concerns; exercise; take a walk outside; volunteer services at a nursing home, hospital, or school).

Older adults often need assistance to develop or enhance skills required to cope with life events. Self-care and personal independence in care choices need to be encouraged. Developing an understanding and a manner of avoiding personal thinking patterns and behaviors that increase depression are important skills to learn as part of managing depression.

Somatoform/Psychosomatic Disorders

The **somatoform** (or **psychosomatic**) disorders are conditions in which physical symptoms occur with no known organic cause. It is believed that the physical symptoms are connected to a psychological conflict. Because the patient is not able to control the symptoms, they are considered to be caused by some unconscious mechanism.

Etiologic Theories

Psychoanalytical theorists believe that the somatoform disorders are rooted in unconscious mechanisms that develop to deny, repress, and displace anxiety. Biologically there is research into the possibility of a genetic predisposition to somatic difficulties.

Types of Somatoform Disorders

There are five separate illnesses within the category of somatoform disorders and two additional illnesses that are closely related. These include **conversion disorder,** hypochondriasis, dysmorphophobia/body dysmorphic disorder, somatization disorder, and somatoform pain disorder. Discussion of all these disorders is beyond the scope of this book. This chapter gives a brief explanation of conversion disorder.

CONVERSION DISORDER. Conversion disorder is an illness that emerges from overuse of the conversion reaction defense mechanism. (See Table 55–2.) In conversion disorder there is a loss or decrease in physical functioning that seems to have a neurological connection. Paralysis and blindness are two common examples of this disorder. Age of onset is usually adolescence and young adulthood, but it can occur later in life as well.

The symptoms, although not caused by organic disease, are very real to the patient. It should not be conveyed to the patient that you think the person is "faking" the illness; this is not true. Patients are truly experiencing the symptoms.

The belief about this disorder is that the symptom is allowing the person to avoid some situation that is unacceptable to him or her. The symptom helps the patient relieve the anxiety. This is called primary gain. Secondary gain results from the extra benefits one may acquire as a result of staying ill, such as extra emotional support, sympathy, love, or financial benefits.

Medical Treatment for Patients with Somatoform Disorders

Because of the physical symptoms, hospitalized patients are usually admitted to a medical unit rather than a psychiatric unit. Treatment is individualized for the patient. Hypnosis and relaxation techniques are used with many patients. Methods of stress management are taught. Behavior management may be effective for some patients. Patients may resist accepting the fact that their problem is psychological or emotional in nature and may feel insulted and become resistant to treatment.

Medications are used sparingly. When they are ordered for a patient, the classifications of choice are usually antidepressants, antianxiety agents, or both.

Nursing Management for Patients with Somatoform Disorders

Nursing management for somatoform disorders includes the following:

- Skillful communication. Honesty and gaining trust encourage the patient to verbalize thoughts and feelings about the physical and emotional aspects of this disorder. An example of a way to be honest about the situation would be to reinforce what the physician has said to the patient, such as, "Ms. Parks, your doctor can find no physical or life-threatening conditions at this time. We will continue to observe and examine you. We will make every attempt to help you improve." Sometimes the practitioner helps the patient find a healthy behavior to substitute for the symptom.
- Therapy. Keeping the patient focused on other topics may help in the recovery.
- Support. When caring for the patient with a somatoform disorder, you must pay attention to the person but must not reinforce the symptom. A thorough head-to-toe assessment should always be done. The patient will see your concern for his or her health, but you will not be focusing on the area of dysfunction or reinforcing the problem. All findings should be documented objectively.

Schizophrenia

Schizophrenia is becoming more widely viewed as a group of illnesses rather than a single condition. It most often strikes adolescents and young adults. The National Institute of Mental Health estimates that nearly 3 million persons in the United States will develop schizophrenia during the course of their lives.

psychosomatic: psych(e)—soul or mind + somatic—body

The term *schizophrenia* (which means "split mind") was first used by a Swiss psychiatrist, Eugene Bleuler.[3] Schizophrenia is a serious psychiatric disorder. People who have schizophrenia have a split between their thoughts and their feelings and between their reality and society's reality. The person may have splits concerning gender identity. People who have schizophrenia may not be able to differentiate between what is "theirs" and what is "everybody else's" in relation to social functioning. Poor self-esteem may be an issue. People with schizophrenia are generally highly intelligent and seem very distractible. It is difficult for them to focus on one topic for any length of time. Schizophrenia is not the same thing as multiple personality disorder.

Schizophrenia has an insidious onset. This means that it develops over time, and symptoms may go unnoticed for a time prior to diagnosis. The early symptoms of quietness and withdrawal in an adolescent person may be shrugged off as normal adolescent behavior. It may be a school nurse or counselor who begins to notice these changes. Grades may begin to suffer. The adolescent may have a change in personality or a change in the way he or she relates to other people. It is easy to misinterpret these new behaviors as part of the adolescent experience.

Eugene Bleuler used a system of four As to define schizophrenia. The four As are associative disturbance, affect, autism, and ambivalence, discussed next.

Associative Disturbance

A patient with associative disturbance (also referred to as associative looseness) typically exhibits three main behaviors: (1) making up words (neologisms); (2) rambling from topic to topic; and (3) using revolving words and syllables that may be associated with a specific word but that are out of context with the conversation. Making up words that rhyme with other words is another behavior that is sometimes observed.

Affect

Affect, as discussed earlier, is the outward expression of emotion. People with schizophrenia generally have what is called a "flat" or "blunted" affect. This means that they rarely show signs of any emotion.

In schizophrenia there may also be inappropriate or incongruent affect. This means the outward expression of the mood does not match the stated feeling (e.g., patients laughing when they say they feel sad or depressed). Exaggerations of affect may also be present in some patients.

Autism

Autism associated with schizophrenia is an emotional detachment. People who display autistic behavior are preoccupied with the self and show little concern for any reality outside their own world.

Ambivalence

Ambivalence means to have opposite feelings about a person or situation at the same time. An example of this is the love/hate relationships sometimes seen in jobs or marriages.

(Caution: Not all people who have love/hate relationships are schizophrenic.)

In addition to the four As, patients with schizophrenia display other common symptoms, such as delusions, hallucinations, and **illusions.** Delusions are fixed, false beliefs that cannot be changed by logic or factual proof. Typically patients exhibit delusions of grandeur, persecutory delusions, or guilt. Hallucinations are false sensory perceptions. They can affect any of the five senses; auditory and visual are most common. Illusions are mistaken perceptions of reality.

Hallucinations and illusions are easily confused. Anyone who has ever watched a magician or seen certain pictures in cloud formations has witnessed an illusion. In illusion, something is there; it is just perceived incorrectly. The card did not magically appear; it was set up to appear when the magician wanted it to. In hallucination there is nothing to misinterpret. The person who sees a lamb in a certain cloud formation is experiencing an illusion; the person who sees a lamb in the sky when there are no clouds is experiencing a hallucination.

Etiologic Theories

There are psychoanalytical and biologic theories of the causes of schizophrenia. The symptoms of schizophrenia are highly debated on both sides of the nature-versus-nurture theories.

The psychoanalytical, or nurture, theories revert to the anal stage of freudian theory. The inability to meet the challenge of oral gratification leaves people in the adolescent and young adult years unable to handle their developing sexuality, according to Freud. Lack of nurturing mother-child relationships can also lead to personalities that are cool or aloof (or indifferent) in their relationships. Freud would also attribute the disruption of effective communication to failure to attain oral gratification.

The role of genetics in schizophrenia have been examined in twins studies, family studies, and adoption studies for more than 75 years. Studies of identical twins (the psychobiologic, or nature, theory) show that if one twin has schizophrenia, the other has about a 50-percent chance of developing it also. In fraternal twins that percentage drops to about 10 percent. It is believed that the more genes twins or family members have in common, the greater the probability of the second twin developing schizophrenia. Other studies have examined the relationship of dopamine and schizophrenia. Patients with schizophrenia generally have elevated amounts of dopamine or a brain that overreacts to the amount of dopamine that is present.

Types of Schizophrenia

There are several categories of schizophrenia. This chapter discusses only the most common: paranoid schizophrenia. Paranoid schizophrenia is defined as schizophrenia in which the person exhibits unusual suspiciousness and fear. People who have paranoid schizophrenia may also be hostile and aggressive in their behavior.

Patients with paranoid schizophrenia tend to have delusions of persecution or grandeur. In persecutory delusions, patients state that they feel tormented and followed by people. Patients often integrate people around them into their delusions. These people could be staff, relatives, or the announcer on the radio or television. In delusions of grandeur, patients may state that they are God or the President of the United States.

Hallucinations often accompany delusions. The hallucinations can affect any of the five senses but are most commonly auditory or visual. Patients with paranoid schizophrenia speak about "the voices." These voices are frightening and derogatory to the patient and are responsible for many of the actions performed by people with paranoid schizophrenia. Patients experience increased fear, anxiety, and suicidal ideation as a result of the voices. You may see or hear patients arguing with what at first appears to be themselves. Actually the patient is arguing with the voices. Describing the voices is difficult, but imagine that you are in a room with six televisions on different stations at the same time. This example comes close to what some patients have described as the voices.

Medical Treatment for Patients with Schizophrenia

Medications, ECT, and psychotherapy are indicated for patients with schizophrenia. Among the classifications of medications that may be prescribed for certain patients are the antipsychotics, which block dopamine action in the brain. Unfortunately, decreased dopamine action creates extrapyramidal (parkinsonian) side effects. Anticholinergic medications such as benztropine (Cogentin) or trihexyphenidyl (Artane) are used to combat the extrapyramidal side effects of the antipsychotics by helping return balance between dopamine, acetylcholine, and other neurotransmitters. Newer antipsychotic medications such as clozapine (Clozaril) and risperidone (Risperdal), with fewer side effects, have been developed specifically for use in schizophrenia.

Psychotherapy may include individual, group, and family therapy. Electroconvulsive therapy is used in some severe cases or in cases that are difficult to treat; ECT is usually not used until other methods of therapy have been exhausted.

Nursing Interventions for Patients with Schizophrenia

■ Develop trust. It is crucial for a trusting relationship to exist between the nurse and the patient. Promises must be kept. Be honest and consistent in all areas of the patient's treatment plan. Allow the patient to vent thoughts and feelings when appropriate to the time and place. Whenever possible, the same nurse should be assigned to the same patients to ensure the best possible continuity of care.

■ Never whisper or laugh when the patient cannot hear the whole conversation. Face the patient when having a conversation, even a personal conversation. Turning away may be interpreted as rejection.

■ Avoid placing the patient in situations of competition or embarrassment.

■ Never reinforce hallucinations, delusions, or illusions. It is necessary to keep the patient in reality as you know it. Some examples of responses to patients who are hallucinating are listed in Table 55–5.

■ Provide a therapeutic milieu. Structure the treatment setting in a way that promotes healthy behaviors and minimizes anxiety. Providing written instructions and information boards can help promote reality and self-responsible behavior.

■ Keep communication simple. Be brief and clear with all directions. State what is acceptable; give the rationale and consequences at the same time. Stating information in positive rather than negative terms is also helpful; for example, "Eat your food calmly" rather than "Do not throw your food!"

ᴠᴧ▶ CRITICAL THINKING

Anne

While preparing to invite Anne, a patient receiving chemotherapy on your oncology unit, to the movie in the day room, you observe her standing in the corner of her room trembling. You ask her what's wrong and she responds that she's talking to the woman in the wall. Your first instinct is to giggle, but you ask her, "What woman?" She tells you that you wouldn't understand and says, "You helped put her there and you told me that it is my job to be sure the woman can't get out." You report this to the charge nurse, who calls the doctor. Tests are run, and it is determined that Anne is not experiencing side effects from the chemotherapy. Further workup delivers the diagnosis of paranoid-type schizophrenia for Anne. What responses are appropriate for the situation above? What special needs might Anne now experience relating to her chemotherapy, if any? How will you get Anne to the movie or to participate in other care activities?

Answers at end of chapter.

Substance Abuse Disorders

Alcoholism and chemical dependency are serious conditions. People start using alcohol and drugs for many reasons, but often it is to feel accepted by a peer group or to feel comfortable in a social situation. People mistake the temporary high as a stimulant. In reality, alcohol is a depressant. You must understand that any chemical can be potentially dangerous.

Physical **dependence** on a substance, **tolerance** (the ability to endure the effects of a drug or the need for increasing amounts of the drug to produce the high), and **withdrawal** (unpleasant physical, psychological, or cognitive effects that result from decreasing or stopping the use of the chemical after regular use) are characteristics of addiction. The general definitions of substance abuse and substance dependence apply to any substance. Substance dependence is a condition in which a person has had several (usually three) of the following symptoms for 1 month or longer:

TABLE 55-5 SUGGESTED INTERVENTIONS FOR PATIENTS WITH SCHIZOPHRENIA WHO ARE HALLUCINATING

Suggested Action	Rationale
1. "Mr. R., I don't see any snakes. It is time for lunch. I will walk to the dining room with you."	1. Lets the patient know you heard him, but brings him immediately into the reality of time of day and need to go to the dining room.
2. "I see a crack in the wall, Mr. R. It is harmless; you are safe. Susan is here to take you down to occupational therapy now."	2. This is in response to a probable illusion. It lets the patient know that you see something. It validates his fear, but it tells him what you see and then moves him into the here and now.
3. "I know that your thoughts seem very real to you, Ms. C., but they do not seem logical to me. I would like for you to come to your room and get dressed now, please."	3. Again, you are validating the patient's concern without exploring and focusing on the delusion.
4. "Ms. C, it appears to me that you are listening to someone. Are you hearing voices other than mine?"	4. This is a method of validating your impression of what you see. This is as far as you will go into exploring what she may be hearing.
5. "Thank you, Ms. C. I want to help you focus away from the other voices. I am real; they are not. Please come with me to the reading room."	5. Responds to her in the present and reinforces her response to you. Attempts to redirect her thinking.

- The patient needs more of the substance and at more frequent intervals to achieve the "high," or the desired effect of the substance.
- The patient spends much time obtaining the substance.
- The patient gives up important social or professional functions to use the substance.
- The patient has tried at least once to quit but still obsesses about the substance.
- Misuse or withdrawal symptoms interfere with job, family, or social activities.
- The patient uses the substance regardless of the problems it causes.
- Tolerance increases greatly (by approximately 50 percent).
- The patient uses the substance to avoid withdrawal symptoms.

Substance abuse is defined differently than dependence and is often diagnosed by a rating system similar to the following:

- Mild—The person meets three of the above criteria, but social functioning is only minimally affected.
- Moderate—Symptoms are somewhere between mild and severe.
- Severe—The patient meets more than three of the above criteria, and social obligations are impaired.
- Partial remission—No substance use for more than 6 months; some symptoms have occurred.
- Full remission—No substance use for more than 6 months; no symptoms have occurred.

Not all patients who have a substance abuse disorder go through all stages, and not everyone goes through them in the same order.

Nurses need to be informed about chemical dependency for several reasons. First, many medical-surgical patients are chemically dependent. This affects their healing and the effect of their medications. Second, as part of the human ex-perience, your chance of being in a close personal relationship with a person who is chemically dependent is great. Third, and perhaps most importantly, you are part of a profession whose members are statistically high users and abusers of drugs and alcohol (Ethical Considerations Box 55-6). Studies indicate that 10 to 20 percent of nurses in the United States will be chemically impaired at some point in their lifetime.[4]

Substance abuse is not a one-person illness; it affects personal and professional relationships with people who are associated with the user. The term **dysfunctional** is often used to refer to the relationships within an alcoholic family or work environment. Dishonesty and inability to discuss the situation are strong components of the disease. Many times, people who live or work in the dysfunctional group begin to cover up for the user's behaviors and lack of responsibility. Family members or significant others often take sides, begin to be dishonest with each other, and erode the bond within that group. Eventually this leads to a condition called **codependence,** which can be as serious as the use and abuse of the substance. In codependence the significant others in the family group begin to lose their own sense of identity and purpose and exist solely for the abuser. Their actions take away the opportunity for the user to take responsibility for his or her own actions. This is called enabling.

Etiologic Theories

Why do some people become addicted or dependent and others do not? Can it be the chemical, or is it the person? Some theorists believe in the existence of an addictive personality, which may begin to explain addictions to food, sex, and gambling, as well as alcohol, chemicals, and other dependencies.

dysfunctional: dys—bad or difficult + functional—performance

BOX 55-6 ETHICAL CONSIDERATION

Ellen and Julie are LPNs who work on a busy medical unit in a large hospital. They were close friends in nursing school and have been working the night shift together for 3 years. Recently Ellen went through a difficult and painful divorce, and Julie has noticed that Ellen's personality has changed. Ellen, usually serious and almost compulsive in the completion of her work, has taken on a vary lackadaisical attitude, laughs or giggles at inappropriate times, and often calls in sick. She also displays hostility toward the hospital administration and seems very irritated when corrected. Julie has observed Ellen taking increasingly large doses of "nerve medication" and fears her friend is becoming a drug abuser.

One evening, Ellen arrives at work with glassy eyes and slurred speech. She asks Julie to watch her patients for her while she takes a little nap to "sleep it off." Julie asks Ellen if she is becoming a drug abuser, and after some initial denial, Ellen admits that she has been taking increasingly large doses of medication to continue functioning from day to day. Ellen pleads with Julie not to tell the head nurse or anyone else about the drug problem. Because of their friendship, Julie consents to cover for Ellen this night.

The next night, Ellen again comes to work with obvious signs of drug intoxication. She again asks Julie to cover for her, swearing that this would be the absolute last time it would happen. Ellen falls asleep while listening to the taped report for the change of shifts. What should Julie do?

The substance-abusing nurse is one of the most common situations nurses may encounter during their careers. It is estimated that about 40,000 nurses who work in the United States are alcoholics, and drug addiction among nurses is reported to be 30 to 100 times greater than among the general population. Nurses have easy access to drugs because they are usually responsible for obtaining and maintaining the supply of controlled drugs in a hospital. Other factors that contribute to the increased drug abuse among nurses include job stress, short staffing, double shifts, unrealistic expectations, frustration and anxiety, personal problems, and lack of autonomy in practice.

Professional and emotional conflicts are felt by the drug-abusing nurse's colleagues, who must use their professional training as promoters of health to try to understand a nurse's abuse of drugs. Underlying the ethical dilemma is the right of the patient to safe and competent treatment versus the nurse's right to self-determination. However, the nurse does not have a totally unrestricted right to self-determination. This right has limits when it comes to endangering others, especially the patients assigned to this nurse for their care and safety. Although the National Association for Practical Nurse Education and Services (NAPNES) Code does not directly address any obligation to protect the public from practitioners who are unsafe, various elements that are related include faithfulness and responsibility with a general admonition to protect patients.

In this particular situation Julie does have an obligation to do something about Ellen's drug problem. Covering for Ellen will not solve the problem. The first thing to do is to get a second opinion about Ellen's drug problem and then to confront her. She can also file a report through the institution's chain of command to the unit manager and head nurse and finally the director of nursing. If the institution hierarchy does not take any action, the nurse should submit the report to the state board of nursing. This report should be well documented and signed, and it should include a request for confidentiality if the nurse submitting the report does not want to be known. To allow Ellen to practice while under the influence of drugs puts Julie in a very serious legal position. If Ellen were to do something wrong and harm a patient and Julie knew that Ellen was under the influence of drugs, Julie could be held liable under the law, depending on the state in which she is practicing.

Psychoanalytical theories state that people who develop addictions to alcohol or other substances are people who failed to successfully pass through the "oral" stage of development.

Biologic theories include numerous studies that imply that there is some sort of genetic metabolic disorder. Many of these studies were done on twins born to an alcoholic parent or parents and who were separated from the parents at birth or shortly after birth. The number of twins who were born of alcoholic parents but raised by nonalcoholic adoptive or foster parents and yet developed alcoholism was consistently elevated.

Cognitive-behavioral theorists suggest the way in which a person perceives being high may influence the act of becoming high. It can be a very innocent beginning: Obtaining relief from the medications given by the doctor can, according to cognitive theory, leave people perceiving that the use of these drugs is a miracle cure. It becomes appealing to want that kind of relief again, and very soon a pattern is formed and other substances may be involved.

Differential Diagnosis

Commonly the alcoholic patient is admitted to the medical unit with primary medical diagnoses including dehydration, hyperemesis, or respiratory infections. Nursing assessment, laboratory test results, patient need for frequent pain medication, or symptoms of withdrawal may lead you or the physician to pursue the possibility of chemical dependency.

Types

ALCOHOL ABUSE AND DEPENDENCE. Use and abuse of alcohol is present in all walks of life, in all economic levels, and in both genders. Sometimes a very fine line exists between a person who is a social drinker and a person who has an abuse condition. One factor used to make that differentiation is the degree of need or compulsion to drink. There is a high incidence of alcohol use and abuse among the elderly, teenagers, and even younger children. Alcoholism either directly or indirectly decreases a person's life expectancy by an average of 10 to 12 years.

Denial is a common defense mechanism used by people who are substance abusers. The alcohol-dependent person often uses statements such as "I can quit anytime I want to" or "I just need a little bump to loosen me up."

Typical symptoms of alcohol dependence include the following:

- Impaired social function and relationships
- Inability to cut down or stop using; daily use is common
- Binges usually lasting 2 days or more
- Blackouts (amnesia while intoxicated)
- Vomiting
- Dehydration
- Disorientation
- Increased vulnerability to infections, accidents, and other injuries

Sometimes patients who are actively using drugs or alcohol when admitted to an inpatient setting, or who are cut off from their alcohol abruptly, experience a condition called **delirium tremens** (DTs). In DTs, hyperexcitability in the sensory activity of the individual can cause visual hallucinations, tremors, and possibly tonic-clonic seizures. Elevated blood pressure and pulse and cardiac dysrhythmias may also occur. This condition occurs within hours or days after the patient has stopped drinking and can last a week or more; hospitalization is necessary.

Treatments for Alcohol Abuse and Dependence. Perhaps the single most effective treatment for alcoholism is involvement in Alcoholics Anonymous (AA). Treatment for and recovery from alcohol dependency and abuse is a slow process. With very few exceptions, an alcoholic who is recovering cannot ever have another drink, or he or she will risk the chance of returning to previous abusive patterns.

Rationale-emotive therapy (RET) is used as an adjunct therapy for control of substance abuse. RET advocates believe that with homework and practice, a person can learn to think differently about the event that led to the drinking. When the person changes the belief system about the activating event and the drinking, the consequences of drinking will be less powerful. AA and RET are group activities.

Psychoanalysis may also be used. Psychoanalysis provides one-on-one therapy. Because addiction affects an entire family, family therapy is important in reinstating honesty in communications.

Medications are used cautiously because of risk for abuse. It is not always wise to substitute another chemical for alcohol. If, however, the anxiety level prohibits participation in therapy, or if a depressive disorder accompanies the abuse, medications may be prescribed. Antidepressant or antianxiety drugs are most often prescribed. Disulfiram (Antabuse) is a medication that is sometimes prescribed as a deterrent to using alcohol. Disulfiram should never be administered without full informed consent of the patient. If the patient taking disulfiram ingests alcohol, a severe reaction causes chest pain, nausea, vomiting, confusion, and other symptoms. Persons taking disulfiram can also be ad-

versely affected if they use products that contain alcohol, such as cologne, mouthwash, aftershave, or cough syrup. The effects of disulfiram last 2 to 3 weeks after the last dose.

Research is being conducted on other medications to assist in treating alcohol addictions. One such drug showing promise is acamprasate, which works on neurotransmitters to alter functions of other brain chemicals. Use of benzodiazepines (Valium, Ativan) can help prevent symptoms of DTs during acute withdrawal.

Milieu can be varied as well. Therapy may range from in-house hospitalizations of 2 weeks or more to independent attempts to help oneself by attending AA. It is not uncommon for patients to seek treatment more than once. This is not to be interpreted as a weakness in the individual or the treatment program. It is only a sign that the person is learning more about the disorder and the need to help himself or herself. People with all kinds of chronic disease experience relapse at times (Home Health Hints).

HOME HEALTH HINTS

- Listen to what your patient says and how he or she says it. listen to what is said by other persons in the home who are familiar with the patient.

- Look for signs of any chemical use, misuse, or abuse by the patient or any caregiver. Count pills and take note of alcohol containers in refrigerators, on counters, or in trash cans.

- Be observant of the health status of the patient's caregivers, with particular attention to signs of exhaustion, depression, or anger. Discuss this with your supervisor.

- Familiarize yourself with the mental health services and professionals in your community. Learn about criteria for admission, cost, and modalities of treatment.

Nursing Care for Patients with Alcohol Abuse and Dependence

- Be honest. Effective therapeutic communication is essential. You need to be in touch with your own thoughts and feelings about addictions.
- Provide group support. Many chemical dependency units provide group support meetings. Twelve-step programs are a popular form of group support.
- Be aware of the use of defense mechanisms. The patient should be confronted immediately if rationalization or denial behaviors are noted.
- Use positive reinforcement. Positive reinforcement for successes is important when helping a person with an addiction. Every step is a big one in this field; every step taken is a new one.
- Provide a safe environment. Patients who are chemically addicted may become suicidal or display other bizarre behavior, especially during DTs. A patient under the influence of alcohol or another chemical may have poor impulse control or judgment. Maintaining a safe milieu and calm demeanor will help the patient through this difficult time. Also, the fact that a patient is hospitalized does not

guarantee that he or she does not have access to the chemical or even use it in your presence. Unfortunately family members or friends sometimes smuggle drugs or alcohol in to the patient. You must be constantly alert. Suspicions should be expressed honestly and nonjudgmentally to the patient. All findings and behaviors that may be potential safety issues for the patient are reported and documented.

■ Practice "tough love." This concept encourages patients to be responsible for their own healing. "Doing for" patients may be tempting, but it is not in the patient's best interest most of the time. Praising (validating) the patient's attempts at self-responsible behavior and using therapeutic communication skills to constructively confront behaviors that are inconsistent with the plan of care are two examples of effective nursing interventions.

DRUG ABUSE AND DEPENDENCE. Many substances can be addictive to humans. Caffeine and nicotine are two that are very readily available. Coffee, tea, soda, and cigarettes are everywhere in our society and yet are very addicting. Many experts believe that the single most difficult addiction to overcome is the addiction to nicotine. Illegal substances such as marijuana, cocaine, crack, and PCP and prescription medications for pain and mental health treatment are also among potentially addictive substances.

It is becoming popular among the youth in the United States to use inhalants such as lighter fluid, paint, paint thinners, and gasoline to get high. The term for this is *huffing*. These are highly toxic, potentially lethal, and usually available in the house or garage.

Signs and Symptoms of Drug Abuse and Dependence. The signs and symptoms of drug abuse and dependence can be very similar to those of people who are alcohol abusers. Additional signs of drug abuse include the following:

■ Red, watery eyes
■ Runny nose
■ Hostile behavior
■ **Paranoia**
■ Needle tracks on arms or legs

Treatments for Patients with Drug Abuse and Dependence

■ Narcotics Anonymous
■ Group therapy
■ Psychotherapy
■ Methadone programs

Methadone acts as a sort of "stepdown" for people addicted to certain drugs. Methadone can be legally prescribed and dispensed. It, too, is potentially addicting, and its critics believe it is only a substitute for heroin. It is typically given once a day. Psychotherapy is also provided for patients in methadone programs.

Nursing Interventions for Patients with Drug Abuse and Dependence. Nursing care for people who are drug dependent is essentially the same as for those who are alcohol dependent. It is important to remember that nurses and doctors cannot "fix" the patient who is chemically dependent. The desire to be chemically free must come from the individual who is addicted.

Caring for patients with mental health disorders is challenging and rewarding. You will learn that there are very few absolutes in the area of mental health nursing. There are many guidelines about the illnesses, but caring for the patients who have the illnesses is as individualized as the patients themselves. It is also important to remember to care for the whole person. The mind and body work together, so be sure to take care of the physical, emotional, cognitive, and behavioral parts of patients.

CRITICAL THINKING

Maria

You are one of the team of school nurses in your local high school. You notice that Maria, a 17-year-old student, is behaving oddly. She has always been rather loud and even has been referred to as "obnoxious" by several of her peers. Lately you have observed her sitting alone, as if waiting for someone, but when you approach her, she barely greets you and then moves away. What are your concerns about Maria? What are some of the possibilities that might be affecting her? How will you approach her more effectively the next time you see her?

Answers at end of chapter.

REVIEW QUESTIONS

1. When assessing mental health, which patient behavior would cause the nurse to be concerned or ask further questions?
    a. Patient is always happy and smiling.
    b. Patient can verbalize emotions.
    c. Patient is able to cope with bad news.
    d. Patient maintains some close, personal relationships.

2. A person who always sounds like he or she is making excuses is displaying which defense mechanism?
    a. Denial
    b. Fantasy
    c. Rationalization
    d. Transference

3. Which of the following is one of the major skills a chemically dependent person and family can learn during treatment?
    a. Honest communication
    b. Codependency
    c. Denial
    d. Scapegoating

4. Which of the following best defines a therapeutic environment (milieu)?
    a. Able to provide for all the patient's needs
    b. Locked and supervised
    c. Structured to decrease stress and encourage learning new behavior
    d. Designed to be homelike for persons who are hospitalized for life

5. Which of the following does not state a goal of psychotherapy?
    a. Decrease emotional pain/discomfort.
    b. Increase social functioning.
    c. Increase ability to behave appropriately.
    d. Allow patient to avoid/deny uncomfortable situations.

6. Psychopharmacology is used for which of the following purposes?
    a. As a cure for mental illness and substance abuse
    b. Only when necessary to control violent behavior
    c. To alter the pain receptors in the brain
    d. To decrease symptoms and facilitate other therapies

7. If an extrapyramidal side effect such as tardive dyskinesia occurs, which of the following is the treatment of choice?
    a. Administer anticholinergic drugs as ordered.
    b. Discontinue the drugs per order.
    c. Increase the dose per order.
    d. Administer antianxiety drugs per order.

8. A patient who is a veteran of the Gulf War has difficulty driving through a parking ramp because "there are people hiding behind the pillars! They have guns! Be careful!" This person is most likely experiencing which of the following?
    a. Auditory hallucinations
    b. Flashbacks
    c. Delusions of grandeur
    d. Free-floating anxiety

REFERENCES

1. Purnell, L, and Paulanka, B (eds): Transcultural Health Care: A Culturally Competent Approach. FA Davis, Philadelphia, 1998.
2. Shives, L: Basic Concepts of Psychiatric Mental Health Nursing, ed 3. JB Lippincott, Philadelphia, 1994.
3. Bleuler, E: Dementia Praecox (Emil Kraepelin) or the Group of Schizophrenias. International Press, New York, 1911, p 26.
4. Skinner, K: The hazards of chemical dependency among nurses. J Pract Nurs 8, December 1993.

Answers to Critical Thinking

Joe

Different people cope in different ways. This may be a healthy way to cope in this gentleman's culture. Gently guide the grieving husband to a room where he can express his emotions without disturbing others. Ask if he would like you to contact someone to come in to support him. Document objectively: "Patient's husband weeping loudly; guided to consultation room for privacy."

Mrs. Beison

Mrs. Beison is using denial to cope with her cancer diagnosis. Although at times denial can be an effective coping mechanism, if Mrs. Beison continues to deny her disease and refuse treatment, her life will be in danger. The nurse can help Mrs. Beison verbalize her fears about cancer and cancer treatment and can provide accurate information to help her make wise choices. If necessary, a psychiatric evaluation can be requested.

Tommy

You can reassure Tommy's mother that his OCD is not her fault. Tommy can learn to control his illness with medications and therapy. The family must be part of the therapy, for both Tommy's sake and the family's sake. Positive communication between Tommy and his family is encouraged. Tommy's mother can also be encouraged to attend a support group herself.

Mr. Zenz

It is important to be supportive of Mrs. Zenz while maintaining Mr. Zenz's confidentiality and privacy. Encouraging Mrs. Zenz to talk to Mr. Zenz's physician is acceptable. Showing empathy with statements such as, "It must be confusing and difficult to watch your husband change moods so quickly" are good tools to use. It may be a bit more challenging to approach him as your boss. You may certainly ask if you can speak frankly and share specific observations. You may share your concern, such as "Mr. Zenz, you are a wonderful boss, but I am frightened when you become loud and boisterous." This may help him to reflect. Chances are, however, that if he is in manic stage, he will not hear your concern. It may require delicate talking to the next in the corporate chain of command. This is a tough one. Good luck!

Anne

Appropriate communication skills are to be positive, reassuring and not reinforcing of the hallucinations. "I don't see or hear a woman, Anne. It is time for the movie. I'd like you to come with me for a while at least" is an example of an appropriate verbal interaction. Reinforcing expectations is also appropriate. "Anne, part of your care plan includes attending one major activity per day. This is the last opportunity for you to meet your care plan objective for today" is also acceptable. At all times, nurses need to be aware of drug interactions. Anne will most likely be medicated for her schizophrenia and those medications may interact unfavorably with her chemotherapy. Good nursing data collection skills are essential.

Maria

A number of options may explain Maria's behavior, including depression, drug use, schizophrenia, anorexia, or bulimia. Next time you see Maria you might try constructively confronting her behavior by saying something like "Maria, you used to be much more outgoing. We always were friendly and now you leave when I'm near. That change in you concerns me. I'm here if you want to talk." Or, "Maria, I see your behavior is changing. You are loud one moment and very quiet the next. That is unusual for you. What is happening?"

UNIT SIXTEEN BIBLIOGRAPHY

Alcohol, Drug Abuse and Mental Health Administration. DHHS pub. No. (ADM) 90-1609, Bethesda, MD, 1990.

American Psychiatric Association. Diagnostic and Statistical Manual of Mental Disorders: DSM-IV, ed 4. Author, Washington, DC, 2000.

Bauer, J: RN news watch: Complementary therapies. Can St. John's wort really alleviate depression? RN 64(9):20, 2001.

Bonner, J: Cancer vaccines. Chemistry & Industry 9:277, 2001.

Bradley-Springer, L: HIV prevention: What works? American Journal of Nursing 101:6, 2001.

Cross, ML, and Gill, HS: Can immunoregulatory lactic acid bacteria be used as dietary supplements to limit allergies? International Archives of Allergy and Immunology 125(2):112, 2001.

D'Amato, G, et al: The role of outdoor air pollution and climatic changes on the rising trends in respiratory allergy. Respiratory Medicine 95(7):606, 2001.

Esch, JF, and Frank, SV: HIV drug resistance and nursing practice. American Journal of Nursing 101:6, 2001.

Fox, BR, et al: Multiple chemical sensitivity. Risk Management 48(5):6, 2001.

Gause, A, and Berek, C: Role of B cells in the pathogenesis of rheumatoid arthritis: Potential implications for treatment. BioDrugs 15(2):73, 2001.

Helpful Facts About Depressive Illness. DHHS-NIH pub. No. 94-3875, Bethesda, MD, 1994.

Johnsen, C: Preventing perinatal HIV transmission. Nursing Profile 2(4), 2001.

Karon, JM, et al: HIV in the United States at the turn of the century: An epidemic in transition. American Journal of Public health 91:1060–1068, 2001.

Kirk, AD: Immunosuppression without immunosuppression? How to be a tolerant individual in a dangerous world. Transplant Infectious Disease 1(1):65, 1999.

Konsman, JP, and Danizer, R: How the immune and nervous systems interact during disease-associated anorexia. Nutrition 17(7):664, 2001.

Osundeko, O, et al: Anticardiolipin antibodies in Hashimoto's disease. Endocrine Practice 7(3):181, 2001.

Pequegnant, W, and Stover, E: Behavioral prevention is today's AIDS vaccine! AIDS 14(Suppl2):S1, 2000.

Porakishvili, N, et al: Recent progress in the understanding of B-cell functions in autoimmunity. Scandinavian Journal of Immunology 54(1–2):30, 2001.

Reith, W, and Mach, B: The bare lymphocyte syndrome and the regulation of MHC expression. Annual Review of Immunology 19:331, 2001.

Sloan, A, and Vernarec, E: Impaired nurses: Reclaiming careers. RN 64(2), 2001.

Smith, GB, and Dandorth, D: Persons living with HIV disease and AIDS. In CA Glod (ed.), Contemporary Psychiatric-Mental Health Nursing: The Brain-Behavior Connection, pp 604–619. FA Davis, Philadelphia, 1998.

Sowell, R: AIDS care nursing: Looking forward, thinking back. Journal of the Association of Nurses in AIDS Care 11:1, 2000.

Townsend, MC: Psychiatric Mental Health Nursing: Concepts of Care. FA Davis, Philadelphia, 2000.

Ungvarski, PJ: The past 20 years of AIDS. American Journal of Nursing 101:6, 2001.

Ungvarski, PJ, and Grossman, AH: Health problems of gay and bisexual men. Nursing Clinics of North America 34:2, 1999.

Weiten, W: Psychology Themes and Variations—Briefer Version, ed 3. Brooks-Cole, Pacific Grove, CA, 1997.

Weitzel, CA: Could you spot this psych emergency? RN 63(9):35–40, 2000.

Williams, A: Adherence to HIV regimens: Ten vital lessons. American Journal of Nursing 101:6, 2001.

Appendix A

North American Nursing Diagnosis Association (NANDA) Nursing Diagnoses

Activity Intolerance
Activity Intolerance, risk for*
Airway Clearance, ineffective
Allergy Response, latex
Type IV reactions [chemical and delayed-type hypersensitivity]: • Delayed onset (hours) • Eczema • Irritation • Reaction to additives (e.g., thiurams, carbamates) causes discomfort • Redness
Anxiety, death
Aspiration, risk for
Attachment, risk for impaired parent/infant/child
Autonomic Dysreflexia
Autonomic Dysreflexia, risk for
Body Image, disturbed
Body Temperature, risk for imbalanced
Bowel Incontinence
Breastfeeding, effective
Breastfeeding, ineffective
Breastfeeding, interrupted
Breathing Pattern, ineffective
Cardiac Output, decreased
Caregiver Role Strain
Caregiver Role Strain, risk for
Communication, impaired verbal
Conflict, decisional (specify)
Confusion, acute
Confusion, chronic
Constipation
Constipation, perceived
Constipation, risk for
Coping, community, ineffective

Coping, community, readiness for enhanced
Coping, defensive
Coping, family: compromised
Coping, family: disabled
Coping, family: readiness for enhanced
Coping, ineffective
Denial, ineffective
Development, risk for delayed
Diarrhea
Disuse Syndrome, risk for
Diversional Activity, deficient
Energy Field, disturbed
Environmental Interpretation Syndrome, impaired
Failure to Thrive, adult
Falls, risk for
Family Processes, dysfunctional: alcoholism
Family Processes, interrupted
Fatigue
Fear [specify focus]
[Fluid Volume, deficient (hyper/hypotonic)]
Fluid Volume, deficient [isotonic]
Fluid Volume, excess
Fluid Volume, risk for deficient
Fluid Volume, risk for imbalanced
Gas Exchange, impaired
Grieving, anticipatory
Grieving, dysfunctional
Growth, risk for disproportionate
Growth and Development, delayed
Health Maintenance, ineffective
Health-Seeking Behaviors (specify)

Home Maintenance, impaired
Hopelessness
Hyperthermia
Hypothermia
Infant Behavior, disorganized
Infant Behavior, risk for disorganized
Infant Behavior, readiness for enhanced organized
Infant Feeding Pattern, ineffective
Infection, risk for
Injury, risk for
Injury, risk for perioperative-positioning
Knowledge, deficient [Learning Need] (specify)
Loneliness, risk for
Memory, impaired
Mobility, impaired bed
Mobility, impaired physical [specify level]
Mobility, impaired wheelchair
Nausea
Noncompliance [Adherence, ineffective] (specify)
Nutrition: imbalanced, less than body requirements
Nutrition: imbalanced, more than body requirements
Nutrition: imbalanced, risk for more than body requirements
Oral Mucous Membrane, impaired
Pain, acute
Pain, chronic
Parental Role Conflict
Parenting, impaired
Parenting, risk for impaired
Peripheral Neurovascular Dysfunction, risk for
Poisoning, risk for
Post-Trauma Syndrome [specify stage]
Post-Trauma Syndrome, risk for
Powerlessness [specify level]
Powerlessness, risk for
Protection, ineffective
Rape-Trauma Syndrome
Rape-Trauma Syndrome: compound reaction
Rape-Trauma Syndrome: silent reaction
Relocation Stress Syndrome
Relocation Stress Syndrome, risk for
Role Performance, ineffective
Self-Care Deficit [specify level] feeding, bathing/hygiene, dressing/grooming, toileting
Self-Esteem, chronic low

Self-Esteem, situational low
Self-Esteem, risk for situational low
Self-Mutilation
Self-Mutilation, risk for
Sexuality, dysfunction
Sexuality Patterns, ineffective
Skin Integrity, impaired
Skin Integrity, risk for impaired
Sleep Deprivation
Sleep Pattern, disturbed
Social Isolation
Sorrow, chronic
Spiritual Distress
Spiritual Distress, risk for
Suffocation, risk for
Suicide, risk for
Surgical Recovery, delayed
Swallowing, impaired
Therapeutic Regimen: Community, ineffective management
Therapeutic Regimen, effective management
Therapeutic Regimen, Family, ineffective management
Therapeutic Regimen, ineffective management
Thermoregulation, ineffective
Thought Processes, disturbed
Tissue Integrity, impaired
Tissue Perfusion, ineffective (specify): renal, cerebral, cardiopulmonary, gastrointestinal, peripheral
Transfer Ability, impaired
Trauma, risk for
Unilateral Neglect
Urinary Elimination, impaired
Urinary Incontinence, functional
Urinary Incontinence, reflex
Urinary Incontinence, stress
Urinary Incontinence, total
Urinary Incontinence, urge
Urinary Incontinence, risk for urge
Urinary Retention [acute/chronic]
Ventilation, impaired spontaneous
Ventilatory Weaning Response, dysfunctional
Violence, [actual]/risk for other-directed
Violence, [actual]/risk for self-directed
Walking, impaired
Wandering

Appendix B
NORMAL REFERENCE LABORATORY VALUES

BLOOD, PLASMA, OR SERUM VALUES

Determination	Conventional	SI
		Reference Range
Aldolase	1.3–8.2 U/L	22–137 nmol · sec⁻¹/L
Ammonia	12–55 μmol/L	12–55 μmol/L
Amylase	4–25 units/mL	4–25 arb. unit
Ascorbic acid	0.4–1.5 mg/100 mL	23–85 μmol/L
Bilirubin	Direct: up to 0.4 mg/100 mL	Up to 7 μmol/L
	Total: up to 1.0 mg/100 mL	Up to 17 μmol/L
Blood volume	8.5–9.0% of body weight in kg	80–85 mL/kg
Calcium	8.5–10.5 mg/100 mL (slightly higher in children)	2.1–2.6 mmol/L
Carbamazepine	4.0–12.0 mg/mL	17–51 μmol/L
Carbon dioxide content	24–30 mEq/L	24–30 mmol/L
Chloride	100–106 mEq/L	100–106 mmol/L
CK isoenzymes	5% MB or less	
Creatine kinase (CK)	Female: 10–79 U/L	167–1317 nmol · sec⁻¹/L
	Male: 17–148 U/L	283–2467 nmol · sec⁻¹/L
Creatinine	0.6–1.5 mg/100 mL	53–133 μmol/L
Ethanol	0 mg/100 mL	0 mmol/L
Glucose	Fasting: 70–110 mg/100 mL	3.9–5.6 mmol/L
Iron	50–150 mg/100 mL (higher in males)	9.0–26.9 μmol/L
Iron-binding capacity	250–410 μg/100 mL	44.8–73.4 μmol/L
Lactic dehydrogenase	45–90 U/L	750–1500 nmol · sec⁻¹/L
Lipase	2 units/mL or less	Up to 2 arb. unit
Lipids		
Cholesterol	120–220 mg/100 mL	3.10–5.69 mmol/L
Very low density lipoprotein	13–32 mg/dL	
Low density lipoprotein	38–40 mg/dL	
High density lipoprotein	20–48 mg/dL	
Triglycerides	40–150 mg/100 mL	0.4–1.5 g/L
Lithium	0.5–1.5 mEq/L	0.5–1.5 mmol/L
Magnesium	1.5–2.0 mEq/L	0.8–1.3 mmol/L
Osmolality	280–296 mOsm/kg water	280–296 mmol/kg
Oxygen saturation (arterial)	96–100%	0.96–1.00
PCO₂	35–45 mm Hg	4.7–6.0 kPa
pH	7.35–7.45	Same
PO₂	75–100 mm Hg (dependent on age) while breathing room air	
	Above 500 mm Hg while on 100% O₂	10.0–13.3 kPa
Phenobarbital	15–50 μg/mL	65–215 μmol/L
Phenytoin (Dilantin)	5–20 μg/mL	20–80 μmol/L
Phosphatase (acid)	Male—Total: 0.13–0.63 sigma U/mL	36–175 nmol · sec⁻¹/L
	Female—Total: 0.01–0.56 sigma U/mL	2.8–156 nmol · sec⁻¹/L
	Prostatic: 0–0.5 Fishman–Lerner U/100 mL	

BLOOD, PLASMA, OR SERUM VALUES—CONT'D

Determination	Reference Range	
	Conventional	SI
Phosphatase (alkaline)	13–39 U/L, infants and adolescents up to 104 U/L	217–650 nmol · sec^{-1}/L, up to 1.26 μmol · sec^{-1}/L
Phosphorus (inorganic)	3.0–4.5 mg/100 mL (infants in first year up to 6.0 mg/100 mL)	1.0–1.5 mmol/L
Potassium	3.5–5.0 mEq/L	3.5–5.0 mmol/L
Protein: Total	6.0–8.4 g/100 mL	60–84 g/L
Albumin	3.5–5.0 g/100 mL	35–50 g/L
Globulin	2.3–3.5 g/100 mL	23–35 g/L
Salicylate:	0	
Therapeutic	20–25 mg/100 mL	1.4–1.8 mmol/L
	25–30 mg/100 mL to age 10 yr 3 hr post dose	1.8–2.2 mmol/L
Sodium	135–145 mEq/L	135–145 mmol/L
Transaminase, aspartate aminotransferase	7–27 U/L	117–450 nmol · sec^{-1}/L
Transaminase, alanine aminotransferase	1–21 U/L	17–350 nmol · sec^{-1}/L
Urea nitrogen (BUN)	8–25 mg/100 mL	2.9–8.9 mmol/L
Uric acid	3.0–7.0 mg/100 mL	0.18–0.42 mmol/L

URINE VALUES

Determination	Reference Range	
	Conventional	SI
Acetone plus acetoacetate (quantitative)	0	0 mg/L
Amylase	24–76 units/mL	24–76 arb. unit
Calcium	300 mg/day or less	7.5 mmol/day or less
Catecholamines	Epinephrine: under 20 mg/day	<109 nmol/day
	Norepinephrine: under 100 μg/day	<590 nmol/day
Creatine	Under 100 mg/day or less than 6% of creatine. In pregnancy: up to 12%. In children younger than 1 yr: may equal creatinine. In older children: up to 30% of creatinine.	<0.75 mmol/day
Creatinine	15–25 mg/kg of body weight/day	0.13–0.22 mmol · kg^{-1}/day
Cystine or cysteine	0	0
Hemoglobin and myoglobin	0	
pH	5–7	5–7
Phosphorus (inorganic)	Varies with intake; average, 1 g/day	32 mmol/day
Protein:		
Quantitative	<150 mg/24 hr	<0.15 g/day
Steroids:		

Steroids:
17-Ketosteroids (per day)

Age	Male	Female		
10	1–4 mg	1–4 mg	3–14 μmol	3–14 μmol
20	6–21	4–16	21–73	14–56
30	8–26	4–14	28–90	14–49
50	5–18	3–9	17–62	10–31
70	2–10	1–7	7–35	3–24

Determination	Conventional	SI
17-Hydroxysteroids	3–8 mg/day (women lower than men)	8–22 μmol/day as tetrahydrocortisol
Sugar:		
Quantitative glucose	0	0 mmol/L
Urobilinogen	Up to 1.0 Ehrlich U	To 1.0 arb. unit

HEMATOLOGIC VALUES

Determination	Reference Range	
	Conventional	SI
Coagulation screening tests:		
Bleeding time (Simplate)	3–9.5 min	180–570 sec
Prothrombin time	Less than 2-sec deviation from control	Less than 2-sec deviation from control
Partial thromboplastin time (activated)	25–38 sec	25–38 sec
Whole-blood clot lysis	No clot lysis in 24 hr	0/day
"Complete" blood count:		
Hematocrit	Male: 45–52%	Male: 0.45–0.52
	Female: 37–48%	Female: 0.37–0.48
Hemoglobin	Male: 13–18 g/100 mL	Male: 8.1–11.2 mmol/L
	Female: 12–16 g/100 mL	Female: 7.4–9.9 mmol/L
Leukocyte count	4300–10,800/mm³	$4.3–10.8 \times 10^9$/L
Erythrocyte count	4.2–5.9 million/mm³	$4.2–5.9 \times 10^{12}$/L
Mean corpuscular volume (MCV)	86–98 μm³/cell	86–98 fl
Mean corpuscular hemoglobin (MCH)	27–32 pg/RBC	1.7–2.0 pg/cell
Mean corpuscular hemoglobin concentration (MCHC)	32–36%	0.32–0.36
Platelet count	150,000–350,000/mm³	$150–350 \times 10^9$/L

MISCELLANEOUS VALUES

Determination	Reference Range	
	Conventional	SI
Carcinoembryonic antigen (CEA)	0–2.5 ng/mL	0–2.5 μg/L
Digoxin	1.2 ± 0.4 ng/mL	1.54 ± 0.5 nmol/L
	1.5 ± 0.4 ng/mL	1.92 ± 0.5 nmol/L
Gastric analysis	Basal:	
	Females: 2.0 ± 1.8 mEq/hr	0.6 ± 0.5 μmol/sec
	Males: 3.0 ± 2.0 mEq/hr	0.8 ± 0.6 μmol/sec
	Maximal (after histalog or gastrin):	
	Females: 16 ± 5 mEq/hr	4.4 ± 1.4 μmol/sec
	Males: 23 ± 5 mEq/hr	6.4 ± 1.4 μmol/sec
Gastrin-I	0–200 pg/mL	0–95 pmol/L
Immunologic tests:		
Alpha-fetoprotein	Undetectable in normal adults	
Alpha-1-antitrypsin	85–213 mg/100 mL	0.85–2.13 g/L
Rheumatoid factor	<60 IU/mL	
Antinuclear antibodies	Negative at a 1:8 dilution of serum	

Reference range values may differ from one institution to another. These data from Scully, Robert E (ed.): Case Records of the Massachusetts General Hospital, New England Journal of Medicine 314:39–49, 1986.

Appendix C
ANSWERS TO
REVIEW QUESTIONS

Chapter 1: 1. c; 2. a; 3. c; 4. d; 5. b; 6. d; 7. c
Chapter 2: 1. d; 2. c; 3. a; 4. c; 5. d
Chapter 3: 1. b; 2. d; 3. b; 4. a; 5. c; 6. a
Chapter 4: 1. c; 2. c; 3. d; 4. a; 5. b
Chapter 5: 1. b; 2. c; 3. a; 4. d; 5. a; 6. b
Chapter 6: 1. b; 2. c; 3. c; 4. d; 5. a; 6. b
Chapter 7: 1. c; 2. d; 3. a; 4. c; 5. c
Chapter 8: 1. c; 2. c; 3. c; 4. b; 5. b
Chapter 9: 1. b; 2. a; 3. c; 4. c; 5. d; 6. d
Chapter 10: 1. b; 2. a; 3. d; 4. a; 5. c; 6. d
Chapter 11: 1. a; 2. b; 3. d; 4. a; 5. b; 6. a; 7. d
Chapter 12: 1. d; 2. b; 3. a; 4. d; 5. b
Chapter 13: 1. d; 2. d; 3. c; 4. c; 5. d
Chapter 14: 1. d; 2. b; 3. a; 4. c; 5. d
Chapter 15: 1. a; 2. c; 3. b; 4. d; 5. b; 6. c; 7. d; 8. a; 9. b; 10. d; 11. a; 12. b
Chapter 16: 1. d; 2. b; 3. b; 4. c; 5. a
Chapter 17: 1. a; 2. d; 3. b; 4. a; 5. a
Chapter 18: 1. b; 2. a; 3. d; 4. c; 5. c
Chapter 19: 1. b; 2. b; 3. b; 4. d; 5. c
Chapter 20: 1. b; 2. c; 3. c; 4. a; 5. b
Chapter 21: 1. b; 2. a; 3. a; 4. c; 5. a
Chapter 22: 1. c; 2. b; 3. a; 4. c; 5. d
Chapter 23: 1. b; 2. c; 3. b; 4. c; 5. b; 6. a
Chapter 24: 1. b; 2. c; 3. d; 4. a; 5. d; 6. a; 7. a
Chapter 25: 1. c; 2. b; 3. b; 4. d; 5. a
Chapter 26: 1. a; 2. c; 3. b; 4. d; 5. a; 6. b; 7. c; 8. d
Chapter 27: 1. d; 2. b; 3. d; 4. a; 5. a; 6. c
Chapter 28: 1. c; 2. d; 3. b; 4. a; 5. a; 6. c; 7. b
Chapter 29: 1. d; 2. b; 3. b; 4. b; 5. c
Chapter 30: 1. b; 2. c; 3. d; 4. c; 5. c
Chapter 31: 1. a; 2. b; 3. a; 4. a; 5. b
Chapter 32: 1. a; 2. b; 3. b; 4. c; 5. a; 6. d
Chapter 33: 1. d; 2. b; 3. b; 4. c; 5. b; 6. d; 7. a
Chapter 34: 1. c; 2. b; 3. d; 4. c; 5. d
Chapter 35: 1. a; 2. b; 3. c; 4. a; 5. b
Chapter 36: 1. a; 2. b; 3. d; 4. c; 5. a; 6. b
Chapter 37: 1. b; 2. c; 3. a; 4. a; 5. d; 6. b; 7. a; 8. b
Chapter 38: 1. a; 2. b; 3. b; 4. d; 5. a; 6. c; 7. a; 8. b

Chapter 39: 1. c; 2. d; 3. d; 4. c; 5. b; 6. c
Chapter 40: 1. a. 2. b; 3. d; 4. c; 5. c; 6. c
Chapter 41: 1. c; 2. b; 3. c; 4. a; 5. d; 6. b
Chapter 42: 1. c; 2. a; 3. d; 4. b; 5. c; 6. b
Chapter 43: 1. b; 2. c; 3. c; 4. c; 5. a
Chapter 44: 1. b; 2. d; 3. b; 4. a; 5. c
Chapter 45: 1. b; 2. b; 3. a; 4. d; 5. c; 6. b; 7. a; 8. b
Chapter 46: 1. c; 2. d; 3. a; 4. b; 5. d; 6. c; 7. a; 8. b
Chapter 47: 1. c; 2. b; 3. a; 4. d; 5. c; 6. c
Chapter 48: 1. d; 2. b; 3. b; 4. a; 5. d
Chapter 49: 1. d; 2. b; 3. c; 4. a; 5. c
Chapter 50: 1. b; 2. c; 3. b; 4. a; 5. d; 6. a
Chapter 51: 1. d; 2. c; 3. a; 4. c; 5. c; 6. c; 7. b; 8. a
Chapter 52: 1. d; 2. a; 3. c; 4. c; 5. a
Chapter 53: 1. d; 2. b; 3. c; 4. a; 5. d
Chapter 54: 1. c; 2. d; 3. c; 4. a; 5. d
Chapter 55: 1. a; 2. c; 3. a; 4. c; 5. d; 6. d; 7. a; 8. b

Appendix D

MEDICAL ABBREVIATIONS

ABG	arterial blood gas
a.c.	before a meal
AD	advance directive
ad lib.	freely; as desired
ALT	alanine aminotransferase
AM	morning
A-P	anterior-posterior
AST	aspartate aminotransferase
AV	atrioventricular
b.i.d.	twice a day
BM	bowel movement
BP	blood pressure
BUN	blood urea nitrogen
$\bar{c}$	with
cap.	a capsule
CBC	complete blood count
cc	cubic centimeter
cm	centimeter
CNS	central nervous system
CSF	cerebrospinal fluid
CV	cardiovascular
D and C	dilatation and curettage
dc	discontinue
DNR	do not resuscitate
DOA	dead on arrival
dr.	dram
Dx	diagnosis
ECF	extracellular fluid
ECG	electrocardiogram
ECT	electroconvulsive therapy
EEG	electroencephalogram
EMG	electromyogram
EMS	emergency medical service
ENT	ear, nose, and throat
EOM	extraocular muscles
ER	Emergency Room
ESR	erythrocyte sedimentation rate
F	Fahrenheit
g, gm	gram
GERD	gastroesophageal reflux disease
GI	gastrointestinal
gr	grain
Gtt, gtt	drops
GYN	gynecology
h, hr	hour
hgb	hemoglobin
hor. som., h.s.	bedtime
IM	intramuscular
IUD	intrauterine device
IV	intravenous
IVP	intravenous pyelogram
J	joule
kg	kilogram
KUB	kidney, ureter, and bladder
L	liter
lb	pound
lmp	last menstrual period
mEq	milliequivalent
mg	milligram
ml	milliliter
mm	millimeter
MRI	magnetic resonance imaging
MS	mitral stenosis; multiple sclerosis
mEq	microequivalent
mg	microgram
n.p.o.	nothing by mouth
NSAID	nonsteroidal anti-inflammatory drug
NSR	normal sinus rhythm
OB	obstetrics
OC	oral contraceptive
OD	right eye
OS	left eye
OU	both eyes
oz	ounce
$\bar{p}$	after
p.c.	after meals
PCO_2	carbon dioxide pressure
PERRLA	pupils equal, regular, react to light and accommodation
pH	hydrogen ion concentration
PM	afternoon/evening
PMI	point of maximal impulse

post.	posterior
p.r.	through the rectum
p.r.n.	as needed
qh	every hour
q2h	every 2 hours
q3h	every 3 hours
q.i.d.	four times a day
q.s.	as much as is needed
RBC	red blood cell; red blood count
s̄	without
SA	sinoatrial
SC, sc, s.c.	subcutaneous(ly)
SOB	shortness of breath
s.o.s.	if necessary
s.q.	subcutaneous(ly)
stat.	immediately

STD	sexually transmitted disease
T	temperature
tab.	medicated tablet
temp.	temperature
t.i.d.	three times a day
top.	topically
URI	upper respiratory infection
USP	United States Pharmacopeia
UTI	urinary tract infection
WBC	white blood cell; white blood count
WF/BF	white female/black female
WM/BM	white male/black male
wt.	weight

Adapted from Thomas, CL: Taber's Cyclopedic Medical Dictionary, 18th ed. FA Davis, Philadelphia, 1997, pp 2224–2226.

Appendix E
PREFIXES, SUFFIXES, AND COMBINING FORMS

a-, an-. Without; away from; not.

ab-, abs-. From; away from; absent.

abdomin-, abdomino-. Abdomen.

-ad. Toward; in the direction of.

aden-, adeno-. Gland.

adip-, adipo-. Fat.

-aemia. Blood.

aer-, aero-. Air.

-algesia, -algia. Suffering; pain.

andro-. Man; male; masculine.

angi-, angio-. Blood or lymph vessels.

aniso-. Unequal; asymmetrical; dissimilar.

ankyl-, ankylo-. Crooked; bent; fusion or growing together of parts.

ante-. Before.

antero-. Anterior; front; before.

ant-, anti-. Against.

arteri-, arterio-. Artery.

arthr-, arthro-. Joint.

-ase. Enzyme.

-asis, esis, -iasis, -isis, -sis. Condition; pathological state.

aut-, auto-. Self.

axo-. Axis; axon.

bacteri-, bacterio-. Bacteria; bacterium.

bi-, bis-. Two; double; twice.

bili-. Bile.

bio-. Life.

blast-, -blast. Germ; bud; embryonic state of development.

blephar-, blepharo-. Eyelid.

brady-. Slow.

bronch-, bronchi-, broncho-. Airway.

cardi-, cardio-. Heart.

cat-, cata-, cath-, kat-, kata-. Down; downward; destructive; against; according to.

cent-. Hundred.

cephal-, cephalo-. Head.

cervic-, cervico-. Head; the neck of an organ.

chrom-, chromo-. Color.

-cide. Causing death.

contra-. Against; opposite.

crani-, cranio-. Skull; cranium.

cry-, cryo-. Cold.

cyan-, cyano-. Blue.

cyst-, cysto-, -cyst. Cyst; urinary bladder.

cyt-, cyto-, -cyte. Cell.

derm-, derma-, dermato-, dermo-. Skin.

di-. Double; twice; two; apart from.

dors-, dorsi-, dorso-. Back.

-dynia. Pain.

dys-. Difficult; bad; painful.

ec-, ecto-. Out; on the outside.

-ectomy. Excision.

ef-, es-, ex-, exo-. Out.

electr-, electro-. Electricity.

-emesis. Vomiting.

-emia. Blood.

en-. In; into.

end-, endo-. Within.

ent-, ento-. Within; inside.

enter-, entero-. Intestine.

ep-, epi-. Upon; over; at; in addition to; after.

erythr-, erythro-. Red.

eury-. Broad.

ex-. Out; away from; completely.

exo-. Out; outside of; without.

extra-. Outside of; in addition; beyond.

-facient. Causing; making happen.

-ferous. Producing.

ferri-, ferro-. Iron.

fluo-. Flow.

fore-. Before; in front of.

-form. Form.

-fuge. To expel; to drive away; fleeing.

gaster-, gastero-, gastr-, gastro-. Stomach.

gen-. Producing; forming.

-gen, -gene, -genesis, -genetic, -genic. Producing; forming.

glosso-. Tongue.

gluc-, gluco-, glyc-, glyco-. Sugar; glycerol or similar substance.

gyn-, gyne-, gyneco-, gyno-. Woman; female.

hem-, hema-, hemato-, hemo-. Blood.

hemi-. Half.

hepat-, hepato-. Liver.

heter-, hetero-. Other; different.

histo-. Tissue.

homo-. Same; likeness.

hydra-, hydro, hydr-. Water.

hyp-, hyph-, hypo-. Less than; below; under.

hyper-. Above; excessive; beyond.

hyster-, hystero-. Uterus.

-ia. Condition, esp. an abnormal state.

-iasis. SEE: *-asis*.

-iatric. Medicine; medical profession; physicians.

in-. In; inside; within; intensive action; negative.

infra-. Below; under; beneath; inferior to; after.

inter-. Between; in the midst.

intra-, intro-. Within; in; into.

ipsi-. Same.

irid-, irido-. Iris.

-ism. Condition; theory.

iso-. Equal.

-itis. Inflammation of.

kera-, kerato-. Horny substance; cornea.

kolp-, kolpo, colp-, colpo-. Vagina.

kypho-. Humped.

leuk-, leuko-. White; colorless; rel. to a leukocyte.

lip-, lipo-. Fat.

-lite, -lith, lith-, litho-. Stone; calculus.

-logia, -logy. Science of; study of.

lumbo-. Loins.

-lysis. 1. Setting free; disintegration. 2. In medicine, reduction of; relief from.

macr-, macro-. Large; long.

mal-. Ill; bad; poor.

med-, medi-, medio-. Middle.

mega-, megal-, megalo-. Large; of great size.

-megalia, -megaly. Enlargement of a body part.

melan-, melano-. Black.

mening-, meningo-. Meninges.

-meter. Measure.

metr-, metra-, metro-. Uterus.

micr-, micro-. Small.

mon-, mono-. Single; one.

muc-, muci-, muco-, myxa-, myxo-. Mucus.

multi-. Many; much.

musculo-, my-, myo-. Muscle.

my-, myo-. SEE *musculo-*.

myel-, myelo-. Spinal cord; bone marrow.

naso-. Nose.

necr-, necro-. Death; necrosis.

neo-. New; recent.

nephr-, nephra-, nephro-. Kidney.

neur-, neuri-, neuro-. Nerve; nervous system.

non-. No.

normo-. Normal; usual.

oculo-. Eye.

-ode, -oid. Form; shape; resemblance.

-odynia, odyno-. Pain.

olig-, oligo-. Few; small.

-ology. Science of; study of.

-oma. Tumor.

onco-. Tumor; swelling; mass.

oo-, ovi-, ovo-. Egg; ovum.

oophor-, oophoro-, oophoron-. Ovary.

ophthalm-, ophthalmo-. Eye.

-opia. Vision.

optico-, opto-. Eye; vision.

orchi-, orchid-, orchido-. Testicle.

orth-, ortho-. Straight; correct; normal; in proper order.

os-. Mouth; bone.

-osis. Condition; status, process; abnormal increase.

oste-, osteo-. Bone.

ot-, oto-. Ear.

-otomy. Cutting.

-ous. 1. Possessing; full of; 2. Pertaining to.

pan-. All; entire.

para-, -para. 1. Prefix: near; alongside of; departure from normal. 2. Suffix: Bearing offspring.

path-, patho-, -path, -pathic, -pathy. Disease; suffering.

ped-, pedi-, pedo-. Foot.

-penia. Decrease from normal; deficiency.

peri-. Around; about.

perineo-. Perineum.

phaco-. Lens of the eye.

phag-, phago-. Eating; ingestion; devouring.

-phil, -philia, -philic. Love for; tendency toward; craving for.

phlebo-. Vein.

-phobia. Abnormal fear or aversion.

photo-. Light.

phren-, phreno-, -phrenia. Mind; diaphgram.

-phylaxis. Protection.

-plasia. Growth; cellular proliferation.

plasm-, -plasm. 1. Prefix: Living substance or tissue. 2. Suffix: To mold.

-plastic. Molded; indicates restoration of lost or badly formed features.

-plegia. Paralysis; stroke.

pneo-. Breath; breathing.

pneum-, pneuma-, pneumato-. Air; gas; respiration.

-poiesis, -poietic. Production; formation.

poly-. Much; many.

post-. After.

pre-. Before; in front of.

presby-. Old age.

pro-. Before; in behalf of.

proct-, procto-. Anus; rectum.

pseud-, pseudo-. False.

psych-, psycho-. Mind; mental processes.

pulmo-. Lung.

py-, pyo-. Pus.

pyro-. Heat; fire.

ren-, reno-. Kidneys.

retro-. Backward; back; behind.

rheo-, -(r)rhea. Current; stream; to flow; to discharge.

rhino-. Nose.

-(r)rhage, -(r)rhagia. Rupture; profuse fluid discharge.

-(r)rhaphy. A suturing or stitching.

salping-, salpingo-. Auditory tube; fallopian tube.

sclero-. Hard; relating to the sclera.

-scopy. Examination.

semi-. Half.

sero-. Serum.

somat-, somato-. Body.

sperma-, spermat-, spermato-. Sperm; spermatozoa.

steno-. Narrow; short.

-stomosis, -stomy. SEE: *-ostomosis*.

sub-. Under; beneath; in small quantity; less than normal.

super-. Above; beyond; superior.

supra-. Above; beyond; on top.

tachy-. Swift; rapid.

tel-, tele-. 1. End. 2. Distant; far.

tendo-, teno-. Tendon.

thorac-, thoraci-, thoraco-. Chest; chest wall.

thrombo-. Blood clot; thrombus.

thyro-. Thyroid gland; oblong; shield.

-tomy. Cutting operation; excision.

top-, topo-. Place; locale.

tox-, toxi-, toxico-, toxo-, -toxic. Toxin; poison; toxic.

tracheo-. Trachea; windpipe.

trans-. Across; over; beyond; through.

-tropin. Stimulation of a target organ by a substance, esp. a hormone.

tympano-. Eardrum; tympanum.

ultra-. Beyond; excess.

-uria. Urine.

uter-, utero-. Uterus.

vaso-. Vessel (e.g., blood vessel).

veno-. Vein.

ventro-, ventr-, ventri-. Abdomen; anterior surface of the body.

vertebro-. Vertebra; vertebrae.

vesico-. Bladder; vesicle.1

Adapted from Thomas, CL: Taber's Cyclopedic Medical Dictionary, 18th ed. FA Davis, Philadelphia, 1997, pp 2214–2218.

Glossary

Ablation: (un-**BLAY**-shun) Removal of part, pathway, or function by surgery, chemical electrocautery, or radio frequency. (Ch. 20)

Abrasion: (a-**BRAY**-zhun) A scraping away of skin or mucus membrane as a result of injury or by mechanical means. (Ch. 12)

Abuse: (uh-**BYOOS**) Misuse; excessive or improper use. May refer to substances or individuals. (Ch. 55)

Acidosis: (ass-i-**DOH**-sis) An actual or relative increase in the acidity of blood caused by an accumulation of acid or a loss of base. (Ch. 5, 8)

Accommodation: (uh-KOM-uh-**DAY**-shun) A reflex action of the eye for focusing. (Ch. 48)

Acquired immunodeficiency syndrome (AIDS): (uh-**KWHY**-erd IM-yoo-noh-de-**FISH**-en-see SIN-drohm) Suppression or deficiency of the cellular immune response, acquired by exposure to human immunodeficiency virus (HIV). (Ch. 54)

Active immunity: (**AK**-tiv im-**YOO**-ni-tee) Acquired immunity attributable to the presence of antibodies or of immune lymphoid cells formed in response to antigenic stimulus. (Ch. 52)

Activities of daily living (ADLs): (ack-**TIV**-i-tees of **DAY**-lee **LIV**-ing) Those activities and behaviors that are performed in the care and maintenance of self (e.g., bathing, dressing, eating). (Ch. 14)

Acupuncture: (ak-yoo-**PUNGK**-chur) Technique using needles inserted at specific points to create anesthesia or treat certain conditions. (Ch. 4)

Acute coronary syndromes: (a-cute **KOR**-un-na-ree sin-**DROMES**) Group of conditions, including unstable angina, non-Q wave myocardial infarction, and ST segment elevation myocardial infarction, caused by a lack of oxygen to the heart muscle. (Ch. 18)

Acute pulmonary hypertension: (ah-**KEWT PULL**-muh-**NAIR**-ee **HIGH**-per-**TEN**-shun) Sudden obstruction of the pulmonary artery causes excessive buildup of pressure in the pulmonary arteries. (Ch. 8)

Adaptation: (ad-dap-**TAY**-shun) Adjustment to changes in internal or external conditions or circumstances; coping. (Ch. 55)

Addiction: (uh-**DIK**-shun) Psychological dependence characterized by drug seeking and craving for an opioid or other substance for effects other than the intended purpose of the substance. (Ch. 9, 55)

Adjunct: (**ADD**-junkt) An addition to the principal procedure or course of therapy. (Ch. 11)

Adjuvant: (ad-**JOO**-vant) Something that assists something else, such as a second form of treatment added to treat a disease. (Ch. 9, 28)

Administrative laws: (ad-MIN-i-**STRAY**-tiv **LAWZ**) Establishes the licensing authority of the state to create, license, and regulate the practice of nursing. (Ch. 2)

Adnexa: (ad-**NECK**-sah) Appendages or accessory organs. (Ch. 39)

Adventitious: (ad-ven-**TI**-shus) Abnormal or extra; often refers to extra breath sounds, such as wheezes or crackles. (Ch. 26)

Aerobic: (air-O-bick) Living only in the presence of oxygen. (Ch. 7)

Afterload: (**AFF**-ter-lohd) The forces impeding the blood flow out of the heart (vascular pressure, aortic compliance, blood mass, and viscosity). (Ch. 21)

Affect: (**AF**-feckt) Emotional tone. (Ch. 36, 55)

Agenesis: (ay-**JEN**-uh-sis) Failure of an organ or part to develop or grow. (Ch. 40)

Agonist: (**AG**-un-ist) A type of opioid that binds to opioid receptors in the central nervous system to relieve pain. (Ch. 9)

Akinesia: (a-ki-**NEE**-zee-ah) Absence or loss of the power of voluntary movement. (Ch. 46)

Alkalosis: (al-ka-**LOH**-sis) An actual or relative decrease in the acidity of blood caused by loss of acid or accumulation of base. (Ch. 5)

Allopathic: (**AL**-oh-**PATH**-ik) Method of treating disease with remedies that produce effects different from those caused by the disease. (Ch. 4)

Alopecia: (**AL**-oh-**PEE**-she-ah) The loss of hair from the body and the scalp. (Ch. 10, 50)

Amenorrhea: (ay-MEN-uh-**REE**-ah) The absence or suppression of menstruation. Amenorrhea is normal before puberty, after menopause, and during pregnancy and lactation. (Ch. 37, 40)

Amputation: (am-pew-**TAY**-shun) The removal of a limb or other appendage or outgrowth of the body. (Ch. 12)

Anaerobic: (**AN**-air-**ROH**-bik) Able to live without oxygen. (Ch. 7, 8)

Analgesic: (**AN**-uhl-**JEE**-zik) A drug that relieves pain. (Ch. 9)

Anaphylactic shock: (**AN**-uh-fi-**LAK**-tik) Systemic reaction that produces life-threatening changes in the circulation and bronchioles. (Ch. 12, 52)

Anaphylaxis: (**AN**-uh-fi-**LAK**-sis) A sudden severe allergic reaction. (Ch. 8, 53)

Anastomose: (uh-NAS-tuh-**MOS**) To surgically connect two parts. (Ch. 22)

Anemia: (uh-**NEE**-mee-yah) A condition in which there is reduced delivery of oxygen to the tissues as a result of reduced numbers of red cells or hemoglobin. (Ch. 10, 24)

Anergy: (**AN**-er-jee) Diminished ability of the immune system to react to an antigen. (Ch. 28)

Anesthesia: (**AN**-es-**THEE**-zee-uh) Lack of feeling or sensation; artificially induced loss of ability to feel pain. (Ch. 11)

Anesthesiologist: (an-es-**THEE**-zee-uhl-la-just) A physician who specializes in anesthesiology. (Ch. 11)

Aneurysm: (**AN**-yur-izm) A sac formed by the localized dilation of the wall of an artery, a vein, or the heart. (Ch. 18)

Angina pectoris: (an-**JIGH**-nah **PEK**-tuh-riss) Severe pain and pressure in the chest caused by insufficient supply of blood and oxygenation to the heart. (Ch. 18)

Angioedema: (**AN**-gee-o-eh-**DEE**-ma) A localized edematous reaction of the deep dermis or subcutaneous or submucosal tissues appearing as giant wheals. (Ch. 53)

Anion: (**AN**-eye-on) Electrolyte that carries a negative electrical charge. (Ch. 5)

Anisocoria: (an-i-soh-**KOH**-ree-ah) Inequality in size of the pupils of the eyes. (Ch. 45)

Ankylosing spondylitis: (**ANG**-ki-**LOH**-sing SPON-da-**LIGHT**-is) Inflammatory disease of the spine causing stiffness and pain. (Ch. 53)

Annuloplasty: (**AN**-yoo-loh-**PLAS**-tee) Repair of a cardiac valve. (Ch. 19, 22)

Anorexia: (**AN**-oh-**REK**-see-ah) Absence or loss of appetite for food. Seen in depression, with illness, and as a side effect of some medications. (Ch. 10, 30)

Anorexia nervosa: (**AN**-oh-**REK**-see-ah ner-**VOH**-sah) Refusal to maintain body weight over a minimal normal weight for age and height. (Ch. 30)

Antagonist: (an-**TAG**-on-ist) Medication used to counteract the effects of an opioid (e.g., naloxone). (Ch. 9)

Anteflexion: (**AN**-tee-**FLECK**-shun) The abnormal bending forward of part of an organ. (Ch. 40)

Anteversion: (**AN**-tee-VER-zhun) A tipping forward of an organ as a whole, without bending. (Ch. 40)

Antibodies: (**AN**-ti-baw-dees) An immunoglobin molecule having a specific amino acid sequence that gives each antibody the ability to adhere to and interact only with the antigen that induced the synthesis. (Ch. 7, 52)

Anticholinesterase: (**AN**-ti-KOH-lin-**ESS**-ter-ays) A substance that breaks down acetylcholinesterase. (Ch. 47)

Antidiuretic: (**AN**-ti-**DYE**-yoo-**RET**-ik) Lessening urine excretion. (Ch. 5)

Antigen: (**AN**-tih-jen) A protein marker on the surface of cells that identifies the type of cell. (Ch. 7, 52)

Antitussive: (an-tee-**TUSS**-iv) An agent that prevents or relieves cough. (Ch. 28)

Anuria: (an-**YOO**-ree-ah) Complete suppression of urine formation by the kidney. (Ch. 35)

Anxiety: (ang-**ZIGH**-uh-tee) The uncomfortable feeling of apprehension or dread that occurs in response to a known or unknown threat. (Ch. 55)

Aphasia: (ah-**FAY**-zee-ah) Defect or loss of the power of expression by speech, writing, or signs, or of comprehension of spoken or written language, caused by disease or injury of the brain centers, such as stroke syndrome. (Ch. 45)

Aphthous stomatitis: (**AF**-thus STOH-mah-**TIGH**-tis) Small, white, painful ulcers (also known as canker sores) that appear on the inner cheeks, lips, gums, tongue, palate, and pharynx. They tend to recur. (Ch. 30)

Apnea: (ap-**NEE**-ah) Temporary absence of breathing. (Ch. 26)

Appendicitis: (uh-PEN-di-**SIGH**-tis) Inflammation of the vermiform appendix. (Ch. 31)

Arcus senilus: (**AR**-kus se-**NILL**-us) A benign white or gray opaque ring in the corneal margin of the eye. (Ch. 48)

Arrhythmia: (uh-**RITH**-mee-yah) Irregular rhythm, especially heartbeat. (Ch. 14)

Arthritis: (are-**THRYE**-tis) Inflammation of a joint. (Ch. 44)

Arthrocentesis: (ar-**THROW**-sen-tee-sis) Puncture of a joint space with a needle to remove fluid accumulated in the joint. (Ch. 43, 44)

Arteriosclerosis: (ar-**TIR**-ee-oh-skle-**ROH**-sis) Term applied to a number of pathological conditions in which there is gradual thickening, hardening, and loss of elasticity of the walls of the arteries. (Ch. 15, 18)

Arthroplasty: (**AR**-throw-**PLAS**-te) Repair of a joint. Also called joint replacement. (Ch. 44)

Arthroscopy: (are-**THROSS**-scop-ee) Examination of the interior of a joint with an arthroscope. (Ch. 43)

Articular: (ar-**TIK**-yoo-lar) Pertaining to a joint. (Ch. 43)

Ascites: (a-**SIGH**-teez) Abnormal accumulation of fluid in the peritoneal cavity. (Ch. 33)

Asepsis: (ah-**SEP**-sis) A condition free from germs, infection, and any form of life. (Ch. 7)

Aseptic: (ah-**SEP**-tik) Free of pathogenic organisms; asepsis. (Ch. 11)

Asphyxia: (as-**FIX**-ee-a) A condition in which there is a deficiency of oxygen in the blood and an increase in carbon dioxide in the blood and tissues. (Ch. 12)

Aspiration: (ASS-pi-**RAY**-shun) Accidental drawing in of foreign substances into the throat or lungs during inspiration. (Ch. 14)

Assessment: (ah-**SESS**-ment) An appraisal or evaluation of a client's condition. (Ch. 1)

Asterixis: (AS-ter-**ICK**-sis) Hand flapping tremor and involuntary movements of tongue and feet; may be present in hepatic encephalopathy. (Ch. 35)

Astigmatism: (uh-**STIG**-mah-**TIZM**) An error of refraction in which a ray of light is not sharply focused on the retina but is spread over a more or less diffuse area. (Ch. 49)

Ataxia: (ah-**TAK**-see-ah) Failure of muscular coordination; irregularity of muscular action. (Ch. 46)

Atelectasis: (AT-e-**LEK**-tah-sis) Collapsed or airless condition of the lung or portion of lung, caused by obstruction or hypoventilation. (Ch. 11, 22, 28)

Atheroma: (ATH-er-**OMA**) Fatty deterioration or thickening of the walls of the larger arteries occurring in atherosclerosis. (Ch. 18)

Atherosclerosis: (ATH-er-oh-skle-**ROH**-sis) A form of arteriosclerosis characterized by accumulation of plaque, blood, and blood products lining the wall of the artery, causing partial or complete blockage of an artery. (Ch. 15, 18)

Atrial depolarization: (**AY**-tree-uhl DE-poh-lahr-i-**ZAY**-shun) Electrical activation of the atria. (Ch. 20)

Atrial systole: (**AY**-tree-uhl **SIS**-tuh-lee) The contraction of the atria. (Ch. 20)

Atrioventricular node: (**AY**-tree-oh-ven-**TRICK**-yoo-lar NOHD) Located in lower right atrium; receives an impulse from the sinoatrial (SA) node and relays it to the ventricles. (Ch. 20)

Atrophy: (**AT**-ruh-fee) Without nourishment; wasting. (Ch. 47)

Atypical: (ay-**TIP**-i-kuhl) Deviating from normal. (Ch. 28)

Augmentation: (AWG-men-**TAY**-shun) The act or process of increasing in size, quantity, degree or severity. (Ch. 40)

Auscultation: (AWS-kul-**TAY**-shun) Process of listening for sounds within the body, usually sounds of thoracic or abdominal viscera, to detect an abnormality. (Ch. 1)

Autoimmune: (AW-toh-im-**YOON**) A condition in which the body does not recognize itself and the immune system attacks normal cells. (Ch. 37)

Avascular necrosis: (a-**VAS**-cue-lur ne-**CROW**-sis) Disruption of blood supply causing tissue death. (Ch. 12, 44)

Ayurvedic: (**AY**-**YUR**-**VAY**-dik) An ancient Hindu system of medicine that improves health by harmonizing mind and body. (Ch. 4)

Azotemia: (**AY**-zoh-**TEE**-me-ah) An increase in nitrogenous bodies in the blood, especially urea, as measured by the serum blood urea nitrogen (BUN) level. (Ch. 35)

Bacteria: (back-**TEER**-e-ah) One-celled organism that can reproduce but needs a host for food and supportive environment. Bacteria can be harmless, normal flora, or disease-producing pathogens. (Ch. 7)

Balanitis: (BAL-uh-**NIGH**-tis) Inflammation of the skin covering the glans penis. (Ch. 40)

Bariatric: (BAR-ry-**AT**-rick) Branch of medicine that deals with the prevention, control, and treatment of obesity. (Ch. 30)

Basal cell secretion test: (**BAY**-zuhl SELL see-**KREE**-shun TEST) Part of a gastric analysis; measures the amount of gastric acid produced in 1 hour. (Ch. 29)

Behavior management: (be-**HAYV**-yer **MAN**-ij-ment) Treatment method that uses positive and negative reinforcement to alter behavior. (Ch. 55)

Belief: (bee-**LEEF**) Something accepted as true. Does not have to be proven. (Ch. 3)

Beneficence: (buh-**NEF**-i-sens) To provide good care; to do good for clients. One of the oldest requirements for health care providers. (Ch. 2)

Benign: (bee-**NINE**) Not progressive; for example, a tumor that is not cancerous. (Ch. 10)

Beta-hemolytic streptococci: (**BAY**-tuh-HEE-moh-**LIT**-ick STREP-toh-**KOCK**-sigh) Gram-positive bacteria that, when grown on blood-agar plates, completely hemolyze the blood and produce a clear zone around the bacteria colony. Group A beta-hemolytic streptococci cause disease in humans. (Ch. 17)

Bigeminy: (bye-**JEM**-i-nee) Occurring every second beat, as in bigeminal premature ventricular contractions. (Ch. 20)

Bilateral salpingo-oophorectomy: (by-**LAT**-er-uhl sal-PINJ-oh-ah-fuh-**RECK**-tuh-mee) Surgical removal of both fallopian tubes and ovaries. (Ch. 40)

Bimanual: (by-**MAN**-yoo-uhl) With both hands. (Ch. 39)

Biofeedback: (BYE-oh-**FEED**-bak) A form of therapy that uses provision of visual or auditory evidence to a person of the status of an autonomic body function such as heart rate, blood pressure, or respiratory rate. (Ch. 55)

Biopsy: (**BY**-ahp-see) A sample of tissue removed for examination. (Ch. 10)

Bipolar: (bye-**POH**-ler) Having two poles or pertaining to both poles. Bipolar disorder is characterized by episodes of manic and depressive behavior. (Ch. 55)

Blanch: (**BLANCH**) To lose color. (Ch. 51)

Bleb: (**BLEB**) An irregularly shaped elevation of the skin, such as a blister. May also occur in lung tissue. (Ch. 28)

Blepharitis: (BLEF-uh-**RIGH**-tis) Inflammation of the glands and lash follicles along the margin of the eyelids. (Ch. 49)

Blindness: (**BLYND**-ness) Lack or loss of ability to see. (Ch. 49)

Bolus: (**BOH**-lus) A dose of intravenous medication injected all at once. (Ch. 6)

Bone: (BOWN) The hard, rigid form of connective tissue constituting most of the skeleton of vertebrates, composed chiefly of calcium salts. (Ch. 43)

Bowel sounds: (BOW'L SOWNDS) Gurgling and clicking sounds heard over the abdomen caused by air and fluid movement from peristaltic action. Normal bowel sounds occur every 5 to 15 seconds at a rate of 5 to 35 sounds per minute. Absent—no bowel sounds heard after 5 minutes of listening in each quadrant. Hyperactivebowel sounds that are rapid, high-pitched, and loud. Hypoactivebowel sounds that occur at a rate of one every minute or longer. (Ch. 29)

Bradycardia: (BRAY-dee-**KAR**-dee-yah) A slow heartbeat characterized by a pulse rate below 60 beats per minute. (Ch. 15, 20)

Bradykinesia: (BRAY-dee-kin-**EE**-zee-ah) Abnormal slowness of movement; sluggishness. (Ch. 46)

Breakthrough: (**BRAYK**-throo) Pain that occurs during use of long-acting analgesic therapy. (Ch. 9)

Bronchiectasis: (BRONG-key-**EK**-tah-sis) Chronic dilation of a bronchus or bronchi, usually associated with secondary infection and excessive sputum production. (Ch. 28)

Bronchitis: (brong-**KIGH**-tis) Inflammation of the mucous membrane of the bronchial airways; may be viral or bacterial. (Ch. 28)

Bronchodilator: (BRONG-koh-**DYE**-lay-ter) A drug that expands the bronchial tubes by relaxing bronchial smooth muscle. (Ch. 28)

Bronchospasm: (**BRONG**-koh-spazm) Spasm of the bronchial smooth muscle resulting in narrowing of the airways; associated with asthma and bronchitis. (Ch. 8, 28)

Bruit: (**BROUT**) A humming heard when auscultating a blood vessel that is caused by turbulent blood flow through the vessel. (Ch. 15)

Bulimia nervosa: (buh-**LEE**-mee-ah ner-**VOH**-sah) Recurrent episodes of binge eating and self-induced vomiting. (Ch. 30)

Bulla: (**BUHL**-ah) A large blister or skin lesion filled with fluid. May also occur in lung tissue. (Ch. 28)

Bundle of His: (**BUN**-duhl of HISS) A bundle of fibers of the impulse-conducting system of the heart. Originates in the atrioventricular (AV) node. (Ch. 20)

Bursae: (**BURR**-sah) A small fluid-filled sac or saclike cavity situated in tissues such as joints where friction would otherwise occur. (Ch. 43)

Calculi: (**KAL**-kyoo-lye) An abnormal concentration, usually composed of mineral salts, occurring within the body, chiefly in the hollow organs or their passages. Called also stones, as in kidney stones and gallstones. (Ch. 35)

Cancer: (**KAN**-sir) A general name for over 100 diseases in which abnormal cells grow out of control; a malignant tumor. (Ch. 10)

Cannula: (**KAN**-yoo-lah) A flexible tube that can be inserted into the body guided by a stiff, pointed rod. For example, an intravenous cannula is guided by a metal needle. (Ch. 6)

Capillary permeability: (**KAP**-i-lar-ee **PER**-me-a-**BILL**-i-tee) The ability of substances to diffuse through capillary walls into tissue spaces. (Ch. 12)

Capillary refill: (**KAP**-i-lar-ee **RE**-fill) The amount of time required for color to return to the nailbed after having been compressed; normally 3 seconds or less. Indicator of peripheral circulation. (Ch. 12)

Caput medusae: (**KAP**-ut mi-**DOO**-see) Dilated veins around the umbilicus, associated with cirrhosis of the liver. (Ch. 32)

Carbuncle: (**KAR**-bung-kull) A necrotizing infection of skin and subcutaneous tissue composed of a cluster of boils. (Ch. 49)

Carcinoembryonic antigens (CEA): (**KAR**-sin-oh-EM-bree-ah-nik **AN**-ti-jens) A class of antigens normally present in fetal cells; CEA level is elevated in many cancers and is measured to guide cancer treatment. (Ch. 29)

Carcinogen: (kar-**SIN**-oh-jen) Specific agents known to promote the cancer process. (Ch. 10)

Cardiac output: (**KAR**-dee-yak **OWT**-put) A measure of the pumping ability of the heart; amount of blood pumped by the heart per minute. (Ch. 8, 16)

Cardiac tamponade: (**KAR**-dee-yak **TAM**-pon-**AID**) The life-threatening compression of the heart by the fluid accumulating in the pericardial sac surrounding the heart. (Ch. 17)

Cardiogenic shock: (**KAR**-dee-o-**JEN**-ick **SHOCK**) Occurs when the heart muscle is unhealthy and contractility is impaired. (Ch. 8, 12)

Cardiomegaly: (**KAR**-dee-oh-**MEG**-ah-lee) Enlargement of the heart. (Ch. 17)

Cardiomyopathy: (**KAR**-dee-oh-my-**AH**-pah-thee) A group of diseases that affect the myocardium's (heart muscle's) structure or function. (Ch. 17)

Cardioplegia: (**KAR**-dee-oh-**PLEE**-jee-ah) Arrest of myocardial contraction, as by use of chemical compounds or cold temperatures in cardiac surgery. (Ch. 22)

Cardioversion: (**KAR**-do-oh-**VER**-zhun) An elective procedure in which a synchronized shock is delivered to attempt to restore the heart to a normal sinus rhythm. (Ch. 20)

Cataract: (**KAT**-uh-rakt) Opacity of the lens of the eye. (Ch. 13, 49)

Cation: (**KAT**-eye-on) Electrolytes that carry a positive electrical charge. (Ch. 5)

Ceiling effect: (**SEE**-ling e-**FEKT**) The dose of medication at which the maximum therapeutic effect is achieved. Increasing the dose beyond the therapeutic dose will not result in increased relief and may result in undesirable side effects. (Ch. 9)

Cell-mediated immunity: (**SELL ME**-dee-ay-ted im-**YOO**-ni-tee) Production of lymphocytes by thymus in response to antigen exposure. (Ch. 52)

Cellulitis: (sell-yoo-**LYE**-tis) Inflammation of cellular or connective tissue. (Ch. 51)

Cerebrovascular: (**SER**-ee-broh-**VAS**-kyoo-lur) Pertaining to the blood vessels of the cerebrum or brain. (Ch. 45)

Chalazion: (kah-**LAY**-zee-on) A small eyelid mass resulting from chronic inflammation of a meibomian gland. (Ch. 49)

Chancre: (**SHANK**-er) A hard, syphilitic primary ulcer, the first sign of syphilis, appearing approximately 2 to 3 weeks after infection. (Ch. 42)

Chemotherapy: (**KEE**-moh-**THER**-uh-pee) The treatment of disease with medication; often refers to cancer therapy. (Ch. 10)

Chiropractic: (ky-**RUH**-prak-tik) Treatment modality that uses manual adjustment of the vertebral column and extremities to remove interference with nerve function. (Ch. 4)

Cholecystitis: (**KOH**-lee-sis-**TIGH**-tis) Inflammation of the gallbladder. (Ch. 33)

Choledocholithiasis: (koh-LED-oh-koh-li-**THIGH**-ah-sis) Gallstones in the common bile duct. (Ch. 33)

Choledochoscopy: (KOH-**LED**-oh-**KOS**-koh-pee) An endoscopic test of the gallbladder and common bile duct. (Ch. 33)

Cholelithiasis: (KOH-lee-li-**THIGH**-ah-sis) Gallstones in the gallbladder. (Ch. 33)

Chorea: (kaw-**REE**-ah) A nervous condition marked by involuntary muscular twitching of the limbs or facial muscles. (Ch. 17)

Chronic illness: (KRAH-nick **ILL**-ness) An illness that is long-lasting or recurring, which usually interferes with a person's ability to perform activities of daily living. Medical care and hospitalization are often required on an ongoing basis. (Ch. 13)

Circumcise: (**SIR**-kuhm-size) Surgical removal of the foreskin covering the head of the penis. (Ch. 39)

Cirrhosis: (si-**ROH**-sis) Chronic disease of the liver, associated with fat infiltration and development of fibrotic tissue. (Ch. 33)

Civil law: (**SIV**-il **LAW**) Provides the rules by which individuals seek to protect their personal and property rights. (Ch. 2)

Claudication: (KLAW-di-**KAY**-shun) Severe pain in the calf muscle from inadequate blood supply. (Ch. 15)

Clubbing: (**KLUB**-ing) A condition in which the ends of the fingers and toes appear bulbous and shiny, most often the result of lung disease. (Ch. 15)

Cochlear implant: (**KOK**-lee-er **IM**-plant) A device consisting of a microphone, signal processor, external transmitter, and implanted receiver to aid hearing. (Ch. 48, 49)

Code of ethics: (**KOHD** of **ETH**-icks) A traditional compilation of ideal behaviors of a professional group. (Ch. 2)

Codependence: (KO-de-**PEN**-dense) A situation in which the significant others in a family group begin to lose their own sense of identity and purpose and exist solely for the abuser. (Ch. 55)

Cognitive: (**KAHG**-ni-tiv). The ability to think rationally and to process thoughts. (Ch. 55)

Colectomy: (koh-**LEK**-tuh-me) Excision of the colon or a portion of it. (Ch. 31)

Colic: (**KAH**-lick) Spasm of a hollow organ or duct, causing pain. (Ch. 33)

Colitis: (koh-**LYE**-tis) Inflammation of the colon. (Ch. 31)

Collateral circulation: (koh-**LA**-ter-al SIR-kew-**LAY**-shun) Small branches off of larger blood vessels that will increase in size and capacity next to a main blood vessel that is obstructed. (Ch. 18)

Colonization: (COLLIN-i-**ZAY**-shun) The presence of pathogenic microbes in the body, without development of a symptomatic infection. (Ch. 7)

Colonoscopy: (KOH-lun-**AHS**-kuh-pee) Examination of the upper portion of the rectum with a colonoscope. (Ch. 29)

Colostomy: (koh-**LAH**-stuh-me) An artificial opening (stoma) created in the large intestine and brought to the surface of the abdomen for evacuating the bowels. (Ch. 31)

Colporrhaphy: (kohl-**POOR**-ah-fee) Surgical repair of the vagina. (Ch. 40)

Colposcopy: (kul-**POS**-koh-pee) Examination of the vulva, vagina, and cervix by means of a magnifying lens and a bright light. (Ch. 39)

Comedone: (**KOH**-me-doh) Skin lesion that occurs in acne vulgaris (closed form: whitehead; open form: blackhead). (Ch. 51)

Commissurotomy: (KOM-i-shur-**AHT**-oh-mee) Surgical incision of any commissure as in cardiac valves to increase the size of the orifice. (Ch. 19, 22)

Compliance: (kom-**PLIGH**-ens) The ability to alter size or shape in response to an outside force; the ability of the lungs to distend. (Ch. 28)

Compulsion: (kum-**PUHL**-shun) A recurrent, unwanted, and distressing urge to perform an act. (Ch. 55)

Conductive hearing loss: (kon-**DUK**-tiv **HEER**-ing **LOSS**) Impaired transmission of sound waves through the external ear canal to the bones of the middle ear. (Ch. 49)

Condylomata acuminata: (KON-di-**LOH**-ma-tah ah-KY-OOM-in-**AH**-tah) Warts in the genital region caused by the human papillomavirus (HPV); a contagious sexually transmitted disease. (Ch. 41)

Condylomatous: (KON-di-**LOH**-ma-tus) Pertaining to a condyloma. (Ch. 41)

Confidentiality: (**KON**-fi-den-she-**AL**-i-tee) Maintaining privacy of client information. Client and client's care can be discussed only in the professional setting. (Ch. 2)

Congestive heart failure: (kon-**JESS**-tive HART **FAIL**-yur) Results from inability of heart to pump sufficient amounts of blood because of impaired pumping function and sodium and water retention. Congestion refers to the buildup of fluid that ranges from mild to life-threatening (pulmonary edema). With right-sided heart failure, the fluid buildup is seen systemically (lower legs/feet, sacral area in bedridden persons, jugular veins, liver, spleen). With left-sided heart failure, the fluid buildup occurs in the lungs and if severe, immediate treatment is required or death can occur.

Conization: (KOH-ni-**ZAY**-shun) The removal of a cone of tissue, as in partial excision of the cervix uteri. (Ch. 39)

Conjunctivitis: (kon-JUNK-ti-**VIGH**-tis) Inflammation of the conjunctiva of the eye. (Ch. 42, 49)

Consensual response: (kon-**SEN**-shoo-uhl ree-**SPONS**) Reaction of both pupils when one eye is exposed to greater intensity of light than the other. (Ch. 48)

Constipation: (KON-sti-**PAY**-shun) A condition of sluggish or difficult bowel action/evacuation. (Ch. 14, 31)

Contraceptive: (KON-truh-**SEP**-tiv) Any process, device, or method that prevents conception. (Ch. 40)

Contracture: (kon-**TRACK**-chur) Abnormal accumulation of fibrosis connective tissue in skin, muscle or joint capsule that prevents normal mobility at that site. (Ch. 14, 46)

Contralateral: (KON-truh-**LAT**-er-uhl) Originating in or affecting the opposite side of the body. (Ch. 46)

Conversion disorder: (kon-**VER**-zhun dis-**OR**-der) An illness that emerges from overuse of the conversion reaction defense mechanism, in which there is impaired physical functioning that appears to be neurological, but no organic disease can be identified. (Ch. 55)

Coping: (KOH-ping) The process of contending with the stresses of daily life in an effort to overcome or work through them. (Ch. 55)

Coronary artery disease: (KOR-uh-na-ree **AR**-tuh-ree di-**ZEEZ**) Narrowing of the coronary arteries sufficient to prevent adequate blood supply to the myocardium. (Ch. 18)

Cor pulmonale: (KOR PUL-mah**NAH**-lee) Hypertrophy or failure of the right ventricle from disorders of the chest wall, lungs, and pulmonary vessels. As with increased pulmonary pressure caused by chronic obstructive pulmonary disease (COPD). (Ch. 21)

Craniectomy: (KRAY-nee-**EK**-tuh-me) Excision of a segment of the skull. (Ch. 46)

Cranioplasty: (**KRAY**-nee-oh-plas-tee) Any plastic repair operation on the skull. (Ch. 46)

Craniotomy: (KRAY-nee-**AHT**-oh-mee) Any incision through the cranium. (Ch. 46)

Crepitation: (crep-i-**TAY**-shun) A dry, crackling sound or sensation, such as that produced by the grating of the ends of a fractured bone. (Ch. 43)

Crepitus: (**KREP**-i-tuss) Crepitation. (Ch. 26)

Criminal law: (**KRIM**-i-nuhl **LAW**) Regulates behaviors for citizens within a country. (Ch. 2)

Critical thinking: (**KRIT**-i-kuhl **THING**-king) Use of knowledge and skills to make the best decisions possible in client care situations. (Ch. 1)

Cryotherapy: (KRY-oh-**THER**-uh-pee) The therapeutic use of cold. (Ch. 40)

Cryptorchidism: (kript-**OR**-ki-dizm) A birth condition in which one or both of the testicles have not descended into the scrotum. (Ch. 41)

Culdocentesis: (KUL-doh-sen-**TEE**-sis) The procedure for obtaining material from the posterior vaginal cul-de-sac by aspiration or surgical incision through the vaginal wall, performed for therapeutic or diagnostic reasons. (Ch. 40)

Culdoscopy: (kul-**DOS**-koh-pee) Direct visual examination of the female viscera through an endoscope introduced into the pelvic cavity through the posterior vaginal fornix. (Ch. 39)

Culdotomy: (KUL-**DOT**-uh-mee) Incision or needle puncture of the cul-de-sac of Douglas through the vagina. (Ch. 40)

Cultural awareness: (KUL-chur-uhl a-**WARE**-ness) Being aware of history and ancestry and having an appreciation of and attention to the crafts, arts, music, foods, and clothing of various cultures. (Ch. 3)

Cultural competence: (KUL-chur-uhl **KOM**-pe-tens) Having an awareness of one's own culture and not letting it have an undue influence over another person's culture. Having the knowledge and skills about a culture that are required to provide care. (Ch. 3)

Cultural diversity: (KUL-chur-uhl di-**VER**-si-tee) Representing two or more cultures; the differences among cultures. For example, the United States includes people from many different countries. (Ch. 3)

Cultural sensitivity: (KUL-chur-uhl SEN-si-**TIV**-i-tee) Being aware of and sensitive to cultural differences. Avoiding behavior or language that may be offensive to another person's cultural beliefs. (Ch. 3)

Culture: (KUL-chur) The socially transmitted behavior patterns, beliefs, values, customs, arts, and all other characteristics of people that guide their worldview. (Ch. 3)

Curet: (kyoo-**RET**) A loop, ring, or spoon-shaped instrument, attached to a handle and having sharp or blunt edges; used to scrape tissue from a surface. (Ch. 39)

Custom: (KUS-tum) A custom is the usual way of acting in a given circumstance or something that an individual or group does out of habit. For example, many people have turkey on Thanksgiving. (Ch. 3)

Cyanosis: (SIGH-uh-**NOH**-sis) Slightly bluish, grayish, or dark purple discoloration of the skin caused by the presence of abnormal amounts of reduced hemoglobin in the blood. (Ch. 8, 21, 26)

Cystic: (**SIS**-tik) Pertaining to cysts or the urinary bladder. (Ch. 39)

Cystitis: (sis-**TIGH**-tis) Inflammation of the urinary bladder. (Ch. 35)

Cystocele: (**SIS**-toh-seel) A bladder hernia that protrudes into the vagina. (Ch. 40)

Cytomegalovirus: (sigh-**TOW**-meg-ul-low-vigh-rus) Species-specific herpesvirus; usually harmless to those with functional immune systems. May cause fatal pneumonia in those who are immunocompromised. Affects retina and may cause blindness in those with acquired immunodeficiency syndrome. (Ch. 54)

Cystoscopy: (sis-**TAHS**-koh-pee) A diagnostic procedure using an instrument (cystoscope) via the urethra to view the bladder. (Ch. 34)

Cytotoxic: (SIGH-toh-**TOCK**-sick) Destructive to cells. (Ch. 10, 42)

Data: (**DAY**-tuh) A group of facts or statistics. (Ch. 1)

Data, objective: SEE objective data. (Ch. 1)

Data, subjective: SEE subjective data. (Ch. 1)

Debridement: (day-breed-MAHNT) The removal of foreign material and contaminated and devitalized tissues from or adjacent to a traumatic or infected area until surrounding healthy tissue is exposed. (Ch. 11)

Decerebrate: (dee-**SER**-e-brayt) Posture of an individual with absence of cerebral function. (Ch. 45)

Decorticate: (dee-**KOR**-ti-kayt) Posture of an individual with a lesion at or above the upper brain stem. (Ch. 45)

Defibrillation: (dee-**FIB**-ri-lay-shun) Use of an electrical device that applies countershock to the heart through electrodes placed on the chest wall to stop fibrillation of the heart. (Ch. 20)

Degeneration: (de-jen-er-**AY**-shun) Deterioration. (Ch. 47)

Dehiscence: (dee-**HISS**-ents) A splitting open (i.e., rupture) of an incision. (Ch. 11)

Dehydration: (**DEE**-high-**DRAY**-shun) A condition resulting from excessive loss of body fluid that occurs when fluid output exceeds intake. (Ch. 5)

Delirium tremens: (dee-LIR-ee-uhm **TREE**-menz) An acute alcohol withdrawal syndrome marked by acute, transient disturbance of consciousness. (Ch. 55)

Delusions: (dee-**LOO**-zhuns) False beliefs that are firmly maintained in spite of incontrovertible proof to the contrary. (Ch. 55)

Dementia: (dee-**MEN**-cha) A broad term that refers to cognitive deficit, including memory impairment. (Ch. 14, 46)

Demyelination: (dee-**MY**-uh-lin-**AY**-shun) Loss of myelin from neurons. (Ch. 47)

Deontology: (DA-on-**TOL**-o-gee) The study of moral obligations and commitments, including medical ethics. (Ch. 2)

Dependence: (di-**PEN**-dens) A state of reliance on something. Psychological craving for a drug that may or may not be accompanied by a physiological need. (Ch. 55)

Depression: (dee-**PRESS**-shun) A mental disorder marked by altered mood with loss of interest. (Ch. 14)

Dermatitis: (DER-mah-**TIGH**-tis) Inflammation of the skin. (Ch. 51)

Dermatophytosis: (DER-mah-toh-fye-**TOH**-sis) A fungal infection of the skin. (Ch. 51)

Dermoid: (**DER**-moyd) Resembling the skin. (Ch. 40)

Developmental stage: (DEE-vell-up-**MEN**-tal STAYJ) An age-defined period with specific psychological tasks that need to be accomplished to maintain ego as proposed by Erik Erikson, a psychoanalyst. (Ch. 13)

Diabetes mellitus: (DYE-ah-**BEE**-tis mel-**LYE**-tus) A chronic disease characterized by impaired production or use of insulin and high blood glucose levels. (Ch. 38)

Diarrhea: (DYE-uh-**REE**-ah) Passage of fluid or unformed stools. (Ch. 31)

Diastolic blood pressure: (dye-ah-**STAH**-lik BLUHD **PRE**-shure) The amount of pressure exerted on the wall of the arteries when the ventricles are at rest. The bottom number in a blood pressure reading. (Ch. 16)

Diffusion: (di-**FEW**-zhun) The tendency of molecules of a substance (gaseous, liquid, or solid) to move from a region of high concentration to one of lower concentration. (Ch. 5)

Dilation and curettage: (DIL-**AY**-shun and kyoor-e-**TAHZH**) A surgical procedure that expands the cervical canal of the uterus (dilation) so that the surface lining of the uterine wall can be scraped (curettage). (Ch. 40)

Displacement: (dis-**PLAYSS**-ment) Transference of emotion from the original idea with which it was associated to a different idea, allowing the client to avoid acknowledging the original source. (Ch. 55)

Disseminated intravascular coagulation: (dis-**SEM**-i-NAY-ted IN-trah-**VAS**-kyoo-lar koh-AG-yoo-**LAY**-shun) A pathological form of coagulation that is diffuse (widespread) rather than localized, as would be the case in normal coagulation. Clotting factors are consumed to such an extent that generalized bleeding may occur. (Ch. 24)

Distributive shock: (dis-**TRIB**-yoo-tiv) Excessive dilation of the venules and arterioles, leading to decreased distribution of blood, resulting in shock. (Ch. 8, 12)

Distributive justice: (dis-**TRIB**-yoo-tiv **JUS**-tiss) The right of individuals to be treated equally regardless of race, sex, marital status, sexual preference, medical diagnosis, social standing, economic level, or religious belief. (Ch. 2)

Diverticulitis: (DYE-ver-tik-yoo-**LYE**-tis) Inflammation of a diverticulum (a sac or pouch in the walls of a canal or organ, usually the colon), especially inflammation involving diverticula of the colon. (Ch. 31)

Diverticulosis: (DYE-ver-tik-yoo-**LOH**-sis) The presence of diverticula in the absence of inflammation. (Ch. 31)

Dormant: (**DOOR**-mant) Condition of greatly reduced metabolic activity permitting long-term survival and possible reactivation of bacterial endospores, protozoan cysts, larval stages of worm parasites, and viruses. (Ch. 7)

Dressler's syndrome: (**DRESS**-lers **SIN**-drohm) Postmyocardial infarction syndrome; pericarditis. (Ch. 17)

Dysarthria: (dis-**AR**-three-ah) Imperfect articulation of speech caused by disturbances of muscular control resulting from central or peripheral nervous system damage. (Ch. 45)

Dysfunctional: (dis-**FUNCK**-shun-uhl) Family or work environment that does not function effectively, sometimes because of other problems of members. (Ch. 55)

Dysmenorrhea: (DIS-men-oh-**REE**-ah) Pain in association with menstruation. (Ch. 40)

Dyspareunia: (DIS-puh-**ROO**-nee-ah) Occurrence of pain in the labia, vagina, or pelvis during or after sexual intercourse. (Ch. 40)

Dysphagia: (dis-**FAYJ**-ee-ah) Inability to swallow or difficulty swallowing. (Ch. 27, 37, 45)

Dysplasia: (dis-**PLAY**-zee-ah) Abnormal development of tissue. (Ch. 40)

Dyspnea: (**DISP**-nee-ah) Subjective sense of labored breathing that occurs because of insufficient oxygenation. (Ch. 26)

Dysreflexia: (DIS-re-**FLEK**-see-ah) State in which an individual with a spinal cord injury at or above T7 experiences an uninhibited sympathetic response to a noxious stumulus. (Ch. 46)

Dysrhythmia: (dis-**RITH**-mee-yah) Abnormal, disordered, or disturbed cardiac rhythm. (Ch. 5, 8, 15, 20)

Dysuria: (dis-**YOO**-ree-ah) Difficult or painful urination. (Ch. 34, 35)

Ecchymoses: (ECK-uh-**MOH**-sis) A bruise of varying size, the color of which may be blue-black, changing to greenish yellow or yellow with time. (Ch. 23, 50)

Ectasia: (ek-**TAY**-zee-ah) Replacement of normal tissue with fibrous tissue. (Ch. 40)

Ectopic: (eck-**TOP**-ick) Ectopic hormones are secreted from sites other than the gland where they would normally be found. (Ch. 28, 37, 39)

Edema: (uh-**DEE**-muh) Collection of excess fluid in body tissues. (Ch. 5, 14, 21)

Ejaculation: (ee-JAK-yoo-**LAY**-shun) The release of semen from the male urethra. (Ch. 39)

Electrocautery: (ee-LECK-troh-**CAW**-tur-ee) Cauterization using platinum wires heated to red or white heat by an electric current, either direct or alternating. (Ch. 42)

Electrocardiogram: (ee-LECK-troh-**KAR**-dee-oh-GRAM) A recording of the electrical activity of the heart. (Ch. 20)

Electrocoagulated: (ee-LECK-troh-coh-**AG**-yoo-LAY-ted) Coagulation of tissue by means of a high-frequency electric current. (Ch. 42)

Electroconvulsive therapy (ECT): (ee-**LEK**-troh-kun-**VUL**-siv **THER**-uh-pee) A type of somatic therapy in which an electric current is used to produce convulsions to treat such conditions as depression. (Ch. 55)

Electroencephalogram: (ee-**LEK**-troh-en-**SEFF**-uh-loh-gram) A record produced by electroencephalography; tracing of the electrical impulses of the brain. (Ch. 45)

Electrolyte: (ee-**LEK**-troh-lite) A substance that when dissolved in water can conduct electricity. (Ch. 5)

Electroretinography: (ee-**LEK**-troh-RET-in-**AHG**-ruh-fee) Measurement of the electrical response of the retina to light stimulation. (Ch. 48)

Emboli: (**EM**-boh-li) Solid, liquid, or gaseous masses of undissolved matter traveling with the fluid current in a blood or lymphatic vessel. (Ch. 17)

Embolism: (**EM**-buh-lizm) Foreign substance or blood clot that travels through the circulatory system until it obstructs a vessel. (Ch. 18, 28)

Empathy: (**EM**-puh-thee) Objective awareness of and insight into the feelings, emotions, and behavior of another person. (Ch. 2)

Emphysema: (**EM**-fi-**SEE**-mah) Distention of interstitial tissue by gas or air; chronic pulmonary disease marked by terminal bronchiole and alveolar destruction and air trapping. (Ch. 28)

Empyema: (**EM**-pigh-**EE**-mah) Pus in a body cavity, especially the pleural space. (Ch. 28)

Encephalitis: (**EN**-seff-uh-**LYE**-tis) Inflammation of the brain. (Ch. 46)

Encephalopathy: (en-**SEFF**-uh-**LAHP**-ah-thee) Dysfunction of the brain. (Ch. 22, 33, 46)

Endarterectomy: (end-AR-tur-**ECK**-tuh-mee) Excision of thickened atheromatous areas of the innermost coat of an artery. (Ch. 22, 46)

Endogenous: (en-**DAH**-jen-us) Produced or originating from within a cell or organism. (Ch. 38, 40)

Endometritis: (**EN**-doh-me-**TRY**-tis) Inflammation of the endometrium of the uterus. (Ch.42)

Endorphins: (en-**DOR**-fins) Naturally occurring opioids in the body, many times more potent than analgesic medications. (Ch. 9)

Endoscope: (**EN**-doh-skohp) A device consisting of a tube and optical system for observing the inside of a hollow organ or cavity. Can be flexible or rigid. (Ch. 29)

Enkephalins: (en-**KEF**-e-lins) One type of endorphin. (Ch. 9)

Enteritis: (en-ter-**EYE**-tis) Inflammation of the intestines, particularly of the mucosa and submucosa of the small intestine. (Ch. 31, 42)

Enucleation: (ee-NEW-klee-**AY**-shun) Removal of an organ or other mass intact from its supporting tissues, as of the eyeball from the orbit. (Ch. 49)

Epidemiological: (EP-i-DEE-me-ah-**LAHJ**-i-kuhl) The study of the distribution and determinants of health-related states and events in populations and the application of this study to the control of health problems. (Ch. 42)

Epididymitis: (EP-i-DID-i-**MY**-tis) Inflammation or infection of the epididymis. (Ch. 41)

Epidural: (EP-i-**DUHR**-uhl) Situated on or outside the dura mater. (Ch. 46)

Epinephrine: (EP-i-**NEFF**-rin) A hormone secreted by the adrenal medulla in response to stimulation of the sympathetic nervous system. (Ch. 8)

Epistaxis: (EP-iss-**TAX**-iss) Nosebleed. (Ch. 27)

Epithelialization: (ep-i-THEE-lee-al-eye-**ZAY**-shun) The growth of skin over a wound. (Ch. 51)

Equianalgesic: (EE-kwee-AN-uhl-**JEE**-zik) Drugs having equal pain killing effect. The same degree of pain relief may require different doses when different medications are given or medications are given by different routes. (Ch. 9)

Erectile dysfunction: (e-RECK-tile dis-**FUNCK**-shun) Inability to have an erection sufficient for sexual intercourse. (Ch. 41)

Erection: (e-**REK**-shun) Enlargement and hardening of the penis caused by engorgement of blood. (Ch. 39)

Erythema: (ER-i-**THEE**-mah) Diffuse redness over the skin. (Ch. 50)

Eschar: (**ESS**-kar) Hard scab or dry crust that results from necrotic tissue. (Ch. 51)

Escharotomy: (ess-kar-**AHT**-oh-mee) Removal of a slough or scab formed on the skin and underlying tissue of severely burned skin. (Ch. 51)

Esophagogastroduodenoscopy: (e-**SOFF**-ah-go-GAS-troh-doo-AH-den-**AHS**-kuh-pee) An endoscopic procedure that allows the physician view the esophagus, stomach, and duodenum. (Ch. 32)

Esophagoscopy: (ee-soff-ah-**GAHS**-kuh-pee) Examination of the esophagus using an endoscope. (Ch. 29)

Esotropia: (ESS-oh-**TROH**-pee-ah) Strabismus in which there is deviation of the visual axis of one eye toward that of the other eye, resulting in diplopia. Also called cross-eyed. (Ch. 48)

Essential hypertension: (e-**SEN**-shul HIGH-per-**TEN**-shun) Chronic elevation of blood pressure resulting from an unknown cause. (Ch. 16)

Ethical: (**ETH**-i-kuhl) Describes behavior guided by a system of moral principles or standards. (Ch. 2)

Ethnic: (**ETH**-nick) Pertaining to a religious, racial, national, or cultural group. For example, individuals may identify with the Jewish, Catholic, or Islamic religions. (Ch. 3)

Ethnocentrism: (**ETH**-noh-SEN-trizm) The tendency to think that one's own ways of thinking, believing, and acting are the only right ways. People who are different are seen as strange or bizarre. An example is one who believes that his or her religious beliefs are the only right beliefs and other religions are wrong. (Ch. 3)

Eustress: (**YOO**-stress) Stress from positive experiences. (Ch. 55)

Euthyroid: (yoo-**THY**-royd) Normal thyroid function. (Ch. 37)

Evaluation: (e-**VAL**-yoo-**AY**-shun) The judgment of anything. (Ch. 1)

Evisceration: (E-**VIS**-sir-a-shun) Extrusion of viscera outside the body, especially through a surgical excision. (Ch. 11)

Exacerbation: (egg-sass-sir-**BAY**-shun) Aggravation of symptoms. (Ch. 47)

Exophthalmos: (ECKS-off-**THAL**-mus) Abnormal protrusion of the eyeball. (Ch. 36, 37)

Exotropia: (EKS-oh-**TROH**-pee-ah) Abnormal turning outward of one or both eyes; divergent strabismus. (Ch. 48)

Expectorant: (ek-**SPEK**-tuh-rant) Agent that promotes removal of pulmonary secretions. (Ch. 28)

Expectorate: (eck-**SPECK**-tuh-RAYT) The act or process of coughing up materials from the air passageways leading to the lungs. (Ch. 14)

External otitis: (eks-**TER**-nuhl oh-**TIGH**-tis) Inflammation of the external ear. (Ch. 49)

Extracardiac: (EX-trah-**KAR**-dee-ack) Outside the heart. (Ch. 8)

Extracellular: (EX-trah-**SELL**-yoo-ler) Outside the cell. (Ch. 5)

Extracorporeal shock wave lithotripsy (ESWL): (ECKS-trah-koar-**POR**-ee-uhl **SHAHK** WAYV LITH-oh-**TRIP**-see) Noninvasive treatment using shock waves to break up gallstones or kidney stones. (Ch. 33, 35)

Extravasation: (eks-**TRA**-vah-**ZAY**-shun) The escape of fluids into surrounding tissue. (Ch. 6)

Extrinsic factors: (eks-**TRIN**-sik **FAK**-ters) External variables. (Ch. 14)

Exudate: (EKS-yoo-dayt) Accumulated fluid in a cavity; oozing of pus or serum; often the result of inflammation. (Ch. 27, 28)

Fasciculation: (fah-SIK-yoo-**LAY**-shun) Twitching. (Ch. 47)

Fasciotomy: (fash-e-**OTT**-oh-me) Incision of fascia. (Ch. 44)

Feminist: (**FEM**-un-nist) A person who advocates for women the same rights as men. (Ch. 2)

Fetor hepaticus: (**FEE**-tor he-**PAT**-i-kus) Foul breath associated with liver disease. (Ch. 33)

Fibrocystic: (FIGH-broh-**SIS**-tik) Consisting of fibrocysts, which are fibrous tumors that have undergone cystic degeneration or accumulated fluid. (Ch. 40)

Fidelity: (fi-**DEL**-i-tee) The obligation to be faithful to commitments made to self and others. (Ch. 2)

Filtration: (fill-**TRAY**-shun) The process of removing particles from a solution by allowing the liquid portion to pass through a membrane or other partial barrier. (Ch. 5)

Fissure: (**FISH**-er) A narrow slit or cleft, especially one of the deeper or more constant furrows separating the gyri of the brain. (Ch. 31)

Fistula: (**FIST**-yoo-lah) Any abnormal, tubelike passage within body tissue, usually between two internal organs, or leading from an internal organ to the body surface. (Ch. 31)

Flaccid: (**FLA**-sid) Weak, lax, soft muscles. (Ch. 46)

Flail chest: (**FLAY**-ul chesst) Condition of the chest wall caused by two or more fractures on each affected rib resulting in a segment of rib that is not attached on either end; the flail portion moves paradoxically in with inspiration and out with expiration. (Ch. 12)

Flora: (**FLOOR**-a) Microbial life adapted for living in a specific environment such as the intestines, skin, or urinary tract. (Ch. 7)

Fluoroscope: (**FLAW**-or-oh-skohp) A device consisting of a fluorescent screen suitably mounted, either separately or in conjunction with an x-ray tube, by means of which the shadows of objects interposed between the tube and the screen are made visible. (Ch. 29)

Fluoroscopy: (fluh-**RAHS**-kuh-pee) The use of a fluoroscope for medical diagnosis or for testing various materials by roentgen rays. (Ch. 20)

Full-thickness burn: (**FUL-THICK**-ness **BERN**) Burn in which all of the epithelializing elements and those lining the sweat glands, hair follicles, and sebaceous glands are destroyed. (Ch. 12, 51)

Fungi: (**FUNG**-guy) A general term for a group of eukaryotic organisms (e.g., mushrooms, yeasts, molds). (Ch. 7)

Furuncle: (**FYOOR**-ung-kull) An acute circumscribed inflammation of the subcutaneous layers of the skin or of a gland or hair follicle. (Ch. 49)

Gastrectomy: (gas-**TREK**-tuh-mee) Any surgery that involves partial or total removal of the stomach. (Ch. 30)

Gastric acid stimulation test: (**GAS**-trik ASS-id STIM-yoo-**LAY**-shun TEST) A test that measures the amount of gastric acid for 1 hour after subcutaneous injection of a drug that stimulates gastric acid secretion. (Ch. 29)

Gastric analysis: (**GAS**-trik ah-**NAL**-i-sis) A test performed to measure secretions of hydrochloric acid and pepsin in the stomach. (Ch. 29)

Gastric lavage: (**GAS**-trik la-**VAHJ**) Washing out of the stomach; used to empty the stomach when the contents are irritating. (Ch. 12)

Gastritis: (gas-**TRY**-tis) Acute—The inflammation of the stomach mucosa; also known as heartburn or indigestion. Chronic—Gastritis that is recurrent; classified as type A (asympotomatic) or type B (symptomatic). (Ch. 30)

Gastroduodenostomy: (**GAS**-troh-**DOO**-oh-den-**AHS**-toh-mee) Excision of the pylorus of the stomach with anastomosis of the upper portion of the stomach to the duodenum. (Ch. 30)

Gastroepiploic: (**GAS**-troh-**EP**-i-**PLOH**-ick) Pertaining to the stomach and greater omentum. (Ch. 22)

Gastrojejunostomy: (**GAS**-troh-JAY-joo-**NAHS**-toh-mee) Subtotal excision of the stomach with closure of the proximal end of the duodenum and side-to-side anastomosis of the jejunum to the remaining portion of the stomach. (Ch. 30)

Gastroparesis: (**GAS**-troh-puh-**REE**-sis) Paralysis of the stomach, resulting in poor emptying. (Ch. 38)

Gastroplasty: (**GAS**-troh-**PLAS**-tee) Plastic surgery of the stomach. Used to decrease the size of the stomach to treat morbid obesity. (Ch. 30)

Gastroscopy: (gas-**TRAHS**-kuh-pee) Examination of the stomach and abdominal cavity by use of a gastroscope. (Ch. 29)

Gastrostomy: (gas-**TRAHS**-toh-mee) Surgical creation of a gastric fistula through the abdominal wall. (Ch. 29)

Gavage: (gah-**VAZH**) Feeding with a stomach tube or with a tube passed through the nares, pharynx, and esophagus into the stomach. The food is in liquid or semiliquid form at room temperature. (Ch. 29)

Generalization: (**JEN**-er-al-i-**ZAY**-shun) An assumption about a group or an individual item or person that leads to seeking additional information to determine if the generalization fits the individual. Whereas generalizations are true for the group, they may not be true for the individual. (Ch. 3)

Glaucoma: (glaw-**KOH**-mah) A group of eye diseases characterized by increased intraocular pressure. (Ch. 14, 49)

Glomerulonephritis: (gloh-**MER**-yoo-loh-ne-**FRY**-tis) A form of nephritis in which the lesions involve primarily the glomeruli. (Ch. 35)

Glossitis: (glah-**SIGH**-tis) An inflammation of the tongue. (Ch. 24)

Glycosuria: (**GLY**-kos-**YOO**-ree-ah) Abnormal amount of glucose in the urine, often associated with diabetes mellitus. (Ch. 38)

Goitrogens: (**GOY**-troh-jenz) Foods or medications that cause a goiter. (Ch. 37)

Gummas: (**GUM**-ahs) A soft granulomatous tumor of the tissues characteristic of the tertiary stage of syphilis. (Ch. 42)

Gynecomastia: (JIN-e-koh-**MASS**-tee-ah) Excessive breast tissue on a male. (Ch. 39)

Hallucinations: (huh-LOO-si-**NAY**-shuns) False perceptions having no relation to reality and not accounted for by any exterior stimuli. (Ch. 55)

Health: (**HELLTH**) A condition in which all functions of the body and mind are normally active. (Ch. 13)

Hearing aid: (**HEER**-ing **AYD**) An instrument to amplify sounds for those with hearing loss. (Ch. 48)

Heatstroke: (**HEET**-strohk) An acute and dangerous reaction to heat exposure, characterized by high body temperature, usually higher than 105° F (40.5° C). (Ch. 12)

Helicobacter pylori: (**HEH**-lick-co-back-tur **PIE**-lori) Bacterium that causes some peptic ulcers. (Ch. 30)

Hemarthrosis: (**HEEM**-ar-**THROH**-sis) Bleeding into a joint. (Ch. 24, 43)

Hematochezia: (HEM-uh-toh-**KEE**-zee-uh) Blood in the feces. (Ch. 31)

Hematoma: (**HEE**-muh-**TOH**-mah) A localized collection of extravasated blood, usually clotted, in an organ, space, or tissue. (Ch. 11)

Hematuria: (HEM-uh-**TYOOR**-ee-ah) Blood in the urine. (Ch. 34)

Hemiparesis: (hem-ee-puh-**REE**-sis) Weakness affecting one side of the body. (Ch. 45, 46)

Hemipelvectomy: (hem-ee-pell-**VEC**-toe-me) The surgical removal of half of the pelvis and the leg. (Ch. 44)

Hemiplegia: (hem-ee-**PLEE**-jee-ah) Paralysis of only one side of the body. (Ch. 46)

Hemodialysis: (**HEE**-moh-dye-**AL**-i-sis) A method for replacing the function of the kidneys by circulating blood through tubes made of semipermeable membranes. (Ch. 35)

Hemolysis: (he-**MAHL**-e-sis) The destruction of the membrane of red blood cells with the liberation of hemoglobin, which diffuses into the surrounding fluid. (Ch. 22, 24)

Hemophilia: (HEE-moh-**FILL**-ee-ah) A hereditary blood disease marked by greatly prolonged coagulation time, with consequent failure of the blood to clot and abnormal bleeding. (Ch. 24)

Hemoptysis: (hee-**MOP**-ti-sis) Coughing up of blood from respiratory tract. (Ch. 28)

Hemorrhoids: (**HEM**-uh-royds) A mass of dilated, tortuous veins in the anorectum involving the venous plexuses of that area. (Ch. 31)

Hemothorax: (HEE-moh-**THAW**-raks) Blood in the pleural space; may be associated with trauma, tuberculosis, or pneumonia. (Ch. 28)

Hepatomegaly: (HEP-uh-toh-**MEG**-ah-lee) Enlargement of the liver. (Ch. 21)

Hepatitis: (HEP-uh-**TIGH**-tis) Inflammation of the liver, most often viral. (Ch. 33)

Hepatorenal syndrome: (hep-**PAT**-oh-REE-nuhl **SIN**-drohm) A deadly kidney failure that sometimes accompanies liver disease. (Ch. 33)

Hepatosplenomegaly: (he-PA-toh-SPLE-noh-**MEG**-ah-lee) Enlargement of the liver and spleen. (Ch. 42)

Hernia: (**HER**-nee-uh) The protrusion or projection of an organ or a part of an organ through the wall of the cavity that normally contains it. (Ch. 31)

Herpetic: (her-**PET**-ick) Pertaining to herpes. (Ch. 42)

Hiatal hernia: (high-**AY**-tuhl **HER**-nee-ah) A condition in which part of the stomach protrudes through and above the diaphragm. (Ch. 30)

High-density lipoprotein (HDL): (HIGH **DEN**-si-tee LIP-oh-**PROH**-teen) Plasma lipids bound to albumin consisting of lipoproteins. It has been found that those with high levels of HDL have less chance of having coronary artery disease. (Ch. 18)

Histamine: (**HISS**-ta-mean) A substance produced in the body that increases gastric secretion, increases capillary permeability, contracts the bronchial smooth muscle. Plays a role in allergic reaction. (Ch. 53)

Homans' sign: (**HOH**-manz SIGHN) An assessment for venous thrombosis in which calf pain with dorsiflexion occurs if thrombosis is present. (Ch. 15)

Homeopathy: (HO-mee-**AH**-pa-thee) System of medicine based on the theory that "like cures like," and uses tiny doses of a substance that create the symptoms of disease. (Ch. 4)

Homeostasis: (HOH-mee-oh-**STAY**-sis) Maintaining a constant balance, especially whenever a change occurs. (Ch. 14)

Hopelessness: (**HOHP**-less-ness) Subjective state in which a person sees limited or unavailable alternatives; lacking energy. (Ch. 13)

Hordeolum: (hor-**DEE**-oh-lum) Sty. (Ch. 49)

Host: (**HOE**-st) The organism from which a parasite obtains its nourishment. (Ch. 7)

Human immunodeficiency virus (HIV): (**HYOO**-man im-YOO-noh-dee-**FISH**-en-see **VIGH**-rus) A retrovirus that causes acquired immunodeficiency syndrome (AIDS). (Ch. 54)

Humoral: (HYOO-mohr-uhl) Pertaining to body fluids or substances contained in them. (Ch. 52)

Hydrocele: (**HIGH**-droh-seel) A collection of fluid in the scrotal sack. (Ch. 39, 41)

Hydrocephalus: (HIGH-droh-**SEF**-uh-luhs) A condition caused by enlargement of the cranium caused by abnormal accumulation of cerebrospinal fluid within the cerebral ventricular system. (Ch. 46)

Hydronephrosis: (HIGH-droh-ne-**FROH**-sis) Abnormal dilation of kidneys caused by obstruction of urine flow. (Ch. 35, 41)

Hydrostatic: (**HIGH**-droh-**STAT**-ik) Pertaining to the pressure of liquids in equilibrium and to the pressure exerted by liquids. (Ch. 5)

Hypercalcemia: (HIGH-per-kal-**SEE**-mee-ah) An excessive amount of calcium in the blood. (Ch. 5)

Hyperglycemia: (HIGH-per-gligh-**SEE**-mee-ah) Excess glucose in the blood. (Ch. 38)

Hyperkalemia: (HIGH-per-kuh-**LEE**-mee-ah) An excessive amount of potassium in the blood. (Ch. 5, 20)

Hyperlipidemia: (HIGH-per-**LIP**-i-**DEE**-mee-ah) Excessive quantity of fat in the blood. (Ch. 18)

Hypermagnesemia: (**HIGH**-per-**MAG**-nuh-**ZEE**-mee-ah) Excess magnesium in the blood. (Ch. 5)

Hypernatremia: (**HIGH**-per-nuh-**TREE**-mee-ah) Excess sodium in the blood. (Ch. 5)

Hyperopia: (HIGH-per-**OH**-pee-ah) Farsightedness. (Ch. 49)

Hyperplasia: (HIGH-per-**PLAY**-zee-ah) Excessive increase in the number of normal cells. (Ch. 37, 41)

Hypertension: (HIGH-per-**TEN**-shun) Abnormally elevated blood pressure. (Ch. 16)

Hypertensive crisis: (HIGH-per-**TEN**-siv **CRY**-sis) Arbitrarily defined as severe elevation in diastolic blood pressure above 120 to 130 mm Hg. (Ch. 16)

Hypertonic: (**HIGH**-per-**TAHN**-ik) Exerts greater osmotic pressure than blood. (Ch. 5, 6)

Hypertrophy: (high-**PER**-truh-fee) An increase in the size of an organ or structure, or of the body, owing to growth rather than tumor formation. (Ch. 16, 40)

Hyperuricemia: (HIGH-per-yoor-a-**SEE**-me-ah) An excess of uric acid or urates in the blood. (Ch. 44)

Hyperventilation: (**HIGH**-per-**VEN**-ti-**LAY**-shun) Increased ventilation that results in a lowered carbon dioxide (CO_2) level (hypocapnia). (Ch. 5)

Hypervolemia: (**HIGH**-per-voh-**LEE**-mee-ah) An abnormal increase in the volume of circulating blood. (Ch. 5)

Hypocalcemia: (**HIGH**-poh-kal-**SEE**-mee-ah) Reduced amount of calcium in the blood. (Ch. 5)

Hypoglycemia: (HIGH-poh-gligh-**SEE**-mee-ah) Below-normal amount of glucose in the blood. (Ch. 38)

Hypokalemia: (**HIGH**-poh-kuh-**LEE**-mee-ah) Reduced amount of potassium in the blood. (Ch. 5)

Hypomagnesemia: (**HIGH**-poh-**MAG**-nuh-**ZEE**-mee-ah) Reduced amount of magnesium in the blood. (Ch. 5, 20)

Hyponatremia: (**HIGH**-poh-nuh-**TREE**-mee-ah) Reduced amount of sodium in the blood. (Ch. 5)

Hypophysectomy: (**HIGH**-pah-fi-**SECK**-tuh-mee) Surgical removal of the pituitary gland. (Ch. 37)

Hypoplasia: (HIGH-poh-**PLAY**-zee-ah) Underdevelopment of a tissue organ or body. (Ch. 40)

Hypoproteinemia: (**HIGH**-poh-pro-teen-**EE**-mee-ah) A decrease in the amount of protein in the blood. (Ch. 12)

Hypospadias: (HIGH-poh-**SPAY**-dee-ahz) A congenital male defect in which the opening of the urethra is on the underside of the penis, instead of the tip. (Ch. 39)

Hypostatic: (HIGH-poh-**STA**-tik) Hypostatic pneumonia occurs from congestion in the lungs associated with lack of activity. (Ch. 28)

Hypotension: (HIGH-poh-**TEN**-shun) Abnormally low blood pressure below 90 mm Hg systolic. (Ch. 8)

Hypothermia: (HIGH-poh-**THER**-mee-ah) Body temperature below 95° F (35° C). (Ch. 11, 22)

Hypotonic: (**HIGH**-poh-**TAHN**-ik) Pertaining to defective muscular tone or tension; having a lower concentration of solute than intracellular or extracellular fluid. (Ch. 5, 6)

Hypovolemia: (**HIGH**-poh-voh-**LEE**-mee-ah) The most common form of dehydration resulting from the loss of fluid from the body; results in decreased blood volume. (Ch. 5)

Hypovolemic: (**HIGH**-poh-voh-**LEEM**-ick) Low volume of blood in the circulatory system. (Ch. 8, 37)

Hypovolemic shock: (**HIGH**-poh-voh-**LEEM**-ick **SHAHK**) Shock that occurs when blood or plasma is lost in such quantities that the remaining blood cannot fill the circulatory system despite constriction of the blood vessels. (Ch. 12)

Hypoxemia: (HIGH-pock-**SEE**-mee-ah) Deficient oxygenation of the blood. (Ch. 22)

Hypoxia: (high-**POCK**-see-ah) Diminished availability of oxygen to the body tissues. (Ch. 22)

Hysterectomy: (HISS-tuh-**RECK**-tuh-mee) Surgical removal of the uterus through the abdominal wall or vagina. (Ch. 40)

Hysterosalpingogram: (HIS-tur-oh-**SAL**-pinj-oh-gram) Radiograph of the uterus and fallopian tubes. (Ch. 39)

Hysteroscopy: (HIS-tur-**AHS**-koh-pee) Endoscopic direct visual examination of the canal of the uterine cervix and the cavity of the uterus. (Ch. 39)

Hysterotomy: (HISS-tuh-**RAH**-tuh-mee) Incision of the uterus. (Ch.40)

Icterus: (**ICK**-ter-us) Yellowing of the skin and the sclera of the eye. (Ch. 32)

Idiopathic thrombocytopenic purpura: (ID-ee-oh-**PATH**-ik THROM-boh-SIGH-toh-**PEE**-nik **PUR**-pew-rah) The total number of circulating platelets is greatly diminished, even though platelet production in the bone marrow is normal, resulting in slowed blood clotting. (Ch. 24)

Ileostomy: (ILL-ee-**AH**-stuh-me) An artificial opening (stoma) created in the small intestine (ileum) and brought to the surface of the abdomen for the purpose of evacuating feces. (Ch. 31)

Illness: (**ILL**-ness) The state of being sick. (Ch. 13)

Illusions: (i-**LOO**-zhuns) Mistaken perceptions of reality. (Ch. 55)

Imagery: (**IM**-ij-ree) The use of the imagination to promote relaxation. (Ch. 55)

Immunocompromised: (**IM**-yoo-noh-**KAHM**-prah-mized) Having an immune system that is not capable of reacting to a pathogen or tissue damage. (Ch. 28)

Impaction: (im-**PAK**-shun) An immovable accumulation of feces in the bowels. (Ch. 29, 31)

Imperforate: (im-**PER**-foh-rate) Without an opening. (Ch. 40)

Induction: (in-DUCK-shun) The process or act of causing to occur, as in anesthesia induction. (Ch. 11)

Induration: (**IN**-dyoo-**RAY**-shun) Area of hardened tissue. (Ch. 28)

Infective endocarditis: (in-**FECK**-tive EN-doh-kar-**DYE**-tis) Inflammation of the heart lining caused by microorganisms. (Ch. 17)

In situ: (in-**SIT**-yoo) Localized, not invading surrounding tissue. (Ch. 10)

Inspection: (in-**SPEK**-shun) Use of observation skills to systematically gather data that can be seen. (Ch. 1)

Insufficiency: (**IN**-suh-**FISH**-en-see) The condition of being inadequate for a given purpose, such as heart valves that do not close properly. (Ch. 19)

Insufflation: (in-suff-**LAY**-shun) Used to inflate the abdomen during laparoscopic or endoscopic procedures to enhance visualization of structures. (Ch. 39, 40)

Intermittent claudication: (**IN**-ter-**MIT**-ent KLAW-di-**KAY**-shun) A symptom associated with arterial occlusive disease. It refers to pain in the calf of a lower extremity, usually brought on by activity or exercise, and ceases with rest. (Ch. 18)

International normalized ratio: (**IN**-ter-**NASH**-uh-nul **NOR**-muh-lized **RAY**-she-oh) The World Health Organization's standardization for reporting the prothrombin time assay test when the thromboplastin reagent developed by the first International Reference Preparation is used. The reagent was developed to prevent variability in prothrombin time testing results and provide uniformity in monitoring therapeutic levels for coagulation during oral anticoagulation therapy. (Ch. 17)

Interstitial: (**IN**-ter-**STISH**-uhl) Fluid between tissues. (Ch. 5)

Intervention: (in-ter-**VEN**-shun) One or more actions taken in order to modify an effect. (Ch. 1)

Intracellular: (**IN**-trah-**SELL**-yoo-ler) Fluids located within the blood cell. (Ch. 5)

Intracranial: (**IN**-trah-**KRAY**-nee-uhl) Within the cranium or skull. (Ch. 5)

Intraoperative: (**IN**-trah-**AHP**-er-uh-tiv) Occurring during a surgical procedure. (Ch. 11)

Intravascular: (**IN**-trah-**VAS**-kyoo-lar) Fluids located within the blood vessels. (Ch. 5)

Intravenous: (**IN**-trah-**VEE**-nus) Within or into a vein. (Ch. 6)

Intrinsic factors: (in-**TRIN**-sik **FAK**-ters) Internal variables. (Ch. 14)

Intussusception: (**IN**-tuh-suh-**SEP**-shun) The slipping of one part of an intestine into another adjacent to it. (Ch. 31)

In vitro fertilization: (in **VEE**-troh FER-ti-li-**ZAY**-shun) Fertilization in a test tube. (Ch. 40)

Ipsilateral: (**IP**-si-**LAT**-er-uhl) On the same side; affecting the same side of the body. (Ch. 46)

Ischemia: (iss-**KEY**-me-ah) Condition of inadequate blood supply. (Ch. 8, 15)

Isoelectric line: (EYE-so-e-**LEK**-trick LINE) The period when the electrical tracing is at zero and is neither positive nor negative. (Ch. 20)

Isolated systolic hypertension: (EYE-suh-lay-ted sis-**TAH**-lik high-per-**TEN**-shun) The systolic pressure is 160 mm Hg or more, but the diastolic pressure is lower than 95 mm Hg. (Ch. 16)

Isotonic: (EYE-so-**TAHN**-ik) A fluid that has the same osmolarity as the blood. (Ch. 5, 6)

Jaundice: (**JAWN**-diss) Yellowing of the skin and the sclera of the eye. (Ch. 32)

Joint: (JOYNT) An articulation. The point of juncture between two bones. (Ch. 43)

Kaposi's sarcoma: (ka-**POE**-sees sar-**CO**-mah) A vascular malignancy that is often first apparent in the skin or mucous membranes but may involve the viscera. (Ch. 54)

Ketoacidosis: (KEE-toh-ass-i-**DOH**-sis) A condition in which fat breakdown produces ketones, which cause an acidic state in the body; may be associated with weight loss or diabetes mellitus. (Ch. 38)

Kussmaul's: (**KOOS**-mahlz) Term describing deep respirations of an individual with ketoacidosis. (Ch. 38)

Laceration: (la-sir-**A**-shun) A wound or irregular tear of the flesh. (Ch. 12)

Lactic acid: (**LAK**-tik **ASS**-id) By-product of anaerobic metabolism. (Ch. 8)

Laminectomy: (LAM-i-**NEK**-toh-mee) The excision of a vertebral posterior arch, usually to remove a lesion or herniated disk. (Ch. 46)

Laparoscopy: (LAP-uh-roh-**SKOP**-ee) Exploration of the abdomen with an endoscope. (Ch. 33)

Laparotomy: (LAP-uh-**RAH**-tuh-mee) The surgical opening of the abdomen; an abdominal operation. (Ch. 40)

Laryngeal edema: (lah-**RIN**-jee-uhl uh-**DEE**-muh) Sudden swelling of the larynx occurring with severe allergic reactions. (Ch. 8)

Laryngectomy: (lar-in-**JEK**-tah-mee) Surgical removal of the larynx. (Ch. 27)

Laryngitis: (lare-in-**JIGH**-tiss) Inflammation of the larynx. (Ch. 27)

Laser ablation: (LAY-zer uh-**BLAY**-shun) Therapeutic destruction of a growth or part of a growth by laser treatment. (Ch. 40)

Lavage: (lah-**VAZH**) Washing out of a cavity. (Ch. 29)

Law: (LAW) The further formalization of moral considerations. (Ch. 2)

Leiomyoma: (LYE-oh-my-**OH**-ma) A myoma consisting principally of smooth muscle tissue. (Ch. 40)

Leukemia: (loo-**KEE**-mee-ah) A malignancy of the blood-forming cells in the bone marrow. (Ch. 24)

Leukocytosis: (LOO-koh-sigh-**TOH**-sis) An increase in the number of leukocytes in the blood, generally caused by presence of infection and usually transient. (Ch. 22)

Leukopenia: (LOO-koh-**PEE**-nee-yah) Abnormal decrease of white blood cells, usually below 5000/mm3. (Ch. 10)

Liability: (LYE-uh-**BIL**-i-tee) The level of responsibility that society places on individuals for their actions. (Ch. 1)

Libido: (li-**BEE**-doh) Sexual drive, conscious or unconscious. (Ch. 39)

Lichenified: (lye-**KEN**-i-fyed) Thickened or hardened from continued irritation. (Ch. 51)

Limitation of liability: (lim-i-**TAY**-shun OF LYE-uh-**BIL**-i-tee) Steps that health care professionals can take to limit their liability. (Ch. 2)

Lobectomy: (loh-**BEK**-tuh-mee) Surgical removal of a lobe of any organ or gland. (Ch. 28)

Lower gastrointestinal series (lower GI): (**LOH**-er GAS-troh-in-**TES**-ti-nuhl **SEER**-ees) The use of barium sulfate as an enema to facilitate x-ray and fluoroscopic examination of the colon. (Ch. 29)

Lymphadenopathy: (lim-FAD-e-**NAH**-puh-thee) Any disorder of the lymph nodes. (Ch. 42)

Lymphangitis: (lim-FAN-je-**EYE**-tis) Inflammation of lymphatic channels or vessels. (Ch. 18)

Lymphedema: (LIMPF-uh-**DEE**-mah) An abnormal accumulation of tissue fluid (potential lymph) in the interstitial space. (Ch. 23)

Lymphocytes: (**LIM**-foh-sites) Cells present in the blood and lymphatic tissue that provide the main means of immunity for the body; white blood cells. (Ch. 52)

Lymphoma: (lim-**FOH**-mah) A usually malignant lymphoid neoplasm. (Ch. 25)

Macular degeneration: (**MACK**-you-lar dee-**JEN**-uh-**RAY**-shun) Age-related breakdown of the macular area of the retina of the eye. (Ch. 14, 49)

Malignant: (muh-**LIG**-nunt) Growing, resisting treatment; used to describe a tumor of cancerous cells. (Ch. 10)

Malpractice: (mal-**PRAK**-tiss) A breach of duty arising out of the relationship that exists between the client and the health care worker. (Ch. 2)

Mammography: (mah-**MOG**-rah-fee) Use of radiography of the breast to diagnose breast cancer. (Ch. 39)

Mammoplasty: (MAM-oh-**PLAS**-tee) Plastic surgery of the breast. (Ch. 40)

Mania: (**MAY**-nee-ah) Mental disorder characterized by excessive excitement. (Ch. 55)

Marsupialization: (mar-SOO-pee-al-i-**ZAY**-shun) Process of raising the borders of an evacuated tumor sac to the edges of the abdominal wound and stitching them there to form a pouch. (Ch. 40)

Mastalgia: (mass-**TAL**-jee-ah) Pain in the breast. (Ch. 40)

Mastectomy: (mass-**TECK**-tuh-mee) Excision of the breast. (Ch. 40)

Mastitis: (mass-**TIGH**-tis) Inflammation of the breast. (Ch. 40)

Mastopexy: (MAS-toh-**PEKS**-ee) Correction of a pendulous breast by surgical fixation and plastic surgery. (Ch. 40)

Mediastinum: (ME-dee-ah-**STYE**-num) A septum or cavity between two principal portions of an organ. (Ch. 22)

Megacolon: (**MEG**-ah-KOH-lun) Extremely dilated colon. (Ch. 31)

Melena: (muh-**LEE**-nah) Black, tarry feces caused by action of intestinal secretions on free blood. (Ch. 31)

Menarche: (me-**NAR**-kee) The initial menstrual period, normally occurring between the ninth and seventeenth year. (Ch. 39)

Meniere's disease: (MAY-nee-**AIRZ** di-**ZEEZ**) A recurrent and usually progressive group of symptoms including progressive deafness, ringing in the ears, dizziness, and a sensation of fullness or pressure in the ears. (Ch. 49)

Meningitis: (men-in-**JIGH**-tis) Inflammation of the membranes of the spinal cord and brain. (Ch. 46)

Menopause: (**MEN**-oh-pawz) The period that marks the permanent cessation of menstrual activity, usually occurring between the ages of 35 and 58. (Ch. 39)

Mental health: (**MEN**-tuhl HELLTH) State of being adjusted to life; able to be flexible, successful, maintain close relationships, solve problems, make appropriate judgments, and cope with daily stresses. (Ch. 55)

Mental illness: (**MEN**-tuhl ILL-ness) Any illness that affects the mind or behavior. (Ch. 55)

Metastasis: (muh-**TASS**-tuh-sis) Movement of bacteria or body cells (especially cancer cells) from one part of the body to another. (Ch. 10)

Milieu: (me-**LYU**) Environment. (Ch. 55)

Miotic: (my-**AH**-tik) An agent that causes the pupil to contract. (Ch. 49)

Morality: (muh-**RAL**-i-tee) A social barometer that dictates what is good or bad in a society. (Ch. 2)

Morbidity: (more-**BID**-it-ee) State of being diseased. (Ch. 7)

Mortality: (more-**TAL**-it-ee) Condition of being mortal; number of deaths in a population. (Ch. 7)

Mucolytic: (MYOO-koh-**LIT**-ik) Agent that liquefies sputum. (Ch. 28)

Mucopurulent cervicitis: (MYOO-koh-**PYOOR**-uh-lent SIR-vi-**SIGH**-tis) Inflammation of the cervix producing mucus and purulent discharge. (Ch. 42)

Mucositis: (MYOO-koh-**SIGH**-tis) Inflammation of a mucous membrane. (Ch. 10)

Multifocal: (MUHL-tee-**FOH**-kuhl) Many foci (areas) or sites. (Ch. 20)

Murmur: (**MUR**-mur) An abnormal sound heard on auscultation of the heart and adjacent large blood vessels. (Ch. 15, 19)

Muscle: (**MUSS**-uhl) A bundle of long slender cells or fibers that have the power to contract and hence to produce movement. (Ch. 43)

Myalgia: (my-**AL**-jee-ah) Muscle pain or tenderness. (Ch. 27)

Myectomy: (my-**ECK**-tuh-mee) Surgical removal of a hypertrophied muscle. (Ch. 17)

Myelogram: (**MY-**e-loh-gram) The film produced by radiography of the spinal cord after injection of a contrast medium into the subarachnoid space. (Ch. 45)

Myocardial infarction: (**MY-**oh-**KAR-**dee-yuhl in-**FARK-**shun) Death of cells of an area of the heart muscle, myocardium, as a result of oxygen deprivation, which in turn is caused by obstruction of the blood supply. Commonly referred to as a heart attack. (Ch. 18)

Myocarditis: (**MY-**oh-kar-**DYE-**tis) The inflammatory process that causes nodules to form in the myocardial tissue; the nodules become scar tissue over time. Inflammation of the heart muscle. (Ch. 8, 17)

Myocardium: (**MY-**oh-**KAR-**dee-um) Heart muscle. (Ch. 8)

Myomectomy: (my-oh-**MECK-**tuh-mee) Removal of a portion of muscle or muscular tissue. (Ch. 40)

Myopia: (my-**OH-**pee-ah) The error of refraction in which rays of light entering the eye parallel to the optic axis are brought to a focus in front of the retina; nearsightedness. (Ch. 49)

Myringoplasty: (mir-**IN-**goh-PLASS-tee) Surgical reconstruction of the tympanic membrane. (Ch. 49)

Myringotomy: (MIR-in-**GOT-**uh-mee) Incision of the tympanic membrane, usually performed to relieve pressure and allow for drainage of either serous or purulent fluid in the middle ear behind the tympanic membrane. (Ch. 49)

Myxedema: (MICK-suh-**DEE-**mah) Condition resulting from hypofunction of the thyroid gland. (Ch. 37)

Nasoseptoplasty: (NAY-zoh-**SEP-**toh-plass-tee) Surgical correction of the nasal septum. (Ch. 27)

Naturopathy: (NAY-chur-**AH-**pa-thee) System of medicine that uses natural therapies such as nutrition, herbs, hydrotherapy, counseling, physical medicine, and homeopathy to treat disease, promote healing, and prevent illness. (Ch. 4)

Negligence: (**NEG-**li-jense) An unintentional tort. (Ch. 2)

Neoplasm: (**NEE-**oh-PLAZ-uhm) New abnormal tissue growth, as in a tumor. (Ch. 10)

Nephrectomy: (ne-**FREK-**tuh-mee) Surgical removal of a kidney. (Ch. 35)

Nephrogenic: (NEFF-roh-**JEN-**ick) Caused by the kidneys. (Ch. 37)

Nephrolithotomy: (NEFF-roh-li-**THOT-**uh-mee) Incision of a kidney for removal of kidney stones. (Ch. 35)

Nephropathy: (ne-**FROP-**uh-thee) Any disease of the kidney. (Ch. 35, 38)

Nephrosclerosis: (NEFF-roh-skle-**ROH-**sis) Hardening of the kidney associated with hypertension and disease of the renal arterioles. (Ch. 35)

Nephrostomy: (ne-**FRAHS-**toh-mee) Creation of a permanent opening into the renal pelvis. (Ch. 35)

Nephrotoxin: (NEFF-roh-**TOCK-**sin) A toxin having a specific destructive effect on kidney tissue. (Ch. 22, 35)

Neuralgia: (new-**RAL-**jee-ah) Nerve pain. (Ch. 47)

Neurogenic: (NEW-roh-**JEN-**ik) Originating in the nervous system. (Ch. 8)

Neuropathic pain: (NEW-roh-**PATH-**ik **PAYN**) Pain resulting from peripheral nerve injury. (Ch. 9)

Neuropathy: (new-**RAH-**puh-thee) A general term denoting functional disturbances and pathologic changes in the peripheral nervous system. (Ch. 38, 47)

Neutrophils: (**NEW-**troh-fils) Granular leukocytes (white blood cells) having a nucleus with three to five lobes connected by threads of chromatin and cytoplasm containing very fine granules. (Ch. 52)

Nociceptive: (NOH-see-**SEP-**tiv) Pain sensitive. (Ch. 9)

Nocturia: (nock-**TYOO-**ree-ah) Excessive urination at night. (Ch. 14, 37, 38)

Nodal or junctional rhythm: (**NOHD-**uhl or **JUNGK-**shun-uhl **RITH-**uhm) A cardiac rhythm with its origin at the atrioventricular (AV) node. (Ch. 20)

Nonmaleficence: (NON-muh-**LEF-**i-sens) The requirement that health care providers do no harm to their clients, either intentionally or unintentionally. (Ch. 2)

Norepinephrine: (NOR-ep-i-**NEFF-**rin) A hormone produced by the adrenal medulla, similar in chemical and pharmacological properties to epinephrine, but chiefly a vasoconstrictor with little effect on cardiac output. (Ch. 8)

Normotensive: (nor-moh-**TEN-**siv) Normal blood pressure. (Ch. 16)

Nosocomial infection: (no-zoh-**KOH-**mee-uhl in-**FECK-**shun) Infection acquired in a health care agency. (Ch. 7)

Nuchal rigidity: (**NEW-**kuhl re-**JID-**i-tee) Rigidity of the nape, or back, of the neck. (Ch. 46)

Nursing diagnosis: (**NER-**sing **DYE-**ag-NOH-sis) A standardized label placed on a client's problem to make it understandable to all nurses. (Ch. 1)

Nursing process: (**NER-**sing **PRAH-**sess) An orderly, logical approach to administering nursing care so that the client's needs for such care are met comprehensively and effectively. (Ch. 1)

Nystagmus: (nis-**TAG-**muss) Involuntary, rapid, rhythmic eye movement. (Ch. 45, 48)

Obesity: (oh-**BEE-**si-tee) Abnormal amount of fat on the body from 20% to 30% over average weight for age, sex, and height. (Ch. 30)

Objective data: (ob-**JEK-**tiv **DAY-**tuh) Factual data obtained through physical assessment and diagnostic tests; objective data are observable or knowable through the five senses. (Ch. 1)

Obsession: (ub-**SESH-**un) Repetitive thought, urge, or emotion. (Ch. 55)

Obstipation: (OB-sti-**PAY-**shun) Intractable constipation. (Ch. 31)

Obstructive shock: (ub-**STRUCK-**tive **SHAHK**) Shock caused by indirect pump failure. (Ch. 12)

Occult blood test: (ah-**KULT** BLUHD TEST) A chemical test or microscopic examination for blood, especially in feces, that is not apparent on visual inspection. (Ch. 29)

Oligura: (AWH-li-**GYOO-**ree-ah) Diminished urination. (Ch. 8, 35)

Oncology: (on-**CAW-**luh-jee) The study of cancer and cancer treatment. (Ch. 10)

Oncovirus: (**ON-**koh-VIGH-russ) Viruses linked to cancer in humans. (Ch. 10)

Onychomycosis: (ON-i-koh-my-**KOH**-sis) Disease of the nails caused by fungus. (Ch. 51)

Ophthalmia neonatorum: (ahf-**THAL**-mee-ah NEE-oh-nuh-**TOR**-uhm) Conjunctivitis in the newborn resulting from exposure to infectious or chemical agents. (Ch. 42)

Ophthalmologist: (AHF-thal-**MAH**-luh-jist) A physician who specializes in the treatment of disorders of the eye. (Ch. 48)

Ophthalmoscope: (ahf-**THAL**-muh-skohp) An instrument used for examining the interior of the eye, especially the retina. (Ch. 48)

Opioid: (**OHP**-ee-OYD) A narcotic drug with morphine-like effects. True opioids are derived from opium. (Ch. 9)

Optician: (ahp-**TISH**-uhn) One who specializes in filling prescriptions for corrective lenses for eyeglasses and contact lenses. (Ch. 48)

Optimum level of functioning: (**OP**-teh-mum **LEV**-uhl of **FUNK**-shun-ing) Highest level of client activity considering the client's condition. (Ch. 14)

Optometrist: (ahp-**TOM**-uh-trist) A doctor of optometry who diagnoses and treats conditions and diseases of the eye per state laws. (Ch. 48)

Orchiectomy: (or-ki-**EK**-toh-mee) Removal of one or both testicles; a treatment for prostate cancer. (Ch. 41)

Orchitis: (or-**KIGH**-tis) Inflammation of a testis. (Ch. 41)

Orgasm: (**OR**-gazm) Pleasurable physical release sensation related to physical, sexual, and psychological stimulation. (Ch. 39, 41)

Orientation: (OR-ee-en-**TAY**-shun) The ability to comprehend and to adjust oneself in an environment with regard to time, location, and identity of persons. (Ch. 55)

Orthopnea: (or-**THOP**-knee-a) Labored breathing that occurs when lying flat; relieved when sitting up; associated with left ventricular heart failure. (Ch. 21)

Osmolality: (ahs-moh-**LAL**-i-tee) Osmotic concentration; ionic concentration of the dissolved substances per unit of solvent. (Ch. 37)

Osmosis: (ahs-**MOH**-sis) The passage of solvent through a semipermeable membrane that separates solutions of different concentrations. (Ch. 5)

Osteomyelitis: (AHS-tee-oh-my-**LIGHT**-tis) Inflammation of bone, especially the marrow, caused by a pathogenic organism. (Ch. 44)

Osteopathy: (**AHS**-tee-ah-**PATH**-ee) System of medicine emphasizing the inter-relationship of the body's nerves, muscles, bones, and organs; involves treating the whole person, and stresses the importance of diet, exercise, and fitness, with a focus on prevention. (Ch. 4)

Osteoporosis: (AHS-tee-oh-por-**OH**-sis) A condition in which there is a reduction in the mass of bone per unit volume. (Ch. 5, 14)

Osteosarcoma: (AHS-tee-oh-sar-**KOH**-mah) A malignant sarcoma of a bone. (Ch. 44)

Otalgia: (oh-**TAL**-jee-ah) Pain in the ear. (Ch. 48)

Otorrhea: (OH-toh-**REE**-ah) Inflammation of the ear with purulent discharge. (Ch. 48)

Otosclerosis: (OH-toh-skle-**ROH**-sis) A condition characterized by chronic, progressive deafness, especially for low tones. (Ch. 49)

Ototoxic: (OH-toh-**TOK**-sik) Having a detrimental effect on the eighth cranial nerve or the organs of hearing. (Ch. 48)

Pain: (PAYN) An unpleasant sensory and emotional experience associated with actual or potential tissue damage, or described in terms of such damage. Is whatever the client says it is whenever the client says it occurs. (Ch. 9)

Palliation: (pal-ee-**AY**-shun) The relief of symptoms without cure. (Ch. 10)

Palpation: (pal-**PAY**-shun) Use of the fingers or hands to feel something. (Ch. 1)

Pancreatectomy: (PAN-kree-uh-**TECK**-tuh-mee) Removal of all or part of the pancreas. (Ch. 33)

Pancreatitis: (PAN-kree-uh-**TIGH**-tis) Inflammation of the pancreas. (Ch. 22)

Pancytopenia: (PAN-sigh-toh-**PEE**-nee-ah) Abnormal depression of all the cellular elements of the blood. (Ch. 24)

Panhysterectomy: (PAN-hiss-tuh-**RECK**-tuh-mee) Excision of the entire uterus, including the cervix uteri. (Ch. 40)

Panmyelosis: (PAN-my-e-**LOH**-sis) Increased level of all bone marrow components, red blood cells, white blood cells, and platelets. (Ch. 24)

Paradoxical respirations: (PAR-uh-**DOK**-si-kuhl RES-pi-**RAY**-shuns) Chest movement on respiration that is opposite to that expected. (Ch. 28)

Paranoia: (PAR-uh-**NOY**-uh) Behavior that is marked by delusions of persecution or delusional jealousy. (Ch. 55)

Paraparesis: (PAR-ah-pah-**REE**-sis) Partial paralysis of the lower extremities. (Ch. 46)

Paraphimosis: (PAR-uh-figh-**MOH**-sis) Uncircumcised foreskin that has swollen and stuck behind the head of the penis. (Ch. 41)

Paraplegia: (PAR-ah-**PLEE**-jah) Paralysis of the lower body, including both legs, resulting from a spinal cord lesion. (Ch. 46)

Paresis: (puh-**REE**-sis) Weakness; incomplete paralysis. (Ch. 45)

Paresthesia: (PAR-es-**THEE**-zee-ah) A heightened sensation, such as burning, prickling, or tingling. (Ch. 22, 45, 46)

Paroxysmal novturnal dyspnea: (PEAR-ox-**IS**-mall knock-TURN-al DISP-knee-a) Sudden attacks of shortness of breath that usually occur during sleep. Person wakes gasping for breath and sits up to relieve symptoms; associated with left ventricular heart failure. (Ch. 21)

Partial-thickness burn: (**PAR**-shul **THICK**-ness **BERN**) Burn in which the epithelializing elements remain intact. (Ch. 12)

Passive immunity: (**PASS**-iv im-**YOO**-ni-tee) Reinforcement of the immune system with immune serum for such conditions as tetanus, diptheria, and venomous snake bite. (Ch. 7, 52)

Paternalism: (puh-**TER**-nuhl-izm) A unilateral and sometimes unreasonable decision by health care providers that implies they know what is best, regardless of the client's wishes. (Ch. 2)

Pathogen: (**PATH**-o-jen) A microorganism or substance capable of producing a disease. (Ch. 7)

Pathological fracture: (PATH-uh-**LAH**-jik-uhl **FRAHK**-chur) Fracture resulting from weakening of the bone structure by pathological processes such as neoplasia or osteomalacia. (Ch. 24, 25)

Patient-controlled analgesia (PCA): (**PAY**-shent kon-**TROHLD** an-uhl-**JEE**-zee-ah) An apparatus that delivers an intravenous analgesic to relieve pain, which is controlled by the client. (Ch. 9)

Pedicle: (**PED**-i-kuhl) The stem that attaches a new growth. (Ch. 40)

Pediculosis: (pe-DIK-yoo-**LOH**-sis) Infestation with lice. (Ch. 51)

Pemphigus: (**PEM**-fi-gus) Acute or chronic serious skin disease characterized by the appearance of bullae (blisters) of various sizes on normal skin and mucous membranes. (Ch. 51)

Peptic ulcer disease: (**PEP**-tick **UL**-sir di-**ZEEZ**) A condition in which the lining of the esophagus, stomach, or duodenum is eroded. (Ch. 30)

Perception: (per-**SEP**-shun) A unique impression of events by an individual. These impressions are strongly influenced by personality, cultural orientation, attitudes, and life experiences. (Ch. 14)

Percussion: (per-**KUSH**-un) A tapping technique used by physicians and advanced practice nurses to determine the consistency of underlying tissues. (Ch. 1)

Percutaneous: (PER-kyoo-**TAY**-nee-us) Through the skin; may refer to an injection, a medication application, or a biopsy. (Ch. 34)

Perfusion: (per-**FEW**-zhun) Supplying an organ or tissue with blood. (Ch. 8)

Pericardial effusion: (PER-ee-**KAR**-dee-uhl ee-**FYOO**-zhun) A buildup of fluid in the pericardial space. (Ch. 17)

Pericardial friction rub: (PER-ee-**KAR**-dee-uhl **FRICK**-shun RUB) Friction sound heard over the fourth left intercostal space near the sternum; a classic sign of pericarditis. (Ch. 15, 17)

Pericardial tamponade: (PER-ee-**KAR**-dee-uhl TAM-pon-**AID**) Compression of the heart by an abnormal filling of the pericardial sac with blood. (Ch. 8, 17)

Pericardiectomy: (PER-ee-kar-dee-**ECK**-tuh-mee) Excision of part or all of the pericardium. (Ch. 17)

Pericardiocentesis: (PER-ee-KAR-dee-oh-sen-**TEE**-sis) Surgical perforation of the pericardium. (Ch. 17, 22)

Pericardiotomy: (PER-ee-KAR-dee-**AH**-tah-mee) Incision of the pericardium. (Ch. 22)

Pericarditis: (PER-ee-kar-**DYE**-tis) Inflammation of the pericardium. (Ch. 17, 22)

Perimenopausal: (PER-ee-MEN-oh-**PAWS**-uhl) The phase before the onset of menopause, during which the cycle of a woman with regular menses changes, perhaps abruptly, to a pattern of irregular cycles and increased periods of amenorrhea. (Ch. 40)

Perinatal: (PAIR-ee-**NAY**-tuhl) Concerning the period beginning after the 28th week of pregnancy and ending 28 days after birth. (Ch. 42)

Perioperative: (PER-ee-**AHP**-er-uh-tiv) Occurring in the period immediately before, during, and after surgery. (Ch. 11)

Peripheral arterial disease: (puh-**RIFF**-uh-ruhl ar-**TIR**-ee-uhl di-ZEEZ) Disease of the peripheral arteries that interferes with adequate flow of blood. (Ch. 18)

Peripheral parenteral nutrition: (puh-**RIFF**-uh-ruhl par-**EN**-te-ruhl new-**TRISH**-un) Nutrition by intravenous injection. (Ch. 29)

Peripheral vascular resistance: (puh-**RIFF**-uh-ruhl **VAS**-kyoo-lar ree-**ZIS**-tense) Opposition to blood flow through the vessels. (Ch. 16, 21)

Peristalsis: (paris-**TALL**-sis) Progressive, wave-like movement that occurs involuntarily in hollow tubes of the body such as the alimentary (digestive) canal; causes contents of tube to be moved onward. (Ch. 29)

Peristomal: (PER-i-**STOH**-muhl) Area around a stoma. (Ch. 31)

Peritoneal dialysis: (PER-i-toh-**NEE**-uhl dye-**AL**-i-sis) The employment of the peritoneum surrounding the abdominal cavity as a dialyzing membrane for the purpose of removing waste products or toxins accumulated as a result of renal failure. (Ch. 35)

Peritonitis: (per-i-toh-**NIGH**-tis) Inflammation of the peritoneum. (Ch. 31)

Petechiae: (pe-**TEE**-kee-ee, puh-**TEE**-kee-eye) Small, purplish, hemorrhagic spots on the skin that appear in certain illnesses and bleeding disorders. (Ch. 17, 23, 50)

Phagocytosis: (fay-go-sigh-**TOH**-sis) Ingestion and digestion of bacteria and particles by phagocytes, cells that have the ability to ingest and destroy particulate substances such as bacteria, protozoa, and cell debris. (Ch. 7)

Pharyngitis: (fair-in-**JIGH**-tiss) Inflammation of the mucous membranes and lymph tissues of the pharynx, usually caused by infection. (Ch. 27)

Pheochromocytoma: (FEE-oh-KROH-moh-sigh-**TOH**-mah) Rare tumor of the adrenal system that secretes catecholamines. (Ch. 37)

Phimosis: (figh-**MOH**-sis) Uncircumcised foreskin that cannot be moved down from the head of the penis. (Ch. 41)

Phlebitis: (fla-**BYE**-tis) Inflammation of a vein; may be due to irritating IV fluids or thrombosis. (Ch. 6)

Phlebotomy: (fle-**BAH**-tuh-mee) Entry into a vein for the removal or withdrawal of blood. (Ch. 24)

Phobia: (**FOH**-bee-ah) A persistent, irrational, intense fear of a specific object, activity, or situation. (Ch. 55)

Photophobia: (FOH-toh-**FOH**-bee-ah) Abnormal visual intolerance to light. (Ch. 46, 49)

Physical dependence: (**FIZ**-ik-uhl dee-**PEN**-dens) A pharmacologic phenomenon characterized by signs and symptoms of withdrawal when medication is withdrawn. (Ch. 9)

Plaque: (PLAK) A deposit of fatty material on the lining of an artery. (Ch. 16, 18)

Plasmapheresis: (PLAS-mah-fer-**EE**-sis) Removal of blood to separate cells from plasma. (Ch. 47)

Pleurodesis: (PLOO-roh-**DEE**-sis) Creation of adhesions between the parietal and visceral pleura to treat recurrent pneumothorax. (Ch. 28)

Pneumocystis carinii pneumonia: (new-mo-**SIS**-tis ca-**RIN**-ee-eye new-**MOH**-nee-ya) An acute pneumonia caused by Pneumosyctis carinii, a fungus. It occurs in immunodeficient adults and is a defining opportunistic infection of AIDS. (Ch. 54)

Pneumonectomy: (NEW-moh-**NEK**-tuh-mee) Surgical removal of all or part of a lung. (Ch. 28)

Pneumothorax: (NEW-moh-**THAW**-raks) Air in the pleural space. (Ch. 28)

Poikilothermy: (POY-ki-loh-**THER**-mee) The absence of sufficient arterial blood flow, causing the extremity to become the temperature of the environment. (Ch. 15)

Point of maximal impulse: (POYNT of **MAKS**-i-muhl **IM**-puls) The area of the chest where the greatest force can be felt with the palm of the hand when the heart contracts or beats. Usually at the fourth to fifth intercostal space in the midclavicular line. (Ch. 15)

Polycythemia: (PAH-lee-sigh-**THEE**-mee-ah) Excessive red cells in the blood. (Ch. 18, 24, 28)

Polydipsia: (PAH-lee-**DIP**-see-ah) Excessive thirst. (Ch. 37, 38)

Polymyositis: (PAH-lee-my-oh-**SIGH**-tis) A rare, inflammatory disease of the skeletal muscle tissue characterized by symmetric weakness of proximal muscles of the limbs, neck, and pharynx. (Ch. 44)

Polyneuropathy: (PAH-lee-new-**RAH**-puh-thee) A disease involving multiple nerves. (Ch. 47)

Polyphagia: (PAH-lee-**FAY**-jee-ah) Excessive eating. (Ch. 38)

Polyuria: (PAH-lee-**YOOR**-ee-ah) Excessive urination. (Ch. 35, 37)

Portal hypertension: (**POR**-tuhl HIGH-per-**TEN**-shun) Persistent blood pressure elevation in the portal circulation of the abdomen. (Ch. 33)

Postcoital: (post-**KOH**-i-tal) Following sexual intercourse. (Ch. 40)

Postictal: (pohst-**IK**-tuhl) Occurring after a sudden attack, such as an epileptic seizure. (Ch. 46)

Postoperative: (post-**AHP**-er-uh-tiv) Following a surgical operation. (Ch. 11)

Powerlessness: (**POW**-er-less-nes) Perceived lack of control over a situation. (Ch. 13)

Preload: (**PREE**-lohd) End-diastolic stretch of cardiac muscle fibers; equals end-diastolic volume. (Ch. 15, 21)

Preoperative: (pre-**AHP**-er-uh-tiv) Preceding an operation. (Ch. 11)

Presbycusis: (PRESS-bee-**KYOO**-sis) Progressive, bilaterally symmetrical perceptive hearing loss occurring with age; usually occurs after age 50 and is caused by structural changes in the organs of hearing. (Ch. 49)

Presbyopia: (PREZ-bee-**OH**-pee-ah) Diminution of accommodation of the lens of the eye occurring normally with aging, and usually resulting in hyperopia, or farsightedness. (Ch. 49)

Pressure ulcer: (PRESS-sure **ULL**-sir) An open sore or lesion of the skin that develops because of prolonged pressure against an area. (Ch. 14)

Priapism: (**PRY**-uh-pizm) Erection that lasts too long. (Ch. 41)

Primary hypertension: (**PRY**-mare-ee HIGH-per-**TEN**-shun) Abnormally elevated blood pressure of unknown cause. Also called essential hypertension. (Ch. 16)

Proctitis: (prock-**TIGH**-tis) Inflammation of the rectum and anus. (Ch. 42)

Proctosigmoidoscopy: (PROK-toh-SIG-moy-**DAHS**-kuh-pee) Visual examination of the rectum and sigmoid colon by use of a sigmoidoscope. (Ch. 29)

Prodrome: (**PROH**-drohm) A symptom indicating the onset of a disease. (Ch. 46)

Prostaglandins: (**PRAHS**-tah-**GLAND**-ins) Chemical neurotransmitters usually associated with pain at the site of an injury, periphery. (Ch. 9)

Prostatectomy: (PRAHS-tuh-**TEK**-tuh-mee) Removal of the prostate gland. (Ch. 41)

Prostatitis: (PRAHS-tuh-**TIGH**-tis) Inflammation or infection of the prostate gland. (Ch. 41)

Protozoa: (pro-tow-**ZOH**-ah) Single-celled parasitic organisms that can move and live mainly in the soil. (Ch. 7)

Pruritis: (proo-**RYE**-tis) Severe itching. (Ch. 51)

Pseudoaddiction: (soo-doh-ad-**DICK**-shun) Syndrome in which behaviors similar to addiction appear as a result of inadequate pain control and clients fear not receiving adequate pain medications. (Ch. 9)

Psoriasis: (suh-**RYE**-ah-sis) Chronic inflammatory skin disorder in which epidermal cells proliferate abnormally fast. (Ch. 51)

Psychoanalysis: (SIGH-koh-uh-**NAL**-i-sis) Form of therapy based on the theories of Sigmund Freud, regarding the dynamics of the unconscious. (Ch. 55)

Psychogenic: (SIGH-koh-**JEN**-ick) Of mental origin. (Ch. 37)

Psychological dependence: (SY-ko-**LAW**-ick-al dee-**PEN**-dens) Obsession of obtaining drugs for use other than medicinal; addiction. (Ch. 9)

Psychopharmacology: (SIGH-koh-FAR-meh-**KAHL**-uh-jee) The study of the action of drugs on psychological functions and mental states. (Ch. 55)

Psychosomatic: (SIGH-koh-soh-**MAT**-ik) Having bodily symptoms of psychological, emotional, or mental origin; illness traceable to an emotional cause. (Ch. 55)

Psychotherapy: (SICH-koh-**THER**-uh-pee) A method of treating disease (especially mental illness) by mental rather than pharmacological means. (Ch. 55)

Ptosis: (**TOH**-sis) Drooping of eyelid. (Ch. 47, 48)

Puerperal: (pyoo-**ER**-per-uhl) Concerning the puerperium, or period of 42 days after childbirth. (Ch. 42)

Pulmonary edema: (**PULL**-muh-NAIR-ee uh-**DEE**-muh) Acute heart failure in which there is severe fluid congestion in the alveoli of the lungs; life-threatening. (Ch. 21)

Pulse deficit: (PULS **DEF**-i-sit) A condition in which the number of pulse beats counted at the radial artery is less than those counted in the same period of time at the apical heart rate. (Ch. 15)

Purpura: (**PUR**-pur-uh) Hemorrhage into the skin, mucous membranes, internal organs, and other tissues. (Ch. 23)

Purulent: (**PURE**-u-lent) Fluid that contains pus. (Ch. 11, 51)

Pyelogram: (**PIE**-loh-**GRAM**) A diagnostic procedure involving x-ray of the kidneys; may be done after injection of a dye into the bloodstream or directly into the kidneys. (Ch. 34)

Pyelonephritis: (PYE-e-loh-ne-**FRY**-tis) Inflammation of the kidney and renal pelvis. (Ch. 35)

Pyoderma: (PYE-oh-**DER**-mah) Any acute, inflammatory, purulent bacterial dermatitis. (Ch. 51)

Quadriparesis: (kwod-ri-par-**E**-sis) Weakness involving all four limbs caused by spinal cord injury. (Ch. 46)

Quadriplegia: (KWA-dri-**PLEE**-jah) Paralysis of all four limbs caused by spinal cord injury. (Ch. 46)

Radiation therapy: (RAY-dee-**AY**-shun **THER**-uh-pee) Cancer treatment with ionizing radiation. (Ch. 10)

Range of motion (ROM): (RANJE of **MOH**-shun) The range of movement of a body joint. (Ch. 14)

Raynaud's disease: (ra-**NOHZ** di-**ZEEZ**) A primary or idiopathic vasospastic disorder characterized by bilateral and symmetrical pallor and cyanosis of the fingers. (Ch. 18)

Reality orientation: (ree-**AL**-i-tee OR-ee-en-**TAY**-shun) A process to orient a person to facts such as names, dates, and time, through the use of verbal and nonverbal repeating messages. (Ch. 14)

Rectocele: (**RECK**-toh-seel) Protrusion or herniation of the posterior vaginal wall with the anterior wall of the rectum through the vagina. (Ch. 40)

Regurgitation: (ree-GUR-ji-**TAY**-shun) A backward flowing, as in the backflow of blood through a defective heart valve. (Ch. 19)

Remyelination: (ree-MY-uh-lin-**AY**-shun) Replacement of myelin or neurons. (Ch. 47)

Replantation: (re-plan-**TAY**-shun) The replacement of an organ or other structure, such as a digit, limb, or tooth, to the site from which it was previously lost or removed. (Ch. 44)

Reservoir: (REZ-er-**VWAR**) A person, animal, arthropod, plant, soil, or substance in which an infectious agent normally lives and multiplies, on which it depends for survival. (Ch. 7)

Respiratory excursion: (**RES**-pi-rah-TOR-ee eks-**KUR**-zhun) Downward movement of diaphragm with inspiration. (Ch. 26)

Respite care: (**RES**-pit CARE) Short-term, intermittent care for the chronically ill; provides rest for the family members or caregivers from the stress of sustained caregiving. (Ch. 13)

Respondeat superior: (ress-**POND**-ee-et sue-**PEER**-ee-or) An institution that employs a worker may be liable for the acts or omissions of its employees. (Ch. 2)

Retinopathy: (RET-i-**NAH**-puh-thee) Disease of the retina of the eye. (Ch. 38, 49)

Retroflexion: (RET-roh-**FLECK**-shun) A bending or flexing backward. (Ch. 40)

Retrograde: (**RET**-roh-grayd) Moving backward; degenerating from a better to a worse state. (Ch. 40, 41)

Retrograde cholangiopancreatography: (**RET**-roh-grayd koh-LAN-jee-oh-PAN-kree-ah-**TOG**-rah-fee) An endoscopic procedure that permits the physician to visualize the liver, gallbladder, and pancreas using an endoscope, dye, and x-ray examinations. (Ch. 32)

Retroversion: (RET-roh-**VER**-zhun) A turning, or a state of being turned back; the tipping of an entire organ. (Ch. 40)

Rheumatic carditis: (roo-**MAT**-ick kar-**DYE**-tis) Serious complication of rheumatic fever in which all layers of the heart become inflamed. (Ch. 17)

Rheumatic fever: (roo-**MAT**-ick **FEE**-ver) A hypersensitivity reaction to antigens of group A beta-hemolytic streptococci. (Ch. 17)

Rhinitis: (rye-**NIGH**-tis) Inflammation of the nasal mucosa, usually associated with congestion, itching, sneezing, and nasal discharge. (Ch. 27)

Rhinoplasty: (**RYE**-noh-plass-tee) Plastic surgery of the nose. (Ch. 27)

Rickettsia: (ra-**KET**-see-ah) A genus of bacteria of the tribe Rickettsiae that multiply only in host cells. (Ch. 7)

Rinne test: (**RIN**-nee **TEST**) A test of hearing made with tuning forks. (Ch. 48)

Romberg's test: (**RAHM**-bergs **TEST**) A test to determine if a person has the ability to maintain body balance when the eyes are shut and the feet are close together. (Ch. 48)

Roux-en-Y: (roo-ehn-**Y**) Gastric bypass surgery. A small stomach pouch the size of a thumb is created with staples, then a **Y**-shaped section of the small intestine is attached to the pouch to allow food to bypass the lower stomach and duodenum. (Ch. 30)

Rule of nines: (**ROOL** of NINES) A formula for estimating percentage of body surface areas, particularly helpful in judging the portion of skin that has been burned. (Ch. 12)

Sacral radiculopathy: (**SAY**-krul ra-DICK-yoo-**LAH**-puh-thee) Pathology of sacral nerve roots. (Ch. 42)

Salpingitis: (SAL-pin-**JIGH**-tis) Inflammation of a fallopian tube. (Ch. 42)

Salpingoscopy: (SAL-ping-**AHS**-koh-pee) Endoscopic visualization of the fallopian tubes. (Ch. 39)

Scleroderma: (SKLER-ah-**DER**-ma) A chronic manifestation of progressive systemic sclerosis in which the skin is taut, firm, and edematous, limiting movement. (Ch. 44)

Sclerosis: (skle-**ROH**-sis) A hardening or induration of an organ or tissue, especially from excessive growth of fibrous tissue. (Ch. 47)

Seborrhea: (SEB-oh-**REE**-ah) Disease of the sebaceous glands marked by increase in the amount and often alteration of the quality of sebaceous secretions. (Ch. 51)

Secondary hypertension: (SEK-un-DAR-ee HIGH-per-**TEN**-shun) High blood pressure that is a symptom of a specific cause, such as a kidney abnormality. (Ch. 16)

Semipermeable: (SEM-ee-**PER**-mee-uh-buhl) Partly permeable; said of a membrane that will allow fluids but not the dissolved substance to pass through it. (Ch. 5)

Sensorineural: (SEN-suh-ree-**NEW**-ruhl) Hearing loss caused by impairment of a sensory nerve. (Ch. 49)

Sensory deprivation: (**SEN**-suh-ree DEP-ri-**VAY**-shun) No or minimal stimulation of the senses that creates the potential for maladaptive coping. (Ch. 14)

Sensory overload: (**SEN**-suh-ree **OH**-ver-lohd) Excessive stimulation of the senses that creates the potential for maladaptive coping. (Ch. 14)

Sepsis: (**SEP**-sis) Systematic infection caused by microorganisms in the bloodstream. (Ch. 7, 8)

Serologic: (SEAR-uh-**LAJ**-ick) Study of substances present in blood serum. (Ch. 42)

Serosanguineous: (SEER-oh-**SANG**-gwin-ee-us) Fluid consisting of serum and blood. (Ch. 11)

Serotonin: (SER-ah-**TOH**-nin) A chemical neurotransmitter important in sleep-wake cycles. Reduced serotonin levels are associated with depression. (Ch. 9)

Shock: (**SHAHK**) A clinical syndrome in which the peripheral blood flow is inadequate to return sufficient blood to the heart for normal function, particularly transport of oxygen to all organs and tissues. (Ch. 12)

Sinusitis: (SINE-u-**SIGH**-tiss) Inflammation of the sinuses; may be due to viral or bacterial infection, or to allergies. (Ch. 27)

Sinoatrial node: (SIGH-noh-**AY**-tree-al NOHD) Node at the junction of the superior vena cava and right atrium, regarded as the starting point of the heartbeat. (Ch. 20)

Snellen's chart: (**SNEL**-ens **CHART**) A chart imprinted with lines of black letters graduating in size from smallest on the bottom to largest on top; used for testing visual acuity. (Ch. 48)

Somatoform: (soh-**MAT**-uh-form) Denoting psychogenic symptoms resembling those of physical disease; psychosomatic. (Ch. 55)

Spider angioma: (**SPY**-der an-jee-**OH**-mah) Thin reddish-purple vein lines close to the skin surface. (Ch. 32)

Spirituality: (SPIHR-it-u-**AL**-it-tee) Sense of connectedness with all of life and the universe. (Ch. 13)

Splenectomy: (sple-**NEK**-tuh-mee) Excision of the spleen. (Ch. 25)

Splenomegaly: (SPLEE-noh-**MEG**-ah-lee) Enlargement of the spleen. (Ch. 21, 25)

Standard of best interest: (**STAND**-erd OF **BEST IN**-ter-est) A type of decision made about clients' health care when they are unable to make an informed decision about their own care. (Ch. 2)

Standard precautions: (**STAN**-derd pre-**KAW**-shuns) Guidelines recommended by the Centers for Disease Control and Prevention to reduce the risk of the spread of infection. (Ch. 7)

Stapedectomy: (stuh-puh-**DEK**-tuh-mee) Excision of the stapes in order to improve hearing, especially in cases of otosclerosis. (Ch. 49)

Staphylococcus: (STAFF-il-oh-**KOCK**-uss) A genus of gram-positive bacteria; they are constantly present on the skin and in the upper respiratory tract and are the most common cause of localized suppurating infections. (Ch. 7)

Status asthmaticus: (**STAT**-us az-**MAT**-i-kus) Prolonged period of unrelieved asthma symptoms. (Ch. 28)

Steatorrhea: (STEE-ah-toh-**REE**-ah) Fat in the stools; may be associated with pancreatic disease. (Ch. 29, 30, 31, 33)

Stenosis: (ste-**NOH**-sis) The constriction or narrowing of a passage or orifice, such as a cardiac valve. (Ch. 19)

Stent: (STENT) Any mold or device used to hold tissue in place or to provide a support, graft, or anastomosis while healing is taking place. (Ch. 35)

Stereotype: (**STER**-ee-oh-**TIGHP**) An opinion or belief about an individual or group that may not be true. (Ch. 3)

Sternotomy: (stir-**NAH**-tuh-mee) The operation of cutting through the sternum. (Ch. 22)

Stoma: (**STOH**-mah) A mouth, small opening, or pore. (Ch. 31)

Stomatitis: (STOH-mah-**TIGH**-tis) Inflammation of the mouth. (Ch. 10, 30)

Stress: (STRESS) The physical (gravity, mechanical, pathogenic, injury) and psychological (fear, anxiety, crisis, joy) forces that are experienced by individuals. (Ch. 55)

Stressor: (**STRESS**-er) Any person or situation that produces an anxiety response. (Ch. 55)

Striae: (**STRIGH**-ee) A line or band of elevated or depressed tissue; may differ in color or texture from surrounding tissue. (Ch.32)

Subarachnoid: (SUB-uh-**RAK**-noyd) Below or under the arachnoid membrane and the pia mater of the covering of the brain and spinal cord. (Ch. 45)

Subdural: (sub-**DUHR**-uhl) Beneath the dura mater. (Ch. 46)

Subjective data: (sub-**JEK**-tiv **DAY**-tuh) Information that is provided verbally by the client. (Ch. 1)

Suffering: (**SUFF**-er-ing) A state of severe distress associated with events that threaten the intactness of the person. Emotional pain associated with real or potential tissue damage. (Ch. 9)

Summons: (**SUM**-muns) A notice of suit. (Ch. 2)

Suprapubic: (SOO-pruh-**PEW**-bik) Bone of the groin (or region) located above the pubic arch. (Ch. 41)

Surgeon: (**SURGE**-on) A medical practitioner who specializes in surgery. (Ch. 11)

Synovitis: (sin-oh-**VIGH**-tis) Inflammation of the synovial membrane that may be the result of an aseptic wound, a subcutaneous injury, irritation, or exposure to cold and dampness. (Ch. 43, 44)

Systolic blood pressure: (sis-**TAL**-ik BLUHD **PRESS**-ur) Maximal pressure exerted on the arteries during contraction of the left ventricle of the heart. The top number of a blood pressure reading. (Ch. 16)

Tachycardia: (TAK-ee-**KAR**-dee-yah) An abnormal rapidity of heart action, usually defined as a heart rate greater than 100 beats per minute in adults. (Ch. 8)

Tachydysrhythmia: (TACK-ee-dis-**RITH**-mee-yah) An abnormal heart rhythm with rate greater than 100 beats per minute in an adult. (Ch. 22)

Tachypnea: (TAK-ip-**NEE**-ah) Abnormally rapid respiratory rate. (Ch. 8, 28)

Tamponade: (TAM-pon-**AYD**) Compression of a part. (Ch. 41)

Tension pneumothorax: (TEN-shun NEW-moh-**THOR**-raks) Abnormal accumulation of air with buildup of pressure in the pleural space. (Ch. 8)

Teratoma: (ter-uh-**TOH**-muh) A congenital tumor containing one or more of the three primary embryonic germ layers. (Ch. 40)

Terminal illness: (TERM-in-al **ILL**-ness) An illness that will probably cause death in 6 months or less. (Ch. 13)

Tetanus: (**TET**-nus) A highly fatal disease caused by the bacillus Clostridium tetani and characterized by muscle spasm and convulsions. (Ch. 12)

Tetany: (**TET**-uh-nee) Muscle spasms, numbness, and tingling caused by changes in pH and low serum calcium. (Ch. 37)

Thoracentesis: (THOR-uh-sen-**TEE**-sis) Insertion of a large-bore needle into the pleural space to remove fluid. (Ch. 26)

Thoracotomy: (THAW-rah-**KAH**-tah-mee) Surgical incision into the chest wall. (Ch. 28)

Thrill: (THRILL) Abnormal vessel that has a bulging or narrowed wall; a vibration is felt. (Ch. 15)

Thrombi: (**THROM**-bye) Blood clots. (Ch. 8)

Thrombocytopenia: (THROM-boh-SIGH-toh-**PEE**-nee-uh) Abnormal decrease in the number of blood platelets. (Ch. 10, 23, 24)

Thrombolytic: (throm-bo-**LIT**-ik) Agent that dissolves or splits up a thrombus, an aggregation of blood factors. (Ch. 46)

Thrombophlebitis: (THROM-boh-fle-**BYE**-tis) The formation of a clot and inflammation within a vein. (Ch. 17)

Thrombosis: (throm-**BOH**-sis) Formation, development, or presence of a thrombus, an aggregation of blood factors. (Ch. 18)

Tidaling: (**TIGH**-dah-ling) Rise and fall; may refer to water in water-seal chamber of a chest drainage system. (Ch. 26)

Titration: (tigh-**TRAY**-shun) Adjustment of medication up or down to meet client needs. (Ch. 9)

Tolerance: (**TALL**-er-ens) The response of the body to medication that requires increased medication administration to achieve the same effect. Often refers to opioids. (Ch. 9, 55)

Torts: (TORTS) Lawsuits involving civil wrongs. (Ch. 2)

Toxemia: (tock-**SEE**-me-ah) Spread of poisonous products of bacteria throughout the body. (Ch. 8)

Tracheostomy: (TRAY-key-**AHS**-tuh-me) A surgical opening in the neck into the trachea to provide an airway when the trachea is obstructed. (Ch. 26)

Tracheotomy: (TRAY-key-**AH**-tuh-me) An opening in the neck into the trachea. (Ch. 26)

Traditions: (tra-**DISH**-uns) Practices and customs handed down through the generations, often by word of mouth. (Ch. 3)

Transcellular: (trans-**SELL**-yoo-lar) Across cell membranes. (Ch. 5)

Transdermal: (trans-**DER**-mal) Entering through the dermis, or skin, as in administration of a drug applied to the skin in ointment or patch form. (Ch. 9)

Transillumination: (TRANS-i-loo-mi-**NAY**-shun) The passage of strong light through a body structure to permit inspection of an observer on the opposite side. (Ch. 39)

Transjugular intrahepatic portosystemic shunt (TIPS): (**TRANZ**-jug-yoo-lar intra-hep-**PAT**-ik POR-toh-sis-**TEM**-ik SHUNT) Shunt that sidetracks venous blood around the liver to the vena cava for treatment of ascites. (Ch. 33)

Transmyocardial (TRANS-my-o-**KAR**-dee-yah) Across all layers of the heart. (Ch. 22)

Trauma: (**TRAW**-mah) Physical injury caused by an external force. (Ch. 8)

Trendelenburg: (tren-**DELL**-en-berg) A position in which the client's head is low and the body and legs are on an elevated and inclined plane. (Ch. 8)

Triage: (**TREE**-ahj) To sort. (Ch. 12)

Trigeminy: (try-**JEM**-i-nee) Occurring every third beat, as in trigeminal premature ventricular contractions. (Ch. 20)

Tropia: (**TROH**-pee-ah) A manifest deviation of an eye from the normal position when both eyes are open and uncovered. (Ch. 48)

T-tube: (**TEE**-toob) A T-shaped tube in the bile duct that allows drainage of bile following gallbladder surgery. (Ch. 33)

Tumor: (**TOO**-mur) An abnormal growth of cells or tissues; tumors may be benign or malignant. (Ch. 10)

Turbid: (**TER**-bid) Cloudy. (Ch. 46)

Turgor: (**TER**-ger) The resistance of the skin to being grasped between the fingers. Dehydration causes poor skin turgor. (Ch. 50)

Unifocal: (YOO-ni-**FOH**-kuhl) Coming or originating from one site or focus. (Ch. 20)

Upper gastrointestinal series (upper GI, UGI): (UH-per GAS-troh-in-**TES**-ti-nuhl SEER-ees) X-ray and fluoroscopic examinations of the stomach and duodenum after the ingestion of a contrast medium. (Ch. 29)

Uremia: (yoo-**REE**-mee-ah) An excess in the blood of urea, creatinine, and other nitrogenous end products of protein and amino acid metabolism. (Ch. 34, 35)

Urethritis: (YOO-ree-**THRIGH**-tis) Inflammation of the urethra. (Ch. 35, 42)

Urethroplasty: (yoo-**REE**-throh-PLAS-tee) Plastic repair of the urethra. (Ch. 35)

Urinary incontinence: (**YOOR**-i-NAR-ee in-**KON**-ti-nents) Inability to control urine excretion creating accidental urinary leakage. (Ch. 14, 34, 35)

Urodynamic: (YOO-roh-dye-**NAM**-ik) The study of the holding or storage of urine in the bladder, the facility with which it empties, and the rate of movement of urine out of the bladder during urination. (Ch. 41)

Urosepsis: (YOO-roh-**SEP**-sis) Septicemia resulting from urinary tract infection. (Ch. 35, 41)

Urticaria: (UR-ti-**CARE**-ee-ah) Hives signifying an allergic reaction. (Ch. 8, 53)

Utilitarian: (yoo-TILL-I-**TAR**-I-en) Consequences or outcomes of a dilemma are the most important element. (Ch. 2)

Vaginosis: (VAJ-i-**NOH**-sis) Inflammation of the vagina caused by Gardnerella vaginalis. (Ch. 40)

Values: (**VAL**-use) Ideals or concepts that give meaning to an individual's life. (Ch. 2, 3)

Valvotomy: (val-**VAH**-tuh-mee) Cutting through a valve. (Ch. 22)

Valvuloplasty: (**VAL**-vyoo-loh-PLAS-tee) Plastic or restorative surgery on a valve, especially a cardiac valve. (Ch. 19)

Varices: (**VAR**-i-seez) Dilated veins. (Ch. 33)

Varicocele: (**VAR**-i-koh-seel) Varicose veins of the scrotum; can lead to infertility. (Ch. 39, 41)

Varicose veins: (**VAR**-i-kohs VAINS) Swollen, distended, and knotted veins, usually in the subcutaneous tissue of the leg. (Ch. 18)

Vasculitis: (VAS-kue-**LIGH**-tis) Inflammation of a vessel. (Ch. 44)

Vasectomy: (va-**SEK**-tuh-mee) Surgically cutting and sealing the vas deferens to prevent sperm from getting outside the body. Used as a birth control method for men. (Ch. 41)

Vector: (**VECK**-tur) Living organism that transmits disease. (Ch. 7)

Venous stasis ulcers: (VEE-nus **STAY**-sis UL-sers) Poorly healing ulcers that result from inadequate venous drainage. (Ch. 18)

Ventricular diastole: (ven-**TRICK**-yoo-lar dye-**AS**-tuh-lee) The period of relaxation of the two ventricles. (Ch. 20)

Ventricular escape rhythm: (ven-**TRICK**-yoo-lar es-**KAYP RITH**-uhm) The naturally occurring rhythm of the ventricles when the rest of the cardiac conduction system fails. (Ch. 20)

Ventricular repolarization: (ven-**TRICK**-yoo-lar RE-pol-lahr-i-**ZAY**-shun) Reestablishment of the polarized state of the muscle after contraction. (Ch. 20)

Ventricular systole: (ven-**TRICK**-yoo-lar **SIS**-tuh-lee) The contraction of the two ventricles. (Ch. 20)

Ventricular tachycardia: (ven-**TRICK**-yoo-lar TACK-ee-**KAR**-dee-yah) A series of at least three beats arising from a ventricular focus at a rate greater than 100 beats per minute. (Ch. 20)

Verrucous: (ve-**ROO**-kus) Wartlike, with raised portions. (Ch. 42)

Vertebrae: (**VER**-te-bray) Any of the 33 bony segments of the spinal column: 7 cervical, 12 thoracic, 5 lumbar, 5 sacral, and 4 coccygeal vertebrae. (Ch. 43)

Vesicant: (**VESS**-i-kant) Agent that causes blistering of tissue. (Ch. 10)

Vesicular: (ve-**SICK**-yoo-ler) Pertaining to vesicles or small blisters. (Ch. 42)

Virulence: (**VEER**-you-lence) The power of an organism to cause disease. (Ch. 7)

Virus: (**VIGH**-rus) The smallest organism identified by use of electron microscopy; intracellular parasites that may cause disease. (Ch. 7)

Viscosity: (vis-**KAH**-si-tee) Thickness, as of the blood. (Ch. 16)

Volvulus: (**VOL**-view-lus) A twisting of the bowel on itself, causing obstruction. (Ch. 31)

Vulvovaginitis: (VUL-voh-VAJ-I-**NIGH**-tis) Inflammation of the vulva and vagina. (Ch. 42)

Weber test: (**VAY**-ber **TEST**) A test for unilateral deafness. (Ch. 48)

Welfare rights: (**WELL**-fare **RIGHTS**) Also called legal rights; rights that are based on a legal entitlement to some good or benefit. (Ch. 2)

White blood cells: (WIGHT BLUHD SELLS) Leukocytes; the body's primary defense against infection. (Ch. 52)

Withdrawal: (with-**DRAW**-ul) Symptoms caused by cessation of administration of a drug, especially a narcotic or alcohol, to which the individual has become either physiologically or psychologically addicted. (Ch. 55)

Worldview: (**WERLD**-vyoo) The way individuals look on the world to form values and beliefs about life and the world around them. (Ch. 3)

Xerostomia: (ZEE-roh-**STOH**-mee-ah) Dry mouth caused by reduction in secretions. (Ch. 10)

Index

Page numbers in italics indicate figures. Page numbers followed by t indicate tables; b, boxes.

A

Abdomen, 479–480
  flat plate of the abdomen, 481, 580
  hernias, 517–518
  preoperative phase of surgery assessments, 151
  trauma, 182
Ablation, 316, 322–323
Abortion, 705–706
  chemical agents, 706–707
  complications, 707
  early pregnancy methods, 707
  ectopic pregnancy, 706
  later pregnancy methods, 707
  methods, 706–707
  nursing care and teaching, 707
  prenatal abnormalities, 706
  risks, 707
Abrasions, 178
Absorption disorders, 518–519
Abuse of drugs, 1039, 1046
Accommodation, 891
Acid-base balance, 58–60
  kidneys and, 576
  renal failure, 604
  respiration and, 406
Acidosis, 55
  shock, 93
Acne vulgaris, 963–964
Acquired chronic illnesses, 201b
Acquired immunodeficiency syndrome. See HIV and AIDS.
Acromegaly, 629–630
Active immunity, 990–991
Activity, 212
  anemias, 378
  chronic heart failure, 332–333, 338
  myocardial infarction, 285
  nervous system therapy, 820–821
Acupuncture, 35
Acute bronchitis, 440
Acute compartment syndrome, 791–792
Acute coronary syndromes, 279
Acute gout, 769
Acute heart failure, 326, 329
Acute hydrocephalus, 848
Acute leukemia, 387
Acute pancreatitis, 556–560, 559b–561b
Acute poststreptococcal glomerulonephritis, 599
Acute pulmonary hypertension, 96
Acute renal failure, 600–602
Acute respiratory distress syndrome, 462–463

Acute respiratory failure, 462–463
Adaptation, 1031, 1036t
Addiction, 105, 1037
Adenoiditis, 433–434
Adjunct agents, 158
Adjunct anesthesia techniques, 159
Adjuvants, 106, 108–109
  lower respiratory tract disorders, 466
Administrative laws, 19
Adnexa, 678
Adrenal cortex, 618, 620t
Adrenal glands, 617–618, 620t, 622t, 639
  adrenocortical insufficiency, 640–641
    Cushing's syndrome, 641–642
    pheochromocytoma, 639–640
Adrenal medulla, 617–618, 620t
Adrenocortical insufficiency, 640–641
Adrenolectomy, 642
Adventitious, 409
Aerobic bacteria, 77
Affect, 620, 1027, 1041
African-Americans/blacks, 28b, 30b, 32
  gastrointestinal system, 478b
  pain, 103b
Afterload, 327
Agenesis, 695
Aging, 210
  and anemias, 376
  and blood, 369, 374
  and cancer, 129
  and cardiovascular system, 210t, 214–215, 233–234
  and chronic illness, 202
  and cognition, 221
  and coping abilities, 221
  and cultural considerations and influences, 23, 24b
  and dehydration, 48
  and dementia, 221–222
  and depression, 221
  and diabetes mellitus, 663
  and diarrhea, 508
  and diverticulosis and diverticulitis, 512b
  and dysrhythmia, 234, 313
  and ears, 896–897
  and endocrine system, 210t, 216, 619
  and eyes, 889, 891, 892b
  and gallbladder, 537, 538
  and gastrointestinal system, 210t, 215–216, 476, 477, 478b
  and genitourinary and reproductive system, 210t, 216–217, 674
  and health promotion, 223
  and HIV and AIDS, 1012